LABORATORY PROTOCOLS
in Applied Life Sciences

LABORATORY PROTOCOLS
in Applied Life Sciences

Prakash S. Bisen

Emeritus Scientist

Defense Research Development Establishment
Defense Research Development Organization
Ministry of Defense, Government of India
Gwalior, India

CRC Press
Taylor & Francis Group
Boca Raton London New York

CRC Press is an imprint of the
Taylor & Francis Group, an **informa** business

Library Resource Center
Renton Technical College
3000 N.E. 4th Street
Renton, WA 98056

CRC Press
Taylor & Francis Group
6000 Broken Sound Parkway NW, Suite 300
Boca Raton, FL 33487-2742

© 2014 by Taylor & Francis Group, LLC
CRC Press is an imprint of Taylor & Francis Group, an Informa business

No claim to original U.S. Government works

Printed on acid-free paper
Version Date: 20140114

International Standard Book Number-13: 978-1-4665-5314-9 (Hardback)

Library of Congress Cataloging-in-Publication Data

Bisen, Prakash S., author.
  Laboratory protocols in applied life sciences / by Prakash Singh Bisen.
    p. ; cm.
  Includes bibliographical references and index.
  ISBN 978-1-4665-5314-9 (hardcover : alk. paper)
  I. Title.
  [DNLM: 1. Clinical Laboratory Techniques. 2. Biological Science Disciplines--methods. QY 25]

  RC71.3
  616.07'5--dc23   2013026167

Visit the Taylor & Francis Web site at
http://www.taylorandfrancis.com

and the CRC Press Web site at
http://www.crcpress.com

I would like to dedicate this work to one of my mentors, late Professor G. L. Farkas, former director of the Institute of Plant Physiology, Biological Research Centre, Hungarian Academy of Sciences, Szeged, Hungary. He was not only a great teacher, scientist, and author but also a wonderful human being who helped me develop my interest in biological research.

Contents

Contents

Contents

Preface

Laboratory Protocols in Applied Life Sciences presents the basic as well as latest techniques in wide areas of life sciences, namely, biotechnology, analytical biochemistry, clinical biochemistry, biophysics, molecular biology, genetic engineering, bioprocess technology, industrial processes, animal, plant, and microbial biology, computational biology, biosensors, etc., in an illustrative fashion. The book includes 27 chapters covered under 12 units with appendices, important links, and a glossary.

This book analyzes information derived through real experiments and focuses on cutting-edge techniques in the field. Up-to-date protocols covering a range of frequently used methods are provided. The main focus in writing this book has been on developing highly doable experiments that will provide students and professionals with a successful learning experience. Basic as well as advanced methods available in wide-ranging areas of biology are covered and compiled in one book. Each experiment is presented with an introduction to the topic, concise objectives, a list of the necessary materials and reagents, and step-by-step, readily reproducible laboratory protocols. Each chapter is self-contained and written in a style that enables the student to progress from elementary concepts to advanced research techniques. *Laboratory Protocols in Applied Life Sciences* serves as a valuable tool for both beginner research workers and experienced professionals. I have taken utmost care to include information to help students to develop concept and to design and execute the experiment in a given field of biology. I hope that this book will be able to promote interest among students, teachers, and encourage beginners to the level of excitement.

Constructive suggestions and comments are welcomed for the improvement of the book.

Prakash S. Bisen
Gwalior, India

About the Book

Life sciences have been qualitative in nature, seeking to identify the mere presence of the molecules but seldom explain them quantitatively. Although the development of instrumentation techniques began as far back as the early part of the twentieth century, these technologies progressed at a slow pace, contributing only minimally to the field; it is only in the middle of the twentieth century that instrumentation began to make rapid strides, increasing its speed and accuracy. With the advent of molecular biology, studies in biological sciences have become increasingly dependent on new and emerging techniques, essentially based on well-known principles of physical and chemical sciences. For a number of years, it has been difficult for the students of biology to grasp the basic principles of these potential emerging techniques, particularly owing to their mathematical bias and an absence of a comprehensive book dealing with all the integrated techniques used for the qualitative as well as quantitative analysis of living systems. *Laboratory Protocols in Applied Life Sciences* identifies several important technologies now used that are critical to fulfilling the potential of life sciences based on the use of increasingly sophisticated instrumentation. The book attempts to present a nonmathematical account of the underlying principles of a variety of experimental techniques, including electron microscopy, spectroscopy, ultracentrifugation, radiation biology, biochemistry, bioinformatics, enzyme technology, separation technology, protein engineering, bioprocess engineering, environmental biology, etc., with the hope that it would be useful to all students of biology at graduate and postgraduate levels. This book is an endeavor toward that end. While standard laboratory books are available for such disciplines as genetics, biochemistry, microbiology, molecular biology, immunology, environmental biology, biochemical engineering, etc., students of interdisciplinary fields such as life sciences, pharmacy, medical and paramedical sciences, biotechnology, etc., find them too exhaustive and terse. Students of these fields require a single volume, integrating all branches with contents arranged and presented in a way comprehensible to them. The book deals with the principles, concepts, techniques, and applications used in life sciences in a very comprehensive and illustrative manner, which is particularly useful for all biology students.

Acknowledgments

I would like to express my gratitude to Dr. G.S. Davar and Puneet Davar, Tropilite Foods Pvt. Ltd, Gwalior, India, and to Rakesh Singh Rathore, Anil Singh, and Richa Verma, Vikrant Institute of Technology and Management, Gwalior, Madhya Pradesh, India, for their invaluable advice and moral support. My sincere thanks to Dr. Dinesh Kumar, senior scientist, Indian Statistical Agricultural Research Institute, Pusa, New Delhi, and Dr. Prem Prakash Dubey, Department of Animal Genetics and Breeding, Guru Angad Dev University of Veterinary and Animal Sciences, Ludhiana, for contributing a full chapter on bioinformatics and to Dr. Divya Shrivastava and Shweta Mittal, School of Life Sciences, Jaipur National University, Jaipur, for helping me in preparing the chapter on genetic engineering.

I am grateful to Professor G.B.K.S. Prasad, School of Studies in Biochemistry; Professor Ishan K. Patro and Dr. Nisha Patro, School of Studies in Neurosciences, Jiwaji University, Gwalior; Dr. Ram P. Tiwari and Dr. Anubhav Jain, Diagnostic Division, RFCL Limited (formerly Ranbaxy Fine Chemicals Limited), Avantor Performance Materials, New Delhi, India; Dr. Ruchika Raghuvanshi, Bhagwan S. Sanodiya, Gulab S. Thakur, Rakesh K. Baghel, and Rohit Sharma, Tropilite Foods Pvt. Ltd., Gwalior; and Sumit Govil and Shailesh Kumar of Jaipur National University, Jaipur, India, for their valuable assistance in the preparation of this book.

My sincere thanks to Devendra Singh, Avinash Dubey, and Rahul Jha, Gwalior, India for all the computational work involved in making this book presentable. The cover page of the book was designed by courtesy of Mr Rahul Jha.

Thanks are also due to Michael Slaughter, Kari Budyk, and Michele Smith of CRC Press/Taylor & Francis Group, for their support in publishing this book in time and for their patience and enthusiasm. Finally, I would like to thank my wife Shashi Bisen for her constant support, cooperation, and understanding. I am grateful to the Council of Scientific and Industrial Research, New Delhi, for bestowing me with the title of Emeritus Scientist.

Prakash S. Bisen
Gwalior, India

Author

Prakash S. Bisen received his PhD in 1972 and his DSc in 1981 from Jabalpur University, Jabalpur, India. He has guided 60 doctoral thesis students and has to his credit 200 publications in indexed international journals of repute as well as 7 international books on contemporary issues in biological sciences. He was a fellow of the UNESCO/UNDP/ICRO and Hungarian Academy of Sciences at the Institute of Plant Physiology, Biological Research Center, Szeged, Hungary; a U.S. National Science Foundation fellow in the Department of Biological Sciences at the University of Illinois at Chicago and in the Department of Bacteriology at the University of California, Davis; UNESCO/WHO fellow at the Institute of Microbiology, Czechoslovak Academy of Sciences, Budejovika, Praha, Czechoslovakia; Marie Curie fellow at the Institute of Environmental and Biological Sciences, University of Lancaster, United Kingdom; and DAAD (German Academic Exchange Fellowship) fellow at the University of Bonn, Germany. He was awarded the Life Time Professorship, the highest honor of Bundhelkhand University, Jhansi, India. Professor Bisen is currently pursuing his research in the field of medical biotechnology as emeritus scientist at Defence Research Development Establishment, Defence Research Development Organization, Ministry of Defence, Government of India, Gwalior, India. He has been elected fellow of the National Academy of Sciences, India, for his enormous contributions to the field of biology. The diagnostic technology developed by Professor Bisen is unique in several respects and has received recognition worldwide. The technology has been patented in India, the United States, Europe, and Japan. Other patents are under active consideration. He has published six international patents in the United States, Europe, Japan, and India. Professor Bisen is the external scientific advisor to a Barcelona-based multinational company, Biokit, which is involved in the manufacturing of diagnostic kits. He also serves as the chief of the research and development center of an upcoming industry, Tropilite Foods, based at Gwalior, which is involved in food biotechnology and develops several innovative products for the food industry.

1

Microscopy

1.1 Introduction

Microscopy includes techniques used to study the cell structure of all organisms. The unaided eye cannot comfortably distinguish two points <0.2 mm apart, as resolving power is inadequate for the study of most cells. Magnification is, therefore, essential to render them visible. Cells that have been fixed and stained can be studied in a conventional light microscope, while antibodies coupled to fluorescent dyes can be used to locate specific molecules in cells using a fluorescence microscope. Living cells can be seen with phase contrast, differential contrast, dark field, or bright field microscopes. All forms of light microscopy are facilitated by electronic image-processing techniques, which enhance sensitivity and refine the image. Confocal microscopy and image deconvolution both provide thin optical sections and can be used to reconstruct three-dimensional images.

1.2 Light Microscopy

Compound light microscopes are no more than optical benches set up in a manner that makes them comfortable for humans to use. Observation of microscopic preparations can be made from a position that suits human posture, rather than the demands of an ordinary optical bench. They comprise three units: an illumination system, an object, and an analytical system. The object to be viewed is mounted on a glass microscope slide that is held on the microscope stage in the object plane. Light is focused on the object by means of the substage condenser. The image of the object is magnified by the combined action of two lens systems situated above the object. These are the objective lens and the eyepiece or ocular lens. Together, the objective and ocular lenses comprise the analytical system of the microscope (Figure 1.1).

Cells are measured in microns (one micron or $\mu m = 0.001$ mm) and millimicrons (one millimicron or $m\mu m = 0.001$ μm or 0.000001 mm). As per the alternate symbols of measurement, the micron is termed a micrometer (μm) and the millimicron is termed a nanometer (nm). The resolving power of the light microscope is limited by the wavelength of visible light used, and even with the best objectives, particles less than about 300 μm cannot be resolved. The particulate nature of the majority of viruses cannot, therefore, be determined with the ordinary light microscope. Useful observations on larger microorganisms, however, can be made.

1

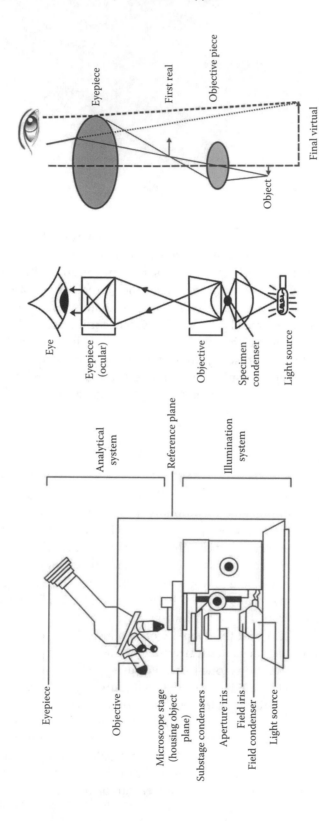

Figure 1.1   A compound microscope.

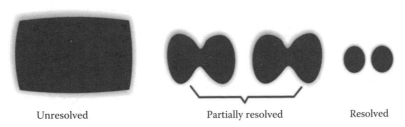

Unresolved   Partially resolved   Resolved

Figure 1.2 Resolution of two points. Allow resolution, structures blur together greater the resolution, the more detail that can be observed.

Commonly, eyepieces magnify 10×, and the objectives multiply 10×, 40×, and 100×. The overall magnification of an object is obtained by multiplying the magnification of the eyepiece lens by the magnification of the objective lens. Thus, objects are magnified 100×, 400×, and 1000× by commonly used combinations. Useful magnification is limited, however, by the phenomenon of resolution. The resolving power of a microscope is its ability to permit objects located close to one another to be distinguished as two separate and distinct entities. The limit of resolution is the smallest distance at which two points appear as separate entities (Figure 1.2). The resolving power of a microscope is governed by the wavelength of light used to illuminate the object and also by the numerical aperture (NA) of the objective lens. The NA of a lens is the measure of its light-gathering capacity. This number is engraved on the objective together with its magnifying power.

The distance between two points that can just be distinguished is called the resolving limit d. The resolving limit is dependent on the wavelength of light (λ) and numerical aperture (NA):

$$d = \frac{0.5\lambda}{\text{NA}}$$

NA is the property of the lens that describes the amount of light that can enter it:

$$\text{NA} = \mu \sin \theta$$

where
    μ is the refractive index of the medium between the specimen and the lens
    θ is half the angular aperture

Angular aperture is the angle between the most divergent rays of the inverted cone of light emerging from the condenser that enters the objective (Figure 1.3). Resolving power can be increased by (a) using shorter wavelength and (b) increasing the refractive index of the medium filling the space between the specimen and the front of the objective.

Due to the physiological restriction of the human retina to perceive light between 400 and 700 nm, in a light microscope, the resolving limit d is approximately 200 µm. Ultraviolet (UV) light, which has shorter wavelength, is preferable for increasing resolution, but because UV light cannot penetrate glass lenses well and because viewing UV light directly results in eye damage, it is normally not

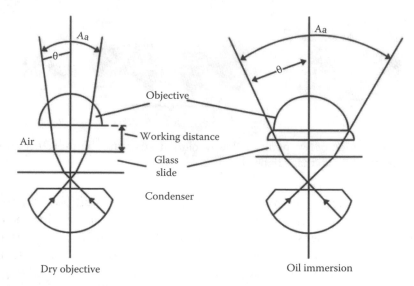

Dry objective   Oil immersion

Figure 1.3 NA is improved by the use of immersion oil to replace the air between the specimen on the glass slide and objective as shown by the wider angular aperture (Aa) obtained using an oil immersion lens. θ = angle between the most divergent rays entering the objective and the optical axis and is equal to 1/2 Aa. The working distance is also reduced in an oil immersion.

possible to depend on the improved resolving power that could be achieved using this shorter λ of light. Also, the advent of the electron microscope that utilizes the wave motion of electrons that have much shorter wavelength has made such UV microscopes obsolete.

The refractive index of the medium filling the space between the specimen and the objective can be increased considerably by using immersion oil between the object and the objective. Cedar oil has a refractive index of 1.52. This was the oil that was traditionally used for oil immersion, but it has been superseded by synthetic oils: the NA of oil immersion lenses is typically about 1.3. This gives a resolution of about 0.2 µm when objects are viewed using oil immersion. Most bacterial preparations are viewed under oil immersion to improve the resolution of the image. However, oil immersion can only be used in conjunction with objectives that are designed for the purpose. Objectives for use with oil immersion are appropriately engraved.

Many of the divergent peripheral rays lost by reflection and refraction at the surface of the condenser, slide, and objective lens are refracted within the angular aperture, thereby increasing the NA and consequently the resolution. It might be recalled that the use of a substage condenser concentrates peripheral light waves on the object, thereby effectively increasing the NA and decreasing the resolving limit, which means increasing the resolution.

Use of immersion oil also effectively decreases the focal length of the lens; so, the specimen has to be very close to the objective in order to be focused. Therefore, there is a short working distance between the lens and the objective. A short focal length also reduces the depth of the field so that only very thin sections can be focused.

The observation of algae, fungi, and protozoa can be achieved with dry objectives, that is, when air occupies the space between the specimen and the objective. The viewing of bacteria, which are smaller in size, normally requires the use of oil immersion lenses. Such lenses are specially designed for use with immersion oil and should never be used without it.

Except for some pigments that are able to absorb light at certain wavelengths, the majority of cell components absorb very little light of the visible region. This means that living cells studied with the light microscope exhibit low contrast. The use of dyes that selectively stain different cell components helps to overcome this limitation by providing some of the necessary contrast. Unfortunately, most staining techniques cannot be used with living cells. Instead, the cells must be fixed, dehydrated, and the larger ones embedded and sectioned prior to staining, and these often prolonged procedures may introduce a variety of changes in both the chemical and morphological makeup of the cell.

Observation of the living cell may be aided by the use of vital stains. The recent use of phase contrast and interference microscope techniques has resulted in further advances in the study of living cells.

Microbes can be visualized using a variety of microscopic techniques, and different types of microscopy are used for different purposes. All the techniques rely, however, on the fact that microorganisms interfere with the path of the light or electron beams. These interference effects can be exploited in different ways. It would be very expensive to have to stock a laboratory with light microscopes dedicated to a particular technique: one for bright field microscopy, one for dark ground microscopy, etc. To overcome this problem, modern compound microscopes have a modular design. It is possible to replace individual components such as objectives or the substage condenser to build a customized system for each particular function.

1.3 Atomic Force Microscopy

Atomic force microscopy (AFM) is a modification of light microscopy that also provides three-dimensional images of live preparations. A tiny stylus is positioned extremely close to the specimen such that weak repulsive atomic forces are established between the probe and the specimen. As the specimen is scanned in both the horizontal and vertical directions, the movement of the stylus up and down the hills and valleys constantly records its interaction with the surface. This pattern is processed by a series of detectors that feed the digital information into a computer that generates an image. No fixatives or coatings are required in AFM and thus allows living and hydrated specimens to be viewed. AFM is useful for mapping plasmids by locating the restriction enzymes bound to specific sites or to see the behavior of living bacteria.

1.4 Bright Field Microscopy

The most commonly used light microscope system in microbiology is bright field microscopy. This technique, however, permits only the examination of dead

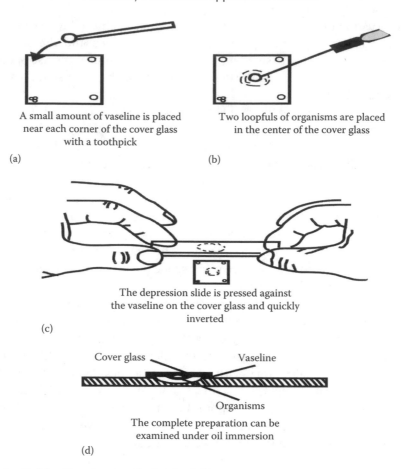

A small amount of vaseline is placed
near each corner of the cover glass
with a toothpick

(a)

Two loopfuls of organisms are placed
in the center of the cover glass

(b)

The depression slide is pressed against
the vaseline on the cover glass and quickly
inverted

(c)

Cover glass ——   —— Vaseline

—— Organisms

The complete preparation can be
examined under oil immersion

(d)

Figure 1.4 Kohler illumination for bright field microscopy.

microorganisms. The entire field is evenly illuminated by using the Köhler illumination system. This involves the use of an extra substage condenser lens that allows light to fill the optical field as illustrated in Figure 1.4. Microbes appear as dark objects against a bright background. Since many bacteria do not produce pigments, stains are used to enhance their visibility in bright field microscopy.

Microscopes require to be set up to optimize the light path through the instrument, and thereby to obtain the best image of the preparation. Many students feel very daunted when first using compound microscopes, but the steps involved in setting up a microscope for bright field illumination are relatively simple. Many modern compound microscopes have an integral lamp that is regulated by a voltage control unit. The lamp acts as the source of illumination, and light from the lamp passes through a field condenser and field iris. This regulates the area of the field of illumination. Above the field iris lies the aperture iris, also known as the substage condenser iris. This iris regulates the angle of one of the light leaving the condenser and consequently entering the objective. This, in turn, controls the resolution of the instrument.

When setting up a compound microscope for bright field microscopy, both the field iris and the aperture iris are fully opened. Using the objective with lowest magnification, an object is placed on the microscope stage and is brought into focus. Even though microbes are viewed at a much higher magnification, the reason an object is focused using the lowest power objective first is to optimize the conditions for correcting the misalignment of the optical path. As the magnification increases, so too does the apparent misalignment, and if the optical path is badly aligned, it is impossible to correct using objectives of high magnification. The next step is to ensure that light from the lamp is focused on the object. To do this, the field iris is closed, and the substage condenser is raised and lowered until the edge of the field iris is brought into sharp focus. The light source must then be centered so that the light beam can pass directly through the object, and then up through the analytical system along the optical axis. When the light on the object is focused and centered, the aperture iris is then adjusted to give the maximum resolution. Bright field microscopy is most frequently used in conjunction with stained preparations, for example, to determine whether a bacterium is Gram-positive or Gram-negative, or in a Ziehl–Neelsen-stained preparation used in the diagnosis of tuberculosis.

1.5 Dark Ground Microscopy

Another method used for observing certain details of cell structure is dark field microscopy. By means of a condenser which in its simplest form is fitted with a central circular stop, the object is illuminated with a cone of light without permitting any direct rays to enter the objective. Only the light reflected or scattered by the object is taken into the objective, with the result that the object appears self-luminous against a black background. The resolving power of the lens system is subject to the same limitations as in the ordinary microscope with direct illumination, but particles below the resolving power of the lenses may be seen as dots of light. The dispersed light is able to enter the objective, causing a brilliantly illuminated image of the specimen to appear against the dark background (Figure 1.5). This method of microscopy is widely used for the examination of bacterial motility and for organisms, which are not readily stained and are too small to be resolved by ordinary light microscopy, for example, spirochetes and larger viruses.

1.6 Phase Contrast Microscopy

By this method, structures within the cells of refractive index slightly different from that of the rest of the cell are rendered visible. Also, differences in refractive index between the whole cell and the surrounding medium make the cell clearly visible. The rays of light passing through the object are retarded by a fraction of a wavelength compared with similar rays passing through the suspending medium. This produces a difference in "phase" between the two emerging types of rays. The phase contrast

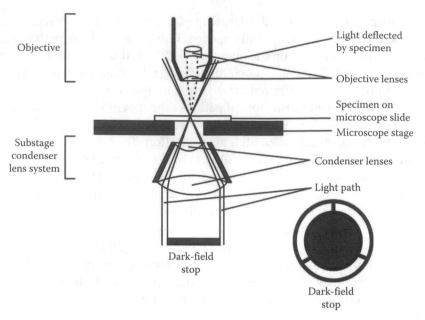

Figure 1.5   Dark ground microscopy.

microscope converts the phase differences into light intensity, producing light and dark contrasts in the image. The transparent objects show contrast where differences in refractive index or thickness occur. Many structures, such as chromatin bodies in bacteria, not usually visible in unstained organisms, are thereby revealed.

This phase difference occurs in bright field illumination, but because of the intensity of the background lighting, it is far too weak to be exploited usefully when the microscope is properly set up. Phase contrast microscopes are set up to enable microscopists to exploit the phase differences. These become translated into differences in light intensity when light passes through a specimen. Differences in phase may result in constructive interference when the object appears brighter than its surroundings, or destructive interference, leading to the production of a darker image. As the light leaves the object, the phase differences cannot be detected by the human eye (Figure 1.6).

Phase contrast microscopy has a great advantage over other optical systems in the examination of microbes. It relies simply on the physical structure of the specimen and thus living cells may be viewed with ease. Phase contrast microscopy has been of great value in the study of dynamic cellular processes such as bacterial spore formation and germination. It avoids the need to examine killed and fixed specimens.

1.7 Confocal Scanning Laser Microscopy

Confocal scanning laser microscopy (CSLM) allows for the generation of three-dimensional digital images by coupling a laser source to a light microscope. A laser beam is bounced off a mirror that directs the beam through a scanning device

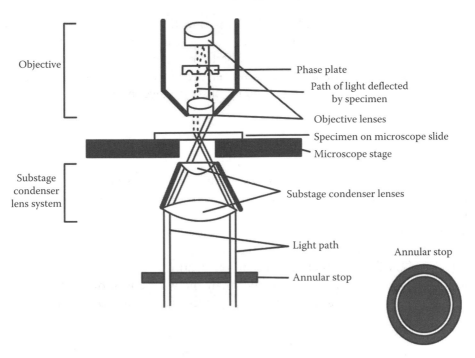

Objective

Phase plate

Path of light deflected
by specimen

Objective lenses

Specimen on microscope slide

Microscope stage

Substage
condenser
lens system

Substage condenser lenses

Light path

Annular stop

Annular stop

Figure 1.6 Phase contrast microscopy.

followed by directing the beam through a pinhole that precisely adjusts the plane of focus of the beam to a given vertical layer within a specimen. Only a single plane of the specimen is illuminated because of which the illumination intensity drops off above and below the plane of focus. Thus, in the case of a microbial biofilm, not only the cells of the surface of the biofilm are apparent, but also the cells in the various layers can be observed by using fluorescent dyes. This type of microscopy has found its vast applications in the field of microbial ecology, especially for identifying phylogenetically distinct population of cells present in a microbial habitat.

1.8 Differential Interference Contrast Microscopy

Differential interference contrast microscopy (DIC) not only reveals cell organelles but enables the cell biologist to make quantitative measurements of such things as the lipid, nucleic acid, and protein contents of the cell. DIC uses two beams of light to create high-contrast, three-dimensional images of live specimens. DIC microscopy is similar to phase contrast in that it creates an image in detecting differences in refractive indices and thicknesses. It is a type of light microscopy that provides a detailed view of unstained live specimens by employing a polarizer to produce polarized light. The polarized light then passes through a prism that generates two distinct beams. These beams traverse the specimen and enter the objective lens where they are recombined into one. An interference effect is produced, because the two beams pass through different substances with slightly different refractive indices and are not totally in phase. Thus, by DIC microscopy, three-dimensional structures of the nucleus of eukaryotic cell, endospores, vacuoles, and granules can be observed. Interference microscope has higher NAs, better contrast, and can produce colored pictures with vivid topographic relief. However, because of their expenses, phase contrast microscope is the preferred method for observing wet mounts.

1.9 Polarization Microscopy

Some cell components behave characteristically when observed with polarized light in that the light is transmitted through them with varying velocity. Such structures or materials are termed birefringent because they possess two different indices of refraction corresponding to the different velocities of light transmission. Until the electron microscope was devised and built, polarization microscopy was an important device in the indirect analysis of cell ultrastructure, since birefringence is dependent on the structural properties smaller than the wavelength of the light. In this respect, it has been particularly useful in the study of fibrillar structures such as the mitotic apparatus.

1.10 Fluorescence Microscopy

Fluorochromes are chemicals that absorb light of short wavelength and emit light of a longer wavelength. This will result in a change in the color of the light emitted by a fluorochrome. If UV light is absorbed, then visible light is emitted in the process of fluorescence. Fluorochromes may be used as fluorescent dyes to stain microorganisms. When fluorescently stained, microbes visualized in fluorescence microscopes appear as brightly colored cells against a dark background. This makes the detection of scanty cells in a stained preparation much easier than when using bright field microscopy. The detection of *Mycobacteria* stained with auramine has greatly enhanced the ability to make a provisional diagnosis of tuberculosis when only scanty *Mycobacteria* are present (Figures 1.7 and 1.8).

1.10.1 Antibodies Can Be Used to Detect Specific Molecules by Fluorescence Microscopy

Antibodies are proteins produced by the vertebrate immune system as a defense against infection. They are unique among proteins because they are made in billions of different forms, each with a different binding site that recognizes a specific target molecule (or antigen). The precise antigen specificity of antibodies makes them powerful tools for the cell biologist. When labeled with fluorescent dyes, they are invaluable for locating specific molecules in cells by fluorescence microscopy labeled with electron-dense particles such as colloidal gold spheres; they are used for similar purposes in the electron microscope.

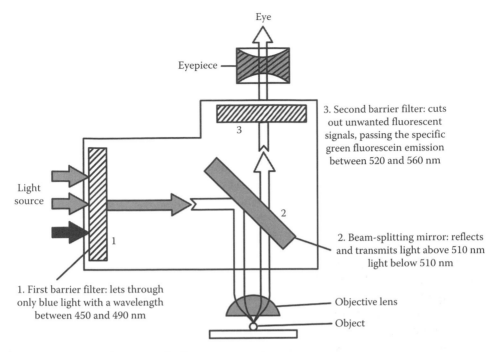

Figure 1.7   Optical system of a fluorescence microscope.

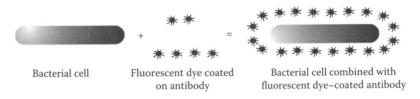

Figure 1.8   Fluorescence staining technique and microscopy.

The sensitivity of antibodies as probes for detecting and assaying specific molecules in cells and tissues is frequently enhanced by chemical methods that amplify the signal. For example, although a marker molecule such as a fluorescent dye can be linked directly to an antibody used for specific recognition, a stronger signal is achieved by using an unlabeled primary antibody and then detecting it with a group of labeled secondary antibodies that bind to it (Figure 1.9).

In addition to being used directly, fluorochromes may be attached to antibodies. Fluorescently tagged antibodies may then be used to flood a microscope preparation to locate the position of antigens. This technique is widely used to locate viruses in infected tissue specimens using antibodies raised against virus antigens. Chlamydial infections are also diagnosed in this manner, because *Chlamydia* can be cultured only with difficulty in mammalian tissue cultures. Furthermore, antibodies from one animal can be used as antigens to raise anti-antibodies in another animal, and these anti-antibodies can be fluorescently labeled. This technique is exploited in the

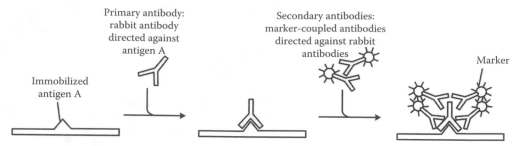

Figure 1.9   Fluorescence immunochemistry.

serological diagnosis of syphilis. Cells of *Treponema pallidum* are fixed on to a slide, and this is flooded with serum taken from the patient. If the patient has syphilis, the serum will contain antibodies that will attach to the fixed *T. pallidum* cells. After the serum is washed off, any antitreponemal antibodies will remain on the slide, attached to the bacterial cells. These antibodies can be detected using antihuman antiserum containing fluorescently labeled antihuman antibodies.

Fluorochrome dyes are not necessarily used to dye whole cells or to tag antibodies. Diamidinophenylindole (DAPI) is a fluorochrome dye that is used to determine the intracellular location of DNA. It has been used in the study of yeasts that have lost their mitochondrial DNA, to demonstrate the absence of this material.

The most sensitive amplification methods are to use an enzyme as a marker molecule attached to the secondary antibody. The enzyme alkaline phosphatase, for example, in the presence of appropriate chemicals, produces inorganic phosphate and leads to the local formation of a colored precipitate. This reveals the location of the secondary antibody that is coupled to the enzyme and hence the location of the antibody–antigen complex to which the secondary antibody is bound. Since each enzyme molecule acts catalytically to generate many thousands of molecules of the product, even tiny amounts of antigen can be detected. An enzyme-linked immunosorbent assay (ELISA) based on this principle is frequently used in medicine as a sensitive test for pregnancy or for various types of infections. Although the enzyme amplification makes enzyme-linked methods very sensitive, diffusion of the colored precipitate away from the enzyme means the spatial resolution of this method for microscopy may be limited, and fluorescent labels are usually used for the most precise optical localization.

Antibodies are made most simply by injecting a sample of the antigen several times into an animal such as a rabbit or a goat and then collecting the antibody-rich serum. This antiserum contains a heterogeneous mixture of antibodies, each produced by a different antibody-secreting cell (β-lymphocyte). The different antibodies recognize various parts of the antigen molecule, as well as impurities in the antigen preparation; the specificity of an antiserum for a particular antigen can sometimes be sharpened by removing the unwanted antibody molecules that bind to other molecules; an antiserum produced against protein X can be passed through an affinity column of antigens Y and Z to remove any contaminating anti-Y and anti-Z antibodies. Even so, the heterogeneity of such antisera sometimes limits their usefulness. This problem is largely overcome by the use of monoclonal antibodies. However, monoclonal antibodies can also have problems. Since they are single-antibody protein species, they show almost perfect specificity for a single site or epitope on the antigen, but the accessibility of the epitope, and thus the usefulness of the antibody, may depend on the specimen preparation. For example, some of the monoclonal antibodies will react only with unfixed antigens, others only after the use of particular fixatives, and still others only with proteins denatured on SDS polyacrylamide gels, and not with the proteins in their native conformation.

This detection method is very sensitive because the primary antibody is itself recognized by many molecules of the secondary antibody. The secondary antibody is covalently coupled to a marker molecule that makes it readily detectable. Commonly used marker molecules include fluorescent dyes (for fluorescence microscopy) (Figure 1.10), the enzyme horseradish peroxidase (for either conventional light microscopy or electron microscopy), colloidal gold spheres (for electron microscopy), and the enzymes alkaline phosphatase or peroxidase (for biochemical detection).

Older fluorescence microscopes were transmission instruments, set up in a fashion similar to that of bright field microscopes, but with a short wave rather than a white light source. Modern instruments employ incident light, and the microscope objective also serves as its own condenser. Modern fluorescence microscopes employ a short wave light source at right angles to the optical axis of the instrument. The light beam is directed onto the specimen by means of a dichroic mirror placed at 45° to the beam of light and within the optical axis of the instrument. This will reflect the short

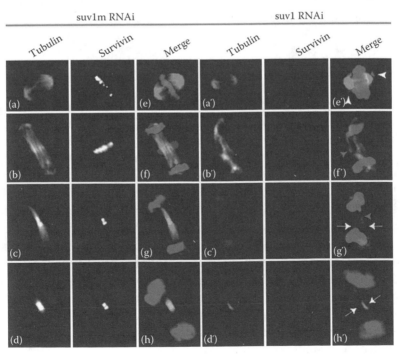

Figure 1.10 (**See color insert.**) Immunoflouresence detection of survivin in cancer tissue.

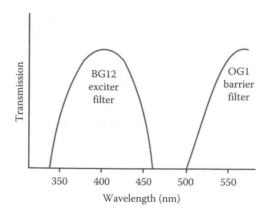

Figure 1.11 Spectral transmissions of BG12 and OG1 filters in fluorescence microscopy.

wave light onto the specimen where it may excite any fluorochrome present to emit long wave light. The dichroic mirror will allow the long wave fluorescent light to pass through to the eyepiece where it may be detected. Light from the specimen passes through a barrier filter that blocks short wave light, and allows detection of long wave light, before entering the eyepiece (Figure 1.11).

1.11 Experiments on Microscopy

Training is required to use the basic microscopes (dissecting and compound oil immersion) safely and to observe various microorganisms.

1.11.1 Introduction

Microscopes are the most important tools of any biological studies. They are needed to observe cell structures that are too small to be seen by the naked eye. Magnification and resolution are the two major steps to achieve quality observation. Most are familiar with magnification, but not all are familiar with resolution. Optic technology in the area of electron microscope allows the achievement of millions of times magnification of a small object, but these same principles limit magnification to about thousands of times in light microscope. Why are the two systems so different? Resolution is the answer to this question. The eye is the ultimate receptor of any image. Eye resolution is determined by the distance between the receptor cells in the retina. Pictures of two objects that are received on the same receptor simultaneously cannot be distinguished from each other. However, if these two pictures are received on adjacent receptors, they may be resolved. The optics of the eye and spacing of receptors on the retina make the optimum focusing distance for any eye. Regular eye can distinguish 25 cm lines as separate lines if they are 60–100 μm apart. Lines <60 μm apart are unresolved and will appear as a single solid line. Eye resolution is an individual phenomenon, and it is different from one eye to another of the same person, and from one person to another. For practical applications, 100 µm will be considered the resolving power of the human eye. Microscope magnification is required to observe an object that is smaller than 100 µm or to distinguish between two structures that are closer together than 100 µm. Microscope is also equipped to resolve mixed light rays. Achieving this mainly depends on the ability of the glass to bend light, its refractive index, and the wavelength of light used to form the image. These factors are quantified by the equation $R = 0.61\lambda/NA$ (where R is the resolution in µm, λ is the wavelength of light in µm, NA is the numerical aperture of the objective lens that includes the refractive index of the lens (η) and the sine of the lens angle (U): $NA = n * \sin U$).

Example:

Using blue light, $\lambda = 45$ µm.

The 45× lens on the microscope (color coded yellow), NA = 0.66.

$R = (0.61 \times 0.45)/0.66 = 0.42$ µm resolution.

The smallest object that can be observed in this lens is 0.42 µm. Microscope magnification is the product of the ocular lens (10×) and the objective lens (45×) or 450× together. In the previous example, the 0.42 µm object will be magnified 450× and will be appeared as $450 \times 0.42 = 190$ µm in size (Figure 1.1).

It is important to know that lens resolution is limited by the refractive index of air space between the objective lens and the cover glass. The materials commonly found in this region and their refractive indices (RI) are listed as follows:

Subject	RI
Air	1.00
Water	1.33
Glass	1.50

Air space that is filled with liquid of higher refractive index will have a large effect on NA. Some specific lens is designed to focus with a specific medium (water immersion or oil immersion) in contact with the lens surface. The degree to which light is bent depends on the ratio between the two materials. Achieving best resolution depends mainly on proper microscope setting, proper adjustment of stage condenser, and proper use of oil immersion with the right lens. Clean lenses and thin specimens are also required to achieve good resolution.

1.11.2 Microscope Operation

A microscope is an expensive and delicate instrument. *Always use two hands to carry* the microscope as to grab the microscope arm with one hand and support the microscope base with the other hand.

Gently remove the microscope from its cabinet. Plug the microscope cord into the electrical outlet on your table. Adjust light setting to give reasonable light illumination. Higher settings shorten the bulb life; bulb will burn out in 30 min. Recent microscopes have a revolving head to change the direction of the ocular lens unit. Find the proper interpupillary distance by adjusting the oculars with both hands until a single image is seen.

Locate the coarse and fine focusing knobs and rotate them back and forth gently while observing the movement of the objective lenses. Rotate the objective lens and notice that there is a positive click into position for each lens. Look at the markings on each lens, colored rings, NA, and magnification. Secure the slide on stage with slide clips. Adjust the condenser under the stage to achieve good resolution. Put the 4× objective into position and adjust the condenser position to see the light coverage in the field of view. Put the 10× objective into position and do adjustments for a better view. Move the iris diaphragm lever back and forth for the best amount of light to pass through. Microscope oculars are focused independently of each other. Close your left eye and bring the object on focus with the fine adjustments. Then close the right eye and focus the left ocular by turning the focusing ring on the left ocular tube. Recheck the right eye for proper focus. This should be a regular practice before any microscope use. When using oil immersion lens, use oil of 1.5 RI; then gradually move from 4×, 10×, 45×, and finally 100×. Excessive oil will obscure large portions of the slide from further examination at lower powers.

1.11.3 Smear Preparation

The success of a staining procedure depends upon the preparation of a suitable smear of the organisms. A good smear is one that, when dried, appears as a thin, whitish layer or film. A properly prepared smear involves one or more washings during staining without loss of organisms. The first step in preparing a bacterial smear differs according to the source of the organisms; those made from broth cultures or cultures from a solid medium require variations in technique (Figure 1.12).

A sample for light microscopy is generally prepared by mounting a sample on a suitable glass slide that is placed on the stage between the condenser and the objective lens. The way in which a mount is prepared depends upon the condition of the sample, either in living or preserved state; the objectives of the biologists, whether to examine overall structure, identifying the cell structure, or for observing the motility of the microorganism; and the type of microscopy available, whether it is bright field, dark field, phase contrast, or fluorescence microscopy. Live samples of microorganisms are placed in wet mounts or in hanging drop mounts so that they can be observed as near to their natural state as possible. The cells are suspended in a suitable fluid (sterile water, broth, saline) that temporarily maintains viability and provides space and a medium for locomotion. A wet mount consists of a drop or two of the culture placed on a slide and overlaid with a cover glass. The wet mount has certain disadvantages. The cover glass can damage larger cells, and the slide is very susceptible to drying and can contaminate the handlers' fingers. The alternative for this method is the hanging drop preparation made with a special concave (depression) slide, a vaseline adhesive or sealant, and a coverslip from which a tiny drop of the sample is suspended. These types of short-term mounts provide a true assessment of the size, shape, arrangement, color,

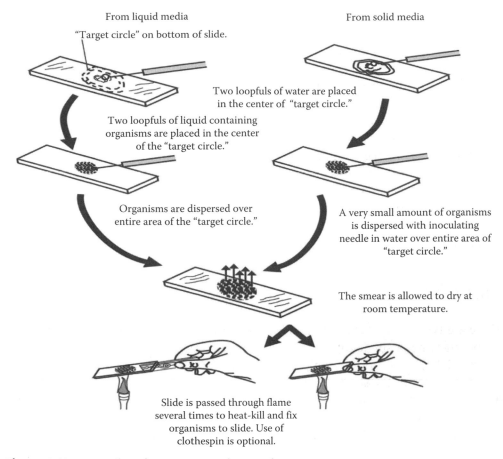

From liquid media

"Target circle" on bottom of slide.

From solid media

Two loopfuls of water are placed
in the center of "target circle."

Two loopfuls of liquid containing
organisms are placed in the center
of the "target circle."

Organisms are dispersed over
entire area of the "target circle."

A very small amount of organisms
is dispersed with inoculating
needle in water over entire area of
"target circle."

The smear is allowed to dry at
room temperature.

Slide is passed through flame
several times to heat-kill and fix
organisms to slide. Use of
clothespin is optional.

Figure 1.12   Procedure for preparing a bacterial smear.

and motility of cells. A more permanent mount for long-term study can be obtained by preparing fixed, stained samples. The smear technique, developed by Robert Koch more than 100 years ago, involves spreading a thin film made from a liquid suspension of cells on a slide and air-drying it. The air-dried smear (thin film of the culture/microorganism is called a smear) is usually heated gently by a process called heat fixation. Heat fixing denatures bacterial proteins (or enzymes), prevents autolysis, and also enhances adherence of bacteria to the slide. Fixation also permits preservation of various cellular components in their natural state with minimal distortion. Fixation of some microorganisms is performed with alcohol and formaline.

Some of the microorganisms cannot be observed properly under light microscopes, because they are transparent, practically colorless, and therefore, difficult to see when suspended in an aqueous medium. Microorganisms, are, therefore, routinely stained to increase visibility and to reveal additional information to aid identifying microbes. The chemical substances commonly employed to stain bacteria are called dyes. Dyes are classified as natural and synthetic. The former are used mainly for histological purposes while the latter are used mainly for bacterial stain preparation. These are coal tar (aniline) dyes.

1.11.4 Dyes

Chemically, a dye (stain) is defined as an organic compound containing a benzene ring plus a chromophore and an auxochrome group. Such dyes are acidic, basic, or neutral. The acidic dyes (e.g., picric acid, rose bengal, acid fuchsin, eosin) possess negatively charged groups and stain the cytoplasmic components of the cells which have a positive charge. On the other hand, the basic dyes (e.g. methylene blue, crystal violet, safranin) possess positively charged groups and combine with those cellular elements which are negatively charged (e.g., nucleic acids). Neutral stains are formed by mixing aqueous solutions of certain acidic and basic dyes together. The coloring matter in neutral stains is present in both negatively and positively charged components. Staining solutions are prepared by dissolving a particular stain in either distilled water or alcohol. The stain is applied to smears for 30–60 s (as per the staining procedure employed), washed, dried, and examined under the microscope.

1.11.5 Staining Procedure

There are two kinds of staining procedures, simple and differential (Figure 1.13). In simple staining, a single dye is used to stain microorganisms (e.g., methylene blue, crystal violet, or corbol fuchsin), and the cells and structures within each cell will attain the color of the stain. Differential staining procedures require more than one dye, and they are used to distinguish between structures within a cell or types of cells by staining them with different colors. These are used for distinguishing between microbial groups or different types of living and dead tissues.

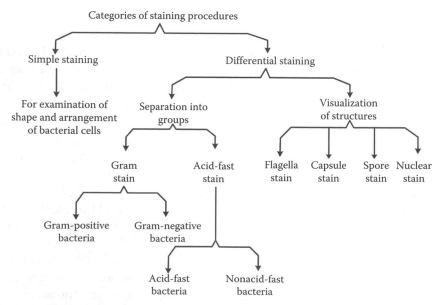

Figure 1.13   Stains used for bacterial staining.

For rapid and routine examination of almost all types of fungi, the material to be examined (usually spores and spore-bearing structures) is teased out on a clean slide in a drop of mounting fluid, with the help of two clean needles, and a coverslip placed over it, and the preparation is then ready for microscopic examination.

Glycerine jelly, lactophenol, lactic acid, Necol, and Melzer's iodine are the commonly used mounting media for fungi. Stains used for visualizing the morphology of fungi include cotton blue (or trypan blue) in lactophenol, India ink, erythrosine, lactofuchsin, periodic acid–Schiff, nigrosin, thionin, and orange. Giemsa stain is used for staining fungal nuclei, whereas, cresyl blue stain is used for spore walls of basidiomycetes.

Two basic types of staining techniques are being employed depending upon how a dye reacts with the specimen: negative and positive staining. Positive staining involves sticking of the dye to the specimen and giving it a color. In negative staining technique, a simple stain is used that stains the background, leaving the test organism unstained, and thereby the stain is termed as a negative stain. An acidic stain like nigrosin or India ink is employed in this type of staining. The acidic stain carries a negative charge, which is repelled by the bacteria that too carry a negative charge on their surfaces; therefore, bacterial cell appears transparent and unstained upon examination. Negative staining is advantageous for two reasons: microbial cells appear less shriveled or distorted as heat fixation is not carried out, and capsulated bacteria that are difficult to stain can also be observed by this technique.

1.11.6 Microscopic Examination of Microorganisms

Bacteria: Bacteria can either be stained through simple or differential staining techniques as described earlier (Figure 15.17).

Fungi: Place the material to be examined onto a clean glass microscopic slide. Add a drop of lactophenol to the material and mix. Place a cover glass over the preparation, and observe under the appropriate microscope (Figure 15.16).

Algae and cyanobacteria: Place the material to be examined onto a clean glass microscopic slide. Place a coverslip over the smear preparation in simple water or 10% glycerol and observe under the appropriate microscope (Figure 16.15).

1.11.7 Hanging Drop Technique

Hanging drop technique enables viewing of the size, shape, arrangement, and motility of live microorganisms in fluid media. It requires the use of special ground slides (Figure 15.25).

In this technique, a loopful of bacterial suspension is placed in the center of a coverslip. In the four corners, tiny droplets of mineral oil are placed. The hollow ground slide is placed over the coverslip with the depression side down, and the slide is inverted quickly so that the water cannot run off to one side. However, the lack of contrast yields limited though valuable information.

1.11.8 Hemocytometer

The hemocytometer devised originally for counting hemocytes can be used for counting the number of bacteria, fungal spores, etc., in a given volume of sample. The hemocytometer is a glass slide with the central area partitioned off by ridges into regular cubicle chambers of exactly known volume (Figure 1.14). By counting the individual cells in each chamber under a microscope and adding them up, the number of organisms living and dead may be computed.

1.11.9 Ocular Meter and Stage Micrometer for Micrometry

This is a technique in which the microscope is calibrated so that the size of objects being viewed can be found out. It involves the use of an ocular meter or eyepiece micrometer and a stage micrometer or object micrometer (Figure 1.15).
The following operations are performed in sequence:

1. The eyepiece is removed from the microscope and ocular meter inserted between the lens and the diaphragm.
2. The stage micrometer is viewed through this eyepiece.
3. The number of divisions of the eyepiece and stage micrometer is noted. Say, x divisions of stage micrometer = y divisions of ocular meter; x/y division of stage micrometer = 1 division of ocular meter; x/y is called the least count.
4. The stage micrometer is then replaced by the slide with the specimen.

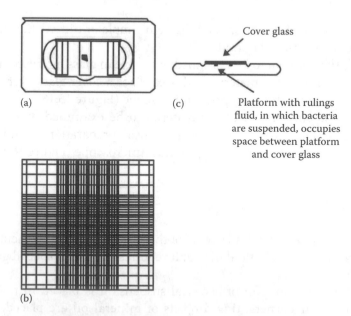

(a)

(c)

Cover glass

Platform with rulings
fluid, in which bacteria
are suspended, occupies
space between platform
and cover glass

(b)

Figure 1.14 A hemocytometer adapted for counting bacteria and other microorganisms: (a) plane view showing dark central square covered by ruled chambers which are seen enlarged in (b) and (c) vertical section with cover glass in place.

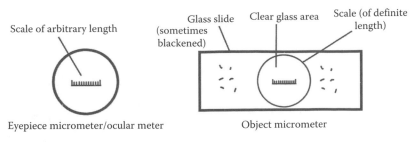

Figure 1.15 Ocular meter and stage micrometer.

5. The number of divisions of the ocular meter which is equal to the length of an object to be measured is observed.
6. This value is then multiplied by the least count to give the size of the object.

1.12 Electron Microscopy

The relationship between the limit of resolution and the wavelength of the illuminating radiation holds true for any form of radiation, whether it is a beam of light or a beam of electrons. With electrons, however, the limit of resolution can be made very small. The wavelength of an electron decreases as its velocity increases. In an electron microscope with an accelerating voltage of 100,000 V, the wavelength of an electron is 0.004 nm. Theoretically, the resolution of such a microscope should be about 0.002 nm, which is 10,000 times that of the light microscope. Because the aberrations of an electron lens are considerably harder to correct than those of a glass lens, however, the practical resolving power of most modern electron microscopes is, at best, 0.1 nm (1 Å). This is because only the very center of the electron lenses can be used, and the effective NA is tiny. Furthermore, problems of specimen preparation, contrast, and radiation damage have generally limited the normal effective resolution for biological objects to 2 nm (20 Å).

1.12.1 Transmission Electron Microscopy

Transmission electron microscopy (TEM) was the first system of electron microscopy to be developed. It operates on theoretical principles identical to those exploited in light microscopy, and is used to visualize objects that are too small to be seen using light microscopes. Viruses fall into this category because of the limitations of resolution imposed by the wavelength of visible light. This restriction is overcome if an electron beam is used in place of visible light, since electron beams have an extremely short wavelength. Using electron microscopy, it is possible to resolve points only 5 Å apart. This is equivalent to a distance of 0.5 nm. In electron microscopy, magnets are used to focus the electron beam rather than the glass lenses used in light microscopy, and the dyes used are electron-dense rather than colored compounds. Heavy metal ions are used to absorb electrons in electron microscopic preparations. Typically, these are provided in the form of lead citrate or uranyl acetate. Osmium tetroxide is used in electron microscopy as a fixative, and the osmium ions also serve

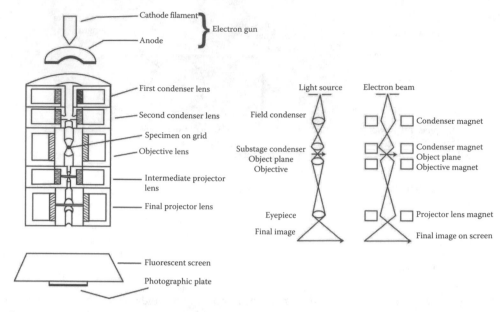

Figure 1.16   TEM allows the visualization of the fine details of the microbial cell.

as dye since they are electron-dense. The electron microscope differs from the optical microscope in using a beam of electrons instead of light rays. Glass is opaque to electrons, and the focusing of the electron beam is brought about by circular electromagnets (analogous to the glass lenses of the light microscope). The focus varies with the strength of the magnetic field applied. The object, by scattering the electrons, produces an image which is focused onto a fluorescent viewing screen. The wavelength of the electrons used in the electron microscope is approximately 0.05 Å (compared with approximately 5000 Å for visible light). In practice, at the present stage of technical proficiency, it is possible to resolve particles as small as 1–2 Å; 1 Angstrom (Å) = 0.1 mμm = 0.1 nm = 10^{-10} m.

TEMs are in many respects analogous to light microscopes (Figure 1.16). They require the electron beam to pass through a vacuum. Since gas molecules scatter the electron beam, it is necessary with the instruments at present available to examine the object *in vacuo*, and hence only killed organisms can be studied at present; these are subjected to fixation, and dried during their preparation; so shrinkage and other artifacts must always be considered. If this is not done, molecules in the air deflect the electron beam, and a sharp image cannot be obtained. Some of the first techniques for preparing biological specimens for examination in the electron microscope involved the use of whole, fixed organisms dried down onto grids and shadowed with a heavy-metal alloy such as gold–palladium. Later, improvements were made in the fixation techniques such that finer detail could be resolved in ultrathin sections of material embedded in epoxy resin and stained with heavy-metal stains such as uranyl acetate and lead citrate. Extremely thin specimens must be used for TEM studies. These may be prepared in a number of different ways. Although TEM produces stunning and often very beautiful pictures, the techniques of specimen

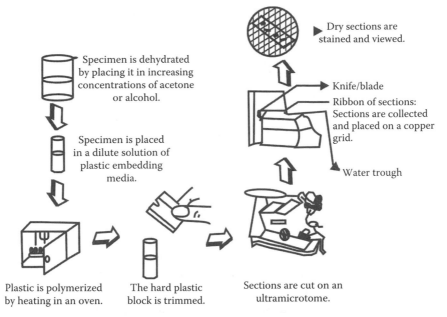

Specimen is dehydrated by placing it in increasing concentrations of acetone or alcohol.

Specimen is placed in a dilute solution of plastic embedding media.

Dry sections are stained and viewed.

Knife/blade

Ribbon of sections: Sections are collected and placed on a copper grid.

Water trough

Plastic is polymerized by heating in an oven.

The hard plastic block is trimmed.

Sections are cut on an ultramicrotome.

Figure 1.17   Preparation of a specimen for viewing by TEM.

preparation employ very harsh conditions, and this may cause considerable damage to specimens. Consequently, the images produced are prone to artifacts, and these make the interpretation of electron microscope images very difficult.

In ultrathin sectioning, staining samples are embedded in a plastic resin and cut into extremely fine sections using a specially prepared glass or diamond knife in an apparatus known as an ultramicrotome. These ultrathin sections are then mounted on a copper grid prior to examination. Ultrathin sections allow the internal structures of cells to be examined. Ferritin-labeled antibodies may be used to detect specific antigens in ultrathin sections, in a process analogous to immune fluorescent microscopy (Figure 1.17).

Negative staining exploits the electron-dense properties of compounds such as phosphotungstic acid (PTA). The heavy-metal ions do not penetrate the specimen, but form thick deposits upon the background material and in crevices within the specimen, enhancing the contrast between these structures and the object to be visualized. This method is routinely used for the detection of viruses in clinical specimens and is useful for studying the external appendages of bacteria, for example, flagella.

Shadowing is a technique used to display the surface topology of a structure. Specimens are mounted on a grid and shadowed with a thin layer of a metal such as platinum or gold. The shower of metal particles is directed at the specimen at an oblique angle so that only the surface facing the shower will be coated. This causes a shadow to be cast in the lee of the particles. This technique is used to study the external structures of viruses. The topology of plasmid DNA is also studied using shadowing. DNA preparations are first coated with a layer of protein to make the molecules easy to be visualized. The cost of the metals used for this preparation limits

its routine use. Attempts have been made to overcome the disadvantages of chemical fixation and dehydration, which, as already pointed out, may cause a certain amount of structural change in the specimen, by the use of freeze-etching. This method involves the deep freezing of specimens in a liquid gas and the subsequent formation of carbon–platinum replicas of fractured surfaces of the material. Since cells frozen in this way remain viable, it is claimed that freeze-etching enables the investigator to observe the ultrastructure as it appears in the living condition.

In freeze-etching, rapidly frozen samples have their superficial layers fractured. A carbon replica is produced of the exposed sample surface, and the replica is then examined. Fracturing occurs along the weakest planes within a structure and frequently along the surface of membranes. Freeze-etching is used to study the internal structures of microbes. The internal structures are not exposed to the distortions caused by other fixative methods, but the results are often more difficult to interpret.

The examination of microorganisms in the electron microscope has also been facilitated by the use of the negative staining technique. In this technique, the particles to be examined are embedded in an electron-dense material, for example, PTA. PTA not only outlines the general shape of the particle but also delineates details of fine structure on the surface. This simple technique has been used very successfully in revealing structures almost at the molecular level, especially in the study of viruses.

A great deal of information has been obtained from observations made on the surfaces of whole or fragmented cells with the scanning electron microscope (SEM). Here, the image is formed by collecting the electrons that are scattered or reflected from the specimen surface which is usually coated with a suitable conductor. The instrument has proved particularly useful in the investigation of such things as fungal spores and diatom cell walls.

1.12.2 Scanning Electron Microscopy

A more recent development is that of SEM. An electron beam is used to scan the surface of a specimen in a manner somewhat similar to the scanning of a television tube. This causes the surface of the specimen to emit secondary electrons and other types of radiation at various angles, producing a three-dimensional image. The emitted radiation is detected in a scintillator, and the signal is amplified to produce an image on a cathode ray tube or other recording device. The closer a surface is to the electron beam, the more the emission of secondary radiation, and the brighter its final image. SEM is used to produce striking images of the surfaces of structures, although it does lack the very fine resolution associated with TEM (Figure 1.18).

Autoradiography, biochemical analysis, fractional centrifugation, and x-ray diffraction used in conjunction with the electron microscope have also yielded a wealth of information on all aspects of cell ultrastructure.

The recent development of very high-voltage electron microscopes (operating at voltages up to 1.5 MV compared with the usual 50–200 kV) may eventually make it practicable to examine living organisms. The use of these high voltages greatly increases the penetrating power of the electron beam, and preliminary

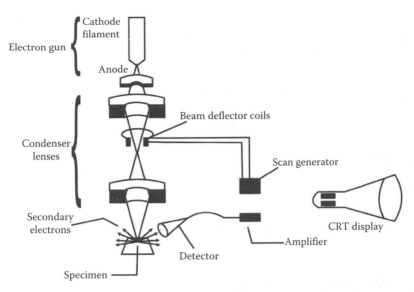

Figure 1.18   SEM is used for viewing surface structures and their three-dimensional spatial relationships.

observations have been made on intact, hydrated bacteria held in a special thin-walled double chamber.

Despite the great advantages of tremendous magnification and resolution, there are several limitations to electron microscopy. The electron microscope operates under vacuum and, therefore, the biological specimen cannot be examined in the living stage. In addition, the drying process may alter some morphological characteristics. Another limitation of the technique is the low penetration power of the electron beam, necessitating the use of thin sections to reveal the internal structures of the cell.

1.12.3 Scanning Tunneling Microscope

Although light and electron microscopes have become quite sophisticated and reached an advanced stage of development, powerful new microscopes are still being created. The scanning tunneling microscope (STM) invented in 1980 is an excellent example. It can achieve magnifications of 100 million and allow scientists to view atoms on the surface of the solid. The electrons surrounding surface atoms tunnel or project out from the surface boundary in a very short distance. STM has a needle-like probe with a point so sharp that often there is only one atom at its tip. The probe is lowered toward the specimen surface until its electron cloud just touches that of the surface atoms. If a small voltage is applied between the tip and the specimen, electrons flow through a narrow channel in the electron clouds. This tunneling current, as it is called, is extraordinarily sensitive to distance and will decrease about a 1000-fold if the probe is moved away from the surface by a distance equivalent to the diameter of an atom.

The arrangement of atoms on the specimen surface is determined by moving the probe tip back and forth over the surface while keeping it at a constant height by adjusting the probe distance to maintain a steady tunneling current. As the tip moves up and down while following the surface contours, its motion is recorded and analyzed by a computer to create an accurate three-dimensional image of the surface atoms. The surface map can be displayed on a computer screen or plotted on paper. The resolution is so great that individual atoms are easily observed. The microscope's inventors Gerd Binnig and Heinrich Rohrer shared the 1986 Noble Prize in Physics for their work. Interestingly, the other recipient of the prize was Ernst Ruska, the designer of the first TEM.

STM will likely have a major impact in biology. Recently, it has been used to view DNA directly. Since the microscope can examine objects when they are immersed in water, it may be particularly useful in studying biological molecules.

1.13 Sample Preparation for Electron Microscopy

The limit of resolution of the light microscope is roughly 2000 Å, which is insufficient to visualize cell organelles, viruses, and macromolecules of current interest. This is possible, however, with the electron microscope for which the limit is less than the diameter of a uranium atom (approximately 5 Å) under special conditions. In electron microscopy, the light is replaced by an electron beam, the sample holder is a wire screen called a grid, and the lenses are electromagnetic rather than glass. A magnetic field is generated by windings of copper wire. Lenses of short wave focal length are needed for high resolution, and the focal length decreases by increasing the magnetic field. The magnetic field could be made larger by increasing the current in the windings, but this would generate a considerable amount of heat. Instead, the field is concentrated by enclosing the wires in a soft iron casing and by inserting two conical pieces of soft iron called pole pieces, each of which contains a small orifice through which the beam passes.

The illumination source consists of a white-hot tungsten filament, which emits electrons. The potential of the anode to which electrons are drawn is normally from 40 to 100 kV greater than that of the filament. The filament and the anode together constitute the electron gun. The anode contains a small orifice through which some of the fastest electrons pass. The hole and a small aperture just below it collimate the electrons to form a beam. The beam is slightly divergent because the electrons are deflected toward the edge of the orifice of the anode owing to its positive potential. The divergent beam is then made to converge onto the specimen by an electromagnetic condenser lens. The beam is rarely focused sharply on the sample, because an intense beam could destroy it.

The image is formed by the subtractive action of the sample, that is, some of the electrons are scattered from the atoms of the object. The pattern of this loss of electrons generates the image pattern (in much the same way that the light intensity is reduced by an absorbing object in the light microscope). The objective lens, which is adjusted so that the sample is precisely at its focal point, then refocuses the beam to produce an image. This image is then magnified in several stages by three electro-magnetic lenses called the diffraction, intermediate, and projector lenses. The final projector lens forms the image on either a fluorescent screen or a photographic plate.

The electron microscope differs in three respects from the light microscope. First, because electrons do not travel very far in the air, the entire microscope column must be in a high vacuum; hence, the object must always be dry. If it were not dry, the water in the sample would boil in the vacuum of the column. This requirement for dryness means that an object cannot be alive. Second, because the magnification of an electromagnetic lens is proportional to the magnetic field, which in turn is proportional to the current in the windings, the magnification can be varied continuously by varying the current through the windings of the lens. In light optics, magnification is fixed by the set shape of the glass lens; hence, many objectives are needed to cover a range of magnification. Third, none of the primary aberrations can be corrected in standard electromagnetic lenses, because the magnetic lenses are always convergent. To reduce spherical aberration and thereby improve image quality, the lenses are operated at very small NAs. The resolution is approximately 500 times as great as that obtained with a light microscope.

1.13.1 Preparation of Specimen Supports

The scattering electrons require that a specimen must be sufficiently thin, otherwise, no beam will get through to form an image. In practice, the maximum thickness is approximately 0.1 μm (1000 Å) for 100 Å resolution, and from approximately 50 to 100 Å for 10 Å resolution for observing viruses, fibrils, or macromolecules, but for most cells which range from 1 to 50 μm in thickness, it is necessary to make thin sections. This requirement clearly means that the sample support (equivalent of a microscopic slide for a light microscope) must also be very thin, uniform in thickness, and without obvious structure at high magnification.

The specimen support or grid for all samples consists of a disk cut from a rigid copper (in some cases, platinum) mesh with openings approximately 75 μm per side, overlaid with a thin "electron-transparent" film called the support film. Electron transparency indicates that its electron-scattering power is both low and uniform. Commonly used films consist of layers from 100 to 200 Å thick of either carbon or various plastics. Unfortunately, no support film is truly structureless, and for high-resolution work with macromolecules whose dimensions are comparable to film thickness, variations in the intensity of the background produce a "granularity," the grains ranging from 5 to 10 Å. This seriously limits the attainable resolution. Films are prepared in one of the following three ways (Figure 1.19):

1. Parlodion films are prepared by placing a drop of a solution of Parlodion in amyl acetate on a water surface (a liquid surface is used, because it is very smooth). The droplet spreads, the solvent evaporates, and a thin film of plastic is formed.
2. Formvar films are prepared by dipping a smooth glass microscope slide into a solution of plastic and then removing it. When dry, the thin film on the glass will slip onto the water surface if the slide is slowly lowered into the water.

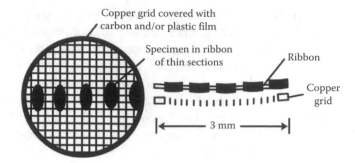

Figure 1.19   A specimen support or grid, which usually consists of a fine copper mesh (the grid itself) overlaid with a thin film of plastic or carbon (the support film). The sample rests on the support film.

3. Carbon films are prepared by evaporating onto freshly cleaved mica (which is a molecularly smooth surface, being a single plane of a crystal), and floating the film onto a water surface as is done in preparing Formvar films.

In all these cases, the film is mounted on the grid in either of the two ways: (i) it can be lowered onto the grids which have been placed on the bottom of the container beforehand by draining the water off, or (ii) the grids can be placed on the film from above and the entire support picked up by touching the surface with a sheet of plastic or absorbant paper. It should be noticed that there is an inherent difference between methods 1 and 2 for attaching the film to the grid. In method 1, the sample is ultimately placed on the side of the film that faces the air; in method 2, the sample is on the water side of the film.

Method 1

A droplet of Parlodion dissolved in isoamyl acetate is placed on a clean surface of water. The droplet spreads and the solvent evaporates leaving a thin film of the plastic (Figure 1.20).

Method 2

A glass microscope slide is coated with a thin film of either Parlodion or Formvar by dipping the slide into a solution of the plastic considerably less concentrated than the solution used with the droplet method.

In some cases, the film is a thin layer of carbon that has been evaporated onto the glass or onto a mica sheet. When the slide is lowered into the water, the film comes off the glass and floats on the water. Several grids are then placed on the surface of the film, and a piece of absorbent paper is placed on the grids. When the paper is lifting up, the grids will adhere to it. An alternative method is to use a vessel with a bottom drain. The grids are placed on a screen platform under the water surface before forming the film. The film is then formed, and the vessel is drained. As the water level drops, the film comes in contact with the grids.

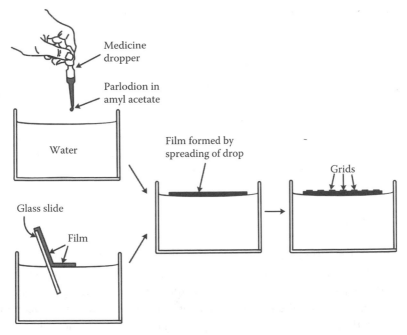

Figure 1.20   Preparation of plastic support film.

The support films are hydrophobic. Since most biological macromolecules are hydrophilic, it is often difficult to transfer the sample from the solution to the support film. Various means of sample preparations are used that overcome this difficulty. A useful alternative is to alter the support film so that it becomes hydrophilic. The most important are the glow discharge and the polylysine procedures.

1.13.1.1 Glow Discharge
The grid holding the support film is placed in a chamber having a partial vacuum, and a high voltage is applied between two metal plates. The air molecules become ionized and produce a glow discharge as in fluorescent lamps. The support film is exposed to the glowing plasma and gradually acquires a charge that is retained for about an hour. The charge film is able to draw proteins from the solution and bind the protein molecule tightly. The charge films also bind nucleic acids but rather inefficiently. It may be the case that the charge predominantly has a negative sign (nucleic acids are negatively charged), because negatively charged proteins are also poorly bound.

1.13.1.2 Polylysine Procedure
The film is glow-discharged and then a droplet of synthetic polypeptide polylysine is placed on each support film for about 30 s. The positively charged polylysine molecule binds tightly. The liquid is then removed, leaving a monomolecular film of polylysine on the support film. The support film in this way acquires a permanent positive charge and can adsorb nucleic acid very strongly.

1.13.2 Sample Preparation and Contrast Enhancement

The intrinsic contrast of biological material is poor because the scattering of the carbon atoms in the support film is of roughly the same magnitude as that of all the principal atoms (C, N, P, O, S) of the material. The usual method of correcting this situation is to deposit heavy metals of very high scattering power on the structure in such a way that the pattern of metal somehow indicates the features of the sample. Useful methods are osmium, platinum, lead, and uranium, although chromium, palladium, tungsten, and gold are sometimes used. Several standard methods for sample preparations and contrast enhancement follow.

1.13.3 Embedding, Sectioning, and Staining

Specimen must be sufficiently thin to be transparent and must possess sufficient contrast to permit the resolution of structural detail. Thinness may be an intrinsic property of the object to be examined. The sample must be made rigid so that it can be cleanly cut. This process, called embedding, consists of the gradual replacement of the aqueous material of the sample with an organic monomer (e.g., methyl methacrylate) that can be hardened by polymerization. The usual procedure is to place the sample in a solution of a fixative, which is usually a dilute formaldehyde solution or a mixture of acetic acid and ethanol. The fixative denatures and often cross-links proteins and other structures and presumably fixes all structures in place so that they will not be moved or disrupted by further handling. The "fixed" sample is then transferred to an ethanol–water mixture, with gradually increasing the concentration of ethanol until the sample is in 100% ethanol. This process is called dehydration. The sample is then transferred to an alcoholic solution of the monomer, which is then stimulated to polymerize.

After the sample has become solid, the plastic containing the supposedly undisrupted sample is sliced with an ultramicrotome (a kind of knife) into layers from 500 to 1000 Å thick. The sections are then stained (although staining is done sometime before embedding) by exposure to solutions of salts of molybdenum, tungsten, lead, or uranium, or to the vapor of osmium tetroxide (the word staining refers to the deposition of a metal by a chemical reaction or the formation of a complex with certain components of the sample, to increase the electron density). These stains react with the proteins and other macromolecules and aggregates, and thereby add electron-dense material in the sample. Stained preparations are beautiful to look at and appear to contain considerable detail, yet, it must be realized that what is being observed is the distribution of metal atoms and, therefore, of the chemical groups that can react with a particular stain. The embedding and sectioning procedures themselves can induce distortion because of uneven permeation of the sample by the organic monomer and because of the cutting itself.

Only a single layer through a sample is observed when looking at a thin section, and this may not always be adequate. To get a picture of the entire sample, a large number of sections are normally examined. An elegant though tedious method is serial sectioning in which successive sections are collected in sequence and examined.

1.13.3.1 Replica Formation

The method used for observing the surface of an electron-opaque or easily destroyed specimen is called replica formation. The specimen is coated first with a thin layer of platinum and then with a supporting layer of carbon (for strength), both deposited by vacuum evaporation or shadow-casting (Figure 1.21).

This bilayer is then floated off onto the water and picked up on a grid. The replica is thus a facsimile of the surface of the object. The contours are the same as those of the sample. This method is used to study the surfaces of viruses, membranes, and certain protein crystals that are immediately destroyed by the electron beam.

1.13.3.2 Freeze-Etching and the Critical Point Technique

In replica formation, the water in the sample must be removed before preparing the replica; production of a film by shadow-casting must be in vacuum. This exposes a problem; that structures usually during air drying as a result of surface tension effects change that occur during evaporation of the solvent. Freeze-etching and the critical point method avoid the production of artifacts due to drying. In freeze-etching, the sample is rapidly frozen, sectioned, or fractured, and placed in a vacuum with conditions of pressure and temperature such that the water sublimes from the surface of the sample. A replica of this surface is then prepared by evaporating platinum or carbon while it is still in the vacuum.

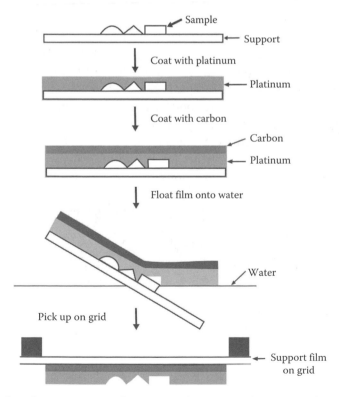

Figure 1.21 Flowchart explaining formation of a carbon–platinum replica.

The critical point method makes use of the fact that no liquid phase can exist above a "critical temperature" characteristic of each substance. The procedure follows.

First, a wet sample is soaked in ethanol; the ethanol is then exchanged with liquid CO_2 under pressure at 15°C. The temperature for the sample is then raised to 31°C (the critical temperature), and the liquid CO_2 becomes a gas. Presumably, all three-dimensional relations are preserved. A replica can then be prepared, or the sample can be observed directly in the microscope if it has been stained beforehand. This method is especially useful in preserving macromolecular structures if molecules are deposited on a film from a solution.

1.13.3.3 Freeze-Fracture

The replica method and freeze-etching have been combined in a technique that allows visualization of the internal structure of extended objects consisting of two or more layers. This method, freeze-fracture electron microscopy, is the best procedure currently available for the study of biological membranes. A sample containing membranes is frozen and then fractured by the impact of a microtome blade (Figure 1.22). Often the cleavage plane of the membrane, which consists of two layers, as shown in the figure lies along the middle of the bilayer. The ice is then sublimed away and a replica is made by successive coating with platinum and carbon (Figure 1.23).

1.13.3.4 Shadow-Casting

The size and other limited information of viruses, phages, ribosomes, and other macromolecules can be studied using shadow-casting. The particles in solution or suspension are applied by spraying of the particles onto a grid overlaid by a support film. The liquid quickly evaporates; the sample is placed in a vacuum, and a heavy metal is applied by evaporation. This requires boiling a metal. A thin metal wire is wrapped around a tungsten wire, or small lumps of metal, or a metal oxide, are placed in a tungsten wire basket. An electric current is passed through the tungsten wire until it becomes white hot. At this temperature, the metal to be evaporated boils away. The metal atoms are projected in all directions, and if the vacuum is good, in straight lines. If evaporation is from an acute angle, metal will pile up on only one side of the sample and will cover the grid except in the shadow of the particle. The amount of

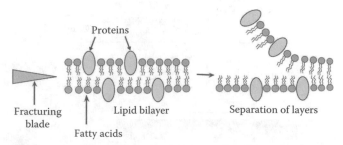

Figure 1.22　Schematic drawing of a lipid bilayer membrane being fractured.

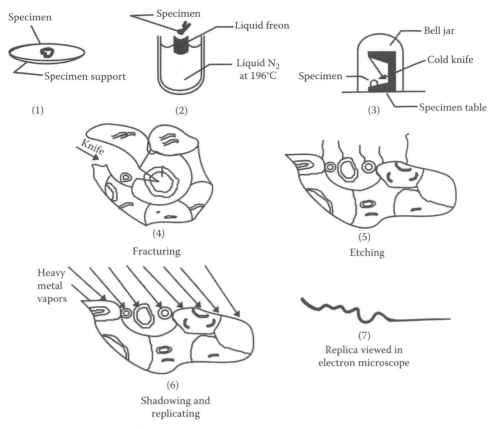

Figure 1.23 Procedure for the formation of freeze-fracture replicas used for visualizing surface structure in conjunction with TEM.

metal deposited on the sample affects the contrast. If there is too little, the sample is not visible, and if there is a great excess, the sample may be totally buried. The correct amount of metal is determined empirically by using metal wire of a particular diameter and counting the number of turns of this wire around the tungsten that, when used, results in optimal contrast. When metal oxide grains are used in a tungsten wire basket, the grains are usually weighed.

1.13.3.4.1 Protocol   The bell jar is evacuated by vacuum and diffusion pumps. The tungsten filament, around which the metal wire to be evaporated is wrapped, is heated to the boiling point of the metal. In 10 or 15 s, the metal is boiled away and forms a film on the sample. The thickness of the film is proportional to the amount of metal put on the filament and inversely proportional to the square of the distance from the filament to the sample (Figure 1.24).

Some metals, for instance, 100% platinum, may yield variable contrast. For platinum, this is a result of formation of a platinum–tungsten alloy. The platinum in the alloy does not evaporate; hence, a little metal is deposited on the sample. If a wire consisting of 80% platinum and 20% palladium is used, this problem does not occur because palladium prevents the formation of the platinum–tungsten alloy.

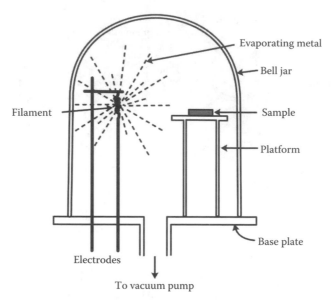

Figure 1.24 Apparatus for vacuum evaporation or shadow-casting.

The magnitude of the vacuum during shadow-casting also affects the picture quality. If the pressure is too high, a significant fraction of the metal atoms collides with air molecules and fails to reach the sample. One can, of course, compensate somewhat for this effect by increasing the amount of metal wire. This is not a particularly effective technique as most of the metal atoms evaporate as single atoms, but some of these are in large clusters. These clusters occur infrequently, but they are less likely to be deflected if they collide with an air molecule. At higher pressure, the clusters have a higher probability of reaching the sample than do individual atoms. Thus, by increasing the amount of metal evaporated when the pressure is too high, one merely enhances the fraction of molecules that arrives as clusters. The sample is then very coarsely shadowed, and most of the detail will be lost.

Shadow-casting has not been too successful for smaller macromolecules because of the small size of the shadow and the granularity of the support film. Many problems can be avoided by using the negative contrast procedure.

1.13.3.5 Negative Contrast Technique

The negative contrast (incorrectly commonly called negative staining) consists of embedding small particles or macromolecules in a continuous stain or electron-opaque film. The stain penetrates the interstices of the particle but not the particle itself. The image is the result of the relative intensity of the beam at every point, which is proportional to the thickness of the opaque material at that point. Hence, contrast is achieved by virtue of the particle, reducing the effective thickness of the opaque film, that is, the particles are seen in outline (Figure 1.25).

The sample is either mixed with the stain and sprayed on the grid or sprayed on the grid first and then sprayed with the stain. The interpretations of negative-stained samples are sometimes difficult because various patterns can be observed depending

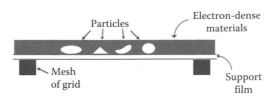

Figure 1.25 The negative contrast method.

on (1) the thickness of the stain; (2) whether it has penetrated the interstices of the particle; (3) whether it lies above or below the particle; and (4) whether any of it has adsorbed specifically to the sample (positive staining). For this reason, it is usually necessary to look at a large number of preparations and particles. Nonetheless, the negative contrast method has been used successfully for a wide variety of phages and viruses. The most useful stains for negative contrast are PTA salts and uranyl acetate, nitrate, or formate. The resolution in the negative contrast method depends on the size of the opaque atoms (i.e., approximately 5 Å).

The negative contrast procedure coupled with glow-discharged support films has become one of the most important techniques for studying the structure of individual protein molecules and the arrangement of subunits in a complex protein. A droplet containing the molecule of interest is placed on a glow-discharged carbon film. After about 30 s, at which time most of the molecules have adsorbed to the film, the liquid is removed and the grid is washed. For best results, uranyl formate, which provides an extremely fine-grained and uniform background, is used. The protein molecules are adsorbed to the film in various orientations, allowing numerous views of the molecule so that the arrangement of different parts of the molecule can be seen. Proteins have also been visualized after binding to the polylysine-coated carbon film, followed by negative staining. However, the simple glow-discharged films usually yield better micrographs.

1.13.3.6 Positive Staining
Positive staining has not had widespread use for most macromolecules, because it is not usually possible to attach sufficiently large numbers of heavy atoms to obtain good contrast, although it has been possible with large molecules and structures such as ribosomes, DNA, RNA polymerase, and collagen.

1.13.3.7 Kleinschmidt Spreading with Positive Staining and Rotary Shadowing
It is the most spectacular method of sample preparation for visualizing DNA. In a single step, artifacts due to drying are eliminated, and extraordinary contrast is obtained. This is a widely used technique nowadays. A drop of a DNA solution in 0.5–1.0 M ammonium acetate containing 0.1 mg/mL of cytochrome c is allowed to flow down a glass slide onto the surface of 0.15–0.25 M ammonium acetate. As the drop touches the surface, a film of denatured cytochrome c spreads across the surface. This film contains somewhat extended DNA molecules to which a thick (100–200Å) layer of denatured cytochrome c binds. If a grid is made to touch the denatured protein film, a drop containing a part of the film is transferred to it.

When the grid with the adhering drop is immersed in alcohol, the aqueous phase is removed and the film adheres tightly to the support film on the grid. The technique includes a preliminary positive staining with uranyl acetate; the protein absorbed to the DNA becomes stained as well as the background film, but owing to the excess protein bound to the DNA, good contrast is achieved. Contrast is enhanced (or created, if staining is not used) by shadow-casting a metal (usually platinum) at a very small angle while the sample is rotating. Because the DNA coated with protein projects above the protein film, metal piles up against the DNA–protein complex like snow drifting against a fence, but on both sides and on all molecules, regardless of orientation, because of the rotation, and the contrast is extraordinary. This method can be used to determine the length of the DNA and also whether it is circular or supercoiled. Under certain conditions (usually by incorporating denatured formamide into all solutions and keeping the salt concentration low), single-stranded polynucleotides become extended and are easily visualized. Single-stranded DNA and RNA are distinguishable from native DNA by their relative thinness and kinkiness (Figure 1.26).

The technique using formamide is called formamide spreading: the usual procedure for preparing double-stranded DNA is called an aqueous spreading. In a variation of this method, the diffusion method, a cytochrome *c* film is formed on the DNA solution, and DNA molecules diffuse upward and adhere to the film. This is a slow process but allows much smaller DNA concentrations to be used.

One important application of the Kleinschmidt method is seen in the heteroduplex method. Here, single strands from two different DNA molecules are allowed to hybridize. Homologous regions show up as double-stranded DNA, but nonhomologous regions remain as single-stranded loops. Special procedures are often needed when attempting to localize protein molecules bound to DNA molecules. This is because often the solutions needed for the spreading contain salt, detergents, or other substances that are not compatible with the stable DNA–protein interaction. Preincubation of the sample with the dialdehyde, glutaraldehyde, stabilizes the interaction. Glutaraldehyde produces protein–protein, protein–nucleic acid, and nucleic acid–nucleic acid covalent cross-links that prevent dissociation of the components. The chemistry of the reaction is not known. Glutaraldehyde is always worth trying when a sample of any kind is found to dissociate during preparation. It has also had wide use in preserving thin sections of cells and tissues of plants and animals.

A variant of the Kleinschmidt technique replaces the cytochrome *c* with benzalkonium chloride, a surfactant that also forms a film to which DNA adheres. The DNA is shadowed with platinum, but because the DNA is not coated with protein, it appears much narrower than when cytochrome *c* is used. Positive staining is still used under conditions that allow the uranium atoms to bind directly to the DNA. This procedure gives better resolution of macromolecules (e.g. RNA polymerase) bound to the DNA; however, it is not as reliable as the cytochrome *c* procedure. Probably the best method for viewing proteins bound to nucleic acids uses adsorption of the complexes to polylysine-coated support films followed by negative staining.

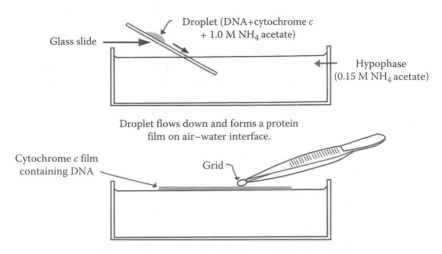

Droplet flows down and forms a protein
film on air–water interface.

Film spreads across surface. Grid is touched to film surface
so that support film on grid is in contact with protein film.

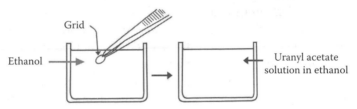

Sample is dehydrated by immersion in ethanol and
then dipped into uranyl acetate solution for staining.

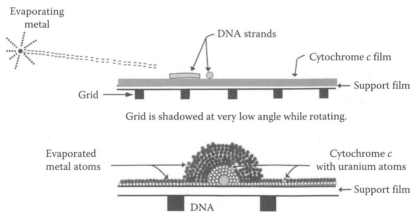

Grid is shadowed at very low angle while rotating.

Enlarged view of a DNA strand coated with
cytochrome *c*, stained and shadowed.

Figure 1.26   Kleinschmidt method for the preparation of DNA for electron microscopy.

1.13.3.8 Special Protocol Used with the Kleinschmidt Technique

There are several procedures that can be used to localize specific regions of a DNA molecule. One of the most important is denaturation mapping. When a solution of DNA molecules is heated, the increased kinetic energy causes disruption of the hydrogen bonds that are responsible for base pairing. The adenine–thymine (A–T) pairs are joined by two hydrogen bonds, whereas there are three hydrogen bonds in a guanine–cytosine (G–C) base pair. Thus, the A–T pairs, and not the G–C pairs, are broken at a lower temperature. It is possible to choose a temperature at which a few particularly $A = T$-rich sequences are totally single-stranded, and the remainder of the molecule is double-stranded. If formaldehyde is added, it reacts nearly irreversibly with the amino groups on the bases that are not hydrogen-bonded; then, the sample can be cooled without reformation of any base pairs. Such a molecule is said to be partially denatured. As an alternative to elevated temperature, high pH (plus formaldehyde) can be used to create partially denatured DNA; in fact, this procedure is used more often than high temperature, despite its requirement for controlling the pH to ±0.05 pH unit.

Partially denatured DNA can be prepared for electron microscopy by a formamide spreading. An electron micrograph of such DNA shows each denatured region as a section of DNA consisting of two separate single strands; the single-stranded regions are usually called denaturation loops or bubbles. It is found that the positions of the bubbles are nearly identical for all molecules in a sample of molecules of the same type. A histogram showing the positions obtained by measuring a large number of molecules is called a denaturation map. The important aspects of this method are that the denaturation maps of two different types of DNA molecules are never the same. Thus, a denaturation map is a unique way to identify a particular type of DNA molecule. Denaturation mapping helps in localizing the end position of the linear DNA molecule in circular form, identifying the long DNA molecule, and determines the direction of replication of DNA, etc.

1.13.4 Visualization of Nucleic Acid Protein Complexes on Polylysine Films

Polylysine procedure is considered to be superior to all other procedures because of its simplicity and reproducibility. A droplet of an aqueous solution containing a nucleic acid–protein complex is placed on a polylysine-coated carbon film. After removal of the liquid, a negative stain is added.

1.13.5 Improving the Quality of the Image and Image Reconstruction

CryoEM (cryoelectron microscopy) is a powerful tool because it permits the structures of biological samples to be preserved in a near native state. Samples are flash-frozen (vitrified) in liquid ethane in a sub-millisecond time frame, which prevents ice crystal formation that would damage the specimen. This technique uses no chemical fixatives or stains and thereby helps preserve the hydrated structure of the virus (or any macromolecular complex). Specimens are then loaded into computer-controlled, transmission electron microscopes that are able to maintain specimens at liquid nitrogen or lower temperature. These microscopes collect upward of tens of thousands or more virus images that are needed to obtain high-resolution 3D reconstructions. Such large numbers are required because each individual virus image is extremely noisy owing to radiation-induced damage to the sample caused by the high-voltage electron beam and because the contrast in unstained samples is extremely low. Computer reconstruction techniques are continuously being developed and tested to more reliably and efficiently extract usable information from the noisy image data. In favorable cases, reconstructions at sub-nanometer resolution in a day or less after the images are recorded. In addition, tools are designed to provide real time, 3D reconstruction feedback (i.e., in a few minutes or less) to the microscopist to further improve success of data collection.

1.13.6 Minimal Beam Exposure

It consists of localizing the electron beam on a small portion of the sample to allow focusing and then moving either the beam or the sample so that an unexpected region of the sample can be photographed.

1.13.7 Three-Dimensional Reconstruction of Molecular Structure from Unstained Sample

Accessories are available by which a sample can be tilted with respect to the electron beam. This makes it possible to photograph a single particle or structure at several angles and a great deal of information can be obtained. For example, a rectangular prism and/or a computer program allows reconstruction of the three-dimensional structure from micrographs taken at many angles. The technique has been used to determine the fine points of the structure of viruses and of membranes.

The principal limitation of the image reconstruction technique is that the sample is progressively damaged by repeated beam exposure. The exposure of kinetic energy of electron beam can rapidly raise the temperature by many hundred degrees which will disrupt the covalent bond and can even cause vaporization of parts of the sample. If the sample is stained or shadowed with a heavy metal, the apparent effect is much less; the biological molecule is probably mostly destroyed after a few seconds, but the pattern of heavy metal atoms remains, and it is this pattern that is observed. However, the loss of the sample certainly causes an alteration of the array of the metal atoms and considerably reduces the information obtained from the micrograph.

1.13.8 Image Rotation and Rotational Filtering

The fine details of organized arrays of the molecules in virus particles are lost in the negative contrast methods because of the minor disarray of the heavy metal atoms on the surface of the particle. When a particle has rotational symmetry, the random variation (noise) of the surface pattern can be smoothed out by image rotation. In this technique, a large number of micrographs of nearly identical objects are super-imposed; the noise will cancel, whereas the details of the structure will be enhanced. It has also been shown that it is not necessary to superimpose many micrographs but only to rotate the image about its axis of rotational symmetry and combine the views obtained at various angles of rotation. The rotational motion can also be accomplished by computer processing using rotational filtration. Here, the blackening of a micrograph is measured. A computer then performs mathematical rotation and provides a printout of a processed image.

1.13.9 Special Mechanisms of Image Formation

With increasing frequency, cellular organelles and nuclear structures are being investigated at high resolution using electron microscopic tomography of thick sections (0.3–1.0 microns). In order to reconstruct the structures in three dimensions accurately from the observed image intensities, it is essential to understand the relationship between the image intensity and the specimen mass density. The imaging of thick specimens is complicated by the large fraction of multiple scattering which gives rise to incoherent and partially coherent image components (Han et al. 1995). Nanostructures fabricated on thick substrates (typically a silicon wafer) are building blocks for high-performance electronic devices. While detailed structural analysis using high-resolution electron microscopy usually requires a thinning process to obtain an electron transparent specimen, a nondestructive approach is also required to analyze the structure. Heat dissipation into the substrate from a nanotube device governs its electronic transport characteristics. In such a system, structural analysis that can be performed without any sample modification is favorable since imaging can be performed before, after, and even during the electrical operation of devices to investigate structural change (Pop et al. 2005).

1.14 Dark Field Electron Microscopy

Dark field electron microscopy allows for substantially increased contrast, as is the case with the light microscopy. A dark field can be obtained either by using hollow-cone illumination, as in light microscopy, or by making the beam fall on the sample at an angle such that it does not enter the imaging system. The image is formed by the scattered and diffracted electrons alone. Dark field electron microscopy has not been used very much, but for the observation of macromolecules it deserves greater attention. It is especially useful if the contrast of a sample is poor, because the contrast is enhanced considerably by dark field microscopy.

1.15 Crewe Microscope

In an ordinary TEM, the incident beam covers the entire sample. As the electrons interact with the sample, they do so in several different ways: (1) essentially no interaction at all (i.e., traveling through the interatomic spaces), the most abundant class of electrons; (2) in elastically scattered (i.e., without loss of energy) by the orbital electrons of the atoms of the sample, the second most abundant; and (3) elastically scattered (i.e., without loss of energy) from atomic nuclei, the least abundant. The ratio of the last two classes is a characteristic of each element because the size of the nuclear target and hence the cross section for inelastic events increases greatly for the larger atoms.

In Crewe electron microscope, the beam is focused into a very small (approximately 5 Å) spot. The spot is swept across the sample as in a television set. As the beam moves, the ratio of the latter two classes of electrons is measured at each point by an electron-energy spectrometer. This ratio is converted into an image on a television screen by suitable electronic circuits; the information attained can also be processed by computer analysis. This important step in electron microscopy gives a new element of analysis to electron microscopy because individual atoms can be identified. In some cases, the limit of resolution is improved, and a picture at 2 Å resolution has been obtained.

1.16 Backscatter Scanning Microscope

The SEM is a device that has produced the many beautiful photographs of cell surfaces seen in the past few years. This microscope is limited to about 200 Å in resolution and operates on a very different principle from the TEM. As in the Crewe electron microscope, the beam is collimated into a small (100 Å) spot, and the spot is swept across the sample surface, which has been coated with a thick (200 Å) layer of gold or other heavy metal. As the beam impinges on the metal and penetrates a short distance into it, electrons are emitted from the gold either as secondary emission or as directly backscattered electrons from the beam. Because of an unexpected angular relationship between the number of electrons emitted and the angle of the surface to the incident beam, which is close to, but not identical with, the way that light reflects from the surface of an object, the image formed by the collected electrons gives dramatic images of the surface being examined.

1.17 Experiment on Scanning Electron Microscopy of Cyanobacteria

A. cylindrica cells are harvested by centrifugation and prefixed in culture medium by the addition of an equal volume of 1% glutaraldehyde in phosphate buffer. Cells are allowed to stand for 30 min on ice, pelleted, suspended in phosphate buffer with 2% glutaraldehyde and incubated for 1 h at room temperature. Samples are washed with phosphate buffer and postfixed in 1% osmium tetraoxide in the same buffer and

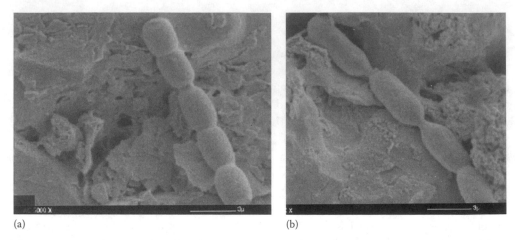

(a) (b)

Figure 1.27 Scanning electron photomicrograph of (a) NaCl-untreated and (b) NaCl (200 mM)-adapted cells of *Anabaena cylindrica*.

washed once in distilled water. After that, the samples are kept on carbon stubs, and gold coating is done with a fine coat ion sputter JFC 1100. Samples are ready to be observed with an electron microscope (JEOL JSM-840) (Figure 1.27).

1.18 Transmission Electron Microscopy of Phages (Bacteriophage and Cyanophage)

1. An aliquot of the clonal phage suspension is absorbed onto carbon-coated copper grids, stained with 2% uranium acetate, and observed at 80 kV using JEOL JEM-1010 TEM.
2. The virion size is estimated from the negatively stained images.
3. Exponentially growing host cells are collected on a 0.8 μm polycarbonate membrane filter.
4. Cells are then suspended in 10 mL of fresh medium and then inoculated with 2.5 mL of the clonal phage suspension.
5. Host cells without inoculation served as a control.
6. After incubation for an appropriate period of postinoculation, 250 μL samples are removed, mixed with 50 μL of fresh medium containing 6% agarose, and solidified at 4°C.
7. The agarose blocks with host cells are cut into 1–2 mm cubes and fixed with 1% glutaraldehyde for 7 days.
8. After the cubes are washed with phosphate-buffered saline (13 mM $NaH_2PO_4 \cdot 2H_2O$, 86.8 mM $Na_2HPO_4 \cdot 12H_2O$, 85.6 mM NaCl; pH 7.4), they are then fixed with 1% osmic acid at 40°C for 3 h.
9. After three washes with phosphate-buffered saline, the cubes were dehydrated in a graded ethanol series (50%–100%) and embedded in Quetol 653 resin.
10. Ultrathin sections are stained with 2% uranium acetate and 3% lead citrate and observed at 80 kV using a JEOL JEM-1010 TEM (Figures 1.28 through 1.31).

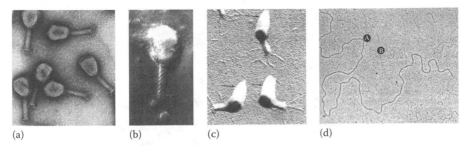

(a) (b) (c) (d)

Figure 1.28 Electron micrograph of *E. coli* bacteriophage T4 prepared by (a) negative contrast, (b) freeze-etching, (c) shadowing, and (d) Kleinschmidt methods using formide to prevent collapse of the single strands: A—double-stranded DNA and B—single-stranded DNA.

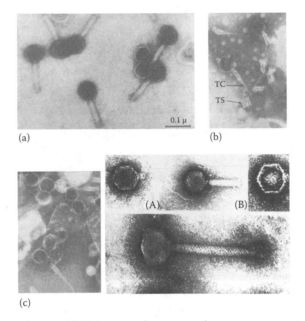

(a) (b)

(c)

Figure 1.29 Cyanophages: (a) N-1 cyanophage uranyl acetate preparation. (b–c) Phosphotungstate preparation and different types of cyanophages: (A) LPP-1 and (B) N-1 and SM-1.

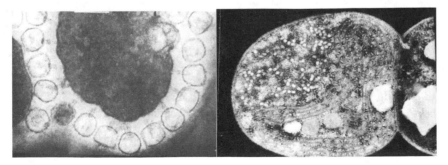

Figure 1.30 Cyanophages' attachment to the host and their development inside the host cell in LPP-1 cyanophage under electron microscopy with negative staining.

Figure 1.31 Electron microscopic view of bacterial (*Spirillum* species) polar flagella, negatively stained with phosphotungstate.

Suggested Readings

Abramowitz, M. 2003. *Microscope Basics and Beyond.* Melville, NY: Olympus of America Inc.

Alberts, B., A. Johnson, J. Lewis et al. 2002. *Molecular Biology of the Cell*, 4th edn. New York: Garland Science.

Andrews, P. D., I. S. Harper, and J. R. Swedlow. 2002. To 5D and beyond: Quantitative fluorescence microscopy in the postgenomic era. *Traffic* 3: 29–36.

Baruch, A., D. A. Jeffery, and M. Bofyo. 2004. Enzyme activity—It's all about image. *Trends Cell Biol* 14: 29–35.

Braga, P. C. and D. Ricci, eds. 2003. *Atomic Force Microscopy—Biomedical Methods and Applications.* In *Methods in Molecular Biology*, vol. 242, New York: Springer.

Crewe, A. V. 1971. A high-resolution scanning electron microscope. *Sci Am* 234: 26–35.

Davis, R. W., M. Simon, and N. G. Davidson. 1971. Electron microscope heteroduplex methods for mapping base sequence homology. In *Nucleic Acid in Methods in Enzymology*, ed. L. Grossman, vol. 21. London, U.K.: Academic Press, pp. 413–428.

Finch, J. T. 1975. Electron microscopy of proteins. In *The Protein*, eds. H. Neurath and R. L. Hill, 3rd edn. London, U.K.: Academic Press, pp. 413–497.

Fisher, H. W. and R. C. Williams. 1979. Electron microscopic visualization of nucleic acids and of their complexes with proteins. *Ann Rev Biochem* 48: 649–680.

Frankel, F. 2002. *Envisioning Science. The Design and Craft of the Science Image.* Cambridge, MA: MIT Press.

Freifelder, D. 1982. *Physical Biochemistry: Applications to Biochemistry and Molecular Biology.* New York: W.H. Freeman and Company, Direct observation, pp. 39–72.

Greenstone, A. 1968. *The Electron Microscope in Biology.* New York: St. Martin's Press.

Haggis, G. H. 1967. *The Electron Microscope in Molecular Biology.* New York: John Wiley & Sons.

Han, K. F., J. W. Sedat, and D. A. Agard. 1995. Mechanism of image formation for thick biological specimens: Exit wavefront reconstruction and electron energy loss spectroscopic imaging. *J Microsc* 178: 107–119.

Hall, C. E. 1966. *Introduction to Electron Microscopy.* New York: McGraw-Hill.

Haugland, R. 2002. *Handbook of Fluorescent Probes and Research Products*, 9th edn. Eugene, OR: Molecular Probes, Inc.

Henderson, R. and P. N. T. Unwin. 1975. Three-dimensional model of purple membrane obtained by electron microscopy. *Nature* 257: 28–32.

Hurtley, S. M. and L. Helmuth. 2003. Biological imaging. *Science* 300: 75–143.

Inoue, S. and K. Spring. 1997. *Video Microscopy, The Fundamentals*, 2nd edn. New York: Plenum Publishing Corp.

Ishijima, A. and T. Yanagida. 2001. Single molecule nanobioscience. *Trends Biochem Sci* 26: 438–444.

Kleinschmidt, A. K. 1968. Monolayer techniques in electron microscopy of nucleic acid molecules. In *Methods in Enzymology*, eds. L. Grossman and K. Moldave, vol. 12B. London, U.K.: Academic Press, pp. 361–376.

Lewis, A., H. Taha, A. Strinkovski et al. 2003. Near-field optics: From subwavelength illumination to nanometric shadowing. *Nat Biotechnol* 21: 1378–1386.

Markham, R., S. Frey, and G. J. Hills. 1963. Methods for the enhancement of image detail and accentuation of structure in electron microscopy. *Virology* 20: 88–102.

Mcintosch, J. R. 2001. Electron microscopy of cells: A new beginning for a new century. *J Cell Biol* 153: F25–F32.

Meek, G. A. 1976. *Practical Electron Microscopy for Biologists.* New York: John Wiley & Sons.

Oliver, R. M. 1973. Negative stain electron microscopy of protein macromolecules. In *Methods in Enzymology*, eds. C. H. W. Hirs and S. N. Timasheff, vol. 27. London, U.K.: Academic Press, pp. 616–672.

Paddock, S. W. 2005. Microscopy. In *Principles and Techniques of Biochemistry and Molecular Biology*, eds. K. Wilson and J. Walker, 6th edn. Cambridge, U.K.: Cambridge University Press, pp. 131–163.

Periasamy, A. 2001. *Methods in Cellular Imaging.* Oxford, U.K.: Oxford University Press.

Pop E., D. Mann, J. Cao, et al. 2005. Negative differential conductance and hot phonons in suspended nanotube molecular wires. *Phys Rev Lett* 95: 155505.

Sedgewick, J. 2002. *Quick Photoshop for Research: A Guide to Digital Imaging for PhotoShop.* New York: Kluwer Academic/Plenum Publishing.

Shapiro, H. M. 2003. *Practical Flow Cytometry*, 4th edn. New York: John Wiley & Sons.

Slayter, E. M. 1970. *Optical Methods in Biology.* New York: John Wiley & Sons.

Spector, D. L. and R. D. Goldman. 2004. *Live Cell Imaging: A Laboratory Manual.* Cold Spring Harbor, NY: Cold Spring Harbor Laboratory Press.

Van Roessel, P. and A. H. Brand. 2002. Imaging into the future: Visualizing gene expression and protein interaction with fluorescent proteins. *Nat Cell Biol* 4: E15–E20.

Wallace, W., L. H. Schaefer, and J. R. Swedlow. 2001. Working person's guide to deconvolution in light microscopy. *Bio Techniques* 31: 1076–1097.

Williams, R. C. 1977. Use of polysine for adsorption of nucleic acids and enzymes to electron microscope specimen films. *Proc Natl Acad Sci USA* 74: 2311–2315.

Williams, R. C. and H. W. Fisher. 1970. Electron microscopy of tobacco mosaic virus under conditions of minimal beam exposure. *J Mol Biol* 52: 121–123.

Wu, M. and N. Davidson. 1978. An electron microscopic method for the mapping of proteins attached to nucleic acids. *Nucl Acids Res* 5: 2825–2846.

Younghusband, H. B. and R. B. Inman. 1974. The electron microscopy of DNA. *Ann Rev Biochem* 43: 605–619.

Zhang, J., R. E. Campbell, A. Y. Ting et al. 2002. Creating new fluorescent probes for cell biology. *Nat Rev Mol Cell Biol* 3: 906–918.

Important Links

General microscopy

http://www.microscopyu.com/
http://microscope.olympus-global.com/en/ga/?ad = mic1
http://www.jenoptik.com/en-digital-cameras-for-microscopy?open&gclid = CIfJhuadt7U
    CFQd76wodIBMAhQ
http://www.microscopy-analysis.com/
http://www.microscopy.org/
http://www.rms.org.uk/
http://east.mesa.k12.co.us/subjects/science/rajnowski/documents/MicroscopyLab.pdf

Fluorescent probes

http://www.clontech.com/
http://www.invitrogen.com/site/us/en/home/brands/Molecular-Probes.html
http://www.bdbiosciences.com/
http://www.probes.zeiss.com/
http://www.jacksonimmuno.com/

Image processing

http://www.apple.com/quicktime/download/
http://rsbweb.nih.gov/ij/docs/examples/stained-sections/index.html
http://rsb.info.nih.gov/ij/developer/source/

2
Microtome

2.1 Introduction

A microtome is a sectioning instrument that allows for the cutting of a material into extremely thin slices known as sections. Microtomes are important devices in microscopy preparation, allowing for the preparation of samples for observation under transmitted light or electron radiation (Figure 2.1a). Microtomes use steel, glass, or diamond blades depending upon the specimen being sliced and the desired thickness of the sections being cut. Steel blades are used to prepare sections of animal or plant tissues for light microscopy histology. Glass knives are used to slice sections for light microscopy and to slice very thin sections for electron microscopy. Industrial grade diamond knives are used to slice hard materials such as bone, teeth, and plant matter for both light microscopy and for electron microscopy. Gem quality diamond knives are used for slicing thin sections for electron microscopy.

2.2 Sectioning

This rather simple device consists of a stationary knife-holder/blade and a specimen-holder which advances by preset intervals with each rotation of the flywheel mounted on the right-hand side. In operation, it is similar to the meat and cheese slicers found within delicatessens. A control knob adjusts the internal cams, which advance the paraffin block with each stroke. It is relatively easy to section paraffin at 10 μm (micrometers) but requires a lot of skill and practice to cut at 5 μm. Since each section comes off of the block serially, it is possible to align all of the sections on a microscope slide and produce a serial section from one end of a tissue to the other (Figure 2.1b).

While virtually anyone can cut a section within minutes of being introduced to the microtome, proper use of the microtome is an art and requires practice and inventiveness. Many a cell biology research project has depended on the skills inherent in the use of this instrument. A microscope is nearly useless without a good, thin, flat, and undistorted section from properly fixed, dehydrated, and embedded tissue.

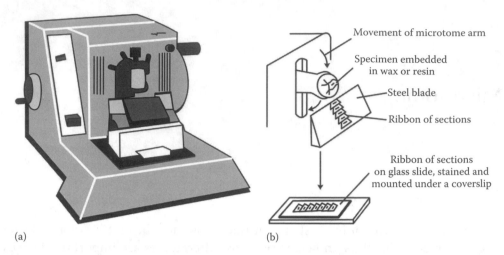

(a)

(b)

Figure 2.1 Illustration showing (a) microtome and (b) embedding and sectioning of preparation with a microtome for examination in the light microscope.

2.3 Traditional Histology

In this technique, tissues are hardened by replacing water with paraffin and then cut in the microtome at varying thicknesses from 2 to 50 µm thick. The tissue sections can be mounted on a microscope slide, stained with appropriate aqueous dye(s) after prior removal of the paraffin, and examined using a light microscope.

2.3.1 Fixation, Dehydration, Embedding, and Staining

The processing involves a series of steps: fixation, dehydration, embedment, and subsequent sectioning with a microtome. These steps are time-consuming and often alter the cell structure in subtle ways. Fixing cells with formaldehyde, for example, will preserve the general organelle structure of the cell, but may destroy enzymes and antigens that are located in the cell.

Pathologists routinely examine tissues that have been fixed in formaldehyde and embedded in paraffin wax prior to sectioning. The process requires a minimum of 24 h, and usually more if automated equipment is not available. This time delay can be crucial when a diagnosis of a benign or malignant tumor is at stake. Valuable time can be saved by skipping the fixation and dehydration steps required for paraffin embedding and freezing the tissue in a modified microtome, the cryostat. Sections can be prepared within minutes and diagnoses made while the patient remains on the operating table. Additionally, frozen sections will more often retain their enzyme and antigen functions. The use of frozen sections can reduce the processing time, but it is not a panacea. Freezing is not adequate for long-term preservation of the tissues, and the formation of ice crystals within the cells destroys subcellular detail. Frozen sections are also thicker, since ice does not section as thin as paraffin.

This results in poor microscopic resolution and poor images of what subcellular structures remain. If time or enzyme function is critical, using frozen sections is the preferred process. If subcellular detail is important, other procedures must be used. Selection of the correct procedure depends on what the cell biologist is looking for, and to a point, becomes an art form. The histologist must choose among hundreds of procedures to prepare tissues in a manner that is most appropriate to the task at hand.

2.3.1.1 Fixation

Since cellular decomposition begins immediately after the death of an organism, one must fix the cells to prevent alterations in their structure through decomposition. Routine fixation involves the chemical cross-linking of proteins (to prevent enzyme action and digestion) and the removal of water to denature the proteins of the cell. Heavy metals may also be used for their denaturing effect. Fixation makes cells permeable to staining reagents and cross-links their macromolecules so that they are stabilized and locked in position. Some of the earliest fixation procedures involved immersion in acids or in organic solvents such as alcohol. Current procedures involve treatment with reactive aldehydes, especially formaldehyde and glutaraldehyde; they form covalent bonds with the free amino groups of proteins and thereby cross-link adjacent protein molecules.

A typical laboratory procedure involves the use of an aldehyde as the primary fixative. Glutaraldehyde is primarily used for transmission electron microscopy (TEM), whereas formaldehyde is used for routine light microscopy. The formaldehyde solution most often employed was originally formulated by Baker in 1944.

Baker's formalin fixative contains

Calcium chloride	1 g
Cadmium chloride	1 g
Formalin concentrated	10 mL
Distilled water	100 mL

Blocks of tissue (liver, kidney, pancreas, etc.) of approximately 1 cm are rapidly removed from a freshly killed organism and placed in the fixative. They are allowed to remain in the fixative for a minimum of 4 h, but usually overnight. The longer the blocks remain in the fixative, the deeper the fixative penetrates into the block and the more protein cross-linking occurs. The fixative is, therefore, termed progressive. Blocks may remain in this fixative indefinitely, although the tissues will become increasingly brittle with long exposures and will be more difficult to section. While it is not recommended, sections have been cut from blocks left for years in formalin.

Formalin has lately been implicated as a causative agent for strong allergy reactions (contact dermatitis with prolonged exposure) and may be a carcinogen. It should be used with care and always in a well-ventilated environment. Formalin is a 39% solution of formaldehyde gas. The fixative is generally used as a 10% formalin or the equivalent 4% formaldehyde solution. The key operative term here is gas formaldehyde, which should be handled in a hood, if possible. As a gas, it is quite capable of fixing nasal passages, lungs, and cornea.

2.3.1.2 Dehydration

Fixatives, such as formaldehyde, have the potential to further react with any staining procedure that may be used later in the process. Consequently, any remaining fixative is washed out by placing the blocks in running water overnight or by successive changes of water and/or a buffer. There are myriad means of washing the tissues (using temperature, pH, and osmotically controlled buffers), but usually simple washing in tap water is sufficient.

If the tissues are to be embedded in paraffin or plastic, all traces of water must be removed; water and paraffin are immiscible. The removal of water is dehydration, which can be accomplished by passing the tissue through a series of increasing alcohol concentrations. The blocks of tissue are transferred sequentially to 30%, 50%, 70%, 80%, 90%, 95%, and 100% alcohol for about 2 h each. The blocks are then placed in a second 100% ethanol solution to ensure that all water is removed. Note that ethanol is hygroscopic and absorbs water vapor from the air. Absolute ethanol is only absolute if steps are taken to ensure that no water has been absorbed.

It is important to distinguish between dehydration and drying. Tissues should never be allowed to air-dry. Dehydration involves slow substitution of water in the tissue with an organic solvent. For comparative purposes, consider the grape. A properly dehydrated grape would still look like a grape. A dried grape is a raisin. It is virtually impossible to make a raisin look like a grape again, and it is equally impossible to make a cell look normal after you allow it to dry.

2.3.1.3 Embedding

After dehydration, the tissues can be embedded in paraffin, nitrocellulose, or various formulations of plastics. Paraffin is the least expensive and, therefore, the most commonly used material. More recently, plastics have come into increased use, primarily because they allow thinner sections (about 1.5 μm compared to 5–7 μm for paraffin). There is a danger that any treatment used for fixation and embedding may alter the structure of the cell or its constituent molecules in undesirable ways. Rapid freezing provides an alternative method of preparation that to some extent avoids this problem by eliminating the need for fixation and embedding. The frozen tissue can be cut directly with a special microtome that is maintained in a cold chamber. Although frozen sections produced in this way avoid some artifacts, they suffer from others; the native structure of individual molecules such as proteins is well-preserved, but the fine structure of the cell is often disrupted by ice crystals.

2.3.1.3.1 Paraffin.   For paraffin embedding, first clear the tissues. Clearing refers to the use of an intermediate fluid that is miscible with ethanol and paraffin, since these two compounds are immiscible. Benzene, chloroform, toluene, and xylol are the most commonly used clearing agents although some histologists prefer mixtures of various oils (cedar wood oil, methyl salicylate, creosote, clove oil, amyl acetate, or cellosolve). Dioxane is frequently used and has the advantage of short preparation time, but it has the distinct disadvantage of inducing liver and kidney damage to the user and should only be used with adequate ventilation and protection. (*Note: Beware of all organic solvents. Most are implicated as carcinogenic agents. Follow all precautions for the proper use of these compounds.*)

The most often used clearing agent is toluene. It is used by moving the blocks into a 50:50 mixture of absolute ethanol:toluene for 2 h. The blocks are then placed into pure toluene and then into a mixture of toluene and paraffin (also 50:50). They are then placed in an oven at 56°C–58°C (the melting temperature of paraffin).

The blocks are transferred to pure paraffin in the oven for 1 h and then into a second pot of melted paraffin for an additional 2–3 h. During this time, the tissue block is completely infiltrated with melted paraffin.

Subsequent to infiltration, the tissue is placed into an embedding mold, and the melted paraffin is poured into the mold to form a block. The blocks are allowed to cool and are then ready for sectioning.

2.3.1.3.2 Plastic.   More recent developments are in the formulation of plastic resins, which have begun to alter the way sections are embedded. For electron microscopy that requires ultrathin sections, paraffin is simply not suitable. Paraffin and nitrocellulose are too soft to yield thin enough sections.

Instead, special formulations of hard plastics are used, and the basic process is similar to that for paraffin. The alterations involve placing a dehydrated tissue sample of about 1 mm^3 into a liquid plastic, which is then polymerized to form a hard block. The plastic block is trimmed and sectioned with an ultramicrotome to obtain sections of a few hundred Angstroms.

Table 2.1 presents a comparison of paraffin embedding with the typical Epon embedment for TEM. Softer plastics are also being used for routine light microscopy. The average reported thickness of a paraffin-sectioned tissue is between 7 and 10 μm, which will often consist of two cell layers, and consequently lack definition for cytoplasmic structures. With a plastic such as Polysciences JB-4, it is possible to section tissues in the 1–3 μm range with increased sharpness. This is particularly helpful if photomicrographs are to be taken. With the decrease in section thickness, however, comes a loss of contrast, and thin sections (1 μm) usually require the use of a phase

TABLE 2.1   Light and Electron Microscopy Preparations

	Light Microscopy	Electron Microscopy
Sample size	1 cm^3	1 mm^3
Fixative	Formaldehyde	Glutaraldehyde
Postfixation	None	Osmium tetroxide
Dehydration	Graded alcohol	Alcohol or acetone
Clearing agent	Xylol/toluene	Propylene oxide
Embedding material	Paraffin	Various plastics
Section thickness	5–10 μm	60–90 nm
Stains	Colored dyes	Heavy metals

contrast microscope as well as special staining procedures. The sharp image makes the effort worthwhile.

These soft plastics can be sectioned with a standard steel microtome blade and do not require glass or diamond knives, as with the harder plastics used for electron microscopy.

2.3.1.4 Staining

Once sections have been cut, by whatever method, the next step is usually to stain them.

There are little contents in the most cells (which are 70% water by weight) to impede the passage of light rays. Thus, most cells in their natural state, even if fixed and sectioned, are almost invisible in an ordinary light microscope. One way to make them visible is to stain them with dyes.

In the early nineteenth century, the demand for dyes to stain textiles led to a fertile period for organic chemistry. Some of the dyes were found to stain biological tissues and, unexpectedly, often showed a preference for particular parts of the cell, the nucleus or mitochondria, for example, making these internal structures clearly visible. Today, a rich variety of organic dyes is available, with such colorful names as Malachite green, Sudan black, and Coomassie blue, each of which has some specific affinity for particular subcellular components. The dye hematoxylin, for example, has an affinity for negatively charged molecules and, therefore, reveals the distribution of DNA, RNA, and acidic proteins in a cell. The chemical basis for the specificity of many dyes, however, is not known. The relative lack of specificity of these dyes at the molecular level has stimulated the design of more rational and selective staining procedures and, in particular, of methods that reveal specific proteins or other macromolecules in cells. It is a problem, however, to achieve adequate sensitivity for this purpose. Since relatively few copies of most macromolecules are present in any given cell, one or two molecules of the stain bound to each macromolecule are often invisible. One way to solve this problem is to increase the number of stain molecules associated with a single macromolecule. Thus, some enzymes can be located in cells through their catalytic activity.

2.4 Ultramicrotome

Operationally, the only difference is that smaller samples are handled, which in turn requires a binocular dissecting microscope mounted over the blade. The tissue sections are too thin to see their thickness with the naked eye; one usually estimates thickness by the color of the diffraction pattern on the section as it floats off the knife onto the surface of a water bath. The sections are also too thin to be handled directly, and they are, therefore, transferred with wire loops, or picked off the water directly onto an electron microscope (EM) grid. This process requires a good light source mounted to cast the light at just the correct angle to see the color pattern.

Since the plastics are hard enough to break steel knives, freshly prepared glass knives or commercially available diamond knives are used. A glass knife costs several dollars each, while a good diamond knife will cost in excess of $3000. Either can be permanently damaged with a single careless stroke by the operator. Diamond knives are used in research laboratories by trained technicians because they have the advantage of a consistent knife edge (unlike glass which varies with each use) and can last for years if treated properly. They can usually be resharpened several times before discarding.

To minimize vibrations (which lead to uneven sections), ultramicrotomes are cast in heavy metal, mounted on shock-absorbent tables and, preferably, kept in the draft-free environments of relatively constant temperature. To minimize vibrations further, some manufacturers have replaced the block's mechanical advance mechanism with a thermal bar, advancing the tissue by heating a metal rod. This can be exquisitely precise and is the ultimate in thin sectioning. Of course, with this advancement comes increased cost and maintenance, and decreased ability to withstand rough treatment. The mechanically advanced ultramicrotome remains as the workhorse of the cell biology laboratory.

2.5 Cryostat

Whether the sectioning is performed with a microtome or an ultramicrotome, one of the major delays in preparing a tissue section is the time required to dehydrate and embed the tissue. This can be overcome by direct sectioning of a frozen tissue. Typically, a piece of tissue can be quick-frozen to about −15°C to −20°C (for light microscopic work) and sectioned immediately in a device termed a cryostat. The cryostat is merely a microtome mounted within a freezer box (Figure 2.2a and b).

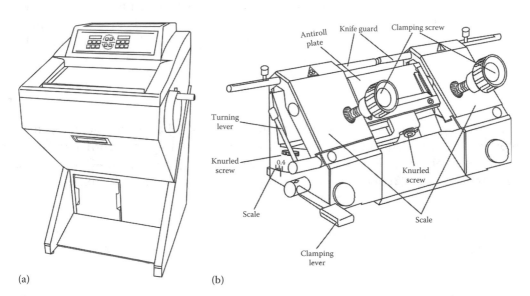

(a)   (b)

Figure 2.2   (a) The cryostat microtome and (b) inside view of the microtome cabinet.

A piece of tissue is removed from an organism, placed onto a metal stub, and covered with a viscous embedding compound to keep it in a form convenient for sectioning. The stud and tissue are placed within the cryostat and quick-frozen.

This method has the advantage of speed, maintenance of most enzyme and immunological functions (fixation is unnecessary), and relative ease of handling (far fewer steps to manipulate). It has the disadvantage that ice crystals formed during the freezing process will distort the image of the cell (e.g., bursting vacuoles and membranes), and the blocks tend to freeze-dry or sublimate. Thus, the blocks must be used immediately, and great care must be taken to guard against induced artifact from the freezing process.

While studying temperature-sensitive (or lipid-soluble) molecules, where speed is of the essence (such as pathological examination during an operation), this is the preferred method. Sectioning operation with the cryostat is similar to that of the microtome, with the exception that one handles single frozen sections, and thus all operations must be handled at reduced temperatures.

2.5.1 Tissue Chilling

Pretreatment: One of the advantages of this procedure is that a wide range of tissues, from many animals and plants, can all be chilled successfully by the same technique. Usually the tissue is chilled directly as it is removed from the body or plant. Delicate structures, however, may benefit from immersion for 10–30 min in a 5% aqueous solution of polyvinyl alcohol (PVA) before being chilled.

2.5.1.1 Apparatus and Materials
The main apparatus required is a chilling bath, which should be heat-insulated by having expanded polystyrene or similar insulating matter put around it.

Also required are a beaker of 50–100 mL capacity which can be covered; a low-temperature thermometer (recording down to $-100°C$); carbon dioxide ice; commercial spirit (or absolute alcohol); hexane (boiling range $67°C–70°C$), which must be free from aromatic hydrocarbons; 3×1 in. corked specimen tubes; Dewer flasks (the 1-gal size is very useful for storage of tubes).

2.5.1.2 Protocol
The chilling bath is prepared by adding small chips of solid carbon dioxide (which cool quicker than do larger pieces) to alcohol in the chilling bath until a saturated solution is obtained. This state is obvious, because (1) the alcohol–CO_2 mixture becomes viscous; (2) addition of more solid CO_2 does not cause bubbling; (3) the thermometer should record about $-70°C$.

The beaker, containing 30–50 mL of hexane, is inserted into the bath, preferably through a close-fitting space cut in the lid. More CO_2 ice is added to the bath to maintain the temperature. The temperature of hexane should be at least $-65°C$ before it is used. Isopentane can also be used in place of hexane.

The tissue is cut into suitable pieces (which can be as large as 5 mm³ in size), and these are dropped into hexane at about $-70°C$. Small pieces can be blown into the hexane from a hollow spatula fitted with a large rubber bulb. Care must be taken to ensure that the tissue is plunged straight into the hexane without touching the sides of the beaker. The specimen is left in the hexane for at least 30 s and preferably not longer than 2 min. It is then transferred, with the use of precooled forceps, to a dry tube at −70°C [*Note: from this stage onward, the chilled tissue should be handled only with cold instruments*]. It is often advisable to shake the surface hexane off the specimen while transferring it. Care must be taken to ensure that the tissue does not warm up during this process; the specimen, in the corked dry tube, should be encased in solid carbon dioxide, in the Dewar flask for storage.

For tissues like lungs which float on the hexane, this can be pressed gently but firmly with a scalpel or forceps to eliminate air pockets against the inside of a glass specimen tube at about −70°C (preferably low down inside the tube to avoid the heating that occurs close to hand). The temperature of the tube can be achieved by immersing it in the CO_2 alcohol bath, or by allowing it to equilibrate with solid CO_2 packed around it in a Dewar flask.

2.5.2 Tissue Mounting

This is the most hazardous operation in the whole process. It should be effective expeditiously, and care must be taken to perform all operations in cold condition at all stages. Prepare another alcohol–solid CO_2 bath at about −70°C. A metal chuck has to be placed in this bath with the top clear of the alcohol; leave it to equilibrate with the bath; place a drop of water on top of the chuck. The water should begin to freeze rapidly; remove the chilled tissue from the 3×1 in. corked tube and place it in a cavity, in a piece of CO_2 ice so that its orientation can be determined before it is mounted, while keeping it chilled. The water on the chuck is left to freeze until there is only a thin film of water, of comparable size to that of the specimen, left unfrozen. Then, expeditiously transfer the specimen (with cold forceps) to this film of water. The residual water will freeze the tissue to the drop of ice on the chuck; remove the chuck (with the specimen) from the bath; wipe its sides free of adherent alcohol and stand it in the refrigerated cabinet of the cryostat.

2.5.3 Preparation for Sectioning

The angle of the knife is critical. On the knife-mounting of the microtome, there are four screws: two threads into lugs at the back and two into lugs at the front. The back pair is marked into eight divisions. Before the knife is placed in the mounting, the two back index screws should be unscrewed far enough to prevent them from protruding into the knife-mounting space. The knife is then placed in the mounting, and the two front screws are screwed up as tight as they will go, forcing the knife flat against the back of the mounting. The two index screws are then screwed forward until they touch the back of the knife (Figure 2.3). The number on each index screw

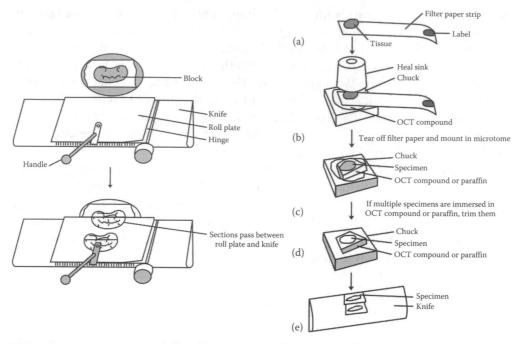

Figure 2.3   Cryosectioning procedure.

opposite the line marked on top of the respective lugs at the back of the mounting should then be read off and recorded.

The front screws are then unscrewed, releasing the knife. Both the index screws are then advanced one complete revolution until the same recorded number of each appears opposite the line on its respective lug. The front screws are then tightened back onto the knife so that it will be held firm and it will now be at the right angle. Once it has been set, the knife must be cooled to −70°C by packing CO_2 ice around the handle. The knife should be left packed around the CO_2 ice for at least 1 h before it is used. The ambient temperature in the cabinet must be maintained at −30°C.

Water-rich tissues are hardened by freezing and cut in the frozen state in this technique with a freezing microtome or microtome cryostat; sections are stained and examined with a light microscope. This technique is much faster than traditional histology (5 min vs 16 h) and is used in conjunction with medical procedures to achieve a quick diagnosis (Figure 2.4).

2.5.4 Cryosectioning

The preparation of sections for histochemistry can be done in the following ways, each of which has its own limitations:

1. *By chemical fixation and embedding in a hard matrix:* The focus on chemical fixation is that the fixative should react with specific active chemical groups. It is these active groups that are stained in the histochemical nature of the tissue. Treatment with

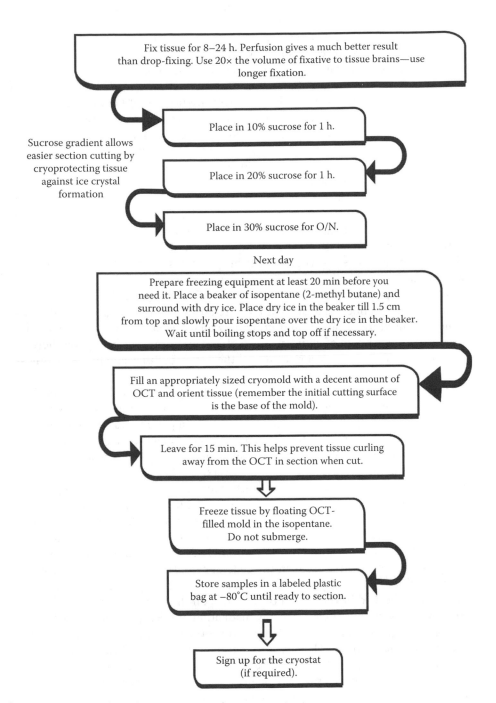

Fix tissue for 8–24 h. Perfusion gives a much better result than drop-fixing. Use 20× the volume of fixative to tissue brains—use longer fixation.

Sucrose gradient allows easier section cutting by cryoprotecting tissue against ice crystal formation

Place in 10% sucrose for 1 h.

Place in 20% sucrose for 1 h.

Place in 30% sucrose for O/N.

Next day

Prepare freezing equipment at least 20 min before you need it. Place a beaker of isopentane (2-methyl butane) and surround with dry ice. Place dry ice in the beaker till 1.5 cm from top and slowly pour isopentane over the dry ice in the beaker. Wait until boiling stops and top off if necessary.

Fill an appropriately sized cryomold with a decent amount of OCT and orient tissue (remember the initial cutting surface is the base of the mold).

Leave for 15 min. This helps prevent tissue curling away from the OCT in section when cut.

Freeze tissue by floating OCT-filled mold in the isopentane. Do not submerge.

Store samples in a labeled plastic bag at −80°C until ready to section.

Sign up for the cryostat (if required).

Figure 2.4   Flowchart for preparation for cryosectioning.

alcohol, or with acetone, whether for fixation or as a part of the process of embedding, will cause denaturation and hence can also change the active chemical sites that are available for the staining reaction. This will also affect the lipid–protein bonds, which stabilize membranes and control enzymatic activities. Other fat solvents, such as xylene or chloroform, will not only remove free fats but will also enhance the splitting of lipid–protein complexes of the protoplasm.

2. *By freeze-drying:* In this technique, autolytic processes are stopped by "quenching" tissues in liquefied gases at very low temperature (e.g., −190°C); this necessitates the use of very small specimens (preferably 1–2 mm² in cross section). Water is removed (at about −40°C, drying proceeding for a few days), and the dried tissue is then embedded in paraffin wax. This sets the tissue in a hard matrix so that sections can be cut with a conventional microtome. The wax is then removed (e.g., with xylene) and the tissue is fixed, in the dried state, with absolute alcohol. The fact that the alcohol acts on dry protoplasm is said to be the reason for the improved preservation of the protoplasm over what would occur when the fresh tissue is plunged into the absolute alcohol. The drawback with this technique is that "quenching" entails the use of liquefied gases, which are not easily available, and the use of very small amounts of specimens. A gas, like nitrogen or air, when liquefied, will boil around the warm specimen, and the gas so liberated will insulate the specimen from the cooling effect of the liquid gas. It is necessary to immerse the specimen in a gas that has a relatively high boiling point, like propane or a mixture of butane and propane which has been cooled to −190°C by means of an outer bath of liquid nitrogen. These hydrocarbons have good thermal conductivity and do not vaporize around the tissue, but they add to the technical difficulties.

3. *By freeze-substitution:* The tissue in this method is quenched, as for freeze-drying, but the water is then removed by substituting methanol for water at low temperatures (such as −70°C). The tissue can be embedded in paraffin wax, or in ester wax, provided that either it is substituted again into butanol and then taken into mixtures of butanol and paraffin wax in the oven at 60°C or if the original substitution is done in acetone and the tissue embedded in ester wax.

4. *By freeze-sectioning techniques:* In this technique, the autolysis can be stopped and tissues hardened simultaneously by the expedient of freezing the specimen and cutting sections of the frozen block. However, the intracellular dehydration produced by the extracellular ice crystals increases the ionic concentration of the cytoplasm to a level that causes denaturation of the protoplasm and ruptures the lipid–protein complexes, whether these occur in the general cytoplasm or in cellular or subcellular membranes. During the process of cutting, the frozen tissue thaws because of the heat liberated by the impact of the block on the knife, and freezes again. In this way, the section is "flash dried" by distillation of the supercooled water over a temperature gradient of nearly 100°C operating over a gap of a few millimeters. Such sections are then stable. Ice damage is undetectable in such tissues and sections. The detecting device may be a thermocouple in the tissue; it may be a simple microscopic examination for ice crystals; it may be dark ground illumination for investigating the degree of denaturation–aggregation of the protoplasm or it may involve a study of lipid–protein associations or of the permeability of subcellular membranes. It is recommended that this procedure be used for all histochemical investigations. For certain reactions, to demonstrate particular substances, it is necessary to fix the

sections specifically for the demonstration of these substances. Other substances are better studied in unfixed sections. Variations in the histochemical reaction are due to true variations in activity (or in manifest activity), and so indicate metabolic differences.

Cryosections can also be used in immunohistochemistry as freezing stops degradation of tissue faster than using a fixative. Also, it does not alter much or mask its chemical composition.

The chuck is locked on the front of the arm of the microtome. It is positioned in such a way that the smallest of the block faces downward. Blocks, which have an epithelium on one edge, should be positioned with this epithelium on one side and not at the top or bottom facing the knife. The screw, which determines the thickness of the sections, is set to the required size. In general, sections 8–10 μm thick are suitable for all histochemical works. However, tissues chilled and prepared in this way can be cut at any required thicknesses from 2 to 20 μm.

On the front of the microtome is the antiroll plate assembly. As its name suggests, it is designed to prevent the sections from curling up as they are cut. These plates can be of two types, being made of either glass or Teflon. The Teflon plates have two disadvantages. They are only translucent so that the microtomist is forced to remove the plate when one wishes to observe the state of the section; also they tend to wrap. Glass plates are preferred which can be made simply by cutting a convenient length from a 3 × 1 in. slide, grinding off the leading edge, and wrapping Sellotape around the vertical sides, hence leaving a gap on the leading face for the section to move down (Figure 2.5).

The angle of this plate to the knife is also critical, but may have to be varied according to the tissue to be cut; consequently, the correct angle setting will be found only by experience. However, whatever angle of the antiroll plate is used, the top of the plate should never be lower than the cutting edge or high enough to touch the block.

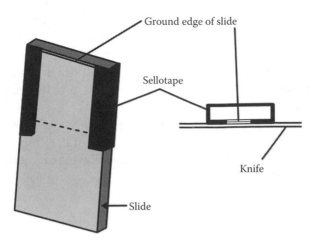

Figure 2.5 Surface and cross-sectional view of the antiroll plate. Note that the Sellotape is placed so as to leave a gap for the section to pass between the knife and the antiroll plate.

2.5.5 Section Picking

The antiroll plate keeps the sections flat against the knife while they are being cut. Then (1) sway the antiroll plate away from the knife; (2) take a glass slide from the ambient temperature of the laboratory and bring it up to, and parallel to, the section on the knife; there should be no need to press the slide on to the section—it should move on to the slide; (3) store the sections, on the slides, in the cryostat until used.

2.6 Incubation

2.6.1 Methods of Incubation

The simplest way is to immerse the section in a Coplin jar, very much as is done in conventional histology. This has the following advantages:

1. The solutions in the Coplin jars can be left in a water bath, or an incubator, or an ice bath, to equilibrate to a constant, predetermined temperature.
2. The sections are immersed in a constant volume of the solution, that is, there is no danger of drying out. Moreover, since there is such a large volume of the solution relative to the volume of the sections, there is no danger that the reactants will be seriously depleted during the incubation, nor that deleterious matter, produced by the sections, could become sufficiently concentrated as to have inhibitory effects on the reaction.
3. It is simple to perform.

The disadvantage of using Coplin jars is that they require a large volume of the incubation medium. Should this contain coenzymes (NAD, NADP, etc.), the cost of these substances would make histochemistry prohibitively expensive. The usual way of overcoming this, in the past, has been to prepare only small volumes of expensive incubating media (e.g., 1 mL) and to add this as a drop, to cover the section that is laid horizontally in some sort of a humidity chamber to stop evaporation of the drop (Figure 2.6).

The advantage of this procedure is that it avoids expense. The disadvantages are that (1) frequently, part of the drop does evaporate, leaving some of the section free of fluid; it is difficult to be sure that a negative reaction in a part of the slide is not due to this; (2) evaporation can change the concentration of the reagents in the drop of the incubation fluid; (3) the drop is not of uniform thickness over the section and so, depending on how it has spread, some parts may have a greater depth of solution above them than have others; and (4) the small volume of the drop could allow inhibitory substances produced by the histochemical reaction to reach a concentration that could seriously affect the result of the reaction.

2.6.2 Microcell Method

It was designed to overcome the objections made to the use of a drop of incubation medium. The disadvantages of the procedure are (1) it is difficult to make the

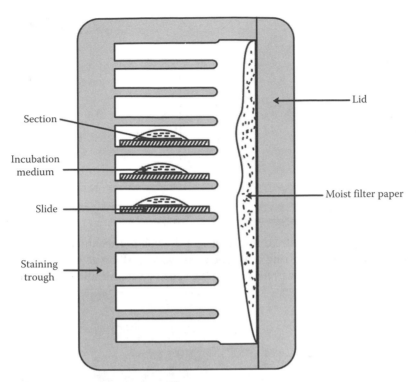

Figure 2.6 Illustration of the cross section of a humidity chamber. The atmosphere is kept humid by a layer of the moist filter paper. An ordinary staining trough stands on edge so that the sections lie horizontally.

microcell, and it is awkward to use; (2) the microcell makes it virtually impossible to control the gaseous atmosphere in which the reaction is performed. The atmosphere can be controlled in both the other methods. Thus, air or selected mixtures of oxygen, nitrogen, or CO_2 can be bubbled into the solution in the Coplin jar, or can be passed through the humidity chamber in which the section lies covered with a thin drop of incubation fluid.

The inability to control the gaseous atmosphere is the main reason to discard the use of microcell procedure.

The method commonly in practice is the open ring technique for costly incubation fluid. In this, the section is picked up normally on to a slide and brought into the open laboratory (or into a room or cabinet kept at 37°C). A Perspex ring just large enough to encircle the section, and of depth of about 3 mm, is set around the section and held on to slide by a thin film of vaseline (Figure 2.7). The incubation medium (a known volume, e.g., 0.25 mL can be pipetted into the space bounded by the ring if this precision is required) is added to the section and confined around the section by the open ring. The gaseous atmosphere can be controlled, especially if the incubation is done in specially designed incubation boxes (Figure 2.8). The rings and the boxes are available commercially, or can be made readily from Perspex. The whole incubation box can be put into a large incubator or into a hot room so that the reaction can be performed at 37°C if necessary. The incubation

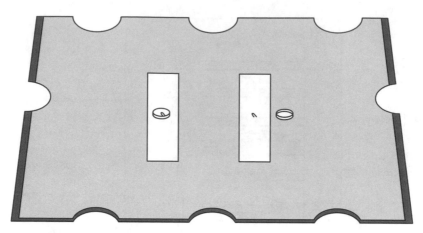

Figure 2.7   Diagram to illustrate the open ring incubation technique. On the right is a section on a glass slide with a ring of Perspex alongside, just big enough to encompass the section. On the left, the ring has been held to the slide by vaseline and can be filled with the incubation medium.

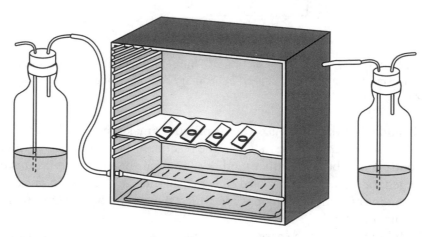

Figure 2.8   Diagram of the incubation box. The slides bearing sections surrounded by Perspex rings are placed on Perspex trays, which can be stacked in layers. The bottom of the box is filled with moist filter paper. The front of the box is closed with a Perspex lid. The gaseous atmosphere is controlled by bubbling appropriate gas through the wash bottle on the left; the moistened gas enters the box through the holes in the tube at the bottom of the box. The flow of gas is checked by the bubbling of the gas through the escape wash bottle on the right.

medium and preferably the atmosphere should be equilibrated to the incubation temperature before they are used.

Cryosections can also be used in immunohistochemistry as freezing stops degradation of tissue faster than that using a fixative. Also, it does not alter much or mask its chemical composition.

2.7 Electron Microscopy Technique

After embedding tissues in epoxy resin, a microtome equipped with a glass or gem-grade diamond knife is used to cut very thin sections (typically 60–100 nm). Sections are stained with an aqueous solution of an appropriate heavy metal salt and examined with a TEM. This instrument is often called an ultramicrotome. The ultramicrotome is also used (with its glass knife or an industrial grade diamond knife) to cut survey sections prior to thin sectioning. These survey sections are generally 0.5–1 μm thick and are mounted on a glass slide and stained to locate areas of interest under a light microscope prior to thin sectioning for TEM. Thin sectioning for the TEM is often done with a gem-quality diamond knife. Ultramicrotomes complement traditional TEM techniques. They are increasingly found mounted inside a scanning electron microscope (SEM) chamber so that the surface of the block face can be imaged at and then removed with the microtome to uncover the next surface which is ready for imaging. This technique is called serial block-face scanning electron microscopy (SBFSEM).

2.7.1 Special Preparation for Electron Microscope

In the early days of its application to biological materials, the electron microscope revealed many previously unimagined structures in cells, but before these discoveries could be made, electron microscopists had to develop new procedures for embedding, cutting, and staining tissues.

Since the specimen is exposed to a very high vacuum in electron microscopy, there is no possibility of viewing it in the living, wet state. Tissues are usually preserved by fixation, first with glutaraldehyde, which covalently cross-links protein molecules to their neighbors, and then with osmium tetroxide, which binds to and stabilizes lipid bilayers, as well as proteins (Figure 2.9). Because electrons have very limited penetrating power, the fixed tissues normally have to be cut into extremely thin sections (50–100 nm thick, about 1/200 the thickness of a single cell) before they are viewed. This is achieved by dehydrating the specimen and permeating it with a monomeric resin that

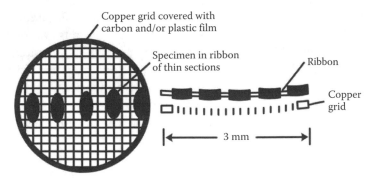

Figure 2.9 Copper grid support for thin sectioning of a specimen in a TEM.

polymerizes to form a solid block of plastic; the block is then cut with a fine glass or diamond knife on a special microtome. These thin sections, free of water and other volatile solvents, are placed on a small circular metal grid for viewing in the microscope.

The steps required to prepare the biological material for viewing in the electron microscope have challenged electron microscopists from the beginning. How can we be sure that the image of the fixed, dehydrated, resin-embedded specimen finally seen bears any relation to the delicate aqueous biological system that was originally present in the living cell? The best current approaches to this problem depend on rapid freezing. If an aqueous system is cooled fast enough to a low enough temperature, the water and other components in it do not have time to rearrange themselves or crystallize into ice. This state can be achieved by slamming the specimen onto a polished copper block cooled by liquid helium by plunging it into or spraying it with a jet of a coolant such as liquid propane, or by cooling it at high pressure.

Some frozen specimens can be examined directly in the electron microscope using a special, cooled specimen-holder. In other cases, the frozen block can be fractured to reveal interior surfaces, or the surrounding ice can be sublimed away to expose external surfaces. However, we often want to examine thin sections and to have them stained to yield adequate contrast in the electron microscope image. A compromise is, therefore, to rapid-freeze the tissue; then replace the water maintained in the vitreous glassy state by organic solvents; and finally embed the tissue in plastic resin, cut sections, and stain. Although technically still difficult, this approach stabilizes and preserves the tissue in a condition very close to its original living state.

Contrast in the electron microscope depends on the atomic number of the atoms in the specimen: the higher the atomic number, the more electrons are mattered and the greater is the contrast. Biological tissues are composed of atoms of very low atomic number (mainly carbon, oxygen, nitrogen, and hydrogen). To make them visible, they are usually impregnated (before or after sectioning) with the salts of heavy metals such as uranium and lead. Different cellular constituents are revealed with various degrees of contrast according to their degree of impregnation, or "staining," with these salts. Lipids, for example, tend to stain darkly after osmium fixation, revealing the location of cell membranes.

2.7.1.1 Metal Shadowing

TEM can also be used to study the surface of a specimen, generally at a higher resolution than that of the SEM in such a way that individual macromolecules can be seen. As in SEM, a thin film of a heavy metal such as platinum is evaporated onto the dried specimen. The metal is sprayed from an oblique angle so as to deposit a coating that is thicker in some places than others—a process known as shadowing, because a shadow effect is created that gives the image a three-dimensional appearance. Some specimens coated in this way are thin enough or small enough for the electron beam to penetrate them directly. This is the case for individual molecules, viruses, and cell walls, all of which can be dried down, before shadowing, onto a flat supporting film made of a material that is relatively transparent to electrons, such as carbon or plastic. For thicker specimens, the organic material of the cell must be dissolved away after shadowing so that only the thin metal replica of the surface of the specimen is

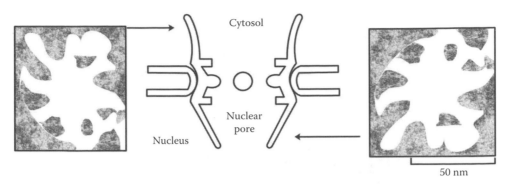

Figure 2.10 Nuclear pore of a frozen nuclear envelope in high-resolution SEM equipped with emission gun as source of elements.

left. The replica is reinforced with a film of carbon; so, it can be placed on a grid and examined in the TEM in the ordinary way (Figure 2.10). These views of each side of a nuclear envelope represents the limit of resolution of SEM. Metal shadowing allows surface features to be examined at high resolution by TEM.

2.7.1.2 Freeze-Fracture and Freeze-Etch Electron Microscopy

Freeze-fracture electron microscopy allows the visualization of the interior of cell membranes. Cells are frozen (as already described), and then the frozen blocks are cracked with a knife blade. The fracture plane often passes through the middle of the hydrophobic lipid bilayers, thereby exposing the interior of cell membranes. The resulting fracture faces are shadowed with platinum; the organic material is dissolved away, and the replicas are floated off and viewed in the electron microscope (Figure 2.11).

Note that the thickness of the metal reflects the surface contours of the original specimen. Such replicas are studded with small bumps, called intramembrane particles, which represent large transmembrane proteins. The technique provides a convenient and dramatic way to visualize the distribution of such proteins in the plane of a membrane.

In this freeze-fracture electron micrograph, the thylakoid membranes performing photosynthesis are stacked up in multiple layers. The plane of the fracture moves from layer to layer, passing through the middle of each lipid bilayer, exposing the transmembrane proteins that have sufficient bulk in the interior of the bilayer to cast a shadow, and show the tip as intramembrane particles in this platinum replica. The largest particles seen in the membrane are the complete photosystem II, a complex of multiple proteins.

Another related replica method is freeze-etch electron microscopy, which can be used to examine either the exterior or interior of the cells. In this technique, the frozen block is cracked with a knife blade as already described, but here the ice level is lowered around the cells (and to a lesser extent within the cells) by the sublimation of ice in a vacuum as the temperature is raised, a process called freeze-drying. The parts of the cell exposed by this etching process are then shadowed as before to make a platinum replica. This technique exposes structures in the interior of the cell and can reveal their three-dimensional organization with exceptional clarity.

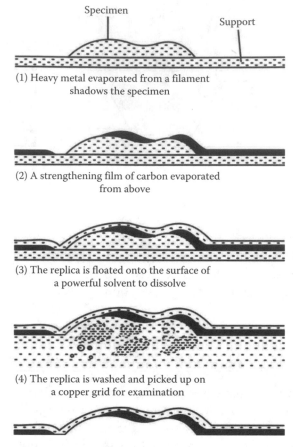

Figure 2.11   Preparation of a metal-shadowed replica of the surface of a specimen.

2.7.1.3 Spectroscopy (Especially FTIR or Infrared Spectroscopy)

Thin polymer sections are needed such that the infrared beam could penetrate the sample under examination. It is normal to cut samples between 20 and 100 μm in thickness. For more detailed analysis of much smaller areas in a thin section, FTIR microscopy can be used for sample inspection.

A recent development is the laser microtome, which cuts the target specimen with a femtosecond laser instead of a mechanical knife. This method is contact-free and does not require sample preparation techniques. The laser microtome has the ability to slice almost every tissue in its native state. Depending on the material being processed, slice thicknesses of 10–100 μm are feasible.

2.8 Botanical Microtomy

Hard materials like wood, bone, and leather require a sledge microtome. These microtomes have heavier blades and cannot cut sections as thin as that by a regular microtome. A few cell types are thin enough to be viewed directly in a microscope (algae, protozoa, blood, tissue cultures), but most tissues (kidney, liver, brain) are too thick to allow light to be transmitted through them. The tissues can be sliced into very thin sections, provided they are first processed to prevent cell damage.

2.9 Processing of Tissue

Stabilized tissues must be adequately supported before they can be sectioned for microscopical examination. While they may be sectioned following a range of preparatory freezing methods, tissues are more commonly taken through a series of reagents and finally infiltrated and embedded in a stable medium, which, when hard, provides the necessary support for microtomy. In order to be examined with a microscope, a specimen must be sufficiently thin to be transparent and must possess sufficient contrast to permit the resolution of structural detail. Thinness may be an intrinsic property of the object to be examined. Thus, small animals and plants, films, smears of cells, tissue cultures, macerated or teased tissues, and spread-out sheets of epithelial or connective tissue are all thin enough to mount on slides directly. It is important to be able to subject sections to some preferred histological stain for the whole range of interests that lie between these extremes. It is quite obvious that the end product of the histological staining method should be a section that resembles, as closely as possible, one prepared by the conventional histological procedures which include fixation, dehydration, and embedding in paraffin wax. To achieve this end, some of the conventional histological techniques have had to be modified in a seemingly irrational way. However, these modifications have not only produced histological preparations, which, by their conventional appearance, make the histology clearer for the histochemist, but they have also had the additional advantage of opening new possibilities for the histopathologist. It is now possible to cut a frozen section for rapid surgical diagnosis; to have it stained within a few minutes of taking the specimen and yet to have a preparation which easily rivals, one would take several hours or even a few days by conventional paraffin-wax histology. The pathologist can use such sections for diagnosis just as if they were conventional preparations, but one can also have additional sections cut for such histochemical procedures as it may aid his work. Another drawback to the use of cryostat sections has been the difficulty of storing the "frozen" blocks over a period of some years, because in a routine pathology laboratory they must be kept, should it be necessary to refer to them at some later stage. This has been overcome by the procedure by which the "frozen" tissue can subsequently be fixed and embedded in paraffin wax; such tissue yields sections in which the histology is indistinguishable from that of similar pieces of tissue which have never been chilled but have been subjected only to fixation and embedding.

2.9.1 Methods of Tissue Processing

Most tissues can be chilled fresh in hexane at −70°C. Muscle may benefit by being treated with 5% PVA for 15–30 min before it is chilled. (Sometimes, the addition of small amounts of carnosine to PVA can improve the histology of the muscle.)

Pretreatment with PVA (5%, with or without about 1%–2% of calcium chloride) is advisable when delicate tissue is to be chilled; retina and necrotic malignant tissue typically benefit from such pretreatment.

Contrary to some belief, tissue that has been fixed in formalin can be chilled and sectioned just as effectively as fresh tissue. Certain precautions should be taken: the temperature of the chilling bath must be below $-65°C$, and the temperature of the cabinet must be $-30°C$ with the knife really well-chilled by prolonged contact with solid CO_2. The slides must be well-albuminized (i.e., smear with egg albumin and leave to dry for 30 min before use).

2.9.2 Fixatives

1. *Picric–formalin fixative*

Formaldehyde (40% technical)	10 mL
Absolute ethyl alcohol	56 mL
Sodium chloride	0.18 g
Picric acid (wet)	0.15 g
Distilled water to make up to	100 mL

[*Note: The picric acid is wet. It is placed on a filter paper to remove excess moisture and then weighed.*]

2. *Picric–acetic acid fixative*
   Same as the picric–formaldehyde, but with the addition of 5% of acetic acid.

3. *Acetic–ethanol fixative*

Either: Glacial acetic acid	10 mL
Absolute ethyl alcohol	30 mL
Mix just prior to use	
Or: Glacial acetic acid	5 mL
Absolute ethyl alcohol	95 mL
Mix just prior to use	

4. *Formol–calcium*

Formaldehyde (40% technical)	10 mL
Calcium chloride ($CaCl_2 \cdot 2H_2O$)	0.18 g
Distilled water to make up to	100 mL

Especially if this is to be kept as a stock solution, it is advisable to add marble chips.

5. *Formol–saline*

Formaldehyde (40% technical)	10 mL
Sodium chloride	0.85 g
Distilled water to make up to	100 mL

6. Heidenhain's Susa (for cryostat sections)

Mercuric chloride	4.0 g
Sodium chloride	0.5 g
Trichloroacetic acid	2.0 g
Acetic acid (glacial)	4.0 mL
Formaldehyde (40% technical)	20 mL
Distilled water	114 mL

2.9.3 Staining Procedures

1. *Normal hematoxylin–eosin method for cryostat sections*
This is done at room temperature; the whole procedure takes about 45 min:

a. Remove fresh section from the cryostat and dry at room temperature for a few minutes.
b. Immerse in picric–formalin fixative (5 min).
c. Wash in 70% alcohol (10 s).
d. Leave in periodic acid solution for 5 min. (Steps 4–8 help to convert the histology to that seen in more conventional fixed and embedded sections).
e. Wash in 70% alcohol (10 s).
f. Immerse in reducing rinse for 5 min.
g. Wash in 70% alcohol (10 s).
h. Leave in absolute alcohol for 1 min.
i. Stain in Ehrlich's hematoxylin for 20 min.
j. Rinse in tap water.
k. Differentiate in acid–alcohol (1% concentrated hydrochloric acid in 70% alcohol) according to requirement (e.g., up to 10 s).
l. "Blue" in alkaline water (5% sodium bicarbonate solution) for about 30 s.
m. Stain in 0.5% aqueous solution of eosin for 40–60 s as per requirement.
n. Rinse briefly in tap water.
o. Dehydrate through a graded series of alcohols, clear in xylol, and mount in DPX or EPX.

Extra solutions required for this method
Periodic acid solution

Dissolve 0.4 g of periodic acid in 15 mL of distilled water. To this, add 35 mL of absolute alcohol in which is dissolved 0.135 g of hydrated crystalline sodium acetate. The final solution should be stored in the dark.

Reducing rinse

To 20 mL of distilled water, add 1 g of potassium iodide and 1 g of sodium thiosulfate. Add, stirring continuously, 30 mL of absolute alcohol and then 0.5 mL of 2 N hydrochloric acid. A precipitate of sulfur will form, which is allowed to settle.

2. *Modified hematoxylin–eosin method*
The special advantages of this procedure are that it enhances the clarity of the nuclei and is quicker than the normal technique. It takes about 15 min to complete.

All stages can be performed at room temperature:

    a. Take the section straight from the cryostat cabinet. Dry it in air for 5 min, preferably with slight warming.
    b. Fix in picro-acetate fixative (3 min).
    c. Wash in 70% alcohol (10 s).
    d. Immerse in celestine blue solution for 3 min.
    e. Rinse briefly in tap water.
    f. Stain with Mayer's hemalum for 3 min.
    g. Rinse in tap water.
    h. Differentiate to taste in acid–alcohol (1% concentrated hydrochloric acid in 70% alcohol).
    i. "Blue" in alkaline water (containing a few drops of a saturated solution of lithium carbonate).
    j. Stain with a 0.5% aqueous solution of eosin (10 s).
    k. Rinse briefly in tap water.
    l. Dehydrate through a graded series of alcohols, clear in xylol, and mount in DPX or EPX.

Extra solutions required for this method

Celestine blue solution

Dissolve 2.5 g of iron alum in 50 mL of distilled water. Add 0.25 g of celestine blue; boil for 3 min; filter when the solution is cool.

Mayer's hemalum

Dissolve 1 g of hematoxylin in 1 L of distilled water (heat if necessary). Add 50 g of ammonium alum; shake to dissolve. Then add 0.2 g of sodium iodate, 1 g of citric acid, and 50 g of chloral hydrate.

    3. *Van Gieson stain for collagen*
      a. Take fresh section from the cryostat cabinet; dry it in air for 5 min.
      b. Fix either in picrate–formalin for 15 min or in 1:3 acetic–alcohol for 10 min.
      c. Rinse in tap water.
      d. Stain with celestine blue for 3 min.
      e. Rinse briefly in tap water.
      f. Stain in Mayer's hemalum for 3 min.
      g. Rinse briefly in tap water.
      h. Differentiate very briefly (if necessary) in 1% acid alcohol. (*Note: Further differentiation occurs during Step l owing to the picric acid in the Van Gieson solution.*)
      i. Wash in tap water.
      j. "Blue" in 5% aqueous solution of sodium bicarbonate.
      k. Wash in distilled water.
      l. Stain with the Van Gieson solution for 1–2 min.
      m. Rinse briefly in distilled water and blot dry.
      n. Dehydrate in absolute ethyl alcohol, clear in xylol, and mount in DPX or EPX.

Result

  Nuclei stain blue or black
  Collagen fibers stain red
  Muscle and other tissue stain yellow

Solutions required for this procedure

Celestine blue, Mayer's hemalum, and acid alcohol as for the modified hematoxylin–eosin method.

Additional solutions

Van Gieson solution

1% aqueous solution of acid fuchsin	10 mL
Saturated aqueous solution of picric acid	90 mL
Distilled water	100 mL
Boil for 3 min	

4. *Gordon and Sweet's silver impregnation method for reticulin*
   a. Take the fresh section from the cryostat cabinet and dry it thoroughly in air.
   b. Fix in picric–formalin for 5 min.
   c. Rinse in distilled water.
   d. To oxidize the tissue, treat for 1–5 min in the acidified permanganate solution.
   e. Rinse in distilled water.
   f. Bleach until white with 1% solution of oxalic acid.
   g. Wash well in several changes of distilled water.
   h. Mordant by immersing for 2–15 min in 2% aqueous iron alum.
   i. Wash well in several changes of distilled water.
   j. Treat for 5–8 s in Wilder's silver bath.
   k. Rinse thoroughly in distilled water.
   l. Reduce in formol–calcium (neutral 10% formaldehyde) for 10–30 s.
   m. Rinse in tap water.
   n. "Fix" the silver in 5% solution of sodium thiosulfate (2–5 min).
   o. Wash well in tap water.
   p. Dehydrate through a graded series of alcohols, clear in xylol, and mount in DePeX.

Result

   Reticulin fibers stain jet-black
   Collagen fibers stain golden brown

Solutions required for this procedure

Acidified permanganate solution

0.5% potassium permanganate in water	47.5 mL
3% sulfuric acid	2.5 mL

Wilder's silver bath

To 5 mL of a 10% solution of silver nitrate in water, add ammonia drop by drop. A precipitate will form. Continue to add ammonia, drop by drop, until the precipitate is just redissolved. Then, add 5 mL of a 3% solution of sodium hydroxide. Again add ammonia, drop by drop, until the solution just becomes clear. Add distilled water to make the final volume to 50 mL.

This solution should be stored in a dark bottle for several months.

5. *Phosphotungstic acid–hematoxylin method (for routine cryostat sections)*
   a. Take the section from the cryostat cabinet. Dry the section in air.
   b. Fix in picric–formalin (5 min).
   c. Rinse in water.
   d. Treat with Lugol's iodine (2–3 min).
   e. Rinse in distilled water.
   f. Treat with a 5% solution of sodium thiosulfate (2–3 min).
   g. Rinse in distilled water.
   h. Immerse in acidified potassium permanganate (2–3 min).
   i. Rinse in tap water.
   j. Bleach in 1% oxalic acid (until white).
   k. Wash for 5 min in running water.
   l. Leave in the phosphotungstic acid-hematoxylin (PTAH) solution (overnight).
   m. Shake off the excess fluid, but do not wash.
   n. Place in 70% alcohol (2 min).
   o. Dehydrate, clear in xylol, and mount in DePeX.

Result

This reaction is particularly good for identifying isolated myoblasts and for cross-striations in striated muscle. It is also good for cellular detail generally.

Solutions required for this method

Lugol's iodine

Iodine	1 g
Potassium iodide	2 g
Distilled water	100 mL

Acidified potassium permanganate

PTAH solution

Dissolve 0.1 g of hematoxylin in distilled water (up to 50 mL). Also, dissolve 2 g of phosphotungstic acid in distilled water (up to 50 mL may also be used). Make up the final volume to 100 mL and leave to ripen for 3 months.

    Rapid ripening can be achieved by adding 0.017 g of potassium permanganate to the solution.

6. *Iron hematoxylin*
   a. Fix dried cryostat sections in picric–formalin (5 min).
   b. Wash with water.
   c. Immerse in iron alum solution (5% ferric ammonium sulfate) for 30–45 min at 56°C. This step mordants the tissue.
   d. Wash well in distilled water.
   e. Stain in the hematoxylin solution at 56°C for 30–45 min.
   f. Wash well in distilled water.
   g. Differentiate in a 2% solution of ferric ammonium sulfate, that is, until the particular structures to be studied are clearly defined.
   h. Wash in running water for 5 min.

i. Dehydrate through a graded series of alcohols, clear in xylol, and mount in DePeX. (Some structures can be seen more clearly if the section is dehydrated and mounted in Euparal.)

Hematoxylin solution

Hematoxylin	0.5 g
Absolute ethyl alcohol	10 mL
Distilled water	90 mL

Allow to ripen for 4–5 weeks. Alternatively, add 0.1 g of sodium iodate to ripen immediately.

Methyl violet method for the demonstration of amyloid

1. Dry a fresh cryostat section in air.
2. Immerse in a 1% aqueous solution of methyl violet (2–5 min).
3. Rinse in distilled water.
4. Differentiate in 0.5%–1% acetic acid until the amyloid is pink and the rest of the tissue remains blue-violet.
5. Wash well in distilled water.
6. Mount in Farrant's medium.

To store "frozen" blocks in paraffin wax: After fresh sections for histology and histochemistry have been cut from correctly chilled tissue, the block of the tissue can be treated as follows:

1. Remove the tissue from the block-holder, place it in a refrigerator at 4°C, and leave for 1 h.
2. Immerse in formol–saline at room temperature. Fixation may take at least 24 h.
3. Dehydrate, clear, and embed in paraffin wax by whatever method is preferred for a normally fixed tissue.

These paraffin blocks can be treated in exactly the same way as those of tissues treated by conventional histological procedures. When required, they can be sectioned, dewaxed, and stained by the usual histological methods, and should yield preparations indistinguishable from tissues that have never been chilled.

2.9.4 Fixation, Dehydration, Embedding, and Staining

2.9.4.1 Introduction
The objective of this module is to learn a general procedure for fixation, dehydration, embedding, and staining of plant and animal materials like leaf, stem, root or callus tissue, liver, kidney, spleen, etc., for examination under light microscope using a microtome. The material is collected and cut into 5–10 nm thick pieces with a sharp

degreased blade, and immediately processed for fixation; pieces should not exceed 1–5 mm for semithin sections.

2.9.4.2 Material Preparation

1. The material should remain submerged at the bottom of the passing tubes or vials. When floating (even after 10 min in the fixing fluid), the bottles with their slightly loosened lids are kept in a vacuum desiccator attached to a vacuum pump. The material is placed under reduced pressure in the vacuum desiccator until it sinks (maximum period should not exceed 30 min: bubbling of fixing fluid must be avoided at all costs).
2. An alternative method (all other fixative mixtures excepting acrolein, glutaraldehyde, osmic acid) is to place small cotton pieces inside the passing tubes or vials at the surface of the fixing fluid so that the material remains completely submerged in the fluid.
3. In fixing fluids that are harmful (e.g., acrolein, etc., being a potent tear gas), the sinking of the material should be carried out inside the fume chamber.
4. Care should be taken to add adequate amount of fixing solution to the material (10:1 ratio).
5. The bottles and vials containing the fixing reagent and material should be tightly capped.
6. Fix the material for the appropriate time and at the required temperature.
7. After fixation, remove the cotton pieces (when used) from the fixing fluid carefully without any loss of material.
   Whenever indicated, the material is washed as per schedule, by transferring the passing tubes at the prescribed time intervals through a series of bottles containing the reagent for washing, or under running water prior to their transfer to the next reagent; the excess fluid is quickly absorbed on a blotting paper by touching the base of the passing tubes. If the material is contained in vials, the fixing fluid is decanted or taken out with a dropper/doctor's syringe and replaced immediately with the washing fluid.
8. Pieces of material are placed in the passing tubes (cylindrical glass tubes at both the ends); muslin cloth is tied on one end. Appropriate-sized labels giving the name of the plant, part or organ or the stage of development, date of fixation, name of the fixative, written with Indian ink or black lead pencil must be placed in these vials. The passing tubes with labels and the material are lowered into large, widemouthed glass bottles (250–500 mL capacity) containing the fixing fluid at the required temperature.

OR

The cut pieces may be directly put in small 5–10 mL glass vials containing the labels and fixing fluid.

2.9.4.3 Material Required

Callus tissue or rat liver, tertiary butyl alcohol, ethyl alcohol, acetone, acetic acid, chloroform, xylol, formalin, glass vials.

Solution preparation

Ethyl alcohol (52% or 70%)	90 mL
Acetic acid (glacial)	5 mL
Formalin (40% commercial)	5 mL
Total	100 mL

2.9.4.4 Protocol

2.9.4.4.1 Fixation. Fix the material for 12 h; the material can be left even for a week or more. Wash it by giving two changes of 50% or 70% alcohol as the case may be (alcohol strength used in fixative). While fixing the delicate tissues from aquatic habitats, lower percentage of alcohol is used. Equally good fixation is obtained when acetic acid is substituted with propionic acid.

2.9.4.4.2 Dehydration. The fixed material is passed through either of the following series:

1. Ethanol–xylene series
   (i) 25%, (ii) 50%, (iii) 70%, (iv) 90%, (v) 100%, (vi) 3:1 ethanol (100%):xylene (100%), (vii) ethanol:xylene (1:1), (viii) ethanol:xylene (1:3), (ix) xylene 1, and (x) xylene 2.
2. Ethanol–TBA series (material kept for 2–4 h at room temperature)
   (i) 25% ethanol, (ii) 50% ethanol, (iii) 95% ethanol:TBA:water (5:4:1) for 2–4 h at room temperature, (iv) 95% ethanol:TBA water (5:3.5:3), (v) 95% ethanol:TBA:water (5:3.5:3), (vi) 95% ethanol:TBA (1:1) for 2–4 h at room temperature, (vii) 100% ethyl alcohol:TBA (1:3), (viii) 100% anhydrous TBA 1, (ix) 100% anhydrous TBA II, and (x) anhydrous TBA III overnight.

OR

3. Acetone series (material kept for 1–2 h at room temperature)
   (i) 7.5%, (ii) 15%, (iii) 30%, (iv) 50%, (v) 70%, (vi) 90%, (vii) 100%, (viii) 100% acetone:100% chloroform (3:1) for 2–4 h at room temperature, (ix) acetone:chloroform (1:1), (x) acetone:chloroform (1:3), and (xi) 100% chloroform for 2 h at room temperature.

2.9.4.4.3 Infiltration. Melted paraffin wax is slowly added along the sides of the vial containing the material in xylene/TBA/chloroform as mentioned earlier in series 1, 2, or 3. The ratio of the solution to paraffin should be nearly 1:1, and the latter should form a distinct layer on the top of the solution; this is achieved by adding paraffin very slowly and carefully, touching the sides of the vials. TBA has a specific gravity lower than that of the paraffin wax, and the wax thus sinks upon the tissues. This can be prevented by adding 25% chloroform to TBA or by adding wax in liquid in very small increments and the solution homogenized to prevent additional wax from sinking and enveloping the tissues.

The vials containing the material are then stoppered and kept for 24 h at room temperature. The following day, the vials are left for another 24 h at 37°C–40°C in an oven, or approximately this temperature can be easily obtained by keeping the vials close to a table lamp with 100 W incandescent bulbs. The vials are shaken at intervals. When paraffin and xylene/TBA/chloroform have completely dissolved, 50% of the solution is poured off, and the same amount of fresh melted paraffin is added; the used wax can be stored separately in a container labeled as impure wax and can be used later to infiltrate another set of materials.

Vials are transferred to an oven at 50°C; paraffin slowly melts at this temperature. At an interval of approximately 3–4 h, half of the paraffin wax–xylene/TBA/chloroform mixture is decanted off and fresh paraffin is added. This is done for one day, and the next day, temperature of the oven is raised to 60°C, and the vials are now uncovered. Changes of fresh, pure, melted paraffin wax are provided as earlier for 3 days, and the vials are shaken well at intervals.

2.9.4.4.4 Embedding. Porcelain trays and "paper folded trays" are generally used to embed the material in paraffin; alternatively, lids of Coplin jars are also found to be useful for the purpose:

1. The inside surface of the lids or embedding trays is smeared with glycerin to facilitate removal of the block subsequently.
2. Hot paraffin wax is poured into the embedding trays, but only to half of its capacity, and it is allowed to solidify undisturbed.
3. The wax is poured off along with the material into a watch glass. If it has solidified, it has to be heated slowly and carefully along the sides of the vials but never at the bottom, as the material lying there might get damaged due to excessive heat.

OR

4. When the wax in the watch glass containing material to be embedded has solidified partly, its top portion is heated by passing a Bunsen burner flame over it in order to melt the upper layer of wax.
5. Quickly the material is transferred from the watch glass to the embedding mold using a round tipped forceps. The material can be oriented using a needle that is heated over a flame. During orienting, care should be taken to leave wide spaces between materials. The material gets lifted easily because of the capillary action of wax; an important precaution, which should be followed is that the procedure should be carried out as quickly as possible to avoid entrapping of any air bubbles.
6. Embedding molds are left undisturbed for 5–10 min.
7. Let the embedding mold, placed on hand, be kept afloat on the surface of cold water in a trough or tray until the surface of paraffin becomes sufficiently firm. The mold is plunged slowly beneath the surface of cold water; cooling of paraffin should be abrupt; otherwise, the blocks would become full of fluffy white patches.
8. The block is loosened by inserting a fine blade/needle between the block and the covering surface and lifting it gradually.
9. A label (containing the details of the material, fixative, date of fixation, embedding, etc.) can be inserted onto the block by heating the scalpel over a flame and melting the top portion of the block and slowly replacing the label with the scalpel.

2.9.4.4.5 Affixing Microtome Sections to Slides and Staining. The sections taken from paraffin blocks are to be transferred onto a slide for subsequent staining.

For transferring paraffin ribbons to the slide, the following steps are taken:

1. It is necessary to soak the slide in alcohol and to dry them with a clean, lint-free cloth. If the slides are reused or dirty, it is necessary to soak the slides in a standard cleaning solution, rinse them well in running water, soak them in alcohol, and allow them to dry. For general use, frosted-end slides work well. The data concerning the ribbon may be written on the slide by a diamond point pencil or a carborundum pencil.
2. The paraffin ribbon is cut into shorter units, which are approximately two-third or three-fourth the length of the coverslip to be used.

3. The upper surface of the slide is covered with a thin coating of Haupt's adhesive. To make Haupt's adhesive, 1 g of gelatin is dissolved in 100 mL of water at 90°C; the mixture is cooled to 30°C, and then 15 mL of glycerin is added. Finally, the adhesive is stirred and filtered and stored in a refrigerator. As a preservative, 2 g of phenol may be added. (Another excellent adhesive may be prepared by dissolving 5 g of gelatin in 1000 mL of warm distilled water, and then adding 0.5 g of chromium potassium sulfate to it.)

4. A few drops of 4% formalin are added to the slide and the ribbon is made to float on the slide by means of a needle, brush, or a toothpick. The ribbon segments are arranged neatly and compactly on the slide.

5. The slide is transferred to a slide warmer, the temperature of which is several degrees lower than the melting point of the paraffin or tissue mat used. More drops of 4% formalin are added to permit the tissue to stretch as it warms. When the tissue is fully extended, the slide is carefully removed from the slide warmer and allowed to cool. Staining material is embedded in the paraffin ribbons.

2.9.4.4.6 Precaution.

1. During the course of handling the paraffin-embedded materials and subsequent staining, there is a great danger that the sections may become loosened from the slide and be lost. This can be avoided by coating the slides with collidine. The collidine coating will not interfere with the sections and will usually keep the sections on the slide. The paraffin is removed from the sections in the usual manner and the slides placed in 95% alcohol.

2. The slide is dipped in 0.5% solution of collidine, alcohol, or ether.

3. The slide is allowed to drain for a moment, and placed in 95% alcohol to harden the collidine.

Affixing the microtome paraffin sections to the slide is one of the important processes. The slides after affixing may be stained with one of the following:

1. Safranin–Fast green
2. Safranin–Aniline blue
3. Delafield's hematoxylin

2.9.4.4.7 Materials. Coplin jars, xylene, slide warmer, HCl, light microscope, collidine, coverslips, Canada balsam, safranin, absolute alcohol, ether, fast green, aniline blue, methyl salicylate, clove oil, hematoxylin, ammonium aluminum sulfate, methyl alcohol, and sodium carbonate.

2.9.4.4.8 Stain Preparation.

1. *1% Safranin:* Weigh 1 g of safranin and dissolve in 35% alcohol and make up the volume to 100 mL.

2. *0.5% Fast green:* Dissolve 500 mg of fast green in 50% clove oil and 50% absolute alcohol in a total of 100 mL.

3. *Aniline blue (saturated solution):* Dissolve aniline blue in absolute alcohol until saturation and dilute to 50% with clove oil.

4. *Delafield's hematoxylin:* Weigh 4 g of hematoxylin and add 25 mL of 95% alcohol, and carefully add this to 400 mL of a standard aqueous solution of ammonium aluminum sulfate. Expose this solution to light and air for days to a week. After exposure, add 10 mL of glycerin and 100 mL of methyl alcohol. Allow to stand to expose to air for several months until the color is of a dark wine shade. Filter and use.

2.9.4.4.9 Staining.

1. When the slide is cool, drain off most of the formalin solution of the slide into a paper to fold well in a Petri dish, and use a needle to prevent the sections from floating off the slide. Arrange the tissue with dissecting needles, remove the remaining liquid from the slide using a clean cloth, and carefully wipe the surface of the slide not covered by sections.
2. Remove the slide from the slide warmer (Figure 2.12) and allow it to dry overnight; the slides may be stored in this condition for an indefinite period.

The materials embedded in paraffin ribbons may be stained with a single stain (Delafield's hematoxylin) or a combination of two stains (Safranin–fast green, Safranin–aniline blue, Crystal violet–fast green, etc.).

2.9.4.4.9.1 Staining Method with Safranin–Fast Green.

1. Remove the paraffin from the sections by placing the sections on the slide in xylene for 5 min and then in a 1:1 mixture of xylol and absolute alcohol for another 5 min. The xylene and all other solutions discussed in these histological procedures are kept in Coplin stain jars or in some other type of container in which the slides are held such that they will not move against one another.
2. Partially hydrate the sections by passing through a series of alcohols of descending concentrations: absolute, 95%, 70%, and 50% (5 min in each).
3. Stain in safranin for about 1–24 h.
4. Wash in water, pass briefly through acidified 70% alcohol to destain excess of safranin, and pass rapidly through 95% and absolute alcohol.

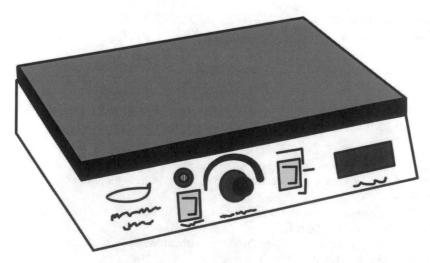

Figure 2.12　Slide warmer.

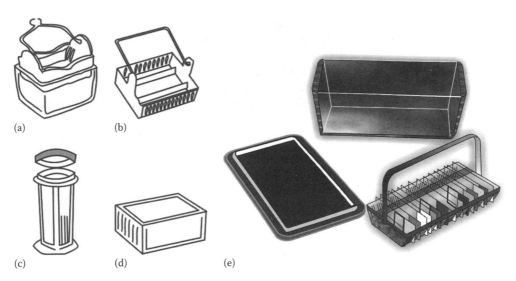

Figure 2.13 Microscope slide-staining dishes: (a) staining dish and slide rack, (b) metal slide-holder rack, (c) Coplin jar, (d) staining dish with slots, and (e) staining chamber.

5. Counterstain with fast green for 30 s.
6. Differentiate the fast green by placing it in a mixture of 50% clove oil, 25% absolute alcohol, and 25% xylene. Use two changes for 5–10 min each.
7. Place the sections in xylene, making three changes at least for 15 min. Mount each section with a coverslip and Canada balsam or any other mounting medium, and examine under light microscope.

Result: Nucleoli, chromosomes, cuticle, and lignified cell walls stain red; the remaining structures stain green. This is an easy and useful stain combination.

(*Note: Figure 2.13 shows the instruments required for the staining procedure.*)

2.9.4.4.9.2 Staining with Safranin–Aniline Blue.
1. Steps 1–3 are identical to the first three of the preceding procedure.
2. Wash sections in water; destain if necessary in acidified water.
3. Counterstain with aniline blue.
4. Place in a mixture of 50% alcohol, 25% alcohol, and 25% xylene or in methyl salicylate.
5. Finally, place in three changes of xylene for 15 min each, and mount the coverslip with Canada balsam and examine under light microscope.

Result: This is a good stain for gymnosperm tissue, although it can be used for most other tissues as a substitute for fast green. The color combination is similar to the safranin–fast green, with blue replacing green. (All soft tissues take blue stain.)

2.9.4.4.9.3 Staining with Delafield's Hematoxylin.
1. Remove the paraffin and hydrate the sections through alcohol series to water.
2. Place in Delafield's hematoxylin for 5–30 min.

3. Wash in running tap water for 5 min.
4. Place in water acidified with a few drops of HCl for 1–2 min. Do not allow sections to remain in this solution for a long time.
5. Return the sections to tap water and wash until they turn purple. If this does not occur, add a small amount of sodium carbonate to the water.
6. Dehydrate rapidly through the alcohol series (a few minutes in each member of the series).
7. Place in three changes of xylene and mount with a coverslip with Canada balsam.

Result

Chromosomes, nuclei, resting nuclei, mitochondria, and other cytoplasmic particles will appear deep blue-black; cytoplasm will appear blue gray.

Suggested Readings

Bltensky, L., R. Elus, A.A. Silcox et al. 1962. Histochemical studies on carbohydrate material in liver. *Ann Histochim* 1: 9–14.

Chayen, J. and E. F. Denby. 1968. *Biophysical Technique: As Applied to Cell Biology*. London, Methuen, U.K.

Gudrun, L. 2006. *Histotechnik. Praxislehrbuch für die Biomedizinische Analytik. (Histology: Practical Textbook for Analytical Biomedicine)*. Wien, Austria: Springer.

Holger, L. 2007. *Laser Microtomy*. Verlag GmbH, Weinheim, Germany: Wiley-VCH.

Klaus, H. 2006. Das Schneiden mit dem Mikrotom. Mikrobiologische Vereinigung München e. V. Accessed February 15, 2009.

Peachey, L. D. 1958. Thin sections: A study of section thickness and physical distortion produced during microtomy. *J Biophys Biochem Cytol* 4: 233–242.

Quekett, J. 1848. *A Practical Treatise on the Use of the Microscope*. London, U.K.: Hippolyte Bailliere.

Roberts, D. R. 1966. A rapid staining method giving sharp nuclear definition in frozen sections. *J Med Lab Tech* 23: 119–120.

Werner, L., L. Werner, and F. Jochen. 1998. *Histologie: Zytologie, Allgemeine Histologie, Mikroskopische Anatomie. (Histology: Cytology, General Histology, Microscopial Anatomy)*. Berlin, Germany: Walter de Gruyter.

Important Links

http://www.archive.org/details/practicaltreatis00quekuoft
http://jcb.rupress.org/cgi/reprint/4/3/233.pdf

3

pH and Buffers

3.1 Introduction

All physiological reactions take place in aqueous acidic or alkali environments, and pH plays a significant role for optimal reactions to occur. pH measurement is one of the most basic and necessary skills in any life science laboratory. Essentially all biochemical reactions depend strongly on pH, in which [H$^+$] is the molar hydrogen ion concentration. The function and physical characteristics of biological molecules are highly sensitive to the pH of the environment. Common biological buffers are used with the appropriate pH, usually close to neutral for most biological applications. pH measurement is accomplished with a commercial pH meter by simply immersing two electrodes into a solution and reading the pH value on a dial. However, it is important to know how the instrument measures pH, because several factors can cause the observed value to differ from the actual value. A pH meter measures the voltage between two electrodes placed in a solution. The heart of the system is an electrode whose potential is pH-dependent. This chapter describes the basic pH measurement theory to critical parameters for the technique and covers the correct use, handling, and storage of pH instruments, and the role of different buffers in biological functions. Experiments have been designed for the basic understanding of the role of hydrogen ion concentrations in biological functions.

3.2 Acids and Bases

3.2.1 Definitions

3.2.1.1 Acids and Bases

An acid is a substance that forms H$^+$ ions as the only positive ion in aqueous solution. Bronsted defined an acid as any substance that can donate a proton and a base as a substance that can accept a proton:

$$HCl \rightarrow H^+ + Cl^-$$

$$CH_3COOH \longrightarrow H^+ + CH_3COO^-$$

$$NH_4 \rightarrow NH_3^- + H^+$$

And the generalized expression would be

$$HA \longrightarrow H^+ + A^-$$

The corresponding bases are now reacting with a proton:

$$Cl^- + H^+ \longrightarrow HCl$$

$$CH_3COO^- + H^+ \longrightarrow CH_3COOH$$

$$NH_3 + H^+ \rightarrow NH_4^+$$

The corresponding base for the generalized weak acid HA is

$$A^- + H^+ \longrightarrow HA$$

Each acid, therefore, has a conjugate base:

$$Acid \rightarrow Base + H^+$$

3.2.1.2 Alkali

Alkalis are all Arrhenius bases, which form hydroxide ions (OH^-) when dissolved in water. Alkalis are normally water-soluble, although some like barium carbonate is only soluble when reacting with an acidic aqueous solution:

$$KOH \rightarrow K^+ + OH^-$$

3.2.1.3 Ampholytes

A molecule, carrying both a positive and a negative charge, is also known as ampho-teric. Amphoteric compounds can act as both acids and bases (Table 3.1).

It is convenient to write acid–base equilibria as shown earlier; the proton does not exist as such, but is usually solvated. In aqueous media, the hydrogen ion exists as the hydronium ion (H_3O^+):

$$H^+ + H_2O \rightarrow H_3O^+$$

TABLE 3.1   Some Examples of Acids and Bases

Acid	Conjugate Base
$HCl \rightarrow H^+ + Cl^-$	
$CH_3COOH \rightarrow H^+ + CH_3COO^-$	
$NH_4^+ \rightarrow H^+ + NH_3$	
$H_2CO_3 \rightarrow H^+ + HCO_3^-$	
$HCO_3^- \rightarrow H^+ + CO_3^-$	
$H_2O \rightarrow H^+ + OH^-$	
$H_3O \rightarrow H^+ + H_2O$	

3.2.2 Strength

3.2.2.1 Strong Acids or Bases

These compounds are completely ionized in solution so that the concentration of free H^+ or OH^- is the same as the concentration of the acid or base:

Strong acid (nitric acid): $HNO_3 \rightarrow H^+ + NO_3^-$
Strong base (sodium hydroxide): $NaOH \rightarrow Na^+ + OH^-$

3.2.2.2 Weak Acids or Bases

These compounds dissociate only to a limited extent, and the concentration of free H^+ and OH^- depends on the value of their dissociation constants. Water acts as a solvent in biological systems:

$$\text{Weak acid (formic acid): } HCOOH \rightarrow H^+ + HCOO^-$$

$$\text{Weak base (aniline): } C_6H_5NH_2 + H^+ \rightarrow C_6H_5NH_3^+$$

3.3 Hydrogen Ion Concentration and pH

3.3.1 Definition of pH

pH can be viewed as an abbreviation for power of hydrogen or more completely, power of the concentration of hydrogen ion. The hydrogen ion concentration of most solutions is extremely low and, in 1909, Sorensen introduced the term pH as a convenient way of expressing hydrogen ion concentration, which avoids the use of cumbersome numbers and defines pH as equal to the negative log of the hydrogen ion concentration, or $pH = -\log [H^+]$. Using the Brønsted–Lowry approach, that would be $pH = -\log [H_3O^+]$. In strict sense, it is the negative logarithm of the hydrogen ion activity, but in practice, the hydrogen ion concentration is usually taken. This is virtually the same as the activity, except in strongly acidic solutions.

The value of using pH can be seen in the case of human blood, which has an extremely low hydrogen ion concentration:

$$\text{Plasma } H^+ = 0.398 \times 10^{-7}$$

$$\text{Plasma } pH = -\log(0.398 \times 10^{-7}) = 7.4$$

3.3.2 Water Dissociation

Derivation of K_w: As per conductivity measurements, water is very weakly ionized, and at 25°C, the concentration of hydrogen ions is only 10^{-7} mol/L:

$$H_2O \rightarrow H^+ + OH^-$$

The equilibrium constant for the dissociation of water is given by

$$K = \frac{[H^+][OH^-]}{[H_2O]}$$

Now the concentration of water to all intents and purposes is constant, so we can write

$$K_w = [H^+][OH^-]$$

The ionic product of water at 25°C is 10^{-14}, so the pH of pure water at 25°C is 7:

$$[H^+] = [OH^-] = 10^{-7}$$

$$pH = -\log_{10}[H^+] = 7$$

3.3.2.1 Temperature and K_w

At other temperatures, pH at neutrality is not 7 since K_w varies with temperature (Table 3.2). Even a small change in temperature from 37°C to 40°C causes a percentage increase in hydrogen and hydroxyl ions; hence, a slight rise or fall in temperature may produce a profound biological change in a living system sensitive to hydrogen ion concentration.

3.3.3 Dissociation of Acids and Bases

3.3.3.1 Strong Acids

These are compounds in which complete dissociation to hydrogen ions and the conjugate base occurs so that the hydrogen ion concentration is the same as that of the acid. The pH of such solutions can, therefore, be very easily calculated:

TABLE 3.2 Ionic Product of Water and pH of Neutrality at Various Temperatures

°C	K_w	pH of Neutrality
0	$0.12 \times 10^{-14} = 10^{-14.94}$	7.97
25	$1.03 \times 10^{-14} = 10^{-14.00}$	7.00
37	$2.51 \times 10^{-14} = 10^{-13.60}$	6.80
40	$2.95 \times 10^{-14} = 10^{-12.77}$	6.77
75	$16.90 \times 10^{-14} = 10^{-12.77}$	6.39
100	$48.00 \times 10^{-14} = 10^{-12.32}$	6.16

Notes: The ionic product of water at 25°C is 10^{-14}; so, the pH of pure water at 25°C is 7.
$[H^+] = [OH^-] = 10^{-7}$.
$pH = -\log_{10}[H^+] = 7$.

(a) 0.01 mol/L HCl: $pH = -\log_{10}(10^{-2}) = 2$

(b) 0.1 mol/L HCl: $pH = -\log_{10}(10^{-1}) = 1$

(c) 0.01 mol/L NaOH: $[H^+] = \dfrac{K_w}{[OH^-]} = \dfrac{10^{-14}}{10^{-2}} = 10^{-12}$

$$pH = -\log_{10}(10^{-12}) = 12$$

3.3.3.2 Weak Acids

The Henderson–Hasselbalch equation: Weak acids are only slightly ionized in solution, and a true equilibrium is established between the acid and the conjugate base. If HA represents a weak acid, then

$$HA \rightarrow H^+ + A^-$$

According to the law of mass action, K_a, the acid dissociation constant is defined as

$$K_a = \frac{[H^+][A^-]}{[HA]}$$

$$[H^+] = \frac{K_a[HA]}{[A^-]}$$

Taking negative logarithms,

$$-\log_{10}[H^+] = -\log_{10} K_a + -\log_{10}\frac{[HA]}{[A^-]}$$

$$pH = pK_a + \log_{10}\frac{[A^-]}{[HA]}$$

In general terms,

$$pH = pK_a + \log_{10}\frac{[\text{Conjugate base}]}{[\text{Acid}]}$$

The activities of A^- and HA are not always known, so it is convenient to express A^- and HA as concentration terms. Thus,

$$pH = pK_a + \log\frac{C_{A^-}}{C_{HA}} + \log\frac{f_{A^-}}{f_{HA}}$$

where f_{A^-} and f_{HA} are the activity coefficients of A^- and HA, respectively. Since $\log(f_A - /f_{HA})$ is constant for a given acid, these activity coefficients can be incorporated into the pK_a term of a given apparent dissociation constant pK_a'.

$$pH = pK_a' + \log \frac{C_A}{C_{HA}}$$

This relationship is known as the Henderson–Hasselbalch equation and is valid over the pH range 4–10, where the hydrogen and hydroxyl ions do not contribute significantly to the total ionic concentration.

pK_a: This, as previously defined, is the negative logarithm of the acid dissociation constant of a weak acid. Another way of defining pK_a is the pH at which the concentrations of the acid and its conjugate base are equal or the pH at which the acid is half-titrated:

$$pH = pK_a + \log 1$$

$$pH = pK_a$$

3.4 Measurement of pH

3.4.1 pH Indicators

An approximate idea of the pH of a solution can be obtained using indicators. These are organic compounds of natural or synthetic origin whose color is dependent upon the pH of the solution. Indicators are usually weak acids that dissociate in solution:

$$\text{Indicator} = \text{Indicator}^- + H^+$$

Applying the Henderson–Hasselbalch equation,

$$pH = pK_{In} + \log_{10} \frac{[\text{Indicator}^-]}{[\text{Indicator}]}$$

The two forms of the indicator have different colors and, as mentioned earlier, the actual color of the solution will depend upon pK_{In} and pH. The greatest color change occurs around pK_{In}, and this is where the indicator is most useful. For example, if a solution has a pH close to 6, then bromocresol purple with a pK_{In} of 6.2 is the best indicator to use. However, this color change occurs over a wide pH range; so, indicators will only give an approximate indication of pH. Another disadvantage is that indicators are affected by oxidizing agents, reducing agents, salt concentration, and protein, so these facts must be borne in mind when using them. A final precaution to be taken when using them is to add only a small quantity of indicator to the solution under examination; otherwise, the acid–base equilibrium of the test solution may be displaced, and the pH is changed. Indicators are probably of more value in determining the end point of a titration, and some of those commonly employed are shown in Table 3.4.

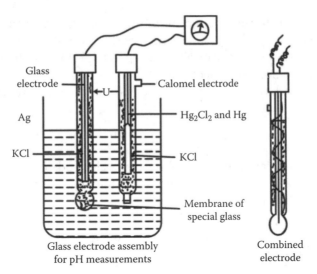

Figure 3.1 pH meter.

3.4.2 Measurement of pH

The most convenient and reliable method for measuring pH is by the use of a pH meter, which measures the electromotive force (emf) of a concentration of cell formed from a reference calomel electrode, the test solution, and a glass electrode sensitive to hydrogen ions (Figure 3.1).

3.4.2.1 Glass Electrode
The pH glass electrode, although somewhat mechanically fragile, resists a variety of sample media and with the exception of hydroxide is largely free from interferences. Moreover, pH-sensitive glass electrodes form the basis of many successful sensors for environmentally sensitive gases. Thus, glass membranes represent an important class of solid membrane ion selective electrodes (ISEs). It consists of a very thin bulb about 0.1 mm thick blown on to a hard glass tube of high resistance. Inside the bulb is a solution of hydrochloric acid (0.1 mol/L) connected to a platinum wire via a silver–silver chloride electrode, which is reversible to hydrogen ions. A potential is developed across the thin glass of the bulb, which depends on the pH of the solution in which it is immersed. This potential is not readily affected by salts, protein, or oxidizing and reducing agents, so the electrode can be used in a wide variety of media.

The glass electrode in the test solution constitutes a half-cell, and the measuring circuit is completed by a reference electrode, which is not sensitive to hydrogen ions.

3.4.2.2 Calomel Electrode
The two most popular types of the reference electrode are the saturated calomel and silver/silver chloride systems. Both types of reference electrodes exhibit many ideal

characteristics, which include maintaining fixed potential over time and temperature, having long-term stability, and returning to the initial potential after being subject to small currents. It may be either a separate probe or built around the glass electrode giving a combination electrode.

pH meter is formed by the linking of these two electrodes; therefore,

$$E = E_{ref} - E_{glass},$$

where E_{ref} is the potential of the calomel reference electrode (which at normal room temperature) is +0.250 V, and is the potential of the glass electrode, which depends on the pH of the solution under test (pH_s):

$$E_{glass} = 0.342 - 0.058 pH_s.$$

So,

$$E = 0.250 - (0.342 - 0.058 pH_s)$$

$$= -0.092 + 0.058 pH_s.$$

This relationship can be changed by the presence of other potentials in the cell, but these are usually constant and can be allowed for when calibrating the instrument.

The glass electrode has a very high resistance (10^6–10^8 Ω), so, a potentiometer of high-input impedance is needed to measure the potential.

3.5 Titration Curves

The titration curve of 0.1 N HCl is also represented in Figure 3.2. When a strong base is mixed with a solution of an acid, and the pH measured, a plot of the base added against pH recorded can be obtained, and this is known as the titration curve. The curves will have the same shape with the exception of the strong acid HCl. Similar curves are obtained when a base is titrated with strong acid.

3.5.1 Strong Acid and a Strong Base

There is little change in pH on adding the base until complete neutralization, when only a slight excess of base causes a large increase in pH. In effect, the strong acid resists a change in pH or acts as a buffer solution until close to the neutralization point. This can be seen when HCl is titrated with NaOH (Figure 3.2).

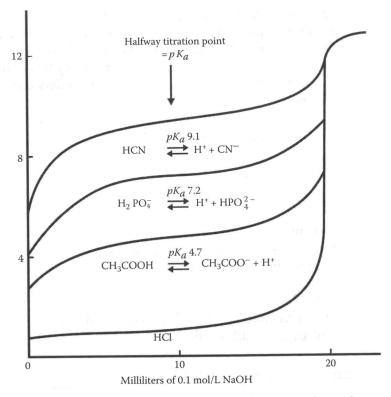

Figure 3.2 pH titration curves of 20 mL solutions of some 0.1 mol/L acids.

3.5.2 Weak Acid and a Strong Base

All titration curves of a weak acid or base titrated with a strong base or acid are of the same type since one has a buffer solution present, whose pH changes according to the Henderson–Hasselbalch equation. The pK_a values are different for each acid, but the general shape of the curve is the same in all cases (Figure 3.2).

3.6 Buffer Solutions

3.6.1 Theory

A buffer solution is one that resists pH change on the addition of acid or alkali. Most commonly, a buffer solution consists of a mixture of a weak Bronsted acid and its conjugate base; for example, mixtures of acetic acid and sodium acetate or of ammonium hydroxide and ammonium chloride are buffer solutions. Such solutions are used in many biochemical experiments where the pH needs to be accurately controlled.

From the Henderson–Hasselbalch equation, the pH of a buffer solution depends on two factors: one is the pK value, and the other is the ratio of the salt to acid. This ratio is considered to be the same as the amount of salt and acid mixed together over the pH range 4–10, where the concentration of hydrogen and hydroxyl ions is very low and can be ignored. Let us take as an example, acetate buffers consisting of a mixture of acetic acid and sodium acetate:

$$CH_3COOH \rightarrow CH_3COO^- + H^+$$

$$CH_3COONa \rightarrow CH_3COO^- + Na^+$$

Since acetic acid is only weakly dissociated, the concentration of acetic acid is almost the same as the amount put in the mixture; likewise, the concentration of acetate ion can be considered to be the same as the concentration of sodium acetate placed in the mixture, since the salt is completely dissociated.

3.7 What Is the pH of a Mixture of 5 mL of 0.1 mol/L Sodium Acetate and 4 mL of 0.1 mol/L Acetic Acid?

Concentration of $CH_3COO^- = \dfrac{5}{9} \times 0.1 \, mol/L$

Concentration of $CH_3COOH = \dfrac{4}{9} \times 0.1 \, mol/L$

pK_a of acetic acid at 25°C $= 4.76$

Therefore,

$$pH = 4.76 + \log\frac{5}{4}$$

$$= 4.76 + (+0.097)$$

$$= 4.86$$

3.8 How the pH Changed on Adding 1 mL of 0.1 mol/L HCl in the Previous Mixture

Addition of HCl provides H^+ that combines with the acetate ion, to give acetic acid. This reduces the amount of acetate ion present and increases the quantity of undissociated acetic acid, leading to an alteration in the salt/acid ratio and hence to a change in pH:

Concentration of $CH_3COO^- = \dfrac{5}{10} \times 0.1 - \dfrac{1}{10} \times 0.1 = 0.04 \, mol/L$

$$\text{Concentration of CH}_3\text{COOH} = \frac{4}{10} \times 0.1 + \frac{1}{10} \times 0.1 = 0.05 \text{ mol/L}$$

Therefore,

$$\text{pH} = 4.76 + \log \frac{0.04}{0.05}$$

$$= 4.76 + (-0.097)$$

$$= 4.66$$

The pH of the solution has been reduced from 4.86 to 4.66, a change of only 0.2 of a unit, whereas if the HCl had been added to distilled water, the pH would be 2. The solution has, therefore, acted as a buffer by resisting the pH change on the addition of acid.

Buffer value β: Buffer solutions vary in the extent to which they resist pH changes, and in order to compare different buffer solutions, Van Slyke introduced the term *buffer value.* When acid or alkali is added to a buffer solution, a titration curve is obtained. The slope of this curve is given by $dB/d(pH)$, where dB is the increment of strong acid or base added in mol/L, and $d(pH)$ is the change in the pH increment. This slope is the buffer value β, which is always positive since dB is negative when acid is added, causing a negative change in pH. As can be seen from the titration curves illustrated in Figure 3.2, the buffer value is maximum at pK_a.

3.9 Buffers Used in Biochemical Experiments

3.9.1 Common Laboratory Buffers

The actual buffer chosen for a particular experiment needs to be selected with care, as experimental results may sometimes be due to specific ion effects and not the pH. For example, borate forms complexes with sugars, and citrate readily combines with calcium.

The pH in many biological experiments often needs to be kept constant in the pH range 6–8, but there are few weak acids or bases that are effective buffers in this part of the pH scale, and particularly around 7.4, the pH of blood.

3.9.1.1 Bicarbonate
This buffer spontaneously liberates carbon dioxide and must, therefore, be maintained in an atmosphere of CO_2. Bicarbonate has a pK_a of 6.1, so the buffering capacity around pH 7.4 is poor.

3.9.1.2 Phosphate
This is probably the most popular buffer, but phosphate readily forms complexes with heavy metals. Phosphate is also inconvenient as it plays an active part in a number of

biochemical reactions where it can act as an activator, inhibitor, or metabolite. The buffering capacity above pH 7.5 is also poor.

3.9.1.3 Tris

This is a popular buffer since it can be used with heavy metals, but it also acts as an inhibitor in some biochemical systems. Tris has a high lipid solubility and penetrates membranes, which can be a disadvantage. However, its greatest problem is the temperature effect, which is often overlooked. A tris buffer of pH 7.8 at room temperature has a pH of 8.4 at 4°C and 7.4 at 37°C; so, the hydrogen ion concentration increases 10-fold from the preparation of material at 4°C to its measurement at 37°C. Tris also has a poor buffering capacity below pH 7.5.

3.9.1.4 Zwitterionic Buffers

These buffers were introduced by Good and coworkers to overcome the disadvantages of the traditional materials. As the name suggests, they contain both negative and positive groups; so, they do not readily penetrate membranes. They are now widely used in many biological experiments, and HEPES has proved to be particularly useful.

$$HO-CH_2-CH_2-\overset{+}{NH}\diagdownN-CH_2-CH_2-SO_3^-$$

3.10 pH and Life

Most cells can only function within very narrow limits of pH and require buffer systems to resist the changes in pH that would otherwise occur in metabolism. The three main buffer systems in living materials are protein, bicarbonate, and phosphate, and the relative importance of each depends on the type of cell and organism.

3.10.1 Animals

The most important buffering system in mammalian plasma is bicarbonate:

$$H_2CO_3 \rightarrow HCO_3^- + H^+$$

From the Henderson–Hasselbalch equation,

$$pH = pK_a + \log_{10}\frac{[HCO_3^-]}{[H_2CO_3]}$$

The plasma pH, therefore, depends on the ratio of bicarbonate to carbonic acid and not on the absolute concentrations. Any tendency for the pH to change is buffered and can be corrected by adjusting this ratio. For example, large quantities of acids formed during normal metabolism react with bicarbonate to form weakly dissociated carbonic acid, so that free hydrogen ions are effectively mopped up. At the same time, carbonic acid is removed at the lungs as carbon dioxide, thus maintaining the pH of the plasma. The kidneys also play an important role in the maintenance of acid–base balance by adjusting the excretion of acid or base in the urine, so that the pH of urine can normally vary from 4.8 to 7.5 in man.

The extracellular fluid is usually slightly alkaline at pH 7.4, while the average cytoplasmic pH of many animal cells is about 6.8, which at 37°C is neutral. It is almost certain that the pH in the organelles will differ from 6.8, and the pH at most membrane surfaces will be lower than this due to the absorption of H^+ on the negatively charged surface.

3.10.2 Plants

The average cytoplasmic pH of plants is similar to that of animals, but the cell sap of the majority of land plants is acidic, and lies within the pH range 5.2–6.5. Some plant juices are very acidic (orange and grapefruit, pH 3), but the acid seems to be present in vacuoles and, therefore, segregate from the rest of the cytoplasm.

3.10.3 Bacteria

The internal pH of a majority of bacteria is near neutrality, although many forms can grow well at pH 6 or 9, a 1000-fold difference in hydrogen ion concentration. *Thiobacilli* sp. will actually grow at pH 0, which is that of 1 mol/L HCl, while some fungi grow at pH 11. Most organisms pathogenic to man have, as might be expected, an optimum rate of growth at pH 7.2–7.6.

Clearly, the simpler forms of life appear to be able to survive a wide pH range of the external medium, though it does not follow that the internal pH varies to any extent.

3.11 Care and Use of the pH Meter

The precise method of operation of a pH meter depends on the particular model; so, detailed instructions here are inappropriate. However, some general guidance on the use and care of the electrode assembly will be useful.

1. *New glass electrode*: New electrodes must be soaked in 0.1 mol/L HCl or distilled water for several hours before use.
2. *Mixing*: The solution must be thoroughly stirred before measuring the pH, and this is ideally carried out with a stream of pure nitrogen, although a magnetic stirrer is probably most convenient for class work.
3. *Temperature*: The beaker should be thermostated, since K_w, pK, and the pH of the standard solution all vary with temperature. The temperature compensator on the instrument does not allow for the previous factors, but only for the change in emf with the temperature of the electrode assembly.
4. *Electrode contamination*: The electrodes should be washed in distilled water before and after use, and must not be touched. In particular, they should be thoroughly washed after measuring the pH of a solution, with a high concentration of biological

TABLE 3.3   Primary Standards for Calibration of a pH Meter

	pH	
	25°C	37°C
1. 0.05 mol/L potassium hydrogen phthalate	4.01	4.02
2. 0.025 mol/L potassium dihydrogen phosphate 0.025 mol/L disodium hydrogen phosphate	6.86	6.84
3. 0.01 mol/L sodium tetraborate	9.18	9.06

macromolecules as these may adhere to the glass and distort subsequent pH measurements unless removed.

5. *Standardization*: The pH meter is calibrated before use by means of a standard solution. The meter should be calibrated with a solution whose pH is close to that under test, and several convenient standards are given in Table 3.3.

6. *Measurements at extreme pH values*: The glass electrode remains accurate down to quite low pH values, but in strong alkali the recorded pH is often too low due to interference by other ions, particularly sodium. When lithium glass is used, this error is considerably reduced, but ordinary glass electrodes cannot be used above pH 10. When titrating at these high pH values, it is best to use KOH and not NaOH, and it is also advisable to use a burette with a soda lime trap to prevent absorption of CO_2.

7. *Storage*: The pH meter should be switched to zero, but the mains switch left on. After use, the electrodes are stored in distilled water and must never be allowed to dry out. If this does occur, then the electrodes need to be soaked in water and calibrated frequently before their measurements can be reliedon.

3.12 Titration Curves

3.12.1 Practical Limits of Titration Curves

If 0.1 mol/L strong acid or base is used in a titration, the curves will asymptotically approach pH 1 or pH 13 after complete neutralization. Likewise, the limits for 0.01 mol/L solutions will be pH 2 or pH 12; so, pK_a values below 2 or above 12 cannot be determined using 0.01 mol/L solutions.

3.12.2 Solvent Correction

Experimental titration curves must be corrected for the amount of acid or base consumed in titrating the solvent, usually distilled water. This is carried out as follows:

1. Plot the titration curves for the same volume of sample and water.
2. Select a pH value on the curve for the sample and note the volume of acid or base used; let this be X mL. Likewise, note the amount of acid or base consumed in order to bring the water to the same pH value; let this be Y mL.

3. The actual amount of acid consumed in the titration of the sample is, therefore, given by $(X–Y)$ mL. Repeat this for a number of pH values and plot the corrected titration curve.

3.12.3 Determination of pK_a Values Can Be Obtained from Titration Data by Three Methods

1. The pH at the point of inflection is the pK_a value, and this may be read directly. A more convenient way is to plot $dB/d(pH)$, the buffer value, against pH when a maximum is obtained at pK_a.
2. By definition, the pK_a value is equal to the pH at which the acid is half-titrated. The pK_a can, therefore, be obtained from the knowledge of the end point of the titration.
3. The ratio of salt/acid can be calculated from the experimental data, and a graph prepared of $\log_{10}$ salt/acid against pH. The intercept on the axis is the pK_a value.

3.13 Determination of pH Using Indicators

3.13.1 Materials

Indicators (solutions in aqueous ethanol of the indicators in Table 3.4; these are available commercially); samples to be tested (saliva, orange juice, lemonade, egg white) diluted 1 in 10.

3.13.2 Procedure

Pipette 1–2 mL of each sample into a test tube; add two drops of methyl red indicator solution and observe the color. Repeat the exercise with other indicators until the approximate pH is found by comparing the colors of the sample with those of the acidic, intermediate, and basic forms of the indicator (Table 3.4). Pipette 2 mL of laboratory distilled water into a test tube and determine the pH using the indicators as described.

Explain your observations.

TABLE 3.4 Color Change and Useful pH Range of Some Common Indicators

Indicator	Color of Indicator		pK_a	Useful pH Range
	Acid	Base		
Thymol blue	Red	Yellow	1.7	1.2–2.8
Bromophenol blue	Yellow	Blue	4.0	3.0–5.0
Methyl red	Red	Yellow	5.0	4.3–6.1
Bromocresol purple	Yellow	Purple	6.3	5.5–7.0
Phenol red	Yellow	Red	7.9	6.8–8.2
Phenolphthalein	Colorless	Red	9.7	8.3–10.0

3.14 Titration of a Mixture of a Strong and a Weak Acid

3.14.1 Principle

The relative amounts of HCl and weak acids are determined by titration with an indicator such as thymol blue that has two pH ranges. The pH ranges and the accompanying color changes for thymol blue are

pH 1.2–2.8—Red–Orange–Yellow

pH 8.0–9.6—Yellow–Green–Blue

Thus, titration to the first color change gives a measure of the amount of free HCl, while titration to pH 9.6 gives the total acidity. If genuine gastric juice is not readily available, then a mock sample can he prepared by mixing hydrochloric and acetic acids.

3.14.2 Materials

Gastric juice or mock sample prepared by mixing 40 mL of 0.1 mol/L hydrochloric acid, 15 mL of 0.1 mol/L acetic acid, and 45 mL of water; potassium hydroxide (0.1 mol/L); thymol blue (0.1% *w/v* in 20% *v/v* ethanol); microburette 10 mL.

3.14.3 Procedure

Place 10 mL of sample in a small conical flask; add four drops of thymol blue indicator and titrate with 0.1 mol/L potassium hydroxide until the reddish orange color changes to orange yellow; carefully note the volume of potassium hydroxide used. Continue the titration until the full blue color forms, and again note the volume of alkali required. Calculate the "free HCl" and total acidity of the sample as mL 0.1 mol/L acid/100 mL of sample. Normal values for genuine gastric juice are

Free HCl 20–30 mL: 0.1 mol/L acid/100 mL

Total acidity 30–70 mL: 0.1 mol/L acid/100 mL

Compare the results obtained on the mock sample with those expected on theoretical grounds.

3.15 Titration of a Strong Acid with a Strong Base

3.15.1 Materials

Hydrochloric acid (0.1 mol/L); potassium hydroxide (0.1 mol/L); burette 25 mL; pH meter.

3.15.2 Procedure

Pipette 20 mL of 0.1 mol/L HCl into a 100 mL beaker and measure the pH. Titrate this solution with 0.1 mol/L KOH and measure the pH after the addition of 1 mL aliquot of alkali. When nearing the end point, reduce the increments of alkali added to give a reasonable change in pH.

Plot the titration curve and compare it with that calculated from theory as previously described.

3.16 Titration of a Weak Acid with a Strong Base

3.16.1 Materials

Acetic acid (0.1 and 0.01 mol/L); potassium hydroxide (0.1 and 0.01 mol/L); burette 25 mL; pH meter.

3.16.2 Procedure

Titrate 20 mL of 0.1 mol/L acetic acid as mentioned earlier. Plot the titration curve and deduce the pK value using the methods given. Repeat the titration using 0.01 mol/L acetic acid and compare both the plots with the theoretical curve obtained from the relationship:

$$pH = pK + \log_{10} \frac{[\text{Salt}]}{[\text{Acid}]}$$

What are the differences between these curves and why?

3.17 Determination of pK_a

3.17.1 Materials

Solutions of the following acids (1.4 g/L)

Acid	Molecular Weight	pK
Acetic	60.05	4.8
Benzoic	122.12	4.2
Hippuric	179.17	3.6
Imidazole	68.08	7.0
Lactic	90.08	3.9
p-Nitrophenol	139.11	7.1

Potassium hydroxide (0.02 mol/L); burette 25 mL; pH meter.

3.17.2 Procedure

Titrate the unknown solution with the standard alkali and deduce its equivalent weight. Add half the amount needed for complete neutralization to another portion of the acid and measure the pH of the solution; this is the *pK* of the acid. From the data, identify the acid as far as possible, using the list provided earlier.

3.18 *pK*$_a$ Values of a Dicarboxylic Acid

3.18.1 Materials

Dicarboxylic acid solution (0.1 mol/L)

Acid	pK_1	pK_2
Malic	3.5	5.0
Oxalic	1.3	4.3
Succinic	4.2	5.6
Tartaric	3.0	4.4

Potassium hydroxide (0.1 mol/L); Burette 25 mL; pH meter.

3.18.2 Procedure

Prepare a titration curve of the acid provided and identify it from the *pK* values obtained.

3.19 Acetate Buffers

3.19.1 Materials

Sodium acetate (0.1 mol/L); Acetic acid (0.1 mol/L); HCl (0.1 mol/L); pH meter.

3.19.2 Procedure

Mix the solutions of sodium acetate and acetic acid as given in the table and record the pH. Calculate what the pH should be from the Henderson–Hasselbalch equation using the examples given earlier.

Add a further 2 mL of 0.1 mol/L HCl to each mixture, measure the pH again, and compare this value with that obtained by calculation.

Tube no.	1	2	3	4	5	6
Sodium acetate (mL)	2	6	10	14	18	18 mL of water
Acetic acid (mL)	16	12	8	4	0	

From the results, explain why some solutions are better buffers than others. The pK_a of acetic acid at 25°C is 4.76.

3.20 Titration Curves of Amino Acids

3.20.1 Principle

Amino acids are present as zwitterions at neutral pH and are amphoteric molecules that can be titrated with both acid and alkali:

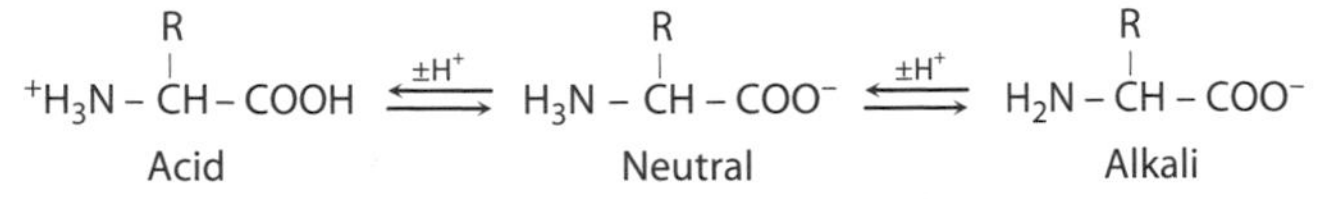

The strong positive charge on the $-NH_3^+$ group induces a tendency for the $-COOH$ to lose a proton; so, amino acids are strong acids. Some amino acids have other ionizable groups in their side chains, and these can also be titrated.

3.20.2 Materials

HCl (100 mmol/L); sodium hydroxide (100 mmol/L); amino acids (100 mmol/L glycine, alanine, histidine, and lysine; 50 mmol/L glutamic acid); pH meter; burette (10 mL).

3.20.3 Protocol

1. Pipette 20 mL of the amino acid solution into a 100 mL beaker. Standardize the pH meter and determine the pH of the solution by dipping the electrode of a pH meter. Now, measure exactly 0.5 mL of 0.001 N HCl; mix well with a glass rod (or with a magnetic stirrer) and measure the pH.
2. Repeat the previous steps till the pH of the solution is about 1. Wash the electrodes in distilled water, restandardize the pH meter, and titrate a further 10 mL with 100 mmol/L NaOH solution to pH 12.5.

3.20.4 Results and Calculations

Interpret the results keeping in view their structure. Deduce the pK and pI values. Some amino acids like glutamic acid, lysine, and histidine show more than two plateaus, indicating the presence of ionizable groups other than the $-COOH$ and $-NH_2$. Table 3.5 lists some of these values. Determine the pK values from your curves and compare them with the values given in Table 3.5.

TABLE 3.5 Ionization Properties of Amino Acids

Amino Acid	pK Values	
	Carboxyl	Amino
Glycine	2.34	9.60
Alanine	2.35	9.69
Leucine	2.36	9.60
Isoleucine	2.21	9.68
Serine	2.09	9.15
Aspartate	2.19	9.82
Glutamate	2.18	9.67
Lysine	1.82	8.95
Histidine	2.17	9.17
Arginine	1.83	9.04
Phenylalanine	1.71	9.13
Cysteine	1.65	10.33
Cystine	2.26	7.85
Proline	1.99	9.85
Hydroxyproline	1.92	10.60

pH calculations

1. Calculate the pH of (a) 0.1 mol/L HCl, (b) 0.1 mol/L KOH, (c) 10 mmol/L HCl, (d) 50 mmol/L H_2SO_4.
2. What are the pK_a values of succinic acid, given that the acid dissociation constants are $K_1 = 6.4 \times 10^{-5}$ and $K_2 = 2.7 \times 10^{-6}$.
3. Calculate the hydrogen ion concentration of blood plasma, pH 7.42. What will be the pH in acidosis if this hydrogen ion concentration is doubled?
4. A solution is prepared by dissolving sodium benzoate and benzoic acid in water to give a 9 mmol/L concentration of each compound. The mixture has a pH of 4.21 at 25°C. Calculate the acid dissociation constant of benzoic acid.
5. Calculate the amount of glycine and HCl required to prepare 1 L of 0.1 mol/L buffer of (a) pH 2.0 and (b) pH 3.0, given that the pK_a value of the glycine carboxyl group is 2.4. Plot HCl equivalents on Y axis and pH on the X axis. Connect all points (Figure 3.3). This is called the titration curve for the amino acid glycine.

3.21 Determination of pH of the Given Water/Soil Sample

3.21.1 Principle

pH is the measure of the relative acidity or alkalinity of the solution and is represented as the negative logarithm of the concentration of free hydrogen ions in a solution. The "p" of pH denotes the power of the hydrogen ion activity in moles per liter:

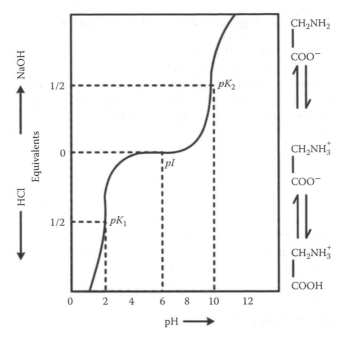

Figure 3.3 Titration curve of an amino acid.

$$pH = -\log 10\,(H^+) = \log 10 \times 1/(H^+).$$

When two electrodes are dipped in two solutions of different pH levels and connected, a potential difference is set up between the two electrodes, which is measured by the potentiometer. This is directly related to the pH of the solution.

pH is measured of H^+ ion concentration in an aqueous medium. Chemically pure water at 25°C is dissociated equally into 10^{-14} g molecules/L of H^+ ions is 10^{-7}. If free H^+ ions are more than OH^- ions, the water is acidic and vice versa. The term pH is expressed as the negative logarithm of H^+ concentration. Thus, pH 7 indicates neutral water, pH below 7 is acidic, and pH 7–14 is alkaline.

3.21.2 Materials

pH meter; standard pH buffers; tissue paper.

3.21.3 Protocol

pH can be measured by two different methods:

1. *Colorimetric method (using indicators)*
   Commercially available pH papers are used. Color change of a paper strip on application of a drop of a test sample indicates the pH, which is matched with the standard color chart.
2. *Electrometric method (using pH meter)*

A pH meter with a combined glass electrode is generally used:

1. Wash the electrode with distilled water.
2. Standardize the electrode with standard buffers ranging from 4 to 9.2.
3. Adjust the temperature knob setting according to that of the sample.
4. Wash the electrode again with distilled water and wipe with tissue paper.
5. Dip the electrode in the test sample and note the pH.

3.21.4 Precautions

1. Soak a new or dried glass electrode for several hours in 0.01 M HCl.
2. There should be no air bubbles in the ends of either electrode or in the salt bridge.
3. Temperature correction should be done before starting.
4. After each use, the electrode should be washed with distilled water and wiped clean with tissue paper.

3.22 Determination of pH of Fruit Juice

3.22.1 Principle

The acidity and alkalinity of a fruit juice is measured as pH.

3.22.2 Protocol

The pH of the fruit juice can be determined colorimetrically by the use of various indicators are as follows:

1. *pH paper*: In this, the color of the pH paper corresponds to a definite pH. A drop of fruit juice is put on the pH paper, and the color of the pH paper is matched with the color of the pH given with the pH paper by the manufacturing company.
2. *Lovibond comparator*: This comparator consists of a Bakelite case with two holes at the top for tubes of standard bore and of colorless glass. Tube A contains water. The hinged door of the case holds a rotatable disk containing a series of standard colored glasses corresponding to various pH values, and each glass can be brought in front of tube A in turn and viewed through aperture A. A solution of the indicator phenol red is added to tube B that contains fruit juice and the pH disk is rotated until a match is obtained. The pH is then read in the aperture at the bottom of the apparatus.
3. *pH meter*: This instrument gives accurate pH reading. The instrument is adjusted with a standard buffer solution of known pH at room temperature, setting its knob. The electrode is then washed with distilled water. Thereafter, the fruit juice is filled in a beaker, and the reading is recorded. The electrodes should be washed with distilled water after every use. When the instrument is not in use, the electrodes should be always immersed in distilled water.

3.23 Measurement of pH of Soil Sample

3.23.1 Principle

Soil pH affects plant nutrient availability, microbial and fungal activity, and plant growth. It is a measurement of hydrogen ion activity in a soil/water solution. Soil pH has many primary determining factors: decomposition of organic material, parent material, rainfall, and anthropogenic alteration. A neutral solution has a pH of 7, while an acidic solution has a pH <7. Note that a basic solution always has a pH >7, but an alkaline does not. This is due to the acid-neutralizing capacity or buffering of a soil by strong bases.

Altering soil pH is a common practice in many fields of plant production, cultivation, and growth. The methods include addition of organic matter, hydrogen/aluminum sulfate, sulfur, fertilizers containing ammonium or urea, and liming agents. In this experiment, we will be measuring pH.

3.23.2 Materials

Prepare all the solutions in previously boiled and cooled distilled water.
Standard buffer solutions:

1. Potassium tetraoxalate ($KH_3C_4O_{82}\cdot H_2O$): Dissolve 3.175 g $KH_3C_4O_{82}\cdot H_2O$ in water and make up to 250 mL (pH 1.68).
2. Potassium tartrate ($KHC_4H_4O_6$): To 250 mL water in a glass stopper bottle, add excess $KHC_4O_4O_6$; shake the bottle vigorously for about 10 min to obtain a saturated solution. Filter and preserve the solution by adding 0.1 g thymol (pH 3.55).
3. Potassium phthalate ($KHC_8H_4O_6$): Dissolve 2.552 g $KHC_8H_4O_6$ in water and make up to 250 mL (pH 4.0).
4. Potassium dihydrogen phosphate (KH_2PO_4): Dissolve 0.30 KH_2PO_4, 0.8875 g Na_2PO_4, and 0.8875 g Na_2HPO_4 in water and make up the solution to 250 mL (pH 6.86).
5. Borax ($Na_2B_4O_7$): Dissolve 0.9502 g $Na_2B_4O_7$ in water and make up to 250 mL (pH 9.18).

3.23.3 Protocol

1. Warm up the instrument for 15 min.
2. Calibrate the instrument with the known buffer solutions. (Calibration is done by a buffer solution whose pH is close to that of the sample.)
3. Immerse the electrode in the unknown sample and stir for 3 min and note the pH.

(*Note: Instead of preparing standard buffer solutions, commercially available buffer tablets can be used for calibrating the pH meter.*)

3.A Appendix A Indicators of Hydrogen Ion Concentration

Table 3.A.1 indicates the pH range of each indicator and the colors that occur. To determine the exact pH within a particular range, one should use a set of standard colorimetric tubes that are available.

3.B Appendix B Buffer Solutions

3.B.1 Introduction

Biological reactions work well only within a narrow pH range. However, many of these reactions themselves generate or consume protons. Buffers are substances that undergo reversible protonation within a particular pH range and therefore maintain the hydrogen ion concentration within acceptable limits. An ideal biological buffer should

1. Have a pK_a between pH 6.0 and pH 8.0
2. Be inert to a wide variety of chemicals and enzymes
3. Be highly polar, so that it is both exquisitely soluble in aqueous solutions and also unlikely to diffuse across biological membranes and thereby affect intracellular pH
4. Be nontoxic
5. Be inexpensive
6. Not be susceptible to salt or temperature
7. Not absorb visible or ultraviolet light

TABLE 3.A.1   pH Range of Indicators and Resultant Colors

Indicator	Full Acid Color	Full Alkaline Color	pH Range
Cresol red	Red	Yellow	0.2–1.8
Metacresol purple (acid range)	Red	Yellow	1.2–2.8
Thymol blue	Red	Yellow	1.2–2.8
Bromophenol blue	Yellow	Blue	3.0–4.6
Bromocresol green	Yellow	Blue	3.8–5.4
Chlorocresol green	Yellow	Blue	4.0–5.6
Methyl red	Red	Yellow	4.4–6.4
Chlorophenol red	Yellow	Red	4.8–6.4
Bromocresol purple	Yellow	Purple	5.2–6.8
Bromothymol blue	Yellow	Blue	6.0–7.6
Neutral red	Red	Amber	6.8–8.0
Phenol red	Yellow	Red	6.8–8.4
Cresol red	Yellow	Red	7.2–8.8
Metacresol purple (alkaline range)	Yellow	Purple	7.4–9.0
Thymol blue (alkaline range)	Yellow	Blue	8.0–9.6
Cresolphthalein	Colorless	Red	8.2–9.8
Phenolphthalein	Colorless	Red	8.3–10.0

None of the buffers used in life sciences meat all of the mentioned criteria. Very few weak acids are known that have dissociation constants between 10^{-7} and 10^{-9}. Among inorganic salts, only borates, bicarbonates, phosphates, and ammonium salts lie within this range. However, they are all incompatible in one way or another with physiological media.

In 1946, George Gomori suggested that organic polyamines could be used to maintain pH between 6.5 and 9.7. One of the three compounds he investigated was Tris (2-amino-2-hydroxymethyl-1, 3-propanediol), which had been first described in 1897 by Piloty and Ruff. Tris turned out to be an extremely satisfactory buffer for many biochemical purposes, and today it is the standard buffer used for most enzymatic reactions in molecular cloning. Tris buffers (0.05 M) of the desired pH can be made by mixing 50 mL of 0.1 M Tris base with the indicated volume of 0.1 N HCl and then adjusting the volume of the mixture to 100 mL with water (Table 3.B.1).

3.B.2 Tris Buffers

One of the first commercial successes, which received wide attention, was the reduction of mortality during handling and hauling of fish in 1940s. Live fish those days were carried to markets in tanks of seawater, and many of the fish died during transit because of the decline in pH resulting from an accumulation of CO_2. This problem was only partially alleviated by adding anesthetics in the water to minimize the fishes' metabolic activities by stabilizing the pH of the seawater but with side effects to the people who ate the fish (McFarland and Norris, 1958). Tris turned out to be an extremely satisfactory buffer for many biochemical purposes, and today it is the standard buffer used for most enzymatic reactions in physiology and molecular cloning.

TABLE 3.B.1   Preparation of Tris Buffers of Various Desired pH Values

Desired pH (25°C)	Volume of 0.1 N HCl (mL)	Desired pH (25°C)	Volume of 0.1 N HCl (mL)
7.10	45.7	8.10	26.2
7.20	44.7	8.20	22.9
7.30	43.4	8.30	19.9
7.40	42.0	8.40	17.2
7.50	40.3	8.50	14.7
7.60	38.5	8.60	12.4
7.70	36.6	8.70	10.3
7.80	34.5	8.80	8.5
7.90	32.0	8.90	7.0
8.00	29.2		

Tris (2-Amino-2-hydroxymethyl-propane-1,3-diol) has a very high buffering capacity, is highly soluble in water, and is inert in a wide variety of enzymatic reactions. However, Tris also has a number of deficiencies:

1. The pK_a of Tris is pH 8.0 (at 20°C), which means that its buffering capacity is very low at pH below 7.5 and above 9.0.
2. Temperature has a significant effect on the dissociation of Tris. The pH of Tris solution decreases by 0.03 pH units for each 1°C increase in temperature. For example, a 0.05 M solution has pH values of 9.5, 8.9, and 8.6 at 5°C, 25°C, and 37°C, respectively. By convention, the pH of Tris solutions given in the scientific literature refers to the pH measured at 25°C. When preparing stock solutions of Tris, it is best to bring the pH into the desired range and then allow the solution to cool to 25°C before making final adjustments to the desired pH.
3. Concentration has a significant effect on the dissociation of Tris. For example, the pH of solutions containing 10 and 100 mM Tris differs by 0.1 of a pH unit, with the more concentrated solution having the higher pH.
4. Tris is toxic to many types of mammalian cells.
5. Tris, a primary amine, cannot be used with fixatives such as glutaraldehyde and formaldehyde. Tris also reacts with glyoxal. Phosphate or MOPS buffer is generally used in place of Tris with these reagents (Table 3.B.2).

3.B.3 Ideal Buffers

Tris is a poor buffer at pH values below 7.5. In the mid-1960s, Norman Good and his colleagues responded to the need for better buffers in this range by developing a series of N-substituted aminosulfonic acids that behave as strong zwitterions at biologically relevant pH values (Good et al., 1966; Ferguson et al., 1980). Without these buffers, several techniques central to molecular cloning either would not exist at all or would work at greatly reduced efficiency. These techniques include high-efficiency transfection of mammalian cells (HEPES, Tricine, and BES), gel electrophoresis of RNA (MOPS), and high-efficiency transformation of bacterial cells (MES).

Data compiled from various sources are presented here which includes data from the Biochemical and Reagents for Life Science Research 1994 (Sigma-Aldrich) and references therein. $pK_a = 9.0$ for the second dissociation stage.

3.B.4 Preparation of Imidazole, MOPS, and HEPES Buffers

For preparing these buffers, the molar ratio of the protonated to nonprotonated species for obtaining a buffer of required pH can be accomplished by mixing the computed amount of free buffer and one of its salts (from Henderson–Hasselbalch equation). Table 3.B.3 depicts the buffer as free acid, and its salt can be mixed in the ratio ranging from 90:10 to 10:90 to obtain the desired pH. For example, for preparing 100 mL 0.1 M HEPES buffer of pH 7.22, mix 70 mL of 0.1 M HEPES and 30 mL of 0.1 M sodium salt of HEPES.

TABLE 3.B.2 Properties of Good Buffers

Acronym	Chemical Name	FW	pK_a	pH Range
MES	2-(N-morpholino) ethanesulfonic acid	195.2	6.1	5.5–6.7
Bis–Tris	Bis (2-hydroxyethy) imino tris (hydroxymethyl) methane	209.2	6.5	5.8–7.2
ADA	N-(2-acetamido)-2-iminodiacetic acid	190.2	6.6	6.0–7.2
ACES	2-(2-amino-2-oxoethyl) amino ethanesulfonic acid	182.2	6.8	6.1–7.5
PIPES	Pipeazine-N,N'-bis (2-ethanesulfonic acid)	302.4	6.8	6.1–7.5
MOPSO	3-(N-mopholino)-2-aminoethanesulfonic acid	225.3	6.9	6.2–7.6
Bis–Tris	1, 3-bis (tris(hydroxyethyl)-2-aminoethanesulfonic acid	282.3	6.8a	6.3–9.5
Propane BES	N,N-bis (2-hydroxyethyl)-2-aminoethanesulfonic acid	213.2	7.1	6.4–7.8
MOPS	3-(N-morpholino) propanesulfonic acid	209.3	7.2	6.5–7.9
HEPES	N-(2-hydroxyethyl) piperazone-N-(2-ethanesulfonic acid)	238.3	7.5	6.8–8.2
TES	N-tris (hydroxymethyl) methyl-2-aminoethanesulfonic acid	229.2	7.4	6.8–8.2
DIPSO	3-(N,N bis (2-hydroxyethyl) amino) 2-hydroxypropanesulfonic acid	243.3	7.6	7.0–8.2
TAPSO	3-(N-tris (hydroxymethyl) methylamino) 2-hydroxypropanesulfonic acid	259.3	7.6	7.0–8.2
TRIZMA	Tris (hydroxymethyl) aminomethane	121.1	8.1	7.0–9.1
HEPPSO	N-(2-hydorxyethyl) piperazine-N (2-hydroxypropanesulfonic acid)	268.3	7.8	7.1–8.5
POPSO	Piperazine-N,N'-bis (2-hydroxypropanesulfonic acid)	362.4	7.8	7.2–8.5
EPPS	N-(2-hydroxyethyl) piperazine-N-(3-propanesulfonic acid)	252.3'	8.0	7.3–8.7
TEA	Triethanolamine	149.2	7.8	7.3–8.3
Tricine	N-tris (hydroxymethyl) methylglycine	179.2	8.1	7.4–8.8
Bicine	N,N-bis (2-hydroxyethyl) glycine	163.2	8.3	7.6–9.0
TAPS	N-tris(hydroxymethyl)methyl–3-aminopropanesulfoni; acid	243.3	8.4	7.7–9.1
AMPSO	3-((1,1-dimethyl-2-hydroxyethyl) amino) 2-hydroxypropanesulfonic acid	227.3	9.0	8.3–9.7
CHES	2-(N-cyclohexylamino)-ethanesulfonic acid	207.3	9.3	8.6–10.0
CAPSO	3-(cyclohexylamino)-2-hydroxy-l-propanesulfonic acid	237.3	9.6	8.9–10.3
AMP	2-amino-2-methyl-1-propanol	89.1	9.7	9.0–10.5
CAPS	3-(cyclohexylamino)-I-propanesulfonic acid	221.3	10.4	9.7–11.1

TABLE 3.B.3   Ratio of Free Acid and Free Base at Different pH with Imidazole, MOPS, and HEPES

% Buffer (as Free Acid)	% Buffer (as Free Base)	pH		
		Imidazole	MOPS	HEPES
90	10	6.09	6.19	6.60
80	20	6.44	6.54	6.95
70	30	6.68	6.78	7.18
60	40	6.87	6.97	7.37
50	50	7.05	7.15	7.55
40	60	7.18	7.32	7.73
30	70	7.22	7.52	7.92
20	80	7.35	7.75	8.15
10	90	8.00	8.05	8.50

3.B.5 Preparation of Phosphate Buffers (Gomori Buffers)

The most commonly used phosphate buffers are named after their inventor Gomori. The buffer consists of a mixture of monobasic dihydrogen phosphate and dibasic monohydrogen phosphate. By varying the amount of each salt, a range of buffers can be prepared between the pH range 5.8 and pH 8.0 (Table 3.B.4). Phosphates have a very high buffering capacity and are highly soluble in water. However, they have a number of potential disadvantages, which are

1. Phosphates inhibit many enzymatic reactions involved with the molecular cloning, including cleavage of DNA by many restriction enzymes, ligation of DNA, and bacterial transformation.
2. Because phosphates precipitate in ethanol, it is not possible to precipitate DNA and RNA from buffers that contain significant quantities of phosphate ions.
3. Phosphates sequester divalent cations, such as Ca^{2+} and Mg^{2+}.

TABLE 3.B.4   Preparation of 0.1 M Potassium Phosphate Buffer

pH	Volume of 1 M K_2HP_4 (mL)	Volume of 1 M KH_2PO_4 (mL)
5.8	8.5	91.5
6.0	13.2	86.8
6.4	27.8	72.2
6.6	38.1	61.9
6.8	49.7	50.3
7.0	61.5	38.5
7.2	71.7	28.3
7.4	80.2	19.8
7.6	86.6	13.4
7.8	90.8	9.2
8.0	94.0	6.0

3.B.6 Acids and Bases

Dilute the combined 1 M stock solutions to 1 L with distilled water. pH is calculated according to the Henderson–Hasselbalch equation:

$$pH = pK' + \log\left\{\frac{(proton\ acceptor)}{proton\ donor}\right\},$$

where $pK' = 6.86$ at 25°C.

With some acids and bases, stock solutions of different molarity/normality are in common use. These are often abbreviated "conc" for concentrated stocks and "dil" for dilute stocks (Table 3.B.5).

3.B.7 Buffers and Stock Solutions

3.B.7.1 Phosphate-Buffered Saline
137 mM NaCl; 2.7 mM KCl; 10 mM Na_2HP_4; 2 mM KH_2PO_4 (Table 3.B.6).

TABLE 3.B.5   Concentrations of Acids and Bases with Common Commercial Strengths

Substance	Formula	M.H.	Moles/L	Grams/L	% by Weight	Specific Gravity	mL/L to Prepare 1 M Solution
Acetic acid	CH_3COOH	60.05	17.4	1045	99.5	1.05	57.5
Glacial acetic acid		60.05	6.27	376	36	1.045	159.5
Formic acid	HCOOH	46.02	23.4	1080	90	1.20	527
Hydrochloric acid	HCl	36.5	11.6	424	36	1.18	86.2
			2.9	105	10	1.05	344.8
Nitric acid		63.02	15.99	1008	71	1.42	62.5
			14.9	938	67	1.40	67.1
			13.3	837	61	1.37	75.2
Perchloric acid	$HCIO_4$	100.5	11.65	1172	70	1.67	85.8
			9.2	923	60	1.54	108.7
Phosphoric acid	H_3PO_4	80.0	18.1	1445	85	1.70	55.2
Sulfuric acid	H_2SO_4	98.1	18.0	1766	96	1.84	55.6
Ammonium hydroxide	NH_4OH	35.0	14.8	251	28	0.898	67.6
Potassium hydroxide	KOH	56.1	13.5	757	50	1.52	74.1
			1.94	109	10	1.09	515.5
Sodium hydroxide	NaOH	40.0	19.1	763	50	1.53	52.4
			2.75	111	10	1.11	363.6

TABLE 3.B.6   Approximate pH Values for Various Concentrations of Stock Solutions

Substance	1 N	0.1 N	0.01 N
Acetic acid	2.4	2.9	3.4
Hydrochloric acid	0.10	1.07	2.02
Sulfuric acid	0.3	1.2	2.1
Citric acid		2.1	2.6
Ammonium hydroxide	11.8	11.3	10.8
Sodium hydroxide	14.05	13.07	10.8
Sodium bicarbonate		8.4	12.12
Sodium carbonate		11.5	11.0

Dissolve 1 g of NaCl, 0.2 g of KCl, 1.44 g of Na_2HPO_4, and 0.24 g of KH_2PO_4 in 800 mL of distilled water. Adjust the pH to 7.4 with HCl. Add distilled water to 1 L. Dispense the solution into aliquots and sterilize them by autoclaving for 20 min at 15 psi (1.05 kg/cm²) on liquid cycle or by filter sterilization. Store the buffer at room temperature.

Phosphate-buffered saline (PBS) is a commonly used reagent that has been adapted for different applications. Note that the recipe presented here lacks divalent cations; if necessary, PBS may be supplemented with 1 mM $CaCl_2$ and 0.5 mM $MgCl_2$.

3.B.7.2 10× Tris EDTA (TE)

pH 7.4
100 mM Tris–HCl (pH 7.4)
10 mM EDTA (pH 8.0)
pH 7.6
100 mM Tris–HCl (pH 7.6)
10 mM EDTA (pH 8.0)
pH 8.0
100 mM Tris–HCl (pH 8.0)
10 mM EDTA (p11 8.0)

Sterilize solutions by autoclaving for 20 min at 15 psi (1.05 kg/cm²) on liquid cycle. Store the buffer at room temperature.

3.B.7.3 Tris–HCl (1 M)

Dissolve 121.1 g of Tris base in 800 mL of H_2O. Adjust the pH to the desired value by adding concentrated HCl.

pH	HCl (mL)
7.4	70
7.6	60
8.0	42

Allow the solution to cool to room temperature before making final adjustments to the pH. Adjust the volume of the solution to 1 L with water. Dispense into aliquots and sterilize by autoclaving. If the 1 M solution has a yellow color, discard it and obtain Tris of better quality. The pH of Tris solutions is temperature-dependent and decreases by 0.03 pH units for each 1°C increase in temperature. For example, a 0.05 M solution has pH values of 9.5, 8.9, and 8.6 at 5°C, 25°C, and 37°C, respectively.

3.B.7.4 Tris Magnesium Buffer (TM)
50 mM Tris–HCl (pH 7.8)
10 mM $MgSO_4$

3.B.7.5 Tris-Buffered Saline (TBS)
Dissolve 8 g of NaCl, 0.2 g of KCl, and 3 g of Tris base in 800 mL of distilled water. Add 0.015 g of phenol red and adjust the pH to 7.4 with HCl. Add distilled water to 1L. Dispense the solution into aliquots and sterilize them by autoclaving for 20 min at 15 psi (1.05 kg/cm^2) on liquid cycle. Store the buffer at room temperature.

Suggested Readings

Akhtar, M. 2007. *Method of Analysis and Assay: Potentiometry. Pharmaceutical Analysis.* New Delhi, India: Jamia University.

Barrow, G. M. 1979. *Physical Chemistry*, Chapter 18, 4th edn. New York: McGraw Hill.

Bates, R. G. 1973. *Determination of pH: Theory and Practice.* New York: John Wiley & Sons.

Bisen, P. S. and A. Sharma. 2012. *Introduction to Instrumentation in Life Sciences.* Boca Raton, FL: CRC Press.

Buck, R. P. 1971. Potentiometry: pH measurements and ion-selective electrodes. In *Techniques in Chemistry*, eds. A. Weissberger and B. W. Rossiter, vol. 1, part IIA, Chapter 2. New York: Wiley Interscience.

Cavins, T. J., J. L. Gibson, B. E. Whipker et al. 2000. *pH and EC Meters—Tools for Substrate Analysis.* Florex.001. Raleigh, NC: North Carolina State University.

Chang, R. 1979. *Physical Chemistry with Application to Biological Systems*, 2nd edn. Oxford, U.K.: Clarendon Press.

Dole, M. 1941. *The Glass Electrode.* New York: John Wiley & Sons.

Durst, R. A. 1973. Ion selective electrodes in science, medicine and technology. *Am Sci* 59: 353–361.

Eisenman, G. R., R. Bates, G. Matlock et al. 1965. *The Glass Electrode.* New York: Interscience.

Ferguson, W. J., K. L. Braumschweiger, W. R. Braumscheiger et al. 1980. Hydrogen ion buffers for biological research. *Anal Biochem* 104: 300–310.

Fischer, R. B. 1974. Ion-selective electrodes. *J Chem Educ* 51: 387–390.

Gardner, M. L. G. 1979. *Medical Acid-Base Balance, The Basic Principles.* London, U.K.: Cassell.

Good, N. E., G. D. Winget, W. Winter et al. 1966. Hydrogen ion buffers for biological research. *Biochemistry* 5: 467–477.

Levine, I. N. 1978. *Physical Chemistry*, Chapter 14. New York: McGraw-Hill.

McFarland, W. N. and K. S. Norris. 1958. A new harbor porpoise of the genus Phocoena from the Gulf of California. *J Mammal* 39: 22–39.

Price, N. C. and R. A. Dwek. 1979. *Principles and Problems in Physical Chemistry for Biochemists.* Oxford, U.K.: Clarendon Press.

Steve, R. 2005. *Principles and Guidelines of pH Measurement*. Research Triangle Park, NC: Instrument Society of America, Technology Publications.

Van Holde, K. E. 1985. *Physical Biochemistry*, 2nd edn. Englewood Cliffs, NJ: Prentice Hall.

Vytras, K. 1995. Potentiometry. In *Encyclopedia of Pharmaceutical Technology*, eds. J. Swarbric and J. C. Boylan, vol. 12. New York: Marcel Dekker.

Williams, V. R., W. L. Mattice, and H. B. Williams. 1978. *Basic Physical Chemistry for the Life Sciences*. San Francisco, CA: Freeman & Company.

Wilson, K. and J. Walker. 2004. *Practical Biochemistry: Principles and Techniques*, 5th edn. Cambridge: Cambridge University Press.

Important Links

http://www.horiba.com/in/application/material-property-characterization/water-analysis/water-quality-electrochemistry-instrumentation/ph-guide/1/details/thick-membrane-electrode-9611–10d-9451/

http://www.enotec.com/en/products-services/products/oxitec/oxitec-5000.Html#c9 35

http://www.ysilifesciences.com/index.php?page = special-methods-and-applications

4

Spectrophotometer

4.1 Introduction

Scientific advances in life science laboratories often depend on technological advances containing instruments and tools to help scientists study living creatures from single-celled organisms to massive plants and mammals. Specialized scientific equipment enables biologists to study cellular processes, genetics, bacteria, and other life forms that are invisible to the naked eye. Scientists need sophisticated instruments to investigate how living organisms function, breathe, eat, reproduce, and evolve. Biology had changed forever when the microscope revealed the complex world of cells and bacteria. Instrumentation increases the sensitivity of the normal senses, quantifies our observations with precision, and expands the domains of these senses in time and space. Knowing the components of cells and how cells work is fundamental to all biological sciences.

A spectrophotometer is an optical device that can determine the concentration of a compound or particles in a solution or suspension. Among the instrumental analytical techniques, the spectrophotometric technique occupies a unique position because of its simplicity, sensitivity, accuracy, and rapidity. Formerly, visible spectrophotometry was often called colorimetry, and even now, such definitions as colorimetric, photometric, or absorptiometric methods are sometimes used in the literature, as equivalents to the term spectrophotometric method. The basis of the spectrophotometric method is the simple relationship between the absorption of radiation by a solution and the concentration of colored species in the solution. In order to determine a species spectrophotometrically, it is usually converted into a colored complex. The color of the determinant itself is utilized much less often. When the determinant is not colored, or didn't form colored compounds, indirect spectrophotometric methods may be used for its determination. Spectrophotometric methods are precise among most of the instrumental methods of analysis. Some spectrophotometers are equipped with a single measuring chamber. For these so-called single-beam instruments, the absorbance of a sample is taken, followed by the absorbance of a control. In double-beam spectrophotometers, the light beam is split into two beams by means of mirrors. One light path goes through the sample chamber and the other light beam passes through what is referred to as the reference cell or chamber. The data of the absorbance between the two chambers is computed and is used to determine sample concentration. Depending on the spectrophotometer, absorbance can be taken at a single wavelength or scanned through a spectrum of wavelengths. The latter can be a useful means of identifying components of the sample, based on their preferential absorption of certain wavelengths of light.

Light of a preselected wavelength is shown through a chamber that houses the sample. The sample particles, bacteria, for example, will absorb some of the light. The amount of light that is absorbed increases with increasing numbers of bacteria in a predictable way. The relationship between absorbance and the number of absorbing sample molecules is expressed mathematically as the Beer–Lambert law. The absorbance of light can also be described as the optical density (O.D.) of the sample solution or suspension.

The percent of light that has been absorbed can be determined and, by comparing this absorption to a graph of the absorption of known concentration in the suspension, can be computed. In a life science laboratory, such measurements are routinely used based on the absorbance of the suspension. A standard curve can be constructed that relates the various measured optical densities to the resulting concentration as determined by the standard from a defined portion of the suspensions.

4.2 Beer–Lambert Law

When a beam of light passes through a solution, a part of it is absorbed by the components of that solution. In addition, certain portions (wavelengths) of the light beam are selectively absorbed. The mathematical relationship between the concentration of a substance and the absorption of light is provided by the Beer–Lambert, or Lambert–Beer, law. The basis of this law comes from the work of Lambert on the transmission of monochromatic light by homogeneous solid substances. Beer applied the law to solutions and found that both the concentration and thickness of the solution affect light transmission through it.

Lambert's law may be expressed as follows:

$$\log\left(\frac{I_0}{I}\right) = K_1 b$$

where
　I_0 is the original intensity of the light beam
　I is the intensity of the beam after passing through the homogeneous substance
　b is the thickness of the layer of solution through which the light has passed (usually expressed in centimeters)
　K_1 is the proportionality constant, the value of which depends on the absorption characteristics of each compound, units of thickness, temperature, and wavelength

Beer's law may be expressed as

$$\log\left(\frac{I_0}{I}\right) = K_2 c$$

where

I_0 is as mentioned earlier
I is the intensity of the beam passing through the solution
c is the concentration of the solution in moles per liter
K_2 is similar to K_1

Combining the two aforementioned formulas,

$$\log\left(\frac{I_0}{I}\right) = Kcb$$

The log quantity earlier is called the O.D., while K is called the extinction coefficient and may be written as E for convenience, or as E. It is specific for a certain wavelength and is large when a particular wavelength is absorbed efficiently by a compound:

$$O.D. = Ecl$$

If the Beer–Lambert law is obeyed and l is kept constant, then a plot of extinction against concentration gives a straight line passing through the origin (Figure 4.1), which is far more convenient than the curve for transmittance. Some colorimeters and spectrophotometers have two scales, a linear one of percent transmittance and a logarithmic one of extinction. It is this latter scale that is related linearly to concentration and is the one used in the construction of a standard curve. With the aid of such a standard curve, the concentration of an unknown solution can easily be determined from its extinction.

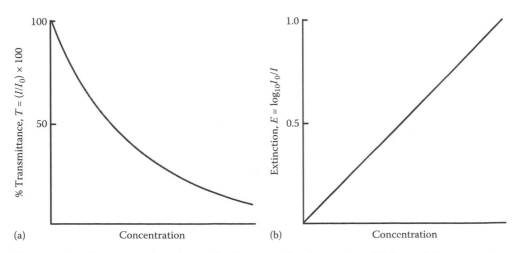

Figure 4.1 Illustrating the relationship between the absorption of light and the concentration of an absorbing solution: (a) transmittance and (b) extinction.

In short, the quantity and quality of light absorbed or transmitted by a solution depends on the absorption characteristics of the solute and solvent and is directly proportional to the concentration of the solution and the thickness of the layer used (length of light path).

The Beer–Lambert law may, therefore, be used to determine the extinction coefficient of substances in solution, as well as the concentration of one or more such substances in a specific solution. Examples 4.1 and 4.2 illustrate this point.

Example 4.1

Given the following absorption data for two compounds, A and B, at the indicated wavelengths, calculate the molar extinction coefficients for the two compounds at the two wavelengths. The concentration of A in both cases is 1.5×10^{-4} mol/L, and the concentration of B in both cases is 8×10^{-4} mol/L. In all cases, the light path is 1 cm.

Calculations

Compound A		Compound B	
Wavelength (μm)	O.D.	Wavelength (μm)	O.D.
400	0.55	400	0.07
600	0.04	600	0.75

Compound A
    At 400 μm,

$$0.55 = E(1.5 \times 10^{-4})\,(1.0)$$

$$0.55 = E(1.5 \times 10^{-4})$$

$$\frac{E \times 0.55}{1.5 \times 10^{-4}} = 3667$$

where
    0.55 is the O.D.
    E is unknown
    1.5×10^{-4} is the concentration c in M
    1.0 is b in centimeters

At 600 μm,

$$0.04 = E(1.5 \times 10^{-4})(1.0)$$

$$0.04 = E(1.5 \times 10^{-4})$$

$$E = \frac{10.04}{1.5 \times 10^{-4}} = 267$$

where
    0.04 is the O.D.
    E is unknown
    1.5×10^{-4} is as previously mentioned
    1.0 is as mentioned earlier

Compound B
At 400 μm,

$$0.07 = E(8.0 \times 10^{-4})(1.0)$$

$$0.07 \times E(8.0 \times 10^{-4})$$

$$E = \frac{0.07}{8.0 \times 10^{-4}} = 87.5$$

where
0.07 is the O.D.
E is unknown
8.0×10^{-4} is the concentration c in M
1.0 is as previously mentioned

At 600 μm,

$$0.75 = E(8.0 \times 10^{-4})(1.0)$$

$$0.75 = E(8.0 \times 10^{-4})$$

$$E = \frac{0.07}{8.0 \times 10^{-4}} = 87.5$$

where
0.75 is the O.D.
E is unknown
8.0×10^{-4} and 1.0 are as mentioned earlier

Example 4.2

Using the molar extinction coefficients obtained earlier, calculate the concentration of A and B in a mixture of the two, which gives the following data (the light path is 1 cm):
O.D. at 400 μm is 0.349
O.D. at 600 μm is 0.7324
Suppose that
Concentration of $A = a$
Concentration of $B = b$
Then at 400 μm,

$$\text{O.D. } (A) = (3667) \ (a) \ (1.0)$$

$$\text{O.D. } (B) = (87.5) \ (b) \ (1.0)$$

$$\text{O.D. (mixture } A + B) = \text{O.D. } (A) + \text{O.D. } (B)$$

Therefore,

$$0.349 = 3667a + 87.5b \tag{4.1}$$

and at 600 μm,

$$\text{O.D. } (A) = (267) \ (a) \ (1.0)$$

$$O.D.\ (B) = (938)\ (b)\ (1.0)$$

$$O.D.\ (\text{mixture } A + B) = O.D.(A) + O.D.(B) \qquad (4.2)$$

Therefore,

$$0.7324 = 267a + 938b \qquad (4.3)$$

Simultaneously solving Equations 4.1 through 4.3,

$$a = 0.000024 \text{ mol/L}$$

$$(2.4 \times 10^{-5} \text{ M})$$

$$b = 0.00077 \text{ mol/L}$$

$$(7.7 \times 10^{-4} \text{ M})$$

4.2.1 Molar Extinction Coefficient

If l is 1 cm and c is 1 mol/L, then the absorbance is equal to k, the molar extinction coefficient, which is characteristic for a compound. The molar extinction coefficient k is thus the extinction given by 1 mol/L in a light path of 1 cm and is usually written as $\Sigma_{1\,cm}^{1\,mol/L}$. It has the dimension of L/mol/cm.

4.2.2 Specific Extinction Coefficient

The molecular weights of some compounds such as proteins or nucleic acids in a mixture are not readily available and, in this case, the specific extinction coefficient is used. This is the extinction of 10 g/L (formerly known as 1% w/v) of the compound in the light path of $\Sigma_{1\,cm}^{10\,g/L}$.

4.3 Limitations of the Beer–Lambert Law

Sometimes, a nonlinear plot is obtained of extinction against concentration, and this is probably due to one or an other of the following conditions not being fulfilled:

1. Light must be of a narrow wavelength range and preferably monochromatic.
2. The wavelength of light used should be at the absorption maximum of the solution: this also gives the greatest sensitivity.
3. There must be no ionization, association, dissociation, or solvation of the solute with concentration or time.
4. The solution is too concentrated, giving an intense color. The law only holds up to a threshold maximum concentration for a given substance.

4.4 Measurement of Extinction

The earliest colorimeters relied on the human eye, to match the color of a solution with that of one of a series of colored discs. The results obtained were too subjective and not particularly accurate. Visual colorimeters are now of historical interest only and are not described here. The photoelectric cell is superior to the human eye in assessing the degree of absorption of a color and is more objective.

4.5 Colorimeter

A colorimeter is used to measure the absorbance of particular wavelengths of light by a specific solution. This instrument is most commonly used to determine the concentration of a known solute in a given solution by the application of the Beer–Lambert law, which states that the concentration of a solute is proportional to the absorbance.

The output from a colorimeter may be displayed by an analogue or digital meter and may be shown as transmittance (a linear scale from 0% to 100%) or as absorbance (a logarithmic scale from zero to infinity). The useful range of the absorbance scale is from 0 to 2, but it is desirable to keep within the range 0–1 because, above 1, the results become unreliable due to scattering of light. In addition, the output may be sent to a chart recorder, data logger, or computer. Changeable optic filters are used in the colorimeter to select the wavelength of light that the solute absorbs the most, in order to maximize accuracy. The usual wavelength range is from 400 to 700 nanometers (nm). If it is necessary to operate in the ultraviolet (UV) range (below 400 nm), then some modifications to the colorimeter are needed. In modern colorimeters, the filament lamp and filters may be replaced by several light-emitting diodes of different colors (Figure 4.2).

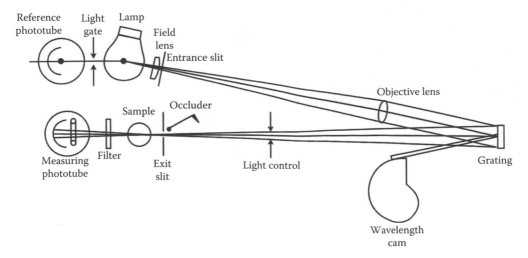

Figure 4.2 Schematic diagram of a colorimeter.

The most important parts of a colorimeter are as follows:

- A light source, which is usually an ordinary tungsten lamp
- An aperture, which can be adjusted
- A set of filters in different colors
- A detector that measures the light that has passed through the solution

4.5.1 Filters

Different filters are used to select the wavelength of light that the solution absorbs the most. This makes the colorimeter more accurate. Solutions are usually placed in glass or plastic cuvettes. The usual wavelengths used are between 400 and 700 nm. If it is necessary to use UV light (below 400 nm), then the lamp and filters must be changed (Table 4.1).

4.5.2 Output

The output of the colorimeter may be shown in graphs or tables, by an analogue or digital meter. The data may be printed on paper or stored in a computer. It either shows the amount of light that is absorbed by the solution or the amount of light that has passed through the solution.

4.5.3 Directions for Use of the Spectronic 20 Spectrocolorimeter

The Spectronic 20 spectrocolorimeter is a routinely used instrument found in most of the laboratories and educational institutions. It is, therefore, described in some detail (Figure 4.3).

4.5.3.1 Description of the Optical System

White light emanating from the tungsten lamp passes through the entrance slit and is focused by the field lens onto the objective lens. The objective lens focuses an image

TABLE 4.1 The Relationship between the Color of the Solution Examined and the Filter Chosen for Colorimetric Analysis

Color of Solution	Filter
Red orange	Blue–blue green
Blue	Red
Green	Red
Purple	Green
Yellow	Violet

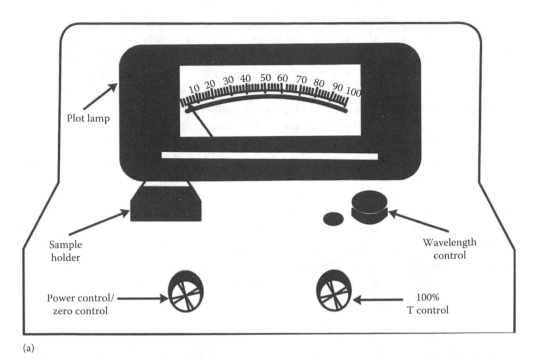

(a)

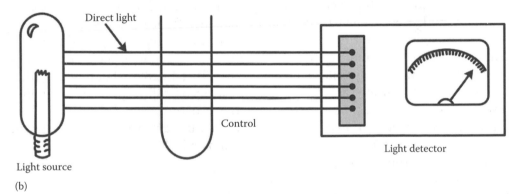

(b)

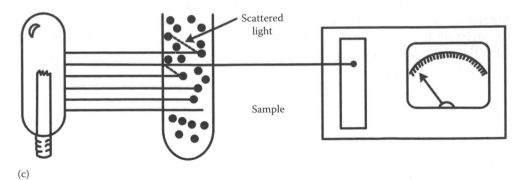

(c)

Figure 4.3 (a) Spectronic 20 spectrocolorimeter, (b) adjustment of instrument with control, and (c) reading on instrument with the sample.

of the entrance slit at the exit slit after it has been reflected and dispersed by the diffraction grating. To obtain various wavelengths, the grating is rotated by means of an arm that rides on the wavelength cam. In setting the wavelengths, the cam rotates the grating so that the desired wavelength passes through the exit slit. The monochromatic light, which passes through the exit slit, is contained in a test tube or cuvette placed in the light path (the light also goes through the red filter in the infrared [IR] wavelength range) and finally terminates at the measuring phototube where the light energy is converted to an electric signal. Whenever the sample is removed from the instrument, an occluder automatically falls into the light beam so that the zero may be set without further manipulation. A light control is provided to set 100% transmittance or zero absorbance with a reference or standard solution in the sample compartment. The optics of the lens tube provides an extended range that goes to 340 nm.

A wall separates the optical system from the electronics, thus preventing dust and dirt from spoiling the efficiency of the optical system. At the same time, the wall shields the meter to prevent erroneous readings and fluctuations from stray light entering through the meter face or permitting error-free operation in bright light or even fluctuating sunlight.

4.5.3.2 Description of the Electrical System

The measuring circuit consists of a dc Wheatstone bridge–type differential amplifier in which variation in characteristics, nonlinearities, and drift of one triode are cancelled by another. A current gain of 5000 V is obtained from this amplifier.

A Wheatstone bridge circuit similar to that in the detection–amplification system is used to balance and compare the monitoring output, with the constant voltage from the power supply voltage regulator tubes. This differential output is further amplified by a transistor cascade (powered by a silicon hill-wave rectifier circuit), transformed into conductance variation of the two transistors operating on the center-tapped secondary transformer. Thus, a change in transformed load is reflected back to the transformer primary and to the 75 ohm and 50 W resistors causing readjustment of the primary voltage. The net effect is that increase or decreases in supply voltage is cancelled out through the feedback system, providing a constant voltage at the primary of the supply transformer. Since this transformer supplies the lamp and amplifier tube filament, the lamp output and filament current are independent of line voltage. In this way not only amplifier stability is achieved by a compensating circuit for variation in tube characteristics but any second-order errors, due to individual tube filament temperature dependence, are removed by keeping the filament at constant operating conditions.

4.5.4 Choice of Instruments for Colorimetry

When the absorption characteristics of a compound are unknown, a spectrophotometer or a recording spectrophotometer should be employed. By scanning the compound through the entire light spectrum, an absorption curve is obtained. This curve or some of its portions may be utilized in identifying the compound or in the development of analytical techniques. When a narrow absorption peak is utilized for such purposes, a spectrophotometer should be employed due to its narrow wavelength "window." In instances where peaks are broad or may shift slightly, a colorimeter with its broader "windows" is a more useful apparatus.

4.5.4.1 Absorbance Curves of Two Dyes

4.5.4.1.1 Principle. Colored compounds have their own characteristic absorption spectra and careful selection of the wavelengths where maximum absorption is found enables a mixture of two colored substances to be analyzed.

4.5.4.1.2 Materials. Colorimeter with a series of filters; bromophenol blue (10 mg/L); methyl orange (10 mg/L); an "unknown" mixture of the two dyes.

4.5.4.1.3 Protocol. Determine the extinction of each dye in turn against the range of filters supplied with the colorimeter. Remember, the instrument must be reset on zero extinction with distilled water in the cuvette for each filter. Carefully note the wavelength of maximum transmission (minimum absorbance) of each filter and plot a graph of the absorbance recorded against this wavelength.

4.5.4.1.4 Observations
What is the wavelength that gives maximum absorbance for each dye?
How does mixing the dyes affect the absorption spectrum?

4.5.4.2 Demonstration of Beer's Law

4.5.4.2.1 Materials
1. As for experiment in Section 4.5.4.1.

4.5.4.2.2 Protocol. Prepare a range of concentrations of one of the dyes by setting up a series of tubes as follows:

Bromophenol blue (10 mg/L) (mL)	1	2	3	4	5
Distilled water (mL)		4	3	2	1	0

Place the filter that gives maximum extinction in the light path and zero the colorimeter with distilled water. Next, record the absorbance of each solution and plot this against the concentration of dye in each tube (mg/L or μg/mL).

1. Repeat the aforementioned experiment using the filter that gave maximum absorption with the methyl orange, one other filter, and, if possible, white light with no filter. How do the curves of extinction against concentration conform to Beer's law?
2. Repeat the whole of the experiment with the dye methyl orange.
3. Finally, use the information gained in these experiments, to determine the concentration of each dye present in the mixture.

4.5.4.3 Colorimetric Estimation of Inorganic Phosphate

4.5.4.3.1 Principle. Unlike the aforementioned dyes, most biochemical compounds are colorless and can only be analyzed colorimetrically after reacting them with a specific chemical reagent to give a colored product. This point is illustrated in the measurement of inorganic phosphate, which is probably one of the commonest determinations carried out in a biological laboratory, and the production of a standard curve now will prove useful in future experiments.

Inorganic phosphate reacts with ammonium molybdate in an acid solution to form phosphomolybdic acid. Addition of a reducing agent reduces the molybdenum in the phosphomolybdate to give a blue color, but does not affect the uncombined molybdic acid. In this method, the reducing agent used is *p*-methylaminophenol sulfate. The presence of copper in the buffer solution increases the rate at which the color develops.

4.5.4.3.2 Materials. Ammonium molybdate (50 g/L) 250 mL; copper acetate buffer pH 4.0 (dissolve 2.5 g of copper sulfate ($CuSO_4 \cdot 5H_2O$) and 46 g of sodium acetate ($CH_3COONa \cdot 3H_2O$) in 1 L of 2 mol/L acetic acid (check the pH and adjust to 4.0 if required); reducing agent (dissolve 20 g of *p*-methylaminophenol sulfate in a 100 g/L solution of sodium sulfite ($Na_2SO_4 \cdot 7H_2O$) and make up to 1 L; store in a dark bottle until required); trichloroacetic acid (100 g/L); stock phosphate solution containing 100 mg phosphorus/100 mL (dissolve 438 mg of potassium dihydrogen phosphate in water and make up to 100 mL; store in the refrigerator); working phosphate solution containing 1 mg phosphorus/100 mL (dilute the stock phosphate solution 100 times with 50 g/L TCA); colorimeter.

4.5.4.3.3 Protocol. Pipette 0.1–1 mL of the standard solution of phosphate into a test tube, and, where necessary, add water to bring the final volume to 1 mL. Then add 3 mL of copper acetate buffer, 0.5 mL of ammonium molybdate, and 0.5 mL of reducing agent mixing thoroughly after each addition. Allow to stand for 10 min and read the extinction at 880 nm. Set up a blank by replacing the phosphate with 1 mL of 50 g/L TCA.

Prepare a graph of the extinction against the concentration of phosphate.

When measuring the concentration of phosphate in a solution containing protein, the test solution is mixed with an equal volume of 20 g/L TCA, the precipitate centrifuged, and an aliquot of the supernatant treated as mentioned earlier.

4.5.4.4 Validity of Beer's Law for the Colorimetric Estimation of Creatinine

4.5.4.4.1 Principle. Creatinine, in the presence of picric acid in alkaline solution, forms a red tautomer of creatinine picrate.

4.5.4.4.2 Materials. Picric acid (saturated aqueous solution); sodium hydroxide (1 mol/L); creatinine standard (2 g/L); colorimeter; volumetric flasks (100 mL).

4.5.4.4.3 Protocol. Prepare a range of creatinine solutions by suitable dilution of the standard and place 1 mL of each solution in a standard 100 mL volumetric flask.

Add 1 mL of 1 mol/L sodium hydroxide and 2 mL of saturated picric acid solution to the flask, mix thoroughly, and stand for 10 min. Make up to the mark with water and measure the extinction at 530 nm. Plot a graph of the extinction against the creatinine concentration.

4.6 Spectrophotometry

Colorimeters are relatively simple instruments, which are designed to function only in the visible range. On the other hand, spectrophotometers are equipped to operate both in the visible and UV range. They are fitted with deuterium or hydrogen (for UV light) or tungsten (visible range) as the light sources. Several models of spectrophotometers of varying degrees of sophistication are available. These include single-beam, double-beam, recording, and multibeam instruments.

In the dual-beam spectrophotometers, the incident light is split into two beams of equal intensity, one of which passes through the reference cuvette and the other through the sample cuvette. The multibeam instrument is designed to record, simultaneously, absorbance changes at two or more predetermined wavelengths. Recording spectrophotometers can both scan the absorption spectrum of the sample in the desired range and also record the change in absorbance with time at a fixed wavelength. Reflectance spectrophotometers are used for determining spectra of pastes and suspensions such as that of microorganisms. The latter type of spectrophotometers enables measurement of the radiation absorbed when the light beam is reflected by a sample, which is too opaque to allow transmission of light (Figure 4.4).

4.6.1 Applications

UV/Vis spectroscopy is routinely used in the quantitative determination of solutions of transition metal ions, highly conjugated organic compounds, and biological macromolecules.

1. Qualitative analysis may be performed in UV and visible regions to identify certain classes of compounds both in the pure state and in biological mixtures, for example, proteins, nucleic acids, cytochromes, and chlorophylls (Figure 4.5a and b). Spectrophotometric analysis has been helpful in identification of the

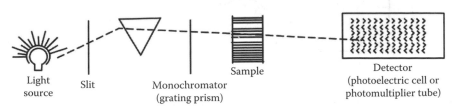

Light source Slit Monochromator (grating prism) Sample Detector (photoelectric cell or photomultiplier tube)

Figure 4.4 Schematic representation of a single-beam spectrophotometer.

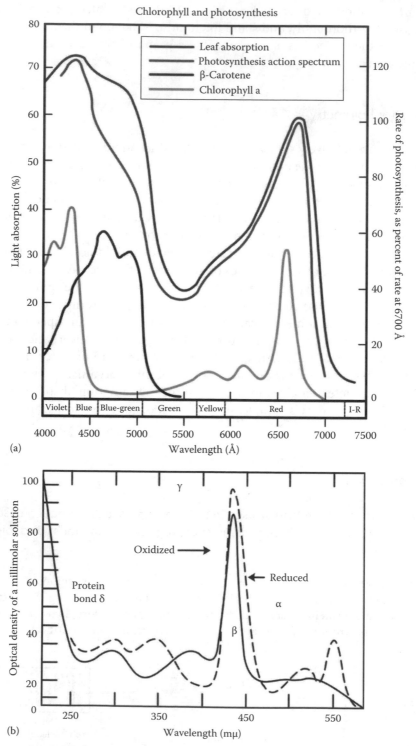

Figure 4.5 Absorption spectra of (a) plant pigments and (b) oxidized and reduced forms of cytochrome "c."

chromophores in light-mediated responses and processes. The rate of the light-dependent response or process is determined at varying wavelengths. The presence of cytochromes, heme, or flavins as prosthetic groups of a number of enzymes is largely deduced from spectral studies. Spectral studies have also been extremely useful in indicating involvement of certain compounds in various complex processes. A notable example of this is in establishing the role of quinones, flavins, and various cytochromes in the mitochondrial electron transport chain and in photophosphorylation.

The amount of substances with overlapping spectra, such as chlorophyll a and b in diethyl ether, may be estimated in their extinction coefficients that are known at two different wavelengths. For n compounds, absorbance data are required at λ_{max} wavelengths (Figure 4.6).

This technique may also be used to indicate chemical structure and intermediates occurring in a system.

2. The most widespread application of the spectrophotometric technique is for continuous assay of enzyme activities. The method can be employed for any enzymatic reaction in which the product, substrate, or cofactor shows absorbance at a unique wavelength.

3. Binding spectra or substrate-binding spectra may also study the extent of interaction between an enzyme and its substrate. An example of this is the binding of a drug (substrate) to liver microsomal monooxygenase, which causes a blue shift of the cytochrome P_{450} component of the enzyme from 420 to 390 nm (a hypsochromic shift).

4. Valuable structural studies may be performed on some particular biological macromolecules such as proteins and nucleic acids.

During enzyme purification, the presence as well as the amount of proteins in different fractions can be monitored simply by measuring their absorbance at 280 nm. Evidently, it is a much quicker, convenient, and nondestructive method of

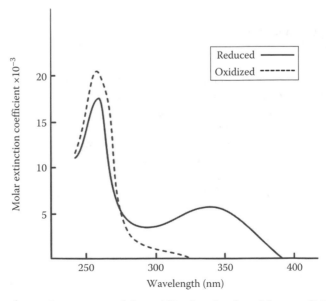

Figure 4.6 The absorption spectra of the oxidized and reduced forms of NAD and NADP.

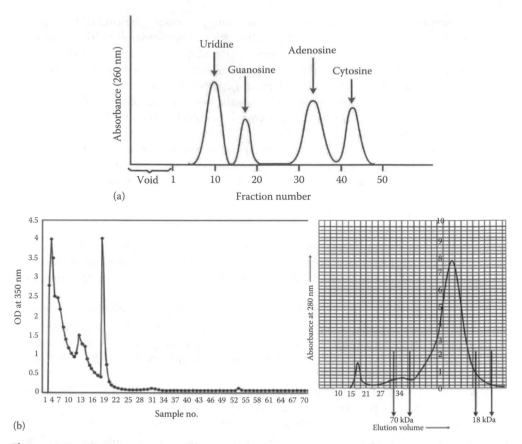

Figure 4.7 Illustrating separation and absorption spectra of (a) RNA nucleosides and (b) protein.

establishing elution profiles of such measurable substances during column chromatography or density gradient centrifugation. Most of the analytical methods for quantitative estimation of chromogenic biomolecules such as chlorophylls, phycocyanin, flavin, carotenoids, anthocyanins, and hemoglobin (Figures 4.7 and 4.8) are colorimetric or spectrophotometric and are based on measurement of absorbance at a specific wavelength known as absorption maxima (λ_{max}).

5. The aromatic amino acids are powerful chromophores in the UV processes such as denaturation (unfolding) of a polypeptide chain by pH, temperature, and ionic strength and can be monitored as more of these residues become exposed to the incident radiation.

6. Many of the processes may be followed, particularly if the amino acid residue tyrosine is involved, for example, protein–protein interaction, protein–metal, or protein–small molecule interaction.

7. In nucleic acid studies, solvent perturbation may be used to estimate the number of unpaired bases in RNA. The denaturation of the helical structure of DNA in solution may be investigated when the double-stranded DNA (ds DNA) is heated through its transition temperature. Effect on the secondary structure of DNA by pH and ionic strength may be studied in a similar way.

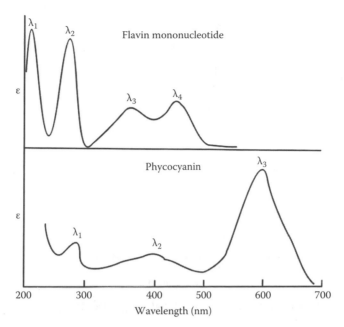

Figure 4.8   Absorption spectra of flavin mononucleotide and phycocyanin.

8. A UV/Vis spectrophotometer may be used as a detector for HPLC. The presence of an analyte gives a response assumed to be proportional to the concentration. For accurate results, the instrument's response to the analyte in the unknown should be compared with the response to a standard; this is very similar to the use of calibration curves. The response (e.g., peak height) for a particular concentration is known as the response factor.

4.6.2 Direction for Operations

The detailed operation of a particular instrument must, of course, be obtained by carefully reading the instruction manual, but a few general points concerning the use and care of colorimeters and spectrophotometers are given in the following:

1. *Cleaning cuvettes*: Cuvettes are cleaned by soaking in 50% v/v nitric acid and then thoroughly rinsed in distilled water.
2. *Using the cuvettes*: First of all, fill the cuvettes with distilled water and check them against each other to correct for any small differences in optical properties. Always wipe the outsides of the cuvettes with soft tissue paper, before placing in the cell holder, and do not handle them by the optical faces. When all the measurements have been taken, wash them with distilled water and leave them in the inverted position to drain.
3. *Absorption of radiation by cuvettes*: All cuvettes absorb radiation and the wavelengths at which significant absorption occurs depend on the material from which the cuvette is made (Figure 4.9). Silica cuvettes are the most transparent to UV light, but they are expensive, so they are generally used in the far UV. Glass cuvettes are

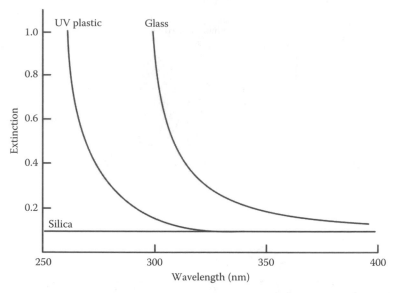

Figure 4.9   Relative absorption of UV light by glass, plastic, and silica cuvettes against air as a blank.

much cheaper than silica and so they are used whenever possible and invariably in the visible region of the spectrum. However, they do absorb UV and cannot be used below 360 nm. Plastic disposable cuvettes also absorb in the UV but much less than glass (Figure 4.9), and they can be conveniently used at 340 nm for the assay of dehydrogenase enzymes that require NAD or NADP as coenzymes.

4. *Light source*: A tungsten lamp produces a broad range of radiant energy down to about 360 nm. To obtain the UV region of the spectrum, a deuterium lamp is used as the light source. If the tungsten lamp is used in the range 360–400 nm, then a blue filter is placed in the light beam.

5. *Blanks*: The extinction of a solution is read against a reagent blank that contains everything except the compound to be measured. This blank is first placed in the instrument and the scale adjusted to zero extinction (100% transmittance) before reading any test solutions. Alternatively, the extinction can be read against distilled water, and the absorbance of the blank subtracted from that of the test solution.

6. *Replicates*: It is essential to prepare all blanks and standard solutions in duplicate so that an accurate standard curve can be constructed. In addition, the test solutions should also be prepared in duplicate wherever possible.

4.7 Infrared Spectroscopy

4.7.1 Introduction

IR spectroscopy deals with the IR region of the electromagnetic spectrum that is light with a longer wavelength and lower frequency than visible light. The technique is mostly based on absorption spectroscopy. As with all spectroscopic techniques, it can be used to identify and study chemicals. The IR portion of the electromagnetic spectrum is usually divided into three regions; the near-, mid-, and far-IR, named for their relation to the visible spectrum. The higher-energy near-IR, approximately 14,000–4,000/cm (0.8–2.5 μm wavelength), can excite overtone or harmonic vibrations. The mid-IR, approximately 4000–400/cm (2.5–25 μm), may be used to study the fundamental vibrations and associated rotational–vibrational structure; mostly useful for analytical purposes. The far-IR, approximately 400–10/cm (25–1000 μm), lying adjacent to the microwave region, has low energy and may be used for rotational spectroscopy. The names and classifications of these subregions are conventions and are only loosely based on the relative molecular or electromagnetic properties (Figure 4.10).

4.7.2 Principle

The absorption of electromagnetic radiations, in the IR region, results in vibrational and rotational transitions of the bonding atoms in the molecule leading to asymmetrical charge distribution. Nonlinear molecules may undergo vibrational motions like stretching and deforming. The energy requirement of stretching vibrations (symmetrical and asymmetrical) is higher and occurs at higher frequencies than deforming or bending (in-plane scissoring and rocking; out-of-plane wagging and twisting) vibrations.

Thus, absorption of IR radiations does not cause excitation of electrons but induces vibrations of bonding atoms, and the vibrations, which bring about alteration in dipole moment or displacement of charge, can be detected. The other types of vibrations are studied by Raman spectroscopy.

The fundamental vibrations are those that correspond to $V_0 - V_1$ transitions. The intensity of a particular absorption depends on the dipole moment of the molecule in the ground state and the vibration excited state. The higher is the difference in dipole moments, the greater will be the intensity of absorption. The IR spectra are observed only in the heteronuclear molecules since homonuclear molecules have no dipole moment. For obtaining an IR spectrum, electromagnetic radiation of increasing wavelength is passed through the sample and percent transmittance or absorbance is measured. The energy of vibration is quantified. The IR spectrum of the functional groups in a molecule is highly specific, and this characteristic spectrum or fingerprint facilitates identification of the compound. In IR spectroscopy, it is usual to work in frequency units (Hz) than wavelength.

4.7.3 Instrumental Components

The spectrum of the sample is recorded using an IR spectrophotometer. Double-beam spectrophotometers are preferred since CO_2 and H_2O show significant absorbance of IR light, and in these instruments, their interference is compensated automatically as half of the splitted beam of light passes through the reference cell and the other through the sample.

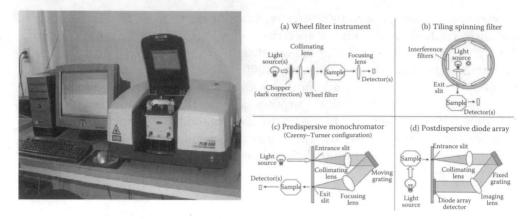

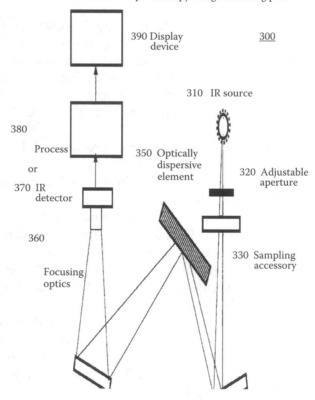

Figure 4.10 Illustrating IR spectrophotometer and operational diagrams.

4.7.4 Sources

An inert solid is electrically heated to a temperature in the range 1500–2000 K. The heated material will then emit IR radiation. *The Nernst glower* is a cylinder (1–2 mm diameter, approximately 20 mm long) of rare earth oxides. Platinum wires are sealed to the ends, and a current passed through the cylinder. The Nernst glower can reach temperatures of 2200 K.

4.7.4.1 Globar Source

It is a silicon carbide rod (5 mm diameter, 50 mm long) that is electrically heated to about 1500 K. Water cooling of the electrical contacts is needed to prevent arcing. The spectral output is comparable with the Nernst glower, except at short wavelengths (less than 5 μm) where its output becomes larger.

4.7.4.2 Incandescent Wire Source

It is a tightly wound coil of nichrome wire, electrically heated to 1100 K. It produces a lower intensity of radiation than the Nernst or globar sources but has a longer working life.

4.7.5 Detectors

There are three categories of detector: (1) thermal, (2) pyroelectric, and (3) photoconducting:

1. *Thermal or thermocouples* consist of a pair of junctions of different metals, for example, two pieces of bismuth fused to either end of a piece of antimony. The potential difference (voltage) between the junctions changes according to the difference in temperature between the junctions.
2. *Pyroelectric detectors* are made from a single-crystalline wafer of a pyroelectric material, such as triglycerine sulfate. The properties of a pyroelectric material are such that when an electric field is applied across it, electric polarization occurs (this happens in any dielectric material). In a pyroelectric material, when the field is removed, the polarization persists. The degree of polarization is temperature dependent. So, by sandwiching the pyroelectric material between two electrodes, a temperature-dependent capacitor is made. The heating effect of incident IR radiation causes a change in the capacitance of the material. Pyroelectric detectors have a fast response time. They are used in most Fourier transform IR instruments.
3. *Photoelectric detectors* such as the mercury cadmium telluride detector comprise of a film of semiconducting material deposited on a glass surface sealed in an evacuated envelope. Absorption of IR promotes nonconducting valence electrons to a higher conducting state. The electrical resistance of the semiconductor decreases. These detectors have better response characteristics than pyroelectric detectors and are used in FT-IR instruments—particularly in GC–FT-IR.

4.7.6 Sample Preparation

Preparation of the sample depends on its physical state. Solid samples are used in the form of solutions dissolved in an appropriate solvent such as CCl_4 or $CHCl_3$ or are prepared in mulls or as solids dispersed in KBr.

Gaseous samples require a sample cell with a long path length to compensate for the dilutions. The path length of the sample cell depends on the concentration of the compound of interest. A simple glass tube with length of 5–10 cm equipped with IR windows at both ends of the tube can be used for concentrations down to several hundred ppm. Sample gas concentrations, well below ppm, can be measured with a White's cell in which the IR light is guided, with mirrors, to travel through the gas. White's cells are available with optical path length starting from 0.5 up to 100 m.

Liquid samples can be sandwiched between two plates of a salt (commonly sodium chloride, or common salt, although a number of other salts such as potassium bromide or calcium fluoride are also used). The plates are transparent to the IR light and do not introduce any lines onto the spectra.

Solid samples can be prepared in a variety of ways. One common method is to crush the sample with an oily mulling agent (usually Nujol) in a marble or agate mortar, with a pestle. A thin film of the mull is smeared onto salt plates and measured. The second method is to grind a quantity of the sample with a specially purified salt (usually potassium bromide) finely (to remove scattering effects from large crystals). This powder mixture is then pressed, in a mechanical press, to form a translucent pellet, through which the beam of the spectrometer can pass. A third technique is the "cast film" technique, which is used mainly for polymeric materials. The sample is first dissolved in a suitable, nonhygroscopic solvent. A drop of this solution is deposited on the surface of a KBr or NaCl cell. The solution is then evaporated to dryness and the film formed on the cell is analyzed directly. Care is important to ensure that the film is not too thick, otherwise light cannot pass through it. This technique is suitable for qualitative analysis. The final method is to use microtomy to cut a thin (20–100 μm) film from a solid sample. This is one of the most important ways of analyzing failed plastic products, for example, because the integrity of the solid is preserved.

In photoacoustic spectroscopy, the need for sample treatment is minimal. The sample, liquid or solid, is placed into the sample cup that is inserted into the photoacoustic cell that is then sealed for the measurement. The sample may be one solid piece, powder, or basically any form for the measurement. For example, a piece of rock can be inserted into the sample cup and the spectrum measured from it. It is important to note that spectra obtained from different sample preparation methods will look slightly different from each other due to differences in the samples' physical states.

4.7.7 Applications

It is used in quality control, dynamic measurement, and monitoring applications such as the long-term unattended measurement of CO_2 concentrations in greenhouses and growth chambers by IR gas analyzers.

The main application of IR spectroscopy in biochemical investigations is in the elucidation of the structure of purified biological molecules of intermediate size such as small peptides, metabolic intermediates, substrates, and drugs. This technique has been employed for examining the secondary structure of proteins.

It is also used in forensic analysis in both criminal and civil cases, for example, in identifying polymer degradation. It can be used in detecting how much alcohol is in the blood of a suspected drunk driver measured as 1/10,000 g/mL = 100 μg/mL.

Interfacing IR spectroscopy with gas chromatography is a powerful technique for analyzing drug metabolites. Its most important application is to study carbon dioxide metabolism during photosynthesis and respiration in plants and microorganisms.

With increasing technology in computer filtering and manipulation of the results, samples in solution can now be measured accurately (water produces a broad absorbance across the range of interest and thus renders the spectra unreadable without this computer treatment).

Some instruments will also automatically tell you what substance is being measured from a store of thousands of reference spectra held in storage.

IR spectroscopy is also useful in measuring the degree of polymerization. Changes in the character or quantity of a particular bond are assessed by measuring at a specific frequency over time. Modern research instruments can take IR measurements across the range of interest as frequently as 32 times a second. This can be done while simultaneous measurements are made using other techniques. This makes the observations of chemical reactions and processes quicker and more accurate.

4.7.7.1 Absorption Spectrum of p-Nitrophenol

4.7.7.1.1 Materials. p-Nitrophenol (10 mmol/L) 50 mL; hydrochloric acid (10 mmol/L) 50 mL; sodium hydroxide (10 mmol/L) 50 mL; volumetric flasks (100 mL) capacity; UV–Vis scanning spectrophotometer.

4.7.7.1.2 Protocol. Dilute the p-nitrophenol solution 0.2–50 mL with (a) 10 mmol/L HCl and (b) 10 mmol/L NaOH. Determine the absorption spectra of each solution by scanning from 250 to 500 nm. Comment on the differences between the two spectra and calculate the molar extinction coefficient at the wavelength for the maximum absorption.

4.7.7.2 Absorption Spectra of Coenzyme Cytochrome c

4.7.7.2.1 Principle. A characteristic absorption spectrum is simply a plot of absorbance of light by a compound at different wavelengths. Absorption spectra of oxidized and reduced forms of cytochrome c are presented (Figure 4.5b). The presence of cytochrome, ham, or flavins as a prosthetic group of a number of enzymes was largely deduced from spectral studies. This technique has also been extremely useful in indicating involvement of certain compounds in various complex processes. A notable example of this is in establishing the role of quinones, flavins, and various cytochromes in the mitochondrial electron transport chain and in the photosynthetic electron transfer chain (Figure 4.11). Special investigations revealed that under anaerobic conditions, the addition of NADH to mitochondrial preparations results in the reduction of cytochrome "c." It undergocs rapid oxidation on the introduction of oxygen, thereby suggesting that it acts as an intermediate carrier of electrons during their transport from NADH to oxygen via the mitochondrial electron transfer

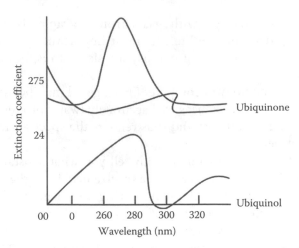

Figure 4.11   Absorption and differences of spectra of ubiquinone and ubiquinol.

system. Such studies are generally carried out by examining the difference spectra because the various states of compounds exhibit qualitatively alteration in their spectral characteristics. As the name denotes, the difference spectra is a graphical representation of difference in absorbance of light at different wavelengths by two forms of a compound.

Spectrophotometric analysis has been helpful in identification of the chromophores in light-mediated responses and processes. First, the rate of the light-dependent response or process is determined at varying wavelengths. A plot of rate of the process (such as photosynthetic oxygen evolution) at various wavelengths is prepared. Such a plot is known as an action spectrum. An attempt is made to isolate the compound, from the tissue that has an absorption spectrum superimposable on the action spectrum. This approach has successfully been employed in establishing the central role of chlorophylls as the primary light-harvesting pigments in photosynthesis and of phytochrome in red-far red-induced responses in plants (Figures 4.5 and 4.8).

4.7.7.2.2 Materials.   1 mM solution of NADH in 0.05 M Tris–HCl buffer (pH 7.5).

4.7.7.2.3 Protocol.   Scan the absorption spectra of the NADH solution in the range of 200–560 nm.

4.7.7.3 Determination of the pK$_a$ Value of p-Nitrophenol

4.7.7.3.1 Principle.   p-Nitrophenol dissociates as shown in Figure 4.12, and the undissociated form that is present in acid solution does not absorb in the visible region, while the quinonoid structure present in alkaline solution absorbs strongly. Now the pK value is the pH at which there is 50% ionization or, in other words, the pH that gives half the absorbance value obtained in alkaline solution assuming 100% ionization in alkali.

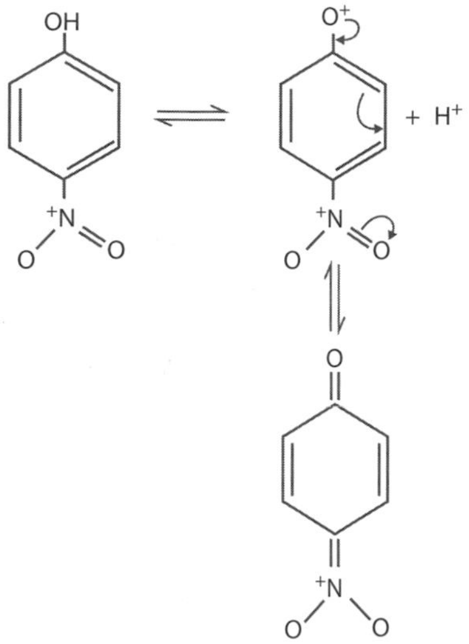

Figure 4.12   Dissociation of p-nitrophenol.

4.7.7.3.2 Materials.   p-Nitrophenol (10 mmol/L) 50 mL; hydrochloric acid (10 mmol/L) 50 mL; sodium hydroxide (10 mmol/L) 50 mL; volumetric flasks (100 mL) capacity; UV–Vis scanning spectrophotometer.

Buffer solutions, 50 mmol/L—(1) citrate phosphate buffer (pH 3.0); (2) citrate phosphate buffer (pH 4.0); (3) citrate phosphate buffer (pH 5.0); (4) citrate phosphate buffer (pH 6.0); (5) Tris–HCl buffer (pH 7.0); (6) Tris–HCl buffer (pH 7.5); (7) Tris–HCl buffer (pH 8.0); (8) Tris–HCl buffer (pH 9.0); (9) carbonate–bicarbonate (pH 10.0); (10) carbonate–bicarbonate (pH 11.0).

4.7.7.3.3 Protocol.   Prepare 0.2 mL in 50 mL dilutions of the p-nitrophenol in the aforementioned buffer solutions and determine the extinction of each solution at a wavelength (405 nm) where the undissociated phenol has zero absorption. Determine the pK_a value. Titrate 25 mL of mmol/L p-nitrophenol with NaOH to a pH of 10.5. Correct the values obtained after titrating distilled water and prepare a titration curve. Determine the pK_a value and compare it with that obtained previously.

4.7.7.4 Estimation of Barbiturates with the UV–Vis Spectrophotometer

4.7.7.4.1 Principle.   The 5, 5′-substituted barbiturates give a characteristic absorption spectrum in the UV with a maximum at 240 nm at pH 10. At pH 13.4, the maximum is shifted to 253 nm, and a minimum is obtained at 235 nm. To detect the presence of barbiturates, the spectrum of a solution is plotted over the range 220–270 nm at pH 10.0 and pH 13.4 (Figure 4.13).

As well as the characteristic maxima and minima, the curves cross at 227 and 250 nm, the isosbestic points (it is a specific wavelength at which two chemical species have the same molar absorptivity). There is also a maximum difference in absorption at 260 nm, which is used to determine the amount of barbiturate present. The reason

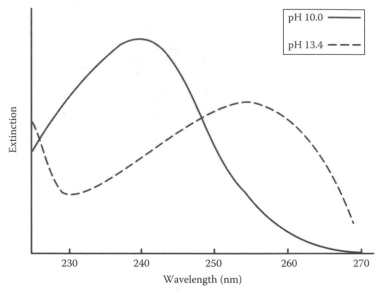

Figure 4.13   Absorption spectra of barbiturates in alkaline solution.

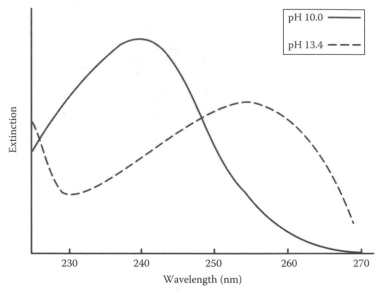

Figure 4.14   Ionized forms of 5,5′-substituted barbiturates.

for this difference in absorption spectra is that these compounds are dibasic acids with pK values of about 8 and 12 so that the single ionized form is present at pH 10 and the double ionized at pH 13.4 (Figure 4.14).

The detection and estimation of barbiturates in blood or urine is quite important in cases of suspected poisoning, and the method can be used with blood, plasma, or urine. Barbiturates present in body fluids are extracted into chloroform, then into alkali.

An "unknown" solution of phenobarbitone, in plasma, can be usefully examined. Concentrations in the plasma in excess of 100 mg/L are usually fatal, and for convenience, the "unknown" should be somewhere in the region of 10–60 mg/L.

4.7.7.4.2 Materials   Chloroform (extra pure grade) 500 mL; phosphate buffer (0.4 mol/L, pH 7.4) 100 mL; sodium hydroxide (50 mmol/L) 100 mL; boric acid–potassium chloride solution (37.2 g boric acid and 100 mL 45 g potassium chloride in 1 L); phenobarbitone standard (20 mg/L in 0.45 mol/L NaOH) 100 mL; Whatman filter papers; sulfuric acid (6 mol/L) 10 mL; separating funnels; spectrophotometer.

4.7.7.4.3 Protocol

4.7.7.4.3.1 Extraction. Place 5–10 mL of the fluid in a separating funnel and extract three times with 30 mL of chloroform. Combine the extracts and filter through a Whatman filter paper. Wash the filter with a little chloroform.

Return the extract to a clean separating funnel and wash twice with 5 mL of phosphate buffer, which is discarded. Add 10 mL of 0.45 mol/L sodium hydroxide and shake vigorously for 1–2 min. Allow the phases to separate and centrifuge the aqueous layer.

4.7.7.4.3.2 Method 1 Detection of Barbiturate. Add 2 mL of the aqueous layer to 2 mL of 0.45 mol/L sodium hydroxide and plot the spectra from 220 to 270 nm against a blank of mol/L sodium hydroxide (pH 13.4).

Add 2 mL of the aqueous layer to 2 mL boric acid solution. Measure this against a blank of 2 mL of 0.45 mol/L sodium hydroxide and 2 mL of boric acid (pH 10.0).

4.7.7.4.3.3 Calculation. The test extinction of the pH 13.4 solution is less than the pH 10.0 solutions at 260 nm.

The standard solution of phenobarbitone is treated the same as the aqueous layer from the chloroform extract, and the standard extinction at 260 nm is calculated as for test:

$$\text{Barbiturate conc. in fluid (mg/L)} = \frac{\text{Test extinction}}{\text{Standard extinction}} \times 20$$

4.7.7.4.3.4 Method 2 Detection of Barbiturate. Record the spectrum at pH 10.0 as mentioned earlier then add four drops of 6 mol/L H_2SO_4 to the sample and blank cuvettes and record the spectrum again, but this time at pH 2. Explain the difference observed and suggest what ionized form is probably present at pH 2. Treat the standard solution of phenobarbitone the same way, and calculate the barbiturate concentration by using the difference in extinction at 240 nm of the two solutions.

4.7.7.5 Estimation of Hemoglobin Spectra

4.7.7.5.1 Principle. The oxygen-carrying protein hemoglobin, when combined with oxygen, there is a shift in the absorption spectrum and the color of the blood is changed from dark to bright red; the reverse change occurs on deoxygenation, and this is the reason for the difference in color of venous and arterial blood. In both hemoglobin and oxyhemoglobin, the iron is present in the ferrous form and is not oxidized on oxygenation. If the ferrous iron is oxidized to ferric with an oxidizing agent such as ferricyanide, then methemoglobin is formed, and the molecule can no longer combine with oxygen or carbon monoxide. Normally, human blood contains only about 1% methemoglobin, but this may be increased following the ingestion of certain drugs.

Hemoglobin combines with carbon monoxide some 200 times more readily than with oxygen to form carboxyhemoglobin; the amount of hemoglobin available for oxygen transport is thereby reduced, and if sufficient CO is present, death ensues from oxygen starvation of the tissues. The absorption spectrum of carboxyhemoglobin is

TABLE 4.2　Absorption Maxima of Different Forms of Hemoglobin

Hemoglobin Derivative	Absorption Maxima (nm)		
Hemoglobin	555	430	—
Oxyhemoglobin	577	541	413
Carboxyhemoglobin	570	535	418
Methemoglobin	630	500	406

only slightly different from that of hemoglobin, but the difference is sufficient to be the basis for detecting the compound in blood and this may have medicolegal implications in cases of death from coal gas poisoning (Table 4.2). If blood is shaken in the air, then oxyhemoglobin is formed, so Stokes' reagent is added to the blood to remove the oxygen completely and form hemoglobin.

4.7.7.5.2 Materials.　Stokes' reagent (20 g/L ferrous sulfate and 30 g/L tartaric acid: just before use, add ammonia until a faint precipitate that forms at first is just dissolved. This is a solution of ammonium ferrotartrate, a reducing agent) 50 mL; potassium ferricyanide (100 g/L) 50 mL; hemoglobin preparation from hemolyzed red cells of human blood 20–50 mL; sodium chloride (0.15 mol/L) 200 mL; direct-vision spectroscope; source of carbon monoxide; UV–Vis spectrophotometer.

4.7.7.5.3 Protocol.　Prepare a solution of hemoglobin in saline at a concentration of about 1 mg/mL and scan the absorption spectrum over the range 400–700 nm. Plot the absorption spectrum after treating the hemoglobin solution as follows:

1. Add two drops of freshly prepared Stokes' solution to 4 mL of the hemoglobin solution.
2. Pass carbon monoxide through the hemoglobin solution (not in the open lab) and seal the top of the cuvette.
3. Repeat (2) after adding two drops of Stokes' reagent.
4. Add two drops of 100 g/L potassium ferricyanide solution to 4 mL of hemoglobin.
5. Repeat (4), but add Stokes' reagent after the ferricyanide.

In addition, examine the visible spectra of the aforementioned solutions using a direct-vision spectroscope.

4.7.7.6　Ultraviolet Absorption of Proteins and Amino Acids

4.7.7.6.1 Principle.　Absorption at 210 nm: Below 230 nm, the extinction of a protein solution rises steeply reaching a maximum at 190 nm; this is mainly due to the peptide bond. In practice, it is more convenient to measure the extinction at 210 nm where the specific extinction coefficient $\Sigma_{1\,cm}^{10\,g/L}$ is about 200 for most proteins. All proteins have a similar specific absorption here since the peptide bond content is similar. A number of compounds such as carboxylic acids, buffer ions, alcohols, bicarbonate, and aromatic compounds also absorb in this region, so the results need to be interpreted with care.

Absorption at 280 nm: Tyrosine and tryptophan absorb at 275 and 280 nm and so proteins containing these amino acids will also absorb in this region. The specific extinction coefficient $\Sigma_{1\,cm}^{10\,g/L}$ varies according to how much of these amino acids are present in the particular protein. The values found in practice range from 6 to 60, although many proteins have a value close to 10: that is, 1 mg/mL of protein gives extinction at 280 nm of about 1 when viewed through a 1 cm light path.

The disadvantage of this method is that many other compounds absorb in this region, particularly nucleic acids that have a peak at 260 nm. Pure proteins have a ratio of absorption (at 280 nm/260 nm) of about 1.8, while nucleic acids have a ratio of 0.5.

4.7.7.6.2 Materials. Proteins (1 g/L albumin and casein); amino acids (0.1 mmol/L tyrosine, tryptophan, and phenylalanine in 10 mL water adjusted to pH 7, in 10 mmol/L HCl, and 10 mmol/L NaOH) scanning UV–Vis spectrophotometer.

4.7.7.6.3 Protocol. Scan the absorption spectra of the aforementioned compounds over the range 190–400 nm. At the lower end of the wavelength, the proteins will need to be diluted about 1 in 200.

4.8 Fluorescence Spectroscopy

4.8.1 Fluorescence

Some compounds not only absorb radiation but also emit some of the energy in the form of fluorescent light. Energy is absorbed in the UV region of the spectrum and molecules are elevated from the ground state to a high energy level. The excited molecules then return to the ground state with the consequent emission of visible light. The wavelength of the emitted light is always higher than that of the absorbed radiation.

The requirements for a compound to fluoresce are an absorbing structure and a high resonance energy. Aromatic compounds in general are often capable of fluorescence, particularly if the substituent in the ring is electron donating.

4.8.2 Quenching

However, fluorescence is not as common as absorption due to quenching. Molecules containing Br, I, NO_2, and azo groups show little fluorescence because of this. Quenching decreases the quantum yield so that the absorbed energy is used for competitive electronic transitions with excited molecules or for breaking weak bonds instead of being emitted as fluorescent light. Quenching can also occur by interaction with the solvent and other molecules in solution. In some cases, the quenching reactions are fairly specific and can be used to identify a particular fluorescent compound.

4.8.3 Applications

Fluorescent compounds are used extensively in biochemical investigations as they can be detected at very low concentrations and with a high degree of selectivity. The absorption and fluorescent spectra of a compound are quite characteristic so that when the maxima are selected by filters or monochromators on the incident and emitted beams, the fluorescent compound can be detected and measured even when other fluorescent compounds are present. Some of the applications include the use of fluorescent compounds as membrane probes, substrates for sensitive enzyme assays, and immunofluorescence.

4.8.4 Intensity of Fluorescence and Concentration

The fluorescence (F) depends on the intensity of light absorbed, so that if the intensity of the incident and emergent beams is I_0 and I, respectively, then

$$F = K(I_0 - I)$$

Now Beer's law states that

$$I = I_0 \, e^{-kcl}$$

So that

$$I_0 - I = I_0 \, (l - e^{-kcl})$$

Therefore,

$$F = KI_0 \, (l - e^{-kcl})$$

Expanding the exponential expression and assuming c to be small so that the higher terms can be ignored,

$$F = KI_0 \, (kcl)$$

The constant K is known as the quantum yield and is the ratio of the number of quanta emitted to the number absorbed. For a particular compound and instrument, when I_0, k, and l are constant, the fluorescence is directly proportional to the concentration:

$$F = K'c$$

This equation holds in practice provided the solution absorbs less than 5% of the exciting radiation so that the greater the intensity of the incident light, the higher is the concentration that gives a linear response. High light intensities also produce problems such as photodecomposition and light scattering. The useful concentration range for the determination of fluorescent compounds is 0.001–10 µg/mL depending on the material under investigation.

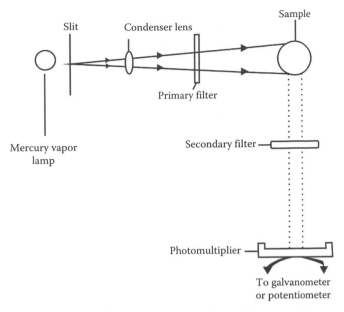

Figure 4.15   Schematic representation of fluorimeter.

4.8.5 Description of the Instrument

4.8.5.1 Light Source

A high-intensity source of UV light is provided by a mercury vapor lamp. UV light is a hazard to be reckoned with so adequate shielding of the lamp must be provided (Figure 4.15).

4.8.5.2 Filters

After passing through a condenser lens, the light meets the primary filter, which isolates a particular region of the UV spectrum, then on through the sample in a circular cuvette to the secondary filter set at right angles to the incident beam. The secondary filter cuts off all the scattered light and passes all the emitted fluorescence. The filters are carefully selected so that their transmission matches the excitation and fluorescent spectra as closely as possible. Ideally, monochromatic light should be used, but this increases the cost of the instrument. A fluorimeter with a monochromator on the secondary side and filters on the primary side is adequate for many purposes.

4.8.5.3 Photomultiplier

The light then passes on to the photomultiplier that produces an electric signal proportional to the intensity of the fluorescent beam. This may then be displayed or recorded on an appropriate measuring device such as a microammeter or flat bed recorder.

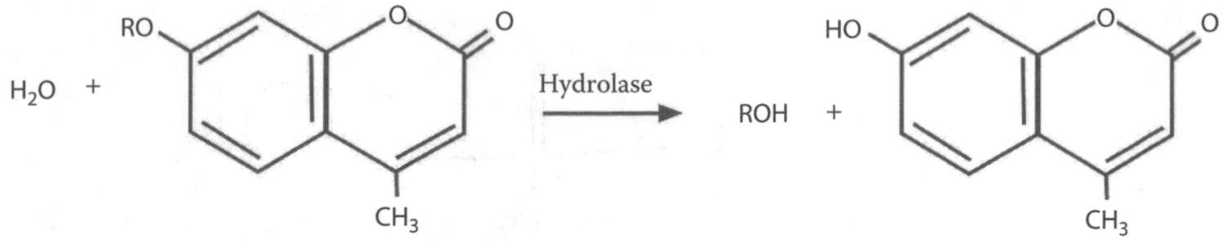

Figure 4.16   Assay of hydrolases by using fluorescent substrate.

4.9 Sensitivity of Fluorescence Assays

4.9.1 Principle

The 4-methylumbelliferyl compounds form the basis of a number of fluorimetric assays of hydrolases (Figure 4.16). The product 4-methylumbelliferone is highly fluorescent and can be distinguished from the lower fluorescence of the substrate by careful selection of the excitation and emission wavelengths.

The sensitivity of such assays is demonstrated by comparing the low concentration of 4-methylumbelliferone that can be detected by absorption spectroscopy with the concentration of the compound that can be detected fluorimetry. Quinine sulfate is included so that maximum reading is obtained with 4 μL.

4.9.2 Materials

Quinine sulfate (4 μg/mL in 0.1 mol/L H_2SO_4) 20 mL; 4-methylumbelliferone primary standard (1 mmol/mL in methanol) 10 mL; glycine buffer (0.05 mol/L, pH 10.4) 100 mL; H_2SO_4 (0.1 mol/L) 50 mL; fluorimeter; UV–Vis spectrophotometer.

4.9.3 Protocol

Adsorption: Plot the absorption spectra against the solvent blank for the 0.1 mmol/L 4-methylumbelliferone and the quinine sulfate. Determine the wavelength that gives the maximum absorption and measure the extinction of a range of concentrations of each compound. Prepare a plot of extinction against concentration for the quinine sulfate and the 4-methylumbelliferone.

Fluorescence: Set the fluorimeter to give a primary wavelength of 360 nm and a secondary wavelength of 420 nm, and adjust the galvanometer to give a maximum deflection with the solution of quinine sulfate. Prepare a series of dilutions down to 0.01 μg/mL of the quinine sulfate and 5 ng/mL of the 4-methylumbelliferone.

Read the fluorescence of each solution and plot a graph of fluorescence against concentration. Compare the relative sensitivities of the absorption and fluorescence assays of the solutions and estimate the minimum concentration that can be assayed by the two methods.

4.10 Fluorescence Quenching

4.10.1 Materials

2,4-Dinitrophenol (DNP) (0.1 mmol/L in 0.05 mol/L glycine buffer, pH 10.4); HCl (0.1 mol/L) 50 mL.

4.10.2 Protocol

1. Repeat the standard curve of 4-methylumbelliferone in glycine buffer but this time include 0.1 mmol/L 2,4-DNP in the solutions and analyze the result.
2. Prepare a standard curve of the quinine sulfate in 0.1 mol/L 2,4 DNP.

4.11 Measurement of α-Naphthyl Phosphatase Activity

4.11.1 Principle

At pH 10, alkaline phosphatase hydrolyzes α-naphthyl phosphate to α-naphthol and phosphate. The solution is then adjusted to pH 12 with sodium hydroxide when the α-naphthol is present largely as the phenolate ion, which is highly fluorescent (Figure 4.17). The substrate is slightly fluorescent, but careful selection of the excitation and emission wavelengths avoids the weak fluorescence of the α-naphthyl phosphate.

4.11.2 Materials

α-Naphthol (1 mmol/L prepared by dissolving the pure compound in a little ethanol and making up to the mark with water) 100 mL; α-naphthyl phosphate (10 mmol/L, stored in a brown bottle) 50 mL; sodium carbonate–bicarbonate buffer (0.1 mol/L, pH 10) 100 mL; sodium hydroxide (0.5 mol/L) 100 mL; fluorimeter: primary wavelength 335 nm and secondary wavelength 455 nm; alkaline phosphatase.

4.11.3 Protocol

Standard curve: Make up a range of concentrations of α-naphthol from 0.1 to 1 μmol/L and prepare the following mixtures (Table 4.3).

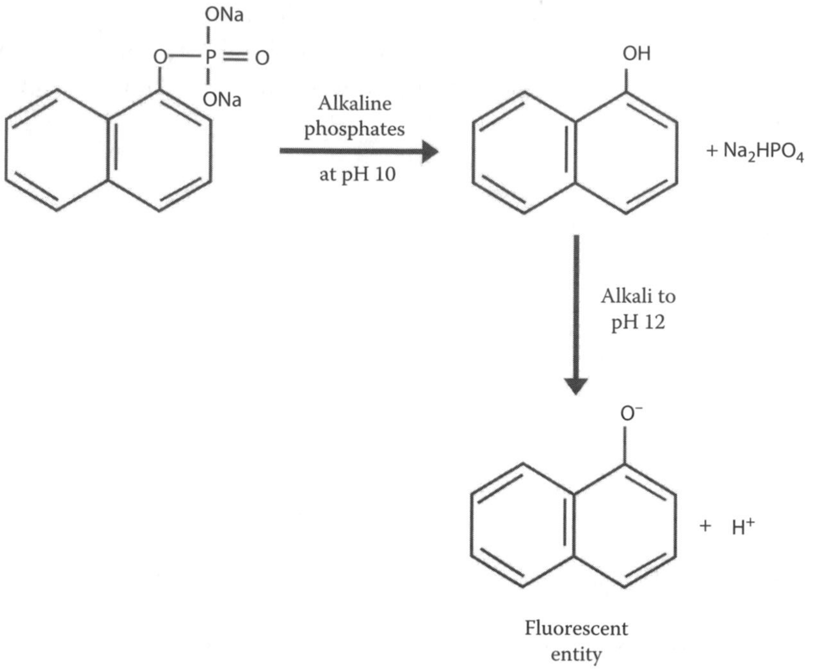

Figure 4.17 Hydrolysis of α-naphthyl phosphate by alkaline phosphatase.

TABLE 4.3   Range of Concentrations of α-Naphthol for Standard Curve

	1 (mL)	2 (mL)	3 (mL)
α-Naphthol (0.1–1 μmol/L)	1	1	1
Sodium carbonate–bicarbonate buffer	1	1	—
Water	1	—	3
Sodium hydroxide (0.5 mol/L)	1	1	—
α-Naphthyl phosphate (10 mmol/L)	—	1	—

Set the potentiometer reading to 100 for the highest concentration of α-naphthol in 1, and prepare standard curves of the potentiometer readings for each concentration of α-naphthol in 1, 2, and 3.

Record the effect of α-naphthol in Tube 2 and what happens when the α-naphthol is prepared in water (Tube 3)?

Enzyme activity: Prepare a mixture of 1 mL of α-naphthyl phosphate and 1 mL of buffer solution, equilibrate at 37°C, and add 1 mL of an aqueous solution of alkaline phosphatase of suitable dilution. After a measured time interval, add 1 mL of 0.5 mol/L sodium hydroxide to stop the reaction and give the pH for maximum fluorescence of the α-naphthol. Read the fluorescence in the fluorimeter and plot a progress curve of the reaction by stopping the reaction after different intervals of time. Express the activity from the initial reaction rate as μmol of α-naphthol produced per mL of enzyme.

4.11.4 Calculations

1. A solution of 10^{-5} mol/L ATP shows a transmittance of 70.2% at 260 nm in a 1 cm cuvette; calculate the following:
   a. The absorbance
   b. The transmittance in a 3 cm cuvette
   c. The absorbance of 50 μmol/L ATP in a 1 cm cuvette
2. The specific extinction coefficient of a glycogen–iodine complex at 450 nm is 0.20. Calculate the concentration of glycogen in a solution of iodine that has an extinction of 0.36 in a 3 cm cuvette.
3. A solution of UTP of 29.3 mg/L has an extinction of 0.25 at 260 nm. If the light path is 1 cm and the molecular weight of the UTP is 586; calculate the following:
   a. The molar extinction coefficient
   b. The transmittance of a 10 μmol/L solution
4. A solution of the amino acids tyrosine and tryptophan has an extinction of 0.65 at 280 nm and 0.5 at 295 nm in a 1 cm cuvette. Given the extinction coefficients of the pure amino acids, calculate the concentration of tyrosine and tryptophan present in the mixture (Table 4.4).
5. A tissue extract (0.3 mL) was diluted with 0.9 mL of water. An aliquot of the diluted solution (0.5 mL) was added to 2.5 mL of biuret reagent and gave an extinction of 0.324 at 540 nm in a 1 cm cuvette. A standard solution of albumin (4 mg/mL) gave an extinction of 0.24 when 0.5 mL was added to 2.5 mL of the biuret reagent. Calculate the concentration of protein in the original tissue extract.
6. A solution containing NAD^+ and NADH has an extinction of 0.316 at 340 nm and 1.11 at 260 nm. Calculate the concentrations of oxidized and reduced forms of the coenzyme in the solution given that both NAD^+ and NADH absorb at 260 nm but only NADH absorbs at 340 nm (Table 4.5).

TABLE 4.4 Molar Extinction Coefficient of Tyrosine and Tryptophan at Two Different Wavelengths

Wavelength (nm)	Molar Extinction Coefficient (L/mol/cm)	
	Tyrosine	Tryptophan
280	1500	5000
295	2500	2500

TABLE 4.5 Molar Extinction Coefficient of NAD and NADH at Two Different Wavelengths

Wavelength (nm)	Molar Extinction Coefficient (L/mol/cm)	
	NAD^+	NADH
260	18,000	15,000
340	0	6,320

Suggested Readings

Bassler, G. C. and T. C. Morrill. 1981. *Spectrometric Identification of Organic Compounds*, 4th edn. New York: John Wiley & Sons.

Bisen, P. S. and A. Sharma. 2012. *Introduction to Instrumentation in Life Sciences*. Boca Raton, FL: CRC Press.

Burrin, D. H. 1986. Spectroscopic techniques. In *A Biologist's Guide to Principles and Techniques of Practical Biochemistry*, eds. K. Wilson and K. H. Goulding, 3rd edn. London, U.K.: Arnold.

Campbell, I. D. and R. A. Dwek. 1984. *Biological Spectroscopy*. Menlo Park, CA: Benjamin/Cummings.

Edisbury, J. R. 1967. *Practical Hints on Absorption Spectrometry*. New York: Plenum Press.

Schwedt, G. 1997. *The Essential Guide to Analytical Chemistry* (Brooks Haderlie, trans.). New York: John Wiley & Sons.

Sharma, B. K. 1991. *Instrumental Methods of Chemical Analysis*, 11th edn. New Delhi, India: Goyal Publishing House.

Snavely, B. B. 1969. Flashlamp-excited organic dye lasers. *Proc IEEE* 57: 1374–1390.

Udenfriend, S. 1969. *Fluorescence Assay in Biology and Medicine*. New York: Academic Press.

Van Holde, K. E. 1985. *Physical Biochemistry*, 2nd edn. Englewood Cliffs, NJ: Prentice-Hall.

Westermeier, R. and T. Naven. 2002. *Proteomics in Practice: A Laboratory Manual of Proteome Analysis*, 3rd edn. Weinheim, Germany: Wiley-VCH Verlag Gmbh.

Wilson, K. and J. Walker. 2003. *Practical Biochemistry; Principle and Techniques*, 5th edn. Cambridge, U.K.: Cambridge University Press.

Wrigglesworth, J. M. 1983. *Biochemical Research Techniques: A Practical Introduction*. Chichester, U.K.: Ellis Horwood.

Important Links

Circular dichroism spectroscopy: http://www.biocompare.com/ProductDetails/665635/FVS-6000-Vibrational-CD-Spectrometer.html

ESR spectroscopy: http://www.jeol.com/PRODUCTS/AnalyticalInstruments/ElectronSpin Resonance/tabid/98/Default.aspx

Fluorescence spectroscopy: http://www.perkinelmer.com/Catalog/Category/ID/Fluorescence %20Spectroscopy

Infrared spectroscopy: http://www.perkinelmer.com/Catalog/Product/ID/L1280002

MALDI-TOF MS: http://www.thermoscientific.com/ecomm/servlet/productsdetail?product Id=11962154&groupType=PRODUCT&searchType=0&storeId=11152&gclid=CLaV oIr-oKsCFYh_6wodyVEXfA

NMR spectroscopy: http://www.magritek.com/kea.html

UV–visible spectroscopy: http://www.uv-groebel.com/pms_spek1.php

5

Centrifugation

5.1 Introduction

A handful of sand is thrown into the water in a beaker; the particles sediment at a different rate depending upon their size and density. If, instead of water, glycerol or castor oil is used, sedimentation rate is considerably slow. All these common observations point out that the rate of sedimentation of a particle is dependent on (1) the density of a particle, (2) the size of the particle, and (3) the viscosity of the medium. The gravitational pull is yet another parameter, which under normal conditions is about 980 cm/s or 1 g unit. If this force is increased, the light particles will also sediment.

The extra gravitational force can be introduced by applying centrifugal force. If a stone tied to a string is swirled around holding the string at the other end, the stone is subjected to centrifugal force, that is, the force is acting in a direction away from the center of the axis (the force can be felt by introducing the finger in the path of the stone and feeling the impact!). This simple experiment will help us to understand several other principles. The faster the speed of rotation measured as the angular velocity or rotation per minute, the greater will be the force. The longer the radius of rotation, the greater is the force.

These can be expressed mathematically as (1) centrifugal force = (angular velocity), and (2) radius angular velocity is related to rotation per minute (rpm) by the following formula:

$$\text{Angular velocity} = \frac{2n \times \text{rpm}}{60} \, \text{rad/s}$$

The centrifugal field is generally expressed as relative centrifugal field. RCF in g units as

$$\text{RCF} = \frac{4\pi^2 (\text{rpm})^2}{3600 \times 980} \, \text{g}$$

$$\text{RCF} = (1.119 \times 10^{5}) \, (\text{rpm}) \, 2(r) \, \text{units}$$

5.2 Centrifuges

A centrifuge is an instrument, which produces centrifugal forces. Basically, it has containers fixed in such a way that they can be rotated around the central axis with the help of electric motors.

The most common centrifuge is the clinical centrifuge, which is sufficient to carry out most of the laboratory experiments like sedimentation of blood cells, chloroplasts, and emulsions. Most of the clinical centrifuges are manufactured to give a maximum speed up to 3000–4000 rpm. The rotor head is usually of swinging buckets (Figure 5.1). There are high-speed centrifuges, which produce speed up to 20,000 rpm. The friction of the rotor (sample containers), with air, produces much heat at high speed and, therefore, they have to be run under refrigeration. Centrifuges producing speed up to 60,000 rpm with gravitational force of the order of 100,000g are called ultra-centrifuges, and here, the rotors are spun, not only under refrigeration but also under vacuum to reduce friction, and the rotors are made up of special metals like titanium to withstand the high force.

5.3 Ultracentrifugation

The technique is useful to investigate the molecular mass, shape, and density of the protein. The basis of this technique is to exert the larger force than does the gravitational force of the earth, thus increasing the rate at which the particle settles (Figure 5.2).

The technique can be divided into the following types.

5.3.1 Preparative Centrifugation

It is applicable for the separation of whole cells, subcellular organelles, plasma membranes, polysome, chromatin, nucleic acids, lipoprotein, and virus.

5.3.2 Analytical Centrifugation

It is devoted to purifying the macromolecules or particles. Rate depends on the centrifugal field G directed radially outward (angular velocity).

5.3.3 Differential Centrifugation

This method is based on the difference in the sedimentation rate of the particles of different sizes or mass. During centrifugation, the larger particle sediments first and the less dense will sediment later. Macromolecules with the same size having

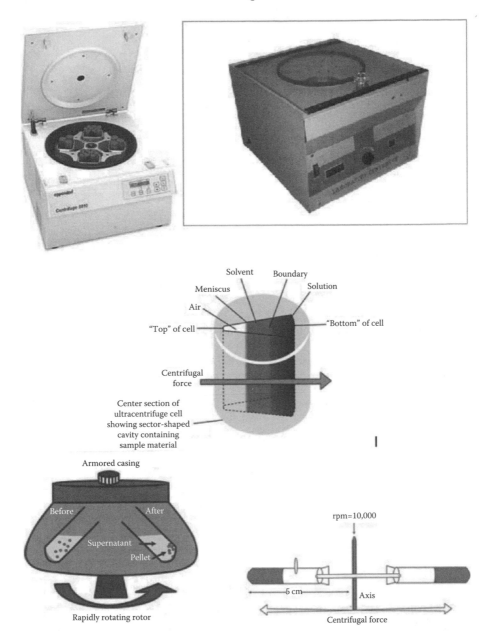

Figure 5.1   Clinical centrifuge and rotors demonstrating operation.

a narrower density range are separated by the rate zonal technique that is based mainly upon differences in their sizes and cannot be separated easily like mitochondria, lysosomes, and peroxisomes. The technique have been used for the separation of enzymes, hormones, RNA and DNA hybrids, ribosomal subunits, and subcellular organelles. In differential centrifugation, the particles to be separated are divided centrifugally into a number of fractions by increasing the applied centrifugal field. After the centrifugation, pellet and supernatant are obtained. Pellet is washed several

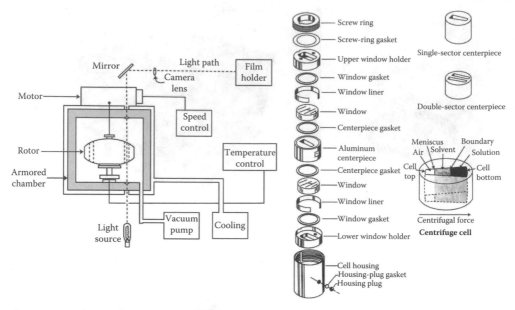

Figure 5.2   Illustrating various components of an ultracentrifuge.

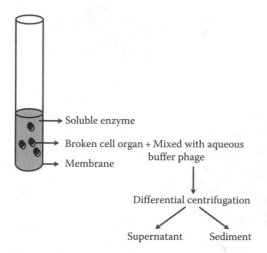

Figure 5.3   Differential centrifugation used for isolation of enzymes.

times and again centrifugation is done. The particle moves against respective sedimentation rates. Centrifugation is continued long enough to obtain pellet and the resulting supernatant (Figure 5.3). A cell homogenate is centrifuged several times, each time at a greater speed and for longer periods of time. In early spins, only the largest particles sediment (pellet) to the bottom of the centrifuge tubes; such particles might be whole cells, fragments of cell wall, nuclear cells, or grains of starch. A higher speed will sediment chloroplast and mitochondrial pellets. Microsomes, ribosomes, and also the dissolved protein molecules can be sedimented at a very high speed, might be tens of thousands of revolution per minute (rpm).

After homogenization, the suspension is separated into a number of fractions by centrifuging at various *g* values. The actual conditions of the fractionation depend on the tissue studied and those for the separation of rat liver mitochondria are not necessarily the same as those for the isolation of mitochondria from the other rat tissue. Also, some fractions, which are more or less homogeneous for one tissue, may be very heterogeneous in others.

5.3.4 Density Gradient Centrifugation

Subcellular particles can also be separated by using differences in their density rather than mass. To do this, the homogenate is placed on top of a discontinuous gradient formed by layering a series of different sucrose concentrations on top of each other. The tubes are then centrifuged, and, at equilibrium, the particles will be found as a band in that concentration of sucrose whose density is close to that of the organelles. The technique has been particularly useful in fractionating brain tissue when nerve endings and myelin can be isolated in a more or less homogenous condition. The alternative to a discontinuous density gradient is a continuous one, and relatively large quantities of material can be fractionated on such a gradient set up in a hollow centrifuge rotor, a technique known as *zonal centrifugation*. In density gradient centrifugation, the particles have to reach sedimentation equilibrium. The technique can yield data about molecular weight, density, shape, and purity of a given group of molecules. The operation is simple in principle, but more complex in practice. There are several materials that can serve as density gradient supports, for the preparation of gradients in the density range of 1–1.4 g/mL. These include cesium chloride, sucrose (sugar), Ficoll, and Percoll. The density gradient is prepared in the centrifuge tube, and the sample is layered and centrifuged (Figure 5.4). The components (macromolecules) migrate until they reach the zone of equilibrium position, where the density equals the density of the medium.

5.3.5 Isopycnic Centrifugation

It depends solely upon the buoyant density of the particles and not its shape or size and is independent of time. The size of the particle affecting only the rate at which it reaches its isopycnic position in the gradient. The technique is useful in separating the particle of the same size but differing in density (Figure 5.5).

In isopycnic centrifugation, aqueous solutions of different densities are prepared with sucrose or other solutes in them. Before centrifugation, these solutions are layered in the centrifuge tube from higher to least dense (bottom to top), and gradients can be prepared in discrete layers (so-called step gradients) or as a single layer of continuously changing density (Figure 5.6). The cell homogenate to be fractionated is then layered on top of the gradient and centrifugation is carried out. Particles suspended in the homogenate move and sediment during the spin as per their density matching with the gradient.

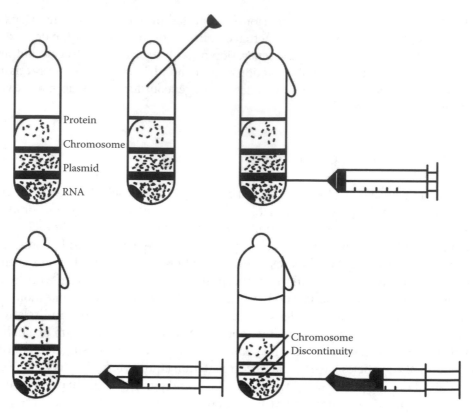

Figure 5.4 Separation of macromolecules by density gradient centrifugation.

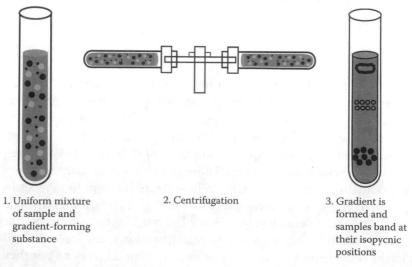

Figure 5.5 Isopycnic centrifugation.

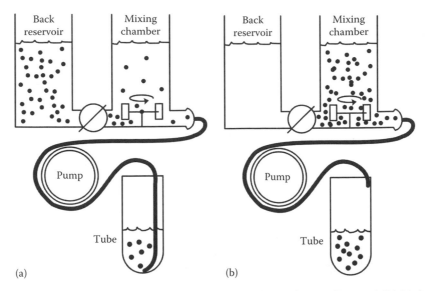

Figure 5.6 Setup for continuous centrifugation: (a) lower density first and (b) higher density first.

5.4 Suspending Medium

The homogenizing medium should be cheap, uncharged, and metabolically inert, and for these reasons, sucrose is the compound most frequently employed. For rat liver, a slightly hypotonic solution of sucrose (0.25 mol/L), buffered with 20 mmol/L Tris to pH 7, has been found to be quite suitable. Ethylene diamine tetraacetic acid (EDTA) adjusted to pH 7 is sometimes incorporated into the medium at a concentration of 0.1 mmol/L. This chelates calcium and other divalent ions that if present in even in trace amounts can cause extensive swelling of the mitochondria. On the other hand, EDTA renders the mitochondrial membrane more permeable to monovalent ions; therefore, it is advisable to use sucrose alone.

5.5 Rotors

There are two types of rotors: fixed angle heads and swinging bucket (Figure 5.7). In the former, the samples are kept at an angle of 30° to the horizontal, whereas in the later, the samples, while spinning, are horizontal. Simple calculations show that for the same radius, the swinging bucket produces more gravitational force. Microcentrifuge (microfuge) with refrigeration is ideally suited for spinning small volumes. Cell organelles differ from one another in mass and density and can be efficiently separated from one another by centrifugation. The larger or denser would fall (sediment) through the solution more quickly than the other in response to gravity and resistance provided by the medium or water. In a centrifugal field, therefore, sedimentation occurs more quickly than in the gravitational field.

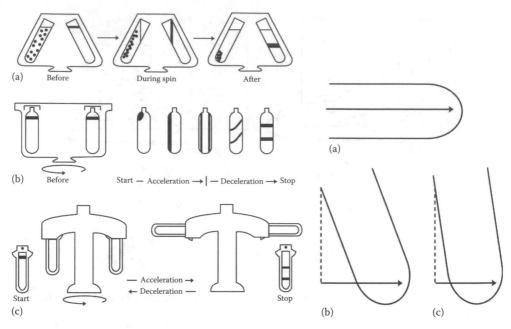

Figure 5.7 Type of rotors and mechanism of separation: (a) fixed-angle rotor, (b) vertical rotor, and (c) swinging-bucket roctor.

5.6 Fractionation of Rat Liver

5.6.1 Principle

Rat liver has been subjected to fractionation for separating the subcellular particles, which is fractionated at different centrifugal force at different period (Table 5.1).

5.6.2 Materials

Isolation medium (0.25 mol/L sucrose; 5 mmol/L Tris–HCl buffer pH 7.4; 0.1 mmol/L EDTA), rat liver, cell homogenizer, ice bath, ultracentrifuge.

TABLE 5.1 Fractionation of Different Components at Different Centrifugal Forces

Centrifugation Conditions		Components Fractionated
g Value	Time (min)	Major Components in Fraction
500	5	Nuclei whole cells, debris
8,000	10	Mitochondria, some lysosomes
15,000	10	Lysosomes, some mitochondria
100,000	60	Microsomes (membrane fragments, largely endoplasmic reticulum) and ribosomes
Final supernatant		Soluble components of the cell

5.6.3 Protocol

Kill a rat, exsanguinate it, and rapidly remove the liver. Wash the tissue free of blood in ice-cold sucrose, lightly blot, and place in a beaker to weigh. Cut the liver into small fragments and homogenize in sucrose (20 g/100 mL) at 2000 rpm by moving the mortar relative to the pestle for 8–10 complete strokes. Centrifuge the suspension in a refrigerated centrifuge according to the scheme shown in Figure 5.8. Ideally, each fraction should be resuspended in sucrose and the washing combined with the supernatants. This has the advantage of producing purer fractions, but the disadvantage of introducing increasing dilution of cellular components.

Carefully resuspend the pellets in about 10 mL of sucrose and store on ice until required.

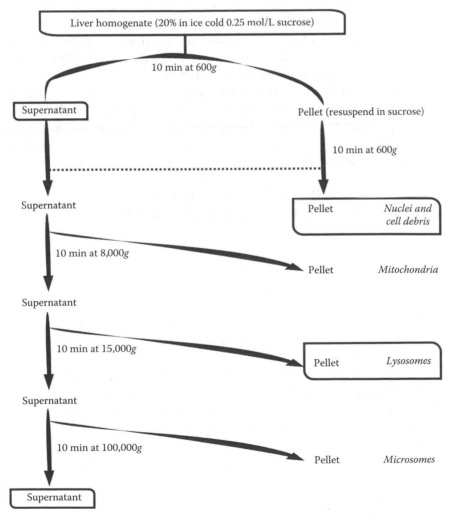

Figure 5.8 Fractionation of different components at different centrifugal forces.

Certain tissues, for example, pig brain tissues, are very heterogeneous in nature so that differential centrifugation gives rise to a fraction of a different composition to those from rat liver. Each fraction contains a very complex mixture of components and density gradient centrifugation is required to purify them further. This is illustrated in a flowchart in Figure 5.8.

5.7 Estimation of DNA, RNA, and Protein in the Isolated Cell Fractions

5.7.1 Principle

Nucleic acids: The tissue fraction is mixed with trichloroacetic acid (TCA) to remove acid-soluble components, then extracted with ethanol to remove phospholipids. The lipid-depleted sediment is incubated overnight with warm alkali, which hydrolyzes the RNA to acid-soluble nucleotides, but does not affect the DNA. On acidification, the DNA is dissolved in hot TCA before estimation. This method has the advantage that DNA and RNA are separated and can be estimated by determining the phosphorous, sugar, or base content. The DNA is measured by the diphenylamine method and the RNA by the orcinol method.

Protein: Sucrose interferes with the biuret assay for protein and so the protein may have to be precipitated from the sample with TCA. If the sample needs to be diluted more than 1 in 50 for the assay, then the TCA precipitation stage can be omitted. The detergent deoxycholate is incorporated into the biuret reagent, to solubilize protein in association with lipid material so that the total protein content of each fraction is obtained.

Presentation of results: The total amount of DNA, RNA, and protein in each fraction is calculated. The specific concentration of the nucleotides is expressed as μg/mg protein.

5.7.2 Materials

TCA (10% w/v), potassium hydroxide (1 mol/L), hydrochloric acid (5 mol/L), ethanol (95% v/v), protein standard (bovine serum albumin [BSA] 5 mg/mL), boiling water bath.

Orcinol reagent for assay of RNA: Dissolve 1 g of ferric chloride ($FeCl_3 \cdot 6H_2O$) in 1 L of concentrated HCl and add 35 mL of 6% w/v orcinol in alcohol.

Diphenylamine reagent for the assay of DNA: Dissolve 10 g of diphenylamine in 1 L of glacial acetic acid and add 25 mL of concentrated sulfuric acid. This solution must be prepared fresh.

Biuret reagent with deoxycholate: Dissolve 1.5 g of $CuSO_4 \cdot 5H_2O$ and 6 g of Na, K tartrate in 100 mL of water. Add 300 mL of 2.5 mol/L NaOH and make up to 900 mL with water. Add 1 g of KI and 15 g of Na deoxycholate and make up to a final volume of 1 L.

5.7.3 Protocol

Nucleic acids: Mix 2 mL of the fraction with 3 mL of 10% TCA; centrifuge the precipitate and wash it with 3 mL of ice-cold TCA. Extract the precipitate three times with 5 mL of 95% ethanol to give a lipid-depleted residue. Suspend the residue in 2 mL of 1 mol/L KOH and incubate at 37°C overnight. The next morning, add 0.5 mL of 5 mol/L HCl and 2 mL of 5% TCA and centrifuge. The supernatant contains the acid-soluble nucleotides from RNA and the precipitate contains the DNA that is then dissolved in TCA by heating at 90°C for 10 min. Take suitable aliquots of the extracts and use them to estimate the nucleic acids with diphenylamine reagent by measuring extinction at 595 nm for DNA and with orcinol reagent at 665 nm for RNA.

Protein: Take a suitable aliquot of the sample and add TCA to a final concentration of 5%. Centrifuge in a bench centrifuge and decant off the supernatant. Add 1 mL of distilled water to resuspend the precipitate and add 2 mL of the biuret reagent; mix and heat in a boiling water bath for 1 min. Cool and read the extinction at 540 nm against a blank prepared by adding the biuret reagent to 1 mL of distilled water. Compare the extinction obtained with a series of protein standards of BSA (0–5 mg/mL) and construct a standard curve. If the sample remains cloudy after incubation, shake the solution with 2 mL of diethyl ether and read the extinction of the aqueous phase.

5.8 Enzyme Distribution in the Cell

5.8.1 Principle

Enzyme markers: Some enzymes are located almost exclusively in one part of the cell and this fact is used to check the "purity" of a particular preparation of cell organelles. Such enzymes are known as marker enzymes, and some of those that can be used for this purpose are shown in Table 5.2.

Fractions may be cross-contaminated due to damaged organelles by adverse physical or chemical conditions. This happens particularly to the acid phosphatase of lysosomes, which is released into the supernatant if the particles are damaged. Please avoid careless handling. Low levels may be obtained for total enzyme activity, unless

TABLE 5.2 Marker Enzymes in Cell Fractions

Fraction	Marker Enzyme
Mitochondria	Glutamate dehydrogenase
Lysosomes	Acid phosphatase
Microsomes	Glucose-6-phosphatase
Supernatant	LDH

the organelle is first rendered permeable to the substrate or even ruptured so that the enzyme is released into the medium.

Temperature of enzyme assays: All the enzyme activity should be assayed preferably at (a constant temperature) 37°C with temperature controlled UV spectrophotometer cell maintained at 37°C or else at room temperature for convenience.

Storage of fractions: It is best to carry out all the estimations on the same day, but if this is not possible, then aliquots of each fraction can be stored in the deep freeze and the dehydrogenases measured later. However, glucose-6-phosphatase should be measured as soon as possible after the preparation of the fractions as this enzyme is unstable. It will probably also be convenient to measure the acid phosphatase on the day of fractionation.

5.8.2 Protocol

Measure the activity of all four enzymes in each of the fractions including the homogenate and also measure the protein content of each fraction. Enzyme assays must be done in duplicate and should be repeated if the two results differ markedly. It is suggested that assays are carried out on a 1:10 and 1:100 dilution of the fractions.

Calculate the total enzyme activity in each of fractions and illustrate results by expressing the activity in each fraction as a percentage of the total (Figure 5.9). In order to do this, you will need to know the dilution of the sample, the volume of this is used in the assay, and the total volume of each fraction. Remember to allow for the fact that a sample of the homogenate is removed at the beginning when you come to calculate the recovery of each enzyme in the subcellular fractions.

5.8.3 Results and Calculations

Also express results by plotting the relative specific activity (RSA) of the enzyme in each fraction against the protein in each fraction expressed as a percentage of the total (Figure 5.9). The RSA is obtained by dividing the specific activity in the fraction by that in the homogenate and is a useful indicator of the location of the enzyme and its "purity" in the fraction.

5.9 Glucose-6-Phosphatase

5.9.1 Principle

Glucose-6-phosphatase catalyzes the hydrolysis of glucose-6-phosphate to glucose and phosphate. The phosphate is then estimated by the method previously described

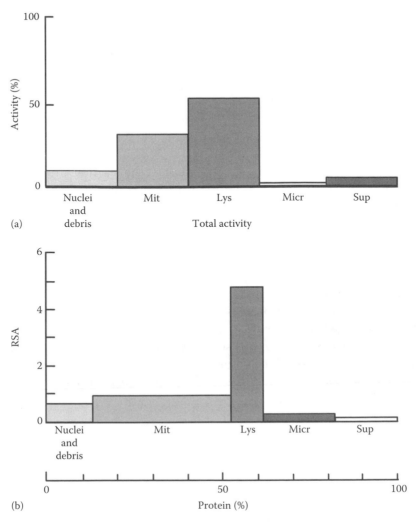

Figure 5.9 The distribution of enzyme activity among fractions obtained from rat liver by differential centrifugation: (a) percent activity and (b) relative specific activity.

in Section 4.5.4.3. EDTA is incorporated into the reaction mixture to chelate Mg^{2+} required for alkaline phosphatase activity.

5.9.2 Materials

Sodium cacodylate buffer (0.1 mol/L, pH 6.5). *Care: poison!* EDTA (10 mmol/L in buffer adjusted to pH 6.5), glucose-6-phosphate (50 mmol/L in buffer), TCA (10% w/v), reagents for phosphate estimation as per Section 4.5.4.3.

5.9.3 Protocol

Prepare the following mixture and equilibrate at room temperature for 10 min:

Content	Volume (mL)
Cacodylate	1.2
EDTA	0.2
Glucose-6-phosphate	0.4

Add 0.2 mL of suitably diluted tissue fraction to start the reaction and incubate for 0, 10, and 20 min. Stop the reaction by adding 1 mL of ice-cold TCA (10%). Centrifuge the precipitate and remove 1 mL of the supernatant for phosphate estimation.

5.9.4 Results and Calculations

Check that the reaction is linear and calculate the activity. The phosphorus content of 0.322 µmol of glucose-6-phosphate is 10 µg, so that 1 µg of phosphorus arises from 0.0322 µmol of glucose-6-phosphate. In each test, 0.2 mL of diluted tissue fraction is taken and made up to 3 mL from this solution; 1 mL is taken for phosphate estimation so that the final dilution of the tissue is as follows: initial dilution × 3/0.2. Always set up suitable blanks and express the final results and micromoles of glucose-6-phosphate hydrolyzed per minute.

5.10 Glutamate Dehydrogenase

5.10.1 Principle

Glutamate dehydrogenase catalyzes the reversible oxidative deamination of glutamate to 2-oxoglutarate. NAD is required as coenzyme so that the reaction can be readily measured by following the change in extinction at 340 nm. EDTA is present in the reaction mixture to remove heavy metal ions that might otherwise inactivate the enzyme. Maximum activity of the enzyme is only obtained after repeated freezing and thawing of the mitochondria or following the addition of detergent Triton X-100.

5.10.2 Materials

Sodium phosphate buffer (0.1 mol/L, pH 7.4), sodium-2-oxoglutarate (0.15 mol/L prepared in the buffer and adjusted to pH 7.4), ammonium acetate (0.75 mol/L in phosphate buffer adjusted to pH 7.4), EDTA (30 mmol/L in buffer adjusted to pH 7.4), NADH (2.5 mg/mL in phosphate buffer prepared fresh), Triton X-100 (10 w/v in buffer), UV–visible spectrophotometer.

5.10.3 Protocol

Prepare the following mixture and equilibrate at room temperature for 10 min:

Content	Volume (mL)
Phosphate buffer	2.1
Tissue fraction	0.2
NADH	0.1
Ammonium acetate	0.2
EDTA	0.2
Triton X-100	0.1

5.10.4 Result and Calculations

Start the reaction by adding 0.1 mL of the substrate 2-oxoglutarate, and follow the rate of change of extinction at 340 nm with time.

Calculate the enzyme activity as micromoles or nanomoles NADH oxidized per minute:

$$\text{Molar extinction coefficient NADH} = 6.3 \times 10^3 \text{ L/mol/cm}$$

5.11 Acid Phosphatase

5.11.1 Principle

Acid phosphatase can only be determined after a fixed incubation time contrary to alkaline phosphatase because the product of the reaction, p-nitrophenol, does not absorb at 405 nm in the acid incubation buffer and is only colored in the alkaline medium. The reaction is, therefore, stopped by the addition of an alkaline buffer containing 0.4 mol/L phosphate. The high phosphate concentration effectively inhibits any alkaline phosphatase activity that may be present giving a stable color. The linearity of the enzymes reaction must be checked. Enzyme assays should never rely on only a single timed incubation with the assumption that linearity occurs. This can be checked by incubating one sample for 10 min and another for 20 min. If the response is linear, then the extinction reading for 20 min should be twice that found at 10 min. If this is not the case, then the experiment should be repeated with a different enzyme dilution until a linear response is obtained.

5.11.2 Materials

Acetate buffer (0.2 mol/L, pH 4.5), p-nitrophenyl phosphate (8 mmol/L, prepared fresh on the day of use), Tris–HCl buffer (1 mol/L, pH 9.0 containing 1 mol/L Na_2CO_3

and 0.4 mol/L K_2HPO_4), p-nitrophenol standard (50 µmol/L; prepare a 5 mmol/L stock solution in the alkaline Tris buffer then dilute this solution 1 in 100 with the buffer), Triton X-100 (10% w/v).

5.11.3 Protocol

Add 0.2 mL of the enzyme solution to 1.2 mL of the acetate buffer and 0.1 mL of Triton X-100 and add 0.5 mL of the substrate solution with thorough mixing. Incubate for 10 and 20 min and stop the reaction by adding 2 mL of the alkaline Tris buffer.

5.11.4 Results and Calculation

Read the extinction at 405 nm and calculate the enzyme activity by reference to a standard curve of p-nitrophenol. If the extinction at 20 min is not doubled at 10 min, then the progress curve is nonlinear, and the assay must be repeated with a lower dilution of the tissue extract. Always set up suitable blanks and express the final results in the calculation of enzyme activity.

5.12 Mitochondrial Swelling

5.12.1 Principle

Changes in the volume of mitochondria suspended in an isotonic solution can be brought about by a wide variety of agents including calcium ions, phosphate, arsenate, thyroxin, and the higher fatty acids. The swelling action of these compounds is complex and depends on a number of factors such as the presence and absence of substrate, adenosine diphosphate (ADP), adenosine triphosphate (ATP), uncouplers, and inhibitors. Some compounds act by stimulating the production of fatty acids from the mitochondrial membrane and BSA that binds fatty acids and, therefore, blocks their swelling action. A large number of compounds can cause mitochondrial swelling, but only ATP in the presence of Mg^{2+} can induce contraction of mitochondria and extrusion of water. Mitochondrial swelling is shown by a fall in extinction and contraction by a rise.

5.12.2 Materials

Rat liver mitochondrial fraction after centrifugation (Section 5.6), potassium chloride–Tris solution (0.125 mol/L KCl–0.02 mol/L Tris–HCl, pH 7.4), BSA (5 g/100 mL in KCl–Tris), 2,4-dinitrophenol (1 mmol/L in KCl–Tris), thyroxin (1 mmol/L in KCl–Tris), ATP–$MgCl_2$ (0.06 mol/L ATP, 0.15 mol/L $MgCl_2$ in KCl–Tris), calcium chloride (15 mmol/L in KCl–Tris), sodium phosphate (0.5 mol/L adjusted to pH 7.4), Spectronic 20 or any UV–Vis spectrophotometer.

5.12.3 Protocol

Fractionate the rat liver and wash the mitochondrial pellet twice with the isolation medium by resuspension and centrifugation at 8000*g* for 10 min. Suspend the final pellet in about 5 mL of sucrose and store on ice until required.

Immediately before use, dilute the suspension in such a way so that 0.1 mL is added to the KCl–Tris solution giving an initial extinction in the range of 0.4–0.7. When a suitable dilution is found, add 0.1 mL to the following mixtures and follow the extinction at 520 nm with time. Take readings every half minute for the first 3 min, then continue to read at convenient time intervals for about 10 min. At the end of this time, add 0.1 mL of the ATP–MgCl$_2$ mixture and continue to read for a further 5 min (Table 5.3).

5.13 Determination of Lysosomal Integrity

5.13.1 Principle

The latency of lysosomal enzymes: Lysosomes are bounded by a single lipoprotein membrane that is normally impermeable to the enzymes and their substrates. If some of the lysosomes are broken by careless handling or other causes, then the enzymes are released into the medium and can be assayed. The degree of damage can be determined by measuring this soluble or free activity after removing the intact lysosomes by centrifugation and comparing this with the total activity measured after deliberately breaking the membrane with detergent. The difference between these values is known as the latent activity, and this is usually expressed as a percentage of the total:

$$\text{Percent latency} = \frac{(\text{Total} - \text{Free})}{\text{Total} \times 100}$$

TABLE 5.3 Effect of Different Agents in mL on Mitochondria

Component	Experiment No.					
	1	2	3	4	5	6
KCl–Tris	2.9	2.8	2.7	2.8	2.8	2.8
Thyroxin	—	0.1	0.1	—	—	—
BSA	—	—	0.1	—	—	0.1
Sodium phosphate	—	—	—	0.1	—	—
Calcium chloride	—	—	—	—	0.1	—
Mitochondria	0.1	0.1	0.1	0.1	0.1	0.1
ATP–MgCl$_2$ added after 10–15 min	0.1	0.1	0.1	0.1	0.1	0.1

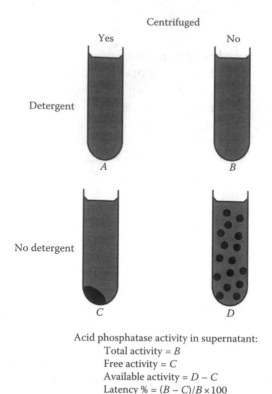

Acid phosphatase activity in supernatant:
  Total activity = B
  Free activity = C
  Available activity = $D - C$
  Latency % = $(B - C)/B \times 100$

Figure 5.10   The effect of Triton X-100 on the lysosomal membrane.

Under certain conditions, the membrane becomes permeable to the substrate but not to the enzymes and can determine the available activity (Figure 5.10).

Rat kidney lysosomes: Kidney lysosomes have been chosen for the study since they can be obtained in a more pure state than lysosomes prepared from rat liver. On high-speed centrifugation, two distinct layers can be seen in the pellet. The upper light-brown layer of a loose consistency is the mitochondrial fraction and the lower layer of a small dark-brown button of a sticky consistency is the lysosomes. The mitochondria can be removed by careful washing and the lysosomes obtained in a reasonable state of purity. Another advantage of kidney lysosomes is that they are tougher than liver lysosomes and can take more rough treatment. However, they are still fragile structures and should be handled carefully and left on ice until required for the experiment.

5.13.2 Materials

Isolation medium (0.35 mol/L sucrose buffered to pH 7 with 10 mmol/L triethanolamine butyrate), rats (about 200 g body weight), reagents for the assay of acid phosphatase (Section 5.11), Triton X-100 (2 g/100 mL), ice baths, refrigerated ultracentrifuge, recording visible spectrophotometers, coaxial homogenizers.

Preparation of lysosomes: Kill two rats and remove their kidneys. Decapsulate them and prepare 8%–10% w/v homogenate in the isolation medium using six up-and-down strokes of the homogenizer rotating at 1500 rpm. Centrifuge the homogenate at 500g for 5 min to sediment nuclei, intact cells, and debris and resuspend the sediment. Centrifuge the suspension at 500g for 5 min and combine the supernatants and centrifuge at 10,000g for 5 min. Remove the supernatant and discard. Add a small volume of the sucrose solution and gently swirl with a Pasteur pipette until the upper mitochondrial layer is suspended, but the lower lysosomal pellet remains intact. Remove the mitochondrial suspension and resuspend the lysosomal pellet in the isolation medium. Centrifuge at 10,000g for 5 min and repeat the washing procedure. Finally, resuspend the lysosomal pellet by gentle homogenization by hand in the buffered sucrose solution.

5.13.3 Protocol

Determination of lysosomal integrity: Take 0.5 mL aliquots of the lysosomal preparation and place into six small-volume high-speed centrifuge tubes. Add 0.5 mL of Triton X-100 or the isolation medium as indicated and centrifuge tubes A, C, and E; centrifuge at 10,000g for 10 min.

Tube	Lysosomes (mL)	2% Triton X-100 (mL)	Isolation Medium (mL)	H$_2$O (mL)	Centrifuged
A	0.5	0.5	—	—	+
B	0.5	0.5	—	—	−
C	0.5	—	0.5	—	+
D	0.5	—	0.5	—	−
E	0.5	—	—	0.5	+
F	0.5	—	—	0.5	−

5.13.4 Calculations

Assay the acid phosphatase activity of either the suspension or the supernatant and determine the latency of the preparation.

5.14 Effect of Detergents on the Stability of the Lysosomal Membrane

5.14.1 Principle

Light scattering: Light scattering measurements can be of use following rapid changes in lysosomes that cannot be conveniently determined by other means. The measurements are taken at 520 nm, and a decrease in the extinction of suspension of lysosomes is taken to mean swelling or rupture of the particles. For light scattering measurements, the lysosomal preparation must contain a minimum of contaminating material, and for kidney lysosomes, this is reasonably true. However, changes in the extinction at 520 nm should be correlated with lysosomal enzyme release whenever possible.

Acid phosphatase: The effect of detergents on the lysosomal membrane is also investigated by measuring the release of the enzyme acid phosphatase from the organelles, as per Section 4.5.4.3.

5.14.2 Materials

Materials for the preparation of rat kidney lysosomes (Section 5.13), detergent solutions prepared in the buffered sucrose solution (20 mmol/L, except Triton X-100, which is 2% w/v).

Detergent	Charge	
Cetyltrimethylammonium bromide	Cationic	−
$CH_3(CH_2)_{15}(CH_3)_3N^+$ $\qquad$ Br^-		
Sodium dodecyl sulfate	Anionic	+
$CH_3(CH_2)_{11}OSO_3^-$ $\qquad$ Na^+		
Lysophosphatidyl choline	Zwitterionic	+−

$$\begin{array}{l} CH_2O-COR_1 \\ | \\ HO-\overset{|}{C}-H \quad O \\ | \qquad \parallel \\ CH_2O-P-O-(CH_2)_2N^+(CH_3)_3 \\ \qquad | \\ \qquad O^- \end{array}$$

Triton X-100	Nonionic	0

$$\begin{array}{l} CH_3 \qquad CH_3 \\ | \qquad\quad | \\ HO-\overset{|}{C}-CH_2-\overset{|}{C}-C-\!\!\bigcirc\!\!-(OCH_2CH_2)_nOH \\ | \qquad\quad | \\ CH_3 \qquad CH_3 \qquad n = 9\ or\ 10 \end{array}$$

5.14.3 Method

Light scattering: Adjust the dilution of part of the lysosome preparation so that the addition of 0.1 mL of the suspension to 2.0 mL of buffered sucrose solution gives an initial absorbance in the range 0.3–0.5 at 520 nm when read against sucrose blank.

5.14.4 Calculations and Result

Examine the effect of a range of concentrations of the detergents on the light scattering properties of the lysosomal suspension. Examine concentrations of detergent over a wide range then select a narrower range where an effect is observed.

Acid phosphatase: From the data obtained, select appropriate concentrations of the detergents to observe the effect on the release of acid phosphatase on the lysosomes. An aliquot of the lysosomal suspension (0.5 mL) is mixed with an equal volume of detergent and the lysosomes removed by centrifugation. The supernatant is then removed and assayed for acid phosphatase activity as in Section 4.5.4.3.

5.15 Density Gradient Centrifugation

5.15.1 Principle

Discontinuous gradient is performed by loading a centrifuge tube with layers of varying densities of the separating medium using Percoll as the density gradient medium. The sample to be separated is added to the top layer of medium and centrifuged at high speed. A continuous gradient is formed when a medium with a uniform density is exposed to centrifugal force (Figure 5.11). The separation medium is a suspension of small particles when placed under a high centrifugal field concentrate toward the bottom of the centrifuge tube forming a gradient. As with the discontinuous gradient, the cells or subcellular particles to be separated are layered above the separating medium, sucrose or Percoll, and centrifuged at a high speed. The particles will sediment at their respective densities in the self-generated density

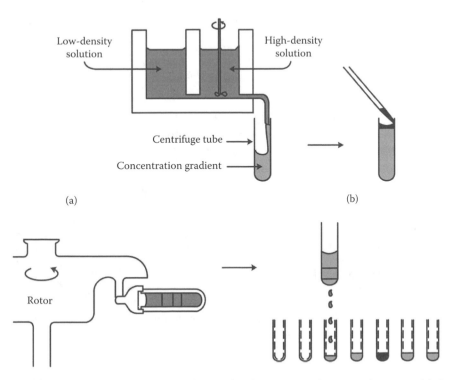

Figure 5.11 Continuous gradient maker and sedimentation by centrifugation: (a) formation of gradient and (b) the sample is layered on the top of the gradient.

gradient. Sucrose cannot be used for cells and some cellular organelles because it alters osmotic pressure.

Percoll is a registered trademark of GE Healthcare and well suited for density gradient experiments because it possesses a low viscosity compared to alternatives, a low osmolarity and no toxicity toward cells and their constituents. Percoll consists of colloidal silica particles of 15–30 nm diameter (23% w/w in water), which have been coated with polyvinylpyrrolidone (PVP). They are used for the separation of cells, subcellular particles, and larger viruses (down to ~70S) under gentle conditions, which preserve viability and morphological integrity. They are nontoxic to cells and adjustable to physiological ionic strength and pH. Gradients can be spontaneously generated by centrifugation at moderate speeds in an angle-head rotor. Gradients are isosmotic throughout and cover a range of densities up to 1.3 g/mL. Density beads may be reused.

5.15.2 Materials

Fifteen mL centrifuge tubes, Percoll solution of different densities, centrifuge, density gradient marker beads.

Preparation of the density gradient

1. The formula used to prepare the desired densities of Percoll is as follows:

$$V_y = V_j \left\{ \frac{(D_i - D)}{(D - D_v)} \right\}$$

   where
   V_y is the volume of diluting media
   V_j is the volume of stock Percoll
   D_i is the density of stock solution (Percoll density = 1.130 g/mL)
   D_v is the density of diluting media (density = 1.002 g/mL)
   D is the density of diluted solution produced

2. Prepare a working Percoll solution at a density of 1.100 g/mL:

$$V_y = V_j \left\{ \frac{(D_i - D)}{(D - D_v)} \right\}$$

$$= 50 \text{ mL} \left\{ \frac{(1.130 \text{ g/mL} - 1.100 \text{ g/mL})}{(1.100 \text{ g/mL} - 1.102 \text{ g/mL})} \right\}$$

   = 50 mL of stock Percoll added to 14.79 mL of media will produce 84.79 mL of working Percoll at a density of 1.100 g/mL

3. Other concentration will be diluted from the working Percoll.
4. Other densities include the following:
   a. 1.05 g/mL 5 mL working Percoll + 5.21 mL media + 1 drop yellow food coloring
   b. 1.0 g/mL 3 mL working Percoll + 3.45 mL media + 1 drop water

c. 1.065 g/mL 5 mL working Percoll + 2.7 mL media + 1 drop green food coloring
d. 1.075 g/mL 5 mL working Percoll + 1.7 mL media + 1 drop red food coloring
e. 1.08 g/mL 5 mL working Percoll + 1.3 mL media + 1 drop blue food coloring
5. Refrigerate the density gradient solution after preparation.

5.15.3 Protocol

Place the denser layer on the bottom using a 1 mL pipette into a 15 mL centrifuge tube. Add the layers on top of each other very slowly, being careful not to mix the different densities. See the interface on the layers (one drop of food coloring is to be added to each solution to facilitate visualization of the various layers). The density to be used will depend on the density of the marker beads:

1. The Percoll densities of 1.080, 1.075, 1.065, and 1.050 g/mL are layered in the centrifuge tube in a ratio of 1:1:1:3:1 mL. The appropriate marker beads for this gradient are as follows: three green (1.048 g/mL), four red (1.062 g/mL), five blue (1.072 g/mL), and six orange (1.088 g/mL).
2. Add approximately 10–50 µL of different density marker beads to the top layer in the tube.
3. To a second centrifuge tube, add the first three of the density layers then add the marker beads and then add the other density layers.
4. Centrifuge both tubes at approximately 1000 rpm for 25 min.

5.15.4 Result

Record results and calculate R_f value for each beads. Measure the total length of the solution in the tube; this is the Y value. Measure the length from the bottom of the tube up to each of the bands; these are the X values. R_f value is determined by dividing x/y for each band.

5.16 Fractionation of Pig Brain by Density Gradient Centrifugation

5.16.1 Principle

Heterogeneity of brain tissue: The cellular structure of the brain is very heterogeneous so that homogenization and differential centrifugation gives rise to subcellular fractions of a different composition to those from rat liver (Figure 5.12). Each of these fractions contain a highly complex mixture of components and density gradient centrifugation is required to purify them further. The mitochondrial pellet obtained by differential centrifugation is resolved into its constituents by centrifugation on a discontinuous gradient of Ficoll (Figure 5.12).

Synaptosomes: During the initial homogenization, the presynaptic nerve endings shear off, and the membranes reseal to form distinct particles known as synaptosomes

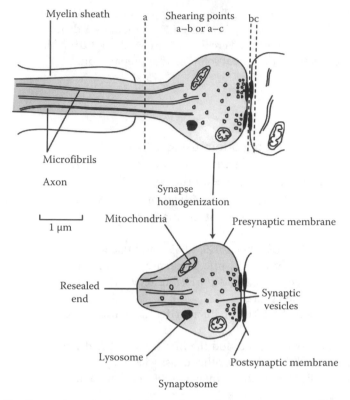

Figure 5.12   The formation of synaptosomes during homogenization of brain.

(Figure 5.12). These nerve-ending particles can be purified by density gradient centrifugation and their contents separated on a sucrose gradient following their disruption by osmotic shock (Figure 5.13).

5.16.2 Materials

Pig brain obtained fresh from the slaughter house, isolation medium (0.32 mol/L sucrose, 5 mmol/L Tris–HCl, pH 7.4), buffered sucrose (0.25 mol/L sucrose, 5 mmol/L Tris–HCl), Ficoll solution (2.8% and 14% w/v in the buffered sucrose solution), sucrose solutions for density gradients (0.4 mol/L, 0.81 mol/L, 1.08 mol/L, and 1.26 mol/L adjusted to pH 7.2–7.4 with 0.1 mol/L NaOH), reagents for electron microscopy, coaxial homogenizers (clearance 0.25 mm), ultracentrifuge, ice baths.

5.16.3 Protocol

Fractionation of the cerebral cortex: Strip the brain of its external membranes, wash with ice-cold isolation medium, and remove the cerebral cortex with a sharp pair of scissors. Remove as much white matter as possible from the cortex by scraping with

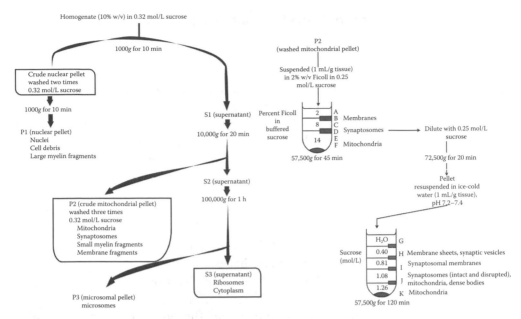

Figure 5.13 Flow diagram of differential centrifugation and the preparation of synaptosomes and synaptosomal plasma membranes by density gradient centrifugation.

a blunt scalpel, weigh, and then finely chop the tissue. Homogenize the cortex in the isolation medium by six or eight complete strokes of the coaxial homogenizer attached to an electric motor rotating at 1500 rpm. Fractionate the homogenate by differential then density gradient centrifugation using the scheme outlined in Figure 5.13.

5.16.4 Precautions

It is essential to take the following precautions to obtain the optimum resolution and purification:

1. *Source of brain*: Obtain the pig brains from freshly killed animals and transport them from the slaughterhouse to the laboratory in a polythene bag placed on crushed ice in a Dewar flask.
2. *Gradients*: Precool all solutions to 0°C and carefully layer the solutions on top of each other. Leave the gradients for 2–4 h at 0°C–5°C before use.
3. *Washing of pellet*: It is essential to wash the mitochondrial pellet (P_2) at least three times before placing it on the Ficoll gradient.
4. *Suspension of pellet*: Suspend the pellet P_2 in 2% w/v Ficoll by gently sucking the mixture up and down in a Pasteur pipette, then homogenize by one complete stroke in a homogenizer with a loose fitting pestle.
5. *Osmotic shock*: Rupture the synaptosomes by suspending them in glass distilled water (1 mL/g tissue) and forcing the suspension through a 13-gauge needle attached to a 25 mL syringe 10 times. Transfer the suspension to a coaxial homogenizer and complete the suspension using five strokes by hand.

5.17 Distribution of Enzymes in the Fractions Obtained from Pig Brain

5.17.1 Principle

The distribution of four enzymes among the subcellular fractions is investigated in terms of their total and relative specific activities.

Lactate dehydrogenase (LDH): This is a cytoplasmic enzyme marker, so little or no activity would be expected in the washed mitochondrial pellet. However, results show that LDH is present in bands B and D where it is occluded in synaptosomes and other membrane fragments.

Glutamate dehydrogenase (GD): This enzyme is found in the mitochondrial matrix and would, therefore, be expected to occur in fraction F, which is the mitochondrial pellet. However, some activity should also be detected in the synaptosomes (D) from the mitochondria present in these particles. Osmotic lysis of these nerve-ending particles and subsequent fractionation can be used to isolate these synaptosomal mitochondria (K). Traces of GDH activity in other fractions are probably due to fragmented mitochondria.

Acetylcholinesterase: The enzyme is present on the postsynaptic membrane of cholinergic neurons and a high activity would be expected in the synaptosomes. However, separation and fragmentation of the postsynaptic membrane mean that this activity is also found in other fractions.

Na^+–K^+ ATPase: This enzyme is a marker for the plasma membrane and a high activity would be expected in the fraction containing the purified synaptosomal membranes (I). This enzyme is assayed by linked enzyme system with the following sequence of reactions:

$$ATP \xrightarrow{\text{ATPase}} ADP + Pi$$

$$ADP + \text{phosphoenol pyruvate} \xrightarrow[\text{kinase}]{\text{Pyruvate}} \text{Pyruvate} + ATP$$

$$\text{Pyruvate} + NADH + H^+ \xrightarrow[\text{dehydrogenase}]{\text{Lactate}} \text{Lactate} + NAD^+$$

In brain, the optimal activity is found in the presence of 100 mmol/L Na^+ and 30 mmol/L K^+ and is inhibited by ouabain.

5.17.2 Materials

Materials for the fractionation of pig brain (Section 5.15), reagents for the assay of glutamate dehydrogenase (Section 5.10), reagents for the assay of LDH (Section 12.12), reagents for the assay of protein (Section 5.7).

Reagents for the assay of ATPase: Tris–HCl buffer (100 mmol/L, pH 7.4), ionic solution (30 mmol/L, $MgSO_4$; 1 mol/L, NaCl; 200 mmol/L, KCl), NADH (10 mg/mL in Tris buffer), pyruvate kinase, LDH, phosphoenolpyruvate (0.1 mol/L in water), KCN (1 mol/L), ouabain (10 mmol/L), Na–ATP (0.1 mol/L, pH 7.4).

5.17.3 Protocol

Assay of Na^+–K^+ ATPase: Prepare several test tubes containing the following reaction mixture and incubate at room temperature for 10 min:

Tris–HCl buffer	2.30 mL
Ionic solution	0.30 mL
NADH	0.05 mL
Pyruvate kinase	0.03 mL
Phosphoenolpyruvate	0.05 mL
KCN	0.03 mL
Water	0.09 mL

Place the mixture in a UV spectrophotometer and establish the baseline by measuring the extinction at 340 nm for 1–2 min. At the end of this time, add 50 µL of ouabain and measure the rate after the addition of the inhibitor. The Na^+–K^+ ATPase is taken to be the difference between these two rates.

Assay of protein, GDH, and LDH: Assay the protein content and the activity of these enzymes in each fraction using the methods described earlier.

5.17.4 Results and Conclusions

Calculate the total activity of the enzymes in each fraction and determine their recoveries, then use the protein measurements to calculate the relative specific activities in each of the fractions. Express the results in terms of their distribution among the subcellular fractions.

5.18 Isolation of Chloroplasts

5.18.1 Protocol

Differential centrifugation and density gradient centrifugation are being used to isolate two plant organelles, namely, chloroplasts and mitochondria, from pea seedling. Chloroplasts are the subcellular sites of photosynthesis, the process by which green plants convert light energy to chemical energy by producing carbohydrate and oxygen using carbon dioxide and water. Under the microscope, chloroplasts are recognized as bean-shaped, membrane-bound, green (chlorophyll-containing) organelles. Experiment describes isolation and assay of chloroplast from a cell homogenate by density gradient centrifugation; saved for the isolation of DNA. Four modules are described herein.

5.18.2 Materials

Homogenization buffer: 10 mM KCl, 1 mM MgCl$_2$, 1% (w/v) dextran T40, 1% (w/v) Ficoll, 0.1% (w/v) BSA; make to volume with 30% (w/w) sucrose in 0.1 M tricine buffer, pH 7.5.

Sucrose solutions (gradient prepared in 0.1 M tricine buffer, pH 7.5) of 60%, 50%, 40%, and 30% (w/w) in 0.1 M tricine buffer, pH 7.5.

Plant material (Pea seedlings): Fresh pea seeds should be used for reproducible result. Pea seeds are relatively inexpensive and easy to grow. Seedlings should be about 7 days old for the laboratory. To plant, soak seed overnight in a large container. Sow the seeds on a layer of about 1.5 in. of wet horticultural grade vermiculite in standard nursery flat (21 × 10 × 2 in.). The seed can be sown thickly nearly touching one another. Cover the seed with 0.5–1 in. of vermiculite and water well. Cover the flat with plastic wrap to hold in moisture until the seedlings begin to emerge. Once the seedling has emerged, plastic wrap is removed, and keep well watered. The seedling can be grown in the lab on a window sill, in a growth chamber, or in the greenhouse. For best results, however, do not grow the seedlings under intense light. Chloroplasts tend to accumulate large granules of starch under bright light, which will damage during blending and centrifugation.

Ice bucket, cheese cloth, centrifuge tubes, Pasteur pipettes, blender (blades of the tissue homogenizer or blender should be as sharp as possible), centrifuges (refrigerated, superspeed centrifuge equipped with a swing bucket rotor), UV–Vis spectrophotometer.

5.18.3 Method

All solutions and samples should be kept on ice:

1. Prepare two sucrose gradients in 50 mL centrifuge tubes:
    a. Pipette 5 mL 60% sucrose solution into the bottom of each tube.
    b. Layer 5 mL 50% sucrose, then 10 mL 40% sucrose into each tube. Layer should be distinct from one another.
    (*Hint: Tip the tube as you add each layer of sucrose. Let the tip of the pipette just touch the surface of liquid in tube.*)
    c. With the tip of a Pasteur pipette or stirring rod, gently mix at the interface of the 50% and 40% layers to diffuse slightly (Figure 5.14).
    d. Layer 5 mL 30% sucrose on top of the gradient.
    e. Keep the gradients on ice while you prepare tissue sample.

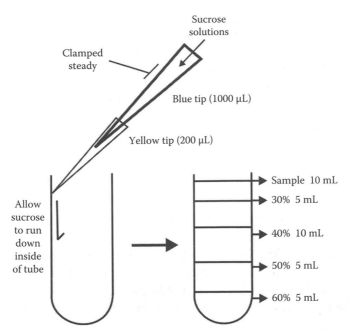

Figure 5.14 Setup for sucrose discontinuous gradients.

2. Harvest 5 g of 7 days old pea seedlings at the soil line with a sharp razor blade.
3. Chop the tissue into small pieces with a razor blade or scissors and transfer them to a chilled blender containing 20 mL ice-cold homogenization buffer.
4. Homogenize with five 2–3-s bursts of the blender at high speed.
5. Filter the homogenate into a breaker (on ice) through four layers of cheesecloth, squeezing the cloth gently to remove most of the liquid (wear gloves).
6. Refilter the first filtrate through one layer of miracloth, moistened in homogenization buffer, by gravity. Do not squeeze. You may want to prepare a wet-mount slide of the residue left in the cheesecloth of miracloth for the microscopic observation.
7. Layer 10 mL of the filtrate onto the top of each of your gradients (prepared in Step 1). Check to see that the two gradients are balanced against one another. If necessary, add homogenization buffer to make the tube balance. Centrifuge at 4°C at 4000 rpm for 5 min, then increase speed to 10,000 rpm for 10 min. Allow the centrifuge to come to a stop. Carefully remove gradients from the rotor.
8. Two green bands appear in the gradient. The green band toward the bottom of the tube is the fraction containing intact chloroplasts. Remove the top of the gradient carefully with a Pasteur pipette. Save the two chlorophyll fractions separately in clean tubes on ice.
9. Prepare wet-mount slides of the chlorophyll-containing fractions and examine microscopically.
10. Assay chlorophyll content of the two green bands. For each sample
    a. Into a clean centrifuge tube, pipette 50 μL of gradient fraction to be assayed and 0.95 mL distilled water.
    b. Add 4 mL acetone.

  c. Centrifuge in a clinical centrifuge (5 min).
  d. Measure the absorbance of the solution using UV–Vis spectrophotometer at 652 nm against 80% acetone in water as blank.
  e. Calculate chlorophyll content:

$$A_{652} \times 29 = \mu g \text{ chlorophyll } 10 \ \mu L \text{ chloroplast fraction}$$

11. Freeze and save the intact chloroplast fraction for DNA isolation.

5.18.4 Result

Make a diagram of gradient. After centrifugation, the gradient should show two bands of chlorophyll. The upper of the two contains broken chloroplasts and membrane fragments. The lower band (about three quarters of the distance of the bottom of the tube) contains intact chloroplasts. Avoid contaminating the intact chloroplast fraction with broken chloroplasts for DNA isolation. Also, try to retrieve the intact chloroplast fraction from the gradient in as small a volume as possible.

5.19 Isolation of Chloroplast DNA

5.19.1 Principle

The DNA of plant cells is found in three distinct genomes. First, there is nuclear DNA familiar as the DNA that makes up the chromosomes, but mitochondria and chloroplasts each have DNAs of their own. These genomes are closed circular DNA molecules encoding many of enzymes necessary for the function of the organelles.

5.19.2 Precautions

Cetyltrimethylammonium bromide (CTAB) is a strong detergent and can cause burns to the skin. Chloroform is toxic by inhalation or contact with skin.

5.19.3 Materials

Extraction buffer (EB): 50 mM Tris pH 8.0; 1% CTAB; 50 mM EDTA; 1 mM 1,10 *o*-phenanthroline; 0.7 M NaCl; 0.1% β-mercaptoethanol; chloroform; isopropyl alcohol; 80% ethanol; 15 mM ammonium acetate pH 7.5.

  TE buffer: 10 mM Tris pH 8.0, 1 mM EDTA, centrifuge tubes, water bath (65°C), centrifuge.

5.19.4 Protocol

1. Start with the frozen chloroplast preparation (from Section 5.18). This sample will have a volume of several mL. For each mL of chloroplasts, add 4 mL EB. If necessary, transfer the mixture to a capped centrifuge tube of at least twice the volume of the chloroplasts and EB.
2. Incubate the mixture at 65°C for 1 h on a water bath.
3. Remove the tube from the water bath and allow cooling on the bench for several minutes to proceed further.
4. Add an approximately equal volume of chloroform to the tube, recap, and mix by inversion.
5. Centrifuge at >3500g for 10 min.
6. The tube contents will be separated into two distinct layers. Transfer the upper (aqueous) layer into a fresh centrifuge tube with the help of an ultra pipette. This tube should be of the same size as that used in the first step. The lower (organic) layer is hazardous water, dispose as per safety guidelines.
7. Add 0.6 mL of isopropanol for each mL of DNA containing extract in the centrifuge tube and mix by inversion.
8. Centrifuge at >10,000g for 20 min.
9. Decant liquid in the tube away from the DNA containing pellet. Stand the tube upside down on a paper towel or "Kimwipe" for several minutes to allow the liquid to drain. The tube's inside can be wiped carefully to remove liquid, but take care not to dislodge the DNA pellet.

5.19.5 Result

The yield of chloroplast DNA is expected to be low and will depend on the quality of the chloroplast preparation produced. Also note that chloroplast DNA may not digest well with restriction endonucleases if not isolated properly.

5.20 Isolation of Mitochondria and Assay of a Marker Enzyme

5.20.1 Principle

Roots of pea seedlings are used to isolate mitochondria by differential centrifugation and assay the activity of a mitochondrial enzyme to assess the success of the isolation protocol. Differential centrifugation separates cell components based on differences in the rate at which they sediment in a centrifugal field. The first centrifugation step will remove whole cells, cell wall fragment, nuclei, starch, etc. Mitochondria, because of their small size, will not be sedimented by this step but will remain suspended in the supernatant. A subsequent spin at a higher speed will then be used to pellet the mitochondria to the bottom of the centrifuge tube.

Assay the activity of a "marker enzyme" in the mitochondrial pellet and in the supernatant from which the mitochondria are isolated in order to assess the efficiency of the isolation protocol. A marker enzyme is any enzyme whose activity is confined to the organelle being isolated. If the isolation protocol is absolutely effective, all marker enzyme activity should appear in mitochondrial fraction. Cytochrome *c* oxidase is the marker enzyme used here for assay. Enzyme activity is measured by determining the rate of oxidation of cytochrome *c*. This can be followed by measuring increase in absorbance at 550 nm using a double-beam UV–Vis spectrophotometer in a reaction mixture during the initial linear phase of the reaction.

5.20.2 Materials

Pea seedlings, cheesecloth/miracloth, centrifuge tubes, Pasteur pipettes, small paint brush, ice bucket, blender (blades of the tissue homogenizer or blender should be as sharp as possible), centrifuges (refrigerated, superspeed centrifuge equipped with a swing bucket rotor), UV–Vis spectrophotometer.

Harvest pea root tissue, prepare extract, isolate mitochondria by differential centrifugation, and assay mitochondrial activity.

Homogenization buffer: 70 mM sucrose, 220 mM mannitol, 0.5 g/L BSA, 1 mM HEPES pH 7.4 homogenization buffer.

Solutions: 0.1 M potassium phosphate buffer, pH 7.4, 0.8 M ascorbic acid, 4% Triton X-100, 5 mg/mL cytochrome *c*, sodium dithionate crystals.

5.20.3 Isolation of Mitochondria

5.20.3.1 Protocol I

1. All solutions and samples should be kept on ice while working.
2. Harvest 5 g of 7-day-old pea roots, shake off vermiculite in which they are growing, and rinse in distilled water in a beaker.
3. Chop roots into small pieces with a razor blade or scissors and put into a chilled blender with 20 mL ice-cold homogenization buffer.
4. Homogenize the tissue with five 2–3 s bursts of the blender at high speed.
5. Filter the homogenate through four layers of cheesecloth plus one layer of miracloth. It may be necessary to squeeze the filtrate through the cloth. Wear gloves.
6. Pour the filtrate into a centrifuge tube balance against a tube of water and centrifuge at 4°C, 700g 10 min (2500 rpm in a Sorvall SS34 rotor).
7. Decant the supernatant into a clean centrifuge tube and centrifuge at 4°C, 10,000g 10 min (9500 rpm in a Sorvall SS34 rotor).
8. Decant the supernatant from the tube into a beaker and save it on ice. Use it for the next experiment. The pellet at the bottom of the centrifuge tube should contain isolated mitochondria. Wash them gently by resuspending in 20 mL of fresh homogenization buffer. This is most easily done by pipetting 1–2 mL of the buffer into the tube and using a small paintbrush to break up the pellet. Once the pellet is resuspended in this small volume, it can be diluted with the remaining 18–19 mL of buffer.
9. Recentrifuge the washed mitochondria as in Step 6 earlier.
10. Discard the supernatant from this spin and resuspend the mitochondrial pellet in 5 mL of homogenization buffer.

5.20.3.2 Protocol II

5.20.3.2.1 Material.

Pea seedlings (use whole seedlings for this experiment); cheese-cloth/miracloth, centrifuge tubes; Pasteur pipettes; dropper bottles containing methylene blue, Janus green, iodine (KI), and vegetable or mineral oil; clinical centrifuge; 37°C water bath; blender (a standard household blender will work well, but the blades should be as sharp as possible).

Homogenization buffer: 70 mM sucrose, 220 mM mannitol, 0.5 g/L BSA, 1 mM HEPES pH 7.4.

Equipment: A refrigerated superspeed centrifuge (Sorvall RC series) equipped with a fixed angle rotor SS34 is used. Other comparable centrifuges with equal performance, for example, Beckman, can also be used but specified speeds and times of centrifugation might have to be changed slightly. The operation manual for particular centrifuge and rotor will explain these changes. If a superspeed centrifuge is not available, consider the alternative protocol for cell fractionation described herein:

The assay of mitochondrial marker enzyme activity requires the use of a spectrophotometer with a bandwidth of less than 5 nm. Before using the cytochrome *c* solution in the assay, check that it is completely reduced by adding 2 crystals of sodium dithionate to 1 mL of cytochrome *c* stock solution. Transfer the solution to the spectrophotometer cuvette and measure its absorbance at 550 and 565 nm. The A_{550}/A_{565} ratio should be between 9 and 10.

5.20.3.2.2 Assay of a Mitochondria Marker Enzyme.

1. From the ice bucket, remove 0.2 mL of the mitochondria preparation to each of two small test tubes and 0.2 mL of the 10,000*g* supernatant to one small test tube and allow them to come to room temperature on the lab bench. To one of the tubes containing mitochondria, add 0.8 mL of potassium phosphate buffer (a 1/5 dilution). To the other tube of mitochondria, add 1.8 mL buffer (a 1/10 dilution).
2. Assemble four reaction mixtures in four small test tubes (Table 5.4).
3. One of the four aforementioned tubes contains no sample to be assayed. Transfer the contents of this tube to a cuvette and use it as a blank in the spectrophotometer.
4. Transfer a reaction mix containing sample to a cuvette and place into the spectrophotometer to perform the assay. Record the absorbance at 550 nm in 20 s intervals for 1 min to determine the rate of the reaction in the absence of substrate. This value (the slope of this line) will be used in Step 7 as a correction factor in calculating the reaction rate. If a double-beam spectrophotometer uses a reference cuvette, then this step is not necessary.
5. Repeat Steps 4 and 5 for each sample to be assayed.

TABLE 5.4   Reaction Mixture Each Tube Contains

Solution	Volume (µL)
0.1 M potassium phosphate buffer pH 7.5	50
4% Triton X-100	25
5 mg/mL cytochrome *c*	50
Distilled water	675
Mitochondrial preparation in suitable dilution	200

5.20.3.2.3 Results and Calculations.   Calculate the rate of cytochrome *c* oxidation by each of the samples:

$$\text{Rate} = \frac{\text{Change in absorbance/min (Step 5)} - \text{Change in absorbance/min (Step 4)}}{\text{Molar absorptivity of cytochrome } c \times \text{Path length of light through the cuvette}}$$

$$\text{Molar absorptivity of cytochrome } c = 18.5 \times 10^6 / \text{M}^{-1}$$

Path length through cuvette is usually 1 cm.

The units attached to rate are moles of cytochrome *c* oxidized per minute. In general, this value would be reported as micromoles cytochrome *c* oxidized per minute; thus, calculation is simplified to

$$\text{Rate} = \frac{\text{Change in absorbance/min (corrected as above)}}{18.5}$$

Note: Enzyme activity is usually reported in the literature as specific activity. This is a way of standardizing the reporting of enzyme activity since the exact molar concentration of enzyme in the preparation being assayed is not known.

$$\text{Specific activity} = \text{Rate/mg protein (total) in the assay mixture}$$

5.20.3.3 Protocol III (by Clinical Centrifuge)
All solutions and samples should be ice-cold:

1. Harvest whole pea seedlings and wash in distilled water to remove planting material. Blot dry with paper towels.
2. Weigh about 50 g of seedlings chop them into small pieces with a razor blade or scissor and transfer to a chilled blender. Add 250 mL ice-cold homogenization buffer.
3. Homogenize tissue with five 2–3-s bursts of the blender at high speed.
4. Filter the homogenate through four layers of cheesecloth plus one layer of miracloth into a beaker on ice. Gently squeeze the cloth to remove most of the liquid. Wear gloves. Save the residue in the cheesecloth for microscopic examination later.
5. Divide filtrate in two centrifuge tubes. Tubes should be filled to the same level to keep centrifuge balanced. Label tubes 1 and 2.
6. Centrifuge for 3 min at 200*g*. Start timing when the centrifuge rotor reaches a top speed. Allow the rotor to coast to a stop when minutes are up.
7. Remove the tubes and decant the supernatant from tube 2 into a clean tube (label the new tube 3). Save the pellet in tube 2 on ice. Add buffer to tube 3 to balance it against tube 1.
8. Return the tubes (1 and 3) to the centrifuge and spin at 700*g* for 10 min and remove the tubes from the rotor.
9. Decant the supernatant from tube 3 into a fresh tube (label 4). Save the pelleted material in tube 3.

10. Prepare wet-mount slides of the cell fractions (residue in cheesecloth and tubes 1–4) and examine microscopically. The residue in the cheesecloth consists largely of unbroken pieces of seedling tissue, cell debris, etc. Tube 2 contains material pelleted at low speed (200g) from the homogenate. Add a drop of iodine solution to the edge of the cover slip of your slide. Any starch grains present should turn blue–black in the presence of iodine. The pellet in tube 3 is separated from the homogenate at higher speed (700g). Compare the components present in the two pellets (2 and 3). The supernatant in tube 4 should be relatively clear. Only the smallest cell components remain in this solution. Prepare a wet-mount slide of this supernatant. Add a drop of Janus green stain. It is used to stain mitochondria. With the stain, at high magnification, mitochondria appear as tiny, dark specks. Examine tube 1, this tube shows a history of the entire experiment and illustrates the principle behind differential centrifugation. The sediments at the bottom of the tube are arranged in layers. The largest and most dense cell components are at the tube's bottom. Additional layers of sediment are added to the pellet by higher centrifugal forces applied for longer periods of time. Only the tiniest of particles remain in solution at the end of the experiment

5.20.3.3.1 Assay the Activity of Mitochondria. To demonstrate that mitochondria remain in the supernatant of the high-speed spin (tube 4) and that they have been separated efficiently from other cell components, perform the following experiment:

Mitochondria are the subcellular sites of respiration. The activities of these organelles thus consume oxygen. Methylene blue dye is blue in the presence of oxygen but is colorless when reduced (i.e., when oxygen is removed from it). In this experiment, the presence of mitochondria is detected by the disappearance of blue color from the reaction mixture. Prepare three test tubes according to Table 5.5.

Mix each of the tubes well and add 1 mL of vegetable or mineral oil to the top of each. Incubate the tubes at 37°C water bath for several hours to overnight. Compare the tubes and record your observations.

5.20.3.3.2 Result and Conclusion. The experiment is generally very reliable. Use of pea roots than shoots avoid possible problem with chloroplast contamination. Fresh tissue gives the best results. Seedlings should be about 7 days old (in any case, not older than 12 days).

TABLE 5.5 Methylene Blue Reaction with Mitochondrial Preparation

	Reaction 1	Reaction 2	Reaction 3
Component (control) buffer pellet in tube 3	6 mL	3 mL	3 mL
(Resuspended in buffer)	3 mL	—	
From tube 4	—	—	3 mL
Methylene blue[a]	2–3 drops	2–3 drops	2–3 drops

[a]The same amount should be added to each of the tubes.

Yield of mitochondria is difficult to predict and depends on a number of factors including the effectiveness of homogenization. For this reason, two dilutions of the mitochondrial fraction are assayed for marker enzyme activity.

5.21 Isolation of Genomic DNA

5.21.1 Principle

Isolation of DNA from plant tissues is at the heart of plant molecular biology. Plants contain three distinct genomes: (a) the nuclear genome, (b) the chloroplast genome, and (c) the mitochondrial genome. There are several protocols developed for DNA isolation from plants, and the one presented here is one of the simplest and most effective for a variety of plant species.

5.21.2 Precautions

CTAB is a strong detergent and can cause burns to the skin. Chloroform is toxic by inhalation or contact with skin. Follow standard safety guidelines.

5.21.3 Materials

Plant tissue: Pea tissue should be air-dried or freeze-dried. Tissue can be air-dried in the laboratory by spreading the cut shoots on a sheet of absorbent paper and turning daily. Tissue treated in this way will typically dry in 5–7 days. Dried tissue can be saved for use at a later date by sealing in a tightly closed jar and freezing.

EB (extraction buffer): 50 mM Tris pH 8.0; 1% CTAB; 50 mM EDTA; 1 mM 1,10 o-phenanthroline; 0.7 M NaCl; 0.1% β-mercaptoethanol; chloroform; isopropyl alcohol; 80% ethanol; 15 mM ammonium acetate pH 7.5.

TE buffer: 10 mM Tris pH 8.0, 1 mM EDTA, centrifuge tubes (50 mL capacity, capped), water bath (65°C), centrifuge.

Chloroform, isopropyl alcohol, 80% ethanol, 15 mM ammonium acetate, pH 7.5.

5.21.4 Method

1. Grind 1 g dried (air-dried or freeze-dried) pea shots to a fine powder in a mortar with pestle.
2. Mix the powder with 25 mL EB in a 50 mL capped centrifuge tube.
3. Place the tube in a 65°C set water bath and incubate for 1 h. Mix the tube's content by inversion several times.
4. Remove the tube from the water bath and allow it to cool for several minutes on the bench.
   Note: Do not skip this step. The chloroform added in the next step will boil out of the tube if added at 65°C.

5. Add 20 mL chloroform to the tube, cap, and mix by inversion until the contents are thoroughly mixed. The EB and the chloroform will form a thick emulsion.
6. Centrifuge at >3500g for 10 min to break the emulsion and separate the tube contents into two phases.
7. The contents of the tube form three distinct layers; the bottom a green layer of chloroform with debris of plant material, the middle whitish or yellowish in color an interphase consisting largely of denatured protein and of debris of leaf tissue. At the top of the tube is the straw-colored aqueous layer. This top layer contains the majority of the DNA in the preparation. Using a pipette, remove this top layer to a small flask or beaker. Avoid transferring any of the interphase material from the centrifuge tube.
8. Interphase and organic (chloroform) layers are hazardous waste left in centrifuge tubes. Follow laboratory safety direction for proper disposal.
9. Add 2/3 volume of isopropyl alcohol to the aqueous phase in the flask or beaker. For example, if you have transferred 18 mL to the flask, add 2/3 × 18 or 12 mL of isopropyl alcohol. Mix by swirling the contents. DNA will precipitate to form a cottony mass.
10. Transfer the DNA to a clean flask with the help of a Pasteur pipette or a glass rod. Add 10 mL of 80% ethanol and 15 mM ammonium acetate and swirl to wash.
11. After about 20 min, transfer the precipitated DNA to a microcentrifuge tube and centrifuge briefly to drive the DNA to the bottom of the tube. Using a Pasteur or capillary pipette, remove the residual ethanol. Allow the DNA to dry in the uncapped tube for about 10 min on the bench top.
12. Add 0.75 mL of TE buffer to the DNA to dissolve the precipitate.
    Note: Large quantities of DNA may require some time to dissolve completely. Leave the capped tube in the refrigerator until required.

5.21.5 DNA Digestion

Cut 2 μL of the DNA with different restriction endonuclease and run the digested product on an agarose gel; compare total DNA (both undigested and digested) with DNA isolated from chloroplasts (both undigested and digested).

DNA prepared by this method forms a flocculent than cottony precipitate. Such precipitates will not spool on a glass rod and must be collected by centrifugation. If desired, RNA can be removed from the DNA preparation by adding 20 μg of RNase A (heat-treated to destroy DNases) along with the TE buffer in Step 12 of the protocol.

Note: If you want to avoid the use of chloroform in this exercise, make the following modifications to the lab protocol:

1. Grind tissue as before, mix with EB, and incubate in the water bath.
2. Centrifuge to pellet undigested plant tissue. Transfer the supernatant to a fresh tube, flask, or beaker and precipitate DNA with isopropanol as before.

Note: The supernatant will be dark green in color in this preparation. The precipitation of DNA somewhat becomes more difficult to observe. Under such circumstances, the supernatant can be diluted with additional EB or TE buffer before the precipitation step. DNA prepared in this way is not ideally suited for DNA digestion.

5.22 Isolation of Chloroplasts from Spinach Leaves by Differential Centrifugation

5.22.1 Materials

Fresh spinach, clean sharp sand, 50 mL 0.5 M sucrose (17% w/v), cheese cloth, 12×12 in. ice, ice bath, 25 mL graduated cylinder, mortar and pestle (or blender), top clinical centrifuge, glass filter funnel, two 16×150 mm test tubes in rack, three 13×100 mm test tubes in rack, plastic-capped 15 mL centrifuge tube, double-pan balance, glass stirring rods.

5.22.2 Protocol

1. Grind 8 g deveined spinach with ½ tsp clean sharp sand in a mortar and pestle to a paste.
2. *Suspend in 0.5 M sucrose*: Measure 16 mL ice-cold 0.5 M sucrose solution in a 25 mL graduated cylinder. Add in 3–4 mL increments, and grind to smooth pulp with each addition.
3. *Filter*: Homogenate through about eight layers of clean cheese cloth in a glass funnel into an ice-cold 16×150 mm test tube.
4. Pour filtrate back into 25 mL cylinder and record volume. Save ~0.5 mL of the filtrate (F1) in a labeled 13×100 mm test tube to examine at 400× under the microscope to determine the composition and illustrate in notebook. Note the appearance of components and degree of heterogeneity (label cells, ghosts, chloroplasts, mitochondria, debris.)
5. *Centrifuge at low speed*: Prepare a balance tube against the filtrate in a 16×150 tube and spin at 50g for 10 min (speed 2 on the clinical centrifuge).
6. *Decant* the top 10 mL into a clean cold centrifuge tube, discard sediment, and record volume. Save ~0.5 mL supernatant (S1) to examine under the microscope to determine composition, illustrate, and label as in Step 2.
7. *Centrifuge* the supernatant from Step 3 opposite a carefully balance tube at 1000g for 10 min (speed 7) to precipitate chloroplasts. Carefully decant all of the supernatant into 16×150 mm tube but save the pellet. Discard supernatant if you have a significant pellet (may lose some soft pellet, but not to worry).
8. *Resuspend* pellet from Step 4 to 1/10th of the volume of the Step 2; filtrate in ice-cold 0.5 M sucrose with a clean, ice-cold stirring rod. Record final volume. Keep on ice at all times. Examine suspended organelles (SOs) under the microscope to determine composition, and illustrate as in Step 2.

5.23 Isolation of Plasmid DNA by Centrifugation

5.23.1 Principle

Isolation of plasmid DNA from *Escherichia coli* is a common routine in research laboratories. A widely practiced procedure that involves alkaline lysis of cells is described herewith. This protocol often referred to as a plasmid "mini-prep" yields fairly clean DNA quickly and easily.

5.23.2 Materials

Solutions: Solution 1 (per 500 mL)

50 mM glucose	9 mL 50% glucose
25 mM Tris–HCl pH 8.0	12.5 mL 1 M Tris–HCl pH 8.0
10 mM EDTA pH 8.0	10 mL 0.5 M EDTA pH 8.0
Add H_2O to 500 mL	

Solution 2 (per 500 mL)

1% SDS	50 mL 10% SDS
0.2 N NaOH	100 mL 1 N NaOH
Add H_2O to 500 mL	

Solution 3 (per 500 mL)

3 M K^+	300 mL 5 M potassium acetate
5 M Acetate	57.5 mL glacial acetic acid
Add H_2O to 500 mL	

TE (per 100 mL)

10 mM Tris–HCl pH 8.0	1 mL 1 M Tris–HCl pH 8.0
1 mM EDTA	0.5 mL 0.5 M EDTA pH 8.0
Add H_2O to 100 mL	

Optional: RNase can be added to TE at a final concentration of 20 µg/mL. Microfuge, microfuge tubes, capped centrifuge tubes, etc.

5.23.3 Protocol

Fill a microcentrifuge tube with saturated bacterial culture grown in LB broth + antibiotic. Spin tube in microcentrifuge for 1 min, and make sure tubes are balanced in microcentrifuge. Dump supernatant and drain tube briefly on paper towel.

Repeat Step 1 in the same tube filling the tube again with more bacterial culture. The purpose of this step is to increase the starting volume of cells, so that more plasmid DNA can be isolated per preparation. Spin tube in microcentrifuge for 1 min. Pour off supernatant and drain tube on paper towel.

Add 0.2 mL ice-cold solution 1 to the cell pellet and resuspend cells as much as possible using disposable transfer pipette.

Solution 1 contains glucose, Tris, and EDTA. Glucose is added to increase the osmotic pressure outside the cells. Tris is a buffering agent used to maintain a constant pH 8.0. EDTA protects the DNA from degradative enzymes (called DNases); EDTA binds divalent cations that are necessary for DNase activity.

Add 0.4 mL Solution 2, cap tubes, and invert five times gently. Let tubes sit at room temperature for 5 min.

Solution 2 contains NaOH and SDS (a detergent). The alkaline mixture ruptures the cells, and the detergent breaks apart the lipid membrane and solubilizes cellular proteins. NaOH also denatures the DNA into single strands.

Add 0.3 mL ice-cold Solution 3, cap tubes, and invert five times gently. Incubate tubes on ice for 10 min.

Solution 3 contains a mixture of acetic acid and potassium acetate. The acetic acid neutralizes the pH, allowing the DNA strands to renature. The potassium acetate also precipitates the SDS from solution, along with the cellular debris. The *E. coli* chromosomal DNA, a partially renatured tangle at this step is also trapped in the precipitate. The plasmid DNA remains in solution.

Centrifuge tubes for 5 min. Using clean disposable transfer pipette, transfer supernatant to fresh microcentrifuge tube. Try to avoid taking any white precipitate during the transfer. It is better leave a little supernatant behind, to avoid accidentally taking the precipitate.

This fractionation step separates the plasmid DNA from the cellular debris and chromosomal DNA in the pellet.

Fill remainder of the centrifuge tube with isopropanol. Let tube sit at room temperature for 2 min.

Isopropanol effectively precipitates nucleic acids but is much less effective with proteins. A quick precipitation can, therefore, purify DNA from protein contaminants.

Centrifuge tubes for 5 min. A milky pellet should be at the bottom of the tube. Pour off supernatant without dumping out the pellet. Drain tube on paper towel.

This fractionation step further purifies the plasmid DNA from contaminants. Cap tubes and store in freezer.

Add 1 mL of ice-cold 70% ethanol. Cap tube and mix by inverting several times. Spin tubes for 1 min. Pour off supernatant (be careful not to dump out pellet) and drain tube on paper towel.

Ethanol helps to remove the remaining salts and SDS from the preparation.

Allow the tube to dry for ~5 min. Add 50 μL TE to tube. If needed, centrifuge tube briefly, to pool TE at the bottom of the tube. DNA is ready for use and can be stored indefinitely in the freezer.

Suggested Readings

Birnie, G. D. and D. Rickwood. 1978. *Centrifugation Separations in Molecular and Cell Biology.* London, U.K.: Butterworth.

Dean, R. T. 1977. *Lysosomes.* London, U.K.: Arnold.

Dingle, J. T. 1977. *Lysosomes: A Laboratory Handbook*, 2nd edn. Amsterdam, the Netherlands: North Holland.

Ford, T. C. and J. M. Graham. 1991. *An Introduction to Centrifugation*. Oxford, U.K.: Bios Scientific.

Griffith, O. M.1983. *Techniques in Preparative, Zonal and Continuous Flow Ultracentrifugation*, 4th edn. Palo Alto, CA: Beckman Instruments Inc.

Ralston, G. 1993. *Introduction to Analytical Ultracentrifugation*. Palo Alto, CA: Beckman Instruments Inc.

Rickwood, D. 1984. *Centrifugation: A Practical Approach*, 2nd edn. Oxford, U.K.: IRL Press.

Rickwood, T. C. and F. J. Steensgard. 1994. *Centrifugation Essential Data*. New York: John Wiley & Sons.

Scheeler, P. 1981. *Centrifugation in Biology and Medical Science*. New York: John Wiley & Sons.

Sharpe, P. T. 1998. Laboratory techniques in biochemical and molecular biology. In *Methods of Cell Separation*, eds. R. H. Burden and P. H. van Knippenberg, vol. 18. Amsterdam, the Netherlands: Elsevier.

Van Holde, K. E., W. Johnson, and P. Shing Ho. 1998. *Principles of Physical Biochemistry*. Upper Saddle River, NJ: Prentice Hall.

Wilson, K. and J. Walker. 2004. *Practical Biochemistry—Principles and Techniques*, 5th edn. Cambridge, U.K.: Cambridge University Press.

Wrigglesworth, J. M. 1983. *Biochemical Research Techniques: A Practical Introduction*. Chichester, U.K.: Ellis Horwood.

Important Links

http://www.sumanasinc.com/webcontent/animations/content/cellfractionation.html

http://www.coleparmer.com/TechLibraryArticle/30

http://www.thestudentroom.co.uk/showthread.php?t = 2130518

http://en.wikibooks.org/wiki/Proteomics/Protein_Separations–Centrifugation/ Differential_Centrifugation

http://www.sigmaaldrich.com/technical-documents/articles/biofiles/centrifugation-separations.html

http://www.filariasiscenter.org/parasite-resources/Protocols/purification-of-microfilariae-by-density-gradient-centrifugation

http://www.google.co.in/search?hl = en&prmd = imvns&tbm = isch&source = univ&sa = X &ei= vs9sUOnHPM2IrAeKjIHQAg&sqi = 2&ved = 0CDAQsAQ&biw = 1024&bih = 648 &q = Important%20links%20on%20density%20gradient%20centrifugation

http://www.ncbi.nlm.nih.gov/pmc/articles/PMC375628/

6

Radiation Biology

6.1 Introduction

Most naturally occurring elements are a mixture of slightly different isotopes. These differ from one another in the mass of their atomic nuclei, but because they have the same number of protons and electrons, they have the same chemical properties.

In radioactive isotopes, or radioisotopes, the nucleus is unstable and undergoes random disintegration to produce a different atom. In the course of these disintegrations, either energetic subatomic particles, such as electrons, or radiations, such as gamma rays, are given off. By using chemical synthesis to incorporate one or more radioactive atoms into a small molecule of interest, such as a sugar or an amino acid, the fate of that molecule can be traced during any biological reaction.

Although naturally occurring radioisotopes are rare (because of their instability), they can be produced in large amounts in nuclear reactors, where stable atoms are bombarded with high-energy particles. As a result, radioisotopes of many biologically important elements are readily available (Table 6.1). The radiation they emit is detected in various ways. Electrons (β particles) can be detected in a Geiger–Müller counter by the ionization they produce in a gas, or they can be measured in a scintillation counter by the small flashes of light they induce in a scintillation fluid. These methods make it possible to measure, accurately, the quantity of particular radioisotopes present in a biological specimen. Using light or electron microscopy, it is also possible to determine the location of a radioisotope in a specimen by autoradiography. All of these methods of detection are extremely sensitive: in favorable circumstances, nearly every disintegration/decay of a radioactive atom can be detected.

Note: The isotopes are arranged in decreasing order of the energy of the β-radiation (electrons) they emit. I^{131} also emits γ-radiation. The half-life is the time required for 50% of the atoms of an isotope to disintegrate.

The chambers labeled A, B, C, and D represent either different compartments in the cell (detected by autoradiography or by cell fractionation experiments) or different chemical compounds (detected by chromatography or other chemical methods). The results of a real pulse–chase experiment can be seen in Figure 6.1.

TABLE 6.1   Some Radioisotopes Commonly Used in Biological Research

Element	Most Abundant Nonradioactive Isotopes Mass Number	Radioisotope	Half-Life	Uses
Hydrogen	1,2	Tritium(3)	12.4 years	Tracers for metabolic studies
Oxygen	16	18(not radioactive)	—	Metabolic studies and determining temperatures of ancient seas by O16/O18 ratios
Carbon	12	14	5745 years	Metabolic studies and dating recent archeological artifacts
Phosphorus	31	32	14.3 days	Metabolic and ecological tracer. Studies of nucleotides, nucleic acids
Sulfur	32, 33, 34	35	87.9 days	Labeling proteins
Cobalt	59	60	5.3 years	Cancer therapy, source of gamma rays

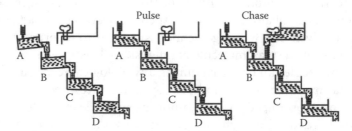

Figure 6.1   The logic of a typical pulse–chase experiment using radioisotopes.

6.2 Radioisotopes

6.2.1 Atomic Structure

An atom is composed of a positively charged nucleus and a surrounding cloud of negatively charged electrons. The mass of an atom is concentrated in the nucleus that is composed of protons and neutrons. Protons are positively charged particles with mass approximately 1850 times greater than that of an orbital electron. The number of orbital electrons in an atom is equal to the number of protons present in the nucleus. This number is called the atomic number (Z). Neutrons are uncharged particles with mass equivalent to that of protons. The sum of the number of protons and neutrons in a given nucleus is the mass number (A). Since the number of neutrons in a given nucleus is not related to its atomic number, it does not affect the chemical properties of the atom. Hence, atoms of a given element may not necessarily contain the same number of neutrons. Atoms that contain the same number of protons (same atomic number, Z value) but vary in their number of neutrons are called isotopes, that is, they are the atoms of the same atomic number but different mass number.

A specific nuclear species symbolically is represented by a subscript number for the atomic number and a superscript number representing the mass number, followed by the alphabetical symbol for the element, for example, $^{12}C_6$, $^{13}C_6$, and $^{14}C_6$. However, in practice, we simply specify the mass number ($^{14}C_6$) as the atomic number is the same in every case. The number of known isotopes for each element varies quite widely, for example, there are three isotopes for hydrogen, seven for carbon C^{10} to C^{16}, and 20 or more for some of the elements of high atomic number.

Only a few combinations of neutrons and protons in a given nucleus give stable isotopes. The number again varies widely from element to element. Though the reasons for stability of certain combinations of neutrons and protons are not fully understood, it is apparent that nuclear stability in general is related to the ratio of neutrons to protons, which, in turn, is related to the binding within the nucleus. Stable isotopes for elements with low atomic number tend to have an equal number of neutrons and protons, whereas for elements of higher atomic number, stability is associated with neutrons/protons in excess of 1. The unstable isotopes tend to attain stability by attaining this ratio. In this process, they emit radiations and are hence called radioisotopes. These isotopes are generally produced artificially but may also occur in nature.

6.2.1.1 Radioisotopes Are Used to Trace Molecules in Cells and Organisms

One of the earliest uses of radioactivity in biology was to trace the chemical pathway of carbon during photosynthesis. Unicellular green algae were maintained in radioactively labeled CO_2 ($^{14}CO_2$)-containing atmosphere, and at various times after they had been exposed to sunlight, their soluble contents were separated by paper chromatography. Small molecules containing ^{14}C atoms derived from CO_2 were detected by a sheet of photographic film placed over the dried paper chromatogram. In this way, most of the principal components in the photosynthetic pathway from CO_2 to sugar were identified.

Radioactive molecules can be used to follow the course of almost any process in cells. In a typical experiment, the cells are supplied with a precursor molecule in radioactive form. The radioactive molecules mix with the preexisting unlabeled ones: both are treated identically by the cell as they differ only in the weight of their atomic nuclei. Changes in the location or chemical form of the radioactive molecules can be followed as a function of time. The resolution of such experiments is often sharpened by using a pulse–chase labeling protocol, in which the radioactive material (the pulse) is added for only a very brief period and then washed away and replaced by nonradioactive molecules (the chase). Samples are taken at regular intervals, and the chemical form or location of the radioactivity is identified for each sample. Pulse–chase experiments, combined with autoradiography, have played an important role in elucidating the pathway, for example, secreted proteins from the ER to the cell exterior.

The results of a pulse–chase experiment in which pancreatic B cells were fed with ^3H-leucine for 5 min (the pulse) were followed by excess unlabeled leucine (the chase). The amino acid is largely incorporated into insulin, which is destined for secretion. After a 10 min chase, the labeled protein has moved from the rough ER to the Golgi stacks, where its position is revealed, by the black silver grains, in the photographic emulsion. After a further 45 min chase, the labeled protein is found in electron-dense secretory granules. The small round silver grains seen here are produced by using a special photographic developer and should not be confused with the similar-looking black dots seen with immune gold labeling methods. Experiments similar to this were important in establishing the intracellular pathway taken by newly synthesized secretory proteins.

Radioisotopic labeling is a uniquely valuable way of distinguishing between molecules that are chemically identical but have different histories, for example, those that differ in their time of synthesis. In this way, for example, it was shown that almost all of the molecules in a living cell are continually being degraded and replaced, even when the cell is not growing and is apparently in a steady state. This "turnover," which sometimes takes place very slowly, would be almost impossible to detect without radioisotopes.

Today, nearly all common small molecules are available in radioactive form from commercial sources, and virtually any biological molecule, no matter how complicated, can be radioactively labeled (Figure 6.2). Compounds can be made with

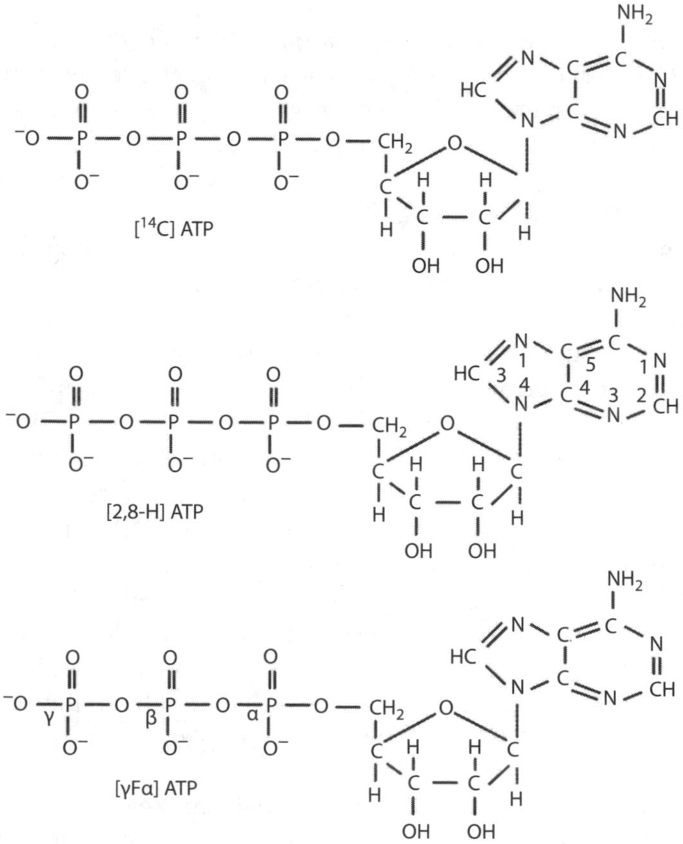

Figure 6.2 Commercially available radioactive forms of ATP with nomenclature and position and type of the radioactive atoms.

radioactive atoms incorporated at particular positions in their structures, enabling the separate fates of different parts of the same molecule to be followed during biological reactions.

One of the important uses of radioactivity in cell biology is to localize a radioactive compound in sections of whole cells or tissues by autoradiography. In this procedure, living cells are briefly exposed to a pulse of a specific radioactive compound and then incubated for a variable period, to allow them time to incorporate the compound, before being fixed and processed for light or electron microscopy. Each preparation is then overlaid with a thin film of photographic emulsion and left in the dark for several days, during which the radioisotopes decay. The emulsion is then developed, and the position of the radioactivity in each cell is indicated by the position of the developed silver grains. If cells are exposed to ^{3}H-thymidine, a radioactive precursor of DNA, it can be shown that DNA is made in the nucleus and remains there. By contrast if cells are exposed to ^{3}H-uridine, a radioactive precursor of RNA, it is found that RNA is initially made in the nucleus and then moves rapidly into the cytoplasm. Radiolabeled molecules can also be detected by autoradiography after they are separated from other molecules by gel electrophoresis; the positions of both proteins and nucleic acid are commonly detected on gels in this way.

6.2.2 Units of Radioactivity

The commonly used unit of radioactivity is the curie (Ci), which is defined as the quantity of radioactive material in which the number of nuclear disintegrations per second (dps) is the same as that in 1 g of the Ra, namely, 3.7×10^{10}. This definition was given in 1950 by the commission of the International Union of Pure and Applied Chemistry and the Union of Pure and Applied Physics. Since this unit is too large, several of its subdivisions in common use are mill curie (mCi), which is equal to 1/1000 of a curie, or 3.7×10^7 dps, and microcurie (μCi), which is 1/1,000,000 of a curie, or 3.7×10^4 dps, or 2.220×10^6 disintegrations per minute (dpm). However, the curie refers to the number of disintegrations actually occurring in a sample, rather than to the disintegrations detected by a radiation counter, which are usually a fraction of the total disintegrations occurring and referred to as counts, that is, counts per second (cps). The International System of Units (SI system) uses becquerel (Bq) as the unit of radioactivity, which is defined as one disintegration per second (1 dps). The bigger units of Bq could be megabecquerel (MBq), 10^6 Bq or 27.027 μCi; gigabecquerel (GBq), 10^9 Bbq or 27.027 mCi; or terabecquerel (TBq), 10^{12} Bq or 27.027 Ci. A carrier of the stable isotope of the element is added to radionuclides while conducting experiments with radioisotopes. It becomes, therefore, necessary to express the amount of radioisotope present per unit mass. This is achieved by the use of the term specific activity, which could be expressed in a number of ways including disintegration rate (dps or dpm), count rate (cps or cpm), or curies (mCi or μCi) per unit of mass of mixture (units of mass are normally either moles or g). It can also be expressed as the amount of radioactivity per given volume or any other expression of quantity.

6.2.3 Measurement of Radioactivity

The three commonly used methods for detecting and quantifying the radioactivity are generally based on the ionization of the gases (proportional counters), on the excitation of solids or solutions (scintillation counters), or on the induction of specific chemical reactions in certain emulsions (autoradiography).

6.2.4 Proportional Counters

These operate on the principle of gas ionization. In gas ionization, a charged particle while passing through the gas dislodges orbital electrons from the atom sufficiently close to its path and causes ionization. The α-particles cause intense ionization (above 10^4–10^5 ion pairs/cm), whereas the β-particles produce rather diffused ionization (10^2–10^3 ion pairs/cm). γ-rays cause little ionization (about 1–10 ion pairs/cm) and that too by secondary mechanism. Accordingly, α- and β-particles may be detected by gas ionization methods while these methods are inadequate for γ-radiation. A simple ionization chamber consists of cylindrical closed metal container filled with air or some other gas and fitted with a central collecting electrode. The chamber wall is connected to the negative (−) pole of a potential source, making it the cathode; the collecting electrode is connected through a resistor to the positive (+) side of the power supply so that it serves as the anode. A pulse (current) flows when ionization occurs, the magnitude of which is related to the applied potential and the number of radiation particles entering the chamber. As a radioactive particle produces only one ion pair per collision, the current produced is low and sensitive measuring devices are necessary. Consequently, this method is of little use in quantitative work. However, various electroscopes, which operate on this principle, are useful in demonstrating the various properties of radioactivity. At voltage levels higher than that of the simple ionization chambers, electrons produced from ionization move much faster toward the anode, causing secondary ionization of the gas in the chamber. These secondary ionization electrons cause further ionization. Consequently, a whole torrent of electrons reaches the anode. Thus from a few primary ion pairs, a geometric increase results in a variable torrent of ions moving toward the chamber electrode. This process is known as gas amplification, and the flood of ions produced is termed as "Townsend avalanche" after its discoverer. As a consequence of this gas amplification, a very large number of electrons are collected at the anode within a microsecond or less from a single β-particle, in the chamber. A strong pulse is thus produced and fed into the external circuit that could be measured directly. However, as the potential gradient is further increased, the number of electrons (mostly secondary) reaching the anode rises sharply and eventually a potential is reached at which the chamber undergoes continuous discharge and no longer can be used as a detector.

There are three distinct potential regions between the saturation current and continuous discharge (Figure 6.3). These regions are the proportional region, the limited proportional region, and the Geiger–Müller region. In the proportional region, the number of ion pairs collected is directly proportional to the applied voltage.

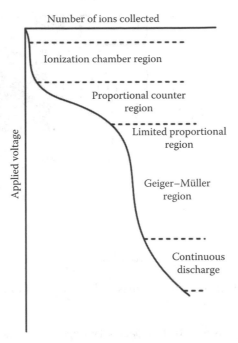

Figure 6.3   Relation of pulse size to potential gradient in an ionization chamber.

The counters that are manufactured to operate in the proportional region require a very stable voltage supply because small fluctuations in voltage result in significant changes in amplification. These counters (proportional counters) are particularly useful for detection and quantification of α-emitting isotopes. The next potential region, that is, the limited proportional region, is similar to the proportional region in operating characteristics. However, in this region, the extent of gas amplification has an upper limit set by the size of the detector chamber, the form of the collecting electrode, and the number of gas atoms present. The limited proportional chamber is not normally used for the purposes of detection. In the Geiger–Müller region, all radiation particles, including even weak β-particles or γ-rays, are able to initiate sufficient ion pair formation, to completely fill the ion space available in the chamber. Consequently, the size of the charge collected on the anode is no longer dependent on the number of primary ions produced. Hence, various types of radiations cannot be distinguished in this region. Ionization chambers operated in this potential region are generally called the Geiger–Müller detectors. Since maximum gas amplification is realized in this region, the size of the output pulse fronting the detector remains the same over a considerable voltage range until continuous discharge occurs (Figure 6.4).

6.2.4.1 Scintillation Counters

Counters based on gas ionization used to be the main instruments employed in quantification of radioisotopes in biological samples. These have now been replaced by scintillation counters. However, small handheld radioactivity monitors based on gas ionization are used for monitoring contamination. They can also be used in situations

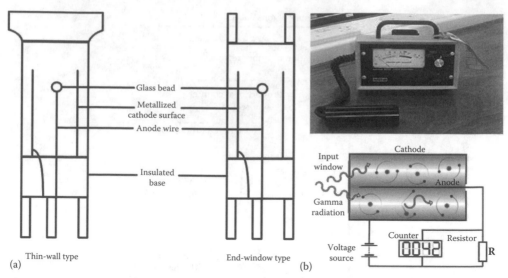

Figure 6.4   (a) Types of Geiger–Müller tubes and (b) Geiger–Müller counter.

where the presence or absence of radioactivity needs to be established. Scintillation detection is based on the interaction of radiation with substances known as fluors (solid or liquid), or scintillation. Excitation of the electron in the fluor leads to the subsequent emission of a flash of light (scintillation), which is converted to an electronic pulse by a photomultiplier tube. The magnitude of the pulse is proportional to the energy loss by the incident radiation and the excitation of the fluor. Accordingly, the number of scintillations is proportional to the rate of decay of the sample, that is, the amount of radioactivity and the intensity of light given out and, therefore, the signal from the photomultiplier is proportional to the energy of radiation. Hence, the counter could be used both for the qualitative and quantitative purposes.

6.2.4.2 Types of Scintillation Counting

There are two types of scintillation counting, that is, solid scintillation counting and liquid scintillation counting. In solid scintillation counting, the sample is placed adjacent to crystals of a fluorescent material. Crystals of sodium iodide, zinc sulfide, and anthracene are the commonly used crystals for γ-isotopes, α-emitters, and β-emitters, respectively. The crystals are placed near to a photomultiplier that in turn is connected to a high-voltage supply and a scanner. These counters are particularly useful for γ-emitting isotopes because γ-rays are electromagnetic radiations and collide only rarely with neighboring atoms to cause ionization or excitation. However, they are not good for weak β-emitting isotopes (^3H and ^{14}C) because even the highest energy negatrons emitted by these isotopes hardly have sufficient energy to penetrate the walls of the counting vials in which the samples are placed for counting. In liquid scintillation counting, the radioactive sample and the fluor are mixed intimately in a homogenous medium. However, the detection sensitivity is such that even the lower energy β-emitters (^3H, ^{14}C, and ^{35}S) can be assayed with significant efficiency (Figure 6.5).

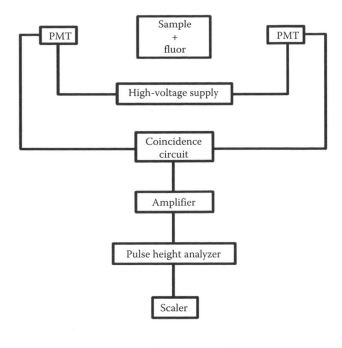

Figure 6.5 Block diagram of a liquid scintillation counter; (inset) a liquid scintillation counter.

6.2.4.3 Mechanism of the Liquid Scintillation Detection

Fluor molecules are directly or indirectly excited by ionizing radiation resulting in the emission of photons, which in turn interact with the photocathode of a photomultiplier to yield photoelectrons from the photocathode surface. These photoelectrons pass through a diode series, resulting in the production of a greatly amplified electron pulse at the photomultiplier anode.

In liquid scintillation, the energy of the nuclear radiations is primarily transferred to the more abundant solvent molecules (toluene) that, in turn, get ionized, dissociated, or excited. The excitation energy of the solvent can either be emitted as photons (in the UV region) or transferred to the molecules of the primary solute. The excited fluor molecules return to the ground state by emitting photons with a wavelength in visible or near-UV region. In the case of para-terphenyl in toluene, the fluorescence peak appears at 3500 A. This may not match the most sensitive range of the

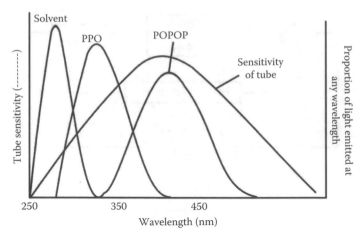

Figure 6.6　Emission spectra of various fluors in relation to sensitivity of phototubes.

photocathode that is used to detect the photons. In such cases, a secondary solute (also a fluor) is used, which absorbs energy from primary fluor and reemits it as light of a longer wavelength. These secondary solutes are, therefore, called wave shifters. POPOP [1, 4-bis-2-(5-phenyl-oxazolyl) benzene] is the commonly used secondary solute that greatly improves detection efficiency. Figure 6.6 illustrates the relationship between fluorescence peaks of primary and secondary solutes and the spectral sensitivity of photocathode.

PPO (2, 5-diphenyloxazole) and POPOP were among the most commonly used primary and secondary fluors, respectively. However, a large number of ready-made scintillation cocktails are now available in the markets that are specifically organically designed for aqueous and organic samples, separately. Accordingly, it is important to make the appropriate selection (Figure 6.6).

Liquid scintillation counting has several advantages over gas ionization counting. The major advantages include much higher counting efficiencies particularly for low-energy β-emitters; ability to accommodate samples of any type including solids, liquids, suspensions, and gels; and the ability to count separately different isotopes in the same sample. However, scintillation counters are highly automated, and hundreds of samples can be estimated automatically. The built-in computer facility can carry out many forms of data analysts, such as efficiency correction, graph plotting, and radioimmunoassay.

The major problem encountered in the use of liquid scintillation counting is that of photomultiplier thermal noise. This is reduced by cooling the photomultipliers. However, some of the latest developments introduced in the instrumentation have taken care of these problems. Another major problem in liquid scintillation detection is the effect known as "quenching," which generally means reduction of efficiency in the energy transfer process in the scintillation solution. Furthermore, the extent of "quenching" varies considerably with different sample materials. Thus, in these experiments, a determination of how much the counting efficiency of every sample has been decreased by quenching is generally determined. Despite the drawbacks mentioned earlier, the liquid scintillation method of detection is the most effective means of assaying low-energy β-emitting samples especially ^{3}H and ^{14}C. Furthermore, the relative ease of sample preparation, the versatility of the sample form possible, and the commercial availability of reliable counting systems with high-capacity automatic sample changers have made this detection method the most useful one where large numbers and types of samples are to be assayed.

6.2.5 Counting of Dual-Labeled Samples

Since the size of electric pulses produced by the conversion of light energy in the photomultiplier is related directly to the energy of the original radioactive event, two different β-emitting isotopes having different energy spectra can be quantified separately in a single sample, for example, ^{3}H and ^{14}C, ^{3}H and ^{35}S, H^3 and P^{32}, ^{14}C and ^{32}P, and ^{35}S and ^{32}P constitute such pairs. The mechanism of estimation is depicted in Figure 6.7. The spectra of two isotopes overlap only slightly, and by setting a pulse height analyzer to reject all pulses of an energy below X (threshold X) and to reject all pulses of energy above Y (window Y) attributable to radioisotope S and also to reject below a threshold of A and a window of B for isotope T, it is possible to separate the pulses generated by these two isotopes completely. A pulse height analyzer set with a threshold and window for particular isotopes is known as a channel (e.g., a ^{3}H channel). Most of the modern counters have multichannel analyzers and are based on an analogue-to-digital converter. The electronic signals from a photomultiplier are converted to digital signals stored in a computer, and the entire energy spectrum is then analyzed simultaneously (Figure 6.7). This type of dual-label counting is of great use in certain fields of molecular biology (nucleic acid hybridization and transcription) and metabolism (steroid synthesis and drug development).

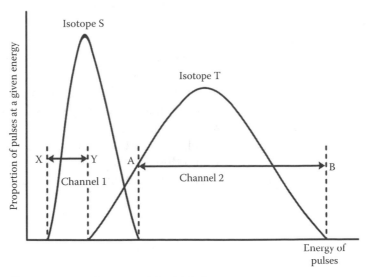

Figure 6.7 Illustrating mechanism of dual-label counting.

6.2.5.1 Determination of Counting Efficiency

Quenching is a major problem encountered in scintillation counting. This necessitates the determination of the counting efficiency of some, if not all, of the sample in a particular experiment. There are several methods by which counting efficiency can be determined.

6.2.5.1.1 Use of Internal Standard.

It is the most commonly employed method and consists of adding a precisely known amount of known isotopes (an internal standard) in a nonquenching form, to a previously counted sample. The resulting mixture is then recounted, and the counting efficiency of the sample is calculated as follows:

$$\text{Counting efficiency} = \frac{\begin{array}{c}[\text{Net count rate (in cpm) of int. std.} + \text{sample}]\\ \times[\text{Net count rate (in cpm) of sample}]\end{array}}{\text{Disintegration rate (in dpm) of int. std.}}$$

Toluene ^{14}C, n-hexadecane ^{14}C, benzoic acid ^{14}C, titrated toluene, titrated water, and titrated n-hexadecane are some of the commonly used internal standards. However, each standard has its limitations. Benzoic acid and water are themselves quenching agents and consequently it is necessary to use these compounds at a very low concentration. Similarly, it is difficult to accurately pipette a small amount of labeled toluene. Furthermore, in case of heterogeneous samples such as gel preparations, it is important to employ an internal standard that can stay in the same phase as the radioactive compound. The method is time consuming because each sample must be counted twice. Moreover, the accuracy of the determination often depends upon accurate measurement of small volumes of the liquid containing the compound used as the standard.

6.2.5.1.2 Channel Ratio.

The efficiency of the scintillation process decreases as less light is produced for a given quantum of radiation energy on quenching of the sample in a scintillation counter. Accordingly, the energy spectrum for a quenched sample is lower than for an unquenched sample. The higher the degree of quenching, the more pronounced is the effect. Channels ratio method for determining counting efficiency is based on this principle and involves preparation of a calibration curve based on counting in two channels that cover different but overlapping parts of the spectrum. The ratio of counts in each channel varies (Figure 6.8).

For preparing standard curves, a set of quenched standards is counted, the absolute amount of radioactivity is known, and the efficiency of counting in each channel is determined. The standard curve (Figure 6.9) is then drawn by plotting counting efficiency against channel ratio. One standard curve applies only to one set of circumstances (one radioisotope, counter, and scintillation fluid). Once the standard curve is drawn, the efficiency of counting experimental samples can be determined. Samples are counted in two channels: ratio calculated and the efficiency read by putting the ratio in the graph. In practice, all data are stored in the counter computers and the corrected values printed automatically.

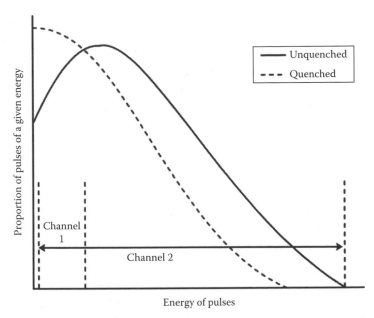

Figure 6.8 The effect of quenching of β-energy spectrum in channel 1 and channel 2.

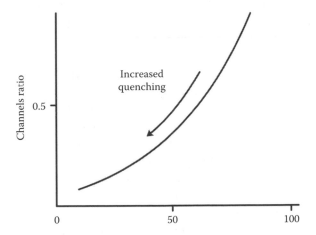

Figure 6.9 Channel ratio quench correction curve.

Multichannel scintillation counters are available in the market, and they operate on the same principle except that, in these instruments, the whole shape and position of the spectrum is analyzed. This is given a digital parameter that relates the counting efficiency. For this reason, these systems have better precision. The channel ratio method is suitable for all types and is less time consuming than either internal or external standardization. However, at low count rates and for very highly quenched samples, the method is inaccurate.

6.2.5.1.3 External Standardization. In this method, each sample to be counted is exposed to the external source of γ-emitting built into the counter. The γ-radiations penetrate the vial and excite the scintillation fluid. The resulting spectrum is unique to the source and is significantly different from the sample in the vial. ^{137}Cs, ^{133}Ba, and ^{226}Ba are some of the commonly used γ-sources. The spectrum obtained is significantly affected by the quenching agents present in the scintillation fluid. The instrument analyzes this spectrum and assigns a quench parameter to it. As for the channel ratio method, a standard curve is required. This method is suitable even for samples with low count rates.

6.3 Sample Preparation

Biological samples in radiotracer experiments are in diverse forms like blood, urine, water, milk, microbes, plant or animal tissue, or respiratory gases. Such varied samples cannot be assayed directly and must be converted to a suitable form for assay. A sample in the solid form has to be mounted in a uniform and reproducible manner. For liquid scintillation counting, the sample material should be suitably incorporated into the fluor medium. Generally, the choice of the counting sample form is determined by the type and energy of the radiation emitted by the sample radioisotope. In samples containing low-energy β-emitters, counting efficiency is poorer because of the problem of self-absorption. This necessitates conversion of the sample to a standard and most suitable form. However, sample containing γ-emitting nuclides can be assayed directly with a minimum of pretreatment.

Sample preparation is easy in solid scintillation counting and only involves transferring the sample to glass or plastic vials compatible with the computer. However, in liquid scintillation counting, sample preparation is more complex. The process starts with the sample vial to be used, which could be glass, low potassium glass, or polyethylene. Each one of these has its own advantages and disadvantages as polyethylene vials give better light transfer and result in slightly higher counting efficiency, but they exhibit more fluorescence than glass vials. The counters used these days accept many types of vials. If possible, vials should be used to save cost.

For scintillation counting, toluene-based cocktails are quite good. However, they cannot be used for aqueous samples as toluene and water are immiscible. Cocktails based on 1,4-bioxane and naphthalene are not used due to toxicity. Aqueous samples are mostly counted with emulsifier-based cocktails that contain Triton X-100 and can accept up to 50% water. However, ready-made cocktails are also available in the market that could be used depending upon the precise sample conditions. Sample volume is also one of the factors determining counting efficiency. Accordingly, sample vials in a given series contain same volume of the sample. In case of color quenching, a sample could be bleached before counting. However, bleaching agents such as hydrogen peroxide may result in chemiluminescence in some of the cocktails.

Plant and animal tissues (solid samples) can be counted best after solubilization by quaternary amines, such as NCS solubilizer toluene. These solutions are highly toxic, and care must be taken in handling them. The sample is digested after adding it to the counting vial containing a small amount of the solubilizer. On completion, scintillation cocktail is added, and the sample is counted. Chemiluminescence is again a problem with tissue solubilization. Combustion techniques are good alternatives to bleaching of colored samples or digestion of tissues. In these methods, the samples are combusted in an apparatus. ^{14}C when combusted is converted to $^{14}CO_2$, which is trapped in a solution of sodium hydroxide and then counted. A ^3H-containing sample is converted to ^3H_2O for counting. However, the sample containing β-emitting isotopes could be prepared for counting in liquid scintillation counters.

6.4 Autoradiography

Ionizing radiation acts upon photographic emulsions, to produce a latent image much as the visible light does. In photography, light initiates a process in the emulsion that produces electrons leading to the reduction of silver halide and metallic silver. This, in turn, acts as a catalyst for further reduction of silver halide in its immediate vicinity. Ionizing radiations produce the same end effect. In autoradiography (also known as radioautography), a sample containing radioactive material is placed in close correct location of the radioactive matter in the sample that is then determined from the pattern of darkening of the film. Thus, the technique of autoradiography is primarily a means of localization of radioisotopes in a given sample. For example, the sites of ^{45}Ca concentration in growing bone tissue, its role in plants and mammalian cells (calcium-regulated proteins), the uptake of sugars and amino acids by living tissues, the relative distribution of ^{32}P in plants, or the localization of ^3H-thymidine in DNA of a cell nucleus can be readily demonstrated by this technique. Even radioactive metabolites separated by chromatographic or electrophoretic techniques could be located on the chromatogram or electrophoretogram. In general, weak β-emitting isotopes like ^3H, ^{14}C, and ^{32}S are more suitable for autoradiography, particularly for cell and tissue localization experiments. This is mainly because of the reason that ionizing tracks of these isotopes are short, which give rise to a discrete image. This is particularly important when radioactivity associated with some subcellular organelles is being located. ^3H is a commonly used radioisotopes for this purpose. Electron microscopy is also used for identifying radioactivity containing zones in the developed image for examining distribution patterns within the whole organism or tissue, either ^{14}C or ^3H is used. In case of very thin gels, even ^{35}S or ^{14}C can also be detected with high resolution. This method is precisely employed in DNA sequencing gels with ^{35}S as the radioisotope.

6.4.1 Film Emulsion

The emulsion is formed by grains of silver halide dispersed in gelatin medium and is backed by the sheet of cellulose acetate or glass. The sensitivity and resolution of the emulsion is governed by the size and concentration of silver halide grains. While grain size is directly related to sensitivity, the grain concentration determines the degree of resolution. Various x-ray-sensitive films are used for gross autoradiography, whereas

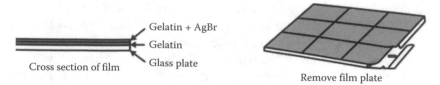

Figure 6.10　Illustrating autoradiography of film cassette.

nuclear emulsions are used for microscopic autoradiography (Figure 6.10). In addition, a number of specialized emulsions are also available for specific autoradiographic purposes. Stripping film has high resolution but low sensitivity and is the emulsion of choice for intracellular localization studies. Liquid emulsions are prepared by melting strips of emulsion by heating them to around 60°C and then either the emulsion is poured over the sample or the sample attached to a support is dipped into it. The emulsion is then allowed to set before being dried. The method is known as dipping film methods and is recommended when a very thin film is required. Accidental exposure to light, chemical in the sample, natural background radioactivity, and even pressure applied during handling and storage can cause a background form on the developed film. It may cause problems particularly in microscopy and care must be taken to minimize this effect. To decrease background form, exposure time should always be kept at a minimum.

6.4.2 Determination of Exposure Time

Exposure time depends upon the isotope, sample type, level of activity, film type, and purpose of the experiment. It has been estimated that 10^6–16^7 β-particles must strike each cm^2 of x-ray film to produce optimum blackening. Accordingly, a rough estimate of exposure time can be made by measuring the activity of the sample/cm^2 with a thin end window Geiger–Müller detector.

6.4.3 Tissue Preparation and Artifacts

While preparing gross samples or tissue sections for autoradiography, extreme care must be taken to avoid leaching of radioactive material. The problem of leaching depends upon tissue and type of radioisotope used, for example, inorganic ^{32}P is easily leached out of many tissues, whereas organically bound ^{32}P is rarely affected. For these reasons, the freeze-drying technique is highly favored for tissue preparation. Where tissue sections and emulsions are permanently mounted, the subsequent staining may result in a reaction with emulsion. Formations of artifacts constitute a major difficulty in interpreting an autoradiograph. Artifacts of the developed film may result from various sources such as vapors from volatile agents in the samples, mechanical pressure and extraneous light during film processing, dust or debris, finger prints, and shrinkage or expansion of either sample or film. To evaluate the extent of artifacts, one may run a parallel sample without radioactivity for comparison.

6.4.4 Specific Autoradiographic Techniques

In general, two basic methods of autoradiography are employed in biological investigations. In the first case, there is contact between the emulsion and the sample only during exposure time, after which the emulsion is removed and developed. This method is employed mostly for gross samples. In the second method, there is permanent contact between emulsion and the sample. This method is used exclusively with thin tissue samples. The first method (temporary contact method) is the most suitable for use with chromatograms of labeled material, leaf and whole plant tissues, gross bone sections, and tissue sections that have well-defined outlines. This requires little pretreatment of sample and any subsequent tissue staining does not affect the film. Because of poor contact, the method gives mediocre resolution and cannot normally be used for cellular localization studies. Several unique adaptations of this method have also been proposed. For better resolution, the second method (permanent contact reasonably) is generally preferred. Three modifications of this method are in common use, which includes mounting method, coating method, and stripping film method. In the mounting method, the tissue section is floated on water and the emulsion, on a glass slide, brought up underneath it so that the tissue lies on the film. Following the exposure, the film is developed and the tissue stained. This is a simple method with reasonably high resolution but suffers from the disadvantage of spotty development due to nonuniform penetration of the developer through the sample. To avoid this problem, the tissue section may be mounted on a glass slide and the emulsion is applied over it. In the coating method, the emulsion is melted and poured onto the tissue sample where it spreads and hardens. This method gives better resolution than the mounting method. In the stripping film method, the films used are the ones that allow the emulsion to be stripped off the base and applied directly on to the tissue section by means of water flotation. This gives better resolution than any of the previous methods and is commonly employed for determination of intracellular localization. In more recent techniques, a monolayer of silver halide crystal is applied on ultrathin tissue section embedded in methacrylate. Following the development of the emulsion, the tissue section is stained with uranyl or lead stains and examined by means of electron microscope.

6.5 Distribution of $^{14}CO_2$ in Different Plant Parts

6.5.1 Principle

The objectives are as follows: (1) feeding plants with $^{14}CO_2$ and (2) determining the radioactivity in different plant parts and fractions. Since ^{14}C (radioactive carbon) and ^{12}C (naturally occurring nonradioactive isotope of carbon) have the same atomic number, they have the same chemical properties. The living organisms, including plants, are unable to differentiate between these two isotopes of carbon. Thus, the plant leaves utilize ^{12}C and ^{14}C carbon dioxide at the same rate. The photosynthetically fixed carbon is then subsequently partitioned between various plant organs, that is, root, shoot, seed, and fruits. The partitioning of assimilates is influenced by a variety of factors such as plant species, age, stage of development, and environmental factors. The partitioning pattern can be quantitatively determined by estimating the radioactivity in different plant parts following exposure of the seedlings to $^{14}CO_2$ in light. The percent distribution of ^{14}C in various parts is then calculated.

6.5.2 Precautions

Radioisotopes emit ionizing radiations and are dangerous if used indiscriminately. Because of the health hazard involved, the use of radioactive materials in laboratories is controlled under license with stringent restrictions on use and waste disposal. Some simple precautions should be followed both for our own safety and that of our colleagues: (1) The place where the isotopes are to be used must be kept very clean. (2) Never pipette by mouth in radioactive laboratory even if a particular liquid is not radioactive. It may be contaminated. (3) Always wear a laboratory coat or apron. (4) Always spread on your work tables some disposable sheets of cellophane or absorbent paper. (5) Wipe all spillage several times using tissue paper. (6) Discard the wiped papers used containers or the soluble waste in their respective containers. (7) Clearly mark all glassware used as "radioactive." (8) Wash all the glassware personally first with "carrier" solution (this is a solution that contains the same nonradioactive compound that was used in the experiment. Normal glucose solution will be the carrier solution for ^{14}C glucose in the experiment). (9) It is advisable to use a pair of rubber gloves. (10) Wash your hands thoroughly after completion of the experiment, before using the counters. If we handle the planchets or planchet holders with contaminated hands, they may increase the counting rate and mess up the background counting. (11) Never prepare the samples in the counting room where the counters are placed. (12) Subscribe for film badge service and get the dosage level reported once in 15 days. (13) Using an ordinary rate meter, check the background activity as often as possible.

6.5.3 Disposal of Radioactive Waste

Keep all glassware, paper towels, etc., used in experiments, in a separate container, and do not mix it with other material in the laboratory. It is advisable to use a foot-operated bin for radioactive waste material (Figure 6.11). It should be periodically emptied and buried. Liquid waste is best kept diluted and stored in carboys that can be disposed off from time to time.

6.5.4 Where to Buy the Isotopes

Compounds labeled with radioactive isotopes are manufactured by regulatory authorities in respective countries. Permission should be obtained for the recognition of the department or laboratory, enclosing details of the space, and other facilities

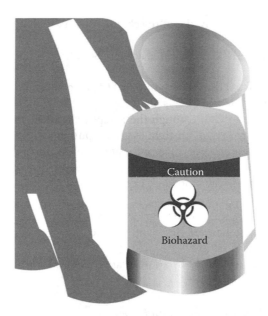

Figure 6.11   Foot-operated bin for radioactive biohazardous wastage.

available. For film badge service, apply to Radiation Protection Service. Isotopes can be obtained from Amersham Inc., England, or various other commercial manufacturers.

6.5.5 Materials

Syringes, double-walled (water-jacketed) Perspex chamber, airtight circulation pump, Geiger–Müller counter, gas proportional counter or liquid scintillation counter, planchets or scintillation vials, 0.1 N HCl (8.3 mL/L of water) and 0.2 N HCl (16.6 mL/L of water) and 0.1 N NaOH (4.12 g/L of water), scintillation fluid, ^{14}C sodium bicarbonate, fully grown plants of wheat, pea, or tomato, reagent bottles, balance, clinical centrifuge, water bath.

Scintillation fluid: Dissolve 4 g of POP (2,5-diphenyloxazole) and 100 g of POPOP (1,4-bis(5-phenyloxazol-2-yl) benzene) in 1 L of toluene.

6.5.6 Protocol

6.5.6.1 Feeding of Plants with $^{14}CO_2$

1. Enclose the plant (e.g., wheat, or tomato or pea plant) in a double-walled Perspex chamber of an appropriate size and place it under natural sunlight.
2. Circulate water between the chamber walls through the water jacket to prevent any rise in temperature in the chamber during the feeding period.
3. Connect outlet of an air tight CO_2 generating tube with a flexible propylene or rubber tubing to the inlet of the air tight circulating pump and outlet of the pump to the

inlet of chamber. Connect outlet of the chamber with different tubing with the inlet of CO_2 generating tube.

4. Place 1–2 μmol of $NaH^{14}CO_3$ (specific activity 50 mCi/rnmol) at the bottom of CO_2 generating tube.
5. Release 0.1 N HCl from the stopper into $NaH^{14}CO_3$ containing CO_2 generating tube so that 50–100 μCi of labeled ^{14}C is liberated.
6. Circulate the evolved $^{14}CO_2$ through the Perspex chamber (in which the plant has been enclosed), using a small air tight electrical pump.
7. Allow the plant to fix $^{14}CO_2$ for an appropriate time (5 min–1 h).
8. Remove excess $^{14}CO_2$ by diverting the circulation of gas through NaOH trap, which is attached via a three-way stop cock in between the CO_2 generating tube and the pump.
9. Return the ^{14}C-fed plants to natural growth conditions.
10. After 24 h of feeding, harvest the plants and separate the parts, for example, leaf, stem, roots, seeds, and fruits.
11. Dry different parts separately in an oven at 65°C to a constant weight and record their weight.

6.5.6.2 Extraction and Measurement of Radioactivity
Radioactivity can be measured by any one of the following methods after suitably processing the plant sample:

6.5.6.2.1 Method 1.
1. Grind the oven-dried sample of different plant parts to a fine powder.
2. Place a known weight of the sample in a planchet having some adhesive material on its surface so that the sample gets stuck to the planchet.
3. Count the dpm either by a gas proportional counter or by Geiger–Müller counter or liquid scintillation counter.

6.5.6.2.2 Method 2.   Prepare the ethanol soluble and insoluble fractions of the powder sample and then determine the radioactivity in each of them individually.

1. Take an appropriate amount of dry sample in a centrifuge tube.
2. Add 5 mL of 80% aqueous ethanol (v/v) and place in a water bath at 40°C for 1 h. Shake tube occasionally.
3. Centrifuge at 5000g for 15 min and decant the supernatant.
4. Extract the residue in the similar manner with 50% (v/v) aqueous ethanol, centrifuge, and collect the supernatant.
5. Finally extract the residue with water, centrifuge, and collect the supernatant.
6. Pool all the supernatants obtained in steps 3–5, and this will constitute ethanol soluble fraction.
7. Suspend the residue in 2 N HCl and autoclave at 15 kg/cm² for 1 h.
8. Centrifuge the hydrolyzed material and wash the pellet twice with 2 N.
9. Pool all the acid fractions representing ethanol insoluble fraction.
10. Evaporate both the fractions (ethanol soluble and insoluble) separately to dryness and redissolve each of them in 2 mL distilled water.
11. Pipette 0.2 mL of the earlier extract on a planchet and dry under the infrared lamp then record the counts in a gas flow proportional counter.

6.5.6.2.3 Method 3.   Measurement of radioactivity by liquid scintillation counter
1. Take a suitable aliquot (1 mL) of the earlier fractions in scintillation vials and dry it.
2. Add 10 mL of scintillation fluid to each vial.
3. Record the counts on a liquid scintillation counter with an appropriate correction for quenching.

6.5.7 Calculations

Determine the total amount of ^{14}C assimilated in the plant by adding the radioactivity (dpm) in both the fractions of all plants. While doing this, take into account weight of the part used for sample preparation, dilution factors (if any in case of fractions), and the total weight of that plant or organ in the plant. For example, if radioactivity is determined in powdered nonfractionated leaf sample, then

$$\text{Total }^{14}\text{C-labeled material} = \frac{\text{dpm} \times \text{total weight of leaves on the plant}}{\text{wt. of powdered sample taken}}$$

For the fractionated sample, take into account the dilution factor as well, that is, the volume of fraction used for determination of radioactivity and the total volume of the fraction.

1. Calculate the total ^{14}C-labeled assimilates in each plant part A, B, C, and D in leaves, stem, seeds, and roots, respectively, from the sum of the counts in both the fractions of an organ.
2. Compute the percent distribution of ^{14}C in different parts as follows:

$$\% \, ^{14}\text{C in leaves} = \frac{A}{X} 100 \text{ dpm}$$

$$\% \, ^{14}\text{C in stem} = \frac{B}{X} 100 \text{ dpm}$$

$$\% \, ^{14}\text{C in seeds} = \frac{C}{X} 100 \text{ dpm}$$

$$\% \, ^{14}\text{C in roots} = \frac{D}{X} 100 \text{ dpm}$$

where X is the total ^{14}C assimilated by the entire plant and a, b, c, and d represent total ^{14}C-labeled material in each of the plant parts, respectively.

6.5.8 Result

Tabulate the results of ^{14}C labeling in each plant part in terms of percent incorporation.

6.6 Photosynthetic Reduction of $^{14}CO_2$ to Primary Metabolic Product

6.6.1 Principle

The mechanism of photosynthetic CO_2 assimilation in plants varies with species and may occur via either C_3 or C_4 pathway. C_3 or C_4 mechanism in a particular plant can be established by identifying the first product of CO_2 fixation reaction. In case of C_3 pathway, 3-phosphoglycerate (3-PGA) is the first product, whereas in C_4 type, the initial product of carboxylation reaction is a C_4 organic acid such as oxaloacetate, which is rapidly converted to either malate or aspartate. Plants are then exposed to $^{14}CO_2$; the label is incorporated into the primary products. Determination of these primary products (viz., 3-PGA or malate and aspartate) for radioactivity provides the confirmatory evidence for operation of a particular photosynthetic pathway in a plant species.

6.6.2 Precautions

Follow all the safety guidelines as described in Section 6.5.2.

6.6.3 Materials

Experimental plants (select one C_3 and one C_4 plant), chamber made of Perspex or glass for feeding the plants with $^{14}CO_2$, gas mixture pump, ^{14}C-labeled sodium bicarbonate (NaH$^{14}CO_3$), chromatographic chamber for descending paper chromatography, Whatman No. 1 filter paper.

Scintillation fluid: Dissolve 4 g of POP (2,5-diphenyloxazole) and 100 g of POPOP (1,4-bis(5-phenyloxazol-2-yl)benzene) in 1 L of toluene.

Authentic standards: Prepare a solution containing 1 mg/mL of authentic sample of 3-PGA, glucose-6-P, aspartic acid, and malic acid.

Solvent system for paper chromatography: Mix n-butanol–propionic acid–water in the ratio of 10:5:7 (v/v) and use as the solvent system.

Detection reagents: Prepare various spraying reagents as given in step II in the procedure.

Buffer: 50 mM tricine–KOH buffer with 0.33 M sorbitol.

6.6.4 Protocol

1. Soak the glass fiber discs in 50 mM tricine–KOH buffer (pH 7.5) containing 0.33 M sorbitol.
2. Quickly transfer 0.1 g plant leaf segments on these glass discs to prevent dehydration of plant tissues. Place these glass fiber discs in steel planchets. Prepare five such planchets and place them on a strip having a well in its center. Add 2 N HCl to this central well and keep the strip in the Perspex chamber ($19 \times 12 \times 37$ cm) with an internal volume of 50 mL.

3. Expose the chamber to direct sunlight for 5 min. Then inject $NaH^{14}CO_3$ through a rubber septum into the acid containing well on the strip such that the concentration of $^{14}CO_2$ in the chamber is 0.1% and circulate the air through the chamber with the help of an air tight gas mixture pump.

4. Remove the samples from the chamber after 20, 40, 60, 120, and 300 s of exposure to $^{14}CO_2$.

5. Immediately kill the tissue by dropping it in boiling ethanol.

6. Extract the plant material with 80% (v/v) ethanol, centrifuge the homogenate at $8000 \times g$, and collect the supernatant.

7. Extract the residue with distilled water, centrifuge again, and carefully decant the supernatant. Pool the obtained supernatants and make to a known final volume of 10 mL with 80% ethanol.

8. Take 1 mL of the extract in a scintillation vial containing 1.5 mL scintillation fluid and evaporate it to dryness. Record the counts in a liquid scintillation counter.

9. Take another aliquot of the extract, evaporate to dryness, and extract chlorophyll from the residue with chloroform. Evaporate off the excess chloroform from this preparation and dissolve the residue in water.

10. Apply a suitable aliquot of the earlier water extract onto Whatman No. 1 filter paper sheet in the form of a spot. Perform descending paper chromatography using n-butanol–propionic acid–water in the ratio of 10:5:7 as the mobile phase.

11. Remove the chromatogram and dry it at room temperature. Spray the chromatogram for developing color using the different spraying reagents as detailed as follows:

Compound Detected	Spray Reagent
Malate sucrose	0.08% (w/v) bromocresol green in ethanol, 1.5 g benzidine + 10 mL glacial acetic acid + 10 mL 40% (w/v) TCA, and ethanol to make the volume to 100 mL
3-PGA and glucose-6-P	5 mL 60% (w/v) $HClO_4$ + 25 mL 4% (w/v) ammonium molybdate + 10 mL N HCl + water to make the final volume to 100 mL
Aspartate	0.4% (w/v) ninhydrin in 95% (v/v) acetone

After spraying, dry at room temperature and then keep it at 80°C–100°C in an oven for 5 min. Colored spots appearing on the chromatogram are due to malate, aspartate, and sucrose.

1. Autoclave the paper for 2 min at 8–10 psi for detecting 3-PGA and glucose-6-P. The paper becomes blue. Now expose the paper to ammonia vapors. Blue color of phosphomolybdate is due to the reaction with phosphorylated compounds while the blue color of the background would disappear.

2. Identify the compounds by comparing their R_f values with those of authentic standards run parallel. Punch the area of paper containing labeled compounds from the developed and sprayed chromatograms and count in a scintillation counter. Make sure that necessary corrections for quenching are made for any colored segments of the chromatogram.

6.6.5 Conclusions

By examining the kinetics of distribution of ^{14}C label in 3-PGA and C_4 dicarboxylic organic acids, the mode of photosynthetic CO_2 assimilation can be deduced. If at the initial stages the ^{14}C label is initially present predominantly in C_4 dicarboxylic acids (oxaloacetate, malate, or aspartate) but little in 3-PGA, and this is followed by a decrease in ^{14}C label in C_4 acids accompanied by a concomitant increase of ^{14}C in 3-PGA, then it is an indication of C_4 pathway of CO_2 fixation. If, in contrast, at the early stages the ^{14}C is largely in 3-PGA and ^{14}C-labeled dicarboxylic acids appear at a much later stage, then it denotes C_3 mechanism of photosynthetic CO_2 assimilation.

6.6.6 Result

Express results as percentage incorporation of radioactivity in each fraction.

6.6.7 Autoradiography

Radioactive isotopes of various elements can be used to follow the specific molecules biochemically; one of the earliest uses of radioactivity was to trace the chemical path way of carbon during photosynthesis. The soluble contents can be separated by paper chromatography (either by ascending, descending, or 2D) and small molecules containing ^{14}C atoms derived from CO_2 can be detected by a sheet of x-ray film placed over the dried chromatogram enclosed in an x-ray cassette. The cassette is kept in the dark room for 48–72 h. Remove the x-ray film and develop the film in the dark room in a manner that is followed for the development of a photographic film. The metabolites formed can be identified. The location of the individual molecules blotted on the x-ray film can be compared with the compounds identified on the chromatographic paper with the help of spray reagents. After identification of the compounds, each area on the chromatogram is cut and eluted in ethanol and added to the scintillation vial containing scintillation fluid and radioactivity of each biomolecule is measured.

6.7 Incorporation of Labeled Acetate into Leucoplastic Fatty Acids of Developing Brassica Seeds

6.7.1 Principle

The process of *de novo* fatty acid biosynthesis, in all types of plastids, occurs by a series of virtually identical enzymatic steps. Starting from acetate, the process involves the enzymes and sequence of repeating reactions. Use of labeled acetate in *in vitro* studies on fatty acid biosynthesis in isolated plastids could provide valuable information about the substrate and cofactor requirement of each of the enzymatic reactions involved in the process of fatty acid biosynthesis. The extent of incorporation of radioactivity into fatty acids serves as an indicator of the rate of fatty acid biosynthesis by isolated plastids in the presence of various exogenously supplied cofactors and substrates.

6.7.2 Precautions

Follow all safety guidelines mentioned in Section 6.5.2.

6.7.3 Materials

Developing mustard seeds (*Brassica campestris*) in sodium salt of ^{14}C acetate.

Extraction medium: 50 mM HEPES-NaOH (pH 7.5) containing 0.4 M sorbitol, 0.4 mM EDTA, 1 mM MgCl$_2$, 1 mM DTT, 1% BSA, and 1% Ficoll.

PBF-Percoll: Prepare a stock solution of Percoll containing PEG 4000 (3% w/v), BSA (1% w/v), and Ficoll (1% w/v).

Resuspension buffer: 50 mM HEPES-KOH (pH 7.5) containing 0.4 M sorbitol, 2 mM EDTA, and 1 mM MgCl$_2$.

Assay buffer: 0.4 M Bis-tris propane buffer (pH 8.5), 0.2 M sorbitol, 8 mM MgCl$_2$, 8 mM ATP, 4 mM MnCl$_2$, 0.8 mM Na^{14}C acetate (40.7 − 51.8 × 10^{-7} Bq/m mol), 0.6 mM CoA, 20 mM NaHCO$_3$, 1 mM NADH, and 1 mM NADPH.

Scintillation fluid: Dissolve 4 g of POP (2,5-diphenyloxazole) and 100 g of POPOP (1,4-bis(5-phenyloxazol-2-yl)benzene) in 1 L of toluene.

Chloroform: Methanol mixture (2: 1, v/v), KOH (0.65 N), and KCl (0.5 M).

6.7.4 Protocol

1. For preparing leucoplasts, hand homogenize 10 g seeds in chilled pestle and mortar using 20 mL of ice cold extraction medium. Filter the homogenate through four layers of cheese cloth and centrifuge the filtrate at 4°C at 500 × g for 5 min.
2. Further centrifuge the resulting supernatant in cold at 6000 × g for 10 min and resuspend the pellet in 5 mL extraction medium.
3. Layer this preparation onto a discontinuous Percoll gradient consisting of 5 mL of 80% (v/v) PBF-Percoll, 7.5 mL of 35% PBF-Percoll, 7.5 mL of 22% PBF-Percoll, and 7.5 mL of 10% PBF-Percoll, all prepared in the resuspension buffer.
4. Centrifuge in a swing-out rotor at 9200 × g for 4 min and collect the band of leucoplasts at 22%–35% Percoll interface and dilute it with 20 volumes of the resuspension buffer.
5. Centrifuge at 10,000*g* for 10 mi, and suspend the pellet in 5 mL of resuspension buffer and use it as leucoplast preparation for further investigations.
6. For examining the rate of fatty acid biosynthesis, take appropriate volume of leucoplasts preparation (containing 50–80 µg protein) in 0.2 mL of the assay buffer containing labeled acetate and cofactors for 2 h at 30°C. Stop the reaction by adding 0.4 mL of 0.65 N KOH.

7. Partition the radiolabeled fatty acids produced by partitioning them into chloroform by adding 750 μL of chloroform–methanol (2:1). Wash the chloroform layer three to four times with 750 μL of 0.5 M KCl.
8. Take the washed chloroform layer into scintillation vial and add 10 mL of scintillation fluid and count using liquid scintillation counter.
9. Calculate the moles of ^{14}C acetate incorporated and record dpm by applying the value of specific activity of the labeled acetate used for the experiment.

6.7.5 Result

Express result in normal acetate incorporated into fatty acids/mg protein/h.

Suggested Readings

Biilington D., G. G. Jayson, and P. J. Maltby. 1992. *Radioisotopes*. Oxford, U.K.: Bios Scientific.
Bisen, P. S. and A. Sharma. 2012. *Introduction to Instrumentation in Life Sciences*. Boca Raton, FL: CRC Press.
Chapman J. M. and G. Avery. 1981. *The Use of Radioisotopes in the Life Sciences*. London, U.K.: George Allen & Unwin Ltd.
Connor K. J. and I. S. Mclintock. 1994. *Radiation Protection Handbook for Laboratory Workers*. Leeds, U.K.: H and H Scientific Consultants.
Gavrilets, S. and A. Vose. 2009. Dynamic patterns of adaptive radiation: Evolution of mating preferences. In *Speciation and Patterns of Diversity*, eds. R. K. Butlin, J. Bridle, and D. Schluter. Cambridge, U.K.: Cambridge University Press.
Slater R. J. 1995. Radioisotopes in molecular biology. In *Molecular Biology and Biotechnology, A Comprehensive Desk Reference*, ed. R. A Myers, New York: VCH.
Slater R. J. 2000. *Radioisotopes in Biology—A Practical Approach*, 2nd edn. Oxford, U.K.: IRL Press
Slater, R. J. 2001. *Radioisotopes in Biology*. Oxford, U.K.: Oxford University Press.

Important Links

Application: http://www.idph.state.ia.us/eh/common/pdf/radiological_health/radioisotopes.pdf; http://bricker.tcnj.edu/tech/BIOL311useofradioisotopes.htm; http://www.edurite.com/kbase/application-of-radioisotopes-in-biology
Autoradiography: http://www.gelifesciences.com/aptrix/upp01077.nsf/Content/na_blotting_site~na_ detection~na_autoradiography_cassettes ?op endo cument&cmpid=ppcaw100
Scintillation counter: http://www.hidex.com/products/hidex-300-sl.aspx

7

Chromatography

7.1 Chromatography

Chromatography is a broad range of physical methods used to separate and/or to analyze complex mixtures. The separation process is achieved by distributing the components of a mixture between two phases, a stationary phase and a mobile phase. Solutes are eluted from the system as local concentrations in the mobile phase, in the order of their increasing distribution coefficients with respect to the stationary phase bed. Chromatography techniques separate molecules on the basis of difference of size, shape, mass, charge, solubility, and adsorption properties. There are several types of chromatography to be separated, a solid phase and a solvent. The magnitude of these interactions depends on the particular method, and the solute–solid interaction is dominant in ion-exchange chromatography, whereas the solvent–solvent interaction is more important in a particular chromatography (Table 7.1).

Depending on the mode by which separation is achieved, there are three types of chromatography:

Paper chromatography in which the stationary phase is supported by cellulose fibers of paper and the mobile phase moves through the interstitial spaces by capillary action. Sometimes the cellulose fibers of paper themselves act as solid stationary phase.

Thin-layer chromatography (TLC) in which the stationary phase is thinly coated onto glass plates and the mobile phase moves along it.

Column chromatography in which the stationary phase is packed into glass or metal columns and the mobile phase percolates through the column.

7.2 Paper Chromatography

7.2.1 Introduction

In paper chromatography, the sample to be analyzed is applied in the form of a spot near one of the edges of a Whatman filter paper. Filter papers are then kept in an atmosphere already saturated with water vapors to form a thin film of water around cellulose fibers of the paper that then act as a stationary phase. An appropriate solvent system, which functions as a mobile phase, is then allowed to flow over the sample spot. While coming in contact with the mobile phase, the various components of the

TABLE 7.1 Solute Property Used for Chromatographic Separation

Technique	Solute Property	Solid Phase	Solvent
GF	Size and shape	Hydrated gel	Usually aqueous
Adsorption	Adsorption	Adsorbent	Nonpolar
Chromatography inorganic material			
Partition chromatography	Solubility	Inert support	Mixture of polar and nonpolar solvent
Ion exchange chromatography	Ionization	Matrix with ionized group	Aqueous buffer

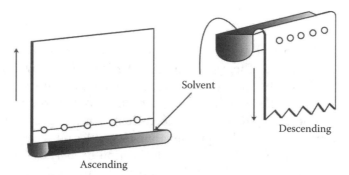

Figure 7.1 Ascending and descending paper chromatography.

sample get partitioned between the stationary and the mobile phases. Those constituents having a higher affinity for the stationary phase move less rapidly as compared to those having higher affinity for the mobile phase. Evidently, the components will get well separated from each other if their K_d values are sufficiently different. Satisfactory separation of components of interest can hence be achieved by judicious selection of the mobile phase.

Depending upon the direction of flow of the mobile phase, the two commonly used systems are ascending and descending paper chromatography. In the ascending method, the solvent is kept at the base of the chamber, and the edge of the paper where the sample has been applied is immersed in the solvent taking care that the sample spots do not get dipped in the solvent but remain just above the surface of the solvent (Figure 7.1). The solvent moves up or ascends the paper by capillary action, and the separation of different components occurs on the basis of differences in their partition coefficients. In the descending method, the paper is hung in such a way that the side where the sample has been spotted dips in a trough that is fitted at the top of the chamber and contains the mobile phase, and this solvent travels down the paper under the force of gravity (Figure 7.2).

7.2.2 Sample Preparation

Biological material should be desalted before chromatography by electrolysis or electrodialysis. Excess salt results in a poor chromatogram with spreading of spots

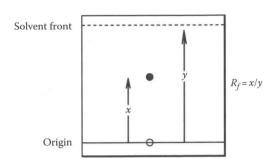

Figure 7.2 R_f determination.

and change in their R_f values (Figure 7.2). It can also affect the chemical reaction used to detect the compounds being separated. Macromolecules such as proteins are removed prior to chromatography by ultrafiltration or gel filtration (GF). The sample (10–20 μL) is then applied to the paper with a micropipette.

7.2.3 Paper

Whatman No. 1 is the paper most frequently used for analytical purposes. Whatman No. 3 MM is a thick paper and is best employed for separating large quantities of material; the resolution is, however, inferior to Whatman No. 1. For a rapid separation, Whatman Nos. 4 and 5 are convenient, although the spots are less well defined. In all cases, the flow rate is faster in the machine direction, which is normally noted on the box containing the papers. The paper may be impregnated with a buffer solution before use or chemically modified by acetylation. Ion-exchange papers are also available commercially. For the separation of lipid and similar hydrophobic molecules, silica-impregnated papers are available commercially.

7.2.4 Solvent

This choice like that of the paper is largely empirical and will depend on the mixture investigated. If the compounds move close to the solvent front in solvent A, then they are too soluble, while if they are crowded around the origin in solvent B, then they are not sufficiently soluble. A suitable solvent for separation, therefore, would be an appropriate mixture of A and B, so that the R_f values of the components of the mixture are spread across the length of the paper. The pH may also be important in a particular separation, and many solvents contain acetic acid or ammonia to create a strongly acidic or basic environment.

7.2.5 Component Detection

After development, the filter paper is dried in air. The location of the compounds under investigation is carried out by making use of their specific chemical, physical, or biological property. For instance, if these components form a colored complex or product with a particular reagent, they can then be conveniently located by spraying the chromatogram with that reagent. In case the compounds absorb ultraviolet (UV) light or show UV fluorescence, the paper can be examined under strong UV light in the darkness. The UV-absorbing compounds would appear as dark spots, while UV fluorescent compounds would show a characteristic fluorescence under UV light. If in metabolic studies a radioactive precursor has been used, then the products derived from it can be detected from the radioactive zones or spots on the chromatogram.

7.2.6 Identification

The separated compounds are identified on the basis of their R_f values, which denotes relative to the front. The R_f value is calculated as follows:

$$R_f = \frac{\text{Distance traveled by the component from baseline}}{\text{Distance traveled by the solvent from baseline}}$$

Since a number of factors such as composition of solvent system, temperature, pH, mode of development, grade of filter paper, and flow rate influence the R_f value, it is always advisable to run authentic standards along with the sample.

7.3 Ascending Chromatography

The sheet of paper is supported on a frame with the bottom edge in contact with a trough filled with solvent. Alternatively, the paper can be rolled into a cylinder, be fastened with a paper clip, and stand in the solvent. The arrangement is constant atmosphere and separations are carried out in a constant temperature room (Figure 7.2).

7.4 Descending Chromatography

This method is convenient for compounds that have similar R_f values since the solvent drips off the bottom of the paper, thus giving a wider separation. In this case, of course, the R_f values cannot be measured, and substances are compared with a standard reference compound such as glucose, for example, in the case of sugars:

$$R_g = \frac{\text{Distance moved by unknown}}{\text{Distance moved by a standard compound}}$$

Different modifications for the development of chromatograms: Quite often, a single run with one solvent system in a single dimension is not sufficient to obtain satisfactory separation of the components. A number of modifications have been employed to achieve better resolution of such overlapping or very closely located compounds. Some of these have been described in the succeeding text.

7.4.1 Multiple Developments

In this procedure, the paper is developed with the same solvent system several times in the same direction. The chromatogram is air-dried between the successive developments. This mode of development of chromatogram can be used advantageously when the sample contains a mixture of components, some of which migrate quite fast and get well separated from each other and a group of other components remain clustered near the origin due to their low and closely similar R_f values. These slow-moving components get further apart with each successive run resulting in their better separation.

7.4.2 Continuous Development

It is also known as overrun chromatography. In this case, the chromatogram is developed, in the same solvent system, in the same direction continuously for a long time even after the solvent has started dripping down from the other end of the paper. Care, however, must be taken that the fastest-moving compound does not get eluted out of the paper. This approach is useful for achieving better separation of compounds having very low and similar mobilities. Evidently, R_f value cannot be calculated as the solvent front cannot be determined due to overflow of the solvent. The identification is hence done on the basis of R_1 value, which is calculated as follows:

$$R_1 = \frac{\text{Distance moved by unknown}}{\text{Distance moved by a standard compound}}$$

7.4.3 2D (Bidimensional) Chromatography

In this method, the chromatogram is first developed in one solvent system, in one direction. After air-drying, the paper is again developed in a second solvent system, in a direction perpendicular to the previous run. The compounds having similar R_f values in the first solvent system might have different mobilities in the second solvent system, and hence, they get separated. However, the limitation of bidimensional chromatography is that only one sample can be spotted or applied for analysis on each filter paper (Figure 7.3).

7.4.4 Reversed-Phase Chromatography

In all the previously mentioned cases, the aqueous phase of the solvent system is used to saturate the internal atmosphere of the chamber, and water absorbed by the cellulose fibers of the paper serves as the stationary phase. Thus, the stationary phase is hydrophilic, whereas the mobile phase is hydrophobic, in nature. For separation

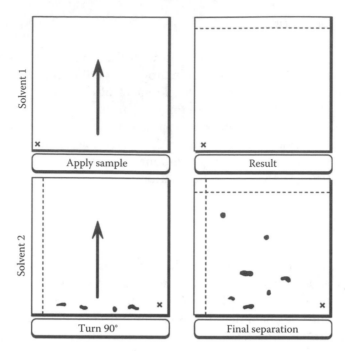

Figure 7.3   2D paper chromatography.

of hydrophobic substances like fatty acids, the filter paper is treated with lipophilic compounds like silicon encase, kerosene oil, paraffin, and vaseline, while the mobile phase is hydrophobic in nature. Since the characteristics of stationary and mobile phases are converse of that in conventional paper chromatography, this modified form is referred to as "reverse phase chromatography (RPC)."

7.4.4.1 Partition Chromatography

The separation of a compound that is readily soluble in organic liquids, but sparingly soluble in water, is carried out by adsorption chromatography, while ionizable water-soluble compounds are best separated by ion-exchange chromatography. Partition chromatography is intermediate between adsorption and ion-exchange chromatography, and compounds that are soluble in both water and organic solvents are readily separated by partition methods.

When a compound is shaken with two immiscible solvents, it will distribute itself unevenly between the two phases, and at equilibrium, the ratio of the concentration of the compound in the two solvents is constant and is known as the partition coefficient (α):

$$\alpha = \frac{\text{Concentration of } x \text{ in solvent 1}}{\text{Concentration of } x \text{ in solvent 2}}$$

In partition chromatography, one solvent, usually water, is held on the stationary supporting phase that is in the form of a column of film of inert material. The other phase consists of a mobile, water-saturated, organic liquid that flows past the

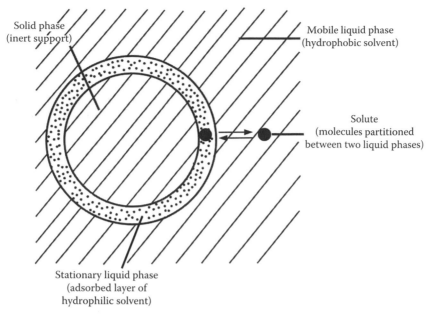

Figure 7.4 Illustrating liquid–liquid partition chromatography principle.

stationary phase. The components of a mixture are separated if their partition coefficients between the solvents are different (Figure 7.4).

7.5 Separation and Identification of Amino Acids by Descending Paper Chromatography

7.5.1 Principle

Amino acids in a given mixture or sample aliquot are separated on the basis of differences in their solubility and hence differential partitioning coefficients in a binary solvent system. The amino acids with higher solubility in the stationary phase move slowly as compared to those with higher solubility in the mobile phase. The separated amino acids are detected by spraying the air-dried chromatogram with ninhydrin reagent. All amino acids give purple or bluish-purple color on reaction with ninhydrin except proline and hydroxyproline that give a yellow-colored product. The reactions leading to the formation of a purple complex are given as follows:

7.5.2 Materials

Whatman filter paper No. 1 (cut it as per the size of the chromatographic chamber); solvent (butanol/glacial acetic acid/water, 4:1:5, in a separating funnel—mix it thoroughly and saturate by incubating for 3–6 h. Allow the phases to separate out completely. Use the lower aqueous phase for saturating the chamber. The upper organic phase is used as mobile phase); ninhydrin location reagent (dissolve 0.2 g in 100 mL of acetone just before use); micropipette/microsyringe; hair drier; sprayer; oven set at 105°C; chromatographic chamber saturated with solvent vapors; standard amino acid solutions (prepare a small volume of mixture of unknown amino acids as 10 mg/mL and as individual amino acids as separate standard solutions containing alanine, aspartic acid, cysteine HCl, cystine, glutamic acid, glycine, histidine, leucine, isoleucine, proline, serine, threonine, tryptophan, valine, and tyrosine in 10% isopropyl alcohol v/v; sometimes a drop of acid or alkali is required to bring the amino acids in solution).

7.5.3 Protocol

1. Take Whatman No. 1 filter paper and lay it on a rough filter paper. Throughout the experiment, care should be taken not to handle the filter paper with naked hands, and for this purpose, either gloves should be used or it should be handled with the help of a folded piece of rough filter paper.
2. Fold the Whatman No. 1 filter paper about 2–2.5 cm from one edge. Reverse the paper and again fold it 2 cm further down from the first fold.
3. Draw a line across the filter paper with a lead pencil at a distance of about 2 cm from the second fold. Put circular marks along this line at a distance of 2.5 cm from each other.
4. With the help of a micropipette or microsyringe, apply 20 μL of solution of each standard amino acid on a separate make. Also apply spots of the sample or mixture to be analyzed, preferably on the mark at the center of this baseline. The size of the spot should be as small as possible so that the developed spots are compact and do not overlap. If necessary, the wet sample spot should be dried with a hair drier before applying additional aliquot.
5. Hang a line filter paper in a chromatographic chamber that has previously been saturated with the aqueous phase of the solvent system. This is done by keeping Petri plates containing the aqueous phase at the bottom of the chamber. The paper is hung from the trough/tray, and a glass rod is kept to hold it in place. Care should be taken to ensure that the baseline should not get submerged when the mobile phase is added to the trough; otherwise, the spotted material would get dissolved in the solvent (Figure 7.5).
6. Close the chamber firmly so that it is airtight. Allow sufficient time for cellulose fibers of the paper to get fully hydrated.
7. Pour the mobile phase through the holes provided on the lid of the chamber into the trough. Replace the rubber bungs in the hole, and allow the mobile phase to run down the paper till the solvent front reaches about 5 cm from the opposite edge.
8. Remove the paper and mark the solvent front with lead pencil and let it dry at room temperature.

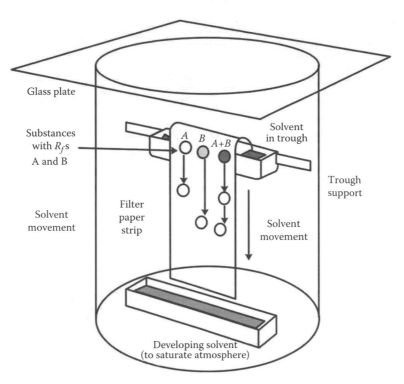

Figure 7.5 Descending paper chromatography.

9. Spray the filter paper (chromatography) with ninhydrin reagent, and after drying it at room temperature, transfer it to an oven at 105°C for 5–10 min.
10. Blue- or purple-colored spots would appear on the paper. Mark the boundary of each spot with lead pencil.
11. Measure the distance between the center of the spots and also the distance of the solvent front from the baseline.
12. Calculate the R_f value of standard amino acids as well as those in the given mixture or sample as follows:

$$R_f = \frac{\text{Distance traveled by unknown amino acid}}{\text{Distance traveled by the solvent system}}$$

13. Identify the amino acids in the mixture or sample by comparing their R_f values with those of the reference standards.

Note: It is advisable to carry out chromatography in three different solvent systems before the identity of the amino acid in the mixture or sample can be established with any degree of certainty.

7.5.4 Observations

Distance traveled by the solvent = x cm front from baseline
Distance traveled by the glycine = a cm from baseline

Distance traveled by alanine $= b$ cm from baseline
Distance traveled by threonine $= c$ cm from baseline
Distance traveled by methionine $= d$ cm from baseline
Distance traveled by spot No. 1 $= a$ cm sample from baseline

Calculation
R_f value of glycine $= a/x$
R_f value of alanine $= b/x$
R_f value of threonine $= c/x$
R_f value of methionine $= d/x$
R_f value of spot No. 1 $= a/x$

7.5.5 Conclusion

The sample contains glycine since the R_f value of spot No. 1 is identical to that of authentic glycine standard.

7.6 Separation and Identification of Amino Acids by Ascending Paper Chromatography

7.6.1 Materials

Whatman filter paper No. 1 (cut it as per the size of the chromatographic chamber); solvent (butanol/glacial acetic acid/water, 4:1:5, in a separating funnel—mix it thoroughly and saturate by incubating for 3–6 h. Allow the phases to separate out completely. Use the lower aqueous phase for saturating the chamber. The upper organic phase is used as mobile phase); ninhydrin location reagent (dissolve 0.2 g in 100 mL of acetone just before use); micropipette/microsyringe; hair drier; sprayer; oven set at 105°C; chromatographic chamber saturated with solvent vapors; standard amino acid solutions (prepare a small volume of mixture of unknown amino acids as 10 mg/mL and as individual amino acids as separate standard solutions containing alanine, aspartic acid, cysteine HCl, cystine, glutamic acid, glycine, histidine, leucine, isoleucine, proline, serine threonine, tryptophan, valine, and tyrosine in 10% isopropyl alcohol v/v; sometimes a drop of acid or alkali is required to bring the amino acids in solution). Cylindrical chromatography chambers are needed for this experiment.

7.6.2 Protocol

1. Take a Whatman No. 1 filter paper sheet of appropriate size so that it can be rolled into a cylinder and can be accommodated in a cylindrical chromatographic jar.
2. Draw a baseline 2 cm from one of the breadth wise edges of the paper. Put small circular marks along the baseline in such a way that the distance from the edge of the paper and the first spot and the distance between the adjacent spot are not less than 2.5 cm.

3. Apply 20 µL aliquots of the standard amino acids and of the sample as different spots. The diameter of the spotted material should be as small as possible, and if required, the applied solution may be dried prior to loading an additional volume.

4. Roll the paper into a cylinder and fasten its edges with a paper clip. Pour a sufficient volume of the mobile phase into the chromatographic jar that has been earlier saturated with water vapors by lining the tank with filter paper saturated with the aqueous phase of the solvent system.

5. Gently place the rolled filter paper upright in the jar ensuring that it does not touch the sides of the chamber and at the same time taking care that the baseline where the spots have been applied does not dip into the solvent.

6. Close the tank with an airtight lid or a glass plate to which a sufficient amount of silicone grease has been applied.

7. Leave the setup undisturbed and allow the solvent to move up till it reaches about 5 cm from the upper edge.

8. Remove the chromatogram from the chamber and air-dry it.

9. Spray the paper with ninhydrin reagent, and let it dry again at room temperature prior to transferring it to an oven at 105°C for 5–10 min. Locate the position of amino acids from the bluish- or purple-colored spots on the chromatogram.

7.6.3 Observations

Calculate the R_f values of the standard amino acids and those in the sample or mixture as described in descending chromatography. Identify the amino acids in the mixture or sample by comparing R_f values with those of applied standard amino acids.

7.7 Separation and Identification of Amino Acids in a Given Mixture by 2D Paper Chromatography

7.7.1 Principle

Amino acids having very close R_f values in a particular solvent system may appear as a single or overlapping spots in a single-dimensional chromatography and may be mistaken as one component. They can be separated into individual components by developing the chromatogram again in a direction perpendicular to the first run in a second solvent system in which they have different R_f values. The main limitation of this method is that only one spot either of the sample or of a standard amino acid can be applied on each filter paper sheet necessitating running of a large number of chromatograms for the standard amino acids.

7.7.2 Materials

Whatman filter paper No. 1 (cut it as per the size of the chromatographic chamber); solvent (butanol/glacial acetic acid/water, 4:1:5, in a separating funnel—mix it thoroughly and saturate by incubating for 3–6 h. Allow the phases to separate out completely. Use the lower aqueous phase for saturating the chamber. The upper organic phase is used as mobile phase); ninhydrin location reagent (dissolve 0.2 g in 100 mL of acetone just before use); micropipette/microsyringe; hair drier; sprayer; oven set at 105°C; chromatographic chamber saturated with solvent vapors; standard amino acid solutions (prepare small volume of mixture of unknown amino acids as 10 mg/mL and as individual amino acids as separate standard solutions containing alanine, aspartic acid, cysteine HCl, cystine, glutamic acid, glycine, histidine, leucine, isoleucine, proline, serine, threonine, tryptophan, valine, and tyrosine in 10% isopropyl alcohol v/v; sometimes a drop of acid or alkali is required to bring the amino acids in solution). Cylindrical chromatography chambers are needed for this experiment. An additional chromatographic chamber for the second solvent system is required. Phenol (distilled)/water, 80:20 w/v, is used as second solvent system. Add 125 mL of water to 500 g of phenol and a few drops of ammonia (0.88%) to this mixture just before use (Caution: Phenol is corrosive and can cause burns on the skin). Standard amino acids—prepare 1% solution of standard amino acids such as asparagine, glycine, serine, and arginine in 10% isopropanol (w/v).

7.7.3 Procedure

1. Lay the chromatographic paper sheet flat on the rough filter paper using gloves.
2. Draw a baseline 5 cm from one of the edges of the paper.
3. Draw another line perpendicular to the first line, again 5 cm away from the adjacent edge.
4. Apply 60 μL of the sample solution or given mixture containing unknown amino acids at the point of intersection of these two lines. The sample should be applied in small volumes at a time with the help of a micropipette, with intermittent drying, to ensure that the zone of applied solution is as small as possible.
5. Repeat the same procedure for a mixture of three standard amino acids using a separate chromatographic sheet for each mixture. The composition of mixture of standard amino acids should be such that each amino acid is present in at least two different mixtures so that its identity can be established (Figure 7.6).
6. Hang the paper in the chromatographic tank whose interior has previously been saturated with the aqueous phase of solvent system No. 1 (butanol/acetic acid/water mixture in the ratio of 4:1:5).
7. After allowing an equilibration period of half an hour, pour solvent number 1 into the trough of the chamber and let it run till it is about 10 cm from the opposite edge of the paper.
8. Take the paper out, air-dry it, turn it at 90°C angle, and now, develop the paper in the second chromatographic chamber using solvent system No. 2 (phenol/ammonia/water).
9. Remove it when the solvent has traveled up to about 10 cm from the opposite end.
10. Dry it at room temperature and spray it with ninhydrin reagent. After air-drying it, keep the chromatogram in an oven at 105°C for 10 min. Mark the blue- and/or purple-colored zones that appeared on the paper.

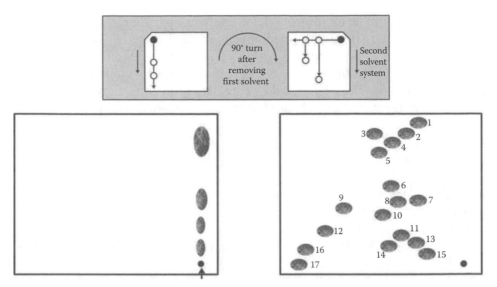

Figure 7.6 2D paper chromatography.

7.7.4 Observations

Calculate the R_f value of the standard amino acids and those in the given mixture as given in both the solvents. From these values identify the amino acids in the given mixture.

7.8 Thin Layer Chromatography

7.8.1 Introduction

This technique is similar to paper chromatography but is more convenient and less time consuming. Here, instead of paper, the supporting material is a glass plate, a plastic sheet, or a piece of metal foil. A thin layer of the stationary phase is laid over this inert support; the layer may be as thin as 250 μm for analytical separations and as thick as 2–5 mm for preparative separations. A binding agent such as calcium sulfate or gypsum may be incorporated into the chromatographic media, to facilitate firm adhesion of adsorbent to the plate. The most widely used adsorbent in TLC is silica gel "G" (Merck) that contains 13% $CaSO_4$. Another commercially available medium is silica gel H (Merck) that does not contain $CaSO_4$. Since silica gel G has a binder, its slurry should not be kept for a long time, and it should be spread immediately after the preparation of its suspension.

7.8.2 Principle

Separation of compounds on a thin layer is similar in many ways to paper chromatography but has the added advantages that a variety of supporting media can be used so that separation can be by adsorption, ion-exchange partition chromatography, or GF depending on the nature of the medium employed. The method is very rapid, and many separations can be completed within an hour. A compound can be detected at a lower concentration than on paper as the spots are very compact. Furthermore, separated compounds can be detected by corrosive sprays and elevated temperature with some thin-layer material, which of course is not possible with paper.

7.8.3 Production of Thin Layer

The R_f value is affected by the thickness of the layer below 200 µm and a depth of 20 µm is suitable for most separations. There are several good spreaders on the market that, when carefully used, can produce an even layer of required thickness. There are now a number of prepared thin-layer plates that are available commercially, and these may be more convenient to use than trying to prepare the plate in the laboratory (Figure 7.7).

7.8.4 Development

It is essential to make sure that the atmosphere of the separation chamber is fully saturated; otherwise, R_f values will vary widely from tank to tank. This can be ensured by using as small a tank as possible and lining the walls with paper soaked in the solvent. The development of the plate is by the ascending technique and is very rapid.

7.8.5 Location

The compounds are located as for paper chromatography by spraying with the appropriate reagent or by scanning in the case of radioactive substances.

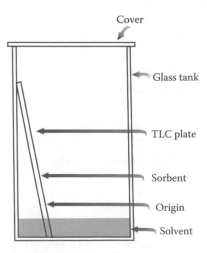

Figure 7.7   Illustration of a TLC tank with ascending chromatography.

TLC has a number of advantages over paper chromatography. Some of these are the following:

1. In paper chromatography, the separation is based on partitioning only, while in TLC, depending upon the nature of the chromatographic media used, the separation can be achieved either by partitioning, adsorption, ion exchange, or molecular sieving phenomenon.
2. This method is extremely rapid as the separation can be achieved within an hour or so, while in paper chromatography, it may take up to about 8–24 h.
3. It is relatively more sensitive since lower concentrations of compounds in the mixture can be successfully separated and detected.
4. Corrosive agents like H_2SO_4 and high temperatures can be used to locate the separated compounds that are not possible in paper chromatography.

Development, location, and identification of components: For development, the plate on which the sample spots have been applied is placed in an airtight glass jar containing the solvent. The location and identification of separated components is carried out in the same manner as in the case of paper chromatography. Also, like paper chromatography, TLC is amenable to a 2D mode of development.

7.9 Thin-Layer Gel Chromatography

Thin-layer gel chromatography has all the advantages of TLC such as quick separation, high sensitivity, simple equipment, and ready elution plus the opportunities afforded by the gels, especially the ability to separate the molecules of high molecular weight (Figure 7.8). Thin-layer gel chromatography is similar to TLC but with the following differences.

In TLC, the support is dried before applying the sample. However, because gels cannot be dried and easily rehydrated, in the thin-layer gel chromatography, the sample is spotted onto a wet layer equilibrated with the appropriate solvent.

TLC is performed with the ascending mobile phase, whereas in the thin-layer gel chromatography; the descending method is used (this is because the capillary moves the eluent in dry material, but gravity is needed for the already wet gel). The plate is

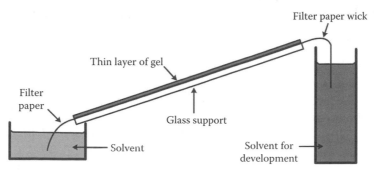

Figure 7.8 Illustrating the setup for thin-layer gel electrophoresis.

put in an airtight chamber, to avoid drying of the gel, and connected to a reservoir at both ends by filter paper bridges. Liquid flows through the layer at a rate determined by the angle (usually 20°). As in TLC, the run terminates before the material of interest enters the lower reservoir. However, there is a continuous flow of liquid so that, unlike TLC, there is no solvent front. Therefore, there is no measurement of R_f values and positions must be measured with respect to added standards.

7.10 Separation and Identification of Sugars by Adsorption Thin Layer Chromatography

7.10.1 Principle

Sugars get separated on the basis of differential adsorption onto silica gel. The sugars, which have higher affinity for the stationary phase, are adsorbed more strongly, and hence, they migrate slowly when the mobile phase moves over them. On the other hand, those having lower affinity for the stationary phase are weekly adsorbed and are more easily carried by the mobile phase. The separated sugars are then located as colored zones by spraying TLC plates with aniline–diphenylamine phosphate reagent.

7.10.2 Materials

TLC chromatographic tank, glass plates (20×20 cm), spreader, micropipettes/microsyringe, oven maintained at 105°C, hair drier, sprayer or atomizer, aniline–diphenylamine phosphate, acetone, standard sugar solution—prepare 1% solution of standard sugars such as glucose, ribose, fructose, and sucrose in 10% isopropanol. Solvent system—ethyl acetate/isopropanol/water/pyridine in the ratio of 26:14:7:2.

7.10.3 Protocol

1. Place thoroughly cleaned and dried glass plates (20×20 cm) on a flat plastic tray side by side with no gap between the two adjacent plates. The chromatographic tank should be airtight, and chromatography should be performed under temperature-controlled conditions.
2. Prepare a slurry of the stationary phase (silica gel G) free of clumps in water or in an appropriate buffer.

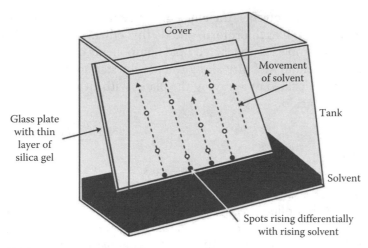

Figure 7.9 TLC.

3. Spread a uniform layer of 250 μm thickness with the help of a spreader or an applicator by moving it from one end of the tray to its other end. Ready-made precoated thick silica gel–coated aluminum or plastic sheets are also available.
4. Activate the plates by keeping them at 105°C for 30 min. Allow the plates to cool in desiccators before use.
5. Gently put marks in a straight line with the help of a pin at a distance about 2 cm from one edge of the plate. The adjacent marks should be taken that silica does not get scratched off while putting these marks.
6. Carefully apply the solution of individual standard sugars and the mixture or alcoholic extract of the sample on the separate marked spots.
7. Gently put marks in a straight line.
8. Develop the TLC plates in the solvent (ethyl acetate/isopropanol/water/pyridine in the ratio of 26:14:7:2 [Figure 7.9]).
9. Take out the plates after the solvent reaches 2 cm from the bottom.
10. Dry them in the air for at least 12 h.

7.10.4 Observations

Identify the sugar species by spraying with aniline–diphenylamine phosphate reagent (mix 5 volumes of 1% aniline and 5 volumes of diphenylamine in acetone with 1 volume of 85% o-phosphoric acid), and calculate the R_f values of each spot.

7.11 Identification of Lipids in a Given Sample by Thin Layer Chromatography

7.11.1 Principle

In biological materials, lipids are found as lipoprotein complexes and these have to be extracted. Lipids being soluble in nonpolar organic solvents and proteins being soluble in polar aqueous solvents, the efficient lipid extraction can be achieved only with an aqueous solvent like chloroform and diethyl ether. This would help in breaking the lipoprotein complexes. Extracted lipid components can be separated on TLC based on their differential mobility along the porous stationary phase such as silica gel, and these can be located by spraying the plates with either 2′,7′-dichlorofluorescein or 50% sulfuric acid.

7.11.2 Materials

TLC tank; glass plates (20 × 20 cm) for TLC; 2′,7′-dichlorofluorescein (prepare 0.2% solution of 2′,7′3-dichlorofluorescein in 95% v/v ethanol) or 50% sulfuric acid as locating reagent; glass plates; spreader; silica gel G; oven set at 110°C; UV lamp; solvent system—petroleum ether (b.p. 60°C–70°C) or hexane/diethyl ether/glacial acetic acid (80:20:1, v/v); lipid standards, various lipids such as cholesterol acetate, vitamin A palmitate, triacylglycerol (e.g., trioleate, tripalmitate, tristearate).

7.11.3 Protocol

1. *Extraction of lipids from sample*: Grind 1 g of the tissue in the extraction solvent (either diethyl ether/ethanol, 3:1, or chloroform/methanol, 2:1) in a pestle and mortar. Transfer the homogenate to a separating funnel. Shake the contents vigorously, and allow it to stand till the two phases have completely separated. Drain out the lower organic layer that contains the lipids. Evaporate the solvent under vacuum, and keep the concentrated lipid extract protected from light under N_2 atmosphere.
2. Prepare the TLC plates using silica gel G as described in the experiment for sugars.
3. Activate the TLC plates at 110°C for 30 min, cool them in a desiccator, and spot the lipid samples, standards, as well as unknown.
4. Develop the plates in the solvent system consisting of petroleum ether (b.p. 60°C–70°C) or hexane/diethyl ether/glacial acetic acid (80:20:1) till the solvent has traveled up to 1 cm from the opposite side of the plate.
5. Remove the plate and allow it to air-dry.
6. Locate the lipid spots by either of the following methods:
   a. Spray the plate with 2′,7′-dichlorofluorescein and examine it under UV light. Lipids show up as green fluorescent regions against the dark background.
   b. Spray the plate carefully with 50% H_2SO_4 and heat it in an oven at 110°C for 10 min. Areas containing lipids get charred and appear as black spots.

7.11.4 Observation and Calculations

Calculate the R_f value of the lipid components in the sample and identify them by comparing their R_f values with lipid standards.

7.12 Identification of Sugars in Fruit Juices Using Thin Layer Chromatography

7.12.1 Materials

Thin-layer plates of silica gel G (prepare a slurry of silica gel G in 0.02 mol/L sodium acetate and pour onto the plate 250 μm thick. Activate before use by heating at 105°C for 30 min); TLC developing chambers; solvent (ethyl acetate/isopropanol/water/ pyridine, 26:14:7:2 v/v); fruit juices from fresh fruit (lemon, orange, grapefruit, pine-apple); absolute alcohol; oven at 105°C; hair drier; spray guns; standard sugar solutions (10 g/L in 10% v/v isopropanol) of glucose, fructose, xylose, ribose, lactose, galactose, and rhamnose; aniline–diphenylamine location reagent (prepare fresh by mixing 5 volumes of 10 g/L aniline and 5 volumes of 10 g/L diphenylamine in acetone with 1 volume of 85% phosphoric acid. Caution: Toxic!).

7.12.2 Protocol

Add 3 mL of ethanol to 1 mL of fruit juice and centrifuge to remove denatured pro-tein. Carefully spot the supernatant onto a thin-layer plate together with some stan-dard sugar solutions. Place the plate in a chamber saturated with solvent, and develop the chromatogram until the solvent front is close to the top of the plate. Draw a line across the plate at this point and remove the chromatogram when the solvent reaches the mark. Dry the plate in a stream of cold air, and locate sugars by spraying the plates with freshly prepared aniline–diphenylamine, in a fume chamber, and heating briefly at 100°C. Note the color of each sugar, and use this and the R_g value to identify the sugars present in the fruit juices.

7.13 Column Chromatography

Paper chromatography and TLC use a flat bed or surface for the separation of mol-ecules. In column chromatography, separation is achieved by the passage of the sam-ple through a vertically fixed tubular glass or polypropylene column that is packed with an appropriate chromatography media. Usually commercially available columns have a porous sintered plate fused at their base, which prevents the stationary phase from flowing out of the column. This sintered base is positioned *as* near the base as possible in order to minimize the dead space to reduce the chances of post-column mixing of the separated compounds. Alternatively, a simple glass burette with a plug of glass wool at the base can be used as a column. At the base, there is a small capil-lary tubing through which the effluent from the column flows into the test tubes in which fractions are collected. At the top of the column, a solvent reservoir with a delivery system is fitted (Figure 7.10).

The stationary phases used in column chromatography are water-insoluble, porous, solid particles, and the resolution of sample components occurs depending upon the principle underlying the separation phenomenon. The flow rate and resolution

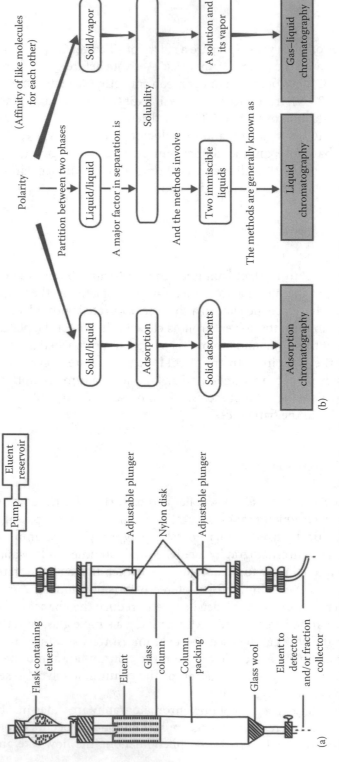

Figure 7.10 (a) Types of columns and (b) procedure for column chromatography with mobile and stationary phases.

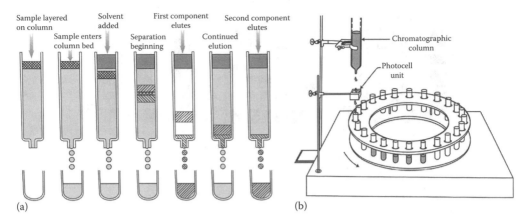

Figure 7.11 Fraction collection by (a) column chromatography manually and (b) by fraction collector.

characteristics are influenced by the size and shape of the stationary phase. Large and coarse particles have higher flow rate but give comparatively poor resolution. Finer particles with large surface to volume ratio have slower flow rate but greater resolution efficiency. Generally, particles of 100–200 mesh are used for routine analysis, but for high resolution, the smaller particles of 200–400 mesh are preferred.

For fractionation and separation of components, the sample is loaded on top of the column and eluted with an appropriate buffer. The effluent emerging from the base of the column is collected in the form of fractions of fixed volume or fixed time in individual test tubes either using an automatic fraction collector (Figure 7.11) or manually. The collected fractions are then analyzed for the presence of the desired substance. The detection technique depends on the physical, chemical, or biological property of the compound. The presence of colored compounds can be identified simply from visual observation, but for colorless compounds, alternative methods of detection are employed. They can be either the color reactions or may be based on its unique physical property such as UV absorption, fluorescence, and refractive index or biological activity such as enzyme activity. Column chromatographic techniques have been classified on the basis of the nature of the interactions occurring between solutes and the stationary phase that ultimately results in their separation. Various types of column chromatography are described in the succeeding text.

7.14 Gradient Preparation

A gradient maker consists either of changing the ratio of two solvents or of increasing the concentration of one or more of the components in the solvent (e.g., the salt concentration). The latter is the most common way of eluting adsorption and ion-exchange columns. Gradients are prepared by means of an apparatus of the type shown in Figure 5.5. It consists of two vessels, a reservoir and a mixing chamber, which are connected at their bases. The reservoir contains the more concentrated solvent.

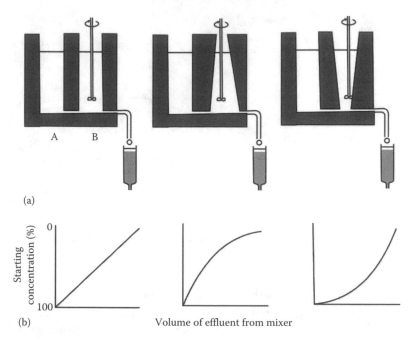

Figure 7.12   (a) Three types of gradient-making apparatuses and (b) the resulting gradients. *Note:* A, added solvent in reservoir; B, starting concentration in mixing chamber.

The liquid leaves the mixing chamber and enters the column. Because the hydrostatic heads (not necessarily the heights because the densities of the liquid in the two vessels may differ) must be equal, the liquid simultaneously flows from the reservoir to the mixing vessel. If the chambers have the same shape, the gradient is linear; concave and convex gradients can also be prepared (Figure 7.12).

7.15 Adsorption Chromatography

Adsorption is a phenomenon in which compounds are held onto the surface of a solid adsorbent, having specific adsorption sites, through weak nonionic interactions such as van der Waals forces and hydrogen bonding. Different compounds bind with varying strengths and hence can selectively be desorbed. For good resolution, selection of the right type of adsorbents includes charcoal, silica, alumina, and hydroxyapatite. Eluent influences the quality of separation since polarity of the mobile phase influences the adsorption considerably. Nonpolar solvents favor maximum adsorption decreases with an increase in polarity of the solvent. In general, the polar solvents are preferred for the substances having polar or hydrophilic groups and nonpolar solvents for substances having hydrophobic or nonpolar groups, for example, alcoholic solvents for OH group containing substances, acetone or ether for substances with carbonyl groups, and hydrocarbons such as toluene or hexane for nonpolar substances. For gradient elution, a mixture of the polar and nonpolar solvents of different ratios can be used to obtain eluent of varying polarities. The mass of solute adsorbed per unit weight of adsorbent (*m*) depends on the concentration of the solute (*c*), and Langmuir derived an equation on the basis that (a) only a monolayer is adsorbed and (b) only a proportion of the molecules in collision will result in adsorption. This is known as the Langmuir adsorption isotherm:

$$m = \frac{K_1 K_2 c}{1 + K_2 c}$$

K_1 is a measure of the number of active adsorption sites per unit weight of adsorbent and depends on the nature of the adsorbent. K_2 is a measure of the affinity of the solute for the adsorbent and is affected by all the components of the system. Langmuir assumed only one binding site, but in practice, there are a number of different sites on the surface of the adsorbent, each with a different affinity, thus giving a series of Langmuir-type isotherms. Hinshelwood, therefore, suggested that the equation gives a more accurate picture of the situation:

$$m = \sum \frac{K_1 K_2 c}{1 + K_2 c}$$

And this, indeed, approximates to the Freundlich adsorption isotherms found in practice:

$$M = Kc^x$$

where *K* and *x* are constants depending on the particular system used.

The difference between these two isotherms is illustrated in Figure 7.13. This mixture of binding sites of differing affinities is the cause of "tailing" observed in the elution profile of compounds (Figure 7.13a and b). This tailing can be overcome by eluting with a suitable gradient of pH, ionic strength, or polarity so that the more strongly adsorbed molecules meet a higher concentration of displacing compound than the more weakly bound molecules.

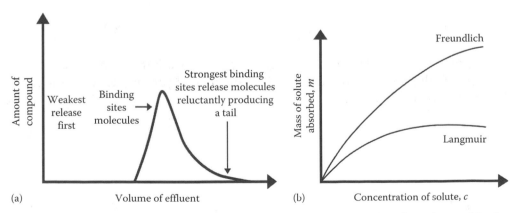

Figure 7.13 (a) Freundlich and Langmuir adsorption isotherms. (b) Tailing observed in the elution of a compound from an adsorption column.

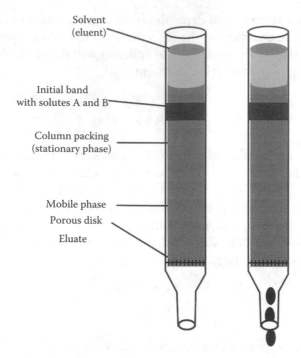

Figure 7.14   Adsorption chromatography.

7.16 Separation of Pigments from Leaves or Flowers by Adsorption Column Chromatography

7.16.1 Principle

Different pigments get adsorbed to alumina to different extents. They can be selectively desorbed by using mobile phases of increasing polarity in a stepwise manner (Figure 7.14).

7.16.2 Materials

Leaves or flowers, pestle and mortar, glass column or a burette, Whatman No. 1 filter paper, alumina or icing sugar, sodium sulfate (anhydrous), acetone. Solvent—benzene/methanol (2:1 v/v).

7.16.3 Protocol

7.16.3.1 Sample Preparation
Homogenize the leaves or flowers (5 g wet weight) in a pestle and mortar, using sand as an abrasive in 20 mL of benzene/methanol (2: 1), adding a small amount of this extractant at a time. Filter the extract through Whatman No.1 filter paper and

transfer the filtrate to a separating funnel. Add 10 mL of water to the filtrate, and after shaking the contents and allowing the phases to separate out, drain out the lower aqueous methanol layer. Repeat this step, avoiding very vigorous shaking. Collect the benzene layer in a beaker, and add a small amount of solid anhydrous Na_2SO_4 to remove the traces of moisture. Decant the clear benzene layer to another beaker and concentrate the extract by evaporating the solvent on a boiling water bath.

7.16.3.2 Column Preparation

Mount a burette or a glass column vertically on a burette stand with the help of clamps. Lightly place a plug of glass wool at the base of the burette and close the stopcock or outlet at the bottom of the column. Take 5 g of alumina or icing sugar (adsorbent) previously dried at 120°C for 8 h and prepare its slurry in benzene. Pour the slurry carefully into the column or burette, by gentle tapping of the column or burette, so that no air bubbles get trapped in the adsorbent. Allow the adsorbent to settle by opening the outlet. After the adsorbent has completely settled, add 20 mL of benzene and let it pass out of the column. Care should be taken not to let the adsorbent get dried.

7.16.3.3 Sample Application

Allow the solvent at the surface of the column to drain out slowly, and transfer the leaf or flower extract with the help of a pipette without disturbing the surface of the column adsorbent. Let it enter into the column, and then add a few drops of benzene to wash the traces of the extract sticking to the wall of the column. Add 20 mL of benzene to wash out the column of any unabsorbed material.

7.16.3.4 Column Development

For desorption of the adsorbed substances, change the polarity of the solvent in a step-wise manner. After 20 mL of benzene has passed through the column, add 10 mL of 5% acetone (v/v) in benzene and let it percolate through the column and collect 1–2 mL fractions of the effluent from the outlet. Continue increasing the concentration of acetone in benzene at even succeeding steps. Finally, pass pure acetone through the column.

7.16.3.5 Result

Note the change in the color of the collected fractions. In case of the leaf extract, the initial fractions are colorless followed by yellow-colored and then by the green-colored ones. The colorless fractions do not contain any pigments, but it is quite possible that these fractions may contain some UV-absorbing materials.

7.17 Ion-Exchange Chromatography

7.17.1 Introduction

Ion-exchange chromatography is a type of adsorption chromatography in which retention of a solute occurs due to its reversible electrostatic interaction with the oppositely charged groups on an ion exchanger. This technique is useful for the separation of compounds that bear a net electric charge such as proteins, amino acids, and nucleic acids. Ion exchangers are prepared from either certain synthetic resins, which are insoluble porous organic molecules, or naturally occurring biopolymers such as cellulose to which various groups known as fixed ions are covalently attached. These fixed ions are balanced by equal and oppositely charged ions from the solution referred to as counterions. Depending upon the nature of the counterions, these ion exchangers are of two types, cation exchangers in which the counterions are cationic or positively charged ions and anion exchangers that have negatively charged counterions. Counterions are mobile and can be easily exchanged by other similarly charged molecules in the sample. The nature of the resin matrix is unchanged during this exchange process. Generally, resin-based ion exchangers are used for the separation of low molecular weight biomolecules, and cellulosic ion exchangers are more suitable for isolation of macromolecules such as proteins and nucleic acids.

7.17.2 Theory

7.17.2.1 Matrix

Ion-exchange materials were earlier based on synthetic resins of an aromatic nature and were suitable for the separation of inorganic ions and small molecules. However, they could not be used for the separation of large molecules such as proteins that cannot penetrate the closely linked structure of the resin and tend to be denatured by the hydrophobic matrix. The first ion exchangers suitable for macromolecules were based on cellulose, but they had low capacities as too much substitution made the cellulose soluble. Since then, other materials have been developed based on dextran and an acrylamide type of polymer that have a much higher capacity.

7.17.2.2 Ionizable Groups

Charged groups are attached to the matrix, and the type of group defines the nature and strength of the ion exchanger. These groups may be either anionic or cationic, according to the nature of their affinity for either negative or positive ions. For example, the cation-exchange materials exchange positive ions, so it is the charge carried by the exchangeable ion that decides whether a material is anionic or cationic and not the charge carried on the matrix (Figure 7.15).

These two types can be further divided into materials that contain strongly ionized groups, such as $-SO_3H$ and $-NR_3$, and the weakly ionized groups, such as $-COOH$, $-OH$, and $-NH_2$. The strong ion-exchange resins are completely ionized and exist in the charged form except at extreme pH values:

$$-SO_3H \rightleftharpoons -SO_3^- + H^+$$
$$-NR_3OH \rightleftharpoons -\overset{+}{N}R_3 + OH^-$$

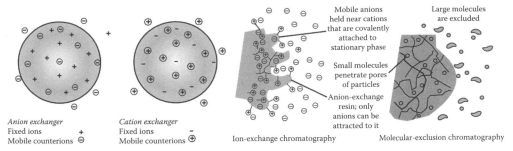

Figure 7.15 Anion- and cation-exchange matrix used in separation of mixtures.

The weak ion-exchange materials, on the other hand, contain groups whose ionization is dependent on the pH, and they can only be used at maximum capacity over a narrow pH range:

$$-COOH \rightleftharpoons -COO^- + H^+$$
$$-NH_3^+ \rightleftharpoons -NH_2 + H^+$$

Normally, resins containing carboxyl groups have a maximum capacity above pH 6, while those with amino groups are effective below pH 6. The number of ionizable groups determines the ion-exchange capacity. The total capacity is the number of ionizable groups per gram of material, whereas the available capacity is the amount of a given molecule that can bind under defined experimental conditions. In the case of some materials, large molecules may be unable to penetrate the matrix and can only react with charged groups on the surface. In this case, the available capacity will be considerably less than the total capacity (Figure 7.16).

7.17.2.3 Ion-Exchange Equilibria

The typical way that an ion-exchange material functions is illustrated in the following example, where an anion-exchange resin containing amino groups is used to separate two negatively charged ions X^- and Y^- (Figure 7.17).

Ion-exchange materials are claimed to be monofunctional, and only the ion-exchange process is used in separation. However, in practice, some molecular sieving and adsorption can occur. The adsorption is small, but it can sometimes be used to separate closely related compounds.

7.17.2.4 Elution of Bound Ions

The bound ions can be removed by changing the pH of the buffer. For example, as the pH of a protein moves toward its isoelectric point, the net charge decreases and the macromolecule is no longer bound. Separation is achieved as other charged proteins remain on the column. Alternatively, ions can be removed by increasing the ionic strength, when high concentrations of ions in the solvent displace the bound ions by increasing competition for the charged groups of the ion-exchange

(a) Cation exchange media	Structure
Strongly acidic, polystyrene resin (Dowex–50)	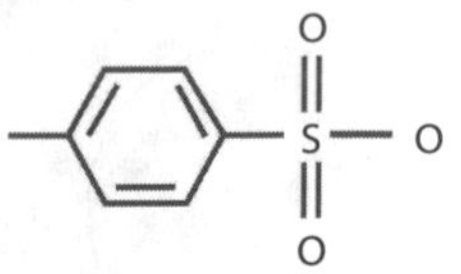
Weakly acidic, carboxymethyl (CM) cellulose	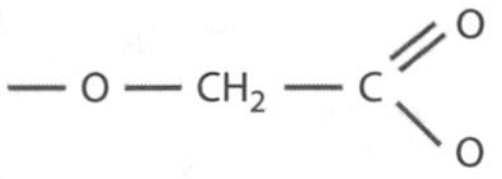
Weakly acidic, chelating, polystyrene resin (Chelex–100)	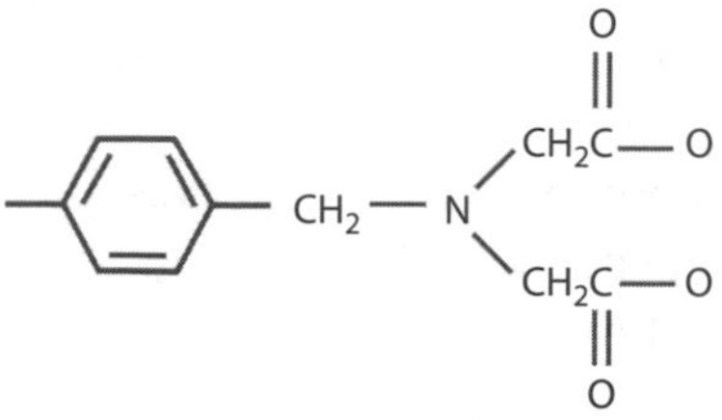

(b) Anion exchange media	Structure
Strongly basic, polystyrene resin (Dowex–1)	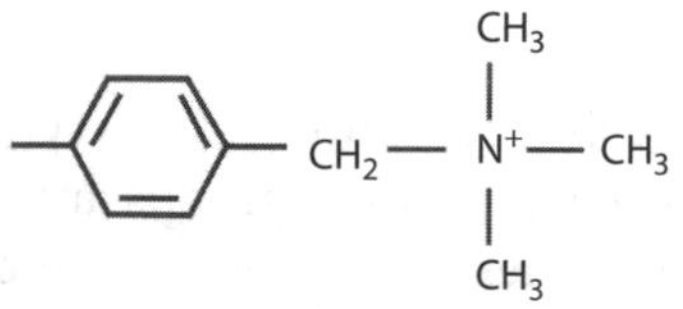
Weakly basic, diethylaminoethyl (DEAE) cellulose	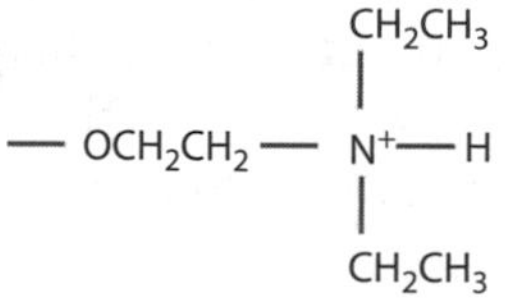

Figure 7.16 (a) Cation and (b) anion matrix and their structure.

material. The pH or ionic strength can be altered sharply, by changing the eluting buffer or gradually by means of a gradient. It is important to realize that it means that the binding of ions is not an all-or-none phenomenon but involves equilibrium between those ions firmly bound to the material and those free in solution. The extent of the binding depends on the nature of the resin and the temperature, ionic strength, and composition of the solvent. In the case of weakly ionized resins, the uptake of ions also depends on the degree of ionization of the resin and materials to be separated.

7.17.2.5 Preparation of Material
Ion-exchange materials are first allowed to swell in the buffer and the fines removed as described earlier. The ion-exchange material is then obtained in the required

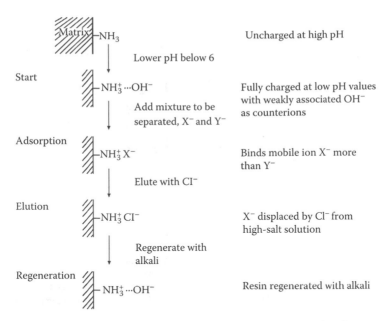

Figure 7.17 Schematic representation of different stages of anion-exchange chromatography.

ionic form by washing with the appropriate solution. For example, the H⁺ form of a cation-exchange resin is obtained by washing the material with hydrochloric acid then water until the washings are neutral. Similarly, the Na⁺ form is prepared by washing the resin with sodium chloride or sodium hydroxide then water. The final stage before preparing the column is to equilibrate the material by stirring with the eluting buffer.

7.18 Separation of Amino Acids by Ion-Exchange Chromatography

7.18.1 Materials

Chromatography column (20 cm × 1.5 cm); strongly acidic resin; hydrochloric acid (4 mol/L); hydrochloric acid (0.1 mol/L); glass wool; amino acid mixture (dissolve aspartic acid, histidine, and lysine in 10 mg mol/L HCl to a final concentration of 2 mg/mL of each); Tris–HCl buffer (0.2 mol/L, pH 8.5); sodium hydroxide (0.1 mol/L); separating funnels (500 mL); acetate buffer (4 mol/L, pH 5.5); ninhydrin reagent (dissolve 20 g of ninhydrin and 3 g of hydrindantin in 750 mL of methyl cellulose and add 250 mL of acetate buffer. Prepare fresh and store in a brown bottle); methyl cellulose (ethylene glycol monomethyl ether); ethanol (50% v/v); ninhydrin (200 mg/100 mL in acetone) (Caution: Carcinogenic!); oven 105°C.

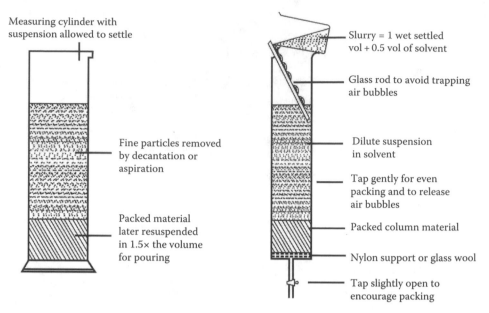

Figure 7.18 Illustrating preparation of column.

7.18.2 Protocol

7.18.2.1 Preparation of the Column

Gently stir the resin with 4 mol/L HCl until fully swollen (15–30 mL/g dry resin). Allow the resin to settle, then decant the acid. Repeat the washing with 0.1 mol/L HCl, resuspend in this solution, and prepare the column (Figure 7.18).

7.18.2.2 Elution of Amino Acid

Carefully apply 0.2 mL of the amino acid mixture to the top of the column, open the tap, and allow the sample to flow into the resin. Add 0.2 mL of 0.1 mol/L HCl, allow to flow into the column as before, and repeat the process twice. Finally, apply 2 mL of 0.1 mol/L HCl to the top of the resin, and connect the reservoir to give a flow rate of about 1 mL/min and collect a total of forty 2 mL fractions. Test five of the tubes at a time for the presence of amino acids by spotting a sample from each tube on a filter paper: dip this in the acetone solution of ninhydrin and heat in an oven at 105°C. If amino acids are present, they will show up as blue spots on the filter paper. When the first amino acid has been eluted, remove the reservoir of 0.1 mol/L HCl and allow the level of acid to fall to just above the resin. Run 2 mL of 0.2 mol/L Tris–HCl buffer (pH 8.5) on to the column, then connect to a reservoir of this buffer, and continue with the elution until the third amino acid is removed from the column (Figure 7.19).

7.18.2.3 Detection of Amino Acids

Adjust the pH of each tube to 5 by the addition of a few drops of acid or alkali. Add 2 mL of the buffered ninhydrin reagent and heat in a boiling water bath for 15 min.

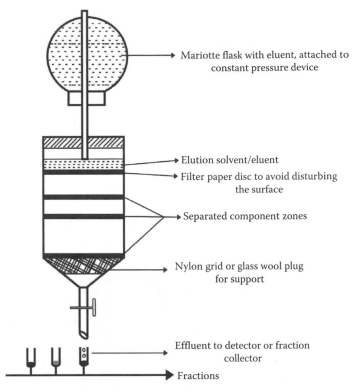

Mariotte flask with eluent, attached to
constant pressure device

Elution solvent/eluent

Filter paper disc to avoid disturbing
the surface

Separated component zones

Nylon grid or glass wool plug
for support

Effluent to detector or fraction
collector

Fractions

Figure 7.19   Ion-exchange separation of amino acid mixture.

Cool the tubes to room temperature, add 3 mL of 50% v/v ethanol, and read the
extinction at 570 nm after allowing the tubes to stand for 10 min.

7.19 Separation of Proteins from Human Serum by Ion-Exchange Chromatography

7.19.1 Principle

The binding and release of compounds to ion-exchange materials is dependent on
pH and salt concentration. At pH 7, most of the serum proteins are bound to diethylaminoethyl (DEAE)-cellulose apart from the γ-globulin that appears in the first
fraction. The bound protein can then be eluted in three more fractions, by stepwise
increase in the ionic strength of the eluting medium. Different compounds are bound
with different strengths. The strength of binding depends upon the degree of charge
and the charge density (amount of charge per unit volume of the molecule) of the
solute. The other solute particles (positively charged and neutral) have no affinity for
the stationary phase and are washed down along with the starting buffer. The bound
solute molecules are then released in succession by altering pH or ionic strength of
the elution buffer. This type of elution is known as gradient elution.

7.19.2 Materials

Ion-exchange column of DEAE-cellulose (3 cm × 1 cm); eluting buffer/sodium phosphate (0.01 mol/L; pH 7.0); eluting buffer containing 0.1 mol/L NaCl 50 mL; eluting buffer containing 0.2 mol/L NaCl; eluting buffer containing 0.3 mol/L NaCl 50 mL; human serum; UV spectrophotometers.

7.19.3 Protocol

Wash the ion-exchange column thoroughly with the elution buffer then carefully pipette 0.5 mL of serum on the top column. Allow the serum to pass into the DEAE-cellulose then elute slowly with 3 mL of the eluting buffer. Collect the eluate in a test tube, repeat the elution with the buffer containing increasing concentrations of salt, and record data:

Fraction	NaCl (mol/L)	Volume (mL)
1	0	3
2	0.1	3
3	0.2	3
4	0.3	3

7.19.3.1 Recovery of Protein
Calculate the recovery of protein by determining the protein content of the serum and the four fractions using the extinction at 280 nm.

7.19.3.2 Characterization of Protein
Identify the protein in the serum and each of the fractions using cellulose acetate electrophoresis. From these results, explain the relative ease of binding of the protein to the DEAE-cellulose.

7.20 Gel Filtration/Size-Exclusion Column Chromatography

In size-exclusion chromatography (SEC), the matrix consists of porous particles and separation is instead achieved according to the size and shape of the molecules. The technique is sometimes also referred to as GF, molecular sieve chromatography, or gel permeation chromatography.

For separation in GF, molecular sieve properties of a variety of porous materials are utilized. SEC matrices consist of a range of beads with slightly different pore sizes. The separation process depends on the different abilities of various proteins to enter all, some, or none of the channels in the porous beads. Molecules running through SEC columns have to solve a maze that becomes more complex the smaller the molecule is, as the small molecules have more potential channels that they can access. Larger molecules, on the other hand, are for steric reasons excluded from the

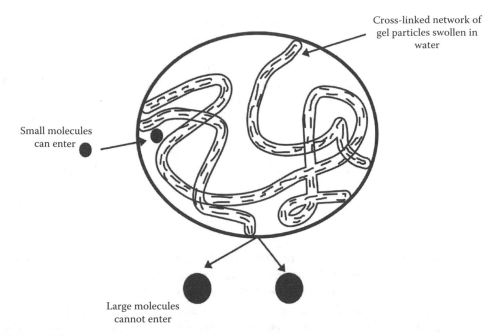

Figure 7.20 Illustration of the GF principle.

channels and pass quickly between the beads. The route through the channels will thus retard smaller molecules in comparison to larger proteins. The matrices used in SEC are often composed of natural polymers such as agarose or dextrans but may also be composed of synthetic polymers such as polyacrylamide. Gels may be formed from these polymers by carefully cross-linking to form a 3D network. Diffcrent pore sizes can be obtained by slightly differing amounts of cross-linking. The degree of cross-linking will define the pore size. The first commercial SEC media, Sephadex, is composed of dextran that was cross-linked with epichlorohydrin (Figure 7.20). Many gels are now commercially available in a broad range of porosity. The surface of these supports contains predominantly hydroxyl groups and provides a good environment for hydrophilic proteins. However, the hydrophilic nature is somewhat reduced by the introduction of cross-linking reagents. Some polymers, like agarose, are able to, spontaneously, form gels under the appropriate conditions. Macroporous silica has also been employed in SEC but must then be coated, with a hydrophilic layer, to prevent denaturation of proteins.

The elution volumes of globular proteins are determined largely by their relative molecular size. Hence, the construction of a calibration curve, with proteins of a similar shape and known molecular weight, enables the molecular size values of other proteins to be estimated, even in crude preparations. In this way, SEC also provides a means of determining differences in the shape of native or denatured globular proteins under a wide variety of conditions. In analytical SEC, the resolution is of utmost interest, and the correct choice of gel and operating conditions is critical if good results are to be obtained. The general expression for the appearance of a solute in an effluent is

$$V = V_0 + K_D V_1$$

where

   V is the elution volume of a substance with a given K_D
   V_0 is the void volume of the total volume of the external water (outside the gel grain)
   V_1 is the internal water volume in the gel grain
   K_D is the distribution coefficient for a solute between the water in the gel grain and
      the surrounding water

A substance with K_D of zero is completely excluded from the gel beads, and substances with K_D volumes between 0 and 1 are partially excluded. If a sample containing a solute with a $K_D = 1$ and another with $K_D = 0$ is introduced in the column, the latter will appear in the effluent after a volume V_0, and the former will appear after a volume $V_0 + K_D V_1$.

The procedure of dialysis can be readily carried out on a suitable Sephadex column. The column is first equilibrated with the new buffer. The protein solution is introduced to the top of the column and eluted with the new buffer. When the volume V_0 has passed, the protein is eluted in the new buffer medium, while the original buffer and small molecular weight compounds, etc., are eluted after a volume of $V_0 + K_D V_1$. The process is very rapid and hence is especially useful when working with labile proteins. Since K_D varies with proteins of different molecular weights, it is possible to fractionate proteins on GF. The biochemist has a choice of several types of Sephadex beads to prepare columns for this procedure. Thus, Sephadex G-25 excludes compounds of molecular weight of 3,500–4,500; Sephadex G-50, 8,000–10,000; and Sephadex G-75, 40,000–50,000. Sephadex G-100 and G-200 gels may be used for higher molecular weight proteins (Figures 7.21 and 7.22).

The ease of diffusion is dependent on the hydrodynamic volume, which is the volume created by the movement of the molecule in water. The difference between hydrodynamic volume and molecular weight is the shape. Proteins tend to be globular molecules, while DNA or polysaccharides tend to be linear molecules. Linear molecules have

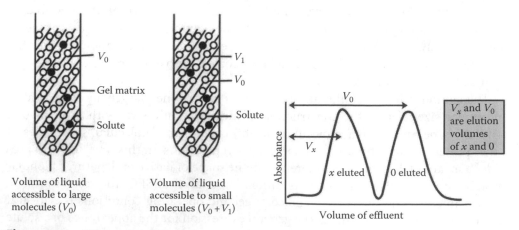

Figure 7.21   GF process.

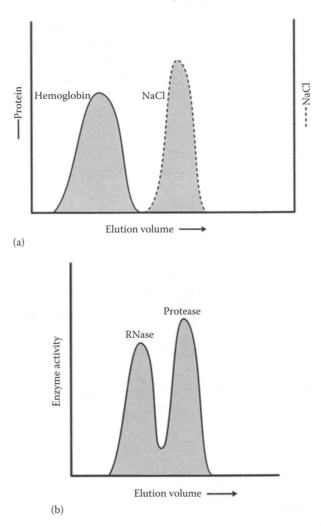

Figure 7.22   Elution pattern of (a) a dialysis on Sephadex G-25 to separate hemoglobin from salt and (b) to show separation of RNase from a protease in pancreatic extract employing a Sephadex G-75 column.

many large hydrodynamic volumes than globular molecules, so a 10,000 M wt. DNA molecule will elute much earlier than a 10,000 M wt. protein. Therefore, focus should be on selectivity rather than efficiency. To obtain maximum resolution, the starting zone must be narrow relative to the length of the column. The resolution of two separated zones in SEC increases as the square root of column length. In general, the diameter of the column is decided by the sample volume and the length by the resolution required.

7.20.1 Molecular Weight Determination

An equally useful application of column GF is its use as a method to determine molecular weights, even if the protein has not been extensively purified. Depending on the possible molecular weight of protein, a suitable gel is chosen. Usually, Sephadex G-100 or G-200 is selected, a column is carefully prepared, and the elution volume of pure proteins with known molecular weights and stabilities is determined to estab-lish a calibration curve. The protein whose molecular weight is to be determined is placed on the same column, and its elution volume is determined under conditions identical to those used to elute the known proteins. The results are plotted, K_D versus log mol. wt., as depicted in Figures 7.23 and 7.24.

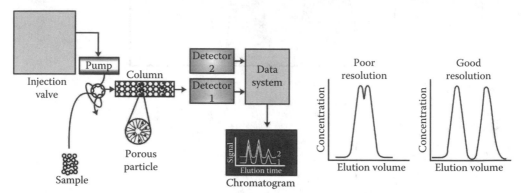

Figure 7.23   Illustration of instrumentation in SEC.

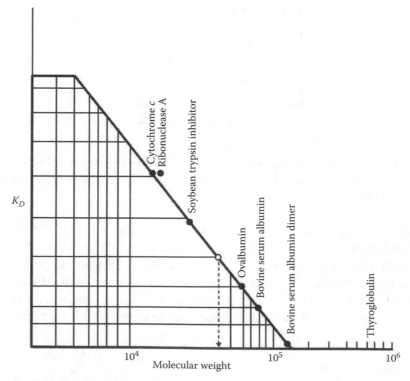

Figure 7.24   Relationship between K_D and log of the molecular weight of proteins as deter-mined by GF on Sephadex G-150.

7.21 Separation of Blue Dextran and Cobalt Chloride on Sephadex G-25

7.21.1 Principle

This experiment explains the separation of molecules on the basis of differences in their size. Blue dextran is a polymer with a molecular weight of several million and is, therefore, excluded from the gel ($K_d = 0$). Cobalt chloride, on the other hand, with a low molecular weight, is freely accessible to the gel particles ($K_d = 1$). The compounds are both colored, so the progress of the filtration can be followed by observing the separation of the colored bands. The completed fractionation is then analyzed by measuring the extinction of each fraction at 625 and 510 nm λmax, respectively: blue dextran = 625 nm and $CoCl_2 \cdot 6H_2O = 510$ nm).

7.21.2 Materials

Sephadex G-25, chromatographic glass column (25 cm × 1 cm), sodium chloride (0.154 mol/L), blue dextran in saline, cobalt chloride in saline, colorimeters with filters or UV–Vis spectrophotometer.

7.21.3 Protocol

Suspend 5 g of Sephadex G-25 in the sodium chloride solution and leave to swell for 3–6 h. During this time, stir the solution and remove any "fines" by decantation. Prepare a column of the gel (20 cm × 1 cm) by pouring the gel suspension into the column and allowing it to settle under gravity while maintaining a slow flow rate through the column. Use a large volume of solvent relative to the amount of the gel to avoid trapping air bubbles in the column.

Add 0.5 mL of the mixture of blue dextran and cobalt chloride to the top of the column and elute with the isotonic saline. Collect 2 mL fractions, until all the cobalt chloride has been eluted. Measure the extinction of each fraction at 510 and 625 nm and plot the elution profile.

7.22 Desalting of Protein Sample by Gel Filtration

7.22.1 Introduction

Separation is based on the fact that proteins are macromolecules, whereas salts are low molecular weight substances. When the sample is passed through a column packed with Sephadex G-10 or G-25, proteins remain totally excluded from the gel and move with the void volume, while salts enter into the gel particles and take a longer time to get eluted.

7.22.2 Materials

Chromatographic glass column (25 cm × 1 cm), Sephadex G-25, bovine serum albumin (BSA), sodium phosphate, 0.1 M Tris–HCl buffer (pH 7.0).

7.22.3 Protocol

7.22.3.1 Column Preparation

Suspend 5 g of Sephadex G-25 in the 0.1 M Tris–HCl buffer (pH 7.0) and leave to swell for 3–6 h at room temperature with intermittent stirring. During this time, stir the solution and remove any "fines" by decantation and obtain a slurry of reasonable thickness. Fix the column upright on a burette stand with the help of a clamp. Prepare the gel (20 cm × 1 cm) by pouring the gel suspension into the column and allowing it to settle under gravity while maintaining a slow flow rate through the column. Keep the outlet of the column closed, place a plug of glass wool at the base of the column, and pour a small volume of the buffer or water into the column. Allow the chromatographic media to settle down evenly and then open the outlet to drain excess liquid from the column. Place a filter paper disc or a nylon gauze on the surface of the packed bed, to prevent disturbance of the upper layer, while loading the sample or feeding the eluent into the column. Use a large volume of solvent relative to the amount of the gel to avoid trapping air bubbles in the column.

7.22.3.2 Running of Column

Prepare a mixture of 10 mg of BSA and 40 mg of sodium phosphate in 2 mL of 0.1 M Tris–HCl buffer (pH 7.0). Apply it to the chromatography column by any one of the following two methods:

1. The mobile phase at the top of the packing is drained out till the bed surface gets exposed. Close the outlet and gently apply the sample uniformly over the bed surface with a pipette and the loaded sample is then just allowed to enter into the column by opening the outlet. A small amount of mobile phase (or buffer) is added to wash the traces of the sample into the column.
2. In the second method, sucrose or glycerol, up to the concentration of 1%, is added in the sample to increase its density. This sample is applied just above the surface of the bed directly through the layer of the buffer in the column bed. Since the sample has higher density, it automatically settles on the surface of the gel. Then, open the outlet to facilitate entry of the sample into the column. When using this procedure, it is advisable to ensure that addition of glycerol or sucrose does not interfere with the separation and subsequent analysis of the separated compounds.

7.22.3.3 Column Separation

Add a sufficient amount of buffer on top of the column and connect it to the buffer reservoir. Collect fractions (2 mL each) either manually or using an automatic fraction collector. Determine the protein content either by monitoring absorbance at 280 nm or by Lowry's method and phosphate ion method in each of the fractions. Plot a graph of concentration of protein and phosphate versus fraction number or elution volume.

7.23 Determination of Molecular Weight of a Given Protein by Gel Filtration

7.23.1 Principle

During GF, solutes are separated primarily on the basis of their molecular size. Due to molecular sieving effect, the large molecules are eluted from the column first followed by compounds of smaller molecular mass. A plot between K_d or elution volume and $\log_{10}$ molecular weight gives a straight line. The molecular weight of a given protein can be established from its elution volume through the GF column that has been previously calibrated with standard marker proteins of known molecular weight.

7.23.2 Materials

Chromatographic glass column (25 cm × 1 cm), Sephadex G-100, $MgCl_2$, dithiothreitol (DTT), glycerol, standard protein markers ((1) β-amylase (200 KD), (2) alcohol dehydrogenase (150 KD), (3) BSA (66 KD), (4) carbonic anhydrase (29 KD), (5) cytochrome (12.4 KD), (6) blue dextran), HEPES–NaOH buffer (20 mM; pH 8.0).

7.23.3 Protocol

Suspend 15 g of Sephadex G-100 in 20 mM HEPES-NaOH (pH 8.0) buffer containing 5 mM $MgCl_2$ and 5 mM DTT for 5 h in a boiling water bath. Allow it to cool and pack it into the glass column. Equilibrate the column by passing buffer equivalent to 2 to 3 volumes of the bed volume. Find out the void volume (V_0) of the column by determining the elution volume of blue dextran solution (2 mg/mL) through the column. Again, pass 2 bed volumes of the starting buffer. Apply the mixture of the standard marker proteins of known molecular weight and elute the column with the buffer. Collect fractions of 2 mL each, and determine the protein content in each of these fractions either by Lowry's method or by recording absorbance at 280 nm.

Determine the elution volume of the standard proteins and prepare a graph of $\log_{10}$ molecular weight V_e or

$$K_d = \frac{V_e - V_0}{V_i}$$

Again, pass 2 volumes of the starting buffer.

7.23.4 Molecular Weight Determination

Calibrate the column by determining the volume at which the blue dextran is eluted (V_0) and the volume when the glucose appears ($V_0 + V_i$). Determine the elution volume (V_e) of the marker proteins and calculate the K_d values. Plot a graph of K_d against $\log_{10}$ mol. wt for each marker and calculate the apparent molecular weight of the unknown sample.

7.24 Concentration of Dilute Protein Solutions Using Sephadex G-25

7.24.1 Principle

When the dry beads of Sephadex G-25 are added to a dilute protein solution, they start swelling and, in the process, absorb water. The proteins being macromolecules are excluded from the swollen gel and hence remain in the solution. Due to the absorption of water, the volume of the solution decreases without affecting the amount of high molecular weight solutes such as proteins in it resulting in the concentration of the solution.

7.24.2 Materials

BSA, Sephadex G-25, Lowry's reagent, UV–Vis spectrophotometer.

7.24.3 Protocol

Dissolve 20 mg of BSA in 100 mL of 1 mM NaCl. Retain 1 mL of this solution for protein estimation by Lowry's method. Add 5 g of dry Sephadex G-25 (coarse) to the remaining protein solution. Let it stand at room temperature for 30 min to swell and then centrifuge at 3000 g for 10 min. Carefully, decant the supernatant into the measuring cylinder and note its volumes. Again retain 1 mL for protein estimation. Subject the supernatant to treat twice or thrice every time, recording the volume of the supernatant and keeping 1 mL aside for protein estimation. Decrease the amount of added Sephadex G-25 progressively (say from 5 g in the first step to 1 g in the final step) at each step. Determine the concentration of protein in 0.5 mL of the supernatant obtained at each step, and express the concentration of protein in terms of the amount of protein/mL of the solution.

7.24.4 Result

In each step, the volume of the protein solution decreases with the increase in concentration of protein (amount of protein/mL of solution).

7.25 Affinity Chromatography

7.25.1 Introduction

Affinity chromatography occupies a unique place in separation technology since it is the only technique that enables purification of almost any biomolecule on the basis of its biological function. Selective separation of a particular group of biopolymers from a complex has undergone a major transformation due to the introduction of affinity chromatography. There occurs a specific interaction between the enzyme and the immobilized ligand. The immobilized ligand is a substrate or competitive inhibitor of the enzyme. Ideally, it should be possible to purify an enzyme from a complex mixture in a single step, and indeed, purification factors of up to several thousandfold have been achieved. These interactions might occur with low molecular weight substances such as substrates or inhibitors. An interacting protein has binding sites with complementary surfaces to its ligand. The binding can involve a combination of electrostatic or hydrophobic interactions as well as short-range molecular interactions such as van der Waals forces and hydrogen bonds. An alternative, equally specific approach is to use an antibody (they are specific in binding because they are raised against the antigen inside an animal) to the enzyme as the ligand. Several requirements must be met for success in affinity chromatography: (1) The matrix should not itself adsorb molecules to any significant extent; (2) the ligand must be coupled without altering its binding properties; (3) a ligand should be chosen whose binding is relatively tight because, although weak binding will enhance retardation, it may not be adequate for separation to result; and (4) it should be possible to elute without destroying the sample. The most useful matrix material is agarose because it exhibits minimal adsorption, maintains good flow properties after coupling, and tolerates the extremes of pH and ionic strength as well as 7 M guanidinium chloride and urea, which are often needed for successful elution.

7.25.2 Principle

Affinity chromatography is a type of adsorption chromatography in which there is a high degree of specificity in the interaction between the adsorbent and the compound to be separated. In the case of enzymes, the ligand attached to the adsorbent is usually a powerful inhibitor with a high affinity constant that will bind only one enzyme in a complex mixture of proteins.

In affinity chromatography, a molecule (usually an antibody) that will bind to the protein to be purified is attached to the glass beads (Figure 7.25). In general, affinity chromatography achieves a higher purification factor than ion-exchange chromatography. A number of synthetic triazine dyes have been explored for protein purification. Dye-ligand chromatography can be the least expensive of those to use, and it is based on the tendency of some dyes (like those used in clothing manufacture) to bind

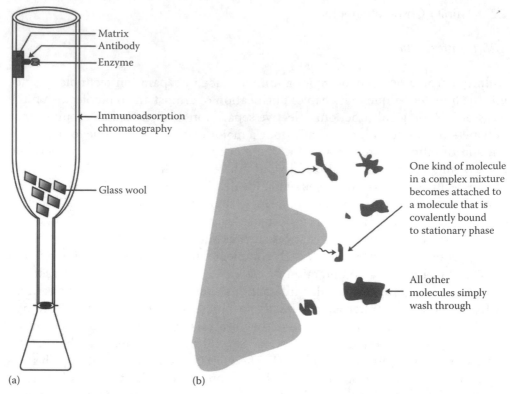

(a)   (b)

Figure 7.25   (a) Affinity chromatography where an antibody is attached to the glass wool (beads) and proteins having matching structure will be captured and this can be eluted later. (b) Mechanism of affinity chromatography where only one molecule is selected specifically and covalently bound to a stationary phase.

fairly specifically and quite solidly to certain types of proteins. Dye-ligand chromatography is often used in the purification of diagnostic enzymes.

7.25.3 Affinity Material

The ligand is covalently attached to a supporting medium so that the chromatographic material can be designed for a specific purification task. The matrix has to be macroporous to allow large molecules access to the binding sites and have good flow properties. It has to be devoid of nonspecific adsorption sites but must contain functional groups to which ligands can be attached. There are a number of commercially available materials that fit these criteria including cross-linked Sepharose (CL-Sepharose). Generally, it is observed that the active site of the biological substance (e.g., enzymes) is located deep within the molecule, and the attachment of the ligand directly to the matrix sometimes interferes with its ability to bind the macromolecule due to steric hindrances between the matrix and the substance to be bound to the ligand. This interference can be prevented by interposing a "spacer arm" between the matrix and the ligand to facilitate effective binding (Figures 7.26 and 7.27).

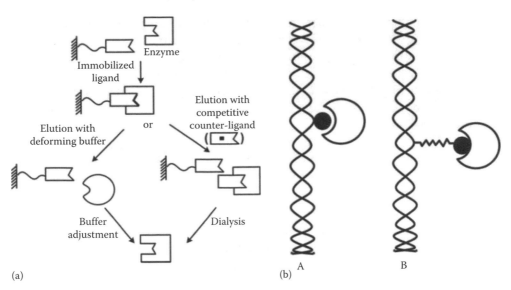

(a)   (b)

Figure 7.26 (a) Illustrating affinity chromatographic principle and (b) role of spacer arm in proper binding of the macromolecules.

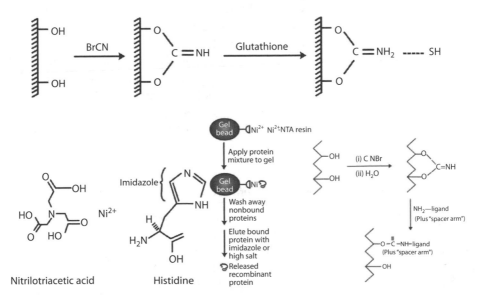

Figure 7.27 NTA used as tag in affinity chromatography and is bound to spacer arm to Sepharose or agarose that can bind to di- or trivalent metal ion, coupling thiol-containing compound to solid matrix.

7.25.4 Spacer Arm

The length of the spacer arm is critical. If it is too short, the arm is ineffective, and the ligand fails to bind the substance in the sample. If it is too long, nonspecific effects become pronounced and reduce the selectivity of separation. The optimum length of the spacer arm is generally 6–10 C atoms or their equivalent.

The chemical nature of the spacer arm (e.g., whether hydrophobic or hydrophilic) is also critical for the success of separation. A spacer arm is nearly always inserted between the matrix and the ligand so that large molecules can gain access to the binding sites. If this spacer arm is too short, then steric hindrance can still occur, while if it is too long, there is an increased risk of nonspecific adsorption, particularly of hydrophobic compounds. In practice, a spacer arm of 2–10 C atoms has been found to be optimal.

There are two approaches that are generally followed for coupling of the ligand. In one case, a spacer arm is first linked to the matrix followed by the coupling of the ligand, whereas the second approach involves the binding of the spacer arm to the ligand that is then linked to the matrix. It is the first approach that is more convenient and is preferred over the second one. For attachment of the ligand with the matrix, the matrix is given preliminary treatment with cyanogen bromide (CNBr) at pH 11. This causes activation of the matrix, and the molecules containing primary amino groups can then easily be coupled to CNBr-activated matrices. Different spacer arms including 1,6-diaminohexane, 6-aminohexanoic acid, and 1,4-bis(epoxy-propoxy) butane have been used to which the ligand can be attached by conventional organo-synthetic procedures involving the use of succinic anhydride and a water-soluble carbodiimide (Figure 7.27). The list in Table 7.2 also includes a number of supports of agarose, dextran, and polyacrylamide type that are commercially available with a variety of spacer arms and ligands preattached.

The CNBr activation has been the most generally used method for coupling ligands. Its application has developed so rapidly that it is now used in almost every laboratory concerned with purification of biological substances. The widespread use of affinity chromatography in such a short time reflects its success in achieving rapid separations, which are otherwise time consuming, difficult, or even impossible using conventional techniques.

TABLE 7.2   Some Characteristics of Affinity Chromatographic Matrices

Type of Matrices	Type of Ligands Attached
CNBr-activated Sepharose 4 B	All types of protein and ligands containing primary amino acids.
AH-Sepharose 4B	Having 6-C-long spacer arm (1,6-diaminohexane) with free amino group. Any ligand with carboxyl group.
CH-Sepharose 4B	Having 6-C long spacer arm (6-aminohexanoic acid) with free carboxyl group. Any ligand having amino group can be attached.
Epoxy-activated Sepharose 6B	Having a hydrophilic spacer and oxirane ring. Ligands having hydroxyl, amino, or thiol groups can be attached.
Activated CH-Sepharose 4B	Activated from the CH-Sepharose 4B formed by esterification of carboxylic group using N-hydroxysuccinimide compounds having free primary amino groups.
Cen A Sepharose	Concanavalin A attached to Sepharose 4B. A group-specific adsorbent for polysaccharides and glycoprotein.
Poly(U) Sepharose 4B	Used for isolation of mRNA having poly (A) sequences.
5'-AMP Sepharose 4B	For adsorption of kinases and dehydrogenases that requires a cofactor having an adenylate residue.

Biospecific affinity chromatography refers to the use of biological ligands (often proteins such as enzymes or antibodies) to grab hold of very specific molecules in a mixture. The ligand may have an affinity for a group of molecules (such as all immunoglobulins) or it can be very specific for one product. When an antibody is bound to a matrix, the resulting medium is called an immunosorbent. As a mixture flows through the column, the antigens bind to the antibody and stay inside, while the rest of the mixture continues through. The antigens are then flushed out of the column and collected separately. This technique has been used to purify interferon.

7.26 Hydrophobic Interaction Chromatography or Affinity Elution

Hydrophobic interaction chromatography (HIC) is very effective, when it was noted that certain proteins were unexpectedly retained on affinity columns containing hydrophobic spacer arms. HIC adsorbents now available include octyl or phenyl groups (Figure 7.28). Hydrophobic interactions are strong at high solution ionic strength, so samples need not be desalted before application to the adsorbent. Elution is achieved by changing the pH or ionic strength or by modifying the dielectric constant of the eluent using, for instance, ethanediol. Cellulose derivative complex is reported to have even more hydroxyl groups. The principle of

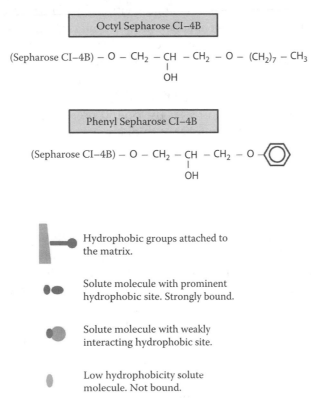

Figure 7.28 Matrix in HIC.

affinity chromatography has been extended to purify a large number of enzymes; other proteins including immunoglobulins and receptor proteins and nucleic acids have contributed considerably to recent developments in the field of molecular biology. Poly(U) Sepharose 4B, poly(A) Sepharose 4B, and immobilized single-stranded DNA nucleotides are quite useful. The technique of affinity chromatography has also been successfully employed for the separation of a mixture of cells into homogenous populations where it relies either on the antigenic properties of the cell surface or on the chemical nature of exposed carbohydrate residue on the cell surface or on a specific membrane receptor–ligand interactions. Useful modifications or methods of affinity chromatography technique have also been developed; for example, a number of proteins including interferon, plasminogen, and restriction endonucleases have been purified by "dye-ligand chromatography". In this technique, immobilized Cibacron blue, one of the triazine dyes, is used as a ligand to purify proteins.

The ionic groups and the conjugated ring system of the ligand bind with catalytic or effector sites of some proteins and so can be exploited to purify such proteins. In metal affinity chromatography, metals such as Zn^{2+}, Cu^{2+}, Cd^{2+}, Hg^{2+}, Co^{2+}, and Ni^{2+} are immobilized by chelation using either iminodiacetic acid or tris(carboxymethyl) ethylenediamine. The binding of proteins with such metals is pH dependent. The proteins having similar molecular weights and isoelectric points can be best separated by this technique if they have the differential binding affinity for metal ions. The bound proteins can be eluted by changing either the pH or ionic strength of the buffer. Proteins such as fibrinogen, superoxide dismutase, and nonhistone nuclear proteins have been purified by this method. Another approach is the covalent chromatography. In this technique, a covalent bond such as disulfide bond is formed between the immobilized ligand and proteins are isolated. Ligands like thiopropyl Sepharose and thiol Sepharose are used to purify proteins having a large number of thiol groups such as papain and urease. The bound protein is then eluted either by DTT or cysteine. The larger the number of thiol groups on the protein to be isolated, the better is the separation.

7.27 Purification by Affinity Chromatography

The principle of affinity chromatography is relatively simple to understand, and this is illustrated in Figure 7.26, which shows the purification of an enzyme from a mixture of other proteins. However, like most things, the practice is not quite so straightforward, and technical problems arise such as unwanted side reactions during the synthesis and attachment of the ligand to the side arm and matrix. This tends to increase the nonspecific adsorption of the affinity material with a consequent lowering of the specificity. In other cases, the binding of the compound to the ligand can be so strong that it becomes quite difficult to remove it from the affinity column.

However, in spite of these and other technical problems, separation by affinity chromatography can lead to a very high degree of purification in a single step.

7.28 Preparation of an Affinity Column

7.28.1 Principle

CNBr reacts with the free hydroxyl groups of CL-Sepharose 4B to give imidocarbonates that can then react with any compound that contains a primary amino group. CNBr-activated Sepharose 4B is used in the present experiment as a starting material that avoids having to handle free CNBr, an unpleasant and hazardous reagent. The spacer arm is introduced next by reacting the activated Sepharose with diaminobutane and then lengthened by the addition of succinic anhydride so that the ligand (3-amino-*N*-methylpyridinium ion) can be linked to the activated spacer arm by a peptide bond using a water-soluble carbodiimide. The affinity material prepared is *N*-methyl-3-aminopyridine agarose abbreviated to MAP-agarose for convenience.

7.28.2 Materials

CNBr-activated Sepharose 4B, HCl (1 mmol/L), sodium borate buffer (0.1 mol/L, pH 9.5), succinic anhydride, NaOH (4 mol/L), diaminobutane dihydrochloride, 3-aminopyridine, iodomethane (Caution: Volatile and possibly carcinogenic!), acetone,1-(3-dimethylaminopropyl)-3-ethyl carbodiimide hydrochloride, sintered glass funnels (G3) to fit Buchner filtration flask, mechanical shakers, elution buffer (0.5% w/v Triton X-100 in 30 mmol/L sodium phosphate buffer, pH 7.4), round-bottomed flasks (250 mL), melting point apparatus, melting point tubes.

7.28.3 Protocol

7.28.3.1 Initial Washing and Swelling
Weigh out 2 g of freeze-dried CNBr–Sepharose and mix with 1 mmol/L HCl on a sintered glass filter (G3). Add about 400 mL of solution in several aliquots over 15–20 min and suck off the supernatant between successive additions. The washing of the gel at low pH preserves the ligand binding of the Sepharose better than washing at pH 7.4. The CNBr-activated Sepharose contains dextran and lactose to preserve its activity during freeze-drying, and these are removed by the washing procedure. During this time, prepare a diaminobutane dihydrochloride solution by dissolving 25 g in 100 mL of the borate buffer. Carefully check that the pH is 9 and adjust if necessary (Figure 7.29).

7.28.3.2 Addition of the Spacer Arm
Give the gel a rapid final wash with cold borate buffer, remove the supernatant by suction, and then immediately add 100 mL of ice-cold diaminobutane dihydrochloride solution. Transfer the mixture to a 250 mL round-bottomed flask and shake gently at 4°C overnight.

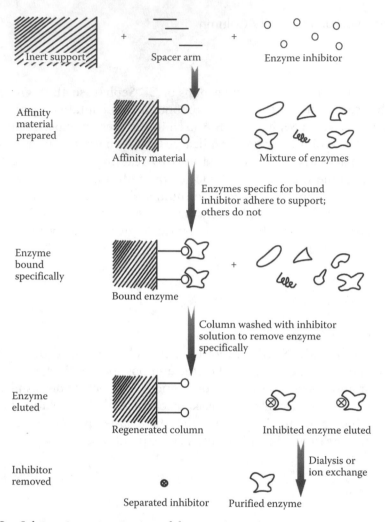

Figure 7.29 Schematic representation of the experiment.

CNBr–activated $-O-\underset{\underset{OH}{|}}{C}=NH + H_2N-(CH_2)_4-NH_2$
Sepharose

$\downarrow$

Sepharose with $-O-\underset{\underset{NH}{\|}}{C}-NH-(CH_2)_4-NH_2-N$-substituted isourea
side arm   (major product)

Shown as: Sepharose $-NH-(CH_2)_4-NH_2$

7.28.3.3 Preparation of the Ligand

Dissolve 1 g of aminopyridine in 15 mL of acetone and stir with 3 mL of iodomethane for 18 h. Filter the precipitated *N*-methyl-3-aminopyridinium iodide (MAP) and wash with acetone. Allow the product to dry and record the yield and melting point. At melting point, 119°C–120°C, the 100% yield is equal to 2.51 g.

7.28.3.4 Coupling of Ligand

Remove the excess amine by filtration on a sintered glass funnel. Wash the Sepharose thoroughly then disperse in 100 mL of water. Transfer the suspension to a flask and cool, on ice, to 4°C. Prepare a saturated solution of succinic anhydride by adding 10 g to 100 mL of warm water, cool to 4°C, decant if necessary, and add to the Sepharose with stirring. Filter the material, wash thoroughly with water to remove the free succinic anhydride, and then add 0.13 g of 3-amino-*N*-methylpyridinium iodide to 6 mL of the succinylated Sepharose in the presence of 0.86 g of 1-(3-dimethylaminopropyl)-3-ethyl carbodiimide hydrochloride. Shake the reaction mixture gently overnight at 4°C and thoroughly wash with the elution buffer before use.

The function of the soluble carbodiimide is to promote the formation of the peptide bond between the –COOH of the spacer arm and the –NH₂ of the ligand. The mechanism is complex, but a simplified representation of the role of the carbodiimide in this process is given as follows:

$$RN = C=NR + R'COOH +R''NH_2 \longrightarrow R'CONHR'' + RNHCONHR$$

Peptide   Substituted
bond   urea

7.28.3.5 Suggested Timetable

There are several stages where the preparation has to be left overnight and four successive days are needed to carry out the experiment in different stages. A convenient timetable is presented as follows:

Day 1

1. Prepare ligand and leave for 18 h overnight.
2. Wash the CNBr–Sepharose and add the spacer arm; leave overnight.

Day 2

3. Couple ligand to affinity material; leave it overnight.
4. Prepare brain extract and store in the deep freeze until day 4.

Day 3

5. Wash MAP-agarose and prepare column.

Day 4

Thaw brain extract (assay for acetylcholinesterase [AChE] and protein) and use affinity column to purify AChE.

7.29 Preparation of Rat Brain Acetylcholinesterase

7.29.1 Principle

7.29.1.1 Extraction

AChE is a membrane-bound enzyme and the first step in any investigation is to bring the enzyme into solution. This can be conveniently achieved by extraction of the tissue with the nonionic detergent Triton X-100. The criterion of solubility is an operational one, and the enzyme is considered to be soluble if it remains in the supernatant following centrifugation at 100,000g for 1 h.

7.29.2 Materials

Rat brain; sodium phosphate buffer (30 mmol/L, pH 7); Triton X-100 (1% w/v in 30 mmol/L; sodium phosphate buffer); bladed homogenizers; refrigerated ultracentrifuges.

7.29.3 Protocol

Remove the brain from a rat and wash with the sodium phosphate buffer. Remove the membrane from the organ, finely chop the tissue, and homogenize in 10 mL of ice-cold sodium phosphate buffer: add 10 mL of 1% w/v Triton X-100 at 4°C and mix thoroughly with gentle stirring. Centrifuge for 1 h at 100,000g and carefully remove the supernatant that is the soluble AChE. This can then be stored in the deep freeze without loss of activity until required.

7.30 Acetylcholinesterase Assay

7.30.1 Principle

The substrate used in the assay system is acetylthiocholine, the ester of thiocholine and acetic acid. The mercaptan formed as a result of the hydrolysis of the ester then reacts with an oxidizing agent, 5,5′-dithiobis-(2-nitrobenzoic acid) (DTNB), that is split into two products, one of which (5-thio-2-nitrobenzoate) absorbs at 412 nm. The activity of the enzyme can thus be measured by following the increase in absorbance at 412 nm in a double-beam spectrophotometer.

7.30.1.1 Enzyme Hydrolysis

$$(CH_3)_3 \overset{+}{N}CH_2CH_2SCOCH_3 \xrightarrow{\ H_2O\ } (CH_3)_3 \overset{+}{N}CH_2CH_2S^- + CH_3COO^- + 2H^+$$

7.30.1.2 Production of Color

7.30.2 Materials

Sodium phosphate buffer (0.1 mol/L, pH 8); DTNB (10 mmol/L; this is unstable at alkaline pH values, and so the stock solution is prepared just before use by dissolving 39.6 mg in 10 mL of sodium phosphate buffer [0.1 mol/L containing 15 mg sodium bicarbonate]); acetylthiocholine iodide (158.5 mmol/L in the sodium phosphate buffer, prepare fresh); UV–Vis double-beam spectrophotometers.

7.30.3 Protocol

Add 50 μL of the enzyme to 3 mL of sodium phosphate buffer (0.1 mol/L, pH 8) and incubate at room temperature for 5 min. Add 10 μL of DTNB (10 mmol/L), followed by 20 μL of acetylthiocholine iodide (158.5 mmol/L) to give a final concentration of 1 mmol/L of the substrate. Record the increase in absorbance at 412 nm on a double-beam spectrophotometer against a blank of the previously mentioned mixture prepared at the same time, but in the latter case, the 50 μL of the enzyme is replaced with 50 μL of the buffer solution.

7.30.4 Calculation of Enzyme Activity

If the change in extinction is ΔE per minute, then the activity in international units (IU) is as follows:

$$AChE \text{ activity} = \frac{(\Delta E \times 1000 \times 3.17)}{(1.36 \times 10^4 \times 0.05)}$$

where
   ΔE is the extraction charge per minute
   1000 is the factor to obtain µmoles
   3.17 is the total volume of the reaction mixture (mL)
   0.05 is the volume of the enzymes (mL)
   1.36×10^4 is the molar extinction coefficient of chromophore at 412 nm (L/mol/cm)

$$AChE \text{ activity} = \Delta E \times 4.66 \text{ L moles/min/mL}$$

In some cases, a larger volume of the enzyme may be needed, and in this case, an appropriate adjustment needs to be made to the previously presented equation.

7.31 Purification of Acetylcholinesterase by Affinity Chromatography

7.31.1 Principle

7.31.1.1 Binding
The ligand attached to the agarose by a spacer arm is a strong inhibitor of AChE so that when the crude extract is added to the column, the AChE molecules are specifically bound to the ligand. The other proteins present in the extract do not bind to the affinity material and are removed from the column by washing. This is achieved by running the solution containing the enzyme through the column followed by the eluting buffer until protein is no longer detected in the eluate. If the column is functioning properly, there should be no enzyme detected in any of the fractions collected up to this stage.

7.31.1.2 Elution
The enzyme is eluted from the column by incorporating an inhibitor (decamethonium bromide) in the elution buffer that has an even higher affinity for AChE than shown by the ligand.

7.31.1.3 Reactivation of Acetylcholinesterase
The reactivation of AChE is carried out by adding a cation-exchange resin (Amberlite CG-120) that strongly binds the decamethonium bromide.

7.31.1.4 Removal of Protein Bound Nonspecifically
In practice, some nonspecific binding of the other enzymes and other proteins always occurs and these are removed before using the column again. This is achieved by

eluting with a strong salt solution (1 mol/L NaCl) that also removes some enzyme activity and by washing with 6 mol/L guanidine HCl that denatures any remaining proteins and removes them from the affinity material. The column is then given a final wash with the elution buffer and stored in the cold room until required again.

7.31.1.5 Protein Assay

The detergent Triton X-100 absorbs strongly the UV at 280 nm, and so measurements at this wavelength cannot be used to estimate protein. The Folin–Lowry method is more sensitive, but a precipitate forms in the presence of detergent and the assay takes some time to complete. The biuret method has a low sensitivity but is probably the best procedure for this experiment when a rapid assay of protein is called for to monitor the fractions. Low extinction values are obtained that can be accurately recorded on a UV–Vis double-beam spectrophotometer.

7.31.2 Materials

MAP-agarose affinity column; small chromatography columns (10 cm × 1 cm); sodium phosphate buffer (30 mmol/L, pH 7); Triton elution buffer (0.5% Triton X-100 in the sodium phosphate buffer); decamethonium bromide (10 mmol/L and 50 mmol/L dissolved in the Triton elution buffer) (*Note: This is quite toxic and will inhibit AChE, so be careful!*); sodium chloride (1 mol/L in the Triton elution buffer); guanidine hydrochloride (6 mol/L); Amberlite CG-120; biuret reagent for protein assay.

7.31.3 Protocol

Thaw the frozen brain extract, and remove 1 mL for the assay of protein and AChE. Add the remainder of the solution to the column and then wash with the Triton elution buffer collecting 3 mL fractions. Assay each fraction for protein using 0.5 mL in duplicate and continue collecting fractions until protein can no longer be detected in the eluate. Store the remainder of each fraction on ice until required for the assay of AChE.

At this stage, elute with 10 mmol/L decamethonium bromide and collect a further five fractions. Increase the concentration of decamethonium bromide to 50 mmol/L and collect another five fractions then wash the column with NaCl (1 mol/L) dissolved in the elution buffer until no further protein or enzyme activity is detected. Assay each of the fractions for AChE activity and select those fractions that contain the enzyme. Add a small quantity of the ion-exchange resin, gently mix, and assay these fractions again for enzyme activity. Finally, plot the total protein and AChE in each fraction against the elution volume. Calculate the specific activity of the purified AChE and compare it with the crude extract.

7.31.3.1 Concentration of the Purified Protein

At the end of the protein purification, the protein often has to be concentrated by the following methods.

7.31.3.1.1 Lyophilization.   If the solution does not contain any other soluble component, then the protein in question can be lyophilized (dried). This is normally done after confirmation with high-performance liquid chromatography (HPLC) analysis. This simply removes all volatile components leaving the proteins behind. Biological materials often need to be dried to stabilize them for storage or distribution. Drying always causes some loss of activity or other damage. Lyophilization, also called freeze-drying, is a method of drying that significantly reduces such damages. Because lyophilization is the most complex and expensive form of drying, its use is usually restricted to delicate, heat-sensitive materials of high value.

The simplest form of lyophilizer would consist of a vacuum chamber into which wet sample material could be placed, together with a means of removing water vapor so as to freeze the sample by evaporative cooling and freezing and then maintaining the water-vapor pressure below the triple point pressure. The temperature of the sample will then continue to fall below the freezing point, and sublimation would slow down until the rate of heat gain in the sample by conduction, convection, and radiation is equal to the rate of heat loss as the more energetic molecules sublimed away are removed. An alternative is to freeze the material before it is placed under vacuum. This is commonly done with small laboratory lyophilizers where material is frozen inside a flask. The flask is then attached to a manifold connected to the ice condenser. To speed the process, the material can be shell-frozen by rotating the flask in a low-temperature bath, giving a large surface area and small thickness of material. The first step in the lyophilization process is to freeze a product to solidify all of its water molecules. Once frozen, the product is placed in a vacuum and gradually heated without melting the product. This process is called sublimation, which transforms the ice directly into water vapor, without first passing through the liquid state. The water vapor given off by the product in the sublimation phase condenses as ice on a collection trap, known as a condenser, within the lyophilizer's vacuum chamber. To be considered stable, a lyophilized product should contain 3% or less of its original moisture content and be properly sealed (Figure 7.30).

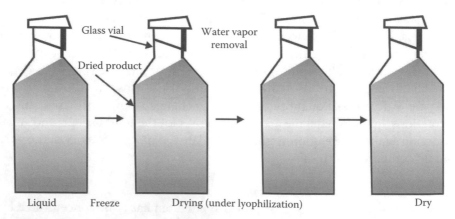

Figure 7.30 Lyophilization process involving freezing of liquid leading to dried product.

7.31.3.1.2 Ultrafiltration. Ultrafiltration is a technique for separating dissolved molecules in solution on the basis of size, which means that molecules larger than the membrane pore size rating will be retained at the surface of the membrane. The accumulation of retained molecules may form a concentrated gel layer. In ultrafiltration, hydrostatic pressure forces a liquid against a semipermeable membrane. It uses asymmetric microporous membranes with a relatively dense, thin skin containing pores supported by a coarse, strong substructure. Membranes are available that possess molecular weight cutoffs from 1,000 to 100,000 and usable at pressure up to 2 MPa. Suspended solids and solutes of high molecular weight are retained, while water and low molecular weight solutes pass through the membrane. This separation process is used for purifying and concentrating macromolecular protein solutions. The solution is forced against the membrane by a mechanical pump or gas pressure or centrifugation. The device is fitted with an ultrafilter membrane of the desired molecular weight, such that proteins of interest will be retained in the cell. The pressure cell is filled with the protein solution, and nitrogen gas is kept at about 50 psi, while the cell is stirred gently at 4°C. After about 1 h, the solution will be decreased in volume usually without loss of activity. The cell is filled with the desired buffer, and the concentration process is repeated. Stirred cells represent the simplest configuration of ultrafiltration cell. The membrane rests on a rigid support at the base of a cylindrical vessel, which is equipped with a magnetic stirrer, to combat concentration polarization. It is suitable for large-scale use but is useful for preliminary studies and for the concentration of laboratory column eluates. Problems associated with the membrane blockage and fouling can usually be overcome by treatment of membranes with detergents or proteases or with care acid or alkali treatment (Figure 7.31). Ultrafiltration differs from conventional filtration to the size being retained (<50 nm diameter).

At each step of protein purification, the following parameters are to be measured:

Total protein: The quantity of protein present in a fraction is obtained by determining the protein concentration of a part of each fraction and multiplying by the fraction's total volume.

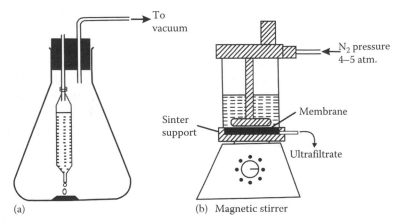

(a)   (b) Magnetic stirrer

Figure 7.31 (a) Ultrafiltration of protein solution in the dialysis bag ($P < 1$) and (b) ultrafiltration at pressure more than 5, forcing filtrate through membrane.

Total activity: The enzyme activity for the fraction is obtained by measuring the enzyme activity in the volume of fraction used in the assay and multiplying by the fraction's total volume.

Specific activity: This parameter is obtained by dividing total activity by total protein.

Yield: This parameter is a measure of the activity retained after each purification step as a percentage of the activity in the crude extract. The amount of activity in the initial extract is taken to be 100%.

Purification level: This parameter is a measure of the increase in purity and is obtained by dividing the specific activity, calculated after each purification step by the specific activity of the initial extract.

7.31.3.1.3 High-Performance Liquid Chromatography. HPLC utilizes a column that holds chromatographic packing material (stationary phase), a pump that moves the mobile phase(s) through the column, and a detector that shows the retention times of the molecules. Retention time varies depending on the interactions between the stationary phase, the molecule being analyzed, and the solvent(s) used. HPLC is a form of chromatography applying high pressure to drive the solutes through the column faster. This means that the diffusion is limited, and the resolution is improved. The most common form is "reversed-phase HPLC," where the column material is hydrophobic. The proteins are eluted by a gradient of increasing amounts of an organic solvent, such as acetonitrile. The protein elutes according to its hydrophobicity. After purification by HPLC, the protein is in a solution that only contains volatile compounds and can easily be lyophilized. HPLC purification frequently results in denaturation of the purified proteins and is thus not applicable to proteins that do not spontaneously refold (Figure 7.32).

In the characterization of biopharmaceutical products, the HPLC is used, along with other analytical techniques, to demonstrate the identity, purity, potency, and stability. Virtually all protein purification processes rely on at least one column chromatography step. The number and sequence of such steps are usually tailored to the protein, the feed material, and the application of the product.

Chromatographic separation is a stepwise process by nature. The actual steps involved are relatively simple, which are column equilibration, sample application (and adsorption), column washing, elution of bound molecules, column regeneration, and re-equilibration. These steps are repeated as long as the column resolution is good, as long as it can be reused.

The mode of elution of a sample usually depends on the scale of the process. Isocratic elution makes use of an elution buffer solution to move the solute through the column. Gradient elution changes their relative affinity by changing the conditions (such as pH or salinity), either stepwise or continuously. Displacement elution uses a substance with greater affinity for the stationary phase to displace the molecule already bound. Chromatographic techniques can be classified by their respective modes of separation: In nonadsorptive chromatography, the rate at which substances pass through the column is controlled by basic physical, kinetic characteristics such as the size and shape of molecules and particles. The choice of chromatographic methods and the stationary phases to be used are important to the design of a purification

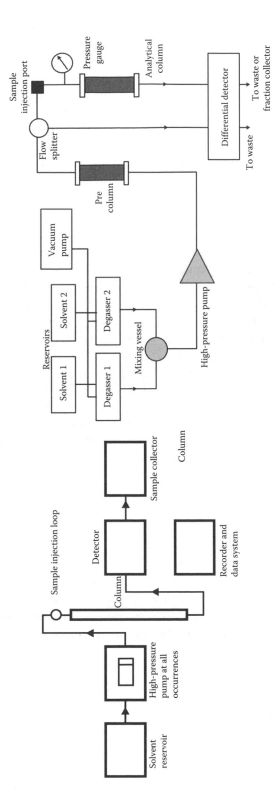

Figure 7.32 Schematic representation of HPLC unit.

system. Equipment is designed and optimized for the necessary conditions to provide robust, reproducible performance. Process design takes into account the economics and efficiency of purification based on scientific parameters such as hydrodynamics and kinetics, mass transfer rates, molecular interactions, and surface chemistry.

There are various types and variations of the chromatographic system that differ in binding property for the enzyme. Volatile components of cells can be analyzed by gas chromatography systems, while hydrophilic or hydrophobic components are best separated by various existing chromatographic systems. Hydrophobic components are mostly soluble in organic solvent, and they can be separated by molecular-exclusion chromatography if they differ in molecular weight or by adsorption chromatography if they have different functional groups. Water-soluble compounds are separated by reversed-phase partition chromatography (RPC) system if they are weakly polar, by ion-exchange chromatography if they are strongly ionic, or by molecular exclusion system if they differ in molecular weight. RPC uses organic solvents for the separation and elution of a protein. Because the conditions are harsh and because the proteins are bound tightly, this method works best with smaller, more stable proteins. RPC is powerful, but it is also limited in its application. The organic solvents used (butanol, isopropanol, ethanol, methanol, acetonitrile, etc.) are sometimes flammable and sometimes require special equipments for their storage and use to prevent fires or explosions. They may denature proteins, and they can be toxic as well, necessitating the use of special ventilation systems. Solvents that provide the best resolutions can also be too difficult to separate from the product after the fact that adding another purification step can negate the advantage of the RPC separation power in the long run. Both HIC and RPC are used sometimes as concentrating steps in the purification process. RPC is less common in protein purification because of the harsher conditions it employs, but it is used quite often in preparative separation of peptides.

7.32 MALDI-TOF

In this technique, protein ions are generated and then accelerated through an electric field. They travel through the flight tube with the smallest traveling fastest and arriving at the detector first. Thus, the time of flight (TOF) in the electric field is a measure of the mass (or more precisely, the mass/charge ratio). Tiny amounts of biomolecules, as small as a few picomoles (pmol) to femtomoles (fmol), can be analyzed in this manner (Figure 7.33).

Matrix-assisted laser desorption–ionization (MALDI)-TOF is indeed an accurate means of determining protein mass. Mass spectrometry has permitted the development of peptide mass fingerprinting. This technique is used for the identification of peptides and has increased the utility of 2D gel electrophoresis. The sample of interest is extracted and cleaved specifically by chemical or enzymatic means after running the 2D gel electrophoresis. The mass of the protein fragment is then determined with the use of mass spectrometry. Finally, the peptide mass, or fingerprint, is matched against the fingerprint found in the database of proteins that have been "electronically

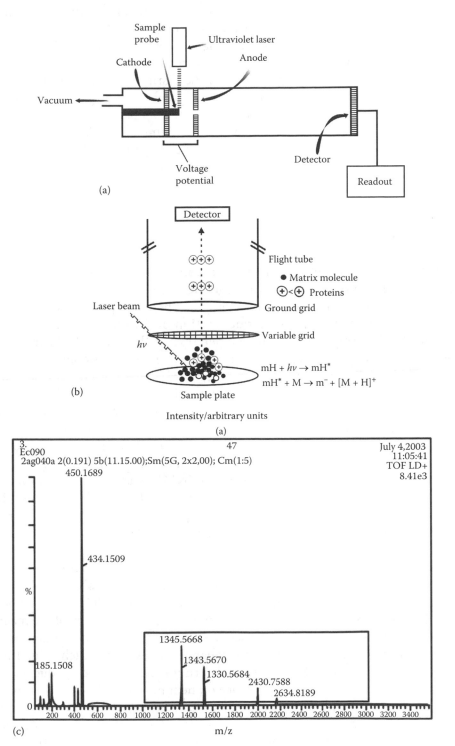

Figure 7.33 (a) Internal structure and (b) mechanism of MALDI. (c) MALDI-TOF analysis of *Mycobacterium* macrophage.

cleaved" by a computer simulating the same fragmentation technique used for the experimental sample. The laser is fired at the crystals in the MALDI spot. The matrix absorbs the laser energy, and it is thought that primarily the matrix is ionized by this event and transfers part of its charge to the analyte molecules (e.g., protein), thus ionizing them while still protecting them from the disruptive energy of the laser. Ions observed after this process consist of a neutral molecule (M) and an added or removed ion. Together, they form a quasi-molecular ion, for example, $[M^+H]^+$ in the case of an added proton, $[M^+Na]^+$ in the case of an added sodium ion, or $[M^-H]^-$ in the case of a removed proton. MALDI is capable of creating single-charged ions, but multiple charged ions ($[M^+nH]n^+$) can also be created, as a function of the matrix, the laser intensity, and/or the voltage used.

7.33 Nuclear Magnetic Resonance Spectroscopy

In this technique, a sample is immersed in a magnetic field and bombarded with radio waves. These radio waves encourage the nuclei of the molecule to resonate or spin. As the positively charged nucleus spins, the moving charge creates what is called a magnetic moment. The thermal motion of the molecule—the movement of the molecule associated with the temperature of the material—further creates a torque or twisting force that makes the magnetic moment "wobble" like a child's top. When the radio waves hit the spinning nuclei, they tilt even more, sometimes flipping over. These resonating nuclei emit a unique signal that is then picked up on a special radio receiver and translated using a detector. This decoder is called the Fourier transform algorithm, a complex equation that translates the language of the nuclei. By measuring the frequencies at which different nuclei flip, one can determine the molecular structure, as well as many of the interesting properties of the molecule. Nuclear magnetic resonance (NMR) has proved as a power alternate to x-ray crystallography for the determination of molecular structure. NMR has the advantage over crystallographic techniques in that the experiments are performed in solution as opposed to the crystal lattice. However, the principles that make NMR possible tend to make this technique very time consuming and limit the application to small- and medium-sized molecules. NMR measures the distances between atomic nuclei, rather than the electron density in a molecule. With NMR, a strong, high-frequency magnetic field stimulates atomic nuclei of the isotopes H^1, C^{13}, or N^{15} (they have a magnetic spin) and measures the frequency of the magnetic field of the atomic nuclei during its oscillation period back to the initial state. The important step is to determine which resonance comes from which spin. The distance and the type of neighboring nuclei determine the resonance frequency of the stimulated atomic nuclei. The dependence on next neighbors is known as chemical shift (or spin–spin coupling constant) and reflects the local electronic environment and the information contained in the 1D NMR spectra. For proteins, NMR usually measures the spin of protons. The following reasons make the H^1 NMR spectroscopy the method of choice for biological H at many sites in proteins, nucleic acids,

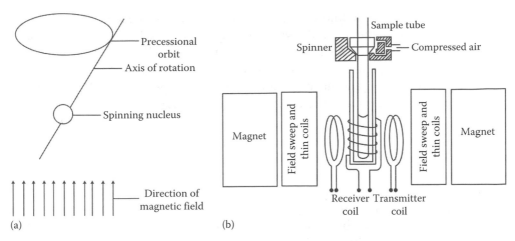

Figure 7.34 (a) Nuclear spinning in external magnetic field and (b) NMR spectrometer.

TABLE 7.3   Comparison of NMR and X-Ray Crystallography

NMR Spectroscopy	X-Ray Crystallography
Short time scale of protein folding	Long time scale, static structure
Solution, purity	Single crystal purity
<20 kD, domain	Any size, domain. Complex
Functional active site	Active or inactive
Domains	Domains
Aromatic nuclei, chemical bonds	Electron density
Resolution limit 2–3.5 Å	Resolution limit 2–3.5 Å
Primary structure must be known	Primary structure must be known (except if resolution is 2 Å or better for every single residue)

and polysaccharides since H has a high abundance for each site and the H nucleus is the most sensitive to detect (Figure 7.34 and Table 7.3).

The 1D spectra contain information about all the chemical shifts of all the H in the protein. The frequency resolution is often not enough to distinguish individual chemical shifts. The 2D NMR solves these problems by containing information about the relative position of H in molecular structures. The 2D NMR spectra contain information about interaction between H that is covalently linked through one or two other atoms (COSY or correlation spectroscopy). Alternatively, pairs of H can be closed in space, even if they are from residues that are not close in sequence (NOE spectra or nuclear overhauser effect). A complete structure can thus be calculated by sequentially assigning cross-peak correlations in the 2D spectra. Currently, the size limit for proteins that are amenable to NMR solution structure analysis is about 200 amino acids. An important feature of the identification of cross-peaks is that regular patterns that stem from secondary structure elements such as alpha helices and parallel or antiparallel beta sheets can be recognized, because they contain typical hydrogen bonding network.

NMR also requires the knowledge of the amino acid sequence, but the protein does not have to be in the ordered crystal, yet high concentrations of solubilized protein must be available (NMR structures are therefore also called solution structures). In biopolymers, the primary structure (sequence) logically breaks up the molecule into groups of coupled spins, normally one or two groups per residue. This is true not only for proteins but also for nucleic acids and polysaccharides.

7.34 Maintenance of Enzyme Activity

The key to maintain the enzyme activity is to maintain the conformation and prevent unfolding, aggregation, and change in covalent structure. This can be done (1) by addition of chaotropes, (2) by the addition of some polyols, and (3) by modification of side chain or by enzyme immobilization.

7.35 Addition of Chaotropes

In general, proteins are stabilized by increasing their concentration and the ionic strength of their environment. Neutral salts compete with proteins for water and bind to charged groups or dipoles. This may result in the interactions between an enzyme's hydrophobic areas being strengthened, causing the enzyme molecules to compress and make them more resistant to thermal unfolding reactions. But not all salts are equally effective in stabilizing hydrophobic interactions. Some are much effective at their destabilization by binding to them since they cause disruption of the localized structure of water.

Chaotrope is a substance that disrupts the 3D structure in macromolecules such as proteins, DNA, or RNA and denatures them. A chaotropic agent interferes with stabilizing intramolecular interactions mediated by noncovalent forces such as hydrogen bonds, van der Waals forces, and hydrophobic effects. Ammonium sulfate and potassium hydrogen phosphate are powerful enzyme stabilizers, whereas sodium thiosulfate and calcium chloride destabilize enzymes.

Many enzymes are specifically stabilized by low concentrations of cations that may or may not form part of the active site, for example, Ca^{2+} stabilizes α-amylases and Co^{2+} stabilizes glucose isomerases. At high concentrations (e.g., 20% NaCl), salt discourages microbial growth due to the osmotic effect. In addition, ions can offer some protection against oxidation to groups such as thiols by salting out the dissolved oxygen from solution (Figure 7.35).

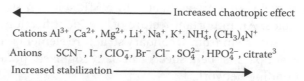

Figure 7.35 Effect of ions on enzyme stabilization.

7.36 Role of Polyols/Hydrophilic Polymers in Stabilizing Structures

Low molecular weight polyols (e.g., glycerol, sorbitol, and mannitol) are also useful for stabilizing enzymes, by repressing microbial growth or by reducing the water activity, or by the formation of protective shells, which prevent the unfolding process. Glycerol may be used to protect enzymes against denaturation due to the ice crystal formation at a subzero temperature. Some hydrophilic polymers (e.g., polyvinyl alcohol, polyvinylpyrrolidone, and hydroxypropyl cellulose) stabilize enzymes by a process of compartmentalization whereby the enzyme–enzyme and enzyme–water interactions are somewhat replaced by less potentially denaturing enzyme–polymer interactions. They may also act by stabilizing the hydrophobic effect within the enzymes.

7.37 Stabilization by Side-Chain Modification

Many specific chemical modifications of amino acid side chains are possible that may (or more commonly, may not) result in stabilization. An example of this is the derivatization of lysine side chains in proteases with N-carboxyamino acid anhydrides, which form polyaminoacylated enzymes with various degrees of substitution and length of amide-linked side chains. This derivatization is sufficient to mask the proteinaceous nature of the protease and prevents its autolysis.

Important lessons about the molecular basis of thermostability have been learned by comparison of enzymes from mesophilic and thermophilic organisms. A frequently found difference is the increase in the proportion of arginine residues at the expense of lysine and histidine residues. This may be possibly explained by noting that arginine is bidentate and has a higher pK_a than lysine or histidine. Therefore, it forms stronger salt links with bidentate aspirate and glutamate side chains that result in more rigid structures. This observation, among others, has given hope that site-specific mutagenesis may lead to enzymes with significantly improved stability. In the meantime, it remains possible to convert lysine residues to arginine-like groups by reaction with activated urea. It should be noted that enzymes stabilized by making them more rigid usually show lower activity (i.e., V_{max}) than the "natural" enzyme. Enzymes are more stable in the dry state than in solution. Solid enzyme preparations sometimes consist of freeze-dried protein, and usually, they are bulked out with inert materials such as lactose, starch, carboxymethyl cellulose, and other polyelectrolytes that protect the enzyme during a cheaper spray-drying stage. Other materials, which are added to enzymes, may consist of substrate thiols to create a reducing environment, antibiotics, inhibitors of contaminating enzyme activities, benzoic acid esters as preservatives for liquid enzyme preparation, and chelating agents. Additives of these types must be compatible with the final use of the enzyme's product.

7.38 Criteria of Purity

Purity of the enzyme can be assayed at a different stage. Therefore, the first step is to determine total protein content in each fraction. There are different methods

TABLE 7.4   Protein Testing Methods

Method	Mode of Action	Sensitivity
Biuret	Cu2+–peptide bond	Complex 1–10 mg protein
Folin (Lowry)	Heavy metal complex	With aromatic amino acid 20–300 µg protein
Bradford		Dye reaction with amino group. Side chains (Lys) 1–100 µg protein

to determine total protein like Lowry's method (combination of biuret and Folin–Ciocalteu). Folin reagent increases the sensitivity of the procedure by forming tetradentate copper complex that gives color according to the presence of amino acid residue taken at A_{280} nm (range 0.1–1 mg/mL). The presence of nucleic acid also interferes in total protein determination; therefore, correction factor is involved after determination of absorbance $= 1.55 \times A_{280} - 0.76 \times A_{260}$ nm. Note that assay should be performed in the presence of buffer. For the first, a standard protein assay is performed using BSA for preparation of calibration graph. It gives the slope for determination of unknown protein quantity in the unknown sample. It helps in the determination of specific activity of protein that is defined as IU per mg protein or as katal per kg protein. IU is defined as loss of 1 µmol of substrate per minute under specified conditions (Table 7.4).

7.39 Biuret

This method is the most linear because its color depends on a direct complex between the peptide bonds of protein and the Cu^{2+} ion. It is highly sensitive since the complex does not have a high extinction coefficient (Figure 7.36).

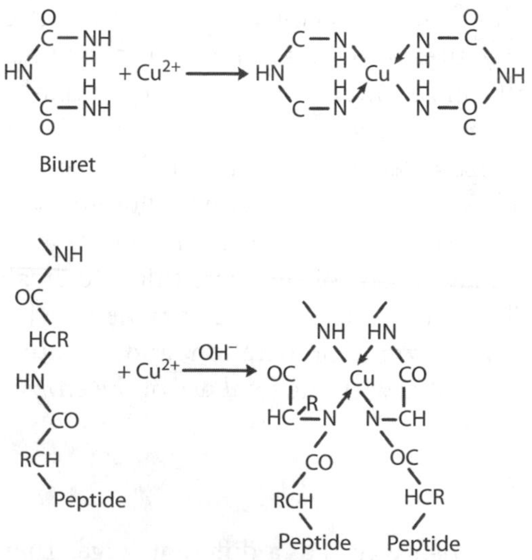

Figure 7.36 Chemical reaction for protein test by biuret method.

7.40 Folin (Lowry's Method)

The Folin assay is dependent on the presence of aromatic amino acids in the protein. First, a cupric/peptide bond complex is formed, and then this is enhanced by a phosphomolybdate complex with the aromatic amino acids. Overall, it is about 10–50 times more sensitive than the biuret method. Many substances sometimes interfere with the Folin assay for protein (Figure 7.37).

7.41 Bradford

Bradford assay is based on a blue dye (Coomassie brilliant blue) that binds to free amino groups in the side chains of amino acids, especially Lys. This assay is as sensitive as the Folin, especially when using the commercial kits available from chemical manufacturers. This method can yield quite a linear standard curve with BSA but often found to be not quite linear. Few substances interfere with this assay. The assay can also be done in the presence of detergent that makes it useful for determining protein concentration for membrane proteins (Figures 7.38 and 7.39).

Figure 7.37   Calibration graph of protein assay.

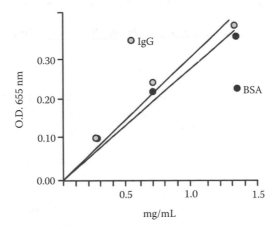

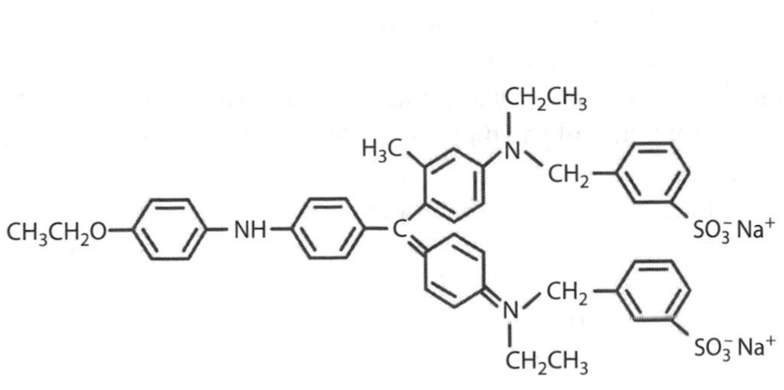

Figure 7.38   Coomassie blue G-250 used in Bradford total protein assay.

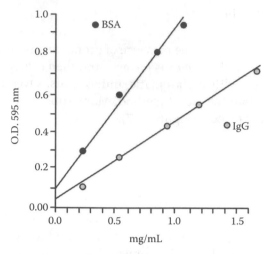

Figure 7.39   Calibration graph of protein assay using BSA.

Suggested Readings

ÄKTA. 2000. *FPLC, System Manual* 18-1140-45, AB edn. Uppsala, Sweden: Amersham Pharmacia Biotech AB.

AKTA. 2006. *Design Purification Method Handbook*. Piscataway, NJ: Amersham Biosciences. Catalog number 18-1124-23.

Bisen, P. S. and A. Sharm. 2012. *Introduction to Instrumentation in Life Sciences*. Boca Raton, FL: CRC Press.

Buffington, R. and M. K. Wilson. 1987. *Detectors for Gas Chromatography—A Practical Primer*. Palo Alto, CA: Hewlett-Packard Corporation. Part No. 5958-9433.

Chaplin, M. F. and C. Bucke. 1990. *Enzyme Technology*. Cambridge, U.K.: Cambridge University Press.

Cooper, E. H., R. Turner, E. A. Johns et al. 1983. Applications of fast protein liquid chromatography TM in the separation of plasma proteins in urine and cerebrospinal fluid. *Clin Chem* 29: 1635–1640.

Desai, S. N., R. B. Colah, and D. Mohanty. 1998. Comparison of FPLC with cellulose acetate electrophoresis for the diagnosis of beta-thalassaemia trait. *Indian J Med Res* 108: 145–148.

Ettre, L. 1993. Nomenclature for chromatography. *Pure Appl Chem* 65: 819–872.

Feinberg, J. G. and I. Smith. 1972. *Paper and Thin Layer Chromatography and Electrophoresis*, 2nd edn. London, U.K.: Longman.

Ferreira, P. O. and M. A. Ferreira. 2002. Solid phase micro-extraction in combination with GC/MS for quantification of the major volatile free fatty acids in ewe cheese. *Anal Chem* 74: 5199–5204.

Gogou, A. I., M. Apostolaki, and E. G. J. Stephanou. 1998. Adsorption chromatography. *J Chromatogr* 799: 215.

Golemis, E. A. and P. D. Adams. 2005. *Protein–Protein Interactions: A Molecular Cloning Manual*, 2nd edn. New York: Cold Spring Harbor Laboratory Press.

2002. *GST Gene Fusion System Handbook*, vol. 18, pp. 1157–1158. Piscataway, NJ: Amersham Biosciences.

Hage, D. S. and P. F. Ruhn. 2006. An introduction to affinity chromatography. In *Handbook of Affinity Chromatography*, 2nd edn., ed. D. S. Hage. New York: Taylor & Francis Group.

Hill, H. H. and D. G. McMinn, Eds. 1992. *Detectors for Capillary Chromatography*. New York: John Wiley & Sons.

Jeppsson, J. O., P. Jerntorp, G. Sundkvist et al. 1986. Measurement of haemoglobin A1c by a new liquid-chromatographic assay: Methodology, clinical utility, and relation to glucose tolerance evaluated. *Clin Chem* 32: 1867–1872.

Lee, W.-C. and K. H. Lee. 2004. Applications of affinity chromatography in proteomics. *Anal Biochem* 324: 1–10.

Marz, W., R. Siekmeier, H. Scharnagl et al. 1993. Fast lipoprotein chromatography: New method of analysis for plasma lipoproteins. *Clin Chem* 39: 2276–2281.

Mehler, A. H. 1993. Glutathione S-transferases function in detoxification reactions. In *Textbook of Biochemistry*, 3rd edn., ed. T. M. Devlin. New York: Wiley-Liss.

Parikh, I. and P. Cuatrecasas. 1985. Affinity chromatography. *Chem Eng News* 63: 17–29.

Per-Erik, G. and L. Per-Olof. 2003. Fast chromatography of proteins. In *Isolation and Purification of Proteins*, eds. R. Hatti-Kaul and B. Mattiasson. New York: Marcel Dekker.

Sandra, J. F. 2002. *Gas Chromatography. Ullmann's Encyclopedia of Industrial Chemistry*. Weinheim, Germany: Wiley-VCH Verlag GmbH.

Scouten, W. H. 1981. *Affinity Chromatography: Bio-Selective Adsorption on Inert Matrices*. New York: Wiley.

Sheehan, D. and S. O'Sullivan. 2003. Fast protein liquid chromatography. *Protein Purification Protocols* 244: 253.

Shibasaki, T., H. Gomi, F. Ishimoto et al. 1990. Urinary N-acetyl-beta-D-glucosaminidase isoenzyme activity as measured by fast protein liquid chromatography in patients with nephrotic syndrome. *Clin Chem* 36: 102–103.

Smith, I. 1960. *Chromatographic and Electrophoretic Techniques*, vol. 1, *Chromatography*, vol. 2, *Zone Electrophoresis*. New York: Interscience Publishers, Inc.

Smith, I., J. Seakins, and T. William. 1976. *Chromatographic and Electrophoretic Techniques*, 4th edn. London, U.K.: William Heinemann Medical Books.

Stahl, E. and M. R. F. Aisworth. 1969. *Thin Layer Chromatography: A Laboratory Handbook*, 2nd edn. New York: Springer.

Turkova, J. 1978. *Affinity Chromatography*. Amsterdam, the Netherlands: Elsevier.

Walters, R. R. 1985. Affinity chromatography. *Anal Chem* 57: 1099A–1114A.

Wilson, K. and J. Walker. 2003. *Practical Biochemistry; Principle and Techniques*, 5th edn. Cambridge, U.K.: Cambridge University Press.

Important Links

Affinity chromatography: http://www.jenabioscience.com/cms/en/1/browse/1450affinitychromatography.html

Chromatography, theories, FPLC, and beyond. http://www.mnstate.edu/biotech/chrom_fplc.pdf

Column chromatography: http://www.repligen.com/bioprocessing/products/prepacked column?gclid = ckogjcjzoksFQl76wodARLC-fw

Exclusion chromatography: http://www.pall.com/main/Laboratory/Literature-Library-Details.page?id 47506

GLC: http://www.makarandelectronics.com/gas_chromatograph.html

HPLC: http://www.hplc.com/Eicom/index.html

HPTLC: http://www.camag.com/v/products/tlc-ms/

Ion-exchange chromatography: http://www.gelifesciences.com/aptrix/upp01077.nsf/Content/
    Products?OpenDocument&parentid = 5179&moduleid = 165889&zone = Labsep

8

Separation Technology

8.1 Electrophoresis

Electrophoresis is the study of movement of charged particles in an electric field and is being used to separate macromolecules on the basis of size, electric charge, and other physical properties. The technique is widely used for separation and analysis of a large number of biomolecules such as amino acids, peptides, proteins, nucleotides, and nucleic acids (Figure 8.1). Gel electrophoresis refers to a technique in which molecules are forced across a span of gel, motivated by an electric current. The driving force for electrophoresis is the voltage applied to electrodes at either end of the gel. The properties of a molecule determine how rapidly an electric field can move it through a gelatinous medium. Many important biological macromolecules (e.g., amino acids, peptides, proteins, nucleotides, and nucleic acids) possess ionizable groups and, at any given pH, exist in solution as electrically charged species either as cations (+) or anions (−). Depending on the nature of the net charge, the charged particles will migrate either to the cathode or to the anode. For example, when an electric field is applied across a gel at neutral pH, the negatively charged phosphate groups of the DNA cause it to migrate toward the anode.

8.2 Principle

Any charged ion or molecule migrates when placed in an electric field. The rate of migration of a compound depends on its net charge, size, shape, and the applied current. This can be represented by the following equation:

$$v = \frac{E \cdot q}{f}$$

where
    v is the velocity of migration of the molecule
    E is the electric field (V/cm)
    q is the net electric charge on the molecule
    f is the frictional coefficient that is a function of the mass and shape of the molecule

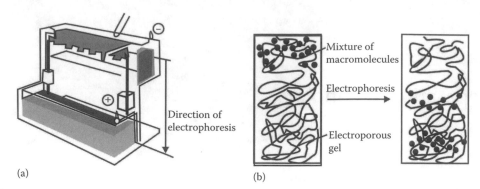

Figure 8.1 Schematic representation of separation of macromolecules by electrophoresis.

The movement of a charged molecule in an electric field is often expressed in terms of electrophoretic mobility (μ), which is defined as the velocity per unit of electric field:

$$\frac{\mu - v}{E} = \frac{E \cdot q}{f \cdot E} = \left[\text{since } v = \frac{E \cdot q}{f} \right] = \frac{q}{f}$$

For molecules with similar conformation, f varies with size but not with the shape. Thus, electrophoretic mobility (μ) of a molecule is directly proportional to the charge density (charge/mass ratio). Molecules with different charge/mass ratios migrate under the electric field at different rates and hence get separated. This is the underlying basic principle for all the electrophoretic techniques.

8.3 Types of Electrophoresis

Depending upon the nature of the support medium, electrophoresis is of different types such as paper, starch, polyacrylamide, and agarose gel electrophoresis.

8.3.1 Paper Electrophoresis

In this form of electrophoresis, the sample is applied as a circular point or spot on a strip of Whatman filter paper or cellulose acetate paper moistened with the buffer solution. The ends of the paper are immersed in separate reservoirs containing buffer and in which the electrodes are fitted. On passing electric current, the ions in the sample migrate toward oppositely charged electrodes at characteristic rates. This method is suitable for separation of low molecular weight (MW) compounds such as amino acids, small peptides, and nucleotides. They are of the following two types: (1) low-voltage paper electrophoresis and (2) high-voltage paper electrophoresis.

8.3.2 Low-Voltage Paper Electrophoresis

It is rarely used today, having been superseded by gel electrophoresis; however, it is useful to begin a discussion of zone electrophoresis by examining this method because it is very simple and illustrates the basic techniques. In this method, the paper strip is first dipped in the buffer solution and then placed in the tank as indicated in Figure 8.2.

The sample is then applied either as a spot or as a line. The paper is enclosed in a tank to prevent evaporation and voltage is applied. When separation is completed, the paper is removed and dried. If the sample contains sufficient material, each component can be located either by its color or fluorescence or by staining with various dyes. If quantitative measurement is required, the dye can be eluted with appropriate eluent and determined spectrophotometrically. Since dye uptake is rarely quantitative, the accuracy is about 20%. If the sample is radioactive, spots can be located by cutting up the paper and counting the radioactivity or by autoradiography of the whole sheet using x-ray film. A particularly useful technique is to combine stains with radioactivity to determine what substances are radioactive. Low-voltage paper electrophoresis was most useful for the analysis of protein mixtures that are poorly separated by chromatography.

Low-voltage electrophoresis is inefficient for small molecules (e.g., amino acids and nucleotides) because their small charge results in low mobility (e.g., slow separation) and their small charge allows considerable diffusional spreading.

8.3.3 High-Voltage Paper Electrophoresis

The speed of separation is increased by the use of a potential gradient of 200 V/cm. The high voltage produces a high current, which heats the paper, so that cooling is necessary. This can be accomplished by immersing the paper in a large volume of an immiscible and nonconducting liquid or by pressing it against cool paper or glass. The immersed-strip technique is not often used because the required coolants (e.g., toluene, carbon tetrachloride, or various oils) are toxic and sometimes inflammable. The enclosed-strip method, illustrated in Figure 8.3, is far more common. In this, this paper is not immersed in the buffer, and after electrophoresis is complete, the paper is dried, and spots are identified as they are in the low-voltage method. High-voltage paper electrophoresis is of great value for resolving amino acids and

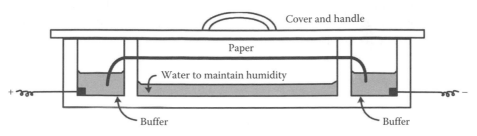

Figure 8.2 Low-voltage paper electrophoresis apparatus.

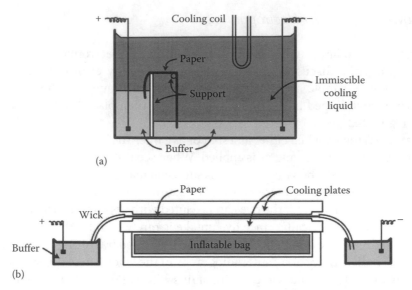

Figure 8.3 High-voltage paper electrophoresis: (a) immersed-strip method and (b) exclosed-strip method.

peptides. Because complete resolution of mixtures is not always possible in a single high-voltage electrophoresis operation, it is frequently coupled with chromatography in the 2D separation technique. The sample is first electrophoresed and then chromatographed at right angles, or vice versa.

8.4 Separation of Amino Acids by Paper Electrophoresis

8.4.1 Principle

The charge carried by a molecule depends on the pH of the medium, and this is illustrated in the case of three amino acids: aspartic acid, histidine, and lysine. Electrophoresis at low voltage is not usually used to separate low MW compounds because of diffusion, but it is easier to illustrate the relationship between charge and pH with amino acids than with proteins or other macromolecules.

At pH 7.6, histidine will carry zero net charge, aspartic acid will be negatively charged, and lysine will be positively charged (Table 8.1).

These three amino acids can, therefore, be readily separated by electrophoresis. Glucose, an uncharged molecule, is included in the mixture, to check for any movement of the origin due to electroosmosis.

8.4.2 Materials

Horizontal electrophoresis apparatus, power pack, amino acid (aspartic acid, histidine, lysine, and a mixture of all three in the Tris–acetate (TAE) buffer containing

TABLE 8.1 pK Values and pH of Zero Net Charge of Three Amino Acids

Amino Acids	pK_1	pK_2	pK_3	Isotonic Point
Aspartic acid	2.0	3.9	10.0	2.9
Histidine	1.8	6.0	9.2	7.6
Lysine	2.2	9.0	10.5	9.7

10 g/L glucose), Tris–acetate buffer (0.07 mol/L, pH 7.6), ninhydrin location reagent (dissolve 0.2 g in 100 mL of acetone just before use), paper strips (10 cm × 2.5 cm), aniline–diphenylamine reagent, citrate buffer (0.07 mol/L, pH 3.0), oven at 110°C.

8.4.3 Protocol

Fill both parts of each electrode compartment with buffer solution to the same level; check this by arranging a siphon between them. Remove the siphon, place five filter paper strips as shown (Figures 8.2 and 8.3), and carefully apply a streak of the amino acid mixture to two of these, avoiding the edge of the paper. Streak three other paper strips with only one amino acid and run all five electrophoresis strips together. Wet the paper from each electrode compartment to within a few centimeters of the point of application, leave the rest to be wetted by capillary attraction, and immediately switch on the current. This way there is minimum spreading of the sample. Carry out electrophoresis for 3 h at 8 V/cm; remove the strips and dry in an oven at 110°C. Develop one of the strips containing the mixture of glucose with aniline–diphenylamine reagent, and dip the remaining four strips rapidly in freshly prepared ninhydrin solution: allow the acetone to evaporate in the air and develop the colors by heating in the oven for a few minutes. Identify the amino acids and check for any electroosmosis.

8.5 Cellulose Acetate Strip Electrophoresis

Many biological macromolecules adsorb to cellulose (e.g., paper) by means of the cellulose hydroxyl groups. Adsorption impedes movement and, therefore, causes the tailing of spots or bands, which reduces resolution. This can be avoided by using a cellulose acetate membrane instead of paper, where most of the hydroxyls have been converted into acetate groups, which are generally nonadsorbing, so that separation is rapider at low voltage. The fact that spots are smaller also means that the material in the spot is more concentrated and easier to detect. The low adsorption of cellulose acetate also reduces background staining, thus improving the sensitivity of detection. Other advantages are that the material is transparent aiding in the spectrophotometric detection of material and is easily dissolved in various solvents, thus allowing simple elution of material. Gel electrophoresis is the method of choice for maximum resolution with variations (Figure 8.4).

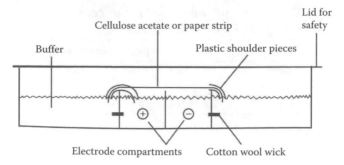

Figure 8.4 Cellulose acetate strip paper electrophoresis apparatus in a horizontal plane.

8.5.1 Separation of Serum Proteins by Electrophoresis on Cellulose Acetate

8.5.1.1 Principle
Cellulose acetate electrophoresis is a rapid and convenient method frequently used in medicine and clinical biochemistry for the analysis of serum for changes in the proteins during disease.

8.5.1.2 Materials
Horizontal electrophoresis apparatus, power pack, barbitone buffer (0.07 mol/L, pH 8.6), ponceau S protein stain (2 g/L in 30 g/L trichloroacetic acid [TCA]), acetic acid (5% v/v), serum, cellulose acetate strips (Oxoid Ltd.), Whatman 3 MM paper, methanol/water (3:2), citrate buffer (0.1 mol/L, pH 6.8).

8.5.1.3 Protocol
Moisten a strip of cellulose acetate (10 cm × 2.5 cm) by placing it on the surface of the buffer in a flat dish, and allow the buffer to soak up from below. Immerse the strip completely by gently rocking the dish, and then remove it with forceps. Lightly blot the strip, place in the apparatus and connect it to the buffer compartments with filter paper wicks. Switch on the current and adjust to 0.4 mA/cm width of the strip. Apply a streak of serum from a microliter pipette or melting point tube about one-third of the length of the strip from the cathode. This is best carried out by guiding the application with a ruler placed across the tank.

Carry out electrophoresis for 2 h, remove the strips, and stain with ponceau S for 10 min. Prior heating is not required as the TCA fixes the proteins for staining. Remove excess dye from the strip by washing repeatedly in 5% v/v acetic acid. Compare your electrophoresis pattern with Figure 8.5.

8.5.2 Identification of Proteins by Cellulose Acetate Electrophoresis Following the Fractionation of Human Blood Plasma

8.5.2.1 Principle
Plasma contains albumin, fibrinogen, and a variety of globulins that can be separated and identified by electrophoresis. These proteins can be fractionated by

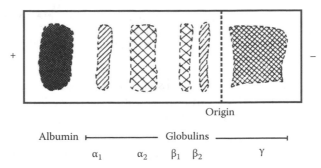

Figure 8.5   Separation of human proteins by electrophoresis on cellulose acetate at pH 8.6.

TABLE 8.2   Proteins in
Human Plasma in g/100 mL

Albumin	3.5–4.5
α-Globulins	0.6–1.2
β-Globulins	0.6–1.3
γ-Globulins	0.6–1.5
Total serum proteins	6.0–7.5
Fibrinogen	0.2–0.4

precipitation with salt or organic solvents, and electrophoresis is used to monitor the separation achieved.

Sodium sulfate is used as the salt as this is very effective at room temperature. This salt has the added advantages that it does not interfere with the assay of protein by the biuret or Folin–Lowry method, unlike the more commonly used ammonium sulfate. The amount of protein in each fraction can, therefore, be measured (Table 8.2). Methanol is a convenient organic solvent as this precipitates the globulins with little or no denaturation.

8.5.2.2 Materials
Sodium sulfite (12.6 g/100 mL), sodium sulfite (15.8 g/100 mL), sodium sulfite (21 g/100 mL), human blood plasma, methanol/water (3:2), citrate buffer (0.1 mol/L, pH 6.8), sodium hydroxide (1 mol/L), sodium chloride (0.89 g/100 mL), materials for the separation of proteins by electrophoresis.

8.5.2.3 Protocol

8.5.2.3.1 Salt Precipitation.   Add 0.5 mL of plasma to 9.5 mL of sodium sulfite solution, and leave to stand for 15 min at room temperature after mixing thoroughly. Use all three concentrations of sodium sulfite and collect the precipitates by centrifugation at 3000*g* for 10 min. Keep the supernatant from the solution containing 20% sodium sulfite, but remove the other supernatants and discard. Turn the centrifuge tubes upside down on filter paper to drain and dissolve the precipitates in a minimum volume of saline. Carry out electrophoresis on the dissolved precipitates and the 20% supernatant, and compare the patterns obtained with that of whole plasma.

8.5.2.3.2 Organic Solvent Precipitation. All steps must be carried out at 0°C. Mix 2 mL of plasma with 1 mL of the citrate buffer in a centrifuge tube. Slowly add 7 mL of ice-cold methanol in water with stirring. Leave on ice for 30 min, and then centrifuge for 10 min at 0°C. Remove the supernatant and dissolve the precipitate in saline. Carry out electrophoresis on these two solutions. Compare this separation with that obtained by salt precipitation.

8.5.2.3.3 Assay of Protein. Assay the amount of protein in each fraction by repeating the experiment and dissolving the precipitates in 2 mL of 1 mol/L NaOH. A suitable aliquot is then used to measure the amount of protein present using the biuret method.

8.6 Gel Electrophoresis

The use of gels such as starch, polyacrylamide, agarose, and agarose–acrylamide as supporting media provides enhanced resolution, particularly for proteins and nucleic acids. The earliest gel electrophoresis of protein was done with starch gels. The gel consists of a paste of potato starch whose grains have been burst by heating in the buffer. When the gel is mounted horizontally, the sample is applied to a slot cut with a razor blade either as a single solution or as a slurry with starch grains. The slot is sealed with wax or grease, and the voltage is applied. After electrophoresis, the semirigid gel is removed and frequently sliced into two or three layers, each to be differentially stained. The various components appear as a series of bands in the gel.

8.6.1 Starch Gel Electrophoresis

Starch gels are prepared by heating and then cooling a solution of partially hydrolyzed starch in an appropriate buffer. The separation is accomplished due to differences in charge as well as in size of the molecules because of the molecular-sieving effect of the gel. Assuming that the charge on the molecule of two compounds is identical, the compound with smaller molecular mass will travel at a faster rate and vice versa.

Starch gel electrophoresis was very useful at one time because of ease of visualization of bands of proteins by chemical stains and efficient elution of enzymes, which facilitated assay of enzymatic activity. Sometimes amounts of enzymes too small to be seen by stains could be assayed merely by placing a slurry of starch grains containing the enzyme into the standard assay medium. Starch gel electrophoresis provided the first evidence for the existence of isozymes, that is, different

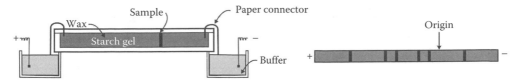

Figure 8.6   A starch gel electrogram of separated proteins.

forms of the same enzymatic activity. It is still the method of choice for analyzing many mixtures of isozymes: however, for general use with proteins, starch gel has been superseded by polyacrylamide gels (Figure 8.6).

8.6.1.1   Isolation of Lactate Dehydrogenase Isoenzymes by Starch Block Electrophoresis

8.6.1.1.1   Principle.   Isoenzymes are proteins that catalyze the same chemical reaction but differ in a number of physicochemical and kinetic properties. Lactate dehydrogenase has up to five isoenzymes, which can be separated by electrophoresis. The advantage of starch block electrophoresis is that the isoenzymes can be separated in sufficient quantities for their properties to be examined. A marker solution of bovine serum albumin stained with bromophenol blue and human hemoglobin is included to check for any drift of the pattern due to electroosmosis.

8.6.1.1.2   Materials.   Extract from leaf lettuce (*Lactuca sativa*) 500 g/500 mL in 50 mM sodium phosphate buffer or commercial dehydrogenase can be used instead.

Sodium phosphate buffer (0.1 mol/L, pH 7.4), potato starch, Perspex former for block (30 cm × 12 cm × 1 cm), absorbent gauze, electrophoresis power pack, marker solution of bovine serum albumin and human hemoglobin (dissolve bovine serum albumin in hemolyzed erythrocytes and add bromophenol blue to stain the albumin), sintered glass funnels.

Reagents for the spectrophotometric assay of lactic dehydrogenase: NADH$_2$ (3.5 mmol/L), phosphate buffer, sodium pyruvate 21 mmol/L prepared fresh in phosphate buffer, ultraviolet (UV)–Vis spectrophotometer with recorder.

8.6.1.1.3   Protocol.   Wash 500 g of potato starch by decantation twice in distilled water and twice in the phosphate buffer. After the final wash, allow the starch to sterile for about 2 h, remove the excess buffer by decantation, and compress the starch into a block using the Perspex former. Remove excess liquid from the block by blotting with Whatman 3 MM paper. Connect the block to the electrode compartment with several thicknesses of absorbent gauze saturated with buffer solution. Apply a potential of 6–8 V/cm to allow the system to equilibrate, then insert 1 mL of the tissue extract into a groove (9 cm), and cut horizontally the block 22 cm from the anodic end and 1 cm from the nearest edge. At the same time, apply the marker solution at the same distance from the anode and 1 cm from the furthest edge. Fill the groove with starch and cover the block with a Perspex plate for safety (Figure 8.7).

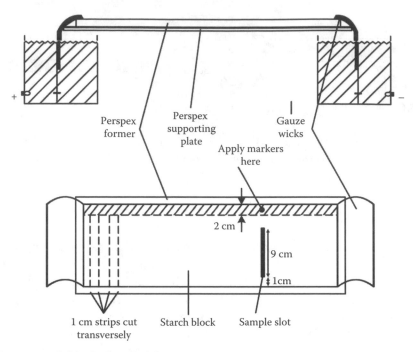

Figure 8.7   Starch block electrophoresis.

Place the whole of the apparatus in the cold room or a refrigerator and apply a potential of 6–8 V/cm for 20 h. After electrophoresis, measure the distance migrated and the spread of the markers, and then remove the part of the block containing the markers and discard it. Cut the rest of the block transversely into strips 1 cm wide and transfer each strip to a sintered glass funnel. Pack the starch tightly with a flattened glass rod, elute twice with 2 mL of ice-cold phosphate buffer pH 7.4, and record the final volume of the eluate. Remove a convenient sample (0.1–1 mL) for the assay of the dehydrogenase activity, and prepare a diagram of the total activity in each fraction against the distance from the cathode. Assay the crude extract and work out the recovery from the block.

Polyacrylamide gels have replaced starch gels because the amount of molecular sieving can be controlled by the concentration of the gel, and the adsorption of protein is negligible. Polyacrylamide is currently the most effective support medium in use for proteins, small RNA molecules, and very small fragments of DNA.

8.6.1.2 Separation of Isoenzymes of Malate Dehydrogenase Starch Gel Electrophoresis

8.6.1.2.1 Introduction.   Different isoenzymic forms of malate dehydrogenase (MDH) get separated due to the difference in their charge density and size. Separation of isoenzymes may be carried out by starch slab gel or native polyacrylamide gel electrophoresis (PAGE) in rods or slab gels. Activity of the enzyme on the gel is detected from the formation of colored formazan due to NADH-linked reduction of nitroblue tetrazolium as per the following reaction:

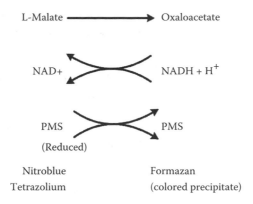

Nitroblue   Formazan
Tetrazolium   (colored precipitate)

8.6.1.2.2 Materials.   Glass plates, two glass plates ($22 \times 12 \times 0.4$ cm), and glass strips of 22 and 14 mm long, 5 mm wide, and 1 and 3 mm thick. Electrophoretic tank and power supply, young leaves, Tris hydroxyl amine, citric acid, starch, boric acid, sodium hydroxide, maleic acid, nicotinamide dinucleotide, nitroblue tetrazolium.

1. *Gel buffer (pH 8.6)*:

Tris–HCl	9.2 g
Citric acid	1.05 g

  Dissolve in distilled water, adjust pH to 8.6, and make final volume to 1 L with distilled water.
2. *Hydrolyzed starch* (electrophoresis grade) (*Note*: Hydrolyzed starch, suitable for electrophoresis, can also be prepared by warming potato starch in acidified acetone).
3. *Electrode buffer (pH 7.9)*:

Boric acid	18.6 g
Sodium hydroxide	2.0 g

  Dissolve in distilled water, adjust pH to 7.9, and then make final volume to 1 L with distilled water.
4. *Enzyme staining solution*:

Tris–HCl pH 8.0	0.1 M
L-malate	0.036 M
NAD	0.3 mg/mL
Nitroblue tetrazolium	0.8 mg/mL
Phenazine methosulfate	0.14 mg/mL

  Adjust pH of the previous mixture to 8.0.
5. *Bromophenol blue* as tracking dye.

8.6.1.2.3 Protocol
1. *Enzyme extract*: Grind 1 g young leaves in 5 mL of 0.05 M Tris–HCl buffer (pH 7.4) in a chilled pestle and mortar. Centrifuge the extract at 10,000g for 30 min at 0°C–4°C. Use the supernatant as the enzyme preparation.

2. Prepare the gel mold by sticking 1 m thick glass edge strips to all sides of one of the glass plates to give a shallow tray of 21 cm length, 14 cm width, and 1 m depth.
3. Prepare starch gel (10%, w/v) by adding 4 g of dry starch to 40 mL of gel buffer in a conical flask and boil the suspension till it becomes a viscous translucent gel.
4. Pour the hot starch solution into the gel mold avoiding any air bubble to trap. Insert a comb in the gel at a distance of about 6 cm from one of the ends and allow the gel to set for 10–15 min.
5. Remove the comb, load 15–20 μL of the leaf extract onto a piece of 3 M Whatman filter paper, and insert it in the slot created by the comb. Similarly, load the dye (0.1% bromophenol blue) in one of the slots.
6. Place the gel plate on the platform of the electrophoretic tank and connect wicks to both ends of the gel. The wicks are prepared by taking several layers of filterer paper cut to the width of the gel, soaked in electrode buffer, and then these are placed in such a way that they overlap 1–2 cm of the gel surface on both ends and dip into the electrode buffer in the buffer chambers of the apparatus.
7. Place a glass plate on the top of the gel to hold the wicks in place as well as to reduce evaporation of buffer from the gel.
8. Run the gel at 5–20 V/cm in the cold room (0°C–40°C).
9. After the electrophoresis is completed, switch *Off* and disconnect the power supply, and remove the glass plate at the top as well as the wicks. Lift the gel plate from electrophoresis apparatus and rinse it in cold 0.1 M Tris–HCl buffer (pH 8.0). Now place the gel in enzyme staining solution and incubate at 37°C for 30 min. Observe for the appearance of dark brown colored bands on the gel.
10. Measure the distance traveled by the dye and the various bands of the enzyme, and calculate the R_m values of the individual enzyme:

$$R_m = \frac{\text{Distance traveled by the protein band}}{\text{Distance traveled by the tracking dye}}$$

8.7 Polyacrylamide Gel Electrophoresis

Electrophoresis in polyacrylamide gel is the most widely used technique for analysis and characterization of proteins and nucleic acids. The use of polyacrylamide gels has several advantages: it is chemically inert, gives superior resolution, and is amenable to the preparation of gels with a wide range of pore sizes and that the gels are stable over a wide range of pH, temperature, and ionic strength.

Acrylamide [$CH_2–CH–CONH_2$] is a white, crystalline substance. It is stable inert and nonreactive with sample molecules. Activation of acrylamide by free radicals causes its polymerization into a transparent viscous material. The ammonium persulfate (APS) is the initiator of the polymerization process. The homolytic rupture of O–O bonds produces two rather stable free radicals, each with an unpaired electron at the oxygen atom:

$$NH_4-O-\overset{\overset{O}{\|}}{\underset{\underset{O}{\|}}{S}}-O-O-\overset{\overset{O}{\|}}{\underset{\underset{O}{\|}}{S}}-O-NH_4 \longrightarrow 2NH_4-O-\overset{\overset{O}{\|}}{\underset{\underset{O}{\|}}{S}}-O$$

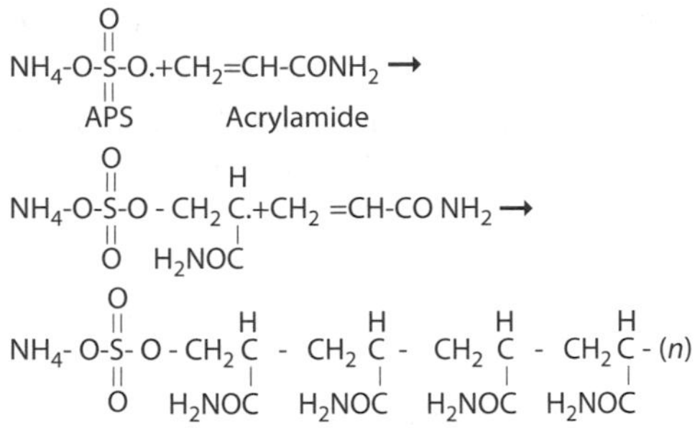

$$O$$
$$\parallel$$
$$NH_4\text{-}O\text{-}S\text{-}O.+CH_2=CH\text{-}CONH_2 \rightarrow$$
$$\parallel$$
$$APS \qquad Acrylamide$$

$$O \qquad\qquad H$$
$$\parallel \qquad\qquad \mid$$
$$NH_4\text{-}O\text{-}S\text{-}O - CH_2 \overset{\mid}{C}.+CH_2 =CH\text{-}CO\ NH_2 \rightarrow$$
$$\parallel \qquad \mid$$
$$O \quad H_2NOC$$

$$O$$
$$\parallel \qquad H \qquad H \qquad H \qquad H$$
$$NH_4\text{-}O\text{-}S\text{-}O - CH_2 \overset{\mid}{C} - CH_2 \overset{\mid}{C} - CH_2 \overset{\mid}{C} - CH_2\overset{\mid}{C}\text{-}(n)$$
$$\parallel \qquad\ \mid \qquad\quad \mid \qquad\quad \mid \qquad\quad \mid$$
$$O \quad H_2NOC \quad H_2NOC \quad H_2NOC \quad H_2NOC$$

Figure 8.8 Acrylamide chain with *n* number of monomeric molecules.

This radical is added to the double bond of the acrylamide molecule to form another free radical at the carbon atom bearing an unpaired electron. This carbon-centered radical in turn attacks the double bond of another acrylamide monomer giving rise to a new free radical, and the process continues until all the available acrylamide molecules are chained together to form polyacrylamide and two free radicals react to form a covalent bond (Figure 8.8).

The cations have no influence on initiation of polymerization and potassium persulfate may be used equally well. The APS-initiated polymerized viscous acrylamide cannot form a suitable gel unless it is hooked together by a cross-linking agent. This is done by *N,N'*-methylene-bisacrylamide [CH$_2$(NHCOCH–CH$_2$)$_2$] commonly known as bis. One of the vinylic groups of bis is incorporated into the growing linear chain of acrylamide, and another (second) can be built into the other linear polymer chain, thus forming a cross-linkage (Figure 8.9).

Both acrylamide and bis are toxic substances affecting the skin and nervous system; hence, they should be handled with care and skin contact must be avoided. Solutions should be prepared under the hood and rubber gloves should be used. Polymerized substance is almost nontoxic.

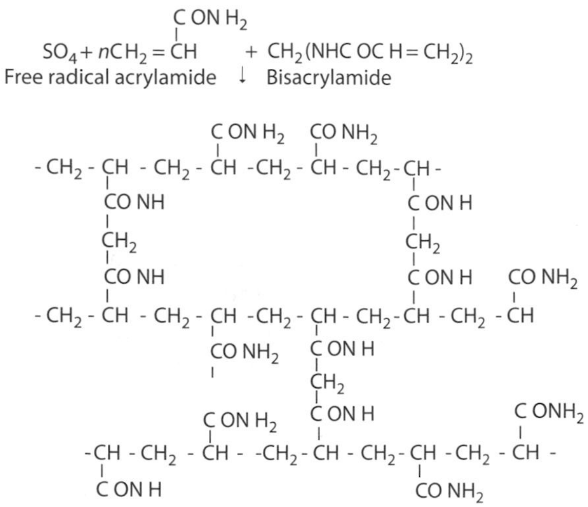

$$\qquad\qquad C\,ON\,H_2$$
$$\qquad\qquad \mid$$
$$SO_4 + nCH_2 = CH \qquad + CH_2\,(NHC\,OC\,H = CH_2)_2$$
$$Free\ radical\ acrylamide \quad \downarrow \quad Bisacrylamide$$

$$\qquad\qquad C\,ON\,H_2 \quad CO\,NH_2$$
$$\qquad\qquad \mid \qquad\quad \mid$$
$$-CH_2 - CH - CH_2 - CH -CH_2 - CH - CH_2\text{-}CH -$$
$$\qquad\quad \mid \qquad\qquad\qquad\qquad \mid$$
$$\qquad\quad CO\,NH \qquad\qquad\qquad C\,ON\,H$$
$$\qquad\quad \mid \qquad\qquad\qquad\qquad \mid$$
$$\qquad\quad CH_2 \qquad\qquad\qquad\quad CH_2$$
$$\qquad\quad \mid \qquad\qquad\qquad\qquad \mid$$
$$\qquad\quad CO\,NH \qquad\qquad\qquad C\,ON\,H \quad CO\,NH_2$$
$$\qquad\quad \mid \qquad\qquad\qquad\qquad \mid \qquad\qquad \mid$$
$$-CH_2 - CH - CH_2 - CH -CH_2 - CH - CH_2\text{-}CH - CH_2 - CH$$
$$\qquad\qquad CO\,NH_2 \quad C\,ON\,H$$
$$\qquad\qquad \mid \qquad\qquad CH_2$$
$$\qquad\qquad\qquad\qquad\qquad \mid$$
$$\qquad\qquad C\,ON\,H_2 \quad C\,ON\,H \qquad\qquad C\,ONH_2$$
$$\qquad\qquad \mid \qquad\qquad \mid \qquad\qquad\qquad \mid$$
$$-CH - CH_2 - CH - \text{-}CH_2\text{-} CH - CH_2\text{-}CH - CH_2 - CH -$$
$$\qquad\quad \mid \qquad\qquad\qquad\qquad\qquad \mid$$
$$\qquad\quad C\,ON\,H \qquad\qquad\qquad\qquad CO\,NH_2$$

Figure 8.9 Vinyl polymerization.

Riboflavin may also be used as a more potent initiator of polymerization gel mixture in place of APS. Riboflavin is illuminated by a fluorescent lamp (445 nm) for 30–45 min. It is reduced to leucoflavin that is oxidized by dissolved oxygen with the formation of hydrogen peroxide. H_2O_2 decomposes to generate free radicals (HO) necessary for chain polymerization reaction.

The pore size in PAG depends on the concentration of the acrylamide and bisacrylamide and on the ratio of bis- to acrylamide. These two parameters can be critically controlled to get the desired pore size. Gels with high bisacrylamide content are fragile and nontransparent. They easily break away from the glass surface, show cracking during gel drying, and stain intensively. Gels ranging from 3% to 30% acrylamide concentration can be made, and these can be used for separation of molecules of size up to 1×10^6 D. A gel with a low percentage has larger pore size and is suitable for separation of high MW compounds, and a high-percentage gel has smaller pore size and is used for separation of relatively low MW compounds. Gradient gels with linear gradients of increasing acrylamide concentration give better resolution and are quite commonly used. Polyacrylamide gels can also be prepared by using (1) ethylene diacrylate (EDA) $[CH_2=CHCOOCH_2=CH_2]$ and (2) N,N'-diallyltartardiamine (DATD) $[CH_2=CHCH_2–NHCO–CHOH–CH_2CH=CH_2]$.

The gels formed by the previous cross-linking agents are free from the drawbacks as observed with high concentrations of bis. However, these gels can be solubilized. In EDA gels, the ester bonds can be cleaved by treatment with aqueous alkali or piperidine. Similarly, DATD gels are soluble in 2% aqueous periodic acid at room temperature within 20–30 min. EDA and DATD gels are used for some specific applications.

Tetramethyl ethylenediamine (TEMED) $[(CH_3)_2NCH_2CH_2N)]$ is a colorless liquid. It is usually added along with APS at a concentration of about 0.4% in gel mixture to serve as a powerful catalyst of the polymerization process because it exists in active free radical form. Dimethylaminopropionitrile (DMPAN) $[(CH_3)2NCH_2CH_2N)]$ is a more potent catalyst than TEMED and requires three to four times less concentration.

8.7.1 Apparatus

8.7.1.1 Rod Gel Electrophoresis

It consists of power pack and an electrophoresis unit. The power pack supplies a stabilized direct current at controlled or required voltage and current output. The electrophoresis unit contains the electrodes, buffer reservoirs, and gel casting assembly. Initially, electrophoresis in polyacrylamide gels was carried out in cylindrical rods of gel in glass tubes (0.7 m in diameter and 10 cm in length) that were then fixed in a specially designed apparatus (Figure 8.10). The stacking gel is of a large pore size, to concentrate the sample so that it enters the separation gel as a tight band. The electrophoresis mobility of the glycinate ion is very much less than the chloride ion in the separation gel so that a sharp boundary is

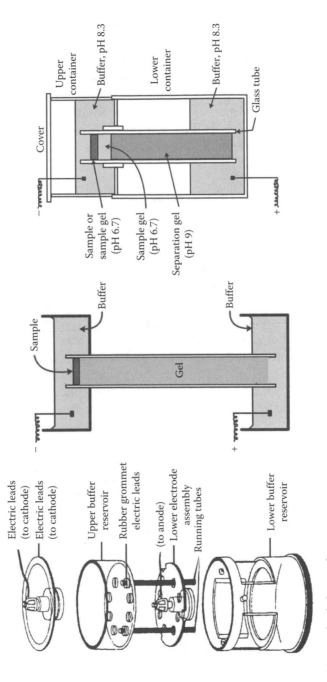

Figure 8.10 Rod gel electrophoresis apparatus.

maintained between these ions with proteins of intermediate mobility sandwiched in between. This sharp boundary is maintained during electrophoresis, and at the pH of the separation gel (pH 8.9), the glycine mobility is greater than that of the protein, so the buffer boundary always runs ahead of the molecules being separated. Bromophenol blue is incorporated into one of the gels as a marker, and this dye marks the boundary between the glycinate and chloride ions. The buffer system is, therefore, a discontinuous one, and this gives rise to another name for this separation method, namely, disc electrophoresis. This particular method of electrophoresis gives very sharp bands. The pore size of the gel can be altered by varying the concentration of monomers in the gel solution. Proteins can be conveniently separated on 7.5% acrylamide, but larger molecules such as ribosomal nucleic acids require a more open gel of 2.5% acrylamide.

8.8 Polyacrylamide Disc Gel Electrophoresis

8.8.1 Principle

Polyacrylamide: The gel is prepared by polymerizing acrylamide ($CH_2=CH\cdot CO\cdot NH_2$) and a small quantity of cross-linking reagent, methylenebisacrylamide ($CH_2=CH\cdot CO\cdot NH)_2\cdot CH_2$ (bis) in the presence of a catalyst, APS. TEMED is also present to initiate and control the polymerization. The gel mixture is allowed to polymerize in small tubes sealed at the bottom with a rubber cap. A layer of water is placed on top of the gel to ensure a flat surface and also to exclude oxygen that inhibits the polymerization.

The pore size of the gel can be altered by varying the concentration of monomers in the gel solution. Proteins can be conveniently separated on 7.5% acrylamide, but larger molecules such as ribosomal nucleic acids require a more open gel of 2.5% acrylamide.

Apparatus: A diagrammatic representation of the apparatus used for PAGE on rods is shown in Figure 8.10. This particular method of electrophoresis gives very sharp bands as can be seen in the following experiments.

Toxicity: Acrylamide is toxic as marked on the reagent bottle and should be handled with care, particularly avoiding contact of the material with the skin.

8.8.2 Materials

8.8.2.1 Stock Solutions
Prepare all the solutions except A and B in 100 mL volumetric flasks with distilled water, filter, and store in brown bottles at 4°C:

   A. *Stock Tris buffer*: Tris, 6 g; glycine, 28.8 g made up to 1 L with water (pH 8.3).
   B. *Working Tris buffer*: Dilute the stock solution 1 in 10 with distilled water.
   C. Tris 36.6 g; 1 mol/L HCl, 48 mL; TEMED, 0.23 mL, and water to 100 mL (pH 8.9).

D. Tris, 6.0; 1 mol/L HCl, 48 mL; TEMED, 0.46 mL, and water to100 mL (pH 6.7, adjust with HCl if required).

E. Acrylamide, 28 g; bis, 0.74 g, water to 100 mL. *Caution*: Toxic.

F. Acrylamide, 10 g; bis, 2.5 g, water to 100 mL. *Caution*: Toxic.

G. Riboflavin, 4 mg, water to 100 mL.

H. Sucrose (40 g/100 mL).

I. Ammonium persulfate (0.14 g/100 mL).

8.8.2.2 *Working Solutions*

Prepare the working solutions by mixing the stock solution in the following proportions:

1. *Small-pore solution*: One part of C, two parts of E, and one part of water, pH 8.9, 40 mL.
2. *Large-pore solution*: One part of D, two parts of F, one part of G, and two parts of H, pH 6.7, 16 mL.

8.8.2.3 *Samples*

Serum: albumin, 5 mg/mL; γ-globulin, 2 mg/mL; transferrin, 2 mg/mL, 2 mL each.

8.8.2.4 *Staining Solutions*

Amido black, 1 g/L in 7% v/v acetic acid.
Coomassie blue, 2.5 g/L in 200 g/L TCA.
Acetic acid, 7% v/v.
Tracking dye (bromophenol blue, 0.1 g/L).

8.8.2.5 *Equipments*

PAGE apparatus
Fluorescent light
Syringe and needle

8.8.3 *Protocol*

Place rubber caps over the bottom of the hollow glass tubes, and prepare the gels as given in the following discussion.

8.8.3.1 *Separation Gel*

(Seven percent polyacrylamide) Mix equal volume of the catalyst solution of ammonium persulfate (J) and the small-pore solution (1), and transfer 0.9 mL to each glass tube. Carefully add a water overlay and leave the gels to set (25–40 min).

8.8.3.2 *Spacer Gel*

(2.5% polyacrylamide) Remove the water overlay from the rods of polyacrylamide and layer 0.15 mL of the large-pore solution over the top. Overlay the gel solution with water and place the tubes under a fluorescent light until gelation is complete (20–50 min.). Finally, cover the gel with a layer of dilute Tris buffer, pH 8.3 (B).

8.8.3.3 Sample Application

Mix the following sample with some of the 40% w/v sucrose (H) so that they will layer on top of the gel underneath the buffer. Add some bromophenol blue to one of the serum samples to mark the ion boundary.

Tube No.	Sample
1 and 2	50 μL of serum
3 and 4	50 μL of albumin (5 mg/mL)
5 and 6	50 μL of γ-globulin (2 mg/mL)
7 and 8	50 μL of transferrin (2 mg/mL)

8.8.3.4 Electrophoresis

Remove the rubber caps at the bottom of the tubes, connect up the buffer reservoirs, and carry out electrophoresis at 5 mA, until the bromophenol blue has migrated almost to the end of the tube.

8.8.4 Staining

Switch off the current. Remove the gels inserting a syringe needle between the gel and the wall of the tube, and carefully discharge water from the syringe while rotating the tube. Collect the gels in test tubes and stain with amido black or Coomassie blue. Destain the gels by repeatedly washing in 7% v/v acetic acid until the background is clear. Make a sketch of separation and compare the results for serum with those obtained for cellulose acetate electrophoresis in an earlier experiment.

8.9 SDS Gel Electrophoresis

8.9.1 Principle

Sodium dodecyl sulfate (SDS) (CH_3 (CH_2)$_{10}$ $CH_2OSO_3^-Na^+$) is a detergent that readily binds to proteins. At pH 7, in the presence of 1% w/v SDS and 2-mercaptoethanol, proteins dissociated into their subunits and bind large quantities of the detergent. Under these conditions, most proteins bind about 1.4 g of SDS/g of protein that completely masks the molecules; therefore, the greater the charge, so the electrophoretic mobility of the complex depends on the size (mol. wt) of protein, and a plot of log mol. wt against relative mobility gives a straight line (Figure 8.11). In this experiment, the MW of a protein is determined by comparing its mobility with a series of protein standards. The sieving effect of the polyacrylamide is important in these techniques, and the range of MWs that can be separated on a particular gel depends on the "pore

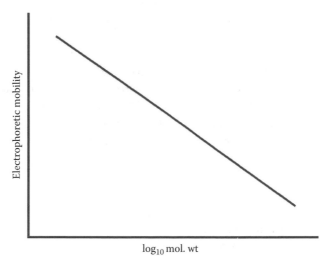

Figure 8.11   Mobility variation with the MW of proteins in SDS gel.

size" of the gel. The amount of cross-linking and hence pore size in a gel can be varied by simply altering the amount of acrylamide to make 5% or 10% gels.

8.9.2 Materials

PAGE disc gel apparatus, SDS buffer (sodium phosphate, 0.2 mol/L, pH 7.1 containing 0.2% w/v SDS), TEMED (5% v/v), acrylamide solution (dissolve 50 g of acrylamide and 1.36 g of bis in water to 200 mL. Filter if necessary and store in a dark bottle. *Caution*: The monomer is highly toxic; avoid inhalation and skin contact), sample buffer (sodium phosphate, 0.1 mol/L, pH 7.1 containing 50 mL 1% w/v SDS and 1% w/v 2-mercaptoethanol), ammonium persulfate (1% w/v, prepare fresh), water baths at 37°C, vacuum flask (200 mL), 2-mercaptoethanol, bromophenol blue (0.25% w/v in the sample buffer), glycerol, Coomassie blue (0.2% w/v in methanol/acetic acid/ water 5:1:5). Filter and store in a dark bottle, acetic acid (7% w/v), unknown protein α-chymotrypsin, standard protein as marker.

Protein	Mol. Wt	$\log_{10}$ Mol. Wt
Insulin	5,700	3.76
Cytochrome *c*	12,400	4.09
Myoglobin	17,200	4.24
Ovalbumin	45,000	4.65
Transferrin	77,000	4.89
Bovine serum albumin		
Monomer	66,000	4.82
Dimer	132,000	5.12
Tetramer	264,000	5.42

8.9.3 Method

8.9.3.1 Preparation of Gel

The gels are prepared by mixing the components to give 5% and 10% polyacrylamide. The first three components are mixed together and deaerated in a vacuum flask for a few minutes. The ammonium sulfate and TEMED are then added and the mixture immediately introduced into the tubes. Distilled water is then added, down the side of the tube, to ensure a flat surface to gel and to exclude air.

Component	5% Gel (mL)	10% Gel (mL)
Distilled water	10.4	0.4
SDS buffer	25.0	25.0
Acrylamide solution	10.0	20.0
Ammonium persulfate	3.6	3.6
TEMED	1.0	1.0
Total volume	50.0	50.0

The gels polymerize in a few hours but are best if kept overnight and used the next day. In order to minimize evaporation and prevent contamination from atmospheric dust, a layer of parafilm is placed over the top of the gels.

8.9.3.2 Preparation of Samples

Dissolve about 1 mg of the proteins in 1 mL of the sample buffer, and incubate for 3 h at 37°C or heat for 2 min at 100°C in a fume chamber. Some proteins are sensitive and require the milder treatment, and a comparison can be made of the two methods of denaturation.

8.9.3.3 Electrophoresis

Remove the water from the surface of the gel and apply the sample of proteins to the top of the gel. For the application, 100 µL of the protein solution is mixed with 45 µL of water, 45 µL of SDS buffer, 5 µL of 2-mercaptoethanol, 10 µL of bromophenol blue, and 15 µL of glycerol. The 2-mercaptoethanol maintains a reducing environment; the bromophenol blue is present as a tracking dye and the glycerol increases the density of the mixture. The rest of the tubes is then filled with half-strength SDS buffer and electrophoresis carried out at 8 mA/tube until the tracking dye is almost at the end of the gel. The buffer used in the electrode compartments is prepared by diluting the SDS buffer with an equal volume of distilled water.

8.9.3.4 Fixing and Staining

Carefully remove the gels from the tubes with a fine hypodermic syringe and wash with distilled water. This procedure requires practice to avoid breaking the gel rods. Measure the length of each gel and the distance migrated by the tracking dye, and then immerse in a solution of Coomassie blue overnight.

Remove the dye solution, wash the gel with water, and destain by washing repeatedly with 7% w/v acetic acid.

8.9.4 Mobility Determination

Measure the distance traveled by each protein and calculate the following mobility relative to the tracking dye:

$$\text{Mobility} = \frac{\text{Distance moved by protein}}{\text{Distance moved by bromophenol blue}}$$

Check for any shrinkage in the gels following the staining and allow for this in the calculation of the mobility.

Plot a curve of mobility against $\log_{10}$ mol. wt for the 5% and the 10% gels, and determine the MW of the chymotrypsin.

8.10 Slab Gel Electrophoresis

It is more commonly used. Slab gel (0.75–1.5 m thick) is a gel that is in the form of a rectangular or square flat slab. There are two types of slab gel apparatus: (1) vertical and (2) horizontal. The vertical type is commonly used for protein separation (Figure 8.12), and the horizontal type is used for nucleic acid separation (Figure 8.13).

Slab gels have the advantage over rod gels since a number of samples can be coelectrophoresed under identical conditions in a single gel so that band patterns can be directly compared. On the other hand, only one sample can be loaded onto each gel rod. Also, the heat produced during electrophoresis is more easily dissipated by thin slab gels as compared to thick rod gels, thus reducing distortion of the band due to heating.

8.10.1 Buffer Systems

Gel electrophoresis can be carried out in buffers ranging from pH 3 to 10. Ionic strength of the buffers is generally kept low (0.01–0.1 M) to reduce heat production during electrophoresis to a minimum. Depending upon the types of buffer systems used, PAGE may be of the following four different kinds:

8.10.1.1 Native Continuous Polyacrylamide Gel Electrophoresis
In native continuous PAGE (Native PAGE), the same buffer ions are present throughout the sample, gel, and electrode vessel reservoirs at constant pH as such

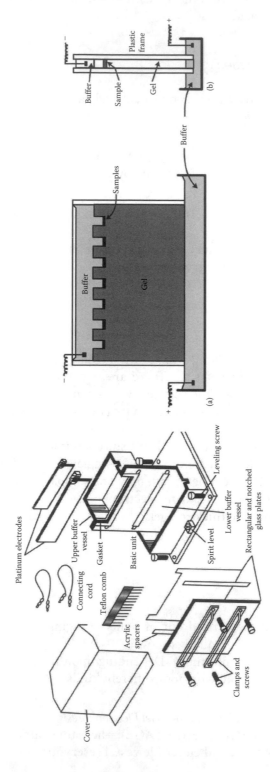

Figure 8.12   Vertical slab gel electrophoresis apparatus (a) front view and (b) side view.

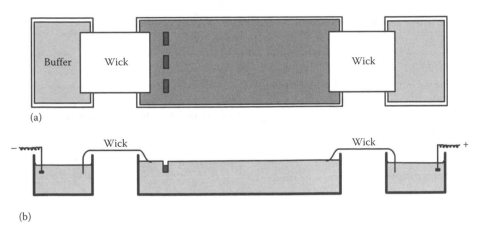

Figure 8.13   Horizontal gel electrophoresis apparatus (a) front view and (b) side view.

that the proteins have native conformation and biological activity, and hence, it is known as native continuous PAGE. In this system, protein sample is loaded directly onto the gel, referred to as resolving gel, in which separation of the sample components occurs.

8.10.1.2 Native Discontinuous Polyacrylamide Gel Electrophoresis (Disc Gel Electrophoresis)
In contrast to continuous system, discontinuous or disc gel electrophoresis embodies (1) contains different buffer ions in the gel than those in the electrode reservoirs (2) consists of two types of gels, a large-pore stacking gel that is polymerized on top of small-pore resolving gel. The buffers used to prepare these two gels are different in ionic strength and pH (3); most discontinuous buffer systems have discontinuities of both buffer composition and pH (Table 8.3).

Discontinuous buffer systems have an advantage over continuous systems as relatively large volumes of dilute solutions of protein samples can be applied onto the gels and yet good resolution of sample components can be obtained. This is because the proteins in applied samples get concentrated into extremely narrow zones or get stacked during migration through large-pored stacking gel prior to their entry and separation in small-pored resolving gel. The process of sample concentration in the gel is as follows.

Let us consider a discontinuous system where electrode buffer is Tris–glycine, pH 8.3, sample and stacking gel contain Tris–HCl buffer of pH 6.8, and the resolving gel contains Tris–HCl buffer of pH 8.8. At the pH of the sample and stacking gel (pH 6.8), glycine is very poorly ionized so that its effective mobility is very low and chloride ions have a much higher mobility at this pH, while mobilities of proteins are intermediate between that of chloride and glycine. The moment voltage is applied, the chloride ions (leading ions) migrate away from the glycine ions (the trailing ions) leaving behind a zone of lower conductivity. Since conductivity is inversely proportional to field strength, this zone attains a higher-voltage gradient that now accelerates glycine molecules so that it keeps up with the chloride ions. A steady state is established where the products of mobility and voltage gradient for

TABLE 8.3   Buffer for Native Discontinuous PAGE System

Types of Buffer	Preparation of Buffer
High-pH buffer system	
Stacking gel buffer (Tris–HCl, pH 6.8)	Tris 6 g 1 N HCl 48 mL Adjust pH to 6.8 and make final volume to 100 mL with water
Resolving gel buffer (Tris–HCl, pH 8.8)	Tris 36.3 g 1 N HCl 48 mL Adjust pH to 8.8 and make final volume to 100 mL with water
Reservoir buffer (Tris–glycine, pH 8.3)	Tris 3 g Glycine 14.4 g Adjust pH to 8.3 and make final volume to 1 L with water
Neutral pH buffer	
Stacking gel buffer (Tris–HCl, pH 5.5)	Tris 6.85 g Water 40 mL Adjust pH to 7.5 with 1 M orthophosphoric acid and make final volume to100 mL with water
Resolving gel buffer (Tris–HCl, pH 7.5)	Tris 6.85 g Water 40 mL Adjust pH to 7.5 with 1 N HCL and make final volume to 100 mL with water
Reservoir buffer (Tris–ethyl barbiturate, pH 7.0)	Diethylbarbituric acid 55.2 g Tris 10.0 g Add water to make up to 1 L
Low-pH buffer	
Stacking gel buffer (acetic acid–KOH, pH 6.8)	1 N KOH 48 mL Glacial acetic acid 2.9 mL Adjust pH to 6.8 and make volume to 100 mL with water
Resolving gel buffer (acetic acid–KOH, pH 4.3)	1 N KOH 48 mL Glacial acetic acid 17.2 mL Adjust pH to 4.3 and make volume to 100 mL with water
Reservoir buffer (acetic acid, β-alanine, pH 4.5)	β-Alanine 31.2 g Acetic acid 8 mL Add water to make up to 1 L

glycine and for chloride are equal, and these charged species now move at the same velocity with a sharp boundary between them. As this glycine/chloride boundary moves through the sample and then the stacking gel, a low-voltage gradient moves before the moving boundary and a high-voltage gradient behind it. Any proteins, in front of the moving boundary, are rapidly overtaken since they have lower velocity than the chloride ions. Behind the moving boundary, in the higher-voltage gradient, the proteins have a higher velocity than glycine. Thus, the moving boundary sweeps up the proteins so that they become concentrated into very thin zones or "stacks."

When the moving boundary reaches the interface of the stacking and resolving gels, the pH of the gel increases markedly, and this leads to a large increase in the degree of dissociation of glycine. Therefore, the effective mobility of glycine increases so that glycine overtakes the proteins and now migrates directly behind the chloride ions. At the same time, the gel pore size decreases, thereby retarding the migration of the proteins because of molecular sieving. These two effects cause the proteins to be unstacked. The proteins move in the zone of uniform-voltage gradient and pH value that are now separated according to their intrinsic charge and size.

8.10.1.3 Native Discontinuous Polyacrylamide Gel Electrophoresis (Nonreducing) (SDS-PAGE)

In this system, proteins are dissociated into their constituent subunits using anionic detergent such as SDS or sodium lauryl sulfate (SDS). The proteins mixture is denatured by heating the sample at 100°C in the presence of excess SDS. Under these conditions, most polypeptides bind SDS in a constant weight ratio (1.4 g SDS/g polypeptide). The intrinsic charges of the polypeptides are insignificant compared to negative charges provided by the bound detergent. Consequently, SDS polypeptide complexes have essentially identical charge densities (charge/mass ratio), and these migrate in polyacrylamide gels of an appropriate porosity strictly according to polypeptide size or their molecular mass. Thus, in addition to the analysis of polypeptide composition of the sample, the MW of the polypeptides of the sample protein can also be determined by using standard MW markers. SDS-PAGE can be carried out using either continuous or discontinuous buffer systems (Table 8.4). Resolution is, of course, superior in a discontinuous system.

Problem arises during the electrophoresis in the settling of the sample and exact tracking of the protein position in the gel. For settling the problem, sucrose or glycerol solution is used, and for tracking the protein, bromophenol blue is used. The protein first of all is loaded in the stalking gel. In this master mix,

TABLE 8.4   Buffers for SDS-PAGE

Buffer System	Buffers

Continuous system	Resolving gel buffer
	  0.1 M sodium phosphate, pH 7.2
	Reservoir buffer
	  0.1 M sodium phosphate, pH 7.2
	  1% SDS

Discontinuous system	Stacking gel buffer
	  0.125 M Tris–HCl, pH 6.8
	Resolving gel buffer
	  0.375 M Tris–HCl, pH 8.8
	Reservoir buffer
	  0.025 M Tris
	  0.192 M Glycine
	  0.1% SDS
	  pH 8.3

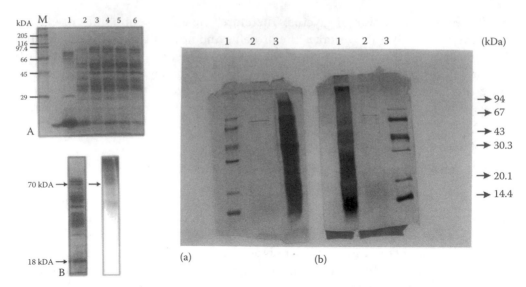

Figure 8.14   (See color insert.) SDS-PAGE analysis of purified proteins with MW markers stained with (a) Coomassie blue and (b) silver stain.

glycine is added. This is done to sharpen the protein band. Please note that glycinate ion has a lower electrophoresis mobility than protein–SDS complex, which has lower mobility than chloride ion of the loading buffer and the stacking gel Cl^- > protein–SDS > glycinate. Therefore, protein–SDS band lies in between the Cl^- and glycinate ion. pH of the stacking gel is kept generally 6.8, and resolving gel has pH of about 8.0. The negatively charged protein is attracted toward the anode. After the protein reaches the bottom, the gel is removed and placed in stain (more generally Coomassie blue). The gel is then placed in the destained solution. Staining of gel is done for 2–3 h and destaining requires overnight. Generally 15% polyacrylamide gel is used in the separating gel. This can allow separation of the protein in the range of 100,000–10,000 kDa. For the protein of MW more than 150,000 kDa, 7.5% gel is used. The MW of the protein can be determined by converting its mobility with those of a number of proteins used as standard (Figure 8.14).

8.10.1.4 Gel Electrophoresis SDS-Polyacrylamide (Reducing)

It is similar to SDS-PAGE described previously except that, in addition to SDS, a thiol reagent such as β-mercaptoethanol is included that cleaves the disulfide bonds. Hence, polypeptides in a native protein that still remain associated even in the presence of SDS get separated. In this method, a protein sample is boiled in the presence of excess SDS and β-mercaptoethanol to denature proteins and their individual polypeptides. SDS polypeptide complexes (1.4 SDS/1 protein) have net negative charge and migrate toward the anode at rates based solely on size or MW of the polypeptides.

8.10.2 Staining of Proteins in Polyacrylamide Gels

After electrophoresis, protein bands in gel can be visualized by staining with dyes such as amido black 10B, Coomassie brilliant blue R-250, or silver staining. Coomassie brilliant blue R-250 is the most commonly used stains though silver stain is about 100 times more sensitive and can detect up to 1 ng protein/band.

8.10.3 Detection, Estimation, and Recovery of Protein Gel

Coomassie blue R-250 is most commonly used for the detection of the protein in the gel. Staining is done in the 0.1% (w/v) CB in methanol/water/glacial acetic acid. This acid–methanol mixture acts as a denaturant to precipitate or fix the protein in the gel. This prevents the protein from being washed out while it is stained. Coomassie blue is highly sensitive, which means it can detect the 0.1 μg of protein. Silver stain is more sensitive. $Ag+$ ion is reduced to metallic silver on the protein, where silver is deposited to give a black band. It can be used immediately after the electrophoresis. It is 100 times more sensitive than Coomassie blue. Glycoprotein can be detected by the stain called as periodic acid Schiff [PAS] stain. It gives pink red bands. It is difficult to observe in the gel. Quantitative analysis of the protein can be done by the method called as scanning densitometry. Protein can be further analyzed, and protein can be purified. Protein bands can be cut out of protein and sequenced by the gas phase sequencer.

8.11 Native Disc Gel Electrophoresis of Proteins

8.11.1 Principle

The separation of proteins in native polyacrylamide gels is based on both charge density and size of the molecules.

8.11.2 Materials

PAGE disc gel apparatus with glass tubes (10 cm × 7 diameter), stand for glass tubes, upper and lower troughs, electrophoresis leads, constant voltage power supply (0–500 V), sucrose (40%, w/v). Weigh 40 g of sucrose and dissolve in 100 mL of water, 0.01% bromophenol blue solution. Take 10 mg of bromophenol blue and dissolve in 100 mL of water as tracking dye.

Acrylamide–bisacrylamide stock solution:

Acrylamide	30.0 g
Bisacrylamide	0.8 g

Dissolve in water and make final volume to 100 mL. Filter the solution through Whatman No. 1 filter paper and store in a brown bottle at 0°C–4°C. This solution is stable for 1 month.

Stacking gel buffer stock (Tris–HCl pH 6.8):

Tris	36.0 g
1 M HCl	48.0 mL

Adjust its pH 8.3 and then make the final volume to 100 mL, filter through Whatman No. 1 filter paper, and store at 0°C–4°C.

Resolving gel buffer stock (Tris–HCl, pH 8.8):

Tris	36.3 g
1 M HCl	48.0 mL

Adjust its pH 8.8 and then make the final volume to 100 mL. Filter through Whatman No. 1 filter paper and store at 0°C–4°C.

1.5% (w/V) APS: Prepare by dissolving 0.15 g of APS in 10 mL water. This reagent should be prepared fresh just before use.

TEMED 5%:

Reservoir Buffer or Electrode	Buffer (Tris–Glycine, pH 8.3)
Tris	3.0 g
Glycine	14.4 g

Adjust pH to 8.3 and make final volume to 1 L with double distilled water (Table 8.3).

Staining solution:

Coomassie brilliant blue R-250	1.25 g
Methanol	200 mL
Glacial acetic acid	35 mL

Make final volume to 500 mL with distilled water. Filter to remove any undissolved material and store at room temperature.

Destaining solution:

Glacial acetic acid	75 mL
Methanol	500 mL

Mix the previously given components and add water, to make its final volume to 1:1.

8.11.3 Sample Preparation

To 0.2 mL of a solution of the sample preparation containing about 3–4 mg protein/mL, add 0.1 mL sucrose (40%, w/v) and 0.1 mL bromophenol blue (0.01% w/v) solution just prior to loading the sample. In place of sucrose, glycerol at a final concentration of 10% (v/v) may be used. Add one drop of glycerol with a pipette.

8.11.4 Preparation of the Gel Rods

1. Take clean glass tubes and place them vertically in the rubber septum in the rack or tube stand so that the tubes fit snuggly in it.
2. Mix all the components of the resolving gel, except TEMED and APS as indicated in the previous table. Degas the mixture for 1 min using a water pump and then add APS and TEMED in the required quantities; mix gently but quickly (Table 8.5).
3. Without any delay, fill the resolving gel in each tube to the mark of tubes. The positions of resolving and stacking gels on all tubes should be marked with the help of a glass marker using Pasteur pipette to avoid trapping of any air bubbles.
4. Overlay the gel with a drop of water as gently as possible to ensure a flat meniscus. Water also avoids contact with oxygen that inhibits polymerization.
5. Allow the gel to stand for 20–30 min for polymerization.
6. When the gel has polymerized, remove water from the top of the gel and rinse surface of the gel with stacking gel buffer.
7. Mix various components of stacking gel as specified in the table except APS and TEMED. Degas the solution as previously given and add indicated amounts of APS and TEMED. Pour the stacking gel up to the mark that should be of about 1 cm in height on top of the resolving or running gel, and allow it to polymerize for 10–15 min when completely polymerized; rinse its surface with reservoir buffer.
8. Remove rubber septum from the base of the glass tubes containing rod gels. Label the tubes and place them in rubber grommets of the electrophoretic system with stacking gel toward the upper side. The lower end of each tube must dip in the buffer into the lower unit, so that upper ends of all the tubes are submerged in buffer. Remove any air bubbles from the tubes that might have gotten entrapped while adding the buffer.
9. Load 0.1 mL of the protein sample with a syringe or a micropipette on the top of glass tubes. Only one sample should be loaded onto each of the tube.

TABLE 8.5 Composition of Stacking and Resolving Gels

Reagents	Stacking Gel 2.5%	Resolving Gel 7.5%
	mL Solution Required	
Acrylamide–bis stock solution	2.5	7.5
Stacking gel buffer stock solution (Tris–HCl, pH 6.8)	5.0	—
Resolving gel buffer stock solution (Tris–HCl, pH 8.8)	—	3.75
1.5% APS	1.50	1.50
Water	11.0	17.25
TEMED	0.015	0.15
Total volume	20.00	30.00

10. Carry out electrophoresis in cold (0°C–4°C). Initially supply a current of 2 mA/tube for 10 min, and then increase it to 3 mA/tube and continue electrophoresis until the tracking dye migrates to about 1 cm from the bottom end of the tubes. This may take 1–2 h. Turn off the power supply and disconnect the electrophoretic apparatus from the power pack, and take out the glass tubes. Remove the gels from the tubes by squirting water with a syringe along the sides between the wall of the glass tube and gel while constantly rotating the tubes.

11. Place the gels in Petri plates or screw-capped tubes. Add sufficient amount of the staining solution to the Petri plates or tubes so as to completely cover the gels, and allow them to stand for 3–4 h for staining (Figure 8.14).

12. For destaining, transfer the gels to destaining solution. Change the destaining solution frequently until the background of the gels is clear.

13. Measure the distance traveled by the tracking dye and also by the different protein bands, and calculate the following given relative mobility (R_m):

$$R_m = \frac{\text{Distance traveled by the protein band}}{\text{Distance traveled by the tracking dye}}$$

14. Preserve the gels 7% acetic acid till these can be photographed for permanent record.

8.12 SDS-Polyacrylamide Slab Gel Electrophoresis of Proteins under Reducing Conditions (SDS-PAGE)

8.12.1 Principle

SDS is a detergent that readily binds to proteins. At pH 7 in the presence of 1% w/v SDS and 2-mercaptoethanol, proteins dissociate into their subunits and bind large quantities of the detergent. Under these conditions, most proteins bind about 1.4 g of SDS/g of protein, which completely masks the molecules; therefore, the greater the charge, the more the electrophoretic mobility of the complex depends on the size (mol. wt) of protein and a plot of log mol. wt against relative mobility gives a straight line (Figure 8.11).

8.12.2 Materials

1. Dry powder of plant or animal origin.
2. Vertical slab gel electrophoresis apparatus: The various items include two glass plates, with one of the glass plates being notched with vertical glass having spacers of 1.5 mm in thickness on its sides gel casting assembly (Figure 8.12). Upper and lower troughs, electrophoresis leads, constant voltage power supply (0–500 V), sucrose (40%, w/v). Weigh 40 g of sucrose and dissolve in 100 mL of water, 0.01% bromophenol blue solution. Take 10 mg of bromophenol blue and dissolve in 100 mL of water as tracking dye.

8.12.2.1 Reagents
Acrylamide–bisacrylamide stock solution:

| Acrylamide | 30 g |
| Bisacrylamide | 0.8 g |

Dissolve in water and make final volume to 100 mL. Filter the solution through Whatman No. 1 filter paper and store in a brown bottle at 0°C–4°C. This solution is stable for 1 month.

Stacking gel buffer stock (Tris–HCl pH 6.8):

| Tris | 36 g |
| 1 M HCl | 48 mL |

Adjust its pH 8.3 and then make the final volume to 100 mL. Filter through Whatman No. 1 filter paper and store at 0°C–4°C.

Resolving gel buffer stock (Tris–HCl, pH 8.8):

| Tris | 36.3 g |
| 1 M HCl | 48.0 mL |

Adjust its pH 8.8 and then make the final volume to 100 mL. Filter through Whatman No. 1 filter paper and store at 0°C–4°C.

1.5% (w/V) APS: Prepare by dissolving 0.15 g of APS in 10 mL water. This reagent should be prepared fresh just before use.

Reservoir buffer (Tris–glycine pH 8.3):

Tris	3 g
Glycine	14.4 g
SDS	1 g

Adjust pH to 8.3 and add water to make final volume to 1 L.

SDS (10% w/v): Dissolve 1 g SDS in 10 mL of distilled water. Store reagent at room temperature.

Sample buffer 2×: The sample preparation buffer that contains a twofold concentration of various components is prepared as per the following composition:

1 M Tris–HCl, pH 6.8	12.5 mL
SDS	4 g
β-mercaptoethanol	10.0 mL
Glycerol	20 g
1% bromophenol blue	4 mL

Add water to make final volume to 100 mL.

Standard MW marker proteins: The standard MW protein kits, having different MW ranges, are commercially available. Dissolve 1 mg of each of the proteins in 1 mL of the sample buffer diluted with water in the ratio of 1:1. About 20 μL of this mixture should be loaded into the one of the wells.

Proteins	Molecular Weight (kDa)
β-lactoglobulin	18.4
Trypsinogen	24.0
Carbonic anhydrase	29.0
Pepsin	34.7
Egg albumin	45.0
Bovine serum albumin	66.0

Stacking and resolving gels: Stacking and resolving gels for SDS-PAGE are prepared according to the particulars given in Table 8.6.

Staining solution:

Coomassie brilliant blue R-250	1.25 g
Methanol	200 mL
Glacial acetic acid	35 mL

Make final volume to 500 mL with distilled water. Filter to remove any undissolved material and store at room temperature.

Destaining solution:

Glacial acetic acid	75 mL
Methanol	500 mL

Mix the previous components and add water to make its final volume to 1:1.

TABLE 8.6 Composition of Stacking and Resolving Gels

Reagents	Stacking Gel 2.5%	Resolving Gel 12.5%
	mL Solution Required	
Acrylamide–bis stock solution	2.5	12.5
Stacking gel buffer stock solution (Tris–HCl, pH 6.8)	5.0	—
Resolving gel buffer stock solution (Tris–HCl, pH 8.8)	—	3.75
10% SDS	0.20	0.30
1.5% APS	1.00	1.50
Water	11.30	11.95
TEMED	0.015	0.015
Total volume	20	30

8.12.3 Protocol

8.12.3.1 Sample Preparation

Mix the protein sample with an equal volume of sample buffer. Boil the mixture for 3 min in a boiling water bath and cool to room temperature. If protein sample is too dilute, precipitate the proteins with 10% TCA (incubate in ice for 30 min). Centrifuge at 10,000*g* for 5 min. Wash the precipitate with ethanol/ether (1:1) to remove TCA. Dissolve the precipitate in the sample buffer diluted with water in 1:1 ratio, boil for 3 min, and cool.

8.12.3.2 Preparation of Slab Gels (Table 8.6)

1. Thoroughly clean and dry the glass plates and assemble them in gel casting assembly. Seal the two glass plates with the help of Teflon tubing, clamp them, and place the whole assembly in an upright position.
2. Mix various components of resolving gel as indicated in the previous table except for SDS, APS, and TEMED. Degas the solution for 1 min using a water pump and then add the remaining components.
3. Mix gently and pour the gel solution into the mold in between the clamped glass plates taking care to avoid entrapment of any air bubbles. Overlay distilled water on the top as gently as possible, and leave for 30 min for setting of the gel.
4. When gel has polymerized, remove the water layer and rinse the gel surface with stacking gel buffer.
5. Mix the stacking gel components in the same way as described previously for the resolving gel.
6. Pour the stacking gel and immediately insert the supplied plastic comb in the stacking gel. Care should be taken that no air bubbles are entrapped. Allow the gel to polymerize for about 20 min.
7. After the stacking gel has polymerized, remove the comb without distorting or damaging the shapes of the wells. Clean the wells by flushing with electrode buffer using a syringe.
8. Remove the Teflon tubing and install the gel plate assembly into the electrophoretic apparatus. Pour reservoir buffer in the lower and upper chambers. Remove any trapped bubbles at the bottom of the gel.

8.12.3.3 Electrophoresis of Sample

1. Load 10–20 µL sample (100–200 µg protein) in the sample wells. Also load MW marker proteins in one or two of the wells.
2. Switch *On* the current maintaining it at 10–15 mA for initial 10–15 min until the samples have traveled through the stacking gel. Then increase the current to 30 mA until the bromophenol blue dye reaches near the bottom of the gel slab. This may require 3–4 h.
3. After the electrophoresis is complete, turn *Off* and disconnect the power supply and carefully remove the gel slab from in between the glass plates.
4. Place the gel in a trough containing staining solution for 3–4 h, or it can be kept for staining overnight. Destain the gel with destaining solution till a clear background of gel is obtained.

5. Record the distance traveled by the dye and the various protein bands, and calculate R_m values.

6. Plot a graph between $\log_{10}$ MW versus R_m values of standard MW markers. A straight line would be obtained. From the R_m values of the sample polypeptides, determine their MWs using the previous standard calibration curve:

$$R_m = \frac{\text{Distance traveled by the protein band}}{\text{Distance traveled by the tracking dye}}$$

8.13 Slab Gel Gradient Electrophoresis of Protein and Multiple Molecular Forms of Acetylcholinesterase

8.13.1 Principle

If electrophoresis is carried out on a concave gradient of polyacrylamide, molecules will migrate through the slab until the pore size becomes too small for them to move any further. Electrophoresis is, therefore, carried out for a prolonged period, usually 24 h so that effectively no further migration takes place. Under these conditions, the distance migrated depends on the size and shape of the molecule and not on its charge. A plot of migration distance against $\log_{10}$ mol. wt gives a straight line if the molecules are approximately spherical and an example of an actual experiment is given in Figure 8.15.

SDS is not used in this method of separation and so the molecular weights obtained are those of the native rather than the denatured protein subunit.

Multiple molecular forms of enzymes can, therefore, be separated by this technique, and their position on the gel can be detected by a suitable stain. In the present experiment, acetylcholinesterase from rat brain is separated into a number of

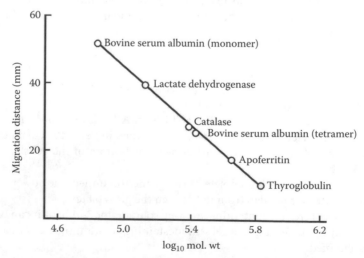

Figure 8.15   Calibration curve for gradients of polyacrylamide slab gel.

MW, and the zones of activity are detected by incubating with acetylthiocholine and the released thiocholine as the copper salt:

$$2CH_3COSCH_2CH_2N^+(CH_3)_3 + (NH_2CH_2COO)_2Cu+H_2O$$

$$\downarrow$$

$$Cu [SCH_2CH_2N(CH_3)_3]_2 + 2NH_2CH_2COOH + 2CH_3COOH$$

Copper thiocholine
(white ppt)

8.13.2 Materials

Slab gel PAGE apparatus for use with slabs of polyacrylamide, polyacrylamide slabs prepared as a concave gradient from 4% to 24% polyacrylamide, protein extract from unknown sample, electrophoresis buffer (88.7 mmol/L Tris, 81.5/L boric acid, 2.5 mmol/L EDTA pH 8.3), Coomassie blue staining solution (0.12% w/v in methanol/acetic acid/water 5:1:5), bromophenol blue as tracking dye and staining solution for acetylcholinesterase (dissolve 200 mg of acetylthiocholine iodide in 16 mL of water), and add 28 mL of 0.1 mol/L copper acetate solution dropwise with stirring. Centrifuge to remove the precipitate and dissolve 120 mg of glycine in the supernatant, and add 2 mol/L sodium acetate to give a final pH of 6.5–7.0.

8.13.3 Protocol

8.13.3.1 Electrophoresis

Place the gels in the electrophoresis apparatus without the sample and connect to the power pack. Set the voltage on 100 V and leave until the current falls to a value between 35 and 40 mA. Apply the samples (30 mL) to the top of the gel in a plastic spacer arm that allows up to 14 samples to be run per gel. Overlay the samples with buffer and carry out electrophoresis at 100 V for 24 h at room temperature. Some means of cooling the apparatus should be employed, and the easiest method is to circulate the electrophoresis buffer through a heat exchanger immersed in a bathe of cold water.

8.13.3.2 Staining

Ideally the gel should be sliced vertically to produce two slabs of the same size but half the thickness, but this may not always be possible and reasonable results can be obtained by staining the whole gel. Remove that part of the gel that contains at room temperature, and then destain with 7% w/v acetic acid. Stain the remainder of the gel for protein as previously described. Plot the distance migrated against the $\log_{10}$ of the MWs of the proteins (Table 8.7), and comment on the range of species obtained for the acetylcholinesterase.

TABLE 8.7   Standard Protein Solution Approximately 1 mg/mL

Protein	Mol. Wt	$\log_{10}$ Mol. Wt
Bovine serum albumin		
Monomer	66,000	4.82
Dimer	132,000	5.12
Tetramer	264,000	5.42
Lactate dehydrogenase	136,000	5.13
Catalase	250,000	5.40
Apoferritin	450,000	5.65
Thyroglobulin	670,000	5.83

8.14 Protein (Western) Blotting

The transfer of proteins from an acrylamide gel onto a more stable and immobilizing support such as nitrocellulose filter is called as protein blotting or Western blotting. It is so named in view of the previously used nomenclature for nucleic acid blotting procedures. Blotting of protein bands allows them to become more accessible for detection and identification using a variety of methods, and also the filters can be stored for much longer periods. A number of support matrices such as nitrocellulose, diazobenzyloxymethyl cellulose, nylon membrane, and hybrid membrane, with high protein-binding capacity, are available, but nitrocellulose that is available in a range of pore sizes (0.2–0.45 g) is sufficiently effective for most purposes and is commonly used. The transfer of proteins from gels can be achieved by any of methods such as simple capillary action, application of vacuum, or electrophoretically. Thus, based on the technique employed, Western blotting is of different types, namely, capillary blotting, vacuum blotting, and electroblotting, the latter being the most efficient and has become a standard method.

8.14.1 Electroblotting (Western Blotting of Proteins from SDS-PAGE)

8.14.1.1 Introduction
When an electric field is applied, proteins migrate from cathode to anode, and as they come in contact with nitrocellulose sheet, which has the property to bind proteins, the proteins get immobilized into the sheet and an exact replica of protein bands are formed on nitrocellulose sheet. The Western blot (sometimes called the protein immunoblot) is a widely accepted analytical technique used to detect specific proteins in the given sample of tissue homogenate or extract. It uses gel electrophoresis to separate native proteins or denatured proteins by the length of the polypeptide. The proteins are then transferred to a membrane (typically nitrocellulose or PVDF), where they are stained with antibodies specific to the target protein.

8.14.1.2 Materials

1. Western blot apparatus consisting of gel holder, sponge, and transfer tank
2. Power pack and electric leads
3. Slab gel containing separated proteins (use gel obtained after electrophoresis of proteins on SDS-PAGE)
4. Nitrocellulose sheet cut to the size of the gel
5. Whatman 3 MM paper cut to the gel size

Transfer buffer consists of the following:

Tris	3 g
Glycine	14.7 g
Methanol	200 mL

After mixing the previously given components, adjust the pH of the solution to 8.3 and make the final volume to 1 L with distilled water.

Staining solution:

Coomassie brilliant blue R-250	1.25 g
Methanol	200 mL
Glacial acetic acid	35 mL

Make final volume to 500 mL with distilled water. Filter to remove any undissolved material and store at room temperature.

Destaining solution:

Glacial acetic acid	75 mL
Methanol	500 mL

Mix the previously given components and add water to make its final volume to 1:1.

8.14.1.3 Protocol

1. Take the gel obtained after electrophoresis in slab gel. Do not stain it and mark it by cutting of the one corner. Place the gel for 30 min in transfer buffer for equilibrium.
2. Take a nitrocellulose sheet, cut to size of the gel, and dip it in the transfer buffer by carefully wetting one edge and then slowly lowering the sheet into the buffer. Leave it under the buffer for 30 min.
3. Soak the sponge in transfer buffer and place the wet sponge on the gel holder. Now keep a sheet of Whatman No. 3 MM paper (presoaked in transfer buffer) on the sponge.
4. Place the equilibrated gel carefully on the filter paper avoiding trapping of any air bubbles.
5. Now lay down carefully the nitrocellulose membrane with shining side toward the gel. Gently roll a sterile 10 mL pipette over the membrane to remove air bubbles for ensuring a good contact between the membrane and the gel.
6. Complete the sandwich by placing wet Whatman No 3 MM filter paper over the membrane and a second sponge on the filter paper. Close the gel holder and place it in the transfer tank containing sufficient transfer buffer, to completely cover the blot.

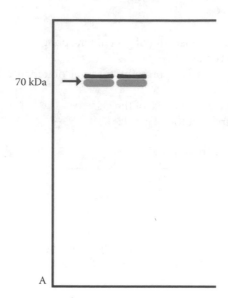

Figure 8.16   Illustrating a Western blot.

7. Connect to the power supply and run for 5 h at 60 V or 30 V over night.
8. When the transfer is complete, lift the membrane from the gel. Stain and destain using Coomassie brilliant blue R-250 stain. Examine the nitrocellulose sheet for the presence of blue bands of the transferred proteins (Figure 8.16).

8.15 Isoelectric Focusing

Proteins are ampholytes, that is, they contain both positively and negatively charged groups. All ampholytes have the property that their charge depends on pH; they are positively charged at low pH and negatively charged at high pH. For every ampholyte, there exists a pH at which it is uncharged, and this is called the isoelectric point. At the isoelectric point, the ampholyte will not move in an electric field. If a protein solution is placed in a pH gradient, the molecules will move until they reach a point in the gradient at which they are uncharged; then they will cease to move. With a mixture of different proteins, each type of molecule will come to rest at a point in the pH gradient corresponding to its own isoelectric point. This method of separating proteins according to their isoelectric points in a pH gradient is called isoelectric focusing. The process of the migration of two different proteins with different isoelectric points is shown schematically in Figure 8.17.

The pH gradient is established in an unusual way. If it were established by simply allowing two buffers at different pH to diffuse into one another or by mixing two buffers in the way that is standard for preparing a concentration gradient, the resulting gradient would not be stable in an electric field because the buffer ions would migrate in the field; fractionation of the macromolecules could not occur because

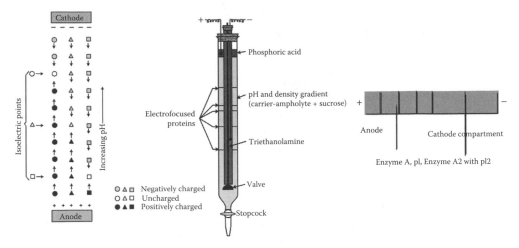

Figure 8.17   Isoelectric focusing.

the macromolecules would migrate much more slowly than the pH gradient is disrupted. The method used to produce a stable pH gradient consists of distributing a mixture of synthetic, low MW (300–600) polyampholytes (multicharged structures) that cover a wide range of isoelectric points (up to 1000 different isoelectric points/interval of 1 pH unit). These polyampholytes are usually mixed polymers of aliphatic amino and either carboxylic or sulfonic acids. They are commercially available as Ampholine, Pharmalyte, and BioLyte. A pH gradient is established by starting with a mixture in distilled water of polyampholytes having isoelectric points covering a range of either 2 or 7 pH units (depending on the resolution required). Before the application of an electric field, the pH throughout the system is constant and is averaged from all the polyampholytes in the solution. When the field is applied, the polyampholytes start to migrate. Because of their own buffering capacities, a pH gradient is gradually established. Soon each particle will come to rest in this self-established gradient at the point corresponding to its own isoelectric point. If the mixture contains proteins of different isoelectric points, they will migrate (but much more slowly) to the positions corresponding to their isoelectric points as long as the concentration and buffering of the protein (which is also a polyampholyte) is not so high that the pH gradient is disrupted.

For purification of a particular type of protein, the pH gradient is formed in a water-cooled glass column containing a cathode tube and a cathode. The tube is filled with a uniform concentration gradient to eliminate convection. The sample material is also contained in the polyampholytes suspension. The cathode tube is filled with a strong base (typically triethanolamine) and the main column is overlaid with phosphoric acid; the anode is in this acid layer. The valve at the bottom of the cathode tube is opened, followed by the application of a few hundred volts between the electrodes. The polyampholyte near the cathode will have a negative charge and will move to the anode. From 1 to 3 days later, the system will be at equilibrium and the proteins distributed throughout the pH gradient according to their own isoelectric points. The tube is then drained and fractionated through the stopcock at the bottom. The various proteins can be detected by spectrophotometry, enzyme activity, or radioactivity.

A substantial improvement in the resolution of individual molecules is obtained by using as a supporting medium a continuous agarose or polyacrylamide gel rather than a sucrose concentration gradient. When this is done, a protein whose isoelectric points differ by 0.001 pH unit is readily separated. The single requirement of gel is that the pore size must be so large that molecular sieving is unimportant, and the isoelectric point is the single factor that determines the position of the band. This means using 2% polyacrylamide or, even better, 1% agarose. Tubular gels are sometimes used but, because of their greater capacity, horizontal slab gels are preferable. The gel initially contains the polyampholyte. The proteins may be present throughout the gel initially, but they are usually applied in a single zone as in ordinary gel electrophoresis. Following focusing, the bands are visualized by staining with Coomassie brilliant blue and Crocein scarlet. Narrow-range amplification can be obtained by using polyampholytes. This mode of isoelectric focusing is used to test the purity or composition of a sample.

8.15.1 Separation of Proteins by Isoelectric Focusing

8.15.1.1 Introduction
When any enzyme mixture is placed in the gel having the solution of different pH, then after passing the current, a potential difference is established that helps in the development of the pH gradient. This pH gradient decides the final movement of the enzyme in the gel because enzyme will stop its movement only after a situation where the pH of enzyme will be equal to the pH of the buffer.

8.15.1.2 Materials
Electrofocusing apparatus, vertical slab gel electrophoresis apparatus, power supply (0–500 V).

8.15.1.2.1 Reagents
Acrylamide–bisacrylamide solution (stock solution):

Acrylamide	30 g
Bisacrylamide	1 g
Distilled water	100 mL

Ampholyte solution (pH 3.5–10 and pH 4–6): Ampholytes are amphoteric compounds, which are supplied as a mixture of molecules with closely spaced isoelectric points. Ampholyte solutions for different pH range are available commercially.

10% APS: Dissolve 100 mg ammonium persulfate in 1 mL distilled water. Prepare fresh just before use.

TEMED.

Prepare gel mixture (for pH gradient 4–6): It consists of the following:

Acrylamide–bisacrylamide solution (stock solution)	2.0 mL
Ampholyte solution, pH 3.5–10	48.0 µL
Ampholyte solution, pH 4–6	240.0 µL
Distilled water	9.7 µL
10% ammonium persulfate	50.0 µL
TEMED	20.9 µL

Note: Add APS and TEMED just before pouring the gel.

Analyte solution (10 mM phosphoric acid): Prepare a stock solution of 1 M phosphoric acid. Dilute 100 times before use.

Catholyte solution (20 mM NaOH): Prepare stock solution of 1 M NaOH. Dilute 50 times before use.

Sample buffer (2×) for pH gradient 4–6:

Glycerol	3 mL
Ampholyte solution (pH 3.5–10)	33 µL
Ampholyte solution (pH 4.6)	33 µL
Distilled water	1.8 mL

10% TCA solution: Weigh 10 g of TCA and dissolve in 100 mL of distilled water.

1% TCA: Weigh 1 g of TCA and dissolve in 100 mL of distilled water.

Staining solution:

Coomassie brilliant blue R-250	1.25 g
Methanol	200 mL
Glacial acetic acid	35 mL

Make final volume to 500 mL with distilled water. Filter to remove any undissolved material and store at room temperature.

Destaining solution:

Glacial acetic acid	75 mL
Methanol	500 mL

Mix the previously given components and add water, to make its final volume to 1:1.

8.15.1.3 Protocol

1. *Sample preparation*: Mix protein sample with an equal volume of 2× sample buffer.
2. *Preparation of slab gel*:
   a. Thoroughly clean and dry the glass plates and assemble them in gel casting assembly. Seal the two glass plates with the help of Teflon tubing, clamp them, and place the whole assembly in an upright position.

    b. Mix various components of the gel mixture and pour the gel mixture into the experiments between two glass plates taking care to avoid entrapment of air bubbles. Fill the gel plates to the top with gel mixture.

    c. Insert comb, taking care not to trap air bubbles in the teeth of the comb.

    d. Allow the gel to polymerize for about 1 h. After the gel has polymerized, remove the comb carefully.

3. *Electrophoresis*:

    a. Attach gel assembly to inner cooling core and insert into electrophoresis tank.

    b. Load about 20 µL sample (50–100 µg protein) in the sample well with the help of a syringe or gel-loading pipette.

    c. Pour catholyte (20 mM NaOH) into the upper buffer chamber and analyte (10 mM phosphoric acid) into the lower buffer chamber.

    d. Attach electrodes and run electrophoresis at room temperature for 30 min at 150 V (constant voltage). Then set the voltage at 200 V for 2.5 h (constant voltage).

    e. After electrophoresis is over, turn "OFF" and disconnect the power supply.

    f. Carefully remove the gel slab from in between the glass plates and place in a trough containing 10% TCA for 10 min to allow fixing of the gel. Replace with 1% TCA and keep the gel overnight in this solution to remove ampholytes.

    g. Place the gel in staining solution for 3–4 h or it can be kept for staining overnight.

    h. Destain the gel with several changes of destaining solution till a clear background of gel is obtained and protein bands become clearly visible and the gel is photographed for permanent record.

8.16 Agarose Gel Electrophoresis

Agarose gels are more porous and have a larger pore size as compared to polyacrylamide gels and are, therefore, used to fractionate large macromolecules such as nucleic acids that cannot readily penetrate into and move through other types of supporting materials.

Agarose is a linear polymer of D-galactose and 3,6-anhydro-L-galactose. Agarose gels are cast by boiling agarose in the presence of a buffer, then poured into a mold, and allowed to harden to form a matrix. Porosity of gel is determined by the concentration of the agarose. The higher the agarose concentration, the smaller the pore size, and the lower the agarose concentration, the larger the pore size. When an electric field is applied across the gel, DNA molecules that are negatively charged at neutral pH migrate toward oppositely charged electrodes at rates determined by their molecular size and conformation. Since charge/mass ratio in nucleic acids is one, the rate of migration of DNA molecules is inversely proportional to $\log_{10}$ of the MWs, that is, smaller DNA molecules will travel faster as compared to the larger ones. Further, DNA molecules of the same size but with different conformations travel at different rates. The order of migration velocity in the increasing order of various forms of DNA is as follows: supercoiled DNA > linear double-stranded DNA > open circular DNA.

8.16.1 Apparatus

Both vertical and horizontal types of apparatus, for casting of agarose gels, are available. However, horizontal submarine or submerged gels (so named because the gel is immersed in the buffer) are more commonly employed (Figure 8.18).

Horizontal slab gels have advantages over vertical ones, in that low agarose concentration can be used as the entire gel is supported from beneath and also the processes of loading, pouring, and handling of gels are more convenient.

8.16.2 Buffers

Commonly used buffers include TAE, Tris–borate (TBE), or Tris–phosphate (TPE) at a concentration of approximately 50 mM, over a pH range of 7.5–8.0. EDTA is almost invariably incorporated in these buffers to suppress the activity of DNase.

8.16.3 Nucleic Acids

The phosphate group of each nucleotide carries a single strong negative charge that is much greater than any of the charges on the bases. Thus, charge-to-mass ratio of all polynucleotides is independent of the base composition. In electrophoresis of nucleic acids, the molecular-sieving effect is also the principal factor in separation because the charge-to-mass ratio is nearly the same for all polynucleotides. Hence, small molecules move faster than large ones. Because naturally occurring nucleic acid molecules are very large, the pore size of the gel must be large, that is, the gel must be dilute. To strengthen the gels, especially for large molecules ($M < 5 \times 10^6$),

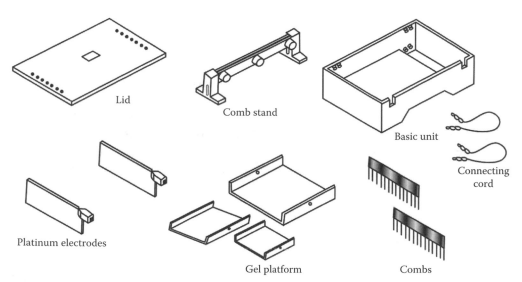

Figure 8.18 Submarine horizontal gel electrophoresis apparatus.

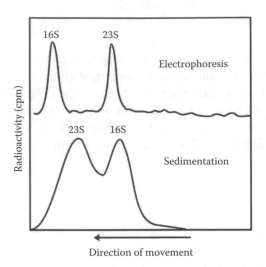

Figure 8.19 Comparative separation of ribosomal RNA species (16S and 23S) by PEG and sedimentation through sucrose gradient.

agarose (a highly porous polysaccharide) is added or sometimes agarose alone is used. Electrophoresis is done in slab gels or column gels. For RNA molecules (e.g., messenger and ribosomal RNA), resolution is far better than in zonal centrifugation and electrophoresis is certainly the method of choice. Furthermore, the distance D migrated is related to the MW M by the following equation:

$$D = a - b \log M \text{ (see Figure 8.19)}$$

where a and b are constants so that M is measurable if two samples of known MW are included. Separation in polyacrylamide and agarose is also very good for single-stranded DNA up to $M = 50 \times 10^6$.

8.16.4 Detection of Nucleic Acids in Gels

Nucleic acid bands can be detected in gels in a variety of ways. Radioactivity is usually measured by slicing the gel, solubilizing it in 0.5 M NaOH or H_2O_2, adding a suitable scintillation cocktail, and counting the sample with a scintillation counter. With material of high MW, slicing can be difficult because the gel is necessarily soft to ensure large pore size. An alternative is to detect radioactivity by autoradiography; in this procedure, the gel is dried and pressed firmly against autoradiographic film. Another procedure uses the dye stains.

8.16.5 Agarose Gel Electrophoresis of Double-Stranded DNA

Naturally occurring DNA molecules usually have MWs that are so high that the molecules cannot penetrate even a weakly cross-linked polyacrylamide gel.

Agarose solves this problem. A 0.8% agarose gel can accept DNA molecules whose MW is as high as 50×10^6, and this gel is still sufficiently rigid that it can be used in the vertical configuration. Concentration as low as 0.2% agarose has been used in the horizontal mode, and DNA molecules having a MW of more than 150×10^6 can be electrophoresed in such a gel.

8.16.6 Separation of DNA by Molecular Weight

Good separation is achieved with molecules whose MWs differ by as little as 1% using agarose gel electrophoresis. Furthermore, the range of applicability is from molecules containing less than ten base pairs to about 3×10^5 base pairs. For the different size ranges, gels of different concentrations must be used: 0.8%–1.5% agarose for molecules up to about 5×10^4 base pairs and 0.2%–0.4% for the very large molecules. This is necessary because very large molecules cannot penetrate the more concentrated gels and very small molecules would pass through the dilute gels so easily that their velocity becomes only weakly dependent on MW.

Two methods of performing agarose gel electrophoresis have become popular. These procedures eliminate the wicks and make running times more reproducible. The most useful setup uses agarose legs and is depicted in Figure 8.20a. In some experiments, electrophoresis must be carried out for a very long time, and the gels

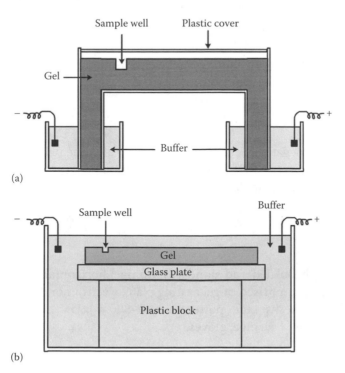

Figure 8.20 Two modes of horizontal gel electrophoresis: (a) agarose legs and (b) submarine gels.

tend to dry and crack. To avoid this, the submarine system depicted in Figure 8.20b is used. In both cases, the technique is very simple. First, the well is filled with buffer. Then a sample of about 25 μL of DNA in buffer containing 50% glycerol is placed in each well. The glycerol raises the density of the sample so that it can be layered beneath the buffer. The DNA concentration of each sample is adjusted, so that there is 0.001–0.01 μg of DNA for each band that is expected. The electric field is then applied. When electrophoresis is complete, the DNA is visualized by soaking the gel in a dilute solution of the fluorescent dye ethidium bromide, whose quantum yield is considerably enhanced when bound to native DNA. The gel is washed to remove unbound dye from the agarose, and then it is illuminated with UV light to excite the fluorescence of the ethidium bromide. The gel with its fluorescent bands is usually photographed.

Sometimes it is valuable to be able to visualize the bands before electrophoresis is complete. The submarine gel system is particularly useful when this is required. Both gel and the buffers are prepared to contain ethidium bromide, and when viewing is desired, the glass plate that supports the gel is lifted out and the gel is illuminated with UV light. If the bands have not migrated far enough to achieve the desired resolution, the plate can be put back into the buffer. The MWs of the DNA of each band are determined from the distance of the band from the origin. The equation that describes the relation between the MW (M) and the distance (D) is $D = a - b \log M$ in which a and b are constants that are a function of the time of electrophoresis, the buffer, and the gel concentration. MWs can be obtained from a plot of D versus $\log M$, as long as at least two values of M are known.

8.17 Determination of Molecular Weight of Plasmid DNA by Agarose Gel Electrophoresis

8.17.1 Introduction

DNA molecules are negatively charged at neutral or alkaline pH and migrate toward the anode when an electric field is applied. Migration occurs largely on the basis of molecular size of the DNA molecules.

8.17.2 Materials

Plasmid DNA (both isolated and standard), minigel horizontal agarose gel electrophoretic unit that is comprised of gel casting plate, electrophoretic tank, comb, electrophoretic leads, adhesive tape, power pack (0–500 V), UV transilluminator with camera, micropipette or syringe, gloves.

8.17.2.1 Reagents
TAE buffer stock solution (5×): A fivefold concentrated TAE buffer stock solution is prepared as per the following composition:

Tris base	24.2 g
Glacial acetic acid	7.71 mL
0.5 M EDTA	10 mL

Adjust pH of the previous solution to pH 9.0 and add water to make 1 L. Dilute five times before use to obtain the working buffer (1× buffer).

One percent agarose in 1× TAE buffer: Dissolve 0.75 g agarose in 75 mL of 1× TAE buffer (working TAE buffer) by boiling, and maintain it at 50°C till it is to be used.

Gel-loading solution: 10% glycerol and 0.025% bromophenol blue in water.

Ethidium bromide: Dissolve 10 mg of ethidium bromide/mL of the 1× TAE buffer. Gloves must be worn while preparing this solution.

Plasmid DNA preparation: To 20 μL of plasmid DNA preparation, add 10 μL of gel-loading solution and mix properly.

Standard DNA marker: Take 20 μL of the λ Hind III DNA digest. Add 10 μL of gel-loading solution and mix them well.

8.17.3 Protocol

1. Take a clean dry gel casting plate and make a gel mold using an adhesive tape along the sides of the plate to prevent running of the material to be poured on the plate.
2. Pour 50 mL of 1% agarose solution kept at 50°C onto the casting plate. Immediately place the comb about 1 cm from one end of the plate ensuring that teeth of the comb do not touch the glass plate. Wait till a firm layer of gel is formed.
3. Remove the comb and the tape surrounding the plate carefully, and transfer the get plate to the electrophoretic tank in such a way that wells are toward the cathode.
4. Pour 1× TAE buffer into the tank, until the gel is completely submerged. Connect the electrodes to the power supply.
5. Load the plasmid DNA preparation and the standard DNA markers into separate wells with the help of a micropipette or a syringe.
6. Turn "ON" the power supply and run at 100 V (10–15 mA). Monitor the progress of fast-running tracking dye (bromophenol blue) during electrophoresis.
7. Turn "OFF" the power supply when the tracking dye has reached near the opposite edge of the gel.
8. Transfer the gel from casting plate onto a UV transparent thick plastic sheet, place it in a staining tray containing ethidium bromide solution, and stain for 20–30 min.
9. For destaining the gel, place it in water for 15–20 min.
10. Now place the gel along with UV transparent sheet on a UV transilluminator, and view the gel in UV light for the presence of orange-colored bands. For a permanent record, the gel may be photographed (*Caution: UV light is extremely injurious for eyes, and wear UV-protective glasses while viewing the gel*).
11. Measure the distance moved by each band front edge of the loading well. Draw a graph between log_{10} of MW of standard DNA markers and the distance traveled by each of them.
12. From the distance traveled by the supplied plasmid DNA preparation, determine its MW using the calibration curve (Figure 8.21).

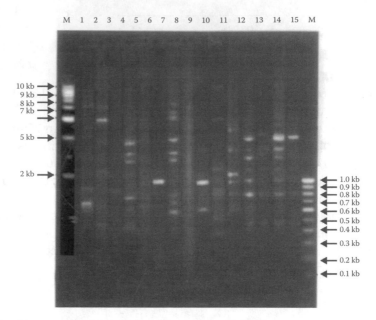

Figure 8.21   Photographed horizontal fluorescence slab gel.

8.18 Electrophoretic Evaluation of Quality of High Molecular Weight DNA Isolated from Plant Material

8.18.1 Introduction

DNA molecules are negatively charged at neutral or alkaline pH and migrate toward the anode when an electric field is applied. The charge/mass ratio in nucleic acids is unity; thus, migration occurs largely on the basis of molecular size of the DNA molecules.

8.18.2 Materials

Isolated DNA from plant material, standard DNA, minigel horizontal agarose gel electrophoretic unit that is comprised of gel casting plate, electrophoretic tank, comb, electrophoretic leads, adhesive tape, power pack (0–500 V), UV transilluminator with camera, micropipette or syringe, gloves, Hind III digest as marker.

8.18.2.1 Reagents

TAE buffer stock solution (5×): A fivefold concentrated TAE buffer stock solution is prepared as per the following composition:

Tris base	24.2 g
Glacial acetic acid	7.71 mL
0.5 M EDTA	10 mL

Adjust pH of the previous solution to pH 9.0 and add water to make 1 L. Dilute five times before use to obtain the working buffer (1× buffer).

0.8% agarose in 1× TAE buffer: Dissolve 0.6 g agarose in 75 mL of 1× TAE buffer (working TAE buffer) by boiling, and maintain it at 50°C until used.

Gel-loading solution: 10% glycerol and 0.025% bromophenol blue in water.

Ethidium bromide stock: (10 mg/mL).

Isolated plant DNA.

Hind III digest as marker.

8.18.3 Protocol

Isolation of DNA from plant tissues: Follow the procedure described in Section 11.5.12.

Electrophoresis of DNA sample: The procedure for electrophoresis of DNA is outlined in experiment described in Section 11.5.12.

8.18.4 Result

The presence of a thick band near the mouth of the wells will indicate that DNA is of high MW, whereas sheared DNA will appear as a smear throughout the gel lane indicating poor quality of the preparation. RNA contamination is indicated by smear near the bottom of the gel.

8.19 Large-Scale Preparative Electrophoresis

8.19.1 Block Electrophoresis

A slurry of starch grains (or glass powder, silica gel, cellulose, polyvinyl chloride, or agar) in a buffer is poured into a large rectangular trough, which is mounted so that there is an electrode at either end. The sample is applied by removing starch from a large groove near one electrode, mixing the protein solution with the starch, and refilling the groove. After electrophoresis, the block is cut up and each section eluted.

8.19.2 Column Electrophoresis with Polyacrylamide

This procedure employs the apparatus illustrated in Figure 8.22. Material continually flows into the elution buffer, which flows through as indicated. The elution buffer is collected by a fraction collector.

8.19.3 Immunoelectrophoresis

A variety of special techniques that combine electrophoresis and immunology have been developed for identifying molecules and for determining the amount of a

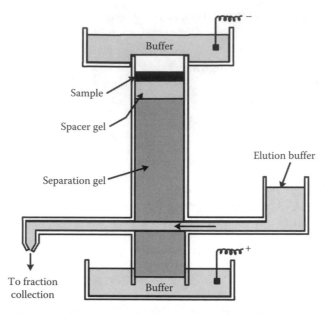

Figure 8.22   Apparatus for large-scale electrophoresis.

particular molecule present in a mixture. These procedures, immunoelectrophoresis and rocket electrophoresis, are described in the section of immunology.

8.19.4 Capillary Electrophoresis

Capillary electrophoresis (CE), also known as capillary zone electrophoresis (CZE), can be used to separate ionic species by their charge and frictional forces. Separation by CE that can be detected by several detection devices is the adaptation of traditional gel electrophoresis into the capillary using polymers in solution to create a molecular sieve also known as replaceable physical gel. This allows analysis having similar charge-to-mass ratios to be resolved by size. This technique is commonly employed in SDS gel MW analysis of proteins and sizing of applications of DNA sequencing and genotyping. CE encompasses a family of related separation techniques that use narrow-bore fused silica capillaries to separate a complex array of large and small molecules. High electric field strengths are used to separate molecules based on differences in charge, size, and hydrophobicity. Sample introduction is accomplished by immersing the end of the capillary into a sample vial and applying pressure, vacuum, or voltage. Depending on the types of capillary and electrolytes used, the technology of CE can be segmented into several separation techniques.

The majority of the commercial systems use UV or UV-B absorbance as their primary mode of detection. In these systems, a section of the capillary itself is used as the detection cell. The use of on tube detection enabled detection of separated analytes with no loss of resolution. In general, capillaries used in CE are coated with a polymer for increased stability. The portion of the capillary used for UV detection,

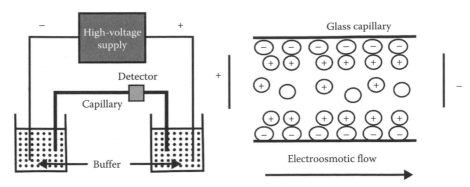

Figure 8.23 Illustrating CE.

however, must be optically transparent. Bare capillaries can break relatively easily, and as a result, capillaries with transparent coatings are available to increase the stability of the cell window. In order to obtain the identity of sample components, CE can be directly coupled with mass spectrometers or surface-enhanced Raman spectroscopy (SERS). In most systems, the capillary outlet is introduced into an ion source that utilizes electrospray ionization (ESI). The resulting ions are then analyzed by the mass spectrometer. This setup requires volatile buffer solutions, which will affect the range of separation modes that can be employed and the degree of resolution that can be achieved. The measurement and analysis are mostly done with specialized gel analysis software. Capillaries are typically of 50 μm inner diameter and 0.5–1 m in length. The applied potential is 20–30 kV. Due to electroosmotic flow, all sample components migrate toward the negative electrode. A small volume of sample (10 μL) is injected at the positive end of the capillary, and the separated components are detected near the negative end of the capillary. CE detection is similar to detectors in HPLC and includes absorbance, fluorescence, electrochemical, and mass spectrometry. The capillary can also be filled with a gel, which eliminates the electroosmotic flow. Separation is accomplished as in conventional gel electrophoresis, but the capillary allows higher resolution, greater sensitivity, and online detection (Figure 8.23).

8.19.5 Two-Dimensional Electrophoresis

Isoelectric focusing and chromatography can be combined with SDS-PAGE, to obtain very high-resolution separations. A single sample is first subjected to isoelectric focusing or chromatography. This single-lane gel is then placed horizontally on top of an SDS-polyacrylamide slab. The proteins are thus spread across the top of the polyacrylamide gel according to how far they migrated during isoelectric focusing. They then undergo electrophoresis again in a perpendicular direction (vertical) to yield a 2D pattern of spots. In such a gel, proteins have been separated in the horizontal direction on the basis of isoelectric point and in the vertical direction on the basis of mass. More than a thousand different proteins can be resolved in a single experiment by 2D electrophoresis (Figures 8.24 and 8.25).

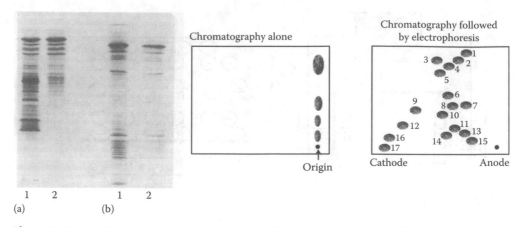

Figure 8.24 (a) Two-protein mixture separated by isoelectric focusing. (b) Chromatography combined with electrophoresis to separate amino acids.

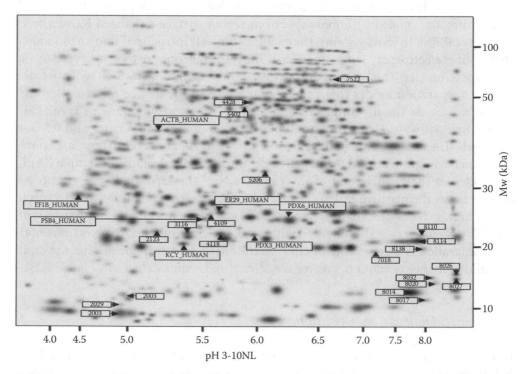

Figure 8.25 A 2D electrophoresis of macrophage proteomes of S1P-stimulated protein with respect to *Mycobacterium*.

Suggested Readings

Amersham. 1999. *Protein Electrophoresis: Technical Manual.* Piscataway, NJ: Amersham Biosciences Inc.

Andreas, M., P. Nicole, and L. I. Dimitri. 2004. *Bioanalytical Chemistry.* London, U.K.: Imperial College Press.

Arakawa, T., L. Hung, V. Pan et al. 1993. Analysis of the heat-induced denaturation of proteins using temperature gradient gel electrophoresis. *Anal Biochem* 208: 255–259.

Baughman, K. 2005. *Principles and Applications of DGGE*. Pittsburgh, PA: Microbac Laboratories, Inc.

Baumstark, T. and D. Riesner. 1995. Only one of four possible secondary structures of the central conserved region of potato spindle tuber viroid is a substrate for processing in a potato nuclear extract. *Nucleid Acids Res* 23: 4246–4254.

Biometra A Whatman Company. 1999. *TGGE System. Manual*. Göttingen, Germany: Biometra biomedizinische Analytik GmbH.

Birmes, A., A. Sattler, K. H. Maurer et al. 1990. Analysis of the conformational transition of proteins by temperature-gradient gel electrophoresis. *Electrophoresis* 11: 795–801.

Birren, B. W., E. Lai, S. M. Clark et al. 1988. Optimized conditions for pulsed-field gel electrophoretic separations of DNA. *Nucleic Acids Res* 16: 7563–7581.

Carle, G. F. and M. V. Olson. 1984. Separation of chromosomal DNA molecules from yeast by orthogonal-field-alternation gel electrophoresis. *Nucleic Acids Res* 14: 5647–5663.

Chrambach, A., M. J. Dunn, and B. J. Radola. 1983. *Advances in Electrophoresis*, vol. 4, pp. 189–195. Weinheim, Germany: VCH Verlagsgesellschaft.

Dempsey, J. A. F., W. Livaker, A. Madhure et al. 1991. Physical map of the chromosome of *Neisseria gonorrhoeae* FA1090 with locations of genetic markers, including opa and pil genes. *J Bacteriol* 173: 5476–5486.

Gardiner, K. 1991. Pulsed-field gel electrophoresis. *Anal Chem* 63: 658–665.

Hecker, R., Z. Wang, G. Steger, and D. Riesner. 1988. Analysis of RNA structure by temperature-gradient gel electrophoresis: Viroid replication and processing. *Gene* 72: 59–74.

Henco, K. and M. Heibey. 1990. Quantitative PCR—The determination of template copy numbers by temperature gradient gel electrophoresis. *Nucleic Acids Res* 18: 6733–6734.

Herschleb, J., G. Ananiev, and D. C. Schwartz. 2007. Protocol: Pulsed-field gel electrophoresis. *Nat Protoc* 2: 677–684.

Horn, D., P. N. Robinson, A. Boddrich et al. 1996. Three novel mutations of the NF1 gene detected by temperature gradient gel electrophoresis of exons 5 and 8. *Electrophoresis* 17: 1559–1563.

Instruction Manual. 2011. *Denaturing Gradient Gel Electrophoresis Systems*. Del Mar, CA: C.B.S. Scientific Company, Inc.

Jain, A., D. S. Gour, P. S. Bisen et al. 2008. Single strand conformation polymorphism detection in alpha-lactalbumin gene of Indian Jakhrana milk goats. *Acta Agr Scand A* 58: 205–208.

Jeroen, H. R. and J. M. P. Dorien. 1998. Denaturing gradient gel electrophoresis (DGGE). In *Medical Biomethods Handbook*, eds. J. M. Walker and R. Rapley. Totowa, NJ: Humana Press Inc.

Joppa, B., S. Li, S. Cole et al. 1992. Pulsed field electrophoresis for separation of large DNA. HIS Laboratories, Hoefer Scientific Instruments, Fall San Francisco, *Probe Volume* 2(3): Fall.

Kappes, S., K. Milde-Langosch, P. Kressin et al. 1995. p53 mutations in ovarian tumors, detected by temperature-gradient gel electrophoresis, direct sequencing and immuno-histochemistry. *Int J Cancer* 64: 52–59.

Kuhn, J. E., T. Wendland, H. J. Eggers et al. 1995. Quantitation of human cytomegalovirus genomes in the brain of AIDS patients. *J Med Virol* 47: 70–82.

Lai, E., B. W. Birren, S. M. Clark et al. 1989. Pulsed-field gel electrophoresis. *Biotechniques* 7: 34–42.

Lessa, E. P. and G. Applebaum. 1993. Screening techniques for detecting allelic variation in DNA sequences. *Mol Ecol* 2: 119–129.

Linke B, J. Pyttlich, M. Tiemann et al. 1995. Identification and structural analysis of rear-ranged immunoglobulin heavy chain genes in lymphomas and leukemia. *Leukemia* 9: 840–847.

Loss, P., M. Schmitz, G. Steger et al. 1991. Formation of a thermodynamically metastable structure containing hairpin II is critical for the potato spindle tuber viroid. *EMBO J* 10: 719–728.

Menke, M. A., M. Tiemann, D. Vogelsang et al. 1995. Temperature gradient gel electrophoresis for analysis of a polymerase chain reaction-based diagnostic clonality assay in the early stages of cutaneous T-cell lymphomas. *Electrophoresis* 16: 733–738.

Milde-Langosch, K., K. Albrecht, S. Joram et al. 1995. Presence and persistence of HPV and p53 mutation in cancer of the cervix uteri and the vulva. *Int J Cancer* 63: 639–645.

Nübel, U., B. Engelen, A. Felske et al. 1996. Sequence heterogeneities of genes encoding 16S rRNAs in *Paenibacillus polymyxa* detected by temperature gradient gel electrophoresis. *J Bacteriol* 178: 5636–5643.

Orbach, M. J., D. Vollrath, R. W. Davis et al. 1998. An electrophoretic karyotype of *Neurospora crassa*. *Mol Cell Biol* 8: 1469–1473.

Prischmann, J. 2011. *Basics and Theory of Electrophoresis*. Diagnostic Laboratories, North Dakota State Seed Department.

Richter, A., L. Plobner, and J. Schumacher. 1997. Quantitatives PCR-Verfahren zur Bestimmung der Plasmidkopienzahl in rekombinanten Expressionssystemen. *BIOFORUM* 20: 545–547.

Riesner, D. 1998. Nucleic acid structures. In *Antisense Technology*. Practical Approach Series. pp. 1–24. Oxford, U.K.: Oxford University Press.

Riesner, D., K. Henco, and G. Steger. 1990. Temperature-gradient gel electrophoresis: A method for the analysis of conformational transitions and mutations in nucleic acids and protein. *Adv Electrophoresis* 4: 169–250.

Sambrook, J., E. F. Fritsch, and T. Maniatis. 1989. Gel electrophoresis of DNA. In *Molecular Cloning: A Laboratory Manual*, eds. J. Sambrook, E. F. Fritsch, and T. Maniatis. Cold Spring Harbor, NY: Cold Spring Harbor Laboratory Press. Chapter 6.

Schwartz, D. C. and C. R. Cantor. 1984. Separation of yeast chromosome-sized DNAs by pulsed field gradient gel electrophoresis. *Cell* 37: 67–75.

Sheffield, V. C., D. R. Cox, L. S. Lerman et al. 1989. Attachment of a 40-base-pair G+C-rich sequence (GC-clamp) to genomic DNA fragments by the polymerase chain reaction results in improved detection of single-base changes. *Proc Natl Acad Sci USA* 86: 232–236.

Skoog, D. A., F. J. Holler, and S. R. Crouch. 2007. *Principles of Instrumental Analysis*, 6th edn. Belmont, CA: Thomson Brooks/Cole Publishing.

Westermeier, R. 1997. *Electrophoresis in Practice: A Guide to Methods and Applications of DNA and Protein Separation*. Weinheim, Germany: Wiley-VCH.

Whatman, Biometra. 2002. TGGE System. Manual. Version 6.0. *Biometra, biomedizinische*. Göttingen, Germany: Analytik GmbH.

Wieland, U., H. Suhr, B. Salzberger et al. 1996. Quantification of HIV-1 proviral DNA and analysis of genomic diversity by polymerase chain reaction and temperature gradient gel electrophoresis. *J Virol Methods* 57: 127–139.

Wiese, U., M. Wulfert, B. Stanley-Prusiner et al. 1995. Scanning for mutations in the human prion protein open reading frame by temporal temperature gradient gel electrophoresis. *Electrophoresis* 16: 1851–1860.

Wilson, K. and J. Walker. 2000. *Practical Biochemistry—Principles and Techniques*, 5th edn. Cambridge, U.K.: Cambridge University Press.

Important Links

Capillary electrophoresis: http://www.biocompare.com/ProductDetails/3177514/Agilent-7100-Capillary-Electrophoresis-System.html?fi=3177514

Density gradient gel electrophoresis: http://www.cbsscientific.com/dgge.aspx

Pulsed field gel electrophoresis: http://www.biolabo.com/Electrophoresis/Pulsed-Field-Gel-Electrophoresis-PFGE-system

Temperature gradient gel electrophoresis: http://www.cbsscientific.com/ttge.aspx

9

Histochemical Techniques

9.1 Introduction

9.1.1 Histochemistry

It deals with the identification of chemical components in cells and tissues. It is performed by examining cells and tissues commonly by sectioning and staining, followed by examination under a light microscope or electron microscope. The ability to visualize and/or differentially identify microscopic structures is frequently enhanced through the use of biological stains. Histochemical techniques are an essential tool of biology and medicine. The microscopic study of diseased tissue is an important tool in anatomical pathology since accurate diagnosis of cancer and other diseases usually requires histochemical examination of samples. It is a histological technique used for studying chemistry of tissues and cells by (1) histochemistry, (2) enzyme histochemistry, (3) immunocytochemistry, and (4) *in situ* hybridization.

The pathologist may see abnormal matter either inside or outside the cells and may be concerned to know what it is composed of: Is it a mass of dead cells (which will contain nucleic acids and proteins); is it an abnormal protein complex, for example, "fibrinoid"; is cholesterol present; or does it contain calcium? Generally the histologist applies a number of tests, selected largely at random. Serial sections will be required so that the tests can be applied, in sequence, to the same matter. The tests are normally performed for four major classes of compounds shown in Figure 9.1.

In this stage, the major classes of compounds present in the material are decided:

1. Test for protein by the dinitrofluorobenzene (DNFB) method or by the tetrazotized dianisidine method (or by both of them used separately). If positive, the nature of the protein will be examined (Figure 9.2).
2. Test for acidic groups, such as nucleic acids. At this stage, all that is required is to stain with toluidine blue (at pH 6) or with the methyl green–pyronin mixture (at pH 4.2). If positive, then tests in Figure 9.3 will have to be done to decide on the nature of the acidic matter present.
3. Test for carbohydrate by the use of a controlled periodic acid–Schiff (PAS) test. If positive, tests for carbohydrate will have to be used (Figure 9.4).

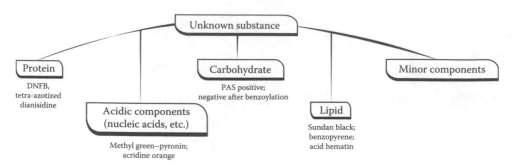

Figure 9.1 The determination of the major classes of compounds present.

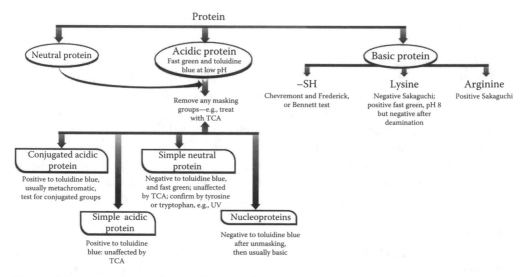

Figure 9.2 Histochemical identification of proteins.

4. Investigate the material for the presence of lipid. Sudan black is a good general indicator of lipid although benzopyrene may be rather better, both as regards sensitivity and because it does not involve the use of a fat solvent. If positive, apply the methods of Figure 9.5.
5. When all these have been tested (1–4 inclusive, earlier), and when the major components have been defined (stages 2–5 inclusive, in the following), minor components can be examined as described in Figure 9.6.

9.2 Protein

Proceed with this stage only if a positive reaction is obtained with DNFB and/or tetrazotized dianisidine method and/or Baker's tyrosine technique. If negative to all of these, it is unlikely to contain protein. If positive, then first decide whether the protein is predominantly basic, acidic, or neutral:

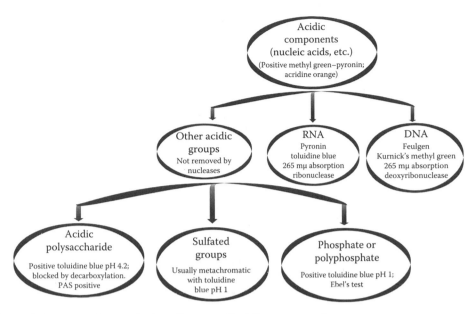

Figure 9.3 Histochemical identification of acidic compounds.

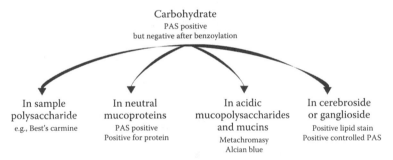

Figure 9.4 Histochemical identification of carbohydrates.

1. It is basic if it stains with acidic dyes (such as fast green) at a high pH value. (*Note*: Above pH 7, the $-NH_2$ groups present in all proteins are relatively unionized and so cannot bind the acidic dye, whereas the more basic groups remain ionized and so stain with fast green.)

2. It is acidic if it stains with basic dyes (such as toluidine blue) at low pH values. (*Note*: Below pH 6, the carboxyl groups present in proteins are relatively unionized so that if the material stains at pH 4·2 it may be taken that it contains conjugated acidic matter such as the nucleic acids or acidic polysaccharide.)

3. It is a neutral protein if it shows neither of these stain characteristics, that is, it may be stained both with fast green and with toluidine blue at about neutral pH, but staining with the basic dye is suppressed at more acidic pH values, while staining with the acidic fast green is suppressed the more basic is the pH.

4. If the protein appears to be basic, this is probably due to its containing a high proportion of basic amino acids. Therefore,

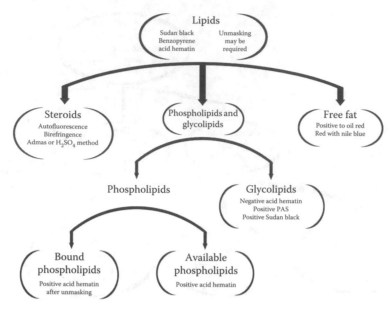

Figure 9.5   Histochemical identification of lipids.

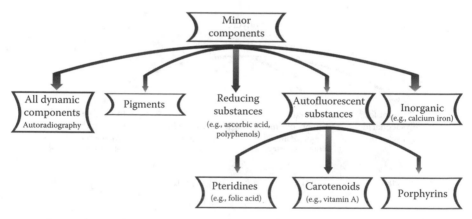

Figure 9.6   Minor compounds.

Test for arginine by the modified Sakaguchi reaction.

Test for lysine: Lysine can be assumed to be present, when the protein stains strongly with fast green at pH 8 but is negative after deamination, provided that the Sakaguchi reaction (for arginine that will respond similarly) is either negative or weak.

Test for –SH groups by the method of Chevremont and Frederick or by Bennett's mercury orange method.

5. If the protein appears to be acidic, this could be due to its containing a high proportion of acidic amino acids like glutamic acid (whose ionization is suppressed at pH 4.2); however, it is more probably due to its being a neutral (or even a basic) protein, which is conjugated to an acidic compound. Hence, both neutral and "acidic" proteins can be tested together as follows:

Treat the section with a 5% aqueous solution of trichloroacetic acid (TCA) at 90°C for 15 min. This will remove the nucleic acids. Then proceed as follows:

Simple neutral protein will stain with fast green and with toluidine blue only close to neutrality (as before); it will be unaffected by the TCA. It is advisable to prove that such material is proteinaceous by testing for tyrosine or for tryptophan.

Simple acidic protein will stain positively with basic dyes, like toluidine blue, but not appreciably with acidic dyes (like fast green) at pH values of between 4.2 and 6 as before; their staining will not have been affected by the TCA (although some intensification may be observed because of the coagulating effect of the TCA and the unmasking of acidic side chains due to the denaturation of the protein).

6. Nucleoproteins will have stained strongly and orthochromatically (blue) with toluidine blue at pH 4.2 and with the methyl green–pyronin method before the TCA treatment; after TCA extraction, these reactions will be abolished, and the residual protein will stain either as a neutral or as a basic protein (e.g., the latter will be stained by fast green at pH 8). If it behaves as a basic protein, then it may be subjected to the tests for basic amino acids detailed previously.

7. Other conjugated acidic proteins may be either lipoprotein or protein conjugated with acidic polysaccharide. If the former, it will behave as a simple protein and will also stain for lipid.

8. If conjugated to acidic polysaccharide, it will stain positively and usually metachromatically (purple to red) with toluidine blue, and this property will not have been lost as a result of the TCA treatment. It should be tested for the acidic conjugate as in Figure 9.4 (for carbohydrate–polysaccharide) or Figure 9.3 (for other acidic groups such as sulfate or phosphate).

9.2.1 Reactions for Protein

9.2.1.1 Dinitrofluorobenzene Method

9.2.1.1.1 Materials. DNFB reacts with amino acid residues and groups such as tyrosine, –NH, and –SH. The color produced is too weak for histochemistry; it is intensified by synthesizing an azo dye from the DNFB protein complex inside the section.

Since the DNFB will react with groups that must occur in proteins, it can be used as a general protein stain. It can probably be made specific for a particular group by blocking the other groups that can react. DNFB will react with tyrosine alone if –NH$_2$ groups are blocked by prior treatment with nitrous acid and –SH groups by treatment with iodoacetamide.

DNFB solution: Dinitrofluorobenzene 0·5 mL; 65% alcohol saturated with sodium bicarbonate 30 mL; add excess of the solid sodium bicarbonate to ensure that alcohol is fully saturated.

Sodium dithionite solution: 20% in distilled water. (*Note*: Dissolve in warm water [e.g., at 40°C]).

Nitrous acid: 2% sodium nitrite 1 volume; 0·1 N hydrochloric acid 1 volume and mix just before use.

H-acid solution: 0·25 g; 1% sodium bicarbonate solution 50 mL.

9.2.1.1.2 Protocol.   The sections need not be fixed for this procedure if fixation is required then fix them in 5% acetic acid in absolute ethanol for 10 min. Take the sections through graded concentrations of alcohol to water before processing.

Place sections for 2 h in the DNFB solution, in a Coplin jar. Stir occasionally to maintain the saturated condition of the bicarbonate solution; wash in 60% alcohol (to remove adsorbed DNFB); place in 20% solution of sodium dithionite at 37°C for 10 min; wash in distilled water at room temperature; place in the nitrous acid solution for 5 min; pass the slides through three consecutive washes, 1 min in each, of acidified distilled water; transfer the slides to the H-acid solution; leave for 15 min; wash in distilled water; and remove from the ice bath. Dehydrate. Mount in Euparal or in DePeX.

9.2.1.1.3 Result.   The reactive groups in protein, and hence proteins generally, stain red-purple.

9.2.1.2 Tetrazotized Dianisidine Method

9.2.1.2.1 Principle.   Diazonium hydroxides react with histidine, with tryptophan, and with tyrosine (also with many phenols), yielding a colored complex. This color can be intensified by coupling another phenol or amine on to the unreacted diazo group of the diazonium hydroxide. Of the various diazonium hydroxides that are available, probably the most suitable is the tetrazotized dianisidine.

This method is specific for one or the other (histidine, tryptophan and/or tyrosine) by pretreating the sections with substances that will block the reactive groups of the other amino acids. Pretreatment with DNFB will block tyrosine and probably histidine so that any reaction with tetrazotized dianisidine would be due to tryptophan alone. Pretreatment with performic acid should eliminate tryptophan and leave only histidine and tyrosine available for the tetrazotized dianisidine reaction. Benzoylation should block all three amino acids. The advantage of this and of the DNFB reaction is that they are stoichiometric reactions, which can be measured quantitatively.

9.2.1.2.2 Materials.   Solutions required for this method are as follows:

Barbiturate solution: Sodium diethyl barbiturate (Veronal) 6 g; make up to 300 mL with distilled water.

Tetrazotized dianisidine: Grind 0.15 g tetrazotized dianisidine with cool barbiturate solution in a cooled mortar until dissolved. Make up to 100 mL with barbiturate solution.

H-acid: H-acid 0·5 g; sodium bicarbonate 1 g; make up to 100 mL with distilled water. The bicarbonate is a convenient way of achieving an alkaline pH.

Pretreatment with performic acid:

Solution: 30% (100 volume) hydrogen peroxide 4 mL; conc. sulfuric acid 0.5 mL; 98% formic acid 40 mL.

The performic acid should be vigorously stirred before use to remove gas and peroxide must be fresh. Place the section in this solution for 45 min at room temperature.

9.2.1.2.3 Protocol. The sections need not be fixed for this procedure if fixation is required then fix the cryostat sections for 10 min in acetic alcohol (5:95 v/v or 1:3 v/v). Take the sections through graded concentrations of alcohol to water before staining:

1. Place slide in freshly prepared ice cool solution of tetrazotized dianisidine.
2. Transfer to another freshly prepared solution of tetrazotized dianisidine for a further 6 min. (*Note*: This solution should be prepared during the first 6 min reaction; the reason for such fresh solution is that the tetrazotized dianisidine decomposes very rapidly so that one can be certain of good activity for only about 6 min.)
3. Wash well in 2% cool barbiturate solution.
4. Wash well in cool distilled water. (*Note*: It is essential to remove adsorbed, unreacted tetrazotized dianisidine because it will also be intensified and yield a strong, but spurious, stain. The reacted compound is held by true chemical linkage and cannot be washed out of the protein.)
5. Stain for 15 min in cold H-acid solution in a Coplin jar.
6. Wash in cold distilled water. Remove from ice bath.
7. Dehydrate and mount in Euparal or DPX.

9.2.1.2.4 Result. Red or brown red.

9.2.1.3 Baker's Method for Tyrosine

9.2.1.3.1 Principle. The phenolic amino acid tyrosine is a useful marker for most proteins (but not for collagen). The most used histochemical procedure for this amino acid has been a modification of the chemical method. The phenol is converted into a nitrosophenol by the sodium nitrite, and a mercury links to the nitrogen atom of this nitroso group to form a red compound.

9.2.1.3.2 Material. *Mercuric sulfate solution*: Add 10 mL conc. sulfuric acid to 90 mL distilled water. Then add 10 g mercuric sulfate, and heat to dissolve. Cool and make up to 200 mL with distilled water. This solution is stable.

Sodium nitrite solution: Sodium nitrite 0.25 g; make up to 100 mL with distilled water.

9.2.1.3.3 Protocol. Fresh cryostat sections may be fixed in acetic alcohol (5% glacial acetic acid in absolute alcohol) for 10 min and then taken through graded concentrations of alcohol to water.

Put 25 mL of mercuric sulfate solution into a 50 mL beaker; add 2.5 mL sodium nitrite solution; put the section in a beaker and heat gently until the mercuric sulfate–sodium nitrite solution is just boiling; remove the sections; wash well in three changes of distilled water; and either mount in glycerine or glycerine jelly or dehydrate and mount in Euparal or in DPX.

9.2.1.3.4 Results. Phenols, especially tyrosine and hence most proteins, stain red, pink, or yellow-red.

9.2.1.4 Fast Green Method for Basic Protein

9.2.1.4.1 Principle.   At pH 8, the main groups, which are still sufficiently ionized to bind acidic dyes, are the guanidine groups of arginine and the ε-amino groups of lysine. They are usually bound to other acidic groups in the tissue, particularly to the nucleic acid. Hence, these have to be removed by treatment with hot TCA. This has the additional advantage of washing out acid-soluble basic matter of low molecular weight, that is, which is not part of the protein molecule.

The basic groups unmasked by the acid treatment tend to bind acidic ions present in buffer solutions, and these will interfere with the ability of the basic groups to bind fast green. Consequently, the dye should be prepared in distilled water and brought to pH 8 by judicious addition of alkali. Citrate buffer is found to be a more convenient way of achieving the required pH and does not produce the competitive inhibition.

9.2.1.4.2 Material.   TCA: 5% in distilled water

Fast green FCF: 0.1% in the citrate phosphate buffer, pH 8.0

Buffer (McIlvaine's): 0.1 M citric acid (0.48 g in 25 mL) 2.75 mL; 0.2 M disodium phosphate (7.1 g Na_2HPO_4. Anhydrous in 250 mL) 97.25 mL

9.2.1.4.3 Protocol.   Fresh cryostat sections should be fixed in acetic alcohol (5:95 v/v in absolute alcohol) for 10 min and then taken through graded concentrations of alcohol to water. Immerse the sections for 15 min in 5% TCA, which is at 95°C, for example, in a Coplin jar standing in a beaker of boiling water); wash well in three changes (10 min each) of 70% alcohol and then in distilled water to remove all the TCA; stain in fast green for 30 min at room temperature; wash briefly in distilled water; transfer the section directly to 95% alcohol; and dehydrate and mount in Euparal or in DPX.

9.2.1.4.4 Results.   Basic protein stains green or green-blue. Histone should be relatively strong colored.

9.2.1.5 Sakaguchi Reaction for Arginine

9.2.1.5.1 Principle.   Dichloro-naphthol gives a strong color and reacts more rapidly to the guanidino–amino group of arginine at a very alkaline pH in the presence of hypochlorite. Hypochlorite must be removed because its prolonged action will reduce the final color, and secondly, the mountant should be alkaline because the color formed is a pH indicator, being straw colored at low pH value but strongly red-brown at higher pH values. Sometimes it may be necessary to use special fixation to unmask and yet retain the arginine in tissue sections.

9.2.1.5.2 Materials.   *Lewitsky's fluid*: 10% formalin (i.e., 10% of the 40% formaldehyde) 1 volume; 1% chromic acid (chromium trioxide) 1 volume; mix just before use. It is advisable to wash well after this fixative (e.g., 5 min in each of three baths of distilled water).

Celloidin solution: Ether (diethyl ether) 50 mL; absolute ethyl alcohol 50 mL; celloidin 1 g. This is used to protect the tissue against the subsequent treatments, which tend to remove the section from the slide.

Reaction medium: 1% aqueous solution of sodium hydroxide 30 mL; 1% solution of 2,4-dichloro-α-naphthol in 70% ethyl alcohol 0.6 mL; sodium hypochlorite 1.2 mL; mix immediately before it is to be applied to the sections.

Urea solution: 5% aqueous solution.

Sodium hydroxide solution: 1% aqueous solution.

Alkaline glycerol: Glycerol 9 mL; 10% aqueous solution of sodium hydroxide 1 mL.

9.2.1.5.3 Protocol. Cryostat sections should be fixed either in Lewitsky's fluid (15 min) or in 10% neutral formalin. For this reaction, the tissue itself may be fixed in either of these fluids and may be embedded in paraffin wax before it is sectioned. Then pass the section up to the alcohol series to absolute alcohol; dip the whole slide into celloidin solution; shake off excess; drain and air-dry; pass the section through graded concentrations of alcohol to water; treat it in freshly prepared reaction medium for 6 min; rinse rapidly in 5% solution of urea (to stop the reaction; please note that prolonged treatment with urea will decolorize the reaction product); place in 1% solution of sodium hydroxide for 5 min; and mount in alkaline glycerol (alkaline is needed to retain the color).

9.2.1.5.4 Results. Orange-red color denotes arginine.

9.2.1.6 Bennett's Mercurial Method for Sulfhydryl Groups

9.2.1.6.1 Principle. The principle of this method is that sulfhydryl groups are such strong reducing agents that they can reduce ferricyanide to ferrocyanide. The latter reacts with ferric ions to produce an intense and insoluble precipitate. Mercurials react specifically with sulfhydryl groups. This is the basis of the specific blockade methods. This reagent should be highly specific for sulfhydryl groups.

9.2.1.6.2 Materials. *Ferricyanide solution (containing ferric ions)*: Solution 1—ferric ammonium sulfate $(NH_4)_2SO_4 \ Fe_2(SO_4)_3 \ 24H_2O$ 2.4 g and distilled water to 100 mL. Solution 2—potassium ferricyanide 0.1 g; distilled water to 100 mL. Prepare this solution just before use.

The reaction medium is prepared by adding 3 volumes of the solution to 1 volume of solution 2 just before it is to be used. Then adjust the pH to 2.4.

Blockade: Either a saturated solution of mercuric chloride in water or a saturated solution of phenylmercuric chloride in butyl alcohol. The advantage of the latter may be that it may minimize the presence of globules of free mercury in the tissue section.

Lugol's iodine: Iodine 1 g; potassium iodide 2 g and distilled water to 100 mL.

Thiosulfate solution: 5% aqueous solution of sodium thiosulfate.

9.2.1.6.3 Protocol. For fixation, a 1% solution of TCA in 80% ethanol is recommended. If cryostat sections are used, they have to be taken to absolute alcohol before proceeding as follows: Immerse for 16–24 h in mercury orange solution at room temperature; the duration of the reaction does not seem to be critical; wash in two changes of absolute ethyl alcohol; mount in Euparal.

9.2.1.6.4 Results.   Orange-red color denotes sulfhydryl groups.

9.3 Acidic Compounds

9.3.1 Introduction

Acidic substances have been shown to be present (as discussed in Figure 9.1). In this stage, their nature is determined:

1. Stain with methyl green–pyronin. If green, or blue-green, test for deoxyribonucleic acid (DNA) by the Feulgen reaction or by Kurnick's methyl green method (or by ultraviolet microscopy, if available). If these tests yield apparently positive results, then use controlled digestion with crystalline deoxyribonuclease (DNase) to confirm that DNA is indeed present.
2. If DNA is not present, or if it is not the sole constituent, all that is known is that acidic matter is present, that is, it stains at pH 4.2 with basic dyes like pyronin (in the methyl green–pyronin test) or toluidine blue.
3. Test for ribonucleic acid (RNA) by observing if the basophilia is decreased after controlled digestion with ribonuclease (RNase). (If ultraviolet microscopy is available, the maximal absorption in the 265 nm region, due to the purines and pyrimidines, is helpful in proving that it is a nucleic acid.)
4. If the basophilia is not removed by digestion with RNase and DNase, tests for other acidic groups must be used. These tests should also be applied where the nucleases are not available. (*Note*: RNA can usually be removed almost as effectively as by RNase by immersing the section in 1 N hydrochloric acid at 60°C for 6 min.)
5. Stain with toluidine blue at pH 1. Only phosphate or sulfate is likely to remain ionized at this pH and so to be capable of staining.
6. Apply Ebel's test for phosphate and polyphosphate. If negative, the acidic group is, probably, sulfate. Usually such substances stain metachromatically (pink or red) with toluidine blue.
7. If the material stains with toluidine blue at pH 4.2, but the staining is abolished at pH 1, it is probably due to acidic groups of the type found in acidic (but not sulfated) polysaccharides. The ability of such groups to stain with basic dyes is lost after decarboxylation; they should be positive with the PAS method.

9.3.1.1 Acidic Compounds (Nucleic Acids)

9.3.1.1.1 Feulgen Reaction for Deoxyribonucleic Acid
9.3.1.1.1.1 Principle.   Neither DNA nor RNA contains free aldehyde groups. However, acid hydrolysis can unmask aldehyde groups in the deoxyribose moiety of DNA, but not in the ribose moiety of RNA, and enables the phosphate to migrate to different sites in the sugar and to allow the ring form of sugar to be opened up to give a straight-chain molecule with a terminal aldehyde (H–C=O) group.

(a) Deoxyribonucleotide   (b) Ribonucleotide

Another chemical difference between ribose and deoxyribose is the fact that only the latter can form levulinic acid when treated with dilute acid. Ketonic group gives a positive reaction with Schiff's reagent.

9.3.1.1.1.2 Materials. *Schiff's reagent*: Dissolve 1 g basic fuchsin in 200 mL of boiling distilled water. Shake and cool to 50°C. Filter and add 30 mL of 1 N hydrochloric acid and 3 g of potassium metabisulfite. This solution is allowed to stand for 24 h in a dark-colored stopper bottle. During this time the bisulfate decolorizes the fuchsin, that is, it forms the colorless SO_2–pararosaniline complex (leuco-basic fuchsin). Impurities and other colored matter are then removed by adsorption on to activated charcoal. Therefore, add 0.5 g of a decolorizing activated charcoal and shake well for about 1 min. Filter rapidly (to avoid recolorizing the leuco-basic fuchsin) through a coarse filter paper (or glass wool). The solution should be clear and colorless; it can be stored for months in a well stopper, dark-colored bottle. However, commercially prepared Schiff's reagents are available with several chemical manufacturers.

SO_2–water: 1 N hydrochloric acid 25 mL; 0.5% aqueous solution of potassium metabisulfite 50 mL; mix immediately before use.

1 N hydrochloric acid: Hydrochloric acid extra pure grade with Sp. Gr. 1.18 89 mL; distilled water 911 mL. *Always add the acid to the water.*

9.3.1.1.1.3 Protocol. Fix cryostat sections in acetic ethanol (1:3 v/v) for 10 min; take through graded concentrations of alcohol to water; and rinse in 1 N hydrochloric acid; plunge into 1 N hydrochloric acid that is at 60°C. Maintain at this temperature for 6 min; transfer to 1 N hydrochloric acid at room temperature; immerse in Schiff's reagent in the dark for 60 min; rinse in three changes of freshly prepared SO_2–water; rinse in distilled water; and dehydrate and mount in Euparal or in DPX.

Control: Fix serial section in 1:3 acetic ethanol for 10 min; take down to water; rinse in 1 N hydrochloric acid and leave in this at room temperature for 15 min; immerse in Schiff's reagent in the dark for 60 min; rinse in three changes of freshly prepared SO_2–water; rinse in distilled water; dehydrate and mount in Euparal or in DPX. Any stain produced by this control procedure is not due to DNA and must be subtracted from the results obtained by the full test procedure.

9.3.1.1.1.4 Result. DNA stains magenta.

9.3.1.1.2 Methyl Green–Pyronin Reaction for DNA and RNA

9.3.1.1.2.1 Principle. The mixture of basic dyes, methyl green, and pyronin stains nuclei and cytoplasm differentially. The mechanism of the reaction is based on the competition between the slow staining but doubly charged methyl green and the more rapidly staining, singly charged pyronin. The pyronin derives the methyl green out of all its sites of linkage except where its double charge gives it some selective advantage, namely, when it responds to an acidic polymer such as DNA. Consequently, the methyl green stains DNA and retains its binding to this substance against the competitive action of the pyronin. On the other hand, the pyronin can stain the less polymerized RNA more rapidly and can displace methyl green from its linkage with smaller polymeric acidic substances (such as RNA, but this applies also to acidic polysaccharides and similar molecules). The reaction is done at pH 4.2, which is the isoelectric point of isolated nucleic acids. This has the advantages of having them in the charged condition and also in their least soluble state.

9.3.1.1.2.2 Materials. *Acetate buffer at pH 4.2*: (A) 0.2 M sodium acetate solution (CH_2 COONa $3H_2O$) 6.8 g made up to 250 mL in distilled water; (B) 0.2 M glacial acetic acid (CH_3COOH) made up to 250 mL with distilled water. Add 70 mL of B to 30 mL of A.

Stain mixture: Prepare 0.25% (w/v) solution of methyl green in 0.2 M acetate buffer at pH 4.2. Extract with chloroform, until almost all the methyl violet impurities have been removed in the chloroform. Separate the aqueous methyl green solution from the chloroform and leave to stand, uncovered, in the dark until the dissolved chloroform has evaporated.

Add an equal volume of 0.5% (w/v) solution of pyronin G in 0.2 M acetate buffer and mix. This solution is stable for some weeks.

9.3.1.1.2.3 Protocol. Fresh cryostat sections can be used. They will be fixed by the acidic acetate buffer used in the staining mixture. The sections can be fixed in acetic alcohol, which is a good precipitant of nucleoprotein. The higher the concentration of the acetic acid used, the more will the nucleic acids be freed from the protein and hence the stronger the stain. Consequently 1:3 (v/v) acetic alcohols may be used, although 5% acetic acid in absolute ethyl alcohol is satisfactory. Fixation should be for 15 min. If the sections have been fixed in an alcoholic fixative, rinse in 70% alcohol and transfer to acetate buffer; immerse in stain mixture for 30 min; rinse very briefly in acetate buffer; and blot dry. This must be done carefully but firmly. (If any moisture is left, the methyl green will be lost from the sections at the next step); immerse in absolute *n*-butyl alcohol for 2 min; mount in Euparal.

9.3.1.1.2.4 Result. Nuclei stain green or blue; nucleoli and cytoplasm stain red. (*Note:* Mast cells stain a characteristic flame orange-red by this method. Plasma cells stain so intensely red that this procedure has been used to identify them.)

Interpretation: Very often, as a first approximation, the red color is said to denote the presence of RNA and the blue or green color that of DNA. This observation is a rough approximation. In practice, however, when nuclei stain blue-green, this is probably a true indication of the presence of DNA; however, should the presence of DNA be in the least critical, this apparent localization should be confirmed by the use of DNase. Alternatively, the use of short hydrolysis with hydrochloric acid may be a useful and less costly way of strengthening the belief that this blue-green-staining matter is DNA. (*Note*: DNA stains red after acid hydrolysis.)

9.3.1.2 Use of Deoxyribonuclease

The final proof that a substance in sections is indeed DNA must be that it is removed specifically by DNase. But even the crystalline enzyme may contain proteolytic activity; the removal of a protein, to which may be attached an acidic substance, could stimulate removal of DNA. To inhibit such proteolytic activity, 0.05 M cysteine or hydroxylamine may be added to the incubation medium along with magnesium to activate the DNase even though most tissues may contain enough of this element. As a control for the specific action of the DNase, a serial section can be treated with the same incubation medium, for the same time, but without the addition of DNase to the solution. A more rigorous control is to use the same incubation medium, containing DNase, but also containing zinc to inhibit the action of the DNase. This ensures the presence of the enzyme but in an inactive form.

9.3.1.2.1 Incubation Medium. The final concentration of the DNase is to be 0.1%. First, prepare a 0.05 M solution of cysteine hydrochloride (or hydroxylamine hydrochloride) in 0.05 M sodium acetate solution at pH 7. Dissolve enough of the DNase in this so that the final concentration of the enzyme will be 0.1% (e.g., dissolve 2 mg of the DNase in 1 mL of the cysteine-acetate buffer solution). Adjust to pH 7 if this should be necessary. Leave the solution to stand at room temperature (or at 37°C) for 5–15 min. (This is to inactivate the proteolytic activity.) Before it is to be used, dilute this solution with an equal volume of Veronal HCl buffer (barbitone) at pH 7. Add 1% of magnesium chloride to this enzyme solution.

9.3.1.2.2 Treatment. Add the complete medium to the section and leave at room temperature. One hour should be sufficient to remove DNA from unfixed sections. Then stain with methyl green as the previous.

9.3.1.2.3 Confirmation of the Specificity of the Deoxyribonuclease. Take three serial sections; treat one with the DNase incubation medium for 1 h and the second for the same time with an exactly similar incubation medium to which has been added zinc sulfate (at a concentration of 0.1–0.01 M; 0.1 M sodium arsenate may be used instead of zinc sulfate). This should inhibit the DNase. Leave the third section untreated (or it may be treated with the incubation medium lacking the enzyme entirely). Then stain all three slides together with the methyl green method.

9.3.1.2.4 Result. DNA should stain green in the third slide; it should not be present in the first slide (having been removed by the active DNase). It should be present and stained green in the second slide (which has been subjected to the inactivated deoxyribonuclease).

Interpretations: Malignant and lymphocyte cells apparently synthesize extra amounts of proteins responding to antibodies. Under these circumstances, it may be gravely misleading to assume that "pyroninophilia" equals RNA. The gravity of this assumption can best be emphasized by an example with which histologically the section contained unusual cells that are mitotically active. When tested by the methyl green–pyronin method, these cells stained vividly red due to neoplastic growth. But, when they were tested either with RNase or by the hydrochloric acid method, the ability of these cells to stain with pyronin was undiminished. Consequently, the "pyroninophilia" was not due to RNA. In fact, these cells were actively growing, but not malignant, bile duct cells in which the basophilia (to pyronin) was due to some acidic mucopolysaccharides matter. In histochemical terms, the interpretation of pyronin-positive matter as RNA depends on the loss of this basophilia after treatment either with RNase or with hydrochloric acid.

9.3.1.3 Use of Ribonuclease

9.3.1.3.1 Method. This enzyme acts optimally at about pH 7.7. Hence, 1 mg of the crystalline preparation is dissolved in 1 mL of 0.2 M Veronal (barbitone) HCl buffer at pH 7.7. (*Note*: It is often far simpler to dissolve 1 mg of the enzyme in an M/40 aqueous solution of sodium bicarbonate: this is a convenient way of achieving a pH of 8.2, which is sufficiently close to the optimum.)

Treat one section with the RNase solution for 1 h at room temperature; treat a control section for the same length of time either with the buffer (or bicarbonate) solution alone or with the RNase solution to which has been added zinc sulfate (at a concentration of 0.1–0.01 M) to activate the enzyme; wash both sections and stain with methyl green–pyronin.

9.3.1.3.2 Result. Pyronin-positive matter, which is lost after treatment with RNase but not after exposure to the control, can be taken to be RNA.

9.3.1.4 Hydrochloric Acid Method for RNA

9.3.1.4.1 Principle. 1 N hydrochloric acid at 60°C extracts all the RNA from sections in 5 min; it also causes sufficient alteration in DNA to make it stain with the pyronin, and not with the methyl green, component of the methyl green–pyronin mixture.

9.3.1.4.2 Method. Take two sections, and fix in acetic alcohol (either 1:3 or 5% as required); take both sections through graded concentrations of alcohol to 0.2 M acetate buffer at pH 4.2; immerse one in 1 N hydrochloric acid at 60°C for 5 min; wash both in 0.2 M acetate buffer; stain both in the methyl green–pyronin mixture; blot dry; immerse in *n*-butyl alcohol for 2 min; and mount in Euparal.

9.3.1.4.3 Result. RNA stains red in the nonhydrolyzed section but is absent from the hydrolyzed section. The DNA should stain green-blue in the nonhydrolyzed section but a reddish or purple color in the hydrolyzed section.

9.3.1.5 Toluidine Blue Method for Basophilia

9.3.1.5.1 Principle. Toluidine blue is a basic dye and so will stain acidic matter. The exact nature of the matter cannot be determined just by the fact that it stains with this dye; all that can be said is that it is acidic. The use of nucleases can show if it is

RNA or DNA. Another diagnostic feature is whether the dye stains orthochromatically, that is, blue, or metachromatically, namely, red or red-purple.

9.3.1.5.2 Materials. *Acetate buffers (0.1 M)*: (A) 0.1 M glacial acetic acid (CH$_3$COOH) 2.85 mL glacial acetic acid made up to 500 mL with distilled water. (B) 0.1 M sodium acetate solution (CH$_2$ COONa 3H$_2$O) 6.8 g sodium acetate made up to 500 mL in distilled water; for pH 4.2 add 140 mL of A to 60 mL of B; for pH 5.0 add 60 mL of A to 140 mL of B; for pH 5.6 add 20 mL of A to 180 mL of B.

Toluidine blue solution: Toluidine blue 0.1 g makes up to 100 mL with the relevant buffer.

9.3.1.5.3 Protocol. Fresh cryostat sections are recommended. Otherwise fix fresh sections in 1:3 (v/v) acetic alcohol (or 5% acetic acid in absolute ethyl alcohol) and take them through graded concentrations of alcohol to acetate buffer. Stain in an aqueous solution of toluidine blue in acetate buffer for 30 min (*Note*: 10 min may be sufficient); wash in acetate buffer; blot dry and place in *n*-butyl alcohol (*n*-butanol) for 2 min; and mount in Euparal.

9.3.1.5.4 Result. Normal basophilia is shown by blue or purple-blue staining (the exact hue depends on the composition of the mixture of the dyes known collectively as "toluidine blue." For this reason, it may be preferred to use azure B in place of toluidine blue). Pink or red staining denotes metachromatic coloration.

9.3.1.6 Metachromatic Staining with Toluidine Blue

9.3.1.6.1 Principle. When the dye reacts with a polymeric molecule in the tissue section, the dye shows its ordinary color, that is, it stains orthochromatically. However, should these dye molecules come to be so close together along the polymer that they can interact with each other, the ionization of the dye becomes suppressed and so gives rise to a change of color. Basic dyes stain acidic substances orthochromatically. They stain metachromatically when the acidic substances are present as long polymeric molecules, with the acidic groups closely packed along the polymer. Consequently, acidic mucopolysaccharides are typically stained metachromatically but certain other acidic polymers may also show this type of coloration. The nature of the acidic groups can be tested by finding at what pH value their ionization is suppressed.

For pH values of 5.6, 5.0, or 4.2, use the acetate buffer recommended for the usual method for toluidine blue. For lower pH values, use distilled water and add a few drops of 1 N HCl to achieve the desired pH. The dye is dissolved (0.1%) either in buffer or in water. The pH of the dye-in-water solution should be adjusted to the required value after the dye has been dissolved, in case the dye itself should affect the pH. (*Note*: Azure can be used in place of toluidine blue. The azure B [0.1–0.2 mg/mL in citrate buffer at pH 4] stains RNA metachromatically.)

9.3.1.6.2 Protocol. Fresh cryostat sections are recommended. Alternatively, the sections can be fixed in the picric–formalin fixative. Immerse in 0.1% solution of toluidine blue at the required pH, stain for 30 min; rinse in the buffer at the same pH, either blot dry or shake free of buffer, dry in air, place in *n*-butanol for 2 min, and mount in Euparal.

9.3.1.6.3 Result. Nucleic acids in sections stain orthochromatically with toluidine blue at pH 4.2. At lower pH values, their ability to bind the dye becomes progressively diminished.

Acidic polymers are stained metachromatically by toluidine blue at pH 5.6. If the acidity is due to carboxyl groups, this staining is abolished at lower pH values (e.g., 4.2). Of the acidic groups found commonly in biological material, only sulfate groups (and polyphosphate) remain stainable at pH 1.0. Phosphate moieties of organic molecules become less stainable between pH values of 4.2 and 1.0.

9.3.1.7 Acridine Orange Method for DNA and RNA

9.3.1.7.1 Principle. Acridine orange produces a different-colored fluorescence when it is bound to low-polymer nucleic acids than when it is linked to high-polymer DNA. It also shows its "high-polymer" color when linked to other polymers, such as keratin. The exact color, which is seen, depends on the excitation wavelength and on the eyepiece (barrier) filters used in the fluorescence equipment. The action of dyes is also very sensitive to the pH at which the staining reaction is done and to the fixative used.

9.3.1.7.2 Materials. *Walpole's acetate buffer at pH 4.2*: 1N sodium acetate (13.6 g/100 mL; $CH_3COONa\ 3H_2O$) 50 mL; 1 N hydrochloric acid 35 mL.

Acridine orange solution: 1:10,000 (w/v) parts of acridine orange in Walpole's acetate buffer (pH 4.2).

9.3.1.7.3 Protocol. Fresh cryostat sections should be used or they may be fixed in acetic alcohol (1:3 v/v). Tissues can be fixed in calcium–formaldehyde, but fixatives containing heavy metals should not be used. Immerse in acetate buffer, pH 4.2 for 5 min. (*Note*: This will also fix fresh cryostat sections.) Stain for 15–30 min in acridine orange solution at pH 4.2, wash in the buffer at pH 4.2, and mount in a drop of the buffer and ring the cover slip with wax and examine with a fluorescence microscope.

9.3.1.7.4 Results. When excited with purple light, the DNA–acridine orange complex fluoresces green-yellow to bright yellow; the RNA–acridine orange complex gives a dull reddish brown to flame orange, depending on its concentration in the cells. With ultraviolet light (365 mμ), DNA appears greenish-yellow and RNA crimson red. "False-positive" mast cell granules and cartilage matrix give red fluorescence. Vascular elastic elements fluoresce yellow and keratin is often green.

9.4 Carbohydrates–Polysaccharides

9.4.1 Introduction

The material is positive with the alcoholic PAS test but is negative (or the reaction is markedly depressed) after benzoylation. Hence, it contains carbohydrate. Is it removed by rinsing in water or in dilute acid (e.g., 0.1 N TCA)? If it is lost after this treatment, it is either free carbohydrate or a simple polymer such as glycogen or starch:

1. Test with Lugol's iodine for starch.
2. Test with Best's carmine for glycogen or use the alcoholic PAS test both with and without prior treatment with diastase or with α- and β-amylases.

If it is not removed by dilute acid, it may be linked to a protein. Hence, apply a test for protein (Figure 9.2). Does the carbohydrate material react metachromatically with toluidine blue? If so, it may be an acidic polysaccharide, also possibly linked with protein. Test with Alcian blue for acidic mucopolysaccharides or with Stoward's reactions to confirm this. The carbohydrate material may comprise part of a cerebroside or ganglioside complex. These will be positive with lipid stains (Figure 9.5). Moreover, the presence of unsaturated fatty acids in such complexes may give an apparent PAS reaction, which will not be abolished by benzoylation but which will require oxidation (e.g., by bromination; this will "saturate" the fatty acids) prior to treatment with periodic acid. The material contains lipid if it stains with benzopyrene or with Sudan black (although the "burnt" Sudan black method or some other unmasking technique may have to be employed).

9.4.1.1 Reactions for Polysaccharides

9.4.1.1.1 Periodic Acid–Schiff Method

9.4.1.1.1.1 Principle. The PAS technique (and its numerous variations) is by far the most commonly performed special stain within the histopathology laboratory. The PAS technique is most commonly used to highlight molecules with a high percentage carbohydrate content such as mucins, glycogen, fungi, and the basement membrane in skin. The PAS method works by exposing the tissue to periodic acid. This acts an oxidizing agent that oxidizes vicinal (neighboring) glycol groups or amino/alkylamino derivatives. This oxidation creates dialdehydes. These dialdehydes, when exposed to Schiff's reagent, create an insoluble magenta compound, which is similar to the basic fuchsin dye within Schiff's reagent.

Tissue sections are first oxidized by periodic acid. The oxidative process results in the formation of aldehyde groupings through carbon-to-carbon bond cleavage. Free hydroxyl groups should be present for oxidation to take place. Oxidation is completed when it reaches the aldehyde stage. The aldehyde groups are detected by the Schiff reagent. A colorless, unstable dialdehyde compound is formed and then transformed to the colored final product by restoration of the quinoid chromophoric grouping. This method is used for detection of glycogen in tissues such as liver, cardiac, and skeletal muscle on formalin-fixed, paraffin-embedded tissue sections and may be used for frozen sections as well. The glycogen, mucin, and fungi will be stained purple, and the nuclei will be stained blue.

9.4.1.1.1.2 Materials. *Picrate–formalin fixative*: 40% formaldehyde 10 mL; ethyl alcohol 56 mL; sodium chloride 0.18 g; picric acid 0.15 g; distilled water to 100 mL.

Periodate solution: Dissolve 400 mg of periodic acid in 15 mL of distilled water. To this solution, add 135 mg of crystalline sodium acetate ($CH_3.COONa.3H_2O$); dissolve in 35 mL of absolute ethyl alcohol. Prepare just before use.

Reducing rinse: Potassium iodide 1 g; sodium thiosulfate 1 g; distilled water 20 mL. When dissolved, add (while stirring the solution) 30 mL of ethyl alcohol and then 0.5 mL 2 N hydrochloric acid. A precipitate of sulfur may form; this is left to settle out.

9.4.1.1.1.3 Protocol.　Fresh cryostat sections should be fixed for 5 min in the picrate–formalin fixative (*Note*: For an estimate of the total amount of glycogen, it may be preferable to fix in absolute alcohol or in the picrate–formalin solution to which 5% acetic acid has been added); wash in 70% alcohol; immerse in the periodate solution for 5 min at room temperature; wash in 70% alcohol; immerse in the reducing rinse for 5 min. (*Note*: Periodate is a very strong oxidizing agent, even traces of it left in the section will recolor Schiff's reagent. Consequently, it is necessary to remove all adsorbed traces of periodate with this reducing solution.) Wash in 70% alcohol; immerse in Schiff's reagent for 30 min; wash in three changes of SO_2–water; dehydrate (counterstain with a hematoxylin for approx. 15 s; differentiate (if necessary) and blue; dehydrate, clear, and mount in the nuclei with Ehrlich's hematoxylin if required); and mount in Euparal or in DPX.

9.4.1.2 Alcian Blue Method

9.4.1.2.1 Principle.　This is an extremely valuable stain for mucins or acidic mucopolysaccharides. "Mucins" and acidic mucopolysaccharides should stain blue.

9.4.1.2.2 Materials.　*0.1% Alcian blue solution*: 0.1% Alcian blue 8GS in 3% acetic acid. If Alcian blue 8GS cannot be obtained, Alcian green 3 BX or Alcian green 2GS may be used, but the concentration of this solution will have to be increased up to 1%.

1% aqueous Alcian blue solution: 1% solution of Alcian blue in distilled water. If this solution is saturated with thymol, it will keep for some weeks but should be filtered before use.

9.4.1.2.3 Protocol.　Fix cryostat sections for 5 min in the picrate–formalin fixative; take the section through graded concentrations of alcohol to water; stain in 0.1% Alcian blue solution for 20 min; stain in the 1% aqueous Alcian blue solution for 2 min; wash in tap water; dehydrate; and mount in Euparal. (*Note*: Taking the sections from water straight into a 0.5% solution of borax in 80% ethyl alcohol and leaving them there for 2 h. During this time the Alcian blue is converted, by the alkaline ethanol, into the very insoluble Monastral blue.)

9.4.1.2.4 Result.　"Mucins" and acidic mucopolysaccharides should stain blue.

9.4.1.3 Lugol's Iodine Method

9.4.1.3.1 Principle.　This is sometimes of use in polysaccharide histochemistry. Fresh cryostat sections may be used, or they may be fixed in the picrate–formalin solution. They are stained for variable periods of time in Lugol's iodine, rinsed, and can be mounted in Farrant's medium or dehydrated and mounted in Euparal or DPX. Glycogen should stain brownish and native starch should be colored deep blue.

9.4.1.3.2 Materials. *Lugol's iodine solution*: Iodine 1 g; potassium iodide 2 g; distilled water to 100 mL.

9.4.1.3.3 Protocol. Fix sections with the same fixative as is used for the PAS reaction. Wash in 70% alcohol, and treat the sections for 3 h at 37°C with 0.5% solution of α-amylase in 0.004 M acetate buffer, pH 5.5. Control sections must be treated, with the buffer alone, to ensure that any loss of PAS-positive material is due to the specific action of the amylase and is not a mere solution of the stainable matter. The sections are then washed in 70% alcohol and subjected to the PAS reaction.

All glycogen and starch must be removed by such treatment. In special investigations, it may be found helpful to distinguish between straight-chain and more complex, branched glycogen. To do this, the sections are treated similarly, but a 0.5% solution of β-amylase in 0.004 M acetate buffer at pH 5.0 is used, instead of the α-amylase. The β-amylase digests only straight-chain glycogen.

9.4.1.3.4 Result. Glycogen should stain brownish and native starch should be colored deep blue.

9.5 Lipids

9.5.1 Introduction

1. Steroids may not respond well to such lipid colorants. Consequently test for auto-fluorescence or for birefringence. Adam's reaction, or the sulfuric acid method, may help to demonstrate steroids.
2. Free fat will take up lipid colorants very readily, even those with lower affinity for lipids that is possessed by Sudan black (which also stains phospholipids that are not as "fatty" as are free triglycerides). Hence, test with oil red. Moreover, neutral triglycerides should yield a red color with Nile blue.
3. Test with the acid hematein method: If positive (but negative after bromination), then the material contains phospholipid. The extent to which the phospholipid is masked can be tested by the unmasking procedure (see following text).
4. If the material stains with benzopyrene or Sudan black (i.e., it contains lipid) but is negative to oil red (hence, not freely available triglyceride) and to the acid hematein test, even after unmasking procedures (i.e., it does not contain phospholipid), and it gives a positive PAS reaction (which is lost after benzoylation), then it contains a glycolipid.

9.5.2 Reactions for Lipids

Histochemically lipids can be defined as fatty matter, which has some of the chemical characteristics of lipid molecules, for example, they contain choline or unsaturated fatty acids or they are phosphatides. The tests are of two types: (1) physical methods, which depend on the readiness of lipophilic molecules to partition between their solvent and the lipid in the section (these are, therefore, tests of the degree of "fattiness" of the material) and (2) chemical methods, like the acid hematein test or the use of Nile blue.

Of these, the physical methods are generally the more important so that, to a large extent, "lipid" is characterized histochemically by its ability to concentrate Sudan black or benzopyrene from semi-aqueous solutions in which these compounds are not very soluble.

9.5.2.1 Sudan Black Method

9.5.2.1.1 Principle.   Sudan black B is a fat soluble colorant, which is soluble in absolute alcohol, but insoluble in water. It is a strong colorant and has a high affinity for fatty material, including phospholipids. It is used as saturated solution in 70% ethyl alcohol in which it is still soluble (it is very much less soluble in 50% alcohol), and the section is exposed to this solution. The colorant partitions between alcohol and the tissue lipids (in which it is more soluble). Because of its intense color, and of its strong affinity for all (or most) classes of lipids, the coloration of tissue components by Sudan black B is often taken as the histochemical criterion for the presence of lipids. The purpose of the various "burnt Sudan black" methods is to increase the sensitivity of tissue lipids, including some masked lipids but also steroids (which, being solid, will stain only at higher temperatures, that is, closer to their melting point).

9.5.2.1.2 Materials
9.5.2.1.2.1 Solutions Required.   A. *Bromine water*: Add 5 mL bromine in 195 mL distilled water. Bromine is caustic, and its brown, pungent vapor is injurious to the respiratory system. Keep it in a fume hood and don't think of pipetting it by mouth! The aqueous solution is stable for about 3 months at room temperature in a glass stoppered bottle and may be used repeatedly. It is nonhazardous in aqueous form.

B. *Sodium metabisulfite*: 0.5% aqueous $Na_2S_2O_3$ freshly prepared. Prepare on the day it is to be used.

C. *Sudan black B*: Add 600 mg Sudan black to 200 mL of 70% ethanol. Place on a magnetic stirrer for 2 h, and then pour into a screw-capped bottle. Leave to stand it for overnight. To use the solution, filter it into a Coplin jar and try not to disturb the sediment of undissolved dye in the bottom of the bottle. The solution of Sudan black B may be kept (and used repeatedly) for only 4 weeks. With older solutions, there is gray nonlipid background staining and weaker coloration of lipids.

D. *Counterstain*: Suitable red nuclear counter stains can be used to differentiate.

9.5.2.1.3 Protocol.   Fresh cryostat sections must be used because most, if not all, fixatives can modify the nature of lipid–protein and lipid–carbohydrate complexes. Immerse slides in bromine water (solution A) for 1 h; rinse in water; immerse in sodium metabisulfite (solution B) for about 1 min until the yellow color of bromine has been removed from the section; wash in four changes of water; rinse in 70% ethanol, 5–10 s with agitation; then wash in water; apply counterstain if desired; wash in water; and mount in an aqueous medium.

9.5.2.1.4 Result. Lipid appears in shades of deep gray, very dark blue, and black.

9.5.2.2 Oil Red O Method for Staining Fats

9.5.2.2.1 Principle. Oil red O is a strong colorant of fats, but it requires a high concentration of fatty matter for it to be drawn out of its alcoholic solution into the section. Triglycerides ("fats") are fattier than are phospholipids, which contain polar groups (i.e., hydrophilic and so lipophobic parts of the molecule); moreover, the former occur more generally in highly concentrated globules of hydrophobic molecules, whereas the latter may be associated with other more hydrophilic molecules. Consequently, oil red O is an excellent colorant of fats but very poor for phospholipids.

9.5.2.2.2 Materials. *Stock solution of oil red O*: A saturated solution (0.5%) of oil red O or of oil red 4B in absolute isopropyl alcohol. This solution can be stored for an indefinite period.

Staining solution: Mix stock solution of oil red O 60 mL with 40 mL distilled water. Leave to stand for 10 min and then filter before use. (*Note*: This solution is oversaturated so as to make the oil red O more likely to dissolve out of this solution and into the fat. It must be filtered just before use.)

9.5.2.2.3 Protocol. Fresh cryostat sections must be used. Rinse the section in 60% isopropyl alcohol, immerse in oil red O solution for 10 min, rinse briefly in 60% isopropyl alcohol, wash in water, then either mount in Farrant's medium or stain in hematoxylin for 30–60 min, wash in water, place in 5% sodium bicarbonate for a few minutes, wash well in distilled water, and mount in Farrant's medium.

9.5.2.2.4 Result. Fats are stained red or yellow-red, depending on their concentration.

9.5.2.3 Benzopyrene Method for Lipids

9.5.2.3.1 Principle. Benzopyrene is an intensely fluorescent molecule and detects relatively very low concentrations in the section. It is a highly sensitive method for all types of lipids (except solid steroids) and can be dispersed in a "hydrotropic solution" in water.

9.5.2.3.2 Materials. *Solutions required*: Saturated distilled water with caffeine at room temperature for 2 days (about 1.5% caffeine, filter and add about 2 mg of 3,4-benzopyrene [also called benzopyrene]) to 100 mL of this caffeine solution. Incubate at about 35°C for 2 days, filter off excess benzopyrene, dilute the solution with an equal volume of distilled water (to prevent the caffeine from crystallizing out the solution), leave to stand for 2 h, and filter. This solution is now ready for use although it can be kept for several months in a dark bottle. It should contain about 0.75 mg of benzopyrene in 100 mL of solution.

Rapid method for preparing the benzopyrene reagent: Prepare a 1% solution of benzopyrene in acetone; add this, drop by drop, to 100 mL of a saturated solution of caffeine, which must be stirred briskly. When a persistent turbidity is obtained, enough benzopyrene has been added (0.1–0.2 mL has been recommended, but we have found rather more to be required). Dilute with an equal volume of distilled water filter twice. The solution is then ready for use. (*Note*: Care should be taken to handle benzopyrene because this substance is carcinogenic.)

9.5.2.3.3 Protocol.　Fresh cryostat sections should be used although the technique works on fixed sections, even if treated with dichromate (as in the hematein procedure). Cover the section with the benzopyrene reagent for 20 min, rinse well, mount in water, and examine with the near ultraviolet light (e.g., 365 mμ) with a fluorescence microscope.

　　For cell suspension, add 0.6 mL of the benzopyrene reagent to about 1 mL of the cell suspension. Leave for 20 min. Make a smear or hanging drop preparation (preferably in fresh suspending medium) and inspect with the fluorescence microscope.

9.5.2.3.4 Result.　Lipid fluoresces silver to blue-white or even yellow, depending on the concentration of the benzopyrene in the lipid. Use the caffeine solution without added benzopyrene.

9.5.2.4 Rhodamine B Method for Lipids

Prepare 1:1,000 or even 1:10,000 solutions in water, or in some Ringer solution that has been much used as a vital fluorescent dye (fluorochrome), particularly for plant cells. It is quite a useful colorant in lipids. The phosphine 3R method is another useful fluorescence method for lipids.

9.6 Minor Compounds

Minor components are described in Figure 9.6.

9.6.1 Reaction with Minor Components

There are a number of well-documented methods for detecting relatively large concentrations of inorganic elements in tissue.

9.6.1.1 Iron

9.6.1.1.1 Unmasked Iron

9.6.1.1.1.1 Protocol.　Immerse the section in freshly prepared, filtered 2% potassium ferrocyanide solution for 5 min; transfer to acidified ferrocyanide solution (equal volume of the 2% potassium ferrocyanide solution and 0.2 N hydrochloric acid); leave for 20 min; wash in distilled water; and nuclei can then be stained with neutral red. Dehydrate and mount in Euparal.

9.6.1.1.1.2 Result. Reactive (unmasked) iron stains blue.

9.6.1.1.2 Masked Iron. Treat the section for at least 30 min with 100 volumes hydrogen peroxide to which have been added a few drops of sodium carbonate solution, to render it slightly alkaline. Then stain for reactive iron as previously discussed.

9.7 Autoradiography

9.7.1 Introduction

The localization of inorganic elements could be found if the elements (1) were radioactive and (2) were maintained at their original site during both the preparation of the tissue and the procedure for marking their location. The first requirement is fairly simply met because radioactive isotopes are treated by living organisms as if they were stable isotopes; consequently if they are injected or otherwise ingested into an animal (or taken up by a plant), they become dispersed throughout the organism, mingled with the nonradioactive elements. Once the radioisotope has been incorporated into a tissue, its presence—even in very low concentration—can be detected by virtue of the radiation that it emits. In histochemistry, this radiation can be detected, and its source inside the tissue can be discovered, by the fact that it affects photographic emulsions. Theoretically the element can be present as the free element itself or as part of a particular radical (e.g., ^{32}P in phosphorus or ^{35}S in sulfate): equally it can be part of a large molecule, in which case the radioactive element may act merely as a means of marking that molecule. Of great value in metabolic studies is the fact that the radioactive isotopes can be used in a molecule, which is a precursor for metabolic processes and so becomes incorporated into specific tissue components. Practically it may be very difficult to ensure that water-soluble radioactive substances do not diffuse from their original location during the various stages of processing the tissue. Certain practical points have to be noticed: (1) The radiations are usually weak and so will not penetrate through much thickness of tissue and (2) the localization of the source of the radiation depends on opposing a photographic emulsion as close to the tissue section as possible (Figure 9.7). (3) The resolution of the method will also depend on the energy of the rays emitted. The stronger radiations will penetrate deeper into the emulsion and may form silver grains at any point of their path; very weak radiation will be stopped close to its entry into the emulsion (Figure 9.8).

9.7.2 Autoradiography Methods and Uses

The various methods of autoradiography are now so valuable and so widely used that a wide range of different techniques is available.

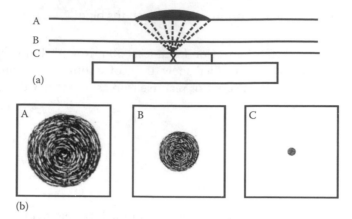

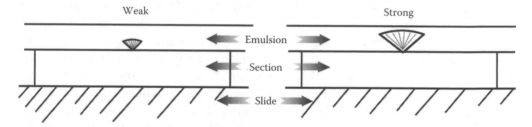

Figure 9.7 (a) Diagrammatic representation of stripping film autoradiography. (b) Diagrammatic representation of the image of the source of radiation on the photographic film. (A) Point source of radiation X in the section will be well localized if the photographic emulsion is closely applied (as in C), less well localized if there is a slight gap (B) but poorly localized if the film and the section are widely separated (A).

Figure 9.8 Diagram to illustrate the formation of silver grains in the photographic emulsion from a point source of radiation. The stronger radiations penetrate deeper into the emulsion and hence give poorer resolution of the point source than do the weaker radiations.

9.7.3 Developer D 19 B

Elon 2.2 g; anhydrous sodium sulfite 72 g; hydroquinone 8 g; anhydrous sodium carbonate 48 g; potassium bromide 4 g; distilled water 1000 mL. This should be filtered before use.

9.7.4 Protocol

The basic stripping film technique is as follows:

1. *Sections*: For most studies, which have been concerned with substances that can be fixed in cells, such as DNA or protein, fixed tissues have been embedded in paraffin and sectioned normally. Certain metals, used in some fixatives, themselves react with photographic emulsion; consequently 1:3 acetic:alcohol or formaldehyde–calcium is usually the recommended fixatives. Cryostat sections of unfixed tissue can be used but should be fixed in acetic alcohol.

2. The slides on which the sections are to be picked up from the knife must be washed well in hot water (but not treated with acid) before being coated (or "subbed") by being dipped briefly in a solution that contains 0.5 g gelatin and 0.05 g chrome alums in 100 mL distilled water. The slide must be dried for 2 days before it is used. Prolonged storage should be at 4°C in a refrigerator. (*Note*: This coating has two advantages: It helps to hold the section on to the slide during all the subsequent processing and it stops the photographic emulsion from slipping once it is in position.)

3. If paraffin sections are used, they are caused to adhere to the coated slide and are then treated with xylol and taken through graded concentrations of alcohol to water.

4. The stripping film is prepared and handled in a dark room, with as little light as possible from a Wratten Series No.1 red safe light. First, the edges of photographic stripping film plate (AR 10 or AR 50) are trimmed by tracing a cut (with a sharp scalpel or razor blade) all around the plate and about ½ in. in from the edge. Other cuts are made as shown in Figure 3.9 to produce eight segments of the film.

   With the scalpel, pick up the edge of each segment of film and very slowly strip that segment off the plate. Hold it with as little contact as possible and float it, face (emulsion) downward, on to the surface of water in a large bowl. The temperature of the water should be maintained at 25°C, to permit the maximum expansion of the film (2½ min).

5. Place the slide, section uppermost, under the water and slowly raise it, under the film. Ideally the slide should be held slightly oblique to the film. As the slide comes out of the water, it should be tilted so as to wrap the film closely about it (Figure 9.9). Care must be taken to ensure a tight fit of the film to slide to avoid pockets of air or of water above the section.

6. Hang the slide (wrapped in film), pegged at one end, on a rack to drain and to dry.

7. When dry, store in the dark in a lighttight, dry container at 4°C. The film is now exposed to radiation. The time of exposure can vary greatly, depending on the radioactive matter used and the concentration, which it has achieved in the tissue; it may be some days or some weeks (Figure 9.10).

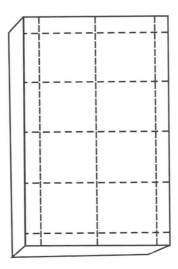

Figure 9.9 Diagrammatic view of a stripping film on its plate-backing. Cut broken lines are made in the photographic film, to produce eight segments of film.

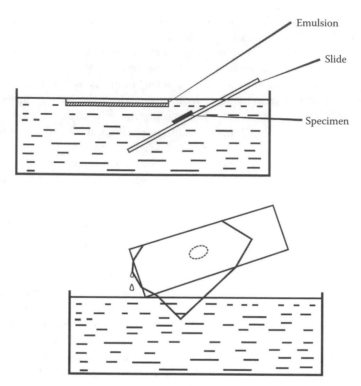

Figure 9.10   The photographic film is floated on the surface of the water, emulsion side downward. When it has expanded, a section mounted on a slide is immersed obliquely into the water, beneath the film. The slide is lifted out of the water and is tilted to wrap the film around the slide.

8. The film is developed on the section, all the process being done at 17°C–18°C, to avoid damaging the film. It is essential to prepare a commercially available developer for photographic fixation.
9. The section can then be stained with hematoxylin and eosin or with methyl green–pyronin. These processes too should be done at 17°C–18°C.

9.7.5 Results

The photographic emulsion acts as a cover slip for the section but contains black regions or dots, caused by the radioactivity in the tissue section below it. Hence, at one focus, the tissue histology can be inspected while at a slightly higher focus, one should be able to observe how much radioactivity has been emitted by each tissue or cellular component. Quantitatively the number of photographic grains in the emulsion bears a definite relationship to the number of radioactive atoms in the tissue below that point in the emulsion. Care must be taken to assess the background count because some photographic grains may be found over those parts of the slide that lack tissue. This background depends on many factors, including the skill of the operator at many of the steps of the procedure; it may also be related to the glass used for making the microscopic slide.

9.8 Enzyme Localization in Embedded Material

9.8.1 Introduction

Identification of enzymes in the cell is one of the important studies in cell biology. The enzymes can be detected by various histochemical techniques. Some of the enzymes are detected in microtomy sections and/or with hand sections employing enzyme-specific stains for the detection of enzyme proteins such as adenosine triphosphatase (ATPase), peroxidases, and dehydrogenases. The module given in this unit deals with the histochemistry of few representative enzymes. The detection of enzymes involves fixation, incubation, and staining.

9.8.1.1 ATPase Localization

ATPases are membrane-bound ion channels (actually transporters, as they are not true ion channels) that couple ion movement through a membrane with the synthesis or hydrolysis of a nucleotide, usually ATP. Different forms of membrane-associated ATPases have evolved over time to meet specific demands of cells. These ATPases have been classified as F-, V-, A-, P-, and E-ATPases based on functional differences. They all catalyze the reaction of ATP synthesis and/or hydrolysis. The driving force for the synthesis of ATP is the H^+ gradient, whereas during ATP hydrolysis the energy from breaking the ATP phosphodiester bond is the driving force for creating an ion gradient. Structurally, these ATPases can differ: F-, V-, and A-ATPases are multi-subunit complexes with a similar architecture and possibly catalytic mechanism, transporting ions using rotary motors. The P-ATPases are quite distinct in their subunit composition and in the ions they transport, and do not appear to use a rotary motor.

9.8.1.1.1 Protocol

9.8.1.1.1.1 Fixation

1. Excised ovaries are cut into two halves—one for ATPase activity and the other for control—and placed in 0.1 M cacodylate buffer at pH 7.2.
2. Fixation is done in 3% buffered p-formaldehyde at room temperature ($25° + 1°C$) 75 min.
3. Material is transferred to 0.05 M Tris–maleate buffer (pH 7.2) at 25°C for 30 min.
4. Wash two times in the same buffer for 30 min each.

9.8.1.1.1.2 Incubation.   Incubate the material in the following incubation medium:

ATP 25 mg in	20 mL distilled water
0.2 M Tris–maleate buffer pH 7.2	20 mL
2% lead nitrate	3 mL
0.25% magnesium sulfate	5 mL
Distilled water	2 mL

40 mg of DNP (2,4-dinitrophenol) can be added to this medium.

The incubation was allowed for 150 min at 25°C. Incubate the control in the same medium without adenosine triphosphate (ATP) or $Pb(NO_3)_2$. Give three successive changes of 1 h duration each in 0.1 M cacodylate buffer (pH 7.2).

9.8.1.1.1.3 Staining.   Ammonium sulfide solution (2%) is placed over the sections for 2 min. The slides are then gently rinsed in distilled water, air-dried, and mounted in distrene plasticizer xylene (DPX) or butyl phthalate plasticized styrene (BPS).

The reaction product of ATPase hydrolyzing activity appears as brown precipitate of lead sulfide.

9.8.2 Peroxidase Localization

Peroxidase belongs to the family of oxidoreductases, to be specific those acting on a peroxide as acceptor (peroxidases). Histochemical study offers the advantage of localization of enzymes in intact cells based on specific activity staining and viewing under different types of microscopy.

9.8.2.1 Protocol
9.8.2.1.1 Fixation.   Excise the tissue and cut in 0.05 M potassium phosphate buffer at pH 5.8 and immediately fix for 2 h in 2.5% buffered glutaraldehyde at 25°C. Subsequent steps are same as for ATPase.

9.8.2.1.2 Incubation.   Incubate the material in a reaction medium containing 0.5% phenylenediamine and 0.5% H_2O_2 in potassium phosphate buffer pH 5.8 at 25°C for 2 h. Give three changes of buffer each of 1 h duration.

Follow the steps as of ATPase localization. The slides are then gently rinsed in distilled water, air-dried, and mounted in DPX or BPS. No staining is required. The reaction product appears reddish brown in color. The control ovary is either boiled after fixation or incubated in the reaction medium lacking H_2O_2. Take photographs for both the enzyme localizations immediately using photomicroscope.

9.8.3 Dehydrogenases

9.8.3.1 Introduction
Dehydrogenases are enzymes that oxidize their substrate by removing hydrogen from the substrate and passing it to a suitable acceptor. They differ from oxidases in that the suitable acceptor cannot be atmospheric oxygen; in almost all cases, it is a coenzyme-like nicotinamide adenine dinucleotide (NAD), nicotinamide adenine dinucleotide phosphate (NADP), or a flavoprotein (FAD). The hydrogen is then passed either through a hydrogen transport system to react ultimately with oxygen or to an enzymatic process in which it is used for biosynthetic reactions.

In histochemistry reactions, fresh cryostat sections are incubated at 37°C in the presence of (1) the substrate that is to be oxidized by the dehydrogenase,

(2) any cofactor that is required as the immediate hydrogen acceptor of the dehydrogenase activity (e.g., NAD or NADP), (3) any activators of the enzyme that may be known from biochemical or histochemical studies, and (4) a tetrazolium salt that acts as the trapping agent for the hydrogen and in becoming reduced by it yields a colored precipitate of formazan.

9.8.3.1.1 Succinate Dehydrogenase. The enzyme is an essential part of Kreb's tricarboxylic acid cycle (TCA) and so is found in all aerobic cells. It is fairly firmly bound to the framework of the mitochondria but has been isolated and purified; it is then very unstable. The isolated enzyme contains a flavin prosthetic group, which is related to FAD; its pH optimum is about 7.6.

9.8.3.1.1.1 Materials. *Reaction medium*: Add 0.1 g of neotetrazolium chloride (or nitroblue tetrazolium) to 100 mL of phosphate buffer (or glycyl glycine) at pH 7.8. Heat gently to dissolve, cool, filter, and adjust to pH 7.8 if necessary.

To 50 mL of this solution, add 0.68 g of sodium succinate ($6H_2O$); this gives a 0.05 M solution. Bubble nitrogen through the solution at 37°C before and during use.

9.8.3.1.1.2 Protocol. For all methods, fresh cryostat sections of unfixed tissue must be used.

9.8.3.1.1.3 Incubation
1. Incubate in the reaction medium in a Coplin jar, or in a ring, at 37°C in an atmosphere of nitrogen. With neotetrazolium, the incubation requires 2 h; with nitroblue tetrazolium about 30 min.
2. Wash well in distilled water, air-dry, and mount in DPX or BPS.
3. Control tissue should be incubated in a 0.05 M solution of sodium malonate in the buffer for 15 min at 37°C, wash, and mount in Ferrant's medium. This treatment will inhibit all true succinate dehydrogenase activity.

9.8.3.1.1.4 Results
1. *With neotetrazolium*: Red-blue granules and red color, which is absent in the control, show the amount of succinate dehydrogenase activity.
2. *With nitroblue tetrazolium*: Red to purple staining, which is not found in the control, indicates both the amount of succinate dehydrogenase activity and the site at which hydrogen, removed from the succinate, has been trapped by the tetrazole.

9.8.3.1.2 Glutamate Dehydrogenase. This dehydrogenase, which is present in mitochondria, has a key role in cellular metabolism, and consequently it is found throughout the animal, plant, and bacterial kingdom. In some tissues, it depends on NAD, in others on NADP.

9.8.3.1.2.1 Materials. *Reaction medium*: Dissolve 0.1 g of neotetrazolium chloride or of nitroblue tetrazolium in 100 mL of 0.5 M phosphate buffer (or 0.05 M glycyl glycine) at pH 7.8. Warm to dissolve, cool, and filter.

Then prepare the reaction medium as follows:

NAD 50 mg; monosodium glutamate 170 mg tetrazole in buffer 10 mL

Control: Incubate as for the test but in a reaction medium lacking glutamate/NAD.

9.8.3.1.2.2 Protocol.　Fresh cryostat sections of unfixed tissue must be used. Care must be taken not to damage the tissue sample mechanically; even pressure by forceps can be detected in the final reaction if it occurs sufficiently long before the chilling is done.

Incubate in the reaction medium in an atmosphere of nitrogen either in a ring or in a Coplin jar. Incubate for 2 h at 37°C if neotetrazolium is used; with nitroblue tetrazolium, 30 min should be sufficient. Wash in distilled water, air-dry, and mount in DPX or BPS.

9.8.3.1.2.3 Result.　Blue, red, or purple formazan (depending on the tetrazole used) indicates glutamate dehydrogenase activity if it is not present in the control.

9.8.3.1.3 Lactate Dehydrogenase

9.8.3.1.3.1 Introduction.　It is found in all cells that are capable of glycolysis. In the formation of lactic acid in muscles, it operates by reducing pyruvate, that is, the equation goes from right to left, with optimal activity around pH 7.4; its optimal oxidation of lactate is said to be close to pH 10. In animal tissues, it can react with a number of α-hydroxy acids as well as lactic acid. In yeast, the lactate dehydrogenase is not dependent on NAD and is almost identical with cytochrome b2.

9.8.3.1.3.2 Materials.　*Reaction medium*: Dissolve 0.2 g of sodium lactate in 25 mL of phosphate–tetrazole solution (as used for succinate dehydrogenase: 1 mg/mL of neotetrazolium or of nitroblue tetrazolium in 0.05 M phosphate buffer at pH 8.0). Check that the pH is still 8.0 after dissolving the lactic acid. To 1 mL of this solution, add 5 mg of NAD.

9.8.3.1.3.3 Protocol.　Use fresh cryostat sections of unfixed tissue. Incubate in the reaction medium at 37°C in the ring (or in a humidity chamber or in a Coplin jar). The duration of the incubation can be decided by inspection of intensity of color produced by the formazan. Generally 2 h are enough when neotetrazolium is used as the hydrogen acceptor and 30 min for nitroblue tetrazolium. Wash in distilled water, air-dry, and mount in DPX or BPS. Control tissue should be incubated excluding first the lactate and then the NAD from the reaction mixture.

9.8.3.1.3.4 Results.　Red, blue, or purple formazan demonstrates lactate dehydrogenase activity.

9.8.3.1.4 Steroid Dehydrogenases

9.8.3.1.4.1 Introduction.　These enzymes are of great importance in the metabolism of steroids, in the role of steroids as hormones, and so in cellular physiology and in endocrinology. In general, these dehydrogenases are more active with NAD than with NADP and so are considered together with other NAD-dependent dehydrogenases.

9.8.3.1.4.2 Materials. Reaction medium: Contains 2 mg/mL of NAD (or of NADP, but all reactions seem to be greater when NAD is used), 0.5 mg/mL of nitroblue tetrazolium, 0.25 mg/mL of the selected steroid (dissolved in dimethyl formamide), all in 0.1 M Tris or phosphate buffer at pH 7.5.

9.8.3.1.4.3 Protocol. The sections (8–15 μ as desired) are incubated for various periods (e.g., 2 h) at 37°C. The control lacks the substrate but should include the same volume of dimethyl formamide. Wash in distilled water, air-dry and mount in DPX or BPS. Control tissue should be incubated excluding first the substrate and then the NAD/NADP from the reaction mixture.

9.8.3.1.4.4 Results. Blue, red, or purple formazan (depending on the tetrazole used) indicates the presence of steroid dehydrogenase activity.

9.8.3.1.5 Glucose-6-Phosphate Dehydrogenase

9.8.3.1.5.1 Introduction. Glucose-6-phosphate dehydrogenase (G6PD or G6PDH) is a cytosolic enzyme in the pentose phosphate pathway, a metabolic pathway that supplies reducing energy to cells (such as erythrocytes) by maintaining the level of the coenzyme nicotinamide adenine dinucleotide phosphate (NADPH). The NADPH, in turn, maintains the level of glutathione in these cells that help protect the red blood cells against oxidative damage. Of greater quantitative importance is the production of NADPH for tissues actively engaged in the biosynthesis of fatty acids and/or isoprenoids, such as the liver, mammary glands, adipose tissue, and the adrenal glands.

9.8.3.1.5.2 Materials. Fresh cryostat sections of unfixed tissue must be used.
Reaction medium: Prepare 0.05 M glycyl glycine/sodium hydroxide buffer solution or 0.05 M Tris–HCl buffer (pH 7.6–8.0) containing 0.1% neotetrazolium chloride (or nitroblue tetrazolium) and 1% of calcium chloride (1.3 g/100 mL of $CaCl_2$ containing 25% moisture). To this add 1 mg/mL of NADP and 1.5 mg/mL of either glucose-6-phosphate or of 6-phosphogluconate (sodium salt), depending on which of the two dehydrogenases is to be tested. Saturate with nitrogen by bubbling the gas through the reaction medium for 2 min at 37°C before use.

9.8.3.1.5.3 Protocol. Incubate the relevant reaction medium in rings, in an atmosphere of nitrogen for 20 min (for active tissue like liver), wash in distilled water, air-dry and mount in DPX or BPS. Control tissue should be incubated excluding first substrate and then the NADP from the reaction mixture.

9.8.3.1.5.4 Results. Activity is shown by the deposition of the formazan (red-purple stain if nitroblue tetrazolium is used; grains and red color if neotetrazolium is used).

Suggested Readings

Bltensky, L., R. Elus, A. A. Silcox, and J. Chayen. 1962. Histochemical studies on carbohydrate material in liver. *Ann Histochim* 1: 9–14.

Chayen, J. and E. F. Denby. 1968. *Biophysical Technique*. London, U.K.: Methuen.

Khan, Z., R. P. Tiwari, R. Mulherkar, N. K. Sah, G. B. K. S. Prasad, B. R. Shrivastava, and P. S. Bisen. 2009. Detection of survivin and p53 in human oral cancer: Correlation with clinicopathological findings. *Head Neck*. DOI:10.1002/hed.21071, 31: 1039–1048.

Kiernan, J. A. 2008. *Histological and Histochemical Methods: Theory and Practice*, 4th edn. Bloxham, U.K.: Scion Publishing Ltd.

Lillie, R. D. and H. M. Fullmer. 1976. *Histopathologic Technic and Practical Histochemistry*, 4th edn. New York: McGraw-Hill, Inc.

Roberts, D. R. 1966. A rapid staining method giving sharp nuclear definition in frozen sections. *J Med Lab Tech* 23: 119–120.

Wick, M. R. 2008. *Diagnostic Histochemistry*. Cambridge, U.K.: Cambridge University Press.

Important Links

http://www.leicabiosystems.com/pathologyleaders/fixation-and-fixatives-1-the-process-of-fixation-and-the-nature-of-fixatives/

http://www.millipore.com/immunodetection/id3/immunostainingtechniques

http://depts.washington.edu/compmed/HIC/files/Berod1981Importance%20of%20fixation%20in%20immunohistochemistry.pdf

http://www.leicabiosystems.com/pathologyleaders/fixation-and-fixatives-3-fixing-agents-other-than-the-common-aldehydes/

http://www.scielo.cl/scielo.php?script=sci_arttext&pid=S0717-95022007000400036

http://www.biologicalstaincommission.org/

http://journals.lww.com/pathologycasereviews/fulltext/2002/11000/the_current_roles_of_immunohistochemistry_and.1.aspx

10

Cytogenetics

10.1 Introduction

Living cells are composed of cells. Some of them consist of only one cell (unicellular), but some, like higher animals and flowering plants, are made up of many cells. The cell is the unit of construction of living organisms just as the atom is the unit of molecules. Knowledge of cell and its contents has progressed hand in hand with the development of the microscope employed for its observation. The study of the cell is known as cytology. Cytology, more commonly known as *cell biology*, studies cell structure, cell composition, and the interaction of cells with other cells and the larger environment in which they exist. The term "cytology" can also refer to *cytopathology*, which analyzes cell structure to diagnose disease. Microscopic and molecular studies of cells can focus on either multi-celled or single-celled organisms.

The fact that we as humans are made up of millions of tiny cells, and that other life forms around us are similarly constituted, now barely needs explanation. The concept of the cell is relatively new. However, the scientific community did not accept the idea of the existence of cells until the late eighteenth century. Cytology became, in the nineteenth century, a way to describe and identify cells and also to diagnose certain medical diseases. Recognizing the similarities and differences of cells is of utmost importance in cytology. Microscopic examination can help identify different types of cells. Looking at the molecules that form a cell, sometimes called molecular biology, helps in further description and identification. All fields of biology depend on the understanding of cellular structure. The field of genetics exists because we understand cell structure and components.

10.2 Study Different Stages in Root Tip Cells of Plant

Flowering plants, which reproduce by sexual means, develop from a single cell called a zygote. Zygote is the product of fertilization and gives rise to a mature individual by repeated divisions. The divisions are not synchronous, only a few cells dividing at any time.

Divisions continue and millions of cells are produced, ultimately giving shape to a young plant. Since all the cells arose from a single cell, the zygote, they have something in common. Each cell has the same hereditary potentialities and, therefore, has the same complement of chromosomes and genes. This continuity of the hereditary material is ensured by a process called mitosis. In mitosis, the division or reproduction of a cell takes place in a very precise and orderly manner. In plants, growing points of stems and roots are the centers of active cell multiplication, and it is possible to examine mitosis in these areas.

Cells, which are not in the process of cell division, are said to be in interphase or a stage between the two consecutive divisions of a particular cell. Each plant cell has its own wall composed mainly of cellulose; its shape is influenced to some extent by the cells of the surrounding tissues. The cell contains a clear or viscous fluid called cytoplasm in which numerous granular and rod-like bodies are present. Within the cytoplasm lies the spherical nucleus with a thin nuclear wall. At this stage, the cell is a working unit with cytoplasm and nucleus making and exchanging chemical substances, some of which are used by the cell for its reproduction and others transported to other cells for the use of the plant as a whole.

Interphase nucleus is composed of a number of chromosomes that are not individually distinguishable. They are highly extended chromatin threads. The nucleus appears as a diffuse network of chromatin. In some plants, there are small regions of chromosomes that remain visible even during interphase. These are called chromocenters. Each nucleus has one or more spherical bodies called nucleoli (singular nucleolus). These bodies are formed close to small regions of some of the chromosomes called nucleolar organizers.

A regular synthesis of nucleic acids, proteins, and other substances take place at interphase. Mitosis begins with a stage called prophase. Chromosomes are long threads, each dividing along its length into two identical halves called chromatids, which are twisted around each other. In the beginning of interphase, the chromosomes are single threads, but at prophase, they are double threads, each thread having reproduced during interphase. One region of a chromosome, the centromere, is single (not replicated yet).

As prophase proceeds, the chromosomes become more distinct due to coiling, which leads to their contraction and thickening. Nucleolus gradually disappears. As the chromosomes reach maximum contraction, the nuclear membrane breaks down. This is the end of the prophase.

The next phase is called metaphase. A new structure called spindle makes its appearance. The extreme ends of the spindle are called poles and the central region the equator. The chromosomes move gradually and become arranged on the equator in a special way. The centromere is in contact with the spindle. Centromeres are always arranged in one plane across the equator, but the arms of the chromosomes are not restricted in any one direction.

The next stage in the mitotic cycle is anaphase. The centromere divides so that each chromatid now has its own centromere. The sister centromeres appear to repel each other, and they move apart. This is a progressive separation that starts at the centromere region and continues along the chromatid arms toward the ends. The sister chromatids attain the status of chromosomes now and move slowly toward opposite poles of the spindle. The spindle also lengthens and becomes narrower at the equator. The daughter chromosomes now form two closely packed groups, one at each spindle pole. This marks the end of anaphase and the beginning of the next stage telophase.

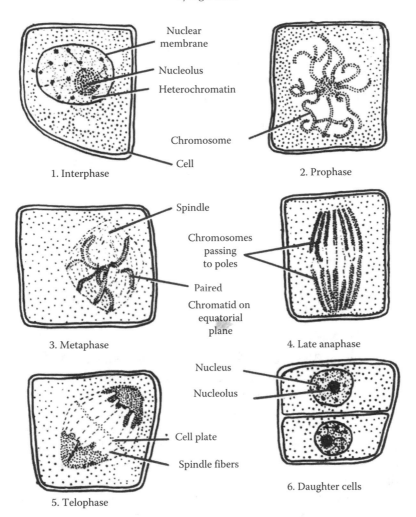

Figure 10.1 Position of chromosome in different stages of mitosis in root tip cells.

In telophase, the chromosomes uncoil and become long thin threads, and nucleoli begin to form. The chromocenters also become evident. Nuclear membrane is reconstituted, and each daughter nucleus becomes an interphase nucleus. This is just opposite to what happens in prophase.

At the equator of the spindle, a new cell wall begins to form that separates the two daughter cells (Figure 10.1).

10.3 Preparation of Root Tips for Mitotic Studies

Root tips of the following plants are normally used for mitotic studies: (1) *Allium cepa*, (2) *Aloe vera*, and (3) *Vicia faba*.

Allium cepa: Bulbs are pealed out of tunica and the dry roots are removed. Gently scraping the portion of the stem with root primordia induces better rooting. The bulb is

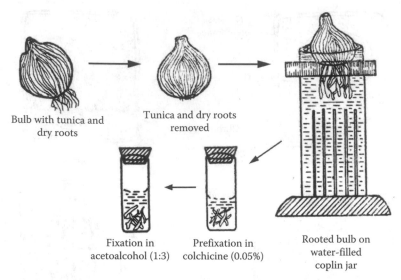

Bulb with tunica and
dry roots

Tunica and dry roots
removed

Fixation in
acetoalcohol (1:3)

Prefixation in
colchicine (0.05%)

Rooted bulb on
water-filled
coplin jar

Figure 10.2   Various steps for preparation of root tips from *Allium cepa*.

then placed over a coupling jar filled with water in such a way that the stem portion (disc) is in contact with water. The setup is placed where light is abundant. The bulb roots arise in 3–5 days depending upon the temperature. One centimeter long roots are cut with scissors or forceps and are fixed or prefixed as the case may be. Various steps involved in the preparation of onion root tips for making squashes are shown in Figure 10.2.

Aloe vera: Healthy potted plants are removed from the pots and washed in running water taking care not to break the tips. One centimeter long secondary roots are cut with scissors and collected in a Petri dish containing water. They are fixed after washing with water.

Vicia faba: *Vicia faba* seeds have hard seed coat. Therefore, before keeping them for germination, they are soaked in water for 3–4 days. After this, the seed is transferred to the germination trays filled with wet sand. When the radicle and plumule are about 1″ long, each seed is taken out of the sand and washed. Each seed (cotyledon) is then pierced with a pin taking care not to damage the radical and plumule. It is then placed on a vial containing water so that the radical dips into the water. When radical grows into primary root, its tip is cut to induce the growth of the secondary roots. When secondary roots are 1 cm long, they are cut with scissors and fixed. The process of preparation of *Vicia faba* root tips is diagrammatically shown in Figure 10.3.

10.4 Preparation and Study of Slides for Mitosis Using Acetocarmine Squash Techniques from Onion Root Tips

10.4.1 Materials

Onion root tips, acetocarmine solution, glass tube, spirit lamp, N HCl, glass slides, microscope.

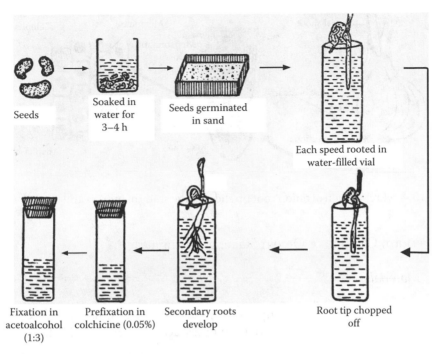

Figure 10.3   Various steps for preparation of root tips from *Vicia faba*.

10.4.2 Protocol

Take nine parts acetocarmine solution and one part N HCl in a small glass tube. Now put onion root tips in the solution and warm the tube slightly on a spirit lamp. Take out the root tips with the help of forceps on a glass slide. Put a drop of acetocarmine solution on the slide and cut the root tip by knife. Prepare a squash and study through a microscope; draw diagrams of different stages and write comments.

10.4.3 Results and Observations

The process of mitosis is based on two events (Figure 10.4): (1) karyokinesis and (2) cytokinesis.

1. *Karyokinesis*: The process of division of the nucleus of a cell into two similar daughter nuclei is called karyokinesis.
2. *Cytokinesis*: The process of division of the cytoplasm of a cell into two similar daughter cells, soon after karyokinesis, is called cytokinesis.

The process of mitosis goes on continuously, that is, once it has started, it continues till the process of cell division completes.

The process of karyokinesis of mitosis is studied under the following five phases: (1) interphase, (2) prophase, (3) metaphase, (4) anaphase, and (5) telophase.

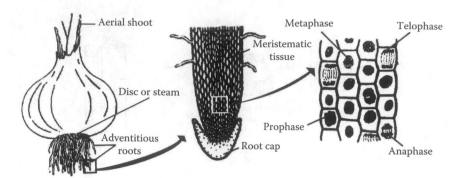

Figure 10.4   Preparation of onion root tip slide for the study of mitosis and its various stages.

10.5 Perform Carmine or Orcein Staining Techniques

10.5.1 Materials

Microscope, glass slides, cover slips, watch glasses, brushes, needles, forceps, spirit lamp, root tips, 2% acetocarmine solution or 2% acetic orcein, 1 N HCl.

10.5.2 Protocol I

1. Warm the root tips in 1 N HCl in watch glass for 5 min.
2. Transfer the root tips to another watch glass containing 2% acetocarmine solution and boil for 2–3 min.
3. Lift a root tip from the watch glass and place it onto a drop of acetocarmine in the center of the slide.
4. Cut off and remove the older part of the root tip with the help of a needle. Place the cover glass on the tip.
5. Squash the root tip by applying uniform pressure on the cover glass with the thumb, through a piece of blotting paper.
6. If necessary, the cells can be spread before Step 5 by tapping with the index finger or with the blunt end of a pencil (a slight heating on the flame improves the staining).
7. Seal the cover slip with sealing wax.
8. Observe the slide under the microscope.

[*Note: If the root tips are fixed in propionic acid–ethanol (1:3), then they can be stained in propionocarmine instead of acetocarmine*].

10.5.3 Protocol II

1. Boil the root tips in 1 N HCl–acetocarmine 1:1 mixture in a watch glass for 5 min.
2. Transfer the root tips in 1 N HCl in a watch glass and keep them in this for 15 min.
3. Squash in 2% acetocarmine as in procedure I.

10.5.4 Protocol III

1. Treat roots with an aqueous solution of α-bromonaphthalene for 30–45 min.
2. Collect root tips and wash them thoroughly with tap water.
3. Fix in 1:1 acetic acid–methanol mixture for 2 h.
4. Transfer into newcomers' fluid propionic acid–chloroform–alcohol (1:4:3) and store for 2–4 h at 5°C–8°C.
5. Hydrolyze in 1 N HCl at 60°C for 10 min.
6. Wash in distilled water giving two to three changes.
7. Stain overnight in 1% acetocarmine.
8. Squash in 1% acetocarmine.

10.5.5 Protocol IV (for Aloe vera)

1. Cut root tips about 1 cm long and after washing, keep in a corked glass vial containing a saturated aqueous solution of p-dichlorobenzene (PDB) for 2–3 h at 12°C–14°C.
2. Fix in glacial acetic acid–ethyl alcohol mixture (1:2) and keep for 30 min to 2 h followed by treatment in 45% acetic acid for 15 min.
3. Place the root tips in a glass vial containing 2% acetic orcein solution and N HCl mixture in the proportion of 9:1 and heat gently over a flame for 5–10 s taking care that the liquid does not boil.
4. Squash the root tip in 1% acetic orcein.
   To intensify the stain, a drop of ferric chloride solution can be added to the acetic orcein–N HCl mixture.

10.6 Feulgen Staining Techniques

10.6.1 Materials

Microscope, glass slides, cover slips, watch glasses, brushes, needles, forceps, spirit lamp, root tips, 0.5% acetocarmine solution, Feulgen stain, 1 N HCl, 45% acetic acid, root tips (fixed in acetic alcohol, stored in 70% alcohol).

10.6.2 Protocol I

1. Hydrolyze the root tips in 1 N HCl at 60°C for 10–15 min in staining vials.
2. Rinse in water.
3. Transfer the root tips to the leuco-basic fuchsin solution for 20 min to 1 h, till the tips are magenta colored.
4. Place a drop of 0.5% acetocarmine solution or 45% acetic acid in the center of a slide and transfer one root tip onto it.
5. Discard the older part of the root and place a cover slip on the rest of the material.
6. Squash as in carmine staining technique and observe under the microscope.

Root tips can be pretreated with chemicals like colchicines, acenaphthene, coumarin, α-bromonaphthalene, and 8-hydroxyquinoline before the fixation.

10.6.3 Protocol II

1. Root tips are pretreated in a saturated solution of α-bromonaphthalene for 30 min.
2. Wash in running water.
3. Hydrolyze in a solution of 22% acetic acid and HCl (12:1) at 60°C for 15 min.
4. Wash under tap water.
5. Hydrolyze in 1 N HCl for 15 min.
6. Wash in tap water.
7. Stain in leuco-basic fuchsin.
8. Squash in 2% acetocarmine.

10.6.4 Results and Observations

10.6.4.1 Interphase
Important characteristics of this phase are as follows:

1. The dividing cell has more conspicuous and large nucleus. The size of the cell and the nucleus reaches to its maximum.
2. For the cell division, the energy is stored in the form of proteins.
3. Nucleolus is very clear.
4. The chromosomes are not visible clearly.

10.6.4.2 Prophase
Important characteristics of this phase are as follows:

1. Nuclear membrane starts to disappear.
2. Nuclear spindles are formed from the nucleoplasm. Continuous spindle fibers join both the centrosomes.
3. Both the chromatids and centromere of the chromosomes are seen distinctly.
4. Nucleolus begins to be deformed in shape and disappears gradually (Figures 10.5 and 10.6).

10.6.4.3 Metaphase
The phase has the following important characteristics:

1. The chromosome comes to the equatorial line.
2. The chromatids of each chromosome separate from each other and centromere divides into two parts. Thus, one chromatid with one centromere forms a new chromosome.
3. On the centromere of the chromatids, spindle fibers are attached (Figures 10.5 and 10.6).

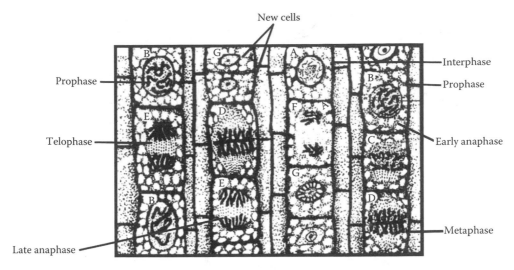

Figure 10.5 Different stages of mitosis in onion root tip.

10.6.4.4 Anaphase

This is a very short and active phase of cell division. In this phase, the following changes are found:

1. The daughter chromosomes (chromatids) of the metaphase begin to move toward opposite poles.
2. The structure of the chromosome becomes "L," "J," or "V" shape.
3. The size of every daughter chromosome becomes small and thick.

10.6.4.5 Telophase

This is the last phase of the mitotic division that shows the following characteristics:

1. In the beginning of this phase, the daughter chromosomes reach on opposite poles.
2. In the daughter chromosomes, DNA coils disappear and become long and thin.
3. The endoplasmic reticulum surrounds the chromosome and forms the nuclear membrane. Thus, the formation of nucleus starts; simultaneously, the formation of nucleolus also starts.
4. After this stage, two daughter nuclei are formed in the mother cell, and the process of karyokinesis is completed and cytokinesis starts.

10.6.4.6 Cytokinesis

After the telophase, a wall is formed inside the cell that divides the cytoplasm into two parts and two daughters are separated by the plate.

10.7 Preparation of Onion Flower Buds for Meiosis Studies

Meiosis is a special type of cell division that occurs in reproductive cells only. Four daughter cells are formed as a result of meiosis, which have half (haploid) the

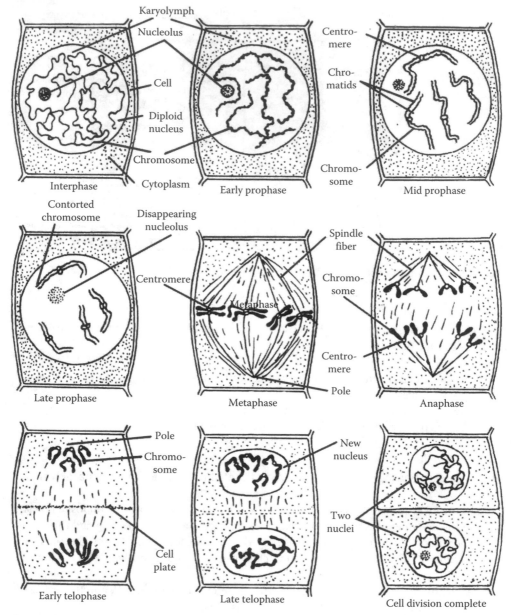

Figure 10.6 Mitosis showing different stages in *Aloe vera*.

chromosome number as that of the mother cell. Gametes or spores are the products of this division. This division is formed only in diploid germ cells. The cells formed after meiosis are haploid. Germ cells in sexual reproduction form sperms and ova that are called male and female gametes that unite together and form a diploid zygote. The nuclear division in meiosis takes place twice as described in the following:

1. Heterotypic or first meiotic division
2. Homotypic or second meiotic division

10.7.1 Preparation and Study of Slides for Meiotic Stages in Onion Flower Buds by the Squash Technique

10.7.1.1 Carmine Staining Technique

10.7.1.1.1 Materials. Microscope, glass slides, cover slips, watch glasses, brushes, needles, forceps, spirit lamp, root tips, 0.2%–2% acetocarmine solution, 1 N HCl, 45% acetic acid, onion flower bud.

10.7.1.1.2 Protocol
1. Take flower buds serially from an inflorescence, starting from the smallest and working up to the largest, until a correct bud having divisional stage is found.
2. Dissect out a single anther from the selected bud with the help of a needle.
3. Place the anther in the center of a clean slide and smear it with the help of a scalpel.
4. Place a drop of acetocarmine on the smeared material. In the case of a large anther, debris should be removed before placing the carmine drop on the material.
5. Heat slightly over a flame.
6. Place a cover glass on the smeared stained material. Tap gently to spread the cells.
7. Take away the extra stain by touching the corners of the cover glass with a blotting paper.
8. Ring the cover glass with paraffin wax or Cutex.

Observe under the microscope.

The slides can be kept safely for 2–3 weeks in a refrigerator. If fresh material is not available, material can be fixed in a suitable fixative (chromic acid 1.5 g–acetic acid 10% aqueous 100 mL–formalin 40% formaldehyde 40 mL–distilled water 60 mL). A crystal of ferric chloride added to acetocarmine improves the stain.

10.7.2 Feulgen Staining Techniques

10.7.2.1 Materials
Microscope, glass slides, cover slips, watch glasses, brushes, needles, forceps, spirit lamp, root tips, 0.2%–2% acetocarmine solution, Feulgen stain, 1 N HCl, 45% acetic acid, onion flower bud.

10.7.2.2 Protocol
Fix the flower buds in acetic alcohol or in Carnoy's fluid I for 2–4 h (material without fixation can also be used).

1. Dissect out the anther with proper divisional stages.
2. Hydrolyze the anthers in 1 N HCl at 60°C for 10 min.
3. Wash in tap water with three changes.
4. Stain in leuco-basic fuchsin.
5. Squash in 45% acetic acid or 2% acetocarmine solution.

6. Single anther should be squashed on one slide. Ovules fixed in Carnoy's fluid can also be squashed following the previous schedule for anther squashing. The success of meiosis depends upon the material from which the samples are taken at the best possible condition. If there are no divisions going on in the sample, not even the most refined staining method gives any good slides.

10.7.2.3 Results and Observations

10.7.2.3.1 First Meiotic Division or Heterotypic Division. This division of meiosis consists of the four phases: (1) first prophase or prophase I, (2) first metaphase or metaphase I, (3) first anaphase or anaphase I, and (4) first telophase or telophase I.

10.7.2.3.1.1 Prophase I. The first meiotic prophase is very long and complex as compared to the first mitotic prophase. This phase is completed in the six subphases: (1) preleptotene, (2) leptotene, (3) zygotene, (4) pachytene, (5) diplotene, and (6) diakinesis (Figures 10.7 and 10.8).

1. *Preleptotene*: The following changes are found in this subphase:
   a. Both cells and nucleus are enlarged in size.
   b. The number of a chromosome is diploid and arranged in chromatin net.
   c. Chromatin is condensed.
2. *Leptotene*: This subphase shows the following characteristics:
   a. The homologous chromosomes are clearly seen in the mother cell.
   b. Chromosomes become straight and long.
   c. Chromomeres are seen on the chromosome; they have a definite number and shape and look like a bead.
   d. The centromere of each chromosome appears clearly.
3. *Zygotene*: In this subphase, the following changes are seen in the mother cell:
   a. The scattered chromosomes are arranged in pairs. First of all, homologous chromosomes come together by attraction force and are arranged parallel to each other, which is called pairing or synapsis. Each pair of homologous

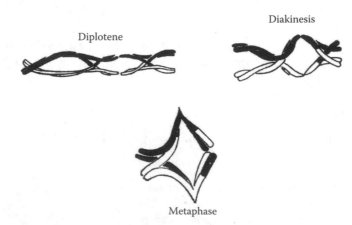

Figure 10.7 Progressive changes in a bivalent with two chiasmata from diplotene to metaphase I of meiosis.

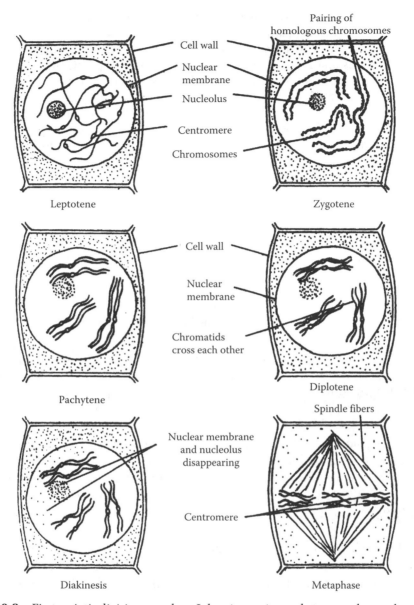

Figure 10.8 First meiotic division: prophase I showing various substages and metaphase I stage.

chromosomes is called bivalent in which one chromosome is paternal and the other one is maternal.

b. At the same time, the size of the nucleolus in the nucleus increases.

4. *Pachytene*: This subphase shows the following changes:

a. The chromosome pairs shrink and begin to become smaller and thicker; the attraction between them increases. This substage takes a long time.

b. Each chromosome of the homologous chromosomes splits longitudinally. The chromatids remain attached on the centromeres. Thus, the bivalent of homologous chromosomes shows four clear chromatids, which is now called as tetravalent.

5. *Diplotene*: This is the most important subphase of the first prophase in which the following changes takes place:
   a. The chromatids of homologous chromosomes cross each other at one or two places and form chiasma or chiasmata. At this point, the exchange of chromatids of the two homologous chromosomes takes place. This is called crossing over.
   b. The piece of a chromatid is joined with the piece of another chromatid, and then they are broken at the chiasma.
   c. The gene exchange occurs as a result of which new gene combination takes place in the resulting chromatids.
   d. The process of crossing over takes place in pachytene substage, but it is seen in diplotene substage.
6. *Diakinesis*: This is the last stage of the first prophase in which the following changes occur:
   a. The attraction between the homologous chromosomes decreases and homologous chromosomes are separated from each other due to repulsion.
   b. After the separation of the homologous chromosomes, their chromatids are shortened and get surrounded by a matrix.
   c. Nucleolus begins to disappear, and the nuclear membrane is ruptured; at the same time, nuclear spindles are formed from the nucleoplasm.

10.7.2.3.1.2 Metaphase I.   At this stage, the nuclear membrane and nucleolus completely disappear, and spindle fiber starts appearing, which seems to be joined along with their length between the two poles. The chromosome pairs collect at the middle plane of the spindle. The centromere of each pair is toward the poles.

10.7.2.3.1.3 Anaphase I.   During this stage, homologous chromosomes are separated. Two chromosomes that come together during synapsis now move toward opposite poles due to repulsive forces. Therefore, this division is known as reductional division. Two halves of chromosomes move toward opposite poles (Figure 10.9).

10.7.2.3.1.4 Telophase I.   Coiled chromosomal groups are reopened in this stage to form chromatin net, and nucleolus and nuclear membrane formation is also initiated simultaneously as cytokinesis also starts. As a result, a parent cell now contains two

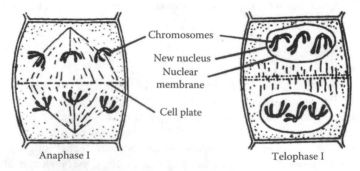

Anaphase I   Telophase I

Chromosomes
New nucleus
Nuclear membrane
Cell plate

Figure 10.9   First meiotic division: showing anaphase I and telophase I stages.

nuclei with (*n*) number of chromosomes. Thus, the first step of meiosis results in the formation of two haploid daughter cells from a single diploid cell (Figure 10.9).

10.7.2.3.1.5 Cytokinesis. This takes place after the telophase. The cytoplasm is divided into two parts, and two daughter cells are formed from a single mother cell.

10.7.2.3.1.6 Interphase. This starts after separation; both the daughter cells undergo interphase stage. After some time, the second meiotic division starts.

10.7.3 Homotypic or Second Meiotic Division

This division resembles with mitosis, during which both the haploid daughter cells pass through the following stages to form four haploid cells (Figures 10.10 and 10.11).

10.7.3.1 Prophase II
Each chromosome divides into two chromatids, which are joined with each other. Chromosome appears like a double thread.

10.7.3.2 Metaphase II
In this stage, chromosomes lie on the equatorial plane of the spindle fiber and are attached with the centromere.

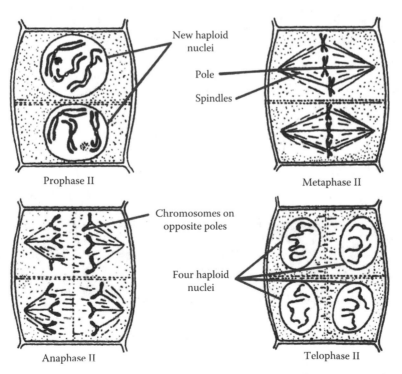

Figure 10.10 Second meiotic division: showing stages of prophase II, metaphase II, anaphase II, and telophase II.

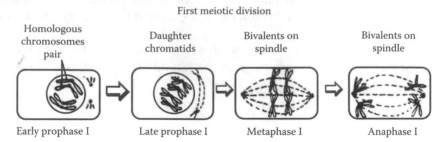

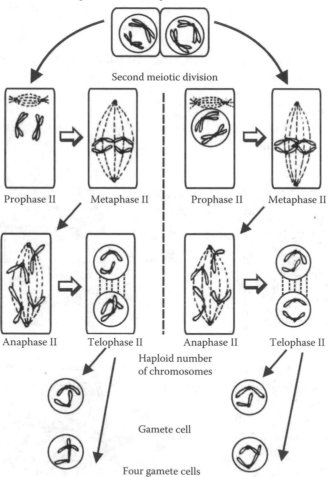

Figure 10.11   Showing stages I and II of the meiotic division.

10.7.3.3 Anaphase II

During this stage, both the chromatids separate with the dividing centromere and move toward opposite poles.

10.7.3.4 Telophase II

In this stage, chromosomes reach the opposite poles and again spread to form chromatin net and nucleus. This is followed by the cell wall formation between the two nuclei of each cell. As a result, four haploid cells are formed.

10.8 Fixation

Coloring a living cell poses problems because it resists being penetrated by foreign matter from outside. This problem is solved by killing the cell taking care to preserve its structures. For this purpose, mixtures of various chemicals are used. This type of killing a cell is known as fixation. Cytology deals with the study of the nucleus, a fairly large, round body near the center of the cell. It is surrounded by a nuclear membrane and is filled with a viscid fluid called karyoplasms or nucleoplasm. Inside the nucleus, there is (are) a smaller round body (bodies), the nucleolus (nucleoli). Floating in the nucleoplasm are the chromosomes, the threadlike structures that change their shape and appearance at different stages of the cell life. Their visibility depends upon the type of chemicals used for staining. These are the most important constituents of the cell and obviously demand a special stain. In order to see the chromosomes clearly under the microscope, the prior treatments are essential, namely, (1) prefixing, (2) fixation, and (3) staining.

10.8.1 Prefixing

The objectives of prefixation are (1) clearing of cytoplasm and softening the tissue, (2) inhibiting the spindle formation, (3) dissolving of middle lamella, (4) contraction and scattering of chromosomes, (5) clearing the surface conditions of cells and tissue for rapid penetration of the fixative and the stain, and (6) removal of extra nuclear structures, separation of chromosomes, and clarification of chromosomal structure.

Chemicals commonly used for pretreatment are colchicines, acenaphthene, chloral hydrate, 8-hydroxyquinoline, and α-bromonaphthalene.

10.8.1.1 Colchicines

The corm of *Colchicum autumnale* is basically an underground tubular part of the plant that has been employed since long in the form of a strong poison. Its chemical formula is $C_{22}H_{25}O_6N$, and the molecular weight is 399.45. Its chemical structure is

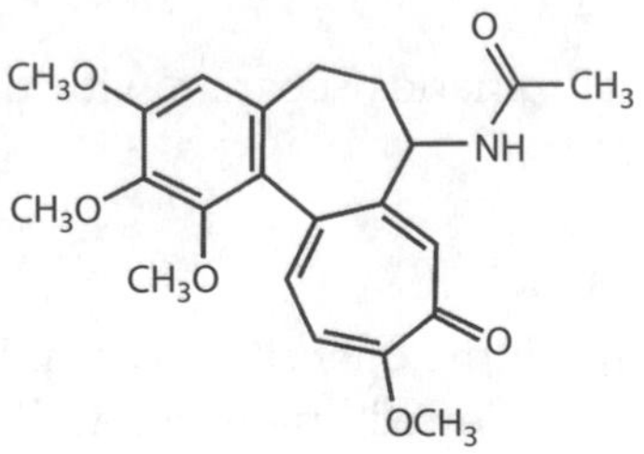

Colchicine

Colchicines are highly soluble in water and are very active even at an extremely low concentration and between a wide range of temperatures 9°C–44°C. It brings about a change in the colloidal state of the cytoplasm causing spindle disturbance by increasing the fluidity of the nuclear substance. The inhibition of spindle formation results in arresting of metaphases. High concentrations with longer treatments cause polyploidy in the cells and heavy contraction of chromosomes. Colchicine is added in the medium in artificial cultures of plants and animals. Coating of colchicines hampers the visibility of chromosomes and the penetration of the fixative. A thorough washing of the material is, therefore, recommended after colchicine treatment. A rapidly penetrating fixative like acetic alcohol is advisable after colchicine treatment.

Fifty milligrams of colchicine powder dissolved in 100 mL of distilled water gives 0.05% colchicine solution. The solution is made fresh before use.

10.8.1.2 Acenaphthene
Acenaphthene is sparingly soluble in water and, therefore, a saturated solution in water is used. Its action is similar to colchicines, that is, it straightens the chromosomes and inhibits the spindle formation. Its chemical structure is

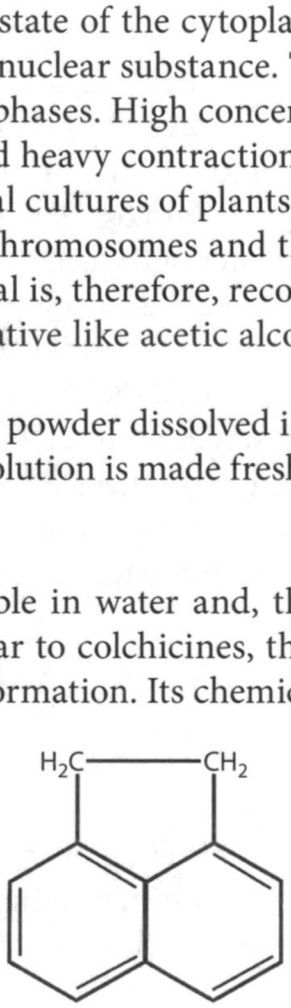

Acenaphthene

Acenaphthene is successfully used in chromosome preparation of the pollen tube.

10.8.1.3 Chloral Hydrate
Its chemical formula is $CCl_3 \cdot CH(OH)_2$. Its effect on chromosomes is similar to that of colchicines, and the effective concentration is also the same as that of colchicines.

10.8.1.4 8-Hydroxyquinoline
8-Hydroxyquinoline demonstrates C-mitotic properties by inactivating the spindle. It contracts the chromosome arms equally. Unlike colchicines, it allows the

metaphase chromosomes to maintain their relative arrangements at the equatorial plane. It also clarifies the primary and secondary constrictions, and the satellite gap is greatly exaggerated. The centrometric structure is easily analyzed after its use. 8-Hydroxyquinoline is specially suitable for plants with long chromosomes.

10.8.1.5 α-Bromonaphthalene
Its chemical structure is

A saturated solution of this, if applied for 15 min to 4 h, inhibits the spindle formation, contracts the chromosomes, and clarifies the constriction regions of the chromosome.

10.8.2 Preparation

One or two drops of the chemical are added to 20 mL of distilled water in a test tube. The test tube is shaken vigorously until a milky precipitate is formed. Root tips are soaked in this mixture for 1–2 h. Material should be thoroughly washed in tap water before fixation.

10.8.3 Fixation

The main objectives of fixation are (1) to kill the cells suddenly and uniformly so that they retain, as near as possible, the same appearance that they possessed in life, (2) to preserve the tissues and cells by the inhibition of putrefaction and autolytic changes, and (3) to induce differences in the refractive indices of certain cell elements. This helps in the differential visibility of the different elements; otherwise, they would be undifferentiated owing to the narrow differences between the refractive index of one and that of the other (4) to set and hold intracellular bodies in the position that they occupied in life. This is achieved by the precipitation of these structures (5) to render the cell constituents resistant to the subsequent processes such as dehydration and staining prior to their examination under the microscope (6) to facilitate proper staining of tissue.

Fixatives are of two types: (1) mordants and (2) inhibitors for certain stains.

It is, therefore, important that a suitable fixative should be employed for a particular stain technique or a particular stain should be chosen to suit material fixed by a particular chemical mixture.

If good results are to be achieved, the following points must be observed:

- Tissue to be fixed should be cut into slices 2–6 mm in thickness to permit penetration of the fixative.

- The containers in which the material is to be fixed should be of sufficient size, to take the pieces of tissue without folding or bending.
- Material should be completely dipped in the fixative.
- Material should not be left in the fixative beyond the necessary time, but should be stored in a fluid suitable for the particular material, until it is required for staining.
- Washing out of the fixative is necessary, except in the case of alcohol that requires no washing out. Liberal quantities of water or any other liquid should be used for washing out.
- A fixative suitable to the material to be examined and compatible with the stains to be employed should be chosen.

For chromosomes, the fixation should

- Increase the visibility of the chromosome
- Clarify the details of chromosome morphology such as chromatic and heterochromatic regions and the primary and secondary constriction
- Maintain the structural integrity of the chromosome intact, and it should be able to enhance the basophilia of the chromosomes

Since all the previous properties are rarely found within a single chemical, a fixative is normally a combination of several compatible fluids that together satisfy all the earlier requirements.

Even with the best fixatives, some of the chemical changes are still evident in the chromosomes. There is always a tendency to shrink on coming into contact with the chemicals. These can be minimized by fixing through freezing at low temperature.

The following are the common fixatives used:

10.8.3.1 Ethanol or Ethyl Alcohol (C_2H_5OH)

1. It precipitates nucleic acid.
2. It has immediate penetration effect.
3. It has dehydrating properties.
4. It causes irreversible denaturation of proteins. It breaks the hydrogen bonds and salt links in protein chains thus revealing several side groups.
5. It has undesirable hardening effect on the tissue. Therefore, it is used in combination with acetic acid, formaldehyde, or chloroform.

10.8.3.2 Acetic Acid (CH_3COOH)

1. It can be combined with other fixatives.
2. It has a remarkable penetrating property. Its smaller ions are responsible for this property.
3. It causes swelling of cell constituents, so it is used when mixed with formalin, alcohol, etc., to counteract their shrinking effect. This preserves the chromosome structure without distortion. It is quite suitable to study the behavior of meiotic chromosomes.
4. It is a good solvent for aniline dyes. It is, therefore, used in staining-cum-fixing mixtures, like acetic carmine, acetic orcein, and acetic lamoids.

10.8.3.3 Formaldehyde (HCHO)

Its commercial form is formalin that is 40% formaldehyde in water:

1. It acts, chiefly, on proteins. It reacts with the amino groups of proteins with the production of water or attaches itself to the amino acid without any liberation of water molecules.
2. Formaldehyde hardens the tissue, so alone it is not a good fixative for chromosomes, but in combination with other fixatives like acetic acid, it serves as a good fixative. Too much hardening by formalin poses difficulties in smearing the tissue.

10.8.3.4 Propionic Acid (C_2H_5COOH)

It is a colorless liquid with an acid odor. It is miscible in water, ethyl alcohol, and ether in all proportions. It is generally used as a substrate for acetic acid:

1. Penetration is not as rapid as that of acetic acid, but it causes much less swelling of the chromosomes.
2. It can be used in staining-cum-fixing mixtures, like propionic carmine and propionic orcein.

10.8.3.5 Chloroform ($CHCl_3$)

It is sparingly soluble in water but miscible in alcohol, ether, and acetone in all proportions. It is used as an anesthetic agent. It is slowly converted into poisonous carbonyl chloride in the presence of air and light. It is a good solvent for fatty acids, and because of this property, it is used as a fixative for chromosomes. In plant chromosomes, it is used in the fixative to dissolve the fatty and waxy secretions from the upper surface, facilitating the penetration of the fixative. Long period of treatment with chloroform is toxic. It should be used mixed with alcohol as alcohol checks the decomposition of the chloroform into carbonyl chloride. It causes brittleness in tissues; therefore, it is not a desirable fixative for squash and smears.

10.8.4 Fixing Mixtures

10.8.4.1 Navashin's Fluid

For smears, 1 h of fixation is sufficient. For mitotic chromosomes, it is food for chromosome counts but not for morphological studies.

Chromic acid 1.5 g, glacial acetic acid (10% aqueous) 100 mL, formalin (40% formaldehyde) 40 mL, distilled water 60 mL.

The formaldehyde should not be added until the solution is required for immediate use. This is because oxidizers are never kept with reducers.

10.8.4.2 Carnoy's Fixative

The addition of chloroform in acetic acid–alcohol mixture aids in rapid penetration and removes fatty substances from chromosomes, thus securing a clear background for the study. Carnoy's fluid is good for animal chromosome studies. Prolonged fixation in Carnoy's fluid leads to DNA extraction; so long duration fixation should be avoided. The addition of formalin is ideal for blood smears. Formalin increases the cell volume, and thus chromosome spreading is better achieved. But hardening caused by formalin makes the fixative unsuitable for smears. The addition of 1% chromic acid and 10% formalin aids in studying somatic chromosomes of plants. The increased proportion of formalin causes wide scattering of chromosomes, which is essential for morphological studies. Chromic acid and formalin are oxidizer and reducer, respectively, but short duration treatment does not pose any problems. Mitotic chromosomes of soft materials like root tips, which need lesser time for fixations, respond well with this mixture.

10.8.4.3 Carnoy's Fluid I

(1) Glacial acetic acid 1 part, (2) absolute ethyl alcohol 3 parts.

It is used for plant material for squash preparations. The time of fixation is 15 min to 24 h in cold or at room temperature. The fluid should be washed out with 70% alcohol. Modifications of Carnoy's fluid I include mixtures in the proportion of 1:1, 1:2, and 3:2.

10.8.4.4 Carnoy's Fluid II

(1) Glacial acetic acid 1 part, (2) chloroform 3 parts, (3) absolute ethyl alcohol 6 parts.

It is used for flower buds and human tissues. The period of treatment is from 15 min to 24 h in cold or at room temperature.

10.8.4.5 Carnoy and Lebrun's Fluid

(1) Glacial acetic acid 1 part, (2) chloroform 1 part, (3) absolute ethyl alcohol 1 part.
It is useful to study the chromosomes of insects.

10.8.4.6 G. S. Sansom's Mixture

(1) Glacial acetic acid 1 part, (2) chloroform 6 parts, (3) absolute ethyl alcohol 13 parts.

It is useful for vertebrate materials. It penetrates very rapidly, and the period of fixation is 10–30 min.

10.8.4.7 Cutter's Mixture

(1) Propionic acid 1 part, (2) 95% ethyl alcohol 3 parts.
It is useful for plant tissue. It should be washed out with 70% alcohol.

10.8.4.8 Mark's Mixture

(1) Propionic acid 100 mL, (2) 95% ethyl alcohol 100 mL, (3) ferric hydroxide 0.4 g and a few drops of carmine.

This is effective for plants with small chromosomes.

10.8.4.9 Newcomer's Mixture

(1) Propionic acid 1 part, (2) chloroform 4 parts, (3) absolute ethyl alcohol 3 parts. This is suitable for plant chromosomes.

10.8.4.10 Iron Acetate–Acetic Alcohol Mixture

(1) Glacial acetic acid 1 part, (2) absolute alcohol 3 parts, (3) iron acetate small quantity.

It is suitable for anthers with small chromosomes. The period of fixation is 12 h followed by keeping for 5–15 min in saturated solution of iron acetate in 45% acetic acid 3 parts, 45% acetic acid 5 parts, and 1% aqueous formaldehyde 2 parts. The tissue is rinsed in 45% acetic acid.

10.8.4.11 Newcomer's Fluid

(1) Isopropyl alcohol 6 parts, (2) propionic acid 3 parts, (3) petroleum ether 1 part, (4) acetone 1 part, (5) dioxane 1 part.

It can be applied to both plants and animals for smears or sections.

10.8.4.12 Original Navashin's Fluid

Solution A—Chromic anhydride 1.5 g, glacial acetic acid 100 mL, distilled water 90 mL.

Solution B—40% aqueous formaldehyde solution 40 mL, distilled water 60 mL.

The two solutions are mixed in equal proportions before use. It is useful for squashes of root tips and flower buds. The period of fixation is 24 h, followed by washing in running water.

10.8.4.13 Karpechenko's Fluid

Ten percent aqueous chromic acid solution 15 parts, glacial acetic acid 1 part, 16% aqueous formaldehyde 3 parts, distilled water 17 parts. It is useful for general purposes.

10.8.4.14 Belling's Fluid

Solution A—Chromic anhydride 5 g, acetic acid 50 mL, distilled water 320 mL.

Solution B_1 (for prophase)—40% aqueous formaldehyde solution 20 mL, distilled water 175 mL.

Solution B_2 (for prophase)—40% aqueous formaldehyde solution 100 mL, distilled water 275 mL.

Solution A is mixed with equal quantity of solution B_1 or B_2 just before use. Fixation period is 3–12 h. It is good for studying meiosis in flower buds.

10.8.4.15 Levitsky's Fixative

Solution A—1% aqueous chromic acid solution.

Solution B—10% aqueous formaldehyde solution.

These two solutions are mixed in different proportions depending upon the material to be fixed.

It is very effective for fixing root tip of plant chromosomes. Fixation is done in 12–24 h at room temperature. After this, root tips are thoroughly washed in tap water.

10.8.4.16 La Cour's Fluid

Glacial acetic acid 2 parts, 40% aqueous formaldehyde solution 7 parts, absolute methyl alcohol 100 parts, distilled water 70 parts.

The mixture is suitable for blood smears.

10.8.5 Staining

Staining is a process of adsorption or it is a result of chemical reaction and physical adsorption. Staining is classified into two groups: vital and nonvital. In vital staining, nontoxic dyes are used, and the living tissue can be studied without being killed. The most common and effective vital stain for chromosomes is methylene blue, a basic dye of thiazine group. It is used in demonstrating cell division in tissue culture. Its chemical formula is $C_{16}H_{18}N_3SCl$. It is soluble in water. Vital staining is not very effective in chromosome studies.

In nonvital staining of chromosomes in the killed tissue, chemical agents insoluble in the chromosome substance are used. There are synthetic organic dyes, derived from coal tar. Certain chemical configuration is called chromophores, which are responsible for the color of the dyes. Quinonoid ring is the best example of a chromophoric group.

Chemical groups in the dye are responsible to retain the color of the dye in the tissue, which is called as auxochromes. These are mostly $-NH_2$ and $-OH$ groups. Auxochromes are responsible for the adherence of the dye to the tissue.

The chemical nature and behavior of the dye classifies it as acidic or basic. Acidic or anionic dye has a net negative charge on its ion, whereas the basic or cationic dye has a net positive charge on its ion. Some dyes are amphoteric, that is, they behave both as acids and bases in a solution.

The use of certain salts before or after staining improves the staining. The process is called mordanting. Mordant forms a double compound, one with stain and the other with the tissue. In carmine staining, iron salts are used as mordant, and they are mixed with the stain itself. Post-mordanting helps to retain the stain for a prolonged period and also clears the cytoplasm.

The intensity of the stain fades away with age. This may be due to the following reasons:

1. The effect of ultraviolet (UV) light by continued exposure to daylight.
2. Progressive acidity of the mounting medium.
3. Retention of contamination of elements during the process of staining.

The following are the commonly used stains for chromosomes:

10.8.5.1 Basic Fuchsin

It has the quinonoid ring responsible for its own color and also has auxochromes (NH_2) that allow the color to be retained in the tissue. Basic fuchsin belongs to the triphenylmethane series and is magenta red in color. The commonly obtained basic fuchsin is a mixture of *p*-rosaniline chloride, basic magenta (rosaniline chloride), and new magenta. The chief constituent is *p*-rosaniline chloride.

Basic fuchsin is easily soluble in water and alcohol. When basic fuchsin solution is treated with sulfurous acid, a colorless fuchsin sulfurous acid is obtained. This is the Schiff's reagent or Feulgen reagent, which stains DNA of the chromosomes. Sulfurous acid is obtained by the action of HCl on potassium metabisulfite.

10.8.5.2 Feulgen Reagent (Fuchsin Sulfurous Acid or Leuco-Basic Fuchsin)

Basic fuchsin 0.5 g, 1 N HCl 10 mL, potassium metabisulfite 0.5 g, activated charcoal 3.5 g, distilled water 100 mL.

Dissolve 0.5 g of basic fuchsin in 100 mL of boiling distilled water. Cool to 60°C and filter. Add 10 mL of 1 N HCl and 0.5 g of potassium metabisulfite to the filtrate. Store the reagent in a dark bottle for 24 h. Add charcoal powder, shake thoroughly, and keep overnight in the refrigerator at 4°C; filter and use.

Precaution: Store in a cool chamber away from light.

The main steps involved in leuco-basic fuchsin staining of plant material are as follows:

1. Hydrolysis of the fixed material in 1 N HCl at 60°C for 5–20 min
2. Staining in leuco-basic fuchsin (within a short time, the chromosomes take up a magenta color)
3. Mount and squash in 45% acetic acid or 0.2% acetic carmine

10.8.5.3 Chemical Basis of the Reaction

1. By hydrolysis, the purine-containing fraction of DNA is separated from the sugar, unmasking the aldehyde groups of the sugar. After removal of the base (purine), carbon atom 1 of the sugar is so arranged so as to form a potential aldehyde, capable of reacting with fuchsin sulfurous acid (Figure 10.12).
2. The reactive aldehyde groups then enter into a combination with fuchsin sulfurous acid (Schiff's reagent) to give the magenta color. Therefore, Feulgen reaction is the Schiff's reaction for aldehydes. The ribose sugar with an –OH in place of –H at carbon 2 is not hydrolyzed by 1 N HCl and does not react with fuchsin sulfurous acid. Therefore, basic fuchsin does not stain RNA.

In the pyrimidine–sugar linkage, on dissociation, the aldehyde groups are not free to react, unlike the open and reactive aldehydes obtained after the breakdown of the purine–sugar linkage.

Sometimes, the term "restoration of color" is used, but this is misnomer as the colored aldehyde–addition complex is entirely different from the original basic fuchsin.

10.8.5.4 Carmine

It is a crimson-colored dye obtained from the ground-up dried bodies of females of *Coccus cacti*, a tropical American Homoptera living on the plant *Opuntia coccinellifera*. The extract of the *C. cacti* is mixed with compounds of aluminum

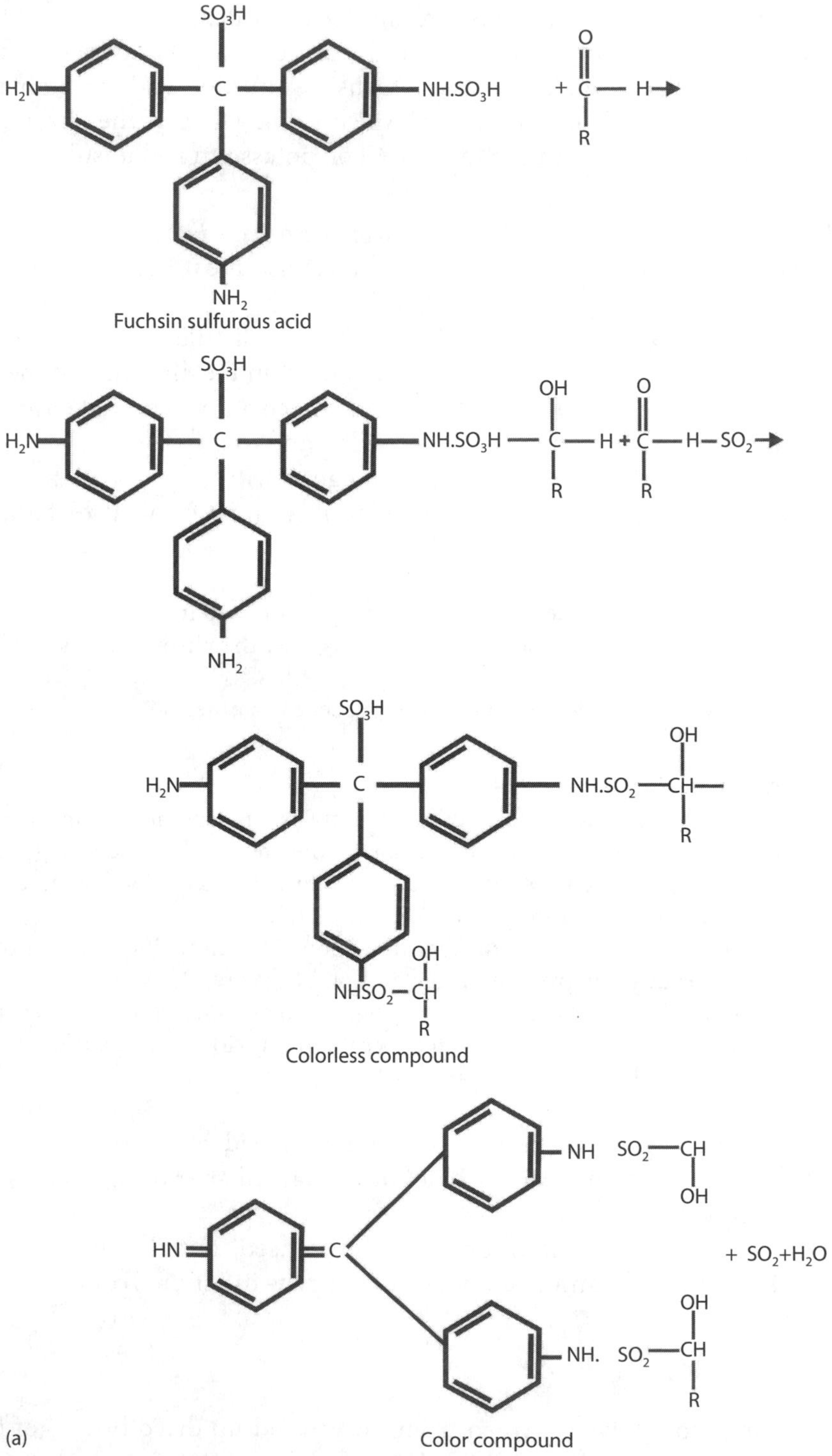

Figure 10.12   (a) The chemical basis of Feulgen reaction.

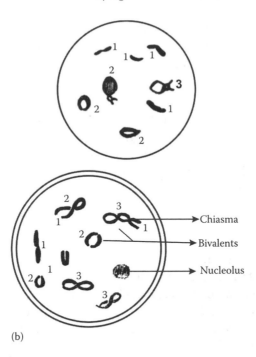

(b)

Figure 10.12 (continued) (b) Showing diplotene stage with chiasma formation in the chromosomes of pollen mother cells of *Allium cepa*.

or calcium. The composition of the carmine dye varies depending upon the method of manufacture. The staining property of all the samples is due to an active principle called carminic acid. Carminic acid can be obtained by extracting cochineal (grounded dried bodies of female *C. cacti*) with boiling water followed by treatment with lead acetate and then decomposition of lead carminate formed with sulfuric acid. Its molecular weight is 492.3 with the following chemical structure:

$$CH_3 \qquad O \qquad OH$$

$$HO \qquad C_6H_{10}O_5$$

$$HOOC \qquad O \qquad OH \qquad OH$$

Carmine

It has quinonoid linkage, which is responsible for the chromophoric property, and also has auxochromes. The dye is soluble in water in all proportions and is a dibasic acid. If it is dissolved in an acid, it behaves like a basic dye and stains chromatin. When used in solution with 45% acetic acid (called acetic carmine), it serves as a fixative as well as a stain. It is the most widely used stain in chromosome studies.

10.8.5.5 Orcein

It is a deep purple-colored dye obtained from the action of hydrogen peroxide and ammonia on the colorless orcinol. Its chemical formula is $C_{28}H_{24}N_2O_7$ with a molecular weight of 500.48. Orcinol is obtained from two species of lichens *Roccella tinctoria* and *Lecanora parella*. It is also prepared synthetically.

Orcinol is soluble in water and in alcohol. In the study of chromosomes, it is used as acetic orcein, in the same way as acetic carmine. Acetic orcein does not need iron mordanting, unlike carmine. It is a very effective stain for salivary gland chromosomes.

Before staining with acetic carmine or acetic orcein, the plant material is heated in a mixture of the stain and normal HCl. This helps in softening the material and facilitates better spreading of the cells. HCl also dissolves the middle lamella, thus avoiding overlapping of the cells and facilitating better spreading of the chromosomes. Too much heating of the slides that are stained with acetic orcein–HCl mixture induces chromosome breakage.

10.8.5.5.1 Preparation (2% Solution).
Carmine or orcein 2 g, glacial acetic acid 45 mL, distilled water 55 mL.

Add distilled water to glacial acetic acid to form a 45% acetic acid solution. Heat the solution in a conical flask to boiling. Add the dye slowly to the boiling solution, stirring with a glass rod. Boil gently till the dye dissolves. Cool down to room temperature. Filter and store in a bottle with a glass stopper.

10.8.5.5.2 Modified Carmine or Orcein Solution.
Iron is added to 1% acetic carmine solution as ferric acetate, ferric chloride, or ferric hydroxide in different proportions.

10.8.5.5.3 Propionic Carmine or Propionic Orcein.
The preparation is similar to acetic carmine or acetic orcein. Instead of acetic acid, propionic acid is used.

10.8.5.6 Giemsa

It is the mixture of several dyes, namely, methylene blue and its oxidative products, the azures, and cosin Y. The quality of the stain varies with regard to the proportion of the dyes used. Giemsa is used for staining different types of bacteria and chromosomes of various eukaryotes. It stains both RNA and DNA. RNA is generally removed by treatment with normal hydrochloric acid or perchloric acid. Stain is prepared by adding Giemsa powder 3.8 g, glycerine 250 mL, and absolute methyl alcohol 250 mL. Dissolve 3.8 g in 250 mL glycerine and 250 mL absolute methyl alcohol.

10.9 Mounting and Making Temporary Slides Permanent

10.9.1 Mounting Medium

The qualities of good mounting are the following:

1. It should have a refractive index same as slightly higher than the glass slide and almost the same as that of the tissue.
2. It should harden quickly in contact with air so as to fix the cover glass firmly to the slide.
3. It should check destaining of the material.
4. It should not decompose with storage.

The medium used for mounting the permanent slides serves the following purposes:

1. It renders the tissue transparent.
2. It increases the visibility of the tissue under the microscope.
3. It holds the tissue with the cover glass firmly in place.
4. It preserves the tissue.

The following mounting media are commonly used:

10.9.1.1 Balsam

It is an oleoresin collected from blisters formed in the bark of a gymnospermic plant *Abies balsamea*. It is an essential oil in which resins are dissolved. Resins are the oxidation products of essential oil itself. Balsam fulfills most of the conditions necessary for a mounting medium. Canada balsam, a light yellow, transparent liquid, is used in the laboratory for cytological studies. It has the optical dispersion and refractive index similar to that of glass. It hardens when in contact with air and fixes the cover glass firmly to the slide. It is 24% essential oil and 76% resin component, which is soluble in xylol and partly soluble in ethyl alcohol. The essential oils are removed on drying. Xylol, in which Canada balsam is dissolved, is converted into toluol and phthalic acid and thus becomes acidic. The acidity weakens the basic stains generally used in smears and squashes.

10.9.2 Procedure for Making Slides Permanent

1. Remove the paraffin seal around the cover slip of the temporary preparation by wiping the top of the cover slip with a piece of cheese cloth soaked in xylol.
2. Invert the slide in a covered Petri plates containing glacial acetic acid–ethyl alcohol mixture 1:1. The slide can be supported on glass rods placed in the Petri plate. Cover slip is detached within half an hour.
3. Transfer the slide and the cover slip to a Petri dish containing acetic ethanol mixture 1:3 and keep in this mixture for 5 min.
4. Give two changes of absolute alcohol of 5 min each.
5. Mount the slides and cover slips separately in Euparal.

If slides are stained with propionate carmine, propionic acid is used instead of acetic acid. An alternative procedure is as follows:

(Three- or four-day-old temporary squashes or smears are ready for making permanent.)

1. Using a blade, remove the wax of the cover slip, taking care not to allow slipping or moving of the cover slip.
2. Place the slide inverted in 1:1 mixture of glacial acetic acid and *n*-butyl alcohol. The slide should rest on a glass rod, and the cover slip should dip into the mixture.
3. When the cover slip slips off the slide, transfer both to another Petri plate containing glacial acetic acid and *n*-butyl alcohol mixture 1:1. The material on the slide and cover slip should touch the mixture. Leave the slide and cover slip in this mixture for 5 min.
4. Transfer to a Petri plate containing *n*-butyl alcohol and keep in it for 5 min.
5. Place a drop of Euparal on the material containing portion of the slide and slowly place a fresh cover slip on it.
6. Place a drop of Euparal on a fresh slide and place the cover slip with the material onto this. Mounting the slide and the cover slip separately avoids overlapping of the cells.
7. Keep the slides in the oven (60°C) for a day. This removes any air bubbles present and also removes the extra Euparal.

10.9.3 Preparation of Albuminized Cover Slip

Place a small spot of glycerine albumen on the cover slip and smear it on the surface with a finger. After smearing, wipe off the albumen with a clean finger. The amount that remains following this is sufficient. Pass the cover slip through a flame until it begins to "smoke." This should not be overdone for the albumen, otherwise it turns brown. The cover slip is now ready for use. Albuminized cover slip should be prepared immediately before use.

10.10 Study the Effect of Colchicines or Acenaphthene or Chloral Hydrate on Chromosomes of *Allium cepa*

10.10.1 Principle

In the majority of the plants, the meristematic region is a heterogeneous mass of cells in which the nuclei are at different stages of division, that is, the divisions are nonsynchronous. The best nuclear phase for chromosome analysis is the metaphase stage. To induce high frequency of metaphase stage, which is low in the heterogeneous mass of cells at different divisional phases, the most suitable chemical is colchicine. Colchicine causes metaphase arrest by inhibiting the operation of the spindle mechanism. This property of arresting the metaphase is an essential prerequisite for inducing cell synchronization. The colchicine is effective in very low concentrations at late prophase of mitosis. The metaphase stages after colchicine induction show clear euchromatid segments and are known as colchicine mitosis or c-mitosis. There are other chemicals also that are used for metaphase arrest. These are chloral hydrate, acenaphthene, gammexane, etc. However, the efficacy of colchicines is decidedly superior to these chemicals.

Colchicine is soluble in water. Due to high solubility in water, it is very active at extremely low concentrations. Colchicine brings about a change in the colloidal state of the cytoplasm, causing spindle disturbance by increasing the fluidity of the nuclear substance. It causes polyploidy. Polyploidy is defined as an increase in gene dosage. Polyploidy has importance in agricultural and horticultural practices. This brings about gigantism in all characters in general. Polyploids are tolerant and more adaptable to changes in the environment. The induction of polyploidy has helped in making certain interspecific and intergeneric crosses fertile. Colchicine inhibits the formation of spindle, and the duplicated chromosomes are confined within one nucleus. The division of the chromosome is unhampered. The tissue recovers as soon as the influence of the chemical is removed. Therefore, polyploidy cells, which are formed by colchicine action, divide normally once the chemical is removed.

Acenaphthene is less soluble in water, so it should be used as saturated solution to get the desired effect. It is most suitable for clarifying the chromosomes of the pollen tubes. It has the same property of arresting metaphase as that of colchicines. It is a derivative of benzene. Benzene itself is ineffective, but the side chains of its derivatives are responsible for c-mitosis property.

Chloral hydrate is similar to colchicines. It is less expensive and the concentrations used are the same as that of colchicines on plant tissues. However, it is unsuitable for critical studies.

10.10.2 Materials

Allium bulbs; different concentrations of colchicine solution, that is 0.1%, 0.2%, 0.3%, 0.4%, and 0.5%; coplin jars; microscope, glass slides, cover slips, watch glasses, brushes, needles, forceps, spirit lamp, root tips, 2% acetocarmine solution or 2% acetic orcein.

10.10.3 Protocol

1. Take healthy young bulbs of *Allium cepa*, denude them of roots, and let them grow in a jar containing tap water until fresh crops of roots are produced.
2. When the roots are just 2–3 mm long, fit the bulbs on the top of jars containing the following: (a) tap water, (b) 0.1% colchicine solution, (c) 0.2% colchicine solution, (d) 0.3% colchicine solution, (e) 0.4% colchicine solution, and (f) 0.5% colchicine solution. Place the bulbs in such a way that the roots dip into the solution.
3. Harvest the roots from the bulbs after different intervals of time and fix them in acetic acid–alcohol mixture 1:3. Label the materials indicating clearly (a) concentration of the colchicines used and (b) duration of treatment.
4. Squash the root tips using acetocarmine stain or leuco-basic fuchsin stain according to the schedule given in the experiment earlier.

TABLE 10.1   Effect of Chemicals in Varying Concentrations on Chromosomes of *Allium cepa*

Treatment		Number of Normal Cells		Number of Polyploidy Cells		% Polyploidy Induced	
Conc. (%)	Time (h)	Control[a]	Treated	Control[a]	Treated	Control[a]	Treated
0.1	1						
	2						
	3						
0.2	1						
	2						
	3						
0.3	1						
	2						
	3						
0.4	1						
	2						
	3						
0.5	1						
	2						
	3						

[a]Control = Roots grown in the tap water for the same interval of time as treated grown in colchicine solution.

5. Observe the slides under the microscope and record the observation according to the following data sheet (Table 10.1).
6. Compare the contraction and straightening of chromosomes in the treated and control cells.

10.11 Estimate the Chiasma Frequency at Diplotene–Diakinesis

10.11.1 Introduction

Meiosis leads to the formation of haploid cells from a diploid cell. Homologous chromosomes replicate at interphase; synapse at zygotene exchanges the genetic material between them at pachytene and forms points of crossover called chiasmata. These chiasmata terminalize at diplotene–diakinesis, and finally the homologous chromosomes separate at metaphase–anaphase.

10.11.2 Chiasma Frequency

Chiasmata are the points of crossover between two nonsister chromatids of homologous chromosomes seen during prophase I of meiosis. They are the cytological expression of genetic exchange at pachytene. Synapsis of homologous chromosomes

TABLE 10.2   Chiasma Formation between Homologous Chromosomes Involving Different Number of Strands

	Type of Crossover Event	Example of Bivalent	Type of Strands after Crossover Event		Frequency of Recombinant Strands	Tetrad Type
A	Two-strand single crossover		A B A b a B a b	Two parental Two crossover types	½	Tetra type
B	Two-strand double crossover		A B A B a b a b	All parental	0	Parental ditype
C₁	Three-strand double crossover		A b A B a B a b	Two parental Two crossover types	½	Tetra type
C₂	Three-strand double crossover		A B A b a b a B	Two parental Two crossover types		
D	Four-strand double crossover		A b A b a B a B	All crossover types	1	Nonparental ditypes

is the prerequisite to crossing over and chiasma formation. According to the number of chiasma involved, the crossing over is one of the following types (Table 10.2):

10.11.3   Single Crossing Over

Only one chiasma occurs at two points in a bivalent. It can be of three types:

1. *Regressive*—where only one strand out of four parental chromatids is involved in reciprocal crossing over. This results in four parental chromatids.
2. *Progressive*—here three strands are involved in forming two chiasmata. One strand is common for both chiasmata. This results in two parental and two recombinant strands.
3. *Digressive*—all the four strands are involved in forming one chiasma and the rest of the two forms the second chiasma. This results in four recombinant chromatids (Table 10.2).

10.11.4   Multiple Crossing Over

Chiasmata occur at three, four, or more points between any two nonsister chromatids.

Chiasmata help to calculate the percentage recombinants. The greater the distance between two genes on a chromosome, the more is the chance of chiasma formation in between the two genes. The occurrence of one chiasma reduces the chance of formation of another chiasma in its near vicinity. This phenomenon is known as "interference" and is measured as the coefficient of coincidence that is equal to the ratio of actual double crossover to expected double crossover.

The frequency with which crossing over occurs between two linked genes is a function of the distance between them (Table 10.2). Thus, one map unit is defined as the length of the chromosome within 1% of crossing over occurs. Percent cross over is calculated using the following formula:

$$\% \text{ Crossing over} = \frac{\text{Total number of recombinants observed}}{\text{Total number of offsprings}} \times 100$$

Even if crossover occurs in every meiotic product, only 50% recombination could be detected. This results from the fact that, from any crossover event, only two parental and two recombinant chromosomes can result. Sex, age, temperature, water content, radiations, chemicals, heterochromatic regions, and proximity of centromere are the factors that affect the frequency of chiasma formation.

10.11.5 Materials

Microscope, glass slides, cover slips, watch glasses, brushes, needles, forceps, spirit lamp, root tips, 0.2%–2% acetocarmine solution or 2% acetic orcein, Allium flower bud.

10.11.6 Protocol

Prepare the slides of pollen mother cells of Allium using acetocarmine. Get well-spread diplotene or diakinesis plates as shown in Figure 10.12b. Count and record the frequency of chiasmata formation as per the following data sheet (Table 10.3):

10.11.7 Observations

Results are observed under the microscope and recorded and calculated as per Table 10.3.

TABLE 10.3 Calculation of Chiasma Frequency

Cell Number	Number of Bivalents	Number of Chiasmata	Chiasmata/Bivalent
1.			
2.			
3.			
4.			
5.			
6.			
7.			
8.			
9.			
10.			
⋮			
25.			

Note: Number of cells studied should be 25 or more. The average chiasmata/bivalent in these cells gives a near accurate picture of chiasma frequency.

10.12 Study the Effect of γ Radiation on Chromosomes

10.12.1 Principle

Ionizing radiations when absorbed by a solid, liquid, or gas, directly or indirectly, produce ion clusters. The effective radiations are x-rays, gamma rays, alpha particles, beta rays, neutrons, and protons. UV rays are the nonionizing radiations with short wave length. The ionizing radiations produce ion clusters and increase local reactivity, but UV rays act by direct or indirect disruption of molecular integrity. The absorption of ionizing radiation by a biological material means absorption of a considerable amount of energy. This energy brings about changes in the molecules as (1) molecular disintegration and (2) production of a variety of reactive radicals, the type of radicals produced depending upon the material exposed.

The biological change after energy absorption and disintegration of the molecules depends upon (1) the degree and kind of disruption and (2) the reaction of the products of disintegration.

10.12.2 Effect of Radiation

The effect of radiation depends upon the following:

1. *The source and nature of the radiation*: This determines degree of penetration and ionizing density, which in turn depends upon wavelength, velocity of the particles, and charge and mass of the particles.
2. The common overall unit of measuring ionizing radiation is "rad," defined as the amount of energy absorbed per gram of target (material), but this is not readily converted into either amount or density of ionization. X- and γ-rays are normally measured in *roentgen* (*r*-unit), which is defined as the number of ion pairs produced per unit volume of a gas or liquid.
3. *Nature of the material exposed*: Different biological materials have different degree of sensitivity to radiations. Man cannot withstand more than 400 to 500 *r* without the lethal effect, while *Drosophila* is not killed by even 50,000 *r*.
4. *Conditions of exposure*: Mutations, chromosome damage, etc., are low under lowered O_2 tension. A wide variety of chemicals provides some protection if present at the time of radiation. Some reducing chemicals act as oxygen removers; others like cyclic amines protect susceptible molecules by simply "bleeding off" the energy.

10.12.3 Physiological Effects

Radiation interferes with a number of biochemical processes by inactivation of essential enzymes: Stickiness of chromosomes so often observed may be due to depolymerization of DNA caused by radiation. Mitosis is generally inhibited even by a small dose.

10.12.4 Mutational Effects

1. Yield of mutations is directly proportional to the dose for a particular material.
2. Increase in ion intensity decreases the rate of mutation for any given total amount of ionization. This reflects target restriction (over kill).
3. Sensitivity of different genes varies. Some genes are mutable by a particular dose; others are not.

10.12.5 Chromosomal Effect

1. *Chromosome breakage*: Results from disruption of bonds for integrity either directly or as a result of strain.
2. *Healing or union of broken ends*: Healing of free bonds takes place, and the resultant rearrangements are observed in the final result of the irradiation effect. The capacity to rejoin the broken ends is retained for some time. In Allium microspores, it is approximately 1 h.

10.12.6 Types of Breaks

1. Irradiation of nuclei in interphase produces chromosome breaks.
2. Irradiation of nuclei in early prophase results in chromatid breaks.
3. Heterochromatic regions are comparatively more sensitive to breaks than euchromatic regions.
4. Dividing cells are more sensitive to breaks than the nondividing ones. Early interphase is least susceptible.

10.12.7 Radiation Hazards

Radiation kills cells and organisms, modifies metabolic systems on the deteriorating side, and induces changes in the genetic material. To sum up, the changes induced by radiations that are cytologically visible are (1) chromosome breaks, (2) chromatid break, (3) chromosomal gaps, (4) single-hit aberrations as observed from chromosome bridges during anaphase, (5) multiple-hit aberrations (fragmentations of chromosome), (6) translocations and other gross morphological changes of chromosomes, and (7) stickiness of the chromosomes.

10.12.8 Mechanism of Changes due to Radiation

Ionizing radiations emitted by γ-rays knock off electrons from the atoms of the material through which they pass. In biological systems, water is the most abundant in the cell. It is ionized by radiations producing hydrogen ions and hydroxyl ions:

$$H_2O + \gamma \text{ rays photon} \rightarrow H_2O + e^-$$

$$\downarrow$$

$$H^+ + OH^- + e^-$$

The resulting H^+ ion reacts with available oxygen to produce a hydroperoxide radical,

$$H_2 + O_2^- = H_2O_2^-$$

These radicals are strong oxidizing agents, and when they react with a DNA of a chromosome, they produce various abnormalities. The breakage in the sugar phosphate backbone of DNA may lead to breaks, deletions, additions, and translocations.

10.12.8.1 Materials

1. Expose barley seed (any other crops seed can be used) to two different concentrations (more than two concentrations can be considered upon the type of experiment) of γ radiation, that is, 75 KR and 100 KR, for a fixed period.
2. Keep one set as control (without exposure).
3. Soak the seeds in water for half an hour.
4. Keep the seeds in Petri plates lined with wet blotting paper for germination.
5. Label the plates for control and for different doses.
6. After 2 or 3 days (different seeds take different time duration for germination), count the number of seeds germinated.
7. After 48 h of germination, measure the length of root and plumule in each treatment.
8. Prefix the roots in 0.5% colchicines for 3 h.
9. Wash thoroughly with water.
10. Fix in (Carnoy's fluid I) for 4 h and preserve in 80% alcohol.
11. Following the Feulgen staining technique, squash the root tips and observe the slides under the microscope.
12. Record the observations in the following data sheet. Record the dose, which gives more drastic effects.
13. Flower buds can be exposed to radiations, and after allowing different recovery periods, meiosis can be studied in pollen mother cells by observing (a) univalents, (b) multivalents, (c) laggards, and (d) bridges.

10.12.8.2 Source of Gamma Rays

Gamma rays are obtained as electromagnetic radiations either from radium itself or from the isotope ^{60}Co. The material can be exposed to these rays by placing it within a ^{60}Co field for known periods.

10.12.8.3 Observations

Record any other abnormalities, if present, like laggards and bridges (Tables 10.4 through 10.6).

TABLE 10.4 Percentage Seed Germination, Root Length, and Chromosome Breakages

% Germination

Treatment	Number of Seeds Germinated	Total Number of Seeds Treated	% Germination
Control			
Dose 1			
Dose 2			

TABLE 10.5 Root Length Measurements

Treatment	Replication I	Replication II	Replication III	Total	Mean
Control					
Dose 1					
Dose 2					

TABLE 10.6 Breakages

Treatment	Total No. of Cells Observed	No. of Normal Cells	Cells with Breakages	% Cell with Breakages	Mean No. of Breaks/Cell
Control					
Dose 1					
Dose 2					

Note: Record any other abnormalities, if present, like laggards and bridges.

10.13 To Prepare Slides Showing Diplotene Stages in Spermatocytes of Desert Locust (*Schistocerca gregaria*) or in Oocytes of the Newt (*Triturus viridescens*) Showing Lampbrush Chromosomes

10.13.1 Principle

Meiosis in the oocytes of most vertebrates is blocked at diplotene, and at the ovulation, the process is resumed. In human females, this stage lasts for about 40 years. For the growth of the egg and for the development of the zygote after fertilization, important biochemical activities are required. Frequently, the bivalent decondenses completely (dictyotene), and transcription is carried out. In some cases, the chromosomes extend tremendously.

In *Triturus* (newt) and Xenopus, the chromomeres of these giant bivalents become very conspicuous, and some of them carry large loops. These loops have DNA core in highly decondensed state and active transcription is carried out on it. The chromosomes have a brushy appearance due to these loops and hence are called lampbrush chromosomes. Some loops can be recognized because of their characteristic shapes. The chiasmata in the lampbrush chromosomes are formed in or between the chromomeres but are never formed in the loops.

10.13.2 Materials

Dissecting kit, *Triturus* (newt) or any other amphibian or *S. gregaria*, fixatives, stain, saline solution, glassware, slides, and cover slips.

10.13.3 Protocol

10.13.3.1 Protocol 1 (for Triturus, Newt, or Any Other Amphibian)
1. Anesthetize the animal with light ether or chloroform.
2. Dissect out the ovary or oocytes by making a small incision at the ventral side of the animal.
3. Dip the oocytes in 0.7% saline NaCl solution for half an hour.
4. Wash the oocytes in distilled water.
5. Fix in Carnoy's fluid for 1 h.
6. Preserve in 70% alcohol.
7. Smear in 1% acetocarmine and observe under the microscope.

10.13.3.2 Protocol 2 (for S. gregaria)
1. Etherize the specimen in chloroform.
2. Dissect out the pair of testes that is present below the third abdominal segment of the male.
3. Dip the testes in 0.7 saline sodium chloride solution (NaCl) for 0.5 h.
4. Wash the testes in distilled water.
5. Fix in Carnoy's fluid II for 1 h. (FeCl$_2$ crystal is added to Carnoy's as a mordant.)
6. Material can be preserved in 70% alcohol.
7. Testis is made up of tubules. Take two or three tubules and smear them in 1% acetocarmine and observe under the microscope.

10.14 Study Karyotypes of *Allium cepa* and *Vicia faba*

10.14.1 Principle

A karyotype is a morphological aspect of the chromosome complement as seen at mitotic metaphase. Five different characteristics of the complement are usually observed and compared in constructing a karyotype. These characteristics are

1. Differences in absolute size of the chromosomes
2. Differences in the position of the centromere
3. Differences in relative chromosome size
4. Differences in basic number
5. Differences in the number and position of satellites

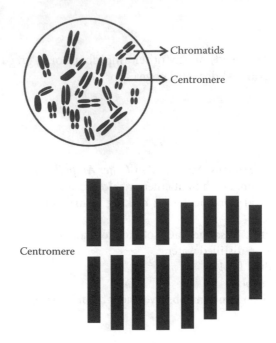

Figure 10.13 Showing symmetrical karyotype of *Allium cepa* and ideogram of *Allium cepa*.

Karyotypes are of two types:

Symmetrical: The chromosomes are all of approximately the same size and have median or submedian centromeres (Figure 10.13).

Asymmetrical: A heterogeneous type of karyotype. This is due to the shifting of the centromere from median to subterminal or terminal position in some chromosomes and also due to the differences in relative sizes between the chromosomes of the complement (Figure 10.14).

Symmetrical karyotypes are regarded as more primitive and heterogeneous and more specialized.

10.14.2 Karyotype Concept

More species of living organisms show a distinct and constant individuality of their somatic chromosomes, and closely related species have more similar chromosomes than those of more distantly related ones.

10.14.3 Idiogram

Diagrammatic representation of chromosome morphology used diagnostically for comparison of the karyotypes of different species and varieties is known as idiogram. The construction of idiograms is based on measurements of total chromosome

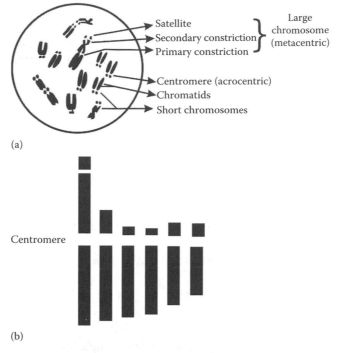

Figure 10.14 (a) Asymmetrical karyotype of *Vicia faba* and (b) ideogram of *Vicia faba* karyotype.

length and arm length ratio long arm/short arm, centromere position and position of secondary constriction (Figure 10.14).

10.15 Micrometry

To measure the length of the chromosomes, ocular micrometer is used. Before using ocular micrometer, it is standardized with the stage micrometer. A stage micrometer has 100 divisions equal to 1 mm. So, one division of stage micrometer is equal to 10 μ (microns). The stage micrometer is placed under the objective lens (oil immersion) that has a power of 100× and the ocular micrometer is placed inside the eyepiece lens with a power of 10× for standardization.

Under the modification of 10× eyepiece and 100× oil immersion objective lens, the number of divisions of ocular micrometer equivalent to the number of divisions of stage micrometer is counted to know the divisions of ocular micrometer in terms of microns.

Suppose 5 divisions of stage micrometer are equal to 30 divisions of ocular micrometer. So, one division of stage micrometer is equal to 6 divisions of ocular micrometer. Hence, one division of ocular is equal to 10/6 = 1.667 μ. The chromosomes are measured using ocular micrometer, and the reading is multiplied by 1.667 to express the length of the chromosome in microns.

10.15.1 Materials

Allium cepa root tips, *Vicia faba* root tips, microscope, glass slides, cover slips, watch glasses, brushes, needles, forceps, spirit lamp, root tips, 2% acetocarmine solution, or 2% acetic orcein.

10.15.2 Protocol

1. Prepare well-spread metaphase plates.
2. Calculate the length of each arm of the haploid number of chromosomes, that is, in Allium 8 chromosomes representing a haploid set and in Vicia 6 chromosomes representing a haploid set.
3. Record the observation as per the following sheet (Table 10.7).

Pool the data as per the following data sheet (Table 10.8).

10.15.3 Idiogram

1. Represent the chromosomes diagrammatically by simple lines.
2. Arrange the chromosomes according to the descending order of length.
3. Show the position of the centromere and the secondary constriction.

TABLE 10.7 Karyotype Classification (Proportion of Chromosomes with Arm Ratio Greater than 2:1)

Ratio Largest/Smallest Chromosome	0.0	0.01–0.50	0.51–0.99	1.00
<2:1	1 A	2A	3A	4A
2:1–4:1	1B	2B	3B	4B
>4:1	1 C	2 C	3 C	4 C

Note: 1 A is the most symmetrical and asymmetry increases in both directions, that is, 1A to 4 A and 1A to 1 C. This means the 4 C type of karyotype is the most asymmetrical type. Similarity in karyotypes between species of genera does not reflect complete similarity in chromosome structure.

TABLE 10.8 Showing Chromosomal Variation as per Karyotype Classification

Absolute Length (μ)			Total Length in Descending Order (μ)	Long/Short Arm Ratio	Largest/Smallest Chromosome Ratio	p^a	Category[b]
L	S	Total					

Note: L denotes for long arm, S denotes for short arm.
[a]Proportion of chromosomes with arm ratio more than 2:1.
[b]Category of karyotype asymmetry.

10.16 To Demonstrate Reciprocal Translocation in Pollen Mother Cells of *Rhoeo discolor* (Tradescantia)

10.16.1 Introduction

Translocation is a process in which a piece of a chromosome is transferred to another chromosome. The process through which two chromosomes are broken and then mutually exchange parts is known as reciprocal translocation or segmental interchange. The two new chromosomes function normally if each has a single centromere. If the reciprocal translocation produces chromosomes with two centromeres (dicentric) or with no centromere (acentric), these are generally eliminated because they would fail to segregate properly at anaphase.

10.16.1.1 Homozygous Translocation
In this process, translocations involve homologous chromosomes that behave like normal chromosomes except that the new linkage groups are formed.

10.16.1.2 Heterozygous Translocation
It involves nonhomologous chromosomes, and they form characteristic pairing configurations during prophase of first meiotic division. The four chromosomes (two pairs of homologous chromosomes) after undergoing reciprocal translocation (between two nonhomologous ones) now are partially homologous to one another. So, at pachytene, a cross-like configuration is produced. This is so because of the synapsis between the homologous parts. From the appearance of the configuration produced, the position of the translocation break can be determined, provided the synaptic behavior is exact. This cross would open out into a ring at diakinesis or early metaphase if chiasmata are formed in each of the paired arm. If one arm fails to form chiasmata, a chain of four chromosomes is formed. Translocation may involve more than two nonhomologous chromosomes. If the arm of one of the translocated chromosomes is being involved in a second interchange with a third nonhomologous chromosome, a ring or chain of six would form at metaphase. A third interchange would give a ring of eight and so on. In *Rhoeo discolor*, multiple translocations involving all 12 chromosomes are observed forming a ring. Normal-appearing translocation heterozygotes may produce abnormal offsprings because of meiotic irregularities. Translocation is used as genetic markers. This leads to rearrangement of genes and may give rise to certain hereditary manifestations.

10.16.2 Materials

Fresh Rhoeo flower buds (fixed material does not give good results), slides, cover glass, needles, forceps, spirit lamp, acetocarmine stain.

10.16.3 Protocol

1. Prepare slides of pollen mother cells of *Rhoeo discolor* as per the smear technique described earlier.
2. Look for diakinesis–metaphase stages. A ring or chain of 12 chromosomes, joined end to end, is seen.

10.17 To Study the Spiral Nature of Chromosomes Using Any Plant Material

10.17.1 Principle

A somatic metaphase chromosome consists of two chromatids. Each chromatid contains helically coiled chromonema. The first metaphase or first anaphase chromosome of meiosis also has the same structure with coiling being longer gyred and looser than the mitotic metaphase chromosome coil. Chemically, the chromosome consists of two major parts, DNA and protein. The length of the double-helix DNA that can be contained in the total amount of DNA present in a diploid human cell turns out to be almost two yards, or in other words, the average length of DNA in a single chromosome is roughly 4 cm. No chromosome of this size is ever seen. Therefore, the chromosome is either composed of many strands of DNA or made up of a single double helix that has been folded on it many times. The multiple-stranded hypotheses are no longer a dominant view now, and each chromosome is considered a single-stranded DNA helix folded many times on itself.

There are two views regarding the single double-stranded DNA helix structure of chromosome:

1. Chromosome is a long continuous single molecule of DNA stretching from one end of the chromatid through the centromere to the other end of the chromatid.
2. A number of different DNA molecules of shorter length held together by protein "linkers."

The evidence so far favors the first view, that is, it is a single long DNA molecule without any protein present as a necessary part of their structure. The digestion of chromosomes with proteolytic enzymes does not result in breakage of the chromosomes. If the different molecules are held together by protein linkers, then these proteolytic enzymes should break the chromosomal thread, which is not the case. On the other hand, DNase destroys the integrity of the chromosome suggesting that it is made up of DNA backbone. Digestion with proteolytic enzymes makes the chromosome thinner. This suggests that proteins of the chromosomes (which make 50%–60% of the dry weight) surround the DNA double helix.

The interphase chromosome is the most extended appearance of the chromosome and has the finest fibrils of 200–300 Å in diameter. The double DNA helix is 20 Å thick, thus supporting the fact that even in the relatively extended condition of the interphase chromosomes, the DNA is supercoiled rather than being a completely extended double helix.

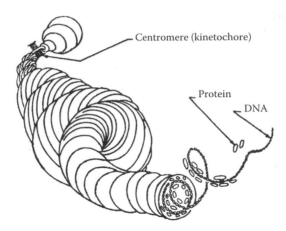

Centromere (kinetochore)

Protein

DNA

Figure 10.15   A model of chromosome showing supercoiling of DNA double helix inside a chromosome strand. Protein present on the DNA surface facilitates the supercoiling.

To achieve this supercoiling, folding up of the double helix is essential. Folding can cause impairing or even occasional breaks in the phosphodiester backbone of one or the other chain of the DNA. This is avoided and the folding is enhanced by protein molecules interacting with two regions of the folded DNA molecule, thus stabilizing the phosphodiester backbone. From interphase to metaphase, the supercoiled regions can themselves be coiled once, or many times, until the final dimensions of the metaphase chromosomes are achieved. The coiling of the DNA is occasioned by its packing with chromosomal proteins. Thus, the entire nucleoprotein fibril is then coiled into a much thicker, highly compacted structure (Figure 10.15). In this way, the presence of large quantities of proteins coating the DNA is justified. To sum up, the chromosomes are chemically made up of deoxyribonucleic acid and proteins. DNA forms the backbone and the protein coating helps in the supercoiling of the DNA, together with forming the compact chromosomes.

The schedules given for the study of the spiral nature of chromosomes are based upon the digestion of protein to expose the spirals of DNA.

10.17.2 Materials (Shock Treatment: For Procedure I and II)

Allium bulbs with fresh roots and Allium buds; coplin jar; concentrated nitric acid or concentrated hydrochloric acid; acetic acid–ethyl alcohol—1:1; 1 N HCl; leuco-basic fuchsin and 2% acetocarmine; 45% acetic acid; micro slides; cover slips; microscope.

10.17.2.1 Protocol I (for Mitotic Chromosome)
1. Hold the bulbs of *Allium cepa* with fresh roots near the mouth of a jar containing concentrated HCl or HNO_3 for 1–2 min.
2. Cut the root tips and fix in acetic alcohol for half an hour.

3. Wash first in 70% alcohol and then in water.
4. Hydrolyze in 1 N HCl at 60°C for 10 min.
5. Wash in water.
6. Stain in leuco-basic fuchsin and squash in 45% acetic acid or 2% acetocarmine.
7. Observe for spirals under the microscope.

10.17.2.2 Protocol II (for Meiotic Chromosome)

1. Smear the tissue (pollen mother cells) and expose the slide on the smeared side to the acid fumes for 1–2 min.
2. Stain in acetocarmine.
3. Observe under the microscope.

Instead of acid fumes, ammonia vapors can also be used.

Root tips, if frozen for 3–5 days, show nucleic acid starvation. After this cold temperature treatment, they can be fixed in acetic alcohol 1:3 and then stain in leuco-basic fuchsin.

10.17.2.3 Protocol III (by Dissolving the Outer Nucleic Envelope)

1. Treat the root tips with ammonia in 30% alcohol (six drops of ammonia in 50 mL of alcohol) for 5–20 s.
2. Wash in water.
3. Fix in acetic alcohol for half an hour.
4. Stain and squash according to leuco-basic fuchsin stain technique.

10.17.2.4 Protocol IV (Precipitation of DNA as a Metallic Salt and then Partially Digest Proteins)

1. Treat the root tips in 1/100 g mol solution of NACN for 30 s.
2. Fix in acetic alcohol for 24 h.
3. Wash in water several times and then keep root tips in 0.1% lanthanum acetate for 12 h to precipitate the nucleic acid as lanthanum salt.
4. Digest in 1% trypsin solution for 24 h at 37°C.
   a. Prepare squashes according to leuco-basic fuchsin technique.

10.17.2.5 Protocol V (Pretreatment)

1. Treat fresh root tips of Allium or Trillium with PDB or 8-oxyquinoline for 4 h at 10°C–15°C.
2. Fix in acetic alcohol for 1 h.
3. Prepare squashes according to leuco-basic fuchsin stain technique.

10.18 To Show the Different Banding Patterns in Eukaryotic Chromosomes

10.18.1 Principle

Different chromosomes show different banding patterns, and this technique led to the identification of human chromosomes and had an impact on clinical cytogenetics.

Based on the banding pattern and the location of the centromere, chromosomes can be readily identified. Banding patterns are important to distinguish different structures of chromosomes, not the gene activity. The staining used here has high preference for some regions, AT-rich domains or GC-rich domains, where in general the AT-rich domains are highly active in gene expression (*in vivo*) and are being used to pinpoint structural abnormalities.

Using different staining techniques, the following types of banding patterns are observed:

1. *Q-bands*: These are produced by quinacrine mustard fluorescence technique. Quinacrine binds preferentially to a certain region of metaphase chromosomes to produce characteristic banding patterns. Good fluorescence preparations revealing band patterns provide sufficient information to identify the chromosomes. This clearly differentiates between the G + Y group chromosomes in humans and provides distinction between C + X group.
2. *G-bands*: Giemsa stain reveals banding patterns, which are characteristic of particular chromosomes, thus helping in the identification of those chromosomes.
3. *R-bands*: These bands are opposite in staining intensity to that of G-bands. The method used for staining is called reverse staining (Giemsa) method.
4. *C-bands*: These bands demonstrate constitutive heterochromatin revealing banding patterns in these areas.

10.18.2 Q-Band Technique

10.18.2.1 Materials
Root tips of *Vicia faba* fixed in acetic alcohol, alcohol series from rehydration, distilled water, phosphate buffer pH 4.1–7.0 (0.05 M), quinacrine mustard stain (QM) 50 µg/mL.

10.18.2.2 Protocol
1. Prepare the slides of root tips without staining and air dry them.
2. Rehydrate the slide in alcohol series and distilled water.
3. Soak in phosphate buffer.
4. Stain in QM for 20 min.
5. Wash the slides in three changes of buffer of the same pH as used before.
6. Mount in the buffer and seal.
7. Investigate the chromosomes in a fluorescence microscope. Acridine orange is also used as a stain instead of QM.

10.18.3 G-Band Technique

10.18.3.1 Materials
Root tips of *Ornithogalum* (*Ornithogalum umbellatum*), α-bromonaphthalene, acetic alcohol, 1 N HCl, 45% acetic acid, saturated aqueous solution of barium hydroxide,

liquid nitrogen, 90% absolute ethyl alcohol, distilled water, 2× SSC solution (0.3 M NaCl + 0.03 M trisodium citrate, pH 7.0), Giemsa stain (2 mL of Giemsa stock solution in 100 mL of 0.1 M Sorensen's phosphate buffer pH 6.8), Euparal.

10.18.3.2 Protocol

1. Pretreat with saturated aqueous solution of a α-bromonaphthalene for 30 min, fixation in 3:1 alcohol–acetic mixture and hydrolysis for 2–5 min at 60°C in 1 N HCl; squash in 45% acetic acid on gelatinized slides.
2. Remove the cover slips by freezing the slides in liquid nitrogen and wash in 90% absolute ethyl alcohol and then in absolute ethanol. Air dry and treat with saturated aqueous solution of barium hydroxide at 45°C–50°C for 6 min for alkali denaturation.
3. Wash in warm distilled water twice for 2 min each and then in running tap water for 30 min.
4. Incubate the slides in 2× SSC for 2 h at 60°C.
5. Stain in Giemsa solution for 1 h at room temperature.
6. Remove the excess stain by repeated washing in distilled water till a clear-differential staining of chromosomes is achieved.
7. Make the preparations permanent by mounting the air dried slides in Euparal.

10.18.4 Modified Technique for Giemsa Staining

1. Seed germination on moist filter paper in dark at 20°C. Roots from seedlings treated with 0.2% colchicines for 5 h and fixed for 4 h in 1:3 acetic acid–alcohol.
2. Hydrolysis in 1 N HCl at 60°C for 25 s followed by thorough washing in distilled water. Further softening by immersion in 45% acetic acid.
3. Root tip squashing in 45% acetic acid on albuminized solution of barium hydroxide for 5 min followed by thorough washing in running water.
4. Incubation at 60°C in 2× SSC solution for 1 h. Rinse in distilled water.
5. Cover slips in "improved" Giemsa R 66 at pH 6.8 for 5–10 min. Rinse in distilled water.
6. Cover slip mounted in Canada balsam after dipping three times in xylene.

10.18.5 Giemsa Staining Technique

1. Prefix the root tips in α-bromonaphthalene for 1–3 h and fix in glacial acetic acid.
2. Soften for 1–2 h in a 50% solution of pectinase (EC 3.2.1.15 polygalacturonase) and cellulase (EC 3.2.1.4) to which are added two to three drops of 1 N HCl for each 5 mL of enzyme solution pH 5.8.
3. Place the material on the slide and place a cover slip. Apply the pressure. Separate the cover slip by CO_2 freezing and immerse the slide in absolute alcohol for 2–3 h and then dry by air blowing.
4. Immerse the slide in a freshly prepared solution of barium hydroxide for 5 min. Wash in distilled water (three changes) for a total duration of 20 min.

5. After air dry, incubate in 0.30 M NaCl–0.03 M sodium citrate solution at 60°C for 1 h. Wash thoroughly in distilled water and again air dry.
6. Stain in Giemsa solution at pH 7 for 1–2 min. Wash quickly in water and air dry. Store in xylene overnight and mount in Canada balsam.

10.19 Fungal Cytogenetics

10.19.1 Principle

Neurospora is an ascomycetes fungus and is commonly called bread mold. The fungus is being widely used as a eukaryotic model to unravel the questions in eukaryotic genetics. It is a haploid eukaryote with seven chromosomes. It can be grown on a chemically defined medium and has a brief morphologically distinct asexual life cycle producing billions of conidia with identical genotype. Controlled genetic analysis is possible in the sexual phase of its life cycle using the mating-type alleles A and a. The fusion of nuclei of the two mating types results in the formation of a diploid nucleus that undergoes meiosis producing eight ascospores after mitotic division of each of the meiotic products (Figure 10.16).

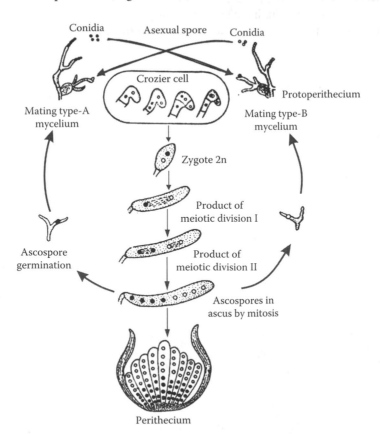

Figure 10.16 Diagrammatic representation of the life history of *Neurospora* sp.

10.20 Harvesting of Neurospora

10.20.1 Flamed-Loop Method

10.20.1.1 Materials
Stock culture, prepared and labeled agar slants, wire loops, Bunsen burner, sterile distilled water.

10.20.1.2 Protocol
1. Place the stock cultures and the prepared agar slants on your left and an empty rack on your right.
2. Hold one stock culture and one agar slant in your left hand. Loosen caps and plugs.
3. Using your right hand, flame the wire loop until red hot. With the fingers of your right hand, remove the caps or plugs and briefly flame the mouths of both test tubes (Figure 10.17).
   Dip the hot loop into the sterile water and carefully touch the wet loop to some of the conidia in the stock culture.
4. Lightly touch the wire loop to the prepared agar and smear the conidia over the surface. Do not break the surface.
5. Reflame the mouths of both the test tubes. Replug or recap.
6. Reflame the loop until red hot. Place both the test tubes in the rack on your right.

10.20.2 Spore Suspension Method

10.20.2.1 Materials
Stock culture, prepared and labeled agar slants, spatula or wire needle, 95% alcohol, Bunsen burner, sterile distilled water, 5 or 10 mL sterile pipettes, 5 mL sterile reverse pipettes (mouth plugged with cotton), sharp tipped tweezers, sterile test tubes or dropper bottles.

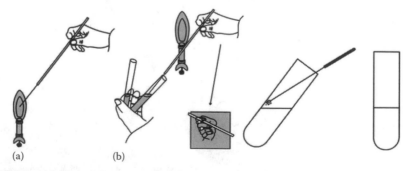

(a)   (b)

Figure 10.17   (a) Flame the inoculation loop. (b) Correct method for holding tubes before and after inoculation by transferring cultures.

10.20.2.2 Protocol

1. Place the tip of the spatula or wire needle and the tweezers into the alcohol for sterilization.
2. Pour a small amount of sterile water into the flasks (or test tubes) of stock cultures, wetting as much of the conidia as possible. Rotate the stock culture to wet the conidia.
3. Flame the spatula to burn off the alcohol and then cool it by immersing the tip in the water in the flask.
4. Use the spatula to break up the conidia, remove the spatula, and place in the alcohol.
5. Place the reverse or plugged end of the reverse pipette into the flask and withdraw the conidial suspension.
6. Remove the pipette from the flask.
7. Flame the tweezers and remove the plug from the pipette with the help of tweezers and place both in the alcohol.
8. Release the liquid suspension into a clean and sterile test tubes or dropper bottle.
9. Repeat the procedure for each culture (Figure 10.18).
10. Store in a cool place until needed for inoculation.
11. To inoculate, add one or two drops of well-shaken conidial suspension to prepared media.
12. Disperse the conidia by rotating the prepared media.
13. Keep the inoculated media (by spore suspension method or by flamed–loop method) in an incubator at 30°C–34°C until the desired amount of growth has been achieved.
14. Autoclave samples for 30 min at 20 psi to kill spores and mycelia when the experiment is completed.

[*Note: Methods of study of* Sordaria, Aspergillus, *and* Saccharomyces *(yeast) are similar to* Neurospora *experiments. They all belong to the same group of fungus, Ascomycetes, as* Neurospora *having similar life cycles.*]

1. Inoculating loop is heated until it is red hot.

2. With free hand, raise the lid of the Petri plates just enough to access a colon to pickup a loopful of organisms.

3. After flaming the mouth of a sterile slant, streak its surface.

4. Flame the mouth of the test tube and recap the tube.

5. Flame the inoculating loop and return it to receptacle.

6. Culture slants after incubation.

Figure 10.18   Demonstration of complete inoculation procedure.

10.21 To Study the Ordered (Tetrad) and Random Ascospore Analysis in *Neurospora* sp.

10.21.1 Principle

The sexual spores (ascospores) are produced from the fusion of two nuclei of opposite mating types through exchange of conidia between the two mating types. Fertilization is followed by immediate meiotic division, and the four products of meiosis undergo one mitotic division resulting in four pairs of ascospores that are linearly arranged in an ascus. The members of each pair are genetically identical and each pair represents one meiotic product. The position of ascospores in the ascus reflects the behavior of the chromosomes during meiosis. The asci are enclosed in a cuplike structure called the perithecium. The ripened ascospores escape through an opening called the ostiole. The formation of the eight ascospores is represented as

$$\text{Zygote} \rightarrow \text{meiosis} \rightarrow 4 \text{ products} \rightarrow \text{mitosis} \rightarrow 8 \text{ ascospores}$$

Ascospore analysis in *Neurospora* is of three distinct types: (1) ordered or linear tetrad analysis, (2) unordered or nonlinear tetrad analysis, and (3) random spore analysis. Ordered tetrad analysis is based upon the first and second division segregation of alleles as detected by the spore position in the ascus.

The first division segregation for any particular gene indicates no crossover between that gene and its respective centromere. This results in 4:4 ascospore patterns. If a gene for black color having an allele causing albino spore is present on two homologous chromosomes and there is no crossover at meiosis 1, the pattern of spore is top four black and bottom four white or top four white and bottom four black, depending upon the pole to which the chromosome with the colored allele passes at first anaphase. The chances for the two arrangements are the same, so the two cases occur with equal frequency (Figure 10.19a and b). The second division segregation reflects a crossover between a particular gene and its centromere, thus producing a variety of ascospore arrangements (Figure 10.19c through f). If a crossover between the locus of the spore color factor and the centromere occurs, both chromosomes have one chromatid with one allele and the other chromatid with the other allele. Depending upon the orientation at second anaphase, one of the four types is produced. All these types have equal probability.

For an ordered tetrad analysis, the perithecia (of 10–12-day-old cross) are broken open. The asci are separated, and the ascospores are carefully leased out of the ascus. Each ascospore, in order, is very carefully implanted on an agar block (placed in a tube), which is then placed at 60°C for 30 min (heat shocked). This initiates germination and colony formation.

Tetrad analysis reveals that crossing over takes place at the four strand stage of meiosis. It also has helped in determining the gene order on the chromosomes and the distribution of the homologous chromosomes in meiosis I and II.

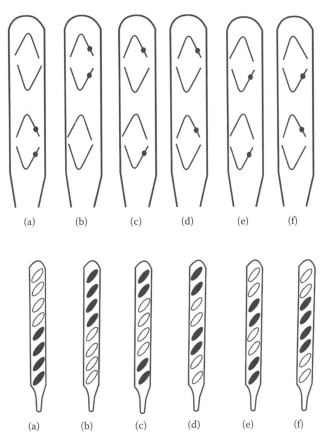

Figure 10.19 Linearly arranged second anaphase of meiosis with linearly arranged ascospores of *Neurospora*. (a and b) First division segregation. (c through f) Different arrangements of ascospores resulting from second division segregation of chromosomes.

Ordered and unordered tetrad analysis does not differ significantly. Mathematical analysis is necessary for unordered tetrad since the type of segregation for any given ascus cannot be determined. If tetrad analysis is not possible, random single ascospores are isolated for study. Perithecia of 1-month-old cross eject the ascospores from the asci onto the wall of the cross tube. Ascospores of many perithecia can be pooled, diluted, and plated on the appropriate medium and placed in separate tubes for growth and subsequent analysis. The class data can be pooled and analyzed statistically. Such analysis is used mainly for calculating the recombination values allowing the study of linkage relationships between genes. A two-factor cross or a three-factor cross is used for such linkage studies. This type of analysis has limited applications where genetic analysis is needed, which is possible in the case of an ordered tetrad analysis. Random spore analysis identifies only one of the four meiotic products whereas with ordered tetrad analysis, all four meiotic products are recognized.

Here, the cross Pan⁻ nic⁻ hist⁻ albino a × ad⁻ A of *Neurospora* is described for tetrad and random analysis. The arrangement of ascospores in an ascus reveals the occurrence of second division segregation, and this allows for the mapping of the centromeres. The distance separating a gene from its centromere is reflected by the frequency of the recombinants in the progeny. More recombinants mean the gene and its centromere are farther part; the closer a gene and its centromere, the fewer times a crossover will occur between the two and hence less number of recombinants in the progeny.

10.22 To Demonstrate the Tetrad and Random Analysis

10.22.1 Crossing of Two Mating Types

Strain a: Phenotype: albino
  Genotype: hist⁻nic⁻pan⁻
Strain A: Phenotype: purple-brown
  Genotype: ad⁻
  hist⁻: histidine requirement
  nic⁻: nicotinic acid requirement
  pan⁻: pantothenic acid requirement
  ad⁻: adenine requirement

10.22.1.1 Materials
Inoculating loops, absolute alcohol, sterilized water, crossing tube (25×150 mm) containing a strip of Whatman No. 1 filter paper (2.5×7.5 cm), one tube containing *Neurospora* strain.

10.22.1.2 Protocol
1. Heat the loop until red hot or dip in alcohol; then flame to sterilize.
2. Cool the loop (if it is heated); dip it briefly into the tube of sterile water. Remove the cap of the cross tube containing strain A.
3. Lightly brush the loop against the conidia of strain A and gently remove the loop and replace the cap.
4. Place the loop with the conidia into the cross tube and swirl the loop in the medium to disperse the conidia.
5. Flame the loop until red hot; dip it into alcohol and flame again.
6. Repeat steps 2 to 5 for strain a.
7. Store the crossing tube at 25°C.

10.23 Tetrad Analysis

After 12 days of the cross.

10.23.1 Materials

Dissecting microscope; wire loop; razor blade; glass slide; 4% agar in Petri dish; agar block 2.5 × 5 cm cut when needed from the 4% agar; 24 tubes each containing 2 mL of solidified Vogel's medium N plus 2% glucose plus supplements pan, nic, hist, and ad; cross tube of cross of section A.

10.23.2 Protocol

1. Remove a plump, dark, fertile perithecium and place it on an agar block. Add one or two drops of household diluted bleach to kill contaminant asexual conidia.
2. With the sharp needles, slice open the perithecium under the dissecting microscope. Perithecium can also be squeezed with the tips of forceps.
3. After squeezing, a bunch of asci, looking like a bunch of bananas, comes out. It normally has 10 asci with 8 spores inside.
4. Separate the asci carefully with needles, scissors, etc.
5. Three asci can be selected (one may select more) and transfer them (by picking with the tip of the needle) to a fresh agar block on the glass slide (Figure 10.20).
6. When the block is dry, squeeze the ascospores from the asci, one at a time, and move the ascospore to clear areas of the agar block. Using a sterile forceps or needle, pick up the agar under the ascospore and place into individual tubes with the medium. Label the tubes and keep them in order.
7. Heat shock the 24 cultures (3 asci × 8 spores = 24) in 60°C water bath for 30 min. Then place them at 25°C for 48 h so that enough conidia should be formed.

10.24 Pigmentation and Nutritional Requirement Analysis

10.24.1 Materials

One Petri plate each of the following special media:

Vogel's minimal + 1% sorbose + 0.2% sucrose + one of the following supplement combination:

hist, nic, ad; hist, nic, pan; hist, ad, pan; ad, pan, nic.

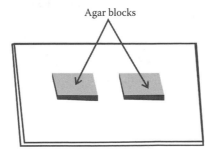

Figure 10.20 Agar block on a slide.

(Histidine hydrochloride 200 mg/L, adenine sulfate 200 mg/L, nicotinamide 20 mg/L, pantothenic acid 20 mg/L, adjust pH of all media to 5–6 and autoclave for 8 min).

24 separate cultures of Step 7 section B, sterile inoculating needle.

10.24.2 Protocol

1. Observe each of the 24 cultures for pigment formation.
2. The expected pigment types are albino, parental; purple-brown, parental; albino-purple, recombinant; and wild, recombinant; wild-type pigment is orange. Orange + purple = brown color; white + purple = albino-purple color.

    A polymer results in the form of Table 10.9.

    Place Petri plates of special media on the corresponding places shown in Figure 10.21.

TABLE 10.9 Tetrad Analysis (Pigment and Nutritional Requirement)

Ascus	Spore Position	Pigment	Growth of Colony on				Genotype
			ad	pan	hist	nic	
1	1						
	2						
	3						
	4						
	5						
	6						
	7						
	8						
2	1						
	2						
	3						
	4						
	5						
	6						
	7						
	8						
3	1						
	2						
	3						
	4						
	5						
	6						
	7						
	8						

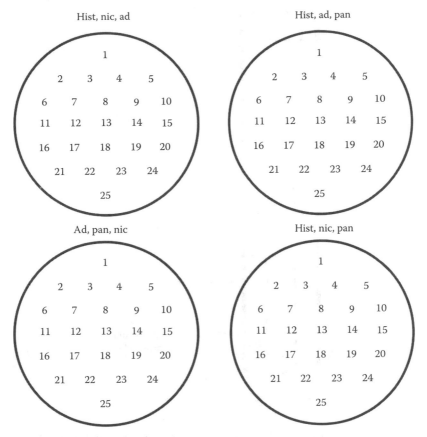

Figure 10.21 Plate pattern for the analysis of nutritional requirements.

Take tube number 1 from 24 tubes and place conidia from this tube on spot number 1 of all the four plates. Repeat this for the rest of the 23 tubes placing conidia on corresponding points.

3. Incubate plates for 48 h at 30°C. Score growth as positive (+) or negative (−) and record data in the form of Table 10.9.

10.25 Random Spore Analysis

Should be done 1 month after the cross.

10.25.1 Materials

Agar block on slides as shown in Figure 10.20.

Sterile tube containing 1 mL of sterile water, crossing tube (25 × 150 mm) containing the original cross of part A, with a black dust on the walls. These black spores are the ascospores ejected out of the asci on the wall of the tube (Figure 10.22). Twenty-five tubes (label 1 to 25) containing solidified Vogel's + 2% glucose medium + hist, nic, ad, pan.

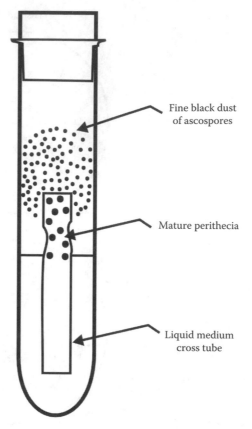

Figure 10.22   The cross tube as it looks 3–4 weeks after the cross is made.

10.25.2 Protocol

1. Flame the loop until red hot and then dip it into diluted household bleach to cool. Place one drop of household bleach onto the agar block.
2. Using the wet loop, scrap some ascospores from the tube wall and spread them in the drop on the agar block. Do not dig into the agar.
3. Allow the agar to dry up and touch the spores with a sterile picking needle.
4. Place a single spore carefully into one of the 25 tubes with agar medium.
5. Repeat Steps 3 and 4 for the rest of the 24 tubes.
6. Heat shock all the tubes in a 60°C water bath for 30 min.
7. Place tubes at 25°C for 1 week.

10.25.3 Scoring Tubes for Pigmentation

Observe the tubes of Step 7 of the preceding section and record the observations in the form of Table 10.10.

TABLE 10.10   Random Spore—Pigmentation Analysis

Color	Number
Albino	
Purple-brown	
Albino-purple	
Wild-type	
No growth	
Total	

10.26 Test for Nutritional Requirement

10.26.1 Materials

One plate each of the four special media listed in Section 10.27.

10.26.2 Protocol

1. Place the first Petri plate on top of Figure 10.21.
2. With a sterile and cooled needle, pick a few conidia from tube 1 and place them onto the agar on spot 1. Small amount of conidia should be placed.
3. Repeat Steps 1 and 2 for the other three Petri plates.
4. Repeat Steps 1, 2, and 3 for the rest of the 24 tubes inoculating all corresponding spots in the four Petri plates. Needle should be sterilized each time it is placed in the tube.
5. Leave the four plates at 30°C for 2 days.

10.26.3 Scoring for Nutritional Requirement

1. Observe the four plates for growth and record the observations in the form of Table 10.11.
2. Write symbol + (positive) for a colony and – (negative) for no growth or poor growth.

10.26.4 Analysis

Tetrad analysis from Table 10.9.

1. Are asci 1, 2, and 3
   a. A tetratype (four genotypes present)?
   b. A parental ditype (two genotypes, both parental)?
   c. Nonparental ditype (two genotypes, both recombinants)?
   Determine the nature of each ascus based upon the previous three types.

TABLE 10.11   Random Score Analysis (Nutritional Requirement)

Strain	Pigment	Growth of Colonies on ad	pan	hist	nic	Genotype
1						
2						
3						
4						
5						
6						
7						
⋮						
25						

2. Are the parental ditypes equal to the number of nonparental ditypes?
3. Make the diagrams for the chromatid exchanges that must have occurred to result in the type of asci obtained in this experiment.
4. If any of the ascospores failed to germinate, infer their genotypes.
5. Did ascus 1 or 2 or 3 show second division segregation for any of the genetic markers that are considered in this cross? Calculate the centromere distance for each marker in the cross using the following formula:

$$\% \text{Crossover} = \frac{\text{Second-division segregation frequency}}{2} \times 100$$

$$1\% = 1 \text{ map unit}$$

Random spore analysis from Table 10.11.
Try to find out the following:

1. Number of genotypic classes present in the asci.
2. Number of each genotype present.
3. Number of genotypic classes missing.
4. Genotypes of each class missing.
5. The most frequent and the rarest genotype occurring in the asci.
6. Since the original cross was pan⁻ nic⁻ hist⁻ al a × ad⁻ A, taking genes pair wise, determine the recombination frequency between them (e.g., for nic and hist, the parental types are nic, hist ad⁺⁺). The recombination frequency can be calculated as

$$\% \text{ Recombinants} = \frac{(\text{Number of nic}^+)+(\text{Number of hist}^+)}{\text{Total number of ascospores}} \times 100$$

Percentage recombination can be calculated similarly for other gene pairs. Enter your data in the form of Table 10.12.

TABLE 10.12   Linkage (Percentage Crossing Over between)

	pan	nic	hist	ad
pan				
nic				
hist				
ad				

7. Find out the genes showing maximum segregation (do not show linkage).
8. Also find out the genes that move together as a unit (show linkage).

10.27 Suggested Media for *Neurospora*, *Sordaria*, and *Aspergillus* Experiments

10.27.1 Vogel's Medium

Trace elements solution: Distilled water 95 mL, citric acid·H_2O 5 g, $ZnSO_4$·$7H_2O$ 5 g, $Fe(NH_4)_2(SO_4)_2$·$6H_2O$ 1 g, $CuSO_4$·$5H_2O$ 0.25 g, $MnSO_4$·H_2O 0.05 g, H_3BO_3 0.05 g, Na_2MoO_4·$2H_2O$ 0.05 g, preservative (chloroform) 0.5 mL.

10.27.2 Biotin Solution

Distilled water 500 mL, 95% absolute ethyl alcohol 500 mL, biotin 10 mg.

10.27.3 Medium N Stock Solution (50X)

Sodium citrate·$2H_2O$ 125 g, KH_2PO_4 anhydrous 250 g, NH_4NO_3 anhydrous 100 g, $MgSO_4$·$2H_2O$ 10 g, $CaCl_2$·$2H_2O$ 5 g, trace element solution 5 mL, biotin solution (0.1 mg/mL) 2.5 mL.

Add the ingredients successively. Each should dissolve before the next is added. Add 3 mL of chloroform as a preservative.

10.27.4 Medium N Final Solution

To prepare 1 L of medium N.

Medium N stock solution 20 mL, sucrose 20 g, Difco bacto agar 20 g, distilled water 980 mL. Autoclave the solution for 15 min at 20 psi. Cool to 50°C, pour into test tubes, and prepare slants. Allow the agar to harden.

10.27.5 Wastegaard–Mitchell Crossing Medium

Trace element solution

$Na_2B_4O_7 \cdot 10H_2O$ 88.4 mg, $CuCl_2 \cdot 2H_2O$ 268 mg, $FeCl_3 \cdot 6H_2O$ 970 mg, $MnCl_2 \cdot 4H_2O$ 72 mg, $(NH_4)_6 \cdot 6Mo_7O_{24} \cdot 4H_2O$ 37 mg, $ZnCl_2$ 42 mg, distilled water to 1 L.

10.27.6 Biotin Solution

95% absolute ethyl alcohol 500 mL, distilled water 500 mL, biotin 10 mg.
   Dissolve the biotin in alcohol, and then add water.

10.27.7 Wastegaard–Mitchell Crossing Medium Stock Solution (20X)

KNO_3 20 g, KH_2PO_4 20 g, $MgSO_4 \cdot 7H_2O$ 20 g, NaCl 2 g, $CaCl_2$ 2 g, trace element solution 20 mL, biotin solution 10 mL, distilled water 1 L.
   The components are added successively with constant stirring. Do not autoclave or heat. To preserve, add 2 mL of chloroform.

10.27.8 Crossing Medium Final Solution

To prepare 1 L of the crossing medium, use the following:
   Distilled water 950 mL, stock solution 50 mL, sucrose 15 g, NaOH enough to adjust pH to 6.5, Difco bacto agar 12 g.
   Follow the same procedure as given for the preparation of 1 L of medium N.

10.28 Study of Heterokaryosis in *Neurospora* from the Nutritionally Deficient Mutants

10.28.1 Principle

Classically, genes are studied, using three methods: recombination, mutation, and analysis of gene products. In this experiment, an attempt is made to explain what a gene is by utilizing a technique for gene analysis called heterokaryosis or complementation.

Heterokaryosis is an act of complementation or the act of coming together to make whole or complete. In genetics, this term is used when two mutant genomes are brought together in a common cytoplasm, allowing for growth or an increase in a specific gene function over that of either mutant type. Thus, the complementary action of homologous sites of genetic material results in wild or near-wild phenotype.

The two genomes may come together in the same cell or cytoplasm by several means. These are the fusion of germ cells or gametes and the formation of dikaryons in fungi.

The terms complementation and recombination are often confused. Recombination is a physical exchange of genetic material between homologous chromosomes resulting in the recombination of genes and physical establishment of gene order. The recombinant individual has the total amount of genetic content equal to the original content, the quality of genetic content being changed by the act of recombination. Complementation is the expression of a wild or near-wild phenotype by an organism containing two mutant genomes. It can be used to determine the order of the mutational sites within a gene and is based on functional rather than recombinational tests. In complementation tests, two → different mutant strains of the same mating type were when jointly inoculated onto to minimal medium, their hyphae would fuse (but not their nuclei), and the mixing of the cytoplasm takes place. Such a mixed strain is called a heterokaryon (hetero = more than one, karyon = nucleus). If two different nonallelic nutritionally requiring auxotrophic mutants, ad^- (adenine requiring) and pro^- (proline requiring), form a heterokaryon. It is described as ad^- pro^+, ad^+ pro^-, because the genome that is ad^- carries the wild-type pro^+ allele and the genome pro^- carries the wild-type ad^+ allele. This heterokaryon is capable of growing on a minimal medium because the nuclei that are ad^+ synthesize adenine and the nuclei that are pro^+ synthesize proline. This way, the mutant nuclei are compensated for in a common cytoplasm, and the complementation of the nuclei results in a near-wild growth. Nonallelic mutants that code for specific enzymes necessary in a biochemical pathway also show complementation. For example, two mutant strains of *Neurospora* requiring riboflavin (rib-1^- and rib-2^-) form heterokaryon, which grows on minimal medium, since each of the two strains has a different enzyme block in the pathway for the production of riboflavin

$$A \rightarrow B \rightarrow C \rightarrow \text{riboflavin}$$

$$rib\text{-}1 \qquad rib\text{-}2$$

The definition of the gene as the smallest unit of mutation, recombination, and function is changed as a result of data obtained from intergenic complementation studies.

(intragenic—within a gene, interallelic—between alleles, intergenic—between different genes)

"Cistron" is now used to describe a functional genetic unit and is formed from the words *cis* and *trans*. In the *cis* configuration, both mutations are positioned on the same chromosome and are obtained as a recombination product from a cross between two mutants. The alternate *trans* configuration contains both mutations but one on each homologue of the diploid.

A_____B	A_____b
A_____b	a_____B
Cis position	*trans* position

In this diagram, a and b are mutant alleles of genes A and B.

10.28.2 Complementation between Genes or Nonallelic Mutants

Since separate genes are responsible for the coding of separate enzymes or polypeptides, when two genomes come together in a common cytoplasm, the wild-type allele of one gene compensates for its homologous mutant gene. A second form of complementation results when each gene is responsible for the coding of a specific polypeptide of one enzyme and both polypeptides interact or polymerize in the cytoplasm to form an enzyme or other protein (hemoglobin). Thus, in either case, each mutant can produce at least one normal product, and a form of complementation occurs when the two genomes come together. The phenomenon of "allelic complementation" (interallelic) has been demonstrated for a number of loci in *Neurospora*.

In short, the act of coming together of two mutants and compensating for the mutant genes of each other is described as complementation, and the formation of hyphae with two different nuclei is known as heterokaryosis.

To obtain stable heterokaryons, it is necessary to inoculate the two strains together on a medium that will permit neither strain to grow by itself. Thus, if one strain requires riboflavin and the other lysin for growth and the two are inoculated together on minimal medium, neither strain can grow alone, but the two can cooperate and grow together since each can synthesize the compound required by the other. Heterokaryons produced in this way are called "balanced" or "forced" heterokaryons.

If conidia from a "balanced" heterokaryon between two auxotrophs (nutrition-requiring mutants) are harvested and plated on minimal medium, no colonies will appear because the conidia are uninucleate and the heterokaryon is dissociated into its two components. The synthesis of new heterokaryons is possible if harvested conidia are plated together on minimal medium in high density. Sometimes a diploid colony, heterozygous for the nutritional characters of the two parents, appears and hence is prototrophic.

10.28.3 Materials

One Petri dish containing complete medium, eight Petri dishes containing minimal medium, long-tipped pipettes, container for used pipettes, conidial suspension of two mutants of the hits$^-$ nic$^-$ pan$^-$ = phenotype-albino = A, ad$^-$ = phenotype-purple-brown = B, 37°C incubator.

10.28.4 Protocol

1. Place the rack of conidial suspension on your left and another empty rack on your right.
2. Using your right hand, grasp a sterile pipette near its mouth (wide end).
3. Using your left hand, pick up the test tube containing suspension A.
4. Remove the cap of the test tube by grasping it between the palm and little finger. Do not place the cap on the desk; it may lead to contamination.

5. Using the pipette, remove a small portion of the conidial suspension. Recap the test tube and place the test tube in the container on your right.

6. Using your left hand, remove the top of the Petri dish with complete medium long enough for the inoculation.

7. Inoculate the Petri dish at different spots. Place the used pipette in the wash container.

8. Select a fresh sterile pipette and repeat Steps 3 to 7 for suspension B.

9. After the liquid from the suspension has dried into the agar, carefully place the plates in a 37°C incubator for 24–48 h.

10. After a sufficient incubation period, round patches of mycelia growth and tufts of conidia appear. Place the plates in a cold room.

11. Transfer small pieces of mycelial mat from the sections showing mixed conidia (white and purple) to plates containing minimal medium at various spots.

12. Incubate the plates at 37°C for 72 h.

13. Select the mycelial sectors showing thick growth, with mixed conidia of both "types of pigmentation" and transfer these to fresh minimal medium plates.

14. Incubate the plates at 37°C for 48 h.

15. Fast-growing mycelia represent heterokaryons.

10.29 General Information and Methods for *Drosophila* Experiments

10.29.1 Introduction

Drosophila means "lover of dew" and *melanogaster* means "black belly." *Drosophila melanogaster* was introduced by T.H. Morgan and W.E. Castle for genetic studies. Since then, it has become a favorite laboratory animal. It is commonly called fruit fly or vinegar fly because it is fond of wineries, being attracted by the smell of fermentation.

The fly is well suited for laboratory studies because it has a short life cycle, is easily and inexpensively raised and handled, and produces much progeny from a single mating. Its giant polytene chromosomes from salivary glands are usefully employed in cytological studies.

The life history consists of four main stages (Figure 10.23).

10.29.2 Life Cycle

(1) Egg (0.5 mm) is white, oblong with a pair of anterior filaments and (2) larva (4.5 mm), also called "maggot," is very active, burrowing into food and a voracious eater. Since the skin does not stretch, a young larva periodically sheds its skin (molting). Like this, there are three larval stages. They are called instars. (3) Pupa (0.3 mm) crawls out of the medium when the larva pupates (after it stops eating). (4) Adult (2 mm) is formed after the metamorphosis of the pupa.

The average life of an adult is 1 month. The female begins to lay eggs about the third day of adult life and continues until she dies. The time from the laying of the

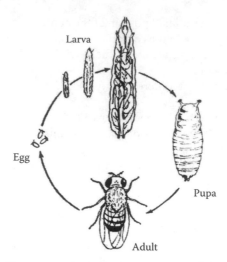

Figure 10.23   Life cycle illustrating the different stages in the life history of *D. melanogaster.*

eggs to the emergence of the adult is about 10 days under good culture conditions and optimal temperature (about 25°C).

10.29.3 Sex Differences

(1) Female is generally larger than the male. (2) Female abdomen curves to a point; the male abdomen is round and much shorter. (3) The last few segments of the male are fused. Alternating dark and light dorsal bands can be seen on the entire rear portion of the female. (4) On males, there is a tiny tuft of hairs on the basal tarsal segment of each leg. This is called sex comb. (5) The ovipositor of the female is located at the tip of the abdomen. (6) During the larval stage, males have a large, white mass of testicular tissue located at the beginning of the posterior third of the larva in the lateral fat bodies. The ovarian tissue of the female is a much smaller mass (Figure 10.24).

10.29.4 Hereditary Traits

Although thousands of mutations in *Drosophila* are known, we will consider a few here.

10.29.4.1 Body Color

Wild type—basically gray, with the pattern of light and dark areas. Mutants—black, yellow.

10.29.4.2 Wings

Wild type—smooth edges, uniform venation, extend beyond the abdomen. Mutants—changes in size and shape, absence of specific veins, change in position of wings.

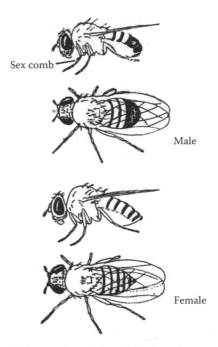

Figure 10.24 Diagram showing male and female *Drosophila*.

10.29.4.3 Bristles
Wild type—fairly long and smooth. Mutant—shortened, thickened, or deformed bristles.

10.29.4.4 Eyes
Wild type—red, oval, and many faceted in shape. Mutants—white, black, apricot, scarlet red, pink, brown, changes in shape and number of facets.

All mutant traits are recessive to the wild type.

10.29.5 Etherization

To experiment with flies, it is necessary to anesthetize them with a light dose of ether. This is called etherization. The procedure is as follows:

Dose the cotton pad that is stapled to the cork of the etherizing bottle with a few drops of ether. Tap the culture bottle lightly on a pad of paper until all the flies have been shaken down away from the mouth of the bottle. Quickly remove the cotton plug from the culture bottle and in its place, insert the mouth of the etherizing bottle. Reverse the position of the bottles so that the etherizing bottle is now at the bottom. Holding the two jars tightly together with one hand, tap the side of the culture bottle with the other hand so that the flies fall into the etherizer. Quickly separate the two bottles and replace the cork on the etherizing bottle. The flies will be anesthetized within 2 min and should be dumped on a white index card. They remain anesthetized for 5–l0 min. Sorting out or counting should be done with a soft camel's hair brush. Dissecting microscope should be used for examination.

10.29.6 How to Get Virgin Female Flies

Males rarely mate with females as early as 8–12 h after emergence. Therefore, if all adult flies are taken out from the culture bottle and the bottle left for 8 h or so, all females removed the second time are virgins.

Keep the culture bottle on its side and carefully slide the etherized flies from the index card to the glass surface. Do not keep the bottle upright until the flies have revived, because etherized flies get permanently stuck to the moist medium and die, so do not let them drop on to the surface of the food.

10.29.7 Study of Random Mating in Population of Drosophila, Heterozygous for a Single Gene Pair

10.29.7.1 Introduction

In 1857, Mendel, an Austrian monk and a teacher of physics, began experiments on heredity using the common garden pea as the experimental material. Work done by Mendel is truly amazing. His article (Experiments in Plant Hybridization) is a work of diligence and clear thinking and the spirit of innovation manifested in several related fields. In all his initial experiments, he crossed plants that differed in only one characteristic and were quite identical in all the others. These pairs of contrasting characteristics were size of the plant, shape and color of the pod, shape and color of fruit, and the color and location of the flowers: pea plant is a self-pollinated plant, and the chance of cross-pollination is completely ruled out with it. For each of the seven traits, Mendel observed an apparent loss of one of the variant types in the next generation. For example, Mendel had crossed varieties with smooth and wrinkled seeds; their progeny were all smooth (called F 1 or filial 1).

Mendel sowed the hybrid seeds and did nothing more and let them pollinate themselves. Harvesting time came, and the results were amazing. Whereas all the plants in the F_1 had been quite uniform, those of the second generation (called F_2) were found to be varied. Both parental types appeared in F_2, and a consistent ratio of 3 smooth seeded to one wrinkle seeded was demonstrated. This 3:1 ratio is often referred to as a monohybrid ratio. "Hybrid" implies heterozygosity here. For each of the seven characteristics, tested same results were obtained. The results of Mendel's experiments can be summed up:

1. For any crossing between two variations of a trait, only one of these variations appears in F_1, and this is called the dominant characteristic and one that does not appear is called the recessive characteristic.
2. The recessive characteristic in F_1 reappears in F_2 and comprises one-fourth of the F_2 population.
3. The results are not affected by which parent variant produced the pollen or the ova.

Mendel's findings suggested that the factors that determine traits (called genes now) were transmitted between generations in a uniform and predictable manner. From this, Mendel postulated the following:

1. Each gene could exist in two alternative states (now called alleles) that are responsible for a given phenotypic character.
2. Plants that produce only one phenotype (dominant or recessive) for several generations of self-fertilization are considered to be homozygous or pure breeding; those producing two phenotypes upon self-fertilization ($F_1 \times F_2$) are heterozygous or hybrids. The diploid generations produce gametes. These are haploid cells produced by a meiotic division of a diploid cell. Gametes are the link between two diploid generations. The homozygous individual produces only one type of gamete, but the heterozygous produces two types of gametes (we are taking only one pair of allele here).

	Homozygous diploid cell	Heterozygous diploid cell
	AA or aa	Aa
Gametes	A or a	A and a

The number of different types of gametes can be calculated using the formula

$$2^n = \text{Number of different types of gametes,}$$

where n is the number of heterozygous pairs of alleles.

Based upon his results on hybridization of peas, Mendel formulated his first law, the law of segregation that states the following.

Hereditary factors are present in the individual in pairs. Each gamete produced carries only one "factor" of each pair (both cannot pass into a gamete). These factors may be alike or different, but even when different, they segregate unchanged by their association with each other to give pure gametes. They do not lose their identity, and there is no blending. Mendel continued his experiments by crossing parents differing in two characteristics or in other words considering two pairs of contrasting characters. For example, he crossed a variety having round yellow seeds with one whose seeds were wrinkled and green. All seeds in the F_1 were round and yellow. When the F_1 was self-fertilized, all four possible phenotypes were observed in F_2 9/16 of the total F_2 population expressed both dominant traits (yellow and round), 3/16 were yellow wrinkled, 3/16 were round green, and 1/16 were green wrinkled (both recessive). This 9:3:3:1 ratio is called the dihybrid ratio, which is defined as being heterozygous with respect to two pairs of alleles.

If all the yellow-seeded peas in F_2 are combined 0001 and all the green seeded are combined 0001, it gives a ratio 0001 = 3:1.

From these ratios, Mendel concluded that gene pairs behave independent of each other and are not changed during their transmission to the next generation. This was Mendel's second law of inheritance—the law of independent assortment.

The traits (now called genes), which Mendel chose for his experiments, are located on different chromosomes, and the chromosomes segregate independent of each other during meiosis. These genes, which are present on different chromosomes, are nonlinked. There are thousands of genes present on each chromosome. Genes present on the same chromosome are called linked genes, and they are inherited together as one linkage group. These genes do not follow Mendel's second law of independent assortment. Linked genes can be recognized by a decrease in the number of phenotypic classes observed. More often, genes on the same chromosome undergo crossing over, and then four phenotypic classes are observed in F_2 in a dihybrid class. The number of parental types would be much more in number than the number of recombinants, which are the results of a physical crossing over occurring between homologous chromosomes at the four-strand stage of meiosis.

The frequency of crossover between two genes depends upon the distance between them on the chromosome. The closer the two genes, the less chance there will be a crossover occurring between them. Percent crossover is calculated using the following formula:

$$\% \text{ Crossing over} = \frac{\text{Total number of recombinants observed}}{\text{Total number of offsprings}} \times 100$$

1% crossover = 1 map unit distance between the two genes

The effect of a gene is not always expressed. A dominant allele can inhibit the expression of a gene at a separate and distinct location. This type of interaction is called epistasis. Suppose gene A makes the effect of gene B, then A is said to be epistatic to B. Epistasis modifies the normal dihybrid ratio. Instead of 9:3:3:1, it may be 12:3:1 or 9:3:4 or 13:3.

To test for dominance or recessiveness where epistatic phenomenon or linkage is suspected, the questionable genotype is crossed with a known homozygous recessive individual. This is called a test cross. In a test cross, the offspring should be in the ratio of 1:1 (two parental types) if the genes are nonlinked or there is no epistasis phenomenon.

Nonlinked

$$AaBb + aabb$$

$$(F_1 \text{ of AABB} \times aabb) \text{ (homozygous recessive parent)}$$

$$1 \text{ A}aBb: 1 \text{ A}abb: 1 \text{ } aaBb: 1 \text{ } aabb$$

Linked

$$\frac{Ab}{aB} \times \frac{ab}{ab}$$

$$1\frac{ab}{Ab} : 1\frac{ab}{aB}$$

Further, it was discovered that sex was determined by chromosomes, and in certain insects (protenor), there is an XX–XO method of sex determination. The females in this group have 14 chromosomes, 6 pairs of autosomes, and 2 X, the sex chromosomes, whereas the males have 13 chromosomes, 6 pairs of autosomes, and 1 sex or X-chromosome.

Males	Females
12+X	12+XX

Gametes

6+X	6+X
6+0	6+X

Sex of the progeny depends upon which type of sperm (6+X or 6+0) fertilizes the egg:

$$(6+X) \times (6+X) - 12+XX$$

Egg   Sperm Female

$$(6+X) \times (6+0) - 12+XO$$

Egg   Sperm Male

In man, the sex determination is of a different type, that is, XX–XY type. The chromosome numbers in male and female are as follows:

Males

44 autosomes + XY (sex chromosomes)

Females

44 autosomes + XX (sex chromosomes)

Gametes of Males	Gametes of Females
Sperms (22+X)	Eggs (22+X)
(22+Y)	(22+X)

Fertilization

(22+X) x	(22+X)	44+XX
Egg	Sperm	Female
(22+X) x	(22+Y)	44+XY
Egg	Sperm	Male

Drosophila has four pairs of chromosomes, namely, three pairs of autosomes and one pair of sex chromosome, XY in males and XX in females. The ratio of X-chromosomes to the autosomes determines maleness or femaleness. The Y-chromosome is necessary for fertility but does not determine the sex of the fly whereas in man it is the sex determiner. In some animals like birds, the males have XX situation and the females XY.

Genes that are present on the X-chromosome but do not influence sex determination are called sex-linked genes. The eye color gene in *Drosophila* is sex linked, that is, present on the X-chromosome. Therefore, the eye color is not transmitted from male to male (in *Drosophila* only) but male to F_2 males through F_1 females, never directly to F_1 males.

In the present experiment, cross two mutant stocks of *D. melanogaster* and observe the offsprings of each generation. After classifying and counting the number of each of the various phenotypes, construct a hypothesis to explain the results and try to answer the following:

1. Are the traits we are observing present on autosome or sex linked?
2. If autosomal, do they show linkage to each other or independent assortment?
3. If X-linked, what is the pattern of transmission?

While analyzing the results, consider the following:

1. Mendel's law of segregation
2. Dominant and recessive alleles
3. Autosomal and X-linkage
4. Crossing over
5. Epistasis

10.29.7.2 Materials

Two vials (cross and its reciprocal) containing adult *D. melanogaster* that are mutant and pure breading for certain genes and are mated 10 days before this experiment, vial of wild-type *D. melanogaster*, camel's hair brush, 21 teasing probes, etherizer, morgue, blank labels, dissecting microscope, and white index cards.

10.29.7.3 Observation and Destruction of Parents (P_1)

10.29.7.3.1 Protocol

1. Etherize the vial of wild-type flies and determine the differences between the sex.
2. Etherize one vial of mutant flies at a time and compare the mutant traits to the wild-type traits.
3. Record the genotype and phenotype of the cross and its reciprocal separately in the form of a data sheet.
4. Destroy all mutant mes.
5. Place the vials at 20°C–25°C, out of direct sunlight.

10.29.7.4 Observation and Crossing of First Filial Generation (F_1)

10.29.7.4.1 Materials.   Two ½ pint jars containing *Drosophila* medium F_1 vials

10.29.7.4.2 Protocol

1. Etherize one vial containing F_1 and compare traits and record phenotypes of F_1s in the form of data sheet 1.

2. Place six males and six females in a ½ pint bottle.
3. Label the bottle with the original genotype of the parents (P_1).
4. Repeat Steps 1 to 3 with the other vial of mutants.
5. Place the jars in a cool place away from sunlight. After about 10 days, F_2 will appear.

10.29.7.4.3 Observation of Filial Generation (F_2). At least 150 F_2 flies of each cross should be tabulated. Flies that appear 2 weeks after the appearance of the first F_2 should not be counted. These may represent the F_3 generation.

10.29.7.4.4 Protocol
1. Etherize the flies of one bottle and identify and record the various F_2 phenotypes in a data sheet.
2. Repeat Step 1 for the second bottle.
3. Destroy the flies that are counted.
4. Repeat this procedure until 150 flies of each cross have been recorded and tabulate the results.

10.29.7.5 Tabulation of Results
1. Compare the totals of the phenotypic classes in F_2.
2. Explain the results of your crosses based upon Mendel's laws of inheritance.
3. Subject your results to statistical analysis using the chi-square method.
4. If results do not support your hypothesis, try to get a new explanation.

Data Sheet 1

Parental Cross (P_1)

	Number	Phenotype	Genotype
Males			
Females			

First Filial Generation (F_1)

	Number	Phenotype	Genotype
Males			
Females			

Second Filial Generation (F_2)

	Number	Phenotype	Genotype
Males			
Females			

10.30 Study of Salivary Gland Chromosomes of *Drosophila melanogaster* for Nature of Polyteny and Regions of Chromosome Puffs

10.30.1 Introduction

Chromosomes composed of many strands are called polytene. These chromosomes are produced by a process called endomitosis. The cell grows larger, chromonemata replicate many times without separation, and there is no cell division. The newly formed chromonemata, along with the old ones, behave as a unit. The two homologues of each chromosome pair remain permanently synapsed and are considered to be in perpetual interphase. It is easy to recognize deletions, inversions, and translocations because of the synapsed nature of the homologous chromosomes.

Polytene chromosomes have localized increased diameter, and these expanded portions are called "puffs" or Balbiani rings (Figure 10.25). They appear at definite physiological stages of development and involve only a single band of 300–400 cross bands. The cross bands are the heterochromatic regions produced by the linear arrangement of chromomeres of individual chromatids. These banding patterns are identical for particular chromosomes in different individuals of *D. melanogaster*. Cytological chromosome maps are constructed based on this consistency of banding patterns. Bands are more deeply stained than the interband regions with Feulgen

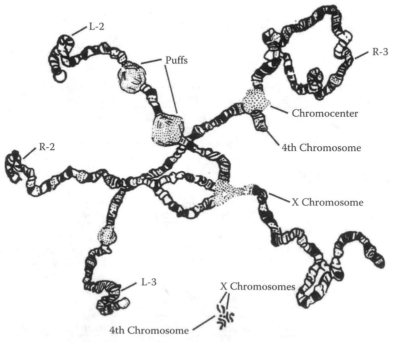

Figure 10.25   Polytene chromosomes of *D. melanogaster* exhibiting "puffed" regions (R-1, R-2, R-3, right arms of chromosomes 1, 2, 3, respectively, and L-1, L-2, L-3, left arms of chromosomes 1, 2, 3, respectively).

stain. Puffs are caused by the uncoiling of the chromatids in a particular band and indicate genetic activity. The size of the puff relates to the amount of RNA produced by it. Uncoiling in the puffed region exposes more and more DNA strands for transcription by RNA polymerase into messenger RNA.

Polytene chromosomes are an excellent material to study chromosome structure and gene action. A simple chromosome squash preparation demonstrates the consistency of the banding patterns and puffs (Figure 10.25).

10.30.2 Materials

Larva of *D. melanogaster* at the third instar stage of development; slides; cover slips; dissecting needles; paper towels; dissection microscope; insect saline solution (Ringer's in a dropper bottle—Ringer's saline solution consists of 0.82% NaCl, 0.02% KCl, 0.02% $CaCl_2 \cdot 2H_2O$ in double distilled water, mixed in equal proportions); acetic carmine stain 2% or acetic orcein stain 2%; five Coplin jars, two filled with 100% ethyl alcohol, one filled with 1:1 ethanol–xylene, and two filled with xylene.

10.30.3 Protocol

1. Transfer full grown (instar-3) larva on a slide in a drop of Ringer's saline solution.
2. Under the dissecting microscope, perform the following:
   a. Insert one dissecting needle through the middle of the larva to hold it in place.
   b. Place the other needle just behind the black mouth parts and gently pull the mouth parts forward. The salivary glands, which are attached to the mouth parts, should be pulled out. Salivary glands appear as two long, sausage-shaped bags with a characteristic fat body along one side.
   c. Detach the salivary glands from the mouth part and fat bodies.
3. Place a drop of 2% acetic carmine or acetic orcein stain on a slide and transfer the salivary glands onto the stain. Keep in the stain for 8–15 min.
4. Place a cover slip over the glands and place the slide in a fold made of paper towelling or a filter paper covering the slide. Press firmly and directly downward with your thumb.
5. Examine the slide under the microscope (low power). If necessary, for better spreading of the chromosomes, squashed glands can be gently tapped with the blunt end of a pencil.
6. Draw the chromosomes under high power of the compound microscope.

Slides can be preserved for several days by simply applying melted paraffin along the edges of the cover slip. If the slides are to be preserved for an indefinite period, the following procedure can be adopted:

1. Place the slide on dry ice until the slide and the cover slip are frosted over and using the needle, flick the cover slip from the slide.
2. Place the slide, cover slip (if cover slip has some material on it, otherwise process only the slide) in two changes of 100% ethanol for 3 min each.

3. Transfer to ethanol–xylene (1:1) mixture for 3 min.
4. Place the slide into two changes of xylene for 3 min each.
5. Mount in Euparal (mount the cover slip separately). The other method, that is, *n*-butyl-acetic acid making slides permanent, can also be used.

10.31 Human Genetics

10.31.1 Introduction

Laws of heredity are universal in character and what is valid for a pea, or *Drosophila*, is fully valid for man. We have often heard people say that such and such person has blue eyes like his mother or has the same blood group as his father. These are some of the characteristics that are inherited generation after generation by the progeny. It is not difficult, however, to give an example illustrating a very common characteristic but of extremely great consequence. Sickle cell anemia is a severe disease in which red blood cells (RBC) assume a sickle shape in low O_2 concentrations. In certain individuals, this causes a severe sickness resulting in death in infancy. In others, the disease is innocuous, and they feel normal but suffer from severe anemia and pains in the joints and muscles. It is a hereditary disease and is associated with a change in a single gene, and the changed gene (mutated) is transmitted from one generation to another strictly according to Mendel's first law of inheritance. Homozygotes with the sickling gene develop severe disease, and in the heterozygotes, the shape of the red blood corpuscles is affected with not very severe effects. This disease is common in the tropics. The biochemical basis of sickle cell anemia is the changed properties of hemoglobin, a protein present in the blood corpuscles. An amino acid glutamic acid present in normal hemoglobin is replaced by valine in the changed gene. Such a small single change leads to such dramatic consequences.

If there is a small cut in a water pipe, the whole water runs out of it. A cut on our skin, which results in the cut in the blood capillaries, results in bleeding, but this bleeding is stopped automatically after some time, and it never occurs to our mind that this cut may lead to loss of all our blood like water in a water pipe. The stoppage of blood flow is due to the clotting of the blood caused by conversion of fibrinogen, a protein in the plasma, to fibrin. This happens because of the presence of a number of blood coagulation factors. In rare individuals, one of these factors is missing, the blood does not clot in them, and the bleeding is very difficult to arrest, which may be fatal. This defect is hereditary, and the disease is known as hemophilia. The afflicted gene is inherited by the son from his mother. It is present on the X-chromosome. If the mother is heterozygous for this gene, she may transfer the X-chromosome with hemophilic gene to her son. The son will be a sufferer in the absence of a normal gene that is dominant because he has only one X-chromosome. Such genes, which are present on the sex chromosomes, are called sex-linked genes and are inherited according to the Mendel's laws.

Now, coming to the blood groups, again the inheritance is strictly according to Mendelian laws. There is a story of King Solomon in the Bible. He was renowned for his wisdom. Once, two women with a baby, each claiming him to be her baby, went to him and demanded justice. The wise king gave his verdict. He ordered to divide the child into two and give half to the one and a half to the other woman. The real mother agreed to hand over the child to the other woman, whereas the other woman insisted on dividing him into two. This determined the real mother and the child were handed over to her. With the modern research progressing, we need not look for the type of decisions taken by King Solomon, but such disputed cases of paternity are now solved with the aid of blood groups, which are inherited like any other trait in Mendelian fashion.

There are other traits like eye color, hair color, baldness, attached ear lobes, webbed toes, hand clasping, ear wax type, color blindness, tongue rolling, phenylthiocarbamide tasting, etc., which obey Mendel's laws of inheritance in human beings.

10.32 Human Sex Chromosomes and Barr Bodies

10.32.1 Study of Barr Body in the Nuclei by Means of Buccal Smear

10.32.1.1 Introduction

In most mammals, the female has two X-chromosomes and the male has only one X-chromosome. This means, all X-chromosome genes in the female are present in a double dose, whereas in males, except for the very few genes that have an allele on the Y-chromosome, X-chromosome genes are present in a single dose.

This difference in dosage can affect gene expression and may lead to genetic imbalance in one of the two sexes. In most mammals, this is corrected by inactivation by means of heterochromatinization of the greater part of one of the X-chromosomes of the females so that the transcription of these genes is not possible. This is called dosage compensation. In late interphase in several tissues, the heterochromatinized X-chromosome is visible as a condensed body of approximately 1 μ diameter. It is positioned on the inside of the nuclear membrane. This sex chromatin body is called *Barr body* after the name of the discoverer M.L. Barr, who first of all recognized it as the heterochromatinized X-chromosome.

Heterochromatinization starts early in the development of the individual, and once it has been determined, which of the two X-chromosome will be heterochromatinized, it remains the same chromosome in all succeeding generations of that cell. In the human female, a light scraping of cells from the inside of the cheek (buccal mucosa) is spread on a slide and fixed and stained to reveal the presence or absence of sex chromatin or Barr body, which distinguishes the female from the male or determines if the individual has more than one X-chromosomes. A normal male (XX) has one Barr body pressed against the inner surface of the nuclear membrane. A normal male (XY) having only one X-chromosome does not have a Barr body. The number of Barr bodies is one less than the number of X-chromosomes. The XXY male has one and the XXX female has two Barr bodies. The determination of the number of Barr bodies by buccal smears is a useful screening procedure for patients in whom an X-chromosome anomaly is suspected. There is no relationship of Barr bodies to the number of Y-chromosomes.

TABLE 10.13   Correlation of Sex, Number of Sex Chromosomes, and Number of Barr Bodies in Humans

Number of Sex Chromosomes	Sex	Number of Barr Bodies Present
XY	Male	None
XO	Female	None
XYY	Male	None
XX	Female	One
XXYY	Male	One
XXXY	Female	Two
XXX	Female	Two

Table 10.13 gives the correlation of sex, number of sex chromosomes, and number of Barr bodies in humans.

If one X-chromosome is completely inactive, then why should patients with the XO Turner syndrome or XXY Klinefelter syndrome have any abnormalities? And why should increasing numbers of X-chromosomes as found in patients with XXX, XXXX, XXXY be accompanied by progressively greater abnormality? Russell has suggested from studies on the mouse that some portions of the "inactivated" X-chromosome may not be inactivated. The inactivated X-chromosome may contain loci that are required to be present in duplicate if normal development and function is to take place. There are loci on Y-chromosome, homologous with certain loci on the X-chromosome in the normal XX female, and normal XY male has these loci in duplicate. The XO Turner syndrome is deficient in these loci and XXY Klinefelter syndrome has these loci in triplicate.

10.32.1.2 Drumsticks

In certain types of blood cells, the polymorphonuclear leukocytes, the sex chromatin sticks out of the body of the nucleus and consists of a head of 1.5 µ in diameter, connected to the nucleus by a filament. It is called drumstick.

10.32.1.3 Lyon's Hypothesis

In 1961, Lyon, on the basis of variegation patterns of X-chromosome genes in female mice, postulated that chance decides which of the two chromosomes remains active. Abnormal chromosomes, however, are generally preferentially heterochromatinized. If the genes are present in heterozygous condition on two X-chromosomes, one of the alleles will be expressed in some cells, the other in other cells. This explains the mosaicism of the coat color of a mouse. The Lyon hypothesis is briefly explained as follows:

1. In each female somatic cell in mammals, one of the two X-chromosomes is inactive, the inactivation occurring early in development.
2. It is a matter of chance whether the maternal or paternal X is inactivated.
3. Once an X-chromosome is inactivated, all progeny of that particular cell maintains the same X as the inactive chromosome.

In XXY, XXXY males, the extra X-chromosomes will be inactive.

10.32.1.4 Amniotic Fluid Examination

Amniocentesis. It is the removal of fluid from the amniotic sac. Desquamated fetal cells from the amniotic fluid are cultured to perform karyotype analysis or biochemical tests. Karyotype analysis of such cells reveals the chromosome number abnormalities. The extra number of Barr bodies reveals X-chromosome anomalies. Timely termination of pregnancy can save the parents and the child from a lot of agony.

10.32.1.4.1 Materials.
Clean glass slide, clean tongue depressor, glass bowl containing 1:1 ether–alcohol, distilled water, absolute alcohol, cresyl violet or acetocarmine or aceto-orcein, xylene.

10.32.1.4.2 Protocol 1
1. Scrap the inside of your cheek with the tongue depressor to obtain a sample of surface epithelial cells.
2. Smear cells thickly onto a clean glass slide.
3. Immediately immerse in 1:1 ether–ethanol.
4. Remove the slide from ether–alcohol solution and immerse in 95% alcohol for 2 h to fix.
5. Immerse the slide in distilled water for 5 min.
6. Immerse in cresyl violet for 10 min.
7. Immerse in 95% alcohol for 5 min—two changes.
8. Immerse in absolute alcohol for 5 min.
9. Dip in xylene for 10 min.
10. Blot dry and observe under the microscope.

10.32.1.4.3 Protocol 2
1. Spread the epithelial cells from the buccal mucosa thickly on albuminized slide.
2. Immerse inversely in 95% ethanol for 15–30 min.
3. Treat with absolute alcohol for 3 min.
4. Treat with ether–ethanol 1:1 mixture for 2 min.
5. Dry in the air for 20 s and pass through 70% ethanol for 5 min and two quick changes in distilled water.
6. Stain in acetocarmine or aceto-orcein.
7. Observe under the microscope.

10.32.2 Study of Blood Groups

10.32.2.1 To Study ABO and Rh Blood Types in Man
Harvey's discovery and study of blood circulation (1616–1628) gave an insight into the blood composition. Landsteiner discovered the ABO blood groups in 1900 and was awarded the Nobel Prize in 1930 for his discovery. Landsteiner proved that the RBC of persons vary with respect to specific proteins (antigens) found on the surface of the RBC. He found that when the RBC of certain individuals was mixed with the

serum from certain others, the cells formed clumps known as the agglutination reaction. Based upon this reaction, he classified the two different kinds of antigens A and B in type A and B, respectively. People with type A blood (antigen A) have a factor in the serum that causes agglutination of type B blood when they are mixed and vice versa. This factor is again another kind of protein called an antibody.

Thus, the four blood types have the following characteristics:

Type A Blood	Type B Blood
Antigen A	Antigen B
Antibody B	Antibody A

Certain people have both antigens A and B (type AB blood).

Some people do not have either of the antigen (Type O).

Type O	Type AB
No antigen	Antigen A + Antigen B
Antibodies A and B	No antibodies

The four phenotypes of blood with their transfusion qualities are listed in the following:

Blood Groups	Genotypes	Has Antigens	Has Antibodies Against	Can Give Blood to	Can Receive from
A	$I^A I^A$, $I^A i$	A	B	A, AB	A, O
B	$I^B I^B$, $I^B i$	B	A	B, AB	B, O
AB	$I^A I^B$	A, B	None	AB	All (universal recipient)
O	ii	None	A,B	All (universal donor)	O

The effect of the antibodies of the donor, especially the type O donor, on the cells of the recipient is not important since the antibodies of the donor are immediately diluted during transfusion.

ABO blood groups are controlled through a series of multiple alleles at a single locus. A and B alleles are inherited without apparent dominance between either type A or B (codominance). However, both A and B alleles are dominant to type O.

The knowledge of the inheritance of blood groups ABO is helpful in solving a variety of legal and social problems like

1. Disputed parentage in case of illegitimate children
2. Establishment of parents for a child after a mix up in the maternity wards
3. Identification of blood stains in cases of hit-and-run driving and murder
4. Establishment of parents in kidnapping cases of children

These cases can be solved because of the Mendelian mode of inheritance of ABO blood group alleles.

Certain diseases are known to be associated with blood group O. The risk of developing duodenal ulcer is greater in people with blood group O than in persons of the other blood types. Apart from ABO blood group system, there are at least 15 more different blood group systems known in man. Each is determined by a separate genetic locus. Some of these are P, Rh, kelI, Duffy, Diego, Lewis, and so on.

10.32.2.2 Rh Factor

The symbol Rh means rhesus, and it has come from a species of monkey "Macaca rhesus." Rh antigen is present on the RBCs of man as well as rhesus monkey.

Men are either Rh^+, that is, contain Rh antigen on their RBCs, or Rh^-, that is, absence of Rh antigen on their RBCs. In the case of blood transfusion, Rh antigen of an Rh positive person will bring forth the production of antibodies in an Rh negative recipient. These antibodies will react with the Rh antigen causing the agglutination of the Rh^+ RBC in the transfused blood.

The first transfusion of Rh^+ blood into Rh^- recipient sensitizes the individual producing antibodies but at low levels. So, the severe reaction docs not take place. Second Rh^+ blood transfusion or subsequent transfusions cause a severe reaction or even death of the individual because the level of antibodies against Rh^+ RBC rises and is sufficient to cause agglutination of transfused Rh^+ blood.

In case of Rh^- mother and Rh^+ father, the fetus inherits Rh^+ from father because Rh^+ is dominant over Rh^- antigen. During pregnancy, Rh^+ RBC of the fetus may cross the placenta into the Rh^- blood of the mother. The mother is sensitized and produces antibodies, but the level of production is low, so it does not severely affect the mother or the child. In subsequent pregnancies, if the fetus is Rh^+, the level of antibodies in the mother's blood rises so much that they cross the placental barrier and destroy the RBC of the fetus. This condition is known as erythroblastosis fetalis and causes the death of the fetus. According to Weiner, there are three Rh factors, R^0, R^1, and R^2, and a single gene is responsible for the production of all of them (multiple allele system).

10.32.2.3 The MN Blood Groups

In 1927, Landsteiner and Levine discovered the MN blood groups after injecting rabbits with human RBC and used the resulting immune serum to distinguish other human red cell samples. They proposed a two-allele mode of codominance. The two alleles, M and N, produce three genotypes, MM, NN, and MN, and three phenotypes, that is, M, MN, and N. The MN antigens do not stimulate the production of antibodies in man, so they do not create problems in blood transfusion cases, but it can be used in solving medicolegal cases.

10.32.2.4 Sickle Cell Anemia

Hemoglobin was the first molecule in which an association was shown between a mutant gene for sickle cell disease and a specific amino acid substitution. Sickle cell disease is a form of chronic hemolytic anemia characterized by the presence of elongated crescent-shaped RBC. The mutant gene is in the recessive form. Heterozygotes show the sickle cell trait. The heterozygous RBC do not show sickling in the normal blood, but when the cells are treated with sodium metabisulfite, sickling would occur.

Ingram demonstrated that sickle cell hemoglobin differs from normal hemoglobin by a single amino acid, a valine substituted for a glutamic acid. The hemoglobin molecule consists of two alpha and two beta chains. The substitution involves the sixth amino acid from the N-terminal end of the beta chain. The sickling of RBCs is most likely caused by structural differences in the folded protein due to the substitution of valine for glutamic acid. Under reduced oxygen, the defective molecules of hemoglobin adhere to each other, causing the various twisted shapes. Patients with sickle cell anemia are unusually susceptible to *Salmonella* bacterial infections, and persons heterozygous with sickling gene are more resistant to malarial infection by *Plasmodium falciparum* than the nonsickled persons. Details on experiments have been described in Chapter 18 with immunology.

10.32.3 PTC Test, Color Blindness, and the Inheritance of These Traits in a Human Population: To Work Out an Inheritance Pattern of a Dominant Gene (PTC) and a Sex-Linked Gene (Color Blindness) in a Human Population

10.32.3.1 Introduction
There are traits that fall into five different groups. The type of inheritance is different in each group. These groups are

1. Dominant and recessive traits
2. Intermediate traits
3. Quantitative traits
4. Sex-linked traits
5. Sex-influenced traits

The inheritance of these traits depends upon the number and position of the genes influencing each trait, although the inheritance of individual genes is strictly according to the Mendelian fashion. In addition to normal morphological traits (like eye color, baldness), a number of interesting physiological traits have been identified. These are the following:

10.32.3.1.1 PTC Taste Threshold.   The ability to taste phenylthiocarbamide shows marked variation between individuals. The majority of people can taste very weak concentrations of the compound, and they are called tasters, but some people can taste much higher concentrations and are called nontasters.

This physiological difference is determined by a single locus. The taster allele is dominant, and the nontaster is the recessive one. If taste thresholds are carefully measured, some individuals fall into the intermediate range or the heterozygotes having somewhat lower taste thresholds for PTC than the homozygous tasters.

Since tasting is dominant, we can denote it by the symbol (T) and the nontasting can be designated (t). These alleles do not have any influence on the ability to taste other substances. A person having an ability to taste PTC has the genetic constitution TT or T*t*. The children of a couple in which mother is nontaster and father is taster can have the following genotype:

Mother, nontaster = tt

Father, taster = Tt or TT

Children's genotypes can be as follows:

1. Parents Tt × tt
   Children 1 Tt: 1 tt (two types, i.e., tasters and nontasters)
2. Parents IT × tt
   Children Tt (only one type, i.e., all are tasters)

10.32.3.1.2 Color Blindness. Lack of the chloroplast pigment in the retinal cone cells results in the inability to discriminate green colors, or deuteranopia. The responsible mutant gene is on the X-chromosome and is recessive to the normal gene. A lack of erythrocyte pigment, necessary for discrimination in the red end of the spectrum, results in protanopia. This gene is also X-linked and appears to be quite close to the deuteranopia locus. It is also recessive to the normal gene.

X- and Y-chromosomes are concerned with the determination of sex. A female has two X-chromosomes and is called homogametic (produces only one type of eggs), and the male having XY sex chromosomes is heterogametic since it produces two types of sperms, one with X-chromosome and the other with Y-chromosome.

Color blindness, the inability to distinguish between green and red colors, is more frequent in males because they do not have another X-chromosome that could carry a dominant normal gene to mask the expression of the recessive gene for color blindness in the other. It is possible for a woman to be color blind if she receives the recessive gene from both the parents, for example,

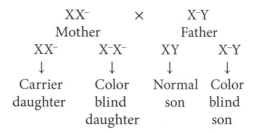

Sex-linked inheritance is also called the "skip generation inheritance." The sex-linked character is inherited from the father to the grandson and not to the son as illustrated in the following:

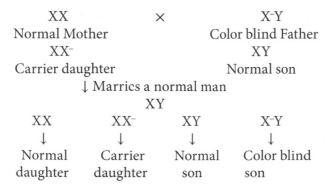

10.32.3.2 PTC Testing

10.32.3.2.1 Materials.   Phenylthiocarbamide; filter paper strips; glassware.

10.32.3.2.2 Protocol
1. Dissolve 1.5 g of PTC in hot water and make up to 1 L.
2. Soak filter papers in the PTC solution and allow them to dry and then cut into strips.
3. From the remaining solution, make up seven more solutions, diluting each time to half strength as before (0.15%, 0.075%, 0.0375%).
4. Soak separate filter papers in these dilutions and allow them to dry and then cut into strips.
5. Put the dried filter paper strips at the back of the tongue and record the number of tasters and nontasters among students.

This test can be applied on all the members of some families and see whether the tasters and nontasters suggest simple dominance.

10.32.3.3 Color Blindness
The color blindness gene is present on the X-chromosome. If a boy is color blind, it shows that the gene has come from the mother. If a girl is color blind, it shows that both the parents have contributed the recessive allele and the father is also color blind. Prepare the following chart for the inheritance of the sex-linked gene in a family:

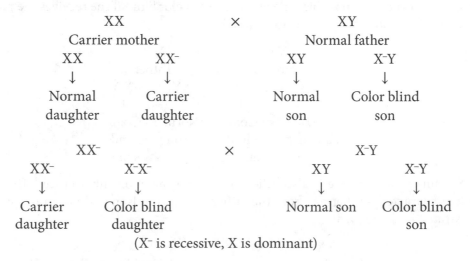

10.33 Autoradiography

It is a photographic technique for detecting radioactive substances. A photographic emulsion is placed in direct contact with the object to be tested and is left for several hours, days, or weeks, depending upon the suspected concentration of radioactive material. The film is then developed, fixed, and washed as in normal photographic processes. At sites where the film was close enough to the radioactive substance, it appears dark because of the presence of silver grains. When the number of grains is insufficient to darken the film to the unaided eye, the film may be examined with the aid of a phase-contrast or light microscope, the former being preferred for unstained preparations. The pattern formed by the grain depends in a large measure on the type of radiation and to some extent on the nature of photographic emulsion. Autoradiographic technique has been extensively used in understanding the basic cell metabolism specially when used in conjunction with other cytochemical methods such as micro-spectrophotometry, enzyme digestion, and some specific cytochemical staining. It has also been used for studying replication of chromosomes.

Autoradiography can be used to detect radioactive materials in almost any object. However, in biological research, the object can be

1. Either a whole plant or animal that can be flattened against a film, the cut surface of the plant or animal, or one of its organs
2. Thin sections of tissue
3. Squashed and well-flattened tissues or cells
4. Sheets of paper on which radioactive substances have been deposited and separated by some technique such as chromatography

The degree of resolution of the radioactive substance, that is, the precision with which it is located, can be greatly improved by preparing the material in smears, squashing tissues or cells against the film, or sections cut at 4–6 µ. In practice, the tissue is often mounted on a glass microscopic slide. The photographic emulsion is then applied in a darkroom with a safe light.

In preparing specimens for autoradiography, it is a common practice to fix them before photographic emulsion is applied. The choice of fixative for sections is usually limited since artifacts such as fogging or desensitization of the emulsion can be produced. Alcohol, acetic alcohol, formal acetic alcohol, formal saline, and freezing substitution methods are satisfactory. The method of fixation depends on the chemical constituents to be studied. For the study of radioisotope incorporation into the nucleoproteins of cell structures such as the chromosomes, Carnoy's acetic alcohol is generally employed. When it is desirable to retain all of the incorporated radioactive material, tissues may be quick frozen and dried by vacuum desiccation. To ensure adherence of the film to glass slides for mounting, slides should be subbed, that is, filmed with egg albumen or dipped in 0.5% aqueous solution of gelatins and 0.1% alum. The subbed slides should be dried for at least 2 days before using, preferably in an incubator at 37°C or less. Heating over flame needs to be avoided.

The smears are prepared in the usual way but on subbed slides. The choice of fixative is once again limited to those recommended. For squashes, roots, anthers, etc., tests are best fixed in acetic alcohol. The tissues can be hydrolyzed for 6–8 min in N HCl at 60°C to facilitate squashing, or plant tissues can be treated with pectinase to avoid loss of RNA. Small pieces of tissue are tapped out in 10% acetic acid in order to separate cells as a single layer, preferably on a cover slip that is then inverted on to the subbed slide. Cells should be flattened by slight pressure on the cover slip and not with finger tips in case they become contaminated with radioactive material. Salivary glands should be squashed on subbed slides in 45% acetic acid. The prepared slides should never be heated. It is preferable to leave them on the bench for 5–10 min after squashing and before inverting the slide in a ridged dish containing 10% acetic acid to separate the slide from the cover slip.

Frequently, it is necessary to stain cells in order to determine whether incorporation of the isotope has occurred in a particular type of cell. Staining of the specimen is most conveniently done before application of the emulsion film; however, some fading of the stain may occur during the developing process. The Feulgen stain works well in this respect as it is not appreciably affected in the photographic processing of the autoradiograph. Toluidine blue is very suitable for staining nuclei if the tissues have previously been hydrolyzed in NHC. Euparal is suitable for mounting stained preparations.

Three methods are commonly used to make contact between the specimen and the photographic emulsion. The first method, often known as Belanger and Leblond's method, involves applying a melted photographic emulsion to the specimen on a dry slide. The second is Evans' and Endicott and Jagoda's method, sometimes called opposition method; the specimen is simply placed in contact with photographic emulsion in the dark. Paraffin sections of tissues are floated on warm water and then picked up on a microscopic slide that has previously been coated with a thin emulsion film. The sections remain attached to the slide during photographic processing and subsequent staining and mounting.

The third method is the stripping film technique of contrast autoradiography. It is a widely used method. A specially prepared thin stripping film is pulled off a glass plate and spread with the emulsion side to the surface of a bowl of water. The wet film is then lifted up by dipping the slide beneath it. The slide is dried, exposed, and then developed with specimen remaining firmly attached. When examined under the microscope, the tissue, cells, or even parts of a single cell are seen with the silver grains above the site of any radioactive substance.

10.34 Photomicrography

Photomicrography is the recording of microscopic images by photographic methods involving the combination of the principles of microscopy and photography. A light micrograph or photomicrograph is a micrograph prepared using a light microscope, a process referred to as photomicroscopy. At a basic level, photomicroscopy may be performed simply by hooking up a regular camera to a microscope, thereby enabling the user to take photographs at reasonably high magnification. Photomicrography is a process of forming a visible image directly or indirectly by the action of light or other forms of radiation on the sensitive surface. When the object is focused under the microscope, the photograph can be taken with the help of simple or specialized cameras fitted on the eyepiece. Photomicrographs are of immense use in studying the cellular structures. They also provide the supporting evidence of the observations described. It is thus of utmost importance that a photograph must be of sound quality and free from any ambiguity. There are different types of photography like x-ray photography, light photomicrography, fluorescence photomicrography, and photography of living cells programmed with different types of software attached with computer with high pixel.

Suggested Readings

Campbell, L. J. 2011. *Cancer Cytogenetics: Methods and Protocols* (Methods in Molecular Biology). New York: Humana Press.

Czepulkowski, B. 2000. *Analyzing Chromosomes* (Basics: From Background to Bench). Oxford, U.K.: BIOS Scientific Publishers.

Fan Yao-Shan. 2010. *Molecular Cytogenetics: Protocols and Applications.* Methods in Molecular Biology. New York: Humana Press.

Gardner, R. J. M., G. R. Sutherland, and L. G. Shaffer. 2011. *Chromosome Abnormalities and Genetic Counseling.* Oxford, U.K.: Oxford University Press.

Gersen, S. L. and M. B. Keagle. 2012. *The Principle of Clinical Cytogenetics.* Dordrecht, the Netherlands: Springer.

Heim, S. and F. Mitelman. 2009. *Cancer Cytogenetics: Chromosomal and Molecular Genetic Aberrations of Tumor Cells.* New York: Wiley Blackwell.

Roy, D. 2009.*Cytogenetics.* Oxford, U.K.: Alpha Science Intl Ltd.

Important Links

http://www.sugarcane.res.in/index.php/about-us/divisions—sections/crop-improvement/genetics-a-cytogenetics/28?phpMyAdmin = 11c501a2a5dt8788ed6

http://www.ehaweb.org/assets/Education-Book-PDF/EHA16-Education-Book.pdf

http://www.cost.eu/domains_actions/bmbs/Actions/B19

http://www.le.ac.uk/biology/phh4/

http://www.britannica.com/EBchecked/topic/228936/genetics/48733/Cytogenetics

11

Biomolecules

11.1 Introduction

All living organisms are composed of numerous organic compounds, most of which fall into distinct classes having different chemical compositions and properties. They are primarily of carbon and hydrogen, along with nitrogen, oxygen, phosphorus, and sulfur. Other elements sometimes are incorporated but these are much less common. Survival of an organism depends on its ability to reorganize the chemicals in its environment into molecules that compose its own cellular material and termed as biomolecules. Biomolecules are necessary for the existence of all known forms of life. For example, the main component of hair is keratin, an agglomeration of proteins that are themselves polymers built from amino acids. Amino acids are some of the most important building blocks used, in nature, to construct larger molecules. Another type of building block is the nucleotides, each of which consists of three components: a purine or pyrimidine base, a pentose sugar, and a phosphate group. These nucleotides, mainly, form the nucleic acids. Besides the polymeric biomolecules, numerous small organic molecules are absorbed or synthesized by living systems. All biomolecules are, therefore, polymers that include carbohydrates, lipids, proteins, and nucleic acids (Table 11.1). They are classified as noninformational (in which sequences of monomers do not carry genetic information) and informational (sequence of monomers carrying genetic information). Carbohydrates and lipids are the examples of noninformational molecules, and protein and nucleic acids are the examples of informational molecules. Proteins are the most abundant class of macromolecules in the cell. Many biomolecules may be useful as important drugs.

11.2 Carbohydrates

11.2.1 Function of Carbohydrates in the Biosphere

Carbohydrates are a major source of energy for living organisms. In man's food, the chief source of carbohydrate is starch, the polysaccharide produced by plants, especially the cereal crops, during photosynthesis. Plants may store relatively large amounts of starch within their own cells in time of abundant supply, to be used later by the plant itself when there is a demand for energy production or to be consumed by animal for food. In simple carbohydrates, the ratio of hydrogen to oxygen is 2:1,

TABLE 11.1   Macromolecules

Macromolecules	Monomeric Units	Examples
Noninformational		
Carbohydrates		
Monosaccharides	3–7 Carbon sugars	Glucose, fructose
Disaccharides	Two monosaccharides	Maltose
Polysaccharides	Chains of monosaccharides	Starch, cellulose, glycogen
Lipids		
Triglycerides	Fatty acids and glycerol	Fats and oils
Phospholipids	Fatty acids, glycerol, and phosphate	Membranes
Waxes	Fatty acids and alcohols	Mycolic acid
Steroids	Ringed structure	Cholesterol, ergosterol
Informational		
Proteins	Amino acids	Enzymes, ribosomes, antibodies
Nucleic acids	Pentose sugars, phosphate, and nitrogenous base	
	Purines: adenine and guanine	
	Pyrimidines: cytosine, thymine, and uracil	
DNA	Deoxyribose sugar and thymine	Chromosomes; genetic material of viruses/inheritance
RNA	Ribose sugar and uracil	Ribosomes, mRNA, tRNA

the same as in water. The atoms are present in H–C–OH configurations. The combination occurs frequently in carbohydrate structures (Figure 11.1).

Carbohydrates are of fundamental importance in living organisms as a source of metabolic energy.

11.2.1.1 Photosynthesis

It is the primary process on which all life on earth is dependent. It is the manner in which the carbohydrates are synthesized in green plants and algae from water and CO_2 using the energy of sunlight. Plants may store relatively large amount of starch within their own cells in time of abundant supply, to be used later by the plant itself when there is a demand for energy production or to be consumed by animals for food. After ingestion, the plant carbohydrates are broken down to glucose that is stored as glycogen in liver, muscle, and other tissues.

11.2.1.2 Respiration

In animals, carbohydrate is stored as glycogen, primarily in the liver (2%–8%) and muscle (0.5%–1%). In the latter, glycogen serves as an important source of energy for contraction for a finite time. In liver, the primary role of glycogen is maintenance of the blood glucose concentration. Simple sugars such as sucrose, glucose, fructose, mannose, and galactose are also encountered in nature and are utilized by living

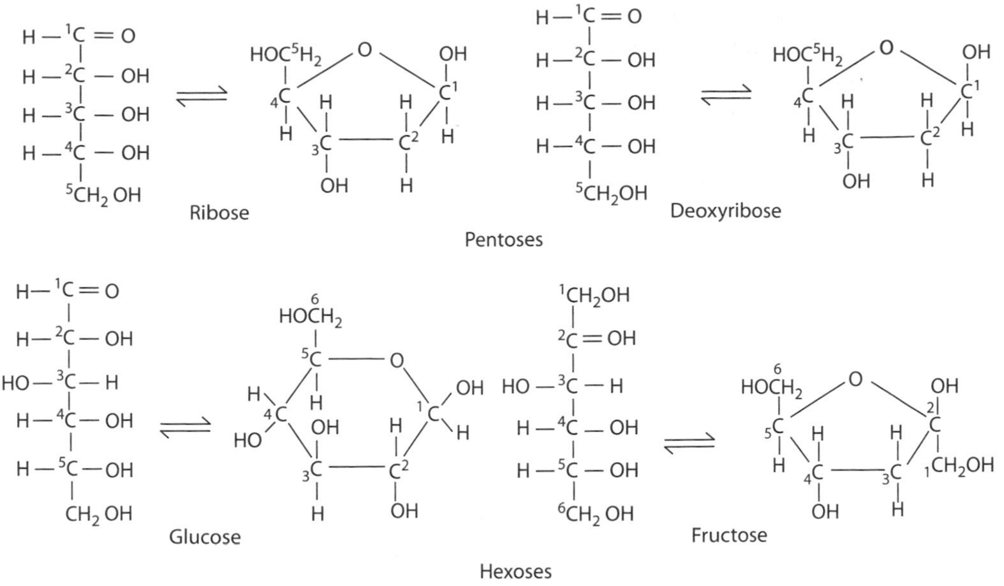

Ribose

Deoxyribose

Pentoses

Glucose

Fructose

Hexoses

Figure 11.1 Chemical structures of four common sugars, open chains, and the closed ring structures.

forms as food. Glucose is the compound formed from both starch and glycogen on metabolism.

The stored carbohydrate is converted to glucose that is then transported to the cells where it is oxidized to CO_2 and O_2 in cellular respiration, in animals. The energy released during this oxidation is then "captured" and used to drive the metabolic machinery of the cells. In the case of humans, more than 60% of their total energy requirements are provided by the oxidation of carbohydrates. In this way, the energy of sunlight is made available to the animal kingdom that cannot carry out photosynthesis (Figure 11.2).

11.2.2 Structure of Cells and Molecules

Complex carbohydrates play not only a structural role in the cell but may serve as a reservoir of chemical energy to be enlarged and depleted as the organism

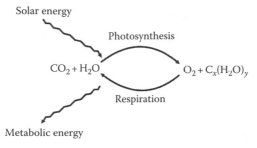

Figure 11.2 Carbohydrate production and oxidation.

wishes. As an example of structural carbohydrates, we may cite cellulose, the major structural component of plant cell walls, and the peptidoglycans of bacterial cell walls and mucopolysaccharides of skin and connective tissue in animals. In addition, monosaccharides are an important part of biochemical compounds such as nucleic acids, coenzymes, and flavoproteins. The storage carbohydrates include the more familiar starch and glycogen, polysaccharides that may be produced and consumed in line with the energy needs of the cell. Carbohydrates are involved in cell recognition, contact inhibition, and the antigenic properties of blood group substances.

11.2.3 Structure of Carbohydrates

11.2.3.1 Introduction
Carbohydrates may be defined as polyhydroxy aldehydes or ketones or as substances that yield one of these compounds on hydrolysis. Many carbohydrates have the empirical formula $(CH_2O)_n$ where n is 3 or larger. This formula obviously contributed to the original belief that this group of compounds could be represented as hydrates of carbon with a ratio of 1:2:1. Basic units of carbohydrates are monosaccharides that cannot be split further by hydrolysis. They are named according to the number of carbon atoms in the chain so that triose contains three, tetrose four, pentose five, and hexoses six carbon atoms.

11.2.3.2 Stereochemistry

11.2.3.2.1 Optical Activity. This is the type of isomerism commonly found in carbohydrates. It is usually encountered when a molecule contains one or more chiral or asymmetric carbon atoms, and because of this, a number of stereoisomers are possible. For example, glyceraldehydes, one of the simplest sugars, have one asymmetric carbon atom, and there are two possible arrangements of the four groups around this carbon. The formulae of these two isomers are given in the following discussion where the asymmetric carbon is placed at the center of a tetrahedron, the four groups being situated at each corner and the dotted line shown as lying below the plane of the paper. The two forms can be likened to mirror images or left- and right-handed gloves. These two configurations are *optical isomers* or *enantiomers* and rotate the plane of polarized light to the same extent but in opposite directions. When the plane of polarized light is rotated to the right, the compound is *dextrorotatory* and is labeled (D) or (+). Likewise, when the plane of polarized light is rotated to the left, the compound is *levorotatory* (L) or (−) (Figure 11.3).

11.2.3.2.2 D and L Forms. Aldoses are derived from the parent compound, glyceraldehydes, by the addition of successive secondary alcohol groups (−CHOH). Since glyceraldehydes can exist in two forms, two distinct families of aldoses emerge: those derived from D-glyceraldehyde known as D-*sugars* and those from L-glyceraldehyde called L-*sugars*. The letters D and L do not give an indication of the optical activity,

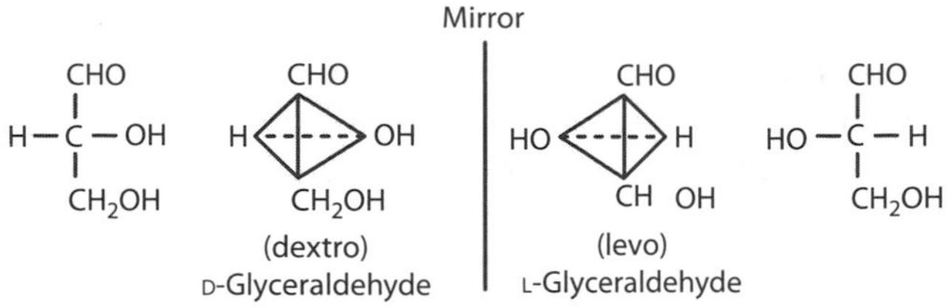

Mirror

CHO   CHO   CHO   CHO

H—C—OH   H⟨⤏⟩OH   HO⟨⤏⟩H   HO—C—H

CH₂OH   CH₂OH   CH OH   CH₂OH

(dextro)   (levo)

D-Glyceraldehyde   L-Glyceraldehyde

Figure 11.3   Optical isomerism.

but refer to the configuration of the carbon atom next but one furthest removed from the end of the chain containing the aldehyde or ketone group. Thus, a D-sugar could be dextrorotatory D (+) or levorotatory L (–) depending on the configuration of the other carbon atoms present:

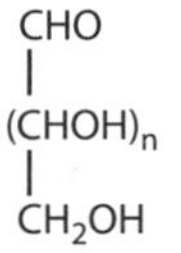

CHO

$|$

$(CHOH)_n$

$|$

CH_2OH

Aldoses

The simplest ketose is the triose, dihydroxyacetone, but this compound is optically inactive, so the tetrose D-erythrulose is the parent compound of the D-ketoses:

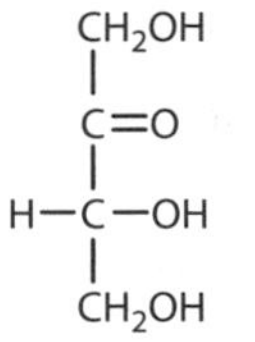

CH_2OH

$|$

$C=O$

$|$

$H—C—OH$

$|$

CH_2OH

D-Erythrulose

Pentoses, hexoses, etc., are formed by the addition of secondary alcohol groups.

11.2.3.2.3 Numbering of Carbon Atoms.   The carbon atoms are numbered from the end of the chain containing the reactive carbonyl group; thus,

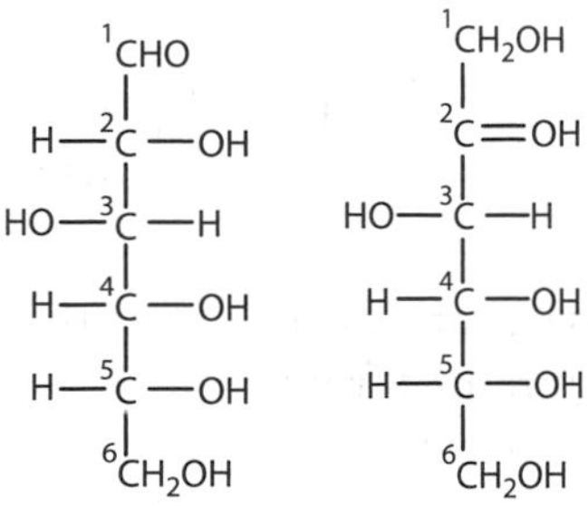

^1CHO $^1CH_2OH$

$H—^2C—OH$ $^2C=OH$

$HO—^3C—H$ $HO—^3C—H$

$H—^4C—OH$ $H—^4C—OH$

$H—^5C—OH$ $H—^5C—OH$

$^6CH_2OH$ $^6CH_2OH$

Glucose   Fructose

11.2.3.2.4 Ring Formulae.   When D(+)-glucose is dissolved in water, a specific rotation of +113° is obtained, but this slowly changes, so that at 24 h the value has become +52.5°. This phenomenon is known as *mutarotation* and is shown by a number of pentoses, hexoses, and reducing disaccharides. The reason for this change in rotation is that glucose exists in solution mainly in the ring form, which is obtained by the formation of an intramolecular hemiacetal, as follows:

α-D-glucose　　　　　　Open chain form　　　　　　β-D-glucose

| *Equilibrium mixture* | 36%, | 0.02%, | 64% |
| *Specific rotation* | + 113° | — | + 19.7° |

This creates another asymmetric carbon atom (C_1) so that two ring forms (α and β) are now possible, and these are known as *anomers*. The open-chain form is in equilibrium with the two ring forms, and this is responsible for the reducing properties of the sugar and shows the ring as a plane with the hydroxyl groups orientated above and below the plane. The hydrogen atoms are not usually shown for the sake of clarity.

The ring formed in the case of glucose contains five carbon atoms and one oxygen and is analogous to pyran; glucose is said, therefore, to exist in the *pyranose* form. Many of the common sugars are present as the pyranose ring, but some exist as a four-carbon ring known as the *furanose* form, after the compound furan Figure 11.4a.

The structures of a number of common monosaccharides (Figure 11.4b) are shown in the following discussion.

11.2.3.2.5 Glycosidic Link. *Glycosides and disaccharides* (lactose, sucrose, maltose): The carbonyl group present in all sugars is very reactive and can form hemiacetals or acetals with other hydroxylic compounds:

Ketone　　　　　　　　　　　Hemiacetal　　　Acetal

Disaccharides are two monosaccharides joined together by a linkage termed a glycosidic linkage. The linkage that results from the reaction between hydroxyl groups of two sugars units, with the loss of a water molecule, is called glycosidic bond, which is an example of dehydration synthesis.

The glycosidic bond can exist in two different geometric orientations, referred to as alpha (α) and beta (β). Maltose is composed of two molecules of glucose linked by α-1,4 bond (Figure 11.5).

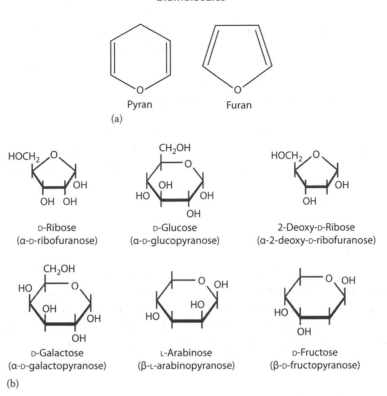

Figure 11.4 (a) Pyran and furan ring. (b) Structural formulae of some common sugars.

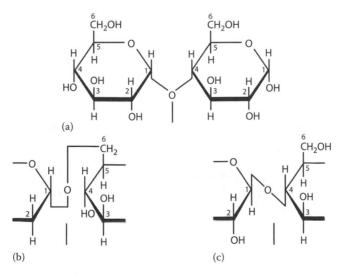

Figure 11.5 Structures of different glycosidic bonds: (a) α-1,4-glycosidic bond, (b) α-1, 6-glycosidic bond, and (c) β-1,4-glycosidic bond.

Maltose occurs in central grains such as barley and is fermented by yeasts to form the alcohol in beer. Lactose, a milk sugar, is composed of the monosaccharides glucose and galactose.

Formation of an internal ring structure is brought about by reaction of the hydroxyl group on C_4 or C_5 with the carbonyl group to give an intramolecular hemiacetal where XOH is the rest of the molecule. The remaining hydroxyl on C_1, known as the *glycosidic hydroxyl*, is very reactive and readily forms a glycosidic link with other hydroxyl groups by the elimination of water. If the hydroxylic compound is not a sugar, then a *glycoside* is formed, as, for example, in the condensation of glucose and methanol to give methyl glucoside (Figure 11.6). Quite often, though, the hydroxyl is from another sugar, and in this case, a *disaccharide* is the product. Provided the bond is not between two C_1 atoms, then a free aldehyde or ketone group is left on the disaccharide and the molecule shows all the reactions associated with these groups as well as mutarotation.

Oligosaccharides and polysaccharides: This process can be repeated so that a second monosaccharide is linked to a third by another glycosidic bond to give a *trisaccharide* and so on to give an *oligosaccharide* of 2–10 units linked as a chain. In the case of *polysaccharides*, chains of 10 to several thousand monosaccharide units are joined together by glycosidic bonds to give a very large molecule. The monomer units

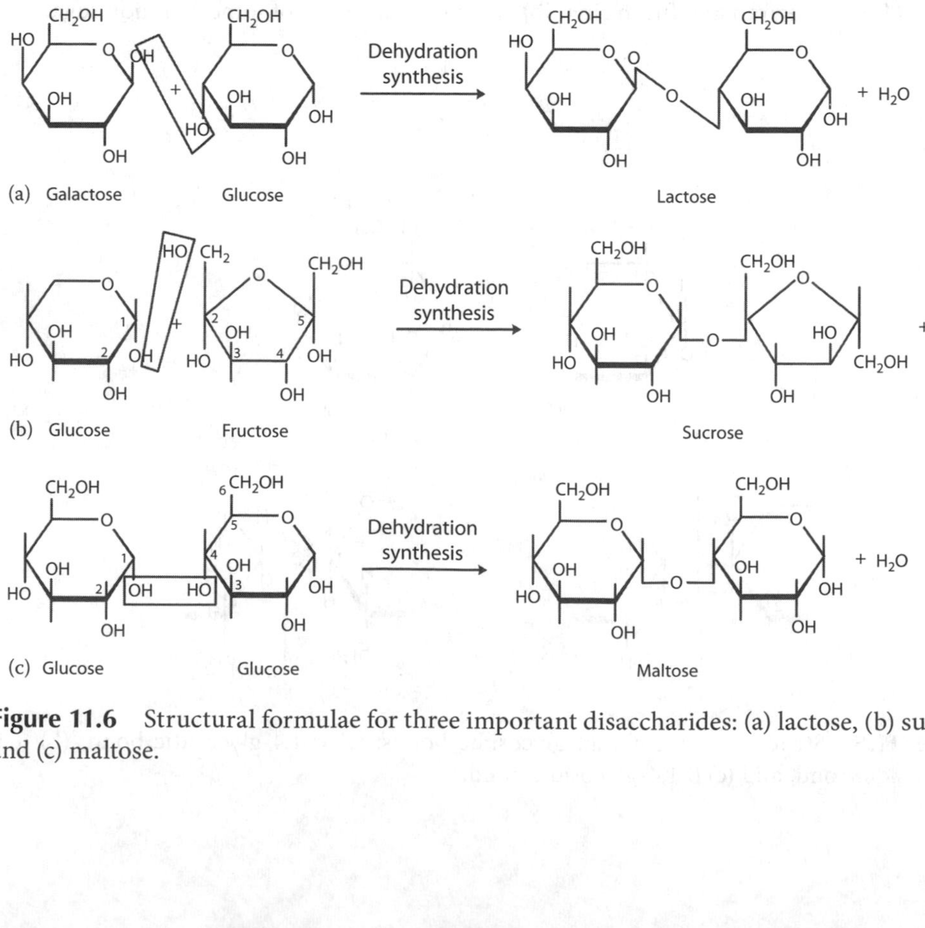

Figure 11.6 Structural formulae for three important disaccharides: (a) lactose, (b) sucrose, and (c) maltose.

are not always the same, and branching can occur as well as substitution to give a quite complex molecule.

11.2.3.2.6 Simple Monosaccharide Derivatives.   *Oxidized products*: The aldehyde and primary alcohol groups of aldoses can be oxidized to the corresponding *aldonic* and *uronic* acids. Further oxidation of these acids yields *saccharic* acid. The oxidation of a monosaccharide is shown in the following text. The names in brackets represent the products when the hexose is the monosaccharide glucose (Figure 11.7).

Reduction: The aldehyde or ketone group present can be reduced to the primary or secondary alcohol with sodium amalgam. For example, fructose and glucose give the hexahydric alcohol *sorbitol* and glyceraldehyde is reduced to *glycerol*:

$$
\begin{array}{ccc}
CHO & & CH_2OH \\
| & & | \\
CHOH & \xrightarrow{\ 2H\ } & CHOH \\
| & & | \\
CH_2OH & & CH_2OH \\
\text{Glyceraldehyde} & & \text{Glycerol}
\end{array}
$$

Esters: Another derivative useful in structural determination is the acetyl derivative of the carbohydrates. Sugars form esters very readily with acids by reaction with the acid chloride or anhydride in the presence of a catalyst. In particular, the phosphate esters of the monosaccharides are of fundamental importance in carbohydrate metabolism. Esters formed from the primary alcohol group of the ultimate carbon (glucose-6-phosphate) are usually the most stable to acid hydrolysis. Hemiacetals formed through the reducing group are, on the other hand, more acid labile (glucose-L-phosphate).

Amino derivatives: The hydroxyl in the 2 position of many sugars can be replaced by an amino group, as, for example, in *galactosamine* and *glucosamine*. These

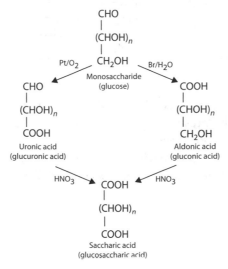

Figure 11.7   Derivatization of monosaccharides.

TABLE 11.2   Some Common Mono- and Disaccharides

Sugar	Occurrence and Function
Monosaccharides	
Pentoses	
L-Arabinose	Present as pentosans in wood gums and fermented by some bacteria. The D-sugar occurs in the glycoside of tubercle bacilli.
D-Ribose	An essential part of RNA, the macromolecule involved in protein synthesis, also present in many coenzymes (NAD, FAD, ATP).
2-Deoxy-D-ribose	An important constituent of the macromolecule DNA, the genetic material of all living organisms.
Hexoses	
D-Glucose	The most widely distributed of the sugars; glucose is transported in the blood and oxidized in the cells to produce energy.
D-Fructose	The sweetest of the sugars found in fruit and honey. It is also present in seminal fluid where it provides the energy source for the spermatozoa.
D-Galactose	This sugar is part of the glycolipids of nervous tissue and chloroplast membranes. Galactosamine is present in blood group substances, cartilage, and tendons.
Disaccharides	
Lactose (α-D-galactosyl-1,4-D-glucose)	The sugar present in the milk of mammals that may also be found in the urine during pregnancy.
Maltose (α-D-glucosyl-1,4-D-glucose)	The sugar produced during the digestion of starch by amylase and present in germinating cereals and malt.
Sucrose (α-D-glucosyl-β-1,2-D-fructose)	This is a nonreducing sugar since the glycoside link is through both of the potentially reducing groups of the monosaccharides. It occurs in plants and is obtained mainly from sugarcane and beet.

compounds also exist as the *N*-acetyl derivatives that are part of the structure of several polysaccharides. The presence of the amino group close to the glycosidic bond renders it more stable to hydrolysis.

Simple sugars: Simple sugars are widespread throughout nature and perform many and varied functions. Some of the common sugars are listed in Table 11.2.

11.2.4 Macromolecules

11.2.4.1 Homopolysaccharides
The addition of one or more monosaccharide units to a disaccharide yields a trisaccharide and several more oligosaccharides. A polysaccharide is a large

TABLE 11.3 Properties of Some Homopolysaccharides of Glucose

Polysaccharide (Monosaccharide Unit)	Straight Chain	Branch Points	Mol. wt.
Storage polysaccharides			
Amylopectin (α-D-Glucose)	α-1–4	α-1–6	$0.2–1 \times 10^6$
The major component of starch as the main carbohydrate store in plants			
Amylose (α-D-Glucose)	α-1–4	None	50,000
Present in most starches making up to 20% of the total			
Glycogen (α-D-Glucose)	α-1–4	α-1–6	$1–3 \times 10^6$
Major reserve carbohydrate in animals present mainly in the liver and muscle			
Inulin (β-D-Fructose)	β-1–2	None	5,000
The reserve carbohydrate of plants such as dahlias and artichokes			
Structural polysaccharides			
Cellulose (β-D-Glucose)	β-1–4	None	—
The major structural component of the cell walls of plants and some algae and bacteria			
Chitin (β-D-N-Acetyl-glucosamine)	β-1–4	None	—
The main component of the exoskeleton of insects and crustacean such as crabs and lobsters.			

compound consisting of a chain of hundreds or thousands of glucose units. Starch, glycogen, cellulose, and pectin are some of the common plant products and constitute the most abundant organic compounds on the earth. These are very large molecules that contain only one monosaccharide component, although several macromolecules can be built up from the same monosaccharide unit. This multiplicity arises from the different sizes of the molecules, the variable degree of branching and cross-linking, and the nature of the glycosidic bond (Table 11.3 and Figure 11.8).

11.2.4.2 Heteropolysaccharides

These molecules are formed from two or more basic units and are frequently associated with protein. Such complexes are known as *proteoglycans* or *mucoproteins* when the polysaccharide component dominates. These materials are gelatinous and viscous in nature and frequently act as lubricants or intracellular cement (Table 11.4).

The carbohydrate–protein complexes are called *glycoproteins* when the carbohydrate is the minor component. The small quantities of short-chain oligosaccharides

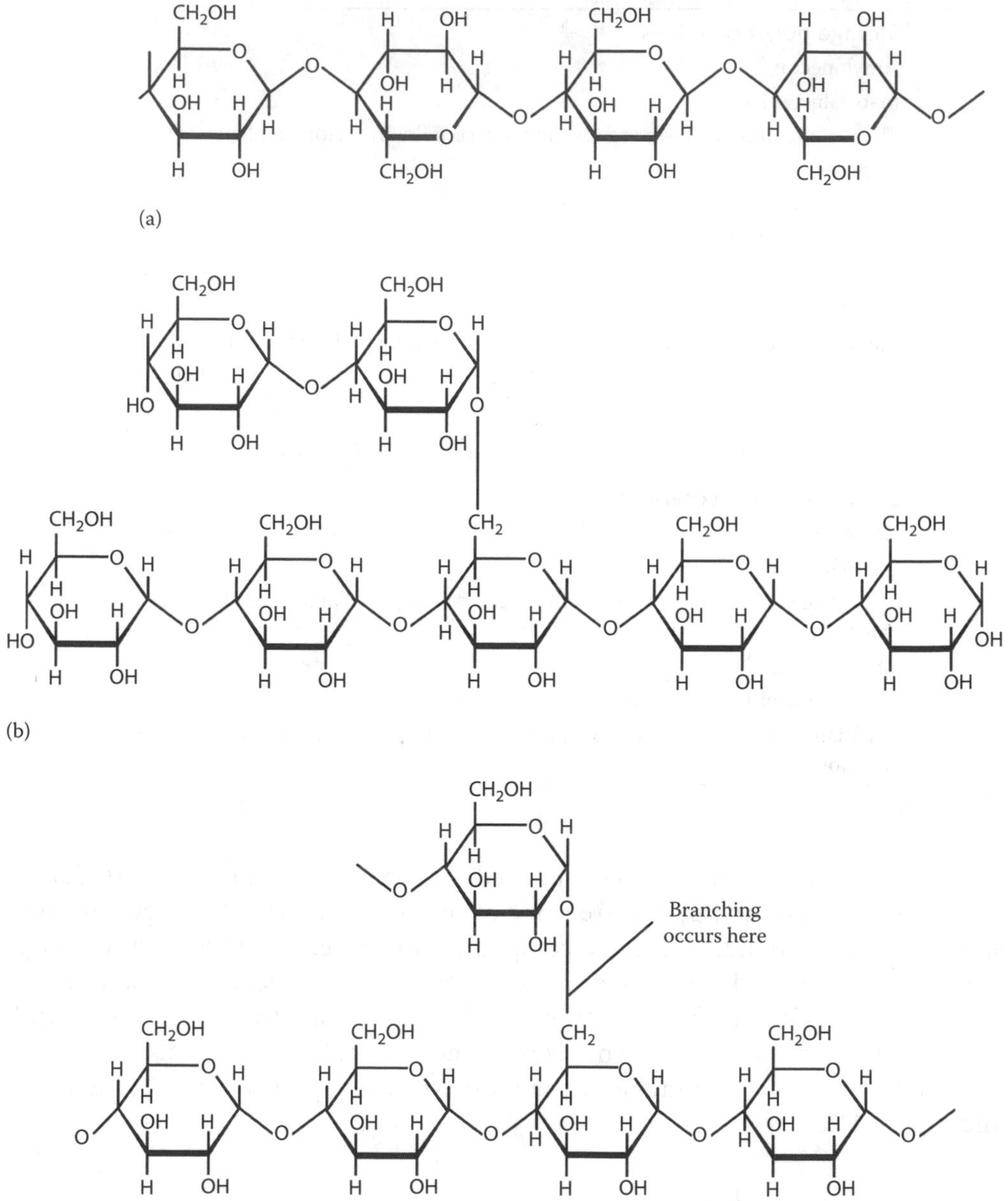

Figure 11.8 Structural formulae of three important homopolysaccharides: (a) cellulose (β-1, 4-glycosidic bond), (b) glycogen (α-1,4- and α-1,6-glycosidic bond), and (c) starch (α-1,4- and α-1,6-glycosidic bond).

TABLE 11.4   Some Common Heteropolysaccharides

Polysaccharide	Monosaccharide Components
Mucopolysaccharides	
Heparin	
A natural anticoagulant in lungs and arteries	D-Glucuronic acid and D-glucosamine sulfate
Hyaluronic acid	
A lubricant in synovial fluid, extracellular cement around joints	D-glucuronic acid and N-acetyl-D-glucosamine
Chondroitin sulfates	
Found in skin and cartilaginous tissue	D-glucuronic acid and N-acetyl-D-glucosamine sulfate
Bacterial polysaccharides	
Murein	N-acetyl-D-glucosamine, N-acetylmuramic acid, and oligopeptide
Backbone of bacterial cell wall	
Capsule	Exact composition depends on the organism
Responsible for antigenic properties of bacteria	

can have a considerable influence on the biological properties of the proteins. For example, the antigenic properties of blood group substances depend on the chemical nature of these carbohydrate chains. The structure and linkages are not always known for certain.

11.2.5 Experiment: Benedict's Test for Reducing Sugars

11.2.5.1 Principle

Alkaline solutions of copper are reduced by sugars that have a free aldehyde or ketone group, with the formation of colored cuprous oxide. Benedict's solution is composed of copper sulfate, sodium carbonate, and sodium citrate (pH 10.5). The citrate will form soluble complex ions with Cu^{++}, preventing the precipitation of $CuCO_3$ in alkaline solutions. Carbohydrates with a free or potentially free aldehyde or ketone group have reducing properties in alkaline solution. In addition, monosaccharides act as reducing agents in weakly acid solution. Benedict modified the original Fehling's test to produce a single solution that is more convenient for tests as well as being more stable than Fehling's reagent. A black cupric oxide is formed when a suspension of copper hydroxide in alkaline solution is heated:

$$Cu(OH)_2 \rightarrow CuO + H_2O$$

However, if a reducing substance is present, then rust-brown cuprous oxide is precipitated:

$$2Cu(OH)_2 \rightarrow Cu_2O + 2H_2O + \tfrac{1}{2}O_2$$

An alkaline solution of a copper salt and an organic compound containing alcoholic –OH is used rather than the previous suspension. Under these conditions, the copper forms a soluble complex, and the reagent is stable.

11.2.5.2 Materials
Benedict's reagent: Dissolve 173 g of sodium citrate and 100 g of sodium carbonate in about 800 mL of warm water. Filter through a fluted filter paper into a 1000 mL measuring cylinder, and make up to 850 mL with water. Meanwhile, dissolve 17.3 g of copper sulfate in about 100 mL of water, and make up to 150 mL. Pour the first solution into a 2 L beaker, and slowly add the copper sulfate solution with stirring.
Glucose solutions: 10 and 1 g/L

11.2.5.3 Protocol
Add five drops of the test solution to 2 mL of Benedict's reagent and place in a boiling water bath for 5 min. Examine the sensitivity of Benedict's test using increasing dilutions of glucose.

11.2.6 Experiment: Iodine Test for Polysaccharides

11.2.6.1 Principle
The use of Lugol's iodine reagent is useful to distinguish starch and glycogen from other polysaccharides. Lugol's iodine yields a blue-black color in the presence of starch. Glycogen reacts with Lugol's reagent to produce a brown-blue color. Other polysaccharides and monosaccharides yield no color change; the test solution remains the characteristic brown yellow of the reagent. It is thought that starch and glycogen form helical coils. Iodine atoms can then fit into the helices to form a starch–iodine or glycogen–iodine complex. Starch in the form of amylose and amylopectin has less branches than glycogen. This means that the helices of starch are longer than glycogen, therefore binding more iodine atoms. The result is that the color produced by a starch–iodine complex is more intense than that obtained with a glycogen–iodine complex.

11.2.6.2 Materials
Iodine solution: 5 mmol/L in KI (30 g/L).
    Cellulose, glycogen, starch, and inulin (10 g/L).

11.2.6.3 Protocol
Acidify the test solution with dilute HCl, then add two drops of iodine, and compare the colors obtained with water and iodine.

11.2.7 Optical Activity

Optical rotation (optical activity) is the turning of the plane of linearly polarized light about the direction of motion as the light travels through certain materials. It occurs in solutions of chiral molecules such as sucrose (sugar), solids with rotated crystal planes such as quartz, and spin-polarized gases of atoms or molecules. It is used in the sugar industry to measure syrup concentration, in optics to manipulate polarization, in chemistry to characterize substances in solution, and in optical mineralogy to help identify certain minerals in thin sections. It is being developed as a method to measure blood sugar concentration in diabetic persons.

11.2.7.1 Polarimeter

Polarimetry is a sensitive, nondestructive technique for measuring the optical activity exhibited by inorganic and organic compounds. A compound is considered to be optically active if plane-polarized light is rotated when passing through it. The amount of optical rotation is determined by the molecular structure and concentration of chiral molecules in the substance. Most sugars contain one or more asymmetric carbon atoms and show optical activity. This rotation of the plane of polarized light can be demonstrated and measured with a polarimeter (Figure 11.9). Monochromatic light passes through a *Nicol prism* and emerges polarized in one plane. This polarized beam then passes through the sugar sample that rotates the plane of the light. The second *Nicol prism* is rotated until its plane of polarization lies at right angles to that of the first prism and the light beam is prevented from passing through the instrument.

Alternatively, the second prism is rotated so that it corresponds with the plane of polarization produced by the first prism to give a field of maximum brightness. The instrument is set at zero on either of these positions with only solvent in the sample chamber. In practice, the emergent light is seen as two semicircular zones, and the zero is obtained when these two halves of the field of view are of uniform darkness or brightness. The solvent is then replaced with the solution to be investigated and the analyzer rotated to restore the situation of minimum or maximum brightness. The degree and direction of the rotation are then recorded. If the angle of rotation is *clockwise*, then the compound is *dextrorotatory* (+), and if *anticlockwise*, then the sugar is *levorotatory* (−).

11.2.7.2 Specific Rotation

The degree of rotation recorded depends on a number of factors including the length of the light path (l, dm) and the concentration of the solute (c, g/mL) as well as the temperature (t, °C) and the wavelength of the light used (λ, nm). The *specific rotation* is

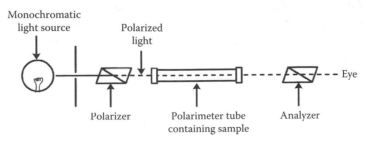

Figure 11.9 The essential features of a polarimeter.

characteristic of a particular compound and is defined as the rotation of monochromatic light caused by 1 g/mL of optically active solute in a 1 dm tube at a fixed temperature:

$$\text{Specific rotation}\left[\alpha\right]_{\lambda}^{t} = \frac{\alpha}{1 \times c}$$

Most simple polarimeters use sodium lamps ($\lambda = 589$ nm), and observations are generally made at 20°C, and in this case, the specific rotation is presented as $\left[\alpha\right]_{D}^{20}$.

11.2.8 Experiment: Mutarotation of Glucose

11.2.8.1 Principle
D-Glucose can be crystallized in either α- or β-form, and freshly prepared solutions of these anomers have specific rotations $\left[\alpha\right]_{D}^{20} + 113°$ and $+ 19°$, respectively. On standing, these solutions show mutarotation and an equilibrium mixture of the α- and β-forms is obtained with a specific rotation of +52.5°.

11.2.8.2 Materials
Polarimeter, α-D-glucose, β-D-glucose, sodium carbonate (0.1 mol/L), stop clocks.

11.2.8.3 Protocol

11.2.8.3.1 Mutarotation in Distilled Water. Rinse the polarimeter tube with distilled water and fill completely with water. Add the last few drops with a Pasteur pipette, and screw on the cap carefully to ensure that no air bubbles are trapped. Adjust the instrument to zero with the tube filled with water in the polarimeter. Empty the polarimeter tube and thoroughly dry it. Carefully transfer 5 g of α-D-glucose to a dry 50 mL flask, add 40 mL of water to dissolve the glucose, and make up to the mark with distilled water. Rapidly mix the solution and fill the polarimeter tube with the glucose: obtain a reading for the rotation as soon as possible and start the stop clock (zero time). Measure the rotation every 10 min for the next 30 min and then at longer time intervals until a constant value is obtained. Repeat the experiment with β-D-glucose and plot a graph of the change in specific rotation with time.

11.2.8.3.2 Mutarotation in Alkali. Repeat the previous experiment with the α- and β-forms of D-glucose, but this time add 1 mL of Na_2CO_3 solution (0.1 mol/L) before making up to the mark. Obtain the first reading as soon as possible, and thereafter, take readings every 2 min for the first 10 min then at longer time intervals as appropriate until no further change is observed.

11.2.8.4 Results
Plot a graph of the change in specific rotation with time. Repeat the experiment in alkali, but this time use sucrose instead of glucose and explain the result.

11.2.9 Experiment: Estimation of Carbohydrates by the Anthrone Method

11.2.9.1 Principle

It is a convenient and rapid colorimetric method for the determination of hexoses, aldopentoses, and hexuronic acids, either free or present in polysaccharides. The blue-green solution shows an absorption maximum at 620 nm, although some carbo-hydrates may give other colors. The reaction is not suitable when proteins containing a large amount of tryptophan are present, since a red color is obtained under these conditions. The extinction depends on the compound investigated but is constant for a particular molecule.

11.2.9.2 Materials

Anthrone reagent: Dissolve anthrone 2 g/L in conc. H_2SO_4.

Stock solution: Glucose (0.1 g/L), glycogen (0.1 g/L), other carbohydrates of the same concentration if desired.

11.2.9.3 Protocol

Add 4 mL of the anthrone reagent to 1 mL of a protein-free carbohydrate solution and mix rapidly (*Caution* Strong acid). Place the tubes in a boiling water bath for 10 min with a marble on top to prevent loss of water by evaporation, cool, and read the extinction at 620 nm against a reagent blank.

11.2.9.4 Results

Prepare standard curves for the glucose and glycogen solutions and compare them. Remember that glucose exists in the glycoside form ($C_6H_{10}O_5$) in glycogen of mol. wt. 162, not 180. Examine the purity of a number of samples of commercial glycogen.

11.2.10 Experiment: Determination of Reducing Sugars Using 3,5-Dinitrosalicylic Acid

11.2.10.1 Principle

Reducing sugars have the property to reduce many of the reagents. One such reagent is 3,5-dinitrosalicylic acid (DNS). 3,5-DNS in alkaline solution is reduced to 3-amino-5-nitrosalicylic acid.

Yellow → Orange-red (Reduction)

The chemistry of the reaction is complicated since standard curves do not always go through the origin, and different sugars give different color yields. The method is therefore not suitable for the determination of a complex mixture of reducing sugars.

11.2.10.2 Materials
Reagents
1. *Sodium potassium tartrate*: Dissolve 300 g sodium potassium tartrate in 500 mL of water.
2. *3,5-DNS*: Dissolve 10 g 3,5-DNS in 200 mL of 2 mol/L sodium hydroxide.
   Preparation of DNS reagent: Prepare this fresh by mixing solutions (1) and (2) and making up to 1 L with water.
3. *Sodium hydroxide*: 2 mol/L.
4. *Stock sugar standards*: Glucose, fructose, and maltose 1 g/L solutions in saturated benzoic acid.
5. *Working sugar standards*: Glucose, fructose, and maltose stock solutions diluted 1:4 before use to give a final dilution to solutions containing 250 µg/mL.
6. Some sugar solutions of unknown concentration, boiling water baths, and marbles.

11.2.10.3 Protocol
Prepare the DNS reagent just before use by mixing the stock solutions as indicated, and add 1 mL of the reagent to 3 mL of the sugar solution in a test tube. Prepare a blank by adding 1 mL of the reagent to 3 mL of distilled water. Cover each tube with a marble and place in a boiling water bath for 5 min; cool to room temperature and read the extinction at 540 nm against the blank.

11.2.10.4 Result
Prepare standard curves of the sugars provided and use them to estimate the concentration of the "unknowns" provided.

11.2.11 Experiment: Determination of Glucose by Means of the Enzyme Glucose Oxidase (β-D-Glucose: Oxygen Oxidoreductase)

11.2.11.1 Principle
Glucose oxidase (EC 1.1.3.4) is an oxidoreductase enzyme that catalyzes the oxidation of glucose to hydrogen peroxide and D-glucono-δ-lactone. It is often extracted from *Aspergillus niger* and catalyzes the oxidation of β-D-glucopyranose to D-glucono-1,5-lactone with the formation of hydrogen peroxide: the lactone is then slowly hydrolyzed to D-gluconic acid. The enzyme is specific for β-D-glucopyranose, but most enzyme preparations contain mutarotatase, which catalyzes the interconversion of the α- and β-forms. D-Mannose and D-xylose are hydrolyzed by the enzyme at about one-hundredth of the rate of D-glucose, and the only other common sugar affected is D-galactose, which is hydrolyzed at only one-thousandth of the rate. The method is therefore highly specific for glucose. Peroxidase is incorporated into the reaction mixture and catalyzes the reaction of hydrogen peroxide with the chromogen ABTS (2.2'-azino-di-[3-ethylbenzthiazoline]-6-sulfonate) to give a color that is read at 437 nm. Test is widely used for the determination of free glucose in body fluids (diagnostics) and in the food industry. It also has many applications in biotechnology, typically enzyme assays for biochemistry including biosensors in nanotechnology.

11.2.11.2 Materials

1. Zinc sulfate: 10 g/100 mL $ZnSO_4 \cdot 7H_2O$ in water.
2. Isotonic sodium sulfate: 93 mmol/L in water.
3. Sodium sulfate–zinc sulfate reagent: Prepare by mixing 55 mL of solutions (1) raising to 1 L with solution (2).
4. Sodium hydroxide: 0.5 mol/L.
5. Sodium phosphate buffer: 0.05 mol/L, pH 7.0.
6. Glucose oxidase.
7. ABTS (2.2′-azino-di-[3-ethylbenzthiazoline]-6-sulfonate).
8. Peroxidase.
9. Glucose oxidase reagent: Prepare this reagent fresh by dissolving 25 mg of glucose oxidase and 190 mg of ABTS in the sodium phosphate buffer. Add a small quantity of peroxidase (2 mg) and make up to 250 mL with the buffer. This solution is active for about 4 weeks if stored in a brown bottle at 4°C.
10. Glucose standard: 0.5 mmol/L freshly prepared.
11. Water baths at 37°C.

11.2.11.3 Protocol

Test: Pipette 0.1 mL of the test solution (blood, etc.) into 1.8 mL of the sodium sulfate–zinc sulfate reagent in a centrifuge tube. Add 0.1 mL of 0.5 mol/L sodium hydroxide, centrifuge, and take 0.5 mL of the supernatant in duplicate.

Blank: Take 0.5 mL of distilled water.

Standards: Use 0.5 mL of a range of glucose solutions suitably diluted from the standard. Add 5 mL of the glucose oxidase reagent, incubate for 1 h at 37°C, and read the extinction at 437 nm against the reagent blank.

Blood sugar: Use the glucose oxidase reagent to determine the glucose concentration in blood obtained from a finger prick.

Sugars in honey: Use the DNS and the glucose oxidase methods to estimate the glucose and fructose content in the mixture of these two sugars as found in honey.

$$\beta\text{-D-Glucopyranose} + FAD \rightleftharpoons \text{D-Glucono-1, 5-lactone} + FADH_2$$

$$\text{D-Glucono-1, 5-lactone} + H_2O \longrightarrow \text{D-Gluconic acid}$$

$$FADH_2 + O_2 \longrightarrow H_2O_2 + FAD$$

$$\overline{\beta\text{-D-Glucopyranose} + H_2O + O_2 \longrightarrow \text{D-Gluconic acid} + H_2O_2}$$

11.2.12 Experiment: Determination of Starch in Plant Tissues

11.2.12.1 Principle

Sugars are first extracted by treating the finely powdered dried grain or leaf sample repeatedly with hot 80% (v/v) alcohol. The residue is then treated with cold perchloric acid to solubilize starch. After filtration, starch in the perchloric acid extract is hydrolyzed to glucose in hot acidic medium that undergoes dehydration to hydroxymethyl furfural. This derivative is then condensed with anthrone to give a blue-colored complex that is determined quantitatively by anthrone sulfuric acid procedure.

11.2.12.2 Materials

Dry leaf powder, test tubes, water bath, refrigerated centrifuge, polypropylene tubes, colorimeter, cuvettes, anthrone, sulfuric acid, ethyl alcohol, perchloric acid, glucose.

Anthrone–sulfuric acid reagent: Dissolve 2 g of anthrone in cold 95% conc. H_2SO_4, store at 4°C. Use freshly prepared reagent.

80% absolute alcohol: Eighty milliliters of ethyl alcohol make up to 100 mL with distilled water.

52% perchloric acid: Add 270 mL of 72% perchloric acid to 100 mL of water.

Glucose standard: Dissolve 0.1 g of anhydrous glucose in 100 mL of water containing 0.001% benzoic acid as preservative. The standard solution contains 1 mg glucose per mL.

11.2.12.3 Protocol

1. For extraction of sugars and starch, take 0.2 g of finely ground sample in a 50 mL centrifuge tube, and add 20 mL of hot 80% alcohol. Shake the tubes for 5–10 min, and after centrifuging at 3000 rpm for 10 min, decant the supernatant. Repeat this extraction with 80% hot ethanol five to six times till the supernatant is free of sugars as judged by a negative test with anthrone reagent.
2. Cool the residue in ice water and add 6.5 mL of 52% perchloric acid while stirring the contents with a glass rod. Let it stand for 15 min with occasional stirring and then centrifuge at 4°C. Collect the supernatant and repeat this extraction Step four to five times. Combine the supernatant fractions and make up the volume to 100 mL with water.
3. Dilute the previous extract so that it contains 5–20 μg of glucose per mL. Take 5 mL aliquot of this diluted extract and place the tubes in a cold water bath. Add 10 mL freshly prepared anthrone reagent. Mix properly and transfer the tubes to boiling water bath for 5–7 min. After cooling the tubes under running tap water, note the absorbance of these solutions at 630 nm.
4. Prepare a standard curve using 0–100 μg glucose and anthrone reagent (Step 3). Calculate the amount of glucose in the sample aliquot.

11.2.12.4 Results and Calculations

1. From the standard curve, determine the amount of glucose in 5 mL of the diluted aliquot of sample extract.
2. Calculate the quantity of glucose in the original extract by applying the dilution factor.
3. Multiply the obtained value by 0.9 for conversion of glucose value to starch.
4. Calculate the amount of starch in 1 g of the powdered sample. Tabulate the results after calculations.

11.2.13 Experiment: Isolation and Assay of Glycogen from the Liver and Skeletal Muscle of Rats

11.2.13.1 Principle

Liver glycogen: The liver glycogen maintains the level of the blood glucose and represents a central reserve of fuel for the body tissues. In a well-fed animal, glucose

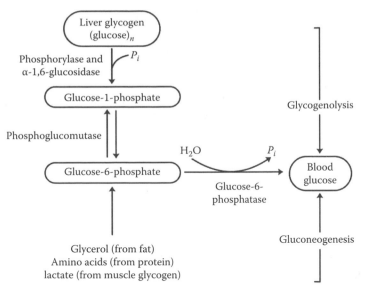

Figure 11.10 Blood glucose maintenance.

is converted to glycogen in the liver (*glycogenesis*), but during starvation, the liver glycogen is broken down to glucose (*glycogenolysis*) and depleted in about 24–48 h. The blood glucose is maintained by the synthesis of glucose from noncarbohydrate sources (*gluconeogenesis*) after incubation period (Figure 11.10).

Muscle glycogen: Muscle glycogen differs from liver glycogen in that it is not particularly affected by the state of the diet and, in the absence of violent exercise, remains fairly constant.

The muscle glycogen cannot contribute directly to the blood glucose as the enzyme glucose-6-phosphatase is missing from muscle, but can do so indirectly when lactate is formed during anaerobic contractions (Figure 11.10).

Isolation of glycogen: Glycogen is released from the tissue by heating with strong alkali and precipitated by the addition of ethanol. Sodium sulfate is added as a coprecipitant to give a quantitative yield of glycogen. The polysaccharide is then hydrolyzed in acid and the glucose released estimated.

11.2.13.2 Materials
Rats of the same age, sex, and weight (fed and starved for 48 h in triplicate), potassium hydroxide (30% in water), calibrated centrifuge tubes, boiling water baths, saturated Na_2SO_4, ethanol (95% v/v), volumetric flasks (100 mL), test tubes calibrated at 10 mL, HCl (1.2 mol/L), marbles, phenol red indicator solution, NaOH (0.5 mol/L), reagents for the estimation of glucose as described in earlier experiment.

11.2.13.3 Protocol

11.2.13.3.1 Isolation of Glycogen. Kill the rat and accurately weigh about 1.5 g of liver and skeletal muscle. Place the tissues into a calibrated centrifuge tube containing

2 mL of KOH (30%), and heat in a boiling water bath for 20 min with occasional shaking. Cool the tubes in ice, add 0.2 mL of saturated Na_2SO_4, and mix thoroughly. Precipitate the glycogen by adding 5 mL of ethanol (95% v/v), stand on ice for 5 min, and remove the precipitate by centrifugation. Discard the supernatant and dissolve the precipitated glycogen in 5 mL of water with gentle warming, then dilute with distilled water to the 10 mL calibration mark, and mix thoroughly. In the case of the fed animals, transfer the liver sample quantitatively to a 100 mL volumetric flask and make up to the mark with water.

11.2.13.3.2 Hydrolysis and Estimation of Glycogen.   Pipette duplicate 1 mL samples of the glycogen solutions into test tubes calibrated at 10 mL, add 1 mL of HCl (1.2 mol/L), place a marble on top of each tube, and heat in a boiling water bath for 2 h. At the end of this period, add one drop of phenol red indicator and neutralize carefully with NaOH (0.5 mol/L) until the indicator changes from pink through orange to a yellow color. Dilute to 5 mL with distilled water and determine the glucose content by the glucose oxidase method.

11.2.13.4 Results
Calculate the grams of glycogen per 100 g of tissue and remember that glucose exists in the glycoside form ($C_6H_{10}O_5$) in glycogen with mol. wt. of 162, not 180.

11.2.14 Experiment: Acid Hydrolysis of Polysaccharides

11.2.14.1 Principle
Acid hydrolysis of the α-1-4 and α-1-6 glycosidic bonds joining the glucose residues in glycogen is quite random so that a whole range of oligosaccharides may be formed as intermediates although the final product is glucose. Other polysaccharides are more resistant or more readily hydrolyzed than glycogen.

11.2.14.2 Materials
Glycogen, other polysaccharides such as cellulose, inulin and chitin, hydrochloric acid (2 mol/L), sodium hydroxide (4 mol/L), weak anion-exchange resin, DNS reagent, iodine solution (5 mol/L) in KI (30 g/L).

11.2.14.3 Protocol
Dissolve 50 mg of glycogen in 10 mL of water and pipette 1 mL into a series of seven test tubes. Add 1 mL of water and 2 mL of 2 mol/L HCl to each tube and place in a boiling water bath. Place a marble on top of each tube to prevent excessive loss due to evaporation and remove a tube at convenient time intervals up to 1 h. Remove one drop of the solution with a Pasteur pipette and mix with a drop of iodine on a white tile. Neutralize the contents of each tube with 1 mL of 4 mol L NaOH, and estimate the total reducing power with the DNS reagent. Plot the progress of the hydrolysis and compare with the color changes seen with iodine with time. Repeat the hydrolysis with one other polysaccharide and compare with

glycogen. Meanwhile add 2 mL of the glycogen solution to 2 mL of 2 mol/L HCl and hydrolyze for 1 h as the previous solution. Cool and then stir with several batches of the weak ion-exchange resin in the hydroxyl form until the mixture has an approximately neutral pH. Decant the supernatant and use it for chromatography. Repeat the hydrolysis with the other polysaccharides using the appropriate optimum conditions.

11.2.15 Experiment: Enzymatic Hydrolysis of Glycogen by α- and β-Amylase

11.2.15.1 Principle
α-Amylase catalyzes the specific hydrolysis of the α-1-4 glycosidic bonds in glycogen but is without effect on the β-linkages in cellulose. The α-1-6 bonds are not affected neither are the α-1-4 bonds linking the glucose units of maltose. Hydrolysis proceeds in a random manner with the formation of a series of intermediates. The final product is mainly maltose with some glucose and maltotriose. The increase in the number of reducing end groups as the hydrolysis proceeds is followed using the dinitrosalicylate reagent, and the final product is identified chromatographically. β-Amylase from higher plants catalyzes the hydrolysis of the α-1-4 bonds to give maltose, but is an exoamylase and therefore splits off maltose units from the free nonreducing end of the chain. When a branch point is reached, the digestion ceases leaving a limit dextrin. The name β-amylase is given to the enzyme as the initial product is β-maltose that later forms an equilibrium mixture of α- and β-maltose. As with α-amylase, the progress and extent of the hydrolysis are followed by determining the amount of reducing sugar produced.

11.2.15.2 Materials
α-Amylase from saliva and β-amylase from plants, rat liver glycogen preparation, phosphate buffer (0.1 mol/L, pH 6.7) containing 0.05 mol/L sodium chloride, dinitrosalicylate (deoxyribonucleic acid [DNA]) reagent, water bath at 37°C.

11.2.15.3 Protocol
Dissolve 50 mg of glycogen in 10 mL of buffered saline and pipette 1 mL into a series of seven test tubes. Add 3 mL of saliva diluted in 1:100 with water and incubate the tubes at 37°C for up to 1 h. Remove a tube at suitable time intervals, add 1 mL of water, and stop the reaction by the addition of 1 mL of dinitrosalicylate reagent. Meanwhile add 6 mL of α-amylase to 2 mL of glycogen solution and incubate for 1 h at 37°C. Desalt the sample with a mixed bed resin and use the supernatant for chromatography. Repeat the previous experiment with a suitably diluted preparation of β-amylase in place of the saliva, and determine the amount of reducing sugar released.

11.2.15.4 Result
Compare the amount of maltose released with that obtained when α-amylase is used, and calculate the percentage degradation of the glycogen by the β-amylase (α-1,4-glucan maltohydrolase 3.2.1.2).

11.2.16 Experiment: Breakdown of Glycogen and the Production of Glucose-ʟ-Phosphate by Muscle Phosphorylase

11.2.16.1 Principle

Glycogen is present in large quantities in the liver and muscles of a well-fed animal. In between meals, this store of carbohydrate is degraded to provide energy, and the first Step in the metabolism is the cleavage of the molecule by phosphorylase. The enzyme catalyzes the removal of one glucose unit at a time from the straight chain parts of the molecule and the formation of glucose-1-phosphate (G-1-P) from inorganic phosphate:

$$(\text{Glucose})_n + P_i \xrightarrow{\text{Phosphorylase}} (\text{Glucose})_{n-1} + \text{glucose-1-phosphate}$$

Cleavage or lysis of the glycosidic link by water is known as hydrolysis and by phosphate as phosphorolysis, hence the name of the enzyme phosphorylase. The reaction is readily reversible but *in vivo* the concentration of inorganic phosphate is such that the equilibrium favors degradation and not synthesis.

The reaction mixture contains cysteine to maintain the thiol groups (–SH) of the enzyme protein in a reduced form, while fluoride is present to inhibit any phosphoglucomutase that would otherwise catalyze the conversion of G-1-P to G-6-P. The AMP is added for maximum enzyme activity as phosphorylase actually exists in two forms *a* and *b*, and the *b* form is only active in the presence of AMP.

The G-1-P produced is isolated as the barium salt and, being a hemiacetal, is readily hydrolyzed by dilute acid (0.5 mol/L H_2SO_4) to glucose and phosphate. The reaction is virtually complete in 10 min at 100°C, and this is used to check the stoichiometry of the product.

11.2.16.2 Materials

Glycogen phosphorylase, sodium-β-glycerophosphate buffer (10 mmol/L pH 6.8 containing 3 mmol/L cysteine and 0.1 mol/L NaF), glycogen (1.6% w/v in 0.26 mol/L phosphate buffer pH 6.8 containing 6 mmol/L cysteine, 2 mol/L AMP, and 0.2 mol/L NaF), amylase.

Barium acetate (200 g/L), phenol red indicator (10 g/L in aqueous ethanol), sodium hydroxide (2 mol/L), sulfuric acid (1 mol/L), absolute ethanol, reagents for phosphate estimation described earlier, boiling water bath, water bath at 37°C, measuring cylinder (25 mL), crushed ice, reagents for the estimation of reducing sugars (DNS reagent), G-1-P (10 mmol/L), glucose-6-phosphate (10 mmol/L).

11.2.16.3 Protocol

11.2.16.3.1 Incubation. Mix 5 mL of the β-glycerophosphate buffer with 5 mL of the glycogen solution in a boiling tube and equilibrate at 37°C for 10 min; start the reaction by adding 0.2 mL of the phosphorylase, suitably diluted in the β-glycerophosphate buffer. Incubate for 30 min at 37°C and stop the reaction by placing the test tube into a boiling water bath for 2 min.

11.2.16.3.2 Isolation.   Cool the tube under running tap water and remove the excess glycogen by adding 1 mL of the amylase solution and leaving at room temperature for 30 min. After the incubation with amylase, add 2 mL of the barium acetate solution and two drops of phenol red indicator followed by 1 mol/L NaOH added dropwise until it becomes pink. Again mix thoroughly and remove the precipitate of barium phosphate by centrifuging at top speed on a bench centrifuge. Carefully remove the supernatant, measure the volume, and place in a flask; now add three times the volume of absolute ethanol, mix well, and leave to stand for 30 min on ice. The precipitate formed under these conditions is the barium salt of G-1-P. Decant the supernatant and centrifuge the suspension for 10 min at top speed on a bench centrifuge. Remove the supernatant and allow the precipitate to drain by inverting the tubes onto filter paper. Repeat the precipitation procedure after adding 4 mL of water to the precipitated sugar phosphate.

11.2.16.3.3 Stoichiometry.   Dissolve the barium salt in 5 mL of water and pipette 2 mL into a test tube containing 2 mL of 1 mol/L H_2SO_4. Mix thoroughly and place in a boiling water bath for 10 min. Cool under running tap water, then add 2 mL of 2 mol/L NaOH to neutralize the reaction, and centrifuge and assay the supernatant for reducing sugar and phosphate. Pipette another 2 mL sample of the phosphate ester into 2 mL of H_2SO_4, but this time add the NaOH immediately after and do not place the tube in a boiling water bath. This nonhydrolyzed control is also assayed for phosphate and reducing sugar. Calculate the yield of G-1-P from the sugar and phosphate assays assuming complete hydrolysis.

11.2.16.4 Results
Compare the results with those obtained after the acid hydrolysis of 1 mmol/L solutions of pure G-1-P and glucose-6-phosphate diluted from 10 mmol/L stock solutions.

11.3 Amino Acids and Proteins

11.3.1 Amino Acids

Amino acids are molecules that contain both amino and carboxylic acid functional groups and are the building blocks of long polymer chains. With 2–10 amino acids, such chains are called peptides; with 10–100, they are often called polypeptides, and longer chains are known as proteins (Figure 11.11). These protein structures have many structural and functional roles in organisms.

There are 20 amino acids that are encoded by the standard genetic code, but there are more than 500 natural amino acids. When amino acids other than the set of 20 are observed in proteins, this is usually the result of modification after translation (protein synthesis). Only 2 amino acids other than the standard 20 are known to be incorporated into proteins during translation.

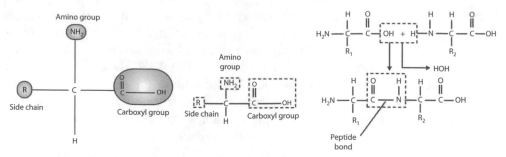

Figure 11.11   Peptide bonding with amino group, side chain, and a carboxyl group.

In certain organisms,

- Selenocysteine is incorporated into some proteins at a UGA codon, which is normally a stop codon.
- Pyrrolysine is incorporated into some proteins at a UAG codon, for instance, in some methanogenic enzymes that is used to produce methane.

Besides those used in protein synthesis, other biologically important amino acids include carnitine (used in lipid transport within a cell), ornithine, GABA, and taurine.

11.3.2 Protein Structure

The particular series of amino acids that form a protein is known as the protein's primary structure. Proteins have several, well-classified, elements of local structure, and these are termed secondary structure. The overall 3D structure of a protein is termed its tertiary structure. Proteins often aggregate into macromolecular structures called quaternary structure (Figure 11.12).

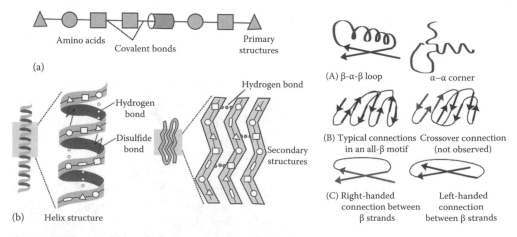

Figure 11.12   (a) Primary and (b) secondary structures of proteins.

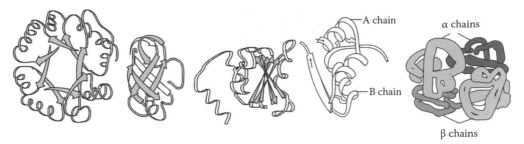

Figure 11.13 Structure details of proteins of different motifs.

11.3.2.1 Metalloproteins

A metalloprotein is a protein that contains a metal cofactor. The metal may be an isolated ion or may be coordinated with a nonprotein organic compound, such as the porphyrin group found in hemoproteins. In some cases, the metal is coordinated with both a side chain of protein and an inorganic nonmetallic ion; this type of protein–metal–nonmetal structure is found in iron–sulfur clusters (Figure 11.13).

11.3.3 Amino Acid Chemistry

11.3.3.1 Acid–Base Properties

As the name suggests, amino acids are organic compounds that contain amino and carboxyl groups and, therefore, possess both acidic and basic properties. There are a large number of chemically possible amino acids, but only a few of these occur naturally. In the case of the 22 or so amino acids found in proteins, nearly all of them are α-amino acids, where the amino group is present on the α-carbon atom (Figure 11.14).

Amino acids are unlike low molecular weight organic compounds in their properties and resemble inorganic salts. In general, they are readily soluble in aqueous media, but only slightly soluble or insoluble in organic solvents. Their melting points are also very high for low molecular weight organic compounds. This is because amino acids exist mostly as *zwitterions* and not as unionized molecules. The high melting points are due to the high energy required to break the ionic bonds of the crystal lattice. The strong positive charge on the $-NH_3^+$

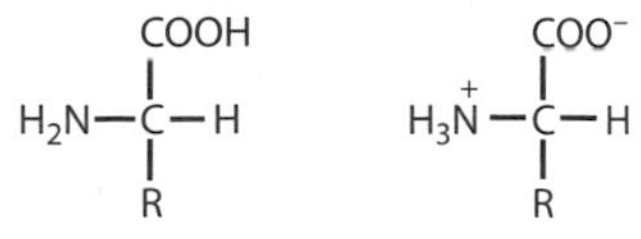

$$
\begin{array}{cc}
\text{COOH} & \text{COO}^- \\
| & | \\
H_2N-C-H & H_3\overset{+}{N}-C-H \\
| & | \\
R & R \\
\text{(Unionized form)} & \text{(Zwitterion)}
\end{array}
$$

Figure 11.14 An α-amino acid.

TABLE 11.5   pK_a for Glycine, For Example, Is Much Lower than the Corresponding Aliphatic Acid

	$\overset{\text{H}}{\underset{}{|}}$ $^+\text{H}_3\text{N}{-}\text{CH}{-}\text{COO}^-$	$\overset{\text{H}}{\underset{}{|}}$ $\text{H}{-}\text{CH}{-}\text{COOH}$	$\overset{\text{H}}{\underset{}{|}}$ $\text{H}_2\text{N}{-}\text{CH}{-}\text{H}$
	Glycine	Acetic Acid	Methylamine
Physical state at 25°C	Solid	Liquid	Gas
Melting point	232°C–236°C	17°C	−94°C
pK –COOH	2.4	4.8	—
pK –NH$_2$	9.7	—	10.7

group induces a tendency for the –COOH group to lose a proton, so that amino acids are strong acids.

Some amino acids contain ionizable groups in the side chain R, and these affect their physical characteristics, whether the amino acids are free in solution or in combination with others in a protein. The charge properties of proteins are determined to a large extent by the ionizable groups of the amino acid side chains. The ionizable groups present in the side chains of amino acids, together with their pK values, are given in Table 11.5.

11.3.3.2 Isoelectric Point

Amino acids migrate in an electric field, and this property is the basis of one method for their separation. The direction and extent of migration depend to a large extent on the predominant ionic form present, and this is determined by the pH of the electrophoresis buffer. The pH at which there is zero net charge and no migration in an electric field is known as the isoelectric point. For amino acids containing only one –COOH and one –NH$_2$ as the ionizable groups, the isoelectric point (*pI*) is halfway between the pK values of these groups, so that in the case of alanine, $pI = \frac{1}{2}\,(2.4 + 9.7) = 6.1$.

When other charged groups are present, the calculation of the *pI* is not so simple, but as a rough guide, the isoelectric point lies midway between the pK values of similar groups (Table 11.6).

11.3.3.3 Stereochemistry

The α-carbon atom is asymmetric and is a chiral center for all amino acids except glycine so that, apart from glycine, all amino acids show optical activity. If serine is taken as the parent compound, then this can be compared with L(−) glyceric acid, the parent compound for the L series of sugars (Figure 11.15).

When –CH$_2$OH is replaced by other groups, two families of amino acids emerge: the D and L series. All of the amino acids present in proteins are of the L configuration although the D form is found in antibiotics and bacterial cell walls. The L and D refers to the absolute configuration about the chiral center and not to the optical activity.

TABLE 11.6 p*K* Values of Ionizable Groups Found in the Side Chains of Some Amino Acids

Ionizing Group		Amino Acid	p*K*
β-Carboxyl		Aspartic acid	3.9
γ-Carboxyl	$-COOH \rightleftharpoons COO^- + H^+$	Glutamic acid Histidine	4.3 6.0
Imidazole	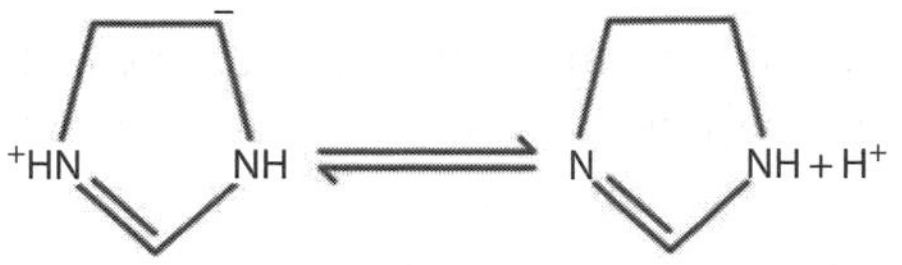	Histidine	6.0
Sulfhydryl	$-CH_2SH \rightleftharpoons CH_2S^- + H^+$	Cysteine	8.3
Phenolic	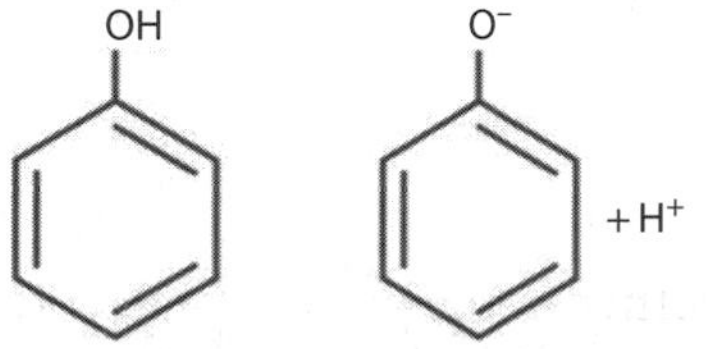	Tyrosine	10.1
ε-Amino	$-NH_3^+ \rightleftharpoons NH_2 + H^+$	Lysine	10.5
Guanidino	$NH_2-C(=NH_2^+)-NH \rightleftharpoons NH_2-C(=NH)-NH + H^+$	Arginine	12.5

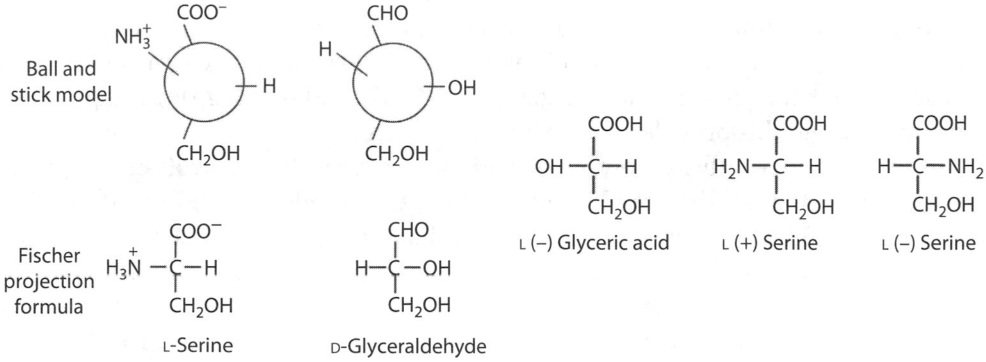

Figure 11.15 Stereochemistry of some amino acids.

Some amino acids such as isoleucine, threonine, and hydroxylysine contain a second chiral center so that more than two forms are possible.

11.3.4 Amino Acid Composition of Proteins

11.3.4.1 Formulae of Amino Acids

Just as monosaccharides are the basic unit of polysaccharides, so amino acids can be thought of as the "bricks" from which the protein "house" is built. Polysaccharides are built up of usually only a few monosaccharide units, but proteins may contain as many as 22 different amino acids. The names and abbreviations of the amino acids commonly found in proteins are given in Figures 11.16 and 11.17. The formulae of all the α-amino acids are the same except for the nature of the side chain R.

11.3.4.2 Peptide Bond

The amino acids are joined together in the protein molecule by peptide ($-CO-NH-$) formed by the condensation of the α-COOH of one amino acid and the α-NH$_2$ group of another one. Low molecular weight polymers of the amino acids are known as polypeptides, while the term proteins is usually reserved for the larger polymers of molecular weight several thousand or more (Figure 11.18).

11.3.5 Protein Structure

11.3.5.1 Primary Structure

Several levels of structural organization can be recognized in proteins, and the first of these is the primary structure, which is the sequence of amino acids. The constituent amino acids of a pure protein can be readily determined by separating them by chromatography or electrophoresis after chemical or enzymatic hydrolysis of the peptide bonds. The quantitative analysis of the protein by automatic ion-exchange chromatography is fairly straightforward, but trying to place the hundreds or thousands of amino acid residues in the right order is more difficult. Fortunately, chemical methods are available for the identification of the free –COOH and –NH, groups of proteins, and peptides, and this provides the key to the problem.

The protein is partially hydrolyzed by acid that gives random hydrolysis and by the use of enzymes that catalyze the hydrolysis of peptide bonds between specific amino acids. A large number of peptides are obtained that are separated by chromatography and electrophoresis and the C- and N-terminal groups identified. The structures of the di- and tripeptides are determined. These are then used to elucidate the amino acid sequence of the large peptide fragments until the complete sequence is known rather like constructing a jigsaw puzzle.

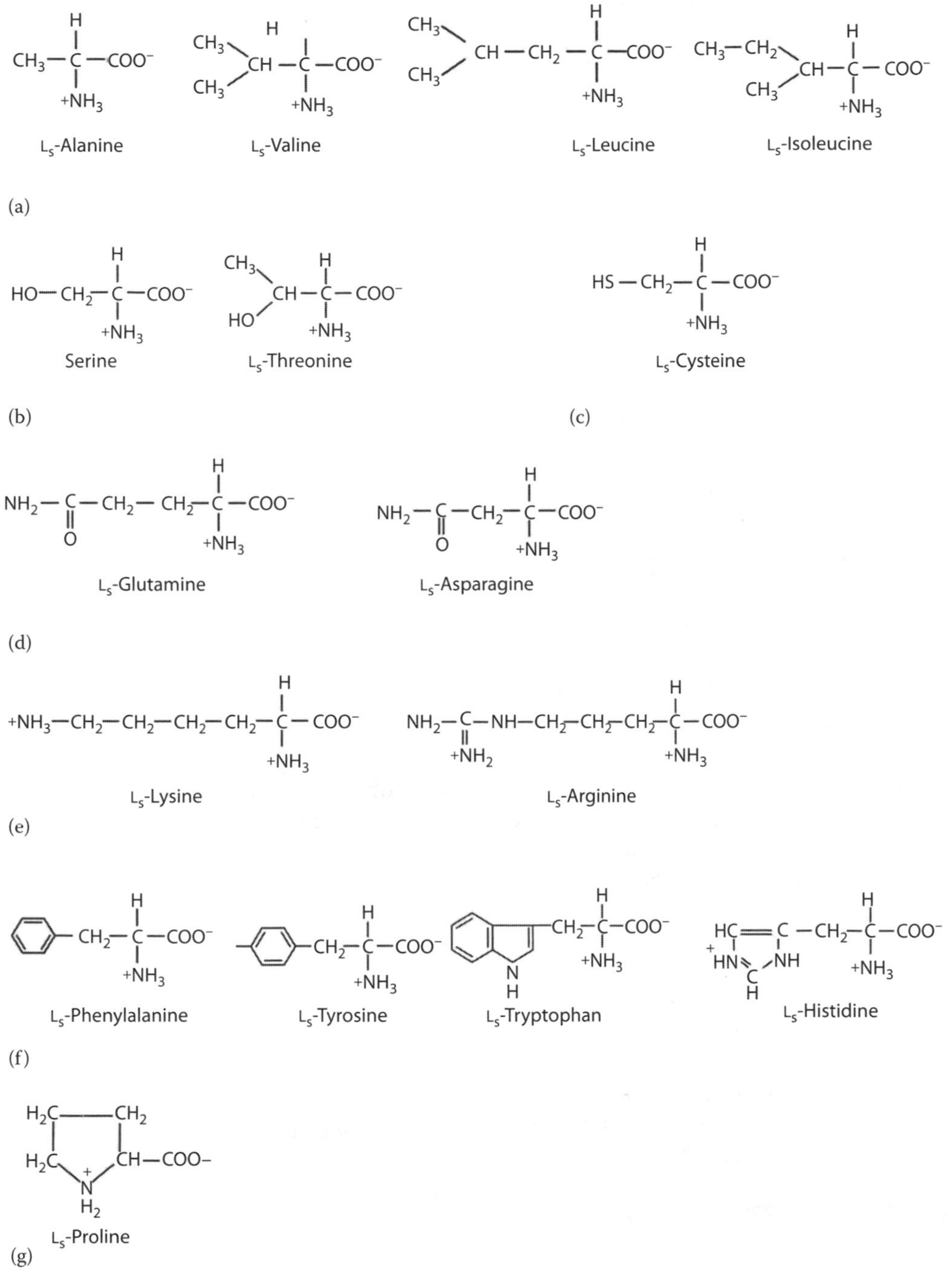

Figure 11.16 Some common amino acids found in proteins: (a) aliphatic amino acids, (b) hydroxylic amino acids, (c) sulfur-containing amino acids, (d) acidic amino acids, (e) basic amino acids, (f) aromatic and heterocyclic amino acids, and (g) imino acids

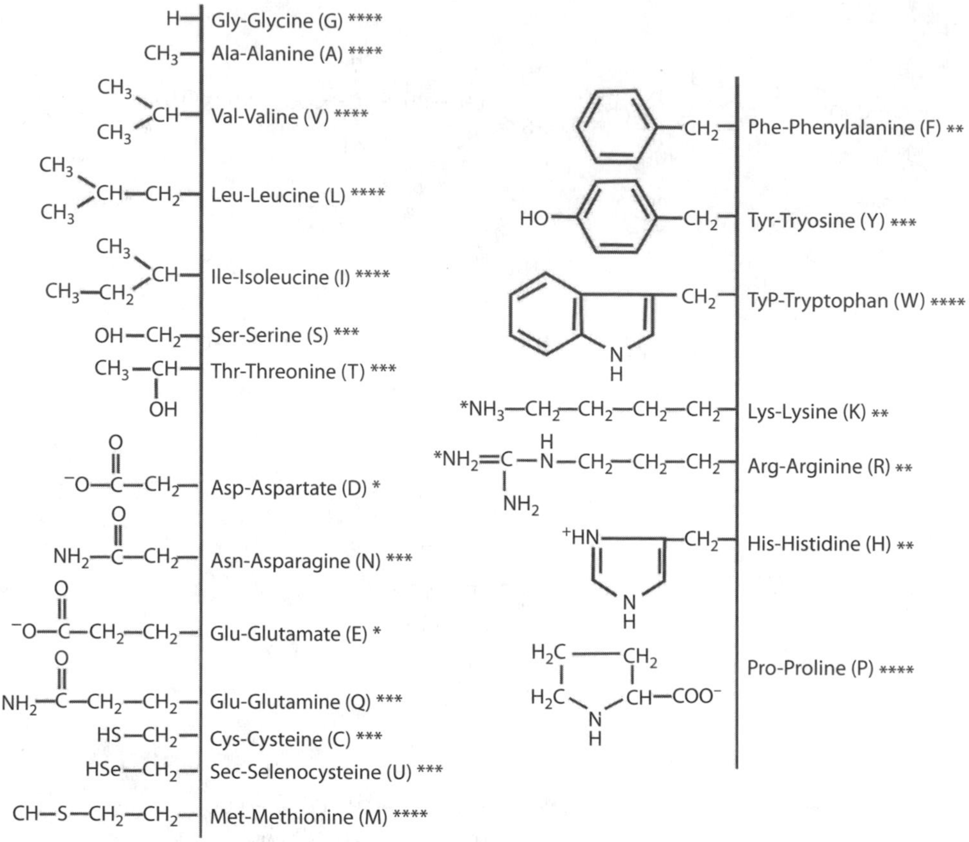

Figure 11.17 Structure of the amino acid "R" groups of 20 amino acids. The entire structure of proline is shown here since proline is an exception in lacking a free amino group; hence, it is termed an imino instead an amino acid. *ionizable acidic, **ionizable basic, ***nonionizable polar, ****nonpolar (hydrophobic).

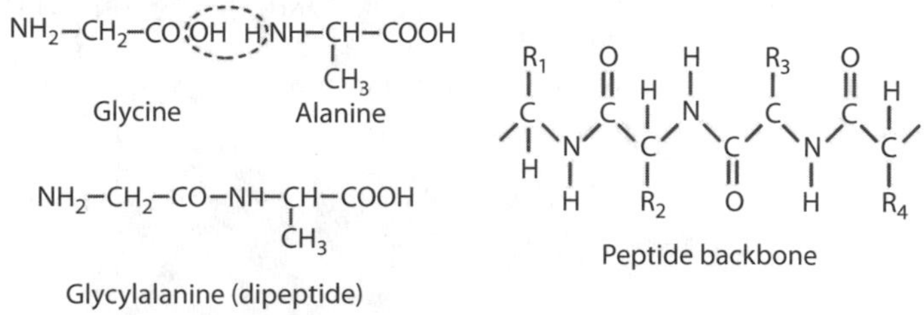

Figure 11.18 Peptide bonding.

11.3.5.2 Secondary Structure

X-ray studies suggest that the peptide chain can exist in the form of a coil or helix. A number of helical forms are known today, but only the α-helix met all the requirements for maximum stability. This helical form has 3.6 amino acid residues per complete turn and a rise along the central axis of 0.15 nm per residue.

The shape of the structure is maintained by intramolecular hydrogen bonds between the carbonyl oxygen and the amide nitrogen three residues apart in the peptide backbone:

$$\text{>C=O--H-N<}$$

The hydrogen bond is fairly weak, but the large number involved in the formation of the α-helix maintains the structure in this stable form. The amino acid side chains can be accommodated in the α-helix since these "stick out" into space away from the coil. The imino acids proline and hydroxyproline, however, do not fit into the normal α-helix, and where these are present a kink or change in the direction of the chain occurs. This is because of the rigid nature of the structure of these imino acids.

Another form of secondary structure is the β-pleated sheet, where hydrogen bonding is between two peptide chains that may be parallel, in that their N atoms point in the same direction, or antiparallel, where alternate chains are orientated the same way. The β-form is found in fibrous proteins such as hair keratin, while the α-helix may be present in both fibrous and globular proteins like albumin and myoglobin.

11.3.5.3 Tertiary Structure

The tertiary structure of a protein is the arrangement in space of the molecular threads or, in other words, the overall shape of the protein molecule. Many protein molecules behave as if they are very compact and are, therefore, known as *globular proteins*. Other proteins are more rigid and form long thin threads; these are the *fibrous proteins*. The tertiary structure is maintained by a number of bonds of the type shown in the following text. The covalent disulfide bond formed between cysteine residues is the strongest and confers a certain amount of rigidity on the protein. The ionic bond occurs when an ionized acidic and basic group are brought into proximity, and this bond of moderate strength is important in the binding of basic proteins with acidic macromolecules as in the formation of nucleoproteins (Table 11.7).

Hydrogen bonds are weak but numerous and can be formed between the amide nitrogen and the carbonyl oxygen of the peptide backbone as well as groups present in the side chain. The side chains of aspartic acid, glutamic acid, tyrosine, histidine, serine, and threonine are all capable of hydrogen bond formation. Hydrophobic bonds arise from a tendency of the nonpolar side chains of amino acid residues to associate with each other. The hydrophilic groups are associated with water and are found on the surface of the protein molecule. This type of association dictates the folding of the polypeptide chain and, therefore, the overall shape of the molecule.

11.3.5.4 Quaternary Structure

A number of proteins are made up of polypeptide units, which are not covalently linked, and the association of these subunits to form the molecule confers quaternary

TABLE 11.7   Ionic Bonding of Basic Protein with Acidic Macromolecules

Amino Acid Residue	Cys	Asp	Ser	Ileu
Amino acid residue	Cys	Lys	His	Phe
Bond strength	Covalent (strong)	Ionic (moderate)	Hydrogen (weak)	Hydrophobic (very weak)

structure on the protein. Probably the best-known example of this is hemoglobin that consists of 2α- and 2β-subunits.

11.3.6 Function in the Living Organism

The name protein is derived from the Greek *proteios* meaning "of primary importance," and the name is fully justified. The main function of proteins is to act as essential components of structural material, in contrast to that of carbohydrates and fats that are to provide energy. This does not mean that proteins are static compounds; on the contrary, they are in a state of continuous flux with regard to their synthesis and degradation.

11.3.6.1 Amino Acids

The amino acids found in proteins arise mainly from the digestion of dietary proteins. Some amino acids can be synthesized by the animal, and these are known as *nonessential* amino acids, while others, the *essential* amino acids, must be supplied in the diet.

The α-amino acids, as well as being concerned in protein synthesis, are also involved in the synthesis of a number of compounds of biological importance, some of which are given in Table 11.8.

11.3.6.2 Peptides

The term peptide is generally used for polymers containing up to 50 amino acid residues or with a molecular weight of less than 5000. Peptide chains are found attached to carbohydrate material to form the peptidoglycans of bacterial cell walls and the glycoprotein of blood group substances. In the former case, many of the

TABLE 11.8   Some Important Amino Acids and Derivatives

Name and Formula		Occurrence and Role
γ-Aminobutyric acid	COOH | $(CH_2)_3$ | NH_2	This is formed in the brain by decarboxylation of glutamic acid, where it may act as a chemical mediator in the transmission of the nerve impulse between some neurons.
Asparagine	COOH | $CHNH_2$ | CH_2CONH_2	Asparagine is present in a number of plant tissues, where it acts as a reservoir of nitrogen for proteins synthesis during germination.
Diaminopimelic acid	COOH | $CHNH_2$ | $(CH_2)_3$ | $CHNH_2$ | COOH	This amino acid is an important constituent of the mucopeptide of bacterial cell walls.
3,4-Dihydroxyphenylalanine	$CH_2CH(NH_2)COOH$ OH OH	This amino acid, known as DOPA for short, is a precursor of the important pigment melanin. This pigment is responsible for the color of hair, skin, and eyes.
Histamine	$CH_2CH_2NH_2$ N HN	Histamine arises from histidine by decarboxylation. It is a vasodilator and is involved in shock and allergic responses.

amino acids are of the D and not the usual L configuration. A number of peptides are also found in the free state and are probably intermediates in the turnover of proteins, although some peptides are synthesized to perform specific biological functions (Table 11.9).

11.3.6.3 Proteins

An adequate intake of protein is essential for higher animals since only the simple forms of life are able to synthesize their protein from other nitrogen sources. Proteins are present in all tissues of the body and make up a large part of the structure of the cell. In addition, a number of proteins have specialized physiological roles.

TABLE 11.9   Some Examples of Biologically Important Peptides

Peptide	Structure
Glutathione	Glu–Cys–Gly

This is a tripeptide with one of the peptide bonds formed through the γ rather than the α-carboxyl group of glutamic acid. Its main function is to protect the thiol groups of molecules and membranes by keeping them in the reduced state.

Gramicidin S

$$D\text{-}Phe \text{--} L\text{-}Leu \text{--} LOrn \text{--} L\text{-}Val\text{--}L\text{-}Pro$$
$$|\qquad\qquad\qquad\qquad\qquad|$$
$$L\text{-}Pro \text{--} L\text{-}Val \text{--} LOrn \text{--} L\text{-}Leu \text{--} D\text{-}Phe$$

This cyclic decapeptide is an antibiotic from the bacterium *Bacillus brevis* and contains the unusual amino acids D-phenylalanine and L-ornithine. A number of other antibiotics are also peptides.

Oxytocin

$$\overbrace{\ S\text{------}S\ }$$
$$Cys \text{--} Tyr \text{--} Ileu \text{--} Gln \text{--} Asn \text{--} Cys \text{--} Pro \text{--} Leu \text{--} Gly$$

This peptide is a hormone synthesized in the posterior lobe of the pituitary and causes uterine contraction and ejection of milk in the female animal.

The numbers and types of proteins present in living matter are vast, and only a few can even be briefly considered.

11.3.6.4 Membranes

The membranes of all cells and organelles contain protein in association with lipid. Lipoprotein membranes act as selective permeability barriers and are involved in the transport of materials into and out of the cell and its compartments.

A number of enzymes, including those involved in the biosynthesis of macromolecules and the detoxification of foreign compounds, are also part of the cell membrane of some tissues.

11.3.6.5 Plasma Proteins

Blood plasma contains large amounts of a number of proteins, each of which has a specialized biological function. Plasma *albumin*, for example, acts as a store of protein, is important in the maintenance of plasma pH and osmotic pressure, and transports a variety of compounds in the blood. The α- and β-*globulins* are associated with the transport of lipids and the γ-*globulins* with antibodies.

Fibrinogen is a soluble protein that is converted into insoluble *fibrin* during blood clotting.

Hormones: Several proteins have hormonal properties. *Insulin*, secreted by the β-cells of the pancreas, controls carbohydrate metabolism by lowering the blood sugar, while glucagon, from the cells, increases the blood sugar. *Gastrin* stimulates acid secretion in the stomach, and the *parathyroid hormone* is concerned with the regulation of calcium and phosphate metabolism.

Enzymes: All enzymes are proteins, and the importance and wide occurrence of these biological catalysts are well known. They owe their high specificity and catalytic activity to the correct arrangement of amino acids in space for the substrate to bind to.

11.3.7 Experiment: Determination of Crude Protein by Micro-Kjeldahl Method

11.3.7.1 Principle
The methods of estimation of proteins vary in their sensitivity. The choice of a method depends on the type of material used. There are inherent limitations in each method. The interference by the other compounds needs consideration. General precautions are detailed in the following discussion, but exact Steps for a particular tissue are to be worked out by trial and error. It is desirable to use acid precipitable material for protein estimation for these samples or extracts.

The amino nitrogen in various nitrogenous compounds in biological samples such as plant material (leaves, seeds, etc.) is converted to ammonium sulfate on digestion with concentrated sulfuric acid in the presence of K_2SO_4 and $CuSO_4$·K_2SO_4 is included in digestion mixture, to raise the boiling temperature, whereas $CuSO_4$ acts as a catalyst. Alternatively, selenium dioxide can be used as a catalyst. On distillation of the digested sample with NaOH, NH_3 is liberated and trapped in boric acid containing a mixture of bromocresol green and methyl red as an indicator. Ammonia reacts, with boric acid, to form ammonium borate, which is then determined volumetrically by titrating against standardized HCl and the amount of nitrogen determined. Since proteins contain about 16% nitrogen, the protein contents of the sample are calculated by multiplying its nitrogen content by 6.25. The value obtained is referred to as crude protein content since this procedure estimates total organic nitrogen rather than true proteins. This method is commonly used for determination of crude protein content of grains forage and animal feeds.

11.3.7.2 Precautions
Caution should be taken that digest is clear without any color before it is distilled.

The pinch cocks should be air tight while performing distillation.

11.3.7.3 Materials
Plant/animal (dry powder), digestion rack, micro-Kjeldahl apparatus also known as Markham distillation apparatus (Figure 11.19), microburette (5 or 10 mL capacity), conc. H_2SO_4, NaOH, $CuSO_4$·$5H_2O$, K_2SO_4, methyl red, ethyl alcohol, 25 mL Erlenmeyer flasks, bromocresol green, selenium dioxide.

Catalyst: Either selenium dioxide powder or mixture of $CuSO_4$·$5H_2O$ and K_2SO_4 in the ratio of 2:1.

NaOH (40%): Dissolve 200 g of NaOH in 500 mL of distilled water.

Boric acid: Dissolve 10 g of reagent grade boric acid in about 470 mL of hot distilled water, cool and add 1 mL of 0.1% alcoholic solution of bromocresol green and 4 mL of 0.1% methyl red solution, and make the final volume to 500 mL.

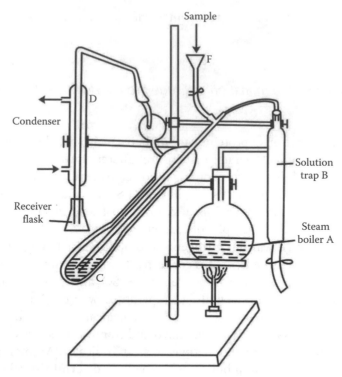

Figure 11.19   Micro-Kjeldahl distillation apparatus.

0.1% Bromocresol green: Dissolve 0.1 g of bromocresol green in 100 mL of absolute alcohol.

0.1% Methyl red: Dissolve 100 mg of methyl red in 100 mL of absolute alcohol.

0.01N HCl: 0.086 mL of HCl makes up the volume to 100 mL with distilled water.

11.3.7.4 Protocol

1. Take 1 g of finely powdered plant or animal sample in a long-necked digestion flask. Add 10 mL of conc. H_2SO_4 and 200 mg of catalyst mixture, and digest the sample on the heater of Kjeldahl's heating unit till the solution becomes clear. Cool it and transfer to 50 mL volumetric flask. Rinse the digestion flask several times with a small amount of water, and pour the washings in the volumetric flask. Finally, make up the volume to 50 mL with distilled water.
2. Take 10 mL of boric acid solution in 100 mL conical flask. Place this receiving flask in such a way that outlet of the condenser of micro-Kjeldahl distillation apparatus dips into the boric acid solution.
3. Transfer 5 mL of acid-digested sample to the steam chamber of micro-Kjeldahl's apparatus. Add about 5–8 mL of 40% NaOH to the aliquot of the digested sample. Immediately close the stopcock and pass the steam through the steam chamber to distill ammonia till about 30–40 mL of distillate is collected in the receiving boric acid-containing flask.
4. Remove the receiving flask; rinse the condenser outlet tip into the receiving flask with water.

5. Titrate the contents of receiving flask against 0.01 N HCl till the bluish green color changes to pink.
6. Run a blank preparation that has been identically prepared, except that it does not contain the sample.

11.3.7.5 Results and Calculations

1 mL of 0.01 N HCl = 0.00014 g nitrogen.

Since average nitrogen content of most proteins is 16%

1 g of nitrogen = 100/16 g of protein

Volume of 0.01 N HCl used for blank = v mL

Volume of 0.01 N HCl used for sample = y mL

Titer volume or sample = $y - v$ mL

Vol. of sample taken for distillation = 5 mL

Total volume made of the digested sample = 50 mL

Nitrogen present in 5 mL of digested sample = $(y - x) \times 0.00014\,g$

$$\text{Nitrogen present in 50 mL digested sample} = \frac{(y-v) \times 0.00014 \times 50\,g}{5}$$

$$\text{Nitrogen present in 1 g sample} = \frac{(y-v) \times 0.00014 \times 50\,g}{5}$$

$$\text{Amount of nitrogen present in 100 g sample} = \frac{(y-v) \times 0.00014 \times 50 \times 100\,g}{5}$$

$$\% \text{ protein content in sample} = \frac{6.25 \times (y-v) \times 0.00014 \times 50 \times 100}{5}$$

11.3.8 Experiment: Quantitative Estimation of Amino Acids Using the Ninhydrin Reaction

11.3.8.1 Principle

Ninhydrin (triketohydrindene hydrate) reacts with α-amino acids between pH 4 and 8 to give a purple-colored compound. Not all amino acids give exactly the same intensity of color, and this must be allowed for in any calculation. The amino acids proline and hydroxyproline give a yellow color, so these are read at 440 nm.

11.3.8.2 Materials

Amino acids (0.1 mmol/L aspartic acid, arginine, leucine, and proline), acetate buffer (4 mol/L, pH 5.5), methyl cellosolve (ethylene glycol monomethyl ether), ethanol (50% v/v), ninhydrin reagent (dissolve 0.8 g of ninhydrin and 0.12 g of hydrindantin in 30 mL of methyl cellosolve, and add 10 mL of acetate buffer; prepare fresh and store in a brown bottle).

(*Caution:* Carcinogenic!)

11.3.8.3 Protocol
Pipette 2 mL of the amino acid solution into a test tube, add 2 mL of the buffered ninhydrin reagent, and heat in a boiling water bath for 15 min. Cool to room temperature, add 3 mL of 50% ethanol, and read the extinction at 570 nm (or 440 nm) after 10 min. Set up the appropriate blanks and compare the color equivalence of the amino acids investigated.

11.3.9 Experiment: Biuret Assay

11.3.9.1 Principle
This method is the most linear because its color depends on a direct complex between the peptide bonds of protein and Cu^{2+} ion. It is highly sensitive since the complex does not have a high extinction coefficient (Figure 11.20).

11.3.9.2 Materials
Protein standard: Freshly prepared 5 mg albumin/mL aqueous solution.
Biuret reagent: Dissolve 3 g of copper sulfate ($CuSO_4 \cdot 5H_2O$) and 9 g of sodium potassium tartrate in 500 mL of 0.2 mol/L sodium hydroxide; add 5 g of potassium iodide and make up to 1 L with 0.2 mol/L sodium hydroxide.
   Water bath at 37°C.

11.3.9.3 Protocol
Add 3 mL of biuret reagent to 2 mL of protein solution, mix, and warm at 37°C for 10 min; cool and read the extinction at 540 nm. Prepare a graph of extinction against albumin concentration.

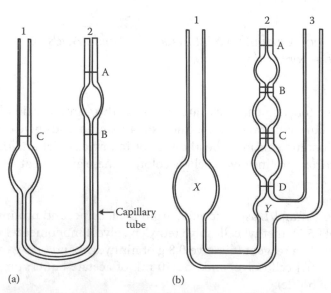

(a)   (b)

Figure 11.20   (a) Ostwald–Fenske viscometer. (b) Ubbelohde viscometer.

11.3.10 Experiment: Folin–Lowry Method of Protein Assay

11.3.10.1 Principle
Protein reacts with the Folin–Ciocalteu reagent to give a colored complex. The color so formed is due to the reaction of the alkaline copper with the protein as in the biuret test and the reduction of phosphomolybdate by tyrosine and tryptophan present in the protein. The intensity of color depends on the amount of these aromatic amino acids present and will thus vary for different proteins. The Folin assay is dependent on the presence of aromatic amino acids in the protein. First, a cupric/peptide bond complex is formed, and then this is enhanced by a phosphomolybdate complex with the aromatic amino acids. Overall, it is about 10–50 times more sensitive than the biuret method (Figure 11.21). Many substances sometimes interfere with the Folin assay for protein.

11.3.10.2 Materials
1. *Alkaline sodium carbonate solution*: 20 g/L Na_2CO_3 in 0.1 mol/L NaOH.
2. *Copper sulfate–sodium potassium tartrate solution*: 5 g/L $CuSO_4 \cdot 5H_2O$ in 10 g/L Na, K tartrate. Prepare fresh by mixing stock solutions.
3. *Alkaline solution*: Prepare on the day of use by mixing 50 mL of (1) and 1 mL of (2).
4. *Folin–Ciocalteu reagent*: Dilute the commercial reagent with an equal volume of water on the day of use. This is a solution of sodium tungstate and sodium molybdate in phosphoric and hydrochloric acids.
5. *Standard protein*: Albumin solution 0.2 mg/mL.

11.3.10.3 Protocol
Add 5 mL of the alkaline solution to 1 mL of the test solution. Mix thoroughly and allow to stand at room temperature for 10 min or longer. Add 0.5 mL of diluted Folin–Ciocalteu reagent rapidly with immediate mixing. After 30 min, read the extinction against the appropriate blank at 750 nm. Estimate the protein concentration of an unknown solution after preparing a standard curve.

11.3.11 Isolation of Proteins

A number of physical techniques are used in isolation and separation of proteins. Some of the experiments detailed on separation methods can be carried out at this stage if required since many of the examples selected involve proteins.

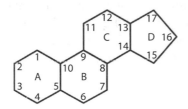

Perhydrocyclopentanophenanthrene

Figure 11.21 Basic steroid structure.

11.3.11.1 Experiment: Isolation of Casein from Milk

11.3.11.1.1 Principle.   Casein is the main protein found in milk and is present at a concentration of about 35 g/L. It is actually a heterogeneous mixture of phosphorus-containing proteins and not a single compound. Most proteins show minimum solubility at their isoelectric point, and this principle is used to isolate the casein by adjusting the pH of milk to 4.8, its isoelectric point. Casein is also insoluble in ethanol, and this property is used to remove unwanted fat from the preparation.

11.3.11.1.2 Materials.   Milk, sodium acetate buffer (0.2 mol/L, pH 4.6), ethanol (95% v/v), ether (*Caution*: Highly inflammable), thermometer 100°C, muslin cloth, Buchner filter equipment, and papers.

11.3.11.1.3 Protocol.   Place 100 mL of milk in a 500 mL beaker and warm to 40°C; also warm 100 mL of the acetate buffer and add slowly with stirring. The final pH of the mixture should be about 4.8, and this can be checked with a pH meter. Cool the suspension to room temperature and then leave to stand for a further 5 min before filtering through muslin. Wash the precipitate several times with a small volume of water, and then suspend it in about 30 mL of ethanol. Filter the suspension on a Buchner funnel and wash the precipitate a second time with a mixture of equal volumes of ethanol and ether. Finally, wash the precipitate on the filter paper with 50 mL of ether, and suck dry. Remove the powder and spread out on a watch glass to allow evaporation of the ether. Weigh the casein and calculate the percentage yield of the protein.

11.3.12 Experiment: Preparation and Properties of Cytochrome c

11.3.12.1 Principle
When hearts are homogenized with trichloroacetic acid, most of the proteins are precipitated, but cytochrome *c* is unusual in this respect and remains in solution. Ammonium sulfate is added to the TCA extract, which completes the precipitation of myoglobin, hemoglobin, and other proteins. Further acidification of the solution with TCA precipitates the cytochrome *c* that is then dialyzed and characterized.

11.3.12.2 Materials
Pig, ox, or goat heart, large mincer, muslin cloth, sodium hydroxide (100 g/L), ammonium sulfate, trichloroacetic acid (200 g/L), trichloroacetic acid (0.145 mol/L), saturated ammonium sulfate solution, dialysis tubing, pH meter, spectrophotometer, Waring blender, potassium ferricyanide (10 mmol/L), sodium dithionite saturated solution.

11.3.12.3 Protocol
11.3.12.3.1 Extraction.   Remove the fat from the hearts and weigh them. Mince 1 kg of the hearts and blend with 1 L of 0.145 mol/L TCA. Allow the suspension to stand

at room temperature for 3 h, and then filter through muslin cloth. Adjust the pH of the cloudy fluid obtained to 7.3 with 100 g/L NaOH. Measure the total volume of the fluid and slowly add solid ammonium sulfate (500 g/L) with stirring. Filter off the precipitate with a large fluted filter paper to obtain a pink filtrate. Measure the volume and add further ammonium sulfate (50 g/L) with stirring. Leave the mixture overnight in the refrigerator. Remove any slight precipitate formed overnight and add 200 g/L TCA (25 mL/L) to precipitate the cytochrome c. Rapidly centrifuge off the cytochrome c (3000g for 15 min) and suspend the brick-red precipitate in saturated ammonium sulfate solution (150 mL/kg original tissue). Place the suspension in a dialysis sac and dialyze against water for 4 h. Centrifuge off the slight, dark brown precipitate of denatured cytochrome c, and store the remaining solution at 15°C.

The cytochrome c can conveniently be characterized in solution or alternatively precipitated by the addition of four volumes of cold acetone. A better product is obtained if the solution is dialyzed free of ammonium sulfate and freeze-dried. Record the final volume of the cytochrome c solution (v).

11.3.12.3.2 Spectrophotometric Standardization. The solution prepared is a mixture of the oxidized and reduced forms of the molecule. The next stage is to oxidize one sample with ferricyanide, to reduce another sample with dithionite, and to record their extinctions at 550 nm. The molar extinction coefficients are known for each form so the purity can be readily checked. If other pigments are present, it is unlikely that they will show the same shift when converted from the oxidized to the reduced form and vice versa. The molar extinction coefficients at 550 nm are as follows:

Oxidized form, $k_1 = 0.9 \times 10^4$ (L/mol/cm)
Reduced form, $k = 2.77 \times 10^4$ (L/mol/cm)

This means that a solution of oxidized cytochrome c of concentration 1 mol/L or 1 mmol/mL has molar extinction of 0.9×10^4 in a light path of 1 cm.

Oxidized form: Prepare the following solution and read the extinction E_1 at 550 nm against a blank of ferricyanide and buffer.

Components:

Sodium phosphate buffer (0.1 mol/L, pH 7.4) 1.9 mL
Cytochrome c preparation suitably diluted 1.0 mL
Potassium ferricyanide (0.01 mol/L) 0.1 mL

Reduced form: Prepare the following mixture and read the extinction E_2 against a blank of dithionite and buffer.

Components:

Sodium phosphate buffer (0.1 mol/L, pH 7.4) 1.9 mL
Cytochrome c preparation 1.0 mL
Sodium dithionite saturated solution 0.1 mL

Use the data obtained to calculate the concentration of cytochrome c present in the cuvette in mg/mL. Multiply this figure by three as there is 3 mL in the cuvette, and remember that this amount of cytochrome c was present in 1 mL of the diluted preparation. The yield of cytochrome c in milligrams (Y) is, therefore, given by the following:

$$Y = \left(\frac{E_1}{k_1}\right) \times \text{mol. wt.} \times 3 \times \text{dilution} \times V \text{ mg}$$

and

$$Y = \left(\frac{E_2}{k_2}\right) \times \text{mol. wt.} \times 3 \times \text{dilution} \times V \text{ mg}$$

Express the final result as milligrams of cytochrome c per kilogram of heart. The yields calculated from the oxidized and reduced forms should be the same if the material is relatively pure. As a final check on purity, determine the total amount of protein present using one of the standard procedures already described.

11.3.12.4 Results and Conclusion

Absorption spectra: Plot the absorption spectra of the oxidized and reduced forms and determine the absorption maxima in the visible (VIS) region.

  Oxidized maxima: 408 and 530 nm.
  Reduced maxima: 415, 520, and 550 nm.

11.3.13 Protein Structure

11.3.13.1 Experiment: Identification of the C-Terminal End of Amino Acid of a Protein

11.3.13.1.1 Principle. The enzyme carboxypeptidase is an exopeptidase that degrades polypeptides by catalyzing the hydrolysis of the peptide bond next to the C-terminal amino acid. This release of the C-terminal amino acid unmasks a new C-terminal residue that is then in turn cleaved by the enzyme. The C-terminal amino acid can thus be identified and also the sequence of several amino acids closes to this. Here, carboxypeptidase A is incubated with several proteins and, during the digestion, samples are removed and the amino acids identified by paper chromatography.

11.3.13.1.2 Materials. Tris–HCl buffer (25 mmol/L, pH 7.5), protein solutions in buffer (muramidase and ribonuclease [RNase], 10 mg/mL), carboxypeptidase A (1 mg/mL in buffer), water baths at 37°C, micropipettes (50 μL), chromatographic chamber, chromatographic paper and solvent for paper chromatography of amino acids, trichloroacetic acid (100 g/L).

11.3.13.1.3 Protocol. Add 0.5 mL of carboxypeptidase A to 0.5 mL of the protein solution, mix thoroughly, and place in a water bath at 37°C. Withdraw 0.2 mL samples at a suitable time intervals (0, 10, 20, 30, and 60 min) and mix with 0.2 mL of

10% w/v TCA. Centrifuge the precipitate and spot 50 μL samples onto Whatman No. 1 chromatography paper. Separate the amino acids and identify them as far as possible; also plot a graph of the number of amino acids appearing with time. The experiment can be shortened by separating the amino acids in one direction only, using the organic phase of butanol/acetic acid/water (4:1:5) as a solvent, and setting up a limited number of standard amino acids (alanine, arginine, leucine, serine, and valine).

11.3.13.1.4 Results and Conclusions. Record results by identification of amino acids with R_f values on paper chromatogram.

11.3.14 Experiment: Determination of the Free Amino End Group of Some Proteins

11.3.14.1 Principle
The N-terminal amino acid can be identified with 1-fluoro-2,4-dinitrobenzene (Sanger's reagent) that reacts in mildly alkaline solution with the free amino group at the end of the chain.

The protein is hydrolyzed in acid solution, and since the DNP amino acid link is resistant to this treatment, the end amino acid is effectively labeled. After hydrolysis, the yellow DNP–amino acid is identified by chromatography.

FDNB also reacts with free amino, imidazole, and phenolic groups on amino acids at neutral to alkaline pH, to give the corresponding DNP derivatives. Fortunately, the nonpolar DNP–amino acids can be readily extracted from the acid hydrolyzate with ether, leaving these charged DNP–amino acid derivatives in the aqueous phase.

11.3.14.2 Materials
Proteins (hemoglobin, muramidase, and RNase), sodium bicarbonate, conc. hydrochloric acid, extra pure ether (peroxide free), 1-fluoro-2,4-dinitrobenzene (5% v/v in ethanol) (*Caution:* Causes blisters), hydrochloric acid (6 mol/L), sodium phthalate buffer (0.1 mol/L, pH 4.6), paper chromatography apparatus, Whatman No. 4 chromatography paper, shakers (capacity 4 tubes), oven at 110°C, acetone, sodium phosphate buffer (0.75 mol/L, pH 6.0), ampoules, ultraviolet (UV) lamp, standard DNP derivatives of glycine, valine, lysine, and phenylalanine. (Keep in the dark.)

Solvent system for paper chromatography: n-butanol–acetic acid–water overnight saturated in a separating funnel; use lower organic phase in a ratio of 4:1:5.

Locating reagent: 0.1% ninhydrin in acetone

11.3.14.3 Protocol

11.3.14.3.1 Preparation of the DNP Amino Acid.   Weigh about 5–20 mg of protein (minimum 0.2 µmol), mix with an equal weight of sodium carbonate, and suspend in 1–2 mL of water. Add twice the volume of FDNB solution and shake for 2 h at room temperature. Maintain the pH in the region of 8–9 by adding more sodium bicarbonate if required. If a large precipitate forms, the pH is too low. Extract the suspension three times with peroxide-free ether to remove dinitrophenol formed from the reaction of the FDNB with water. Adjust the pH to about 1 using strong acid and extract three times with an equal volume of peroxide-free ether; combine the ether extracts and evaporate to dryness in a fume chamber. Add 0.2 mL of acetone to the dried DNP derivative and transfer to a hydrolysis vial. Remove the acetone by evaporation in a stream of air and add 1 mL of 6 mol/L HCl. Seal the ampoule and place in an oven at 110°C for 18 h.

Carefully open the vial after cooling it to room temperature, add 1 mL of water, and extract three times with about 2 mL of ether. Concentrate the combined ether extracts and evaporate to dryness. Dissolve the DNP–amino acid in a little acetone and chromatograph as follows:

11.3.14.3.2 Chromatography of DNP Amino Acids.   Apply 10–20 µL of the test to Whatman No. 4 paper previously saturated with phthalate buffer, and repeat this until the yellow spots are clearly visible. Apply the standard DNP amino acid solutions and develop the ascending chromatogram with 0.75 mol/L phosphate buffer, pH 6.0. Develop the chromatogram by using lower organic phase n-butanol/acetic acid/water in a ratio of 4:1:5. Locate the chromatogram with ninhydrin reagent in acetone (0.1%), and then heat it at 105°C for 5 min.

11.3.14.4 Results and Conclusion
Identify the amino acids on paper chromatograms by measuring R_f values against standard.

11.3.15 Experiment: Detection of Changes in the Conformation of Bovine Serum Albumin by Viscosity Measurements

11.3.15.1 Principle
High concentrations of urea cause an unfolding of proteins by weakening the hydrophobic bonds that maintain the tertiary structure. This change in protein conformation leads to a less compact molecule with a larger viscosity than the native protein. Such changes in tertiary structure can be readily followed using an Ostwald–Fenske viscometer (Figure 11.20). This, essentially, consists of a capillary tube down in which a known volume of protein solution is allowed to flow under gravity. The time taken for this flow is measured (t_1) and also that of the solvent (t_0); the relative viscosity is then given by the following:

$$\eta_{rel} = \left(\frac{\eta_1}{\eta_0}\right) = \left(\frac{t_1}{t_0}\right) \times \left(\frac{p_1}{p_0}\right)$$

where

η_1 is the viscosity of the protein solution of density p_1

η_0 is the viscosity of the solvent of density p_0

If the densities are taken to be the same, then the expression simplifies to

$$\eta_{rel} = \frac{t_1}{t_0}$$

Einstein has shown that, for spherical molecules, the relative viscosity is related to the concentration of the molecule (c) and the partial specific volume (V), which is the volume occupied by the molecule and its bound water:

$$\eta_{rel} = 1 + 2.5cV$$

11.3.15.2 Materials

Viscosity is very sensitive to temperature, so all solutions and the viscometer must be kept at 30°C in the water bath. Ostwald–Fenske viscometer (Figure 11.20), water bath at 30°C, potassium chloride (100 mmol/L), urea solutions (0.5, 1, 2,4, 6, and 8 mol/L in 100 mmol/L KCl), bovine serum albumin (10 g/L in 100 mmol/L KCl and the previous urea solutions), stop watch accurate to at least 0.1 s.

In the Ubbelohde viscometer, B and C are pairs of scratches (Figure 11.20). The upper scratch is for timing the movement of the meniscus from A to B and the lower from B to C. The double scratch is to allow time to restart a stopwatch. The relative shear gradients are usually calculated from constants provided by the manufacturer.

11.3.15.3 Protocol

Always handle the viscometer by one limb only and never squeeze the two arms together. Rinse the viscometer with KCl solution and place it in position in a water bath, by carefully clamping one limb. Check that it is vertical using a plumb line and introduce exactly 20 mL (or the volume marked red on the viscometer) of KCl solution at 30°C into the bulb A with a syringe or pipette. Leave for 5 min to equilibrate, and then either apply positive pressure to the wide limb (1) or gentle suction to the other limb (2) until the meniscus rises above the upper graduation mark A. Release the pressure and measure the time (to the nearest 0.1 s) for the liquid to flow between the two graduation marks A and B. Repeat the experiment until the flow times agree within 0.2 s, and calculate the average flow time. Repeat the whole procedure with the urea solutions alone (t_0), which are the solvents, and then with the bovine serum albumin dissolved in the urea (t_1). Plot the values of t_0 and t_1 against the concentration of urea and join up the points with smooth curves. Select convenient concentrations of urea and calculate the relative viscosities (t_1/t_0) using the values from the curves.

This ensures that any slight errors involved in the determination of t_1 and t_0 are not magnified on taking the ratios.

11.3.15.4 Results and Conclusion

Prepare a graph of the relative viscosity against the concentration of urea and comment on the results. In addition, calculate the partial specific volume of serum albumin in 10 mmol/L KCl and in 8 mol/L urea. Assume that the molecule remains spherical so that Einstein's equation is valid.

11.3.16 Experiment: Effect of pH on the Conformation of Bovine Serum Albumin

11.3.16.1 Materials

Viscosity is very sensitive to temperature, so all solutions and the viscometer must be kept at 30°C in the water bath. Ostwald–Fenske viscometer, water bath at 30°C, potassium chloride (100 mmol/L), urea solutions (0.5, 1, 2,4, 6, and 8 mol/L in 100 mmol/L KCl), bovine serum albumin (10 g/L in 100 mmol/L KCl and the previous urea solutions), stop watch accurate to at least 0.1 s, pH meter.

11.3.16.2 Method

Using the Ostwald viscometer, follow the structural changes in the albumin dissolved in 100 mmol/KCl and distilled water as the pH is varied over the range of 2–12.

11.3.16.3 Results and Conclusions

Comment on the results.

11.4 Lipids

11.4.1 Introduction

Lipids are poorer in oxygen but made up of carbon, hydrogen, and oxygen like carbohydrates, lipids (or) fats, and fatlike substances. They are found almost in every living cell. Lipids form a heterogeneous group of compounds. They dissolve only in organic solvents such as alcohol, acetone, chloroform, benzene, ether, and hexane. The hydrophobicity is the unique feature attributed to their structure. The specific solubility characteristic is being exploited in extracting lipids from tissues, free from any water-soluble matter. The subsequent analytical methods are largely used for individual compounds. A mixture of ethanol and ethyl ether or a mixture of chloroform and ethanol is normally used for lipid extraction from the biological sample. They are generally bound to proteins (as lipoprotein) in the biological samples and cannot be extracted efficiently by nonpolar organic solvents alone. The inclusion of methanol or ethanol helps in breaking bonds between the lipids and proteins.

11.4.2 Classification and Biological Role of Lipids

Lipids are naturally occurring compounds that are esters of long-chain fatty acids. They are insoluble in water but soluble in "fat solvents" such as acetone, alcohol, chloroform, or ether. Alkaline hydrolysis (known as saponification) gives rise to alcohol and the sodium or potassium salts of the constituent fatty acids: these products of hydrolysis may be water soluble. Chemically, lipids can be divided into two main groups: simple lipids and compound lipids. Steroids and the fat-soluble vitamins are also considered as lipids because of their similar solubility characteristic: they are known as derived lipids. However, many of these latter compounds are alcohols and not esters and hence cannot be saponified.

11.4.2.1 Simple Lipids

11.4.2.1.1 Acylglycerols. Esters of glycerol and fatty acids are known as *acylglycerols* or *glycerides*. The trihydric alcohol glycerol can be esterified to give mono-, di-, and triglycerides. The fatty acids may be the same or different, and on saponification, free glycerol and fatty acids are obtained:

$$
\begin{array}{lll}
CH_2O\cdot COR_1 & CH_2OH & R_1COOK \\
| & & \\
CH\cdot OCOR_2 + 3KOH \longrightarrow & CHOH \quad + & R_2COOK \\
| & & \\
CH_2O\cdot COR_3 & CH_2OH & R_3COOK \\
\text{Triglyceride} & \text{Glycerol} & \text{K salts of fatty acids}
\end{array}
$$

Triglycerides are the predominant form in nature, although mono- and diglycerides are known. The acylglycerols are uncharged molecules and for this reason are also known as *neutral lipids*. They are called *fats* or *oils* depending on whether they are solid or liquid at room temperature. If the fatty acids substituted at positions 1 and 3 are different, then C-2 becomes a chiral center and two stereoisomers are possible, although most triglycerides in nature are of the L form.

11.4.2.1.2 Fatty Acids. A wide range of fatty acids are found in nature, and some of the commonest ones are given in the following discussion (Table 11.10). Most of the natural fatty acids have straight chains with an even number of carbon atoms, although branch chain and cyclic fatty acids are not unknown. Many of the fatty acids are unsaturated molecules and the introduction of a double bond into the fatty acid part of an acylglycerol lowers the melting point of the compound; thus, animal fats, which consist largely of triglycerides with fully saturated fatty acids, are solid at room temperature, while vegetable and fish oils, which contain a high proportion of unsaturated fatty acids, are liquid at room temperature. Furthermore, the introduction of a double bond gives rise to two possible geometric forms, the *cis* and *trans* isomers, although the trans form is rare in nature.

TABLE 11.10   Some Common Fatty Acids Found in Living Organisms

Trivial Name	Formula	Symbol
Saturated fatty acids		
Lauric	$CH_3(CH_2) COOH$	12:0
Myristic	$CH_3(CH_2)_{12}COOH$	14:0
Palmitic	$CH_3(CH_2)_{14} COOH$	16:0
Stearic	$CH_3(CH_2)_{16} COOH$	18:0
Unsaturated fatty acids		
Oleic	$CH_3(CH_2)_7CH=CH(CH_2)_7COOH$	$18:1^{\Delta,9}$
Linoleic	$CH_3(CH_2)_4CH=CH·CH_2·CH=(CH_2)_7COOH$	$18:2^{\Delta,9,12}$
Linolenic	$CH_3(CH_2)_4CH=CH·CH_2·CH=CH(CH_2)·CH= CH(CH_2)_7 COOH$	$18:3^{\Delta,9,12,15}$
Arachidonic	$CH_3(CH_2)_4(CH=CH·CH_2)_3 CH=CH(CH_2)_3COOH$	$20:4^{\Delta,5,8,11,14}$

Δ^9 – Octadecanoic acid

$$
\begin{array}{cc}
\underset{\|}{CH(CH_2)_7CH_3} & CH_3(CH_2)_7CH \\
CH(CH_2)_7COOH & \underset{\|}{} \\
 & CH(CH_2)_7COOH \\
cis & trans \\
\text{Oleic acid} & \text{Elaidic acid}
\end{array}
$$

A simple notation for fatty acids is given in Table 11.10. The first number denotes how many carbon atoms are in the chain and the second one the numbers of double bonds. The superscript numbers following Δ indicate the position of any double bonds in the molecule. Stearic acid, for example, is shown as 18:0, which indicates 18 carbon atoms and no double bonds. Linoleic acid is given as $18:2^{\Delta,9,12}$, which shows 18 carbon atoms and two double bonds at positions 9–10 and 12–13.

11.4.2.1.3 Triacylglycerol (Triglycerides). These compounds make up the bulk of ingested lipids. They are partially degraded by lipases in the gut and then re-esterified in the gut mucosa. The fat is then transported to the blood via the lymphatic system in the form of chylomicrons; these are fat droplets from 0.1 to 1 μm in diameter made up largely of triglycerides with some cholesterol and a lipoprotein skin. This lipid may then be oxidized, in the liver, to provide energy or deposited as depot fat, in characteristic regions of the animal, where it acts as a long-term food store and heat insulator. In some seeds, the triglycerides are stored in the form of oil.

11.4.2.2 Compound Lipids
Complete hydrolysis of a compound lipid yields at least one other component as well as the usual alcohol and fatty acids. These compounds are essential structural components of cell membranes.

11.4.2.2.1 Phosphoglycerides (Glycerol Phosphatides). These compounds are also known as *phospholipids* and are very abundant in living organisms. They are chemically similar to the triacylglycerols, being fatty acid esters of glycerol, but in addition, they contain phosphoric acid esterified with an alcohol (X). The general formula of these compounds (L phosphoglyceride) is as follows:

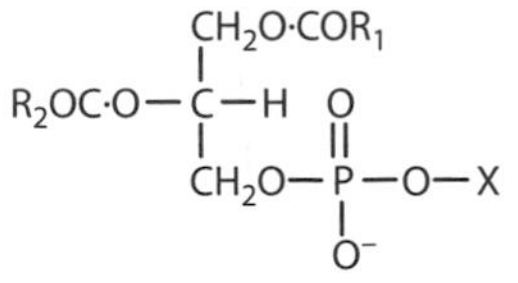

$$CH_2O\cdot COR_1$$
$$R_2OC\cdot O-\underset{|}{C}-H \quad O$$
$$CH_2O-\underset{|}{\overset{\|}{P}}-O-X$$
$$O^-$$

Carbon-2 of the glycerol is a chiral centers and the naturally occurring compounds are part of the L series.

Four of the commonest phospholipids are shown in Table 11.11. Their names are derived from the alcohol X linked to the phosphoric acid residue. (The recommended name is the alcohol followed by phosphoglyceride or phosphatidyl linked to the name of the alcohol.) One of the most abundant phospholipids contains choline, so this is known as *choline phosphoglyceride* or *phosphatidylcholine*.

11.4.2.2.2 Plasmalogens. The plasmalogens have a very similar structure to the phosphoglycerides except that the fatty acid in the 1 position is joined to the glycerol through a vinyl ether bond and not the usual ester link:

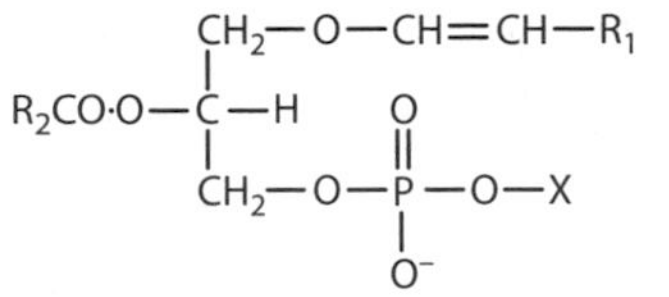

$$CH_2-O-CH=CH-R_1$$
$$R_2CO\cdot O-\underset{|}{C}-H \quad O$$
$$CH_2-O-\underset{|}{\overset{\|}{P}}-O-X$$
$$O^-$$

A plasmalogen

11.4.2.2.3 Sphingolipids. The sphingolipids contain the amino alcohol sphingosine or a related compound as a backbone instead of glycerol:

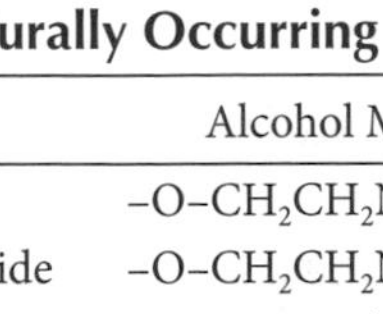

$$CH-CH(CH_2)_{12}CH_3$$
$$CHOH$$
$$CHNH_2$$
$$CH_2OH_2$$

Sphingosine

TABLE 11.11 Some Naturally Occurring Phospholipids

Recommended Name	Alcohol Moiety X	Trivial Name
Choline phosphoglyceride	$-O-CH_2CH_2N^+(CH_3)_3$	Lecithin
Ethanolamine phosphoglyceride	$-O-CH_2CH_2NH_2$	Cephalin
Serine phosphoglyceride	$-O-CH_2CH(NH_2)COOH$	—
Inositol phosphoglyceride	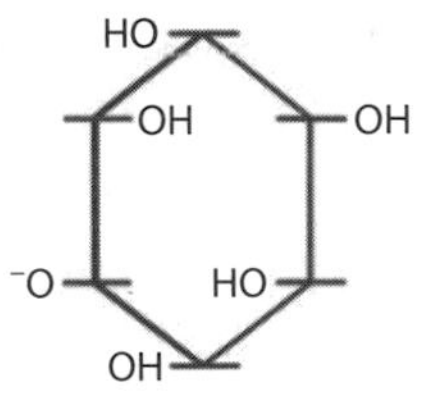	—

In the case of the *sphingomyelin,* the primary alcohol group of *sphingosine* is esterified with phosphatidyl ethanolamine, and the fatty acid residue is present as the acyl derivative of the amino group. Sphingolipids are generally metabolized more slowly than the phospholipids and appear to make up more stable structures of cells. For example, the sphingomyelin of the myelin sheath that acts as an insulator for nerve fibers is not metabolized in adult life.

11.4.2.2.4 Glycolipids. As the name suggests, these compounds contain both carbohydrate and lipid moieties. *Cerebrosides* have a similar structure to sphingomyelin except that the sugar galactose replaces the phosphoryl choline. The gangliosides are again derivatives of sphingosine, but in this case an oligosaccharide chain containing *N*-acetylneuraminic acid is joined to the primary alcohol group of sphingosine. Cerebrosides are present in high concentrations in nervous tissue, particularly in the white matter of brain, while gangliosides, on the other hand, are found in high concentrations in the gray matter.

A sphingomyelin   A cerebroside

11.4.2.2.5 Lipoproteins. Lipids in biological material are frequently associated with proteins as lipoprotein complexes. The soluble lipoproteins are concerned with the transport of lipids in the blood, while insoluble lipoproteins constitute the main part of many biological membranes such as the endoplasmic reticulum, the lamellae of chloroplasts, the cristae of mitochondria, and the myelin sheath of nerves.

11.4.2.3 Derived Lipids

11.4.2.3.1 Steroids. These compounds are soluble in the usual lipid solvents, but most of them are not saponified. They are usually considered along with lipids because of their similar solubility characteristics. All steroids have the 17-carbon perhydrocyclopentanophenanthrene ring as the basis of their structure (Figure 11.22).

When a substituent lies above the plane of the ring, it is known as β and is indicated by a full line, whereas if it is situated below the ring, the letter α is used and the position of the group is denoted by a dotted line.

Steroids are of widespread occurrence in higher animals where they perform a variety of functions. For example, *cholesterol* that is the most abundant of steroids is present in many animal cell membranes and serves as an important precursor of many other steroids.

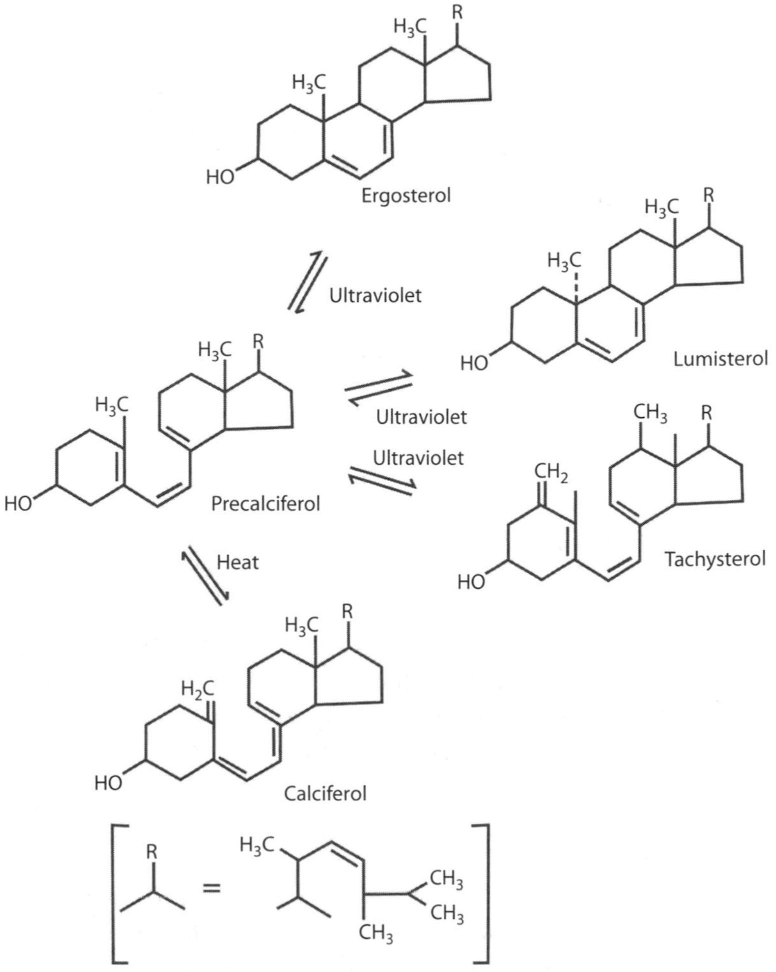

Figure 11.22 The conversion of ergosterol to vitamin D_2.

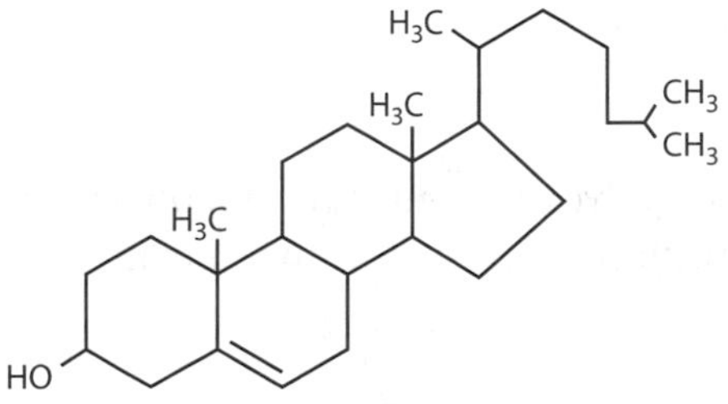

Cholesterol

The bile acids *cholic acid* and *deoxycholic acid*, involved in the digestion and adsorption of fats, are steroids and so is the hormone *aldosterone*, from the adrenal cortex, which regulates water and electrolyte balance in mammals. Many other hormones are steroids and only slight changes in their chemical structure profoundly affect their biological activity, as in the case of the sex hormones shown as follows.

Testosterone
(a male sex hormone)

Progesterone
(a female sex hormone)

11.4.2.3.2 Fat-Soluble Vitamins. These compounds are found in association with natural lipid foods and are included with lipids because of their solubility properties (Table 11.12).

11.4.3 Qualitative Test of Lipids

11.4.3.1 Solubility Test

Take 5 mL of water, dilute acetic acid, and dilute KOH, ethanol, benzene, ether, acetone, chloroform, and carbon tetrachloride in seven test tubes separately. Add a few drops of cottonseed oil or olive oil or any other natural food oil in each test tube. Shake well, allow the tube to stand, and observe that the mixture becomes homogeneous.

11.4.3.2 Emulsification Test

Take 5 mL of distilled water; add a few drops of the emulsifier (bile salt) and the same amount of oil in a test tube. Shake well to mix the contents. The lipid is dispersed forming an emulsion in the form of micelles, that is, fat broken up into minute droplets in water. This is brought about as a result of the lowering of the surface tension between fat and water. The bile salts or other substances that bring about emulsification are referred to as emulsifiers.

11.4.3.3 Paper Spot Test

Rub a bit of fat sample, say olive oil or mustard oil, on a piece of paper. If the spot on the paper assumes a translucent appearance, that means the sample under test is a fat or oil, that is, lipid.

11.4.3.4 Saponification Test

Take 100 mL of distilled water and 50 mL of 10% ethanolic solution of NaOH or KOH, and add about 10 g of fat or oil in a 250 mL flask or beaker. Reflux this over a water bath under a reflux condenser for an hour or so, till the saponification is complete. Place few drops of the saponified mixture on the surface of the water. If oily droplets do not separate, then the reaction is complete. The mixture

TABLE 11.12 Some Fat-Soluble Vitamins

Vitamin	Main Sources	Biological Role
Vitamin A (retinol)	Fish liver, vegetables,	Lack of vitamin A in the diet causes deficient night vision and eventually blindness. The vitamin is also found in dairy products, is needed for the maintenance of a healthy epithelium and the normal formation of bones and teeth.
Vitamin D₃ (cholecalciferol)	Fish liver, dairy products	Vitamin D is needed for normal calcium absorption in the gut and the formation of bone from calcium phosphate. Lack of the vitamin causes rickets in children.
Vitamin E (α-tocopherol)	Cereals, green plants	The tocopherols are needed in animals other than man for normal reproduction. They probably act as antioxidants to protect cell constituents from oxidation.
Vitamin K₁	Green vegetables	Several forms of vitamin K are known that appear to be required for normal blood coagulation. A deficiency of this vitamin causes an increase in the clotting time of the blood.

consists of glycerol, NaOH, water, and sodium soap. Now proceed further with this soap solution:

1. Add powdered sodium chloride to the point of saturation with constant stirring to 20 mL of soap solution in a beaker. The soap is separated from its sodium component and rises to the surface.
2. Take a few milliliters of the saponified mixture in a test tube, and add a few drops of calcium chloride solution (10%) to it and mix well. If a white precipitate that does not dissolve readily on dilution is formed, that means the lipids are present. The precipitate corresponds to the soap of calcium, which is hard. This can also be performed with 10% solution of magnesium sulfate, where the insoluble soap of magnesium is formed. Calcium soaps are formed with hard water where the hardness is due to the calcium ions.

11.4.4 Unsaturated Lipids

Take a few milliliters of ethanol in a test tube and add a few drops of olive oil or any other edible oil to it. To this, add a few drops of 0.4% bromine water drop by drop. An orange color is developed. The intensity of the color developed corresponds to the degree of unsaturation. If the same test is performed with acetic acid, no color will be developed as it is a saturated acid.

11.4.4.1 Liebermann–Burchard Test
It is a specific test for cholesterol. Take 2 mL of chloroform in a test tube and dissolve a pinch of any sterol, for example, cholesterol, in it. To this, add at first 8–10 drops of acetic anhydride and then a drop or two of concentrated H_2SO_4 from the sides of the test tube. A greenish or greenish blue color will be developed indicating the presence of cholesterol.

11.4.4.2 Odor Test
Place 0.5–1 g of fine-powdered potassium bisulfate ($KHSO_4$) in an air-dried test tube, and add three to four drops of olive oil to it. Heat it gently. The heating should be carried out progressively. A characteristic color of acrolein will be felt, which occurs on account of the dehydration of glycerol moiety resulting in the production of acrylic aldehyde or acrolein.

11.4.4.3 Sudan Black B Test
Take a few milliliters of distilled water in a test tube and add a drop or two of the olive oil to it. Shake the tube so that the mixture is converted into an emulsion. Place a small drop of this emulsion on a microslide, and add to it a little bit of Sudan black B stain; mix and put a cover glass. Examine under the microscope and note that only oily droplets are stained; this happens because the stain is very specific for lipids.

11.4.4.4 Sudan III Test

Take a few milliliters of either olive oil or its emulsion in a test tube, and add a few drops of Sudan III solution (in ethanol) to it; appearance of red color indicates the presence of lipids.

11.4.5 Quantitative Analysis of Lipids

11.4.5.1 Introduction

A complete chemical analysis of a naturally occurring fat is quite a lengthy procedure, but there are a number of measurements such as the acid value, the saponification number, and the iodine number that give useful information on the composition and purity of a particular fat.

11.4.6 Experiment: Determination of the Acid Value of a Fat

11.4.6.1 Principle

During storage, fats may become rancid as a result of peroxide formation at the double bonds by atmospheric oxygen and hydrolysis by microorganisms with the liberation of free acid. The amount of free acid present, therefore, gives an indication of age and quality of the fat. The acid value is the number of milligrams of KOH required to neutralize the free fatty acid present in 1 g of fat.

11.4.6.2 Materials

Olive oil, butter and margarine, fat solvent (equal volumes of 95% v/v alcohol and ether neutralized to phenolphthalein), phenolphthalein (10 g/L in alcohol), KOH (0.1 mol/L), burettes (5 and 25 mL).

11.4.6.3 Protocol

Accurately weigh out 10 g of the test compound and suspend the melted fat in about 50 mL of fat solvent. Add 1 mL of phenolphthalein solution, mix thoroughly, and titrate with 0.1 mol/L KOH until the faint pink color persists for 20–30 s. Note the number of milliliters of standard alkali required and calculate the acid value of the fat.

11.4.6.4 Results and Conclusion

   Note: 0.1 mol/L KOH contains 5.6 g/L or 5.6 mg/mL.

11.4.7 Experiment: Saponification Value of Fat

11.4.7.1 Principle

On refluxing with alkali, glyceryl esters are hydrolyzed to give glycerol and the potassium salts of the fatty acids (soaps). The saponification value is the number of milligrams of KOH required to neutralize the fatty acids resulting from the complete hydrolysis of 1 g of fat. The saponification value gives an indication of the nature of the fatty acids in the fat since the longer the carbon chain, the less acid is liberated per gram of fat hydrolyzed.

11.4.7.2 Materials
Fats and oils (tristearin, coconut oil, corn oil, and butter), fat solvent (equal volumes of 95% ethanol and ether), alcoholic KOH (0.5 mol/L), reflux condenser, boiling water bath, phenolphthalein (10 g/L in alcohol), HCl (0.5 mol/L), burettes (10 and 25 mL), conical flasks (250 mL).

11.4.7.3 Protocol
Weigh 1 g of fat in a tared beaker and dissolve in about 3 mL of the fat solvent. Quantitatively transfer the contents of the beaker to a 250 mL conical flask by rinsing the beaker three times with a further addition of solvent; add 25 mL of 0.5 mol/L alcoholic KOH and attach to a reflux condenser (Figure 11.23). Set up another reflux condenser as blank with everything present except the fat, and heat both flasks on a boiling water bath for 30 min (Figure 11.23). Leave to cool to room temperature and titrate with 0.5 mol/L HCl and phenolphthalein indicator. The difference between the blank and test reading gives the number of milliliters of 0.5 mol/L KOH required to saponify 1 g of fat.

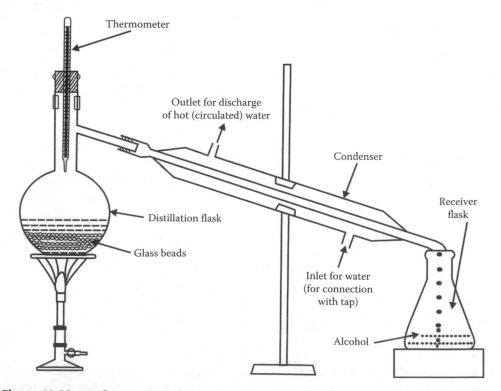

Figure 11.23 Reflex assembly for extraction.

11.4.7.4 Results and Conclusion

The molecular weight of KOH is 56, and since three molecules of fatty acid are released from a triglyceride, then

$$\text{Saponification value } (S) = \frac{3 \times 56 \times 1000}{\text{Average mol. wt. of fat}}$$

Therefore, average mol. wt. of fat $= 3 \times 56 \times 1000/S$.

11.4.8 Experiment: Iodine Number of a Fat

11.4.8.1 Principle

Halogens add across the double bonds of unsaturated fatty acids, to form addition compounds. Iodine monochloride (ICl) is allowed to react with the fat in the dark. The amount of iodine consumed is then determined by titrating the iodine released (after adding KI) with standard thiosulfate and comparing with a blank in which the fat is omitted:

$$-CH{=}CH + ICl \longrightarrow \begin{array}{c} H \quad H \\ | \quad\; | \\ -C-C- \\ | \quad\; | \\ I \quad\; Cl \end{array}$$

$$ICl + KI \longrightarrow HCl + I_2$$
$$I_2 + 2Na_2S_2O_3 \longrightarrow 2NaI + Na_2S_4O_6$$

The reaction mixture is kept in the dark, and the titration is carried out as quickly as possible since halogens are oxidized in the light. The iodine number is the number of grams of iodine taken up by 100 g of fat.

11.4.8.2 Materials

Fats (20 g/L solutions of corn oil, olive oil, linseed oil, and butter in chloroform), ICl (0.2 mol/L approx.), potassium iodide (100 g/L), sodium thiosulfate (0.1 mol/L), starch indicator (10 g/L), stoppered bottles (250 mL), burette (25 mL), chloroform.

11.4.8.3 Protocol

Pipette 10 mL of the fat solution into a stoppered bottle; add 25 mL of the ICl solution, stopper the bottle, and leave to stand in the dark for 1 h, after shaking thoroughly. At the same time, set up a blank in which the fat solution is replaced by 10 mL of chloroform. Rinse the stoppers and necks of the bottles with about 50 mL of water, add 10 mL of the KI solution, and titrate the liberated iodine with the standard thiosulfate. When the solution is pale straw in color, add about 1 mL of starch solution and continue titrating until the blue color disappears. The bottles must be shaken thoroughly throughout the titration to ensure that all the iodine is removed from the chloroform layer.

11.4.8.4 Results and Conclusion

The difference between the blank and test readings ($Bl - T$) gives the number of milliliters of 0.1 mol/L thiosulfate needed to react with the equivalent volume of iodine. This is ($Bl - T$)/2 mL of 0.1 mol/L iodine since two molecules of thiosulfate are needed for each iodine. The mol. wt. of iodine is 2×127, so the weight of iodine in ($Bl - T$)/2 mL of 0.1 mol/L iodine is

$$\frac{(Bl-T)}{2} \times 0.1 \times 2 \times \frac{127}{1000} \, g$$

The amount of fat taken was 0.2 g so the iodine number is

$$\left(Bl-T\right) \times \frac{12.7}{1000} \times \frac{100}{0.2}$$

Iodine number $= (Bl - T) \times 6.35$ g per 100 g of fat

11.4.9 Experiment: Estimation of Blood Cholesterol

11.4.9.1 Principle

Acetic anhydride reacts with cholesterol in a chloroform solution to produce a characteristic blue-green color. The exact nature of the chromophore is not known, but the reaction probably includes esterification of the hydroxyl group in the 3 position as well as other rearrangements in the molecule. Blood or serum is extracted with an alcohol–acetone mixture that removes cholesterol and other lipids and precipitates protein. The organic solvent is removed by evaporation on a boiling water bath and the dry residue dissolved in chloroform. The cholesterol is then determined colorimetrically using the Liebermann–Burchard reaction. Free cholesterol is equally distributed between the cells and plasma, while the esterified form occurs only in plasma.

It is essential to use absolutely dry glassware for this estimation.

11.4.9.2 Materials

Serum or blood, alcohol–acetone mixture (1:1), chloroform, acetic anhydride–sulfuric acid mixture (30:1; mix just before use with *caution!*), stock cholesterol solution (2 mg/mL in chloroform), working cholesterol solution (dilute the previous solution in 1:5 ratios in chloroform to give a strength of 0.4 mg/mL).

11.4.9.3 Protocol

Place 10 mL of the alcohol–acetone solvent in a centrifuge tube, and add 0.2 mL of serum or blood. Immerse the tube in a boiling water bath with shaking until the solvent begins to boil. Remove the tube and continue shaking the mixture for a further 5 min. Cool to room temperature and centrifuge. Decant the supernatant fluid into a test tube and evaporate to dryness on a boiling water bath. Cool and dissolve the residue in 2 mL of chloroform. At the same time, set up a series of standard tubes containing cholesterol and a blank with 2 mL of chloroform. Add 2 mL of acetic anhydride–sulfuric acid mixture to all tubes and thoroughly mix. Leave the tubes in the dark at room temperature and read the extinction at 680 nm.

11.4.9.4 Results and Conclusion

The normal serum cholesterol lies within the range of 100–250 mg/100 mL. The average serum cholesterol is about 200 mg/100 mL at 25 in men and rises slowly with age, reaching a peak figure at age 40–50 and then declining. Women in general show a lower cholesterol level than men until the menopause, when the value rises above that found in men of the same age.

High serum cholesterol of 300 mg/100 mL or more in young adults is a serious indication of coronary disease. High levels of as much as 25% above normal are also found in nephritis, diabetes, myxedema, and xanthomatosis.

11.4.10 Separation and Isolation of Lipids

11.4.10.1 Experiment: Lipid Composition of Wheat Grain

11.4.10.1.1 Principle.   The predominant food reserve in the wheat grain is starch. Lipid accounts for only 2%–4% of the dry weight of the whole grain, and most of this is concentrated in the germ or embryo. The major lipid component is triglyceride, and this is hydrolyzed during germination by the enzyme lipase. The products of enzymatic hydrolysis are then utilized by the growing shoot. Estimation of lipid content in grains includes extraction, separation, and composition of the wheat lipids using thin-layer chromatography (TLC) to identify the main components. The comparative investigation of lipid compositions of whole-wheat grain, germinating wheat seeds and a vegetable oil (olive, sunflower, etc.), is demonstrated.

11.4.10.1.2 Materials.   Wheat grain, wheat germinated seeds, pestle and mortar, chloroform–methanol extraction mixture (2:1), water bath at 50°C, nitrogen cylinders, olive oil, sunflower oil, mustard oil, TLC plate, TLC chamber, etc.

11.4.10.1.3 Protocol.   *Extraction*: Crush 20–25 wheat grains and extract the lipids by grinding in a pestle and mortar for several minutes with 2 mL of the chloroform–methanol mixture. Remove the debris by centrifugation and transfer the supernatant to a test tube in a water bath at 50°C and evaporate to a small volume (0.2 mL) under a stream of nitrogen in a fume chamber. Extract the lipids in a similar way from a commercial sample of wheat germinated seeds.

TLC: Spot samples of whole-wheat lipids, wheat-germ lipids, and olive or sunflower oil onto the thin-layer plates together with the standards. Run the plates in the chromatography solvent and identify the lipids as far as possible and determine the R_f value using the TLC procedure.

Further experiments: Examine the lipid composition of wheat grain after germination for a week. Compare the lipid composition of wheat seedlings grown in the dark with those grown in the light.

11.4.11 Experiment: Preparation of Cholesterol from Brain

11.4.11.1 Principle
Cholesterol is readily soluble in acetone, while most complex lipids are insoluble in this solvent.

11.4.11.2 Materials
Brain tissue from calf or pig or goat, acetone, Warring blender, large Buchner filtration apparatus, distillation equipment, large filter funnel and fluted paper, boiling water bath, ethanol.

11.4.11.3 Protocol
Add 400 mL of acetone to 100 g of brain tissue and blend for 1 min. Rinse out the blender with a little acetone and stir the combined homogenates for 10 min. Filter the suspension on a Buchner funnel, blend the residue with a further 200 mL of acetone as before, and filter and combine the filtrates. After removing most of the acetone by distillation under reduced pressure, cool the flask with tap water and collect the crude cholesterol on a Buchner funnel. Dissolve the crude material in the minimum volume of hot ethanol and filter while still hot using a fluted filter paper. For this, the collecting flask is placed in a boiling water bath with the filter funnel in position. Air-dry the cholesterol and record the yield and melting point. If necessary, the cholesterol may be crystallized from hot ethanol as the previous one. Determine the purity of product by preparing a standard solution in chloroform and estimating the amount of cholesterol present by the Liebermann–Burchard reaction.

11.4.12 Fat-Soluble Vitamins

Vitamins are necessary food factors required for the maintenance of health, and their absence from the diet leads to a number of deficiency diseases (Table 11.12). In some ways, vitamins are similar to hormones in that only small quantities may be needed to produce a large physiological effect. As with hormones, the vitamins are chemically diverse and perform a wide range of functions. Two of the fat-soluble vitamins (A and D) and their interaction with UV radiation are dealt with in the next two experiments.

11.4.12.1 Experiment: Effect of Ultraviolet Light on Vitamin A

11.4.12.1.1 Principle.
Vitamin A absorbs in the UV region of the spectrum with a maximum at 325 nm, and this property can be used for the estimation of the vitamin. The method is rapid and sensitive but suffers from the disadvantage of low specificity.

A number of other substances absorb in the same region of the spectrum as vitamin A, and allowance has to be made for these interfering compounds. One method is to apply a correction formula to check for the presence of interfering compounds and to correct for their absorption in the region of 325 nm. For example, pure vitamin A in isopropanol alcohol gives six-sevenths of the maximum extinction at 310 and 334 nm so that if the extinction at 325 nm is scaled to 1.00, then the absorbance at 310 and 334 nm becomes 0.857. The corrected extinction at 325 nm (E_{corr}) can be obtained by the application of the following formula:

$$E_{corr} + 7E_{325} - 2.625F_{310} - 4.375E_{334}$$

Another way of detecting and allowing for these compounds is to measure the extinction at 325 nm before and after irradiating with UV light. Vitamin A is destroyed by UV light, and it is assumed that the interfering chromogens are unaffected by this treatment. The fall in extinction following irradiation is, therefore, a direct measure of the vitamin A content of the sample.

11.4.12.1.2 Materials. Isopropanol, vitamin A standard solution (1 µg/mL in isopropanol stored in a sealed bottle under nitrogen in the dark), cod liver oil and halibut liver oil, UV–VIS spectrophotometer, UV lamp.

11.4.12.1.3 Protocol. Samples such as fish oils, which contain high concentrations of the vitamin, can be diluted directly with isopropanol although probably more accurate values are obtained when the material is saponified, and the vitamin A, which is unsaponifiable, is extracted. Measure the extinction of a freshly diluted sample of the fish oil in isopropanol at 325 nm against a solvent blank (T_i), remove the cuvette, expose to UV light until the extinction no longer falls with time, and again record the absorbance (T_2). Treat the standard vitamin A solution in the same way (St_i and St_2), and calculate the vitamin A concentration of the following sample:

$$\text{Vitamin A concentration } (\mu g / mL) \frac{T_1 - T_2}{St_1 - St_2} \times 1 \times \text{Dilution factor}$$

11.4.12.1.4 Results and Conclusion. Record the absorption spectrum of the freshly diluted test sample from 280 to 380 nm, and compare this with the absorption spectrum of the standard vitamin A solution by plotting E/E_{325} for the standard and test samples.

11.4.13 Experiment: Preparation of the D Vitamins by Irradiation of Their Precursors with Ultraviolet Light

11.4.13.1 Principle

Ergosterol occupies a similar place in plants to that of cholesterol in animals. It is an important precursor of vitamin D_2. Irradiation of ergosterol (provitamin D_2) with

UV light causes a number of rearrangements to take place in the molecule forming provitamin D and finally calciferol (vitamin D_2). Lumisterol and tachysterol are also formed as by-products of this photochemical reaction (Figure 11.22).

Similar rearrangements take place when 7-dehydrocholesterol is irradiated giving provitamin D_3 and cholecalciferol (vitamin D_3). This reaction takes place in the skin of man and other mammals exposed to sunlight or UV radiation. Gentle warming of a solution of calciferol in the absence of light and in an atmosphere of nitrogen produces an equilibrium mixture of provitamin D_3 and calciferol.

11.4.13.2 Materials
Silica gel G-coated TLC plates, TLC glass chamber, chloroform as chromatography solvent, standards of vitamin D_2 (ergosterol, provitamin D_2, tachysterol, and lumisterol in chloroform, approx. 20 mg/mL) for chromatography, chromatographic standards related to vitamin D_3 (dehydrocholesterol, provitamin D_3, vitamin D_3, and cholesterol in chloroform, approx. 20 mg/mL), ergosterol and dehydrocholesterol, nitrogen cylinder, fluorimeter developing reagent (5% v/v sulfuric acid), apparatus for spraying thin-layer chromatograms, oven set at 140°C, silica cuvette, UV lamp.

11.4.13.3 Protocol
UV irradiation: Prepare a solution of ergosterol in chloroform (20–30 mg/mL), transfer an aliquot to a silica cuvette of a fluorimeter, flush out with nitrogen, and seal. Place the cuvette in the fluorimeter and irradiate with UV light of short wavelength for 20 min. Remove 10 μg of the test solution and spot this onto a thin-layer plate of silica gel G. Rapidly spot out the standards related to vitamin D_2 on the thin-layer plate, and place immediately in the chloroform solvent. Repeat the previous test with dehydrocholesterol instead of ergosterol, and identify the products formed.

Heat: Examine the effect of heating solutions of vitamin D_2 and vitamin D_3 in chloroform at 60°C in the absence of light and air for periods of time up to 4 h.

Detection of vitamins: Examine the plates under UV light, when some of the compounds show up as fluorescent spots. Alternatively, the plates can be sprayed with dilute sulfuric acid (*Caution*), heated at 140°C for 5–10 min, and then viewed in VIS or UV light. Compare the sensitivity of the two methods of detection.

11.4.13.4 Results and Conclusion
Compare the product formed after UV treatment on the TLC plates with the respective standards, and determine the R_f value.

11.5 Nucleic Acids

11.5.1 Introduction

Knowledge concerning the chemistry of the components of nucleic acid has vastly expanded the horizons of biology. The nucleic acids are intimately involved in all life processes. The chemistry of nucleosides and their phosphorylated derivatives play key roles in many of the biological reactions. The nucleic acids DNA and ribonucleic acid (RNA) are the informational macromolecules composed of subunits called nucleotides joined by phosphorus bonds; they are, in fact, polynucleotides. A nucleotide is composed of three units: a five-carbon sugar, ribose (in RNA) or deoxyribose (in DNA), and a nitrogenous base (purine or pyrimidine), covalently bonded to a molecule of phosphate (PO_4). Because of the sugar, the subunit of DNA is called deoxyribonucleotide and RNA is ribonucleotide.

11.5.2 Chemical Composition of the Nucleic Acids

Nucleic acids are nitrogen-containing compounds of high molecular weight found in association with proteins in the cell. They are not only responsible for the storage and transmission of genetic information, but also translate this information for a precise synthesis of protein characteristics of the individual cell. The nucleic acid–protein complexes are known as nucleoproteins, and these can be separated into the component proteins and nucleic acids by treatment with acid or high salt concentration. The proteins are basic in character and the nucleic acids, as the name suggests, are acidic. Two main groups of nucleic acids are known: RNA and DNA. Hydrolysis of DNA and RNA under controlled conditions yields *nucleotides*, which can be regarded as the basic unit of nucleic acids, just as amino acids are the basic unit of proteins and monosaccharides of polysaccharides. Further hydrolysis of the nucleotides yields *nucleosides* and eventually phosphate, a sugar, and a number of purine and pyrimidine bases. The relationship between these components and the nucleoprotein is shown for DNA (Figures 11.24 through 11.26).

RNA has a similar composition, except that the sugar ribose is present instead of deoxyribose and uracil instead of thymine.

The formulae of the main nitrogen bases found in DNA and RNA are shown in Figure 11.27.

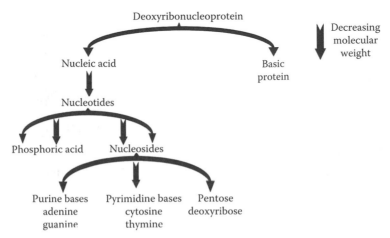

Figure 11.24 The relationship between deoxyribonucleoprotein and its low molecular weight components.

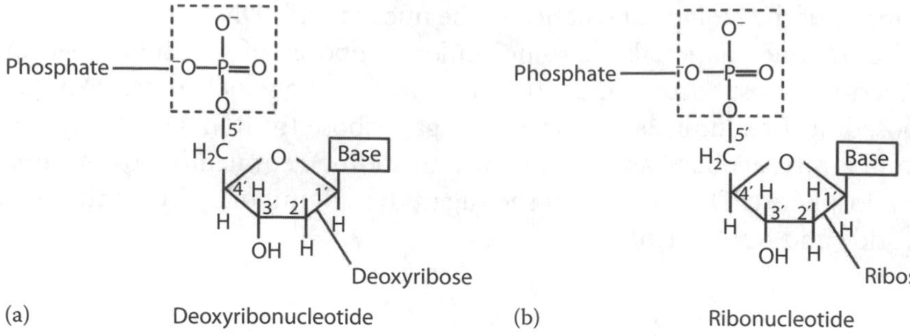

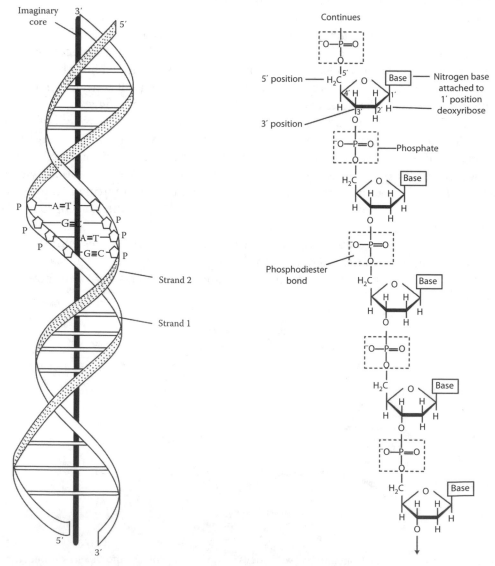

Figure 11.25 Nucleotides of (a) DNA and (b) RNA. Both contains a 5′ phosphate bond.

Figure 11.26 Schematic representation of a common form of DNA double helix.

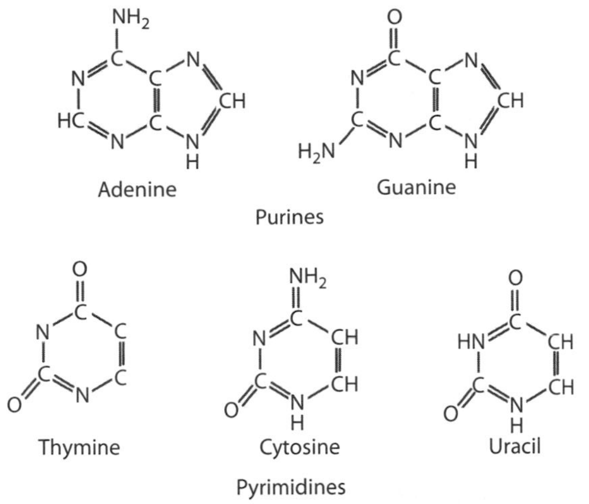

Figure 11.27 The main nitrogen bases found in DNA and RNA.

Small quantities of other bases have been detected in nucleic acids from some sources. It should be noted that both purines and pyrimidine bases can exist in the keto or enol form (Figure 11.28).

Pentose sugars: The two main groups of nucleic acids derive their names from the sugar present, which is either ribose or deoxyribose (Figure 11.29).

The carbon atoms on the sugars are denoted as $1'$, $2'$, etc., in order to differentiate them from the atoms of the bases.

Nucleosides: The C-1 of the sugar is linked to the nitrogen at the 9 position for purines or 1 position for pyrimidines to form a nucleoside. Most of the bonds linking the sugar and the base are as shown in Figure 11.30, but *transfer RNA* (tRNA) does contain an unusual nucleotide, pseudouridine, in which ribose is linked to uracil via the 5 position.

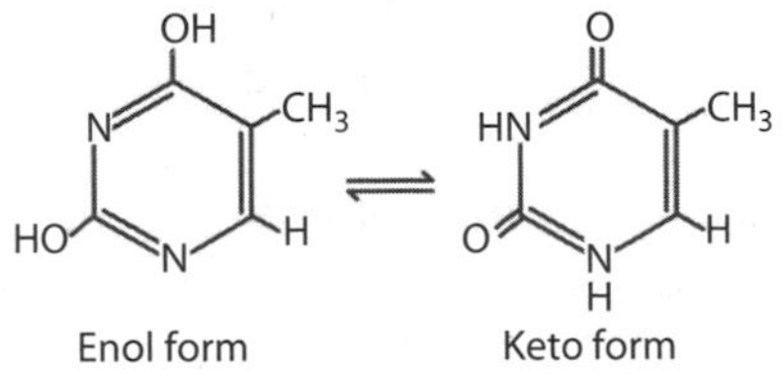

Figure 11.28 The keto and enol forms of thymine.

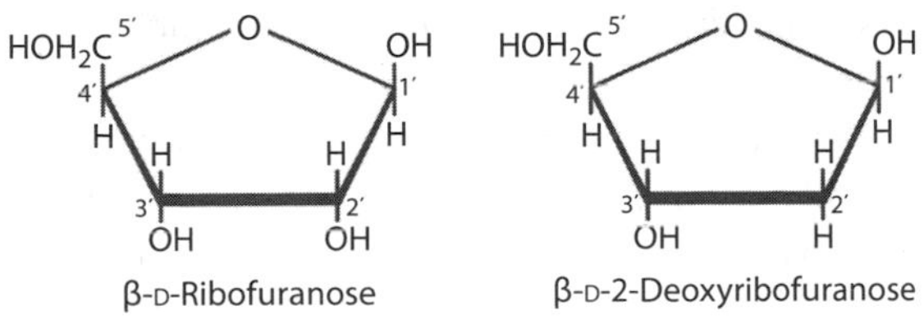

Figure 11.29 Pentose sugars.

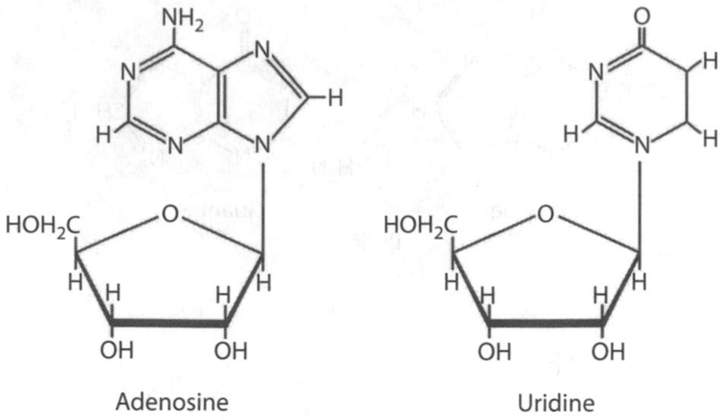

Figure 11.30   Bonds linking the sugar and the base.

Nucleotide: The hydroxyl groups on positions 2′, 3′, and 5′ of ribose can be esterified with phosphoric acid, and all these esters are known. Similarly, the 3′ and 5′ positions of deoxyribose can be esterified, and these esters are also known to exist. Nucleotides and nucleosides are named after the bases contained in their structure, as follows:

Base	Nucleoside	Nucleotide
Adenine	Adenosine	Adenylic acid
Guanine	Guanosine	Guanylic acid
Uracil	Uridine	Uridylic acid
Cytosine	Cytidine	Cytidylic acid
Thymine	Thymidine	Thymidylic acid

If the base is linked to deoxyribose, then the names are modified so that a nucleoside consisting of adenine and deoxyribose would be called deoxyadenosine. As well as the nucleotides indicated, a number of biologically important nucleotides such as adenosine di- and triphosphate (ADP and ATP), guanosine di- and triphosphate (GDP, GTP), and nicotinamide adenine dinucleotide (NAD) occur in the free state (Figure 11.31).

Nucleic acids: Nucleic acids are macromolecules in which the nucleotides are linked by phosphodiester bonds between the 3′ and 5′ positions of the sugars.

A portion of a molecule of RNA, therefore, has the structure given in Figure 11.32 where the base is either a purine or pyrimidine. Most nucleic acids are very large molecules so that to show the complete formulae would be rather cumbersome. A useful form of shorthand for the structure shows the bases present by using their first letter. The sequence of bases is extremely important so that a nucleic acid may be shown by the first letters of the bases only (Figure 11.33).

DNA: *It* consists of two chains of polynucleotides interwoven in the form of a spiral structure that is stabilized by hydrogen bonding between particular base pairs. The stereochemistry of the bases is such that adenine pairs with thymine and guanine with cytosine so that the ratio of A/T and G/C is unity (Figures 11.34 and 11.35). Most DNA molecules are of the double-helical type, although some viruses contain only

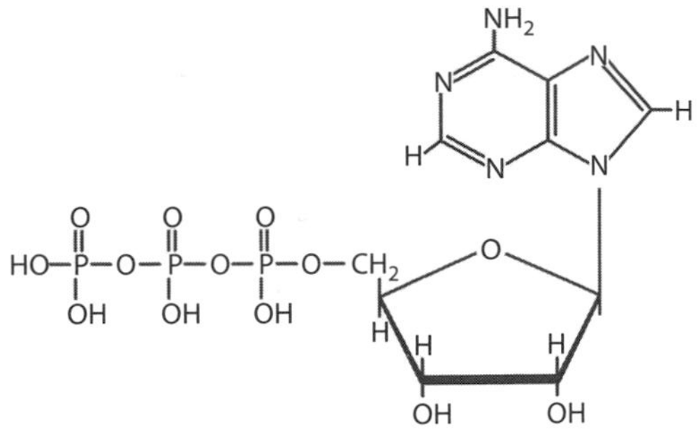

Adenosine triphosphate (ATP)

Figure 11.31   ATP in free state.

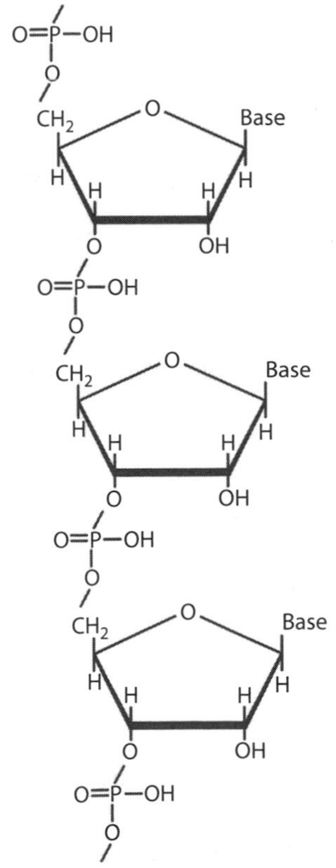

Figure 11.32   Structure of RNA molecule.

single-stranded DNA and mitochondrial DNA as a closed loop. It is difficult to arrive at an accurate value for the molecular weight of DNA since the methods used for its isolation may result in breaking of the molecule, but values of 10^9 have been obtained.

RNA: The simple relationship of A/U = G/C = 1 does not hold for most forms of RNA since the RNA molecule consists of a single strand of nucleic acid in the form of a random coil with only limited regions of base pairing. The molecular weight

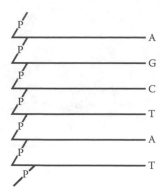

Figure 11.33 Shorthand notation to show the base sequence of a nucleic acid.

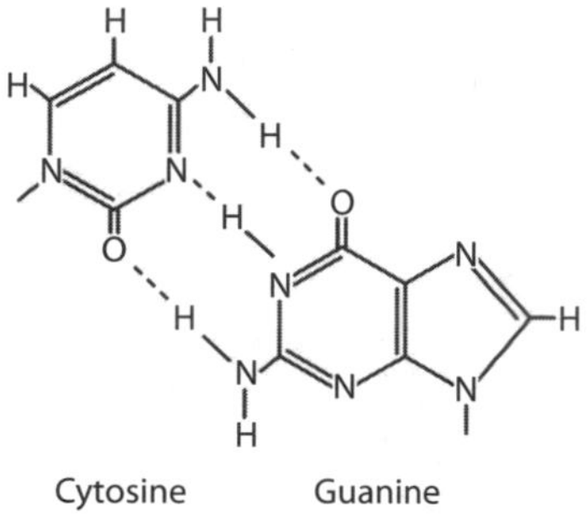

Cytosine   Guanine

Figure 11.34 The pairing of cytosine and guanine by hydrogen bonding.

Thymine   Adenine

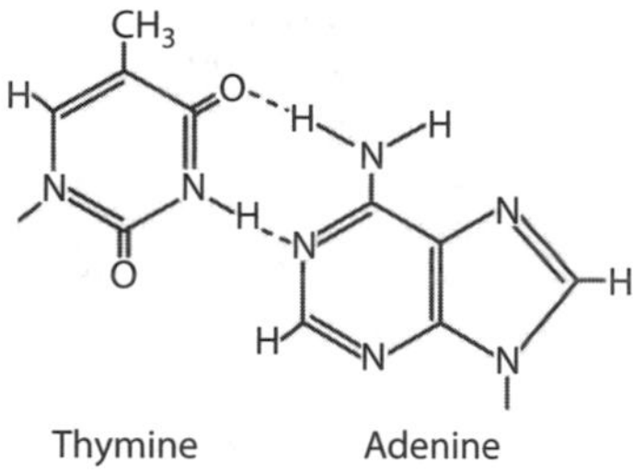

Figure 11.35 The pairing or thymine and adenine by hydrogen bonding (–) as in DNA.

of tRNA is about 25,000, but other forms of RNA have a high molecular weight of a million or so.

11.5.3 Biological Significance of Nucleic Acids

Nucleic acids are large molecules that carry tons of small details: all the genetic information. Nucleic acids are found in every living thing—plants, animals, bacteria, viruses, fungi—that uses and converts energy. Every single living thing has something in common.

11.5.4 DNA

DNA is an extremely long chemical thread made up of two strands, but unlike most threads, the strands are not twisted round one another but wound together round an imaginary core to form a double spiral or more accurately "double helix" (Figure 11.26). Each thread of the molecule consists of a chain of deoxyribose and phosphate residues arranged alternatively.

11.5.4.1 Occurrence
DNA is intimately associated with the genetic material of the cell. In some microorganisms, a single strand of DNA seems to be the store of the genetic information, but in higher organisms, the DNA is present as nucleoprotein in the chromosomes. The amount of DNA in the cell of a particular species is constant, whereas the germ cells, with half the number of chromosomes, contain half the amount of DNA present in other cells. The DNA occurs almost exclusively in the nucleus with trace amounts in the mitochondria and chloroplasts.

11.5.4.2 Function
The main function of DNA is to act as a store of genetic information. Hereditary characteristics are passed on to daughter cells through replication of DNA that appears to take place in a series of short Steps as segments of the double helix are unwound, and the new DNA is formed on each strand simultaneously from free nucleotides. The newly synthesized strands are built up according to the rules of base pairing so that eventually two identical molecules are formed from the original DNA. DNA is also the template for the synthesis of proteins in the cell with a triplet of three bases providing the *genetic code* for each amino acid.

11.5.5 RNA

Superficially RNA has considerable structural similarity to DNA. The molecule has a long chain of alternating sugar and phosphate residues, although the monosaccharide residue is ribose and not deoxyribose as in DNA. As in DNA, the sugar residues are double-substituted: one bond being to the 3' position of the molecule and the other to the 5'. Once again, the ends of an RNA strand can be distinguished: one by having a free 5' phosphate and the other by a free 3' phosphate (Figure 11.32).

11.5.5.1 Occurrence
Whereas DNA is found almost exclusively in the nucleus, RNA is distributed throughout the cell. Most of the RNA is present in the cytoplasm as soluble and ribosomal RNA (rRNA), but about 10% is found in the nucleus with trace amounts also present in the mitochondria. There are three types of RNA present in the cells of higher organisms: *messenger RNA (mRNA)*, *rRNA*, and *tRNA*, all of which are actively involved in the synthesis of proteins.

11.5.5.2 Protein Synthesis

The genetic message is first passed from the nuclear DNA to mRNA through the bases, a process known as *transcription*. The mRNA then migrates into the cytoplasm, and it is here in association with the ribosomes and tRNA that the 4-letter base code is translated into the 20-letter amino acid code of proteins: a process known as *translation*. Each amino acid has its own tRNA whose function is to transfer the activated amino acid to the site of protein synthesis.

11.5.6 Nucleic Acids of Viruses

All viruses contain nucleic acids, which may account for as much as 50% of the particle. The nucleic acid may be single or double stranded and may be present as a linear or a circular molecule. The nucleic acid has a well-defined tertiary structure and is surrounded by a protective coat of protein. The nucleic acid is the infective part of the virus, while the protein accounts for its immunological specificity. The virus attaches itself to the cell and injects its own DNA into the host, thus directing the infected cell to synthesize virus proteins and nucleic acids in order to multiply. Plant viruses contain RNA but no DNA, and in these cases, the RNA acts as the genetic material as well as performs its usual function in protein synthesis.

11.5.7 Experiment: Isolation of RNA from Yeast

11.5.7.1 Principle

Total yeast RNA is obtained by extracting a whole-cell homogenate with phenol. The concentrated solution of phenol disrupts hydrogen bonding in the macromolecules, causing denaturation of the protein. The turbid suspension is centrifuged, and two phases appear: the lower phenol phase contains DNA, and the upper aqueous phase contains carbohydrate and RNA. Denatured protein, which is present in both phases, is removed by centrifugation. The RNA is then precipitated with alcohol. The product obtained is free of DNA but usually contaminated with polysaccharide. Further purification can be made by treating the preparation with amylase.

11.5.7.2 Materials

Dried yeast, phenol solution (90%), potassium acetate (20%, pH 5), absolute ethanol, diethyl ether, water bath at 37°C.

11.5.7.3 Protocol

Suspend 30 g of dried yeast in 120 mL of water previously heated to 37°C. Leave for 15 min at this temperature and add 160 mL of concentrated phenol solution (*Caution*: Corrosive). Stir the suspension mechanically for 30 min at room temperature, and then centrifuge at 3000g for 15 min in the cold to break the emulsion. Carefully remove the upper aqueous layer with a Pasteur pipette, and centrifuge at 10,000g for 5 min in a refrigerated centrifuge to sediment denatured protein. Add potassium acetate to the supernatant to a final concentration of 20 g/L, and precipitate the RNA by adding two volumes of ethanol. Cool the solution in ice and leave to stand for 1 h. Collect the precipitate by centrifuging at 2000g for 5 min in the cold. Wash the RNA with ethanol–water (3:1), ethanol, and, finally, ether; air-dry and weigh. (*Note*: Yeast contains about 4% RNA by dry weight.) Compare with a commercial preparation by measuring the pentose, phosphorus, and DNA content and by determining the absorption spectrum. Keep your preparation for use in later experiments.

11.5.8 Experiment: Isolation of Total RNA from Plant Tissue by SDS–Phenol Method

11.5.8.1 Principle
Procedure is relatively simple, economical, and an all-purpose for the isolation of RNA from leaves, shoots, roots, and other plant tissues. SDS and phenol dissociate nucleoprotein complex and denature proteins. Further treatment with phenol–chloroform–isoamyl alcohol removes denatured proteins, and RNA is recovered by precipitation with ethanol from the aqueous phase, while DNA remains confined in the phenol phase.

11.5.8.2 Precaution
Rinse glassware and plasticware with 1% diethylpyrocarbonate (DEPC) solution to inactivate RNase. *Caution: DEPC is a suspected carcinogen and should be handled carefully.*

11.5.8.3 Materials
Refrigerated centrifuge, plant tissues such as leaves and embryos, phenol, ethanol.
   Phenol reagent: Redistilled phenol 500 g, 8-hydroxyquinoline 0.5 g, water 150 mL.
   SDS solution: 1% SDS, 0.15 M sodium acetate, 5 mM EDTA, 50 mM Tris–HCl (pH 9.0).
   Deproteinizing solution: Mix phenol reagent with chloroform and isoamyl alcohol in a ratio of 25:24:1 at room temperature.
   Chloroform/isoamyl alcohol: 24:1.
   Proteinase solution: 20 mg/mL proteinase K. Incubate for 15 min at 37°C.
   Sodium acetate buffer: Prepare 0.15 M solution of sodium acetate and adjust its pH to 6.0 with acetic acid, and then add SDS so that its concentration is 0.5% (w/v). Store this reagent at room temperature.

11.5.8.4 Protocol
   1. Homogenize the plant tissue in a pestle and mortar in SDS solution at room temperature in a ratio of approximately 1:10. Add 2.5 μL of 20 mg/mL proteinase K. Incubate for 15 min at 37°C.
      Note: With protease treatment, there should be only a small amount of precipitate at the interface between the two phases, although this can vary depending on the cell type. For some cells, the protease Step can be safely omitted.
   2. Transfer the homogenate to a polypropylene centrifuge tube (do not use polycarbonate tubes). Add an equal volume of deproteinizing solution and mix for 5–10 min at room temperature.

3. Centrifuge the tubes at 10,000g for 10 min at 4°C. Three layers will be formed, with the upper aqueous phase containing RNA, the middle layer of denatured proteins, and the lower layer of the organic solvents.

4. Carefully remove the aqueous phase, taking caution not to include the protein interphase, and transfer it to a clean centrifuge tube. If the aqueous phase contains a significant amount of proteins, re-extract RNA by repeating Steps 2 and 3.

5. Transfer the aqueous phase to a clean tube. Add 400 µL of 24:1 chloroform/isoamyl alcohol. Vortex 15–30 s and microcentrifuge 1 min at maximum speed, at room temperature.

6. Again, transfer the aqueous (upper) phase to a clean tube.

Precipitate RNA:

7. Add 40 µL of 3M DEPC-treated sodium acetate, pH 5.2, and 1 mL of 100% ethanol. Mix by inversion. Incubate 15–30 min on ice or store at −20°C overnight.

8. Recover the RNA by microcentrifuging for 15 min at maximum speed, at 4°C.

9. If necessary, remove contaminating DNA (by deoxyribonuclease [DNase] treatment).

10. Rinse the pellet with 1 mL of 75% ethanol/25% 0.1 M sodium acetate, pH 5.2 solution.

Analyze purity:

11. Dry and resuspend in 100 µL DEPC-treated water. Dilute 10 µL into 1 mL water to determine the $A260$ and $A280$. Store the remaining RNA at −70°C.

11.5.9 Experiment: Electrophoresis of RNA Nucleotides

11.5.9.1 Principle

The ester bonds of RNA are readily hydrolyzed due to a free 2′ hydroxyl group on the ribose. This allows the formation of 2′-, 3′-phosphodiester nucleotides that then break down to a mixture of 2′ and 3′ nucleotides. DNA, of course, does not contain a free hydroxyl group on the 2′ position and is, therefore, relatively stable to dilute alkali. The constituent nucleotides are then separated by electrophoresis in citrate buffer at pH 3.5 where they carry quite different negative charges.

11.5.9.2 Materials

Potassium hydroxide (0.3 mol/L), RNA, perchloric acid (20%), citrate buffer (0.02 mol/L, pH 3.5), HCl (0.01 mol/L), incubator set at 37°C, UV spectrophotometer, standard nucleotides (AMP, GMP, CMP, and UMP), equipment for cellulose acetate electrophoresis, UV lamp.

11.5.9.3 Protocol

Dissolve the RNA in 5 mL of 0.3 mol/L potassium hydroxide (20–30 mg/mL), and incubate at 37°C for 18 h. The following day, place the solution in an ice bath and titrate to pH 3.5 with 20% perchloric acid. Remove any precipitate by centrifugation and use the supernatant for the electrophoretic separation. Soak the cellulose acetate in the buffer, gently blot, and place across the electrodes. Apply the nucleotides close to the cathode and carry out the electrophoresis at 10 V/cm.

The progress of the separation can be readily followed by examining the strip in UV light. The cellulose acetate fluoresces under UV irradiation, while the nucleotides, which absorb strongly in the UV, show up as dark spots. Elute each of the nucleotides with 4 mL of 0.01 mol/L HCl, and plot the absorption spectra of the eluates in the region 220–320 nm. As a blank, use similar eluates from a strip run at the same time with no added nucleotides.

11.5.9.4 Results and Conclusion

Identify the nucleotides as far as possible from the extinction data given in Table 11.13.

Use the pK values of the ionic groups on the nucleotides as given in Table 11.14 to predict their order of separation.

11.5.10 Experiment: Separation of RNA Nucleotides by Ion-Exchange Chromatography

11.5.10.1 Principle

Isolated RNA from the previous experiment is hydrolyzed to its constituent nucleotides by dilute alkali. The products of hydrolysis are then separated on a strongly acidic ion-exchange column and identified by their characteristic UV absorption spectra (Table 11.12). UMP and GMP are eluted from the column as distinct peaks, but the AMP and CMP are combined and, in this case, an empirical formula is used to calculate the relative amounts of the two nucleotides.

TABLE 11.13 Absorption of RNA Nucleotides in the Ultraviolet at pH 2

Nucleotide	Millimolar Extinction Coefficient at 260 nm E_{260}	E_{250}/E_{260}	E_{280}/E_{260}
Adenylic acid	14.3	0.85	0.22
Guanylic acid	11.8	0.92	0.68
Cytidylic acid	6.8	0.47	1.90
Uridylic acid	9.8	0.78	0.30

TABLE 11.14 pK Values of the Ionizable Groups on the RNA Nucleotides

RNA Nucleotides	Primary Phosphate	Amino	Secondary Phosphate	Hydroxyl
Adenylic acid	0.9	3.7	5.9	—
Guanylic acid	0.7	2.3	5.9	9.5
Cytidylic acid	0.8	4.3	6.0	13.2
Uridylic acid	1.0	—	5.9	9.4

11.5.10.2 Materials

RNAs from animal and bacterial sources, KOH (0.3 mol/L), perchloric acid (20%), HCl (0.05 mol/L), HCl (1 mol/L), Dowex 50 W chromatography column (15 cm × 1.5 cm), UV spectrophotometer, volumetric flasks (10 mL), volumetric flasks (25 mL).

11.5.10.3 Protocol

Hydrolyze 0.2 g of RNA with 0.3 mol/L KOH and neutralize with perchloric acid as described in the previous experiment. Pipette 3.8 mL of the solution into a test tube, and add 0.2 mL of 1 mol/L HCl to give a final concentration of acid of 0.05 mol/L. Measure the extinction of the solution at 260 nm by diluting a portion with 0.05 mol/L HCl, and apply from 5–12 extinction units to the top of a Dowex 50 W column previously equilibrated with 0.05 mol/L HCl. Elute with 0.05 mol/L HCl and collect the effluent in a 10 mL volumetric flask. Monitor the effluent by measuring the extinction at 260 nm, and when the first nucleotide has emerged from the column, place a second 10 mL flask containing 0.5 mL of 1 mol/L HCl under the column, and elute with water until all the GMP has been eluted. At this stage, increase the flow rate down the column and collect the AMP and CMP into a 25 mL flask containing 1.25 mL of 1 mol/L HCl.

Make up the first flask with 0.05 mol/L HCl and the other two flasks with water. Mix thoroughly and plot the UV spectra of a sample from each flask from 220 to 320 nm.

11.5.10.4 Results and Conclusion

The amount of each eluted nucleotides is calculated by using the following formulae. Does the simple relationship AMP/UMP = GMP/CMP = 1 hold for RNA?

Peak 1: UMP (μmol) = $(E_{260} \times \text{vol})/9.8 = (E_{260} \times 10)/9.8$

Peak 2: GMP (μmol) = $(E_{257} \times \text{vol})/11.8 = (E_{257} \times 10)/11.8$

Peak 3: If $y = (2.32E_{257} - E_{279})/2.08$

$$\text{CMP} (\mu\text{mol}) = (E_{279} - 0.238y) \times \text{vol}/13.2$$

$$= (E_{279} - 0.238y) \times 25/13.2$$

$$\text{AMP} (\mu\text{mol}) = (y \times \text{vol})/14.3 = (y \times 25)/14.3$$

11.5.11 Experiment: Base Composition of RNA

11.5.11.1 Principle

RNA and DNA can be hydrolyzed to the constituent bases by treatment with 72% perchloric acid for 1 h. The method is not completely quantitative since some thymine is lost. The resulting bases are then separated by paper chromatography and detected with UV light.

11.5.11.2 Materials

RNA, perchloric acid (72%), marker bases (adenine, guanine, cytosine, uracil, and thymine, 5 mmol/L), boiling water bath, UV lamp, HCl (0.1 mol/L), UV spectropho-tometer, conc. HCl, isopropanol, paper chromatography apparatus, chromatography solvent (isopropanol–water–cone. HCl, 130:37:33).

11.5.11.3 Protocol

Mix the nucleic acid (100 mg) with 1 mL of perchloric acid, and heat on a boiling water bath for 1 h. Place a marble or ampoule on top of the tube to reduce the loss by evaporation. *Note: This hydrolysis must be carried out* in *a fume chamber behind a protective screen since there is a risk of an explosion. Do not heat to dryness.* Cool the tube, add 1 mL of water, and centrifuge the contents: use the clear supernatant for descending chromatography overnight on Whatman No. 1 paper. Marker spots of the bases should be run at the same time. Dry the chromatogram in a current of cold air for 4 h, and then locate the bases with UV light. Elute the spots with 5 mL of 0.1 mol/L HCl overnight, and plot the absorption spectra. Identify the bases by comparing the absorption spectra with standard purines and pyrimidines, and determine the relative amounts present in the nucleic acid from the absorption coefficients given in Table 11.15. Compare the base ratios with those obtained by alkaline hydrolysis followed by electrophoresis of the nucleotides.

11.5.12 Experiment: Isolation of DNA from Coconut

11.5.12.1 Principle

Nucleic acids are usually present in the cell in combination with proteins and are called nucleoproteins. During isolation, this complex is broken, and nucleic acids are precipitated with alcohol.

11.5.12.2 Precautions

Nucleic acids are more fragile; hence, isolation of nucleic acids with biological activity requires experience and skill. Even mere sucking through a pipette will break the DNA to pieces.

TABLE 11.15 Absorption of Purine and
Pyrimidine Bases in 0.1 mol/L HCl

Nucleotide Bases	Wavelength Maximum	Millimolar Extinction Coefficient
Adenine	260	13.0
Guanine	250	11.2
Uracil	260	7.9
Thymine	265	7.9
Cytosine	275	10.5

11.5.12.3 Material

Coconut endosperm, pestle and mortar measuring cylinder, centrifuge tubes, centrifuge and incubator, phenol, chloroform, sodium acetate, Tris–HCl, EDTA and absolute alcohol, sodium chloride, sodium citrate.

Lysis buffer: 10 mM Tris–HCl buffer of pH 8.0 containing 0.372 g EDTA and 0.5 g sodium dodecyl sulfate.

10 mM Tris–HCl buffer: 0.1211 g of Tris dissolved in 100 mL of distilled water and adjusted to pH 8.0 by adding 0.1 N HCl.

3 mM sodium acetate (pH 7.0): 0.0820 g was dissolved in 100 mL of distilled water.

Saturated phenol: Chloroform mixture (1: 1).

11.5.12.4 Protocol

Homogenize tender coconut endosperm with homogenizing buffer. Extract with phenol chloroform. Dissolve in 014 M solution of NaCl containing 3 mM sodium acetate buffer (pH 7.0) as shown in Figure 11.36.

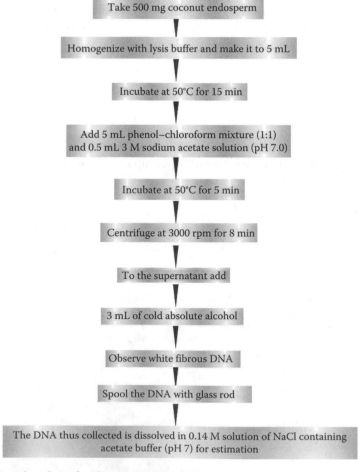

Figure 11.36 Flowchart for the extraction of DNA from endosperm.

11.5.13 Experiment: Isolation of DNA from Pig Spleen

11.5.13.1 Principle

Almost all cells contain DNA, but the amount present in some tissues is quite small so that they are not a particularly convenient source. In addition, some tissues contain high DNase activity so that the DNA is broken down into smaller fragments. A convenient source for the isolation of DNA should, therefore, contain a high quantity of the material and have low DNase activity. Lymphoid tissue is very good in this respect and thymus is the best source, with spleen as a good alternative. DNA is readily denatured, and extreme caution must be taken in order to obtain a product that is structurally related to that found in the cell. Mechanical stress and extreme physical and chemical conditions must be avoided, and nucleases must be inhibited. Sodium citrate is, therefore, present in the solution to bind Ca^{2+} and Mg^{2+} that are cofactors for DNase and the stages until the removal of protein is carried out as rapidly as possible in the cold. The nucleoprotein is soluble in water and solutions of high ionic strength, but is insoluble in solutions of low ionic strength (0.05–0.25 mol/L), and use is made of this property in the initial extraction. The tissue is first homogenized in isotonic saline buffered with sodium citrate, pH 7, when most other macromolecules pass into solution, leaving the insoluble deoxyribonucleoprotein that is then dissolved in 2 mol/L saline. The protein is removed by treatment with a chloroform/amyl alcohol mixture and the DNA precipitated with ethanol. The product is then dissolved in dilute buffered saline and stored frozen. It is stable in this form for several months.

11.5.13.2 Materials

Pig spleen, buffered saline (0.15 mol/L NaCl buffered with 0.015 mol/L sodium citrate, pH 7), sodium chloride (2 mol/L), chloroform–amyl alcohol (6:1), absolute ethanol, ether, Waring blender.

11.5.13.3 Protocol

Chop 50 g of pig spleen into small fragments and homogenize with 200 mL of buffered saline for 1 min. Centrifuge the suspension at 5000g for 15 min and re-homogenize the precipitate in further 200 mL of buffered saline. Discard the supernatant and suspend the combined sediments uniformly in 2 mol/L NaCl to a final volume of 1 L, when most of the material should dissolve. Remove any sediment by centrifugation and stir the solution continuously with a glass rod while adding an equal volume of distilled water. Spool the fibrous precipitate onto a glass rod and leave it to stand in a beaker for 30 min. During this time, the clot will shrink, and the liquid expressed should be removed with filter paper. Dissolve the deoxyribonucleoprotein in about 100 mL of 2 mol/L NaCl, add an equal volume of the chloroform/amyl alcohol mixture (6:1), and blend for 30 s. Centrifuge the emulsion at 5000g for 10–15 min and collect the upper (opalescent) aqueous layer containing the DNA. This is best carried out by gentle suction into a suitable container so that the denatured protein at the interface of the two liquids is not disturbed. Repeat the treatment with organic solvent twice more and collect the supernatant in a 500 mL beaker.

Precipitate the DNA by slowly stirring two volumes of ice-cold ethanol with the supernatant, and collect the mass of fibers on the glass stirring rod. Carefully remove the rod and gently press the fibrous DNA against the side of the beaker to expel the solvent. Finally, wash the precipitate by dipping the rod into a series of solvents and expelling the solvent as described. Four solvents are used: 70% v/v ethanol, 80% v/v ethanol, absolute ethanol, and ether. Remove the last traces of ether by standing the DNA in a fume cupboard for about 10 min. Weigh the dry DNA and dissolve by continuously stirring in buffered saline diluted 1 in 10, with distilled water (2 mg/mL); store frozen until required.

11.5.14 Experiment: Ultraviolet Absorption of the Nucleic Acids

11.5.14.1 Principle
The nucleic acids absorb strongly in the UV region of the spectrum due to the conjugated double-bond systems of the constituent purines and pyrimidines. They show characteristic maxima at 260 nm and minima at 230 nm. The water content of nucleic acids is not usually known, so extinction coefficients cannot be reliably based on weight. The extinction is conveniently expressed as $E\rho$, the extinction of a solution containing 1 mol/L of phosphorus. Even so, the extinction coefficient of a nucleic acid is not a constant but depends on the previous treatment of the material, as well as the pH and ionic strength of the medium. Typical values for $E\rho$ are given as follows:

$$E\rho \text{ for DNA} = 6000\text{--}8000$$

$$E\rho \text{ for RNA} = 7{,}000\text{--}10{,}000$$

The extinction coefficient of a nucleic acid may increase by up to 40% on degradation or hydrolysis, and this is known as the *hyperchromic effect*. In the macromolecule, hydrogen bonding and π–π interactions alter the resonance behavior of the bases are so that the extinction of the nucleic acid is less than that of the constituent nucleotides.

When a solution of double-stranded DNA is slowly heated, there is little change in extinction until the "melting temperature" (T_m) is reached; at this stage, the absorbance increases rapidly to a higher value, which is not significantly changed by further heating. At the melting temperature, the hydrogen bonds between base pairs on opposite strands are broken, and the two DNA threads are separated. If the hot DNA solution is then cooled slowly, the two threads recombine and the "cooling curve" should be superimposed on the "melting curve." If, however, the DNA is cooled rapidly, then some recombination of the two strands takes place, but in a more random manner so that the extinction of the solution at room temperature is higher than that of the original DNA solution before heating (Figure 11.37).

11.5.14.2 Materials
DNA, RNA, DNase, acetate buffer (0.5 mol/L, pH 5.5), acetate buffer–magnesium sulfate solution (prepare a 0.1 mol/L solution of magnesium sulfate in the previous

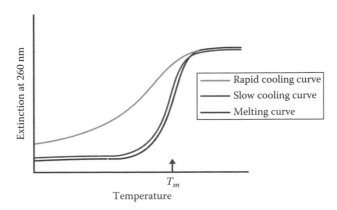

Figure 11.37   The effect of temperature on the extinction of DNA.

buffer solution), buffered saline (0.15 mol/L NaCl, 0.105 mol/L sodium citrate, pH 7), thermo-controlled cell housing, UV spectrophotometer, water bath for use at temperatures up to 95°C.

11.5.14.3 Protocol
Spectrophotometric assay: Dissolve 10 mg of RNA and DNA separately in the buffered saline, and make up to 100 mL. Plot the absorption spectra of these two solutions between 220 and 320 nm, and express the final results as $E\rho$ after determining the phosphorus content. Compare the results obtained with commercial samples.

Action of DNase (deoxyribonucleate–oligonucleotidehydrolase): Prepare a solution of DNA in the acetate buffer–magnesium sulfate solution (10 mg/100 mL), and pipette 2.5 mL of this into a cuvette. Add 0.5 mL of enzyme solution, suitably diluted, and follow the change in extinction at 260 nm. Compare this with a blank containing distilled water instead of the enzyme. Repeat the experiment with RNA.

Effect of heat on DNA: Prepare three solutions of DNA in dilute (0.1), standard (1), and strong (×5) buffered saline to give an initial extinction value of about 0.3–0.5 (0.1 mg/mL). Place the solutions in three cuvettes in a spectrophotometer with a thermostatically controlled cell housing, and record the change in extinction as the temperature is increased to 90°C. The temperature of the water bath heating the cell container is usually higher than that of the contents of the cuvettes, so the temperature of the DNA solutions should be measured directly if possible.

In addition, measure the extinction change as the cuvettes are slowly cooled from 90°C to room temperature. Repeat the experiment, but this time cool the solutions rapidly. Compare the cooling and melting curves for the two experiments.

11.5.15 Experiment: Viscosity of DNA Solutions

11.5.15.1 Principle
The separation of the complementary strands of DNA involves a change from the rigid linear double helix to a random coil formation and leads to a drop in viscosity. When using the Ostwald–Fenske viscometer, slow rates of flow must be used and

very dilute solutions to make sure that the molecules are randomly orientated and not aligned by fluid flow.

The relative viscosity is obtained from viscometer timings of the test solution (t_1) and the solvent (t_0) so that

$$\text{Relative viscosity } \eta_{rel} = \frac{t_1}{t_0}$$

$$\text{Specific viscosity } \eta_{sp} = \eta_{rel} - 1 = \frac{(t_1 - t_0)}{t_0}$$

$$\text{Reduced viscosity } \eta_{red} = \frac{(t_1 - t_0)}{ct_0}$$

where c is the DNA concentration in mg/mL.

The *intrinsic* viscosity η is the viscosity at infinite dilution and is obtained by plotting η_{red} against concentration and reading the intercept when $c = 0$. The intrinsic viscosity depends on the properties of the molecule and is related to the mol. wt. of the DNA so that

$$0.665 \log M = 3.86 + \log (\eta + 0.5)$$

11.5.15.2 Materials
Ostwald–Fenske viscometer and 25°C water baths, buffered saline (0.15 mol/L NaCl, 0.015 mol/L sodium citrate, pH 7), bacterial DNA or calf thymus DNA, stock DNA solution (1 mg/mL in buffered saline).

11.5.15.3 Protocol
The details for the operation of the Ostwald viscometer and the determination of relative viscosity are given in protein section 11.3.15.

Determination of mol. wt.: Prepare a range of concentrations of DNA in the buffered saline (0, 0.05, 0.1, 0.2, 0.4 mg/mL). Equilibrate in the water bath at 25°C and measure the flow times of each of the solutions. Determine the intrinsic viscosity and the mol. wt. of the DNA.

Effect of denaturation: Take the highest concentration of DNA (0.4 mg/mL) and boil for 5 min; cool to 25°C and measure the viscosity.

Hydrolysis by DNase: Investigate the effects of DNase on the viscosity of the DNA under varying conditions.

11.5.15.4 Result and Conclusion
Interpret data by plotting the graph.

11.5.16 Experiment: Estimation of DNA by the Diphenylamine Reaction

11.5.16.1 Principle
When DNA is treated with diphenylamine under acid conditions, a blue compound is formed with a sharp absorption maximum at 595 nm. This reaction is given by 2-deoxypentoses in general and is not specific for DNA. In acid solution, the straight chain formed of a deoxypentose is converted to the highly reactive β-hydroxylevulinaldehyde that reacts with diphenylamine, to give a blue complex. In DNA, only the deoxyribose of the purine nucleotides reacts so that the value obtained represents half of the total deoxyribose present.

11.5.16.2 Materials
DNA (commercial), RNA (commercial), isolated pig spleen DNA solution, isolated yeast RNA solution, buffered saline (0.15 mol/L NaCl; 0.015 mol/L, sodium citrate, pH 7), diphenylamine reagent (dissolve 1 g of pure diphenylamine in 100 mL of glacial acetic acid, add 1.5 mL of concentrated sulfuric acid. This solution must be prepared fresh), boiling water bath.

11.5.16.3 Protocol
Dissolve 10 mg of the nucleic acid in 50 mL of buffered saline, remove 2 mL, and add 4 mL of diphenylamine reagent. Heat on a boiling water bath for 10 min, cool, and read the extinction at 595 nm. Read the test and standards against a water blank. Assay the isolated nucleic acids and the commercial samples for DNA.

11.5.17 Experiment: Estimation of RNA by Means of the Orcinol Reaction

11.5.17.1 Principle
This is a general reaction for pentoses and depends on the formation of furfural when the pentose is heated with concentrated hydrochloric acid. Orcinol reacts with the furfural in the presence of ferric chloride as a catalyst to give a green color. Only the purine nucleotides give any significant reaction.

11.5.17.2 Materials
DNA (commercial), RNA (commercial), isolated pig spleen DNA solution, isolated yeast RNA solution, buffered saline (0.15 mol/L NaCl; 0.015 mol/L, sodium citrate, pH 7), (0.2 mg/mL), orcinol reagent (dissolve 100 mg of ferric chloride ($FeCl_3 \cdot 6H_2O$) in 100 mL of concentrated HCl, and add 3.5 mL of 6% w/v orcinol in alcohol), boiling water bath.

11.5.17.3 Protocol
Mix 2 mL of the nucleic acid solution with 3 mL of orcinol reagent. Heat on a boiling water bath for 20 min cool, and determine the extinction at 665 nm against an orcinol blank.

11.5.18 Experiment: Determination of the Phosphorus Content of a Nucleic Acid

11.5.18.1 Principle
The nucleic acid is oxidized with perchloric acid to give inorganic phosphate that is then estimated by the usual methods described in carbohydrates.

11.5.18.2 Materials
Reagents for the estimation of inorganic phosphate, perchloric acid (60% v/v), digestion flasks and racks.

11.5.18.3 Protocol
Pipette 2 mL of a solution of nucleic acid (0.1 mg/mL) into a digestion flask. Add 0.5 mL of perchloric acid and digest over a low flame for 1 h until all the inorganic matter has disappeared. (*Caution: Explosion risk*) After cooling, add 1 mL of distilled water to each flask and measure the phosphate content as described in spectrophotometer assay section 4.2.4.3.

Suggested Readings

Adams, R. C. P., R. H. Burdon, A. M. Campbell, D. P. Leader, and R. M. S. Smellie. 1981. *The Biochemistry of the Nucleic Acids*, 9th edn. London, U.K.: Chapman & Hall.

Akoh, C. C. and D. B. Min. 2002. *Food Lipids—Chemistry, Nutrition and Biotechnology*, 2nd edn. New York: Marcel Dekker, Inc.

Alexander, P. and H. P. Lundgren. 1966–1969. *A Laboratory Manual of Analytical Methods of Protein Chemistry*, Vols, 1–5. Oxford, U.K.: Pergamon Press.

Bisen, P. S. and G. P. Agarwal. 1980. In vitro production of pectolytic enzymes by *Aspergillus niger* Van Tiegh causing soft rot in apples. *Phytopath Z* 97: 317–326.

Bisen, P. S., K. Ghosh, and G. P. Agarwal. 1982. Induction and inhibition of cellulase complex in *Fusarium solani. Biochem Physiol Pfl* 117: 593–599.

Creighton, T. E. 1983. *Proteins: Structure and Molecular Principles*. New York: W. H. Freeman.

Devlin, T. M. 1997. *Textbook of Biochemistry*. New York: Wiley-Liss, Inc.

Fromm, H. J. and M. Hargrove. 2012. *Essentials of Biochemistry*. New York: Springer.

Furth, A. J. 1980. *Lipids and Polysaccharides in Biology*. London, U.K.: Arnold Press.

Ginsburg, V. 1981. *Biology of Carbohydrates*, Vol. 1. New York: Wiley.

Ginsburg, V. and P. W. Robbins. 1984. *Biology of Carbohydrate*, Vol. 2. New York: Wiley.

Gurr, M., J. Harwood, and F. Keith. 2002. *Lipid Biochemistry: An Introduction*. New York: Wiley Blackwell.

Guthrie, R. D. 1974. *Introduction to Carbohydrate Chemistry*. Oxford, U.K.: Clarendon Press.

Hughe, R. C. 1983. *Glycoproteins*. London, U.K.: Chapman & Hall.

Kerese, I. 1984. *Methods of Protein Analysis*. New York: Wiley.

Lehninger, A. L., D. L. Nelson, and M. M. Cox. 2000. *Principles of Biochemistry*, 3rd edn. Miami, FL: Worth Publishers.

Mainwaring, W. I. P., J. H. Parish, L. D. Pickering, and N. H. Mann. 1982. *Nucleic Acid: Biochemistry and Molecular Biology*. Oxford, U.K.: Blackwell Scientific Publications.

Mead, J. F., R. B. Alfin-Slater, D. R. Howton, and G. Popjak. 1986. *Lipids: Chemistry, Biochemistry and Nutrition*. New York: Plenum Press.

Meeks, J. C., E. Campbell, and P. S. Bisen. 1994. Elements interrupting nitrogen fixation genes in cyanobacteria: Presence and absence of *nifD* element in clones of *Nostoc* sp. strain Mac. *Microbiology* 140: 3225–3232.

Plummer, D. T. 1988. *An Introduction to Practical Biochemistry*. New York: McGraw-Hill.

Scopes, R. 1982. *Protein Purification: Principles and Practice*. New York: Springer-Verlag.

Singh, B. B., I. Curdt, C. Jakobs, D. Schomburg, P. S. Bisen, and H. Böhme. 1999. Identification of amino acids responsible for the oxygen sensitivity of ferredoxins from *Anabaena variabilis* using site directed mutagenesis. *Biochim Biophys Acta* 1412: 288–294.

Singh, B. B., I. Curdt, D. Shomburg, P. S. Bisen, and H. Bohme. 2001. Valine 77 of heterocystous ferredoxin *FdxH2* in *Anabaena variabilis* strain ATCC 29413 is critical for its oxygen sensitivity. *Mol Cell Biochem* 217: 137–142.

Stumph, P. K. and E. E. Conn. 1987. *Lipid Structure and Function*. London, U.K.: Academic Press.

Van Holde, K. E. 1985. *Physical Biochemistry*, 2nd edn. Englewood Cliffs, NJ: Prentice Hall.

Watson, J. D. 1968. *The Double Helix*. London, U.K.: Weidenfeld & Nicolson.

Important Links

http://chemistry.about.com/od/biochemistry/a/carbohydrates.htm.
http://carb.sites.acs.org/links.htm.
http://www.visionlearning.com/library/module_viewer.php?mid=61.
http://www.infoplease.com/cig/biology/lipids.html.
http://link.springer.com/article/10.1007%2FBF02496327?LI=true.
http://www.cell.com/chemistry-biology/abstract/S1074-5521(11)00402-9?switch=standard.
http://www.biology.arizona.edu/biochemistry/problem_sets/aa/aa.html.
http://www.ncbi.nlm.nih.gov/books/NBK22364/.
http://www.veganhealth.org/articles/protein.
http://library.thinkquest.org/04apr/00217/en/biology/dna/index.html.
http://www.photobiology.info/Smith_Shetlar.html.
http://carcin.oxfordjournals.org/content/21/3/461.full.
http://www2.chemistry.msu.edu/faculty/reusch/VirtTxtJml/nucacids.htm.

12

Enzymology

12.1 Introduction

Applied life sciences offer an increasing potential for the production of goods to meet various human needs. In enzyme technology—a subfield of biotechnology—new processes have been and are being developed to manufacture both bulk and high-added-value products utilizing enzymes as biocatalysts, in order to meet needs such as food (e.g., bread, cheese, beer, vinegar), fine chemicals (e.g., amino acids, vitamins), and pharmaceuticals. Enzymes are also used to provide services, as in washing and environmental processes or for analytical and diagnostic purposes. The goal of these approaches is to design innovative products and processes that are not only competitive but also meet criteria of sustainability.

Enzymes are catalytic proteins that speed up the rate of biochemical reactions without themselves being changed in the process. In the absence of an enzyme, most reactions in living organisms would not occur at appreciable rates, whereas in its presence, the rate can be increased up to 10^7-fold. Enzyme-catalyzed reactions usually take place under relatively mild conditions (temperature well below 100°C, atmospheric pressure, and neutral pH) as compared to the corresponding chemical reactions. Each enzyme catalyzes only a single type of chemical reaction. In addition, enzyme activity can be regulated, varying in response to the concentration of substrates or other molecules (Figure 12.1).

All enzymes are proteins. However, without the presence of a nonprotein component, an enzyme is called as cofactor. Many enzyme proteins lack catalytic activity. When this is the case, the inactive protein component of an enzyme is termed as apoenzyme and the active enzyme, including cofactor, the holoenzyme. The cofactor may be an organic molecule, when it is known as a coenzyme, or it may be a metal ion. Some enzyme binds cofactors more tightly and becomes difficult to remove without damaging the enzyme; it is sometimes called prosthetic factor (Figures 12.2).

12.2 Enzymes as Catalysts

12.2.1 Catalysis

Living organisms are able to obtain and use energy very rapidly because of the presence of biological catalysts called enzymes. As with inorganic catalysts, enzymes change the rate of a chemical reaction but do not affect the final equilibrium; also

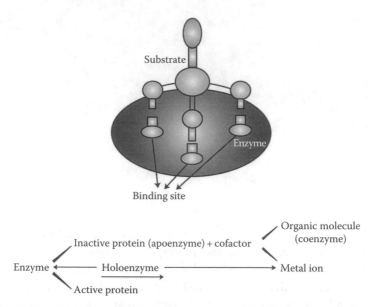

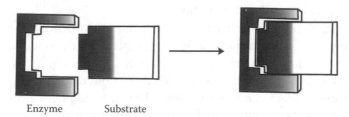

Figure 12.1   Asymmetric binding site in the enzyme with three points: one is catalytic and others are binding sites.

Figure 12.2   Schematic illustration of the lock and key model of enzyme–substrate interactions.

only small quantities are needed to bring the transformation of a large number of molecules. However, unlike most inorganic catalysts, enzymes have a very narrow specificity, that is, they will only catalyze a comparatively small range of reactions or, in some cases, only one reaction. Enzymes will also only function under certain well-defined conditions of pH, temperature, substrate concentration, cofactors, etc. These properties are illustrated in some of the following experiments.

12.3 Classification

Enzymes are named and classified into six main groups according to the type of reaction catalyzed as follows: (1) oxidoreductases, (2) transferases, (3) hydrolases, (4) lyases, (5) isomerases, and (6) ligases (synthetases) (Table 12.1).

TABLE 12.1 Enzyme Classification—Six Major Classes of Enzyme Based on the Type of Reaction Catalyzed

Enzyme	Reaction Catalyzed	Example
Oxidoreductase	Transfer of electron	Alcohol dehydrogenase
Transferase	Transfer of functional groups	Hexokinase
Hydrolases	Hydrolysis of molecules	Trypsin
Lyases	Add groups or remove groups	Pyruvate decarboxylase
Isomerases	Change of substrate isomeric forms	Malate isomerase
Ligases or synthetases	Joining of two molecules	Pyruvate carboxylase

Each enzyme has a systematic name and number that identifies it by indicating its group and subgroups. Enzymes are named either for the substrate acted upon (Figure 12.3) or the type of chemical reaction catalyzed, or both (Figure 12.4), followed by the ending "*ase*." Thus, urease is an enzyme that catalyzes the hydrolysis of urea, cellulase is an enzyme that attacks cellulose, ribonuclease is an enzyme that decomposes ribonucleic acid (RNA), and glucose oxidase is an enzyme that catalyzes the oxidation of glucose and so on. To rationalize the name of enzymes, a system of enzyme nomenclature has been internationally accepted. This system places all enzymes into one of six major classes based on the type of reaction catalyzed (Table 12.1). Each enzyme is then uniquely identified with a four-digit

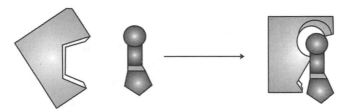

Figure 12.3 Induced fit model in which enzyme conformation changes due to substrate.

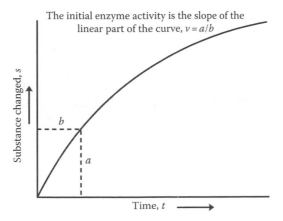

Figure 12.4 Enzyme action progress curve.

classification number. For example, trypsin has the Enzyme Commission (EC) number 3.4.21.4, where the first number (3) denotes that it is a hydrolase, the second number (4) indicates that it is a protease that hydrolyzes peptide bonds, the third number (21) denotes that it is a serine protease with a critical serine residue at the active site, and the fourth number (4) indicates that it was the fourth enzyme to be assigned to this class.

12.4 Measurement of Enzyme Activity

12.4.1 Enzyme Assay

Enzymes are assayed by following the disappearance of the substrate or the appearance of the reaction products with time.

12.4.1.1 Controls
In addition to the "test" mixture, controls are always prepared: one containing no enzymes and the other containing the enzymes but no substrate. Other control mixtures may be required when the enzymes have several cofactors. The object of these controls is to allow for nonspecific and spontaneous chemical reactions, not catalyzed by the enzyme.

12.4.1.2 Progress Curve
A plot of the amount of substrate changed or product formed with time is known as a progress curve (Figure 12.4). This is linear at first but then falls as the reaction proceeds.

The enzyme's activity (v) is obtained from the linear part of the curve that is the initial reaction velocity ($v = a/b$).

12.4.1.3 Enzyme Units
The enzyme activity is most frequently expressed in terms of units (U) such that one unit is the amount of the enzymes that catalyze the conversion of 1 μmole of substrate per minute under defined conditions. In some cases, the unit is too large, and the activity can be more conveniently expressed in terms of nmol/min or pmol/min.

The SI unit of enzyme activity is the *katal* (*kat*) that represents the transformation of 1 mole of substrate per second. This unit is big and more manageable figures are obtained by expressing activities in microkatals (μ*kat*), nanokatals (*nkat*), or picokatals (*pkat*):

1 U = 1 μmol/min
1 kat = 1 mol/s
1 U = μkat/60 = 16.67 nkat

The purity of an enzyme is expressed in terms of the specific activity, which is the number of enzyme units (U) per milligram of protein. The specific activity in SI units is given as katals per kilogram of protein.

12.4.2 Enzyme Activity and Substrate Concentration

12.4.2.1 Michaelis–Menten Enzymes

The rate of an enzyme-catalyzed reaction depends directly on the concentration of the enzyme (Figure 12.5). There exists a relation between the rate of reaction and increasing enzyme concentration in the presence of an excess of the compound that is being transformed (also called the substrate).

With a fixed concentration of enzyme and with increasing substrate concentration, a second important relationship is obtained. A typical curve is shown in Figures 12.6 and 12.7.

With fixed enzyme concentration, an increase of substrate will result at first in a very rapid rise in velocity or reaction rate. As the substrate concentration continues to increase, however, the increase in the rate of reaction begins to slow down until, with a large substrate concentration, no further change in velocity is observed. Michaelis and others correctly suggested that an enzyme—catalyzed reaction at

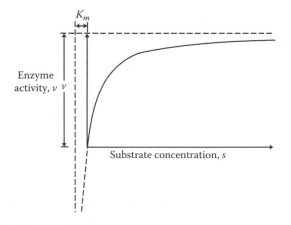

Figure 12.5   The effect of substrate concentration on enzyme activity.

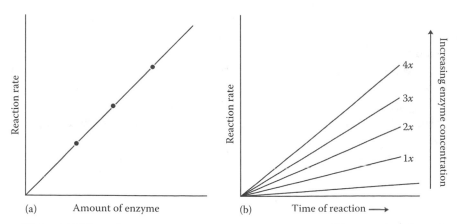

Figure 12.6   (a) Effect of enzyme concentration on reaction rate (b), assuming that substrate concentration is in saturating amounts.

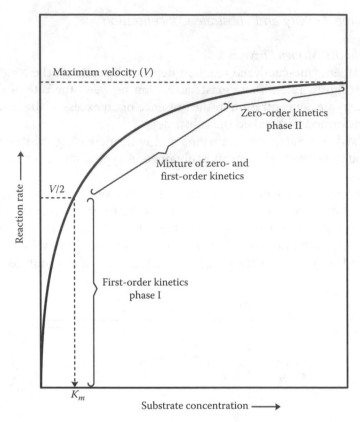

Figure 12.7 Effect of substrate concentration on reaction rate, assuming that enzyme concentration is constant.

varying substrate concentrations—is diphasic; that is, at low substrate concentrations, the active sites on the enzyme molecules are not saturated by substrate, and thus, the enzyme rate varies with substrate concentration (phase 1). As the number of substrate molecules increases, the sites are covered to a greater degree until at saturation no more sites are available, the enzyme is working at full capacity, and now the rate is independent of substrate concentration (phase II). The mathematical equation that defines the quantitative relationship between the rate of an enzyme reaction and the substrate concentration and thus fulfills the requirement of the rectangular hyperbolic curve is the Michaelis–Menten equation.

Kinetics: If the activity of an enzyme is determined over a range of substrate concentration, a curve similar to the rectangular hyperbola of Figure 12.7 is often obtained.

At low substrate concentration, *v* varies linearly with *s*, giving first-order kinetics:

$$v = -\frac{ds}{dt} = ks$$

where *k* is the rate constant.

At high substrate concentration, there is a mixture of first- and zero-order kinetics.

At intermediate substrate concentrations, there is a mixture of first- and zero-order kinetics.

12.4.2.2 Michaelis Equation

The following equation v and s can be obtained for the whole curve and was first derived by Michaelis and Menten in 1913:

$$v = \frac{V}{1 + K_m / s} \quad \text{or} \quad v = \frac{Vs}{s + K_m}$$

where V and K_m are constants.

The following equation can be rearranged to give the form usually shown for the equation of a rectangular hyperbola (Figure 12.2):

$$(V - v)(K_m + s) = V K_m$$

The basic assumption used was that the enzymes and substrate form a complex that then breaks down to give the enzymes and products.

The whole process can be represented by the following general equation:

$$E + S \underset{k_1}{\overset{k_{+1}}{\rightleftarrows}} ES \xrightarrow{k_{+2}} E + \text{Products}$$

$$(e - \rho)(s) \qquad \qquad (\rho)$$

where

e is the enzyme (E) concentration

s is the substrate (S) concentration

ρ is the enzyme–substrate complex (ES) concentration

k_1, k_{+1}, k_{+2}, and k_{-1} are the velocity constants

12.4.2.3 Kinetic Constants K_m and V

If $s = K_m$, then $v = V/2$ so that the Michaelis constant is the substrate concentration that gives half the maximum velocity.

If the equilibrium condition suggested by Michaelis and Menten holds, then k_{+2} is small compared with k_{+1} and can be ignored so that $K_m = k_{-1}/k_{+1}$, which is the dissociation constant for the enzyme–substrate complex. A large K_m therefore means a large dissociation constant or a small association constant ($1/K_m$). Conversely, a small K_m means a small dissociation constant or a large association constant ($1/K_m$). The Michaelis constant therefore gives a measure of the enzyme–substrate affinity:

Large K_m is the low enzyme–substrate affinity.

Small K_m is the high enzyme–substrate affinity.

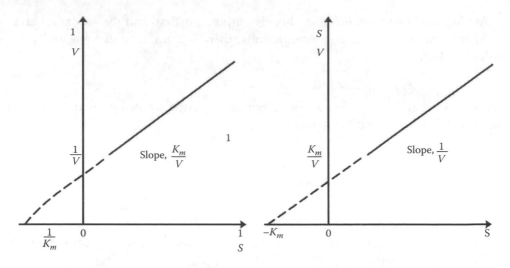

Figure 12.8 Determination of the Michaelis constant.

The kinetic constants K_m and V are most conveniently determined from a linear transformation of the Michaelis equation, obtained by taking the following reciprocals:

$$\frac{1}{v} = \frac{1}{V} + \frac{K_m}{V} \times \frac{1}{s}$$

A plot of $1/v$ against $1/s$ therefore gives a straight line of slope K_m/V (Figure 12.8). The reciprocals of the kinetic constants can then be determined from the intercepts on the axis. Since when

$$\frac{1}{s} = 0, \quad \frac{1}{v} = \frac{1}{V}$$

and when

$$\frac{1}{v} = 0, \quad \frac{1}{s} = \frac{1}{K_m}$$

An alternative plot is of s/v against s (Figure 12.8). This is obtained by multiplying the reciprocal equation by s:

$$\frac{s}{v} = \frac{s}{V} + \frac{K_m}{V}$$

When

$$s = 0, \quad \frac{s}{v} = \frac{K_m}{V}$$

And

$$\frac{s}{v} = 0, \quad s = -K_m$$

The graph of $(1/v)$ against $(1/s)$ is known as the Lineweaver–Burk plot and is the method most frequently used to calculate K_m. However, the slope of the line is most influenced by the activities measured at low substrate concentration, which are the least accurate, and for this reason, the plot of (s/v) against s is to be preferred.

12.4.2.4 Validity of the Michaelis Equation

Most enzymes give a rectangular hyperbola when v is plotted against s in support of the Michaelis–Menten theory. There are some exceptions that are dealt with an allosteric enzymes.

12.4.3 Allosteric Enzymes

12.4.3.1 Effect of Substrate

There are some enzymes that give sigmoid curve when the activity is plotted against substrate concentration (Figure 12.9) and do not obey the usual Michaelis–Menten kinetics. These enzymes are made up of subunits and contain more than one active site per molecule. The sigmoid curve of v versus s can be explained on the basis that when one molecule binds, conformational change takes place in the molecule enabling the next molecules of substrate to bind more readily. The rate of substrate binding therefore depends on the number of active sites already occupied by substrate molecules. Hemoglobin, although not an enzyme, is a good example of an allosteric protein with four binding sites for oxygen. The relative

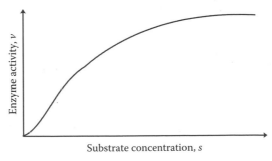

Figure 12.9   Effect of substrate concentration on the activity of an allosteric enzyme.

ease of binding of the oxygen atoms from the first to the fourth is approximately in the ratio of 1:4:24:9.

12.4.3.2 Allosteric Kinetics

The derivation of an accurate equation for the sigmoid curve of allosteric enzymes is complicated, but a simplified equation can be obtained, as suggested by Atkinson, if the following assumptions are made:

1. The intermediates ES, ES_2, ES_3, etc., have only a transient existence.
2. Equilibrium is rapidly attained between the enzymes and the substrate molecules.

If there are n binding sites for the substrate S, then

$$E + nS \xrightleftharpoons[k_1]{k_{+1}} ES_n \xrightarrow{k_{+2}} E + \text{Products}$$

$$\frac{v}{V-v} = \frac{[S]^n}{K}$$

Rearranging the equation,

$$v = \frac{V[S]^n}{K + [S]^n}$$

If there is only one binding site for the substrate ($n = 1$), then this equation becomes the Michaelis equation and a hyperbolic plot is obtained.

12.4.3.3 Kinetic Constants

A convenient linear transformation can be obtained by taking the logarithms of above equation. The kinetic constants can then be read off the straight-line plot (Figure 12.10):

$$\log \frac{v}{(V-v)} = n \log[S] - \log K$$

where n is a measure of the number of binding sites if the original assumptions are correct, but in practice, n is usually less than this since the existence of the intermediates is not transient. n is the degree of cooperativity: the greater the value of n, the more sigmoid the plot of v versus $[S]$.

K is actually a complex steady-state constant and for allosteric enzymes, when $[S] = K$, v is not $V/2$, but if $v = V/2$, then from Equation 2,

$$2[S]^n = K$$

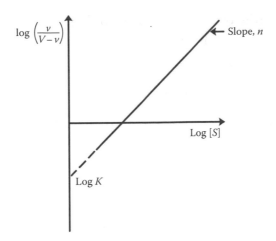

Figure 12.10 Determination of the kinetic constant for an allosteric enzyme.

The substrate concentration that gives 50% of the maximum activity is thus related to the constant K so that

$$K = [S]_{50}^{n}$$

or taking logs

$$\log K = n \log[S]_{50}$$

12.4.3.4 Metabolic Control
Allosteric enzymes have specific sites for the binding of activators and inhibitors that act by causing a change in the conformation of the protein. This enables compounds that are chemically unrelated to the substrate to affect the enzyme activity, and allosteric enzymes are found at many points of metabolic control.

12.4.4 Factors Affecting Enzyme Activity

12.4.4.1 Cofactors
In many cases, if an enzyme is mixed with its substrate under the appropriate conditions, either no catalysis occurs or there is only a slight activity. This is often due to the absence of a coenzyme or activator.

12.4.4.2 Coenzymes
These are low-molecular weight (mol. wt.) organic compounds that are actively involved in the catalysis. They often act as acceptors or donors of specific chemical groups. Nicotinamide adenine dinucleotide (NAD), for example, accepts and donates hydrogen atom and is a coenzyme for many dehydrogenases. The name coenzyme is

reserved for soluble cofactors, while the term prosthetic group is used for coenzymes that are firmly attached to the protein.

12.4.4.3 Activators

These are of a simple chemical nature and are not as specific as coenzymes. They appear to function by activating the enzyme–substrate complex. Several metal ions are known to be activators of a wide range of enzymes: Mg^{2+} for alkaline phosphatase and the kinases.

12.4.4.4 Inhibitors

Many compounds react with enzymes and reduce the measured activity. This property of enzyme is used in designing drugs and insecticides that selectively inhibit enzymes in the infective bacteria or insects, but do not affect the animal or plant. Two classical types of inhibition are recognized: competitive and noncompetitive.

12.4.4.4.1 Competitive Inhibition. In this case, the inhibitor reacts with the enzymes by competing with the substrate for the active site. The degree of inhibition depends on the relative concentrations of substrate and inhibitor, and almost maximal velocity may be found in the presence of the inhibitors if the substrate concentration is high enough.

Competitive inhibitors are of a similar chemical structure to the natural substrate and are fairly specific. This is well illustrated in the case of the enzyme succinate dehydrogenase, which catalyzes the conversion of succinate to fumarate. Malonate and maleate both act as competitive inhibitors of the enzyme.

$$
\begin{array}{ccc}
CH_2COOH & & CHCOOH \\
| & + FAD \rightleftharpoons & \| \\
CH_2COOH & & HOOCCH \quad + FADH_2 \\
& & \text{Fumarate} \\
\text{Succinate} & & \\
\end{array}
$$

$$
\begin{array}{cc}
COOH & \\
| & CH.COOH \\
CH_2 & | \\
| & CH.COOH \\
COOH & \\
\text{Malonate} & \text{Maleate} \\
\end{array}
$$

12.4.4.4.2 Noncompetitive Inhibition. In the noncompetitive-type inhibition, the inhibitor combines with the enzymes but not at the active site, so that enzymes can bind both substrate and inhibitor at the same time. The binding site of the inhibitors is usually sufficiently far removed from the active center so that the binding of the substrate is unaffected. The enzyme–substrate–inhibitor complex formed is unable to break down, and inhibition effectively occurs by the reduction of the amount of enzymes available. Increase of the substrate concentration has no effect on the degree of inhibition.

Most noncompetitive inhibitors are not related chemically to the substrate, and the same inhibitor may affect a number of enzymes.

Examples of noncompetitive inhibition are the action of thiol-blocking agents such as ρ-chloromercuribenzoate, heavy metal ions such as Ag^+ and Cu^{2+}, and the reaction of cyanide with the iron–porphyrin enzymes.

12.4.4.4.3 Determination of Inhibitor Constant (K_i)

In the presence of an inhibitor, the Michaelis–Menten equation is altered as shown in the following discussion, where i is the concentration of the inhibitor and K_i the inhibition constant:

No inhibitor	$v = \dfrac{V}{1+(K_m/S)}$
Competitive inhibitor	$v = \dfrac{V}{1+K_m/s(1+i/K_i)}$
Noncompetitive inhibitor	$v = \dfrac{V}{(1+K_m/s)(1+i/K_i)}$

Taking reciprocals of the previous equations, we obtain the following:

No inhibitor	$\dfrac{1}{v} = \dfrac{1}{V} + \dfrac{K_m}{V}\dfrac{1}{s}$
Competitive inhibitor	$\dfrac{1}{v} = \dfrac{1}{V} + \dfrac{K_m}{V}\left(1+\dfrac{i}{K_1}\right)\dfrac{1}{s}$
Noncompetitive inhibitor	$\dfrac{1}{v} = \dfrac{1}{V}\left(1+\dfrac{i}{K_i}\right) + \dfrac{K_m}{V}\left(1+\dfrac{i}{K_i}\right)\dfrac{1}{s}$

A graph of ($1/v$) against ($1/s$) can be used to determine K_i and the type of inhibition as shown in the following text (Figures 12.11 and 12.12). A competitive type alters K_m, while the noncompetitive type causes a change in V:

$$K_x = K_m\left(1+\frac{i}{K_i}\right)$$

$$V_x = \frac{V}{(1+i/K_i)}$$

12.4.5 Temperature

12.4.5.1 Effect of the Enzyme Reaction

Molecules must possess certain energy of activation (E) before they can react, and enzymes function as catalysts by lowering this energy of activation, thereby enabling the reaction to proceed more rapidly. The overall change in the free energy (ΔG) is unaffected by the enzyme.

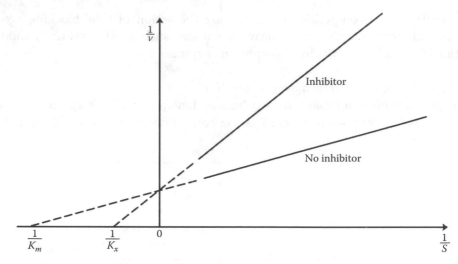

Figure 12.11   Effect of a competitive inhibitor on the Lineweaver–Burk plot.

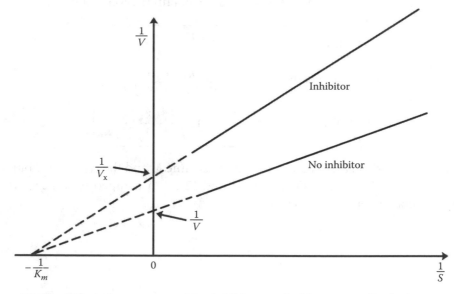

Figure 12.12   Effect of a noncompetitive inhibitor on the Lineweaver–Burk plot.

12.4.5.2 Energy of Activation of an Enzyme

The enzyme-catalyzed reaction can be determined by measuring the maximum velocity (V) at different temperatures and plotting $\log_{10} V$ against $1/T$. The slope of the line is given by $-E/2.303\,R$. This relationship is obtained from the empirical equation of Arrhenius:

$$\frac{d\ln k}{dT} = \frac{E}{RT^2}$$

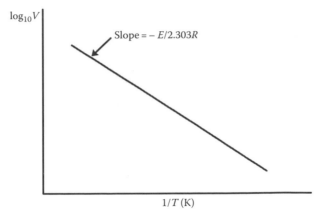

Figure 12.13   An Arrhenius plot.

Integrating above equation, one obtains

$$\log_{10} k = C - \frac{E}{2.303RT}$$

where
   C is a constant
   k is the velocity constant reaction
   T is the temperature
   R is the gas constant $= 8.32$ J/mol/K
   E is the energy of activation (J/mol)

The velocity constant is not easy to obtain, and since V is directly proportional to k, the maximum velocity is plotted against the reciprocal of the temperature for the Arrhenius plot (Figure 12.13).

12.4.5.3 Effect of Temperature on Denaturation
This can be determined by exposing the enzymes to a high temperature for a different period of time and then measuring the activity at a temperature at which the enzymes are stable. This will tell us how much of the enzyme has been destroyed, and a graph can be prepared of the rate of loss of active enzymes with time. This is repeated for a number of different temperatures and the initial rate plotted against $1/T$. The value of E from this plot is the energy of inactivation that is usually quite high due to the large positive entropy change resulting from the unfolding of the molecules during denaturation.

12.4.6 pH

12.4.6.1 pH Optimum
Enzymes are active over a limited pH range only, and a plot of activity against pH usually gives a bell-shaped curve of the type shown in Figure 12.14. The pH value

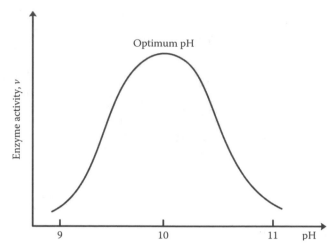

Figure 12.14   pH optimum of an enzyme illustrated by rat alkaline phosphatase.

of maximum activity is known as the optimum pH, and this is a characteristic of the enzymes, provided that the enzymes are stable under the conditions studied. The variation of activity with pH is due to the change in the state of ionization of the enzyme protein and other components of the reaction mixture. Michaelis and Davidson suggested in 1911 that only one of the large number of ionized forms of the protein is active so that a change in pH on either side of the optimum produces a decrease of this form and hence a fall in the activity.

K_m and V changing the pH alter the enzyme activity by affecting V, K_m, or the stability of the enzyme protein. Most published plots of pH optima are the result of changes in V and K_m and are not, therefore, readily amenable to mathematical treatment. Ideally the effect of pH on K_m and V should be studied separately.

12.4.6.2 Enzyme Stability

If the enzyme is unstable at certain pH values, then the optimum pH is no longer a characteristic of the enzyme. The stability can be checked by exposing the enzyme to the appropriate pH value for the time of the experiment and then adjusting the pH to a value at which the enzyme is stable and measuring the activity.

12.5 Time Course Studies of the Reaction Catalyzed by Alkaline Phosphatase (EC 3.1.3.1)

12.5.1 Principle

Phosphatase is a broad term used for nonspecific phosphomonoesterases that hydrolyze organic phosphate esters liberating an alcohol derivative of the substrate and inorganic phosphate (P_i). These enzymes catalyze the following reaction:

$$\text{Orthophosphoric monoester} + H_2O \rightarrow H_2O \rightarrow \text{Alcohol} + Pi$$

Phosphatases have been classified into two groups, namely, acid phosphatases and alkaline phosphatases depending on their pH optima. Acid phosphatases function optimally at acidic pH (4.0–5.5), whereas alkaline phosphatases give maximum activity at alkaline pH (8–10). For assaying phosphatases, p-nitrophenyl phosphate can be used as substrate that is hydrolyzed to p-nitrophenol and Pi, p-nitrophenol is colorless at acidic or neutral pH, but at alkaline pH of 11, it gives yellowish color with absorbance maxima at 410 nm. Hence, the activity of alkaline or acid phosphatases can be conveniently determined colorimetrically by determining the amount of p-nitrophenol produced. The reaction catalyzed by alkaline phosphatase is as follows:

12.5.2 Materials

Test tubes, water bath, colorimeter, refrigerated centrifuge, glycine, NaOH, $MgCl_2$, p-nitrophenyl phosphate, 7-day-old germinating green gram seeds.

Reagents

1. *Glycine–NaOH* buffer (0.05 M, pH 10.5): Dissolve 375 mg glycine in small volume of water, add 42 mL of 0.1 N NaOH solution, and adjust the pH to 10.5.
2. *NaOH* solution (0.085N): Dissolve 340 mg of NaOH in 100 mL of distilled water.
3. *$MgCl_2$* solution (10.5 mM): Dissolve 10 mg of $MgCl_2$ in 10 mL of distilled water. 0.1 mL of this solution is used in 3.5 mL of reaction mixture so that the final conc. of Mg^{2+} during the assay is 0.3 mM.
4. p-*nitrophenyl phosphate* (35 mM): Dissolve 38.8 mg of p-nitrophenyl phosphate in 5 mL of 0.005 M glycine–NaOH buffer pH 10.5.
5. *Standard solution of* p-*nitrophenol* (100 mM): Dissolve 69.75 mg of p-nitrophenol and in 5 mL of distilled water. This solution contains 100 mmoles p-nitrophenol/mL.
6. *Plant material*: Use 7-day-old germinating gram seeds. Ensure that the petri plates contain sufficient amount of water throughout germinating period of seeds.

12.5.3 Protocol

1. Tissue extraction is to be carried out in cold room at 4°C. Grind 1 g of the germinating seeds with chilled 10 mL of glycine–NaOH buffer (0.05 M, pH 10.5) in chilled pestle and mortar in the presence of small amount of acid-washed river sand as an abrasive to facilitate complete breakage of the cells.

2. Centrifuge the homogenate in a refrigerated centrifuge at 10,000g for 20 min. Decant the supernatant and use it as the enzyme preparation.
3. Take nine numbered test tubes and add 3.0 mL of glycine–NaOH buffer, 0.1 mL of $MgCl_2$, and 0.3 mL of the enzyme preparation into each of them.
4. Transfer these tubes to a water bath maintained at 37°C. After 3 min, start the reaction in seven of the previous tubes by adding 0.1 mL p-nitrophenyl phosphate. Note down the time of starting of the reaction for each tube.
5. Exactly after 5, 10, 15, 20, 25, 30, and 45 min, stop the reaction by adding 9.5 mL of 0.085 N NaOH to each tube. In the eighth tube, add NaOH followed by 0.1 mL of p-nitrophenyl phosphate. This represents 0 min control. In the ninth tube, instead of p-nitrophenyl phosphate, add 0.1 mL of 0.05 M glycine–NaOH buffer (pH 10.5), and this serves as the reagent.
6. Adjust the colorimeter at 410 nm to 100% transmission with the reagent blank (tube 9), and record absorbance of the other eight tubes.
7. To prepare a standard curve, take 0–1 mL (0–100 µmoles) of p-nitrophenol. Add 3 mL of 0.05 M glycine–NaOH buffer (pH 10.5) to all the tubes, and make the final volume to 3.5 mL with distilled water. Pipette 9.5 mL of 0.035 N NaOH into each tube, mix the contents, and record the absorbance at 410 using the tube without p-nitrophenol for setting the instrument to 100% transmission or zero absorbance.
8. Plot a graph of A_{410} versus µmoles of p-nitrophenol to obtain a standard curve.
9. Determine the amount of pmoles of p-nitrophenol in the tubes 1 through 8 of step 5.
10. Draw a graph of moles of p-nitrophenol produced versus reaction time.
11. From this graph, note down the maximum assay period up to which the production of p-nitrophenol is linear. In all the subsequent experiments, the duration of reaction should be well within this time limit.
12. Using the experimental data, plot another graph of velocity of reaction (i.e., µ moles of p-nitrophenol produced/min) versus reaction time.

12.5.4 Result

Prepare the graphs (prepared according to steps 11 and 12 of procedure) and interpret the data in terms of reaction velocity and reaction time.

12.6 Effect of Enzyme Concentration on the Rate of Enzyme-Catalyzed Reaction

12.6.1 Principle

Concentration of enzyme in the assay mixture is one of the important factors that determine the velocity of the reaction. Generally, the velocity increases proportionally with increasing concentration of the enzyme, but starts to decline beyond a certain concentration. For obtaining quantitatively reliable values for the enzyme activity in a sample, it is imperative that the amount of the enzyme in the assay mixture should be within the range where proportional increase in activity is obtained. The optimum concentration of the enzyme that should be used can be determined by measuring the activity at varying concentrations of the enzyme in the reaction mixture.

This experiment is designed to study the effect of different amounts of extract on alkaline phosphatase activity on velocity of the reaction.

12.6.2 Materials

All the requirements are as per Section 12.5.

12.6.3 Protocol

1. Prepare the cell-free enzyme extract of germinating gram seeds as described in the earlier experiment 6 (steps 1 and 2).
2. Take two sets of numbered test tubes and pipette 0, 0.05, 0.10, 0.15, 0.20, 0.25, 0.30, 0.35, and 0.40 mL of the enzyme preparation to successive tubes of both the sets. Add appropriate amount of 0.05 M glycine–NaOH buffer (pH 10.5) to each tube so that the combined volume of the extract and the buffer is 3.4 mL. Then add 0.1 mL of $MgCl_2$ into each tube.
3. After 3 min incubation of the tubes at 37°C in a water bath, start the reaction by adding 0.1 mL of p-nitrophenyl phosphate to each tube of the first set. To the second set of tubes, in place of p-nitrophenyl phosphate, add 0.1 mL of 0.05 M glycine–NaOH buffer (pH 10.5), and these tubes serve as minus substrate blanks for the corresponding tubes of the first set. Perform the assay for 30 min and then stop the reaction with 9.5 mL of 0.085 N NaOH.
4. Note down the absorbance at 410 nm of all the tubes after setting the colorimeter to 100% transmission using the tube that did not contain the enzyme preparation.
5. Deduct the absorbance values of the minus substrate blanks from their corresponding tubes of the first set of tubes.
6. Using the standard curve, determine the amount of p-nitrophenol formed in the assay tubes.

12.6.4 Result and Observations

1. Draw a graph of μmoles p-nitrophenol produced versus volume of the enzyme extract in the reaction mixture. Determine the maximum volume of extract required to form a linear curve of p-nitrophenol formation.
2. Plot another graph for velocity of reaction (μmoles of p-nitrophenol formed/min) versus volume of the enzyme preparation. Note down the maximum volume of extract beyond which the velocity starts to decline. Interpret the data obtained.

12.7 Determination of Temperature Optima for Alkaline Phosphatase

12.7.1 Principle

All the enzymes have a narrow temperature range for their efficient functioning. Assays should be carried out at optimum temperature for determining the enzyme activity. Temperature optima ought to be determined in a preliminary experiment for undertaking further investigations of any enzyme. It is essential that appropriate controls must be run simultaneously while carrying out any enzyme assay. It would be important, here, to determine in this experiment whether any p-nitrophenol is produced due to nonenzymatic hydrolysis of p-nitrophenyl phosphate particularly at a higher temperature. The control tubes contain all the components of the reaction mixture except the enzyme preparation and are incubated under identical conditions. After the stipulated period, the reaction is terminated in all the tubes with the reagent that stops the reactions (0.085 N NaOH in the present case) followed by addition of the enzyme preparation to the control tubes. These controls will provide the requisite information about the extent of nonenzymatic formation of the product under the experimental conditions being employed for assaying the enzyme activity.

12.7.2 Materials

All the requirements are as per Section 12.5.

Water baths maintained at the required temperatures (20°C, 25°C, 30°C, 35°C, 40°C, 45°C, and 50°C).

12.7.3 Protocol

1. Prepare the cell-free enzyme extract of germinating gram seeds as described in the earlier experiment 6 (steps 1 and 2).
2. Take two sets of numbered test tubes and pipette 0, 0.05, 0.10, 0.15, 0.20, 0.25, 0.30, 0.35, and 0.40 mL of the enzyme preparation to successive tubes of both the sets. Add appropriate amount of 0.05 M glycine–NaOH buffer (pH 10.5) to each tube so that the combined volume of the extract and the buffer is 3.4 mL. Then add 0.1 mL of $MgCl_2$ into each tube.
3. Add 0.3 mL of the enzyme extract to the assay tubes. Keep one assay tube and one control tube at the temperature at which the enzyme activity is to be determined 20°C, 25°C, 30°C, 35°C, 40°C, 45°C, and 50°C, and after 3 min, add 0.1 mL of p-nitrophenyl phosphate to all the tubes.
4. Allow the reaction to proceed for 30 min or any other suitable time period that falls within linear region of curve of the graph constructed in earlier experiment, stop the reaction with 9.5 mL of 0.085 N NaOH, and add 0.3 mL of the enzyme preparation to control tubes.
5. Record the absorbance of all the tubes at 410 nm.
6. Deduct the obtained A_{410} for the controls from A_{410} of the corresponding assay tubes, and from the standard curve, determine the amount of p-nitrophenol formed in the assay tubes.
7. Draw a graph of amount of p-nitrophenol produced at different temperatures.
8. From the graph, determine the optimum temperature for the activity of alkaline phosphatase.

12.7.4 Result and Observations

Draw a graph of amount of p-nitrophenol produced at different temperatures, and determine the temperature optima for the activity of alkaline phosphatase.

12.8 Determination of Effect of pH on the Activity of Alkaline Phosphatase

12.8.1 Principle

pH is an important parameter for determining enzyme activity. Enzyme assays should be conducted with appropriate control for pH to make corrections for any nonenzymatic formation of the product under varying experimental conditions employed. As in the case of the effect of temperature, zero minute controls may be used here also. The results obtained in earlier experiments should be used in selecting the optimal conditions (i.e., reaction time, amount of enzyme, and temperature) for assay of alkaline phosphatase from germinating gram seeds.

For determining the pH optima, the enzyme activity is measured at varying pH of the reaction mixture. Each buffer has a limited range over which it can effectively fulfill the buffering role. In many instances, two different buffers may have to be tried to ascertain effect over a broader range of pH. In such cases, it is advisable to select at least one common pH for both these buffers so that from this overlapping pH, the variation in enzyme activity due to the nature of the buffer can be distinguished from the actual effect of pH.

12.8.2 Materials

All the requirements are as per Section 12.5.

Prepare 0.05 M glycine–NaOH of different pH (8.6, 9.0, 9.6, 10.0, 10.6, 11.0, 11.5) by adjusting the pH to the desired level with increasing higher volumes of 0.1 M NaOH to 50 mL 0.1 M solution of glycine and then making the final volume to 100 mL with water.

Prepare 0.05 M Tris–HCl buffer with 0.05 M Tris by adjusting to pH 7.2, 7.6, 8.0, 8.6, and 9.0 with 0.1 M HCl.

12.8.3 Protocol

1. Prepare the cell-free enzyme extract of germinating gram seeds as described in the earlier experiment 6 (steps 1 and 2).
2. Take two sets of numbered test tubes and pipette 0, 0.05, 0.10, 0.15, 0.20, 0.25, 0.30, 0.35, and 0.40 mL of the enzyme preparation to successive tubes of both the sets. Add appropriate amount of 0.05 M glycine–NaOH buffer (pH 10.5) to each tube so that the combined volume of the extract and the buffer is 3.4 mL. Then add 0.1 mL of $MgCl_2$ into each tube.

3. Add 0.3 mL of the enzyme extract to the assay tubes. Keep one assay tube and one control tube at the temperature at which the enzyme activity is to be determined at different pH, and after 3 min, add 0.1 mL of p-nitrophenyl phosphate to all the tubes.
4. Note the absorbance at 410 nm against distilled water.
5. Deduct absorbance of the controls from the corresponding assay mixture, and using the standard curve, determine the amount of p-nitrophenol formed in the assay tubes.
6. Since the enzyme extract was prepared in a buffer of pH 10.5, the actual pH of the reaction mixture might deviate from that of the assay buffer. Hence, prepare a mixture containing extraction buffer and assay buffer in the same ratio as used for measuring enzyme activity (1:10 or 1.5 mL of extraction buffer and 15 mL of assay buffer) and using pH meter to determine pH.
7. Plot a graph of µmoles of p-nitrophenol formed/min versus pH of the reaction mixture, and find out the optimum pH for enzyme activity. Also observe whether at a given pH of Tris–HCl and glycine–NaOH buffers the enzyme activity is identical or different.

12.9 Effect of Substrate Concentration on Activity of Alkaline Phosphate and Determination of the K_m and V_{max} of the Reaction

12.9.1 Principle

The enzyme activity is measured at varying concentrations of the substrate under optimal conditions on the basis of the information obtained in earlier experiments. The K_m and V_{max} of the reaction are then determined by processing the data and drawing Lineweaver–Burk ($1/v$ vs. $1/[S]$), Eadie–Hofstee (v vs. $v/[S]$), Hanes ($[S]/v$ vs. v), or Eisenthal and Cornish–Bowden (v vs. $[S]$) plots.

12.9.2 Materials

All the requirements are as per Section 12.5.

12.9.3 Protocol

1. Prepare the cell-free enzyme extract of germinating gram seeds as described in the earlier experiment 6 (steps 1 and 2).
2. Take eight numbered test tubes and add 0.0, 0.05, 0.10, 0.15, 0.20, 0.25, 0.30, 0.40, and 0.50 mL of p-nitrophenyl phosphate (which corresponds to 0–2.5 µmol) in successive tubes.
3. Add calculated amount of 0.05 M glycine–NaOH (pH 10.5) so that the total volume of p-nitrophenyl phosphate and the buffer in each tube is 3.0 mL. Add 0.1 mL of 10 mM $MgCl_2$ to the tubes and keep them in water bath at 37°C.
4. After 3 min warm-up, start the reaction with 0.3 mL of the enzyme extract in each tube.

5. Stop the reaction after 30 min with 9.5 mL of 0.085 N NaOH.
6. Record the absorbance of the color formed at 410 rim against distilled water.

12.9.4 Results and Calculations

1. Determine the amount of *p*-nitrophenol formed in each tube from the standard curve.
2. Plot a graph of μmoles of *p*-nitrophenol produced/30 min against molar conc. of the substrate (ρ-nitrophenyl phosphate) in the assay mixture.

12.10 Progress Curve of Serum Alkaline Phosphatase (Orthophosphoric Acid Monoester Phosphohydrolase, 3.1.3.1)

12.10.1 Principle

Alkaline phosphatase (pH 9–10 optimum) is found in bone, kidney, liver, and intestine and acts on phosphoric esters with the liberation of inorganic phosphate.

$$R.O.P.{=}0 + H_2O \longrightarrow R.OH + HO-\overset{\overset{\displaystyle OH}{|}}{\underset{\underset{\displaystyle OH}{|}}{P}}{=}O$$

p-Nitrophenyl phosphate is used as substrate and the *p*-nitrophenol released by enzymic hydrolysis is measured colorimetrically. In alkaline solution, *p*-nitrophenol absorbs at 405 nm. The substrate *p*-nitrophenyl phosphate does not absorb at this wavelength, so the progress of the enzyme-catalyzed reaction can be readily followed by measuring the change in extinction at 405 nm.

12.10.2 Materials

Sodium carbonate–bicarbonate buffer (0.1 mmol/L in the alkaline buffer), *p*-nitrophenyl phosphate substrate solution (5 mmol/L in the alkaline buffer), *p*-nitrophenol standard (50 μmol/L). This is prepared by dissolving 69.6 mg *p*-nitrophenol in 100 mL of alkaline buffer and then diluting this solution 1:100 with the buffer. Solutions should be freshly prepared; serum, colorimeter.

12.10.3 Protocol

Test: Add 0.8 mL of the carbonate–bicarbonate buffer to 2 mL of the substrate solution and mix thoroughly. At zero time, add 0.2 mL of serum and follow the change in extinction at 405 nm.

Blank: This is the same as the test except that 0.2 mL of buffer replaces the serum.

Standard: Prepare a range of *p*-nitrophenol solutions and plot a graph of extinction at 405 nm against concentration.

12.10.4 Results and Conclusion

Subtract the extinction of the blank from that of the test and calculate the amount of *p*-nitrophenol released from the standard curve. Express the activity of the serum alkaline phosphatase in terms of units per milliliter of serum remembering that 0.2 mL is used in the assay.

12.11 Effect of Variation of Serum Alkaline Phosphatase Activity with Enzyme Concentration

12.11.1 Principle

A plot of enzyme activity against enzyme concentration should give a straight line unless an inhibitor or activator is present in the sample or the activity is wrongly calculated from the nonlinear part of the progress curve.

12.11.2 Materials

Sodium carbonate–bicarbonate buffer (0.1 mmol/L in the alkaline buffer), *p*-nitrophenyl phosphate substrate solution (5 mmol/L in the alkaline buffer), *p*-nitrophenol standard (50 µmol/L). This is prepared by dissolving 69.6 mg *p*-nitrophenol in 100 mL of alkaline buffer and then diluting this solution 1:100 with the buffer. Solutions should be freshly prepared; serum, colorimeter.

12.11.3 Protocol

Repeat the previous experiment with volumes of serum from 0.1 to 0.6 mL. The volume of the substrate is kept constant at 2 mL and the volume of buffer adjusted so that the final mixture is 3 mL as previously. A progress curve for each volume of serum is prepared as previously described, and the initial reaction velocity is determined. A graph is then plotted of enzyme activity and enzyme concentration expressed in milliliters of serum.

12.11.4 Results and Conclusion

Choose a suitable time, which is beyond the linear portion of the progress curve, and express the activity as the average amount of *p*-nitrophenol liberated per minute.

Plot this activity against enzyme concentration and compare the result with the previous curve when initial rates were used to express the activity.

12.12 Effect of Substrate Concentration and Inhibitors on Oxheart Lactate Dehydrogenase

12.12.1 Principle

Lactate dehydrogenase catalyzes the reversible reduction of pyruvate to lactate with $NADH_2$ as the coenzyme. The reduced coenzymes ($NADH_2$) absorb strongly at 340 nm, while the oxidized form (NAD) do not, so the progress of the reaction can be followed by measuring the decrease in extinction at 340 nm with pyruvate as substrate.

$$
\begin{array}{ccc}
CH_3 & & CH_3 \\
| & & | \\
C{=}O \;\; +NADH_2 & \rightleftharpoons & H{-}C{-}OH + NAD \\
| & & | \\
COO^- & & COO^-
\end{array}
$$

Substrates: Pyruvate   L-Lactate

$$
\begin{array}{ccc}
OH & & NH_2 \\
| & & | \\
C{=}O & \text{and} & C{=}O \\
| & & | \\
COO^- & & COO^-
\end{array}
$$

Inhibitors: Oxalate   Oxamate

In this experiment, two compounds with similar structure to the natural substrate pyruvate are examined to see if they are competitive or noncompetitive inhibitors.

12.12.2 Materials

Phosphate buffer (0.1 mol/L, pH 7.4), sodium pyruvate (21 mmol/L, prepared fresh in phosphate buffer), $NADH_2$ (3.5 mmol/L, prepared fresh), oxheart lactate dehydrogenase diluted in phosphate buffer when required, sodium oxalate (3 mmol/L, in phosphate buffer), sodium oxamate (15 mmol/L, in phosphate buffer), water bath at 37°C, UV–Vis recording spectrophotometer.

12.12.3 Protocol

Initial assay: Prepare the following mixture in duplicate and equilibrate at 37°C for 10 min.

Component	mL
Phosphate buffer	2.5
$NADH_2$ (3.5 mmol/L)	0.1
Oxheart lactate dehydrogenase (diluted)	0.3

Rapidly add 0.1 mL of the sodium pyruvate solution also at 37°C, transfer to a cuvette in a thermostatically heated cell housing of an ultraviolet spectrophotometer, and observe the change in extinction at 340 nm with time. Dilute the enzyme preparation so that an extinction change in the region of 0.05–0.10 per minute is obtained.

12.12.4 Results and Conclusion

Enzymes activities are expressed most conveniently in nmoles per minute per milliliter of enzyme or nanomoles per minute per milligram of protein. The molar extinction coefficient of $NADH_2$ at 340 nm is 6.3×10^3 L/mol so that a solution of 1 μmol/mL has an absorption of 6.3.

For the 3 mL reaction mixture used in the assay,

Enzyme activity (μmol/min) = Extinction change per min/6.3 × 3

Since 0.3 mL of enzyme solution was used in the assay,

$$\text{Enzyme activity} \left(\mu\text{mol/min/mL} \right) = \frac{\text{Extinction change per minute}}{6.3} \times \frac{1}{0.3}$$

$$= \text{Extinction change per minute} \times 1.61$$

In general terms, the enzyme activity (μmol/min/mL) = E/min × 1000/extinction coefficient NADH × volume in cuvette × 1.0/volume used for the assay.

The final activity is obtained by multiplying by any dilution factor for micromoles per minute per milliliter and dividing by the protein concentration in milligrams per milliliter for micromolecules per minute per milligram.

12.12.4.1 Determination of the Michaelis Constant
Dilute the sodium pyruvate solution 10 times and repeat the previous assay using volumes from 0.1–1.0 mL of the substrate to start the reaction. Adjust the initial volume of buffer in each case so as to give a final reaction mixture of 3 mL. Calculate the activity and plot graphs of v against s, $1/v$ against $1/s$, and s/v against s, and determine K_m and V.

12.12.4.2 Effect of Inhibitors
1. Repeat the previous experiment, but this time incorporate 0.1 mL of the inhibitor into the reaction mixture, again adjusting the volume of buffer to give a final reaction volume of 3 mL. Plot a graph of $1/v$ against $1/s$, and determine the inhibitor constant K_i.
2. Repeat the initial assay, but this time determines the activity at two fixed substrate concentrations and a range of concentration of the inhibitor, up to a final concentration of 0.1 mmol/L oxalate and 0.5 mmol/L oxamate. If v_i is the activity in the presence of the inhibitor and v is the activity in the absence of the inhibitor, plot a graph of v/v_i against i. Use these graphs to decide what type of inhibition is given by oxalate and oxamate, and determine the inhibitor constants.

12.12.4.3 High Substrate Concentration

Examine the effect of high substrate concentration on a plot of v against s and s/v against s. This is carried out by using volumes up to 1 mL of 21 mmol/L sodium pyruvate to trigger the enzyme reaction.

12.13 Effect of Temperature on the Activity of α-Amylase

12.13.1 Principle

α-Amylase catalyzes the hydrolysis of α 1–4 links of starch with the production of reducing sugars. The reaction is followed by measuring the increase in reducing sugars using the 3,5 dinitrosalicylate reagent when an alkaline solution is followed by measuring the extinction at 540 nm.

12.13.2 Materials

Sodium or potassium phosphate buffer (0.1 mol/L, pH 6.7), buffered starch substrate (5 g/L in phosphate buffer, mix 500 mg of soluble starch to a smooth paste with about 5 mL of buffer solution. Add this quantitatively to 50 mL of boiling phosphate buffer solution, continue to boil for 1 min, then cool to room temperature, and dilute to 100 mL with buffer solution), sodium chloride (10 g/L), α-amylase suitably diluted, sodium hydroxide (2 mol/L), colorimeter, water baths set at a range of temperatures up to 75°C.

DNS reagent

1. *Sodium potassium tartrate*: Dissolve 300 g sodium potassium tartrate in 500 mL of water.
2. *3,5-Dinitrosalicylic acid*: Dissolve 10 g 3,5-dinitrosalicylic acid in 200 mL of 2 mol/L sodium hydroxide.

    Preparation of dinitrosalicylic acid reagent: Prepare this fresh by mixing solutions (1) and (2) and making up to 1 L with water.
3. *Sodium hydroxide*: 2 mol/L.

12.13.3 Protocol

12.13.3.1 Determination of the Energy of Activation
Set up eight tubes containing the following reaction mixture:

Contents	mL
Starch (5 g/L)	2.5
Phosphate buffer (0.1 mol/L, pH 6.7)	1.0
Sodium chloride (10 g/L)	0.5

Place the tube in a water bath at a fixed temperature and equilibrate the enzyme at the same temperature. After 10 min, add 0.5 mL of the enzyme to seven of the tubes and 0.5 mL of water to the blank. Incubate the tubes for 0.5, 10, 15, 20, 30, and 40 min and stop the reaction by adding 0.5 mL of 2 mol/L NaOH to all tubes. Add 0.5 mL of the dinitrosalicylate reagent and heat the tubes for 5 min in a boiling water bath, cool, and read the extinction at 540 nm against a blank. (*Note: The tubes must all be cooled to room temperature before reading since the extinction is sensitive to temperature change.*)

Plot the extinction of each solution against the time of incubation and prepare a progress curve of the reaction. Collect data from the other groups and plot a graph of the change in the initial reaction rate against temperature and also the substrate changes per minute at 10 and 40 min against temperature. Compare these graphs and explain them.

Plot a graph of $\log_{10} v$ against $1/T$ and determine the heat of activation. For convenience, the activity can be expressed in terms of extinction change at 540 nm/min.

12.13.3.2 Temperature and Enzyme Stability

Prepare a suitable dilution of the α-amylase so as to give a good activity when incubated at 37°C for 5 min. Place the enzyme in a boiling tube and equilibrate for 10 min. Pipette 0.5 mL of the α-amylase into four test tubes in a water bath maintained at a fixed temperature from 50°C to 75°C. Remove the tubes after 5, 10, 15, and 20 min (or other suitable time intervals) and rapidly cool under running tap water. Incubate the tubes at 37°C and assay for enzyme activity after the addition of the substrate. The mixture is incubated for 5 min and the activity is determined with 3,5-dinitrosalicylate. The rate of inactivation of the enzyme can then be determined. Collect data at other temperature and prepare a graph of the rate of inactivation (v) against $1/T$ and determine the heat of inactivation. Compare this figure with that obtained for the heat of activation from the last experiment.

12.14 Determination of the Molecular Weight and Purity of Chymotrypsin from the Enzyme Kinetics

12.14.1 Principle

The hydrolysis of certain carboxylic esters by chymotrypsin displays anomalous kinetics. Chymotrypsin catalyzes the hydrolysis of ρ-nitrophenyl acetate, and the enzyme reacts initially with the substrate by becoming acetylated at a serine hydroxyl at the active site. This first stage is accompanied by an initial burst of *p*-nitrophenol after which the release of the acetylchymotrypsin proceeds at a much slower rate.

The amount of *p*-nitrophenol released in the initial burst is stoichiometrically related to the amount of enzyme present and can be used to calculate the minimum mol. wt. of chymotrypsin.

1. Release of p-nitrophenol

$$E - OH + CH_3 COO - \langle \bigcirc \rangle - NO_2 \longrightarrow E - OOC.CH_3 + HO - \langle \bigcirc \rangle - NO_2$$

2. Hydrolysis of acetylated enzyme

$$E - OOC. CH_3 + H_2O \longrightarrow E - OH + CH_3 COO^- + H^+$$

12.14.2 Materials

Tris–H_2SO_4 buffer (0.1 mol/L, pH 7.6), chymotrypsin (0.8 mg/mL in the Tris buffer), stock solution of ρ-nitrophenyl acetate (1 mmol/L in ethanol, freshly prepared), working ρ-nitrophenyl phosphate (0.1 mmol/L in Tris buffer, prepared by diluting the stock solution 1 in 10 with Tris buffer), p-nitrophenol (50 μmol/L in Tris buffer), UV–Vis recording spectrophotometers.

12.14.3 Protocol

Prepare two cuvettes as follows:

Substrate	Reference	Reaction
p-Nitrophenyl acetate (0.1 mmol/L)	1.5 mL	1.5 mL
Chymotrypsin (0.8 mg/mL)	—	1.5 mL
Water	0.5 mL	0.5 mL

Immediately after adding the enzymes, mix thoroughly and record the increase in extinction at 400 nm, to obtain the linear part of the progress curve. Extrapolate the curve back to zero time and record the extinction.

Repeat the experiment with varying concentration of chymotrypsin. Meanwhile, prepare a standard curve of p-nitrophenol (0–50 μmol/L) and use this to plot the μmoles of p-nitrophenol released at zero time against the concentration of chymotrypsin. From this, calculate the minimum mol. wt. of chymotrypsin.

12.14.4 Results and Conclusion

1. Calculate the μmoles of p-nitrophenol in the cuvette.
2. Calculate the μg of chymotrypsin in the cuvette.
3. Mol. wt. = mass/molarity expressed in the same units, that is, mg and mmoles or μg and μmoles.

The true mol. wt. of pure chymotrypsin is 24,500, but it is very unlikely that the value obtained, by this method, will be the same, and it could be almost certainly larger than this. The actual value obtained depends on how good is the sample of the enzymes. Use the data to determine the purity of the enzyme preparation.

12.15 Yeast Isocitrate Dehydrogenase: Allosteric Enzyme

12.15.1 Principle

Yeast isocitrate dehydrogenase catalyzes the conversion of L-isocitrate to α-oxoglutarate and carbon dioxide with NAD as the coenzymes. It is a major control point in the tricarboxylic acid cycle and is modified by a number of effectors including NAD, AMP, and Mg^{2+}. The enzyme is assayed by following the increase in extinction at 340 nm as NAD is reduced to $NADH_2$.

$$
\begin{array}{l}
COO^- \\
|\\
CH_2 \\
|\\
H-C-COOH + NAD \rightleftharpoons \\
|\\
HO-C-H \\
|\\
COO^- \\
\text{L-Isocitrate}
\end{array}
\qquad
\begin{array}{l}
COO^- \\
|\\
CH_2 \\
|\\
CH_2 + NADH_2 + CO_2 \\
|\\
O=C \\
|\\
COO^- \\
\text{α-Oxoglutarate}
\end{array}
$$

12.15.2 Materials

Fresh baker's yeast, acid-washed sand, sodium bicarbonate (0.1 mol/L), Tris–HCL buffer (30 mmol/L, pH 7.4), sodium D,L-isocitrate in Tris buffer, pH 7.4, NAD (2 mmol/L) in Tris buffer, pH 7.4, AMP (3 mmol/L) in Tris buffer, pH 7.4, $MgCl_2$ (0.1 mol/L), pestle and mortar, a recoding UV–Vis spectrophotometer, circulating water bath at 30°C.

12.15.3 Protocol

12.15.3.1 Preparation of the Enzyme

Prepare a crude extract of the enzyme by grinding some freshly grown yeast with washed sand and sodium bicarbonate solution (0.1 mol/L) in a pestle and mortar. The yeast cell wall is extremely tough, and considerable grinding is required to break it. The proportions of yeast and sand are adjusted so that there is plenty of abrasive action by the sand, yet sufficient sodium bicarbonate present in which the released enzyme can dissolve. The grinding requires a little practical, and it may be advisable for a technician to prepare a fresh extract for the class. After grinding, transfer the mixture to centrifuge tubes and remove the sand and large cell debris by centrifuge at 1000 g for 10 min; collect the supernatant and recentrifuge at 30,000 g for 1 h. Discard the precipitate and store the supernatant on ice until required.

12.15.3.2 Enzyme Assay

Prepare a series of tubes containing the following reaction mixture, and incubate at 37°C for 10 min.

Substrate	Volume in mL
NAD (2 mmol/L)	1.5
MgCl$_2$ (0.1 mol/L)	0.1
Diluted enzymes	0.2

The amount of enzymes required has to be determined by experiment, and if the preparation is too active, it should be diluted with Tris buffer.

Start reaction by adding 1.2 mL of substrate previously incubated at 37°C, and follow the increase in extinction at 340 nm. Set up the appropriate blanks and controls to allow for any nonenzymatic changes. Measure the enzyme activity over a range of substrate concentrations from 0.1–1.2 mmol/L. This is achieved by triggering the reaction with 0.1–1.2 mL of the substrate and adding Tris buffer to give a final volume of 1.2 mL. The substrate used is a mixture of D and L forms so that L-isocitrate concentration ranges from 0.05 to 0.6 mmol/L.

Repeat the experiment, but this time include AMP in the reaction mixture at final concentrations of 0.2 and 0.5 mmol/L.

Express the enzyme activity in terms of nanomoles of NAD reduced per minute, and prepare graphs of v against the s, $1/v$ against $1/s$, and log $v/(V - v)$ against log s.

Repeat the previous experiment after heating the enzymes at 60°C for 5 min. Another possible investigation is to examine the effect of Mg^{++} concentration on the reaction.

12.16 Purification and Characterization of Enzymes

12.16.1 Introduction

An enzyme synthesized in a living cell catalyzes or speeds up a thermodynamically possible reaction so that the rate of the reaction is compatible with the biochemical process essential for the maintenance of a cell. The most important requirement for commercial venture would be to use a source that enables large amount of a suitable enzyme extracted by some convenient procedure. Enzyme purification gives the maximum understanding of the behavior of the enzyme in a complex system and is an important investigation for the characterization of the function, structure, and interactions of the protein of interest. Enzyme is, therefore, isolated and purified. Enzymes usually have several applications in food, beverage, and pharmaceutical industries and a wide range of sources are used for commercial enzyme production from different microbes, plant, and animal cells. Of the hundred or so enzymes that are being used industrially, over a half is from fungi and yeast and nearly one-third from bacteria with the remainder from plant and animal sources. Microbes are

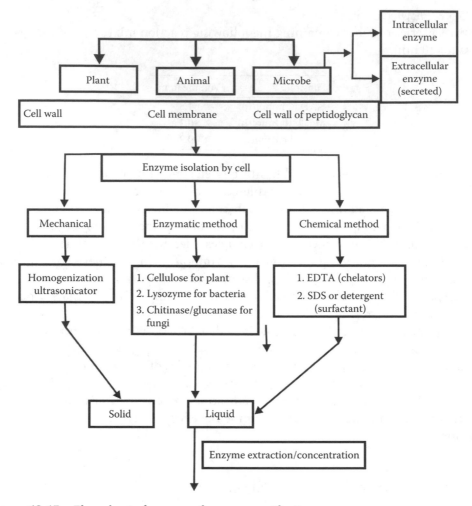

Figure 12.15 Flow chart of process of enzyme purification.

preferred to plants and animals as sources of enzymes because (1) they are generally cheaper to produce, (2) their enzyme contents are more predictable and controllable, (3) they are reliable supplies of raw material of constant composition and are more easily arranged, and (4) plant and animal tissues contain more potentially harmful material than microbes, including phenolics (from plants), endogenous enzyme inhibitors, and proteases (Figure 12.15).

12.16.2 Method of Enzyme Isolation

There are various methods of enzyme isolation after cell lysis by osmotic and enzymic process, sonication, or homogenization method depending on the source and location of the enzyme. Soluble enzymes are easy to extract by simple breaking cell membrane, but isolation of enzymes from intracellular membrane is very difficult and requires

TABLE 12.2 Factors Affecting Enzymes during Extraction

Heat	All mechanical methods require a large input of energy, generating heat. Cooling is essential for most enzymes. The presence of substrates, substrate analogues, or polyols may also help stabilize the enzyme.
Shear	Shear force is needed to disrupt cells and may damage enzymes, particularly in the presence of heavy metal ions and/or air interface.
Proteases	Disruption of cell will release degradative enzymes that may cause serious loss of enzyme activity. Such action may be minimized by increased speed of processing with as much cooling as possible. This may be improved by the presence of an excess of alternative substrates (e.g., inexpensive protein) or inhibitors in the extraction medium.
pH	Buffered solutions may be necessary. The presence of substrate, substrate analogues, or polyols may also help to stabilize the enzyme.
Chemical	Some enzymes suffer conformational changes in the presence of detergent and/or solvent. Polyphenolics derived from plants are protein inhibitors of enzymes. This problem may be overcome by the use of adsorbents, such as polyvinylpyrrolidone, and by the use of ascorbic acid or reduce polyphenol oxidase action.
Oxidation	Reducing agent (e.g., ascorbic acid, mercaptoethanol, and dithiothreitol) may be used.
Foaming	The gas–liquid phase interfaces present in foams may disrupt enzyme conformation.
Heavy metal toxicity	Heavy metal ions (e.g., iron, copper, and nickel) may be introduced by leaching from the homogenization apparatus. Enzymes may be protected from irreversible inactivation by the use of chelating reagents, such as EDTA.

quite harsh conditions. Many of the cells undergo lysis or autolysis in the absence of cold environment. Another major difficulty is to maintain the native state of enzyme. In plant cell, cell wall offers a major barrier in enzyme isolation. Besides cell wall, there are various phenolic compounds and phenol oxidases that bind to proteins and convert them into quinine. Therefore, they require proper filtration and presence of antioxidant. Majority of commercial enzymes are produced by the microbes extracellularly or must be released from the cells into solution for further processing. Solid/liquid separation is generally required for the initial separation of cell achieved by filtration, centrifugation, or aqueous biphasic partition. In general, filtration or aqueous biphasic systems are used to remove unwanted cells or cell debris, whereas centrifugation is the preferred method for the collection of required solid material.

It is the most important in choosing cell-disrupting strategies to avoid damaging the enzymes. The particular hazards to enzyme activity are summarized in Table 12.2.

12.16.2.1 Cell Lysis Method
The methods that are available include osmotic shock; freezing followed by thawing, cold shock, desiccation, enzymic lysis, and chemical lysis; homogenization; and ultracentrifugation. Each method has drawbacks but may be particularly useful under certain specific circumstances. Osmotic shock method is cheap, gentle, and convenient of releasing enzymes but has not been used apparently on a large scale.

Osmotic technique may be used in conjunction with mechanical means of disrupting cells in the presence of a buffer with the osmotic pressure. Often in the laboratory, 20% sucrose is used to lower down the osmotic concentration.

12.16.2.2 Ultrasonic Cell Disruption

Application of high-frequency waves is an effective method of cell breakage that can be applied for extraction from microbes. Ultrasonication utilizes the rapid sinusoidal movement of a probe within the liquid. It is characterized by high frequency (12 kHz–1 MHz), small displacement (less than about 50 m), moderate velocities (a few meters per second), steep transverse velocity gradients (up to 4000/s), and very high acceleration (up to about 20,000 g).

12.16.2.3 Grinding

Several types of apparatus are available for grinding. Cells are broken by impact, tearing, and maceration between the hard surfaces. To avoid heating, cooling liquid is circulated through the jacket.

12.16.2.4 Cell Lysis by High-Pressure Homogenizers

Various types of high-pressure homogenizers are available for use in the food and chemical industries. Two most common cell disruptors are Manton–Gaulin homogenizer and French press. In the Manton–Gaulin homogenizer, a positive displacement pump draws cell suspension (about 12% w/v) through a check valve into the pump cylinder and forces it, at high pressures of up to 150 MPs (10 ton/in.^{2}) and flow rate of up to 10,000 L/h, through an adjustable discharge valve that has a restricted orifice. Cells are subjected to impact, shear, and a severe pressure drop across the valve. Unbound intracellular enzymes may be released by a single pass, whereas membrane-bound enzymes require several passes for reasonable yields to be obtained. High-pressure homogenizers are acceptably good for the disruption of unicellular organisms provided the enzymes needed are not heat labile. The shear forces produced are not capable of damaging enzymes free in solution (Figure 12.16).

12.16.2.5 French Pressure Cell

Cells are broken through a narrow orifice at a pressure of up to 8000 psi disrupting the cell plasma membrane by passing them through a narrow valve under high pressure. The pressure uses an external hydraulic pump to drive a piston within a larger cylinder that contains the sample. The highly pressurized solution is then squeezed past a needle valve. Once past the valve, the pressure drops to atmospheric pressure and generates shear stress that disrupts the cell. A French press is commonly used to break the resilient plasma membrane and cell walls of bacteria during protein isolation (Figure 12.17).

12.16.2.6 Waring Blender and Virtis Homogenizer

These devices consist of a high-speed stirrer with cutting blades, mounted in a glass vessel, the walls of which are indented from top to bottom, forming a clover-leaf cross

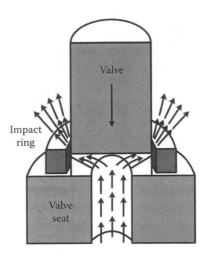

Figure 12.16 Cross section through the Manton–Gaulin homogenizer valve showing the flow of material.

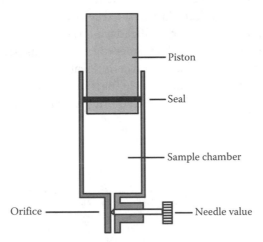

Figure 12.17 Illustration of French pressure cell.

section. The speed of the blades' motion generates strong shear forces, due to laminar flow, while the irregular outline of the vessel gives good overall mixing of the solution. The degree of disruption depends upon the speed of rotation of the blades. At high speed, a blender will disrupt mitochondria and nuclei and may even denature proteins. It is mostly used with plant and animal tissues but is less effective with microbes (Figure 12.18).

12.16.2.7 Bead Mills

When cell suspensions are agitated in the presence of small steel or glass beads (usually 0.2–1.0 mm in diameter), they are broken by the high liquid shear gradients and collision with the beads. The rate and effectiveness of enzyme release can be modified by changing the rates of agitation and the size of the beads, as well as the dimensions

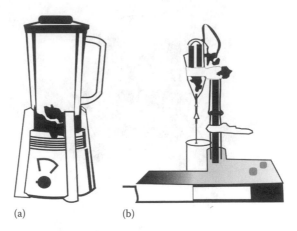

(a)   (b)

Figure 12.18   (a) Waring blender and (b) Virtis homogenizer.

of the equipment. Any type of biomass, filamentous or unicellular, may be disrupted by bead milling, but in general, the larger-sized cells will be broken more readily than small bacteria. For the same volume of beads, a large number of small beads will be more effective than a relatively small number of larger beads because of the increased likelihood of collisions between beads and cells. Heat generation is the major problem in the use of bead mills for enzyme release, particularly on a large scale (e.g., 20 L). However, if cooling is effective through a cooling jacket, there is little damage to the enzyme released (Figure 12.19).

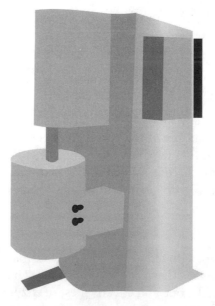

Figure 12.19   Bead mills.

12.16.2.8 Enzymatic Lytic Method

The breakage of cells using nonmechanical methods is attractive as it offers the prospects of releasing enzymes under conditions that are gentle, do not subject the enzyme to heat or shear, and may be very cheap. Some types of cells may be caused to autolysis, in particular yeasts and bacillus species. Autolysis is a slow process compared to mechanical methods and can be used on a very large scale if necessary. Where applicable, desiccation may be very useful in the preparation of enzymes on a large scale. The rate of drying is very important in these cases, slow methods being preferred to rapid ones like freeze-drying (lyophilization). Enzymic lysis using added enzymes has been used widely on the laboratory scale but is less popular for industrial purpose. Lysozyme has been often used to lyse the Gram-positive bacteria but the large-scale use is a costly affair. Sometimes addition of ethylenediaminetetraacetic acid (EDTA) helps in release of lipopolysaccharide from Gram-negative cell envelops. Lysozyme method requires gentle environmental conditions like gentle pH and temperature. Similarly, plant cell walls may be digested with cellulases and fungal cell walls with chitinases.

12.16.2.9 Detergent

Triton X-100 is used alone or in combination with certain chaotropic agents, such as guanidine HCl, which are effective in releasing membrane-bound enzyme.

12.16.3 Clarification of the Extract

The cellular extract prepared by any of the methods described previously needs to be clarified by filtration through a nylon mesh or cheesecloth to remove large cell debris and centrifuged at relatively low speed to remove insoluble cell components.

12.16.3.1 Aqueous Biphasic Separation

Aqueous biphasic separation provides the opportunity for the rapid separation of biological materials with little possibility of denaturation. Aqueous dextran–polyethylene glycol (PEG) system (10% PEG 4000 2% dextran T500) is the most thoroughly investigated system where dextran forms the more hydrophilic, denser, lower phase and PEG the more hydrophobic, less dense, upper phase. Aqueous three-phase systems are also known. Separation by this process may be achieved in a few minutes, minimizing the harmful action of endogenous protease. A powerful modification of this technique is to combine phase partitioning and affinity partitioning. Affinity ligands (e.g., triazine dyes) may be coupled to either polymer in an aqueous biphasic system and thus greatly increase the specificity of the extraction.

12.16.4 Enzyme Concentration

12.16.4.1 Ammonium Sulfate Precipitation

Precipitation of enzyme is a useful method of concentrating the protein as an initial step. It can be used on a large scale and is less affected by the presence of interfering materials than any of the other methods. There are methods of salting in and salting out. Salting in is the method in which ammonium sulfate is used for dissolving the protein. Protein dissolves least at its isoelectric point. However, as the salt concentration is increased, a point of maximum protein solubility is usually reached. Further increase in the salt concentration implies that there is less and less water available to solubilize protein. Finally, protein starts to precipitate when there are not sufficient water molecules to interact with protein molecules. This phenomenon is called as salting out. All this process must be done in the ice-cold solution so that no protein can denature. Hydrophobic proteins precipitate at lower concentration than hydrophilic proteins. Ammonium sulfate is convenient and effective because of high solubility, cheapness, lack of toxicity to most of the enzymes, and its stabilizing effect on some enzymes. Some enzymes do not survive ammonium sulfate precipitation. Most favored alternative is to use organic solvents such as methanol, ethanol, propanol, and acetone. Organic solvents act by reducing the dielectric of the medium and consequently reducing the solubility of the proteins by favoring protein–protein rather than protein–solvent interactions. Organic solvents are not widely used on a large scale because of their cost and their inflammable property. After ammonium sulfate precipitation, the precipitating salt or solvent must be removed. This may be done by dialysis, by ultrafiltration, or by using a desalting column such as Sephadex G-25.

Intracellular enzyme preparations contain nucleic acids that increase the viscosity, interfering with enzyme purification procedures (in particular ultrafiltration). Some organisms particularly plant cells contain sufficient nuclease activity. The nucleic acids must be removed. Various precipitants usually positively charged materials such as polyethyleneimine, the cation detergent cetyltrimethyl ammonium bromide, streptomycin sulfate, and protamine sulfate that form complexes with the negatively charged phosphate residues of the nucleic acids have been recommended for precipitation or degraded by the addition of exogenous nucleases.

12.16.4.2 Dialysis of Protein

After a protein has undergone ammonium sulfate precipitation and has taken back in buffer at a much greater concentration than before precipitation, this solution will now contain a lot of residual ammonium sulfate bound to the protein. One way to remove this excess salt is to dialyze the protein against a buffer low in salt concentration. The concentrated protein solutions are placed in a dialysis bag, allowing water and salt to pass out of the bag while protein is retained. The dialysis bag is placed in a large volume of buffer and stirred for many hours (16–24 h), in a cold room at 4°C that allows the solution inside the bag to equilibrate with the solution outside the bag with respect to salt concentration. This process of equilibrium is repeated several times. This process is performed in the presence of buffer to prevent the protein from denaturing (Figure 12.20a). After the dialysis process, ion-exchange chromatography step is done to remove all the salt, and protein is ready for the next step. Reverse dialysis also helps in purifying the protein (Figure 12.20b).

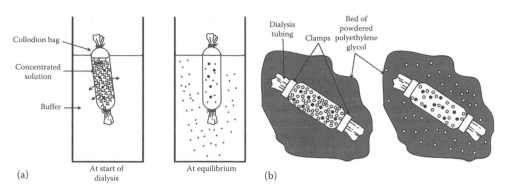

Figure 12.20 (a) Dialysis—only small molecules (dots) diffuse through the collodion (dialysis) membrane, and at equilibrium, the concentration of small molecules is the same inside and outside the membrane. Macromolecule remains in the bag. (b) Reverse dialysis—a solution of macromolecules (solid circle) and solvent molecules (open circle) is placed in a dialysis bag and enters the polyethylene glycol phase. Neither the polyethylene glycol nor the macromolecules can pass through the membrane, and so the solution is concentrated.

12.16.4.3 Enzyme Purification and Characterization

Purification and characterization is broadly based on size and mass (centrifugation, gel permeation chromatography, dialysis, and ultracentrifugation), polarity (ion exchange, electrophoresis, isoelectric focusing, and hydrophobic interaction chromatography), changes in solubility (by change in pH, change in ionic strength salting in or salting out), and change in dielectric strength by adding organic solvent and on specific binding site (affinity chromatography, affinity elution, dye-ligand chromatography, immune-adsorption chromatography, covalent chromatography). Various methods (Figure 12.21) used are described in Chapters 7 and 8.

12.16.4.4 Purification Testing

Purification is tested by ultracentrifugation (impurities <5%), SDS-PAGE electrophoresis (an ideal method for impurity detection and dissociation of nonidentical subunit), capillary electrophoresis, isoelectric focusing (highly sensitive method), and mass spectrometry (powerful method).

12.16.4.5 Ultracentrifugation

The technique is useful to investigate the molecular mass, shape, and density of the protein. The basis of this technique is to exert the larger force than does the gravitational force of the earth, thus increasing the rate at which the particle settles. The technique can be divided into two types.

12.16.4.5.1 Preparative Centrifugation
It is applicable for the separation of whole cell, subcellular organelles, plasma membranes, polysome, chromatin, nucleic acids, lipoprotein, and virus.

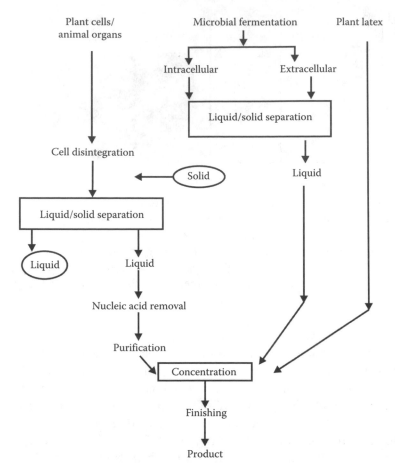

Figure 12.21 Flowchart of purification process of enzymes from different sources.

12.16.4.5.2 Analytical Centrifugation   It is devoted to purify the macromolecules or particles. Rate depends on the centrifugal field G directed radially outward (angular velocity).

12.16.4.6 Differential Centrifugation
This method is based on the difference in the sedimentation rate of the particles of different sizes and density (Figure 12.22). During centrifugation, the larger particle sediments first and less dense will sediment later. Particles having similar density can be separated by the differential centrifugation or the rate zonal method. Proteins having same size are having narrower density range are separated by the rate zonal technique based mainly upon differences in their sizes and cannot be separated easily.

In differential centrifugation, the particle to be separated is divided centrifugally into a number of fraction by increasing the applied centrifugal field. After centrifugation, pellet and supernatant are obtained. Pellet is washed several times

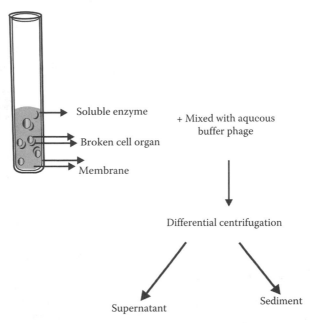

Figure 12.22   Differential centrifugation used for isolation of enzymes.

and again centrifugation is done. The particle moves against respective sedimentation rates. Centrifugation is continued long enough to obtain pellet and the resulting supernatant.

12.16.4.7 Isopycnic Centrifugation

It depends solely upon the buoyant density of the particles and not its shape or size and is independent of time the size of the particle affecting only the rate at which it reaches its isopycnic position in the gradient. The technique is useful in separating the particle of same size but differing in density (Figure 12.23).

12.17 Isolation of Muramidase (Mucopeptide *N*-Acetylmuramyl Hydrolase, 3.21.17) from Egg White

12.17.1 Principle

The enzyme is a potent antibacterial agent and acts by catalyzing the hydrolysis of β 1–4 links between *N-acetylmuramic* acid and *N-acetylglucosamine* residues present in the mucopeptide of the cell walls. This particular property forms the basis of the assay method when the enzyme is mixed with a turbid suspension of freeze-dried bacteria. As the hydrolysis proceeds, the turbidity of the suspension decreases and this can be readily followed in a spectrophotometer at 450 nm.

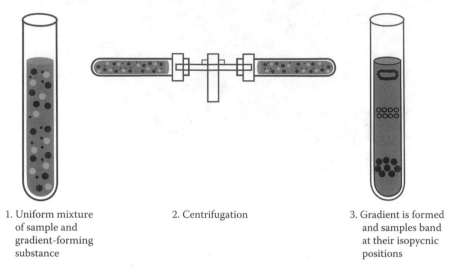

1. Uniform mixture
   of sample and
   gradient-forming
   substance

2. Centrifugation

3. Gradient is formed
   and samples band
   at their isopycnic
   positions

Figure 12.23   Isopycnic centrifugation.

N-Acetylglucosamine   N-Acetylmuramic acid

Muramidase has a low mol. wt. (14, 300) and a high isoelectric point (pH 10.5) that is being used to separate the enzymes from other proteins. Muramidase was known as lysozyme for many years, but this trivial name is no longer recommended.

12.17.2 Materials

Egg, muslin, HCl (1 mol/L and 0.1 mol/L), glass wool, concentrated buffered saline (NaCl, 0.5 mol/L, and Tris-EDTA, 0.5 mol/L, pH 8.2), cation-exchange resin, CM-cellulose, sodium carbonate–bicarbonate buffer (0.2 mol/L, pH 10.5), NaCl, acetic acid (1 mmol/L), standard albumin for protein assays (1 mg/mL) 100 mL, pH meter, fraction collector, UV–Vis spectrophotometer, water bath at 25°C or 37°C, sodium or potassium phosphate buffer (0.1 mol/L, pH 7), suspension of freeze-dried *Micrococcus lysodeikticus* in phosphate buffer (freshly prepared at 0.3 mg/mL).

12.17.3 Protocol

12.17.3.1 Enzymes and Protein Assay

Muramidase: Equilibrate the suspension of freeze-dried bacteria at 37°C and shake before use. Pipette 2.8 mL into a cuvette and start the reaction by adding 0.2 mL of the muramidase solution. Mix thoroughly and follow the decrease in absorption at 450 nm in a recording spectrophotometer. Calculate the activity in terms of microorganism of bacteria hydrolyzed per minute.

Protein: Prepare a standard curve of bovine serum albumin measuring the absorbance at 280 nm of a solution up to 1 mg/mL. The protein content at each stage is then determined by measuring the extinction of the solution at 280 nm and reading the protein value off the standard curve.

12.17.3.2 Purification

12.17.3.2.1 Preparation of Chromatography Column. Equilibrate 10 g CM-cellulose with the buffered saline dilute 1 in 10, by gently stirring for 24 h; then prepare a chromatography column of this material as described in Section 7.17.

Purification table: At each stage of the purification, record the following values in the form of a table:

1. Volume (mL)
2. Enzyme activity (units/mL)
3. Total muramidase activity (unit/mL × vol)
4. Total protein content (mg)
5. Specific activity (total muramidase activity/mg protein)
6. Purification factor for muramidase (specific activity divided by the initial specific activity)
7. Yield (percent of the original total activity)

12.17.3.2.2 Extraction. Collect the white from three eggs, filter through muslin to remove the chalazae, collect 50 mL filtrate, and add this to 100 mL of water. Stir the mixture carefully without whipping air into the whites to prevent denaturation. Set aside 2 mL on ice for enzymes and protein assay.

12.17.3.2.3 pH Precipitation. Adjust of the extract to pH 7.5 by slowly adding 1 mol/L and 0.1 mol/L HCl over a 10 min period. Take care not to overshoot. Remove the protein that precipitates by filtration through a glass wool plug in a filter funnel, and add concentrated buffered saline to the filtrate, to give a final concentration of 0.05 mol/L NaCl and 0.05 mol/L Tris-EDTA, pH 8.2. Measure the pH and adjust to 8.2 if necessary. Record the volume and again set aside 2 mL on ice for the assay of muramidase and protein.

12.17.3.2.4 Column Chromatography. Carefully apply the extract to the column of CM-Sephadex and collect 10 mL fraction on a fraction collector. When the extract has run into the column, wash with further 50 mL of dilute buffered saline, before eluting with 0.2 mol/L sodium carbonate–bicarbonate buffer. Continue collecting fraction until all the muramidase has come off the column.

Determine the protein and enzymes content of each fraction and plot their elution profile. Calculate the percent recoveries from the column.

Crystallization: Muramidase is a strongly basic protein and readily forms crystalline salts with chloride, iodide, carbonate, etc. Under optimal condition, crystallization proceeds slowly and reaches a maximum yield after 72–96 h.

Adjust the pH of the pooled enzyme peak to 10.5 and slowly add NaCl to a final concentration of 0.3 mol/L. Leave to stand in the refrigerator for 3–4 days.

If crystallization has occurred, decant as much as possible of the supernatant and sediment the crystals from the remaining solution by centrifugation. Dissolve the precipitate in 1 mL of mmol/L acetic acid and remove any insoluble materials by centrifugation. Determine the protein and muramidase content and record the final purification factor and yield.

12.18 Separation of the Isoenzymes of Lactate Dehydrogenase by Electrophoresis on Polyacrylamide Gel

12.18.1 Principle

Enzymes that catalyze the same chemical reaction but differ in certain of their physicochemical properties are known as multiple molecular forms, and electrophoresis or chromatography is frequently used to separate and characterize the different forms. If the multiple molecular forms are genetically determined, they are given the name of isoenzymes. Most animal tissues contain up to five isoenzymes of lactate dehydrogenase, which can be readily separated by electrophoresis. They are numbered according to the speed of migration to the anode so that LD_1 is the faster-migrating species and LD_5 the form with the lowest mobility. The isoenzymes arise from various combinations of two subunits, to form five possible tetramers. LD_1 contains only H subunits and so named as this is the predominant form in the heart, while LD_5 contains only M subunits and is the main form present in skeletal muscle. The other hybrid (LD_2, LD_3, and LD_4) arises from the combination of these two subunits to form hybrid molecules (Figure 12.24). The amount and distribution of these enzyme forms are characteristic of the tissue of origin and is a useful way of fingerprinting the tissue.

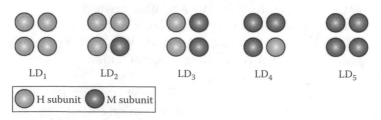

Figure 12.24 Composition of lactate dehydrogenase isoenzyme.

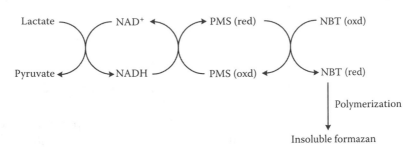

Figure 12.25 Reaction involved in the detection of lactate dehydrogenase isoenzymes where NAD is the nicotinamide adenine dinucleotide, PMS is the phenazine methosulfate, NBT is the nitroblue tetrazolium, oxd is oxidized, and red is reduced.

The isoenzymes of LDH are visualized by incubating the gels with a mixture containing the normal substrate and coenzymes, phenazine methosulfate (PMS) (an artificial electron acceptor), and a tetrazolium dye. On reduction, this dye forms a highly colored and sparingly soluble formazan that precipitates at the sites of enzyme activity (Figure 12.25).

12.18.2 Materials

Solution and equipment for polyacrylamide gel electrophoresis (described in Section 8.6), sodium or potassium phosphate buffer (0.05 mol/L, pH 7.4), rat (200–300 g weight) oxheart and rabbit muscle lactate dehydrogenase instead of rat tissues, Tris–HCl buffer (0.1 mol/L, pH 9.2), lithium lactate, NAD, nitroblue tetrazolium (NBT), PMS, developing reagent (0.1 mol/L lithium lactate, 50 mg NAD and 5 mg, NBT in 100 mL of Tris–HCl buffer, pH 9.2). The solution is stable for a week in the dark at 4°C. Immediately before use, add PMS (approx. 0.5 mg), incubator set at 37°C.

12.18.3 Protocol

Prepare rods of polyacrylamide gel and allow setting. Meanwhile, kill a rat and prepare homogenates of liver, kidney, skeletal muscle, heart, and other tissues in 0.05 mol/L phosphate buffer (0.5–1 g/10 mL). Remove the debris by centrifugation and store the supernatant on the ice until required.

Add 50 μL of each sample in duplicate to the polyacrylamide rods and carry out electrophoresis as described earlier. At the end of the run, incubate the gels with the developing reagent at 37°C for up to 10 min in subdued light. Photograph the electrophorogram or make a careful scale drawing of the position of the bands. Carefully note the rate at which each band develops the final intensity of the stain and the degree of spacing between the isoenzymes.

Repeat the exercise with blood serum, hemolyzed erythrocytes, small intestine, and lung tissue.

12.19 Enzyme Engineering

12.19.1 Introduction

Proteins have a wide application in industry (enzymes) and medicine (drugs). Deliberate design and production of enzyme with novel or altered structure and properties are not found in natural enzyme. The efficient application of enzyme requires the availability of suitable enzymes with high activity and stability under process conditions, desired substrate selectivity, and high enantioselectivity. There are two general strategies for protein engineering. The first as rational design in which detailed knowledge of the structure and function of the protein have to be known to make desired changes. Such studies are advantageous for being inexpensive and easy since site-directed mutagenesis (SDM) is already well developed. However, there is a major drawback in that detailed structural knowledge of a protein is often unavailable, and even when it is available, it can be extremely difficult to predict the effect of various mutations. The second strategy is known as directed evolution where random mutagenesis is applied to enzyme and a specific selection regime is used to pick out variants that have the desired qualities, and in further rounds or mutation and selection, final enzyme with specific property is selected. This method mimics natural evolution and generally produces superior results as compared to rational design. An additional technique known as DNA shuffling mixes and matches pieces of successful variants in order to produce better results. This method has been applied for vanillyl-alcohol oxidase, vanillin 2-hydroxybiphenyl monooxygenase, large-scale production of substituted catechols, galactose oxidase, and production of new oligosaccharide.

The major assumptions are as follows: (1) mutation does not significantly change the structure of the folded state and (2) the target groups do not make new interactions with new partners during the course of the reaction.

12.19.2 Enzyme Production by Gene Manipulation

The production of enzyme by microorganisms may be enhanced by increasing the number of gene copies that code for it by inserting a portion of gene for specific enzyme into the suitable vector. If the vector is expression type, then large amount of protein will be obtained after the translation of vector. The product can be tested by its specific function. The DNA having a gene of interest for specific enzyme can be isolated by partial digestion of a gene or complete digestion of gene (incomplete or complete fragmentation of gene), and selection of specific gene can be done with the help of probe. This principle has been used to increase the activity of penicillin G-amidase in *Escherichia coli*. The cellular DNA from a production strain is selectively cleaved by the restriction endonuclease Hind III. This hydrolyzes the DNA at relatively rare site containing the 5′-AAGCTT-3′ base sequence to give identical "staggered" ends (Figure 12.26). Single DNA strand is rarely cut by restriction endonucleases.

The total DNA is cleaved into about 1000 fragments, only one of which contains the required genetic information. These fragments are individually cloned into a cosmid

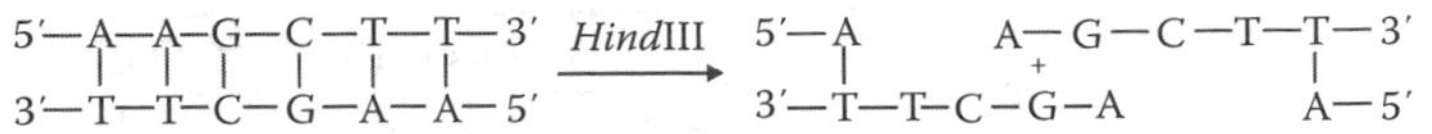

$$5'\text{—A—A—G—C—T—T—}3' \quad \xrightarrow{\textit{Hind}\text{III}} \quad 5'\text{—A} \qquad \text{A—G—C—T—T—}3'$$
$$3'\text{—T—T—C—G—A—A—}5' \qquad\qquad\qquad 3'\text{—T—T—C—G—A} \qquad \text{A—}5'$$

Figure 12.26 Staggered cut of DNA double strand by restriction endonuclease.

vector and thereby returned to *E. coli*. These colonies containing the active gene are identified by their inhibition of a 6-amino penicillanic acid-sensitive organism. Such colonies are isolated and the penicillin G-amidase gene transferred on to pBR 322 plasmids and recloned back into *E. coli*. The engineered cells, obtained by the plasmid amplification at around 50 copies per cell, produce penicillin G-amidase constitutively and in considerably higher quantities than does the fully induced parental strain. Such increased yields are economically relevant, not just for the increased volumetric productivity but also because of reduced downstream processing costs, the resulting crude enzyme being much purer. Another extremely promising area of genetic engineering is protein engineering. New enzyme structures may be designed and produced in order to improve on existing enzymes or create new activities. An outline of the process of protein engineering is shown in Figure 12.27. Such factitious enzymes are produced by SDM to change in the future.

The preferred pathway for creating new enzymes is by the stepwise substitution of only one or two amino acid residues out of the total protein structure. Although

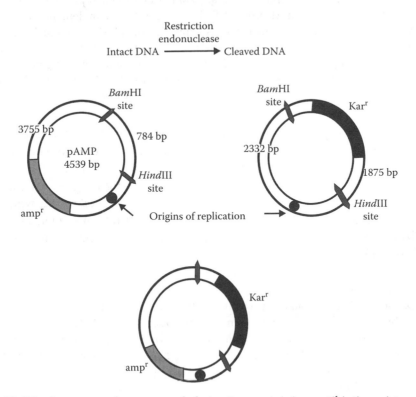

Figure 12.27 Structure of vectors and their sites containing antibiotic resistance genes as a marker.

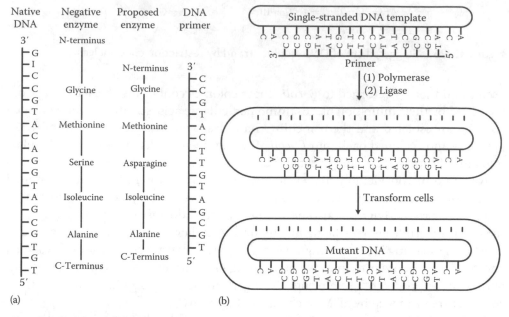

Figure 12.28 An outline of the process of SDM using a hypothecated example. (a) The primary structure of the enzyme is derived from the DNA sequence. A putative enzyme primary structure is proposed with an asparagine residue replacing the serine present in the native enzyme. A short piece of DNA (the primer), complementary to a section of the gene apart from the base mismatch, is synthesized. (b) The oligonucleotide primer is annealed to a single-stranded copy of the gene and is extended with enzymes and nucleotide triphosphate to give a double-stranded gene. On reproduction, the gene fives rise to both mutant and wild-type clones. The mutant DNA may be identified by hybridization with radioactively labeled oligonucleotides of complementary structure.

a large database of sequence-structure correlations is available, US funded National Center for Biotechnology Information (NCBI) grew rapidly together with the necessary software. It is presently insufficient accurately to predict 3D changes as a result of such solutions. The main problem is assessing the long-range effects, including solvent interactions, on the new structure. As the many reported results would attest, the science is at a stage where it can explain the structural consequences of amino acid substitutions after they have been determined but cannot accurately predict them (Figure 12.28).

12.19.2.1 Future Uses

1. Enzymes are being exploited as electrocatalyst for synthesizing specific biosensors. In a biosensor, the enzymes are simply entrapped with a dialysis membrane on the sensitive top of the detector or immobilized on it by covalent linking or cross-linking. At a given concentration of the substance, the enzymatic reaction gives a corresponding reaction rate that alters the local physicochemical environment of the detector. This alteration results in a response of the biosensors.
2. Enzymes are also being used as analytical tools to measure specific compounds or for the regeneration of specific metabolites, which are used in general molecular

biology techniques, for example, polymerase chain reaction (PCR) for gene amplification and sequencing, native and SDS gel electrophoresis, zymogram for activity detection on gels, isoelectric focusing, TLC chromatography to analyze hydrolysis products, measure absorbance to determine enzyme activity, and informatics software for enzyme structure analysis.

a. Enzymes can be utilized in the synthesis of bulk organic materials and in the production of fragrances and cosmetics.
b. Enzymes can be utilized in the formation of food flavors and aroma compounds.
c. The enzymes can be used as tools for the detoxification of pesticide residue.
d. Enzymes can be used for monitoring toxic chemical levels in food and water.

12.19.3 Advantages of Genetic Engineering

Many enzymes are now produced by fermentation of genetically modified microorganisms (GMOs). There are several advantages of using GMOs for the production of enzymes including the following:

1. It is possible to produce enzymes with a higher specificity and purity.
2. It is possible to obtain enzymes, which otherwise are not available for economical, occupational health, or environmental reasons.
3. Due to higher production and functional efficiency, there is an additional environmental benefit through reducing energy consumption and waste from the production plants.
4. In the food industry, particular benefits are, for example, a better use of raw materials (juice industry), better keeping quality of final food and thereby less wastage of food (baking industry), and a reduced use for chemicals in the production process (starch industry).
5. In the feed industry, particular benefits include a significant reduction in the amount of phosphorus released to the environment from farming.
6. Genetic engineering can be used to improve production and purification of recombinant enzymes, for example, targeting enzymes for extracellular secretion to allow for their easy isolation and purification.

Some of the industrially useful enzymes are (1) proteases–subtilisin (detergent enzyme) and rennet (cheese processing); (2) carbohydrases–amylases (starch hydrolysis), cellulases (bioconversions by Stake Technology), xylanases (bioconversions and biopulping), and glucose isomerase (HFCS); (3) enzymes used in pharmaceutical industries—urokinase (plasminogen activator), a protease for the treatment of thrombosis and embolism; and (4) ligninases and Mn peroxidases—pulp processing.

12.19.4 Manipulation of Enzyme Design

Rational protein design was the earliest approach to engineering enzymes and is still widely employed to introduce the desired characteristics into a target protein.

Enzymes evolve and adapt, at the molecular level. Amino acid sequences can vary to such an extent that evolutionary relationships may no longer be apparent from sequences alone. The genetic diversity for evolution is created by mutagenesis and/ or by recombination of one or more parent sequences. These altered genes are cloned back into a plasmid for expression in a suitable host organism (bacteria or yeast). Clones expressing improved enzymes are identified in a high-throughput screen or, in some cases, by selection, and the gene(s) encoding those improved enzymes are isolated and recycled to the next round of directed evolution.

12.19.4.1 Rational Protein Design

It requires detail knowledge of protein structure, planning of mutants (created by SDM or random mutagenesis also called as saturation mutagenesis), preparation of specific vectors containing mutated genes, their transformation in *E. coli* and protein expression and purification, and finally mutant enzyme analysis. Generally rational design strategy is applied for enzyme adaptation, molecular breeding, and metabolic engineering and for enzymes used in industrial applications.

The complete process is inexpensive and easy since mutagenesis technique and SDM are well developed. However, the major drawback is that the detailed knowledge of protein structure is often unavailable, and even when it is available, it can be extremely difficult to predict the effect of various mutations. Creating tailored enzymes with opposite enantiopreference may also be possible with this approach.

12.19.4.2 Site-Directed Mutagenesis

Mutation is created at a defined site in DNA molecule. SDM requires that the wild-type gene sequence be known. The technique is also known as site-specific mutagenesis or oligonucleotide-directed mutagenesis. Sometimes multiple mutageneses are performed by cassette mutagenesis. The benefit is limited but creates site-specific mutagenesis. Amino acid substitutions are often selected by sequence comparison with homologous sequences. Minor changes by a single-point mutation may cause significant structural disturbance. Comparison of 3D structure of mutant and wild-type enzyme is necessary. Structural analysis of the mutant enzyme reveals that the mutation is really site directed. Figure 12.29 reveals where human pancreatic RNase was improved after SDM.

12.19.4.2.1 Method of Site-Directed Mutagenesis.

Gene-encoding proteins under investigation need to be isolated or created first. Nucleotide residues to be mutated need to be first identified by using information from the 3D structure, homology comparison, etc. Oligonucleotide of defined sequences can be synthesized (solid phase synthesis). Nucleotide and amino acid residues can be replaced, deleted, or added (Figure 12.30).

The basic procedure starts with synthesizing a short DNA primer, containing the desired base change and a single DNA strand and a cloning vector like M-13 that makes the process easy. This synthetic primer has to hybridize with a single-stranded DNA containing the gene of interest. The single-stranded fragment is then extended using DNA polymerase, which copies the rest of the gene. The obtained double-stranded molecule is introduced into a host cell and cloned. Mutants are selected

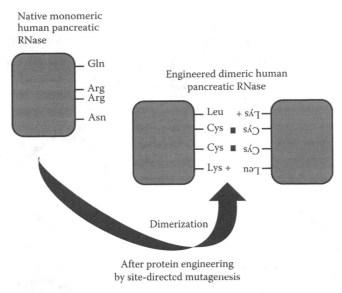

Figure 12.29 Improved dimeric human pancreatic RNase after SDM.

and products purified. Old method of SDM includes uracil-based mutagenesis and modern method is based on PCR mutagenesis.

12.19.4.2.2 Cassette Mutagenesis. It is possible to change two or three adjacent nucleotides so that every possible amino acid substitution is made at a site of interest. This generates a requirement for 19 different mutagenic oligonucleotides assuming only one codon will be used for each substitution. An alternative way of changing one amino acid to all the alternatives is cassette mutagenesis. This involves replacing a fragment of the gene with different fragments containing the desired codon changes. It is a simple method for which the efficiency of mutagenesis is close to 100%. However, if it is desired to change the amino acids at two sites to all the possible alternatives, then 400 different oligos or fragments would be required, and the practicality of the method becomes questionable (Figure 12.31).

12.19.4.2.3 Site-Directed Mutagenesis Using Uracil-Containing Templates. The plasmid to be mutated can be transformed into an *E. coli* deficient in two genes: dUTPase and uracil deglycosidase. The former would prevent the breakdown of dUTP, a nucleotide that replaces dTTP in RNA, resulting in an abundance of the molecule, and deficiency in the latter would prevent the removal of dUTP from newly synthesized DNA. As the double mutant *E. coli* replicates the uptaken plasmid, its enzymatic machinery incorporates the dUTP, resulting in a distinguishable copy. This copy is then extracted and incubated with an oligonucleotide containing the desired mutation, which attaches by base pair hydrogen bonding to the complementary wild-type gene sequence, as well as the Klenow enzyme, dNTPs, and DNA ligase. The reaction essentially replicates the dUTP-containing plasmid using as primer the oligonucleotide, giving a nearly identical copy. The essential differences being that

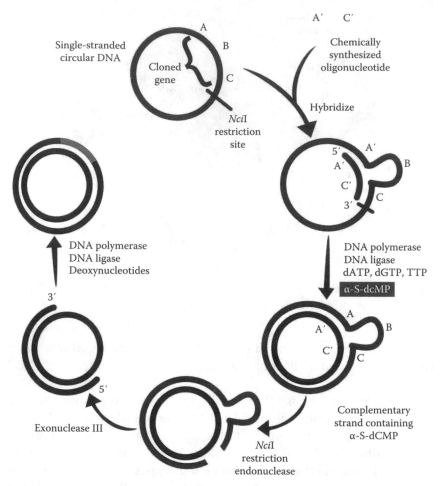

Figure 12.30 Oligonuceotide mutagenesis in which chemically synthesized strand is hybridized, and by using polymerase enzyme, they can be incorporated and copied into new strand, while template strand is digested by exonuclease III. Some digested part acts as a primer and used to make double strand again.

the copy contains dTTP rather than dUTP, as well as the desired mutation. When the chimeric double-stranded plasmid, containing the dUTP (unmutated) strand and the dTTP (mutated) strand, is inserted into a normal, wild-type *E. coli*, the dUTP-containing strand is broken down, whereas the mutation-containing strand is replicated.

Sodium bisulfate reacts with cysteine of sDNA and converts them into uracil, a thymine analogue that base pairs with adenine. Single-stranded DNA is treated with sodium bisulfate to modify a small number of cysteine residues in the molecule. An oligo primer is annealed to the DNA and serves as a primer for synthesis by DNA polymerase. When the polymerase encounters a uracil in the template strand, it incorporates an adenine into the newly synthesized DNA. Since the vector sequence is also damaged by bisulfate treatment, it is necessary to excise the double-stranded

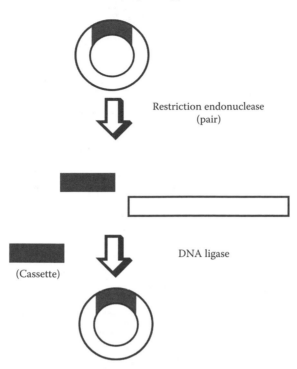

Figure 12.31 Cassette mutagenesis.

DNA fragment by restriction endonuclease cleavage and reclone it into an undamaged vector. Following transformation into *E. coli*, a library of mutant plasmid can be isolated, or individual plasmid can be purified and tested. The average number of substitution in the DNA fragment can be controlled by altering the conditions of bisulfate treatment (Figure 12.32).

12.19.4.2.4 PCR Site-Directed Mutagenesis. The same result can be accomplished using PCR with oligonucleotide primers that contain the desired mutation (Figure 12.33). As the primers are the ends of newly synthesized strands, by engineering a mismatch during the first cycle in binding the template DNA strand, a mutation can be introduced. Because PCR employs exponential growth, after a sufficient number of cycles, the mutated fragment will be amplified sufficiently to separate from the original, unmutated plasmid by a technique such as gel electrophoresis and reinstalled in the original context using standard recombinant molecular techniques.

For plasmid manipulation, this technique has largely been supplemented by PCR-like technique where a pair of complementary mutagenic primers is used to amplify the entire plasmid. This generates a nicked, circular DNA that can undergo repair by endogenous bacterial machinery. However, this process does not amplify the DNA, which must undergo the nick repair and is not super coiled, resulting in lowered efficiency of transformation in bacteria. Finally, the product DNA is of the same size as the plasmid. Therefore, the template DNA must be eliminated by enzymatic digestion with a restriction enzymes specific for methylated DNA. The template, for which

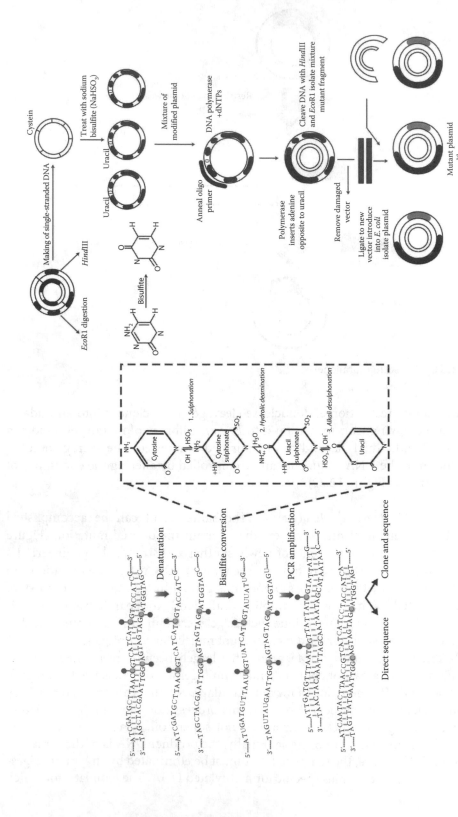

Figure 12.32 Chemical mutagenesis using sodium bisulfate.

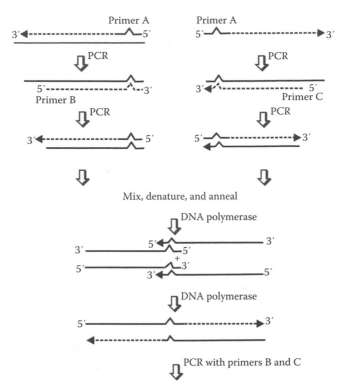

Figure 12.33 SDM with the help of PCR.

this technique should be biosynthesized, will be digested, but the mutated plasmid is preserved because it was generated *in vitro* and is, therefore, unmethylated.

12.19.4.3 Random Mutagenesis

DNA in cells is randomly mutated by chemical mutagens (e.g., hydroxylamine, sodium bisulfate), enzymatic synthesis, and mutagenic strains of bacteria (with deficient repairing systems). There is no need to know the detailed structure of the proteins. This technique can be applied when the current theories are inadequate to predict which structural changes will give improvement on certain property. Appropriate procedures for screening or selecting for desired properties are needed, and "directed evolution" was developed based on this concept (Figure 12.34).

12.19.4.3.1 Genetic Modification (In Vivo Engineering). In irrational engineering (random mutagenesis) and rational engineering (site-directed mutagenesis) the gene (DNA) encoding proteins are mutagenized.

12.19.4.3.2 Chemical Modification (In Vitro Engineering). PEG modification of protein surface results in amino groups, which reduces immunoreactivity, prolongs clearance times, improves biostability, and increases the solubility and activity of enzymes in organic solvents (Figure 12.35).

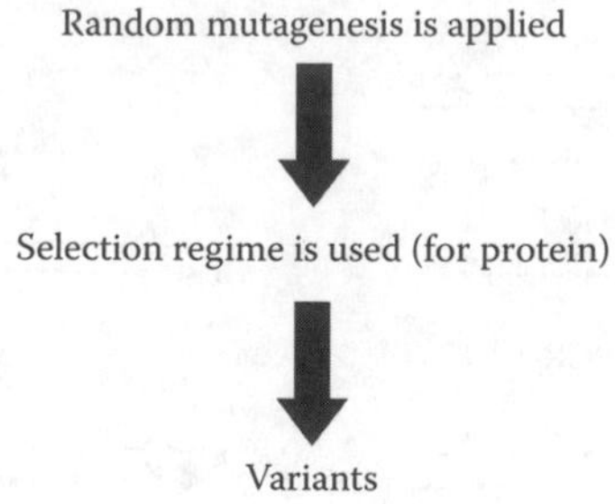

Random mutagenesis is applied

⬇

Selection regime is used (for protein)

⬇

Variants

Figure 12.34 Flow chart to get variants by random mutagenesis and mutant undergoes various selection pressure like temperature and pH. This strategy combines random mutagenesis of the target gene with screening or selection for the desired property.

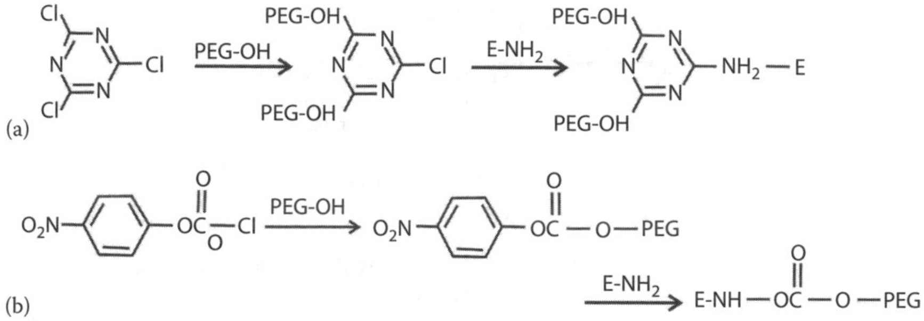

Figure 12.35 Enzyme modification with PEG activated with (a) cyanuric acid and (b) p-phenyl chloroformate. (From Chabala, J.C., *Curr. Opin. Biotechnol.*, 6, 632, 1995.)

12.19.4.4 Combinatorial Mutagenesis

Generally, simple mutations (to be introduced by site-directed mutagenesis, for example) are not expected to have as drastic effect as altering an enzyme's substrate recognition pattern as many amino acid residues in the enzyme (often not close to one another in the primary structure of the protein) affect the binding pocket of the substrate in the enzyme. Obviously, several amino acid residues may need to be altered simultaneously to achieve the goal of altering substrate specificities. However, as any amino acid residue may be altered into 19 other ones, the number of amino acid combinations that can be made if mutation is introduced at various residues simultaneously can become very large. For example, if four amino acid residues are altered simultaneously, there are 19^4 different combinations in which this can occur. It is generally unknown which of these combinations is what one is looking for as it is difficult to predict on the basis of a primary sequence what the 3D structure of a protein (and of the active site) will be in detail. Therefore, all different combinations made at the DNA level in different plasmids (can be done using degenerated oligonucleotides) can be used to identify the desired property.

Combinatorial mutagenesis does not limit itself to applications involving DNA. Peptides can also be synthesized from a degenerate mix of amino acid analogues, and the resulting mix of peptides, which can be screened for desired properties, in particular pharmaceutical applications. Moreover, RNA can be synthesized by combinatorial, and degenerate RNA mixtures have been used to study features that are needed to provide RNA with catalytic properties. In any case, combinatorial mutagenesis provides virtually limitless possibilities for genetic engineering and has become an important tool in biotechnology. The efficient application of biocatalyst requires the availability of suitable enzymes with high activity and stability under process conditions, desired substrate selectivity, and high enantioselectivity.

12.19.5 Directed Evolution

Directed evolution involves basically three steps: (1) generation of the gene pool, (2) recombination of the gene pool, and (3) application of evolutionary pressure by screening and then selection or by using high throughput to screen variants for a desired property.

Selection is done under increasing selective pressure like antibiotics, pH, and organic solvent. *In vitro* homologous recombination of pools of selected genes takes place by random fragmentation and PCR assembly. There are three different means to obtain the desired diversity to mimic the natural design process and speed it up by directed selection *in vitro* toward a simple specific goal: (1) introducing point mutations (and sometimes, deletion and insertion) at random, (2) randomization that is restricted to specific positions or limited regions within the gene (this approach often includes randomization of a limited number of codons, also referred to as saturation mutagenesis), and (3) the use of recombination techniques applied to gene pools derived from the two first approaches and/ or from nature (Table 12.3).

TABLE 12.3   Some Examples of Protein Engineering

Example	Method	Reference
Increased rate and extent of biodesulfurization of diesel by modification of dibenzothiophene monooxygenase	RACHITT	Coco et al. (2001)
Generation of a subtilisin with a half-life at 65°C that is 50 times greater than wild type by recombining segments from five different subtilisin variants	StEP	Zhao et al. (1998)
Conversion of a galactosidase into fucosidase	Suffling	Zhang et al. (1997)
Enhanced activity of amylosucrase	Random mutagenesis plus shuffling	Van der Veen et al. (2004)
Generation of novel DNA polymerase from a combination of rat DNA polymerase beta and African swine fever virus DNA polymerase X	SCOPE	O'Maille et al. (2002)
Generation of novel β-lactamase by recombining two genes with 40% amino acid identity and 49% nucleotide sequence identity	SISDC	Hiraga and Arnold (2003)

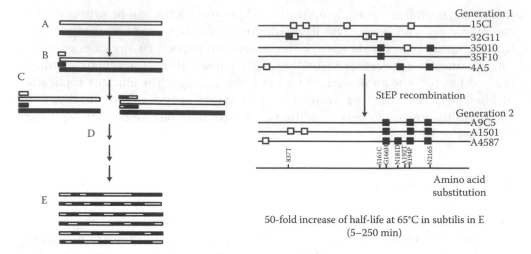

Figure 12.36   StEP method for generating hybrid proteins. Here, a hybrid gene will be constructed from two homologous genes. Cloning of the hybrid gene will result in the product of a hybrid protein. Here, only changed strand is shown.

DNA libraries can be created by the other fast method-like error-prone PCR, combinatorial oligonucleotide mutagenesis, DNA shuffling, exon shuffling, random-priming recombination, random chimeragenesis on transient templates (RACHITT), staggered extension process (StEP) recombination, heteroduplex recombination (Figure 12.36), incremental truncation for the creation of hybrid (ITCHY) (Figure 12.37), recombined extension on truncated templates (RETT), degenerate oligonucleotide gene shuffling (DOGS), and *in vivo* recombination.

12.19.5.1 Gene Shuffling

Gene shuffling refers to a process in which DNase is used to digest and break apart a DNA segment (e.g., a gene), shuffle (i.e., changes the order) the order of the relevant nucleotide within that sequence, and then recombine those nucleotides into an intact DNA segment. In this method of generating genetic diversity, the gene of interest is cleaved with DNase I into many short double-stranded fragments (10–50 base pairs) that are then purified and recombined in a PCR-like process without primers. When the genes are partially recombined, terminal primers are added, and full-length sequences are amplified. This process is usually used after an error-prone PCR step, where different mutants of the gene are generated. Alternatively, the condition of DNA shuffling can be adjusted to introduce mutations and to avoid regeneration of the original sequence. Using a powerful derivative technique called family shuffling, homologous genes pooled from different organisms are recombined.

When repeated DNA shuffling is coupled to a gene expression and assessment/improvement process (e.g., each new shuffled DNA segment is expressed, and

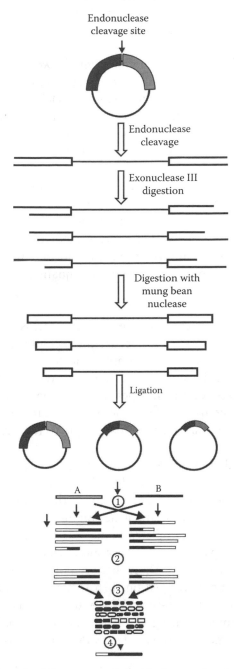

Figure 12.37 ITCHY method for creating hybrid of two related proteins, gene 1 and gene 2, and finally a hybrid gene is obtained, gene 1 at 5′ end and gene 2 at 3′ end. (1) Plasmids A and B (exonuclease digestion). (2) Restriction digestion on B gene. (3) Byligation into A. (4) Two ITCHY libraries. Restriction digestion or PCR amplification. DNase I digestion and Reassembling.

the resultant protein is evaluated against a desired goal), incorporated feedback of the DNA-shuffling process, the process is sometimes called directed evolution. For example, increase the activity of a GAT enzyme 10,000-fold glyphosate *N*-acetyltransferase.

Gene shuffling involves *in vitro* recombination of two or more homologous genes (<70%) for creating engineered variants of the enzyme that one wants to engineer. Recently, by the use of truncation and ligation of DNA fragments, recombination of genes is possible with low sequence similarity that has been described. Synthetic shuffling is possible by the use of specific primers or by using the insertion of tag sequences to induce specific similarity-independent recombination events between sequences (Figure 12.38).

DNA shuffling used related genes from different species, or genes with related function fragments them and reassembles through recombination. Recombined genes are then placed into *E. coli* to identify which new genes produce usable or potentially interesting products. Those genes that express a potentially interesting protein or enzyme are again fragmented and reassembled to form new recombinant genes. The process continues until a protein with the desired qualities is found.

12.19.5.2 Family Shuffling

Here, recombination of two or more parent genes yields a chimeric gene library for evolution of the desired features. Because the recombined sequences are related through divergent evolution from a common ancestor of similar structure and function—and, therefore, the sequence differences are to some extent neutral with respect to structure and function—it appears that very large jumps in sequence space can yield functional proteins. *In vivo* recombination can also yield interesting new chimeric enzymes; the diversity may be more limited than that obtained by *in vitro* methods (Figure 12.39).

Gene shuffling is used to increase DNA library size rapidly. DNA shuffling, in contrast, is PCR without synthetic primers. In this process, a family of related genes—say, the ones that codes for the surface proteins of three different HIV isolates—are digested with restriction enzymes. The gene fragments then are heated up to separate them into single-stranded templates. Some of these fragments will bind to other fragments that share complementary DNA region, which in some cases will be from other family members. Regions of DNA that are noncomplementary hang over the ends of the templates. The PCR reaction then treats the complementary regions as primers and builds the new double-helical DNA, but PCR also adds bases to the overhanging piece of the primer, forming a double helix there, too. This ultimately creates a mixed structure or chimera. In the final step, PCR reassembles these chimeras into full-length, shuffled gene.

The mathematical model, which provides a predictive framework for DNA shuffling, looked into how fragment length, annealing, temperature, sequence identity, and the number of shuffled parent sequences affect the number, type, and distribution of crossovers along the length of reassembled sequences. The more similar genes are the more potential for crossover exists.

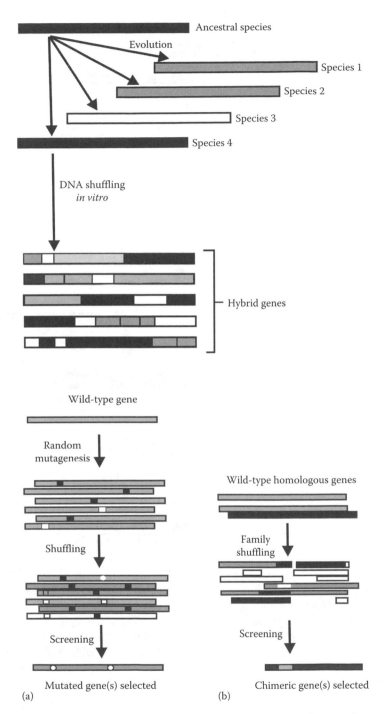

Figure 12.38 Schematic representation of method of gene shuffling and comparison of gene shuffling and family shuffling (white dot = advantageous; black dot = deleterious mutation). Gene shuffling involves recombination of different genes from same species (differ in location and property), and family shuffling involves gene recombination of different species.

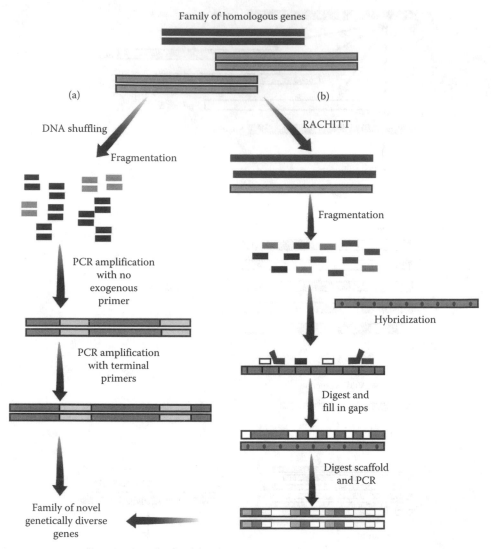

Figure 12.39   The classical DNA-shuffling strategy begins by fragmenting a pool of double-stranded parent genes with DNase I. (a) Selection of small fragments by size fractionation maximizes the probability of multiple recombination events occurring, as the fragments cross-prime each other in a round of PCR amplification. The full-length, diversified products are then obtained by PCR amplification with terminal primers. (b) RACHITT begins with DNase I fragmentation and size fractionation of single-stranded DNA and hybridization—in the absence of polymerase—to a complementary single-stranded scaffold. Any overlapping fragments leave single-stranded overhangs that are trimmed down. The gaps between fragments are filled in; the fragments are then ligated, yielding a pool of full-length, diversified single strands hybridized to the scaffold. This scaffold, which is synthesized so as to include uracil (U), can be efficiently fragmented so as to preclude its amplification. With PCR, it is replaced by a new strand that is complementary to the diversified strand, and the whole is amplified.

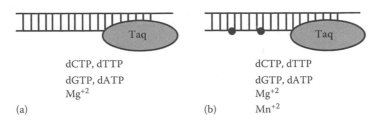

Figure 12.40 Difference in method of (a) PCR and (b) error-prone PCR. From the figure, it is clear that epPCR uses Mn^{2+} in place of Mg^{2+}.

12.19.5.3 Error-Prone PCR

Error-prone PCR uses altered reaction conditions in order to increase the rate of production of copying error. Taq polymerase, when used *in vitro*, has an intrinsic error rate of about one noncomplementary nucleotide for every 10,000–100,000 bases (an error rate of 1 in 10^4 or 10^5). EpPCR is based on the principle that *Taq polymerase* is capable of annealing incompatible base pairs to each other during PCR amplification under imperfect PCR conditions. The errors can be increased by replacing Mg^{2+} in place of Mn^{2+} and to include excess of dGTP and dTTP relative to other two nucleotides (Figure 12.40). Even higher rate of mutagenesis can be created by using nucleotide triphosphate analogues. The main aim is to introduce the incorrect base pair so that random change in amino acid can happen to occur. Even more mutation can be done by ligating an insertion or deleting cassette at random locations within genes. Naturally unusual amino acid can be introduced during protein synthesis.

12.19.6 Selection Parameters

12.19.6.1 In Vivo *Display in Bacteria*

In this method, evolved genes are incorporated into a bacterial plasmid using standard molecular biology techniques. Plasmids containing different evolved genes are used to transform different bacterial cells; the plasmid is incorporated into the bacterial cytoplasm and its genetic code is translated into proteins. Cells are plated on agar, and single colonies (resulting from single cells) are isolated and replated. Each colony produces only one genetic (evolved) variant. The protein is expressed without a direct linkage (bond) with the plasmid and may be secreted or remain in the cytoplasm as shown in Figure 12.41. The number of clones that can be screened using plates or liquid multiwell assays is usually limited to 10^5 clones.

12.19.6.2 Phage Display or Phagemid Display Can Be Used to
Facilitate the Selection of Mutant Peptides

Phage display begins with the ligation of each member of the genetic library to a section of viral DNA that codes for one of the coat proteins of the virus. The most commonly used virus is the filamentous M13 phage, a common lab strain, and the

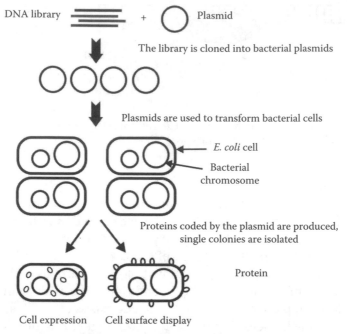

Figure 12.41 *In vivo* methods allow connection between phenotype and genotype using cellular machinery. Synthesized proteins remain in the cytoplasm (or are secreted) in the nondisplay method. Cell surface display is a more recently introduced the technique that uses special plasmids in order to produce proteins capable of transport into and retention by the cell membrane.

coat protein is the gene III protein (gIIIp). Thus, a novel protein can be engineered by fusing the protein-coding sequence of one gene to the protein-coding sequence of a viral gene and can be selected by phase display method. In phase display, a segment of foreign DNA is inserted into either a phagemid or an infectious filamentous phase genome and expressed as a fusion protein. The population of viruses is used to infect a bacterial culture; the resulting daughter phages display the evolved proteins and contain the genetic code responsible for their displayed phenotypic variants. For example, M-13 phase consists of single-stranded DNA surrounded by a coat protein (P8). A mutant can be selected by phase display more than 10 residues since DNA attached with phase DNA can be expressed and packed. The use of phagemid can increase even more residue and attachment of amber codon help in truncation immediately downstream from the foreign DNA and upstream from P8 or P3 (gene-encoding protein). This fusion protein can be attached to signal peptide to allow them to go inside the bacterial membrane especially in Gram-negative bacteria for display where fusion protein remains outside.

12.19.6.3 Ribosome Display (In Vitro)

Ribosome display uses the unmodified DNA library, which is transcribed and translated *in vitro*. The link occurs between the evolved protein and mRNA and is achieved by stalling the translating ribosome at the end of the mRNA, which lacks

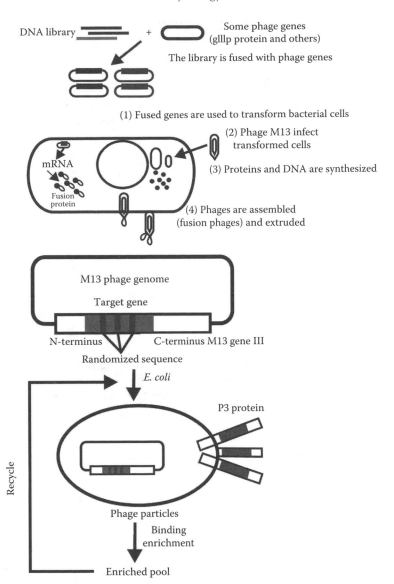

Figure 12.42 Phage display method. When the phages are extruded, binding to immobilized ligands can be used to select evolved variants.

a stop codon. Without a stop codon, the protein is not released by the ribosome, and the complex formed by the mRNA, protein (usually correctly folded), and ribosome is used directly for selection against an immobilized target (Figure 12.42). Later the selected mRNA complexes are separated from the immobilized target and are dissociated with EDTA. Using reverse transcription PCR, the corresponding DNA is synthesized using the purified mRNA as the template. In another technique, mRNA-peptide fusion is done *in vitro* and is also called mRNA display; this method was derived from ribosome display technology. Puromycin-tagged mRNA and several additional steps must be used to achieve covalent coupling between

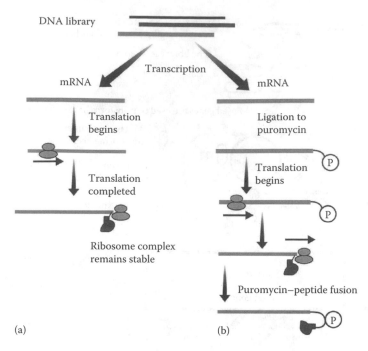

Figure 12.43 Ribosome display is used to display binding molecules: the noncovalent complex (ribosomes, protein, and mRNA) is stabilized by changing the medium composition, and lowering the temperature (a) in mRNA–peptide fusion, the link between genotype (mRNA) and phenotype (protein) is covalent (b).

the protein product and its mRNA (Figure 12.43). Puromycin, which mimics aminoacyl (tRNA ¼ transfer RNA), can be attached covalently to mRNA by the ribosomal machinery.

12.19.6.4 High-Throughput Product Analysis

Using robotics, data processing and control software, and sensitive detectors, high-throughput screening (HTS) allows to conduct millions of biochemical, genetic, or pharmaceutical tests quickly. Through this process, one can rapidly identify active compounds, antibodies, or genes that modulate a particular biomolecular pathway. The results of these experiments provide starting points for drug design and for understanding the interaction or role of a particular biochemical process in biology. Most high-throughput assays are based on chromogenic and flurogenic substrate or sensors. Umbelliferyl and nitrophenyl esters, phosphates, sulfates, and glycosides are classical fluorogenic and chromogenic substrates for the corresponding hydrolytic enzymes; fluorescein and often fluorescein isothiocyanate (FITC) are used to label proteins and, in particular, antibodies. FITC casein is a classical fluorogenic protease substrate; autoquenching occurs between different fluorescein chromatophores coupled to the same casein molecule, but quenching is released by proteolysis to produce a fluorescence increase. Flourescein-labeled peptides have also been used to follow the activity of proteases, kinases, and phosphatases by following changes in the fluorescence polarization induced by complexation of the phosphorylated peptides with polyarginine.

Large gain in the high-throughput efficiency of enzyme assay is possible by miniaturization in the form of microarrays. Microarrays, featuring substrates or enzymes in a surface-bound or surface-displayed format, are relevant to enzyme screening for biocatalysis, inhibitor screening for drug discovery, and enzyme fingerprinting. Peptide microarrays have been prepared for fingerprinting proteases and have also found use in the profiling and inhibitor screening of kinases. A similar approach was demonstrated for a general enzyme profiling microarray.

There is one limitation of this technique that it can be used for checking thermostability of the enzyme since microplate reader can have maximum 50°C and also high temperature causes evaporation, pH change, and substrate instability. For such enzyme, alternative technique is better like stability selection based on phase display, and this technique is also called as protein stability increased by directed evolution (proside).

12.19.6.5 Agar Plate-Based Assay

For testing enzyme property like thermo and pH, stability and specificity can be checked by using agar plate-based assay for colonies obtained.

12.19.7 Application of Enzyme Engineering

12.19.7.1 Subtilisin

Much protein engineering has been directed at subtilisin from *Bacillus amyloliquefaciens*, the principle enzyme in the detergent enzyme preparation as Alcalase. Most of the attempts were on improvements to have concerned alterations to hold (1) the amino acid on the carbonyl side of the targeted peptide bond, (2) stabilization of the tetrahedral intermediate, (3) the neighborhood of the catalytic histidyl residue (*His* 64) that has a general base role, and (4) the methionine residue (*Met*222) that causes subtilisin's liability to oxidation (Table 12.4).

Mutagenesis is very powerful for development of stable enzymes, presumably because it includes useful mutations that do not naturally occur in nature and

TABLE 12.4 Enzyme and Their Improved Properties

Enzymes	Specificity Improvement
Lipase (*Pseudomonas aeruginosa*)	Enantioselectivity toward 2-methyldecanoate
Hydantoinase	Enantioselectivity toward D- and L-form
Aldolase (KDPG)	Enantioselectivity toward D- and L-glyceraldehydes
β-Galactosidase	Changing activity of β-galactosidase
Lipase (*Staphylococcus aureus*)	Changing activity of phospholipase
Esterase (*Pseudomonas fluorescens*)	Activity on sterically hindered substrate
Activity and stability improvement	
Lipase (*Candida antarctica*)	Chimeric lipase more stable and active
D-Amino amidase	Enhanced thermostability and activity

thus are found with low frequency in, for example, family shuffling. Site-saturation mutagenesis exchanges complete codon, whereas in *Error-Prone Polymerase Chain Reaction* epPCR, usually only one nucleotide per codon is exchanged. This site-saturation mutagenesis does not have the limitations of epPCR with respect to the amount of variability that may be introduced at each codon. Interesting examples of stability engineering based on this approach include the stabilization of a haloalkane dehalogenase and a phytase. Some other examples of directed evolution addressing other parameters than temperature include studies on stability/activity at extremes of pH, in oxidative environments, in organic solvents, and in the presence of surfactants. Rational approach has been quite successful in the past, but now-a-days only combinatorial approaches are useful with directed evolution.

12.19.7.2 Protein Engineering of Laundry Detergent

The enzyme subtilisin is a protease produced by bacteria having broad specificity for proteins. Detergent manufacturers add subtilisin to improve the efficiency of laundry detergents. However, subtilisin is inactivated by bleach, and this inactivation was due to oxidation of the amino acid methionine at position 22 of the subtilisin molecule. Using SDM of the subtilisin gene in *E. coli*, this methionine was changed to a variety of other amino acids, and the subtilisin activity and bleach resistance of the mutated protein was tested. The substitution of methionine by alanine was the best in terms of activity and stability, and now many laundry detergents contain cloned genetically engineered subtilisin. The mutant enzyme dramatically increased stability and the specific activity with much better results compared to wild type.

Trypsin: It is another enzyme where the protein engineering has been employed to improve the activity and stability of the enzyme.

12.20 Artificial Enzymes

12.20.1 Synzymes

Several possibilities now exist for the construction of artificial enzymes. Normally, they are synthetic polymers or oligomers with enzyme-like activities prepared to recreate the active site of an enzyme, often called synzymes. The catalysis of chemical reactions occurs with high selectivity and rate in a small portion of the enzyme macromolecule known as the active site. The artificial enzyme must possess two structural entities: a substrate binding site and a catalytically effective site. The facility for producing a substrate binding site is relatively straightforward, but catalytic sites are somewhat more difficult. Both sites may be designed separately but it appears that, if the synzyme has a binding site for the reaction transition state, this often achieves both functions. Synzymes generally obey the saturation Michaelis–Menten kinetics. For a one-substrate reaction, the reaction sequence is given by

$$\text{Synzyme} + \text{S} \rightleftharpoons (\text{Synzyme–S complex}) \rightarrow \text{Synzyme} + \text{P}$$

New approaches based on amino acids or peptides as characteristic molecular moieties have led to a significant expansion of the field of artificial enzymes or enzyme mimics. Some synzymes are simply derivatized proteins, although covalently immobilized enzymes are not considered here. An example is the derivatization of myoglobin, the oxygen carrier in muscle, by attaching $(Ru(NH_3)_5)^{3+}$ to three surface histidine residues. This converts it from an oxygen carrier to an oxidase, oxidizing ascorbic acid while reducing molecular oxygen. The synzyme is almost as effective as natural ascorbate oxidases.

12.20.2 Abzymes

Abzymes are also categorized as artificial enzyme or synzyme. In abzymes, antibodies act as enzymes. The abzymes are also called as catmab that means catalytic monoclonal antibodies. Abzymes are mostly artificial enzymes; however, they are also found in humans with asthma as antivasoactive intestinal peptide autoantibodies and in patients with the autoimmune disease systemic lupus erythematosus where these bind and hydrolyze DNA. Abzymes have a low K_m indicating that they bind a target molecule. However, abzymes also show low V_{max} indicating a slow reaction rate. Abzymes are found to have a potential role in therapeutics. Anti-cocaine abzyme may be used to treat patients who are addicted to cocaine. They may also be used to reverse the lethal effects of a cocaine overdose. Research workers are actively engaged in exploiting the abzyme technology for specific targeting of cancer cells for destruction. Abzymes have also been isolated that cleave viral coat proteins of human immunodeficiency virus and also destroy the viral genome.

12.20.3 Promises of Enzyme Technology

The amount of enzyme produced by a microorganism may be increased by increasing the number of gene copies that code for it. Such increased yields are economically relevant not just for the increased volumetric productivity but also because of reduced downstream processing costs, the resulting crude enzyme being that much purer. New enzyme structures may be designed and produced in order to improve on existing enzymes or create new activities. Much protein engineering has been directed at subtilisin (from *B. amyloliquefaciens*), the principal enzyme in the detergent enzyme preparation, Alcalase. This has been aimed at the improvement of its activity in detergents by stabilizing it at even higher temperatures, pH, and oxidant strength. A number of possibilities now exist for the construction of artificial enzymes synzymes. Enzymes can be immobilized, that is, an enzyme can be linked to an inert support material without loss of activity that facilitates reuse and recycling of the enzyme. Use of engineered enzyme to form biosensor for the analytical use is also a recent activity. Some enzymes make use in disease diagnosis so they can be genetically engineered to make the task easier. The protein engineering technique involves genetic modification by means of recombinant DNA technology of the enzyme-producing microorganism, in particular the enzyme-encoding gene, resulting in substitution of one or more amino acids in the amino acid sequence of the enzyme protein. Strategies for making such amino acid substitutions and developing protein-engineered enzymes are based on the knowledge of the structure/function relationships of enzymes, computer modeling, and techniques for creating and testing enzyme variants. Thus, it is obvious that there is a huge scope of the enzyme technology.

Suggested Readings

Bergmeyer, H. U. 1978. *Principles of Enzymatic Analysis*. New York: Verlag Chemie.

Coco, W. M., L. P. Encell, W. E. Levinson et al. 2002. Growth factor engineering by degenerate homoduplex gene family recombination. *Nat Biotechnol* 20: 1246–1250.

Coco, W. M., W. E. Levinson, M. J. Crist et al. 2001. DNA shuffling method for generating highly recombined genes and evolved enzymes. *Nat Biotechnol* 19: 354–359.

Collins, C. H., Y. Yokobayashi, and D. Umeno. 2003. Engineering proteins that bind, move, make and break DNA. *Curr Opin Biotechnol* 14: 371–378.

Hiraga, K. and F. H. Arnold. 2003. General method for sequence-independent site-directed chimeragenesis. *J Mol Biol* 330: 287–296.

International Union of Biochemistry. 1984. *Enzyme Nomenclature*. New York: Academic Press.

Lutz, S. and W. M. Patrick. 2004. Novel methods for directed evolution of enzymes: Quality not quantity. *Curr Opin Biotechnol* 15: 291–297.

Moss, D. W. 1982. *Isoenzymes*. London, U.K.: Chapman & Hall.

Neylon, C. 2004. Chemical and biochemical strategies for the randomisation of protein encoding DNA sequences: Library construction methods for directed evolution. *Nucl Acids Res* 32: 1448–1459.

O'Maille, P. E., M. Bakhtina, and M. D. Tsai. 2002. Structure-based combinatorial protein engineering (SCOPE). *J Mol Biol* 321: 677–691.

Price, N. C. and L. Stevens. 1982. *Fundamentals of Enzymology*. Oxford, U.K.: Oxford University Press.

van der Veen, B. A., G. Potocki-Véronèse, C. Albenne et al. 2004. Combinatorial engineering to enhance amylosucrase performance: Construction, selection, and screening of variant libraries for increased activity. *FEBS Lett* 560: 91–97.

Zhang, J.-H., G. Dawes, and W. P. C. Stemmer. 1997. Directed evolution of a fucosidase from a galactosidase by DNA shuffling and screening. *Proc Natl Acad Sci USA* 94: 4504–4509.

Zhao, H., L. Giver, Z. Shao et al. 1998. Molecular evolution by staggered extension process (StEP) in vitro recombination. *Nat Biotechnol* 16: 258–262.

Important Links

http://www.lpscience.fatcow.com/jwanamaker/animations/Enzyme%20activity.html
http://silvaggi.com/index.php/links
http://www.unilorin.edu.ng/courseware/BCH/BCH401.pdf
http://www.mendeley.com/catalog/computational-enzymology-insights-enzyme-mechanism-catalysis-modelling-19/

http://www.cliffsnotes.com/study_guide/Enzymes-Are-Catalysts.topicArticleId-24998,articleId-24969.html

http://www.bnl.gov/pubweb/alistairrogers/linkable_files/pdf/Rogers_&_Gibon_2009.pdf

http://www-users.med.cornell.edu/~jawagne/proteins_&_purification.html

http://www.ncbi.nlm.nih.gov/books/NBK22410/

http://www.pharmatutor.org/pharmacognosy/enzyme-biotech.html

http://en.wikibooks.org/wiki/Structural_Biochemistry/Proteins/Purification/Affinity_chromatography

http://www.lsbu.ac.uk/water/enztech/engineering.html

http://cdn.intechopen.com/pdfs/29172/InTech-Protein_engineering_methods_and_applications.pdf

http://www.sbj.or.jp/e/news/news_eei_2013022-26.html

13

Membranes

13.1 Introduction

Cytoplasmic or plasma membrane is present just beneath the cell wall in a cell. It is a thin, flexible sheet-molded structure surrounding the cytoplasm. It acts like a barrier separating the cell from its environment. The cytoplasmic membrane is also a selective permeability barrier, enabling the cell to concentrate specific metabolites and excrete waste materials. It is a phospholipid–protein bilayer. The phospholipids contain both hydrophobic (e.g., fatty acid) and hydrophilic (e.g., glycerol phosphate) components (Figure 13.1). The cytoplasmic membrane appears as two light-colored lines separated by a darker area when observed under the electron microscope.

13.2 Composition of Membranes

Membranes are clearly visible under the electron microscope and form an essential part of the cell structure and may be studied in a number of different ways: for example, by looking at them with high-powered microscopes or by observing their various transports or signaling properties with electrical instruments and enzymatic assays. Plasma membranes from quite different cells have similar compositions. The outer envelope consists mainly of protein and lipid, with very little carbohydrate, which is covalently linked to either protein or lipid, making these molecules, as glycoprotein or glycolipid, respectively. Prokaryotic cells are seen to have a single membrane surrounding the cell. Eukaryotic cells present in higher organisms also have a cell membrane and, in addition, contain other membrane-bound structures such as a nucleus, mitochondria, and lysosomes. The plasma membranes of most eukaryotic cells contain equal weights of lipid and protein (with and without attached carbohydrate), but there are many more lipid than protein molecules in these membranes simply because lipids are much, much smaller than proteins. Inner mitochondrial membranes (and the plasma membranes of many bacteria) are relatively rich in protein, containing three times more protein than lipid on a weight basis. Even so, bacterial and mitochondrial membranes still have about 30 lipid molecules for every protein. These membranes effectively separate cells from each other and also divide

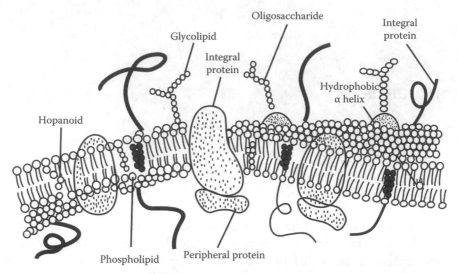

Figure 13.1 Structure of cytoplasmic membrane fluid mosaic model of bacterial membrane structure showing internal proteins floating in the lipid bilayer.

them into distinct aqueous regions. Membranes are also a basic structure that consist of about 40% lipid and 60% protein together with some carbohydrates and ions.

13.2.1 Phospholipids

Phospholipids are a major component of cell membranes, which enclose the cytoplasm and other contents of a cell. A phospholipid is composed of two fatty acids, a glycerol unit, a phosphate group, and a polar molecule. The phosphate group and polar head region of the molecule are hydrophilic (attracted to water), while the fatty acid tail is hydrophobic (repelled by water). Such structural organization arises from the fact that phospholipids have hydrophobic and hydrophilic regions in the same molecules, which is known as amphipathic (Figure 13.2). When placed in water, phospholipids will orient themselves into a bilayer in which the nonpolar tail region faces the inner area of the bilayer. The polar head region faces outward and interacts with the water. Phospholipids form a lipid bilayer in which their hydrophilic head areas spontaneously arrange to face the aqueous cytosol and the extracellular fluid (ECF), while their hydrophobic tail areas face away from the cytosol and ECF. Phospholipids thus orient themselves on the surface of an aqueous solution so that the polar region lies in the water and alkyl side chain in the air (Figure 13.2a). A similar type of association occurs in membranes with the alkyl side chains associating to form a lipid bilayer (Figure 13.2b). The lipid bilayer is semipermeable, allowing only certain molecules to diffuse across the membrane to enter or exit the cell.

Most phospholipids are also zwitterions since the phosphate group carries a negative charge at neutral pH and the base a positive charge (choline phosphoglyceride, ethanolamine phosphoglyceride) or a positive and negative charge

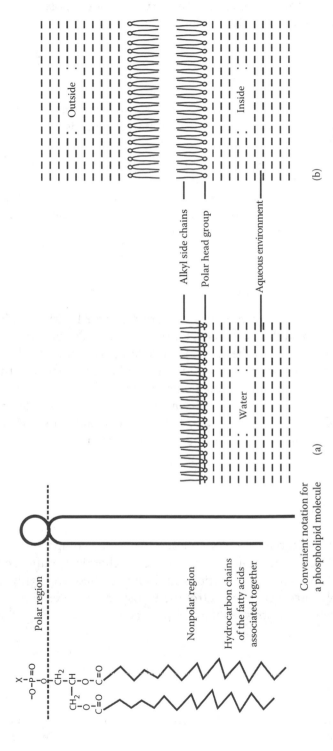

Figure 13.2 Amphipathic nature of phospholipids and association of phospholipid molecules at an (a) air–water interface and (b) in a membrane.

(serine phosphoglyceride). The other common polar group linked to the phosphate residue is inositol, but this is uncharged. The structure of the polar head group thus determines the magnitude and distribution of charge on the membrane.

13.2.2 Cholesterol and Other Lipids

Cholesterol is a fat-like substance with a waxy consistency that is produced in the livers of humans and other animals. Every human cell needs a minute amount of cholesterol in order to function properly. While some cholesterol is essential, too much can be harmful. Because it is waxy, cholesterol can stick to the interior of blood vessel walls. Most of the lipids found in membranes are also amphipathic molecules and may confer particular properties on the membrane. Many glycolipids appear to determine cell and tissue immunological specificity and to be responsible for the major part of the charge carried on the cell surface.

13.2.3 Proteins

Proteins are also present in membranes and are associated with the phospholipid bilayer in a variety of different ways (Figure 13.3). Membrane proteins perform a variety of functions vital to the survival of organisms. They act like membrane receptor, transporter protein, enzymes, and/or cell adhesion molecules. They are also the target of over 50% of all modern medicinal drugs.

A few proteins found on the membrane surface are bound electrostatically to the bilayer and can be removed with high salt concentration. Most proteins, however, are more strongly bound to the membrane structure with detergents or organic solvents.

13.3 Membrane Transport

In membrane transport, the flow of substances from one compartment to another can occur in the direction of a concentration or electrochemical gradient or against it. Membranes carry out a variety of functions, and some of these are highly specialized. Their principal role, however, is to control the transport of material into and out of the cell and between the various cellular compartments.

13.3.1 Diffusion

13.3.1.1 Simple Passive Diffusion

Diffusion is a form of passive transport in which molecules move naturally from a higher to a lower concentration. Membrane proteins are not involved in passive diffusion. The diffusion velocity of a pure phospholipid membrane will depend on

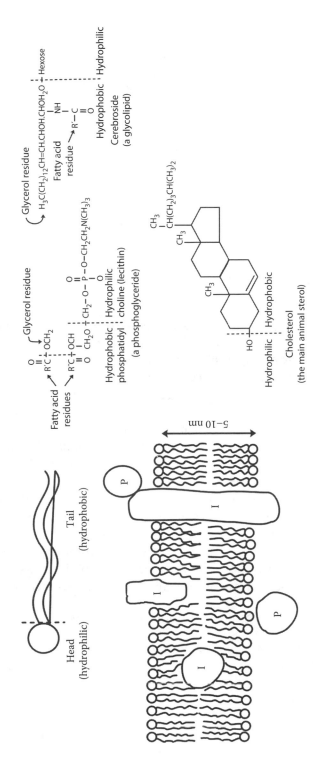

Figure 13.3 The association of protein with the phospholipid bilayer in membrane (P = peripheral protein and I = internal protein) and important examples of lipid found in membranes.

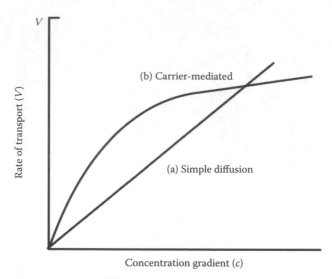

Figure 13.4 Dependence of (a) simple diffusion and (b) carrier-mediated transport on the concentration gradient across the membrane.

concentration gradient, hydrophobicity, size, and charge, if the molecule has a net charge. The rate of transport is linear with concentration and does not show saturation (Figure 13.4a). Some examples of simple diffusion are the transport of ethanol, oxygen, and carbon dioxide across the cell membrane.

13.3.1.2 Facilitated Diffusion

In facilitated diffusion, substances move into or out of cells down their concentration gradient through protein channels in the cell membrane. Simple diffusion and facilitated diffusion are similar in that both involve movement down the concentration gradient. The difference is how the substance gets through the cell membrane. This is also a form of passive transport down a concentration gradient, but the rate of diffusion is greater than that expected from simple diffusion. Transport occurs by means of a carrier across the membrane, and the process is also known as carrier-mediated transport. The rate of transport is hyperbolic with concentration and reaches a maximum value as the carrier becomes saturated (Figure 13.4b). Examples of facilitated diffusion include the passage of glucose and amino acids across membranes:

$$\text{Simple diffusion: } v = kc$$

$$\text{Carrier-mediated transport: } v = Vc/K + c$$

13.3.2 Active Transport

The type of membrane transport discussed so far always involve substances moving down their concentration gradient. It is also possible to move substances across membranes against their concentration gradient (from areas of low concentration to areas of high concentration). Since this is an energetically unfavorable reaction, energy is needed for this movement. The source of energy is the breakdown of adenosine triphosphate (ATP). If the energy of ATP is directly used to pump molecules against their concentration gradient, the transport is called primary active transport. Active transport is similar to facilitated diffusion in that it involves a carrier, shows a high degree of specificity, and becomes saturated at high concentrations. The best known example of active transport is that of Na^+. The concentration of Na^+ inside the cell is much lower than the extracellular fluid (ECF), and this concentration gradient is maintained by the Na^+ pump, which removes Na^+ from the cell to the ECF using the energy obtained by the hydrolysis of ATP.

13.4 Effect of Lipid Composition on the Permeability of a Lipid Monolayer

13.4.1 Principle

The lipid composition of a membrane has a considerable effect on its permeability, and in these experiments, a lipid monolayer is used as a membrane model. If butanol is layered on top of water, then two distinct phases are formed; if amphipathic lipids are present, they will move into the boundary region. The polar part of the molecules will associate with the top aqueous layer and the hydrophobic region with the organic phase. Methylene blue is a highly colored molecule, and its passage across the boundary can be readily followed by the eye. Unlike biological membranes, this model does not contain protein, but useful information can be obtained from this simple experiment as passive diffusion does depend on the lipid composition of the membrane.

13.4.2 Materials

Fatty acids (stearic acid, oleic acid), acylglycerols (triolein, tripalmitin), phospholipid (egg lecithin), sterol (cholesterol), n-butanol, boiling tubes, methylene blue in butanol (0.25 g/L).

13.4.3 Protocol

Set up boiling tubes each containing 5 mL of water. Carefully pipette 5 mL of butanol containing methylene blue and 200 mg of lipid down the side of each tube to form two distinct layers. Leave the tubes to stand at room temperature for 1–2 h and compare the results with the one obtained using a control tube containing water, but no lipid in the butanol. A rough estimate of the effectiveness of the lipids as permeability barriers can be obtained by measuring the extinction of the methylene blue in the aqueous phase.

13.5 Effect of Detergents and Other Membrane-Active Reagents on the Erythrocyte Membrane

13.5.1 Principle

Several detergent-like molecules disrupt membranes by "dissolving" the phospholipid components. In the case of erythrocytes, this effect can be readily followed in a colorimeter by measuring the absorbance of hemoglobin released from the disrupted cells.

13.5.2 Materials

Fresh rat blood or time-expired human blood used for transfusion, isotonic saline 8.9 g/L; *detergents* (1%): Triton X-100 (*neutral*), cetyltrimethylammonium bromide (*cationic*), sodium dodecyl sulfate (*anionic*).

Lysophosphatidyl choline (lysolecithin, 10 mmol/L), progesterone (100 mmol/L in ethanol), hydrocortisone (saturated solution in ethanol, approximately 5 mmol/L), centrifuge tubes (10 mL), incubator at 37°C, colorimeter or spectrophotometer.

13.5.3 Protocol

Use human blood from a transfusion bottle or collect the blood from a freshly killed rat into a tube containing 0.5 mL of an anticoagulant (40 g/L trisodium citrate). Centrifuge the cells, wash them twice with isotonic saline, and resuspend them in the same volume as the original blood. Dilute the erythrocyte suspension with saline so that when 0.5 mL is added to 4.5 mL of Triton X-100, an extinction of about 0.8–0.9 is obtained at 540 nm after centrifugation. This represents 100% lysis and all subsequent extinction values should be expressed as percent lysis. Pipette in duplicate 0.5 mL samples of the diluted erythrocytes into 10 mL centrifuge tubes containing 4.5 mL of saline. Mix thoroughly and add 50 μL of the test compound, mix by gentle swirling, and place in an incubator at 37°C for 20 min. Separate any unbroken cells by centrifugation on a bench centrifuge and measure the absorbance of the supernatant solution at 540 nm.

13.5.4 Results and Conclusion

First, examine a range of concentrations of the three detergents and prepare a graph of percent hemolysis against concentration of the reagent. Repeat the experiment with lysolecithin, progesterone, and hydrocortisone and observe the relative lytic effect of these compounds, and examine the protective effect of hydrocortisone by incubating the erythrocytes with this steroid before exposing them to the lytic agent.

13.6 Permeability of Model Membrane (Liposome)

13.6.1 Principle

13.6.1.1 Liposomes

Dry films of phospholipid swell spontaneously when in contact with aqueous solutions to form multilayered structures known as liposome. These structures are made up of concentric bimolecular leaflets of phospholipid separated by small aqueous spaces (Figure 13.5).

Sucrose is trapped within the liposomes as they form, and as this molecule cannot usually be passed out across the liposome membrane, it acts as an effective osmotic support. If the liposomes are suspended in an isosmotic solution of ions that cannot penetrate the phospholipid bilayer, then no change in the volume of the liposomes occurs. However, if the phospholipids are permeable to the ions, then swelling occurs as water moves into the liposomes with the ions to maintain osmotic equilibrium. The permeability of a solute can, therefore, be determined by following the light scattering of liposome suspension. When liposomes swell the absorbance decreases and when they contract the absorbance rises.

13.6.1.2 Membrane Permeability

Liposomes are a useful model for studying the permeability of the phospholipid part of membranes to anions. The aim of this experiment is to get the ideas of how anions may cross membranes from the results obtained. A few general points should be borne in mind when discussing the results.

13.6.1.3 Ionophores

Many compounds modify ion transport, and the antibiotic valinomycin is particularly interesting as it acts as a specific carrier of K^+. The molecule has a doughnut-like structure with the K^+ fitting neatly into the hydrophilic center of the ring. The rest of the molecule is hydrophobic and, therefore, readily crosses the phospholipid of the membrane carrying the K^+ with it. Addition of valinomycin can, therefore, alter the

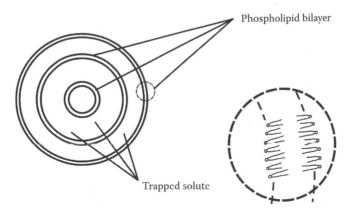

Figure 13.5 Liposomes.

course of liposome swelling in some cases. Molecules like this that facilitate the transport of specific ions across membranes are known as ionophores. 2,4-Dinitrophenol and other uncoupling agents act by discharging the proton gradient across membranes so that they become permeable to H^+.

13.6.2 Materials

Choline phosphoglyceride, dicetyl phosphate, chloroform, sucrose (100 mmol/L), round-bottom flask, rotary evaporator, valinomycin, 2,4-dinitrophenol (5 mol/L in ethanol), nitrogen cylinders, UV–VIS recording spectrophotometer.
Test solutions (50 mmol/L):

NaCl	KCl	NH_4Cl
NaI	KI	NH_4I
NaCNS	KCNS	—
CH_3COONa	CH_3COOK	CH_3COONH_4

13.6.3 Protocol

13.6.3.1 Preparation of Liposomes

Mix 20 μmol of choline phosphoglyceride with 2 μmol of dicetyl phosphate in about 3 mL of chloroform in a 100 mL round-bottom flask (Figure 13.6). The dicetyl phosphate is present to give the liposomes a net negative charge. Remove the chloroform by rotary evaporation and add 3 mL of 100 mmol/L sucrose to the dry lipid film. It is essential to remove all the chloroform before adding the sucrose and this is best checked by any odor in the flask. When the flask appears free of the solvent, flush it out with nitrogen, add the sucrose, and agitate the flask gently by hand to disperse the lipids.

Liposome swelling: The total absorbance change is small, so the condition for the swelling should be optimized first. This is done by adding 100 μL of the liposome suspension to 3 mL of distilled water and following the decrease in extinction at 450 nm in a recording spectrophotometer as the liposomes swell. Make any adjustments necessary to the chart speed, sensitivity, etc., and then proceed with the investigation of anion permeability.

13.6.4 Results and Conclusions

Add 100 μL of the liposomes suspension to a cuvette containing 3 mL of one of the test solutions. Follow any change of absorbance with time. If no swelling occurs, make the liposomes specifically permeable to K+ by adding 1 μg of valinomycin and continue to follow the absorbance at 450 nm and discuss the results obtained

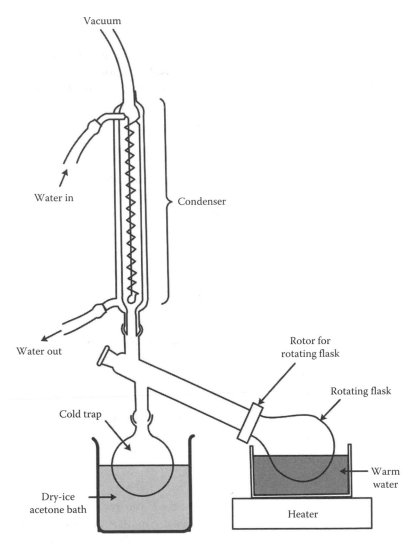

Figure 13.6 Rotary evaporator with round-bottomed flask.

in view of the possible mechanisms by which the various anions may cross phospholipid membranes.

13.7 Effect of Cholesterol on the Anion Permeability of a Phospholipid Membrane

13.7.1 Principle

Cholesterol modifies the packing of phospholipid and affects their mobility in the membrane. Cholesterol, therefore, changes the permeability properties of liposomes.

13.7.2 Materials

Cholesterol, choline phosphoglyceride, dicetyl phosphate, chloroform, sucrose (100 mmol/L), round-bottom flask, rotary evaporator, valinomycin, 2,4-dinitrophenol (5 mol/L in ethanol), nitrogen cylinders, UV–VIS recording spectrophotometer.

13.7.3 Protocol

Prepare another batch of liposomes, but this time, containing 10 μmol of cholesterol as well as 20 μmol of choline phosphoglyceride and 2 μmol of dicetyl phosphate. Select those salts, which were permeable to the liposomes, under the condition of the experiment and repeat the experiment using the cholesterol-containing liposomes.

13.7.4 Results and Conclusion

Discuss and suggest the effect of cholesterol on the permeability of the liposomal membrane in view of the result obtained.

13.8 Effect of Insulin on the Transport into Isolated Fat Cell

13.8.1 Principle

13.8.1.1 Glucose Uptake
Insulin increases the rate of glucose transport across many cell membranes, and this is illustrated in the following experiment by measuring the rate of removal of glucose from the incubation medium by isolated fats cells. In this case, membrane transport is the rate-limiting step in glucose metabolism by the cells, and therefore, there is an increase in the rate of removal of glucose from the medium is taken as indicator for stimulation of glucose transport across the fat cell membrane.

13.8.1.2 Isolated Fat Cells
Isolate fat cells from adipose tissue of the epididymal fat pads from male rats by digestion with collagenase. These enzymes catalyze the hydrolysis of the intercellular matrix and liberate isolated due to their high triglyceride content, float on gentle centrifugation. This procedure effectively isolates the fat cells from other cell types present in the tissue, and a homogeneous population of cells can thus be obtained. These cells are metabolically active and respond to a wide range of physiological and pharmacological stimuli.

13.8.2 Materials

Sodium chloride (0.154 mol/L), potassium chloride (0.154 mol/L), calcium chloride (0.110 mol/L), potassium phosphate (0.154 mol/L), magnesium sulfate (0.154 mol/L), sodium bicarbonate (0.154 mol/L).

Krebs–Ringer bicarbonate (KRB) with reduced calcium concentration: Mix previous solution 1–6 in the following proportion, then gas with 95% O_2/5% CO_2 for 20 min.

1000 mL of solution 1

40 mL of solution 2

15 mL of solution 3

10 mL of solution 4

10 mL of solution 5

210 mL of solution 6

KRB–albumin solution: Prepare a 30% w/v solution of albumin in KRB and, dialyze overnight against KRB in the cold.

The following day dilute to 4% w/v albumin with KRB and gas to pH 7.4 with 95% O_2/5% CO_2. Store at 37°C in a sealed container under an atmosphere of 95% O_2/5% for CO_2.

Incubation medium (KRB–albumin containing 3 mmol/L glucose)

Gas cylinder of 95% O_2/5% CO_2

Male rats (200–300 g)

Plastic centrifuge tubes (10–15 mL)

Collagenase

Insulin stock solution (dissolve 8 mg in 1 mL of mmol/L HCl)

Reagents for the assay of glucose using glucose oxidase described in Section 11.2.11

Shaking water bath at 37°C

Note: Plastic or siliconized glassware must be used throughout for all manipulations of the cells.

13.8.3 Protocol

13.8.3.1 Preparation of Isolated Fats Cells

Kill two male rats and remove the four epididymal fat pads and place in 10 mL of KRB–albumin containing 5 mg of collagenase in a plastic tube. Gas briefly with 95% O_2/5% CO_2, seal the tube, and incubate by shaking for 1 h in a 37°C water bath. Remove any tissue fragments with forceps and centrifuge the cell suspension in a plastic centrifuge tube at 400*g* for 1 min. Penetrate the layer of fat cells on the surface with a Pasteur pipette and remove the cells below (infranatant) and sediment cells by aspiration. Resuspend the fat cells in 10 mL of KBR–albumin and wash by gentle stirring with a plastic rod. Centrifuge at 400g for 1 min and again remove and discard the infranatant. Repeat the washing procedure twice more and finally resuspend the cells in a suitable volume of the incubation medium (KRB–albumin–glucose).

13.8.3.2 Glucose Uptake

First, determine the dose–response curve for the stimulation of glucose uptake by insulin. This is carried out by incubating 1 mL of the fat cell suspension in triplicate with the following final concentration of insulin: 0, 1, 10, 100, and 1000 µU/mL (1 mg = 25 units). The insulin is added in a small volume (5 µL) to the plastic incubation tubes followed by 1 mL aliquots of the stirred cell suspension using a wide-bore 1 mL plastic syringe. Gas with 95% O_2/5% CO_2 for 2 min and incubate in sealed plastic tubes for 2 h at 37°C with gentle shaking.

After incubation, centrifuge the cells and determine the glucose remaining in the incubation medium using the glucose assay described in Section 11.2.11.

13.8.4 Results and Conclusion

Express the results as a mean number of micromoles of glucose taken up per milliliter of suspension per hour ± standard error of the three incubations. Plot a graph of glucose uptake against log 10 insulin concentration to the dose–response curve. Calculate the concentration of insulin giving half of the maximum stimulation of glucose uptake.

13.9 Transport of Amino Acids across the Small Intestine

13.9.1 Principle

13.9.1.1 Transport of Metabolites

After the digestion of food in the gut, the amino acids and other small molecules produced are absorbed from the intestine into the blood stream. In many cases, the absorption is against a concentration gradient and thus involves active transport. The energy needed to drive such process is provided by the generation of ATP, so that metabolic poisons that reduce or block ATP production will have an adverse effect on these transport processes.

The type transport (active or passive) depends very much on the geometry of the molecules, and this is seen in the absorption of the optical isomers of amino acids and sugars.

13.9.1.2 Everted Sac

A convenient method for studying transport in the gut is to use a section of the small intestine, which is turned out and tied at each end. This everted sac is immersed in buffered ionic medium containing the metabolite, and changes in the concentration of the metabolite are measured after incubation. By turning the intestine inside out, transport now from a large volume of the incubation medium into a small volume inside the everted gut. This, therefore, magnifies any absorption that occurs and is a more sensitive preparation than the intestine in the normal way.

13.9.1.3 Histidine Estimation
Histidine is assayed by reading the color produced when the amino acid reacts with diazotized sulfanilic acid.

13.9.2 Materials

Sodium chloride (0.154 mol/L), potassium chloride (0.154 mol/L), magnesium sulfate, (0.154 mol/L), potassium phosphate buffer (0.1 mol/L, pH 7.4), Krebs–Ringer phosphate (KRP) medium with calcium omitted. Mix the previous solutions in the following proportions and gas out with 95% O_2/5% CO_2:

Volume (mL)	Solution
1000	1
40	2
10	3
200	4

13.9.2.1 Incubation Medium (KRP Containing 18 mmol/L Glucose)
Gas cylinder (95% O_2/5% CO_2, rats, glucose–saline (0.154 mol/L NaCl containing 18 mmol/L Glucose), L-histidine (5 mol/L in the KRP medium containing 18 mmol/L glucose).

13.9.2.2 Inhibitor Solutions
Prepare in the incubation medium and gas with 95% O_2/5% CO_2 (*Caution: Poisonous*):

1. Sodium cyanide (2 mmol/L)
2. Sodium iodoacetate (10 mmol/L)
3. Sodium iodoacetate (40 mmol/L)
4. Sodium malonate (5 mmol/L)

Acetic acid for deproteinization (0.35 mmol/L).
Sulfanilic acid (10 g/L in 1 mol/L HCl). If the blank is colored, recrystallize the reagent.
Sodium nitrite (50 g/L, prepare fresh).
Sodium carbonate (75 g/L of the anhydrous salt)
Ethanol (20% v/v), standard L-histidine (0.15 mmol/L), syringes (1 mL capacity with blunted needles), colorimeter, shaking water bath set at 37°C, boiling water bath, hardened filter paper Whatman No. 54, thread for ligatures, glass rods with specially thickened ends (Figure 13.7).

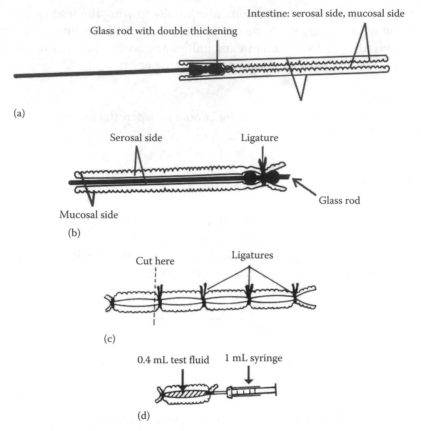

Figure 13.7 Preparation of everted sacs from rat small intestine: (a) glass rod inserted, (b) intestine everted, (c) everted sac tied 2–3 cm lengths, and (d) final taking away of fluid with syringe.

13.9.3 Protocol

13.9.3.1 Preparation of the Everted Sac

Kill the rat and open the abdomen by a midline incision. Remove the small intestine by cutting each end and manually stripping the mesenteries. Wash the entire length of the small intestine with glucose–saline at room temperature to remove blood, debris, etc., and prepare the everted sac. Insert a narrow glass rod with thickening into one end of the intestine (Figure 13.7a). Tie a ligature over the thickened part of the glass rod and evert the sac by gently pushing the rod through the whole length of the intestine (Figure 13.7b). Carefully ligate both ends of the everted sac, remove the rod, and place the intestine in a glucose–saline solution at room temperature. Tie off 2–3 cm lengths of intestine with thread and cut an open sac from the main length (Figure 13.7c). Place a second ligature loosely round the open end of the sac and introduce a blunt needle attached to a 1 mL syringe. Tighten the loose ligature over the needle and inject 0.4 mL of the KRP–glucose solution for experiments measuring absorption rates or 0.4 mL of the amino acid solution for the concentration gradient experiments into the sac (Figure 13.7d); tighten the ligature and withdraw the needle. All ligatures have to be firm enough to prevent leaks, but not too tight so as to damage the tissue.

13.9.3.2 Amino Acid Transport

Immerse the sac in 15 mL of the 5 mmol/L histidine solution, gas briefly with 95% O_2/5% CO_2, seal the flask, and shake for 10 min in a 37°C water bath. At the end of this time, analyze 0.2 mL of the solution inside the everted sac (*serosal side*) and outside the sac (*mucosal side*) for L-histidine after deproteinization.

At the end of the experiment, empty the sac, blot the hardened filter paper (Whatman No. 54), and weigh the tissue after removing the ligatures.

13.9.3.3 Determination of Histidine

Mix 2 mL of the weak acetic acid solution with 0.2 mL of the test solution in a test tube, cover with a marble, and place the tubes in a boiling water bath for 10 min. Cool and add distilled water to give a final volume of 5 mL. Filter or centrifuge as required.

Take 2 mL of the deproteinized solution, add 0.4 mL sulfanilic acid (10 g/L in 1 mol/L HCl), mix thoroughly, and add 0.4 mL of sodium nitrite (50 g/L). Shake the tube and leave to stand for 5 min with occasional shaking. Add 1.2 mL of sodium carbonate solution (75 g/L) and shake vigorously for about 10 s. Add 4 mL of ethanol (20% v/v) and 2 mL of water, mix thoroughly, and read the extinction at 498 nm against distilled water after 30 min. Calculate the concentration of histidine present by reference to a standard curve of histidine concentration using 0–2 mL of 0.15 mmol/L histidine instead of the 2 mL of deproteinized fluid.

13.9.4 Results and Calculation

1. Calculate the concentration gradient, which is the ratio of amino acid concentration in the serosal fluid to that in the mucosal fluid.
2. Express the absorption rate in micromoles of amino acid per gram of wet tissue per hour (μmol/g/h).
3. Calculate the percentage recovery by comparing the amount found in the mucosal and serosal fluids with the amount taken originally.

Further, examine the effect of various inhibitors upon the absorption of L-histidine, and compare the absorption of the L and D forms of histidine and rate of absorption of various sugars using an assay for total carbohydrates.

13.10 Absorption of Xylose from the Gut in Man

13.10.1 Principle

13.10.1.1 Absorption of Xylose

The following test is used, in clinical chemistry, to check for impaired absorption from the upper small intestine. The pentose sugar D-xylose is selected as this compound is not metabolized to any extent. Furthermore, there is no special mechanism for its reabsorption in the kidney, so the sugar is excreted in the urine. The test is carried out by determining the total xylose excreted in the 5 h following an oral dose of 5 g of sugar. During this period, normal persons excrete 23%–50% of the dose, whereas those with steatorrhea due to malabsorption excrete less than 20%.

13.10.1.2 Xylose Estimation
When xylose is heated in an acid solution, furfural is produced, which reacts with p-bromoaniline to give a pink complex. The extinction of this complex is read in a colorimeter.

13.10.2 Materials

13.10.2.1 p-Bromoaniline Reagent
(Dissolve 20 g of p-bromoaniline in glacial acetic acid saturated with 4 g of thiourea then make up to 1 L with acetic acid; shake well and filter.)

Pure D-xylose for human consumption, standard D-xylose for human consumption (0.1 g/L prepared fresh), water bath set at 70°C, UV–VIS spectrophotometer.

13.10.3 Protocol

13.10.3.1 Test
Fast overnight, empty the bladder, and discard the urine, then drink a solution of 5 g of D-xylose dissolved in about 300 mL of water. Collect urine specimens every hour for 5 h and measure the volume and concentration of D-xylose.

13.10.3.2 Estimation of D-Xylose
Dilute the urine 1 to 10 and 1 in 50 and take 1 mL of each dilution and mix with 5 mL of the color reagent in a test tube. Incubate for 10 min in a water bath at 70°C, cool to room temperature, and allow the color to develop in the dark for 70 min. Repeat this using 1 mL of the standard D-xylose solution in place of the diluted urine, also set up "test control" and a "standard control" by adding the 1 mL of diluted urine or 1 mL of standard to the p-bromoaniline solution and read the color immediately.

Carry out all the estimations in duplicate and read the extinctions at 524 nm.

13.10.4 Results and Calculation

The concentration of the D-xylose solution is 0.1 mg/mL; therefore, if V is the volume of urine sample in milliliters and D the dilution, then the amount of xylose in the sample is given by

$$\frac{(E_{test} - E_{test\,control})}{(E_{std} - E_{std\,control})} \times 0.1 \times V \times D \text{ (mg)}$$

Plot histograms of the D-xylose excreted over each hour and calculate the percentage of xylose excreted over the test period.

Suggested Readings

Bangham, A. D., M. M. Standish, and J. C. Watkins. 1965. Diffusion of univalent ions across the lamellae of swollen phospholipids. *J Mol Biol* 13: 238–252.

Bisen, P. S. and G. P. Agarwal. 1984. Respiratory and permeability changes in bean leaves (*Phaseolus vulgaris* L.) infected with *Curvularia lunata* var. *aeria* or with the isolated toxin. *Biochem Physiol Pfl* 179: 13–19.

Bisen, P. S., S. K. Garg, R. P. Tiwari, P. R. Tagore, D. Tiwari, R. Chandra, R. Karnik et al. 2003. Analysis of shotgun expression library of *Mycobacterium tuberculosis* genome for immunodominant polypeptide: Potential use in serodiagnosis. *Clin Diagn Lab Immunol* 6: 1051–1058.

Carafoli, E. and G. Semenza. 1979. *Membrane Biochemistry: A Laboratory Manual on Transport and Bioenergetics*. Berlin, Germany: Springer-Verlag.

Chan, J., T. Fujiwara, P. Brannan, M. McNeil, S. J. Turco, J. C. Sibille, M. Snapper, P. Aisen, and B. R. Bloom. 1989. Microbial glycolipids: Possible virulence factors that scavenge oxygen radicals. *Proc Natl Acad Sci USA* 86: 2453–2457.

Finean, J. B., R. Coleman, and R. H. Michell. 1984. *Membranes and Their Cellular Function*, 3rd edn. Oxford, U.K.: Blackwell Scientific Publication.

Harrison, R. and G. G. Lunt. 1980. *Biological Membranes*, 2nd edn. Glasgow, U.K.: Blackie.

Plummer, D. T. 1988. *An Introduction to Practical Biochemistry*. New York: McGraw-Hill.

Singh, S. K. and P. S. Bisen. 2006. Adjuvanticity of stealth liposomes on the immunogenicity of synthetic gp41 epitope of HIV-1. *Vaccine* 24: 4161–4166.

Tiwari, R. P., D. Tiwari, S. K. Garg, R. Chandra, and P. S. Bisen. 2005. Glycolipids of *Mycobacterium tuberculosis* strain H37Rv are potential serological markers for diagnosis of active tuberculosis. *Clin Diagn Lab Immunol* 12: 465–473.

West, I. C. 1983. *The Biochemistry of Membrane Transport*. London, U.K.: Chapman & Hall.

Work, T. S. and E. Work. 1978. *Laboratory Techniques in Biochemistry and Molecular Biology*. Amsterdam, the Netherlands: Elsevier.

Important Links

http://chemwiki.ucdavis.edu/Analytical_Chemistry/Electrochemistry/Case_Studies/Membrane_Potentials

http://physrev.physiology.org/content/87/4/1175.full

https://biblio.ugent.be/publication/2964832

http://etap.org/demo/biology1/instruction1tutor.html

http://www.ncbi.nlm.nih.gov/books/NBK21570/

14

Photosynthesis and Respiration

14.1 Introduction

14.1.1 Bioenergetics

The process of trapping solar energy (light) and its conversion to chemical energy, which is then used to fix carbon dioxide into organic compounds, is called photosynthesis. The study of production and use of energy by cells is called bioenergetics, including catabolic routes that degrade nutrients and anabolic routes that are involved in cell synthesis. Energy is defined as the ability to do work, and a microbial cell performs a wide and extensive chemical reaction. Organisms require energy for function such as movement, active transport of nutrients into the cell, and the biosynthesis of cell components such as nucleotides, RNA, DNA, proteins, and membranes. In other words, energy is required to derive various biosynthetic chemical reactions and do mechanical work. A simplified view of the cell metabolism is depicted in Figure 14.1.

14.2 Measurement

14.2.1 Principle

Electron flow to oxygen as a result of oxidative phosphorylation can be demonstrated using an oxygen electrode by monitoring the reactions involved in oxygen exchange. Oxygen electrode systems for photosynthesis and respiration measurements take one of two forms, liquid or gas phase. The measurement techniques for each form are quite different; however, the underlying principles of the measurement of oxygen remain the same. Oxygen dissolved in the reaction vessel of liquid-phase systems or that which accumulates in the sample chamber of gas-phase systems is detected polarographically. The electrode compartment is isolated from the reaction chamber by a thin Teflon membrane; the membrane is permeable to molecular oxygen and allows this gas to reach the cathode, where it is electrolytically reduced. The reduction allows a current to flow; this creates a potential difference that is recorded on a flat-bed chart recorder. The trace is thus a measure of the oxygen activity of the reaction mixture by an electrode.

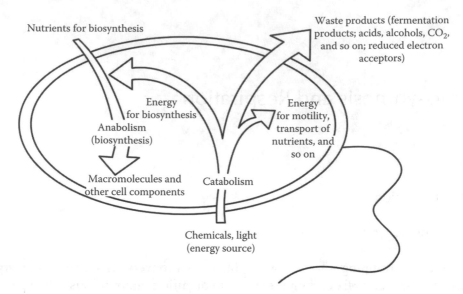

Figure 14.1   Simplified view of bioenergetics.

There are several types of oxygen electrodes differing in design, but the most common is the Clark-type electrode. It utilizes an annular silver reference electrode (anode) and a glass-coated platinum electrode (cathode) connected to a small external voltage source to charge the circuit with a potential difference of 500–700 mV, and then the current generated is proportional to the oxygen concentration in the medium. This is called polarizing voltage. Oxygen gets reduced at the cathode giving rise to a current that is proportional to the concentration of oxygen in the solution provided that the polarizing voltage remains approximately 700 mV. Oxygen from the solution diffuses in through a thin semipermeable membrane that separates the electrodes from the test solution (Figure 14.2). The amplified current is fed to a chart recorder that gives a trace of the change in oxygen concentration with time. Zero oxygen concentration is obtained by adding a crystal of sodium dithionite to the test solution and adjusting the pen to the baseline. The air-saturated buffer is taken to be 100% oxygen and the pen adjusted accordingly. In practice, 100% oxygen is assumed to be 240 µmol of dissolved oxygen per liter, which is the solubility of oxygen in an aqueous solution at 26°C. If 4 mL is present in the reaction vessel, then the total oxygen content is $4 \times 240 = 960$ nmol, when saturated. For this reason, it is best to adjust

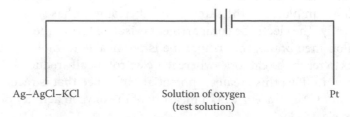

Ag–AgCl–KCl   Solution of oxygen   Pt
                        (test solution)

Figure 14.2   Diagrammatic view of cell formed by oxygen electrode and test solution.

the pen on the recorder to 96 rather than 100. In these circumstances, one division on the chart recorder is then equivalent to 10 nmol of O_2. Both electrodes are immersed in the same solution at saturated KCl.

The oxygen electrode consists of a platinum cathode and silver anode in saturated potassium chloride solution, and when potential is applied across the "cell" formed by these electrodes dipping in the test solution, oxygen is electrolytically reduced (Figure 14.2). Four electrons are generated at the anode, which are then used to reduce a molecule of oxygen at the cathode:

$$\text{Anode} \quad 4Ag + 4Cl^- \longrightarrow 4AgCl + 4e$$

$$\text{Cathode} \quad 4H^+ + 4e + O_2 \longrightarrow 2H_2O$$

$$\overline{4Ag + 4Cl^- 4H^+ + O_2 \longrightarrow 4AgCl + 2H_2O}$$

The current flowing is proportional to the activity of oxygen provided the solution is stirred constantly (stir bar) to minimize the formation of an unstirred layer next to the membrane. A typical field of application is closed chamber respirometry. The evolution of oxygen by illuminated chloroplasts or the utilization of oxygen by respiring organisms and tissues can be readily monitored with the equipment.

14.2.2 Practice

There are a number of electrode designs and the one illustrated in the figure is based on that of Clark type (Figure 14.3). The electrode is separated from the bulk of the solution by a membrane to prevent deposition of materials on the electrode surfaces that would otherwise interfere with the oxygen determination.

The whole unit is mounted on a magnetic stirrer and a magnetic follower (flea) placed inside the reaction vessel. This keeps the test solution constantly stirred so as

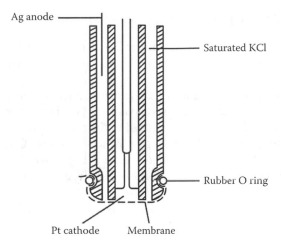

Figure 14.3 Simplified view of Clark oxygen electrode.

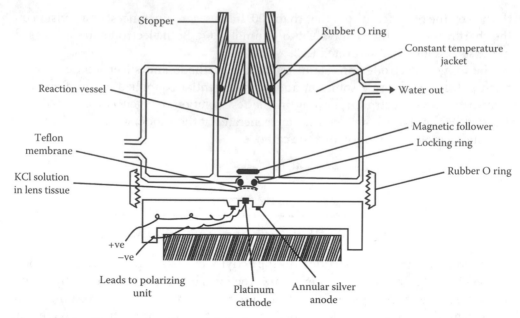

Figure 14.4 Cross-sectional view of Clark-type oxygen electrode.

to obtain correct values of change in dissolved oxygen levels. Since the rate of dissolution of oxygen is temperature dependent, it is important to keep temperature constant. This is achieved by means of circulation of water in a water jacket around the reaction vessel. Both the water jacket and the reaction vessel are made of special type of transparent plastic so that photosynthetic studies requiring the use of light can be carried out. The unit is connected to a control box that takes electric signals from the electrode and conveys it to a chart recorder (Figure 14.4).

14.2.3 Calibration of the Electrode

An oxygen electrode is calibrated in the following manner:

1. Calibration is made with a liquid standard or gas standard that has been equilibrated with water vapor. Either standard must be equilibrated to the measuring temperature.
2. The proper scale is selected, and certain instruments have more than one scale to accommodate wide ranges sometimes encountered during analysis.
3. A standard usually distilled water is added to the reaction vessel; the magnetic follower is introduced and the instrument is switched on for 30 min warm-up time. During this time, the lid of the reaction vessel is kept open. The pen recorder stabilizes at a particular point. This is taken as dissolved oxygen concentration of the standard whose value can be obtained from Table 14.1.
4. If now a few crystals of sodium dithionite ($Na_2S_2O_4$) are added, all the oxygen is used up in accordance with the following reaction:

$$Na_2S_2O_4 + O_2 + H_2O \longrightarrow NaHSO_4 + NaHSO_3$$

TABLE 14.1 Standard Value of Dissolved Oxygen Concentration

Temperature (°C)	O_2 (ppm)	O_2 (μmol/mL)
0	14.16	0.442
5	12.37	0.386
10	10.92	0.341
15	9.76	0.305
20	8.84	0.276
25	8.11	0.253
30	7.52	0.30
35	7.02	0.219

The pen recorder moves a certain distance and stabilizes at a particular point on the chart. This is the zero point of dissolved oxygen concentration. This point is adjusted with the zero calibration control of the recorder.

5. This process is repeated several times till the zero concentration and standard concentration need no further adjustment.

6. The number of divisions between the zero and standard corresponds to the standard oxygen concentration. Hence, one division would represent a certain dissolved oxygen concentration:

$$10 \text{ divisions} = x \text{ μmol/L oxygen}$$

$$1 \text{ division} = 0. \, x \text{ μmol/L oxygen}$$

7. The instrument is thus calibrated and ready to determine unknowns.

14.2.4 Applications

The oxygen electrode is being put to increasingly wide uses because of easy portability of the instrument that is also inexpensive and does not require much expertise in handling.

Chloroplast and mitochondrial studies: Oxygen evolution in cyanobacteria, algae, and chloroplasts, that is, those containing photosystem II (PS II), can be studied using a suitably illuminated Clark oxygen electrode. The oxygen content of the suspension medium is normally reduced below 100% oxygen by bubbling nitrogen through it so that oxygen produced stays in solution and is recorded. In marine biology or limnology, oxygen measurements are usually done in order to measure respiration of a community or an organism, but have also been used to measure primary production of algae. There are however commercially available oxygen sensors that measure the oxygen concentration in liquids with great accuracy. There are two types of oxygen sensors available: electrodes (electrochemical sensors) and optodes (optical sensors). Respiratory control studies and the effect of various inhibitors on mitochondrial respiration can also be studied using the oxygen electrode. Inhibitors of respiration slow down the rate of respiration.

Enzyme assays: Activity of enzymes that require oxygen for reaction, for example, glucose oxidase, catalase, etc., can be studied using the Clark-type oxygen electrode.

Microorganism studies: Bacteria that use oxygen as a terminal electron acceptor can be studied using an oxygen electrode and the effect of electron inhibitors determined. The efficiency of different carbohydrates as respiratory substrates can be studied by comparing the rate of oxygen uptake. In soil respiration studies, oxygen sensors can be used in conjunction with carbon dioxide sensors to help improve the characterization of soil respiration. Typically, soil oxygen sensors use a galvanic cell to produce a current flow that is proportional to the oxygen concentration being measured. These sensors are buried at various depths to monitor oxygen depletion over time, which is then used to predict soil respiration rates. Generally, these soil sensors are equipped with a built-in heater to prevent condensation from forming on the permeable membrane, as relative humidity can reach 100% in soil.

14.3 Photosynthesis

Photosynthesis is the process by which light energy is used to synthesize carbohydrate in plants, algae, and cyanobacteria from CO_2 and H_2O. The synthesis of carbohydrate, therefore, can go on only in the light, and it has two basic aspects: (1) the conversion of light energy into chemical energy and (2) the conversion of CO_2 into organic compounds, which is called fixation of CO_2 also known as light and dark reaction (Figure 14.5).

14.3.1 Light Reaction

The light reaction only takes place in the presence of visible radiant energy that is absorbed by the green pigment chlorophyll present in chloroplasts. The light reaction consists essentially of the removal of electrons from water (photolysis), and these are then used to reduce nicotinamide adenine dinucleotide phosphate (NADP)$^+$ and generate adenosine triphosphate (ATP). There are two light-driven reactions that take place at the reaction centers of photosystem I (PS I) and PS II and operate in series. They consist of electron transport chains connected in the classical zigzag or Z scheme (Figure 14.6).

14.3.2 Dark Reaction

The second stage of photosynthesis does not require the presence of light and is, therefore, referred to as the dark reaction. This stage involves the utilization of NADPH and ATP, generated by the light reaction, to fix carbon dioxide. The complex

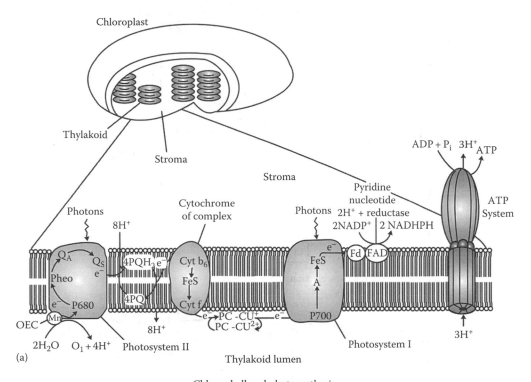

(a)

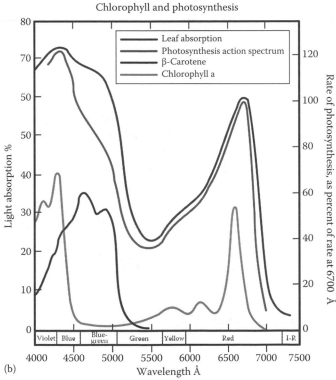

(b)

Figure 14.5 (a) The mechanism of photosynthesis and (b) light absorption spectra of a leaf.

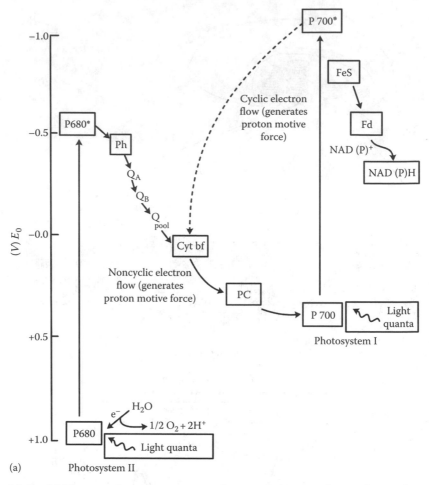

Figure 14.6 (a) Electron flow in oxygenic eukaryotic photosynthesis, the z scheme. Two photosynthesis (PS) are involved: PS I and PS II. Ph, pheophytin; Q, quinine; Cyt, cytochrome; PC, plastocyanin; FeS, nonheme iron sulfur protein; Fd, ferredoxin; P680 and P700 are the reaction center of chlorophyll PS II and PS I, respectively.

metabolic sequence whereby CO_2 is effectively converted to sugars was elucidated by Calvin and his associates and is sometimes known as the Calvin cycle (Figure 14.7):

$$6CO_2 + 12NADPH + 18ATP + 12H^+ + 12H_2O \longrightarrow$$
$$C_6H_{12}O_6 + 18ADP + 18Pi + 12NADP^+$$

14.4 Isolation of Chloroplasts from Spinach Leaves

14.4.1 Principle

The leaves are placed under strong illumination before homogenization as this reduces the starch grain content of the leaves and hence damage to the chloroplasts

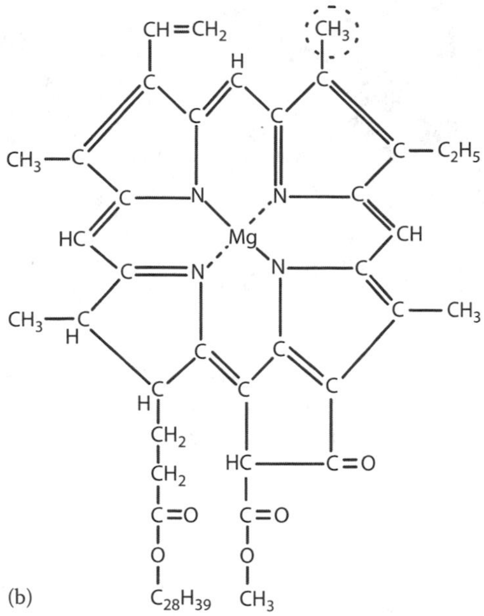

Figure 14.6 (continued) (b) Structure of chlorophyll a. Chlorophyll b differs in having a –CHO group in place of –CH$_3$ group enclosed by the dotted circle.

during isolation. The leaves are then homogenized in buffered hypertonic saline and the chloroplasts isolated by centrifugation.

14.4.2 Materials

Fresh spinach leaves, chopping board and knife, chilled mortar and pestle, cheese-cloth, refrigerated preparative centrifuge, hemocytometer, and microscope.

Grinding solution: 0.33 M sorbitol, 10 mM sodium pyrophosphate (Na$_4$P$_2$O$_7$), 4 mM MgCl$_2$, 2 mM ascorbic acid, adjust pH to 6.5 with HCl.

Suspension solution: 0.33 M sorbitol, 2 mM EDTA, 1 mM MgCl$_2$, 50 mM HEPES, adjust pH to 7.6 with NaOH.

14.4.3 Protocol

1. Prepare an ice bath and precool all glassware to be used.
2. Select several fresh spinach leaves and remove the large veins by tearing them loose from the leaves. Weigh out 4.0 g of deveined leaf tissue.
3. Chop the tissue as fine as possible. Add the tissue to an ice-cold mortar containing 15 mL of grinding solution and grind to a fine paste.
4. Filter the solution through double-layered cheesecloth into a beaker, and squeeze the tissue pulp to recover all of the suspension.
5. Transfer the green suspension to a cold 50 mL centrifuge tube, and centrifuge at 400 g for 1 min at 4°C to pellet the unbroken cells and fragments.

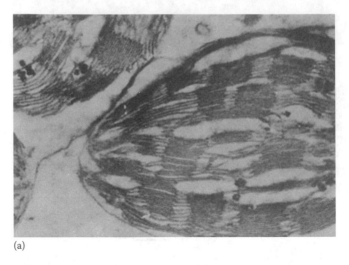

(a)

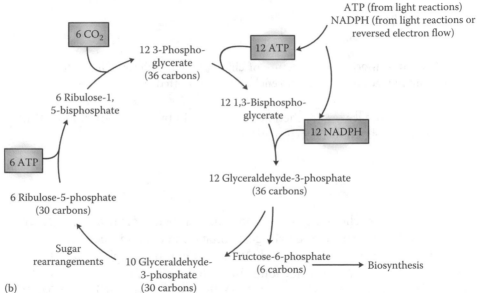

(b)

Figure 14.7   (a) An electron micrograph of a cross section of a chloroplast showing stack of lamellae and stroma. (b) Calvin cycle (or reductive pentose phosphate cycle). For every six molecules of CO_2 incorporated, one fructose-6-phosphate is produced.

6. Decant the supernatant into a clean centrifuge tube, and recentrifuge at 1000 g for 7 min. The pellet formed during this centrifugation contains chloroplasts. Decant and discard the supernatant.
7. Resuspend the chloroplast pellet in 5.0 mL of cold suspension solution or 0.035 M NaCl. Use a glass stirring rod to gently disrupt the packed pellet. This is the chloroplast suspension for use in subsequent procedures.
8. Enclose the tube in aluminum foil and place it in an ice bucket.

14.4.4 Results and Observations

Determine the number of chloroplasts/mL of suspension media using a hemocytometer.

14.5 Determination of Chlorophyll Concentration in the Chloroplast Suspension

14.5.1 Principle

Pigments absorb light to different extents depending on the exposure of light of varying wavelength of the light spectrum. The ideal wavelength to select is the peak of the pigment's absorption spectrum for quantification of a pigment. For example, for chlorophyll a, 663 nm is generally used, but for chlorophyll b, 645 nm is a more appropriate wavelength to choose. The greater the concentration of a pigment in solution, the larger the proportion of light absorbed by the sample at that wavelength. The quantitative relationship is expressed as per Beer–Lambert law $A = Ecd$

14.5.2 Protocol

1. Take 1 mL aliquot of the chloroplast suspension into a 10 mL graduated cylinder, and dilute to 10 mL with 80% acetone. Cover the cylinder with parafilm and mix by inverting.
2. Prepare the spectrophotometer to read the absorbance of the diluted chlorophyll extract.
3. Adjust the wavelength to read 652 nm (why is this wavelength chosen?). Without a cuvette in the machine, adjust to 0% transmittance (left-hand knob).
4. Blank the spectrophotometer with the reagent blank (80% acetone) to read 0 absorbance (right-hand knob). Transfer some of diluted chlorophyll extract to a cuvette and read the absorbance. Record the absorbance value.
5. Chlorophyll absorbance $(A_{652}) = OD$
6. Calculate the chlorophyll content in the diluted sample using the following equation. Record the chlorophyll concentration of the diluted sample.
7. $A = ECd$ where A is the observed absorbance, E is a proportionality constant (extinction coefficient) (=36 mL/cm), C is the chlorophyll concentration (mg/mL), and d is the distance of the light path (=1 cm).
8. Calculate the chlorophyll concentration in the original chloroplast suspension (undiluted) by adjusting for the dilution factor. In order to determine the concentration of chlorophyll in the original suspension, you must multiply the chlorophyll concentration in the diluted sample by the dilution factor.
9. Knowing the chlorophyll content of the undiluted chloroplasts, prepare 10 mL of chloroplast suspension containing approximately 0.02 mg/mL chlorophyll by diluting an appropriate aliquot of original chloroplast suspension with cold 0.5 M sucrose. *Keep this on ice.*

14.6 Evolution of Oxygen by Isolated Chloroplast Using Hill Oxidants

14.6.1 Principle

Unless special precautions are undertaken when chloroplasts are isolated, the stroma and its enzymes are lost. The thylakoid membranes, however, are still capable of electron transport and photosynthetic phosphorylation if an appropriate electron acceptor and substrates for phosphorylation are provided. Electron flows from PS II toward PS I or from ferredoxin toward NADP$^+$ can be intercepted by artificial electron acceptors. This process is known as the Hill reaction and can be used to measure the photochemical activity of PS I and PS I acting in series or of PS II alone. The dye 2,6-dichlorophenolindophenol (DCPIP) can be used as a substitute for the naturally occurring terminal electron acceptor. DCPIP is blue when oxidized (quinone form) but becomes colorless when reduced to a phenolic compound. The photochemical activity of PS I can be measured if the PS II activity is blocked by the powerful herbicide dichlorodimethylurea (DCMU) and the electron flow provided by an artificial electron donor.

1. Ferricyanide

$$2H_2O + 4Fe(CN)_6^{3-} \longrightarrow 4H^+ + 4Fe(CN)_6^{4-} + O_2$$

2. 2,6-DCPIP

2,6-Dichlorophenolindophenol

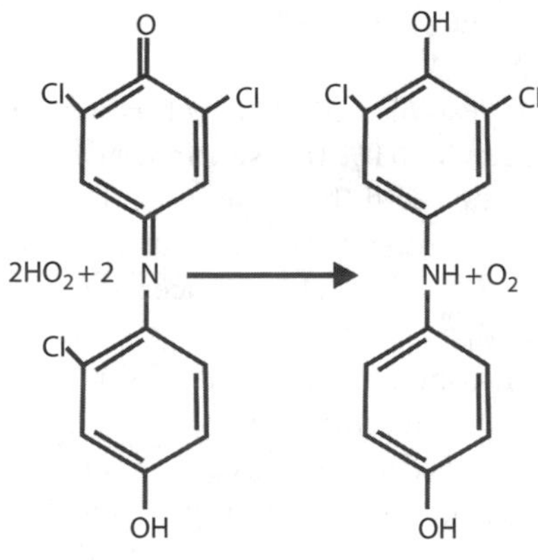

3. Herbicide (DCMU) 3-(3,4-dichloro)-1,1-dimethylurea

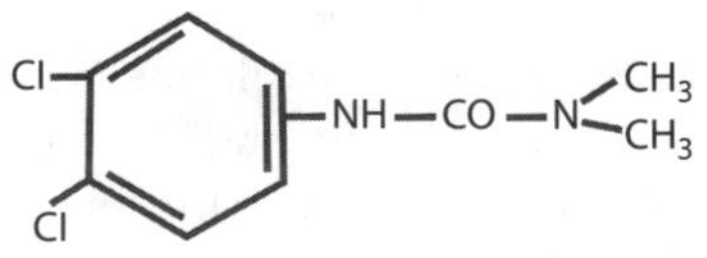

Herbicide (DCMU)

14.6.2 Materials

Materials for the isolation of chloroplasts, potassium ferricyanide (10 mmol/L, prepared fresh), DCPIP (1 mmol/L, prepared fresh), DCMU (250 mol/L), sodium dithionite (solid), oxygen electrode.

14.6.3 Protocol

14.6.3.1 Oxygen Electrode Assays

Set up and calibrate the oxygen electrode in the usual way, and after calibration, move the pen to the center of the chart paper with the zero control as oxygen evolution is being measured.

Start each experiment by adding 2 mL of double strength assay medium to the electrode and sufficient chloroplasts to give an apple-green suspension and a good rate of oxygen evolution in the experiment. This amount of chloroplasts should then be maintained throughout the investigation. Add sufficient distilled water and additions to give a final volume of 4 mL of single strength assay medium.

The assay should be set up in the dark by surrounding the electrode with aluminum foil; the rate of oxygen evolution is monitored in the light following the addition of cofactors, inhibitors, etc. The chloroplast suspension can be conveniently illuminated by a 100 W bulb "focused" roughly through a large round-bottomed flask filled with water.

14.6.3.2 Hill Oxidants

Test the effectiveness of ferricyanide and DCPIP as Hill oxidants. Add increasing amounts of the oxidants recording both the rate and the total amount of O_2 evolved at each concentration. Plot the results as they are recorded. Finally, add sufficient DCPIP to ensure sustained O_2 evolution and test the effect of 50 μL of DCMU.

14.6.4 Results and Conclusion

Observe the rate of oxygen evolution and role of inhibitors on photosynthesis.

14.7 Methyl Viologen as a Terminal Electron Acceptor

14.7.1 Principle

Methyl viologen (MV) reacts as an electron acceptor in redox and radical reactions. As an herbicide, MV acts by inhibiting photosynthesis. In light-exposed plants, it accepts electrons from PS I (more specifically Fd, which is presented with electrons from PS I) and transfers them to molecular oxygen. In this manner, destructive reactive oxygen species are produced. In forming these reactive oxygen species, the oxidized form of MV is regenerated and is again available to shunt electrons from PS I to start the cycle again:

$$H_2O + MV_{oxd} \longrightarrow MV_{red} + \frac{1}{2}O_2$$

$$O_2 + MV_{red} \longrightarrow MV_{oxd} + H_2O_2$$

$$\overline{}$$

$$H_2O + O_2 \longrightarrow H_2O_2 + \frac{1}{2}O_2$$

Therefore, 0.5 mol O_2 is taken up per pair of electrons transferred.

14.7.2 Materials

Materials for the isolation of chloroplasts: potassium ferricyanide (10 mmol/L, prepare fresh), DCPIP (1 mmol/L, prepare fresh), DCMU (250 mol/L), sodium dithionite (solid), oxygen electrode, MV (10 mmol/L, prepared fresh), sodium azide (20 mmol/L), sodium ascorbate (50 mmol/L, pH 7.6 prepare fresh).

14.7.3 Protocol

14.7.3.1 Oxygen Uptake with Methyl Viologen as Electron Acceptor
Measure the rates of oxygen uptake in the presence of MV with, and without, adding sodium azide (0.5 mmol/L) to block catalase activity. Check the effects of varying the amount of MV.

14.7.3.2 Activity of PSI Measured with Methyl Viologen
Measure the rate of oxygen evolution of the sample in the presence of DCPIP, and let the reaction continue until the DCPIP is used up. Add DCMU to block PS II activity and carefully note what happens, and then add sufficient sodium ascorbate to reduce the DCPIP so that the blue color disappears. Measure the rates of oxygen uptake, add MV, and compare the two rates of oxygen uptake. Repeat the experiment but this time omit the second DCPIP addition.

14.7.4 Results and Conclusion

Record and interpret the results.

14.8 Spectrophotometric Assay of the Hill Reaction and the Estimation of Chlorophyll

14.8.1 Principle

Hill reaction: DCPIP loses its blue color on reduction and this can be followed in a spectrophotometer at 600 nm.

Assay of chlorophyll: The chlorophyll is extracted from the chloroplasts by shaking with acetone, and the absorption of the solution is plotted in the visible region of the spectrum.

14.8.2 Materials

Fresh spinach leaves, isolation medium (0.3 mol/L NaCl, 3 mmol/L $MgCl_2$ 0.2 mol/L tricine, pH 7.6), double strength assay medium (0.2 mol/L sorbitol, 6 mmol/L $MgCl_2$, 0.4 mol/L tricine, pH 7.6), acetone (80% v/v), Waring blender and muslin, centrifuge in the cold room, UV–Vis recording spectrophotometer.

14.8.3 Protocol

14.8.3.1 Assay of Hill Reaction

Instead of the oxygen electrode, prepare the sample in a spectrophotometer cuvette and measure the rate of DCPIP reduction by exposing the sample to successive 15 and 30 s of illumination. Adjust the chlorophyll and DCPIP concentration, to yield a rate of reduction that can be conveniently measured.

Taking into account possible variations in the concentrations of the various components, how does the rate measured compare with the rate of oxygen evolution observed with DCPIP in the oxygen electrode?

14.8.3.2 Assay of Chlorophyll Content

Add a suitable volume of chlorophyll suspension (0.2 mL) to 10 mL of the solvent, shake thoroughly, and filter through Whatman No 1 filter paper into a 25 mL volumetric flask. Rinse out the test tube with a further 5 mL of the aqueous acetone and use this to wash the filter paper. Repeat the washing and make up to 25 mL with the solvent. Determine the extinction of the green solution against a solvent blank:

$$\text{Chlorophyll (mg/mL)} = \text{Extinction at 625 nm} \times 5.8$$

14.8.3.3 Oxidation and Reduction in Living Cells

Living organisms require a continuous supply of energy to maintain varied functions charactcristic of the living matter (Figure 14.8). In most cases, this energy is obtained by the oxidation of metabolites from the digestion of food. There are three ways that

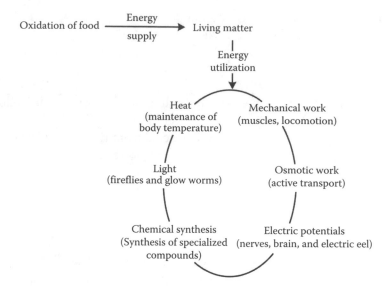

Figure 14.8 Energy cycle.

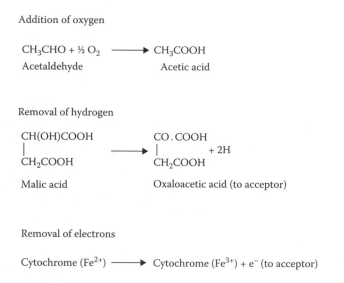

Figure 14.9 Illustrating biological oxidation.

such oxidation can take place, although they are all essentially involved in the loss of electrons from the compound being oxidized (Figure 14.9).

The chemical energy in food that is released during oxidation comes ultimately from the light energy of the sun during photosynthesis. The overall reaction in this process is one of reduction where electrons are added to the compound being formed. Oxidation and reduction are always linked together rather than the formation of an acid and its conjugate base. It is, therefore, more accurate to discuss a redox reaction.

$$AH \rightleftharpoons A^- + H^+$$

Acid   Conjugated   Proton
base

$$AH \rightleftharpoons B^+ + e^-$$

Oxidant reductant
electron

14.8.3.4 Electron Transport Chain

Nearly all biological oxidations take place by the removal of hydrogen from the substrate. Hydrogen atoms thus removed are then passed on to an acceptor, usually one of the pyridine nucleotides nicotinamide adenine dinucleotide (NAD)$^+$ or NADP$^+$. In the aerobic cell, the reduction of NAD is followed by a series of electron transfers with the eventual formation of water and the generation of three molecules of ATP. This sequence of redox reactions are known as the electron transport chain or the respiration chain and like most other oxidative processes takes place in mitochondria:

1. Adjust the wavelength of the spectrophotometer to 600 nm, and then blank the machine with 0.1 M phosphate buffer.
2. Label six cuvettes and add the appropriate volumes of phosphate buffer, DCPIP, DCMU, and 0.02 mg/mL chloroplast suspension as outlined in Table 14.1. Be sure to add the chloroplast suspension last.
3. Once all solutions have been added to each tube, invert to mix and quickly measure the absorbance of each at time 0 and record this in Table 14.2.
4. Place tubes 1, 2, 3, and 5 in a test tube rack 30 cm from a 100 W light source. Place tube 4 in a rack 60 cm from the same light source. Wrap tube six in aluminum foil, leaving only the top open for access when reading absorbance, and place in the rack with tube 4.
5. Measure the absorbance of each tube at 5 min intervals for 30 min. Record results in Table 14.2.
6. To determine the change in absorbance over time in each tube, subtract the interval absorbance from the initial absorbance and record these results in Table 14.3.

TABLE 14.2 Reagent Volumes Used in Hill Reaction Investigation

Tube #	0.1 M Phosphate Buffer (mL)	0.05 mM DCPIP (mL)	0.05 mM DCMU (mL)	Chloroplast Suspension (mL)
1	4	4	0	0
2	4	4	0	0.5
3	4	4	0	1
4	4	4	0	1
5	0	4	4	1
6	4	4	0	1

TABLE 14.3   Absorbance (A_{600}) for the Treatments Investigating the Effect of Chlorophyll Concentration, Light Intensity, and DCMU on Hill Reaction

Tube #	0 min	5 min	10 min	15 min	20 min	25 min	30 min
1							
2							
3							
4							
5							
6							

TABLE 14.4   Change in Absorbance (A_{600}) for the Treatments Investigating the Effect of Chlorophyll Concentration, Light Intensity, and DCMU on the Hill Reaction

Tube #	0 min	0–5 min	0–10 min	0–15 min	0–20 min	0–25 min	0–30 min
1	0						
2	0						
3	0						
4	0						
5	0						
6	0						

14.8.4 Results and Interpretation

Record data as per Tables 14.2 through 14.4 and give interpretation.

14.9 Respiration of Mitochondria and Oxidative Phosphorylation

14.9.1 Principle

When mitochondria are carefully isolated and suspended in an isotonic medium in the presence of substrate, only a slow rate of respiration is observed. On adding ADP, the respiration rate increases until all the ADP is phosphorylated; when the rate of respiration returns to the original slow rate, the quantity of ADP that is added is known, and the amount of oxygen consumed is measured (X), so it is possible to arrive at a P/O ratio for the particular substrate used.

A typical trace of oxygen consumption with time is shown in Figure 14.11 where A is the rate of respiration in the presence of substrate and B is the rate of phosphorylating respiration. The ratio of these rates is known as the respiratory control ratio and is a measure of the degree of coupling of respiration and phosphorylation. A low ratio indicates loose coupling, but freshly prepared mitochondria should show a high respiratory control of 4 or more.

Calculations

Respiratory control ratio = Rate b/Rate A

P/O ratio = μmol ADP added/μatoms O_2 utilized (X)

The phosphorylation sites in the electron transport chain are shown in Figures 14.10 and 14.11, so that substrates using $NADH_2$ as coenzyme should give a P/O ratio of 3, while succinate, which bypasses the first phosphorylation site, should give a ratio of 2. The mixture of ascorbate and tetramethylphenylenediamine (TMPD) uses cytochrome c as electron acceptor and thereby misses two of the phosphorylation sites so that the P/O ratio for this substrate is 1.

Even in freshly prepared mitochondria, the P/O ratios actually found are always less than the whole number values given previously due to partial uncoupling and the action of membrane ATPases.

14.9.2 Materials

Isolation of rat liver mitochondria:

Isolation medium: (0.3 mol/L sucrose; 2.5 mmol/L Tris–HCl, pH 7.4; 0.5 mmol/L EDTA).

Incubation medium: Sucrose (150 mmol/L), potassium chloride (20 mmol/L), magnesium chloride (20 mmol/L), potassium phosphate (1 mmol/L, pH 7.4).

Solutions: Sodium malate (200 mmol/L, pH 7.4), sodium succinate (200 mmol/L, pH 7.4), ascorbate–TMPD (200 mmol/L ascorbic acid, pH 7.4 containing 5 mmol/L tetramethylphenylenediamine [TMPD] prepared fresh and stored in the dark), ADP (20 mmol/L, pH 7.4, prepared fresh).

14.9.3 Protocol

By using the oxygen electrode, make any addition to the mitochondrial suspension in small volumes only (20–100 μL) so that the total volume in the electrode compartment remains close to 4 mL.

Prepare a fresh suspension of rat liver mitochondria and store on ice until required. Set up and calibrate the oxygen electrode as described earlier, and add a small volume of the mitochondrial suspension (50–100 μL) followed by succinate and ADP in that order. Calculate the P/O ratio and the respiration control index; repeat the experiment using malate and ascorbate/TMPD as the substrate.

14.10 Effect of Inhibitors on the Respiratory Chain

14.10.1 Principle

Many compounds are known to inhibit electron flow in the respiratory chain and the sites at which they act can be identified by examining the effect of these inhibitors on

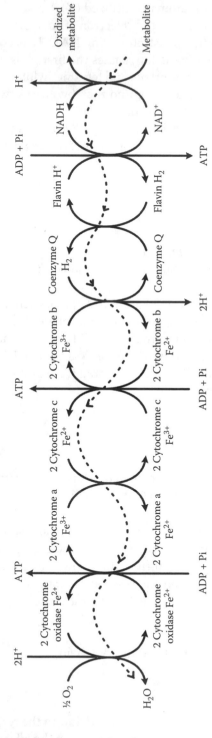

Figure 14.10 Electron transport chain. The dashed line indicates the flow of electrons.

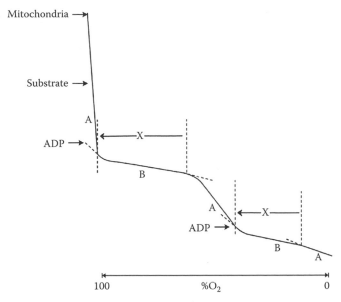

Figure 14.11 A typical curve of oxygen uptake by mitochondria during substrate-induced and phosphorylating respiration.

respiration induced by the substrates. The three inhibitors examined are rotenone, antimycin A, and cyanide. Rotenone blocks electron flow between the nicotinamide nucleotides and flavoproteins, antimycin A acts at the cytochrome b/cytochrome c step, and cyanide inhibits cytochrome oxidase at the end of the respiratory chain (Figure 14.12).

All these inhibitors are highly toxic and must be treated with extreme care.

14.10.2 Materials

Isolation medium: (0.3 mol/L sucrose; 2.5 mmol/L Tris–HCl, pH 7.4; 0.5 mmol/L EDTA).

Incubation medium: Sucrose (150 mmol/L), potassium chloride (20 mmol/L), magnesium chloride (20 mmol/L), potassium phosphate (1 mmol/L, pH 7.4).

Inhibitor solutions: Rotenone (1 mmol/L in 95% v/v ethanol), antimycin A (0.1 mg/mL in 95% v/v ethanol), potassium cyanide (10 mmol/L).

14.10.3 Protocol

Repeat the previous experiment and examine the effect on the oxygen uptake of adding a small volume (10–50 μL) of one of the inhibitor solutions. Take each of the three substrates in turn and record the changes in mitochondrial respiration induced by each of these inhibitors.

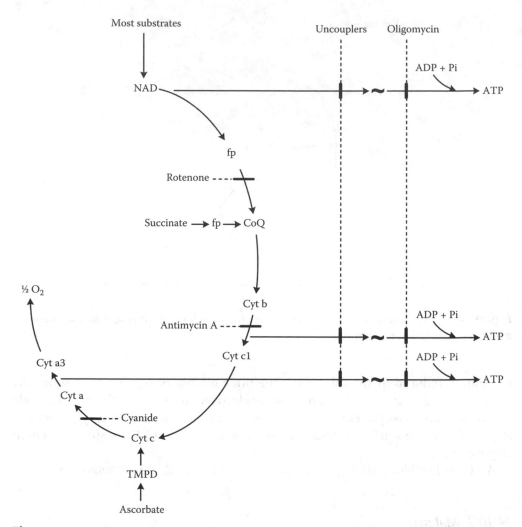

Figure 14.12 The site of action of inhibitors and uncoupler of the respiratory chains.

14.10.4 Results and Conclusion

Record the results on site of action of different inhibitors.

14.11 Compounds That Affect the High-Energy State of Mitochondria

14.11.1 Principle

The rate of respiration in undamaged mitochondria is well below the maximum possible, but in the presence of uncouplers, this restriction is removed, and electron flow is stimulated. The electron transport chain sets up a proton gradient across the mitochondrial inner membrane, and it is the discharge of this gradient via a reversed

ATPase that generates the formation of ATP. 2,4-DNP discharges the proton gradient across the membrane so that the two processes become uncoupled. Respiration continues, but ATP is no longer produced. The practical effect of an uncoupling agent is to cause an increase in the rate of respiration and loss of respiratory control.

Oligomycin does not discharge the proton gradient but inhibits the mitochondrial ATPase so that ATP cannot be formed. Oligomycin, therefore, inhibits the electron flow indirectly by preventing the discharge of the proton gradient by the ATPase. As expected, this inhibition can be released by 2,4-DNP.

14.11.2 Materials

Materials for the isolation of rat liver mitochondria:
Isolation medium: (0.3 mol/L sucrose; 2.5 mmol/L Tris–HCl, pH 7.4; 0.5 mmol/L EDTA).
Incubation medium: Sucrose (150 mmol/L), potassium chloride (20 mmol/L), magnesium chloride (20 mmol/L), potassium phosphate (1 mmol/L, pH 7.4).
Inhibitor solutions: 2,4-Dinitrophenol (5 mmol/L in ethanol), sodium salicylate (0.2 mol/L), oligomycin (0.5 mg/mL).

14.11.3 Protocol

Measure the respiration of freshly prepared mitochondria from rat liver, and then examine the effect on the respiration of adding 20–50 µL of each of the previous solutions to the reaction mixture.

14.11.4 Results and Conclusion

Record the data of the effect of these compounds on substrate-induced and phosphorylating respiration. Determine the alteration of respiratory control.

Suggested Reading

Bisen, P. S. and S. Shanthy. 1993a. Characterization of a DCMU resistant mutant of the filamentous, diazotrophic cyanobacterium *Anabaena doliolum*. *J Plant Physiol* 142: 557–563.
Bisen, P. S. and S. Shanthy. 1993b. Regulation of Hill reaction coupled to photoreduction of nitrate and nitrite in *Anabaena doliolum*. *Curr Sci* 65: 85–86.
Hall, D. O. and K. K. Rao. 1981. *Photosynthesis*, 3rd edn. London, U.K.: Edward Arnold.
Halliwell, B. 1984. *Chloroplast Metabolism*. Oxford, U.K.: Clarendon Press.
Nicholls, D. G. 1982. *Bioenergetics—An Introduction to the Chemiosmotic Theory*. London, U.K.: Academic Press.
Plummer, D. T. 1988. *An Introduction to Practical Biochemistry*. New York: McGraw-Hill.

Sharma, S. K. and P. S. Bisen. 1992. Hg^{2+} and Cd^{2+} induced inhibition of light induced proton efflux in the cyanobacterium *Anabaena flos-aquae*. *Biometals* 5: 163–167.

Sharma, S. K., D. P. Singh, H. D. Shukla, A. Ahmad, and P. S. Bisen. 2001. Influence of sodium ion on heavy metal-induced inhibition of light-regulated proton efflux and active carbon uptake in the cyanobacterium *Anabaena flos-aquae*. *World J Microbiol Biotechnol* 17: 707–711.

Singh, D. P., R. Gothalwal, and P. S. Bisen. 1989a. Toxicity of sodium diethyl-dithiocarbamate (NaDDC) on photoautotrophic growth of *Anacystis nidulans* IU 625. *J Basic Microbiol* 29: 685–694.

Singh, D. P., R. Gothalwal, and P. S. Bisen. 1992. Action of sodium diethyldithiocarbamate (NaDDC) on the Hill activity and chlorophyll fluorescence in *Anacystis nidulans* IU 625. *J Basic Microbiol* 32: 119–128.

Singh, D. P., P. Khare, and P. S. Bisen. 1989b. Effect of Ni^{2+}, Hg^{2+} and Cu^{2+} on growth, oxygen evolution and photosynthetic electron transport in *Cylindrospermum* IU 942. *J Plant Physiol* 134: 406–412.

Singh, D. P., S. K. Sharma, and P. S. Bisen. 1993. Differential action of Hg^{2+} and Cd^{2+} on the photosynthetic apparatus of *Anabaena flos-aquae*. *Biometals* 6: 125–132.

Tzagaloff, A. 1982. *Mitochondria*. New York: Plenum Press.

Important Links

http://bioenergy.asu.edu/photosyn/photoweb/education.html
http://www.helium.com/items/2115700-relationship-between-photosynthesis-and-cellular-respiration
http://www.windows2universe.org/earth/Life/photosynthesis.html
http://www.dnr.sc.gov/ael/personals/pjpb/lecture/lecture.html
http://www.mansfield.ohio-state.edu/~sabedon/campbl10.htm
http://serendip.brynmawr.edu/exchange/bioactivities/cellrespiration
http://scienceline.ucsb.edu/getkey.php?key=153
http://baskt.tripod.com/id9.html
http://sepuplhs.org/pdfs/Shuffle_Teacher_Pages.pdf

15
Microbiology

15.1 Introduction

Microbiology has emerged as the key to biological science. Microorganisms provide the models used in molecular biology for research. This research at the molecular level has provided, and continues to provide, the answers to numerous fundamental questions in genetics, metabolism, and cell forms and functions. Microorganisms are found in almost every habitat present in nature. These are microscopic organisms that are very diverse and include bacteria, fungi, archaea, green algae, and also animals such as plankton and planarian. The true masters of life on earth are not humans but microbes. Microorganisms are the most versatile and adaptable forms of life on Earth as they have existed here for some 3.5 billion years. Microorganisms also provide model systems for studying the relationships between species in mixed populations. There is growing recognition for the potential of microorganisms in many applied areas. The ability of microorganisms to decompose materials such as herbicides, pesticides, and oils in oil spills; the potential of microorganisms as food supplements; the exploitation of microbial activity to produce energy such as methane gas for rural consumption; and the potential of new therapeutic substances produced by microorganisms. Several other uses of microorganisms are becoming increasingly attractive such as recombinant DNA technology, commonly referred to as genetic engineering, is one of the principal thrusts of the emerging high technologies in the biological sciences. Recombinant DNA technology makes it feasible to consider genetically manipulated (engineered) microorganisms for commercial production of new and valuable products for a variety of purposes viz. medicinal, fuel, and food.

15.2 Good Microbiological Practice

15.2.1 Laboratory Safety

During an experiment, it is important to observe good microbiological practice. Fire is a major hazard in microbiology laboratories. Lighted Bunsen burners must never be left unattended, and long hair must be restrained. This avoids both fire risk and the risk of accidental contamination by or of cultures. Bacteriological staining often employs flammable solvents such as acetone or alcohol. Great care must be employed when handling these, particularly near Bunsen burners. Clothing should be protected at all times by a clean laboratory coat, which must be worn properly fastened. An open laboratory coat protects only the wearer's back. Laboratory coat should be preferably white so that contamination can be easily spotted. Laboratory coats should not be allowed to leave the laboratory without being sterilized. They should even be sterilized before laundering. Many microbiological experiments involve the use of potentially dangerous organisms. These only become dangerous when handled incorrectly or carelessly. However, all cultures should be regarded as potentially dangerous.

If accidental spills of culture do occur, any liquid should be quickly and carefully prevented from spreading by the application of absorbent paper towels or cotton wool. This contaminated material should be then placed in a suitable receptacle for later sterilization. The affected work surface should be thoroughly cleaned using a suitable disinfectant, active against the microbe being manipulated. When spills occur, it is important that thorough hand washing is practiced immediately after the spilled material is cleared up and before work is resumed.

Mouth is an important route of infection. Eating, drinking, and smoking are forbidden in microbiology laboratories. Fingers, pens, etc., should never be placed in the mouth, and lipsticks should not be applied in laboratories, to minimize the risk of orally acquired infection or chemical poisoning. The application of other cosmetics in microbiology laboratories is also forbidden to reduce the risk of infection through the skin.

The skin provides a major anatomical barrier to infection. Any cuts and grazes breach this defense and so should be covered before microbiological manipulations are started. The use of rubber gloves helps to protect against accidental infection. The skin provides a major anatomical barrier to infection if cuts or grazes occur. Any cuts and grazes breach this defense and so should be covered before microbiological manipulations are started. The use of rubber gloves helps to protect against accidental infection. If cuts or grazes occur during experimental work, medical advice must be obtained. An open cut or graze provides an unusual portal of entry for infection, and a resulting infection may well not show typical clinical symptoms because the infectious agent has been allowed access to the body through an unusual route. At the end of laboratory work and after laboratory coats have been removed, it is important to wash hands thoroughly to remove contaminating microorganisms, preferably using sinks equipped with knee-controlled taps that do not require manual operation. These reduce the risk of contamination and cross-infection. Outdoor clothes and bags should not be brought into the laboratory, and notebooks, etc., should be kept to a minimum. Live cultures should never be placed on or manipulated over notebooks that will be taken out of the laboratory.

The concept of universal precautions is to treat all human/primate blood and other body fluids, tissues, and cells as if they were known to be infectious. Along with frequent hand washing, no mouth pipetting, no food or drink in the laboratory, and proper disposal of biohazardous/medical waste are the inclusions of engineering controls and personal protective equipment (PPE). Engineering controls include items such as biosafety cabinets, ventilation systems, closed top centrifuge rotors, etc.—these are the primary methods to control exposure. PPE such as gloves, laboratory coats, eye protection, face shields, or others must be selected and used as appropriate.

PPE will be provided without cost to all individuals who are at risk of occupational exposure to pathogens. PPE will be chosen based on the anticipated exposure to potentially infectious materials. It is the responsibility of the laboratory director (the person who has immediate responsibility for the laboratory) to ensure the development and adoption of a biosafety management plan and a safety or operations manual. The laboratory supervisor (reporting to the laboratory director) should ensure that regular training in laboratory safety is provided.

Personnel should be advised of special hazards, and required to read the safety or operations manual and follow standard practices and procedures. The laboratory supervisor should make sure that all personnel understand these. A copy of the safety or operation manual should be available in the laboratory. There should be an arthropod and rodent control program. Appropriate medical evaluation, surveillance, and treatment should be provided for all personnel in case of need, and adequate medical records should be maintained.

There is a tiny minority of microorganisms and viruses that are highly infectious and whose infections are potentially fatal. Cultures of these microbes require much greater degrees of containment than used in most microbiology laboratories. Handling such agents requires the use of safety cabinets and more protective clothing that may cover the entire body. It may even require special isolation facilities in purpose built containment laboratories.

Cultures should be manipulated using a good aseptic technique. This comprises practices designed to minimize the risk of contamination. Liquid cultures should be handled with the minimum of shaking to reduce the risk of aerosols forming and dispensing. Culture should be only exposed to the air for as long as it is necessary to make observations or perform manipulations. Agar cultures in Petri dishes should always be kept upside-down so that any condensation forming on the lid does not fall onto the surface of the culture. Mouth pipetting is never permitted in microbiology laboratories, even for dispensing sterile distilled water. Bottle tops, test-tube caps, and flask stoppers such as cotton wool bungs must never be placed on the work surface where they may easily become contaminated. With a little practice, even large bottle tops can be easily manipulated in the crook of the little finger, leaving the rest of the hand free for manipulation of cultures, loops, and so on (Figure 15.1).

Open cultures should be manipulated close to the flame of a Bunsen burner, and they should be held so that they face away from any convection currents that may carry aerial organisms. The neck of culture vessels and reagent bottles should be passed through a hot Bunsen burner flame upon removal of the stopper, bung, or bottle top and again after the manipulation is complete and before the cap or stopper is replaced. This will kill any contaminating organisms on the neck of the vessel before the manipulation or any of the culture contaminating the mouth of the vessel

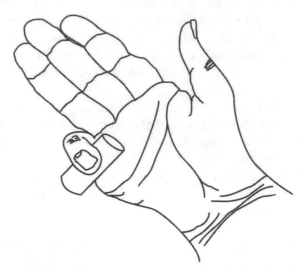

Figure 15.1 Handling the stopper from a tube culture in the crook of the little finger. This leaves the other finger and thumbs free to manipulate other objects.

after manipulation and before replacing the cap or stopper. When cotton wool is used as a bung, care must be taken because the material is flammable.

15.2.1.1 Preparation of Temporary Cotton Plugs

15.2.1.1.1 Principle. Microorganisms are omnipresent. We need cotton plugs to create aerobic environment and to make the growing cultures free from undesired microbes. Cotton plugs have cotton fibers' and air pours. Air can pass through air pours but not the microbial containment. They are adhered on the surface of cotton fibers and, therefore, prevented to enter inside the flasks, tubes, etc. Air facilitates the growth of microorganisms in glass apparatus.

15.2.1.1.2 Materials. Nonabsorbent cotton rolls, culture tubes, flasks, and scissors.

15.2.1.1.3 Protocol
1. Procure a nonabsorbent cotton roll and remove its packing.
2. Open the roll as per number, size, and type of glassware.
3. Cut a small piece of cotton sheet with the help of scissors as per requirement, that is, flask/culture tubes, etc. (Figure 15.1).
4. Fold the cut piece from the center.
5. Roll the sheet from any of the ends. Opposite to the folded side have cotton fibers that emerge out freely.
6. Insert the folded and rolled side in the neck/mouth of a flask/culture tube and the other end of plug bears cotton fibers. If it is tight, open the plug and remove some cotton from one side and readjust as earlier (Figure 15.2).

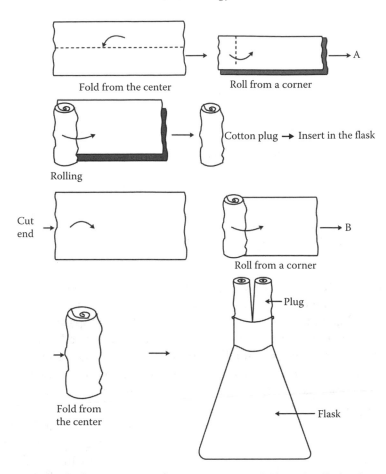

Figure 15.2 Method of preparation of cotton plug (cut, fold, and roll plugs).

15.2.1.2 Preparation of Permanent Cotton Plugs

15.2.1.2.1 Principle. Temporary cotton plugs should not be used repeatedly as they get contaminated by different types of microbes due to adherence of growth media. Permanent cotton plugs should be used for frequent use.

15.2.1.2.2 Materials. Nonabsorbent cotton rolls, culture tubes, flasks, scissors, and muslin cloth.

15.2.1.2.3 Protocol
1. Prepare temporary cotton plugs as described earlier.
2. Cut a small square shaped pieces (4–6 in.) of muslin cloth.
3. Put the folded end of the plug in the center of a single piece of muslin cloth
4. Tightly tie the four corners of muslin cloth at unfolded free end of the plug. This is a permanent cotton plug, which can be used repeatedly.

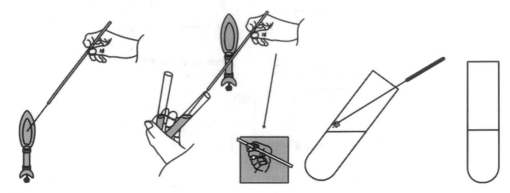

Figure 15.3   Aseptic transfer: loop is heated until re-hot, tube is uncapped, and loop is cooled in air briefly, sample is removed and tube is recapped. Sample is transferred to a sterile tube. Loop is reheated before being taken out of service.

A very widely used tool in microbiology is the metal microbiological loop. Before this is employed, it is heated to red heat in a Bunsen burner flame to sterilize it. It is then allowed to cool, before use, to avoid damaging the culture or spattering live microbes into the atmosphere. It is essential that the loop is flame sterilized after use as well, in order to kill any culture cells that adhere to it. When heating a contaminated loop, care must be taken to avoid dispensing live organisms. To do this, loop is first introduced into the cooler part of the flame and only when it has warmed up, it is placed in the hottest part of the flame (Figure 15.3).

When a liquid culture is to be inoculated, a charged loop should not be placed directly into the broth. Rather, the culture should be tilted slightly and the inoculums rubbed on the side of the vessel. The inoculum should be emulsified in the liquid that clings to the side of the tube in a position where it will be submerged when the culture is placed in its upright position for incubation.

Microorganisms require a favorable nutritional and physical environment to grow and multiply. A combination of nutritional requirement is referred to as a medium. Different organisms are specific in their nutritional requirements and need a specific medium for their growth. There are media, known as selective media, that specifically allow a particular group of bacteria to grow. Culture techniques are available to isolate and purify microorganisms. The main principle behind the isolation and purification technique involves dilution of the sample, on the plate or in water blanks, so as to obtain well-separated discrete colonies (Figure 15.4).

15.3 Culture Media

Media is defined as any substrate or material that will enable microorganisms to grow and multiply. The medium, which is prepared, with several nutrients in the laboratory for the sustenance of microbial growth is also known as culture medium or nutrient medium. All the microbes have the same basic requirements, but there is diversity in the use of organic and inorganic compounds. Some media contain solutions

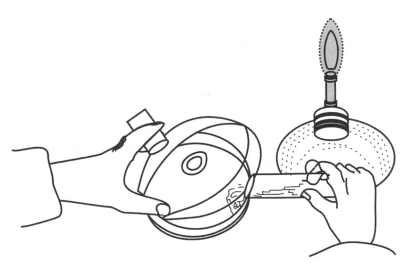

Figure 15.4   Pouring an agar plate.

of inorganic salts and supplemented with organic compounds, while other media contain complex ingredients, hence all media should provide carbon and nitrogen sources, in addition to minerals and other growth factors. There are no media available that can support the growth of all microorganisms.

15.4 Types of Media

Media can be classified into three categories on the basis of chemical components. They are (1) synthetic media, (2) complex media, and (3) natural media.

15.4.1 Synthetic Media

The media in which all the constituents are chemically defined. They are generally used to study the specific nutritional requirements of different microbes (e.g., minimal medium).

15.4.2 Complex Media

The media in which several organic substances are added for special purposes. For example, beef extract, peptone, etc.

15.4.3 Natural Media

Natural media contain the substrates of natural origin that favors microbial growth (e.g., milk). Basically, all culture media are liquid, semisolid, or solid. A liquid medium lacks a solidifying agent and is called a broth medium.

Basically, all culture media are liquid or semisolid. A liquid medium lacks a solidifying agent and is called a *broth medium*. A broth medium supplemented with a solidifying agent called as agar-agar results in a *solid* or *semi solid medium*. A completely solid medium requires an agar concentration of about 1.5%–1.8%, whereas semi solid media require a concentration of less than 1%. Agar is a neutral solidifying agent, which makes the slants easy in handling.

Two other valuable types of media are selective and differential media. Media permitting certain organisms to be distinguished from others by the appearance of their colonies are called *differential media*. Media formulated to permit the growth of certain bacteria but not others are called *selective medium*.

15.5 Preparation of Culture Media

Most essential culture media are available commercially in readymade dehydrated form. If organisms are to be isolated successfully, culture media must be prepared carefully. Each of the following must be performed correctly:

(a) Weighing and dissolving, (b) addition of heat sensitive ingredients, (c) dispensing, (d) sterilization and sterility testing, (e) pH testing, (f) quality control, and (g) storage

The reliable performance of each of the aforementioned stages requires training and experience. To achieve standardization in preparation of media, to avoid errors due to poor media performance, and to prevent the waste of resources a "Media manual" should be prepared and maintained in the laboratory containing full details.

15.5.1 Procedure

1. Dissolve each ingredient of the medium except agar-agar in an appropriate volume of distilled water.
2. Check the pH of the medium and adjust it to desired pH (normally 6.5–7.0) by adding few drops of 1 N HCl or 1 N NaOH as required.
3. Now add agar to the flask and plug it with cotton.
4. Cover the cotton plug with a paper wrap.
5. Sterilize the medium in an autoclave at 15 lb psi for 20 min.
6. Prior to sterilization, the flask should be labeled indicating the type of medium, the date, and your name.

15.6 Maintenance and Preservation of Cultures

Pure cultures of microorganisms need to be preserved so they can be used as and when required for research purpose. The preservation can be short term or long term. Different methods are employed for preservation of microorganisms, and all of them conserve all the characteristics of the organism. Some of the methods of preservation are the following.

15.6.1 Transfer on Fresh Media

Microbial cultures can be maintained by periodic transfer on fresh, sterile media in tubes. The frequency of transfer, however, varies with an organism. The agar-based slants should be stored at 4°C for long-term storage.

15.6.2 Overlaying with Mineral Oil

Many bacteria and fungi can be preserved by covering the fresh growth in agar slants with sterile mineral oil. The oil must be above the tip of the slanted surface. Mineral oil covered cultures are stored at room temperature or preferably at 4°C. Some microorganisms have been preserved satisfactorily for more than 15–20 years by this method.

15.6.3 Freeze-Drying (Lyophilization)

It is a process used for long-term preservation of cultures of certain types of microorganisms. In this process, the cell suspensions are placed in small vials, which are then frozen by immersing in a mixture of dry ice and acetone or liquid nitrogen. The vials are then evacuated and dried under vacuum, sealed, and stored at low temperature. The cells are dehydrated under vacuum (water passing directly from the solid to the vapor phase). Freeze-dried cultures can be stored for a long time at room temperature. These cultures can be reconstituted with distilled water.

15.6.4 Storage at Low Temperature

In this method, cultures are frozen in the presence of cryoprotectants added to the liquid or medium in which cells are suspended. Some of the cryoprotectants used are glycerol, ethylene glycol, and dimethyl sulfoxide (DMSO). Glycerol and ethylene glycol appear to be more widely used for bacterial preservation, while DMSO has been extensively used for the cryopreservation of protozoans. The cultures are preserved in 80°C deep freezers or in liquid nitrogen containers.

15.7 Preparation of Basic Liquid Media (Broth) for the Cultivation of Bacteria

15.7.1 Principle

Bacteria, in contrast to fungi, are often cultured in a liquid broth. The most common constituents of basic media are beef extract (a beef derivative, which is a source of organic carbon, nitrogen, vitamins, and inorganic salts) and peptone (a semi-digested protein). These may be modified in a variety of ways by supplementing with some specific materials for the cultivation of specific types of groups of bacteria. Growth of the organisms in liquid shows diffused growth. Nutrient broth and glucose broth are basic liquid media, but they have many disadvantages, such as identification becomes difficult due to specific characters are not properly exhibited. However, the advantages are, for example, obtaining bacterial growth from blood or water when large volumes have to be tested for preparing bulk cultures for physiological, genetical, and industrial investigations or for characterization of antigens or vaccines.

15.7.2 Materials

Peptone, beef extract, glucose, sodium chloride, HCl, NaOH, distilled water, test tubes, conical flasks, measuring cylinder, glass rods, pH meter, pressure cooker or autoclave, incubator, and nonabsorbent cotton.

15.7.2.1 Nutrient Broth

Peptone: 5 g and beef extract: 3 g. Dissolve in 500 mL distilled water and gently heat to dissolve completely. Make the solution to 1 L. Adjust the pH to 7.0 by adding 1 N HCl or 1 N NaOH as the case may be. Preparation of 1 N NaOH—40 g of NaOH in 1000 mL of H_2O. 1 N HCl—10 mL of concentrated HCl in 90 mL H_2O.

15.7.2.2 Glucose Broth

Peptone: 10 g, glucose: 5 g, and sodium chloride: 5 g. Dissolve the contents in 1000 mL of water. Adjust pH to 7.0.

15.7.3 Protocol

1. Arrange the test tubes in a test tube stand.
2. Pour 10 mL of broth per tube or 30 mL in 250 mL Erlenmeyer flasks; 60 mL in 250 mL Erlenmeyer flasks.
3. Close the test tubes with cotton plugs.
4. Autoclave at 121°C, 15 lb pressure for 15 min.
5. Allow the autoclave to cool.
6. Remove the nutrient broth tubes and store at room temperature (covered with butter paper) for future use. Glucose broth also can also be prepared and stored in the same way (Figure 15.5).

15.8 Solid Media for Growth of Microorganisms

15.8.1 Introduction

Liquid media containing different nutrients are usually solidified by the addition of agar-agar, often simply called as agar. Agar is a complex polysaccharide

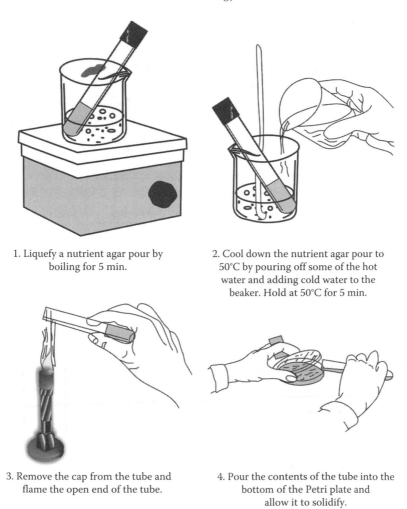

1. Liquefy a nutrient agar pour by boiling for 5 min.

2. Cool down the nutrient agar pour to 50°C by pouring off some of the hot water and adding cold water to the beaker. Hold at 50°C for 5 min.

3. Remove the cap from the tube and flame the open end of the tube.

4. Pour the contents of the tube into the bottom of the Petri plate and allow it to solidify.

Figure 15.5 Pouring an agar plate for streaking.

consisting of 3,6-anhydro-L-galactose, D-galactopyranose, and free of nitrogen, produced from various algae belonging to *Gelidium*, *Gracilaria*, *Gigartina* and *Pterocladia*. It liquefies on heating to 96°C and hardens into a jelly on cooling to 40°C–45°C. Agar can be replaced with gelatin (10–16% w/v), though this cannot be used with incubation temperatures exceeding 20°C or with proteolytic (protein degrading) microbes. The solidified medium kept in a Petri dish called as agar plates provides an artificial environment suitable for the rapid growth of fungi. The medium is taken into a test tube kept in slanting position and allowed to cool. The test tube with nutrient agar in hardened state in the upright position is called agar tube slant. Potato dextrose agar (PDA) and Czapek-Dox agar (CDA) are routinely used for the isolation and maintenance of common fungi. PDA acidified with 25% lactic acid (3–5 drops per 100 mL melted agar) is used for isolation of fungi from plant parts. Nutrient agar medium is used for common bacterial culture (Figure 15.6).

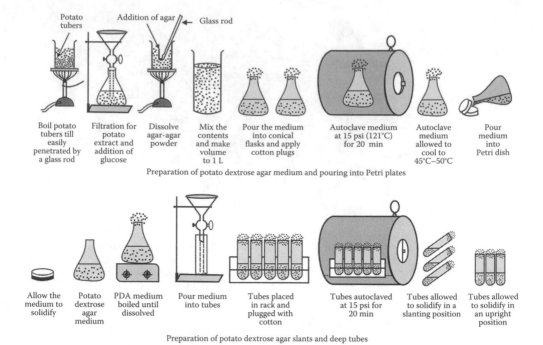

Potato tubers   **Addition of agar**   Glass rod

| Boil potato tubers till easily penetrated by a glass rod | Filtration for potato extract and addition of glucose | Dissolve agar-agar powder | Mix the contents and make volume to 1 L | Pour the medium into conical flasks and apply cotton plugs | Autoclave medium at 15 psi (121°C) for 20 min | Autoclave medium allowed to cool to 45°C–50°C | Pour medium into Petri dish |

Preparation of potato dextrose agar medium and pouring into Petri plates

| Allow the medium to solidify | Potato dextrose agar medium | PDA medium boiled until dissolved | Pour medium into tubes | Tubes placed in rack and plugged with cotton | Tubes autoclaved at 15 psi for 20 min | Tubes allowed to solidify in a slanting position | Tubes allowed to solidify in an upright position |

Preparation of potato dextrose agar slants and deep tubes

Figure 15.6   Illustration for the preparation of potato dextrose agar medium.

15.8.2 Common Culture Media Used in Laboratories

There are a large number of natural, semi-synthetic, and synthetic media that can be used to grow pathogenic and nonpathogenic microorganisms. Some of them are described in the subsequent sections.

15.8.2.1 Natural Media

Natural media is one that is composed of complex natural materials of unknown composition. Among natural substances are plant parts, malt, yeast, peptone, fruits, and vegetables. The following are a few natural media used in the isolation of fungi.

15.8.2.1.1 Cooked Vegetable Agar.
Broth or decoction of the desired vegetative plant parts can be prepared by adding 10%–20% of the tissue in water, steamed for 30 min, and the contents mashed and squeezed through a muslin cloth. To this 1.5%–2% agar-agar may be added for preparing the slants. pH of the medium may be adjusted to 5.5 or to desired level before autoclaving.

15.8.2.1.2 Lima Bean Agar.
The medium is used to cultivate Phytophthora species. Blend 50 g lima beans (*Phaseolus lunatus*), previously soaked in water for 10 h with 500 mL of water and raised to 1 L. Add 1%–2% agar-agar to prepare slants.

15.8.2.1.3 Oatmeal Agar. 60 g of rolled oats are blended in the commercial blender with 500–600 mL water and then heated to 55°C–60°C for 1 h and the volume was adjusted to 1 L with water. Add 1.5%–2% agar-agar to prepare slants. Media is used for general cultivation of fungi.

15.8.2.2 Semisynthetic Media

A semi-synthetic medium is made up of natural substances of unknown composition and by adding some chemical compounds of known composition.

15.8.2.2.1 Potato Dextrose Agar. Boil 200 g of peeled and sliced potato in 1 L of water until the potatoes are soft. Strain through a cheesecloth and adjust the filtrate to 1 L. Add 20 g dextrose and 15–20 g agar-agar for preparing slants and Petri plates and autoclave.

15.8.2.2.2 Malt Extract Peptone Dextrose Agar. Add 20 g malt extract, 20 g dextrose, 1 g peptone, agar-agar 15–20 g (for slants and plates), and distilled water to 1 L and autoclave.

15.8.2.2.3 Martin Rose Bengal Streptomycin Agar. Dextrose 10 g, peptone 5 g, KH_2PO_4 1 g, $MgSO_4 \cdot 7H_2O$ 0.5g, rose bengal 0.05 g, streptomycin 0.03 g, agar-agar 15–20 g, and distilled water to 1 L. Dissolve 1 g of streptomycin sulfate in 100 mL sterile distilled water; after opening the vial aseptically, add 0.3 mL of streptomycin in solution to each 100 mL of the basal medium after it is cooled. This is a general purpose medium to isolate fungi.

15.8.2.2.4 Nutrient Agar. Beef extract 3 g, peptone 5 g, NaCl 8 g, and distilled water to 1 L and autoclave. Add 15–20 g agar-agar for preparing slants or Petri plates. This is a general purpose medium for isolation and culturing of bacteria.

15.8.2.2.5 Soil Extract Agar. Take 500 g fertile soil in 500 mL distilled water, autoclave, cool, and filter the water extract. Add 0.2 g K_2HPO_4 and 15–20 g agar-agar (for slants and Petri plates) and adjust the volume to 1 L and autoclave. The medium is commonly used to enumerate bacteria and fungi.

15.8.2.2.6 Yeast Extract Medium. Yeast extract 1 g, K_2HPO_4 10 g, $MgSO_4 \cdot 7H_2O$ 1 g, Tween 80 0.1%, and distilled water to 1 L. *Erwinia amylovora* grows well in this medium.

15.8.2.3 Synthetic Media

Synthetic medium is of known composition and concentration. Chemicals used for preparing synthetic medium should be of pure quality.

15.8.2.3.1 Brown's Medium. Glucose 2 g, K_2HPO_4 1.25 g, asparagine 2 g, $MgSO_4 \cdot 7H_2O$ 0.75 g, agar-agar 15–20 g (for slants and Petri plates), and distilled water to 1 L.

15.8.2.3.2 Czapek-Dox Agar. $NaNO_3$ 2 g, K_2HPO_4 1 g, $MgSO_4 \cdot 7H_2O$ 0.5 g, KCl 0.5 g, $FeSO_4$ 0.01 g, sucrose 30 g, agar-agar 15–20 g (for slants and Petri plates), and distilled water to 1 L.

If glass distilled water is used, add 1 mL of 1% $ZnSO_4$ and 0.5% $CuSO_4$. Heat all chemical solution with sucrose in a water bath for 15 min. After cooling, add sucrose and agar-agar and autoclave.

15.8.2.3.3 Galactose Nitrate Agar.   Medium is commonly used to isolate *Fusarium oxysporum* from soil. Galactose 10 g, $NaNO_3$ 2 g, KH_2PO_4 1 g, $MgSO_4 \cdot 7H_2O$ 0.5 g, $K_2S_2O_5$ 0.3 g, agar-agar 15–20 g (for slants and Petri plates), and distilled water to 1 L and autoclave.

15.8.2.4 Some Special Media for Cultivation of Members of Actinomycetes

15.8.2.4.1 Basal Mineral Salts Agar.   Glucose 10 g, $(NH_4)_2SO_4$ 2.64 g, KH_2PO_4 anhydrous 2.38 g, $K_2HPO_3 \cdot 3H_2O$ 5.65 g, $MgSO_4 \cdot 7H_2O$ 1 g, distilled water 1000 mL, and agar-agar 20 g. Adjust pH to 6.8–7.0 with 1 N NaOH or 1 N HCl. Media can be used for culture of members of actinomycetes particularly *Streptomyces* with the addition of sterile carbon source at a concentration of 1%.

15.8.2.4.2 Oatmeal Agar (Shirling and Gottlieb No. 3).   Oatmeal 20 g, agar-agar 18 g, distilled water 1000 mL; ($FeSO_4 \cdot 7H_2O$ 0.1 g, $MnCl_2 \cdot 4H_2O$ 0.1 g, and $ZnSO_4 \cdot 7H_2O$ 0.1 g. Dissolve in 100 mL of distilled water and add 1 mL to oatmeal agar solution) steam oatmeal in a liter of water for 10 min, filter through cheesecloth, add trace salts, and make up volume. Adjust to pH 7.2 with NaOH. Add agar and autoclave for 15 min. This medium is good for culture of *Streptomyces*.

15.8.2.5 Media for Soil Bacteria

15.8.2.5.1 Asparagine Lactose Medium.   This medium is used to isolate *Rhizobium* sp. Lactose 10 g, asparagine 0.5 g, sodium taurocholate 5 g, K_2HPO_4 1 g, $MgSO_4 \cdot 7H_2O$ 0.2 g, neutral red 3.5 mL (2/40 mL ethyl alcohol), agar-agar 20 g, and distilled water 1000 mL, pH 7.5. This medium is suitable for the growth of all the Rhizobial species and especially for isolating *Rhizobium* from the nodules of ground nut. This medium can be used for distinguishing the *Agrobacterium* and *Rhizobium*. *Agrobacterium tumefaciens* shows quick growth within 24 h with pink color colony on congo red (10 mL of l/4000 aqueous solution) incorporated in the medium, while *Rhizobium* strains grows late after 72 h showing no absorption of Congo red.

15.8.2.5.2 Ashby's Mannitol Agar Medium.   Used to cultivate *Azotobacter* sp. Potassium acid phosphate 1 g, $MgSO_4 \cdot 7H_2O$ 0.2 g, ammonium sulfate 0.2 g, NaCl 0.2 g, mannitol 10 g, calcium sulfate 0. 1 g, calcium carbonate 5 g, agar-agar 20 g, and distilled water 1000 mL. Dissolve phosphate in 500 mL and make alkaline to phenolphthalein. Melt agar and other ingredients in the remaining 500 mL, mix both the lots, and sterilize by autoclaving for 20 min.

15.8.2.6 Media for Soil Fungi

15.8.2.6.1 V-8 Juice Agar.   Used to support growth and induce sporulation of many fungi. It is preferred over PDA for isolation of sporulating fungi from soil. V-8 juice 200 mL, $CaCO_3$ 3 g, agar-agar 15–20 g, distilled water 1000 mL, and pH 7–7.5 with variation in amount of juice or $CaCO_3$.

15.8.2.6.2 Cassava-Dextrose Agar.   Used in place of PDA for routine work of fungi: peel cassava (*Manihot utilissima*) tubers and cut into chips. Dry overnight at 55°C and grid to 30 mesh. Soak 135 g of the powder in 500 mL of water for 15 min at 60°C, filter through cheesecloth, add 20 g glucose and 12 g agar-agar, adjust the volume to 1 L, and autoclave.

Apart from the aforementioned media for fungi, there are some basic points that are to be kept in mind while preparing media.

1. Fungi usually grow best in a carbohydrate rich medium.
2. Fungi usually prefer acidic zone, pH 5–6.
3. Agar is slow to dissolve thoroughly. It is advisable to dissolve the agar in half the water and nutrients in the other half and then mix. Agar does not solidify satisfactorily in very acid or alkaline solutions.
4. Peptone may generally be omitted from fungus culture.
5. Tap water is often preferable to distilled water as it contains useful trace elements.

15.8.2.7 Selective Media for the Isolation of Plant Pathogens
There are several selective media for the isolation of plant pathogenic fungi. For example, P10VP medium is very useful in isolating *Phytophthora* species from plant tissues and soil. The medium contains Difco cornmeal agar-agar (17–20 g/L), which is autoclaved (cooled to 45°C) and amended with low dosage of pimaricin (10 ppm), vancomycin (200 ppm), and pentachloronitrobenzene (100 ppm). Plates are stored in the dark because of the sensitivity of pimaricin to light. After the usual sterilization procedure of the diseased tissues, the tissues are plated on this selective medium. Infested soil or suspected soil containing the *Phytophthora* propagules or mycelium can also be placed on this medium to isolate the pathogen. Hymexazol can be added after autoclaving to suppress species of *Pythium*. There are several other selective media for the isolation of *Phytophthora* species.

15.9 Isolation and Maintenance of Organisms

15.9.1 Isolation and Maintenance of Organisms by Streak Plate Method

15.9.1.1 Principle
In nature, microbial populations do not segregate themselves by species but exist with a mixture of many other cell types. In laboratory, these populations can be separated into pure cultures. A culture that contains more than one kind of microbes is called a mixed culture, and if it contains only one kind of microorganism is called a pure

culture. Pure cultures are essential in order to study the colony character, biochemical properties, morphology, immunological studies staining, antibiotic sensitivity, resistance, etc. In this process, two kinds of operations are involved: (1) isolation (separation of a particular microbe from the mixed population) and (2) cultivation (growth on culture media under laboratory conditions). Pure cultures of microorganisms that form discrete colonies on solid media may be most simply obtained by (1) streak-plate, (2) pour-plate, and (3) spread plate methods. This method offers a most practical method of obtaining discrete colonies and pure cultures. In this method, a sterilized loop or transfer needle is dipped into a suitable diluted suspension of organisms, which is then streaked on the surface of an already solidified agar plate, to make a series of parallel, nonoverlapping streaks. The microbial population in air, soil, and water is large and include several species of bacteria, fungi, and algae. From this mixed population, pure cultures are prepared. Streak plate method helps to obtain colonies of microorganisms that are pure, that is, growth derived from a single cell/spore (Figures 15.7 and 15.8).

15.9.1.2 Materials
Petri plates, conical flasks, test tubes, inoculation loop, pressure cooker autoclave, incubator, measuring cylinder, pH meter, balance, laminar flow, beef extract, peptone, NaCl, agar-agar, distilled water, and mixed culture.

Nutrient agar medium contains peptone: 5 g, beef extract: 3 g, NaCl: 5 g, and agar: 20 g dissolved in distilled water by gently heating until all contents are completely dissolved and finally adjust to 1 L. The medium is poured in Petri plates gently and kept aside for hardening. These agar plates are used for inoculation (Figure 15.6).

15.9.1.3 Protocol
1. Label the Petri plates.
2. Hold the tube containing the culture in the left hand.
3. Sterilize the loop holding in the right hand, remove the cotton plug using the little finger of the right hand and immediately flame the mouth of the tube (Figure 15.3).
4. Take a loopful of culture and close the tube, keep it aside.
5. Inoculate the agar plate starting from one side, that is, streak the inoculum from side to side in parallel lines across the surface of the area. Label it as area 1 (Figure 15.4).
6. Reflame and cool the loop and turn Petri plate to 90° angles.
7. Repeat the same process, and label it as area 3 and area 4.
8. Close the Petri plate after completing the streaking and sterilize the loop by flaming.
9. Incubate the agar plates at 25°C, in an inverted position, for 48–72 h (Figure 15.5).

15.9.1.4 Results and Observations
Examine the agar plates after incubation. A confluent growth will appear where the initial streak was made; the growth is less dense away from the streak, and discrete colonies are clearly seen (Figure 15.9) at the end of last streak (fourth area). Any colony not growing on the streak is regarded as a contaminant. If discrete pigmented and nonpigmented colonies are observed on the plates inoculated with mixed cultures, it shows that the components of mixed broth have been successfully separated. Select a well-isolated colony from agar plate and record their features.

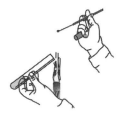

1. Shake the culture tube from side to side to suspend organisms. Do not moisten cap.

2. Heat the loop and wire to redhot. Flame the handle slightly also.

3. Remove the cap and flame the neck of the tube. Do not place the cap down on the table.

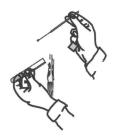

4. After allowing the loop to cool for at least 5 s, remove a loopful of organisms. Avoid touching the slides of the tube.

5. Flame the mouth of the culture tube again.

6. Return the cap to the tube and place the tube in a testtube rack.

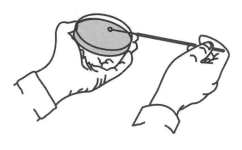

7. Streak the plate, holding it as shown. Do not gouge into the medium with the loop.

8. Flame the loop before placing it down.

Figure 15.7   Petri plate inoculation procedure.

15.10 Isolation of Organisms by Pour Plate Method and Serial Dilation Method

15.10.1 Introduction

Robert Koch developed this procedure in his laboratory. In this procedure, successive dilutions of the inoculum (serially diluting the original specimen) are added into sterile Petri plates to which is poured melted and cooled agar and thoroughly mixed by rotating the plates, which is then allowed to solidify. After

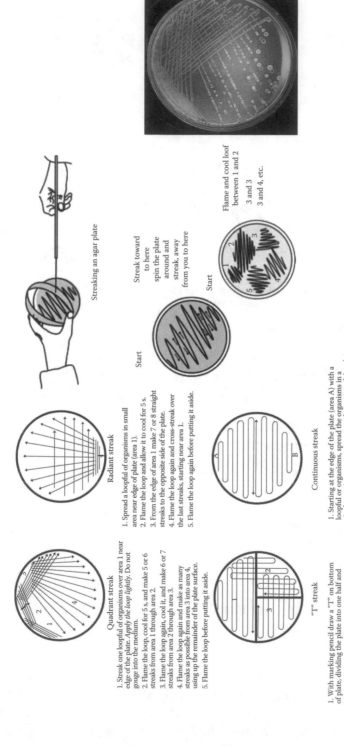

Figure 15.8   Streak plate methods.

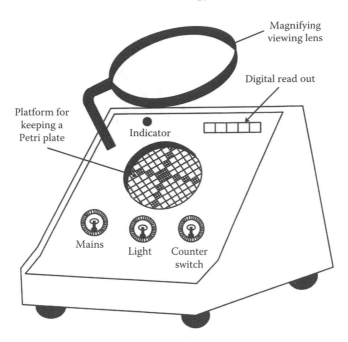

Figure 15.9   Quebec colony counter.

incubation, the plates are examined for the presence of individual colonies growing throughout the medium. The pure colonies, which are of different size, shape, and color, may be isolated or transferred into the test tube culture media for making pure cultures.

In this method, the original sample of inoculum is diluted several times to reduce the microbial population in the sample. This successive dilution ensures the development of separate colonies upon plating on an agar surface. As the medium solidifies, the microbial cells distributed throughout the medium get fixed in position. These trapped cells develop into individual colonies on incubation. Growth of colonies occurs both at the surface and subsurface areas as the medium is soft enough to permit the growth of cells. This pour plate method is also useful for quantitative studies to enumerate the microbial cells in the sample (Figure 15.10).

15.10.2 Materials

Test tubes, test tube stands, Petri plates, water bath, Bunsen burner, incubator, pipettes, nonabsorbent cotton, marking pen or labels, Quebec colony counter, 20 mL nutrient agar, 20 mL CDA for fungi, and mixed culture of bacteria or fungi.

Nutrient agar medium is prepared as described earlier.

CDA: Contains sodium nitrate (KNO_3) 2.0 g, dipotassium hydrogen phosphate (K_2HPO_4) 1.0 g, magnesium sulfate ($MgSO_4 \cdot 7H_2O$) 0.5 g, potassium chloride (KCl) 0.5 g, ferrous sulfate ($FeSO_4 \cdot 5H_2O$) 0.01 g, sucrose 30 g, agar-agar 15 g, and distilled

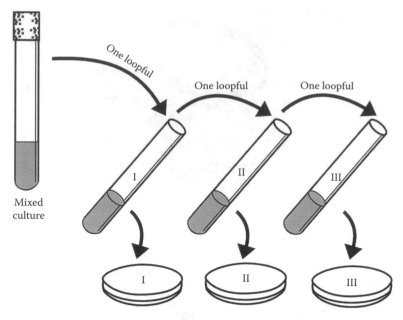

Figure 15.10   Pour plate method—Loop dilution technique.

water to 1 L. Dissolve all the ingredients except phosphate in half of the water and add sucrose. Dissolve phosphate separately and add to the rest. Make volume to 1 L. Sterilize by autoclaving at 121°C for 15 min.

15.10.3 Protocol

Nutrient agar medium is used for bacterial isolation, and CDA is used for fungal isolation.

1. Arrange seven test tubes in a test tube rack.
2. Label tubes from 1 to 7.
3. In the first tube, bacterial culture is taken (if fungi fungal cultures on Czapek-Dox). It acts as a stock.
4. The rest of the tubes contain 9 mL distilled water, respectively,
5. From the culture, 1 mL of the culture solution is transferred to test tube no. 2 with a fresh pipette (10^{-1})
6. From second test tube, again 1 mL is taken into tube no. 3 with another pipette, the process is continued up to the seventh tube. 1 mL is discarded from the seventh tube, so that all tubes contain equal volume of solution.
7. Arrange in the same way the Petri plates with proper labels coinciding with the concentration of the tube dilutions.
8. Take 1 mL of culture from 1 to 7 and transfer to respective Petri plates in laminar flow.
9. Immediately add nutrient agar (CDA) to Petri plates. For example, pour the medium into plate 1 and rotate the plate gently to ensure uniform distribution of the cells in the medium.

10. The same method is applied to plates 2–7.
11. Allow the medium to solidify.
12. Incubate the inoculated plates for 24–48 h at 37°C in an inverted position (Figure 15.11).

15.10.4 Results and Observations

Examine the plates for appearance of individual colonies growing throughout the agar medium. Using a colony counter observe all colonies. Plates with more than 300 colonies cannot be counted and designated as too numerous to count; those with fewer than 30 colonies are designated as too few to count. Count only plates with 30–300 colonies. The number of organisms (viable cells) per mL of the original culture is calculated by multiplying the number of colonies counted by the dilution factor.

$$\text{Number of cells mL}^{-1} = \text{number of colonies} \times \text{dilution factor.}$$

15.11 Spread Plate Technique

15.11.1 Introduction

It is an easy and direct method for achieving isolated colonies on the agar surface. In this method, microorganisms are spread over the solidified agar medium with a sterile L-shaped glass rod while the pad dish is placed on a working or turntable. The principle behind this technique is that, at some stage, single cells will be deposited with the bent glass rod on to the agar surface. Some of these cells will be separated from each other by a distance sufficient to allow the colonies that develop to be free from each other. This method is useful in quantitative studies (Figure 15.12).

15.11.2 Materials

Nutrient agar plates, lazy Susan turntable, L-shaped glass spreader, inoculation loop (or fine capillary tube), 95% alcohol, beaker, Bunsen burner, glass marker, and bacterial culture.

Nutrient broth is prepared as described earlier. Labeled nutrient agar plates are prepared.

15.11.3 Protocol

1. Label agar plates with a marker pen the bacteria used in the experiment.
2. The "L" spreader is sterilized with alcohol.
3. Aseptically transfer a drop or a loopful bacterial culture in the center of the nutrient agar plate, inside the laminar flow.

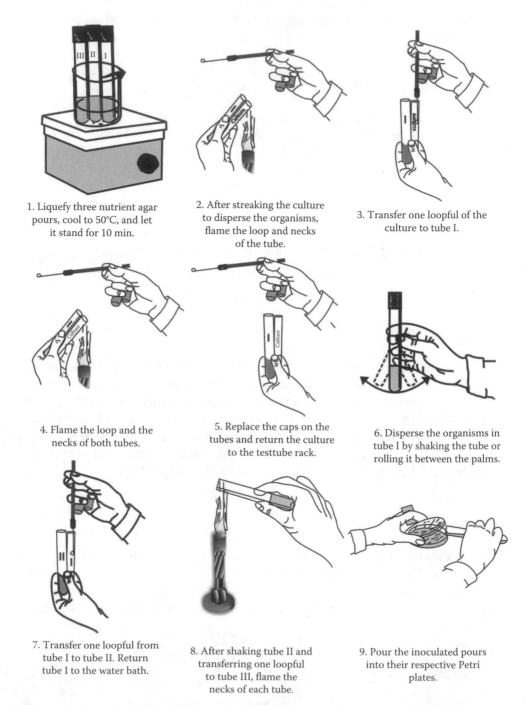

1. Liquefy three nutrient agar pours, cool to 50°C, and let it stand for 10 min.

2. After streaking the culture to disperse the organisms, flame the loop and necks of the tube.

3. Transfer one loopful of the culture to tube I.

4. Flame the loop and the necks of both tubes.

5. Replace the caps on the tubes and return the culture to the testtube rack.

6. Disperse the organisms in tube I by shaking the tube or rolling it between the palms.

7. Transfer one loopful from tube I to tube II. Return tube I to the water bath.

8. After shaking tube II and transferring one loopful to tube III, flame the necks of each tube.

9. Pour the inoculated pours into their respective Petri plates.

Figure 15.11   Procedure to handle inoculations for pour plate technique.

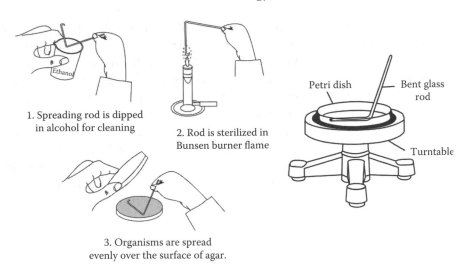

1. Spreading rod is dipped in alcohol for cleaning

2. Rod is sterilized in Bunsen burner flame

Petri dish

Bent glass rod

Turntable

3. Organisms are spread evenly over the surface of agar.

Figure 15.12   Spread plate technique; lazy Susan turn table spreader.

4. Take the "L" spreader gently expose to the flame and cool.
5. Place the Petri plate on turntable (Figure 15.12), gently open the Petri plate and lightly touch "L" spreader to the agar surface and move it back and forth while the turntable is spinning for spreading the culture over the agar surface. Cover the Petri plate with lid and incubate for 24–48 h in an inverted position at 25°C. Repeat the same with the required number of agar plates.

15.11.4 Result

1. Observe the distribution of colonies on different agar plates.
2. Note the size, pigmentation, and color of the colony.

15.12 Isolation of *Vibrio parahaemolyticus* from Seafood Sample

15.12.1 Introduction

Vibrio parahaemolyticus is common microflora of seafood such as prawns, crabs, shellfishes, saltwater fishes, fresh water bodies, plankton, etc. It is Gram-negative, actively motile, capsulated, halophilic short rod bacterium. *V. parahaemolyticus* has an affinity to the shell surface of seafood. Improper shelling, cross contamination at the time of shelling, consumption of raw seafoods, and undercooking of shelled foods are the primary cause of *V. parahaemolyticus* infections. During 1973–1998, Centers of Disease Control and Prevention reported 40 outbreaks of *V. parahaemolyticus*. These outbreaks involved more than 1000 illnesses.

For isolation of *V. parahaemolyticus*, freshly collected samples of prawns, crabs, and salt water fish are enriched in thiosulfate citrate bile sucrose (TCBS) broth and subsequently isolated on TCBS agar. Identification is on the basis of growth in high levels of NaCl, oxidase production, and sucrose fermentation. Pathogenic strains are differentiated on the basis of Kanagawa test lysis of red blood cells. The high concentrations of thiosulfate and citrate and the strong alkalinity of this medium largely inhibit the growth of Enterobacteriaceae. Ox bile and cholate suppress primarily enterococci. Any coliform bacteria, which may grow, cannot metabolize sucrose. Only few sucrose-positive Proteus strains can grow to form vibrid-like yellow colonies. The mixed indicator thymol blue–bromothymol blue changes its color to yellow when acid is formed, even in this strongly alkaline medium.

15.12.2 Materials

Sea food sample, TCBS agar and broth medium, inoculating loop, sterile Petri plates, sterile test tube, Bunsen burner, and glass wear marking pencil.

15.12.2.1 TCBS Media Composition (g/L)
Peptone from casein, 5; peptone from meat, 5; yeast extract, 5; sodium citrate, 10; sodium thiosulfate, 10; ox bile, 5; sodium cholate, 3; sucrose, 20; sodium chloride, 10; iron(III) citrate, 1; thymol blue, 0.04; bromothymol blue, 0.04; and agar-agar, 14.

15.12.3 Protocol

1. Inoculate different tubes containing sterile 10 mL TCBS broth with samples of whole prawns with and without shells.
2. Incubate the tubes at 37°C 24 h for selective enrichment of *Vibrio* sp.
3. Streak loopful of enriched broth on sterile TCBS agar plates and incubate at 37°C for 24–48 h.

15.12.4 Results and Observation

V. parahaemolyticus produce colonies with blue to green centers. Observe the biochemical parameters of the test sample and interpret the results (Figure 15.13).

15.13 Isolation of Lactic Acid Bacteria from Milk and Buttermilk Sample

15.13.1 Introduction

Lactic acid bacterium (LAB) is a group of Gram-positive, aero-tolerant, acid-tolerant, and spherical or rod shaped anaerobic bacteria. LAB is isolated from

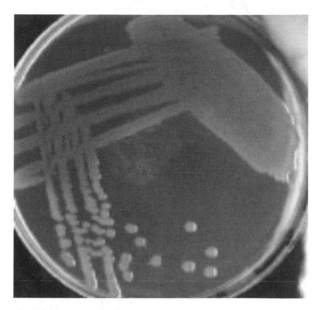

Figure 15.13 Twenty-four hour old colonies of *Vibrio parahaemolyticus* on TCBS.

natural sources such as fermented milks, cheese, sauerkraut, pickles, and human intestine. The name lactic acid bacteria are derived from the fact that adenosine triphosphate is synthesized through fermentation of carbohydrates, which yield lactic acid as a major end product. Lactic acid bacteria are divided into two biochemical subgroups.

a. Homofermenters convert glucose almost quantitatively to lactic acid
b. Heterofermenters, produce equimolar mixture of lactic acid, ethanol, and CO_2

LAB is distinguished from one another by temperature requirements and by fermentation reaction with various carbohydrates. Various species of LAB group are being used in probiotic products; most common are *Lactobacillus casei*, *Lactobacillus rhamnosus*, *Bifidobacterium* spp., *Propionibacterium*, etc.

Lactic acid bacteria (LAB) selectively grows on de Man Rogosa and Sharpe (MRS) agar medium due to the low pH (pH 4.58–5.20). Cycloheximide in the medium suppresses growth of yeast, molds, Gram-negative bacteria and a few Gram-positive bacteria other than LAB. Moreover, growth of lactic culture is enhanced in the presence of 5%–10% CO_2. Use of CO_2 cylinder may not be feasible for every laboratory. Hence, a desiccator jar with a lit candle is used to incubate the plate. The candle flame uses the O_2 as long as it grows. As a result, the relative concentration of CO_2 in the jar is raised. This method, therefore, serves as a cheap alternative to the CO_2 incubator.

15.13.2 Materials

Milk or buttermilk sample, MRS agar medium, beef, inoculating loop, sterile Petri dish, sterile test tube, Bunsen burner, and glass wear marking pencil.

15.13.3 Protocol

1. Weigh 1 g fresh mutton and beef samples and suspend in separate tubes containing 9 mL sterile buffered phosphate saline (pH 7.0).
2. Shake the tubes by vortex mixing and allow standing undisturbed until the solids settle. Consider the supernatant as sample for isolation.
3. Transfer 1 mL freshly prepared buttermilk sample in a sterile test tube and allow it to stand undisturbed until the solids settle. Consider the supernatant as sample for isolation.
4. Inoculate 1 mL supernatant from the mutton and beef samples in 9 mL sterile MRS broth. Incubate the tubes at 35°C for 24 h for selective enrichment of LAB. The buttermilk samples are already enriched preparations of LAB and do not need enrichment.
5. Streak loopful of enriched mutton and beef samples, and buttermilk on sterile MRS agar plates.
6. Incubation
   a. Clean a desiccator jar (or any closed glass chamber). Place few moist filter papers at the bottom. Stand a lighted candle in the jar at one side.
   b. Place the inoculated agar plates in an inverted position in the jar.
   c. Grease the rim of the jar with petroleum jelly/Vaseline and slide the lid slowly to make the jar airtight (Figure 15.14).
   d. The candle in the jar will get extinguished after some time due to depletion of O_2. Such a jar is said to have a 3%–5% CO_2 environment, which favors growth of LAB (do not open the lid thereafter).
   e. Incubate the entire jar at 35°C for 3–4 days.

15.13.4 Results and Observation

Study the colonies that develop on the media after incubation. Determine the motility and gram nature of each variety of colonies isolated.

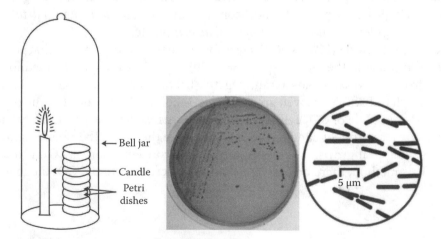

Figure 15.14 Illustrating *Lactobacillus* isolation.

15.14 Isolation of *Staphylococcus aureus* from Food Sample

15.14.1 Introduction

Staphylococcus aureus is a Gram-positive, nonmotile bacterium and a normal inhabitant of skin and nasal mucosa of man and animals. Improperly handled foods are frequently contaminated with *S. aureus*. Some strains of *S. aureus* are toxigenic and cause gastroenteritis (food poisoning). *S. aureus* food poisoning ranks as the second most reported of all types of food borne disease caused by enterotoxin. Enterotoxin is an exotoxin secreted in foods that are heat resistant and may escape in the cooking process, since the symptoms are restricted to the intestinal tract.

The ability of *S. aureus* to grow at a higher concentration of salt is used for its selective enrichment in salt mannitol broth. Selective isolation of *S. aureus* in food sample is carried out Baird–Parker agar medium (BPA). This medium contains

1. Lithium chloride to prevent the growth of many other contaminants
2. Egg yolk to demonstrate lecithinase activity
3. Potassium tellurite, which is reduced to tellurium by *S. aureus* producing pitch black colonies

15.14.2 Materials

Food sample, mannitol broth and BPA, phosphate buffered saline, sterile Petri plates, sterile test tube, inoculating loop, distilled water, Bunsen burner, glassware marking pencil, etc.

15.14.3 Protocol

1. Prepare suspension of the given sample by weighing 10 g and transferring in a flask containing 90 mL sterile phosphate buffered saline.
2. Collect 10 mL of supernatant in a separate sterile tube and label this as 10^{-1}.
3. Carry out serial dilution up to 10^{-4}.
4. Inoculate 1 mL of the supernatant in a tube containing sterile salt mannitol broth and incubate at 35°C for 18 h.
5. Streak loopful of the enriched culture on sterile BPA plate at 37°C for 24 h.

15.14.4 Results and Observations

Observe for typical lecithinase positive colonies of *S. aureus*. Typical colonies are pitch black in color, and a clear zone of lecithin hydrolysis is produced around the colony against a cream yellow background (Figure 15.15).

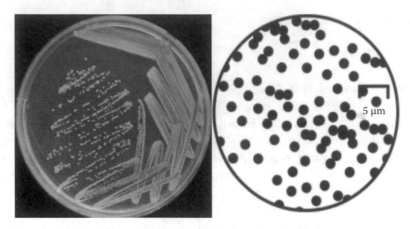

Figure 15.15   Isolation of *Staphylococcus aureus* from food sample.

15.15 Sauerkraut Fermentation

15.15.1 Introduction

Sauerkraut is defined as a clean product with characteristic flavor, obtained by full fermentation of properly prepared and shredded cabbage in presence of not less than 2%–3% salt, finally containing not less than 1%–1.5% lactic acid. Refrigeration and pasteurization is not required, although these treatments prolong storage life. German sauerkraut is often flavored with juniper berries. In simple words, sauerkraut or sour cabbage fermentation is initiated by lactobacilli, which are introduced naturally, as these air-borne bacteria culture on raw cabbage leaves where they grow. NaCl inhibits the growth of initial spoilage organisms like *Pseudomonas* and extracts moisture from the shredded cabbage by osmosis to form the brine in which fermentation takes place and helps to maintain the crisp texture of the cabbage by withdrawing water and inhibiting endogenous pectolytic enzymes enabling the product to soften.

Leuconostoc mesenteroides is a species of bacteria sometimes associated with fermentation, under conditions of salinity and low temperatures. It produces lactic acid and acetic acid, which rapidly lower the pH thereby inhibiting the growth of spoilage microorganisms. The CO_2 produced by them generate anaerobic conditions within the product and prevent the oxidation of vitamin C and loss of color.

Fermentation is then continued by *Lactobacillus brevis* and *Lactobacillus plantarum*, which is mainly responsible for the final high levels of acidity. The lactic acid produces during fermentation inputs characteristic flavor to the product and act as a preservative by inhibiting the growth of food spoilage microorganisms.

15.15.2 Materials

Heads of cabbage (1 kg), NaOH (0.1 N), uniodized NaCl, knife, pH paper, cheese cloth, two wide mouthed jars, 8% phenolphthalein, and methylene blue.

15.15.3 Protocol

1. Remove the outer leaves and all bruised tissues from each of the cabbage heads.
2. Divide cabbage into two equal parts and wash the heads in tap water, and shed the cabbage.
3. Weigh the shredded cabbage on a pan balance and separate into two equal portions.
4. Weigh out the table salt in amounts equal to 3% of the weight of each of the portions of shredded cabbage.
5. Place the shredded cabbage and salt in the alternating layers in the two wide-mouthed jars.
6. Place a wooden board over each of the mixtures and press gently to squeeze out a layer of juice from the cabbage.
7. Place a weight on each of the boards, and cover the jars with cheese cloth.
8. Incubate the jars for 21 days at 30°C.

15.15.4 Results and Observations

Examine the sauerkraut preparation on days 2, 17, 14, and 21 of incubation.

1. Examine the fermenting cabbage for aroma, color, and texture.
2. Remove 10 mL of the fermenting juice and analyze as follows:
   a. Prepare a slide by using methylene blue for microscopic examination
   b. Determine the pH of the juice.
   c. Perform titration against 0.1 N NaOH to know the percentage of lactic acid present
3. *Odor*: Acid, earthy, spicy, or putrid.
4. *Color*: Brown, pink, straw yellow, pale yellow, or colorless.
5. *Texture*: Soft (*L. plantarum*), slimy (*Lactobacillus cucumeris*), or rotted (bacteria, yeast, or mold spoilage).
6. *pH*: The pH of the finished product should be 3.1–3.7.
7. *Total acidity*: (1) Dispense 10 mL fermentation juice and 10 mL distilled water into an Erlenmeyer flask. Boil to remove CO_2, (2) cool and add five drops of 1% phenolphthalein to the juice, and (3) titrate the pink color reaction mix with 0.1 N NaOH.

$$\% \text{ Lactic acid} = \frac{\text{mL of alkali} \times \text{normality of alkali} \times 9}{\text{weight of sample in g } (1 \text{ mL} = 1 \text{ g})}$$

8. Examination of microflora

	Sauerkraut Preparation			
Results	2 days	7 days	14 days	21 days
Odor				
Color				
Texture				
Lactic acid %				
pH				
Draw microflora				

15.16 Isolation of Microorganisms from Air

15.16.1 Materials

Nutrient agar medium, PDA medium, colony counter, sterile Petri plates, sterile test tube, inoculating loop, distilled water, Bunsen burner, glass wear marking pencil, etc.

15.16.2 Protocol

1. Prepare PDA medium and nutrient agar medium by adding a pinch of antibacterial and antifungal, respectively.
2. Pour both the aforementioned molten agar medium on 4–5 Petri plates.
3. Allow the plates to solidify.
4. Open the plates in the air for 5–8 min by removing its top cover.
5. Incubate the exposed plates at 25°C for 48–72 h for PDA plates and 37°C for 24–48 h for nutrient agar plates.

15.16.3 Observation

Observe the plates for microbial growth. Determine the distribution by using colony counter.

15.17 Phyllosphere/Phylloplane Microbiology

15.17.1 Introduction

Leaf surface microbes may perform an effective function in controlling the spread of airborne microorganisms inciting plant diseases. Presence of fungal spores on the surface of leaves incites the formation of a chemical substance referred to as phytoalexin, which are active in host defense mechanisms. The name elicitor has been commonly used to denote the compounds that induce the synthesis of phytoalexins. These are biotic elicitors such as polysaccharides from fungal cell walls, lipids, microbial enzymes, and polypeptides.

15.17.2 Isolation of Microbial Flora from Phyllosphere and Phylloplane

15.17.2.1 Principle
The phyllosphere and phylloplane microflora are of special interest as some of them have antagonistic effect against fungal parasites, some degrade plant surface wax and cuticle and produce phytoalexins. Any change in phyllosphere affects plant growth, which in turn, affects the physiological activity of the root system. Such changes in

the root result in an altered pH and spectrum of chemical exudation causing a change in rhizosphere microflora. Thus, there is a link between phyllosphere microflora and rhizosphere microflora. There is a continuous diffusion of plant metabolites from the leaves, which support the microbial growth, and in turn these microbes protect the plant from pathogens. There are certain methods for isolation of phyllospheric and phylloplane microflora such as direct observations, scanning microscopy, cultural spore plating, etc.

15.17.2.2 Materials
Fresh leaves, CDA medium (fugal isolation), nutrient agar medium (bacterial and actinomycete isolation), distilled water, pipette, sterile Petri plates, sterile test tube, inoculating loop, distilled water, Bunsen burner, disinfectant for desktop, sponge, wax marking pencil, etc.

15.17.2.3 Protocol
1. Collect the leaf samples in a sterile container and take it to the laboratory.
2. Cut small pieces of leaf with sterile cork borer.
3. Transfer the leaf discs in 10 mL distilled water.
4. Prepare serial dilution up to 10^{-6} in the test tube.
5. Transfer 1 mL suspension from each dilution in separate plates and pour molten and cool medium on it.
6. Incubate at 27°C for 2–6 days in CDA plates (fungi) and at 37°C for 24–48 h for nutrient agar plates (bacteria).

15.17.2.4 Results and Observation
Observe the plates after incubation for growth and identify the organisms present in the phyllosphere. Note the results for further studies (Figures 15.16 and 15.17).

15.18 Measurement of Microbial Growth

15.18.1 Principle

Microbial growth is defined as an increase in number of microbial cell in a population, which can also be measured as an increase in microbial mass. Growth rate is the change in cell number or mass per unit time. All the structural components of the cell double during cell-division cycle. The interval for the formation of two cells from one is called a generation, and the time required for this to occur is called the generation time. The generation time is, thus, the time required for the cell number to double. It is also referred some time as the doubling time. Cell numbers in a growing population of bacteria increase in geometric progression. This is often referred to as exponential growth. $1 \rightarrow 2 \rightarrow 4 \rightarrow 8 \rightarrow 16 \rightarrow 32 \rightarrow 64 \rightarrow 128$ and so on, may be written as 2^0 cells $\rightarrow 2^1$ cells $\rightarrow 2^2$ cells $\rightarrow$, and so on (Figure 15.18).

Microorganisms are cultured in water with appropriate nutrients is called as culture medium. The nutrients present in the culture medium provide the microbial cell,

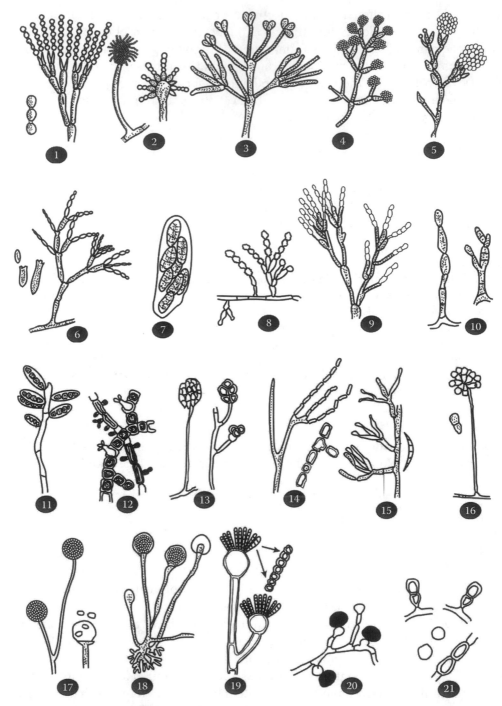

Figure 15.16 Microscopic appearance of some more common molds: (1) *Penicillium*, (2) *Aspergillus*, (3) *Verticillium*, (4) *Trichoderma*, (5) *Gliocladium*, (6) *Cladosporium*, (7) *Pleospora*, (8) *Scopulariopsis*, (9) *Paecilomyces*, (10) *Alternaria*, (11) *Bipolaris*, (12) *Pullularia*, (13) *Diplosporium*, (14) *Oospora*, (15) *Fusarium*, (16) *Trichothecium*, (17) *Mucor*, (18) *Rhizopus*, (19) *Syncephalastrum*, (20) *Nigrospora*, (21) *Montospora*.

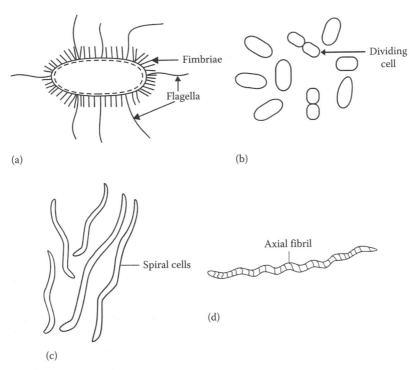

Figure 15.17 Diagrammatic view of some common bacteria on phyllosphere/phylloplane. (a) *Escherichia coli*, (b) *Spirochaete*, (c) *Spirillum*, and (d) *Azotobacter*.

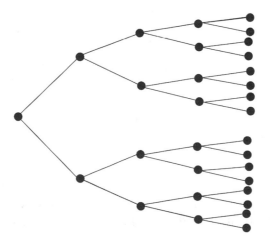

Figure 15.18 Representation of exponential microbial growth by binary fission.

with those ingredients required for the cell to produce more cells like itself. Besides an energy source, which could be an organic or inorganic chemicals, or light, a culture medium must have a source of carbon, nitrogen, and other macro- and micronutrients. Growth is measured by the following changes in number of cell or weight of cell mass by counting cell number, by observing turbidity in liquid broth or estimating cell mass suited to different organisms or different problems. Microbial control by chemical and physical means involves the use of antiseptics, disinfectants, antibiotics, ultraviolet light, and many other agents. The experiments explained on growth are related to these aspects primarily to demonstrate methods of measurement with in-depth evaluation. The complete killing of the microorganism is called as sterilization and is brought about by use of heat, radiation, or chemicals.

15.18.2 Bacterial Growth

Microbiologists involved with industries are concerned, on the one hand, with providing optimum growth conditions to optimize growth for biomass or industrial products. The physician, nurse, and other members of the medical profession, on the other hand, are concerned with the limitation of microbial populations in disease prevention and treatment. An understanding of one of these facets of microbial existence enhances the other. Selection of media for microbial growth that contain all the essential nutritional needs is very important for any microbiological experiment. Temperature, oxygen, and/or hydrogen ion concentration are other limiting factors. An organism provided with all its nutritional needs may fail to grow if one or more be sustained to achieve the desired growth of microorganism. Some techniques are qualitative in their intent. That is, they provide a "yes or no" answer. Other techniques are quantitative in their intent. These techniques provide numerical information about a sample.

Bacterial growth is defined as the division of one bacterium into two cells, taking place by a process called binary fission. One initial bacterium cell eventually becomes millions. Each cell divides into two cells in minutes, making for the exponential rate at which bacteria population increases. Binary fission is the asexual method of bacterial reproduction. The DNA of each cell is first replicated then attaches itself to the plasma membrane. Once the DNA is attached to the membrane, the cell elongates by growing inward and splits into two separate entities, and this process is referred to as cytokinesis. Bacterial growth undergoes four stages. The first two stages are the lag phase, where growth is slow at first, bacteria are becoming "acclimatized" to the new environmental conditions to which they have been introduced (pH, temperature, nutrients, etc.). There is no significant increase in numbers with time. The log phase is defined as the phase where the amount of bacteria doubles every few minutes. The final two stages of growth are the stationary phase, where growth stabilizes due to competition for nutrients among a high number of bacteria, and the death phase, where waste builds up, and nutrients run out, allowing for the bacteria to die off (Figure 15.19).

15.19 Methods to Measure Bacterial Growth

Different methods are available for population counts. The selection of method depends on the purpose of the study. It is possible to do a population count by preparing the dilution and counting the organisms in a number of microscopic fields on a slide without involvement of much of the equipments. Direct examination of milk

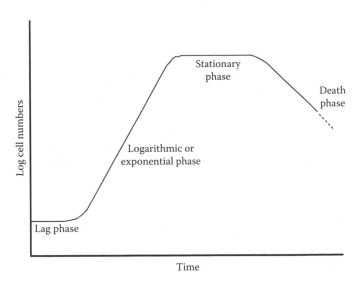

Figure 15.19 Phase of microbial growth in batch culture.

samples with this technique can be performed very quickly, and the results obtained are quite reliable. A technique similar to this can be performed on a Petrof-Hauser counting chamber.

Bacterial counts of gas-forming bacteria can be made by inoculating a series of tubes of lactose broth and using statistical probability tables to estimate bacterial numbers. This method, which we use, in exercise, to estimate numbers of coliform bacteria in water samples, is easy to use, works well in water testing, but is limited to water, milk, and food testing. Quantitative plating (standard plate count or SPC) and turbidity measurements to determine the number of bacteria in a culture sample are being commonly used. The two methods are somewhat parallel in the results they yield; however, there are distinct differences. The SPC reveals information only as related to viable organisms; that is, colonies that are seen on the plates after incubation represent only living organisms, not dead ones. Turbidimetry results, on the other hand, reflect the presence of all organisms in a culture, dead and living.

15.19.1 Turbidimetry Determinations

Following inoculation, the bacterial culture is stored or incubated in an environment suitable for growth in liquid broth or semisolid agar slants. The mass of daughter cells are visible to the naked eye either as cloudiness or turbidity in liquid broth or as an isolated colony on solid media. The quantitative plate count method becomes a rather cumbersome tool while you are handling a large number of cultures. It is time-consuming, besides takes a considerable amount of glassware and media. Measurement of turbidity of the culture is a much faster with a spectrophotometer and translate this into the number of organisms. To accomplish this, however, the plate count must be used to establish the count for one culture of known turbidity.

The bacterial cells act as a colloidal suspension, which will intercept the lights as it passes through the spectrophotometer. The amount of light that is absorbed is directly proportional to the concentration of cells within certain limits.

Figure 15.20 illustrates the path of light through a spectrophotometer. A beam of white light passes through two lenses and an entrance slit into horizontal beams of all colors of the spectrum. Short wavelengths (violet and ultraviolet) are at one end, and long wavelengths (red and infrared) are at the other end. The spectrum of lights falls on a dark screen with a slit (exit slit) cut in it. Only that portion of the spectrum that happens to fall on the slit goes through into the sample. It will be a monochromatic beam of light. By turning a wavelength control Knob on the instrument, the diffraction grating can be reoriented to allow the slit. The light that passes through the culture activates a phototube, which, in turn, registers *percent transmittance (1%T)* on a galvanometer. The higher the 1%T, the fewer are the cells in suspension (Figure 15.21).

There should be a direct proportional relationship between the concentration of bacterial cells and the absorbance (optical density; OD) of the culture. These values will be converted to OD and plotted on a graph as a function of culture dilution. There is a linear relationship between concentration of cells and OD values, the relationship may not be linear, that is, for a doubling in cell concentration, there may be less than a doubling in OD.

15.20 Growth Curve Measurement of Bacterial Population by Serial Dilution Method

15.20.1 Principle

Bacterial population growth studies require inoculation of viable cells into a sterile broth medium and incubation under optimum temperature, pH, and gaseous conditions. Cells will reproduce rapidly, and the dynamics of the microbial growth can be charted by means of a population growth curve, which is constructed by plotting the increase in all numbers versus time of incubation. The graph expresses the generation time, that is, the time required for a microbial population to double. The stages of typical growth curve are (1) lag phase, (2) logarithmic (log) phase, (3) stationary phase, and (4) decline or death phase.

1. *Lag phase*: Cellular metabolism is accelerated, resulting in rapid biosynthesis of cellular macromolecules, primarily enzymes, in preparation for the next phase of the cycle, but no cell division occurs and, therefore, no increase in numbers.
2. *Log phase*: Under favorable growth conditions, the physiologically robust cells reproduce at a uniform and rapid rate by binary fission. There is a rapid exponential increase, which doubles regularly until a maximum number of cells are reached. The time taken for the population to double is the generation time. The length of log phase varies, depending on the organism and the medium, the average is 6–12 h.
3. *Stationary phase*: Here, no further increase in cell number, and the population are maintained at its maximum level for a period of time, but in this phase starts the depletion of metabolites and accumulation of toxic end products.

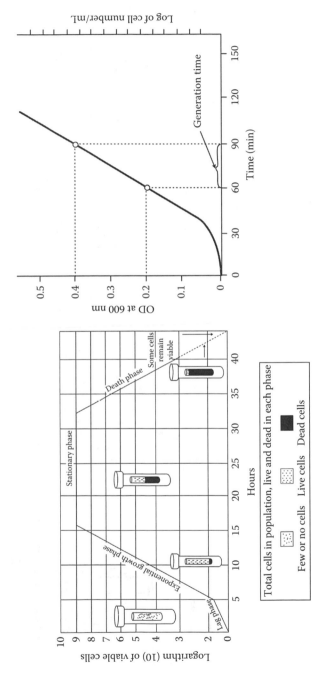

Figure 15.20 Bacterial growth curve illustrating generation time.

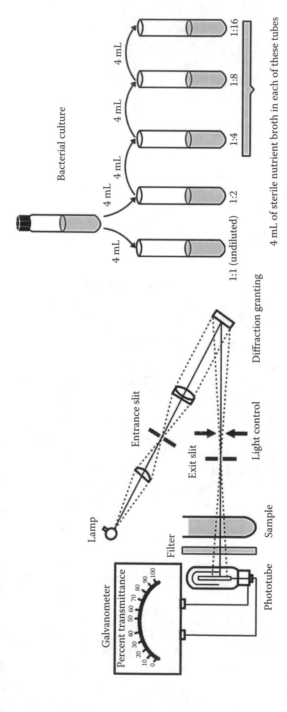

Figure 15.21 Colorimetric measurement of bacterial/turbidity growth.

4. *Decline/death phase*: Due to the earlier activity, that is, accumulation of metabolic wastes, and lack of nutrients, microbes dies at a rapid and uniform rate. This closely parallels its increase during the log phase. However, a small number of highly resistant organisms remain for on indeterminate length of time.

The microbial growth may be measured by a variety of techniques, viz. dry weight determination (determination of total N_2 by Kjeldahl method), direct counting of cells under light microscope (or indirectly by viable count method), which detects living organisms by their ability to form colonies on agar, and turbidity measurements, which relate cell number to the turbidity (cloudiness) of a broth culture. It is not a direct measure of bacterial numbers but increase in turbidity does indicate bacterial growth OD, which is directly proportional to cell concentration. It is measured by spectrophotometer. With the increase of bacterial numbers, the broth becomes more turbid, causing the light to scatter and allowing less light to reach the photoelectric cell.

15.20.2 Materials

Test tubes, conical flasks, test tube rack, non absorbent cotton, and pipettes. Petri plates, spectrophotometer, water bath, autoclave, incubator, laminar flow, pH meter, Bunsen burner, peptone, disodium phosphate, sodium chloride, dextrose, agar, distilled water, and *Escherichia coli* cultures.

Prepare *nutrient broth medium* as described earlier.

15.20.3 Protocol

1. Procure 1 mL of fully grown bacterial culture from broth prepared earlier.
2. By serial dilution technique, dilute the culture to get nearly 1×10^6 cells/mL (it is approximately 1×10^9 cells/mL, which is equal to the fully grown culture).
3. Aseptically transfer 5 mL of the broth culture to 100 mL of nutrient broth flask.
3. Determine the initial OD at 600 nm, which should be 0.08–0.01.
4. Place the inoculated culture flask in the shaker set at 120 rpm at 37°C for 6 h.
5. Aseptically transfer after 30 min incubation 5 mL of the culture to a cuvette.
6. Determine the OD of the sample at 600 nm.
7. Repeat the Steps 5 and 6 at each 30 min interval for a period of 6 h.

15.20.4 Results and Observations

Record optical densities of the broth culture suspension at 30 min time intervals for 6 h and record the OD as per Table 15.1.

Absorbance $\left[\left(A = -\log \dfrac{\%T}{100} \right) \right]$ is a logarithmic value, more linear than %T and is used to plot bacterial growth curve on graphs %T = OD and read the values on the log scale (absorbance) and consider these to be OD values.

TABLE 15.1   Growth of Bacteria Against Optical Density at 600 nm

Culture (Incubation Period in min)	Optical Density at 600 nm[a]
0	
30	
60	
90	
120	
210	
360	

[a]Optical density (OD = 2−log of percent transmittance).

Example: If the 1%T of one of the sample is 53.5, the OD would be

OD = 2 − log of 53.5

= 2 − 1.7284 (0.7284 is the value found in the log table is the mantissa

= 0.272

Plot a graph between OD versus incubation time to prepare the growth curve and calculate the generation time for the three bacterial cultures.

15.21 Serial Dilutions

15.21.1 Principle

SPC is universally used to count the number of bacteria present in water, milk, and food. It is relatively easy to perform and gives excellent results.

15.21.2 Materials

Test tubes, conical flasks, test tube rack, non absorbent cotton, pipettes, Petri plates, spectrophotometer, water bath, autoclave, incubator, laminar flow, pH meter, Bunsen burner peptone, disodium phosphate, sodium chloride, dextrose, agar, distilled water infusion from calf brain and beef heart, and *E. coli* cultures.

Nutrient broth medium is prepared as described earlier.

15.21.3 Protocol

1. Separate 18 9 mL sterile water blank into six sets of 3 mL water each. Label each set as to time of inoculation (t_0, t_{30}, t_{60}, t_{90}, t_{120}, and t_{150}) and the dilution to be effected in each water blank (10^{-2}, 10^{-4}, 10^{-6}, and 10^{-7}).
2. Label six sets of four Petri dishes as to time of inoculation and dilution to be plated (10^{-4}, 10^{-5}, and 10^{-7}).

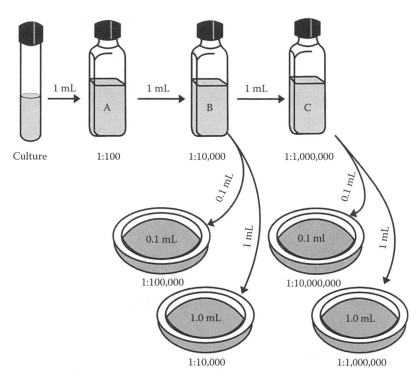

Figure 15.22 Method of preparing proper dilution of samples.

3. Liquify the four bottles of nutrient agar in a water bath. Cool and maintain at 45°C.
4. Take t_0 OD. Shake the culture flask and aseptically transfer 1 mL to the 99 mL water blank labeled t_0 10^{-2} and continue to dilute serially to 10^{-4} and 10^{-6}.
5. Plate t_0 dilutions on the appropriately label to plates. Aseptically pour 15 mL of the molten agar into each plate and mix by gentle rotation.
6. The same procedure is continued in serial dilution, and plate in the respectively labeled Petri plates.
7. When the pour-plate cultures harden, incubate them in an inverted position for 24 h at 37°C (Figure 15.22).

15.21.4 Results and Observations

Perform cell counts on all plates. Because each single cell in the plate becomes visible as a colony, which can then be counted (Table 15.2).

15.22 Measurement of Fungal Growth by Colony Diameter Method

15.22.1 Principle

The fungi grow on semi-synthetic or synthetic media forming colony with radial growth of different size, shape, and color. Size of colonies enhances every day

TABLE 15.2   Qualitative Estimation of Bacterial Count

Incubation Time (min)	Plate Counts (Cells/mL)	Log of Cells/mL
0		
30		
69		
90		
120		
150		

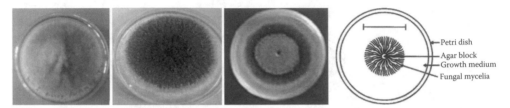

Figure 15.23   (See color insert.) Radial growth of two test fungus on PDA agar plates and schematic diagram.

attaining the full growth at a definite period. However, the fungal growth can be estimated by measuring the size of the colonies per day (Figure 15.23).

15.22.2 Materials

PDA, culture tube of a test organism, sterile Petri dishes, cork borer (95 mm diameter), measuring scale or vernier caliper, parafilm, incubator, and Bunsen burner.

15.22.3 Protocol

1. Grow the fungus on the PDA plates in three replicates, cut agar blocks of 5 mm diameter from actively growing margin of test fungus and transfer block in the center of agar plates.
2. Seal the margin of the plates with parafilm to avoid contamination, and incubate the inoculated plates at 28°C ± 1°C for 5 days in inverted position.
3. At regular intervals of 24 h, measure colony diameter of the fungus from the four different corners growing on the medium and tabulate the results.

15.22.4 Results and Observations

Take the average value of the three readings and calculate radial growth of the test fungus/day by dividing this value by total number of days incubated. For example, a fungus attained 60 mm of its colony in 5 days, then its mean growth rate will be 12 mm/day.

15.23 Biomass Measurement of Fungal Growth (Dry Weight Method)

15.23.1 Principle

Radial growth measurement method does not give the correct picture of growth of some of the fungi. Several fungi form mycelia on the upper surface of agar showing colony, whereas a few penetrate inside the agar and form mycelia network. Therefore, those growing inside semisolid agar could not be taken into account. In addition, growing edges of hyaline mycelia are not readily observed with naked eyes. Therefore, mycelia dry weight method may be used to estimate fungal growth.

15.23.2 Materials

Czapek-Dox broth medium, pure culture of the test fungus, cork borer (5 mm diameter), inoculation needle, Whatman No. 42 filter paper, and incubator.

CDA: $NaNO_3$ 2 g, K_2HPO_4 1 g, $MgSO_4 \cdot 7H_2O$ 0.5 g, KCl 0.5 g, FeSO4 0.01 g, sucrose 30 g, agar-agar 15–20 g (for slants and Petri plates), and distilled water to 1 L.

If glass distilled water is used, add 1 mL of 1% $ZnSO_4$ and 0.5% $CuSO_4$. Heat all chemical solution with sucrose in a water bath for 15 min. After cooling, add sucrose and agar-agar and autoclave.

15.23.3 Protocol

1. Prepare Czapek-Dox broth medium and sterilize at 15 psi/in.², cut agar block of 5 mm diameter from actively growing margin of the test fungus by using a sterile cork borer, and transfer one block into each flask with the help of a sterile inoculation needle, and incubate the flask at 28°C ± 1°C in an incubator for 15 days depending on the fungal growth.
2. Filter the mycelia mat through pre weighed Whatman No. 42 filter paper and dry at 80°C for 24 h in an oven, weigh dry mycelia mat and record dry weight (Figure 15.24).

Figure 15.24   Fungal mat of three different test fungus in culture flasks.

15.23.4 Results and Observations

Measure average mycelia dry weight (average of three readings) and estimate growth rate of the test fungus by dividing the value by the total number of the incubation period.

15.24 Effect of Incubation Temperature on Growth

15.24.1 Materials

Culture of *Curvularia lunata*, *Trichoderma viride*, *E. coli*, *Bacillus globisporus*, and *Bacillus thermophilus*.

Nutrient agar slants, CDA plates, Bunsen burner, inoculating loop, inoculating needle, and four incubators set at 15°C, 35°C, 55°C, and 75°C.

15.24.2 Protocol

1. Label each of the nutrient agar slants with the test organisms (bacteria) and the temperature of incubation (0°C, 4°C, 15°C, 35°C, 55°C, and 75°C).
2. Inoculate six nutrient agar slants (labeled 0°C, 4°C, 15°C, 35°C, 55°C, and 75°C) with *E. coli* by streaking the surface of each slope.
3. Repeat Step 2 for inoculation of the remaining cultures.
4. Label each of the CDA plates with *C. lunata* and *T. viride* and the temperature of incubation (0°C, 5°C, 15°C, 35°C, 55°C, and 75°C).
5. Inoculate six CDA plates with *C. lunata* and other six plates with *T. viride* by transferring a disc on the center of each plate.
6. Incubate one inoculated slant and plate from each set at 0°C (ice box of refrigerator), 4°C (refrigerator) for 10–14 days, and other slants and plates at 15°C, 35°C, 55°C, and 75°C (incubators set at these temperatures) for 24–48 h (bacterial) and 7–14 days (fungal cultures).

15.24.3 Results and Observations

Examine the cultures after 2 days incubation (incubated at 15°C, 35°C, 55°C, and 75°C) and after 10–14 days incubation (at 0°C and 4°C) for the presence (+) or absence (−) of growth and the degree of growth [i.e., minimal growth (+), moderate growth (2+), heavy growth (3+), very heavy (maximum growth (4+))]. Express the results in the form of a table and determine the minimum, maximum, and optimum temperature ranges for each microorganism and classify each organism as psychrophile, mesophile, and thermophile (Table 15.3).

TABLE 15.3   Growth Observed in Four Test Organisms at Various Incubation Temperature

Species	Growth at 0°C, 4°C, 15°C, 35°C, 55°C, and 75°C	Cardinal Temperature (Min, Max, and Opt)	Tem. Classification (T, M, and P)
C. lunata			
T. viride			
E. coli			
B. globisporus			
B. thermophilus			

Note: Growth is indicated as −, +, 2+, 3+, or 4+.
T = thermophile, M = mesophile, and P = psychrophile.

15.25 Effect of pH on Microbial Growth

15.25.1 Introduction

Growth and survival of microorganism are greatly influenced by the pH of the environment, and all microbes differ as to their requirement. Each species has the ability to grow within a specific pH range, which may be broad or limited, with the most rapid growth occurring within a narrow optimum range.

pH is the symbol for the logarithm of reciprocal of hydrogen ion concentration. For example, a pH 4 indicates a concentration of 0.0001 or 10^{-4} g atoms of hydrogen ions in 1 L of solution. It is the negative exponent of the concentration. The measure of the relative acidity or alkalinity of a solution is called pH and is expressed from 0 to 14 range. A more acidic solution has a lower pH value and alkaline solution has a higher pH. The pH values above and below which an organism fails to grow are turned as the minimum and maximum pH concentrations. Enzymes differ greatly in their pH optima.

15.25.2 Materials

Test tubes, pH meter, inoculation loop, pipettes, Bunsen burner, incubator, autoclave, test tube stand, laminar flow, peptone, beef extract, NaCl, distilled water, cultures of C. lunata, Saccharomyces cerevisiae, E. coli, and S. aureus.

Nutrient agar medium g/L in distilled water pH 7.0: Peptone 5, beef extract 3, and NaCl 5.

15.25.3 Protocol

1. Tubes containing sterile, buffered nutrient broth adjusted to pH 2.5, 3.5, 4.5, 5.5, 6.5, 7.5, 8.5, and 9.5 (4 for each pH, i.e., total 32 tubes) are taken in a test tube stand.
2. Inoculate each of a series of the tubes with E. coli by adding 0.1 mL of the culture to each tube.

3. Repeat Step 2 for the inoculation of *S. aureus* and *S. cerevisiae*, to various labeled tubes.
4. Incubate the bacterial inoculated tubes for 24–48 h at 35°C and fungal inoculated tubes for 72–120 h at 25°C.

15.25.4 Result and Observations

Record the growth of each organism by using the symbols −, no growth; +, slight growth; 2+, moderate; 3+, good; or 4+, optimal growth for each microorganisms.

15.26 Microscopic Examination of Bacteria

15.26.1 Hanging Drop Preparation or Motility Test

15.26.1.1 Introduction
Visualization of microorganisms in the living state is most difficult because they are minute, transparent, and practically colorless when suspended in an aqueous medium. Bacteria are quite colorless and transparent; and unless the diaphragm is carefully adjusted, usually there is considerable difficulty in bringing the organisms into focus. To study their properties and to divide microorganisms into specific groups for diagnostic purpose, biological stains and staining procedures in conjunction with light microscopy have become major tools in microbiology. The study of bacterial morphology is performed in two ways:

1. Observing unstained cells alive by hanging drop preparation.
2. Observing dead cells by making use of the chemical nature of their unicellular body. This is achieved by staining.

Hanging drop slide technique is used for nonpathogens, whereas the one tube one plate method can be used for pathogens. Each method has its advantage and limitations. The selection of method will depend on which one is most suitable for the situation at hand.

15.26.1.2 Materials
Coverslip, bacterial broth culture, inoculation loop, microscopic cavity slide, microscope, Vaseline or petroleum jelly, inoculation needle, and Bunsen burner.

15.26.1.3 Protocol
1. Apply Vaseline around the depression of the hanging-drop slide.
2. Using the inoculation loop, aseptically transfer one drop of the culture to the center of a clean cover slip.
3. Invert the hanging-drop slide and center its well over the drop of the culture. Press down on the edge of the cover slip so that the Vaseline makes a firm seal.
4. Quickly and carefully turn the slide right side up so as to suspend the hanging drop in the well. Do not let the drop fall or touch the bottom of the well.

15.26.1.4 Results and Observations

To examine, first locate its edge in center of the microscopic field with low power objective and markedly lower the light. The edge will be visible, as a bright wavy like against a dark background. Now the slide can be focused under oil immersion.

15.27 Microscopic Examination of Pathogens by One Tube One Plate Method

15.27.1 Introduction

When working with pathogenic microorganisms such as the typhoid bacillus, it is too dangerous to attempt to determine motility with slide techniques. A much safer method is to culture the organisms in a special medium that can demonstrate the presence of motility. The procedure is to inoculate a tube of semisolid or sulfur reduction/indole production/motility media (SIM) medium that can demonstrate the presence of motility. Media have a very soft consistency that allows motile bacteria to migrate readily through them causing cloudiness. Figure 15.25 illustrates the inoculation procedure.

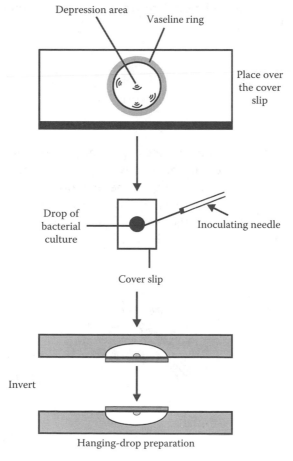

Figure 15.25 Illustrating hanging drop method.

15.27.2 Materials

SIM medium g/L in distilled water: Casein peptone 20, meat peptone 6.10, ferric ammonium sulfate 0.2, sodium thiosulfate 0.2, and bacteriological agar 3.5. Final pH 7.3 at 25°C. Dispense into appropriate containers and sterilize in an autoclave at 121°C for 15 min. The prepared medium should be stored at 2°C–8°C. The color is amber and slightly opalescent.

15.27.3 Protocol

Inoculate the pure culture by stabbing to a depth of three-quarters of the tube. Incubate at 35°C ± 2°C for 18–24 h and read the results.

15.27.4 Results and Observation

Darkening indicates the production of H_2S. Motility is indicated by a diffuse turbidity away from the line of inoculation. Growth only along the inoculation line indicates nonmotility (Figure 15.26).

15.28 Smear Preparation

15.28.1 Introduction

The success of a staining procedure depends upon the preparation of a suitable smear of the organisms. A good smear is one that, when dried, appears as a thin, whitish layer or film. A properly prepared smear is one or more washings during staining without loss of organisms. The first step in preparing a bacterial smear differs according to the source of the organisms, those made from broth cultures, or cultures from a solid medium require variations in technique.

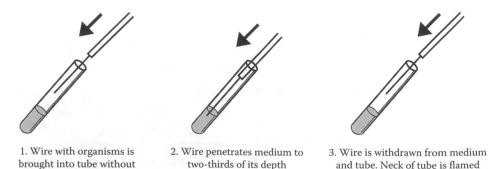

1. Wire with organisms is brought into tube without touching the walls of tube

2. Wire penetrates medium to two-thirds of its depth

3. Wire is withdrawn from medium and tube. Neck of tube is flamed and plugged

Figure 15.26 Stab technique for motility test.

15.28.2 Protocol

15.28.2.1 From Liquid Media (Figure 15.27)
1. Apply one or two loopful of suspended cells to a clean glass slide.
2. Spread it evenly over a small area on a slide.
3. Allow the slide to dry by normal evaporation of the water. Do not apply heat.
4. After the smear has become completely dry, pass the slide over a Bunsen burner, flame to heat-kill and fix the organisms to the slide.

15.28.2.2 From Solid Medium (Figure 15.27)
1. Place a loopful of water on the slide.
2. Flame an inoculating needle, let it cool, pick up a very small amount of the organism, and mix them into the water on the slide.
3. Disperse the mixture over a small area in the center of the slide.
4. Allow the slide to dry by normal evaporation of the water.
5. Once the smear is completely dry, pass the slide over the flame of a Bunsen burner to heat-kill and fix the organism to the slide.

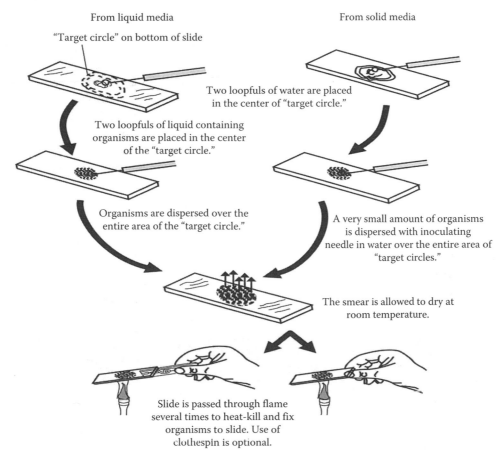

From liquid media

"Target circle" on bottom of slide

Two loopfuls of water are placed in the center of "target circle."

Two loopfuls of liquid containing organisms are placed in the center of the "target circle."

Organisms are dispersed over the entire area of the "target circle."

From solid media

A very small amount of organisms is dispersed with inoculating needle in water over the entire area of "target circles."

The smear is allowed to dry at room temperature.

Slide is passed through flame several times to heat-kill and fix organisms to slide. Use of clothespin is optional.

Figure 15.27 Procedure for preparation of bacterial smear.

15.29 Staining

It is a chemical or a physical union between the dye and the component of a cell. If it is a chemical reaction, a new compound is formed and a simple washing with water does not liberate the bound dye, but if purely physical, it is easy to decolorize such stained organism. Usually, it is a combination of chemical and physical reactions. The main advantages of staining are as follows:

1. Provides contrast between microorganisms and their backgrounds, permitting differentiation among various morphological types
2. Permits study of internal structures of the bacterial cell, such as the cell wall, vacuoles or nuclear bodies, and other cellular structures
3. Enables the bacteriologist to use higher magnifications

15.29.1 Type of Stains

15.29.1.1 Simple Stains

15.29.1.1.1 Principle. The use of a single stain to color a bacterial organism is commonly referred to as simple staining. All these dyes work well on bacteria as they have color bearing ions (chromatophores) and are positively charged. Bacteria are slightly negatively charged when the pH of the surrounding is near neutrality and produces a pronounced attraction between these cationic chromatophores and the organism so that the cell is stained. Such dyes are classified as basic dyes. Commonly used basic stains are methylene blue, crystal blue, and carbol fuchsin. Those dyes that have anionic chromatophores are called acidic dyes. Eosine (sodium$^+$ eosine$^-$) is such a dye. The anionic chromatophores, eosine$^-$, will not stain bacteria because of the electrostatic repelling forces that are involved.

The staining times for most simple stains are relatively short, usually from 30 s to 2 min, depending on the affinity of the dye. After a smear has been stained for the required time, it is washed off gently, blotted dry, and examined directly under oil immersion. Such a slide is useful in determining basic morphology and the presence or absence of certain kinds of granules (Figure 15.28).

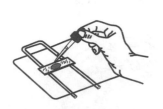 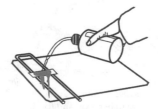 

1. A bacterial smear is stained with methylene blue for 1 min.	2. Stain is briefly washed off slide with water.	3. Water drops are carefully blotted off slide with bibulous paper.

Figure 15.28 Simple staining procedure.

15.29.1.1.2 Protocol
1. Prepare bacterial smear as described earlier.
2. Place the slide on the staining tray and flood with a required simple stain, using the appropriate exposure time (carbol fuchsin requires 15–30 s, crystal violet 20–60 s, and methylene blue 1–2 min).
3. Wash the smear with tap water to remove excess stain. During this step, hold the slide parallel to the stream of water to reduce the loss of cells.
4. Blot-dry the slide and observe under light microscope.

15.29.1.2 Negative Stain

15.29.1.2.1 Principle. A better way to observe bacteria for the first time is to prepare a slide by a process called negative or background staining. Negative staining requires the use of an acidic stain such as eosin or nigrosin. The acidic stain, with its negatively charged chromogen, will not penetrate the cells because of the negative charge on the surface. Therefore, the unstained cells are easily visible against the colored background. This method consists of mixing the microorganisms in a small amount of nigrosine or India ink and spreading the mixture over the surface of the slide (nigrosine is far superior to India ink; Figure 15.29). Since these two pigments are not really bacterial stains, they do not penetrate the microorganisms; instead they obliterate the background, leaving the organisms transparent and visible in a darkened field. Although this technique has limitations, it can be useful for determining cell morphology and size. Since no heat is applied to the slide, there is no shrinkage of the cell, and, consequently more

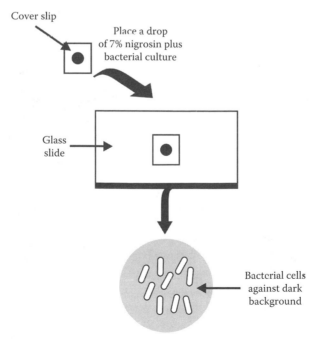

Figure 15.29 Demonstrating negative staining procedure.

accurate cell size determination result than with some other methods. This method is also useful for studying spirochaetes that do not stain readily with ordinary dyes.

15.29.1.2.2 Protocol
1. Place a small drop of nigrosin close to one end of a clean slide.
2. Using sterile technique, place a loopful of inoculums from the culture in the drop of nigrosin and mix.
3. With the edge of a second slide held at 30° angle and placed in front of the bacterial suspension, push the mixture to form a thin smear.
4. Air dry. Do not heat and fix the smear.
5. Examine the slide under oil immersion.

15.29.1.2.2.1 Spreader Slide Technique for Negative Staining. Negative staining can be done by one of the following methods. Figure 15.30 illustrates the more commonly used method in which the organisms are mixed in a drop of nigrosine and spread over the slide with another slide in order to prepare a smear that is thick at one end and feather thin at the other end. Somewhere between the too thick and too thin areas will be an ideal spot to study an ideal spot to study the organisms.

15.29.1.2.2.2 Nonspreader Slide Method for Negative Staining. In the second method, the organisms are mixed in only a loopful of nigrosine instead of a full drop. The organisms are spread over a smaller area in the center of the slide with an inoculating needle. No spreader slide is used in this method. It gives more accurate view of the bacterial cell (Figure 15.31).

While negative staining is a simple enough process to make bacteria more visible with a brightfield microscope, it is of little help when one attempts to observe

1. Organisms are dispersed into a small drop of nigrosine or India ink. Drop should not exceed 1/8 in. diameter and should be near one end of the slide.

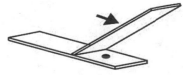

2. Spreader slide is moved toward drop of suspension until it contacts the drop causing the liquid to be spread along its spreading edge.

3. Once the spreader slide contacts the drop on the bottom slide, the suspension will spread out along the spreading edge as shown.

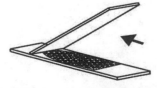

4. Spreader slide is pushed to the left, dragging the suspension over the bottom slide. After the slide has air dried, it may be examined under oil immersion.

Figure 15.30 Spreader slide technique for negative staining.

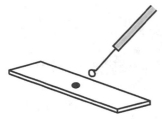

1. A loopful of nigrosine or India ink is placed in the center of a clean microscope slide.

2. A sterile inoculating wire is used to transfer the organisms to the liquid and mix the organisms into the stain.

3. Suspension of bacteria is spread evenly over an area of 1–2 cm with the straight wire.

4. Once the preparation has completely air-dried, it can be examined under oil immersion. No heat should be used to hasten drying.

Figure 15.31   Nonspreader slide method for negative staining.

anatomical microstructures such as flagella, granule, and enodospore. Only by applying specific bacteriological stains to organisms can such organelles be seen. They are of (a) acidic dyes, (b) basic dyes, and (c) indifferent dyes.

15.29.1.3 Differential Stains

15.29.1.3.1 Gram Stain
  15.29.1.3.1.1 Principle.   The most important differential stain used in bacteriology is the Gram stain, developed by Christian Gram in 1884. By using this procedure, bacteria have been classified into two broad groups—Gram-positive and Gram-negative bacteria, which make it an essential tool for classification and differentiation of microorganisms. In the Gram stain, a bacterial smear is dried and then heat fixed. It is then stained with crystal violet (primary stain), which is rinsed off and replaced with an iodine solution. The iodine acts as mordant, that is, it binds the dye to the cell. The smear is then decolorized with alcohol and counterstained with safranin. In Gram-positive organisms, the purple crystal violet dye, complexed with the iodine solution, is not removed by the alcohol and thus the organisms remain purple (Figure 15.32). On the other hand, the purple stain is removed from the Gram-negative organism by alcohol and the colorless cells take up the red color of the safranin counterstain. Gram staining requires four different solutions. (i) A *basic dye*, crystal violet here; (ii) *a mordant*, which increases the affinity or attraction between the cell and the dye, for example, iodine; (iii) *a decolorizing agent*, which removes the dye from a stained cell, for example, alcohol, acetone, or ether; and (iv) *counter stain*, which is a basic dye of a different color than the initial one, for example, safranin (Table 15.4 and Figure 15.33).

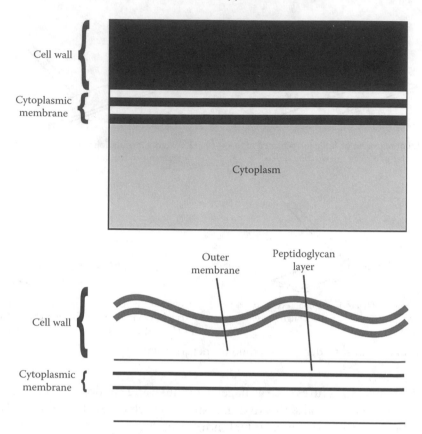

Figure 15.32 Cell wall of Gram-positive and Gram-negative bacteria.

TABLE 15.4 Steps in Gram Staining

Steps	Process	Results Gram⁺	Gram⁻
Initial stain Mordant	CV for 30 s	Stains purple	Stains purple
Decolorization	12 for 30 s	Remains purple	Remains purple
Counterstain	95% alcohol for 10–20 s	Remains purple	Becomes colorless
	Safranin for 20–30 s	Remains purple	Stains pink

15.29.1.3.1.2 Materials

Stain preparation

1. *Crystal violet*: Dissolve 2 g of crystal violet in 20 mL of 95% ethanol. Dissolve 0.8 g of ammonium oxalate in 80 mL of distilled water. Mix these 2 solutions, stand for 24 h, and filter.
2. *Gram's iodine*: Dissolve 2 g of potassium iodide and 1 g of iodine in 300 mL of distilled water.
3. *Safranin*: Grind 0.25 g of safranin in a mortar with 10 mL of 95% ethanol.

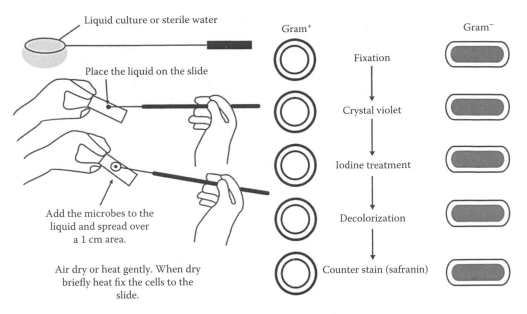

Figure 15.33 Illustrating differential (Gram) staining in bacteria.

15.29.1.3.1.3 Protocol

1. Prepare smear on a clean slide, stain with crystal violet for 30 s, and rinse with water.
2. Flood the film with Gram's iodine and allow it to act for 30 s and rinse with water.
3. Decolorize with 95% alcohol, rinse with water and counter stain with safranin for 20–30 s, and then rinse with water and blot dry, examine under oil immersion objective immediately (Figure 15.34).

Note

1. Use fresh culture and preferably culture for making smear.
2. Old cultures of Gram-positive organisms lose their ability to retain the crystal violet.
3. Do not over decolorize with alcohol. The Gram-positive organisms may also appear Gram-negative.
4. Use thin smears. It is difficult for the dyes to penetrate properly if thick smears are used.

15.29.1.3.2 Acid Fast Stain

15.29.1.3.2.1 Principle. The acid fast stain is useful for identifying bacteria with a waxy lipid cell wall. Most of these organisms are members of a group of the genus *Mycobacterium*. These organisms have a Gram-positive cell wall, but the lipid in the cell wall prevents staining with the Gram-stain dyes.

15.29.1.3.2.2 Materials

Stain preparation: In the Ziehl–Neelsen method, three different reagents are used:

1. *Primary stain*—Carbol fuchsin, a phenolic stain is driven into the waxy cell wall with steam.

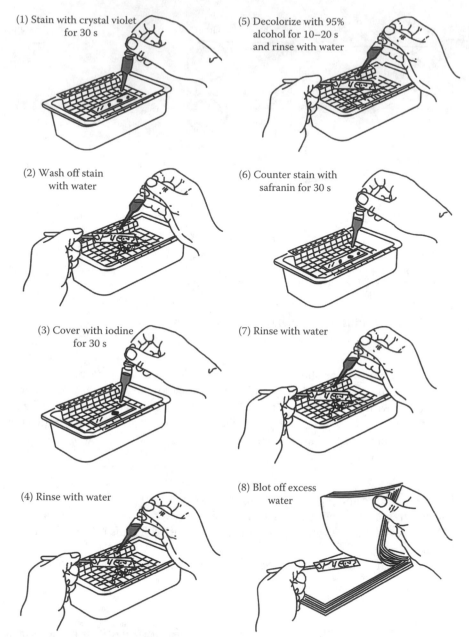

(1) Stain with crystal violet for 30 s

(2) Wash off stain with water

(3) Cover with iodine for 30 s

(4) Rinse with water

(5) Decolorize with 95% alcohol for 10–20 s and rinse with water

(6) Counter stain with safranin for 30 s

(7) Rinse with water

(8) Blot off excess water

Figure 15.34 Gram staining procedure.

2. *Decolorizing agent*—Acid alcohol (3% HCl + 9.5% ethanol) is used a decolorizing agent. The mycobacterial cells are acid fast are not decolorized and thus retain the primary stain. Non mycobacteria are, however, decolorized by acid alcohol.

3. *Counterstain*—Methylene blue is used as the final reagent to stain previously decolorized cells. The colorless, non mycobacteria will take up the blue color, so they contrast with the pink acid-fast bacteria that were not decolorized.

15.29.1.3.2.3 Protocol

1. Prepare a smear of the material and heat fix it.
2. Place the slide over a beaker of boiling water and cover the slide with carbol fuchsin. Cover it with a paper towel to prevent the dye from flowing out.
3. Keep the slide covered with stain and steam for 5 min.
4. Remove paper and wash off carbol fuchsin in tap water.
5. Flush all freely removable stain with acid alcohol.
6. Flood the entire slide with acid alcohol, and allow it to stand for 1–2 min.
7. Wash with water for 5 s.
8. Counterstain for about 30–45 s with methylene blue.
9. Wash with water
10. Blot dry carefully and microscopically examine under the oil immersion lens (Figures 15.35 and 15.36).

15.29.1.4 Structural Stains

15.29.1.4.1 Endospore Stain

15.29.1.4.1.1 Principle. Species of bacteria, belonging principally to the genera *Bacillus* and *Clostridium* produce extremely heat resistant structures called endospores. In addition to being heat-resistant, they are also resistant to many chemicals that destroy nonspore forming bacteria. This resistance to heat and chemicals is due primarily to a thick, tough spore coat. They resist staining and, once stained, they resist decolorization and counterstaining.

Several methods are available that employ heat to provide stain penetration. However, since the malachite green—Schaeffer and Fulton and Dorner methods—are

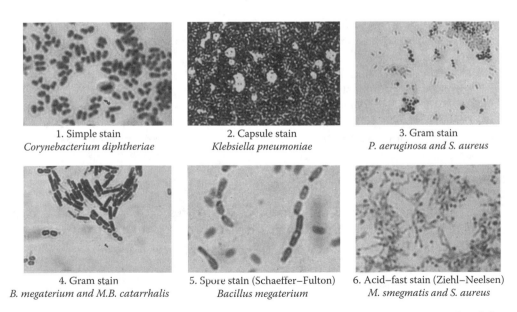

| 1. Simple stain *Corynebacterium diphtheriae* | 2. Capsule stain *Klebsiella pneumoniae* | 3. Gram stain *P. aeruginosa and S. aureus* |

| 4. Gram stain *B. megaterium and M.B. catarrhalis* | 5. Spore stain (Schaeffer–Fulton) *Bacillus megaterium* | 6. Acid–fast stain (Ziehl–Neelsen) *M. smegmatis and S. aureus* |

Figure 15.35 **(See color insert.)** Different colors as observed under microscope of acid fast bacteria.

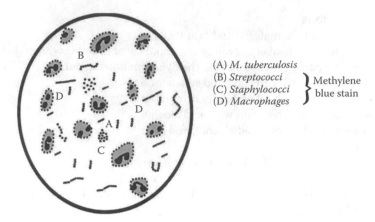

(A) *M. tuberculosis*
(B) *Streptococci*
(C) *Staphylococci*   } Methylene
(D) *Macrophages*   blue stain

Figure 15.36   Various forms of acid fast stain bacteria.

commonly used by most bacteriologists. Thus, endospore stains green, but the rest of the cell or a cell without endospore stains light red (Figure 15.37).

15.29.1.4.1.2 Malachite Green—Schaeffer and Fulton—Method
Principle:

This method utilizes malachite green to stain the endospore and safranin to stain the vegetative portion of the cell. A proper spore former will have a green endospore contained in a pink sporangium.

Protocol:

1. Prepare a smear on a clean slide and heat fix.
2. Take a beaker with about an inch of water and bring it to boil.
3. Place the slide on the beaker.
4. Flood the slide with malachite green, steam over boiling water bath for 5 min, and add additional stain if stain runs off.
5. After the slide has cooled sufficiently, rinse with water for 30 s.
6. Counterstain with safranin for about 30 s and then wash with tap water for 30 s.
7. Blot dry and microscopically examine slide under oil immersion (Figure 15.38).

15.29.1.4.1.3 Dorner Method
Principle:

The Dorner method for staining endospores produces a red spore within a colorless sporangium. Nigrosine is used to provide a dark background for contrast.

Protocol:

1. Make a heavy suspension of bacteria by dispersing several loopfuls of bacteria in five drops of sterile water.
2. Add five drops of carbol fuchsin to the bacterial suspension.
3. Heat the carbol fuchsin suspension of bacteria in a beaker of boiling water for 10 min.

1. Place piece of paper towel on heat fixed smear

2. Add malachite green

3. Remove from heat and discard paper

4. Rinse with water

5. Add safranin

6. Rinse with water

7. Blot dry

Figure 15.37 Steps in the endospore staining.

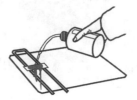

1. Cover smear with small piece of paper toweling and saturate it with malachite green. Steam over boiling water for 5 min. Add additional stain if stain boils off.

2. After the slide has cooled sufficiently, remove the paper toweling and rinsewith water for 30 s.

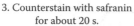
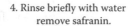

3. Counterstain with safranin for about 20 s.

4. Rinse briefly with water to remove safranin.

5. Hot dry with bibulous paper and examine slide under oil immersion

Figure 15.38　Malachite green—Schaeffer and Fulton stain—method.

4. Mix several loopfuls of bacteria in a drop of nigrosine on the slide.
5. Spread the nigrosine bacteria mixture on the slide.
6. Allow the smear to air dry. Examine the slide under oil immersion (Figure 15.39).

15.29.1.4.2 Capsule Stain
15.29.1.4.2.1 Principle
Some bacterial cells are surrounded by a pronounced gelatinous or slimy layer called a capsule. Capsules appear to be made up of glycoprotein or polypeptides. It can be observed by differential staining.

15.29.1.4.2.2 Protocol
1. Make the suspension of the organism in a drop of water on a clean slide.
2. Put a small drop of India ink next to it.
3. Spread the ink suspension of bacteria over slide and air dry it. Caution: Do not heat fix.
4. Stain the smear with crystal violet for 1 min.
5. Wash off the crystal violet gently with water.
6. Gently blot dry and examine the slide with an optical microscope under oil immersion immediately.

For dry preparations, mix one loopful of Indian ink with one loopful of suspension of an organism in 5% dextrose solutions at one end of a slide. Allow to dry and pour few drops of methyl alcohol on it, and keep the slide over the flame to fix. Stain for a few seconds with 0.5% aqueous solution of methyl violet. The capsule will appear as haloes in blue cell of bacterium under microscope (Figure 15.40).

1. Make a heavy suspension of bacteria
by dispersing several loopfuls of
bacteria in five drops of sterile water.

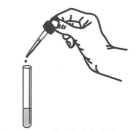

2. Add five drops of carbol fuchsin to the
bacterial suspension.

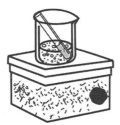

3. Heat the carbol fuchsin suspension of
bacteria in beaker of boiling water
for 10 min.

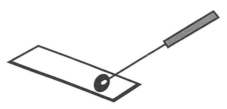

4. Mix several loopfuls of bacteria in
a drop of nigrosine on the slide.

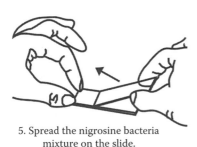

5. Spread the nigrosine bacteria
mixture on the slide.

6. Allow the smear to air-dry. Examine
the slide under oil immersion.

Figure 15.39 Illustrating Dorner method of staining.

15.29.1.4.3 Flagella Stain

15.29.1.4.3.1 Principle.   Some bacteria have flagella for motility. Their width is below the resolving power of the microscope so they cannot be seen in a light microscope. Flagella can be viewed if they are dyed with a special stain that precipitates on them, making them appear thicker. Leifson's method accomplishes this by using a single staining reagent that utilizes pararosaniline as a staining reagent and tannic acid as a mordant. The arrangement of flagella on bacteria is usually characteristic of the organism and can aid in identification.

1. Two loopfuls of the organism are mixed in a small drop of India ink.

2. The ink suspension of bacteria is spread over the slide and air-dried.

3. The slide is gently heat-dried to fix the organisms to the slide.

4. Smear is stained with crystal violet for 1 min.

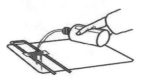

5. Crystal violet is gently washed off with water.

6. Slide is blotted dry with bibulous paper, and examined with oil immersion objective.

Figure 15.40   Illustrating six steps for dry preparation in capsule staining.

15.29.1.4.3.2 Materials
1. *Mordant*: 20% tannic acid solution three parts, 5% tartar emetic solution two parts, and distilled water five parts. Mix and redissolve the heavy precipitate by boiling.
2. *Silver solution*: Dilute a saturated solution of silver sulfate with an equal volume of distilled water. Add 33% monoethylamine till the precipitate formed is redissolved. Solution can be stored for long term; heat gently before using.

15.29.1.4.3.3 Protocol
A. Culture preparation:
1. Inoculate nutrient broth with the organism and incubate at room temperature for 18–20 h.
2. Add 0.25 mL of formalin to the culture, mix by shaking, and let it stand for 15 min.
3. Fill the tube to within 1 cm of top with distilled water, mix, and centrifuge for 3 min.
4. Pour off the supernatant fluid without disturbing the pellet.
5. Resuspend pellet in about 2 mL of distilled water.
6. Dilute the suspension with additional distilled water until suspension is barely turbid.
B. Staining procedure:
1. Heat a clean slide in Bunsen burner flame.
2. While the slide is still hot, mark an oval outline on the slide with pencil.
3. Place several loopfuls of organism at right end of cooled slide.
4. Tilt slide to allow organism to flow down over the surface of the slide.
5. Allow the smear to completely air dry. Do not apply any heat.
6. Cover the smear with Leifson's stain and leave it on the slide until all the alcohol has evaporated.
7. Wash gently to remove the stain from the slide.
8. Allow the stained organisms to air dry and microscopically examine under oil immersion (Figure 15.41).

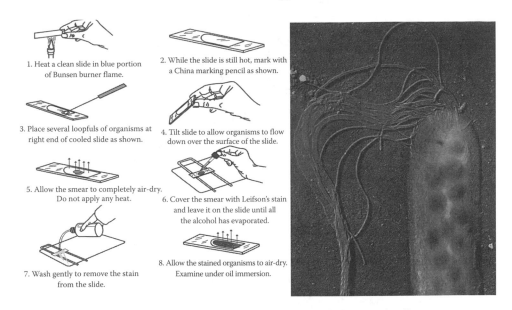

1. Heat a clean slide in blue portion of Bunsen burner flame.

2. While the slide is still hot, mark with a China marking pencil as shown.

3. Place several loopfuls of organisms at right end of cooled slide as shown.

4. Tilt slide to allow organisms to flow down over the surface of the slide.

5. Allow the smear to completely air-dry. Do not apply any heat.

6. Cover the smear with Leifson's stain and leave it on the slide until all the alcohol has evaporated.

7. Wash gently to remove the stain from the slide.

8. Allow the stained organisms to air-dry. Examine under oil immersion.

Figure 15.41   Eight-step procedure for flagella staining of a bacterial cell.

15.30 Biochemical Tests

15.30.1 Introduction

A diverse range of biochemical reagents are known for the identification of certain metabolisms and to differentiate between bacteria. Classical biochemical tests are often used to identify microorganisms; the results are seen by color change. In most cases, detection is based on the reaction of an enzyme with a certain substrate. Furthermore, methods to detect certain metabolites by chemical reaction and or complex building techniques are used. At the end, a color change gives a result that leads to greater recognition of the unknown organism.

15.30.1.1 Amylase Activity

15.30.1.1.1 Principle.   Amylase is an enzyme confined to the starch hydrolyzing organisms. It can be tested using Lugol's iodine. Starch is made up of two polysaccharides:

1. Amylose, soluble in water and a polymer of 200–1000 D-glucose molecules linked by 1–4 glycosidic linkages; it dissolves in hot water giving blue color with iodine.
2. Amylopectin is insoluble in hot water and is a polymer of 100 chains of D-glucose molecules united by 1–4 glucosidic linkages. The chains are united by 1–6 glycosidic linkage (Figure 15.42).

15.30.1.1.2 Mode of Action of Amylase.   It is a complex of α and β amylases. Both degrade amylopectin to maltose unit; the latter is acted upon by β-glucosidase to

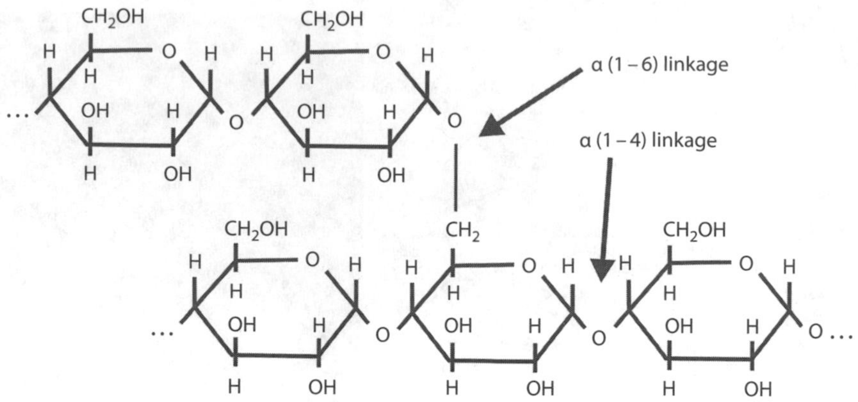

Figure 15.42   An α (1→6) branch point in amylopectin.

give glucose. While α-amylase alone degrades amylose to yield monomer glucose (Figure 15.44).

15.30.1.1.3 Materials
   A. Starch agar medium:
   1. Prepare a 10% solution of soluble starch in water and steam for 1 h.
   2. Add 20 mL of this solution to 100 mL of melted nutrient agar and pour plates. *Composition of nutrient agar in g/L in distilled water*: Beef extract 3, peptone 5, NaCl 5, NaNO$_3$ 5, and pH adjusted to 7.0.
   B. *Lugol's iodine*: Iodine 1 g, potassium iodide 2 g, and distilled water to 300 mL.

15.30.1.1.4 Protocol.
Inoculate starch agar and incubate for 24 h. Flood the plate with Lugol's iodine and observe for the clear zones (Figure 15.43).

15.30.1.1.5 Result and Observations.
Formation of yellow zone around the bacterial colony indicates the production of extracellular β-amylase on amylopectin (Figure 15.44).

15.30.1.2 Caseinase

15.30.1.2.1 Introduction.
Casein is the principal protein of milk. It exists as a colloidal suspension that gives milk its opaque whiteness. Many bacteria are equipped with caseinase enzyme that hydrolyses this protein. This reaction of peptonization is useful for identification of microorganisms (Figure 15.45).

15.30.1.2.2 Material

15.30.1.2.3 Protocol
   1. Reconstitute milk powder and mix equal volumes of this and double strength nutrient agar melted at 55°C, mix and pour plates.
   2. Inoculate the plates with a loop of cells at the center only.
   3. Incubate at 37°C for 24–48 h.

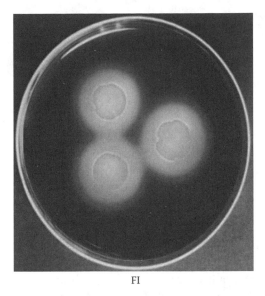

FI

Figure 15.43 (See color insert.) Test for production of amylase after flooding with Lugol's iodine solution.

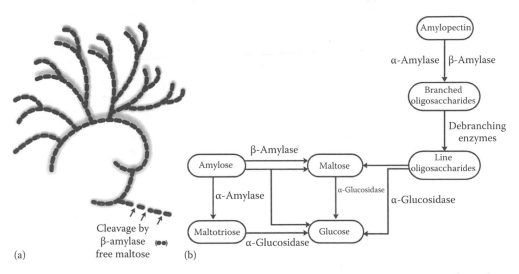

Figure 15.44 (a) Action of β-amylase on amylopectin. (b) Enzyme conversion of starch to glucose initiated by α- and β-amylase.

15.30.1.2.4 Result and Observations. Colonies of organisms that produce caseinase will be surrounded by clear zone. Areas where caseinase has not been produced will remain opaque.

15.30.1.3 Carbohydrate Digestion

15.30.1.3.1 Introduction. Microorganisms obtain their energy through a series of enzymatic reactions leading to the biooxidation of carbohydrate. Organisms

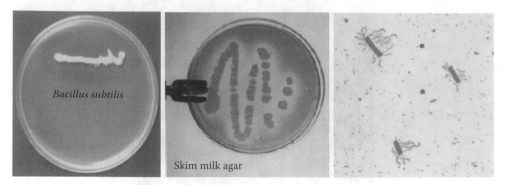

Figure 15.45   **(See color insert.)** Caseinase production by *Bacillus subtilis.*

use carbohydrates differently depending upon their enzyme constitution. Some of the organisms are capable of fermenting sugars such as glucose anaerobically while others use aerobic pathways. Still, the facultative anaerobes are enzymatically competent to use both aerobic and anaerobic pathways. In fermentation, substrates such as carbohydrates and alcohol undergo anaerobic dissimilation and produce an organic acid, for example, lactic acid and may be accompanied by the production of gases like hydrogen or carbon dioxide. Following incubation, carbohydrates that have been fermented and produce acid wastes turn the phenol from red to yellow, thereby indicating the production of acids. In some cases, acid production is accompanied by the evolution of gas (CO_2) that is visible as bubbles in the inverted (Durham) tube. The acid produced can be detected using a pH indicator, phenol red, which is red at pH 8.5 and yellow at pH 6.9. Gas formation is detected through the use of a Vasper (or Vaseline) seal or an inverted vial Durham tube. Gas formation in agar media is achieved by bubble formation. Carbohydrate is digested either by oxidation or fermentation, which in Gram-negative rods is an important identification feature (Figure 15.46). Three simple methods are described here.

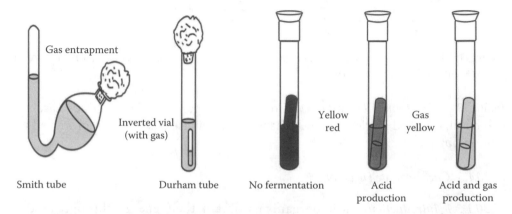

Figure 15.46   Illustrations of carbohydrate utilization with or without gas formation.

15.30.1.3.1.1 Protocol I (Inverted Tube Method). Composition of peptone water broth medium g/L in distilled water: Peptone 0.8, NaCl 1.4, NaHCO₃ 0.02, KCl 0.04 g, CaCl₂ 0.04, KH₂PO₄ 0.24, and Na₂HPO₄ · 2H₂O 0.88.

Steam the ingredients at 100°C for 20 min and filter through double fold Whatman No.1 filter paper. The pH is adjusted to 7.2. Autoclave at 121°C for 15 min.

Addition of sugar: Prepare 1% solution of the desired sugar. Autoclave at low pressure and add to the aforementioned solution in order to make to a final concentration of 10% by volume.

Addition of indicator (phenol red): Dissolve 0.1 g of phenol sulfonephthalein in 14.1 mL of 0.02 N NaOH and dilute to 250 mL with water, autoclave and add to basal medium making a final concentration of 0.4% by volume.

Phenol red is yellow at pH 6.8 (acid production) and red at pH 8.4 (no acid).

1. Transfer media in test tubes and invert a Durham tube in each of these tubes.
2. Inoculate each tube with the culture to be identified.
3. Incubate at 37°C.
4. Observe the change in color, if any

15.30.1.3.1.2 Method II (Disc Method). It differs from the earlier method in that the sugar is not added to medium directly but sterilized filter paper discs are impregnated with the sugar solution and then placed on the inoculated agar plates aseptically. Observe after 48 h (Figure 15.47).

15.30.1.3.1.3 Method III

Principle

It illustrates whether the carbohydrate is utilized oxidatively or by fermentation. Here, oxidation fermentation (OF) medium is used with bromothymol blue as a pH indicator.

The purpose of this test is to determine whether an organism attacks sugars (in this case glucose is used) by fermentation or oxidation. Two tubes of Hugh and Leifson's medium are used. In one tube, the medium is covered with Vaseline. If the organism is an oxidizer, it will produce acid only in the open tube; if it is a fermenter, it will produce acid in the Vaseline-covered tube and in the open tube. Some aerobic

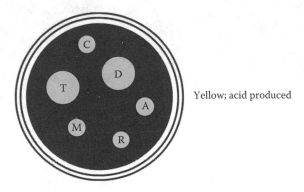

Yellow; acid produced

Figure 15.47 The disc method for determining carbohydrate fermentation.

bacteria may use the peptone in the medium, producing ammonia, with resulting alkalinity (blue) in the top part of the open tube. The indicator used is bromothymol blue. Motile bacteria will show diffuse growth away from the stab line. Nonmotile bacteria will only show growth confined to the stab line. In the examples given here, *E. coli* is motile and *Klebsiella* is nonmotile. Tubes are inoculated by stabbing (Figure 15.26). One tube is incubated aerobically, while other is kept in anaerobic condition. Fermented cultures will turn both tubes yellow, while oxidative culture will turn only one tube yellow. In some oxidative culture tubes, no acid end product will be detected. If the color production is homogeneous, it also confirms the motility of the organism.

Materials

Oxidation-fermentation (OF) agar medium g/L in distilled water: Peptone 2.0, sodium chloride 1.5, K_2HPO_4 1.0, and agar 2 g. Heat to dissolve, adjust pH to 7.2, and add bromothymol blue.

Bromothymol blue: Dissolve 0.1 g dibromo-cresol sulfonephthalein in 8 mL of 0.02 N NaOH and make up to 250 mL with water. This is yellow at pH 6 and blue at pH 7.6. Sterilize and add to OF medium making a final concentration of bromothymol blue solution as 0.4%; sugar to be added is dextrose (dextrose sugar can be replaced by any other desired sugar), which is autoclaved separately or filter sterilized, if possible. Make the final concentration of sugar as 0.1% by volume.

Protocol

1. Inoculate the slants of OF medium by stabbing (Figure 15.26) once.
2. Cover one tube with a thin layer of vaspar (Vaseline in one tube and paraffin in other tube. Melt and mix.)
3. Incubate for 48 h at 37°C.
4. Observe for production of acid and gas from dextrose. Fermentative cultures turn both tubes yellow, while oxidative cultures will turn only the open tube yellow (Figures 15.48 and 15.49).

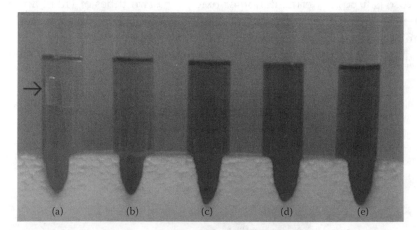

Figure 15.48   **(See color insert.)** Possible phenol red tube results include (a) formation of acid and gas (bubble is indicated by arrow), (b) formation of acid, (c) uninoculated control, (d) alkaline byproducts, and (e) no acid or gas formation.

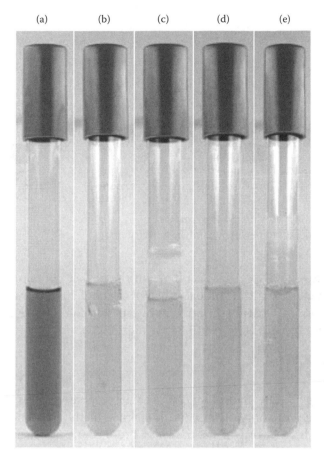

Figure 15.49 **(See color insert.)** (a) Uninoculated medium, (b) and (c) fermentative reaction with gas production in closed tubes (*Escherichia coli*), and (d) and (e) fermentative reaction with gas-production in closed tubes (*Klebsiella*).

15.30.1.4 Catalase Activity

15.30.1.4.1 Principle. Catalase is an enzyme that catalyses the breakdown of hydrogen peroxide to release free oxygen.

$$2H_2O_2 \xrightarrow{\text{catalase}} 2H_2O + O_2$$

The enzyme contains the hemeporphyrin structure. This porphyrin ring structure is characteristic of not only catalase but also of cytochromes.

15.30.1.4.2 Method I Add a few drops of 3% H_2O_2 to the slope and broth cultures and observe closely for the appearance of oxygen bubbles and a surface-broth accumulation (Figure 15.50).

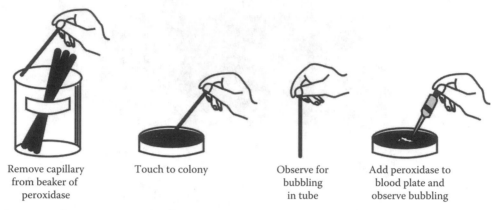

| Remove capillary from beaker of peroxidase | Touch to colony | Observe for bubbling in tube | Add peroxidase to blood plate and observe bubbling |

Figure 15.50   Protocol to perform catalase activity.

15.30.1.4.3 Method II.   Dip a glass capillary in the beaker containing 3% H_2O_2 and allow capillary action to tilt the tube to a height of about 20 mm. Touch the capillary on one of the colonies. Observe the bubbles in capillary for positive catalase reaction, for example, *Lactobacillus* (Figures 15.51 through 15.53).

15.30.1.5 Urease

15.30.1.5.1 Principle.   The hydrolysis of urea by urease releases NH_3. Urease is produced by some microorganisms and is helpful in the identification of *Proteus vulgaris*. Although other organisms may also produce urease, their action is slower than with *Proteus* sp.

Since the NH_3 production raises pH of the medium, urease activity can be detected by the change of the phenol red indicator to purple.

$$\underset{\text{Urea}}{\overset{\displaystyle \underset{NH_2}{\overset{NH_2}{\big|}}}{C=O}} + H_2O \xrightarrow{\text{Urease}} \underset{\text{Ammonia}}{2NH_3 + CO_2}$$

(a)   (b)   (c)

Figure 15.51   Confirmation of catalase activity: (a) showing catalase with bubbles, (b) no bubbles, catalase negative, and (c) control.

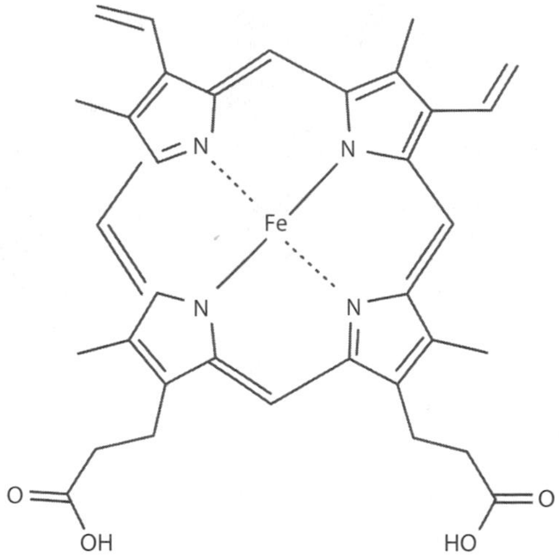

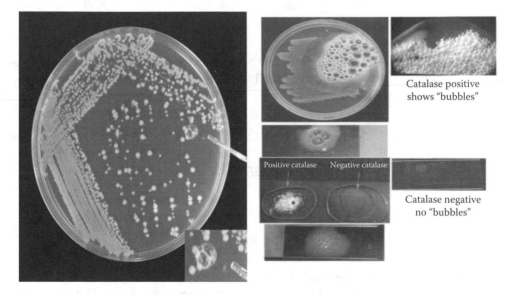

Figure 15.52 Heme, an example of an iron chelate porphyrin.

Catalase positive
shows "bubbles"

Positive catalase   Negative catalase

Catalase negative
no "bubbles"

Figure 15.53 Catalase activity.

15.30.1.5.2 Materials. *Composition of urea agar slant in g/L in distilled water*: Peptone 1.0, glucose 1, NaCl 5, K_2HPO_4 5, phenol red 6 mL of 1:500 aqueous solution, and agar 20, pH 6.8–6.9.

15.30.1.5.3 Protocol
  1. Prepare agar base and sterilize.
  2. Prepare urea solution 40% separately. Sterilize using bacteriological filter.
  3. Add urea solution to agar base so as to make a final concentration of 20% of urea. Dispense in tubes to prepare slopes.

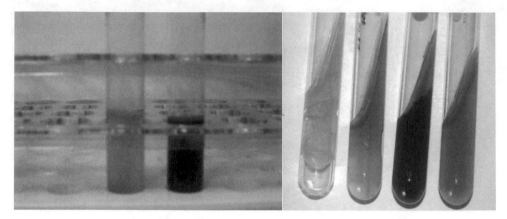

Figure 15.54 (See color insert.) Triple sugar iron agar test.

4. Inoculate and incubate.
5. Observe for color change of phenol red indicator to purple, for example, *P. vulgaris*.

15.30.1.6 H₂S Production

15.30.1.6.1 Principle. The activity of some bacteria on S-containing amino acids frequently results in the liberation of H_2S. This is best observed in a triple sugar iron agar slant (Figure 15.54).

15.30.1.6.2 Material. *TSI agar medium g/L in distilled water*: Trypticase 10, peptone 10, sodium chloride 5, lactose 10, sucrose 10, dextrose 1, ferrous ammonium sulfate 0.2, sodium thiosulfate 0.2, phenol red 0.025, and agar 12, pH 7.3.

15.30.1.6.3 Protocol. See Table 15.5.

15.30.1.7 IMViC Reactions
Introduction: IMViC reactions are a set of four useful reactions that are commonly employed in the identification of members of family enterobacteriaceae. The four reactions are: Indole test, methyl red (MR) test, Voges–Proskauer (VP) test, and citrate utilization test. The letter "i" is only for rhyming purpose.

15.30.1.7.1 Indole Test
15.30.1.7.1.1 Principle. Some bacteria can produce indole from amino acid tryptophan using the enzyme tryptophanase. Production of indole is detected using Ehrlich's reagent or Kovac's reagent. Indole reacts with the aldehyde in the reagent to give a red color. An alcoholic layer concentrates the red color as a ring at the top (Figure 15.55).

Indole is a nitrogenous compound and a degradation product of amino acid tryptophan of various bacteria.

Indole can be detected as follows.

TABLE 15.5　Key Reaction for TSI Agar

Reaction in Butts and Slants	Sugar Fermentation	Possible Organism
A and G in butt; A on slant H$_2$S not produced[a]	Glucose: AG Lactose and/or sucrose: AG	*Escherichia, Klebsiella,* intermediate coliform bacteria, *Proteus,* and *Paracolons*
A and G in butt Alk. on slant: H$_2$S produced	Glucose: AG; lactose and sucrose: not fermented	*Salmonella, Proteus,* and *Paracolons*
A only in butt Alk. on slant: H$_2$S not produced[b]	Glucose: AG; lactose and sucrose: not fermented	*Salmonella, Shigella,* and *Paracolons*
Alk. or neutral on butt Alk. on slant: H$_2$S not produced	None	*Alcaligenes* and *Pseudomonas*
No change in butt in 24 h A on butt in 48 h A on slant: H$_2$S produced	Glucose and sucrose: slight A; lactose: not fermented	*Pasteurella*[c]

Note: A = acid (yellow), Alk = alkaline (red), G = gas.

[a] *Citrobacter freundii* produced H$_2$S.

[b] *Salmonella typhi* produced a small amount of H$_2$S in the butt, and occasional strain may be encountered to give a small amount of gas in the butt.

[c] *Pasteurella multocida* is rarely encountered. *Enterococci* may give a similar type of reaction.

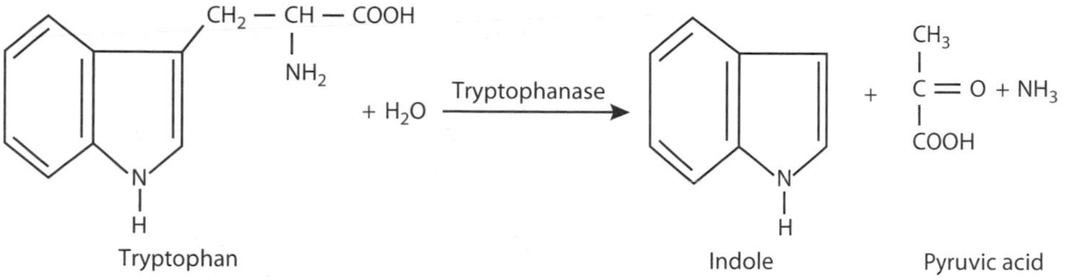

Figure 15.55　Bacterial conversion of tryptophan to indole.

15.30.1.7.1.2 Materials.　*Composition of peptone water broth medium g/L in distilled water*: Peptone 0.8, NaCl 1.4, NaHCO$_3$ 0.02, KCl 0.04 g, CaCl$_2$ 0.04, KH$_2$PO$_4$ 0.24, and Na$_2$HPO$_4 \cdot$ 2H$_2$O 0.88.

Steam the ingredients at 100°C for 20 min and filter through double fold Whatman No.1 filter paper. The pH is adjusted to 7.2. Autoclave at 121°C for 15 min.

15.30.1.7.1.3 Protocol.　Bacterium to be tested is inoculated in peptone water, which contains amino acid tryptophan and incubated overnight at 37°C. Following incubation, few drops of Kovac's reagent is added. Kovac's reagent consists of para-dimethylaminobenzaldehyde, isoamyl alcohol, and concentrated HCl. Ehrlich's reagent is more sensitive in detecting indole production in anaerobes and nonfermenters. Formation of a red or pink colored ring at the top is taken as positive. For example: *E. coli*: positive and *Klebsiella pneumoniae*: negative.

1. Put a strip of filter paper (soaked in a saturated solution of oxalic acid and dried) inside the overnight grown culture tube. Keep the filter paper suspended by the cotton plug. The paper becomes pink when the test is positive.
2. Kovac's reagent method: p-dimethylaminobenzaldehyde 1 g and iso-amyl alcohol 15 mL.

To an overnight peptone medium grown culture, add Kovac's reagent to layer it, shake and allow the reagent to float. Indole is detected by the appearance of pink color at the top (Figures 15.56 and 15.57).

15.30.1.7.2 MR Test

15.30.1.7.2.1 Principle. This is to detect the ability of an organism to produce and maintain stable acid end products from glucose fermentation. Some bacteria produce large amounts of acids from glucose fermentation that they overcome the buffering action of the system. MR is a pH indicator, which remains red in color at a pH of 4.4 or less. It detects organisms that do not convert acidic products to neutral products and produce a final pH lower than that of organisms producing neutral products. Because of the lower pH, the MR indicator changes to a red color indicating positive test.

15.30.1.7.2.2 Materials. *Composition of glucose-phosphate broth g/L in distilled water*: Glucose 5, proteose peptone 5, and K_2HPO_4 5.

MR: 0.1 g of MR in 300 mL alcohol. Dissolve and make volume up to 500 mL.

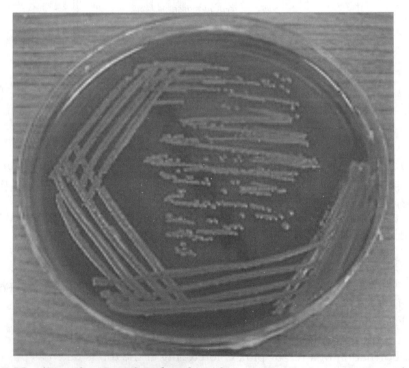

Figure 15.56 **(See color insert.)** *Escherichia coli* produce red colony on MacConkey agar.

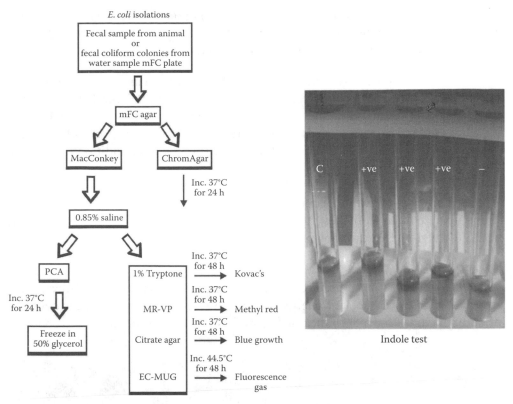

Figure 15.57 (**See color insert.**) Protocol for indole test for *E. coli*.

15.30.1.7.2.3 Protocol. The bacterium to be tested is inoculated into glucose phosphate broth, which contains glucose and a phosphate buffer and incubated at 37°C for 48 h. Over the 48 h, the mixed acid producing organism must produce sufficient acid to overcome the phosphate buffer and remain acid. The pH of the medium is tested by the addition of five drops of MR reagent. Development of red color is taken as positive. MR-negative organisms produce yellow color.

For example: *E. coli*: positive and *K. pneumoniae*: negative.

15.30.1.7.3 VP Test

15.30.1.7.3.1 Principle. While MR test is useful in detecting mixed acid producers, VP test detects butylene glycol producers. Acetyl-methyl carbinol (acetoin) is an intermediate in the production of butylene glycol. In this test, two reagents, 40% KOH and alpha-naphthol, are added to test broth after incubation and exposed to atmospheric oxygen. If acetoin is present, it is oxidized in the presence of air and KOH to diacetyl. Diacetyl then reacts with guanidine components of peptone, in the presence of alpha-naphthol, to produce red color. Role of alpha-naphthol is that of a catalyst and a color intensifier. In some bacteria, for example, *Bacillus* instead of accumulating mostly, acidic products from the fermentation of glucose, it is converted to metabolic intermediate, pyruvic acid and then to neutral products (acetyl-methylcarbinol—$CH_3COCHOHCH_3$) plus CO_2.

15.30.1.7.3.2 Material. Composition of glucose-phosphate broth g/L in distilled water: Glucose 5, proteose peptone 5, K_2HPO_4 5, α-naphthol solution (5% solution in alcohol), and 40% KOH solution containing 0.3% creatine.

15.30.1.7.3.3 Protocol. Bacterium to be tested is inoculated into glucose phosphate broth and incubated for at least 48 h, 0.6 mL of alpha-naphthol and 0.2 mL of 40% KOH are added to the broth and shaken. The tube is allowed to stand for 15 min. Appearance of red color is taken as a positive test. The negative tubes must be held for 1 h, since maximum color development occurs within 1 h after addition of reagents (Figures 15.58 and 15.59).

For example: *E. coli*: negative and *K. pneumoniae*: positive.

Note: The tubes of medium contain only 5 mL solution. This establishes a ratio of a large volume of air in the test tube to a small volume of medium a condition that promote acetyl-methyl carbinol formation.

15.30.1.7.4 Citrate Utilization Test

15.30.1.7.4.1 Principle. This test detects the ability of an organism to utilize citrate as the sole source of carbon and energy. Bacteria are inoculated on a medium containing sodium citrate and a pH indicator, bromothymol blue. The medium also contains

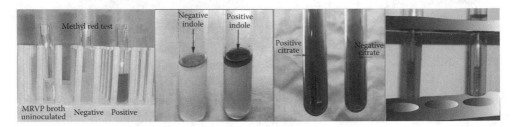

Figure 15.58 **(See color insert.)** Complete demonstration for IMVIC test.

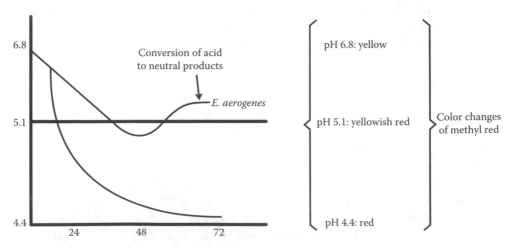

Figure 15.59 Relation between acid and acetylmethylcarbinol production by *Enterobacter aerogenes* and *Escherichia coli* as a function of time.

inorganic ammonium salts, which is utilized as sole source of nitrogen. Utilization of citrate involves the enzyme citritase, which breaks down citrate to oxaloacetate and acetate. Oxaloacetate is further broken down to pyruvate and CO_2.

The ability to utilize citrate as a sole source of carbon and energy can be used to distinguish certain Gram-negative rods. Simmon's citrate agar contains citrate as the only carbon and energy source. Growth on this medium is a positive test for citrate utilization. Production of Na_2CO_3 as well as NH_3 from utilization of sodium citrate and ammonium salt, respectively, results in alkaline pH. This results in change of medium's color from green to blue.

15.30.1.7.4.2 Materials. *Simmon's citrate agar*: $MgSO_4 \cdot 7H_2O$ 0.2 g, $(NH_4)_2HPO_4$ 1 g, K_2HPO_4 1 g, sodium citrate 2 g, NaCl 5 g, agar 15 g, bromothymol blue 0.08 g, and distilled water to 1000 mL. Prepare slants and inoculate.

Procedure: Bacterial colonies are picked up from a straight wire and inoculated into slope of Simmon's citrate agar and incubated overnight at 37°C. If the organism has the ability to utilize citrate, the medium changes its color from green to blue.

For example: *E. coli*: negative and *K. pneumoniae*: positive.

15.30.1.7.4.3 Results and Observation of IMVIC Test

Bacterium	Indole	MR	VP	Citrate
E. coli	+	+	–	–
K. pneumoniae	–	–	+	+

15.30.1.8 Litmus-Milk Reaction

15.30.1.8.1 Principle. Litmus milk is an undefined medium consisting of skim milk and the pH indicator azolitmin.

It is used to differentiate members within the genus *Clostridium*. It differentiates *Enterobacteriaceae* from other Gram-negative bacilli based on enterics' ability to reduce litmus.

The skim milk provides nutrients for growth. The protein is casein, and the lactose is for fermentation. Azolitmin is purple between pH of 4.6 and 8.2. It turns pink when pH reaches 4.5 and blue at a pH of 8.3. Litmus milk has four major reactions: lactose fermentation, reduction of litmus, casein coagulation, and casein hydrolysis. Lactose fermentation turns the litmus pink due to acidity. If litmus is reduced during lactose fermentation, it will turn the medium white. Acid may cause an acid clot; this is due to the casein precipitating. Heavy gas production can cause stormy fermentation; this is due to the breaking up of the clot (Figure 15.60).

15.30.1.8.2 Materials. Dissolve 150 g of skimmed milk powder to a liter of water. Add sufficient amount of 2.5% aqueous solution of azolitmin (a dark red nitrogenous coloring matter obtained from litmus and used as an acid–base indicator).

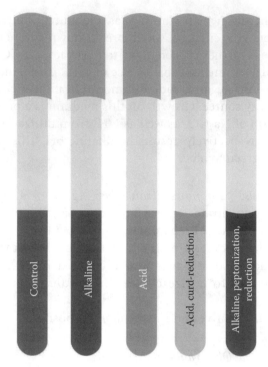

Figure 15.60 (See color insert.) Litmus milk reaction.

15.30.1.8.3 Protocol
1. Inoculate and observe the change in the reaction of milk, it will turn red if acid is produced and blue if alkali is produced.
2. Clotting is followed by peptonization of curd.

15.30.1.9 Nitrate Reduction

15.30.1.9.1 Principle. Bacterial species may be differentiated on the basis of their ability to reduce nitrate to nitrite or nitrogenous gases. The reduction of nitrate may be coupled to anaerobic respiration in some species. The reduction of nitrate is brought about by several soil inhabiting bacteria—nitrate, ammonia, nitrous oxide, nitrogen gas, etc., are end products of nitrate reduction. They serve as H_2 acceptors for these species in the absence of oxygen.

15.30.1.9.2 Materials. *Nutrient broth g/L in distilled water*: Beef extract 3, peptone 5, NaCl 5, and $NaNO_3$ 5, pH 7.0.

Sulfanilic acid solution (reagent A): Dissolve 8 g of sulfanilic acid in 1 L 5 N acetic acid. Store reagent A at room temperature for up to 3 months in the dark. Reagents may be stored in dark brown glass containers; bottles may be wrapped in aluminum foil to ensure darkness.

α-naphthylamine solution (reagent B): Dissolve 6 g of N,N-Dimethyl-1-naphthylamine in 1 L 5 N acetic acid. Store reagent B at 2°C–8°C for up to 3 months

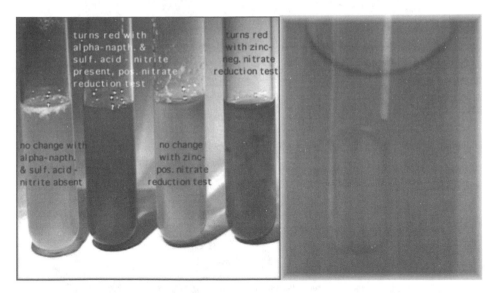

Figure 15.61 (See color insert.) Nitrate reduction test.

in the dark. Reagents may be stored in dark brown glass containers; bottles may be wrapped in aluminum foil to ensure darkness.

15.30.1.9.3 Protocol. Dissolve 9 g of nitrate broth in 1 L distilled water. Dispense 10 mL aliquots of the broth into tubes fitted with Durham tubes. Sterilize by autoclaving at 121°C for 15 min. Inoculate the tubes heavily with a fresh culture of the suspect organism. Inoculate at least 1 mL sample in a tube or take a big part of a colony with an inoculating loop. Do not forget a negative control without any bacteria. Nitrate reaction occurs only under anaerobic conditions. The medium is dispensed in tubes to give a low-surface area to depth ratio, which limits the diffusion of oxygen into the medium. Most bacteria use the oxygen in the medium and rapidly produce anaerobic conditions. It is recommended to give about 1 cm of paraffin oil on the surface of the media or over-gassing to maintain an anaerobic condition faster, for example, with carbon dioxide, and seal the tube with parafilm. Incubate at 37°C for 24 h. Observe for gas formation; if there is no gas, incubate further. Add 1 mL of α-naphthylamine reagent and 1 mL of sulfanilamide reagent to the culture tubes after incubation. A red color within 80 s shows nitrate reduction (Figure 15.61).

$$NO_3 \longrightarrow NO_2 \longrightarrow NH_3 \longrightarrow N_2O$$

15.30.1.10 Oxidase Test

15.30.1.10.1 Principle. The test is used to distinguish *Neisseria gonorrhoeae* (oxidase positive) from *Staphylococcus* spp. and *Streptococcus* spp (oxidase negative). The sensitivity of the oxidase test was increased when Kovács found that a tetramethyl-*p*-phenylenediamine dihydrochloride solution gave a quicker reaction. Gaby and Hadley developed a modified oxidase test using ρ-aminodimethylaniline oxalate

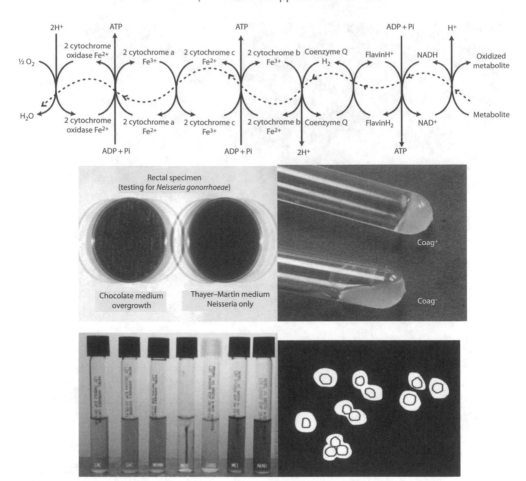

Figure 15.62 **(See color insert.)** The electron transport system; the dashed line indicates the flow of electrons in *Neisseria* sp.

with α-naphthol to detect oxidase in test tube cultures. The oxidase test is a biochemical reaction that assays for the presence of cytochrome oxidase, an enzyme sometimes called indophenol oxidase. In the presence of an organism that contains the cytochrome oxidase enzyme (Figure 15.62), the reduced colorless reagent becomes an oxidized colored product. It measures the ability of a microbe to oxidize aromatic amines, for example, ρ-amino methyl aniline to yield colored end products. Oxidation on one hand correlates with high cytochrome oxidase activity of some bacteria, for example, *Pseudomonas* and on the other a negative oxidase reaction shows the presence of the enteric bacteria, for example, *E. coli*.

15.30.1.10.2 Materials
1. *Kovács oxidase reagent*: 1% tetramethyl-*p*-phenylenediamine dihydrochloride in water. Store refrigerated in a dark bottle no longer than 1 week.
2. *Gordon and McLeod reagent*: 1% dimethyl-*p*-phenylenediamine dihydrochloride in water. Store refrigerated in a dark bottle no longer than 1 week.

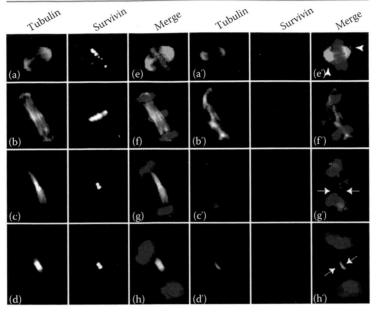

Figure 1.10   Immunoflouresence detection of survivin in cancer tissue.

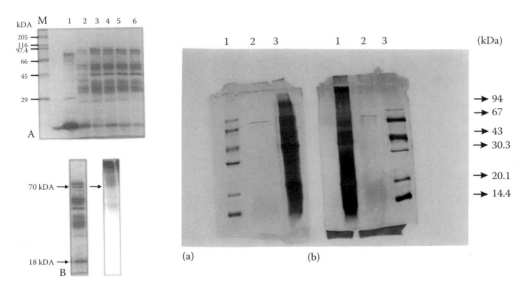

Figure 8.14   SDS-PAGE analysis of purified proteins with MW markers stained with (a) Coomassie blue and (b) silver stain.

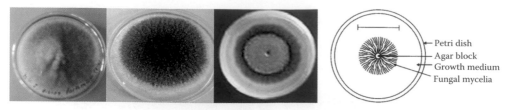

Figure 15.23   Radial growth of two test fungus on PDA agar plates and schematic diagram.

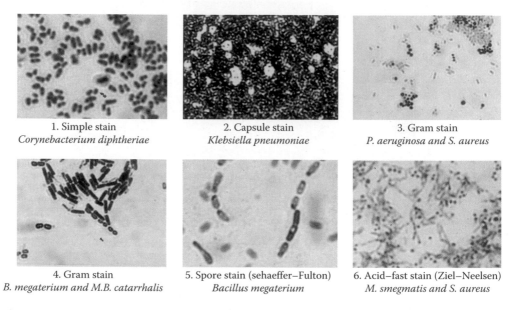

1. Simple stain
Corynebacterium diphtheriae

2. Capsule stain
Klebsiella pneumoniae

3. Gram stain
P. aeruginosa and S. aureus

4. Gram stain
B. megaterium and M.B. catarrhalis

5. Spore stain (sehaeffer–Fulton)
Bacillus megaterium

6. Acid–fast stain (Ziel–Neelsen)
M. smegmatis and S. aureus

Figure 15.35   Different colors as observed under microscope of acid fast bacteria.

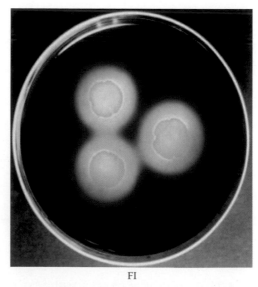

FI

Figure 15.43   Test for production of amylase after flooding with Lugol's iodine solution.

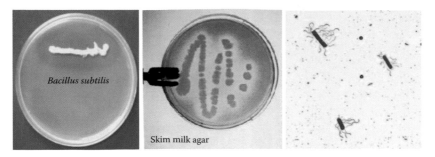

Figure 15.45 Caseinase production by *Bacillus subtilis*.

Figure 15.48 Possible phenol red tube results include (a) formation of acid and gas (bubble is indicated by arrow), (b) formation of acid, (c) uninoculated control, (d) alkaline byproducts, and (e) no acid or gas formation.

Figure 15.49 (a) Uninoculated medium, (b) and (c) fermentative reaction with gas production in closed tubes (*Escherichia coli*), and (d) and (e) fermentative reaction with gas-production in closed tubes (*Klebsiella*).

Figure 15.54 Triple sugar iron agar test.

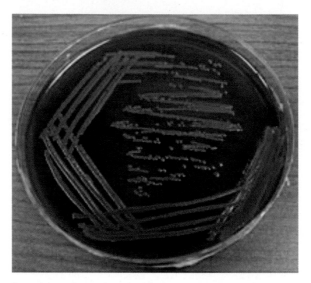

Figure 15.56 *Escherichia coli* produce red colony on MacConkey agar.

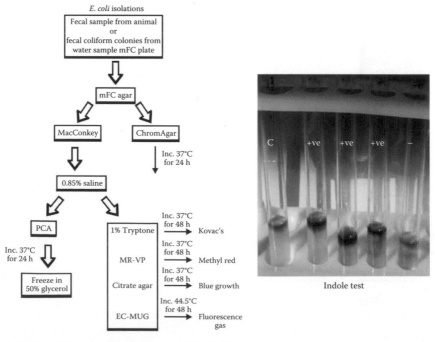

Figure 15.57 Protocol for indole test for *E. coli*.

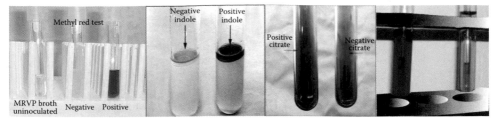

Figure 15.58 Complete demonstration for IMVIC test.

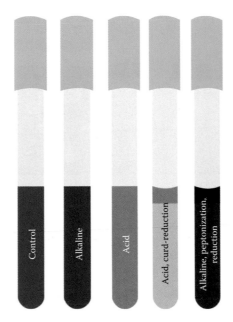

Figure 15.60 Litmus milk reaction.

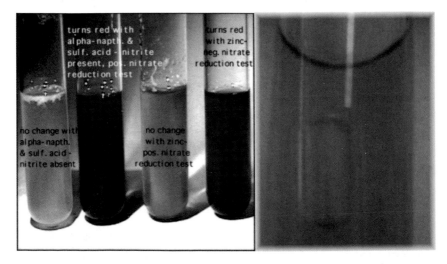

Figure 15.61 Nitrate reduction test.

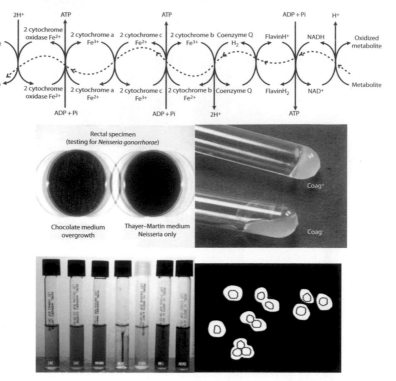

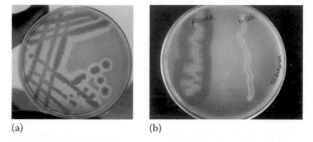

Figure 15.62 The electron transport system; the dashed line indicates the flow of electrons in *Neisseria* sp.

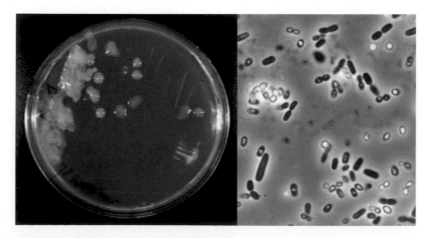

(a) (b)

Figure 15.66 Bacterial lipase hydrolysis test. (a) Soil isolate of *Pseudomonas aeruginosa* and (b) lipid hydrolysis test.

Figure 16.20 Isolated *Azotobacter* sp. from soil sample.

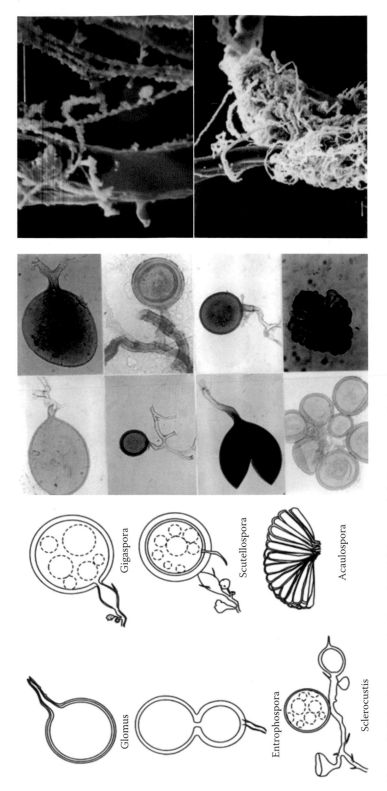

Figure 16.26 Microscopic view of VAM fungi and their root association.

Gigaspora

Scutellospora

Acaulospora

Glomus

Entrophospora

Sclerocustis

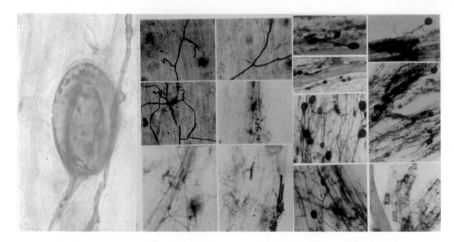

Figure 16.29   Histochemical visualization of various stages of total AMF mycelium in roots.

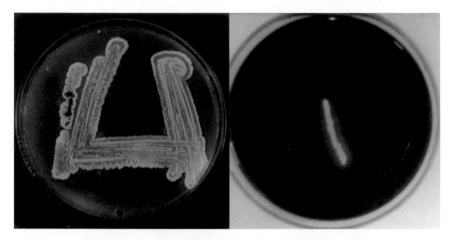

Figure 16.34   Growth of an isolated actinomycete from soil and its amylase positive test.

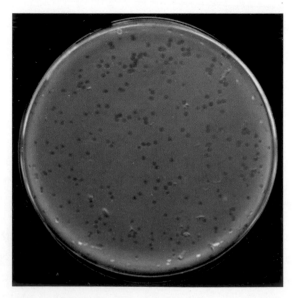

Figure 17.6   Plaques of uniform size of bacteriophages.

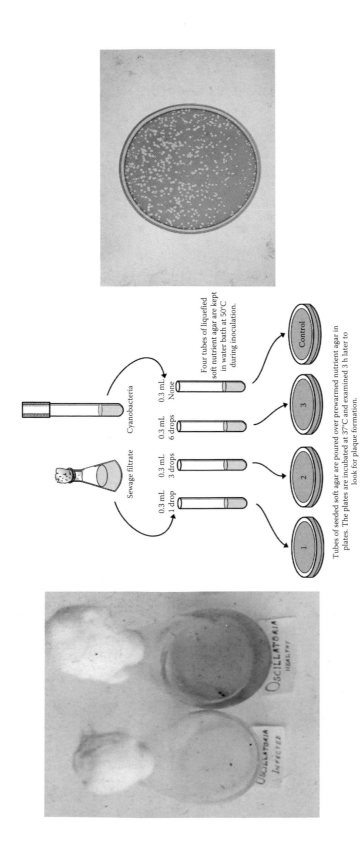

Four tubes of liquefied soft nutrient agar are kept in water bath at 50°C during inoculation.

Cyanobacteria

Sewage filtrate

0.3 mL
1 drop

0.3 mL
3 drops

0.3 mL
6 drops

0.3 mL
None

1

2

3

Control

Tubes of seeded soft agar are poured over prewarmed nutrient agar in plates. The plates are incubated at 37°C and examined 3 h later to look for plaque formation.

OSCILLATORIA
HEALTHY

OSCILLATORIA
INFECTED

Figure 17.12   Procedure for the plaque forming unit assay.

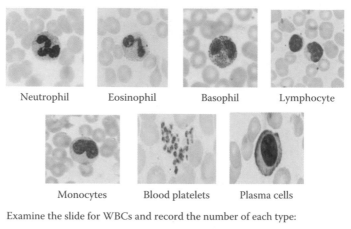

| Neutrophil | Eosinophil | Basophil | Lymphocyte |

| Monocytes | Blood platelets | Plasma cells |

Examine the slide for WBCs and record the number of each type:

Path to follow when observing under microscope

Zone of morphology

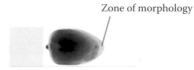

Figure 18.47 Blood morphology containing different types of cells.

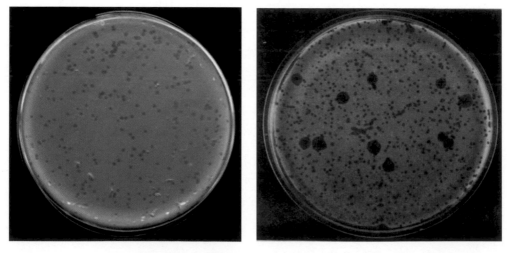

Figure 19.11 λ Coliphage plaques of different sizes caused by at least two bacteriophages on a carpet of actively multiplying bacteria.

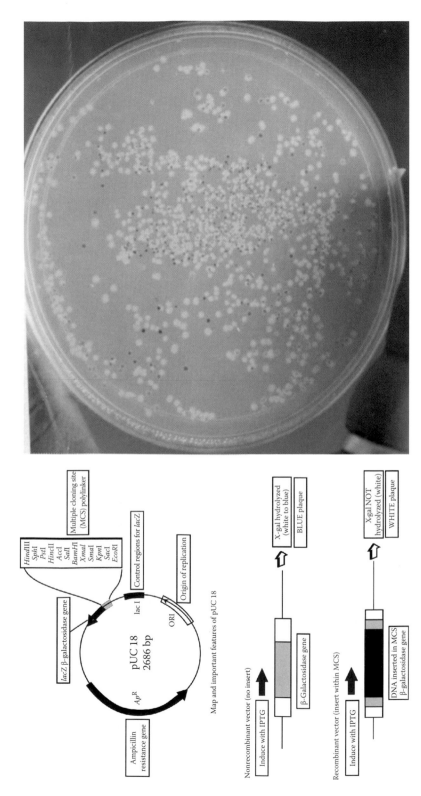

Figure 19.16 Plate showing blue and white recombinant bacterial colonies.

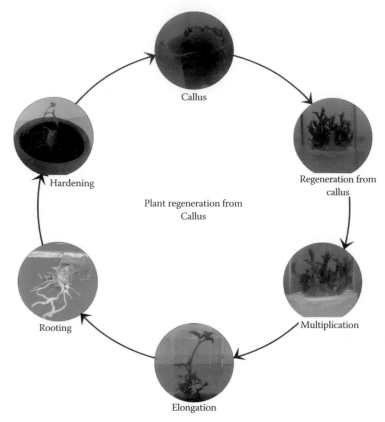

Figure 23.6 Plant regeneration from callus.

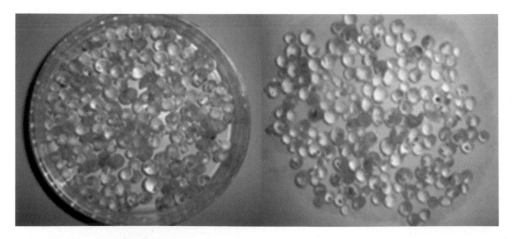

Figure 23.28 Artificial seed.

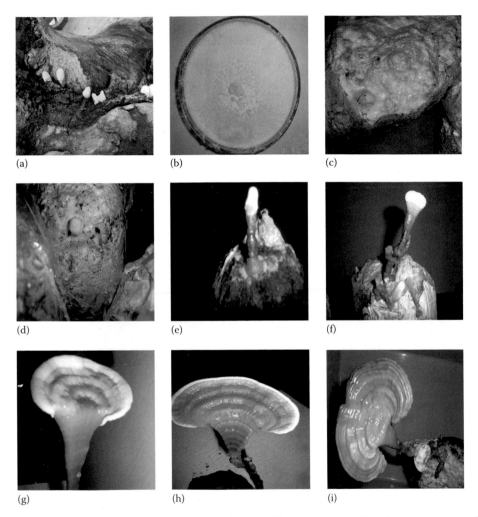

Figure 25.35 Different growth stages of *G. lucidum* during artificial cultivation under solid substrate state fermentation on polypropylene bags: (a) source of strain isolation, (b) culture plate (growing in potato dextrose agar medium), (c) mycelial colonization on solid substrate, (d) primordia formation, (e) elongation of primordia, (f) cap formation, (g) flattening and growth of cap, (h) thickening of cap, and (i) maturation of fruit body.

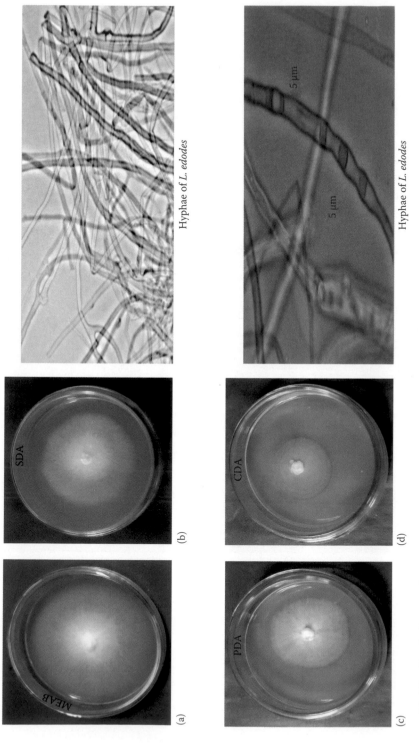

Hyphae of *L. edodes*

Hyphae of *L. edodes*

Figure 25.38 *L. edodes* mycelium growing on different agar media on plates and microscopic view of hyphae of *L. edodes*. (a–d) Different degrees of growth on potato dextrose agar medium on Petri plates.

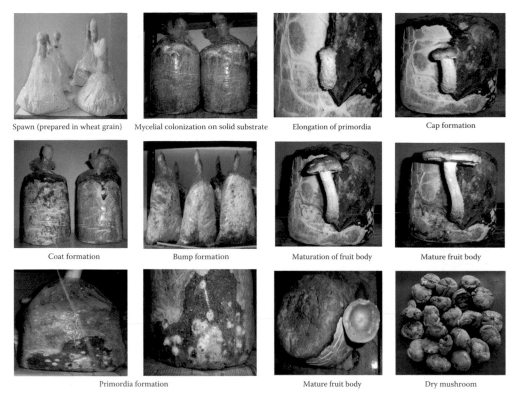

Spawn (prepared in wheat grain)   Mycelial colonization on solid substrate   Elongation of primordia   Cap formation

Coat formation   Bump formation   Maturation of fruit body   Mature fruit body

Primordia formation   Mature fruit body   Dry mushroom

Figure 25.39   Different stages of production of *L. edodes* (shiitake) mushroom.

Pleurotus ostreatus   *Pleurotus sajorcaju*

Figure 25.42   Oyster mushroom growing on solid substrate fermentation.

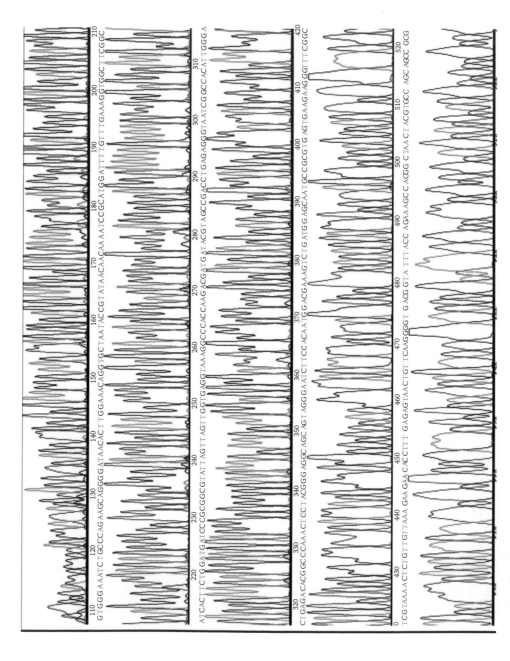

Figure 27.3 *Lactobacillus* identification by rDNA fingerprinting.

3. *Gaby and Hadley oxidase test*: 1% α-naphthol in 95% ethanol and 1% ρ-aminodimethylaniline oxalate. Store refrigerated in dark bottles no longer than 1 week.

Oxidase reagents are also available commercially in droppers, impregnated disks, and test strips.

15.30.1.10.3 Protocol

15.30.1.10.3.1 Test Strip Method.   1% solution of tetramethyl-*p*-phenylenediamine hydrochloride.

Using a sterilized loop, pick growth from a well-isolated colony and rub into area 1 of the test strip (a filter paper disc, impregnated with oxidase reagent 1% solution of tetramethyl-*p*-phenylenediamine hydrochloride), wait for 30 s, and observe the inoculated area. A blue or deep purple color indicates a positive test.

Instead of using paper strips, solution can directly be poured on to the agar plate. However, a test strip keeps the reagent stable for a longer time (Figure 15.63).

15.30.1.10.3.2 Direct Plate Method

1. Grow a fresh culture (18–24 h) of bacteria on nutrient agar using the streak plate method so that well-isolated colonies are present.
2. Place 1 or 2 drops of 1% Kovács oxidase reagent or 1% Gordon and McLeod reagent on the organisms. Do not invert or flood plate.
3. Observe for color changes.
4. When using Kovács oxidase reagent, microorganisms are oxidase positive when the color changes to dark purple within 5–10 s. Microorganisms are delayed oxidase positive when the color changes to purple within 60–90 s. Microorganisms are oxidase negative if the color does not change or it takes longer than 2 min.
5. When using Gordon and McLeod reagent, microorganisms are oxidase positive when the color changes to red within 10–30 min or to black within 60 min. Microorganisms are oxidase negative if the color does not change.

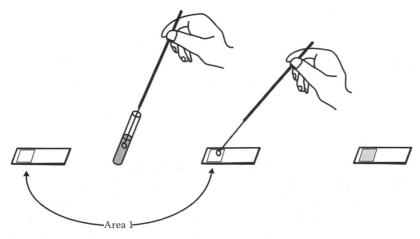

Figure 15.63   Demonstrating oxidase by test strip method.

15.30.1.10.3.3 Test Tube Method

1. Grow a fresh culture (18–24 h) of bacteria in 4.5 mL of nutrient broth (or standard media) that does not contain a high concentration of sugar.
2. Add 0.2 mL of 1% α-naphthol, then add 0.3 mL of 1% ρ-aminodimethylaniline oxalate (Gaby and Hadley reagents).
3. Observe for color changes.
4. Microorganisms are oxidase positive when the color changes to blue within 15–30 s. Microorganisms are delayed oxidase positive when the color changes to purple within 2–3 min. Microorganisms are oxidase negative if the color does not change.

15.30.1.11 Pectinase

15.30.1.11.1 Principle.   Pectinolytic organisms are a constant threat with regard to spoilage of vegetables because of their extensive host range and widespread distribution. It is estimated that between 10% and 30% of fresh vegetables are wasted, mainly due to three factors: mechanical injuries, physiological decays, and microbial spoilage. The role of microorganisms in this wastage is significant, hence the need to better understand the microbial diversity is responsible for soft rot spoilage in vegetables.

15.30.1.11.2 Materials.   Nutrient broth with 1% pectin in the medium as inducer for the synthesis of pectinase, 0.02 N NaOH, and 1% pectin solution.

15.30.1.11.3 Protocol.   Grow the fresh culture of the microorganism in the medium. Take the culture filtrate and perform the test for pectin esterase enzyme.

Pectin methyl esterase is an enzyme used in hydrolysis of pectin, which is expressed in micro equivalent methoxyl group removed by the enzyme sample. To 5 mL of pectin (1%) adjusted to pH 6, add 2 mL of enzyme preparation (culture broth) mix and allow to react at 30°C. Note the pH and again adjust to 6 at 0 h with the help of 0.02 N NaOH. Use the boiled sample of enzyme as control. Allow the enzyme substrate to react for about 10 min and readjust pH to 6 using 0.02 N NaOH.

15.30.1.11.4 Results and Observation.   Express the amount of activity in terms of unit ml^{-1} NaOH used to adjust the pH.

Micro equivalent of NaOH required to maintain the original pH h^{-1} mL^{-1} enzyme sample. The aforementioned method gives a quantitative estimation of pectinase.

Note: About 1% of pectin should be added to culture broth so as to induce pectinase synthesis.

15.30.1.12 Gelatinase

15.30.1.12.1 Principle.   The protein gelatin is obtained by the hydrolysis of collagen a component of connective tissue and tendons of animals. It also demonstrates a test for proteolytic enzymes. Water solutions of gelatin of concentration to be used here are liquid at room temperature and solidify in an acid bath. If the gelatin has been hydrolyzed by the action of microorganism being tested, the medium will remain liquid (Figure 15.64).

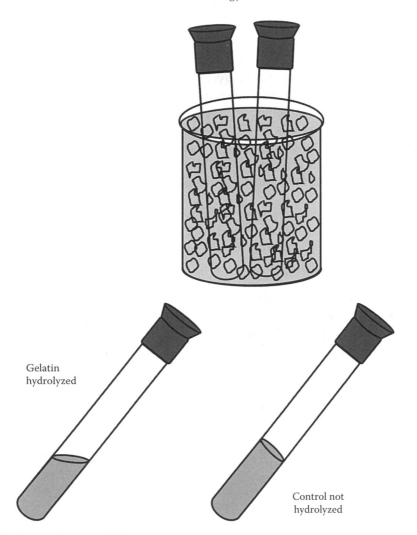

Figure 15.64 Gelatin hydrolysis.

15.30.1.12.2 Materials.   *Preparation of nutrient gelatin*: Add 4% of gelatin in nutrient broth.

Composition of nutrient broth g/L in distilled water: Beef extract 3, peptone 5, and NaCl 3, pH 7.

15.30.1.12.3 Protocol.   Inoculate the tubes of nutrient gelatin with the unknown culture. Incubate at 37°C together with a sterilized tube of nutrient gelatin that will serve as a control. Test after 2 days and up to 7 days or until a positive reaction is obtained.

To examine hydrolysis, chill the tubes in ice water. The control tubes and tubes in which no hydrolysis has taken place will solidify. Hydrolyzed gelatin will remain fluid.

15.30.1.13 Lipase

15.30.1.13.1 Principle. Two common lipids decomposed by microorganisms are triglycerides and phospholipids. The triglycerides are esters of glycerol and fatty acids. The reaction involved in their hydrolysis occurs as given below (Figure 15.65).

Phospholipids are major components of all cell membranes; the ability to hydrolyze host cell phospholipids is an important factor in the spread of virulent microorganisms. It can be tested by growing the culture in egg yolk agar, the cleavage of phosphate ester bonds form a water insoluble lipid. The enzymatic activity is detected by a zone of opalescence in the medium surrounding the cell mass (Figure 15.66).

15.30.1.13.2 Materials. *Composition of the plate count agar medium g/L in distilled water*: Tryptone 5, yeast extract 2.5, glucose 1, and agar 15.

Mix 100 mL of 5% egg yolk with 900 mL of the plate count medium with 15 g agar, autoclave, and pour in Petri plates.

15.30.1.13.3 Protocol. Inoculate egg yolk agar with the unknown culture, incubate at 37°C for 48 h. Observe for an opalescent precipitate. For the egg yolk agar, the growth must have a white halo around the colony growth if it utilizes the lipids therefore having the enzyme lipase. For example, *Bacillus* spp.[+], *E. coli*[-], and *Alcaligenesfaecalis*[-].

15.30.1.14 Gluconate Test

15.30.1.14.1 Principle. Gluconate test aids in identifying the fluorescent group of pseudomonads, especially in the absence of pigment production. These organisms

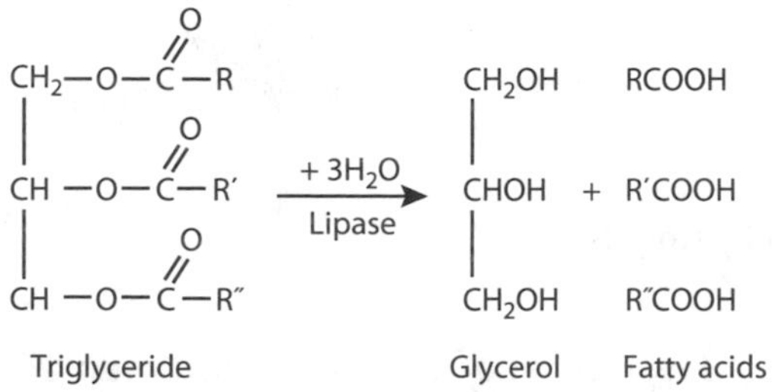

$$
\begin{array}{ccc}
\underset{\text{Triglyceride}}{
\begin{array}{l}
CH_2-O-\overset{\overset{\displaystyle O}{\|}}{C}-R \\[2pt]
| \qquad\; \overset{\displaystyle O}{\|} \\
CH\;-O-\overset{}{C}-R' \\[2pt]
| \qquad\; \overset{\displaystyle O}{\|} \\
CH\;-O-\overset{}{C}-R''
\end{array}}
&
\xrightarrow[\text{Lipase}]{+\,3H_2O}
&
\underset{\text{Glycerol \qquad Fatty acids}}{
\begin{array}{ll}
CH_2OH & RCOOH \\[4pt]
| \\
CHOH & +\ R'COOH \\[4pt]
| \\
CH_2OH & R''COOH
\end{array}}
\end{array}
$$

Figure 15.65 Biochemical reaction of lipase.

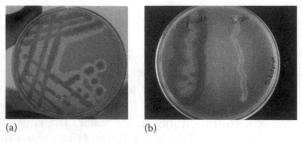

(a)   (b)

Figure 15.66 (See color insert.) Bacterial lipase hydrolysis test. (a) Soil isolate of *Pseudomonas aeruginosa* and (b) lipid hydrolysis test.

(*Pseudomonas aeruginosa*, *Pseudomonas fluorescens*, and *Pseudomonas putida*) are able to oxidize gluconate to keto-gluconate, which will test positive for reducing sugars, while gluconate in the original form will not. Benedict's reagent will produce a color change in the presence of keto-gluconate when added to the test and heated.

15.30.1.14.2 Materials. *Gluconate broth in g/L in distilled water*: Yeast extract 1, peptone 1.5, K_2HPO_4 1, and potassium gluconate 40 or sodium gluconate 37.25. Dissolve by heating and adjust pH to 7.0. Filter if necessary.

Benedict's reagent: Sodium citrate 50 g, sodium carbonate 18.75 g, and potassium thiocyanate 31.25 g. Dissolve in 159 mL of hot distilled water.

This is mixed with 25 mL of an 8.38% (w/v) of $CuSO_4 \cdot 5H_2O$ solutions. Then 1.5 mL of a 5% solution of potassium ferrocyanide is added to the mixture with thorough mixing.

The resultant solution is then made up to 250 mL by adding distilled water (prepared Benedict's reagent are also available commercially).

15.30.1.14.3 Protocol
1. Inoculate gluconate broth.
2. Incubate for 48 h at 37°C.
3. Add an equal volume of Benedict's reagent and place in a boiling water bath for 10 min.
4. An orange or brown precipitate indicates gluconate oxidation. A deep blue color is the negative response.

15.31 Microbial Genetics

15.31.1 Introduction

Bacteria have proved to be the essential organisms for genetic studies because of their extremely rapid rate of growth, haploid genetic state, the availability of large test population, and low cost of maintenance and propagation. Genes on the nucleic acid code for specific peptides such as enzymes, antibodies, or structural proteins. Variations in bacteria are either temporary or permanent. Temporary variations may be morphological or physiological and disappear as soon as the environmental changes that brought them about to disappear. For example, as a culture of *E. coli* becomes old and the nutrients within the tube become depleted, the new cells that form become so short that they appear coccoidal. Variations in bacteria that involve the DNA macromolecule are designated as permanent variations. It is because they survive a large number of transfers that they are so named. Such variations are due to mutations. Variations of this type occur spontaneously. They also might be induced by physical and chemical methods. Some permanent variations also are caused by the transfer of DNA from one organism to another, either directly by conjugation or indirectly by phage. Chemicals in the form of industrial pollutants, pesticides, food additives, hair dyes, and cigarette smoke play a significant role in the induction of mutations in somatic cells. Emphasis is being laid to develop the method for the detection of environmental carcinogens. Although animal testing is available, a more rapid and widely used screening method is based on a bacterial test system, the Ames test. Ames test was developed by Bruce Ames in the 1970s at the University of California. Mutations are mostly due to base substitutions and frame shift mutations. The premise of Ames test is that any chemical capable of mutating bacterial DNA can similarly mutate human DNA and is, therefore, potentially hazardous. The Ames test has proved invaluable for screening an assortment of environmental carcinogens.

15.31.1.1 Mutant Isolation by Gradient Plate Method

15.31.1.1.1 Principle. An excellent way to determine the ability of organisms to produce mutants that are resistant to antibiotics is to grow them on a gradient plate of a particular antibiotic. Such a plate consists of two different wedgelike layers of media: a bottom layer of plain nutrient agar and a top layer of nutrient agar with the antibiotic. Since the antibiotic is only in the top layer, it tends to diffuse into the lower layer, producing a gradient of antibiotic concentration from low to high.

A gradient plate using streptomycin in the medium is used. *E. coli*, which is normally sensitive to this antibiotic, will be spread over the surface of the plate and incubated for 4–7 days (Figure 15.67). Any colonies that develop in the high concentration area will be streptomycin-resistant mutants.

15.31.1.1.2 Materials. Sterile Petri plate, nutrient agar (10 mL per tube), tube of streptomycin solution (1%), 1 wood spacer ($^1/_8 \times \frac{1}{2} \times 2$ in.), and 1 mL pipette.

15.31.1.1.3 Protocol
15.31.1.1.3.1 Plate Preparation. The gradient plate used in this experiment will have a high concentration of 100 mcg of streptomycin per milliliter of medium. The concentration is 10 times the strength used in sensitivity disks in the Kirby–Bauer test method. Prepare a gradient plate as follows:

1. Liquefy two pours of nutrient agar and cool to 50°C.
2. With wood spacer under one edge of Petri plate, pour contents of one agar pour into plate. Let stand until solidifications has occurred.
3. Remove the wood spacer from under the plate.
4. Pipette 0.1 mL of streptomycin into second agar pour, mix tube between palms, and pour contents over medium of the plate that is now resting level on the table.
5. Label the low and high concentration areas on the bottom of the plate.

| 1. Plain nutrient agar is poured into Petri dish with plate in slant position. | 2. Streptomycin agar is poured over plain agar with plate in normal position. |

Figure 15.67 Gradient plate method for the isolation of mutant.

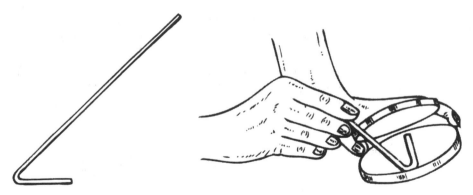

Figure 15.68 Spreading organisms on gradient plates.

15.31.1.1.3.2 Inoculation. The inoculation procedure is illustrated in Figure 15.68. The technique involves spreading a measured amount of culture on the surface of the medium with a glass bent rod to provide optimum distribution.

15.31.1.1.3.2.1 Materials. Beaker with 95% ethanol, glass rod spreader, nutrient broth culture of *E. coli*, and 1 mL pipette.

15.31.1.1.3.2.2 Protocol
1. Pipette 0.1 mL of *E. coli* suspension onto the surface of medium in Petri plate.
2. Sterilize glass spreading rod by dipping it in alcohol first and then passing it quickly through the flame of a Bunsen burner. Cool the rod by placing against sterile medium in plate before contacting organisms.
3. Spread the culture evenly over the surface with the glass rod.
4. Invert and incubate the plate at 37°C for 4–7 days in a closed canister or plastic bag. Unless incubated in this manner, exercise dehydration might occur.

First evaluation: After 4–7 days, look for colonies of *E. coli* in the area of high streptomycin concentration. Count the colonies that appear to be resistant mutants and record your count on the laboratory report.

Select a well-isolated colony in the high concentration area and, with a sterile loop, smear the colony over the surface of the medium toward the higher concentration portion of the plate. Do this with two or three colonies. Return the plate to the incubator for another 2 or 3 days.

Final evaluation: Examine the plate again to note what effect the spreading of the colonies had on their growth.

15.31.1.2 Demonstration of Streaking Plates Using Loop/Toothpick

15.31.1.2.1 Principle. Bacterial culture streaking allows bacteria to reproduce on a culture medium in a controlled environment. The process involves spreading bacteria across an agar plate and allowing them to incubate at a certain temperature for a period of time. Bacterial streaking can be used to identify and isolate pure bacterial colonies from a mixed population. Microbiologists use bacterial and other microbial culture streaking methods to identify microorganisms and to diagnose infection.

15.31.1.2.2 Materials. Mixed culture plate, autoclave sterilized toothpicks, Bunsen burner, laminar flow, fresh Petri plates containing nutrient agar medium, and glass marker.

15.31.1.2.3 Protocol
1. While wearing gloves, sterilize an inoculating loop by placing it at an angle over a flame. The loop should turn orange before you remove it from the flame. A sterile toothpick may be substituted for the inoculating loop. Do not place toothpicks over a flame.
2. Remove the lid from a culture plate containing the desired microorganism.
3. Cool the inoculating loop by stabbing it into the agar in a spot that does not contain a bacterial colony.
4. Pick a colony and scrape off a little of the bacteria using the loop. Be sure to close the lid.
5. Using a new agar plate, lift the lid just enough to insert the loop.
6. Streak the loop containing the bacteria at the top end of the agar plate moving in a zig-zag horizontal pattern until one-third of the plate is covered.
7. Sterilize the loop again in the flame and cool it at the edge of the agar away from the bacteria in the plate that you just streaked.
8. Rotate the plate about 60° and spread the bacteria from the end of the first streak into a second area using the same motion in Step 6.
9. Sterilize the loop again using the procedure in Step 7.
10. Rotate the plate about 60° and spread the bacteria from the end of the second streak into a new area in the same pattern.
11. Sterilize the loop again.
12. Replace the lid and invert the plate. Incubate the plate overnight at 37°C (98.6°F).
13. You should see bacterial cells growing in streaks and in isolated areas.

15.31.1.2.3.1 Streaking Plates. When picking and streaking lots of bacterial colonies, it is often quicker to use sterile toothpicks and sticks instead of using a wire loop that must be sterilized between each colony by passing it through a flame. Using the technique shown next, eight colonies can be streaked for isolation on a single plate (Figure 15.69) (save the used toothpicks and sticks to be re-autoclaved).

15.31.1.2.3.2 Picking and Patching Colonies. When checking out bacterial strains, it is often useful to patch many colonies on a single Petri plate so they can be tested simultaneously by replica printing onto a series of media. To do this, place a fresh master plate over a patching grid. Touch the top of each colony with a sterile toothpick and draw a small "x" on the new master plate. (Only use each toothpick once. Save the used toothpicks to be reautoclaved.) Many patches can be placed on a single plate. After patching, incubate the plate overnight to let the patches grow up. The next day this plate can be used as a master for replica-printing. Always mark each plate at the top of the patch grid since the patch grids are symmetrical (Figure 15.70).

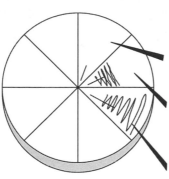

1. Use a sharpie to draw sectors on the bottom of an agar plate. Follow Steps 2–4 for each colony to be streaked out.

2. Pick bacteria with sharp toothpick and make a single line in the center of plate sector.

3. Streak from the middle of the line with the edge of sterile stick. Keep streak closely spaced and stay within the border of plate sector.

4. Using another sterile stick, streak out from the middle of the previous streak. Keep streak closely spaced and stay within the border of plate sector.

Figure 15.69   Spreading on plates by toothpick.

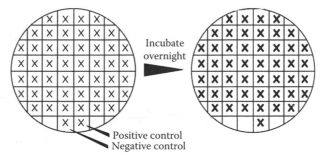

Incubate overnight

Positive control
Negative control

Figure 15.70   Picking and patching colonies.

15.31.1.3 Isolation of Streptomycin-Resistant Mutant by Replica Plating Method

15.31.1.3.1 Principle.   E. coli could develop mutant strains that are streptomycin/ampicillin-resistant. The replica plating technique is applied in order to test the nature of mutation in the bacterium, that is, whether induced by the antibiotic or occurred spontaneously. To prove whether or not such a colony exists on a plain agar plate having between 500 and 1000 colonies could be a difficult task as the transfer of organisms from each colony to medium supplemented with streptomycin would be required. The task of transferring several hundred colonies to test one mutant is simplified by the use of replica plating technique in one step (Figure 15.71).

This experiment deals with the ampicillin-resistant mutant isolation of *E. coli* by the replica plating technique, where simply a velveteen covered colony transfer device is used for making the transfer on nutrient agar medium supplemented with and without ampicillin (Figure 15.72).

15.31.1.3.2 Materials.   Twenty-four hours old actively growing broth of *E. coli* culture, 10 mL nutrient agar deep tubes, streptomycin sulfate (10 mg/100 mL of sterile water),

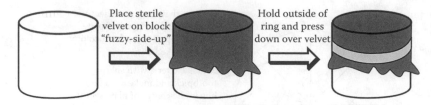

Figure 15.71 Preparation of velveteen covered colony transfer.

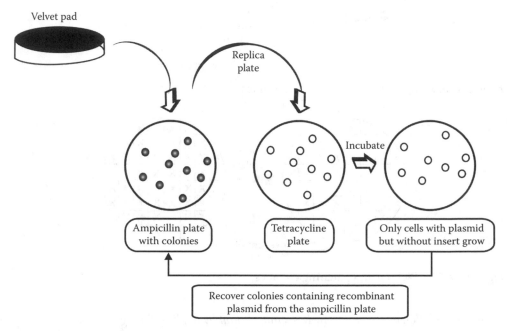

Figure 15.72 Replica plating to detect recombinant plasmids.

sterile Petri dishes, sterile velveteen colony carrier, beaker with 95% ethanol, bent glass rod, 1 mL serological pipette, and Quebec colony counter.

15.31.1.3.3 Protocol
First day:

1. Melt the nutrient agar deep tubes in a hot water bath (96°C) and allow molten medium to cool to 50°C.
2. Pour the molten agar medium to the two sterile plates and allow solidifying in the horizontal position.
3. Using a sterile 1 mL pipette, add 0.1% of streptomycin into the third tube of molten agar (50°C), mix by rotating tube between the palms of your hands, and pour the contents into a sterile plate.
4. Allow to solidify the plate in the horizontal position.
5. Using the sterile 1 mL pipette, add 0.2 mL of the *E. coli* test culture to the surface of nutrient agar plate.

6. Using an alcohol dipped and flamed bent glass rod, spread the inoculums over the entire agar surface by rotating the plate.
7. Incubate the plate in an inverted position for 24–48 h at 37°C.

Observe the incubated plate for the growth of *E. coli* culture on the agar surface. Second day:

1. Carefully lower the sterile velveteen colony carrier on to the colonies of *E. coli* growing on the nutrient agar plates.
2. Inoculate the nutrient agar plate by lightly pressing the carrier onto the medium.
3. Without returning the carrier to the original culture plate, inoculate the streptomycin agar plate by the procedure as followed earlier.
4. Incubate both the nutrient agar and streptomycin agar inoculated plates in an inverted position for 48–72 h at 37°C.

15.31.1.3.4 Results and Observations. Observe both the nutrient agar and streptomycin agar plates for the appearance of discrete colonies. Using Quebec colony counter, count the colonies that occur on both plates and record the results (Figure 15.73).

Nutrient agar plate will show the presence of the higher number of colonies as compared to the streptomycin agar plate, indicating the sensitivity of *E. coli* to these antibiotics. Moreover, the colonies isolated on the antibiotic agar will be of antibiotic-resistant mutants.

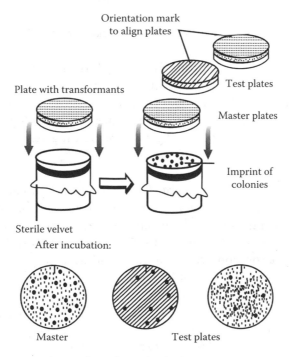

Figure 15.73 Demonstrating replica plating technique.

15.31.1.4 Bioassay for Evaluating the Mutagen for Carcinogen by Ames Test

15.31.1.4.1 Principle.   The Ames test, developed by Bruce Ames at the University of California, Berkeley, has been widely used for screening chemical compounds for possible carcinogen using bacteria as test organism. The chemicals in the form of industrial pollutants, pesticides, food additives, hair dyes, and cigarette smoke play a significant role in the induction of malignant changes in human beings. Such chemicals cause mutations in somatic cells. Mutations are mostly due to base substitutions and/or due to frame shift mutations. It is a mutational reversion assay employing several special stains of *Salmonella typhimurium*, each of which has a different mutation in the amino acid histidine biosynthesis operon. *S. typhimurium*, the indicator organism used in the test is the mutant strain that cannot synthesize histidine (his⁻) lacks the enzyme to repair DNA, so that mutation shows rapidly and has leaky cell walls that permit the rapid entry of chemicals. In addition, his⁻ *S. typhimurium* strain carries mutation-enhanced plasmid; the combined effects of these three characters increase the sensitivity of the organism to mutagenic treatments. A test agent is considered a mutagen if it enhances the rate of back-mutation beyond levels that would occur spontaneously. Since highly potential carcinogens, such as benzanthracene and aflatoxins, are mutagenic agents after being acted on by mammalian liver enzymes, an extract of these enzymes can be added to the test medium. The premise of Ames test is that any chemical capable of mutating bacterial DNA can similarly mutate human DNA and is, therefore, potentially hazardous. The Ames test has proved invaluable for screening an assortment of environmental (e.g., pesticides) and industrial pollutants and food additives for mutagenesis and carcinogenicity.

15.31.1.4.1.1 Ames Test by Plate Incorporation Method (Pour Plate Assay)

15.31.1.4.2 Material.   Twenty-four hours actively grown cultures of *S. typhimurium* Ames strain TA98 or TA1538 (ATCC), three minimal agar (MA) plates (30 mL/plate), three 2 mL top agar plates (2 mL/tube), 2-nitrofluorene dissolved in ethanol (10 μg/mL), commercial hair dye, sterile Pasteur pipettes, 1 mL serological pipettes, water bath, mechanical pipetting device, and glassware marking pencil.

15.31.1.4.3 Protocol
  15.31.1.4.3.1. Preparation of Modified E MA (Standard Minimal Medium for S. typhimurium).   A stock solution of 50× E medium is prepared by dissolving each of the following ingredients in the order given: distilled water 670 mL, $Mg.SO_4 \cdot 7H_2O$ 10 g, citric acid 100 g, K_2HPO_4 (anhydrous) 500 g, and $NaHPO_4 \cdot 4H_2O$ 175 g.
  Store the 50× stock solution over chloroform and sterilize when diluted to 1×. Add 2% glucose and 1.5% agar per liter of 1× E medium.

  15.31.1.4.3.2 Top Agar.   NaCl 5 g, agar 6.5 g, and distilled water 1000 mL.
  Dissolve agar in 500 mL of distilled water and add NaCl dissolved in 100 mL and make volume to 1 L. Dispense 100 mL into bottles, autoclave at 121°C for 15 min. Add 10 mL of a sterile stock mixture of L-histidine HCl (0.05 μM, i.e., 10.5 mg/100 mL) to

100 mL top agar and maintain at 45°C until used. Pour 2 mL of top agar in sterile test tubes (10 × 75 mm).

1. Prepare E MA plates by pouring 30 mL of the medium into each Petri plate.
2. Label the bottoms of the three minimal E agar plates as follows: positive control, negative control, and unknown.
3. Liquify three tubes of top agar and maintain the molten agar at 45°C.
4. To each molten top agar tube, aseptically add 0.1 mL of an overnight grown culture of *S. typhimurium His⁻* auxotroph.
5. Thoroughly mix the bacterium into the top agar by rotating the tube between the palms of your hands.
6. Add 0.5 mL of the test compound (dissolved in water, DMSO, ethanol, or formaldehyde depending upon its solubility—10 µg/mL).
7. Mix the contents immediately by rotating the tube between the palms and pour to contents over the surface of a minimal E glucose agar plate (unknown).
8. Tilt the plate all sides for even distribution of the top agar layer. The entire operation (Steps 4–7) should be carried out within 20 s.
9. Repeat Steps 4–7 for each of the other two tubes of top agar, one plate with control plates of bacteria alone (negative control) and other plate with bacteria plus the mutagen (i.e., 2-nitrofluorene dissolved in alcohol, 10 µg/mL) (positive control).
10. Allow the agar to harden in the dark for a few minutes.
11. Incubate the plates in an inverted position in the dark at 37°C for 24–48 h.

15.31.1.4.4 Results and Observations. Examine the negative control plate for the appearance of the colonies of histidine revertants (phototrophs) and the test plate for the his⁺ colonies induced by the chemical. Count the number of colonies in all the three plates and record the data in a tabular form (Figure 15.74).

The positive test plate, that is, added with the known, and strong mutagenic agent will have a high density of revertants, that is, his⁺ colonies induced by the chemical. The negative control plate will either have no or very few revertants. It is due to spontaneous back mutations (i.e., *his⁻→his⁺*), which always occurs. By comparing the number of colonies growing on the control plate with the test plate, the degree of mutagenicity of the chemical agent can be calculated. Chemicals that produce an increased incidence of back mutation are considered carcinogens (i.e., more number of *his⁺* revertants).

15.31.1.4.4.1 Ames Test by Spot Method
Principle: In this method, the mutagen is omitted from the top agar mixture and is applied after the top agar containing the bacterial inoculum has been poured and hardened. The mutagen can be added directly as a few crystals or 6 mm sterile filter disc impregnated with the test chemical dissolved either in water, alcohol, or DMSO solution is placed in the center of the test plate and the plates are incubated. Following diffusion of the test compound from the disc, a concentration gradient of chemicals is established. A qualitative indication of the test compound can be determined by noting the number of colonies present on the plate. Spot test is widely used for screening of chemical compounds for mutagenicity.

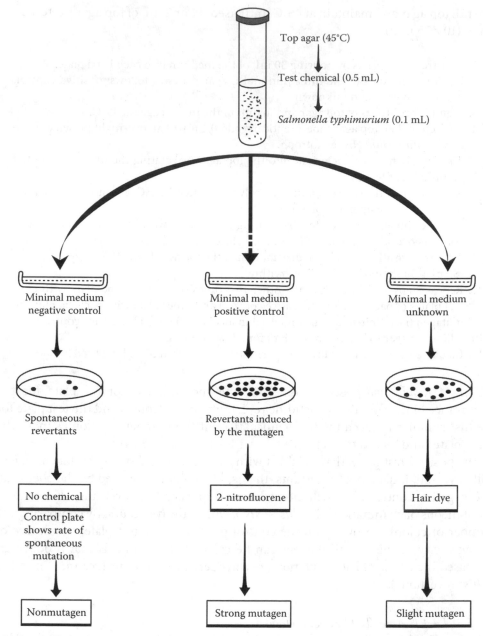

Figure 15.74   Pour plate method for Ames test.

Materials: Twenty-four hours actively grown cultures of *S. typhimurium* Ames strain TA98 or TA1538 (ATCC), three MA plates (30 mL/plate), three 2 mL top agar plates (2 mL/tube), 2-nitrofluorene dissolved in ethanol (10 µg/mL), commercial hair dye, sterile Pasteur pipettes, 1 mL serological pipettes, water bath, mechanical pipetting device, glassware marking pencil, sterile filter paper discs, and forceps.

Protocol:

1. Label the bottoms of three minimal E glucose agar plates as negative control, positive control, and hair dye.
2. Molt the three tubes of top agar in a water bath and maintain at 45°C.
3. Aseptically add 0.1 mL of *S. typhimurium* test culture into molten top agar tube and thoroughly mix the contents by rotating the test tube between the palms of your hands or by vortexing for 3 s.
4. Immediately pour the top agar cultures onto the E MA plate (control positive) and allow to solidify.
5. Repeat Steps 3 and 4 for each of the other two top agar tubes.
6. With a sterile forceps, dip each disc into its respective test chemical solution and drain by touching the side of the container, or position a filter paper disc on the edge of a MA plate with sterile forceps. Saturate the disc with a test compound.
7. Place the chemical impregnated disc in the center of the respectively labeled E minimal glucose agar plates. Gently press down on the discs.
8. Insert a sterile disc dipped in sterile water on the negative control plates.
9. Incubate all the three plates for 24–48 h at 37°C.
10. Examine all the three plates for the appearance of colonies around the disc. Count the number of colonies in each plate and record the result in a tabular form (Figure 15.75).

Results: On examination after 48 h of incubation, the positive test plates, that is, the plates having the disc impregnated with 2-nitrofluorene, a strong mutagen, will have a high density of revertants around the disc. If the hair dye used is mutagenic, only then revertant colonies will be seen around the disc, and the negative control plate will have either no or a few scattered revertants produced as a result of spontaneous back-mutations.

15.32 Bacterial Recombination

15.32.1 Introduction

Recombination is the process in which a recombinant chromosome, one with a genotype different from either parent is formed by combining genetic material from two organisms. It results in a new arrangement of genes or parts of genes and normally accompanied by a phenotypic change. Recombination results in genetic variability, which is essential for the evolutionary success of all organisms. In most eukaryotic organisms, who exhibit a complete sexual cycle, the process of crossing over, exchange of genetic material between homologous chromosomes, and meiosis contribute to this variability. In bacteria, there is no true fusion of male and female gametes to produce a diploid zygote; instead, there is transfer of only some genes from the donor cell to produce a partial diploid. The actual transfer of genetic material between bacteria usually takes place in one of the three ways: (1) conjugation—direct transfer between two bacteria temporarily in physical contact; (2) transformation—transfer of naked DNA fragment; and (3) transduction—transport of a bacterial DNA by bacteriophages (Figure 15.76).

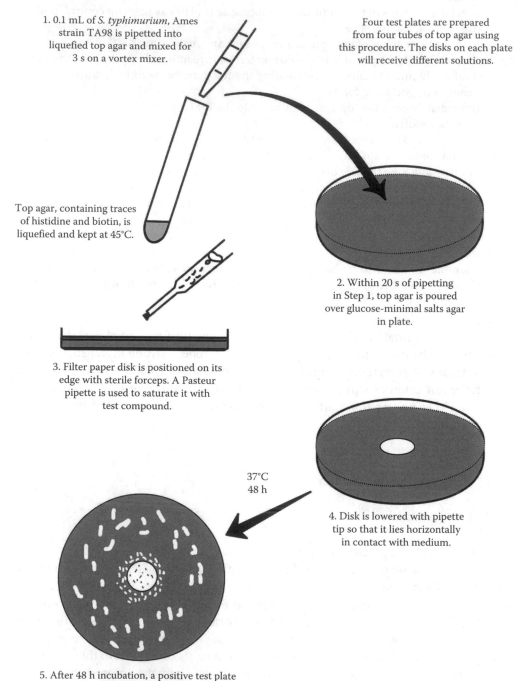

1. 0.1 mL of *S. typhimurium*, Ames strain TA98 is pipetted into liquefied top agar and mixed for 3 s on a vortex mixer.

Four test plates are prepared from four tubes of top agar using this procedure. The disks on each plate will receive different solutions.

Top agar, containing traces of histidine and biotin, is liquefied and kept at 45°C.

2. Within 20 s of pipetting in Step 1, top agar is poured over glucose-minimal salts agar in plate.

3. Filter paper disk is positioned on its edge with sterile forceps. A Pasteur pipette is used to saturate it with test compound.

37°C
48 h

4. Disk is lowered with pipette tip so that it lies horizontally in contact with medium.

5. After 48 h incubation, a positive test plate will have a halo of high-density revertants growing around the disk. The large colonies beyond the halo are spontaneous back mutations.

Figure 15.75 Protocol for Ames test by spot method.

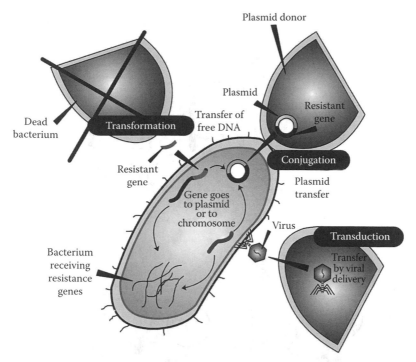

Figure 15.76 Demonstrating three different modes of genetic transfer in bacteria.

Conjugation may be defined as unidirectional transfer of genetic material (DNA) by one organism to another. To enable the DNA to pass from the donor (male) to the recipient (female), a "conjugation tube" forms between the two organisms. Only a fractional segment of the donor's chromosome passes into its single, circular chromosome. With this new acquisition of genes, the recipient becomes a different organism (Figure 15.77).

15.32.1.1 To Demonstrate Transformation in E. coli *or in Any Other Bacterium*

15.32.1.1.1 Principle.   Transformations are an important source of bacterial variability and origin of new types in nature. In short, transformation in bacteria is one of the simplest recombination systems, involving single donor DNA particles interacting with single cell genomes. The donor DNA can be initiated, carried out, and terminated under controlled conditions. The process of transformation essentially includes removal of deoxyribonucleic acid from one strain of a particular bacterium and addition to another strain of that bacterium. Some of the cells in the second strain are able to transmit characteristics of the first to their progeny (Figure 15.78).

The different steps involved in the process of transformation are as follows:

1. Donor DNA segment is absorbed by the recipient cell.
2. Absorbed DNA becomes irreversibly bound.

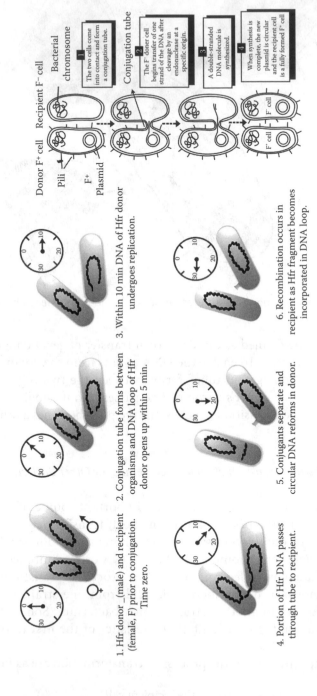

Figure 15.77 Demonstrating mechanism of bacterial conjugation.

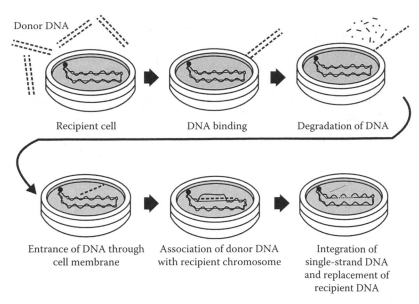

Figure 15.78   Transformation in bacteria.

3. The DNA begins the process of integration. A part of the donor DNA, possibly only one strand, is integrated, probably by substitution into the genome of the recipient. Demonstration of the process of transformation includes two steps: (1) DNA extraction and (2) transformation.

15.32.1.1.1.1 DNA Extraction

Materials: *E. coli* culture at log phase, centrifuge and centrifuge bottles (250 mL), conical centrifuge tubes, saline-EDTA buffer, lysozyme, sodium dodecyl sulfate, 5 M perchlorate, chloroform-isoamyl alcohol, and 95% ethyl alcohol.

Function and Preparation of Chemicals Used: *Saline EDTA (ethylenediaminetetraacetate)*: It inhibits DNase activity. Dissolve 37.22 g of EDTA and 8.77 g of NaCl in 600 mL of distilled water, adjust pH to 8.0, and add distilled water to bring volume to 1 L.

Lysozyme: It lyses cells resistant to detergent. Dissolve 2 g of crystalline lysozyme in 100 mL of distilled water.

Sodium dodecyl sulfate ($NaC_{12}H_{26}SO_4$): It is an ionic detergent and will lyse most of the nonmetabolizing cells. It inhibits enzyme action and denatures some proteins. Dilute 25 mL of SDS in 75 mL of distilled water.

Sodium perchlorate: It dissociates proteins from nucleic acids. Dissolve 141 g of sodium perchlorate in 200 mL of distilled water.

Chloroform-isoamyl alcohol: It causes denaturation of proteins. It aids in separation of proteins and DNA. Add 288 mL of chloroform with 12 mL isoamyl alcohol per 300 mL.

Ethyl alcohol: It precipitates nucleic acid.

Protocol:
1. Grow 200 mL of *E. coli* (any strain) in Penassay broth to log phase (12–24 h in standing, unaerated culture at room temperature) and pour off this 200 mL culture into the 250 mL centrifuge bottle and centrifuge at 8000 rpm for 5 min.
2. Discard the supernate and resuspend the pellet in 25 mL of saline-EDTA buffer.
3. Pour off into screw-cap conical centrifuge tube and centrifuge at 5000 rpm for 5 min.
4. Pour off the supernate and resuspend the pellet in 13 mL of saline-EDTA buffer.
5. Add 1 mL of lysozyme and place in a 37°C water bath for 30 min.
6. Add 2 mL of sodium dodecyl sulfate and place in a 50°C water bath for 10 min.
7. Add 4 mL of 5 M perchlorate to suspension to dilute perchlorate to a final concentration of 1 M.
8. Add 15 mL of chloroform-isoamyl alcohol.
9. Tighten the cap and shake vigorously.
10. Centrifuge at 5000 rpm for 3 min.
11. Decant the top layer (contains DNA) into 30 mL of cool 95% ethyl alcohol.
12. Gently stir the solution with a glass rod. DNA is at the bottom in the form of strands.

15.32.1.1.2 Transformation

15.32.1.1.2.1 Materials. Solution of DNA from a prototrophic strain of *E. coli*; *E. coli* culture of auxotrophic strain for valine and leucine, that is, Val⁻ Leu⁻. DNase solution.

15.32.1.1.2.2 Protocol
1. Add DNase solution to the DNA extract of the prototrophic strain.
2. Inoculate two plates having minimal medium with the auxotrophic strain.
3. Spread DNA solution (100 µg/mL) of Step 1 over one cultured plate of Step 2 with a glass spreader. The other plate serves as the control.
4. Incubate both the plates for 48 h at 37°C.
5. Count the colonies in both the plates and interpret the results.

15.32.1.1.2.3 Results and Observations
1. No colonies are expected in the control dish since auxotrophs cannot grow on minimal medium.
2. Presence of colonies in the experimental dish indicates transformation of auxotrophs to prototrophs by recombination of genes between the donor DNA and auxotrophic genome.

15.32.1.1.3 Calcium Chloride (CaCl₂) Induced Transformation Procedure in E. coli

15.32.1.1.3.1 Introduction. A genetic alteration in a cell resulting from the introduction of free DNA from the environment across the cell membrane is known as transformation. Genetic transformation is a process by which a cell takes up naked DNA from the surrounding medium and incorporates it to acquire an altered genotype that is heritable. Bacteria are the only organisms (except yeast) known to transform in nature although cells of higher organisms including that of mammals are transformable by artificial techniques. The overall process in naturally transformable organisms may be divided into five common steps: (1) development of competence, (2) binding DNA, (3) integration of DNA, (4) formation of pre-integration complex, and (v) integration of DNA into the recipient cell chromosomes or exist as an extrachromosomal DNA (plasmid).

The competence of a bacterial cell for transformation could be artificially induced by exposing cells to calcium chloride prior to the addition of DNA. It involves the exposure of growing bacterial cells to a hypotonic solution of calcium chloride at 0°C, causing cells to swell (spheroplast formation). DNA added to the transformation mixture forms a DNase-resistant complex of hydroxyl-calcium phosphate that adheres to the cell surface. This complex can be taken up by the cell during a brief 42°C heat pulse. After a few hours of growth in rich medium to allow the spheroplast to recover and the transformed genes to be expressed, transformants can be isolated by plating on selective medium. For example, cells transformed by pBR322 DNA can be selected on medium containing the antibiotics ampicillin and tetracycline. With minor modifications, this basic transformation protocol has been made to work for a variety of bacteria that do not take up exogenous DNA normally. It is suggested further that magnesium ions may play an important role in the stability of the DNA during uptake, and many transformation methods now include $MgCl2$ during the treatment of the bacteria (Figure 15.79).

15.32.1.1.3.2 Materials. Conical flasks, test tubes, centrifuge, pipettes, incubator, Petri plates, autoclave, Eppendorf tubes, colorimeter, tryptone, yeast extract, sodium chloride, NaOH, distilled water, $CaCl_2$, DNA, Tris EDTA buffer, ampicillin, tetracycline, distilled water, *E. coli*, and plasmid DNA (pBR 322).

LB medium g/L in distilled water: Casein enzymic hydrolysate 10, yeast extract 5, sodium chloride 5, and final pH (at 25°C) 7.0 ± 0.2.

Tris-EDTA buffer: 10 mM Tris 1 mM EDTA dissolved in distilled water.

15.32.1.1.3.3 Protocol
1. From the overnight grown culture, take 0.1 mL and inoculate into the 25 mL of fresh LB medium in 100 mL flask.
2. Allow it to grow till the OD reaches 0.5–0.6 at 540 nm.
3. Chill the culture by keeping it in ice for 30 min.
4. Spin the culture at 5000 rpm for 10 min and discard the supernatant.
5. Suspend the pellet in 5 mL of 50 mM or 10 mM $CaCl_2$.
6. Keep it in the ice (0°C) for 30–60 min.
7. Spin down the culture and discard the supernatant.
8. Suspend the pellet in 1 mL of 10 mM $CaCl_2$, take 0.2 mL of culture and add 50 mg to 500 mg of DNA in Tris-EDTA buffer and keep in the ice for 30 min.
9. In the control experiment, to 0.2 mL of the culture, add only buffer without DNA.
10. Give a heat shock at 42°C for 2 min and transfer it to ice and keep again for 5 min.
11. Now add 0.8 mL of sterile LB broth and incubate at 37°C for 45 min, for the cells to multiply and express the marker.
12. Take an aliquot of 0.1 mL each of these cells and plate on LB agar containing ampicillin (50 mg/mL) or tetracycline (15 μg/mL) or both and incubate the plates at 37°C for overnight.
13. Furthermore, plate 0.1 mL of the control culture (step 9) on the selective plate and incubate at 37°C along with plates of step 12 (after incubation no colonies should appear in the control plates).

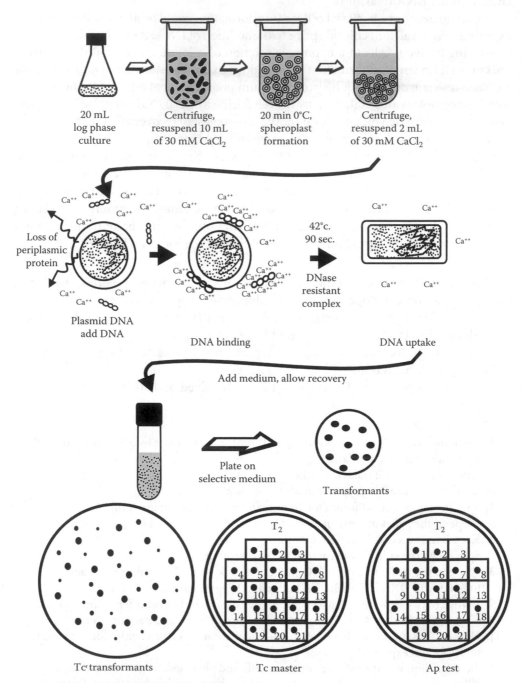

Figure 15.79 A diagrammatic representation of the $CaCl_2$-induced transformation procedure for *E. coli*.

15.32.1.1.4 Demonstration of Genetic Recombination in Bacteria by Conjugation

15.32.1.1.4.1 Principle. In this experiment, an attempt has been made to demonstrate the existence of the phenomenon of genetic recombination by using two auxotrophs of *E. coli*. An auxotroph is a bacterial mutant that requires one or more growth factors that the wild strain, or prototroph, can synthesize. Auxotrophs are produced in the laboratory by subjecting prototrophs to mutagenic agents, such as ultraviolet irradiation or mitomycin C. One of the auxotrophs used in this experiment is able to synthesize methionine, but not threonine, from a minimal medium containing only glucose, ammonia, and inorganic salts. In other words, this auxotroph has lost the gene that exists in the prototroph that would enable it to synthesize threonine in the presence of a minimal medium. Thus, it cannot grow on such a medium. Its genetic composition is designated as thr⁻met+. The other auxotroph in this experiment is able to synthesize threonine, but not methionine, and also is unable to grow on the minimal medium. It is designated as thr⁺met⁻. If these two auxotrophs are able to conjugate to produce a strain that is thr⁺met⁺ to have a genetic combination, like the original prototroph, that will grow on the minimal medium. The experiment demonstrates that the conjugation can occur to bring such a genetic recombination of genes. The donor in this experiment is thr⁺met⁻. The recipient is thr⁻met⁺.

15.32.1.1.4.2 Materials. Twelve hours actively growing (10^8 cells per mL) broth of auxotroph strain of *E. coli* (thr⁺ met⁻) male (M) strain 1, 12 h actively growing (10^8 cells per mL) broth of auxotroph strain of *E. coli* (thr⁻ met⁺) female (F) strain 2, two Petri plate of trypticase soy agar (TSA) complete medium, five Petri plates of MA, three sterile 1 mL pipettes, three tubes of sterile distilled water (9 mL per tube), one sterile serological tube (13 × 100 mm), beaker of 95% alcohol, one bent glass rod, and glass marker pencil.

15.32.1.1.4.3 Protocol. *Control inoculations*: To establish that the two auxotrophic strains of *E. coli* grow well on a complete (TSA) but not at all on MA, inoculate two plates of each of these media as follows:

1. Label one TSA plate "TSA-Male" and the other "TSA-Female." Label one MA plate "MA-Male" and the other "MA-Female."
2. With a good isolation technique, streak all four plates with the appropriate organisms.

Recombination inoculations: The procedure illustrated in Figure 15.79 bring about conjugation of the two auxotrophs of *E. coli*. One milliliter of each strain is placed in an empty test tube and is allowed to stand for 30 min. After this period, when conjugation has presumably occurred, the conjugants are diluted out onto MA plates in dilutions of 1:100, 1:1000, and 1:10,000. Any colonies that appear on this medium are organisms that can synthesize both threonine and methionine (thr⁺met⁺). By counting the colonies on these plates, it is possible to determine the frequency of recombination (Figure 15.80).

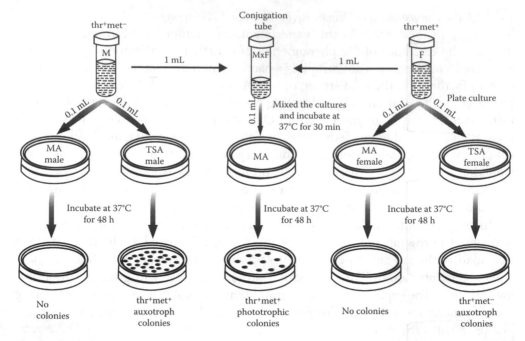

Figure 15.80 Demonstrating conjugation in bacteria using single auxotrophs of *E. coli*.

1. Label a sterile serological tube "M × F." This will be used for conjugation of the two strains of *E. coli*.
2. With separate sterile 1 mL pipettes, aseptically transfer 1 mL from each culture into a sterile 13 × 100 mm test tubes labeled as conjugation tube. Let it stand for 30 min.
3. Mix the two cultures by gently rotating the tube between the palms of hands.
3. While conjugation is taking place, label the three water blanks: I, II, and III. Furthermore, label the three plates of MA: I, 1:100; II, 1:1000; and III, 1:10,000.
4. With a sterile pipette, transfer 1 mL from the conjugation tube into water blank I. Mix the contents in tube I by drawing the dilution up into the pipette three times and discharging slowly into the tube.
5. With the same pipette, transfer 0.1 mL from tube I to plate I. Do not spread at this time.
6. Still using the same pipette, transfer 1 mL from tube I to tube II and mix contents in tube II as mentioned earlier. Now transfer 0.1 mL from tube II to plate II. Do not spread at this time.
7. Still using the same pipette, transfer 1 mL from tube II to tube III and repeat all aforementioned procedures to inoculate plate III from tube III with 0.1 mL of diluted organisms.
8. With a sterile bent glass rod, spread the organisms over each plate. Be sure to sterilize the rod before spreading organism on each plate.
9. Incubate the two TSA and five MA plates, inverted, for 2–4 days at 30° C.

15.32.1.1.4.4 Results and Observations. Observe the two TSA and two MA control plates inoculated with the individual *E. coli* strain as well as minimal plate inoculated with the mixed culture. Count the number of colonies on all the five plates and record the results. Examine the control streak plates.

TSA control plates are expected to exhibit the growth of the two auxotrophic strain of *E. coli* (i.e., thr⁺met⁻ and thr⁻met⁺) on a complete medium but no growth of these strains on the MA (control) plates. Recombinant phototrophic colonies of *E. coli* are observed on the experimental MA plate as a result of conjugation of two auxotrophic strains of *E. coli*. During conjugation, the chromosomes of two auxotrophs get associated and undergo recombination thus producing a recombinant strain with genetic composition of thr⁺ met⁺. The number of colonies counted on the experimental MA plate be compared to controls in order to estimate the per cent of recombinants produced.

15.32.1.1.5 Demonstration of Transduction

15.32.1.1.5.1 Introduction. Transduction is a phenomenon in which a bacteriophage transfers genetic material from one cell to another. This is a process in which a bacterial trait is transmitted from donor cell to a recipient bacterial cell by means of a temperate bacteriophage. Transducing phage may originate from cells that are lysed or may originate as the prophage particle from lysogenized cells following induction. Transduction system is of two types. Generalized and specialized. In the former, the phage may exchange genetic material at any site on the bacterial genome, and in the latter, the phage shows specificity for a single site on the bacterial genome.

15.32.1.1.5.2 Materials. Test tubes, cooling centrifuge, conical flasks, pH meter, pipettes, Petri plates, incubator, tryptone, yeast extract, sodium chloride, NaOH, $MgSO_4 \cdot 7H_2O$, $CaCl_2 \cdot 2H_2O$, HCl, Na_2HPO_4, KH_2PO_4, NH_4Cl, agar, distilled water, bacterial culture of *E. coli* RR$_1$ (Pro⁻), and *E. coli* (C 600, Pro⁺) P1 phage lysate.

LB agar plates (g/L): Tryptone 10 g, agar 10 g, yeast extract 5 g, and NaCl 10 g, pH adjusted to 7.2–7.4 with 10 N NaOH. Agar medium is autoclaved and cooled, to pour in to Petri plates. Soft agar plates are prepared with 0.5% agar.

MA medium (M63): Na_2HPO_4 5.8 g, agar 15 g, KH_2PO_4 3g, NaCl 0.5 g, and NH_4Cl 1 g. Dissolve all the chemicals in 1 L of distilled water, and pH is adjusted to 7.0 and sterilized.

The following sterilized ingredients are added just before use to complete the medium: 10% glucose or glycerol 50 mL, 1 M $MgSO_4 \cdot 7H_2O$ 1 mL, 0.01 M $FeCl_3$ 1 mL, 1.5% agar, and poured to plates.

MC buffer: $MgSO_4 \cdot 7H_2O$ 0.1 M and $CaCl_2 \cdot 2H_2O$ 0.005.

TM buffer: 10 m M Tris-HCl (pH 8) containing 10 mM $MgSO_4 \cdot 7H_2O$.

15.32.1.1.5.3 Protocol

1. Grow the *E. coli* RR$_1$ Pro-culture in LB medium (5 mL) over night.
2. Centrifuge and pellets are suspended in 5 mL of MC buffer.
3. Aerate the suspension at 37°C for 15 min.
4. Add 0.8 mL of the suspended cells to each of "6" test tubes and label them 10^0, 10^{-1}, 10^{-2}, 10^{-3}, and 10^{-4}.

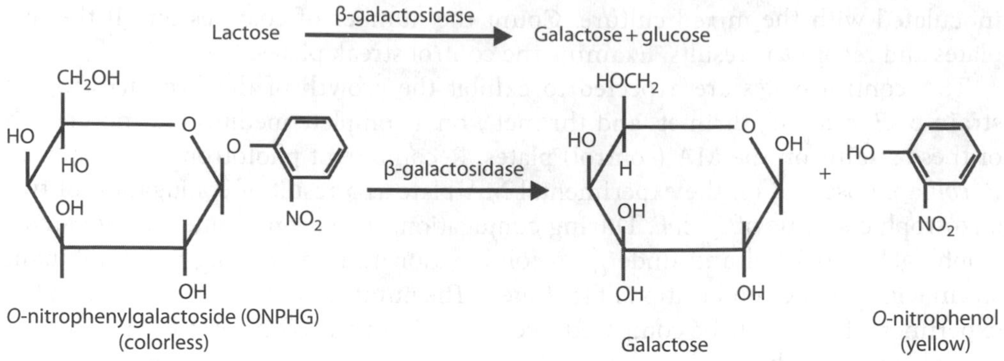

Figure 15.81　Chemical reaction of β-galactosidase.

5. Add to each tube 0.2 mL of diluted (10^{-1} to 10^{-4}) P1 lysate obtained from *E. coli* C 600 Pro$^+$. Add only buffer in one of the tube as control, which has no lysate.
6. Allow for preabsorption of phage particles on cells by incubating the tubes at 37°C for 20 min.
7. An aliquot (0.1 mL) of the aforementioned medium is mixed with 10 mL of top agar and kept at 45°C.
8. Mix well and pour the mixture on LB bottom agar plates.
9. Incubate at 37°C for 24–48 h.
10. Observe the appearance of colonies on LB agar.

15.32.1.1.5.4　Result.　Note the colonies on LB agar plates. Then replica plate the colonies on MA and score for prototroph. Calculate the transduction frequency and express it in terms of the number of prototrophs (transductant) per recipient cell (Figure 15.81).

Suggested Readings

Alexopoulos, C. J., C. W. Mims, and M. Blackwell. 1996. *Introductory Mycology*, 4th edn. New York: John Wiley & Sons.

Benson, H. J. 1990. *Microbiological Applications—A Laboratory Manual in General Microbiology*. Dubuque, IA: Win. C. Brown Publishers.

Bisen, P. S., G. B. K. S. Prasad, and M. Debnath. 2012. *Microbes: Concepts and Applications*. 1st edn. New York: John Wiley & Sons, p. 699.

Bisen, P. S. and A. Sharma. 2012. *Introduction to Instrumentation in Life Sciences*, 1st edn. Boca Raton, FL: CRC Press, p. 3553.

Cappuccino, J. G. and N. Sherman. 1991. *Microbiology: A Laboratory Manual*. Amsterdam, the Netherlands: Benjamin/Climmings Publishing Company.

Cochran, W. G. 1950. Estimation of bacterial densities by means of the most probable number. *Biometrics* 6: 105–116.

Corliss, J. O. 1991. Introduction to the protozoa. In *Microscopic Anatomy of Invertebrates*, eds. F. W. Harrison and J. J. Corliss, vol. 1. New York:Wiley-Liss, pp. 1–12.

Hleyn, J., M. Bicknell, and M. Gilstrap. 1995. *Microbiology Experiments: A Health Science Perspective*. London, U.K.: Wm. C. Brown Publishers, p. 354.

Ingraham, J. L. and C. A. Ingraham. 2000. *Introduction to Microbiology: A Case-History Study*. Pacific Grove, CA: Brooks Cole.

Marinšek, L. R. and M. Vodovnik. 2007. The applications of microbes in environmental monitoring. In *The Applications of Microbes in Environmental Monitoring*. Current Research and Educational Topics and Trends in Applied Microbiology, eds. A. Méndez-Vilas. ©FORMATEX, Badajoz, Spain.

Pollack, R. A., L. Findlay, W. Mondschein, and R. Ronald Modesto. 2004. *Laboratory Exercises in Microbiology*. New York: John Wiley & Sons, p. 280.

Prsecott, L. M., J. P. Harley, and D. A. Klein. 2005. *Microbiology*, 6th edn. New York: McGraw Hill, p. 992.

Seeley, H. W., Jr. and P. J. Denmark. 1975. *Microbes in Action: A Laboratory Manual of Microbiology*. Mumbai, India: D.B. Taraporevala Sons & Co. Pvt. Ltd, p. 361.

Shirling, E. B. and D. Gottlieb. 1966. *Methods, Classification, Identification and Description of Genera and Species*, vol. 2. Baltimore, MD: The Williams & Wilkins Company.

Tortora, G. J., B. R. Funke, and C. L. Case. 2009. *Microbiology: An Introduction*, 10th edn. San Francisco, CA: Benjamin Cummings, p. 960.

Important Links

http://www.microbiologyonline.org.uk/links
http://www.asm.org/
http://clevelandcc.edu/uploads/catalog/bio-175— -general-microbiology.htm
http://www.tcd.ie/Microbiology/news/
http://books.google.co.in/books/about/General_Microbiology.html?id=DrHQtIbiunkC
http://www.fems-microbiology.org/website/nl/page16.asp
http://www.dartmouth.edu/~ehs/biological/
http://www.foodsafetysite.com/educators/competencies/general/microbiology/mic6.html

16

Soil Biology

16.1 Introduction

The diversity of the soil biology is so high that inevitably some selection must be made. Soil biology plays an important role in nutrient transformations and in the reclamation of waste lands. The health of soil is often described in terms of the microbial activity and the organism content of the soil. The decomposition of organic matter is an important soil process for organically managed farms and gardens. Organic matter includes a vast array of compounds that can be biologically decomposed at various rates, depending on their physical and chemical complexity. Environmental factors such as temperature and moisture also determine decomposition rate. Soil respiration is an indicator of biological activity (i.e., microbial and root) or soil life. This activity is as important to the soil ecosystem as healthy lungs are to us. However, more activity is not always better because, in some circumstances, it may indicate an unstable system undergoing net carbon loss (i.e., after the tillage). Soil respiration is the production of carbon dioxide (CO_2) as a result of biological activity in the soil by microorganisms, live roots, and macroorganisms such as earthworms, nematodes, and insects (Figure 16.1). CO_2 emitted from soil is a colorless and odorless gas that enters the atmosphere and annually exceeds the amount emitted by all human activities. The activity of organisms in the soil is considered to be a positive attribute for soil quality. Soil respiration can be limited by moisture, temperature, oxygen, soil reaction (i.e., pH), and the availability of decomposable organic substrates. Optimum respiration usually occurs at around 60% of water-filled pore space. Soil respiration will decrease under saturated or dry conditions. Biological activity doubles for every 18°F rise in temperature until the optimum temperature is reached (varies for different organisms). Activity declines as temperature rises above optimum. The most efficient soil organic matter decomposers are aerobic; thus, soil respiration rates decline as soil oxygen concentration decreases. The organism fraction of the soil, often referred as humus, is a product of synthetic and decomposing activities of microflora. Since it contains the organic carbon and nitrogen needed for microbial development, it is the dominant food reservoir. Because humus is both a product of microbial metabolism and an important food source, the organic fraction is of special interest to the microbiologist. Estimation of microbial counts can be enumerated by simple plate count technique or most probable number (MPN) technique; the microbial

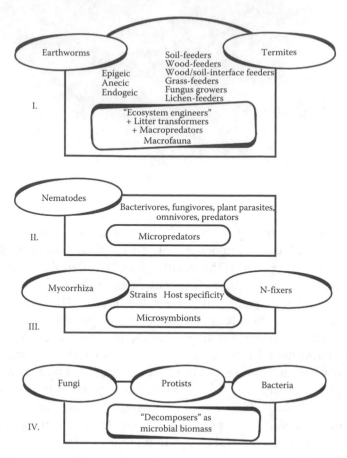

Figure 16.1   Illustrating soil diversity.

activity can only be determined by assaying certain microbial enzymes like dehy-drogenases and ureases or by estimating the amount of CO_2 liberated. Several ben-eficial microorganisms like *Rhizobium*, *Azotobacter*, *Azospirillum*, phosphorus (P) solubilizing microorganisms, and plant growth promoting microorganisms can be isolated by employing specific media and technique. The efficiency of nitrogen fix-ers can be estimated by assay of nitrogenase enzyme, while for P solubilizers, avail-able P(AM) fungi forms association with the roots of higher plants and mobilizes P and other micronutrients like Zn and Cu from the soil into the plants.

16.2 Estimation of Humus in Soil/Manure

16.2.1 Principle

Humus is a complex mixture of heterogeneous organic compounds derived from plant and animal residues and is formed by the reaction of decomposition, synthesis, and polymerization. It is an integral part of soil organic matter and can be extracted by neutral salts or alkali separating humic and nonhumic fraction.

16.2.2 Materials

Extracting solution: Dissolve 44.6 g of sodium pyrophosphate and 4 g NaOH in 1 L water to maintain 0.1 M strength of the solution.

16.2.3 Protocol

1. Take 20 g soil or manure in a conical flask and add 100 mL of extractant (1:5 ratio). Shake well and allow standing for 18 h.
2. Filter through Whatman filter paper no. 1 (the residue portion on filter paper is called humin and the filtrate is humus. It is dark brown in color with pH 13).
3. To get rid of sodium, the humus is dialyzed in dialysis tubes or parchment paper against running tap water for 24 h.
4. It is free of sodium. Measure the volume, and a known volume (10 mL) is taken in a pre-weighed 100 mL beaker. The beaker is dried in an oven at 70°C to a constant weight. The difference in weight gives the amount of humus in 10 mL. Accordingly, it is calculated on a percentage basis.

16.3 Determination of Microbial Carbon Biomass

16.3.1 Principle

The content of soil organic matter affects the stability of agro-ecosystems and is controlled by many factors such as cultivation. Over short periods, changes in microbial biomass C can be a sensitive index of changes in the content of soil organic matter. The soil microbial biomass acts as the transformation agent of the organic matter in soil. As such, the biomass is both a source and sink of the nutrients C, N, P, and S contained in the organic matter. It is the center of the majority of biological activity in soil. Soil microorganisms are important for the release and retention of nutrients, and soil microbial biomass does play an important role while assessing nutrient and energy flow in the soil system. If a soil is fumigated, the respiration rate of soil immediately after removal of the fumigant will be less than that of nonfumigated soil. The respiration rate of the fumigated soil will increase sharply to a value exceeding that of the nonfumigated soil and then subside after a short period of incubation. There is a temporary flush in CO_2 evolution in the previously fumigated soil. This flush of decomposition (CO_2 evolution) is the result of the decomposition of microbial cells killed by the fumigation. The measurement of this flush in respiration could indicate the approximate size of the soil biomass. The fumigation–incubation method helps in determining the soil microbial biomass carbon (MBC).

16.3.2 Materials

Soil sample, desiccators, vacuum pump, Millipore filtration assembly, filter paper, centrifuge, test tubes, beakers, conical flasks, chloroform, and glass beads.
Reagents freshly prepared (quantities suitable for about 60 samples):

2 M KCl solution, 1 L

50% analytical grade ethanol in distilled water (dH$_2$O), 1 L

4 N acetate buffer (544 g sodium acetate to 400 mL water and stir while heating until dissolved). Stabilize at room temperature and add 100 mL of glacial acetic acid and make up to 1 L; adjust the pH to 5.51 ± 0.03.

Ninhydrin reagent: Dissolve 2 g ninhydrin and 0.2 g hydrindantin in 50 mL 2-methoxy ethanol, add 50 mL acetate buffer and mix; store in a dark glass reagent bottle.

2.5 × 10^{-4} M leucine solution: Prepare a stock solution of 2.5 × 1^{-2} M by dissolving 0.3279 g of L-leucine in 100 mL 2 M KCl. For the assay, dilute 1 mL of the stock solution with 99 mL of 2 M KCl; this standard contains 7 μg N per 2 mL.

2.5 × 10^{-4} M ammonium sulfate solution: Prepare a stock solution of 2.5 × 10^{-2} M by dissolving 0.1652 g analytical grade ammonium sulfate in 100 mL 2 M KCl. For the assay, dilute 1 mL of stock solution with 99 mL of 2 M KCl; this standard contains 7 μg N per 2 mL solution.

16.3.3 Protocol

16.3.3.1 Sample Collection

In the field, soil should be sampled with a spade from the top, 1–10 cm, after the clearance of dead wood and leaf litter. The number of samples taken should be sufficient to calculate confidence limits for the data produced, that is, at least five samples per transect, but two per 5 m section is a better number, making 16 in total, or two bulked samples per section, making a total of 8. At each sampling point, about 500 g of soil should be placed in a clean plastic bag and taken immediately to the laboratory and brushed firmly through a coarse sieve to remove stones and pieces of wood. The soil should be then air-dried for about 24 h, subsampled for the determination of moisture content (by evaporation to dryness at 105°C) and rebagged for further processing, which should begin as soon as possible and not more than 48 h later. Soil waiting for processing can be stored at 4°C, but must not be frozen or oven dried.

16.3.3.2 Analysis

The first step is to determine water-holding capacity (WHC). This need not be done on more than one or two samples per transect, as long as the site is pedologically homogeneous.

1. Place about 50 g of soil in a large filter funnel plugged loosely with glass wool and clamp to stand in a beaker. Pour about 100 mL of water onto the soil and leave overnight.

2. Weigh the moist soil and funnel, then dry in an oven at 105°C for 24 h and weigh again.
   Remove the dry soil from the funnel and weigh this alone.

$$\text{WHC} = \frac{\text{Weight of wet soil} - \text{Weight of dry soil}}{\text{Weight of dry soil}} \text{ g H}_2\text{O/g dry soil}$$

   WHC is then used to adjust all soil samples subsequently weighed out to 40% WHC (= 8.0 g H$_2$O per 20 g soil), taking account of the moisture already present (from the earlier determination of moisture content). Then, 20 g subsamples adjusted in this way are then ready for fumigation and/or extraction.

3. Weigh two subsamples of 20 g each of air-dried soil into small glass beakers and adjust to 40% WHC by the addition of water. One subsample will be fumigated with chloroform and then extracted with KCl solution and the other, which is a control, will be extracted with KCl solution without exposure to chloroform. Mark each beaker accordingly using pencil and tape (not felt-tip pens).

4. Place the samples for fumigation in large vacuum desiccators containing a beaker with about 50 mL of chloroform. Add a few glass beads to the chloroform, close the desiccators and evacuate (using either an electrical or water-jet vacuum pump) until the chloroform starts boiling. Seal the desiccators and leave in the dark for exactly 10 days at 25°C. Leave the control soil samples, loosely covered with foil, under the same conditions, but without exposure to chloroform.

5. After 10 days, extract each soil sample with 63 mL of 2 M KCl for 1 h, using an end-over-end (or rotary) shaker. Centrifuge for 2–5 min at 2500 × g to separate soil and supernatant. Carefully remove 10 mL of supernatant with a graduated plastic syringe, fit a Millipore prefilter (AP20 013 00) to the syringe and expel the supernatant through the prefilter into a clean 10 mL screw-cap plastic or glass vial. This filtrate can then be frozen pending further processing and analysis, but note that until this point is reached the timetable of the procedure must be adhered to strictly.

6. Use the standard leucine and ammonium sulfate solutions to check the linearity of the color development with ninhydrin and calibrate the reagent as follows:
   Make up four tubes containing 1 mL of solution with 0, 0.7, 1.4, 2.1, and 2.8 µg N, respectively (i.e., use 0, 0.2, 0.4, 0.6, and 0.8 mL of diluted stock and make up to 1 mL with 2 M KCl). Put 1 mL of each of these standards in each of three tubes. Add 1 mL of freshly prepared ninhydrin to each, mix with a vortex and boil in a water bath for 15 min. Cool the tubes in cold water and add 5 mL of 2 M KCl to each. Seal with parafilm, shake vigorously for 5 min and immediately read optical density (OD) at 570 nm against a blank tube of 2 M KCl. Average OD for the three replicates.

7. For each sub sample filtrate, prepare three tubes:
   A blank containing 1 mL 2 M KCl.
   Sample 1, containing 0.5 mL of filtrate from the fumigated samples + 0.5 mL 2 M KCl.
   Sample 2, containing 0.5 mL of filtrate from the unfumigated samples + 0.5 mL 2 M KCl.

Add 1 mL of freshly prepared ninhydrin to each, mix with a vortex, and boil in a water bath for 15 min. Cool the tubes in cold water and add 5 mL of 2 M KCl to each. Seal with parafilm, shake vigorously for 5 min and immediately read OD at 570 nm against a blank tube of 2 M KCl.

16.3.4 Results and Observations

MBC is defined as 21×, the release of ninhydrin reactive N from soils fumigated for 10 days at 25°C. The calculation of results is therefore

MBC = ([t_{10}abs – $t_0$abs – y intercept]/slope) × (sample dilution) × (extraction volume/dry soil mass) × (21). It should be expressed as µg C per g dry weight of soil.

16.4 Estimation of Algal (Photosynthetic) Biomass in Soil Cores

16.4.1 Introduction

Among the cellular constituents that can be used as a measure of biomass, pigments, primarily chlorophylls are more widely accepted. All chlorophylls are extractable in 80%–90% acetone and can be measured spectrophotometrically in the spectrum of 665 nm. Addition of dimethyl sulfoxide (DMSO) is known to improve the extent of extraction, especially in soil samples. Soil chlorophyll estimation provides a good indicator of growth rate of algal populations and has been extensively utilized in soil ecological studies.

16.4.2 Materials

Fresh soil cores, glass vials, acetone, DMSO, and centrifuge.

16.4.3 Protocol

1. Fresh soil cores (0–3 or 0–5 cm) are collected with the help of tube auger, and placed in 55 mL glass vials and tightly sealed using Suba-Seal stoppers.
2. Acetone:DMSO (1:1) mixture is added (4 mL/g) to the soil and vortexed to allow proper mixing of soil and extractant.
3. The vials are stored in the dark at room temperature until all the pigments get extracted (48–96 h).
4. The samples are thoroughly shaken; and the colored solvent is removed, centrifuged to remove debris, and OD values taken at 665 nm with a spectrocolorimeter against a blank and expressed in terms of per g soil.

16.4.4 Result and Observations

Quantification of *chlorophyll a* is made by employing the absorption coefficient 12.5.

Concentration of chlorophyll a (µg/mL) = 12.5 × A_{665} where A_{665} = absorption at 665 nm.

16.5 Soil Decomposition and Microbial Community Structure (Winogradsky Technique)

16.5.1 Principle

Soil decomposition is a necessary and important process in every ecosystem. It is the process of recycling nutrients that all organism need for survival. Bacteria and fungi are the main component of this process that involves the breaking down of detritus and dead organic materials. Humus (decomposed material) is the final step in this process that supply soil with nutrient such as calcium, phosphate, potassium, and other ions. In any given condition, natural environment teems with microorganisms to provide a specific combination of nutrients and oxygen that allows only certain microorganisms to survive. One can use their own designed composed media and inoculum from any water body to create and study a mini-ecosystem, called a Winogradsky column, in the laboratory. The objective of this experiment is to demonstrate the impact of eutrophication and plant nutrients on soil decomposition and microbial community construction in the forest and agriculture environments using Winogradsky technique.

16.5.2 Materials

Water sample, agriculture soil, forest soil, mixing container, grass cuttings, pine needles, calcium sulfate, and calcium carbonate.

16.5.3 Protocol

1. Obtain a clean and clear container and label your initial, date, and soil type.
2. Place about 200 mL compost soils (agriculture or forest soil) to a mixing container.
3. Add pond water and stir until it is about the consistency of apple sauce.
4. Add 5 g grass cutting to the agriculture soil and 5 g pine needle cutting to the forest soil as a carbon source (carbon source is in the form of cellulose).
5. Add an equal amount of calcium carbonate and calcium sulfate and mix until the mixture become drier (source of carbon and sulfur).
6. Pour or spoon this mixture into the decomposition chamber to approximately 3–4 cm in depth.
7. Mix it well with a spoon or stirring rod to remove any air pockets.
8. Add plain agriculture or forest soil to the respective chamber until the depth of the soil mixture reaches between 6 and 8 cm. DO NOT STIR!
9. Add pond water, leaving about 4 cm of headspace.
10. Insert a thermometer into the soil and record the starting temperature in °C.
11. The soil column in each decomposition chamber should be covered with aluminum foil, seal with parafilm, and place next to the window (Figure 16.2).

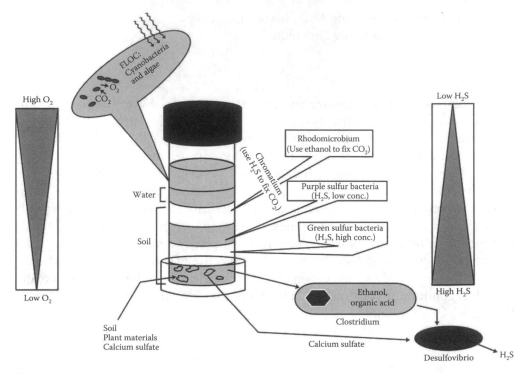

Figure 16.2   Demonstration of Winogradsky technique.

16.5.4 Observation and Results

Create data sheet for weekly observations and discussion of the developed soil layers.

16.6 Soil Algae Enumeration

16.6.1 Algae Test

16.6.2 Introduction

Since algae are photoautotrophs, they can be grown on selective culture media devoid of organic carbon. This eliminates competing microbes, which would outgrow the algae and possibly inhibit their development. One can make these media further selective for cyanobacteria (blue-green algae) by eliminating all sources of fixed nitrogen (ammonium or nitrate, etc.). Enumeration of soil algae is most frequently accomplished by a dilution technique involving inoculation of replicate tubes of an appropriate liquid medium with subsamples from the dilution tubes. This technique is known as the MPN technique and represents a statistical probability approach to counting microorganisms. It is employed when the group of organisms being investigated is not readily cultured on solid media.

16.6.3 Materials

Modified Bristol's solution (9 mL per test tube) g/L in dH$_2$O: (NaNO$_3$, 0.25; CaCl$_2$, 0.025; MgSO$_4$·7H$_2$O, 0.075 g; K$_2$HPO4, 0.075; KH$_2$PO$_4$, 0.018; NaCl, 0.025; FeCl$_3$, 0.01; NaMoO$_4$·2H$_2$O, 0.002; and trace minerals, 1 mL. Combine and bring the volume with dH$_2$O to 1 L.

Trace element stock solution: MnSO$_4$, 2.1; H$_3$BO$_3$, 2.8; Cu (NO$_3$)$_2$·3H$_2$O, 0.4; and ZnSO$_4$·7H$_2$O, 0.24 g. Combine and bring the volume with dH$_2$O to 1 L. Use 1 mL stock solution per liter of medium. Sterile water blanks 90 and 9 mL.

16.6.4 Protocol

Prepare a dilution series of a soil sample from 10^{-1} (1/10) to 10^{-5} (1/100,000) as in earlier exercises (remember that the 10^{-1} dilution is prepared by adding 10 g [approximately 5 mL] in a 95 mL dilution blank). Transfer 1 mL of this suspension to a 9 mL sterile water blank to obtain a 10^{-2} dilution and repeat until you have reached the 10^{-5} dilution. Inoculate three tubes of modified Bristol's solution with 1 mL aliquots from each soil dilution and label the tubes. After inoculating all tubes, incubate them in diffuse light on a windowsill, in a greenhouse, or in a growth chamber. Examine occasionally and make a final observation after 30 days. To determine the number of algae in your soil samples, determine the number of tubes at each dilution, which exhibit noticeable growth. To facilitate this, hold the tubes against a white background or up to the light and look for green coloration. Record the number of positive tubes at each dilution and refer to Table 16.1. This table indicates the MPN of algae per gram of soil based on the number of positive tubes. Locate the series of numbers, which correlated to your positive tubes, at the highest dilutions showing growth. For example, consider the following results:

Dilution 10^{-1}, 10^{-2}, 10^{-3}, 10^{-4}, and 10^{-5}. Positive tubes 3, 3, 2, 1, and 0.

The code (series) relating to these results is 3, 2, and 1 (10^{-2}, 10^{-3}, and 10^{-4}) or 2, 1, and 0 (10^{-3}, 10^{-4}, and 10^{-5}).

Refer to Table 16.1 and find these "codes" by reading across the columns. MPN number for 3, 2, and 1 is 1.50 and for 2, 1, and 0 is 0.15. The MPN number associated with the proper code is multiplied by the reciprocal of the center dilution of the series (series B in this table) to obtain the MPN per gram of the original soil. Thus, the 3, 2, and 1 code yields 1.5 × 10^3/g and 2, 1, and 0 yields 0.15 × 10^{-4} or 1.5 × 10^3/g also. It should be stressed here that the three-tube MPN provides minimal accuracy and is used in this study primarily to introduce the concept and the technique. After you have calculated the MPN of algae in your soil samples, observe the contents of some tubes (low dilution and high dilution) microscopically using simple wet mounts. Note the pigmentation and morphology of the organisms. Try to classify one or more of these using Prescott's book *How to Know the Fresh Water Algae* available in the laboratory. The MPN technique is a statistical approach to quantification and, as in the plate count technique, involves the preparation of a decimal dilution series. The sample must be diluted to extinction. An aliquot of each dilution (usually 1.0 or 0.1 mL) is

TABLE 16.1 Three-Tubes Most Probable Number (MPN) Table

Number of Positive Tubes				Number of Positive Tubes			
Series A	Series B	Series C	MPN	Series A	Series B	Series C	MPN
0	0	0	<0.03	2	0	0	0.09
0	0	1	0.03	2	0	1	0.14
0	0	2	0.06	2	0	2	0.20
0	0	3	0.09	2	0	3	0.26
0	1	0	0.03	2	1	0	0.15
0	1	1	0.06	2	1	1	0.20
0	1	2	0.09	2	1	2	0.27
0	1	3	0.12	2	1	3	0.34
0	2	0	0.06	2	2	0	0.21
0	2	1	0.09	2	2	1	0.28
0	2	2	0.12	2	2	2	0.35
0	2	3	0.16	2	2	3	0.42
0	3	0	0.09	2	3	0	0.29
0	3	1	0.13	2	3	1	0.36
0	3	2	0.16	2	3	2	0.44
0	3	3	0.19	2	3	3	0.53
1	0	0	0.04	3	0	0	0.23
1	0	1	0.07	3	0	1	0.39
1	0	2	0.11	3	0	2	0.64
1	0	3	0.15	3	0	3	0.95
1	1	0	0.07	3	1	0	0.43
1	1	1	0.11	3	1	1	0.75
1	1	2	0.15	3	1	2	1.20
1	1	3	0.19	3	1	3	1.60
1	2	0	0.11	3	2	0	0.93
1	2	1	0.15	3	2	1	1.50
1	2	2	0.20	3	2	2	2.10
1	2	3	0.24	3	2	3	2.90
1	3	0	0.16	3	3	0	2.40
1	3	1	0.20	3	3	1	4.60
1	3	2	0.24	3	3	2	11.00
1	3	3	0.29	3	3	3	>24.00

then added to a series of tubes containing a growth medium. The number of replicate tubes employed dictates the degree of accuracy. Tables are available for 3, 5, 8, and 10 tube determinations. Table 16.1 is a three-tube table. As an example, consider the following: a sample is decimally diluted to extinction. Three 1 mL aliquots from each dilutions is dispensed to three tubes of nutrient broth. After incubation, growth is determined by turbidity. Any turbidity is considered a positive test. At a dilution of 10^{-2}, all three tubes are turbid, 10^{-3} has two turbid tubes, 10^{-4} has one turbid tube, and 10^{-5} has none. The combination is 3, 2, 1, and 0 or 2, 1, and 0. Referring to the table, the combination 3, 2, and 1 gives a MPN value of 1.50 for the center dilution and 2, 1, and 0 gives a value of 0.15. The MPN value is multiplied by the reciprocal of the dilution. In these examples, the MPN is (1.5×10^3).

16.6.4.1 Soil Metabolic Assessment

16.6.4.1.1 Enzyme–Phosphatase Activity

16.6.4.1.1.1 Principle. While enzymes in soil are sometimes associated with roots, many are of microbial origin.

Soil enzymes may retain their activity long after their release from microbial cells primarily because the enzymes form complexes with humic and clay colloids. Complexing with these colloids renders the enzymes highly resistant to denaturation and degradation. Thus, soil can have significant enzymatic activity independent of a native microbial community.

Phosphatase mediates the generalized reaction:

Phosphatase

$$C - PO_4 + H_2O \rightarrow C - OH + HPO_4$$

Phosphatase activity in soil is estimated by using p-nitrophenyl phosphate as a substrate. As the phosphate is cleaved, p-nitrophenol is left. This hydrolyzed product is light sensitive. Samples should, therefore, be kept covered with aluminum foil, to prevent photolysis.

16.6.4.1.1.2 Materials. Soil samples (agriculture or forest), 2 mL Eppendorf microcentrifuge tubes, p-nitrophenyl phosphate solution (100 mM) g/L = 1 M, 0.7422 g/20 mL = 100 mM, and $CaCl_2$ solution (0.5 M).

NaOH solution (0.5 M), p-nitrophenol standard stock solution (100 μM), 0.013911 g/100 mL = 1 mM solution. To make 100 mL of 1 μM, take 10 mL of 1 mM solution + 90 mL water. Use the 1 μM solution to make five standard solutions between 0 μM and 10 μM (10, 5, 2.5, 1.25, and 0).

Microplate reader (vertical path length photometer with 405 nm interference filter), two flat bottom 300 μL wells, 96-well microplate reader rack, aluminum foil, micropipettes and micropipette tips, spatula, marking pen, and tape.

16.6.4.1.1.3 Protocol

1. Label the microcentrifuge tubes as enzyme assay treatment and control.
2. Wrap each microcentrifuge with aluminum foil.
3. Weigh out 0.25 g soil (dry mass basis).
4. Add 1 mL distilled H_2O to the soil and swirl gently.
5. Initiate the reaction by adding 0.25 mL of 100 mM p-nitrophenyl-phosphate to the treatment microcentrifuge tube but NOT to control.
6. Place microcentrifuge tubes in the rack and shake at 175 rpm for 30 min.
   Note: The algae MPN assay should be started during this incubation time.

7. After the incubation period, add 0.25 mL of the *p*-nitrophenyl-phosphate solution to control.
8. Terminate the reaction by adding 0.25 mL of each 0.5 M $CaCl_2$ and 0.5 M NaOH to all tubes.
9. Place the treatment and the control tubes of each soil opposite each other in the microcentrifuge and centrifuge at 10,000 rpm for 10 min.
10 Remove tubes from the microcentrifuge and cover with aluminum foil.
11. Run samples against standard solutions of 0–10 µM of *p*-nitrophenol.
12. Samples may require dilution to fit the standard curve. Use dH_2O to make 1:3 dilutions (0.1 mL sample to 0.2 mL distilled H_2O).
13. Place sample wells in the microplate reader rack and mark their locations.
14. The Microplate Program Manager will calculate the concentration of *p*-nitrophenol in each well based on the standard curve.

16.6.4.1.1.4 Results and Observations. Subtract the control *p*-nitrophenol concentration from the treatment *p*-nitrophenol concentration to get the net reaction result. Because each mole of *p*-nitrophenol produces one mole of phosphate, the final data can be directly expressed as µmol phosphate released. Multiply net results by the dilution factor to express the total as µmol phosphate released per gram soil per hour.

16.6.4.1.2 Dehydrogenase Activity of the Soil

16.6.4.1.2.1 Principle. Biological oxidation of organic compounds is generally through a dehydrogenation process, and there are many dehydrogenases (enzymes catalyzing dehydrogenation), which are highly specific. The overall process of dehydrogenation may be presented as follows:

$$XH_2 + A \rightarrow X - AH_2$$

where XH_2 is an organic compound (hydrogen donor) and A is a hydrogen acceptor.

Tetrazolium salts are representatives of a unique class of compounds. These compounds have combinations of desirable properties. They are quaternary NH_4^+ salts and, as such, possess a high degree of water solubility. The method described is based on extraction with methanol and colorimetric determination of triphenylformazan (TPF) produced from the reduction of 2,3,5-triphenyltetrazolium chloride (TTC) in soils. The transformation occurs through rupture of the ring of TTC.

16.6.4.1.2.2 Materials
1. *Three percent solution of TTC*: Dissolve 3 g of TTC in about 80 mL of water and finally make up the volume to 100 mL with water.
2. *TPF standard solution*: Dissolve 100 mg of TPF in about 80 mL of methanol and adjust the volume to 100 mL with methanol, mix thoroughly.

16.6.4.1.2.3 Protocol
1. Weight 6 g soil sample for each of the three sterile screw cap tubes.
2. Add 2.5 mL sterile dH_2O in two tubes and 1.0 mL of 3% aqueous solution of triphenyltetrazolium chloride.

3. To the third tube (control), add 3.5 mL sterile dH_2O.
4. Mix thoroughly with a mixer and incubate at 30°C for 24 h.
5. Methanol extraction: Remove the sample from tubes by shaking with methanol onto a Buchner funnel fitted with Whatman no. 5 paper. Wash the soil with methanol until no more color could be extracted.
   [Precaution: During this procedure, it is necessary to keep the sample wet at all times until extraction is complete to avoid air drawn through the soil].
6. Pour the filtrate in a 100 mL volumetric flask and make up to volume with methanol.
7. After mixing, determine the absorbance at 485 nm.
8. If suspended soil particles are present in tubes, centrifuge the filtrate at 14,000 rpm for 10 min before reading absorbance.

16.6.4.1.2.4 Result and Observations. Calculate the amount of TPF produced by reference to a calibration graph prepared from TPF standards. (To prepare this graph, dilute 10 mL of TPF standard solution to 100 mL with methanol (100 μg of TPF/mL.) Pipette 5, 10, 15, and 20 mL aliquots of this solution into 100 mL with methanol and mix thoroughly. Measure the intensity of the red color of TPF as described for samples. Plot the absorbance readings against the amount of TPF in 100 mL standard solutions).

16.6.4.1.3 Respiration/Biomass Tests

16.6.4.1.3.1 Principle. Respiration of microorganisms in soil is one of the earliest and still is one of the most frequently used indices of soil microbial activity. Measurements of respiration have been found to be well correlated with other parameters of microbial activity such as nitrogen or P transformations, metabolic intermediates, and average microbial numbers.

Carbon comprises about 45%–50% of the dry matter of plant and animal tissues. When these tissues or residues are metabolized by microorganisms, O_2 is consumed and CO_2 is evolved in accordance with the following generalized reaction:

$$(CH_2O) x + O_2 \rightarrow CO_2 + H_2O + Intermediates + Cell\ material + Energy$$

In this reaction, all of the organic carbon should eventually be released as CO_2. In actual practice, under normal aerobic conditions, only 60%–80% of the carbon is evolved as CO_2 because of incomplete oxidation and synthesis of cellular and intermediary substances. The quantities of CO_2 evolved depend on the type of carbon substrate, the environmental conditions, and the types and numbers of microorganisms involved. When a portion of the soil is fumigated (or otherwise sterilized—e.g., microwave irradiation), the difference in CO_2 evolved between fumigated and non fumigated soil can be used to estimate microbial biomass. The experiment introduces a simple titrimetric method for measuring CO_2 evolution from soils and demonstrates the effects that different carbonaceous substrates have on the rate of CO_2 evolution (microbial respiration). In addition, comparisons between the fumigated and non fumigated samples will provide data for estimating the microbial biomass of the original soil.

16.6.4.1.3.2 Materials. Soil samples, glucose, cellulose, NaOH, NH_4NO_3, beaker, and $BaCl_2$.

16.6.4.1.3.3 Protocol.   Response to carbon addition: Treatments are agriculture soil (alone, with glucose, or with cellulose additions) and forest soil (alone, with glucose, or with cellulose additions).

Weigh 50 g (dry mass basis) of soil (soil should be at ca. 10% moisture) and place in one pint Mason jar. Add 2 mL of NH_4NO_3 solution, 250 mg glucose plus NH_4NO_3 or 250 mg cellulose powder plus NH_4NO_3 (note: add dry ingredients and mix before adding N-solution). Place a 50 mL beaker on the soil. Add exactly 10 mL of 1 N NaOH to the 50 mL beaker and seal with the Mason jar lid. CO_2 produced by the soil is absorbed by the alkali. After 1 week of incubation, remove NaOH with a syringe, place in the beaker, and recharge beaker with fresh NaOH for the second week reading. To determine the amount of CO_2 produced:

1. Add 1 mL of phenolphthalein and 1 mL of 50% $BaCl_2$ to the beaker of NaOH to precipitate the carbonate as an insoluble barium carbonate.
2. Titrate to neutralize the unused alkali with 1 N HCl and record the exact used acid volume.

16.6.4.1.3.4 Results and Observations.   Calculate the amount (mg) of C or $CO_2 = (B - V)$ NE

Note: $(B - V)$ is a simplification of $(T - V) - (T - B)$
where

T is the total volume of NaOH at the start of the experiment

V is the volume (mL) of acid to titrate the alkali in the CO_2 collectors from treatments (i.e., with glucose and cellulose) to the endpoint

B is the volume (mL) of acid to titrate the alkali in the CO_2 collectors from controls (i.e., without C addition) to the endpoint

N is the normality of the acid

E is the equivalent weight

If data are expressed in terms of carbon, $E = 6$ and if expressed as CO_2, $E = 22$. Express the results as mg CO_2 produced per gram soil.

16.6.4.1.4 Demonstrate the Biomass by a Biometer Vessel

16.6.4.1.4.1 Materials.   Soil sample, Erlenmeyer flask, microwave oven, NaOH, and HCl.

16.6.4.1.4.2 Protocol.   Weigh 50 g dry mass of well-mixed soil (soil should contain about 10% moisture) and place in Erlenmeyer flask of a biometer vessel. One vessel from each soil is microwaved for 30 s (high power, 700 W, 2450 MHz microwave). Note this is sufficient to kill approximately 90% of the cell. After microwaving, close the vessel with rubber stopper. A similar vessel is assembled with nonmicrowaved soil. Charge side tubes with exactly 10 mL of 1.0 N NaOH and seal. After 10 days, titrate the NaOH as given earlier to determine the CO_2 evolved (Figure 16.3).

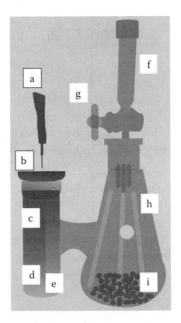

Figure 16.3 Biometer vessel: (a) Rubber policeman, (b) 15 gauge needle, (c) 50 mL tube, (d) alkali solution, (e) polyethylene tubing, (f) ascarite filter, (g) stopcock, (h) 250 mL flask, and (i) 50 g soil sample.

16.6.4.1.4.3 Result and Observations. Calculate the mg C biomass by the formula (HCl used in control – HCl used in microwaved soil)/0.5 = mg C biomass (assuming 50% conversion of biomass to C in 10 days).

16.7 Phosphate Solubilizing Bacteria

P is one of the major essential macronutrients for plants and is applied to soil in the form of phosphatic fertilizers. However, a large portion of soluble inorganic phosphate applied to the soil as chemical fertilizer is immobilized rapidly and becomes unavailable to plants. Microorganisms are involved in a range of processes that affect the transformation of soil P and are thus an integral part of the soil P cycle. Soil bacteria are effective in releasing P from inorganic and organic pools of total soil P through solubilization. Genus *Bacillus*, *Rhodococcus*, and *Arthrobacter* are some important genus of bacteria, which are actively involved in phosphate solubilization (Figures 16.4 and 16.5).

16.7.1 Isolation Phosphate Solubilizing Bacteria from Soil

16.7.1.1 Principle
P is one of the vital nutrients for microorganism next to nitrogen. Several species of bacteria such as *Bacillus* sp. and *Pseudomonas* sp. degrade and solubilize the insoluble

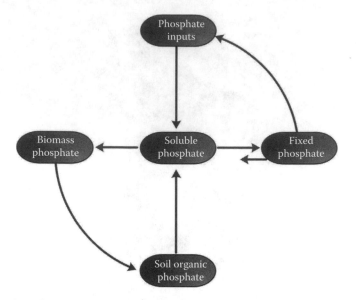

Figure 16.4   Phosphorus cycle.

phosphates into soluble forms through the mechanism of secretion of organic acids such as glycolic acid, acetic acid, etc. This acid decreases pH and causes dissolution of bound form of phosphate. The source of phosphates in soil is both minerals and organic phosphates. For isolation of these bacteria, a special type of Pikovskaya's medium is used.

16.7.1.2 Materials
Soil sample, sterile Petri plates, sterile test tube, inoculating loop, dH_2O, Bunsen burner, glassware marking pencil, etc.

Pikovskaya's medium g/L in dH_2O: Glucose, 10; yeast extract, 0.5; $Ca_3(PO_4)_2$, 2 5*; $(NH_4)_2SO_4$, 0.5; KCl, 0.2; NaCl, 0.2; $MgSO_4 \cdot 7H_2O$, 0.1; yeast extract, 0.5 g; $FeSO_4 \cdot 7H_2O$, trace; $MnSO_4$, trace; agar, 15.0 g; pH 7.0 ± 0.2.

16.7.1.3 Protocol
Prepare soil sample by serial dilution method up to the dilution of 10^{-4} described earlier, transfer 1 mL of suspension from 10^{-4} dilution into agar plates containing *Pikovskaya's medium* supplemented with phosphate; incubate in Petri plates at 25°C for 4–5 days.

16.7.1.4 Result and Observation
Transparent zones of clearing around the colonies of microorganisms indicate that phosphate has been solubilized in clearing zones, isolate the culture and identify.

---

* Stock suspension of 2.5% $Ca_3(PO_4)_2$ was prepared in dH_2O and was autoclaved for preparation of plates or broth; 10 mL of stock suspension was added aseptically to the 90 mL of sterilized medium.

Figure 16.5 Schematic illustration of phosphate mineralization in soil by different bacterial genus and the phosphate-solubilizing zone formed on assay medium containing (a) $Ca_3(PO_4)_2$ and (b) lecithin.

16.8 Isolation of Sulfur Oxidizing Bacteria from Soil

16.8.1 Principle

Sulfur is considered the fourth major plant nutrient after nitrogen, P, and potassium and is one of the 16th nutrient elements, which are essential for the growth and development of plants, especially in the agricultural crop production. The importance of sulfur is equal to that of nitrogen in terms of protein synthesis. Plant roots take sulfur in the form of sulfate (SO_4), which undergoes a series of transformation prior to its incorporation into the original compounds. The soil bacteria like *Thiobacillus thiooxidans* play a major role in sulfur transformation. Sulfur is accumulated in the soil mainly as a constituent of organic compounds and has to be converted to sulfate to become readily available to the plants.

Elemental sulfur is accumulated in the soil as constituents of organic compounds, which cannot be utilized by the plants. It is oxidized to sulfates by the action of chemolithotrophic bacteria of the family *Thiobacteriaceae* (*T. thiooxidans*).

$$2S + H_2O + 3O_2 \rightarrow 2H_2SO_4$$

Sulfates are the compounds that can readily be taken by the plants and are beneficial to agriculture in the following three ways:

1. It is the most suitable sources of sulfur and readily available to plants
2. Accumulation of sulfate solubilizing organic salt that contain plant nutrients such as phosphates and metals.
3. Sulfate is the anion of a strong mineral acid (H_2SO_4) and prevents excessive alkalinity due to ammonia formation by soil microorganisms. Sulfates is assimilated by plants and is incorporated into sulfur amino acids and then into proteins. Animals fulfill their sulfur demand by feeding of plants and plant products.

16.8.2 Materials

Soil sample, thiosulfate mineral salt medium (MSM), flask, sterile Petri dish, sterile test tube, dH_2O, Bunsen burner, inoculating loop, wax marking pencil, etc.

Thiosulfate MSM in dH_2O g/L: KNO_3, 2; NH_4Cl, 1; KH_2PO_4, 2; $NaHCO_3$, 2; $MgSO_4 \cdot 7H_2O$, 0.8; $Na_2S_2O_3 \cdot 5H_2O$, 5; and 1.0 mL trace element solution with the pH adjusted to 6 with 1 N potassium hydroxide (KOH). The trace element solution contained in 1 L of dH_2O g/L: Na_2-ethylenediaminetetraacetic acid (EDTA), 50; $CaCl_2 \cdot 2H_2O$, 7.34; $FeSO_4 \cdot 7H_2O$, 5; $MnCl_2 \cdot 4H_2O$, 2.5; $ZnSO_4 \cdot 7H_2O$, 2.2; $(NH_4)_6Mo_7O_{24} \cdot 4H_2O$, 0.5; $CaSO_4 \cdot 5H_2O$, 0.2; and 11.0 g NaOH. Fifteen grams of agar was added to solidify the medium.

16.8.3 Protocol

Five milliliters of the soil/wastewater samples were added to a shaking flask containing 50 mL of the MSM and incubated in a rotary shaker (100 rpm) at 30°C for 7 days. The development of turbidity in the medium was assumed to be due to bacterial growth and any flask that also showed a pH drop indicating the growth of Thiobacilli which is chosen for purification by streaking onto the thiosulfate MSM solid medium, also incubated at 30°C for 7 days. The purified cultures are initially screened by growing in the MSM containing Na_2S instead of $Na_2S_2O_3$.

16.8.4 Results and Observation

Observe pH reduction of thiosulfate mineral salt broth and on thiosulfate mineral salt agar medium. Observe for round, smooth, white colonies with an average diameter of 0.5 mm. Microscopic and biochemical examination of sulfur oxidizing bacteria are Gram-negative, nonmotile, short rod catalase, and oxidase positive.

16.9 Nitrogen Cycle

The activity of the organisms involved in the series of processes that constitute the nitrogen cycle results in nitrogen being made available to plants in an assimilable form. These organisms, therefore, act in an analogous way to those involved in the carbon cycle and which make CO_2 available as a result of oxidation of carbon-containing compounds.

An important difference between the two cyclic systems is that nitrogen turnover determines productivity in most agricultural situations, hence the activity of those soil organisms metabolizing nitrogen compounds in such a way as to remove nitrogen in a form assimilable by higher plants from the soil, may be very important with respect to loss of soil fertility. The various transformations involving nitrogen can conveniently be summarized in diagrammatic form (Figure 16.6). The process of the production of ammonia from organic compounds is called ammonification. The illustration only includes the main biological processes involved in nitrogen conversion. Nitrogen is essential for the synthesis of structural and functional protein for all living organisms. Protein is decomposed to amino acids that, in turn, are deaminated to liberate ammonia when the organisms die. The process of production of ammonia from organic compounds is called as ammonification. Majority of bacteria in soil participate in the process and is a very important step for bacteria and plants to assimilate it.

The process of the production of ammonia from organic compounds is called ammonification (Figure 16.7). In addition to the ammonification of amino acids, other compounds such as nucleic acids, urea, and uric acid go through the ammonification process. The bacteria that accomplish it (*Bacillus, Clostridium, Proteus, Pseudomonas,* and *Streptomyces*) are called ammonifying bacteria. Ammonification of organic compounds is a very important step in the cycling of nitrogen in soils, since most autotrophs are unable to assimilate amino acids, nucleic acids, urea, and uric acid and use them for their own enzyme and protoplasm construction.

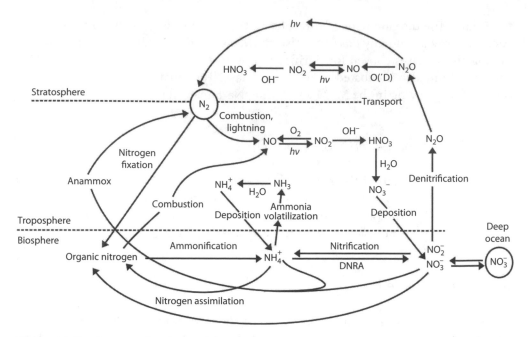

Figure 16.6 Ammonification cycle.

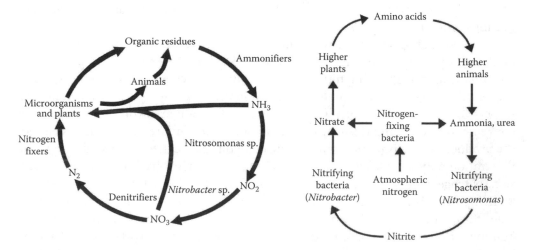

Figure 16.7 Nitrogen cycles.

16.10 Demonstrate the Process of Ammonification in Soil

16.10.1 Principle

To demonstrate the existence of the ammonification process, inoculate peptone broth with a sample of soil, incubate it for a week, and test for ammonia production. After a total of 7 days incubation, it will be tested to see if the amount of ammonia has increased.

16.10.2 Material

Peptone broth, garden soil, inoculation needle, test tubes, Nessler's reagent (commercially available), spot plate, bromothymol blue, and indicator chart.

16.10.3 Protocol

1. Weigh out 0.1 g of soil.
2. Add the soil to a tube of peptone broth (ammonia forming broth).
3. Label the tube with the soil description and incubate at room temperature with the cap loosened for 1 week.
4. After 7 days, test the medium for ammonia with the following procedure.
5. Deposit a drop of Nessler's reagent into a depression of a spot plate (Figure 16.8).
6. Add a drop of the inoculated peptone broth/soil culture to the Nessler's reagent in the depression. Repeat steps 1–5 using uninoculated peptone broth.
7. Check the pH of the two tubes by placing 50 μL of each in separate depressions on the spot plate and adding one drop of bromothymol blue to each one and compare the color with a color chart or set of indicator tubes to determine the pH.

16.10.4 Result and Observation

Interpret the amount of ammonia present as follows:

A. No color—no ammonia (–)
B. Pale orange color—small amount of ammonia (+)
C. Orange—more ammonia (++)
D. Brown precipitate—large amount of ammonia (+++)

Compare results with the uninoculated control. Record results.

	Incubated Soil Culture	Uninoculated Broth
Ammonia produced		

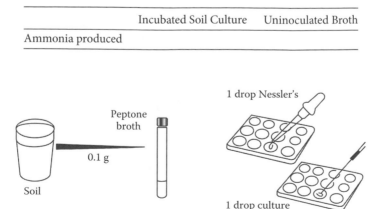

Figure 16.8   Test for ammonification.

16.11 Demonstrate the Process of Nitrification in Soil

16.11.1 Principle

Some microbes can utilize nitrates as electron acceptors and metabolize organic substances without the need for oxygen. This process reduces nitrates to nitrites and free nitrous oxide and nitrogen gases ($NH_4^+ \rightarrow NO_2^- \rightarrow NO_3^-$). The nitrate released into the soil is highly soluble and easily assimilated by photoautotrophic bacteria, algae, and plants that convert it into the amino acids needed for their own enzyme and protoplasm construction. This process is called nitrification (Figure 16.9). The chemoautotrophic bacteria that accomplish it (*Nitrobacter, Nitrococcus, Nitrosococcus,* and *Nitrosomonas*) are called nitrifying bacteria. These chemoautotrophs use the energy of reduced nitrogen compounds, such as ammonium and nitrite, as an energy source for the autotrophic production of organic compounds. Nitrification requires oxygen as a reactant and occurs in aerobic soils.

Nitrosomonas and *Nitrobacter* are autotrophic and appear to be the principal organisms in soil that can perform these reactions. There are some heterotrophs that can produce nitrites and nitrates, but the amount is insignificant. These organisms are very small Gram-negative rods and do best in a completely inorganic medium using CO_2 as their carbon source. Many kinds of organic matter are toxic to them, especially the substances that have free amino groups.

To demonstrate the existence of nitrification, inoculate two broths with a soil sample, incubate them for a week, and test for nitrite and nitrate production. Ammonium sulfate broth (nitrite forming broth) will be observed to see if ammonium has been converted to nitrite ($NH_4^+ \rightarrow NO_2^-$) in the tube. Nitrate forming broth will be observed to see if nitrite has been converted to nitrate ($NO_2^- \rightarrow NO_3^-$) in

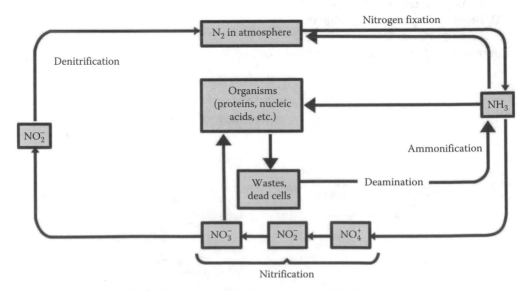

Figure 16.9 Biological conversion of ammonia to nitrate by bacteria.

$$NH_4^+ \xrightarrow[\text{Nitrosomonas}]{O_2} NO_2^- \xrightarrow[\text{Nitrobacter}]{O_2} NO_3^-$$

Ammonium ion   Nitrite ion   Nitrate ion

Figure 16.10 Nitrification process by bacteria in soil.

the tube (Figure 16.10). After a total of 7 day incubation, the tubes will be observed to see if nitrite and nitrate has been released.

16.11.2 Materials

Composition of Medium (g/L)

Inorganic Compounds	Ammonium	Nitrite Medium
Ammonium sulfate [$(NH_4)_2SO_4$]	2	—
Sodium nitrite [$NaNO_2$]	—	1
Magnesium sulfate [$MgSO_4 \cdot H_2O$]	0.5	0.5
Ferrous sulfate [$FeSO_4 \cdot 7H_2O$]	0.03	0.03
Sodium chloride [NaCl]	0.3	0.3
Magnesium carbonate [$MgCO_3$]	10	—
Sodium carbonate [Na_2CO_3]	—	1
Dipotassium phosphate [K_2HPO_4]	1	1

16.11.2.1 Reagents

Trommsdorf's solution: Add slowly, with constant stirring, a boiling solution of 20 g of zinc chloride in 100 mL of dH_2O to a mixture of 4 g of starch in water. Continue heating until the starch is dissolved as much as possible and the solution is nearly clear. Dilute with water and add 2 g of zinc iodide (potassium iodide will do). Dilute to 1 L and filter. Alternatively, the reagent is commercially available.

Nessler's reagent (for ammonia): Dissolve 50 g of potassium iodide (KI) in the smallest possible quantity of cold water (50 mL). Add a saturated solution of mercuric chloride (about 22 g in 350 mL of water will be needed) until an excess is indicated by the formation of a precipitate. Then add 200 mL of 5 N sodium hydroxide (NaOH) and dilute to 1 L. Let settle and draw off the clear liquid. Alternatively, the reagent is commercially available.

Diphenylamine: Dissolve 0.2 g in 100 mL of concentrated sulfuric acid.

Ammonium sulfate and sodium nitrate is separately sterilized by bacterial filter or at low pressure 5–8 psi for 30 min so as to protect the chemical nature of salt and added to maintain the required concentration.

Ammonium ion containing medium as a nitrogen source is to be used to detect the presence of nitrite producing (*Nitrosomonas)* bacteria, and a medium containing nitrate is to be used for isolating nitrate producing *(Nitrobacter)* bacteria.

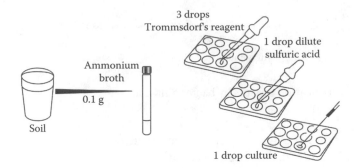

Figure 16.11   Test for nitrification (nitrite production by *Nitrosomonas*).

16.11.3 Protocol

16.11.3.1 Test for Nitrosomonas
Take 25 mL broth of the medium in 6 oz flat bottle separately for *Nitrosomonas* and *Nitrobacter* bacterial isolation, inoculate with 100–200 µL soil suspension and incubate at room temperature for 1 week, and measure the nitrite production in ammonium containing medium and nitrate production in nitrite medium (Figure 16.11).

16.11.3.1.1 Measurement of Nitrite Production.   Mix 3–5 drops of *Trommsdorf's reagent* with 1 drop of dilute sulfuric acid (1:3 concentration) on a spot plate and transfer 2–3 drops of the culture from the ammonium medium to this reagent, mix to avoid false result. Appearance of a blue black color confirms the production of conversion of ammonium to nitrite. Subsequently perform the test of the residual ammonium ions in the medium by *Nessler's reagent*. Negative test for absence of ammonium ions confirms the complete conversion of ammonium to nitrite. Perform the Gram stain test each time Gram-negative cocci confirm for the presence of *Nitrosomonas*.

16.11.3.2 Test for Nitrobacter
Diphenylamine test is performed for the production of nitrate by *Nitrobacter* bacteria. This reagent produces a blue–black color in the presence of nitrate or nitrites. It is, therefore, necessary to make sure that no nitrite is present when it is used as a test for nitrates (Figure 16.12).

First, test the nitrite medium culture with *Trommsdorf's reagent* to establish the absence of nitrites. Mix one drop of diphenylamine reagent, two drops of concentrated sulfuric acid, and one drop of culture on the plate. The blue–black color will be evidence of nitrate production. Perform the Gram stain test each time Gram-negative cocci confirm for the presence of *Nitrobacter*.

16.11.4 Results

	Incubated Soil Culture	Uninoculated Broth
Nitrite		
Nitrate		

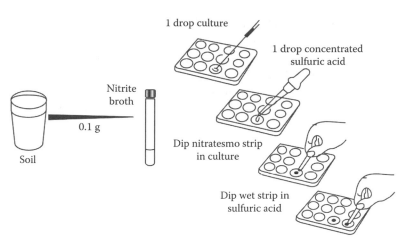

Figure 16.12   Test for nitrate production by *Nitrobacter.*

16.12 MPN Method for *Nitrosomonas* and *Nitrobacter*

16.12.1 Principle

The soil bacteria that oxidize ammonium to nitrite and further to nitrate are chemoautotrophic. They are able to use an inorganic material as their sole source of energy and CO_2 as their sole source of carbon. This property is taken to enumerate the nitrifying bacteria. In the case of *Nitrosomonas*, dilutions of soil are inoculated into an inorganic medium containing ammonium as the source of nitrogen. If *Nitrosomonas* is present in viable form in the inoculums, growth will occur, and nitrite will be produced. Hence, a positive test for nitrite in the inoculated medium but not in the controls indicates the presence of *Nitrosomonas*. To enumerate and isolate the *Nitrobacter* group, the medium employed is free of organic material and contains nitrite as an energy source. The presence of *Nitrobacter* in the inoculums is indicated by a negative test for nitrite will have been oxidized to nitrate if *Nitrobacter* group are present but not if they are absent.

16.12.2 Materials

Erlenmeyer flasks 250 mL capacity, culture tubes, shaker, sterile pipettes, Griess–Ilosvay reagent (Merck).

Medium for *Nitrosomonas* and *Nitrobacter* g/L in dH_2O: $(NH_4)SO_4$, 0.235; KH_2PO_4, 0.200; $CaCl_2 \cdot 2H_2O$, 0.040; $MgSO_4 \cdot 7H_2O$, 0.040; $FeSO_4 \cdot 7H_2O$, 0.005; $NaEDTA \cdot 7H_2O$, 0.005; and phenol red, 0.005; pH is adjusted to 8.0. The same medium is used for both *Nitrobacter* and *Nitrosomonas*; but for the cultivation of *Nitrobacter*, along with the aforementioned ingredients, $NaNO_2$ (247.0 mg/mL) is added.

Griess–Ilosvay reagent: (a) Dissolve 0.6 g sulfanilic acid in 70 mL hot (90°C) dH_2O; cool the solution, add 20 mL of concentrated hydrochloric acid (HCl), dilute the mixture to 100 mL with dH_2O, and mix; (b) dissolve 0.6 g of alpha naphthylamine in 10–20 mL of the dH_2O containing 1 mL of concentrated hydrochloric acid (HCl); dilute to 100 mL with dH_2O and mix, and (c) dissolve 16.4 g of sodium acetate ($CH_3COONa \sim 3H_2O$) in dH_2O, dilute to 100 mL with dH_2O and mix.

Zinc–copper–manganese dioxide mixture: Mix together 1 g of powdered zinc metal (Zn), 1 g of powdered manganese dioxide (MnO_2), and 0.1 g of powdered copper (Cu).

16.12.3 Protocol

1. Collect the soil sample and prepare serial dilutions.
2. From the highest serial dilution prepared, transfer 1 mL aliquots to each of five tubes containing sterile medium for *Nitrobacter* and *Nitrosomonas*.
3. Make a similar set of inoculations using the four next lower and consecutive serial dilutions.
4. Incubate tubes for 3 weeks at 30°C. Also, include a set of uninoculated controls.
5. After incubation, test each tube using Griess–Ilosvay reagent. Development of pink color indicates the presence of nitrite.
6. To all those tubes that yield a negative test, add a small pinch of Zn–Cu–MnO_2 mixture. Find the MPN from MPN table.

16.12.4 Results and Observations

The presence of *Nitrobacter* in the inoculums is indicated by a negative test for nitrite will have been oxidized to nitrate if *Nitrobacter* group are present but not if they are absent.

16.13 Isolation of *Azospirillum* sp. from Soil and Plant Root

16.13.1 Introduction

Azospirillum appear to have a worldwide distribution and occur in a large number in rhizospheric soils in association with the roots of a variety of C3 and C4 plants. The *Azospirillum* are common habitant of tropical and temperate regions. *Azospirillum* have well-adapted and highly versatile metabolism that allow them to proliferate in carbon rich, nitrogen poor rhizosphere, rhizoplane, and the endorhizosphere environments. *Azospirillum* prefers organic acids like malate, succinate, and lactate. There are significant interspecies differences for the utilization of carbohydrates. It contains five species named *Azospirillum brasilense*, *Azospirillum lipoferum*, *Azospirillum amazonense*, *Azospirillum halopraeference*, and *Azospirillum irakense*. All species were proliferating in the rhizosphere and host root of numerous plants of many families. After pertaining in the rhizosphere, in sufficient numbers, they usually, but not always, promote the growth of the host plant, and are commonly known as plant-growth promoting bacteria.

Members of the genus *Azospirillum* are versatile toward Gram's nature, curved-rod shape, motile, oxidase positive, and exhibit acetylene reduction activity under microaerophilic conditions. They are capable of nitrogen synthesis since they can also grow on minimal media like nitrogen-free semisolid malate (NFb).

16.13.2 Materials

Soil sample, NFb medium, root sample, dH_2O, sterile Petri plates and test tube, inoculating loop, dH_2O, Bunsen burner, glassware marking pencil, etc.

N-free malate medium for Azospirillum g/L: Malic acid 5, K_2HPO_4 0.5, KOH 4, $MgSO_4 \cdot 7H_2O$ 0.1, NaCl 0.02, $CaCl_2$ 0.01, $FeSO_4 \cdot 7H_2O$ 0.05, Na_2MoO_4 0.002, $MnSO_4$ 0.01, and 0.5% Alc. Bromophenol thymol blue (0.5% alcoholic solution) 2 ml, NH_4Cl 1, agar 1.8 (solid), pH 6.9–7.3, to 1 L dH_2O.

16.13.3 Protocol

1. Collect the soil sample and prepare a 10-fold dilution series up to 10^{-6}/plant root from the field.
2. Cut 0.5 cm long root pieces and wash it with sterile dH_2O.
3. Prepare semi solid agar medium and autoclave at 15 psi for 30 min.
4. When temperature cool down, transfer the medium into small screw cap bottles.
5. Place small pieces of washed roots/soil sample (a few microgram) on a semi-solid agar medium containing Na/Ca-malate as carbon source.
6. Incubate the plates for 2 days at 28°C–30°C. Pellicles of *Azospirillum* are seen 1–2 mm below the upper surface of the medium. The bacteria grow in the semi-solid medium at low partial pressure of O_2, which favors the organism to grow at a state of N_2 fixation.
7. Transfer *Azospirillum* serially from thin pellicle into fresh semi-solid medium to obtain pure culture.

16.13.4 Result and Observations

Development of white, dense, and undulating fine pellicle on the semi-solid malate medium is very characteristic of *Azospirillum*. It is polymorphic, varying in size, Gram-positive, containing β-hydroxybutyrate granules. Cells have one-half spiral turn and show spiral movement. Lateral and single polar flagella are present.

16.14 Enumeration of *Azospirillum* by MPN Method

16.14.1 Introduction

The MPN method permits estimation of population density without an actual count of single cells or colonies. It is sometimes called the method of ultimate or extinction dilution or end point dilution (Alexander 1965). The MPN technique is based on a determination of the presence or absence of microorganisms in several individual aliquots of each of several consecutive dilutions of soil. A prerequisite of the method is that microorganisms whose population is to be determined must be able to bring about some characteristic and readily recognizable transformation in the medium into which it is inoculated, or else the microorganism itself after it has undergone multiplication, must be easily recognizable in positive and negative tubes receiving a certain quantity of inoculum, the MPN of organisms in that quantity of inoculums. Multiplying the result by the appropriate dilution factor gives the MPN for the sample. *Azospirillum* is a microaerophilic nitrogen fixer. A semi-solid N-free malate medium with bromothymol blue dye is used for enumeration. The presence of *Azospirillum* in the tube is indicated by

1. Change in dye color from green to blue
2. Formation of pellicle at the subsurface of medium
3. Reduction of acetylene to ethylene

16.14.2 Materials

Glass culture tubes, soil sample, sterile pipettes, sterilized dH_2O, and Petri plates.

N-free malate medium for Azospirillum g/L: Malic acid 5, K_2HPO_4 0.5, KOH 4, $MgSO_4 \cdot 7H_2O$ 0.1, NaCl 0.02, $CaCl_2$ 0.01, $FeSO_4 \cdot 7H_2O$ 0.05, Na_2MoO_4 0.002, $MnSO_4$ 0.01, 0.5% alcoholic bromophenol thymol blue (0.5% alcoholic solution) 2 mL, NH_4Cl 1, and agar 1.8 (solid) pH 6.9–7.3, to 1 L dH_2O.

16.14.3 Protocol

Follow from the earlier experiment.

16.14.4 Result and Observations

To calculate the MPN in the original sample, select as p1: The number of positive tubes in the least concentrated dilution in which the greatest number of tubes is positive; and let p2 and p3 represent the numbers of positive tubes in the next two higher dilutions.

Use the table of MPN for use with 10-fold dilution and five tubes per dilution. Then find the row of numbers in MPN table in which p1 and p2 correspond to the values observed experimentally. Follow that row of numbers across the table to the column headed by the observed value of p3. The figure at the point of intersection is the MPN of organisms in the quantity of the original sample represented in the inoculums added in the second dilution (p2). Multiply this figure by the appropriate dilution factor to obtain the MPN for the original sample.

Example: Suppose following observations were made

Dilution +ve
10^{-3}: 5
10^{-4}: 3
10^{-5}: 1
10^{-6}: 0

In this series, p1 = 5, p2 = 3, and p3 = 1. For this combination of p1, p2, and p3, the MPN table gives 1.1 as the MPN of organisms in the quantity of inoculums applied in the 10^{-5} (p2) dilution. Multiplying this number with dilution factor 10^5 gives 1.1×10^5 as the MPN for the original sample.

16.15 Demonstration of Microbial Denitrification in Soil

16.15.1 Principle

Some microbes can utilize nitrates as electron acceptors and metabolize organic substances without the need for oxygen. This process reduces nitrates to nitrites and free nitrous oxide and nitrogen gases ($NO_3^- \rightarrow NO_2^- \rightarrow N_2O \rightarrow N_2$). The gases diffuse out the soil and into the atmosphere removing their nitrogen from the nitrogen exchange pool. This process is called denitrification (Figure 16.13).

The bacteria that accomplish it (*Alcaligenes*, *Bacillus*, *Paracoccus*, and *Pseudomonas*) are called denitrifying bacteria. Denitrification happens most frequently in waterlogged, anaerobic soils. It can be minimized by aeration of the soil through cultivation.

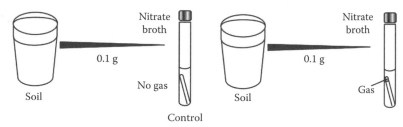

Figure 16.13 Test for denitrification.

To demonstrate the existence of this process, inoculate nitrate broth in a Durham tube with a sample of soil, incubate it for a week, and test for gas production. After a total of 7 days incubation, observe to see if nitrogen gas has collected in the Durham tube.

16.15.2 Materials

Two tubes of nitrate broth (nitrogen gas forming broth), Durham tube (containing an inverted vial), and soil sample.

16.15.3 Protocol

1. Weigh 0.1 g of soil and add into a tube containing nitrate broth (nitrogen gas forming broth).
2. Label the tube with the soil description. If any gas is present in the inverted vial, remove it by inverting the entire tube and tapping lightly.
3. Incubate at room temperature with the cap tight for 1 week.
4. After 7 days, test the medium for nitrogen gas.

16.15.4 Results and Observation

1. Observe the inoculated and incubated nitrate broth/soil culture in the Durham tube. Look for a gas bubble inside the inverted vial.
2. Repeat step 1 using uninoculated nitrate broth in Durham tube.
3. Interpretation of nitrogen gas presence is as follows:
   a. Gas present in inverted vial: +
   b. No gas in inverted vial: –

Compare results with the uninoculated control. Record results as per following tabulated form.

	Incubated Soil Culture	Uninoculated Broth
Nitrogen gas		

16.16 Isolation of Plant Growth Promoting Rhizobacteria

16.16.1 Principle

Plant growth promoting rhizobacteria (PGPR) are free living bacteria that enhance plant growth by direct and indirect means, but the specific mechanisms involved have not all been well characterized, Direct mechanisms of plant growth promotion by PGPR can be demonstrated in the absence of plant pathogens or other rhizosphere microorganisms, while indirect mechanisms involve the ability of PGPR to reduce the harmful effects of plant pathogens on crop yield. PGPR have been reported to, directly, enhance plant growth by a variety of mechanisms: fixation of atmospheric nitrogen that is transferred to the plant, production of siderophores that chelate iron and make it available to the plant root, solubilization of minerals such as P, and synthesis of phytohormones. PGPR strains may use one or more of these mechanisms in the rhizosphere. PGPR also includes that synthesize auxins and cytokinins or that interfere with plant ethylene synthesis. A number of different nitrogen fixing and phosphate solubilizing bacteria may be considered to be PGPR, including *Azotobacter*, *Azospirillum*, and *Rhizobium*. Other bacterial genera, for example, *Arthrobacter*, *Bacillus*, *Burkholderia*, *Enterobacter*, *Klebsiella*, *Pseudomonas*, *Xanthomonas*, and *Serratia* are also reported as PGPR. These PGPR can be isolated on different types of selective and nonselective media.

16.16.2 Protocol

1. Pour plates of different media and allow them to solidify. Keep them inverted until use.
2. Take 10 g soil from the rhizosphere of host plant and prepare 10-fold dilution series up to 10^{-6}.
3. Withdraw by sterile pipette 0.1 mL aliquots from dilutions (10^{-2} to 10^{-6}) and pour on plates containing respective media.
4. Dip the spreader in alcohol and sterilize it in the flame. Cool the spreader and use it for spreading the diluted aliquot by rotating the plate.
5. Incubate at 30°C for 2–5 days.
6. Record visual observation after development of colonies.

16.16.2.1 Characterization of PGPR

16.16.2.1.1 Introduction. PGPR has been reported to influence the growth and yield of many plants. The effect of PGPR on plant growth can be mediated by direct or indirect mechanisms. The direct effect have been most commonly attributed to the production of plant hormones such as auxins, gibberellins, and cytokinins; or by supplying biologically fixed nitrogen. These PGPR also affects growth by indirect mechanism such as suppression of bacterial, fungal, and nematode pathogens by production of siderophores, HCN, ammonia, antibiotics, volatile metabolites, etc., by induced systemic resistance and/or by competing with the pathogen for nutrients or for colonization space.

The isolates can be screened for their plant growth promotion activities by assaying the following attributes:

- Seed germination
- Production of ammonia
- Production of indole acetic acid (IAA)
- Phosphate solubilization in liquid medium
- Nitrogen fixation

- HCN production
- Production of antibiotics
- Production of antifungal metabolites

16.16.2.1.1.1 Seed Germination Test.   Protocol: The selected rhizobium strains can be bioassayed for their ability to promote/inhibit seedling growth using the following method. Surface sterilize the seeds with 0.1% $HgCl_2$ for 3 min followed by successive washing with sterile dH_2O.

1. Decant the water.
2. Add the seeds to cultures grown in their respective medium for 48 h, containing at least 10^6 cells/mL. Keep for 10 min in culture medium and decant the medium.
3. Pour the plates with 0.1% sterile agar.
4. Keep the seeds on soft agar plates and incubate at 30°C for 2–3 days. Keep at least three replicates.
5. In control plates, place the seeds treated with sterilized medium alone.
6. After 3 days, record root and shoot length.

16.16.2.2 Production of Ammonia

16.16.2.2.1 Protocol
1. Grow the isolates in peptone water in tubes.
2. Incubate the tubes at 30°C for 4 days.
3. Add 1 mL Nessler's reagent in each tube.
4. Presence of a faint yellow color indicates small amount of ammonia and deep yellow to brownish color indicate production of higher concentration of ammonia.

16.16.2.3 IAA

16.16.2.3.1 Principle.   A pink color develops when a mineral acid is added to the solution containing IAA in the presence of ferric chloride. Different mineral acids, HCl, phosphoric acid, nitric acid, sulfuric acid, and perchloric acid can be used for the development of the color. $FeCl$-$HClO_4$ reagent is the most sensitive and shows least interference from the other indole compounds, for example, tryptophan, skatole, acetyltryptamine, allylindole, indole, and indole aldehyde show the slight interference with this colorimetric estimation. Since Beer's law is not followed at high concentration of IAA, absorbances obtained are converted to IAA concentration by a standard curve. Such a curve does not vary appreciably with reagent stored in light at room temperature.

16.16.2.3.2 Materials

16.16.2.3.2.1 Medium.   Luria broth g/L in dH_2O: Tryptone 10, yeast extract 5, and sodium chloride 10; final pH 7.0.

16.16.2.3.2.2 Reagents
1. Reagent A: 1 mL of 0.5 M $FeCl_3$ in 50 mL of 35% $HClO_4$
2. IAA stock solution: 100 µg/mL in 50% ethanol.

16.16.2.3.3 Protocol
1. Inoculate a loopful of culture in 25 mL broth of Luria–Bertani (LB) medium and LB amended with 50 μg/mL tryptophan.
2. Incubate for 24 h at 28°C on rotary shaker.
3. Centrifuge at 10,000 g for 15 min.
4. Take 2 mL of supernatant and add 2–3 drops of O-phosphoric acid.
5. Add 4 mL of reagent to the aliquot.
6. Incubate the samples for 25 min at room temperature.
7. Read absorbance at 530 nm
8. Record auxin quantification values by preparing calibration curve made by using IAA as standard (10–100 μg/mL).

16.16.2.4 HCN Production
HCN production by the isolates may be detected by the following method.

16.16.2.4.1 Principle. Cyanogenesis from glycine results in the production of HCN, which is volatile in nature. Reaction of HCN with picric acid in the presence of Na_2CO_3, results in the color change of the filter paper from deep yellow to orange and finally to orange brown to dark brown. In the case of negative test, the deep yellow color of the filter paper remained unchanged after growth of bacteria.

16.16.2.4.2 Material. *King's B medium g/L in dH₂O*: Proteose peptone 0.2, glycerol 0.1, K_2HPO_4 1.5, $MgSO_4 \cdot 7H_2O$ 1.5, and agar 0.15; pH adjusted to 7.2.
Steam sterilize for 30 min at 15 psi at 121°C.

16.16.2.4.3 Protocol
1. Prepare King's B medium amended with 4.4 g/L glycine and sterilize it.
2. Pour 25 mL of this medium in each plate.
3. After solidification, streak the isolate in plates, only single isolate has to be streaked on each plate.
4. Soak Whatman no. 1 filter paper disc (9 cm in diameter) in 0.5% picric acid in 2% sodium carbonate.
5. Place the soaked disc in the lid of each Petri plate.
6. Seal the Petri plate with parafilm and incubate at 28°C ± 2°C for 4 days.
7. Keep an uninoculated control for comparison of the results.

16.16.2.4.4 Results and Observations. Incubate at 30°C for 48 h. Observe for yellow to orange brown to dark brown.

16.16.2.5 Antibiotic Production

16.16.2.5.1 Principle. Certain bacteria produce antibodies that inhibit the growth of other bacteria. This antagonistic property of the bacteria can be used against pathogenic microorganisms.

16.16.2.5.2 Protocol
1. Weigh 10 g of soil and place it into 90 mL dilution medium (water) as blank. Shake well and prepare dilutions up to 10^{-2}.

2. Take one mL aliquot from 10^{-2} dilution and pour plate it along with TY medium in Petri dishes.
3. After solidification of the plates, spot the test cultures on these plates.
4. Incubate for the plates at 30°C ± 2°C for 48 h.
5. Observe for the appearance of inhibition zones around the test cultures.

16.16.2.6 Antifungal Activity

16.16.2.6.1 Principle. Certain bacteria produce antifungal metabolites like NH_3, HCN, antifungal antibiotic, and fungal cell wall degrading enzymes like chitinase and glucanase that attribute to their antifungal properties. Bacteria that can inhibit the fungal pathogens can be used as a bio control agent.

16.16.2.6.2 Protocol
1. Grow the fungal cultures on the potato dextrose agar (PDA) medium.
2. Raise broth cultures of rhizobacterial isolates in nutrient broth or their specific medium.
3. Cut 2–3 mm agar disc containing fungal growth from previously grown plates and place it at the center of the fresh PDA plate.
4. Streak the test bacteria on the periphery of the plate.
5. Incubate the plates at 25°C ± 2°C.
6. Observe for the inhibition of the radial growth of fungus.

16.17 Isolation of Photosynthetic Bacteria

16.17.1 Principle

Photosynthetic bacteria are some of the most widely distributed microorganisms in the environment, playing an important role in our food chain among all animals and plants. This genera of microbes receives solar energy with bacteriochlorophyll, which produce purple carotenoid pigments. At present, they are represented in three families with around 20 genera. They are all Gram-negative. Cell suspension of various species occurs as purple, red, orange brown, brown, or green in color. The genera are subdivided into two main groups: those that are purple and those that are green. They are all aquatic, being abundant in the ooze of ponds. Photosynthetic bacteria differ from the algae; and green plants in that the bacteria are anaerobic and utilize bacteriochlorophyll instead of chlorophyll a for photosynthesis.

16.17.2 Materials

Sample of pond mud, Petri dish, Erlenmeyer flasks, inoculation loop, illuminated growth chamber, and cotton plugs.

Beneck's broth: KNO_3 0.2 g, $MgSO_4 \cdot 7H_2O$ 0.2 g, K_2HPO_4 0.2 g, $CaCO_3$ 0.1 g, $FeCl_3$ (1%) two drops, dH_2O to 1 L, pH adjusted to 7.5, and 2% agar is added to prepare agar medium).

16.17.3 Protocol

1. Prepare Beneck's broth medium, transfer 250 mL in 500 mL Erlenmeyer flasks, plug, and autoclave them.
2. When the broth is cool, add 1 g of pond mud in each flask, shake well, and incubate the flasks at 35°C for 15 days in illuminated growth chamber. The lamp should remain on 24 h/day. Examine the flasks each day to look for color change.
3. After incubation, algal colonies appear in flasks in broth. Make a hanging drop slide of organisms from broth and examine it with phase optics of brightfield oil immersion.
4. Prepare Beneck's agar medium by adding 2% agar in Beneck's broth and autoclave at 15 psi.
5. To get pure culture, transfer a few cells of algal colonies with the help of the inoculation loop on the surface of the Beneck's agar medium in Petri plates.
6. Incubate the plates at 37°C for a week in illumination chamber (Figure 16.14).

16.17.4 Results and Observations

Photosynthetic bacteria appear on the plates, identify them with the help of manual.

16.18 Cyanobacteria

16.18.1 Introduction

Cyanobacteria include unicellular and colonial species. Colonies may form filaments, sheets, or even hollow balls. Some filamentous colonies show the ability to differentiate into several different cell types: vegetative cells; the normal, photosynthetic cells that are formed under favorable growing conditions; akinetes, the climate-resistant spores that may form when environmental conditions become harsh; and thick-walled heterocysts, which contain the enzyme nitrogenase, vital for nitrogen fixation (Figure 16.15 and Table 16.2). Heterocysts may also form under the appropriate environmental conditions (anoxic) when fixed nitrogen is scarce. Heterocyst-forming species are specialized for nitrogen fixation and are able to fix nitrogen gas into ammonia (NH_3), nitrites (NO_2), or nitrates (NO_3), which can be absorbed by plants and converted to protein and nucleic acids [atmospheric nitrogen cannot be used by plants directly].

Often initial isolations of cyanobacteria from the natural environment give mixed cultures. Therefore, it is essential to purify the individual types of cyanobacteria from the mixture. This could be achieved easily by streak or pour plate techniques. These are essentially dilution techniques resulting in the physical separation of individual cyanobacterium from the mixture allowing them to form distinct colonies that can then be picked up to make pure cultures. Furthermore, for obtaining clonal cultures of filamentous forms, these techniques prove extremely useful.

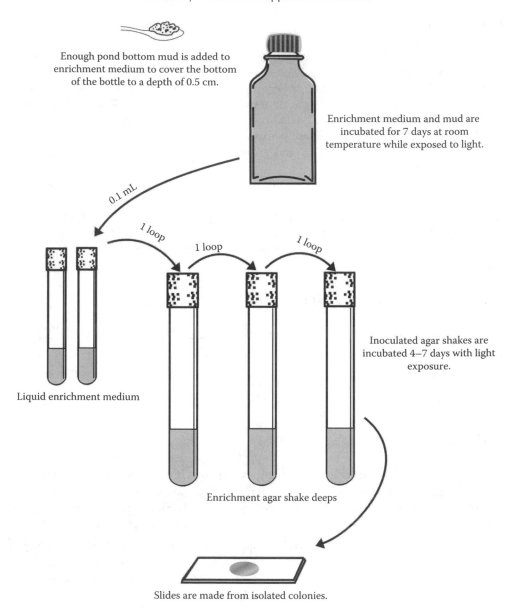

Enough pond bottom mud is added to enrichment medium to cover the bottom of the bottle to a depth of 0.5 cm.

Enrichment medium and mud are incubated for 7 days at room temperature while exposed to light.

0.1 mL

1 loop

1 loop

1 loop

Inoculated agar shakes are incubated 4–7 days with light exposure.

Liquid enrichment medium

Enrichment agar shake deeps

Slides are made from isolated colonies.

Figure 16.14 Enrichment and isolation procedure of photosynthetic bacteria.

16.18.2 Isolation and Purification of Cyanobacteria

16.18.2.1 Isolation

16.18.2.1.1 Streak Plate

16.18.2.1.1.1 Principle. A loop containing varying numbers of cyanobacteria, when touched to the agar surface in a Petri dish, allows to adhere maximum number of cyanobacteria, as the streak progresses further on the surface of the medium; and

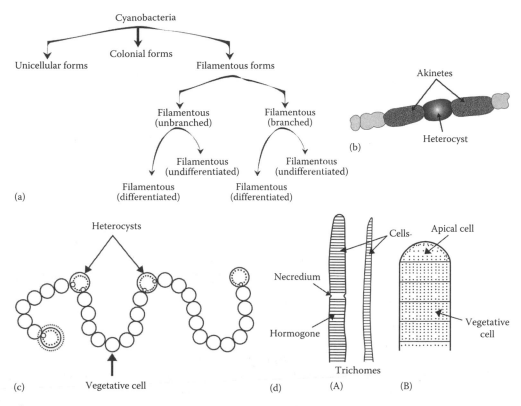

Figure 16.15 (a) Classification of cyanobacteria with filament showing (b) akinetes and (c) heterocyst.

TABLE 16.2   Broad Classification of Cyanobacteria

Cyanobacteria

Unicellular; cells single or forming colonical aggregates held together by additional outer cell wall layers	Reproduction by binary fission or by budding		Section I
	Reproduction by multiple fission giving rise to small daughter cells (baeocytes), or by both multiple fission and binary fission		Section II
Filamentous; a tri-chome (chain if cells) that grows by intercalary cell division	Reproduction by ran-dom trichome break-age, by formation of hormogonia, and (Sections IV and V only) sometimes by germination of akinetes	Trichome always composed only of vegetative cells	Division in only one plane Section III
		In the absence of combined nitrogen, trichome contains heterocysts; some also produce akinetes	Division in only one plane Section IV Division in more than one plane Section V

toward the end of streak, the number gets so much reduced that on incubation form separate colonies.

16.18.2.1.1.2 Protocol
1. Prepare agar plates containing suitable medium.
2. Flame the inoculation needle and cool it by jabbing it into the edge of the agar medium.
3. With the drop on the edge away from the body, streak the culture back and forth, edge to edge in parallel lines, moving toward the body.
4. When the needle reaches the centre of the plate, spin it around 180° and continue streaking, now moving away from the body. This reversal avoids interference with the needle.

16.18.2.1.2 Spread Plate

16.18.2.1.2.1 *Protocol.* In this method, the cyanobacteria in liquid medium is directly spread over the entire surface of the solid medium resulting in separation at many places.

16.18.2.1.2.2 Protocol
1. Prepare suitable agar medium.
2. Sterilize the medium, Petri plates, tissue grinder, etc.
3. Grind a small portion of cyanobacteria with the help of tissue grinder homogenously.
4. Take 1 mL of ground material and make serial dilutions.
5. Pour the sterilized agar medium and allow it to solidify.
6. Inoculate 0.1 mL of diluted cyanobacterial sample on to the surface of agar and spread it thoroughly with the L-shaped glass rod.
7. Label the Petri plate before incubation.

16.18.2.1.3 Pour Plate

16.18.2.1.3.1 *Principle.* In pour plate technique, there is direct dilution of cyanobacteria while being suspended in molten agar medium, resulting in separation at the time of plating. Solidifying substance traps the individual cells in place and produces a fixed colony of the cell or filament.

16.18.2.1.3.2 Procedure
1. Prepare a set of sterile 90 mL water blanks in 250 mL conical flasks.
2. Add 10 mL of cyanobacterial sample in water blank and shake the contents thoroughly. Label it as 10^{-1} dilution.
3. Transfer 10 mL from 10^{-1} dilution to 90 mL water blank and mix thoroughly. Label it as 10^{-2} dilution.
4. Transfer 0.1 or 1 mL of the diluted suspension into sterile Petri dish and pour molten agar medium cooled to 47°C.
5. Mix the contents by swirl action and allow solidification.
6. Prepare 10, 100, and 1000 dilutions following similar procedure and transfer 0.1 mL suspension from each dilution in separate Petri dish and pour the medium to prepare poured plates from each dilution.

Comments: There is no accurate way of predicting the number of viable cells in a given sample by this method. Therefore, one must always make several dilutions of the sample and pour several plates. However, experience with a certain type of samples will minimize wasteful dilutions. The success of this method depends upon thorough mixing and even distribution of cells.

16.18.2.2 Inoculation Procedure
In order to grow an organism in a sterilized medium, a number of cells or filaments (inoculum) are transferred (inoculated) into the medium with special precautions to maintain the purity of the culture. In the inoculation procedure, the needle or loop, when used to transfer cyanobacteria, should be heated to redness by flaming immediately, before and the lower part of the handle. The flaming destroys any living forms on the surface of needle or loop. During transfer, hold the tube in the left hand and hold the plug between the fingers of the right hand. Never lay a plug down. Hold the tube as nearly horizontal as feasible during transfer and do not leave it open longer than necessary. The mouths of the tubes from which cultures are taken and into which they are transferred should also be passed through the burner flame immediately before and after the needle is introduced and removed. In addition to destroying any organism on the tip of the tube, flaming tends to create outward convection currents, thus decreasing the chance of contamination.

16.18.2.3 Incubation Conditions
The cultures are normally incubated at 29°C ± 2°C under continuous illumination (2000–3000 lux) unless otherwise required. Because of the high concentration of water in agar, condensation of water may occur in Petri dishes during incubation, and moisture is likely to drip from the cover into the surface of the agar and spread out resulting in a confluent mass of growth and ruining individual colony formation. To avoid this, the Petri dishes are incubated bottom side up. The slant cultures should be placed in a slanting position, with the streak toward the light source. Always label all Petri dishes, tubes, flasks, and culture bottles with name, date, and identification of contents.

16.18.2.4 Culture Room
A culture room (growth room or incubation chamber) for cyanobacteria can be designed depending upon the space and nature of work. In general, a growth room should have a provision of controlled illumination, temperature, and aeration facilities for the culture vessels.

16.18.2.5 Isolation Procedures

16.18.2.5.1 Unicellular Cyanobacteria
1. Transfer sample from primary enrichment tubes to fresh liquid medium and incubate.
2. From the secondary tubes, streak the cyanobacteria onto the mineral agar plates and incubate.

3. After incubation, isolate colonies with a sterile loop and inoculate in fresh medium.
4. Successive transfers from liquid to solid media usually result in unicyanobacterial cultures.

16.18.2.5.2 Filamentous Cyanobacteria. Pure cultures of filamentous forms usually can be obtained simply by repeated liquid transfer of small amounts of material. Phototactic response of the cyanobacteria is also utilized to separate them from other cyanobacteria.

1. Place a small mass of cells on a slide in a liquid medium.
2. Cover with cover glass and seal with Vaseline, paraffin, or similar compounds.
3. Incubator slide in unidirectional light for 2–10 h. The filament moves from the centre so that single filament is observed at the outer edge.
4. Isolate with a sterile Pasteur-type capillary pipette and transfer it to a tube of liquid medium or on the surface of solidified medium (1%) in a Petri dish (a low agar concentration less than 1.5% is important for obtaining good growth and the filament rapidly migrates through 1% agar).
5. Incubate in unidirectional light.
6. Isolate filaments that are visibly free of bacterial contamination from the rapidly moving front on the illuminated side of the plate. By several passages through agar, axenic filamentous strains can be obtained (Figure 16.16).

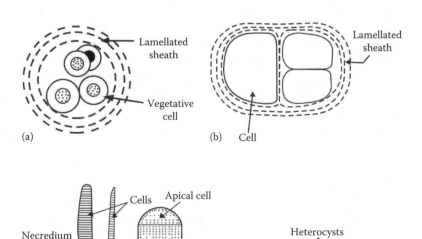

Figure 16.16 Diagrammatic view of some common cyanobacteria. (a) *Gloeocapsa* sp., (b) *Chroococcus* sp., (c) *Oscillatoria*, and (d) *Nostoc.* sp.

16.18.2.6 Purification Methods

It is often possible to obtain axenic cultures from field material by performing isolation and the subsequent purification steps on plates as described. Sometimes, however, this is not the case, particularly if the natural sample is highly contaminated by bacteria and cyanobacterial species are immobile and incapable of self-purification by gliding away from their contaminants, or if the single filaments or micro colonies do not grow on solid media. Purification of such cyanobacteria should be attempted after the establishment of unicyanobacterial isolates in liquid cultures.

16.18.2.6.1 Repeated Liquid Subculture. This technique has been successfully used when a natural collection is particularly rich in a specific cyanobacterium.

16.18.2.6.1.1 Protocol
1. Prepare suitable liquid medium and distribute it to flask and sterilize.
2. Make frequent subcultures to the fresh medium to get rid of the other contaminating bacteria.

16.18.2.6.2 Fragmentation
16.18.2.6.2.1 Protocol
1. Homogenize filamentous forms with a glass homogenizer for 5–10 min or using a sonicator to break mucilaginous colonies allowing short filaments (four to eight cells) to be obtained.
2. Then follow either streak/spread/pour plate method to obtain individual colonies.

16.18.2.6.3 Antibiotic Treatments. It may be difficult to remove certain contaminants by repeated subculturing and/or ultrasonic treatment. In such instances, use of a chemical method rather than a physical method is preferred, but physical methods such as washing and ultrasonic treatment may proceed. One such chemical method uses antibiotics, singly or in combination, to kill or inhibit the growth of tenaciously attached contaminants. Purification though antibiotic treatment has been adopted by many workers.

16.18.2.6.4 Protocol
1. Dissolve 100 mg penicillin G (K or Na salt) and 50 mg streptomycin sulfate, together in 10 mL dH_2O: add 10 mg chloramphenicol (dissolved in 1 mL 95% ethanol) to the penicillin–streptomycin solution and mix well.
2. Filter the triple antibiotic solution quickly, using a membrane or Seitz filter.
3. Place 1 mL of cyanobacterial suspension to be purified in each of six 125 mL Erlenmeyer flasks, each containing 50 mL culture medium.
4. Add one of the following volumes of antibiotic solution to each of the flasks: 0.125, 0.25, 0.5, 1.0, 2.0, and 3.0 mL. This provides penicillin levels ranging from approximately 20–500 mg and corresponding levels of two other antibiotics.
5. Place the culture flasks in dark conditions suitable for bacterial growth.

6. After 12–24 h, aseptically transfer some cyanobacterial cells from each flask to tubes of sterile, antibiotic free culture medium. Prepare tubes in triplicates at specific intervals and incubate.
7. Check culture tubes for bacterial contamination after 2–3 days using nutrient agar plates or nutrient broth.

Some of the bacteriostatic compounds like potassium tellurite (K_2IeO_3–10 µg/mL) have also been used to obtain bacteria-free cyanobacterial algal cultures. Purification can also be done by treatment with cycloserine (5 µg/mL).

16.18.2.6.5 Ultraviolet Irradiation.
This method has been widely used to obtain pure cultures of cyanobacteria.

16.18.2.6.5.1 Protocol.
Place a dilute cyanobacterial suspension in the quartz-windowed chamber and irradiate for 20–30 min at 275 nm. Agitate the suspension by continuous stirring during the irradiation period. Remove the cyanobacterial samples from the irradiation chamber at 5 min intervals and prepare a large number of dilution cultures from each. Inoculate the diluted cyanobacterial suspension into fresh medium and incubate in a culture room. Usually cultures from long exposure times fail to grow, while those from shorter exposure are contaminated. However, certain subcultures, particularly from samples with intermediate exposures, will be free of bacteria.

16.18.2.7 Test for Purity
Cyanobacterial cultures grown in liquid medium under stationary phase are examined critically under the microscope using phase contrast objectives and oil immersion. Sometimes, the medium is supplemented with casamino acids (0.02–0.05% w/v) and glucose (0.5% w/v).

The plates are normally incubated in the dark for several days at a temperature typical for growth of the cyanobacteria to be tested since any contaminants still present should grow at that temperature. The contaminants will then be easily detected by their superior growth, color, and other typical appearances (e.g., fungal hyphae). If the culture is heavily contaminated, the entire area of the test drop will be covered with bacterial or fungal growth: a low degree of contamination will reveal itself in the form of micro colonies on the cyanobacterial lawn. Some contaminants (e.g., gliding bacteria) do not form discrete colonies on the surface of the agar. Their presence (or, its hoped absence) can only be confirmed by additional microscopy, after screening small sections from the agar occupied by the cyanobacterial deposit.

16.18.2.8 Culture Media
To isolate and maintain the cyanobacteria, appropriate media are used, and congenial environmental conditions are maintained. Phosphate, nitrate, magnesium, and calcium are the macronutrients generally required by them. The essential micronutrients are iron, zinc, manganese, copper, and molybdenum. Good growth is obtained at 2000–3000 lux light intensity. The optimal temperature required for their growth

is 29°C ± 1°C. A number of culture media for growing cyanobacteria are available. For the isolation and maintenance of heterocystous cyanobacteria, the source of combined nitrogen ($NaNO_3$ or KNO_3) in the synthetic medium should be omitted and possibly be replaced by the corresponding chlorides.

16.18.3 Isolation of Cyanobacteria from Soil/Water

16.18.3.1 Introduction
Cyanobacteria are photoautotrophic, nitrogen fixing prokaryotes occurring everywhere in a diversified habitat. Some of the members of cyanobacteria are being commercially exploited for mass production of biofertilizer, pharmaceuticals, and value added products. They have been classified on the basis of their morphology from unicellular to filamentous and differentiated structures.

16.18.3.2 Materials
Soil sample, 10 mL sterile pipettes, Erlenmeyer's flasks (250 mL capacity), illuminated growth chamber, inoculation needle, polythene bags, Bunsen burner, and sterile dH_2O.

Modified Chu-10 medium: $Ca(NO_3)_2 \cdot 4H_2O$ 40 mg, $K_2HPO_4 \cdot 3H_2O$ 10 mg, $MgSO_4 \cdot 7H_2O$ 25 mg, Na_2CO_3 20 mg, $NaSiO_3 \cdot 9H_2O$ 25 mg, ferric citrate 3 mg, citric acid 3 mg, and glass dH_2O to 1 L.

To this medium 1 mL/L of *Fogg's micronutrient solution* is to be added. The composition of the solution is as follows: $MnCl_2 \cdot 4H_2O$ 10 mg, $Na_2MoO_4 2H_2O$ 10 mg, H_3BO_3 10 mg, $CuSO_4 \cdot 5H_2O$ 1 mg, $ZnSO_4 \cdot 7H_2O$ 1 mg, and dH_2O 100 mL; pH of this medium should be adjusted to 7.5 before autoclaving.

16.18.3.3 Protocol
1. Collect the sample in a polythene bag or colonies of cyanobacteria floating on the water surface in the rice fields in a glass bottle and bring it to the laboratory.
2. Prepare Chu-10 medium, transfer 20 mL into 100 mL Erlenmeyer flasks and autoclave at 15 psi for 15 min.
3. Serially dilute the soil/water sample (10^{-2} to 10^{-7}) in sterile dH_2O.
4. Pour aliquots of 1 mL of various dilution of soil/algal sample into 20 mL of sterilized medium in flasks.
5. Incubate the flasks at 30°C ± 1°C for 2–3 weeks in an alternate light and dark regime of 12 h. Cyanobacterial colonies appear in broth medium.
6. Soon, pick up individual colonies and transfer onto fresh agar slants and incubate as mentioned earlier.

16.18.3.4 Results and Observations
Several colonies of cyanobacteria grow on slants. They secrete copious amount of mucilage. Observe them under the microscope and identify following literature. Filaments of various color and shape are observed under the microscope. The filament may be heterocystous or nonheterocystous and branched or unbranched.

16.18.4 Measurement of Specific Growth Rate and Generation Time in a Given Cyanobacterium

16.18.4.1 Introduction

Microorganism in a culture will increase exponentially until an essential nutrient is exhausted. Typically the first organism splits into two daughter organisms, which is then each split to form four, split to form eight, and so on. Consequently, the organisms continuously add in their growth and biomass. There are several parameters to measure the growth of a given cyanobacterium such as cell number, chlorophyll content, DNA and RNA content, protein contents, absorbance (OD), and dry weight. Of all the growth process, a sigmoidal curve (s-shaped growth) is a typical one (Figure 16.17). The cyanobacteria show four distinct types of growth phases as are observed in bacteria. In addition, the generation time is the time required by any of the initial population to double its population. It is designed as G. Any two points are selected in the log phase of growth and generation time can be calculated by using the following formula:

$$G = \frac{T_2 - T_1}{2.302 \, (\log N_2 - \log N_1)}$$

where
  G is the generation time
  T_1 is the initial time of growth
  T_2 is the final time of growth
  N_1 is the initial number of cell mass or any other parameter
  N_2 is the final cell number or cell mass or any other parameter

16.18.4.2 Materials

Cyanobacterial culture, Erlenmeyer flasks, pipettes, culture tubes, spectrocolorimeter, and water bath.

  BG-11 growth medium (g/L) in dH$_2$O: NaNO$_3$ 1.5, K$_2$HPO$_4$·3H$_2$O 0.04, MgSO$_4$·7H$_2$O 0.75, CaCl$_2$·2H$_2$O 0.036, citric acid 0.006, ferric ammonium citrate 0.006, EDTA (Na$_2$Mg salt) 0.001, and Na$_2$CO$_3$ 0.02.

  Micronutrient solution (g/L) in dH$_2$O: H$_3$BO$_3$ 2.86, MnCl$_2$·4H$_2$O 1.81, ZnSO$_4$·7H$_2$O 0.222, Na$_2$MoO$_4$·2H$_2$O 0.390, CuSO$_4$·5H$_2$O 0.79, and Co(NO$_3$)$_2$·6H$_2$O 0.0494.

  One milliliter of the micronutrient solution is added to BG-11 growth medium.

  Note: K$_2$HPO$_4$·3H$_2$O is autoclaved separately and prepared fresh when added to the medium as it precipitates when autoclaved with other ingredients.

  After autoclaving and cooling, the pH of the medium is 7.1.

16.18.4.3 Protocol

1. Autoclave the BG-11 medium at 121°C for 30 min, inoculate the medium with actively growing pure culture of the cyanobacterium and incubate at 30°C–35°C for 10–15 days, harvest the exponentially growing cells by centrifugation at 5000×g.
2. Wash the pellets twice with sterile dH$_2$O, re suspend the pellets in BG-11 medium, and measure immediately OD at 663 nm after resuspending the culture in fresh medium.

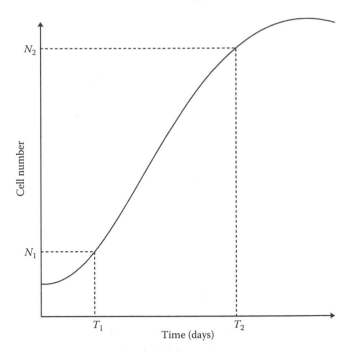

Figure 16.17 A typical sigmoidal growth curve.

3. Again take the sample after each day interval and measure OD at 663 nm.
4. Plot the graph between incubation time and OD (Figure 16.17).

16.18.4.3.1 Growth Measurement by Measuring Chlorophyll a. Three milliliters of culture is centrifuged at $3000 \times g$ for 5 min; pellet is washed and suspended in equal volume (3 mL) of methanol 90% v/v. After thorough mixing, the resulting suspension is heated at 60°C for 4–5 min in a water bath. Thereafter, it is centrifuged to collect the cell free supernatant, and the pellet is discarded. Absorbance of the methanol extract is estimated at 665 nm with a spectrocolorimeter against 90% methanol as a blank. Quantification of chlorophyll a is made by employing the absorption coefficient 12.5.

Concentration of chlorophyll *a* (μg/mL) $= 12.5 \times A_{665}$ where $A_{665} =$ absorption at 665 nm.

16.18.4.4 Results and Observations
Calculate the specific growth rate (K) by using the following formula:

$$K = \frac{T_2 - T_1}{2.302 \, (\log N_2 - \log N_1)}$$

where
N_1 is the OD at time T_1
N_2 is the OD at time T_2

16.19 Nitrogen Fixation

The nitrogen cycle involves a series of rather complex processes, most of which are mediated by soil microorganisms. Information about the role of microorganisms on maintaining the nitrogen cycle can be gathered by conducting a series of microbiological and biochemical tests on a soil sample. The process is illustrated in Figure 16.18.

16.19.1 Free Living

N_2 fixation is central to our understanding of ecosystem function and to predicting future ecosystem responses to global environmental change. Free-living N_2 fixation-also commonly called asymbiotic N_2 fixation or nonsymbiotic. N_2 fixation is nearly ubiquitous in terrestrial ecosystems, occurring on the surface of plants and in their leaves, on leaf litter and decaying wood, and in soil. Nevertheless, free-living N_2 fixation represents a critical N input to most terrestrial ecosystems, particularly those lacking large numbers of symbiotic N_2-fixing plants. N_2 fixers are unique in their ability to exploit an essentially limitless supply of N_2 in the atmosphere. Biological N_2 fixation occurs via two primary pathways: symbiotic and free living. However, the line separating these two pathways is not as clearly defined as it may seem, as relationships between N_2-fixing microorganisms and plants occur in a diversity of forms.

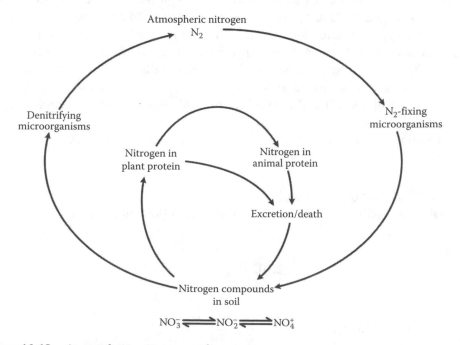

Figure 16.18   Atmospheric nitrogen cycle.

16.19.1.1 Introduction

Soil contains several microorganisms, which may live freely as well as in association with some plants. Numerous microorganisms occur in soil subsisting on traces of nutrients exuding from the plants. These species do not penetrate the plant. They are able to survive under severe climate conditions, many of the bacteria being protected by a mucous sheath. Comparatively few species are represented, but the number of individual organisms may be very large. Among the free living microorganism, *Azotobacter* sp. fix atmospheric nitrogen into ammonia, which is utilized by plants as a nitrogen sources. *Azotobacter* is Gram-negative, polymorphic, motile, oval or spherical bacteria that form thick-walled cysts and may produce large quantities of capsular slime. Old population of bacteria includes encapsulated forms and have enhanced resistant to heat, desiccation, and adverse conditions. The cyst germinates under favorable conditions to give vegetative cells. They also produce polysaccharides. Azotobacter sp. is sensitive to acidic pH, high salts, and temperature above 35°C. There are four important species of Azotobacter viz. *Azotobacter chroococcum, Azotobacter agilis, Azotobacter paspali,* and *Azotobacter vinelandii* of which is most commonly found in our soils.

Azotobacter sp. can be isolated by the soil dilution method, direct isolation, or soil plate method on in any of the medium for Azotobacter viz. Ashby's medium, Jensen's medium, and Beijerinck medium. In direct isolation, lumps of soil are spread out on a nitrogen-free medium. After incubation of 3 days at 28°C, Azotobacter colonies appear on the agar medium.

16.19.1.2 Isolation of Free-Living Nitrogen-Fixing Bacteria from Soil

16.19.1.2.1 Principle. *Azotobacter* is a genus of free-living diazotrophic bacteria, which have the highest metabolic rate compared to any other microorganisms. *Azotobacter* is capable of nitrogen fixation, which uses mannitol as a carbon source and atmospheric nitrogen as nitrogen source. It does not require organic nitrogen source. Hence, a medium comprising of mineral salts and an easily metabolizable sugar mannitol allows selective growth of *Azotobacter* sp. A selective enrichment step prior to the isolation is always recommended (Table 16.3).

16.19.1.2.2 Materials. Soil sample, ingredients of mannitol and Ashby's media, dH$_2$O, sterile Petri dish and test tube, inoculating loop, glassware marking pencil, etc.

16.19.1.2.3 Protocol. Prepare 50 mL nitrogen-free mannitol broth, inoculate 1 mL of finely ground and sieved soil sample, and incubate for 7 days at room temperature; a thin film of growth appears on the surface of the broth. Touch a sterile loop to lift a little amount of the film and transfer it on a sterile nitrogen-free mannitol agar plate (Figure 16.19).

1. Streak as usual for isolation of the culture.
2. Incubate the plate in an inverted position for 3 days at room temperature.

TABLE 16.3 Differential Characters in Members of Azotobacteraceae

Microbes	Cysts	Motility	Long Filaments	Water-Soluble Pigments					
				Brown–Black	Brown–Black to Red–Violet	Red Violet	Green	Yellow–Green Fluorescent	Blue–White Fluorescent
Azotobacter									
A. chroococcum	+	+	–	–	–	–	–	–	–
A. vinelandii	+	+	–	–	–	d	d	+	–
A. beijerinckii	+	–	–	–	–	–	–	–	–
A. nigricans	+	–	–	d	+	d	–	–	–
A. armeniacus	+	+	–	–	+	+	–	–	–
A. paspali	+	+	+	–	–	+	–	+	–
Azomonas									
A. agilis	–	+	–	–	–	–	–	+	+
A. insignis	–	+	–	d[1]	–	d	–	d	–
A. macrocytogenes	–	+	–	–	–	–	–	d	d

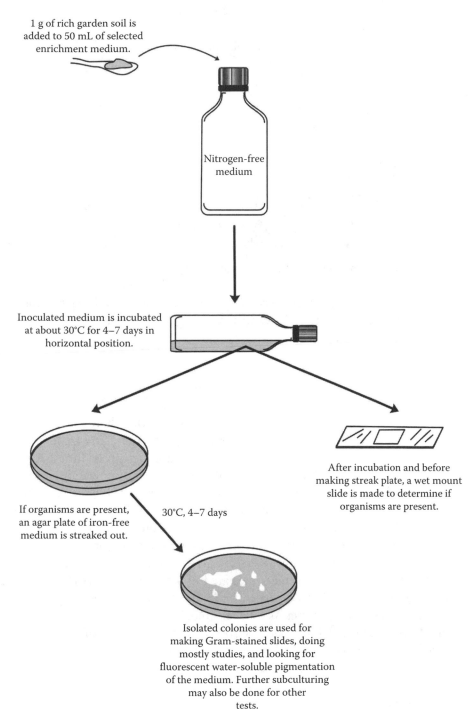

1 g of rich garden soil is added to 50 mL of selected enrichment medium.

Nitrogen-free medium

Inoculated medium is incubated at about 30°C for 4–7 days in horizontal position.

If organisms are present, an agar plate of iron-free medium is streaked out.

30°C, 4–7 days

After incubation and before making streak plate, a wet mount slide is made to determine if organisms are present.

Isolated colonies are used for making Gram-stained slides, doing mostly studies, and looking for fluorescent water-soluble pigmentation of the medium. Further subculturing may also be done for other tests.

Figure 16.19 Isolation of free-living nitrogen-fixing bacteria.

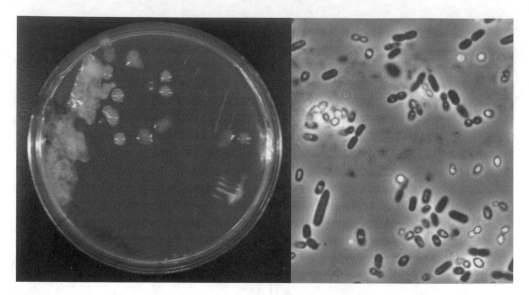

Figure 16.20   (See color insert.) Isolated *Azotobacter* sp. from soil sample.

16.19.1.2.4 Results and Observation   *Azotobacter* sp. produce small watery colonies, which are brown black or colorless. They are Gram-negative aerobic rods (Figure 16.20).

16.19.2 Symbiotic Nitrogen-Fixing Bacteria

16.19.2.1 Introduction

The rhizobia are soil microorganisms that inhabit the rhizosphere of legumes and other plants. They are a rather more diverse group of organisms than might be supposed, but are united by their ability to produce nodules on legumes. A proper understanding of the Rhizobium legume association requires knowledge of the relationship of rhizobia one to another and to other members of the Rhizobiaceae. The Leguminosae is a large family of about 750 genera and 20,000 species (Figure 16.21). However, not all species or genera associate with Rhizobium to form root nodules. The legumes are important in agriculture not only for the fact that they fix nitrogen and can thus conserve nitrogen fertilizers but also because the plants and seeds are high in protein. The seeds of grain legumes are important nutritionally for humans, and the forage legumes (clovers, lucerne or alfalfa, and vetches) also supply a high-protein diet to livestock. Legumes also have ecological significance as they provide nitrogen inflow to natural habitats such as tropical forests. Some tropical legume trees are grown for their wood, and the nitrogen-enriched soil is exploited when the wood is removed. There are several other uses of legumes, but even so only a tiny fraction of the total number of species in this family is exploited by man. Because there is such a wide range of plants in this family, nodule form and modes of infection will differ in detail according to the host species and rhizobial strain.

Leguminous plants play an important role in the maintenance of soil fertility. M.W Beijerinck succeeded in 1988 in isolation and cultivation of root nodule bacteria and

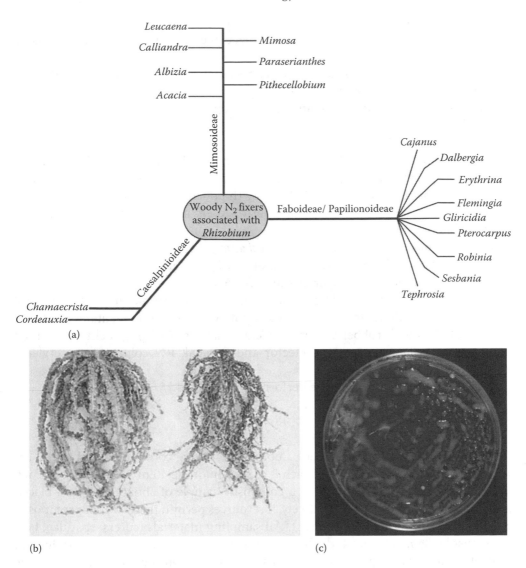

Figure 16.21 (a) Woody N_2-fixer taxa associated with *Rhizobium*, (b) root nodules, and (c) rhizobial growth on Petri plates.

named it *Bacillus radicicola*, now placed under the genus Rhizobium. Rhizobia are soil bacteria belonging to genus *Rhizobium*. They live freely in soil and root region of both leguminous and nonleguminous plants and form root nodules. They require a plant host and cannot independently fix nitrogen. In young nodule, the bacteria occur mostly as rods but subsequently acquire irregular shapes. These bacteria take nitrogen from the air and convert into ammonium (NH_4^+), which is used by plants as a nitrogen source. This process is known as nitrogen fixation and bacteria are often called nitrogen fixers. The nitrogenase is an enzyme that controls the process of nitrogen fixation. Morphologically, they are generally Gram-negative, motile, nonsporulating rods. Rhizobia grow well on laboratory culture media enriched with yeast extract agar.

16.19.2.2 Sampling

Fieldwork: Alcohol, insulated cold-box, sterilized plastic bags (300 mL), spatula, large plastic bags, and small soil corer.

Nodule sampling: Small scissors, spade, hoe, mattock, forceps, shovel, and screw cap tubes with silica gel or anhydrous $CaCl_2$.

Plant vouchers: Alcohol, press, and old newspapers.

16.19.2.3 Laboratory Work

Rhizobia enumeration: 1 and 5 mL pipettes, diluent solution, 1 and 125 mL Erlenmeyer flasks, orbital shaker, sterile plastic bags (125 mL, as growth pouches) or glass tubes (150×20 mL or 200×30 mm), racks for growth pouches or tubes, nutrient solution, seeds of promiscuous host plants, and controlled environment room (temperature, light, and humidity).

Rhizobia isolation and culture characterization: Petri dishes, 95% alcohol, 0.1% $HgCl_2$ (acidified with concentrated HCl at 5 mL/L), sterilized water, forceps, and yeast-mannitol-mineral salts agar medium, pH 6.8.

Nitrogenase assay: (by acetylene reduction) Kitasato Erlenmeyers, rubber ball (of the type used inside footballs), 1 mL gas-tight syringes, 5 mL vacutainers, 10 mL (or larger) vials with rubber stoppers, calcium carbide (CaC_2), gas chromatograph equipped with flame ionization detector and Porapak RN column for acetylene/ethylene determinations.

Note: Nitrogenase assays can be performed in the field.

16.19.2.3.1 Protocol

16.19.2.3.1.1 Soil Sampling. Small amounts of soil are cored to a depth of 20 cm from 20 points distributed within each 8 m section of the 40 m × 5 m transect. Each set of 20 samples is bulked to form a composite sample of about 300 g and placed inside a sterile plastic bag. Alternatively, if resources permit, three or more composite samples can be collected per transect. All sampling materials (corers, spatulae, hoe, etc.) must be flamed before and after sampling at each site to avoid the introduction of exotic Rhizobia. Steps inside the transect should be limited, and litter must be removed just before sampling takes place. Soil samples should be transferred to the laboratory in an insulated container (preferably at 4°C) as soon as possible. If time and resources are short, a single bulked sample can be collected from 20 cores distributed across the whole transect. A second bulk sample of about 200 g should be collected in a nonsterile plastic bag for soil physical and chemical analysis.

16.19.2.3.1.2 Nodule Sampling. Leguminous species inside the transect should be identified and collected. It is helpful if those, which are able to nodulate, are known in advance, in which case the collection can be confined to these species. For herbaceous plants, the whole root system can be removed from the soil (using hoe, spade, or mattock as required) with care not to accidentally sever existing nodules. Nodules of woody plants must be discovered by excavation of the roots, taking care to explore the finer ramifications where nodulation is more commonly found. The nodules are then excised (leaving a piece of root to facilitate manipulation) and stored individually in screw-cap tubes containing desiccant. At least 50 nodules should be collected per site and be representative of all nodulating species in the transect. Occasional nodules may be too large for the ordinary screw-cap tubes and should be stored in a larger container.

16.19.2.3.1.3 Nitrogenase Activity. Nitrogenase activity can be measured in the field on individual nodules, just after sampling, or in the laboratory (Figure 16.22).

The nodule is put in a 10 mL (or larger if needed) vial with a rubber stopper. Acetylene is produced in a Kitasato Erlenmeyer by the reaction of CaC_2 with water (Figure 16.23), and 1 mL of this gas injected into the nodule containing vial. After 1 h (or less), 1 mL of headspace gas is removed and transferred to a vacutainer for the analysis of ethylene in the laboratory by gas chromatography.

16.19.2.3.1.4 Rhizobia Counting. Soil samples are submitted to serial dilutions before candidate host plants are inoculated (Figure 16.24). The dilution ratios vary between 2.0 and 14.5, depending on the expected concentration of cells in the soil sample, that is, greater dilution for soils with more Rhizobia. However, it is still necessary to inoculate host plants at all dilutions (see subsequent sections). Furthermore, at

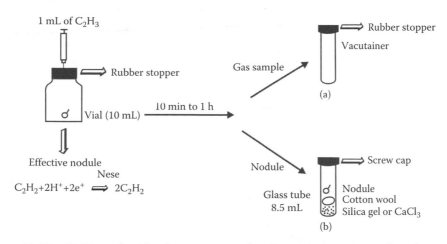

Figure 16.22 Field work: (a) taking gas samples from nitrogenase-mediated acetylene reduction and (b) storage of nodules until isolation in the laboratory.

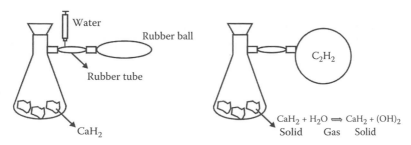

Figure 16.23 Acetylene production in the field or laboratory.

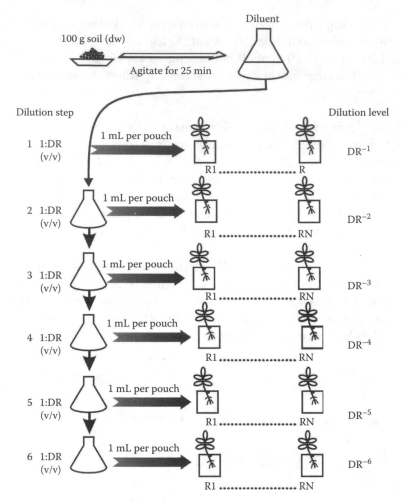

Figure 16.24 Method for calculation of most probable number of rhizobia cells in soil by the plant infection technique. Base dilution rates (DR) can vary from 2 to 14.5 and replicate numbers per dilution (N) from 2 to 5.

each dilution used, replication of the bioassay (two to five times) should be employed. Plants are grown under controlled environment conditions and examined for nodule formation after 15 days. Populations of Rhizobia are estimated by the MPN method (Figure 16.24).

16.19.2.3.1.5 Rhizobia Isolation and Characterization. Rhizobia are isolated from nodules collected in the field and from those obtained under laboratory bioassay. In the latter case, it is necessary to use nodules obtained at each inoculum dilution level, in order to address those strains that are rare in the soil sample, as well as common ones. The first step is to surface sterilize the nodules by a brief immersion in 95% alcohol, followed a longer immersion up to 3–4 min in $HgCl_2$ (Na or Ca hypochlorite, or H_2O_2 can be substituted) and washing in several rinses of sterile water. The nodule is then crushed in a few drops of sterile water, using forceps, and a loopful of this suspension is streaked onto an agar medium. In the case of desiccated nodules, they should first be soaked in sterile water, to improve their wettability by the sterilants. Immersion times in $HgCl_2$ should be adjusted to nodule size (shorter for smaller nodules). Composition of the yeast-mannitol-mineral salts agar medium (especially pH and carbohydrate source) can be varied to take account of a particular soil conditions. Bromothymol blue can be included as an indicator as pH changes caused by rhizobial growth may be useful in genus identification. Other characters include growth rate (time of appearance of isolated colonies; TAIC), the extent of extracellular polysaccharide deposition, colony shape, and colony color. The main generic descriptors are as follows:

Allorhizobium, Rhizobium, and *Sinorhizobium*: Colonies are circular, 2–4 mm in diameter, but usually coalesce due to copious extracellular polysaccharide production, convex, semi-translucent, raised, and mucilaginous, most with a yellowish center (due to pH indicator), and fast growers (TAIC: 3 days).

Mesorhizobium: Same as *Rhizobium*, but intermediate growers (TAIC: 4–5 days).

Bradyrhizobium: Colonies are circular, do not exceed 1 mm in diameter, extracellular polysaccharide production from abundant to little (the latter generally in those strains taking >10 days to grow), opaque, rarely translucent, white and convex, granular in texture, produce an alkaline pH shift, and slow or very slow growers (TAIC: 6 or more days).

Azorhizobium: Colonies are circular, 0.5 mm in diameter with a creamy color, very little extracellular polysaccharide production (much less than in *Bradyrhizobium*), produce an alkaline pH shift, and are fast to intermediate growers (TAIC: 3–4 days).

Rhizobia generally are nonspore forming Gram-negative rods usually containing poly-β-hydroxybutyrate granules refractile under phase contrast microscopy. Isolates with these characteristics must be reconfirmed as rhizobia by demonstration that they will again nodulate a test host plant under bacteriologically controlled conditions.

16.20 Mycorrhizal Symbiotic Association

16.20.1 Introduction

Arbuscular mycorrhizal (AM) fungi are of considerable interest because of their archaic existence, ability to form symbiotic associations with 85% of plant taxa, and their potential use as a biofertilizer to increase yield of crop and tree species. AM fungi are obligate symbionts and can be multiplied only on the roots of host plant. This mutualistic association provides the fungus with relatively constant and direct access to carbohydrates, such as glucose and sucrose supplied by the plant. The carbohydrates are translocated from their source (usually leaves) to root tissue and on to fungal partners. In return, the plant gains the benefits of the mycelium's higher absorptive capacity for water and mineral nutrients (due to comparatively large surface area of mycelium:root ratio), thus improving the plant's mineral absorption capabilities. Plant roots alone may be incapable of taking up phosphate ions that are demineralized, for example, in soils with a basic pH. The mycelium of the mycorrhizal fungus can, however, access these P sources, and make them available to the plants they colonize. Mycorrhizal plants are often more resistant to diseases, such as those caused by microbial soil-borne pathogens and are also more resistant to the effects of drought. These effects are perhaps due to the improved water and mineral uptake in mycorrhizal plants.

16.20.2 Sampling

Twenty random small auger cores of about 50 g each should be taken to a depth of 20 cm in each 5 m section of the transect. The samples are bulked together and mixed thoroughly after collection and aggregates hand-broken. The transect as a whole, therefore, should yield eight bulked samples of about 1 kg per sample. Each sample is then placed in a plastic bag and air-dried (with the tops of the bags rolled down) for 24 h under cover, then brushed through a 2 mm mesh and stored at 4°C. Spores are separated from soil by a wet sieving method using a graded series of sieving baskets (45 through 710 mesh). Three replicates of ca. 100 g each should be processed from each sample. Collected spores are normally identified under a stereomicroscope, by attention to color, size, and shape, following the key of Schenck and Perez (1990). Fine roots on the sieves are also collected and stained with acid fuchsin. Characterization of mycorrhizal infection follows evidence of vesicles, arbuscules, and surface hyphae (Figure 16.25). Further evaluation of the variety of mycorrhizal infection potential in soil samples can be obtained by bioassay. Three-hundred grams of soil sample is used as growth medium for the host plants *Vigna unguiculata* and *Pennisetum americanum*. Mycorrhizal infection and sporulation are evaluated (as described earlier) after 3 months growth in a greenhouse. At present, the best available measure of mycorrhizal diversity is the number of types present per soil, that is, the sum of spore types from soil extraction and bioassay infections.

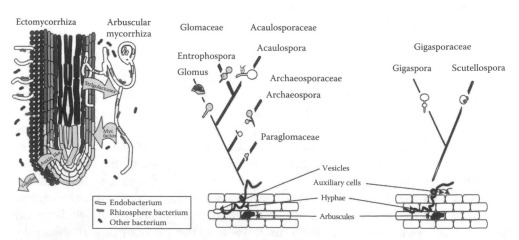

Figure 16.25 Mycorrhizal association of plant roots.

16.20.3 Manipulation and Staining of Spores and Roots

16.20.3.1 Determination of Percent Mycorrhizal Colonization in Root

Plant roots are collected, washed thoroughly with tap water, and kept in KOH (10%) for 2–3 days depending on the thickness of the roots boiled at 90°C for 1 h; washed with dH_2O and acidified by immersing in 5 N HCL for 5 min. Root pieces are stained with trypan blue (0.1% in lactophenol). For pigmented roots, after KOH treatment, they are then immersed in 6% hydrogen peroxide until bleached (Figure 16.26). Excess stain in the roots is removed with lactophenol. Root segments are mounted temporarily on slides containing acetic acid:glycerol (1:1 v/v); edges of the cover slips are sealed with DPX mountant and observed under an optical microscope. A root segment is considered to be mycorrhizal if showing the presence of either vesicular-arbuscular mycorrhizal (VAM) fungal hyphae, arbuscule, or vesicle. About 100 segments are mounted and calculated the percent mycorrhizal colonization using the formula

$$\text{Percent mycorrhiza colonization} = \frac{\text{Number of root segments colonized}}{\text{Total number of root segments examined}} \times 100$$

16.20.3.2 Spore and Root Extraction from Pot Cultures/Field Samples

Remove soil sample from the rhizosphere of the host plant growing in the pot with a 10–20 mm diameter core borer. If the sample is taken from the field, larger quantities should be sieved (100–200 g) and mixed into a 1 L beaker of water before pouring through the sieves. Clay-based soils will block the finer sieve quickly, and care must be taken to tap the base of that sieve to encourage excess water to drain through. The same procedure is used for pot culture material should then be followed:

1. Wash the soil through 710 and 45 μm pore sieves with running water.
2. Remove root material trapped on the 710 μm sieve to check for attached mycelium of AMF with spores or for staining of roots (trypan blue, chlorazole black E, alkaline phosphatase [ALP], acid fuchsin, etc.) if required.
3. Backwash the contents of the 45 μm sieve into a small beaker. Try to keep the volume to a minimum.
4. Swirl the beaker contents and quickly decant the contents into 50 mL centrifuge tubes up to a maximum half way up the tube.
5. Gently inject an equal amount of a 60% (w/v) commercial sugar (sucrose) solution into the pellet at the bottom of each tube using a syringe with a plastic tube extension. There should be a clear interface visible between the water (above) and sugar phase (below).
6. Centrifuge the capped tubes at approximately 3000 rpm for 2 min in a bench centrifuge.
7. Remove the spores caught at the interface of the two layers with the syringe and tube attachment. Start above the interface and work down into the sugar phase using a circular motion as some species produce spores that can sink in the sugar solution while others can float just above the interface (Figure 16.27).

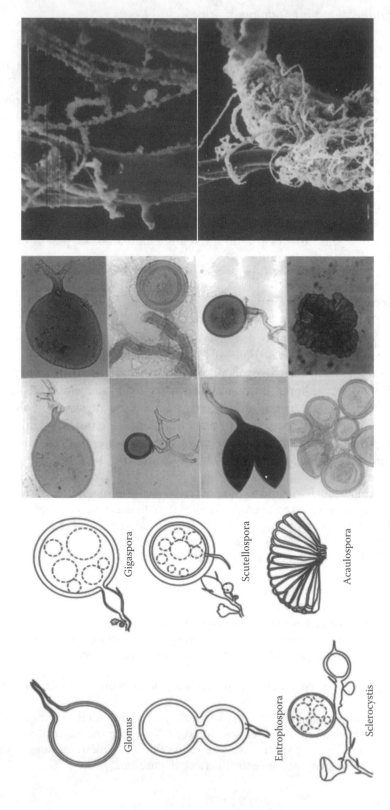

Figure 16.26   (See color insert.) Microscopic view of VAM fungi and their root association.

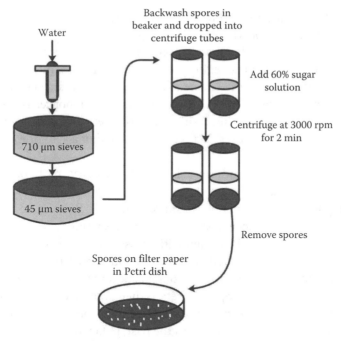

Figure 16.27 Schematic view of root and spore extraction from field sample.

a. Pour the contents of the syringe into a clean 45 μm sieve, and wash thoroughly to remove traces of sugar solution.
b. Backwash contents into a Petri dish and view under a stereomicroscope.

16.20.3.3 Preparation of Permanent Slide Mount for Reference

16.20.3.3.1 Reagents

16.20.3.3.1.1 Polyvinyl-Lacto-Glycerol.   Polyvinyl-lacto-glycerol (PVLG) is used to permanently mount whole or broken spores on glass slides. For best results, mounted specimens should not be studied for 2–3 days after they were mounted to give time for spore contents to clear. Whole spores will change color, generally darkening to varying degrees, and shrink or collapse with plasmolysis of spore contents. Discrete layers of the spore wall or flexible inner walls of broken spores will swell to varying degrees and appear fused after long storage in some instances.

Ingredient quantity: dH$_2$O 100 mL, lactic acid 100 mL, and polyvinyl alcohol (PVA) 16.6 g.

It is most important to mix all ingredients in a dark bottle; *BEFORE* adding PVA. The PVA should have the following properties: 50%–75% hydrolyside, and a viscosity of 20–25 centipoise in a 4% aqueous solution at 20°C. The PVA is added as a powder with the other mixed ingredients and then placed in a hot water bath to dissolve (70°C–80°C), which takes between 4 and 6 h. PVLG stores well in dark bottles for approximately 1 year.

16.20.3.3.1.2 Melzer's Reagent.   Ingredient quantity: Chloral hydrate 100 g, dH_2O 100 mL, iodine 1.5 g, and potassium iodide 5.0 g.

Melzer's reagent can be used alone to mount spores and look for diagnostic iodine staining reactions (to hydrophobic regions of structures), but the mounts are temporary and subject to drying out within 1–2 years of storage. For permanence, Melzer's reagent is mixed in equal proportions with PVLG in a separate dark bottle. There is no diminishing of a staining reaction with the 1:1 dilution. However, the reaction will fade (or disappear in lightly staining structures) in prepared slides after a year or longer of storage.

16.20.3.3.1.3 Sodium Azide.   Sodium azide is a respiratory inhibitor and, therefore, should be handled with care (wearing gloves) in the preparation of stock solutions (2.5 g in 50 mL of dH_2O). A 1 mL aliquot of the stock is added to 90 mL of dH_2O for a 0.05% working solution. For vial vouchers, spores are collected and added to 2 mL vials in a minimum of water. The vial is then filled with the sodium azide working solution and labeled. Solutions and vials are stored at 4°C as an added precaution to optimize safety of the workplace. Spores will darken, and contents become cloudy after long-term storage, but sub cellular structural properties retain their integrity to a great extent. Other preservative solutions such as formalin + acetic acid + alcohol and lactophenol (lactic acid + phenol) have been used extensively in the past, but evidence from type specimens indicates they can cause major changes or degradation of subcellular structure of spores.

16.20.3.3.2 Protocol

1. After extracting spores from a fresh pot culture, isolate a minimum of 10–20 spores.
2. On two clean microscope slides place one drop each of the mountant polyvinyl lactoglycerol and Melzer's PVLG. Transfer half the spores to the first drop of mountant and the second half to the second drop using fine-tip forceps.
3. Try and orientate the spores so that distinguishing features will be apparent once the coverslip is added (Figure 16.28).
4. Carefully place a clean coverslip over each drop, making sure to lower the coverslip at an angle to prevent air bubbles being trapped.
5. Gently apply a pressure to the coverslips of one of the slides to break open the spores. Wait for 30 s and then apply gentle pressure in a circular motion with a soft (B) pencil to break spore walls open further (the pressure will depend on the species of AMF). This should be done under a stereomicroscope.
6. If using PVLG, remember to allow the mountant to polymerize and top-up as necessary before sealing with clear nail varnish or white/silver car paint.
7. Label the slide at one end with the species name and reference code, date, your name, and the mountant used.

16.20.4 Histochemical Visualization of Total AMF Mycelium in Roots

16.20.4.1 Principle

The presence of AM fungi in roots is not visible without appropriate staining. Different nonvital strains are available (e.g., trypan blue, chlorazole black, and fuchsin) to detect intraradical mycelium and they enable an estimation of the abundance of AM fungi

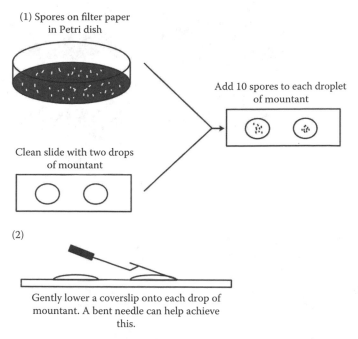

(1) Spores on filter paper
in Petri dish

Add 10 spores to each droplet
of mountant

Clean slide with two drops
of mountant

(2)

Gently lower a coverslip onto each drop of
mountant. A bent needle can help achieve
this.

Figure 16.28 Protocol for the isolation of spores.

within a root system. However, they stain both dead and living fungal structures. Complete understanding of AM functioning requires consideration of the metabolic states of both internal and external hyphae, and the relationship between these, because the physiological interactions will necessitate the presence of an active symbiotic fungus. Activity of succinate dehydrogenase (SDH), a mitochondrial enzyme, is considered as an indicator of viability of mycorrhiza but does not appear to reflect mycorrhizal efficiency for plant growth enhancement. ALP activity, located within the phosphate-accumulating vacuoles of AM hyphae, has been proposed as a physiological marker for analyzing the efficiency of mycorrhiza. Measurements of these two enzyme activities make it easy to directly compare the total production of fungal tissue with the proportion that is living or functional, and to compare, simultaneously, the production of mycelium within roots and in soil in order to determine whether (1) biomass produced in the two compartments is interdependent and (2) the proportion of metabolically active hyphae differs with time.

16.20.4.2 Materials
Root sample, razor blade, KOH, HCl, trypan blue, and lactoglycerol.

16.20.4.3 Protocol
Wash the roots, make them free from soil, and cut roots into 1 cm long segments.
  Trypan blue staining of total mycelium:

1. Clear roots in 2% (w/v) KOH (10% can be used for highly pigmented tree roots) for 15 min at 120°C in a pressure cooker (1 h at 90°C in a water bath or oven) (do not use samples that are more than 2 g).

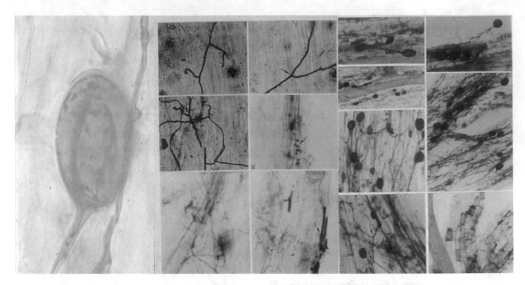

Figure 16.29 **(See color insert.)** Histochemical visualization of various stages of total AMF mycelium in roots.

2. Rinse roots with water three times on a fine sieve or using a mesh and forceps.
3. Cover roots with 2% (v/v) HCl for at least 30 min and preferably longer.
4. Throw away the HCl and cover roots with 0.05% (w/v) trypan blue in lactoglycerol (1:1:1 lactic acid, glycerol, and water; 5:1:1 may be used if tree roots are to be stained) for 15 min at 120°C in a pressure cooker or 15 min to 1 h at 90°C in a water bath or oven.
5. Place roots into Petri dish with 50% (v/v) glycerol for destaining and viewing under stereomicroscope (Figures 16.29 and 16.30).

16.20.4.4 Results and Observations
Microscopic examination of roots.

16.20.5 Histochemical Localization of Lipids in Mycorrhizal Roots

16.20.5.1 Principle
Most of the chemicals are visualized by stains. They make complex inside the tissue, and these insoluble complexes are visible under the microscope. The chemicals have specific reactions for their localization.

16.20.5.2 Materials
Root sample, KOH, sudan III, trypton blue, lactophenol, razor blade, microscopic slides, cover slip, compound microscope, and glycerol.

16.20.5.3 Protocol
1. Select a root showing endomycorrhizal association, cut root sections with a sharp razor or blade and stain with a 0.1% trypan blue plus 0.1% sudan III in lactophenol.
2. Mount the section on slides in glycerol and examine under the microscope. The red stain indicates the localization of lipid.

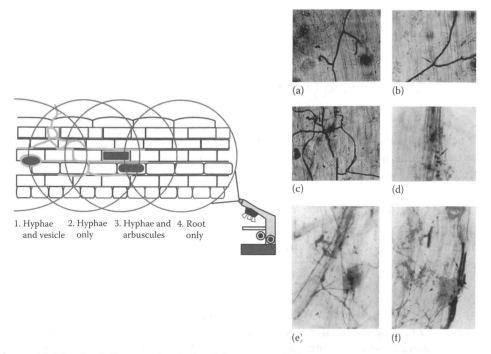

1. Hyphae   2. Hyphae   3. Hyphae and   4. Root
   and vesicle   only   arbuscules   only

Figure 16.30   (a–f) Showing hyphal infection on root of *Zea mays* and degree of colonization of mycelia of *G. mosseae*.

16.20.5.4 Results and Observations
Lipids are the main component of cell; hence red color of varying intensity reveals the qualitative presence of lipid.

16.20.6 Histochemical Localization of SDH

16.20.6.1 Principle
The localization of enzymes require fresh root for sectioning, because in fixed root, the enzyme may become inactive. Effective VAM infection and active external hyphae can be quantified through histochemical staining of this enzyme and measuring metabolic activity of VAM fungus.

16.20.6.2 Materials
Root sample, razor blade, Tris, HCl, $MgCl_2$, nitro blue tetrazolium (NBT), pipette, measuring cylinder, balance, sodium succinate, KOH, sudan III, trypton blue, lactophenol, microscopic slides, cover slip, compound microscope, and glycerol.

16.20.6.3 Protocol
1. Incubate the root sections overnight in 50 mM Tris-HCl buffer (pH 7.4) containing 0.5 mM $MgCl_2$, 1 mg/mL NBT and 0.25 M sodium succinate and rinsed with water.
2. Mount the sections in glycerol. Appearance of the dark purple color indicates the presence of SDH.

16.20.6.4 Results and Observations
Red stain of cytoplasmic contents shows the presence of enzyme SDH.

16.20.7 Histochemical Localization of ALP and Acid Phosphatase

16.20.7.1 Principle
ALP activity, located within the phosphate-accumulating vacuoles of AM hyphae has been proposed as a physiological marker for analyzing the efficiency of mycorrhiza. The violet–black stain indicates the presence of enzymes.

16.20.7.2 Materials
Root sample, razor blade, cellulase-pectinase solution, sodium alpha-naphthyl acid phosphate, $MgCl_2$, $MnCl_2$, glycerol, microscopic slides, cover slip, and compound microscope.

16.20.7.3 Protocol
1. Incubate the root sections in cellulase-pectinase solution for 4 h to digest and then transfer to incubate further overnight in 0.5 M Tris citric acid buffer (pH 8.5–9.2) containing 1 mg/mL sodium alpha-naphthyl acid phosphate, 1 mg/mL fast blue, 0.5% M $MgCl_2$, and 5% $MnCl_2$.
2. Mount all the sections in glycerol and examine for the presence of black–violet color.
3. Appearance of color indicates the presence of alkaline and acid phosphatase.

16.20.7.4 Results and Observation
Darkening (blackish brown) of the cell content reveals the presence of enzyme ALP and acid phosphatase.

16.20.8 Histochemical Localization of Peroxidase

16.20.8.1 Principle
Mitochondrial enzyme is considered as an indicator of viability of mycorrhiza but does not appear to reflect mycorrhizal efficiency for plant growth enhancement Peroxidases, a group of specific enzymes such as NAD-peroxidase, NADP-peroxidase, etc., and nonspecific enzymes such as peroxidases, catalyze dehydrogenation of a large number of organic compounds present both in plant and microbes.

16.20.8.2 Materials
Root sample, razor blade, ammonium chloride, EDTA, benzidine, H_2O_2, glycerol, microslide, coverslip, and compound microscope.

16.20.8.3 Protocol
1. Incubate the root sections at room temperature in a solution containing one part NH_4Cl, one part 5% EDTA, six parts saturated benzidine, and one part 3% H_2O_2.

2. Rinse the section in water.
3. Mount in glycerol for microscopic analysis. The presence of blue–black stain indicates the enzyme peroxidase.
4. Appearance of color indicates the presence of alkaline and acid phosphatase.

16.20.8.4 Results and Observation

The young cells show dark blue–black stain, whereas the dead cells did not show any enzymatic activity.

16.20.9 Estimation of Mycorrhizal Colonization

16.20.9.1 Protocol

1. Mount 15 root fragments on one slide; prepare two slides (30 root fragments in total).
2. Observe these fragments under the microscope and rate according to the range of classes indicated in Figure 16.31.
3. These classes give a rapid estimation of the level of mycorrhizal colonization of each root fragment and the abundance of arbuscules.
4. Put the values into the computer program "Mycocalc" to calculate the parameters: %F, %M, %m, %a, and %A, according to Trouvelot et al. (1986). Date: F% =, soil: M% =, plant: A% = fungus: m% =, treatment: a% = replication:
5. Frequency of mycorrhiza in the root system F% = (nb of fragments myco/ total nb) × 100.

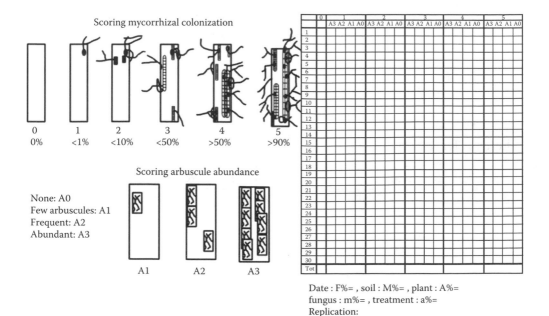

Figure 16.31 Illustrating quantification of mycorrhizal colonization.

6. Intensity of the mycorrhizal colonization in the root system M% = (95n5 + 70n4 + 30n3 + 5n2 + n1)/(nb total) where n5 = number of fragments rated 5; n4 = number of fragments 4, etc.
7. Intensity of the mycorrhizal colonization in the root fragments m% = M × (nb total)/(nb myco).
8. Arbuscule abundance in mycorrhizal parts of root fragments a% = (100mA3 + 50mA2 + 10mA1)/100 where mA3, mA2, and mA1 are the % of m, rated A3, A2, A1, respectively, with mA3 = ((95n5A3 + 70n4A3 + 30n3A3 + 5n2A3 + n1A3)/nb myco) × 100/m and the same for A2 and A1.

16.20.9.2 Results and Observations

Calculate the arbuscule abundance in the root system A% = a*(M/100).

16.20.10 Extraction and Measurement of AM Fungal Hyphae in Soil

16.20.10.1 Protocol

1. Take soil cores (1 cm × 6 cm) randomly from pots; mix the soil sample well and then put a 2 g sample in a 500 mL beaker, suspend the soil in 250 mL dH$_2$O, and filter the soil suspension through a 300 μm mesh sieve.
2. The washings are collected and blended for 30 s at high speed in a blender, transfer the suspension to a flask, shake by hand and then stand on the bench for 1 min, pipette 10 mL (5 mL × 2) aliquots onto a millipore filter (1 mm pore size) and filter under vacuum using the filter holder.
3. Place the filter on a microscope slide and let it dry, stain the hyphae on the filter in lactic glycerol–trypan blue (0.05% v/v) for 5 min and observe the stained filter under a coverslip at 200 × magnification.
4. Examine 30 random fields and estimate hyphal length by using a grid line interception method as used for evaluating mycorrhizal root lengths (Figure 16.32).

16.20.11 Estimation of SDH and ALP Active Hyphae in Soil

16.20.11.1 Materials

Solution A for SDH staining

Chemical	Concentration	Volume (mL)
Tris-HCl buffer pH 7.4	0.2 mol/L	5
MgCl$_2$	5 mmol/L	2
NBT; freshly prepared[a]	4 mg mL/L	5
H$_2$O	—	6
Sodium succinate	2.5 mol/L	2

[a]NBT: nitroblue tetrazolium.

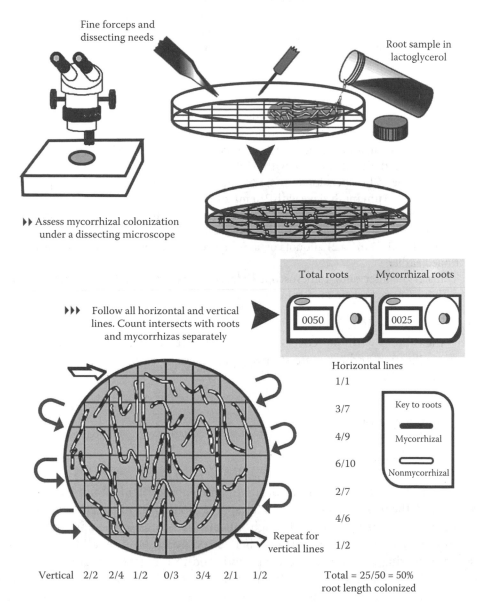

Figure 16.32 Protocol for the measurement of AM fungal hyphae in soil by gridline intersection method.

Solution B for ALP staining

Chemical	Concentration	Volume (ml)
Tris–citric acid buffer pH 9.2	0.05 mol/L	18
α-Naphthyl acid phosphate	1 mg/mL	20
Fast blue RR salt	1 mg/mL	20
$MgCl_2$	0.5 mg/mL	1
$MnCl_2.4H_2O$	0.8 mg/mL	1

16.20.11.2 Protocol

1. Take soil cores and put in a beaker on ice, mix the soil sample well and take two 2 g subsamples, add the subsamples in two bottles marked with SDH and ALP separately, and cover the soil with 20 mL ice-cold water immediately.
2. Add 20 mL incubation solution A in the bottle marked with SDH and add 20 mL solution B in the bottle marked with ALP.
3. Incubate the soil suspension at room temperature for 3 h and filter the suspension through a 300 μm mesh sieve with 210 mL dH_2O.
4. The collected washings are blended at high speed for 25 min, transfer the suspension to a flask and leave to stand on the bench for 1 min.
5. Pipette 10 mL (5 mL × 2) aliquots on a millipore filter, filter under vacuum as mentioned earlier.
6. Transfer the filter onto a microscope slide, counterstain AMF hyphae on the filter with 0.1% basic fuchsin for 5 min.

16.20.11.3 Results and Observations

Cover the filter with a coverslip and observe under the microscope at 200 × magnification and estimate stained hyphal length using the gridline intersect method (Figure 16.32).

16.21 Actinomycetes

16.21.1 Introduction

Actinomycetes are aerobic, Gram-positive sporulating filamentous bacteria, and reproduce by the formation of unicellular spores. Some actinomycetes develop only in mycelial state and reproduce through unicellular spores termed as sporoactinomycetes or euactinomycetes; while in other actinomycetes, mycelial development is less and occurring only during active growth and immediately fragmented during slow growth rate, in extreme conditions appearing as branched single cell. These actinomycetes are termed as proactinomycetes. The number and types of actinomycetes present in particular soil is influenced by geographical location such as soil, temperature, pH, organic matter, aeration, and moisture content. About 100 genera of actinomycetes exists in soil, which provided valuable products like enzyme and antimicrobial product such as streptomycin from *Streptomyces griseus* used for the treatment of tuberculosis caused by *Mycobacterium* and immunosuppressive drug, tacrolimus produced by *Streptomyces tsukubaensis*.

16.21.2 Isolation of Actinomycetes from Garden Soil on Agar Plates

16.21.2.1 Materials

Garden soil sample, starch casein agar medium, sterile Petri plates, sterile test tube, dH_2O, normal saline, Bunsen burner, glass L shaped spreader, inoculating loop, wax marking pencil, etc.

Starch casein agar medium: starch 10 g, casein (vitamin free) 0.3 g, KNO_3 2 g, NaCl 2 g, K_2HPO_4 2 g, $MgSO_4 \cdot 7H_2O$ 0.5 g, $CaCO_3$ 0.2 g, $FeSO_4 \cdot 7H_2O$ 0.01 g, $ZnSO_4 \cdot 7H_2O$ 1 mg, agar-agar 18 g, and dH_2O to 1 L. pH is adjusted to 7–7.2 before autoclaving and nystatin and actidine (50 mg/mL each) are to be added to autoclaved medium; cool (45°C) before being poured in Petri dish.

16.21.2.2 Protocol

1. Prepare starch-casein agar medium and autoclave after adjusting the pH to 7–7.2.
2. Collect garden soil, dry at 45°C for 30 min so that most of bacteria and fungi may be killed.
3. Take 1 g garden soil sample, prepare serial dilution of 10^{-1}, 10^{-2}, 10^{-3}, and 10^{-4} dilutions in sterile dH_2O or sterile normal saline.
4. Soil suspension is heated at 50°C for 10 min in a water bath.
5. Take 1 mL suspension in sterile Petri dish and pour 20 mL pre-autoclaved starch casein agar medium.
6. Incubate the plates in the inverted position at 30°C–35°C for 14 days or more as the member of actinomycetes are slow growing microbes.
7. Check for the growth of typical actinomycetes colonies up to 15th day.

16.21.2.3 Results and Observation

Actinomycetes look like chain bacteria producing thin slender mycelium and conidia in chains. Observe mycelium structure, color, and arrangement of conidiospore and arthrospore through the oil immersion microscope. Mostly they are white–grey or black powdery surface with radiating edge (Figure 16.33).

16.21.3 Qualitative Determination of Antagonistic Properties of the Isolated Streptomyces from Soil

16.21.3.1 Introduction

Soil is the source of all natural antibiotics so far discovered, which are being used for treatment of various infectious diseases. Most of the antibiotics isolated belong to the members of Actinomyces that can be used for manufacture of antibiotics commercially. Although, many organisms in soil produce antibiotics, only a small portion of new antibiotics are suitable for medical use (Figures 16.33 and 16.34).

16.21.3.2 Material

Soil sample, Waksman's medium, nystatin, spreader, dH_2O, culture tubes, and incubator.

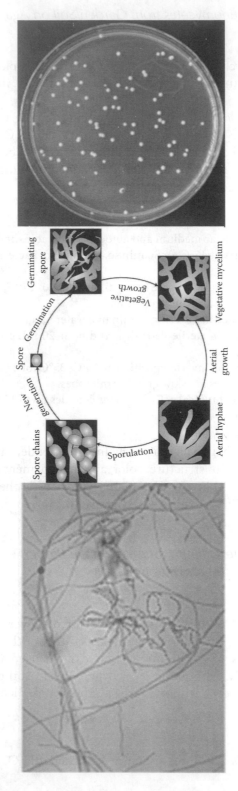

Figure 16.33 Actinomycetes isolated from soil growing on agar plates.

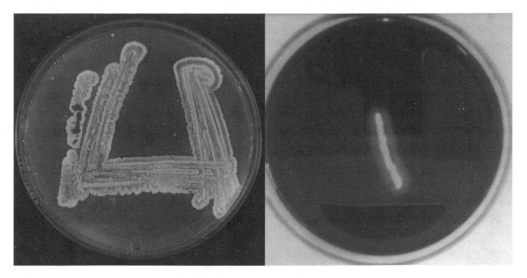

Figure 16.34   **(See color insert.)** Growth of an isolated actinomycete from soil and its amylase positive test.

Waksman's medium g/L: Glucose 10, peptone 5, K_2HPO_4 1, $MgSO_4 \cdot 7H_2O$ 0.5, agar 20, and dH_2O 1 L, pH 4.0.

16.21.3.3 Protocol

16.21.3.3.1 Isolation from Soil

Take 1 g of farm soil in a test tube, vortex mix to form a homogeneous suspension, and prepare a serial dilution from tube 1 to 6 by transferring 1 mL from tube to tube; and from each of the last three tubes, transfer 1 mL to a plate of Walksman's agar medium. Spread the organisms over the agar surfaces on each plate with L-shaped sterilized glass rod. Always sterilize the glass rod in alcohol and open flame. Incubate the plates for 7–12 days. Colonies of the actinomycetes will appear on the plate, which can be identified on the basis of spore character, arrangement of spore, pigment formation, color of the mycelia, color of the pigments, biochemical characters, etc.

The organisms will be streaked in nutrient agar plates that have been seeded with *Staphylococcus* sp. After incubation, the inhibitory zones formed are measured against the tested Actinomycetes. Place four liquefied agar pours in a water bath (40°C–45°C) to prevent solidification, inoculate each one with 1 mL spore suspension of *Staphylococcus*, pour each of the inoculated tube into Petri plates. Allow agar to cool and solidify. Take out the actinomycetes from the primary isolation plates from the Actinomycetes colonies from Walksman's plates. They may have dusty appearance due to the presence of spores and over growth of the actinomycetes. Inoculate each *Staphylococcus* sp. plates with the actinomycetes spores and incubate at 30°C for 10–12 days and observe the inhibition zone (Figure 16.35).

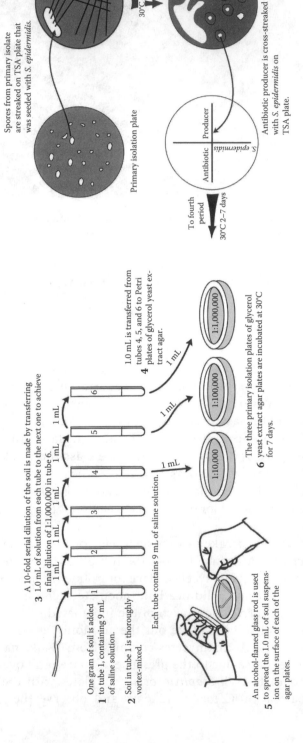

Figure 16.35   Test for the antagonistic properties of the isolated actinomycetes.

16.22 Aquatic Molds

16.22.1 Introduction

Aquatic molds require water to disperse throughout their environment. Unlike other members of fungi, they produce flagellated, asexual reproductive cells known as zoospores. They complete their life cycle in water and solely are of aquatic habitats. Besides being commonly found in lakes, streams, ponds, roadside ditches, and coastal marine environments, they also are present in the soil. As members of terrestrial and aquatic microbial communities, they play an important ecological role in the degradation of recalcitrant materials like keratin, chitin, and cellulose. They live saprobically or as parasites in, or on, a number of different organisms and substrates such as pollen grains, insect exoskeletons, protists and small invertebrates, amphibian skin, other fungi, pieces of plants, fruits, and waterlogged twigs.

They can be isolated by baiting in soil and water samples. When baiting water samples, place the sample into a dish with some of the accompanying organic material and then add a few pieces of bait. The type of bait used will depend on the kind of aquatic molds you are seeking. The size of the bait should be big enough to be retrieved but small enough to be examined on a microscope slide. For soil samples, to avoid the culture from being overrun by bacteria, add only 1–2 g of soil to a dish half filled with sterilized water. The type of baits commonly used include purified shrimp exoskeleton (chitin), snake skin, defatted blond baby hair (keratin), cellophane, lens paper, boiled or autoclaved dialysis tubing, white onion skin, hemp rope, corn straw (cellulose), and pollen (pine pollen, spruce pollen, and sweetgum pollen). Another way to bait is to use bait bags. Bait bags are made from mosquito netting and can be used to determine what molds are active in aquatic habitats. Several pieces of bait and two or three marbles are put into the bags. The bags are then placed into the lake, pond, stream, bog, or ditch for a week or more before being re-collected. The length of time the bait bags remain depends on the time of the year, and a good rule of thumb is a week when water temperature is above 20°C and longer at lower temperatures. The baits are then examined microscopically for aquatic molds back in the laboratory.

Another good thing to do is to record the temperature of the water. This is important later when culturing because some cultures may grow better at the temperature of the environment from which they were isolated versus room temperature. When examining baits from bait bags or gross cultures in the laboratory, allow a couple of weeks before discarding them as some will take longer to appear than others. This is especially true with members of the *Nowakowskiella* clade, which can take up to 3 weeks to appear on onion skin and corn straw baits.

16.22.2 Isolation and Purification of Aquatic Molds

16.22.2.1 Materials

Water sample in sterile bottles, baits (hemp, mustard seed, grass leaves, house or drosophila flies, termite, wings, cellophane, pollen grains of conifers, snake skin, and white human hair), sterile Petri dishes, sterile glass dH$_2$O, forceps, beakers, and Bunsen burner.

PmTG medium with antibiotics: 1 g peptonized milk, 1 g tryptone, 5 g dextrose, 8 g agar, 0.5 g streptomycin sulfate, 0.5 g penicillin-G, and 1 L deionized water.

mPmTG agar (modified) for plates and slants: 0.4 g peptonized milk, 0.4 g tryptone, 2 g glucose, 10 g agar for plates, 12 g for slants, and 1 L deionized water (1000 mL distilled).

T/2 broth for flasks and Petri dishes with cover slips: 5 g tryptone and 1 L deionized water.

16.22.2.2 Protocol

1. Collect water samples, soil, and plant and animal debris in a sterile bottle.
2. Put a few mL of collected water or a small amount of soil or debris in sterile Petri dishes (5–10 mL), which are half filled with cool, sterilized glass dH$_2$O.
3. Add suitable baits (sterilized or boiled hemp seeds, brassica seeds, tomato seeds, filter paper strips, etc.) to the above Petri dishes (one bait/three Petri dishes).
4. Incubate the plates at 5°C–20°C for 2–3 weeks in the dark (to discourage algal growth).
5. Change the water in Petri dishes frequently (first change after 2–3 days) with sterile glass dH$_2$O to avoid a profusion of bacteria and protozoans.
6. As soon as the growth of a fungus is seen on a bait, transfer the bait to a new Petri dish containing fresh bait and sterile water.
7. Transfer the bait with fungus on a PmTG agar plates with antibiotics and incubate the Petri dishes at 20°C.
8. Renew the culture in PmTG plates or slants and maintain for identification.

16.22.2.3 Results and Observations

Observe the baits, with fungus growth, under low-power magnification for their identification. The baits will be found to have growth of various aquatic fungi. The genera most frequently observed in fresh water include *Achlya*, *Allomyces*, *Saprolegnia*, *Dictyuchus*, and *Monoblepharis*, which can be identified by consulting the monographs and available books on aquatic fungi.

16.22.3 Isolation of Cellular Slime Molds

16.22.3.1 Introduction

Slime moulds inhabit forest soil and consume bacteria and yeast, which they track by chemotaxis—chemical gradient sensing (of folic-acid in this case). Starvation, however, prompts the solitary cells to aggregate and develop as a true multicellular organism, producing a fruiting body comprised of cellular, cellulosic stalk supporting a bolus of spores. Thus, slime mould has evolved mechanisms that direct the differentiation of a homogeneous population of cells into distinct cell types, regulate the proportions between tissues, and orchestrate the construction of an effective structure of the dispersal of spores. Many of the genes necessary for these processes in slime mould were inherited by Metazoa (animal with specialized cells) and fashioned through evolution for use within many different modes of development. Analysis of the proteins show that slime mould diverged from the animal–fungal lineage after the plant–animal split, but it seems to have retained more of the diversity of the ancestral genome than have plants, animals or fungi, that is, it possesses a level of complexity that is greater than the yeast but much simpler than plants or animals. It represents one of the earliest branches from the last common ancestor of all eukaryotes. As described earlier, cellular slime molds are not true fungi. A typical example of this group is *Dictyostelium discoideum*. It starts its life cycle with the dispersal of spores from a sporangium. Upon the absorption of water, the wall of the spore cracks open and a single amoeba like cell, a *Myxamoeba*, emerges. These cells live independently; as they move about feeding on bacteria, they grow and reproduce by fission. This, free-living, feeding stage ends following depletion of nutrients, the myxamoeba stream to an aggregation centre and forms a *Pseudoplasmodium*. The pseudoplasmodium migrates as a unit, leaving behind a trail of slime. When the migration ceases, the cells of the pseudoplasmodium differentiate to form a sporangium with spores. The cells from the front, or apical, region of the slug organize to form a stalk. Cells from posterior region move up the stalk as a ball of cells and become tip of the stalk. These cells transform into spores, each capable of initiating another life cycle of this slime mold (Figure 16.36).

16.22.3.2 Material
Sterile Petri dishes, Bunsen burner, and soil sample. They can be isolated from the sample using the following medium.

Peptone yeast extract medium: Proteose peptone 10 g, yeast extract 5 g, glucose 10 g, sodium acetate* 0.35 g, and KH_2PO_4 0.35 g, pH 6.4–6.6.

16.22.3.3 Protocol
1. Pour the melted and cooled (45°C) medium to Petri dish.
2. Allow the medium to solidify.

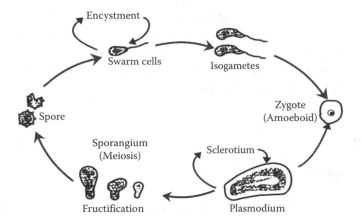

Figure 16.36 Life cycle of slime mold.

---

* In the place of sodium acetate, 2.3 mL of glacial acetic acid can be used.

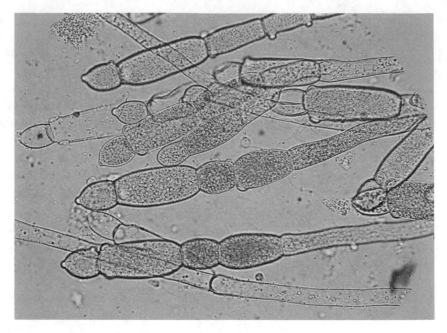

Figure 16.37 Sporangium and sporangiospore in a slime mold.

3. Sprinkle soil particles or finely divided decomposing leaf litter on the surface of the medium.
4. Incubates the plates at room temperature (20°C–24°C) for 3–7 days.
5. After incubation, examine the growth on the plate with the naked eye and with a hard lens.

16.22.3.4 Results and Observations

Slime molds grow forming translucent patches and producing heel-like aggregating pseudoplasmodia (Figure 16.37). The pseudoplasmodia will show a characteristic protoplasm movement. It is advisable to observe the plate on the stage of a microscope and examine under the 4 and 16 mm objectives. If an aggregation is visible, disrupt the aggregate with a needle to observe the component cells. Repeat the procedure with a pseudoplasmodium. If no fruiting bodies are detected, reincubate the culture for 2 days and then re-examine.

16.23 Fungi

16.23.1 Introduction

A large number of fungi of different groups are found in soil. The isolation, culture, and microscopic examination of fungi require the use of suitable selective media and special microscopic slide techniques. Different techniques are used for the isolation of fungi from the soil, which include serial dilution agar plate, Warcup soil plate, syringe inoculation, immersion tube method, screened immersion plates, plate profile, hyphal tip isolation, soil washing, partial pasteurization, soil sieving, floating baiting technique, etc. A single method cannot be used to count all the different types of fungi present in a given sample. A sample, therefore, is to be processed by a variety of technique. The dilution plates and soil plate methods are the two most widely used methods for the fungal isolation from soil. Direct soil plate or Warcup soil plate method was developed by Warcup in 1950 and is superior to dilution plate method as it is easier to use, less time consuming, and requires less glassware and more qualitative information.

16.23.2 Isolation and Microscopic Examination of Fungi

16.23.2.1 Materials
Martin's agar medium: KH_2PO_4 1 g, $MgSO_4 \cdot 7H_2O$ 0.5 g, peptone 5 g, dextrose 10 g, rose bengal (1% aqueous solution) 3.3 mL, streptomycin/streptopenicillin 30 mg (to be added after autoclaving and just before pouring), agar-agar 15 g, and dH_2O to 1 L. The medium is also known as peptone-dextrose-rose bengal agar medium.

Soil sample, pipettes of 10 mL and 1 mL capacity, Erlenmeyer flasks 250 mL capacity, incubator, cotton blue, lactophenol, syringe with needle, and spirit lamp/ Bunsen burner.

16.23.2.2 Protocol
1. Collect soil at random in sterilized polythene bags and make a composite sample by mixing the above samples.
2. Add 0.005–0.15 g (approximately) air dried soil each in 5 sterile Petri plates with the help of a sterilized cooled loop or transfer needle.
3. Add 15–20 mL of melted, cooled (45°C) medium, supplemented with streptomycin/ streptopenicillin and rose Bengal, to each soil inoculated Petri plate.
4. Dispense soil particles throughout the medium by gentle rotation of the Petri dishes.
5. Allow the plates to solidify.
6. Incubate the plates at 25°C in an inverted position for 15 days.

16.23.2.3 Results and Observations
Examine selected colonies using the dissecting microscopes and low power objective of the compound microscope. Look especially for fruiting structures (e.g., spores). Aseptically remove small portions of fungal and an actinomycete culture from dilution plates and make "squash" mounts in various stains. Observe the plates for the appearance of the fungal colonies after 3 days of incubation and continue till 15 days. Purify various fungal isolates for identification up to a specific level and future use (Figure 16.38).

16.23.3 Serial Dilution Plate Technique

16.23.3.1 Materials
As mentioned earlier.

Figure 16.38 Mixed culture of fungi from soil in a Petri plate, *Trichoderma viride* on culture plate and slants, and microscopic examination of conidial attachment of *Trichoderma viride*.

16.23.3.2 Protocol

1. Take 3 Erlenmeyer flasks (250 mL capacity), transfer 90 mL dH$_2$O in each flask, plug them properly, label 1–3, and autoclave at 15 psi for 15 min.
2. Collect a small amount of soil from five different places of the desired field and mix to make one lot.
3. Weigh 10 g of soil sample and transfer into flask 1 containing 90 mL sterilized water. It gives the dilution 1:10 (10^{-1}).
4. Shake the flask gently for 10 min on shaker to get homogeneous soil suspension.
5. Transfer 10 mL soil suspension from 10^{-1} dilution into flask 2 containing 90 mL sterilized water to get dilution 10^{-2}. Mix the suspension gently.
6. Similarly, serially transfer 10 mL soil suspension from 10^{-2} dilution into flask 3 containing 90 mL sterilized water to get the final dilution 10^{-3}. Mix the suspension gently.
7. Aseptically pour 1 mL soil suspension from 10^{-3} dilution into Martin's agar plates supplemented with streptomycin/streptopenicillin. Gently rotate the plates so as to spread the suspension on medium.
8. Incubate the plates at 25°C for 4–5 days.

16.23.3.3 Results and Observations

Fungal colonies of different size and color grow on medium. For qualitative study, pick up the small amount of mycelia growth with a sterile needle, transfer on a glass slide containing one drop of cotton blue plus lactophenol. Observe under a microscope and identify them following mycological literature. Measure occurrence of each fungal species by using the following formula:

$$\text{Occurrence (\%)} = \frac{\text{Average number of colonies of a species}}{\text{Average number of colonies of all the fungal species}} \times 100$$

Quantitative analysis: Take 10 g soil from the same lot of soil, transfer in a pre-weighed beaker (100 mL capacity), and put in an oven. Make the soil oven dry at 85°C for 24 h. Count all the colony forming units (CFUs), take an average of three plates, and calculate average number of CFUs/g dry soil by using the following formula:

$$\text{CFUs/g dry soil} = \frac{\text{Average number of colonies}}{\text{Dry weight of soil}} \times \text{Dilution factor}$$

If the number of colonies are innumerable and difficult to count, further dilute the soil sample and plate it as done before.

Many soil fungi belong to the order "Hyphomycetes" of the fungi imperfecti—that is, they produce conidia (asexual spores) on conidiophores. These structures are very fragile and require special techniques for observation. The slide culture mount is an excellent technique to observe these fungi.

16.23.4 Slide Culture Mount (Riddell Mount)

16.23.4.1 Introduction

Members of fungi are viewed by picking a small portion of colony growing on Petri plates or slants. This disturbs the arrangement of spores or conidia and the cells developing conidia. Therefore, the slide culture technique is used to observe these structures of fungi without disturbing the natural shape and arrangement of conidia, spores which helps in identification.

16.23.4.2 Materials

Petri dish, sterile filter paper, glass U-shaped rods, pure fungal culture plate to be studied, Czapek Dox agar plates, scalpel, inoculating loop, sterile water, microscopic glass slides, cover slips, and forceps.

16.23.4.3 Protocol

1. Prepare the moist chamber by placing the sterilized circular filter paper discs as per the diameter of the Petri dish on its bottom half.
2. Pour dH_2O in the plate and see that the lower end of the U-tube sinks in water as shown in Figure 16.39.
3. With forceps, place a sterile slide on the U-shaped rod.
4. Gently flame a scalpel to sterilize and cut a 5 mm square block of the medium from the plate of Czapek Dox agar plates.
5. Pick up the block of agar by inserting the scalpel into one side as illustrated in Figure 16.39.
6. A needlepoint inoculum is placed on the corners of the agar and cover with a sterile cover slip. Be sure to flame and cool the loop prior to picking up spores.
7. The slide is incubated in a moist chamber on glass rods for 1–2 weeks.
8. Observe periodically; and once sporulation has occurred, transfer cover slip to a drop of cotton blue stain on a new slide and observe conidiophores arrangement.

16.23.4.4 Results and Observation

After 48 h, examine the slide under low power in a compound microscope. If growth has occurred, one should be able to see hyphae and spores. If growth is inadequate

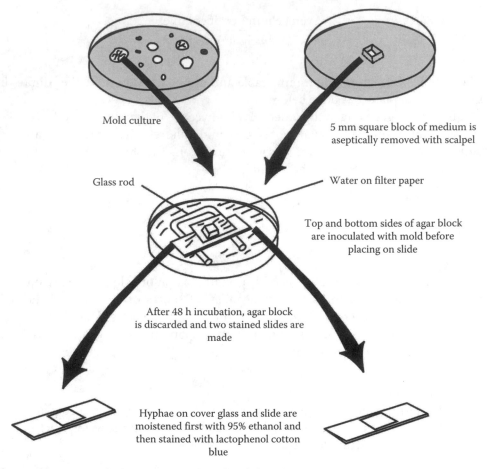

Mold culture

5 mm square block of medium is aseptically removed with scalpel

Glass rod

Water on filter paper

Top and bottom sides of agar block are inoculated with mold before placing on slide

After 48 h incubation, agar block is discarded and two stained slides are made

Hyphae on cover glass and slide are moistened first with 95% ethanol and then stained with lactophenol cotton blue

Figure 16.39　Displaying the procedure for slide culture.

and spores are not evident, allow the mold to grow another 48 h before making the stained slides. Attempt to identify to genus using the classification key of fungi.

16.23.5 Isolation of Pure Culture

16.23.5.1 Principle
Once discrete, well-separated colonies develop on the surface of the agar plates, each may be picked up with a sterile needle and transfer to new agar slants. Each slant will represent a single bacterial species and is designated as pure or stock culture.

16.23.5.2 Protocol
1. Mark the characteristics, discrete, well separated colony.
2. In the laminar bench, using a sterile needle pick up the colony.
3. Streak the loopful of selected culture on a slant.
4. Incubate the slants for 48–72 h at optimum temperature.

16.23.6 Isolation of Microorganisms from Soil and Water

16.23.6.1 Introduction

Microorganisms in soil are important because they affect the structure and fertility of different soils. Soil microorganisms exist in large numbers in the soil as long as there is a carbon source for energy. A large number of bacteria in the soil exist, but because of their small size, they have a smaller biomass. Actinomycetes are a factor of 10 times smaller in number but are larger in size, so they are similar in biomass to bacteria. Fungus population numbers are smaller, but they dominate the soil biomass when the soil is not disturbed. Bacteria, actinomycetes, and protozoa are hardy and can tolerate more soil disturbance than fungal populations, so they dominate in tilled soils, while fungal and nematode populations tend to dominate in untilled or no-till soils. Despite this diversity, bacteria, including the mold-like actinomycetes and fungi, are the most prevalent as shown in Table 16.4.

16.23.6.2 Methods for the Isolation of Pure Cultures from Soil/Water

16.23.6.2.1 Materials.   Following media are being used

Organisms	Media
1. Bacteria	Nutrient agar
2. Actinomycetes	Glycerol yeast extract agar
3. Cyanobacteria/algae	Chu10/A and A agar
4. Molds	Glucose peptone acid agar/PDA with rose bengal and antibiotics to check bacterial growth

Though various methods are available to isolate microbes from soil, the serial dilution agar plating method or viable plate count method is one of the commonly used one.

TABLE 16.4   Relative Number and Biomass of Microbial Species at 0–6 in. (0–15 cm) Depth in Soil

Microorganisms	Number/g of Soil	Biomass (g/m^2)
Bacteria	10^8–10^9	40–500
Actinomycetes	10^7–10^8	40–500
Fungi	10^5–10^6	100–1500
Algae	10^4–10^5	1–50
Protozoa	10^3–10^4	Varies
Nematodes	10^2–10^3	Varies

Note: Microbes are in the mixed population and should be separated into individual species for study. Several techniques are applied to isolate and study microorganisms in pure culture. Different media are employed to support the growth of these microbes.

16.23.6.2.2 Protocol.   The method is based upon the principle that when material containing microorganisms is cultured, each viable microorganism will develop into a colony; hence, the number of colonies appearing on the plates represent the number of living organisms present in the sample. In serial dilution agar-plate method, a known amount (1 mL or 1 g) of material is suspended or agitated in a known volume of sterile NaCl water blank (9 mL to make the total volume to 10 mL) to make a microbial suspension. Serial dilutions 10^{-2} to 10^{-7} are made by pipetting measured volumes (1–9 mL) of sterile water. Finally 0.1 mL aliquot of various dilutions is added to sterile Petri dishes to which 15 mL of sterile, cool, molten media is added to each one. If it is fungi or actinomycetes, the glycerol yeast agar and Sabouraud's agar supplemented with 10 µg of chlortetracycline per millimeter of medium to inhibit the growth of bacteria are used. The dilutions of 10^{-2} to 10^{-5} are selected for enumeration of fungi, 10^{-3} to 10^{-6} for actinomycetes, and 10^{-4} to 10^{-7} for bacteria. Upon solidification, the plates are incubated in an inverted position for 3–7 days at 25°C (Figure 16.40).

16.23.6.2.3 Results and Observations.   The number of colonies appearing on dilution plates are counted, averaged, and multiplied by the dilution factor to find the number of cells/spores per gram of the sample

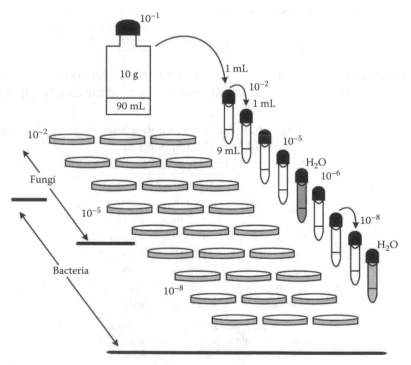

Figure 16.40 Dilution plate technique is commonly used for counting the microorganisms.

$$\text{Number (average of three replicates) of cells / mL or g} = \frac{\text{Colonies dilution factor}}{\text{Dry weight of soil}}$$

Dilution factor = Reciprocal of the dilution (e.g., $10^7 = 10^7$)

16.23.6.3 Experiment Isolation of Pure Cultures from Water/Soil

16.23.6.3.1 Materials. Petri plates, conical flasks, test tubes, colony counter, Bunsen burner, glass marker, pipettes, test tube stand, incubator, pH meter, peptone, dextrose, agar, beef extract, glycerol, yeast extract dipotassium hydrogen phosphate, dH_2O, aureomycin, streptopenicillin, NaCl, and soil samples.

Nutrient agar medium (for the isolation of bacteria): Prepared as described earlier.

Glycerol yeast extract: Glycerol 5 mL, yeast extract 2 g, dipotassium hydrogen phosphate 1 g, agar-agar 15 g, aureomycin 0.01 g, and dH_2O 1000 mL. Dissolve the chemicals in dH_2O, adjust the pH to 7.0 and raise to 1 L, autoclave at 15 psi for 15 min, and add aureomycin to the sterile, cooled, and molten agar, aseptically.

Sabouraud's agar (pH 5.6): Peptone 10 g, dextrose 40 g, agar-agar 15 g, and dH_2O 1 L.

Dissolve all the chemicals in 1000 mL of water autoclave, cool, and then add streptopenicillin aseptically. Soil dilution is prepared in 0.85% NaCl solution.

16.23.6.3.2 Protocol

1. Collect soil samples at random, minimum five, from a field, mix thoroughly to make a composite sample for microbiological analysis.
2. Label 90 mL sterile 0.85% NaCl blanks as 1–7 and sterile Petri dishes as 10^{-2} (three plates), 10^{-3} (six plates), 10^{-4} (nine plates), 10^{-5} (nine plates), 10^{-6} (six plates), and 10^{-7} (three plates) with a marker pen.
3. Add 1 g sample of finely pulverized, air dried soil into numbered NaCl (0.85% blank to make 10 dilution (10^{-1})).
4. Vigorously, shake the dilution on a magnetic shaker for 20–30 min to obtain uniform suspension of microbes.
5. Transfer 1 mL of suspension from test tube number 1 into NaCl blank number 2 with a sterile pipette under aseptic conditions to make 1:100 (10^{-2}) dilution and shake it well for about 5 min.
6. Prepare another dilution 1:1000 (10^{-3}) by pipetting 1 mL of the suspension into NaCl blank number 3, using a fresh sterile pipette and shake it.
7. Make further dilutions 10^{-4} to 10^{-7} by pipetting 1 mL suspension into additional NaCl blanks (4–7) as prepared earlier.
8. Transfer 0.1 nil aliquots each from 10^{-2} dilution blank into three sterile Petri dishes, from 10^{-3} dilution blank to six sterile Petri dishes, from 10 to nine sterile Petri dishes from, 10^{-6} to six Petri dishes, and from 10^{-7} to three petri dishes.

9. Add 15 mL of the cooled medium (45°C) to each Petri dish and mix the inoculum by gentle rotation of the Petri dish. The three media can be added to various dilutions as follows:

   a. For bacteria—nutrient agar medium to 12 plates with 10^{-4} to 10^{-7} dilutions.
   b. For actinomycetes—glycerol yeast agar medium supplemented with aureomycin to plates with 10^{-3} to 10^{-6} dilutions.
   c. For fungi—Sabouraud's agar medium supplemented with streptopenicillin to plates with 10^{-2} to 10^{-5} dilutions.

10. Upon solidification of the media, incubate all the plates in an inverted position at 25°C for 2–7 days.

16.23.6.3.3 Result and Observations. Observe the plates for the number and distribution of colonies of bacteria, fungi, and actinomycetes from each dilution. Calculate the number of organisms (bacteria, actinomycetes, and fungi) per gram of the soil by applying the following formula:

$$\text{Viable cells/g dry soil} = \frac{\text{Mean plate count} \times \text{Dilution factor}}{\text{Dry weight of soil}}$$

For example, if 160 bacterial colonies were counted (average of three replicate) in 1:100000 dilution of a soil sample (8 g dry wt. basis), the number of colonies per g of the soil would be

$$\frac{160 \times 100,000}{8} 20,00,000 = 2 \times 10^6$$

16.23.6.4 Isolation of Yeast from Soil Ripened Grape

16.23.6.4.1 Introduction. Yeast is a group of unicellular fungi, which are cosmopolitan in distribution. Most yeast are placed in Ascomycotina of the kingdom Mycota. They have been isolated from a natural substance like leaves, flower, sweet fruits, grains, bakery product, and soil. Yeast plays a major role in many industry viz. fermentation and food industries (*Saccharomyces cerevisiae*), production of single cell protein (*Candida succiphila*), animal feed industries (*Candida utilis*), and natural decomposition of organic matter in soil and water. While it is common colonizer of human mucosal surfaces and considered being nonpathogenic for immunocompetent hosts and may cause infections particularly in immunocompromised patients. Morphologically yeast are Gram-positive, nonfilamentous, unicellular fungi, which are typically spherical or oval in shape. They are reproduced by budding, a process in which new cells are formed as protuberance (bud) from the parent cell and separates from it upon reaching maturity. Sometimes, daughter cell fail to detach themselves, a short chain of cells is formed which is termed as pseudomycelium or pseudohyphae (Figure 16.41).

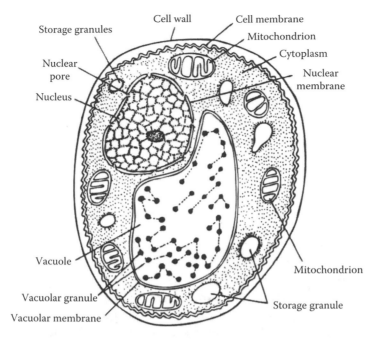

Figure 16.41   Diagrammatic view of electron micrograph of a yeast cell.

16.23.6.4.2 Materials.   Glucose yeast extract (GYE) agar medium, sterile Petri plates, sterile test tube, dH$_2$O, glass rod, Bunsen burner, inoculating loop, wax marking pencil, etc.

16.23.6.4.3 Protocol
   1. Place a ripened grape into sterile dH$_2$O and crush with the help of a sterile glass rod.
   2. Perform Gram's staining of the suspension and observe for yeast cell in smear.
   3. Take loopful suspension and streak on GYE agar plate and incubate the plate at 27°C for 24–48 h.
   4. After incubation, observe for colonies of yeast and note down cultural characteristics.
   5. Pick up a well isolated colony of yeast and prepare the suspension in sterile dH$_2$O.
   6. Perform the Gram's staining of the suspension and verify the purity of the yeast culture.
   7. If necessary, preserve the culture by transferring the culture on GYE agar slant.

16.23.6.4.4 Results and Observations.   Since yeasts are unicellular microorganisms, its growth resembles a bacterial colony rather than fungal colony. Growth is rapid, and colony is formed within 24–48 h. Colony size ranges from 2–3 mm and have flat, slightly raised, or convex elevation. Most colonies are smooth, dull, and have soft pasty consistency with characteristic alcoholic smell; the colony appears to have white, creamy, dirty white, or light brown pigments.

16.24 Rhizosphere

16.24.1 Introduction

The region in the vicinity of roots can be distinguished into many microhabitats. The term rhizosphere was introduced in 1904 by the German scientist Hiltner to denote that region of the soil, which is subject to the influence of plant roots. Rhizosphere is characterized by greater microbiological activity than the soil away from plant roots. Rhizosphere is the region of the soil immediately surrounding the roots of a plant. From an operational viewpoint, rhizosphere can be defined as the region extending a few millimeters from the root surface in which the microbial population of soil is influenced by the chemical activities of plant. Root secrete a wide range of organic materials (root exudates) into the soil, which greatly influence the development of a rhizosphere microbial community. Microbial populations differ both quantitatively and qualitatively from those in the bulk soil. It is now clearly established that the greater number of bacteria, fungi, and actinomycetes are present in the rhizospheric soil than in nonrhizospheric soil. Several factors such as soil type, its moisture, pH, temperature, age, and condition of the plants are known to influence the rhizosphere effect. For isolation of microorganisms from rhizosphere, the rhizospheric soil can be separated, and suspension is obtained by shaking roots in sterile water from which subsequent dilutions are made further. One milliliter of the appropriate dilutions is plated on suitable agar media for enumeration of bacteria, actinomycetes, and fungi (Figure 16.42).

16.24.2 Isolation of Microorganisms from Rhizosphere

16.24.2.1 Materials
Rhizospheric soil, PDA medium, nutrient agar medium, dH_2O, pipette, sterile Petri plates, sterile test tube, inoculating loop, dH_2O, Bunsen burner, wax marking pencil, etc.

16.24.2.2 Protocol
1. Prepare nutrient agar medium and PDA medium in sterile flasks.
2. Separate rhizospheric soil from four to five roots and collect in a Petri plates.
3. Add about 1 g soil in 10 mL dH_2O in a test tube and mark it as 10^0 or crude.
4. Now serially dilute the soil sample by transferring 1 mL of previous soil solution from test tube to 9 mL dH_2O in next tube.
5. Prepare dilution up to 10^{-6} (Figure 16.43).
6. Transfer 1 mL of soil suspension from each dilution to sterile Petri plates.
7. Now pour the molten and cooled PDA medium and nutrient agar medium in various plates.
8. Incubate the plates for 24–48 h at 37°C for bacterial growth and 2–6 days at 27°C for fungal growth.

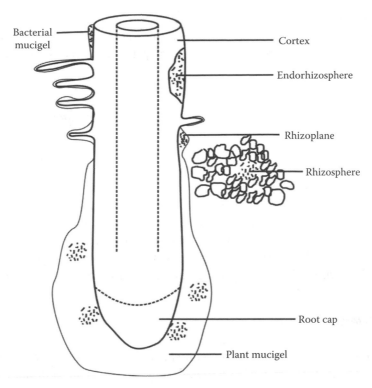

Bacterial
mucigel

Cortex

Endorhizosphere

Rhizoplane

Rhizosphere

Root cap

Plant mucigel

Figure 16.42   Distribution of microorganisms in the rhizosphere.

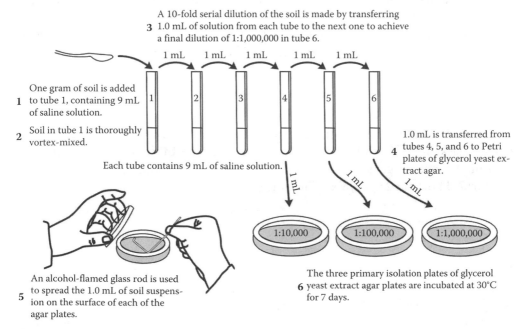

3 A 10-fold serial dilution of the soil is made by transferring 1.0 mL of solution from each tube to the next one to achieve a final dilution of 1:1,000,000 in tube 6.

1 mL   1 mL   1 mL   1 mL   1 mL

1 One gram of soil is added to tube 1, containing 9 mL of saline solution.

2 Soil in tube 1 is thoroughly vortex-mixed.

Each tube contains 9 mL of saline solution.

4 1.0 mL is transferred from tubes 4, 5, and 6 to Petri plates of glycerol yeast extract agar.

1 mL   1 mL   1 mL

1:10,000   1:100,000   1:1,000,000

5 An alcohol-flamed glass rod is used to spread the 1.0 mL of soil suspension on the surface of each of the agar plates.

6 The three primary isolation plates of glycerol yeast extract agar plates are incubated at 30°C for 7 days.

Figure 16.43   Diagrammatic view for preparing serial dilution of soil sample.

16.24.2.3 Observations

Observe the plates for growth and note different type of colonies on the plates.

Record the colony morphology and identify the organisms by following the literature (Figure 16.44).

16.25 Microbial Production of Citric Acid by *Aspergillus niger*

16.25.1 Introduction

Citric acid is used in food, beverage, and in pharmaceutical industries as a preservative. Ninety-nine percent of the world's output of citric acid is produced by microbial fermentation. It is a primary metabolic product formed through TCA cycle. The chemical name of citric acid is 2-hydroxy propane 1,2,3 tricarboxylic acid. It is produced by fungus like *Aspergillus niger, Aspergillus wentii, Aspergillus clavatus, Penicillium luteum, Penicillium citrinum, Candida, Saccharomyces*, and *Mucor*.

16.25.2 Materials

Actively growing culture of *A. niger*, conical flasks 250 mL capacity, Petri plates, pipettes, test tubes, autoclave, dH_2O, inoculating needle, Whatman no. 1 chromatographic paper, n-butanol, acetic acid, λ pipette, chromatographic chamber, auto glass sprayer, hair dryer, xylose, aniline, and ethanol.

Czepek Dox agar medium (g/L): Sucrose 30, sodium nitrate 2, dipotassium phosphate 1, magnesium sulfate 0.5, potassium chloride 0.5, ferrous sulfate 0.01, and agar 15, final pH (at 25°C) 7.3 ± 0.2.

Aniline xylose reagent: 5 mL aniline, 5 g xylose, dissolved in 100 mL of 50% ethanol. This reagent is used to locate the citric acid on paper chromatogram.

16.25.3 Protocol

Prepare Dox medium as per the composition, pour 50 mL of the medium into a conical flask, and adjust pH to 4.5–5.0. Sterilize the flask by autoclaving at 121°C for 15 min. After cooling, aseptically inoculate the medium with *A. niger* culture, incubate for 7–14 days at 25°C–28°C. Filter the broth.

16.25.4 Result

Check the pH of the broth with pH meter. The pH of the broth should be acidic and analyze the culture filtrate for citric acid by descending paper chromatography using organic phase of n-butanol:acetic acid:water (4:1:5) as solvent. The citric acid is located by spraying the overnight run chromatogram with aniline xylose reagent and heated at 100°C in an oven.

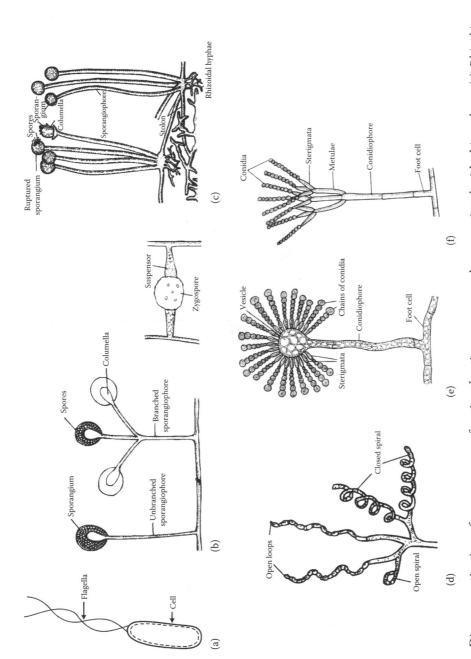

Figure 16.44 Diagrammatic view of some common fungi and actinomycete members associated with rhizosphere. (a) *Rhizobium* with two polar flagella, (b) *Mucor* sp., (c) *Rhizopus* sp., (d) *Streptomyces* sp., (e) *Aspergillus* sp., and (f) *Penicillium* sp.

(continued)

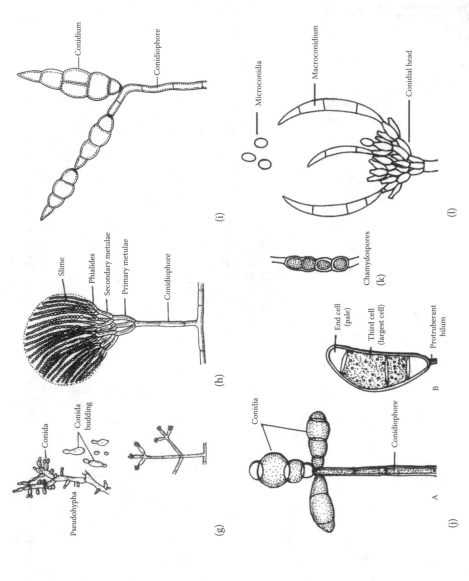

Figure 16.44 (continued) Diagrammatic view of some common fungi and actinomycete members associated with rhizosphere. (g) *Candida* sp., (h) *Trichoderma* sp., (i) *Gliocladium* sp., (j) *Alternaria* sp., (k) *Curvularia* sp., and (l) *Fusarium* sp.

16.26 Isolation of Industrially Important Antibiotic Producing Microorganisms

16.26.1 Introduction

Antibiotics are microbial products or their derivatives that kill susceptible microorganism or inhibit their growth even when used at low concentration. Alexander Fleming first discovered penicillin from *Penicillium notatum*. This mold inhibited the growth of *Staphylococcus aureus*. The fungus subsequently was identified as *P. notatum*, and the crude antimicrobial extract obtained from it was named Penicillin. It was the first of many effective antibiotics for treating infectious diseases. It is being produced on a commercial scale from *Penicillium chrysogenum*.

16.26.2 Materials

Czepek Dox agar medium (g/L): Sucrose 30, sodium nitrate 2, dipotassium phosphate 1, magnesium sulfate 0.5, potassium chloride 0.5, ferrous sulfate 0.01, agar 15, final pH (at 25°C) 7.3 ± 0.2, conical flasks, Petri plates, pipettes, test tubes, autoclave, dH$_2$O, and spreader.

16.26.3 Methods

Take 1 g of soil sample. Make suspension with 10 mL of sterile dH$_2$O. Prepare different dilutions with dH$_2$O and label them as 10^{-4} to 10^{-3}. Prepare Czapek Dox agar medium as per the composition and sterilize by autoclaving at 15 psi for 15 min, pour the medium into sterile Petri plates. Add 1 mL of 10^{-3} soil dilution over the surface of solidified agar medium. Spread the inoculum with a bent glass rod. Incubate the plates at 37°C for 24 days.

16.26.4 Results and Observations

Isolate the colonies with the zones of inhibition having an antimicrobial activity.

16.27 Assay of Antibiotics and Demonstration of Antibiotic Resistance

16.27.1 Introduction

Antibiotic is a chemical substance, such as penicillin or streptomycin, produced by or derived from certain fungi, actinomycetes and bacteria, and other organisms that can destroy or inhibit the growth of other microorganisms. Antibiotics are widely used in the prevention and treatment of infectious diseases. They are now being synthesized by chemical methods as well. These are low-molecular-weight compounds. They have a variety of chemical structures, elemental composition, and physical chemical properties. Once the causative organism is identified for specific disease and is isolated, it is important to test the sensitivity of the organism to the effective antibiotics. In the clinical laboratories, antibiotic-impregnated discs are commonly used to identify the antibiotic sensitivity of the causative organism. The effectiveness of the antibiotic in this test is based on the size of zone of inhibition. The zone of inhibition also depends on the diffusibility of antibiotic, the size of the inoculum, type of medium, and other factors. The concentration of the antibiotic at the edge of zone of inhibition represents minimal inhibitory concentration (MIC) of antibiotic. Alternatively, the antibiotic disc with different concentration of antibiotic could be employed in the test. The MIC is the lowest concentration of antibiotic that exhibits the zone of inhibition on the assay plate.

16.27.2 Materials

Petri plates, test tubes, autoclave, incubator, refrigerator, forceps, swabs, pH meter, beef extract, peptone, NaCl, agar, dH$_2$O, and antibiotic discs.

Nutrient agar plates g/L in dH$_2$O pH 7.2: Peptone 5, beef extract 3, and NaCl 5.

Fresh bacterial cultures of *Escherichia coli*, *Pseudomonas aeruginosa*, and *S. aureus*.

16.27.3 Antibiotic Sensitivity Test

16.27.3.1 Protocol (Kirby–Bauer Method)

1. Select and label the cultures that are to be used for antibiotic sensitivity assay (*E. coli*, *P. aeruginosa*, and *S. aureus*).
2. Prepare nutrient agar plates (at least three plates for each set).
3. Aseptically plate the culture on the entire surface of the agar plate swabbed with organism to be tested, or a bacterial lawn is prepared on the plate.
4. Allow at least 5 min for the agar surface to dry before applying disc.
5. Take forceps and sterilize the tip by dipping in alcohol and then flame it. Allow to cool.
6. Carefully take the antibiotic disc and place over the agar plate at least 15 m from the edge of the plate. Gently press the disc to give a better contact with the agar.
7. Place at least six different antibiotic discs at the same distance apart from each other in the agar plates.
8. Incubate the plates in the inverted position for 16–18 h at 37°C.

16.27.3.2 Results and Observation

1. Observe the zone of inhibition around the antibiotic discs (Figure 16.45).
2. Indicate whether test organism is resistant (no zone of inhibition) or sensitive (clear zone of inhibition) to the antibiotic.

16.27.4 Determination of MIC of Antibiotic

16.27.4.1 Protocol

1. Prepare four nutrient agar plates.

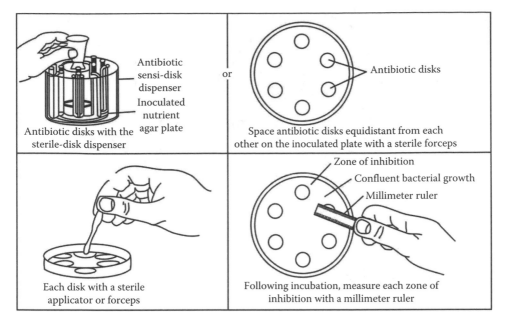

Figure 16.45 Antibiotic sensitivity test.

2. Select a test organism (*E. coli* or *P. aeruginosa*) and an antibiotic from the previous experiment that is showing good inhibition zone.
3. Inoculate the nutrient agar plate with the selected test organisms as lawn culture. Allow the plate to dry for 5 min.
4. Place a disc impregnated with different concentrations of antibiotic (10–500 µg/mL). Incubate the plates for 12–18 h.

16.27.4.2 Result and Observations

1. Measure the zone of inhibition after incubation on the plate by a ruler. Tabulate the result—concentration of antibiotic versus diameter of zone of inhibition (Figure 16.46).
2. Plot a graph taking concentration of antibiotic in X-axis and square of the diameter of zone on the Y-axis.
3. The straight line intercepting the X-axis is the minimal concentration of antibiotic required for inhibition of the growth of test organisms.

16.28 Characterize and Identify Aflatoxin by Thin-Layer Chromatography

16.28.1 Introduction

Wide spectrum of fungi infests most of the agricultural produce including food grains and feedstuffs under warm humid conditions, especially when the moisture content of the stuffs is high. Although the grains are high in carbohydrates and protein, microbes do not grow on these because of low water activity, if harvested and stored properly. The growth of the molds and other microorganisms is stimulated

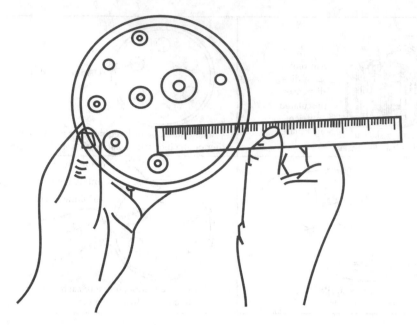

Figure 16.46   Antimicrobial sensitivity testing by filter paper disc method.

and cause spoilage when these are stored under moist conditions. In tropical climates, the commonest molds responsible for spoilage of grains belong to *Rhizopus stolonifer, Penicillium, Aspergillus, Alternaria,* and *Fusarium.* Some of these molds produce extensive mycelium with the sporulating structures on the surface. Certain species of *Aspergillus* (e.g., *Aspergillus flavus*), infesting peanuts and some grains, produce aflatoxin. Of the 18 aflatoxins known, aflatoxin B_1 is the most common and the most potent mutagen and carcinogen (Figure 16.47). Continuous intake of fungus-contaminated materials leads to the damage of liver and kidney by the toxic substances. *Aspergillus* species is the most widely contaminating fungi producing aflatoxins B_1, B_2, G_1, and G_2. Because of the potential health hazards produced by these toxins even in minute amounts, a watch over their presence in various food materials is very much inevitable.

Aflatoxin is known for causing liver and colon cancer in humans. Fungal contamination of corn (maize) by *Fusarium moniliforme,* when improperly stored, is very serious. The mold on corn produces fumonisins (Figure 16.48), another potent carcinogenic mycotoxin that causes leukoencephalomalacia in horses, pulmonary edema in pigeon, and esophageal cancer in humans.

Thin-layer chromatography (TLC) method is simple and can be performed without any sophisticated equipment.

16.28.2 Materials

TLC Kit, ultraviolet (UV)-chamber, mechanical shaker, "Quick-fit" distillation set, toluene, ethyl acetate, formic acid, chloroform, and silica gel G (TLC Grade).

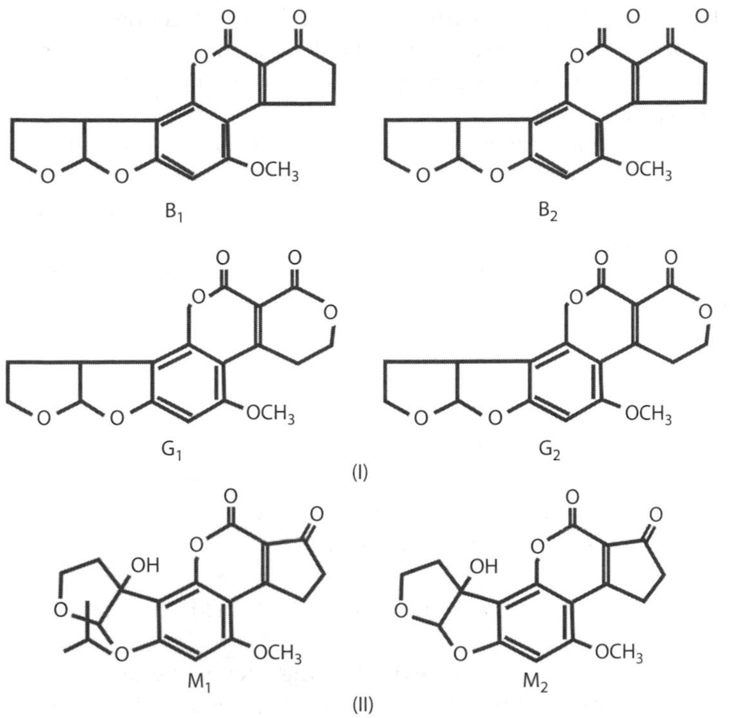

Figure 16.47 Four basic structures of common aflatoxins B_1, B_2, G_1, G_2, and M_1 and M_2. When viewed under UV light, the six mycotoxins give fluorescence B_1 and B_2: blue, G_1 and G_2: green–blue, M_1: blue violet, and M_2: violet.

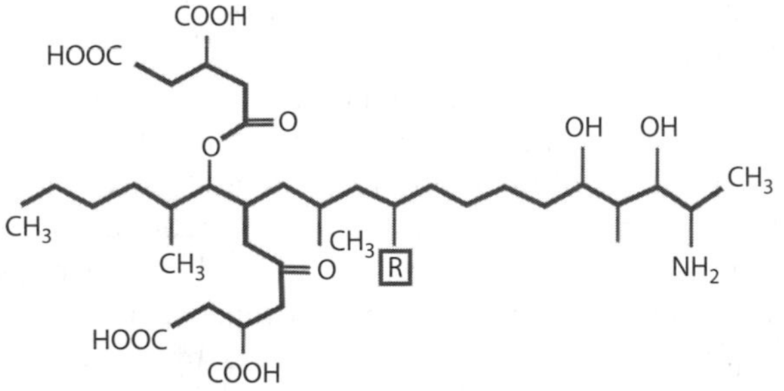

Figure 16.48 The basic structure of toxin fumonisins produced by *Fusarium moniliforme*, a fungal contaminant of stored corn.

16.28.3 Procedure

16.28.3.1 Extraction of Toxins

1. Weigh exactly 50 g of ground sample material and transfer it into a 250 mL conical flask.
2. Moist the material uniformly by adding 10–15 mL of dH_2O and add about 200 mL chloroform; stopper the mouth with a cotton plug in aluminum foil.
3. Shake the flask for 1 h mechanically (it is important that the oil-containing materials are defatted prior to extraction).

4. Filter the slurry through a Buchner funnel under mild suction. Equal amount of a filtering aid such as celite may be mixed before filtering in order to ease filtration. Wash the flask and the slurry thoroughly with additional chloroform (25 mL) and collect the filtrate.

5. Transfer the filtrate quantitatively to a separatory funnel and shake with water one-half volume of chloroform. After the phases separate, drain the bottom (chloroform) phase into a flask containing about 10 g sodium sulfate (anhydrous) to absorb any water.

6. Concentrate the clear, chloroform extract "under vacuum" over a warm water bath using quick fit distillation set. Make up the concentrate to a known volume with chloroform and store in amber-colored vials under refrigeration until analysis.

16.28.3.2 Preparation of TLC Plates

1. Place 30 g silica gel G (with $CaSO_4$ as a binder) in a stoppered flask, shake vigorously with 60–65 mL dH_2O for about 1 min, transfer to the applicator and spread uniformly on five clean glass plates (20 cm × 20 cm). The exact quantity of water required to get a good slurry will vary from batch to batch of silica gel G. The thickness of layer should be usually 0.25 mm.

2. Allow the plates to dry for 1–3 h in dust-free conditions. Activate the gel, prior to use, for 30 min at 110°C in a hot-air oven. The activated gel plates should be stored in a desiccator chamber.

3. Divide the gel into a number of lanes by drawing lines on the gel with a sharp needle.

4. Spot different known volumes (5 mL, 10mL, etc.) of the sample extract in various lanes carefully with a microsyringe on an imaginary line 2.5 cm away from one end of the plate. Similarly, spot standard aflatoxins (B_1, B_2, G_1, and G_2) mixture in the concentration range 0.0025–0.0125 mg in parallel lanes.

5. Develop the plate in a solvent system of toluene:ethyl acetate:formic acid (6:3:1) in a chromatographic tank for about 50 min. By then, the solvent front might have moved up to 20 mm below the top end of the plate.

6. Dry the plate at room temperature to remove the solvent. Visualize the fluorescing spots of toxins under UV light in a cabinet. Protective glass should be worn while viewing under UV light; otherwise, eye sight will be affected.

7. Identify each fluorescing spot of the sample extract by comparing with the authentic toxin spot co-chromatographed. Determine the R_f value of each spot.

8. For quantitative estimation of the toxins in the sample extract, match the intensity of spots of the sample with that of standard toxin spots, by diluting both to extinction. Calculate the amount of toxin in a kg sample material.

Notes

1. *Care to collect a representative sample of the experimental material.* Conveniently, small portions of samples from various points are collected, mixed thoroughly, and quartered. Sampling procedure varies from commodity to commodity (for details read the reference in the subsequent sections).

2. *Pre-run of activated plates in diethyl ether* is useful to eliminate UV fluorescing substances, if any, in the gel.

3. *A variety of solvent systems are used for developing the plates.* Choose the system to get all the spots clearly resolved.

4. The R_f value is in the order of $B_1 > B_2 > G_1 > G_2$.

5. Under UV light, B_1 and B_2 fluoresce blue and G_1 and G_2 green. B_1 content is usually greater than the other toxins.

6. Dilution to extinction: The standard or sample is diluted serially. Each dilution is spotted in equal volume on the plate. After developing, at one dilution, a particular toxin (say B_1) will be visible under UV light while, at its next high dilution, the spot will not be visible. The dilution at which the spot is visible is termed "dilution to extinction". For each toxin, the dilution to extinction is different from the other.

16.29 Protozoa

16.29.1 Introduction

Protozoa are the eukaryotic, unicellular, and nonphotosynthetic microorganisms devoid of cell wall. They fall into the division Protista. Total number of species of protozoa account for nearly 4000, of which some of them are human pathogens and the others nonpathogens. The nonpathogenic species are free-living in soil and involved in the establishment of relationships with the other microorganisms. They become a source of nutrient release for other microorganisms. The human pathogens are associated with certain diseases such as amoebic dysentery (*Entamoeba histolytica*), malaria (*Plasmodium vivax*, *Plasmodium falciparum*, *Plasmodium ovale*, and *Plasmodium malariae*), giardiasis (*Giardia lamblia*), balantidiasis (*Balantidium coli*), trypanosomiasis (*Trypanosoma gambiense*), etc.

On the basis of locomotion in the mature stage, protozoa have been classified in to three phylum (Figure 16.49).

16.29.2 Isolation of Protozoa from Soil

16.29.2.1 Materials
Culture *of Aeromonas* sp. (2–3 days old), glass bottle with sterile tap water, inoculation loop, microscope, Petri dishes, pipettes (1 mL capacity), soil sample, and water agar medium (2%).

16.29.2.2 Protocol
1. Sterilize Petri dishes, pipettes, glass bottle, water agar medium 2% agar, etc.
2. Pour 20 mL of sterilized water agar medium into three sterile Petri dishes.

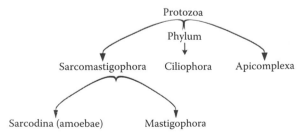

Figure 16.49 Classification of protozoa.

3. Inoculate the water agar plates at three different places with a thick suspension of *Aeromonas* sp. With inoculation loop, spread gently so a circular patch of 2.5 cm diameter may be formed.
4. Add a small amount of soil (50 mg) in the centre of each circle.
5. Pour a few drops of sterile tap water on each circle so that soil may be moistened.
6. Incubate the plates for about 14 days at room temperature.
7. Add more water on the soil to keep it always moist, as and when required.

16.29.2.3 Results and Observations
Observe the plates under the microscope for the presence of protozoa, which move on an agar surface (Figure 16.50).

16.29.3 Staining of Free-Living (Nonpathogenic) Protozoa

16.29.3.1 Introduction
The free-living protozoa are generally found in the stagnant pond water and soil. Unlike parasitic group, they take nutrients by ingesting bacteria, yeast, and algae, for example, *Amoeba proteus, Euglena, Paramecium*, etc.

16.29.3.2 Materials
Stagnant pond water, methyl cellulose, microscope, glass microslides, and pipettes (1 mL capacity).

16.29.3.3 Protocol
1. Take a drop of pond water from the bottom of sample bottle and place in the center of a clean slide.
2. Pour a drop of methylcellulose sample (methylcellulose slow down the movement of the protozoa).
3. Place a clean cover slip in such a way that no bubble forms.
4. Spread the sample by the edge of the cover slip and gently lower the cover slip onto the slide.
5. Examine this slide under the microscope.

16.29.3.4 Results and Observations
Observe the slide under the microscope and note the types of locomotory parts, nucleus, food vacuoles, flagella, cilia, etc., and identify the following standard literature (Figure 16.51).

16.30 Observation of Microscopic Invertebrates from Pond Water

16.30.1 Introduction

A large number of invertebrates, including single-celled organisms, are seen in the aquatic environment. They are all multicellular with organ system. Collectively, these microscopic forms are designated as "invertebrates." There is tissue differentiation in

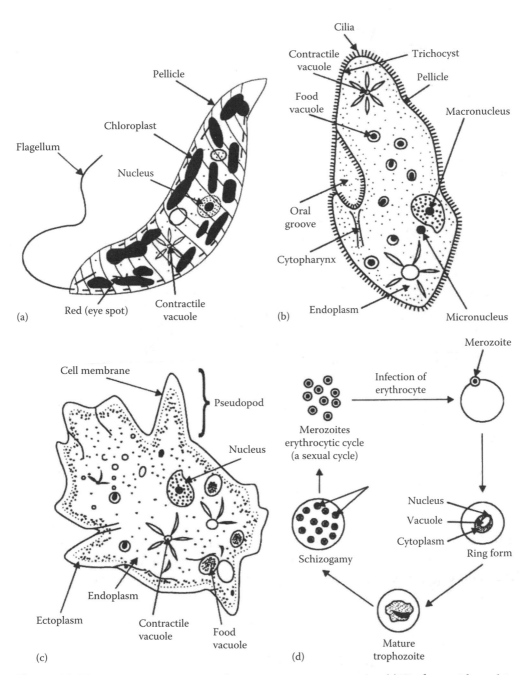

Figure 16.50 Diagrammatic view of some common protozoans. (a) *Euglena* with a whip like flagellum for movement, (b) *Paramecium* with cilia over its surface, (c) *Amoeba* with pseudopods (lob like projections), and (d) *Plasmodium sporozoites* (nonmotile) are released into the blood when an infected mosquito bites a person.

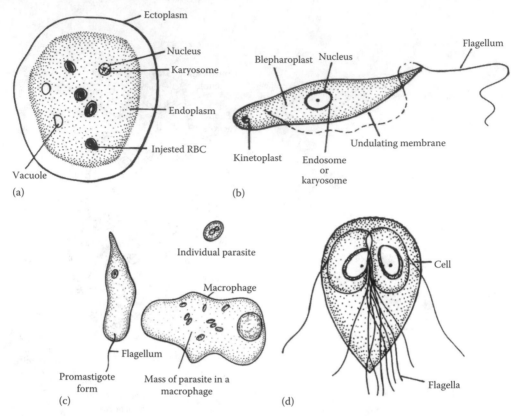

Figure 16.51 Diagrammatic view of some common free living protozoans (a) *Entamoeba histolytica*, (b) *Trypanosoma gambiense*, (c) *Leismania* sp. different stages, and (d) *Giardia intestinalis*.

microscopic invertebrates and, therefore, they cannot be included amongst Protista. Many tissues from multicellular invertebrates do not require supplemental oxygen or complex media, unlike vertebrate tissues. Their central nervous systems are not as well-developed as those of vertebrates; therefore, invertebrates are probably not self-aware. Invertebrates are the eukaryotic, unicellular, and nonphotosynthetic microorganisms devoid of cell wall. They have been divided into seven phyla according to the degree of complexity, the simplest being the Coelenterata (Figure 16.52).

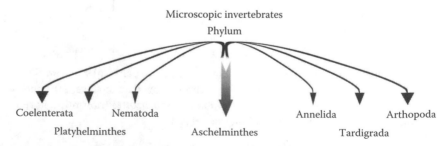

Figure 16.52 Classification of microscopic invertebrates.

16.30.2 Materials

Pond water in a glass bottle, microscope, Petri dishes, and pipettes (1 mL capacity).

16.30.3 Protocol

One or two drops of sample water is removed with the help of Pasteur pipet and placed in the center of a clean glass microscope slide, observe with the aid of a dissecting microscope to locate individual cells. Invertebrates are then examined at low magnifications (40× and 100×). They can be seen under direct observations. Some of the common microscopic invertebrates are illustrated in Figure 16.53.

16.30.4 Results and Observations

Observe the microslides with samples under the microscope for the presence of invertebrates.

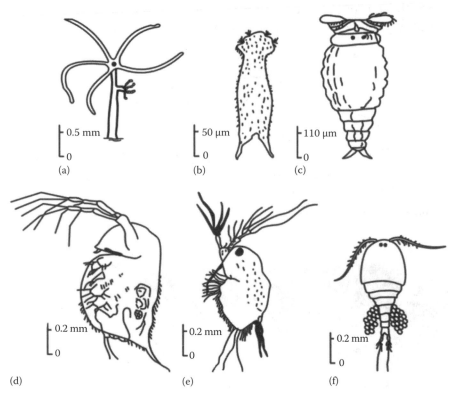

Figure 16.53 Diagrammatic view of some common microscopic invertebrates: (a) *Hydra*, (b) *Lepidodermella*, (c) *Philodina*, (d) *Daphnia*, (e) *Latonopsis*, (f) and *Cyclops*.

16.31 Observation of Algae from Pond Water

16.31.1 Introduction

Algae are all photosynthetic differing from plant in that they lack tissue differentiation. They may be unicellular, colonial form, or filamentous. Undifferentiated algal structures are referred to as a thallus. They are universally present and are classified on the basis of pigments present in them (Figure 16.54).

16.31.2 Protocol

They can be seen in a soil or water sample directly under a compound microscope on a microscopic slide. Some common aquatic members of algae are shown in Figure 16.55.

16.32 Microscopic Examination of Soil Microbes

16.32.1 Observe Soil Bacteria

Because of the transparency of unstained bacteria, it is very difficult to observe them without either staining them or using special techniques of microscopy to view them.

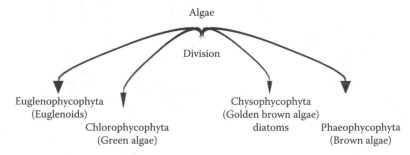

Figure 16.54 Classification of algae.

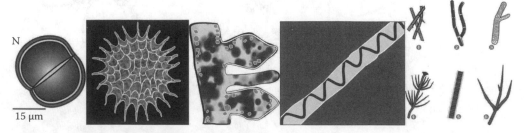

Figure 16.55 Schematic diagram of different forms of algae from unicellular to filamentous and microscopic to macroscopic.

To make the bacteria visible through the microscope, they must be stained with dyes that have an affinity for the bacterial cytoplasm or other cell constituents such as the cell wall. Many commonly used dyes are positively charged (cationic) molecules and hence combine strongly with negatively charged cell constituents such as nucleic acids and acidic polysaccharides. Cationic dyes include methylene blue, crystal violet, and safranin. Other dyes are negatively charged (anionic) molecules and combine with positively charged cell constituents such as many proteins. Anionic dyes include eosin, acid fuchsin, and Congo red. Still another group of dyes is called fat soluble. Dyes in this group combine with fatty materials in the cell and are often utilized to reveal the location of fat droplets or deposits. A common fat soluble dye is Sudan black. Methylene blue is a good, simple stain that works well on many bacterial cells and does not produce such an intense staining that cellular details are obscured. Another attribute of methylene blue is that it does not combine well with most non-cellular materials so that it is especially useful in examining natural samples for the presence of bacteria.

16.32.2 Preparation of a Smear Mount

16.32.2.1 Materials
Soil isolates, microscope slides, inoculating loop, staining solutions—methylene blue, and bibulous paper or Kim wipes.

16.32.2.2 Protocol
1. Place a small drop of clear water on a clean microscope slide using the inoculating loop.
2. Flame the inoculating loop, let it cool, and transfer a small portion of a bacterial colony to the drop of water.
3. Using a circular motion, mix and spread the resulting cell suspension to cover an area about the size of a dime. Flame the loop again immediately.
4. Allow this smear to air dry and then heat fix it by passing it briefly through the flame of the burner several times. Add a few drops of methylene blue to cover the smear. Use sparingly to avoid making a mess with spillage.
5. After the dye has been on the smear for 4 min, gently rinse it in a stream of water in the sink.
6. Blot the slide dry using a pad of bibulous paper. Examine the stained smear at the different magnifications on the microscope. Be sure to use the oil immersion objective to view the preparation.

16.32.3 Preparation of Wet Mounts

16.32.3.1 Protocol
It is not always desirable to view only stained preparations of bacteria since no information is gained as to whether or not the organisms are motile, that is, if they have

flagella (the organelle of locomotion in bacteria: singular = flagellum). To determine whether an organism is motile, it is usually observed in a wet mount.

Place a small drop of water on a microscope slide using a flamed inoculating loop. Flame the loop, let it cool, and add some bacterial from a culture to the drop of water. Mix but do not spread the drop out. Alternatively, if you have a broth culture, simply place several loop-full onto the center of the slide to form a drop. After you have placed the cell suspension on the slide, simply place a cover slip over the drop to obtain as few air bubbles as possible. This is most easily done by bringing one edge of the cover slip into contact with the edge of the drop and then laying the cover slip down on an angle, so the fluid flows evenly across the slide.

16.32.3.2 Gram Stain for Bacteria

There are basically two types of bacteria, Gram-positive (purple color) and Gram-negative (no purple color) when stained with crystal violet and iodine. This is mainly due to the chemical and structural composition of the bacterial cell wall. The following protocol is to be followed to differentiate two different types of bacteria.

16.32.3.3 Protocol

1. Place a loop full of bacteria on a slide (mix very well).
2. Add a drop of water, spread, and allow them to dry.
3. Hold the slide with a clothespin.
4. Quickly pass the slide through a flame to fix most of the bacteria cells.
5. Once the slide cool down, flood with crystal violet and let sit for 30 s.
6. Rinse with water.
7. Flood the slide with iodine and let sit for 1 min.
8. Rinse the slide with 95% alcohol until no purple color drips off the slide.
9. Rinse the slide with water.
10. Flood the slide with safranin for 1 min.
11. Rinse the slide with water.
12. Place a cover slide, observe the slide under a compound microscope and sketch.

16.33 Bacteria Frequently Isolated from Soil

16.33.1 Gram-Negative Chemolithotrophs

16.33.1.1 Nitrobacteraceae

Nitrobacter—Short rods, reproduce by budding, yellow pigment, oxidize nitrite to nitrate and fix CO_2 to fulfill energy and carbon needs, and strict aerobes.

Nitrosomonas—Ellipsoidal to short rods, obligately chemolithotrophic, oxidize ammonia to nitrite and fix CO_2 to fulfill energy and carbon needs, and strictly aerobic.

16.33.1.2 Sulfur Metabolizing

Thiobacillus—Small rod-shaped, energy derived from the oxidation of one or more reduced sulfur compounds, and mostly autotrophic.

16.33.2 Gram-Negative Aerobic Rods and Cocci

16.33.2.1 Pseudomonadaceae
Pseudomonas—Straight or curved rods, motile by polar flagella, chemoorganotrophs, most are strict aerobes (a few species can denitrify), and abundant in rhizosphere.

Xanthomonas—Straight rods, motile by polar flagellum, growth on agar: yellow, chemoorganotrophs, strict aerobes, and mostly plant pathogens.

16.33.2.2 Azotobacteraceae
Azotobacter—Large ovoid to coccoid cells, marked pleomorphism, form thick-walled cysts and capsular slime, motile with peritrichous flagella or nonmotile, Gram-negative or variable, and fix atmospheric nitrogen.

Beijerinckia—Straight to pear-shaped with rounded ends, up to 6 μm with occasional branching, cysts and capsules in some species, produces copious slime in culture, fixes atmospheric nitrogen, and acid tolerant.

16.33.2.3 Rhizobiaceae
Rhizobium—Rods but pleomorphic under adverse conditions, motile by two to six peritrichous flagella, nonsporing, copious extracellular slime in culture, chemoorganotrophs, aerobic to microaerophilic, and able to invade root hairs of leguminous plants.

Agrobacterium—Rods, motile by one to four peritrichous flagella, nonsporing, slime production, chemoorganotrophs, aerobic, and most initiate plant hypertrophies.

16.33.2.4 Uncertain Affiliation
Alcaligenes—Rods to cocci, motile by up to four peritrichous flagella, chemoorganotrophs, strict aerobes, and saprophytes in animals and soil.

16.33.3 Gram-Negative Facultative Anaerobic Rods

Flavobacterium—Coccobacilli to slender rods, motile with peritrichous flagella or nonmotile, pigmented in culture, chemoorganotrophs, and fastidious as to requirement.

16.33.4 Gram-Negative Cocci and Coccobacilli

Acinetobacter—Short rods to cocci, large irregular cells and filaments in culture, no spores or flagella, chemoorganotrophic, strict aerobes, and resistant to penicillin.

16.33.5 Gram-Positive Cocci

Micrococci—Spherical, nonmotile, chemoorganotrophs, and aerobic.

Staphylococcus—Spherical, chemoorganotrophs, metabolism respiratory or fermentative, produce extracellular enzymes and toxins, and facultative anaerobes.

Streptococcus—Spherical to ovoid, chemoorganotrophs, metabolism fermentative, and facultative anaerobes.

Sarcina—Nearly spherical, nonmotile, chemoorganotrophs, strictly fermentative metabolism, and strict anaerobes.

16.33.6 Endospore-Forming Rods

Bacillus—Rod-shaped, motile, flagella lateral, heat-resistant endospore, chemoorganotrophs, strict aerobes to facultative anaerobes, and Gram-positive.

Clostridium—Rods, peritrichous flagella, form spherical to ovoid spores, Gram-positive, chemoorganotrophs, and most strictly anaerobic.

16.33.7 Budding and/or Appendaged

Hyphomicrobium—Rod-shaped with pointed ends, produce mono- or bipolar filamentous outgrowths, Gram stain unknown, multiply by budding at tip, growth in liquid culture on surface, chemoorganotrophic, aerobic, and temp 15°C–30°C.

Pedomicrobium—Spherical to rod-shaped, multiply by budding at tip of cellular extension producing uniflagellate swarmers, Gram-negative, microaerophilic to aerobic, chemoheterotrophic, and mesophilic.

Caulobacter—Rod-shaped to vibrioid, typically with stalk extending from one pole, cells may adhere to each other in rosettes, cell division by asymmetrical fission, Gram-negative, chemoorganotrophic, and aerobic.

Metallogenium—Coccoid, attached to surfaces, may form flexible filaments, multiply by budding, manganese and iron oxides deposited on filaments, and heterotrophic.

16.33.8 Coryneform Group

Corynebacterium—Irregular shape with club-like swellings, nonmotile, Gram-positive, chemoorganotrophs, aerobic, and facultatively anaerobic.

Arthrobacter—Old culture coccoid, on fresh media swellings from coccoid cells giving rise to irregular rods, Gram-positive, chemoorganotrophs, and strict aerobes.

Cellulomonas—Irregular rods, motile by one or more flagella, Gram-positive, chemoorganotrophs, decompose cellulose, and aerobic.

16.33.9 Mycobacteriaceae

Mycobacterium—Curved to straight rods, filamentous growth may occur. Acid-fast reaction, nonmotile, lipid content in wall high, and aerobic.

16.34 Fungi Frequently Isolated from Soil

Examine selected colonies using the dissecting microscopes and low power objective of the compound microscope. Look especially for fruiting structures (e.g., spores). Aseptically remove small portions of fungal and an actinomycete culture from dilution plates and make "squash" mounts in various stains. Many soil fungi belong to the order "Hyphomycetes" of the Fungi imperfecti—that is, they produce conidia (asexual spores) on conidiophores. These structures are very fragile and require special techniques for observation. The slide culture mount is an excellent technique to observe these fungi. Attempt to identify to genus using the key in Barron (1968). *Classification of Fungi* (Adapted from Cavalier-Smith 1998; Alexander, 1965; Alexopoulos and Mims 1983; Alexopoulos et al., 1996).

16.34.1 Kingdom Protista (Protozoans)

Phagotrophic, organisms with somatic structures devoid of cell walls, we are including them here because mycologists traditionally study them. These organisms are the cellular slime molds and the true slime molds. The cellular slime molds have a reproductive stalk that consist of walled cells and is simple. The most prevalent form of the organisms is the myxamoeba that feeds by engulfing bacteria (Alexopoulos and Mims 1983). The true slime molds have the plasmodial somatic phase but produce spores with definite walls from elaborate sporophores.

16.34.2 Kingdom Stramenophila

Eukaryotic organisms with either tubular ciliary mastigonemes or with chloroplast bounded by an envelope of two membranes and surrounded by two chloroplast endoplasmic reticulum membranes and a periplastidal space containing the periplastidal reticulum; mitochondrial cristae are rounded tubules or flattened finger-like projections. Pseudofungal organisms typically produce flagellate cells.

16.34.3 Phylum Heterokonta

Anterior cilium with tubular retronemes; posterior cilium smooth or absent.

16.34.4 Class Hyphochytridiomycetes

A very small group of aquatic fungi with motile anteriorly uniflagellate cells each with a tinsel flagellum.

16.34.5 Class Oomycetes

Soma varied but usually filamentous, consisting of a coenocytic, walled mycelium; hyphal wall containing glucans and cellulose, with chitin also present in one order (Leptomitales); zoospores each bearing one whiplash and one tinsel flagellum; and sexual reproduction, oogamous resulting in the formation of oospores. Root parasitic species, *Pythium* and *Phytophthora*, belong to this class.

16.34.6 Kingdom Eumycota (Fungi)

Eukaryotic organisms without chloroplasts or phagocytosis, but with saprobic or parasitic nutrition and typically with chitinous walls and plate-like mitochondrial cristae; develop from spores; cilium, when present, single posterior without rigid, tubular mastigonemes. The kingdom has four phyla: Chytridiomycota, Zygomycota, Ascomycota, and Basidiomycota (Figure 16.56).

16.34.7 Phylum Chytridiomycota

Zoospores with single posterior whiplash cilium; perfect state spores are oospores or zygospores; and have a coenocytic thallus of chitinous walls. These fungi are

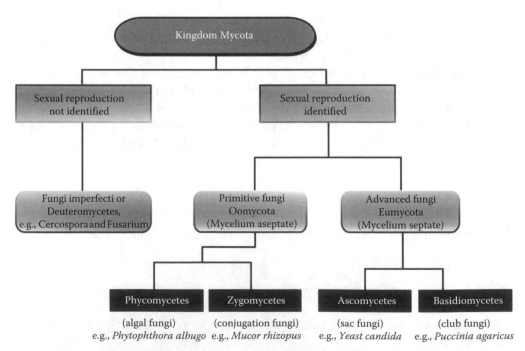

Figure 16.56 Illustrating broad classification of fungi.

prevalent in aquatic habitats but many inhabit the soil. Some parasitize and destroy algae and thus form a link in the food chain.

16.34.8 Phylum Zygomycota

Sexual reproductions are by the fusion of usually equal gametangia resulting in the formation of a zygospore. Asexual reproduction is by the aplanospores, yeast cells, arthrospores, or chlamydospores. Motile spores are absent. The phylum consists of two classes, the Trichomycetes (arthropod parasites) and Zygomycetes.

16.34.9 Class Zygomycetes

Mainly terrestrial saprobes or parasites of plants or mammals, or predators of microscopic animals; if parasitic, mycelium immersed in host tissue; asexual reproduction by aplanospores, borne singly or in groups within sporangial sacs; and sexual reproduction by fusion of usually equal gametangia resulting in the formation of a zygosporangium containing a zygospore.

16.34.10 Phylum Ascomycota

Unicellular or more generally with septate mycelium; sexual reproduction by formation of meiospores (ascospores) in sac-like cells (asci) by free cell formation. Three Subphyla based on ascus formation.

16.34.11 Subphylum Euascomycotina

Ascomata and ascogenous hyphae present; thallus mycelial. These classes include most fungi imperfecti and lichen forming groups. The teleomorphs of Aspergillus, Penicillium, Thielaviopsis, and Fusarium belong to orders in this subphylum.

16.34.12 Subphylum Laboulbeniomycotina

Thallus reduced; ascoma a perithecium. These are exoparasites of athropods and can survive in the soil as resting structures.

16.34.13 Subphylum Saccharomycotina

These fungi lack ascogenous hyphae and have a yeast-like thallus or mycelial. They are the budding yeast and their filamentous relatives.

16.34.14 Phylum Basidiomycotina

They are saprobic, symbiotic, or parasitic fungi. Morphologically are unicellular (yeastlike) or more typically, with a septate mycelium with a vegetative heterokaryophase, sexual reproduction by producing meiospores (basidiospores) on the surface of various types of basidia.

16.34.15 Classes Urediniomycetes and Ustomycetes

Basidiocarps are lacking and resting spore germination results in formation of basidiospores. These fungi cause rust and smuts of plants and their resting spores may survive in the soil for decades.

16.34.16 Class Gelimycetes

Basidia transversely or longitudinally septate (phragmobasidia) produced on various types of sporophores or directly on the mycelium. These are mainly decomposing fungi found on litter but *Thanatephorus cucumeris* (teleomorph of *Rhizoctonia solani*) also belongs here.

16.34.17 Class Homobasidiomycetes

Basidia nonseptate (holobasidia), produced on persistent hymenia on various types of open sporophores or, rarely, directly on the mycelium; or inside closed sporophores opening, if at all, after the spores are mature. These are the more commonly seen mushrooms and wood rot fungi of which many species form ectomycorrhizal associations with trees.

16.34.18 Form Phylum Deuteromycotina

Generally are called the imperfecti fungi. Their main characteristic are teleomorph absent, saprobic, symbiotic, parasitic, or predatory fungi, unicellular or, more typically, with a septate mycelium, usually producing conidia from various types of conidiogenous cells. Sexual reproduction is unknown but a parasexual cycle may operate. A few species produce no spores of any kind.

16.34.19 Form Class Blastomycetes

Soma consisting of yeast cells with or without pseudomycelium; true mycelium, if present, not well developed.

16.34.20 Form Class Coelomycetes

True mycelium present; conidia produced in pycnidia or acervuli.

16.34.21 Form Class Hyphomycetes

True mycelium present; conidia produced on special conidiogenous hyphae (conidiophores) arising in various ways other than in pycnidia or acervuli. A few species do not produce spores of any kind.

16.34.22 Series Aleuriospore

Spores develop terminally as blown-out ends of the sporogenous cells and usually thick walled and pigmented.

16.34.23 Series Annellosporae

First spore produced terminally with each new spore blown out through the scar left by the previous. A succession of proliferations is accompanied by increased length of sporogenous cells.

16.34.24 Series Arthrosporae

Conidia produced after separation and breaking up of sporogenous hyphae.

16.34.25 Series Blastosporae

Develop in acropetal succession as blown out ends of conidiophore.

16.34.26 Series Botryoblastosporae

Conidia produced on well differentiated, swollen sporogenous cells.

16.34.27 Series Meristem Blastosporae

Conidia borne singly at apex in irregular whorls, which elongate from the base.

16.34.28 Series Phialosporae

Sporogenous cells stay constant in length and conidia are abstricted successively in basipetal succession from an opening.

16.A Appendix

16.A.1 Common Culture Media for Soil Microorganisms

There are a large number of natural, semisynthetic and synthetic media that can be used to grow pathogenic and nonpathogenic microorganisms.

1. *Natural media*: A natural media is one that is composed of complex natural materials of unknown composition. Among natural substances are plant parts, malt, yeast, peptone, fruits, and vegetables. The following are a few natural media used in the isolation of fungi:
   a. *Cooked vegetable agar*: Broth or decoction of the desired vegetative plant parts can be prepared by 10%–20% of the tissue in water, steamed for 30 min, and the contents mashed and squeezed through muslin cloth. To this 1.5%–2% agar-agar may be added for preparing the slants. pH of the medium may be adjusted to 5.5 or to desired level before autoclaving.
   b. *Lima bean agar*: Used to cultivate Phytophthora species. Blend 50 g lima beans (*Phaseolus lunatus*), previously soaked in water for 10 h with 500 mL of water and raised to 1 L. Add 1%–2% agar-agar to prepare slants.
   c. *Oatmeal agar*: 60 g of rolled oats are blended in commercial blender with 500–600 mL water and then heat to 55°C–60°C for 1 h and adjust the volume to 1 L with water. Add 1.5%–2% agar-agar to prepare slants. Media is used for general cultivation of fungi.
2. *Semisynthetic media*: A semisynthetic medium is made up of natural substances of unknown composition and by adding some chemical compounds of known composition.
   a. *Potato-dextrose agar (PDA)*: Boil 200 g of peeled and sliced potato in 1 L of water until the potatoes are soft. Strain through cheesecloth and adjust the filtrate to 1 L. Add 20 g dextrose and 15–20 g agar-agar for preparing slants and Petri plates and autoclave.
   b. *Malt extract peptone dextrose agar*: 20 g malt extract, 20 g dextrose, 1 g peptone and agar-agar 15–20 g (for slants and plates) and dH_2O 1 L and autoclave.
   c. *Martin_rose Bengal streptomycin agar*: Dextrose 10 g, peptone 5 g, KH_2PO_4 1 g, $MgSO_4 \cdot 7H_2O$ 0.5 g, Rose bengal 0.05 g, streptomycin 0.03 g, agar-agar 15–20 g, and dH_2O to 1 L. Dissolve 1 g of streptomycin sulfate in 100 mL sterile dH_2O; after opening the vial aseptically, add 0.3 mL of streptomycin in solution to each 100 mL of the basal medium after it is cooled. This is a general purpose medium to isolate soil fungi.
   d. *Nutrient agar*: Beef extract 3 g, peptone 5 g, NaCl 8 g, and dH_2O to 1 L, and autoclave. Add 15–20 g agar-agar for preparing slants or Petri plates. This is general purpose medium for isolation and culturing of bacteria from soil.

e. *Soil extract agar*: Take 500 g fertile soil in 500 mL dH_2O, autoclave, cool, and filter the water extract. Add 0.2 g K_2HPO_4 and 15–20 g agar-agar (for slants and Petri plates) and adjust the volume to 1 L and autoclave. The medium is commonly used to enumerate bacteria and fungi from soil.

f. *Yeast extract medium*: Yeast extract 1 g, K_2HPO_4 10 g, $MgSO_4 \cdot 7H_2O$ 1 g, Tween-80 0.1%, and dH_2O to 1 L. *Erwinia amylovora* grows well in this medium.

3. *Synthetic media*: Synthetic medium is of known composition and concentration. Chemicals used for preparing synthetic medium should be of pure quality.

   a. *Brown's medium*: Glucose 2 g, K_2HPO_4 1.25 g, asparagine 2 g, $MgSO_4 \cdot 7H_2O$ 0.75 g, agar-agar 15–20 g (for slants and Petri plates) and dH_2O to 1 L.

   b. *Czapek Dox agar*: $NaNO_3$ 2 g, K_2HPO_4 1 g, $MgSO_4 \cdot 7H_2O$ 0.5 g, KCl 0.5 g, FeSO4 0.01 g, sucrose 30 g, agar-agar 15–20 g (for slants and Petri plates) and dH_2O to 1 L.

      If glass dH_2O is used, add 1 mL of 1% $ZnSO_4$ and 0.5% $CuSO_4$. Heat all chemical solution with sucrose in a water bath for 15 min. After cooling, add sucrose and agar-agar and autoclave.

   c. *Galactose nitrate agar*: Medium is commonly used to isolate *Fusarium oxysporum* from soil. Galactose 10 g, $NaNO_3$ 2 g, KH_2PO_4 1 g, $MgSO4 \cdot 7H_2O$ 0.5 g, $K_2S_2O_5$ 0.3 g, agar-agar 15–20 g (for slants and Petri plates), and dH_2O to 1 L and autoclave.

4. *Some special media for cultivation*

   a. Media for actinomycetes

      i. *Basal mineral salts agar*: Glucose 10 g, $(NH_4)2SO_4$ 2.64 g, KH_2PO_4 anhydrous 2.38 g, $K_2HPO_3 \cdot 3H_2O$ 5.65 g, $MgSO_4 \cdot 7H_2O$ 1 g, dH_2O 1000 mL, agar-agar 20 g. Adjust pH to 6.8–7.0 with 1 N NaOH or 1 N HCl. Media can be used for culture of *Streptomyces* with addition of sterile carbon source at a concentration of 1%.

      ii. *Oatmeal agar (Shirling and Gottlieb No.3)*: Oatmeal 20 g, agar-agar 18 g, dH_2O 1000 mL; ($FeSO_4 \cdot 7H_2O$ 0.1 g, $MnCl_2 \cdot 4H_2O$ 0.1 g, $ZnSO_4 \cdot 7H_2O$ 0.1 g. Dissolve in 100 mL of dH_2O and add 1 mL to oatmeal agar solution) steam oatmeal in a liter of water for 10 min, filter through cheesecloth, add trace salts, and make up volume. Adjust to pH 7.2 with NaOH. Add agar and autoclave for 15 min. This medium is good for culture of *Streptomyces*.

   b. Media for soil bacteria

      i. *Asparagine lactose medium*: This medium is used to isolate *Rhizobium* sp. lactose 10 g, asparagine 0.5 g, sodium taurocholate 5 g, K_2HPO_4 1 g, $MgSO_4 \cdot 7H_2O$ 0.2 g, neutral red 3.5 mL (2/40 mL ethyl alcohol), agar-agar 20 g, and dH_2O 1000 mL, pH 7.5. This medium is suitable for the growth of all the Rhizobial species and especially for isolating Rhizobium from the nodules of ground nut. This medium can be used for distinguishing the *Agrobacterium* and *Rhizobium*. *Agrobacterium tumefaciens* shows quick growth within 24 h with pink color colony on Congo red (10 mL of l/4000 aqueous solution) incorporated in the medium, while *Rhizobium* strains grown late after 72 h showing no absorption of Congo red.

      ii. *Ashby's mannitol agar medium*: Used to cultivate *Azotobacter* sp. potassium acid phosphate 1 g, $MgSO_4 \cdot 7H_2O$ 0.2 g, ammonium sulfate 0.2 g, NaCl 0.2 g, mannitol 10 g, calcium sulfate 0.1 g, calcium carbonate 5 g, agar-agar 20 g, and dH_2O 1000 mL. Dissolve phosphate in 500 mL and make alkaline to phenolphthalein. Melt agar and other ingredients in the remaining 500 mL, mix both the lots and sterilize by autoclaving for 20 min.

c. Media for soil fungi

i. *V-8 juice agar*: Used to support growth and induce sporulation of many fungi. It is preferred over PDA for isolation of sporulating fungi from soil. V-8 juice 200 mL, $CaCO_3$ 3 g, agar-agar 15–20 g, and dH_2O 1000 mL, pH 7–7.5 with variation in amount of juice or $CaCO_3$.

ii. *Cassava-dextrose agar*: Used in place of PDA for routine work of fungi: Peel cassava (*Manihot utilissima*) tubers and cut into chips. Dry overnight at 55°C and grid to 30 mesh. Soak 135 g of the powder in 500 mL of water for 15 min at 60°C. Filter through cheesecloth. Add 20 g glucose and 12 g agar-agar. Adjust the volume to 1 L and autoclave.

Apart from the aforementioned media for fungi, there are some basic points that are to be kept in mind while preparing media:

1. Fungi usually grow best in a carbohydrate rich medium.
2. Fungi usually prefer an acidic zone, pH 5–6.
3. Agar is slow to dissolve thoroughly. It is advisable to dissolve the agar in half the water and nutrients in the other half and then mix. Agar does not solidify satisfactorily in very acid or alkaline solutions.
4. Peptone may generally be omitted from fungus culture.
5. Tap water is often preferable to dH_2O as it contains useful trace elements.

16.A.2 Some Common Media for Cyanobacteria

1. *Modified Chu-10 medium* (Gerloff et al. 1950) *g/L*: $Ca(NO_3)_2 \cdot 4H_2O$ 0.04, $K_2HPO_4 \cdot 3H_2O$ 0.01, $MgSO_4 \cdot 7H_2O$ 0.025, $NaCO_3$ 0.02, $NaSiO_3 \cdot 9H_2O$ 0.025, ferric citrate 0.003, citric acid 0.003, and dH_2O to 1 L. To this medium, add 1 mL/L Fogg's micronutrient solution containing $MnCl_2 \cdot 4H_2O$ 0.010, $Na_2MoO_4 \cdot 2H_2O$ 0.010, H_3BO_3 0.010, $CuSO_4 \cdot 5H_2O$ 0.001, and $ZnSO_4 \cdot 7H2O$ 0.001. pH of this medium should be adjusted to 7.5 before autoclaving.

2. *Allen and Arnon's medium (Allen and Arnon 1955) g/L*: KNO_3 2.020, $K_2HPO_4 \cdot 3H_2O$ 0.456, $MgSO_4 \cdot 7H_2O$ 0.246, NaCl 0.232, and $CaCl_2 \cdot 2H_2O$ 0.074. To this medium 1 mL/L Fogg's micronutrient solution and 4 mg/L of Fe (as EDTA complex) should be added. pH of this medium should be adjusted to 7.5 before autoclaving.

3. *BG-11 medium (Rippka et al. 1979) g/L*: $NaNO_3$ 0.0015, $K_2HPO_4 \cdot 3H_2O$ 0.04, $MgSO_4 \cdot 7H_2O$ 0.075, $CaCl_2 \cdot 2H_2O$ 0.036, citric acid 0.006, ferric ammonium citrate 0.006, EDTA (Na_2Mg salt) 0.001, Na_2CO_3 0.02, and dH_2O to 1 L. Trace metal solution 1 mL. For preparing N-free medium $NaNO_3$ may be omitted. Trace element solution g/L: H_3BO_3 2.86, $MnCl_2 \cdot 4H_2O$ 1.81, $ZnSO_4 \cdot 7H_2O$ 0.22, $Na_2MoO_4 \cdot 2H_2O$ 2.86, $CuSO_4 \cdot 5H_2O$ 1.81, and $Co (NO_3)_2 \cdot 6H_2O$ 0.494.
   After autoclaving and cooling, the pH of the medium is 7.1.

4. *KM modified after Kratz and Myers (1955) g/L*: $MgSO_4 \cdot 7H_2O$ 0.25, $K_2HPO_4 \cdot 3H_2O$ 1, $Ca(NO_3)_2 \cdot 4H_2O$ 0.025, $NaNO_3$ 1, Na_2 EDTA 0.031, Fe $(SO_4) \cdot 6H_2O$ 0.004, and NH_4Cl 0.1. To this medium, add 0.6 mL/L trace element solution, which contains $ZnSO_4 \cdot 7H_2O$ 8.82, $MnCl_2 \cdot 4H_2O$ 1.44, MoO_3 0.71, $CuSO_4 \cdot 5H_2O$ 1.57, and $Co(NO_3)_2 \cdot 6H_2O$ 0.49.

This medium is employed for unicellular coccoid cyanobacteria especially for the quantitative surface plating of *Aspergillus nidulans* to accurate estimation of viable counts. After autoclaving and cooling, the pH of the medium is 7.1.

Suggested Reading

Alexander, M. 1961. *Introduction to Soil Microbiology*. p. 314. New York: John Wiley & Sons.

Alexander, M. 1965. Most-probable-number method for microbial population. In *Methods of Soil Analysis*, ed. C. A. Black, pp. 1467–1472. Part 2. New York: Academic Press.

Alexopoulos, C. J., C. W. Mims, and M. Blackwell. 1996. *Introductory Mycology*. p. 868. 4th edn. New York: John Wiley and Sons.

Allen, M. B and D. I. Arnon. 1955. Studies on nitrogen-fixing blue-green algae. I. Growth and nitrogen fixation by *Anabaena cylindrica* Lemm. *Plant Physiol* 30: 366–372.

Bakker, A.W. and B. Schippers. 1987. Microbial cyanide production in the rhizosphere in relation to potato yield reduction and *Pseudomonas* spp. mediated plant growth simulation. *Soil Biol. Biochem* 19: 451–457.

Barron, G. L. 1968. *The Genera of Hyphomycetes from Soil*. p. 364. Baltimore: Williams & Wilkins.

Bartha, R. and D. Pramer. 1965. Features of a flask and method for measuring the persistence and biological effects of pesticides in soil. *Soil Sci* 100: 68–70.

Benson, H. J. 1990. *Microbiological Applications—A Laboratory Manual in General Microbiology*. p. 459. Dubuque, IA: Win. C. Brown Publishers.

Bric, J. M., R. M. Bostock, and S. E. Silversone. 1991. Rapid in situ assay for indole acetic acid production by bacteria immobilized on a nitrocellulose membrane. *Appl Environ Microbiol* 57: 535–538.

Brockwell, J. 1963. Accuracy of plant infection technique for counting populations of *Rhizobium trifolii*. *Appl Microbiol* 11: 377–383.

Bushby, H. V. A. 1981. Quantitative estimation of rhizobia in non-sterile soil using antibiotics and fungicides. *Soil Biol Biochem* 13: 237–239.

Cappuccino, J. G. and N. Sherman, N. 1991. *Microbiology: A Laboratory Manual*. p. 462. Amsterdam, the Netherlands: The Benjamin/Climmings Publishing Company.

Carr, N. G. and B. A. Whitton. 1982. *The Biology of Cyanobacteria*. p. 688. Berkeley, CA: University of California Press.

Casida, L. E. 1968. Methods for the isolation and estimation of activity of soil bacteria. In *The Ecology of Soil Bacteria*, eds. T. R. G. Gray and D. Parkinson. Liverpool, U.K.: Liverpool University Press.

Casida, L. E., Jr., D. A. Klein, and T. Sautoro. 1964. Soil dehydrogenase activity. *Soil Sci* 98: 371–376.

Cavalier-Smith, T. 1998. A revised six-kingdom system of life. *Biol Rev* 73: 203–266.

Cochran, W. G. 1950. Estimation of bacterial densities by means of the most probable number. *Biometrics* 6: 105–116.

de Lorenzo, V., M. Herrero, U. Jakubzik et al. 1960. Mini-Tn5 tranposon derivatives for insertion mutagenesis, promoter probing, and chromosomal insertion of cloned DNA in gram negative Eubactria. *J Bacteriol.* 172: 285–294.

Dick, W. A. and M. A. Tabatabai. 1992. Significance and uses of soil enzymes. In *Handbook of Soil Sciences Properties and Processes*, eds. P. M. Huang, Y. Li, and M. E. Sumner, 2nd edn., pp. 95–127. New York: CRC Press.

Domsch, K. H., W. Gams, and T. H. Anderson. 1980. *Compendium of Soil Fungi*. New York: Academic Press.

Drahos, D. J., B. C. Hemming, and S. McPherson. 1986. Tracking recombinant organisms in the environment: β-Galactosidase as a selectable, non-antibiotic marker for fluorescent pseudomonads. *Bio/Technology* 4: 439–443.

Dye, D. W. 1962. The inadequacy of the usual determinative tests for identification of *Xanthomonas* spp. *N Z J Sci* 5: 393–416.

Elliot, L. F. and J. M. Lynch. 1984. *Pseudomonas* as a factor in the growth of winter wheat (*Triticum aestivum L.*) *Soil Biol. Biochem* 16: 69–71.

Fred, E. B., I. L. Baldwin, and F. Mc Coy. 1932. *Root Nodule Bacteria and Leguminous Plants*. Madison, WI: University of Wisconsin Press.

Gerloff, G. C., G. P. Fitzgerald, and F. Skoog. 1950. The mineral nutrition of *Coccochloris peniocystis*. *Amer J Bot* 37: 835–840.

Gibson A. H. 1963. Physical environment and symbiotic nitrogen fixation. I. The effect of root temperature on recently nodulated *Trifolium subterraneum L.* plants. *Aust J Biol Sci* 16: 28–42.

Glick, B. R. 1995. The enhancement of plant growth of free living bacteria. *Can J Microbiol* 41: 109–117.

Gordon A. S. and R. P. Weber. 1951. Colorimetric estimation of indole acetic acid. *Plant Physiol* 26: 192–195.

Gould, W. D., C. Hagedorn, T. R. Bardinelli, and R. M. Zablotowiez. 1985. New selective medium for enumeration and recovery for fluorescent pseudomonads from various habitats. *Apl Environ Microbiol* 49: 28–32.

Hagedorn, C. and J. G. Holt. 1975. Ecology of soil arthrobactrers in Clarion-Websters toposequence of Iowa. *Appl. Microbiol* 29: 211–218.

Halvorson, H. O. and N. R. Ziegler. 1933. Application of statistics to problems in bacteriology: I. A means of determining bacterial population by the dilution method. *J Bacteriol* 25: 101–121.

Hardy R. W. F., R. C. Burns, and G. W. Parshall. 1971. The biochemistry of N_2 fixation. In *Bioinorganic Chemistry*. Advances in Chemistry, pp. 219–247. Series 100. Washington, D.C.: American Chemical Society.

Hardy R. W. F., R. D. Holsten, E. K. Jackson et al. 1968. The acetylene-ethylene assay for N_2 fixation: Laboratory and field evaluation. *Plant Physiol* 43: 1185–1207.

Hartel, P. G., J. J. Fuhrmann, Jr. W. F. Johnson et al. 1994. Survival of a *lacZY* containing *Pseudomonas putida* strain under stressful abiotic soil conditions. *Soil Sci Soc Am J* 58: 770–776.

Hattemer-Frey, H. A., E. J. Brandt, and C. C. Travis. 1990. Small-scale field test of the genetically engineered *lacZY* marker. *Reg Toxicol Pharmacol* 11: 253–261.

Hendricks, C. W. and N. Pascoe. 1988. Soil microbial biomass estimates using 2450 MHz microwave irradiation. *Plant Soil* 110: 39–47.

Herrero, M., V. de Lorenzo, and K. T. Timmis. 1990. Transposon vectors containing non-antibiotic resistance selection markers for cloning and stable chromosome insertion of foreign genes in gram-negative bacteria. *J Bacteriol* 172: 6557–6567.

Hleyn, J., M. Bicknell, and M. Gilstrap. 1995. *Microbiology Experiments: A Health Science Perspective*. p. 354. London,U.K: Wm. C. Brown Publishers.

Horwath, W. R. and E. A. Paul. 1994. Microbial biomass. In *Methods of Soil Analysis*, eds. R. W. Weaver, S. Angle, P. Bottomley et al., pp. 753–773. Part 2. Madison, WI: SSSA (Soil Science Society of America Series, 5).

Jackson, M. L. 1967. *Soil Chemical Analysis*. New Delhi, India: Prentice Hall of India Pvt Ltd.

Jensen, H. L. 1942. Nitrogen fixation in leguminous plants. I. General characters of root-nodule bacteria isolated from species of *Medicago* and *Trifolium* in Australia. *Proc Linn Soc NSW* 66: 98–108.

Jensen, H. L. 1954. The Azotobacteriaceae. *Bact Rev* 18: 195–214.

Johnson, L. F. and E. A. Curl. 1972. Methods for research on the ecology of soil-borne plant pathogens. 247. Minneapolis, MN: Burgess Publishing Co.

Johnston, A. W. B. 1975. Identification of *Rhizobium* strains in pea root nodules using genetic markers. *J Gen Microbiol* 87: 343–350.

Josey, D. P., J. L. Beynon, A. W. B. Johnston et al. 1979. Strain identification in *Rhizobium* using Intrinsic antibiotic resistance. *J Appl Bacteriol* 46: 333–350.

Kapulnik, Y. 1996. Plant growth promoting rhizosphere bacteria. In: *Plant Roots: The Hidden Half*, eds. Y. Waisel, A. Eshel, and U. Kafkafi, pp. 769–781. New York: Marcel Dekker.

King, E. O. K., M. K.Ward, and D. E. Rancy. 1954. Two simple media for the demonstration of pyocyanin and fluorescein. *J Lab Clin Med* 44: 301–307.

Kirk, P. M., P. F. Cannon, D. W. Minter, and J. A. Stalpers. 2008. *Ainsworth and Bisby's Dictionary of Fungi*. Oxon, U.K.: CAB International.

Klein, D. A., T. C. Loh, and R. L. Goulding. 1971. A rapid procedure to evaluate dehydrogenase activity of soils low in organic matter. *Soil Biol Biochem* 3: 385–387.

Kloepper, J. W., J. Leong, M. Teintze et al. 1980. Enhanced plant growth by siderophores produced by plant growth promoting rhizobacteria. *Nature* 286: 885–886.

Kloepper, J. W. 1993. Plant growth-promoting rhizobacteria as biological control agents. In *Soil Microbial Ecology: Applications in Agricultural and Environmental Management*. ed. F. B. Metting, Jr., pp. 255–274. New York: Marcel Dekker, Inc.

Kormanik, P. P. and A. C. McGraw. 1982. Quantification of vesicular-arbuscular Mycorrhizae in plant roots. In *Methods and Principles of Mycorrhizal Research*, ed. N. C. Schenck, pp. 37–36. St. Paul, MN: The American Phytopathological Society.

Laguerre, G., M. R. Allard, F. Revoy et al. 1994. Rapid identification of rhizobia by restriction fragment length polymorphism analysis of PCR amplified 16S rRNA genes. *Appl Environ Microbiol* 60: 56–63.

Mishustin, E. N. and V. K. Shilinkova. 1969. The biological fixation of atmospheric nitrogen by free-living bacteria. In *Review of Research, Soil Biology*. Paris, France UNESCO, pp. 72–124.

Mullis, K. B. 1990. The unusual origin of the polymerase chain reaction. *Sci Am* 262: 56–61.

Nautiyal, C. S. (1999). An efficient microbiological growth medium for screening phosphate solubilizing microorganisms. *FEMS Microbiol. Lett.* 170: 265–270.

Norris, J. R. and H. M. Chapman. 1968. Classification of *Azotobacter*. In *Identification Methods for Microbiologists*, eds. B. M. Gibbs and D. A. Shapton, pp. 19–27. New York: Academic Press.

Okon, Y., S. L. Albrecht, and R. H. Burris. 1977. Methods for growing *Spirillum lipoferum* and for counting it in pure culture and in association with plants. *Appl Environ Microbiol* 33: 85–87.

Pagan, J. D., J. J. Child, W. R. Scowcroft et al. 1975. Nitrogen fixation by *Rhizobium* cultured on a defined medium. *Nature* 256: 406–407.

Phillips, J. M. and D. S. Hayman. 1970. Improved procedures for clearing roots and staining parasitic and vesicular-arbuscular mycorrhizal fungi for rapid assessment of infection. *Trans Brit Mycol Soc* 55: 158–161.

Pikovskaya, R. E. 1948. Mobilisation of phosphorus in soil in connection with vital activity of some microbial species. *Mikrobiologiya* 17: 362–370.

Powlson, D. S. 1976. The effect of biological treatment on metabolism in soil. V.A method of measuring soil biomass. *Soil Biol Biochem* 8: 209–213.

Riddell, R. S. 1950. Permanent stained mycological preparations obtained by slide culture. *Mycologia* 42: 265–270.

Rodriguez-Caceres, E. A. 1982. Improved medium for isolation of *Azospirillum sp. Appl Environ Microbiol* 44: 990–991.

Schollhorn, R. and Burris, R. H. 1967. Acetylene as a competitive inhibition of N_2 fixation. *Proc Natl Acad Sci USA* 58: 213–216.

Schwyn, B. and J. B. Neilands. 1987. Universal chemical assay of the detection and determination of siderophore. *Anal Biochem* 60: 47–56.

Seeley, H. W., Jr. and P. J. Denmark. 1975. *Microbes in Action: A Laboratory Manual of Microbiology*, p. 361. Mumbai, India: D.B. Taraporevala Sons & Co. Pvt. Ltd.

Shende, S. T., R. G. Apte, and T. Singh. 1977. Influence of *Azotobacter* on germinatin of rice and cotton seeds. *Curr Sci* 46: 675–676.

Smith, S. E. and D. J. Read. 1997. Mycorrhizal symbiosis. In *Soil Science Society of America* (SSSA book series) 5, 2nd edn., p. 605. Madison, WI: Academic Press.

Stewart W. D. P., G. P. Fitzgerald, and R. H. Burris. 1967. In situ studies on N_2 fixation using the acetylene reduction technique. *Proc Natl Acad Sci USA* 58: 2071–2078.

Sylvia, D. M. 1994. Vesicular-arbuscular mycorrhizal (VAM) fungi. In *Methods of Soil Analysis,* Vol. 5, pp. 351–378. Madison, WI.

Tabatabai, M. A. 1994. Soil enzymes. In: *Microbiological and Biochemical Properties*, eds. R. W. Weaver, J. S. Angle, and P. S. Bottomley. Part 2, pp. 775–833. Madison, WI: Soil Science Society of America (SSSA book series).

Trouvelot, A., J. L. Kough, and V. Gianinazzi-Pearson. 1986. Estimation of mycorrhizal colonization according to arbuscular mycorrhizal fungi of assessment methods. In: *Physiological and Genetical Aspects of Mycorrhizae*, V. Gianinazzi-Pearson and S. Gianinazzi (eds.). Paris, France: INRA Press, pp. 217–221.

Vance, E. D., P. C. Brookes, and D. S. Jenkinson. 1987. A extraction method for measuring soil microbial biomass C. *Soil Boil Biochem* 19: 703–707.

Vela, G. R. and O. Wyss. 1964. Improved stain for visualization of *Azotobacter* encystment. *J Bacteriol* 87: 476–477.

Vincent, J. M. 1970. *A Manual for the Practical Study of the Root Nodule Bacteria*, pp. 1–13. Oxford, U.K.: Blackwell Scientific Publication.

Voorhorst, W. G. B., R. I. L. Eggen, E. J. Luesink et al. 1995. Characterization of the *cel*/B gene coding for β-glucosidase from the hyper thermophicli archaeon *Pyrococcus furiosus* and its expression and site directed mutation in *Escherichia coli*. *J Bacteriol* 177: 7105–7111.

Watanabe, T. 1994. *Pictorial Atlas of Soil and Seed Fungi: Morphologies of Cultured Fungi.* Boca Raton, FL: Lewis Publishers.

Willems, A. and M. D. Collins. 1993. Phylogenetic analysis of rhizobia and agrobacteria based on 16S ribosomal RNA gene sequences. *Int J Sys Bacteriol* 43: 305–313.

Woese, C. R. 1987. Bacterial evolution. *Microbiol Rev* 51: 221–271.

Zuberer, D. A. 1994. Recovery and enumeration of viable bacteria. In *Methods of Soil Analysis. Microbiological and Biochemical Properties,* eds. R.W. Weaver, J. S. Angle, and P. S. Bottomley, Part 2, pp. 119–144. Madison, WI: Soil Science Society of America (SSSA book series).

Important Links

http://www.tfrec.wsu.edu/pdfs/P2326.pdf
http://www.dpi.nsw.gov.au/_data/assets/pdf_file/0018/41643/Soil_biology_testing.pdf
http://soilsmart.com.au/diagnistic_services.php
http://vro.dpi.vic.gov.au/dpi/vro/vrosite.nsf/pages/soilhealth_biology_tests
http://www.scribd.com/doc/34781068/Manual-of-Soil-Analysis-%E2%80%93-Soil-Bioremediation

17

Virology

17.1 Introduction

Viruses are ultramicroscopic particulate macromolecules composed primarily of a nucleic acid genome, either DNA or RNA, and protein. These obligate parasites, whose nucleic acid or genome controls and utilizes the synthetic capacity of their host cells for replication, occur morphologically either as rod-shaped, cubical, or polyhedral virions and those possessing more complex structures. Basically, a viral particle, or virion, is a nucleic acid core, surrounded by a protein coat or capsid, composed of protein subunits or capsomers. Viruses that infect bacteria are referred to as bacteriophages and can be isolated from natural environments. Animal and plant viruses can be isolated from infected tissues. Bacteriophages are also called bacterial viruses. Schematic procedure for the isolation of bacteriophage is illustrated in Figure 17.1.

These viruses exist in many shapes and sizes. Some of the simplest ones exist as a single-stranded DNA virion. Most of them are tadpole-like, with a head and a tail. The head, or capsid, may be round, oval, or polyhedral and is composed of protein. It forms a protective envelope for the DNA of the organism. The tail structure is hollow and provides an exit for the DNA from the capsid into the cytoplasm of the bacterial cell. The extreme end of the tail has the ability to become attached to specific receptor sites on the surface of phage-sensitive bacteria. Once the tail of the virus attaches itself to a cell, it literally digests its way through the wall of the host cell.

With the invasion of a bacterial cell by the DNA, one of two things occur: lysis or lysogeny. In the event that lysis occurs, as illustrated in Figure 17.2, the metabolism of the bacterial cell becomes reoriented to the synthesis of new viral DNA and protein to produce mature phage particles. Once all the cellular material is used up, the cell bursts to release phage virions that, in turn, are prepared to invade other cells. Phage that causes lysis is said to be virulent. If the phage does not cause lysis, however, it is termed as temperate and establishes a relationship with the bacterial cell known as lysogeny. In these cells, the DNA of the phage becomes an integral part of the bacterial chromosome. Lysogenic bacteria grow normally, but their cultures always contain some phage. Periodically, however, phage virion is released by lysogenized cells in lytic bursts similar to that seen in the lytic cycle.

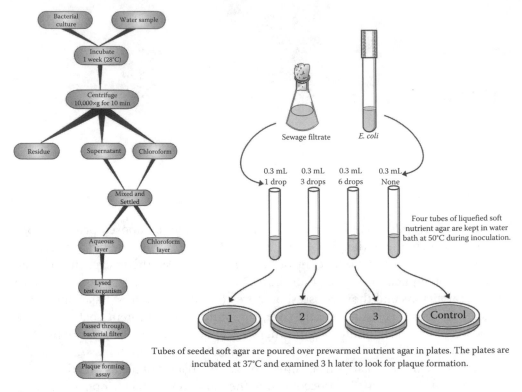

Four tubes of liquefied soft nutrient agar are kept in water bath at 50°C during inoculation.

Tubes of seeded soft agar are poured over prewarmed nutrient agar in plates. The plates are incubated at 37°C and examined 3 h later to look for plaque formation.

Figure 17.1 Schematic procedure for the isolation of bacteriophage.

17.2 Isolation of Bacteriophage from Sewage

17.2.1 Principle

Visual evidence of lysis is demonstrated by mixing a culture of bacteria with phage and growing the mixture on nutrient agar. Areas where the phages are active will show up as clear spots called plaques. The most thoroughly studied bacterial viruses are those that parasitize *Escherichia coli*. They are collectively called as coliphages. They are readily isolated from raw sewage.

17.2.2 Materials

Raw sewage, phage broth, pure culture of *E. coli*, incubator, centrifuge, membrane filtration unit, Erlenmeyer flask, test tubes, nutrient broth, soft nutrient agar, hard agar, Petri dishes, graduate and 5 mL pipettes, and water bath.

17.2.2.1 Media
Nutrient broth (1 g peptone, 0.5 g yeast extract, 0.25 g NaCl, 0.8 g potassium phosphate, and dibasic in 100 mL distilled water), 10× strength nutrient broth (peptone 20 g,

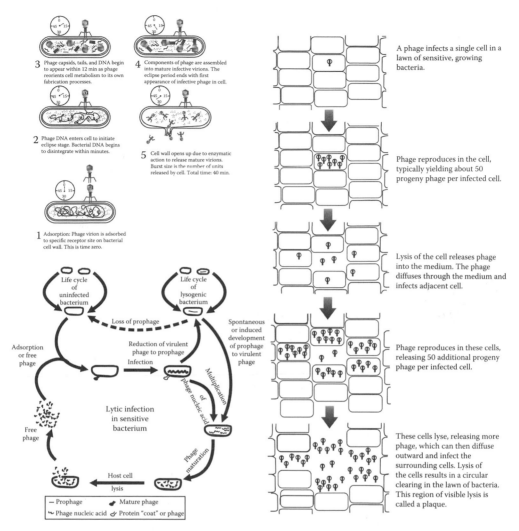

Figure 17.2   Life cycle of lytic and lysogenic bacteriophage.

yeast extract 10 g, NaCl 5 g, and potassium phosphate dibasic 16 g in 200 mL distilled water), and warm nutrient agar plates (6100 × 15 mm plates) (12–15 g agar/liter nutrient medium). Tubes containing 3 mL each of warm, top agarose (one per plate) (7.5 g agarose/L nutrient broth, molten, cooled to 45°C).

17.2.3 Protocol

17.2.3.1 First Step: Filtration and Amplification of Bacterial Viruses (Figure 17.3)

1. Collect the raw sewage from sewage treatment.
2. Mix the sewage (100 mL) with 10 mL of 10× strength nutrient phage broth and 10 mL of *E. coli* suspension.

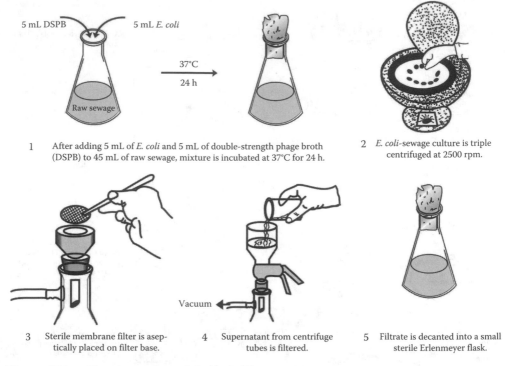

1   After adding 5 mL of *E. coli* and 5 mL of double-strength phage broth   2   *E. coli*-sewage culture is triple
    (DSPB) to 45 mL of raw sewage, mixture is incubated at 37°C for 24 h.   centrifuged at 2500 rpm.

3   Sterile membrane filter is asep-   4   Supernatant from centrifuge   5   Filtrate is decanted into a small
    tically placed on filter base.   tubes is filtered.   sterile Erlenmeyer flask.

Figure 17.3   Filtration and amplification of bacterial viruses.

3. Incubate the mixture at 37°C for 24 h.
4. Centrifuge the sewage at 2500 rpm for 10 min so that the bacteria and other solid matrix are separated. Decant the supernatant.
5. Repeat the process three times to obtain the filter.
6. Filter the supernatant through the membrane filter (0.45 μm) to remove the bacterial cells. It is quite possible that some time, filter gets clogged or choked. It is, therefore, desirable to change the filter and pass the filtrate through the new filter.
7. Connect the filtration assembly through a vacuum pump to quick the filtration process.
8. Keep vacuum pump on so that filtrate may accumulate the flask of filtration unit.
9. Transfer aseptically the final filtrate so obtained into a presterilized Erlenmeyer flask.
10. Incubate both cultures at 37°C, shaking for 24 h.

17.2.3.2 Second Step: Bacteriophage Isolation and Plating

1. Procure four tubes of soft nutrient agar of jelly like consistency and keep on a water bath at 50°C. Label them as 1, 2, 3, and 4.
2. Pour 1 mL filtrate in tube 1, 2 mL filtrate in tube B, and 4 mL in tube C and keep tube D as blank having no filtrate.

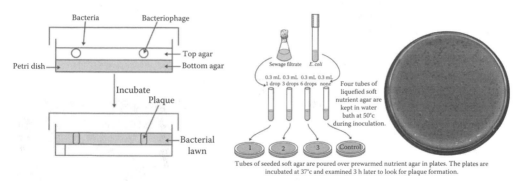

Figure 17.4 Double agar layer method of seeding *Escherichia coli* culture with phage.

3. Inoculate all the tubes with 0.5 mL in *E. coli* and pour into Petri dishes previously containing hard agar and label them 1, 2, 3, and 4, respectively, to the tubes.
4. After the medium is solidified, invert the plates, keep them in the incubator for several hours, and examine for plaque formation (Figure 17.4).

17.2.4 Results and Observations

Record the results in triplicate and take the average of the readings as plaque forming units (PFUs).

17.2.4.1 Phage Typing

Phage typing is a method used for detecting single strains of bacteria. It is used to trace the source of outbreaks of infections. The viruses that infect bacteria are called bacteriophages ("phages" for short), and some of these can only infect a single strain of bacteria. These phages are used to identify different strains of bacteria within a single species.

A culture of the strain is grown in the agar and dried. A grid is drawn on the base of the Petri dish to mark out different regions. Inoculation of each square of the grid is done by a different phage. The phage drops are allowed to dry and are incubated. The susceptible phage regions will show a circular clearing where the bacteria have been lysed, and this is used in differentiation (Figure 17.5).

17.3 Preparation of High Titer of Bacteriophage of Isolated Bacterial Virus

17.3.1 Principle

The titer is defined as high titer of phage, which is required for plaque bioassay. In this procedure, initially, a single plaque is taken from a plate by sterilized inoculating needle or sterilized tooth picks and suspended in tryptone broth. After vortex

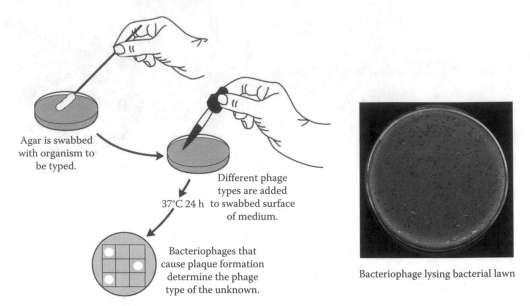

Agar is swabbed with organism to be typed.

37°C 24 h

Different phage types are added to swabbed surface of medium.

Bacteriophages that cause plaque formation determine the phage type of the unknown.

Bacteriophage lysing bacterial lawn

Figure 17.5 Illustrating method for phage typing.

for 2–3 min, it is transferred to an overnight grown culture of specific host bacteria, and the lysate is subjected to plaque forming assay as described in the earlier experiment.

17.3.2 Materials

Tryptone broth, host bacterial culture, nutrient agar plates with plaque, sterile needle or sterile tooth picks, sterile pipettes, and incubator shaker.

17.3.3 Protocol

1. Procure a plate containing plaque.
2. Scoop out a plaque from the plate by sterilized inoculating needle or sterilized tooth picks.
3. Meanwhile, prepare an overnight grown bacterial culture of the host cell.
4. Mix 0.1 mL plaque broth in 10 mL culture of the host cell.
5. Incubate the mixture at 37°C for 6–8 h.
6. Centrifuge the mixture at 10,000 rpm for 20 min to remove the bacterial cells.
7. Add one to two drops of chloroform for lysing the bacterial cell, if any, present in the suspension.
8. Pour the suspension (1–2 mL) on each plate containing nutrient agar.
9. Incubate the plates at 37°C for 24 h in the inverted position.
10. Observe the plate for plaque formation.

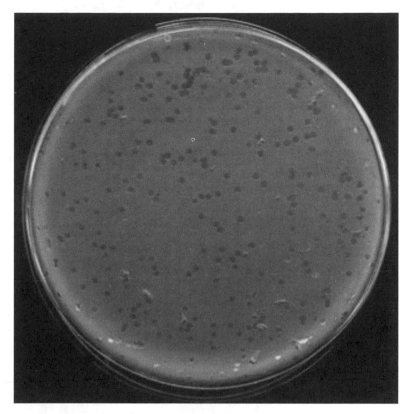

Figure 17.6   (See color insert.) Plaques of uniform size of bacteriophages.

17.3.4 Results and Observations

Numerous plaques of uniform size are visible against the dark background of agar (Figure 17.6).

17.4 One-Step Growth Curve of Coliphage and Determining the Number of Phage Particles Released by the Lysis of the Single Infected Bacterial Cell (Burst Size)

17.4.1 Introduction

The growth cycle of the bacteriophage begins with the adsorption of the phages on the surface of the sensitive bacterial host cells. The phage genome is injected into the bacterial cells with the functions of the tail of the phages piercing the cell wall. Inside the cell, the phage organizes its synthetic capacity by utilizing the host metabolism and rapidly reproduces itself. Proteins of the host cells are rapidly reduced. Proteins are synthesized to assemble the viral genome giving rise to proper viral particles.

The host cell then lyses releasing up to several hundred bacteriophages. The period of time from the adsorption of the phage to its host cell until the beginning of the release of progeny is called "latent period or eclipse period". The period between the end of the latent period and the *plateau phase* (where maximum PFU number is observed and remained constant) is called "rise period." The number of phage particles released per cell at the end of this period is called "Burst size" (Figure 17.7). Multiplicity of infection (MOI) is a frequently used term in virology, which refers to the number of virions that are added per cell during infection. If one million virions are added to one million cells, the MOI is one. If 10 million virions are added, the MOI is 10. Add 100,000 virions, and the MOI is 0.1. The concept is straightforward. High MOI is used when the experiment requires that every cell in the culture is infected. By contrast, low MOI is used when multiple cycles of infection are required. However, it is not possible to calculate the MOI unless the virus titer can be determined—for example, by plaque assay or any other method of quantifying infectivity. In other words, it may be defined as the ratio of the number of infectious virus particles to the number of target cells present in a defined space.

Many bacteriophages require divalent cations such as Mg^{++} and Ca^{++} for attachment to bacterial host cells. All host cells do not recognize the viral particles, and some viral particles are aborted due to over infection of the single cell in the adsorption mixture. The number of viral particles injecting the host cells in real term is defined as the input multiplicity.

17.4.2 Materials

Raw sewage, phage broth, pure culture of *E. coli*, incubator, centrifuge, membrane filtration unit, Erlenmeyer flask, test tubes, nutrient broth, soft nutrient agar, hard agar, Petri dishes, graduate and 5 mL pipettes, water bath, semi log graph papers pH meter, test tube stand, beef extract, peptone, agar, *E. coli* culture, and bacteriophage lysate.

17.4.3 Protocol

17.4.3.1 Bacteriophage Production Burst Size Determination (Single-Step Growth Curve)

1. Plate out an *E. coli* strain B culture at dilutions of 10^{-5}, 10^{-6}, and 10^{-7}.
2. Pipette 0.9 mL of *E. coli* into a sterilized test tube.
3. Pipette out 0.1 mL of a bacteriophage suspension in 0.9 mL of *E. coli* wait for 5 min so that the phage particles are absorbed by 80% of the bacterial cells.
4. Pipette the phage and *E. coli* mixture in a sterile centrifuge tube and centrifuge at 5000 rpm for 20 min. Decant the supernatant, resuspend the pellet of cells in 0.9 mL of nutrient broth and mix well by swirling. This removes unabsorbed phage particles.
5. Add 0.1 mL of the aforementioned cell suspension to 9.9 mL of nutrient broth and mix well by swirling.

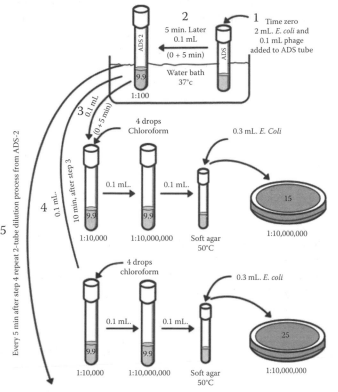

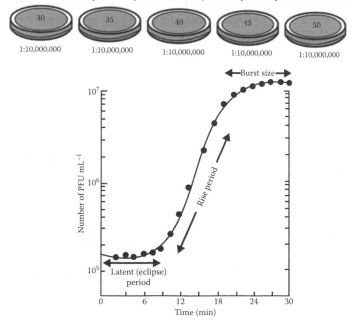

At proper time intervals (every 5 min) 0.1 mL. is pipetted from ADS-2 tube through two tubes of tryptophan broth and into soft agar (as in steps 3 and 4) to overlay five more plates. All plates are incubated at 37°C

Figure 17.7 Single step growth curve of coliphage number of plaque forming unit as a function of time (min).

6. Inoculate 0.1 mL of the aforementioned diluted cell suspension to 9.9 mL of nutrient broth. Mix well by swirling. Incubate at 37°C in the water bath provided.
7. Melt five tubes of regular nutrient agar and five tubes of soft nutrient agar (0.7% agar), temper the agar to 45°C–50°C, and pour a tube of regular nutrient agar into each of five Petri plates, and allow them to harden.
8. Temper the soft agar tubes to 45°C, and add three to four drops of original *E. coli* strain B culture to each tube.
9. Consider the step 3 to be zero time. Add 0.1 mL sample from diluted culture of step 6 to a tube of the inoculated soft agar at intervals of 20, 25, 30, 35, and 40 min. Mix each tube and pour on the surface of one of the nutrient agar plates prepared in step 7. Allow the soft agar to solidify and incubate the plates (inverted) at 37°C.
10. Determine the count of *E. coli* from the plates made in step 1. Examine the plates from step 8, and count and record number of plaques at various time intervals.
11. Determine the *E. coil* count from step 2.
12. Examine the plaque number and calculate PFU in the original infected culture taking into consideration of dilution factor.
13. Plot the number of PFU versus time (use semi-log graph sheet). Calculate the burst size of the bacteriophage in one-step growth.

17.4.4 Result

17.4.4.1 Calculation for Example

Assume that the 0.1 mL of the phage added contained 10^8 PFU. Therefore, the PFU in the 2 mL of the bacterial culture would contain 5×10^7 PFU/mL. Consider the bacterial population (in 2 mL) in the beginning of the experiment is 1×10^8/mL, but the number of bacterial cells infected by the phage is only 5×10^7 cells/mL (IM – input multiplicity).

$$\text{Therefore, the MOI is} \frac{\text{Number of phages infecting bacterial cells}}{\text{Number of host cells}} = \frac{5 \times 10^7}{1 \times 10^8} = 0.2$$

This means that every 2 cells of 10 bacterial cells are infected with the phages.

If the total PFU obtained at the plateau is 5×10^9 PFU/mL, then the burst size can be calculated as follows:

$$\text{The burst size} = \frac{\text{Number of PFU released at the plate}}{\text{Number of infected bacterial cells}}$$

$$\frac{\text{Final PFU} - \text{Initial PFU}}{\text{Number of infected bacterial cells}} = \frac{5 \times 10^9 - 5 \times 10^7}{5 \times 10^7} \times \frac{4.95 \times 10^9}{5 \times 10^7} = 99$$

The burst size is 99 indicating that each cell after lysis releases about 99 phage particles.

17.5 Demonstration of Lysogeny and Isolation of Rhizobium Phage

17.5.1 Introduction

With the invasion of bacterial cell by the viral DNA, one of the two things occur; lysis or lysogeny. In the event that lysis occurs, the metabolism of the bacterial cell becomes reoriented to the synthesis of new viral DNA and protein to produce mature phage particles. Once all the cellular material is used up, the cell bursts to release phage virions that, in turn, are prepared to invade other cells. Phage that causes lysis is said to be virulent. If the phage does not cause lysis, however, it is termed as temperate and establishes a relationship with the bacterial cell known as lysogeny. In these cells, the DNA of the phage becomes an integral part of the bacterial chromosome becoming a prophage and behaves as if it is a bacterial chromosome. Lysogenic bacteria grow normally, but their cultures always contain some phage. Periodically, however, phage virions are released by lysogenized cells in lytic bursts similar to that seen in the lytic cycle. Upon exposure to physical or chemical mutagens such as ultraviolet (UV) exposure or mitomycin c, NTG treatment, the viral genome excises and reverts back.

17.5.2 Materials

Rhizobial culture, indicator strain of rhizobia, incubator, shaker, sterilized pipettes, Petri dishes, test tubes, and media.

Medium A (yeast sucrose phage broth) g/L in distilled water: Sucrose, 2.5; $K_2HPO_4 \cdot 7H_2O$, 0.5; $MgSO_4 \cdot 7H_2O$, 0.2; $CaSO_4$, 0.16; NaCl, 0.10; $FeCl_3 \cdot 6H_2O$, 0.02; yeast extract, 0.5; and distilled water, 1 liter.

Medium B (yeast sucrose agar plates): Yeast sucrose broth with 7 g agar per liter.

Medium C (defined agar medium).

Solution 1 g/L in distilled water: Mannitol, 10; $Na_2HPO_4 \cdot 12H_2O$, 0.45; $Na_2SO_4 \cdot 10H_2O$, 0.6; KNO_3, 0.6; Thiamine HCl, 0.1; biotin, 0.5; agar, 7.5; and distilled water to 1 L.

Solution 2 g/L in distilled water: $MgCl_2 \cdot 6H_2O$, 0.1; $CaCl_2 \cdot 6H_2O$, 0.1, and distilled water to 1 L.

Sterilize solutions 1 and 2 separately and pour 0.5 mL solution 2 in to Petri dishes. After the medium is solidified, pour 5 mL of melted agar solution 1 over the surface of solution 2.

Medium D (diluting broth) g/L in distilled water: Sucrose, 2.5; K_2HPO_4, 0.5; NaCl, 0.1; and distilled water to 1 L.

17.5.3 Protocol

1. Prepare medium A, B or the defined medium C, inoculate 5 mL of any of the medium with desired rhizobial strain and incubate for 12 h.
2. Pour 5 mL of the inoculated prepared media on a presterilized Petri dish (9 cm diameter), incubate to get translucent background of colonial growth.

3. Inoculate a flask containing 100 mL sterilized medium B with a loopful of young culture and incubate at 26°C in a shaker until it contains 2×10^8 rhizobia/mL. Pour 1 mL of medium over the first one and incubate, and confluent growth of bacterium is obtained.

4. Mix 1 mL of 24 h old culture of the same bacterium with 0.5 mL indicator suspension in 5 mL of melted agar medium, add to the Petri dishes to form the second layer of agar.

5. Incubate the Petri dish for 48 h at 26°C and then observe the development of the plaques.

17.5.4 Results and Observations

Note the plaque size. Free phage particles from plaques (i.e., spontaneous lysis or rhizobia) are obtained due to the induction of the tested culture by the indicator strain demonstrating lysogeny.

Avoid too heavy bacterial growth. Use low salt medium D to check complete lysis followed by rapid growth due to aggregation.

17.6 Isolation of Cyanophage LPP from Pond Water

17.6.1 Introduction

Cyanophages are the viruses that attack the members of cyanobacteria. They are known to be abundant and dynamic component in the aquatic microbial communities that can regulate the biomass production and species composition of varied microorganisms and other aquatic life forms like bacteria and phytoplankton. They are also known to mediate gene transfer between microorganisms in aquatic ecosystems. Cyanophages are known to play a significant role in the ecology of cyanobacteria and that they are very possibly affecting natural control of cyanobacterial blooms (Figure 17.8).

Distributions of cyanophages are affected immensely by the change in environmental conditions. Temperature, salinity, pH, solar radiation, depth, and seasonal variation greatly affect the attachment and multiplication of cyanophage population (Figure 17.9).

Cyanophages are capable of entering into a lysogenic relationship with their host. However, induction of the lytic state under lysogenic conditions appears to be more difficult with cyanophage than with bacteriophage. High MOI is known to favor lysogeny. Also, the nutritional status of the cells is also an important factor. In nutrient rich media, the lytic course predominates, whereas growing cells in poor media increases the chances of the lysogenic pathway after single infection. Cells raised in nutrient rich media have a higher concentration of the host global regulator ribonuclease (RNase) III, which leads to elevated rates of expression of the protein N favoring lytic growth. In carbon-starved cells, on the other hand, RNase III and consequently N concentrations are low. Under these

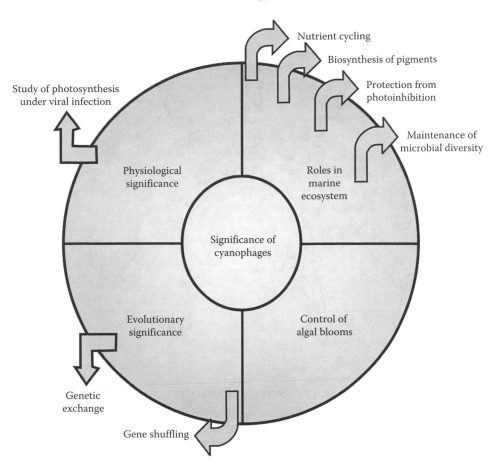

Figure 17.8 Significance of cyanophages.

conditions, N translation is repressed. This reduction of N concentration would reduce Q expression to a level that provides more opportunity for lysogenic response. When the phage DNA enters the host cell, then there are options for the operation of both lytic and the lysogenic cycle. Both the processes require the synthesis and also the expression of the immediate early and the delayed early genes. But after this point, there occurs the point of deviation of both the processes: lytic process occurs if the late genes are expressed, and lysogeny happens if the repressor gets expressed.

The cyanobacterial–cyanophage association represents an ideal model system for studying the mechanism of oxygenic photosynthesis under viral infection and also the correlation between the viral abundance and the surrounding environment. The cyanophages also infect marine cyanobacteria. Due to the resemblance between cyanophages and the bacteriophages, as well as the presence of common type of photosynthetic activity in cyanobacteria and higher plants, they provide an ideal and excellent model system to study the photosynthetic activity under viral infection. The cyanophages show extreme diversity in their structure, habitat, and host range. On the structural basis, they are differentiated into three major groups (Figure 17.10).

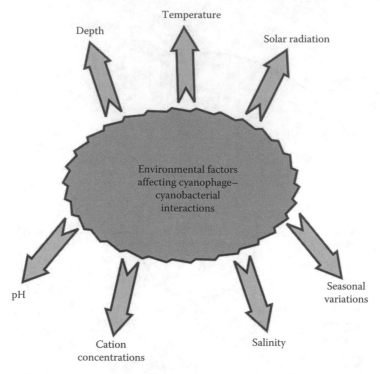

Figure 17.9   Environmental factors affecting cyanobacterial interactions.

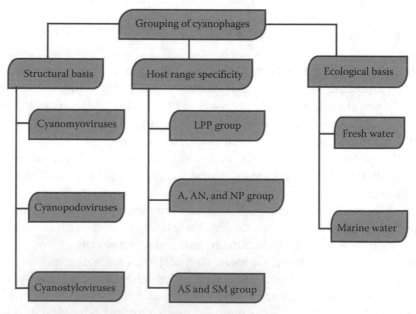

Figure 17.10   Cyanophage grouping based on different parameters.

17.6.2 Materials

Fresh culture of cyanobacteria (Lyngbya, Plectonema, and Phormidium), modified Chu 10 agar slants, pond water, chloroform, sterile Petri dishes, illuminated incubator, sterile culture tubes, centrifuge, sterile membrane filtering assembly, and membrane filter.

17.6.2.1 Modified Chu 10 Medium

g/L in distilled water: $Ca(NO_3)_2 \cdot 4H_2O$, 0.04; $K_2HPO_4 \cdot 3H_2O$, 0.01; $MgSO_4 \cdot 7H_2O$, 0.025; $NaCO_3$, 0.02; $NaSiO_3 \cdot 9H_2O$, 0.025; ferric citrate, 0.003; citric acid, 0.003; and glass distilled water to 1 L.

To the aforementioned medium add 1 mL/L of Fogg's micronutrient solution. The composition of the solution: $MnCl_2 \cdot 4H_2O$, 0.01; $Na_2MoO_4 2H_2O$, 0.01; H_3BO_3, 0.01; $CuSO_4 \cdot 5H_2O$, 0.001; $ZnSO_4 \cdot 7H_2O$, 0.001; and distilled water to 100 mL. pH of this medium should be adjusted to 7.5 before autoclaving. Then, 1.5% agar is added for preparing agar slants for culturing cyanobacteria.

17.6.2.2 Magnesium Saline Composition g/L in Distilled Water

0.2 $MgCl_2 \cdot 6H_2O$, 5.85 NaCl, and distilled water to 1 L.

17.6.3 Protocol

1. Collect two liters of water sample from ponds, add two to three drops of chloroform, shake vigorously and filter through glass filter paper (Whatman GF/C) to remove extraneous matter.
2. The filtrate is concentrated down to 20 mL through dialysis against a solution of high molecular weight (20,000) polyethylene glycol in magnesium saline. Such a solution virtually soaks up water in the filtrate.
3. The concentrate is then filtered through an ultrafine sintered glass filter that allows only virus size particles to pass through. Prepare dilution as per Figure 17.11.
4. Five milliliters of this ultrafiltrate is incubated with the host culture for 10–15 days. There is complete lysis of the cyanobacterial culture with the disappearance of green color in the culture flask.

17.6.3.1 PFU Assay

5. Plaque counts are determined on plates in which 5 mL of an inoculated agar had been evenly distributed over a 15 mL solidified layer of 1.5% modified Chu 10 modified agar medium.
6. The surface layer is prepared in test tubes, which is consisted of 0.5 mL of an appropriately diluted virus suspension, 2 mL concentrated cyanobacterial filament of 3–4 weeks old culture of host cells, and 2.5 mL of melted agar medium (medium with 1% agar) plated on agar plates having a base of 0.8% agar containing medium (30 mL in 100 × 17 mm Petri plates).

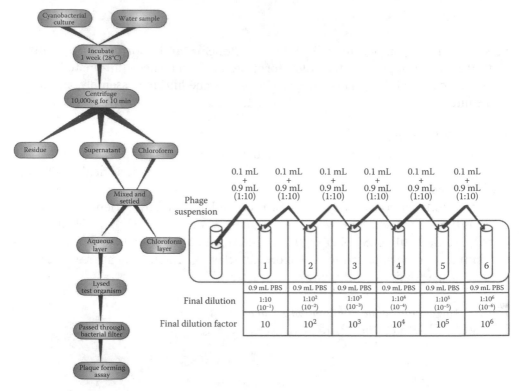

Figure 17.11 Schematic demonstrations for the isolation of cyanophage and preparation of dilution for plaque assay count.

7. The inverted plates are incubated in an illuminated incubation chamber at 30°C. Plaques start appearing after 4 days of incubation (Figures 17.11 and 17.12).
8. The clonal population of the virus is raised from a single plaque isolate, and lysate is always filtered through a sintered glass filter (2.5 cm GF/C; Whatman, England) before being used in the experiment.
9. Single step growth curve and burst size of the cyanophage is determined similar to the bacteriophage experiment except for that medium and illumination is essential for cyanophage multiplication as the host is a photosynthetic autotroph.

17.6.4 Results and Observations

Clear plaque formation on the green lawn of the host cells shows the presence of cyanophages in the pond water infecting Lyngbya, Plectonema, and Phormidium. Examine the bacterial green lawn for clear areas, that is, plaques that represent for the presence of cyanophages released after bursting the cyanobacterial cells. Record the number, shape, and size of the plaques.

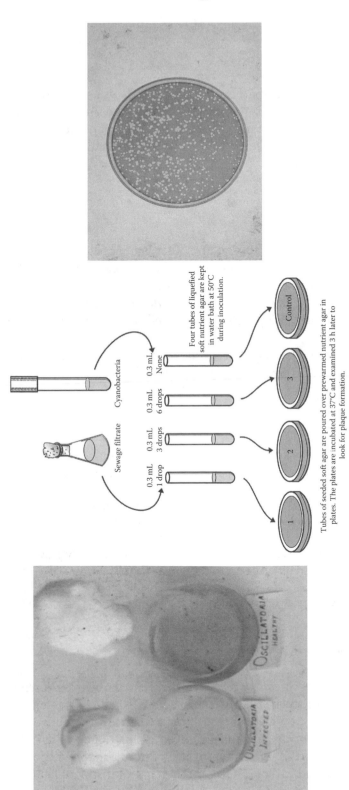

Figure 17.12 (See color insert.) Procedure for the plaque forming unit assay.

17.7 Demonstration for the Presence of Actinophages in Soil

17.7.1 Introduction

Actinophages are the viruses that lyse the members of actinomycetes. There are several actinomycetes that reside in soil and infect plants and animals and cause disease besides their beneficial role in nitrogen fixation and organic matter decomposition. Actinophages play a vital role in soil ecosystem by maintaining the balance of actinomycete population.

17.7.2 Materials

Pure culture of Streptomyces sp. and any other genera belonging to actinomycetes, 20 g garden soil sample, peptone corn agar medium, meat peptone broth, Erlenmeyer flask, membrane filter, and membrane filter assembly.

Peptone-corn agar medium g/L in distilled water: Glucose, 10; NaCl, 5; $CaCO_3$, 0.5; peptone, 5; corn extract, 5; agar, 20; and distilled water to 1 L. pH of the medium is adjusted to 7–7.2.

Beef extract peptone broth g/L in distilled water: Beef extract, 3; peptone 5, and distilled water to 1 L. pH of the medium is adjusted to 7–7.2.

17.7.3 Protocol

1. Prepare corn agar plates and inoculate with the pure culture of *Streptomyces* sp.
2. Collect 20 g garden soil sample and transfer into a 70 mL beef extract peptone broth in a 250 mL conical flask.
3. Incubate the flask at 25°C–30°C for 24 h on a shaker at low speed.
4. Decant the water and filter through the membrane filter aseptically under vacuum, and store the filtrate in a sterilized screw cap tubes.
5. Place one to two drops of filtrate in the centre of young actively growing cultures of *Streptomyces* sp. and tilt the plate so as to spread the filtrate over a definite part of the agar surface.
6. Incubate the culture at 25°C–26°C for 24–72 h.

17.7.4 Result and Observations

Observe the growth of *Streptomyces* sp. in the plate. Growth of *Streptomyces* sp. is inhibited in filtrate inoculated areas, and plaques (small clear lysed areas) are also observed. It can also be demonstrated by subculturing the active agent. Prepare the fresh culture of *Streptomyces* sp. again. Cut a small piece of agar from plaques/inhibition zone and spread it over the surface of pure culture with spatula. Incubate the plates as mentioned earlier. Appearance of several distinct plaques confirms the presence of actinophages.

17.8 Isolation of Phage from Flies

17.8.1 Introduction

Coprophagous insects contain various kinds of bacterial viruses, and house flies fall into coprophagous category. They deposit their eggs in fecal material where the young larvae feed, grow, pupate, and emerge as adult flies. This type of environment is heavily populated by *E. coli* and its inseparable parasitic phages. The method adopted here for isolation of phage is similar to the one used for the isolation of phages from raw sewage. The procedure is schematically illustrated in Figure 17.13.

17.8.2 Materials

Nutrient broth medium, culture of *E. coli*, 40–50 flies, 10 tubes of soft nutrient agar (5 mL per tube) with metal cap, 10 plates of nutrient agar (15 mL per plate and prewarmed at 37°C), and 1 mL serological sterile pipette.

Phage growth medium g/L in distilled water: KH_2PO_4, 1.5; Na_2HPO_4, 3; NH_4Cl, 1; $MgSO_4 \cdot 7H_2O$, 0.2; glycerol, 10; acid-hydrolyzed casein, 5; DL-tryptophan, 0.01; gelatin, 0.02; Tween–80, 0.2; and distilled water to 1 L.

Lysis buffer: 50 mM Tris–HCl, pH 7.4, 5 mM ethylenediaminetetraacetic acid (EDTA), 1% sodium dodecyl sulfate, 200 mM NaCl, 0.5 mg/mL proteinase K (add freshly before use; stock solution: 50 mg/mL in 10 mM Tris–HCl, pH 8.0, and 1 mM EDTA, pH 8.0).

Nutrient broth: (1 g peptone, 0.5 g yeast extract, 0.25 g NaCl, 0.8 g potassium phosphate, and dibasic in 100 mL distilled water), 10× strength nutrient broth (peptone 20 g, yeast extract 10 g, NaCl 5 g, and potassium phosphate-dibasic 16 g in 200 mL distilled water), warm nutrient agar plates (6, 100 × 15 mm plates) (12–15 g agar/L nutrient medium), and tubes containing 3 mL each of warm, top agarose (one per plate) (7.5 g agarose/L nutrient broth, molten, cooled to 45°C).

17.8.3 Protocol

17.8.3.1 Extraction

1. Pour half of the growth medium into the mortar and grind the flies to a fine pulp with the pestle.
2. Transfer the fly–broth mixture to an empty flask.
3. Wash the mortar to an empty flask. Use the remainder of the growth medium to rinse out the mortar and pestle, pouring the entire medium into the flask.
4. Incubate the fly–broth mixture for 48 h at 37°C. Add 50 mL of the lysing buffer to the fly–broth mixture. Incubate the mixture for 6 h.
5. Centrifuge the mixture at 5000 rpm for 10 min, and decant the liquid from the tube to another clean tube without disturbing the bottom of the tube. Again centrifuge at 5000 rpm for 10 min and pour off the top two-third of each tube in a clean set of tubes and centrifuge again in the same manner.

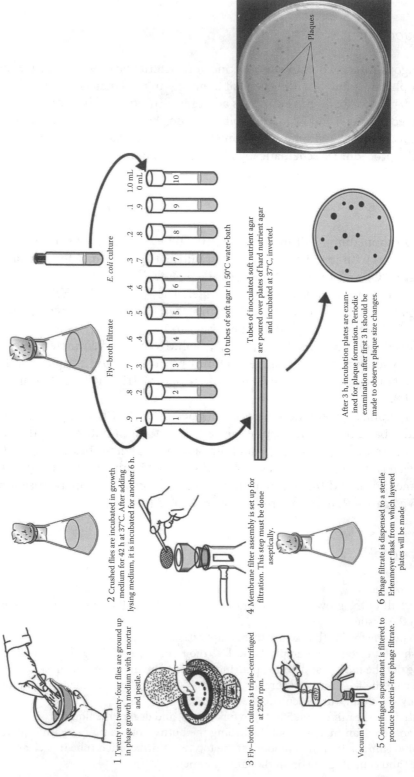

Figure 17.13 Flow diagram for the isolation of phage from flies.

6. Extracted liquid is filtered through a membrane filter assembly using 22 μm cellulose nitrate filter aseptically using a vacuum pump (Figure 17.13).

17.8.3.2 PFU Assay

1. Liquify 10 tubes of soft nutrient agar and cool to 50°C. Keep tubes in a water bath to prevent solidification.
2. Arrange tubes in order in test tubes and label them. Label 10 plates of prewarmed nutrient agar 1 through 10. In addition, label plate 10 as negative control. Pre warming these plates will allow the soft agar to solidify more evenly.
3. With a 1 mL serological pipette, deliver 0.1 mL of the fly broth filtrate to tube 1, 0.2 mL of tube 2, etc., until 0.9 mL has been delivered to tube 9 (Figure 17.12). Tube 10 will be a negative control without fly–broth filtrate.
4. With a fresh 1 mL pipette, deliver 0.9 mL of *E coli* to tube 1, 0.8 mL to tube 2, etc., as illustrated in Figure 17.13.
5. After flaming the necks of the each of the tubes, pour them in to similarly numbered plates.
6. When the agar has cooled completely, put the plates, inverted, into a 37°C incubator.

17.8.4 Results and Observations

After about 3–6 h incubation, examine the plates and look for plaques. Measure the visible plaques and record their diameter. If there are no plaques, then incubate further for 3–6 h and record the number. Check the plaque size after 12 h and keep on recording up to 24 h.

17.9 Virus Culture in Embryonating Chicken Eggs

17.9.1 Introduction

Before the development of cell culture, many viruses were propagated in embryonated chicken eggs. Today, this method is most commonly used for growth of the virus. The excellent yield of virus from chicken eggs has led to their widespread use in research laboratories and for vaccine production. Avian-embryo culture is economical and convenient compared to laboratory animals. This is commonly used for the isolation, identification, titrating, and maintenance of many animal viruses as well as for the production of vaccines.

17.9.2 Materials

Chicken eggs, candling apparatus, syringe, sterile normal saline solution, incubator, sterile pipettes, and disinfectant solution.

17.9.3 Protocol

1. Candle two embryonated eggs with the long axis in the horizontal plane. Mark the location of the air sac. Swab the marked area with the iodine–alcohol disinfectant. After sterilizing a 18-gauge needle by dipping it into the disinfectant solution and flaming, use the needle to puncture the shell of an egg at the upper extremity of the shell over the air sac (Figure 17.14).
2. Using a 1 mL syringe, with a 27 gauge, 3/4″ needle, inoculate the allantoic cavity with the dilution of disease Newcastle viral disease or any other virus by inserting the needle perpendicularly through the puncture in the shell, passing the needle into the egg along its full length parallel to the long axis of the egg. Inject 0.2 mL of the virus preparation and remove the syringe needle. Seal the puncture hole with Duco cement or any other suitable cell cementing material (Figure 17.15).
3. As in step 2, inoculate the second embryonated egg with 0.2 mL of sterile saline using the aforementioned steps.
4. Incubate the inoculated eggs at 37°C in an incubator having coaster trays in order to maintain the humidity.
5. Candle the inoculated eggs each succeeding laboratory period; examining for the death of embryos as evidenced by the cessation of the movement or disappearance of veins from the egg shell. Newcastle disease virus causes death of embryo within 3 or 4 days following inoculation.
6. Crack the shell and empty the contents into a Petri plate when the embryo is dead. Compare the appearance of the two embryos.

Note: Newcastle disease virus can cause conjunctivitis in humans, and hence it is advisable to wash hands carefully with soap before and after the experiment.

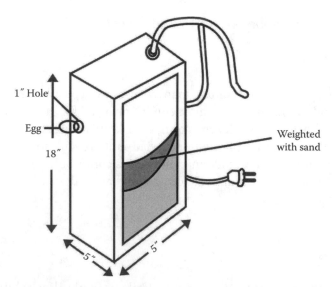

Figure 17.14   Candling an egg.

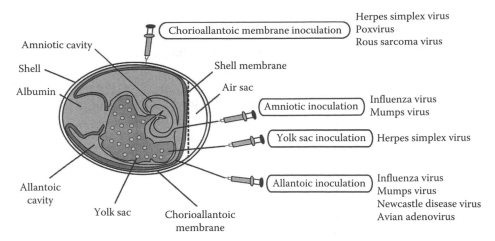

Herpes simplex virus
Poxvirus
Rous sarcoma virus

Chorioallantoic membrane inoculation

Amniotic cavity

Shell

Albumin

Shell membrane

Air sac

Amniotic inoculation · Influenza virus
Mumps virus

Yolk sac inoculation · Herpes simplex virus

Allantoic inoculation · Influenza virus
Mumps virus
Newcastle disease virus
Avian adenovirus

Allantoic cavity

Yolk sac

Chorioallantoic membrane

Figure 17.15 Inoculation of virus into the allantoic cavity.

17.10 Extraction and Isolation of Tobacco Mosaic Virus

17.10.1 Introduction

Tobacco mosaic virus (TMV) is a very well studied single-stranded RNA virus that infects plants, especially tobacco, and other members of the family Solanaceae. The infection causes characteristic patterns (such as mottling and discoloration) on the leaves (hence the name). TMV was the first virus to be discovered. TMV has a rod-like appearance. Its capsid is made from 2130 molecules of coat protein and one molecule of genomic single-strand RNA, 6400 bases long. The coat protein self-assembles into the rod like helical structure (16.3 proteins per helix turn) around the RNA, which forms a hairpin loop structure (Figure 17.16).

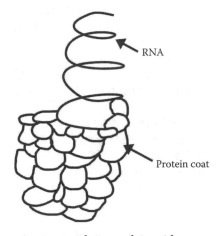

RNA

Protein coat

Figure 17.16 Tobacco mosaic virus with its nucleic acid.

17.10.2 Materials

Diseased (TMV) tobacco leaves, Waring blender, sterilized water, low-speed centrifuge, ultracentrifuge, cheese cloth, test tubes, buffer solution, sodium borate (0.02 M) pH 8.2, and sodium ascorbate (0.01 M). Triton X-100, Tris–HCl (10 mM) pH 8.0, EDTA (1 mM), sucrose solutions of various concentrations (10%–65%), syringe, and dialysis tubing.

17.10.3 Protocol

17.10.3.1 Extraction and Pathogenicity Test

1. Take a few virus-infected leaves; grind the diseased leaves in a mixture of 0.02 M borate buffer pH 8.2 containing 0.01 M sodium ascorbate as a reducing agent to avoid any oxidation of the virus in a Waring blender.
2. Filter the homogenate through several layers of cheese cloth for collection of sap with virus (Figure 17.16).
3. Filter the particles by pouring the contents of the blender through a filter paper.
4. Pass the filtrate through a membrane filter (pore size 0.45 μm) using membrane filtration assembly.
5. Keep a healthy tobacco plant beside and select the leaves that are to be inoculated. Wash them with carborundum using a wash bottle.
6. Rub the leaf with a piece of cotton that has been soaked in the filtrate from step 4.
7. Wash the carborundum and excess virus from the leaf with distilled water from the wash bottle.
8. Keep aside a control set.
9. Check the plants up to 3 weeks. Look for both localized and systemic symptoms.

TMV can also be isolated from some commercial products. Russian and Turkish brand of cigarettes are good sources of these viruses (Figure 17.17).

17.10.3.2 Purification of TMV Virus

1. Take the filtrate from step 3, add 0.5% Triton X-100 and stir for 1 h.
2. Centrifuge the sap at low speed (10,000 rpm) for 10 min, discard pellet, and recover the supernatant.
3. Layer the supernatant over cushion of 5 mL of 30% sucrose in 10 mM Tris–HCl, 1 mM EDTA (pH 8.0), and centrifuge at 27,000 rpm for 3 h.
4. Discard the supernatant and collect the tiny pellets of virus sediments at the bottom of the tube.
5. Resuspend the virus pellet in the buffer and keep overnight.
6. Repeat steps 2–5 two or three times, and further purification of the virus is subjected to density gradient centrifugation.
7. Furthermore, the semipurified preparation is submitted to a linear density sucrose gradient (10%–40%), and ultracentrifugation at 80,000g for 2.5 h or prepare sucrose solution gradient (10%, 20%, 30%, and 45%) in centrifuge tubes for layering (Figure 17.18) of virus suspension on sucrose gradient.
8. Spin the tubes at high speed at 80,000g for 2.5 h in an ultracentrifuge in order to form a band of virus particles at the top of the centrifuge tube.

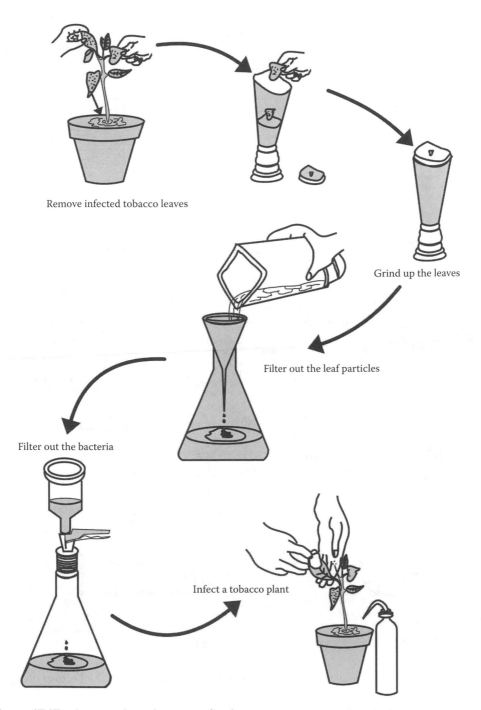

Remove infected tobacco leaves

Grind up the leaves

Filter out the leaf particles

Filter out the bacteria

Infect a tobacco plant

Figure 17.17 Steps in the infection with tobacco mosaic virus.

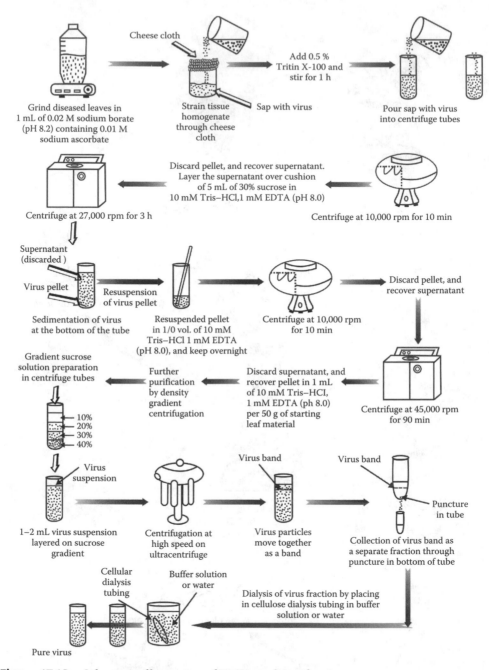

Figure 17.18 Schematic illustration of TMV viral purification.

9. Collect virus band as separate fraction through a puncture in the bottom of the centrifuge tube or by injecting a syringe into plastic centrifuge tube.

10. Transfer virus fraction in dialysis tubing and remove impurities by dialysis against extraction buffer (0.02 M sodium borate pH 8.2 with 0.01 M sodium ascorbate) or against water for 24 h at 4°C.

17.10.4 Result and Observations

Observe the dialysis tubing for virus. Pure virus particles are obtained in the dialysis tubing, which can be used for further studies.

17.11 Cultivation of Viruses in Tissue Culture

17.11.1 Introduction

As the viruses do not reproduce independent of the living host cells, they cannot be cultured in the same as bacteria and eukaryotic microorganisms. Tissue culture is the *in vitro* propagation of cells. A convenient source for tissue is the chick embryo. *in vitro* cultivation of animal viruses has eliminated the need to kill the animals. This technique has become possible by the development of growth media for animal cells and by the availability of antibiotics, which prevent bacterial and fungal contaminations in cultures. Some viruses cause the formation of plaques in tissue culture procedure. The clonal cell lines suspended in suitable media are infected with any desired virus that replicates inside the multiplying cells. If the viruses are virulent, they cause lysis of cells, and virus particles are released in the surrounding medium. These newly produced virus particles (virions) infect the adjacent cells. As a result, localized areas of cellular destruction and lysis (called *plaques*) often are formed. Plaques may be detected if stained with dyes, such as neutral red or trypan blue that can distinguish living from dead cells. Viral growth does not always result in the lysis of cells to form a plaque. Animal viruses, in particular, can cause microscopic or macroscopic degenerative changes or abnormalities in host cells and in tissues called *cytopathic effects*, cytopathic effects may be lethal, but plaque formation from cell lysis does not always occur.

17.11.2 Experiment Study of Morphology of Chicken Fibroblast Monolayers (Figure 17.19)

17.11.2.1 Protocol
1. Place the chicken embryo in phosphate buffered saline (PBS) with the following composition:
   a. NaCl, 8 g; KCl, 0.2 g; $CaCl_2 \cdot 2H_2O$, 0.13 g; $MgCl_2 \cdot 6H_2O$, 0.1 g; and distilled water to 800 mL.
   b. Na_2HPO_4, 1.15 g; KH_2PO_4, 0.2 g; and distilled water to 200 mL.
   Autoclave separately and mix when cool.
2. Cut tissue into pieces of about 1 mm in size
3. Wash fragmented sections several times in fresh PBS.
4. Mix the fragments in trypsin to break the clumps into individual cells.
5. Allow the cell clumps to pass through gauze fitted on a centrifuge tube, add serum, and centrifuge at 1000 rpm for 10 min.

(1) Place chicken embryo in buffered saline (PBS).

(2) Cut tissue into piece of about 1 mm in size.

(3) Wash fragmented sections several times in fresh PBS.

(4) Mix fragments in trypsin to break down clumps of cells into individual cells.

(5) Strain cell clumps through gauze into a centrifuge tube, add serum and centrifuge at 1000 rpm for 10 min.

(6) Pour off supernatant liquid and resuspend packed cells in a measured amount of tissue culture medium.

(7) Using a hemocytometer calculate the number of cells/mL of your suspension from step 6.

(8) Transfer cells to a suitable bottle and dilute with enough tissue culture medium to obtain a final concentration of about 10^6 cells/mL.

(9) Plant each Petri plate (60 × 15 mm) with 5 mL of the cell suspension and incubate at 37°C.

(10) Examine microscopically at intervals.

Shortly after planting   24 h   48 h

Figure 17.19 Demonstration for the preparation of primary tissue for monolayer cultures experiment study of morphology of chicken fibroblast monolayers.

6. Decant the supernatant and resuspend the cells in a measured amount of tissue culture medium.

7. Calculate the number of cells per mL of the aforementioned suspension using a hemocytometer.

8. Transfer the cells to a flask and dilute it with enough tissue culture medium to obtain a final concentration of about 10^6 cells/mL.

9. Inoculate each Petri plate with 5 mL of the cell suspension and incubate at 37°C.

10. Examine the plates microscopically at different intervals.
11. Aseptically aspirate the media from each of the tissue cultures, without damaging the monolayer.
12. Inoculate 0.5 mL of Newcastle disease virus to one monolayer and 0.5 mL of sterilized PBS to the other.
13. Rotate the cultures gently so as to mix the substances thoroughly.
14. Incubate the monolayers keeping the cell side down at 37°C for 1 h.
15. Add 5 mL of tissue culture medium having methyl cellulose to each monolayer after 1 h incubation.
16. Incubate the tissue culture for 2 days at 37°C.
17. Observe the monolayers under the microscope, watch for cytopathic effects in the inoculated set.
18. Aspirate the media from each monolayer.
19. Add 4 mL of 5% formalin to each monolayer culture (to preserve it) for 10 min. Aspirate the formation.
20. Add 1 mL crystal violet solution to each monolayer for 5 s. Aspirate the crystal violet. Allow the stained cells to dry and look for plaques in the monolayers.

17.12 Viral Hemagglutination Test for the Presence of Antigens

17.12.1 Introduction

Red blood cells tend to agglutinate in the presence of antisera against them. Red blood cells can be easily enveloped with several antigens, the agglutination occurs in the presence of antisera against these antigens; the red cells act as passive carriers of the antigens and aid the recognition of the antigen antibody reaction. The technique of using coated red blood cells in this way is called hemagglutination. Antigen may be any chemical, bacterial cell wall, or secondary metabolites or even viral particles show antigenicity in blood. The antigenic properties of antigens can be tested following hemagglutination test. Moreover, when a particular antigen is mixed with its specific antibody in the presence of electrolytes at a suitable temperature and pH, the particles are clumped or agglutinated. Agglutination of erythrocytes is called hemagglutination. Therefore, agglutination serves as a convenient method for the detection and assay of antigens.

17.12.2 Materials

Permanent Lucite "U" or "V" plates, microtiter dropper (0.025 mL), microtiter diluters (0.025 mL), diluents (such as complement fixation test diluents is suitable for influenza), antigen (e.g., hemagglutinin), serum (heated to 56°C for 30 min and suitably treated to remove nonspecific inhibitors; filter sterilized), and RBCs of optimum concentration of a mammal like sheep, goat, mouse, etc.

17.12.3 Protocol

1. Place one drop of diluents in about 11 wells of the microliter plate. Use the last well as control.
2. Place one drop of antigen and one drop of diluents in the first well so as to get 1:2 dilution.

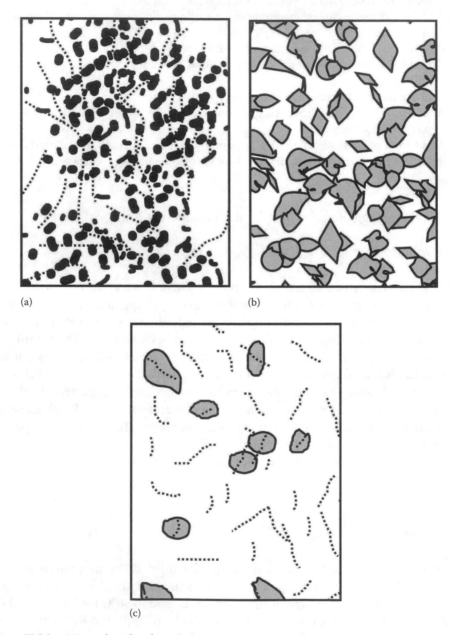

(a)

(b)

(c)

Figure 17.20 Normal and infected chicken fibroblast cells: (a) cells that are not infected, (b) 24 h infection, and (c) 72 h infection.

3. Furthermore, serially dilute the antigen in the subsequent wells and discard the last drop (0.025 mL) from the test well.
4. Add one drop of diluents to each well used.
5. Add one drop of RBCs to each well including the control well.
6. Incubate the well at appropriate temperature so as to settle the cells of control well to the bottom.

17.12.4 Results

The tests should be recorded as follows:

1. Cells distributed evenly in the bottom of the well—complete agglutination (+).
2. Cells compacted in a small red button at the bottom—no agglutination (–).
3. Nonagglutinated cells overlying a thin layer of agglutinated cells—partial agglutination (+).

Read the titer of the antigen as the highest dilution that gives hemagglutination (Figure 17.20).

Suggested Readings

Abedon, S. T., S. Duffy, and P. E. Turner. 2009. Bacteriophage ecology. In *Encyclopedia of Microbiology,* Schaecter, M, ed. Oxford, UK: Elsevier, pp. 42–57.

Ackermann, H. W., and M. S. DuBow. 1987a. *Viruses of Prokaryotes: Natural Groups of Bacteriophages*, Vol. II. Boca Raton, FL: CRC Press.

Ackermann, H. W., and M. S. DuBow. 1987b. *Viruses of Prokaryotes: General Properties of Bacteriophages*, Vol. I. Boca Raton, FL: CRC Press.

Adams, M. H. 1959. *Bacteriophages*. New York: Wiley-Interscience.

Bailey, S., M. R. Clokie, A. Millard et al. 2004. Cyanophage infection and photoinhibition in marine cyanobacteria. *Res Microbiol* 155: 720–725.

Baker, A. C., V. J. Goddard, J. Davy et al. 2006. Identification of a diagnostic marker to detect freshwater cyanophages of filamentous cyanobacteria. *Appl Environ Microbiol* 72: 5713–5719.

Bisen, P. S., S. Audholia, and A. K. Bhatnagar. 1985. Mutation to resistance for virus AS-1 in the cyanobacterium *Anacystis nidulans*. *Microbiol Lett* 29: 7–13.

Desjardins, P. R., and G. B. Olson. 1983. *Viral Control of Nuisance Cyanobacteria (Blue-Green Algae). II. Cyanophage Strains, Stability on Phages and Hosts, and Effects of Environmental Factors on Phage-Host Interactions*. Davis, CA: California Water Resource Center, University of California.

Safferman, R. S., R. E. Cannon, P. R. Desjardins et al. 1983. Classification and nomenclature of viruses of cyanobacteria. *Intervirology* 19: 61–66.

Safferman, R. S., and M. E. Morris. 1963. Algal virus: Isolation. *Science* 140: 679–680.

Shilo, M. 1972. The ecology of cyanophages. *Bamidgeh* 24: 76–82.

Singh, P., S. S. Singh, A. Srivastava et al. 2012. Structural, functional and molecular basis of cyanophage-cyanobacterial interactions and its significance. *Afr J Biotechnol* 11: 2591–2608.

Wilson, W. H., I. R. Joint, N. G. Carr et al. 1993. Isolation and molecular characterization of five marine cyanophages propagated on Synechococcus spp. Strain WH7803. *Appl Environ Microbiol* 59: 3736–3743.

Young, R. 1992. Bacteriophage lysis: Mechanism and regulation. *Microbiol Rev* 56: 430–481.

Young, R., I. N. Wang, and W. D. Roof. 2000. Phages will out: Strategies of host cell lysis. *Trends Microbiol* 8: 120–128.

Important Links

http://www2.warwick.ac.uk/fac/sci/lifesci/study/ug/courses/micvir/studyvirus/curriculum/

http://www.manchester.ac.uk/postgraduate/taughtdegrees/courses/atoz/course/?code=08965&pg=2

http://www.virology.ws/

http://www.ncbi.nlm.nih.gov/pmc/articles/PMC3327398/

http://www.clfs.umd.edu/cbmg/research/virology/links.html

http://www.amazon.com/Plant-Virology-Protocols-Transgenic-Resistance/dp/0896033856

http://www.bioinfo.de/isb/2008/08/0008/main.html

http://www.helmholtz-muenchen.de/en/viro/research/working-groups/viral-vectors-and-vaccines/viral-vector-vaccines/index.html

18

Immunochemical Methods

18.1 Introduction

Immunology is a broad branch of life science that covers the study of all aspects of the immune system in all organisms. Living organisms are able to defend themselves against invading foreign substances, and the immune system is the means whereby they are able to recognize and reject foreign cells and their products. It deals with the physiological functioning of the immune system in states of both health and disease; malfunctions of the immune system in immunological disorders (autoimmune diseases, hypersensitivities, immune deficiency, and transplant rejection); and the physical, chemical, and physiological characteristics of the components of the immune system *in vitro*, *in situ*, and *in vivo*. The relentless expansion of immunological knowledge and the growth of both experimental and clinical immunology continue to accelerate as immunology has become one of the most inclusive and integrative of the science, dependent on insight and information from many other fields of biology.

18.2 Immunity

The innate immune system recognizes and binds to nonself molecules-pathogen associated molecular patterns (PAMPs) by pattern recognition receptors (PRRs). They are limited in number (hundred or so) and are directly encoded (hard-wired) in the genome, common to all normal individuals. They specifically detect molecules produced by a wide variety of other organisms such as those found on bacterial cells but not on human cells. They are a genetically stable set of receptors and are transmitted across generations and expressed in each individual within a species in an essentially identical form. Some PRRs (toll-like receptors; TLRs) are cell-surface, while others (complement molecules) exist in soluble form. The innate immune system treats each encounter with a foreign agent as if they were meeting it for the first time. TLRs are PRRs that bind specific PAMPs. This binding signals the synthesis and secretion of the cytokines to promote inflammation and recruit leucocytes to the site of infection. Scavenger receptors are PRRs that are involved in the internalization of bacteria and in the phagocytosis of host cells undergoing apoptosis. Immunity can, therefore, be defined as the ability to resist damage from foreign substances such as microorganisms and harmful chemicals such as toxins released by microorganisms, and the

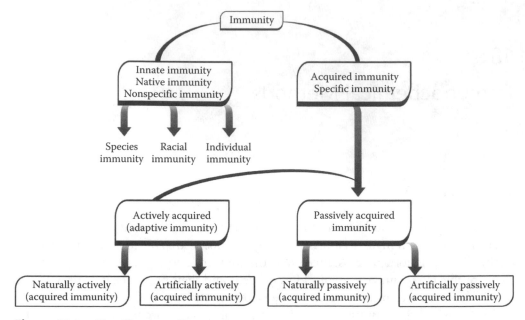

Figure 18.1   Classification of immunity.

ability to ward off disease through our defenses is called resistance. It can be divided into two broad subdivisions, namely, innate immunity (nonspecific immunity) and acquired immunity (Figure 18.1).

18.2.1 Innate Immunity/Native Immunity/Nonspecific Immunity

It is the resistance that an individual possesses by virtue of his/her genetic and body constitution. Thus, it is inborn.

18.2.1.1 Species Immunity
Resistance to infection varies with the species. Species immunity is the resistance exhibited by all members of a species toward a pathogen. Basic anatomical, physiological, and biochemical characteristics of a species can determine whether a microorganism can be pathogenic for that species, or not.

18.2.1.2 Racial Immunity
With a species, different races may exhibit differences in their resistance, due to genetic factors, such as races with sickle cell anemia prevalent in Mediterranean coast are resistant to infection by the malaria parasite, *Plasmodium falciparum*.

18.2.1.3 Individual Immunity
Different individuals in a race exhibit differences in innate immunity and is known as individual immunity, such as, some people have many more colds during the winter than do others. Several factors such as age, malnutrition, and hormonal disturbance influence the level of innate immunity.

TABLE 18.1   Differences between Active and Passive Immunity

Actively Acquired Immunity	Passively Acquired Immunity
1. Produced by active functioning of the immune system	1. Received passively; no involvement of immune system
2. Stimulated by infections, antigens (vaccines and allergens)	2. Acquired by administration of antibodies of sensitized T-cells
3. Provides effective and long-lasting immunity	3. Immunity is less effective and transient
4. There is a lag phase (latent period) required for the formation of antibodies	4. Offers immediate protection
5. Immunological memory is present. Subsequent contact with antigen will have a booster effect	5. No memory. Subsequent administration of antibodies is less effective due to immune elimination
6. Negative phase may occur due to the combination of antigen with preexisting antibody	6. No negative phase
7. Not applicable in immunodeficient hosts	7. Applicable in immunodeficient hosts
8. Active immunization is mainly employed in prophylaxis	8. Passive immunization has therapeutic and prophylactic uses

18.2.2 Acquired Immunity

The resistance that an individual acquires during life is known as acquired immunity. It is of two types—actively acquired and passively acquired (Table 18.1).

Actively acquired immunity may be natural or artificial. Similarly, passively acquired immunity may be natural or artificial. Thus, there are four ways to acquire acquired immunity.

18.2.2.1 Active Natural Immunity
It is due to natural exposure to antigen (Ag) from infections (disease or subclinical infections). For example, persons who recover from viral diseases such as small pox, chicken pox, and measles will develop life-long natural active immunity.

18.2.2.2 Active Artificial Immunity
When an Ag is deliberately introduced into an individual to stimulate his/her immune system, the immunity developed is known as active artificial immunity. Active artificial immunity is the preferred method of acquiring immunity because it produces long-lasting immunity without any disease such as vaccination (Table 18.2).

18.2.3 Passively Acquired Immunity

Passive immunity occurs when another person or animal develops antibodies and the antibodies are transferred to a nonimmune individual, who did not produce antibodies, and there is no involvement of the immune system. Passive immunity is preferred in some situations when immediate protection is required.

TABLE 18.2   Some Common Vaccines

Bacterial vaccines	
Live	BCG for TB, typhoid oral (Type 21 a) for typhoid fever.
Killed	TAB for enteric fever, cholera vaccine for cholera, pertussis vaccine, and anthrax vaccine
Bacterial product	Tetanus toxoid and diphtheria toxoid
Sub unit	Vi antigen for typhoid and hemophilus influenza type b (Hib)
Viral vaccines	
Live	Sabin's OPV for poliomyelitis, varicella vaccine for chicken pox, MMR for measles, mumps, rubella, 17 D Vaccine for yellow fever, influenza vaccine, vaccinia for small pox
Killed	Salk's vaccine for poliomyelitis Hepatitis A vaccine, rabies (semple) vaccine, KFD vaccine, and Japanese B encephalitis vaccine
Sub unit	Hepatitis B vaccine and subunit influenza vaccine

18.2.3.1 Passive Natural Immunity

Immunity results when antibodies are transferred from mother to child. During her life, the mother has been exposed to many Ags. These antibodies protect the mother and the developing fetus against disease. The antibody (Ab) immunoglobulin (Ig)G can cross the placenta from mother to the fetus and provides natural passive immunity to the fetus. If the mother nurses her child, IgA in the mother's milk, particularly in colostrum, provides natural passive immunity to the child.

18.2.3.2 Passive Artificial Immunity

It is achieved by administration of specific antibodies or sensitized T- lymphocytes, for example, tetanus antitoxin (antitetanus serum), diphtheria antitoxin, pooled human gamma globulin against Hepatitis B, convalescent sera, and antirabic serum. Passive immunization may also be used to suppress active immunity in Rh negative women with Rh positive baby by the administration of anti-Rh Ab (IgG). Achieving passive artificial immunity usually begins with vaccinating an animal such as horse. After the animal's immune system responds to the Ag, antibodies (sometimes T-cells) are removed from the animal and injected into the individual requiring immunity. In some cases, a human who has developed immunity is used as a source. However, this technique provides only temporary immunity because the antibodies are metabolized and eliminated (immune elimination).

18.3 Ags

These are foreign compounds that provoke the formation of antibodies. The chemical nature of Ags is quite diverse, and they are usually proteins or polysaccharides, although lipoproteins and lipopolysaccharides are also known. These macromolecules may occur free in solution, or they may be bound to the surface of cells or particles such as viruses, bacteria, pollens, or tissues such as erythrocytes or kidney from another individual.

Ags are usually macromolecules although low-molecular-weight compounds, such as drugs, can become antigenic when bound to a protein. These small molecules, which in themselves are not antigenic, are known as haptens. Classically, an Ag is defined as an organism, a molecule, or part of a molecule that is recognized by the immune system. Often, the term Ag is associated primarily with those molecules recognized by the extremely diverse receptors found on T and B lymphocytes. Thus, as per the definition, an Ag can perform two important functions: (1) inducing an immune response, known as immunogenicity and (2) specific reaction with antibodies or sensitized T-cells, known as immunological reactivity.

18.4 Antigenic Determinant or Epitope

Epitope is the smallest part of an Ag that is "seen" by somatically generated B and T-cell receptors. Ags contain one or more epitopes or antigenic determinants; the basic units recognized by immune receptors and antigenic specificity depend on epitopes (Figure 18.2). The smallest individually identifiable part of an Ag usually consist of four or five monosaccharide or aminoacid residues possessing a specific chemical structure, spatial (steric) configuration and electric charge, capable of sensitizing lymphocytes and of reacting with its complementary site on the specific Ab or T-cell is called as receptor. Antigenic specificity of the Ag depends on epitopes.

Ags may be classified as immunogens if they stimulate an immune response, as haptens if they induce an immune response only when coupled to an immunogenic carrier molecule, or as toleragens if they cause cells of the immune system to become selectively unresponsive to re-exposure to the same molecules.

18.5 Haptens

Haptens are substances that cannot induce the formation of antibodies (not immunogenic) by themselves, but can react with specific antibodies, for example, capsular polysaccharide of pneumococci and cardiolipin. Penicillin is a hapten. It can breakdown and bind to serum proteins to form a combined molecule that can produce an allergic reaction such as rash and fever, rarely death. Haptens may be converted to complete Ags by combining with a large inert carrier molecule.

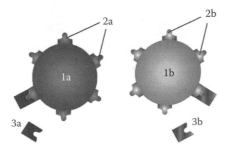

Figure 18.2 Antigens with antigenic determinant.

18.6 Specificity

An Ab produced in response to a foreign substance reacts with the Ag to form an immune complex (Ab–Ag) and is thereby inactivated. In most cases, the immune complex is insoluble and is removed from the circulation:

$$Ab + Ag \rightarrow (Ab - Ag)_n$$

This reaction is highly specific, and the Ab is bound by close range noncovalent van der Waals' forces to a small site on the Ag known as the antigenic determinant. A single antigenic molecule may contain a number of different antigenic determinants; and bovine serum albumin (BSA) has several hundred antigenic determinants per molecule (Figure 18.3).

Antibodies are also specific in that they normally react only with foreign substances and not with the plasma and tissues of the individual. This ability to recognize "self" as opposed to "nonself" is extremely important, and problems occur when antibodies are made against one's own tissue giving rise to autoimmune diseases such as rheumatoid arthritis and ulcerative colitis.

18.7 T-Cell Dependent (TD) and T-Cell Independent (TI) Ags

Based on the ability to induce Ab formation by B-lymphocytes, Ags are classified as T-cell dependent (TD) and T-cell independent (TI) Ags (Table 18.3).

TD Ags require the participation of T lymphocytes, whereas TI Ags do not require the cooperation of T lymphocytes for Ab formation.

18.8 Immunoglobulins

WHO introduced the term Ig, defined as "proteins of animal origin possessing Ab activity and or certain other proteins related to them by chemical structure." As per this definition, besides Ab globulins, the abnormal proteins found in myeloma,

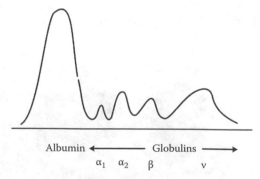

Figure 18.3 Scan of a stained cellulose acetate electrophoresis strip of human serum protein.

TABLE 18.3   Difference between TD and TI Antigens

T-Cell Dependent (TD) Antigens	T-Cell Independent (TI) Antigens
Structurally complex	Structurally simple, contain limited number of epitopes
Antigenic over a wide dose range	Antigenicity is critically dose dependent
Do not cause immunological tolerance	Too much dose may result in immunological tolerance, whereas too little is nonantigenic
They induce full gamut of Ig types: IgM, IgG, IgA, and IgE	Antibody response is limited to IgM and IgG
Produce immunological memory	Do not produce immunological memory
Require preliminary processing	Do not require preliminary processing
Rapidly metabolized in the body, for example, RBC, plasma proteins, and several protein-hapton complexes	Metabolized very slowly, for example, pneumococcal capsular polysaccharide, bacterial LPS, and flagellar protein (flagellin)

macroglobulinemia, and L and H chains (subunits of Igs) are also considered as Igs. The term Ig indicates a structural and chemical concept, while the term Ab is a biological and functional concept. All antibodies are Igs, but all Igs may not be antibodies. The term Ab is applied to an Ig molecule with specificity for an epitope of the molecules that make up Ags.

18.8.1 Classification

Chemically antibodies are proteins that are globulins with an immunological activity, hence the name Igs. Plasma proteins can be separated into albumin, globulins (alpha, beta, and gamma globulins). Albumin is the protein present in highest concentration contributing to the colloidal osmotic pressure of plasma and also act as a nonspecific transport protein. Globulins include specific transport protein (e.g., transferrin for iron), clotting factors, the complement system, and inactive precursors of certain hormones (e.g., angiotensinogen). Gamma globulins act as circulating antibodies important in specific immune process, also known as Igs (Figure 18.4).

Igs are produced from B lymphocytes and are both synthesized and secreted by plasma cells, following antigenic stimulation. The Igs are all found in the y-globulin region, but this band is quite diffuse and includes several different types of Ab each with a different structure and function (Table 18.4).

18.8.2 Structure

Igs constitute 20%–25% of the total serum proteins. Igs are glycoproteins, composed of four polypeptide chains, two identical light-chains (L-chains) and two identical heavy chains (H-chains) linked by disulfide bonds, to form a monomeric unit arranged in the form of a flexible Y with a hinge region (Figure 18.5).

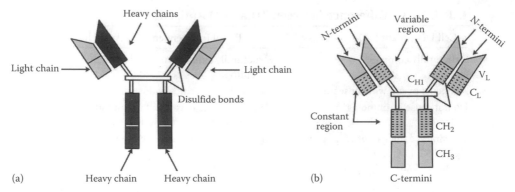

Figure 18.4 Structure of Ig molecule. (a) Immunoglobulin molecule with light chains shown in white and heavy chains shown in black and (b) immunoglobulin molecule with variable shown in white and constant region shown in textured gray.

TABLE 18.4   Classification of the Human Immunoglobulins

Class	Serum (mg/mL)	Molecular Weight	Light Chain	Heavy Chain	Structure[a]
IgG	12	150,000	k or λ	γ	$k_2 \gamma_2$
IgA	3	>300,000	k or λ	α	$(k_2 \alpha_2)_n$
IgM	1	950,000	k or λ	μ	$(k_2 \mu_2)_5$
IgD	0.1	160,000	k or λ	δ	$k_2 \delta_2$
IgE	0.001	200,000	k or λ	ε	$k_2 \varepsilon_2$

[a]Alternative structures are possible with the k chains replaced by λ chains.

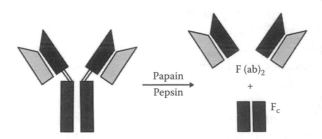

Figure 18.5 Cleavage of Ig molecule at hinge region by proteolysis.

The smaller chains are called L-chains (molecular weight [MW]: 20,000 each), containing 214 amino acids each. The larger chains are called H-chains (MW: 50,000 each), containing 446 amino acids each. The L-chains occur in two forms called kappa (k) and lambda (λ) encoded on chromosomes 2 and 22. A molecule of Ig may have either two identical kappa (k) or two identical lambda (λ) chains, but never both together. The H chains are structurally distinct for each Ig class. There are five types of H chains; all encoded on chromosome 14, termed gamma (γ), alpha (α), mu (μ), delta (δ), and epsilon (ε). Similarly, H-chain carries carbohydrate moiety, which is

TABLE 18.5 Five Different Classes of Immunoglobulins

Ig Class	H-Chain
IgG	Gamma (γ)
IgA	Alpha (α)
IgM	Mu (μ)
IgD	Delta (δ)
IgE	Epsilon (ε)

distinct for each Ig class. Based on physicochemical and antigenic differences, five classes of Igs have been recognized—IgG, IgA, IgM, IgD, and IgE. Both L and H chains contain two different regions (Tables 18.5 and 18.6).

Constant regions (C_L and C_H) have amino acid sequences that do not vary between antibodies of the same subclass.

Variable regions (V_L and V_H) have amino acid sequences that are different for all antibodies. Variable regions are so named for their variation in amino acid sequences between Igs synthesized by different B-cells.

The stalk of Y is termed F_C or constant (crystallizable) fragment. It is composed of only H-chains with carboxy terminal end. It contains the site at which the Ab molecule can bind to a cell. It determines the biological properties of Ig molecule such as complement fixation, placental transfer, skin fixation, and catabolic rate.

A L-chain variable region and a H-chain variable region together at amino terminus form a pocket that constitutes two Ag (epitope)-binding fragments $F(ab)_2$ or paratopes, that can bind to antigenic determinant sites (epitopes). Because an Ig monomer contains two identical L-chains and two identical H-chains, the two binding sites found in each monomeric Ig molecule are also identical.

The F_C fragments are composed of only constant regions, whereas $F(ab)_2$ fragments have both constant and variable regions. The portion of H chain, present in $F(ab)_2$ fragment, is called F_d piece. F_d piece is the H-chain (V_H and C_{H1}) portion of Fab. $F_{d'}$ is a H-chain (V_H and C_{H1}) portion of Fab. The prime (′) mark denotes extra amino acids due to a pepsin cleavage site. Each Ig chain has internal disulfide links or intrachain disulfide bonds (in addition to interchain disulfide bonds) forming loops, within each chain known as domain. These domains can be called as V_{L1}, V_{L2}, etc., and V_{H1}, V_{H2}, etc., for those present in variable regions of L and H chains, respectively. Similarly, C_{L1}, C_{L2}, etc., C_{H1}, C_{H2}, V_{L1}, etc., for those present in constant regions. The FC portion contains C_{H2}, C_{H3}, and sometimes C_{H4} regions of the Ig molecule. It is responsible for many biological activities that occur following engagement of an epitope.

The region between C_{H1} and C_{H2} domains is called hinge region. Disulfide bonds join the H-chains at or near a proline-rich hinge region, which confers flexibility on the Ig molecule. It is highly fragile when exposed to enzymes, such as pepsin or papain, Ig molecule will be cleaved at the hinge region, to produce two Fab fragments and one Fe fragment (Figure 18.5).

TABLE 18.6　Physicochemical Properties of Human Immunoglobulins

Property	IgG[a]	gM	IgA[b]	IgD	IgE
Heavy chain	γ_1	μ	α_1	δ	ε
Mean serum (mg/mL)	9	1.5	3.0	0.03	0.00005
% total antibody	80–85	5–10	5–15	<1	0.002–0.05
Valency	2	5 (10)	2–4	2	2
Mass of heavy chain (kDa)	51	65	56	70	72
Total molecule mass (kDa)	146	970	160[c]	184	188
Placental transfer	+	−	−	−	−
Half-life in serum (days)[d]	23	5	6	3	2
Compliment activation Classical pathway	++	+++	−	−	−
Alternative pathway	−	−	+	−	−
Induction of mast cell degranulation	−	−	−	−	+
Major characteristics	Abundant Ig in body fluids; neutralizes toxins, opsonizes bacteria, activates complement, maternal antibody	First to appear after antigen stimulation; very effective agglutination; expressed as membrane bound antibody on B-cells.	Secretory antibody; protects external surfaces	Present in B-cell surface; B-cell recognition of antigen	Anaphylactic mediating antibody; resistance to helminthes
% carbohydrate	3	7–10	7	12	11

[a]Properties of IgG subclass 1.
[b]Properties of IgA subclass 1.
[c]sIgA = 360–400 kDa.
[d]Time required for half of the antibodies to disappear.

The amino acid sequences of variable regions of L and H chains are not uniformly variable along their length but consist of relatively invariable and some highly variable regions, known as hypervariable regions or hot spots and are involved in the formation of Ag-binding sites. They are complementarity determining regions that can make contact with specific epitope.

Within two of the major Ig classes, there are subclasses. For example, IgG can be grouped into four subclasses (IgG$_1$, IgG$_2$, IgG$_3$, and IgG$_4$) and IgA into two subclasses (IgA$_1$ and IgA$_2$).

18.9 IG Determinants

Each Ig subclass has differences in the amino acid sequences, which determine the Ag specificity.

18.9.1 Isotypes

Differences between constant regions as a result of the use of heavy and L-chain constant region genes are called isotypes. In other words, the genetically different forms of L-chains (k and λ) and of H-chains (γ, α, μ, δ, and ε) are known as isotypes. Ig class or subclass is determined by the H-chain isotype. These determinants are common to or shared by all members of a species. These isotypic markers distinguish H chain classes (lgG-γ, IgA-α, IgM-μ, IgD-δ, and IgE-ε) and subclasses (IgG1-γ_1, IgG$_2$,-γ_2, IgG$_3$-γ_3, IgG$_4$-γ_4, IgA$_1$-α_1, and IgA$_2$-α_2) as well as k and λ L chains. They are located in the C domains of Ig chains in all individuals. Each isotype is encoded by a gene that is characteristic for a particular species and is present in all the members of that species. Antibodies to these determinants may be raised by injecting Igs from one species to another species (Figure 18.6).

18.9.2 Allotypes

Differences owing to different alleles of the same constant region gene are called isotypes. Allotypic markers occur for Ig molecules, with the same isotype but with amino acid sequences that differ from one person to another based on genetic differences among individuals. Every one (allo) of them cannot have the same IgG or IgM. Allotypes are variations that exist between individuals of same species. They are distinct amino acid residues located primarily in γ and α H chains and k L chains. Allotypic determinants reflect genetic polymorphism of Igs within one species. Allotype markers are useful in testing paternity and population genetics (Figure 18.6).

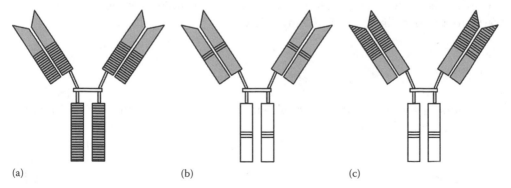

(a)   (b)   (c)

Figure 18.6 Three different determinants of Ig molecules: (a) isotypes, (b) allotypes, and (c) idiotypes.

18.9.3 Idiotypes

Within a given isotype (e.g., IgG), differences in particular rearranged VH and VL genes are called idiotypes. These are located in the hypervariable regions (V regions) of L and H chains and are analogous to epitopes of Ags. Idiotypes are specific for each Ig molecule and are sometimes useful in identifying marker for specific Ig molecules produced by a single clone of cells. Anti-idiotypic antibodies produced against Fab fragments prevents Ag–Ab interaction. Variations in Ig structure reflect the diversity of antibodies generated by an immune response (Figure 18.6).

18.10 Ig Classes

There are five different classes of Igs—IgG, IgA, IgM, IgD, and IgE (Table 18.5).

18.10.1 IgG

It is the major Ig in human serum, accounts for about 80% of the total Ig pool in serum. It is present in blood plasma and tissue fluids. It has a MW 150,000 (7S). Normal serum concentration is 8–16 mg/mL. There are four subclasses, namely IgG_1, IgG_2, IgG_3, and IgG_4; numbered according to their decreasing concentration in serum. The production of IgG occurs later in an immune response. Its presence in serum indicates remote or chronic infection. It has a half life of approximately 23 days. IgG is the major line of defense and neutralizes toxins, microorganisms, and viruses. They are effective in activating complement and are the major cause of lysis of bacteria, enveloped viruses, and infected or aged cells. They also act against bacteria and viruses by opsonization to promote phagocytosis and efficient in immobilizing various motile bacteria. IgG initiates Ab-dependent cell mediated cytotoxicity. They are the only maternal Ig, which is transported across the placenta and provides passive immunity to the child. Anti-Rh (anti-D) antibodies are IgG type. When IgG is administered passively, it suppresses the homologous Ab synthesis by a feedback process. This property is utilized in the isoimmunization of women by the administration of anti-Rh (anti-D) IgG during delivery.

18.10.2 IgM

It accounts for about 10% of the Ig pool, with a serum concentration of 0.5–2 mg/mL. It has a half life of about 5 days. It is a polymer (pentamer) of five molecules, held together by a special joining (J) chain. It is a large molecule (MW: 100,000) and, therefore, called as millionaire molecule (Figure 18.7). It is the first class of Ab produced in immune response, about 4 days after immunization (antigenic stimulus), and reaches peak concentrations by about the first week and then rapidly falls because of its short life span. Hence, their demonstration in serum indicates recent or acute infection.

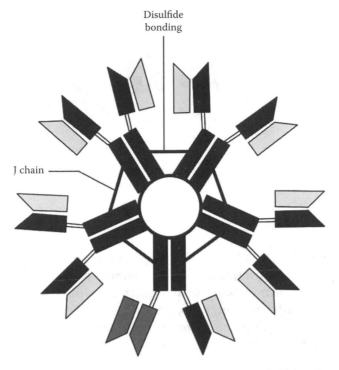

Figure 18.7   IgM molecule.

This class contains special antibodies such as blood group antibodies (anti-A and anti-B) and heterophile antibodies. It is largely confined to intravascular space and responsible for protection against blood invasion by microorganisms. Treatment of serum with mercaptoethanol selectively destroys IgM without affecting IgG antibodies. As IgM is not transported across the placenta, its presence in the fetus or newborn indicates congenital infection. IgM is effective both at immobilizing Ag and in activating the classical pathway of the complement. IgM can fix complement extremely well (30 times better than IgG) and, therefore, can effectively cause the lysis of bacteria, enveloped viruses, and infected or aged cells. Monomeric IgM is found in the membranes of B lymphocytes and acts as a specific receptor for Ag. IgM also enhances phagocytosis.

18.10.3 IgA

It accounts for about 15% of the Ig pool. In serum, IgA is present primarily as a monomer, but in mucus secretions as a dimer held together by a J-chain (Figure 18.8), normal serum concentration is 0.64.2 mg/mL. It has a half life of 6–8 days.

It provides humoral immunity (local immunity) in mucus secretions such as tears, saliva, intestinal mucus, semen, and colostrum (first breast milk). During the transport of IgA from the mucosa associated lymphoid tissues to mucosal surfaces, it acquires a protein called secretory (S) piece, hence it is known as secretory

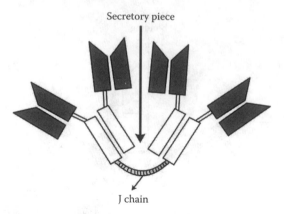

Figure 18.8   IgA secretory molecule.

IgA (sIgA). The plasma cells that produce sIgA are located in the secretory piece predominantly in the connective tissue called lamina propria, which lies immediately below the basement membrane of many surface epithelia. It is found in mucus, tears, saliva, breast milk, and respiratory and gastrointestinal secretions. It is the primary Ig of secretory immune system. It plays a major role in protecting surface tissues in the respiratory tract, intestine, and urogenital system against infectious microorganisms and viruses.

IgA is resistant to proteases and neutralizes toxins, allergens, bacteria, and viruses before they can enter the body through the mucus membranes. IgA activates complement by alternative pathway but not by the classical pathway. IgA has been shown to possess bactericidal activity against Gram negative organisms in the presence of lysozyme, which is also presenting in the same secretions that contain sIgA. sIgA present in breast milk provides natural passive immunity against enteric pathogens to the intestine of the new born. In the intestine, sIgA attaches to viruses, bacteria, and protozoan parasites such as *Entamoeba histolytica* and prevents their adhesion and entry through mucus membrane surfaces. IgA, also activates alternate complement pathway and promotes phagocytosis. IgA has two subclasses—IgA$_1$ and IgA$_2$. Serum IgA is a monomeric 7S molecule (MW about 160,000). sIgA is a much larger molecule than serum IgA (11S and MW about 400,000).

18.10.4 IgD

It is found in trace amounts in the blood serum, 0.03 mg/100 mL. It is mostly intravascular. It is found on the surface of B lymphocytes, along with IgM and may act as a receptor or recognition molecule for Ags to stimulate Ab production. It has a half life of 3 days. While the function of IgD has not been fully elucidated, expression of membrane IgD appears to correlate with the elimination of B-cells that generate self-reactive antibodies. IgD comprises less than 1% of serum Igs.

18.10.5 IgE

It is the Ab that is responsible for hypersensitivity to several Ags—from mosquito bites to pollens. It is a 8S molecule (MW about 190,000) with a half life of 2 days, IgE is also known as reagin Ab and found only in trace amounts in serum, but it is found attached to the surface of basophils, mast cells, monocytes, and eosinophils. When two IgE molecules on the surface of these cells are cross-linked by binding to the same Ag, the cells are activated to release granules, which contain pharmacologically active substances such as histamine, which cause allergic conditions such as asthma, anaphylaxis, eczema, etc. It also stimulates eosinophilia and gut motility that help in the elimination of helminthic parasites. IgE also causes skin sensitization. IgE differs from other Igs in the following aspects:

1. IgE can be detected by radioallergosorbent test, enzyme-linked immunosorbent assay (ELISA), and passive agglutination tests but cannot be detected by conventional serological tests.
2. IgE is species specific (homocytotropic) only human IgE can fix to the surface of human cells.
3. IgE is heat sensitive (unlike other antibodies) and is inactivated at 56°C in 1 h.
4. It can fix complement and does not pass through the placenta.
5. It can be destroyed by β-mercaptoethanol.

IgE is mainly produced in the linings of respiratory and intestinal tracts. IgE levels will be increased in Type 1 hypersensitivity (anaphylaxis and atopy) and in children with a high load of intestinal parasitic infections.

18.10.6 Abnormal Igs

Bence-Jones proteins are so called because of their discovery by Henry Bence Jones (1847) and are typically found in multiple myeloma. These are L-chains (k or λ) of Igs that are synthesized in excess. In about 20% of the patients of multiple myeloma, they are excreted in urine and damage the renal tubules and can be identified in urine by its characteristic property of coagulation when heated to 50°C but redissolve at 70°C. H-chain disease is a lymphoid neoplasia, characterized by the overproduction of the Fc portions of Ig H-chains.

Cryoglobulinemia is a condition in which there is the formation of gel or precipitation on cooling the serum, which redissolves on warming. It may be found in myelomas, macroglobulinemias, and autoimmune diseases such as systemic lupus erythematosus (SLE). The cryoglobulins are either IgG, IgM, or their mixed precipitates.

Waldenstrom's macroglobulinemia is a condition in which there is an excess production of myeloma proteins (M proteins) and of their L-chains (Bence-Jones protein).

18.11 Monoclonal Antibodies

Antibodies produced by a single clone (a single Ab forming cell or a single B lymphocyte or plasma cell) and directed against a single antigenic determinant are called monoclonal antibodies.

However, antibodies produced by infection or immunization are polyclonal because natural Ags possess several epitopes. Polyclonal antibodies are heterologous and are formed by several different clones of plasma cells in response to Ag.

18.11.1 Production

Monoclonal antibodies against any desired Ag can be produced on a large scale by hybridoma technology. Hybridomas are somatic cell hybrids produced by fusing Ab forming spleen cells with myeloma cells. The resultant hybrid retains the Ab producing capacity of the spleen cell and the ability of the myeloma cells to multiply indefinitely.

Animals (mice) are immunized with the desired Ag. Once the animal starts producing antibodies, its spleen is removed, and the spleen cells are separated from each other and fused with myeloma cells (myeloma is cancer or continuous proliferation of Ab forming cell) by the addition of polyethylene glycol, which promotes membrane fusion. These fused cells derived from spleen cells and myeloma cells are called hybridomas.

The mutant myeloma cells are grown in a culture, which do not form Igs and are deficient in the enzyme hypoxanthine guanine phosphoribosyl transferase.

The fused cells are transferred to a culture medium containing hypoxanthine, aminopterin, and thymidine (HAT medium). Aminopterin is a poison that blocks a specific metabolic pathway in cells. Myeloma cells lack HPRT enzyme that allows their growth in the presence of aminopterin. However, the pathway is by-passed in spleen cells provided with the intermediate metabolites hypoxanthine and thymidine. As a result, the hybridomas grow in HAT medium, but the myeloma cells die due to deficiency of HPRT enzyme and metabolic defect and cannot employ the bypass or salvage pathway.

When the culture is initially established using the HAT medium, it contains lymphocytes, myeloma cell, and hybridomas. The unfused splenic lymphoid cells die naturally in culture within a week or two as they cannot replicate indefinitely and the myeloma cells die in the HAT as just described. In contrast, the fused cells survive because they have the immortality of the myeloma and the metabolic bypass or the splenic lymphoid cells (Figure 18.9).

Hybridomas are then separated into individual wells of plastic dishes and allowed to grow into clones. Each hybridoma clone may produce monoclonal Ab (MAb); the antibodies in each well are screened to see which hybridoma is producing the desired Ab.

Once the particular hybridoma is identified, it may be recloned by cell culture or it may be maintained *in vivo* by injection into the peritoneal (abdominal) cavity of living animals. Cell culture yields about 100 mg of MAb; *in vivo culture* yields

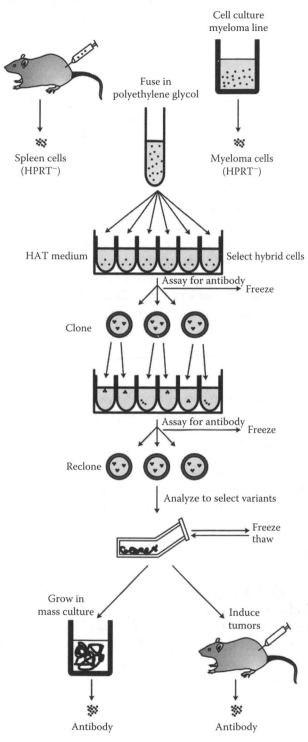

Figure 18.9 Production of monoclonal antibodies.

about 1000 mg (micrograms) or more of MAb per mL of ascitic fluid from the peritoneal cavity. Hybridomas may be frozen for prolonged storage.

18.11.2 Uses

Monoclonal antibodies are important because they are homogenous, highly specific and can be produced readily in large quantities. They bind to one and only one Ag and are, therefore, monospecific. Monoclonal antibodies against several Ags are now available commercially. They have several diagnostic, therapeutic, and research applications. They are useful in tissue typing, blood grouping, identification and epidemiological study of infectious microorganisms, identification of tumor Ags and other surface Ags, in the classification of leukemias, and identification of functional populations of different type of T-cells.

In the future, they can be used in passive immunization against infectious agents and toxic drugs, tissue and organ graft protection, stimulation of tumor rejection and elimination, manipulation of immune response, preparation of more specific and sensitive diagnostic procedures, and to deliver antitumor agents such as lethal doses of radiation and anticancer drugs to tumor cells to destroy them.

18.12 Polyclonal Antibodies

Polyclonal antibodies are effective immunosuppressants and are generally prepared by immunizing animals with human lymphocytes, for example, antilymphocyte serum or antilymphocyte globulin. Their primary effect is to inhibit cell mediated immune responses. Polyclonal antibodies are used clinically to treat organ graft rejection reactions, GVH reactions, and to promote remissions in some cases of aplastic anemia. The use of polyclonal antibodies is associated with certain problems, such as inducing serum sickness and thrombocytopenic purpura.

18.13 Ag–Ab Reaction

The reaction between Ags and antibodies *in vivo* form the basis of Ab mediated immunity in infectious diseases or of tissue injury in some types of hypersensitivity and autoimmune diseases. In the laboratory (*in vitro*) Ag–Ab reactions are useful in the diagnosis of infectious diseases, in the identification of infectious organisms, Ag and enzymes in serotyping, and in epidemiological surveys.

Ag–Ab reactions occur in various stages. Primary stage is the initial interaction between Ags and antibodies, which is rapid and reversible but without any visible effects. The weaker intermolecular forces such as hydrogen bonds, van der Waals forces and hydrophobic interactions are responsible for this primary stage. The primary stage is followed by secondary stage, an irreversible interaction leading to visible effects such as precipitation, flocculation, and agglutination and is due to covalent bonding.

18.13.1 Affinity and Avidity

18.13.1.1 Affinity (Intrinsic Affinity)
It is the intensity of attraction between Ag and Ab molecules. Low affinity antibodies bind Ag weakly and tend to dissociate readily, whereas high affinity antibodies bind Ag more tightly and remain bound longer.

18.13.1.2 Avidity (Functional Affinity)
It is the binding strength of an Ag with many epitopes and multivalent antibodies. The avidity of an Ag–Ab reaction is dependent on the valencies of both Ags and antibodies and is greater than the sum total of individual affinities. The avidity of Ag–Ab reactions is a better indicator of the strength of interactions in biological systems than affinity.

18.13.2 Sensitivity and Specificity

Sensitivity refers to the ability of a serological test to detect even very minute quantities of Ag or Ab. When a serological test is highly sensitive, false negative results will be absent or minimum. Specificity refers to the ability of a serological test to detect reactions between homologous Ags and antibodies only, and not with other. When a serological test is highly specific, false positive reactions will be absent or minimum.

18.13.3 Titer

Titer is a unit for the measurement of antibodies. Titer can be defined as the highest dilution (or lowest) quantity of the serum that gives a visible reaction with the Ag in a particular test.

18.13.4 Zeta Phenomenon

The surfaces of certain particulate Ags (e.g., red blood cell; RBC) possess an electrical charge, as for example, the net negative charge on the surface of RBC caused by the presence of sialic acid. When such charged particles are suspended in saline solution, an electrical potential, termed the zeta potential, is created between particles (e.g., RBC), preventing them from getting very close to each other. This introduces a difficulty in agglutinating charged particles (e.g., RBC) by antibodies, in particular RBC by IgG antibodies. The distance between Fab arms of the IgG molecule, even in its most extended form, is too short to allow effective bridging between two RBCs across the zeta potential. Thus, although IgG antibodies may be directed against Ags on the charged RBC, agglutination may not occur because of the repulsion by the zeta potential. On the other hand, some of the Fab areas of IgM pentamers are far enough apart and can bridge RBCs separated by the zeta potential. This property of IgM antibodies, together with their pentavalent, is a major reason for their effectiveness as agglutinating antibodies.

18.13.5 Mechanism of Ag–Ab Reactions (Marrack's Hypothesis)

The mechanism of Ag–Ab reaction can be explained according to Marrack's lattice hypothesis (Figure 18.10).

The multivalent Ag reacts with bivalent antibodies in varying quantities. But agglutination or precipitation occurs only when a large lattice of Ag–Ab complex is formed, which occurs when Ag and Ab are found in equal concentrations (zone of equivalence) (Figure 18.11).

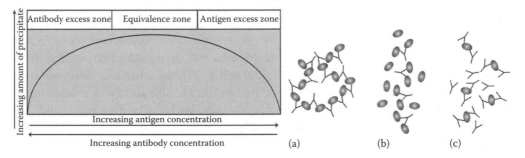

Figure 18.10   Mechanism of antigen–antibody reaction and zone phenomenon: (a) equivalence, (b) antigen excess, and (c) antibody excess.

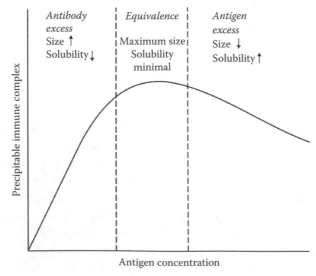

Figure 18.11   A typical precipitation curve for an antiserum titrated against a fixed concentration of antibody.

18.13.6 Zone Phenomenon

If to the same amount of Ab in a series of tubes, increasing quantities of Ags are added; precipitation or agglutination occurs best in one of the middle tubes where Ag and Ab are found in equal concentrations. It is called zone of equivalence. In the initial tubes, where Ab is in excess, the precipitation or agglutination will be absent. It is called prozone. Prozone is of importance in clinical serology as sometimes sera rich in Ab may give a false negative result, unless several dilutions are tested. In the later tubes, where the Ag is in excess, the precipitation or agglutination will be absent. It is called postzone. If the amounts of Ag–Ab complexes in different tubes are plotted on a graph, the resulting curve will have three phases: (1) an ascending zone, prozone or zone of Ab excess, (2) a peak, zone of equivalence, and (3) a descending zone, post-zone, or zone of Ag excess.

This phenomenon is called zone phenomenon. The prozone and postzone phenomena are taken into account in the interpretation of serological tests because false negative reactions can occur in either of these conditions. To avoid prozone phenomenon in clinical serology, several serum dilutions should be tested. When postzone phenomenon is suspected, the test can be repeated about a week later by taking a fresh serum sample. This would give time for the further production of antibodies.

18.14 Determination of Ag–Ab Reactions

Several laboratory tests are available to detect the Ag–Ab reactions. These laboratory tests are commonly used in diagnostic procedures because the presence of a certain Ab indicates the presence of a certain disease. These are precipitation, agglutination, coagglutination, complement fixation, immunolabeling, immunoelectroblotting, immunoelectrochromatography, immunoelectron microscopy, neutralization, opsonization, and chemiluminescence.

18.14.1 Precipitation

18.14.1.1 Definition
When soluble Ag combines with its Ab in the presence of normal saline (NaCl) at a suitable temperature and pH, the Ag–Ab complex forms an insoluble precipitate, as a sediment, or deposit. Such reactions are called precipitation reactions, and the antibodies participating in these reactions are called precipitins and the Ags, agglutinogens. Precipitation can take place in liquid media or in gels such as agar, agarose, or polyacrylamide.

18.14.1.2 Applications
Precipitation reactions are highly sensitive for the detection of Ags. Therefore, they find applications in (1) diagnosis of infectious diseases, (2) forensic medicine in the

identification of blood and semen stains on weapons and clothing or in other crime investigations, and (3) detection of food adulterants.

Precipitation tests may be carried out either as a qualitative or quantitative test. Precipitation can be allowed to occur either in solution or in agar gel or in agar gel with an electric force. Following tests are commonly used.

18.14.1.3 Ring Test

Ag solution is taken over a column of antiserum in a test tube. A precipitate forms at the junction of two liquids.

18.14.1.4 Slide Test

A drop of Ag and Ab are placed on a slide and mixed by rotation, floccules appear.

Venereal Disease Research Laboratory (VDRL) test for syphilis is a slide flocculation test: 0.05 mL of inactivated patient's serum is taken on special slide with depression (VDRL slides). One drop of freshly prepared cardiolipin Ag is added with a syringe that delivers 60 drops per mL. The slide is rotated at 180 rpm, in a VDRL rotator, for 4 min and examined under the microscope with low power objective. Uniform distribution of crystals in the drop indicates that the serum is nonreactive, while the formation of floccules indicates that the serum is reactive. With reactive sera, serial dilutions can be tested to obtain the reactive titer.

Rapid Plasma Reagin (RPR) card test utilizes the cardiolipin Ag coated with fine carbon particles. It makes the test result clearer and evident to the naked eye, and there is no need of microscope. RPR test can be done with unheated serum or plasma. A drop of patient's serum or plasma and a drop of Ag is added on a card in the circle and rotated for a few minutes. If black particles are aggregated in the center, the test is nonreactive; if the black particles are seen throughout the circle, the test is reactive.

18.14.1.5 Tube Test

The Ag solution and antiserum (patient's serum) are taken in a test tube, mixed, incubated, and then observed for precipitation. Usually several tubes with different dilutions of Ag are set up since precipitation may be inhibited in the post zone region.

Kahn test is a tube flocculation test used for the diagnosis of syphilis: 0.15 mL of inactivated patient's serum is taken in three tubes and mixed with different amounts (0.05, 0.25, and 0.0125 mL) of freshly prepared Ag, shaken in the Kahn shaker at 280 oscillations per minute, and examined after addition of saline. Floccules appear in a positive test. Negative test shows uniform opalescence. In Kahn verification test, the test is observed at both 1°C and 37°C. The Ab due to syphilis gives a strong reaction at 37°C and the nonspecific Ab at 1°C.

18.14.1.6 Immunodiffusion (or) Precipitation in Gel (or) Gel Precipitation

When precipitation is allowed to occur in a gel, the reaction is visible as a distinct band, which is stable and can be stained for preservation. An agar concentration of 0.3%–1.5% allows for diffusion of most reactants. Agarose gel may be preferred to agar because agar has a strong negative charge, while agarose has no charge. The rate of diffusion is affected by the size of particles, temperature, gel viscosity,

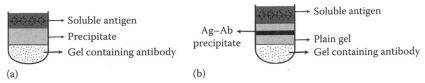

Figure 18.12 (a) Single diffusion in one dimension (Oudin's method). (b) Double diffusion in one dimension (Oakley method).

amount of hydration, and interactions between the matrix and reactants. As each Ag–Ab reaction gives rise to a line of precipitation, the number of different Ags in the sample can be made out.

Immunodiffusion also indicates identity, cross-reaction, and nonidentity between different Ags. Gel precipitation can be done by several ways.

18.14.1.6.1 Single Diffusion in One Dimension (Oudin's Method). In this method, Ab is incorporated into gel in a test tube, and the soluble Ag is directly pored over the Ab-containing gel. The Ag diffuses into the gel and forms a line of precipitation (Figure 18.12 a).

18.14.1.6.2 Double Diffusion in One Dimension (Oakley–Fulthorpe Method). In this method, the Ab is incorporated into gel in a test tube, above which a layer of plain agar gel is placed. The soluble Ag is then poured over on top of this plain gel. This results in the diffusion of both soluble Ag and Ab toward each other into intervening layer of plain agar and interact to form a precipitation ring (Figure 18.12 b).

18.14.1.6.3 Single Diffusion in Two Dimensions (Mancini Method). It is a quantitative immunodiffusion test used to quantitate serum proteins, Igs, complement factors, measure antibodies to influenza virus in the sera, and estimate serum transferring and alpha fetoprotein (Figure 18.13).

Monospecific Ab is added to molten agar and poured on slides or Petri dishes and allowed to set. Wells are cut in the gel and filled with known amounts of various dilutions of reference standard Ag solutions. The unknown test Ag is added to a separate well and allowed to diffuse. A ring of precipitate forms around each well, and the diameter of the ring is proportional to the Ag concentration in the well. The wider the ring, the greater will be the Ag concentration. The diameters of the rings formed by the standard Ag concentrations are measured and plotted as a function of Ag concentration. Concentrations values for the test Ags are then read from the plotted graph.

18.14.1.6.4 Double Diffusion in Two Dimensions (Ouchterlony Method). It is used to detect enterotoxin, in cases of suspected food poisoning, to detect antibodies against specific pathogens in the diagnosis of infectious diseases, widely used for the diagnosis of small pox. Elek's test for toxigenicity of diphtheria bacilli is a special variety of double diffusion in two dimensions, to identify fungal Ags, and to detect antibodies to extractable nuclear Ags (Figure 18.14).

The test is performed by pouring molten agar on to glass slides or Petri dishes and allowing it to settle. The antiserum is placed in the central well, and different Ags

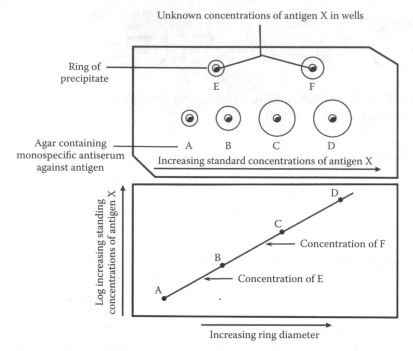

Figure 18.13 Single radial immunodiffusion test (Mancini method).

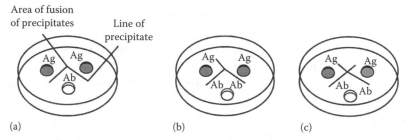

Figure 18.14 Ouchterlony test: (a) identity (precipitate lines fuse), (b) partial identity (spur [arrow]) is due to Ab, and (c) nonidentity (precipitate lines cross).

are added to wells surrounding the central well punched in agar. The Ag and Ab diffuse from the wells independently through the gel in two dimensions, horizontally and vertically. After incubation for 12–48 h in a moist chamber, lines of precipitate are formed at the sites of Ag–Ab combination. If the two adjacent precipitation lines fuse forming an arc (V-shaped), it indicates reaction of identity of the Ags. If the two adjacent precipitation lines fuse and form a spur or projection (Y-shaped, the stem of Y called spur), it indicates reaction of partial identity (cross-reactivity) of the Ags. If the two adjacent precipitation lines cross each other (X-shaped), it indicates reaction of nonidentity (Figure 18.15).

18.14.1.6.5 Uses. It is used for the comparison of various Ags and antisera or for comparison of several Ags at a time on one agar gel plate.

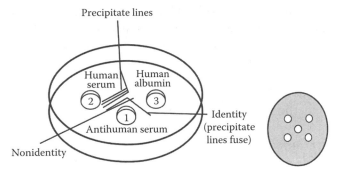

Figure 18.15 Distinguishing a multicomponent antigen–antibody system by the Ouchterlony method.

18.14.1.7 Immunoelectrophoresis

It was devised by Grabar and Williams. It combines the techniques of immunodiffusion and electrophoresis. It is a method in which different Ags in serum are separated according to their charge under an electric field. It has the advantage of greater resolution by its ability to separate components of complex Ag mixtures. It can resolve as many as 39 individual Ags in a mixture.

This technique is used to analyze serum proteins, Igs, and abnormal proteins such as myeloma proteins (Figure 18.16).

The test is performed in two stages:

1. Agar gel is taken on a slide, and a well is cut and filled with an Ag mixture such as human serum. Electric current is applied in an electrophoresis chamber. As a result, Ags are separated and distributed as separate spots based on their electrical charge. Positively charged proteins move to the negative electrode and negatively charged proteins move toward the positive electrode.

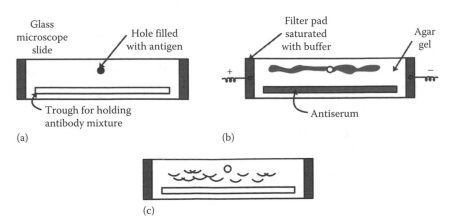

Figure 18.16 Arrangement for immune electrophoresis. (a) Hole is filled with protein, (b) electrophoresis causes migration of proteins; voltage removed and antiserum added to trough, and (c) immunodiffusion forms precipition lines.

2. A rectangular trough is then cut parallel to the electrophoretic separation and filled with appropriate Ab, and incubated. The Ags and antibodies diffuse toward each other into the gel and form a series of complex pattern of arc shaped precipitation bands that can be better visualized by staining.

18.14.1.8 Electroimmunodiffusion (Electroimmunoassay)
There are several methods to perform these tests.

18.14.1.8.1 Counter Current Immunoelectrophoresis.
This test depends on the diffusion of Ags toward anode and antibodies toward the cathode through the gel under electric field. The test is performed on a glass slides with agarose gel in which a pair of wells are punched. One well is filled with Ag solution and the other with antibodies, and electric field applied in electrophoresis chamber. The Ag and Ab, which have opposite charge, diffuse toward each other in an electric field by forming a line of precipitation where they meet in equivalent amounts in about 30–60 min (Figure 18.17).

Uses: It is a rapid and highly specific method for detection of both Ags and antibodies in serum, cerebrospinal fluid (CSF), and other body fluids in the diagnosis of many infectious diseases including bacterial, viral, fungal, and parasitic. It is commonly used for detection of HBsAg in patient serum, alpha fetoprotein in serum, *Cryptococcus* Ags in CSF, meningococcal Ags in CSF, and detection of hydatid and amoebic Ags in serum.

18.14.1.8.2 Rocket Electrophoresis.
This method is an adaptation of radial immunodiffusion. It is used for the quantitative estimation of human plasma Ags and viral Ags. It combines the speed of electrophoresis with quantification of Ag of radial immunodiffusion (Figure 18.18).

The Ab is added to molten agar gel and poured on a glass slide and allowed to set; wells are cut and filled with Ag solution (standard concentrations and unknowns) and current is applied in an electrophoresis chamber, allowing little time for diffusion. When Ags migrate into gel, it forms precipitation lines with antibodies in the gel. As more Ag arrives, the precipitate at the advancing front dissolves, while the precipitate at the sides remains; this gives a rocket like configuration. Hence, the procedure is commonly called rocket immunoelectrophoresis.

The area of each rocket or its height is directly proportional to the initial concentrations of Ag.

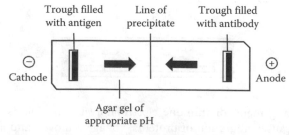

Figure 18.17   Counter immune electrophoresis.

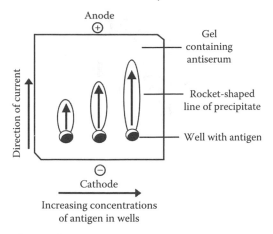

Figure 18.18 Rocket electrophoresis.

18.14.1.8.3 Laurell's Electroimmunodiffusion in Two Dimensions. It is used to quantify each of several components in an Ag mixture. In this method, the Ag mixture is first electrophoretically separated into different components, followed by immunodiffusion into gel containing Ab so that rocket-like precipitation lines are produced. Thus, this technique is similar to immunoelectrophoresis and is a variant of rocket electrophoresis (Figure 18.19).

1. Ags are separated by their electrophoretic mobility on the first run.
2. The second run at right angle to the first drives the Ags into the antiserum containing gel to form precipitation rockets.

18.14.2 Agglutination Reactions

Agglutination means clumping of the particles. In most of the agglutination reactions, the Ag is particulate, and Ab is in the soluble form. However, if antibodies are coated on particles, agglutination can also occur with soluble Ags.

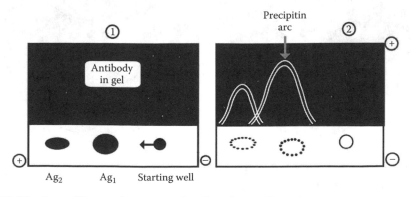

Figure 18.19 Laurell's two-dimensional rocket electrophoresis.

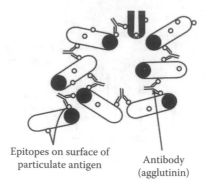

Epitopes on surface of
particulate antigen

Antibody
(agglutinin)

Figure 18.20 Agglutination reaction.

18.14.2.1 Definition

When a particulate Ag combines with its corresponding Ab in the presence of an electrolyte at a suitable temperature and pH, the Ag–Ab complexes form large clumps/aggregates/agglutinates that are visible and can be seen with the unaided eye. Such reactions are called agglutination reactions (Figure 18.20). The antibodies participating in these reactions are called *agglutinins* and Ags referred to as *agglutinogens*. Agglutination tests are more sensitive for both either qualitative or quantitative assay of antibodies.

18.14.2.2 Slide Agglutination Test

This is the direct agglutination reaction, mainly utilized for the identification of bacterial isolates from clinical specimens as well as for blood grouping (to type RBC) and cross-matching.

In this test, a drop of specific antiserum is added to a smooth, uniform saline suspension of a particulate Ag on a slide, agglutination takes place within a few seconds, which is usually visible to the naked eye (Figure 18.21). A control

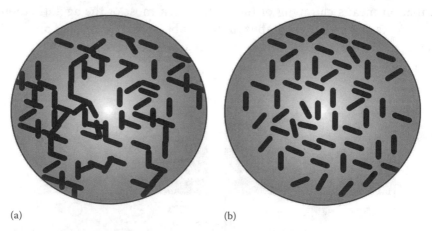

(a)   (b)

Figure 18.21 Illustration of slide agglutination test. (a) Agglutination and (b) no agglutination (antiserum not specific for antigen).

consisting of Ag suspension in normal saline without antiserum is put up to check auto agglutination of the Ag.

18.14.2.3 Tube (Direct) Agglutination Test

This is the most widely used standard quantitative method for the estimation of antibodies in serum specimens.

The test is performed by adding a fixed amount of particulate Ag to a series of tubes containing doubling dilutions of the antiserum (patient's serum); for example, tube 1 = 1/20, tube 2 = 1/40, tube 3 = 1/80, tube 4 = 1/160, tube 5 = 1/320, and so on. The reciprocal of the highest dilution in which agglutination occurs is taken as the titer of the antiserum (Figure 18.22). The test is used viz. Widal test for the serodiagnosis of typhoid and paratyphoid fevers (enteric fever), tube agglutination test for brucellosis, Weil–Felix reaction is a heterophile agglutination test for the serodiagnosis of typhus fevers, Paul–Bunnell test is a heterophile agglutination test for the serodiagnosis of infectious mononucleosis, cold agglutination and *Streptococcus* MG agglutination tests are heterophile agglutination tests for the serodiagnosis of primary atypical pneumonia.

18.14.2.4 Passive (Indirect) Agglutination Test

In this test, soluble Ags are detected by first being made to adhere to the surfaces of inert particulate carrier substances such as latex beads, bentonite particles, and RBC. When specific antiserum is added to such Ag-coated particles, large clumps or aggregates are formed that are easily visible to the naked eye. Thus, a precipitation test can be converted into an agglutination test (Figure 18.23). Passive agglutination tests are more sensitive for detecting antibodies.

18.14.2.4.1 Latex Agglutination Tests.   These are widely employed for the detection of antistreptolysin O antibodies, cardioreactive protein, rheumatoid arthritis factor, etc. Latex particles are cheap, relatively stable, and do not cross react with other antibodies.

18.14.2.4.2 Rose–Waaler Test.   This is a passive hemagglutination test (sensitized sheep RBC agglutination test) used to detect RA factor. In passive agglutination, soluble Ags are first made to adhere to particulate carriers before reaction with antibodies. When the carrier is RBCs, the method is termed passive hemagglutination.

18.14.2.4.3 Indirect Hemagglutination Test.   When RBCs are coated with Ag to detect antibodies in the serum, the test is called the indirect hemagglutination (IHA). This is most commonly used for the serodiagnosis of many parasitic diseases such as extraintestinal amoebiasis, hydatid disease, and toxoplasmosis.

When the Ag is a natural constituent of a particle (e.g., RBC and bacterial cell), the agglutination reaction, is referred to as direct agglutination. When the agglutination reaction takes place between antibodies and soluble Ag that has been attached to an insoluble particle, the reaction is referred to as passive agglutination.

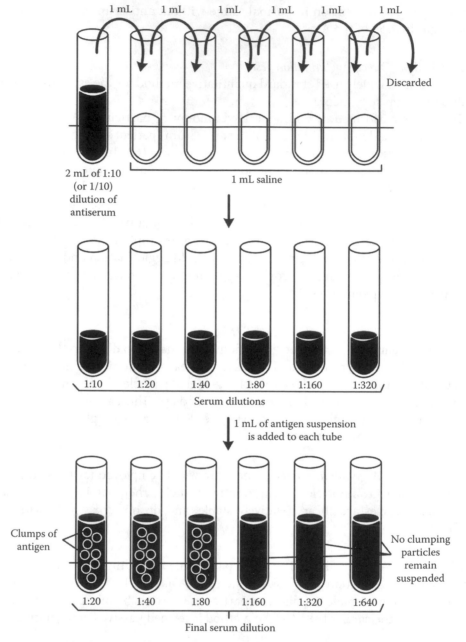

Figure 18.22 Protocol for preparing doubling dilution for tube agglutination test. Passive (indirect) agglutination test.

18.14.2.4.4 Latex Agglutination Test for Pregnancy. The modern pregnancy test detects the increased levels of human chorionic gonadotropin (HCG) hormone that occurs in female urine and blood early in pregnancy.

The urine from the female containing HCG is mixed with an Ab specific for HCG. Then, latex particles coated with HCG are added. If HCG is present, it binds to HCG

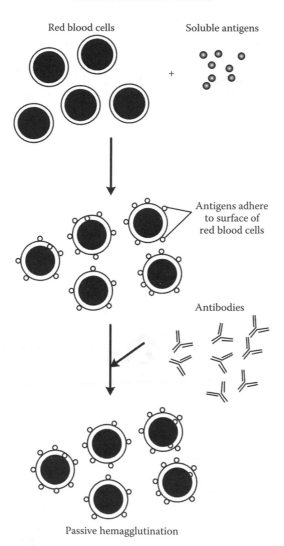

Figure 18.23 Passive agglutination test.

specific antibodies, thereby preventing them from agglutinating the latex particles coated with HCG. Therefore, in a positive test, there will be no agglutination. In a negative test, the latex particles coated with HCG are agglutinated by Ab to HCG.

18.14.2.5 Reverse Passive Agglutination Test
When instead of the Ag, antibodies are attached to the RBCs to detect Ags, it is known as reverse passive hemagglutination, for example, detection of HBsAg in patient's serum.

18.14.2.6 Coagglutination
In this method, *Staphylococcus aureus* (Cowan I strain) cells are used as carrier particles to coat antibodies because protein-A on their cell wall can react with

F$_c$ region of Ig molecules while leaving the Fab region free to combine with a specific Ag. The advantage of the test is that these cells show greater stability than latex particles.

Thus, *Staphylococcus aureus* (Cowan I strain) cells with attached IgG antibodies on their protein-A of cell wall against a specific Ag will agglutinate bacteria that have that specific Ag. This is called co-agglutination (Figure 18.24).

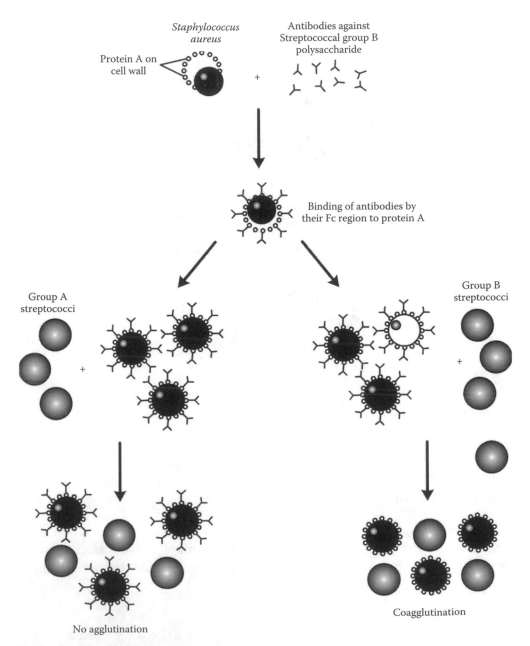

Figure 18.24   Illustrating coagglutination.

It has several important diagnostic applications such as streptococcal grouping, gonococcal and mycobacterial typing, detection of typhoid Ag from enteric fever, detection of cryptococcal Ag from CSF in the diagnosis of cryptococcal meningitis, detection of amoebic and hydatid Ags in the serum for the diagnosis of amoebiasis and hydatid disease, respectively.

18.14.2.6.1 Coombs Test (Antiglobulin Test). It is used to detect nonagglutinating, antierythrocyte (Anti-Rh IgG) antibodies. The major applications of Coombs test are in evaluating hemolytic disease of the new born, diagnosis of autoimmune hemolytic anemia, and erythrocyte (RBC) typing in blood banks.

18.14.2.6.1.1 Principle. Anti-Rh IgG antibodies when react with RBC (coated on RBC) fail to produce agglutination. The addition of an antiglobulin antiserum (Coombs serum) produced in heterologous species (e.g., rabbit antihuman globulin) produces marked agglutination.

The test may be direct or indirect (Figure 18.25). The direct Coombs test measures bound Ab, whereas the indirect Coombs test measures serum Ab.

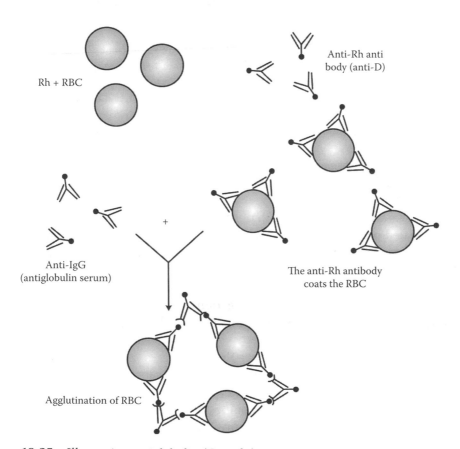

Figure 18.25 Illustrating antiglobulin (Coombs) test.

18.14.2.6.1.2 Direct Coombs Test. In the direct Coombs test, anti-Igs are added to RBC that are suspected of having antibodies bound to Ags on their surfaces. For example, a newborn baby is suspected of having hemolytic disease of the newborn caused by maternal anti-Rh IgG antibodies that are bound to the baby's RBC. Then, the addition of anti-Ig to suspension of baby's RBC causes binding of the anti-Ig to the maternal IgG on the surface of the RBC leading to agglutination.

18.14.2.6.1.3 Indirect Coombs test. It is a two-stage reaction for detection of incomplete (anti-Rh IgG) antibodies in patient's serum. The indirect Coombs test is used to detect the presence of antibodies in the serum, specific to Ags on RBC. The serum antibodies (Anti-Rh IgG), when added to the RBC, may fail to cause agglutination because of the zeta potential. The subsequent addition of anti-Ig will cause agglutination. A common application of the indirect Coombs test is to detect anti-Rh IgG antibodies in the blood of a Rh negative woman.

The woman's (patient's) serum is first incubated with Rh positive RBC so that Ab coats RBC. In the second step, Coombs antiglobulin serum (anti-Ig) is added, as in the direct Coombs test so that Ab-coated RBC are agglutinated.

18.14.2.6.2 Hemagglutination. Different types of hemagglutination with examples are as follows:

1. Viral hemagglutination, for example, influenza virus agglutinates RBC.
2. Passive hemagglutination, for example, Rose–Waaler test is a sensitized sheep RBC agglutination test, used to detect RA factor.
3. IHA test used in the serodiagnosis of several parasitic, bacterial diseases, for example, IHA for malaria, TPHA for syphilis.
4. Fimbrial hemagglutination is used to detect the presence of fimbriae.

18.14.2.6.2.1 Viral Hemagglutination. Certain viruses have the ability to agglutinate RBC from chicken, guinea pig, and humans and is known as viral hemagglutination (Figure 18.26).

When RBCs are added to serial dilutions of the viral suspension (or of serum from a virus-infected patient), in tubes or plastic trays, RBCs which are not agglutinated settle at the bottom in the form of a button, while the agglutinated cells spread as an even coating in a shield like pattern (Figure 18.27).

18.14.2.6.2.2 Uses. Test is used for detection and assay of influenza virus in culture fluids and for titration and standardization of killed influenza vaccines.

18.14.2.6.3 Hemagglutination Inhibition. The antibodies against viruses that can cause hemagglutination can bind to viruses and prevent them from agglutinating RBC. This inhibition phenomenon is called viral neutralization, and the test based on it is called a viral hemagglutination inhibition (HI) test, which is used for detection and quantitation of Ab against virus (Figure 18.28).

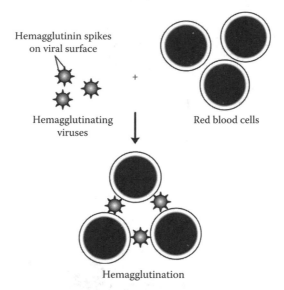

Figure 18.26 Illustration for viral hemagglutination test.

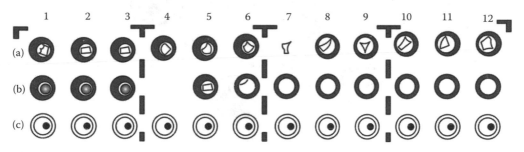

Figure 18.27 Hemagglutination reactions seen in the wells of a microtiter plate. (a) complete hemagglutination has occurred in all of the wells, (b) only wells 3–6 show complete hemagglutination, and (c) no hemagglutination in any of the wells.

18.14.3 Complement Fixation Test

Complement fixation test (CFT) detects the presence of complement fixing antibodies (IgM and IgG) present in a patient's serum. Complement-fixation tests are used in the diagnosis of certain viral, fungal, and bacterial diseases, for example, Wassermann test for the diagnosis of syphilis.

18.14.3.1 Principle
The ability of Ag–Ab complexes to fix complement is made use of in CFT. The test requires five reagents which are as follows:

1. Ag, may be soluble or particulate.
2. Ab (usually patient's serum).

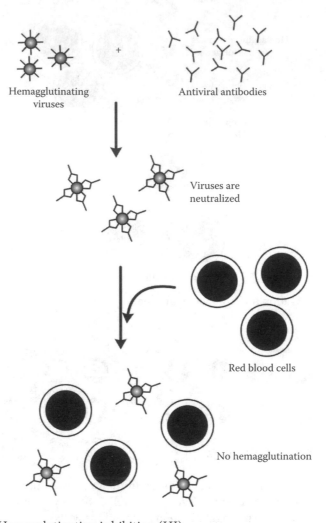

Hemagglutinating viruses

Antiviral antibodies

Viruses are neutralized

Red blood cells

No hemagglutination

Figure 18.28   Hemagglutination inhibition (HI).

3. Complement (C): The source of complement is Guinea pig serum. As C is heat labile, the serum should be freshly drawn or preserved either in lyophilized or frozen state or by Richardson's method.
4. Sheep RBC.
5. Amboceptor, an Ab to sheep RBC raised in rabbits. Sheep RBC and amboceptor together are known as hemolytic system.

18.14.3.2 Procedure

The test may be carried out in small tubes, in plastic microtitration plates or even in the automated system (Figure 18.29).

The test is performed in two stages:

The test serum from the patient must first be inactivated at 56°C for 30 min to destroy its complement activity.

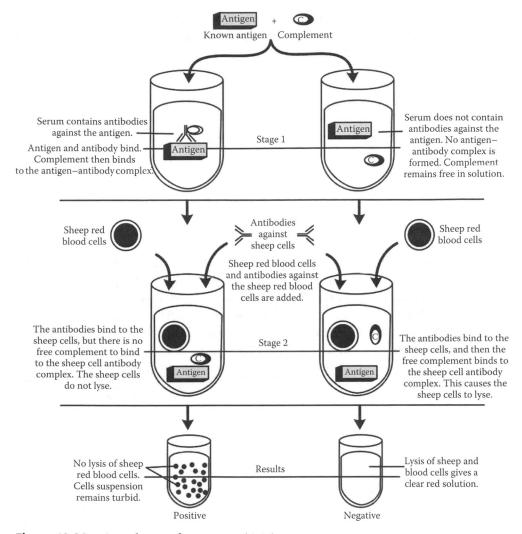

Figure 18.29 Complement fixation test (CFT).

I Stage: Known Ag + test serum + complement mixture is incubated at 37°C for 30 min or longer. The Ag–Ab reaction cannot be seen at the end of this stage since all components are soluble.

II Stage: Sheep RBC + amboceptor are added.

18.14.3.3 Results and Observations

In the first stage, if complement has been fixed because of an Ag–Ab reaction, it is no longer available for the lysis of sheep RBC in stage II. Therefore, if there is no hemolysis, it indicates a positive result because the complement fixing antibodies in the test serum had reacted with Ag in I stage and utilized (fixed) the complement.

If there is hemolysis, it indicates a negative result because complement is available for hemolyzing the sheep RBC in stage II; therefore, no antibodies specific for the known Ag were present in the patient's serum.

18.14.4 Immunolabeled Tests

There are several serological tests in which either antigen or antibody is labeled:

Radioimmunoassay (RIA): ^{125}I and ^{32}P are conjugated/labeled with antigen or antibody.

Immunofluorescent tests: Conjugates used are fluorescein isothiocyanate, lissamine, and rhodamine conjugated with antigen or antibody.

18.14.4.1 Immunofluorescence

Conjugates used are fluorescein isothiocyanate (FITC), lissamine, and rhodamine conjugated with Ag or Ab. Fluorescence is the property of absorbing light rays of one particular wavelength and emitting rays with different wavelengths. Fluorescent dyes show up brightly under ultraviolet (UV) light as they convert UV light (invisible light) into visible light. Such fluorescent dyes can be conjugated to antibodies and such labeled antibodies can be used to detect and identify Ags in tissues. Immunofluorescence is a method for localizing an Ag by the use of fluorescent labeled antibodies. The most commonly used fluorescent dye in immunology is FITC, which fluoresces with a visible greenish color when excited by UV light. The technique is of two types.

18.14.4.1.1 Direct Immunofluorescence.
It is used to directly detect Ag and involves reacting the target tissue (or microorganism) with fluorescently labeled Ag specific antibodies. The specimen (cells or microorganisms) containing the Ag are fixed on to a slide. Then specific fluorescein labeled antibodies are added to the slide and incubated. The slide is washed to remove any unbound Ab and examined with the fluorescence microscope for a yellow–green fluorescence, which reveals the Ag location. Direct fluorescent Ab technique is usually used to identify a particular kind of microorganism in a clinical specimen, for example, *Streptococcus pyogenes* in a throat swab from a patent. Direct immunofluorescence is used to identify Ags found on the surface of Group A *Streptococci* and to diagnose enteropathogenic *E. coli*, *Neisseria meningitidis*, *Salmonella typhi*, *Shigella sonnei*, *Listeria monocytogenes*, *Haemophilus influenzae*, and the rabies virus.

Direct immunofluorescence is also used clinically for identifying lymphocytic subsets and for demonstrating the presence of specific protein deposition in certain tissues such as kidney and skin in cases of SLE. The major disadvantage of the direct immunofluorescence is the preparation of separately labeled Ab for each pathogen.

18.14.4.1.2 Indirect Immunofluorescence.
In this technique, a known Ag is fixed on to a slide. The test antiserum (patient's serum) is then added, and if the specific Ab is present, it reacts with the Ag to form a complex. When fluorescent-labeled anti-Ig is added, it binds to the specific Ab that has already reacted with the Ag in the smear.

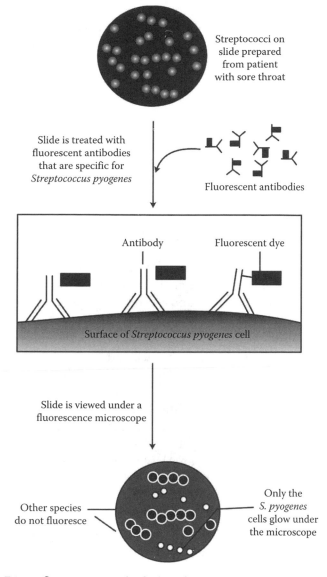

Figure 18.30 Direct fluorescent antibody (DFA) test.

After incubation and washing, the slide is examined with the fluorescence microscope. The occurrence of fluorescence shows that Ab specific to the test Ag is present in the serum (Figure 18.30).

The indirect fluorescent Ab (IFA) is used to detect the presence of antibodies to a specific microorganism in the body fluid of patient, for example, (1) the presence of IgG antibodies to an oral Spirochete in the serum or saliva of a patient with gum disease, (2) detection of *Treponema pallidum* antibodies in the diagnosis of syphilis (FTA-ABS and FTA-200). The IFA is more sensitive and has the advantage of using a single labeled antiglobulin (Ab to IgG) reagent to detect many different specific Ag–Ab interactions.

18.14.4.2 ELISA

ELISA is highly sensitive, highly specific, and less expensive technique used in serology to detect Ags or antibodies. This test involves the linking of various label enzymes to either Ags or antibodies. Horseradish peroxidase or alkaline phosphatase is labeled with either Ag or Ab. The ELISA is used for the detection of many infectious viruses, bacteria, fungi, parasites (protozoa) and drugs, toxins, food allergens, etc. ELISA is based on the following two principles:

1. Antibodies and some Ags can absorb to a solid support such as polystyrene, polyvinyl, or polycarbonate wells or membranes (discs) of polyacrylamide paper or plastic.
2. Ags and antibodies can be linked to enzymes, with the resulting complexes still fully functional, both immunologically and enzymatically.

Enzyme activity is used to measure the quantity of Ag or Ab present in the test sample. Enzymes used in ELISA include β-galactosidase, glucose oxidase, peroxidase, and alkaline phosphatase. The test is usually performed by using microtiter plates (96 well) suitable for automation. There are three basic ELISA techniques viz. double-Ab sandwich technique, indirect immunosorbent technique, and competitive ELISA.

18.14.4.2.1 Double Ab Sandwich Assay (Sandwich ELISA).

It is used for detection of Ags (e.g., HBsAg in patient's serum, rotavirus, and enterotoxin of *E. coli* in feces) (Figure 18.31). In this technique, specific Ab is placed in wells of a microtiter plate (or it may be attached to a membrane) and the Ab adheres to the inner surface of each well. The test Ag is added to each well. If the Ag reacts with the Ab, the Ag is retained when the well is washed to remove unbound Ag (Figure 18.32).

An Ab–enzyme conjugate specific for the Ag is then added; it will bind to the Ag already fixed by the first Ab, forming an Ab (with enzyme)—Ag–Ab sandwich.

A substrate that the enzyme will convert to a colored product is then added, and any resulting product is quantitatively measured by optical density scanning with an ELISA reader.

If the Ag has reacted with the adsorbed antibodies in the first step, the ELISA test is positive. If the Ag is not reacted with the adsorbed Ab, the ELISA test is negative.

18.14.4.2.2 Indirect Immunosorbent Assay (Indirect ELISA).

It is used for the detection of antibodies (e.g., antibodies to HIV, dengue, Japanese encephalitis, and rubella virus in the serum sample) (Figure 18.33).

In this technique, the Ag is adsorbed onto the inner surface of well wall. Test antiserum (usually patient's serum) is added and incubated. If any antibodies in the test antiserum have bound to the Ag, their presence is detected by adding an enzyme-linked anti-Ig (anti-Ab or Ab against Ab). Enzyme substrate is then added, and the color change can be detected by ELISA reader in terms of optical density.

18.14.4.2.3 Competitive ELISA.

In competitive ELISA, two specific antibodies, one conjugated with enzyme and other present in patient's serum are used. Two antibodies compete for the same Ag. Appearance of color indicates a negative test (absence of

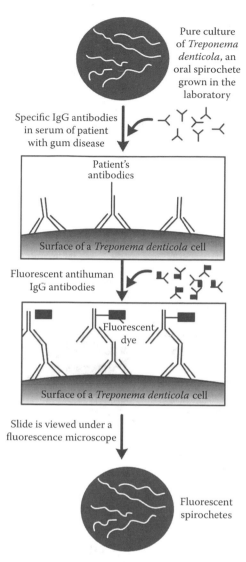

Figure 18.31 Indirect fluorescent antibody (IFA) technique.

antibodies), while the absence of color indicates a positive test (presence of antibodies). It is most commonly used to detect antibodies to HIV. The microtiter plate wells are coated with HIV Ag. The patient's serum to be tested is added and incubated at 37°C and then washed. If antibodies are present in patient's serum, react with Ag. This Ag–Ab complex can be detected by adding enzyme labeled specific antibodies to HIV.

In a positive test, no Ag is left for these antibodies to react. Hence, these antibodies remain free and are removed during washing. When substrate is added, no enzyme is available to act on it. Therefore, positive test shows no color. In a negative test, no antibodies are present in patient's serum, Ag in the coated wells is available to combine with enzyme labeled antibodies and the enzyme acts on the substrate to form color.

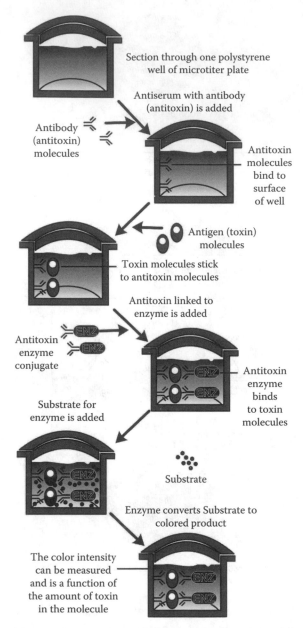

Figure 18.32 Double antibody sandwich ELISA technique for the detection and measurement of antigen (against *Staphylococcus* enterotoxin in a food extract). The presence of as little as 0.4 ng of toxin per mL of food extract can be detected.

18.14.4.3 Radioimmunoassay

Immunological assays using radioactive-labeled reagents are called RIAs. RIAs are highly sensitive but costly and associated with radioactive hazard. The technique was discovered by Rosalyn Yalow to detect insulin concentration in plasma, for which she won a Noble Prize in 1977. RIAs are available for all hormones, drugs in

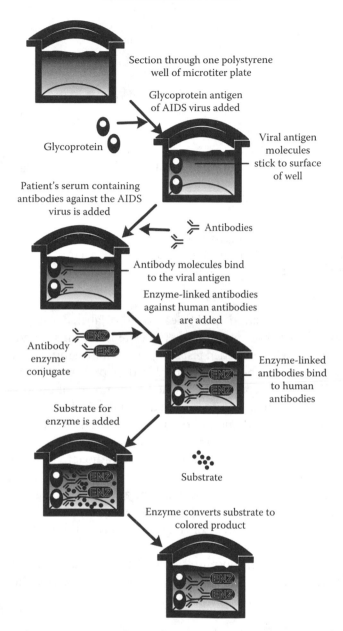

Figure 18.33 Indirect immunosorbent ELISA technique (against HIV) for the detection and measurement of antibody. (*Note*: necessary washing after each step is not shown.)

body fluids, Igs, tumor Ags, and viral Ags (e.g., HBsAg) and it has become an important tool in biomedical research and clinical practice.

18.14.4.3.1 Principle. The assay is based on competition between radioactive-labeled Ag (e.g., ^{125}I) and unlabeled Ags to bind to specific antibodies. This competition is determined by the level of the unlabeled (test) Ag present in reacting system (Figure 18.34).

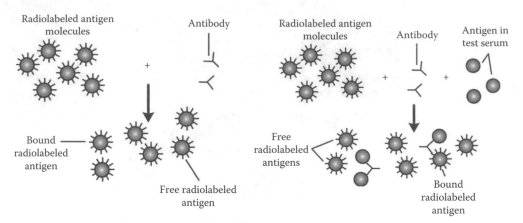

Figure 18.34 Radiolabeled antigens bind to specific antibodies and unlabeled antigen (variable) amount in test sample) competes with labeled antigen (fixed amount) for antibody-combining sites (fixed amount of antibodies).

18.14.4.3.2 Procedure Measured quantities of radio labeled Ag (of the same kind being tested) and Ab (specific to the Ag being tested) are mixed and incubated, one mixture with and one without added test sample. At increasing concentrations of test Ag, (unlabeled Ag), more labeled Ag is displaced or prevented from binding to the Ab molecules. The Ag–Ab complex is washed to remove unbound radiolabeled Ag from the mixture. The radioactivity associated with the Ab is then detected by means of radioisotope analyzers and autoradiography (photographic emulsions that show areas of radioactivity). A little amount of bound radioactivity indicates that there is a large amount of Ag, and a large amount of bound radioactivity indicates that there is little Ag in the sample.

A standard calibration curve is plotted on the graph by taking the ratio of bound radiolabeled Ag to free radiolabeled Ag against varying known amounts of unlabeled Ag and the Ag concentration of the test sample can be read from this curve (Figure 18.35).

To determine an unknown concentration of Ag in a solution, a sample of the solution is mixed with predetermined amounts of labeled Ag and Ab. The ratio of bound

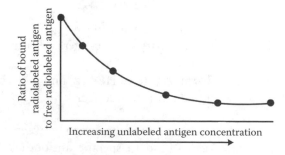

Figure 18.35 Curve showing the ratio of antibody-bound labeled antigen to unbound labeled antigen decreases with the increase of amount of unlabeled antigen.

to free radioactivity is compared with that obtained in the absence of unlabeled Ag (the latter value is set at 100%).

18.14.4.4 Immunoelectroblotting
In immunoblot techniques, antibodies detect individual proteins (Ags) in complex mixtures. The technique combines the sensitivity of enzyme immunoassay with much greater specificity viz. Western blot.

18.14.4.4.1 Western Blotting.
It is used as a confirmatory test for the serodiagnosis of HIV infection. This technique is used to identify proteins (Ags) by using specific antibodies as analytical probes. The sources of Ab probes may be serum from patients with known infections or purified polyclonal Ab preparations or monoclonal antibodies.

Protein Ag mixture is first separated into its components by polyacrylamide and then blotted onto the nitrocellulose membrane. It is then incubated with specific Ab (patient's serum). This primary Ab binds to its target protein Ag and is immobilized on the nitrocellulose blotting support. This Ag–Ab complex can then be identified with radioactive or enzyme labeled antibodies (antiantibodies or antihuman gamma globulin) to primary antibodies, and can then be visualized by autoradiography. If required, the labeled anti-Ab can be removed, and the protein Ags can be studied further.

When enzyme tagged antihuman gamma globulin is added to detect primary Ab, enzyme substrate is added subsequently. The substrate changes color in the presence of enzyme and permanently stains the nitrocellulose paper.

18.14.4.5 Immunochromatographic Technique
Immunochromatographic technique has wide application in serodiagnosis due to its simplicity, economy, and reliability. One-step qualitative immunochromatographic test system for HBsAg contains a small cassette with a membrane impregnated with anti HBsAg Ab-colloidal gold dye conjugate. The membrane is exposed at three windows on the cassette. The test serum is dropped into the first window. As the serum travels upstream by capillary action, a colored band should appear in every case at the third window, which forms an inbuilt control, in the absence of which the test is invalid.

18.14.4.6 Immunoelectron Microscopy
Ag–Ab reactions can be visualized by electron microscope. When viral particles are mixed with specific antisera and observed under the electron microscope, they are seen to be clumped, for example, detection of Hepatitis A virus and rotavirus from fecal samples.

18.14.4.6.1 Immunoferritin Test.
Ag–Ab reactions can be visualized by electron microscope with the help of appropriate markers. Ferritin, an electron dense substance from horse spleen can be conjugated with Ab, and such labeled Ab reacting with an Ag can be visualized under the electron microscope.

18.14.4.6.2 Immunoenzyme Test. Ag–Ab reactions can be visualized by electron microscope with the help of appropriate markers such as enzymes, for example, peroxidases, glucose oxidase, phosphatases, tyrosinase, etc.

Some stable enzymes, such as peroxidases, can be conjugated with antibodies. Tissue sections carrying the corresponding Ags are treated with peroxidase labeled antisera. The peroxidase bound to the Ag can be visualized under the electron microscope, by treatment with enzyme substrate.

18.14.5 Neutralization

Neutralization tests are Ag–Ab reactions that determine whether the activity of a toxin or virus has been neutralized by Ab. Laboratory animals or tissue culture cells are used as indicator systems in these tests. The toxin or virus to be assayed has known effects on the indicator system. The effect in the animals might be death, paralysis, or skin lesions.

18.14.5.1 Toxin Neutralization Tests

The test is based on the principle that the biological action of toxin is neutralized on reacting with specific neutralizing antibodies called antitoxins, for example, Nagler's reaction and Schick test. For example, when the exotoxin of *Clostridium botulinum* is suspected of causing food poisoning in a person, a sample of the suspect food or the serum, stools, or vomits of the patient is collected. Two groups of indicator mice are used. The control group receives the botulinum antitoxin and the experimental group does not. Filtrates of the samples are injected into both groups of mice. If the toxin is present, all mice except those receiving the antitoxin will die, and the test is positive for botulism.

Neutralization of viruses by their specific antibodies is called *virus neutralization* tests. Viral neutralization is frequently used to detect viral infections. Suspected serum containing viral antibodies can be introduced into tissue culture cells or embryonated eggs. If antibodies are present against the virus, viral neutralization will occur and prevent the virus from infecting the culture resulting in no cytopathic effect. Viral HI test is an example of virus neutralization test commonly used in the diagnosis of viral infections such as influenza, mumps, and measles. The mechanism of neutralization of toxins is illustrated in Figure 18.36.

18.14.6 Opsonization

Opsonization (Greek; Opson = prepare victims for) is a process in which microorganisms or other particles are coated by Ab and/or complement, thereby being prepared for recognition and ingestion by phagocytic cells.

Opsonins are substances that make an Ag more susceptible to phagocytosis. Antibodies (IgG) act as opsonins by connecting to the Ag at $F(ab)_2$ site and to the macrophages by FC receptor. This binding forms a bridge between the phagocyte and Ag; the macrophage then phagocytoses the Ag and Ab (Figure 18.37).

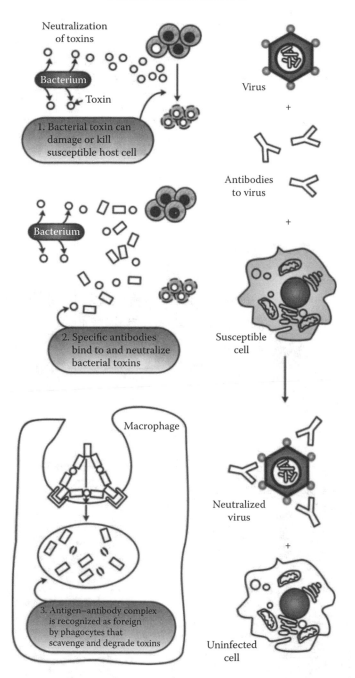

Figure 18.36 Flow diagram showing neutralization of toxin.

When Ag–Ab complexes activate complement pathway, C3b fragment assembles on the surface of microorganisms. Both neutrophils and macrophages possess surface receptors for C3b. A bridge is formed between microorganisms and phagocytic cells via complement that aids in phagocytosis (Figure 18.37). If both Ab and C3b opsonize, binding is greatly enhanced (Figure 18.37). The opsonic index is defined

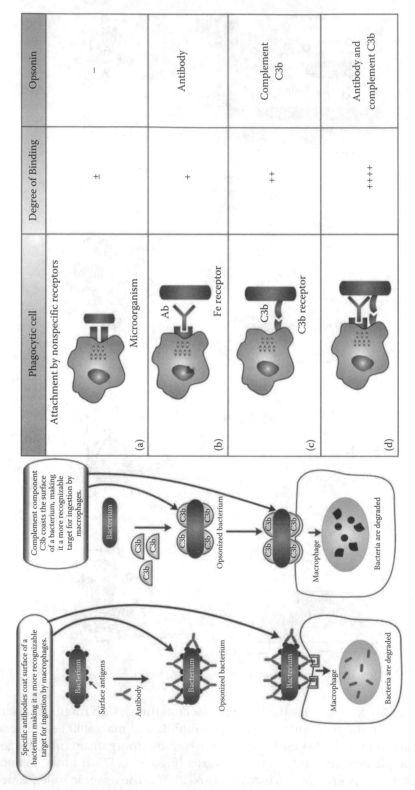

Figure 18.37   Opsonization for neutralization of toxin and bacterium.

as the ratio of the phagocytic activity of patient's blood for a given bacterium, to the phagocytic activity of blood from a normal individual. It is measured by incubating fresh blood with bacterial suspension at 37°C for 15 min and estimating the average number of phagocytosed bacteria per neutrophils (phagocytic index) from stained blood films. It is used to study the progress of resistance during the course of the diseases.

18.14.7 Chemiluminescence Assay

The phenomenon chemiluminescence refers to a chemical reaction in which energy is emitted in the form of light. The chemiluminescence assay uses chemiluminescent compounds that emit energy in the form of a light, during Ag–Ab reactions.

The emitted light is measured, and the concentration of the analyte is calculated. The assay has been fully automated. This property, chemiluminescence, has also been applied for drug sensitivity testing of *Mycobacterium tuberculosis*.

18.15 Collection of Blood Sample

The collection of the blood and plasma, which are frequently used in the experiments, must be carried out with precautions. Collect the blood into a glass container and leave the blood to clot at 37°C for 1 h. Cool on ice to allow the clot to retract, and separate the serum by decanting the liquid from the clot and centrifugation on a bench centrifuge. Store the serum at 0°C until required then warm to room temperature just before use.

Immunochemical assays are now used extensively in the biological sciences, and some examples of those currently in use are described herein as the series of experiments.

18.16 Experiment: To Determine the ABO Blood Group by Using Slide Agglutination Test

18.16.1 Introduction

The interaction between Ab and a particulate Ag results in visible clumping called agglutination. Antibodies that produce such reactions are called agglutinins. Agglutination reactions are similar in principle to precipitation reactions; they depend on the crosslinking of polyvalent Ags. Just as, an excess of Ab inhibits precipitation reactions, such excess can also inhibit agglutination reactions; this inhibition is called the prozone effect. Because prozone effects can be encountered in many types of immunoassays, understanding the basis of this phenomenon is of general importance. Agglutination reactions are routinely performed to type RBCs. In typing for the ABO Ags, RBCs are mixed on a slide with antisera to the A or B blood-group Ags. If the Ag is present on the cells, it agglutinates, forming a visible clump on

TABLE 18.7 Blood Grouping

Blood Group	Antigen on RBC	Antibodies in Plasma	Can Donate to Recipient	Can Receive from Donor
A	A	B (anti B)	A, AB	O, A
B	B	A (anti A)	B, AB	O, B
AB	A and B	Nil	AB,	A, B, AB, O
O	Nil	A and B	A, B, AB, O	O

the slide. Determination of which Ags are present on donor and recipient RBCs is the basis for matching blood types for transfusions (Figure 18.40 and Table 18.7). When RBCs are mixed with various reagent antisera (soluble Ab), agglutination will occur on the slides containing cells positive for (possessing the Ag) the corresponding Ag. No agglutination will occur when the red cells do not contain the corresponding Ag. One primary application of this principle is blood typing (Figure 18.38).

18.16.2 Materials

Ab A, Ab B, RBCs, microscopic slides, applicator sticks, pipettes, and blood lancet (Figure 18.39).

18.16.3 Protocol

On the section of slide label anti-A, place one drop of Ab A over it.
On the section of slide label anti-B, place one drop of Ab B over it.
Place one drop of cells in each Ab containing circle.
Carefully mix each solution with a separate applicator stick.
Tilt slowly for 1 min.
Record results (Table 18.8)

18.16.4 Interpretation

Agglutination (clumping) of the RBCs is positive. No agglutination is negative. It is critical to read the results immediately as false positives can occur when the mixture begins to dry on the slide (Figure 18.40).

18.17 Experiment: Determination of Rh Factor in Human Being

18.17.1 Introduction

Besides the blood groups discussed earlier, the human blood may also contain another factor that is quite independent of the blood groups. This factor has enabled

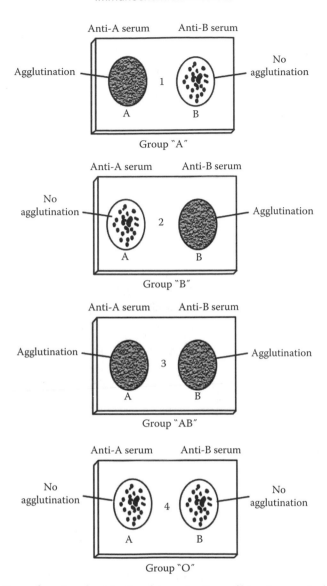

Figure 18.38 Procedure involved in the determination of blood group in human being.

Figure 18.39 Blood lancet.

TABLE 18.8 Result on Slide Agglutination Test

Sample	Anti-A	Anti-B	Interpretation
Number 1			
Number 2			

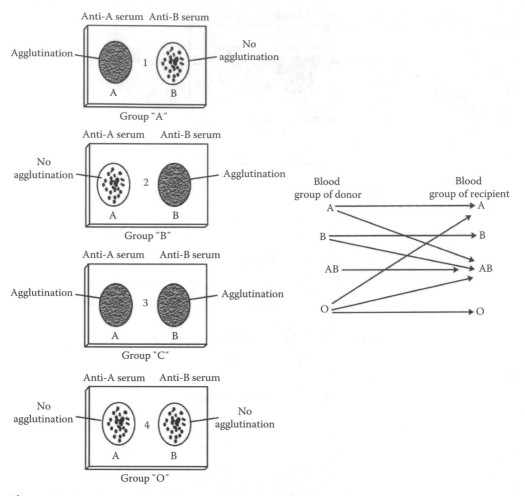

Figure 18.40 Compatible blood groups for transfusion.

to differentiate the human blood further into two groups namely Rh-positive and Rh-negative. This factor is of great significance during blood transfusion. Rh stands for rhesus. When the rabbits are injected with erythrocytes of rhesus monkey called *Muraceous rhesus*, an anti-rhesus serum (or anti-Rh) is developed. This anti-rhesus serum on being tested brings about the agglutination of the erythrocytes of monkey as well as a great percentage of human population. About 97% Indians possess Rh positive factor and the rest Rh negative factor.

18.17.2 Materials

Blood lancet, clean slide, and anti-Rh serum (obtain from a standard company).

18.17.3 Protocol

1. Take a neat and clean air-dried microslide and mark an oval area with the help of a wax pencil in the center of the slide.
2. Prick your sterilized finger with a blood lancet and after discarding the first one or two drops, place the drop of blood in the marked area.
3. Add a drop of anti-rhesus serum (anti-Rh procured from market) to it and mix it gently with a toothpick.
4. Gently warm it over a low flame of burner avoiding overheating or in a warming box (Figure 18.41).

18.17.4 Observation and Result

If the agglutination takes place, the person concerned with the blood sample is called (Rh+), otherwise (Rh−).

18.18 Experiment: Demonstration of Agglutination Reaction of Unknown Bacterial Culture by Slide Agglutination Technique

18.18.1 Introduction

Many closely related microorganisms that are identical physiologically can be differentiated only by determining their antigenic nature. The method of determining

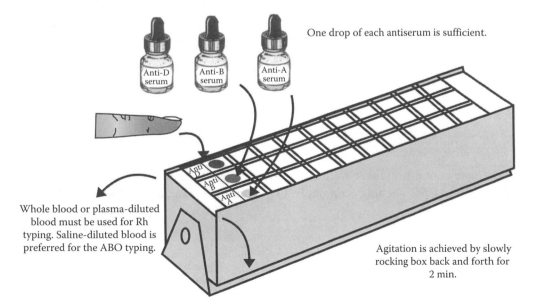

Figure 18.41 Blood typing with warming box.

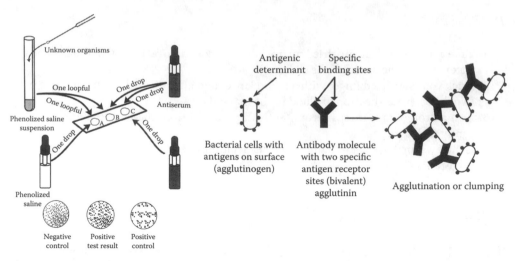

Figure 18.42   Slide agglutination test for serotyping.

the presence of specific Ag in a microorganism is called serological typing (serotyping). Organisms of different species not only differ morphologically and physiology but they also differ in protein makeup. The different proteins of bacterial cells that are able to stimulate Ab production when injected into an animal are Ags. The Ag structure of each species of bacteria is unique to that species and, like the fingerprint of an individual, can be used to identify the organism. Many closely related microorganisms that are identical physiologically can be differentiated only by determining their antigenic nature. Serotyping consists of applying suspension of the organism to antiserum, which is specific for the known Ags. If the Ags are present, the antibodies in the antiserum will combine with the Ags, causing agglutination or clumping of the bacterial cell. Serotyping is playing a key role in vaccine research against several infectious agents including viruses and bacteria. Method used for the serotyping is illustrated in Figure 18.42.

18.18.2 Materials

Slide/plate containing three wells, serological tube, phenolized saline (0.85% sodium chloride and 0.5% phenol), *Salmonella* O polyvalent serum, and *Salmonella* O Ag.

18.18.3 Protocol

1. Take a clean depression slide or spot plate and label the three depressions as A, B, and C.
2. Prepare a phenolized saline suspension of each culture in separate serological tubes by transferring one or more loopfuls of organisms in 1 mL of phenolized saline. Mix it thoroughly and observe that it should be turbid.

3. Transfer 0.05 mL from the phenolized saline suspension of one tube to cavity A and B.
4. Add one drop of *Salmonella* antiserum inside each of the two depressions B and C.
5. Add one drop of phenolized saline to depression A.
6. Add one drop of Ag to depression C.
7. Separately mix the cells with serum in each depression with an unused end of the application stick.
8. Gently rock the slide back and forth, holding it horizontally for 2–3 min.
9. Repeat this process on another depression slide for the other organisms. To cavity B and C, add one drop of *Salmonella* O polyvalent serum.
10. Compare all the three mixtures for agglutination reaction.

18.18.4 Results and Observations

If agglutination appears in cavity B, the bacterial sample is *Salmonella*. Slide agglutination technique is shown in Figure 18.42.

18.19 Experiment: Estimate the Total Leukocyte Count by Hemocytometer

18.19.1 Introduction

The glacial acetic acid lyses the red cells, while the gentian violet slightly stains the nuclei of the leukocytes. The blood specimen is diluted 1:20 in a white blood cell (WBC) pipette with the diluting fluid, and the cells are counted under low power of the micro scope by using a counting chamber. The number of cells in undiluted blood is reported per cubic mm (cu mm) (μL) of whole blood.

Increase in total leukocyte count of more than 10,000/cu mm (μL) is known as leukocytosis and decrease of less than 4000 cu mm (μL) as leukopenia.

18.19.1.1 Causes of Leukocytosis

18.19.1.1.1 Pathological. Leukocytosis is a common disease when there is a transient period of infections. The degrees of the rise in leukocytes depend on the type and severity of the infection and the response of the body. The infection may be (1) bacterial, (2) viral, (3) protozoal (malaria), or (4) parasitic (filaria and hookworm infection). Leukocytosis is also observed in severe hemorrhage and in leukemia.

18.19.1.1.2 Physiological. Age: At birth, the total leukocyte count is about 18,000/cu mm (μL). It drops gradually to adult level

Pregnancy: At "full term," the total count tends to be about 12,000–15,000/cu mm (μL). It rises soon after delivery and then gradually returns to normal

High temperature
Severe pain
Muscular exercise

18.19.1.2 Causes of Leukopenia
Certain viral and bacterial infections (typhoid) lead to leukopenia rather than leukocytosis.

Infections

> Bacterial (typhoid, paratyphoid, tuberculosis, etc.)
> Viral (hepatitis, influenza, measles, etc.)
> Protozoal (malaria)
> Some cases of leukemia
> Primary bone marrow depression (aplastic anemia)
> Secondary bone marrow depression (due to drugs, radiation, etc.)
> Anemia (iron deficiency megaloblastic, etc.)

18.19.1.3 Normal Values
- Adults 4000–10,000/cu mm (µL)
- At birth 10,000–25,000/cu mm (µL)
- 1–3 years 6000–18,000/cu mm (µL)
- 4–7 years 6000–15,000/cu mm (µL)
- 8–12 years 4500–13,500/cu mm (µL)

18.19.2 Materials

Microscope
Improved Neubauer chamber (Figure 18.43)
WBC pipette (Figure 18.44)
WBC diluting fluid containing glacial acetic acid 2.0 mL, 1% (w/v) gentian violet: 1.0 mL and distilled water 97 mL. This solution is stable at room temperature (25°C ± 5°C). A pinch of thymol may be added as a preservative.

18.19.2.1 Specimens
Double oxalated or ethylenediaminetetraacetic acid (EDTA) blood
Capillary blood (specimen need not be a fasting sample)

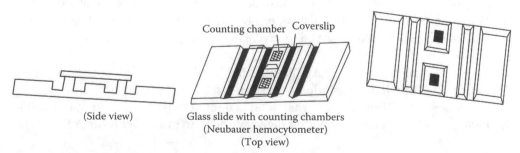

Figure 18.43 Hemocytometer with Neubauer counting chamber with coverslip over it.

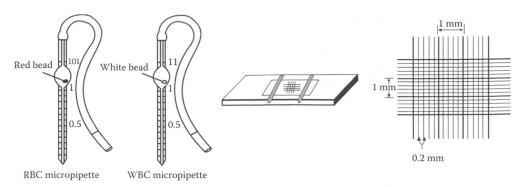

Figure 18.44 Micropipettes and Neubauer ruling on a blood counting chamber.

18.19.3 Protocol

1. Draw blood up to 0.5 mark of a WBC pipette.
2. Carefully, wipe excess blood outside the pipette by using cotton. Draw diluting fluid up to 11 mark.
3. Mix the contents in the pipette and after 5 min by discarding few drops, fill the counting chamber and allow the cells to settle for 2–3 min.
4. Since bulb pipettes are not recommended, the following procedure is performed:
   a. Make a 1:20 dilution of blood with diluting fluid by with 20 µL of blood in a glass tube (10 mm × 75 mm). Cork the tube tightly and mix the suspension by rotating in a cell-suspension mixer for at least 1 min. Fill the Neubauer counting chamber by means of a Pasteur pipette or glass capillary.
   b. Focus on one of the "W" marked areas (each having 16 small squares) by turning objective to low power. Count cells in all four W marked corner.

18.19.4 Calculations

Number of white cells/cu mm (µL) of whole blood = (number of white cells counted × dilution)/(area counted × depth of fluid) (where: dilution = 20)

Area counted 4 × 1 sq. mm = 4 sq mm

Depth of fluid = 0.1 mm (constant)

Hence number of white cells per cu mm (µL) of whole blood = (no. of cells counted × 20)/(4 × 0.1) = no of cells counted × 50

Note: The precautions taken are exactly the same as for recombinant counting technique. The sources of error are also same as for RBC counting technique. However, in the case of WBC counting, extra care is taken during the preparation and storage of WBC diluting fluid. It should be perfectly free from dust particles and yeast cells; otherwise falsely high counts are obtained due to the presence of yeast cells and dust particles.

18.19.5 Error of the Total White Cell Count

1. The inherent distribution error $= \lambda 1/2$, here $\lambda =$ total number of cells in each area.
2. The error as high as 20% may make the difference between 5.0 and 6.0×10^9 cells per liter, which is of little practical significance.
3. The error can be reduced by counting more cells. If 400 cells are counted, the error is reduced to 5%.
4. Error may also be caused due to dirt clumped RBC debris or due to clumping of leukocytes.

The degree of the rise in leukocytes depend on the type and severity of the infection and the response of the body.

18.20 Experiment: Determine the Differential Count of WBCs (Leucocyte) by Staining Method

18.20.1 Introduction

A leucocyte count is of great help to a physician in the diagnosis of disease, like appendicitis, where in acute cases the leucocyte count often reaches 15,000–20,000. The disease in which the leucocyte count increases to the greatest extent is leukemia in which the total counts to the extent of 20,000 to 500,000 are reported. This disorder of blood is referred to as leucocytosis.

The leucocytes are counted in much the same manner as the RBCs except that the diluting fluid is 1.5% acetic acid or Turk's fluid, which destroys the RBC's and make the leucocytes more readily and easily visible. Because of the smaller number of WBCs, the dilution is 1:20 instead of 1:200 as in the ease of RBC count. The count is made under the low power of the microscope and an area of 1 mm² is the unit counted.

In the blood of a normal person, white cells are present to the extent of 5000–10,000 per cu mm. There are several types of WBCs, most of which are larger than RBCs. All WBC's are nucleated.

Blood platelets are normally present to the extent of 350,000/mm³. They are much smaller than red cells and are of irregular shape. They are devoid of nuclei, and in a stained preparation, they are visible in groups or clusters (Figure 18.45) shows the general appearance, relative size, and other characteristics of different types of cells as revealed in the microscopic examination.

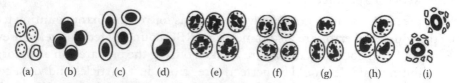

Figure 18.45   (a) RBC, (b) small lymphocyte, (c) large lymphocyte, (d) monocyte or large mono nuclear, (e) polymorphonuclears (neutrophils), (f) eosinophils, (g) basophils (mast cells), (h) transitionals, and (i) RBC surrounded by blood platelets.

18.20.1.1 Hypersegmented Neutrophil

A hypersegmented neutrophil is a mature neutrophil. The nucleus of a hyperseg-mented neutrophil is divided into six or more segments or lobes.

18.20.1.2 Eosinophil

Eosinophils aid in detoxification. They also break down and remove protein material. The cytoplasm of an eosinophil contains numerous coarse, reddish-orange granules, which are lighter colored than the nucleus.

18.20.1.3 Basophil

The function of basophilic cells is unknown. It is believed, however, that basophilic cells keep the blood from clotting in inflamed tissue. Scattered large, dark-blue granules that are darker than the nucleus, characterize the cell as a basophil. Granules may overlay the nucleus as well as the cytoplasm.

18.20.1.4 Lymphocyte

The function of lymphocytes is also unknown, but it is believed that they produce antibodies and destroy the toxic products of protein metabolism. The cytoplasm of a lymphocyte is clear sky blue, scanty, with few unevenly distributed, azurophilic granules with a halo around them. The nucleus is generally round, oval, or slightly indented, and the chromatin (a network of fibers within the nucleus) is lumpy and condensed at the periphery.

18.20.1.5 Monocyte

The monocyte, the largest of the normal WBCs, destroys bacteria, foreign particles, and protozoa. Its color resembles that of a lymphocyte, but its cytoplasm is a muddy gray-blue. The nucleus is lobulated, deeply indented, or horseshoe-shaped and has a relatively fine chromatin structure. Occasionally, the cytoplasm is more abundant than in the lymphocyte.

18.20.2 Protocol

The procedure for the differential white cell count is completed in four steps: Step 1: Making the blood smear, Step 2: Staining the cells, Step 3: Counting the cells, and Step 4: Reporting the count.

18.20.2.1 Preparation of Blood Smears

The simplest way to count the different types of white cells is to spread them out on a glass slide. The preparation is called a blood smear. There are two methods of making a blood smear: the slide method and the cover glass method.

It is very important to make a good blood smear. If it is made poorly, the cells may be so distorted that it will be impossible to recognize them. You should make at least two smears for each patient as the additional smear should be examined to verify any abnormal findings.

The steps to prepare a blood smear for a differential count are as follows:

1. Using a capillary tube, collect anticoagulated blood from a venous blood sample.
2. Deposit a drop of blood from capillary tube onto a clean, grease-free slide. Then place the slide on a flat surface, blood side up.
3. Hold a second slide between your thumb and forefinger and place the edge at a 23 degree angle against the top of the slide that holds the drop of blood. Back the second slide down until it touches the drop of blood. The blood will distribute itself along the edge of the slide in a formed angle.
4. Push the second slide along the surface of the other slide, drawing the blood across the surface in a thin, even smear. If this is done in a smooth, uniform manner, a gradual tapering effect (or "feathering") of the blood will occur on the slide (Figure 18.46).
5. A small drop of blood is placed about three-quarters inch away from one end of the slide. The drop should not exceed one-eighth diameter.
6. The spreader slide is moved in the direction of arrow, allowing drop of blood to spread along slide's back edge.
7. The spreader slide is pushed along the slide, dragging the blood over the surface of the slide.
8. A pencil is used to mark off both ends of the smear to retain the staining solution on the slide.

18.20.2.2 Staining the Cells

1. Fix for at least 30 s in absolute methanol.
2. Remove methanol by tilting the slide or by simply removing from the fixing jar.
3. Apply staining solution I freshly diluted with an equal part of buffer for 5 min on a horizontally positioned slide or in a jar.
4. Transfer slide from the jar without washing (or remove staining solution by holding slide vertically) into staining solution II that has been freshly diluted with nine parts buffer for 10–15 min.
5. Transfer slide to jar with buffer for one rinse after removing stain.
6. Wash slide with ample water.
7. Transfer slide to a jar containing water for 2–5 min.
8. Dry the slide in a tilted position; do not blot dry.
9. Mount a coverglass if desired and observe different types of cells under oil immersion lens (Figure 18.47).
10. Examine the slide for WBC's and record the number of each type

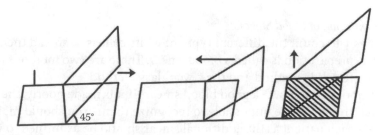

Figure 18.46 Preparation of blood smear.

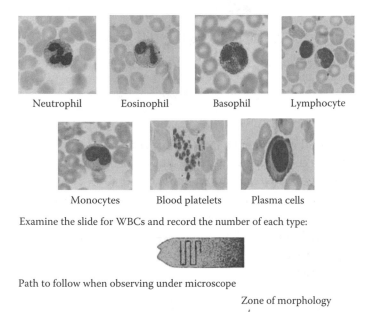

Figure 18.47 **(See color insert.)** Blood morphology containing different types of cells.

18.20.3 Observation

Record results as per Table 18.9.

18.21 Experiment: Determine the Total Erythrocyte Count by Hemocytometer

18.21.1 Introduction

The blood specimen is diluted 1:200 with the RBC diluting fluid and cells are counted under high power (40× objective) by using a counting chamber. The number of cells in undiluted blood are calculated and reported as the number of red cells per cu mm (μL) of whole blood.

TABLE 18.9 Differential Count of WBC

Cells	Counts
% Segmented neutrophils	
% Banded neutrophils	
% Eosinophil	
% Lymphocytes	
% Monocytes	

18.21.1.1 Specimens

Double oxalated or EDTA blood or capillary blood (the specimen need not be a fasting sample).

18.21.2 Materials

Microscope, improved Neubauer chamber, and RBC pipette.

RBC diluting fluid containing sodium citrate 3 g, formalin 1 mL, and distilled water to 100 mL. This solution is stable at room temperature (25°C ± 5°C) for at least 1 year.

Note: RBC diluting fluid is isotonic with blood hence hemolysis does not take place. Normal saline can also be used, but it causes slight deformation of RBCs and allows rouleaux formation. Formalin acts (as a preservative and checks bacterial and fungal growth. Sodium citrate prevents coagulation of blood and provides correct osmotic pressure.

18.21.3 Protocol

1. Mix the anticoagulated blood carefully by swirling the bulb.
2. In the case of capillary blood, the lancet stab should be sufficiently deep to allow free flow of blood. It is drawn quickly in the RBC pipette
3. Draw blood up to 0.5 mark.
4. Carefully wipe the excess blood outside the pipette by using cotton or a gauze, draw diluting fluid up to 101 mark.
5. The pipette is rotated rapidly by keeping it horizontal during mixing.
6. After 5 min, by discarding few drops from the pipette and holding it slightly inclined small volume of the fluid is introduced under the covers which are placed on the counting chamber.
7. Allow the cells to settle for 2–3 min.
8. Place the counting chamber on the stage of the microscope.
9. Switch to low power objective. Adjust light and locate the large square in the center with 25 small squares. Now switch to high power objective.

18.21.4 Calculations

The RBCs in the four corner squares and in the center square (marked in the diagrams as "R") are counted.

Use the following formula for the calculation of RBCs,

- Total RBCs per liter of blood = RBCs/cu mm (μL) × 10^6 or
  use the following formula—Red cell count (per liter) = [(no. of cell counted)/(volume counted μL)] × dilution × 10^6

    Total RBCs/cu mm (μL) = [number of red cells counted × dilution]/[area counted × depth of fluid] Where (1) Dilution = 1:200 (i.e., 200)

Area counted = 80/400 = 1/5 mm^2
Since cell are counted in five bigger squares and such square is further divided into 16 small squares.
Each small square = 1/400 mm^2
Hence area of (5 × 16) = 80 such areas = 80/400 mm^2 = 1/5 mm^2
Depth of fluid = 1/10 mm
Number of red cells counted = N
Hence, total RBCs, cu mm = [(N × 200)]/[(1/5) × (1/10)] = N × 200 × 50 = N × 10000

18.21.5 Observations

18.21.5.1 Additional Information
For dilution of blood, instead of using an RBC pipette the following steps are followed:

1. By using a 20 μL (standardized) pipette, mix 20 μL of blood with 4 mL of RBC diluting fluid in a glass test tube (10 mm × 75 mm). It is mixed well and then introduce small volume of this fluid into a counting chamber.
2. The errors of blood counting by visual means due to the random distribution of the cells in the counting chamber should be avoided. The movement of the cells in the chamber during the filling process causes them to collide, and this influences their distribution. The inherent distribution error can be reduced by counting more cells.
3. To confirm specific results, it is better to repeat a count using a second chamber.

18.21.5.2 Sources of Error

18.21.5.2.1 Falsely High Counts
1. Due to collection of blood from the area where there is hemoconcentration.
2. Due to inadequate wiping of the pipette.
3. Due to improper pipetting of blood as well as the fluid.
4. Because of improper mixing.
5. Due to uneven distribution in the counting chamber.
6. Due to errors in calculations.

18.21.5.2.2 Falsely Low Counts
1. Blood dilution with tissue fluid due to edema or squeezing.
2. Improper pipetting and dilution (when blood draw is less and if dilution is above the mark).
3. Dilution of the contents in the pipette by saliva.
4. Uneven distribution of the cells in the counting chamber.
5. Error in calculations.
6. Use of improperly standardized counting chamber and bad adjustment of the cover slip.

18.21.6 Clinical Significance

At birth, the total erythrocyte count varies from 6.5 million to 7.25 million per mm^3. There is the steady decline after a few hours, and at the end of 15 days to 1 month, there is a small rise to normal adult levels. An increase in total erythrocyte count is observed in conditions such as hem concentration due to burns, cholera, etc., in central cyanotic states as seen in chronic heart disease, conditions of decreased lung function such as emphysema, and in polycythemia. Decrease in erythrocyte count is observed in old age, during pregnancy, and group of diseases classified under anemia.

Normal values
Male: $4.5–6.0 \times 106$ cells/mm³ (µL).
Female: $4.0–4.5 \times 106$ cells/mm³ (µL).

18.21.7 Results and Observations

Report results in view of aforementioned points.

18.22 Experiment: Lymphoid Cells Identification in Blood Smears

18.22.1 Introduction

Blood cells undergo progressive differentiation from stem cells originating in the bone marrow. In the presence of differentiation inducing growth factors, B lymphocytes mature in the bone marrow and T lymphocytes mature in the thymus: both of these lymphoid organs are classified as primary lymphoid organs. Peripheral blood, lymph nodes, spleen, and tonsils are classified as secondary lymphoid organs as they provide battle ground for mature, fully functional lymphocytes. Cytological differences exist between immature stem cells in bone marrow and mature lymphocytes in peripheral blood and tissue sections. The bone marrow contains lymphoid and myeloid cells at various stages of differentiation. The undifferentiated cells are nonfunctional, and differences between lymphoid and various myeloid cell types are more difficult to distinguish.

Peripheral blood smear shows well defined different cell types. Functional myeloid cells include erythrocytes, granulocytes, monocytes, and platelets. Lymphocytes show a circular dark staining nucleus under normal conditions. It is not possible to distinguish T lymphocytes from B lymphocytes in peripheral blood using standard Wright's-Giemsa stain. Distinct circular region of cells in the lymph nodes called germinal centers are present near the outer, cortical edge of the organ.

The differential white cell count is performed to determine the relative number of each type of white cell present in the blood. This provides valuable information concerning infections and other disease processes. Performing the differential smear after counting the cells allows the smear to be used as a double check of the white cell count and platelet count.

The typical color of cell nuclei, namely purple, is due to molecular interaction between eosin Y and an azure B-DNA complex. Both dyes build up the complex later. The intensity of the staining depends on the azure B content and on the ratio azure

B/eosin Y. The staining result can be influenced by several factors such as the pH of the solutions and buffer solution, buffer substances, fixation, and staining time.

18.22.2 Materials

Leishman stain (commercial), glass slide, 70% ethanol, cotton, sterile needle, and compound microscope.

18.22.3 Protocol

1. Scrub the middle finger with a piece of cotton saturated with 70% ethanol and pierce it with a sterile needle.
2. A drop of blood is placed toward one end of each slide about half an inch from the end.
3. The edge of a spreader slide is placed in front of the blood drop at an angle of 35°–40° and spread evenly to get a smear.
4. The smear is dried in air until the blood is fixed to the slide.
5. The thin film is stained with Leishman stain for 3–4 min after that add an equal volume of distilled water and allow to stand for 6–7 min.
6. The slide is washed with distilled water until the film appears light pink in color.
7. The film is air dried and viewed under microscope, different types of WBCs are noted by their stained nucleus shape.
8. The total of 100 cells is counted, taking care not to count the same area again. Neutrophils: 62%–72%, eosinophils: 3%–5%, and basophils: 0%–1%.

18.22.4 Results

Prepare the drawings and compare from the three different lymphoid organs.

18.23 Experiment: Spleen Cell Preparation from Mouse

18.23.1 Introduction

The preparation of lymphocytes from mouse spleen is a technique that readily provides a source of functional T-cells, B-cells, and macrophages. Spleen is classified as a secondary lymphoid organ along with the lymph nodes and mucosa associated lymphatic tissue and contains active differentiated cells. The spleen is easily dissociated into a single cell suspension and provides a high yield (approximately $1-2 \times 10^8$ cells per spleen).

Cells obtained from the spleen are an excellent starting material for primary cell cultures and may be fractionated into T lymphocytes, B lymphocytes, and macrophages. Splenic lymphocytes from immunized animals are key starting cells for the production of hybridomas and monoclonal antibodies.

18.23.2 Materials

Mice, balanced salt solution (BSS), acetic acid counting solution, trypan blue vital dye solution, beaker containing 70% ethanol, squeeze bottle containing 70% ethanol, surgical scissors, curved hemostat, curved forceps, stainless steel wire mesh, rubber policeman, glass centrifuge tube 30 mL, glass tubes 5 mL, glass Petri dish, hemocytometer, and Pasteur pipettes with rubber bulb.

18.23.2.1 Reagents

18.23.2.1.1 Balanced Salt Solution.   Make up $10 \times$ stocks of
Stock 1: Dextrose 10 gm, K_2HPO_4 0.6 gm, 0.5% phenol red solution 20 mL, and dissolve and bring up to 1000 mL with distilled water.
Stock 2: $CaCl_2 \cdot H_2O$ 1.86 g, KCl 4.0 g, NaCl 80.0 g, $MgCl_2$ 1.04 g, $MgSO_4 \cdot 7H_2O$ 2.0 g. Dissolve and bring up to 1000 mL with distilled water.

1. Mix 10 mL of stock 1 and 10 mL of stock 2, bring up to 100 mL with distilled water. The pH should be 7.2–7.4.
2. Sterilize all instruments and wire mesh by storing in 70% ethanol. Place in a loosely covered sterile Petri dish and allow to air dry before use. Pasteur pipettes and glass tubes should also be sterilized.
3. Make up the following solutions for cell counting: 0.5% acetic acid in distilled water (v/v) for lysing RBCs trypan blue $10 \times$ stock, % (w/v) in distilled water.
4. Sacrifice mice just before the laboratory in a closed chamber containing 30% CO_2. An alternative method is to use cervical dislocation.

18.23.3 Protocol

1. After the mouse skin has been wet with 70% ethanol, place the left side facing up on the paper towels.
2. Make a cut through the loose skin by pulling gently upwards and using the blunt scissors on the skin flap. This will expose the peritoneal wall.
3. Pull gently in opposite directions on the two sides of the skin incision to expose a wider area of the peritoneal wall.
4. With the small surgical scissors, make a cut over the spleen to expose it through the peritoneum.
5. Grasp the spleen with the forceps. Pull up gently and cut away attached connective tissue, which appears white using the small surgical scissors.
6. Grasp a folded corner of the stainless steel wire mesh with the hemostat. Insert the mesh into the Petri dish and add about 10 mL of BSS.
7. Put the removed spleen on the wire mesh. With a rubber policeman, rub the spleen back and forth over the mesh. Continue until all the spleen dissociates leaving only white connective tissue on the mesh surface.
8. With a sterile Pasteur pipette, bring the cell suspension up and down several times to dissociate large cell clumps pipette the cell suspension into the glass centrifuge tube.

9. Allow the cells to settle, for 5 min, to remove larger debris. Use the Pasteur pipette to transfer roughly 2 mL volumes of cell suspensions to 5 mL glass tubes.

10. Centrifuge at $200 \times g$ for 10 min. Pour off the supernatant.

11. Resuspend the cell pellet in 2 mL of BSS by vortex mixing.

12. Add 0.1 mL of trypan blue solution to 1.8 mL of acetic acid solution. Add 0.1 mL of the cell suspension and mix thoroughly.

13. Fill a hemocytometer chamber with a drop of cell mixture and proceed as if doing a WBC count. Count both total cells and the fraction, which are dead (those that have taken up the blue dye).

14. Remember that a dilution factor of 20 and a volume factor of 10,000 needs to be multiplied by the average count of a 4×4 square grid. For example, 50 cells in a 4×4 grid would yield a cell count of 10 million.

18.23.4 Result

Record grid counts by averaging four separate grids. Record the number and per cent of dead cells per grid, and average against trypan blue positive counts.

18.24 Experiment: Identification of Lymphocyte Populations

18.24.1 General View

Lymphocytes are one of the major classes of WBCs involved in the immune response. These cells are derived from stem cells in the bone marrow, which migrate to the lymphoid organs and develop into either B lymphocytes or one of the subclasses of T lymphocytes. Each type of lymphocyte plays a distinct and critical role in immunity. These functions are mediated, in part, by unique proteins with specific binding properties, called receptors, present on the surface of each type of lymphocyte. For example, B lymphocytes have cell surface Ig, or Ab, molecules, which bind with a specific Ag. This binding stimulates activation and multiplication of the B lymphocyte. T lymphocytes also have cell surface, Ag-specific receptors, which are distinct from Ig. Another example is the Fc receptor, found on certain lymphocytes, which binds the constant region of Ab molecules, called as Fc region.

If one examines a mixed population of lymphocytes microscopically, the cells are generally indistinguishable with respect to size, shape, etc. However, monoclonal antibodies have been developed, which will identify the specific cell surface receptors so that they are visible by microscopy, and, therefore, allow visual identification of specific cell types. The two experiments described in the subsequent text demonstrate different reagents used to identify specific subpopulations of lymphocytes within a heterogeneous population of lymphocytes, such as blood cells or spleen cells.

18.24.2 Experiment: Identification of FC Receptor-Bearing Cells

18.24.2.1 Introduction

Most mature B lymphocytes, as well as other accessory cells such as macrophages, possess cell surface receptors for the Fc portion of Ig or Ab, molecules of the IgG class. Other cell types have receptors for the Fc region of other Ig classes. The role of Fc receptors in the immune function of B lymphocytes remains unclear. On other immune cell types, however, Fc receptors may facilitate the destruction of cells or particles coated by antibodies during an immune response. It has been demonstrated that Fc receptors on macrophages promote binding and subsequent engulfing and destruction of Ab–Ag complexes, which would damage host tissues if allowed to accumulate. Another important example is the Fc receptors for IgE antibodies found on basophils and mast cells, which are probably involved in allergic reactions (Figure 18.48).

Cells bearing Fc receptors on their surface can be identified easily by the formation of rosettes with Ab-coated erythrocytes (RBCs). This procedure involves mixing suspension of lymphocytes, for example, spleen cells, with erythrocytes which have been coated with IgG antibodies produced against the erythrocytes. The Fab portion of the antibodies will attach to the erythrocyte surface. The Fc portion of the antibodies will be exposed and will bind with the IgG. Fc receptors on the B lymphocytes and other cells bearing Fc receptors are then identified with the help of microscope by the cluster of erythrocytes around them, called a rosette (Figure 18.49).

Rosette cells can be separated from cells which have not formed rosettes by differential centrifugation techniques. Removal of the erythrocytes then yields a cell population enriched for Fc receptor-positive cells. Variations on this procedure are commonly used to isolate human lymphocyte populations in clinical laboratories and to isolate or deplete various WBC populations in research laboratories.

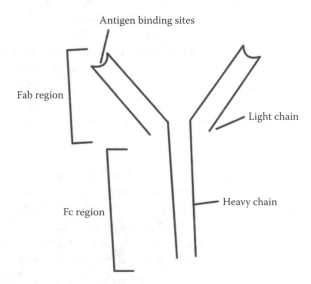

Figure 18.48   Structure of antibody molecule.

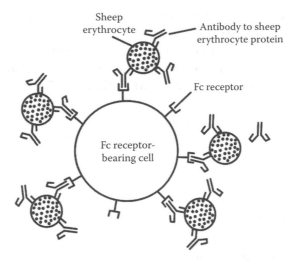

Figure 18.49 Rosettes formation with antibody-coated erythrocytes and Fc receptor-bearing cells.

In this experiment, sheep erythrocytes, which have been coated with IgG antibodies, produced against sheep erythrocyte stroma proteins, are used to identify IgG Fc receptor-bearing cells in a population of cells isolated from mouse spleen. The mouse Fc receptors for IgG will bind with the Fc portions of either mouse or rabbit IgG molecules. The percentage of mouse spleen cells, which have Fc receptors for IgG on their surface can then be calculated.

The sheep erythrocytes should be prepared within 1 week prior to experiment and store in the refrigerator.

18.24.2.2 Materials
Antiserum produced in rabbits or mice against sheep erythrocytes (1 mL), sheep blood, or glutaraldehyde fixed sheep erythrocytes (4 mL), Hank's balanced salt solution (HBSS) (250 mL), lymphocytes isolated from mouse spleen, purified by acetic acid lysis to remove erythrocytes 1×10^6 cells/mL (see subsequent sections for alternative protocol for removing erythrocytes: hypotonic lysis), 5% suspension in HBSS of sheep erythrocytes coated with IgG antibodies, crystal violet 0.5% w/v in distilled water, 17×100 mm clear tubes either with caps or with parafilm for sealing, microscopic slide and cover slips, 5 mL pipettes, 1 mL pipettes, 37°C water bath, 12×75 mm glass tubes, compound microscope, Pasteur pipettes and bulbs, goat Ab to mouse IgM, FITC-labeled rabbit Ab to goat immunization, alkaline phosphatase-labeled rabbit Ab to goat Ig for direct fluorescent labeling, and antibodies against IgM from other species if cells other than mouse are used.

18.24.2.2.1 Hank's Balanced Salt Solution
The medium is also commercially available 1× or 10× solution or in powder form.

Recipe to prepare per liter 1× HBBS solution in the laboratory:

Solution A—1 g glucose, 0.1 g phenol red, 0.6 g KH_2PO_4, and 0.09 g $Na_2HPO_4.7H_2O$.

Solution B—0.14 g $CaCl_2$, 0.4 g KCl, 1.0 g NaCl, 0.1 g $MgCl_2 \cdot 6H_2O$, 0.1 g $MgSO_4 \cdot 7H_2O$, and 0.35 g $NaHCO_3$.

Separately dissolve the components for solution A and solution B, each in 300 mL distilled water. Mix solution A and solution B together. Make sure that all ingredients are in solution, and then add sodium bicarbonate. Adjust the pH to 7.4 and store in refrigerator.

18.24.2.2.2 Sheep Erythrocytes Preparation

1. In a conical 50 mL centrifuge tube, mix 4 mL of sheep blood or reconstituted erythrocytes in 16 mL HBSS.
2. Centrifuge for 10 min, at $400 \times g$.
3. Discard supernatant and resuspend cells in 20 mL HBSS.
4. Repeat steps (2) and (3) twice.
5. Store washed erythrocytes in the refrigerator for no more than 1 week.

18.24.2.2.3 Ab Coating of Erythrocytes

1. Dilute the antiserum to sheep erythrocytes by mixing 1 mL antiserum with 19 mL HBSS.
2. Mix antiserum with washed sheep erythrocytes in conical centrifuge tube(s).
3. Incubate mixture for 30 min at 37°C.
4. Centrifuge mixture for 10 min at 400 g.
5. Discard supernatant and resuspend cells in 40 mL HBSS.
6. Repeat steps 4 and 5 once.
7. Store coated erythrocytes in the refrigerator for 1 week.

18.24.2.2.4 Hypotonic Lysis to Remove Erythrocytes from Spleen Cells (Alternate Protocol)

1. Centrifuge spleen cells for 10 min at $400 \times g$ and decant supernatant.
2. Tap cell pellet to resuspend cells in a small amount of supernatant remaining.
3. Add 9 mL distilled water to the cell pellet. After cells have been in water for 2 s, quickly add 1 mL of $10 \times$ concentrated HBSS or phosphate buffered saline (PBS) and mix thoroughly with a pipette.
   Note: The timing of this step is important. The lymphocytes are somewhat more resistant to the hypotonic conditions than the erythrocytes, but the lymphocytes will also be lost if exposed to the water for too long.
4. Wash the lymphocytes by centrifuging as in step 1, resuspend the pellet in $1 \times$ HBSS and centrifuge again. The cell pellet should be white rather than red at this point. If there is still red in the pellet, repeat steps 1–5 to remove additional erythrocytes.
5. Resuspend the cells in $1 \times$ HBSS and count.

18.24.2.3 Protocol

1. Mix 2.5 mL of lymphocyte suspension with 2.5 mL Ab-coated erythrocytes in a 17 mm × 10 mm tube.
2. Seal the tube tightly with a cap. Use parafilm to seal the cap.
3. Warm the tube to 37°C by placing it in the water bath for 5 min.
4. Remove the tube from the water bath and gently mix it for 10 min by inverting it back and forth. Keep your hand around the tube during mixing to keep it warm.
5. Place the tube on ice until ready to count the cells. For counting the cells, stain the WBCs with crystal violet for easier visualization as follows:

a. Remove 0.1 mL from the tube containing lymphocytes and coated erythrocytes and transfer to a 12 mm × 75 mm tube.
b. Dilute sample with 0.4 mL HBSS.
c. Add one drop of 0.5% crystal violet to sample using Pasteur pipette.
d. Let the tube sit 5 min at room temperature to allow the lymphocytes to take up the stain.

6. Prepare the slide and place on a microscope and focus on the cells, first using the 10× objective, and then using the 20× objective. Stained naked cell and WBCs, which are surrounded by a cluster of smaller cells, can be seen with the microscope. These clusters are the rosettes.

18.24.2.4 Calculations

Count the number of rosettes visible in the field. Also count the total number of WBCs visible. Record these two numbers. Count large enough sample to obtain a total of at least 50 WBCs. If necessary, count more than one field and add the numbers obtained.

Note: 10× objectives allow counting more cells in a single field, but it may be easier to see the cells using the 20× objective. Choose whichever objective you find easier to use to carry out the cell counting.

18.24.2.5 Result

Use the cell counts obtained to calculate the percent of lymphocytes in the spleen cell population that are Fc receptor positive.

If the concentration of erythrocytes in the mixture is too high, they may obscure visualization of the lymphocytes and rosettes. If this seems to be a problem, dilute the sample further in HBSS.

Washing the erythrocytes after Ab coating is an important step, since residual excess Ab will inhibit rosette formation.

18.24.3 Experiment: Identification of B Lymphocytes by Detection of Cell Surface Ig

18.24.3.1 Introduction

The B lymphocytes are the cells that produce specific antibodies when activated during an immune response. While antibodies are secreted during an active immune response, most B lymphocytes contain Ab that remains attached to the B lymphocyte surface. This Ig, which is of the IgM class is inactivated B-cells, is produced when the B-cells develop in the bone marrow. Each B-cell produces IgM antibodies with unique Ag specificity. Binding of the specific Ag with the IgM on a particular B-cell stimulates that cell to become activated and to begin dividing as part of the immune response to the Ag. The presence of surface IgM can be used to identify B lymphocytes. As with many cell surface proteins, specific antibodies against the IgM class of Ig can be used as a tool to mark IgM bearing cells. If these anti-antibodies are labeled, such as with a fluorescent dye or with an enzyme that will produce a color when it acts on an appropriate substrate, the B lymphocytes can be distinguished

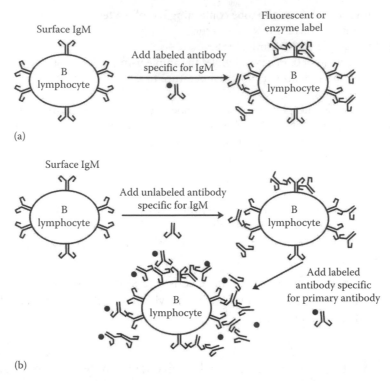

Figure 18.50 (a) Direct and (b) indirect immune labeling of cell surface immunoglobulin.

microscopically from the rest of the cells in a population. The IgM antibodies may be labeled directly or, alternatively, a stronger signal can be achieved by using unlabeled Ab against the IgM (called the primary Ab) in the first step followed by labeled Ab produced against the primary antibodies (called the secondary Ab) in a second step. This indirect labeling method is used frequently in clinical and research laboratories to detect the presence of many types of cell surface proteins on subpopulations of cells. This detection system can be used to determine the percent of cells bearing the cell surface protein (Figure 18.50). The procedure describes the detection of IgM on B lymphocytes isolated from mouse spleen.

18.24.3.2 Precautions
If using the fluorescence microscope, the eyes should be protected from the UV light used to excite the fluorescent dye. Most fluorescent microscope is equipped with protective shields that are positioned so that eyes of the person looking into the objectives are protected.

18.24.3.3 Materials

18.24.3.3.1 Reagents
1. *Preparation of PBS-BSA*: To 100 mL PBS, add 0.1 g BSA, stir to mix. Prepare PBS by dissolving 2.19 g NaCl, 1.28 g KH_2PO_4, and 2.63 g $Na_2HPO_4 \cdot 7H_2O$; adjust pH to 7.2, add distilled water to bring 250 mL.

2. *Hypotonic lysis to remove erythrocytes from spleen cells*:
   a. Centrifuge isolated spleen cells for 10 min at $400 \times g$. Decant supernatant.
   b. Tap cell pellet and resuspend in a small amount of supernatant remaining.
   c. Add 9 mL of distilled water to the cell pellet and quickly add 1 mL of $10 \times$ HBSS or PBS and mix thoroughly with a pipette.
      Note: The timing of this step is important. The lymphocytes are somewhat more resistant to the hypotonic conditions than the erythrocytes, but the lymphocytes will also be lost if exposed to the water for too long.
   d. Centrifuge isolated spleen cells for 10 min at $400 \times g$, decant supernatant.
   e. Tap cell pellet and resuspend in a small amount of supernatant remaining.
   f. Add 9 mL of distilled water to the cell pellet and quickly add 1 mL of $10 \times$ HBSS or PBS and mix thoroughly with a pipette.
3. *Dilution of primary and secondary antibodies*: Antibodies should be diluted fresh, no more than 12 h before the experiment to commence. If desired, aliquot diluted antibodies into labeled tubes for groups. FITC labeled antibodies should be protected from light as much as possible.
4. *Preparation of cells*: Cells should be washed once in PBS-BSA by centrifuging 10 min at $25 \times g$, resuspending in PBS-BSA, and centrifuge again. Resuspend cell pellet in PBS-BSA so that cells are at a concentration of 2×10^6/mL.
5. *Alkaline phosphatase label-preparation of substrate*: Both substrate components are available as tablets from Sigma Chemical Co., which are dissolved in the indicated amount of distilled water just before use. Make substrate solution fresh and protect from light until use.

Note: It is important to keep the cells and reagent on ice throughout the procedure. If the cells become warm, the IgM-Ab complexes may be internalized and will be difficult to visualize. The addition of 0.2% sodium azide to the PBS-BSA would inhibit internalization, but since azide is highly toxic, it is not recommended for use in a control environment.

18.24.3.3.2 Sample Preparation. Mice or spleen from different sources, antibodies against IgM, bovine spleen or alternatively, cultured lymphocytes cell lines available from American Type Culture Collection (ATCC) or from research laboratories, mixture of cultured IgM positive, B lymphocytic cell and IgM negative cells (such as a T lymphocytic cell line), mixed in approximately 50:50 proportions. Wash cultured cells twice with HBSS, count, and adjust concentration to 1×10^6/mL, fluorescent microscope, PBS with 0.1% BSA (PBS-BSA) 100 mL, lymphocytes isolated from mouse spleen and purified by acetic acid lysis or hypotonic lysis, described earlier, 2×10^6/mL in PBS-BSA (2×10^7 cells total), 12 mm $\times$ 75 mm centrifuge tubes 10 mL, microscope slides, cover slip, Pasteur pipettes, bulbs, micropipettes, pipette pumps, gloves, ice, 90% glycerol in PBS-BSA, and goat antibodies against mouse IgM, diluted appropriately in PBS-BSA 2 mL.

Fluorescence detection: FITC labeled rabbit antibodies against goat Ig, diluted appropriately in PBS-BSA 2 mL.

Enzyme-linked color detection: Alkaline phosphatase labeled rabbit antibodies to goat Ig, diluted appropriately in PBS-BSA 2 mL; alkaline phosphatase substrate (nitroblue tetrazolium; NBT) and 5-bromo-4-chloro-3-indolyphosphate (BCIP).

Fluorescence detection: FITC labeled goat antibodies against rabbit Ig (secondary Ab).

Enzyme-linked color detection: Alkaline phosphatase labeled goat antibodies to rabbit Ig (secondary Ab), alkaline phosphatase substrate (NBT), and BCIP, made fresh by adding 6.5 μL BCIP and 33 μL NBT to 5 mL PBS-BSA.

18.24.3.4 Protocol

Transfer 1 mL of lymphocyte suspension to a 12 mm × 75 mm centrifuge tube. Centrifuge for 10 min at 250 × g.

Carefully remove supernatant with Pasteur pipette and discard.

Add 0.1 mL diluted Ab against IgM and gently resuspend the cell pellet by tapping the bottom of the tube or by using pipette to break up the pellet.

Incubate on ice for 15 min.

Add 1 mL PBS-BSA and mix, centrifuge for 10 min, 250 × g.

Remove supernatant and discard.

Add 1 mL PBS-BSA and resuspend the cell pellet, centrifuge for 10 min 250 × g and discard supernatant and discard again.

Perform the following test for enzyme-linked fluorescent color detection:

1. Add 0.1 mL FITC labeled goat Ab against rabbit Ig and resuspend the cell pellet.
2. Incubate for 20 min on ice.
3. Wash cells as in twice with HBSS.
4. Add 0.1 mL PBS-BSA and resuspend the cell pellet.
5. Add 0.1 mL peroxidase substrate and resuspend the cell pellet.
6. Place a drop of 90% glycerol on a microscope slide. Add a drop of stained cell suspension. Place a cover slip over the cells, taking care to avoid air bubbles.

Fluorescence detection: Observe cells using microscope equipped with UV light source and filters for fluorescence detection use a 20 × or 40 × objectives.

First count the total number of lymphocytes using normal light. The lymphocytes will be small, round cells with little cytoplasm.

Then, in the same field, use the UV light to count the number of fluorescent cells. The fluorescence should have a "patchy" appearance on the lymphocyte surface. Enzyme-linked color detection: Observe cells using 20 × or 40 × objective.

Count the total number of lymphocytes that have stained in the field. The lymphocytes are small, round cells with little cytoplasm.

Count the total number of lymphocytes which have stained with peroxidase. These cells will have a purple stain over the surface.

18.24.3.5 Result

1. Use the cell counts obtained to calculate the percent of B-lymphocytes in the lymphocyte sample.
2. The critical parameter is Ab dilution. No positive cells may indicate that one of the antibodies is too dilute. A high degree of background staining (resulting in no distinction between positive and negative cells) may indicate the antibodies are too concentrated, and nonspecific binding is occurring.

3. One may want to include a control sample that is treated with the secondary Ab, but not the primary Ab. This may help to determine the degree of nonspecific binding of the secondary Ab.
4. Another way to reduce nonspecific staining is to add one more washing step between each Ab addition.

18.25 Experiment: Isolation of Monocytes by Adherence

18.25.1 Introduction

Monocytes are produced by the bone marrow from hematopoietic stem cell precursors called monoblasts. Monocytes circulate in the bloodstream for about 1–3 days and then typically move into tissues throughout the body. They constitute between 3% and 8% of the leukocytes in the blood. Half of them are stored as a reserve in the spleen, in clusters. In the tissues, monocytes mature into different types of macrophages at different anatomical locations. Monocytes are the largest corpuscles in the blood. Monocytes, which migrate from the bloodstream to other tissues, will then differentiate into tissue resident macrophages or dendritic cells. Macrophages are responsible for protecting tissues from foreign substances, but are also suspected to be important in the formation of important organs like the heart and brain. They are cells that possess a large smooth nucleus, a large area of cytoplasm, and many internal vesicles for processing foreign material. In this protocol, monocytes are isolated from mononuclear cells by exploiting their ability to adhere to glass or plastic. This is a quick and easy procedure, but it can induce cell activation.

18.25.2 Materials

Peripheral blood mononuclear cells (PBMC), serum-free DMEM supplemented with 2 mM L-glutamine and 50 µg/mL gentamicin, 0.02% EDTA in PBS, 75 cm² tissue culture flasks, Beckman GPR centrifuge or equivalent with GH-3.7 horizontal rotor, 15 mL conical polypropylene centrifuge tube, and additional reagents and equipment for cell counting.

18.25.3 Protocol

1. Suspend PBMC in supplemented serum-free DMEM at 2×10^6 cells/mL.
2. Add 10 mL of the cell suspension to each 75 cm² flask and incubate for 1 h in a humidified 37°C, 5% CO_2 incubator.
3. Decant the medium that contains nonadherent cells, wash twice with 10 mL supplemented serum-free DMEM to remove any residual nonadherent cells, and replace with 10 mL fresh DMEM.

4. Remove adherent monocytes by gently scraping with a plastic cell scraper or by incubating the cells in ice-cold 0.02% EDTA/PBS solution for 10 min followed by firmly tapping the flask. Transfer cells to a 15 mL conical tube and centrifuge 10 min in GH-3.7 rotor at 1400 rpm (300g) to remove the EDTA/PBS solution.
5. Resuspend in supplemented serum-free DMEM and count monocytes. Alternatively, monocytes can be adhered in the appropriate culture plate and use directly.

18.25.4 Result

Count the monocyte with hemocytometer. A high monocyte is a generally a sign of infection. Monocytes are a type of WBCs. It could indicate an infection with the mononucleosis virus. Normal range is 2%–8% of WBCs. Decrease in monocyte indicates aplastic or lymphocytic anemia.

18.26 Experiment: Isolation of Neutrophils

18.26.1 Introduction

Neutrophil or polymorphonuclear granulocytes are the most abundant leukocytes in humans and among the first cells to arrive on the site of inflammatory immune response. They have the key role in inflammation, neutrophil functions such as locomotion, cytokine production, phagocytosis, and tumor cell combat. The neutrophil response to infection *in vivo* is initiated by the adherence of the neutrophils to vascular endothelia and progress into the extravascular tissue space, and finally culminates in neutrophil-mediated phagocytosis and intracellular killing of the invading microorganisms. Protocol comprises of sequential sedimentation in dextran, density centrifugation in Ficoll-Hypaque, and lysis of contaminating RBC. The number and quality of cells obtained are suitable for carrying out most available assays of function. This protocol is the standard method used for isolating neutrophils from blood in high purity from large volumes (≥200 mL) of peripheral blood, but with minor modifications, it is suitable for other species and sources. An alternate protocol that reverses the order of the purification steps may be useful in some situations.

18.26.2 Materials

Heparin, 3% dextran T-500 (Pharmacia) in 0.9% NaCl (dextran/saline solution), 0.9% NaCl, Ficoll-Hypaque solution, 0.2% and 1.6% NaCl ice-cold, phosphate-buffered saline with 10 mM D-glucose (PBS/glucose) ice-cold, 50 mL conical plastic centrifuge tubes, additional reagents and equipment for drawing blood and counting cells, and Ficoll-Hypaque density gradient centrifugation.

18.26.3 Protocol

18.26.3.1 Enrichment for Neutrophils by Dextran Sedimentation

1. Draw blood in a syringe containing preservative-free heparin (20 U/mL final). Heparin is the recommended anticoagulant, although substitution of EDTA or citrate may be acceptable for some applications.

2. Mix blood with an equal volume of dextran/saline solution in a plastic cylinder or 50 mL centrifuge tube and incubate in upright position ~20 min at room temperature. Dextran promotes erythrocyte rouleaux formation, which results in differential sedimentation of RBC and the formation of a leukocyte-rich, RBC-poor plasma layer above the sedimented RBC. The RBC sedimentation rate is proportional to temperature; and at room temperature, it takes about 20 min for a clean interface between plasma and RBC to form.

3. Aspirate and save the leukocyte-rich plasma (upper layer) using a pipette and rubber bulb or tubing and a syringe. Pellet cells from the plasma by centrifuging for 10 min at 1000 rpm, 5°C. Aspirate and discard supernatant.

This cell pellet still contains many RBC. The proportion of neutrophils versus other leukocytes is increased in comparison with blood, but many lymphocytes and monocytes remain.

18.26.3.2 Remove Mononuclear Cells by Ficoll-Hypaque Centrifugation

Repeat steps 1–3 if a high level of neutrophil purity is required. Otherwise, move further.

4. Immediately resuspend cell pellet in a volume of 0.9% NaCl equal to the starting volume of blood.

5. Put ≤40 mL of the cell suspension into one or more 50 mL conical plastic centrifuge tubes. Layer 10 mL Ficoll-Hypaque solution beneath the cell suspension using a pipette or a syringe fitted with a long blunt-end needle. Take care to preserve a sharp interface between the Ficoll-Hypaque solution and the cell suspension.

6. Centrifuge 40 min at 1400 rpm ($400 \times g$), 20°C with no brake. The 20°C temperature is important for achieving the best cell separation. It is most dependably maintained if the refrigeration remains off overnight and is turned on at the 20°C setting just before centrifugation.

7. Aspirate the top (saline) layer, as well as the Ficoll-Hypaque layer, leaving the neutrophil/RBC pellet. If mononuclear cells are desired, aspirate the cloudy band of mononuclear cells at the saline/Ficoll-Hypaque interface prior to aspiration of saline and Ficoll-Hypaque layers.

18.26.3.3 Remove RBC and Count Neutrophils

8. To remove residual RBC, subject cells to hypotonic lysis by resuspending each neutrophil/RBC pellet in 20 mL cold 0.2% NaCl for exactly 30 s. At the end of this period, restore isotonicity by adding 20 mL ice-cold 1.6% NaCl.

   The hypotonic lysis step is based on the high sensitivity of RBC to hypotonicity in comparison with neutrophils. However, the 30 s limit must be carefully observed since a more prolonged period of hypotonicity will result in neutrophil damage.

9. Centrifuge 6 min at 1000 rpm (250 × g), 5°C, discard supernatant. Repeat steps 8 and 9 once or twice until the cell pellets appear relatively free of RBC.
10. Resuspend cells in ice-cold PBS/glucose and combine tubes.
11. Determine cell concentration using an automated counter (e.g., coulter counter). Alternatively, count cells using a hemocytometer, using a 3% acetic acid diluent. This allows visualization of nuclear morphology and thereby facilitates differentiation of neutrophils from other contaminating leukocytes.
12. Adjust cell concentration in ice-cold PBS/glucose as desired and keep the suspension in an ice bath until ready for functional assays.

18.27 Experiment: Isolation of Human Basophils

18.27.1 Introduction

Human basophils are normally the rarest of all circulating leukocytes and thus are among the most difficult of the blood elements to purify. In this unit, basophils are isolated from leukocyte populations, resulting in preparations of cells of which 5%–50% are basophils, and the rest are lymphocytes along with neutrophils and monocytes. This technique is advantageous because of its speed and it minimizes basophil stimulation and histamine release.

18.27.2 Materials

Venous blood, freshly drawn, 0.1 M EDTA, pH 7.7, Percoll gradients in 15 mL polystyrene conical tube, HBSS, 50 mL polypropylene conical tubes, 1 mL disposable polypropylene pipettes, 15 mL polystyrene conical tubes, additional reagents, and equipment for blood collection and Wright-Giemsa cell staining.

18.27.3 Protocol

1. Place 40 mL unheparinized venous blood (immediately after it is drawn) into a 50 mL polypropylene conical tube containing 1:10 the blood volume of 0.1 M EDTA, pH 7.7. Alternatively, the blood may be diluted in a similar manner with HBSS containing 10 U heparin/mL (preservative-free).
2. Layer 4 mL blood mixture over each of 10 Percoll gradients and centrifuge 25 min at 1200 rpm (300 × g), room temperature, with no brake.
3. Gently remove the gradients from the centrifuge. Remove each cell band with a 1 mL polypropylene disposable pipette and pool similar bands from each gradient into a clean 15 mL polystyrene conical tube. Once each band is pooled, remove 3.3 mL aliquots and place in separate 15 mL conical tubes.
4. Wash cells of each aliquot twice as follows: add 10 mL HBSS, mix, centrifuge 7 min at 1200 rpm (300 × g), 4°C, and discard supernatant.
5. Stain an aliquot from each preparation with Wright-Giemsa stain and examine microscopically to establish identity of basophils.

18.27.4 Result

Most basophils stain dark reddish purple.

18.28 Experiment: Isolation of Tissue Mast Cells

18.28.1 Introduction

Mast cells derive from mononuclear precursor cells, which undergo their final phase of differentiation in the tissues. Mast cells express a unique set of proteases and display functional diversity depending on the tissue in which they differentiate—a phenomenon often referred to as mast cell heterogeneity. Mast cells, unlike other immune effectors cell, are not found in circulation and thus must be obtained from accessible tissues such as gastrointestinal and lung tissue. In the following protocol, a combination of mechanical fragmentation, enzyme digestion, and centrifugation procedures are used to obtain suspensions enriched for mast cells from gastrointestinal tissue.

18.28.2 Materials

Small intestine segments, 3–5 g, HBSS containing 5% fetal calf serum (FCS), 0.9% saline, HBSS containing 20% FCS and 1 mg/mL collagenase, HBSS containing 20% FCS and 0.5 mg/mL collagenase, HBSS containing 0.13 mM EDTA (HBSS/EDTA), 50 mM N-acetyl-L-cysteine (prepared in HBSS), 100 mm × 15 mm petri dishes, tweezers, scissors, and scalpels, kept in sterile beaker with 70% ethanol, 50 mL graduated conical tubes, 10 mL syringe, 4 × 4 in. 12 ply gauze sponges, and column 4 mL with nylon wool.

18.28.3 Protocol

18.28.3.1 Prepare Tissue Fragments

1. Obtain small intestine segments surgically and place in several Petri dishes each containing 25 mL HBSS.
2. Remove adherent mesentery and fat from tissue using tweezers and scissors. Keep tissue fragments moist.
3. Wash tissue fragments three times in 100 mL of 0.9% saline in a glass beaker with gentle agitation at room temperature to remove fecal material. Cut the washed material with scissors into ~0.3 cm^3 cubes.
4. Place tissue cubes in a series of plastic Petri dishes containing 25 mL of the following washing solutions with gentle agitation at 37°C: HBSS/EDTA-two times for 5 min each, 50 mM N-acetyl-L-cysteine—one time for 1 min, and HBSS-5 two times for 10 min each.

5. Discard the final wash and mince the tissue with a scalpel into ~1 mm³ pieces. Place tissue pieces in several Petri dishes.
6. Add 25 mL HBSS containing 20% FCS and 1 mg/mL collagenase to the dishes. Incubate for 90 min at 37°C with gentle agitation.
7. Decant liquid into 50 mL conical tubes and centrifuge 10 min at 1000 rpm (150 × g), 4°C. Discard supernatant, resuspend cell pellet in 5 mL HBSS-5, and keep on ice.
8. Incubate remaining tissue fragments an additional 60 min with 25 mL HBSS containing 20% FCS and 0.5 mg/mL collagenase. Mechanically disperse the residual tissue by pulling up and down through a syringe. Remove any remaining undispersed tissue by filtering the suspension through gauze sponges.

18.28.3.2 Recover Mast Cells

9. Pool the cell-containing filtrate with the cell suspension from step 7 and centrifuge 10 min at 1000 rpm (150 × g), 4°C. Discard supernatant, resuspend cell pellet in 20 mL HBSS-5, and centrifuge 10 min at 1000 rpm (150 × g), 4°C. Discard supernatant.
10. Resuspend cell pellet in 5 mL HBSS-5 and filter through a 5 mL nylon wool column to remove debris and clumps of cells. Wash column with 5 mL HBSS-5 and combine with first eluate, for a total volume of 8–9 mL cell suspension. Aseptically remove nylon wool from Leuko-Pak filter and gently pack a 5 mL syringe to the 4 mL mark. If other sources of nylon wool are used, the nylon wool-packed syringe should be autoclaved before use.
11. Determine percent of mast cells by acid toluidine blue staining.

18.28.3.3 Acid Toluidine Blue Staining of Mast Cells

Both connective tissue and mucosal mast cells contain granules within their cytoplasm that exhibit blue metachromasia upon fixation of the mast cells in Mota's fixative followed by staining with acid toluidine blue. The internal structure of mast cells is best visualized when slide preparations are made using a cytocentrifuge.

18.28.3.3.1 Materials.
Mota's fixative, 66% and 100% ethanol, acid toluidine solution, and cytocentrifuge.

18.28.3.3.1.1 Reagents and Solutions

Acid Toluidine Solution: Mix 100 mg toluidine blue O in 100 mL of 30% alcohol, adjust to pH 1 by dropwise addition of 1 N HCl and store at room temperature.

Mota's Fixative: Mix 8 g lead acetate (basic) with 100 mL H_2O using a magnetic stirrer, adding sufficient glacial acetic acid (2–4 mL) to dissolve the salt, and raise to 100 mL with 100% ethanol.

18.28.3.3.2 Method

1. Cytocentrifuge cells into a microscope slide.
2. Flood slide with Mota's fixative and let sit for 10 min.
3. Flood slide with 66% ethanol to remove fixative.
4. Wash slide by dipping it into a beaker of distilled water.
5. Flood slide with acid toluidine solution and let sit for 10 min. Repeat washing as in step 4.

6. Flood slide with 66% ethanol and repeat washing as in step 4.
7. Flood slide with 100% ethanol. Repeat washing as in step 4 and allow the slide to air dry.

18.29 Experiment: Measurement of Percent β-Glucuronidase Secretion

18.29.1 Introduction

A major problem of tumor gene therapy is the low transduction efficiency of the currently available vectors. One way to circumvent this problem is the delivery of therapeutic genes encoding intracellular enzymes for the conversion of a pro-drug to a cytotoxic drug, which can then spread to neighboring nontransduced cells (Bystander effect). One possibility to improve the bystander effect could be the extracellular conversion of a hydrophilic pro-drug to a lipophilic, cell-permeable cytotoxic drug. The physiological function of β-glucuronidase is the degradation of glucuronic acid containing glycosaminoglycans. In higher eukaryotes, the enzyme undergoes a series of post-translational modifications. Mast cell and basophil granules contain several preformed enzymes that are released from secretory granules in parallel to serotonin and histamine. The following protocol is a nonradioactive chromogenic method for assaying β-glucuronidase concentration.

18.29.2 Materials

0.1 M sodium acetate buffer pH 4.5, 0–50 μg dilution of phenolphthalein standard solution (optional), 0.01 M phenolphthalein glucuronic acid, 0.4 M glycine stop solution pH 10.5, and 12 mm × 75 mm polypropylene test tubes.

Note: All incubations are performed in a humidifier at 37°C, 5% CO_2 incubator unless otherwise specified.

18.29.2.1 Reagents and Solutions

18.29.2.1.1 0.4 M Glycine Stop Solution pH 10.5.   3.26 g glycine, 2.53 g NaCl, 3.62 mL 10 N NaOH, and H_2O to 200 mL.
   Sterilize by filtration and store at 4°C. Solution is stable at least for 3 months.

18.29.2.1.2 0.1 M Sodium Acetate Buffer, pH 4.5.   1.158 g sodium acetate, 0.65 mL glacial acetic acid, and H_2O to 200 mL.
   Sterilize by filtration and store at 4°C. Solution is stable at least for 3 months.

18.29.2.1.3 Sodium PIPES Buffer 2× Stock.   8.66 g disodium Piperazine-N,N′-bis (2-ethanesulfonic acid) (PIPES), 25 mL 4 N NaCl, 2.5 mL 2 N KCl, 800 μL 500 mM $MgCl_2$, and 1 g glucose.

Add distilled water to the aforementioned ingredients to just under 500 mL and adjust pH to 7.0 with ~0.6 mL concentrated HCl. Bring volume to 500 mL with distilled water. Filter-sterilize and store at 4°C for ≤6–9 months.

18.29.2.1.4 Sodium PIPES Buffer, 1× Working Solution. 50 mL 2× sodium PIPES stock, 1 mL 10% (w/v) BSA, 100 μL 1 M CaCl$_2$, and H$_2$O to 100 mL. Make fresh each day.

18.29.3 Protocol

1. Prepare RBL cells and set up assay as described in steps 1–7 of the first basic protocol for the histamine assay. Stop each reaction by transferring supernatant to a 12 mm × 75 mm plastic test tube.
2. Add 500 μL of 0.1 M sodium acetate buffer, pH 4.5, to each well and incubate for 30 min at room temperature. Transfer buffer from wells to separate 12 mm × 75 mm plastic test tubes. Repeat extraction with another 500 μL sodium acetate buffer and combine second extract with the first.
3. Add 50 μL of each sample-direct supernatant (from step 1), residual pellet extract (from step 2), or phenolphthalein standard solution to a 12 mm × 75 mm plastic test tube containing 350 μL sodium acetate buffer, pH 4.5, and 100 μL of 0.01 M phenolphthalein glucuronic acid. Prepare two tubes containing 400 μL buffer, 100 μL of 0.1 M phenolphthalein glucuronic acid, and no sample as blanks. Phenolphthalein standard solutions can be used to quantify the reaction products if desired.
4. Vortex, cover with plastic wrap, and incubate overnight.
5. Add 500 μL of 0.4 M glycine stop solution, pH 10.5.

18.29.4 Calculations

Read A$_{540}$ in a standard spectrophotometer. Multiply absorbance value for residual pellet extract by 5 to account for dilution by sodium acetate buffer. Calculate the percent of β-glucuronidase release using the formula:

$$\% \text{ Release} = \frac{100 \times A_{540} \text{ of supernatant}}{A_{540} \text{ of supernatant} + \text{Adjusted } A_{540} \text{ of residual extract}}$$

18.30 Experiment: Generation of Polyclonal Antibodies in Rabbits

18.30.1 Introduction

Polyclonal antibodies (or antisera) are a combination of Ig molecules secreted against a specific Ag, each identifying a different epitope. They are obtained from different B-cell resources. Production of antiprotein antibodies exploits the ability of an animal to protect itself from infection. Macrophages or dendritic cells engulf and break down foreign Ags. Also known as Ag presenting cells (APCs), these cells degrade foreign protein into peptides, which are then presented on the surface of the cells after having been complexed with major histocompatibility complex (MHC) class II protein. Helper T-cells have a receptor that recognizes MHC class II with bound peptide. Each helper T-Cell receptor is highly specific for the MHC class II protein/bound peptide shape, and if the helper T-Cell receptor can bind the complex, interleukin 2 is secreted by the cell, which results in proliferation of that class of T-cell. At the same time as this is happening with the T-cell, the B-cells are also engulfing and degrading the Ag, but these cells have an Ab like receptor that recognizes the particular epitope. The B-cells then present the Ag fragments in a similar way to the APCs. When a Helper T-cell binds via its receptor to the B-cells MHC II peptide receptor, an Ab response ensues. After binding, the helper T-cell secretes interleukin-4 and -5, which cause the B-cells to differentiate into plasma and memory cells. It is the plasma cells that secrete the antibodies whilst the memory cells retain the surface Ab, and are able to mount a more rapid response in the event of being challenged with the same immunogen again (the so-called secondary response). It is this secondary response, which produces far more Ab than the initial challenge that will be exploited for our purposes.

If the Ag is too small, or the presented Ag fragments are not recognized by a helper T-cell, there will be no immune response. Peptides are too small to generate an immune response by themselves and, therefore, need to be coupled to larger molecules in order to become immunogenic. This is because the Ag needs to be sufficiently large in order to be recognized by the APCs (macrophages). In such a case, the Ag is said to be hapten and needs to be coupled to a carrier protein to make it immunogenic. The protein carrier molecules have sequences that are helper T-cell epitopes and, therefore, help helper T-cells to proliferate and interact with the B-cells. This is essential if there is to be a strong immune response. As the carrier protein is chosen to be "foreign" to the animal in which the antibodies will be raised, the carrier protein will enhance the immune response.

The choice of which carrier protein to conjugate will depend on the animal to be used for immunization. Proteins that are least similar to the host animal proteins will generate the best immune response. It is important to remember that antibodies will also be made against the carrier protein, as well as the peptide, so this may have consequences when using the Ab.

Therefore, it is best to avoid using a carrier protein that features in the assay in which the Ab will be used. Keyhole limpet hemocyanin (KLH) and BSA are the two carrier proteins most commonly used, but ovalbumin, rabbit serum albumin, and tetanus toxoid have also been used.

18.30.2 Protocol

The rabbit is the ideal species for the production of polyclonal antibodies and is, therefore, the most widely used animal. It produces sufficient volumes of antiserum with a high concentration and a strong affinity toward the injected Ag.

18.30.2.1 Adjuvants
Adjuvants are used to enhance the recipient's immune response to a supplied Ag while keeping the injected foreign material to a minimum. They are nonspecific

stimulators of the immune system, but how they actually work is not entirely known. However, they are thought to play a multifunctional role. They may slow down the rate of release of the Ag from the site of injection, which is known to be beneficial for a good immune response. Furthermore, the adjuvant may protect the Ag from premature breakdown; it could also be that adjuvant stimulates the release of lymphokines, which would increase the activity of the macrophages (APC cells). There are many kinds of adjuvant, the most famous of which (historically speaking) is Freund's complete adjuvant, which is made up of mineral oil and killed mycobacterial cells (Freunds incomplete adjuvant lacks the killer cells). However, the toxic nature of this adjuvant has led to its use being barred in some countries and in turn has stimulated the development of alternatives, such as produced by RIBI, Titer Max, and many others.

18.30.2.2 Protein Ags
Proteins can be used for immunization purposes from a number of sources viz., whole proteins or portions of proteins, fusion proteins, etc.

18.30.2.3 Expressed or Purified Proteins
These should be preferably in an aqueous buffer. Where the protein is not soluble, detergents, urea, and guanidine hydrochloride can be used as necessary but use in minimal amounts. If the protein is denatured, the Ab generated may only recognize the denatured protein and not the native form.

18.30.2.4 Protein from Bands in a Gel Slice
Bands should be cut from the gel and briefly wash, in water, to reduce the concentration of methanol and acetic acid. For injection purposes, the gel slice should be taken up in a small amount of PBS and homogenized by passing through a large gauge needle. This can then be used for immunization directly. Protein is released from the gel slice slowly, which is good for generating a good Ab response. On the down side, the protein is denatured, and the antibodies may not later recognize the protein in the native state.

Peptides coupled to carriers such as KLH, BSA, etc., can also be used to raise antibodies.

18.30.2.5 Housing of Rabbits
Rabbits are housed in stainless steel cages and allowed free access to complete pelleted rabbit diet and water and with 12 h photoperiod. After no less than 7 days acclimation to the environment and observation for signs of disease, rabbits are given the primary immunization.

18.30.2.6 Rabbit Injection Protocol
When raising polyclonal antibodies in rabbits, it is needed to give an initial injection of approximately 100–200 μg protein/coupled peptide along with the adjuvant. For subsequent boosts, only 70–150 μg protein/coupled peptide are sufficient. A total volume of 1 mL of Ag/adjuvant mix should be injected per rabbit.

18.30.2.7 Immunization/Boost

The Ag/adjuvant mixture is inoculated multiple site subcutaneously along the back and intramuscularly in the hind limbs. We use no more than 0.1 mL of the mixture per site. If we have 1 mL of Ag, we mix it with 1 mL of complete Freund's adjuvant for initial inoculation. Booster doses of Ag are given with Freund's incomplete adjuvant. Two booster immunizations are given at 21 day intervals with the emulsion prepared to use incomplete Freund's adjuvant administered subcutaneously in two locations. Rabbits are checked daily for signs of discomfort or local vaccine reactions.

18.30.2.8 Handling of Blood from Rabbits

Rabbits will have a pre-immune blood sample (1–3 mL) drawn prior to the first immunization from the central auricular artery using a 3 mL syringe and 20 gauge needle. Up to three production blood samples may be drawn from the central auricular artery at two week intervals prior to the terminal bleed.

Once the rabbit is bled, the whole blood is allowed to set at room temperature for about 2 h. It is then placed in a refrigerator for holding overnight. The following day, the whole blood is processed into serum. This process includes centrifuging the whole blood at 5000 rpm × 15 min. The serum is poured off the clot into a clean tube and centrifuged again at 5000 rpm × 5 min. The final serum is poured into a vial, which has the appropriate label attached. A stopper is placed on the vial and then a crimp seal is attached. The vial of serum is then placed in a freezer, which is maintained at below −20°C.

18.30.2.9 Determination of Ab Titer in Rabbit Antiserum

The Ab titer is determined by Ouchterlony immunodiffusion as described in the next exercise.

18.30.2.10 Storage of Antiserum

Made to 0.1% with sodium azide and store at 4°C. Caution! The presence of azide may interfere in assays (for example those involving cell extracts) so beware when using the antibodies for immunodepletion experiments.

Made to 50% with glycerol and store at −20°C

Aliquoted and store at −80°C

18.31 Experiment: Gel Diffusion/Ouchterlony Immunodiffusion

18.31.1 Introduction

The specific combination of an Ab with an Ag is the fundamental reaction of humoral immune response. An Ab is a large protein molecule, which contains two or more identical sites for combining with a specific Ag. A single Ab molecule can bind simultaneously to more than one molecule of Ag. Most Ags are also macromolecules, such as proteins, and each Ag usually contains multiple binding sites and can be bound simultaneously by more than one Ab. Under appropriate conditions of

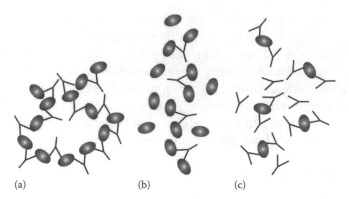

(a)   (b)   (c)

Figure 18.51 Antibody–antigen interactions: (a) equivalence, (b) antigen excess, and (c) antibody excess.

Ab and Ag concentration, antibodies and Ags can form large complexes, or lattices, which are insoluble and precipitate from solution (Figure 18.51). The ability from precipitate makes it possible to perform qualitative and quantitative assays using Ag–Ab systems. Such precipitation assays are very sensitive and relatively quick and easy to perform.

Precipitation assays can be performed either in solution or in a gel such as agarose. When a protein is placed in a well, which has been cut in an agarose gel, the protein will diffuse away from the well, through the gel, in all directions, to create a gradient of the molecule where the concentration decreases with increasing distance from the well.

This occurs because proteins can move freely in the gel, which microscopically is more like a matrix with many holes than like a solid mass, and the kinetic energy of the molecules will cause them to move randomly. If antibodies and Ags are placed in wells in different areas of a gel, they will move toward each other, as well as all other directions, and will form opaque bands of precipitate (precipitation lines) where diffusion fronts meet. Maximum precipitation will occur in the area of gel where concentrations of Ab and Ag molecules are at an optimal ratio. Closer to the Ab well, Ab concentration will be too high for precipitate formation (Ab excess zone). Closer to the Ag well, the Ag concentration will be too high (Ag excess zone).

A major advantage of this type of gel diffusion assay is that more than one Ag–Ab system in a mixture can be detected. It is important to remember that when antibodies are produced by injecting an Ag into an animal, many different Ab molecules reacting with different parts of the Ag (e.g., different proteins in a mixture, or antigenically different parts of a single protein) will be found in the serum of the animal. It is this serum, referred to as antiserum, which is usually used as a source of antibodies in experiments such as these. A single Ag will combine with the Ab it induces (its homologous Ab) to from a single precipitin line in a gel diffusion assay. When two Ags are present, each behaves independently of the other. The positions of the precipitin lines will depend primarily on the sizes of the Ags and their rates of diffusion in the gel. Thus, the number of precipitin lines indicates the number of Ag–Ab systems present. This is actually the minimum number of Ag–Ab systems

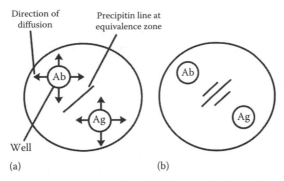

Figure 18.52 Diffusion of antigen and antibody and the formation of precipitin lines: (a) Formation of a precipitin line in gel and (b) multiple precipitin lines from complex mixtures of antigens and antibodies.

since different precipitin lines often develop near each other and appear as a single line (Figure 18.52).

Double diffusion, referring to the process described earlier in which both Ag and Ab molecules diffuse from a well, is a simple but very useful procedure invented by the Swedish scientist, Ouchterlony. It is used to compare Ags for the number of identical or cross-reacting determinants. If a solution of Ag is placed into two adjacent wells, and the Ab is placed equidistant from the two Ag wells, the two precipitin lines formed will join at their closest ends and fuse (Figure 18.53).

This is known as a reaction of identity. When unrelated Ags are placed in adjacent wells, and the center well is filled with Ab reacting with both Ags, the precipitin lines will form independently of each other and will cross. This is known as a reaction of nonidentity. If one of the Ags was used to produce the Ab sample (i.e., it is the homologous Ag) and the second Ag shares with the homologous Ag some, but not all,

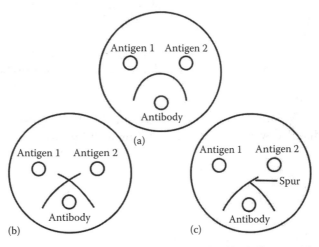

Figure 18.53 Detection of antigenic identity by the double diffusion: (a) identity, (b) nonidentity, and (c) partial identity.

of the specificities recognized by the Ab (i.e., it is a cross-reacting Ag), the two pre-cipitin lines will fuse. However, an additional spur will form, which projects toward the cross-reacting Ag. The spur represents the reaction between the homologous Ag and those Ab molecules that do not bind to the cross-reacting Ag.

18.31.2 Materials

1% agarose dissolved in buffered saline solution pH 7.4 and kept in 60°C water bath, 4 microscope slides, 95% ethanol, 5 mL pipette, pipette pump, well cutter, template for cut-ting wells, transfer pipettes with bulb, practice loading solution (optional), tube A con-taining antibodies produced against whole rabbit serum (Ab), tube B containing whole rabbit serum (Ag B, or AB), tube C containing rabbit serum albumin (Ag C, or AgC), tube D containing rabbit serum Ig (Ag D, or AgD). Antiserum containing antibodies against the Ag being tested, Ag, Phosphate-buffered saline (200 mL), agarose 1.5 g in 250 mL, plastic or lid for pans, toothpicks or spatulas for removing agarose plugs from wells.

18.31.2.1 Staining and Destaining Solution
0.1M Tris buffer pH 7.5, destaining solution 50:40:10 water:methanol:acetic acid, staining solution 0.1% Coomassie blue in 50:40:10 of water:methanol:acetic acid.

18.31.2.2 Preparation of Solutions
1. For phosphate-buffered saline, dissolve the following in 150 mL distilled water: 2.19 g NaCl, 1.28 g KH_2PO_4, 1.63 g $Na_2HPO_4\cdot 7H_2O$ adjust pH to 7.2, add distilled water to bring final volume to 250 mL.
2. 1% agarose solution
   a. Add 180 mL phosphate-buffered saline to 250 mL flask. Add 1.8 g agarose. Swirl to disperse large clumps.
   b. Heat to dissolve agarose using a hot plate to bring mixture to boiling with occa-sional careful swirling. Continue heating until all particles have dissolved, and the final solution is clear.
   c. Keep melted agarose in 60°C water bath to prevent solidification, If the agarose does solidify, remelt it in a hot water bath or in a microwave oven.
3. Tris buffer

Tris solution (0.01 M.), dissolve 1.21 g Tris base in 700 mL distilled water. Adjust pH to 7.5 and add distilled water to bring the volume to 1000 mL.

18.31.3 Protocol

18.31.3.1 Preparation of Slides with Sample Wells
1. Clean all slides by placing a small amount of alcohol. After cleaning, do not touch the clean surface with fingers.
2. Label the slides to be used for the experiment at one end and number them consecutively 1, 2, 3. Be careful not to touch the clean surface with fingers.

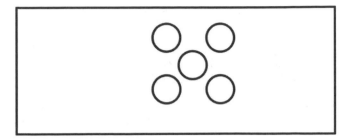

Figure 18.54   Template for cutting wells.

3. Using a 5 mL pipette, carefully pipette 3.5 mL of melted agarose, allow to cool slightly, onto each slide by moving the pipette tip along the center length of the slide and dispensing the agarose. The melted agarose should spread evenly over the surface of the slide, without running over the edges. Repeat with the remaining two slides.

   Note: If you are not successful at covering the slide with agarose, wipe off the slide, clean the slide again with alcohol as described in step 1, relabel the slide, and try again.

4. Allow the agarose to solidify. This will take approximately 10 min at which time the gel will appear slightly opaque.

5. For the three slides to be used for the experiment, place the template under one of the slides so that the pattern is in the center of the slide. The distance between the wells is important. Try to follow the template as accurately as possible (Figure 18.54).

6. Cut the five wells with the well cutter, using a gentle punching motion. Remove the agarose plugs by lifting the edges with a toothpick or spatula.

   Note: If well placement is not accurate, there should be enough room on the slide to re-cut the wells after repositioning the template under the slide.

7. Repeat steps 5 and 6 with the remaining two slides.

18.31.3.2 Loading the Samples

1. Load samples by the following:
   a. Squeeze the pipette bulb to draw slowly, the liquid up into the pipette. The sample should remain in the lower portion of the pipette.
   b. Place the pipette tip just over, not inside, the sample well. Maintain steady pressure on the pipette bulb to prevent sample from being drawn back up into the pipette.
   c. Slowly squeeze the pipette bulb to eject three drops of sample. The well should appear full. Be careful not to overfill the well and cause spillage on the agarose surface. Such spillage would affect your results (Figure 18.55).

2. Fill the center wells of all three slides with 3 drops (approximately 30 µl) of antiserum (Ab, tube A), using a transfer pipette.

3. Fill the outer wells with 3 drops of Ag (tubes B. C. and D), using a clean transfer pipette for each Ag as follows in Figures 18.54 through 18.56.

4. Carefully place the slides in the incubation chamber on top of the wet paper towel layer. Cover the chamber with plastic wrap and let allow precipitin lines to form. The precipitin lines should be visible in 24 h.

5. Predict the patterns that will be formed by the precipitin lines, based on the relationships between the Ags and the information given in the introduction. Sketch your predictions on the diagrams of the slides.

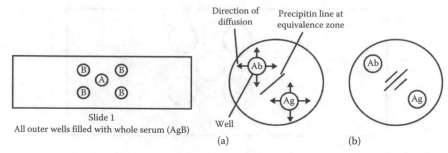

Figure 18.55   Filling of well: (a) formation of a precipitin line in gel and (b) multiple precipitin lines from complex mixtures of antigens and antibodies.

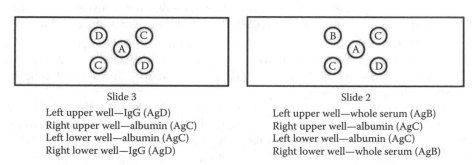

Figure 18.56   Complete test.

18.31.3.3 (Optional) Staining Slides

1. Cover the slides with Iris buffer. 0.01 M pH 7.5 and incubate at room temperature for at least an hour, change the tris buffer once.
2. Remove buffer and cover the slides with 0.1% Coomassie blue stain and let sit 30–60 min.
3. Remove stain and cover slides with destain. Change destain when it becomes blue (about 1 h), add fresh destain, and repeat until the gel is clear enough to visualize blue precipitin lines.
4. Remove destain, cover dish and store slides in the refrigerator until results are observed.

18.31.4 Results and Conclusions

1. The precipitin lines will be visible as opaque white arch on each slide where Ab and Ag complexes precipitated.
2. Draw the patterns of the precipitin lines observed on each slide. The precipitin lines should be seen easily. The expected results are shown in Figure 18.57. However, the intersections of the lines are often faint and more difficult to see, making interpretation difficult. The Coomassie blue staining is recommended to increase visibility of the lines.

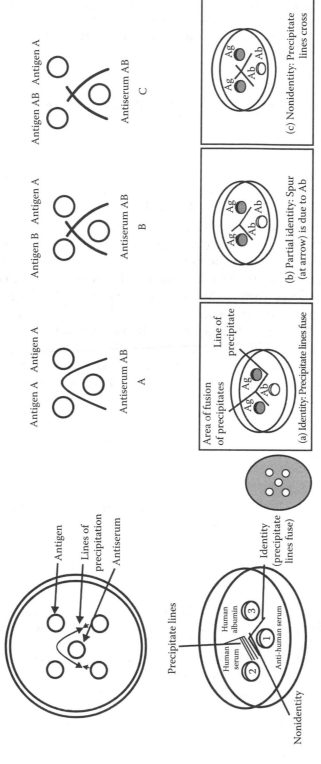

Figure 18.57 Predicted results for double diffusion experiment.

3. Compare the patterns obtained to those illustrated in Figure 18.57. For each slide, determine the relationship between the Ags which is indicated by the results. The results indicate identity, partial identity, or nonidentity between the Ags.
4. The relationships indicate the line patterns against Ag should be is consistent with what you predict.

18.32 Experiment: Demonstration of Radial Immunodiffusion

18.32.1 Introduction

Radial immunodiffusion is a technique that can quantitatively determine the level of an unknown Ag. This is also known as Mancini technique and is similar to Ouchterlony immunodiffusion, but in this method the antiserum is incorporated into the gel. Small wells cut into the agar dish are filled with solutions of known concentration Ag and an unknown solution sample. The Ag solution diffuses outwards front the well in a circular pattern. In the endpoint or Mancini method used in this experiment, Ab is present in excess amount and diffusion continues until a stable ring of Ag–Ab precipitate forms, generally within 24–48 hours. For each standard Ag concentration, an endpoint precipitation ring of a certain diameter is formed. The diameter squared measurements of the rings plotted against the known concentration of each starting standard yields a straight line. From this linear calibration curve, the concentration of the unknown Ag sample may be determined. Unlike many gel and liquid precipitation techniques which qualitatively detect Ag, RID is a sensitive quantitative technique that is often used clinically to detect patient levels of blood proteins.

18.32.2 Materials

RID plates which contain set of reference standard for Ag, calibrated ruler mm, pipette gun 5 μL or plastic micropipettes, graph paper, A kit containing critical reagents required for the experiment is commercially available.

18.32.3 Protocol

1. Perform the test using the samples, following the directions provided with the kit, add 5 μL of Ag solution standards to each of four wells. Rinse micropipettes or change tips (Figure 18.58).

   To a fifth well add 5 μL of the unknown Ag solution. Snap the cover of the Petri dish back on. Place the dish into a zip lock bag and seal.

   Allow the dish to sit undisturbed for 24–28 h at room temperature on a flat surface.

   After precipitin rings have reached equilibrium, remove the plate from the bag and place on a light box or back lit plate.
2. Record the result of the test.

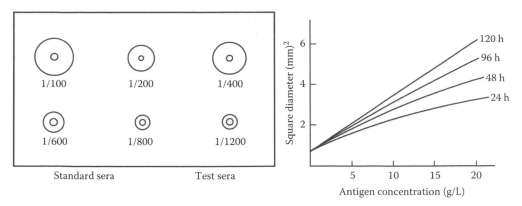

Figure 18.58   Albumin measurement by single radial diffusion method.

3. Measure the diameter of each precipitin ring with the ruler to the nearest tenth of an mm.
4. Square the ring diameters and plot these values on the Y-axis of the graph paper versus the known starting Ag concentrations and draw the best lit line through the standard points.
5. Using the standard line determine the value of the unknown Ag.

18.32.4 Result

1. Interpret the result of the test by measuring ring diameters considering any control sample run.
2. Plot the values of precipitin ring diameter squared versus known Ag concentrations. Draw the best fit line through these points. Calculate the value of the unknown Ag concentration from this line (Figure 18.58).

18.33 Experiment: Counter Current Immunoelectrophoresis

18.33.1 Principle

In counter current immunoelectrophoresis, two parallel wells in agar gel are filled with Ag and Ab move toward each other at a faster rate. The Ag migrates toward the anode and the Ab moves in opposite direction as a result of electro-endosmosis and a precipitin line is formed where they meet (Figure 18.59).

18.33.2 Materials

Noble agar, 0.01 M Veronol buffer pH 8.6, normal saline, amido black (1%), 2% acetic acid, glass microslides, humid chamber, electrophoretic tank with power supply, Whatman No. 3 filter paper, template and gel punch.

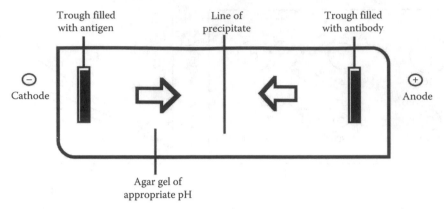

Figure 18.59 Schematic illustration for counter immunoelectrophoresis.

18.33.3 Protocol

1. 1% Nobel agar in veronol sodium acetate buffer (0.01 M, pH 8.6) is layered over glass slides and well are made as described under double diffusion technique.
2. Fill the buffer tank with veronol buffer and, see that level of buffer in the two compartments is same.
3. Arrange the agar slide in electrophoretic chamber and connect to the buffer with Whatman-3 filter paper wicks.
4. Make pre run for 15 min at a constant current of 7.5 mA per slide.
   Disconnect the power supply and 5 μL of Ab in the anode well and 5 μL Ag in the cathode well are added.
6. Connect the electrode to the power supply and run electrophoresis for 1 h at a constant current of 7.5 mA per slide.
7. After the run, disconnect the power supply and incubate the slide in a humid chamber at room temperature for 12 h.

18.33.4 Result and Conclusion

Observe the slides for precipitin bands and the slides may be stained for improved sensitivity in 1% amido black as described under double diffusion.

18.34 Experiment: Serum Electrophoresis

18.34.1 Principle

Serum protein electrophoresis is used to identify patients with serum protein disorders. Electrophoresis separates proteins based on their physical properties, and the subsets of these proteins are used in interpreting the results. Plasma protein levels display reasonably predictable changes in response to acute inflammation, malignancy, trauma, necrosis, infarction, burns, and chemical injury. A homogeneous spike-like peak in a focal region of the gamma-globulin zone indicates a monoclonal gammopathy. Monoclonal gammopathies are associated with a clonal process that is malignant or potentially malignant, including multiple myeloma, Waldenström's macroglobulinemia, solitary plasmacytoma, smoldering multiple myeloma, monoclonal gammopathy of undetermined significance, plasma cell leukemia, H-chain disease, and amyloidosis. In contrast, polyclonal gammopathies may be caused by any reactive or inflammatory process.

Electrophoretic pattern of serum protein depends on the fractions of two major types of the protein: albumin and globulins. Albumin, the major protein component of serum, is produced by the liver under normal physiologic conditions. Globulins comprise a much smaller fraction of the total serum protein content. The subsets of these proteins and their relative quantity are the primary focus of the interpretation of serum protein electrophoresis.

Albumin, the largest peak, lies closest to the positive electrode. The next five components (globulins) are labeled α_1, α_2, β_1, β_2, and γ. The peaks for these components lie toward the negative electrode, with the gamma peak being closest to that electrode. Figure 18.60 shows a typical normal pattern for the distribution of proteins as determined by serum protein electrophoresis.

18.34.2 Materials

Glass slides, agarose, electrophoresis apparatus, power pack,

1 × immune buffer
Prepare from 10 ×: 1 mL 10 × immunobuffer + 9 mL distilled water.
Veronal Buffer 0.1 M pH 8.6

1. Na-barbitone = 10 g/L (5 g/500 mL)
2. Na-Acetate = 6.5 g/L (3.28 g/500 mL)
3. HCl 0.1 N = 34.2 μL/L (17.1 μL/500 mL)

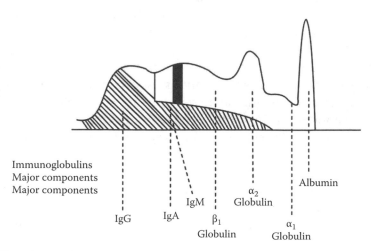

Figure 18.60 Electrophoretic pattern of plasma proteins.

18.34.3 Protocol

1. Prepare 1% agarose in 1×immuno buffer.
2. Prepare Gel plate by pouring 5–6 mL of agarose on to the slide.
3. After the solidification, cut a trough on the Gel with the help of blade.
4. Take an electrophoretic tank and add Veronal buffer.
5. Transfer agarose slide to the electrophoretic chamber.
6. Connect the slides to anodic and cathodic buffer chambers with Whatmann filter paper wicks.
7. Make pre run for 5 min at 7.5 mA.
8. Load 20 µL serum containing few drops of Bromo phenol blue dye in the trough.
9. Run electrophoresis for about 180 min.
10. Remove slides from electrophoretic chamber
11. Wash slides in normal saline for about 30 min.
12. Immerse slides in 2% acetone solution for about 15 min.
13. Dry the slides in an incubator at 40°C.
14. Stain the slides in 1% amido black or Coomassie blue stain.

18.34.4 Results and Conclusion

Observe the slides for bands.

18.35 Experiment: Demonstration of Immunoelectrophoresis

18.35.1 Introduction

In Immunoelectrophoresis, two methods are combined, namely agar-gel electrophoresis and immunodiffusion. Method is used in both clinical and research laboratories for separating and identifying proteins on the basis of their electrophoretic behavior and their immunological properties. Immunoelectrophoresis is used to separate and characterize a mixture of serum proteins as well as to examine the specificity of the Ag–Ab interaction.

In the clinical laboratory, this technique is used for diagnosis. It is utilized in examining certain serum abnormalities especially those involving Igs, and also semi-quantitative proteins analysis of urine, CSF, pleural fluids, etc. In the research laboratory, this technique may be used to monitor Ag and/or Ab purifications, to detect impurities, to analyze soluble Ags of plant and animal tissues and microbial extracts.

In immunoelectrophoresis, the proteins are first subjected to an electric field and separated by zone electrophoresis on the basis of their different charge-to-mass ratios. The electric field is then shut off, and antibodies to the proteins are introduced into the system. Diffusion of both Ag and Ab takes place and at a particular locus, the equivalence point is reached resulting precipitation. In classical immunoelectrophoresis, the Ab is added in a trough cut out of the electrophoretic gel after electrophoresis.

18.35.2 Materials

2 g agarose, 500 mL TAE (Tris-acetate-EDTA) buffer pH 8.0, 10 µL whole serum, 200 µL goat anti-rabbit IgG, horizontal electrophoresis apparatus

TAE (50 × stock, 1 Liter): 24.2 g Tris base, 57.1 mL glacial acetic acid, 37.2 g Na₂EDTA. H₂O pH 8.5.

Stock solution: dissolve 12.1 g tris(hydroxymethyl)aminomethane (Tris) and 17 g of ethylenediaminetetraacetate, disodium salt (EDTA) in 450 mL of ionized water. Adjust pH to 8.5 with glacial acetic acid (25 mL). Make up to 500 mL with water.

Working buffer: dilute stock 50 times

Dilute antiserum in TAE if necessary

Prepare agarose gel: Mix 0.5 g of agarose and 50 ml of diluted TAE buffer in a 250 ml Erlenmeyer flask. Dissolve by boiling in a microwave oven or on a hot plate. Allow to cool to 50°C. Temperature of molten gel is critical to be at 50°C.

18.35.3 Protocol

1. Coat a glass plate by applying 2% agar solution with a small brush.
2. Cast the gel on to the glass plate by pouring the molten gel to a thickness of 1–2 mm and allow cooling.
3. Cut a small hole/trough in the gel by punching the agarose gel with a Pasteur pipette.
4. Transfer the gel to the electrophoresis apparatus and add enough buffers, to just cover the gel.
5. Add 2 drops of a bromophenol blue solution to the serum.
6. Load whole serum into the well with a micropipette or capillary tubing.
7. Elevate slide 2–3 mm above electrophoresis-buffer solution. Make contact between the buffer and agarose by using a wick of filter paper pre-soaked in electrophoresis buffer. One end of the wick should overlap the end of the gel by 3–4 mm and the other end submerged in electrophoresis buffer.
8. Electrophorese at 50 V and 30 mA for 1 h.
9. Carefully remove the gel from the electrophoresis chamber. Cut two troughs, parallel to electrophoretic run. Use a sharp blade to delineate the troughs and a Pasteur pipette tip to remove the cut gel.
10. Load the left through with goat anti-rabbit whole serum
11. Place the gel in a closed humidifying chamber containing moistened paper towels and allow diffusion to take place over a 24 h period or until a visible precipitate is formed (Figure 18.61).

18.35.4 Result

1. Note the formation of arcs of precipitate in the gel.
2. Identify the number of proteins in the rabbit whole serum from the number of arcs of precipitate,
3. The actual number of arcs of precipitate will depend on the number of Ags in serum and the efficiency of separation.

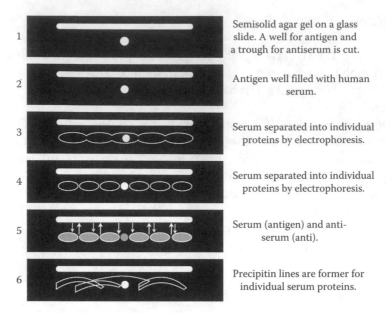

1 — Semisolid agar gel on a glass slide. A well for antigen and a trough for antiserum is cut.

2 — Antigen well filled with human serum.

3 — Serum separated into individual proteins by electrophoresis.

4 — Serum separated into individual proteins by electrophoresis.

5 — Serum (antigen) and anti-serum (anti).

6 — Precipitin lines are former for individual serum proteins.

Figure 18.61   Flow diagram illustrating immunoelectrophoresis.

18.36 Experiment: Determination of Albumin by Laurell Rocket Immunoelectrophoresis

18.36.1 Principle

Rocket immunoelectrophoresis is very similar to radial immunodiffusion in that the Ag is incorporated into the agar-gel. This time, however, the samples move through the gel by electrophoresis rather than by simple diffusion.

The antiserum is incorporated into the agar, and holes are cut along the origin, to hold the test material and dilutions of a standard Ag for calibration. When potential is applied across the gel, the Ag migrates toward the anode while the Ab in the gel moves toward the cathode and immune complex forms where they meet. Initially there is an excess of Ag so that the immune complex is soluble but as the experiment proceeds the Ag continues to migrate and becomes more dilute as more Ag is present in the immune complex. Later on equivalence is reached, and an insoluble immune complex is formed. The complex itself migrates but eventually becomes stationary when the entire Ag is present in the immune complex. The precipitate is rocket shaped and, for a given Ab concentration, there is a linear relationship between the distance moved by the precipitate and the concentration of the Ag (Figure 18.62).

18.36.2 Materials

Materials described for immunoelectrophoresis and immunodiffusion in earlier experiments, Coomassie blue (0.25% w/v in methanol:acetic acid: water, 1 L 5:1:5), destaining solution (methanol:acetic acid:water, 5:1:5), hair drier.

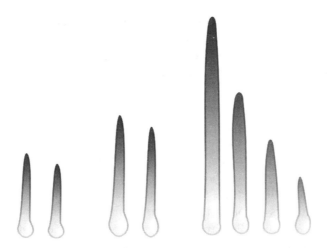

Figure 18.62   Demonstration of Laurell rocket immunoelectrophoresis.

18.36.3 Protocol

Prepare the plates as for the immunoelectrophoresis but add 0.6 μL of rabbit anti-human albumin per cm² of the plate area. Cut 12 wells into the gel in a straight line 1.5 cm from one edge of the plate and place in the electrophoresis apparatus. Apply the diluted standards to four of the wells and 1/400 dilutions of the tests to the other wells. All dilutions are made in saline as in the previous experiments. Pipette 3 μL of the samples into each well as quickly as possible so as to minimize the amount of radial diffusion. When all the samples have been added to the plate, which should be set up in duplicate, place the lid of the tank on the apparatus and turn on the power. Carry out electrophoresis at 5–10 V/cm with the cooling system activated. At the end of this time, turn off the power and remove the plate. Place the slides under a number of filter papers with a light weight on top, leave for 10 min then remove the filter papers. Dry the plate completely in a stream of hot air from a hair drier then immerse the dry plate in Coomassie blue for 10 min. Remove the excess stain with the destaining solution and measure the heights of the rockets from the anodal edge of the well to the peak of the rocket.

18.36.4 Results and Conclusion

Plot the rocket height against albumin concentration to produce a standard curve, and then use this curve to estimate the albumin concentrations in the eight serum samples.

18.37 Experiment: Purification of Igs by Ion Exchange Chromatography

18.37.1 Principle

When human serum is added to an ion exchange column such as DEAE-Sephadex, most of the serum proteins become bound to the column. However, by using a buffer with a pH greater than 6.5, IgG is not absorbed and is eluted in the first fractions. The other proteins are then removed from the column by a stepwise elution with increasing concentrations of salt. The isolated fractions are then analysed by cellulose acetate electrophoresis and immunoelectrophoresis.

18.37.2 Materials

DEAE-Sephadex A50–120, disposable syringes (10 mL), Tris-HCl buffer (10 mmol/L, pH 8.0), Buchner filtration apparatus, NaOH (1 mol/L), HCl (1 mol/L), Glass wool or plastic sinter, Peristaltic pumps that can pump less than 60 mL/h, human serum, UV spectrophotometers or column monitoring equipment, solid polyethylene glycol

Buffered saline solutions: (Prepare the different concentrations of 250 mL saline in the Tris-HCl buffer (a) 20 mmol/L, (b) 50 mmol/L, (c) 100 mmol/L, (d) 150 mmol/L, (e) 200 mmol/L, (f) 300 mmol/L.

Electrophoresis: horizontal electrophoresis apparatus, power pack, barbitone buffer (0.07 mol/L pH 8.6), ponceau S stain (2 g/L in 30 g/L TCA), acetic acid (5% v/v), cellulose acetate strips, Whatman 3 MM paper, methanol: water (3: 2), citrate buffer (0.1 mol/L pH 6.8).

Immunoelectrophoresis: 2 g agarose, 500 mL TAE (Tris-acetate-EDTA) buffer pH 8.0, 10 μL whole serum, 200 μL goat anti-rabbit IgG, horizontal electrophoresis apparatus

TAE (50 × stock, 1 L): 24.2 g Tris base, 57.1 mL glacial acetic acid, 37.2 g Na_2EDTA. H_2O pH 8.5.

Stock solution: dissolve 12.1 g tris(hydroxymethyl)aminomethane (Tris) and 17 g of ethylenediaminetetraacetate, disodium salt (EDTA) in 450 mL of deionized water. Adjust pH to 8.5 with glacial acetic acid (25 mL). Make up to 500 mL with water.

Working buffer: dilute stock 50 times. Dilute antiserum in TAE if necessary

Prepare agarose gel: Mix 0.5 g of agarose and 50 mL of diluted TAE buffer in a 250 mL Erlenmeyer flask. Dissolve by boiling in a microwave oven or on a hot plate. Allow to cool to 50°C. Temperature of molten gel is critical to be at 50°C.

18.37.3 Protocol

Preparation of column: Add 2 g of DEAE-Sephadex A50–120 to 70 mL of water and leave until swollen. Degas the suspension under vacuum and leave to equilibrate to room temperature. Place some glass wool in the bottom of a 10 ml disposable syringe as a support and prepare a column of the ion exchange material. Wash the column with 1 mol/L NaOH then repeatedly with water until the pH falls to about 7. Wash the saline then 1 mol/L HCl and finally with the tris-HCl buffer until the eluate registers pH 8.0. Alternatively the material can be washed on a Buchner funnel.

Separation of Igs: Load 0.5–1.0 mL serum on to the DEAE-Sephadex and collect five fractions of about 2.5 mL each by pumping the tris-HCl buffer through the column. In the meantime prepare 25 mL of a series of fluids containing sodium chloride at the following concentrations in the buffer: 20 mmol/L, 50 mmol/L, 100 mmol/L, 150 mmol/L, 200 mmol/L, and 300 mmol/L. Pump these solutions through the column in turn starting with the lowest salt concentration and collect fractions until all the protein has been eluted with that solution, then move on to a higher concentration of salt. Carefully monitor the proteins eluted by measuring the absorbance of each fraction at 280 nm and plot the extinction fraction number.

18.37.4 Result

Concentrate each fraction between 5 and 10 times by dialysing against solid polyethylene glycol for about 3 h and examine the protein concentrates by cellulose acetate electrophoresis, immunodiffusion and immunoelectrophoresis. Compare the electrophoresis pattern of fractions.

18.38 Experiment: The Detection and Assay of a Myeloma Protein

18.38.1 Principle

Multiple myeloma is a malignant disorder of the bone marrow where single cell multiplies to produce a very large number of cells of one kind. The resultant colony of cells is a clone as they are all descended from one cell. The cells, therefore, have identical properties and produce large quantities of homogeneous Ig. The detection and quantitation of myeloma protein is important in the diagnosis of this condition, and the following experiment shows how electrophoretic and immunological techniques are used in the clinical biochemistry laboratory.

Samples of serum from myeloma patients can be obtained by arrangement with the medical staff of a local chemical pathology laboratory and must be treated even more carefully than usual when handling body fluids. The actual type of myeloma will of course depend on what samples are available at the time.

18.38.2 Materials

Sera from myeloma patients, materials for cellulose acetate electrophoresis, materials for immunoelectrophoresis, Anti-IgA serum, Anti-IgG serum, materials for single radial immunodiffusion.

18.38.3 Protocol

Detection: Use cellulose acetate electrophoresis to detect myeloma protein in the samples provided. Stain the cellulose acetate strips and compare the pattern obtained with normal human serum.

Demonstration: The class of the Ig, which constitutes the myeloma protein, can be demonstrated by immunoelectrophoresis in agarose-gel. Use the anti-IgA and the anti-IgG sera to identify the Ig.

Quantitation: Single radial immunodiffusion is a technique that is frequently used to measure the amount of Igs present in the serum. Use two plates in duplicate for the experiment with anti-IgA in the gel of one pair and anti-IgG in the other pair.

18.38.4 Result and Conclusion

Observe the separation of human serum proteins by electrophoresis and interpret the result obtained by immunodiffusion and immunoelectrophoresis.

18.39 Experiment: ELISA

18.39.1 Introduction

Enzyme-linked immunosorbent assay (ELISA) is a very sensitive and safe technique used in the detection of Ags and antibodies. The sensitivity of this technique is comparable with that of radio immunoassay with an added advantage of safer use of non-radioisotopic reagents and longer shelf life of the same. It also eliminates the requirement of sophisticated isotope counters. ELISA can be done in smaller laboratories and is adaptable to field conditions, as well.

Although initially developed for Ab measurement, enzyme immunoassays have also been adapted successfully to detect and quantitate Ag in unknown samples. For example, antibodies specific for HIV Ags can be produce by injecting the Ags into laboratory animals (e.g., rabbits) and the specific antibodies obtained can then be used to develop immunoassays for measuring HIV Ags in infected individuals. Due to their high specificity and sensitivity, immunoassays are commonly used for.a great variety of measurements of both antibodies and antigenic analytes (e.g., hormones, drugs) in research, analytical and diagnostic laboratories.

The basis of all Immunoassays is an interaction between an Ab and a corresponding Ag. The interaction can in turn be conveniently detected by conjugating a measurable label to either the Ab or the Ag, such as a radioisotope, a fluorescent compound or an enzyme. An enzyme label is generally favored over either fluorescent compounds that require expensive equipment for detection or radioisotopes which is a biohazard and unstable. Thus, in recent years, the enzyme-based immunoassay, generally referred to as enzyme-linked immunosorbent assay (ELISA) has seen increasing popularity in a wide range of applications. There are two basic types of ELISA:

1. Direct binding assays primarily for detections and quantification of antibodies.
2. Competition enzyme immunoassays for the detection and quantification of Ags.

18.39.1.1 Direct ELISA

1. Observe a specific interaction between an Ag and its Ab.
2. Determine by titration the amount (i.e., titer) of Ab in a serum preparation from a rabbit immunized with the Ag.

   Experiment can be completed in 2–3 h and the experimental protocol is divided into the following parts:

   Prepare Ag solutions and coat the Ags to a 98 well micro titer plate.

   Prepare serial dilutions of rabbit serum solutions.
3. Add Ab solutions to Ags coated on the plate to allow Ag–Ab interaction.
4. Add second Ab conjugated with an enzyme (e.g., horseradish peroxidase enzyme conjugated with goat antibodies specific for rabbit antibodies); this second Ab, referred to as anti-Ag which will react with the Ag–Ab complex formed earlier.
5. Add a substrate solution that will react with the enzyme conjugate to form a colored product. Theoretically, the intensity of color developed is proportional to the quantity of Ab in the antiserum sample. No significant color change will be observed if no antibodies are present in the rabbit serum.

Various components involved in ELISA are a solid phase to which specific Ag/Ab is coated, an Ag or Ab enzyme conjugate as a probe as the case may be and enzyme substrate. Solid supports used are polystyrene or PVC microtiter plates, tubes or beads. Nitro cellulose paper or cellulose acetate membrane is also used as solid phases. Enzymes used for conjugation to anti-Igs include peroxidase, β- galactosidase, alkaline phosphatase, penicillinase, urease, glucose oxidase. The influence of the solid phase on the enzyme should be minimal. Conjugation to anti-Ig should be easy, and conjugates should be active and stable. These are the reasons for frequent selection of peroxidase in commercial diagnostics (Figure 18.63).

The enzyme immunoassay basically involves interaction between Ag and Ab on a solid phase and can be modified in several different assays.

18.39.1.2 Indirect ELISA

Indirect ELISA is useful for the detection of Ab using specific Ag. In this assay, the PVC plate is coated with Ag and the test sample (serum/blood, hydrocele fluid, urine, etc) added to the plate. Any Ab specific for the Ag will bind to the available sites. The bound Ab is detected by incubation with an enzyme labeled specific anti-Ig followed by the enzyme substrate (Figure 18.64).

18.39.1.3 Sandwich ELISA

In sandwich assay, two antibodies are used primarily to determine the Ag concentration in unknown samples. This assay is quick and accurate. In this assay system, the Ag to be detected is a sandwich between two similar or different antibodies of which one is labeled with an enzyme. The assay requires two antibodies that bind to non-overlapping regions of the same Ag. One Ab is immobilized to the solid phase, and the Ag in the test sample is allowed to bind. Unbound material is washed away, and a labeled, second Ab is added which will bind to any Ag bound to the immobilized Ab.

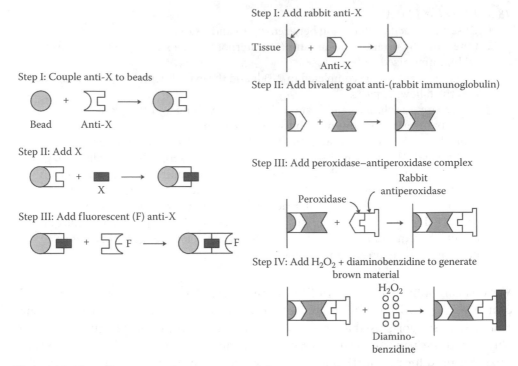

Figure 18.63 Illustrations for preparing solid support and enzymes used for conjugation.

The major advantage of this technique is that it is very sensitive and specific, and crude Ag samples can be used (Figure 18.65).

18.39.1.4 Competitive ELISA

In competitive ELISA, two specific antibodies, one conjugated with enzyme and other present in patient's serum are used. Two antibodies compete for the same Ag. Appearance of color indicates a negative test (absence of antibodies) while the absence of color indicates a positive test (presence of antibodies). It is most commonly used to detect antibodies to HIV. The microtiter plate wells are coated with HIV Ag. The patient's serum to be tested is added and incubated at 37°C and then washed. If antibodies are present in patient's serum, react with Ag. This Ag–Ab complex can be detected by adding enzyme labeled specific antibodies to HIV. In a positive test, no Ag is left for these antibodies to react. Hence these antibodies remain free and are washed away during washing. When substrate is added, no enzyme is available to act on it. Therefore, positive test shows no color. In a negative test, no antibodies are present in patient's serum, Ag in the coated wells is available to combine with enzyme labeled antibodies and the enzyme acts on the substrate to form (Figure 18.66).

18.39.1.5 Inhibition ELISA

Inhibition ELISA works similar to competitive ELISA, but in this system the two Ags (Ag in the test sample and enzyme labeled Ag) are added one after another. This is useful

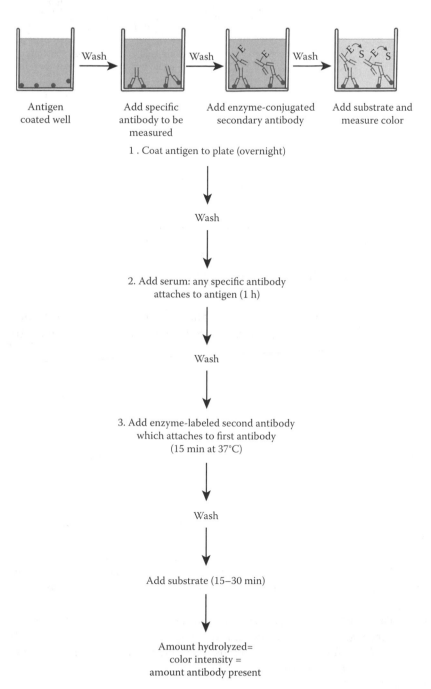

Antigen coated well — Wash → Add specific antibody to be measured — Wash → Add enzyme-conjugated secondary antibody — Wash → Add substrate and measure color

1 . Coat antigen to plate (overnight)

↓

Wash

↓

2. Add serum: any specific antibody attaches to antigen (1 h)

↓

Wash

↓

3. Add enzyme-labeled second antibody which attaches to first antibody (15 min at 37°C)

↓

Wash

↓

Add substrate (15–30 min)

↓

Amount hydrolyzed= color intensity = amount antibody present

Figure 18.64 Flow chart of indirect ELISA.

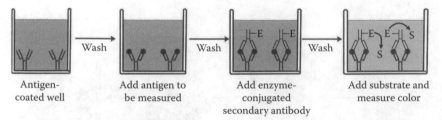

Figure 18.65   Sandwich ELISA.

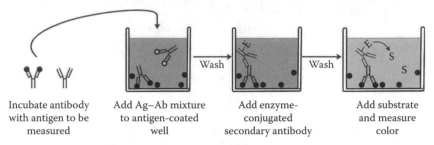

Figure 18.66   Illustrations for competitive ELISA.

especially when test serum contains both Ag and Ab of interest in the immune reaction. Inhibition ELISA is also useful in determining the identity of specific Ag or Ab.

18.39.2 Precautions

Standard lab safety guidelines for handling chemicals and biological samples are adequate here. Wear a lab coat (or apron), gloves and goggles in performing the experiment. The substrate solution which contains TMB, a potential mutagen, and organic solvent should be handled with caution.

The four basic steps of the assay are as follows:

1. *Binding the Ag*: BSA or gelatin is added to separate wells of the microtiter plate. Allow time for the Ag to bind to the plate.
2. *Ab addition*: Plates are washed. Rabbit serum A containing Ab to BSA and rabbit serum B (normal rabbit serum without antibodies to BSA) are serially diluted and added to specific wells of the ELISA plate. This is incubated to allow for the Ag–Ab interaction.
3. *Conjugate addition*: The plates are again washed. This will wash away Ab that has not reacted with the Ag on the plate. The conjugate (enzyme coupled to an anti-1g or second Ab) is added. Conjugate A binds rabbit Ab. Conjugate B binds to mouse Ab. Both conjugates are linked to horseradish peroxidase in this case. Again the plates are incubated.
4. *Addition of substrate*: Again, wash away unbound material. An enzyme substrate (a color forming chromogen) is added. This reacts with the conjugate to produce a color indicating an Ag–Ab reaction.

18.39.3 Materials

ELISA reader
96 well microtiter plate, Multichannel pipette, Disposable pipette tips, Plastic squirt bottles, Glass Test tubes, Plastic Test tubes, 15 and 50 mL Test tube racks, 5 mL pipettes, Vortex mixer (optional).
Reagents
PBS: PBS 10× solution (Commercially available).
Add 50 mL to a 500 mL graduated cylinder. Fill to the 500 mL mark with distilled water for a working solution. (Optional: add 0.5 mL of Tween 20 per L of 1×PBS).
PBS (3% milk): Blocking diluent is made by adding 3 g nonfat dry milk to 100 mL 1×PBS,
Carbonate buffer 0.05 M: Sodium carbonate buffer, pH 9.6, containing 0.02% sodium azide.
Ag I 1% BSA: A vial containing 0.1 g of BSA. This should be mixed with 10 mL of carbonate buffer for a working solution of Ag I.
Ag II 1% gelatin: Prepare 1% solution by dissolving 0.1 g into 10 mL of carbonate buffer. This should be heated until the gelatin is dissolved. Cool before using.
Rabbit serum A: Contains antibodies against BSA provided in vial with 0.1 mL of 1:10 dilution. Dilute this with 9.9 mL of PBS (3% milk) as follows:
Measure out 9.9 mL of PBS (3% milk) in a 15 mL test tube and raise with diluent to 15 mL and mix. Label tube # 1 rabbit serum A 1: 1000
Rabbit serum B Normal rabbit serum: It does not contain antibodies against BSA. Prepare as above for rabbit serum A. label #8 rabbit serum B 1:1000
Anti-rabbit Ig conjugate: Keep frozen until ready to use (0.2 mL). Dilute in 19.8 mL PBS (3% milk) in 50 mL tube. Label rabbit conjugate.
Anti-mouse Ig conjugate: Horseradish peroxidase conjugated Ab specific for mouse. Store and dilute as for anti rabbit Ig conjugate. Label mouse conjugate.
TMB substrate: This is provided as solution A and B. Do not prepare until just before use. Mix 30 mL solution A with 30 mL of solution B. Label TMB. Keep away from light.
Phosphate-buffered saline (PBS): 1.9 mM $NaHPO_4$ (anhydrous), 8.1 mM Na_2HPO_4 (anhydrous), 154 mM NaCl, distilled water to 800 mL, adjust to desired pH (7.2 to 7.4) using 1 M NaOH and make up the volume up to 1 L.
Carbonate buffer: 1.6 g Na_2CO_3 (15 mM), 2.9 g $NaHCO_3$ (35 mM), 0.2 g NaN_3 (3.1 mM), distilled water to 1 L, adjust to pH 9.0. *[Caution: Sodium azide is poisonous; wear gloves]*
PBS with 3% milk: dissolve 3 g nonfat milk powder in 100 mL PBS. Prepare during the day of use.
1% BSA, (Ag I): dissolve 0.1 g BSA in 10 mL carbonate buffer.
1% gelatin solution, (Ag II): Dissolve. 0.1 g gelatin in 10 mL carbonate buffer.
Rabbit anti-BSA antiserum 1: 10 (Rabbit, serum A) mix 0.1 mL of rabbit anti-BSA antiserum with 0.9 mL of 1× PBS. Prepare 0.1 mL aliquots in vials and freeze.

Normal rabbit serum 1: 10 (Rabbit serum B) Prepare as mentioned earlier. Keep aliquots of 0.1 mL and freeze.

Anti-rabbit Ig conjugate 1:50: Obtain goat anti-rabbit Ig conjugated with horseradish peroxidise and mix 0.1 mL of it with 4.9 mL $1 \times$ PBS with 3% milk. Prepare 0.2 mL aliquots and freeze.

Anti-mouse Ig conjugate 1:50: Obtain rabbit anti-mouse Ig conjugated with horseradish peroxidase and prepare dilution as above. Keep aliquots (0.2 mL) and freeze.

TMB substrate solutions

Solution A: In one litre volumetric flask, dissolve 1 g tetramethylbenzidine (TMB) in 600 mL methanol. Bring to one liter with glycerol. Store at 4°C in foil wrapped bottle.

Solution B: In one liter volumetric flask, dissolve following in 500 mL distilled water, 22.82 g K_2HPO_4, 19.2 g citric acid, 1.34 mL 30% H_2O_2, make up the volume up to 1 L with distilled water.

18.39.4 Protocol

18.39.4.1 Ag Application to Plate

1. Prepare Ag I (BSA) and Ag II (gelatin) as described.
2. Copy the chart below (Record chart). The numbers and letters on the chart correspond to the wells in the ELISA test plate. Along the sides of the chart record, the materials added to each well as continue with the procedure.
3. Using a pipette, add 100 µL of Ag I to wells in rows A, B, C and D, columns 1 through 5 and 8 through 12 of the ELISA plate. Leave columns 6 and 7 empty to separate the two sections.
4. Using a clean pipette tip, add 100 µL of Ag II to wells in rows E, F, 0, and H, columns 1 through 5 and 8 through 12 of the ELISA plate. Record on your chart what was added to each row (Table 18.10).
5. Seal the ELISA plate with plastic wrap and incubate at 37°C for 1 h. (Room temperature incubation is acceptable). Refrigerate overnight.

TABLE 18.10   Record Charts

	1	2	3	4	5	6	7	8	9	10	11	12
A												
B												
C												
D												
E												
F												
G												
H												

18.39.4.2 Ab Dilution in Tubes

6. Prepare stock solutions rabbit serum A (Label tube #1) and rabbit serum B (Label tube #2).
7. Label 8 15 mL test tubes as 2, 3, 4, 5, 9, 10, 11, 12.
8. Using a 5 mL pipette, add 4 mL of PBS with dry milk (PBS 3% milk) to each of the 8 test tubes.
9. Using test tube 1 containing rabbit serum A, perform a serial dilution using a 5 mL pipette as follows:
   a. Transfer 4 mL of rabbit serum A from tube 1 to tube 2. Mix well by drawing liquid up and down several times.
   b. Transfer 4 mL from tube 2 to tube 3 and mix well.
   c. Transfer 4 mL from tube 3 to tube 4 and mix well.
   d. Transfer 4 mL from tube 4 to tube 5 and mix well.
10. Using tube #8 containing rabbit serum B, repeat the serial dilution as described earlier for tubes 9 through 12. Seal all of the tubes and refrigerate until needed (step 14).

18.39.4.3 Ab Addition to Plate

11. Empty contents of ELISA plates into a sink by turning upside down and flicking. Blot on paper towels.
12. Fill each well on the ELISA plate with PBS: milk to the top of the wells. Let set for 15 min. Empty as mentioned earlier. (This step serves the purpose of saturating the nonspecific binding sites in each well).
13. Wash by flicking the plate and refilling wells with PBS only. Let sit for 2 min. Empty and blot as mentioned earlier. Repeat.
14. Using a pipette, add 100 µL of diluted rabbit serum to each well according to the chart below. The tube number corresponds to the row on the ELISA plate. Use a clean pipette tip for each rabbit serum (A and B). Record all information on the chart in Table 18.11. (If a pipette is not available, use a dropping pipette and add 2 drops of each serum preparation.)
15. Cover plate and incubate for 1 h at room temperature.
16. Empty ELISA plate into the sink and blot on paper towels.

TABLE 18.11 ELISA Plate Setup

Tube #	Rabbit Serum	Dilution	Column	Row
1	A	I 1,000	1	A–H
2	A	I 2,000	2	A–H
3	A	I 4,000	3	A–H
4	A	I 8,000	4	A–H
5	A	I 16,000	5	A–H
8	B	I 1,000	8	A–H
9	B	I 2,000	9	A–H
10	B	I 4,000	10	A–H
11	B	I 8,000	11	A–H
12	B	I 1,6000	12	A–H

17. Wash as in step 13. Wash a total of three times.
18. Add 100 mL of rabbit conjugate to wells A and B, E and F, rows 1 through 5 and 8 through 12. Add 100 mL of mouse conjugate to wells C and D, G and H, rows 1 through 5 and 8 through 12. Cover plates and record information on the chart.
19. Incubate for 15 min at 37°C.

18.39.4.4 Substrate Addition to Plate

20. Prepare TMB substrate as described. Working solution of substrate should be freshly prepared right before use.
21. Empty plates wash as described in steps 16 and 17.
22. Add 200 μL of fresh substrate to wells in rows A, B, C, D, E, F, G, and H, columns 1 through 5 and 8 through 12. A color change (colorless to blue) after few minutes indicates a positive Ag–Ab reaction.
23. Read optical density or absorbance at wavelength 650 nm on a ELISA plate reader after 15 and 30 min. Alternatively, transfer the reaction product from each well to a micro cuvette and read OD_{650} in a spectrophotometer. If neither a plate reader nor a spectrophotometer is available, place ELISA plate on an index card or sheet of white paper and observe the intensity of any color change. Make observations between 15 and 30 min and score according to the [+++] very reactive, [++]moderately reactive, [+]slightly reactive, [-] no reaction.
24. Record results on the chart.

18.39.5 Result

1. Record and summarize your result as per the following chart (Table 18.12).
2. Using the data in the aforementioned table, prepare a graph to compare the activities of anti-BSA and normal serum control and interpret the data.

TABLE 18.12   Summary of ELISA

	BSA		Gelatin	
Treatments	Rabbit Conjugate	Mouse Conjugate	Rabbit Conjugate	Mouse Conjugate
Anti-BSA				
I: 1,000				
Surface of				
I 4,000				
I 8,000				
I 16,000				
Normal				
Serum				
I 1,000				
I 2,000				
I 4,000				
I 8,000				
I 16,000				

18.40 Experiment: Pregnancy Testing by Using the Immunological Methods

18.40.1 Introduction

Pregnancy is a stage of conception in women that can be detected in the early stage by using the immunological methods. However, this test is based on the presence of a hormone *human chorionic gonadotropin* (hCG). It is a glycoprotein produced by trophoblastic cells of the placenta that begin about 10 days after the conception, and found in the urine of only pregnant women. The hCG is a dimer consisting of two sub-units alpha and beta. The molecular structure of hCG is similar to the other glycoprotein hormones found in urine such as LH (luteinizing hormone), FSH (follicle stimulating hormone) and TSH (thyroid stimulating hormone).

After five weeks of LMP (last menstrual period), the concentration of hCG rises sharply in urine and attains peak level at 10 weeks of gestation (LMP). During the first trimester, it reaches 1,00,000 IU/mL. The IU (international unit) is more accurate and related to specific gonadotropic activity of 0.1 mg of dried standard kept at the National Institute for Medical Research (London). There are several methods developed for a pregnancy test, but the most accurate and rapid ones are the immunological methods. Such as latex particle agglutination inhibition test, RIA for the α-subunit of hCG, ELISA test, etc.

18.40.1.1 Latex Agglutination Inhibition Test
The presence of hCG in urine depends on the specificity of an Ag–Ab reaction. The heat-coagulation test of proteinuria is performed before proceeding for this test. A false pregnancy test can be anticipated if this test is positive. Standard technique for latex hemagglutination inhibition test is given in this experiment. The immunological test can be performed in two ways. It differs in carrier used for an external source of hCG. A positive and a negative control should be prepared.

18.40.1.1.1 Materials. Urine of pregnant women (collected in a clean glass container), glass slides, dropper, anti-hCG serum, a wooden stick, Ag.

18.40.1.1.2 Protocol
1. Collect urine of pregnant women in a thoroughly clean glass container free from any residual soap or detergent.
2. Take anti-hCG serum in a clean dropper and put on a clean glass slide.
3. Add one drop of urine to anti-hCG serum and mix thoroughly.
4. Add 1–2 drop of Ag (hCG-coated latex particle suspension).
5. Mix with a clean wooden stick.
6. Rock the slide very slowly and gently for no longer than 90 s and observe agglutination.

18.40.1.1.3 Results and Observations. The added antiserum will not be neutralized if the urine does not contain hCG. Still active antiserum reacts with hCG causing agglutination when the hCG coated latex suspension is added. Evenly suspended latex particles show a positive result and the clumping formation shows a negative result. In case of proteinuria, a false positive result may occur.

Note: In some of the methods the latex particles are coated with hCG Ab, therefore, reversal of interpretation and the result should be considered, that is, agglutination implies for a positive result.

18.40.1.2 Hemagglutination Inhibition Test

In this test anti-hCG serum and sensitized red cells are used. The anti-hCG serum is neutralized having no effect of the sensitized red cells. The red cells are gathered at the bottom of the tube giving the positive test, and if the hCG is absent in urine the reactions will not take place.

18.40.1.2.1 Materials.
Small size test tubes, urine sample, sensitized red cells, anti-hCG serum

18.40.1.2.2 Protocol
1. Collect the urine sample in a clean glass vial and appropriately dilute it with distilled water.
2. Take one to two drops of anti-hCG serum in the glass tube.
3. Add one to two drops of urine with anti-hCG serum and mix properly.
4. Add one to two drops of sensitized red cells in the suspension.

18.40.1.2.3 Results and Observations.
The anti-hCG serum will be neutralized if urine contains hCG and the sensitized red cells will not be affected. The red cells will gather at the bottom of the tube and form a ring giving the positive evenly dispersed layer in the test tube showing the negative result. This test is not affected by proteinuria.

18.40.1.3 Pregnancy Testing through Commercially Available Kits

There are several pregnancy test kits available commercially such as beta VISIPREG latex agglutination pregnancy test kit (sensitivity = 0.5 IU/mL), VISIPREG STRIP pregnancy test kit (sensitivity = 25 IU/mL), VISIPREG CARD pregnancy test kit, etc.

18.40.1.3.1 VISIPREG STRIP Method

18.40.1.3.1.1 Introduction. The VISIPREG STRIP is a two-site sandwich immune-assay in which antiserum, that is, anti-hCG (purple color) is used both in conjugate as well as in native forms. The conjugates are dried on the strip (Figure 18.67). Its working is based on the principle of ELISA which determines hCG qualitatively present in the pregnant women. When the strip is dipped in urine, the hCG if present migrates through capillary action and reacts with dried conjugate (anti-hCG) and form a complex "anti-hCG – conjugate hCG" complex. This complex again migrates forward on the membrane which is predispersed with antiserum (anti-hCG). Antibodies of this serum can bind the hCG molecules present in "anti-hCG - conjugate-hCG" complex. Where ever the antibodies of hCG serum are present, it forms a visible purple band along the exact location of immobilized anti-hCG conjugate resulting in a positive

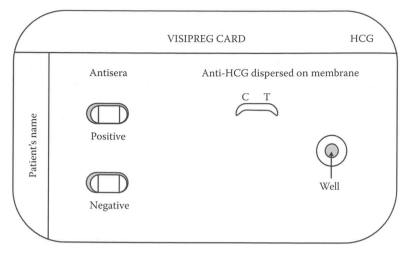

Figure 18.67 Diagrammatic illustration of VISIPREG CARD used for pregnancy testing.

test. If no hCG is present in urine, the anti-hCG conjugate will pass through immobilized anti-hCG band giving no colored line. This shows the negative test.

Also, the strip contains a control region near the holding end. It consists of a band of immobilized antiserum. This band will bind all encountered free conjugate and forms a colored line irrespective of whether hCG is present in urine or not. Appearance of control line indicates the correct testing procedure only.

The VISIPREG STRIP kept in a pouch should be stored at 4°C–8°C so that the activity of antiserum may be preserved. The testing work should be done at room temperature around 25°C.

18.40.1.3.1.2 Materials.   Urine sample, VISIPREG STRIP.

18.40.1.3.1.3 Protocol

1. Collect urine in a clean container at any time in a day and store at 2°C–8°C for about 72 h before testing.
2. Bring both urine sample and pouch of the VISIPREG STRIP at room temperature.
3. Take 0.5 mL urine in a clean vial.
4. Hold the strip at unprinted end and dip the printed end in urine up to the level marked "MAX" (Figure 18.67) (do not dip the strip in urine above the level marked "MAX" as the conjugate may be dissolved into the urine instead of migrating up).
5. Wait for 5 min and see color development.

18.40.1.3.1.4 Results and Observations.   Appearance of purple line near the dipping end of the strip indicates a positive result. Absence of this line indicates the negative result. The control line will appear purple near the holding end in case the finding is negative. However, it should be noted that in a positive test some time two lines also appear (i.e., the positive test line and the control line). It is because when the concentration of hCG is very high in urine, only the positive line will appear. The control line may appear very faint which sometimes may not be visible. If none of these lines appears, the test is wrong. Repeat the test taking the other VISIPREG STRIP.

18.40.1.3.2 Experiment: VISIPREG CARD Test

18.40.1.3.2.1 Introduction. VISIPREG CARD pregnancy test is a qualitative laboratory immunoassay device used for determination of hCG in urine. The VISIPREG CARD is a two-site sandwich assay where antiserum anti-hCG is dispersed on a membrane both in conjugate form as well as in native form. It is based on the technological innovation over the principles of ELISA (Figure 18.67).

The VISIPREG CARD has an efficiency of detecting 25 mIU/mL hCG or greater. This card should be stored at 4°C–5°C but must not be freezed. An anti-hCG conjugate of purple color is immobilized over a pad. When a urine sample is applied, the hCG present in urine migrates and reconstitute the conjugate and forms a complex of anti-hCG conjugate -hCG. This complex migrates forward on the membrane onto which an antiserum anti-hCG has already been dispersed. This immobilized anti-hCG binds with the hCG present in the migrated "anti-hCG conjugate—hCG" complex. This causes agglutination of the antiserum resulting in formation of a purple band along the site where the antiserum anti-hCG has been immobilized (Figure 18.67). Formation of purple band is a positive finding and absence of hCG in urine will not form any colored band. This is a negative result.

Also in the membrane, there is a control region consisting of a band of immobilized antiserum anti-hCG. This band binds all the encountered free conjugate and develops a colored line irrespective of the presence or absence of hCG in the urine. Appearance of this colored line represents the correct procedure and viability of the reagents in the test device.

18.40.1.3.2.2 Materials. Test device, dropper, desiccants pouch (not for use in the test).

18.40.1.3.2.3 Protocol.
1. Collect urine of pregnant women in a clean glass vial. The urine collected in morning contains the highest concentration of hCG, but it may be collected any time in a day.
2. Before testing bring the urine sample, test device, control/reference materials to the room temperature, 25°C–30°C.
3. Write the patient's name or reference number on the space marked on the card.
4. Put three drops (about 0.1 mL) of urine into the center of the sample well with the help of a dropper holding vertically. Care should be taken to allow each drop to soak before adding the next drop. Wait for 5–10 min.

18.40.1.3.2.4 Results and Observations. Appearance of Two purple colored lines (one in the result region and the second in the control region) is suggestive of a positive result (Figure 18.68). However, appearance of only one purple colored line in the control region is indicative of the negative result. If there are no colored lines neither in the test region nor in the control region, does this show the improper handling of the test or due to improper storage and exposure of the test device after opening the pouch and/or expiry of the card?

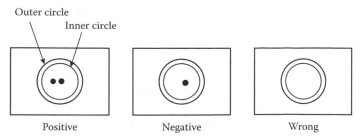

Figure 18.68   HIV DOT test of the suspected AIDS patient.

18.41 Experiment: HIV DOT Test of AIDS Patients

18.41.1 Introduction

HIV 1 and HIV 2 are the causal agent of AIDS (acquired immuno-deficiency syndrome). After entering in the blood of humans, HIV disrupts the immune system leading to multiple complex diseases called "disease syndrome". HIV kits are now available commercially. The HIV DOT kit has in-built procedural control combined test for HIV 1 and HIV 2. The HIV DOT test is a rapid (requires 3–5 min) visual screening test against human serum or plasma. It is an accurate immunoassay of HIV 1 and HIV 2 antibodies in human serum/plasma and does not require any instrumentation. All reagents are ready to be used in the kit. This test is quite cheap, absolutely safe and extremely reliable. Each reagent is individually packed in dropper vials eliminating cross contamination. The test device is such that it contains a porous membrane having two visible circles. The membrane is supported with absorbent. The HIV peptide Ags (toxin) are immobilized onto a porous membrane. The test sample of a patient and solutions can pass through the membrane and absorb into the underlying absorbent.

18.41.2 Materials

Colloidal gold conjugate solution, Protein A, HIV antibodies (Ag Ag Ag), Membrane, Absorbent

Suspected HIV patient, a clean and dry sterile vial, centrifuge, the test device, buffer solution, wash solution, colloidal gold conjugate, positive control (no risk of infection synthetic recombinant peptide used), sample dropper (disposable pipette).

18.41.3 Protocol

1. Collect blood from the suspected HIV patient(s) in a clean, dry and sterile glass vial and allow to clot.
2. Centrifuge the blood at room temperature to separate the serum. Assay the serum immediately, however, it may be stored at 2°C–8°C.

3. Bring all the solution, test device and sample to room temperature before the start of the test.
4. Add two drops of buffer solution into the inner circle of the device (Figure 18.68) and wait for a few second to get it soaked.
5. Add one drop of the patient's serum/plasma with the help of a dropper provided with the kit and wait to get the serum soaked by the membrane.
6. Add two drops of buffer solution and allow it to soak in.
7. Again add two drops of wash solution and wait to soak in the membrane.
8. Add two drops of colloidal gold conjugate solution and wait to soak in.
9. Finally, add three drops of wash solution and record the result.

18.41.4 Results

If the patient's serum contains the HIV antibodies, it will bind to the HIV peptide Ags (toxin). The conjugate complex will bind to the Fe portion of the HIV antibodies to give a distinct red color against a white background. Therefore, the positive result will show two visible dots in the inner circle. However, if only one dot is visible, it shows that the patient's serum does not contain detectable HIV antibodies giving the negative result. On the other hand if none of the dots appears, it shows the procedural error or deterioration of reagents. In such cases, re-test the specimen with a new test device.

For the assessment of positive procedure, a control dot appears in the inner circle. It is because a rabbit IgG is immobilized on the membrane surface. This control verifies that the test reagents (colloidal gold conjugate, wash buffer and the test device) are functioning well, and reagents have been added in the correct order.

18.42 Experiment: Demonstration of Immunofluorescence Assay

18.42.1 Introduction

Target Ags in parasite, tissue or cells against specific antibodies can be localized by incubating them directly with fluorochrome labeled specific antibodies or directly using second antibodies labeled with fluorochromes. Fluorochromes emit fluorescence when excited with UV radiation. Fluorescein is the most widely used fluorochromes. The protocol describes indirect fluorescent Ab test (IFAT) on intact filarial parasite.

18.42.2 Materials

Brugia malayi infective (L3) larvae/microfilariae (mf)/adult worms, anti-human IgG, test sera, FITC, PBS 0.01 M, pH 7.4 and 0.05 M pH 7.4, 0.145 M sodium chloride solution (850 mg in 100 mL distilled water), 0.2 M $Na_2HPO_4 \cdot 2H_2O$ solution (3.5 g in 100 mL distilled water), 0.1 M $Na_2HPO_4 \cdot 2H_2O$ solution (1.75 g in 100 mL distilled water),

0.1 M Na_3PO_4 solution (3.8 g in 100 mL distilled water), Sephadex G-25 column, mounting solution (50% glycerol, 50% PBS, and 0.1% sodium azide), and Evan's blue (0.1%).

18.42.3 Protocol

18.42.3.1 Anti-Human IgG Ab–FITC Conjugate

1. Take 250 µL (2.5 mg protein) of anti-human IgG antibodies in plastic vial and dilute to 1 mL with 0.145 M sodium chloride solution.
2. Add 500 µL of 0.2 M Na_2HPO_4 dropwise while stirring.
3. Add 1 mL FITC solution (12.5 µg FITC/mg protein) in 0.1 M Na_2HPO_4.
4. Adjust the pH to 9.5 using 0.1 M Na_2PO_4.
5. Incubate the mixture at 25°C for 45 min.
6. Apply the conjugate mixture and pass through Sephadex G- 25 column (2 cm × 30 cm) and elute with 0.01 M PBS (pH 7.2) at a flow rate of about 30 mL/mm. Collect the first elute yellow colored fractions that contain anti-human IgG antibodies conjugated with FITC.
7. Concentrate FITC conjugate by ultra-membrane filtration, add preservative sodium azide (0.1%), and store at 4°C.

18.42.3.2 Immunofluorescence Assay

1. Incubate about 10 L3 larvae/100 mf/two adult worms with 200 µL of optimally diluted (1:10) test sera in PBS (0.05 M) in a conical vial at room temperature for 1 h.
2. Wash the parasites with PBS two times using a centrifuge at 500 RPM for 2 min.
3. Add 20 µL of optimally diluted (1:10 in 0.05 M PBS) anti-human IgG-FITC conjugate and incubate for 30 min at 40°C.
4. Wash the parasite with 0.05 M PBS once (500 rpm for 2 min).
5. Add 5 µL of 0.1% Evan's blue.
6. Wash the parasites with 0.05 M PBS once (500 rpm for 2 min).
7. Add 5 µL mounting buffer, transfer the parasite onto a glass slide, place cover slip, and seal with nail polish.

18.42.4 Results and Observations

Examine under a fluorescence microscope and compare the yellow–green fluorescence with the parasite incubated with positive sera compared to those incubated with normal sera.

18.43 Experiment: Identification of Bacteria by Using Fluorescent Ab Technique (FAT)

18.43.1 Introduction

The antibodies produced by bacteria combine with Ag and form Ag–Ab complex. This complex can be observed after fluorescing it against a dark background with UV light by using a fluorochrome, that is, fluorescent dye.

18.43.2 Materials

Microslide, coverslip, spirit lamp, Petri dish, Whatman no. 1 filter paper, lint-free paper, fluorescent microscope, PBS (phosphate buffer saline) (pH 7.4–7.6), specific conjugate, distilled water, low fluorescence immersion oil, mounting fluid (9 mL glycerine plus 1 mL of PBS pH 9), fluorochromes (such as rhodamine, FITC, acridine orange, auramine, etc.).

18.43.3 Protocol

1. Inoculate the bacterial colonies into a few drops of PBS having pH 7.4–7.6 and make the suspension.
2. Clean slide and put a drop of suspension on the test area. Prepare smear, air dry, and heat fix the bacterial cells on a glass slide by gently warming. Do not make the slide hot over the flame of the spirit lamp.
3. Place a drop of the bacterial conjugate over the smear and distribute it evenly over the smear. Care should be taken not to dry the conjugate, otherwise it will give a false result.
4. Place the slide in Petri dish containing a moist filter paper for 1 h or incubate in a humid chamber at 37°C.
5. Rinse the slide in a jar containing PBS for 10 min. Finally, rinse the slide with distilled water.
6. Gently blot dry the slide using lint-free paper. For each test, use separate slide.
7. Place a drop of mounting fluid and apply the cover slip.
8. Clean the objective lenses and condenser lenses of the microscope using lens paper.
9. Place one to two drops of low-fluorescence immersion oil on the condenser after lowering it.
10. Place the slide on the stage of the microscope and put a drop of immersion oil on the cover slip. Adjust the microscope to view the required area.
11. Examine the field under low power and concentrate the light in a small area by adjusting the condenser.
12. Now, switch over to oil immersion objective and observe under the microscope using UV light (Figure 18.69).

18.43.4 Result

The bacterial cells that produce specific antibodies will show fluorescence.

18.44 Experiment: Immunofluorescence Labeling of Cultured Cells

18.44.1 Introduction

The following is a basic "generic" method for localizing proteins and other Ags by indirect immunofluorescence. The method relies on proper fixation of cells to retain cellular distribution of Ag and to preserve cellular morphology. After fixation, the cells are exposed to the primary Ab directed against the protein of interest, in the

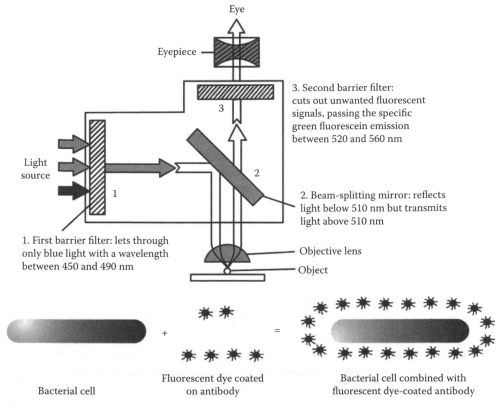

Figure 18.69 Fluorescence staining technique and microscopy.

presence of permeabilizing reagents to ensure Ab access to the epitope. Following incubation with the primary Ab, the unbound Ab is removed, and the bound primary Ab is then labeled by incubation with a fluorescently tagged secondary Ab directed against the primary Ab host species. For example, incubation with a mouse IgG primary Ab might be followed by incubation with a rhodamine isothiocyanate (RITC)-labeled goat anti-mouse IgG secondary Ab. After removal of the secondary Ab, the specimen is ready for viewing on the fluorescence microscope. Once the conditions for observing specific immunolocalization have been identified for a given Ab and cell type, double labeling with two antibodies can be employed to compare localizations. To do this, primary Ab incubation can contain two antibodies generated in two species (e.g., mouse and rabbit), followed by incubation with two secondary antibodies coupled to different fluorophores. Care should be taken, however, that the two Ab combinations, especially the secondary antibodies, should not cross-react.

18.44.2 Materials

Cells of interest growing in tissue culture, Trypsin/EDTA solution, 2% formaldehyde, phosphate-buffered saline pH 7.4, PBS/FBS:PBS, pH 7.4, containing 10% fetal bovine serum (FBS), and primary Ab.

0.1% (w/v) saponin in PBS/FBS: Prepare fresh from 10% (w/v) saponin stock solution.

Controls: Preimmune serum (if using rabbit polyclonal Ab) or Ag add in excess to the primary Ab.

Secondary antibodies (against IgG of species from which primary Ab was obtained) coupled to fluorophore: for example, RITC, FITC (fluorescence isothiocyanate) Cy3, Texas Red, or Alexa 488 and 594.

Mounting medium, 10 cm diameter tissue culture dishes, 12 mm no. 1 round glass coverslips, sterilized by autoclaving or soaking in 70% ethanol, 12 well tissue culture plates, 150 mm Petri dishes, watchmaker's forceps, microscope slides, nail polish, and fluorescence microscope with oil-immersion lens.

18.44.2.1 Reagents and Solutions

Formaldehyde 2%: In a chemical fume hood, dilute 2 mL 37% reagent-grade formaldehyde into 35 mL PBS pH 7.4.

Mounting medium: Use fluoromount G to prepare 50% (w/v) glycerol and 0.1% (w/v) *p*-phenylenediamine in PBS, pH 8.0.

Phosphate-buffered saline (PBS): $0.144\,g\,KH_2PO_4$, $9.0\,g\,NaCl$, $0.795\,g\,Na_2HPO_4 \cdot 7H_2O$, and distilled water to 1 L. Adjust to desired pH with 1 M NaOH or 1 M HCl. Store indefinitely at room temperature.

Poly-L-lysine-coated coverslips: Apply 25 µL of 1 mg/mL poly-L-lysine to each sterilized no. 1 coverslip in a hood and allow to stand ~10 min. Carefully rinse cover slips three times with water, then allow to air dry.

Note: All solutions and equipment coming into contact with live cells must be sterile, and aseptic technique should be used accordingly. All culture incubations should be performed in a humidified 37°C, 5% CO_2 incubator unless otherwise specified.

18.44.3 Protocol

1. For adherent cells, 1–2 days prior to experiment, trypsinize cells and seed onto 10 cm culture dishes, each containing 15–20 sterilized cover slips so that on the day of experiment cells are 20%–50% confluent. Nonadherent cells can be coaxed into adhering to the cover slips by precoating the coverslip with poly-L-lysine. Apply 10–20 µL of suspended cells to each cover slip, let sit 10 min, then proceed with fixation (step 3). Alternatively, cells can be attached to cover slips using a cytocentrifuge by following the manufacturer's instructions.
2. On the day of experiment, transfer each cover slip individually to a well of a 12 well tissue culture dish containing 1 mL culture medium. Subject cells to the desired experimental conditions (e.g., treat with various drugs, inhibitors, or temperatures prior to fixation and immunostaining).
3. Aspirate medium and add 1 mL of 2% formaldehyde to each well. Allow cells to fix at room temperature for 10 min.
4. Aspirate the formaldehyde fixative and wash cover slips twice, each time by adding 1 mL PBS, pH 7.4, let it stand for 5 min, then aspirating the PBS. Add 1 mL PBS/FBS

to the fixed cover slips and let stand for 10–20 min to block nonspecific sites of Ab adsorption.

Note: Throughout the procedure, do not let cells dry out.

5. In 1.5 mL microcentrifuge tubes, dilute primary antibodies in 0.1% saponin/PBS/FBS. Typically affinity-purified antibodies are diluted in the range of 1–10 µg/mL, and rabbit antisera are diluted between 1:100 and 1:1000. If using a commercial Ab, follow suggested dilutions from manufacturer. During initial characterization, it is wise to try a range of dilutions of Ab.

6. Prepare controls containing only 0.1% saponin/PBS/FBS or (if available) containing preimmune antiserum (if rabbit polyclonal Ab is being used) or specific (primary) Ab with the Ag added in excess. Controls are often the most important part of an immunofluorescence experiment.

7. Microcentrifuge Ab dilutions and control solutions 5 min at maximum speed, room temperature, to bring down aggregates in pellet. Pipette the Ab solution from above the aggregate pellet.

8. Place a 10 cm × 10 cm piece of parafilm in the bottom of a 150 mm Petri dish. In a grid pattern that replicates the 12 wells used to incubate the coverslips, label the appropriate place on the parafilm for each coverslip with a marker.

9. Apply a 25 µL drop of appropriate primary Ab solution to each numbered section. Carefully remove each coverslip from the 12 well plate with watchmaker's forceps, blot the excess fluid by touching the edge to a Kimwipe, then invert the coverslip over the appropriate 25 µL drop, making sure that the side with the cells is facing down. Place the top on the Petri dish and incubate for 1 h at room temperature.

   Sometimes proper labeling will require a longer incubation time or the Petri dish will be incubated for >1 h for convenience. For longer incubations, add some wetted Kimwipes to the dish to maintain a humid atmosphere. The incubation can be extended overnight at 4°C, if necessary.

Note: It is important to be always aware of which side of the cover slip the cells are on. Cells should be facing up when in the 12 well plate, but facing down when placed on the Ab. Picking up cover slips with the forceps is awkward at first, but becomes easier with practice.

10. Carefully pick up each inverted coverslip and flip it over so that it is cell-side-up, then place in a well of a 12 well plate. Wash each coverslip three times to remove unbound Ab, each time by adding 1 mL PBS/FBS, letting it stand 5 min, then aspirating the solution.

11. Dilute fluorophore-conjugated secondary antibodies in 0.1% saponin/PBS/FBS. Mix, then microfuge as in step 7 to remove aggregates. Typically, commercial preparations are diluted between 1:100 and 1:500.

12. Prepare an incubation chamber as in step 8. Apply 25 µL of appropriate secondary Ab solution to each numbered section and invert coverslip over drop as in step 8. Cover Petri dish and protect from light with aluminum foil or place chamber in drawer. Incubate for 1 h at room temperature.

13. Wash cover slips as in step 10. After removal of last PBS/FBS wash, add 1 mL PBS.

14. Label slides and place 1 drop of mounting medium onto slide. Pick up coverslip from well, gently blot off excess PBS by touching the edge to a Kimwipe, then invert cover slip, cell-side-down, onto drop. Gently blot mounted coverslip with paper towel, then seal edge of coverslip onto the slide by painting the edge with a rim of nail polish. Let it dry. The fixed, mounted, and nail polish sealed coverslips can be stored in the dark for 6 months to 1 year at 4°C.

18.44.4 Results and Observations

View specimen on fluorescence microscope using a $63 \times$ oil immersion objective.

18.45 Experiment: Isolation of IgG from Serum

18.45.1 Introduction

IgG is the main Ab isotype found in blood and extracellular fluid allowing it to control infection of body tissues by binding many kinds of pathogen-representing viruses, bacteria, and fungi-IgG protects the body from infection. IgG is secreted as a monomer that is small in size allowing it to easily perfuse tissues. It is the only isotype that can pass through the human placenta, thereby providing protection to the fetus in the uterus. Along with IgA secreted in the breast milk, residual IgG absorbed through the placenta provides the neonate with humoral immunity before its own immune system develops. Colostrum contains a high percentage of IgG, especially bovine colostrum. In individuals with prior immunity to a pathogen, IgG appears about 24–48 h after antigenic stimulation.

Isolation of IgG class of Ig is carried out in two steps. In the first phase A, Ig rich fraction is prepared by salt precipitation, and in the second phase B, IgG is separated from other classes of Igs by ion-exchange chromatography.

18.45.1.1 Isolation of Ig-Rich Fraction from Serum by Salt Precipitation

18.45.1.1.1 Principle.   Different protein can be gradually precipitated out from their aqueous solutions using highly soluble salts such as ammonium sulfate. As the concentration of the salts is increased, they compete with protein for water molecules and lead to their gradual precipitation based upon their requirement for water molecules to be in the soluble form. By convention, the final concentration of ammonium sulfate required for precipitation is expressed in term of percentage of saturation. Ig rich fraction can be precipitated out from serum by 33% saturation with ammonium sulfate.

18.45.1.1.2 Materials.   Serum, 0.01 M phosphate buffer pH 7.2, saturated ammonium sulfate solution (add about 75 g of ammonium sulfate to 100 mL distilled water and stir. Heat the solution and filter through Whatman no. 1 filter paper while it is hot. After the solution comes to room temperature, adjust the pH to 6.8 using concentrated ammonium hydroxide and store. Ammonium sulfate crystals should be formed in the container after few hours), magnetic stirrer, and refrigerated centrifuge.

18.45.1.1.3 Protocol
1. Add 5 mL of saturated ammonium sulfate solution gradually to 10 mL of serum under constant stirring at ice cold temperature.
2. Keep the suspension in ice for 15 min with occasional stirring.
3. Centrifuge at 3000 rpm for 30 min at 4°C.

4. Discard the supernatant and wash the precipitate twice with 40% ammonium sulfate solution (mix thoroughly with a glass rod and centrifuge at 3000 rpm for 30 min at 4°C).

5. Reconstitute the precipitate in small volume of 0.01 M sodium phosphate buffer (SPB) (pH 7.2) and dialyze against the same buffer for 24 h with frequent changes of buffer.

18.45.1.2 Removal of Salt from Protein Fraction

The ammonium sulfate is removed from protein fraction by dialysis and the protein concentrated by filtered by ultrafiltration.

18.45.1.2.1 Principle.

Dialysis is the separation process that depends on the differential transport of solutes of different sizes across a porous barrier separating to liquid where the driving force is concentration gradient only. It is usually used to separate solutes too large to diffuse through the barrier from those small enough to diffuse through it (Figure 18.70).

18.45.1.2.2 Materials.

Dialysis membrane, closures for dialysis membrane, SPB (0.01 M SPB pH 7.2), Ab to be desalted (human Ig), magnetic stirrer, and magnetic bar.

18.45.1.2.3 Protocol.

Cut the required length of dialysis membrane and place in distilled water.

Open the membrane by gently seizing from one end to another and wash the inner side of membrane by passing the distilled water gently without damaging the membrane. Close one end of the membrane using closure or thread. Pour the Ab solution through another end and close the same. Place the filled bag in a beaker containing 0.01 M SPB and magnetic bar and keep the beaker on a magnetic stirrer and adjust

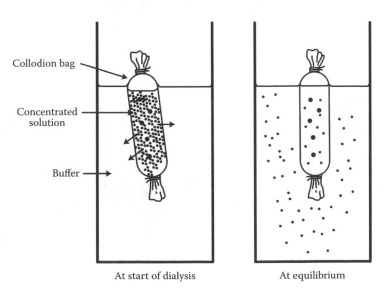

At start of dialysis   At equilibrium

Figure 18.70 Process of dialysis.

the speed (at 4°C). Change the pre-cooled buffer after every 4–6 h for three times (1 L each time). Collect the Ab.

Note: Choice of the pore size of the membrane depends on the substance to be dialyzed. Extensive cleaning of the membrane is necessary before use. The membrane should not dry because it may damage the membrane. It is advisable to test the system for leaks using distilled water before adding the protein solution to the tubing.

18.45.1.3 Ultra-Membrane Filtration

18.45.1.3.1 Principle. Ultrafiltration is a device that can withstand high pressure and concentrates a protein solution using selective permeable membranes. The function of the membrane is to let the water and small molecules pass through while retaining the protein. Ultra-filtration is sometimes called as reverse osmosis. It is a process in which the solutes and solvent, up to a certain critical size, are forced to the barrier by high pressure on one side of the porous barrier than the other. Since there is always a flow of solvent moving along the solute, it can be used for the concentration process.

18.45.1.3.2 Materials. Ultrafiltration tubing, closure for the tubing, conical flask with side arm, funnel and single hold rubber cork, and sample to be concentrated.

18.45.1.3.3 Protocol. Cut 10–15 cm ultra filtration tubing and wash it as mentioned in dialysis.

Insert the tip of the funnel at one end of the tubing and pass the end of the funnel with membrane through the rubber cork.

Close the other end of tube with closure or with thread.

Pour the sample to be concentrated through the funnel.

Apply vacuum pressure through the side arm and close it. Leave the flask at 4°C.

Collect the sample by opening the tied end and dialyze, if necessary (Figure 18.71).

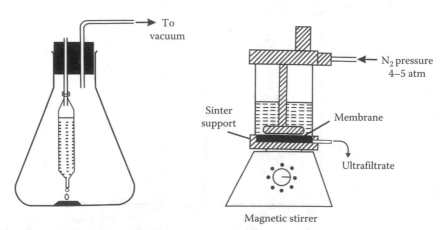

Figure 18.71   Ultrafiltration of IgG solution in a dialyzing sac.

18.45.1.4 Isolation of IgG from Ig-Rich Fraction

18.45.1.4.1 Principle.   Fractionation of protein by ion exchange chromatography exploits differences in their net charges. The Ig rich fraction prepared by ammonium sulfate precipitation is applied on DEAE cellulose column at pH 7.2. At this pH, DEAE cellulose is positively charged, while all classes of Igs except IgG carry negative charges. This situation allows only IgG to be eluted out while the other classes of Igs interact with the matrix and hence are retained in the column.

18.45.1.4.2 Materials.   DEAE cellulose (Whatman DE-52), 0.01 M SBP, pH 7.2, Whatman filter paper no. 1, chromatography set up, spectrophotometer, and Ig (γ- globulin fraction)

18.45.1.4.3 Protocol
1. Suspend about 3 g of DEAE cellulose in 0.01 M SPB (pH 7.2) and wash the slurry by changing the buffer five or six times to remove fine particles.
2. Take a glass column (10 cm × 2 cm) with sintered glass disc or glass-wool plug at the bottom and mount it vertically.
3. Inject about 10 mL of 0.01 M SPB, pH 7.2 through the outlet of column with the help of the syringe and avoid trapping of air bubbles below the sintered glass disc or glass wool plug.
4. Close the outlet and pour thick slurry of DEAE cellulose along the sides of the column and allow the buffer to run about 25 mL/h. Disturb the settled bed slightly with a glass rod if some slurry is to be added.
5. Cover the gel bed with a Whatman no. 1 filter paper circle (having a diameter slightly lesser than that of the column) and equilibrate the column by pumping 0.01 M SPB, pH 7.2, at a flow rate of about 25 mL/h.

18.45.1.4.4 Sample Application
1. Close the outlet and remove the excess buffer above the gel bed.
2. Carefully layer on 25 mg of Ig fraction in 1 mL volume of 0.01 M SPB pH 7.2 and allow this buffer to enter the gel.
3. Wash the sample from sides of column with 1 mL of buffer and allow it to enter the gel.
4. Close the outlet, fill up the column with buffer, and connect to the buffer reservoir.
5. Open the outlet and pump about 30 mL of buffer through the gel at a flow rate of 25 mL/h. Collect 2 mL fractions and read their absorbance at 280 nm wave length. Alternatively estimate protein by Lowry's method in a small sample of each fraction.
6. Test the protein fraction for IgG in Ouchterlony double diffusion using monospecific anti-human IgG. Pool positive fractions, concentrate to 2 mL by ultrafiltration and store at −20°C with sodium azide (0.01%) as preservative.

18.46 Experiment: Purification of IgM

18.46.1 Introduction

IgM forms polymers where multiple Igs are covalently linked together with disulfide bonds, mostly as a pentamer but also as a hexamer. IgM is primarily found in serum;

however, because of the J chain, it is also important as a secretory Ig. Many IgM antibodies will precipitate when dialyzed against distilled water. The IgM produced by this method is of sufficient purity to be used for fragmentation, FITC or biotin labeling, or radiolabeling. The protocol involves centrifugation of ascites fluid or MAb supernatant, followed by simple dialysis against water. IgM is separated from contaminating proteins (such as a normal mouse proteins or proteins from fetal bovine serum) by use of size-exclusion (SE) chromatography. The purity of IgM is checked by sodium dodecyl sulfate polyacrylamide gel electrophoresis (SDS-PAGE), and the pure IgM is stored in borate buffer.

18.46.2 Materials

Ascites fluid or MAb supernatant, PBS, centrifuge, dialysis tubing, 26 mm × 900 mm column with appropriate SE resin.

18.46.3 Protocol

1. Clarify ascites fluid by centrifugation and remove lipid. If using MAb supernatant, centrifuge 30 min at $15,000 \times g$ (11,000 rpm in SS-34 rotor), room temperature or 4°C, to remove cell debris. Save the supernatant.
2. Transfer clarified ascites fluid or MAb supernatant to dialysis tubing and dialyze against ≥ 10 times the sample volume of distilled water for 24 h at 4°C.
3. Centrifuge dialyzed material 1 h at $15,000–20,000 \times g$ (11,000–13,000 rpm in SS-34 rotor), room temperature or 4°C, and discard supernatant.
4. Prepare a 26 mm × 900 mm column with SE resin and equilibrate in PBS. Dissolve pellet from step 3 in ≤ 5 mL PBS and apply to column. Elute protein with PBS and collect 100 fractions (1% of column volume). IgM will be in the first protein peak to elute from the column.
5. Assess IgM concentration spectrophotometrically at A_{280}. Pool the eluates containing pure IgM.
6. Check the purity of the IgM on a 10% SDS-PAGE under reducing conditions. Load 1–80 µL on the gel, depending on the apparatus used. On a reduced gel, IgM shows two bands, one at 78 kDa, representing the H-chain, and the other at 25 kDa, representing the L-chain.
7. Store IgM at 1–20 mg/mL in PBS at 4°C. IgM is stable under these conditions for several months. Some IgM preparations will come out of solution more easily than others; this is unavoidable. Each IgM must be observed for precipitation.

Suggested Reading

Brewer, J. M., A. J. Pesce, and R. B. Ashworth. 1974. *Experimental Techniques in Biochemistry*. Upper Saddle River, NJ: Prentice-Hall.

Dabbas, D. J. 2002. *Diagnostic Immunochemistry*. New York: Churchill Livingstone.

Hayat, M. A. 2002. *Microscopy, Immunochemistry and Antigen Retrieval Methods: For Light and Electron Microscopy*. London, U.K.: Plenum Press.

Hermanson, G. T. 1996. *Bioconjugate Technique*. London, U.K.: Academic Press.

Janeway, C. A., P. Travers, M. Walport et al. 2005. *The Immune System in Health and Disease*, 6th edn. Oxford, U.K.: Garland Science.

Johnstone, A. P. and R. Thorpe. 1996. *Immunochemistry in Practice*, 3rd edn. Oxford, U.K.: Blackwell Science.

Johnstone, A. P. and M. W. Turner (eds.). 1997. *Immunochemistry—A Practical Approach*. Oxford, U.K.: IRL Press.

Lachmann, P. I., D. R. Peters, and P. S. Rosens (eds.). 1992. *Clinical Aspects of Immunology*, 5th edn. Oxford, U.K.: Blackwell Science.

Plummer, D. T. 1988. *An Introduction to Practical Biochemistry*. New York: McGraw-Hill.

Roitt, I. M. 2001. *Essential Immunology*, 10th edn. Oxford, U.K.: Blackwell Science.

Tiwari, R. P., A. Jain, Z. Khan, et al. 2013. Development of novel antigenic peptide cocktail for the detection of antibodies to HIV 1/2 by ELISA. *J Immunol Methods* 387: 157–166.

William, E. and W. E. Paul. 2008. *Fundamental Immunology*. New York: Lippincott Williams & Wilkins.

Wilson, K. and J. Walker. 2003. *Practical Biochemistry: Principle and Techniques*, 5th edn. Cambridge, U.K.: Cambridge University Press.

Important Links

http://www.inoncology.com/news_release/Textbook_of_Gynaecological_Oncology. html?gclid = CMfZ_tTToLMCFQV66wodEgYAdA

http://www.intertek.com/pharmaceutical/immunochemistry/quantitative-immunoassays/?gc lid = CLXEgufToLMCFQob6wodWC4A2A

http://www.ncbi.nlm.nih.gov/pmc/articles/PMC2247169/

http://www.intertek.com/pharmaceutical/immunochemistry/quantitative-immunoassays/?gc lid = CLXvjtvUoLMCFch66wod0ikAIw

http://www.crcnetbase.com/doi/abs/10.1201//9781420082692-c8

19

Genetic Engineering

19.1 Introduction

Occasionally, technical developments in science occur, which enable us to leap forward in our knowledge, thus increasing the potential for innovation. The rapid and fascinating advances in genetic engineering have opened boundless opportunities to harness the already existing technology for the developments in various areas of biological sciences to further benefit mankind. Remarkable success has already been achieved in the unraveling of genome organization in diverse groups of organisms, regulation and molecular mechanism of gene expression, transferring of desirable genes from one organism to another, and thus construct transgenic with a view to improve the commercial performance of the organisms (Figure 19.1).

Furthermore, inherent genes can be altered very minutely through site-directed mutagenesis to enhance biological efficacy of the gene products. All these remarkable accomplishments owe their success to development of highly innovative experimental techniques. Gene cloning encompasses a number of experimental protocols leading to the transfer of genetic information from one organism to another (Figure 19.2), which includes (a) extraction of DNA from a donor organism, (b) cleavage of DNA, (c) ligation to another DNA entity, that is, cloning vector to form a new recombinant molecule, (d) introduction of cloning vector—insert DNA construct into a bacterial host cell, known as transformation, (e) selection of the transformed cells, and (f) expression of the DNA construct for the desired protein product.

Identification of genes and regulation of their expression forms an integral and important aspect of cellular development and metabolic adaptation. There are qualitative and quantitative changes in expression of various genes at different developmental stages, in different organs, and in response to various chemical and physical stimuli. This well-programmed and stringently controlled expression of genetic information is central and of immense importance in cell differentiation and modulation of metabolism requirements of an organism under fluctuating internal and external environment. The entire genetic information of an organism is not expressed at a constant rate at all times during its life cycle. Which genes are being expressed at a given time is usually established by estimating translational products viz., protein by western blotting or measurement of transcripts by northern blotting. These techniques along with complementary DNA (cDNA) cloning and southern blotting also provide insights into the characterization and organization of gene and gene families,

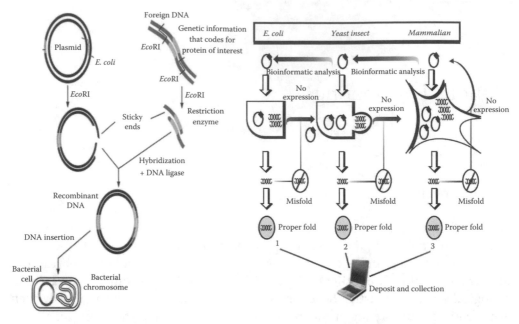

Figure 19.1 Illustration of various steps involved in genetic engineering in prokaryotic, eukaryotic, and mammalian cell.

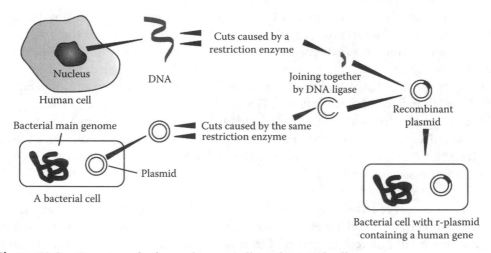

Figure 19.2 Gene transfer from a human cell to a bacterial cell.

estimation of gene copy numbers and for determination of chromosomal location of structural genes. Polymerase chain reaction (PCR) is another revolutionary technique that amplifies DNA *in vitro* and has wide applications in diagnostics, fingerprinting, sequencing, mutagenesis, and so on. Another group of techniques such as restriction fragment length polymorphism and randomly amplified polymorphic DNA provide a way to directly follow chromosomal segments during recombination and are finding applications in construction of molecular maps and map-based cloning of genes. This section deals with laboratory protocols on genetic engineering.

19.2 Tools of Genetic Engineering

Tools used in genetic engineering are a group of enzymes and reagents that are used to isolate, purify, cut, synthesize, modify, join together, and transfer required DNA segments among organisms. Another group of important tools is that of vectors required as intermediate carriers to transfer a gene from one source to another. Chief enzymes that are required for the purpose are tools of genetic engineering, which includes the following.

19.2.1 Enzymes

19.2.1.1 Restriction Endonucleases

Restriction endonucleases or restriction enzymes (RE) are special types of endonucleases that cut DNA at specific sites or motifs. They are able to recognize their site in DNA and act there. REs are obtained from bacteria, which use them to prevent foreign phage DNA from parasitizing by digesting it on its specific sites. In genetic engineering experiments, type II enzymes of this group are used, as they are very specific and cut DNA within recognized motifs, which are mostly 4–7 bases long. The motifs have twofold symmetry called palindromes. For example, the enzyme *Eco*RI is able to identify 5'GAATTC3' motif of DNA molecules and cut between G and A. As such, a double stranded DNA (dsDNA) would have staggered cuts, that is, at two different points on the two strands.

Likewise, BamHI acts on the motif 5'GGATCC3' causes cut in both DNA strands between GG.

Some REs cut both strands of DNA at the same point, for example, *Hind*II, which acts at the motif 5'GTCGAC3' and cuts the strands between and CG (Figure 19.3). The naming of a RE is based on the scientific name of the bacterium from which it has been obtained, its strain, and the serial number of the enzyme from that source. *Eco*RI was obtained from *Escherichia coli* strain R before any other enzyme was obtained from this bacterial strain (hence, serial number I). Some other common REs are *Sal*I, *Eco*RV, *Hae*I, etc., with their specific restriction sites (Table 19.1).

19.2.1.2 DNA Ligase

DNA ligase is the enzyme used for joining two DNA lengths (Figure 19.4). This enzyme forms 5'-3' phosphodiester bond from the 5'-phosphate end of a chain to the 3'-OH end of another chain. DNA ligases are essential for the joining of Okazaki fragments during replication and for completing short-patch DNA synthesis occurring in DNA repair process. There are two classes of DNA ligases. The first uses NAD+ as a cofactor and only found in bacteria. The second uses adenosine triphosphate (ATP) as a cofactor and found in eukaryotes, viruses, and bacteriophages. The smallest known ATP-dependent DNA ligase is the one from the bacteriophage T7 (at 41 kDa). Eukaryotic DNA ligases may be much larger (human DNA ligase I is >100 kDa), but they all appear to share some common sequences and probably structural motifs.

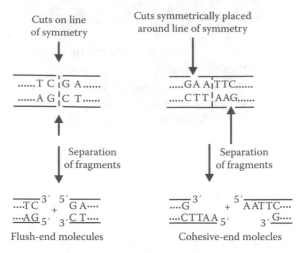

Figure 19.3 Formation of blunt-end and cohesive-end DNA molecules.

TABLE 19.1 Restriction Endonucleases from Various Sources

Microorganism	Restriction Enzyme	Sequence	Remarks
Haemophilus aegyptius	*Hae*III	5'...GG CC...3' 3'...CC GG...5'	Blunt ends are produced
Thermus aquaticus	*Taq*I	5'...T CGA...3' 3'...AGC T...5'	Single strand is the 5' strand
Escherichia coli	*Eco*RI	5'...G AATAC...3' 3'...CTTAA G...5'	Single strand is the 5' strand
Providencia stuartii	*Pst*I	5'...CTGCA G...3' 3'...GA CGTC...5'	Single strand is the 3' strand
Microcoleus	*Mst*II	5'...CC TNAGG...3' 3'...GGANT CC...5'	Base pair N can be any purine or pyrimidine

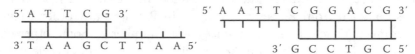

Figure 19.4 Fragments of dsDNA with sticky tails created by *Eco*RI in two DNA. These can be made to join together.

19.2.1.3 Exonucleases

They catalyze the hydrolysis of single nucleotides from the end of a DNA or RNA chain (Figure 19.5). Exonucleases differ from endonucleases in digesting nucleotides of a DNA chain one by one either from its 5' or from 3' end. These enzymes are used to suitably modify DNA chain ends. Some exonucleases have multiple activities. For example, exonuclease III obtained from phage-infected *E. coli* cells removes nucleotides of a DNA strand one by one starting from its 3' end. It can also be used to add a phosphate group to the 3' end of a single-stranded DNA (ssDNA) or dsDNA. It is also able to digest DNA–RNA hybrid molecules. DNA polymerases are used to synthesize small or large DNA lengths as per requirement.

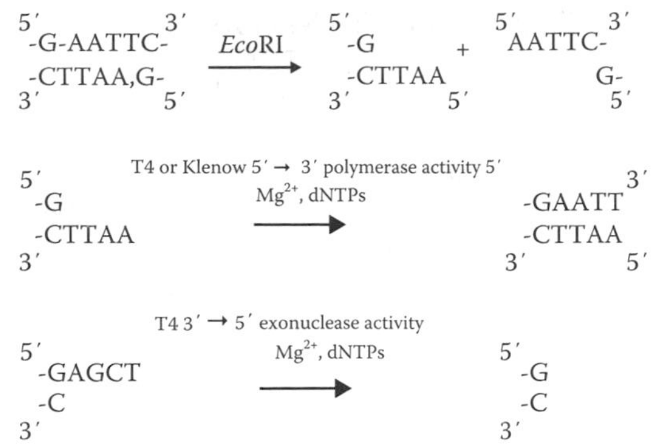

$$5' \quad 3'$$
-G-AATTC- $\xrightarrow{\textit{EcoRI}}$ $5'$ -G + $5'$ AATTC- $3'$
-CTTAA,G- -CTTAA G-
$$3' \quad 5' \quad 3' \quad 5' \quad 5'$$

$5'$ T4 or Klenow $5' \to 3'$ polymerase activity $5'$ $3'$
-G $\xrightarrow[\text{Mg}^{2+}, \text{dNTPs}]{}$ -GAATT
-CTTAA -CTTAA
$$3' \quad 3' \quad 5'$$

T4 $3' \to 5'$ exonuclease activity
$5'$ -GAGCT $\xrightarrow[\text{Mg}^{2+}, \text{dNTPs}]{}$ $5'$ -G
-C -C
$$3' \quad 3'$$

Figure 19.5 Restriction and cloning analysis using exonuclease and to make 3′ overhanging end blunt.

19.2.1.4 DNA Polymerases
DNA polymerases are used to synthesize small or large DNA lengths as per requirement. DNA polymerase I and T4 DNA polymerase are commonly used polymerases. Klenow fragment is the name of the DNA polymerase I from *E. coli* without exonuclease activity, which this enzyme utilizes to replace RNA primers during DNA replication in bacterial cells.

19.2.1.5 Alkaline Phosphatase
Alkaline phosphatase (AP) is used to remove phosphates from the 5′ end of an ssDNA length or from an RNA length.

19.2.1.6 Polynucleotide Kinase
Polynucleotide kinase is the enzyme that can transfer a phosphate group from an ATP molecule to a DNA or RNA. It is obtained from phage-infected *E. coli*.

19.2.1.7 Terminal Deoxynucleotidyl Transferase
Terminal transferase is a kind of DNA polymerase, which is capable to add given nucleotides to a 3′-OH end of a DNA length. It does not need a template for complementary copying as generally required by DNA polymerases.

19.2.1.8 Reverse Transcriptase
Reverse transcriptase, which is also a kind of DNA polymerase but RNA dependent, is used when cDNA is to be produced from a given messenger RNA (mRNA) length.

19.2.2 Cloning Vectors

Vectors are vehicles by whose help genes are transferred between organisms. Their suitability depends on the type of organism to which a gene is to be transferred. Several vectors have been designed by genetic engineers for bacteria, animal species, and plants. For gene transfer into bacterial cells, suitable vectors are plasmids and

TABLE 19.2   Some Common Cloning Vectors

Cloning Vector	Natural Occurrence	Size (kb)[a]	Selective Marker[b]
Plasmid			
pACYC177	*Escherichia coli*	3.7	Amp^r and Kan^r
pBR322	*E. coli*	4.0	Amp^r and tct^r
pBR324	*E. coli*	8.3	Amp^r, tet^r, and El imm.
pMB9	*R. coli*	8.3	Tcf
pRK646	*E. coli*	3.4	Amp^r
pC194	*S. aureus*	3.6	Ery^r
pSA0501	*S. aureus*	4.2	Str^r
pBS161-1	*Bacillus subtilis*	3.65	Tet^r
pWWO	*Pseudomonas putida*	117	Kan^r
Cosmids			
pJC74	Derived plasmid from ColE1	16	Amp^l and El imm.
pJC720	-do-	24	El imn./Rif^r
pHC79	Derivative of pBR322	6	Amp^r, Tet^r
Viruses			
SV40	Mammalian cells	5.2	—
Phage M13$^+$	*E. coli*	6.4	—
Phage^l	*E. coli*	4.9	—

[a] 1 kb (kilobase pairs) = 1000 base pairs = 0.66 mega Dalton.

[b] Resistance to ampicillin (Amp^r), tetracycline (Tet^l), erythromycin (EryO* streptomy-cin (Strr), kanamycin (Kan^r), rifampicin (Rif^r), and colicin production (EL imm.).

bacteriophages. Bacteriophages are used due to their natural ability to transfer their genetic material to bacteria. If you tag your desirable foreign gene to the DNA of the phage, it would easily be delivered into its host bacterial cell. Several advanced and more efficient plasmid vectors have been artificially developed by modifying original bacterial plasmids through genetic engineering tools themselves. The cloning vectors must possess the following: (a) it can be easily isolated from the cells, (b) it must possess at least a single restriction site for one or more REs, (c) insertion of a linear molecule at one of these sites should not alter its replication properties, and (d) it can be reintroduced into a bacterial cell, and cells carrying the plasmid with or without the insert can be selected or identified. Some such plasmid vectors are pBR322, pCR1, pUC, pMB9, etc. (Table 19.2). Based on the nature and source, the vectors are grouped into bacterial plasmids, bacteriophages, cosmids, and phagemids.

19.2.2.1 *Plasmid Cloning Vectors*
Plasmids are widely used as cloning vehicles. They are extra-chromosomal, self-replicating, ds, circular DNA molecules present in the bacterial cell containing sufficient genetic information for their own replication. Their number varies from one per cell to as many as several hundred in each bacterial cell. This is referred to as copy number. A number of host properties are specified by plasmids, such as antibiotic and heavy metal resistance, nitrogen fixation, pollutant degradation, bacteriocin,

and toxin colicin. The small size of plasmids and the presence of antibiotic resistance make them ideal vectors for cloning. They are not easily broken by physical manipulation or by cycles of freeze–thawing during storage in the laboratory. Plasmid vectors can carry DNA fragments ranging from hundreds of base pairs to much longer fragments of many thousand base pairs.

Naturally occurring plasmids often lack several important features required for a high-quality cloning vector like a small size, unique restriction endonuclease recognition sites, and one or more selectable genetic markers for identifying recipient cells that carry the cloning vector-insert DNA construct. They can be modified by *in vitro* digestion and ligation reactions. Most plasmid vectors containing multiple cloning sites (MCSs) are in common use. They have recognition sequences for many different REs. It is almost always possible to find a vector carrying restriction sites that are compatible with the terminus of a particular fragment of foreign DNA.

19.2.2.1.1 pBR322 Plasmid Cloning Vectors. This is an *E. coli* cloning vector, which carries genes conferring resistance to ampicillin and tetracycline. The origin of replication or *ori* site in this plasmid is pMB1 (a close relative of ColE1). The ori encodes two RNAs (RNAI and RNAII) and one protein (called Rom or Rop). The vector size is 4.36 kilobase pairs (kb); and within the ampicillin and tetracycline genes, there are a number of recognition sites for restriction endonucleases. For example, *Pst*I cuts within the ampicillin resistance gene, whereas *Bam*HI, *Hin*dIII, *Sal*I, and *Cla*I cut within the tetracycline resistance gene. These genes are therefore suitable as selection markers in cloning experiments. A physical map of plasmid pBR322 is shown in Figure 19.6. The pBR322 is constructed from the plasmids of *E. coli*, pBR318 and pBR320. It contains origin of replication *(ori)* that was derived from a plasmid related to naturally occurring ring plasmid ColE1. Therefore, its replication may be faster than bacterial DNA. It also possesses genes conferring resistance to antibiotics, for example, ampicillin *(amp^l)* and tetracycline *(tet^r)*, and unique recognition sites for 20 restriction endonucleases. Certain restriction sites, for example, *Bam*HI in the *tet^r* genes of the plasmid are present within the gene in such a way that the insertion of foreign segment of DNA will inactivate the *tet^r* gene. The recombinant plasmid will allow the cells to grow only in the presence of ampicillin but will not protect them against tetracycline. The presence of an antibiotic-resistant gene in a plasmid of bacterium will confer resistance to that antibiotic.

The construction of pBR322 from pMB1, pSC101, and pRSF2124 is shown in Figure 19.7.

19.2.2.1.2 pUC Plasmid Cloning Vectors. The valuable features of pBR322 were enhanced by the construction of a series of plasmids termed as pUC (produced at the University of California) (Figure 19.8). The important features of pUC are the following:

- Presence of antibiotic resistance gene for ampicillin and origin of replication from *E. coli*.
- Most popular restriction sites concentrated into a region termed as MCS.
- MCS is part of a gene coding for a portion of a polypeptide called β-galactosidase.

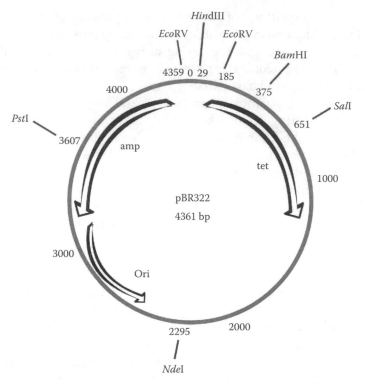

Figure 19.6   Physical map of pBR322.

19.2.2.1.3 pBluescript KS(+) A.   pBluescript KS(+) A is a 2.9 kb derivative of the pUC plasmid series. It contains a MCS (polylinker) with recognition sites for 21 restriction endonucleases (Figure 19.9). The recognition sites are so arranged that those for enzymes creating 5′-overhangs are flanked on both sides by recognition site for endonucleases creating 3′-overhangs. Furthermore, the presence of T3 and T7 promoters on either side of the polylinker allows *in vitro* transcription with either T3 or T7 RNA polymerases. Yet another advantage of this plasmid is that it carries an intergenic sequence of M13 permitting the isolation of ssDNA with the aid of a suitable helper phage.

KS (+) describes the orientation of the polylinker region (KpnI SacI) with respect to the direction of transcription of the *lacZ* gene. The vector pBluescript KS (+) and the pUC series of vectors contain part of the *lacI* gene and the first 492 nucleotides of the *lacZ* gene (laclOPZ) as well as the gene-encoding β-lactamase (*amp*ʳ) and an origin of replication from *E. coli*. The polylinker is inserted into the *lacZ* gene so that the reading frame is not affected. In addition to *E. coli*, some Gram-negative and Gram-positive bacteria have also been investigated and used in medical and agricultural studies. A broad host range cloning vehicle has been developed from RK2, which is a plasmid of P-1 group of Gram-negative bacteria. RK2 plasmid contains a single restriction site for *Eco*RI, *Hin*dIII, *Bam*HI and *Bgl*II. However, a broad host range cloning system has been developed from it by incorporating the transfer and replication regions of this plasmid into two different plasmids. The three regions of

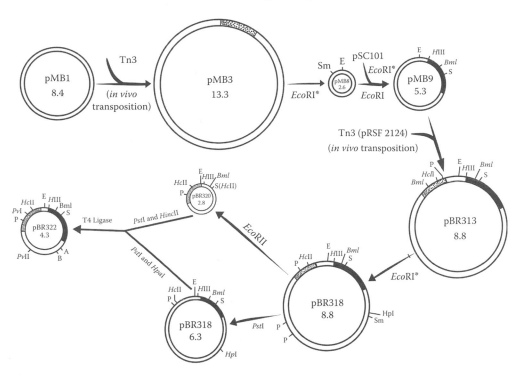

Figure 19.7 Schematic representation showing the steps in construction of a pBR322 from pMB1, pSC101, and pRSF2124 where solid bar = Tc^r gene; stippled bar = *Ap^r* gene; and hatched bar = *Cm^r* gene.

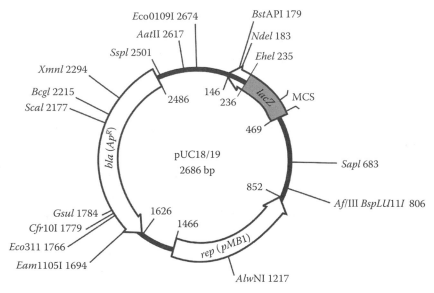

Figure 19.8 Physical map of pUC.

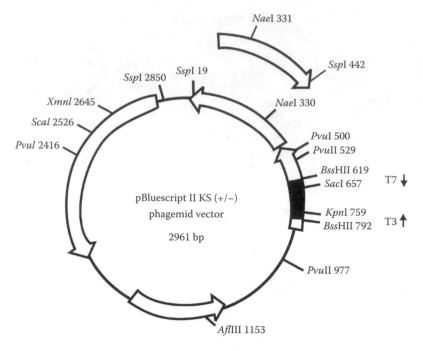

Figure 19.9   Physical map of pBluescript KS(+) A.

RK2 plasmid essential for replication (e.g., *OriV, trfA,* and *trfB*) have also been incorporated into the plasmid cloning vector, the pRK290.

19.2.2.2 Bacteriophages

A bacteriophage is a virus that eats upon bacteria. The plasmid-based vectors used for cloning DNA molecules generally carry up to 10 kb of inserted DNA. For the formation of the library, it is desirable to maintain larger pieces of DNA. A bacteriophage is a virus that eats upon bacteria. A number of such viruses of different genetic material have been reported, for example, ΦX174, phage λ, M_{13}, Fd^{11}, R209, etc. The process of infection and replication of phage in *E. coli* cell are shown in Figure 19.10.

Coliphage lambda (λ) DNA is widely used as a vector for recombinant DNA. This virus is temperate and may reside within the genome of its host through lysogeny. Phage lambda contains a proteinaceous head and a long tail attached to the head. In the head, it possesses 50 genes in its 49 kb genome of which about half of genes are essential. On attachment with tail to the cell wall of *E. coli*, it injects its linear DNA into the cell. The linear dsDNA molecule circularizes through the single strand of 12 nucleotides commonly known as *cos* sites at its end. The *cos* sites are the key feature of the DNA. Replication cycle of phage λ is accomplished into two pathways: the lytic and lysogenic pathways. In the lytic pathway, early in the infection, the circular DNA replicates as theta (θ) forms. By a rolling circle mechanism, it produces the long concatemeric molecules joined end to end and composed of several linearly arranged genomes. At the same time, phage DNA directs the synthesis of many proteins

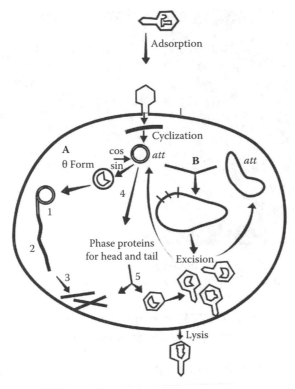

Figure 19.10 Replication of phage lambda in an *E. coli* cell. (A) Lytic cycle: (1) rolling circle replication, (2) production of concatemers, (3) cleavage at cos site, (4) transcription and translation, and (5) packaging. (B) Lysogenic cycle.

required to produce empty heads where DNA is packed after the cleavage of concatemeric DNA at its cos site to yield fragments of such sizes as to fit in their heads. Eventually, a tail is attached to the head, and finally the mature phage particles are released out the bacterial cell. As λ replicates to create more viral particles, infected *E. coli* cells are lysed by λ, creating zones of dead bacteria called plaques. Plaques appear as cleared spots on bacterial lawns (Figure 19.11).

The lysogenic pathway for replication is another alternative mode of propagation where it becomes stably integrated into the host chromosome and replicate along with the bacterial chromosome. Phage genome integrates by an attachment site (*att*) with a partially homologous site on the *E. coli* chromosome, where it replicates as a chromosomal DNA segment. In this case, a protein is produced by *cl* gene, which represses all the genes responsible for lytic pathways. In this pathway, no phage structural proteins are synthesized. The interaction of two proteins, the *cl* genes expressed protein (by phage genome), and *cro* gene expressed protein (by *E. coli* chromosomes) decide between the events of the lytic and lysogenic pathways. The bacteriophage λ DNA is about 50 kb in length. About 20 kb is essential for the integration–excision events. For preparing genomic libraries, this 20 kb of DNA could be replaced with 20 kb of DNA to be cloned. The construct then can be perpetuated as a "recombinant" bacteriophage λ t with no detrimental effect to its replication or packaging. The usual

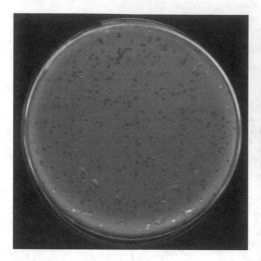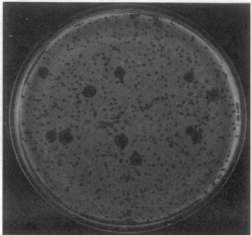

Figure 19.11 **(See color insert.)** λ Coliphage plaques of different sizes caused by at least two bacteriophages on a carpet of actively multiplying bacteria.

recombinant lambda DNA (λ DNA) contains 80% lambda vector DNA and 20% insert. Following are the advantages of phage cloning system over the plasmids:

1. DNA can be packed *in vitro* into phage particles and transduced into *E. coli* with high efficiency.
2. Foreign DNA up to 25 kb in length can be inserted into phage vector.
3. Screening and storage of recombinant DNA is easier.

Two types of phage cloning vectors have been constructed: the insertion vector and the replacement vector. Cloning procedures of both the vectors are shown in Figure 19.12.

19.2.2.2.1 Insertion Vector (λgt10 and λ charon 16A). Insertion vectors have unique cleavage site into which relatively small piece of foreign DNA is inserted into the region of a phage genome with appropriate restriction sites. Foreign DNA fragment does not affect the function of phage. The upper and lower limit for foreign DNA that may be packed into phage particles is between 35 and 53 kb. Therefore, the minimum size of vector must be above 35 kb. It means the maximum size of a foreign DNA to be inserted is about 18 kb.

19.2.2.2.2 Replacement Vector (λEMBL and λZAP). The replacement vectors have cleavage sites present on either side of a length of nonessential DNA of phage. As a result of cleavage, left and right arms are formed, each arm has a terminal cos site and a stuffer region, the nonessential region, which can be substituted by foreign DNA fragment.

The maximum size of inserted DNA fragment depends on how much of the phage DNA is nonessential. It has been found that about 25–30 kb of genome codes for essential products for lytic cycle. The remaining 20–25 kb of genome could be replaced with the foreign DNA fragments of known essential products.

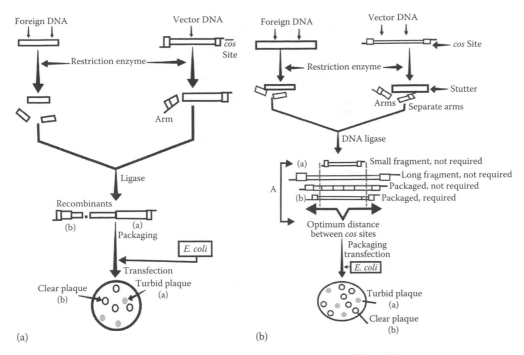

Figure 19.12   (a) Cloning of an insertion vector. (b) Cloning of a replacement vector.

Nonessential part of the λ genome can be separated from the arms by electrophoresis or velocity gradient ultracentrifugation that facilitates to make size differences. Formation of multiple inserts can be checked by using AP before ligation with insert fragment. Recombinant DNA formed by multiple inserts has too large genome to be packed in viral head. However, after ligation the recombinant DNA molecules have left arm plus large insert plus right arm linked by their cos sites at the arms.

Optimum distance from *cos* sites governs efficiency for packaging in *E. coli*. As a result of ligation, the size of recombinant DNA fragment may be less or more than the required size or may have more than two *cos* sites of many small fragments of foreign DNA. Those having the range of viral head can be packed *in vitro* using a preparation of head and tail proteins. The viruses thus constructed are allowed to multiply in *E. coli*. Development of plaque turbidity is a useful criterion for the selection of recombinant phages. Plaque turbidity is determined by the presence of nonlysed bacteria. The recombinant phages give clear plaques due to inactivation of *cl* gene. By using these methods, the clones containing recombinant DNA can be isolated from the wild type clones, that is, turbid plaques. The constructed genome has all the information required for DNA replication and synthesis of the viral protein. The inserts can be identified by colony hybridization technique or immunological assays.

19.2.2.3 M13 Phage Vectors

Bacteriophage M13 is a filamentous bacteriophage containing single-stranded circular DNA genome having 6407 bp (Figure 19.13).

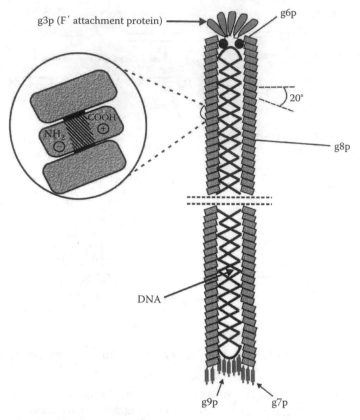

Figure 19.13 Structure of M13 phage.

It is packaged inside rod-shaped protein capsids. Phage particles bind to F pilus, and then ssDNA genome enters the cell designated as "+" strand. The double-stranded replicative form (RF) is formed that contains "+" and "−" strands. "−" strand is template for mRNA synthesis for production of new "+" strands. "+" strands are packaged in phage coat protein that exit in cell as phage particle.

19.2.2.3.1 Salient Characters of M13 as Cloning Vector
1. M13 occurs in both single- and double-stranded forms.
2. RF can be digested with restriction endonucleases.
3. Inserts can be cloned in it.
4. Convenient source of ssDNA.
5. Used for sequencing and site-directed mutagenesis.
6. Different-sized DNA molecules packaged as phage particle.
7. M13 does not kill host.
8. Collected by growing M13 infected cells in culture.
9. Production of ssDNA requires: M13 origin of replication—in DNA molecule and M13 gene products—in cell containing DNA molecule.
10. Phage proteins can be provided in trans by helper phage.
11. Allows "phagemid" vectors to be used (plasmid + M13 origin of replication).

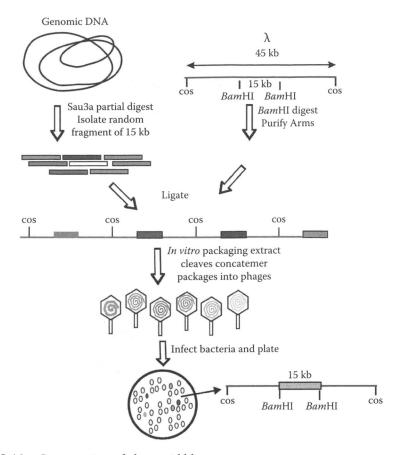

Figure 19.14   Construction of phagemid library.

19.2.2.3.1.1 Phagemid Vectors. Phagemids are filamentous-phage-derived vectors containing the replication origin of a plasmid. They are hybrid vectors. Phagemids usually encode no or only one kind of coat proteins. Other structural and functional proteins necessary to accomplish the life cycle of phagemid are provided by the helper phage. In addition, other elements such as molecular tags and selective markers are introduced into the phagemids to facilitate the subsequent operations, such as gene manipulation and protein purification. One example of phagemid vector is pBluescript KS(+) A. Construction of phagemid library is shown in Figure 19.14.

19.2.2.3.1.2 Cosmids.   A group of Japanese workers showed that the presence of a small segment of phage λ DNA containing cohesive end on the plasmid molecule is a sufficient prerequisite for *in vitro* packaging of this DNA into infectious particles. The cosmids can be defined as the hybrid vectors derived from plasmids, which contain *cos* site of phage λ. Cosmids lack genes encoding viral proteins; therefore, neither viral particles are formed within the host cell nor cell lysis occurs. Cosmid DNA contains 10% vector and 90% insert.

Special features of cosmids similar to plasmids are the presence of

- Origin of replication
- A marker gene coding for antibiotic resistance
- A special cleavage site for the insertion of foreign DNA
- The small size

Character different to plasmid is the presence of extra phage DNA, the *cos* site, which has about 12 bases. It helps the whole genome in circularization and ligation.

The cosmids have a length of about 5 kb, the upper size limit of the foreign DNA fragment that may be inserted in cosmids and packed into phage particles is, therefore, approximately 45 kb, much larger than it would be possible to clone in phage λ or plasmid vector (Dahl et al., 1981). According to the size of *cos* sites and upper size limit that can be packaged in the head of phage, the recombinant DNA molecules can be packed into bacteriophage particles in *in vitro* packaging system consisting of packaging enzymes, head, and tail proteins. Figure 19.15 demonstrates the cloning of a cosmid vector.

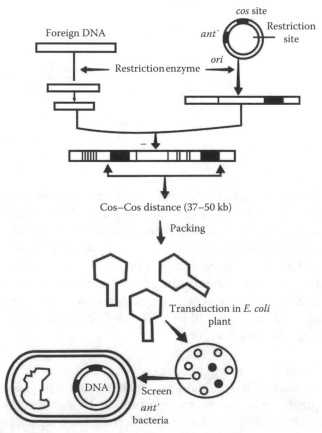

Figure 19.15 Cloning of a cosmid vector. Transduced bacteria contain a cosmid that shows resistance to specific antibiotic. Such bacteria can be screened. Ant[r], antibiotic resistance; ori, origin of replication.

Upon infection of *E. coli* by bacteriophage, the recombinant DNA cycle moves through *cos* sites and then replicates as a plasmid and expresses the drug resistance marker. The antibiotic resistance gene allows colonies carrying the cosmid to grow in the presence of the antibiotic. However, nontransformed cells are sensitive to antibiotic and die. A number of cosmid vectors have been obtained from *E. coli*, yeast, and mammalian cells, and gene bank has been constructed. Other vectors used in cloning are bacteriophage M13, virus SV40, Ti-plasmid, etc., but none of them are useful for building gene bank. The advantages of using cosmids as cloning vectors are the following:

- As the cloning capacity of cosmids is greater than that of plasmids, gene clusters and large genes are easier to clone.
- Fewer clones have to be screened as the insert size is large.

19.2.3 Transformation

The cloned DNA is to be taken up by the host cells. The uptake of cloned DNA into a bacterial cell is called transformation. Bacteria that are able to uptake DNA are called "competent" and are made so by treatment with calcium chloride in the early log phase of growth. The bacterial cell membrane is permeable to chloride ions, but is nonpermeable to calcium ions. As the chloride ions enter the cell, water molecules accompany the charged particle. This influx of water causes the cells to swell and is necessary for the uptake of DNA. Generally, cells can be made competent by suspending them in a solution of ice-cold divalent cations (typically Ca^{++}, although Mg^{++} and Mn^{++} have also been used), followed by a brief heat shock. Actively growing cells seem to give the highest percentage of transformants.

The mechanism of DNA uptake is not known. One theory suggests that the presence of pores in the cell wall allows the formation of channels through which DNA molecules can pass. These pores are present only in actively growing cells, which is consistent with the observation that they make the best competent cells. At 0°C, the membrane is "frozen," stabilizing the structure of the pores. DNA molecules and the phospholipids of cell membranes are both negatively charged. Electrostatic forces are therefore likely to prevent entry of the DNA into the cells. However, the divalent cations present in the cell suspension may interact with the negatively charged phospholipids, coating the pores. Channels into the cells are thereby created, lined with positive charges that attract the negatively charged DNA molecules. The brief heat shock is believed to create a thermal imbalance between the inside of the cell and its external environment, which helps to "pump" the DNA into the cell. Cells are then grown at 37°C, which allows them to recover from the treatment, to make DNA, RNA, and proteins, and to divide.

In many cases, the principal objective of cloning experiment is the insertion of a particular restriction fragment into a suitable plasmid vector and its amplification. Amplification is a process of increasing the number of plasmid in a bacterial cell. The host system used for the propagation of plasmid DNA is the bacterium *E. coli*. Other bacteria such as *Bacillus subtilis* and *Agrobacterium tumefaciens* are often used as host cells.

In this process, a cell containing a relaxed plasmid is treated with a drug to inhibit protein synthesis. Consequently, cells stop replicating. The relaxed plasmid continues to replicate despite drug treatment. Replication of relaxed plasmid neither depends on cell replication nor requires protein synthesis. For example, addition of chloramphenicol causes pBR322 to get increased about 3000 per cell. Finally, the ratio of plasmid DNA is increased to chromosomal DNA, which makes easy to isolate the plasmid DNA.

19.2.3.1 Selection of Recombinants

The following methods are commonly used to identify bacterial colonies that contain recombinant plasmids:

1. *α-Complementation (blue/white selection)*: Vectors used in cloning experiments carry a short segment of *E. coli* DNA that contains the regulatory sequences and the coding fragment for the first 146 amino acids of the β-galactosidase (β-gal) gene (*lacZ*). This coding region contains a polycloning site that does not disrupt the reading frame but results in the harmless incorporation of a small number of amino acids into the amino terminal of β galactosidase. These vectors are used in host cells that code for the carboxy terminal of β-gal. Both these host cells as well as vector encoded fragments are inactive separately, but when associated they form an enzymatically active protein. This phenomenon is called α-complementation. The bacteria that result from α-complementation are easily recognized because they form blue colonies in the presence of chromogenic substrate, 5-bromo-4-chloro-3-indolyl-β-D-galactoside (X-gal) and inducer, isopropylthio-β-D-galactoside (Figure 19.16).

   However, insertion of a fragment of foreign DNA into the polycloning site of the plasmid results in production of an amino-terminal fragment that is not capable of α-complementation. Therefore, bacteria carrying recombinant plasmids form white colonies. This method of identification simplifies the selection of recombinants. This property of β-gal is used as the selection system for pBS and pUC vectors.

2. *Insertional inactivation*: It is a useful selection method for identification of recombinant vectors having the insert. This can be used with vectors that carry two or more antibiotic resistance genes and appropriate distribution of RE cleavage sites. In a pBR32 system where the target DNA is inserted into *Bam*HI site, the identification is done in two steps:

Step I: The cells from the transformation mixture are plated onto medium containing the antibiotic ampicillin. Cells having either intact pBR32 or pBR32-cloned DNA constructs, both having intact *Amp*ʳ gene will grow under these conditions. The nontransformed cells, sensitive to ampicillin will not survive. As the *Bam*HI site of pBR32 is within the tetracycline resistance gene, the insertion of DNA into this gene disrupts the coding sequence resulting in the loss of tetracycline resistance. Therefore, the cells with these plasmid-cloned DNA constructs are resistant to ampicillin but are sensitive to tetracycline. Cells with recircularized pBR322 are resistant to both antibiotics.

Step II: In this step, the cells that grow on the ampicillin-containing medium are transferred to tetracycline-containing medium. Those cells that fail to grow on the tetracycline-containing medium are sensitive to tetracycline and carry the construct.

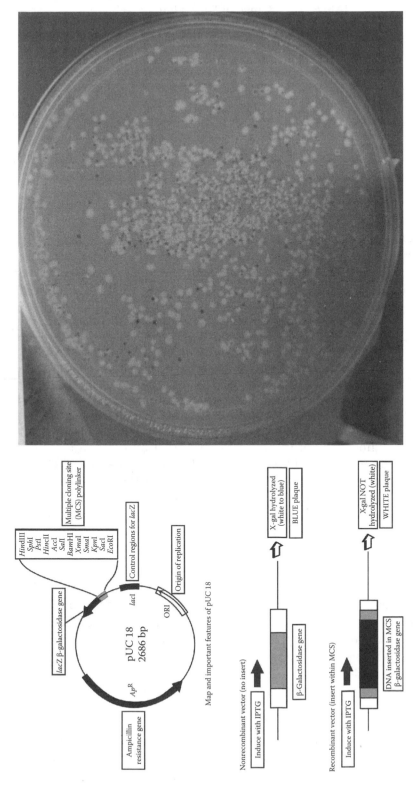

Figure 19.16   (See color insert.) Plate showing blue and white recombinant bacterial colonies.

TABLE 19.3   Range of Separation in Gels Containing Different Amounts of Agarose

% Concentration of Agarose in TAE (w/v)	Efficient Range of Separation of Linear DNA (kb)
0.3	5–60
0.6	1–20
0.7	0.8–10
0.9	0.5–7
1.2	0.4–6
1.5	0.2–3
2.0	0.1–2

19.2.4 Electrophoretic Techniques

Nucleic acid fragments produced by REs can be analyzed by electrophoresis. It has become the principal means for analyzing and characterizing recombinant DNA molecules. Molecules differing in electrical charge, size, and shape move at different rates in an electric field and therefore can be separated under conditions in which they retain their biological activity. The small fragments move much faster than large fragments.

Different supports are used depending on the absolute sizes of the fragments to be separated as well as on the relative differences in the sizes of these fragments. DNA fragments ranging in size from 100 bp to several thousand base pairs can be efficiently separated in a gel varying from 0.3% to 2% agarose (Table 19.3), provided these fragments differ in size from each other by a few hundred base pairs. Polyacrylamide gels are used for resolution of DNA fragments differing from one another by only a single nucleotide. There is a linear relationship between agarose concentration and the logarithm of the molecular weight (MW) of the DNA. The migration of the DNA molecule also depends on its conformation. The DNA molecule can be superhelical, nicked circular, or linear. Depending on the electrophoretic conditions (ionic strength of the buffer and intensity of the electric field), the superhelical can migrate faster than the linear form. Generally the DNA molecule is visualized after electrophoresis by staining with ethidium bromide (EtBr).

Two types of apparatus—horizontal and vertical—are used for gel electrophoresis (Figure 19.17). In both the vertical and horizontal modes, the gel is allowed to polymerize or harden in place. An appropriately shaped mold called a comb is suspended on the top of a frame (vertical mode) or in the horizontal gel during polymerization or hardening in order to make notches to serve as sample wells.

Agarose consisting of polymerized galactose residues dissolves on heating to give a clear solution that can be poured into a gel casting form and upon cooling solidifies to give a gel matrix. The density of the gel matrix is determined by the initial agarose concentration. In an electric field, DNA, which is a polyanion, migrates through the gel matrix toward the anode. The length of time required to

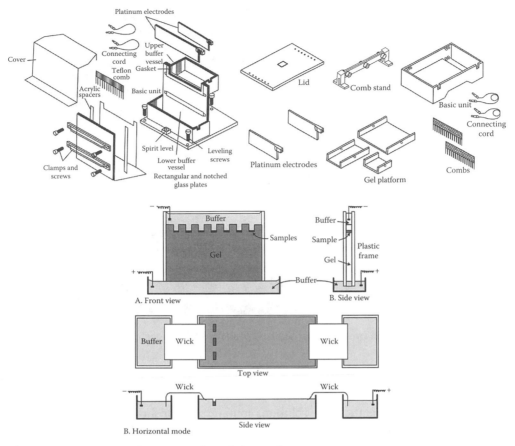

Figure 19.17   Types of apparatus for gel electrophoresis.

separate DNA fragments by electrophoresis is dependent on a number of parameters, for example, agarose concentration, volume, and composition of the buffer.
On completion of the run, the DNA fragments can be visualized by staining the
gel with ethidium bromide (EtBr). EtBr is a fluorescent dye, which intercalates
between the bases of DNA. After irradiation with ultraviolet (UV) light, the bound
dye retransmits the light at 590 nm. Through this staining, which can be done during or after the electrophoresis, small amounts of DNA (<10 ng) can be detected
[note: be careful when handling EtBr. It is a powerful mutagen] (Figure 19.18). Each
band corresponds to a restriction fragment of specific size. The size of each fragment can be determined by calibration with DNA molecules of known sizes. Bands
containing as little as 0.005 µg of DNA can be visualized. The distance travelled by
a linear DNA fragment is inversely proportional to $\log_{10}$ of the size the fragment in
kb or Dalton (g/mol). The two dyes, xylene cyanol and bromophenol blue, can be
used as "markers" for the migration of DNA fragments of 0.7–9.0 kb. For instance,
in a 0.85% agarose gel, the "ideal" position of the two marker dyes at the end of the
electrophoresis are at one-third and two-third of the total length of the gel.

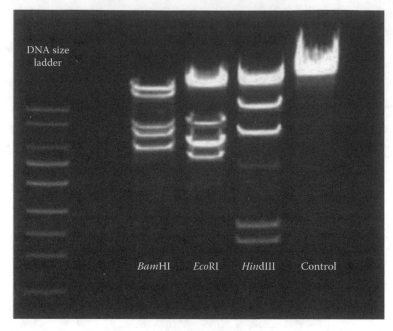

Figure 19.18 Ethidium bromide stained gel of ladder DNA digested with different restriction endonucleases.

19.2.4.1 Decontamination of EtBr Containing Solutions

EtBr is carcinogenic, and therefore solutions containing EtBr should be decontaminated before discarding them. Add 1.0 g active charcoal per 1.0 mL of EtBr-containing buffer Mix thoroughly for an hour. Filter the suspension through Whatman no. I filter paper and discard the filtrate. The active charcoal can be reused several times. Dispose the used active charcoal and filters as per recommendations for halogen-containing waste.

19.3 DNA Sequencing

Two methods of DNA sequencing have been discussed in this section, viz chemical-based method and Sanger's method.

19.3.1 Chemical Cleavage and Modification Method

This method can be used for sequencing both ss and dsDNA. The first hurdle in chemical cleavage method is to radioactively label one end, usually the 5′ end, with ^{32}P. Four sets of reactions are set up in separate aliquots of the nucleic acid samples as follows:

19.3.1.1 G Cleavage

DNA is treated with dimethyl sulfate (DMS), which methylates the G residues of the base thereby making the glycosidic bond of the methylated residue susceptible to hydrolysis. Subsequent treatment with piperidine cleaves the polynucleotide chain at a site just before the G residue.

19.3.1.2 A + G Cleavage

In the second set, DNA is methylated at A and G bases using DMS. Under acidic conditions, both A and G is released resulting in depurinated product. Piperidine treatment then breaks the strand at a point just before both A and G residues. The A residues are then identified by comparing the positions of the G and A + G cleavages.

19.3.1.3 C + T Cleavage

In the third sample aliquot, DNA is made to react with hydrazine followed by piperidine treatment, and this cleaves the DNA molecule before both its C and T residues.

19.3.1.4 C Cleavage

If DNA is treated with hydrazine in presence of 1.5 M NaCl, then only its C residues reacts appreciably and thus the piperidine cleavage occurs at C residues only. The comparison of C + T cleavage positions facilitates identification of the T residues.

In all the four reactions, the conditions are varied so that the strands are cleaved at an average of one randomly located position. These four sets of reaction mixtures are simultaneously electrophoresed in adjacent lanes on a long, thin denaturating, polyacrylamide slab gel (0.1 mm thick × 40 cm long). The DNA fragments separate according to their size. After electrophoresis is over, the gel is subjected to autoradiography, and sequence of the DNA can then directly be read from the autoradiogram in 5′–3′ direction (as bands from bottom to top). Various steps involved in sequencing are summarized in Figure 19.19.

19.3.2 Dideoxy Chain Termination Method

This method involves a primed synthesis of a complementary radioactive copy of a ssDNA template by a DNA polymerase. The base specific ends of the oligonucleotides thus synthesized are produced by incorporating 2′, 3′-dideoxynucleoside triphosphates (2′,3′ ddNTPs) into the newly synthesized DNA. ddNTPs differ from the normal deoxynucleotide triphosphates (dNTPs) that they lack a hydroxyl residue at the 3′ position of deoxyribose. They can be incorporated by DNA polymerase (Klenow enzyme) into a growing DNA chain through their 5′-triphosphate groups. However, the absence of a 3′-OH residue prevents formation of a phosphodiester bond with the succeeding dNTP. Further extension of the growing chain is performed in four different sets each containing a different dideoxynucleotide, and this results in production of nested sets of fragments each terminating with dideoxynucleotide being used. The four reaction mixtures are run simultaneously in different lanes on denaturating polyacrylamide sequencing gels. The fragments get separated according to their size, producing an autoradiogram. This will indicate the sequence of complementary strand and from this the sequence of the original (template) strand can be computed. Several advancements have been made in Sanger's method such as development of

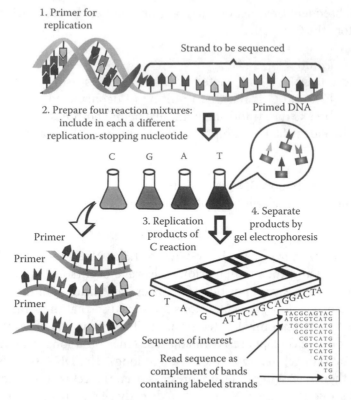

Figure 19.19   Steps involved in DNA sequencing.

sequencing vectors, use of nonradioactive fluorescent dideoxy analogues, and PCR technology, which have led to its automation.

19.3.2.1 PCR

19.3.2.1.1 Introduction.   PCR is based on the DNA polymerization reaction and is an *in vitro* technique enabling chemical amplification of DNA. Using the random primers, the entire sequence of a genome can be amplified in pieces. A primer and dNTPs are added along with a DNA template and the DNA polymerase (in this case, Taq). With the improvement brought by the use of the heat stable *Taq* DNA polymerase of *Thermus aquaticus* and automation, it is possible to obtain quick amplification even of single copy genes, starting from minute amounts of material. The efficiency of PCR amplification is dependent on the purity of the target DNA, *Taq* DNA polymerase is less sensitive to template purity than other molecular biology techniques so that partially purified nucleic acid can be used and is a better alternative to gene cloning. The main concern with PCR is that, in addition to using a primer that sits on the 5′ end of the gene and makes a new strand in that direction, a primer is made to the opposite strand to go in the other direction. The original template is melted (94°C), the primers anneal (45°C–55°C), and the polymerase makes two new strands (72°C), doubling the amount of DNA present. This provides two new templates for the next cycle.

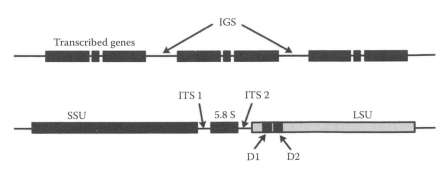

Figure 19.20 Illustrating inter nontranscribed region (ITS) separating small (SSU) and large subunits (LSU).

The DNA is again melted, primers anneal, and the Taq makes four new strands. A wrong temperature during the annealing step can result in primers not binding to the template DNA at all or binding at random. The elongation temperature depends on the DNA polymerase. The time for this step depends both on the DNA polymerase itself and on the length of the DNA fragment to be amplified. Ribosomal genes are multicopy genes tandemly organized in the genome. Each ribosomal genes encodes for three subunits (18S [small subunit], 5.8S, and 28S [large subunit]) separated from each other by inter nontranscribed region (ITS). The genes themselves are separated from each other by an intergenic spacer (IGS) (Figure 19.20).

The various characteristics of rRNA and rDNA have made them a choice target for phylogenetic and taxonomic studies, and comparative studies of the nucleotide sequences in ribosomal genes has provided data for the analysis of phylogenetic relationships over a wide taxonomic range of organisms. The nucleotidic polymorphism is not evenly distributed throughout the ribosomal genes, and the three regions evolve at different rates. ITS and IGS are variable regions, which mutate more frequently than the three conserved coding subunit regions (18S, 5.8S, and 25S). This generally makes the former more informative for analyses of closely related genomes, whereas the coding regions of the small and the large ribosomal subunit are considered to be more useful for understanding more distant relationships at the species/order level.

19.3.2.1.2 Principle. The PCR is an *in vitro* technique that allows the amplification of a specific region of DNA located between two known sequences. After each cycle of denaturation, annealing, and extension, the amount of DNA is double. Potentially, after 20 cycles of PCR, there will be a 2^{20}-fold amplification (or 1×10^6). This illustrates the sensitivity of this method, and the potential artifactual amplification of DNA, as any traces of DNA can be amplified (Figure 19.21).

Before the discovery of thermostable polymerase, DNA polymerases such as the Klenow fragment of *E. coli* DNA polymerase I or T4 DNA polymerase were used. Due to their heat liability, fresh aliquots of enzymes had to be added after each denaturation cycle. The first heat stable DNA polymerase (*Taq* polymerase) was purified from *Thermus aquaticus*. Today, several heat stable polymerase are available, they are of natural or recombinant origin and vary in their biochemical properties such as extension rate, thermal stability, and 5′–3′ or 3′–5′ exonuclease

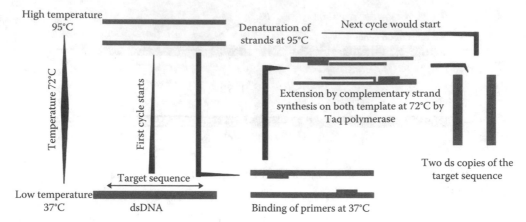

Figure 19.21 Principle of PCR technology.

TABLE 19.4 Thermostable DNA Polymerases and Their Sources

DNA Polymerase	Natural/Recombinant	Sources
Taq	Natural	*Thermus aquaticus*
Amplitaq®	Recombinant	*Thermus aquaticus*
Amplitaq® (Stoffel fragment)	Recombinant	*Thermus aquaticus*
Hot Tub™	Natural	*Thermus flavus*
Pyrostase™	Natural	*Thermus flavus*
Vent™	Recombinant	*Thermococcus litoralis*
DeepVent™	Recombinant	*Pyrococcus GB-D*
Tth	Recombinant	*Thermus thermophilus*
Pfu	Natural	*Pyrococcus furiosus*
Pfu	Cloned	*Pyrococcus furiosus*
Exo⁻*PFU*	Recombinant	*Pyrococcus furiosus*
UITma™	Recombinant	*Thermotoga maritima*

activity (Tables 19.4 and 19.5). The specificity and activity of the same enzymes is also very dependent on the producer. Some enzymes such as *Tth* have a reverse transcriptase activity, they cannot therefore be used for the synthesis of cDNA. Beside the enzyme, the other factors that can affect the PCR reaction are: primers, $MgCl_2$ concentration, primer concentration, primer sequence, reaction stringency, length of the amplification product, number of PCR cycles, and other unknown factors. For each PCR reaction, the optimal conditions can vary depending mainly on the primer–DNA combination. The dNTPs are generally used at a concentration of 100 μM, although at lower concentrations (10–100 μM) *Taq* polymerase has a higher fidelity. The most common buffer used with the Taq polymerase is

- 10 mM Tris/HCl, pH 8.3
- 50 mM KCl
- 1.5 mM $MgCl_2$
- 0.1% (w/v) gelatin

TABLE 19.5 Properties of DNA Polymerases Commonly Used in PCR

Properties	Taq/Amplitaq®	Stoffel Fragment	Vent™	Deep-Vent™	Pfu	Tth	UlTma™
Thermostability—half-life at 95°C	40	80	400	1380	>120	20	>50
5′–3′ exonuclease activity	Yes	No	No	No	No	Yes	No
3′–5′ exonuclease activity	No	No	Yes	Yes	Yes	No	Yes
Extension rate (nt/s)	75	>50	>80	—	—	30–40	—
Reverse transcriptase activity	Weak	Weak	—	—	—	Yes	—
Resulting DNA ends	3′A	3′A	>95% blunt	>95% blunt	—	3′A	Blunt
Molecular weight (kDa)	94	61	—	—	92	94	70

Source: Newton, C.R. and A. Graham, *PCR.* Oxford, U.K.: BIOS Scientific Publishers, Ltd., 1994, p. 13.

The $MgCl_2$ concentration affects the specificity of the PCR reaction. A too low concentration affects the final yield, whereas a too high concentration reduces the specificity of the reaction. Other components often present in DNA extraction buffer can affect the enzyme activity. Sodium dodecyl sulfate (SDS) at a concentration >0.01% inhibits the polymerase. The inhibition of SDS (0.01%) can be reversed by some nonionic detergents (0.5% (v/v) Tween 20, and NP 40.

The primer working concentration is generally of 0.5–1 µM. If the primer concentration is too high, primer dimerization can occur. The primer composition is very important. In most PCR applications, the primers are designed to be exactly complementary to the template DNA. The general rules for the primer design are a length of about 20–30 nucleotides, shorter primers can be used with success and primers longer than 30 do not increase the specificity of the binding the GC content should be about 50%; the 3′ ends should not be complementary, as primer dimerization will occur; the 3′ of the primer should be as homologous as possible; the 5′ can be modified to add a restriction site or a GC clamp, in this case, both primers should be equivalent in their melting temperatures. The number of the cycles can be increased to increase the amount of product recovered, but this will also increase nonspecific amplification.

Beside all these factors, some primer combinations will work very well, and others not. As so many factors affect the PCR reaction, it is very important to have a positive and negative control in a PCR reaction.

19.3.2.1.3 Contamination. As the PCR reaction is so sensitive, precautions have to be taken to avoid undesirable amplifications such as using DNA-free water and negative controls with every set of amplifications.

19.3.2.1.4 Steps Involved with PCR. Because DNA polymerase can add a nucleotide only onto a preexisting 3′-OH group, it needs a primer to which it can add the first nucleotide. This requirement makes it possible to delineate a specific region of template sequence to amplify. At the end of the PCR reaction, the specific sequence will

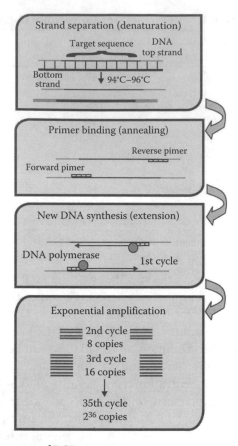

Figure 19.22 Different steps of PCR reaction.

be accumulated in billions of copies known as amplicons. There are three basic steps in a PCR reaction (Figure 19.22).

1. Denaturation (94°C)—It separates the dsDNA template into ssDNA.
2. Annealing (60°C)—Primers bind to the 3′ end of the template with base complementarity.
3. Extension (72°C)—Thermostable DNA polymerase (Taq polymerase) extends the primer complementary to the given template.

19.3.2.1.5 Components of the PCR Reaction. DNA template—It is the sample DNA that contains the target sequence. At the beginning of the reaction, high temperature is applied to the original dsDNA molecule to separate the strands from each other.

DNA polymerase—It is a type of enzyme that synthesizes new strands of DNA complementary to the target sequence. The first and most commonly used of these enzymes is *Taq* DNA polymerase (from *T. aquaticus*), whereas *Pfu* DNA polymerase (from *Pyrococcus furiosus*) is used widely because of its higher fidelity when copying DNA. Although these enzymes are subtly different, they both have two capabilities that make them suitable for PCR: (1) they can generate new strands of DNA using a DNA template and primers and (2) they are heat resistant.

Primers—It is a short piece of ssDNA that are complementary to the target sequence. The polymerase begins synthesizing new DNA from the end of the primer.

Nucleotides (dNTPs)—These are single units of the bases A, T, G, and C, which are essentially "building blocks" for new DNA strands.

Optimizing a PCR requires the adjustment of many parameters, including temperature, salt concentrations, cycle durations, and the primers. However, primers are the key factors. The specificity of the primer binding depends on the length of the primers and their nucleotide composition oligotides between 18 and 24 nucleotides long are highly specific under standard conditions where annealing temperature is close to the melting temperature. Increasing primer length results in an increase in the specificity. As a result, the *Tm* also increases. At too high *Tm*, the required annealing temperature will exceed the temperature for polymerase extension resulting in inaccurate priming.

Various protocols commonly used for gene cloning are summarized in the subsequent sections.

19.4 To Isolate Genomic DNA from *E. coli*

19.4.1 Principle

Genomic DNA and plasmid DNA can be isolated from cells by disrupting the cell wall and membrane and by exploiting the difference in properties of DNA, protein, and other constituents. Bacterial cell wall is destroyed by digestion with lysozyme, and plasma membrane is removed by a detergent, SDS. Proteins are precipitated using phenol. DNA in aqueous phase is collected by centrifugation, and any trace of proteins and phenols remaining in the aqueous phase are removed through extraction by chloroform–isoamylalcohol. DNA is then precipitated using chilled ethanol.

19.4.2 Precautions

1. Phenol should be handled with care as it is an explosive. Acid proof gloves must be worn while working with phenol. Its vapors should not be inhaled.
2. Micropipettes should be properly handled so as to avoid pipetting errors.
3. All equipment must be handled with proper care.
4. All glassware, plastic ware, and solution (except SDS, enzymes, and organic solvents) should be sterilized before use.

19.4.3 Materials

Luria broth; Tris–HCI 1M, pH 8.0; 0.5 M ethylenediaminetetraacetic acid (EDTA), pH 8; 10% SDS; 3M NaCl; TE buffer; lysis buffer (TNE buffer); TE saturated phenol; chloroform–isoamylalcohol; lysozyme solution; absolute alcohol (chilled); 70% ethanol; sterile distilled water (SDW); test tubes (25 mL); beaker (250 mL); conical flask (250 mL); inoculation needle; spirit lamp; Eppendorf tubes (1.5 mL): autoclaved;

microtips (all capacities): autoclave; pH paper; disposable gloves; acid proof gloves; laminar air flow chamber; BOD incubator/shaker; magnetic stirrer (with magnet); cooling centrifuge; water bath; vortex mixer; and micropipettes (variable): 100–1000 μL, 20–200 μL, and 0.5–10 μL.

19.4.4 Protocol

1. Inoculate a loopful of bacterial cells in 2 mL of LB, and grow the culture overnight on a rotary shaker at 150 rotations per minute (rpm) at 37°C in the dark.
2. The next day inoculate the actively growing culture in conical flasks containing 98 mL LB, and keep the flasks in same conditions as mentioned previously.
3. Take bacterial culture in 1.5 mL Eppendorf tube and centrifuge the tubes at 10,000 rpm for 2 min.
4. Discard the supernatant and resuspend the pellet in the residual medium using the finger flicking method or vortex.
5. Add another 1.5 mL of culture to the tube.
6. Repeat the steps 3 through 5 two to three times to obtain more number of cells from about 1.5 mL of culture in the same Eppendorf tube.
7. After discarding the supernatant (spent medium) finally, add 500 μL of TNE buffer to the tube and the vortex the tubes to break off any flocculated cells.
8. To the tube add 10 μL of lysozyme and 40 μL of SDS. Incubate at 37°C for 15 min.
9. To the lysate obtained earlier, add equal volumes (550 μL) of TE saturated phenol, and mix the contents of the tubes by mild inversions.
10. Centrifuge the mixture at 6000 rpm for 15 min.
11. Collect the upper aqueous phase (500 μL) in a fresh Eppendorf and add equal volumes of chloroform–isoamylalcohol to it.
12. Mix the contents and centrifuge at 6000 rpm for 10 min.
13. Collect about 350 μL of the upper aqueous phase in a fresh Eppendorf, and add two volumes of chilled absolute alcohol to it.
14. The DNA fibers can be seen at this stage following spooling by gentle mixing.
15. Centrifuging the tubes at 5000 rpm for 3–4 min to pellet out DNA and discard the alcohol.
16. Wash the pellet with 70% alcohol and then air-dry for about 20 min.
17. Resuspend the pellet in 50 μL of SDW.

19.4.5 Result

A white colored thick cloud of DNA is observed.

19.5 Rapid Method for Isolating Plasmid DNA from Yeast

19.5.1 Principle

Plasmid isolated by this method can be used to transform *E. coli*; however, the nucleic acid is not concentrated or pure enough to perform RE analysis.

19.5.2 Precautions

Gloves must be worn for these procedures, and the procedure should be performed in a fume hood. Phenol is VERY nasty stuff!

19.5.3 Materials

YS medium: 5 mL of yeast minimal medium supplemented with amino acids and vitamins as required.

$10 \times$ salts, 100 mL; 20% glucose, 100 mL; $1000 \times$ metals, 1 mL; $1000 \times$ vitamins, 1 mL; $100 \times$ supplements, 10 mL; distilled water to 1 L volume, pH adjusted to 6–6.5; and agar (for solid medium 2%) [*Note: Add after autoclaving other ingredients*].

$10 \times$ *YS medium (yeast salts (per liter)*: $CaCl_2.2H_2O$, 1 g; NaCl, 1 g; $MgSO_4.7H_2O$, 5 g; KH_2PO_4, 10 g; and $(NH_4)_2SO_4$, 50 g.

Trace metals $1000 \times$ stock solution: Boric acid, 50 mg; $CuSO_4.5H_2O$, 4 mg; KI, 10 mg; $FeCl_3.6H_2O$, 20 mg; $MnSO_4.H_2O$, 40 mg; $Na_2MoO_4 \cdot 2H_2O$, 20 mg; and $ZnSO_4.7H_2O$, 40 mg. Dissolve chemicals in order listed in 100 mL double distilled water. There will be a precipitate that will not dissolve. Shake bottle before using.

Vitamins $1000 \times$ stock solution: Biotin, 0.1 mg; calcium pantothenate, 20 mg; folic acid, 0.1 mg; inositol, 100 mg; niacin, 20 mg; ρ-aminobenzoic acid, 10 mg; pyridoxine HCl, 20 mg; riboflavin, 10 mg; and thiamine HCl, 20 mg. Put indicated amount into 100 mL flask, add 50 mL deionized water to dissolve (riboflavin will not entirely dissolve at room temperature). Make 5 mL aliquots and store frozen.

Supplement for minimal medium ($100 \times$ stock solution): Amino acids (mg/mL): L-alanine, 2; L-arginine, 2; L-asparagine, 2; L-aspartic acid, 10; L-cysteine, 2; glycine, 2; L-glutamic acid, 10; L-histidine (free base), 2; L-isoleucine, 3; L-leucine, 3; L-lysine, 3; L-methionine, 2; L-phenylalanine, 5; L-proline, 2; L-serine, 37.5; L-threonine, 20; L-tryptophane, 2; L-tyrosine, 3; and L-valine, 10. Bases (mg/mL): adenine sulfate, 2 and uracil, 1. Other supplements (mg/mL): cycloheximide (actidione), 1.4 and canavanine H_2SO_4, 6.

YNBD (yeast minimal) medium) (g/L): Difco yeast nitrogen base without amino acids, 7; dextrose, 2; $100 \times$ supplements, 10 mL, double distilled water to 1 liter; and agar (for solid medium 2%).

YEPD (yeast)medium (g/L): Difco peptone, 20; Difco yeast extract, 10; 20% dextrose, 100 mL (add after autoclaving other ingredients); double distilled water to 1 liter; and agar (for solid medium 2%).

This medium supports the growth of most auxotrophic strains without additional supplements and is used as a general growth medium. For storage slants, the amount of each ingredient (except for the agar) is doubled.

Spheroplast buffer (SB): 0.2 M Tris–HCl, pH 9.0; 1 M sorbitol; 0.1 M EDTA; and 0.1 M 2-mercaptoethanol.

Zymolase 60,000: 1 mg/mL in TEN buffer.

SET buffer: 20% sucrose; 50 mM Tris–HCl, pH 7.6; and 50 mM EDTA.

Na (sodium) acetate buffer: 3.0 M, pH 4.8.

RNase buffer: Pancreatic ribonuclease A, 10 mg/mL in 0.1 M Na-acetate, 0.3 mM EDTA, pH 4.8, preheated to 80°C for 10 min.

Isopropanol, ethanol 70%, and sterile double distilled water.

19.5.4 Protocol

1. Inoculate 5 mL of YM with a plasmid containing strain of yeast. Grow the culture to log phase (5×10^7 cells/mL) by incubating at 30°C with aeration for about 36 h.
2. Check the A_{600} value for the culture and adjust the cell density to give an A_{600} of 1.5.
3. Harvest 5 mL of yeast cells by centrifugation in a table top centrifuge.
4. Wash the cells twice with distilled water and pellet the cells as in step 3.
5. Resuspend the cell pellet in 5 mL of SB and add 5 μL of Zymolyase. Incubate for 1 h at 30°C with gentle agitation.
6. Collect cells as in step 3 and resuspend the cell pellets in 150 μL of SET buffer.
7. Add 5 μL of RNase buffer, 350 μL of lytic mixture (at room temperature), and vortex briefly. Place on ice for 10 min.
8. Add 150 μL sodium acetate buffer and invert the tube several times. Incubate for 30 min on ice.
9. Centrifuge the solution for 5 min at 4°C in a microcentrifuge. Pipette the supernatant to a clean microfuge tube (approximately 700 μL).
10. Add an equal volume of isopropanol, invert tubes several times and centrifuge for five more min at room temperature. Invert tubes and drain to remove the isopropanol.
11. Wash the DNA pellet with 1 mL 70% ethanol, and centrifuge for 5 min at room temperature. Vacuum dry the tubes for 10 min. Resuspend in 20 μL of sterile double distilled water.

19.5.5 Results and Observations

Use 5–10 μL of the "miniscreen" DNA for each *E. coli* transformation.

19.6 Isolation of Plasmid DNA from Yeast (Protocol 1)

19.6.1 Introduction

This method works well all the time, and restriction analysis can be performed. The method takes 2 h to complete the extraction process with a yield of 2–3 μg/μL DNA.

19.6.2 Precautions

Gloves must be worn for these procedures, and the procedure should be performed in a fume hood. Phenol is VERY nasty stuff!

19.6.3 Materials

As mentioned in the preceding text.

19.6.4 Protocol

1. Inoculate 5 mL of YM with a plasmid containing strain of yeast. Grow the culture to log phase (5×10^7 cells/mL) by incubating at 30°C with aeration for about 36 h.
2. Check the A_{600} value for the culture, and adjust the cell density to give an A_{600} of 1.5.
3. Harvest 5 mL of yeast cells by centrifugation in a table top centrifuge.
4. Wash the cells twice with distilled water and pellet the cells as in step 3.
5. 50–100 μL packed volume of yeast cells, resuspend in 200 μL TES (TE + 1% SDS).
6. Add 200 μL of glass beads, add 400 μL of phenol:chloroform:isoamyl alcohol (PCI; 25:24:1).
7. Vortex at max speed for 1 min, spin at 13,000 rpm at room temperature for 10 min, take the aqueous layer.
8. Add 1 mL absolute ethanol and incubate for 1 min in liquid nitrogen.
9. Spin at 13,000 rpm for 10 min at 4°C and recover the pellet.
10. Resuspend the pellet in 200 μL TE (add TE and mix with a pipette tip after 5 min), add 500 μL absolute ethanol and keep in liquid nitrogen for 1 min.
11. Spin at 13,000 rpm at 4°C for 20 min and keep the pellet.
12. Add 1 mL of 75% ethanol, invert three to four times, and spin at 13,000 at room temperature for 5 min (repeat the washing step twice).
13. Dry the pellet and add 50–100 μL of TE-RNase (TE+ RNAse A) and incubate at 50°C for 30–40 min.

19.6.5 Results and Observations

Check 2–3 μL on 0.8% agarose gel for purity of the DNA.

19.7 Isolation of Yeast Plasmid DNA (Protocol 2)

19.7.1 Introduction

This method is very effective to isolate plasmid DNA from yeast.

19.7.2 Precautions

Gloves must be worn for these procedures, and the procedure should be performed in a fume hood. Phenol is VERY nasty stuff!

19.7.3 Material

Composition of the culture media as given earlier.
Lysis buffer: stock solution per 100 mL

Lysis Buffer	Stock Solution	Per 100 mL
2% Triton X-100	10% TX-100	20 mL
1% SDS	10% SDS	10 mL
100 mM NaCl	2 M	5 mL
10 mM Tris, pH 8.0	1 M Tris, pH 8.0	1 mL
1 mM EDTA	100 mM EDTA	1 mL
Distilled water		63 mL

TE: 10 mM Tris, pH 8.0, 1 mM EDTA.

19.7.4 Protocol

1. Grow 10 mL of a yeast culture to saturation. A 50 mL Falcon tube works well for this.
2. Collect cells by centrifugation for 5 min at 13,000 rpm. Resuspend cells in 0.5 mL distilled water. Transfer cells to a 1.5 mL microcentrifuge tube and collect them by a 5 s centrifugation.
3. Decant the supernatant and briefly vortex the tube to resuspend the pellet in residual liquid.
4. Add 0.2 mL lysis solution. Add 0.2 mL phenol–chloroform–isoamyl alcohol (25:24:1; isoamyl alcohol is optional). Add 0.3 g acid-washed glass beads (0.45–0.52 mm).
5. Vortex for 3–4 min to disrupt cells. Be careful not to overdo the vortexing as DNA is very sensitive to mechanical damage. Add 0.2 mL TE.
6. Spin for 5 min in a microcentrifuge. Transfer the aqueous layer to a new tube. Add 1.0 mL 100% ethanol to precipitate the DNA. Invert the tube to mix.
7. Spin for 2 min in a microcentrifuge. Resuspend the pellet in 0.4 mL TE plus 30 µg RNase A. Incubate for 5–30 min at 37°C. Add 10 µL 4 M ammonium acetate plus 1.0 mL 100% ethanol. Invert tube to mix. Place on ice for 10 min.
8. Spin for 2 min in a microcentrifuge, air dry the pellet, and resuspend in 50 µL TE.

19.7.5 Results and Observations

Estimated concentration is 0.2–0.4 µg/µL. Concentrations can be checked by electrophoresis on a 0.8% agarose gel in tris-acetate EDTA buffer (TAE). High quality DNA will appear as a single band with a high MW ~22 kbp.

19.8 Isolation of *Drosophila* Genomic DNA

19.8.1 Introduction

Drosophila melanogaster is a small, common fly found near unripe and rotted fruit. It has been in use for over a century to study genetics and lends itself well to behavioral studies. It is one of the few organisms whose entire genome is known and many genes have been identified. Due to its small size, ease of culture, and short generation time, geneticists have been using Drosophila ever since.

19.8.2 Precautions

Gloves must be worn for these procedures, and the procedure should be performed in a fume hood. Phenol is VERY nasty stuff!

19.8.3 Materials

Drosophila flies, ethanol, phenol, Tris-HCl, and LiCl.

Buffer A: 100 mM Tris-HCl pH 7.5, 100 mM EDTA, 100 mM NaCl, and 0.5% SDS.

Tris-buffered phenol (with 500 mM tris-HCl buffer pH 8.0 and 0.1% 8-hydroxyquinoline).

LiCl/potassium acetate (KAc) solution: Mix 1 part 5 M KAc to 2 parts 6 M LiCl.

19.8.4 Protocol

19.8.4.1 Extraction

1. Obtain 30 healthy, freshly closed flies and freeze them at −80°C for 5 min, add 200 μL Buffer A (see subsequent sections) and grind with tissue grinder (blue, plastic), ~5 min (tissue grinder should be cleaned with detergent before use).
2. Add 200 μL Buffer A and continue to grind until only pieces of cuticle remain, incubate at 65°C for 30 min.
3. Add 800 μL 1:2.5 [5 M]KOAc:[6 M]LiCl, precipitate on ice for 10 min, centrifuge at 14,000 rpm for 15 min and then transfer supernatant to two Eppendorf tubes noting volume in each tube.
4. Add 700 μL Isopropanol per mL supernatant (7/10 volume), centrifuge at 14,000 rpm for 15 min, remove supernatant.
5. Wash with 1 mL cold ethanol, centrifuge at 14,000 rpm for 5 min and remove supernatant and resuspend in 100 μL TE.

19.8.4.2 Crude Purification and Precipitation/Resuspension

1. Wash with 150 μL phenol (tris-buffered), transfer aqueous (top) layer to new Eppendorf tube (be careful to avoid transferring material at the meniscus as unwanted protein (junk) will be concentrated here, between the aqueous and organic layers).

2. Wash with 150 µL PCI phenol:chloroform:isoamyl alcohol (25:24:1).
3. Transfer aqueous (top) layer to new Eppendorf tube as before.
4. Wash with 150 µL chloroform:isoamyl alcohol (24:1).
5. Transfer aqueous (top) layer to new Eppendorf tube.
6. Use 200 µL Micropipette to note exact volume of solution.
7. Add 1/10 volume [3 M] NaOAc (pH 5.2) and 2× volume 100% ethanol.
8. Mix well and chill at −80°C for 15 min.
9. Centrifuge at 14,000 rpm for 15 min.
10. Remove ethanol and add 1 mL cold 70% ethanol and mix.
11. Centrifuge at 14,000 rpm for 5 min, remove 70% ethanol and dry.
12. Resuspend in 100 µL TE buffer.

19.8.5 Results and Observation

Estimated concentration is 0.2–0.4 µg per µL. Concentrations can be checked by electrophoresis on a 0.8% agarose gel in TAE. High quality DNA will appear as a single band

19.9 To Determine λ_{max} Value of the Isolated DNA Sample

19.9.1 Principle

UV-spectrophotometer is used for enumeration of DNA. The DNA has maximum and minimum absorption at 260 and 234 nm, respectively. The λ_{max} is used to assess purity of the DNA and contamination present in the sample. The λ_{max} for different compounds that may be present in the sample as contaminant is given in Table 19.6.

19.9.2 Materials

Given DNA sample, UV spectrophotometer, quartz cuvette, distilled water, and wash bottle.

TABLE 19.6 λ_{max} for Different Compounds

S. No.	λ_{max}	Compound/Molecule
1.	230	EDTA, polysaccharide, and ethanol
2.	260	DNA and RNA
3.	270	Phenol
4.	280	Proteins
5.	320	Cell debris

19.9.3 Protocol

1. Make the dilution 1:20 of the DNA sample.
2. Take the absorbance from 200 to 300 nm.
3. Note the readings and plot the graph of absorbance against the wavelength and make the conclusions.

19.9.4 Results

Nucleic acids DNA/RNA absorbs maximally at 260 nm. Hence, maximum absorbance at 260 nm indicates presence of nucleic acids.

19.10 To Determine the Purity of the Isolated DNA Sample

19.10.1 Principle

Once the contamination has been removed from DNA preparation, the concentration of the DNA in the solution can be determined. The method most commonly used to determine DNA concentration involves the use of UV absorption spectroscopy. Just as all organic compounds have characteristic absorption spectra, the nitrogenous bases of dsDNA exhibit strong absorption maxima at a wavelength of 260 nm. At this wavelength, the extinction coefficient of DNA, $E_{260} = 20$, indicates that DNA at a concentration of 1 mg/mL will have an absorption $(A_{260}) = 20$.

As the relationship between DNA concentration and A_{260} is linear to an $A_{260} = 2$, the concentration of DNA in a solution can be determined. For example, $A_{260} = 0.5$ corresponds to 25 µg/mL, $A_{260} = 0.1$ corresponds to 5 µg/mL, and so on. Use of the conversion factor, 50 µg/mL = 1 A_{260} unit, enables the concentration of most DNA solutions to be determining easily. However, this relationship only applies to purify dsDNA with a G + C content of 50%. The presence of RNA, proteins, detergents, and organic solvents will also contribute to absorbance at this wavelength. Since the absorption maxima for DNA and protein are 260 nm and 280 nm, respectively, an approximate measure of the purity of the isolated DNA can be obtained by determining the A_{260}/A_{280} ratio. Pure *E. coli* DNA has $A_{260}/A_{280} = 1.95$. This ratio, however, is dependent on the overall base composition of the DNA and will vary with different organisms.

19.10.2 Materials

Given DNA sample, UV spectrophotometer, Quartz cuvette, distilled water, and wash bottle.

19.10.3 Protocol

1. Make the dilution 1.20 of the DNA sample.
2. Take the optical density (OD) at 260 nm and 280 nm.

19.10.4 Result

Calculate the ratio of A 260 nm/A 280 nm.

19.11 Determine the Quantity of DNA in the Sample

19.11.1 Principle

UV-spectrophotometer is used for enumeration of DNA. The DNA concentration can be calculated by the formula DNA conc. (μg/mL) = $A_{260}/\in_{260}$.

Where, A_{260} is absorbance at 260 nm and ε_{260} is the DNA extinction coefficient. For dsDNA, the ε_{260} is 0.02 (μg/mL)$^{-1}$ cm $^{-1}$. Thus, an absorbance of 1 at 260 nm gives a DNA concentration of 50 μg mL^{-1} (1/0.2 = 50), the ε_{260} is 0.027 (μg/mL)$^{-1}$ cm^{-1} for ssDNA at absorbance of 1 (1/0.027 = 37).

19.11.2 Materials

Given DNA sample, UV spectrophotometer, Quartz cuvette, distilled water, and wash bottle.

19.11.3 Protocol

1. Make the dilution 1.20 of the DNA sample.
2. Take the OD at 260 nm.

19.11.4 Result

Calculate the quantity of the DNA by the formula
DNA conc. (g/mL) = OD 260 nm × 50 × dilution factor (i.e., 20).

19.12 To Perform Agarose Gel Electrophoresis for the Given DNA Sample

19.12.1 Principle

19.12.1.1 Electrophoresis
When a molecule is placed in an electric field, it will migrate to appropriate electrode with a velocity or free electrophoretic mobility (m_0), which is calculated by

$$m_0 = \frac{\in \cdot q}{d \cdot 6\pi\eta}$$

where
   $\in$ is the potential difference between electrodes measured (V)
   q is the net charge of the molecule
   d is the net distance between the electrodes (cm)
   η is the viscosity of the solution
   R is the stocks radius of the molecule
   $\in/d$ is the field strength

19.12.1.2 Agarose Gel

Agarose is the polysaccharide consisting of basic repeated agarobiose units of 1,3 linked β-D-galactopyranose and 1,4-linked-3,6 anhydro-α-L-galactopyranose. These units form long chains of approximately 400 repeats having MW of about 120,000 Da. The migration rate of DNA through agarose gel during electrophoresis depends upon the size of the DNA molecule, the concentration of agarose, the voltage applied, the conformation of DNA, and buffer used for the electrophoresis. Mobility of the DNA depends on the agarose gel concentration of voltage used. A DNA fragment of the given size migrates at different rates in gel containing different concentrations of agarose and different voltage applied. Greater the agarose concentration slower will be the migration rate of DNA and vice-versa.

19.12.1.3 Electrophoresis Buffer

Generally buffers used in agarose gel electrophoresis are TAE buffer, tris-borate EDTA (TBE) buffer, and tris-phosphate EDTA buffer. For general purpose, TAE is used, which has low buffering capacity and is suitable for small gel run, TBE buffer has a very high buffering capacity and gives superior results as compared with TAE buffer. For sharper DNA bands, TBE buffer is used. TPE buffer has still higher buffering capacity and therefore takes much long time to run the gel as compared to other buffers.

19.12.1.4 Gel Size

In general, small gels are used but for better DNA band resolution, the gel size should be of 20 × 15 cm. The thickness of the gel should be 4 mm. The size of the well should be between 0.5 cm and 1 cm, and width 1–2 cm, and the depth of the well should be 3.0–3.5 mm.

19.12.1.5 Sample Concentration

The DNA concentration should not exceed 10 μg/well.

19.12.1.6 Sample Load Solution

The DNA samples are loaded in the well after mixing with loading dye solution. This serves three functions:

1. It is used to terminate enzymatic reactions before electrophoresis.
2. It provides the density for loading the samples into the wells.
3. Helps to monitor the progress of electrophoresis.

Loading dye solution contains EDTA, urea, and glycerol. EDTA stops the enzymatic reactions, urea denatures proteins without affecting DNA mobility, and glycerol provides density to the samples. Ficoll 400 can be added to increase the sharpness of the bands.

19.12.1.7 Gel Staining

Agarose gels are usually stained with EtBr, which is both intercalating and fluorescent dye. The best results are obtained by adding EtBr into the gel at concentration 0.5 µg/mL. This allows direct observation during electrophoresis.

19.12.2 Materials

Agarose, TAE buffer, EtBr, electrophoresis unit, and UV-transilluminator.

19.12.3 Protocol

1. Prepare (1%) agarose by dissolving it in $1 \times$ TAE buffer.
2. Boil to get a clear homogenous solution.
3. Cool upto 55°C.
4. Add EtBr to get 0.5 µg/mL.
5. Mix properly.
6. Pour the gel in the gel mould and place the comb.
7. Allow it to settle at least for 45 min.
8. Remove comb carefully.
9. Pour $1 \times$ TAE buffer to the tank, so that the gel is completely submerged in the buffer.
10. Connect the electrodes, that is, black to the negative (black) terminal and red to the positive (red) terminal.
11. Add one-fifth volume of BPB to the DNA sample, mix by pipetting, and load it carefully in the wells. Also load a suitable DNA MW marker in the last lane.
12. Run at 100 V for initial 10 min and then at 50 V, till the sample reaches the end of the gel.
13. View the gel on an UV-transilluminator.

19.12.4 Results

DNA appears as orange fluorescent bands.

19.13 To Isolate Plasmid DNA from Given Bacterial Culture

19.13.1 Principle

Two methods are routinely used in the laboratory for the isolation of plasmid DNA:

1. Isolation of plasmid DNA on a preparative scale and purification by column chromatography or CsCl density gradient centrifugation.
2. The so-called quick prep on an analytical scale is useful when only the presence or absence of a specific plasmid is to be determined. The basic steps are shown in Figure 19.23.

19.13.1.1 Alkaline Lysis

19.13.1.1.1 Principle. In this method, the cells are lysed with sodium hydroxide (NaOH) and SDS. This highly alkaline solution gave rise to the name of this technique. SDS, a detergent, create holes in the cell membranes. NaOH loosens the cell walls and releases the plasmid DNA and sheared cellular DNA. NaOH denatures the DNA. Cellular DNA becomes linearized and the strands are separated. Plasmid DNA is circular and remains topologically constrained. Addition of KAc results in renaturation of the circular DNA. However, sheared cellular DNA remains denatured as ssDNA. The ssDNA is precipitated, since large ssDNA molecules are insoluble at high salt concentrations. Furthermore, addition of sodium acetate to the SDS forms KDS, which is insoluble. This allows easy removal of the SDS from plasmid DNA. The contaminants (cell debris, KDS, and cellular ssDNA) are removed by centrifugation. Plasmid DNA is retained in the supernatant. For further purification, the plasmid DNA is precipitated by alcohol precipitation (ethanol or isopropanol) and a salt (such as ammonium acetate, lithium chloride, sodium chloride, or sodium acetate). Addition of a salt masks the negative charges on DNA and allows DNA to precipitate.

19.13.1.1.2 Materials. Microcentrifuge, microfuge tubes, *E. coli* DH5α cells containing plasmid pUC19 or any other strain possessing plasmid DNA, yeast extract, tryptone, NaCl, NaOH, sodium acetate, Tris base, HCl, glucose, lysozyme, SDS, and EDTA.

19.13.1.1.3 Protocol
1. Grow *E. coli* DH5α cells overnight in LB medium at 37°C.
2. Transfer 1 mL of the aforementioned culture in an Eppendorf tube and collect cell pellet after centrifugation at 5000 rpm for 5 min in a microfuge.
3. Resuspend bacterial pellet in a total of 1 mL ice-cooled solution I (50 mM). Pipette up and down or vortex as necessary to fully resuspend the bacteria.
4. Add 2 mL solution II at room temperature to the suspension. Mix thoroughly by repeated gentle inversion. *Do not vortex.*
5. Add 1.5 mL ice-cold solution III to the lysate. Mix thoroughly by repeated gentle inversion. *Do not vortex.*
6. Centrifuge at 15,500*g* for 30 min at 4°C.

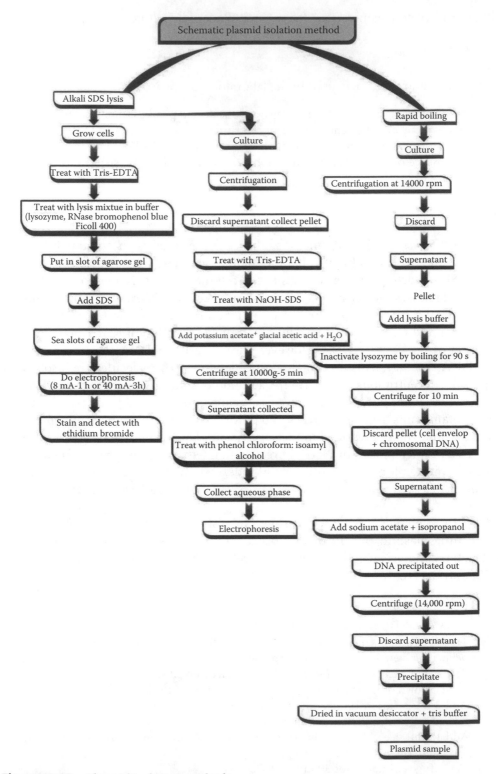

Figure 19.23 Plasmid isolation methods.

7. Recover resulting supernatant.
8. Add 2.5 volumes of chilled isopropanol to precipitate the plasmid DNA. Mix thoroughly by repeated gentle inversion. *Do not vortex.*
9. Centrifuge at 15,500 g for 30 min at 4°C.
10. Discard the supernatant. The pellet is plasmid DNA.
11. Rinse the pellet in ice-cold 70% ethanol and air-dry for about 10 min to allow the ethanol to evaporate.
12. Add double distilled H_2O or TE to dissolve the pellet. After addition of 2 µL RNase A (10 mg/mL), incubate the mixture for 20 min at room temperature to remove RNA.

19.13.1.1.4 Result. Determine the purity of DNA and perform electrophoresis.

19.13.1.2 Rapid Boiling Method: For Small Plasmids in E. coli

19.13.1.2.1 Principle. Boiling lysis plasmid is a quick procedure to perform and is especially suitable for screening large numbers of colonies for plasmid content. The method is suitable for the isolation of plasmids ranging in size from 3.2 to 39.2 kb. The method has been used for isolating plasmids from *E. coli* and *Neisseria* sp. The bacteria are lysed by treatment with lysozyme, detergent, and heat. The chromosomal DNA denatures and remains associated with the bacterial membrane instead of renaturing. It is then removed by centrifugation. The plasmid DNA remains in the supernatant from which it is then precipitated with isopropanol. The quality of the isolated plasmid DNA is lower as compared to the one isolated by alkaline lysis. This is due to the presence of chromosomal DNA.

19.13.1.2.2 Materials. Eppendorf tubes, sucrose, Triton X-100, EDTA, Tris-HCl, lysozyme, $ZnCl_2$, isopropanol, ammonium acetate, and ethanol.

19.13.1.2.3 Protocol
1. Centrifuge 1.5 mL of 24 h old culture in Eppendorf tube.
2. Resuspend pellet in 200 µL of STET buffer (pH 8.0) containing 10 µL of lysozyme (20 mg/mL, freshly dissolved in H_2O) and 20 µL $ZnCl_2$ (1% in H_2O).
3. Incubate at about 100°C for 45–55 s then cool on ice.
4. Centrifuge for 5 min and add supernatant to Eppendorf tube containing 480 µL of IS mix. Incubate at room temperature for 20–30 min.
5. Centrifuge for 5 min, wash DNA pellet with 70% cold ethanol twice, and dry in a vacuum desiccator.

19.13.1.2.4 Result. Resuspend pellet in 20 µL of TE buffer before using for agarose gel electrophoresis.

19.13.1.3 Plasmid Isolation Using Concert High Purity Plasmid Purification Systems

19.13.1.3.1 Protocol
1. Inoculate *E. coli* DH5α cells with pBR322 plasmid into 30 mL of Luria broth media.
2. Grow the cells overnight to obtain turbid cultures.
3. Add equilibrium buffer to the columns and allow the columns to drain by gravity.

4. Centrifuge 15 mL of culture broth 5000 g. Discard the supernatant.
5. Wash the pellet with cell suspension buffer.
6. Add cell lysis solution and incubate for 5 min.
7. Mix neutralization buffer to the sample and centrifuge at 12,000 g for 10 min at room temperature.
8. Transfer the supernatant to the columns and allow to drain through. Wash the columns with wash buffer twice, elute the DNA in the column by adding elution buffer. Mix absolute isopropanol and centrifuge the sample at 12,000 g for 30 min at 4°C to precipitate the plasmid DNA.
9. Wash the plasmid pellet with 70% ethanol and centrifuge at 12,000 g for 5 min at 4°C.
10. Air dry the pellet for 10 min.
11. Dissolve the plasmid pellet in 50 μL of tris-EDTA buffer.
12. Add sample buffer, electrophorese the sample on a 0.8% agarose gel with 1 mg/mL EtBr for 1.5 h at 100 V. Add linear DNA ladder and a supercoiled DNA ladder into the agarose gel for electrophoresis.

19.13.1.3.2 Result. After completion of the run, visualize the DNA.

19.14 Electrophoresis of DNA: Linear, Circular, and Super Coiled

19.14.1 Principle

Several techniques have been developed for the analysis of nucleic acids (both RNA and DNA), but the agarose gel electrophoresis is an ideal technique for analysis of nucleic acids. The mobility of nucleic acids in agarose gel is influenced by the agarose concentration and the molecular size and molecular conformation of the nucleic acid. Agarose concentrations of 0.3%–2% are most effective for nucleic acid separation. The lower limit of 0.3% agarose in the gel allows analysis of linear dsDNA within the range of 5–60 kb (up to $150 \times$ in MW). Gels with an agarose concentration of 0.8% can separate DNA in the range of 0.5–10 kb, and 2% agarose gels are used to separate smaller DNA fragments (0.1–3 kb). For most routine analysis of RE fragments, agarose concentrations in the range of 0.7%–1% are most appropriate. Nucleic acids migrate in agarose medium at a rate that is inversely proportional to their size (kb or MWs). Linear relationship exists between mobility and logarithm of kb (or MW) of a DNA fragment. The DNAs conformations most frequently encountered are superhelical circular (form-I), nicked circular (form-II), and linear (form-III). The small, compact, supercoiled form-I molecules usually have the greatest mobility, followed by the rod-like, linear form-III molecules. The extended circular form-II molecules migrate more slowly. The relative electrophoretic mobility of the three forms of DNA also depends on the ionic strength. Agarose gel electrophoresis is a valuable tool for separation and characterization of nucleic acids for plasmid mapping and DNA recombination experiments. The method is simple, rapid, and inexpensive; ease of operation, sensitive staining procedures, high resolution and a wide range of MW s ($0.6–100 \times 10^6$) can be analyzed.

19.14.2 Materials

19.14.2.1 Reagents

Tris-acetate buffer (TAE) stock solution (5 ×): A 5-fold concentrated TAE buffer stock solution is prepared, which contains Tris-base 24.2 g, Glacial acetic acid 531 mL, and 0.5 M EDTA 10 mL. Adjust to pH 8 and add water to make 1 liter. Dilute five times before use to obtain the working buffer (1 × buffer).

1% agarose in 1 × TAE buffer: Dissolve 0.75 g agarose in 75 mL of 1 × TAE buffer (working TAE buffer) by boiling and maintain at 50°C until use.

Gel loading solution: 10% glycerol and 0.025% bromophenol blue in water.

EtBr: Dissolve 10 mg/mL EtBr in 1 × TAE buffer.

Plasmid DNA preparation: Add 10 µL loading solution to 20 µL plasmid DNA preparation and mix properly.

Standard DNA marker: Add 10 µL of gel loading solution in 20 µL of the λ *Hind*III DNA, digest, and mix them well.

19.14.3 Protocol

1. Take a clean, dry, gel casting plate and make a gel mould using an adhesive tape along the sides of the plate to prevent running off of the material to be poured on the plate.
2. Pour 50 mL of 1% agarose solution kept at 50°C on to the casting plate. Immediately place the supplied comb about 1 cm from one end of the plate ensuring that teeth of the comb do not touch the glass plate. Wait till a firm layer of gel is formed.
3. Remove the comb and the tape surrounding the plate carefully, and transfer the gel plate to the electrophoretic tank in such a way that wells are toward the cathode.
4. Pour 1 × TAE buffer into the tank until the gel is completely submerged. Connect the electrodes to the power supply.
5. Load the plasmid DNA preparation and the standard DNA markers into separate wells with the help of a micropipette or a syringe.
6. Turn "ON" the power supply and run at 100 V (10–15 mA). Monitor the progress of fast running tracking dye (bromophenol blue) during electrophoresis.
7. Turn "OFF" power supply when the tracking dye has reached the opposite edge of the gel.
8. Transfer the gel from casting plate onto a UV-transparent thick plastic sheet and place it in a staining tray containing EtBr solution. Stain for 20–30 min.
9. The gel is then placed in water for 15–20 min for destaining.
10. Now place the gel along with UV-transparent sheet on a UV-transilluminator and view the gel in UV light for presence of orange colored bands. For a permanent record, the gel may be photographed (caution: UV-light is extremely injurious for eyes. Wear UV protection glasses while viewing gel).
11. Measure the distance moved by each band from edge of the loading well. Draw a graph between $\log_{10}$ of MW of standard DNA marks vs. the distance travelled by each of them.

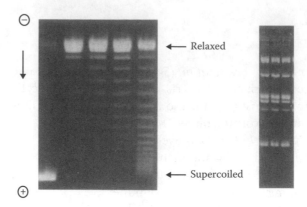

Figure 19.24   Agarose gel electrophoretic pattern of circular DNA in 1% agarose.

19.14.4 Result

From the distance travelled by the supplied plasmid DNA preparation, determine its MW using the calibration curve prepared in step II (Figure 19.24).

19.15 To Perform Restriction Digestion of the Given DNA Sample

19.15.1 Principle

Gene cloning requires DNA molecule to be cut in a very precise manner for a specified sequence in both genomic and vector DNA. The vector (plasmid) is a covalently closed circular DNA molecule, which should be cut at a single position to open up the circle for insertion of the desired DNA fragment.

A unit of RE activity is defined as the amount of enzyme needed to digest 1 μg of DNA completely in 60 min under appropriate reaction conditions. Three types of RE are used—RE I, II, and III. Out of which RE II are used most commonly in molecular biology/genetic engineering experiments. To stop restriction digestion reaction, the method of choice depends on the samples used. If it is just for running and confirmation, the reaction can be stopped by addition of gel loading buffer. If additional manipulations are required, the reaction can be stopped by elevating temperature or by adding phenol or EDTA.

Normally end nucleases are supplied in glycerol that contains storage buffers, which are different for each enzyme. They should be stored at –20°C. Most of the end nucleases will function at pH 7.4. But different enzymes vary in their pH requirements.

19.15.2 Materials

DNA, RE and its buffer, agarose, TAE buffer, EtBr, electrophoresis unit, and UV-transilluminator.

19.15.3 Protocol

1. Take 5–10 μg of DNA dissolved in 10 μL of SDW.
2. Add 5 μL of appropriate 10 × RE buffer.
3. Add a threefold excess (75–100 units) of appropriate RE (approximately 5–10 μL). Add SDW up to a final volume of 50 μL.
4. Incubate the digestion mixture for overnight (12–16 h) at the temperature recommended by the manufacturer.
5. Run the samples under proper electric current.

19.15.3.1 Preparation of 1% Agarose Gel and Electrophoresis
1. Prepare 1 × TAE.
2. Take 50 mL of TAE and add 0.5 g of agarose. Boil to dissolve agarose. Cool to approximately 60°C.
3. Add EtBr.
4. Place the combs of electrophoresis set such that the combs are 2 cm away from the cathode.
5. Pour the agarose solution slowly.
6. Once the gel has solidified, pour 1 × TAE buffer into the gel tank.
7. Lift the comb gently, keeping the wells intact.
8. After loading the ligated and control samples, put on the power supply
9. After the run is complete, switch off the power supply and disconnect the cords.
10. Remove the buffer into a flask.
11. Remove the gel slowly from the tank and place it on the UV-transilluminator.

19.15.4 Result

DNA can be seen as orange band under UV.

Note: If RE digestion is complete, then no DNA band will be seen at the position of the undigested control DNA band. Rather some smaller DNA fragments will be seen on gel as separate bands or a smear.

19.16 To Demonstrate Ligation

19.16.1 Principle

Ligation of the DNA means the joining of the two or more DNA fragments with each other. Ligation of DNA fragments is done by DNA ligases, which catalyze the formation of phosphodiester bonds between 5′ phosphate and 3′ hydroxyl terminus of dsDNA. Two types of ligases are in use—*E. coli* ligase and T_4 DNA ligase. T_4 DNA ligase is the product of gene 30 of phage T_4. Using ATP as a cofactor, T_4 DNA ligase catalyses the repair of ss nicks in dsDNA and joins dsDNA restricted fragments having either blunt or cohesive ends.

Ligation is carried out at 12°C to 15°C. High temperature makes it difficult for the ends to anneal, whereas lower temperature diminishes ligase activity. Blunt-end ligation usually requires 10–100 times more enzymes than cohesive end ligations.

19.16.2 Materials

T4 DNA ligase, DNA sample, ligase buffer (2×), agarose, gel loading buffer, gel running buffer (TAE 50×), EtBr , electrophoresis unit, and UV-transilluminator.

19.16.3 Protocol

1. Thaw the ligase buffer and DNA EcoRI digest. Place on ice.
2. Add 10 μL of EcoRI digest of T4 DNA ligase. Place on ice.
3. Add 10 μL of 2× ligase assay buffer.
4. Incubate for 2 h at 16°C in a water bath.
5. Add 5 μL of gel loading buffer.
6. Add 5 mL of gel loading buffer to 10 μL of DNA digest (control).
7. Prepare 1% agarose gel and load the samples in the wells.
8. After loading the ligation and control samples, put on the power supply.
9. After the run is complete, switch of the power supply and disconnect the cords.
10. Remove the buffer into a flask.
11. Remove the gel slowly from the tank and place it on the UV-transilluminator.

19.16.4 Result

DNA can be seen as orange band under UV.
Lanes
Lane 1—λ/*Eco*RI digest
Lane 2—λ/*Eco*RI digest
After ligation
Ligation of λ/*Eco*RI fragments

19.17 Elution of DNA from Agarose Gel

19.17.1 Principle

DNA is isolated from gel based on its binding properties to silica (glass) at different salt concentrations. At high salt concentrations, sodium cations break hydrogen bonds between the hydrogens in water and the negatively charged oxygen ions in silica. Sodium serves as a cation bridge and attracts the negatively charged oxygen in the phosphate backbone of the DNA molecule (Figure 19.25).

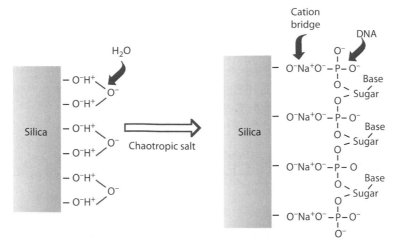

Figure 19.25 DNA binds to silica in the presence of high salt concentration.

Once the salt has been removed by rinsing the DNA–silica pellet several times, the DNA detaches from the silica and can be removed in solution. The DNA is isolated in TE or water and does not need to be purified further.

Glass milk is made from silica 325 mesh, a powdered pottery glaze. Do not use mesh of other sizes, otherwise the glass milk will contain particles that are too large or too fine for efficient purification of DNA fragments from agarose gel. Such fragments can then be used for cloning, PCR, or other molecular biology purposes.

19.17.2 Materials

DNA sample, NaI solution, sodium sulfite, ethanol, tris-HCl, EDTA, Eppendorf tube, vortex, centrifuge, and glass milk.

19.17.3 Protocol

1. Weigh the gel slice in a 1.5 mL microcentrifuge tube.
2. Add 3 volumes of NaI solution per gram of gel.
3. Incubate at 45°C–55°C to melt the gel slice (mix the contents of the tube occasionally until the entire slice is melted—approximately 5 min).
4. Vortex glass milk to resuspend.
5. Add 1 μL glass milk per 1 μg DNA and mix the tube well.
6. Incubate for 5 min at room temperature, mixing occasionally to keep the glass milk in suspension.
7. Quick spin the tube in a microfuge to pellet the glass milk (approximately 5 s at full speed).
8. Wash pellet three times with 500 μL wash solution, spinning after each wash step as in step 7.

Note: The glass pellet in the wash solution is more difficult to resuspend than in the NaI solution. If pellet is resistant to resuspension, swirl the pipette tip in the pellet while pipetting up and down to get the pellet into solution.

9. Dry the pellet at room temperature for 5–10 min, or for a few minutes at 55°C, with the cap open.
10. To elute the DNA, resuspend the glass pellet in an equal volume of TE or water.
11. Spin tube for 30 s at top speed in a microfuge to pellet glass milk.
12. Carefully remove supernatant with eluted DNA into a fresh tube, making sure to avoid transfer of glass pellet.
13. Run aliquot on gel to see proportion recovered, which varies with the size of the fragment. Larger fragments (>54 to 5 kb) and very small (<0.3 kb) give poor yields.

Notes

1. Borate in TBE (and similar) buffers inhibits DNA binding to glass particles. If purifying DNA from TBE gels, weigh the gel slice and add 1/10 volume of 1 M sodium phosphate, pH 6.5, and 1 M mannitol to a final concentration of 0.1 M. Continue with the protocol as mentioned earlier (include the volume of the sodium phosphate and mannitol in the volume of the gel slice when determining how much NaI solution to add).
2. This procedure is probably not appropriate where a high proportion of fully intact, very large DNA is required.

19.18 Purification of Phage λ DNA

19.18.1 Principle

The host for lambda is *E. coli* C600 and its subsequent variants. Traditionally, the *E. coli* host for lambda is grown on NZY medium. This medium is not as rich as LB (and many other) medium, and *E. coli* growth is slower, making it even less likely that *E. coli* will overgrow the phage.

Furthermore, the lambda receptor on the bacterial cell outer surface is part of the maltose-usage pathway. To induce the receptor to high levels (10^2–10^3 receptors per cell), maltose is added to a final concentration of 0.2% in the medium.

When growing lambda on plates to prepare DNA (as opposed to picking plaques or titering), agarose is substituted for agar. Agarose is much more expensive than agar, but does not contain impurities that inhibit enzymes. λ DNA prepared from agar plates may not be digestible by REs because of the presence of inhibitors that have leeched from the agar.

19.18.2 Materials

E. coli C600 ($\lambda cI_{857}S_7$), Tris salt, EDTA, SDS, ethyl alcohol, chloroform, $MgSO_4 \cdot 7H_2O$, DNase I (free of RNase) RNase, casein hydrolysate, yeast extract, NaCl, maltose, gelatin, and agarose.

19.18.3 Protocol

19.18.3.1 Preparation of E. coli Cells

1. Inoculate single colony of actively growing *E. coli* culture into 5 mL of NZY medium containing 0.2% maltose.
2. Place the tube on a shaker incubator at 37°C for approximately 4 h (this time will be much longer if the colony is from an old plate or has been stored at 4°C or frozen).
3. Under aseptic conditions, transfer the culture to a sterile 15 mL conical flask and incubate the flask in a shaker incubator at 37°C. Pellet the *E. coli* cells by centrifugation in an SS34 rotor with adaptors in an RC-5B centrifuge or equivalent. Pour off the medium gently.
4. Resuspend the bacterial pellet in 4–6 mL of 0.01 M $MgSO_4$ by pipetting up and down (under sterile condition), not by vortexing.
5. Adjust the concentration of bacteria to an OD of 2.0 at 600 nm wavelength. *E. coli* cells prepared in this way can be stored at room temperature and used for up to one week.
6. To resuspend, gently pipette the *E. coli* cells. Do not vortex. Cells prepared in this manner are employed at values of 10 μL per small plate or 20 μL per large plate.

19.18.3.2 Isolation/Preparation of λ Phage Stock

1. Pick a single, well-isolated plaque (a lambda plaque contains phage and lysed bacteria and appears relatively clear against the bacterial lawn) using sterile tooth pick and add to the 10 μL of *E. coli* cells suspended in 0.2 mL SM buffer.
2. Place the tubes at 37°C on wheel for 20–30 min. The 20 min incubation allows time for attachment reaction to occur.
3. Add 3 mL of top agar to the mixture of phage and bacteria.
4. Vortex for 10 s.
5. Pour on to an agar plate. Rock the plate by hand to reach an even distribution of top agar.
6. After the top agar hardens (approximately 10 min), invert the plates and incubate at 37°C overnight.
7. Next day, at least one plate should have well isolated plaques. If there are too many plaques, make further dilutions.
8. Punch out a plaque with the wide mouth end of a sterilized Pasteur pipette. The agar stays in the pipette and can be expelled in to a sterile Eppendorf tube by a wrist flick.
9. Add 0.5 mL SM buffer and two drops of chloroform.
10. To elute the phage, allow the Eppendorf tube to stand for 2–4 h at room temperature or at 4°C over night. The chloroform will evaporate, and phage stock is prepared from the isolated phage.

19.18.3.3 Phage Cultivation

1. Prepare NZY agar plates.
2. Add 0.2 mL SM buffer and 10 μL *E. coli* cells. Place on a shaker for 20–30 min.
3. Add 3 mL of NZY top agar and vortex for 10 s.
4. Pour onto NZY agar plates. After the top agar hardens, invert the plates and incubate them at 37°C overnight.
5. Next day, the plates infected and disrupted *E. coli* lawn.

6. Add 5 mL sterile SM buffer to cover the surface of the plates, rotate gently or rock the plates at room temperature for 1–2 h.
7. Harvest the buffer using a sterile pipette and transfer in to a sterile 15 mL conical tube. The milky appearance is caused by the presence of bacterial debris.
8. Add 0.2 mL of chloroform and vortex for 10 s. Spin the tube in an SS34 rotor with adapters; RC-5B centrifuge or equivalent, at 4000 rpm for 8 min.
9. Decant the clear yellow supernatant in to a fresh, sterile 15 mL conical tube.
10. Add a drop of chloroform (to stop the growth of bacteria) and store at 4°C. These stocks are stable for several months at 4°C. Frozen (–70°C) stocks can be made by mixing 0.1 mL dimethyl sulfoxide with 1.4 mL of lambda phage stock.

19.18.3.4 Preparation of Phage Titer from Single Plaque
1. Dilute the lambda stocks tenfold (10^{-2}, 10^{-3}, etc.) in SM buffer supplemented with gelatin.
2. Equilibrate the NZY agar plates at 37°C, melt top agar aliquots and maintained at 50°C on a water bath.
3. Add 0.2 mL of SM buffer and 10 µL of E. coli in small tubes. Add appropriate amounts (e.g., 1, 10, and 100 µL) of each phage dilution to the bacteria culture.
4. Place the tubes in a shaker incubator at 37°C for 20–30 min.
5. Add 3 mL of top agar to the phage and bacteria mix.
6. Vortex the tube for 10 s and pour the contents onto a warm agar plate.
7. Distribute the top agar evenly on the plate while rocking (rotating and swirling) by hand.
8. Allow the top agar to harden at room temperature (approximately 10 min). Invert the plates and incubate at 37°C overnight.
9. The next morning, look at the plaques. They are usually of a reasonable size to count. However, if the plaques are very small, leave the plates at 37°C and check the size of the plaques hourly.
10. Count the plaques on plates where the number is between 50 and 300. A minimum of 100 is needed for reasonable statistical reliability; a plate with over 200 plaques may yield an incorrectly low number, as plaques overlap. Calculate the titer of the stocks. Phage titer is now derived from a single plaque.

19.18.3.5 Preparation of λ DNA
1. Equilibrate the large NZY agarose plates at 37°C.
2. Maintain NZY top agarose at 50°C. Place 0.2 µL of SM buffer in a small tube and add 20 µL of E coli (double the amount for a small plate). The goal is to achieve confluent lysis, maximum phage production, without blowing away the E. coli prematurely.
3. Add the lambda to the bacteria in SM buffer and place the tube on the 37°C shaker incubator for 20–30 min.
4. Add 5 mL (an increased amount for the large plates) of top agarose, vortex for 10 s, and pour onto the warm 150 mm plate. Rotate (rock and swirl) the plate by hand to achieve an even distribution of top agarose.
5. Allow the top agarose to harden (approximately 10 min) and invert the plate at 37°C overnight. The next morning, the plates should have confluent lysis. Confluently lysed large plates will yield adequate amount of DNA even for poorly growing lambda.
6. Add 8 mL of SM buffer to each plate. Make sure the buffer covers the entire plate. Sterile conditions are no longer necessary, but neatness always counts. Rotate gently at room temperature for 1–2 h. The phage will elute into the buffer.

7. Decant (pipette) the buffer into a 15 mL conical tube (or equivalent); 5–6 mL/plate. Centrifuge the tube in an SS34 rotor with adapters, RC-5B centrifuge or equivalent for 8 min at 4000 rpm.

8. The debris will pellet, and most of the phage will remain in the supernatant. Decant the supernatant (9–12 mL) to an Oak Ridge centrifuge tube.

9. Add DNase I to digest contaminating *E. coli* DNA. λ DNA, inside the phage particle is protected from digestion. DNase activity requires magnesium ions, which are provided in SM buffer.

10. Add 0.1 mL of DNase I (100 μg per mL; RNase-free) and mix gently. Place the tube at 37°C for 1–2 h (no attempt is made to remove RNA, and in fact, RNA is used as a carrier for the DNA).

11. Add 0.5 mL of 0.5 M EDTA, pH 8, and 0.5 mL of 10% SDS and mix gently. DNase I is inactivated and phage particles are disrupted. The removal of magnesium ions by the EDTA also causes the polysomes and ribosomes to fall apart, which helps in phage DNA purification.

12. Add 10 mL of phenol equilibrated with TF buffer. Vortex hard for 20 s to remove residual protein and SDS.

13. Centrifuge in an SS34 rotor, RC-5B centrifuge or equivalent for 10 min at 10,000 rpm at 10°C. Decant the upper, aqueous phase to a fresh Oak Ridge centrifuge tube.

14. Add 10 mL of chloroform, vortex hard for 20 s centrifuge in an SS34, RC-5B centrifuge or equivalent for 10 min at 10,000 rpm at 10°C. Decant the upper, aqueous phase to a fresh Oak Ridge centrifuge tube.

19.18.3.6 DNA Precipitation

1. Remove traces of phenol and chloroform by alcohol precipitation. Add two-volumes of cold (95%–100%) ethanol and mix by vortex for 10 s.

2. Centrifuge in an SS34 rotor, RC-5B centrifuge or equivalent for at least 30 min at 12,000 rpm at 10°C.

3. Gently decant the supernatant (pour the alcohol off carefully while watching that the precipitate does not move).

4. To remove excess salt, gently add 10 mL of cold 70% ethanol. Do not disturb the nucleic acid precipitate. Centrifuge for 10 min (or more) at 12,000 rpm at 10°C.

5. Gently decant the supernatant, add 10 mL of cold (95%–100%) ethanol, and centrifuge for 10 min at 12,000 rpm at 10°C. Gently decant the supernatant, drain, and air dry briefly. Most of the precipitate is RNA.

6. Add 2.5 mL TE and incubate for 1 h at room temperature to dissolve the precipitate then gently mix with 0.1 mL RNase (1 mg/mL DNase-free), incubate at 37°C for 1–2 h (to remove RNA).

7. Add 0.1 mL of proteinase K (1 mg/mL, nuclease-free) and incubate at 37°C for 1–2 h to remove contaminating proteins, including RNase and proteinase K itself by digestion. Cool to room temperature.

19.18.3.7 DNA Purification

1. Add 2.5 mL of phenol:chloroform (1:1; the phenol has previously been equilibrated with TE buffer), vortex hard for 10 s, and centrifuge for 10 min at 10 K rpm, 10°C, in an SS34, RC-5B centrifuge (or equivalent).

2. Decant the upper, aqueous layer to a Centricon-30. Centricon takes the place of alcohol precipitation and for dialysis to change buffer and concentrate the DNA. The Centricon works annoyingly slow, but the recoveries are excellent. The plastic that composes the Centricon is resistant to traces of phenol and chloroform but is easily damaged by isoamyl alcohol. Do not use those recipes containing isoamyl alcohol.

3. Use a Centricon-30 spun at 5 K rpm in an SS34, RC-5B centrifuge at 10°C. Other rotor/centrifuge combinations may have a different maximum speed; check Amicon's specifications.

4. The choice of 10°C, rather than 4°C, is to avoid icing. Under these conditions, a 2.5 mL sample will pass through the filter in about 30–60 min. Load the sample, wash three times with 2.5 mL TE, and then spin for an additional 2 h to reach minimum volume.

19.19 To Perform Restriction Digestion of Plasmid DNA with Different Restriction Endonucleases and to Determine the Position of Restriction Site Using Restriction Mapping

19.19.1 Principle

Restriction mapping involves digesting DNA with a series of REs and then separating the resultant DNA fragments by agarose gel electrophoresis. Several bands will be visible on the gel after run, and the lowest one corresponds to the supercoiled form of the plasmid, which has migrated farthest in the gel; the band immediately above this is the "open" circle form. Any bands higher up in the gel are due to concatemer forms of the plasmid. In some plasmid preparations, it is possible that "shearing" of the DNA has occurred; and in this case, bands corresponding to linear plasmids appear between the supercoiled (closed circle) and "open" circle plasmid species.

The distance between RE sites can be determined by the patterns of fragments that are produced by the RE digestion, and information about the structure of an unknown piece of DNA can be obtained. By testing the insert for the presence and location of sites of many different REs, a "restriction map" of the clone is made.

19.19.2 Materials

Electrophoresis chambers (75 × 75 mm or 225 × 115 mm) and accessories; Polaroid MP 4 Land camera, power supply, UV-transilluminator; sterile reaction tubes, agarose (electrophoresis grade), 5 × TBE, 1 × TBE, loading buffer, DNAs: bacteriophage λ. DNA, K acetate, tris-HCl, Mg acetate 0.1 M, dithiothreitol (DTT) 5 mM, bovine serum albumin, EtBr, plasmids, REs (*Hind*III, *Eco*RI, *Bam*HI, and *Pst*I) and their buffers.

19.19.3 Protocol

19.19.3.1 RE Digest

1. Pipette the following into a sterile reaction tube:

   1 µL DNA (1–2 µg)

   2 µL of 10 × TBE (the final concentration of the buffer is 1×)

   1 µL of enzyme (2–5U)

   20 µL of H_2O

   Control tube contains no REs.

2. *Digest the plasmid DNA with different REs* (e.g., *Hind*III + *Eco*RI, *Hind*III + *Bam*HI, *Hind*III + *Pst*I, etc.).

3. λ DNA should be used as marker cut with *Hind*III *or Hind*III/*Eco*RI. An aliquot of one or both of these standards will be loaded onto the gel to allow the sizes of the restriction fragments obtained with the plasmid DNAs to be determined. The lengths of the fragments (in kb) of λ DNA obtained are (a) *Hind*III-restricted λ DNA: 23.1, 9.4, 6.5, 4.3, 2.3, 2.0, 0.56, and 0.125 and (b) *Hind*III/*Eco*RI-restricted λ DNA: 21.2, 5.1, 4.9, 4.3, 2,1,1.6, 1.3, 0.98, 0.83, 0.56, and 0.125.

4. Incubate the samples at 37°C for 1–2 h.

5. Stop the reaction by placing the samples at 65°C for 10 min.

6. Add 1/10 volume loading buffer to each sample and place again at 65°C for 10 min (this second incubation at the higher temperature is particularly important for the λ DNA digest to disrupt any concatemers, which may have formed). Keep samples on ice until the gel is to be loaded.

7. Subject the samples to electrophoresis at 40–120 V either in horizontal "Minigels" (73 75 mm) with 14 wells or in larger gels measuring 225 × 155 mm with 14 or 32 wells. The buffer used for the electrophoresis is 1 × TBE, and the agarose concentration is 0.8% in 1 × TBE.

8. Dissolve the agarose by heating on a Bunsen burner or in a microwave oven. Allow the agarose to cool to approximately 55°C before pouring into the gel form and inserting the comb to form the wells. The gel should be about 0.5 cm in depth.

9. After the gel has set, remove the comb carefully and place the gel gently in the electrophoresis chamber. Pour in sufficient 1 × TBE to cover the gel to a depth of 2–3 mm.

10. When the electrophoresis has been completed, place the gel in a 0.1% solution of EtBr for 5 min to stain the DNA in the gel. DNA is visualized by exposing the gel to UV light. This is done by placing the gel on a transilluminator. The result of the electrophoresis is documented by photographing the gel in UV light using a MP4 Land camera and Polaroid 667 film (wear a protective face shield when using the transilluminator).

19.19.3.2 Restriction Mapping

The determination of the positions or restriction sites in the plasmid DNA used in the restriction digests described earlier is termed restriction mapping and can be carried out. The plasmid DNA is digested with various REs, and the fragments obtained separated by electrophoresis as described. With the help of appropriate double digests, the positions of the restriction sites can be determined.

19.19.3.3 Result

The lengths of the DNA fragments obtained can be determined from the standard curve drawn by plotting the $\log_{10}$ of the distance migrated for each of the λ DNA restriction fragments obtained with *Hind*III and *Eco*RI or *Hind*III/*Eco*RI against the length of the fragment.

19.20 To Perform Transformation

19.20.1 Principle

The bacterial cells treated with ice cold solution of $CaCl_2$ and briefly heating could transfect the cells with bacteriophage λ DNA so is the case with plasmid DNA. Apparently, the treatment induces a transient state of "competence" in the recipient bacteria, during which they are able to take up DNA derived from a variety of sources. The *E. coli* strain DH5α is used as host bacteria.

19.20.2 Materials

Culture of *E. coli* cells for use as transformation host, sterile L-broth, sterile 0.1 M $MgCl_2$, sterile 0.1 M $CaCl_2$, L-agar, ampicillin, X-gal, plasmid pBluescript, and Petri plates.

19.20.2.1 Preparation of Competent Cells

1. Inoculate 10 mL of the saturated culture into 100 mL L-broth in a sterile 500 mL flask and incubate in an orbital shaker at 37°C until the OD of the culture reaches an absorbance (A600) of between 0.4 and 0.5 (measure and record the OD every 30 min). To measure the absorbance the flask should be removed from the incubator, a sample taken aseptically and then the flask returned to the incubator as quickly as possible.
2. Once the culture has reached A600 of 0.4–0.5, chill it on ice for ~10 min. For all subsequent steps, solutions and bottles that have been pre-chilled should be used.
3. Transfer 10 mL of the culture to a sterile round-bottom plastic tube, and pellet the cells at 3500 rpm for 5 min in a benchtop centrifuge. Discard the supernatant.
4. Thoroughly resuspend the cell pellet in 5 mL 0.1 M $MgCl_2$ (0°C) and then pellet the cells at 3500 rpm for 5 min. Discard the supernatant.
5. Carefully resuspend the cell pellet in 5 mL 0.1 M $CaCl_2$ (0°C) and hold on ice for 15 min.
6. Pellet the cells at 3500 rpm for 5 min. Discard the supernatant.
7. Resuspend the cell pellet in 0.5 mL of 0.1 M $CaCl_2$ (0°C) and store on ice.

These are now competent cells. Their cell walls are somewhat fragile and they should be treated gently.

19.20.2.2 Preparation of Plates

1. Cool L-agar slightly (~65°C) before adding the antibiotic.
2. The final concentration of ampicillin should be 80 μg/mL.
3. Calculate the volume of stock solution of ampicillin (10 mg/mL), which has to be added to 200 mL of L-agar to give 80 μg/mL of ampicillin.
4. Prepare 200 mL of molten agar containing ampicillin and pour nine plates with L-agar ampicillin. Allow the agar to set.
5. Spread 40 μL of X-gal on top of the plate with a sterile spreader. Then, let the plates dry. This should take 30 min.

19.20.3 Protocol

19.20.3.1 Transformation

1. Label three sterile round-bottom plastic tubes and place them on ice (N, negative control; P, positive control (known *nonrecombinant* plasmid); L, ligation reaction (potentially recombinant plasmid).
2. To each tube add 120 μL of competent cells.
3. Then to tube N add 10 μL of sterile water; to tube P add 10 μL of control plasmid (pBluescript); and to tube L, 10 μL of the ligation reaction.
4. Mix the tube contents by gentle shaking and stand them on ice for 10 min, occasionally shaking them to mix.
5. Transfer the tubes to a 42°C waterbath for 2 min.
6. Return the tubes to ice for 5 min.
7. Add 1 mL of fresh L-broth to each tube and incubate at 37°C in an orbital shaker for 40 min.
8. Spread 5 μL and 50 μL aliquots of the contents of each of the three tubes on agar/ampicillin/X-gal plates and incubate at 37°C for 48 h.

19.20.4 Result

Observe blue and white colonies and determine the frequency of recombinants.

19.21 Directional Cloning into Plasmid Vectors

19.21.1 Principle

Directional cloning usually requires incompatible terminus at the opposite ends of both vector and target DNAs. However, in certain circumstances, directional cloning can be achieved when both the target and plasmid DNAs carry identical terminus at both ends. For example, the REs *Bam*HI and *Bgl*II, which recognize different hexanucleotide sequences (GGATCC and AGATCT, respectively), generate restriction fragments with identical 3′ protruding terminus. If a DNA fragment carrying *Bam*HI and *Bgl*II terminus is ligated into a vector that has been cleaved with the same two enzymes, then the foreign DNA can be inserted in either orientation. However, if one of the two REs is included in the ligation mixture, or if the enzyme is used to digest the ligated DNA before transformation, then only those ligation events in which the *Bam*HI end is joined to the *Bgl*II end and vice versa (which destroys the recognition sites of both enzymes) will give rise to recombinant products in *E. coli*. This strategy takes advantage of the observation that closed circular DNAs transform bacterial cells with a much higher frequency than linear DNAs. Occasionally, it is impossible to find a suitable combination of vector, target DNA, and REs that will allow directional cloning. There are several solutions to this problem:

1. Synthetic linkers or adaptors can be ligated to the terminus of the linearized plasmid and/or fragment of foreign DNA.
2. The fragment of foreign DNA can be amplified by PCR using oligonucleotide primers that add the desired restriction sites to one or both terminus.
3. DNA fragments with recessed 3′ terminus can be partially filled on controlled reactions using the Klenow fragment of *E. coli* DNA polymerase I. The procedure often generates complementary terminus from restriction sites that are otherwise incompatible, thus facilitating ligation of the vector and foreign DNAs. Because partial filling eliminates the ability of terminus on the same molecule to pair with one another, the frequencies of circularization and self oligomerization during the ligation reaction are also reduced. The micromole concentration of dATP can inhibit bacteriophage T4 DNA ligase. Thus, if dATP is used as a substrate in partial end-filling reaction, the modified DNA product should be purified by spun column chromatography or by two rounds of ethanol precipitation in the presence of the ammonium acetate. This removes incorporated dATP from the DNA preparation. The protocol describes a standard procedure for cloning DNA fragments with protruding ends.

19.21.2 Materials

ATP, ethanol, phenol, chloroform, sodium acetate, bacteriophage T4 DNA ligase, restriction endonucleases, agarose, polyacrylamide gels, vector DNA, foreign or target DNA, equipment for spun column chromatography, and water bath pre set to 16°C.

19.21.3 Protocol

1. Perform restriction digestion of the gene of interest and the vector using compatible REs as per earlier experiment.
2. Ligate the restricted foreign DNA and the vector to obtain a recombinant molecule.
3. Transform competent *E. coli* as per earlier experiment. Controls include known amounts of a standard preparation of superhelical plasmid DNA to check the efficiency of transformation. Transformants arising from ligation of vector DNA alone are due to 1. Failure of one or both restriction endonucleases to digest the DNA to completion and/or 2. Ligation of the vector to residual amounts of the small fragment of the MCSs (multiple cloning sites).

19.22 To Assay Reporter Chloramphenicol Acetyltransferase Gene

19.22.1 Principle

Reporter genes are "markers" widely used for analysis of mutationally altered genes as well as gene regulation. The expressed reporter genes are detected by biochemical activity assays, immunological analysis, or by histochemical staining of tissue sections or cells. Reporter gene assay is an invaluable tool for both biomedical and pharmaceutical researches to monitor cellular events associated with gene expression, regulation, and signal transduction. They represent a gene whose phenotypic expression is easy to monitor and are used to study promoter activity in different tissues and developmental stages. Most reporter genes are placed downstream to the promoter region, but close to the gene under study. This ensures that these genes are expressed together and are not separated during cell division by crossover events. Reporter genes are used extensively for both, *in vitro* and *in vivo* applications. In particular, reporter systems are employed to study the promoter and enhancer sequences or trans-acting mediators for the transcription, mRNA processing, and translation. They can also be utilized to monitor the transfection efficiencies, protein–protein interactions, protein subcellular localization, and recombination events as well as to screen genome-wide libraries for novel genetic regulatory elements. Several commonly used reporter genes are chloramphenicol acetyltransferase (CAT), AP, β-gal, luciferases, green fluorescent protein, and β-lactamase.

CAT, encoded by a bacterial drug-resistance gene, inactivates chloramphenicol by acetylating the drug at one or both of its two hydroxyl groups. This gene is not found in eukaryotes, thus eukaryotic cells produce no background CAT activity. This characteristic, along with the ease and sensitivity of the assay for CAT activity, made the CAT gene one of the first reporter genes used to study mammalian gene expression. The product of the reaction, 3-acetoxychloramphenicol, neither binds to the peptidyl transferase center of 70S ribosomes nor inhibits peptidyl transferase. In members of Enterobacteriaceae and other Gram-negative bacteria, the *CAT* gene is constitutively expressed and is usually carried on plasmids that confer multiple drug resistance.

Several different assays are available to measure CAT activity. CAT activity may be monitored by two alternative methods in the CAT Enzyme Assay System. The most rapid, sensitive, and convenient of these assays is based on liquid scintillation counting (LSC) of CAT reaction products. Cell extracts are incubated in a reaction mix containing ^{14}C- or ^{3}H-labeled chloramphenicol and n-butyryl coenzyme A. CAT transfers the n-butyryl moiety of the cofactor to chloramphenicol. For the LSC assay, the reaction products are extracted with a small volume of xylene. The n-butyryl chloramphenicol partitions mainly into the xylene phase, while unmodified chloramphenicol remains predominantly in the aqueous phase. The xylene phase is mixed with scintillant, and radioactive product is measured with a scintillation counter.

CAT activity also can be analyzed using thin-layer chromatography (TLC). This method is more time-consuming than the LSC assay but allows visual confirmation of the data. After the completion of the CAT reactions, the products are placed on

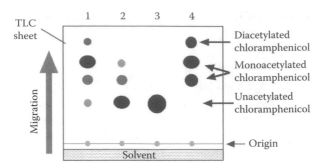

Figure 19.26 CAT assay results. Lane 1: promoter 1 construct; promoter 2 construct; negative control; positive control.

thin-layer chromatographic sheets, placed in the appropriate solvent (a mixture of chloroform and methanol) and allowed to migrate up the TLC sheet, which is exposed to x-ray film. Each product (two forms of chloramphenicol with one acetyl group and one with two acetyl groups added) and all the unused substrate (acetyl CoA, chloramphenicol, and CAT) will migrate on the TLC surface according to their ability to interact with surface of the TLC. Since the only radioactive molecule is chloramphenicol, the only molecule visible on the x-ray film is chloramphenicol and all its acetylated forms.

The amount of acetylated chloramphenicol can be measured either by scanning the density of the dark spots on the x-ray film or by measuring the radioactivity in each spot on the TLC sheet. Either way, CAT assays offer an indirect but quantitative way to measure the amount of transcription driven from any given promoter. In Figure 19.26, two different strength promoters were tested in addition to the two controls. The amount of acetylated chloramphenicol is greater in lane 1 than lane 2, and therefore promoter number 1 was able to produce more mRNA than promoter number 2.

19.22.2 Materials

Tris-HCl, ^{14}C chloramphenicol, acetyl CoA, ethyl acetate, Triton X-100, phosphate buffer saline, chloroform, methanol, dry ice/ethanol bath, rotary vacuum evaporator, and cultured mammalian cells transfected with pCAT vectors carrying DNA of interest.

19.22.3 Protocol

19.22.3.1 Preparation of Transfected Cell Pellets
1. Remove gently the medium from transfected monolayers of cells growing in 90 mm tissue culture dishes. Wash the monolayers three times with 5 mL of PBS without calcium and magnesium salts.
2. Stand the dishes at an angle for 2–3 min to allow the last traces of PBS to drain to one side. Remove the last trace of PBS. Add 1 mL of PBS to each plate and use a

rubber policeman to scrape the cells into microfuge tubes. Store the tubes in ice until all the plates have been processed.

3. Recover the cells by centrifugation at maximum speed for 10 s at room temperature in a microfuge. Gently resuspend the cell pellets in 1 mL of ice-cold PBS, and again recover the cells by centrifugation. Remove the last traces of PBS from the cell pellets and from the walls of the tubes. Store the cell pellets at –20°C for further analysis.

4. The PBS can conveniently be removed with a disposable pipette tip attached to a vacuum line. Use gentle suction, and touch the tip to the surface of the liquid. Keep the tip as far as away from the cell pellet as possible while the fluid is withdrawn from the tube and then be used to vacuum the walls of the tube to remove any adherent droplets of fluid.

19.22.3.2 Preparation of Cell Extracts

Lyse the cells either by repeated cycles of freezing and thawing or by incubating the cells in detergent-containing buffers.

19.22.3.2.1 Lysis of Cells by Repeated Freezing and Thawing

1. Resuspend the cell pellet in 100 μL of 0.25 M Tris-Cl (pH 7.8) per 90 mm dish of cells. Vortex the suspension vigorously to break up clumps of cells.

2. Disrupt the cells by three cycles of freezing in a dry ice/ethanol bath and thawing at 37°C. Make sure that the tubes have been marked with ethanol insoluble ink.

3. Centrifuge the suspension of disrupted cells at maximum speed for 5 min at 4°C in a microfuge. Transfer the supernatant to a fresh microfuge tube and store the extract at –20°C for CAT assay.

19.22.3.2.2 Lysis of Cells Using Detergent Containing Buffers

1. To lyse cells with detergent, resuspend the cell pellets from step 3 in 500 μL of lysis buffer and incubate the mixture for 15 min at 37°C.

2. Use 100 μL of this lysis buffer per cell pellet for extracts prepared from cells grown in 35 mm dishes.

3. Remove the cellular debris by centrifuging the tubes at maximum speed for 10 min in a microfuge. Recover the supernatant. Assay CAT activity using one of the methods described. Snap freeze the remainder of the clear lysates in nitrogen and store them at –70°C.

19.22.3.3 Detection of CAT Activity Using TLC

1. Incubate a 50 μL aliquot of the cell extract for 10 min at 65°C to inactivate endogenous deacetylases. If the extract is cloudy or opaque at this stage, remove the particulate material by centrifugation at maximum speed for 2 min at 4°C in a microfuge.

2. Mix each of the samples to be assayed with 80 μL of CAT reaction mixture 1, and incubate the reactions at 37°C. The length or the incubation depends on the concentration of CAT in the cell extract, which in turn depends on the strength of the promoter and the cell type and investigation. In most cases, incubation for 30 min to 2 h is sufficient.

3. The reactions can be incubated for longer periods of time (up to 145 h) when the expression of the *CAT* gene is low in the transfected cells. However, in this case, it is advisable to add an additional 10 μg aliquot of acetyl-CoA to each reaction after 2 h of incubation.

4. To terminate the reaction add 1 mL of ethyl acetate to each sample, and mix the solutions thoroughly by vortexing three times of 10 s each. Centrifuge the mixture at maximum speed for 5 min at room temperature in a microfuge. The acetylated forms of chloramphenicol partition into the organic (upper) phase; unacetylated chloramphenicol remains in the aqueous phase.

5. Use a pipette to transfer exactly 900 μL of the upper phase to a fresh tube, carefully avoiding the lower phase and the interface. Discard the tube containing the lower phase in the radioactive waste.

6. Evaporate the ethyl acetate under vacuum by placing the tubes in a rotary evaporator for −1 h.

7. Add 25 μL of ethyl acetate to each tube and dissolve the reaction products by gentle vortexing. Apply 10–15 μL of the dissolved reaction products to the origin of a 25 mm silica gel plate. The origin on the plate can be marked with a soft-lead pencil. Apply 5 μL at a time, and evaporate the sample to dryness with a hair dryer after each application.

8. Prepare a TLC tank containing 200 mL of TLC solvent chloroform:methanol in the ratio 95:5. Place the TLC plate in tank, close the chamber, and allow the solvent front to move to 75% of the distance to the top of the TLC plate.

9. Remove the TLC plate from the tank and allow it to dry at room temperature.

10. Place adhesive dot labels marked with radioactive ink on the TLC plate to align the plate with the film, and then expose the plate to x-ray film.

11. Alternatively enclose the plate in a phosphors imaging cassette. Store the cassette at room temperature for an appropriate period of time then expose the plate to x-ray film. Develop the x-ray film and align it with the plate. Alternatively, expose the chromatogram, the imager plate of a phosphors imager device, or subject the plate to scanning (Figure 19.27).

12. Typically, three radioactive spots are detected. The spot that has migrated the least distance from the origin consists of nonacetylated chloramphenicol migrating that partitioned into ethyl acetate. The two faster-migrating spots modified forms are of chloramphenicol that have been acetylated at one or the other or two potential diacetylated chloramphenicol may be detected as a third, even faster migrating spot only when high concentrations of CAT or lengthy incubations with copious amounts of extract protein are used.

13. To quantitate CAT activity, cut the radioactive spots from the TLC plate and measure the amount of radioactivity they contain in a liquid scintillation counter. Use another aliquot of the cell extract (from step 3 mentioned earlier) to determine the concentration of protein in the extract, using a rapid colorimetric assay, such as the Bradford assay. Reduce the concentration of Triton X-100 from 0.2% to 0.1% by dilution before determining the concentration of protein to prevent interference with the assay. Express the CAT activity as pmoles of acetylated product formed per unit time per milligram of cell extract protein.

14. TLC on silica gels followed by scraping and scintillation counting of the modified chloramphenicol products can become quite tedious when large number of samples are prepared. An alternate method for quantitating the amount or product in these cases is to use a phosphorimager to scan the TLC plates. "Intrinsic" acetylation of chloramphenicol in the absence of cellular extracts can contribute to background. If this is a problem, try decreasing the concentration of chloramphenicol by two- to fourfold in the assays.

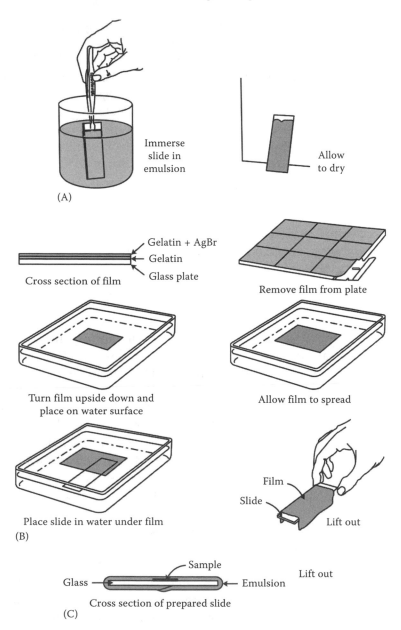

Immerse
slide in
emulsion

Allow
to dry

(A)

Gelatin + AgBr
Gelatin
Glass plate

Cross section of film

Remove film from plate

Turn film upside down and
place on water surface

Allow film to spread

Film
Slide

Lift out

Place slide in water under film

(B)

Lift out

Sample
Glass
Emulsion

Cross section of prepared slide

(C)

Figure 19.27 Diagram showing placing x-ray film on TLC plate.

19.23 To Assay β-Gal in Extracts of Mammalian Cells

19.23.1 Principle

The β-gal gene from *E. coli*, *LacZ*, is often used as a reporter gene in eukaryotic transfection. The enzyme catalyzes the hydrolysis of various β-galactosides. Substrates designed to produce chromogenic, fluorescent, or chemiluminescent products are used for β-gal activity detection. O-nitrophenyl-β-D-galactopyranoside (ONPG) is a widely used chromogenic substrate for determination of β-gal activity in transfected cells or tissues. The standard assay is performed by adding a diluted sample to an equal volume of assay buffer that contains the substrate ONPG. Samples are incubated for 30 min, during which time the β-gal hydrolyzes the colorless substrate to o-nitrophenol, which is yellow. The reaction is terminated by addition of sodium carbonate, and the absorbance is read at 420 nm with a spectrophotometer.

The *E. coli* β-gal gene is often used as an internal reference in transfection studies. A test plasmid containing a eukaryotic gene promoter linked to CAT or luciferase is cotransfected into a mammalian cell line with a small amount of a plasmid containing β-gal gene linked to a strong constitutive promoter. After a period of time to allow expression, enzyme activities arising from the expression of test and control plasmids are determined in cell lysates. By dividing the amount of CAT or luciferase activity by the β-gal activity, a normalized expression value can be obtained. Constructs are available that carry the β-gal gene downstream from promoters that express strongly in a wide variety of eukaryotic cell types (e.g., SV 40 early promoter, the Rous sarcoma virus long terminal repeat promoter, or the immediate early region promoter of cytomegalovirus).

Summarized here are the features of pβ-gal reporter vectors that carry the coding sequence for β-gal. All vectors in the series contain (i) the origin of replication derived from filamentous phage (flori), (ii) an origin of plasmid replication in *E. coli* (ori), (iii) the origin of replication for SV40 that allows replication in mammalian cells, (iv) an ampicillin resistance marker (Amp^R) for selection in prokaryotic cells, and (v) a poly (A) addition signal located 5' of *lacZ* to reduce the background of read through transcription from upstream sequences. The sequence of interest is cloned into the MCS at the 5' end of the *lacZ* gene. (B) pβ-gal-Basic lacks eukaryotic promoter and enhancer sequences. This vector may be used as a negative control or as a vehicle to characterize cloned promoters. (vi) pβ-gal-enhancer lacks the SV 40 promoter, but contains the SV40 enhancer. This vector can be used to study cloned promoter sequences, (C) pβ-gal-promoter lacks the SV40 enhancer, but contains the SV40 promoter. This vector can be used to study cloned enhancer sequences. (D) pβ-gal-control contains the SV40 early promoter and enhancer. This vector can be used as a positive control or as reference, comparing the activities of different promoter and enhancer elements. Extracts of most types of cultured mammalian cells contain relatively low levels of endogenous β-gal activity and an increase in enzyme activity of up to 100-fold can usually be detected during the course of a transfection. With effort, β-gal can also be used as an internal control in certain specialized cells (e.g., gut epithelial cells and human embryonic kidney 293 cells) that express high endogenous levels of β-gal activity. Because endogenous β-gal activity is usually more heat-labile than the bacterial enzyme, a heating step can be used to eliminate the endogenous β-gal activity while preserving the bacterial activity expressed from the control plasmid. In addition, most mammalian β-gals are associated with the lysosome and therefore have an acidic pH optimum. The *E. coli* enzyme has a neutral to slightly alkaline pH optimum. The contribution of the mammalian enzyme can therefore be reduced by carrying out the β-gal assay at pH 7.5.

Several different approaches are used to normalize CAT or other reporter enzyme activities to β-gal activity. In one approach, the amount of protein is first measured in individual extracts prepared from a series of transfected cells, and CAT and β-gal are then assayed separately using a standard amount of protein in each assay. Finally, the CAT activity is normalized to the β-gal activity. In another method, the β-gal activity in a constant volume of extract is first measured, and CAT assays are then carried out using amounts of extract containing the same amount of β-gal activity. Alternatively, both enzymatic assays are carried out in a constant volume of extract, and the results are then normalized to a defined level of β -galactosidase activity (i.e., the amount of CAT activity is divided by the amount of β-gal activity). With sonic marker genes, such as luciferase, the amounts of β-gal and luciferase that are present in the same aliquot of cell lysate can be determined by using different luminescent substrates simultaneously.

This protocol describes the detection of β-gal expressed from reporter vectors transfected into mammalian cells. The assay described is both simple and rapid and can be carried out using a visible light spectrophotometer.

19.23.2 Materials

$100\times$ MgCl$_2$, 1 M sodium bicarbonate solution, 1× ONPG, 0.1 M sodium phosphate (pH 7.5), 1 M tris-HCl (pH 7.8), *E. coli* β-gal, and cultured mammalian cells transfected with the DNA of interest.

19.23.3 Protocol

1. Prepare cell extracts from the transfected cells as described earlier.
2. Set aside ~30 μL of the extract for the β-gal assay. The exact amount of extract required will depend on the strength of the promoter driving the expression of the β-gal gene, the efficiency of transfection, and the incubation time of the assay. If a heat treatment is to be used to inactivate endogenous β-gals, incubate the cell lysates for 45–60 min at 50°C before assay. Luciferase activity is also inactivated by preheating. Assay luciferase and β-gal activities in separate aliquots of cell lysate if a preheating step has been used.
1. For each sample of transfected cell lysate to be assayed, mix

   $100\times$ Mg^{2+} solution 3 μL
   $1\times$ ONPG 66 μL
   Cell extract 30 μL
   0.1 M sodium phosphate (pH 7.5) 201 μL
2. It is essential to include positive and negative controls. These assays check for the presence of endogenous inhibitors and β-gal, respectively. All controls should contain 30 μL cell extract from mock-transfected cells. In addition, the positive control should include 1 μL of a commercial preparation of *E. coli* β-gal (50 units/mL). The commercial enzyme preparation should be dissolved at a concentration of 3000 units/mL in 0.1 M sodium phosphate (pH 7.5). One unit of *E. coli* β-gal is

defined as the amount of enzyme that will hydrolyze 1 µmole of ONPG substrate in 1 min at 37°C.

3. Incubate the reaction mixtures for 30 min at 37°C or until a faint yellow color has developed. In most cell types, the background of endogenous β-gal activity is very low, allowing incubation times as long as 4–6 h to be used.

4. Stop the reactions by adding 500 µL of 1 M Na$_2$CO$_3$ to each tube. Read the OD of the solutions at a wavelength of 420 nm in a spectrophotometer. The linear range of the assay is 0.2–0.8 OD at 420 nm. If the assay is above or below the range, repeat the experiment with the adjustment of protein concentration. The extract can be diluted in 0.25 M Tris-HCl (pH 7.8) to adjust the protein concentration.

19.23.4 Result

The specific activity of the β-gal enzyme is expressed as units of enzyme activity per milligram of cell protein, where 1 unit of *E. coli* β-gal is defined as the amount of enzyme that will hydrolyze 1 µmole of ONPG substrate in 1 min at 37°C. This value can then be used to normalize expression of the marker gene, whose specific activity has also been determined. Sometimes, it is calculated simply by dividing the amount of CAT or luciferase activity by the amount of β-gal activity present in a given volume to normalize for transfection efficiency, thus ignoring the calculation of specific activity.

19.24 Determination of Nucleotide Sequence of DNA by Dideoxy Chain Termination Method

19.24.1 Principle

The chain-termination method (or Sanger method after its developer Frederick Sanger) is more efficient and uses fewer toxic chemicals and lower amounts of radio-activity than the method of Maxam and Gilbert is the method of choice. The key principle of the Sanger method is the use of 2′,3′-dideoxynucleotide triphosphates (ddNTPs) as DNA chain terminators. These ddNTPs are incorporated normally in to a growing DNA chain through their 5′-triphosphate groups. However, they cannot join with the next incoming dNTP as they lack 3′-OH group needed to make the phosphodiester bond.

The classical chain-termination method requires a ssDNA template, a DNA primer, a DNA polymerase, normal dNTPs, and modified dideoxynucleotides (dideoxyNTPs) that terminate DNA strand elongation. These ddNTPs are radio-actively or fluorescently labeled for detection in automated sequencing machines. The DNA sample is divided into four separate sequencing reactions, containing all four of the standard deoxynucleotides (dATP, dGTP, dCTP, and dTTP) and the DNA polymerase. To each reaction is added only one of the four dideoxynucleo-tides (ddATP, ddGTP, ddCTP, or ddTTP), which are the chain-terminating nucleo-tides, lacking a 3′-OH group required for the formation of a phosphodiester bond between two nucleotides, thus terminating DNA strand extension and resulting in DNA fragments of varying length. The fragments thus synthesized are separated by high resolution polyacrylamide gel electrophoresis that allows fragments differing by a single nucleotide to be resolved and the sequence can he directly read from the autoradiogram of the gel as dark bands (Figure 19.28).

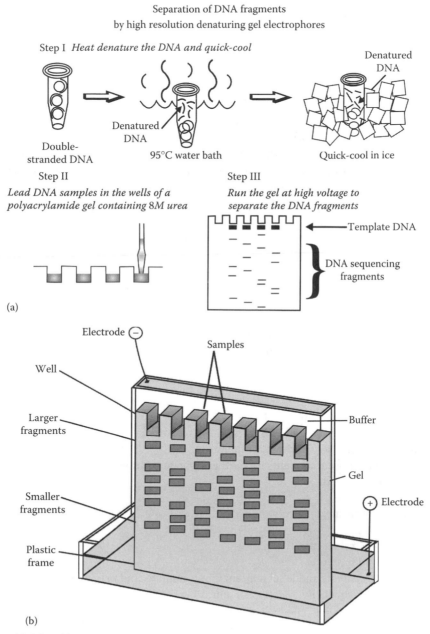

Figure 19.28 (a) Steps involved for high resolution denaturing gel electrophoresis. (b) Manual DNA sequencer.

19.24.2 Materials

19.24.2.1 Reagents
Template DNA, TE buffer, universal primer: commercially available M13 sequencing primer, Tris-HCl, $MgCl_2$, NaCl, ^{35}S dATP (8 mCi/mL), 0.1 M DTT, Klenow enzyme, dNTP stock solutions (dATP, dCTP, dTTP, and dGTP), dNTPs working solutions, ddNTPs (ddX) working solutions, formamide, xylene cyanol FF, bromophenol blue, EDTA, methanol, acetic acid, x-ray developer, x-ray fixer, acrylamide, bisacrylamide, tris, boric acid, urea, TEMED, ammonium persulfate, siliconizing fluid, sealing tape, and paraffin (liquid).

19.24.2.2 DNA Sequencing Apparatus
It consists of two glass plates (one notched and the other unnotched), 40×20 cm each; three spacers: two 50 cm long and one 30 cm long, each 1 cm wide and 0.4 mm thick, shark's tooth comb, electrophoretic chamber; power pack 3000 volts capacity; gel loading pipette or syringe; clamps, Whatman filter paper, sealing tape, liquid paraffin, template DNA, TE buffer, universal primer, ^{35}S dATP, DTT, Klenow enzyme, dNTPs, ddNTPs, formamide, xylene cyanol, bromophenol blue, EDTA, methanol, acetic acid, acrylamide, bisacrylamide, Tris, boric acid, urea, TEMED, and ammonium persulfate

19.24.3 Protocol

19.24.3.1 Sequencing Reaction
1. Add the following to an Eppendorf tube to make a template primer mixture:
   Template DNA (700 ng) 7.0 μL
   Primer DNA (5 ng) 2.0 μL
   $10 \times$ polymerase reaction buffer
   pH 7.5 1.0 μL
   Paraffin (liquid) 15.0 μL
   Keep in boiling water bath for 5 min. Switch off the water bath and allow the contents to cool slowly to room temperature.
2. Add the following components to another Eppendorf tube:
   $[^{35}S]$ dATP 1.0 μL
   0.1 M DTT 1.0 μL
   Klenow enzyme 2–3 units
   SDW 1.0 μL
   Template primer mix 10.0 μL
   (Prepared in step 1)
   Mix the contents and label it as enzyme, template, and primer mix.
3. Carry out the sequencing reaction in four different Eppendorf tubes labeled as G, A, T, and C.
4. Incubate the tubes at room temperature for 20 min.

5. Add 1 mL 0.5 µM dATP to each tube and allow the tubes to stand for another 20 min at room temperature.

6. Add 10 µL formamide dye mix (gel loading solution) to all the tubes and keep at −15°C till use.

19.24.3.2 Electrophoresis

1. Wash the two plates (notched and unnotched or smooth) with detergent in hot water and rinse with deionized water. Dry the surface of the plates with tissue paper. Rinse with distilled water and again wipe dry the surface with tissue paper.

2. Put 1 mL of silane (5% solution of dichlorodimethyl silane in chloroform) on a smooth plate, spread it evenly and keep at room temperature for 30 min, Rinse with ethanol and dry thoroughly with tissue paper (caution: ensure complete removal of all silane).

3. Place the spacers on the notched plate and align properly the smooth plate with the notched one and fix tape at the bottom and on the sides (3–4 in.) of both the plates together. Clamp the plates.

4. Make 8% of denaturing polyacrylamide-bisacrylamide solution with 5 mL 1 × TBE buffer and 25 g urea. Keep in boiling water bath to dissolve urea and make the volume to 50 mL. Filter the solution through nitrocellulose paper (0.45 p) and deareate.

5. Cool the solution and add 50 µL TEMED and 200 µL 10% ammonium persulfate.

6. Immediately transfer the solution between the two plates slowly and continuously using a big syringe, keeping the gel mould slightly slanting at an angle of 30°.

7. Insert the comb. Refill the acrylamide solution, if the gel retracts.

8. Allow the gel to polymerize for about 1 h.

9. Remove the tapes, and clamp the plates to the electrophoretic unit.

10. Run the gel at 20–30 mA, ~2000 V for 30 min.

11. Heat samples at 90°C for 3 min and keep on ice, load 1.5 µL of each sample on the gel in different lanes.

12. Run the gel at 20–30 mA, ~2000 V, till bromophenol blue runs out (90 min).

19.24.3.3 Autoradiography

1. Take out the plates from the electrophoretic chamber and keep them on table top. Remove both spacers, insert a spatula at one corner and separate the plates carefully. Make sure that the gel is sticking evenly to the bottom plate.

2. Keep the gel in methanol-acetic acid solution for 10 min (this step is used when [35]S dATP is used as an isotope).

3. Transfer the gel onto a Whatman filter paper no. 3 and dry it at 80°C for 30 min.

4. Expose the dried gel to a x-ray film for 48–72 h at room temperature.

5. Develop the film for 3 run in x-ray developer in dark. Wash with water and keep for 2 min in fixer. Air dry and label the autoradiogram with date and name of the template. Mark each set of sequencing reactions clearly. Read the sequence from bottom to the top of the autoradiogram as dark bands, in different lanes (Figure 19.29). This will indicate the sequence of the complementary strand from 5′ direction. From this sequence, deduce the sequence of the template strand.

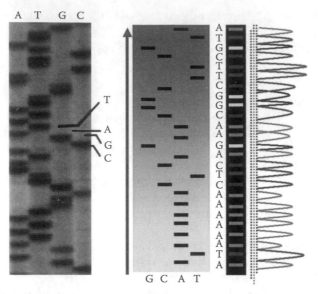

Figure 19.29 The resulting labeled fragments are separated by size on an acrylamide gel and autoradiography is performed. The pattern of the fragments gives the sequence of DNA.

19.25 Amplification of 16S rDNA Genes by the Polymerase Chain

19.25.1 Principle

The PCR technique was devised by Kary Mullis. In the mid-1980s. Like DNA sequencing, the technique has revolutionized the practice of molecular genetics. The PCR can be used to amplify a specific nucleotide present in nearly any environment. This includes DNA in samples such as body fluids, soil, food, and water. This technique can be used to detect organisms that are present in extremely small numbers as well as those that cannot be grown in cultures.

The DNA is denatured and specific primers and dNTPs are added along with the DNA polymerase. Using the random primers, the entire sequence of a genome can be amplified in pieces. Both DNA strands of a double helix can serve as template for synthesis, provided an oligonucleotide primer is supplied for each strand. The primers are chosen to flank the region of DNA that is to be amplified. The newly synthesized strands of DNA, starting at each primer, extend beyond the position of the primer on the opposite strand. The reaction mixture is heated to separate the strands. The primers anneal to the newly available binding sites and new chains are synthesized. These cycles of heating, primer binding, and extension is repeated so that at the end of n cycles, the reaction mixture theoretically has maximum of 2^n dsDNA molecules (Figure 19.30).

In most of the cases, the PCR amplification results in a sufficient quantity of the amplified fragment that can be easily visualized on an agarose gel stained with EtBr.

Factors affecting optimization of PCR includes temperature, salt concentrations, cycle durations, and primers. Primers are the key factors. The specificity of primer

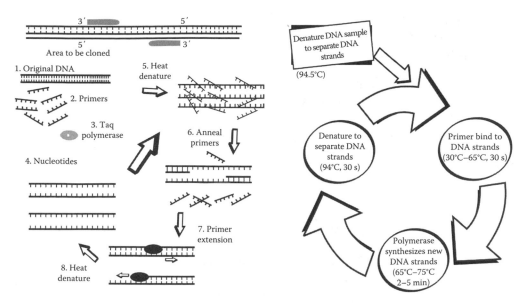

Figure 19.30 Amplifying specific DNA sequences using PCR and PCR cycle.

binding depends on the length of the primers and their nucleotide composition. Oligonucleotides 18–24 nucleotides long tend to be highly specific under standard conditions where the annealing temperature is close to the melting temperature. It is also important to select primer pairs with similar T_ms, so that both anneal at the chosen temperature. Increasing primer length increases specificity but results in an increase in the T_m. If the T_m becomes too high, the required annealing temperature will exceed the temperature for polymerase extension and may result in inaccurate priming.

The most commonly used DNA sequence for bacterial phylogenetics is the highly conserved 16S rRNA gene sequence, and primers have been designed to selectively amplify bacterial 16S rRNA genes.

19.25.2 Materials

Taq DNA polymerase, 0.1 M Tris, 0.5 M KCl, 1.5 mM $MgCl_2$, 0.1% gelatin, dATP, dCTP, dGTP, dTTP, λ DNA (*Hind*III digest) MW marker, 100 bp ladder, Tris-HCl, boric acid, 0.5 EDTA (pH 8.0), primers: PA (5′ AGAGTTTGATCCTGGCTCAG 3′) and PH (5′AAGGAGGTGATCCAGCCGCA 3′). Thermocycler (PCR machine), Eppendorf tubes (0.5 mL capacity), micropipettes, and microfuge.

19.25.3 Protocol

19.25.3.1 Preparation of Template DNA

1. Grow bacteria at 30°C in 25 mL medium on a shaker until OD is 0.3–0.5.
2. Take 3.0 mL of broth culture, centrifuge at 6000 rpm for 5 min.

3. Resuspend pellet in 0.4 mL TE.
4. Add 40 μL of 10% SDS.
5. Add 0.4 mL of phenol.
6. Mix well.
7. Centrifuge for 5 min at 8000 rpm.
8. Transfer the water phase to new tube.
9. Add 0.2 mL phenol and 0.1 mL chloroform.
10. Mix well (do not vortex).
11. Centrifuge for 5 min at 8000 rpm.
12. Transfer water phase to new tube.
13. Add 0.4 mL chloroform.
14. Mix well (do not vortex).
15. Centrifuge for 5 min at 8000 rpm.
16. Transfer water phase to clean tube.
17. Estimate volume.
18. Add 0.1 vol of 3 M sodium acetate, pH 5.2, and mix.
19. Add 2 volume of 96% ethanol and keep it at −20°C for 1 h.
20. Centrifuge DNA for 10 min at 8000 rpm.
21. Wash pellet with 70% ethanol (−20°C).
22. Centrifuge DNA for 10 min at 8000 rpm. Discard the supernatant.
23. Dry pellet and dissolve in 20 μL of water and keep it at 4°C.

19.25.3.2 Preparation of the PCR Amplifications

1. DNA will be amplified by mixing template DNA (50 ng) with the polymerase reaction buffer, dNTP mix (100 micro moles), primers PA and PH (100 ng each) and 0.5U *Taq* polymerase. The reaction master mix should be prepared using the following reagents in the desired concentrations.

Components	Volumes (μL)
Deionized water	79.5
10× *Taq* buffer	10.0
Taq polymerase (0.5U)	0.5
dNTPs (10 mM)	1.0
$MgCl_2$ (50 mM)	6.0
Primer PA (100 ng)	1.0
Primer PH (100 ng)	1.0
Target DNA (50 ng)	1.0

2. Add the reagents in the required volumes, and all the operations have to be carried out on ice. In addition to the samples, negative control (i.e., without template DNA) has to be maintained simultaneously to check for possible contamination.
3. Cap all the PCR tubes and vortex to mix. Pulse the tubes for 2–3 s in a microcentrifuge to pool the contents at the bottom.
4. If you are using a thermal cycler with a heated lid, the tubes are ready to go into the thermal cycler. If not, overlay the reaction mixture with 50 μL of sterile mineral oil and cap the tubes.

5. DNA amplification will be done in Thermal Cycler with the following temperature profile—an initial denaturation at 94°C for 5 min, followed by 30 cycles of denaturation at 94°C for 30s, annealing at 50°C for 40 s, and extension at 74°C for 90 s, and a final extension at 74°C for 7 min.

19.25.4 Result

Examine the amplified DNA by horizontal electrophoresis on 0.8% agarose.

19.26 PCR from Fungal Spores/AMF (Arbuscular Mycorrhizal Fungi)

19.26.1 Introduction

The protocol described herein can be used to amplify the 5′ end of the large ribosomal unit of fungal spore mycorrhiza as starting material. This method can be applied to other types of biological material, like plant roots as well or under symbiotic association with lichens, cycad root and azolla.

19.26.2 Materials

Fungal spores, Eppendorf tube, binocular microscope, forceps, micropestle, glass Pasteur pipette, Tris, HCl, NaOH, Chelex® 100 (chelating material is styrene divinylbenzene ion exchange resin), and PCR machine or thermocycler.

19.26.3 Protocol

19.26.3.1 Preparation of Template DNA

1. Collect clean and shiny Glomus spores (1–10) with forceps under a binocular microscope and rinse with distilled water.
2. Transfer the spores to 1.5 mL Eppendorf tube containing 10 μL water and crush by means of a micropestle, or a glass Pasteur pipette. Disposable micropestles are available from many laboratory suppliers and can be reused after incubation for several hours in 0.1 N NaOH to digest any remaining DNA.
3. Add 30 μL 100 mM Tris/HCl pH 8.0 and 10 μL of 20% Chelex 100 (Bio Rad) to the crushed spores. Vortex this suspension and then bring to 95°C for 5 min. Cool on ice.
4. Clear the suspension by centrifugation for 1 min and discard the pellet. The supernatant contains the nucleic acids for the PCR reactions. Depending on the nature of the species analyzed, and especially its DNA content, the supernatant obtained can be directly used as template for PCR amplification, or be diluted up to 1/100 before use. This DNA preparation should be stored at –20°C until use.

19.26.3.2 Preparation of the PCR Amplifications

1. Each PCR reaction is performed in a final volume of 100 µL.
2. For each set, (a) amplification reaction and (b) negative control, without DNA, has to be made. For convenience and to minimize the risk of contamination, a master mixture is prepared with all the reagents except the template DNA.
3. For PCR machine without heated lid, 25–50 µL mineral oil is laid over the mixture, and quickly spun down onto the surface.
4. DNA will be amplified by mixing template DNA (50 ng) with the polymerase reaction buffer, dNTP mix (100 micro moles), primers PA and PH (100 ng each) and 0.5U *Taq* polymerase. The reaction master mix should be prepared using the following reagents in the desired concentrations.

Components	Volumes (µL)
Deionized water	79.5
10× *Taq* buffer	10.0
Taq polymerase (0.5U)	0.5
dNTPs (10 mM)	1.0
$MgCl_2$ (50 mM)	6.0
Primer PA (100 ng)	1.0
Primer PH (100 ng)	1.0
Target DNA (50 ng)	1.0

Twenty-five to 35 PCR cycles are performed as follows (Figure 19.31).

Denaturation: at 95°C for 3 min for the initial denaturation, 93°C and 45 s for the remaining cycles.

Annealing: at 62°C for 45 s.

Extension: at 72°C for 45 s (1 min per 1 kb is an average polymerization speed of Taq DNA polymerase).

A final extension of 5 min is performed at the end of the cycles.

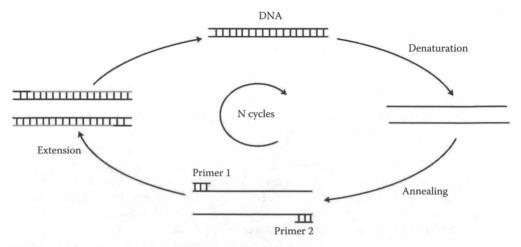

Figure 19.31 PCR reaction for fungal spores.

19.26.4 Result

Examine the amplified DNA by horizontal electrophoresis on 0.8% agarose.

19.27 To Demonstrate the Nested PCR Reaction

19.27.1 Introduction

The aim of the nested PCR reaction is to increase the specificity of the amplification reaction by performing two PCR amplifications one after the other. The first PCR reaction is performed as previously described; but for the second reaction, the amplification products obtained in the first amplification cycles are used as template, after a dilution of up to 10^3, an internal primer. In this way, the specificity of the amplification is increased as the target DNA to be amplified requires to possess the three primer binding, the efficiency of the amplification is increased as the number of cycles can be increased, without loss of specificity.

19.27.2 Materials

19.27.2.1 Solutions and Reagents
- TE lysis solution (100 mM Tris pH 8.0, 10 mM EDTA pH, 10 mM NaCl, and 2% SDS) or use commercially available solution
- Buffered phenol (pH 7) commercially available
- Chloroform/isoamyl alcohol (24:1)
- Ethanol or isopropanol
- 70% ethanol
- TE buffer (1 ×: 10 mM Tris, 1 mM EDTA)
- Glycogen

19.27.3 Protocol

19.27.3.1 Extraction of DNA from Fungal Spores

1. Take approximately 400 mg of fungal mycelium from the outermost area of a colony and place in a screw-capped tube containing 600 μL of DNA lysis TE solution with 300 mg of 0.5 mm diameter glass beads.
2. Triturate the mycelium by beating in a minibead-beater for 4 × 30 s at 5000 beats/min. Cool the tube on ice between each cycle.
3. After the final beating, incubate the disrupted fungal mycelium at 65°C for one hour, then extract with buffered phenol (pH 7) until there is no evidence of protein at the phenol/lysis buffer interface.
4. Extract with chloroform/isoamyl alcohol (24:1) to remove any remaining traces of phenol.

5. Finally, precipitate the DNA with alcohol (ethanol or isopropanol), wash the pellet with 70% ethanol, and resuspend in 50–100 μL TE buffer. DNA concentration can be established using a bench spectrophotometer.

19.27.3.2 Extraction of DNA from Plant

19.27.3.2.1 Protocol

1. Individually brush root pieces in water to remove large pieces of adhering debris, then transfer to tubes containing deionized water and place in a sonicating water bath at 47 kHz for 3 min. This step removes hyphal fragments adhering to or protruding from the root cortex and should be confirmed by subsequent microscopic examination.
2. Place the root pieces individually in screw-capped tubes containing 600 μL of DNA TE lysis solution, 300 mg of 0.5 mm diameter glass beads and 500 mg 1.0 mm diameter zirconium beads.
3. Triturate the root material by beating in a minibead beater for 4×30 s at 5000 beats/min and cool on ice between each cycle. After the final beating, incubate the disrupted root tissue at 65°C for one hour, beat once more and treat as described for the extraction of DNA from fungal isolates.

19.27.3.3 DNA Extraction from AMF Spores

1. Extract spores from the soil or soil medium by wet sieving. DNA can be extracted from single or multiple spores in a similar manner.
2. Place the spores in a sonicating water bath at 47 kHz for 3 min to remove surface debris then transfer to a 0.5 mL tube and crush with a micropestle in 50 μL DNA lysis solution.
3. Centrifuge the crushed spore solution through glass wool to remove fragments of the spore wall and incubate the aqueous phase at 65°C for 1 h.
4. Extract the solution once with chloroform to remove protein and precipitate the DNA with one volume of isopropanol and 20 mg of glycogen acting as an inert carrier.
5. Wash the nucleic acid pellet in 70% isopropanol and resuspend in 20 μL TE.

19.27.3.3.1 PCR Parameters
19.27.3.3.1.1 Solutions and Reagents
- 10× PCR buffer (constituents differ depending on enzyme type)
- 4 mM stock dNTPs
- DNA polymerase
- Sterile water
- PCR primers
- Template at correct concentration

The PCR parameters used are as follows: 96°C for 55 s, 61°C for 55 s, 72°C for 45 s—10 cycles; next 20 cycles—anneal temperature reduced to 59°C and extension time increased to 2 min; and final 15 cycles—anneal temperature reduced to 58°C and extension time increased to 3 min.

All amplifications should be carried out in a volume of 20 µL using 5 ng template DNA, 20 µM dNTPs, 0.4U DNA polymerase, and 20 pmol of each primer. The thermocycler with heated lid did not require any oil overlay.

19.27.4 Result

The quality of PCR products should always be checked by agarose gel electrophoresis.

Suggested Readings

Acquaah, G. 2004. *Understanding Biotechnology*. Upper Saddle River, NJ: Prentice Hall.

Bloom, M., G. Freyer, and D. Micklos. 1996. *Laboratory DNA Science*. San Francisco, CA: Benjamin Cummings.

Brown, T. 2006. *Gene Cloning and DNA Analysis: An Introduction*. Cambridge, MA: Blackwell Publishers.

Dahl, H. H., R. A. Flavell, and F. G. Grosveld. 1981. The use of genomic libraries for the isolation and study of eukaryotic genes. In: *Genetic Engineering*, Volume 2, Williamson, R. (ed.), London: Academic Press, pp. 49–127.

Jacquot-Plumey, E., D. van Tuinen, S. Gianinazzi, and V. Gianinazzi-Pearson. 2000. Monitoring species of arbuscular mycorrhizal fungi in planta and in soil by nested PCR: Application to the study of the impact of sewage sludge. *Plant Soil* 226: 179–188.

Nathans, D., and H. O. Smith. 1975. Restriction endonucleases in the analysis and restructuring of DNA molecules. *Annu Rev Biochem* 44: 273–293.

Newton, C.R. and A. Graham. 1994. In: *PCR*. Oxford, UK: BIOS Scientific Publishers, Ltd. p. 13.

Patten, C. L., B. R. Glick, and J. Pasternak. 2009. *Molecular Biotechnology: Principles and Applications of Recombinant DNA*. Washington, DC: ASM Press.

Russell, D. W., and J. Sambrook. 2001. *Molecular Cloning: A Laboratory Manual*. Cold Spring Harbor, NY: Cold Spring Harbor Laboratory.

Seidman, L., and Moore, C. 2000. *Basic Laboratory Methods for Biotechnology*. Upper Saddle River, NJ: Prentice Hall.

Turnau, K., P. Ryszka, V. Gianinazzi-Pearson, and D. van Tuinen. 2001. Identification of arbuscular mycorrhizal fungi in soils and roots of plants colonizing zinc wastes in southern Poland. *Mycorrhiza* 10: 169–174.

van Tuinen, D., B. Zhao, and V. Gianinazzi-Pearson. 1998. PCR in studies of AM fungi: From primers to application. In *Mycorrhiza Manual*, ed. A. K. Varma. Heidelberg, Germany: Springer-Verlag, pp. 387–400.

van Tuinen, D., E. Jacquot, B. Zhao, A. Golotte, and V. Gianinazzi-Pearson. 1998. Characterization of root colonization profiles by a microcosm community of arbuscular mycorrhizal fungi using 25S rDNA-targeted nested PCR. *Mol Ecol* 7: 103–111.

Watson, J. D. 2007. *Recombinant DNA: Genes and Genomes: A Short Course*. San Francisco, CA: W.H. Freeman.

Wirth, R., A. Friesenegger, and S. Fiedler. 1989. Transformation of various species of gram-negative bacteria belonging to 11 different genera by electroporation. *Mol Gen Genet* 216: 175–177.

Important Links

https://www.neb.com/products/N8082-pGLuc-Basic-Vector
http://www.ornl.gov/sci/techresources/Human_Genome/elsi/cloning.shtml
http://agbiosafety.unl.edu/education/clone.htm
http://learn.genetics.utah.edu/content/tech/cloning/
http://www.addgene.org/plasmid_protocols/PCR_cloning/

20

Molecular Biology

20.1 Introduction

The rapid and fascinating advances in molecular biology have ushered in an exciting era with boundless opportunities of harnessing the already existing technology and also the expected developments in biological sciences, for further benefit to mankind. Molecular biology in the broadest sense can be used to describe the study of biology at the molecular level. Remarkable success has already been achieved in unraveling the genome organization in diverse groups of organisms, regulation and molecular mechanisms of gene expression, transferring desirable genes from one organism to another and thus construct transgenic with a view to improve the commercial performance of the organism, and also alter very minutely the inherent genes through site-directed mutagenesis to enhance biological efficacy of the gene products. All these remarkable accomplishments owe their success to development of highly innovative experimental techniques that has been used in a variety of ways. It is a well-established fact that the entire genetic information is embedded in the chromosomes and is not expressed at a constant rate at all the time during the life cycle of an organism (Figure 20.1).

There are quantitative and qualitative changes in expression of various genes at different developmental stages in different organs and in response to various chemical and physical stimuli. This well-programmed and stringently controlled expression of genetic information is central and of immense importance in cell differentiation and modulation of metabolism to requirements of an organism under fluctuating internal and external environment. Identification of genes and regulation of their expression form an integral and important aspect of cellular development and metabolic adaptation. Which genes are being expressed at a given time is usually established by estimating translational products viz. protein by Western blotting or measurement of transcripts by Northern blotting. These techniques along with cDNA cloning and Southern blotting also provide insights into the characterization and organization of gene and gene families, estimation of gene copy numbers and for determination of chromosomal location of structural genes. Polymerase chain reaction (PCR) is another revolutionary technique which amplifies DNA *in vitro* and has wide applications in diagnostics, fingerprinting, sequencing, mutagenesis, and so on. Another group of techniques such as

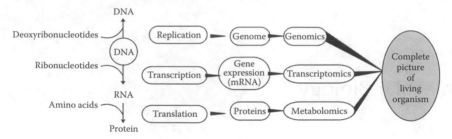

Figure 20.1   Fundamentals of molecular biology.

restriction fragment length polymorphism (RFLP) and randomly amplified polymorphic DNA (RAPD) provide a way to directly follow chromosomal segments during recombination and are finding applications in construction of molecular maps and map-based cloning of genes. Molecular biology can be considered as a discipline at the interface of genetics and biochemistry or is a way of defining an area of biochemistry, rather like enzymology, immunochemistry, or intermediary metabolism. Experiments on molecular biology are a rapidly expanding area of biology and the protocols described here are only a small selection from this field of study, and this chapter deals with some protocols on these techniques.

20.2 Isolation of Nucleic Acids from Biological Samples

Nucleic acids are present as nucleoprotein complexes in cells. The major problem encountered in isolation of pure and intact DNA or RNA molecules are degradation of high-molecular-weight nucleic acids by mechanical damage or by hydrolytic action of nucleases, contamination of DNA preparations with RNA and vice versa, and contamination of nucleic acids with proteins, polysaccharides, and other high-molecular-weight compounds. Methods have been devised for isolation of nucleic acids from different sources taking adequate precautions to eliminate or minimize the earlier problems. The main steps involved in isolation of nucleic acids are described in the following sections.

20.2.1 Homogenization or Disruption of Cells

The procedure employed for breakage of cells should be as gentle as possible. In the case of plant material, this is generally done by grinding it in a pestle and mortar or a Waring blender at low temperature in buffers containing EDTA when isolating DNA for chelating Mg^{2+} ions which are required for DNase activity. For isolation of RNA, an inhibitor of RNase such as bentonite, diethylpyrocarbonate, and placental RNase inhibitor is included in the extraction buffer. Disintegration of bacterial cells can be achieved by treating them with cell coat–hydrolyzing enzyme, that is, lysozyme in the presence of a detergent.

20.2.2 Dissociation of Nucleoprotein Complexes

The approach employed is such that the proteins either get dissociated or degraded while nucleic acids remain unaffected and intact. This is generally achieved by using detergents like SDS, phenol, or broad-spectrum proteolytic enzymes such as pronase or proteinase K. Alkaline pH and high concentration of salts improve efficiency of the process.

20.2.3 Removal of Contaminating Materials and Precipitation of Nucleic Acids

Proteins are removed by treatment of tissue homogenate with phenol or mixture of chloroform–isoamyl alcohol or phenol–chloroform. Upon centrifugation, the denatured proteins form a layer at the interface between aqueous and organic phases, lipids and other contaminants remain in the organic phase while nucleic acids are recovered in the aqueous phase from which they are precipitated with ethanol. For isolation of DNA, the contaminating RNA is removed by selective salt precipitation or treatment with DNase-free RNase, Conversely, while isolating RNA, the preparation is incubated with DNase for eliminating DNA as an impurity.

20.3 Demonstration of Southern Blotting

20.3.1 Introduction

The Southern blot is used to verify the presence or absence of a specific nucleotide sequence in the DNA from the different sources and to identify the size of the restriction fragment that contains the sequence. Southern blotting combines transfer of electrophoresis-separated DNA fragments to a filter membrane and subsequent fragment detection by probe hybridization. The transfer step of the DNA from the electrophoresis gel to a nitrocellulose membrane permits easy binding of the labeled hybridization probe to the size-fractionated DNA. It also allows for the fixation of the target-probe hybrids, required for analysis by autoradiography or other detection methods. Blotting technique was first discovered by Southern in 1975 and hence called Southern blotting (Figure 20.2).

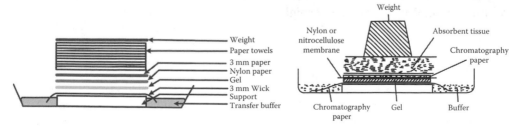

Figure 20.2 Demonstrating the transfer of electrophoresis-separated DNA fragments to a filter membrane.

20.3.2 Materials

Stable (room temperature [RT])

1. H_2O: Double distilled water, sterile.
2. Stop buffer (for labeling): EDTA, 0.2 M, pH 8.0, sterile.
3. Maleic acid buffer (1×), washing buffer (1×), detection buffer (1×) (for detecting): Shake the stock solution to suspend the buffer components; dilute them 10-fold with double distilled water.
4. Depurination solution: 250 mM HCl.
5. Denaturation solution: 0.5 M NaOH, 1.5 M NaCl.
6. Neutralization solution: 0.5 M Tris–HCl (pH 7.5), 1.5 M NaCl.
7. 10× SSC, 2× SSC: dilute 20× SSC with double distilled water.
8. Low stringency buffer: 2× SSC containing 0.1% SDS, filter sterile.
9. High stringency buffer: 0.5× SSC containing 0.1% SDS.
10. TE buffer: 10 mM Tris, 1 mM EDTA, pH ~8.
11. Stripping buffer: 0.2 M NaOH, 0.1% SDS.
12. TAE buffer: 1 × 40 mM Tris acetate with 1 mM EDTA.
13. Ethidium bromide (EtBr): Dissolve 10 mg of EtBr per mL of the 1× TAE buffer. Gloves must be worn while preparing this solution.

Prepare fresh

1. Blocking solution (1×): Dilute 10× blocking solution 1:10 with maleic acid buffer.
2. Antibody solution (150 mU/mL): Centrifuge anti-digoxigenin (DIG)–alkaline phosphatase (AP) for 5 min at 10,000 rpm in the original vial prior to each use, and pipette the necessary amount carefully from the surface. Dilute anti-DIG–AP (750 units/mL) 1:5000 in blocking solution (add 4 µL to 20 mL blocking solution) (stability: 2 h, 4°C).
3. Color substrate solution: Add 200 µL of the NBT/BCIP stock solution to 10 mL of detection buffer. Keep away from light.
   Note: Do not allow the membrane to dry at any time in following procedures.

20.3.3 Protocol

Day 1

20.3.3.1 Digestion of DNA
Digest 10 µg of genomic DNA with 1 µL of the appropriate enzyme making the final volume 40 µL. Run the digest overnight.

Day 2

20.3.3.2 Agarose Gel Construction and Preparation
1. Make sure the gel box sits level (use paper towels to adjust until level).
2. Cast a 0.8% agarose gel in 1× TAE (for the minigel boxes, 100 mL are needed and the regular size gel boxes need 150 mL). Heat the solution in the microwave for ~3 min until all the agarose is dissolved BUT the solution does not boil. Cool off with cold tap water until the bottom is cool enough to touch with a bare hand. Pour into

appropriate gel box and wait for 30 min. Gel will be cloudy when it is finished. Move gel (along with the U-shaped caster) to the gel box. Fill the box with 1× TAE. Make sure to remove all bubbles below the gel caster.

20.3.3.3 Loading, Running, and Staining of the Gel
1. Remove samples from the 37°C water bath.
2. Add 10 µL of loading dye to sample.
3. Heat samples for 1–2 min at 65°C.
4. Spin the samples down and load 50 µL into each well. Remember to load a ladder (13 µL TE pH 8.0; 2 µL 1 kb ladder; 5 µL loading dye) in one of the wells.
5. Run gel at 64 V for about 5 h (make sure that the mA do not stray from 50). The dye front should be at least two-third of the way into the gel before stopping.
6. Remove gel from box and caster and put into *staining solution* (50 µL of EtBr in 200 mL of TAE).
7. Rock at speed 10 for 30 min.

20.3.3.4 Preparation of the Gel for the Pictures and Transfer
1. Look at the gel under the UV box.
2. Cut off the wells and orientate a glow ruler between the ladder and samples to show the running distances of all the bands. Take a Polaroid picture of the gel under the high setting of the UV box. Once the picture shows the gel in a decent exposure, cut off the well and nick the upper left corner (orientation later).
3. Place in *acid nicking and destaining solution* (5 mL of 10 M HCl in 200 mL of distilled water) for 10 min rocking.
4. Remove *acid nicking/destaining solution* and add enough *neutralize/transfer (N/T) buffer* to cover the gel and rock solution at speed 10 for 30 min. While that is occurring, cut out two (9 × 14 cm) membranes. Wet the filter paper first in dH₂O for 10 min followed by 10 min in NT buffer.

Day 3

20.3.3.5 Membrane Preparation
1. Create neutralization solution and make sure that the stack of paper towels are not soaked completely through. Add more paper towels if so.
2. Place blot in neutralization solution for 10 min, but do not rock on rocker.
3. Place blot in 2× SSC and write the sample set along with the enzyme used.
4. Store at 4°C until needed.

20.3.3.6 Neutralization Solution
2 M Tris (pH 7.5) 25 mL
5 M NaCl 20 mL
Distilled water 25 mL
Titrate with 6 M HCl carefully to pH 7.0
Bring volume to 100 mL

20.3.3.7 Clean Up the Probe
1. Invert a G-50 quick spin column several times to resuspend the medium.
2. Remove the top cap and finally the bottom cap.

3. Place column in an empty collection tube.
4. Let drain by gravity and dump the eluate.
5. Put column back into collection tube.
6. Spin in the Beckman tabletop centrifuge for 2 min at maximum setting. Discard eluate and spin again for 2 min.
7. Discard tube and replace with a new one.
8. Behind the radioactive shield, load the probe to the center of the column.
9. Carefully transport the tube with shielding (use the top of the radioactive vessel) to the Beckman centrifuge. Spin at max for 5 min.
10. Transport the tube back to the hood (use shielding).
11. Discard the column properly as radiation waste. Make sure that the probe is clear (a pink color is bad = your probe still in column).
12. Use the scintillation counter to determine the amount of radiation present. Add 1 μL of probe to 4 mL of scintillation fluid. Invert to mix and place in counter. Run the machine. The machine will give you the amount of counts per minute per microliter.

20.3.3.8 Preparation of the Prehybridization/Hybridization of the Membrane

1. For a medium membrane (11.5 cm × 8.5 cm), place 160 μL into a 1.5 mL screw-cap tube.
2. Boil for 10 min; place on ice for 10 min.
3. Add 150 μL of salmon sperm to 15 mL of hybridization/prehybridization solution.
4. Mix well by inverting the tube.
5. Place the membrane on a similar sized mesh, roll, and place into a hybridization tube.
6. Rotate the tube so that the membrane adheres to the side of the tube.
7. Pour in the hybridization/prehybridization solution with salmon sperm.
8. Place in hybridization oven. Make sure to orient the tube so that the rotation of the oven does not reroll the membrane upon itself.
9. Prehybridize at 65°C for 90 min.

20.3.3.9 Preparation of the Labeled Probe

1. Standard Southern probe
   a. Aliquot out enough labeled probe that when added to the hybridization/prehybridization solution will bring the concentration between 7.5×10^5 and 1×10^6 counts per mL of solution.
   b. Place in labeled screw top tube.
   c. Boil for 10 min; ice for 10 min.
   d. Pour 200 μL of hybridization/prehybridization solution into the aliquoted probe and pour back into hybridization tube. Be careful not to touch the sides of the membrane while doing this.
   e. Roll overnight at 65°C.
2. Tough Southern probes
   Aliquot out enough labeled probe that when added to the hybridization/prehybridization solution will bring the concentration to 10 million counts per 15 mL solution. Then add the following in order:
   24 μL dH$_2$O
   25 μL COT DNA
   25 μL 20× SSC

25 µL probe (add TE pH 7.4 if needed)
1 µL SDS

Boil the solution for 10 min, place in hybridization oven at 65°C for 4 h. Pour 200 µL of hybridization/prehybridization solution into the probe and pour back into hybridization tube. Be careful not to touch the sides of the membrane while doing this. Roll overnight at 65°C.

Day 4

20.3.3.10 Washing of Blot
1. Dump the hybridization solution in the radioactive waste container being careful not to drip.
2. Use a paper towel to catch the drips off the lip of the tube.
3. Add 30 mL of 2× SSC to hybridization tube.
4. Roll in hybridization tube for 10 min at RT.
5. Dump solution in radioactive waste.
6. Wash with 30 mL of 2× SSC at RT for 10 min.
7. Dump solution in radioactive waste.
8. Listen to the blot with Geiger counter to check activity. The counts should be around 10–50 counts per second and there should be bands of irradiation peaks.
9. If the counts are higher than 50 per second or no bands can be heard, more washing is needed as given in the following list (remember that it is important to go slow so not to completely wash off all the probes):
    a. 30 mL of 2× SSC + 1% SDS at 65°C for 15 min
    b. 30 mL of 2× SSC + 1% SDS at 65°C for 15 min
    c. 30 mL of 2× SSC + 0.1% SDS at 65°C for 10 min

20.3.3.11 Wrapping and Preparing of the Blot for Developing Film
1. Place a new Saran wrap down on cart.
2. Place blot face down and make sure there is no bubbles.
3. Carefully wrap the blot trimming excess from the edges.
4. Tape blot to the filter paper (the size of the film cassette—8 × 10 in.).
5. Take to dark room and place film on blot making sure to crease the top right corner of the film for reference.
6. Place cassette with film and blot into the −70°C freezer overnight.

20.3.3.12 Develop Film
1. Take film and develop.
2. Mark on the film the samples/enzyme/date/time of the exposure.
3. If needed, place another sheet of film down on the blot and return it to the freezer and let expose over the weekend.

20.3.4 Result and Observation

The DNA-digested fragments transferred from gel to nitrocellulose membrane are visualized on the film making replica of gel pattern.

20.4 Demonstration of Experiment on RFLP Analysis

20.4.1 Introduction

An RFLP is defined by the existence of alternative alleles associated with restriction fragments that differ in size from each other. RFLPs are visualized by digesting DNA from different individuals with a restriction enzyme, followed by gel electrophoresis to separate fragments according to size, then blotting, and hybridization to a labeled probe that identifies the locus under investigation. An RFLP is demonstrated whenever the Southern blot pattern obtained with one individual is different from the one obtained with another individual. RFLPs were the predominant form of DNA variation used for linkage analysis until the advent of PCR. Even now, in the PCR age, RFLPs provide a convenient means for turning an uncharacterized DNA clone into a reagent for the detection of a genetic marker. The main advantage of RFLP analysis over PCR-based protocols is that no prior sequence information, nor oligonucleotide synthesis, is required. Furthermore, in some cases, it may not be feasible to develop a PCR protocol to detect a particular form of allelic variation, DNA isolation and purification. A typical RFLP analysis comprises following five major steps: (i) DNA isolation and purification, (ii) PCR amplification and restriction enzyme digestion, (iii) separation and detection of the digested products via electrophoresis, (iv) analysis of data to generate the fragment profile for each sample, and (v) clustering analysis based on the profile of samples from step (iv). The actual identification of the resistant cultivars came into picture with the evolution of the molecular markers.

20.4.2 Types of Markers

20.4.2.1 Morphological Markers
These are usually dominant or recessive. These are generally confined to the qualitative traits and can be scored visually. They include the genes for dwarfism and leaf morphology, etc. But these are highly influenced by the environment. These will mask the effect of minor genes, hence not advisable in breeding programs.

20.4.2.2 Biochemical Markers
In recent years, isozymes are used as the biochemical marker. These markers are developmental and environmental-specific. These are highly influenced by the environment. The isozyme banding pattern reveal the genetic diversity among the different genotype and to identify the disease/pest-resistant susceptible cultivars with the use of polymorphism obtained in isozyme profiles.

20.4.2.3 Molecular Markers
It is a DNA sequence that is readily detected and whose inheritance can be easily monitored. It depends upon the polymorphism present in the DNA.

20.5 Restriction Fragment Length Polymorphism

20.5.1 Experimental Outline

RFLP variation is environmentally independent. RFLP technically includes the following steps: (1) DNA isolation; (2) production of DNA fragments using restriction enzyme(s); (3) the enzyme is used in a range of 4 units/mg of DNA and the reaction can be proceeded for about 3 h or preferably oil. Digestion of the DNA will be carried at optimum conditions depending upon the enzyme as specified by the respective company; (4) choice of the enzymes depends on the ability to detect the polymorphism, for example, *EcoRI*, *EcoRII*, and *HindIII*, etc.; (5) DNA fragment separation on agarose gel; (6) transfer of DNA fragments onto nylon/nitrocellulose membrane; (7) visualization of DNA fragments using labeled probes; and (8) analysis of the results.

20.5.2 Materials

10× TBE buffer 108 g Tris base, 55 g boric acid, 40 mL 0.5 M EDTA (pH 8.0), dissolve in 1 L working solution 1×.

10× Gel loading buffer: 0.25% bromophenol blue (BPB), 0.25% xylene cyanol, 40% sucrose, 0.25% xylene cyanol, 5 L of loading/25 µL/sample (1:5 ratio).

20.5.3 Protocol

1. Isolate DNA method already described elsewhere.
2. Choose appropriate restriction enzyme for digestion.
3. Total volume of the reaction mixture is 25 mL.
4. Calculate DNA stock required for 10 mg DNA to be used.
5. Four units of enzyme/mg of DNA is used.
6. All the components are added in the following order.
7. Spermidine: To inhibit the ability of polysaccharides that hinder the digestion: (a) water, (b) buffer (2.5 µL), (c) DNA, and (d) enzyme.
8. Mix them well by shaking and incubate at 37°C in an incubator.
9. Analyze the samples on 0.8% agarose gel.

20.5.3.1 DNA Fragment Separation on Agarose Gel

1. Sides of the gel casting plate are sealed with a cellophane tape.
2. Select a comb that fixes into the grooves of the gel plate.
3. Add 0.8 g of agarose for 100 mL TBE buffer.
4. Heat to dissolve the agarose and allow it to cool.
5. Add 5 mL (stock solution) to the 0.5× TBE buffer into the gel and remove the comb gently.
6. Remove the cellophane tape and transfer the gel plate to the electrophoresis unit.
7. Add the sufficient quantity of 0.5× TBE buffer to cover the gel completely.

8. Restricted DNA is loaded into wells by mixing it with 1/10 volume of gel loading buffer (BPB). Molecular weight marker, for example, *Hind*III digest can be used.
9. Apply 30 V and run allow it to till the BPB dye reaches to 1 cm away from the end of the gel.
10. In a dark room (with red safety light), place the membrane in a film cassette with the DNA containing surface facing upward and place a sheet x-ray film on top of it. Carefully close the cassette and keep at 70°C for 2–3 days.
11. Develop the film in the dark room tinder and red safety light for 3 min in x-ray developer and rinse briefly with deionized water light for 3 min in x-ray fixer. Rinse with deionized water for 5 min and let the film air-dry.
12. Cut the nitrocellulose/nylon membrane to the same size as the gel and wet it in 20× SSC. Two Whatman papers for the bridge (15×35 cm) and 2×3 Whatman paper (15×19 cm).

If using Hybord^T N$^+$, it can be transferred by 0.4 M NaOH and there is no need to denaturate and renaturate, after washing by 2× SSC membrane which can be used immediately for hybridization.

Gels can be denatured before transfer by incubating in 4% HCl for 20 min and neutralized by 0.4 M NaOH for 20 min.

20.5.3.1.1 Materials
Reagents

1. Denaturation solution
   1.5 M NaCl and 0.5 M NaOH
   20 g of NaOH/L
   87 g of NaCl/L
2. Neutralizing solution
   1.5 M NaCl and 0.5 M Tris–HCl (pH 7.5)
   1 M Tris = 121.14 g/L
   0.5 M Tris = 60.57 g/L
3. 20× SSC
   3 M NaCl
   0.3 M sodium citrate (pH 7.2)
   NaCl: 175.3 g/L
   Sodium citrate: 83.23 g/L

20.5.3.2 Hybridization

Temperature 65°C

Composition of Church buffer: $Na_2HPO_4 \cdot 2H_2O$ 31.8 g/L, $NaH_2PO_4 \cdot H_2O$ 7.4 g/L, SDS 70 g/L, BSA 10.0 g/L, EDTA 370 mg/L.
   Washing 2× SSC + 0.1% SDS at 65°C for 15 min.

20.5.3.3 Southern Blotting (Described Earlier)
1. After electrophoresis, incubate the gel with slow shaking for 15 min in 0.5 M HCl to depurinate the DNA.
2. Incubate the gel twice for 30 min (two times in 20× SSC).

3. Incubate the gel twice in neutralizing solution for 20 min.
4. Cut the nitrocellulose nylon membrane to the same size as the gel and wet it in 20× SSC. Two Whatman papers will act as the bridge (15×35 cm) and 2×3 Whatman paper (15×19 cm).
5. The gel is put upside down on the first three Whatman papers using two glass plates.
6. Make sure that the tissues above the Whatman papers are dry.
7. Capillary blot to the membrane over input with 20× SSC.
8. After blotting, the membrane is placed on UV light for the DNA face down. Then the membrane is shortly washed in 2× SSC and packed in Saran wrap (thus can be stored for few days at 4°C).

20.5.3.4 Probe Labeling

Probe labeling is carried out to produce the radioactively labeled DNA. This is dependent on DNA polymerase I (from *Escherichia coli*) to add nucleotide residues to the 3′-hydroxyl terminus (which can be created) when one strand of double-stranded DNA molecule is nicked. By replacing the preexisting nucleotides with radioactive nucleotides, it is possible to prepare ^{32}P-labeled DNA. The radioactively labeled DNA can be used to detect the presence of homologous sequences.

20.5.3.4.1 Protocol

1. Take approximately 25–50 ng of DNA and add 8 μL of sterile distilled water.
2. Boil the probe for 5–10 min to denature the DNA and keep it on ice immediately.
3. Add 1 μL of each dGTP, dTTP, dATP and 5 mL of dCTP ^{32}P.
4. Add 2 μL of buffer and add 1 μL of Klenow enzyme (enzyme should be added at the end).
5. Incubate at 37°C for 30 min.
6. Check the reaction for proper labeling with Gieger Müller (GM) counter.
7. Precipitate the DNA by adding 3 μL of sodium acetate and 100 mL of ethyl alcohol (absolute) and incubate at 70°C for 30 min.
8. Centrifuge and dissolve the precipitate in 100 μL of hybridization buffer (Church buffer).
9. Boil the probe and add to the filter.

Notes: (a) Concentration of washing solution and time vary depending upon the probe used. That is homologous or heterologous probe. (b) After washing, the filter is exposed to the autoradiography to observe the signals.

20.5.3.5 Autoradiography

Radioactive nucleic acids and proteins can be detected by autoradiography. The method is being used extensively in molecular biology because of the high sensitivity and much better resolution.

20.5.3.5.1 Materials

1. *X-ray film developer*: Water at 52°C 500 mL, Metol 1 g, sodium sulfite 75 g, hydroquinone 9 g, sodium carbonate (monohydrate) 25 g, potassium bromide 5 g. Chemicals are dissolved in the order given earlier and make up the volume to 1 L with cold water.

2. *Fixer*: Sodium thiosulfate 80 g, potassium metabisulfate 5 g. Make up the volume to 250 mL with water and filter.
3. *Stop bath*: Water containing 2% acetic acid.

20.5.3.5.2 Protocol
1. Remove the film after exposing and develop it in developer for 3–5 min in dark.
2. Wash the film in stop bath for 1 min and fix the film in fixer for 5 min in dark.
3. Wash the film under running water for 30 min and dry it at RT.
4. Avoid reloading the moist or cold cassette. ^{32}P samples are covered with Saran wrap to protect the cassette from contamination with penetrating radioactivity.

20.5.4 Result

Observe the polymorphic bands on the x-ray film and document the variations.

20.6 Isolation of RNA from Biological Sample

20.6.1 Introduction

Ribonucleic acid (RNA) is made of large biological molecules performing multiple vital roles in the coding, decoding, regulation, and expression of genes. Some RNA molecules play an active role within cells by catalyzing biological reactions, controlling gene expression, or sensing and communicating responses to cellular signals. During the RNA extraction procedure, RNA should be protected against endogenous RNase. Most plant material contains relatively high levels of RNase activity which is normally located in the vacuoles. It has been found that in both RNA extraction procedures the addition of RNase inhibitors is unnecessary, thereby omitted the step. The procedure is suitable for all types of tissues from wide variety of animal (and blood) and plant species. All steps are performed at weak acid pH (MOPS or MES free acids) and at RT (without ice) and without DEPC-treated water. RNA precipitates with lithium chloride (LiCl) for increased stability of the RNA preparation and improvement of cDNA synthesis. The following protocol is designed for small and large tissue samples (tissue volume 10–200 µL), which normally yield about 10–500 µg of total RNA.

20.6.2 Materials

GuTC extraction buffer: 2.5 M guanidine thiocyanate, 1% *N*-lauroylsarcosine (Na salt, Sarkosyl), 0.1 M LiCl, 10 mM EDTA, 0.1 M MOPS, pH 4.6.

Distilled phenol, pH 4.5–6.6.

Chloroform–isoamyl alcohol mix (24:1); 100% isopropanol (isopropyl alcohol, 2-propanol); 70% ethanol; 10 M LiCl; Fresh Milli-Q water (or ultrapure water) or autoclaved 1× TE (0.1 mM EDTA, 10 mM Tris–HCl, pH 7.0) or 1× THE (0.1 mM

EDTA, 2 mM Tris, 8 mM HEPES, pH 7.0). When an ultrafiltration cartridge is utilized at the point of use, the water is suitable for genomics applications (quality at least equivalent to DEPC-treated water) and cell culture.

20.6.3 Protocol

1. Two milliliter Eppendorf Safe-Lock Tube with tissue sample and glass boll freeze at −80°C, grind in the MM300 Mixer Mill for 5 min at 30 Hz.
2. In 2 mL tube with mechanically disrupted tissue sample, add fresh 1.5 mL GuTC extraction buffer, vortex very well, and incubate the samples at 60°C for 10–30 min. Spin at maximum speed on table microcentrifuge for 5–10 min.
3. Transfer 1 mL of the supernatant (the pellet contains polysaccharides and high-molecular-weight DNA) to a fresh tube with 500 μL of phenol, vortex very well, and incubate for 5 min.
4. Add 400 μL of chloroform–isoamyl alcohol, vortex very well for 1 min creating an emulsion (or in the MM300 Mixer Mill at 30 Hz). Spin at maximum speed on table microcentrifuge for 3 min at RT.
5. Transfer the aqueous phase to a fresh microcentrifuge 2 mL tube with 700 μL of chloroform–isoamyl alcohol and vortex well (in the MM300 Mixer Mill for 2 min at 30 Hz). Spin at maximum speed on table microcentrifuge for 2 min at RT.
6. Transfer the aqueous phase to a fresh microcentrifuge 2 mL tube with an equal volume of 2-propanol and mix well. Spin at maximum speed on table microcentrifuge at RT for 2 min. Wash the pellet once with 1.5 mL 70% ethanol. Spin immediately at maximum speed on table microcentrifuge at RT for 2 min.
7. Dissolve the pellet (do not dry) in 400 μL 1× TE at 55°C about 10–20 min, with vortex.

20.6.3.1 Optional

Add an equal volume of 10 M LiCl, mix well, and chill the solution at −20°C for several hours (overnight). Spin at maximum speed on table microcentrifuge for 10 min. Carefully remove and discard supernatant (contains small RNA <200 nt and DNA). Wash pellet with 1 mL 70% ethanol, vortex well, microcentrifuge, discard the ethanol, and do not dry the pellet. Dissolve the pellet in 200–400 μL fresh Milli-Q water or 1× TE.

Load 5 μL of the solution onto a standard (nondenaturing) 1.5% agarose gel with 1× THE buffer to check the amount and integrity of the RNA. Add EtBr to the gel to avoid the additional (potentially RNase-prone) step of gel staining. Load a known amount of DNA in a neighboring lane to use as standard for determining the RNA concentration. Intact RNA should exhibit sharp band(s) of ribosomal RNA.

20.6.4 Results and Observations

1. There is a widespread belief that RNA is very unstable, and therefore all the reagents and materials for its handling should be specially treated to remove possible RNase activity. It is now found that purified RNA is rather stable and, ironically, too much anti-RNase treatment can become a source of problems. This especially applies to

DEPC treating of aqueous solutions, which often leads to RNA preparations that are very stable but completely unsuitable for cDNA synthesis. It is found that simple precautions such as wearing gloves (only for your protection from chemicals), avoiding speech over open tubes, using aerosol-barrier tips, and using fresh 1× TE (or 1× THE) solution or ultrapure Milli-Q water for all solutions are sufficient to obtain stable RNA preparations.

2. When an ultrafiltration cartridge is utilized at the point of use, the water is suitable for genomics applications (quality at least equivalent to DEPC-treated water) and cell culture. The BioPak cartridges has been validated in Millipore laboratories to warrant the production of pyrogen-free (<0.001 Eu/mL), RNase-free (<0.01 ng/mL), and DNase-free (<4 pg/µL) ultrapure water, while maintaining both the resistivity and total organic carbon (TOC) of the treated water, it replaces the lengthy diethylpyrocarbonate (DEPC) treatment process to remove nucleases from purified water. All organic liquids (phenol, chloroform, and ethanol) can be considered essentially RNase free by definition, as is the dispersion buffer containing guanidine thiocyanate.

3. The final concentration of guanidine thiocyanate may need to be optimized for certain plant tissue from 2 to 4 M.

4. The volume of tissue should not exceed 1/5 of the extraction buffer volume. To avoid RNA degradation, tissue dispersion should be carried out as quickly and completely as possible, ensuring that cells do not die slowly on their own. To adequately disperse a piece of tissue usually takes 2–3 min of triturating using a pipette, taking all or nearly all volume of buffer into the tip each time. The piece being dissolved must go up and down the tip, so it is sometimes helpful to cut the tip to increase the diameter of the opening for larger tissue pieces. Tissue dispersion can be performed at RT. The tissue dispersed in extraction buffer produces a highly viscous solution. The viscosity is usually due to genomic DNA. This normally has no effect on the RNA isolation (except for dictating longer periods of spinning at the phenol–chloroform extraction steps), unless the amount of dissolved tissue was indeed too great.

5. Note that isolating genomic DNA and RNA not requires very gentle mixing because the DNA and RNA should not be sheared by vortexing.

6. RNA degradation can be assessed using nondenaturing electrophoresis. The first sign of RNA degradation on the nondenaturing gel is a slight smear starting from the rRNA bands and extending to the area of shorter fragments. RNA showing this extent of degradation is still good for further procedures. However, if the downward smearing is so pronounced that the rRNA bands do not have a discernible lower edge, the RNA preparation should be discarded. The amount of RNA can be roughly estimated from the intensity of the rRNA staining by EtBr in the gel, assuming that the dye incorporation efficiency is the same as for DNA (the ribosomal RNA may be considered a double-stranded molecule due to its extensive secondary structure). The rule for vertebrate rRNA—that in intact total RNA the upper (28S) rRNA band should be twice as intense as the lower (18S) band—does not apply to invertebrates. The overwhelming majority have 28S rRNA with the so-called "hidden break." It is actually a true break right in the middle of the 28S rRNA molecule, which is called hidden because under nondenaturing conditions the rRNA molecule is held in one piece by the hydrogen bonding between its secondary structure elements. The two halves, should they separate, are each equivalent in electrophoretic mobility to 18S rRNA. In some organisms the interaction between the halves is rather weak,

so the total RNA preparation exhibits a single 18S-like rRNA band even on nondenaturing gel. In others the 28S rRNA is more robust, so it is still visible as a second band, but it rarely has twice the intensity of the lower one.

20.7 Extraction of Polysomal RNA of Nuclear Sequences

20.7.1 Introduction

Polysomes are a group of ribosomes joined by a molecule of mRNA containing a portion of the genetic code that is to be translated. Polysomes are found in the cytoplasm during protein synthesis. It may also be defined as an mRNA which has many ribosomes translating it at the same time called a polysome (polysomal RNA). This protocol is optimized for isolation of polysomal RNA from leaves of mature *Arabidopsis thaliana*. If you isolate polysomes from different types of tissue, you may need to optimize the condition. This protocol has been used for maize and tobacco tissues. Polysome isolation from seedlings, and other tissues with low cytoplasmic density, may require concentration of polysomes by centrifugation through a sucrose cushion. The RNA isolated by this protocol can be used for DNA microarray hybridizations, RT-PCR, or RNA blot hybridizations.

20.7.2 Materials

Day 1

Before you start:

1. Cool rotor (Beckman SW55Ti) at 4°C at least 1 h prior to use.
2. Warm 20% detergent mix at 42°C–45°C.
3. Thaw gradients (20%–60% sucrose*) at 37°C for 1 h in a rotor bucket, then cool at 4°C for 1 h (see Sections 5.3.4 and 5.15 for how to make gradients).
4. Label collection tubes (1.7 mL Eppendorf tubes) for each gradient.
5. Wash Corex tubes (30 mL) and rinse with 0.1% DEPC, then autoclave at 121°C for 30 min.

20.7.3 Protocol

20.7.3.1 Polysome Extraction
1. Extraction buffer: 2 M Tris (pH 9.0)* 1 mL, 2 M KCl* 1 mL, 0.5 M EGTA (pH 8.3)* 0.36 mL, double distilled water to 8 mL, mix well and then add β-mercaptoethanol 80 μL, 50 mg/mL cycloheximide in ethanol 10 μL, 50 mg/mL chloramphenicol in ethanol 10 μL, 20% detergent mix,** 0.5 mL mix and add 2% PTE 10% DOC 1 mL,*** heparin 10 mg and double distilled water to 10 mL. *Need to prepare stock solutions and autoclave prior to use. Store at −20°C in aliquots.

2. **20% detergent mix: Triton X-100 10 mL, Brij 35 10 g, Tween-40 10 mL, NP-40 10 mL, double distilled water to 50 mL.
3. ***20% PTE and 10% DOC: 4 mL of poly, 2 g deoxycholine, add double distilled water to 20 mL. No need to autoclave chill microtubes (2.0 mL) and spatulas in liquid nitrogen. Place 750 µL of packed tissue volume of frozen leaf tissue (ground) into a microtube (*do not thaw*), then *immediately* add 1250 µL extraction buffer.

1. Mix the extract with spatula well and place on ice for 10 min. Occasionally mix by inverting tubes (*do not vortex*).
2. Spin samples down for 2 min at 14,000 rpm (4°C).
3. If there is debris that remains in the supernatant, it can be removed by the following optional step. Place the supernatant onto a QIA shredder (QIAGEN) (700 µL/column). Spin the column for 1 min at 14,000 rpm (4°C). Combine the flow through (usually from two columns) into a new 1.7 mL microtube. Briefly mix.
4. Carefully layer 700 µL of sample onto a sucrose gradient. Balance tubes and buckets to 0.03 g by adding either sample or extraction buffer.
5. Centrifuge for 90 min at 50,000 rpm (275,000 g) in Beckman LM-80 centrifuge (Figure 20.3).
6. Turn on fraction collector system at least 30 min prior to polysome analysis.
7. Run blanks and samples on ISCO with sensitivity = 1.0, flow rate = 1.5 mL/min.
8. Fractionate polysomes into 14 microtubes (16 drops/tube).

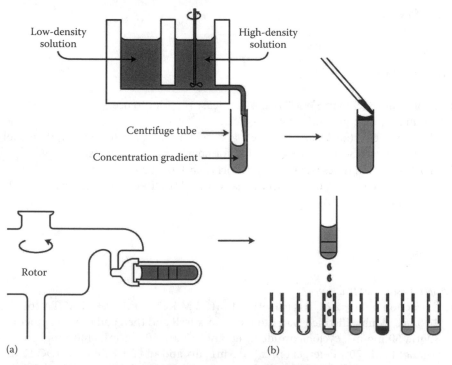

Figure 20.3 Fractionation of the polysomal RNA: (a) formation of gradient and (b) the sample is layered on the top of the gradient.

9. Combine fractions 1–7 (nonpolysomal) and fractions 8–13 (polysomal) in separate 30 mL Corex tubes. (If you wish to perform analyses on individual fractions, you will not want to combine samples at this step.)

10. Add 7 mL (nonpolysomal) or 5 mL (polysomal) 8 M guanidine HCl (0.1% DEPC, filtered, not autoclaved). Vortex for 3 min (seal the tube with parafilm to avoid spilling). (*Note*: If samples were not combined in Step 12, then an equal volume of 8 M guanidine HCl is added to each fraction. RNA precipitation can be performed in 2 mL Eppendorf tubes; add as close to 2 volumes of EtOH, spin at ≥14,000 rpm to pellet sample. Resuspend each pellet in 20 μL DEPC-treated water. This RNA obtained by this method should be pure enough for RT-PCR.)

11. Add 10.5 mL (to nonpolysomal sample) or 7.5 mL (to polysomal sample) 100% ethanol. Vortex for 1 min.

12. Precipitate RNA at –20°C overnight.

Day 2

Before you start:

1. Cool Beckman JA-20 rotor at 4°C.
2. Wash a Corex tube insert for centrifuge rotor with 0.1% DEPC water and autoclave for 30 min.

20.7.3.2 RNA Purification

1. Centrifuge samples at 10,000 rpm for 45 min.
2. Remove supernatant *carefully*. Remove the residual solution by pipette.
3. Invert the Corex tube and let the pellet dry for 20 min.
4. Prepare extraction buffer (EB) by adding β-mercaptoethanol to RLT (QIAGEN) (10 μL/1 mL RLT).
5. Add 450 μL EB to the sample and vortex vigorously.
6. Add 225 μL 100% EtOH to the sample. Mix well by pipetting (*do not vortex*).
7. Apply sample (700 μL) to an RNeasy mini spin column (pink). Let it sit for 3 min.
8. Spin the column for 15 s at 14,000 rpm.
9. Take the flow through and put it back to the column again (double loading). Let it sit for 3 min.
10. Spin the column for 15 s at 14,000 rpm.
11. Add 700 μL RW1 and spin for 15 s at 14,000 rpm. Discard the flow through.
12. Transfer the column into a new 2 mL microtube (without a cap). Add 500 μL RPE (4 volumes of EtOH is already added. See QIAGEN protocol) onto the column and spin for 15 s at 14,000 rpm.
13. Add 500 RPE to the column and spin for 2 min at 14,000 rpm.
14. Place the column in a new 2 mL microtube and spin for 1 min at 14,000 rpm.
15. Transfer the column in a new 1.7 mL microtube and add 50 μL RNase-free water. Let it sit for 5 min.
16. Elute RNA by centrifuge for 1 min at 14,000 rpm (repeat Steps 15 and 16).
17. Take 5 μL RNA solution into a new tube and add 95 μL TE (pH 8.0). Take a spectrophotometer reading at A260 and A280 and calculate RNA concentration.
18. Add 9.5 μL 3 M sodium acetate (pH 5.3) and 190 μL 100% EtOH to the remaining RNA solution (95 μL). Mix briefly and precipitate at –20°C overnight.

Day 3

20.7.3.3 RNA Wash and Final Concentration Determination
1. Centrifuge microtubes for 30 min at 14,000 rpm at 4°C.
2. Wash the pellet with 0.8 mL 75% ethanol.
3. Centrifuge the tube for 10 min at 14,000 rpm at 4°C (repeat Steps 2 and 3 one more time).
4. Quick spin and remove residual ethanol by pipette.
5. Invert tubes and let the pellet dry for 20 min.
6. Add appropriate quantity of RNase-free water (assuming 90% recovery from Day 2) to get approximately final concentration of 1.0 µg/µL.
7. Mix well by pipetting. Let it sit for 20 min at 4°C and vortex 1 min.
8. Take spectral reading (1 µL RNA solution + 99 µL TE8.0) in triplicate and determine the final concentration.
9. Store samples at −80°C until use.

20%–60% Sucrose gradients (5 mL gradients)

Prepare stock solutions

10× Sucrose salts: Trizma base 2.43 g, KCl 0.75 g, $MgCl_2$ 1.02 g, double distilled water 50 mL, adjust pH with HCl to 8.4, autoclave 10 min.

2 M (68.5% Sucrose): Sucrose 171.2 g double distilled water to 250 mL (approximately for 70 gradients).

Make up gradient layers as proportioned as follows. Then make up gradients in 5 mL polycarbonate SW55.1 centrifuge tubes. This is best done by freezing each layer at −80°C before adding the next layer.

Add 1 µL per 10 mL volume of both cyclohexamide and chloramphenicol to each layer.

Layer	2 M 68.5% Sucrose (mL)	10× Sucrose Salt (mL)	Double-Distilled Water (mL)	Volume per Gradient (mL)	% Sucrose
1 (Bottom)	0.88 (44)	0.1 (5)	0.02 (1)	0.75	60
2	1.32 (66)	0.2 (10)	0.48 (24)	1.50	45
3	0.88 (44)	0.2 (10)	0.92 (46)	1.50	30
4 (Top)	0.29 (14.5)	0.1 (5)	0.61 (30.5)	0.75	20

The first value is for one 5 mL gradient. The value in parentheses is for 50 gradients.

20.7.3.4 Electrophoresis on Polyacrylamide Gel
RNA sample of 2–10 µg can be subjected to electrophoresis through either cylindrical or slab of 2.5% polyacrylamide gels in 40 mM Tris–HCl, 40 mM NaH_2PO_4, 2 mM EDTA pH 7.5, 0.2% SDS, 10% glycerol and subsequently scanned at 260 nm gel scanner. Although laborious, though, it is the most reliable method for checking RNA integrity.

20.7.3.5 Electrophoresis on Agarose Gel
RNA sample of ~1 µg can be subjected to electrophoresis through 1%–1.5% agarose gels (in 40 mM Tris, 20 mM hydrogen acetate, 2 mM EDTA pH 8.1) and conveniently stained with EtBr (1 µg/mL).

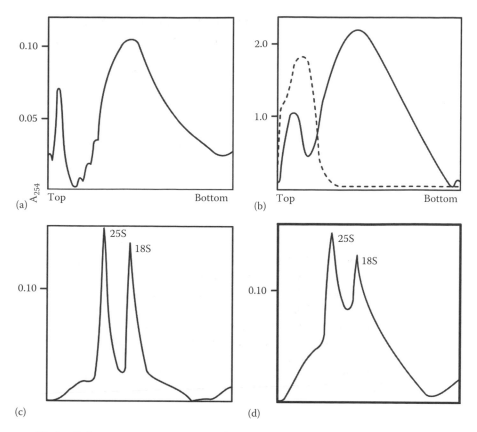

Figure 20.4 Polysome preparation and RNA profiles from etiolated shoots of pea (*P. sativum* L.). (a) Analytical gradient centrifugation of polysomes, (b) preparative sucrose-gradient centrifugation of polysomes, (c) polysomal RNA profile after gel electrophoresis in 2.4% polyacrylamide gels, and (d) same as (c) after enrichment for poly(a) RNA.

20.7.4 Result and Observations

This method is very fast and allows a gross indication of the presence of both ribosomal RNA peaks (Figure 20.4).

20.8 Separation of RNA Species by Ion-Exchange Column Chromatography on Methylated Albumin Kiesselguher Columns

20.8.1 Introduction

Nucleic acids bind to an anion exchanger due to their negatively charged phosphate groups and these can then selectively be eluted by changing the ionic strength of the elution buffer. The method is commonly used for fractionation of different types of RNA. Kiesselguher acts as an inert support material for albumin. The negative charges on albumin are masked by methylation while the positive charges remain exposed and are thus available for binding nucleic acids that are negatively charged.

20.8.2 Materials

Gradient maker, spectrophotometer, glass column, albumin, Kiesselguher, HCl, KOH, NaCl, RNA.

20.8.2.1 Column Preparation

1. Methylated albumin: Prepare by suspending 5 g albumin in 500 mL absolute methyl alcohol and add 1.2 mL of 12 N HCl. The protein dissolves and eventually precipitates again. Allow the mixture to stand in the dark for 3 days or more with occasional shaking. Collect the precipitate in 250 mL centrifuge bottles and wash twice with methyl alcohol and twice with anhydrous ether in the centrifuge. Evaporate most of the ether in air and then *in vaccuo* over KOH till a dry powder is obtained and store it in desiccators over KOH. Failure to remove promptly the residual acid lowers the basicity of the final product. The material is readily soluble in water and is stable both in solution and in dry form. Reproducible batches can easily be made. Make 1% solution of dry powder in water.
2. NaCl solutions (0.1, 0.4, and 2 M).
3. Nucleic acid sample containing ~500 µg RNA in 1–2 mL of 0.4 M NaCl solution.

20.8.3 Protocol

1. Suspend 20 g of Kiesselguher in 100 mL of 0.1 M NaCl (pH 7.0) in a beaker and boil the suspension briefly to expel air. Cover the beaker, cool the solution, and then add 5 mL of 1% (w/v) solution of methylated albumin and stir well. Wash the material thoroughly with 0.4 M NaCl, suspend it in 0.4 M NaCl, and label it as methylated albumin Kiesselguher (MAK) suspension.
2. Boil and then cool a suspension of 6 g Kiesselguher in 50 mL of 0.1 M NaCl. Allow it to cool and then pour it down the column whose base is plugged with glass wool. When Kiesselguher settles down, add 10 mL of MAK suspension (prepared in Step 1) to form a second layer. The third layer in the column is obtained by suspension of 1 g Kiesselguher in 10 mL of 0.4 M NaCl; wash the column with 0.4 M NaCl.
3. Load the nucleic acid sample on surface of the ion exchange and elute the column with a 0.4–2 M linear gradient of NaCl solution. Collect 5 mL fractions and record absorbance of each fraction.
4. Pool fractions showing absorbance at 260 nm of the individual peaks and calculate the amount of nucleic acid in the pooled fractions.

20.8.4 Result and Observations

The nucleic acids are eluted in the following order: rRNA, tRNA, and finally mRNA.

20.9 Northern Hybridization of RNA Fractionated by Agarose Formaldehyde Gel Electrophoresis

20.9.1 Introduction

Northern blots allow to determine the molecular weight of an mRNA and to measure relative amounts of the mRNA present in different samples. RNA (either total RNA or just mRNA) is separated by gel electrophoresis, usually an agarose gel. Because there are so many different RNA molecules on the gel, it usually appears as a smear rather than discrete bands. The RNA is transferred to a sheet of special blotting paper called nitrocellulose, though other types of paper, or membranes, can be used. The RNA molecules retain the same pattern of separation they had on the gel. The blot is incubated with a probe which is single-stranded DNA. This probe will form base pairs with its complementary RNA sequence and bind to form a double-stranded RNA–DNA molecule. The probe cannot be seen but either it is radioactive or it has an enzyme bound to it (e.g., AP or horseradish peroxidase). The location of the probe is revealed by incubating it with a colorless substrate that the attached enzyme converts to a colored product that can be seen or gives off light which will expose x-ray film. If the probe was labeled with radioactivity, it can expose x-ray film directly (Figure 20.5).

20.9.2 Material

Water bath set at 60°C, oblong sponge slightly larger than the gel being blotted, Whatman No. 3 filter paper, nylon membrane, UV transilluminator, UV transparent

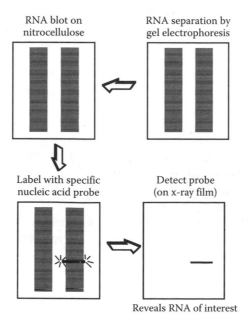

Figure 20.5 Schematic view depicting the Northern blotting.

plastic wrap (e.g., Saran wrap or other polyvinylidene wrap), hybridization oven (e.g., Hybridizer HB-1).

DEPC: 10× MOPS buffer, 1× MOPS buffer.

12.3 M (37%) formaldehyde pH >4

RNA sample: Total cellular RNA or poly(A) RNA, RNA formamide formaldehyde loading buffer, 0.5 M ammonium acetate and 0.8 µg/mL EtBr in 0.5 M ammonium acetate or 10 mM sodium phosphate (pH 7.0) and 0.1 M formaldehyde with 10 ng/mL acridine orange.

0.05 NaOH/1.5 N NaCl (optional).

0.5 M Tris–HCl (pH 7.4) or 1.5 M NaCl (optional).

20×, 2×, and 6× SSC, 0.03% (w/v) ethylene blue in 0.3 M sodium acetate, pH 5.2 (optional).

Prehybridization or hybridization solution: SSC/0.1% (w/v) SDS, 0.2× SSC (w/v) SDS at RT and 42°C, 0.1× SSC, 0.1% (w/v) SDS at 68°C.

20.9.3 Precautions and Preparation

1. Everything must be thoroughly cleaned of RNases, and gloves must be worn at all times.
2. Use RNase–ZAP or some other commercially available RNase cleaner to clean the work area and all equipment that may come into contact, directly or indirectly, with the RNA. If none are available, use a dilute solution of bleach.
3. Clean everything with 95% ethanol. This will also help any residual RNase–ZAP evaporate.
4. Put sign(s) around the work area that stress(es) the importance of a lack of contamination.
5. Also, use only DEPC-treated chemicals.
6. Chemicals can be DEPC-treated by adding 2 mL of DEPC per liter of solution, followed by autoclaving.

20.9.4 Protocol

Following steps are to be followed

1. Pouring the formaldehyde gel (for a 1% gel)
   a. Put 0.50 g agarose into an Erlenmeyer flask.
   b. Add 36 mL of DEPC–water.
   c. Add 5 mL of DEPC–10× MOPS.
   d. Add 9 mL of 12.3 M (37%) formaldehyde (does not need DEPC treatment).
   e. Boil in microwave to dissolve agarose. *Be careful* to inhale as little of the formaldehyde fumes as possible.
   f. Pour gel into setup box with comb in it. If using 66 µL loading samples as described here, three wells must be taped together to form an I using the eight-welled comb in order for the sample to fit. For any sized loading sample, *do not* use the outer wells.
2. Loading the gel (66 µL sample described, but ratios are what is important)

    a. Dilute 30 µg of quantified tRNA to 12.1 µL in DEPC–water.

    b. Add 5.5 µL of 10× MOPS.

    c. Add 9.9 µL of 12.3 M formaldehyde.

    d. Add 27.5 µL of formamide.

    e. Incubate in a PCR machine for 15 min at 55°C.

    f. Add 11 µL of formaldehyde running buffer (RNA-loading dye).

    g. Add to appropriate well, avoiding outer wells.

3. Running the gel

    a. Perform electrophoresis at 100 V for 1 h or until loading dye has run 2/3–3/4 of the way through the gel.

    b. If you wish to stain, use EtBr in DEPC–0.5 M ammonium acetate.

    c. For destaining, use DEPC–0.5 M ammonium acetate.

    d. Before moving gel, let gel cool a little, or it may break because of formaldehyde, gels are much weaker than normal agarose gels. Even after cooling, be exceedingly careful.

4. Vacuum blotting

    a. Rinse the gel in a dish with DEPC–water.

    b. While rinsing, set up the vacuum blotter (clean it for RNases first).

    c. Attach a water trap (side-arm flask) to the vacuum blotter with a tube.

    d. Attach the water trap to the provided regulator port (thing with the pressure gauge) with a tube.

    e. Attach the regulator port to the vacuum (huge object on wheels [actually a freeze-drier]) with a tube.

    f. Cut a piece of Whatman 3 mm paper and a piece of Hybond N⁺ nylon membrane paper on one of the gel patterns. Be especially careful to keep these RNase-free.

    g. Wet the membrane in DEPC–water.

    h. Wet the membrane and the filter paper in DEPC 10× SSC.

    i. Put the porous vacuum plate on the main stage of the vacuum blotter.

    j. Put the wetted filter paper on the porous vacuum plate in the center with long ends of each rectangle parallel to each other.

    k. Put the wetted nylon membrane on top of the filter paper.

    l. Wet the seal o-ring (on the main stage of the vacuum blotter) with DEPC–water.

    m. Put the window gasket that has been precut for the appropriate sized gel on next. The window gasket must overlap the nylon membrane by at least 5 mm, and it should be consistent on all sides. You will probably have to adjust the positioning of the filter paper and the nylon membrane. The window gasket must also cover the entire reservoir seal o-ring to ensure a vacuum.

    n. Place the gel on the nylon membrane such that the edges are overlapped by the window gasket. Be certain that all of the area containing RNA is touching the nylon membrane, but do not move the gel once it touches the nylon membrane.

    o. Place the sealing frame over the entire setup.

    p. Unscrew the bleeder valve on the vacuum regulator.

    q. Hold the gel in place with your gloved hands, while another person turns on the vacuum.

    r. Have them adjust the bleeder until the vacuum regulator reads a pressure of 5 in. Hg, while you hold the gel in place.

    s. While holding the gel in place, add 1 L of DEPC 10× SSC to the transfer reservoir so that the gel is completely covered by the transfer solution.

t. Put the lid on.

u. Allow transfer to occur at 5 in. Hg for 1.5–2.5 h. Effective transfer has not been established outside this pressure or time of transfer range, but the instruction manual contains information suggesting deviation from this protocol is detrimental to effective transfer.

5. Preparation for Probing

a. Disconnect the vacuum from the vacuum blotter, and turn the vacuum off.

b. Remove the sealing frame and allow the transfer solution to drain off and remove the gel.

c. If you wish to estimate transfer efficiency and determine whether RNA was present in the gel to begin with (all of it will not have transferred), you may stain it with EtBr in DEPC–0.5 M acetate and destain it in DEPC–0.5 M ammonium acetate.

d. Remove the window gasket, and remove the nylon membrane.

e. Soak the membrane in DEPC–2× SSC for 5 min or just rinse in DEPC 2× SSC.

f. Wrap the blot in Saran wrap.

g. UV cross-link both sides of the blot in the UV cross-linker. After cross-linking, RNase precautions are no longer necessary for the blot.

h. Rinse all of the blotting equipment with DEPC H20 and dry it. Store all the equipment in an out-of-the-way place not likely to be disturbed by RNase-laden people. Put these things in plastic if possible.

6. Probing

a. For the most part, the Northern blot can be probed as normal.

b. In order to be certain of transfer, each Northern blot should first be probed with a probe of the "gene" encoding 18S rRNA. These probes will work for at least 1 month.

c. Perform all washes as normal (first wash at RT and second wash at 65°C). Put on film for about 1 h for a fresh probe and proportionately less for older probes.

d. After developing, trace the position of the Northern blot onto the developed film, so that you will know where the ribosomal bands are for future probes. There should be one visible band.

e. The darkness of these bands can be used as a loading control.

f. Stop the Northern blot with boiling 0.5% SDS, and allow it to cool to RT.

g. Put it on film for 3 days to ascertain the efficiency of stripping.

h. Probe with a probe of the desired gene as normal.

i. Only perform the first wash 2× at RT.

j. Put on film for 1 day to 1 week depending on circumstances. Generally, 3 days is optimal.

k. Develop and trace the position of the Northern blot onto the film, and compare it with the film when the Northern blot was probed with the ribosomal RNA probe. Some nonspecific binding will probably occur at the location of the ribosomal band, so ignore that and look for other bands. There will also probably be a band toward the lower end of mRNA length. It is not yet known what this band corresponds to, but it has been present every time Northern blots have been probed with cDNA probes.

l. The Northern blot may be probed with other probes after stripping it. But after stripping it four times, there is a significant loss of efficiency of probe binding, probably because some of the RNA is removed each time it is stripped.

20.9.5 Results and Observation

Blot is ready for autoradiography or for further studies.

20.10 Preparation of Radiolabeled Probe by Random Primer Method

20.10.1 Introduction

The random primer method is one of the techniques employed for preparing a labeled DNA probe. The original DNA is first denatured by boiling to obtain a single-stranded DNA template. Synthetic hexanucleotide primers of random sequence are added and annealed with the template. The Klenow enzyme then extends the primers and synthesizes a radioactively labeled complementary strand after incorporation of ^{32}P dCTP from the reaction mixture into elongating chain.

20.10.2 Precautions

1. To keep probe DNA in denatured form, after boiling keep the Eppendorf tube immediately on ice without any undue delay.
2. Radioactive substances are hazardous to health. Hence, necessary precautions such as use of gloves and protective clothing must be taken while handling these materials. Never handle radioactive materials with bare hands.
3. Carry out radioactive work only behind radioactive shield or screen.
4. All glassware plastic ware used for work with radioisotopes should be clearly marked with a radiation symbol.
5. Always wash your hands and monitor yourself before leaving radioactive laboratory.
6. Dispose off radioactive wastes in approved containers provided for the purpose.

20.10.3 Preparation

1. DNA probe: Plasmid DNA containing plant DNA insert (5 ng/μL).
2. Random primer labeling kit: These kits are available commercially and usually contain the following components: random primer mix (10 μL) of random primers in a buffer and mixture of dNTP (dATP, dGTP, dCTP, and dTTP at 100 mM concentration each).
3. λ-DNA *Hind*III digest: It is available commercially or prepared in the laboratory.
4. TE buffer (pH 8.0): Add 0.4 mL of 0.25 M EDTA solution to 1.0 mL of 1 M Tris–HCl buffer (pH 8.0). Make volume to 100 mL with distilled water.
5. SDS (10%): Dissolve 10 g of SDS in 100 mL of water.

Denature the DNA in the Eppendorf tubes for obtaining single-stranded DNA. Incorporate the denatured single-stranded DNA with ^{32}P dCTP in the presence of Klenow enzyme.

20.10.4 Materials

Boiling water bath, water bath which maintains constant temperature (37°C), Eppendorf tubes (1.5 mL), micropipettes (0–10 and 100 μL).

DNA probe: Random primer labeling kit, known enzyme, DNA digest, ^{32}P dCTP (1 μCi/μL) TE buffer, SDS, Klenow enzyme.

20.10.5 Protocol

1. Take 5 μL of probe DNA (25 mg) in an Eppendorf tube with a hole poked in the cap, and add 20 μL water and 8 μL TE buffer to it.
2. Place the Eppendorf tube in a boiling water bath in a floater for 5 min to denature DNA. After this, immediately keep the tube on ice.
3. Add the following reagents, while the tube is on ice (or follow the instructions supplied by the manufacturer of the labeling kit): dNTP mix (6 μL), random primer mix (5 μL), Klenow enzyme (1 μL), and dCTP (5 μL).
4. Incubate the tubes at 37°C for 60 min.
5. Stop the reaction with 2.5 μL of 10% SDS.
6. Add 500 μL of TE buffer and keep in a boiling water bath for 5 min.
7. Cool in ice for 5 min.
8. Label λ-DNA (HindIII digest) in a similar way.

20.10.6 Result and Observation

Probe is now ready for further studies.

20.11 Detection of Specific Fragment of Plant DNA from Restriction Digest by Hybridization with a Labeled Probe

20.11.1 Introduction

Complementary strands of DNA hybridize with each other under suitable conditions of temperature and ionic strength. After blotting, the membrane is treated with radioactive-labeled probe which hybridizes with the specific plant DNA fragment bearing sequence homology with the probe and the radioactive spot on the membrane is detected by autoradiography. Prior to hybridization, the membrane is prehybridized with salmon sperm DNA to block the nonspecific binding sites where the probe may otherwise get bound spuriously. The objective of the experiment is to demonstrate the hybridization of complementary strands of DNA under suitable conditions of temperature and ionic strength.

20.11.2 Materials

X-ray cassette, x-ray film, plastic bags, and boxes; boiling water bath, shaking water bath at 65°C, ultrafreezer (−70°C), Geiger Müller counter, HSB (5×), carrier DNA, Denhardt's III, TES solution, 10× SSC, prehybridization solution.

20.11.3 Precautions

While handling radioisotopes, follow the safety guidelines.

20.11.4 Preparation

1. HSB (5×): This mixture comprises of 3 M NaCl, 100 mM PIPES, and 25 mM Na_2–EDTA. It is prepared by dissolving 175.3 g of NaCl, 30.3 g of PIPES, 7.5 g of Na_2–EDTA in distilled water and adjusting the pH to 6.8 with 4 M NaOH. Make the final volume to 1 L with distilled water.
2. Carrier DNA: To 0.5 g salmon sperm DNA, add 100 mL distilled water and autoclave. Store in refrigerator.
3. Denhardt's III: It contains 2 g gelatin, 2 g Ficoll, 2 g polyvinylpyrolidine, 10 g SDS, and 5 g tetrasodium pyrophosphate. Dissolve these components in water and make the volume to 100 mL.
4. TES solution. Prepare 10 mM Tris–HCl containing 1 mM EDTA and 0.1% SDS.
5. 10× SSC: It comprises of 1.5 M NaCl and 0.15 M sodium citrate. Make solutions of different strength by diluting accordingly.
6. Prehybridization solution: 5× HSB 8 mL, Denhardt's III 4 mL, salmon sperm DNA 4 mL (freshly denatured by boiling for 5 min), sterile distilled water 4.0 mL. Mix well and incubate at 68°C for 5 min.
7. Hybridization buffer: 5× HSB 4 mL, Denhardt's III 2 mL, 25% dextran sulfate 8 mL, salmon sperm DNA 0.5 mL (freshly denatured by boiling for 5 min), sterile distilled water 5.5 mL.
8. Nylon membrane containing DNA fragments: Labeled probe obtained from the last experiment.

20.11.5 Protocol

The experimental procedure consists of the following two steps:

1. Hybridization during which blotted DNA on nylon membranes are incubated with labeled probe:
   a. Place the nylon membrane containing the immobilized DNA fragments on it in a plastic bag, pour 20 mL prehybridization solution, and seal the bag without leaving any entrapped air bubbles.
   b. Incubate the bag at 65°C in a water bath with gentle shaking for at least 2 h.

   c. Denature the probe by incubating in a boiling water bath for 5 min and then, quickly chill it on ice.

   d. Cut open the plastic bag and drain prehybridization solution. Add labeled probe (freshly denatured by boiling for 5 min) in 20 mL hybridization solution and pour the entrapped.

   e. Incubate at 65°C overnight with gentle shaking for hybridization to take place.

   f. Next day, cut open the bag, remove the membrane with forceps, and place it in a box with a lid. Dispose off the plastic bag and the solution in radioactive waste reservoirs.

   g. Wash the membrane successively with 200 mL of the following buffers in a vigorously shaking water bath.
First wash: 2× SSC, 0.1% SDS (20 min at RT)
Second wash: 1× SSC, 0.1% SDS (20 min at 65°C)
Third wash: 0.5× SSC, 0.1% SDS (20 min at 65°C)

2. Autoradiography for location of any radioactive zone on the membrane

   a. For autoradiography, dry the washed membrane between folds of filter paper and then wrap in Saran wrap. Measure the strength of radioactivity signal with a hand Geiger Müller counter.

   b. In a dark room (with red safety light), place the membrane in a film cassette with the DNA containing surface facing upward and place a sheet of x-ray film on top of it. Carefully close the cassette and keep at −70°C for 2–3 days.

   c. Develop the film in the dark room under red safety light for 3 min in x-ray developer, rinse briefly with deionized water, and fix for 3 min in x-ray fixer. Rinse with deionized water for 5 min and let the film air-dry.

20.11.6 Result and Observation

Observe the spots on the film.

20.12 In Vitro Transcription

20.12.1 Introduction

Plasmids are used in *in vitro* transcription. Many plasmid and phagemid vectors are commercially available that contain various combinations of bacteriophage promoters and polycloning sites (e.g., the PGEN series and the PSP series). Some of the vectors also encode the lack of complementing fragment, which allows selection by color of recombinations on plates containing X-gal (5-bromo-4-chloro-3-indolyl-β-D-galactopyranoside). The choice among these vectors is largely a matter of personal preference. However if transcripts of both strands of the template are required, it is better to use a vector carrying two different bacteriophage promoters than to clone the template in two orientations in a vector carrying a single promoter. It is also important to consider the disposition of restriction sites within the template DNA and downstream from it. The 5′ terminus of the transcript is fixed by the bacteriophage promoter but the 3′ terminus is defined by the downstream site of cleavage by the restriction enzyme. By using different restriction enzymes, RNAs of various lengths can be synthesized from a series of linear templates generated from the same polylinker plasmid.

The DNA sequence to be transcribed is cloned into a restriction site in a polycloning region that is flanked by promoters for two bacteriophage polymerases (usually T7 and SP 6). The DNA is then linearized with the restriction enzyme that cleaves downstream from the region to be transcribed. This cleavage prevents synthesis of multimeric transcripts from circular template DNA. Transcription begins from the promoter proximal to the insert when the appropriate polymerase and rNTPs are added in a suitable buffer. The RNA products that are complementary to one strand of the template are equal in length to the distance between the active promoter and the end of the linear DNA distal to the target sequence. At the end of the reaction, the template is removed by digestion with RNase-free DNase. Unincorporated nucleotides and oligodeoxy nucleotides produced by digestion with DNase can be removed by chromatography through Sephadex G-75. However, the presence of plasmid sequences in RNA probes can increase the level of background hybridization to levels that are unacceptable if the probe is to be used to screen a plasmid or cosmid library.

During generation of template DNAs, complete cleavage of superhelical plasmid DNA by a restriction enzyme is essential. Small amounts of circular plasmid DNAs will dramatically reduce yields by producing multimeric transcripts. Restriction enzymes that generate blunt or 5′ protruding termini produce the best linear templates. However, both these types of termini yield RNA products that show heterogeneity at their 3′ ends. Transcription of templates with 3′ protruding termini results in the synthesis of significant amounts of RNA molecules that are aberrantly initiated at the termini of the templates and thus in the production of double-stranded RNA molecules. Restriction enzymes that generate protruding 3′ termini should therefore be avoided. In addition to plasmids, some bacteriophage and cosmid vectors also contain bacteriophage promoters, usually arranged in opposite orientations on either side of the cloning site for foreign DNA. When a recombinant constructed in a vector of this type is digested with the restriction enzyme that cleaves many times within the foreign DNA, a large number of fragments are generated, one of which contains a particular bacteriophage promoter and the foreign sequences that lie immediately adjacent to it. If the fragments do not carry pretending 3′ termini, only the fragment bearing the bacteriophage promoter serves as a template for *in vitro* transcription. The resulting radiolabeled RNA, which is complementary to sequences located at one end of the original segment of foreign DNA, can then be used as a probe to isolate overlapping clones from genomic DNA or cDNA libraries. These vectors greatly simplify the task of "walking" from one recombinant clone to another along the chromosome. The following are two highly efficient methods available to generate strand-specific RNA probes:

1. The relevant DNA fragment may be cloned or subcloned into specialized plasmids that contain promoters for bacteriophage-encoded DNA-dependent RNA polymerases. The recombinant plasmids are a source of double-stranded templates that can be transcribed *in vitro* into single-stranded RNAs of defined length and strand specificity.

2. The DNA fragment to be transcribed may be amplified in PCRs with primers whose 5′ ends encode synthetic promoters for bacteriophage-encoded DNA-dependent RNA polymerase. Following purification, the products of the PCRs are used as double-stranded templates for *in vitro* transcription reaction.

In both cases, the synthesis of RNA is remarkably efficient. When *in vitro* transcription reactions are saturated with ribonucleoside triphosphates (rNTPs), the templates can be transcribed many times, yielding a mass of RNA that exceeds the weight of the template several folds. In addition, because bacteriophage-encoded DNA-dependent RNA polymerase functions efficiently *in vitro* in the presence of relatively low concentrations of rNTPs (1–20 µM), full-length probes of high specific activity can be synthesized relatively inexpensively. Finally, RNA probes can be freed from template DNA by treating the reaction products with RNase-free DNase probes and usually do not need to be purified by gel electrophoresis. However, when using RNA probes of high specific activity to detect rare mRNA transcripts, background hybridization can be kept to a minimum using probes purified by denaturing gel electrophoresis.

The protocol describes procedures for synthesizing RNA probes of high specific activity from plasmids containing promoters for bacteriophage-encoded RNA polymerases.

20.12.2 Safety Guidelines

Clean everything with 95% ethanol. Put sign(s) around the work area that stress(es) the importance of a lack of contamination. Everything must be thoroughly cleaned of RNases, and gloves must be worn at all times.

20.12.3 Troubleshooting: No RNA Is Synthesized

The most common cause of an apparent lack of RNA synthesis is contamination of tubes or reagents with RNase. This contamination can be avoided by taking the precautions.

Less frequently, failure to synthesize RNA is a consequence of precipitation or the DNA template by the spermidine in the 10× transcription buffer. Make sure that the components of the reaction are assembled at RT and in the stated order. If necessary, the presence of soluble template can be confirmed by analyzing an aliquant of the reaction by electrophoresis through an agarose gel.

Transcription is from the wrong strand of DNA. Usually, >99.8% of transcripts synthesized *in vitro* are bacteriophage DNA-dependent. RNA polymerases are derived from the correct DNA strand (Melton et al. 1984). However, this high degree of specificity is only achieved provided templates are both linear and lacking protruding 3′ termini. Contamination of the template with superhelical plasmid DNA causes an increase in aberrant initiation of RNA chains on both strands of the DNA. The presence of a protruding 3′ terminus downstream from the bacteriophage promoter leads to the synthesis of transcripts complementary to the full length of the wrong strand of DNA, that is, four-turn transcripts. Both of these problems can be avoided by careful preparation of the template.

Synthesized transcripts are shorter than the desired length. Synthesis of abbreviated transcripts can be due to the chance occurrence in the template of sequences that terminate transcription by the particular DNA-dependent RNA polymerase being used. Another contributing factor can be limiting concentration of precursor (usually the radiolabeled rNTP). The first of these obstacles may be resolved by constructing a new plasmid in which transcription of the desired sequence is driven by the polymerase from a different bacteriophage. Transcription terminator sequences are not always recognized equally by the various bacteriophage DNA-dependent RNA polymerases.

Although the strengths of transcription termination signals vary greatly, all but the strongest of them can be overcome, at least in part, by increasing the concentration of the limiting rNTP. In most cases, it is impractical to increase the concentration of the rNTP that is radiolabeled in the reaction, since the improvement in yield of full-length product is gained at the expense of reducing the specific activity of the probe. In this case, a number of additional steps can be taken:

1. Lowering the temperature of incubation of the transcription reaction to 30°C.
2. For the sequence to be transcribed to the minimum. In this way, it may be possible to eliminate the termination sequences from the clone.
3. Purity of the desired product is checked by electrophoresis through a polyacrylamide or agarose gel. The transcription reactions are so efficient that it is often possible to purify sufficient quantities of the desired RNA, even if it is only a relatively minor proportion of the total RNA synthesized in the reaction.

20.12.4 Experimental Outline

The DNA used as a template in the *in vitro* transcription reaction need not be highly purified—crude mini preparations work well. The essential requirement is that the template be free of RNase, a criterion that can usually be fulfilled by extracting the preparation of plasmid DNA twice with phenol:chloroform. However, if RNase is added to the plasmid at a later stage in the purification process, that is, after deproteinization, it should be removed by treatment with proteinase K, as follows:

1. Add to the plasmid DNA preparation 0.1 volume of 10× proteinase K buffer (100 mM Tris–HCl—pH 8.0/50 mM EDTA—pH 8.0, 500 mM NaCl), 0.1 volume of 5% (w/v) SDS, and proteinase K (20 mg/mL to a final concentration of 100 µg/mL).
2. Incubate the reaction for 1 h at 37°C.
3. Extract the DNA with phenol:chloroform and recover the DNA by standard precipitation with ethanol.
4. Resuspend the DNA in RNase-free TE (pH 7.6) at a concentration of >100 µg/mL.

20.12.5 Materials

Ammonium acetate (10 N), bovine serum albumin (2 mg/mL fraction V, Sigma), dithiothreitol (1 N), ethanol, phenol:chloroform (1:1 v/v), placental RNase inhibitor (20 units/μL), sodium acetate (3 M, pH 5.2), 10× transcription buffer—(400 mM Tris–HCl [pH 7.5 at 37°C], 60 mM $MgCl_2$, 20 mM spermidine HCl, 50 mM NaCl).

Sterilize 10× buffer by filtration and then store in small aliquots at −20°C. Discard each aliquot after use.

Enzymes and buffers

1. Appropriate restriction enzymes.
2. *DNA-dependent RNA polymerase* of bacteriophage or SP_6. These enzymes are available from several companies and are usually supplied at concentrations of 10–20 units/μL. Most manufacturers also supply a 10× transcription buffer that has presumably been optimized for their particular preparation of the DNA-dependent RNA polymerase. The generic 10× transcription buffer can be used if the manufacturer's buffer is in short supply.
3. *RNase-free pancreatic DNase I (1 mg/mL)*. This enzyme is available from several manufacturers. Alternatively, contaminating RNase can be removed from standard preparation of DNase I by affinity chromatography on agarose 5′-(4-aminophenyl phosphoryl)-uridine-2′(3′)-phosphate.

Gels: Agarose gel (0.8%–1%).

Nucleic acids and oligonucleotides

rNTP solution containing rATP, rCTP, and rUTP each at 5 mM.

rGTP (0.5 mM).

Template DNA

The DNA fragment to be transcribed should be cloned into one of the commercially available plasmids containing bacteriophage RNA polymerase promoters on both sides of the polycloning sequence (e.g., PGEM from Promega or P blue scripts from Strategene).

Radioactive compounds

(^{32}P) rGTP (10 mCi/mL, specific activity 400–3000 Ci/mmol) to minimize problems caused by radiolysis of the precursor. It is best, whenever possible, to prepare radiolabeled probes on the day the (^{32}P) dNTP arrives in the laboratory.

Special equipment

Microfuge tubes (0.5 mL), Sephadex G-50 spun column equilibrated with 10 mM Tris–HCl (pH 7.5), water bath set at 40°C.

20.12.6 Protocol

1. Prepare 5 pmol of linear template DNA by digestion of superhelical plasmid DNA with a suitable restriction enzyme. Analyze an aliquot (100 ng) of the digested DNA by agarose gel electrophoresis. If necessary, add more restriction enzyme and incubate until there is no trace of the undigested DNA.

   Note: Approximately 2 mg of a plasmid 3 kb in length is ~1 pmol.

Important: It is essential that plasmid DNA templates be cleaved to completion, since trace amounts of closed circular plasmid DNA result in the generation of extremely long transcripts that include plasmid sequences. Because of their length, these transcripts can account for a substantial proportion of the incorporated radiolabeled rNTP,

2. If restriction enzymes, such as PSTI or SSTI, that generate protruding 3′ termini must be used, treat the digested DNA with bacteriophage T_4 DNA polymerase in the presence of all four dNTPs to remove the 3′ protrusion.

   Note: Protruding 3′ termini create an opportunity for the DNA-dependent RNA polymerase to transfer to the complementary strand of the template and to generate long-term transcripts with extensive secondary structure.

3. Purify the template DNA by extraction with phenol:chloroform and standard precipitation with ethanol. Dissolve the DNA in double distilled water at a concentration of 100 nm (i.e., 200 µg/mL for a 3 kb plasmid).

4. Warm the first six components listed as follows to RT in a sterile 0.5 mL microfuge tube, and mix in the following order at RT:

   Template DNA: 0.2 mol (400 mg for 3 kb plasmic)

   RNase-free water 10.6 µL, 100 mM dithiothreitol 2 µL, 5 mM rNTP solution 2 µL, 10× transcription buffer 2 µL, 2 mg/mL bovine serum albumin 1 µL, 10 mCi/mL ^{32}P or GTP 5 µL (specific activity 400–3000 Ci/mmol).

   Mix the components or the mixture by gently tapping the outside of the tube. Then add placental RNase inhibitor (10 units) 1 µL, bacteriophage DNA-dependent RNA polymerase 1 µL.

   The components are added in the order shown and at RT to avoid the possibility that the template DNA may be precipitated by the high concentration of spermidine in the transcription buffer.

   Mix the reagents by gently tapping the outside of the tube. Centrifuge the tube for 1–2 s to transfer all of the liquid to the bottom. Incubate the reaction for 1–2 h at 37°C (bacteriophages T_3 and T_7 DNA-dependent RNA polymerases) or 40°C (bacteriophages SP_6 DNA-dependent RNA polymerase).

   Note: The reaction may be scaled from 20 to 5 µL to accumulate more dilute reagents.

   When the reaction is carried out as described earlier, 80%–90% of the radiolabel will be incorporated into RNA. The yield of RNA will be 20 ng (specific activity 4.7×10^9 dpm/mg) when the specific activity of the (^{32}P) GTP is 3000 Ci/mol and ~150 ng (specific activity 6.2×10^8 dpm/µg) when the specific activity of the precursor is 400 Ci/mmol.

5. If full-length transcripts are desired, add 2 µL of 0.5 mM rGTP and incubate the reaction mixture for an additional 60 min at the temperature appropriate for the polymerase.

6. Terminate the *in vitro* transcription reaction by adding 1 µL of 1 mg/mL RNase-free pancreatic DNase I to the reaction tube. Mix the reagents by gently tapping the outside of the tube. Incubate the reaction mixture for 15 min at 37°C.

7. Add 100 µL of RNase-free water, and purify the RNA by extraction with phenol:chloroform.

   Notes: If the probe will be used in experiments where length is important (e.g., RNase protection), purify the radiolabeled RNA by polyethylamide gel electrophoresis. The aim of this extra step is to eliminate truncated radiolabeled molecules from the preparation.

8. Transfer the aqueous phase to a fresh microfuge tube, and separate the radiolabeled RNA from undesired small RNAs and rNTPs by one of the following three methods:

Purification of RNA by ethanol precipitation

1. Add 30 µL of 10 M ammonium acetate to the aqueous phase. Mix, and then add 250 µL of ice-cold ethanol to the tube. After storage for 30 min on ice, collect the RNA by centrifugation at maximum speed for 10 min at 4°C in a microfuge.
2. Remove as much of the ethanol as possible by gentle aspiration and leave the open tube on the bench for a few minutes to allow the last visible traces of ethanol to evaporate. Dissolve the RNA in 100 µL of RNase-free water.
3. Add 2 volumes of ice-cold ethanol to the tube and store the RNA at 70°C until needed.

To recover the RNA, transfer an aliquot of the ethanolic solution to fresh microfuge tube. Add 0.25 volume of 10 M ammonium acetate, mix, and then store the tube for at least 15 min at 20°C. Centrifuge the solution at maximum speed for 10 min at 40°C in a microfuge. Remove the ethanol by aspiration and dissolve the RNA in the desired volume in the appropriate RNase-free buffer.

20.13 *In Vitro* Translation of RNA

20.13.1 Introduction

mRNA molecules are read and translated into protein by complex RNA–protein particles termed ribosomes. The ribosomes are termed as 70S or 80S, depending on their sedimentation coefficient. Ribosomes are composed of two subunits that are held apart by ribosomal binding proteins until translation proceeds. There are sites on the ribosome for the binding of one mRNA and two tRNA molecules and the translation process is in the following three stages:

1. Initiation, involving the assembly of the ribosome subunits and the binding of the mRNA
2. Elongation, where specific amino acids is used to form polypeptides, being directed by the codon sequence in the mRNA
3. Termination, which involves the disassembly of the components of translation following production of a polypeptide

tRNA molecules are also essential for translation. Each of these are covalently linked to a specific amino acid, forming an aminoacyl tRNA and each has an exposed triplet of bases that is complementary to the codon for that amino acid. This exposed triplet is known as the anticodon, and allows the tRNA to act as an "adaptor" molecule, bringing together a codon and its corresponding amino acid. The process of linking an amino acid to its specific tRNA is termed charging and is carried out by the enzyme aminoacyl tRNA synthetase.

In prokaryotic cells, the ribosome binds to the 5′ end of the mRNA at a sequence known as ribosome-binding site or sometimes termed as Shine–Dalgarno sequence, so named after the discoverer of the sequence. In eukaryotes, the situation is similar, although less well understood, but is thought to involve the so-called Kozak sequence. The ribosome appears to scan the mRNA for this sequence before translation. Following translation initiation, the ribosome moves toward the 3′ end of the mRNA, allowing an aminoacyl tRNA molecule to base pair with each successive codon, thereby, carrying in amino acids in the correct order for protein synthesis. There are two sites for tRNA molecules in the ribosome (the A site and the B site) and when these sites are occupied, directed by the sequence of codons in the mRNA, the ribosome allows the formation of a peptide bond between the amino acids. The process is also under the control of the enzyme peptidyl transferase. When a ribosome encounters a termination codon (UAA, UGA, or UAG), a release factor binds to the complex and translation stops, the polypeptide and its corresponding mRNA are released, and the ribosome divides into its two subunits (Figure 20.6).

Since the mRNA base sequence is read in triplets, an error of one or two nucleotides in the positioning of the ribosome will result in the synthesis of an incorrect polypeptide. Thus, it is essential for the correct reading frame to be used during translation. This is ensured in prokaryotes by base pairing between the Shine-Dalgarno sequence and a complementary sequence of one of the ribosomal rRNAs, thus establishing the correct starting point for movement of the ribosome along the mRNA. However if a mutation such as a deletion/insertion takes place within the coding sequence, it will also cause a shift of the reading frame and result in an aberrant polypeptide.

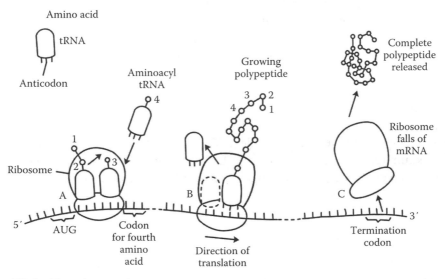

Figure 20.6   Translation ribosome.

20.14 Reticulocytes and Their Use in the Study of Protein Synthesis

20.14.1 Introduction

Reticulocytes are immature red blood cells containing the normal protein-synthesizing machinery typical of eukaryotic cells. They are produced in bone marrow and appear in the blood in large numbers in response to phenylhydrazine-induced anemia. The intact cells will readily incorporate amino acids from a suitable medium and, if the amino acids are tagged with a radioactive atom, protein synthesis can be monitored. As with most eukaryotic cells, the mRNA molecules found in reticulocytes are relatively stable compared to their bacterial counterparts. Thus this cell, whose protein-synthesizing machinery is mainly geared to the synthesis of large amounts of globin, yields a lysate which contains globin mRNA (species for α- and β-chains are present) in addition to ribosomes, tRNA, and other factors. Also present, but in small amounts, are mRNA species for proteins other than globin. Again, protein synthesis may be studied by adding radioactive amino acids to the lysate. The translation process can therefore be studied *in vitro* using the natural endogenous messenger rather than employing exogenous templates.

The reticulocyte lysate system is readily available and has also been widely used for assessing the template activity of exogenous mRNA. However, in this case the very presence of globin mRNA creates a problem, as a high background incorporation of amino acids into globin occurs. Therefore, a measure of overall protein synthesis in terms of incorporation of labeled amino acids into total protein cannot be used as an activity index for added mRNA although this problem can be overcome if product analysis is performed by antiserum precipitation or SDS electrophoresis. An alternative approach is to try and eliminate the endogenous mRNA using ribonuclease, without damaging other types of RNA. Exogenous mRNA is then added as the sole template for translation.

During maturation, *in vivo* reticulocytes lose their ability to initiate globin chains and eventually degrade all mRNA and ribosomes. Mature animal erythrocytes are therefore incapable of protein synthesis, although this ability is retained by avian erythrocytes. Collect blood after 9 days of induction of reticulocytosis and store the reticulocytes for the use in the next experiment after induction of reticulocytosis.

20.14.2 Materials

Rabbits, phenylhydrazine hydrochloride (2.5% w/v adjusted to pH 7 with dilute NaOH), wash medium (0.14 mol/L NaCl, 0.05 mol/L KCl, 5 mmol/L Mg acetate) 250 mL, sterile glassware and centrifuge tubes, syringes, and needles for blood collection.

20.14.3 Protocol

"Reticulocytosis" is induced in rabbits by subcutaneous injections of a 2.5% w/v solution of phenylhydrazine hydrochloride using the protocol: Day 1—1 mL; Day 2—1 mL; Day 3—1 mL; Day 4—1 mL; and Day 5—0.5 mL.

Collect 5–10 mL of blood from an ear on Day 9 where a reticulocyte content of >80% should be obtained. Centrifuge the blood on a bench centrifuge for 7 min at 4°C. Remove supernatant with a Pasteur pipette (autoclaved), taking care to remove any buffy coat which appears at the interface between the packed cells and the plasma. Resuspend the cells in the medium with a glass rod and centrifuge again. Repeat the washing procedure twice or more, and then give a final spin for 15 min.

20.14.4 Results

Discard the supernatant and store the cell on ice until required. If a lysate is required, then retain 0.5 mL of the cells and lyse the remainder detailed later.

20.15 Protein Synthesis in Intact Rabbit Reticulocytes

20.15.1 Experimental Outline

To study the protein synthesis by using labeled leucine, use rabbit reticulocyte.

20.15.2 Precautions

While handling radioisotopes, follow the safety guidelines.

20.15.3 Materials and Prelab Preparations

Rabbit reticulocytes from previous experiment
   Saline solution (0.13 mol/L NaCl, 0.005 mol/L KCl sterile)
   Ferrous ammonium sulfate (0.0105% w/v prepare fresh), trichloroacetic acid (5% w/v), trichloroacetic acid (10% w/v), glass fiber disks (Whatman SF/A 2.1 cm), shaking water baths set at 80°C and 37°C, ethanol:acetone (1:1), oven set at 50°C, scintillation fluid, scintillation vials.
   Amino acid mixture (the mixture is in 100 mL of the saline solution) prepared by weighing out each of the following in mg of each amino acid and dissolved in 100 mL of the saline solution:

Amino Acids	mg/100 mL	Amino Acids	mg/100 mL
Alanine	23	Glycine	50
Methionine	6	Arginine	16
Histidine	63	Phenylalanine	33
Aspartic acid	48	Isoleucine	5
Proline	20	Cysteine	6
Leucine	23	Serine	22
Glutamine	50	Lysine HCl	41
Tyrosine	19	Tryptophan	8
Valine	46	Control	Nil

Master mix 2.70 mL, amino acid mixture 0.15 mL, $MgCl_2$ (0.25 mol/L in 10% w/v glucose)—1.45 mL, Tris–HCl (0.15 mol/L pH 7.6–1.10 mL, trisodium citrate (0.01 mol/L)—1.60 mL, total volume 7 mL.

Radioactive master mix: Dissolve leucine in the master mix to give a final concentration of 1 mmol/L and add 20 µL of ^{14}C leucine with an activity of 50 mCi/mL.

20.15.4 Protocol

Protein synthesis: Mix 0.25 mL of washed reticulocytes with 0.75 µL of radioactive master mix and add 3.75 µL/mL of freshly prepared ferrous ammonium sulfate solution and incubate the mixture at 37°C in a shaking water bath. Remove duplicate 50 µL aliquots at 0, 10, 20, 30, 40, and 60 min (gently mix before sampling), then spot each sample on to a glass fiber disk which is then placed in a scintillation vial. Cover the disks with 4 mL of 5% w/v TCA. When all the samples are collected, wash the disks as follows:

1. 5 mL 10% w/v TCA for 10 min
2. 5 mL 10% w/v TCA at 80°C for 10 min
3. 5 mL ethanol:acetone for 5 min

Please note that the supernatants will be radioactive, so remove with a Pasteur pipette into an appropriate counter.

Dry in an oven at 50°C for 0–15 min, add scintillation fluid, and count the radioactivity in a scintillation counter.

20.15.5 Results and Observations

1. Determine the specific activity of ^{14}C leucine in the incubation mixture.
2. Plot the counts per minute as a function of time and comment on the shape of the graph.
3. Determine the rate of globin synthesis (pmol/min packed cells) assuming 17 leucine residues per globin chain.

Note: 1 µCi $= 2.22 \times 10^6$ disintegrations per minute (dpm).

20.16 Metabolic Labeling of Proteins and Immunoprecipitation

20.16.1 Introduction

Plants and animals have developed mechanisms to withstand physical and chemical stress conditions such as high and low temperatures, heavy metal, radiations, and other pollutants. Many of the adaptive mechanisms are a consequence of stress perception and are likely to be mediated through the stress-induced expression leading to the synthesis of specific stress responsive genes both in plants and animals. Protein turnover is a common mechanism during developmental program of both plants and animals. The protein synthesis can be studied normally by using labeled amino acids. This experiment demonstrates the synthesis of stress protein using methionine as tracer amino acid.

20.16.2 Safety Guidelines

Clean everything with 95% ethanol. Put sign(s) around the work area that stress(es) the importance of a lack of contamination. Everything must be thoroughly cleaned of RNases, and gloves must be worn at all times.

20.16.3 Precautions

While handling radioisotopes, follow the safety guidelines.

20.16.4 Materials

Germinating pea seedlings, Tris–HCl, SDS, phenozene methosulfate (PMSF), β-mercaptoethanol, N,N'-methylene bisacrylamide, TEMED, ammonium persulfate, ^{35}S-methionine, scintillation solution 2,5-diphenyloxazole (PPO), BPB, x-ray sheet (Kodak), deep freezer, homogenizer, high-speed centrifuge, electrophoretic unit, seed germinator.

Extraction buffer (EB): 50 mM Tris pH 8.0, 1% cetyltrimethyl ammonium bromide (CTAB), 50 mM EDTA, 1 mM O-phenanthroline, 0.7 M NaCl, 1% β-mercaptoethanol.

20.16.5 Protocol

1. Germinate pea (*Pisum sativum* L.) on a moist filter paper in a tray at 30°C in a seed germinator. After 48 h of germination, remove the seedling and wash thoroughly. Incubate the seedling with 200 μCi of ^{35}S-methionine with specific activity 500 mCi/mm with activity of 100 μCi/mL for 3 h.
2. After pretreatment with ^{35}S-methionine, wash the seedling with cold methionine followed by water and extraction buffer.
3. Homogenize in 50 mm Tris–HCl buffer (pH 8.0), and centrifuge the homogenate at 10,000g for 10 min. Use the supernatant for further studies.
4. Take an aliquot of supernatant proteins with cold acetone in the ratio of (1:5), and measure the radioactivity of the precipitated protein with scintillation system using a scintillation liquid to quantify the protein yield.
5. Subject another aliquot of protein for SDS-PAGE. Stain with either Coomassie blue or use silver stain for locating the protein bands and determine the molecular weights of individual proteins with the help of molecular markers.

6. Treat the gels with PPO, dry the gels, expose to Kodak x-ray film at −70°C for 48 h, and develop the film in dark.
7. View the newly synthesized protein bands. Carry out immunoprecipitation of these proteins using Western blots.

20.16.6 Western Blots

1. Separate proteins by SDS-PAGE using 12% gel, and transfer to a polyvinylidene difluoride (DVDF) membrane by electroblotting.
2. Carry out Western blotting using polyclonal antiserum containing antibodies for a protein.
3. Detect the new proteins polyclonal antibodies specific to a particular protein and secondary antibody conjugated with AP using 5-bromo-4-chloro-3-indolylphosphate (BCIP) as substrate.

20.17 Induction of β-Galactosidase in Strains of *E. coli*

20.17.1 Introduction

The enzyme β-galactosidase derived from the bacterium *E. coli* is an important catalyst in the breakdown of the 12-carbon sugar lactose into two 6-carbon sugars, glucose and galactose. Bacteria are very adaptable organisms and an alteration in the nutritional environment of some cells leads to a change in the enzyme production enabling the cells to grow under the new conditions. The breakdown of lactose provides an important source of energy for the bacteria. When *E. coli* is grown in a medium lacking lactose, β-galactosidase is either absent or present only in minute quantities. If lactose is added to the growth medium, there is a gradual accumulation of β-galactosidase within the bacterium. The production of β-galactosidase is genetically controlled through the pathway shown later (Figure 20.7).

Many genetically controlled systems are energetically conservative. They remain in an "off" state unless the enzyme is required for a cellular process. Systems of this type are used when a relative slow response is adequate. Other genetically controlled systems are capable of a more rapid response and are in an "on" state unless switched off. The production of β-galactosidase involves a number of steps. Each one of these can be probed to try to determine how the cell regulates production. There are two hypotheses to explain the phenomenon:

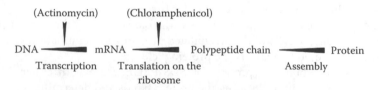

Figure 20.7 β-Galactosidase pathway in *E. coli*.

Hypothesis 1: Lactose stimulates the synthesis of β-galactosidase by activating the DNA–RNA–protein pathway

This hypothesis suggests that the enzyme is not present, or is present in only very low concentration, in the bacterial cells until lactose is added to the medium. Therefore, the amount of β-galactosidase present in the cell will depend on how long the DNA–protein pathway is active (i.e., it is time-dependent).

Hypothesis 2: *E. coli* contains a form of the enzyme β-galactosidase which is present all the time but is inactive in its current configuration

The presence of lactose serves to activate the enzyme molecule. According to this hypothesis, lactose has no influence on the rate of β-galactosidase synthesis, but rather serves as an activator for the already present enzyme molecule. Enzymes that are rapidly synthesized in increased amounts in response to the substrate in the medium are known as *inducible enzymes*. This is in contrast to the *constitutive enzymes* that are found in constant amounts irrespective of the metabolic state of the cell. *lac operon* in prokaryotic cells such as *E. coli* the cellular DNA is contained in a single circular chromosome which contains genes for the production of mRNA (structural genes) and genes associated with transcriptional control (control genes), so that the structural genes are only transcribed when the control genes allow transcription to occur and is known as the *operon model,* which was proposed by Jacob and Monod as illustrated in Figure 20.8.

The lac operon is the gene cluster responsible for the regulation of the three enzymes needed for the metabolism of galactose. It consists of three structural genes and a control region that contains a promoter and operator site. RNA polymerase binds to the promoter site and this is enhanced by a cyclic AMP receptor protein (CAP) which also binds in this region. Immediately adjacent to this area is the operator which is the binding site for a repressor coded by the regulatory gene.

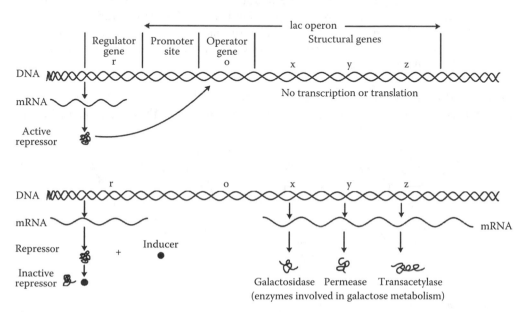

Figure 20.8   Operon model explaining regulation of the lac operon in *E. coli.*

In the absence of an inducer, the repressor binds to the operator site and blocks RNA polymerase so that the structural genes cannot be expressed. However, when the inducer is present, it binds to the repressor protein converting it to an inactive form that cannot bind to the operator site, so that the lac operon can be transcribed.

20.17.2 Experimental Design

To differentiate between these two hypotheses which stimulate the production/activation of β-galactosidase, use an antibiotic at successive time intervals to block protein synthesis, and assay for the amount of β-galactosidase present.

1. Stimulation of the bacteria
   When bacteria are growing on lactose, the lactose in the medium is eventually used up and the level of intracellular β-galactosidase drops. The design of the experiment depends on having a constant level of lactose (the activator). Fortunately, there are several lactose analogs or "look-alikes" that also cause the production of β-galactosidase, but unlike lactose they are not broken down by the enzyme. One such molecule is isopropyl-β-D-thiogalactoside (IPTG). Its structure and that of lactose are shown in Figure 20.9. Note the regions of similarity.
2. Regulation of the DNA pathway
   In this experiment, the antibiotic chloramphenicol is being used to prevent translation of the message encoded in the mRNA. Chloramphenicol prevents mRNA from attaching to the ribosomes. Once chloramphenicol is added to the bacterial culture no additional β-galactosidase will be produced, but any already present will continue to function normally. If β-galactosidase is produced via the DNA–mRNA–protein pathway, then the amount of β-galactosidase in the *E. coli* culture will depend on how long the pathway is active before chloramphenicol is added (i.e., shows a time dependency). If an inactive form of β-galactosidase is always present, then the chloramphenicol should have no effect on the amount of β-galactosidase present.

Bacterial strains. In this experiment, two strains of *E. coli* are investigated. In the *wild type* ($i^+ x^+ y^+ z^+$), all the proteins are inducible and are only synthesized when an

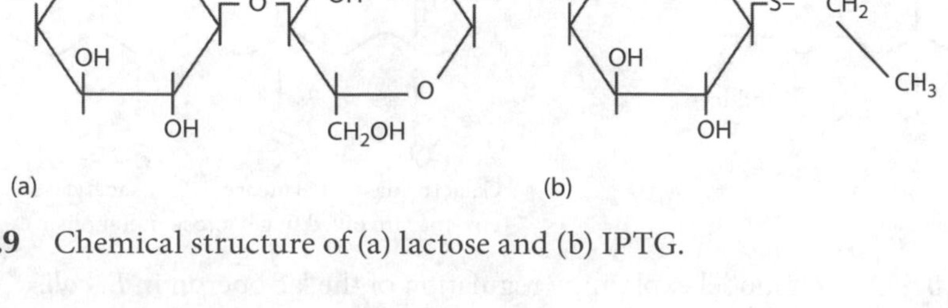

(a)   (b)

Figure 20.9   Chemical structure of (a) lactose and (b) IPTG.

inducer is present. In contrast, the *constitutive mutant* ($i^- x^+ y^+ z^+$) contains high levels of the three enzymes in the absence of an inducer.

Enzyme assay. The substrate used to measure the β-galactosidase activity is O-nitrophenyl-β-D-galactoside that is hydrolyzed by the enzyme to O-nitrophenol and galactose. The activity is assayed calorimetrically by adding alkali that stops the reaction and develops the full yellow color of O-nitrophenol.

20.17.3 Materials

Bacterial culture: *E. coli* (wild type), *E. coli* (constitutive mutant) ATCC 15224.

Bacterial growth medium: Buffered salts medium containing a carbon source (10 g/L sodium succinate), amino acids for protein synthesis (1 g/L vitamin-free casamino acids), and thiamine (10 mg/L). The inducer or inhibitor is added to the medium in appropriate concentration.

CTAB (0.8 mg/mL), O-nitrophenyl-β-D-galactoside (1 mg/mL in buffer prepared just before use), Na_2CO_3 (1 mol/L), standard O-nitrophenol (100 μmol/L in 0.1 mol/L Na_2CO_3), sterile pipettes plugged with cotton wool, lysol, or similar solution for sterilizing glassware after use, spectrocolorimeter, large container for bacterial growth, incubator shaker set at 30°C, lactose (200 mmol/L), chloramphenicol (1 mg/mL).

20.17.4 Protocol

Bacterial growth: Grow the two strains of bacteria overnight in the growth medium, but in the absence of any inducer. Suspend the organisms in 5 mL final volume of growth medium at a concentration of 5 mg/mL and set up five test tubes containing 4 mL of growth medium for each strain of *E. coli*. Add 0.5 mL of the bacterial suspension, warm to 30°C, and at zero time add 0.5 mL of lactose (200 mmol/L) as inducer. Stop one tube of each strain after incubation for 0, 30, 60, 90, and 120 min by adding 0.5 mL of chloramphenicol. Include a control by incubating each strain for 120 min in the absence of inducer. Centrifuge each tube, wash the pellet once with growth medium to remove inducer, and resuspend in 5 mL of growth medium and relate the number of bacteria to the turbidity. Determine the cell growth by measuring the turbidity at 600 nm and assay the β-galactosidase activity in each sample. Deduce a suitable incubation time for subsequent experiment for the induction of β-galactosidase activity.

Assay of β-galactosidase: This enzyme is assayed in disrupted cells because transport across the cell membrane may be rate-limiting in the intact cells. The cell membrane is disrupted without loss of activity by adding the detergent CTAB at a final concentration of 0.1 mg/mL.

Shake 1 mL of the washed bacterial suspension in a centrifuge tube with 0.5 mL of CTAB (0.8 mg/mL). Add 1.5 mL of freshly prepared O-nitrophenyl-β-D-galactoside and incubate at 30°C for 5 min or longer if necessary. Include a control with 1 mL of growth medium but no organisms. Stop the reaction by adding 1 mL of Na_2CO_3 (1 mol/L) and centrifuge to remove the bacteria. Decant the supernatant and determine its O-nitrophenol concentration by comparing its absorbance at 420 nm with a standard curve of O-nitrophenol (0–100 μmol/L) prepared in 0.1 mol/L Na_2CO_3.

20.17.5 Results and Observations

Express the activity in nmole of O-nitrophenyl-β-D-galactoside hydrolyzed per minute per milligram cell weight.

20.18 Effect of Different Inducers on the Induction of β-Galactosidase

20.18.1 Introduction

In some cases, β-galactoside analogs may stimulate induction but they do not act as substrates for the induced enzyme. This nonmetabolizable induction is called *gratuitous induction*. An example of a gratuitous inducer is β-thiogalactoside. Gratuitous inducers enable induction to be studied without interference from substrate or products but they may competitively inhibit β-galactosidase (Figure 20.10).

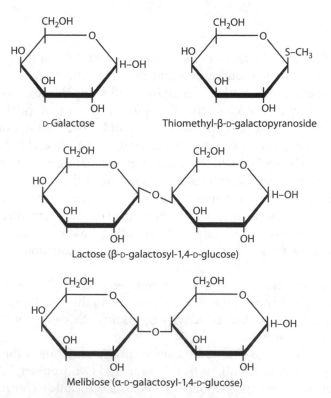

Figure 20.10 Inducers of β-galactosidase activity in *E. coli*.

20.18.2 Materials and Protocol

Using the previous experiment as a model, investigate, with controls, the efficiency of the following compounds as inducers of β-galactosidase. Use each compound at several concentrations and estimate the concentration of inducer that gives half maximal stimulation:

1. Thiomethyl-β-D-galactopyranoside (up to 10 mmol/L)
2. Melibiose (0.5 mol/L)
3. Lactose (0.5 mol/L)
4. Galactose (0.5 mol/L)

Stock solutions of the earlier compounds should be prepared fresh.

20.19 Effect of Protein Synthesis Inhibitors on the Induction of β-Galactosidase

20.19.1 Introduction

Chloramphenicol inhibits protein synthesis by binding to the 70S ribosomes found in prokaryotic cells and mitochondria and blocking peptidyl transferase. However, this antibiotic does not affect protein synthesis by the 80S ribosomes of eukaryotic cells. Cycloheximide on the other hand prevents peptide bond formation by binding to the 80S ribosomes of eukaryotes but is without effect on the 70S ribosomes of prokaryotes.

20.19.2 Materials and Protocol

Investigate the effects of chloramphenicol and cycloheximide on the induction of the enzyme by lactose using concentrations of inhibitor up to 1 mg/mL.

20.20 Turnover of β-Galactosidase

20.20.1 Protocol

Induce the *E. coli* for 60 min with lactose as in previous experiments, then add chloramphenicol and follow the rate of decay of β-galactosidase activity over the next 2 h. Include a control with chloramphenicol added before the lactose.

Suggested Readings

Alberts, B., D. Bray, I. Lewis et al. 1983. *Molecular Biology of the Cell.* New York: Garland Publishing.

Bisen, P. S., M. Debnath, and G. B. K. S. Prasad. 2012. *Microbes: Concepts and Applications.* Hoboken, NJ: Wiley-Blackwell.

Debnath, M., G. B. K. S. Prasad, and P. S. Bisen. 2010. *Molecular Diagnostics: Promises and Possibilities*. Dordrecht, the Netherlands: Springer.

Freifelder, D. 1982. *Physical Biochemistry: Applications to Biochemistry and Molecular Biology*. New York: W.H. Freeman & Company.

Glover, D. M. 1984. *Gene Cloning*. London, U.K.: Chapman & Hall.

Levine, L. 1980. *Biology of the Gene*, 3rd edn. St. Louis, MO: C.V. Mosby.

Lewin, B. 1997. *Genes*. Oxford, U.K.: Oxford University Press.

Lodish, H. 2007. *Molecular Cell Biology Solutions Manual*, 6th edn. New York: W.H. Freeman & Company.

Lodish, H., A. Berk, A. Chris et al. 2012. *Molecular Cell Biology*, 7th edn. New York: W.H. Freeman & Company.

Macleod, A. and K. Sikora. 1984. *Molecular Biology and Human Disease*. Oxford, U.K.: Blackwell Scientific Publications.

Melton, D.A., P. A. Krieg, M. R. Rebagliati, T. Maniatis, K. Zinn, and M. R. Green.1984. Efficient in vitro synthesis of biologically active RNA and RNA hybridization probes from plasmids containing a bacteriophage SP6 promoter. *Nucleic Acids Res* 12: 7035–7056.

Plummer, D. T. 2001. *An Introduction to Practical Biochemistry*, 3rd edn. London, U.K.: McGraw-Hill Book Company (UK) Ltd.

Rees, A. R. and M. J. E. Sternberg. 1984. *From Cells to Atoms: An Illustrated Introduction to Molecular Biology*. Oxford, U.K.: Blackwell Scientific Publications.

Rodriguez, R. L. and R. C. Tait. 1983. *Recombinant DNA Techniques: An Introduction*. London, U.K.: The Benjamin/Cummings Publishing Company, Inc.

Twyman, R. M. 1998. *Advanced Molecular Biology*, 8th edn. Mumbai, India: Bios Scientific Publishers.

Wilson, K. and J. Walker. 2005. *Principles and Techniques of Biochemistry and Molecular Biology*, 6th edn. Cambridge, U.K.: Cambridge University Press.

Important Links

http://www.britannica.com//EBchecked//topic//388110//molecular-biology
http://www.bioinformatics.nl//webportal//background//timeline.html
http://www.scielo.br//scielo.php?pid=S1415-47572008000400002&script=sci_arttext
http://www.biochemweb.org//general.shtml
http://bld.msu.edu//Online%20Education//mldcert.html
http://www.carroll.edu//academics//majors//biochem//
http://cels.uri.edu//cmb//CMB_FacultyView.aspx?fname=Niall&lname=Howlett

21

Biosensors

21.1 Introduction

A biosensor is made from a biological sensing element attached to a signal transducer. The sensing element can be enzymes, antibodies (as in immunosensors), DNA, or microorganisms, and the transducer may be electrochemical, optical, or acoustic. Most commercial biosensors developed are focused on clinical applications, such as for glucose and lactate. The most widespread example is the blood glucose biosensor, which uses an enzyme to break blood glucose down. Electrochemical transducers measure changes in current or voltage; optical transducers measure changes in fluorescence, absorbance, or reflectance; and acoustic transducers measure changes in frequency resulting from small changes in mass bound to their surface. Biosensors are also known as immunosensors, optrodes, chemical canaries, resonant mirrors, glucometers, biochips, biocomputers, and so on. Clark and Lyons first demonstrated the concept of biosensors, in which an enzyme was immobilized into an electrode to form a biosensor. A signal transducer is an essential component of a biosensor. It converts the recognition event into a measurable signal. A biosensor is an analytical device that converts a biological response into an electrical signal in order to determine the concentration of substances and other parameters of biological interest even where they do not utilize a biological system directly (Figure 21.1).

A commonly cited definition is "a biosensor is a chemical sensing device in which a biologically derived recognition entity is coupled to a transducer, to allow the quantitative development of some complex biochemical parameter" or "a biosensor is an analytical device incorporating a deliberate and intimate combination of a specific biological element (that creates a recognition event) and a physical element (that transduces the recognition event)."

The basic requirement of a biosensor is that the biological material should bring the physicochemical changes in close proximity of a transducer. In this direction, immobilization technology has played a major role. Immobilization not only helps in forming the required close proximity between the biomaterial and the transducer, but also helps in stabilizing it for reuse. The biological material is immobilized directly on the transducer or in most cases, in membranes, which can subsequently be mounted on the transducer. Selection of a technique and/or support would depend on the nature of the biomaterial and the substrate and configuration of the transducer used. Biosensors are suitable for detecting a wide variety of analytes including

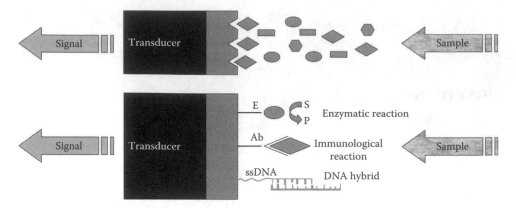

Figure 21.1 Biosensor concepts.

pollutants, explosives, viruses, biochemical and pharmaceutical products, vitamins, amino acids, heavy metals, ions, gases, etc.

Biosensors represent a rapidly expanding field, at the present time, with an estimated 60% annual growth rate; the major impetus is coming from healthcare industry with some contribution from other areas, such as food quality appraisal and environmental monitoring. The estimated analytical market is about US $11 billion per year of which 30% is in the healthcare area. There is clearly a vast market expansion potential as <0.1% of this market is currently using biosensors.

Research and development in this field is wide and multidisciplinary, spanning biochemistry, bioreactor science, physical chemistry, electrochemistry, electronics, and software engineering. Most of the current endeavor concerns potentiometric and amperometric biosensors and colorimetric paper enzyme strips.

21.2 Biosensor

A biosensor consists of a bioelement and a sensor element. The "bio" and "sensor" elements can be coupled together in one of the four possible ways, viz., membrane entrapment, physical adsorption, covalent bonding, and cross-linking (Figure 21.2).

21.2.1 Membrane Entrapment

It refers to mixture of the biomaterial with monomer solution and then polymerized to a gel, trapping the biomaterial. A semipermeable membrane separates the analyte and the bioelement, and the sensor is attached to the bioelement.

21.2.2 Physical Adsorption

The method is dependent on a combination of van der Waals forces, hydrophobic forces, hydrogen bonds, and ionic forces to attach the biomaterial to the surface of the

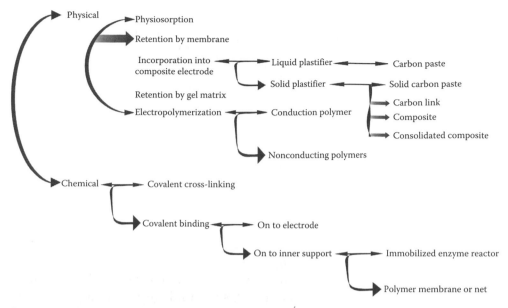

Figure 21.2 Immobilization techniques available for making biosensors.

sensor. The porous entrapment scheme is based on forming a porous encapsulation matrix around the biological material that helps in binding it to the sensor.

21.2.3 Covalent Bonding

In this method, the bond occurs between a functional group in the biomaterial and the support matrix. The sensor surface is treated as a reactive group to which the biological material can bind. Covalent binding is a commonly used technique for the immobilization of enzymes and antibiotics but not so much useful for the immobilization of cells.

21.2.4 Cross-Linking

In this method, biomaterial is chemically bonded to solid supports or to another supporting material such as cross-linking agent to significantly increase the attachment. Cross-linking using bifunctional reagents like glutaraldehyde has been successfully used for the immobilization of cells in various supports. This cross-linking technique is useful in obtaining immobilized nonviable cell preparations containing active intracellular enzymes.

Entrapment and adsorption techniques are more useful when viable cells are used. The synthetic polymers used for microbial biosensor applications include polyacrylamide, polyurethane-based hydrogels, photo-cross-linkable resins, and polyvinyl alcohol. Natural polymers used for the entrapment of the cells include alginate,

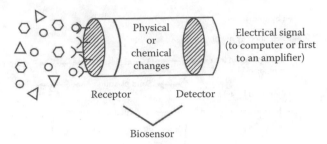

Figure 21.3   Working principle of a biosensor.

carrageenan, low melting agarose, chitosan, etc. But entrapment technique adds another diffusion barrier.

Passive trapping of cells into the pores or adhesion onto the surface of cellulose or other synthetic membranes has the major advantage of direct contact between the liquid phase and the cell, thus reducing or eliminating the problem of mass transfer. The biological material should bring the physicochemical changes in close proximity of a transducer. Immobilization not only helps in forming the required close proximity between the biomaterial and the transducer, but also helps in stabilizing it for reuse. The selection of a technique and/or support material would depend on the nature of the biomaterial and the substrate and configuration of the transducer used. Thus, a typical biosensor consists of mainly three parts (Figure 21.3):

1. *Biological material*: The biologically derived material or biomimic includes tissue, microorganisms, organelles, cell receptors, enzymes, antibodies, and nucleic acid. The sensitive elements can be created by biological engineering.
2. *Transducer*: It converts the recognition signal event into a measurable signal.
3. *Signal processors*: Primarily responsible for the display of the results in a user-friendly way.

21.3 Classification of Biosensors

A specific "bio" element recognizes a specific analyte and the "sensor." Element transduces the change in the biomolecule into an electric signal. The bioelement is very specific to the analyte to which it is sensitive. It does not recognize other analytes. Depending on the transducing mechanism used, the biosensors can be of many types such as (1) resonant biosensors, (2) optical detection biosensors, (3) thermal detection biosensors, (4) ion-sensitive field-effect transistor (ISFET) biosensors, and (5) electrochemical biosensors. The electrochemical biosensors based on the parameter measured can be further classified as (1) conductometric, (2) amperometric, and (3) potentiometric. Therefore, biosensors can be divided into the following types based on type of detection (Figure 21.4):

1. Calorimetric/thermal biosensors
2. Optical biosensors
3. Piezoelectric biosensors

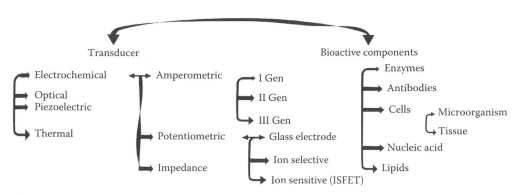

Figure 21.4 Different types of transducers and bioactive components.

4. Electrochemical biosensors
   a. Potentiometric biosensors
   b. Amperometric biosensors

Biosensors may be classified according to several criteria, such as transducers, bioactive components, or immobilization techniques used. There are three so-called "generation" of biosensors:

1. First-generation biosensors where the normal product of the reaction diffuses to the transducer and causes the electrical response.
2. Second-generation biosensors which involve specific "mediators" between the reaction and the transducer in order to generate improved response.
3. Third-generation biosensors where the reaction itself causes the response and no product or mediator diffusion is directly involved.

21.4 Characteristics of Biosensors

Some of the notable characteristics of a good biosensor are as follows (Table 21.1):

1. *Specificity*: Biosensor should be substrate-specific and should have the ability to distinguish between the analyte of interest and other similar substances. With biosensor, an analyte can be detected with great accuracy.
2. *Linearity*: Should have maximum linear value of the sensor calibration curve. Linearity of the sensor must be high for the detection of high substrate concentration.
3. *Response time*: Analytic tracer or catalytic product can be detected directly and instantaneously with the help of biosensors. Thus, they are very much suitable for online monitoring.
4. *Simplicity*: The sensing molecule and the transducer are integrated onto a single probe. Thus, handlings of biosensors are very easy.
5. *Continuous monitoring ability*: Biosensors can be regenerated and reused. Thus, they are suitable for continuous monitoring purposes that can never be done with conventional analysis method.
6. *Reproducibility*: Biosensors are very reliable, the reproducibility of the values are also of great advantage.

TABLE 21.1   Various Biosensor Transducers, Principles, and Applications

Transducer System	Principle	Applications
Enzyme electrode	Amperometric	Enzyme substrate and immunological system
Conductometer	Conductance	Enzyme substrate
Piezoelectric crystal	Mass change	Volatile gases and vapors
Thermistor	Calorimetric	Enzyme, organelle, whole cell or tissue sensors for substrate, products, gases, pollutants, antibiotics, vitamins, etc.
Optoelectronic/wave guide and fiber-optic device	Optical pH	Enzyme substrates and immunological systems
Ion-sensitive electrode (ISE)	Potentiometric	Ions in biological media, enzyme electrodes, enzyme immunosensors
Field-effect transistor (FET)	Potentiometric	Ions, gases, enzyme substrates, and immunological analytes

7. *Portability*: Biosensors are often portable. Portability of such sensors adds to the advantages that can never be the case with a spectrophotometer.
8. *Cost*: The cost of the biosensor has to some extent limited its application. It should be cheap and economical.

21.5 Components of Biosensors

21.5.1 Bioactive Components

This is the biological part of the biosensor that specifically reacts with the analyte of interest sparking a signal that is detectable by the attached transducer.

21.5.1.1 Enzymes

Purified enzymes have been commonly used in the construction of biosensors due to their high specific activities as well as high analytical specificity. They are mostly used in catalytic type biosensors. Enzymes have been immobilized at the surface of the transducer by adsorption, covalent attachment, entrapment in a gel or an electrochemically generated polymer, in bilipid membranes, or in solution behind a selective membrane. Purified enzymes, however, are expensive and unstable, thus limiting their applications in the field of biosensors. As most of the enzymes being employed are intracellular, isolation and purification becomes tough. Their main limitations are that pH, ionic strength, chemical inhibitors, and temperature affect their activity.

21.5.1.2 Antibodies

The binding between an antigen and its corresponding antibody is very specific. This property of antibody is exploited while designing biosensors based on antibodies. Antibodies are usually immobilized on the surface of the transducer by covalent attachment by conjugation of amino, carboxyl, aldehyde, or sulfhydryl group.

The binding reaction between the antibody and antigen can be monitored as a time-dependent change of fluorescence signal which is proportional to the reaction ratio of antibody to analyte. Though antibodies have similar limitations with enzymes, potential advantage of immunosensors over traditional immunoassays is that they could allow faster and in-field measurements. Immunosensors usually employ optical or acoustic transducers.

21.5.1.3 Cells

Either whole microorganisms or tissues can be used as the biocomponent. Whole cells can be used in either a viable or a nonviable form. Viable microbes metabolize various organic compounds either anaerobically or aerobically resulting in various end products like ammonia, carbon dioxide, acids, etc., that can be monitored using a variety of transducers. Viable cells are mainly used when the overall substrate assimilation capacity of microorganism is taken as an index of respiratory metabolic activity, as in the case of estimation of BOD, vitamin, sugars, organic acids, etc. Another mechanism used for the viable microbial biosensor involves the inhibition of microbial respiration by the analyte of interest, like environmental pollutants. Microbial cells have the advantage of being cheaper than enzymes or antibodies, can be more stable, and can carry out several complex reactions involving enzymes and cofactors. However, they are less selective than enzymes, and the major limitation is the diffusion of substrate and products through the cell wall resulting in a slow response as compared to enzyme-based sensors.

21.5.1.4 Nucleic Acids

The ability of a single-stranded nucleic acid to hydrolyze with another fragment of DNA by complementary base pairing is the principle behind the nucleic acid sensors. Technological innovation is introduced in the manner in which the nucleic acid oligomers is attached to the surface of the detector and the manner in which the hybridized nucleic acid is detected by attaching to glass surfaces such as filter optics cable, glass beads, or microscopic slices through covalent bonding with a chemical linker.

21.5.1.5 Lipids

Active biological receptors can be immobilized and stabilized in a polymeric film for determining an analyte of interest in a sample.

21.5.2 Sensing Device

21.5.2.1 Optical Biosensor

The output transduced signal measured is light for this type of biosensor. The biosensor is based on optical diffraction or electrochemoluminescence properties. Optical diffraction-based devices use a silicon wafer coated with a protein via covalent bonds. When the wafer is exposed to UV light through a photomask, the antibodies become inactive to the exposed regions. When the diced wafer chips are incubated in an analyte, antigen–antibody bindings are formed in the active regions, thus creating a diffraction grating. The grating produces a diffraction signal when illuminated with a light source such as laser. The resulting signal can be measured or can be further amplified before measuring for improved sensitivity. Since they are nonelectrical, optical biosensors have the advantages of lending themselves to *in vivo* applications and allowing multiple analytes to be detected by using different monitoring wave lengths. The versatility of fiber-optic probes is due to their capacity to transmit signals that reports on changes in wavelength, wave propagation, time, and intensity, distribution of spectrum, or polarity of light.

21.5.2.2 Piezoelectric Biosensor

In this mode, sensing molecules are attached to a piezoelectric surface (a mass to frequency transducer) in which interactions between the analyte and the sensing molecules set up mechanical vibrations that can be translated into an electrical signal proportional to the analyte such as quartz crystal.

21.5.2.3 Resonant Biosensor

In this type of biosensor, an acoustic wave transducer is coupled with an antibody (bioelement). The mass of the membrane changes when the analyte molecule (or antigen) gets attached to the membrane. The resulting change in the mass subsequently changes the resonant frequency of the transducer. This frequency change is then measured.

21.5.2.4 Thermal Detection Biosensor

This type of biosensor is exploiting one of the fundamental properties of biological reactions, namely absorption or production of heat, which in turn changes the temperature of the medium in which the reaction takes place. They are constructed by combining immobilized enzyme molecules with temperature sensors. When the analyte comes in contact with the enzyme, the heat reaction of the enzyme is measured and is calibrated against the analyte concentration. The total heat produced or absorbed is proportional to the molar enthalpy and the total number of molecule in the reaction. The measurement of the temperature is typically accomplished via a thermistor, and such devices are known as enzyme thermistors. Their high sensitivity to thermal changes make thermistors ideal for such applications. Unlike other transducers, thermal biosensors do not need frequent recalibration and are insensitive to the optical and electrochemical properties of the sample. Common applications of this type of biosensor include the detection of pesticides and the pathogenic bacteria. The use of sophisticated and expensive instrumentation is the major drawback of this technique.

21.5.2.5 Ion-Sensitive Biosensor

There are semiconductor-based devices having an ion-sensitive surface. This is based on the fact that the surface electrical potential changes when the ions and the semiconductor interact. This change in the potential can be subsequently measured. The ISFET can be constructed by covering the sensor electrode with a polymer layer.

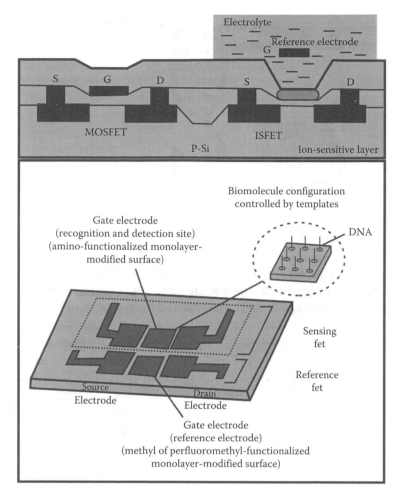

Figure 21.5 Schematic view of ion-sensitive field-effect transistor (ISFET) and enzyme field-effect transistor.

This polymer layer is selectively permeable to analyte ions. The ions diffuse through the polymer layer and in turn cause a change in the FET surface potential. This type of biosensor is also called an enzyme field-effect transistor (ENFET) and is primarily used for pH detection (Figure 21.5).

21.5.3 Electrochemical Biosensor

The underlying principle for this class of biosensors is that many chemical reactions produce or consume ions or electrons which in turn cause some changes in the electrical properties of the solution which can be sensed out and used as measuring parameter. The sensing molecule reacts specifically with compounds to be detected, sparking an electrical signal proportional to the concentration of the analyte (Figure 21.6).

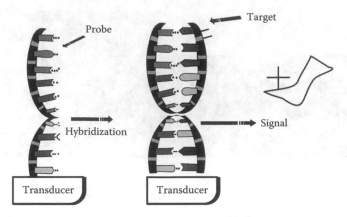

Figure 21.6   Electrochemical DNA biosensor.

In this configuration, sensing molecules are coated onto a covalently bonded to a probe surface. A membrane holds the sensing molecule in place, excluding interfering species from the analyte solution. The common detection method for electrochemical biosensors involve measurement of voltage, current, capacitance, conductance, and impedance. Among such sensors, amperometric (e.g., oxygen or hydrogen peroxide) and potentiometric (e.g., pH and carbon dioxide) transducers have found the widest applications. High sensitivity, selectivity, and ability to operate in turbid solutions are advantages of electrochemical biosensors.

Electrochemical biosensors are mainly used for the detection of hybridized DNA, DNA-binding drugs, glucose concentration, etc. Electrochemical biosensors can be classified based on measuring electrical parameters as (1) conductometric, (2) amperometric, and (3) potentiometric.

21.5.3.1 Conductometric Biosensor

The measured parameter is the electrical conductance/resistance of the solution. The overall conductivity or resistivity of the solution changes when electrochemical reactions produce ions or electrons. This change is measured and calibrated to a proper scale. Conductance measurements have relatively low sensitivity. The electric field is generated using a sinusoidal voltage (AC) which helps in minimizing undesirable effects such as Faradaic processes, double-layer charging, and concentration polarization.

21.5.3.2 Amperometric Biosensor (Measurement of the Current Resulting from a Redox Reaction)

Amperometric detection is based on measuring the oxidation or reduction of an electroactive compound at a working electrode (sensor). A potentiostat is used to apply a constant potential to the working electrode with respect to a second electrode (reference electrode). A potentiostat is a simple electronic circuit that can be constructed using a battery, two operational amplifiers, and several resistors. The applied potential is an electrobiochemical driving force that causes the oxidation or reduction reaction. The electrons required for this electroreduction are produced at the reference electrode (anode) as follows:

$$Ag^0 (solid) + Cl^- \rightarrow AgCl (solid) + e^-$$

Provided Cl⁻ concentration is fixed, the subsequent reaction produces a stable potential. The reaction rate depends on both the rate of electron transfer at the electrode surface and analyte mass transport. Use of amperometric enzyme electrodes is complicated by the enzymatic generation of the electroactive component from the analyte (enzyme substrate). The concentration of the electroactive component at the surface of the electrode is affected by diffusion of the product through the enzyme layer, the activity of the enzyme, and diffusion of analyte.

21.5.3.2.1 In Vivo Sensing. Amperometric biosensors have also been applied for in vivo sensing since their size may be reduced. The sensing probe can be either implanted or connected online with a microdialysis sampling unit for the monitoring of localized biochemical events and for the determination of physiological parameters (glucose, glutamate lactate, hydrogen peroxide, acetylcholine). By the appropriate casting of these membranes onto the biosensors tip, high selectivity and biocompatibility may be achieved (Figure 21.7 and Table 21.2).

21.5.3.3 Potentiometric Biosensor
The measured parameter in this type of sensor is oxidation or reduction potential of an electrochemical reaction. The working principle relies on the fact that when a ramp voltage is applied to an electrode in solution, a current flow occurs because of electrochemical reactions. The voltage at which these reactions occur indicates a particular reaction and particular species.

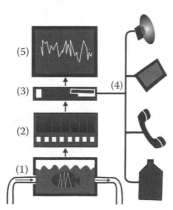

Figure 21.7 Bioelectrical biosensor for detecting water. (1) Water reservoir, (2) transducer with immobilized cells or enzyme, (3) signal detector with amplifier, (4) analytical device, and (5) display board.

TABLE 21.2   Biological Elements Used in Different Transducer

Biochemical Recognition	Physical Transducers
Enzyme	Electrochemical detection
	pH electrode
	Gas sensor O_2, NH_3, CO_2
	Platinum H_2O_2
	Surface-modified electrode
	Ion-sensitive electrode
	Chem FET
	Optical detection
	NADH florescence absorption
	Chromospheres reaction product
	Other
	Conductivity
	Thermistor calorimetric
Antibody	Optical detection
	NADH fluorescence
	Surface plasmon resonance
	Elipsometry
	Electrochemical detection
	Platinum (O_2 and H_2O_2)
	Potentiometric
	Other
	Piezoelectric (mass)
Plant and animal tissues	Amperometric/potentiometric
Microorganisms	Amperometric/potentiometric

21.6 Biochip

Biochip technology is highly effective method that allows monitoring of thousands of genes/alleles at a time in computerized automatic operations with minimal volumes of necessary reagents. Biochips promise an important shift in molecular biology, DNA diagnostics, and pharmacology, research in carcinogenesis and other diseases, and also the possibility of a better understanding of the world of biology in its globality.

A biochip is a collection of miniaturized test sites (microarrays) arranged on a solid substrate that permits many test to be performed at the same time in order to achieve higher output and speed. Biochips can also be used to perform techniques such as electrophoresis or polymerase chain reaction (PCR) using microfluidic technology. For example, DNA biochip is a solid surface to which tiny strands of DNA are attached. It allows thousands of biological reactions to occur within a few seconds. Biochip is a small device capable of performing rapid, small-scale biochemical reactions for the purpose of identifying gene sequences, environmental pollutants, airborne toxins, or other biochemical constituents (Figures 21.8 and 21.9).

A variety of industries currently desire the ability to simultaneously screen for a wide range of chemical and biological agents, with purposes ranging from testing

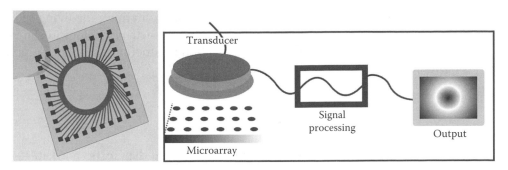

Figure 21.8 Biochip device where biological components are immobilized with an electronic sensor.

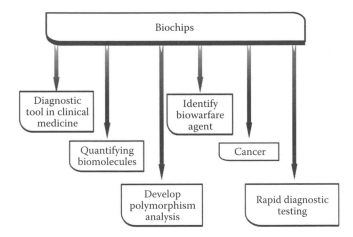

Figure 21.9 Application of biochip.

public water system for disease agents to screening airline cargo for explosives. To achieve these ends, DNA, RNA, proteins, and even living cells are being employed as sensing mediators on biochips. Numerous transduction methods can be employed including surface plasmon resonance (SPR), fluorescence, and chemiluminescence. The particular sensing and transduction techniques chosen depend on factors such as price, sensitivity, and reusability.

The development of microsensor chips comprises different microelectronic sensors namely ISFET, oxygen microelectrodes, etc., and immobilized living cells. Such multiparametric systems allow changes in the extracellular acidification rate to be measured online and noninvasively. Variations of cell adhesion, cell morphology, and intracellular junctions are detectable by impedance measurements with interdigitated microelectrodes. Such microchips may be useful for the detection of both cell metabolic and cell physiological responses to drugs.

A very unique application of a potentiometric sensor exploits living cells or DNA retained on an ISFET where DNA is being used as biosensor (Figure 21.9). Here the phenomenon of extracellular acidification is monitored as a measure of the viability and growth of the cells. The biosensor is based on a combination of a light addressable potentiometric sensor (LAPS) with a microflow chamber containing the living cells. The environmental acidification by the cells is related to their physiological status. This biosensor allows the rapid and sensitive detection of functional responses upon receptor stimulation in real time.

21.6.1 DNA Chip

The DNA probe technology is the fastest growing *in vitro* diagnostic market. It is expected to grow at the rate of 25% per year during the next 5 years. Infectious disease detection is the largest used tests for sexually transmitted diseases, tuberculosis, and pneumonia. Tests for human immunodeficiency virus and hepatitis virus are available now.

DNA microarray (also known as biochip, DNA chip, or gene chip) is a collection of microscopic DNA spots attached to a solid surface, such as glass, plastic, or silicon chip forming an array for the purpose of expression profiling, monitoring expression levels for thousands of genes simultaneously. Several immobilization techniques such as adsorption, covalent attachment, or immobilization involving avidin–biotin complexation were adopted for a DNA probe to the surface of an electrochemical transducer. The transducer was made from carbon, gold, or conducting polymer. In oligonucleotide microarray (or single channel microarrays), the probes are designed to match parts of the sequence of known or predicted mRNA. These microarrays give estimations of the absolute value of gene expression and therefore the comparison of two conditions requires the use of two separate microarrays. In addition to its high sensitivity, the testing time has decreased from more than a day to a few hours. Advances in DNA sensors are especially attracting much interest using chips coupled to fiber-optic waveguides or CCD camera detectors. Commercial devices exploit generally a fluorescent probe for monitoring the hybridization reaction. Considerable progress is observed in portal optical devices for drug screening, sequencing by hybridization, cell screening, epitope mapping.

In DNA chip, when a single immobilized DNA strand encounters a complementary partner in a sample, it will hybridize. This event can be directly detected by electro-oxidation (or reduction) of a redox label accumulated at the immobilized DNA duplex or by monitoring a change of the electrical property at the biosensor interface (capacitance, impedance). For example, conductivity changes can be observed during hybridization with the nucleic acid probe immobilized within a conducting polymer. Other configuration uses lipid bilayers onto gold electrodes with entrapped ion-channel protein acting as an ion-channel switching biosensor. Such biomimetic devices may serve both for immunosensors and for DNA biochip development. Protein and DNA microarrays are generally based on gold electrode since gold electrode is readily shaped into different microconfigurations and functionalized with thiomolecules. The latter, chemisorbed as a monolayer on gold, serve for subsequent oligonucleotide (protein) attachment and for minimizing surface fouling phenomena by large molecules during the assay. ECDNA chips are also suited for studying DNA base damage *in vitro* by measuring the effect of the xenobiotic on the oxidation signal of guanine.

21.6.2 Protein Chip

Protein chips, also referred to as protein arrays or protein microarrays, are modeled after DNA microarrays. The success of DNA microarrays in large-scale genomic experiments inspired researchers to develop similar technology to enable large-scale, high-throughput proteomic experiments. The types of protein chips available include "lab on a chip," antibody arrays, and antigen arrays, as well as a wide range of chips containing "alternative capture agents" such as proteins, substrates, and nucleic acids. In protein array chips, revealing the analyte retained on the chip by the use of surface-enhanced laser desorption/ionization (MALDI) for determining its molecular weight. Such systems allow for comparison of protein profiles from multiple samples simultaneously to rapidly detect changes in protein expression levels.

There are three general types of protein arrays:

1. *Target protein arrays*: Constructed by immobilizing large numbers of purified proteins and used to assay a wide range of biochemical functions, such as protein–protein, protein–DNA, protein–small molecule interactions and enzyme activity, and to detect antibodies and demonstrate their specificity. These protein chips are used to study the biochemical activities of the entire proteome in a single experiment.

2. *Analytical capture arrays*: Carry affinity reagents, primarily antibodies, but may also be alternative protein scaffolds, peptides, or nucleic acid aptamers, and are used to detect and quantitate analytes in complex mixtures such as plasma/serum or tissue extracts. This type of microarray is especially useful in comparing protein expression in different solutions.

3. *Reverse phase protein arrays*: These arrays involve complex samples, such as tissue lysates, which are printed on the surface and targets then detected with antibodies overlaid on them. These allow measurement of protein expression levels in a large number of biological samples simultaneously in a quantitative manner.

Simultaneously from the biosensor definition, DNA and protein chips are not a biosensor since the biocomponent is not directly linked to the physical transducer. Yet, research is underway to focus on immobilization microarray of oligonucleotides directly onto the optical fiber tip.

21.7 Immunobiosensor

Immunobiosensors are applied in the field of medical diagnostics, food and environmental analytics. For instance, medically relevant antigens in blood can be detected (e.g., heart attack markers), as well as antibiotics in milk, or pesticides and other contaminants in the environment. Immunosensors are also utilized to detect explosives or drugs in security sectors. This is based on the fact that antibodies are proteins that are produced by the immune system and belong to the class of globulins when they come in contact with exogenous substances. Antibodies can be produced by immunization of test animals against the target substance and subsequent isolation of the produced antibodies from their blood. When immobilized

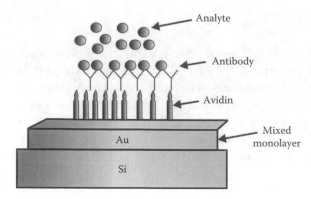

Figure 21.10   Immunosensor design.

antibodies are the receptor components in biosensors, these are referred to as immunosensors. The affinity of the antibodies to their respective antigens is used for detection (Figure 21.10).

Antibodies and aptamers both show high affinities for their target substance. Signal formation originates from the bonding to the target molecule and is detected mainly by optical or microgravimetric transducers. Optical transducers are based on the principles of photometry, where usual changes in color or luminescence are recorded, or they detect changes in the film thickness, for example, by SPR or reflectometric interference spectroscopy (RIfS). The principle of microgravimetry forms the basis of the so-called QCM sensors (quartz crystal microbalance), which detect changes in mass.

21.8 Aptamers

Aptamers are oligonucleic acid or peptide molecules that bind to a specific target molecule. They are the novel biomolecular recognition elements, which can be utilized as receptors in biosensors (amongst other applications). Aptamer-based biosensors are able to detect analytes which were, up to now (using antibody technology), only very difficult to measure or even not at all, like toxic or nonimmunogenic substances (Figures 21.11 and 21.12).

Immunobiosensor with EC, optical (SPR fluorescence), or mass-sensitive (piezo-electric microbalance) transducers are gaining extensive research interest. As for enzymes, antibodies or antigens may be readily immobilized onto transducers. The highly selective molecular recognition may be monitored directly due to a change in the physicochemical parameter at the sensing tip (affinity probe) or indirectly by detection of the labeled immunoagent. Alkaline phosphatase, horseradish peroxidase, and glucose oxidase are the most popular enzyme labels for immunoassays. Immunosensors for protein A, dogoxin, theophylline, salmonellas, IgG, etc., have been described. A general problem with immunosensors is the difficulty to reversibly regenerate the sensing surface. To solve such limitation, single-use strip immunosensors have also been used for field portable devices.

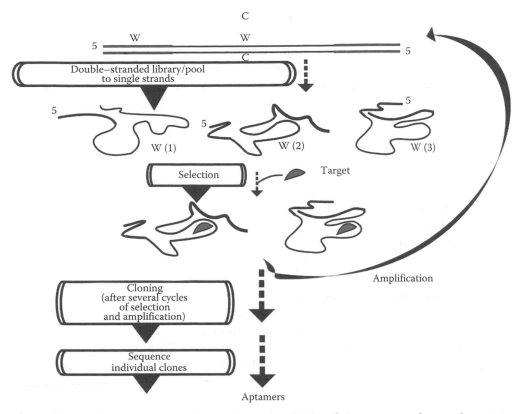

Figure 21.11 The approach can be used to isolate high-affinity aptamers for a wide variety of proteins and small molecules.

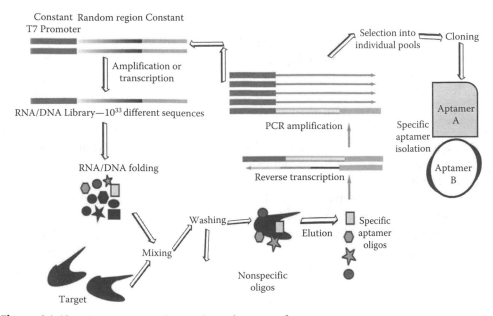

Figure 21.12 Aptamers can be used to select specific regions.

21.9 Surface Plasmon Resonance (SPR) Technology

SPR is an optical–electrical phenomenon involving the interaction of light with the electrons of a metal. The optical–electronic basis of SPR is the transfer of the energy carried by photons of light to a group of electrons (a plasmon) at the surface of a metal. SPR is based on total internal reflection; phenomenon occurs when light waves travel from a high to a low index material at an angle of incidence above the critical angle, where the waves are no longer transmitted. This can be achieved with plane-polarized light directed onto the glass/gold film interface. At a sharply defined angle of incidence, resonance occurs between the incident light and electrons in the gold which can be observed as a dip in reflectance. The light propagates, to some extent, back to the high refractive index material. Coupling between oscillating electrons (plasmons) in the metal film and the incident light produces an evanescent field that penetrates into the low refractive index medium. Within this electric field, a few hundred nanometers from the surface, interactions between the biomolecules can be measured. The next-generation microarray-based SPR systems are designed to help in profiling and characterizing biomolecular interactions in a parallel format. Miniature optical sensors that specifically identify low concentrations of environmental and biological substances are in high demand. Currently, there is no optical sensor that provides identification of the aforementioned species without amplification techniques at naturally occurring concentrations. Triangular silver nanoparticles have remarkable optical properties and their enhanced sensitivity to their nanoenvironment has been used to develop a new class of optical sensors using localized SPR spectroscopy.

21.10 Working Principle of Biosensors

The key of a biosensor is the transducer (Figures 21.13 and 21.14) that makes use of a physical change accompanying the reaction. This may be the heat output (or absorbed) by the reaction (calorimetric biosensors), changes in the distribution of charges causing an electrical potential to be produced (potentiometric biosensors), movement of electrons produced in a redox reaction (amperometric biosensors), light output during the reaction or a light absorbance difference between the reactants and products (optical biosensors), or effects due to the mass of the reactants or products (piezoelectric biosensors). Electrochemical biosensors are normally based on enzymatic catalysis of a reaction that produces ions. The sensor substrate contains three electrodes: a reference electrode, an active electrode, and a sink electrode.

The essence of biosensors is matching appropriate biological and electrical components to relevant signal during analysis. Isolation of the biological component is necessary that only the molecule of interest is bound or immobilized on the electronic component or transducer. The stability of the biological component is critical as it is used outside its normal biological environment. The sensor works by converting the signal produced by the biological sensing element on response to a specific analyte to a measurable electrical signal with the help of the transducer. The amplifier increases

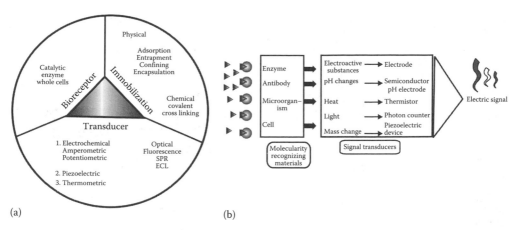

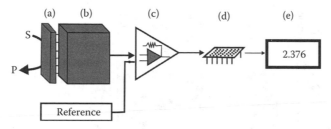

Figure 21.13 (a) Integration of bioreceptor immobilization and transducer to make biosensor. (b) Use of bioactive compound, signal transducer to evoke signal for detection.

Figure 21.14 Different components of an ideal biosensor. The biocatalyst (a) converts the substrate to a product. This reaction is determined by the transducer (b) which converts it to an electrical signal. The output from the transducer is amplified (c), processed (d), and displayed (e).

the intensity of the signal so that it can be readily measured. The digital display then displays the reading in a suitable unit. All these components are generally integrated onto a single probe to make the handling easier (Figure 21.15).

The biosensor is designed with many membranes (cellulose acetate, polyurethane, polycarbonate) specialized in sieving unwanted analyte and products. In glucose biosensor, glucose is converted within the biolayer, containing glucose oxidase enzyme, to hydrogen peroxide and gluconic acid. Hydrogen peroxide passes all the membranes to be finally detected at the platinum electrode. The resulting electrical current is amplified and recorded. Other compounds may give an artificial signal or foul the electrode.

21.11 Immunosensors

Immunosensors act on the principle that the immune response against certain biological species (usually bacteria) or contaminants will produce antibodies, which in turn can be measured. The recognition and amplification abilities of living immune cells are being exploited by measuring or detecting a wide variety of analytes like

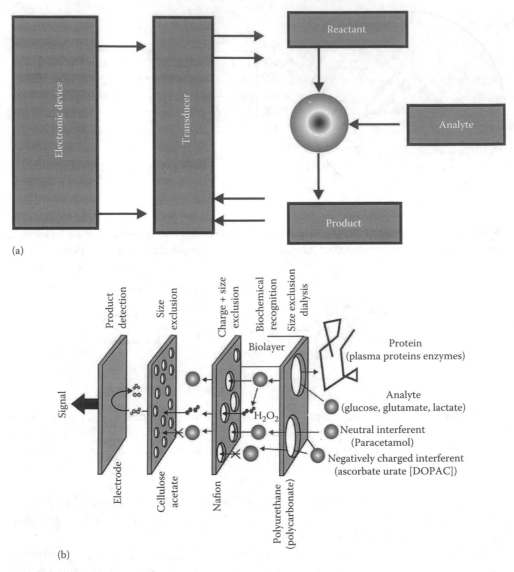

Figure 21.15 (a) Working model of a biosensor. (b) Working principle of glucose biosensor.

explosives, low-molecular-weight compounds, environmental organic pollutants, well-known proteins like CRP, insulin, pesticides, antibiotics, etc. Immunosensors can also be employed for ecological monitoring and control of biotechnological processes (Tables 21.3 and 21.4).

Immunosensors often work on the principle of enzyme-linked immunosorbent assay (ELISA) which measures the amount of enzyme-linked antigen bound to the immobilized antibody, by determining the relative concentration of the free and conjugated antigen and that can be quantified by the rate of enzyme reaction. Enzyme with high turnover numbers is used in order to achieve rapid response. The sensitivity

TABLE 21.3   Different Examples of Available Biosensors

Analyte	Microorganism	Transducer/Immobilization	Detection Limit
BOD	*Trichosporon cutaneum*	Miniature O_2 electrode	0.2–28 mg/L
BOD	Activated sludge (mixed microbes)	O_2 electrode/flow injection system	>3.5 mg/L
Phenolic compound	*Pseudomonas putida*	O_2 electrode/reactor with cells absorbed on PFI glass	100 μM
Nitrite	*Nitrobacter vulgaris*	O_2 electrode (adsorption on Whatman paper)	>10 μM
Cyanide	*Saccharomyces cerevisiae*	O_2 electrode (PVA)	0.15–15 nM
Organophosphate nerve agent	GEM *Escherichia coli*	Potentiometric	0.055–1.8 mM
Mercuric chloride	*Synechococcus* sp. PCC 7942	Photoelectrochemical	0.2 and 0.06 μM
Alcohol	*Candida vini*	O_2 electrode (porous acetyl cellulose filter)	0.2–0.02 mM
Glucose	*Aspergillus niger*	O_2 electrode (entrapment in dialysis membrane)	>1.75 mM
Glucose, sucrose, and lactose	*Gluconobacter oxydans, S. cerevisiae, Kluyveromyces marxianus*	O_2 electrode (gelatin)	Up to 0–0.8 mM
Sugars (glucose)	Psychophilic *Deinococcus radiodurans*	O_2 electrode (agarose)	0.03–0.55 mM
Short chain fatty acids in milk	*Arthrobacter nicotianae*	O_2 electrode (polyvinyl alcohol)	0.11–1.7 mM
CO_2	CO_2-utilizing autotrophic bacteria	O_2 electrode (bound on cellulose nitrate membrane)	0.2–5 mM
Vitamin B6	*Saccharomyces uvarum*	O_2 electrode (adsorption on cellulose nitrate membrane)	0.5–2.5 ng/mL
Vitamin B12	*E. coli*	O_2 electrode (trapped in porous acetyl cellulose membrane)	5.25×10^{-9} mM
Peptides	*Bacillus subtilis*	O_2 electrode (filter paper strip and dialysis membrane)	0.07–0.6 mM
Phenyl alanine	*Proteus vulgaris*	Amperometric O_2 electrode	2.5×10^{-2}–2.5 mM
Pyruvate	*Streptococcus faecium*	CO_2 gas-sensing electrode	0.22–32 mM
Tyrosine	*Aeromonas phenologenes*	NH_3 gas-sensing electrode	8.2×10^{-2}–1 mM
Monitoring toxicity of compounds to eukaryotes	*S. cerevisiae* was genetically modified to express firefly luciferase		
Online monitoring of microbial growth	*E. coli* engineered for constitutive bioluminescence		

(*continued*)

TABLE 21.3 (continued)   Different Examples of Available Biosensors

Analyte	Microorganism	Transducer/Immobilization	Detection Limit
Toxicity of Zn, Cu, and Cd, alone or in combination	*E. coli* HB 101 and *Pseudomonas fluorescens* 10586 genetically modified with *lux*CDABE		
Polycyclic aromatic hydrocarbons	*Pseudomonas fluorescens* HK44 genetically modified with *lux*CDABE		
Ecotoxicity assessment of organotoxins and their breakdown products	Microtox and *lux*CDABE-modified *Pseudomonas fluorescens*		
Ethanol as a model toxicant	*E. coli* TV1061, harboring the plasmid *pGrpELux5*		
Monitoring of biocides	Bioluminescent strain of *E. coli* produced by rDNA technology		
Metals, solvent crop protection chemicals	*E. coli* heat shock promoters *dnaK* and *grpE* were fused with *lux* gene of *Vibrio fischeri*		
Assessment of the toxicity of metals in soils amended with sewage sludge	*lux*CDABE-modified *Pseudomonas fluorescens*		

of such assays may be further enhanced by utilizing enzyme-catalyzed reactions that give intrinsically greater response; for instance, those giving rise to highly colored, fluorescent, or bioluminescent products (Table 21.4):

1. Antibody, specific for the antigen of interest, is immobilized on the surface of a tube. A mixture of a known amount of antigen–enzyme conjugate plus unknown concentration of sample antigen is placed in the tube and allowed to equilibrate.
2. After a suitable period, the antigen and antigen–enzyme conjugate will be distributed between the bound and free states dependent upon their relative concentrations.
3. Unbound material is washed off and discarded. The amount of antigen–enzyme conjugate that is bound may be determined by the rate of the subsequent enzyme reaction.

21.12 Calorimetric Biosensors

Many enzyme-catalyzed reactions are exothermic, generating heat (Table 21.5) and the temperature changes are determined by means of thermistors at the entrance and exit of small packed bed columns containing immobilized enzymes within a constant

TABLE 21.4 Types of Immunosensors

Analyte	Transducer/Immobilization	Characteristics
Low-molecular-weight analytes	Surface plasmon resonance	A monoclonal antibody against HBP is used. Lowest detection limit is 0.1 ng/mL, response time is 15 min
DDT	Surface plasmon resonance, covalent immobilization through an alkane-thiol self-assembled monolayer (SAM)	Lowest detection limit is 20 ng/L
Organophosphorous	Surface plasmon resonance, covalent immobilization through an alkane-thiol self-assembled monolayer (SAM)	Lowest retection limit is 50 ng/L
Carbamate	Surface plasmon resonance, covalent immobilization through an alkane-thiol self-assembled monolayer (SAM)	Lowest detection limit is 0.9 ng/L
CRP	Surface plasmon resonance	Linear detection level is 2.5 g/mL using two different anti-CPR antibodies
Insulin with femtomole detection antibiotics	Liposomal immunosensors; fluorescence dye is used. Poly and monoclonal antibodies are used	As low as 136 attomoles. Total assay time is <30 min
Antibiotics	Poly and monoclonal antibodies are used	

TABLE 21.5 Heat Output (Molar Enthalpies) of Enzyme-Catalyzed Reactions

Reactant	Enzyme	Heat Output (ΔH) (kJ/mol)
Cholesterol	Cholesterol oxidase	53
Esters	Chymotrypsin	4–16
Glucose	Glucose oxidase	80
Hydrogen peroxide	Catalase	100
Penicillin G	Penicillinase	67
Peptides	Trypsin	10–30
Starch	Amylase	8
Sucrose	Invertase	20
Urea	Urease	81
Uric acid	Uricase	49

temperature environment (Figure 21.16). At 80% efficiency, a temperature resolution of 0.0001°C can be attained which is for the biosensor to be generally useful.

A major problem with this biosensor is the difficulty encountered in environmental temperature change that should be avoided. An equal movement of only 1°C in the background temperature of both thermistors commonly causes an apparent change in the relative resistances of the thermistors equivalent to 0.01°C and equal to the full-scale change due to the reaction.

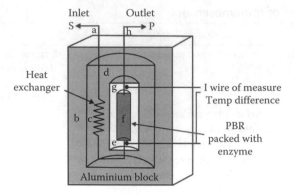

Figure 21.16 Schematic diagram of a calorimetric biosensor.

Both the sensitivity (10^{-4} M) and range (10^{-4}–10^{-2} M) of thermistor biosensors are quite low and it can be increased by using the more exothermic reactions (e.g., catalase) or by increasing the heat output by the reaction. In the simplest case, this can be achieved by linking together several reactions in a reaction pathway, all of which contribute to the heat output. Thus the sensitivity of the glucose analysis using glucose oxidase can be more than doubled by the coimmobilization of catalase within the column reactor in order to disproportionate the hydrogen peroxidase produced. An extreme case of this amplification is shown in the recycle scheme in Figure 21.17 for the deduction of ADP.

ADP is the added analytes and excess glucose, phosphoenol pyruvate, NADH, and oxygen are present to ensure maximum reaction. For enzymes (hexokinase, pyruvate kinase, lactate dehydrogenase, and lactate oxidase) are generally coimmobilized within the packed bed reactor. In spite of the positive enthalpy of the pyruvate kinase reaction, the overall process results in a 1000-fold increase in sensitivity, primarily due to the recycling between pyruvate and lactate. Reaction

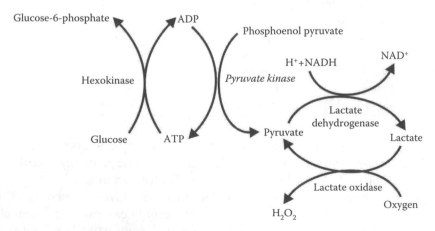

Figure 21.17 Coupling of many heat increasing enzyme in exothermic reaction.

limitation is due to low oxygen solubility which may be overcome by replacing it with benzoquinone, which is reduced to hydroquinone by flavor enzymes. Such reaction systems do, however, have the serious disadvantage in that they increase the probability of the occurrence of interference in the determination of the analyte of interest. Reactions involving the generation of hydrogen ions can be made more sensitive by the inclusion of a base having a high heat of protonation. For example, the heat output by the penicillinase reaction may be almost doubled by the use of Tris (Tris hydroxymethyl aminomethane) as the buffer. In conclusion, the main advantages of the thermistor biosensor are its general applicability and the possibility for its use on turbid or strongly colored solutions. The most important disadvantage is the difficulty in ensuring that the temperature of the sample stream remains constant (±0.01°C).

21.13 Potentiometric Biosensors

Potentiometric biosensors make use of ion-selective electrodes in order to transduce the biological reaction into an electrical signal. This consists of an immobilized enzyme membrane surrounding the probe from a pH meter (Figure 21.18), where the catalyzed reaction generates or absorbs hydrogen ion. The reaction occurring next to the thin sensing glass membrane causes a change in pH which may be read directly from the pH meter's display. The use of such electrodes is that the electrical potential is determined at very high impedance allowing effectively zero current flow and causing no interference with the reaction.

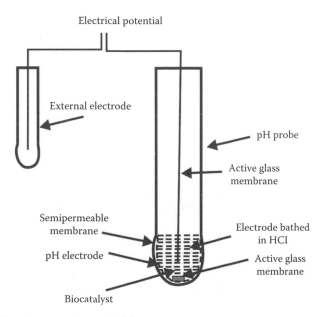

Figure 21.18   Simple potentiometric biosensors.

There are three types of ion-selective electrodes that are of use in biosensors:

1. *Glass electrodes for cations (normal pH electrodes)*: The sensing element in such system is a very thin hydrated glass membrane which generates a transverse electrical potential due to the concentration-dependent competition between the cations for specific binding sites. The selectivity of this membrane is detrimental by the composition of the glass. The sensitivity to H^+ is greater than that achievable for NH_4^+.
2. *Glass pH electrodes coated with a gas-permeable membrane*: They are selective for CO_2, NH_3, or H_2S. The diffusion of the gas through this membrane causes a change in pH of a sensing solution between the membrane and the electrode which is then determined.
3. *Solid-state electrode*: In this, the glass membrane is replaced by a thin membrane of a specific ion conductor made from a mixture of silver sulfide and a silver halide. The iodide electrode is useful for the determination of I^- in the peroxidase reaction and also responds to cyanide ions.

A recent development from ion-sensitive electrodes is the production of ISFETs (ion-sensitive field-effect transistor) and their biosensors used as enzyme-linked field-effect transistors (Figure 21.19).

The main advantage of such devices is of their extremely small size (<0.1 mm²) which allows cheap mass-produced fabrication using integrated circuit technology. As an example, a urea-sensitive FET (field-effect transistors) (ENFET [enzyme field effect transistors] containing bound urease with a reference electrode containing bound glycine) has shown only a 15% variation in response to urea (0.05–10 mg/mL) during its active lifetime of a month. Several analytes may be determined by miniaturized biosensors containing arrays of ISFETs and ENFETs. The sensitivity of FETs, however, may be affected by the composition, ionic strength, and concentrations of the solutions analyzed.

Actual dimension of the ENFET active area is of about 500 m long by 50 μm wide and 300 μm thick. The main body of the biosensor is a p-type silicon chip with two n-type silicon areas: the negative source and the positive drain. The chip is insulated by a thin layer (0.1 μm thick) of silica (SiO_2) which forms the gate of the FET.

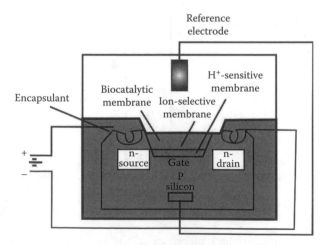

Figure 21.19 Schematic diagram of the section across the width of an ENFET.

Above this gate is an equally thin layer of H^+-sensitive material (e.g., tantalum oxide), a protective ion-selective membrane, the biocatalyst, and the analyte solution, which is separated from sensitive parts of the FET by an insert encapsulating polyimide photopolymer. When a potential is applied between the electrodes, a current flows through the FET dependent upon the positive potential detected at the ion-selective gate and its consequent attraction of electrons into the depletion layer. This current (I) is compared with that from a similar but noncatalytic ISFET immersed in the same solution.

21.14 Amperometric Biosensors

Amperometric biosensors that have been utilized for glucose, fructose, ethanol, malate, and lactate analysis in drinks using the carbon composite concept are commercially available (Biofutura, Torino, Italy). Glucose biosensors are dominating the biosensor market and this can be explained by the commercial availability and good stability of the enzyme glucose oxidase (GO_x), by the relatively high physiological concentration of glucose and of course by the huge market for glucose sensing. About 90% of the consumer diagnostic market is blood glucose monitoring, the rest is largely home testing for pregnancy and blood coagulation.

Electrochemical biosensors for personal diabetes management using test strips allowing accurate blood glucose determination are now-a-days flourishing. Even though blood samples as low as 3 µL can be analyzed with test strips, noninvasive biosensors for glucose have attracted substantial research efforts. The Glucowatch (Cygnus Inc., Redwood City, CA) painlessly measures blood glucose every 20 min for up to 12 h at a time. Glucose is collected through intact skin by ionophoresis via application of a direct electric current. Once in the gel disk at the biosensor, the glucose reacts with GO_x to form hydrogen peroxide, which is measured amperometrically. Currently, the device cannot be perfected for frequent glucose monitoring.

Amperometric biosensors function by the production of a current when a potential is applied between two electrodes. They generally have response times, dynamic ranges, and sensitivities similar to the potentiometric biosensors. This consists of a platinum cathode at which oxygen is reduced and a silver/silver chloride reference electrode. When a potential of −0.6 V relative to the Ag/AgCl electrode is applied to the platinum cathode, a current proportional to the oxygen concentration is produced. Normally both electrodes are bathed in a solution of saturated potassium chloride and separated from the bulk solution by an oxygen-permeable plastic membrane (e.g., Teflon, polytetrafluoroethylene). The following reactions occur:

$$Ag\ anode \qquad 4Ag^0 \rightarrow 4AgCl + 4e$$

$$Pt\ cathode \qquad O_2 + 4H^+ + 4e^- \rightarrow 2H_2O$$

A typical application for this type of biosensor is the determination of glucose concentrations by the use of an immobilized glucose oxidase membrane. The reaction

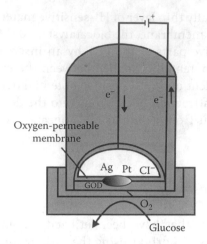

Figure 21.20   Schematic illustration of a simple amperometric biosensor.

results in a reduction of oxygen concentration as it diffuses through the biocatalytic membrane to the cathode, this being detected by a reduction in the current between the electrodes (Figure 21.20).

Other oxidases may be used in a similar manner for the analysis of their substrates (e.g., alcohol oxidase, D- and L-amino acid oxidases, cholesterol oxidase, galactose oxidase, and urate oxidase). An alternative method for determining the rate of this reaction is to measure the production of hydrogen peroxide directly by applying a potential of +0.68 V to the platinum electrode, relative to the Ag/AgCl electrode, and causing the reaction:

$$\text{Pt anode} \quad H_2O_2 \rightarrow O_2 + 2H^+ + 2e^-$$

$$\text{Ag cathode} \quad 2AgCl + 2e^- \rightarrow 2Ag^0 + 2Cl^-$$

The major problem with these biosensors is their dependence on the dissolved oxygen concentration. This may be overcome by the use of "mediators" that transfer the electrons directly to the electrode bypassing the reduction of the oxygen cosubstrate. The ferrocenes represent a commonly used family of mediators (Figure 21.21).

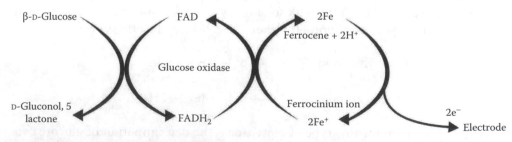

Figure 21.21   Ferrocenes as mediators.

Electrodes have now been developed that can remove the electrons directly from the reduced enzymes, without the necessity for such mediators. They neutralize a coating of electrically conducting organic salts, such as N-methylphenazinium cation (NMP$^+$) with tetracyanoquinodimethane radical anion (TCNQ$^-$). Many flavoenzymes are strongly absorbed by such organic conductors due to the formation of salt links, utilizing the alternate positive and negative charges, within their hydrophobic environment. Such enzyme electrodes can be prepared by simply dipping the electrode into a solution of the enzyme and they may remain stable for several months. These electrodes can also be used for reactions involving NAD(P)$^+$-dependent dehydrogenases as they also allow the electrochemical oxidation of the reduced forms of these coenzymes. The three types of amperometric biosensor utilizing product, mediator, or organic conductors represent the three generations in biosensor development. The reduction in oxidation potential, found when mediators are used, greatly reduces the problem of interference by extraneous material (Figures 21.22 and 21.23).

All electrode potentials (E_0) are relative to the Cl$^-$/AgCl$^-$, Ag^0 electrode (Figure 21.23):

$$\text{Substrate } (2H) + FAD\text{-oxidase} \rightarrow \text{Product} + FADH_2\text{-oxidase}$$

This is followed by the processes:

$$FADH_2\text{-oxidase} + O_2 \rightarrow \text{Biocatalyst} \rightarrow FAD^-\text{oxidase} + H_2O_2$$

$$H_2O_2 \rightarrow \text{Electrode} \rightarrow O_2 + 2H^+ + 2e^-$$

$$FADH_2^-\text{oxidase} + 2\text{Ferricinium}^+ \rightarrow \text{Biocatalyst} \rightarrow FAD^-\text{oxidase} + 2\text{Ferrocene} + 2H^+$$

$$2\text{Ferrocene} \rightarrow \text{Electrode} \rightarrow 2\text{Ferricinium}^+ + 2e^-$$

$$FADH_2^-\text{oxidase} \rightarrow \text{Biocatalyst/electrode} \rightarrow FAD^-\text{oxidase} + 2H^+ + 2e^-$$

The current (i) produced by such amperometric biosensor is related to the rate of reaction (vA) by the expression: $I = \eta \ FAvA$ where η represents the number of

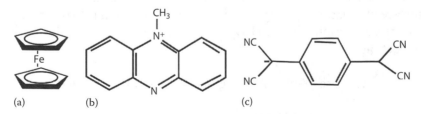

(a) (b) (c)

Figure 21.22 (a) Ferrocene (5-bis-cyclopentadienyl iron), the parent compound of a number of mediators. (b) TMP$^+$ the cationic part of conducting organic crystals. (c) TCNQ$^-$, the anionic part of conducting organic crystals. It is a resonance-stabilized radical formed by the one-electron oxidation of TCNQH$_2$.

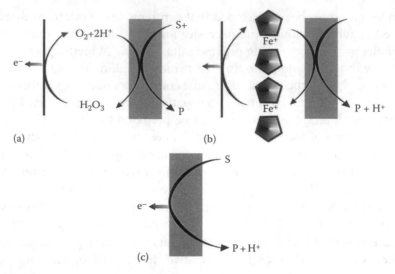

(a)   (b)

(c)

Figure 21.23 Amperometric biosensors for flavin oxidase enzymes illustrating the three generations in the development of a biosensors. (a) First-generation electrode utilizing the H_2O_2 produced by the reaction ($E_0 = +0.68$ V). (b) Second-generation electrode utilizing a mediator (ferrocene) to transfer the electrons, produced by the reaction, to the electrode ($E_0 = +0.19$ V). (c) Third-generation electrode directly utilizing the electrons produced by the reaction ($E_0 = +0.10$ V).

electrons transferred, A is the electrode area, and F is the Faraday. Usually, the rate of reaction is made diffusionally controlled by use of external membranes. Under these circumstances, the electric current produced is proportional to the analyte concentration and independent both of the enzyme and electrochemical kinetics.

21.14.1 Generation of Amperometric Biosensor

Amperometric biosensors have been divided into three generations. The first-generation biosensors were proposed by Clark and Lyons and implemented by Updike and Hicks, who coined the term enzyme electrode. Typically, an oxidative enzyme, that is, glucose oxidase (GO_x), is immobilized behind a dialysis membrane at the surface of a platinum electrode. The enzyme's function is to selectively oxidize analyte by the reduction of O_2 to H_2O_2, that is, GO_x selectively catalyzes the following two reactions:

$$\text{Glucose} + GO_x\text{-FAD} \leftrightarrow \text{glucolactone} + GO_x\text{-FADH}_2$$

$$GO_x\text{-FADH}_2 + O_2 \leftrightarrow GO_x\text{-FAD} + H_2O_2$$

GO_x-FAD and GO_x-FADH$_2$ represent the oxidized and reduced states of the glucose oxidase enzyme's flavin active site. The consumption of O_2 is subsequently measured at a platinum electrode. Most of the commercial benchtop amperometric biosensors rely on reactions catalyzed by the oxidase enzyme and subsequent detection of H_2O_2 on platinum electrodes. A universal problem with this sensor arrangement is the loss in selectivity between the biorecognition events and the amperometric H_2O_2 oxidation results in substantial interference from the oxidation of other compounds in complex matrices.

21.15 Optical Biosensors

There are two main areas of development in optical biosensors. These involve determining changes in light absorption between the reactants and products of a reaction, or measuring the light output by a luminescent process. The former usually involve the widely established, if rather low technology, use of colorimetric test strips. These are disposable single-use cellulose pads impregnated with enzyme and reagents. The most common use of this technology is for whole-blood monitoring in diabetes control. In this case, the strips include glucose oxidase, horseradish peroxidase, and a chromogen (e.g., o-toluidine or 3,3',5,5'-tetramethylbenzidine). The hydrogen peroxide is produced by the aerobic oxidation of glucose by oxidizing the weakly colored chromogen to a highly colored dye:

$$Chromogen\,(2H) + H_2O_2 \xrightarrow{\text{Peroxidase}} dye + 2H_2O$$

The evaluation of the dyed strips is best achieved by the use of portable reflectance meters, although direct visual comparison with a colored chart is often used. A wide variety of test strips involving other enzymes are commercially available at the present time. A most promising biosensor involving luminescence uses firefly luciferase (Photinus-luciferin 4-monooxygenase [ATP-hydrolyzing] to detect the presence of bacteria in food or clinical samples. Bacteria are specifically lysed and the ATP released (roughly proportional to the number of bacteria present) reacted with D-luciferin and oxygen in a reaction which produces yellow light in high quantum yield:

$$ATP + D\text{-luciferin} + O_2 \xrightarrow[\text{luciferase}]{} oxyluciferin + AMP + pyrophosphate$$

$$+ CO_2 + light\,(562\ nm)$$

The light produced may be detected photometrically by use of high-voltage, and expensive, photomultiplier tubes or low-voltage cheap photodiode system. The sensitivity of the photomultiplier containing system is, at present, somewhat greater ($<10^4$ cells/mL, $<10^{-12}$ MATP) than the simpler photon detectors which use photodiodes. Firefly luciferase is a very expensive enzyme, only obtainable from the tails of wild fireflies. Use of immobilized luciferase greatly reduces the cost of these analyses.

21.16 Piezoelectric Biosensors

Piezoelectric crystals (e.g., quartz) vibrate under the influence of an electric field. The frequency of this oscillation (f) depends on their thickness and cut, each crystal having a characteristic resonant frequency. This resonant frequency changes as molecules adsorb or desorbs from the surface of the crystal, obeying the relationship

$$\Delta f = \frac{kf2\Delta m}{A}$$

where
Δf is the change in resonant frequency (Hz)
Δm is the change in mass of adsorbed material (g)
k is a constant for the particular crystal dependent on such factors as its density and cut
A is the adsorbing surface area (cm^2)

In any piezoelectric crystal, the change in frequency is proportional to relatively unsophisticated electronic circuits. A simple use of such a transducer is a formaldehyde biosensor, utilizing a formaldehyde dehydrogenase coating immobilized to a quartz crystal and sensitive to gaseous formaldehyde. The major drawback of these devices is the interference from atmospheric humidity and the difficulty in using them for the determination of material in solution; they are however, inexpensive, small, and robust, and capable of giving a rapid response.

21.17 Second-Generation Biosensors

They use an artificial electron mediator, which replaces O_2 as the electron shuttle. Ferrocene, quinones, quinoid-like dyes, organic conducting salts, and viologens have been used as mediators. Most oxidase enzymes are not selective with respect to oxidizing agent, allowing substitution of a variety of artificial oxidizing agents, as in the following reaction:

$$GO_x - FADH_2 + Mediator_{oxid} \leftrightarrow GO_x = FAD + Mediator_{red}$$

Eliminating the O_2 dependence of the first-generation method facilitated control of the enzymatic reaction and sensor performance. Specifically, the selection of mediators with appropriate redox potentials allow poising of the working electrodes in a potential range where other components in the sample matrix are not oxidized or reduced.

Low O_2 solubility in aqueous solutions and the difficulty associated with controlling the O_2 partial pressure were disadvantages of biosensors based on the O_2/H_2O_2 reaction. When a highly soluble artificial mediator is used, the enzyme turnover rate is not limited by the cosubstrate (O_2) concentration. Use of mediators other than

O_2 allows exploitation of other oxidoreductase enzymes, including peroxidases and dehydrogenases. Unlike oxidases, these enzymes cannot use O_2 as an electron-accepting cosubstrate.

Second-generation biosensors have been commercialized, mostly in single-use testing format. MediSense (Waltham, MA) was the first company to launch a second-generation product. Again the application was blood glucose monitoring, but this device was for home use. The mediation was provided by a ferrocene species.

21.18 Third-Generation Biosensors

They are marked by the progression from use of a freely diffusing mediator (O_2 or artificial) to a system where enzyme and mediator are coimmobilized at an electrode surface, making the biorecognition component an integral part of the electrode transducer. Coimmobilization of enzyme and mediator can be accomplished by redox mediator labeling of the enzyme followed by enzyme immobilization, enzyme immobilization in a redox polymer, or enzyme and mediator immobilization in a conducting polymer. There are even reported cases of direct electrical contact of enzyme to electrode (Figure 21.24). Whether this is direct electrical connection or mediation by surface functionalities is a matter of debate.

Third-generation biosensors offer additional benefits other than second-generation sensors. The new benefits arise from the self-contained nature of the sensor. Since neither mediator nor enzyme must be added, this design facilitates repeated measurements. Sensor use for multiple analyses minimizes cost pressures on sensor design. It also follows that such a sensor could allow for continuous analyte monitoring. TheraSense, Inc. (Alameda, CA) is researching continuous blood glucose monitoring using wired enzyme technology.

Enzymes with covalently attached redox mediators are termed as wired enzymes. The enzyme is in effect wired by the mediator to an electrode. The wired enzymes are able to transfer redox equivalents from the enzyme's active site through the

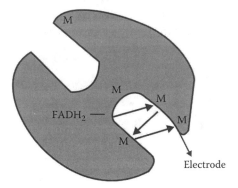

Figure 21.24 Mediated electron transfer. Enzyme-immobilized mediators (M) allow electron transport between an enzyme active site and an electrode surface by shortening the electron tunneling steps. $FADH_2$, flavin adenine dinucleotide.

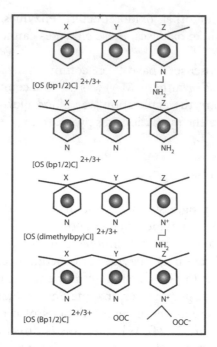

Figure 21.25 Structure of enzyme "wiring" polymers. Variation of X, Y, and Z allows the properties of the polymer to be customized.

mediator to an electrode. This principle resulted in subsequent development of enzyme-immobilizing redox polymers (Figure 21.25).

These polymers effectively transfer electrons from glucose-reduced GO_x flavin sites to polymer-bound redox centers. A series of chain redox reactions within and between polymers transfer the equivalents to an electrode surface (Figure 21.26). The redox enzyme and wire are immobilized by cross-linking to form three-dimensional redox epoxy hydrogels. A large fraction of enzymes bound in the three-dimensional redox epoxy gel are wired to the electrode. These wires provided a general approach to third-generation biosensors, sensitive not only to glucose but also to sarcosine, ʟ-lactate, ᴅ-amino acids, ʟ-glycerophosphate, cellobiose, and choline. This high-sensitivity biosensor can detect electroactive species present in biological

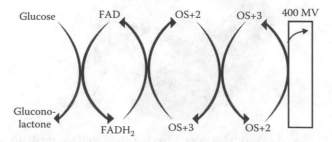

Figure 21.26 The redox cycles occurring at a three-dimensional redox epoxy-wired enzyme electrode. The wired enzyme is a flavin (FAD) containing oxidase enzyme.

test samples. Since the biological test samples may not be intrinsically electroactive, enzymes are needed to catalyze the production of radioactive species. In this case, the measured parameter is current.

21.19 Biosensor for Healthcare

21.19.1 Glucose Biosensor

The most commercially successful biosensors are amperometric glucose biosensors. These biosensors have been made available in the market in various shapes and forms such as glucose pens, glucose displays, etc. The first historic experiment that served as the origin of glucose biosensors was carried out by Clark. He used platinum (Pt) electrodes to detect oxygen. The enzyme glucose oxidase (GOD) was placed very close to the surface of platinum by physically trapping it against the electrodes with a piece of dialysis membrane. The enzyme activity changes depending on the surrounding oxygen concentration (Figure 21.27). Glucose reacts with GOD to form gluconic acid while producing two electrons and two protons, thus reducing GOD. The reduced GOD, surrounding oxygen, electrons, and protons react to form hydrogen peroxide and oxidized GOD (the original form). This GOD can again react with more glucose. The higher the glucose content, more oxygen is consumed. On the other hand, lower glucose content results in more hydrogen peroxide. Hence, either the consumption of oxygen or the production of hydrogen peroxide can be detected by the help of platinum electrodes and this can serve as a measure for glucose concentration. The most commercially successful biosensors are amperometric glucose biosensors.

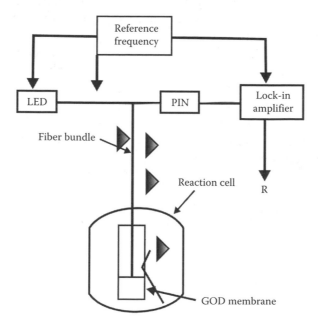

Figure 21.27 Fiber-optic glucose biosensor.

These biosensors have been made available in the market in various shapes and forms such as glucose pens, glucose displays, etc.

The first experiment that served as the origin of glucose biosensors was carried out by Clark. He used platinum (Pt) electrodes to detect oxygen. The enzyme GOD was placed very close to the surface of platinum by physically trapping it against the electrodes with a piece of dialysis membrane. The enzyme activity changes depending on the surrounding oxygen concentration.

Disposable amperometric biosensors for the detection of glucose are also available. The typical configuration is a button-shaped biosensor consisting of the following layers: metallic substrate, graphite layer, isolating layer, mediator-modified membrane, immobilized enzyme membrane (GOD), and a cellulose acetate membrane. This biosensor uses graphite electrodes instead of platinum electrodes. The isolating layer is placed on the graphite electrodes which can filter certain interfering substances (ascorbic acid, uric acid) while allowing the passage of hydrogen peroxide and oxygen. The mediator-modified membrane helps in keeping the GOD membrane attached to the graphite electrode when the electrochemical reaction takes place at a specific applied potential. The cellulose acetate outer layer placed over the GOD membrane also provides a barrier for interfering substances. The amperometric reading of the biosensor (current versus glucose concentration) shows that the relationship is linear up to a specific glucose concentration. In other words, current increases linearly with glucose concentration; hence, it can be used for detection.

The current and future applications of glucose biosensors are very broad due to their immediate use in diabetic self-monitoring of capillary blood glucose. These types of monitoring devices comprise one of the largest markets for biosensors today and their existence has dramatically improved the quality of life of diabetics.

A number of glucose biosensors have been reported which are based on conducting polymers such as the one developed covalently attached glucose oxidase on poly(o-aminobenzoic acid) and fabricated the screen-printed electrodes made of this material. These electrodes have been shown to be useful for glucose estimation from 1 to 40 mM and stability of about 6 days. The i-STAT portable clinical analyzer, which is a significant commercially available biosensor, can measure a range of parameters: sodium, chloride, potassium, glucose, blood urea nitrogen (BUN), and hematocrit. The sensors are normally fabricated using thin-film microfabrication technology on a disposable cartridge (Figure 21.28).

21.19.2 Lactate Biosensors

Lactate measurement is helpful in respiratory insufficiencies, shocks, heart failure, metabolic disorder, and monitoring the physical condition of athletes. Many biosensors have been reported to date. Two different technologies have been approached for the development of miniaturized systems. Thin-film electrodes have been developed, which can be used as either implantable, catheter-type devices or for *in vivo* monitoring in combination with microdialysis system. Second, disposable-type sensors are

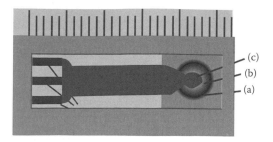

Figure 21.28 Three-electrode screen-printed sensor body made from ceramics (a). Gold working electrode is surrounded by an Ag/AgCl reference electrode (b) and gold auxiliary electrode (c).

developed for the purpose of online analysis. Disposable lactate sensor has a linear dynamic range from 0.2 to 1 mM of lactate and stability of about 3 weeks. The sensor is found to have a diminished enzyme activity (about 10%) and leaching of the enzyme from the matrix.

21.19.3 Urea and Creatinine Biosensors

Urea estimation is of utmost importance in monitoring kidney functions and disorders associated with it. Most of the urea biosensors available in literature are based on detection of NH_4^+- or HCO_3^--sensitive electrodes. A highly sensitive and rapid flow injection type has been developed for urea analysis with a composite film of electropolymerized inactive polypyrrole and a polyion complex. A coimmobilized urease and glutamate dehydrogenase on electrochemically prepared polypyrrole/polyvinyl sulfonate for the fabrication of urea biosensor has been developed.

21.19.4 Cholesterol Biosensor

Determination of cholesterol is clinically important because abnormal concentrations of cholesterol are related with hyperthyroidism, hypertension, anemia, and coronary artery diseases.

Determination based on the inherent specificity of an enzymatic reaction provides the most accurate means for obtaining true blood cholesterol concentration. An amperometric biosensor for cholesterol determination by a layer-by-layer self-assembly using cholesterol oxidase (CHO_x) and poly(styrenesulfonate) on a monolayer of microperoxidase covalently immobilized on Au–alkanethiolate electrodes has been developed. The sensor was found to be responsive even in the presence of potential electrical interferents, L-ascorbic acid, pyruvic acid, and uric acid. A cholesterol biosensor has been presented based on coimmobilization of cholesterol oxidase and peroxidase on sol–gel films and utilized these films for estimation of cholesterol.

21.19.5 Uric Acid Biosensor

Uric acid is one of the major products of purine breakdown in humans and therefore its determination serves as a market for the detection of range of disorders associated with altered purine metabolism, notably gout, hyperuricemia, and Lesch–Nyhan syndrome. Elevated levels of uric acid are observed in a wide range of conditions such as leukemia, pneumonia, kidney injury, hypertension, ischemia, etc. Additionally as a reducing agent uric acid scavenges free oxygen radicals, preventing their destructive action toward tissues and cells. Various attempts have been made to develop a biosensor for the estimation of uric acid.

21.20 Conclusion and Future Challenges

There are interesting possibilities within the field of biosensors. The next-generation technology should integrate biology with nanofabrication. Those technologies that directly convert a biological component to an electrical signal are needed to be explored extensively. In doing so, we may be able to avoid the expense and time usually involved in biological amplification of the signal and the use of special readout material such as fluorescent dyes and then the corresponding complexity of the detection device. The technologies under development must either allow to address biological questions that were previously considered impossible or impractical or make significant improvements in existing technology for speed, accuracy, throughput, or cost. Most commercial biosensors developed are focused in clinical applications, such as for glucose and lactate. Prospective biosensor market for food, pharmaceutical, agriculture, military, veterinary, and environment are still to be explored. Given the existing advances in biological sciences, coupled with advances in various other scientific and engineering disciplines, it is imminent that many analytical applications will be replaced by biosensors. A fruitful fusion between biological sciences and other disciplines will help to realize the full potential of this technology in the future.

Suggested Readings

Bianchi, N., C. Rutigliano, M. Tomassetti et al. 1997. Biosensor technology and surface plasmon resonance for real-time detection of HIV-1 genomic sequences amplified by polymerase chain reaction. *Clin Diagn Virol* 8: 199–208.

Caoili, J. C., A. Mayorova, D. Sikes, et al. 2006. Evaluation of the TB-biochip oligonucleotide microarray system for rapid detection of rifampin resistance in *Mycobacterium tuberculosis*. *J Clin Microbiol* 44: 2378–2381.

Chen, C., L. Q. Chen, G. L. Yang et al. 2008. The application of C12 biochip in the diagnosis and monitoring of colorectal cancer: Systematic evaluation and suggestion for improvement. *Clin Report* 54: 186–190.

Cooper, M. A. 2002. Optical biosensors in drug discovery. *Nat Rev Drug Discov* 1: 515–528.

Du, P., H. Li, Z. Mei et al. 2009. Electrochemical DNA biosensor for the detection of DNA hybridization with the amplification of Au nanoparticles and CdS nanoparticles. *Bioelectrochemistry* 75: 37–43.

Ercole, C., M. Del Gallo, M. Pantalone et al. 2003. Rapid identification of viable pathogens *Escherichia coli* subspecies with an electrochemical screen-printed biosensor array. *Biosens Bioelectron* 18: 907–916.

Evangelyn, C., S. Alocilja, and M. Radke. 2003. Market analysis of biosensors for food safety. *Biosens Bioelectron* 18: 841–846.

Homola, J. 2003. Present and future of surface plasmon resonance biosensors. *Anal Bioanal Chem* 377: 528–539.

Lei, Y., W. Chenb, and A. Mulchandani. 2006. Microbial biosensors. *Anal Chim Acta* 568: 200–210.

Malhotra, B. D. and A. Chaubey. 2003. Biosensors for clinical diagnostics industry. *Sens Actuat B* 91: 117–127.

Sook, J. K., K. G. Vengatajalabathy, H. H. I. Ryohei et al. 2005. Miniaturized portable surface plasmon resonance immunosensor applicable for on-site detection of low-molecular-weight analytes. *Biosens Bioelectron* 15: 2276–2282.

Tuan, V.-D. 2002. Nanobiosensors: Probing the sanctuary of individual living cells. *J Cell Biochem* 87: 154–161.

Turner, A. P. F. 2000. Biosensors—Sense and sensitivity. *Science* 290: 1315–1317.

Wang, J. 2002. Electrochemical nucleic acid biosensors. *Anal Chim Acta* 469: 63–71.

Zhang, Y. Q., Y. F. Wang, and X. D. Jiang. 2009. The application of nanoparticles in biochips. *Recent Pat Biotechnol* 2: 55–59.

Important Links

http://www.biotechnology4u.com/biotechnology_environment_use_biosensors_detect_environmental_pollutants.html

http://www.biosensoracademy.com/eng/readarticle.php?article_id=11

http://www.bbc.co.uk/news/business-17097958

http://pubs.rsc.org/en/Content/ArticleLanding/2006/AN/b603402k

http://www.sensorsmag.com/specialty-markets/medical/strong-growth-predicted-biosensors-market-7640

http://www.imiplex.com/page2/page1/index.html

http://www.cranfield.ac.uk/health/researchareas/biosensorsdiagnostics/page18795.html

22

Enzyme Immobilization

22.1 Introduction

The word "immobilized enzyme" was coined by Katchalski-Katzir in 1971. The first immobilization was done in 1916 by immobilizing invertase on animal charcoal and it was observed that enzyme retained most of the activity over a long period of time. Immobilization is specially needed when enzymes are soluble as they are costly and unstable due to denaturation and this has restricted their analytical involvement. Immobilization has solved many problems. The technology of immobilization was started initially to solve certain analysis problem and after many years of research it was found that for immobilization, it is necessary that support must be stable and should not swell after immobilization. Today, besides using enzyme for the purpose of analysis, it has been largely needed for the industrial application for production of many enzyme-catalyzed products. Enzymes may be immobilized on solid carriers by various techniques such as carrier binding, cross-linking, or entrapment. Immobilization requires carriers or support to bind the enzymes.

A variety of carriers have been utilized like alginate, magnetite, κ-carrageenan, and polyurethane. It is generally believed that the best support for one enzyme may not work for another enzyme, and that each enzyme ought to be evaluated to determine the optimal carrier matrix system and conditions. Generally, the decision will depend on (a) the various characteristics of the enzymes, (b) requirements of specific application, i.e., the operational conditions, and (c) the properties and limitations of the support system.

Immobilization often stabilizes structure of the enzymes, thereby allowing their applications even under harsh environmental conditions of pH, temperature, and organic solvents, and thus enables their uses at high temperatures in nonaqueous enzymology, and in the fabrication of biosensor probes. In future, techniques for the immobilization of multienzymes along with cofactor regeneration and retention system can be gainfully exploited in developing biochemical processes involving complex chemical conversions.

Because of their excellent functional properties (activity, selectivity, specificity), enzymes have a great potential as industrial catalysts in a number of areas of chemical industry: fine chemistry, food chemistry, analysis, and so on. For technical and economical reasons, most chemical processes catalyzed by enzymes require the reuse or the continuous use of the biocatalyst for a very long time. In this context, immobilization of enzymes may be defined as any technique that allows the reuse or continuous use of the biocatalyst. Immobilization of enzymes may be the most relevant approach for stabilization and recovery of enzymes. From an industrial perspective, simplicity and cost-effectiveness are key properties of immobilization techniques:

> Enzyme immobilization is aimed to restrict the freedom of movement of an enzyme in a certain defined region or space with retention of its catalytic activity which can be repeatedly used continuously.

22.2 Selection of Carrier and Support before Immobilization

Desirable properties of the carrier are a high surface-to-volume ratio, high protein-binding capacity, compatibility and insolubility in the reaction medium, high mechanical and chemical stability, and recoverability after use and conformational flexibility. There is no material that fulfills all these requirements. Many of the materials tested at laboratory scale are, however, not well suited to perform at productive scale, because of either their intrinsic properties or their high costs. At the end, availability and cost are key factors in determining the carrier to be used. Therefore, before selection of carrier for immobilization, the following properties should be kept in mind.

22.2.1 Physical Properties

Physical property affects various properties like strength, available surface area, shape or form (e.g., beads, sheets, fibers), and compressibility of carriers. Porosity, pore volume, permeability, density, flow rate, and pressure drop are the main requirements of the immobilization.

22.2.2 Chemical Properties

Certain chemical properties like hydrophilicity; inertness to enzyme(s), substrate(s), or cofactor(s); available functional groups for modification; ability to be regenerated or reused; and compatibility with certain buffers are necessary for the support that has to be used for the immobilization.

22.2.3 Stability

Immobilized enzymes must be stable on storage, with mechanical stability against pressure or water flow. The stability of the enzymes might be expected to either increase or decrease insolubilization, depending on whether the carrier provides a microenvironment capable of denaturing the enzymatic protein or of stabilizing it.

22.2.4 Resistance

Immobilized enzymes must be resistant against bacterial or fungal attack, disruption by chemicals, pH, temperature, organic solvents, and enzymes such as proteases. Inactivation due to autodigestion of proteolytic enzymes should be reduced by isolating the enzyme molecules from mutual attack by immobilizing them on a matrix. It has been found that enzymes coupled to inorganic carriers were generally more stable than those attached to organic polymers when stored at 4°C or 23°C. Stability to denaturing agents may also be changed upon insolubilization.

22.2.5 Safety

Immobilized enzymes must be safe from toxicity of component reagents, health and safety for process workers, and end-product users. Carriers must be "safe" if the end product is to be used for food, medical, or pharmaceutical applications. For example, acrylamide is toxic.

22.2.6 Economic

1. Availability and cost of carrier materials, chemicals, special equipment, reagents, and technical skills are required. Besides this, industrial-scale chemical preparation, feasibility for scale-up, continuous processing, effective working life, and reusability must not be very costly.
2. Immobilized enzymes must not show any diffusion limitations on mass transfer of cofactors, substrates, or products and side reactions. Immobilized support must be biodegradable in certain cases, for example, in waste treatment polyurethane is not naturally occurring and is not easily degraded and therefore its use may be undesirable for cleanup.
3. A substantial saving in costs occurs where the carrier may be regenerated after the useful lifetime of the immobilized enzyme.
4. The surface density of binding sites together with the volumetric surface area determines the maximum binding capacity. The actual capacity will be affected by the number of potential coupling sites in the enzyme molecules and the electrostatic charge distribution and surface polarity on both the enzyme and support.

22.2.7 Kinetic Properties

1. The ideal support should be cheap, inert, physically strong, and stable. It will increase the enzyme specificity (k_{cat}/K_m) whilst reducing product inhibition, shift the pH optimum to the desired value for the process, and discourage microbial growth and nonspecific adsorption. The Michaelis constant has been found to decrease by more than one order of magnitude when substrate of opposite charge to the carrier matrix is used. Again, this only happened at low ionic strengths, and when neutral substrates were used.

2. Some matrices possess other properties that are useful for particular purposes such as ferromagnetism (e.g., magnetic iron oxide, enabling transfer of the biocatalyst by means of magnetic fields), a catalytic surface (e.g., manganese dioxide, which catalytically removes the inactivating hydrogen peroxide produced by most oxidases), or a reductive surface environment (e.g., titanium, for enzymes inactivated by oxidation). Clearly most supports possess only some of these features, but a thorough understanding of the properties of immobilized enzymes does allow suitable engineering of the system to approach these optimal qualities.

3. There is usually a decrease in specific activity of an enzyme upon insolubilization, and this can be attributed to denaturation of the enzymatic protein caused by the coupling process. Once an enzyme has been insolubilized, however, it finds itself in a microenvironment that may be drastically different from that existing in free solution. The new microenvironment may be a result of the physical and chemical characters of the support matrix alone, or it may result from interactions of the matrix with substrates or products involved in the enzymatic reaction.

4. The diffusion of substrate from the bulk solution to the microenvironment of an immobilized enzyme can limit the rate of the enzymatic reaction. The rate at which substrate passes over the insoluble particle affects the thickness of the diffusion film, which in turn determines the concentration of substrate in the vicinity of the enzyme and hence the rate of reaction.

22.3 Carriers for Enzyme Immobilization

22.3.1 Alginates

Alginate is a naturally occurring polymer extracted on industrial scale from various species of brown algae (Phaeophyceae), including Ascophyllum, Laminaria, Lessonia, Ecklonia, Durvillaea, and Macrocystis. Alginate consists of monomeric residues of D-mannuronic acid and L-gluronic acid. The gels are formed by ionic network formation in the presence of cations such as calcium ions or other multivalent counterions. This method qualifies as safe, mild, fast, and cheap. However, high concentrations of alginate are difficult to work with. Chelating agents such as phosphates and citrates are best avoided as they disrupt the gel structure by binding calcium ions. Microspheres of alginate were produced using an encapsulation technique to immobilize glucocerebrosidase as an enzyme delivery matrix for treatment of Gaucher's disease.

22.3.2 Chitin and Chitosan

Chitin and chitosan are polysaccharides containing amino groups. They are attractive and widely studied as they are inexpensive. Chitin is an abundant by-product of the fishing and the fermentation industries. It is composed of (1,4)-linked 2-acetamido-2-deoxy-β-D-glucopyranosyl residues. Chitosan is composed of (1,4)-linked 2-amino-2-deoxy-D-glucose and is obtained from chitin by deacetylation, at high pH. It is more soluble than chitin and can be gelled by either cross-linking agents such as glutaraldehyde or multivalent anionic counterions such as $[Fe(CN)_6]^{4-}$ or polyphosphates. Chitin and chitosan can be prepared in the form of beads and capsules and enzymes can be immobilized using ionotropic gelation methods. Chitosan is of importance because of its primary amino groups that are susceptible for coupling reactions. Furthermore, porous spherical chitosan particles are commercially available allowing noncovalent or covalent attachment of enzymes. This support matrix can be easily prepared.

22.3.3 Carrageenans

Carrageenans are produced from red seaweeds (Rhodophyceae). They are water-soluble, sulfated galactans that are isolated from red seaweed and contain hydroxyl and sulfate groups. There are three forms of carrageenans: κ-carrageenan, ι-carrageenan, and λ-carrageenan and they differ in their ratios of sulfate to hydroxyl groups. Carrageenan has been used for the immobilization of enzymes and cells using entrapment techniques. It is inexpensive, but suffers from weak mechanical and thermal stability. In the field of immobilization of enzymes, κ-carrageenan is one of the main supports used for cell and enzyme immobilization via entrapment. One of the main disadvantages of these biopolymers is that they are usually used for immobilization of enzymes using noncovalent bonds (mainly entrapment/encapsulation) due to the lack of functionalities.

22.3.4 Cellulose

Cellulose is the most widely spread natural polymer, comprised of 1,4-linked β-D-glucopyranosyl chains, additionally bound by hydrogen bonds. The binding capacity for enzymes is generally lower as compared to agarose but it is inexpensive and commercially available in fibrous and granular forms. Some drawbacks are the low particle sizes, which impairs the use in rapid high-pressure applications, and its susceptibility to microbial cellulases.

22.4 Limitations and Advantages of Enzyme Immobilization

22.4.1 Limitations

An important factor in determining the use of enzymes in a technological process is their expense. Several hundred enzymes are commercially available that are very costly, although some are much cheaper and many are much more expensive. As enzymes are catalytic molecules, they are not directly used up by the processes in which they are used. Their high initial cost, therefore, should only be incidental to their use. Limitations have been attributed to the use of immobilized enzymes in biomedical applications, such as mass transfer resistances (substrate in and product out), adverse biological responses of enzyme support surfaces (*in vivo* or *ex vivo*), fouling by other biomolecules, greater potential for product inhibition, and sterilization difficulties.

22.4.2 Advantages

1. The most important benefit derived from immobilization is the easy separation of the enzyme from the products of the catalyzed reaction. This prevents the enzyme contaminating the product, minimizes downstream processing costs and possible effluent handling problems, particularly if the enzyme is noticeably toxic or antigenic.
2. It also allows continuous processes to be practicable, with a considerable saving in enzyme, labor, and overhead costs.
3. The productivity of an enzyme is greatly increased at higher substrate concentrations for longer periods than the free enzyme.
4. Other important advantages of using therapeutic immobilized enzymes are the prolonged blood circulation lifetime without the loss of specific activity and the lower immunogenicity. This advantage is particularly important for delivering enzymes or other biomolecules and may constitute an alternative and suitable method for the enzyme replacement therapy (Table 22.1).

22.5 Methods of Immobilizations

There are three principal methods available for immobilizing enzymes (Figures 22.1 and 22.2): (a) carrier binding, (b) cross-linking, and (c) entrapment.

TABLE 22.1 Examples of Various Supports for Immobilizing Enzyme

Synthetic Polymers	Natural Polymers	Semisynthetic Carrier	Active Synthetic Carrier	Inorganic Carriers
Poly(AAc-MAAn)	DEAE cellulose	Collodion	Halogen	Carbon
PAAM	Sephadex	Nitrocellulose	Epoxy ring	Clay
PVA	Sepharose group with monohalogen	Epoxy ring grafted	Aldehyde	Silica gel
Nylon coupling			Anhydride (carbonyldiim idazole)	Hydroxyapatite
Polystyrene				
	CLEC			
	Starch			
	Gelatin		Acylazide hydroxyl groups	Kaolinite
	Alginate		Spacer	
	Agarose		Carbonate	
	Collagen		Isocyanate	

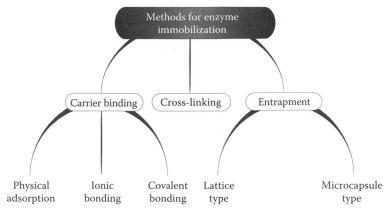

Figure 22.1 Different methods of immobilization.

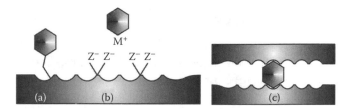

Figure 22.2 (a) Covalent binding, (b) ionic binding, and (c) entrapment.

22.5.1 Carrier-Binding Method

The method includes binding of enzyme covalently (with soluble polymer conjugate, solid surface, and hydrogel) or noncovalently like adsorption, ionic (with polysaccharides and synthetic polymers) ion-exchange centers, and physical binding via van der Waal forces (with silica, carbon nanotube, cellulose, etc.) to inert and water-insoluble support, for example, porous glass polyacrylamide, cellulose, magnetic particles.

22.5.2 Adsorption of Enzymes

In adsorption, the enzyme is bound to the carrier material via reversible surface interactions. The forces involved are electrostatic, van der Waal forces, ionic, H-bonding interaction, and possibly hydrophobic forces. The forces are generally weak, but they are sufficiently large to allow reasonable binding. Adsorption utilizes existing surface interactions between enzyme and carrier, and does not require chemical activation or modification as shown in Figure 22.3. Normally, one does not see any damage to the enzyme by the carriers. The carriers with adsorption properties are selected on the basis of knowledge of their compatibility with the enzyme. The driving force causing this binding is usually due to a combination of hydrophobic effects and the formation of several salt links per enzyme molecule. Although the physical links between the enzyme molecules and the support are often very strong, they may be reduced by many

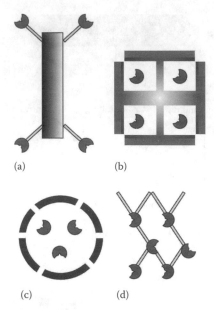

(a)   (b)

(c)   (d)

Figure 22.3 Various types of immobilization techniques: (a) Carrier binding, (b) lattice type, (c) microcapsule type, and (d) cross-linking type.

factors including the introduction of the substrate. Care must be taken that the binding forces are not weakened during use by inappropriate changes in pH or ionic strength.

Adsorption is a simple method, and the carrier can be easily recovered after use by promoting proteins desorption, and also immobilization yields are usually high and no obnoxious reagents are involved. However, its main drawback is that the enzyme can be easily desorbed from its carrier by subtle changes in the reaction medium in the case of aqueous systems. It has been reported that there is a much lesser tendency to desorption in the case of nonaqueous medium, like organic solvents, where simple immobilization by adsorption can be a good option. A typical example is the immobilization of soybean β-amylase on phenyl-boronate agarose, which can be reversed for the recovery of the support using sorbitol. Varieties of biospecific interactions have also been investigated for the reversible immobilization of enzymes by adsorption. Enzymes like acetyl choline esterase, ascorbic acid oxidase, invertase, peroxidase, glucose oxidase, etc., have been immobilized by biospecific-reversible immobilization on lectin-bound supports and invertase, using polyclonal anti-invertase antibodies.

Adsorption of enzymes has high enzyme loading (about 1 g/g matrix) and under appropriate conditions of pH and ionic strength, followed, after a sufficient incubation period, by washing off loosely bound and unbound enzyme will produce the immobilized enzyme in a directly usable form (Figure 22.4). The particular choice of adsorbent depends principally upon minimizing leakage of the enzyme during use. Examples of suitable adsorbents are ion-exchange matrices, porous carbon, clays, hydrous metal oxides, glasses, and polymeric aromatic resins. Ion-exchange matrices, are although more expensive than these other supports, may be used economically

Figure 22.4 Adsorbed enzyme on the carrier.

TABLE 22.2 Preparation of Immobilized Invertase by Adsorption

% Bound at	Support DEAE-Sephadex Anion Exchanger	Type CM-Sephadex Cation Exchanger
pH 2.5	0	100
pH 4.7	100	75
pH 7.0	100	34

due to the ease with which they may be regenerated when their bound enzyme has come to the end of its active life, a process that may simply involve washing off the used enzyme with concentrated salt solutions and resuspending the ion exchange in a solution of active enzyme (Table 22.2). Useful techniques have also been developed for the immobilization of nonviable cells to be used as an enzyme source for simple chemical conversions. Notable among them include treating the cells or the support with trivalent metal ions like Al^{3+} or Fe^{3+} or charged colloidal particles, and use of polycationic polymers like chitosan. Novel techniques have been developed to adhere cells strongly on a variety of polymeric surfaces including glass, cotton cloth, cotton threads, and other synthetic and inorganic surfaces using polyethylenimine. This technique has also been used for the simultaneous filtration and immobilization of cells from a flowing suspension, thus integrating downstream processing with bioprocessing. This technique may have future potentials for the immobilization of cells for food applications. Polyethylenimine is nontoxic and the US Food and Drug Administration has permitted its use as a direct food additive under the Food Drug and Cosmetic Act.

Some advantages of adsorption include

- Little or no damage to the enzyme
- Simple, cheap, and immobilization can be done quickly
- No chemical changes to carriers or enzyme
- Easily reversible

Some disadvantages of adsorption include the following:

- *Desorption or leakage of enzyme from support.* Desorption may occur on changing environmental conditions (such as pH, temperature, ionic strength) or on conformational changes arising from substrate/cofactor binding, contaminant binding, etc. Some physical factors such as flow rate, agitation by stirring, collision, or abrasion can cause desorption.

- *Diffusion limitations and mass transfer problems.* Nonspecific binding by substrate, cofactor, or contaminants to the carrier may result in diffusion limitations and mass transfer problems. This may in turn change the kinetic properties of the reactions. Binding of protons to the support may change the pH of the microenvironment and change the reaction rate.
- *Reduced catalytic activity.* Overloading of carrier is possible that may result in reduced catalytic activity.

22.5.3 Covalent Binding/Coupling

This involves the formation of a covalent bond between the enzyme and the carrier material. The bond is normally formed between functional groups on the carrier and the enzyme. Usually amino acid residues such as amino (NH_2) group from Lys or Arg, carboxyl (COOH) group from Asp, Glu, and hydroxyl (OH) group from Ser, Thr, and sulfhydryl (SH) group from Cys on the enzymes are involved in formation of covalent bond. A covalent bond is established between the functional groups (OH, SH, NH_2, and COOH) in the activated carrier and the functional groups in the amino acid residues of the enzyme. For example, chemical modification of –OH group can give rise to AE cellulose (aminoethyl), CM cellulose (carboxymethyl), and DEAE cellulose (diethylaminoethyl). This increases the range of immobilization methods that can be used for a given carrier (Figure 22.5).

Only small amounts of enzymes can be immobilized by this method (about 0.02 g/g matrix), although in exceptional cases as much as 0.3 g/g of matrix has been reported. The strength of binding is very strong and thus very little leakage of enzyme from the support occurs. However, it should be ensured that the formation of covalent bonds will not inactivate the enzyme. The functional groups of enzyme suitable for covalent binding under mild conditions include (i) the α-amino groups of the chain and the ε (epsilon)-amino groups of lysine and arginine, (ii) the α-carboxyl group of the chain end and the β- and γ-carboxyl groups of aspartic and glutamic acids, (iii) the phenol ring of tyrosine, (iv) the thiol group of cysteine, (v) the hydroxyl groups of serine and threonine, (vi) the imidazile group of histidine, and (vii) the indole group of tryptophan. Lysine residues are found to be the most generally useful groups for covalent bonding of enzymes to insoluble supports due to their widespread surface exposure and high reactivity, especially in slightly alkaline solutions. They also appear to be only very rarely involved in the active sites of enzymes.

The main disadvantages are that the immobilization yield is rather low and the kinetic properties of the enzyme can be severely altered. However, operational stability is high and it is quite flexible, so that directed immobilization can be done to suit the particular characteristics of the process. Besides this, covalent bonding should provide stable, insolubilized enzyme derivatives that do not leach enzyme into the surrounding solution. Multipoint covalent attachment has allowed to significantly increase the stability of a large number of enzymes like ±α-chymotrypsin,

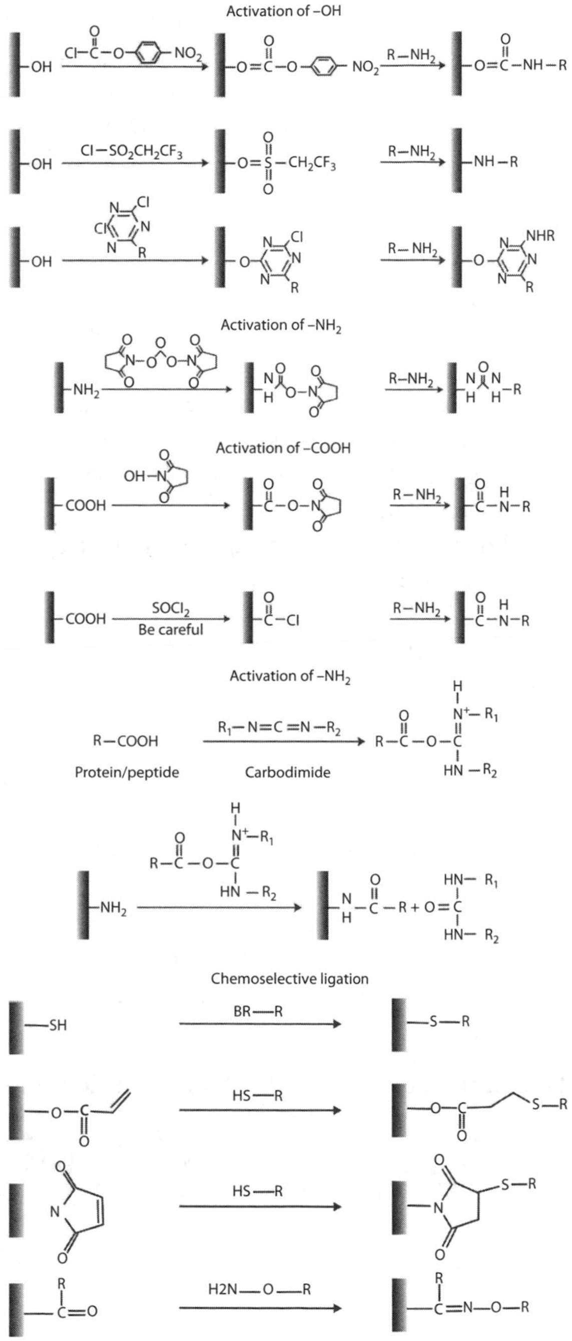

Figure 22.5 Chemical reaction between enzyme and support.

trypsin, carboxypeptidase A, D-amino acid oxidases, ferredoxin NADP oxidoreductase, esterases, rennin, and penicillin G acylase from *Escherichia coli* and *Kluyvera citrophila* (Figure 22.5). Enzymes like glucose oxidase, peroxidase, invertase, etc. have been immobilized using this technique. Covalent binding has been extensively investigated using inorganic supports. Enzymes covalently bound to inorganic supports have been used in the industry. Enzymes have also been bound to synthetic membranes, thus integrating bioconversion and downstream processing. Large-scale processes using such an approach have been demonstrated for the preparation of invert sugar using invertase.

For covalent binding to an inert solid polymer surface, the surface must first be chemically modified to provide reactive groups for the subsequent immobilization step. It is possible in some cases to increase the number of reactive residues of an enzyme in order to increase the yield of insolubilized enzyme and to provide alternative reaction sites to those essential for enzymatic activity. The wide variety of binding reactions and insoluble carriers with functional groups capable of covalent coupling, or being activated to give such groups, makes this a generally applicable method of insolubilization, even if very little is known about the enzyme structure or active site of the enzyme to be coupled (Table 22.3).

There are many procedures for covalent coupling of an enzyme and most of them fall into the following categories:

1. Formation of isourea linkage
2. Formation of diazo linkage
3. Formation of peptide bond
4. An alkylation reaction

22.5.4 Modification of Some Other Chemical Groups

The native enzyme might not be suitable for a desired process, because of its poor performance such as lower activity or selectivity. Thus improvement-by-immobilization is focused mainly on utilization of available immobilization

TABLE 22.3   Stabilization of Enzymes by Multipoint Covalent Attachment to Glyoxyl Agarose

Enzyme	Activity Expressed	(%) Stabilization Factor
Chymotrypsin	70	60,000
Trypsin	75	10,000
Penicillin G acylase (*E. coli*)	70	8000
Penicillin G acylase (*K. citrophila*)	70	1000
Glutamate racemase	70	1000
Esterase (*Bacillus stearothermophilus*)	70	1000
Lipase (*Candida rugosa*)	70	150
Thermolysin (*Bacillus thermoproteolyticus*)	100	100

techniques to alter (or improve) enzyme performance, to suit the desired application. Chemical modification affects covalent bonding (such as formation of an extra hydrogen bond as a result of chemical modification in the covalent immobilization process). Chemical modification of enzymes enhances its functionality and can also improve the performance of the carrier-bound immobilized enzymes. For instance, the stabilization of enzymes by chemical modification can usually be achieved by two major approaches—rigidification of enzyme scaffold with the use of a bifunctional cross-linker and engineering the microenvironment by introduction of new functional groups that favor the hydrophobic interaction (by hydrophobization of the enzyme surface) or hydrophilization of the enzyme surface (because of mitigation of unfavorable hydrophobic interaction) or formation of new salt bridges or hydrogen bonds (because of the introduction of polar groups). Similarly, these two principles have also been increasingly applied to improve the enzyme performance, for instance, the stability, selectivity, and activity.

It is very important that the immobilized enzyme retains as much catalytic activity as possible after the reaction. This can, in part, be ensured by reducing the amount of enzyme bound in noncatalytic conformations. Immobilization of the enzyme in the presence of saturating conditions of substrate, product, or a competitive inhibitor ensures that the active site remains unreacted during the covalent coupling and reduces the occurrence of binding in unproductive conformations. The activity of the immobilized enzyme is then simply restored by washing the immobilized enzyme to remove these molecules.

22.6 Cross-Linking

Immobilization of enzymes has been achieved by intermolecular cross-linking by a chemical group to functional groups on an insoluble support matrix. Cross-linking of an enzyme is both an expensive and insufficient, as some of the protein material will inevitably be acting mainly as a support, resulting in relatively low enzymatic activity. Generally, cross-linking is best used in conjunction with one of the other methods. In order to prevent leakage polyacrylamide gels have been used, after cross-linking as a means of stabilizing adsorbed enzymes, since the enzyme is covalently linked to the support matrix and has very little desorption chances.

Enzymes with low pI, like invertase, urease, glucose oxidase, catalase, and other enzymes, have been bound through adsorption followed by cross-linking on polyethyleneimine-coated supports. This method joins the enzymes to each other to form a large 3D structure without any support. This can be done through chemical (covalent bonding) or physical means (flocculation). Intermolecular cross-linking of enzyme by means of bifunctional or multifunctional reagents (aggregate or crystal), for example, glutaraldehyde, is possible. Immobilization via cross-linking of enzyme molecules with a bifunctional cross-linking agent is a carrier-free method and the resulting biocatalyst essentially comprises of 100% active enzyme. Binding to a carrier inevitably leads to dilution of catalytic activity

TABLE 22.4 Relative Usefulness of Enzyme Residues for Covalent Coupling

Residue	Content	Exposure	Reactivity	Stability	Usefulness
Aspartate	+	++	+	+	+
Arginine	+	++	−	±	−
Cysteine	−	±	++	−	−
Cystine	+	−	±	±	−
Glutamate	+	++	+	+	+
Histidine	±	++	+	+	+
Lysine	++	++	++	++	++
Methionine	−	−	±	−	−
Serine	++	+	±	+	±
Threonine	++	±	±	+	±
Tryptophan	−	−	−	±	−
Tyrosine	+	−	+	±	+
C-Terminus	−	++	++	++	+
Carbohydrate	− to ~++	++	+	+	±
Others	− to ++	−	−	− to ++	−

Note: + denotes for usefulness, ++ higher degree of usefulness, ± for may or may not be, and − of no use.

resulting from the introduction of a large proportion (90%–99% of the total) of noncatalytic mass.

There are numerous other methods available for the covalent attachment of enzyme (e.g., the attachment of tyrosine groups through diazo linkage, and lysine groups through amide formation with acylchlorides or anhydrides).

The technique of protein cross-linking, via reaction of, for example, glutaraldehyde with reactive NH_2 groups on the protein surface, was originally developed >40 years ago. However, the cross-linked enzymes exhibited low activity retention, poor reproducibility, and low mechanical stability and, owing to their gelatinous nature, were difficult to handle. Consequently, binding to a carrier became the most widely used methodology for enzyme immobilization (Table 22.4).

22.6.1 Glutaraldehyde

Glutaraldehyde is a bifunctional reagent that may be used to cross-link enzymes or link them to supports (Figure 22.6). It is particularly useful for producing immobilized enzyme membranes, for use in biosensors, by cross-linking the enzyme plus a noncatalytic diluent protein within a porous sheet (e.g., lens tissue paper or nylon net fabric).

Enzyme immobilization on amine-activated supports activated with glutaraldehyde is a simple process. Amino supports are chemically very stable and can be stored at 4°C for prolonged period of time and its activation is simpler. Glutaraldehyde is a nontoxic reagent and frequently used for enzyme immobilization. This method also

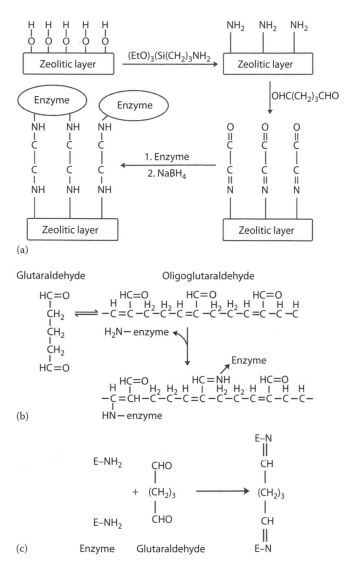

Figure 22.6 (a) Example of multipoint covalent formation in enzyme. (b) Glutaraldehyde is used to cross-link enzymes or link them to support. It usually consists of an equilibrium mixture of monomer and oligomers. The product of the condensation of enzyme and glutaraldehyde may be stabilized against dissociation by reduction with sodium borohydride. (c) Two aldehyde groups of glutaraldehyde form Schiff base linkage with free amino groups.

has some drawbacks, for example, multipoint covalent attachment is not very strong nor is chemically stable and the reactivity of the lysine groups cannot be increased further by using more alkaline pH because of the instability of the glutaraldehyde groups. The use of trialkoxysilanes allows even such apparently inert materials as glass to be coupled to enzymes (Figure 22.7).

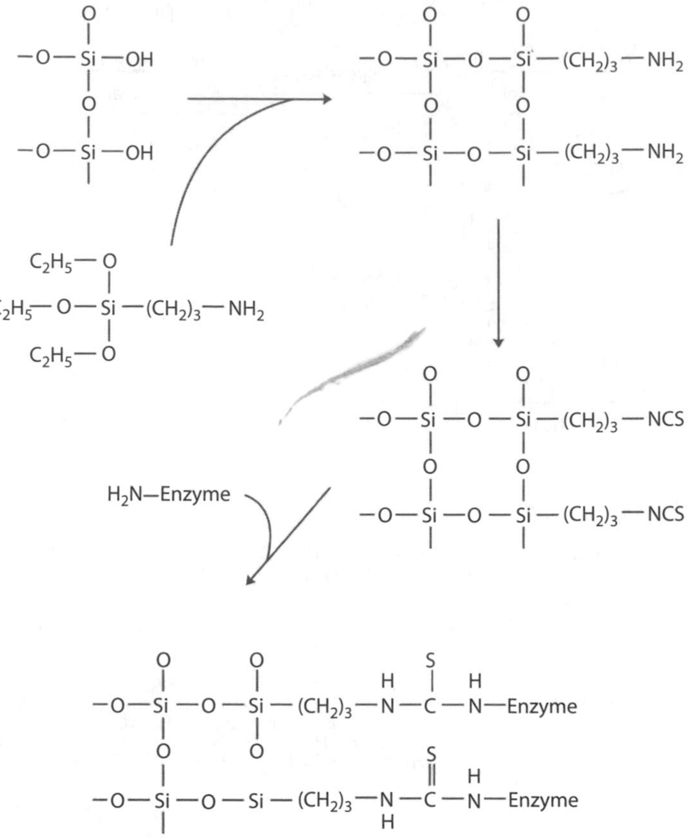

Figure 22.7 The use of trialkoxysilane (3-aminopropyltriethoxysilane) to derivatized glass. The reactive glass may be linked to enzymes by a number of methods including the use of thiophosgene.

22.6.2 Cyanogen Bromide

The most commonly used method for immobilizing enzymes on the research scale (i.e., using <1 g of enzyme) involves sepharose, activated by cyanogen bromide (CNBr). The hydroxyl groups of this polysaccharide combine with cyanogen bromide to give the reactive cyclic imidocarbonate. This reacts with primary amino groups (i.e., mainly lysine residues) on the enzyme under mildly basic conditions (pH 9–11.5). This is a simple, mild, and often successful method of wide applicability. Sepharose is a commercially available beaded polymer that is highly hydrophilic and generally inert to microbial attack. Chemically it is an agarose (poly-{β1,3-D-galactose-α-1,4-(3,6-anhydro)-L-galactose}) gel.

CNBr is often used to activate the hydroxyl group present in polysaccharide support material. In this method, enzyme and support are joined via isourea linkage (Figure 22.8a). The high toxicity of cyanogen bromide has led to the commercial, if rather expensive, production of ready-activated sepharose and the investigation of

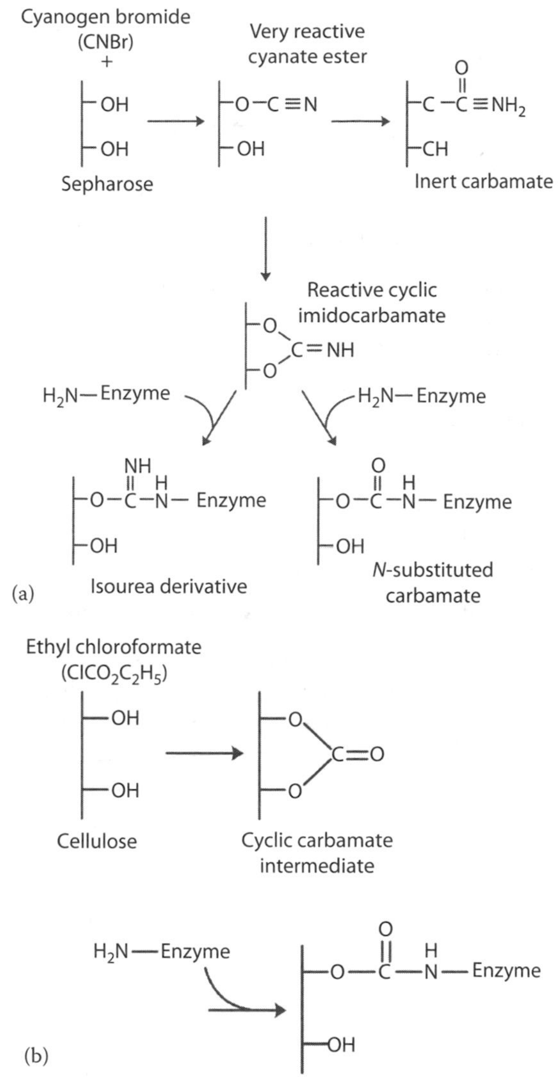

Figure 22.8 (a) Activation of sepharose by cyanogen bromide. (b) Chloroformates may be used to produce similar intermediates to those produced by cyanogen bromide but without its inherent toxicity.

alternative methods, often involving chloroformates, to produce similar intermediates (Figure 22.8b).

22.6.3 Carbodiimides

Carbodiimides (Figures 22.9 through 22.13) are very useful bifunctional reagents as they allow the coupling of amines to carboxylic acids. Careful control of the reaction conditions and choice of carbodiimide allow a great degree of selectivity in this

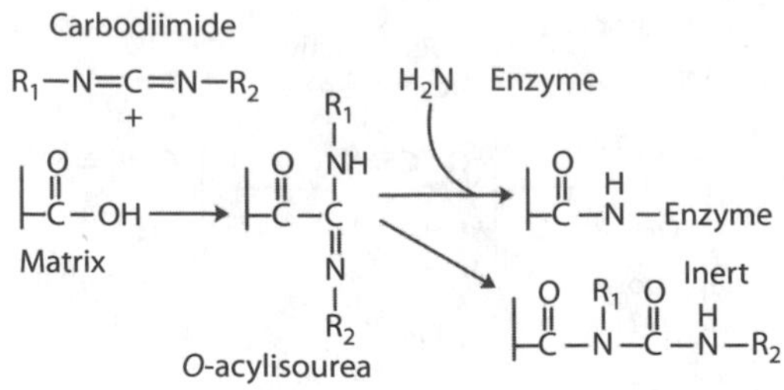

Figure 22.9 Carbodiimides may be used to attach amino groups on the enzyme to carboxylate groups on the support or carboxylate groups on the enzyme to amino groups on the support. Conditions are chosen to minimize the formation of the inert substituted urea.

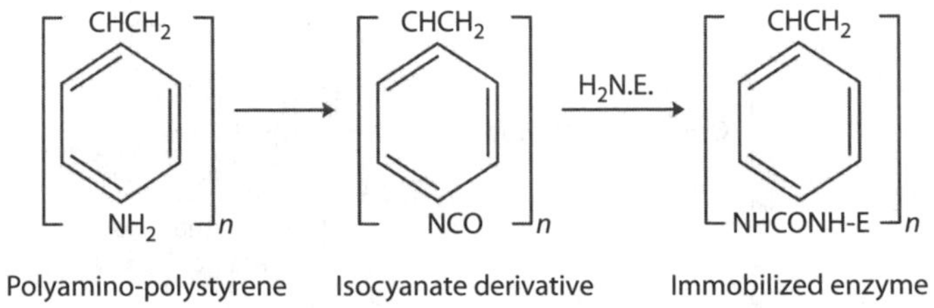

Figure 22.10 Demonstration of peptide bond formation between a free amino group in an enzyme and a carboxyl group, for example, treating polyamino-polystyrene with phosgene $CoCl_2$ for converting it into isocyanate derivative so that enzyme H_2NE can be linked with.

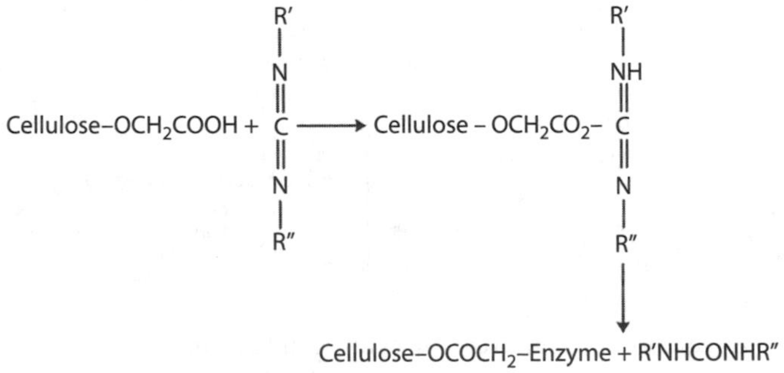

Figure 22.11 Peptide bond formation between enzyme and carried CM cellulose using carbodiimide.

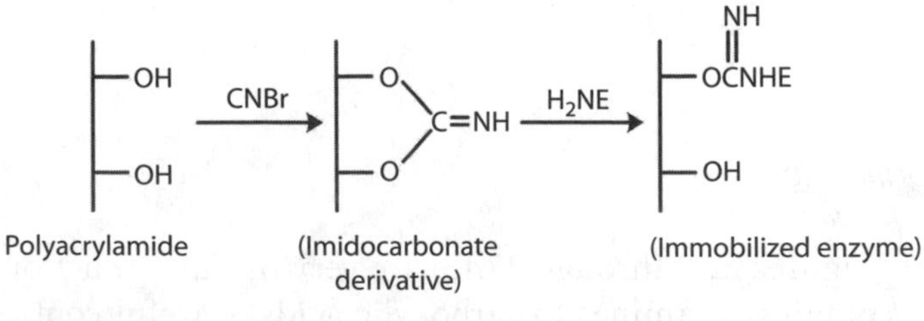

Figure 22.12 Activation of polyacrylamide by CNBr for immobilization of ligands for affinity chromatography.

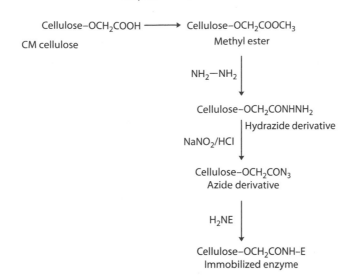

Figure 22.13 Conversion of CM cellulose into azide derivative for attachment to enzyme.

reaction. In the case of carbodiimide activation, the support material should have a carboxyl functional group and the enzyme and support are joined via peptide bond. If the support material contains an aromatic amino functional group, it can be diazotized using nitrous acid. Subsequent addition of enzyme leads to formation of a diazo linkage between the reactive diazo group on the support and the ring structure of an aromatic amino acid such as tyrosine.

22.6.4 Cross-Linked Enzyme Crystals

Another approach that has been explored is by chemical cross-linking of enzyme crystals, thereby stabilizing the crystalline lattice and its constituent enzyme molecules, which result in forming highly concentrated immobilized enzyme particles that can be lyophilized and stored indefinitely at room temperature. Cross-linked enzyme crystals (CLECs) retain catalytic activity in harsh conditions, including temperature and pH extremes, exogenous proteases, and exposure to organic or aqueous solvents. Lyophilized CLECs can be reconstituted easily in these solvents as active monodisperse suspensions. CLECs have shown promise in a variety of applications like the synthesis of aspartame, using thermolysin, and for the resolution of chiral esters. The techniques of immobilization can also be extended to obtain organic solvent–soluble enzymes by treating them with hydrophobic molecules like lipids.

The use of CLECs as industrial biocatalysts was introduced in the early 1990s and subsequently commercialized by Altus Biologics. An inherent disadvantage of CLECs is the need to crystallize the enzyme, a laborious procedure requiring an enzyme of high purity. On the other hand, it is well known that the addition of salts, or water-miscible organic solvents or nonionic polymers, to aqueous solutions of proteins lead to their precipitation as physical aggregates, held together by noncovalent bonding without perturbation of their tertiary structure. This is because of cross-linking of these physical aggregates which would render them permanently insoluble while maintaining the preorganized superstructure of the aggregates and, hence, their catalytic activity.

22.6.5 Cross-Linked Enzyme Aggregates

Cross-linked enzyme aggregate (CLEA) methodology essentially combines purification and immobilization into a single-unit operation. CLEAs are commercially available from CLEA Technologies (http://www.cleatechnologies.com). Initially the synthesis of CLEAs from penicillin amidase (EC 3.5.1.11), which is an industrially important enzyme, used in the synthesis of semisynthetic penicillin and cephalosporin antibiotics. The free enzyme has limited thermal stability and a low tolerance to organic solvents, which makes it an ideal candidate for stabilization as a CLEA. Indeed, penicillin G amidase CLEAs, prepared by precipitation with ammonium sulfate or tertiary butanol, proved to be effective catalysts for the synthesis of ampicillin (Figure 22.14).

Remarkably, the productivity of the CLEA was even higher than the free enzyme from which it was made and substantially higher than that of the CLEC. Not surprisingly, the productivity of the commercial catalyst was much lower, a reflection of the fact that it mainly consists of noncatalytic ballast in the form of the polyacrylamide carrier. Analogous to the corresponding CLECs, the penicillin G amidase CLEAs also maintained their high activity in organic solvents.

The CLEA technology has many advantages in the context of industrial applications. The method is exquisitely simple and amenable to rapid optimization, which translates to low costs and short time-to-market. It is applicable to a wide variety of enzymes, including crude preparations, affording stable, recyclable catalysts with high retention of activity and tolerance to organic solvents. The technique is applicable to the preparation of combi-CLEAs containing two or more enzymes, which can be advantageously used in catalytic cascade processes. CLEAs have also been used in microchannel reactors.

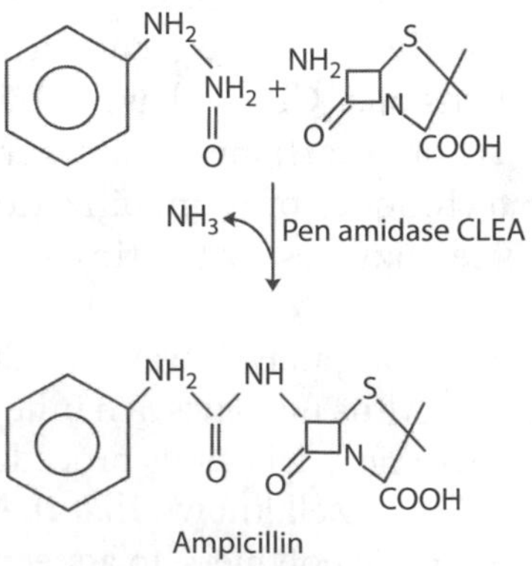

Figure 22.14   Ampicillin synthesis (pen amidase—penicillin G amidase).

22.6.6 Molecular Imprinting Technique

Molecular imprinting techniques (MITs) are based on the hypothesis that the conformation induced by a ligand can be frozen by physical or chemical means such as lyophilization or cross-linking or molecular confinement. One possible explanation is that the population of some enzyme conformers is enhanced by the conformer selectors used and, consequently, enzyme selectivity toward some substrates can be improved, as is exemplified by the so-called MIT.

22.7 Entrapment Method

Entrapment immobilization is based on the localization of an enzyme within the lattice of a polymer matrix or membrane. Immobilization by entrapment differs from adsorption and covalent binding in that enzyme molecules are free in solution, but restricted in movement by the lattice structure of a gel. Entrapment has been extensively used for the immobilization of cells, but not for enzymes. The major limitation of this technique for the immobilization of enzymes is the possible slow leakage during continuous use in view of the small molecular size compared to the cells. Biocatalysts have been entrapped in natural polymers like agar, agarose, and gelatin through thermoreversal polymerization, but in alginate and carrageenan by ionotropic gelation. A number of synthetic polymers have also been investigated. Notable among them are the photo-cross-linkable resins, polyurethane prepolymers, and acrylic polymers like polyacrylamide. Among these, the most widespread matrix made from monomeric precursors is the polyacrylamide gel (Figure 22.15).

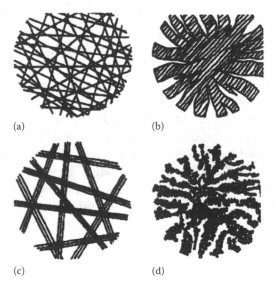

(a)

(b)

(c)

(d)

Figure 22.15 Entrapment carriers for enzyme: (a) Polyacrylamide, (b) porous glass, (c) agarose, and (d) porous silica.

Of the three gels, polyacrylamide is the most widely used matrix for entrapping enzymes. It has the advantage that it is nonionic. The consequence is that the properties of the enzymes are only minimally modified in the presence of the gel matrix. At the same time, the diffusion of the charged substrate and products is not affected. However, dimethylaminopropionitrile, the polymerization initiator, is highly toxic and must be handled with great care. The requirement to purge the monomer solution with nitrogen is also troublesome, although not totally crippling.

Calcium alginate is just as widely used as polyacrylamide. Unlike polyacrylamide gels, gelation of calcium alginate does not depend on the formation of more permanent covalent bonds between polymer chains. Rather, polymer molecules are cross-linked by calcium ions. Because of this, calcium alginate beads can be formed in extremely mild conditions ensuring enzyme activity yield of over 80%. However, just as easily as calcium ions can be exchanged for sodium ions, they can also be displaced by other ions. This property can be both advantageous and disadvantageous. If needed, enzymes or microbial cells can be easily recovered by dissolving the gel in a sodium solution. On the other hand, proper caution must be exercised to ensure that the substrate solution does not contain high concentrations of those ions that can disintegrate the gel.

The main attraction of using gelatin as the immobilization media is that the gel formation process requires only simple equipment and that the reagents are relatively inexpensive and nontoxic. The retention of enzymatic activities for immobilization with a gelatin gel is typically 25%–50% of the original free enzyme. Gelatin gel has the advantage that the mass transfer resistance is relatively low compared to other entrapment methods, but the rate of enzyme loss due to leakage is high. Incorporating enzyme into lattice of semipermeable gel or enclosing the enzyme into semipermeable membrane (microcapsule), for example, collagen often improves the operational stability and sometimes efficiency of enzyme catalysis under industrial conditions, and also allows their easy separation from the medium. This allows for easier recovery of the enzymes for repeat use. The improved stability and ease of separation of immobilized enzymes make them suitable to be used in a variety of bioreactors. They are also easier to scale up. In addition, encapsulation allows for ease of handling and storage of enzymes. Sometimes, other properties of the enzyme may be altered by immobilization, for example, altered pH activity profile. Polyacrylamide may not be a useful support for use in food industry in view of its toxicity, but can have potential in the treatment of waste and in the fabrication of analytical devices containing biocatalysts. One of the major limitations of entrapment technique is the diffusion limitation as well as the steric hindrance, especially when the macromolecular substrates like starch and proteins are used. Diffusion problem can be minimized by entrapment in fine fibers of cellulose acetate or other synthetic materials or by using an open pore matrix. Recently, the development of the so-called hydrogels and thermoreactive water-soluble polymers, like the albumin–poly(ethylene glycol) hydrogel, has attracted attention in the field of biotechnology. In the area of healthcare, they offer new avenues for enzyme immobilization. Such gels with a water content of about 96% provide a microenvironment for the immobilized enzyme close to that of the soluble enzyme with minimal diffusion restrictions.

Entrapment of enzymes within gels or fibers is a convenient method for use in processes involving low-molecular-weight substrates and products. Amounts in excess of 1 g of enzyme per gram of gel or fiber may be entrapped. However, the difficulty which large molecules have in approaching the catalytic sites of entrapped enzymes precludes the use of entrapped enzymes with high-molecular-weight substrates. The entrapment process may be a purely physical caging or involves covalent binding. As an example, the enzymes' surface lysine residues may be derivatized by reaction with acryloyl chloride ($CH_2=CH-CO-Cl$) to give the acryloyl amides. This product may then be copolymerized and cross-linked with acrylamide ($CH_2=CH-CO-NH_2$) and bisacrylamide ($H_2N-CO-CH=CH-CH=CH...CO-NH_2$) to form a gel. Enzymes may be entrapped in cellulose acetate fibers by, for example, making up an emulsion of the enzyme plus cellulose acetate in methylene chloride, followed by extrusion through a spinneret into a solution of an aqueous precipitant. Entrapment is the method of choice for the immobilization of microbial, animal, and plant cells, where calcium alginate is widely used.

22.7.1 Membrane Confinement

Membrane confinement of enzymes may be achieved by a number of quite different methods, all of which depend for their utility on the semipermeable nature of the membrane. This must confine the enzyme whilst allowing free passage for the reaction products, and in most configurations, the substrates. The simplest of these methods is achieved by placing the enzyme on one side of the semipermeable membrane whilst the reactant and product stream is present on the other side. Hollow fiber membrane units are available commercially with large surface areas relative to their contained volumes (>20 m^2/L) and permeable only to substances of molecular weight substantially less than the enzymes. Although costly, these are very easy to use for a wide variety of enzymes without the additional research and development costs associated with other immobilization methods.

22.7.2 Occlusion

Many entrapment methods are used today, and all are based on the physical occlusion of enzyme molecules within a "caged" gel structure such that the diffusion of enzyme molecules to the surrounding medium is severely limited, if not rendered totally impossible. What creates the "wires" of the cage is the cross-linking of polymers.

A highly cross-linked gel has a fine "wire-mesh" structure and can more effectively hold smaller enzymes in its cages. The degree of cross-linking depends on the condition at which polymerization is carried out. Because there is a statistical variation in the mesh size, some of the enzyme molecules gradually diffuse toward the outer shell of the gel and eventually leak into the surrounding medium. Thus, even in the absence of loss in the intrinsic enzyme activity, there is a need to replenish continually the lost enzymes to compensate for the loss of apparent activity. In addition, because an

TABLE 22.5   Generalized Comparisons of Different Enzyme Immobilization Techniques

Properties	Adsorption	Covalent Binding	Entrapment	Membrane Confinement
Preparation	Simple	Difficult	Difficult	Simple
Cost	Low	High	Moderate	High
Binding force	Variable	Strong	Weak	Strong
Enzyme leakage	Yes	No	Yes	No
Applicability	Wide	Selective	Wide	Very wide
Running problems	High	Low	High	High
Matrix effects	Yes	Yes	Yes	No
Large diffusion barriers	No	No	Yes	Yes
Microbial protection	No	No	Yes	Yes

immobilized enzyme preparation is used for a prolonged period of operation, there is also a gradual, but noticeable, decline in the intrinsic enzyme activity even for the best method (Table 22.5). Eventually, the entire immobilized enzyme packing must be replaced.

As there is no bond formation between the enzyme and the polymer matrix, occlusion provides a generally applicable method that, in theory, involves no disruption of the protein molecules. However, free radicals generated on the course of the polymerization may affect the activity of entrapped enzymes. Another disadvantage is that only low-molecular-weight substrates can diffuse rapidly to the enzyme, thus making the method unsuitable for enzymes that act on macromolecular substrates, such as ribonuclease, trypsin, dextranase, etc.

22.7.3 Entrapment and Encapsulation

Enzymes, encapsulated within small membrane-bound droplets or liposomes, may also be used within such reactors. As an example of the former, the enzyme is dissolved in an aqueous solution of 1,6-diaminohexane. This is then dispersed in a solution of hexanedioic acid in the immiscible solvent, chloroform. The resultant reaction forms a thin polymeric (Nylon-6,6) shell around the aqueous droplets that traps the enzyme. Liposomes are concentric spheres of lipid membranes, surrounding the soluble enzyme. They are formed by the addition of phospholipids to enzyme solutions. The microcapsules and liposomes are washed free of nonconfined enzyme and transferred back to aqueous solution before use.

In entrapment and encapsulation methods, the enzyme remains free in solution, but restricted in movement by the lattice structure of a gel. The porosity of the gel lattice is controlled to ensure that the structure is tight enough to prevent leakage of enzyme, while allowing free movement of substrates, cofactors, and products. Encapsulation generally refers to larger capsules that allow for coimmobilization of different combinations of enzymes for selected applications. Notably, red blood cells can be used as encapsulation capsules.

22.8 Whole-Cell Immobilization

The immobilized whole-cell system is an alternative to enzyme immobilization. Unlike enzyme immobilization, where the enzyme is attached to a solid support, in immobilized whole-cell systems, the target cell is immobilized. This technique is used especially with eukaryotic cells where the whole metabolic machinery is often required for their specific application. Such methods may be implemented when the enzymes required are difficult or expensive to extract, such as intracellular enzymes. The reactions catalyzed by immobilized whole-cell biocatalysts can be classified as reactions involving single enzymes (bioconversions), reactions involving multienzyme systems with or without cofactors, reactions involving a complete metabolic pathway yielding primary or secondary metabolites. The cells can be carrier-bound, cross-linked, or entrapped with the suitable matrix depending upon the nature of the cells to be immobilized. The ideal support for immobilized cell culture should fulfill as many as possible the following requirements: high cell mass–loading capacity, easy access to nutrient media, simple nontoxic immobilization procedure, optimum diffusion distance from flowing media to center of support, sterilization, mechanical stability, reusable, easy separation of cells and carrier from media, and suitable for conventional reactor systems.

The major advantage of immobilized cells, in contrast to immobilized enzymes or free cells, is reduction of the cost of bioprocessing. This should follow from the repeated and continuous use of biocatalyst, the maintenance of a high cell density, and the provision of a system with minimal cost of cell separation. In addition, immobilization may provide resistance to shear for shear-sensitive cells. The use of immobilized microbial cells is advantageous when the microorganism does not contain interfering enzymes or when such enzymes can be inactivated without loss of desired catalytic activity and also when substrates and products do not have a high molecular mass and can diffuse through the cell membrane. However, growth of live cells, when immobilized, can be disastrous in that unrestricted growth can burst soft or thin-walled containment structures to release previously entrapped organisms which then either lost or contaminate the product. Disadvantages may also arise from the diffusion barrier created by immobilization matrix as well as by the high cell density. Also, some enzymes may be used for the metabolic needs of the cell, leading to reduce yield.

22.9 Industrial Application of Enzymes

The first industrial use of an immobilized enzyme is amino acid acylase by Tanabe Seiyaku Company, Japan, for the resolution of racemic mixtures of chemically synthesized amino acids. Amino acid acylase catalyzes the deacetylation of the L-form of the N-acetyl amino acids leaving unaltered the N-acetyl-D-amino acid that can be easily separated, racemized, and recycled. Some of the immobilized preparations used for this purpose include enzyme immobilized by ionic binding to DEAE-sephadex and the enzyme entrapped as microdroplets of its aqueous solution into fibers of cellulose triacetate by means of fiber wet spinning developed by Snam Progetti, Rohm GmbH has immobilized this enzyme on macroporous beads made of flexiglass-like material.

By far, the most important application of immobilized enzymes in industry is for the conversion of glucose syrups to high fructose syrups by the enzyme glucose isomerase. It is evident that most of the commercial preparations use either the adsorption or the cross-linking technique. Application of glucose isomerase enzyme technology has gained considerable importance, especially in nontropical countries that have abundant starch raw material. Unlike these countries, in tropical countries like India, where sugarcane cultivation is abundant, the high fructose syrups can be obtained by a simpler process of hydrolysis of sucrose using invertase. Compared to sucrose, invert sugar has a higher humectancy, higher solubility, and osmotic pressure. Historically, invertase is perhaps the first reported enzyme in an immobilized form. A large number of immobilized invertase systems have been patented:

$$\text{Starch (Gn) + orthophosphate} \xrightarrow{\text{Phosphorylase}} \text{Starch (Gn-1) + } \alpha\text{-glucose-1-phosphate}$$

$$\text{Glucose} \xrightarrow{\text{Glucose isomerase}} \text{Fructose}$$

$$\text{Glucose-1-phosphate + fructose} \xrightarrow{\text{Sucrose phosphorylase}} \text{Sucrose + orthophosphate}$$

Currently used immobilized enzyme processes are given in Table 22.6.

Methods compatible with the enzyme, substrate, or cofactor should be selected. For example, the entrapment method may not work well with cellulose substrate because the substrate itself is large and will not readily go into the entrapment matrix to reach the enzyme.

TABLE 22.6   Some Important Industrial Uses of Immobilized Enzymes

Enzyme	EC Number	Product
Aminoacylase	3.5.1.14	L-Amino acids
Aspartate ammonia lyase	4.3.1.1	L-Aspartic acid
Aspartate 4-decarboxylase	4.1.1.12	L-Alanine
Cyanidase	3.5.5.x	Formic acid from waste cyanide
Glucoamylase	3.2.1.3	D-Glucose
Glucose isomerase	5.3.1.5	High fructose corn syrup
Histidine ammonia lyase	4.3.1.3	Urocanic acid
Hydantoinase	3.5.2.2	D- and L-Amino acids
Invertase	3.2.1.26	Invert sugar
Lactase	3.2.1.23	Lactose-free milk and whey
Lipase	3.1.1.3	Cocoa butter substitutes
Nitrile hydratase	4.2.1.x	Acrylamide
Penicillin amidases	3.5.1.11	Penicillins
Raffinase	3.2.1.22	Raffinose-free solutions
Thermolysin	3.2.24.4	Aspartame

Immobilize enzyme systems are used where they offer cost advantages to users on the basis of total manufacturing costs. The plant size needed for continuous processes is two orders of magnitude smaller than that required for batch processes using free enzymes. The capital costs are, therefore, considerably smaller and the plant may be prefabricated cheaply off-site. Immobilized enzymes offer greatly increased productivity on an enzyme weight basis and also often provide process advantages.

L-Aspartic acid is widely used in medicines and as a food additive. The enzyme aspartase catalyzes a one-step stereo-specific addition of ammonia to the double bond of fumaric acid. The enzymes have been immobilized using the whole cells of *E. coli*. This is considered as the first industrial application of an immobilized microbial cell. The initial process made use of polyacrylamide entrapment that was later substituted with the carrageenan treated with glutaraldehyde and hexamethylenediamine. Immobilized fumarase has also been used widely for the production of malic acid for pharmaceutical use. These processes make use of immobilized nonviable cells of *Brevibacterium ammoniagenes* or *Brevibacterium flavum* as a source of fumarase. Malic acid is becoming of greater market interest as food acidulant in competition with citric acid.

A large population of lactose intolerance can consume lactose-hydrolyzed milk. This is of great significance in countries where lactose intolerance is quite prevalent. Therefore preparation of lactose-hydrolyzed milk and whey, using β-galactosidase, helped in lactose hydrolysis and also enhances the sweetness and solubility of the sugars. Lactose-hydrolyzed whey may be used as a component of whey-based beverages, leavening agents, feed stuffs, or may be fermented to produce ethanol and yeast, thus converting an inexpensive by-product into a highly nutritious, good-quality food ingredient. The process makes use of a neutral lactase from yeast entrapped in synthetic fibers or immobilized β-galactosidase for the production of lactose-hydrolyzed whey. Unlike the milk, the acidic β-galactosidase of fungal origin has been used for this purpose. An immobilized preparation obtained by cross-linking α-galactosidase in hen egg white (lyophilized dry powder) has been used for the hydrolysis of lactose. A major problem in the large-scale continuous processing of milk using immobilized enzyme is the microbial contamination that has necessitated the introduction of intermittent sanitation steps. A co-immobilizate obtained by binding of glucose oxidase on the microbial cell wall using Con A has been used to minimize the bacterial contamination during the continuous hydrolysis of lactose by the initiation of the natural lactoperoxidase system in milk. A novel technique for the removal of lactose by heterogeneous fermentation of the milk using immobilized viable cells of *K. fragilis* has also been developed.

More than 50% of 6-aminopenicillinic acid (6-APA) produced today is enzymatically using the immobilized route by the deacylation of the side chain in either penicillin G or V, using penicillin acylase (penicillin amidase). A number of immobilized systems have been patented or commercially produced for penicillin acylase which make use of a variety of techniques using either the isolated enzyme or the whole cells. Similar approach has also been used for the production of 7-aminodeacetoxy-cephalosporanic acid, an intermediate in the production of semisynthetic cephalosporins.

Immobilized oxidoreductases are gaining considerable importance in biotechnology to carry out synthetic transformations of mediated asymmetric synthesis of amino acids, steroids, and other pharmaceuticals and a host of specialty chemicals. They play a major role in clinical diagnosis and other analytical applications like the biosensors. Future applications for oxidoreductases can be in areas as diverse as polymer synthesis, pollution control, and oxygenation of hydrocarbons. Immobilized glucose oxidase can find application in the production of gluconic acid, removal of oxygen from beverages, and removal of glucose from eggs prior to dehydration in order to prevent Maillard reaction.

Immobilized D-amino acid oxidase has been investigated for the production of keto acid analogs of the amino acids, which find application in the management of chronic uremia. Keto acids can be obtained using either L- or D-amino acid oxidase. The use of D-amino acid oxidase has the advantage of simultaneous separation of natural L-isomer from DL-racemates along with the conversion of D-isomer to the corresponding keto acid which can then be transaminated in the body to give the L-amino acid. Of the several microorganisms screened, the triangular yeast *T. variabilis* was found to be the most potent source of D-amino acid oxidase with the ability to deaminate most of the D-amino acids. The permeabilized cells entrapped in either radiation-polymerized acrylamide Ca alginate or gelatin have shown promise in the preparation of α-keto acids. Another interesting enzyme that can be used profitably in immobilized form is catalase for the destruction of hydrogen peroxide employed in the cold sterilization of milk.

Lipase catalyzes a series of different reactions. Although they were designed by nature to cleave the ester bonds of triacylglycerols (hydrolysis), lipase are also able to catalyze the reverse reaction under microaqueous conditions, viz. formation of ester bonds between alcohol and carboxylic acid moieties. These two basic processes can be combined in a sequential fashion to give rise to a set of reactions generally termed as interesterification. Immobilized lipases have been investigated for both these processes. Lipases possess a variety of industrial potentials starting from use in detergents; leather treatment–controlled hydrolysis of milk fat for acceleration of cheese ripening; hydrolysis, glycerolysis, and alcoholysis of bulk fats and oils; production of optically pure compounds, flavors, etc. Lipases are spontaneously soluble in aqueous phase but their natural substrates (lipids) are not. Although use of proper organic solvents as an emulsifier helps in overcoming the problem of intimate contact between the substrate and enzyme, the practical use of lipases in such pseudo-homogeneous reactions poses technological difficulties. Significant research has also been carried out on the immobilization and use of glucoamylase. This is an example of an immobilized enzyme that probably is not competitive with the free enzyme and hence has not found large-scale industrial application. This is mainly because soluble enzyme is cheap and has been used for over two decades in a much optimized process without technical problems. Immobilization has also not found to significantly enhance the thermostability of amylase.

Immobilized renin or other proteases might allow for the continuous coagulation of milk for cheese manufacture. One of the major limitations in the use of enzymes which act on macromolecular substrates or particulate or colloidal substrates like starch or cellulose pectin or proteins has been the low retention of their realistic activities with natural substrates due to the steric hindrance.

Efforts have been made to minimize these problems by attaching enzymes through spacer arms. In this direction, application of tris(hydroxymethyl)phosphine as a coupling agent may have future potentials for the immobilization of enzymes which act on macromolecular substrates.

Other problem, when particulate materials are used as the substrates for an enzyme, is difficulty in the separation of the immobilized enzyme from the final mixture. Efforts have been made in this direction to magnetize the biocatalyst either by directly binding the enzyme on magnetic materials (magnetite or stainless steel powder) or by coentrapping magnetic material so that they can be recovered using an external magnet. Magnetized biocatalysts also help in the fabrication of magneto-fluidized bed reactor.

A variety of biologically active peptides are gaining importance in various fields including in pharmaceutical industries and in food industries as sweeteners, flavorings, antioxidants, and nutritional supplements. Proteases have emerged over the last three decades as powerful catalysts for the synthesis and modification of peptides.

The field of immobilized proteases may have a future role in this area. One of the important large-scale applications will be in the synthesis of peptide sweetener using immobilized enzymes like the thermolysin. Proteolytic enzymes, such as subtilisin, α-chymotrypsin, papain, ficin, or bromelain, which have been immobilized by covalent binding, adsorption, or cross-linking to polymeric supports, are used to resolve *N*-acyl-DL-phenylglycine ester racemate, yielding *N*-acyl-D-esters or *N*-acyl-D-amides and *N*-acyl-L-acids. Immobilized aminopeptidases have been used to separate DL-phenylglycinamide racemates.

22.10 Cell Immobilization by Using Alginate

22.10.1 Principle

Alginate is commercially available as sodium salt of alginic acid commonly called sodium alginate. It is a linear polysaccharide normally isolated from many strains of marine brown sea weed and algae. The immobilization of cells is a technique in which the cells are retained in restricted space or surface of certain matrices, but retaining their catalytic activity. In immobilized system, the cells can he retained at a physiological state at which product formation is maximum. The system is less sensitive to change in operational conditions. Among the various techniques employed for whole-cell immobilization, entrapment in sodium alginate cell is becoming very attractive because of the simplicity and low cost.

The cell immobilization technique can broadly be of two types:

1. Attachment on insoluble support
2. Entrapment in a matrix

Here, the yeast extract medium for seed culture was prepared and inoculated with cells like yeast (*Saccharomyces diastaticus*) and trapped by using matrix-like sodium alginate.

22.10.2 Materials

Organism: *S. diastaticus*
   Culture medium
   Yeast extract medium: Yeast extract 10 g, KH_2PO_4 10 g, glucose 50 g, pH 6.0, distilled water 1 L.
   3.6% w/v sodium alginate in a solution containing 0.1% NaCl, $CaCl_2$ 4% w/v, yeast cell 15 g/L; EDTA 4% w/v.

22.10.3 Protocol

1. Yeast extract medium was prepared, sterilized, and inoculated with *S. diastaticus* and incubated at 37°C for 12–18 h under shaking.
2. Add 3.6% w/v sodium alginate in 0.1% NaCl slowly with continuous stirring.
3. Leave it for 6–8 h to allow the air bubbles to escape.
4. Add 15 mg/L of yeast cells to alginate slum and mix well.
5. Take the slurry in a syringe and exclude the slurry drop wise into the 4% $CaCl_2$. The drops formed into spherical beads of 2–3 mm in size.
6. Keep the beads in $CaCl_2$ solution for 30 min for gelation.
7. Wash the beads in sterile distilled water repeatedly till there is no leakage.

25.10.4 Result

Several beads were observed with immobilized yeast culture in sodium alginate solution when transferred drop wise into $CaCl_2$ solution.

22.11 Immobilization of Enzyme Horseradish Peroxidase (Donor: H_2O_2 Oxidoreductase; EC 1.11.1.7) by Aminoantipyrine–Phenol Assay

22.11.1 Principle

This method has been chosen because it is rapid, inexpensive, convenient, accurate and allows kinetic characterization of the immobilized enzyme. Immobilized peroxidase catalyzes a reaction that has commercial potential and interest. There is a reductive cleavage of H_2O_2, by an electron donor, AH_2.
   In the presence of peroxidase, H_2O_2 rapidly reacts with phenol and 4-aminoantipyrine, which forms a chromogen complex peroxidase quinone imine that is intensely colored with absorbance maxima at 510 nm:

$$H_2O_2 + phenol + 4\text{-aminoantipyrine} \rightarrow Peroxidase\ quinone\ imine + 2H_2O$$

The amount of peroxidase present will influence the amount of quinone imine product formed. There is a linear relationship between the quantity of peroxidase present in the gel or solution and the intensity of the color produced. The intensity of color or absorbance can be measured with a spectrophotometer. This assay is commonly used to measure the rate of the enzyme-catalyzed reaction which allows calculation of important kinetic constants including the number of enzyme activity units (U), the Michaelis constant (K_m), and the maximum velocity (V_{max}).

22.11.2 Materials

KH_2PO_4, $NaOHPO_4$, acrylamide, methylene bisacrylamide, ammonium persulfate, tetramethylethylene diamine (TEMED); horseradish peroxidase, spectrophotometer with cuvettes, clinical centrifuge, 4-aminoantipyrine, phenol, hydrogen peroxide, syringes (5 mL) with 0.8 μm filter system, constant temperature water bath (60°C), scintillation vials.

Reagents

1. Potassium phosphate buffer 0.2 M, pH 7.0.
2. 30% acrylamide and 0.8% methylene bisacrylamide: Dissolve 30 g of acrylamide and 0.8 g of methylene bisacrylamide in small quantity of potassium phosphate buffer (0.2 M pH 7.0) and make the volume to 100 mL with buffer.
3. Ammonium persulfate (10%): Dissolve 10 g of ammonium persulfate in small quantity of phosphate buffer and make the total volume to 100 mL with buffer (prepare fresh just before use).
4. 4-Aminoantipyrine–phenol solution: Dissolve 810 mg phenol in 40 mL of glass distilled water and add 25 mg of 4-aminoantipyrine. Dilute to a final volume of 50 mL with glass distilled water.
5. Hydrogen peroxide 0.0077 M in water: Prepare by mixing 1 mL of 30% H_2O_2 with 99 mL of glass distilled water. Further, dilute 1 mL of this solution to 50 mL with 0.2 M phosphate buffer, pH 7.0 (prepare fresh before use).
6. TEMED 2% in the final volume of the gel.

22.11.3 Protocol

1. *Preparation of immobilized peroxidase.* To prepare the acrylamide gel, add the following to a 20 mL vial: 3.25 mL potassium phosphate buffer, 2.7 mL acrylamide solution (solution 2), and 80 μL ammonium persulfate (solution 3). Mix and add 1 mL of 0.1 mg/μL peroxidase solution. Add 10 μL of TEMED reagent and mix with a vortex mixer. The solution should become opaque within a few minutes and completely polymerized within 20–30 min. With a spatula, transfer the gel to a vacuum filtration system to remove most of the solution. To wash the gel, transfer it to a test tube containing 5 mL of water. Break up the gel by aspirating with a Pasteur pipette. Centrifuge the gel mixture for 5 min at 1000–1200 rpm. Decant supernatant and add 10 mL of water to the gel. Again, break up the gel by aspirating with a Pasteur pipette and centrifuge as before. Repeat this washing process two more times. The final centrifugation should be done for 5 min at 1200–1500 rpm. Dry the gel by vacuum filtration for a minute and weigh on a balance. Approximately 2–3 g of semiwet gel is obtained. Proceed for further processing.

2. *Assay of immobilized peroxidase.* A 3 min fixed time is used to measure peroxidase activity. Two tubes are set up for each assay: one tube for the zero point and the other tube for the 3 min assay. The amount of gel-immobilized peroxidase will be different for each assay. Obtain six test tubes for reaction vessels and label them as 1–6. Weigh two samples of 0.05 g of immobilized peroxidase and transfer one sample to tube 1 and the other sample to tube 2. Add 2.5 mL of the aminoantipyrine–phenol solution to tube 1 and mix well. To tube 1 (zero point), add 2.5 mL of the H_2O_2 solution, rapidly mix well, and immediately quench the reaction by pouring into the barrel of syringe with filter system. Use the plunger to force the solution through the filter into glass cuvette. This whole procedure from mixing to transfer should take no more than 10 s. Read and record the absorbance of the reaction mixture at 510 nm.

Add 2.5 mL of the solution to tube 2 (3 min point), immediately mix, and note the time. Gently and continuously mix the reaction mixture. At the end of exactly 3 min, transfer the reaction mixture to the barrel of a syringe with filter system. Force the solution through the filter into a glass cuvette. Read the absorbance of the reaction mixture at 510 nm.

The 0.1 g of immobilized peroxidase is used for tubes 3 and 4. Tube 3 will represent the zero point and tube 4 the 3 min point. Weigh two samples of 0.1 g of immobilized peroxidase and transfer to tubes 3 and 4. Add 2.5 mL of aminoantipyrine–phenol solution to each tube and mix well. Measure the peroxidase activity in each tube by adding 2.5 mL of the H_2O_2 solution as before; tube 3 will be treated like tube 1 and tube 4 like tube 2. Read the A_{510} for tubes 3 and 4. Repeat the procedure with tubes 5 (zero point) and 6 (3 min point), each containing 0.2 g of immobilized peroxidase.

3. *Thermal stability of immobilized peroxidase.* The thermal stability of acrylamide gel–immobilized peroxidase is compared to the free enzyme.

The free enzyme is assayed in the following manner:

Dilute 1 mL of the stock horseradish peroxidase (0.1 mg/mL, 15 units/mL) with 299 mL of glass distilled water. Add 10 mL of this diluted enzyme to each of two test tubes. Place one of the tubes in a 60°C set water bath for exactly 4 min, and allow the other tube to sit at room temperature for the same time interval. Cool the higher temperature tube to room temperature. To each tube add 2.0 mL of aminoantipyrine–phenol stock solution and 2 mL of the H_2O_2 solution and mix well. Allow the tubes to sit at room temperature for exactly 3 min; then immediately read the absorbance A_{510} for each tube.

The immobilized enzyme is assayed in the following manner:

Weigh two samples of 0.1 g of the gel and transfer to two separate test tubes each containing 0.5 mL of phosphate buffer. Place one of the tubes in the 60°C water bath and allow the other to remain at room temperature. At the end of 4 min, remove the tube from the water bath, cool to room temperature, and add 2.0 mL aminoantipyrine–phenol solution and 2 mL of H_2O_2 solution. Gently mix the reaction mixture for exactly 3 min and remove the gel by passing the mixture through the syringe filter system. Read the absorbance of the reaction mixture at 510 nm.

To the room temperature tube, add 2.25 mL of aminoantipyrine–phenol solution and 2.25 mL of H_2O_2. Gently mix the reaction mixture for exactly 3 min and separate the gel with a syringe filter system. Read the absorbance of the reaction mixture at 510 nm.

22.11.4 Results and Observations

1. *Assay of immobilized peroxidase.* The activity of the *immobilized* peroxidase can be calculated from the change of absorbance for each reaction mixture. The absorbance change is calculated as follows:

$$\Delta A = A_{3\,min} - A_{0\,min}$$

where
   ΔA is overall absorbance change
   $\Delta A_{3\,min}$ is absorbance at 510 nm of tubes 1, 3, and 5

Calculate the $\Delta A/min$ for each of the three sets of conditions (0.05, 0.1, and 0.2 g of gel). Prepare a plot of $\Delta A/min$ (Y-axis) vs. mg of gel (X-axis). Read the shape of the graph.

Use the following equation to calculate the units of activity per mg of gel. The number 6.58 in the denominator is the absorption coefficient for the quinone imine chromogen assay product:

$$Units/mg = \frac{\Delta A/min}{6.58 \times mg \text{ of gel}}$$

Repeat this calculation for all three amounts of gel (0.05, 0.1, 0.2 g) and compare the three results.

2. *Thermal stability of immobilized peroxidase.* Determine the reaction rate ($\Delta A/min$) for each of the four reaction conditions. Assure that $A_{0\,min} = 0$ for each assay.
   ΔA_1 is absorbance change for free peroxidase at room temperature
   ΔA_2 is absorbance change for free peroxidase at 60°C
   ΔA_3 is absorbance change for immobilized peroxidase at room temperature
   ΔA_4 is absorbance change for immobilized peroxidase at 60°C
   Calculate the % activity remaining after heating the free and immobilized enzymes:

$$\% \text{ Activity remaining (free)} = \frac{\Delta A_2}{\Delta A_1} \times 100$$

$$\% \text{ Activity remaining (immobilized)} = \frac{\Delta A_4}{\Delta A_3} \times 100$$

Compare the two final results and find out which one is more stable to heat, free or immobilized enzyme.

Calculate the specific activity of the free enzyme in units/mg using the following equation:

$$\text{Units/mg} = \frac{\Delta A_{510/\text{min}}}{6.58 \times \text{mg enzyme/mL reaction mixture}}$$

Compare this number with known specific activity of the enzyme listed on the reagent bottle.

Suggested Readings

Mateo, C., J. M. Palomo, G. Fernandez-Lorente, J. M. Guisan, and R. Fernandez-Lafuente. 2007. Improvement of enzyme activity, stability and selectivity via immobilization techniques. *Enzyme Microb Technol* 40: 1451–1463.

Pandey, P. K., R. K. Saxena, and P. S. Bisen. 2002. Immobilization results in sustained calcium transport in *Nostoc calcicola* Bréb. *Curr Microbiol* 44: 173–177.

Sheldon, R. A. 2007. Enzyme immobilization: The quest for optimum performance. *Adv Synth Catal* 349: 1289–1307.

Spahn, C. and S. D. Minteer. 2008. Enzyme immobilization in biotechnology. *Recent Pat Eng* 2: 195–200.

Important Links

http://www.lsbu.ac.uk/water/enztech/immethod.html
http://onlinelibrary.wiley.com/doi/10.1111/j.1751-7915.2010.00227.x/pdf
http://www.rpi.edu/dept/chem-eng/Biotech-Environ/IMMOB/methods.htm
http://www.biotecharticles.com/Biotechnology-products-Article/Immobilized-Enzyme-and-its-Applications-1041.html
http://link.springer.com/protocol/10.1007%2F978-1-59745-053-9_22?LI=true
http://link.springer.com/article/10.1007%2FBF02293003?LI=true

23

Plant Tissue Culture

23.1 Introduction

Plant cell culture has found wide applications ranging from studies of basic plant biochemistry and molecular biology to mass propagation and genetic engineering of crop species. The underlying principle of the science of plant cells and tissue culture is based on the concept of totipotency. All living cells of a plant are genetically alike and can behave like zygotic cells when they are relieved of the constraints of surrounding cells. If proper nutritional and hormonal conditions are provided, it is possible to raise complete plants from a single cell or group of cells. Various techniques are used routinely to facilitate the regeneration of new plants from fully differentiated cells. This is achieved by inducing cells with various agents, including hormones (auxins, cytokinins), to undergo a new developmental program and regenerate an entire new individual plant. Breakthroughs have been made in the transformation of important crop species such as corn, wheat, and soybean, and transgenic crops are now being marketed. With good sterile technique and some patience for the comparatively slow growth, much can be accomplished today with plant cell culture.

23.2 Facilities

As with all other tissue culture procedures, plant tissue culture requires a sterile environment, aseptic manipulation of specimens, and defined growth conditions. The general laboratory requirements are similar to those required for the animal cell but with certain modifications specific to plant tissue culture. For instance, a plant growth chamber or a controlled environment room may be used instead of the standard CO_2 incubators required for animal cells. Other environmental conditions required for optimum growth will vary depending on the species but, in general, consideration should be given to the diurnal temperature variations, light quality, and intensity and the relative length of light–dark cycles required by the plants. Successful plant cell culture requires extremely close attention to proper sterile technique and temperature control. The growth of plant cell cultures is relatively slow compared to common microbial cultures, necessitating very long culture times for experiments of a low level of contamination which would be outgrown and perhaps unnoticed in a bacterial culture would have time to overwhelm plant cultures. It may take months to establish a high-producing cell line or to regenerate trait plants from the original explants.

23.2.1 Sterile Transfer Facilities

Laminar flow hoods are strongly recommended. An even flow of high-efficiency particulate air (HEPA) filter-sterilized air comes from the back of the hood, protecting the cultures when the vessels are opened. Unlike primary animal cell cultures, no biohazard containment of cultures is required. Some laboratories prefer to use containment hoods when introducing microbes into the plant cell cultures, such as during genetic transformation or inoculation with plant pathogens. Each hood should be equipped with a natural gas outlet for a Bunsen burner and lighting (usually overhead). A UV germicidal lamp is strongly recommended for use in decontaminating the hood work surface after the HEPA filter is replaced if the hood is switched off for >30 min or if the hood is accidentally contaminated. A selection of stainless steel forceps, spatulas, and scalpels with replaceable blades are required for culture initiation, callus subculture, transformation, and regeneration. For quantitative transfers, an electronic (digital) top-loading balance (400 g to 4 kg numerical capacity) with auto-tare feature is essential. The good clean bench and laboratory practice and stringent housekeeping procedures have become essential requirements in a modern tissue culture laboratory.

23.2.2 Laboratory Layout

The proper layout of the laboratory is essential for an effective housekeeping practice of any tissue culture laboratory (Figure 23.1). The main components of any tissue culture laboratory include

1. Area for glassware cleaning and media preparation
2. Facilities for aseptic manipulations of the material
3. Incubation room with a controlled environment

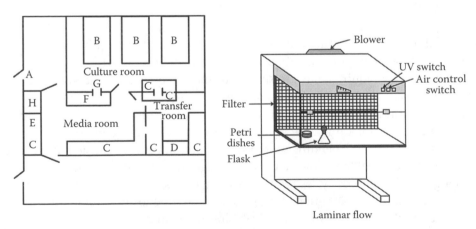

Figure 23.1   Design of a plant tissue culture laboratory: (A) entrance, (B) shelve, (C) counter, (D) laminar air flow cabinet, (E) sinks, (F) gas outlet/burner, (G) window, and (H) refrigerator and deep freezer.

Special facilities, for example, microscope for cytological studies, chemicals, microtome, and shaker facility for suspension cultures, are necessary. The basic organization and facilities of most tissue culture laboratories are as follows:

- Large sinks (some lead-lined to tolerate strong acids and alkali) and draining areas
- Cabinet and shelf space for storage of chemicals and dust-free storage space for clean glassware
- Transfer areas/chamber/laminar airflow set for aseptic manipulations if possible fitted with UV lighting
- Culture rooms or incubators with controlled light, temperature, air, and vacuum in the working areas
- Essential services such as electricity, water, gas, and compressed air and vacuum in the working areas
- Washing machines to wash glassware in bulk
- Hot air cabinets to dry-washed glassware

23.2.3 Instruments

1. Balances one to weigh small quantities (electronic analytical to five decimal places) and the other to weigh comparatively large quantities (to two decimal places)
2. Pressure cooker and/or autoclave for steam sterilization of media and apparatuses
3. Oven for dry heat sterilization of glassware
4. Hot plate-cum-magnetic stirrer to dissolve chemicals
5. Refrigerator to store chemicals, stock solutions, and plant materials
6. Heater or heat-regulated hot plate for heating purposes
7. Exhaust pump to facilitate filter sterilization
8. pH meter to adjust pH of solutions and media
9. Shaker to grow suspension cultures
10. Steamer to dissolve agar and melt media
11. Low-speed centrifuge to sediment cells and cleaning of protoplasts
12. Atomizer to spray spirit in the inoculation chamber
13. Deep freeze to store stock solutions, certain enzymes, coconut milk, and so on
14. Water distillation apparatus or demineralization apparatus to obtain distilled or deionized water
15. Filter membranes and their holders to filter sterilize solutions
16. Instruments stand to keep sterilized instruments
17. Metal trays and carts for transport of culture flasks, racks of tubes, and so on
18. Dissecting (low-power) microscope, for dissecting out microscopic explants and overall observation of tissue development
19. Bright-field (compound) microscope to observe cells and tissues

23.2.4 Tools

1. Spirit lamps or Bunsen burners, to flame instruments
2. Large forceps with blunt ends to inoculate and subculture

3. Forceps with fine tips to peel leaves
4. Fine needles for dissections, autopipettes of different capacity
5. Scalpels to shred the tissues
6. Cork borer to remove tissue cylinders of a standardized size
7. Hemocytometer to determine cell counts
8. Cavity slides for hanging-drop cultures
9. Microscope slides and coverslips
10. Plastic or steel buckets to soak labware for washing
11. Large plastic carboys (e.g., 10 or 25 L) for storing high-quality water
12. Hypodermic syringes to filter sterilize solution
13. Screw-cap bottles to sterilize plant material
14. Culture tube rack slanted to prepare medium slants
15. Culture tube rack, upright to store cultures
16. Stainless steel or Teflon sieve of various pore sizes (40–100 mesh screens, i.e., with 0.38 openings), to separate cell clumps of different dimensions

23.2.5 Glassware for Media

1. Erlenmeyer flasks (50, 100, 125, 500 mL, 1, 5 L capacity)
2. Volumetric flasks (500 mL, 1–3 L capacity)
3. Measuring cylinders (10, 25, 100, 500 mL and 1 L capacity)
4. Graduated pipettes (1, 2, 5, and 10 mL capacity)
5. Pasteur pipettes and teats for them
6. Culture vials (Petri dishes [50×10, 100×17, 500×20 mm], culture tubes [25×150 mm], screw-cap bottles of various sizes)
7. Buchner funnels

23.2.6 Culture Vessels

Successful plant cell culture requires extremely close attention to proper sterile technique and temperature control. Plant material is inoculated in glass vessels containing the nutrient media. All glassware used are of high-quality glass. Depending upon the convenience, one can use culture tubes, Erlenmeyer flasks, or Petri plates. Polythene vessels and wide mouth screw bottles are also used. The vessels are plugged with nonabsorbent cotton.

The cultures are maintained under sterilized conditions. Therefore, it is necessary that glassware, nutrient medium, explants, and place of inoculation should be properly sterilized.

23.3 Glassware Preparation

It is extremely important to use clean glassware for the growth of tissues *in vitro*. Following steps are generally adopted for the purpose. Glassware should be free of residual detergents. All traces of chromium should be removed by treating the glassware with ethylenediaminetetraacetic acid (EDTA) or any metal chelators if glassware are cleaned with chromic acid. In fermenters or bioreactors, all parts in contact with the medium must be made up of high-quality stainless steel, glass, or plastic, to work with sensitive cell lines:

- Immerse glassware in 10% aqueous washing soda (Na_2CO_3) and boil for 1 h.
- Glass pipettes should be soaked in 10% Na_2CO_3 for 24 h. Do not heat or boil.
- Rinse the glassware thoroughly with tap water.
- Soak in 1 N HCl or chromic acid for 2 h.
- Wash thoroughly with tap water to remove traces of acid.
- Rinse with double distilled water and dry overnight at room temperature.

23.3.1 Sterilization

The objective of sterilization is to make glassware and instrument free of microorganisms such as bacteria, fungi, etc. Sterilization of glassware is usually accomplished by wet heat method and dry heat method.

23.3.2 Wet Heat Sterilization

- Plug glassware such as conical flasks, test tubes, etc., with nonabsorbent cotton.
- Wrap Petri dishes in a clean wrapping paper or aluminum foil or keep them in a Petri can.
- Keep forceps and scalpels in test tubes, plug the tubes with nonabsorbent cotton, and cover with wrapping paper.
- Plug the mouth end of the pipettes with cotton. Wrap individually in wrapping paper or aluminum foil or keep them in a pipette can.
- Autoclave glassware and instruments in an autoclave at 121°C for 1 h.

23.3.3 Dry Sterilization

- Plug and wrap the glassware and instruments according to the procedure described earlier.
- Keep glassware and instruments in an oven at 140°C–160°C for 2 h.

23.3.4 Filter Sterilization

- Compounds such as amino acids, vitamins, number of plant growth hormones, etc., are usually destroyed during autoclaving. Such thermolabile compounds are sterilized by filtration through Millipore filtration membranes of 0.22 or 0.45 μm porosity (Figure 23.2).
- Plug 100 mL or 500 filtration flask with cotton.
- Assemble Millipore filtration unit with bacteriological membranes (0.22 or 0.45 μm) in place.

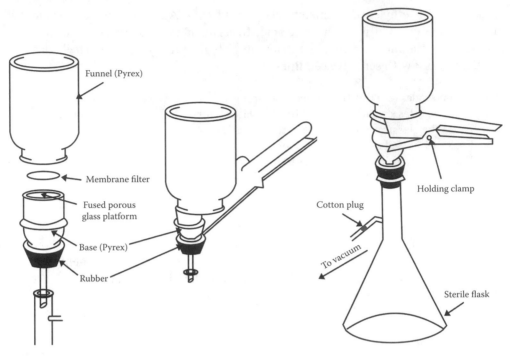

Figure 23.2 Millipore filtration assembly.

- Wrap filtration unit with wrapping paper.
- Autoclave filtration flask and filtration unit at 121°C for 1 h.
- Do not sterilize membrane filter by dry heat method as it will damage the membrane.
- Fix filtration unit to the filtration flask in a sterilized cabinet.
- Pour solution to be sterilized into the filtration unit.
- Apply slight air pressure to commence filtration. Do not exceed 5 lb/in.² pressure.
- Under sterile conditions, transfer filter-sterilized solution to a sterile flask.
- Using a sterile pipette, add filter-sterilized solution to the autoclaved medium under aseptic conditions.

23.4 Media Preparation

Only high-purity (reagent-grade) chemicals should be used for plant cell culture. Water should be of extra pure quality viz. double distilled (glass stills preferred) or deionized, such as 18 MΩ water produced by Milli-Q devices (Millipore, Befoul, MA). Success of any plant tissue culture depends on the appropriate composition of the medium. Different plant species have different nutritional requirements. It is often necessary to modify and standardize composition of medium to suit a particular tissue. Several basal media formulations are commonly employed for plant tissue culture. In general, the components of media used for plant tissue can be classified into six groups: (1) major inorganic nutrients; (2) trace elements; (3) carbon source; (4) iron source; (5) organic supplements, for example, vitamins, plant growth regulators; and (6) complex organic substances such as yeast extract, malt extract, casein hydrolysates, coconut milk, etc. are sometimes added to improve growth of the cultures. Several plant cell culture media are now available commercially in the form of dry powders containing all components except growth regulators, sucrose, and agar. They are convenient for routine culture work but suffer from the disadvantage of expense and the fact that they are of limited use for studies involving manipulation of levels of media components.

A general method adopted for preparation of a standard medium is given in the following sections:

23.4.1 Preparation of Stock Solution

- Prepare a list of chemicals to be weighed as per the requirement of composition of standard culture media.
- Weigh the chemicals in a glass container or on a butter paper.
- Dissolve each component in glass double distilled water.
- Major inorganic and trace elements are kept ready in the form of 10× stock solution whereas trace elements as 100× each Na_2MoO_4, $CuSO_4 \cdot 5H_2O$, and $CoCl_2 \cdot 6H_2O$.
- Stock solutions should be stored under refrigerated conditions.
- Stock solutions of plant growth regulators and vitamins commonly used for culture media are prepared as given in Tables 23.1 through 23.4.
- Powdered vitamin mixtures are hygroscopic and must be protected from moisture.
- Precipitation in the vitamin products may occur after a prolonged storage. This can be brought back into solution by warming the solution in a water bath (35°C–37°C) for a short period of time.
- To prepare 1 L of a 10× iron chelate solution dissolve 0.278 g of ferrous sulfate heptahydrate in 350 mL using DOW. Heating may be required. Dissolve 0.378 g disodium EDTA in 350 mL using DOW. When both components are dissolved, combine the solutions and bring up to the final volume 1 L.

23.4.2 Protocol

- Transfer appropriate amounts of stock solution of salts as per Tables 23.1 through 23.4 to 1 L flask.
- Add vitamin and iron stock solution.
- Add correct amounts of auxins, cytokinins, and organic supplements.
- Add sucrose, glucose, or any other carbohydrate source (2%–5%) as the case may be.
- Adjust the pH of the medium to 5.8–6.0 using pH indicator or pH meter.
- Make up the desired volume of the medium using double distilled water.
- Add agar powder at a concentration of 0.8%–1% for preparation of static media, and keep on a hot plate or heater for melting the agar under continuous stirring.
- Shake the flask well for uniform distribution of agar.
- Dispense medium in sterile tubes or flasks as per requirement.
- Use the prepared media within a week of its preparation or within a month if stored under refrigeration (10°C).

TABLE 23.1 Inorganic Macronutrients in Selected Plant Tissue Culture Media

Constituent	Helter (1953)	Nitsch and Nitsch (1956)	White (1963)	Hildebrandt et al. (1946)	Murashige and Skoog (1962)	Gautheret (1942)	White (1943)	Wood and Hraun (1961)	Eriksson (1965)	Gamberg et al. (1968)
KCl	750	1500	65	65	—	—	65	910	—	—
$NaNO_3$	600	—						1800		
$MgSO_4 \cdot 7H_2O$	250	250	720	190	370	125	3(×)	3(×)	370	250
$NaH_2PO_4 \cdot H_2O$	125	250	16.5	33			16.5	300		150
$CaCl_2 \cdot 2H_2O$	75				440				440	150
KNO_3		2000	80	80	1900	125	80	80	1900	2500
$CaCl_2$		25								
Na_2SO_4			200	800			88	88		
$(NH_4)_2SO_4$								790		134
NH_4NO_3					16.50	125			1200	
KH_2PO_4					170				340	
$Ca(NH_3)_2 \cdot 4H_2O$			300	400		500	200	200		

Note: Concentrations are expressed as mg/mL.

TABLE 23.2 Inorganic Micronutrients in Selected Plant Tissue Culture Media

Constituent	Helter (1953)	Nitsch and Nitsch (1956)	White (1963)	Hildebrandt et al. (1946)	Murashige and Skoog (1962)	Gautheret (1942)	White (1943)	Wood and Hraun (1961)	Eriksson (1965)	Gamberg et al. (1968)
$NiSO_4$						0.05				
$FeSO_4 \cdot 7H_2O$		0.3			27.8	0.05			27.8	27.8
$MnSO_4 \cdot H_2O$										10.0
$MnSO_4 \cdot 4H_2O$	0.01		7	4.5	22.3	3			2.23	
$MnCl_2 \cdot 4H_2O$							4.5	4.5		
KI	0.01		0.75	3.0	0.83	0.5	0.75	0.75		0.75
$NiCl_2 \cdot 6H_2O$								0.03		
$CoCl_2 \cdot 6H_2O$					0.025				0.0025	0.0025
$Ti(SO_4)_3$						0.2				
$ZnSO_4 \cdot 7H_2O$	1.0	0.5	3	6.0	8.6	0.18	1.5	1.5		20
$ZnNa_2EDTA$									15	
$CuSO_4 \cdot 5H_2O$	0.03	0.025			0.025	0.05	0.013	0.013	0.013	0.0025
$BeSO_4$						0.1				
H_3BO_3	1.0	0.5	1.5	0.38	6.2	0.05	1.5	1.5	0.63	3.0
H_2SO_4						1.0				
$FeCl_3 \cdot 6H_2O$	1.0						2.5	2.5		
$Na_2MoO_4 \cdot 2H_2O$		0.025			0.25				0.025	0.25
H_2MoO_4							0.0017	0.0017		
$AlCl_3$	0.03									
$Fe(SO_4)_3$			2.5							
Ferric tartrate				40.0						

Note: Concentrations are expressed as mg/mL.

TABLE 23.3   Organic Constituents Present in Selected Plant Tissue Culture Media

Constituent	Helter (1953)	Nitsch and Nitsch (1956)	White (1963)	Hildebrandt et al. (1946)	Murashige and Skoog (1962)	Gautheret (1942)	White (1943)	Wood and Hraun (1961)	Eriksson (1965)	Gamberg et al. (1968)
Sucrose	20,000	34,000	20,000	20,000	30,000	30,000	20,000	20,000	20,000	20,000
Glycine			3	3	2	3	3	3	2	0
Indole-3-acetic acid		0.18–0.18			100		100	100		100
Myo-inositol					1.30					
Cysteine			1.0			10				
Vitamin B1	1.0		0.1	0.1	0.1	0.1	0.1	0.1	0.5	10
Vitamin B6		25	0.1		0.5	0.1	0.1	0.1	0.5	1.0
Nicotinic acid			0.5		0.5	0.5	0.5	0.5		
EDTA (disodium salt)					37.3				37.3	37.3
Cad-panthothenic acid			1.0							
2,4-Dichlorophenoxyacetic acid										0.1–0.1
1-Naphthalene-acetic acid								1.0	1.0	
Kinetin			6		0.04–10			0.5	0.02	0.1

Note: Concentrations are expressed as mg/mL.

TABLE 23.4 Plant Growth Regulators Commonly Used in Cell Culture Media

Regulator	Concentration		Preparation of Stock Solution	Comments
	M_w	(μM)		
Auxins				
p-Chloroacetic acid (pCPA)	186.6	0.1–10	Auxins are usually titrated into solution with NaOH used as the sole auxin in a culture medium	IAA may be readily oxidized by plant cells, it is seldom
2,4-Dichlorophenoxyacetic acid (2,4-D)	221.0	0.1–10		
Indole-3-acetic acid (IAA)	175.2	0.1–10		
Indole-3-butyric acid (IBA)	203.2	0.1–10		
1-Naphthaleneacetic acid (NAA)	186.2	0.1–10		
β-Naphthoxyacetic acid (NOA)	202.2	0.1–10		
Cytokinins				
6-Benzylaminopurine (BAP)	225.2	0.1–10	Cytokinins are normally dissolved in dilute NaOH or in aqueous EtOH	Zeatin is thermolabile and should not be autoclaved (21P)
n-Isopentenylaminopurine	203.2	0.1–10		
6-Furfurylaminopurine (kinetin K)	215.2	0.1–10		
Zeatin (*Zea*)	219.2	0.1–10		
Gibberellins				
Gibberellic acid (GA₃)	346.4	0.1–5	Freely soluble in H_2O	GA_3 is thermolabile, must not be autoclaved. GA_3 is seldom necessary for initiation or maintenance of cultures. Sometimes necessary for plantlet regeneration

Source: Dixon, R.A., *Plant Cell Culture. A Practical Approach*. IRL Press, Oxford, London, 1985.

23.5 Sterilization of Media

Plant tissue culture media are generally sterilized by autoclaving at 121°C and 1.05 kg/cm² (15–20 psi). The time required for sterilization depends upon the volume of the medium in the vessel. The minimum time required for sterilization of different volumes of media is listed as follows:

Minimum autoclaving time for plant tissue culture media

Volume of Media per Vessel (mL)	Minimum Autoclaving Time
25	20
50	25
100	28

Volume of Media per Vessel (mL)	Minimum Autoclaving Time
250	31
500	35
1000	40
2000	48
4000	68

It is advisable to dispense media in small aliquots whenever possible because many media components are broken down on prolonged exposure to heat and pressure. There is evidence that media exposed to temperatures in excess of 121°C may not support any growth or may result in poor cell growth.

Several media components are considered thermolabile and should not be autoclaved. Stock solutions of the heat-labile components are prepared and then filter sterilized through a 0.22 μm filter into a sterile container. The filtered solution is then aseptically added to the culture medium that has been autoclaved and allowed to cool to ~45°C. The medium is then dispensed under sterile conditions.

23.6 Aseptic Manipulation

All aseptic manipulations are carried out in a laminar flow bench. The working surface is cleaned with 70% aqueous industrial methylated spirit or 70% aqueous ethanol and irradiated with UV germicidal tubes for 15–30 min before use.

23.7 Sterile Transfer Facilities

Laminar flow hoods are strongly recommended. An even flow of HEPA filter-sterilized air comes from the back of the hood, protecting the cultures when the vessels are opened. Unlike primary animal cell cultures, no biohazard containment of cultures is required. Some laboratories prefer to use containment hoods when introducing microbes into the plant cell cultures, such as during genetic transformation or inoculation with plant pathogens. Each hood should be equipped with a natural gas outlet for a Bunsen burner and lighting (usually overhead). A UV germicidal lamp is strongly recommended for use in decontaminating the hood work surface after the HEPA filter is replaced if the hood is switched off for >30 min or if the hood is accidentally contaminated. A selection of stainless steel forceps, spatulas, and scalpels with replaceable blades are required for culture initiation, callus subculture, and transformation and regeneration. For quantitative transfers, an electronic (digital) top-loading balance (400 g to 4 kg numerical capacity) with auto-tare feature is essential.

23.8 Temperature

The most common growth temperature is 25°C. Some plant species, especially heat-tolerant plants, can continue to grow at temperatures of 30°C and above while heat-sensitive cultures will cease to grow above 27°C and will die at higher temperatures. Cyclic temperature changes may occur during the day/night cycles used for some culture procedures. Localized heating of shelves occurs during the day, and they then cool during the night. Although not directly lethal to cultures, this contributes to the formation of condensation inside the lids of Petri dishes and other culture vessels, which can dry out the cultures or contribute to contamination. Stacking an uninoculated Petri dish containing water agar on top of the important cultures will greatly reduce condensation yet still allow sufficient light for regeneration.

23.9 Light

Callus and shake flask cultures generally do not require light for growth. Usually they are grown in dark rooms, equipped with lights that are turned on when inspecting or subculturing the cultures. Absolute darkness is not usually required. High light intensities or low levels of continuous illumination can cause cultures to turn green (form small numbers of chloroplasts) or brown (accumulate polymerized phenolics). Enough light to induce either color change can be sufficient to inhibit the secondary-metabolite accumulation, and browning may greatly slow the growth of cultures. If callus and shake flasks must be incubated in lighted rooms (such as rooms shared with regenerating cultures or aseptically grown plants), they can be shielded with foil or boxes (Figures 23.3 and 23.4).

For regeneration, cool white fluorescent lights are recommended. Interspersing incandescent bulbs will greatly increase the quality of the light, providing a spectrum close to that of sunlight. Sufficient distance must be maintained between the lights and vessels, to allow cooler air to circulate and avoid overheating the plates. Wire shelving, positioned a few inches from the culture room walls, is often preferred providing good air circulation at a reasonable cost. The shelves can be partially covered with aluminum foil or Plexiglas to provide a more stable, easily cleaned surface. If several layers of closely lit shelves are used, heat from a lower shelf may warm the shelf above plates can partially be insulated by placing them on foil-covered sheets of styrofoam, or lights should be suspended, by chains, to allow heat dissipation. Excessive light can be detrimental to culture growth and regeneration. One indication of a high light level is the accumulation of anthocyanins (red to purple pigments) in callus or differentiated tissues. Continuous light is sometimes detrimental to whole-plant growth and is rarely used in plant cell cultures. The use of household timers is adequate for generating day–night cycles (16–18 h of light, 6–8 h of dark).

23.10 Aeration

Under most conditions, shake flask cultures of plant suspension have a lower volumetric oxygen demand than do microbial cultures. The respiration rates (QO_2) reported for plant cell cultures range from 0.2 to 3.6 mmol of O_2/h/g (dry weighed), compared with 2–16 mmol/h/g for *Escherichia coli*, 39–570 mmol/h/g for *Bacillus* spp., and

Sterilized lab materials in
laminar airflow

Labwares used in plant tissue
culture laboratory

Sterilized materials and medium required
for micropropagation

Performing experimental work in laminar
airflow

Maintenance of cultures in plant tissue culture room

Figure 23.3 An ideal illuminated culture facilities for plant tissue culture.

1.9–2.8 mmol/h/g for *Saccharomyces* spp. Shaker speeds (100–150 rpm with a 1 in. throw) are generally lower than those used for microbial cultures, and the fill ratios in the flasks (volume of medium/volume of flask generally 0.2–0.4) are generally higher. Baffled flasks are rarely used to increase aeration but are occasionally used for breaking up clumps of cells, especially when suspension is first being initiated. A wide variety of closures has been used for shake flask cultures. Traditional cotton plugs in Erlenmeyer flasks work well. A preferred option is straight-walled Delong neck shaker flasks with metal culture tube closures; a small piece of folded cheesecloth or

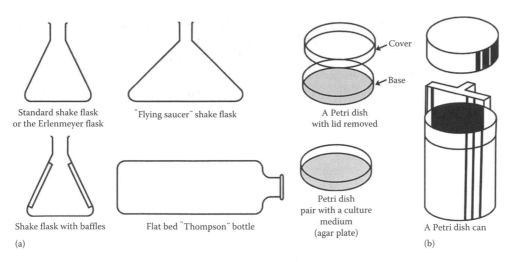

Figure 23.4 (a) Shaker flasks and (b) Petri dish and Petri can.

gauze is placed, in the top of the closure, to act as a sterile filter and the prongs on the closures hold it securely on the flask (Figure 23.4). These caps have the advantage of providing the same sterility barrier and good aeration as cotton plugs while covering the rim and upper edge of the flasks. Any dust from the shaker rooms or medium storage areas can be easily removed by a brief flaming before the flask is opened.

Petri dishes containing agar media and callus cultures or regenerating plants are incubated agar side down, not in inverted position like microbial cultures. Plates are always sealed with parafilm or porous medical tape in order to avoid contamination and also to protect from dehydration of agar media during prolong incubation period. Dedicated growth rooms provide the most efficient and flexible use of space. Commercial growth chambers equipped with lights temperature control and (optional) humidity control are available; several growth chambers allow many different conditions to be maintained, which may be important if diverse species are to be cultivated. For important cultures, it is recommended that several separate cultures be maintained, preferably in separate culture rooms or incubators.

Temperature recorders are strongly recommended in diagnosing possible problems in temperature control. Motion sensors and temperature monitors with audible alarms can also be used to minimize the damage caused by heating and cooling and by shaker equipment breakdowns.

23.11 Temperature

Most cell cultures grow well only in a narrow temperature range. The most common growth temperature is 25°C. Some plant species, especially heat-tolerant plants such as tobacco or *Catharanthus roscus*, can continue to grow at temperatures of 30°C and above, while heat-sensitive cultures like *opium poppy* will cease to grow above 27°C and will die at higher temperatures.

Cyclic temperature changes may occur during the day–night cycles used for some culture procedures. Localized heating of shelves occurs during the day, and they then cool during the night. Although not directly lethal to cultures, this contributes to the formation of condensation inside the lids of Petri dishes and other culture vessels, which can dry out the cultures or contribute to contamination. Stacking an uninoculated Petri dish containing water agar on top of the important cultures will greatly reduce condensation yet still allow sufficient light for regeneration.

23.12 Plant Cell Culture System

Setting up a plant cell or tissue culture begins with the excision and surface sterilization of explants, which may be chosen from any part of the plant, depending on the objective of the study. Leaves are frequently preferred for protoplast isolation (cells without walls), anthers for production of haploids, shoot meristems for proliferating shoot cultures, and root tips for root cultures. Plant tissue culture technology can be divided into five classes, based on the type of materials used:

1. *Callus culture*: The culture of callus (cell masses) on agar media produced from one explant of seedling or other plant source.
2. *Cell culture*: The culture of cells in liquid media, usually aerated by agitation.
3. *Organ culture*: The aseptic culture of embryos, anthers (spores), ovaries, roots, shoots, or other organs on nutrient media.
4. *Meristem culture*: The aseptic culture of shoot meristems or other explant tissue on nutrient media to get complete plants.
5. *Protoplast culture*: The aseptic isolation and culture of plant protoplasts from cultured cells or plant tissue.

23.13 Explants

Explants may be excised either in the field or from glasshouse grown plants or seedlings incubated under aseptic conditions. Surface sterilization most frequently employs several minute exposure to a 1%–10% (w/v) sodium hypochlorite solution in which chlorine gas acts as a biocide. After the set time (of the order of 5 min), excess hypochlorite must be removed immediately by copious washing in sterile distilled water.

Explants are most likely to survive *in vivo* when the tissue chosen is physiologically active (i.e., not dormant). Many types of explants contain meristematic tissue capable of cell division. Undifferentiated callus is formed soon after excision and comprises cells with a small amount of cytoplasm but large vacuoles. Such tissue may develop localized growth centers called meristemoids from which caulogenesis (shoot indication), rhizogenesis (root initiation), or both may ensure. The ability of callus to undergo such organogenesis is genetically controlled but may be encouraged *in vitro* by manipulating the cytokinin to auxin ratio in the medium. A high cytokinin to auxin ratio usually favors shoot proliferation, whereas a low cytokinin to auxin ration usually promotes rooting. In practical terms, whole plantlets can be

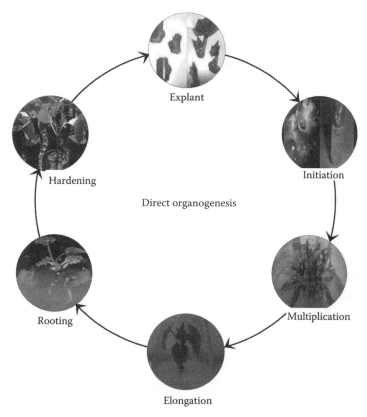

Figure 23.5 Direct organogenesis from explant.

produced *in vitro* from organogenic callus by first encouraging shoot proliferation on a cytokinin-rich medium and then transferring leafy shoots to auxin-rich medium for root induction (Figure 23.5).

Roots may also be developed *in vivo*, provided plants are protected from desiccation. Organogenesis from callus is not generally recommended for plantlet multiplication *in vitro* (micropropagation) because there is considerable evidence that it results in genetically aberrant plants being recovered. A greater degree of chromosomal stability can be achieved by using shoot meristems as initial explants for reasons that are as yet ill defined. Meristem cultures are consequently a convenient starting material for micropropagation. Meristem culture can also be used sometime to cure plants infected with viruses, following the use of high temperature treatment (thermotherapy) and/or chemicals (chemotherapy).

23.14 Calluses

In general, plant tissue culture begins with the initiation of callus cultures, that is, dedifferentiated masses of rapidly dividing cell cultures, on solid, agar-based nutrient media. The callus is generated by exposing sterile pieces of a plant to plant growth

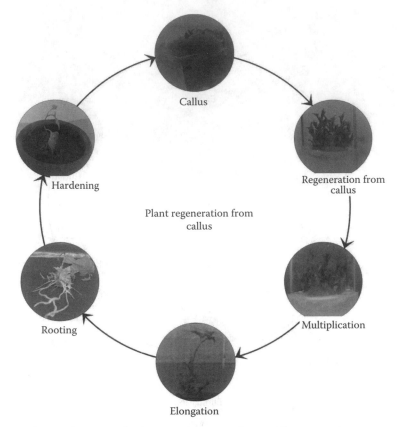

Figure 23.6   (See color insert.) Plant regeneration from callus.

regulators, Normally, cell division in an intact plant is restricted to various meri-stematic regions (e.g., root and shoot tip meristems) by the plant regulating natural hormone distributing system. The plant growth regulators are either natural plant hormones or synthetic analogs which bind to receptors in the plant cells, causing all cells to resume division and growth (Figure 23.6).

Callus cultures can be maintained indefinitely by passing clumps of cells to fresh agar media at regular intervals. Alternatively, callus pieces can be placed in shake flasks containing liquid media, to eventually form suspension cultures. They grow faster than callus cultures, probably due to better nutrient exchange with the medium. Plant cell suspensions consist of some single cells and many clumps of a few to hundreds of cells; the size of the clumps depends upon the species, the medium, and the length of time since the culture was initiated. Suspension cultures can be used as a uniform, year-round source of plant cells for biochemical and plant physiological studies or as a source of valuable plant secondary metabolites following appropriate treatments (Figure 23.7).

The callus cultures can also be allowed to differentiate (regenerate) back into plants; this process is often accelerated by the omission or changes in the types of plant growth regulators. Each piece of callus can give rise to one to hundreds of plantlets, each with the same genome as the parent, giving rise to the industry of micropropa-gation. Either early or late in the callus stage, foreign DNA can be introduced into

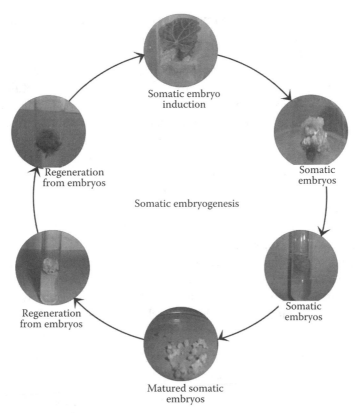

Somatic embryogenesis

Figure 23.7   Somatic embryogenesis.

the dividing plant cells. If the cells are then allowed to regenerate, a portion of the plantlets will be transgenic and can be selected for in a number of ways. At one time, the main way of introducing foreign DNA and organelles into plant cells involved the production of protoplasts (wall-less plant cells released by the action of cellulases and pectinase). Currently, several methods of DNA delivery are used, including delivery by the "tamed" plant pathogen *Agrobatecrium tumefaciens* or delivery by direct shooting of DNA-coated particles into the cells.

23.15 Cell Suspension

Cell suspension may grow as batch, fed-batch, or continuous culture on a small, medium, or large scale. For example, single individual cells may be grown in incubated microscope slide chambers where knowledge of the clonal origin of the plant is a prerequisite for the study. The wide-necked Erlenmeyer flask, shaken on a horizontal platform orbital shaker, is a favorite container for small-scale batch culture. Fermenters with capacities of 1–50,000 dm^3 have been designed for large-scale cultures.

Cell suspension provides an excellent model system for studies on cell division, cell expansion, cell differentiation, and intermediary metabolism, because of the ease of adding test compounds and of harvesting cell and medium for analysis. The physiological properties of plant suspensions, however, render them more difficult to exploit than microbial cells. Moreover, the higher cost of plant culture medium, longer fermentation time (usually of the order of weeks), higher downstream processing costs, and potential changes in the number of sets of chromosomes (i.e., ploidy instability) expressed in cells on prolonged subculture significantly detract from their usefulness for commercial exploitation (i.e., as important drugs, food additives, perfumes, biocides, etc.). Despite these difficulties, research is being conducted into using cell fermenters for plant secondary product synthesis because such compounds are either not possible or economically impractical to synthesize chemically. Since many desirable compounds are derived from plants found only in environmentally sensitive areas, there is added pressure on chemical manufacturers to turn to alternative sources of supply.

Plant cell culture suspensions are useful in experiments in which mutants are selected by growing colonies in the presence of the selecting agent. Ideally, suspensions comprising only single haploid cells should be sued for mutant selection. Single-cell suspensions may be obtained by sequential filtration through a graded series of filter down to 50 μm pore size. Single-cell colonies selected for superior yields are usually transferred to a production medium, possibly involving mild stress, which suppresses cell division whilst promoting some degree of differentiation.

23.16 Protoplast

A protoplast is a spherical, osmotically sensitive cell with an intact cell membrane but lacking a cell wall. This can be prepared from whole cells following the removal of the cell wall either enzymatically using pectinase in conjunction with cellulases, or mechanically by dispersing plasmolyse tissues in which the protoplasts have shrunk away from the cell wall following incubations in a concentrated osmoticum such as mannitol (at ~500 mM). Of the two protocols, enzymatic isolations give much higher yields of uniform protoplast. However, incubation conditions must be carefully defined and monitored to avoid overexposure to the wall-degrading enzymes that may alter metabolism with deleterious consequences.

Protoplasts may be isolated from healthy leaves that have initially been sterilized in a solution of hypochlorite for ~10 min. The lower epidermis is peeled off, and the leaf cut into small sections and incubated with the digesting enzyme mixture in the dark at 25°C overnight. Protoplasts released into the suspension can then be separated from cell debris and purified by a combination of filtration using a fine nylon mesh, centrifugation at low speed, and washing. The quality of the yield may be verified by checking their viability using dyes such as fluorescein diacetate (FDA) or Evans blue. The former is accumulated by viable protoplasts and subsequently converted by endogenous esterases to fluorescein, which can be visualized using ultraviolet microscopy. In contrast, Evans blue is excluded by viable protoplasts and thus can be used to distinguish these from dead isolates.

Protoplasts will start to undergo division and form new calluses within 24 h of culture, following the regeneration of a new cell wall. To facilitate this, freshly isolated protoplast should be counted to find effective inoculum density for growth

(10^4 cells/cm^3) prior to plating. As with previous counting procedures, this can be performed using a hemocytometer. For certain experimental purposes, protoplast fusion may be required. This may be particularly valuable in plant breeding experiments where sexual incompatibility between genera may be a limiting factor. Protoplast fusion can be induced by either chemical or physical means (electrofusion). Chemically, fusogens such as polyethylene glycol (PEG; ~30%) or a higher pH calcium solution (1.1% (w/v) $CaCl_2 \cdot 6H_2O$ in 10% (w/v) mannitol, pH 10.4) may be used to initiate this process, which can be achieved in <30 min.

Electrofusion is carried out in two steps. In the first step, protoplasts are placed in a medium of low conductivity between two electrodes (platinum wires arranged in parallel on a microscope slide). A high-alternating field (0.5–1.5 MHz) is applied between the electrodes, which causes the protoplasts to align in a process known as dielectrophoresis. In the second step, one or more short (10–200 µs) direct current pulses (of 1–3 kV/cm) are applied, which causes pores to form in the membranes of protoplasts and allows fusion to take place where there is close membrane contact. This technique allows a higher degree of control over the fusion process than do chemical methods.

23.17 Surface Sterilization

There are two basic sources for obtaining surface-sterilized plant tissues. One is to surface sterilize seeds, which are dormant and often dried and can, therefore, tolerate very harsh sterilization conditions. The seeds are then germinated under sterile conditions and the various tissues are placed on callus-inducing media. The other is to surface sterilize undamaged tissue from a growing plant which may be the only option for vegetatively propagated species or tree species that are slow to set seed. The same sterilization strategies can be used in each case. Most commonly, a brief ethanol wash serves to remove and kill spores and bacteria on the plant surface and to thoroughly wet the tissue; it is followed by a hypochlorite soak to kill any remaining microbes.

23.18 Common Culture Media

Refer Tables 23.1 through 23.4.

23.19 Sterilization of Plant Materials

23.19.1 Introduction

Surface of plant parts carries a wide range of microbial contaminants, fungi and bacteria being the common. To avoid the microbial growth that is detrimental to culture growth, explants must be surface-sterilized in disinfectant solutions before planting these in the nutrient medium. The most common surface sterilants are listed in

TABLE 23.5   Sterilization Procedures for Different Plant Organs

Tissue	Procedure Presterilization	Procedure Sterilization	Procedure Poststerilization	Remarks
Seeds	Submerge in absolute EtOH for 10 s and rinse in sterile distilled water	Seeds with intact testas submerge for 20–30 min in 10% (w/v) calcium hypochlorite or for 5 min in a 1% (w/v) solution of bromine water	Wash three times in sterile water and germinate in sterile water. Wash five times with sterile distilled water and germinate on damp sterile filter paper	Root and shoot tissue for callus culture. Excellent for tomato seeds
Fruits	Rinse briefly with absolute EtOH	Submerge for 10 min in 2% (w/v) sodium hypochloride	Wash repeatedly with sterile water, dissect out seeds or interior tissue	Good source of sterile seedlings
Pieces of stem	Scrub in running tap water and rinse with absolute EtOH	Immerse for 15–30 min in 2% (w/v) sodium hypochlorite: remove ends	Wash three times in sterile water	Plant vertically in agar medium or dissect tissue and culture in isolation
Storage organs	Scrub in running tap water	Submerge for 20–30 min in 2% (w/v) sodium hypochloride	Wash repeatedly with sterile water. Dry with sterile tissue paper	Remove tissue from the inside of the structure
Leaves	Rub surfaces briefly with absolute EtOH	Immerse for about 1 min in 0.1% (w/v) mercuric chloride	Wash repeatedly with sterile water. Dry with sterile tissue paper	Difficult material to sterilize. Choose very young leaves. Lamina laid in agar or petiole inserted in agar

Source: Yeoman, M.M. and A.J. Macleod, In: *Plant Tissue and Cell Culture*, Blackwell Scientific Publications, Oxford, U.K., 1977, pp 31–59.

Tables 23.5 and 23.6 with concentration and exposure times that preserve the explant but at the same time kill any microbial contaminant.

23.19.2 Materials

Explant materials, commercial bleach/calcium hypochlorite, 70% ethanol, surfactant (Triton-X/Tween-20 or -80), mold detergent, screw-cap bottles, sterile forceps, scalpels, and sterile distilled water.

23.19.3 Protocol

1. Make small pieces of explants materials (seedlings, swelling buds, stem or storage organs, and leaf materials) using scalpel.
2. Wash explants in a mild detergent (herbaceous materials may not be treated).

TABLE 23.6   Concentration and Exposure Times of Different Disinfectants

Disinfectants	Concentration (%)	Exposure Time (min)	Effectiveness
Calcium hypochlorite	9–10	5–30	Very good
Sodium hypochlorite	0.5–1.5[a]	5–30	Very good
Hydrogen peroxide	3–12	5–15	Good
Ethyl alcohol	70–95	0.1–5	Good
Bromine water	1–2	2–10	Very good
Silver nitrate	1.0	5–30	Good
Mercuric chloride[b]	0.1–1.0	2–10	Satisfactory
Antibiotics	4–50 mg/L	30–60	Fairly good

[a]Commercial bleach contains about 5% sodium hypochlorite and thus may be used at a concentration of 10%–20% (v/v) of the commercial preparation equivalent to 0.5%–1.05% sodium hypochlorite.

[b]Usage to be avoided because of its poisonous nature.

3. Rinse explants under running tap water for 10–30 min.
4. Rinse explants in 70% ethanol for 30 s and material left exposed in the sterile hood for evaporation of alcohol.
5. Aseptically transfer explants into a vial containing a mixture of wetting agent (surfactant) to reduce surface tension and for better surface contact and 20% commercial bleach (5% sodium hypochlorite), a disinfectant.
6. Keep the explants submerged in the aforementioned solution for 5–30 min and shake the vial two to three times for sterilization.
7. Decant the liquid.
8. Pour an adequate amount of sterile distilled water into the vial and replace the cap.
9. Shake the vial a few times and discard water.
10. Repeat Steps 8 and 9 to rinse explants four to five times.

23.19.4 Observation and Result

Explant, a material that has been surface-sterilized, is ready for inoculation on a nutrient medium.

23.20 Preparation of Tissue Culture Media

23.20.1 Introduction

The nutrient preparation on or in which a culture (i.e., a population of cells) is grown in the laboratory is called a culture medium. Nutritional requirements for optimal growth of a tissue *in vitro* may vary with the species. Even tissues from different parts of a plant may have different requirements for satisfactory optimal growth of a tissue *in vitro*. The nutritional requirement consists of essential and optional components. The essential nutrients consist of inorganic salts, a carbon and energy source, and vitamins. A medium with these constituents referred as basal medium. Some growth regulators such as auxins, gibberellins, and cytokinins may also be added to the basal medium either alone or in various combinations. Some tissues, in addition to the ingredients mentioned, require natural plant extracts, for example, coconut milk, various fruit juices, casein hydrolysate, and yeast extract, for their growth; agar, which liquefies on heating to 96°C, and hardens into a jelly on cooling to 40°C–45°C, is used, at the rate of 0.8% to solidify liquid media. The pH of the medium is adjusted to 5.8 (slightly acidic).

The choice of a medium (identified by its mineral salt composition) for growing cell in callus culture depends on the plant species and to a degree upon the intended use of the culture. Of the various tissue culture media, for example, modified White's medium, Murashige and Skoog's (MS) medium, Nitsch's medium, and Takebe's medium, MS medium (Table 23.7) is widely used for plant tissue culture because it has proven effective for growth promotion of both monocotyledons and dicotyledons. The medium is characterized by high concentrations of mineral salts, nitrate (NO_3^-), and ammonium (NH_4^+) which are preferred by cells of some species.

The most simple method of preparing tissue culture media today is to use commercially available (Sigma Chemicals Company, St. Louis, MO; Carolina Biological Supply Company, Burlington, NC; Irvine, Ayrshire, Scotland; GIBCO, Grand Island, New York; Hi Media Laboratories Pvt. Ltd., Mumbai, India) dry powder that is dissolved in distilled water. After adding agar, sugar, and other constituents, the final volume is made up with distilled water. The desired pH is adjusted, and the medium is autoclaved.

23.20.2 Materials

Constituents of the MS medium, Erlenmeyer flasks of various capacity (100, 250, 500 mL, and 1 L), measuring cylinders (100 and 1000 mL), pipettes (1, 5, and 10 mL), distilled or demineralized water, pH meter, 1 N NaOH, HCl, or KOH, autoclave.

23.20.3 Protocol

1. Prepare macronutrients solution in 100 mL distilled water.
2. Prepare micronutrients solution in 100 mL distilled water following the stock solution dilution chart (Table 23.3).
3. Add macronutrient and micronutrient solutions while stirring into 700 mL distilled water taken in 1 L Erlenmeyer flask.
4. Add the other heat stable constituents (e.g., sucrose, vitamins, and hormones) and agar powder (if desired at a concentration of 0.8%–1.0%) (vitamins and auxins can be added after autoclaving for better results).
5. Make the final volume of the medium by the addition of more distilled water.
6. Adjust pH of the medium to 5.7, while stirring, using 0.1 N NaOH or 0.1 N HCl.

TABLE 23.7 Composition of Murashige–Skoog Medium for Plant Tissue and Cell Culture

Constituents	mg/L
Macronutrients	
NH_4NO_3	1650
KNO_3	1900
$CaCl_2 \cdot 2H_2O$	440
$MgSO_4 \cdot 7H_2O$	370
KH_2PO_4	170
Micronutrients	
KI	0.83
H_3BO_3	6.2
$MnSO_4 \cdot 4H_2O$	22.3
$ZnSO_4 \cdot 7H_2O$	8.6
$Na_2MoO_4 \cdot 2H_2O$	0.25
$CuSO_4 \cdot 5H_2O$	0.025
$CaCl_2 \cdot 6H_2O$	0.025
Fe versenate (EDTA)	43.0
Vitamins and hormones	
Inositol	100
Nicotinic acid or niacin (vitamin B_3)	0.5
Pyridoxine–HCl	0.5
Thiamine–HCl	0.1
IAA	1–30
Kinetin	0.04–10
Carbon source	
Sucrose	30,000
pH	5.7

7. If solid medium is desired, agar is used, heat the solution while stirring until agar is dissolved.
8. Pour the medium into the desired culture vessels (15 mL in a 25×150 mm culture tube and 50 mL in a 250 mL flask).
9. Plug the culture vessels with nonabsorbent cotton wool wrapped in cheese cloth, or with any other suitable closure.
10. Transfer the culture vessels to appropriate baskets covered with aluminum foil to check wetting of plugs during autoclaving.
11. Transfer the baskets to autoclave.
12. Sterilize the medium by autoclaving at 121°C (1.05 kg/cm²) (15 psi) for the time period depending upon the volume of the medium in the vessel (e.g., for 25, 50, 100, 250, 500, 1000, 2000, and 4000 mL capacity time required is 20, 25, 28, 31, 35, 40, 48, and 63 min, respectively).
13. The medium is allowed to cool at room temperature.

23.20.4 Observation and Result

The cooled medium is ready for inoculation and should be stored at 4°C for future use.

23.21 Organ Culture

23.21.1 Introduction

Regeneration of plants from cell culture offers a practical strategy for plant cloning in order to regenerate genetically identical cultures, which is otherwise difficult, costly, or inefficient to propagate by cuttings or other sexual means for commercially valuable plants. Cell culture sometimes offers the only practical means of propagation. Plant cultures are also being investigated as sources of valuable plant products like drugs, flavors, and fragrances. The amounts of growth regulators added to the medium are variable, depending on whether the culture is maintained as callus or made to regenerate whole plants.

23.21.2 Material

1. Plant material: Young leaves, internodes, nodes, and hypocotyls.
2. Chemicals: MS basal medium (preparation of medium is described earlier supplemented with following growth regulators)
3. IAA (0.5 mg/L) in combination with BA (1–4 mg/L), NAA (1–4 mg/L), and 2,4-D (1–8 mg/L), 5% Teepol or any other detergent, 0.05% mercuric chloride ($HgCl_2$).
4. Equipment and others: Scalpel, forceps, disposable Petri dishes (commercially sterilized) or tissue culture bottles, and laminar flow chamber.

23.21.3 Protocol

1. Prepare MS medium supplemented with different hormones.
2. Using a razor blade or scalpel, cut into small pieces of young leaves collected from the field.
3. Wash the leaves first with 5% Teepol, and constantly shake the container for 3–4 min.
4. Wash thoroughly with running tap water.
5. Surface sterilize with 0.05% $HgCl_2$ for 3–5 min.
6. Rinse these with sterile distilled water.
7. Put all the pieces of explant in a sterile Petri dish and inoculate onto MS medium supplemented with different hormones.

23.21.4 Observations and Results

Auxin (NAA and 2,4-D) induces callus from the explant, low-level NAA (1–4 mg/L) induces root, and high-level 2,4-D (5–8 mg/L) induce callus formation. Auxin (IAA) in combination with a cytokinin (BA) induces shoot. IAA (0.5 mg/L) and BA (1–4 mg/L) induce shoot and maximum number of shoots produced at 0.5 mg/L IAA in combination with 3 mg/L BA. Wash ~50 mm tall plantlets with tap water and transfer them into a small pot containing autoclaved potting compost cover, each plantlet with a glass beaker to prevent desiccation and maintain the plants in a humid greenhouse. After 1 week, remove the beaker and transfer to a large pot.

23.22 Callus Propagation

23.22.1 Introduction

The MS medium is the most suitable for plant regeneration from tissue and callus. The hormones are the important compounds in the plant-regenerating medium. The capacity for plant regeneration of tissues varies widely in different species. In some species, the morphogenesis is readily induced (e.g., carrot, coffee, celery) and the species develop into complete plants, while in others it fails to occur. Callus tissue can be induced from different plant tissues of many plant species. Carrot is a standardized material. Callus culture from excised tap root of carrot is described in Figure 23.8.

23.22.2 Materials

Carrot roots, MS liquid medium supplemented with NAA (4 mg/L) and 2,4-D (5 mg/L), microchamber (a sterile microscope slide enclosing single cell within the

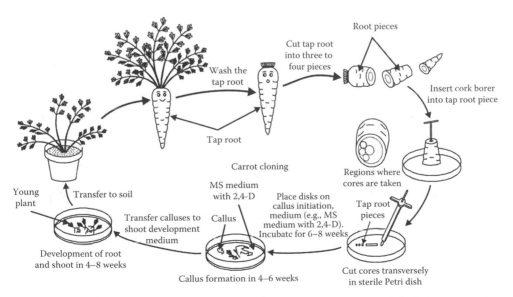

Figure 23.8 Protocol for somatic embryogenesis in carrot. (Modified from Carolina Biological Supply Company, Burlington, NC.)

mineral oil, surrounded by two cover glasses, placed on oil drop on each side, and finally covered with a third cover glass bridging the two cover glasses), culture bottles with screw caps, gyratory shaker (that moves in a circle or spiral), sterile forceps, sterile Petri plates, glass tubes or jars, pots, greenhouse or growth chamber, laminar flow, 5% Teepol or any other detergent, 0.05% $HgCl_2$.

23.22.3 Protocol

1. MS medium supplemented with NAA (4 mg/L) and 2,4-D (0.5 mg/L) is prepared as per instructions described earlier.
2. A fresh tap root of carrot is taken and washed thoroughly under running tap water to remove all surface detritus.
3. The tap root is then dipped into 5% "Teepol" for 10 min and then the root is washed.
4. The tap root is surface-sterilized by immersing in 70% v/v ethanol for 60 s, followed by 20–25 min in sodium hypochlorite (0.8% available chlorine).
5. The root is washed three times with sterile distilled water to remove hypochlorite completely.
6. The carrot is then transferred to a sterilized Petri dish containing a filter paper. A series of transverse slice 1 mm in thickness is cut from the tap root using a sharp scalpel.
7. Each piece from cambial tissue of carrot is transferred to another sterile Petri dish. Each piece contains a whitish circular ring of cambium around the pith. An area of 4 mm² across the cambium is cut from each piece so that each small piece contains part of the phloem, cambium, and xylem. Size and thickness of the explants should be uniform.
8. Always replace the lid of Petri dish after each manipulation.
9. The closure (cotton plug) from a culture tube is removed and flamed the uppermost 20 mm of the open end. While holding the tube at an angle of 45°, an explant is transferred using forceps onto the surface of agarified nutrient medium. Nutrient medium is Gamborg's B 5 or MS medium supplemented with 0.5 mg/L 2,4-D.
10. The closure is immediately placed on the open mouth of each tube. The forceps are always flamed before and after use. Date, medium, and name of the plant are written on the culture tube by a glass marking pen or pencil.
11. Culture tubes after inoculation are taken to the culture room where they are placed in the racks. Cultures are incubated in dark at 25°C.
12. Usually, after 4 weeks in culture the explants incubated on medium with 2,4-D will form a substantial callus. The whole callus mass is taken out aseptically on a sterile Petri dish and should be divided into two or three pieces.
13. Each piece of callus tissue is transferred to a tube containing fresh same medium.
14. Prolonged culture of carrot tissue produces large calli.

23.22.4 Observation and Result

Appearance of callus and shoot takes place after 4–6 weeks of incubation on MS medium supplemented with hormones.

23.23 Plant Regeneration from Callus or Plant Tissue

23.23.1 Introduction

The MS medium is the most suitable for plant regeneration from tissue and callus. The hormones are the important compounds in the plant-regenerating medium. The capacity for plant regeneration of tissues varies widely in different species. In some species, the morphogenesis is readily induced (e.g., carrot, coffee, celery) and the species develop into complete plants while in others it fails to occur (Figure 23.9).

23.23.2 Materials

Callus (raised from cells), MS liquid medium, microchamber (a sterile microscope slide enclosing single cell within the mineral oil, surrounded by two cover glasses, placed on oil drop on each side, and finally covered with a third cover glass bridging the two cover glasses), gyratory shaker (that moves in a circle or spiral), forceps, sterile Petri plates, glass tubes or jars, pots, greenhouse or growth chamber.

23.23.3 Protocol

1. Transfer callus into the flask each containing 20 mL MS medium.
2. Shake the culture flasks on a gyratory shaker at 150 rpm to dissociate callus into single cells.
3. Transfer a cell from the flask and place it in a drop of MS medium in a microchamber.

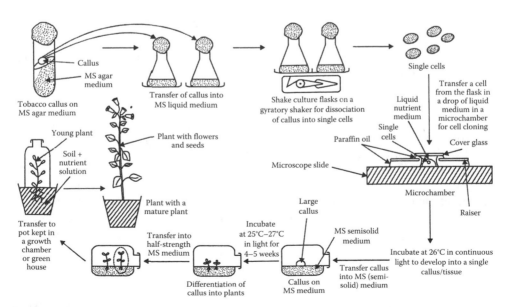

Figure 23.9 Protocol for regeneration of a tobacco plant from a single cell.

4. Place the microchamber in a sterile Petri dish.
5. Incubate the Petri plate at 26°C in continuous light for the development of tissue/callus.
6. Remove the cover glass of the microchamber and aseptically the small callus (tissue) onto MS medium with 4.5 μm² 2,4-D and 1–2 g/L of casein hydrolysate.
7. Incubate at 25°C–27°C in light (~1000 lux) to develop the tissue into large callus and shoot for 4–5 weeks.
8. When shoot appears, transplant these on half-strength MS medium or hormone-free MS medium.
9. Transfer the rooted plants to pots containing soil and water with a nutrient solution.
10. Keep the pots in a greenhouse or growth chamber where high humidity is maintained.

23.23.4 Observations and Results

Observe the microchamber and culture tubes for the developing calluses and plant-lets and pots for the mature plants.

Appearance of callus and shoot takes place after 4 weeks of incubation on MS medium supplemented with hormones. Root appears later from the callus with shoots in half-strength MS medium. The plants in pots will grow to maturity, flower, and set seeds to the plants growing in nature.

23.24 Single Cell Culture

23.24.1 Introduction

Various methods have been developed to grow isolated individual cells on a nutrient medium. Free cells are normally suspended in a liquid medium, later mixed with an equal volume of molten, cooled agar medium, and plated in a Petri plates and can be cultured using the following techniques:

1. Filter paper raft nurse tissue technique (Figure 23.10)
2. Microchamber technique (Figure 23.11)
3. Microdrop method (Figure 23.12)
4. Bergmann's plating technique (Figure 23.13)
5. Thin layer liquid medium

Isolated single cells fail to divide in normal tissue culture media. Therefore, either a nurse tissue or a conditioned medium is used for their culture. A conditioned medium is a medium in which plant cells have been grown for about 48 h (cells are then filtered out).

23.24.1.1 Filter Paper Raft Nurse Tissue Technique
Single cells are placed on small pieces (8 × 8 mm) of filter paper (sterilized), which themselves are placed on top of established callus cultures several days in advance.

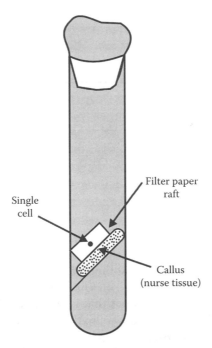

Figure 23.10   Filter paper raft nurse tissue technique.

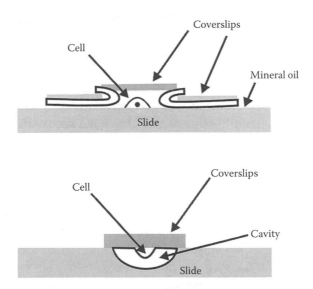

Figure 23.11   Microchamber techniques.

This allows the filter papers to be wetted by the exudates from callus tissues. The single cells placed on the filter papers (Figure 23.10) derive their nutrition from the callus exudates diffusing through the filters.

The cells divide and form macroscopic colonies on the filter; the colonies are isolated and cultured.

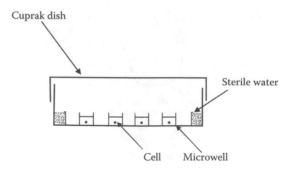

Figure 23.12   Microdrop technique.

23.24.1.2 Microchamber Technique

A microchamber can be created either by using a microscope slide and coverslips (latter held in place by sterile mineral oil), or by a cavity slide (Figure 23.11). Single cells are suspended in the conditioned medium, and a drop of medium having a single cell is placed in the microchamber that is covered with a coverslip. In case of cavity slide, the drop is placed onto a coverslip that is then inverted into the slide cavity. Microchambers allow microscopic observation, and they can be kept in a Petri dish for incubation.

23.24.1.3 Microdrop Method

A specially designed dish, Cuprak dish, having a smaller outer chamber (to be filled with sterile distilled water to avoid desiccation of cells) and a larger chamber (having several numbered microwells) is employed. Microdrops of 0.25–0.5 mL are distributed in the microwells, and the dish is sealed with parafilm (Figure 23.12). Cell density in the medium is so adjusted as to give, on an average, one cell per droplet (it works out as $2–4 \times 10^3$ cells/mL). This method has been successfully used for protoplasts and should work with single cells as well.

23.24.1.4 Bergmann's Plating Technique

In this widely used technique, cells are suspended in a liquid medium at a cell density that is twice the desired density in the plate. Sterilized agar (1%) medium is kept melted in a water bath at 35°C. Equal volumes of the liquid and agar media are mixed thoroughly and quickly spread in 1 mm thick layer in a Petri dish. The cells that remain embedded in the soft agar medium are observable under a microscope, and when macroscopic colonies develop they are isolated and cultured separately (Figure 23.13).

23.24.1.5 Thin Layer Liquid Medium Culture

Cells can be placed in a thin layer of liquid medium to allow adequate aeration. Since cells are not fixed in position, it is not possible to follow up individual cells during culture. This technique is common for protoplast culture.

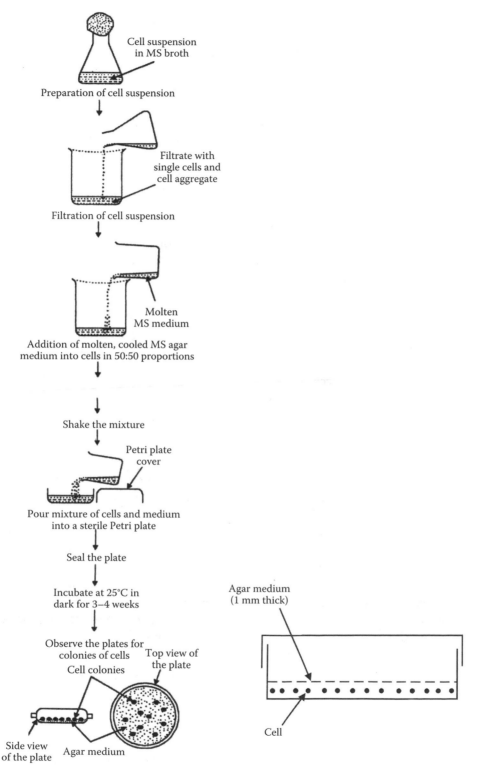

Figure 23.13 Bergmann's plating technique.

23.24.2 Culture Requirements

It is important (i) to culture the single cells in dark and keep microscopic observations to the minimum since light has a detrimental effect on cell proliferation and (ii) either a conditioned medium or a suitably enriched medium should be used since standard tissue culture media are unsuitable.

Some highly enriched synthetic media permit the culture of as few as 25–50 cells/mL while some media supplemented with casamino acids and coconut milk support cell division at 1–2 cells/mL. It is postulated that some biochemicals essential for cell division leach into the culture medium when they are subcultured; this produces the lag phase (already described in the previous section). Cells begin to divide only when equilibrium is established for these metabolites between the medium and the cells; this happens much later at lower cell densities and below a critical cell density it may not be achieved. For this reason, conditioned or specially enriched media are required for culture of cells at low densities.

23.24.3 Cell Viability Test

Cell viability should be monitored by one of the following approaches:

1. Phase contrast microscopy
2. Staining with 2,3,5-triphenyltetrazolium chloride (TTC)
3. FDA (fluorescein diacetate assay)
4. Evan's blue

Live cells show cytoplasmic streaming and a well-defined healthy nucleus which are easily observable with a phase contrast microscope or even a light microscope. Cell masses can be stained with 1%–2% solution of TTC which is reduced by living cells to formazan which yields red color. Formazan can be extracted and measured with a spectrophotometer to give a quantitative estimate of viability, but it is not suitable for single or few cells. Cell is treated with 0.01% solution of fluorescein diacetate. Live cells cleave FDA by esterase activity and produce fluorescein, which cannot cross the plasma membrane. With UV exposure, fluorescein gives green fluorescence so that live cells appear green while dead cells do not fluoresce. Evan's blue (0.025%) is not taken up by live cells while it freely enters into damaged dead cells. Therefore, all cells that take up stain are dead. Evan's blue is usually used in conjunction with fluorescein diacetate.

23.25 Preparation and Fusion of Protoplasts

23.25.1 Introduction

Protoplast (cell minus cell wall) is the biologically active and most significant material of cells. When cell wall is mechanically or enzymatically removed, the isolated protoplast is known as "naked plant cell" on which most of the recent researches on manipulation, microinjection, and fusion are based. Plant cell wall acts as a physical barrier and protects cytoplasm from microbial invasion and environmental stress. It consists of a complex mixture of cellulose, hemicellulose, pectin, lignin, lipids, protein, etc. For dissolution of different components of the cell wall, it is essential to have the respective enzymes. Until suitable methods were developed, protoplasts were isolated by cutting the plasmolysed plant tissues and releasing protoplast through deplasmolysis of cells. Because naked plant cells are fragile and can burst in solution of low osmotic potential, mannitol is included in the digestion mixture as an osmoticum. Other ingredients included improve the quality of the protoplasts.

23.25.2 Isolation

23.25.2.1 Isolation of Single Cell from Intact Plant Organ

Cells from any plant species can be cultured aseptically in a nutrient medium. The cultures are initiated by planting (inoculating) a portion of sterile tissue (explant) on an agar medium which after 2–4 weeks incubation gives rise to callus. Such a callus can be subcultured by transferring a small piece to fresh agar medium. To get a liquid suspension culture, a callus mass is transferred to liquid medium, and the vessel is incubated on a shaker. A liquid suspension culture results after several weeks' incubation. Whole plants can be raised *in vitro* from these cultured cells. Leaf tissue is the most suitable material for the isolation of single cell. Isolation of mesophyll cell in the laboratory can be achieved by either mechanical method or enzymatic methods.

23.25.2.1.1 Mechanical Method. The procedure has been successfully used to isolate mesophyll cells, active in photosynthesis and respiration, from mature leaves of several species of dicots and monocots including the grasses. Even metabolically active single cells from the bundle sheath of crab grass (*Digitaria sanguinalis*) can be isolated using a similar procedure.

23.25.2.1.1.1 Materials. Fresh leaves of any plant, maceration medium with 0.8% agar and liquid medium (20 μL mol sucrose, 10 μmol $MgCl_2$, 20 μmol Tris–HCl buffer, pH 7.8), mortar and pestle, 90% ethanol, 7% calcium hypochlorite, scalpel, sterile metal Tyler filters (38 and 61 μm mesh diameters), centrifuge (low speed), culture tubes/vials, sterile distilled water

23.25.2.1.1.2 Protocol
1. Immerse the leaves rapidly in 95% ethyl alcohol.
2. Rinse the leaves in filter-sterilized 7% solution of calcium hypochlorite for 15 min.
3. Wash the leaves in sterile distilled water two to three times.
4. Cut the leaves into small pieces (0.5–1 cm²) with a sterile scalpel.
5. Transfer 1.5 g of leaves into a mortar and pestle glass homogenizer tube.

6. Add 10 mL of maceration culture medium to the tube.
7. Homogenize the leaves.
8. Filter the homogenate through two leaves of sterile metal Tyler filter, with mesh diameters of 61 μm (upper) and 38 μm (lower filter).
9. Centrifuge the filtrate by low-speed centrifugation to remove fine debris.
10. Discard the supernatant.
11. Suspend the sediment consisting of free cells, in a volume of the maceration liquid medium sufficient to achieve the required density.
12. Inoculate the free cells into an agar plate or into the liquid medium for culturing.
13. Incubate the plates or vials at 26°C in the dark or light.

23.25.2.1.1.3 Observation and Result. Free mesophyll cells would sediment in the centrifuge tubes during centrifugation which could be used for culturing on a suitable medium.

23.25.2.1.2 Enzymatic Method. Isolation of single cells by the enzymatic method has been found convenient for several herbaceous species as it is possible to obtain high yields from preparations of spongy parenchyma with minimum damage or injury to the cells. This can be accomplished by providing osmotic protection to the cells while the enzyme macerozyme degrades the middle lamella and cell wall of the parenchymatous tissue. Applying the enzymatic method to cereals (*Hordeum vulgare*, *Triticum vulgare*, *Zea mays*) has proven rather difficult since the mesophyll cells of these plants are apparently elongated with a number of interlocking constrictions, thereby preventing their isolation. Addition of potassium dextran sulfate in the enzyme mixture improves the yield of free cells.

23.25.2.1.2.1 Materials. Tobacco leaves are collected either 1–2 h after sunrise or from dark pretreated plants that are 60–80 days old. Enzymes are dissolved in 0.4 M mannitol containing 0.1% (w/v) $CaCl_2 \cdot 2H_2O$ and 0.5% (w/v) potassium dextran sulfate (sulfur content 18.1%, inherent viscosity 0.013 dL/g) and pH adjusted to 5.7, 1.5% enzyme mixture (1.5% cellulase, 0.2% macerase, pectinase 0.4%), or enzymes solution 80 mL (0.5% macerozyme + 0.8% mannitol + 1% potassium dextran sulfate), 70% ethanol, 3% sodium hypochlorite, Teepol, sterile distilled water, culture vials, fine forceps, scalpel, vacuum pump, reciprocating shaker (stroke 4–5 cm at the speed of 120 cycles/min), spirit lamp/bunsen burner.

23.25.2.1.2.2 Protocol
1. *Surface sterilization of leaf samples*: Mature leaves are collected from healthy plants which are washed in tap water to remove adhering soil particles and sterilized with sodium hypochlorite solution.
2. *Rinsing in suitable osmoticum*: After 10 min, sample is properly washed with sterile distilled water or MS medium adjusted to a suitable pH and buffer to maintain osmotic pressure. Washing should be done for about six times, to remove the traces of sodium hypochlorite.
3. *Plasmolysis of cells*: The lower epidermis covered by thin wax cuticle is removed with a forceps. Stripping should be done from midrib to margin of lamina. The stripped

surface of the leaf is kept in mannitol solution (13% w/v) for 3 h to allow plasmolysis of cells.

4. *Peeling of lower epidermis*: Thereafter, about 1 g leaves are peeled off and transferred into enzyme mixture already sterilized through a Seitz filter (0.45 mm). This facilitates the penetration of enzyme into tissue within 12–18 h at 25°C.

5. *Isolation and purification of protoplasts*: Leaf debris are removed with forceps, and enzyme solution containing protoplasts is filtered with a nylon mesh (45 mm). Filtrate is centrifuged at 75g for 5 min, and the supernatant is decanted. Again a fresh MS medium plus 13% mannitol is added to centrifuge. Repeated washing with nutrient medium, centrifugation, and decantation are done for about three times. Finally, specific concentration of protoplast suspension is prepared. Observe under the microscope (Figure 23.14).

23.25.2.1.2.3 Observation and Result. Predominantly appeared palisade cells after incubation in flask could be used for culturing on a suitable medium.

23.25.3 Protoplast Fusion and Somatic Hybridization

With the development of techniques for enzymatic isolation of protoplasts and subsequent regeneration, a new tool of genetic manipulation of plants has now become available. Moreover, the fusion of protoplasts of genetically different lines or species has also been possible. For example, some plants that show physical or chemical incompatibility in normal sexual crosses may be produced by the fusion of protoplasts obtained from two cultures of different species. Somatic hybridization of crop plants represents a new challenge to plant breeding and crop improvement. In the field of pest and disease resistance and transfer of C_3 photosystems into C_4 crop plants, somatic crosses show most interesting promises.

23.25.4 Mechanism of Protoplast Fusion

The mechanism of protoplast fusion is not fully known. Several explanations have been put forward to understand the mechanism of protoplast fusion. Some are explained here: When the protoplasts are brought into close proximity, this is followed by an induction phase thereby induced changes in electrostatic potential of the membrane results in fusion. After the fusion, the membranes stabilize and the surface potential returns to its former state. Other literature showed when the protoplasts are closely adhered, the external fusogens cause disturbance in the intramembranous proteins and glycoproteins. This increases membrane fluidity and creates a region where lipid molecules intermix, allowing coalescence of adjacent membranes. The negative charge carried by protoplast is mainly due to intramembranous phosphate groups. The addition of Ca^{2+} ions causes a reduction in the zeta potential of plasma membrane and under this situation protoplasts are fused. The high-molecular-weight polymer (1000–6000) of PEG acts as a molecular bridge connecting the protoplasts. Calcium ions linked the negatively charged PEG and

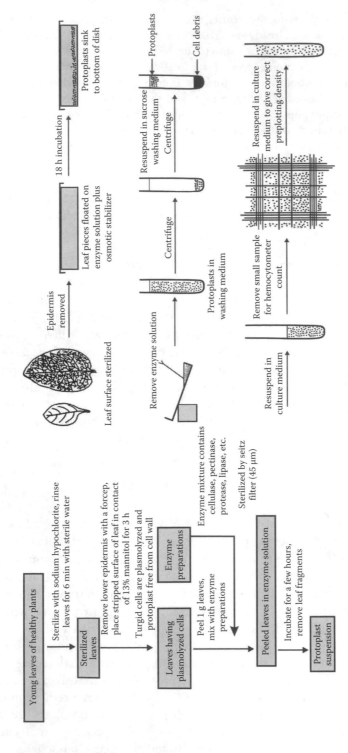

Figure 23.14   Flowchart and diagrammatic method of protoplast isolation.

membrane surface. The surface potential is disturbed, leading to intramembrane contact and subsequent fusion on elution of the PEG. Besides this, the strong affinity of PEG for water may cause local dehydration of the membrane and increase fluidity, thus inducing fusion. Protoplast fusion takes place when the molecular distance between the protoplasts is 10 Å or less indicating that protoplast fusion is highly a traumatic event (Figure 23.15).

23.26 Protocols on Protoplast Fusion

Protoplast fusion can be broadly classified into two categories.

23.26.1 Spontaneous Fusion

Protoplasts during isolation often fuse spontaneously and this phenomenon is called spontaneous fusion. During the enzyme treatment, protoplast from adjoining cells fuses through their plasmodesmata, to form multinucleate protoplasts.

23.26.2 Induced Fusion

Fusion of freely isolated protoplasts from different sources with the help of fusion-inducing chemical agent is known as induced fusion. Normally, isolated protoplasts do not fuse with each other because the surface of isolated protoplast carries negative charges (−10 to −30 mV) on and around outside of the plasma membrane. Thus, there is a strong tendency in the protoplast to repel each other due to their same charges. Such type of fusion needs a fusion-inducing chemical to reduce the electronegativity of the isolated protoplasts and allows them to fuse with each other. The isolated protoplasts can be induced to fuse by three ways.

23.26.2.1 Mechanical Fusion
In this process, the isolated protoplasts are brought into intimate physical contact mechanically under microscope using micromanipulator or perfusion micropipette.

23.26.2.2 Chemofusion
Several chemicals have been used to induce protoplast fusion such as sodium nitrate, PEG, calcium ions (Ca^{2+}). Chemical fusogens cause the isolated protoplast to adhere each other and lead to tight agglutination followed by fusion of protoplast. Chemofusion is a nonspecific, inexpensive, can cause massive fusion product which can be cytotoxic and nonselective having less fusion frequency.

The results obtained so far from somatic hybridization represent that it is possible to recover fertile and stable amphidiploid somatic hybrids after protoplast fusion. There are four major aspects of protoplast fusion as (i) production of fertile amphidiploid somatic hybrids of sexually incompatible species is achieved. Induced

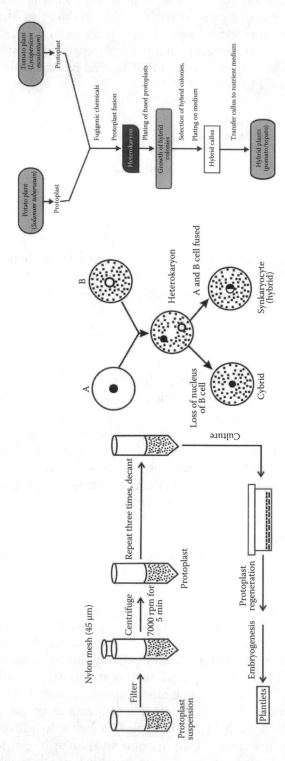

Figure 23.15 Fusion of the protoplasts of potato and tomato, and the production of hybrid plant (pomato).

fusion of protoplasts from two genetically different lines of species must result in a variety of homokaryotic as well as heterokaryotic fusion products (heterokaryon or heterokaryocytes). Selection of a few true somatic hybrid colonies from the mixed population of regeneration protoplast is a key step in successful somatic hybridization technique; (ii) production of heterozygous lines within one plant species which normally will be propagated only vegetatively, for example, potato; (iii) the transfer of only a part of genetic information from one species to another using phenomenon of chromosome elimination; and (iv) the transfer of cytoplasmic genetic information from one to a second line or species. It has been possible to transfer useful genes (e.g., disease resistance genes, rapid growth genes) from one species to another, thereby to widen the genetic base for plant breeding.

23.26.2.3 Electrofusion

The electrofusion technique offers greater advantage in terms of fusion and transformation of cells. Technique offers high frequencies with greater experimental reproducibility. The process can be monitored microscopically while standardizing various parameters. The fusion is synchronous, and the number of fusing protoplasts can be controlled to some extent. Electrofusion does not require any severe chemical treatment to cells, and it is not cytotoxic. This technology allows recombination to take place not only between related species but also between unrelated genera and is of great potential in the breeding and improvement of strains (Figure 23.16).

Electrofusion consists of two steps. In the first step, close membrane contact is established by the action of electrical treatment which can also be achieved by mechanical, chemical, and other treatments but with lesser efficiency. At a high frequency, alternating electric field (ac current) dipoles are generated within protoplast. Dipolar attraction between them results in their pearl chain orientation, in a process called dielectrophoresis. In the second step, fusion is triggered between adjacent cells by the application of an intense field pulse of very short duration leading to reversible electrical breakdown of the membrane. Application of momentary direct current pulse creates pores in the aligned membranes, resulting in fusion of protoplasts. A slight modification of this technique makes it possible to form transient pores in the plasma membrane without the induction of fusion. The pores formed by a transient depolarization of the membrane lipid–protein interface have stability in the microsecond range which can be increased up to several seconds by an appropriate combination of Ca^{2+}/phosphate. Pores formed in cell membranes can facilitate the transfer of exogenous macromolecules into cells. This high-voltage pulse transfection/ electroinjection termed electroporation was used to transfer DNA/RNA into a different type of cells of microbes including mammals and plants (Figure 23.17).

The membrane is regarded as a layer of dielectric (the hydrocarbon region and intrinsic proteins) zone with an aqueous solution on either side serving as capacitor plates. The presence of an external electric field leads to charge separation at the membrane capacitor owing to ion displacement. As a result, a transmembrane potential is built up, and a dipole is generated. In a nonuniform field, the field strengths on both sides of the cell are different giving a net force that results in a translational motion of the cell toward the region of higher field intensity. Thus, translational motion in one

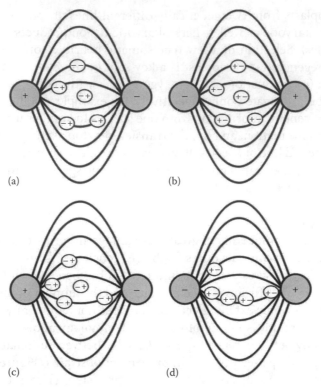

Figure 23.16 Dielectrophoretic collection of cells in an ac field. Protoplast is represented by circles and opposing electrodes are shown in cross section. Frames (a) through (d) represent states of transient dipole generation in protoplast in ac field resulting to formation of long pearl chain of protoplast which provides the requisite membrane–membrane contact for subsequent fusion.

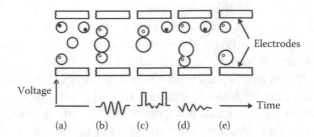

Figure 23.17 Schematic representation of the sequence of stages in electrofusion of protoplast: (a) ac field. (b) Pearl chain formation by protoplasts in ac field as a result of dielectrophoresis. (c–e) Pulse induces fusion of membranes of adenating protoplasts. (d) and (e) Fusion is completed and spherical shape of fusion product regained (ready for culture).

Figure 23.18 Fusion chamber, the interelectrode distance is 500 μm. The + and − designated as two electrodes, and the arrow indicates the path of the protoplast through the chamber.

direction is accomplished. Protoplast comes in intimate contact by dielectrophoresis as a result of the effect one polarized cell approaching another during the movement toward the region of high field intensity. The cells are drawn to whichever electrode they are closer when the ac field is turned on (Figure 23.18). The frequency and voltage of the ac field required to cause this effect depend on the membrane composition. As the cells approach the electrodes, they tend to form chains of cell because the electrical conductivity of the cell is greater than the conductivity of the medium leading to locally higher electric fields at the poles of the cell. This pearl chain formation brings the plasma membranes of neighboring cells into intimate contact, a condition required for subsequent fusion. Once the cells are aligned, fusion is achieved, with the application of short dc pulse of sufficient voltage, to cause membrane breakdown. The cell membrane breakdown occurs preferentially at the points of contact between neighboring cells as the cells are better conductors than the surrounding medium as the high conductivity of the cell interior causes the current to move through the cell rather than through the medium. Strong dc pulse results in perforation of the plasmalemma inducing the fusion process between adenating cells (Figure 23.18). Since the membranes break down at the point of contact, the dc pulse leads to cell fusion rather than cell lysis. The fusion pulse must be in the form of a square wave, with the rise and decay time of <2 μs, usually pulse duration of 10–50 μs is given for fusion. The dc voltage necessary for fusion depends on the distance between the electrodes and the density of cells in the fusion chamber. Normally, a single square wave of pulse (400–1000 V/cm) will cause fusion. Long pulse durations and supraoptimal voltage cause cell lysis.

23.26.2.3.1 Fusion Chamber.   The schematic equipment for fusion chamber used for various electrode arrangement setup for electrofusion is illustrated in Figure 23.19. The chamber is constructed on a standard glass microscope slide. The electrodes may be of any good electrical conductor (gold, platinum–irridium, silver, etc.) that resists oxidation. It should be ensured that the surface of the electrodes is highly polished. Electrodes should not be too thick (thin wire works well). Gold-plated copper can be used (0.5 mm diameter). The wire or rods that serve as electrodes can then be affixed to the edge of rectangular pieces of glass or Plexiglas and glued to the surface of glass microscope slide. Epoxy cement or cyanoacrylate glue works well for cementing these chambers together. It is important that the electrodes be spaced 200–500 μm apart, and be kept as parallel as possible. Similarly, the area between two electrodes should

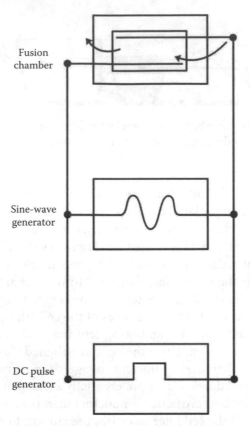

Figure 23.19 Schematic representation of equipment setup used for the electrofusion. Fusion chamber is connected in parallel to sine-wave generator (which provides ac) and a square-wave pulse generator (supplies the dc fusion pulse while the cells are held in place by the ac field).

be glue-free. A thin glass coverslip can then be glued over the chamber and thin polyethylene tubing (0.3 mm internal diameter) can be attached to each end of this chamber to create a flow-through chamber (Figure 23.19). The end is sealed carefully. The chamber is connected to two syringe pumps to have protoplasts to be fused in one and sterile osmoticum (for washing and flushing the chamber clean) in the other. An appropriate three-way valve connected to these pumps and the chamber can be used to precisely regulate the introduction and density of protoplasts in the fusion chamber.

The entire assembly can be sterilized by first pumping through 70% ethanol and then by washing with a large volume of sterile water. Sterilization can also be achieved by UV treatment but with longer exposure periods. Any waveform generator, which produces an ac signal from 0 to 100 V at frequencies from 100 to several million Hz, can be used. A square-wave pulse generator, capable of delivering a dc square-wave pulse from 1 to 200 μs in duration and at voltage from 0 to at least 150 V, would be adequate. The square-wave pulse should be very clean with the rise and decay time 2 μs. An oscilloscope is required to monitor the performance of the electrical system, and calibrate pulse frequencies duration and field strength, etc. The signal generators are connected in parallel to a fusion chamber as outlined in schematic fashion (Figure 23.19).

23.27 Fusion Products: Hybrids and Cybrids

Fusion of cytoplasm of two protoplasts results in coalescence of cytoplasms. The nuclei of two protoplasts may or may not fuse together even after fusion of cytoplasms. The binucleate cells are known as heterokaryon or heterocyte. When nuclei are fused the cells are known as hybrid or synkaryocyte, and when only cytoplasms fuse, and genetic information from one of the two nuclei is lost, they are known as cybrid, that is, cytoplasmic hybrid or heteroplast. There are some genetic factors that are carried in cytoplasmic inheritance, instead of nuclear genes, for example, male sterility in some plants. Susceptibility and resistance to some of the pathotoxins and drug are controlled by cytoplasmic genes. Therefore, production of cybrids which contain the mixture of cytoplasms but only one nuclear genome can help in the transfer of cytoplasmic genetic information from one plant to another. Thus, information of cybrid can be applicable in plant breeding experiments. In China, cybrid technology in rice is a great success. Such plants are very useful in producing hybrid seeds without *emasculation*. However, cybrid technology has successfully been applied to carrot, *Brassica* sp., *Citrus*, tobacco, and sugar beet.

23.28 Methods of Somatic Hybridization

Procedure for successful somatic hybridization is as follows:

1. Isolation of protoplasts from suitable plants.
2. Mixing of protoplasts in centrifuge tube containing fugigenic chemicals, that is, chemicals promoting protoplast fusion, such as PEG (20%, w/v), sodium nitrate ($NaNO_3$), maintenance of high pH 10.5 and temperature 37°C (as a result of fusion of protoplasts viable heterokaryons are produced). PEG induces fusion of plant protoplasts and animal cells and produces heterokaryon.
3. Wall regeneration by heterokaryotic cells.
4. Fusion of nuclei of heterokaryon to produce hybrid cells.
5. Plating and production of colonies of hybrid cells.
6. Selection of hybrid, subculture and induction of organogenesis in the hybrid colonies.
7. Transfer of mature plants from the regenerated callus.

Intergeneric hybrids are difficult to achieve by sexual crosses. However, they have been obtained through protoplast fusion, for example, tomato + potato, *Datura* + *Atropa*, barley + wheat, barley + rice, wheat + oat, and sugarcane + sorghum. Potato + tomato cross is shown in Figure 23.20. The interfamilial hybrids of plant cells are genetically

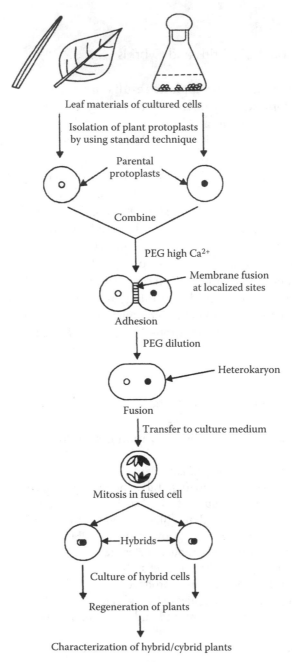

Figure 23.20 Flow diagram for the steps in production of hybrid plant by protoplast fusion technique.

unstable and show species-specific elimination/reconstruction of chromosome belonging to one of the parents; this was usually the parent represented by the mesophyll cells in hybridization process. These hybrids are incapable of regeneration and morphogenesis into plants. Intertribal hybrids are more widely used due to their higher genetic stability. The first intertribal cell hybrid was obtained from callus cells of *Arabidopsis thaliana* crossed with mesophyll protoplasts of turnip (*Brassica campestris*).

23.29 Protoplast Fusion (Somatic Cell Hybridization)

The technique of hybrid production through the fusion of body (somatic) cell eliminating sex altogether is called somatic hybridization. Protoplasts from different varieties or species of plants can be fused when protoplasts come into contact with other protoplasts. Fusion of cell membranes is induced and then the cytosols of both protoplasts are mixed. The resulting fused protoplast regenerates the cell wall forming a cell, the hybrid cell, by the fusion process that retains the genetic properties of both protoplasts. Somatic hybridization has been considered of special significance in improvement of crop such as banana, cassava, potato, sweet potato, sugarcane, and yam in which sexual reproduction is either weak or absent. Successful somatic hybrids have been obtained between the species of *Brassica*, *Nicotiana*, *Petunia*, and *Solanum*. Pomato, the somatic hybrid between potato (*Solanum tuberosum*) and tomato (*Lycopersicon esculentum*), is an example of intergeneric hybrid.

The protoplast fusion technique involves the use of a high-molecular-weight compound, PEG at a concentration of 28%–30% (w/v). The procedure (Figure 23.20) for the somatic hybridization consists of the mixing of the two protoplasts which are prepared, separated, and collected by centrifugation, in the desirable ratios of 1:1 to 1:4 in a solution containing osmotic stabilizer(s) and a calcium salt. The protoplasts are dispended in droplets and allowed to settle at the bottom of a Petri dish. Protoplasts form adhesion bodies on addition of PEG solution (at the rate of 5 g/10 mL solution). Removal of PEG is initiated by adding droplets of an alkaline high calcium solution followed by washing with a culture medium. The protoplasts and the fusion products adhere to the surface. After washing, a thin layer of the medium is retained and Petri dishes are incubated for growth/culture.

23.29.1 Materials

Protoplasts from two parents (in enzyme solution), protoplast culture medium (as used in previous experiment), washing solution (500 mmol/L glucose, 0.7 mmol/L $KH_2PO_4 \cdot H_2O$ and 3.5 mmol/L $CaCl_2 \cdot 2H_2O$, pH 5.5), eluting solution (50 mmol/L glycine, 50 mmol/L $CaCl_2 \cdot 2H_2O$, 300 mmol/L glucose, pH 9–10.5), PEG solution (50% PEG, 10.5 mmol/L $CaCl_2 \cdot 2H_2O$, 0.7 mmol/L $KH_2PO_4 \cdot 4H_2O$, Calcoflour (Calcoflour white MZR), 0.1% in 0.4 M sorbitol, silicon 200 fluid, sterile 60 μm pore size filter, humidified chamber (covered plastic box lined by a moist blotting paper in a beaker with 1% $CuSO_4$ solution), Pasteur pipettes, Petri dishes (60 × 15 mm), sterile cover-slips (22 × 22 cm), centrifuge, centrifuge tubes with screw cap, inverted microscope, fluorescence microscope

23.29.2 Protocol

1. Take freshly isolated protoplast of the two desired parents.
2. Mix the protoplast in 1:1 ratio.
3. Filter the suspension/mixture of protoplasts, through a 62 μm pore size filter.
4. Transfer the filtrate into a screw-cap centrifuge tube.
5. Close the mouth of the tube.
6. Centrifuge at 50g for 6 min.
7. Remove the supernatant with a sterile Pasteur pipette so that protoplasts remain at the bottom of the centrifuge tube.
8. Wash the protoplasts with 10 mL washing solution.
9. Resuspend the protoplasts in washing solution to make a suspension with 4%–5% (v/v) protoplasts per milliliter.
10. Place a 3 mL drop of silicon 200 fluid in the center of a 60 × 15 mm sterile Petri dishes using a 1 mL sterile syringe.
11. Put a sterile coverslip (22 × 22 mm) on the drop using a sterile forceps.
12. Using a sterile Pasteur pipette, pour 150 μL of the mixed protoplast suspension onto the coverslip.
13. Allow the protoplasts to settle on the coverslip for 5 min to form a thin layer.
14. Slowly add drop by drop, 450 μL (i.e., ~6 drops) of PEG solution to the protoplast suspension.
15. Incubate the protoplasts in PEG solution for 10–25 min at room temperature (24°C).
16. Gently add 0.5 mL of eluting solution to the mixture.
17. After another 10 min, add additional 1 mL of eluting solution.
18. Wash the protoplasts five times at 5 min intervals, every time using 1–2 mL (16–20 drops) of fresh protoplast culture medium. Allow the protoplasts to remain in the culture medium between washings.
19. Culture both the protoplasts (fused and unfused) on the same coverslip in a thin layer of 500 μL of the culture medium.
20. Put additional 0.5–1.0 mL of culture medium in droplets in the Petri dish to maintain humidity.
21. Seal the Petri dish with parafilm.
22. Incubate the sealed Petri dish at 25°C in diffused light (60–100 lux) in a humidified chamber.

23.29.3 Results and Observations

Observe the protoplast agglutination, after the addition of PEG solution on the stage of an inverted microscope (after Step 15).

Examine the protoplasts for digestion and regeneration of cell wall (fused and unfused protoplasts) after attaining with Calcoflour in fluorescence microscope, nuclear condition of protoplasts (heterokaryons or multinucleated) by staining with carbol–fuchsin stain, and division in hybrid cells and formation of cellular colonies (after 1–2 weeks). Cellulose layers will fluoresce when irradiated with UV light at 366 nm where regeneration of cell walls had taken in protoplasts (fused and unfused).

23.30 Electrofusion Isolation of Oat and Corn Protoplast

23.30.1 Introduction

Electrofusion of plant protoplast. The upper epidermis of leaves from light brown, 6 day old oat seedlings (*Avena sativa*) and 8 day old seedlings of maize (*Z. mays*) were removed with fine forceps. The pealed leaves are floated, upside down, on a solution containing 2% (w/v) Cellulysin (Calbiochem), 0.5 M mannitol, 3 mM $CaCl_2$, 1 mM KCl, and 3 mM morpholinoethanesulfonic acid (MES) at pH 5.6. Digestion is completed after 3–4 h at 30°C in the dark. The protoplasts are filtered through a nylon screen (pore diameter 80 µm), layered onto a 17% (w/v) sucrose pad, and centrifuged for 10 min at 100g. The protoplast at the interface is collected, resuspended in 12 mL of 0.5 M mannitol, and centrifuged for 3 min at 70g. The pellet is washed once with 0.5 M mannitol.

Isolation and culture of Vigna *protoplast.* After fusion, the protoplasts are concentrated by centrifugation and cultured in 25 µL sitting drops on the bottom of Petri dishes. The culture dishes are sealed with parafilm and maintained at 26°C under day light fluorescent tubes at a photon flux density of 30–50 µE/m² s. The culture medium contained MS macro- and micronutrients with vitamins and organic acids and other additives.

23.30.2 Materials

Chemicals: Enzymes are dissolved in 0.4 M mannitol containing 0.1% (w/v) $CaCl_2 \cdot 2H_2O$ and 0.5% (w/v) potassium dextran sulfate (sulfur content 18.1%, inherent viscosity 0.013 dL/g, and pH adjusted to 5.7).

23.30.3 Protocol

Isolated protoplast is subjected to electrofusion as per procedure described earlier.

23.30.4 Observations and Results

Observe the protoplast agglutination. Examine the protoplasts for digestion and regeneration of cell wall (fused and unfused protoplasts) after attaining with

Calcoflour in fluorescence microscope, nuclear condition of protoplasts (heterokary-ons or multinucleated) by staining with carbol–fuchsin stain, and division in hybrid cells and formation of cellular colonies (after 1–2 weeks). Cellulose layers will fluo-resce when irradiated with UV light at 366 nm where regeneration of cell walls had taken in protoplasts (fused and unfused). Electrofusion is found to be easy to control having fusion frequency up to 100%, gives reproducibility, less cytotoxic, but equip-ment is sophisticated and expensive.

23.31 Culture of Protoplasts

23.31.1 Introduction

Isolated protoplast can be cultured in several ways of which agar-embedding tech-nique in small Petri dish is commonly followed. In this technique, protoplast suspen-sion is mixed with equal volume of melted 1.6% "Difco" agarified medium (37°C) and the protoplast agar mixture is poured into small Petri dish. The plated protoplast can be handled very easily, and the agar medium provides a good support to the proto-plast. *In situ*, developmental stages of embedded protoplast and also separated clones from individual protoplast can be monitored by employing a compound microscope (Figure 23.21).

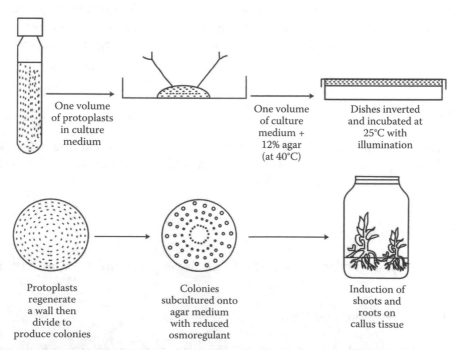

One volume of protoplasts in culture medium

One volume of culture medium + 12% agar (at 40°C)

Dishes inverted and incubated at 25°C with illumination

Protoplasts regenerate a wall then divide to produce colonies

Colonies subcultured onto agar medium with reduced osmoregulant

Induction of shoots and roots on callus tissue

Figure 23.21   Protoplast culture by agar-embedding technique in small Petri dishes.

23.31.2 Materials

Plant material: Tobacco leaf protoplast.

Chemicals: Macronutrients, micronutrients, iron source, vitamins, carbohydrate source, growth regulators, mannitol. Composition of Nagata and Takebe (NT) medium (1971) for protoplast culture.

Equipments and glassware: Laminar flow, culture bottles, Petri dishes, culture racks fitted with lighting system, pipettes.

23.31.3 Protocol

1. The protoplasts in liquid NT medium* are counted with the help of hemocytometer. The protoplast density is adjusted to 1×10^5–2×10^5 protoplast/mL.
2. Solidified (1.6% Difco agar) NT agar medium is melted.
3. The tight lid of Falcon plastic Petri dish (35 mm diameter, 5 mm thickness) is opened; 1.5 mL of protoplast suspension is taken out; and an equal amount of melted agar medium is added maintaining at 37°C–40°C temperature (Figure 23.22).
4. The lid is quickly replaced tightly, and the whole dish is swirled gently to disperse the protoplast agar medium mixture uniformly throughout the dish.

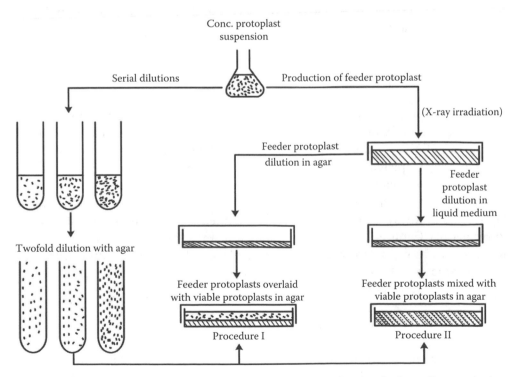

Figure 23.22 The technique using x-ray–irradiated protoplasts as feeder cells to stimulate growth of viable protoplasts at low density.

5. The medium is allowed to solidify in Petri dish and incubated under inverted position at 25°C, 500 lux illumination (16 h light).
6. The cultures are subcultured periodically in the same solid medium (0.8% agar) with gradually reducing mannitol.

23.32 Other Protocols

23.32.1 Droplet Culture

Suspending protoplasts in liquid culture media are placed on Petri dishes in the form of droplet with the help of a micropipette. This method enables the subsequent microscopic examination of protoplast development. In this method, cultured protoplast clumps together at the center of droplets.

23.31.2 Coculture

Sometimes, a reliable fast-growing protoplast is mixed in varying ratio with the less fast-growing protoplast. The mixed protoplasts are plated in solid medium as described earlier. The fast-growing protoplast presumably provides some growth factors which induces the growth and development of the desirable protoplasts. This is known as a coculture technique.

23.32.3 Feeder Layer Techniques

Fast-growing protoplasts are sometimes made mitotically blocked protoplast by low doses (1–2 krad) of x-ray treatment. Such irradiated protoplasts are plated with agar medium. Upon this thin solidified layer of irradiated protoplast, desirable protoplasts are again plated at a low density with agar medium. As a result, it makes two agar layers containing irradiated protoplast in the lower layer and desirable protoplast in the upper layer. The lower irradiated protoplast is known as feeder layer, which improves the growth and development of normal protoplasts, even at lower density (Figure 23.23). The technique of using x-ray-irradiated protoplasts as feeder cell to stimulate growth of viable protoplasts at low density is known as feeder layer technique.

23.32.4 Hanging Droplet Method

Culture of protoplasts in an inverted liquid droplet (0.25–0.50 µL) is known as hanging droplet method. Each droplet contains very small group of protoplasts. A number of droplets are generally placed on the inner surface of the lid of a Petri dish. Very thin layer of water is generally kept on the lower part of the Petri dish to make

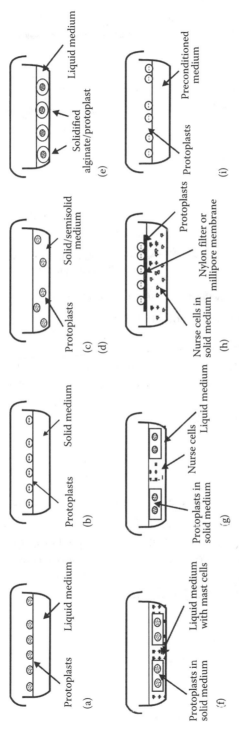

Figure 23.23 Illustration of various ways of growing cultures: (a) Culture in liquid medium, (b) culture in liquid medium, (c) culture in solid medium, (d) culture in solid medium, (e) alginate beads culture, (f) nurse culture, (g) nurse culture with millicell, (h) feeder layer culture, and (i) culture on preconditioned medium.

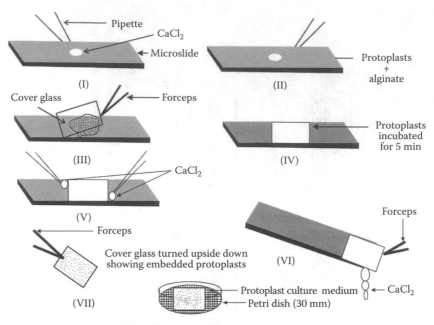

Extra thin alginatic film technique

Figure 23.24　Different methods of protoplast fusion.

a humid condition inside the Petri dish as well as to prevent the desiccation of the droplets. This technique facilitates to observe the development of protoplast under microscope. Protoplast also gets better aeration as they go down to the hanging surface of the droplets (Figure 23.24).

23.32.5 Bead Culture

Sometimes, protoplast suspension is mixed with several polymers like alginate, carageenan, etc. as well as melted standard Difco agar. Small beads are made by dripping the mixture into the liquid medium and then beads in liquid medium are put on moving shaker. Entrapped protoplasts culture has shown several advantages over static liquid culture or slowly moving liquid culture where the protoplast suffers the mechanical breakage (Figures 23.23 and 23.24).

23.33 Somatic Embryogenesis

23.33.1 Introduction

Embryo production is a characteristic feature of the flowering plants. However, such structures (embryoids) have also been artificially induced in cultured plant tissues, besides zygote. Somatic embryogenesis was first induced in suspension culture and callus culture of carrot. More than 30 plant families are known so far where somatic embryoids have been induced. The one singular event that was demonstrated during the century was cell totipotency that is the development of a whole organism starting from a single asexual cell via embryogenesis, a finding which revolutionized plant breeding methods. This potency has been found to persist even in generative cells of pollen grains. Somatic tissues of a plant represent products of mitotic divisions and each cell possessed the built-in capacity to regenerate the phenotype of the whole organism from which it is derived through embryogenesis when grown under appropriate culture conditions. The phenomenon is designated "somatic embryogenesis" which occurs both in diploid and haploid cells without the intervention of a sexual fusion. Embryogenesis in both somatic and reproductive cells corresponds at least in early stages to cell divisions characteristic of zygotic development, though with differing chromosome numbers and occurs in the absence of an auxin. The adventive embryos are bipolar in the organization with an integrated root–shoot axis arising in culture, unlike the development of shoots or roots (organogenesis) which are monopolar, originating independently and later integrating into one axis. The morphogenetic events of the somatic cells showed a striking resemblance to developmental sequences occurring in the fertilized egg cell, acquiring early bipolarity passing through proembryonal, globular, heart and torpedo stages exemplified by carrot cell cultures grown as both suspensions and callus. The plane of cell division, may, however, be irregular. The cell totipotency concept implies that somatic cells retain their functional nature free from neighboring tissues and the DNA is conserved during differentiation processes in several zygotes. Carrot has served as a model system for the study of somatic embryogenesis in all its manifestations. The sequence of events has been confirmed to be similar in both wild and cultivated carrots in the production of the enormous number of embryoids and plantlets from free cells in suspension, and in isolated tobacco cells for mass propagation. Somatic embryogenesis can be initiated in two ways: (i) by inducing embryogenic cells within the preformed callus and (ii) directly from preembryonic determined cell (without callus), which are ready to differentiate into embryoids. In the first case, embryoids are initiated in callus from superficial cell aggregates where cells contain a large vacuole, dense cytoplasm, large starch granules, and nucleus. Two nutritional media of different composition are required to obtain embryoids. First medium contains auxin to initiate embryogenic cells. Second medium lacks auxin or reduced level of auxin is needed for subsequent development of the embryonic cells into embryoids and plantlets. In both the cases, reduced amount of nitrogen is required. The embryogenic cells pass through three different stages, for example, globular, heart-shaped, and torpedo-shaped, to form embryoids (Figure 23.25). These embryoids can be separated and isolated mechanically by using glass beads. When embryoids reach torpedo stage, they are transferred to filter paper bridge (a sterile and plugged culture tube containing about 10 mL MS liquid medium supplemented with kinetin [0.2 mg/L] and sucrose [2% w/v] on which Whatman No. 1 filter paper is placed to make a bridge). Some plants in which somatic embryogencsis has been induced *in vitro* are *Atropa belladonna*, *Brassica oleracea*, *Carica papaya*, *Coffea arabica*, *Citrus sinensis*, *Daucus carota*, *Nicotiana tabacum*, *Pinus ponderosa*, and *Saccharum officinarum*.

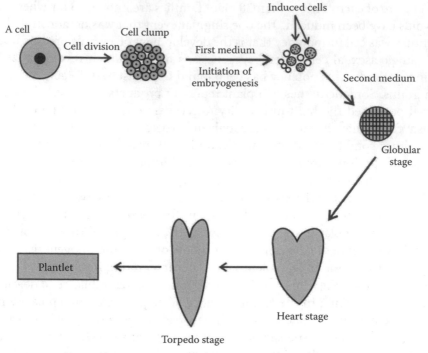

Figure 23.25   Events of somatic embryogenesis.

23.33.2 Material

Media

1. Suspension culture of carrot (*D. carota*) cells growing in MS medium supplemented with 2,4-D (1 mg/L) and sucrose (3% w/v).
2. Culture medium A. Sucrose containing MS salts zeatin (0.2 mg/L), 2,4-D (0.1 mg/L) sucrose (2% w/v) and agar.
3. Culture medium B. Sucrose containing MS salts zeatin (0.2 mg/L), sucrose (2% w/v) and agar (0.8% or 1% w/v).
4. Culture medium C. Disposable Petri dishes commercially sterilized MS salts kinetin (0.2 mg/L) and sucrose (2% w/v).

Equipments and accessories

Black cord with lent grid lines, scalpel, forceps, 2 cm³ pipettes (six), parafilm sealing tape, microscope slides, coverslips, light microscope (100× magnification), dissection microscope (optical).

23.33.3 Protocol

Prepare suspension culture from taproot of carrot. Take healthy taproot of fresh carrot, remove the external 1–2 mm of tissue with the help of vegetable scraper, cut the carrot transversely into l cm slices, surface sterilize in the usual manner with the help of sterile cork borer. Remove cylindrical tissue sample in the vicinity of the vascular cambium. Culture the explant on MS basal medium supplemented with myoinositol (100 mg/L), sucrose (3.0% w/v), gelrite (0.2% w/v), IAA (10 mg/L), and kinetin (0.1 mg/L). Incubate the culture (25°C–29°C) for 4–6 weeks, maintain carefully, so they are suitable for the induction of somatic embryogenesis:

1. Prepare five Petri plates containing culture medium A (with auxin).
2. Prepare five Petri plates containing culture medium B (without auxin).
   a. Add to each of the Petri dish aliquot (2 cm³) of carrot suspension culture, seal the dishes with parafilm, and incubate at 25°C in the dark for 2–3 weeks.
   b. The test for embryogenic potential is based on a visual count of the embryoids. Suspend the callus in double distilled H_2O and determine the number of embryoids present by placing the Petri dish over a black card marked with 1 cm² gridlines.
   c. Place small aliquots of the dispersed sample on a microscope slide and examine under the microscope (visualize various stages of somatic embryogenesis).
   d. Place the somatic embryos on agar medium devoid of 2,4-D (culture medium) for plantlet development.
   e. Transfer the plantlets to jiffy pots or vermiculite for subsequent development.

23.33.4 Results

In the Petri dish containing culture medium A, the cells develop into a callus, embryoids, however, are not formed. In the Petri dishes containing culture medium B, a large number of embryoids are formed (with a wide range of developmental stages).

23.34 Production of Haploid Plants

23.34.1 Introduction

Anther, a male reproductive organ, is diploid in chromosome numbers. As a result of microsporogenesis, tetrads of microspores are formed from a single spore mother cell. They are known as pollen grains after release from tetrads. The aim of anther and pollen culture is to get haploid plants by induction of embryogenesis. Haploid plants have single complete set of chromosomes that in turn may be useful for the improvement of many crop plants. Moreover, chromosome set of these haploids can be doubled by mutagenic chemicals (e.g., colchicine) or regeneration technique to obtain fertile homozygous diploid. In premitotic anthers, where the microspores have completed meiosis but not started first pollen division, the best response is achieved, for example, *H. vulgare*. In mitotic anthers where first pollen division has started the optimum responses are achieved, for example, *N. tabacum* and *Datura innoxia*. In postmitotic anthers, the early bicellular stage of pollen development is the best time to culture, for example, *A. belladonna*. Haploid plants are very useful in (i) direct screening of recessive mutation because in diploid or polyploid screening of recessive mutation is not possible and (ii) development of homozygous diploid plants following chromosome doubling of haploid plant cells. In China, the most widely grown wheat is a doubled haploid produced through homozygous diploid lines. Anther culture of rice is also successfully grown. Haploid plants have been produced in tobacco, wheat, and rice through pollen culture. These are used for the development of disease-resistant and superior diploid lines. At present, >247 plant species and hybrids belonging to 38 genera and 34 families of dicots and monocots have been regenerated using anther culture technique (Figures 23.26 and 23.27).

23.34.2 Anther Culture

23.34.2.1 Materials
Flower bud of *Nicotiana tobacum*, scalpel, forceps, dissecting needles, Petri dishes (10), pipettes, double distilled water, culture tube, MS medium supplemented with sucrose (2% w/v) and agar (0.6%–0.8% w/v), 5% Teepol and 10% sodium hypochlorite, 80% ethanol, 70% ethanol, spirit lamp, microscope, microscope slides and coverslip, acetocarmine stain, refrigeration unit (7°C–8°C), growth chamber (light intensity with 300 and 5000 lux), colchicines (0.5% w/v), parafilm.

23.34.2.2 Protocol
1. Select the lower buds with corolla length of 21–23 mm and prechill it for 12 days at 7°C–8°C.
2. Surface sterilize the buds with 10% sodium hypochlorite and 5% Teepol for 10 min.
3. Wash the bud again with sterile double distilled H_2O.
4. With the help of forceps and dissecting needle, tear open the buds and remove the anthers.
5. Take one anther of the group, prepare a squash in acetocarmine, and observe it under microscope; the pollen should exhibit the pollen division.
6. The remaining anthers will also be at the same stage. The filaments are removed and inoculated onto the medium for culturing at 25°C and incubated in darkness for 2–4 weeks.

23.34.2.3 Results
1. After plantlet formation is initiated, light is essential to keep them under continuous illumination.
2. Separate the plantlets (3 mm in length) with forceps and inoculate in the medium identical to the anther culture (agar 0.5% w/v) and other compounds are provided in half strength.
3. Illuminate the plants to day light region at about 2000 lux with the help of white fluorescent lamps.

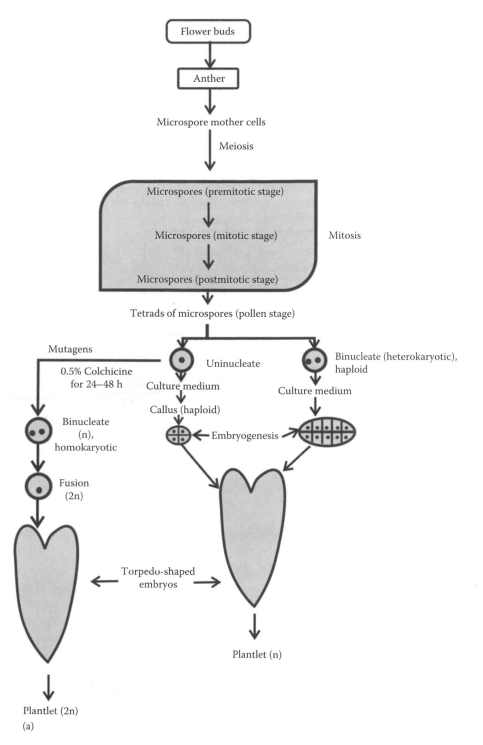

Figure 23.26 (a) Experimental outline of anther culture, and production of haploid and diploid plants *in vitro*.

(*continued*)

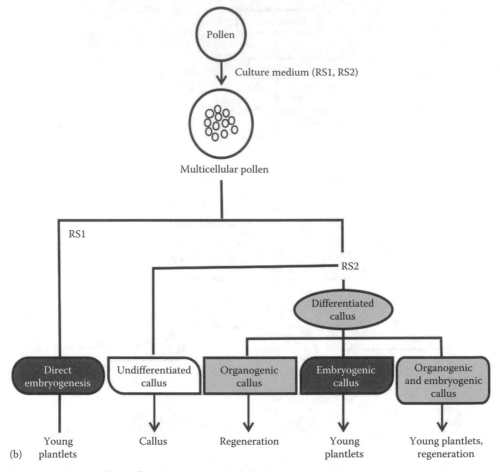

(b)

Figure 23.26 (continued) (b) Types of *in vitro pollen* morphogenesis in *Datura*.

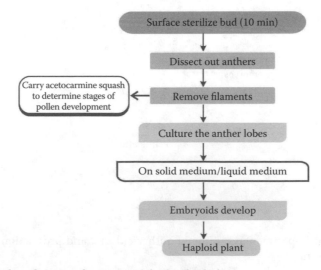

Figure 23.27 Flow diagram for protocol for haploid plant.

4. Examine the root tip of the plantlets with acetocarmine in order to verify the haploid chromosome number.

5. Wash ~50 mm tall plantlets with tap water and transfer it into a small pot containing autoclaved potting compost. Cover each plantlet with a glass beaker to prevent desiccation and maintain the plants in a humid greenhouse. After 1 week, remove the beaker and transfer to larger pots. Observe the zygotic embryos through stereomicroscope.

23.35 Embryo Encapsulation: Production of Artificial Seeds

23.35.1 Introduction

The potential application and importance of *in vitro* somatic embryogenesis and organogenesis are more or less similar. The mass production of adventitious embryos in cell culture is still regarded by many as an ideal propagating system. Adventitious embryo is a structure that develops directly into a complete plantlet, and there is no need for rooting phase as is done in shoot culture. Somatic embryos do not possess the food reserves of their own; therefore, suitable nutrients could be packaged by coating or encapsulation to form some kind of artificial seeds. Such artificial seeds produce the plantlets in the field, unlike organogenesis. Somatic embryos may arise from single cells. It is, therefore, of special significance in mutagenic studies. Plants derived from asexual embryos may in some cases be free of viral and other pathogens and offer as an alternative approach for the production of disease-free plants (Figure 23.28).

23.35.2 Materials

Embryos of *Crotalaria verrucosa*, field beans and cluster beans, etc., weighing balance, volumetric flasks, sodium alginate, calcium chloride.

Figure 23.28 **(See color insert.)** Artificial seed.

4% Sodium alginate: Dissolve 4 g of sodium alginate in 100 mL distilled water. Steam the mixture to fully dissolve the alginate.

Calcium chloride solution: Dissolve 1.47 g of $CaCl_2 \cdot 2H_2O$ in 100 mL distilled water.

23.35.3 Protocol

1. Embryos from seeds of *C. verrucosa* or cyanosis were isolated and stored in distilled water until use.
2. The embryos were collected and mixed with sodium alginate solution.
3. A Pasteur pipette was filled with embryo suspended in sodium alginate solution.
4. The suspended embryos were dipped in a beaker containing $CaCl_2$ solution.
5. Care is taken to see that at least one embryo is dropped along the sodium alginate.
6. The embryos were allowed to get encapsulated by placing them to remain in calcium chloride solution for 30 min.

23.35.4 Results

After 1 h, the gel becomes hard, and it forms a bead-like structure inside the $CaCl_2$ solution enclosing an embryo. This is taken out from the solution and washed with distilled water and stored for about a year or more and is ready for farmers. Water solution of soluble hydrogels has been found suitable for making artificial seeds with complexing agents like calcium chloride or ammonium chloride. Hydrogels include sodium alginate, sodium alginate with gelatin, carageenan with locust beam gum or gel rite.

23.36 Cytological Examination of Regenerated Plants

Histology may be studied to scrutinize the development of the shoot from leaf explant.

23.36.1 Microtomy

Regenerating tissue was cut into small pieces (1 × 1 cm) and fixed in a FAA solution. FAA (Formaldehyde:acetic acid:ethyl alcohol)

Ethyl alcohol (50%–70%)	90 mL
Acetic acid (glacial)	5 mL
Formaldehyde	5 mL

The tissue was allowed to fix for 12 h and then stored in 70% ethyl alcohol after washing two to three times.

23.36.2 Dehydration

Dehydration removes free water from the fixed and washed tissue and simultaneous replace with suitable organic solvent. The tissue is gradually dehydrated in a mixture of tertiary butyl alcohol (TBA) and ethyl alcohol in ascending and descending series, respectively. The earlier process involves transference of fixed material from ethyl alcohol to tertiary butanol.

The concentration of ethyl alcohol and TBA during the process of dehydration and the period of dehydration is as follows:

Ethyl Alcohol	TBA (mL)	Period of Incubation
90	10	2 h
80	20	Overnight
65	35	1 h
45	55	1 h
25	75	1 h with pinch erythrocin
—	100	

The tissue was transferred into pure TBA mixed with paraffin oil.

23.36.3 Infiltration, Embedding, and Sectioning

Infiltration is the process of replacement of a dehydrating agent from the fixed and dehydrated tissue with embedding medium. The infiltration is carried at 60°C in a hot air oven for an hour, and the mixture of TBA and paraffin oil was decanted and replaced with molten paraffin wax (melting point 50°C) and incubated until the smell of TBA was not felt. The tissue was embedded in paper boats/trays containing molten wax (melting point 50°) and allowed to cool over a tray of water. The molten blocks were cut into smaller blocks suitable for block holder of a rocking microtome. The blocks were trimmed and fixed to the block holder. Sections of 5–10 µm were cut and placed on a microscopic slide smeared with egg albumin.

TBA–xylene series:

1. 25%
2. 50%
3. 70%
4. 90%
5. 100%
6. 3:1 TBA (100%):xylene (100%)
7. TBA:xylene (1:1)
8. TBA:xylene (1:3)
9. Xylene 1
10. Xylene 2

Ribbons with serial sections were kept in coplin jars (Figure 23.29) containing xylene for 1 h to relieve the wax from the sections. The wax-free slides were subjected to

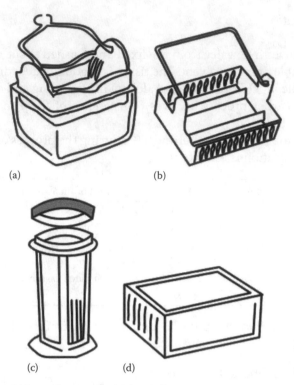

Figure 23.29   (a and b) Coplin jars and (c and d) staining trays.

reverse process of dehydration to bring the tissue to the aqueous medium. The sections were stained with safranin for 1 min. After removing the excess of safranin from the sections, they were dehydrated with TBA–xylene series reducing the TBA at the end. Finally, the sections were brought into xylene. The sections are placed over the glass slide and then covered with a coverslip with a drop of DPX mount-out.

23.36.4 Result

Examine the cells of the tissues like the nature of parenchymatous cells, xylem, phloem, and cambium.

23.37 Demonstration of Root Hair Culture of Different Explants Source of Cereal Using *Agrobacterium rhizogenes*

23.37.1 Introduction

A. rhizogenes is a Gram-negative soil-borne phytopathogenic bacterium, which causes the neoplastic plant disease syndrome known as "hairy root" disease. The typical disease syndrome is characterized by the sprouting of numerous adventitious roots at the wound site upon infection by the bacterium. The development of hairy toots is a consequence of the transfer and integration of the T-DNA from a portion of the *A. rhizogenes* root-inducing plasmid of Ri plasmid.

The transformation process can be arbitrarily divided into four steps: (i) bacterial colonization and attachment to the plant cell, (ii) T-DNA processing and delivery, (iii) T-DNA integration, and (iv) T-DNA expression.

Metabolically active wounded plant cells produce an abundance of low-molecular-weight signal compounds such as acetosyringone or hydroxyl acetosyringone, which act as chemical attractants for the bacterium. The T-DNA of the Ri plasmid also bears genes responsible for the synthesis of opines (agropine, mannopine, cucumpine, mikimopine) which are amino acid derivatives and serve as the carbon and nitrogen source for the bacterium. The ability of the specific region of the pathogenic bacterium persists, and since these compounds are not normally found in plants, their presence in plant tissue is indicative of transformation.

The roots produced at the infection site can be removed and cultured in simple hormone-free media where they exhibit a high rate of growth, genetic stability, and high levels of production of secondary metabolites. Genetically transformed "hairy roots" are being investigated worldwide as a tool for commercial production of plant chemicals. So far, several plant species have been transformed by *A. rhizogenes*. Over the past few years, the number of reports about the successful culture of hairy roots in bioreactors have considerably increased. Six different types of reactors, namely air lift, bubble column, ebb and flow, trickle bed, stirred tank, and turbine blade, have so far been attempted for hairy root cultures. However, a need is raised to focus further efforts for scale-up of technology to have adequate replication, improved inoculation, and tissue harvest.

23.37.2 Materials and Protocol

23.37.2.1 Seed Sterilization and Germination

Maize seeds (*Z. mays*) were thoroughly washed and soaked in water overnight and then surface sterilized by repeated washing with detergents (Extran) and soaking in 95% (v/v) ethanol for 10 s. Seeds were thoroughly rinsed with sterile distilled water and were placed in 0.01% mercuric chloride for 3 min and again rinsed several times with sterile distilled water to remove the last traces of chemical. Washed seeds were placed on 1% water agar for germination and incubated at 25°C for 5 days. Five seeds were placed on 25 mL of the basal MS agar-solidified culture medium in Petri dishes (100×15 mm). The MS medium was supplemented with sucrose (30 g/L) and agar (8 g/L), adjusted to pH 5.8 and autoclaved at 121°C for 25 min. The seeds were germinated in the growth chamber under standard cool white fluorescent tubes with a flux rate of 35 μmol/s/m^2 and a 16 h photoperiod and at 25°C.

23.37.2.2 Preparation of A. rhizogenes

The culture of *A. rhizogenes* was initiated from stock and maintained on MYA-solid medium (5.0 g/L yeast extract, 0.5 g/L casamino acids, 8.0 g/L mannitol, 2.0 g/L ammonium sulfate, 5.0 g/L NaCl, and 15 g/L agar) for 48 h at 28°C in the dark. The single clones were grown for 24 h in 20 mL MYA-liquid medium at 28°C on a rotary shaker at 100 rpm in the dark.

23.37.2.3 Callus Formation

After 5 days of the seed germination, 1 cm long shoot was cut and placed on Petri plates containing MS media fortified with growth regulator 6 µg/L 2,4-D and 0.8% agar. Petri plates were incubated at 25°C for 10 days in an inverted position.

23.37.2.4 Establishment of Hairy Root Cultures

Excised stems, leaves, and roots were collected from plants grown *in vitro* (21 day old) and used as the explant materials for cocultivation with *A. rhizogenes*; stems (2 cm), leaves with petioles, and the roots (2–3 cm length and ~3 mm separated from the root tip). Then, each explant was immersed in bacterial suspension separately for 5 min. The explants were blotted dry on sterile filter paper to remove excess bacteria and incubated in the dark at 28°C on 200 mL Erlenmeyer flask with 50 mL of liquid hormone-free MS medium with 30 g/L sucrose on a rotary shaker at 100 rpm. Uninfected explants (control) were cultured under the same conditions. After 24 h of cocultivation, the explant tissues were transferred to new liquid hormone-free MS medium with 30 g/L sucrose containing 500 mg/L cefotaxime to eliminate bacteria and then incubated in the dark. Numerous hairy roots were observed emerging from the wound sites within 2 weeks. The hairy roots were separated from the explant tissues and subcultured in the dark at 25°C on liquid hormone-free MS medium. After repeated transfer to fresh medium, rapidly growing hairy root cultures were obtained. Isolated roots (300 mg) were transferred to 50 mL of MS liquid medium, containing 30 g/L sucrose, in 200 mL flasks. The root cultures were maintained at 25°C on a rotary shaker at 100 rpm in a growth chamber under standard cool white fluorescent tubes with a flux rate of 35 µmol/s/m² and a 16 h photoperiod. After 3 weeks of culture, the hairy roots were harvested and analyzed biochemically for desired biochemical parameter. Six flasks were used for each culture condition, and the experiments were repeated twice.

23.37.3 Results and Observations

Hairy roots of cereals were initiated with three different explant sources (roots, leaves, and stems) inoculated with *A. rhizogenes* strain. After 2 days of cocultivation, explant tissues were transferred to free hormone liquid MS medium containing 500 mg/L cefotaxime to eliminate *A. rhizogenes*. Visible roots were formed after 5–7 days at the site of bacterial inoculation of different explants. After 10–14 days, the roots began to grow more rapidly. Observe the prolific growth of hairy root organ on MS medium. The rate of root elongation followed a distinct pattern over the first 20 days of culture. Sometimes, the growth of roots was further enhanced by adding growth regulating substances like auxins to the culture media. It is postulated that the first 15 days were taken for physiological adjustment period during which the lateral roots assumed the functions of the main roots which, however, produced several new lateral roots (Figures 23.30 and 23.31).

23.38 Experiment Demonstration of Root Hair Culture of Different Explants Using Arbuscular Mycorrhizal Fungi

23.38.1 Introduction

The term mycorrhiza refers to the association between fungi and roots of higher plants. This association is usually considered as a mutualistic symbiosis because of the highly independent, and commonly beneficial relationships established between both partners, in which the host plant receives mineral nutrients via fungal mycelium (mycotrophism) while the heterotrophic fungus obtain carbon compounds from the host photosynthesis. Despite the scarcity of experimental knowledge, it can be assumed that symbiosis must be the result of continuous molecular dialogue between plant and fungus as carried out through the both recognition and acceptance signals. One of the major advantages of the root organ cultures system is that many observations on extrametrical fungal development can be made nondestructively using a binocular lens. Biological studies are made difficult by the obligate biotrophic relationship they form with vascular plants. Empirical test of many chemical substances and conditions has failed to provide sustained hyphal growth from germinated spores in the absence of host root. For a critical study of the effect of arbuscular mycorrhiza on plant growth, typical infections must be produced under controlled microbiological condition. Present experiment demonstrates that root organ culture has potential for growing AM fungus *in vitro*. However, the method does not result in the reproducible germination of the entire life cycle of the fungus including the production of viable spores. The two most widely used species in this system are *Glomus* sp. but the species list is expanding at a good pace (Figures 23.32 and 23.33).

23.38.2 Materials and Protocols

23.38.2.1 Seed Sterilization and Germination

Maize seeds (*Z. mays*) were thoroughly washed and soaked in water overnight and then surface-sterilized by repeated washings with detergents (Extran) and soaking in 95% (v/v) ethanol for 10 s. Seeds were thoroughly rinsed with sterile distilled water and were placed in 0.01% mercuric chloride for 3 min and again rinsed several times with sterile distilled water to remove the last traces of chemical. Washed seeds were placed on 1% water agar for germination and incubated at 25°C for 5 days. Five seeds

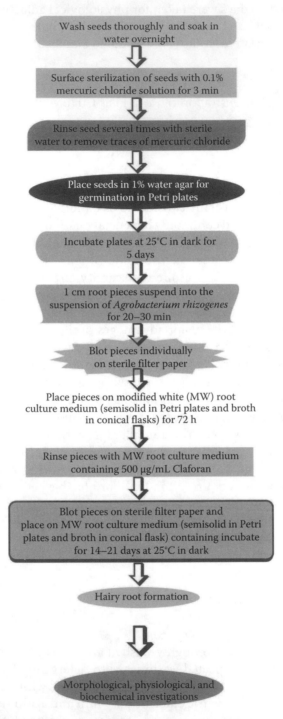

Figure 23.30 Flow diagram for root hair culture demonstration.

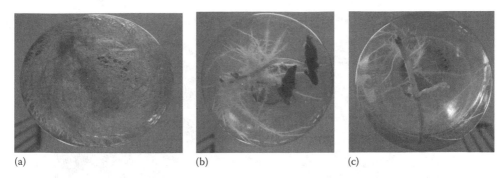

(a)   (b)   (c)

Figure 23.31 Development of hairy roots from different explant sources of a cereal within 10 days after inoculation with *A. rhizogenes* strain A4: (a) roots, (b) leaves, and (c) stems.

were placed on 25 mL of the basal MS (Murashige and Skoog 1962) agar-solidified culture medium in Petri dishes (100 × 15 mm). The MS medium was supplemented with sucrose (30 g/L) and agar (8 g/L), adjusted to pH 5.8 and autoclaved at 121°C for 25 min. The seeds were germinated in the growth chamber under standard cool white fluorescent tubes with a flux rate of 35 μmol/s/m^2 and a 16 h photoperiod and at 25°C.

23.38.2.2 Callus Formation
After 5 days of the seed germination, 1 cm long shoot was cut and placed on Petri plates containing MS media (Murashige and Skoog 1962) fortified with growth regulator 6 μg/L 2,4-D and 0.8% agar. Petri plates were incubated at 25°C for 10 days in an inverted position.

23.38.2.3 Root Regeneration
Ten days old calli were transferred on MS medium containing growth regulator NAA at a concentration of 1 μg/L. Plates were incubated at 25°C for 10 days. After several transfers, a clonal culture derived from a single root was established.

23.38.2.4 Culture Media
MS medium solidified with 0.8% agar containing 1 μg/L of NAA is used for routine maintenance.

23.38.2.5 Mycorrhizal Inoculum
A monoculture of *Glomus mosseae* Schenck and Perez is to be used to inoculate maize (*Z. mays*) maintained in sterile earthen ware pots in a greenhouse.

23.38.2.6 Minimal Agarose Medium
Refer Table 23.8.

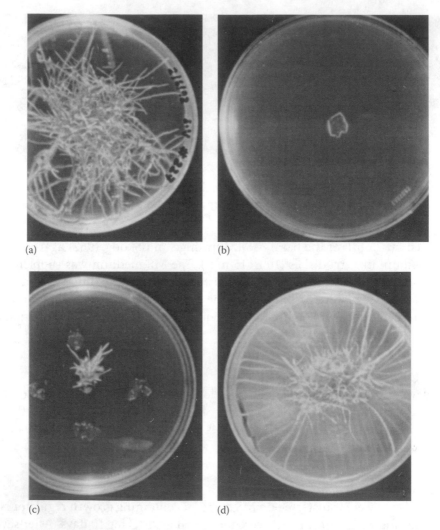

Figure 23.32 Establishment of dual culture. Regenerated roots of *Z. mays* grown on MS media. One square cm agarose disk mycelia was transferred on to a fresh minimal agarose media and incubated 25°C for 20 days: (a) Regenerated root of *Z. mays*. (b) Axenic endosymbiont growing on to minimal agarose media. (c) Early stage of establishment of dual culture. (d) Regenerated roots of *Z. mays* colonized with AM fungus.

23.38.3 Observation and Result

Obtain result and interpret as per Table 23.9.

Control: Without mycorrhizal spore or hyphae.

Agar disk: Mycelia of *G. mosseae* grown in minimal agarose media.

Broth: Mycelia of *G. mosseae* grown in minimal broth.

Spore: Spores obtained from pure soil culture (*G. mosseae*).

Z. mays seeds were inoculated with mycelia and spore culture. Plants were incubated in environmentally controlled growth chamber for 3 months.

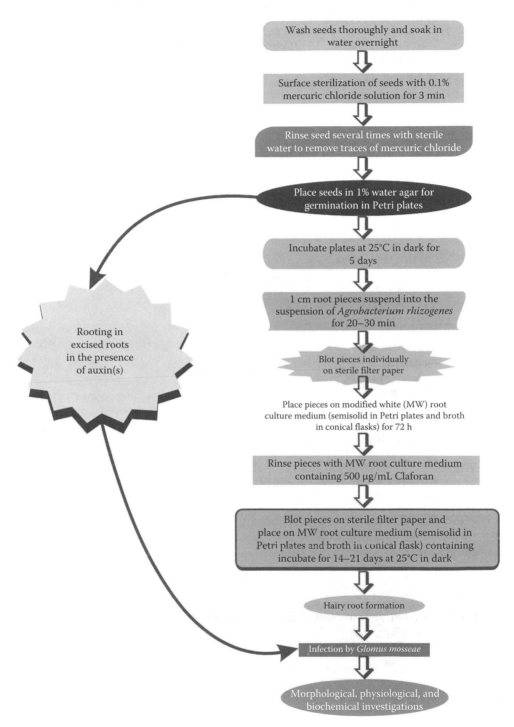

Figure 23.33 Flow diagram for root hair culture with arbuscular mycorrhizal (AM) infection.

TABLE 23.8   Composition and Preparation of Media for Spore Germination

Constituents	Concentration of Solution Stock (g/L)	Volume of Stock per Liter of Medium (mL)
Major inorganic nutrients		
$MgSO_4 \cdot 7H_2O$	7.3	
$Na_2SO_4 \cdot 10H_2O$	4.5	
$Ca(NO_3)_2 \cdot 4H_2O$	2.9	100
KNO_3	0.8	
KCl	0.65	
$NaH_2PO_4 \cdot 2H_2O$	0.195	
Trace elements		
$MnCl_2 \cdot 4H_2O$	0.60	
$ZnSO_4 \cdot 7H_2O$	0.26	
H_3BO_3	0.15	20
KI	0.075	
$CuSO_4 \cdot 5H_2O$	0.002	
$Na_2MoO_4 \cdot 2H_2O$	0.17	
Iron source		
EDTA (trisodium salt)	0.80	10
$FeCl_3 \cdot 6H_2O$	0.31	
Organic supplements		
Glycine	0.003	
Nicotinic acid	0.005	
Thiamine–HCl	0.001	1
Pyridoxine–HCl	0.001	
Carbon source		
Sucrose	20,000	
Agarose	7.0	

TABLE 23.9   Growth Parameters of *Z. mays* in Environmentally Controlled Conditions

Inoculum Used	Plant Height (cm)	Leaf Area (cm^2)	Leaf Length (cm)	Shoot Biomass (g)	Root Biomass (g)	No. of Spores/100 g Soil
Control						
Agar disk						
Broth						
Spore						

23.39 Useful Secondary-Metabolite Production from Plant Tissue Culture

23.39.1 Introduction

The use of plant tissue cultures in bioproduction of useful secondary metabolites of pharmaceutical significance holds an interesting alternative for controlled production of plant constituents. It has become difficult to maintain ample supply of medicinal plants due to their over exploitation, lack of conservation, and problems associated with the cultivation of medicinal plants. Production of secondary metabolites by plant cell cultures is particularly useful when these compounds (a) cannot be chemically synthesized, (b) are produced by the plant in small quantities, (c) are produced in an unharvestable stage of the plant development, (d) are produced in plants when they are not amenable to agriculture, and (e) have high commercial value.

Nearly four decades of research have resulted in the large-scale cultivation of few systems. There is, however, a wide gap in our understanding of the basic aspects of cellular metabolism, control mechanism, and genetic basis of controls involved in the production/biotransformation by plant cell cultures. The need for new plant products effective against AIDS, cancer, drug-resistant pathogen and effective in the treatment of circulatory and heart diseases is ever increasing. With this background vis-à-vis the progress made in the area of plant tissue and organ cultures, there is great optimism in the next decade for plant cell culture systems which would give us certain macro- and megalevel technologies leading to the synthesis of novel plant constituents (Figure 23.34).

Study of secondary metabolites involves extraction from dried plant samples in organic solvents (cold or luke warm) for a few hours, purification of extract to remove unwanted chlorophylls, lipids, carbohydrates, protein, and other metabolites through acid, basic, phase and extracting with organic solvents, and then separation by TLC. Different solvents are used depending upon the polarity of compound. Quantification of isolated products is done by densitometer, spectrophotometer, spectrofluorimeter, HPTLC, HPLC, FPLC and properties were determined by NMR and mass spectrophotometer.

The preliminary examination of cell extracts can be carried out by either TLC or paper chromatography. Table 23.10 summarizes the methods adopted for the detection of a range of chemical compounds in cell extracts.

23.39.2 Tropane Alkaloid Production from Plant Tissue Culture

Tropane alkaloids are a class of alkaloids and secondary metabolites, sharing the tropane skeleton as a common structural feature, and are sharply divided into two classes: tropine and ecgonine derivatives. The first group, represented by well-known alkaloids, atropine and scopolamine that are considered to be model anticholinergic drugs, continues to provide inspiration in the search for more selective muscarinic receptor antagonists. The second class accommodates one of the principal drugs of abuse, cocaine. Many of these alkaloids are glycosides that occur in aereal parts of *Solanum* spp. About 1400 alkaloids of this group are known. Many of the plants that contain these alkaloids are of economic importance, for example, potatoes, tomatoes,

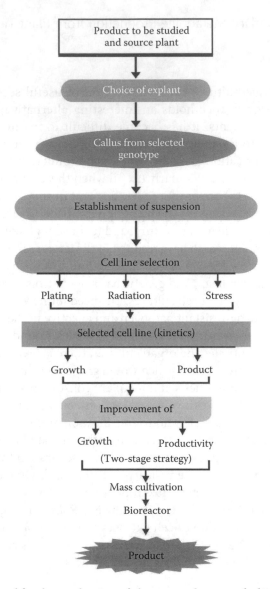

Figure 23.34   Protocol for the production of plant secondary metabolites using tissue culture.

tomatillos, and eggplant. Species of this genus are often involved in animal-poisoning problems. Tissue culture has been shown to be a potential tool to produce a great variety of secondary metabolites of commercial importance.

23.39.3 Materials

Equipments: Laminar flow, usual tissue culture facilities, shaker, oven 50°C–250°C, precoated silica gel G TLC plates, TLC sprayer, separating funnels, conical flasks, HPLC, and flash evaporator.

TABLE 23.10 Examination of Cell Extracts for Different Classes of Secondary Products

Compound	Chromatography System	Solvent System	Detection Method	Color Reaction	Examples
Anthocyanins	TLC cellulose	Forestal	White light, $AlCl_3$	Red, blue, purple	Cyanidin, petunidin
Flavans	Paper chromatography Whatman No. 1	BAW	White light, $AlCl_3$	Red, blue, purple	Cyanidin, petunidin
	TLC cellulose				
Flavonols and flavones	TLC cellulose	BAW	$UV + NH_3$ fumes	Pink	Catechin, epicatechin
	Paper chromatography Whatman No. 1	Vanillin/conc. HCl			
			$AlCl_3$	Yellow fluorescence	Myricetin, querectin
		BAW	Lead acetate	Yellow fluorescence	Apigenin, kaempferol
	Paper chromatography Whatman No. 1	HCl/acetic acid/H_2O (3:30:10)			
		Forestal	Vanillin/HCl	Yellow fluorescence, red/ purple	Apigenin, kaempferol
		Forestal			
Hydroxycinnamic acids and hydroxycoumarins	Paper chromatography Whatman No. 1	n-BuOH/acetic acid/H_2O (4:1:5)	$UV + NH_3$ fumes	Blue, green, yellow fluorescence	p-Coumaric acid, caffeic acid, ferulic acid
	TLC cellulose	(Top phase) BAW	General phenol sprays	Blue, green, yellow fluorescence	p-Coumaric acid, caffeic acid, ferulic acid
Phenylpropenes	TLC silica gel G	n-Hexane/chloroform (3:2)	Vanillin/conc. HCl	Red/pink	Thymol, eugenol
Simple phenols and phenolic acids	TLC silica gel G	Chloroform/acetic acid (9:1)	Folin reagent	Blue	Orcinol
	TLC cellulose	6% Acetic acid	Vanillin/HCl	Pink	Catechol, vanillic acid
	TLC cellulose	6% Acetic acid	$FeCl_3/K_3Fe(CN)_6$	Blue	Catechol, vanillic acid

(continued)

TABLE 23.10 (continued)   Examination of Cell Extracts for Different Classes of Secondary Products

Compound	Chromatography System	Solvent System	Detection Method	Color Reaction	Examples
Terpenoids, monoterpenes	TLC silica gel G	Benzene/chloroform (1:1)	$KMnO_4$, $SbCl_3$, vanillin/H_2SO_4	Various colors	Limonene, loganin, menthol, apinene
Sesquiterpenes	TLC silica gel G	Chloroform/MeOH (99:1)	Iodine	Yellow/brown	Farnesol
	TLC silica gel G	Chloroform/MeOH (99:1)	H_2SO_4	Green, brown, red, blue	Abscisic acid
Triterpenoids, cardenolides	TLC silica gel G ± $AgNO_3$	Ethyl acetate/MeOH/H_2O (16:1:1)	Chloramine/ trichloroacetic acid	Yellow/green fluorescence	Digoxin, digitoxin
Sapogenins	TLC silica gel G ± $AgNO_3$	Chloroform/acetone (4:1)	20% $SbCl_5$ in chloroform	Pink/purple	Diosgenin, yarnogenin
Steroids	TLC silica gel G ± $AgNO_3$	Chloroform	Acetic anhydride/ H_2SO_4	Various colors	Sitosterol, stigmasterol
Triterpenes	TLC silica gel G ± $AgNO_3$	Chloroform/MeOH (99:1)	20% $SbCl_5$ in chloroform	Yellow/green fluorescence	Amycin, cholestone
Alkaloids, Rauwolfia	TLC silica gel G	Ethyl acetate/EtOH (3:1)	Cerric ammonium sulfate	Range of colors	Ajmalicine, reserpine
	TLC silica gel G	Chloroform/NH_4OH (9:1)	Dragendorff's reagent	Orange brown	Serpentine
Catharanthus	TLC silica gel G	Ethyl acetate/EtOH (3:1)	Cerric ammonium sulfate	Range of colors	Vindoline, catharanthine
	TLC silica gel G	Chloroform/acetone/ diethylamine (5:4:1)	Cerric ammonium sulfate	Range of colors	Vinblastine, vindoline, catharanthine, vinblastine

	Stationary phase	Solvent system	Detection reagent	Color	Compounds
Cinchona	TLC silica gel G	Chloroform/ethyl acetate/i.PrOH/diethylamine (20:70:4:6)	Dragendorff's reagent	Orange	Quinine
	TLC silica gel G				
	TLC silica gel G	Chloroform/ethyl acetate/i.PrOH/diethylamine (20:70:4:6)	Lodoplatanate reagent	Blue/violet	Quinidine
			Formic acid	Blue fluorescence	Cinchonine
Opium	TLC silica gel G	Acetone/EtOH (4:1)	Iodoplatanate reagent	Purple/brown	Morphine, codeine
	TLC silica gel G	Xylene/butanone/MeOH/diethylamine (40:40:4:6.2)	Dragendorff's reagent	Orange brown	Thebaine, papaverine
Tobacco	TLC silica gel G	Chloroform/MeOH/NH$_4$OH (60:10:1)	Iodoplatanate reagent	Purple/blue/pink	Nicotine, anabasine
Tropane	TLC silica gel G 0.5 M KOH	70% EtOH/25% MeOH/NH$_4$OH	Dragendorff's reagent	Orange	Atropine, scopolamine
Amines, aliphatic	TLC cellulose	BAW	Ninhydrin	Pink, blue, purple	Putroscino Spermidine
Aromatic	TLC cellulose	BAW	Ninhydrin	Blue, purple	Tryptamine
	TLC cellulose	Chloroform/ethyl acetate/acetic acid (6:3:1)	Ninhydrin Ehrlich's reagent	Pink	Mescaline
Phenolics, anthraquinones	TLC silica gel G ± 4% tartaric acid	Chloroform/acetic acid (9:1)	White light	Yellow/orange	Juglone, chrysophanol, emodin
	TLC silica gel G ± 4% tartaric acid	Toluene/MeOH/acetate/acetic acid (9:1)	UV + NH, fumes	Orange, red	Juglone, chrysophanol, emodin
	TLC silica gel G ± 4% tartaric acid	Benzene/ethyl acetate/acetic acid (75:24:1)	10% MeOH/KOH	Red, violet	Juglone, chrysophanol, emodin

Chemicals: Chloroform, methanol, sodium carbonate, sodium bicarbonate, anhydrous sodium sulfate, HCl, bismuth subnitrate, acetic acid, and potassium iodide.

Plant source: Well-established tissue cultures (callus, shoots, or roots and suspension cultures).

23.39.4 Protocol

1. Harvest tissue, dry at 105°C for 15 min to inactivate enzymes and then at 60°C, until a constant weight is reached.
2. Grind each sample to a fine powder in a pestle mortar.
3. Immerse powdered sample (500 mg) in 20 mL ethanol–25% NaOH (19:1) mixture and keep overnight at room temperature.
4. Decant the ethanol solution and evaporate in vacuum.
5. Dissolve residue in 5–10 mL 0.1 N HCl and keep it for 1 h.
6. Basify acid extract with sodium carbonate/bicarbonate buffer (pH 9.6).
7. Transfer alkaline phase into separating funnel and extract with 10–15 mL $CHCl_3$.
8. Repeat earlier step for three times.
9. Combine chloroform extract and pass through anhydrous sodium sulfate.
10. Evaporate to dryness.
11. Dissolve extract in $CHCl_3:NH_4$ (10:1) and apply to TLC.
12. For HPLC, dissolve the extract in methanol (pH 2.5), centrifuge, and clear extract is subjected to HPLC analysis.

23.39.5 Thin Layer Chromatography

Take silica gel G precoated TLC plates (20 × 20 cm) and activate at 100°C for 30 min before use:

1. Prepare the standard solution for hyoscyamine (tropane alkaloids) by dissolving in $CHCl_3$.
2. Apply 20 µL of standard sample as spot. Similarly, apply sample extract (50–100 µL) as a spot at a distance of 2 cm.
3. Develop the plate in presaturated TLC chamber using $CHCl_3$:MeOH:ammonia (85:1:14) as solvent system.
4. Remove the plate and allow it to air dry.
5. Spray the plate with modified Dragendorff's reagent.

23.39.6 HPLC

The alkaloid mixture can be separated by HPLC on RP Nucleosil CB column using acetonitrile CH_3CN (pH 2.5):H_2O (pH 2.5) with a flow rate of 2 mL/min and UV detector at 254 nm.

Dragendorff's reagent

Solution A: Bismuth nitrate (0.17 g) in glacial acetic acid (2 mL) and H_2O (8 mL)

Solution B: Potassium iodide (4 g) in glacial acetic acid (10 mL) and H_2O (20 mL)

Mix solutions A and B and dilute to 100 mL with H_2O.

23.40 Genetic Transformation of Plants through *Agrobacterium tumefaciens* as Cloning Vector

23.40.1 Introduction

A. tumefaciens, a soil bacterium, has the ability to infect most dicotyledonous plants, usually at a wound site. The tissue around the infected wound develops a neoplastic growth known as a crown on hormone-free medium independent of the bacterium. In cultures, the crown gall tissue produces a set of metabolites termed opines (amino acids or sugar derivatives which plant cells do not metabolize normally but can be utilized as a carbon and nitrogen source by *A. tumefaciens* responsible for induction of the tumor). *Agrobacterium* harbors the Ti plasmid, which is directly responsible for tumor induction. The transfer of small DNA segments from this plasmid and their integration into the genome of host cells induces tumor formation in the plants. The property of the Ti plasmid DNA segment (T-DNA integrating with the genome of higher plants) has attracted worldwide attention for using *Agrobacterium* in genetic transformation techniques.

A. rhizogenes is another soil bacterium that produces hairy root disease in dicotyledonous plants used in plant genetic manipulations. Unlike *A. tumefaciens* tumors, the tumors produced by *A. rhizogenes* have the capability for regenerating mature fertile plants. *A. rhizogenes* induces the formation of proliferative multi-branched adventitious roots at the site of infection, so-called "hairy roots." In the rhizosphere, plants may suffer from wounds by soil pathogens or other sources. This leads to the secretion of phenolic compounds like acetosyringone that have chemotactic effects that attract the bacteria. Under such conditions, certain bacterial genes are turned on leading to the transfer of its T-DNA (TL-DNA and TR-DNA) from its root-inducing plasmid (Ri plasmid) into the plant through the wound. After integration and expression, *in vitro* or under natural conditions, the hairy root phenotype is observed, which typically includes overdevelopment of a root system that is not completely geotropic, and altered (wrinkled) leaf morphology if leaves are present.

Genes within the T-DNA are transcribed and translated by the plant to produce tumor growth. There are genes for the production of cytokinins affecting cell division. Transformation of the plant with cytokinin biosynthetic genes produces the unregulated cell division resulting in the development of a tumor. There are few genes for the synthesis of nonprotein amino acids called opines used as a source of carbon by bacteria and probably also influence bacterial mating.

The objective of this experiment is

1. The induction of crown gall tumor by infecting a plant with wild-type *A. tumefaciens*.
2. Assay opine production in tumor tissue.

Notes: *A. tumefaciens* is a plant pathogen and a serious pest on some crop plants. All cultures and contaminated materials (including syringes, blades, swabs, etc.) should be autoclaved before disposal.

23.40.2 Materials

Plant tumor tissue, razor blade or scalpel, microfuge tube 1.5 mL (one for each tumor assayed), glass rod (3 mm diameter), sterile sand, paper chromatography apparatus, ninhydrin spray, microfuge, hair dryer, Nopaline standard solution, lambda pipette, Whatman No. 1 chromatographic paper/TLC plate, *n*-butanol, glacial acetic acid, distilled water, sprayer, YEP medium (10 g bactoyeast extract, 5 g NaCl per liter).

23.40.3 Protocol

1. *Experiment (week 1)*: Inoculate tobacco plant (*Nicotiana* species) with bacterium using a syringe or razor blade, make shallow wounds on the leaf, petiole, or stem, and observe tumor formation over a period of weeks. Cover the inoculation site with a piece of clear tape. Attach a label bearing the date and your name to the inoculated site. Return the plant to the growth chamber, greenhouse, or window and observe tumor formation over a period of 2–3 weeks.

2. *Assay of opine production*: Remove a small piece of each tumor to be assayed from the plant (not more than 0.1 g) and transfer each to microfuge tube. Also include an uninfected plant sample as a control. For each sample, moisten the tip of a glass rod with water, touch on the sand to pick up a small amount, and use it to grind tumor or control tissue in its tube. Spot the supernatant on paper chromatogram/TLC plate and air dry along with a standard of nopaline (2–5 µL of a 500 µg/mL stock). Nopaline is a chemical compound derived from the amino acids glutamic acid and arginine. It is classified as an opine. Develop the chromatogram by descending paper chromatography on Whatman No. 1 chromatographic paper or through TLC with a solvent system in a chromatography tank of 4:1:1 of butanol:acetic acid:water. Remove the chromatogram/TLC plate, air dry it, and spray with 0.25% ninhydrin in acetone to reveal the production of nopaline.

23.41 Assay of β-Glucuronidase in Transformed Plant Tissue

23.41.1 Introduction

This is a continuation of the earlier experiment. In order to identify transformed cells or plants that have been growing on a selective medium, it is necessary to have an easily assayable reporter gene. The most useful reporter genes encode an enzyme activity not found in the organism being studied. A number of genes currently are being used; however, one of the most popular is the β-glucuronidase (*GUS*).

23.41.2 Taming Agrobacterium for Use in Biotechnology

A. tumefaciens C58 is a nopaline-producing strain. The Ti plasmid has been replaced in this strain with two smaller plasmids. This type of construction is referred to as a "binary vector." One of the binary vector plasmids (here called pTiC58) is a shortened version of the Ti plasmid. The second of the binary plasmids (here, pMSG) contains T-DNA which has been altered by the insertion of a marker gene, a transformation. The marker gene in pMSG is the *GUS* gene. GUS has several properties that make it useful as a marker. First, the enzyme GUS is not normally found in plants. This means that false positive results are a rare event. Second, different substrates are available for the fake enzyme that produce either a visible histochemical stain (used in this experiment) or fluorescent product detectable at lower levels upon cleavage by the enzyme. In addition to the features already mentioned, each of the plasmids in the binary system carries genes for resistance to antibiotics. pTiC58 makes the bacterium resistant to spectinomycin while pMSG conditions resistance to tetracycline. The inclusion of the two antibiotics in the bacterial culture medium ensures that only bacterial cells containing both plasmids will grow. During the transformation process, only the T-DNA and the *GUS* gene cloned into it are transferred to the plant cell and integrated into the plant genome.

23.41.3 Materials and Protocols

A. tumefaciens culture-strain C58 containing the plasmids pTiC58 and pMSG, radish roots, razor blade or scalpel, Petri dish of plant medium, and X-gluc stain.

Note: This strain of *Agrobacterium* can be obtained by contacting Dr. Mark Holland, Department of Biology, Salisbury State University, Salisbury, MD. Before requesting for the bacteria, you must contact your state Department of Agriculture for information on regulations that apply to distribution and use of this plant pathogen.

Equipment: Autoclave, spectrophotometer, shakers, and incubator at 37°C.

Plant material: Radish root, A. tumefaciens strain, bacterial media, YEP (recipe earlier), M9 + (per liter), 200 mL 5 × salts (recipe later), 50 mL 20% glucose, 12.5 mL 20% casamino acids, 1 mL 1 M $MgSO_4 \cdot 7H_2O$, and 0.1 mL 0.5% thiamine HCl.

Filter sterilize glucose and thiamine through a 0.45 mm filter, autoclave the rest separately, mix the components, adjust pH to 5.5, filter, and sterilize.

5× salts (per 200 mL), 2.7 g Na_2HPO_4, 15 g KH_2PO_4, 25 g Na_2HPO_4, 5 g NH_4Cl, and 2.5 g NaCl.

X-gluc stain (50 mL) make fresh before use: 25 mL 0.2 M Na_2HPO_4 buffer pH 7.0, 23.5 mL sterile water, 0.25 mL 0.1 M $K_3Fe(CN)_6$, 0.25 mL 0.1 M $K_4Fe(CN)_6 \cdot 3H_2O$, 1 mL 0.5 M Na_2EDTA, 50 mL X-gluc stock solution, and store at 4°C in the dark.

Make a stock solution of 0.1 mg/mL glue in *N,N*-dimethylformamide.

Other solutions: 0.5 M Acetosyringone in DMSO antibiotics for plasmid selection 70% Ethanol, 5% bleach, sterile water, MS salts, 0.75% agar, 0.5× MS plates.

23.41.4 Observation

Inoculate radish slices with *A. tumefaciens* and observe. Stain radish slices of GUS activity.

23.42 Random Amplified Polymorphic DNA Analysis

Polymorphism obtained by random amplified polymorphic DNA (RAPD) can be used as the Mendelian genetic markers. The amplified bands can be used for genomic finger printing. RAPD is also helpful in genetic diversity studies.

23.42.1 Introduction

RAPD is a PCR-based technique in which the random primers are used. The primers will bind to a specific sequence of DNA and the amplification takes place randomly. The binding of primers to DNA differs from genotype to genotype. The banding pattern outline after amplification can be used as markers in characterization of different genotypes.

The different steps involved are denaturation (94°C for 1 min), annealing of primers (36°C for 2 min), and complementary strand synthesis (72°C for 1.5 min).

23.42.2 Materials

1. RAPD primers can be purchased from the company, diluted to 15 mg/μL with TE, and store at 20°C
2. 10× PCR buffer (explained for PCR)
3. 50 mM $MgCl_2$
4. 5 mM dNTPs aliquot 50 μL each of dNTPs (dATP, dCTP, dTTP) from 100 mM stocks and add 800 mL of sterilized water
5. 10 mg/mL ethidium bromide (store in a brown bottle or wrapped with aluminum foil at 4°C)
6. 10× TBE stock solution (108 g Tris base, 55 g boric acid, 40 mL 0.5 M EDTA pH 8.0, distilled water/liter)
7. 50× TAE stock solution (242 g Tris base, 57.1 mL glacial acetic acid, 100 mL 0.5 M EDTA pH 8.0 diluted to get 1 L)

23.42.3 Protocol

RAPD is performed under two main categories:

1. DNA isolation
2. RAPD analysis

DNA isolation has been discussed in Chapters 11 and 19. The isolated DNA is stored at 20°C.

23.42.4 RAPD Analysis

1. Take out the template DNA from –20°C and keep it on ice.
2. Store all the stocks used for PCR at –20°C. Take out these stocks and allow them to be thawed by keeping them on ice.
3. Prepare a total volume of 15 μL of the reaction mixture by adding the following components:

Components	Volume (μL) (If Using Lab Taq)	Volume (μL) (If Using Company Taq)
Primer (15 ng/μL)	1.2	1.2
dNTPs (5 mM)	0.75	0.75
PCR buffer (10×)	1.5	1.5
MgCl$_2$	0.9	0.9
Taq polymerase	0.3	0.12
Sterile water	9.35	9.53
DNA (25 mg/μL)	1.0	1.0
Total volume	15	15

4. Add Taq polymerase at the end and do not spin the mixture after adding the enzyme.
5. Add 1.0 μL of template DNA (25 mg/μL) to the tube.
6. Take 14.0 μL reaction mixture into 0.5 mL tube.
7. Likewise, prepare the required number of the samples with the 0.5 mL Eppendorf tubes.
8. Label the tubes with unerasable marker the name of the sample and the primer.
9. Program for RAPD machine.

File #	Step	Temperature	Time	No. of Cycles
50	Initial denaturation	94°C	2 min	1
51	Denaturation	94°C	30 s	45
	Annealing	37°C	30 s	45
	Extention	72°C	1 min 6 s	45
52	Extended extention	72°C	7 min	1
53	Soak	4°C	Until recovery	

10. To run the program:
    a. Place the tubes in the wells.
    b. Press: Enter.
    c. Enter: File # 100 (all the files 50, 51, 52, 53 are linked sequentially in the program no. 100).
    d. Press: Enter.
    e. Enter: 15 vol.
    f. Press Enter (machine will start from file no. 50).

11. Resolve the amplified PCR production under 1.5% agarose gel.
12. Carry out the electrophoresis in either 1× TAE or TBE buffer.
13. Take out the samples from thermocycler and add 3 mL of loading dye (sample buffer) and spin it for uniform mixing with the dye.
14. Load the samples on the agarose gel and the electrophoresis is carried out at 100 V to 4–5 h.
15. Stain the gel by immersing in a liter of the staining solution which consists of 20 mL ethidium bromide (10 mg/mL) in a liter of water.
16. The gel is incubated for 25 min, and destain gel with distilled water for 20–25 min.
17. Visualize the bands under UV light and can be photographed.
18. Follow scoring for the presence or the absence of a particular band for the data analysis.

23.43 Efficiency of Rhizobia for Nodulation in Leguminous Crops

23.43.1 Introduction

Interaction between microorganisms and higher plants is of varied type, ranging from the usually beneficial effects of soil microorganisms on soil fertility and hence on plant growth, to destructive parasitism, resulting in damage or death of the plant. The Leguminosae is a large family of about 750 genera and 20,000 species. However, not all species or genera associate with *Rhizobium* to form root nodules. The legumes are important in agriculture not only for the fact that they fix nitrogen and can thus conserve nitrogen fertilizers, but also because the plants and seeds are high in protein. The seeds of grain legumes are important nutritionally for humans, and the forage legumes (clovers, lucerne, or alfalfa and vetches) also supply a high-protein diet to livestock. Legumes also have ecological significance as they provide nitrogen inflow to natural habitats such as tropical forests (Figure 23.35). Some tropical legume trees are grown for their wood, and the nitrogen-enriched soil is exploited when the wood is removed. Because there is such a wide range of plants in this family, nodule form and modes of infection will differ in detail according to the host species and rhizobial strain (Figure 23.36).

Legumes that resist infection and nodulation are inhibited if the nitrogen content of the soil is high. When the soil is depleted of nitrogen, rhizobia are recognized, and infection occurs. The recognition process is thus controlled by the nitrogen status of the environment through its interaction with the plant. Rhizobia have a restricted host range; each strain can only infect and nodulate a limited number of legume species so that there must be a high degree of specificity in the recognition process, and the specificity is expressed by suitable molecules, which are proteins and polysaccharides. The range of sugars in the surface polysaccharides of both plant and bacterium is wide enough to provide sufficient diversity. If the sugars are the recognition signals, then there must be a mechanism to distinguish specific sugars and to mediate the recognition response between plant and bacterium. Lectins, proteins that have several sites, will bind sugar molecules could act in this way, since different

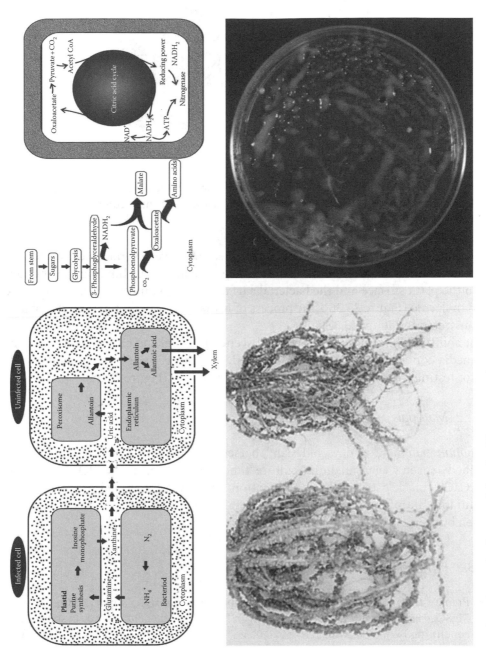

Figure 23.35 Physiology of infection of rhizobial/legume symbiotic association.

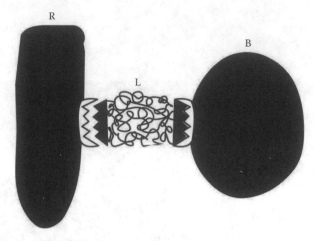

Figure 23.36 The mechanism of lectin binding. The specific binding sites on the lectin attach to sugars on the *Rhizobium* and root hair surfaces. L, lectin; R, root; B, bacterium.

lectins show specificity for different sugars. Lectins have several binding sites for sugars and can thus attach the bacterium to the root surface by binding specific sugars on the plant surface and the bacterial surface through binding sites on the same lectin molecule (Figure 23.36). The process of invasion of the root hair by rhizobia is demonstrated in Figure 23.37.

The ability of *Rhizobium* to infect small-seeded legumes can be tested on nitrogen-free media containing agar, vermiculite, sand, etc. There are several methods available for testing *Rhizobium* isolates for nodulation ability (Figure 23.38).

23.43.2 Materials

Rhizobium trifolii (2–4 days old slant, berseem [*Trifolium alexandrinum*] seeds, modified Jensen's agar medium, culture tube (150 × 20 mm), $NaClO_3$, dessicator with $CaCl_2$, black paper, sterile water, cotton, specimen tubes, sterile forceps, aluminum foil, illuminated growth room maintained at 28°C ± 2°C, inoculating loop, bunsen burner.

23.43.3 Protocol

1. Prepare the medium of the following composition: $CaHPO_4$ 1 g, K_2HPO_4 0.2 g, $MgSO_4 \cdot 7H_2O$, NaCl 0.2 g, $FeCl_2$ 0.1 g, agar 15 g, trace elements 10 mL (a solution containing Bo 0.05%, Mn 0.05%, Zn 0.005%, Mo 0.005%, Cu 0.002%), and distilled water to 1 L.
2. Weigh the constituents and dissolve (except K_2HPO_4 and agar) in 200 mL distilled water, dissolve K_2HPO_4 in 100 mL of distilled water separately. Dissolve agar in 400 mL of warm distilled water. Mix all the constituents and raise the volume to 1 L by the addition of distilled water. Adjust pH to 6.5–7.

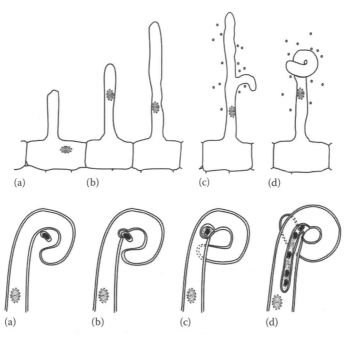

Figure 23.37 The invasion of the root hair by *Rhizobium*: (a) the bacterium becomes enclosed in the curled root hair, (b) the bacterium penetrates the root hair and is surrounded by cell wall material, and (c) initial infection thread growth. (d) The bacteria divide within the infection thread.

3. Place the test tubes and seeds in a desiccator over $CaCl_2$ for 24 h.
4. Surface sterilize the seeds with 1% $NaClO_3$ by incubating for 10–20 min in solution and then wash with sterile water with quick changes four to five times to avoid any damage of the seed coat.
5. Pour 13–15 mL of agar medium into each culture tube.
6. Plug the tubes with loose cotton plugs.
7. Sterilize the plugged tubes at 15 lb/in.^{2} pressure (121°C) for 20 min.
8. Keep the tubes in a slanting position in such a way that the slope of the agar reaches the top of the tube and allow the medium to solidify.
9. Cover the top of each tube by thin circles of aluminum foil held in place with a rubber band.
10. Aseptically make a small aperture on the aluminum foil with a sterile knife and fit the aperture with cotton wool for aeration and for replenishing nutrient solution in tubes.
11. Transfer aseptically, using a pair of sterile forceps, grown seedlings in a second tube (opposite to the hole carrying the cotton plug) of the test tube in such a way that the root system lies on the agar slope, and the shoot system comes out of the tube.
12. Cover the tubes carrying seedlings with large specimen tubes containing moist sterile cotton to prevent desiccation of the seedlings.
13. Allow the seedlings to establish in the agar slopes.
14. Remove the cover tube after the establishment of the seedlings.

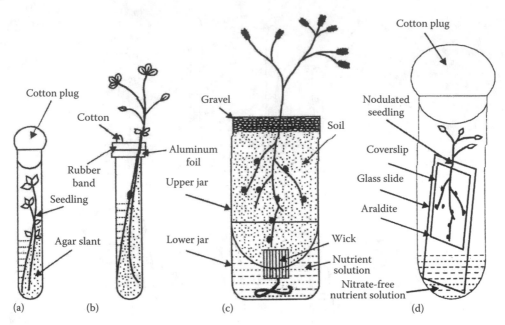

Figure 23.38 Four commonly used techniques for testing nodulation ability by rhizobia: (a) Jensen's seedling agar method, (b) agar seedling method modified by Gibson, (c) Leonard jar assembly for growing large-seeded legumes, and (d) Fahraeus seedling tube for studying infection of legume root by rhizobia.

15. Cover the lower portion of the tube with black paper to avoid light from falling on the root system.
16. Make suspension of *Rhizobium* from scrapings of growth from 2 to 4 day old fresh slants.
17. Mix the bacterial suspension with ¼ strength nitrogen-free nutrient solution (medium without agar).
18. Pour 5–10 mL of the *Rhizobium* nutrient solution mixture into each tube from the second hole.
19. Incubate the inoculated tube in growth cabinets for 3–4 weeks.

23.43.4 Results and Observations

Remove the seedlings from the inoculated tubes periodically for the development of nodules on the roots. Record the time taken for the development of nodules postinoculation and number of nodules formed per plant. Cut off a nodule from a legume, stain the bacterium with methylene blue for 30 s, and observe under oil immersion for the cells of *Rhizobium*.

A loopful of the crushed nodule may be streaked on mannitol–yeast agar and incubate at room temperature for 7 days. Compare the growth and stained preparation of the isolate with the original culture used for inoculation.

23.44 Seed Inoculation with Rhizobia

23.44.1 Introduction

Several properties of agriculturally important microbes are being explored to enhance crop yields. Rhizobia are present in root nodules of legumes, because of their ability to fix nitrogen, they are being used as biofertilizers to reduce the need for expensive chemical fertilizers. Seed inoculation of rhizobia is done by mixing the rhizobia culture in 10% sugar and 40% gum Arabic to form a slurry to which seeds are added. With the result, a uniform coat of the *Rhizobium* culture is formed around the seeds. The inoculated seeds are dried in the shade and sown immediately.

23.44.2 Materials

Culture of *Rhizobium japonicum*, *Glycine max* (soybean) seeds, yeast extract mannitol agar medium, 10% sugar solution, 40% gum Arabic solution, and needle or scalpel

23.44.3 Protocol

1. Inoculate yeast extract mannitol agar medium plates with *Rhizobium* culture.
2. Incubate the plates at 26°C for 10 days.
3. Using a scalpel or needle, scrap the growth of the bacterium from the plates and suspend in water.
4. Take 50 g of sugar in 500 mL of water.
5. Heat the sugar solution for 15 min.
6. Add 200 g of gum Arabic into hot sugar solution.
7. Allow the solution to cool at room temperature.
8. Mix the inoculum prepared in Step 3 to form a slurry.
9. Add soybean seeds to inoculum slurry.
10. Mix the seeds with inoculum slurry by hand so as to uniformly coat the seeds with the inoculants.
11. Transfer the seeds on sheet of paper.
12. Sow the seeds in the fields (Figure 23.39).

23.44.4 Observations and Results

Observe the roots of rhizobia inoculated and noninoculated (control) plants for the production of nodules after 10–15 days of sowing of seeds in the fields.

Compare the yields of rhizobia inoculated and noninoculated fields and record the per cent increase in yields due to the addition of *Rhizobium* as biofertilizer.

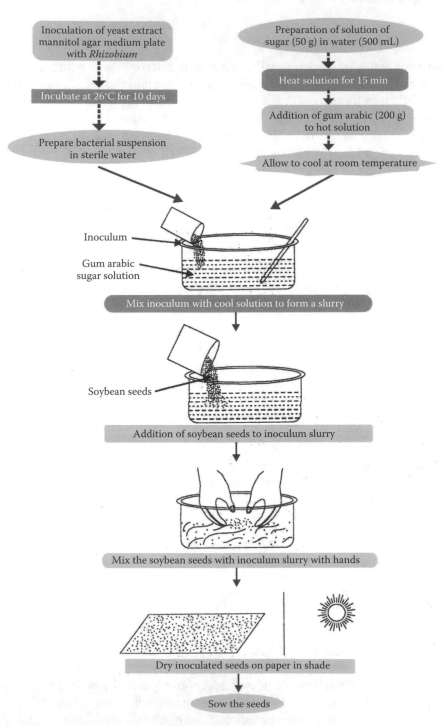

Figure 23.39 Protocol for inoculating soybean seeds with rhizobia.

Suggested Readings

Bhojwani, S.S. and M. K. Razdan. 1996. *Plant Tissue Culture: Theory and Practice*. Oxford, U.K.: Elsevier.

Debnath, M., C. P. Malik, and P. S. Bisen. 2006. Micropropagation: A tool for the production of high quality plant-based medicines. *Curr Pharm Biotechnol* 7: 33–49.

Dixon, R.A. 1985. *Plant Cell Culture. A Practical Approach*. Oxford, U.K.: IRL Press.

Edwin, F., M. A. Hall, and D. K. Geert-Jan, eds. 2008. *Plant Propagation by Tissue Culture*. Dordrecht, the Netherlands: Springer.

Murashige, T. and F. Skoog. 1962. A revised medium for rapid growth and bioassay with tobacco tissue culture. *Physiol Plant* 15: 473–497.

Neuman, K.-H., A. Kumar, and J. Imani. 2009. *Plant Cell and Tissue Culture: A Tool in Biotechnology*. Dordrecht, the Netherlands: Springer.

Razdan, M. K. 2003. *Introduction to Plant Tissue Culture*. New Delhi, India: Science Publishers.

Sathyanarayana, B. N. 2007. *Plant Tissue Culture: Practices and New Experimental Protocols*. New Delhi, India: I.K. International Pvt. Ltd.

Smith, R. H. 2000. *Plant Tissue Culture, Techniques and Experiments*, 2nd edn. London, U.K.: Academic Press.

Vasil, I. K. and T. A. Thorpe. 1994. *Plant Cell and Tissue Culture*. Dordrecht, the Netherlands: Kluwer Academic Press.

Yeoman, M. M. and A. J. Macleod. 1977. Tissue (callus) culture techniques. In: *Plant Tissue and Cell Culture*, H. E. Street, ed. Oxford, U.K.: Blackwell Scientific Publications, pp. 31–59.

Important Links

http://www.ncbe.reading.ac.uk/ncbe/protocols/pdf/ptc2002.pdf

http://www.scribd.com/doc/35044234/Importance-of-Plant-Tissue-Culture

http://www.btfskarnataka.org/courses/courses-offered/plant-tissue-culture-and-micropropagation.html

http://link.springer.com/protocol/10.1007/978-1-61779-818-4_19

http://www.agriquest.info/index.php/micropropagation/127-micropropagation

http://www.iisc.ernet.in/currsci/nov25/articles6.htm

http://link.springer.com/protocol/10.1007/978-1-61779-818-4_2

24

Animal Cell Science and Technology

24.1 Introduction

Cell culture is a technique that involves the isolation and maintenance *in vitro* of cells isolated from the tissues or whole organs derived from animals, microbes, or plants. The cells are made free from tissue by mechanical, enzymatic, or chemical means. The dispersed cells are resuspended in an appropriate nutrient medium at optimum conditions. These cells are inoculated into culture vessels and then incubated in a stationary position. After settling out, the cells adhere to and stretch out on the surface of the vessel and transform into morphology very close to *in vivo*. These cells multiply until a confluent sheet layer is formed. In general, animal cells have more complex nutritional requirements and usually need more stringent condition for growth and maintenance. By comparison, microbes and plants require less rigorous conditions and grow effectively with the minimum of needs. Regardless of the sources of material used, in practice, cell culture is governed by the same general principles, requiring a sterile pure culture of cells, needs to adopt appropriate aseptic techniques and the utilization of suitable conditions for optimal viable growth of cells. The primary objective in cell culture procedures is to disrupt/digest intercellular material that binds cells together. The deaggregation of cells of some organ can be achieved by perfusion of organ with an enzyme (e.g., trypsin, collagenase) or a chemical. However, some organs are not easily perfused and, therefore, alternate procedures are adopted viz. mincing the tissue into small fragments with sharp scissors, scalpels, razor blades, or cataract knives. Mechanical dispersion involves pressure homogenization and passage through a nylon or stainless mesh is applicable when soft tissues especially those from young embryo are employed. Animal cell culture techniques have vast applications in the field of pharmaceutical biotechnology, virology and vaccine research, and cancer biology. Cell culture models are used for examining intestinal absorption of drugs or peptide absorption across the blood–brain barrier (BBB), hepatic absorption of drugs for *in vitro* experimentation, etc. Once established, cells in culture can be exploited in many different ways. For instance, they are ideal for studying intracellular processes including protein synthesis, signal transduction mechanisms, and drug metabolism. They have also been widely used to understand the mechanisms of drug actions, cell–cell interaction, and genetics. Additionally, cell culture technology has been adopted in medicine, where genetic abnormalities can be determined by chromosomal analysis of cells derived, for example, from expectant mothers. Similarly, viral infections can be assayed both qualitatively and quantitatively on isolated cells in culture. In industry, cultured cells are used routinely to test both the pharmacological and toxicological effects of pharmaceutical compounds. This technology thus provides a valuable tool to scientists, offering a user-friendly system that is relatively cheap to run and the exploitation of which avoids the legal, moral, and ethical questions generally associated with animal experimentation.

24.2 Chronology of Cell Culture Development

The cell types used in cell culture fall into two categories generally referred to as either a primary culture or a cell line.

24.2.1 Primary Cell Culture

Primary cultures are cells derived directly from tissues following enzymatic dissociation or from tissue fragments referred to as explants. These are usually the cells of preference since it is argued that primary cultures retain their characteristics and reflect the true activity of the cell type *in vivo*. The disadvantage in using primary cultures, however, is that their isolation can be labor-intensive and may produce a heterogeneous population of cells. Moreover, primary cultures have a relatively limited life span and can be used over only a limited period of time in culture. Cells are derived from the tissues and organs of animals with a diploid karyotype. These cells when culture *in vitro* have the tendency to attach onto the surface of the vessel and after a definite time they attain the *in vivo* morphology. Such cells divide mitotically and remain identical to their progenitors. A progenitor cell is a biological cell that, like a stem cell, has a tendency to differentiate into a specific type of cell, but is already more specific than a stem cell and is pushed to differentiate into its "target" cell. The most important difference between stem cells and progenitor cells is that stem cells can replicate indefinitely, whereas progenitor cells can divide only a limited number of times. These cells retain their ability to recognize each other when they come in contact. This is attributed to their identical subcomponents; genetically identical cells show a ruffling movement toward each other when they come in contact, ruffling, stops which is termed as coinhibition of movement. A chronological pattern of growth of human cells cultured *in vitro* is shown in Figure 24.1, in which phase I terminates with the formation of the first confirm culture. Phase II shows the continuous subdivision of cells while phase III represents the degeneration of cell culture.

Primary cultures can be obtained from many different tissues and the source of tissue used generally defines the cell type isolated. For instance, cells isolated from the endothelium of blood vessels are referred to as endothelial cells whilst those isolated from the medial layer of the blood vessels and other similar tissues are smooth

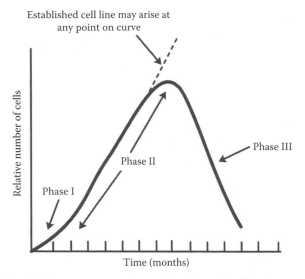

Figure 24.1   Graphical representation of growth of animal cell culture.

muscle cells. Although both can be obtained from the same vessels, endothelial cells are different in morphology and function, generally growing as a single monolayer characterized by a cobble-stoned morphology. Smooth muscle cells, on the other hand, are elongated, with spindle-like projections at either end and grow in layers even when maintained in culture. In addition to these cell types, there are several other widely used primary cultures derived from a diverse range of tissues, including fibroblasts from connective tissue, lymphocytes from blood, neurons from nervous tissues, and hepatocytes from liver tissue.

24.2.2 Continuous Cell Lines

Cell lines consist of a single cell type that has gained the ability for infinite growth. This usually occurs after transformation of cells by one of several means that include treatment with carcinogens or exposure to viruses such as the monkey Simon virus 40 (SV40), Epstein-Barr virus (EBV), or Abelson murine leukemia virus (A-MuLV) among others. These treatments cause the cell to lose their ability to regulate growth. As a result, transformed cells grow continuously and, unlike primary culture, have an infinite life span (become "immortalized"). The drawback to this is that transformed cells generally lose some of their original *in vivo* characteristics. For instance, certain established cell lines do not express particular tissue-specific genes. One good example of this is the inability of liver cell lines to produce clotting factors. Continuous cell lines, however, have several advantages over primary cultures, not least because they are immortalized. In addition, they require less serum for growth, have a shorter doubling time, and can grow without necessarily needing to attach or adhere to the surface of the flask.

Many different cell lines are currently available from various cell banks, which make it easier to obtain these cells without having to generate them. One of the largest organizations that supply cell lines is the European Collection of Animal Cell Cultures (ECACC) based in Salisbury, United Kingdom.

24.2.2.1 Established Cell Lines

Established cell lines are like tumor cells. Such cell lines are generated spontaneously as a rare occurrence in diploid cells. This process is called as transformation. Transformation can be induced in cell culture by the following ways:

1. A diploid culture cell transformed by some spontaneous event
2. A haploid cell transformed by adding a carcinogen to the culture
3. Transformation in haploid cells induced by virus

Cell cultures are grown by monolayer cultures and suspension cultures.

24.2.3 Monolayer Cultures

When cultures are grown on a supporting surface, they are called monolayer cultures, e.g., cultures grown in glass or plastic bottles, flasks, tubes, Petri dishes, microtiter plates, etc. After dispensing the cells in these vessels, they are incubated in a stationary condition so that the cells form a confluent monolayer.

24.2.4 Suspension Cultures

Cells are grown in suspension and are not allowed to attach to any support surface in suspension culture. Cell suspension is agitated continuously so that the cells remain in suspension. Such cultures are termed as suspension culture.

24.3 Cell Culture Laboratory

The design and maintenance of the cell culture laboratory is perhaps the most important aspect of cell culture since a sterile surrounding is critical for handling of cells and culture media, which should be free from contaminating microorganisms. Such organisms, if left unchecked, would outgrow the cells being cultured, eventually resulting in the culture-cell demise owing to the release of toxins and/or depletion of nutrient from the culture medium.

Where possible, a cell culture laboratory should be designed in such a way that it facilitates preparation of media and allows for the isolation, examination, evaluation, and maintenance of cultures under controlled sterile conditions. In an ideal situation, there should be a room dedicated to each of the above tasks. However, many cell culture facilities, especially in academia, form part of an open plan laboratory and as such are limited in space. It is not unusual, therefore, to find an open plan area where places are designated for each of the earlier functions. This is not a serious problem as long as a few basic guidelines are adopted. For instance, good aseptic techniques should be used at all times. There should also be adequate facilities for media preparation, sterilization and all cell culture materials should be maintained under sterile condition until used. In addition, all surfaces within the culture area should be nonporous to prevent adsorption of media and other materials that may provide a good breeding ground for microorganism, resulting in the infection of the cultures. Surfaces should also be easy to clean, and all waste generated should be disposed off immediately. The disposal procedure may require prior autoclaving of the waste, which can be carried out using pressurized steam at 121°C under 105 kPa for a defined period of time. These conditions are required to destroy microorganisms.

For smooth running of the facilities, daily checks should be made of the temperature in incubators and of the gas supply to the incubators, by checking the CO_2 cylinder pressure. Water baths should be kept clean at all times and areas under the work surfaces of the flow cabinets cleaned of any spills.

Cell culture is specifically conducted in a well-defined laboratory premises where near absolute sterility is maintained by all means and modes are available to avoid any possible cross contamination. A well self-contained laboratory (Figure 24.2)

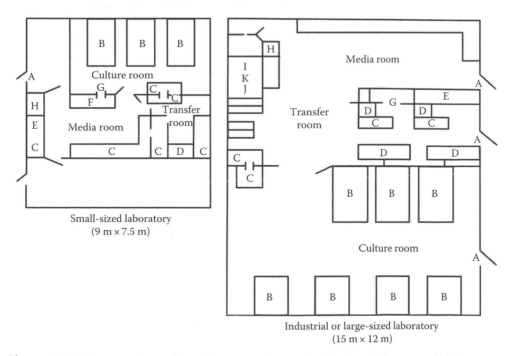

Figure 24.2 Design of a small- and large-sized animal cell culture laboratory. (A) Entrance, (B) shelve, (C) counter, (D) laminar airflow cabinet, (E) sinks, (F) gas outlet/burner, (G) window, (H) refrigerator and deep freezer, (I) store, (J) conference room, and (K) library.

for the purpose houses a range of equipments used commonly. Some important equipment with their principles and uses are discussed in the following sections.

24.3.1 Laminar Flow/Cell Culture Hoods

The cell culture hood is the central piece of equipment where all the cell handling is carried out and is designed not only to protect the cultures from the operator but in some cases to protect the operator from the cultures. These hoods are generally referred to as laminar flow hoods as they generate a smooth uninterrupted stream-lined flow (laminar flow) of sterile air that has been filtered through a high-efficiency particulate air (HEPA) filter. There are two types of laminar flow hood classified as either vertical or horizontal. The horizontal hoods allow air to flow directly at the operator and, as a result, are generally used for media preparation or when one is working with noninfectious materials including those derived from plants. The vertical hoods (also known as biological safety cabinets) are best for working with hazardous organisms since air within the hood is filtered before it passes into the surrounding environment.

Currently, there are at least three different classes of laminar flow used which offers various levels of protection to the cultures, the operator, or both and these are described in the following sections.

24.3.1.1 Class I Hoods

These hoods, as with the class II type, have a screen at the front that provides a barrier between the operator and the cells but yet allows access into the hood through an opening at the bottom of the screen (Figure 24.3). This barrier prevents too much turbulence to airflow from the outside and more importantly, provides good protection

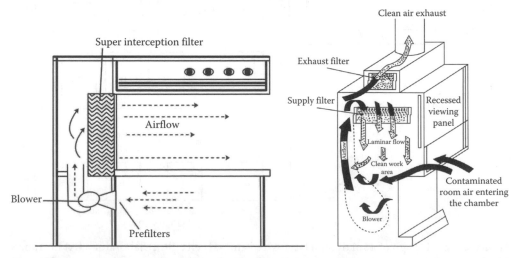

Figure 24.3   Diagrammatic representation of horizontal laminar airflow cabinet.

for the operator. Cultures are also protected but to a lesser extent when compared to the class II hoods as the air drawn in from the outside is sucked through the inner cabinet to the top of the hood. These hoods are suitable for use with low-risk organisms and when operator protection only is required.

24.3.1.2 Class II Hoods

Class II hoods are the most common units found in tissue culture laboratories. These hoods offer good protection to both the operator and the cell culture. Unlike with class I foods, air drawn from outsides passed through the grill in front of the work area and filtered through the HEPA filter at the hood before streaming down over the tissue culture (Figure 24.3). This mechanism protects the operator and ensures that the air over the cultures is largely sterile. These hoods are adequate for an animal cell culture that involves low to moderate toxic or infectious agents, but are not suitable for use with high-risk pathogens that may require a higher level of containment.

24.3.1.3 Class III Hoods

Class III safety cabinets are required when the highest levels of operator and product protection are required. These hoods are completely sealed, providing two glove pockets through which the operator can work with material inside the cabin. Thus, the operator is completely shielded, making class III hoods suitable for work with highly pathogenic organisms including tissue samples carrying known human pathogens.

24.3.1.4 Safety Guidelines

All hoods must be maintained in a clutter-free and clean state at all times as too much clutter may affect airflow and contamination will introduce infections. It is advisable to place only items that are required inside the cabinet and clean all work surfaces before and after use with industrial methylated spirit (IMS). The latter is used at an effective concentration of 70% (prepared by adding 70% v/v IMS to 30% Milli-Q water), which acts against bacterial and fungal spores by dehydrating and fixing cells, thus preventing contamination of cultures.

Some cabinets may be equipped with a short-wave ultraviolet light that can be used to irradiate the interior of the hood to kill microorganisms. When present, switch on the ultraviolet light for at least 15 min to sterilize the inside of the cabinet, including the work area. Ultraviolet radiation can cause adverse damage to the skin and eyes and precaution should be taken always to ensure that the operator is not in direct contact with the ultraviolet light when using this option to sterilize the hood. Once finished, ensure that the front panel door (class I and II hoods) is replaced securely after use. In addition, always turn the hood on for at least 10 min before starting work to allow the flow of air to stabilize. During this period, monitor the airflow and check all dials in the control panel at the front of the hood to ensure that they are within the safe margin.

24.3.2 CO₂ Incubators

CO_2 incubators are required to facilitate optimal cell growth under regulated conditions, normally requiring a constant temperature of 37°C and an atmosphere of 5%–10% CO_2 plus air and are water-jacketed. CO_2 maintains the required physiological pH (usually pH 7.2–7.4) of the culture medium. CO_2 is supplied from a gas cylinder into the incubator through a valve triggered to draw in CO_2 whenever the levels fall below the set value of 5% or 10%. The CO_2 enters the inner chamber of the incubator and dissolves into the culture medium containing bicarbonate. The latter reacts with H^+ (generated from cellular metabolism), forming carbonic acid, which is in equilibrium with water and CO_2, thereby maintaining the pH in the medium at pH 7.2:

$$HCO_3^- + H^+ \rightarrow H_2CO_3 \rightarrow CO_2 + H_2O$$

These incubators are generally humidified by the inclusion of a tray of sterile water to the bottom deck. The evaporation of water creates a highly humidified atmosphere and prevents evaporation of medium from the cultures.

An alternative to humidified incubators is the dry nongassed, not humidified unit and relies on the use of alternative buffering systems such as 4(2-hydroxyethyl)-1-piperazine-ethanesulfonic acid (HEPES) or morpholinopropane sulfonic acid (MOPS) for maintaining a balanced pH within the culture medium. The advantage of this system is that it eliminates the risk from infections that can be posed by the tray of water in the humidified unit. The disadvantage, however, is that the culture medium will evaporate rapidly, thereby stressing the cells. One way round this problem is to place the cell culture plate in a sandwich box containing little pots of sterile water. With the sandwich box lid partially closed, evaporation of water from the pots will create a humidified atmosphere within the sandwich box, thus reducing the risk of evaporation of medium from the culture plate.

24.3.2.1 Safety Guidelines

The incubator should be maintained at 37°C and supplied with 5% CO_2 at all times by keeping a thermometer in the incubator, preferably on the inside of the inner glass door, checked on a regular basis, and adjustments made as required. CO_2 level inside the unit can be monitored and adjusted by using a gas analyzer such as the Fryrite Reader, and checked on a regular basis on the levels of CO_2 in the gas cylinders and replaced when levels are very low. Most incubators are designed with an in-built alarm that sounds when the CO_2 level inside the chamber drops. At this point, the gas cylinder must be replaced immediately to avoid stressing or killing the cultures. It is now possible to connect two gas cylinder change over a unit that switches automatically to the second source of gas supply when the first is empty. It is, therefore, advisable to use this device where possible.

In a humidified incubator, it is essential that the water tray is maintained and kept free from microorganisms by adding various agents to the water such as the antimicrobial agent 1% (w/v) Roccal, or Thimerosal or Sigma Clean from Sigma Aldrich. The interior of the incubator should be cleaned regularly with 70% IMS. Copper-coated incubators reduce the microbial contamination due to the antimicrobial properties of copper.

24.3.3 Microscopes

Inverted phase contrast microscope is routinely used to visualize cells in culture. They are easy to operate with a light source located above and the objective lenses below the stage on which the cells are placed. Visualization of cells by microscopy can provide useful information about the morphology and state of the cells. Early signs of cells stress may be easily identified, and appropriate action is taken to prevent loss of cultures.

24.3.4 Centrifuges

The centrifuges used in cell culture laboratory range from bench models for sedimenting cells to ultracentrifuges for sedimenting small viruses. The performance of the centrifuges is rated as the relative centrifugal force (RCF) expressed in terms of G, which is calculated from the following formula:

$$RCF(G) = 1.118 \times 10^{-5} \times R \times N^2$$

where
   R is the distance in centimeters from the center of the centrifuge shaft to the extended tip of the centrifuge tube
   N is the revolutions of the centrifuge head per minute (rpm)

It is obvious from the earlier formula that although the speed of the centrifuge (rpm) is the most important factor, the radius (R) of the centrifuge head also influences the rate of sedimentation. It is, therefore, advisable to state the speed of the centrifuge in G rather than in rpm to avoid errors in techniques caused by using centrifuges of different sizes (Figure 24.4).

24.3.5 Filters

A variety of filters are available, but the most commonly used in the laboratory are the Seitz filters with ESK grade. Seitz filter pads are used for sterilization by filtration of reagents and media. The Millipore filters consist of Millipore filter, filter holder, and filter pads with varying pore diameters. They are suitable for the sterilization by filtration of the reagents and media as well as for removing the microbial contamination in viral samples and also for determining the size of virus particles in a viral sample (Figure 24.5).

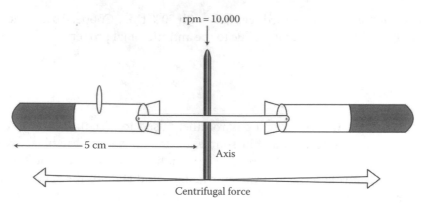

Figure 24.4 Running centrifuge.

24.3.6 Other Common Equipment

Several other pieces of equipment are required in cell culture: a water bath for thawing frozen sample of cells and warming media to 37°C before use, and a fridge and freezer for storage of media and other materials required for cell culture, balances, incubator (Figure 24.6), distillation apparatus, pH meter, roller cell culture apparatus, liquid nitrogen containers, and surgical kit. Glassware like pipette, culture tube, inoculation needle, and the container for oven sterilizing the glassware are required in an ideal cell culture laboratory (Figure 24.7).

24.3.6.1 Laboratory Accessories
Some cells need to attach onto a surface in order to grow and are, therefore, referred to as adherent. These cells are cultured in nontoxic polystyrene plastic that contains a biologically inert surface on which the cells attach and grow. Various types of plastics are available for this purpose and include Petri dishes, multiwell plates (with either 96, 24, 12, or 6 wells per plate), and screw cap flasks classified according to their surface area: T-25, T-75, T-225 (cm^2 of surface area). A selection of these plastics is shown in Figures 24.6 through 24.8.

24.3.7 Washing of Glassware

24.3.7.1 Washing of New Glassware
New glassware contain toxic products derived from the manufacturing process and packing and hence they should be carefully washed as suggested as follows:

1. Autoclave at 15 lb/in.^{2} pressure for 30 min to 1 h.
2. Wash several times with tap water.
3. Immerse in a 4% solution of hydrochloric acid in water overnight.
4. Rinse in tap water.
5. Rinse once in double distilled water.
6. Put immediately for drying in a hot air cabinet at 63°C–70°C.

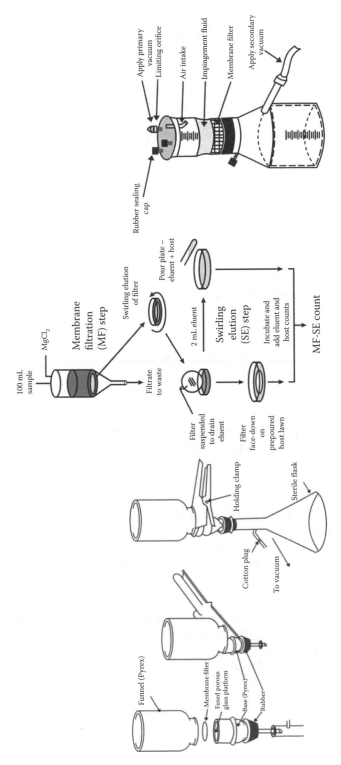

Figure 24.5 Microfiltration assemblies and microfiltration.

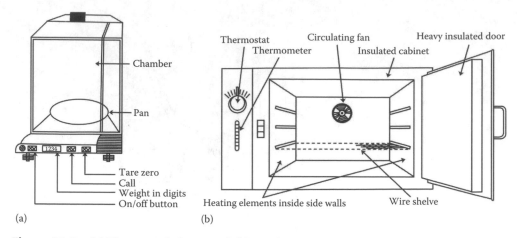

Figure 24.6 (a) Electronic balance and (b) incubator.

24.3.7.2 Washing of Used Glassware

1. Scrub the glassware with 10% soap solution.
2. Wash with tap water thoroughly.
3. Rinse in 0.05% HCl solution to neutralize the alkalinity of detergent and tap water.
4. Thoroughly wash with tap water.
5. Rinse several times with triple glass distilled water.
6. Put for drying in a hot air cabinet at 63°C–70°C.

24.3.7.3 Washing of Rubber Bungs

The earlier mentioned procedure is followed for rubber bungs, which are in use. New rubber bungs are subjected to the following treatment after autoclaving at 15 lb/in.^{2} pressure for 30 min:

1. Rinse in hot tap water.
2. Boil in 20% sodium hydroxide solution for 20 min.
3. Rinse in hot tap water.
4. Boil in 20% hydrochloric acid solution for 20 min.
5. Rinse in hot tap water. Rinse thoroughly in glass distilled water.

24.3.7.4 Sintered Glass Filters

1. Sintered glass filters should be cleaned with strong acids after use. Concentrated sulfuric acid with a few crystals of sodium nitrate and sodium chlorate is passed through the filters.
2. After this, the filter should be rinsed with a large volume of glass distilled water.

24.3.7.5 Glass Pipettes

1. Wash in tap water.
2. Boil in appropriate detergent solution for 10 min.
3. Wash in continuous flow pipette washer for 4 h.
4. Rinse in deionized water. Dry in oven.

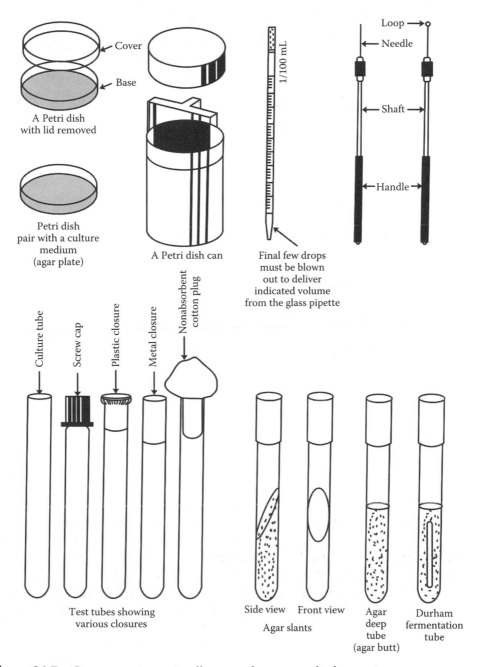

Figure 24.7 Common minor miscellaneous glassware and other equipments.

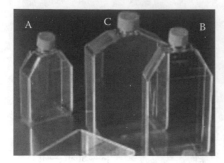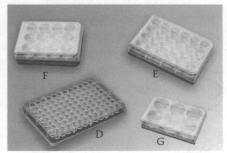

Figure 24.8 Tissue culture plastic ware used for cell culture (A–C) T flasks and (D–G) multiwell plates.

24.3.8 Disposable Syringes

Always use disposable sterile syringes.

24.3.9 Silicone Glassware

1. Remove the silicone grease from the glassware using benzene.
2. Place the glassware in grease remover for a few hours. Grease remover contains 150 mL of 50% potassium hydroxide and 850 mL methanol.
3. Proceed with the washing procedure in the normal way as described for used glassware.

24.4 Culture Medium

24.4.1 Cell Culture Media and Growth Requirements for Animal Cells

The cell culture medium used for an animal cell growth is a complex mixture of nutrients (amino acids, a carbohydrate such as glucose, and vitamins), inorganic salts (e.g., containing magnesium, sodium, potassium, calcium, phosphate, chloride, sulfate, and bicarbonate ions), and broad-spectrum antibiotics. In certain situations, it may be essential to include a fungicide such as amphotericin B, although this may not always be necessary. For convenience and ease of monitoring the status of the medium, the pH indicator phenol red may also be included. This will change from red at pH 7.2–7.4 to yellow or fuchsia as the pH becomes either acidic or alkaline, respectively.

The other key basic ingredient in the cell culture medium is serum, usually bovine or fetal calf. This is used to provide a buffer for the culture medium, but, more importantly, enhances cell attachment and provides additional nutrients and hormones like growth factors that promote healthy growth of cells. An attempt to culture cells in the absence of serum does not usually result in successful or healthy cultures, even though cells can produce growth factors of their own. However, despite these benefits, the use of serum is increasingly being questioned not least because of many of the other unknowns that can be introduced including infectious agents such as viruses and *Mycoplasma*. The more recent resurgence of "mad cow disease" (bovine spongiform encephalitis) introduced an additional drawback, posed a particular risk for the cell culturist, and increased the need for alternative products in this regard; several cell culture reagent manufacturers have now developed serum-free medium supplemented with various components including albumin, transferring, insulin, growth factors, and other essential elements required for optimal cell growth. This is proving very useful, particularly for the pharmaceutical and biotechnology companies involved in the manufacture of drugs or biological products for human and animal consumption.

Most commonly the balanced salt solutions (BSS) are used for washing cells and tissue fragments, for suspending enzymes that are used to deaggregate cells from tissue/organ and for separation of cells from glass and plastic surfaces and as a basic unit in formulating cell culture media. Some commonly used media used in animal cell culture are given in Table 24.1.

24.4.2 Preparation of an Animal Cell Culture Medium

Preparation of the culture medium is perhaps taken for granted as a simple straightforward procedure that is often not given due care and attention. As a result, most infections in cell culture laboratories originate from infected media. Following the simple yet effective procedures should prevent or minimize the risk of infecting the media when they are being prepared.

Preparation of the medium itself should also be carried out inside the culture cabinet and usually involves adding a required amount of serum together with antibiotics to a fixed volume of medium. The amount of serum used will depend on the cell type

TABLE 24.1 Composition of Various Mediums Used in Animal Cell Culture

Component	Mol. W.	Quantity (mg/L)					
		Ringer's	Earle's	Hank's	Dulbecco's	Eagle's MEM	CMF-PBS
NaCl	58.44	6500	6800	8000	8000	6800	8000
KCl	75.56	140	400	400	200	400	300
$CaCl_2 \cdot 2H_2O$	146.99		265	186			
$MgSO_4 \cdot 7H_2O$	110.99	120	200	140	100		
$MgCl_2 \cdot 6H_2O$	246.50		200	200			
Na_2HPO_4	203.31					200	
$Na_2HPO_4 \cdot 7H_2O$	141.96				1150		
$NaH_2PO_4 \cdot 7H_2O$	267.96			90			730
$NaH_2PO_4 \cdot H_2O$	138.0		140	60		1400	
KH_2PO_4	136.09		2200	350	200		200
$NaHCO_3$	84.01	200		1000		2000	
Glucose	180.16			10		1000	2000
Phenol red						10	

but usually varies between 10% and 20%. The most common antibiotics used are penicillin and streptomycin, which inhibit a wide spectrum of Gram-positive and Gram-negative bacteria. Penicillin acts by inhibiting the last step in bacterial cell wall synthesis whilst streptomycin blocks protein synthesis.

Once prepared, the mixture, which is referred to as complete growth medium, should be kept at 4°C until used. To minimize wastage and risk of contamination, it is advisable to make just the required volume of medium and use this within a short period of time. As an added precaution, it is also advisable always to check the clarity of the medium before use. Any infected medium, which will appear cloudy or turbid, should be discarded immediately. In addition to checking the clarity, a close eye should also be kept on the color of the medium, which should be red at physiological pH owing to the presence of phenol red. Media that looks acidic (yellow) or alkaline (fuchsia) should be discarded as these extremes will affect the viability and thus growth of the cells.

24.5 Cell Subculturing

Subculturing is the process by which cells are harvested, diluted in fresh growth medium, and replaced in a new culture flask to promote further growth. This process, also known as passaging, is essential if the cells are to be maintained in a healthy and viable state; otherwise, they may die after a certain period in continuous culture. The reason for this is that adherent cells grow in a continuous layer that eventually occupies the whole surface of the culture dish and at this point they are said to be confluent. Once confluent, the cells stop dividing and going into a resting state where they stop growing (senesce) and eventually die. Thus, to keep cells viable and facilitate efficient transformation, they must be subcultured before they reach full contact inhibition. Ideally, cells should be harvested just before they reach a confluent state.

Cells can be harvested and subcultured using one of the several techniques. The precise method used is dependent to a large extent on whether the cells are adherent or in suspension.

24.5.1 Subculture of Adherent Cells

Adherent cells can be harvested either mechanically using a rubber spatula (also referred to as a rubber policeman) or enzymatically using proteolytic enzymes. Cells in suspension are simply diluted in fresh medium by taking a given volume of cell suspension and adding an equal volume of medium.

24.5.2 Harvesting of Cells Mechanically

The method is simple and easy. It involves gently scraping cells from the growth surface into the culture medium using a rubber spatula that has a rigid polystyrene

Figure 24.9 Cell scrapers.

handle with a soft polyethylene scraping blade (Figure 24.9). This method is not suitable for all cell types as the scraping may result in membrane damage and significant cell death. Before adopting this approach, it is important to carry out some test runs where cell viability and growth are monitored in a small sample of cells following harvesting.

24.5.3 Harvesting of Cells Using Proteolytic Enzymes

Several different proteolytic enzymes can be exploited including trypsin, a proteolytic enzyme that destroys proteinaceous connections between cells and between cells and the surface of the flask in which they grow. As a result, harvesting of cells using this enzyme results in the release of single cells, which is ideal for subculturing as each cell will then divide and grow, thus enhancing the propagation of the cultures.

Trypsin is commonly used in combination with EDTA, which enhances the action of the enzyme. EDTA alone can also be effective in detaching adherent cells as it chelates the Ca^{2+} required by some adhesion molecules that facilitate cell–cell or cell–matrix interaction. Although EDTA alone is much gentler on the cells than trypsin, some cell types may adhere strongly to the plastic, requiring trypsin to detach.

The standard procedure for detaching adherent cells using trypsin and EDTA in Ca^{2+}/Mg^{2+} free phosphate-buffered saline (PBS). The growth medium is aspirated from confluent cultures and washed at least twice with a serum-free medium such as Ca^{2+}- or Mg^{2+}-free PBS to remove traces of serum that may inactivate the trypsin. The trypsin–EDTA solution (~1 cm^3/25 cm^2 of surface area) is then added to the cell monolayer and swirled around for a few seconds. Excess trypsin–EDTA is aspirated, leaving just enough to form a thin film over the monolayer. The flask is then incubated at 37°C in a cell culture incubator for 2–5 min but monitored under an inverted light microscope at intervals to detect when the cells are beginning to roundup and detach. This is to ensure that the cells are not overexposed to trypsin as this may result in extensive damage to the cell surface, eventually resulting in cell death. It is important, therefore, that the proteolysis reaction is quickly terminated by the addition of complete medium containing serum that will inactivate the trypsin. The suspension of cells is collected into a sterile centrifuge tube and spun at 1000 rpm for 10 min to pellet the cells, which are then resuspended in a known volume of fresh complete culture medium to give a required density of cells per cubic centimeter volume.

As with all tissue culture procedure, aseptic techniques should be adopted at all times. This means that all the earlier procedures should be carried out in a tissue culture cabinet under sterile conditions. Other precautions worth nothing include the handling of the trypsin stock. This should be stored frozen at −20°C and, when needed, placed in a water bath just to the point where it thaws. Any additional time in the 37°C water bath will inactivate the enzymatic activity of the trypsin. The working solution should be kept at 4°C once made and can be stored for up to 3 months.

24.5.4 Subculture of Cells in Suspension

For cells in suspension, it is important initially to examine an aliquot of cells under a microscope to establish whether cultures are growing as single cells or clumps. If cultures are growing as single cells, an aliquot is counted as described in Section 24.6 and then reseeded at the desired seeding density in a new flask by simply diluting the cell suspension with fresh medium, provided the original medium in which the cells were growing is not spent. However, if the medium is spent and appears acidic, then the cells must be centrifuged at 1000 rpm for 10 min, resuspended in fresh medium, and transferred into a new flask. Cells that grow in clumps should be centrifuged first and resuspended in fresh medium as single cells using a glass Pasteur or fine bore pipette.

24.6 Cell Quantification

It is essential that when cells are suspended they are seeded at the appropriate seeding density that will facilitate optimum growth. If cells are seeded at a lower seeding density, they may take longer to reach confluency and some may expire before getting to this point. On the other hand, if seeded at a high density, cells will reach confluency too quickly, resulting in irreproducible experimental results. This is because trypsin can digest surface proteins, including receptors for drugs, and these will need time (sometimes several days) to renew. Failure to allow these proteins to be regenerated on the cell surface may, therefore, result in variable responses to drugs specific for such receptors.

Several techniques are now available for quantification of cells, and of these the most common method involves the use of a hemocytometer. This has the added advantage of being simple and cheap to use. The hemocytometer itself is a thickened glass slide that has a small chamber of grids cut into the glass. The chamber has a fixed volume and is etched into 9 large squares, of which the large corner squares contain 16 small squares each; each large square measures 1 mm × 1 mm and is 0.1 mm deep (Figure 24.10).

Thus, with a coverslip in place, each square represents a volume of 0.1 mm^3 (1.00 mm^2 area × 0.1 mm depth) or 10^{-4} cm^3. Knowing this, the cell concentration (and the total number of cells) can, therefore, be determined and expressed per cubic centimeter. The general procedure involves loading ~10 µL of cell suspension into a clean hemocytometer chamber and counting the cells within the four corner squares

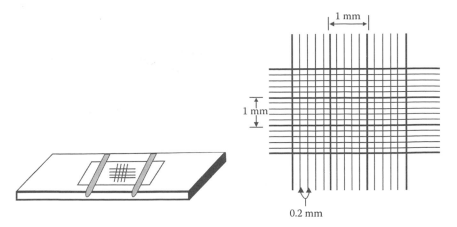

Figure 24.10   Hemocytometer.

with the aid of a microscope set at 20× magnification. The count is mathematically converted to the number of cells/cm³ of suspension.

To ensure the accuracy, the coverslip must be firmly in place, and this can be achieved by moistening a coverslip with exhaled breath and gently sliding it over the hemocytometer chamber, pressing firmly until Newton's refraction rings (usually rainbow like) appear under the coverslip. The total number of cells in each of the four 1 mm³ corner squares should be counted, with the proviso that only cells touching the top or left borders but not those touching the bottom and right borders are counted. Moreover, cells outside the large squares, even if they are within the field of view, should not be counted. When present, clumps should be counted as one cell. Ideally, 100 cells should be counted to ensure a high degree of accuracy in counting. If the total cell count is <100 or if >10% of the cells counted appear to be clustered, then the original cell suspension should be thoroughly mixed and the counting procedure repeated. Similarly, if the total cell count is greater than 400, the suspension should be diluted further to get counts of between 100 and 400 cells.

Since some cells may not survive the trypsinization procedure, it is usually advisable to add an equal volume of the dye trypan blue to a small aliquot of the cell suspension before counting. This dye is excluded by viable cells but taken up by dead cells. Thus, when viewed under the microscope, viable cells will appear as bright translucent structures while dead cells will stain blue. The number of dead cells can, therefore, be excluded from the total cell count, ensuring that the seeding density accurately reflects viable cells.

24.6.1 Calculating Cell Number

Cell number is usually expressed per cubic centimeter and is determined by multiplying the average of the number of cells counted by a conversion factor that is constant for each hemocytometer. The conversion factor is estimated at 1000, based on the fact that each large square counted represents a total volume of 10^{-4} cm³.

Thus if

$$\text{Cells/cm}^3 = \frac{\text{Number of cells counted}}{\text{Number of squares counted}} \times \text{Conversion factor}$$

were diluted before counting, then the dilution factor should also be taken into account. Therefore,

$$\text{Cells/cm}^3 = \frac{\text{Number of cells counted}}{\text{Number of squares counted}} \times \text{Conversion factor} \times \text{Dilution factor}$$

finds the total number of cells harvested. The number of cells determined per cubic centimeter should be multiplied by the original volume of fluid from which the cell sample was removed, that is,

$$\text{Total cells} = \text{Cells/cm}^3 \times \text{Total volume of cells}$$

24.6.2 Alternative Methods for Determination of Cell Number

Several other methods are available for quantifying cells in culture, including direct measurement using an electronic Coulter counter. This is an automated method of counting and measuring the size of microscopic particles. The instrument itself consists of a glass probe with an electrode that is connected to an oscilloscope (Figure 24.11). The probe has a small aperture of fixed diameter near its bottom end. When immersed in a solution of cell suspension, cells are flushed through the aperture causing a brief increase in resistance owing to a partial interruption of current flow. This will result in spikes being recorded on the oscilloscope, and each spike is counted as a cell. One disadvantage of this method, however, is that it does not distinguish between viable and dead cells.

24.6.3 Calculation of Cell Number

Calculate the total number of cells suspended in a final volume 5 cm³, taking into account that the cells were diluted 1:2 before counting and the number of cells counted with the hemocytometer was 400.
  Answer:

$$\text{Cells/cm}^3 = \frac{\text{Number of cells counted}}{\text{Number of squares counted}} \times \text{Conversion factor}$$

$$= \frac{400}{4} \times 1000 = 100{,}000 \text{ cells/cm}^3$$

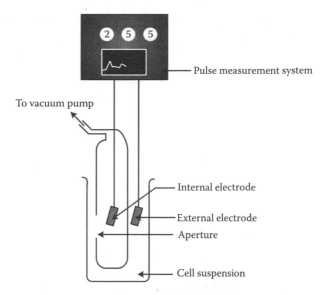

Figure 24.11 Coulter counter—cells entering the aperture create a poles of resistance between the internal and external electrodes that is recorded on the oscilloscope.

Because there is a dilution factor of 2, the correct number of cells/cm³ is given as

$$100,000 \times 2 = 200,000 \text{ cells/cm}^3$$

Thus, in a final volume of 5 cm⁻³, the total number of cells present is

$$200,000 \times 5 = 1,000,000 \text{ cells}$$

Indirectly, cells can be counted by determining total cell protein and using a protein versus cell number standard curve to determine cell number in test samples. However, protein content per cell can vary during culture and may not give a true reflection of cell number. Alternatively, the DNA count of cells may be used as an indicator of cell number since the DNA content of diploid cells is usually constant. However, DNA content of cells may change during the cell cycle and therefore not give an accurate estimate of cell number.

24.7 Seeding Cells onto Culture Plates

Once counted, cells should then be seeded at a density that promotes optimal cell growth. It is essential therefore that when cells are subcultured they are seeded at the appropriate seeding density. If cells are seeded at a lower density, they may take longer to reach confluency and some may die before getting to this point. On the other hand, if seeded at too high a density cells will reach confluency too quickly, resulting in irreproducible experimental results as already discussed earlier. The seeding density will vary depending on the cell type and on the surface area of the culture flask into which the cells will be placed. These factors therefore should be taken into account when deciding on the seeding density of any given cell type and the purpose of the experiments carried out.

24.8 Maintenance of Cells in Culture

It is important that after seeding, flasks are clearly labeled with the date, cell type, and the number of times the cells have been subcultured or passaged. Moreover, a strict regime of feeding and subculturing should be established that permits cells to be fed at regular intervals without allowing the medium to be depleted of nutrients or the cells to overgrow or become superconfluent. This can be achieved by following a standard but routine procedure for maintaining cells in a variable state under optimum growth conditions. In addition, cultures should be examined daily under an inverted microscope, looking particularly for changes in morphology and cell density. Cell shape can be an important guide when determining the status of growing cultures. Round or floating cells in subconfluent cultures are not usually a good sign and may indicate distressed or dying cells. The presence of abnormally large cells can also be useful in determining the well-being of the cells since the number of such cells increases as a culture ages or becomes less visible. Extremes in pH should be avoided by regularly replacing spent medium with fresh medium. This may be carried out on alternate days until the cultures are ~90% confluent, at which point the cells are either used for experimentation or trypsinized and subcultured.

The volume of medium added to the cultures will depend on the confluency of the cells and the surface area of the flasks in which the cells are grown. As a guide, cells, which are under 25% confluent, may be cultured in ~1 cm^3 of medium per 5 cm^2 and those between 25% and 40% or ≥45% confluency should be supplemented with 1.5 or 2 cm^3 culture medium per 5 cm^2, respectively. When changing the medium, it is advisable to pipette the latter either to the sides or on the opposite surface of the flask from where the cells are attached. This is to avoid making direct contact with the monolayers as this will damage or dislodge the cell.

24.9 Growth Kinetics of Animal Cells in Culture

When maintained under optimum culture conditions, cells follow a characteristic growth pattern (Figure 24.12), exhibiting an initial lag phase in which there is enhanced cellular activity but no apparent increase in cell growth. The duration of this phase is dependent on several factors including the viability of the cells, the density at which the cells are plated, and the media component.

The lag phase is followed by a log phase in which there is an exponential increase in cell number with high metabolic activity. These cells eventually reach a

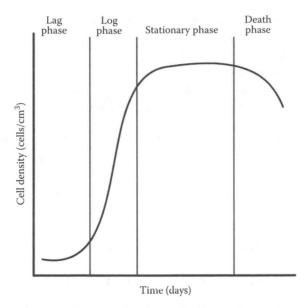

Figure 24.12 Growth curve showing the phase of cell growth in culture.

stationary phase where there is no further increase in growth due to depletion of nutrients in the medium, accumulation of toxic metabolic waste, or a limitation in available growth space.

24.10 Cryopreservation of Cells

Cells can be preserved for later use by freezing stocks in liquid nitrogen. This process is referred to as cryopreservation and is an effective way of sustaining stocks. Indeed, it is advisable that, when good cultures are available, aliquots of cells should be stored in the frozen state. This provides a renewable source of cells that could be used in the future without one necessarily to culture new batches from tissues. Freezing can, however, result in several lethal changes within the cells, including formation of ice crystals and changes in the concentration of electrolytes and in pH. To minimize these risks, a cryoprotective agent such as dimethyl sulfoxide (DMSO) is usually added to the cells prior to freezing in order to lower the freezing point and prevent ice crystals from forming inside the cells. In addition, the freezing process is carried out in stages, allowing the cells initially to cool down slowly from room temperature to −80°C at a rate of 1°C/min–3°C/min. This initial stage can be carried out using a freezing chamber or alternatively a cryofreezing container ("Mr. Frosty") filled with isopropanol, which provides the critical repeatable −1°C/min cooling rate required for successful cell cryopreservation. When this process is complete, the cryogenic vials, which are polypropylene tubes that can withstand temperatures as low as −190°C, are removed and immediately placed in a liquid nitrogen storage tank where they can remain for an indefinite period or until required.

The actual cryogenic procedure is itself relatively straightforward and involves harvesting cells and resuspended them in 1 cm³ of freezing medium, which is basically culture medium containing 40% serum. The cell suspension is counted and appropriately diluted to give a final cell count of between 106 and 107 cells/cm³. A 0.9 cm³ aliquot is transferred into a cryogenic vial labeled with the cell type; passage number, and date harvested. This is then made up to 1 cm³ by adding 100 mm³ of DMSO to give a final concentration of 10%. The cells should be mixed gently by rotating or inverting the vial and placed in a "Mr. Frosty" cryofreezing container. Then the container and cell should be placed in a −80°C freezer and allowed to freeze overnight. The frozen vials may then be transferred into a liquid nitrogen storage container. At this stage, cells can be stored frozen until required for use.

All procedures should be carried out under sterile conditions to avoid contaminating cultures as this will appear once the frozen stocks are recultured. As an added precaution, it is advisable to replace the growth medium in the 24 h period prior to harvesting cells for freezing. Moreover, cells used for freezing should be in the log phase of growth and not too confluent in case they may already be in growth arrest.

24.11 Resuscitation of Frozen Cells

When required, frozen stocks of cells may be revived by removing the cryogenic vial from storage in liquid nitrogen and placing in a water bath at 37°C for 1–2 min or until the ice crystals melt. It is important that the vials are not allowed to warm up to 37°C as this may cause the cells to die rapidly. The thawed cell suspension may then be transferred into a centrifuge tube, to which fresh medium is added and centrifuged at 1000 rpm for 10 min. The supernatant should be discarded to remove the DMSO used in the freezing process and the cell pellet resuspended in 1 cm³ of fresh medium, ensuring that clumps are dispersed into single cells or much smaller clusters using a glass Pasteur pipette. The required amount of fresh prewarmed growth medium is placed in a culture flask and the cells pipetted into the flask, which is then placed in a cell culture incubator and the cells allowed adhering and growing.

24.12 Determination of Cell Viability

Determination of cell viability is extremely important since the survival and growth of the cells may depend on the density at which they are seeded. The degree of viability is most commonly determined by differentiating living from dead cells using the dye exclusion method. Basically, living cells exclude certain dyes that are readily taken up by dead cells. As a result, dead cells stain the color of the dye used whilst living cells remain refractile owing to the inability of the dye to penetrate into the cytoplasm. One of the most commonly used dyes in such assays is trypan blue. This is incubated at a concentration of 0.4% with cells in suspension and applied to a hemocytometer. The hemocytometer is then viewed under an inverted microscope set at 100× magnification and the cells counted, keeping separate counts for viable and nonviable cells.

The total number of cells is calculated using the following equation as described previously:

$$\text{Cells/cm} = \frac{\text{Number of cells counted}}{\text{Number of squares counted}} \times \text{Conversion factor} \times \text{Dilution factor}$$

and the percentage of viable cells determined using the following formula:

$$\% \text{ Viability} = \frac{\text{Number of unstained cells counted}}{\text{Total number of cells counted}} \times 100$$

To avoid underestimating cell viability, it is important that the cells are not exposed to the dye for >5 min before counting. This is because uptake of trypan blue is time-sensitive and the dye may be taken up by viable cells during prolonged incubation period. Additionally, trypan blue has a high affinity for serum proteins and as such may produce a high background staining. The cells, therefore, should be free from serum, which can be achieved by washing the cells with PBS before counting.

24.13 Safety Considerations in Cell Culture

Safety in the cell culture laboratory must be of a major concern particularly when one is working with pathogenic microbes or with fresh primate or human tissues or working with fresh human lymphocytes, which may contain infectious agents such as the human immunodeficiency virus (HIV) and/or hepatitis B virus. It is, therefore, essential that the infection status of the donor is determined in advance of use, and all necessary precautions are taken to eliminate or limit the risks of exposure. A recirculation class II cabinet would be a minimum requirement for this type of cell culture work and the operator should be provided with protective clothing including latex gloves and a face mask if required. Such work should also be carried out under the guidelines laid down by the Advisory Committee on Dangerous Pathogens (ACDP).

The operator should also be aware of his or her work environment and be fairly conversant with the equipment being used as these may also pose a serious hazard. The culture cabinet should be serviced routinely and checked (approximately every 6 months) to ensure its safety to the operator. Additionally, one should adopt some common precautionary measures such as refrain from eating or drinking whilst working in the cabinet and using a pipette aid as opposed to mouth pipetting to prevent ingestion of unwanted substances. Gloves and adequate protective clothing such as a clean laboratory coat should be worn at all times and gloves must be discarded after handling of nonsterile or contaminated material.

24.14 Aseptic Techniques and Good Cell Culture Practice

24.14.1 Good Practice

1. Adequate aseptic or sterile technique should be adopted at all times by working under conditions that prevent contaminating microorganisms from the environment from entering the cultures by washing hands with antiseptic soap and ensuring that all work surfaces are kept clean and sterile by swabbing with 70% IMS before starting work. Moreover, all procedures, including media preparation and cell handling, should be carried out in a sterile condition in clean cell culture cabinet.
2. Other essential precautions should include avoiding talking, sneezing, or coughing into the cabinet or over the cultures. A clean pipette should be used for each different procedure and the same pipette should not be used under any circumstance between different bottles of media as this will significantly increase the risk of cross contamination.
3. All spillages must be cleaned quickly to avoid contamination from microorganisms that may be present in the air.
4. In the event of cultures becoming contaminated, these should be removed immediately from the laboratory, disinfected, and autoclaved to prevent the contamination spreading. Under no circumstance can an infected culture be opened inside the cell culture cabinet or incubator.
5. Moreover, all waste generated must be decontaminated and disposed off immediately after completing the work in accordance with the national legislative requirements, which state that cell culture waste including media can be inactivated using a disinfectant before disposal and that all contaminated material and waste be autoclaved before being discarded or incinerated.
6. In animal cell cultures, bacterial and fungal infections are relatively easy to identify and isolate. The other most common contamination originates from *Mycoplasma* that are the smallest (~0.3 μm in diameter) self-replicating prokaryotes in existence. They lack a rigid cell wall and generally infect the cytoplasm of mammalian cells. There are at least five species known to contaminate cells in culture: *Mycoplasma hyorhinis*, *Mycoplasma arginini*, *Mycoplasma orale*, *Mycoplasma fermentans*, and *Acholeplasma laidlawii*. Infections caused by these organisms are more problematic and not easily identified or eliminated. Moreover, if left unchecked, *Mycoplasma* contamination will cause subtle but adverse effects on culture, including changes in metabolism, DNA, RNA, and protein synthesis, morphology, and growth. This can lead to nonreproducible, unreliable experimental results, and unsafe biological products.
7. Maintaining a clean environment and adopting good laboratory practice and aseptic techniques, therefore, should help to reduce the risk of infection. However, should infection occur, it is advisable to address this immediately and eradicate the problem.

24.14.2 Identification and Eradication of Bacterial and Fungal Infections

Bacterial and fungal contaminations are easily identified as the infective agents are readily visible to the naked eye in the early stages usually noticeable by the increase in turbidity and the change in color of the culture medium owing to the change in pH caused by infection. In addition, bacteria can be easily identified under microscopic examination as motile round bodies. Fungi, on the other hand, are distinctive by their long hyphal growth and by the fuzzy colonies they form in the medium. In most cases, the simplest solution to these infections is to remove and dispose off the contaminated cultures. In the early stages of an infection, attempts can be made to eliminate the infecting microorganism using repeated washes and incubations with antibodies or antifungal agents but this is not advisable as handling infected cultures in the sterile work environment increases the chances of spreading infection.

As a part of the good laboratory practice, sterile testing of culture should be carried out regularly to ensure that cultures are free from microbial organisms. This is particularly important when preparing cell culture products or generating cells for storage. Generally, the presence of these organisms can be detected much earlier, and necessary precautions should be taken to avoid a full-blown contamination crisis in the laboratory. The testing procedure usually involves culturing a suspension of cells or products in an appropriate medium such as tryptone soya broth (TSB) for bacterial or thioglycollate medium (TGM) for fungal detection. The mixture is incubated for up to 14 days but examined daily for turbidity, which is used as an indication of microbial growth. It is essential that both positive and negative controls are set up in parallel with the sample to be tested. Uninoculated flasks containing only the growth medium are used as negative controls. Any contamination in the cell cultures will result in the broth appearing turbid, as would the positive controls. The negative controls should remain clear. Infected cultures should be discarded, whilst clear cultures would be safe to use or keep.

24.14.3 Identification of Mycoplasma Infections

Mycoplasma contaminations are more prevalent in cell culture. *Mycoplasma* contaminations are not evident under light microscopy nor do they result in a turbid growth in culture. Changes induced are more subtle and manifest themselves mainly as a slow down in growth and in change, in cellular metabolism and functions. However, cells generally return to their native morphology and normal proliferation rates relatively rapidly after eradication of *Mycoplasma*.

Mycoplasma contamination in culture has, until recently, been difficult to determine, and samples had to be analyzed by specialist laboratories. Improved techniques are now available for detection of *Mycoplasma* in cell culture laboratories. These involve either microbiological cultures or infected cells or an indirect DNA staining technique using the fluorochrome dye Hoechst 33258. With the former technique, cell in suspension is inoculated into liquid broth and then incubated under aerobic conditions at 37°C for 14 days. A noninoculated flask of broth is used as a negative control. Aliquots of broth are taken every 3 days and inoculated onto an agar plate, which is incubated anaerobically. All plates are then examined under an inverted microscope at a magnification of 300× after 14 days of incubation. Positive cultures will show the typical *Mycoplasma* colony formation, which has an opaque granular central zone surrounded by a translucent border, giving a "fried egg" appearance. Always set up positive controls in parallel having plates and broth inoculated with a known strain of *Mycoplasma* such as *M. orale* or *Mycoplasma pneumoniae*.

The DNA-binding method offers a rapid alternative for detecting *Mycoplasma* and works on the principle that Hoechst 33258 fluoresces under ultraviolet light once bound to DNA. Thus, in contaminated cells, the fluorescence will be fairly dispersed in the cytoplasm of the cells owing to the presence of *Mycoplasma*. In contrast, uncontaminated cells will show localized fluorescence in their nucleus only.

The Hoechst 33258 assay, although rapid, is relatively less sensitive compared with the culture technique. For this assay, an aliquot of the culture to be tested is placed on a sterile coverslip in 35 mm culture dish and incubated at 37°C in a cell culture incubator to allow cells to adhere. The coverslip is then fixed by adding a fixative consisting of 1 part glacial acetic acid and 3 parts methanol, prepared fresh on the day. A freshly prepared solution of Hoechst 33258 stain is added to the fixed coverslip, incubated in the dark at room temperature to allow the dye to bind to the DNA and then viewed under ultraviolet fluorescence at 1000×. *Mycoplasma* DNA appearing as small cocci or filaments in the cytoplasm of the contaminated cells will show fluorescence. Negative cultures will show only fluorescing nuclei of uncontaminated cells against a dark cytoplasmic background. However, this technique is prone to errors, including false negative results. To avoid the latter, cells should be cultured in antibiotic-free medium for two or three passages before being used. A positive control using a strain of *Mycoplasma* seeded onto a coverslip is essential. Such controls should be handled away from the cell culture laboratory to avoid contaminating clean cultures of cells. It is also important to ensure that the fluorescence detected is not due to the presence of bacterial contamination or debris embedded into the plastics during manufacture. The former normally appear larger than the fluorescing cocci or filaments of *Mycoplasma*. Debris, on the other hand, would show a nonuniform fluorescence owing to the variation in size of the particles usually found in plastic.

24.14.3.1 Eradication of Mycoplasma

Bacterial antibiotic Plasmocin™ has been shown to be effective against *Mycoplasma* even at relatively low, noncytotoxic concentrations. The antibiotics contained in this product are actively transported into cells, thus facilitating killing of intracellular *Mycoplasma* but without any adverse effects on actual cellular metabolism. Apart from antibiotics, other products have also been introduced into the cell culture market for eradicating *Mycoplasma* efficiently and quickly without causing any adverse effects to the cells. One such product is Mynox®, a biological agent that integrates into the membrane of *Mycoplasma*, compromising its integrity and eventually initiating its disintegration. This process apparently occurs within an hour of applying Mynox and may have the added advantage that being not an antibiotic will not lead to the development of resistant strains. It is safe to culture and eliminated once the medium has been replaced. Moreover, this reagent is highly sensitive, detecting as little as 1–5 fg of *Mycoplasma* DNA, corresponds to 2–5 *Mycoplasma* per sample and is effective against many of the common *Mycoplasma* contaminations encountered in cell culture.

24.15 Preparation of Tissue Culture Medium

24.15.1 Introduction

Choice of a suitable culture medium for establishing cell lines is essential. The medium used should be predetermined. A large number of different types of medium and BSS has been formulated and published; however, a vast majority of cells are cultivated in one of the seven or eight different media. Several formulations have been developed specially for the cultivation of cells with more specialized nutritional requirements. The type of medium used for primary cell culture is the choice of the requirement of the experiment and researcher. Wear gloves, lab coat, and other safety procedures should be followed.

24.15.2 Functions of the Main Ingredients in Culture Media

24.15.2.1 Balanced Salt Solution

The BSS is designed to provide a selection of metabolically important inorganic salts in concentrations sufficient not to reach depletion during the passage period of the cell culture and balance to provide a combined osmotic pressure similar to that experienced by the cell type *in vivo*. It also contains sodium bicarbonate, which in most cases acts primarily as a buffer but also functions as an important metabolite. The phosphates also function as a secondary source of buffering.

24.15.2.2 Amino Acids and Vitamins

These are essential nutrients required by mammalian cells for growth and division.

24.15.2.3 Other Ingredients

The selection of the other ingredients varies according to the type of medium, which in turn will depend on the type of cells and function for which they are being grown. There is always a prime source of energy, usually in the form of glucose, and an indicator to aid visual assessment of pH of the medium, usually phenol red.

24.15.2.4 Buffers

Cell metabolism produces comparatively large amount of waste products that are usually acidic in nature. All cell culture medium includes at least one chemical compound whose primary function is to act as a buffer by reacting with and to some extent neutralize the acidity. Secondary buffering is supplied by the phosphate salts in the BSS and to a lesser extent by any serum present.

The ideal buffer should have the following properties:

1. Consistently to maintain a given pH in a particular medium
2. No interference with chemical or biochemical processes occurring in the medium
3. No impedance of measurements or observations made on the system

Most widely used system for buffering cells culture media has traditionally been sodium carbonate/carbon dioxide:

$$NaHCO_3 + H_2O \rightarrow Na^{2+}HCO_3^- + H_2O \rightarrow Na^{2+}H_2CO_3 + OH^-$$

$$\rightarrow Na^{2+}OH^- + H_2O + CO_2$$

Sodium bicarbonate is helpful in maintaining pH, osmotic pressure and provides a source of energy. However, there are two main disadvantages to use sodium bicarbonate:

1. It has a pK_a of 6.3 at 37°C that gives suboptional buffering in the pH 7.0–7.5 range.
2. It requires a special gas mixture of 5% or 10% CO_2 in 95% or 90%, air in which to equilibrate. To overcome the equilibrium problem, Hanks formulated his BSS. It contains a lower concentration of sodium bicarbonate to eliminate the need for a special mixture. Earle's BSS is still more popular.

Another alternate to bicarbonate as a buffer was Tris (Tris hydroxy methyl amino methane).

The following are the disadvantages of Tris buffer:

1. It has a high chemical reactivity as a primary amine.
2. It often acts as an inhibitor.
3. It permeates biological membranes.
4. It has a high efficiency on pH dependence with temperature.

24.15.2.5 Antibiotics

Antibiotics are being routinely included in most cell culture media. They reduce the incidence of opportunistic contamination by microorganisms. Penicillin is being used most widely as antibiotic supplement.

24.16 Preparation of Single Cell Suspension of Spleen and Thymus

24.16.1 Introduction

Suspension cell culture is derived from cells that can survive and proliferate without attachment. Cell cultures may be derived from primary explants or dispersed cell suspension. A cell line arises from a primary culture at the time of the first successful subculture. The term cell line implies that cultures from it consist of lineages of cells originally present in the primary culture. The terms finite or continuous are used as prefixes if the status of the culture is known. The propagation of cell lines becomes feasible as often cell proliferation is found in such cultures. Cell suspensions may simply be diluted and reseeded or subcultured into fresh vessels to constitute a passage and the daughter cultures so formed are the beginnings of a cell.

The formation of cell line from a primarily culture implies

1. An increase in the total number of cells over several generations
2. The ultimate predominance of cells or cell lineages with the capacity for high growth
3. A degree of uniformity in the cell population

A cell strain is derived from either a primary culture or a cell line by the selection or cloning of cells having specific properties or markers. In describing a cell strain, its specific features must be defined. Cell strains are propagated in the form of a mono-layer or suspension.

For the purpose of cell suspension, the suspension is agitated continuously so that cells remain suspended in the liquid medium.

They are useful in the production of vaccines, interferons, etc.

24.16.2 Materials

Hank's buffered salt solution (HBSS) medium, centrifuge, CO_2 incubator, Ficoll column, and Tris ammonium buffer

Hanks BSS medium to 1 L of double distilled water: $CaCl_2$ 0.14 g, KCl 0.4 g, KH_2PO_4 0.06 g, $MgCl_2·6H_2O$ 0.1 g, $MgSO_4·7H_2O$ 0.1 g, NaCl 8 g, $NaHCO_3$ 0.35 g, $Na_2HPO_4·7H_2O$ 0.09 g, HEPES or NaCl 2.08 g + D-glucose 1 g, phenol red 0.1 g. The pH should be adjusted to 6.5.

24.16.3 Protocol

1. Collect spleen and thymus from a young animal of desired species in a Petri dish containing BSS.
2. Remove the adherent at and wash the spleen and thymus several times.
3. Mince the spleen with the help of a pair of scissors so as to release the cells.
4. Pass the suspension through muslin cloth layer.
5. As the cell suspension would contain erythrocytes also, to remove them, treat with Iris ammonium buffer or separate by using Isopaque Ficoll.
6. Adjust the cell density at 2×10^6 cells/mL and dispense in suitable culture vessels and incubate at 37°C in 5% CO_2 atmosphere.

24.17 Cell Counting

24.17.1 Introduction

The viable cell density is one of the most important parameters to monitor in any cell culture process. Quantification is required in order to determine the cell number, cellular constituents, and cell predefinition kinetics. Proper quantification should be performed to maintain for consistency and for analysis of the response of cells to environmental variations or experimental conditions. Cell counting must be done before and after each experiment for standardization of culture conditions and for proper quantitative experiments.

24.17.2 Hemocytometer

The concentration of cell suspension may be determined by placing the cells in an optically flat chamber under a microscope. The cell number within a defined area of known depth is counted, and the cell concentration is derived from the count (Figure 24.13).

24.17.3 Materials

Sterilized at 15 psi for 15 min PBSA medium (KCl 0.2 g, KH_2PO_4 0.2 g, NaCl 8 g, Na_2HPO_4 2.2 g, HEPES salt 2.02 g dissolved in 1 L of distilled water); 0.25% aqueous crude trypsin; 1 mM EDTA (3.72 g/L in PBSA medium); growth medium; pipettor adjustable 20 μL and yellow tips, hemocytometer (improved Neubauer), Tally counter, microscope.

24.17.4 Protocol

24.17.4.1 Cell Sampling
1. Monolayers
    a. Trypsinize the monolayer and resuspend in medium to give an estimated 1×10^6 cells/mL.

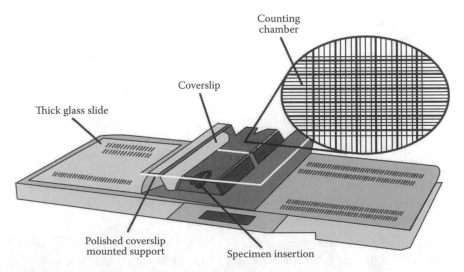

Figure 24.13   Hemocytometer.

    b. Mix the suspension thoroughly to disperse the cells, and transfer a small sample to a vial or universal container.

2. Suspension culture
    a. Mix the suspension thoroughly to disperse any clumps.
    b. Transfer 1 mL of the suspension to a vial or universal container.

Minimum of ~1 × 10^6 cells/mL is required for this method so the suspension may need to be concentrated by centrifuging and resuspending in a measured smaller volume.

24.17.4.2 Slide Preparation

1. Clean the surface of the slide with 70% alcohol without scratching the semisurface.
2. Clean the edges of the coverslip and press it down over the grooves in semisilvered count area. The appearance of interference patterns indicates that the coverslip is properly attached and thereby determined the depth of the counting chamber.
3. Mix the cell sample thoroughly by vigorous pipetting to disperse any clumps and collect about 20 μL with the help of Pasteur pipette or pipettor.
4. Transfer the cell suspension immediately to the edge of the hemocytometer, and let the suspension run out of the pipette and be drawn under the coverslip. Do not overfill or underfill the chamber or else its dimension may change due to alterations in surface tension, the fluid should run only to the edges of the groves.
5. Blot off any surplus fluid and transfer the slide to microscope stage.
6. Select a 10× objective and focus on the grid in the chamber.
7. Move the slide so that the fluid is seen in the central area of the grid and the largest area is bounded by parallel lines. This area is 1 mm^2 with a standard 10× objective.
8. Count the cells lying within this 1 mm^2 area, using the subdivisions and single grid lines.
9. Attempt to count between 100 and 300 cells/mm^3, more the cells are counted, the result becomes more accurate and precise for quantitative experiment; 200–1000 cells should be counted:
    a. If there are very few cells (<100/mm^2), count one or more additional squares surrounding the central square.
    b. If there are too many (>1000/m^2), count only five small squares across the diagonal of the larger (1 mm^2) square.

24.17.5 Results

The concentration of the sample using the formula

$$C = \frac{n}{v}$$

where
    C is the cell concentration (cells/mL)
    n is the number of cells counted
    v is volume counted from improved membrane slide, the depth of the chamber is 0.1 mm, and only central 1 mm^2 is used

Therefore,

$$v = 0.1 \, \text{mm}^3 \text{ or } 1.0 \times 10^4 \, \text{mL}$$

That formula for concentration

$$C = \frac{n}{10^{-4}} = C = n \times 10^4$$

If the cell concentration is high and only five diagonal squares were counted within the central 1 mm² (i.e., 1/4 of the total), then

$$C = n \times 5 \times 10^4$$

If the cell concentration is low and only eight 1 mm² squares were counted, in addition to the central 1 mm², this expression becomes

$$C = n \times \frac{10^4}{9}$$

24.18 Cell Viability Assay

24.18.1 Introduction

Cell viability is a determination of living or dead cells, based on a total cell sample. Cell viability is an important parameter to measure when determining cell responses to endogenous factors, such as hormones and growth factors, cytotoxic drugs, and environmental stresses. Evaluation of cell viability is essential in cell biology and drug discovery. Cell viability measurements may be used to evaluate the death or life of cancerous cells and the rejection of implanted organs. In other applications, cell viability tests might calculate the effectiveness of a pesticide or insecticide, or evaluate environmental damage due to toxins. Cell viability counts have a tremendous number of applications. Testing for cell viability usually involves looking at a sample cell population and staining the cells or applying chemicals to show which are living and which are dead. There are numerous tests and methods for measuring cell viability. Viable cells are impermeable to naphthalene black, Trypsin blue, and a number of other dyes.

24.18.2 Materials

Experimental cells, growth medium, 0.25% aqueous crude trypsin, sterilized PBSA (described in an earlier experiment), hemocytometer, viability testing stain (e.g., 0.4% trypsin blue or 1% naphthalene black in PBSA), Pasteur pipettes, microscope, tally counter.

24.18.3 Protocol

1. Prepare cell suspension ($1\text{--}5 \times 10^6$ cells/mL concentration) by trypsinization or by centrifugation and resuspension.
2. Take a clean hemocytometer slide and fix the coverslip in place.
3. Mix one drop of cell suspension with one drop (trypsin blue) or four drops (naphthalene black) of stain.
4. Load the counting chamber of the hemocytometer.
5. Leave the slide for 1–2 min (do not leave them for a longer period of time, or else, viable cells will deteriorate and take up the stain).
6. Place the slide on the microscope, and use a 10× objective to look at the counting grid.
7. Count the total number of cells and the number of stained cells.
8. Wash the hemocytometer and return it to the box.

24.18.4 Result

Calculate the percentage of unstained cells. This figure is the percentage viability.

24.19 Estimation of Viability by Dye Uptake

24.19.1 Introduction

Viable cells take up diacetate fluorescein and hydrolyze it to fluorescein to which the cell membrane of live cells is impermeable. Live cells fluoresce green, dead cells do not. Nonviable cells may be stained with propidium iodide and subsequently fluoresce red. Viability is expressed as the percentage of cells fluorescing green. This method can be applied to CCD analysis or flow cytometry.

24.19.2 Materials

Single cell suspension, fluorescein diacetate 10 μg/mL in HBSS, propidium iodide, fluorescence microscope with fluorescein excitation 450/590 nm, emission LP 515 nm and propidium iodide: excitation 488 nm, emission 615 nm filters.

24.19.3 Protocol

1. Prepare the cell suspension as for dye exclusion but in medium without phenol red.
2. Add the fluorescent dye mixture at a proportion of 1:10 to give a final concentration of 1 μL/mL of diacetyl fluorescein and 50 μg/mL of propidium iodide.
3. Incubate the cells at 37°C for 10 min.
4. Place a drop of cell on a microscope slide, add a coverslip, and examine the cells by fluorescence microscopy.

24.19.4 Results

Cells that fluoresce green are viable while those that fluoresce red are nonviable. Viability may be expressed as the percentage of cells that fluoresce green as a proportion to the total number of cells.

24.20 Macrophage Monolayer from PEE and Measurement of Phagocytic Activity

24.20.1 Introduction

Phagocytosis by polymorphonuclear neutrophils and monocytes constitutes an essential part of host defense against bacterial or fungal infection. The phagocytic process can be separated into several major stages: chemotaxis, attachment of particles to the cell surface of phagocytes ingestion, and intracellular killing by oxygen-dependent and oxygen-independent mechanism. Phagotest is a complete diagnostic kit for the investigation of phagocytic function of granulocytes and monocytes by determining the quantitative estimation of leukocyte phagocytosis in heparinized whole blood. It contains fluorescein-labeled opsonized bacteria and necessary reagents that measure the overall percentage of monocytes and granulocytes showing phagocytosis in general and the individual cellular phagocytic activity. The evaluation can be made with the help of flow cytometry or by fluorescence microscopy.

24.20.2 Materials

Whole-blood phagokit comprised of fluorescent-labeled *E. coli* cells, quenching solution, washing solution, lysing solution, DNA staining solution, flow cytometer, and heparinized whole blood.

24.20.3 Protocol

1. Incubate the heparinized whole blood with fluorescein-labeled *E. coli* bacteria at 37°C.
2. Negative control should be kept on ice.
3. Phagocytosis is stopped by placing the sample in ice and adding quenching solution. (This enables discrimination between attachment and internalization of bacteria by quenching the FTC fluorescence on surface of bacteria and leaving the fluorescence of internalized particles unaltered.)
4. After two steps of washing with washing solution, the erythrocytes are removed by addition of lysing solution.
5. DNA staining solution is added just prior to flow cytometric analysis, excludes aggregation artifacts of bacteria or cells.
6. The flow cytometer enables to enumerate the cells that have internalized the fluorescent dye.

24.20.4 Applications

1. Phagotest is a kit used to evaluate the effect of drugs and various clinical disorders with respect to phagocytic activity. The defects can be associated with the neutrophil itself or with an immunoglobulin or complement defect.
2. Reduced phagocytosis is seen in patients with recurrent bacterial skin and sinopulmonary infections, in patients with wound infections, in patients with wound infections from burns, in AIDS patients, and in neonates or in elderly people.

24.21 Trypsinization of Monolayer and Subculturing

24.21.1 Introduction

The need to subculture implies that the primary culture has increased to occupy all the available substrates. The primary culture may have a variable growth fraction, depending on the type of cells present in the culture, after the first subculture, the growth fraction is usually high. A more homogenous cell line emerges from a very heterogeneous primary culture, containing many of the cell types present in the original tissue. Subculturing, in addition to its biological significance, has considerable practical importance, as the culture can be propagated, characterized, and stored. Once a primary culture is subcultured, it becomes a cell line with several cell lineages of similar or distinct phenotypes. If one cell lineage is selected by cloning, by physical cell separation or by any other selection technique, to have certain specific properties that have been identified in the bulk of the cells, in the culture, this cell line becomes cell strain. *In vitro* transformation of cell line gives rise to a continuous cell line and if selected, cloned, characterized, and stored, gives rise to a continuous cell strain. The first subculture gives secondary culture and secondary to tertiary and so on. The number of times the subculture is done is termed as passage number and number of doubling cell population is generation number.

24.21.2 Materials

Sterilized at 15 psi for 15 min PBSA medium (KCl 0.2 g, KH_2PO_4 0.2 g, NaCl 8 g, Na_2HPO_4 2.2 g, HEPES salt 2.02 g dissolved in 1 L of distilled water); 0.25% aqueous crude trypsin; 1 mM EDTA (3.72 g/L in PBSA medium); growth medium; pipettor adjustable 20 µL and yellow tips, hemocytometer (improved Neubauer), tally counter, microscope, laminar flow hoods.

24.21.3 Protocol

1. Bring the reagents and materials to the working laminar flow hood.
2. Examine the culture carefully for signs of contamination or deterioration.
3. Take the culture to a sterile work area, remove, and discard the medium.

4. Add PBSA prewash to the side of the flask opposite the cells so as to avoid dislodging cells, rinse the cells, and discard the rinse. This step is designed to remove traces of serum that will otherwise inhibit the action of the trypsin.

5. Add trypsin, to the sides of the flask, opposite the cells turn the flask over and lay it down.
   Ensure that the monolayer is completely covered. Leave the flask stationary for 15–30 s and then withdraw all but a few drops of the trypsin, making sure that the monolayer has not detached. Using trypsin at 4°C helps to prevent premature detachment.

6. Incubate with the flask lying flat, until the cells round up, when the bottle is tilted, the monolayer should slide down the surface (usually 5–15 min). Do not leave the flask longer than necessary, but on the other hand, do not force the cells to detach before they are ready to do so, or else "clumping" may result.

7. Add medium (0.1–0.2 mL/cm²) and disperse the cells by repeated pipette resting over the surface bearing the monolayer. Finally, pipette the cell suspension up and down a few times with the tip of the pipette resting on the bottom corner of bottle, taking care not to create foam. The degree of pipetting required will vary from one cell line to another. Some cell lines disperse easily while others require vigorous pipetting in order to disperse them. Almost all cells incur mechanical damage from shearing forces if pipetted too vigorously. Primary suspension and early passage cultures are particularly prone to damage, partly due to their greater fragility and partly due to their larger size. Continuous cell lines are usually more resistant and require vigorous pipetting for complete disaggregation. Pipette the suspension up and down sufficiently to disperse the cells into a single cell suspension. Apply a more aggressive dissociating agent if finding this stage is difficult.

8. A single cell suspension is desirable at subculture to ensure an accurate cell count and uniform growth on reseeding. It is essential if quantitative estimates of cell proliferation or of platting efficiency are being made and if cells are to be isolated as clones.

9. Count the cells using a hemocytometer or an electronic particle counter.

10. Dilute the suspension to the appropriate seeding concentration:
    a. By adding the appropriate volume of cells to a premeasured volume of medium in a culture flask
    b. By diluting the cells to the total volume required and distributing that volume among several flasks.

11. If the cells are grown in elevated CO_2, gas the flask by blowing the correct gas mixture from a premixed cylinder, through a filtered line into the flask above the medium. Do not bubble gas through the medium, as doing so will generate bubbles, which can denature some constituents of the medium and increase the risk of contamination.

12. Cap the flask and return them to the incubator, check after about 1 h for a change in the pH. If the pH rises in a medium with a gas phase affair, then return the flask to the aseptic area and gas the culture briefly (1–2 s) with 5% CO_2. Since each culture will behave predictably in the same medium. One will eventually know which cells to gas when they are reseeded, without having to incubate them first. If the pH rises in the medium that already has a 5% CO_2 gas either increase the CO_2 to 7%–10% or add sterile 0.1 N of HCl.

The expansion of air inside plastic flasks causes larger flasks to swell and prevent them from lying flat. Release the pressure 30 min after placing the flask in the incubator by briefly slackening the cap, or alternatively, this problem may be prevented by compressing the top and bottom of large flasks before sealing them to restore the correct shape.

24.22 Cryopreservation

24.22.1 Introduction

The cell line becomes a valuable resource, replacement of which would be expensive and time consuming. If unique, it might be impossible to replace. Therefore, it is essential to protect this considerable investment, by preserving the cell line.

24.22.2 Importance

Cell lines in a continuous culture are prone to variation due to genetic instability, senescence, transformation, phenotypic instability due to selection and dedifferentiation, microbial contamination, cross contamination by other cell lines selection, incubator failure in early passage culture, and senescence in finite cell lines and genetic instability in continuous cell lines. In addition, there is a need for distribution to other users. Therefore, the freezing down a stock of the cell is necessary in order to preserve the valuable cell lines.

24.22.3 Preservation

24.22.3.1 Cell Line Selection
A cell line is selected with the following properties:

1. It must be a finite cell line.
2. It must be grown to around the fifth population doubling in order to create a sufficient bulk of cells for freezing.
3. Continuous cell lines should be cloned and an appropriate clone selected, and grown up to sufficient bulk to freeze.
4. Before freezing, the cells should be maintained under standardized culture conditions, characterized, and checked for contamination.
5. Continuous cell lines have the advantages that they survive indefinitely, grow more rapidly, and can be cloned more easily, but they may be less stable genetically.
6. Finite cell lines are usually diploid or close to it and stable between certain passage levels, but they are harder to clone, grow more slowly, and eventually die out or transform.
7. Storage in liquid nitrogen is preferred for preserving cultures.

24.22.4 Materials

Cultures to be frozen, if monolayer: Sterilize at 15 psi for 15 min PBSA medium (KCl 0.2 g, KH_2PO_4 0.2 g, NaCl 8 g, Na_2HPO_4 2.2 g, HEPES salt 2.02 g dissolved in 1 L of distilled water); 0.25% aqueous crude trypsin; 1 mM EDTA (3.72 g/L in PBSA medium); growth medium; pipettor adjustable 20 μL and yellow tips, hemocytometer (improved Neubauer), tally counter, microscope, laminar flow hoods, syringe 1–5 mL for dispensing glycerol, plastic ampoules 1–2 mL, curved forceps, protective gloves, canes or racks for storage, insulated container for freezing, polystyrene box lined with cotton wool or plastic foam insulation tube.

Growth medium: Serum improves survival of the cells after freezing, up to 50% or even more. If serum is being used with serum-free cultures, it should be washed off after thawing).

Preservative, free of impurities: DMSO or glycerol (DMSO should be colorless and store in glass or polypropylene as it causes impurities with rubber and plastics). Glycerol should not be >1 year old as it may become toxic after prolonged storage.

24.22.5 Protocol

1. Check the culture for the following:
   a. Healthy growth
   b. Free from contamination
   c. Specific characteristic
2. Grow the culture up to the late log phase. If using a monolayer trypsinized cell, then count the cells. If the suspension is used, count and centrifuge to sediment the cells.
3. Dilute with one of the following preservation in growth medium to make freezing medium:
   a. Add DMSO to 5%–10%.
   b. Add glycerol of 10%–15%.
4. Resuspend the cells in freezing medium at $\sim 1 \times 10^6$–1×10^7 cells/mL. It is not advisable to place ampoules on ice in an attempt to minimize deterioration of the cells. It can withstand up to 30 min at room temperature and is not harmful when using DMSO and is beneficial when using glycerol.
5. Dispense the cell suspension into ampoules and seal the ampoules.
6. Place the ampoules on canes for canister storage.
7. Freeze the ampoules by one of the following methods:
   a. Lay the ampoules on cotton wool in a polystyrene foam box with a wall thickness of ~15 mm. This box plus the cotton wool should provide sufficient insulation in such a way that the ampoules will cool at 1°C/min when the box is placed at −70°C or −90°C in a regular deep freeze or insulated container with solid CO_2.
   b. Insert the canes in tubular foam pipe insulation with a wall thickness of ~15 mm and place the insulation at −70°C or −90°C in a regular deep freeze or insulated container with solid CO_2.
   c. Place the cells in a freezer.

8. When the ampoules have reached −70°C (a minimum of 4–6 h after placing them at −70°C if starting from 20°C ambient), transfer them to a liquid N_2 freezer. This transfer must be done quickly as the ampoules will reheat at ~10°C/min and the cells will deteriorate rapidly if the temperature rises above −50°C.

24.23 Thawing

24.23.1 Introduction

Thawing is usually carried out by putting the ampoule containing the sample in warm water (35°C–48°C) for thawing. Frozen tips of the samples in tubes or ampoules are plunged into warm water with a vigorous swirling wrist action just to the point of the appearance. It is important for survival of the tissue that the tubes should not be left in the warm bath after the ice melts. Just at the point of thawing, quickly transfer the tubes to a water bath maintained at room temperature (20°C–25°C) and continue the swirling action for 15 s, to cool the warm walls of the tubes. Then continue with washing and culture transfer procedures. It is necessary to avoid excessive damage to the fragile thawed tissues, or cells by minimizing the amount of handling at pregrowth stage.

24.23.2 Precautions

A face shield and gloves must be worn. Ampoules stored under liquid nitrogen may imbibe the liquid and thawing will explode violently. A plastic bucket with a lid is, therefore, necessary to contain any explosion.

24.23.3 Materials

Protective gloves and face mask, bucket of water at 37°C with lid, forceps, 70% alcohol Swab, naphthalene black (amido black) 1% or trypan blue 0.4%, culture flask, centrifuge tube, growth medium, pipettes 1–10 mL, syringe, and 19 gauge needle

24.23.4 Protocol

1. Check the index for the location of the ampoule to be thawed.
2. Collect all materials, prepare the medium, and label the culture flask.
3. Retrieve the ampoule from the freezer, check that it is the correct one, and place it in 10 cm level in water at 37°C in a bucket with a lid.
4. When the ampoule is thawed, then swab the ampoule thoroughly with 70% alcohol and open it.
5. Transfer the contents of the ampoule to a culture flask.

6. Add medium slowly to the cell suspension, 10 mL over about 2 min added drop wise at the start and then a little faster, gradually diluting the cells. This gradual process is particularly important with DMSO, with which sudden dilution can cause severe osmotic damage and reduce cell survival by half.
7. Cell viability may be determined by staining with naphthalene black or trypan blue.

The number of cells frozen should be sufficient to allow for 1:10 or 1:20 dilution on thawing to dilute the preservative but still keep the cell concentration higher than at normal passage, for example for cells subcultured normally at 1×10^5–1×10^7 cells/mL should be frozen in 1 mL of medium after thawing the cells. Then 1 mL should be diluted to 20 mL of medium. This dilutes the preservative from 10% to 0.5% and is likely to be less toxic. Residual preservative may further be diluted as soon as the cells start to grow, or medium is changed as soon as the cells have attached. The proportion of cells attached to those still floating after 24 h should be recorded by employing dye exclusion viability test. One ampoule should be thawed from each batch as it is frozen to check success rate.

24.24 Role of Serum in Cell Culture

24.24.1 Introduction

Serum is an extremely complex mixture of many small and large biomolecules with different growth-promoting and growth-inhibiting activities. Three major classes of factors are known to be important in the serum that includes hormonal factors stimulating cell growth and functions, attachments, spreading factors, and transport of proteins carrying hormones, minerals, lipids, etc. Although many attempts have been made to produce completely defined media for cell culture for optimum cell growth, most cells still usually require the addition of 5%–10% serum to the media.

24.24.2 Precautions

A face shield and gloves must be worn. Ampoules stored under liquid nitrogen may imbibe the liquid and thawing will explode violently. A plastic bucket with a lid is, therefore, necessary to contain any explosion.

24.24.3 Materials

Hemocytometer, trypan blue, media, and serum.

24.24.4 Protocol

Grow cells in the culture media and then count the cells.

1. Counting of cells
   a. Remove 0.2 mL of the cell suspension and place it into a microfuge tube and take it to the lab for counting.
   b. Add 0.2 mL of trypan blue solution to the tube and thoroughly mix the contents by pipetting up and down.
2. Place a coverslip, use a Pasteur pipette to transfer a small amount of trypan blue/cell suspension to both chambers of a hemocytometer by carefully touching the edge of the coverslip with the pipette tip, and allow capillary action to fill the counting chambers. Do not overfill or underfill the chambers.
3. Take the culture tubes with the desired amount of cells, centrifuge at 1500 rpm for 5 min at 4°C.
4. Decant the supernatant and immediately resuspend cells in the tube by gentle shaking.
5. To one tube add 12 mL of medium w/10% serum, to the second tube add 12 mL of medium w/5% serum, and to the third tube 12 mL of medium w/2% serum.
6. Mix the contents by carefully inverting the tube several times. From each tube, pipette 1 mL into each of the wells on a 12-well plate. Be sure to label the plate with initials, the start date, time, and serum concentration.
7. Place the 12-well plates into the incubator at ~24 h intervals for 4 days. Perform cell counts of the contents.

24.25 Metaphase Chromosome Preparation from Cultured Cells

24.25.1 Introduction

Chromosome content is well-defined criteria for identifying cell lines and relating them to the species and sex from which they were derived. Metaphase chromosomes give a clear picture of the chromosome and the chromosomal number. Metaphase chromosome analysis also distinguishes between normal and malignant cells, since the chromosome number is more stable in normal cells (except in mice where the chromosome complement of the normal cells can change quite rapidly after explanation into culture).

24.25.2 Materials

Cell culture in log phase, colcemid 10^{-5} M in PBSA, PBSA sterilized at 15 psi for 15 min (KCl 0.2 g, KH_2PO_4 0.2 g, NaCl 8.0 g, Na_2HPO_4 2.2 g, HEPES salt 2.02 g dissolved in 1 L of distilled water); 0.25% aqueous crude trypsin in PBSA, hypotonic solution 0.04 M KCl; 0.025 M sodium citrate, acetic methanol fixative: 1 part glacial acetic add + 3 parts anhydrous methanol or ethanol, made up fresh, and kept on ice, Giemsa stain, DPX or permount mountant, ice, centrifuge tubes, Pasteur pipettes, slides and coverslips, slide dishes, low-speed centrifuge, vortex mixer.

24.25.3 Protocol

1. Set up a 75 cm^2 flask culture at between 2×10^4 and 5×10^4 cells/mL in 20 mL.
2. Approximately 3–5 days later, when cells are in log phase of growth, add 1:100 v/v, 1×10^{-5} M colcemid (1×10^{-7} M, final) to the medium already in the flask.
3. Remove medium gently after 4–6 h, add 5 mL of 0.25% trypsin, and incubate the culture for 10 min.
4. Centrifuge the cells in trypsin and discard the supernatant.
5. Resuspend the cells in 5 mL of a hypotonic solution and leave them for 20 min at 37°C.
6. Add an equal volume of freshly prepared, ice cold acetic methanol, mix constantly, and then centrifuge the cells at 100g for 2 min.
7. Discard the supernatant mixture, "buzz" the pellet on a vortex mixer, and slowly add fresh acetic methanol with constant mixing.
8. Leave the cells for 10 min on ice.
9. Centrifuge the cells for 2 min at 100g.
10. Discard the supernatant acetic methanol and resuspend the pellet by "buzzing" in 0.2 mL of acetic methanol, to give a finally dispersed cell suspension.
11. Draw one drop of the suspension into the tip of a Pasteur pipette and drop onto a microslide. Tilt the slide and let the drop run down the slide as it spreads.
12. Dry off the slide rapidly and examine it on the microscope with phase contrast to see that cells are evenly spread. Prepare more slides at the same cell concentration. If the cells are piled up and overlapped, then dilute the suspension two- or fourfold and make a further drop preparation.
13. Stain the cells with Giemsa. (i) Immerse the slide in stain for 2 min, (ii) place the dish in the sink and ~10 volume of water, allowing surplus stain to over flow from the top of the slide dish, (iii) leave the slides for a further 2 min, (iv) displace the remaining stain with running water and finish by running the slide individually under tap water to remove precipitated stain, and (v) check the staining under the microscope. If satisfactory, then dry the slide thoroughly and mount the coverslip in DPX or permount.

24.25.4 Results

1. Count the chromosome number per spread and between 50 and 100 spread:
   a. Count all the mitoses seen and classify them by chromosomal number.
   b. If counting is impossible, then classify them under near diploid unclassifiable or polyploidy uncountable.
2. Karyotype: Photograph or save about 10–20 good spreads of banded chromosome and print on 20×25 cm high contrast paper. If the image is recorded with a CCD camera or scanned from a slide or print, then chromosome sorting may be easier.

24.26 Isolation of DNA and Demonstration of Apoptosis of DNA Laddering

24.26.1 Introduction

The thymus gland and spleen apart from being soft tissues contain relatively large amount of nucleic acids. They are a good source of sample for DNA isolation. The addition of detergent in the mixed tissue causes the rupture of cell membrane and dissolves the lipid and proteins of the cells. Cell wall and nuclear membrane are to be ruptured, deproteinized, and DNA is precipitated after adding cold ethanol for DNA isolation.

24.26.2 Materials

Thymus/spleen, 2 M NaCl, 0.14 M NaCl, 0.02 M sodium citrate, extraction medium (pH 1.4) 0.14 M NaCl and 0.02 M sodium citrate, blender, centrifuge, ethanol.

24.26.3 Protocol

1. Suspend about 25 g of the tissue in about 100 mL of extraction medium.
2. Homogenize in a blender.
3. Centrifuge at 3000 rpm for 10 min and discard the supernatant.
4. Rehomogenize the precipitate with the extraction medium and discard the supernatant.
5. Suspend the sediment in about 50 mL of NaCl (2 M).
6. Centrifuge at 10,000 rpm for 10 min at room temperature.
7. Carefully pipette out the clear aqueous phase in a beaker. This phase contains nucleic acid.
8. Gently stir the nucleic acid solution with a sterilized glass rod while slowly adding two volumes of ice-cold ethanol down the side of the beaker so that ethanol is layered over the viscous aqueous phase and spool all of the gelatinous, threadlike DNA-rich precipitate on the glass rod.
9. Drain off excess fluid from the spooled crude DNA by pressing the rod against the wall of the beaker until no further fluid can be squeezed from the spooled preparation.
10. Dissolve the crude DNA in 5 mL of extraction medium and reprecipitate with ethanol.
11. The DNA is best stored in extraction medium in cold.

24.27 Demonstration of Apoptosis of DNA Laddering

24.27.1 Introduction

Apoptosis or programmed cell death is one of the major control mechanisms by which a cell undergoes self-destruction which is important in controlling cell number and proliferation as part of normal growth and development and also plays an important role in tissue homeostasis. The process is characterized by nuclear chromatin condensation, cytoplasmic shrinking, dilated endoplasmic reticulum, and membrane bleeding. Mitochondria remain unchanged morphologically. Apoptosis is recognized as one of the important processes in all biological systems, including embryonic development, cell turnover, and immune response against tumorigenic or virus-infected cells. The procedure adopted is simple and uses nontoxic reagents. The cells are prefixed with

70% ethanol and if desired can be stored for several weeks before analysis. After fixing, the cells are extracted with 0.2 M phosphate–citrate buffer at pH 7.8. Under these conditions, the partially degraded, oligonucleosomal DNA is extracted selectively from the cells whereas the higher molecular DNA stays associated with the nuclei. The extracts are treated with RNase A and proteinase K and then loaded to the agarose gel. DNA laddering can be detected from samples with only 8% apoptotic cells.

24.27.2 Materials

HBSS, ethanol, phosphate–citrate buffer 192 part of 0.2 M Na_2HPO_4 and 8 parts of 0.1 M citric acid (pH 7.8), cells, trypsin, proteinase, loading buffer 0.25% bromophenol blue and 30% glycerol are mixed together, agarose gel, ethidium bromide, electrophoretic unit, UV transilluminator, centrifuge, speed vacuum concentrator.

24.27.3 Protocol

1. Harvest the cells by trypsination and spin down by centrifugation at 200g for 5 min.
2. Resuspend cells in 1 mL of HBSS.
3. Transfer the cells into 10 mL of ice-cold 70% ethanol and store at –20°C for 24 h or longer.
4. Spin down fixed cells by centrifugation at 800g for 5 min and remove the ethanol thoroughly.
5. Resuspend the cell pellets in 40 µL phosphate–citrate buffer consisting of 192 parts of 0.2 M Na_2HPO_4 and 8 parts of 0.1 M citric acid (pH 7.8) and incubate for 30 min.
6. Spin down cells by centrifugation at 1000g for 5 min.
7. Transfer supernatant to new tubes. Optionally, the samples can be concentrated by vacuum SpeedVac concentrator.
8. Add 3 µL of 0.25% Nonidet P-40 (NP-40) (in water) and 3 µL of RNase (1 mg/mL in water) to the samples and incubate for 30 min at 37°C.
9. Add 3 µL of proteinase K (1 mg/mL) to the samples and incubate for another 10 min at 37°C.
10. Add 12 µL of loading buffer to the samples and load on a 1.5% agarose gel.
11. Run the gel at 4 V/cm for about 4 h and detect DNA by ethidium bromide 0.5 µg/mL in loading buffer under UV light.

24.28 MTT Assay for Cell Viability and Growth

24.28.1 Introduction

Cells in the exponential phase of growth are exposed to a cytotoxic drug and the duration of exposure is usually determined by the time required for maximal damage. The cells are allowed to proliferate for two or three population doubling times (PDT) after removal of drugs in order to distinguish between cells that remain viable but cannot proliferate and those capable of proliferation. The number of surviving cells is then determined indirectly by MTT dye reduction. The amount of MTT formazan produced can be determined spectrophotometrically, once the MTT formazan has been dissolved in a suitable solvent.

24.28.2 Materials

Sterile: Growth medium, trypsin (0.25%) + ETTA 1 mM in PBSA, MTT (3,4,5-dimethylthiazol-2yl) 2,5-diphenyl tetrazolium bromide (Sigma) 50 mg/mL, filter-sterilized Sorensen's glycine buffer, microtitration plates, pipettor tips, preferably in an autoclavable tip box, Petri dishes (non-TC-treated) 5 and 9 cm, universal containers—30 and 100 mL.

Nonsterile: Plastic box (clear polystyrene to hold plates), multichannel pipettor, DMSO, DMSO dispenser, ELISA plate reader, plate carrier for centrifuge (for cells growing in suspension)

24.28.3 Protocol

1. Cell plating: Trypsinize a subconfluent monolayer culture and collect the cells in growth medium containing serum.
2. Centrifuge the suspension (5 min at 200*g*) to pellet the cells. Resuspend the cells in growth medium and count them.
3. Dilute the cells to $2.5-50 \times 10^3$ cells/mL depending on the growth rate of the cell line and allowing 20 mL of cell suspension per microtitration plate.
4. Transfer the cell suspension to a 9 cm Petri dish and with a multichannel pipette, add 200 µL of the suspension into each well of the central 10 columns of a plate followed by 96 others in the plate starting with column 2 and ending with column 11 and placing $0.5-10 \times 10^{-3}$ cells into each well.
5. Add 200 µL of growth medium to the light wells in columns 1 and 12. Column 1 will be used to blank to plate reader and column 12 helps to maintain the humidity for column 11 and minimizes the edge effect.
6. Put the plates in a plastic lunch box and incubate in a humidified atmosphere at 37°C for 1–3 days such that the cells are in an exponential phase of growth at the time the drug is added.
7. For nonadherent cells, prepare suspension in fresh growth medium. Dilute the cells to $5-100 \times 10^3$ cells/mL and plate out only 100 µL of the suspension into round-bottomed 96-well plates. Add drug immediately to these plates.
8. Drug addition: Prepare a serial fivefold dilution of the cytotoxic drug in growth medium to give eight concentrations. This set of concentrations should be selected in such a way that the highest concentration kills the cells and lowest kills none of the cells. Once the toxicity of the drug is known, a small range of concentration can be used. Normally, three plates are used for each drug to give triplicate determinations within one experiment.
9. For adherent cells, remove the medium from the wells in columns 2–11. This can be achieved with a hypodermic needle attached to a suction line. Feed the cells such that eight cells in columns 2 and 11 with 200 µL of fresh growth medium. These cells are the controls.

10. Add the cytotoxic drugs to the cells in columns 3–10. Only four wells are needed for each drug concentration in such a way that rows A–D can be used for one drug and rows E–H for a second drug.

11. Transfer the drug solution to 5 cm Petri dishes and add 200 μL to each group of four wells with a four-tip pipettor.

12. Return the plates to the plastic box and incubate them for a defined exposure period for nonadherent cells, prepare the drug dilution as twice the desired final concentration, and add 100 μL of dilution to 100 μL of cells already in the wells.

13. Growth period. At the end of the drug exposure period, remove the medium from all of the wells containing cells and feed the cells with 200 μL of fresh medium. Centrifuge plates containing nonadherent cells to pellet the cells: Then remove the medium, using a fine gauge needle to prevent disturbance of the cell pellet.

14. Feed the plates daily for two to three PDTs.

15. Estimation of surviving numbers: Feed the plate with 200 μL of fresh medium; at the end of the growth period, add 50 μL of MTT to all the wells in columns 1–11.

16. Wrap the plates in aluminum foil and incubate them for 4 h in a humidified atmosphere at 37°C. Note that 4 h is a minimum incubation time and plates can be left for up to 8 h.

17. Remove the medium and MTT from the wells, and dissolve the remaining MTT formazan crystals by adding 200 μL of DMSO to all the wells in columns 1–11.

18. Add glycine buffer (25 μL per well) to all the well containing DMSO.

19. Record absorbance at 570 nm immediately, since the product is unstable use the well in column 1, which contain medium and MTT but no cells, as blank.

24.28.4 Analysis

1. Plot a graph of absorbance (y-axis) against drug concentration (x-axis).
2. The mean absorbance reading from the wells in columns 2 and 11 is used as control absorbance.
3. The IC_{50} concentration of the drug is determined, which is required to reduce the absorbance to half that of the control,
4. The absolute value of the absorbance should be plotted so that control values may be compared and convert the data to percentage.

Note: The absorbance in columns 2 and 11 must be same if not uneven plating of cells across the plate has taken place.

24.29 Cell Fusion by PEG

24.29.1 Introduction

Many cell lines undergo spontaneous fusion and the frequency of such events is very low. In order to produce hybrids in significant numbers, the cells are treated with either the Sendai virus or most commonly, the chemical fusogen polyethylene glycol (PEG). Selection systems that kill parental cells but not hybrids are then used to isolate clones of hybrid cells.

24.29.2 Materials

Sterile

PEG: 1000 (Merck), complete growth medium, serum-free growth medium, NaOH 1.0 M, Petri plates 50 mm, universal containers.

24.29.3 Methods

24.29.3.1 Monolayer Fusion

Inoculate equal numbers of the two types of cells to be fused into 50 mm tissue culture dishes. Between 2.5×10^5 and 2.5×10^6 of each parental cell line per dish is usually sufficient. Incubate the mixed culture overnight:

1. Warm the PEG solution to 37°C which may be necessary at this point to readjust the pH, using NaOH.
2. Remove the medium thoroughly from the culture medium and wash them once with serum-free medium. Add 3.0 mL of the PEG solution and spread it over the monolayer of cells.
3. Remove the PEG solution after exactly 1 min and rinse the monolayer three times with 10 mL of serum-free medium before returning the cells to complete medium.
4. Culture the cells overnight before adding selection medium.

24.29.3.2 Suspension Fusion

Centrifuge a mixture of 4×10^6 cells of each of two parental cells at $150g$ for 5 min at room temperature, and carry out centrifugation and subsequent fusion in 30 mL plastic universal containers or centrifuge tubes:

1. Resupend the pellet in 15 mL of serum-free medium and centrifuge again.
2. Aspirate off the medium and resuspend the cells in 1 mL of PEG solution by gently pipetting.
3. After 1 min, dilute the suspension with 9 mL of serum-free medium and transfer half of the suspension to each of two universal containers or centrifuge tubes containing a further 15 mL of serum-free medium.
4. Centrifuge the suspensions at $150g$ for 5 min, remove the supernatant, and resuspend the cells in complete medium.
5. Incubate overnight and clone the cells in selection medium.

Suggested Readings

Bisen, P. S. and A. Sharma. 2012. *Introduction to Instrumentation in Life Sciences.* Boca Raton, FL: CRC Press.

Davis, J. M. 2002. *Basic Cell Culture: A Practical Approach*, 2nd edn. Oxford, U.K.: Oxford University Press.

Freshney, R. I. 2000. *Culture of Animal Cells: A Manual of Basic Technique*, 4th edn. New York: John Wiley & Sons.

Furr, A. K., ed. 1995. *CRC Handbook of Laboratory Safety*, 4th edn. Boca Raton, FL: CRC Press.

HSC Advisory Committee on Dangerous Pathogens. 2001. *The Management Design and Operation of Microbiological Containment Laboratories*. Sudbury, Ontario, Canada: HSE Books.

Khan, Z., N. Khan, A. K. Varma, R. P. Tiwari, M. Shahul, G. B. K. S. Prasad, and P. S. Bisen. 2010. Oxaliplatin-mediated inhibition of survivin increase sensitivity of head and neck squamous cell carcinoma cell lines to paclitaxel. *Curr Cancer Drug Target* 10: 660–669.

Khan, Z., N. Khan, R. P. Tiwari, G. B. K. S. Prasad, and P. S. Bisen. 2012. Induction of apoptosis and sensitization of head and neck squamous carcinoma cells to cisplatin by targeting survivin gene expression. *Curr Gene Therapy* 12: 444–453.

Khan, Z., K. Noor, R. P. Tiwari, I. K. Patro, G. B. K. S. Prasad, and P. S. Bisen. 2010. Down-regulation of survivin by oxaliplatin diminishes radioresistance of head and neck squamous carcinoma cells. *Radiother Oncol* 96: 267–273.

Parekh, S. R. and V. A. Vingi. 2003. *Handbook of Industrial Cell Culture: Mammalian, Microbial and Plant Cells*. Totowa, NJ: Humana Press.

Wilson, K. and J. Walker. 2006. *Principles and Techniques of Biochemistry and Molecular Biology*. Cambridge, U.K.: Cambridge University Press.

Important Links

http://www.level.com.tw/html/ezcatfiles/vipweb20/img/img/20297/intro_animal_cell_culture.pdf

http://www.biosafety.be/CU/animalcellcultures/mainpage.html

http://www.ncbi.nlm.nih.gov/pubmed/11248856

http://www.ncbi.nlm.nih.gov/books/NBK21682/

http://www.opsdiagnostics.com/notes/ranpri/rpcryoprescell.htm

http://www.vetelib.com/threads/20350-Animal-Cell-Culture-and-Technology-The-Basics-2nd-Edition

25

Bioprocess Engineering

25.1 Introduction

Bioprocess engineering deals with studying various biotechnological processes used in industries for large-scale production of biological product for optimization of yield in the end product and the quality of the end product. It deals with the design and development of equipment and processes for the manufacturing of products such as food, feed, pharmaceuticals, nutraceuticals, chemicals, and polymers and paper from biological materials. The commercialization of new products and processes is made possible through application of life sciences technologies requiring a coordinated coupling of unit operations in order to develop efficient processes. The usual objective of a process is to convert relatively inexpensive raw materials into more valuable products or services. While discoveries and developments in fundamental molecular biology and molecular genetics today have catalyzed much of the excitements in life science technologies, it is the role of the process engineer to translate these discoveries and observations into usable processes. The schematic diagram of a typical biochemical process illustrated in Figure 25.1 easily visualizes this synthesis.

The central point of this process is the bioreactor using biological cells of microbial, plant, or animal origin for process development (Figure 25.2). Value is added through biocatalysis, fermentation, or cell culture to make valuable products and services. The bioreactor, however, does not exist in isolation; its efficiency and successful operation depend upon adequate upstream processing. Furthermore, recovery of the final product requires a series of steps referred to, collectively, as downstream processing.

25.2 Isolation of Industrially Important Microorganisms

25.2.1 Introduction

Several properties of microorganisms are increasingly important in industry which is being exploited in a variety of ways, namely, as a source of high-quality food and other products such as enzymes for the industrial applications degrading pollutants and toxic wastes to clean environment, as vectors to treat disease and enhance agricultural productivity. In industry, microorganisms are used to make products such as antibiotics, vaccines, steroids, alcohol and other solvents, vitamins, amino acids, and enzymes. Microorganisms can even leach valuable minerals from low-grade ores.

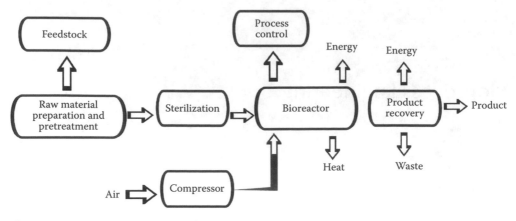

Figure 25.1 Flow sheet for a biochemical process.

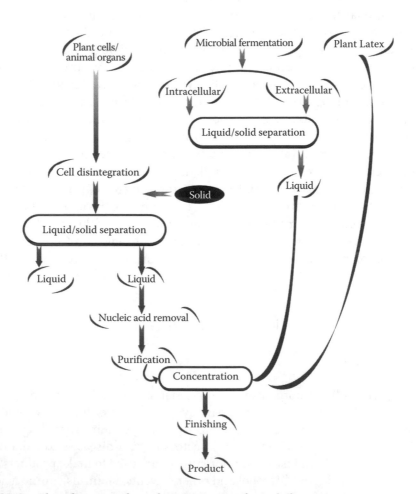

Figure 25.2 Flow diagram of purification process from different sources.

25.2.2 Materials

Shake flasks, Petri plates, agar medium, and yeast extract.

25.2.3 Protocol

Isolation of pure or mixed culture involves the following criteria for selecting a microorganism of industrially important processes:

1. *The nutritional characteristics of the organism*: It is frequently required that a process should be carried out by employing an inexpensive medium or a predetermined one.
2. *The optimum temperature of the organisms*: The use of organisms having an optimum temperature above 40°C considerably reduces the cooling costs of a large-scale fermentation; therefore, the use of such a temperature in the isolation procedure may be beneficial.
3. The reaction of the organisms with the equipment to be employed and the suitability of the organism to the type of process to be used.
4. The stability of the organism and its amenability to genetic manipulation.
5. The productivity of the organism, measured in its ability to convert substrate into product and to give a high yield of product per unit time.
6. *The ease of product recovery from the culture*: The ideal isolation procedure commences with an environmental source (frequently soil) which is highly probable to be rich in the desired types, as to favor the growth of those organisms possessing the industrially important characteristic. Alternatively, the isolation procedure may be designed to include certain microbial "weeds" to encourage the growth of more novel types.

25.2.4 Isolation Methods Utilizing Selection of the Desired Characteristics as Selective Factors

Enrichment culture is a technique which involves a mixed population and provides conditions either suitable for the growth of the desired type or unsuitable for the growth of the other organisms, for example, by the provision of particular substrates or the inclusion of certain inhibitors prior to the culture stage. It is often advantageous to subject the environment source (normally soil) to conditions which favors the survival of the organisms.

25.2.5 Enrichment Liquid Culture

1. Frequently carried out in shake flasks.
2. Growth of the desired type from a mixed inoculum results in the modification of the medium and therefore changes the selective force which may allow the growth of other organisms, viable from the initial inoculum.

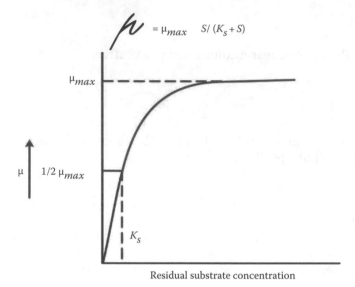

Figure 25.3 The effect of substrate concentration on the specific growth rate of a microorganism.

3. The selective force may be reestablished by inoculating the enriched culture into identical fresh medium (subculturing). Such subculture may be repeated several times.

4. The time of subculture in an enrichment process is critical and should correspond to the point at which the desired organisms are dominant.

5. The problems of time of transfer and selection on the basis of maximum specific growth rate in batch culture may be overcome by the use of a continuous process where fresh medium is added to the culture at a constant rate.

6. In continuous culture, growth rate is controlled by the dilution rate and is related to the limiting substrate concentration by the equation (Figure 25.3).

7. In continuous culture the specific growth rate is determined by the substrate concentration which is equal to the dilution rate, so that at dilution rates below point Y strain B would be able to maintain a higher growth rate than strain A, whereas at rates above Y strain A would be able to maintain a higher growth rate (Figure 25.4).

8. Thus the organisms which are isolated by continuous enrichment culture will depend on the dilution rate employed.

9. The enrichment medium employs organic carbon as a sole source of carbon and should be carbon-limited. The inclusion of other organic carbon sources, such as vitamins or yeast extract, may result in the isolation of strains adapted to using these, rather than the principal carbon source, as energy sources.

25.2.6 Disadvantages

The main difficulty in using a continuous enrichment process is the washout of the inoculum before an adopted culture is established. It is, therefore, advisable to start the isolation process in batch culture using 20% inoculum and as soon

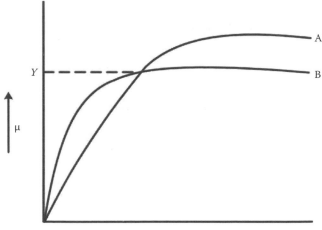

μ

Residual substrate concentration

Figure 25.4 The effect of substrate concentration on the specific growth rate of two microorganisms A and B.

as growth is observed, the culture should be transferred to fresh medium and the subsequent purification and stabilization of the enrichment is performed in continuous culture.

25.2.7 Solutions to the Problem of Early Washout

Use a turbidostat and/or two-stage chemostat. In the turbidostat, the cell density of the culture is continually monitored photometrically and whenever the turbidity of the culture reaches a critical point, the addition of fresh medium to the growth chamber is triggered. In the chemostat, fresh medium is continuously being added to the culture chamber and the rate of growth of the microorganism being cultured is regulated by the rate of flow of fresh medium into the vessel. Growth is regulated by the concentration of a limiting nutrient in this system.

25.2.7.1 Turbidostat
1. It is a continuous-flow system provided with photoelectric cell to determine the turbidity of the culture and maintain the turbidity between set points by initiating or terminating the addition of medium.
2. Thus, washout is avoided as the medium supply will be switched on/off if the biomass falls below the lower fixed point.
3. The use of turbidostat will result in selection on the basis of maximum specific growth rate as it operates at high levels of limiting substrate.

25.2.7.2 Chemostat
The first stage of the system was used as a continuous addition of inoculum and the second stage consisted of a large bottle containing basic medium inoculated with a soil infusion (Figure 25.5).

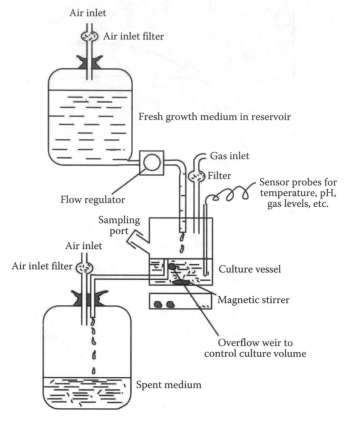

Figure 25.5 Continuous culture system (chemostat).

25.2.8 Enrichment Cultures Using Solidified Media

1. Soils of various pH were used as the initial inoculum and to a certain extent, the number of isolates correlated with the alkalinity of the soil sample.
2. The soil samples were pasteurized to eliminate nonsporulating organisms and then the sample is spread onto the surface of agar media at pH 9–10, containing an insoluble protein.
3. The size of the clearing zone could not be used quantitatively to select high products as there was not an absolute correlation between the size of the clearing zone and the production of alkaline protease in submerged culture.

25.2.9 Isolation Methods Not Utilizing Selection of the Desired Characteristic

The synthesis of some products does not give the producing organism any selective advantages which may be exploited directly in an isolation procedure. Examples include the production of antibiotics and growth promoters. Therefore, a pool of organisms has to be isolated and subsequently tested for the desired characteristic. A major problem faced by industrial microbiologists in this situation is the resolution of strains which have already been screened many times before:

1. Developing procedures to favor the isolation of unusual taxa which are less likely to have been screened previously
2. Identifying selectable features correlated with the undetectable industrial trait thus enabling an enrichment process to be developed
3. Incorporation of particular antibiotics into isolation media may result in the selection of the resistant taxa

25.2.10 Screening Methods

1. The evolution of antibiotic screens serves as an excellent illustration.
2. Antibiotics were initially detected by growing the potential producer on an agar plate in the presence of an organism against which antimicrobial action was required.
3. Production of the antibiotic was detected by inhibition of the test organisms.
4. Alternatively, the microbial isolate could be grown in liquid culture and the cell-free broth is tested for activity.
5. Antibiotics that inhibit cell wall biosynthesis and the mode of action resulted in a very significant increase in the discoveries of new β-lactose antibiotics. Increasing frequency of penicillin and cephalosporin resistance amongst clinical bacteria led to the development of mechanism-based screens for the isolation of more effective antibacterial antibiotics.

25.2.11 Screening Mechanism

The progress in molecular biology, genetics, and immunology has also contributed extensively to the development of innovative screens by enabling the construction of specific detector strains, increasing the availability of enzymes, receptors, and developing extremely sensitive assays. Molecular probes for particular gene sequences may enable the detection of organisms capable of producing certain product groups. The development of reporter gene assays has been used in the search for metabolites that disrupt viral replication.

25.3 Determination of Thermal Death Point and Thermal Death Time of Microorganisms for Design of a Sterilizer

Temperature is one of the most important physical factors affecting microbial growth. Bacteria are different from higher plants and animals as they lack homeostatic mechanism and cannot regulate against heat generated by metabolism and are, therefore, directly and readily affected by temperature. Over a limited temperature range, there is a twofold increase in the rate of enzyme-catalyzed reactions for every 10°C rise in temperature.

Bacteria may be divided into three major groups with respect to their temperature requirements: (i) *psychrophiles*, those with optimum temperature between 0°C and

20°C; (ii) *mesophiles*, those with optimum temperature between 20°C and 40°C; and (iii) *thermophiles*, those with optimum temperature between 40°C and 80°C. Thermophiles are of two types: (i) *facultative thermophiles* with an optimum temperature of growth between 45°C and 60°C and (ii) *obligate thermophiles* with optimum temperature above 60°C. Temperature in the range between 50°C and 100°C are normally in the lethal range for bacterial cells and spores. The range of temperature preferred by bacteria is genetically determined resulting in enzyme expression with different temperature requirements.

Each organism grows within a particular temperature range (i.e., cardinal temperature points). The minimum growth temperature is the lowest temperature at which growth of a species will occur. The highest temperature is the temperature at which a species can attain its maximum growth temperature and a species grows faster at its optimum growth temperature.

25.3.1 Effect of Incubation Temperatures on Growth

25.3.1.1 Precautions
Culture incubated at higher temperatures (55°C and above) should be kept in moist chamber in order to avoid drying of the slants.

25.3.1.2 Materials
Cultures of *Trichoderma viride*, *Fusarium solani*, *Escherichia coli*, and *Bacillus cereus*, nutrient agar slants (bacterial culture); Czapek–Dox agar plates (fungal culture); thermometer; glass marker; pipettes; spreader; water bath; bunsen burner; inoculating loop; inoculating needle; refrigerator set at 4°C; and four incubators set at 15°C, 35°C, 55°C, and 75°C.

25.3.1.3 Protocol
1. Label nutrient agar slants with the test organisms and the temperature of incubation (0°C, 4°C, 15°C, 35°C, 55°C, and 75°C).
2. Inoculate six nutrient agar slants each (label 0°C, 4°C, 15°C, 35°C, 55°C, and 75°C) with *E. coli* and *B.cereus* by streaking the surface of each slope.
3. Repeat Step 2 for inoculating the remaining cultures.
4. Label each of the Czapek–Dox agar plates with *F. solani* and *T. viride* and the temperature of incubation (0°C, 5°C, 15°C, 35°C, 55°C, and 75°C).
5. Inoculate six Czapek–Dox agar plates with *F. solani* and other six plates with *T. viride* by transferring a disk on the center of each plate.
6. Incubate one inoculated slant and plate from each set at 0°C (ice box of refrigerator), 4°C (refrigerator) for 10–14 days, and other slants and plates at 15°C, 35°C, 55°C, and 75°C (incubators set at required temperature) for 24–48 h (bacterial) and 7–14 days (fungal cultures).

25.3.1.4 Observations and Results
Examine cultures after 2 days incubation (incubated at 15°C, 35°C, 55°C, and 75°C) and after 10–14 days incubation (at 0°C and 4°C) for the presence (+) or absence (−)

TABLE 25.1 Minimum, Maximum, and Optimum Temperatures of Different Microorganisms

	Temperature (°C)						Cardinal Temperature			Classification as T, M, and P
Organism	0	4	15	35	55	75	Min.	Max.	Opt.	
T. viride										
F. solani										
E. coli										
B. cereus										

Notes: Growth expressed as – for nil, and 1+, 2+, and 3+ for degrees of growth. T, M, and P expresses for *thermophilic, mesophylic,* and *psychrophilic*.

of growth and the degree of growth (i.e., minimal growth [+], moderate growth [2+], heavy growth [3+], very heavy [maximum] growth [4+]). Express the results in the form of a table and determine the minimum, maximum, and optimum temperature range for each microorganism and classify each organism as *psychrophile, mesophile,* and *thermophile* (Table 25.1).

25.3.2 Lethal Effects of Temperature on Microorganism: Thermal Death Point

25.3.2.1 Introduction

Time of exposure is a vital factor in assessing the lethal effect of high temperature on bacterial cell. For this purpose, two methods are useful: (i) the thermal death point (TDP), the temperature at which an organism is killed in 10 min of exposure, and (ii) thermal death time (TDT) is the time required to kill cells/spores suspension at a given temperature. It is necessary to compare the susceptibility of different organisms to rising temperatures. However, some factors such as pH, moisture, composition of media, and age of cells influence the TDP. Enzymatic activities of any organism operate well at their optimum temperature which varies with the organism. Increase/decrease in temperature influences the microbial growth and survival leading to death also. The degree of tolerance is measured by exposing the microbes to gradually increasing temperature for a given period, for example, 10 min and determining their survival (Figure 25.6).

25.3.2.2 Precautions

1. The inoculum should be at least 1–2 cm away from the plate center where the streak sector converges on Petri plates.
2. Constant attention should be given to the temperature of water bath so that adjustment can be made before a serious fluctuation in temperature occurs.

25.3.2.3 Materials

Nutrient broth cultures of *E. coli* and *Bacillus subtilis* and *Staphylococcus aureus* (30–48 h old culture in flasks), hot water bath, nutrient agar in Erlenmeyer flask 100 mL, sterile empty tubes with cap (15), sterile cotton-plugged 10 mL pipettes (3), culture tube with 10 mL water (1), centigrade thermometer, wax glass marking

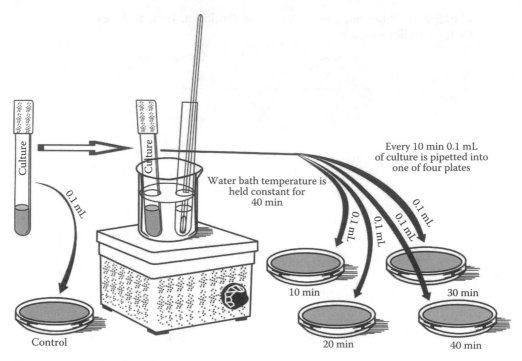

Figure 25.6 Flowchart for determining thermal death point.

pencil, culture tube stand, inoculating loop/needle, and bunsen burner. Set the water bath up to 60°C, 70°C, 80°C, 90°C, and 100°C.

25.3.2.4 Protocol

1. Liquefy the nutrient agar medium and cool to 50°C.
2. Label five Petri plates: control, 10, 20, 30, and 40 min.
3. Shake the culture of organisms and transfer 0.1 mL of organisms with a 1 mL pipette to the control plate.
4. Place the culture and a tube of sterile nutrient broth into the water broth. Remove the cap from the tube of nutrient broth and insert a thermometer into the tube.
5. As soon as the temperature of the nutrient broth reaches the desired temperature, record the time. Watch the temperature carefully to make sure it does not vary appreciably.
6. After 10 min, transfer 0.1 mL from the culture to the 10 min plate with a fresh 1 mL pipette. Repeat this operation at 10 min intervals until all the plates have been inoculated. Use fresh pipette each time and be sure to shake the culture before each delivery.
7. Pour liquefied nutrient agar (50°C) into each plate, rotate, and cool.
8. Incubate at 37°C for 24–48 h. After evaluating plates, record results (Table 25.2).

25.3.2.5 Observations and Results

Examine the plates for presence (+) or absence (−) of growth in all the sectors and compare with control.

Represent your results in a tabular form in the following table and find the TDP (at which death of an organism occurs) of each organism.

TABLE 25.2 Growth of Different Microorganisms at Different Temperatures

Organism	Temperature (°C)				
	60	70	80	90	100
B. megaterium					
E. coli					
S. aureus					

25.3.3 Determination of Thermal Death Time of an Organism

25.3.3.1 Introduction

TDT is the time required to kill the cells/spores/propagules of an organism at a given temperature. TDT differs from organism to organism. The degree of heat tolerance by a microorganism is determined first by exposing cells, etc., at a gradually increasing time, and then assessing its survival by subculturing on respective nutrient medium (Figure 25.7).

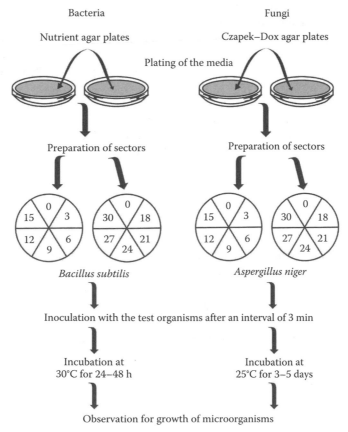

Figure 25.7 Flowchart for determination of TDT of microorganisms.

25.3.3.2 Materials

Nutrient broth cultures of *E. coli* and *B. subtilis* (30–48 h old culture in flasks), hot water bath, nutrient agar in Erlenmeyer flask 100 mL, sterile empty tubes with cap (15), sterile cotton-plugged 10 mL pipettes (3), culture tube with 10 mL water (1), centigrade thermometer, wax glass marking pencil, culture tube stand, inoculating loop/needle, and bunsen burner.

25.3.3.3 Protocol

1. Prepare nutrient broth, pour 2 mL into 11 small-sized tubes, plug properly, and autoclave at 121°C for 20 min.
2. Cool down and inoculate each tube with a loopful suspension of the bacterial culture and incubate at 30°C for 24 h.
3. Simultaneously, prepare nutrient agar medium and pour sterilized and cooled down medium into two Petri dishes.
4. Take nutrient agar plates and divide into six sectors on the back of the lower half with the help of a glass marker (pen or pencil).
5. Label the sector of one plate with time intervals to be treated as 0, 3, 6, 9, 12, and 15 min and the other with 0, 18, 21, 24, 27, and 30 min where 0 refers control.
6. Inoculate the 0 sector with the test bacterium suspension.
7. Note the TDP of the test bacterium from the experiment as described in the earlier experiment (see Section 25.3.3).
8. Prepare a water bath and maintain the temperature at 60°C.
9. Place 10 tubes containing 24 h old bacterial culture into the water bath maintained at 60°C.
10. After an exposure to 3 min, take out one tube and quickly cool under running tap water.
11. Make streak inoculation of the suspension on sector 3 of the plate.
12. Repeat the process at every 3 min interval, that is, exposure to 6, 9, 12, 15, 18, 21, 24, 27, and 30 min and make streak inoculated on sectors 6–30, respectively.
13. Incubate and inoculate plates at 30°C for 24–48 h.
14. Tabulate the results as per Table 25.3.

25.3.3.4 Observations and Results

Observe the inoculated plates for the presence of growth in various sectors and compare with the "0" time sector and record the growth as + (growth present) and − (no growth) and determine the TDT of each organism, that is, the time to kill the test organism at the given temperature.

TABLE 25.3 Determination of TDT of the Microorganisms

Test Organisms	Time of Exposure (min) at 60°C											Thermal Death Time (TDT) (min)
	0	3	6	9	12	15	18	21	24	27	30	
E. coli												
B. subtilis												
S. aureus												

Notes: Identical method can be adopted to determine the TDT of fungi by using conidial/hyphal suspension. Medium used for fungal should be Czapek–Dox agar and Czapek–Dox broth. The inoculated plates should be incubated at 28°C for about 1 week.

25.4 Determination of Ethanol Production Using Different Substrates

25.4.1 Introduction

Fermentation is the conversion of a carbohydrate such as sugar into an acid or an alcohol through microbes. Fermentation occurs naturally in many different foods given the right conditions, and humans have intentionally made use of it for several years. Fermentation is a biological and enzymatic process of conversion of carbohydrates into alcohols, lactic acid, carbon dioxide, etc. by some bacteria and fungi. These organisms use fermentation as a method of obtaining energy in the form of adenosine triphosphate (ATP). It is a cellular process that transfers the energy in glucose bonds to bonds in ATP:

$$C_6H_{12}O_6 \rightarrow 2CH_3CH_2OH + 2CO_2$$

A series of intermediate steps involving enzymes result in the breakdown of $C_6H_{12}O_6$ to pyruvic acid which is then converted to ethyl alcohol and CO_2:

$$CH_3CCOCOH \xrightarrow{\text{Pyruvate decarboxylase}} CH_3CHO + CO_2$$

$$CH_3CHO + NADH_2 \xrightarrow{\text{Alcohol dehydrogenase}} CH_3CH_2OH + NAD$$

The three main by-products of alcohol industry are CO_2, yeast, and distillery effluent.

25.4.2 Materials

Grape juice broth culture of *Saccharomyces cerevisiae* var. *ellipsoideus*, pasteurized commercial grape juice (500 mL), 1% phenolphthalein solution, 0.1 N sodium hydroxide, sucrose 20 g, 1 L sterile Erlenmeyer flask, rubber stopper (fitted with a 5 cm glass tube plugged with cotton), sterile 10 mL pipettes, burette, spatula, alcoholometer, distillation apparatus, wire gauze, glass beads, ring stand, and bunsen burner.

25.4.3 Protocol

1. Take 500 mL of the grape juice into a 1 L sterile flask.
2. Add 20 mL of sucrose and 50 mL of culture of the yeast (10% starter culture) to the flask. Size of the inoculum ranges from 8% to 10% with an average of 4% v/v.

3. High degree of aeration and agitation is necessary for preparing the inoculum since this favors rapid growth of yeast, producing a large amount of cells mass. pH is maintained at 4.8–5 while temperature at 28°C–30°C.
4. Close the flask with the rubber stopper.
5. Inoculate the flask at 25°C for 21 days and determine total acidity, volatile acidity, and alcohol at different intervals.
6. Add 20 g of sucrose to the fermenting juice after 2 and 4 days of incubation.
7. Determine total acidity and volatile acidity by titration method at 1–7 day interval, as follows:
   a. Take 10 mL of the fermenting juice to a flask.
   b. Add 10 mL of distilled water and five drops of 1% phenolphthalein solution as indicator.
   c. Mix the contents.
   d. Titrate against 0.1 N sodium hydroxide taken in a burette to the first persistent pink color.
   e. Note the volume of sodium hydroxide used.
   f. Repeat Steps (a) through (e) taking fermenting juice after 14 and 21 days.
   g. Repeat Steps (a) through (e) using uninoculated (control) grape juice.
8. Determine the % alcohol by dipping alcoholometer in the fermenting juice at 7 day intervals.
9. Assemble the distillation apparatus as illustrated in Figure 25.8.
10. Pour 100 mL of fermented grape juice in the distilling flask, add a few glass beads, and put thermometer in the hole.
11. Start circulation of water through the condenser.
12. Light the bunsen burner and allow the liquid to drip in the receiver flask.
13. Adjust the heat in such a way that the drops come at the rate of one per second.

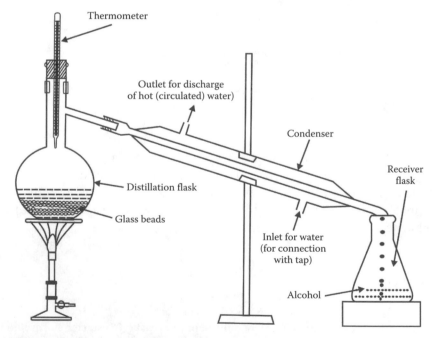

Figure 25.8   Distillation assembly for recovery of alcohol.

25.4.4 Observations and Results

Record and note the appearance of odor and presence of turbidity or sediment of the content of the flask.

Total acidity and volatile acidity can be calculated by applying the following formula:

Total acidity (expressed as % tartaric acid)

$$= \frac{\text{Volume (mL) of alkali used} \times \text{Normality of alkali} \times 7.5}{\text{Weight of sample in grams (1 mL} = 1 \text{ g})}$$

Volatile acidity (expressed as % acetic acid)

$$= \frac{\text{Volume of alkali} \times \text{Normality of alkali} \times 60}{\text{Weight of sample in grams}}$$

Record total acidity, volatile acidity, and % alcohol of the fermenting juice at 7 day intervals. Determine the amount of alcohol per 100 mL of the grape juice produced under laboratory conditions at 12 h, 24 h, or 7 day intervals.

25.5 Sugar Fermentation by Yeast (*S. cerevisiae*) for Production of Ethanol and Cell Biomass Using Laboratory Fermenter of 2 L Capacity

25.5.1 Introduction

Yeast can metabolize sugar in two ways, aerobically, with the aid of oxygen, or anaerobically, without oxygen. When the yeast respires aerobically, oxygen gas is consumed at the same rate that CO_2 is produced; there would be no change in the gas pressure in the test tube. When yeast ferments the sugars anaerobically, CO_2 production will cause a change in the pressure of a closed test tube, since no oxygen is being consumed. The use of this pressure change monitors the fermentation rate and metabolic activity of the organism (Figures 25.9 and 25.10).

25.5.2 Material

A small laboratory-grade fermenter of 2 L capacity, 50 mL centrifuge tubes, glucose reagent, yeast inoculum, and distilled water

25.5.3 Protocol

1. Take two centrifuge tubes and add glucose to each centrifuge tubes. Add 1, 2, 3, and 4 g glucose to achieve 5%, 10%, 15%, and 20% w/v glucose concentrations, respectively.
2. Add distilled water to make up the volume to 20 mL.

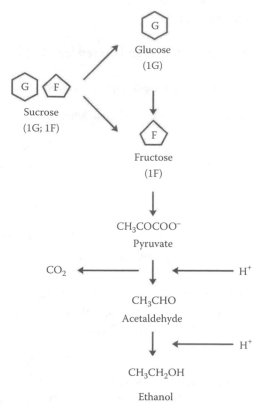

Figure 25.9 Fermentation reaction.

Figure 25.10 Fermenter after 24 h of incubation. The visible color in the media shows high cell density of *S. cerevisiae*.

3. Add 150 μL yeast inoculum to glucose solution. Yeast inoculum is prepared by incubating active dry yeast with distilled water for 20 min in 32°C water bath at 120 rpm.
4. Weigh the tubes. This is the 0 h reading.
5. Loosen the cap of centrifuge tubes. Keep them in 32°C water bath with shaking at 100 rpm.
6. Weigh the tubes again and record the time.
7. Next day, weigh the tubes after 24 h (according to the time you started the fermentation). Introduce the procedures for the laboratory.

Necessary data, see Table 25.4.
  Necessary equations

$$W_{CO_2} = W_t - W_0, \text{ refer Table 25.4} \tag{25.1}$$

$$m_{EtOH} = m_{CO_2} = W_{CO_2}/M_{CO_2} \tag{25.2}$$

$$A_{EtOH} = m_{EtOH} * M_{EtOH} \tag{25.3}$$

W_{CO_2} is mass of CO_2 produced at a particular time period; m_{CO2} is moles of CO_2 produced; m_{EtOH} is moles of ethanol produced; M_{CO2} is molecular weight of CO_2; M_{EtOH} is molecular weight of ethanol; A_{EtOH} is amount of ethanol produced (g) (Table 25.4).

TABLE 25.4 Result on the Effect of Different Concentrations of Glucose on Fermentation

Glucose Concentration (% w/v)	Time (h)	Weight of Centrifuge Tube (W_{time})	Calculate Loss in Weight ($W_t - W_0$)	Amount of Ethanol Produced (g)
5	0	$W_0 =$		
	8	$W_8 =$		
	24	$W_{24} =$		
	40	$W_{40} =$		
10	0	$W_0 =$		
	8	$W_8 =$		
	24	$W_{24} =$		
	40	$W_{40} =$		
15	0	$W_0 =$		
	8	$W_8 =$		
	24	$W_{24} =$		
	40	$W_{40} =$		
20	0	$W_0 =$		
	8	$W_8 =$		
	24	$W_{24} =$		
	40	$W_{40} =$		

25.5.4 Data Analysis Procedure

1. Determine amount of ethanol produced (in grams) at respective time periods for four different glucose concentrations in the fermentation media.
2. Plot amount of ethanol produced versus fermentation time for four glucose concentrations.
3. Discuss correlation of total ethanol produced with glucose concentrations.

25.6 Measurement of Yeast Biomass Using Hemocytometer Technique

25.6.1 Materials

Dry yeast, distilled water, and methylene blue solution.

25.6.2 Protocol

1. Yeast inoculum is prepared by adding 5 g dry yeast and 25 mL distilled water.
2. This solution is incubated in 32°C water bath at 120 rpm for 20 min.
3. After incubation, yeast solution is diluted 1000 times.
4. To 1000 µL of diluted yeast solution, methylene blue dye is added. Dead yeast cells become blue in color while live cells retain original mustard yellow color.
5. Pour 100 µL of the earlier solution into hemocytometer well.
6. Count yeast cells in the four corner squares of the hemocytometer.
7. Yeast concentration is determined based on dimensions of the hemocytometer square and cells counted.

25.6.2.1 Calculations

$$\text{Total live cell concentration, cells/mL} = \frac{\text{Number of live cells} \times \text{Dilution factor}}{\text{Volume of corner square}}$$

$$\text{Total dead cell concentration, cells/mL} = \frac{\text{Number of dead cells} \times \text{Dilution factor}}{\text{Volume of corner square}}$$

Volume of the corner square, mL = 16 × volume of smallest square.

25.6.2.2 Experimental Procedures
1. Determining cell concentration:
   a. *Zeroing the spectrophotometer.* Set the wavelength to 630 nm. Using the knob on the left, set the reading to 0% transmission when the chamber is empty; using the knob on the right, set the reading to 100% transmission when the chamber contains a test tube with about 4 mL of pure medium.
   b. *Determining cell concentration.* First flush out the sample tube by taking an 8–10 mL fermenting sample. Take about 4 mL from your sample, mix by shaking the

tube, and transfer it to a glass test tube. Clean the outside of the test tube with ethanol, insert it into the spectrophotometer, and record the absorbance reading. If the absorbance reading is greater than 0.25, a typical limit of linear correlation between the absorbance and cell mass concentration, dilute the sample with a known amount of pure medium, and measure the absorbance again to check if the absorbance reading is on the linear portion of the calibration curve. Record the time you take the sample along with the absorbance reading in the linear range as well as the dilution factor.

2. Glucose measurement:

Take undiluted yeast sample tube and insert the thin plastic sample tubing from the glucose analyzer into it. If the analyzer is in the "Standby" mode, hit the "Run" key; after the machine has calibrated, hit the "Sample" key. If the analyzer is not in the "Standby" mode, just hit the "Sample" key. The glucose concentration in the medium from your sample will print out in about 1 min. Record this value.

3. Ethanol assay:

   a. Take 0.5 mL from your undiluted sample, pipette it into an Eppendorf tube, and spin at 14,000 rpm for 5 min.

   b. Add 2 mL of ethanol assay reagent to each of three cuvettes.

   c. Add 10 µL of pure medium to the first cuvette; this is your blank. Cover with parafilm and mix gently.

   d. Add 10 µL of the ethanol standard (0.08% w/v) to the second cuvette; this is your standard. Cover with parafilm and mix gently.

   e. Add 10 µL of sample (from the supernatant from Step a) to the third cuvette; this is your sample. Cover with parafilm and mix gently.

   f. Incubate cuvettes at room temperature for 10 min.

   g. Clean the outside of the cuvette, and then read the absorbance of each at 340 nm; use the blank to zero the spectrophotometer. If the sample absorbance is >1.7, dilute the supernatant from your sample 1:4 and perform the assay again using this diluted supernatant. Don't redo the blank and standard.
   Record this value.

   h. Calculate the ethanol concentration for your sample using the following formula:

$$[\text{Ethanol}](\text{g/L}) = \frac{A(\text{Sample})}{A(\text{Standard})} \times 8.0 \left(\text{for the undiluted samples} \right)$$

25.7 Study Fermentation Process for Lactic Acid Production by *Streptococcus thermophilus* and Cell Biomass

25.7.1 Introduction

Fermentation is an anaerobic (without oxygen) process; cellular respiration is aerobic (utilizing oxygen). All living organisms, including bacteria, protists, plants, and animals, produce ATP in fermentation or cellular respiration and then use ATP in their

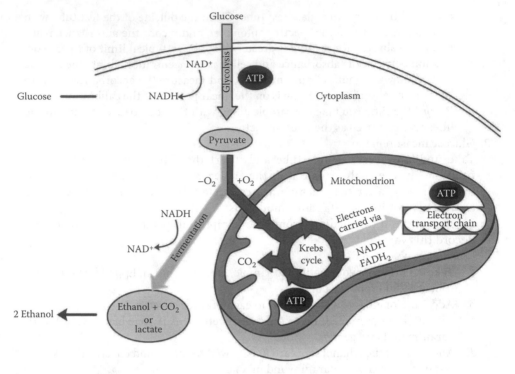

Figure 25.11 Stages of cellular respiration and fermentation.

metabolism. The process of fermentation followed by *S. thermophilus* is homolactic fermentation in which 1 mol of glucose gets converted into 2 mol of lactic acid:

$$C_6H_{12}O_6 \rightarrow 2CH_3CHOHCOOH$$

Before the fermentation starts, molecule of glucose must split into 2 mol of pyruvate and the process is known as glycolysis. Fermentation and cellular respiration involve redox reactions (oxidation–reduction reactions). Redox reactions are defined in terms of electron transfers, being reduction the gain of electrons and oxidation the loss of electrons. In cellular respiration, two hydrogen atoms are removed from glucose (oxidation) and transferred to a coenzyme called nicotinamide adenine dinucleotide (NAD^+), reducing this compound to NADH (Figure 25.11).

25.7.2 Material

Small laboratory-grade fermenter, balance, autoclave, cooling centrifuge, pH meter, and hot plate
Media composition

Ingredients	g/L
Universal peptone	10
Meat extract	5

Yeast extract	5
Glucose	20
Dipotassium hydrogen phosphate	2
Diammonium hydrogen citrate	2
Sodium acetate	5
Magnesium sulfate	0.1
Manganous sulfate	0.05

Final pH 6.5 ± 0.2 at 37°C.

25.7.3 Protocol

1. Prepare 1 L MRS broth and pour into fermenter vessel.
2. Autoclave media at 121 psi for 20 min in vessel and allow it to cool overnight.
3. Inoculate actively growing culture of *S. thermophilus* to fermenter vessel and initiate fermentation process.
4. Allow the process to run for 24 h.
5. Harvest the culture from the fermenter by centrifugation at 4000 rpm.

25.7.3.1 Fermentation Process

The inoculation of *S. thermophilus* should always be done carefully under total aseptic conditions through the inoculation port on the top plate of fermenter with the help of inoculation syringe (Figure 25.12).

After inoculation, adjust pH to 7 and allow the fermentative process to run for 24 h for complete cell multiplication leading to high cell density or mass production. The required conditions for *S. thermophilus* to grow in a uniform manner are the following:

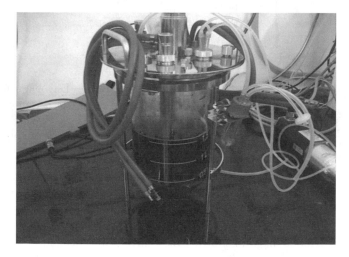

Figure 25.12   A laboratory-grade fermenter 2 L capacity with MRS broth.

- Temperature should not be less than 36°C and not more than 41°C
- pH of the media should be in between 6.5 and 7.2
- Very low amount of oxygen can be supplied as the organism is facultative anaerobe

The pH of the medium starts dripping as the microbe enters the log phase of the bacterial growth curve because of the production of lactic acid which is an end product of the homofermentative process of the microbe to gain energy for survival.

25.8 Microbial Production of Citric Acid Using *Aspergillus niger*

25.8.1 Introduction

Citric acid ($CH_2COOH\ COH\ COOH\ CH_2\ COOH$), a tricarboxylic acid, is a natural constituent of a variety of fruits (Figure 25.13). Citric acid is a commodity chemical, and more than a million tones are produced every year by fermentation. It is used mainly as an acidifier, as a flavoring, preservative, and as a chelating agent. In this production technique, which is still the major industrial route to citric acid used today, cultures of *A. niger* are fed on a sucrose- or glucose-containing medium to produce citric acid. The source of sugar is corn steep liquor, molasses, hydrolyzed corn starch, or other inexpensive sugary solutions. After the mold is filtered out of the resulting solution, citric acid is isolated by precipitating it with lime (calcium hydroxide) to yield calcium citrate salt, from which citric acid is regenerated by treatment with sulfuric acid.

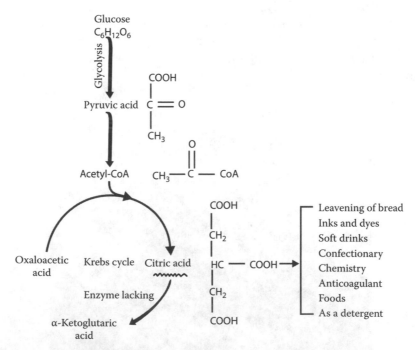

Figure 25.13 Schematic diagram showing mechanism of citric acid production through glycolytic pathway and Krebs cycle in *A. niger.*

25.8.2 Material

25.8.2.1 Organism and Culture Maintenance

Soil isolate of *A. niger* is used. The strain is developed by alternate treatment of ultraviolet irradiations (1.6×10^2 J/m²/s) and *N*-methyl-*N'*-nitro-*N*-nitrosoguanidine (NTG—100 mg/mL) for different time intervals (5–45 min) and maintained on sterilized potato dextrose agar (PDA) medium (diced potato 200 g/L, dextrose 20 g/L, and agar 15 g/L), pH 4.5, and stored at 5°C in the refrigerator. All the culture media, unless otherwise stated, are sterilized in autoclave at 15 lb/in.² pressure (121°C) for 15 min.

25.8.2.2 Preparation of Fermentation Medium for Citric Acid Production

Sugar (molasses/sucrose)	14%–15%
Nitrogen source	
$(NH_4)_2SO_4$, NH_4NO_3, $NaNO_3$, KNO_3, urea, etc.	0.25%
KH_2PO_4	0.10%–0.15%
$MgSO_4 \cdot 7H_2O$	0.02%–0.025%
Trace element	
Iron	0.2 ppm
Copper sulfate	4.7 mg%
Zinc	1–2 µM
pH molasses medium	5–6
Sucrose medium	2–3

25.8.2.3 Pretreatment of Molasses

Cane molasses obtained from sugar mills is used as an inexpensive carbon substrate. Cane molasses contains water 20%, sugar contents 62%, nonsugar contents 10%, and inorganic salts (ash contents) 8%, making a blackish homogenous liquid with high viscosity. Ash contents include ions such as Mg, Mn, Al, Fe, and Zn in variable ratio. Sugar content is diluted to about 25% sugar level. The molasses solution, after adding 35 mL of 1 N H_2SO_4 per liter, is boiled for half an hour, cooled, neutralized with limewater (CaO), and is left to stand overnight for clarification. The clear supernatant liquid is diluted to 15% sugar level. When molasses is used, it is diluted to a sugar concentration of 15%–20% with dilute sulfuric acid and to a pH 5.5–6.5. Other nutrients for fungal growth are then added and the mixture is sterilized for 30 min to reduce or eliminate the inhibitory action of metals.

25.8.2.4 Inoculum Preparation

Hundred milliliters of molasses medium (sugar 15%, pH 6.0) containing glass beads, in 1 L cotton wool–plugged conical flask, is sterilized. One milliliter of conidial suspension (6.5×10^6 conidia) from the slant culture is aseptically transferred. The conidial count is made by hemocytometer slide bridge. The flask is then incubated at 30°C in an incubator shaker at 200 rpm for 24 h.

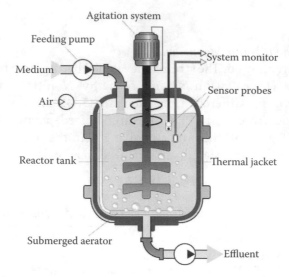

Figure 25.14   External view of an ideal stainless steel fermenter of 15 L capacity.

25.8.3 Method

25.8.3.1 Fermentation Technique

Stainless steel fermenter of 15 L capacity with working volume of 9 L (60%) is employed for citric acid fermentation. Prepared inoculum is transferred to the production medium at a level of 5% (v/v). The incubation temperature is kept at 30°C ± 1°C throughout the fermentation period of 144 h. Agitation speed of the stirrer is kept at 200 rpm while aeration rate is maintained at 1.0–4.0 L/L/min. Sterilized silicone oil is used to control foaming during fermentation (Figure 25.14).

25.8.3.2 Product Recovery

The crude-fermented liquor containing citric acid, obtained either from surface or submerged, or solid state, is subjected to the following method for product recovery:

1. Filter crude-fermented mother liquor to remove mycelia or other suspended impurities. Heat the mother liquor to 80°C–90°C and add small amount of hydrated lime to precipitate oxalic acid.
2. Critic acid is then precipitated as calcium citrate using 1 part of hydrated lime for every 2 parts of liquor added over a 1 h period, while the temperature raised to 95°C. Precipitation of citric acid can be carried at a temperature of 50°C for 20 min.
3. The precipitated calcium citrate is filtered, washed with water several times, and transferred to acidulate. Treat with H_2SO_4 and filter the solution again to remove $CaSO_4$: Calcium citrate + $H_2SO_4 \rightarrow CaSO_4$ + citrate. Mother liquor containing citric acid is decolorized by passing through ion-exchange column with activated charcoal.
4. Concentrate the liquor under vacuum and finally run into low-temperature crystallizer. Citric acid is crystallized as monohydrate.

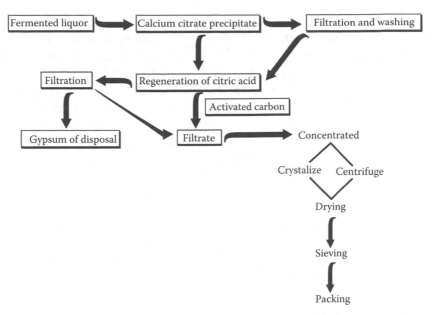

Figure 25.15 Flowchart for the downstream processing of citric acid production by *A. niger*.

5. Quality of citric acid can further be improved by treating with metal-free $Ca(OH)_2$. Also add 9%–12% calcium ferrocyanide to mother liquor at 95°–97°C for 5–8 min to improve the quality of citric acid (Figure 25.15).

25.8.4 Result

Citrate obtained from *A. niger* recovered by this method is recommended for food, drug, and cosmetic uses.

25.9 Microbial Production of Antibiotics (Penicillin)

25.9.1 Introduction

Antibiotics are secondary metabolites produced by microorganisms that inhibit or kill a wide spectrum of other microorganisms. These are produced by molds (*Penicillium, Cephalosporium, Sordaria*) and bacteria of the genus *Streptomyces* and a few *Bacillus* and *Paenibacillus* species. Antibiotics are produced by the submerged culture method.

25.9.1.1 Biotransformation of Antibiotics
There has been an emergence in the production of new and improved antibiotics by the microbial transformation of existing compounds. The objective is the

development of modified antibiotics with improved effectiveness, reduced toxicity, wider antimicrobial spectrum, improved oral absorption, decreased development of resistance, or lower allergic effects. In most cases, any transformation step causes a partial or complete inactivation of the antibiotic. The addition of inhibitors or modified precursors to the medium may result in the synthesis of altered antibiotics via controlled biosynthesis. For example, in the presence of cis-4-methylproline, *Streptomyces parvulus* produces two new actinomycins in which proline is replaced by this proline analog. New compounds have been found when mutants blocked in the synthesis of a particular antibiotic were used. A few improved antibiotics have been isolated after mutational synthesis, of which only 5-episisomycin has proved of sufficient utility to undergo clinical trials. Semisynthetic antibiotics, for example, semisynthetic penicillin and cepalosporins, are produced by chemical modification of the penicillin, 6-aminopenicillinic acid. The penicillin produced penicillin G and penicillin V.

The industrial production of penicillin was broadly classified into two processes, namely

- Upstream processing
- Downstream processing

Upstream processing encompasses any technology that leads to the synthesis of a product. Upstream includes the exploration, development, and production. The antibiotic penicillin is produced from high-yielding strain of the fungus *Penicillium chrysogenum* for the commercial production. Six major steps involved in the commercial production of penicillin are (1) inoculum preparation, (2) preparation of the medium (broth), (3) inoculation of the medium in the fermenter, (4) incubation with forced aeration during fermentation, (5) separation of fungal biomass, and (6) extraction, purification, and recovery of the penicillin.

25.9.2 Material

25.9.2.1 Strain Development
It is highly desirable to choose a high-yielding strain which could be obtained by sequential genetic selection. In other words, such a strain can be obtained by stepwise development with the help of a series of mutagenic treatments, each followed by the selection of mutants. Mutations in *P. chrysogenum* are induced by UV. Mutant possesses a far greater capacity for antibiotic production rather than the wild strain.

Mutagenically improved strains of *P. chrysogenum* are being used for the commercial production of penicillin in excess of 30,000 mg/L. The strain is grown on barley seed for preparation of the inoculums. The flask containing 15 g barley seeds are inoculated with the culture from the slant and incubated at 25°C for 3 days for the conidia to develop on the seeds. Conidial suspension is made in sterile distilled water and is used to inoculate the seed tank (Figure 25.16).

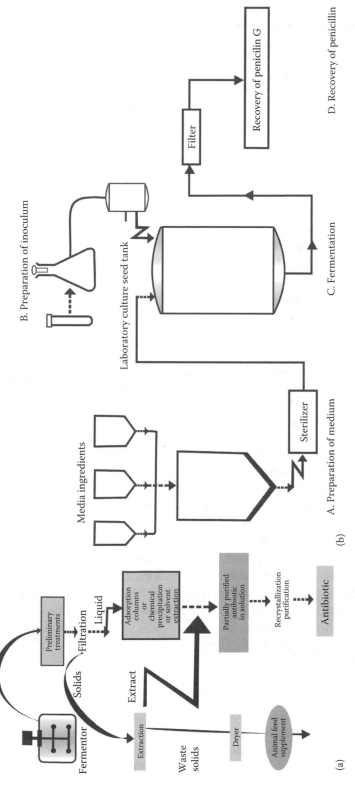

Figure 25.16 Flowcharts showing production of (a) antibiotics and (b) penicillin.

25.9.2.2 Storage of Strains

Production strains are stored in a dormant form by any one of the following standard culture preservation techniques:

- A spore suspension may be mixed with a sterile, finely divided inert support and desiccated.
- Spore suspension can be lyophilized in appropriate media.
- Spore suspension can be stored under liquid nitrogen, that is, in a frozen state.

25.9.2.3 Inoculum Preparation

The aim is to develop pure inoculums in sufficient volume and in the fast growing phase for the production-stage fermenter. The time taken for each stage is measured in days, and it decreases as the sequence progresses. The medium is designed to provide the organism with all the nutrients that it requires. For example, Moyer and Coghill (1946) suggested the following sporulation medium (g/L) glycerol 7.5; cane molasses 7.5; corn-steep liquor 2.5; $MgSO_4 \cdot 7H_2O$ 0.05; KH_2PO_4 0.06; peptone 5.0; NaCl 4.0; Fe tartrate 0.005; $CuSO_4 \cdot 5H_2O$ 0.004; agar 2%; distilled water to 1 L.

The flask containing the seeded medium is incubated on a rotary shaker to effect aeration and agitation. Adequate oxygen is supplied in the form of sterile air. In addition, the temperature is controlled at 24°C.

25.9.2.4 Fermentation Medium

It contains corn steep liquor solids 3.5%; $CaCO_3$ 1%; lactose 3.5%; KH_2PO_4 0.4%; edible oil 0.25%, glucose 1%; *Penicillium* precursor with slow feeding rate.

The medium of a typical fed-batch culture may vary depending on the strain and usually consists of corn steep liquor (4%–5%, dry weight), which may be replaced by soy meal, yeast extract, or whey; a carbon source (lactose acts as a very satisfactory carbon compound, provided that is used in a concentration of 6%, others such as glucose and sucrose may also be used); ammonium sulfate and ammonium acetate can be used as nitrogenous sources; and various buffers. The pH is kept constant at 6.5. The elements namely potassium, phosphorus, magnesium, sulfur, zinc, and copper are essential for penicillin production. Some of these are applied by corn steep liquor. Phenylacetic acid or phenoxyacetic acid is fed continuously as a precursor. Lactose, the carbon source, is obtained from whey and is the initial carbon source for the mold. Inoculation broth consisting of corn steep liquor (a by-product of turning corn into starch) which contains nitrogen and several growth factors, lactose, ammonium sulfate, calcium carbonate, calcium hydroxide, sodium hydrogen phosphate, and phenylacetic acid (pH 5.2) is prepared, sterilized, cooled, and pumped into the fermenter.

25.9.3 Methods

25.9.3.1 Inoculation Methods

Any one of the following methods may be used to inoculate the fermentation medium in the submerged culture production of penicillin:

- Dry spores may be used to seed the fermentation medium.
- The fermentation medium may be seeded by pellet inocula obtained by the germination of spores, with the formation of mycelia growth, under submerged conditions. It is a common practice to seed the fermentation medium with pellets 2 or 3 days after the medium has been seeded with pairs.

25.9.3.2 Fermentation Process

The fermentation process is usually carried out in a large fermenter. The fermentation medium is formulated and fed into the fermenter. The inoculum is also maintained properly. Then a small amount of inoculum is seeded into the fermenter. The fermentation conditions such as temperature, pH, mineral contents should also be maintained. The process starts immediately after the addition of the inoculum strain. After the required amount of time, the required antibiotic is produced in the fermenter. It is followed by the inoculation of the medium in the fermenter from the seed tank; sterile air is forced to aerate the culture and incubation to be run at 23°C–25°C for 7 days. Penicillin yield can be increased by the addition of glucose along with a low level of nitrogen in the fermentation broth at a time when most of the lactose has been metabolized by the mold. Synthesis of penicillin begins when nitrogen from ammonia becomes limiting in the broth. Once the fermentation is complete, that is, the maximum yield of penicillin has taken place, the fungal biomass is removed by filtration, and the penicillin is recovered in pure form by a series of manipulations which include absorption, precipitation, redissolving and filtration, and crystallization to yield the final product, that is, penicillin G. The basic product can be modified by chemical procedures to yield semisynthetic penicillins such as ampicillin and amoxicillin.

25.9.3.3 Downstream Processing

The extraction and purification of a biotechnological product from fermentation is referred as downstream processing.

25.9.3.3.1 Stages in Downstream Processing.   *Solid–liquid separation*: The first step in product recovery is the separation of whole cells and other insoluble ingredients from the culture broth. Several methods such as floatation, flocculation, etc. are used to serve this purpose.

Concentration: This is a step followed to concentrate the desired product. The methods used to concentrate include adsorption, precipitation, etc.

Purification: Chromatographic techniques are generally used to purify the product. As concerned with the production of antibiotics, ion-exchange chromatography seems to be the better option.

25.9.3.3.2 Extraction and Purification of Penicillin.   There are some special techniques used for the extraction of penicillin from the fermentation medium. They are as follows:

Removal of mycelium: At harvest, the penicillin is in solution exocellularly, together with a range of other metabolites and medium components. First remove the mycelium by filtration. A rotary vacuum filter is employed for the filtration of fermented production medium. This stage is carried to avoid contamination of the filtrate with penicillinase-producing microorganisms which may cause serious or total loss of an antibiotic.

Countercurrent solvent extraction of penicillin: The next stage is to extract the penicillin. The pH is adjusted to 2–2.5 with the help of phosphoric or sulfuric acids. In aqueous solution at low pH values, there is a partition coefficient in favor of certain organic solvents. This step has to be carried out quickly for penicillin as is very unstable at low pH values. Podbielniak countercurrent solvent extractor is used for this purpose. Antibiotic is then extracted back into an aqueous buffer at a pH of 7–7.5, the partition coefficient now being strongly in favor of the aqueous phase. The resulting aqueous solution is again acidified and reextracted with an organic solvent. These shifts between the water and solvent help in the purification of penicillin. The spent solvent is rediscovered by distillation for reuse.

Treatment of crude extract: The treatment of the crude penicillin extract varies according to the objective, but involves the formation of an appropriate penicillin salt. The solvent extract recovered in the previous stage is carefully extracted back with aqueous sodium hydroxide. This is followed by charcoal treatment to eliminate pyrogens and by sterilization. Pure metal salts of penicillin can be safely sterilized by dry heat, if desired, and the aqueous solution of penicillin is subjected to crystallization. For parental use, the antibiotic is packed as a powder or suspension, in sterile vials. For oral use, it is tabletted usually with a film coating. Searching tests (e.g., for purity, potency) are performed on the appreciable number of random samples of the finished product. It must satisfy fully all the strict government standards before being marketed.

25.9.4 Production of Steroid Hormones through Biotransformation

The derivatives of sterols are called steroids. Steroids are important animal hormones involved in regulating various metabolic processes and are also used as life-saving drugs by humans, such as in the treatment of arthritis and control of fertility.

Steroids can be obtained by chemical synthesis, but this is a complicated and extensive process. Some fungi have been discovered that can bring about precise changes in steroid molecules, and used to carry out some of the steps involved in the conversion of readily available plant sterols into therapeutically useful compounds that would be difficult to accomplish by purely chemical means. The process used is called biotransformation or bioconversion, a reaction in which one compound is converted to another by enzyme in cells.

Two well-known examples of biotransformation are use of a zygomycetous fungus, *Rhizopus nigricans* to hydroxylate progesterone (progesterone is obtained by a series of chemical steps from stigmosterol, extracted from soybeans) for the synthesis of cortisone hormone (Figure 25.17). This conversion involves three reactions. The first reaction is a typical microbial transformation carried out by *R. nigricans* and the other two reactions are performed chemically as follows:

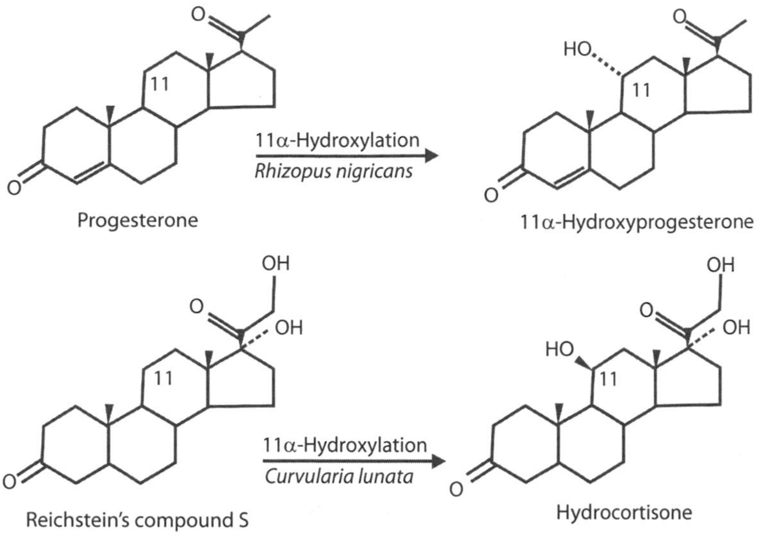

Figure 25.17 Two common examples of steroid biotransformation by employing *R. nigricans* and *C. lunata*.

$$\text{Progesterone} \xrightarrow[\text{R. nigricans}]{\text{Biotransformation}} 11\alpha\text{-Hydroxyprogesterone}$$

$$\xrightarrow{\text{Chemical reaction}} \text{Hydrocortisone} \xrightarrow{\text{Chemical reaction}} \text{Cortisone}$$

Another steroid hormone, prednisolone is being produced by using the fungus *Curvularia lunata* which bring out β-hydroxylation to Reichstein compound to hydrocortisone (Figure 25.17). Hydrocortisone is chemically converted to the therapeutic useful compound, the prednisolone.

Currently, biotransformation process is being used for the industrial production of most sterols. The process followed is the growth of the mold in large fermenters, followed by the addition of a sterol precursor at an appropriate time. Fermentation is allowed to proceed till the sterol precursor is transformed in the fermentation broth and the sterol is further concentrated and purified. Other hormones that are being produced industrially include insulin, human growth hormone, or somatostatin by using recombinant DNA technology through modified strains of *E. coli*.

25.10 Production of Vinegar (Acetic Acid)

25.10.1 Introduction

The acetic acid (vinegar) fermentation is one of the oldest known to mankind, occurring naturally as an unwanted spoilage of wine. The active ingredient in vinegar is acetic acid, which is produced by acetic acid bacteria oxidizing an alcohol containing fruit juice. Vinegar is actually produced in two stages: the first stage utilizing

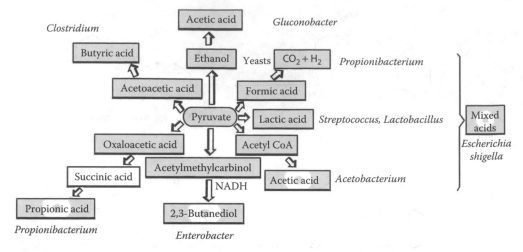

Figure 25.18 Schematic representation of production of various organic acids by different bacteria.

a yeast such as *Saccharomyces cerevisiae* to produce ethanol from plant sugar. The second stage involves an aerobic fermentation carried out by acetic acid bacteria such as *Acetobacter* and *Gluconobacter* (Figure 25.18). These bacteria oxidize ethanol through acetaldehyde to acetic acid as shown below:

$$2C_2H_5OH + \frac{1}{2}O_2 \xrightarrow{\text{Alcohol dehydrogenase}} \underset{\text{Acetaldehyde}}{CH_3CHO} + H_2O$$

$$CH_3CHO + \frac{1}{2}O_2 \xrightarrow{\text{Aldehyde dehydrogenase}} \underset{\text{Acetic acid}}{CH_3COOH}$$

25.10.2 Material

25.10.2.1 Substrate

The substrate for the first fermentation can be any natural material that can be fermented to yield alcohol, including various fruits, berries, honey, wine, malt, etc. Based on the substrate used, names of vinegar are given: (i) sugar vinegar (alcoholic fermentation of sugar by yeast followed by oxidation of alcohol to acetic acid), (ii) malt vinegar (alcoholic fermentation of an infusion of barley malt followed by its acetic acid fermentation), (iii) cider vinegar (fermentation of apple cider to hard cider [alcohol] followed by acetic fermentation), and (iv) wine vinegar (oxidation of wine by *Acetobacter*), etc. (Figure 25.19). The final volume is diluted so as to contain 4% acetic acid in vinegar. In addition to acetic acid, vinegar also contains traces of other compounds such as glycerol, ethanol, esters, salts, etc. Sulfur dioxide is added to the fermentation vessel to control the bacterial growth. At the end of the alcohol fermentation, the yeast cells and various sediments are allowed to settle, and the supernatant alcoholic broth is withdrawn for second fermentation to be carried out by acetic acid bacterium.

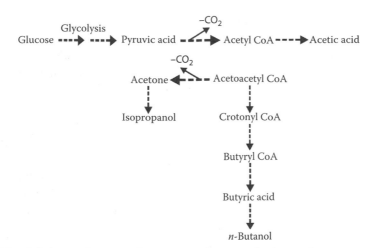

Figure 25.19 Schematic diagram showing production of *n*-butanol, acetone, isopropanol, and acetic acid during acetone butanol fermentation.

Vinegar is produced by several techniques but the most common is inoculation of cane juice yeast sugar. The sugar at 4% level are most suitable for fermentation. The high concentration inhibits the growth of yeast.

Erlenmeyer flask (500 mL), Brewer's jar, palladium crystal, sealing wax, large size glass sheet, filter paper, Buchner funnel, horizontal shaker, and incubator.

25.10.2.2 Microorganisms
Pure cultures of *Saccharomyces cerevisiae, Acetobacter* species.

25.10.3 Process

There are three different processes for the production of vinegar: the open-vat method, trickle method, and the bubble method. The *open-vat method* or *Orleans method* is the original process which is still used in France where it was first used. The fermentation vessel for the Orleans process consists of large barrels or casks placed either upright or on their sides. Holes in the walls of these vessels above the liquid level admit air. The *Acetobacter* bacteria produce considerable slime and grow as a film on the surface of the alcoholic broth, being supported at the surface on a floating raft composed of wooden grafting. The oxidation of ethanol to acetic acid by this process is quite slow, however, requiring an incubation of 1–3 months, and during this period other nonacetobacter bacteria also are active, producing lactic and propionic acids which, are esters, impart a unique flavor and aroma to the vinegar.

The *trickle method* (or the packed generator or quick vinegar method), developed in Germany, is a more rapid method for the microbial oxidation of ethanol to acetic acid. The vinegar generator employed in this process is a large vertical tank, either opened or closed at the top, and loosely packed with beech-wood shavings, twigs, corn cobs, bamboo-stick bundles, or other packing agent, and the alcoholic broth, which is added intermittently to the top of the generator, trickles down through the shavings so as to provide contact of the alcohol with the cells and consequent oxidation of the alcohol. The alcoholic broth is recycled through the fermenter until the alcohol becomes virtually completely oxidized, or it passes through several generators in series to successively oxidize the alcohol at each of the generators.

The *bubble method* is basically a submerged fermentation process. Fruit wines and mashes with low concentrations of alcohol were first used in submerged processes. With such low-yielding processes, aeration was not critical, but in present day high-yielding processes, which produce 13% acetic acid in quantities of up to 50 m^3, aeration must be highly regulated. The fermenter resembles other bioreactors. The tanks are constructed of stainless steel and are stirred from the bottom. The aeration apparatus consists of a suction rotor with the incoming air coming down through a pipe from the top of the fermenter. Heat exchangers control the temperature and mechanical foam eliminators must usually be installed. Household vinegar is produced in a semicontinuous, fully automatic process, under continuous stirring and aeration, beginning with a starting material that contains 7–10 g acetic acid and 5% ethanol. The ethanol concentration is measured continuously during the process and when the concentration sinks to 0.05%–0.3% (about 35 h), 50%–60% of the solution is removed and replaced with a new mash containing 0–2 g acetic acid/100 mL and 10%–15% ethanol. The acetic acid obtained in the submerged process is turbid due to the efficiency of the bubble method is quite high and 90%–98% of the ethanol is converted to acetic acid.

25.10.4 Protocol

1. Take 100 mL of substrate juice (sugar cane juice) in a 500 mL Erlenmeyer flask and sterilize through Millipore filter assembly. Incubate this juice with 1 mL suspension of pure culture of *S. cerevisiae*. Place the Brewer's jar (a special jar) over the flask. Tightly seal the jar with wax to warrant the entry of air. Introduce H$_2$ or mixture of gases (95% N$_2$ and 5% CO$_2$) into the apparatus after proper sealing. Heat the palladium crystal to activate for reaction of residual O$_2$ and H$_2$ and formation of water. This results in anaerobic condition (anaerobic condition may also be created by candle jar method) (Figure 25.20).

2. Incubate this assembly at 30°C ± 1°C for 7–10 days. During this period, yeast anaerobically ferments sugar into alcohol.

3. After incubation, remove the jar and filter the fermented product first through sterile filter paper and then through bacterial membrane filter.

4. Acidify the fermented product by mixing three parts of alcoholic filtrate and two parts of acetic acid. It is done to speed up the oxidation process.

5. Inoculate it with 1 mL broth of pure culture of *Acetobacter* species.

6. Place it on horizontal shaker for proper agitation and air supply at room temperature (28°C–35°C) for a week. This content of bacteria, medium, and air results in higher rate of oxidation. This fermented product is called as vinegar.

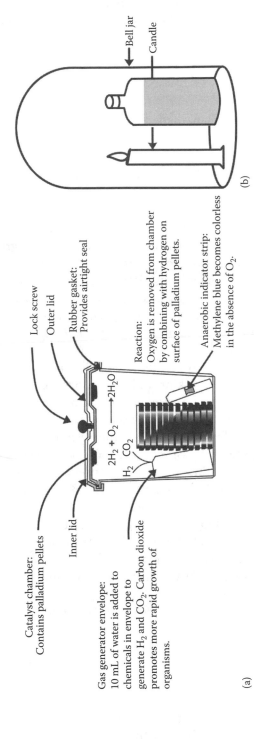

Figure 25.20 (a) Brewer's jar and (b) candle jar created anaerobic condition.

Test for acetic acid

1. Take 1 mL of solution and add a few drops of neutral ferric chloride solution.
2. There appears blood red color which indicated the presence of acetic acid in the solution.

25.10.5 Result

Filter the mixture to get clear supernatant of vinegar.

25.11 Demonstration of Wine Production Using Grape Juice

25.11.1 Introduction

Wine is an alcoholic beverage made from the fermentation of grape juice. The production of wine is referred to as enology and the industries producing wines are called wineries. Wine can be made by using the naturally occurring microorganisms on the grapes surface or by artificial inoculation of *Saccharomyces ellipsoideus* in the grape juice of *Vitis vinifera* and allowing the fermentation to take place between 20°C and 28°C for 3–5 days for red wine and 10°C–21°C for 7–12 days or for several weeks for white wine.

In general, grape wines are of two types—red wine and white wine. For red wine, black grapes are used, including skins and sometimes, the stems. White wine, in contrast, is made only from the juice of either white or red grapes, that is, without their skins or stems. Steps involved in wine production are shown in flowchart (Figure 25.21).

Commercially, wine making begins with the selection of good-quality grapes which are crushed in a grape processing unit that breaks and separates the stems and releases the juice. In making white wines, gentle pressing follows to release further juice from the pulp, after which the skins and pips are discarded. The unfermented grape juice, called must, is rich in sugars and other nutrients and is strongly acidic (pH 2.9–3.9) and is used as a substrate during fermentation. The must is sterilized by the addition of sulfur dioxide (at the rate of 100 ppm) or sulfite such as potassium or sodium metabisulfite in the fermentation vat to retard the growth of natural microbial population such as acetic acid bacteria, wild yeasts, and molds.

25.11.2 Materials

Grape juice, yeast culture, flasks, balloons, and lead acetate strip.

25.11.2.1 Inoculation of Must
Sterilized must is inoculated with a selected wine strain of *S. ellipsoideus* (or *S. cerevisiae* strain), grown on sterilized grape juice, at the rate of 2%–5%.

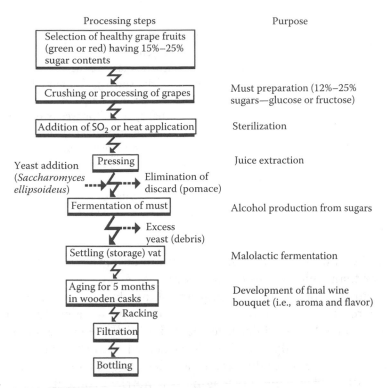

Figure 25.21   Flowchart for different steps involved in the wine production.

S. ellipsoideus is usually more resistant than most other yeasts to the antimicrobial action of sulfur dioxide.

25.11.3 Protocol

25.11.3.1 Fermentation

1. Collect fresh and healthy grapes and squeeze to get the juice. Sterilize the grape juice passing through bacterial membrane filter. Collect the grape juice which is subjected to fermentation.
2. The fermentation is carried out in vats of various sizes ranging from 200 L casks to 200,000 L tanks made of cement, stone, oak, or glass-lined metal. The fermenters used are fitted with a special one-way valve so that carbon dioxide produced during fermentation can escape but air cannot enter into it. Rapid growth of the yeast starts and fermentation proceeds immediately which soon results in anaerobic conditions. The fermentation is allowed to continue at 10°C–21°C for white wine for 7–12 days or for several weeks, until an alcohol concentration of 10%–12% (v/v) is reached. This results in partial inhibition of the yeast, and cell numbers and metabolic activity fall. Although CO_2 and ethanol are the main products of yeast fermentation, hundreds of other compounds are produced in smaller amounts. These, which include higher alcohols, organic acids, and ethyl esters, are responsible for distinctive characteristics of different wines.

25.11.3.2 Aging

1. When the first phase of fermentation is over, the wine is pasteurized with sulfur dioxide. During pasteurization, proteins are removed. It is cooled, filtered, and transferred to wooden casks and other containers, plastic concrete tanks for maturation and storage. During this time, most of the sugar is fermented into ethanol and CO_2, generally resulting in a final alcohol concentration of <14%.

2. During the aging process in wooden casks that may go for weeks or months, the wine develops unique flavor, aroma due to the production of alcohols, organic acids, aldehydes, amino acids, and pigments by the action of yeasts' enzymes. Finally, the wine is clarified, filtered through diatomaceous earth, asbestos, or membrane filters (i.e., sterile filtration, cold sterilization) before bottling to eliminate bacteria that could spoil wine, for example, by converting the ethanol into acetic acid, and the yeasts (*Candida valida*). Finally, the wine is canned, barreled, or bottled. The final alcohol content of wine varies from 6% to 15%.

Red wines. They vary in color from a faint red to deep, rich burgundy. They have strong flavor than white due to the presence of large amount of tannins that are extracted into the juice from the red grape skins, during the fermentation. The protocol for red wine preparation is similar to white wine except that in white wine preparation only grape juice is fermented, while in red wine making crushed grape must (including pulp and skin) is fermented with the yeast at 20°C–28°C for 3–5 days. Red wine is usually aged in wooden casks for several years but white wine is typically sold without much aging. The nutrient content of wine is shown in Table 25.5.

Champagne. A carbonated wine is an example of sparkling wine, which is made by a secondary fermentation of wine in the sealed bottles. It is produced by adding sugar, citric acid, and the yeast to the bottles containing table wine. The inoculated bottles are corked, clamped, and stirred horizontally for 6 months for fermentation to take place. They are then agitated and aged for an additional period of up to 4 years. The final sedimentation of tartrates and yeast cells is accelerated by reducing the temperature of the wine to 25°C and holding for 1–2 weeks. Sediments are collected on the necks of inverted champagne bottles after the bottles have been carefully turned over a period of 2–6 weeks by frequent rotation of the bottle. Finally, the sediment is frozen and disgorged upon removal of the cork. The bottles

TABLE 25.5 Nutritive Value of Red Table Wine per 100 g (3.527 oz) Energy 80 kcal 360 kJ

Carbohydrates	2.6 g
Sugars	0.6 g
Fat	0
Protein	0.1
Alcohol	10.6 g

are refilled with clear champagne from another disgorged bottle, and the product is ready for final packaging and labeling.

25.11.4 Results

The total acidity is denoted by % tartaric acid. For calculation, take 10 mL fermenting juice and add 10 mL distilled water with five drops of 1% phenolphthalein solution. Mix them and titrate with 0.1 N NaOH. Pink color appears first which is an indication of ethanol production. The total acidity is calculated by putting the values in the following formula:

$$\% \text{ Tartaric acid} = \frac{\text{Alkali(mL)} \times \text{Normality of alkali} \times 7.5}{\text{Weight of sample in grams}}$$

whereas volatile acidity is expressed as % acetic acid using the following formula:

$$\% \text{ Acetic acid} = \frac{\text{Alkali(mL)} \times \text{Normality of alkali} \times 6}{\text{Weight of sample in grams}}$$

Weight of 1 mL = 1 g.

25.11.4.1 Types of Wines

Depending upon the level of free sugar present at the end of the fermentation, wines are of two types:

1. *Dry wines.* Those wines in which all the sugar present in the juice is metabolized during fermentation.
2. *Sweet wines.* Those wines in which smoke of the sugar is left due to the stopping of the fermentation process before completion, or additional sugar is added after fermentation.

On the basis of the presence of carbon dioxide, wines are of two types:

1. *Still wines.* Those wines from which all the gas (CO_2) is released before the wine is bottled.
2. *Sparkling wine* is the carbonated wines and is produced by continuing the process of secondary fermentation carried out in the bottles.

Distillation or addition of other alcoholic spirit to wine after fermentation is used for production of other wines such as the following:

1. *Brandy*: It is distilled wine in which alcoholic content is increased up to 21%.
2. *Fortified wines*: Those wines to which brandy or other spirit is added. *Sherry* and *port* are the examples of fortified wines. The alcohol content of distilled liquors is rated by proof, a measurement that is usually two times the alcohol content; thus, 80 proof vodka consists of 40% ethyl alcohol.

25.12 Production and Manufacture of Beer

25.12.1 Introduction

Beer is one of the world's oldest prepared beverages and is a product of starch fermentation from germinated barley grain (malt) by yeasts. The process of making beer is known as brewing. The carbohydrates of grains exist loosely in the form of starch which must be converted to simple sugars by the naturally occurring enzymes before they can serve as the substrate for the yeast since the brewing yeasts are unable to digest the starch.

25.12.2 Protocol

Brewing consists of seven basic steps: malting, mashing, wort boiling, fermentation, aging, membrane filtration or pasteurization, and bottling (Figure 25.22).

25.12.2.1 Malting

The barley (*Hordeum vulgare*) grains are malted (i.e., partially germinated for 1–2 days by soaking in water) to increase the concentration of naturally occurring starch-digesting enzymes (amylase) that digest the starch to maltose sugar for fermentation (starch → maltose). This process of conversion of barley starch to fermentable sugars is called as malting. Malt not only acts as a source of fermentable sugars and other nutrients for yeasts but is also a source of amylolytic and proteolytic enzymes for hydrolysis of any additional unmalted cereal in the recipe.

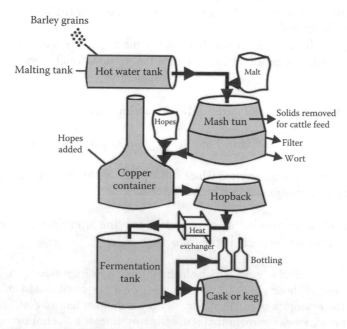

Figure 25.22   The steps involved in commercial production of lager beer.

25.12.2.2 Mashing Process

Malted grain (malt) is crushed (called milling) and mixed with hot water (65°C) to produce mash and the process is called as mashing. Mashing results in extraction of sugars, amino acids, and other yeast nutrients and enzymes to yield sweet wort. Sugar and starch supplements are then added to the mash mixture and heated between 65°C and 70°C. During this process, further digestion of the starch takes place. Mashing is followed by the separation of the liquid extract, called wort from the mixture. Dried petals of the vine *Humulus lupulus* called hops are added as a source of pyrogallol and catechol tannins, resins, essential oils, and other constituents for the purpose of precipitating unstable proteins during boiling of wort and to provide for biological stability, bitterness, and aroma. The addition of hops also prevents contamination of the wort because of the presence of antimicrobial substances. The wort is boiled for 1.5–2.5 h, usually in large copper kettles to stop enzyme action, to precipitate proteins and sterilization. It is followed by filtration, cooling, and transferred to the fermentation vessel for the addition of yeast (i.e., fermentation).

25.12.2.3 Fermentation

Wort is inoculated with *Saccharomyces carlsbergensis* (*Saccharomyces uvarum*), a yeast especially selected for making beer. It is a bottom-fermenting yeast that functions deep in the fermentation vat, settles at the bottom, and gives rise to lager and other beers with pH values of 4.1–4.2. Fermentation takes place at 6°C–12°C and wort sugar is converted to CO_2 and ethanol in a period of 7–12 days. Fermentation is self-limiting and ceases when ethanol concentration of 3%–6% is reached in the fermentation vat.

25.12.2.4 Lagering or Aging

After 2 weeks of fermentation, the beer from fermentation tanks is pumped off in large tanks—primary aging tanks and secondary aging tanks for lagering (to store or age) or aging. When it is stored at a cold temperature (0°C–1°C) for several weeks to months for maturation, it results in settling of yeasts, protein, resin, and other nutrients and the clear mellow fluid (i.e., lager beer) is pasteurized at 60°C to destroy the spoilage organisms or membrane-filtered. The alcoholic content is about 4%. Finally, it is carbonated (0.45%–0.52%) with CO_2 collected during fermentation, and packaged in bottles, cans, or other packaging. This process is termed finishing.

25.12.3 Kinds of Beers

Breweries also produce several other malt products which are as follows:

Ale is a beer that is prepared by using the top yeast *S. cerevisiae* that carries out fermentation to the top of the vat at high temperature (14°C–23°C) within 5–7 days with pH of 3.8.

Bock beer is made from roasted germinated barley seeds and is dark in color with higher alcoholic content.

Sake (rice beer or wine) is an oriental beer derived from rice starch. It is made by using *Aspergillus oryzae* and *Saccharomyces sake.*

Sonti—Rice beer or wine in India is called sonti. It is manufactured in India using the mold *Rhizopus sonti* and yeast.

Pilsener—Lager-type beer in which fermentation is carried out at low temperature. It is light in color and contains little fermentable carbohydrate with alcohol content varying from 12% to 15%.

Ginger beer—It is made by the fermentation of a sugar solution flavored with ginger by using a yeast (*Saccharomyces pyriformis*) and bacterium (*Lactobacillus vermiformis*).

25.13 Production and Estimation of Alkaline Protease

25.13.1 Introduction

Proteases are one of the most important industrial enzymes accounting for nearly 60% of the total worldwide enzyme sales. Of these, alkaline proteases are employed primarily as cleansing additives. Among proteases, bacterial proteases are the most significant compared to animal and fungal proteases.

Alkaline proteases are produced by many bacteria like *Bacillus licheniformis, B. amyloliquefaciens, B. megaterium,* and *B. pumilus; Streptomyces* strains such as *S. fradiae, S. griseus,* and *S. rectus; Pseudomonas* sp.; *Proteus;* and *Serratia* and by fungi like *A. niger, A. sojae, A. oryzae,* and *A. flavus.* However, the enzymes associated with these microbes are actually mixtures of proteinases and peptidases, with the proteinases usually being excreted out to the fermentation liquid during growth, while the peptidases often liberated only on autolysis of cells. At present, proteases are of more commercial significance than peptidases.

Alkaline proteases are employed in the leather and textile industry to remove proteinaceous sizing and in the silk industry to liberate the silk fibers from the naturally occurring proteinaceous material in which they are embedded. In addition, these enzymes are used in tenderizing of meat and also as the active ingredients in the spot-remover preparations for removing food spots in the dry-cleaning industry.

25.13.2 Materials

B. subtilis, glucose, peptone, KH_2PO_4, $MgSO_4 \cdot 7H_2O$, $FeSO_4 \cdot 7H_2O$, tyrosine, casein, Na_2CO_3, sodium alginate, $CaCl_2$, and Folin–Ciocalteu reagent.

25.13.2.1 Medium
Bacterial culture medium: Glucose 5 g/L, peptone 5 g/L, KH_2PO_4 0.25 g/L, $MgSO_4 \cdot 7H_2O$ 0.25 g/L, and $FeSO_4 \cdot 7H_2O$ 0.01 g/L

25.13.2.2 Solutions
Carbonate buffer (pH 10): 0.2 M Na_2CO_3 27.5 mL + 0.2 M $NaHCO_3$ 22.5 mL. Adjust the pH of the medium to 10.

Tyrosine: Dissolve 100 mg in 100 mL of glass distilled water.

Sodium alginate (3% w/v): Weigh 3 g of sodium alginate, suspend in distilled water, and make up the volume to 100 mL.

0.2 M CaCl$_2$·2H$_2$O: Prepare 0.2 M CaCl$_2$·2H$_2$O solution in glass distilled water.

25.13.3 Protocol

1. Culture *B. subtilis* in glucose–peptone medium adjusted to pH 10.
2. Subculture the bacterium on nutrient agar and incubate the subculture for 24 h at 37°C.
3. After the bacteria attain maximum growth, immobilize the cells with 20 mL of 3% sodium alginate and 0.2 M CaCl$_2$·2H$_2$O. The beads are formed. Wash the beads for two or three times.
4. Keep the beads in culture broth consisting of 2% casein solution in 0.2 carbonate buffer at pH 10 (substrate).
5. Centrifuge the culture broth at 3000 rpm for 15 min and supernatant was taken as enzyme source.
6. Take 0.5 mL of enzyme source and add 0.5 mL of substrate.
7. After 10 min, the reaction was terminated by the addition of 1 mL 10% trichloroacetic acid.
8. The mixture is centrifuged at 4000 rpm for 15 min.
9. One milliliter of twofold-diluted Folin–Ciocalteu reagent was added to the supernatant.
10. After 30 min, the color developed is read at 660 nm against a reagent blank prepared in the same manner.
11. One protease unit (PU) is defined as the amount of enzyme that releases 1 µg of tyrosine/mL/min—under the earlier assay conditions:

    1 PU = 1 µg tyrosine/min/mL of enzyme–substrate mixture
    That means,
    10 PU = 10 µg tyrosine/min/mL
    That implies,
    10 PU = 100 µg tyrosine/10 min/mL.

25.13.4 Result

Prepare standard graph with different concentrations of tyrosine.

25.14 Mushroom Production Technology

25.14.1 Introduction

Mushrooms can be found in forests around the world. Given the proper environment, mushrooms will grow and can offer a good source of natural vitamins and minerals. Mushrooms can also bring illness and even death to people who are unaware of certain types of wild mushrooms. Cultivated mushrooms are therefore the preferred and most reliable source of supply. Mushrooms are commonly used for various dishes in different shapes and forms. The most commonly and easily cultivated mushrooms around the world are white button (*Agaricus brunnescens* [*A. bisporus*]), oyster mushrooms (*Pleurotus ostreatus*), ear mushrooms (*Auricularia polytricha*), straw mushrooms (*Volvariella volvacea*), shiitake (*Lentinula edodes*), and Reishi mushroom (*Ganoderma lucidum*). *Macrocybe* sp., *Agrocybe* sp. types can also be cultivated successfully but will require more attention and knowledge. Mushrooms (often refers to the fruiting body of the gill fungi), yeasts, and algal foods are frequently mentioned as alternative sources of food. Of these, mushrooms are the most preferred. Modern mushroom culture produces more protein per unit area of land than any other kind of agriculture and technology at present available. Mushroom farming is becoming successful because of its very low inputs. It is estimated that about 300 million tons of fresh mushroom can be produced from just one-fourth of world's annual yield of straw (2325 million tons). It was calculated that ~317 million metric tons of fresh mushroom could be produced annually that would provide 197 g of fresh mushrooms daily to each person in the world.

25.14.1.1 Morphology

Mushrooms can be defined as "a macrofungus with distinctive fruiting bodies, epigeous, or hypogeous, large enough to be seen with naked eyes and picked up by the hands." The mushroom fruiting body may be umbrella-like or of various other shapes, size, and color. Commonly, it consists of a cap or pileus and a stalk or stipe but others have additional structures like veil or annulus, a cup or volva. Cap or pileus is the expanded portion of the carpophore (fruit body) which may be thick, fleshy, membranous, or corky. On the underside of the pileus, gills are situated. These gills bear spores on their surface and exhibit a change in color corresponding to that of the spores. The attachment of the gills to the stipe helps in the identification of the mushroom.

Mushroom as a vegetable has become popular due to its food value in terms of protein and medicinal contents. They also show potential for use in waste management. However, the successful operation of mushroom cultivation depends on the proper knowledge of laboratory techniques and practices involved in spawn production.

Mushrooms are the fleshy fungi which constitute a major group of lower plant kingdom. The mushroom is a common fungal fruit body that produces basidiospores at the tip of club-like structures, called *basidia*, which are arranged along the gills of the mushroom. Beneath the mushroom, in the soil, is the mold colony itself, consisting of a mat of intertwined hyphae, sometimes several feet in diameter. The mushrooms first appear as white tiny balls consisting of short stem (stipe) and a cap (pileus), which begin to open up like an umbrella (Figures 25.23 and 25.24). The delicate membrane or veil (velum) enveloping the cap tears off, if allowed to develop fully, and lamellae (gills) radiating from the stalk into the cap become visible. These gills become darkened as the basidiospores (seeds) develop into millions and fall to the ground for starting their life cycle once again for second generation of mushrooms. Since mushrooms grow independently of sunlight, they can be grown in complete darkness but the darkness is not an essential prerequisite. They are relatively fast growing,

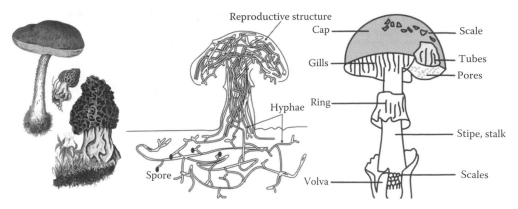

Figure 25.23 Morphology of mushrooms.

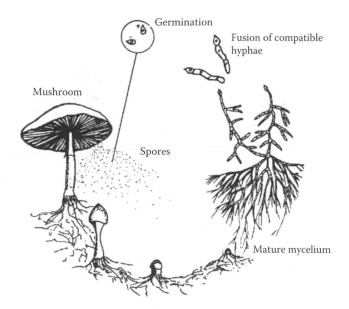

Figure 25.24 Life cycle of mushroom in natural environment.

do not require fertile soil, since grown on composted or uncomposted agrowastes and their culture can be concentrated within a relatively small space. In addition to floor, air space is also utilized resulting in higher production. It is a labor-intensive indoor activity which can help the landless, small, and marginal farmers to raise their income, diversify economic activity, and can create gainful employment especially for unemployed/underemployed youths, weaker section of the society, and women folk. *Agaricus campestris* is one among several types of mushroom seen growing wild (Table 25.6). It is popularly called temperate mushroom or "*Khumb*" and grows on dead organic matter under suitable environmental conditions. It derives its carbonaceous food by decomposing lignin, cellulose, and hemicellulose present in agrowastes with the help of extracellular enzymes secreted by the mycelium. Microbic protein available in compost is the chief source of organic nitrogen for its assimilation.

TABLE 25.6 Incubation Period for the Cultivation of Few Selected Tropical Mushrooms

| | Incubation Period | | |
Type of Mushroom	Incubation Time (Weeks)	Mushroom Flushes[a]	Production Time[b] (Weeks)
Oyster mushrooms (*P. ostreatus*)	4	first	5
		second	8
		third	11
		fourth	14
		fifth	17
Ear mushrooms (*A. polytricha*)	4	first	4–5
		second	8–9
		third	11–12
		fourth	14–15
		fifth	Beware of mites
Lentinus squarrosulus	4–5 or more	first	5–6
		second	8–9
		third	11–12
		fourth	14–15
		fifth	17–18
Lentinus polychrous	4–5 or more	first	5–6
		second	8–9
		third	11–12
		fourth	14–15
		fifth	17–18
Straw mushrooms (*V. volvacea*)	3–4 days for mycelium and 4–5 days for fruiting body	first	7–9 days
		second	14–16 days
		third	21–23 days

[a] Flushes means harvesting time or number of harvests.

[b] Production time is the number of weeks following inoculation. This will depend on the season and on the amount of care given by farmers.

25.14.1.2 *Food Value of Edible Mushrooms*

The use of mushrooms as food is probably as old as civilization itself. Mushrooms have been a delicacy since ancient times. The Egyptians regarded them as a food for Pharaohs. The Greeks and Romans described them as "*food for the Gods*" and were served only on festive occasions. These were earlier preferred for their flavor and taste while their nutritive value was recognized later. Mushrooms provide a rich addition to the diet in the form of proteins, carbohydrates, minerals, and vitamins. The protein content of fresh mushrooms (3 g/100 g) is about 3.7%. It is twice as high as that in most vegetables except green peas, Brussels sprouts, and pulses and is much lower to meat, egg, fish, and cheese. They have a high percentage of all the nine essential amino acids with traces of sugar and without cholesterol having low caloric value (<35 kcal/100 g), richer in vitamins (B_1, B_2, niacin, B_{12}, pantothenic acid, and C) than most vegetables and almost free from fat (0.2 g/100 g), and a good source of minerals P, K, Fe, and Cu recognizing them as a valuable source of nutritive and protective food.

25.14.1.3 Cultivated Edible Mushrooms

Presently. about a dozen fungi are cultivated in over 100 countries with a production of 2.2 million tons. Five genera, *Agaricus*, *Lentinus*, *Volvariella*, *Pleurotus*, and *Flammulina*, contributed about 91% of total production. White button (*A. brunnescens* [*A. bisporus*]) has the largest share (56%) followed by shiitake (*L. edodes*) (14%), paddy straw (*V. volvacea*) (8%), oyster (*Pleurotus* spp.) (7.7%), and others (13%). The United States produces about one-fourth of total white button mushroom production (0.35 million metric tons) followed by China (0.31), France (0.2), the Netherlands (0.15), United Kingdom (0.12), and Taiwan (0.062). It is now grown in almost all the continents. Technology for its cultivation has advanced to a state in which this mushroom can be grown profitably in locations with climatic conditions quite different from those in Europe and America where mushroom culture had its beginning and greatest development.

25.14.1.4 Nutraceutical Properties of Mushrooms

Studies show that certain types of mushrooms have a direct impact on body activities (Table 25.7).

25.14.1.5 Important Medicinal Mushrooms

Mushrooms have a long history of use in traditional Chinese medicine. In fact, it is estimated that in China >270 species of mushrooms are believed to have medicinal properties with 25% of them thought to have antitumor capability. *G. lucidum* (reishi mushroom), *Coriolus versicolour* (turkey tail), *Grifola frondosa* (maitake), *L. edodes* (shiitake), *Cordyceps* species (keera ghas), *Tremella fuciformis*, *Poria cocos*, and *Pleurotus* species (oyster or dhingri) have gained importance in modern medicine for their various pharmacological, immunological, anticancerous, and other medicinal values.

TABLE 25.7   Mushrooms as Nutraceutical

Termitomite sp.	Good for brain and memory
V. volvaceae	Heal wounds
Coprinus sp.	Help the digest and decrease phlegm
Auricularia sp.	Clean lungs
Agaricus sp.	Increase mother's milk
Hericium erinacius	Heal wounds in intestine
Pleurotus sp.	Decrease muscle malpighia
L. edodes	Good for baby's cartilage
Flammulina velutipes	Good for liver
Agrocybe cylindraceae	Good for kidney and urine
Schizophyllum commune	Decrease leucorrhea
Dictiophora sp.	Cure dysentery and decrease rotting
T. fuciformis	Good for sperm, semen, and kidney
Lentinus sp.	Control the whole body system

25.14.1.6 Poisonous Mushrooms

The order Agaricales, commonly called *gill fungi*, contains over 270 genera. These include the mushrooms, the toadstools, and the boletes. The edible species are called "mushrooms" and the poisonous ones, toadstools. These fungi are incapable of causing infectious diseases but produce toxic substances that poison a person who ingests them. These poisonous substances are collectively known as mycotoxins (*myco* = fungus + *toxin* = poison) and result in *mycetismus* (mushroom poisoning) following ingestion of poisonous mushrooms. These mushrooms contain lethal substances that destroy liver cells and excite the nervous system. The most deadly mushrooms are the *death cap—Amanita phalloides*, the closely related *destroying angles—Amanita virona*, the *fool's cap—Amanita verna*, and the *fly-agaric—Amanita muscaria*. In the *death cap*, toxic principle is a mixture of α- and β-*amanitin* and *phalloidin*—both complex cyclic polypeptides containing sulfur. Cooking does not destroy the toxin nor is it affected by the human digestive juices. Symptoms of poisoning appear only after 8–24 h of ingestion and by time the toxin is absorbed by the body; neither vomiting nor a stomach pump can help then. Eating poisonous mushrooms may result in various types of reaction like nervous disorder, gastric disorder, hemolytic disorder, and muscular disorder. Sometimes, even edible mushrooms can cause indigestion in healthy people and some people may be allergic to a species which is harmless to others. Causes of discomfort and indigestion may be due to eating too much, or eating mushrooms with indigestible food, or having been incorrectly cooked. Mushrooms may also cause illness if they are taken with alcohol.

25.14.1.7 Cultivation of Mushrooms

The most common mushrooms that can be cultivated using straw as the principal ingredient are *A. brunnescens* (= *A. bisporus*) and *V. volvacea*. Other edible species that can be artificially cultivated using straw as main substrate are mostly species of *Pleurotus* and *Stropharia*. Another quite popular mushroom, *L. edodes* can be grown on freshly cut wood of several tree species. Usually, the wheat and rice straw are used for the cultivation of species of *Agaricus* and *Volvariella*, respectively.

25.15 Identification of Mushrooms by Spore Print Method

25.15.1 Introduction

Several methods are used to identify mushroom and one of the most common methods adopted is the spore colors. One of the generic features of mushroom is the color of spores which varies in different genera, for example, white, green, orange, brown, cream, etc. Though *Agaricus* and *Amanita* look alike from outer appearance, the spores of *Agaricus* are chocolate brown and that of *Amanita* white to pale. Spores are produced on basidia formed in gills of the basidiocarp.

The spore print is a simple method used for identification of gilled fungi. When the cap of mushroom is held stationary on a paper in a closed chamber for a long period,

TABLE 25.8 Specific Spore Prints of Mushrooms

Mushroom Type	Genus	Spore Color
Gilled mushroom	Amanita	White to pale
	Chlorophyllum	Green
	Clitocybe	Fleshy color
	Hebeloma	Rusty to yellow brown
	Hygropharus	Cream
	Inocybe	Brown or cinnamon
	Lepiota	Green
	Naematoloma	Purple brown
	Panaceolus	Black
	Pleurotus	Lavender
	Russula	White, cream, or yellow
	Schizophyllum	Pink, salmon, or flesh
Tubed mushroom	Boletus	Yellow to olive brown
	Cyrodon	Yellow to yellow brown
	Suillus	Yellow brown to cinnamon
Puffball mushroom	Scleroderma	Purple black

the impression of pattern of gills is printed, and the spores get accumulated onto the paper and invisible spores appear colored (Table 25.8).

25.15.2 Materials

Fresh mushroom, sharp knife, half black and half white paper, and glass bowl.

25.15.3 Method

1. Take a fresh mushroom for preparing the spore print for identification. Carefully cut the gill with the help of sharp knife.
2. Place gill side down of the cap on a half black and half white paper in such a way that half of gill should be toward black side and half toward white side.
3. Place a drop of water on the cap of mushroom and cover it by a glass bowl.
4. Leave this setup for about 8 h or overnight undisturbed (Figure 25.25).

25.15.4 Results

Gently remove the glass bowl and cap from the paper. This area shows the print of gills and spore deposits. Carefully study the gill pattern and spore color and identify mushrooms studied by this method. Match color with the color dictionary and identify the mushroom.

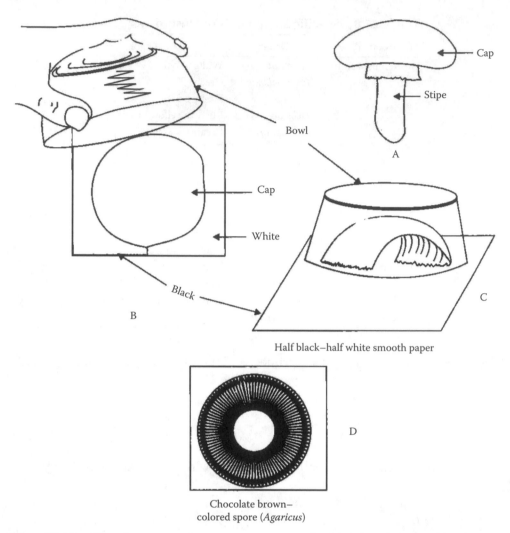

Half black–half white smooth paper

Chocolate brown–
colored spore (*Agaricus*)

Figure 25.25 Flow diagram explaining spore print method for identification of mushroom.

25.16 Isolation and Maintenance of Mushroom Culture

25.16.1 Materials

Fresh mushroom, inoculation needle, sterilized PDA medium (diced potato 200 g/L, dextrose 20 g/L and agar 15 g/L), pH 4.5, Petri plates, laminar airflow, bunsen burner, and sterilized PDA slants.

25.16.2 Method

1. Take a fresh mushroom for tissue culture, clean the surface with 70% aqueous ethyl alcohol, transfer a small piece with the help of inoculation needle under laminar airflow.

2. Clean both hands and bottles with alcohol and insert hands into the laminar air cabinet.

3. Hold needle with two fingers in a 45° angle, flame needle to disinfect until the needle turns red. Make sure it does not touch any surface after flaming.

4. While needle cools down (15–20 s—hold needle not to touch anything or place it on the clean surface of a glass).

5. Using other fingers, tear mushroom lengthwise (do not use knife to cut).

6. With the needle, cut a small piece (2 mm × 2 mm) of fleshy tissue from inside the mushroom (in the middle between the cap and the stalk). Make sure that it is clean and did not touch the outside of the mushroom.

7. Flame around the PDA (potato dextrose agar) plate. Remove lid of PDA plate in front of flame to secure against contamination using other fingers. Inoculate the mushroom flesh on to the plates. Close the plate immediately near the flame (Figure 25.26).

8. Insert the needle in the bottle and inoculate by placing small piece of cut mushroom in the middle of the PDA's surface. Make sure the piece of mushroom does not touch anything before entering the PDA bottle. Close the mouth of the bottle before the flame with cotton plug. Label plates and indicate: date, type of mushroom, and culture #.

9. Incubate plates and culture bottles at 28°C ± 20°C in an incubator for 7 days and observe growth.

25.16.3 Observation and Result

Observe cottony radial growth on agar media.

25.17 Production of Spawn for White Button Mushroom (*A. brunnescens* syn. *A. bisporus*)

25.17.1 Introduction

For the successful cultivation of any mushroom on a small scale or commercial scale, one of the most important requirements is the seed of that species/variety. *Spawn—a pure culture of the mycelium grown on a special medium*—is the mushroom seed, comparable to the vegetative seed in crop plants. The production of spawn is done by professionals in the laboratory under controlled conditions of temperature, light, and humidity. Spawn can be produced either by germinating basidiospores or by culturing small pieces of vegetative mycelium of a mushroom on a suitable substrate. The success of mushroom cultivation and its yield depend to a large extent on the purity and quality of the spawn used. Mushrooms for spawn production can be grown on sterilized cereal grain (wheat, rye, sorghum, or bajra), but usually grain colonized with mycelium (*grain–spawn*) is used as an inoculum for composts. Grain is preferred as a substrate for mushroom spawn because grain gives a large number of inoculation sites, each with a high inoculum potential derived from the nutrient base the outgrowing fungi can utilize. This helps to ensure that the compost is rapidly

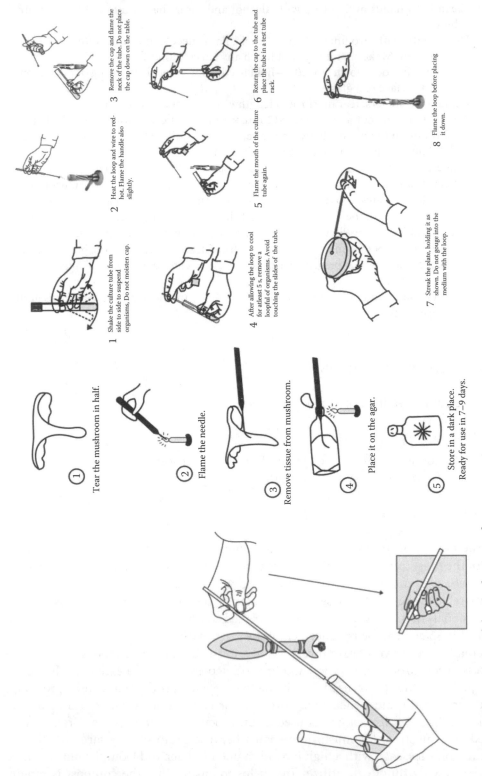

1 Shake the culture tube from side to side to suspend organisms. Do not moisten cap.

2 Heat the loop and wire to red-hot. Flame the handle also slightly.

3 Remove the cap and flame the neck of the tube. Do not place the cap down on the table.

4 After allowing the loop to cool for atleast 5 s, remove a loopful of organisms. Avoid touching the slides of the tube.

5 Flame the mouth of the culture tube again.

6 Return the cap to the tube and place the tube in a test tube rack.

7 Streak the plate, holding it as shown. Do not gouge into the medium with the loop.

8 Flame the loop before placing it down.

① Tear the mushroom in half.

② Flame the needle.

③ Remove tissue from mushroom.

④ Place it on the agar.

⑤ Store in a dark place. Ready for use in 7–9 days.

Figure 25.26   Isolation of the mushroom culture.

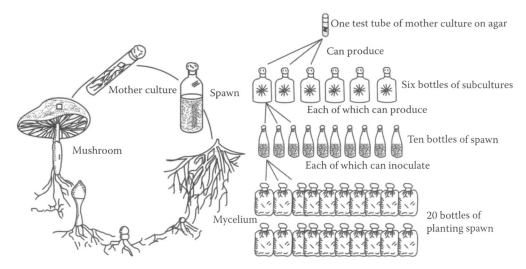

One test tube of mother culture on agar

Can produce

Mother culture Spawn

Six bottles of subcultures

Each of which can produce

Mushroom

Ten bottles of spawn

Each of which can inoculate

Mycelium

20 bottles of planting spawn

Figure 25.27 Life cycle from mushroom to spawn and multiplication of spawn.

permeated, which is important for the exclusion of competitors as well as for the rapid production of fruit bodies. The media used for maintenance, multiplication, and preservation of mushroom culture are PDA, yeast PDA (YPDA), malt extract agar, and rice bran decoction medium.

Spawn production mainly consists of three steps: (i) substrate preparation, (ii) substrate inoculation, and (iii) incubation of the inoculated substrate for spawn production or growth of the mycelium on the substrate. Preferably, fresh spawn should be used for mixing with compost for better results. Protocol for the spawn production of *A. bisporus* is given in Figure 25.27.

25.17.2 Materials

Organism: Pure culture of *A. brunnescens* syn. *A. bisporus*, cereal grains (sorghum, wheat, rye, etc.), bottles (flask type), cotton (gauze), paper squares 7×7 cm, calcium sulfate (gypsum), calcium carbonate (chalk), glucose bottles/milk bottles/polypropylene bags, cotton, alkathene sheets, autoclave, laminar flow cabinet, incubator/storage room, wire gauge balance, bunsen burner, and water.

25.17.3 Method

1. Substrate preparation:
    a. Soak cereal grains for one night; 2 L of water per 1 kg of grain, wash and strain cereal grains to remove all water.
    b. Steam sorghum seeds for 30–45 min to soften grains, drain water, and spread cereal grains to cool down and decrease moisture.
    c. Fill three-fourth of bottle with cereal grains, carefully prepare cotton plug, tightly plug mouth of bottle with cotton, and leave out for ventilation.

d. Allow the grain to surface dry by spreading over alkathene sheets, in shade, for a few hours.

e. Mix the grain thoroughly with chemicals (e.g., calcium sulfate and calcium carbonate at 2% and 0.5%, respectively, on dry weight basis of the grain), to adjust pH of the grain at 7–7.8. The grain must not be coagulated at this stage.

f. Fill the grain–chemical mixture in 500 mL glucose/milk bottles/polypropylene bags (300–350 g boiled grains/container). However, the first generation spawn (master spawn) must be prepared in glass bottles due to their convenience in handling for further subculturing.

g. Plug the bottles/containers with nonabsorbent cotton (Figure 25.28).

h. Sterilize the substrate by autoclaving at 121°C (15 psi) for 30 min.

i. Repeat the process of sterilization after 24 h of first autoclaving.

j. Allow the substrate container to come to room temperature for making the substrate ready for inoculation.

2. Inoculation of substrate:

a. Inoculate the substrate (grain in containers) with the mycelium of the mushroom grown on a specific medium by transferring mycelium in agar on the grain under aseptic (sterile) conditions.

b. Shake the containers, after plugging, to distribute fragments of the mycelium (Figure 25.29).

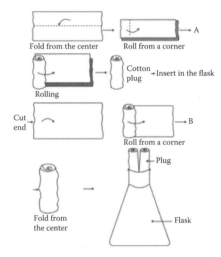

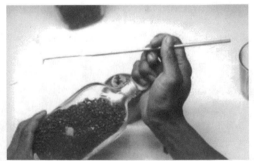

Figure 25.28 Plug preparation and method of inoculation for spawn preparation.

Figure 25.29 Plugging after inoculation.

3. Incubation:
   a. Store (incubate) the inoculated containers at 20°C–25°C in darkness for 3 weeks.
   b. Shake the containers for an even distribution of mycelium, after a few days of incubation or as soon as mycelium is visible on grain.

At the time of boiling of grain in water, observe the grain for their intact nature. Observe the inoculated grain during incubation at regular intervals for the appearance of white mycelium on the grain as well as for the appearance of contaminants.

25.17.4 Result

Appearance of silky whitish growth completely covering the grain indicates the preparation of spawn of white button mushroom.

D. Storage of spawn:
   Store the spawn, if not needed immediately, at 0°C–4°C in a refrigerator for a maximum period of 6 months. Spawn, if stored at low temperature, should be allowed to attain room temperature (25°C) before being used for spawning the compost.
E. Transportation of spawn:
   Transport the spawn in refrigerated vans as higher temperature (above 32°C) is detrimental to mushroom mycelium.

25.17.5 Precautions

1. Always use healthy and whole grain for spawn making.
2. Remove the damaged grain, weed seeds, and inert material before boiling of grain.
3. Never boil the grain for a longer period.
4. Check the master spawn for contamination; contaminated as well as suspected to be contaminated bottles should be discarded immediately.
5. Always keep a check for microbial contamination of the mushroom culture during incubation.
6. Always use cotton plugs to maintain aerobic conditions in the spawn containers.

25.18 Cultivation of White Button Mushroom (*A. brunnescens* = *A. bisporus*)

25.18.1 Introduction

A. brunnescens = *A. bisporus* is a temperate mushroom which can grow best only under temperate conditions and 24°C is the optimum temperature required for its development and spread of mycelium in the compost (Figure 25.30). However, for the formation of fruit bodies, it requires temperature ranging from 14°C to 18°C. Temperature, moisture, ventilation, and good spawn (seed) are the essential conditions for the optimal growth of the mushroom. They require nearly a saturated atmosphere with moisture yet the direct application of water on beds is harmful to the growing crop. Good ventilation maintains congenial environmental conditions for the exchange of toxic gas with the introduction of adequate fresh air. Carbon dioxide level of 0.10%–0.15% volume is necessary during crop production that can be achieved by giving 4–6 air changes/h or introducing 10 ft^3 fresh air/ft^2 bed area/h. Spawn, a pure culture of the mycelium from a selected mushroom grown on a convenient medium, should be of good quality, that is, it should be a strain originating from a single specimen or of a perfect crop; the substratum must be covered with the white mycelium, it must be uniform and at the time of removal from the container, be absolutely free from molds and other microbial contaminants. Commercial and amateur mushroom cultivation is done indoors. The various installations required for mushroom cultivation vary with the size of a mushroom house. The space requirements necessary for successful production of *A. bisporus* are (i) the location should be easily accessible so that the manure and casing soil can be brought in and removed conveniently, (ii) the room should be well-ventilated, (iii) direct sunlight should not fall on the mushroom bed, (iv) the room temperature should not exceed 15°C during the growing period, (v) the room should be fitted with heating facility, and (vi) location for growing mushroom should not be too moist and should be at such a site where there should be good fresh

Figure 25.30　*A. bisporus* growing wild on field.

water supply, availability of the fertilizers for making compost, and a good market for the sale of the mushrooms. Mushrooms have been grown successfully in cellars, garages, and in abandoned rooms.

Protocol for cultivation of A. bisporus: Preparation of compost, filling of tray beds with compost, spawning (inoculation) of beds, casing, watering of beds, harvesting of mushrooms, and storage.

25.18.2 Materials

Cement concrete platform (14×7 m) for preparing compost, tray beds (wooden trays made of pinus or deodar wood, $100 \times 50 \times 22$ cm) (Figure 25.31), compost ingredients, spawn (mushroom seed) of *A. bisporus* 10 bottles/25 beds, casing soil, sprayer with a fine nozzle, box for gathering mushrooms (Figure 25.30), sterilization facility, and wooden mold. It consists of three wooden boards: one end board and two side boards, fastened with a clamp with notches at the top, wooden board (25×12 cm) with handle for compressing the compost, filter paper sheets, lime or carbonate of lime, and pH meter.

25.18.3 Protocol

1. Preparation of compost:

   Compost is the substrate on which mushroom grows. The microbial degradation of organic wastes by several microorganisms makes the substrate selective for the growth of *A. bisporus*. During composting, distinct changes occur in the physical, chemical, and biological characteristics of the straw, all of which influence the productivity of *A. bisporus* subsequently. Synthetic compost, made out of wheat straw for growing *A. bisporus*, consists of the following constituents for 25 tray beds:

   Chopped wheat straw (3–6 cm long) 300 kg, wheat bran 15 kg, calcium ammonium nitrate (or ammonium sulfate) 6 kg, superphosphate 7.5 kg, urea 2.4 kg, potassium sulfate 3 kg, gypsum 30 kg, and saw dust 10–12 kg.

   Procedure:

   1. Wet the saw dust by spraying water or leaving it overnight after mixing all the constituents except wheat straw.
   2. Spread the wheat straw over the cement floor on the following day and wet it thoroughly by sprinkling water.
   3. Spread the premixed constituents over the wheat straw surface and mix thoroughly.
   4. Stack this mixture into a pile of 1.30 m wide and 1.30 m high, using indigenously fabricated wooden mold (Figure 25.31).
   5. Allow the compost to decompose for 28–30 days under aerobic conditions in the compost pile.
   6. Dismantle the heap (pile) repeatedly and prepare pile again and again at periodic intervals by placing the outer compost inside and inner compost layers in the

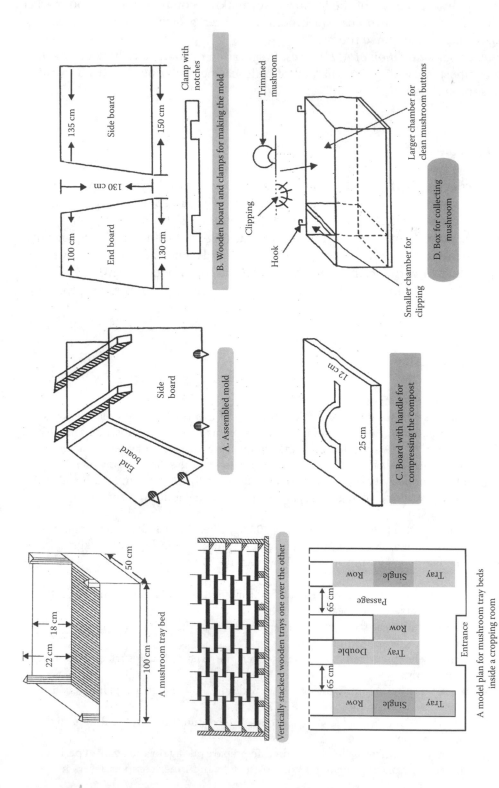

Figure 25.31 Details of mushroom tray bed and mode of stacking of trays and appliances for cropping of white button mushroom.

outer periphery, the process called turning of the compost pile, to obtain uniform fermentation of the entire pile. Watering is to be done in the first two turnings. For good results, periodic turning of the pile should be done according to the following schedule:

First turning: Sixth day, add more fertilizers (CAN or ammonium sulfate, 3 kg and urea, 1.2 kg) and wheat bran (15 kg).

Second turning: Tenth day, add gypsum (30 kg) and more water.

Third to seventh turning: After every 3 days. Twenty-fifth day should normally be the last day for turning the compost.

Observations: Well-prepared compost will be brown to dark brown in color and free from ammonia in color.

2. Filling of tray beds:
   a. Spread the prepared compost on the platform.
   b. Mix 3 kg of calcium carbonate to it.
   c. Fill the compost in all corners and edges of a tray.
   d. Compress fairly the compost in the tray using a wooden board (Figure 25.31) leaving 1 cm clear space on the top of the tray.

3. Spawning:
   Spawning means planting mushroom mycelium, growing on a suitable substrate, in the compost. *A. bisporus* spawn is the mycelium especially grown on wheat, sorghum, or other cereal grains.
   a. Perform the spawning by spreading the spawn on tray beds when half filled with compost and again after the tray is filled completely. During spawning, the spawn is gently mixed with forefingers and pressed uniformly each time.
   b. Cover the trays with filter paper/newspaper sheets.
   c. Sprinkle water on newspaper sheets, to provide humidity.
   d. Stack the inoculated trays vertically, one over the other, depending on the height of the room as shown in Figure 25.31.
   e. Continue water spraying twice a day or less depending upon available humidity in the atmosphere throughout the spawn running and cropping period.

4. *Maintain temperature* of the room between 24°C and 25°C for 12–15 days for running of the spawn, that is, formation of mycelium strands all over the tray beds.

   Observations: Observe for the white cottony mycelium over the compost surface and color of the compost that changes from dark to light brown which is indicative of successful completion of spawn running period.

5. Casing:
   Casing means covering the compost with a thin layer of soil or soil-like material after the spawn has spread in the compost (spawn run). Soil has been the universal casing material. But all soils as such cannot be used as "casing soil" with advantage, so it is especially prepared soil that can be used for casing:
   a. Well-rotten cow dung, mixed with light soil in 3:1 ratio
   b. Soil and sand in the ratio of 1:1
   c. Farm yard manure and gravel, ratio 4:1
   d. Farm yard manure and loam ratio 1:1
   e. Soil peat mixture 2:1
   f. Spent compost, sand, and slaked lime (4:1:1) and nematicide mixture
   Of these, the last one is most often used.

A. Preparation of casing soil:
   a. Mix four parts of spent compost with one part of sand to which 5 kg of slaked lime per cubic meter of compost is added.
   b. Treat the mixture with nemagon, a nematicide, by spraying or sprinkling it at the rate of 5 mL/m^3.
   c. Leave the material in a pile (1.20 M × 1.0 M) under a tree shade for a period of 1 year, turning it every 4 months.
   d. Sieve the spent compost, which turns into black soil in 1 year, for removal of undecomposed lumps.

   Observations: Observe the casing soil for black color and pH. A pH between 8.0 and 8.5 is the most suitable for mushroom yield. Adjust pH, if desired, by the addition of lime or carbonate of lime or free stone to the casing mixture.

B. Sterilization of casing soil:
   Sterilize the casing material either by the chemicals (formalin, chloropicrin, methyl bromide) or by heating or by steam from boiler through perforated pipes (Figure 25.32) and temperature raised to 70°C–75°C and maintained for 6 h. The purpose of sterilization is to kill harmful microorganisms (fungi, nematodes, insects).

C. Casing the beds:
   a. Remove the newspaper sheets from the spawned beds after 3 weeks.
   b. Gently press the compost with the help of wooden board.
   c. Cover the beds with 2–2.5 cm thick layer of sterilized casing material.

6. Watering of beds:
   Spray the beds over the casing soil with a fine nozzle of a sprayer to maintain relative humidity (RH) between 70% and 80%.

   Observations: Observe the beds for mushroom crop which can be expected after 5–20 days. Mushrooms mostly appear *in-"flushes"* and at a temperature of 15°C; it

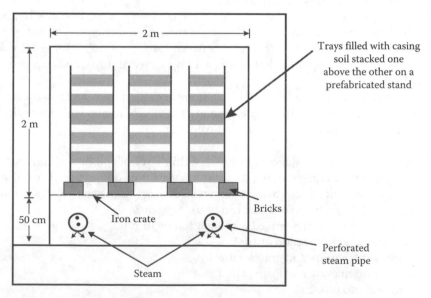

Figure 25.32 Outline plan of a specially equipped room for pasteurization (sterilization) of casing soil.

generally takes 7–8 days to come to the button stage from the first appearance of the formation of a pinhead. There is an interval of 8–10 days between flushes.

7. Harvesting of mushrooms:
   Harvest the crop everyday (in the morning and evening depending upon the market demand) to get a good quality of mushrooms (the cap still being tight to its stalk is the right stage to harvest the mushrooms). Harvesting is done by holding the cap with forefingers slightly pressed against the soil and twisted out. The mycelial strands and soil particles clinging to the base of the stalk (stem) are cut off with a knife. Specially designed wooden box is used for collection of mushrooms from multistoried trays, each provided with the hooks for resting it against the side board of the mushroom tray.

8. Storage:
   Store the mushrooms at 4°C in a refrigerator for a few days to avoid the quality deterioration, because mushrooms are a highly perishable commodity. The white color turns brown and then black in a couple of hours at high summer temperature. Soon after water oozes out, the mushrooms become unfit for cooking.

25.18.4 Precautions

1. Always prepare compost with wheat straw over 300 kg in quantity, as the heat generated in smaller piles will not be sufficient for fermentation.
2. The compost pile should not become hard pressed to avoid anaerobic conditions as decomposition takes place only under aerobic conditions.
3. If ammonia smell is noticeable, more turnings of the compost should be done to free the compost from ammonia smell.
4. Care should be taken to wet the composting material to the extent that water drops do not ooze out, if the compost is squeezed in the palm of hand.
5. Water should never be poured over the spawn to avoid the failure of the crop.
6. Proper care should be taken not to allow the temperature to rise above 75°C so that useful microorganisms present in the soil are not killed.
7. Care should be taken to leave at least 1 m clear space in between the last tray and the ceiling, so that air circulation is not impaired.

25.19 Cultivation of G. lucidum (Ling Zhi)

25.19.1 Introduction

G. lucidum (Ling Zhi) is a basidiomycete white rot macrofungi which has been used extensively as "the mushroom of immortality" in China, Japan, Korea, and other Asian countries for 2000 years. A great deal of work has been carried out on therapeutic potential of *G. lucidum*. The basidiocarp, mycelia, and spores of *G. lucidum* contain ~400 different bioactive compounds that include triterpenoids, polysaccharides, nucleotides, sterols, steroids, fatty acids, proteins/peptides (Figure 25.34), and trace elements that have been reported to have a number of pharmacological effects including immunomodulation, antiatherosclerotic, anti-inflammatory, analgesic, chemopreventive, antitumor, chemo- and radioprotective, sleep promoting, antibacterial, antiviral (including anti-HIV), hypolipidemic, antifibrotic, hepatoprotective, antidiabetic, antiandrogenic, antiangiogenic, antiherpetic, antioxidative and radical scavenging, antiaging, hypoglycemic, estrogenic activity, and antiulcer properties. *G. lucidum* has now become recognized as an alternative adjuvant in the treatment of leukemia, carcinoma, hepatitis, and diabetes. The macrofungus is very rare in nature rather not sufficient for commercial exploitation for vital therapeutic emergencies; therefore, the cultivation on solid substrates, stationary liquid medium, or submerged cultivation has become an essential aspect to meet the driving force toward the increasing demands in the international market. It can be grown under solid substrate and submerged fermentation and cultivation methods are described, keeping in mind that *G. lucidum* is a therapeutic fungal biofactory (Figures 25.33 and 25.34).

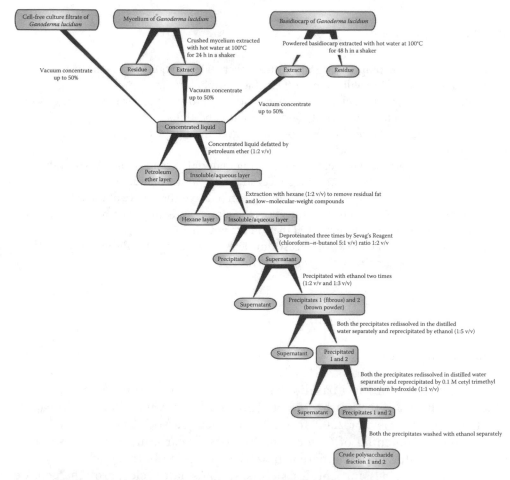

Figure 25.33 Flowchart for the isolation and characterization of polysaccharides from *G. lucidum*.

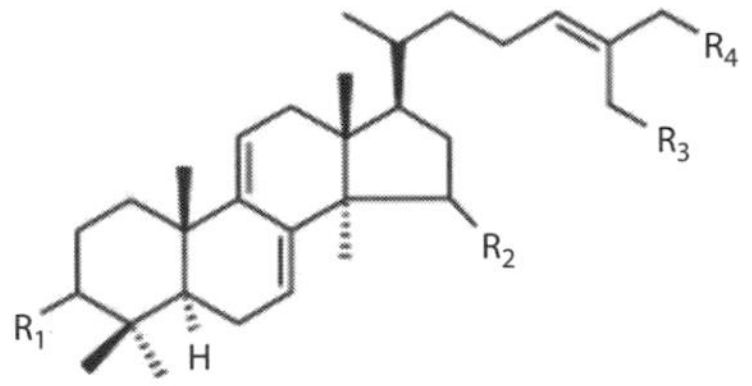

Ganoderiol F R_1=O, R_2=H, R_3=R_4=OH
Ganodermadiol R_1=β-OH, R_2=R_3=H, R_4=OH
5A-lanosta-7,9(11), 24-triene-15A-26-dihydroxy-
       3-one R_1=O, R_2=α-OH, R_3=H, R_4=OH

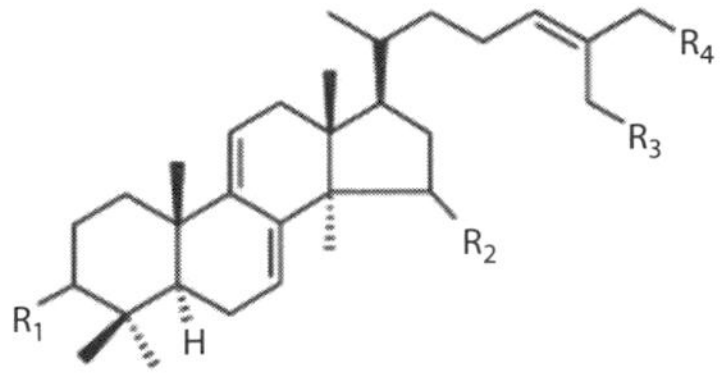

Ganoderiol F R_1=O, R_2=H, R_3=R_4=OH
Ganodermadiol R_1=β-OH, R_2=R_3=H, R_4=OH
5A-lanosta-7,9(11), 24-triene-15A-26-dihydroxy-
       3-one R_1=O, R_2=α-OH, R_3=H, R_4=OH

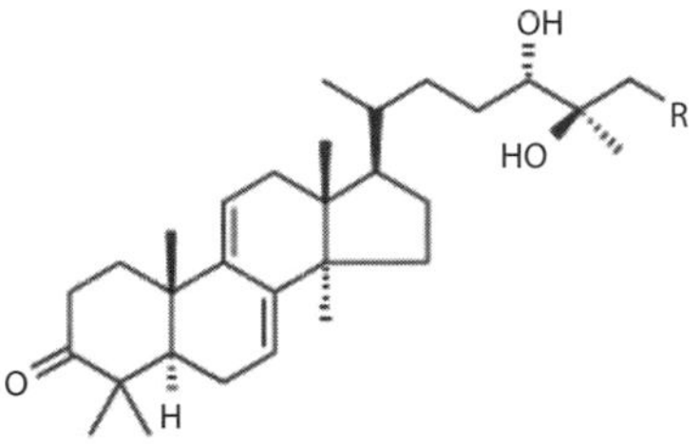

Ganodermanontriol   R=OH
Ganodermanondiol   R=H

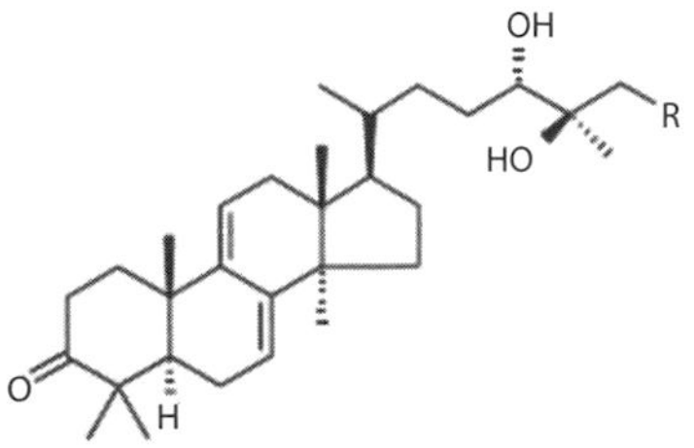

Ganodermanontriol   R=OH
Ganodermanondiol   R=H

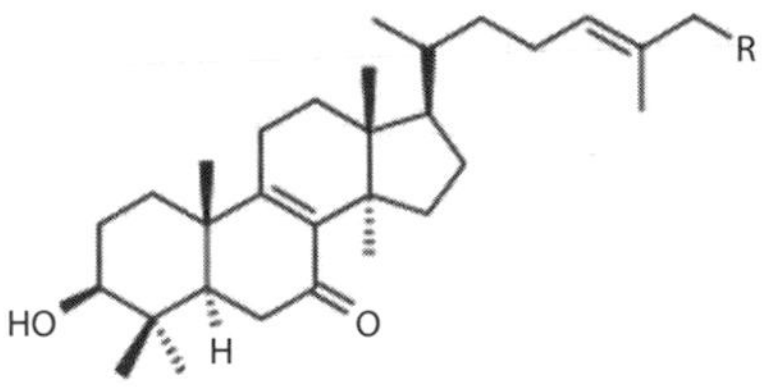

Lucidadiol   R=OH
Lucialdehyde C   R=O

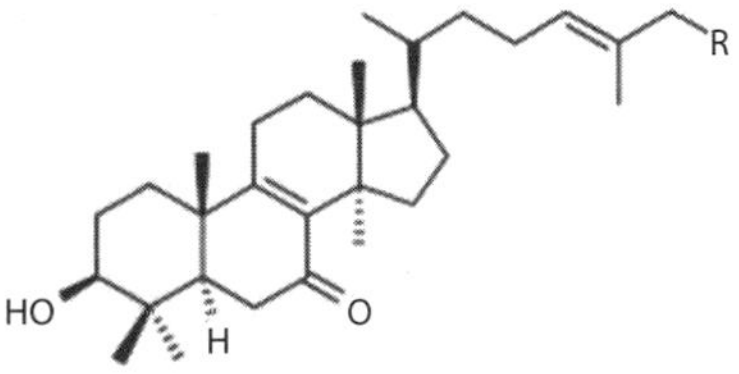

Lucidadiol   R=OH
Lucialdehyde C   R=O

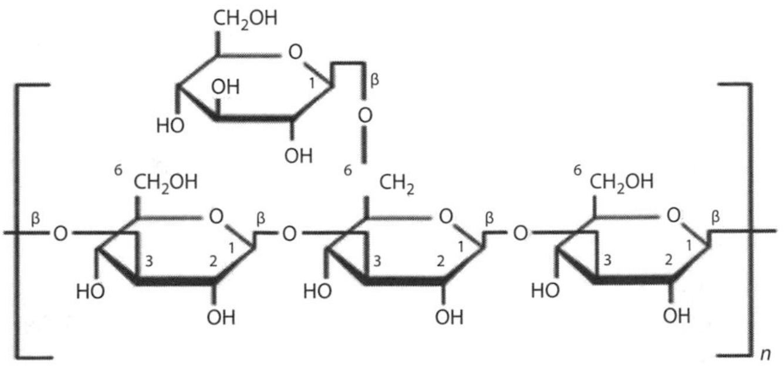

(a)

Figure 25.34 Structure of triterpenoids of *G. lucidum* and structure of β-D-glucans of *G. lucidum*: (a) Primary molecular diagram.

(*continued*)

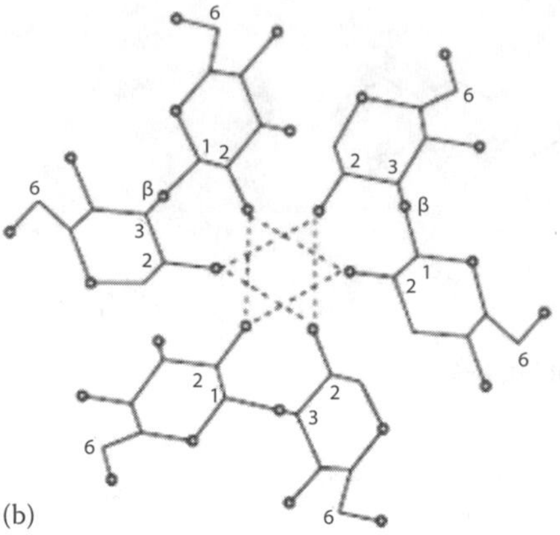

(b)

Figure 25.34 (continued) Structure of triterpenoids of *G. lucidum* and structure of β-D-glucans of *G. lucidum*: (b) Higher-level molecular diagram.

25.19.2 Materials

Organism: Pure culture of *G. lucidum* (Ling Zhi), cereal grains (sorghum, wheat, rye, etc.), bottles (flask type), cotton (gauze), saw dust of the broad leaf plant *Tectona grandis* (teak), narrow leaf plant *Acacia* spp. as substrate, wheat bran, rice bran, $CaSO_4$, $CaCO_3$, plastic bag, autoclave, laminar flow cabinet, incubator/storage room, wire gage balance, bunsen burner, and water.

Culture maintenance: The media used for maintenance, multiplication, and preservation of mushroom culture are PDA and YPDA.

Spawn production: Spawn is produced and maintained in spawn bottle using wheat grain with calcium sulfate and calcium carbonate at 2% and 0.5%, respectively, on dry weight basis of the grain, to adjust pH of the grain at 4.5–5.0 pH. The grain must not be coagulated at this stage. After full colonization of spawn packets, a thick mycelial coat is formed on the outer surface of colonized substrate. Clumps of mycelia appear as blister-like bumps of various sizes on the surface of the mycelial coat in each packet. Bumping usually start when colonization of white mycelia changed to brown. The data on different parameters are recorded regularly.

Preparation of substrate: The substrate prepared contains saw dust (teak) + 20% wheat bran + 2% $CaSO_4$ + 0.6% $CaCO_3$; saw dust (teak) + 20% rice bran + 2% $CaSO_4$ + 0.6% $CaCO_3$; saw dust (*Acacia* spp.) + 20% wheat bran + 1.5% $CaSO_4$ + 0.5% $CaCO_3$; saw dust (*Acacia* spp.) + 20% rice bran + 1.5% $CaSO_4$ + 0.5% $CaCO_3$ and are named as substrates I, II, III, and IV, respectively. $CaSO_4$ and $CaCO_3$ at different ratios are also added to substrate to maintain the pH 4.5. The moisture level of the substrate is maintained at 65% and bag system is adopted for cultivation. The 500 g substrate is filled in polypropylene bags and plugged with nonabsorbent cotton after putting a plastic (PVC) ring at the neck. Successive sterilization of substrate is done in 3 days at 15 psi for 60 min each day.

25.19.3 Protocol

Upon cooling, the sterilized bags are inoculated with wheat grain spawn at the rate of 4% and incubated in dark. After complete mycelial colonization (bags are white all over), the bags are opened and rolled back to expose upper surface and proper conditions (temperature, RH, and light intensity) for different stages of primordia and fruit body formation are optimized and provided. After harvesting the first flush, conditions for primordia formation are again switched on (i.e., temperature 28°C ± 2°C, RH 95%, light 800 lux) for starting and completing the second flush and the same procedure has been repeated for the successive flush. The optimum temperature for mycelial colonization is 30°C ± 2°C.

25.19.4 Observations and Results

For the complete mycelial colonization, substrate I normally takes 18 days; substrate II takes 16 days; substrate III takes 26 days; and substrate IV takes 23 days. Further, primordia is formed in 12–14, 10–12, 16–19, and 14–16 days in substrates I, II, III, and IV, respectively, soon after mycelial colonization of G. lucidum. The optimum light intensity (800 lux), RH (95%), and temperature (30°C ± 2°C) are suitable for primordia formation (Figure 25.35).

Data record. The following data is taken to account for record during the experiment:

1. *Days for completion of running of mycelium*: Time is recorded in days for completion of spawn running and mycelial cover on each substrate.
2. *Days for primordia formation*: Record the time in days for primordia formation on each substrate.
3. *Fruiting body formation*: Record the time in days for fruiting body formation on each substrate.

25.19.5 Submerged Cultivation of G. lucidum

Submerged culture by fermentation is an alternative approach for efficient production of polysaccharides of G. lucidum. There are several advantages of submerged culture over solid culture for polysaccharides production, viz. high productivity, low costs, availability of convenient control, and easy downstream processing. The production of fruit body takes at least 3–5 months, while reasonable amount of product of interest can be obtained by submerged fermentation only after 2–3 weeks.

The advantage of using bioreactors for submerge culture of G. lucidum is that it is easier to control environmental conditions such as temperature, dissolved oxygen, and pH. These fermentations were done with different objectives. Some aimed simply to produce biomass, with no concern for its composition. Others aimed to maximize the production of polysaccharides and to understand how different variables affect their production (Figure 25.36).

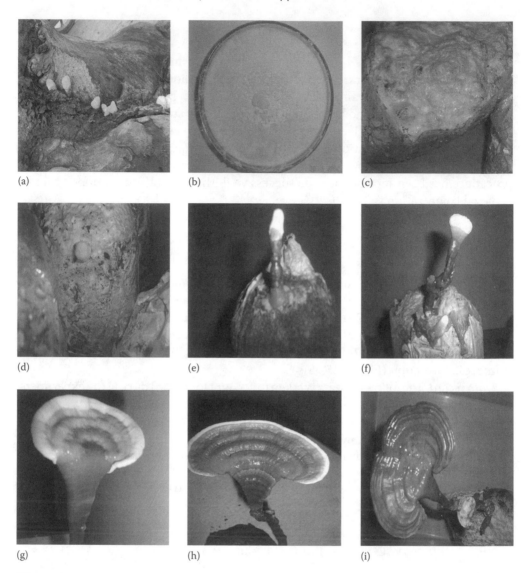

Figure 25.35 **(See color insert.)** Different growth stages of *G. lucidum* during artificial cultivation under solid substrate state fermentation on polypropylene bags: (a) source of strain isolation, (b) culture plate (growing in potato dextrose agar medium), (c) mycelial colonization on solid substrate, (d) primordia formation, (e) elongation of primordia, (f) cap formation, (g) flattening and growth of cap, (h) thickening of cap, and (i) maturation of fruit body.

Inoculum preparation: For inoculum preparation, culture disk from PDA medium is inoculated into Erlenmeyer flasks containing inoculation culture broth (20% potato extract, 2% glucose, and 2% olive oil) and incubated at 30°C for 5 days in a shaking incubator at 120 rpm. After 5 days, culture broth is ready to inoculate into bigger bioreactor. This inoculum culture broth contains *G. lucidum* mycelium of very high cell density.

Figure 25.36 Large-scale submerged cultivation of *G. lucidum*.

Production medium: Following production medium can be used for submerged cultivation of *G. lucidum*:

Ingredients	Medium 1 (%)	Medium 2 (%)	Medium 3 (%)	Medium 4 (%)
Potato extract	20.00	—	30.00	20.00
Malt extract	2.00	—	—	2.00
Yeast extract	—	0.10	—	1.00
Glucose	2.00	5.00	2.00	2.00
Olive oil	—	—	2.00	1.00
KH_2PO_4	0.10	0.05	0.05	0.10
K_2HPO_4	—	0.05	0.05	—
$MgSO_4$	0.05	0.05	0.05	0.05
NH_4Cl	—	0.40	—	—

Production: Desired quantity of production medium can be prepared in bioreactor and sterilize for 30 min at 121°C temperature and 15 psi pressure. Then after cooling of medium bioreactor is inoculated by 5% inoculum and fermentation carried out in optimum cultivation conditions such as temperature, $T = 30°C$, shaking 120 rpm, pH = 5.5. During fermentation, sampling is to be done at different time intervals from the bioreactor and mycelium growth is observed by OD. After 7–8 days, mycelium is separated from the cultivation broth by vacuum filtration and filtered cultured medium is concentrated one-fourth at 50°C. Extracellular polysaccharides (EPS) is isolated from concentrated filtrate and intracellular polysaccharides (IPS) is isolated from separated mycelium. The best yields reported till date are 22.1 g/L for mycelium, 1.71 g/L for EPS, and 2.49 g/L for IPS.

25.20 Cultivation of *L. edodes*

25.20.1 Introduction

L. edodes (shiitake) is the world's second largest cultivated and most popular medicinal and edible mushroom used as "functional foods." *L. edodes* is among the most valuable medicinal mushroom. This mushroom is known in China as Xingu and as shiitake in Japan, which has been renowned for thousands of years as both food and medicine. It is a source of two well-studied and widely approved polysaccharide medicines: LEM (an acronym for *L. edodes* mycelia), a protein-bound polysaccharide derived only from the mycelium, and lentinan—a cell wall–branched—β-D-glucan extracted from both the fruiting body and mycelium. Both compounds are immune system enhancers that demonstrate anticancer activity (Figure 25.37). Additionally, *L. edodes* contains other compounds that inhibit blood aggregation, reduce cholesterol levels, and exhibit antibacterial and antiviral effects. A lignan-rich compound derived from *L. edodes* mycelium holds a promising factor for treating both hepatitis B and AIDS. Another antitumor active polysaccharide, KS-2, has been isolated from shiitake mycelia (LEM).

  L. edodes mushroom-derived polysaccharides, immunomodulating and anticancer compounds are used in clinical applications as adjuvant to standard chemotherapy. Another potential use for *L. edodes* is as foodstuff, consumed whole or in concentrated extracts or as dietary supplements. There are several types of dietary supplements and medicinal formulations derived from *L. edodes*: dried and pulverized fruiting bodies, hot water and alcohol extracts of fruiting bodies, biomass or extracts of mycelia, or broth harvested from submerged liquid cultures. Commercial preparations are available as tablets, capsules, or elixirs, and are available in most Asian countries, and are increasingly available in the United States, New Zealand, Australia, and Europe.

25.20.2 Materials

Organism: Pure culture of *L. edodes*, cereal grains (sorghum, wheat, rye, etc.), bottles (flask type), cotton (gauze), saw dust of the broad leaf plant *T. grandis* (teak), wheat bran, rice bran, $CaSO_4$, $CaCO_3$, plastic bag, autoclave, laminar flow cabinet, incubator/storage room, wire gage balance, bunsen burner, and water.

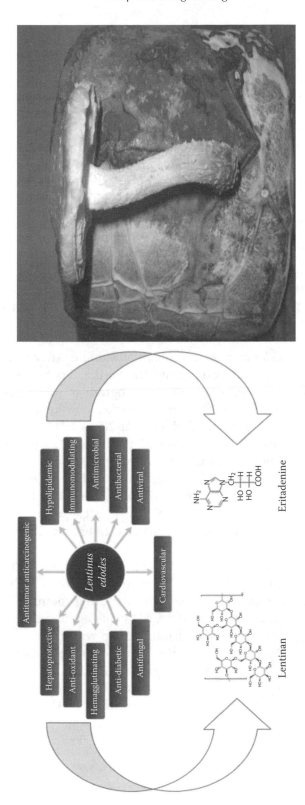

Figure 25.37 Chemotherapeutic compounds and activity of *L. edodes*.

25.20.3 Protocol

Culture maintenance: The media used for maintenance, multiplication, and preservation of mushroom culture are PDA, and YPDA (Figure 25.38).

Spawn production mainly consists of three steps: (i) substrate preparation, (ii) substrate inoculation, and (iii) incubation of the inoculated substrate for spawn production or growth of the mycelium on the substrate.

Preparation of substrate: Shiitake mushroom was cultivated by using the polybag method of cultivation. A substrate used is normally saw dust of the broad leaf plant *T. grandis* (teak). Substrates is supplemented with wheat bran and rice bran. $CaSO_4$ and $CaCO_3$ at different ratios are also added to substrate to maintain the pH 5.5. The substrate prepared contains 78% saw dust (teak) + 20% wheat bran + 1.5% $CaSO_4$ + 0.5% $CaCO_3$; 78% saw dust (teak) + 20% rice bran + 1.5% $CaSO_4$ + 0.5% $CaCO_3$ and are named as substrates I and II, respectively. The moisture level of the substrate is maintained at 60%–65% and bag-log system is adopted for cultivation. One kilogram substrate is filled in polypropylene (heat-resistant) bags and plugged with nonabsorbent cotton after putting a plastic (PVC) ring at the neck. Successive sterilization of substrate is done three times for 1.30 h at 121°C (Figure 25.39).

Spawning, mycelial colonization, and bump formation: After sterilization, these bags are kept to cool for 24 h. Upon cooling, the sterilized bags are inoculated with wheat grain spawn at the rate of 2%–5% per kilogram of substrate on dry weight basis under aseptic condition. The inoculated packets are kept in iron rack in an incubation room in a 4 h/20 h light/dark cycle at 20°C–25°C temperature and RH 65%–75%. During incubation period, whitish mycelia started to grow in the substrate. Mycelium running rate of each type of substrate is measured after 10 days. After full colonization of spawn packets, a thick mycelial coat is formed on the outer surface of colonized substrate. Clumps of mycelia appears as blister-like bumps of various sizes on the surface of the mycelial coat in each packet. Bumping usually starts when colonization of white mycelia changed to brown. The data on different parameters are recorded regularly.

Fructification: A hard and thick brown coat is on the surface of the bags, and the bags are then opened and dipped into cold water for 2 min. Then the bags of each type are placed separately side by side on the floor of fruiting room. The moisture level of fruiting room is maintained at 80%–90% RH by using humidifier. Light is either totally absent or slightly present in the fruiting room with maintained temperature (19°C–20°C) and ventilation at 10–15 min, at four times per day for proper oxygen management. The first primordia appears 7–10 days after depending upon type of substrate. The harvesting data also varied depending upon different types of substrates. During the fruiting process, temperatures of 15°C–20°C, humidity of 90%–95% is maintained in almost total darkness. The water is used in humidification. Fruit body formation is optimized and provided. Fruit body is oven dried at <50°C–60°C after maturation.

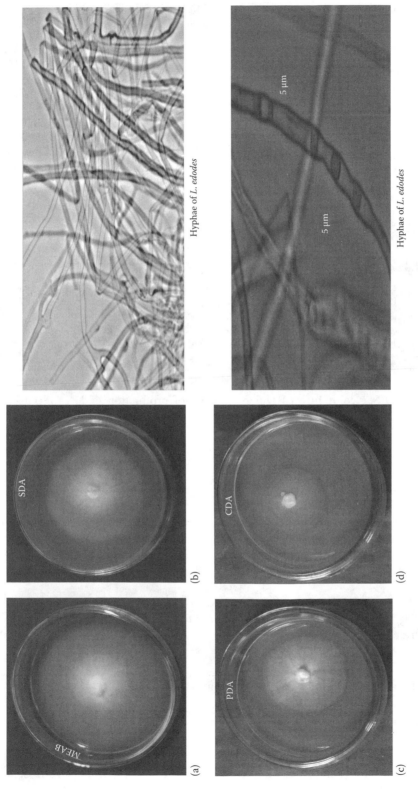

Figure 25.38   (See color insert.) *L. edodes* mycelium growing on different agar media on plates and microscopic view of hyphae of *L. edodes*. (a–d) Different degrees of growth on potato dextrose agar medium on Petri plates.

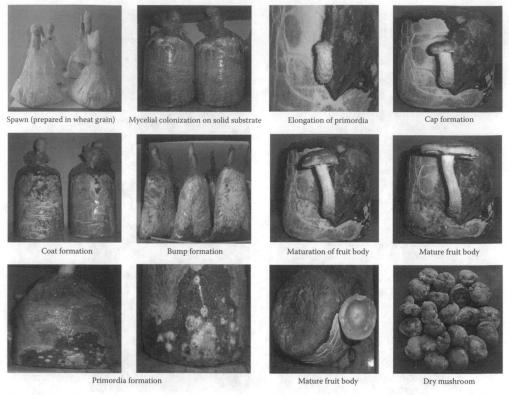

Spawn (prepared in wheat grain)	Mycelial colonization on solid substrate	Elongation of primordia	Cap formation
Coat formation	Bump formation	Maturation of fruit body	Mature fruit body
Primordia formation		Mature fruit body	Dry mushroom

Figure 25.39 **(See color insert.)** Different stages of production of *L. edodes* (shiitake) mushroom.

25.20.4 Observations and Results

The following data are taken to account for record:

a. *Days for completion of mycelium running*: Record time in days for completion of spawn running and mycelial cover on each substrate
b. *Days for bump formation*: Record time in days for bump formation on each substrate
c. *Days for pinheads formation*: Record time in days for pinhead formation on each substrate
d. *Days for cap formation*: Record time in days for cap formation on each substrate
e. *Fruiting body formation*: Record time in days for fruiting body formation on each substrate (Figure 25.38)

25.21 Cultivation of Paddy Straw Mushroom (*V. volvacea*)

25.21.1 Introduction

Paddy straw mushrooms are a species belonging to the genus *Volvariella* of the family Pluteaceae, order Agaricales in Basidiomycetes. It differs from *Agaricus* (*temperate* or *European mushroom*) in always having a cup-shaped persistent volva at the base of the stipe but never an annulus, and pink rather than sepia colored spore print. It differs from *Amanita*, a deadly poisonous mushroom, in having pink gills (against white) and lack of annulus. Of the over 25 species known, *V. volvacea*, *V. diplasia*, and *V. esculenta* are edible and are well-known table delicacy in many parts of the world. They are very delicate and must be consumed fresh. These are commercially grown in China, Indonesia, Burma, Philippines, Malaya, Madagascar, India, and Nigeria. The white to dull brown fruit bodies of *V. volvacea* are generally found growing solitarily or in groups on the soil in gardens, on compost heaps, at roadsides, occasionally on the mortar of brick work and hot houses and cellars. *V. diplasia* differs from *V. volvacea*, by its lack of pigments. *V. esculenta* grows in the hollow trunks of rotting oil-palm trees. It is rather fascinating to watch the outer skin of the basidiocarp crack and the cap emerges leaving the skin as a volva at the base of the stipe. Cultivation methods for this mushroom vary in different countries according to the availability of local waste materials and traditional practices (Figure 25.40).

A variety of materials have been tried for its cultivation, for example, paddy straw, cotton waste, water hyacinth, oil-palm bunch (inflorescence stalk), oil-palm pericarp waste, banana leaves and saw dust, sugarcane thrash (bagasse), etc. All these substrates have cellulose, hemicellulose and lignin as the main constituents. It grows at high temperatures between 30°C and 45°C. It can, therefore, also be cultivated during the summer months. In case of *Volvariella*, composting of straw is not required, merely a short period of straw wetting without the addition of any supplements is sufficient. It is generally believed that in areas where minimum temperature during the growing period is lower than 30°C, cultivation of straw mushroom is not practicable.

The traditional method of growing *V. volvacea* involves the use of bridge beds (20–80 cm wide) under partial shade or in the open field. The most commonly used substrate, the paddy straw, is soaked in water by complete immersion in tanks, before being folded and arranged into a stack on a soil base. In case of *Volvariella*, no casing material is required to initiate the formation of fruiting bodies. The fruiting body is formed within 14 days following spawning. However, light requirement is a must for the formation of a fruit body. Peak-heated or pasteurized compost is used and the

(a)   (b)

Figure 25.40 (a) Preparation of straw bundle and (b) paddy straw mushroom *V. volvacea*.

mushroom beds are cased with a 2 cm thick layer of day loam. With this method of cultivation, the yield is just doubled than that observed by the outdoor method. Mushroom yield is adversely affected by the contamination of beds with other fungi which act as weeds and compete for nutrients. Important fungi which act as weeds are *Coprinus aratus*, *Psathyrella* spp. and molds like *Aspergillus* spp., *Rhizopus stolonifer*, *Sclerotinia rolfsii*, and several others. Spraying with captan or zineb suppresses the growth of the weed fungi without affecting the crop of paddy straw mushroom. Protocol for growing *V. volvacea*:

1. Preparing paddy straw bundles
2. Wetting the straw bundles for 20–24 h
3. Laying of beds and spawning
4. Picking of mushrooms 12–26 day
5. Removing used straw and cleaning the site 27–28 day

25.21.2 Materials

Paddy straw, small water tanks, bricks, bamboo frame, Gram powder, spawn of *V. volvacea* (350–400 g of grain spawn/bed), thermometer (0°C–100°C), and polythene sheets.

25.21.3 Protocol

1. Collect paddy straw which is not very leafy and should not be more than one year old and not uncrumbled.
2. Store the straw at a protected place where it does not get wet.
3. Make the bundles of paddy straw.
4. Soak the bundles in water for 18–24 h in small tanks, keeping in view that the bundles are completely immersed in water, by putting weight on the bundles.
5. Take out the bundles from the tanks.
6. Drain off the excess water.
7. Laying of beds and spawning:
   a. Make square beds (1 m × 1 m × 1 m or 1 m × 0.75 m × 1 m) of the soaked straw bundles with their butt ends on one side placed lengthwise close to each other on a bamboo frame supported on bricks. The number of bundles are to be placed in such a way that it approximates to the length of the straw, to make a square bed.
   b. Make a second layer over the first layer by placing the bundles having the butt ends on the opposite side. An arrangement of this type makes one layer.
   c. Place small bits of spawn 7.5–10 cm inside the margin, leaving a space of 5–5.5 cm from each other.
   d. Place the third layer of straw bundles at right angles to the previous layer, that is, in a criss-cross fashion.
   e. Place the fourth layer on the third layer with the opposite butt ends.
   f. Spawn this layer too as done earlier (in Step c).
   g. Place another layer of bundles with butt ends at right angles to the previous one.
   h. Spawn this layer all over.

      i. Cover the inoculated layer with loose straw.
      j. Press down the bed.
  8. Cover the bed with polythene sheets.
  9. Maintain temperature and moisture in the beds.

Observe the beds for watering and temperature of beds. The bed temperature should remain between 30°C and 35°C after spawning and should not go below 30°C during growing season.

25.21.4 Observations

Remove the polythene sheet after 7–10 days of spawning for the appearance of small buttons.

Harvesting of mushrooms: Harvest the mushrooms, when the volva is about to rupture or is just ruptured, by gently twisting the fruiting bodies.

Preservation: The paddy straw mushrooms are consumed fresh but can be preserved either by dehydration or canning. Dry the mushrooms either in the sun or at a temperature of 50°C–60°C and pack either in aluminum foil or in polythene bags.

25.22 Cultivation Method of Oyster Mushroom

25.22.1 Materials

Wheat grains, calcium carbonate, calcium sulfate, polypropylene, Petri plates, culture tubes, rice straw, wheat straw, sugarcane baggas and cotton waste, sharp knife, and blades.

Organism and growth conditions: Cultures of Oyster mushroom *Pleurotus djamor*, *P. florida*, and *P. sajorcaju* are cultivated in a growth room and maintained on potato dextrose medium at 25°C–30°C ± 2°C in slants and Petri plates. The culture are sub cultured after every 3 weeks (Figure 25.41).

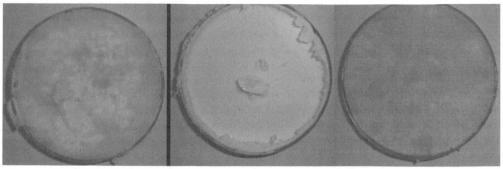

    Pleurotus ostreatus   *Pleurotus sajorcaju*   *Pleurotus djamor*

Figure 25.41 Three different oyster (*Pleurotus* sp.) mushroom growing on PDA Petri plates.

25.22.2 Protocol

Preparation of spawn: The seed of mushroom called spawn is prepared on wheat grain. The wheat grain is softened by soaking in boiling water for 20 min and 0.5% calcium carbonate and 2% calcium sulfate are added in relation to their mass. Moisture level of the mixture is maintained at 65%. Then the mixture is transferred in to 9×12 cm polypropylene bags and packed tightly. The neck of the bag is prepared by using heat-resistant PVC tube. The neck is plugged with cotton and covered with a sterile brown paper and tied with a rubber band. The bags are sterilized by autoclaving them at $120°C \pm 2°C$ at 15 psi for 30 min. After sterilization, the bags are allowed to cool for 24 h and subcultured by a 9 mm disc of 15 days old oyster mushroom growing on Petri plate culture cut with sterile cork borer and aseptically transferred to polypropylene bags and incubated at $25°C-30°C \pm 2°C$ for 15 days.

Preparation of substrate: Rice straw, wheat straw, sugarcane baggas and cotton waste are used as substrate. They are chopped into small pieces of 1–3 cm. These substrate (dry and soaked) are in boiling water for 15 min and extra water present in the substrates is drained off and the substrate are spread on the surface of a clean blotting paper and air-dried for 15 min to remove the excess water. After cooling of the substrates (when the moisture content were left around 65%–70%), 1 kg substrates are filled in the polythene bags of 18×12 cm in size. All the substrates are sterilized in an autoclave at 15 psi for 1 h.

Spawning and fructification: After sterilization and cooling, next day the bags are inoculated with the spawn grain of oyster mushroom at 10 g/kg of substrate on dry weight basis under aseptic condition. The bags are then inoculated for spawn running under complete darkness at controlled temperature of $25°C-30°C$ and the mycelium is allowed to ramify the substrate. The RH of the room is maintained between 65% and 85% with the help of a humidifier. After bags are completely colonized by the mycelium for 30 days, they are moved to the fruiting room and large holes are made in the polythene bags to allow the normal development of pinheads. Photoperiod of 9 h light/15 h darkness is given. The air in the cultivation room is renewed six times per hour. When such bags become full of growth and pinheads started appearing, the bags are mouth opened to facilitate the development of fruiting bodies. As soon as the fruiting bodies developed (oyster mushroom) and attained their full size, they are cut just above surface of the substrate with sharp knife or blade (Figures 25.42 and 25.43).

25.22.3 Observations and Results

The following data should be recorded during the experimentation:

- *Completion of spawn running*: The days required for completion of one-fourth, one-half, three-fourth, and full growth of mycelium on substrate.
- *Appearance of pinheads*: Days required for the appearance of pinhead formation.
- *Maturation of fruiting body*: Days for fruiting body formation on each substrate.

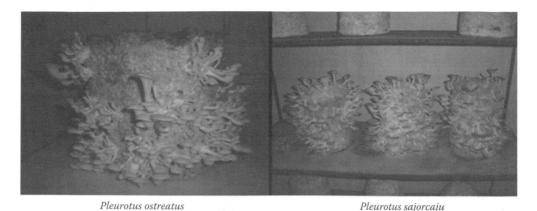

Pleurotus ostreatus   *Pleurotus sajorcaju*

Figure 25.42 **(See color insert.)** Oyster mushroom growing on solid substrate fermentation.

Figure 25.43 *Pleurotus florida* growing under solid substrate fermentation.

- *Harvesting dates and total yield of mushroom*: Harvesting of first, second, and third crops. The weight of different crops is added to get the total yield of the crop.
- *Temperature*: The temperature of the mushroom growing room should be recorded daily at 7.00 a.m., 1.00 p.m., and 7.00 p.m. The average of 7 days is to be worked out.
- *Relative humidity*: The RH of the mushroom growing room is to be maintained with humidifier and noted daily at 9.00 a.m., 1.00 p.m., and 6.00 p.m. with the help of humidity recorder.

Suggested Readings

Bioprocess Engineering
Doran, P. M. 2012. *Bioprocess Engineering Principles*. Waltham, MA: Academic/Elsevier.

Gyun, J. K., C. Jeong, J. S. Kwang, and Y. Y. Je Kim. 2011. Biotechnology and bioprocess engineering 16 gene cloning and expression of a 3-ketovalidoxylamine C-N-lyase from *Flavobacterium saccharophilum* IFO 13984. *Biotechnol Bioprocess Eng* 16: 366–373.

Moyer, A.J., and R.D. Coghill. 1946. Penicillin: VIII. Production of penicillin in surface cultures. *J Bacteriol* 51:57–78.

Rao, D. G. 2010. *Introduction to Biochemical Engineering*. New Delhi, India: Tata McGraw-Hill Education.

Shuler, M. L. and F. Kargi. 2012. *Bioprocess Engineering: Basic Concepts*. New Delhi, India: Pearson Education.

Stanbury, P. F. and A. Whitaker. 1984. *Principles of Fermentation Technology*. New York: Pergamon Press.

Mushrooms

Antonio, C. 2005. *The Complete Mushroom Book*. London, U.K.: The Quiet Hunt Quadrille Publishing Ltd.

Arora, D. 1986. *Mushrooms Demystified*. Berkeley, CA: Ten Speed Press.

Bisen, P. S., R. K. Baghel, B. S. Sanodiya, G. S. Thakur, and G. B. K. S. Prasad. 2010. *Lentinus edodes*: A macrofungus with pharmacological activities. *Curr Med Chem* 17: 2419–2430.

Christopher, H. 2002. *Medicinal Mushrooms* (Herbs and Health Series). Redondo Beach, CA: Botanica Press.

Preuss, H. G. and S. Konno. 2002. *Maitake Magic*. Evanston, IL: Freedom Publishing Company.

Sanodiya, B. S., G. S. Thakur, R. K. Baghel, G. B. K. S. Prasad, and P. S. Bisen. 2009. *Ganoderma lucidum*: A potent pharmacological macrofungus. *Curr Pharm Biotechnol* 10: 717–742.

Stamets, P. 1983. *The Mushroom Cultivator: A Practical Guide to Growing Mushrooms at Home*. Seattle, WA: Agarikon Press.

Stamets, P. 1996. *Psilocybin Mushrooms of the World: An Identification Guide*. Berkeley, CA: Ten Speed Press.

Stamets, P. 2000. *Growing Gourmet and Medicinal Mushrooms*. Berkeley, CA: Ten Speed Press.

Important Links

http://www.bioh2iitkgp.in/links.php
http://www.nae.edu/Publications/Bridge/BiotechnologyRevolution/TheRoleof BioprocessEngineeringinBiotechnology.aspx
http://www.shef.ac.uk/cbe/prospectivepg/taught/bbe/whatisit
http://www.wisegeekedu.com/what-does-a-bioprocess-engineer-do.htm
http://www.montereymushrooms.com/our-mushrooms/quality-assurance/
http://www.windows2universe.org/earth/Life/fungi.html
http://www.integral-health-guide.com/types-of-mushrooms/
http://www.bios.niu.edu/calvo/calvo_lab.shtml

26

Environmental Biology

Environmental biology is a multidisciplinary with its constituent areas of life sciences. All living and nonliving components present on this planet interact among themselves and exist together in a dynamic state. Such an organization or assemblage is called an environment. For that matter, the entire earth with all its components is considered as a giant environment or giant biosphere (bio = life forms, sphere = an area occupied). Living organisms and nonliving matter in any ecosystem do not exist in isolation and function all independent. Most of the abiotic constituents provide raw materials, energy sources, etc. for the living organisms to consume and produce the organic matter. The functional components are mainly the processes involved in the flow of energy (solar energy), from abiotic components (including nutrients) to biotic components (as biomass), from one biotic to another biotic system, and lastly from biotic back to abiotic system. The success of an environment mainly depends upon the longevity (or half-life) of the bioenergy retained within the biomass. The half-life of the bioenergy in a biomass in turn is controlled by the rate of producer's activity, the rate of consumer's activity, the rate of detritivores activity, and rate at which these three interact with each other.

The major functional process of an environment is autotrophic mechanism, by which solar energy is converted into chemical energy as the main capital. Using such energy and other abiotic ingredients biomass is built up by various respiratory and intermediary metabolic processes, responsible for the growth of biomass. Lastly, the biomass (after death) is converted by various oxidative processes into basic abiotic ingredients and there is a net loss of energy in the form of heat. All biogeochemical cycles are involved. In all these energy transformation, there is a loss of energy in one or the other form; thus, they obey the second law of thermodynamics. It is the functional process that ultimately determines the success of biosystem in an environment. The study of environmental biology involves specific application of applied life sciences to the management of environment and related socioeconomic and developmental issues, keeping in view the concept of sustainable development. Environmental biology encompasses issues like

- Environmental monitoring
- Restoration of environmental quality
- Resource/residue/water-recovery utilization/treatment through application of rDNA technology

- Global changes
- Biological diversity
- Substitution of nonrenewable resources by renewable ones
- Strain improvement for degradation of highly toxic pollutants with the production of chemicals
- Risk management

The market for environmental biology is developing fast with a growth rate of 17% in environmental cleanup areas against a growth rate of 7% in the market of food and pharmaceutical applications. The use of environmental biotechnology for pollution control is well-documented. With the advancement of bioreactor designs, the use of genetically engineered or adapted cultures, biodegradation technology has been successful in making its impact feel on pollution abatement efforts.

Commercial exploitation of life science applications needs more technical evaluation, demonstration of processes, performance data, and wide literature publicity. Life science technology is effective in the removal of many xenobiotic compounds from the aqueous, solid, and gaseous wastes. Polynuclear aromatics, halogenated aliphatic, heterocyclic, polar nonhalogenated nitrocompounds, halogenated compounds, cyanides, phenolic, urea, etc. can be removed with the use of microorganisms or their enzymes. Anaerobic biodegradation rather than the aerobic one is the major change in the trend and this not only saves energy expenditures on treatment but generates energy in the form of methane or ethanol. Bioremediation of land sites, ground water polluted by spillage, drainage, and leakage is also getting established as a successful remedial measure to remove hazardous chemicals like C_6H_6, styrenes, xylene, vinyl chloride (PAHs), and nitrates.

Single cell protein (SCP) production and mushroom cultivation are relatively old applications of biotechnology. But now treating organic-rich effluents from dairy, brewery, confectionery, fruit pulp processing, and starch production units gives product formation from wastes. There are now a dozen of companies active in developing and marketing engineered microbes for effluent treatment schemes for wastewater of dairy, distillery, tannery, sugar, antibiotic industries including bioremediation of contaminated soil and wastelands. The environmental biology also deals with disposal of solid waste by composting vermiculture and methane production. The global environmental problems like ozone layer depletion, UVB radiation, greenhouse effect, and acid rain, their impact and biotechnological approaches for management are always warranted.

26.1 Wastewater Treatment Process

Wastewater is the used water supply of a community and consists of domestic wastewater or sewage, including human excrement and wash water that is drained into a sewage system, industrial waterborne wastes such as acids, oil, greases, and animal and vegetable matter discharged by factories, and ground, surface, and atmospheric waters that enter the sewage system. Domestic waste-water or sewage consists of ~99.9% water and 0.2%–0.03% suspended solids, and other soluble organic and inorganic substances. Wastewater and sewage treatment involves a large-scale use of microorganisms for bioconversion on an industrial scale. Wastewater treatment is necessary before wastewater can be disposed off without producing significant undesirable or even harmful effects. Disposal of inadequately treated wastewater leads to greater possibility for dissemination of pathogenic microorganisms and toxic chemicals that endanger ecosystems and threaten public health. Wastewater and sewage treatment from individual dwellings or households can be accomplished by anaerobic digestion and/or by aerobic metabolic processes. One of the commonly used structures is the septic tank, which is an aerobic digesting system.

A septic tank (Figure 26.1) is an enclosed sewage settling tank designed to retain the solids of the sewage entering the tank long enough to permit adequate decomposition of the sludge (collected particulate matter). As sewage enters this type of tank, sedimentation occurs from the upper portion, permitting a liquid with fewer suspended solids to be discharged from the tank. The sedimented solids are subject to degradation by anaerobic bacteria; hence, the end products are still very high in biological oxygen demand (BOD) and odorous. The effluent from the septic tank is distributed under the soil surface through a disposal field.

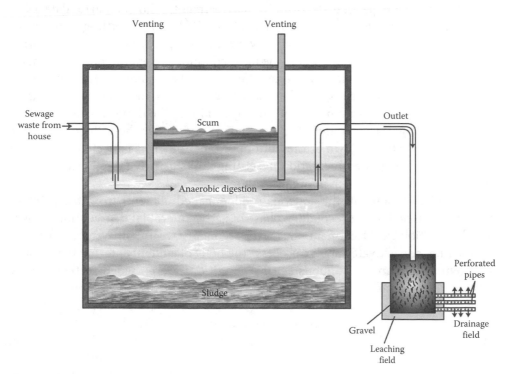

Figure 26.1   Cross section of a septic tank.

26.1.1 Artificial Wetlands

In artificial wetlands, the land can be subdivided into small lots with a common portion set aside for a series of ponds. Sewage is channeled into successive ponds, which carry out both aerobic and anaerobic stabilization. By the time the water reaches the final pond, it is suitable for decorative or recreational purpose.

26.1.2 Municipal Treatment Processes

Complete municipal wastewater treatment consists of a series of steps (Figure 26.2): primary, secondary, tertiary, and final treatments and solid processing. In primary treatment, physical means are used to remove solid waste from wastewater. In secondary treatment, biological means such as action of decomposers are used to remove solid wastes that remain after primary treatment. In tertiary treatment, chemical and physical means are used to produce an effluent of water pure enough to drink. Final treatment is carried out to disinfect and dispose of liquid effluent. Solid processing is done to stabilize solids removed from liquid processes, to dewater solids, and ultimately to dispose of solids (by land application, landfill, and incineration).

Primary wastewater treatment separates solid and particulate organic and inorganic materials from freshwater. Primary treatment can physically remove 20%–30% of the BOD present in particulate form. In this treatment, solid and particulate waste is removed by screening, precipitation of small particulate, and settling in basins or tanks. The most commonly used primary treatment processes are sedimentation and froth floatation. Sedimentation is the most commonly used method for primary treatment of wastewater as it is very easy, effective, and economical. The sedimentation units include tank, basins, or other mechanical devices, in which sedimentation is allowed to take place for a period of time by retaining the liquid waste undistributed in it. During sedimentation, the suspended solid under the influence of force of gravity starts settling down at the bottom of the tank. This collected particulate matter is called sludge. Sedimentation in settling tanks usually takes between 2 and 10 h and leaves a mixture rich in organic matter.

In secondary treatment, the aqueous residue (called effluent) from the primary treatment is carried into a secondary phase of active microbial degradation. This is also called as biological treatment phase. The secondary treatment system consists of trickling filter and activated sludge process. The BOD of effluent entering the secondary system is high and continuous aerobic conditions are provided so that aerobic microorganisms could act at a fast speed to decompose the organic matter.

The trickling filter consists of a bed of crushed stone, gravel, slag, or synthetic material with drains at the bottom of the tank. The waste effluent is passed over the surface of the bed (either by a rotating arm or through nozzles; Figure 26.3). Spraying oxygenates the sewage so that the aerobes can decompose organic matter in it. The microbial film consists of bacteria (*Pseudomonas, Beggiatoa, Alcaligenes, Nitrobacter, Sphaerotilus*), fungi (*Penicillium, Fusarium, Mucor, Geotrichum, Sporotrichum*),

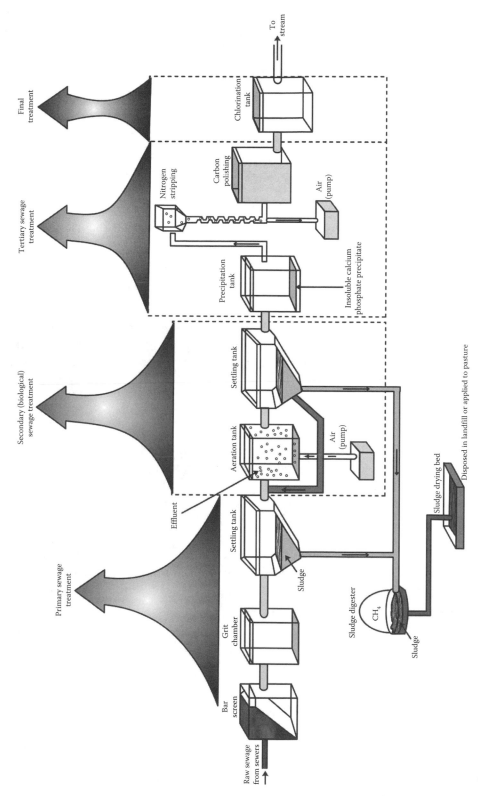

Figure 26.2 Illustrating municipal wastewater or sewage treatment plant.

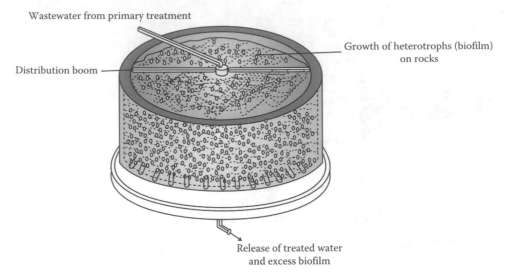

Figure 26.3   The trickling filter.

protozoa (*Opercularia, Amoeba, Paramecium, Treponema*), and algae (*Phormidium, Chlorella*, and *Ulothrix*). The microbial community in these films degrades the organic waste. Such a system is less efficient but less subject to operational problems.

26.1.3 Activated Sludge Process

The mechanisms by which populations of aerobic organisms stabilize sewage at most secondary treatment plants are called the activated sludge process (Figure 26.4). In this treatment, the sewage serves as a nutrient source for mixed populations of aerobic organisms like bacteria, yeasts, molds, and protozoa adapted to grow in it. A high amount of oxygen is supplied by mixing the sewage in an aerator. Most of the biologically degradable organic material is then converted into gases or oxidized product, and a very small percentage is incorporated into the cell material of the organisms growing on the sewage. This process is often the most satisfactory way to treat domestic sewage, but it can be rendered totally ineffective by the presence of toxic industrial wastes, which poison the microbial population.

26.1.4 Oxidation Ponds

They are also referred to as lagoons or stabilization ponds. They are shallow ponds, 2–4 ft in depth, and are used for treating sewage on a small scale. The sewage remains in the pond for several days during which settling occurs and sewage materials are stabilized by anaerobic/aerobic microorganisms. Oxygen requirements for the oxidation of nutrients are generally fulfilled by the atmospheric air, but the release of oxygen during photosynthesis of algae provides additional oxygen.

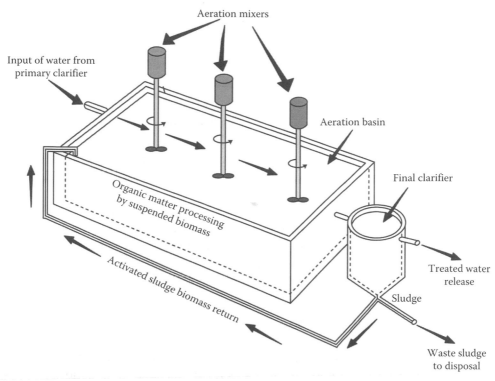

Figure 26.4 Activated sludge with microbial biomass recycling.

26.1.5 Tertiary Treatment

The separation of solid sludge leaves a certain amount of water that may be further processed in tertiary treatment by purifying the water. Tertiary wastewater treatment is a physicochemical or biological process involving bioreactors, precipitation, filtration, or chlorination procedures similar to those employed for drinking water purification. It sharply reduces the levels of inorganic nutrients, especially phosphate, nitrate, and nitrite from the final effluent.

26.1.6 Final Treatment

The liquid effluent is disinfected and usually discharged into a body of water upon completion of other treatments. Disinfection of wastewaters is necessary to protect public health when the receiving waters are used for purposes such as downstream water supply, recreation, or for irrigation. Chlorination is the most common method of disinfection. However, the chlorination has been proved adverse to the aquatic life, and hence alternative disinfection procedures such as the use of ozone and ultraviolet light are becoming more prevalent. Dissolved oxygen (DO) may also be added to the treated wastewater prior to final discharge, a process called postaeration, which minimizes the decrease in receiving water–DO that normally occurs when treated wastewater effluent is discharged.

26.1.7 Solid Processing

Solids are processed during the primary, secondary, and tertiary stages of the treatment and generally involve thickening, stabilization, dewatering, and disposal of sludge. Thickening is generally used to further concentrate the solids or sludge prior to stabilization. Thickening is achieved by gravity thickening or by dissolved air floatation. Stabilization is achieved by aerobic and anaerobic digestion, composting, chemical addition, and heat treatment, the most common being anaerobic digestion. Dewatering is achieved by employing physical methods, and the equipment used for dewatering includes vacuum filters, belt filter presses, plate and frame presses, and centrifuges.

26.1.8 Anaerobic Sludge Digestion

The treatment involves a series of digestive and fermentative reactions carried out by a number of bacterial and archaeal species under anaerobic conditions. The anaerobic degradation process is carried out in large enclosed tanks called sludge digesters or bioreactors. Within a digester, anaerobic organisms act on the solids remaining in sewage after its aerobic treatment. The digestor provides anaerobic stabilization and removes water from the sewage so that a minimum of solid matter remains in it. Various microbial populations act sequentially. In the sewage digester, the anaerobic methane-forming bacteria can perform their role of converting the simple organic acids in sewage into a useful end product methane (CH_4), which can be further used as a biogas. Undigested matters dry into cakes, which are carried by conveyor belts to incinerators. The incinerator reduces the sludge cake to ash for final disposal in a landfill.

26.1.9 Composting

Composting is the natural decomposition of organic solid material. The dewatered sludge is mixed with a bulking agent such as wood chips. The bulking material is added to enhance circulation of air throughout the sludge to improve the stabilization process. The mixture of sludge and bulking material is placed in aerated piles and allowed to decompose biologically for a period of time. The stabilization is achieved in 21 days. The bulking agent is separated thereafter from the sludge and the sludge is allowed to decompose further. The end result of composting is the formation of humus-type material and is used as a soil conditioner.

26.1.10 Landfills

Landfills are used to dispose of solid wastes near towns and cities. The area of land chosen for this purpose is generally not valuable. Since piling up the solid wastes on the ground attracts insects and rodents causing health and aesthetic problems, sanitary landfills are used in which a layer of dirt is put over the wastes at the end of each day. When the landfill is over, it can be used for recreation and for construction purposes.

26.2 Measurement of pH

26.2.1 Principle

pH is the measure of the relative acidity or alkalinity of the solution and is represented as the negative logarithm of the concentration of free hydrogen ions in a solution. The "p" of pH denotes the power of the hydrogen ion activity in moles per liter:

$$pH = -\log 10(H^+) = \log 10 \times \frac{1}{H^+}$$

When two electrodes are dipped in two solutions of different pH levels and connected, a potential difference is set up between the two electrodes, which is measured by the potentiometer. This is directly related to the pH of the solution.

26.2.2 Materials

Prepare all the solutions in previously boiled and cooled distilled water.
Standard buffer solutions

1. Potassium tetraoxalate ($KH_3C_4O_{82}H_2O$): Dissolve 3.175 g $KH_3C_4O_{82}H_2O$ in water and make up to 250 mL (pH 1.68).
2. Potassium tartrate ($KHC_4H_4O_6$): To 250 mL water in a glass-stoppered bottle, add excess $KHC_4O_4O_6$ and shake the bottle vigorously for about 10 min to obtain a saturate solution. Filter and preserve the solution by adding 0.1 g thymol (pH 3.55).
3. Potassium phthalate ($KHC_8H_4O_6$): Dissolve 2.552 g $KHC_8H_4O_6$ in water and make up to 250 mL (pH 4.0).
4. Potassium dihydrogen phosphate (KH_2PO_4): Dissolve 0.30 KH_2PO_4, 0.8875 g Na_2PO_4, and 008875 g Na_2HPO_4 in water and make up the solution to 250 mL (pH 6.86).
5. Borax ($Na_2B_4O_7$): Dissolve 0.9502 g $Na_2B_4O_7$ in water and make up to 250 mL (pH 9.18).

26.2.3 Protocol

1. Warm up the instrument for 15 min.
2. Calibrate the instrument with the known buffer solutions. (Calibration is done by a buffer solution whose pH is close to that of the sample.)
3. Immerse the electrode in the unknown sample, stir for 3 min, and note the pH.

Note: Instead of preparing standard buffer solutions, commercially available *buffer tablets* can be used for calibrating the pH meter.

26.3 Measurement of Alkalinity of Water

26.3.1 Principle

In most natural waters, bicarbonates and sometimes carbonates are present in appreciable amounts. Their salts get hydrolyzed in solution to produce hydroxyl ions, consequently raising the pH.

Alkalinity is determined by titrating the sample with a standard solution of strong acid. Alkalinity due to hydroxide and carbonate is determined to the first end point (pH 8) using phenolphthalein indicator and bicarbonate alkalinity is determined to the second end point (pH 4.5) using methyl orange indicator.

26.3.2 Materials

1. Sulfuric acid titrant (0.02 N): Dilute 0.6 mL of sulfuric acid to 1 L to obtain 0.02 N acid titrant.
2. Phenolphthalein indicator: Dissolve 1.25 g phenolphthalein in 125 mL ethyl alcohol and add 125 mL distilled water. Add 0.02 N NaOH drop wise until a faint pink color appears.
3. Methyl orange indicator: Dissolve 0.1 g methyl orange in 200 mL distilled water.

26.3.3 Protocol

Pipette out 50 mL sample to an Erlenmeyer flask and two drops of phenolphthalein indicator:

1. If a slight pink color appears, titrate with acid titrant to a colorless end point and note the reading as "*P*" (mL of titrant used for phenolphthalein alkalinity).
2. Now add two drops of methyl orange to the same flask and continue to titrate further till the color changes from yellow to orange. Note this reading as "*T*" (total value of the titrant used for both the titrations).

TABLE 26.1 Relationship between Different Forms of Alkalinity

Result of Titration	Value of Radical Expressed in Terms of CaCO₃	
	CO_3^{2-}	HCO_3^-
$P=0$	0	T
$P < 1/2T$	$2P$	$T-2P$
$P = 1/2T$	$2P$	0
$P > 1/2T$	$2(T-P)$	0
$P = T$	0	0

Note: T, Total alkalinity; *P*, phenolphthalein alkalinity.

26.3.4 Calculation

Calculate the phenolphthalein and total alkalinity by the given formulae

$$\text{Phenolphthalein alkalinity } (p) \text{ as mg/L CaCO}_3 = \frac{\text{mL of titrant ''}p\text{''} \times 1000}{\text{mL of sample}}$$

$$\text{Total alkalinity } (t) \text{ as mg/L CaCO}_3 = \frac{\text{mL of titrant ''}t\text{''} \times 1000}{\text{mL of sample}}$$

For the computation of amounts of carbonates and bicarbonates, refer to Table 26.1.

26.3.5 Result

Express the total alkalinity and contribution of bicarbonates (HCO_3^-) and carbonates (CO_3^{2-}) as mg/L of $CaCO_3^-$.

26.4 Measurement of DO of Water

26.4.1 Principle

This method depends upon the oxidation of manganese dioxide (bivalent manganese) by the oxygen dissolved in the water, resulting in the formation of a tetravalent compound. When the water containing the tetravalent compound is acidified, free iodine is liberated from the oxidation of potassium iodide. The free iodine is chemically equivalent to the amount of DO present in the sample and is determined by titration with a standard solution of sodium thiosulfate:

$$MnSO_4 + 2KOH \rightarrow Mn(OH)_2 + K_2SO_4$$

If the precipitate is white, there is no DO in the sample. A brown precipitate indicates that oxygen was present and reacted with the manganous hydroxide forming manganic oxide:

$$2Mn(OH)_2 \rightarrow 2MnO(OH)_2$$

On addition of sulfuric acid, the precipitate is dissolved forming manganic sulfate:

$$MnO(OH)_2 + 2H_2SO_4 \rightarrow Mn(SO_4)_2 + 3H_2$$

There is an immediate reaction between this compound and the potassium iodide previously added, liberating iodine and resulting in the typical iodine coloration (brown) of the water:

$$Mn(SO_4)_2 + KI \rightarrow MnSO_4 + K_2SO_4 + I_2$$

The number of molecules of iodine liberated by the reaction is equivalent to the number of molecules of oxygen present in the sample. The quality of iodine can be determined by titrating a portion of the solution with a standard solution of sodium thiosulfate:

$$2Na_2S_2O_3 + I_2 \rightarrow Na_2S_4O_6 + 2NaI$$

Since 1 mL of 0.025 N sodium thiosulfate is equivalent to 0.2 mg oxygen, the number of mL of sodium thiosulfate used is numerically equivalent to the concentration in mg/L of DO if 200 mL of original sample was titrated.

Similarly if a 100 mL sample is titrated with 0.005 N sodium thiosulfate, the number of mL of titrant times 0.4 equals the number of mg of O_2 per liter.

The iodine should be uniformly distributed throughout the bottle; a portion is decanted for titrating.

26.4.2 Materials

1. Manganous sulfate solution: Dissolve 364 g of manganous sulfate ($MnSO_4H_2O$) in distilled water, filter, and dilute to 1 L.
2. Alkaline iodide azide solution: Dissolve separately 700 g of potassium hydroxide (KOH) and 150 g potassium iodide (KI) in distilled water. Mix them and make the volume up to 1 L. Dissolve separately 10 g sodium azide (NaN_3) in 40 mL of distilled water. Add this solution to 960 mL of alkaline iodide reagent.
3. Sodium thiosulfate titrant (0.025 N): Dissolve 6.20 g sodium thiosulfate ($Na_2S_2O_3$) in freshly boiled and cooled distilled water and dilute to 500 mL. Add one pellet of sodium hydroxide (NaOH) as preservative.

4. Starch indicator: Dissolve 1 g of starch (soluble) in 200 mL hot distilled water and add few drops of toluene as preservative.
5. Concentrated sulfuric acid.

26.4.3 Protocol

1. Rinse clean BOD bottles (125 mL) twice with sample water before sampling. Introduce samples through a long tube into the bottom of the bottle so as to avoid bubbles while filling. Allow the water to overflow the bottle before stoppering.
2. Winklerize (pretreat at site) the sample at site, that is, after sample is collected in the BOD bottle, remove the stopper, add 1 mL of manganous sulfate solution followed by 1 mL of alkaline iodine solution.
3. Restopper the bottle. Shake thoroughly to mix the reagents and disperse the precipitate formed evenly. Allow the precipitate to settle half way down the bottle.
4. Back in the laboratory, add 1 mL of conc. H_2SO_4. Stopper and mix thoroughly. The entire precipitate should redissolve after the addition of acid.
5. Transfer 50 mL of acidified sample into a conical flask, 3–5 min after acidification.
6. Titrate at once with standard 0.025 N thiosulfate solution until the brown iodine color becomes pale or straw yellow.
7. Add 1 mL of starch indicator. Blue color appears.
8. Continue titration until the blue color disappears. Note the titer value.

Calculation

$$O_2(\text{mg/L}) = \frac{\text{CD} \times M \times E \times 100 \times 0.698 \times v_t}{v_s}$$

where CD is the correction for displacement of sample when reagents are added:

$$\text{CD} = \frac{\text{Volume of bottle}}{\text{Volume of bottle} - \text{Volume of reagents}}$$

M is molarity of thiosulfate (0.025)
E is equivalent weight of oxygen (8)
100 is to express per liter
0.698 is to convert ppm to mg of oxygen
v_t is titer value
v_s is volume of the sample used in titration

Comments: Rapid titration is recommended. After the end point (disappearance of blue color), exposure to air may cause liberation of more iodine and hence color will reappear. This is caused by interference of nitrites in the sample.

26.4.4 Result

Express oxygen as mg/L of sample.

26.5 Measurement of Chloride in Water

26.5.1 Principle

The chloride present in the water sample is titrated with silver nitrate solution:

$$Cl^- + Ag^+ \rightarrow AgCl$$

Potassium chromate is used as indicator. At the end point, the concentration of chloride ion in solution reaches zero, the silver ion concentration increases to a level at which the solubility product of silver chromate exceeds. Then, silver chromate is precipitated as a reddish brown product:

$$2Ag^+ + CrO_4^{2-} \rightarrow Ag_2CrO_4$$

26.5.2 Materials

1. Silver nitrate ($AgNO_3$) (0.014 N): Dissolve 2.395 g of $AgNO_3$ in distilled water and make up this solution to 1000 mL and store it in a brown bottle.
2. Potassium chromate (K_2CrO_4) indicator: Dissolve 10 g of K_2CrO_4 in a little distilled water. Add $AgNO_3$ solution in drops till a red precipitate is formed. Allow it to stand overnight, filter, and dilute to 200 mL with distilled water.

26.5.3 Protocol

1. Pipette 50 mL of the water sample (if sea water, take 0.5 mL of sample) into a conical flask.
2. Pipette into it, 0.5 mL of K_2CrO_4 indicator. This gives yellow color to the sample.
3. Titrate the solution with shaking, against standard silver nitrate solution till the appearance of reddish brown color.
4. Perform a duplicate titration in an identical manner.
5. Carry out a blank titration using 50 mL of deionized, chloride-free water and 0.5 mL of the indicator. Subtract this titer value from that obtained for the water sample.

Calculation

$$\text{Chloride in mg/mL} = \frac{\text{mL of titrant used} \times N \times 35.46 \times 1000}{\text{mL of sample}}$$

where N is normality of titrant

Comments: The pH of the water sample should be in the 7–8 range: when it is less, CrO_4^{2-} is converted into CrO_7^{2-}; when it is more, Ag^+ is precipitated as AgOH. For all the titrations, a definite volume of the indicator should be accurately pipetted into the analyte. Use double distilled water for the preparation of all solutions.

26.5.4 Result

Express chlorides as mg/L.

Salinity
The salinity of the water is detected from the chlorinity of the water sample.

Chlorinity of the sample (mg/L)

$$= \frac{\text{Chlorinity of standard water} \times \text{Volumes of titrant used for chloride}}{\text{Volume of AgNO}_3 \text{ used for standard chloride}}$$

Salinity ppt (%) = 0.03 + (1.805 × chlorinity)

26.6 Measurement of Nitrate in Water

26.6.1 Principle

1,2,4-Phenoldisulfonic acid produces 6-nitro-1,2,4-phenoldisulfonic acid (an alkaline salt) with nitrates yielding yellow color. The color is conveniently read at 410 nm.

26.6.2 Materials

1. Standard nitrate solution: Dissolve 13.7 mg sodium nitrate ($NaNO_3$) in 100 mL distilled water (conc. NO_3 100 µg/mL, as N-22 µg/mL).
2. Brucine reagent: Take 50 mL of water and 3 mL of conc. HCl in a beaker. Heat this just to boiling and add 1 g brucine sulfate and 0.1 g sulfanilic acid with stirring. Cool and dilute this solution to 100 mL.
3. Sulfuric acid: Mix carefully 500 mL conc. sulfuric acid with 100 mL distilled water.

26.6.3 Protocol

1. Prepare a series of 50 mL standard solutions of nitrate from the standard nitrate solution to obtain a range of concentration (10–100 µg).
2. Pipette 2 mL aliquots of the standard solutions into different dry 100 mL beakers provided with glass rods.
3. To each beaker, add 1 mL brucine–sulfanilic acid into different dry 100 mL beakers provided with glass rods.
4. Cover the beakers with watch glasses and keep them in the dark for 10 min during which time a yellow color develops.

5. Add 10 mL distilled water to each of beakers and incubate in the dark for 30 min. The absorbance of each solution is then measured at 410 nm.
6. A graph is drawn by plotting the absorbance values against the nitrate concentrations of the standard solutions.

Treat a suitable volume of the water sample with the same amount of reagents in an identical manner and read the absorbance of this solution. Then, using the standard graph find out the nitrate concentration of the water sample.

26.6.4 Result

Express NO_3 nitrogen in mg/L.

26.7 Measurement of Nitrite in Water

26.7.1 Principle

In acid solution, nitrite yields nitrous acid, which diazotizes the sulfanilamide. The diazonium salt on reacting with aromatic amine, N-1-naphthylethylene diamine dihydrochloride (NED), forms a red azo dye, which is determined spectrophotometrically at 543 nm.

26.7.2 Materials

1. Standard nitrite solution: Dissolve 15 mg of $NaNO_2$ in distilled water and make up to 100 mL (conc. NO_2 100 μg/mL).
2. Sulfanilamide solution: Dissolve 1 g sulfanilamide in 10 mL conc. hydrochloric acid and 60 mL distilled water. Dilute this solution to 100 mL.
3. NED: Dissolve 0.200 mg of NED salt in 100 mL distilled water and store in an amber-colored bottle (prepare freshly every month.)

26.7.3 Protocol

1. Take a known volume of sample in a 50 mL volumetric flask.
2. To it add 1 mL of sulfanilamide reagent and mix.
3. After 5 min, add 1 mL NED reagent and mix well. Make up the contents to 50 mL by adding distilled water.
4. Shake thoroughly and measure the absorbance at 543 nm against distilled water blank.
5. Pipette out known concentration from the standard solution (10–100 μg).
6. Add reagent as earlier and draw a standard graph. From the standard, deduce the amount of nitrite content.

26.7.4 Result

Express NO_2 nitrogen in mg/L.

26.8 Measurement of Ammonia in Water

26.8.1 Principle

Ammonia reacts with phenol and alkaline hypochlorite to form indophenol blue. The reaction is catalyzed by nitroprusside or ferrocyanide. The resulting absorbance is proportional to the concentration of ammonia and is measured spectrophotometrically at 640 nm.

26.8.2 Materials

1. Standard ammonia solution: Prepare a standard solution by dissolving 3.1 mg NH_4Cl in 100 mL distilled water in a standard flask (concentration 10 µg/mL).
2. Hypochlorite stock: Commercially available chlorine solution (5.5% available chlorine). Normality should be at least 1.6.
3. Alkaline stock: Dissolve 100 g trisodium citrate and 5 g of sodium hydroxide in distilled water and make up to 500 mL.
4. Nitroprusside reagent: Dissolve 1 g of sodium nitroprusside and make up to 200 mL with distilled water.
5. Oxidizing reagent: Freshly prepare during use with 4 parts alkaline stock (3) and 1 part of hypochlorite stock (2) and keep in a stoppered bottle.
6. Phenol reagent: Dissolve 100 g phenol in 50 mL of 65% ethyl alcohol and make up to 100 mL in distilled water.

26.8.3 Protocol

1. To the sample, add 0.4 mL of phenol reagent (6) and 0.4 mL of nitroprusside reagent (4) and mix well.
2. To it add 1 mL of the oxidizing reagent (5) and stopper the tubes immediately.
3. Vortex and incubate for hours at room temperature in the dark.
4. Measure the absorbance at 640 nm in spectrophotometer.
5. Prepare a standard graph using different dilutions of the standard (1) (concentration 1–10 µg). From this find out the ammonia concentration of the sample.

26.8.4 Result

Express the resulting NH_3^- nitrogen mg/L.

26.9 Measurement of Total Phosphorus in Water

26.9.1 Principle

Phosphorus occurring as orthophosphate can be easily estimated. The organically bound phosphorus is converted into inorganic phosphorus on oxidation with potassium persulfate because of the production of sulfuric acid. Phosphorus in the presence of a reducing agent (ascorbic acid) is reduced to molybdenum blue with ammonium phosphomolybdate reagent.

26.9.2 Materials

1. Standard phosphate solution: Dissolve 14.3 mg of KH_2PO_4 and make up to 100 mL with distilled water (concentration 100 µg/mL).
2. 5% potassium persulfate (K_2SO_4): Dissolve 500 mg of potassium persulfate in 100 mL of distilled water.
3. 5 N Sulfuric acid (H_2SO_4): Dilute 70 mL of conc. H_2SO_4 to 500 mL with distilled water.
4. Ammonium molybdate: Dissolve 1.32 g of ammonium molybdate in 100 mL of distilled water and make up to 500 mL with distilled water and store in a polythene bottle.
5. Ascorbic acid solution: Dissolve 0.264 g of ascorbic acid in 15 mL of distilled water (prepare this solution freshly).
6. Potassium antimonyl tartrate solution: Dissolve 0.2473 g of potassium antimonyl tartrate in 100 mL of distilled water.
7. Mixed reagent: Mix 25 mL of 5 N H_2SO_4 thoroughly with 7.5 mL of molybdate reagent. To this add 15 mL of ascorbic acid solution and 2.5 mL of potassium antimonyl tartrate solution. Then mix the contents thoroughly. Prepare this reagent as and when required.

26.9.3 Protocol

1. Take 50 mL of the sample and heat the contents until the volume is reduced to 15 mL.
2. Add 1 mL of perchloric acid and heat it until the volume is reduced to 5 mL.
3. Add 2 mL of phenolphthalein indicator solution. Then add saturated NaOH solutions drop by drop until the solution turns pink in color.
4. Make up the solution to 50 mL with distilled water. Use this sample for the estimation of total phosphorous.
5. Pipette out a known volume of standard solutions to test tubes (10–100 µg). Then add 2 mL of mixed reagent followed by 2 mL of potassium persulfate reagent and mix well. Incubate for 10 min at room temperature and read the absorbance of the solution at 882 nm.
6. Pipette out a known volume of water sample and add reagents as earlier and measure the absorbance. Find out the concentration of phosphorus of the known sample from the standard curve.

26.9.4 Result

Express total phosphorus in mg/L.

26.10 Measurement of Inorganic Phosphate in Water

26.10.1 Principle

Phosphate and ammonium molybdate react in an acid solution to give phosphomolybdic acid, which can be reduced by a number of reagents to molybdenum blue. In this method, ascorbic acid is used as the reducing agent. Antimony salt is also needed for the rapid color development of molybdenum blue complex.

26.10.2 Materials

1. Standard phosphate solution: Dissolve 14.3 mg of K_2HPO_4 and make up to 100 mL with distilled water (100 μL/mL).
2. 5 N Sulfuric acid (H_2SO_4): Dilute 70 mL of conc. H_2SO_4 to 500 mL with distilled water.
3. Ammonium molybdate: Dissolve 1.32 g of ammonium molybdate in 100 mL of distilled water and make up to 500 mL with distilled water and store in a polythene bottle.
4. Ascorbic acid solution: Dissolve 0.264 g of ascorbic acid in 15 mL distilled water. (Prepare this solution freshly.)
5. Potassium antimonyl tartrate solution: Dissolve 0.273 g of potassium antimonyl tartrate in 100 mL of distilled water.
6. Mixed reagent: Mix 25 mL of 5 N H_2SO_4 thoroughly with 7.5 mL of molybdate reagent. To this add 15 mL of ascorbic acid solution and 2.5 mL of potassium antimonyl tartrate solution. Then mix the contents thoroughly. Prepare this reagent as and when required.

26.10.3 Protocol

Pipette 10 mL of the water sample to a test tube:

1. To it add 2 mL of mixed reagent and make it up to 15 mL with distilled water. Vortex the contents.
2. After 10 min, measure the absorbance at 882 nm in a spectrophotometer.
3. Estimate the amount of inorganic phosphorus from the standard curve.

26.10.4 Result

Express inorganic phosphorus in mg/L.

26.11 Measurement of Sulfate in Water

26.11.1 Principle

Sulfate gets precipitated with barium in an acid solution to form $BaSO_4$ crystals of uniform size (when glycerol–ethanol solution is added as stabilizer). The quantity of $BaSO_4$ can easily be estimated spectrophotometrically at 420 nm.

26.11.2 Materials

1. Standard sulfate solution–sodium sulfate (Na_2SO_4): Dissolve 1.42 g of Na_2SO_4 in 10 mL distilled water (concentration 100 mg/mL).
2. NaCl–HCl solution: Dissolve 60 g of NaCl in 200 mL of distilled water. To this add 5 mL of conc. HCl. Make up the solution to 250 mL with distilled water.
3. Glycerol–ethanol solution: Add 50 mL of glycerol to 100 mL of ethyl alcohol and mix thoroughly.
4. Barium chloride, dry crystals ($BaCl_2$): 0.15 g.

26.11.3 Protocol

1. To 50 mL of sample, add 10 mL of NaCl–HCl solution followed by 10 mL of glycerol–ethanol solution.
2. Mix and keep it on the stirrer. When the stirrer is on, add 0.15 g of barium chloride and mix for 60 s.
3. Immediately read the absorbance at 420 nm against a suitable blank.
4. Run a similar experiment with standard sulfate solutions and plot (concentration 10–100 mg).
5. Measure the amount of sulfate from the standard curve drawn with sodium sulfate.

26.11.4 Result

Express sulfates in mg/L.

26.12 Measurement of Sulfide in Water

26.12.1 Principle

Sulfides react with phenanthroline in the acid solution to give an orange color, due to the reduction of phenanthroline.

26.12.2 Materials

1. Standard sulfide solution–sodium sulfide (Na_2S): Dissolve 2.4 g of Na_2S in 100 mL of distilled water (concentration 10 mg/mL).
2. Phenanthroline monohydrate solution (0.1%): Dissolve 100 mg of phenanthroline monohydrate in 100 mL of distilled water.
3. Acetate buffer (3.5 pH):
   a. Sodium acetate (0.1 M) solution: Dissolve 1.36 g of sodium acetate in 100 mL of distilled water.
   b. Acetic acid (0.1 M) solution: Mix 0.6 mL of acetic acid in 100 mL of distilled water.

Acetic acid (0.1 M) solution is added dropwise to 100 mL of sodium acetate (0.1 M) solution until 3.5 pH is reached.

26.12.3 Protocol

1. To 7 mL of acetate buffer (3.5 pH), add 3 mL of phenanthroline monohydrate (0.1%).
2. To it add 10 mL of water sample. Then make up to 25 mL with distilled water and incubate at 25°C for 1 h.
3. Run a parallel experiment with standard sulfide solution (10–50 mg).
4. Read the developed color at 510 nm using suitable blank. Calculate the amount of sulfide by using standard curve drawn with sodium sulfide.

26.12.4 Result

Express sulfide in mg/L.

26.13 Measurement of Calcium and Magnesium in Water

26.13.1 Principle

Eriochrome black T forms wine-red complex with metal ions (Ca^{2+} and Mg^{2+}). The disodium salt of ethylenediaminetetraacetic acid (EDTA) chelates the metal ions from the dye–metal ion complex leaving a blue color of the dye. Murexide indicator forms pink-colored complex with Ca^{2+} ions. With the addition of the disodium salt of EDTA, the Ca^{2+} forms a colorless chelate complex leaving behind a purple solution of the dye.

26.13.2 Materials

1. Disodium salt EDTA 0.1 N: Dissolve 37.224 g of EDTA (disodium salt) in 1 L of distilled water and store in a polythene bottle.

2. Ammonium buffer: Dissolve 70 g of ammonium chloride in 570 mL of ammonium solution.
3. Sodium hydroxide solution: Dissolve 80 g of NaOH in 1 L of distilled water.
4. Eriochrome black T indicator: Prepare a fine mixture of Eriochrome black T (Solochrome black T) using sodium chloride in a pestle and mortar in the ratio 1:200 (w/w).
5. Ammonium purpurate indicator: Prepare a fine mixture of ammonium purpurate (murexide) using sodium chloride in a pestle and mortar in the ratio of 1:100 (w/w).

26.13.3 Protocol

1. Pipette 5 mL of water sample to a 250 mL conical flask.
2. To this add 5 mL of ammonium buffer and dilute to 100 mL with distilled water.
3. Add a pinch of Eriochrome black T and warm the solution to 60°C.
4. Titrate against EDTA until the red color turns to blue. Note the end point as "A."

To another flask pipette 5 mL of water sample and add 5 mL of NaOH solution, and dilute it to 100 mL with distilled water. Add a pinch of murexide indicator and shake well. Titrate against EDTA until the pink color turns to blue. Note the end point "B."

Calculations
mL of consumption of EDTA by Ca and Mg $= A$
mL of consumption of EDTA by Ca alone $= B$
mL of consumption of EDTA by Mg alone $= A - B$

Amount of Ca in sample $= \dfrac{F \times B \times 100}{\text{Volume of sample}}$ (mg/L); Factor value for calcium

$(F) = 2$ volume of sample

Amount of Mg in sample $= \dfrac{F \times A - B \times 100}{\text{Volume of sample}}$ (mg/L); Factor value for Mg $(F) = 1.2$

volume of sample

26.13.4 Result

Express Ca and Mg in mg/L of sample.

26.14 Detection of Coliforms for the Determination of the Purity of Potable Water

26.14.1 Principle

The microbiological examination of water is a direct measurement of deleterious effects of pollution of human health. The microorganisms in water include several harmless and harmful bacteria. Pathogenic bacteria are often removed during water

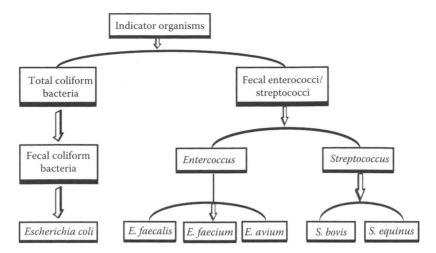

Figure 26.5   Illustration of various coliform bacteria as water pollution indicator.

treatment. Water pollution is generally indicated by the presence of coliform bacteria. The coliform bacteria include the genera *Escherichia, Citrobacter, Enterobacter*, and *Klebsiella* (Figure 26.5).

The *Escherichia coli* is entirely of human origin but their exclusive estimation is difficult and hence the entire coliforms are used as indicator. In routine tests to confirm the water quality to drinking water standards, the actual number of coliforms is not reported but they are reported as an approximate count—most probable number (MPN). In this experiment, membrane filtration method is explained. Coliform is not a taxonomic classification but rather a working definition used to describe a group of Gram-negative, facultative anaerobic rod-shaped bacteria that ferments lactose to produce acid and gas within 48 h at 35°C. In 1914, the US Public Health Service adopted the enumeration of coliforms as a more convenient standard of sanitary significance.

26.14.2 *Experimental Outline*

The membrane filter (MF) technique is an effective, accepted technique for testing fluid samples for microbiological contamination. It involves less preparation than many traditional methods, and is one of a few methods that will allow the isolation and enumeration of microorganisms. The MF technique also provides presence or absence information within 24 h. The MF technique was introduced in the late 1950s as an alternative to the MPN procedure for microbiological analysis of water samples. The MF technique offers the advantage of isolating discrete colonies of bacteria, whereas the MPN procedure only indicates the presence or absence of an approximate number or organisms. In the membrane filtration method, water sample is passed through a thin sterile MF (pore size 0.45 μm) that is kept in a special filter apparatus contained in a suction flask (Figure 26.6). The filter disk that contains

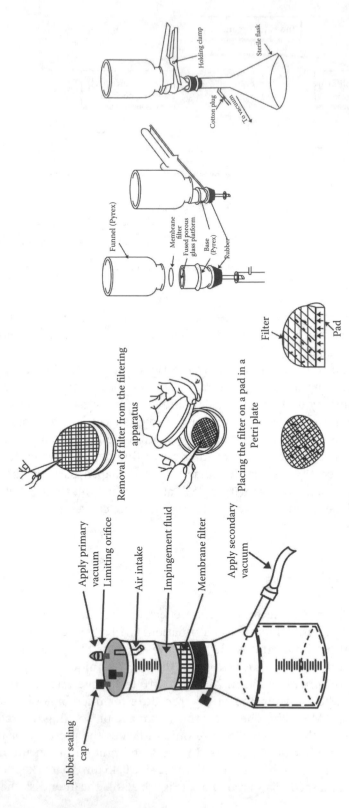

Figure 26.6   Microfiltration assemblies and microfiltration.

the trapped microorganisms is aseptically transferred to a sterile Petri dish having an absorbents pad saturated with a selective, differential liquid medium and the colonies that develop following incubation are counted. This method enables large volume of water to be tested more economically. The various steps are described.

26.14.3 Materials

Water sample, 20 mL tube of M-endobroth, sterile Millipore membrane apparatus (base, funnel, and clamps), 1 L suction flask, sterile 0.45 absorbent pad, 50 mL alcohol, forceps, sterile rinse water, dissecting microscope, colony counter, Petri dishes.

26.14.4 Protocol

Collect the sample and make necessary dilutions (Figure 26.7):

1. *Select the appropriate nutrient or culture medium. Dispense the broth into a sterile Petri dish, evenly saturating the absorbent pad.*
2. Flame the forceps, and remove the membrane from the sterile package.
3. Place the MF into the funnel assembly.
4. Flame the pouring lid of the sample container and pour the sample into the funnel.
5. Turn on the vacuum and allow the sample to draw completely through the filter.
6. Rinse funnel with sterile buffered water. Turn on vacuum and allow the liquid to draw completely through the filter.
7. Flame the forceps and remove the MF from the funnel.
8. Place the MF into the prepared Petri dish.
9. Incubate the plates in inverted position at 37°C for 24 h.
10. Count the colonies under 10–15× magnification with colony counter.
11. Remove the filter disks from the Petri dish and allow drying on absorbent paper for 1 h.
12. Confirm the colonies and report the results.

26.14.5 Results

Examine filter paper disk under a dissecting microscope for the presence of coliforms and count their number. On endomedium, coliforms appear as pink to red colonies and some may have a green metallic screen. Express results per 100 mL of water sample by applying the formula

$$100 \text{ mL water} + \frac{\text{Number of coliforms per colony count}}{\text{Volume of sample used}} \times 100$$

Water is considered safe for drinking if it contains fewer than four coliforms per 100 mL of water.

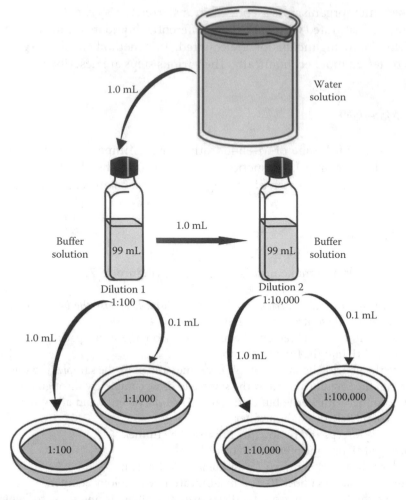

Figure 26.7 Method of preparing proper dilution.

26.15 Determination of Total Dissolved Solids of Water

26.15.1 Principle

Water, the universal solvent, has a large number of salts dissolved in it. These salts largely govern the physicochemical properties of water and in turn have an indirect effect on the flora and fauna. Total dissolved solids (TDS) are determined as the residue left after evaporation of the filtered sample.

26.15.2 Materials

Water sample to be tested, evaporating dish, hot water bath, desiccators, Whatman filter paper No. 4, weighing balance

26.15.3 Protocol

1. Weigh the evaporating dish.
2. Filter the sample in suitable quantity (250–500 mL) through Whatman filter paper No. 4.
3. Transfer the sample to the evaporating dish.
4. Evaporate on water bath.
5. Note the weight of the dish along with the contents after cooling in a desiccator.

26.15.4 Result

Calculate TDS and express in mg/L by using the formula

$$TDS\ (mg/L) = \frac{(B-A)}{V} \times 10^6$$

where
  A is initial weight of the dish (g)
  B is final weight of the dish (g)
  V is volume of the water sample taken (mL)

26.16 Determination of DO Concentration of Water by Winkler's Iodometric Method

26.16.1 Principle

DO of water is of paramount importance to all the living organisms and is considered to be the lone factor that is to a great extent can reveal the nature of the whole aquatic system at a glance, even when information on other chemical physical and biological parameters are not available. DO concentration in water explains two distinct phenomena: (i) direct diffusion from the air and (ii) photosynthetic evolution by aquatic autotrophs. Eutrophic water bodies have a wide range of DO and show clinograde oxygen curve in which O_2 is more at the surface and depletes fast with the depth. The oligotrophic lakes have narrow range of DO and show orthograde O_2 curve in which the O_2 content increases with the depth to saturation levels.

26.16.2 Experimental Outline

The manganous sulfate reacts with the alkali (KOH/NaOH) to form a white precipitate of manganous hydroxide that in the presence of O_2 gets oxidized to a brown color compound. In the strong acid medium, manganic ions are reduced by iodide ions that get converted to iodine equivalent to the original concentration of O_2 in the sample. The iodine can be titrated against thiosulfate using starch as an indicator:

$$MnSO_4 + 2KOH \rightarrow Mn(OH)_2 + K_2SO_4$$

$$Mn(OH)_2 + O \rightarrow MnO(OH)_2$$

$$MnO(OH)_2 + 2H_2SO_4 + 2KI \rightarrow MnSO_4 + K_2SO_4 + 3H_2O + I_2$$

26.16.3 Materials

Water samples, potassium iodide, KOH, conc. sulfuric acid, manganese sulfate solution, formaldehyde solution, NaOH or borax, pipettes, sodium thiosulfate (0.025 N), narrow mouth 250 mL BOD bottles

1. Sodium thiosulfate 0.025 N: Dissolve 24.82 g of $Na_2S_2O_3\cdot5H_2O$ in boiled distilled water and make up to 1 L. Add 0.4 g of borax or NaOH as stabilizer. This is 0.1 N stock. Dilute it four times to prepare 0.025 N. Keep in a brown glass-stoppered bottle.
2. Alkaline KI solution: Dissolve 100 g of KOH and 50 g of KI in 200 mL of boiled distilled water.
3. Manganous sulfate solution: Dissolve 100 g of $MnSO_4\cdot H_2O$ in 200 mL of boiled distilled water and filter.
4. Starch solution: Dissolve 1 g of starch in 100 mL of warm (80°C–90°C) distilled water and add a few drops of formaldehyde solution.

26.16.4 Protocol

1. Collect water sample without bubbling in the 250 mL (BOD) glass bottle.
2. Add 2 mL each of manganous sulfate and alkaline iodide solutions in succession right at the bottom of the bottle with separate pipettes and replace the stopper.
3. Shake the bottle in the upside down direction at six times.
4. Allow the brown precipitate to settle.
5. Add 2 mL of conc. sulfuric acid and shake the stoppered bottle to dissolve the brown precipitate.
6. Take 50 mL in a flask and titrate with thiosulfate solution (taken in the burette) till the color changes to pale straw.
7. Add "two" drops of starch solution to the aforementioned flask that changes the color of the contents from pale to blue.
8. Titrate again with thiosulfate solution till the blue color disappears.

26.16.5 Result

Find out the total amount of sodium thiosulfate solution (titrant) used and calculate the dissolved O_2 content of water (mg/L) by applying the equation

$$DO\ (mg/L) = \frac{8^* \times 100 \times N}{v} \times V$$

where
  V is volume of the sample taken (mL)
  v is volume of titrant used
  N is normality of titrant
  $*$ = 8 is the constant since 1 mL of 0.025 sodium thiosulfate solution is equivalent
    to 0.2 mg of O_2

26.17 Determination of Biochemical Oxygen Demand of Sewage Water

26.17.1 Principle

The biochemical oxygen demand (BOD) is a way of expressing the amount of organic compounds in sewage as measured by the volume of O_2 required by bacteria to metabolize it under aerobic conditions. It is a good index of the organic pollution. If organic matter is high, more O_2 will be utilized by bacteria. Sewage that contains high BOD increases the concentration of soluble organic compounds in the aquatic body where it is discharged. Digestion in natural ecosystems depletes available O_2 and results in asphyxiation of fish.

26.17.2 Experimental Outline

The main principle involved is the amount of O_2 required by the microorganisms in stabilizing the biologically degradable organic matter under aerobic conditions. Water sample is incubated at 20°C for 5 days in the dark under aerobic conditions (BOD_5 at 20°C). In tropical and subtropical belts where metabolic activities are higher, samples are incubated at 27°C for 3 days (BOD_3 at 27°C).

26.17.3 Materials

Water samples, BOD-free water (deionized glass distilled water passed through a column of activated carbon and redistilled), allylthiourea solution (0.5%), phosphate buffer—KH_2PO_4, K_2HPO_4, $Na_2HP_4 \cdot 7H_2O$, NH_4Cl, distilled water, sulfuric acid, NaOH, reagents for DO estimation, BOD bottles (6), flask, Pipettes, BOD incubator, pH meter

  1. Phosphate buffer pH 7.2: Dissolve 8.5 g KH_2PO_4, 21.75 g K_2HPO_4, 33.4 g $Na_2HPO_4 \cdot 7H_2O$, and 1.7 g NH_4Cl in distilled water to prepare 1 L of solution and adjust pH 7.2.

2. Sulfuric acid (1 N): 2.8 mL of conc. H_2SO_4 added to 100 mL of BOD-free distilled water.
3. NaOH (1 N): 4 g of NaOH dissolved in 100 mL of distilled water.
4. Allylthiourea solution: 0.5%.

26.17.4 Protocol

1. Adjust the pH of the water to neutrality using 1 N acid or 1 N alkali solutions.
2. Fill the water sample in six BOD bottles without bubbling.
3. Add 1 mL of allylthiourea to each bottle.
4. Determine dissolved O_2 content in three of the bottles by the titration method as in DO method.
5. Take the mean of the three readings (D_1).
6. Incubate the rest BOD bottles (3) at 27°C in a BOD incubator for 3 days.
7. Estimate the O_2 concentration in all three incubated samples.
8. Take the mean of three readings (D_2).

26.17.5 Result

Calculate the BOD of the water in mg/L by applying the formula

$$BOD_3 \text{ in mg/L} = D_1 - D_2$$

where
D_1 is initial DO in sample (mg/L)
D_2 is DO after 3 days incubation (mg/L)

26.18 Determination of Chemical Oxygen Demand of Sewage Sample

26.18.1 Principle

A large amount of several chemically oxidizable organic substances of different nature are being discharged in the aquatic system leading to severe aquatic pollution by different industries. BOD alone does not give a clear picture of the organic matter content of the water sample. In addition, the presence of various toxicants in the sample may severely affect the validity of the BOD test. Therefore, chemical oxygen demand (COD) is a better estimate of the organic matter, which needs no sophistication and is time saving. However, COD, that is, the O_2 consumed (OC) does not differentiate the stable organic matter from the unstable form and the COD values are not directly comparable to BOD.

26.18.2 Experimental Outline

Chemical O_2 demand is the measure of OC during the oxidation of the oxidizable organic matter by a strong oxidizing agent. Potassium dichromate in the presence of sulfuric acid is generally used as an oxidizing agent in the determination of COD. The sample is refluxed with $K_2Cr_2O_7$ and H_2SO_4 in the presence of mercuric sulfate to neutralize the effect of chlorides and silver sulfate (catalyst). The excess of potassium dichromate is titrated against ferrous ammonium sulfate using ferroin as an indicator. The amount of $K_2Cr_2O_7$ used is proportional to the oxidizable organic matter present in the sample.

26.18.3 Materials

Water sample, silver sulfate, phenonthroline, sulfuric acid, potassium dichromate solution, ferrous ammonium sulfate, mercuric sulfate

1. Potassium dichromate solution (0.025 N): Dissolve 12.259 g of dried $K_2Cr_2O_7$ (analytical grade) in distilled water and make up to 1 L. Dilute 100 mL of the aforementioned solution to 1000 mL, which gives 0.025 N strength.
2. Ferrous ammonium sulfate (0.1 N): Dissolve 39.2 g of $Fe(NH_4)_2 (SO_4)_2 \cdot 6H_2O$ in water adding 20 mL conc. H_2SO_4 to make it 1 L. Standardize this solution with $K_2Cr_2O_7$ by diluting 10 mL of $K_2Cr_2O_7$ to 100 mL, add 30 mL conc. H_2SO_4, and titrate with ferrous ammonia sulfate using ferroin as an indicator. Dilute 0.1 N ferrous ammonium sulfate to 10 times (100 → 1000 mL) to get 0.01 N.
3. Ferro indicator: Dissolve 1.485 g of 1,10-phenonthroline and 0.695 g of ferrous sulfate $FeSO_4 \cdot 7H_2O$ in distilled water to make 100 mL solution.

26.18.4 Protocol

1. Take 20 mL of sample in a 250–500 mL COD flask.
2. Add 10 mL of 0.025 N potassium dichromate solution and a pinch of Ag_2SO_4 and H_2SO_4 (30 mL).
3. Reflex at least for 2 h on a water bath or a hot plate. Remove the flask, cool, and add distilled water to make the final volume to 140 mL.
4. Add two to three drops of ferroin indicator mix thoroughly and titrate with 0.1 N ferrous ammonium sulfate (with 0.01 N ferrous ammonium sulfate if 0.025 N $K_2Cr_2O_7$ has been used).
5. Run a blank with distilled water using the same quantity or the chemicals.

26.18.5 Result

$$COD \ (/mL) \ of \ the \ water \ sample = \frac{8 \times C \times (B - A)}{Volume \ of \ water \ sample \ in \ mL}$$

where
A is mL of titrant with sample
B is mL of titrant with blank
C is concentration of titrant (mmol/L)
8 is the constant

26.19 Determination of the Efficiency of Removal of Air Pollutant, Using Fibrous Air Filter

Bioremediation of Air Pollutants: Air emissions from various industrial or waste treatment processes release into the atmosphere substances that may be noxious or hazardous to humans and may contribute to smog formation or to depletion of the atmosphere ozone layer. The use of microorganisms for control of air emissions is a relatively new technique. Three types of microbial devices, namely, biofilters, trickling biofilters, and bioscrubbers are used for air emission control. Air pollutants, especially odor-causing compounds, are bioremediated from air near farms, sewage treatment facilities, and various industrial operations. These bioreactors have biofilms of microorganisms that can remove over 99% of volatile pollutants from air. Toxic compounds such as hydrogen sulfide, toluene, chlorobenzene, and nitrobenzene can also be removed in this way (Figure 26.8).

Biofilter: Biofilter is a device that consists of an immobilized microbial community as a biofilm through which air is passed to detoxify contamination and is used for the bioremediation of air pollutants such as odors and volatile organic compounds. Biofilteration systems are used to remove formaldehyde from air released from plywood production factories and phenol from resin producers, to remove solvents from indoor air at paint production facilities and for removal of odorous gases from the cattle feed extrusion plants and from ceramic factories for the removal of ethanol and isopropyl alcohol released into the air from the drying ceramics.

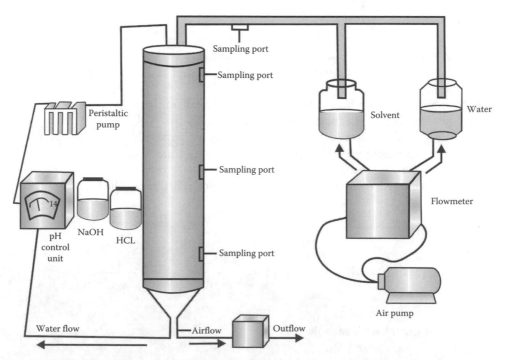

Figure 26.8 Laboratory-scale trickling air biofilter used for the removal of chlorobenzene and nitrobenzene from air.

Bioscrubber: It is a device in which air moves through a fine spray of a microbial suspension in order to remove pollutants from air. The actual degradation takes place in a stirred reactor into which the collected spray is channeled. The device contains no solid phase. It is effective only for highly water-soluble volatile organic chemicals (VOCs).

26.19.1 Principle

The function of an air filter is to reduce the concentration of solid particles in the air stream to a level that can be tolerated by the process or space occupancy purpose. The situation influences the selection of the equipment that varies with the degree of cleanliness required and economics. Solid particles in the air stream range in size from 0.01 μm up to things that can be caught by ordinary fly screens, such as lint, feathers, and insects. The particles generally include soot, ash, soil, lint, and smoke, and may also include almost any organic or inorganic material, bacteria, and mold spores.

Following types of air filters are in common use today:

Viscous impingement dry and electronic device: The principles employed by these filters in removing airborne solids are viscous impingement interception, impaction, diffusion, and electrostatic precipitation. Some filters utilize only one of these principles. Others employ combinations. A fourth method, inertial separation, is finding increasing use as a result of the construction boom throughout the world.

Fabric filtration: By collection of particles on a fabric makes use of inertial impaction, direct interception, and diffusion capture. The filtration fabric may be configured as bag envelopes of cartridges. Bag filters are the most important for use with processes having high volumes of emissions.

26.19.2 Materials

Fabric filtration bag houses (Figure 26.9).

Inside-to-outside bags are cleaned by reversing the gas flow momentarily, shaking the bags or a combination of both method.

26.19.3 Description

Inside-to-outside bags (Figure 26.10) are usually made from woven fibers, whereas outside-to-inside bags use felted fibers. In both cases the particulate matter falls into a hopper, from where it is removed. The bags are attached to a tube sheet that separates the clean side of the bag house from the dirty side. Outside-to-inside flow bags contain an internal cage to prevent their collapse during filtration. Inside-to-outside bags are normally used in industrial applications. They are as large as 36 ft (10.8 m)

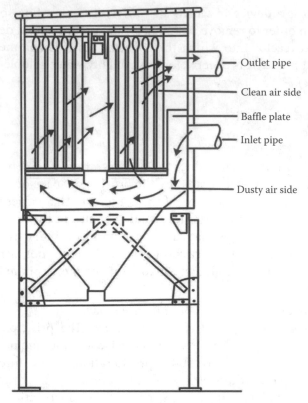

Figure 26.9   Cloth collector.

in length with a diameter of 12 in. (360 mm); outside-to-inside bags range from 8 to 24 ft (2.5–7.2 m) in length with diameters from 4 to 8 in. (100–200 mm). Depending upon the size of the process to be controlled, a bag house might contain from fewer than 100 bags to a quantity as large as several thousand bags. Power required for operation is needed predominantly to remove the gas through the device. The pressure drop for filtration is greater than for *electrostatic precipitator pressure* (ESP), but less than for scrubbers.

26.19.4 Protocol

The dust-laden gas is passed through a woven or felted fabric upon which the gradual deposition of dust form a precoat that then serves as a filter for the subsequent dust. These units are analogs to those used in liquid filtration and represent a special type of packet bed. The resistance to gas flow gradually increases due to continuous dust accumulation. The cloth must therefore be periodically vibrated or flexed or back-flushed with a stream or pulse of air to dislodge accumulated dust. Cotton or wool sateen or felts are usually used for temperature below 212°F (100°C). Some of the synthetic fibers may be used at temperature up to 500°F (260°C). Glass and asbestos

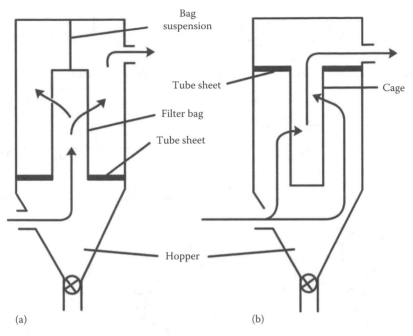

Figure 26.10 Basic configuration of fabric filtration of bag houses. (a) Inside-to-outside flow. (b) Outside-to-inside flow. Arrow indicates flow.

or combinations thereof have been employed for temperature up to 650°F (343°C) for special high temperature applications such as metallic screen and porous ceramics or stainless steel. Collection efficiencies of over 98% are attained readily with cloth collectors, even with very fine dust.

26.20 Isolation of Xenobiotic-Degrading Bacteria

26.20.1 Bioremediation of Xenobiotic Compounds

Xenobiotics are the synthetic or human-made compounds that are resistant or recalcitrant to biodegradation and/or decomposition (Figure 26.11). The xenobiotics have molecular structure and chemical bond sequences not recognized by the existing degradative enzymes.

The compounds that resist biodegradation and thereby persist in the environment are called recalcitrant. Recalcitrant molecules are fossil organic matter (humus), polyaromatic compounds (tannins and lignins), persistent microorganisms (endospores and melanin-rich fungi), synthetic molecules (fungicides, nematicides, herbicides, insecticides), polyhalogenated biphenyls (flame retardants and solvents), plastics, and detergents. The persistence of xenobiotics ranges from days to years and minor alterations in biodegradable compound can render them recalcitrant (Table 26.2).

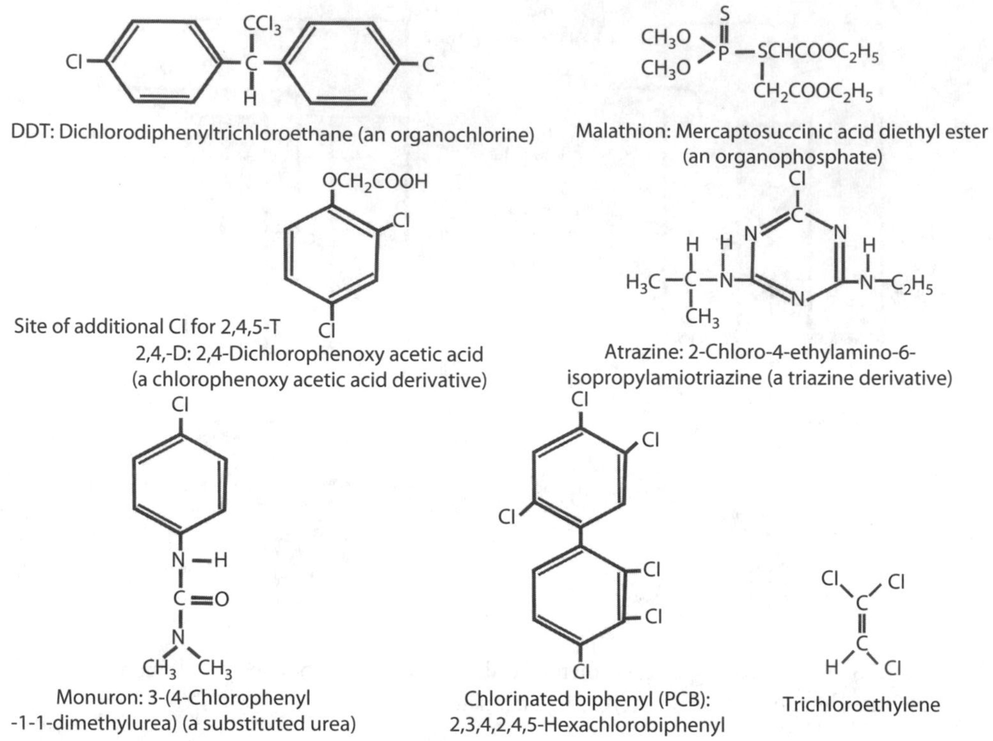

Figure 26.11   Common examples of xenobiotic compounds.

TABLE 26.2   Persistence of Xenobiotic Compounds in Soil

Xenobiotic Compounds	Duration for 75%–100% Disappearance
Chlorinated insecticides	
DDT	4 years
Aldrin	3 years
Chlordane	5 years
Heptachlor	2 years
Lindane	3 years
Organophosphate insecticides	
Diazonin	12 weeks
Malathion	1 week
Parathion	1 week
Herbicides	
2,4-D (2,4-dichlorophenoxyacetic acid)	4 weeks
2,4,5-T (2,4,5-trichlorophenoxyacetic acid)	20 weeks
Dalapon	8 weeks
Atrazine	40 weeks
Simazine	48 weeks
Propazine	1.5 years

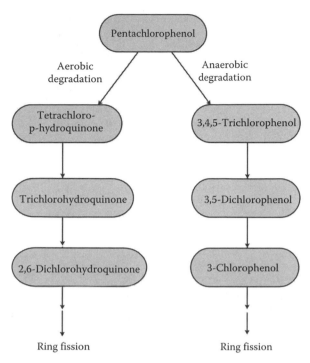

Figure 26.12   Outline of aerobic and anaerobic degradation of pentachlorophenol.

Autoxidation, degradation by sunlight, and microbial actions are three major forces responsible for the breakdown of the compounds in the environment. Majority of the xenobiotics after their release go to soil and aquatic sediments (Figure 26.12).

26.20.2 Principle

Use of pesticides and artificial chemical compounds has benefited the modern society by improving the quantity and quality of human comforts. Gradually, usage of such compounds has become an integral part of modern system. Many of the artificially made complex compounds, that is, xenobiotics persist in environment and do not undergo biological transformation. Some of xenobiotic halo-substituted and nitro-substituted organics such as propellants, solvents, refrigerants, plastics, detergents, explosions, pesticides are completely resistant to biodegradation. Xenobiotics are defined as a synthetic product not formed by natural biosynthetic processes; a foreign substance; or a poison. Microorganisms play an important role in the degradation of xenobiotics, and maintaining of steady-state concentrations of chemicals in the environment. The complete degradation of a pesticide molecule to its inorganic components that can be eventually used in an oxidative cycle removes its potential toxicity from the environment. However, there are two objectives in relation to biodegradation of xenobiotics: (i) how biodegradation activity arises, evolves, and transferred among the members of soil micro flora and (ii) to device bioremediation methods for removing or detoxifying high concentration of dangerous pesticide residues. The characters of pesticide degradation by microorganisms are located on plasmids and transposons that are grouped in clusters on chromosome. Understanding of the characters provide clues to the evolution of derivative pathways and makes the task of gene manipulation easier to construct the genetically engineered microbes capable of degrading the pollutants. Pollution problems caused by xenobiotics have necessitated an elaborate environmental safety testing program for newly introduced chemical. Bacteria are isolated from the enrichments and roseto axenic cultures, maintained in nutrient agar medium and urea agar medium.

26.20.3 Materials

Collect soil from gas station and store at 4°C until needed.

Mineral salts medium (MSM) contains KH_2PO_4 12 g; K_2HPO_4 12 g; NH_4Cl 4 g; $FeSO_4 \cdot 7H_2O$ 0.01 g; $MgSO_4 \cdot 7H_2O$ 0.1 g; and 20 g agar per 1 L of water (pH 6.8).

Xylene, ethyl ether, and *m*-octane are used as carbon sources for MSM. Xylene, ethyl ether, and *m*-octane are diluted to 0.1%, 1.0%, and 10.0% (density of xylene = 0.864 g/mL, density of ethyl ether = 0.713 g/mL, density of octane = 0.70 g/mL). Dilution and bacterial exposure are completed according to established guidelines for toluene 0.1%.

26.20.4 Protocol

26.20.4.1 Regulation of Toluene Concentration

26.20.4.1.1 Protocol 1. Toluene vapors are used as a sole carbon source in MSM. Plates are incubated at room temperature (25°C) in glass desiccators containing varying concentrations of toluene. Vacuum pump oil (VPO) is used to dilute toluene because of its lower vapor pressure. Desiccator opening is avoided unless growth is being checked. The appropriate concentration of toluene is calculated using the density of toluene 0.8669: 0.1% (58 µL toluene, 49.95 g VPO), 1.0% (577 µL toluene, 49.5 g VPO), and 10.0% (5.8 mL toluene, 45.0 g VPO).

Isolation of toluene-degrading bacteria: Tryptic soy agar (TSA) plates are used to spread with suspensions of soil bacteria in sterile H_2O at final dilutions of 10^{-2}, 10^{-4}, and 10^{-6}. Single colonies were picked up and transferred to MSM with a sterile tooth pick and exposed to 1.0% toluene in a desiccator. After 96 h of incubation, plates were removed and examined. TSA slants were inoculated with growth spots in order to maintain stock cultures.

Growth confirmation: Growth with toluene as a carbon source was confirmed by reexposure to 1.0% toluene exposure to 100% VPO, and inoculation of MSM without supplement of carbon source.

26.20.4.1.2 Protocol 2.   Enrichment cultures were developed by adding 1 mL of 1000 ppm aqueous solution of xenobiotics into 20 g portion of the soil samples, maintained at 50% water-holding capacity (WHC). After five additions of the xenobiotics at 10 day intervals, the soil samples were withdrawn for isolation of predominant bacteria following MPN method. Bacterial cultures were isolated from the enrichment, raised to axenic cultures, and maintained in nutrient agar medium/ urea agar medium. The ability of selected bacterial species to utilize the xenobiotics as a source of C and N is tested by streaking cultures on agar growth medium.

26.20.5 Result and Observation

The plates are observed for visual growth of the cultures over a period of 3–4 days.

26.21 Degradation of Aromatic Hydrocarbons by Bacteria

26.21.1 Principle

The most widely distributed marine organic pollutants are hydrocarbons/crude oil, spilled into marine environment by tanker accidents, oil spillage, anthropogenic activities, ship-breaking yard activities, etc. Microbial degradation of crude oil is one of the major routes in the natural decontamination process and microorganisms play a key role in the degradation of petroleum hydrocarbons/crude oil in both terrestrial as well as aquatic environment. One of the recent examples of oil pollution is of Jessica oil spill that discharged ~600 tons of diesel and 300 tons of bunker fuel at Galapagos Island on January 16, 2001. Bioremediation was used to clean up oil-contaminated sites by spraying N.P. fertilizer to accelerate degradative abilities of indigenous microorganisms. Most of the bioremediation techniques currently used are based on enhancement of oil-degrading abilities of indigenous microbes by adding nutrients.

Bioremediation is emerging as most ideal alternative technique for reducing the different pollutants from the contaminated location and preventing toxic impacts. This ecofriendly technology is expanding with the addition of range of organisms for pollution cleanup.

26.21.2 Experimental Outline

Bioremediation is the use of living organisms (primarily microorganisms) to degrade environmental pollutants or to prevent pollution through waste treatment. Bioremediation can occur naturally by microbial and plant processes or can be accelerated by addition of nutrients to support the indigenous microbial flora or by application of *genetically engineered microorganisms* (GEM) or by manipulating physical, chemical, or hydrological processes. Identify the isolated microbes by performing various morphological examination tests specified in Chapter 15, which includes Gram staining, biochemical, motility, capsule staining, and growth investigations versus degradative potential using aromatic hydrocarbon as a sole source of carbon in the medium.

26.21.3 Materials

Oil-polluted seawater samples, mineral salt medium (MSM), incubators, Bushnell–Haas medium (BHM), luminescent spectrofluorometer

For isolation of oil-degrading bacteria, crude oil enrichment technique is used. The 0.1 mL of diluted oil-contaminated sea water sample collected from different ship-breaking plots is streaked on MSM containing $(NH_4)_2SO_4$, 18 mM; $FeSO_4 \cdot 7H_2O$, 1 µM; $CaCl_2 \cdot 2H_2O$ 100 µm; $MgSO_4 \cdot 7H_2O$, 1 mM; and NaCl, 8.5 mM in 10 mM Na_2HPO_4–K_2HPO_4 buffer adjusted to pH 7.0. The 1.5% agar is added in MSM before autoclaving. An ethereal solution of crude oil (10% w/v) was uniformly sprayed over the surface of agar plate. The ether immediately vaporizes at ambient temperature and then layer of oil remained on the entire surface. The plates are then incubated at 25°C for 16 days. The growth of bacteria is compared to control plate without crude oil.

The oil-degrading bacteria are characterized by Gram's reaction, IMViC (indole, methyl red, Voges Proskauer, and citrate utilization), sugar fermentation, and catalase tests described in Chapter 15 to facilitate identification of organism.

26.21.4 Protocol

26.21.4.1 Preculture Preparation
BHM contains NH_4NO_3, 1 g; $Mg_5O_4 \cdot 7H_2O$, 0.2 g; $CaCl_2 \cdot 2H_2O$, 0.02 g; K_2HPO_4, 1.0 g; KH_2PO_4, 1.0 g; $FeCl_3 \cdot 6H_2O$, 0.05 g; and distilled water to 1 L. The medium is supplemented with 5000 mg/L of crude oil. Fifty milliliters of medium is dispensed in 250 mL Erlenmeyer flask and inoculated with oil-degrading bacteria, The flasks are incubated at 25°C for 7 days on a shaker at 175 rpm.

For biodegradation experiment and observing emulsifying activity, 10^6 cells/mL is used as inoculums. BHM supplemented with 5000 mg/L of crude oil inoculated with the isolated bacteria are incubated at 25°C for 34 days on a shaker at 175 rpm. The control comprises of uninoculated BHM with 5000 mg/L crude oil. Degradative abilities are examiner after every 4, 10, 16, 22, 28, and 34 days of incubation by gravimetric and spectrometric analysis.

26.21.4.2 Gravimetric Analysis
Residual oil was quantified at mentioned time intervals in inoculated and uninoculated flasks by adding 2 mL of sulfuric acid, 10 mL of petroleum ether, and 1 mL of ethyl alcohol (C_2H_5OH) to the aforementioned flasks and shaken well. The mixture was taken in separating funnel. The upper layer of petroleum ether contains residual

oil that was separated. The residual petroleum ether was drained out from the separating funnel, through a filter paper soaked in petroleum ether, in preweighed glass beaker. The beaker was kept in water bath at 60°C to evaporate petroleum ether. The breaker with residual oil was weighed. The oil-degrading activity was evaluated by measuring decrease in weight compared to control.

26.21.4.3 Spectrofluorometric Analysis

Oil degradation by spectrofluorometric analysis: Five milliliters of *n*-hexane (HPLC grade) was added in BHM medium in Erlenmeyer flask inoculated with isolates, incubated for 34 days at 25°C with 5000 mg/L of crude oil as sole carbon source. Uninoculated flasks with the same quantity of crude oil served as control. The mixture was transferred to separating funnel and extracted. Extraction was carried out twice. The extract was transferred in clean test tube containing 0.4 g of anhydrous sodium sulfate. The extract was decanted into another test tube leaving behind sodium sulfate and evaporated to dryness in rotary evaporator under reduced pressure. Extract was transferred in volumetric flask and volume was made up to 10 mL with *n*-hexane.

Fluorescence was measured using luminescent spectrophotometer. Crude oil gives maximum fluorescence at 310 nm excitation and 360 nm emission wavelengths. Calibration curves were constructed using oil dissolved in *n*-hexane in concentrations ranging from 1 to 20 μg/L.

Synchronous excitation/emission spectra were taken of each sample. Oil degradation rates were extrapolated from the calibration curve. Since the aromatics in crude oil fluoresce, this technique provides an estimation of the aromatic fraction in the oil.

26.21.5 Results

Isolated bacterium still needs to be examined further for degradation rates if the growth continues to increase even after 34 days of incubation. The isolates showing higher degradation rate also showed high emulsification activity. Bacteria produce extracellular emulsifiers that convert large oil droplets to smaller ones in size, in increasing the surface area. The increase in surface area of emulsified oil droplets corresponds to higher bacterial growth rates and hence high biological degradation rates; smaller the oil droplets, higher the biodegradation rates.

26.22 Survey of Degradative Plasmids in Microbes Growing in Polluted Environment

26.22.1 Introduction

Plasmid refers to a circular piece of DNA, primarily independent of the host chromosome, found in bacteria. Degradative plasmids allow the host bacterium to metabolize unusual molecules such as toluene and salicylic acid, for example,

TOL (toluene-degrading plasmid of) *Pseudomonas putida*. Mobile genetic elements such as plasmids, bacteriophage, transposons, and integrons allow genes to move within the genome of a single bacterium, and between genomes of different bacteria. Such elements allow rapid acquisition of genes in response to environmental pressure, with profound consequences for evolution. The impact of mobile DNA has been felt most strongly in the clinical setting, with the development of multiple antibiotic-resistant bacteria due to our intensive use of antibiotics in human therapy and agriculture. However, it is increasingly realized that mobile DNA plays a more general role influencing diverse processes such as for the development of metabolic pathways for pollutant biodegradation and for the establishment of both symbiotic and pathogenic relationships between bacteria and plants. Integrons can be thought of as nature's little genetic engineers that are capable of capturing shuffling and expressing their associated modular coding units known as gene cassettes. Integrons are notorious for their ability to rapidly accumulate arrays of resistance gene cassette, making host bacterium difficult to kill with antibiotics.

26.22.2 Catabolic Plasmids of Alkene Degradation

Alkenes are hydrocarbons that contain a double bond between two of the carbon atoms. The simplest alkene is ethylene ($H_2C=CH_2$), which is produced by plants as a hormone. Alkenes are important industrial chemical—ethylene, vinyl chloride, and propylene used in huge quantities as the starting materials for plastic manufacture. Bacteria from the genera *Mycobacterium*, *Xanthobacter*, *Nocardioides*, and *Pseudomonas* can grow on alkenes as sole sources of carbon and energy.

26.23 Effect of SO_2 on Crop Plants

26.23.1 Principle

Agricultural crops can be injured when exposed to high concentrations of various air pollutants including SO_2. Injury ranges from visible markings on the foliage, to reduced growth and yield, to premature death of the plant. The development and severity of the injury depends not only on the concentration of the particular pollutant, but also on a number of other factors. These include the length of exposure to the pollutant, the plant species and its stage of development as well as the environmental factors conducive to a buildup of the pollutant and to the preconditioning of the plant, which make it either susceptible or resistant to injury.

Major sources of sulfur dioxide are coal-burning operations, especially those providing electric power and space heating. Sulfur dioxide emissions can also result from the burning of petroleum and the smelting of sulfur-containing ores (Figure 26.13). Sulfur dioxide in the atmosphere is produced mainly by combustion of coal and oil. Atmospheric concentration of sulfur dioxide is generally expressed in micrograms

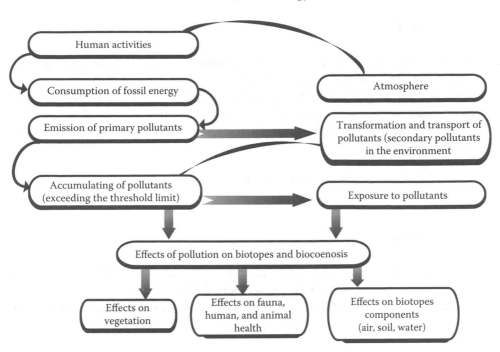

Figure 26.13 Schematic representation of effect of atmospheric pollution on different human activities.

of SO_2 per cubic meter of air ($\mu g/m^3$) or volume of SO_2 per hundred million volume of the air (1 vphm = 28.6 $\mu g/m^3$). Monthly mean concentration of SO_2 in urban areas ranges between 150 and 400 $\mu g/m^3$ in winter and 50–100 $\mu g/m^3$ in summer, but for periods of up to a few days under very stable weather conditions, concentrations may exceed 1000 $\mu g/m^3$. The adverse effect of SO_2 on plant growth consists mainly of visible reactions such as leaf necrosis, chlorosis, and discoloration. These effects occur only at SO_2 concentration above 3000 $\mu g/m^3$.

Sulfur dioxide enters the leaves mainly through the stomata (microscopic openings) and the resultant injury is classified as either acute or chronic. Acute injury is caused by absorption of high concentrations of sulfur dioxide in a relatively short time. The symptoms appear as two-sided (bifacial) lesions that usually occur between the veins and occasionally along the margins of the leaves. The color of the necrotic area can vary from a light tan or near white to an orange-red or brown depending on the time of year, the plant species affected, and weather conditions. Recently expanded leaves usually are the most sensitive to acute sulfur dioxide injury, the very youngest and oldest being somewhat more resistant.

26.23.2 Experimental Outline

The effect of SO_2 on crop plants is studied with the help of an exposure chamber. Follow the instructions of the manufacturer.

26.23.3 Materials

Exposure chamber, gas generation assembly, solution of Na_2SO_3, H_2SO_4 (10%), handy air sampler, spectrophotometer, plants (in pots)

Sulfur dioxide is generated by the action of H_2SO_4 on Na_2SO_3 under control reaction conditions in a gas generator. The concentration of SO_2 (0.05 ppm, 0.1 ppm) are determined by sampling the air with handy air sampler and analyzed by spectrophotometer:

$$Na_2SO_3 + H_2SO_4 \rightarrow SO_2 + Na_2SO_4 + H_2O$$

26.23.4 Protocol

Plants in pots are exposed in varying concentrations of SO_2 ranging from 0.05 to 0.1 ppm for 3 h on alternate days for a given period (30 or 60 days). The experiment is continued for 2–3 months. After every exposure, pots should be transferred to glass house.

26.23.5 Results and Observations

The effect of SO_2 on the plants is studied by taking a suitable parameter depending upon the objective of the experiment. For instance, data may be recorded for the treated and control plants for plant height, visual changes, photosynthetic activity, physiology and morphology of stomata color of leaves, pigment composition of leaves, other suitable parameter for specific studies.

26.24 Estimation of Heavy Metals in Water/Soil by Atomic Absorption Spectrophotometry

26.24.1 Principle

Living organisms require trace amounts of some metals including cobalt, copper, iron, manganese, molybdenum, vanadium, strontium, and zinc. Excessive levels of these essential metals are detrimental to the organisms. Nonessential metals like cadmium, chromium, mercury, lead, arsenic, and antimony are of more concern to surface water system because these metals produce undesirable effects on human and animal life. Once these metals enter in a system, they remain for relatively longer periods. Once absorbed, inorganic metals are capable of reacting with a variety of binding sites in the human body and have strong attraction to biological tissues. Natural water contains toxic metals in traces. Industrial wastes containing metals have aggravated the problem of metal pollution. Electroplating, metallurgical industry, galvanizing plants, tanneries, and thermal power stations are few of the major contributors of metal pollution in surface water. All metals exist in surface water in colloidal, particulate, and dissolved forms, although dissolved concentrations are generally low. The soluble forms are generally ions or unionized organometallic chelates or complexes. The solubility of trace metals in surface water is predominately controlled by pH, the type, and concentration of legends on which the metal can adsorb and the oxidation state of the mineral components. Atomic spectroscopic methods are rapid, convenient, and usually of high selectivity. Absorption spectrophotometer is being used for the qualitative and quantitative determination of >70 elements in water, soil, and plants. Typically, atomic absorption spectrophotometry (AAS) methods can detect parts per million to parts per billion (ppm to ppb) amounts, and in some cases, even smaller concentration.

26.24.2 Sampling and Preservation

26.24.2.1 Sample Containers
The best containers are quartz or Teflon (TFE). The container made up of polypropylene (PP) or linear polyethylene; borosilicate transparent glass capacity 500 mL is preferred. The containers with screw cap (single or double) without rubber or cork liner in the inner side are suited to avoid leakage of water.

26.24.2.2 Precaution
- Avoid contamination of metals from containers, distilled water, or membrane filter.
- Thoroughly clean sample container with a metal-free nonionic detergent solution. Rinse with tap water, soak in acid, and then rinse with metal-free water.
- Use 1:1 HNO_3 or aqua regia conc. HNO_3 (3 parts of conc. $HCl + 1$ part of conc. $HNO_3 + 4$ parts of water).

Metal concentration can be determined by the following methods:

- AAS
- Inductively coupled plasma (ICP) analyzer
- Colorimetric methods

ICP techniques and AAS are applicable over a broad linear range and are especially sensitive for refractory elements. In general, detection limits for ICP methods are higher than AAS. Colorimetric methods are applicable when interferences are known to be within the limit of the particular method:

- For mercury analysis, use 2 mL/L 20% w/v $K_2Cr_2O_7$ solution prepared in $(1 + 1)$. Store in a refrigerator not contaminated with mercury.
- Labeling of sample containers: The sample containers should have label.

Analysis: Drinking water sample for metal analysis.

The label should have removable sticker with the information written in permanent black ink (Table 26.3). The sample containers should be packed in cartons,

TABLE 26.3 Coding for Water Sampling

Code
Place
Source
Date
Time
Drawn by
Sealed by

sealed, and labeled. Types of metals are categorized; their treatment and preservation of collected samples of water are depicted in Table 26.4.

26.24.2.3 Sampling Error

The sources of sampling error, which is a combination of quantifiable components and qualitative factors, can be divided into four groups:

1. Errors originating from inherent sample variability
2. Errors originating from population variability
3. Sampling design error
4. Field procedure error

TABLE 26.4 Types of Metals, Their Treatment, and Preservation of Collected Samples of Water

Step	Dissolved Metals	Suspended Metals	Total Metals	Acid-Extractable Metals
I	Filter known volume of sample water through a 0.45 μM membrane[a]		Transfer the sample in sample container	
II	Filtered water	Residue over membrane filter	Add conc. HNO_3 toward pH <2 (2–5 mL/L)	
III	Transfer filtered water to sample container	Soak the filter paper in 1 N HNO_3	Label the container	
IV	Add conc. HNO_3 having pH <2 and transfer to sample container and then seal it	Transfer the filter paper and collect HNO_3 in sample container	Send the sample to laboratory	
V	Label the container	Label the container	Store the sample at 4°C to avoid evaporation	
VI	Send the sample to laboratory	Send the sample to laboratory	—	
VII	Store the sample at 4°C to avoid evaporation losses			

[a]*Membrane filter*: Polycarbonate or cellulose esters.

The quantifiable components of sampling error are the natural variability within a sample itself and the variability between sample populations that may be randomly selected from the same sampled area. Both of these errors can be evaluated quantitatively in the form of statistical variance.

Suspended metals: Metal fraction of a nonacidified sample retained on a 0.45 mm membrane filter.

Total metals: Metal fraction determined in an unfiltered sample after vigorous digestion or sum of metal concentrations both dissolved and suspended fractions.

Acid-extractable metals: Metal fraction after treatment of an unfiltered sample with hot dilute mineral acid and controlled to some extent through the application of appropriate sampling designs.

The quantitative components of sampling error are the following errors in the planning process and in the implementation of field procedure:

- *Sampling design error*—The collection of excessive or insufficient amount of data; the unrepresentative selection of sampling points; wrong time of sampling that may affect data relevancy; the choice of improper analysis.
- *Field procedure error*—Misidentified or missed sampling points in the field; failure to use data consistent to sampling procedure; gaps in field documentation; the use of incorrect sampling equipment and sample containers; incorrect sample preservation techniques or storage.

26.25 Preliminary Digestion for Metals

To reduce interference by organic matter and to convert metal associated with particulate to a form (usually the free metal) that can be determined by ICP spectroscopy, use one of the digestion techniques. Use the least rigorous digestion method required providing complete and consistent recovery compatible with the analytical method and the metal being analyzed:

- Open digestion
- Closed system digestion (microwave-assisted digestion)

Nitric acid digests most samples adequately. Nitrate is an acceptable matrix for both flame and electrothermal atomic absorption. Some samples may require addition of perchloric, hydrochloric, or sulfuric acid for complete digestion. Confirm metal recovery for each digestion and analytical procedure used. As a general rule, HNO_3 alone is adequate for clean samples or easily oxidized materials; HNO_3–H_2SO_4 or HNO_3–HCl digestion is adequate for readily oxidizable organic matter; HNO_3–$HClO_4$ or HNO_3–$HClO_4$–HF digestion is necessary for difficult-to-oxidize organic matter or minerals. Dry ash formation is helpful if large amounts of organic matter are present.

26.25.1 Open Digestion

26.25.1.1 Nitric Acid Digestion Apparatus

Materials: Hot plate, conical (Erlenmeyer) flasks, 125 mL or Griffin beakers 150 mL, acid-washed and rinsed with double distilled water; volumetric flasks, 100 mL.

    Reagents: Nitric acid, concentrated analytical grade, or trace metal grade.

    Procedure: Transfer a measured volume (50 mL) of well-mixed, acid-preserved sample to a flask or beaker. Add 5 mL conc. HNO_3 and a few boiling chips or glass beads. Bring to a slow boil and evaporate on a hot plate to the lowest volume possible (about 10–20 mL). Continue heating and adding conc. HNO_3 as necessary until digestion is complete as shown by a light-colored, clear solution. Do not let sample dry during digestion. Wash down flask or beaker walls with water and then filter if necessary. Transfer filtrate to a 100 mL volumetric flask with two 5 mL portions of water, adding these rinsing to the volumetric flask. Cool, dilute to mark, and mix thoroughly. Take portions of this solution for required metal determinations.

26.25.1.2 Nitric Acid–Hydrochloric Acid Digestion Apparatus

Materials: Hot plate, conical (Erlenmeyer) flasks 125 mL or Griffin beakers 150 mL, acid-washed and rinsed with double distilled water, volumetric flasks 100 mL, watch glasses, steam bath.

    Reagents: (a) Concentrated nitric acid (HNO_3) and (b) concentrated hydrochloric acid. Mix in the ratio of 1:1.

Protocol:

Total metals by digestion with HNO_3/HCl: Transfer a measured volume of well-mixed acid-preserved sample appropriate for the expected metal concentrations to a flask or beaker. Add 3 mL conc. HNO_3. Place flask or beaker on a hot plate and cautiously evaporate to <5 mL, making sure that the sample should not boil and that no area of the bottom of the container is allowed to go dry. Cool and add 5 mL conc. HNO_3. Cover container with a watch glass and return to hot plate. Increase temperature of hot plate so that a gentle reflux action occurs. Continue heating, adding additional acid as necessary, until digestion is complete. Evaporate to <5 mL and cool. Add 10 mL (1 + 1) HCl and 15 mL water per 100 mL and make up the final volume. Heat the sample for an additional 15 min to dissolve the precipitate or residue. Cool, wash down beaker walls and watch glass with water, filter to remove insoluble material that could clog the nebulizer, and transfer filtrate to a 100 mL volumetric flask with rinsing. Alternatively centrifuge or let settle overnight. Adjust to volume and mix thoroughly.

    Recoverable HNO_3/HCl: For this, transfer a measured volume of well-mixed, acid-preserved sample to a flask or beaker. Add 2 mL (1 + 1 HNO_3) and 10 mL (1 + 1 HCl) and heat on a steam bath or hot plate until volume has been reduced to near 25 mL, making certain sample does not boil. Cool and filter to remove insoluble material or alternatively centrifuge or let settle overnight. Transfer sample to volumetric flask, adjust volume to 100 mL, and mix.

26.25.1.3 Nitric Acid–Sulfuric Acid Digestion Apparatus
Materials: Hot plate, conical (Erlenmeyer) flasks 125 mL or Griffin beakers 150 mL, acid-washed and rinsed with double distilled water, volumetric flasks 100 mL, watch glasses, steam bath.

Reagents: (a) Concentrated nitric acid (HNO_3) and (b) concentrated sulfuric acid (H_2SO_4).

Protocol: Transfer a measured volume of well-mixed, acid-preserved sample appropriate for the expected metal concentrations to a flask or beaker. Add 5 mL conc. HNO_3 and a few boiling chips, glass beads. Bring to slow boil on hot plate and evaporate to 15–20 mL. Add 5 mL conc. HNO_3 and 10 mL conc. H_2SO_4. Evaporate on a hot plate until dense white fumes of SO_3 just appear. If solution does not clear, add 10 mL conc. HNO_3 and repeat evaporation to fumes of SO_3. Heat to remove all HNO_3 before continuing treatment. All HNO_3 will be removed when the solution is clear and no brownish fumes are evident. Do not let sample dry during digestion. Cool and dilute to about 50 mL with water. Heat to almost boiling slowly to dissolve soluble salts, filter if necessary, and then complete procedure as described earlier.

26.25.1.4 Nitric Acid–Perchloric Acid Digestion
Materials: Hot plate, conical (Erlenmeyer) flasks 125 mL or Griffin beakers 150 mL, acid-washed and rinsed with double distilled water, volumetric flasks 100 mL, watch glasses, steam bath, safety shields, safety goggles, watch glasses.

Reagents: (a) Concentrated nitric acid (HNO_3), (b) perchloric acid ($HClO_4$), and (c) ammonium acetate solution. Dissolve 500 g $NH_4C_2H_3O_2$ in 600 mL water.

Caution: Heated mixtures of $HClO_4$ and organic matter may explode violently. Avoid this hazard by taking the following precautions:

- Do not add $HClO_4$ to a hot solution containing organic matter.
- Always pretreat samples containing organic matter with HNO_3 before adding $HClO_4$.
- Avoid repeated fuming with $HClO_4$ in ordinary hoods (for routine operations, use a water pump attached to glass fume eradicator. Stainless steel fume hoods with adequate water wash-down facilities are available commercially and are acceptable for use with $HClO_4$).
- Never let samples being digested with $HClO_4$ evaporate to dryness.

Protocol: Transfer a measured volume of well-mixed, acid-preserved sample appropriate for the expected metal concentrations to a flask or beaker. Add 5 mL conc. HNO_3 and a few boiling chips, glass beads, or Hengar granules, and evaporate on a hot plate to 15–20 mL. Add 10 mL each of conc. HNO_3 and $HClO_4$ cooling flask or beaker between additions. Evaporate gently on a hot plate until dense white fumes of $HClO_4$ just appear. If solution is not clear, cover container with a watch glass and keep solution just boiling until it is clear. If necessary, add 10 mL of conc. HNO_3 to complete digestion. Cool, dilute to about 50 mL with water, and boil to expel any chlorine or oxides of nitrogen. Filter.

26.25.1.5 Nitric Acid–Perchloric Acid–Hydrofluoric Acid Digestion

Materials: Hot plate, TFE beakers, 250 mL, acid-washed and rinsed with water, volumetric flasks 100 mL, PP or other suitable plastic ware.

Reagents: (a) Concentrated nitric acid (HNO_3), (b) (1 + 1) perchloric acid ($HClO_4$), and (c) hydrofluoric acid (HF), 48%–51%.

Caution: Handle HF with extreme care and provide adequate ventilation, especially for the heated solution. Avoid all contact with exposed skin. Provide medical attention for HF burns.

Protocol: Transfer a measured volume of well-mixed, acid-preserved sample appropriate for the expected metal concentration into a 250 mL TFE beaker. Evaporate on a hot plate to 15–20 mL. Add 12 mL conc. HNO_3 and evaporate to near dryness. Repeat HNO_3 addition and evaporation. Let solution cool; add 20 mL $HClO_4$ and 1 mL HF, and boil until solution is clear and white fumes of $HClO_4$ have appeared. Cool, add about 50 mL water, filter, and make up the volume to known concentration using volumetric flask.

26.25.1.6 Dry Ashing

Materials: Evaporating dishes: Dishes of 100 mL capacity made up of one of the following materials: porcelain (90 mm diameter), platinum—generally satisfactory for all purposes, high-silica glass. Muffle furnace for operation at 550°C + 50°C; steam bath; desiccator provided with a desiccant containing a color indicator of moisture concentration; drying oven for operation at 103°C–105°C; analytical balance capable of weighing up to 0.1 mg.

Protocol: Mix sample and transfer to a known volume into a platinum or high-silica glass-evaporating dish. Evaporate to dryness on a steam bath. Transfer dish to a muffle furnace and heat sample to a white ash. If volatile elements are to be determined, keep temperature at 400°C–450°C. If sodium is to be determined alone, ash the sample at a temperature of 600°C. Dissolve ash in a minimum quantity of conc. HNO_3 and warm water. Filter diluted sample and adjust to a known volume preferably so that the final HNO_3 concentration is about 1%. Take portions of this solution for metal determination.

26.25.2 Microwave Digestion System Scope and Application

The microwave region of the electromagnetic spectrum lies between infrared radiation and radio frequencies and corresponds to wavelength of 1 cm to 1 m (frequencies of 30 GHz to 300 MHz, respectively, used for RADAR transmission and the remaining wavelength range is used for telecommunication). In order not to interfere with these uses, domestic and industrial microwave heaters are required to operate at either 12.2 cm (2.45 GHz) or 33.3 cm (900 MHz) unless the apparatus is shielded in such a way that no radiation losses occur. Laboratory/domestic microwave ovens generally operate at 2.45 GHz.

26.25.2.1 Microwave-Assisted Digestion

Microwave system is designed with time-controlled programmable power supply, having a corrosion-resistant, well-ventilated cavity and having all electronics protected against corrosion for safe operation. Use a unit having a rotating turntable with a minimum speed of 3 rpm to ensure homogeneous distribution of microwave radiation. Only laboratory-grade microwave equipment and closed digestion containers with pressure release that are specifically designed for hot acid may be used.

Open-vessel and closed-vessel microwave system: Microwave sample preparation is now a standard analytical tool employing a variety of microwave equipment, including both low to high pressure closed vessels and atmospheric pressure open vessels. Both open-vessel and closed-vessel microwave systems use direct absorption of microwave radiation through essentially microwave transparent vessel materials. Atmospheric pressure microwave systems can generate more stable temperature condition. In comparison, closed-vessel microwave dissolution systems are limited by the temperature and pressure safety tolerances of the reaction vessel and the microwave absorption characteristics of the solution. Microwave digestion rotor (MDR) technology provides analysts with unsurpassed performance capabilities and the highest standard of safety in closed-vessel microwave digestion. The rotor consists of high-density PP core. The compact core has the required strength to withstand the extreme pressure generated inside the vessel during digestion. Special plastic compression screws are located on the top of the rotor next to each niche. The vessel cover is also made of PFA (TFE). Microwave digestion of organic samples frequently involves exothermic reactions, which instantaneously generate large amount of decomposition gases (CO_2 and NO_x) as the sample is oxidized (Table 26.5). The relief valve is capable of instantaneously venting excess decomposition gases and resealing the vessel for completion of the digestion procedure. The acid used in microwave digestion may be classified into two main groups:

1. Nonoxidizing acids such as hydrochloric acid, HF, phosphoric acid, dilute sulfuric acid, and perchloric acid
2. Oxidizing acids, such as nitric acid, hot concentrated perchloric acid, concentrated sulfuric acid, and hydrogen peroxide

TABLE 26.5   Sample Program for Microwave-Assisted Digestion

Step	Time	Power (W)	Pressure (psi)	Internal Temperature (°C)	External Temperature (°C)
1	00:05:00	250	3	150	70
2	00:05:00	400	6	180	80
3	00:04:00	500	10	210	80
Ventilation	00:03:00				

26.26 Estimation of Metals

26.26.1 ICP Method

Scope and application: Emission spectroscopy using ICP is rapid, sensitive, and conventional method for the determination of metals in water and wastewater samples. Dissolved metals are determined in filtered and acidified samples. Total metals are determined after appropriate digestion. Care must be taken to ensure that potential interference is dealt with, especially when dissolved solids exceed 1500 mg/L.

Equipment

1. *ICP source*: The ICP source consists of a radio frequency (RF) generator capable of generating at least 1.1 kW of power, torch, tesla coil, impedance matching network, nebulizer, spray chamber, and drain (Figure 26.14). High-quality flow regulators are required for both the nebulizer argon and the plasma support gas flow. A peristaltic pump is recommended to regulate sample flow to the nebulizer. The type of nebulizer and spray chamber used may depend on the sample to be analyzed as well as on the equipment manufacture. In general, pneumatic nebulizer of the concentric or cross-flow design are used. Viscous samples and samples containing particulates or high dissolved solids content may require nebulizers of the Babington type.

2. *Spectrometer*: The spectrometer may be of the simultaneous or sequential type with air-part, inert gas purged, or vacuum optics. A spectral bandpass of 0.05 nm or less is required. The instrument should permit examination of the spectral background

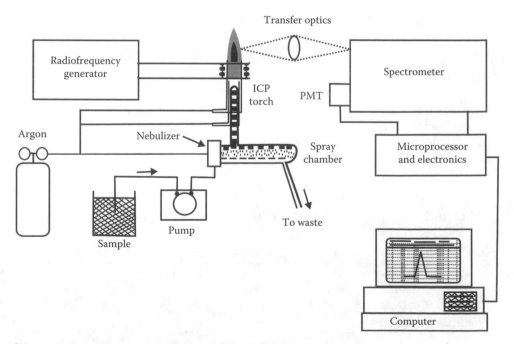

Figure 26.14 Schematic diagram of ICP for metal analysis.

surrounding the emission lines used for metals determination. It is necessary to be able to measure and correct for spectral background at one or more positions on either side of the analytical lines.

Reagents and standards: Use reagents that are of ultrahigh-purity grade or equivalent. Redistilled acids are acceptable. Dry all salts at 105°C for 1 h and store in a desiccator before weighing. Use deionized water prepared by passing water through at least two stages of deionization with mixed bed cation/anion-exchange resins. Use deionized water for preparing all calibration standards, reagents and for dilution (Table 26.6):

1. Concentrated hydrochloric acid (HCl) and (1 + 1)
2. Concentrated nitric acid (HNO_3)
3. Nitric acid (HNO_3) (1 + 1): add 500 mL conc. HNO_3 to 400 mL water and dilute to make 1 L

TABLE 26.6   Standard Solutions for ICP

Element	Standard Solution for ICP
Aluminum	Dissolve 0.100 g aluminum metal in an acid mixture of 4 mL (1 + 1) HCl and 1 mL conc. HNO_3 in a beaker. Warm gently to effect solution. Transfer to 1 L flask, add 10 mL (1 + 1) HCl, and dilute to 1000 mL with water; 1.00 mL = 100 μg Al.
Arsenic	Dissolve 1.534 g arsenic pentoxide As_2O_5 in distilled water containing 4 g NaOH. Dilute to 1000 mL; 1.00 mL = 1.00 mg As(V).
Cadmium	Dissolve 0.100 g cadmium metal in 4 mL conc. HNO_3. Add 8 mL conc. HNO_3 and dilute to 1000 mL with water; 1.00 mL = 100 μg Cd.
Chromium	Dissolve 0.1923 g CrO_3 in water. When solution is complete, acidify with 10 mL conc. HNO_3 in a beaker and dilute to 1000 mL with water; 1.00 mL = 100 μg Cr.
Copper	Dissolve 0.100 g copper metal in 2 mL conc. HNO_3, add 10.0 mL (1 + 1) HCl, and dilute to 1000 mL with water; 1.00 mL = 100 μg Cu.
Iron	Dissolve 0.100 g iron wire in a mixture of 10 mL (1 + 1) of HCl and 3 mL conc. HNO_3. Add 5 mL conc. HNO_3 and dilute to 1000 mL with water; 1.00 mL = 100 μg Fe.
Lead	Dissolve 0.1598 g lead nitrate, $Pb(NO_3)_2$, in a minimum amount (1 + 1) of HNO_3, add 10 mL of conc. HNO_3 and dilute to 1000 mL with water; 1.00 mL = 100 μg Pb.
Manganese	Dissolve 0.1000 g manganese metal in 10 mL conc. HCl mixed with 1 mL of conc. HNO_3. Dilute to 1000 mL with water; 1.00 mL = 100 μg Mn.
Nickel	Dissolve 0.1 g Ni metal in 10 mL hot conc. HNO_3, cool and dilute to 1000 mL water; 1.00 mL = 100 μg Ni.
Potassium	Dissolve 0.1907 g potassium chloride, KCl (dried at 110°C) in water and make up to 1000 mL; 1.00 mL = 100 μg K.
Sodium	Dissolve 0.2542 g sodium chloride, NaCl, dried at 140°C, in water and at 10 mL conc. HNO_3 and make up to 1000 mL; 1.00 mL = 100 μg Na.
Selenium	Dissolve 2.393 g sodium selenate, Na_2SeO_4, in water containing 10 mL conc. HNO_3. Dilute to 1 L; 1.00 mL = 1.00 mg Se (VI).
Silver	Dissolve 0.1575 g silver nitrate, $AgNO_3$, in 100 mL water, add 10 mL conc. HNO_3 and make up to 1000 mL; 1.00 mL = 100 μg Ag.
Zinc	Dissolve 0.10 g zinc metal in 20 mL (1 + 1) of HCl and dilute to 1000 mL with water; 1 mL = 100 μg Zn.

Secondary standard: The standard solutions for metals are available with supplier, or otherwise arrange the standard prepared by recognized laboratory. Use these standards to verify the prepared standards.

Calibration standards: In 100 mL volumetric flasks, add 2 mL of (1 + 1) HNO_3 and 10 mL of (1 + 1) HCl and dilute to 100 mL with water. Before preparing mixed standards, analyze each stock solution separately to determine possible spectral interference or the presence of impurities. When preparing mixed standards take care that the elements are compatible and stable. Store mixed standard solutions in an FEP fluorocarbon (TFE) or unused polyethylene bottle. Verify calibration standards initially using the quality control standard; monitor weekly for stability.

Mixed standard solution V Silver: If addition of silver results in an initial precipitation, add 15 mL water and warm flask until solution clears. Cool and dilute to 100 mL with water. For this acid combination, limit the silver concentration to 2 mg/L. Silver under these conditions is stable in a tap water matrix for 30 days. Higher concentrations of silver require additional HCl.

Caution: Many metal salts are extremely toxic and may be fatal, if swallowed. Wash hands thoroughly after handling.

Calibration blank: Dilute 2 mL of (1 + 1) HNO_3 and 10 mL of (1 + 1) HCl to 100 mL with water. Prepare a sufficient quantity to be used to flush the system between standard and samples.

Method blank: Carry a reagent blank through entire sample preparation procedure. Prepare method blank to contain the same acid types and concentrations as the sample solution.

Instrument check standard: Prepare instrument check standard by combining compatible elements as a concentration of 2 mg/L.

Instrument quality control sample: Certified aqueous reference standards are available. Use the same standards for the calibration.

Argon: Use technical or welder's grade.

Protocol

Sample preparation: Use digested sample as described in Section 26.25.

Operating conditions: Establish instrumental detection limit, precision, optimum background correction positions, linear dynamic range, and interference for each analytical line. Verify that the instrument configuration and operating conditions satisfy the analytical requirements. An atom-to-ion emission intensity ratio (Cu(I) 324.75 nm/Mn(II) 257.61 nm) can be used to reproduce optimum conditions for multielement analysis precisely. The Cu/Mn intensity ratio may be incorporated into the calibration procedure, including specifications for sensitivity and for precision.

Instrument calibration: Warm up for 30 min. For polychromators, perform an optical alignment using the profile lamp or solution. Check alignment of plasma torch and spectrometer entrance slit. Make Cu/Mn or similar intensity ratio adjustment. Calibrate instrument using calibration standards and blank.

Aspirate the standard or blank for a minimum of 15 s after reaching the plasma before beginning signal integration. Rinse with calibration blank or similar solution for at least 60 s between each standard to eliminate any carryover from the previous

TABLE 26.7   Suggested Wavelength, Detection Limit, and Upper Limit Concentration for Elements

Element	Suggested Wavelength (nm)	Estimated Detection Level (µg/L)	Alternate Wavelength (nm)	Calibration Concentration (mg/L)	Upper Limit Concentration (mg/L)
Aluminum	308.22	40	237.32	10.0	100
Arsenic	193.70	50	189.04	10.0	100
Cadmium	226.50	4	214.44	2.0	50
Chromium	267.72	7	206.15	5.0	50
Copper	324.75	6	219.6	1.0	50
Iron	259.94	7	238.20	10.0	100
Lead	220.35	40	217.0	10.0	100
Manganese	257.61	2	294.92	2.0	50
Nickel	231.60	15	221.65	2.0	50
Potassium	766.49	100	269.90	10.0	
Sodium	589.0	30	589.59	10.0	100
Selenium	196.03	75	203.99	5.0	100
Silver	328.07	7	338.29	2.0	50
Zinc	213.86	2	206.20	5.0	100

standards. Use average intensity of multiple integration of standard or sample to reduce random error. Before analyzing samples, ensure the instrument check that concentration values obtained should not deviate from the actual values by more than +5%. The wavelength for the element is as suggested in Table 26.7.

26.26.2 Analysis of Samples

Analyze the samples using calibration blank. This permits a check of the sample preparation reagents and procedures for contamination. Analyze samples alternately with analyses of calibration blank. Rinse for at least 60 s with dilute acid between samples and blanks. After introducing each sample or blank, let system equilibrate before starting signal integration. Examine each analysis of the calibration blank to verify that no carryover memory effect has occurred. If carryover is observed, repeat rinsing until proper blank values are obtained. Make appropriate dilutions of the sample to determine concentrations beyond the linear calibration.

Spike test: To the known volume of digested sample, add known volume of the standard. Shake well and aspirate through ICP. Check the increase in concentration of metal for added quality of standards.

Instrument quality control: Analyze instrument check standard to confirm proper recalibration. Reanalyze one or more samples analyzed just before termination of the analytical run. Results should agree within 15% error. Analyze instrument quality control sample within every run. Use this analysis to verify accuracy and validity of the calibration standards.

26.26.3 Data Analysis and Calculations

Blank correction: Subtract result of an adjacent calibration blank from each sample result to make a baseline drift correction. Use the result of the method blank analysis to correct for reagent contamination.

Dilution correction: If the sample was diluted or concentrated in preparation, multiply result by a dilution factor (DF) calculated as follows:

$$DF = \frac{\text{Final weight of volume}}{\text{Initial weight of volume}}$$

Calculations: Report analytical data in concentration units of mg/L using up to three final weights or volume DF = Initial weight or volume significant figures.

26.27 Flame AAS

26.27.1 Scope and Application

In flame AAS, a sample is aspirated into a flame and atomized. A light beam is directed through the flame, into a monochromator, and onto a detector that measures the amount of light absorbed by the atomized element in the flame. For some metals, atomic absorption exhibits superior sensitivity over flame emission. Because each metal has got its own characteristic absorption wavelength a source lamp composed of the elements is used; this marks the method relatively free from spectral or radiation interference. The amount of energy at the characteristic wavelength absorbed in the flame is proportional to the concentration of the element in the sample over a limited concentration range. Most atomic absorption instruments are also equipped for operation in an emission mode, which may provide better linearity for some elements.

26.27.2 Interferences

26.27.2.1 Chemical Interference

Many metals can be determined by direct aspiration of sample into an air–acetylene flame. The most troublesome type of interference is termed "chemical" and results from the lack of absorption by atoms bound in molecular combination in the flame. This can occur when the flame is not hot enough to dissociate the molecules or when the dissociated atom is oxidized immediately to a compound that will not dissociate further at the flame temperature.

26.27.2.2 Background Correction

Molecular absorption and light scattering caused by solid particles in the flame can cause erroneously high absorption values resulting in positive errors. When such phenomena occur, use background correction to obtain accurate values. Use any one of three types of background correction:

1. *Continuum-source background correction*: Continuum-source background corrector utilizes either a hydrogen-filled hollow-cathode lamp with metal cathode or a deuterium arc lamp. When both line sources are placed in the same optical path and are time-shared, the broadband background from the elemental signal is subtracted electronically, and the resultant signal will be background-compensated.
2. *Zeeman background correction*: This correction is based on the principle that a magnetic field splits the spectral line in to two linearly polarized light beams parallel and perpendicular to the magnetic field. One is called the pi component and other the sigma component. These two line beams have exactly the same wavelength and differ only in the plane of polarization. Zeeman background correction provides accurate background correction at much higher absorption levels than is possible with continuum-source background correction systems.
3. *Smith–Hieftje background correction*: This correction is based on the principle that absorbance measured for a specific element is reduced as the hollow-cathode lamp is increased while absorption of nonspecific absorption substance remains identical at all current levels. When this method is applied, the absorption at a high-current mode is subtracted from the absorption at a low current mode. Under these conditions, any absorbance due to nonspecific background is subtracted out and corrected for.

26.27.3 Apparatus and Equipment

1. *Atomic absorption spectrometer*: It consists of a light source emitting the line spectrum of an element, a device for vaporizing the sample, a means of isolating an absorption line, and a photoelectric detector with its associated electronic amplifying and measuring equipment (Figure 26.15).
2. *Burner*: The most common type of burner is a premix, which introduces the spray into a condensing chamber for removal of large droplets. The burner may be fitted with a conventional head containing a single slot; a tree-slot bowling head that may be preferred for direct aspiration with an air–acetylene flame; or a special head for use with nitrous oxide and acetylene.
3. *Readout*: Most instruments are equipped with either a digital or null meter readout mechanism. Most modern instruments are with microprocessor or stand-alone control computers capable of integrating absorption signals over at high concentrations.
4. *Lamps*: Use either a hollow-cathode lamp or electrodeless discharge lamp (EDL). Use one lamp for each element being measured. Multielement hollow-cathode lamps generally provide lower sensitivity than single-element lamps. EDLs take a longer time to warm up and stabilize.
5. *Pressure-reducing valves*: Maintain supplies of fuel and oxidant at pressures somewhat higher than the controlled operating pressure of the instrument by using suitable reducing valves. Use a separate reducing value for each gas.
6. *Vent*: Place a vent about 15–30 cm above the burner to remove fumes and vapors from the flame. This precaution protects laboratory personnel from vapors, protects the instrument from corrosive vapors, and prevents flame stability from being affected by room drafts. A damper or variable speed blower is desirable for modulating airflow and preventing flame disturbance. Select blower size to provide the airflow recommended by the instrument manufacturer. In laboratory locations with heavy particulate air pollution, use clean laboratory facilities.

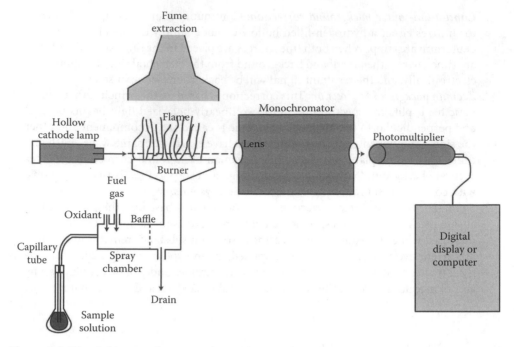

Figure 26.15 Schematic diagram of AAS for metal analysis.

26.27.4 Reagents and Standards

Air: Air should be cleaned and dried through a suitable filter to remove oil, water, and other foreign substances. The sources may be a compressor or commercially bottled gas.

Acetylene: Standard commercial grade in which acetone is always present in acetylene cylinders. This prevents the entering and damaging the burner head by replacing a cylinder when its pressure has fallen to 689 kPa (100 psi) acetylene.

Caution: Acetylene gas represents an explosive hazard in the laboratory. Follow instrument manufacturer's directions in plumbing and using this gas. Do not allow gas contact with copper, brass with >65% copper, silver, or liquid mercury; do not use copper or brass tubing, regulators, or fittings.

Nitrous oxide (for aluminum): The gas is commercially available in cylinders. Fit nitrous oxide cylinder with a special nonfreezable regulator or wrap a heating coil around an ordinary regulator to prevent flashback at the burner caused by regulation in nitrous oxide flow through a frozen regulator.

Caution: Use nitrous oxide with strict adherence to manufacturer's directions. Improper sequencing of gas flow at start-up and shutdown of instrument can produce explosions from flashback.

Metal-free water: Use metal-free water for preparing all reagents and calibration standards and as dilution waste. Prepare metal-free water by deionizing tap water and/or by using one of the following processes, depending on the metal concentration in the sample: single distillation, reinstallation, or subboiling.

Standard solution: Prepare standard solutions of known metal concentrations. Stock standard solution can be obtained from several commercial sources that should be used as secondary standard solution for calibration of instrument and the prepared standards.

26.27.5 Protocol for Sample Preparation

Use digested sample as described in Section 26.25.

26.27.5.1 Operating Conditions

Install a hollow-cathode lamp for the desired metal in the instrument and roughly set the wavelength dial according to Table 26.8. Set slit width according to manufacturer-suggested setting for the element being measured. Turn on instrument; apply to the hollow-cathode lamp the current suggested by the manufacturer; and let instrument warm-up until energy source stabilizes, generally about 10–20 min. Readjust current as necessary after warm-up. Optimize wavelength by adjusting wavelength dial until optimum energy gain is obtained. Align lamp in accordance with manufacturer's instruction. Install suitable burner head and adjust burner head position. Turn on air and adjust flow rate to that specified by manufacturer to give maximum sensitivity for the metal being measured. Turn on acetylene, adjust flow rate to value specified, and ignite flame. Let flame stabilize for a few minutes. Aspirate a blank consisting of deionized water containing the same concentration of acid in standard and samples. Adjust the instrument to zero, aspirate a standard solution, and adjust aspiration rate of the nebulizer to obtain maximum sensitivity. Adjust burner both vertically and horizontally to obtain maximum response. Aspirate blank again and check zero reading of the instrument. Aspirate a standard near the middle of the liner range. Record absorbance of this standard when freshly prepared and with a new hollow-cathode lamp. Refer to the data on subsequent determination of the same element to

TABLE 26.8 Wavelength Instrument Detection Level Sensitivity and Optimum Concentration Range for Elements

Element	Wavelength (nm)	Flame Gases	Instrument Detection Level (mg/L)	Sensitivity	Optimum Concentration Range (mg/L)
Al	328.1	N-Ac	0.1	1	5–100
Cd	228.8	A-Ac	0.002	0.025	0.05–2
Cr	357.6	A-Ac	0.02	0.1	0.2–10
Cu	324.7	A-Ac	0.01	0.1	0.2–10
Fe	248.3	A-Ac	0.02	0.12	0.3–10
Mn	279.5	A-Ac	0.01	0.05	0.1–10
Ni	232.0	A-Ac	0.02	0.15	0.3–10
Pb	283.3	A-Ac	0.05	0.5	1–20
Zn	213.9	A-Ac	0.005	0.02	0.05–2

check consistency of instrument setup and aging of hollow-cathode lamp and standard. The instrument now is ready to operate. When analysis is finished, extinguish flame by turning off first acetylene and then air (Table 26.8).

26.27.5.2 Standardization

Select at least three concentrations of each standard metal solution to bracket the expected metals concentration of a sample. Aspirate blank and adjust zero of the instrument. Then aspirate each standard in turn into flame record absorbance. Prepare a calibration curve by plotting on linear graph paper absorbance of standard versus their concentrations. For instruments equipped with direct concentration readout, this step is unnecessary. With some instrument, it may be necessary to convert percent absorption by using a table generally provided by the manufacturer.

26.27.5.3 Data Analysis and Calculations

Calculate concentration of each ion, in mg/L for trace elements, and in mg/L for more common metals, by referring to the appropriate calibration curve. Alternatively read concentration directly from the instrument readout if the instrument is so equipped. If the sample has high values, multiply it by the appropriate DF.

Spike test: To the known volume of digested sample, add known volume of the standard. Shake well and aspirate through ICP. Check the increase in concentration of metal for added quantity of standards.

26.28 Measurement of Aluminum in Soil/Water Sample

26.28.1 Scope and Application

Aluminum is the third most abundant element on the earth's crust. The presence of aluminum in all natural water is in the form of soluble salt, a colloid, or an insoluble compound. Soluble, colloidal, and insoluble aluminum may be present in treated water in residual form of coagulation with aluminum-containing material. The USEPA water standard for Al is 0.05 mg/L max. BIS desirable limit is 0.03 mg Al/L.

26.28.2 Methods for Analysis

26.28.2.1 ICP Method

Described in Section 26.26.1.

26.28.2.2 Eriochrome Cyanine R Method

26.28.2.2.1 Principle. With Eriochrome cyanine R dye, dilute aluminum solutions buffered to a pH 6.0 produce a red to pink complex that exhibits maximum absorption at 535 nm. The aluminum concentration, reaction time, temperature, pH, alkalinity, and concentration of other ions in the sample influence the intensity of the complex. To compensate for color and turbidity, the aluminum in one portion of sample is complexed with EDTA to provide a blank. Adding ascorbic acid eliminates the interference of iron and manganese. The optimum aluminum range lies between 20 and 300 µg/L but can be extended by sample dilution.

26.28.2.2.2 Interference. Both fluoride and polyphosphates cause negative errors. When the fluoride concentration is constant, the percentage error decreases with increasing amount of aluminum. Because the fluoride concentration often is known or can be determined readily, fairly accurate results can be obtained by adding the known amount of fluoride to a set of standards. Orthophosphate in concentrations under 10 mg/L does not interfere. The interference caused by even small amount of alkalinity is removed by acidifying the sample just beyond the neutralization point of methyl orange. Sulfate does not interfere up to a concentration of 2000 mg/L.

26.28.2.2.3 Minimum Detectable Concentration. The minimum aluminum concentration detectable by this method in the absence of fluorides and complex phosphates is ~6 µg/L.

26.28.2.2.4 Apparatus and Equipment. Spectrophotometer, wavelength at 535 nm, with a light path of 1 cm.

Glassware: Treat all glassware with warm (1 + 1) HCl and rinse with aluminum-free distilled water to avoid errors due to materials absorbed on the glass. Rinse sufficiently to remove all acid.

26.28.2.2.5 Reagents and Standards. Use reagents and distilled water free from aluminum contamination:

1. Stock aluminum solution: Use either the metal (1) or soluble salt (2) for preparing stock solution; 1 mL = 500 µg Al: (1) Dissolve 500 mg aluminum metal in 10 mL conc. HCl by heating gently. Dilute to 1000 mL with distilled water. (2) Dissolve 8.791 g aluminum potassium sulfate, $AlK(SO_4)_2 \cdot 12H_2O$, in water and dilute to 1000 mL. Correct this weight by dividing by the decimal fraction of analyzed $AlK(SO_4)_2 \cdot 12H_2O$ in the reagent used.
2. Standard aluminum solution: Dilute 10.0 mL stock aluminum solution to 1000 mL with water; 1 mL = 5.00 µg Al. Prepare fresh stock.
3. Sulfuric acid (H_2SO_4): 0.02 and 6 N.
4. Ascorbic acid solution: Dissolve 0.1 g ascorbic acid in water and make up to 100 mL in volumetric flask. Prepare fresh daily.
5. Buffer reagent: Dissolve 136 g sodium acetate, $NaC_2H_3O_2 \cdot 3H_2O$, in water, add 40 mL 1 N acetic acid, and dilute to 1 L.
6. Stock dye solution: Eriochrome cyanine R: Dissolve 100 mg in water and dilute to 100 mL in volumetric flask. This solution should have a pH of about 9–2.9. Stock solutions have excellent stability and can be kept for at least a year.
7. Working dye solution: Dilute 10 mL of selected stock dye solution to 100 mL in volumetric flask with water. Working solutions are stable for at least 6 months.

8. Methyl orange indicator solution or bromocresol green indicator solution: As in alkalinity test.
9. EDTA (sodium salt of EDTA dihydrate) (0.01 M): Dissolve 3.7 g in water and dilute to 1 L.
10. Sodium hydroxide (NaOH): 1 and 0.1 N.

26.28.3 Protocol

26.28.3.1 Preparation of Calibration Curve

Prepare a series of aluminum standards from 0 to 7 μg (0–280 μg/L) based on a 25 mL sample by accurately measuring the calculated volumes of standard aluminum solution into 50 mL volumetric flasks or Nessler tubes. Add water to a total volume of ~25 mL. Add 1 mL of 0.02 N H_2SO_4 to each standard and mix. Add 1 mL ascorbic acid solution and mix. Add 10 mL buffer solution and mix. With a volumetric pipette, add 5 mL working dye reagent and mix. Immediately make up to 50 mL with distilled water. Mix and let it stand for 5–10 min. The color begins to fade after 15 min. Read transmittance or absorbance on a spectrophotometer, using a wavelength of 535 nm. Adjust instruments to zero absorbance with the standard containing no aluminum. Plot concentration of Al (micrograms of Al in 50 mL final volume) against absorbance.

26.28.3.2 Sample Treatment in Absence of Fluorides and Complex Phosphate

Place 25 mL sample, or a portion diluted to 25 mL, in a porcelain dish or flask, add a few drops of methyl orange indicator, and titrate with 0.02 N H_2SO_4 to faint pink color. Record reading and discard sample. To two similar samples at room temperature, add the same amount of 0.02 N H_2SO_4 used in the titration and 1 mL in excess. To one sample, add 1 mL EDTA solution. This will serve as a blank by complexing any aluminum present and compensating for color and turbidity. To both samples add 1 mL ascorbic acid, 10 mL buffer reagent, and 5 mL working dye reagent. Set instrument to zero absorbance or 100% transmittance using the EDTA blank. After 5–10 min contact time, read transmittance or absorbance and determine aluminum concentration from the calibration curve previously prepared.

26.28.3.3 Removal of Phosphate Interference

Add 1.7 mL 6 N H_2SO_4 to 100 mL sample in a 200 mL Erlenmeyer flask. Heat on a hot plate for at least 90 min, keeping solution temperature just below the boiling point. At the end of the heating period, the solution volume should be about 25 mL. Add water if necessary to keep it at or above that volume. After cooling, neutralize to a pH of 4.3–4.5 with NaOH, using 1 N NaOH at the start and 0.1 N for the final time adjustment. Monitor with a pH meter. Make up to 100 mL with water, mix, and use a 25 mL portion for the aluminum test. Run a blank in the same manner, using 100 mL distilled water and 1.7 mL 6 N H_2SO_4. Subtract blank reading from sample reading or use it to set instrument to zero absorbance before reading the sample.

26.28.3.4 Correction for Samples Containing Fluoride
Add the same amount of fluoride as in the sample to each aluminum standard and draw calibration curve for aluminum standard.

26.28.4 Data Analysis and Calculation

$$mg\ Al/L = \frac{mg\ Al\ (in\ 50\ mL\ final\ volume)}{mL\ sample}$$

26.29 Measurement of Arsenic in Soil/Water Sample

26.29.1 Scope and Application

Severe poisoning can arise from the ingestion of as little as 100 mg of arsenic trioxide. Chronic effects may result from the accumulation of arsenic compounds in the body at low intake levels. Carcinogenic properties have also been observed to arsenic compounds. For the protection of aquatic life, the average concentration of As^{3+} in water should not exceed 72 µg/L. And maximum should not exceed 140 µg/L. FAO-recommended maximum level for irrigation water is 100 µg/L. USEPA water standard is 0.05 mg/L. BIS desirable limit is 0.05 mg As/L.

26.29.2 Methods for Analysis

1. Kit (determination of arsenic by kit method)
2. Silver diethyldithiocarbamate method
3. ICP method

26.29.2.1 Kit Method
Arsenic is liberated as arsine (AsH_3) by zinc in acid solution in a conical flask–type arsine generator. The generated arsine is passed through a mercuric bromide test paper. The generated arsine produces yellow to brown stain on test paper impregnated with mercuric bromide. The developed color of the stain is proportional to the amount of arsenic present. The field kit has the following salient features:

- Aesthetic
- Sturdy
- Lightweight
- Free from occupational hazards
- Contains arsenic-free reagents
- Convenient to carry in the field
- Useful for rapid on-site screening of water sources for arsenic levels

The field kit has the following components:

Glass assembly comprising arsine generator, sulfide trap, trap cover (all having standard glass joint arrangements to prevent any escape of arsenic):

- Two polythene bottles containing Reagents 1 and 2*
- One small polythene bottle containing lead acetate solution
- One small plastic bottle containing cotton plugs
- One small plastic bottle containing cotton $HgBr_2$ strips (most effective for 1 month)
- One elongated box with color comparator chart on the top (inside forceps, plastic rod, spatula)
- Adhesive tape
- Tissue paper
- Record pad
- Instruction for use

The field kit can detect arsenic in the range 0.01–0.1 mg/L (10–100 ppb). It has low precision and accuracy at arsenic level <2 ppb but for arsenic levels >50 ppb the kit is precise and accurate.

The field kit has been subjected to extensive "within-laboratory" tests of reliability and reproducibility at six different arsenic concentrations (Table 26.9).

One field kit contains sufficient chemicals to conduct 100 field tests.

Instructions for use:

- Take test water up to 20 mL mark in the arsine generator.
- Push using the plastic rod one cotton plug into the sulfide trap and soak it with three to four drops of lead acetate solution using the dropper bottle.
- Place with the help of forceps one mercuric bromide test strip on top of the sulfide trap.
- Carefully put the trap cover over the sulfide trap so that the mercuric bromide test strip is held between the sulfide trap and the trap cover.
- Add two spatula of Reagent 1 to the test water.
- Add 2 mL of Reagent 2 to the test water using the dropper.
- Immediately fix the lower end of the sulfide trap (joined with trap cover) to the arsine generator.
- Allow the reaction to proceed for 20 min.
- Take out the mercuric bromide strip with the help of forceps and compare the developed color with the color comparator chart.
- Stick the mercuric bromide strip to the record pad with the adhesive tape.

TABLE 26.9   Details of the Field Kit for Arsenic Concentration

Arsenic Concentration (ppb)	10 (%)	20 (%)	50 (%)	100 (%)	500 (%)	1000 (%)
Sensitivity to detect the concentration at A	≤85	≤85	≤95	≤95	≤95	≤95
Reproducibility (precision)	≤85	≤85	≤95	≤95	≤95	≤95

---

* Reagent 2 is hydrochloric acid and should be filled in the polythene bottle on the day of field visit.

26.29.2.2 Silver Diethyldithiocarbamate Method

26.29.2.2.1 Summary of Method

26.29.2.2.1.1 Principle.   Arsenite, containing trivalent arsenic, is reduced selectively by aqueous sodium borohydride solution to arsine, AsH_3, in an aqueous medium of pH 6. Arsenate, methyl arsenic acid, and dimethyl arsenic acid are not reduced under these conditions. The generated arsine is swept by a stream of oxygen-free nitrogen from the reduction vessel through a scrubber containing glass wool or cotton impregnated with lead acetate solution into an absorber tube containing silver diethyldithiocarbamate and morpholine dissolved in chloroform. A red color intensity that develops is measured at 520 nm. To determine total inorganic arsenic in the absence of methyl arsenic compounds, reduce another sample portion at a pH of about 1.0. Alternatively, determine arsenate in a sample from which arsenite has been removed by reduction at pH 6. The sample is acidified with hydrochloric acid and another portion of sodium borohydride solution is added. Arsine formed from arsenate is collected in fresh absorber solution.

26.29.2.2.2 Interferences.   Although certain metals, viz. chromium, cobalt, copper, mercury, molybdenum, nickel, platinum, silver, and selenium influence the generation of arsine, their concentrations in water seldom are high enough to interfere. H_2S interferes, but the interference is removed with lead acetate. Antimony is reduced to stibine, which forms a colored complex with absorption maximum at 510 nm and interferes with the arsenic determination. Methyl arsenic compounds are reduced at pH 1 to methyl arsines, which form colored complexes with the absorber solution. If methyl arsenic compounds are present, measurements of total arsenic and arsenate are unreliable. Methyl arsenic compounds do not influence the results for arsenite. Minimum detectable quantity: 1 μg arsenic.

26.29.2.2.3 Apparatus and Equipment.   *Arsine generator, scrubber, and absorption tube*: Apparatus is shown in Figure 26.16. Use a 200 mL three-necked flask with a side arm (19/22 or similar size female ground-glass joint) through which the inert gas delivery tube reaching almost to the bottom of the flask is inserted; a 24/40 female ground-glass joint to carry the scrubber; and a second side arm closed with a rubber septum, or preferably by a screw cap with a hole in its top for insertion of a TFE-faced silicon septum. Place a small magnetic stirring bar in the flask. Fit absorber tube (20 mL capacity) to the scrubber and fill with silver diethyldithiocarbamate solution. Do not use rubber or cork stoppers because they may absorb arsine. Clean glass equipment with concentrated nitric acid.

Fume hood: Use apparatus in a well-ventilated hood with flask secured on top of a magnetic stirrer.

Spectrophotometer: For use at 520 nm with current 1 cm path.

26.29.2.2.4 Reagents and Standards.   Use reagents and distilled water free from arsenic contamination:

1. Acetate buffer: pH 5.5. Mix 428 mL 0.2 M sodium acetate, $NaC_2H_3O_2$ and 72 mL of 0.2 mL/L acetic acid, CH_3COOH.

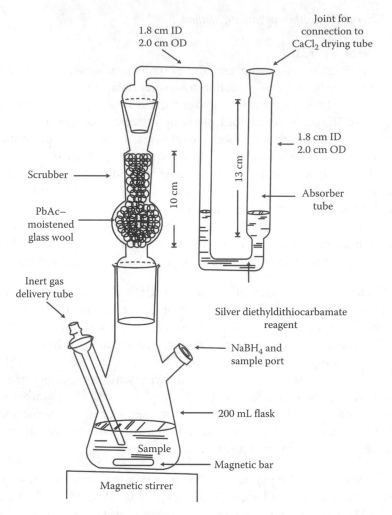

Figure 26.16 Arsenic generation.

2. Sodium acetate (0.2 M): Dissolve 16.46 g anhydrous sodium acetate or 27.36 g sodium acetate trihydrate, $NaC_2H_3O_2 \cdot 3H_2O$ in water. Dilute to 1000 mL with water.

3. Acetic acid (0.2 M): Dissolve 11.5 mL glacial acetic acid in water. Dilute to 1000 mL.

4. Sodium borohydride solution (1%): Dissolve 0.4 g sodium hydroxide, NaOH (4 pellets), in 400 mL water. Add 4.0 g sodium borohydride, $NaBH_4$ (check for absence of arsenic). Shake to dissolve and to mix. Prepare fresh every few days.

5. Hydrochloric acid (HCl) (2 M): Dilute 165 mL conc. HCl to 1000 mL with water.

6. Lead acetate solution: Dissolve 10.0 g Pb $(CH_3COO)_2$ $3H_2O$ in 100 mL water.

7. Silver diethyldithiocarbamate solution: Dissolve 1.0 mL morpholine (caution: corrosive—avoid contact with skin) in 70 mL chloroform, $CHCl_3$. Add 0.30 g silver diethyldithiocarbamate, $AgSCSN(C_2H_5)_2$ and shake in a stoppered flask until most is dissolved. Dilute to 100 mL with chloroform. Filter and store in a tightly closed brown bottle in a refrigerator.

8. Stock arsenite solution: Dissolve 0.1734 g $NaAsO_2$ in water and dilute to 1000 mL with water; 1.00 mL = 100 μg As.

Caution: Toxic—avoid contact with skin and do not ingest.

9. Intermediate arsenite solution: Dilute 10.0 mL stock solution to 100 mL with water; 1.00 mL = 10.0 μg As.

26.29.2.2.5 Standard Arsenite Solution. Dilute 10.0 mL intermediate solution to 100 mL with water; 1.00 mL = 1.00 μg As (9). Standard arsenate solution: Dissolve 0.416 g $Na_2HAsO_4 \cdot 7H_2O$ in water and dilute to 1000 mL. Dilute 10.0–100 mL with water; dilute 10 mL of this intermediate solution to 100 mL; 1.00 mL = 1.00 μg As.

26.29.2.2.6 Protocol. *Arsenite*: Preparation of scrubber and absorber. Dip glass wool into lead acetate solution and remove excess by squeezing glass wool. Press glass wool between pieces of filter paper, and then fluff it. Alternatively, if cotton is used, treat it similarly but dry in a desiccator and fluff thoroughly when dry. Place a plug of loose glass wool or cotton in scrubber tube. Add 4 mL silver diethyldithiocarbamate solution to absorber tube (5 mL may be used to provide enough volume to rinse spectrophotometer cell).

Loading of arsine generator: Pipette not more than 70 mL sample containing not more than 20 μg As (arsenite) into the generator flask. Add 10 mL acetate buffer. If necessary, adjust total volume of liquid to 80 mL. Flush flask with nitrogen at the rate of 60 mL/min.

Arsine generation and measurement: While nitrogen is passing through the system, use a 30 mL syringe to inject through the septum 15 mL 1% sodium borohydride solution within 2 min. Stir vigorously with magnetic stirrer. Pass nitrogen through system for an additional 15 min to flush arsine into absorber solution. Pour absorber solution into a clean and dry spectrophotometric cell and measure absorbance at 520 nm against chloroform. Determine concentration from a calibration curve obtained with arsenite standards. If arsenate also is to be determined in the same sample portion, save liquid in the generator flask.

Preparation of standard curves: Treat standard arsenite solution containing 0.0, 1.0, 2.0, 5.0, 10.0, and 20.0 μg as described earlier. Plot absorbance versus micrograms arsenic in the standard.

Arsenate: After removal of arsenite as arsine, treat sample to convert arsenate to arsine. If the lead acetate–impregnated glass wool has become ineffective in removing hydrogen sulfide (if it has become gray to black), replace glass wool. Pass nitrogen through system at the rate of 60 mL/min. Cautiously add 10 mL 2 N HCl. Generate arsine with all precautions and prepare standard curves with standard solutions of arsenate.

Total inorganic arsenic: Prepare scrubber and absorber for arsine. Load arsine generator using 10 mL of 2 N HCl instead of acetate buffer. Generate arsine and measure the absorbance as directed earlier. Prepare standard curves. Curves obtained with standard arsenite are almost identical to those obtained with arsenate standard solutions. Therefore, use either arsenite or arsenate standards.

26.29.2.2.7 Calculation.   Calculate arsenite, arsenate, and total inorganic arsenic from readings and calibration curves obtained from standard

$$\text{mg As/L} = \frac{\text{mg As (from calibration curve)}}{\text{mL sample in generator flask}}$$

26.29.2.3 ICP Method
Described in Section 26.26.1.

26.30 Measurement of Cadmium in Soil/Water Sample

26.30.1 Scope and Application

Cadmium occurs in sulfide minerals that also contain zinc, lead, or copper. The metal is used in electroplating, batteries, paint pigments and in alloys with various other metals. Cadmium is usually associated with zinc. Cadmium is highly toxic and has been implicated in some cases of poisoning through food. Minute quantities of cadmium are suspected of being responsible for adverse changes in arteries of human kidneys. Cadmium also causes generalized cancers in laboratory animals and has been linked epidemiologically with certain human cancers. A cadmium concentration of 200 µg/L is toxic to certain fishes. Cadmium may enter water as a result of industrial discharges or the deterioration of galvanized pipe. The FAO-recommended maximum level for cadmium for irrigation water is 10 µg/L. USEPA drinking water standard for cadmium is 0.005 mg/L. BIS desirable limit is 1 mg/L.

26.30.2 Methods for Analysis

1. ICP method/AAS
2. Dithizone method

26.30.2.1 ICP Method
Described in Section 26.26.1.

26.30.2.2 Dithizone Method
26.30.2.2.1 Principle.   Cadmium ions under suitable conditions react with dithizone to form a pink to red color that can be extracted with chloroform ($CHCl_3$). Chloroform extracts are measured photometrically and the cadmium concentration is obtained from a calibration curve prepared from a standard cadmium solution treated in the same manner as the sample.

Interference: Under the specified conditions, concentrations of metal ions normally found in water do not interfere. Lead up to 6 mg, zinc up to 3 mg, and copper up to 1 mg in the portion analyzed do not interfere. Ordinary room lighting (incandescent or fluorescent) does not affect the cadmium dithizonate color. Minimum detectable concentration: 0.5 µg/L Cd with 1 cm light path.

26.30.2.2.2 Apparatus

1. *Spectrophotometer*: Use at 518 nm with minimum light path of 1 cm.
2. *Separatory funnels*: 125 mL, preferably with TFE stopcocks.
3. *Glassware*: Clean all glassware, including sample bottles, with (1 + 1) HCl and rinse thoroughly with tap water and distilled water.

26.30.2.2.3 Reagents and Standards.   Use reagents and distilled water free from cadmium contamination:

1. Stock cadmium solution: Weigh 100 mg pure Cd metal and dissolve in a solution composed of 20 mL water + 5 mL conc. HCl. Use heat to assist metal dissolution. Transfer quantitatively to a 1 L volumetric flask and dilute to 1000 mL; 1.00 mL = 100 µg Cd. Store in a polyethylene container.
2. Standard cadmium solution: Pipette 1 mL stock cadmium solution into a 100 mL volumetric flask, add 1 mL conc. HCl, and dilute to 100 mL with water. Prepare as required and use on the same day; 1 mL = 1 µg Cd.
3. Sodium potassium tartrate solution: Dissolve 250 g, $NaKC_4H_4O_6 \cdot 4H_2O$ in water and make up to 1 L.
4. Sodium hydroxide–potassium cyanide solutions:
   Solution I: Dissolve 400 g NaOH and 10 g KCN in water and make up to 1 L. Store in a polyethylene bottle. This solution is stable for 1 month.
   Solution II: Dissolve 400 g NaOH and 0.5 g KCN in water and make up to 1 L. Store in a polyethylene bottle. This solution is stable for 1–2 months.
   Caution: Potassium cyanide is extremely poisonous. Be especially cautious when handling it. Never use mouth pipettes to deliver cyanide solutions.
5. Hydroxylamine hydrochloride solution: Dissolve 20 g, NH_2OH HCl in water and make up to 100 mL.
6. Stock dithizone solution: In a fume hood, dissolve 100 mg dithizone in 50 mL $CHCl_3$ in a 150 mL beaker and filter through a 7 cm diameter paper. Receive filtrate in a 500 mL separatory funnel. Wash the paper with three 5 mL portion of $CHCl_3$. Transfer with $CHCl_3$ to 500 mL separatory funnel. Add 100 mL (1 + 99) NH_4OH to a separatory funnel and shake moderately for 1 min. Let layers separate. Transfer $CHCl_3$ layer to a 250 mL separatory funnel retaining the orange red aqueous layer in the 500 mL funnel. Repeat extraction receiving $CHCl_3$ layer in another 250 mL separatory funnel and transferring aqueous layer using (1 + 99) NH_4OH to the 500 mL funnel holding the first extract. Repeat extraction transferring the aqueous layer to 500 mL funnel. Discard $CHCl_3$ layer. To combine extracts in the 500 mL separatory funnel add (1 + 1) HCl in 2 mL portions mixing after each addition, until dithizone precipitates and solution is no longer red. Extract precipitated dithizone with three 25 mL portions $CHCl_3$. Dilute combined extracts to 1000 mL with $CHCl_3$; 1 mL = 100 mg.

7. Working dithizone solution: Dilute stock dithizone solution with $CHCl_3$ to produce a working solution of 10 µg/mL. Prepare daily.
8. Chloroform: Test for a satisfactory $CHCl_3$ by adding a minute amount of dithizone to a portion of the $CHCl_3$ in a stoppered test tube so that a faint green color is produced; the green color should be stable for a day.
9. Tartaric acid solution: Dissolve 20 g, $H_2C_4H_4O_6$ in water and make up to 1 L. Store in refrigerator and use while still cold.
10. Hydrochloric acid (conc. HCl).
11. Thymol blue indicator solution: Dissolve 0.4 g thymolsulfonephthalein sodium salt in 100 mL water.
12. Sodium hydroxide (NaOH): 6 N.

26.30.2.2.4 Protocol. *Preparation of standard curve*: Prepare a blank and a series of standards by pipetting the appropriate amounts of standard Cd solution into separatory funnels. Dilute to 25 mL and pipette 1 mL stock cadmium solution into a 100 mL volumetric flask, add 1 mL conc. HCl, and dilute to 100 mL with water. Prepare as needed and use the same day; 1 mL = 1 µg Cd. Plot a calibration curve.

Treatment of samples: Pipette a volume of digested sample containing 1–10 µg Cd to a separatory funnel and dilute to 25 mL as necessary. Add three drops thymol blue and adjust with 6 N NaOH to the first permanent yellow color, pH 2.8.

Color development, extraction, and measurement: Add reagents in the following order, mixing after each addition: 1 mL sodium potassium tartrate solution, 5 mL NaOH–KCN solution, 1 mL NH_2OH HCl solution, and 15 mL stock dithizone solution. Stopper the funnels and shake for 1 min, relieving vapor pressure in the funnels through the stopper rather than the stopcock. Drain $CHCl_3$ layer into a second funnel containing 25 mL cold tartaric acid solution. Add 10 mL $CHCl_3$ to first funnel, shake for 1 min, and drain into second funnel. Do not permit aqueous layer to enter second funnel. Because time of contact of $CHCl_3$ with the strong alkali must be kept to a minimum, make the two extractions immediately after adding dithizone (cadmium dithizonate decomposes on prolonged contact with strong alkali saturated with $CHCl_3$).

Shake second funnel for 2 min and discard $CHCl_3$ layer. Add 5 mL $CHCl_3$, shake 1 min, and discard $CHCl_3$ layer, making as close a separation as possible. In the following order, add 0.25 mL NH_2OH HCl solution and 15.0 mL working dithizone solution. Add 5 mL NaOH–KCN solution II, immediately shake for 1 min, and transfer $CHCl_3$ layer into a dry cuvette. Read absorbance at 518 nm against the blank. Obtain Cd concentration from the calibration curve.

26.30.2.2.5 Calculation

$$\text{mg Cd/L} = \frac{\text{µg Cd (in } \sim 15 \text{ mL final volume)}}{\text{mL sample}}$$

26.31 Measurement of Mercury in Soil/Water Sample

26.31.1 Introduction

Mercury is a liquid metal found in natural deposits as ores containing other elements. Electrical products such as dry-cell batteries, fluorescent light bulbs, switches, and other control equipment account for 50% of mercury used. Large amounts of mercury are released naturally from the earth's crust. Combustion of fossil fuels, metal smelters, cement manufacture, municipal landfills, sewage, metal refining operations, from chloralkali plants are important sources of mercury release. Mercury is unique among metals in that it can evaporate when released to water or soil. Mercury is toxic to microorganisms and to survive in its presence, they transform Hg^0 into methylated compounds such as CH_3Hg (methyl mercury) and CH_3HgCH_3 (dimethyl mercury). These methylated forms of Hg are volatile and harmless to microorganisms since the Hg concentration decreases. Also, microbes can convert inorganic forms of mercury to organic forms, which can be accumulated by aquatic life (Figure 26.17).

26.31.2 Scope and Application

Naturally occurring levels of mercury in groundwater and surface water are <0.5 μg/L, although local mineral deposits may produce higher levels in groundwater. Mercury detection in drinking water is predominantly in the form of inorganic mercury. The principal target organ of inorganic mercury is kidney with neurological and renal disturbances. Methyl mercury compounds are very toxic to the central nervous system; they are also the major source of environmental contamination. The average daily intake of mercury from food is in the range 2–20 μg, but may be much higher in regions where ambient waters have become contaminated with mercury and where

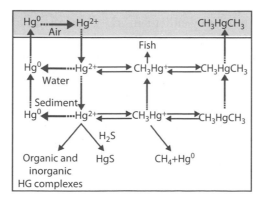

Figure 26.17 Mercury cycling in nature. The major reservoirs of mercury are in water and in sediments where it can be concentrated in animal tissue or precipitated out as Hg.

fish constitute a high proportion of the diet. On the assumption of an ambient air level of 10 ng/m^3, the average daily intake of inorganic mercury by inhalation would amount to about 0.2 µg. If a level in drinking water of 0.5 µg/L is assumed, the average daily intake of inorganic mercury from this source would amount to about 1 µg. The average daily intake of mercury from food is in the range 2–20 µg. Mercury in drinking water is considered to be a minor source of exposure to mercury except in circumstances of significant pollution. Standard methods of estimation of Hg advise to prepare samples by treatment with HNO_3 to reduce the pH to <2. Atomic absorption spectroscopy is the method of choice, with the "dithizone method" as the method to be used in potable water for high levels of Hg.

26.31.3 Methods for Analysis

26.31.3.1 Sampling
A 100–2000 mL sample is collected directly into a cleaned, pretested fluoropolymer or glass bottle using sample handling techniques designed for collection of mercury at trace levels. For dissolved Hg, the sample is filtered through a 0.45 µm capsule filter prior to preservation. Samples are collected into rigorously cleaned fluoropolymer bottles with fluoropolymer or fluoropolymer-lined caps. Glass bottles may be used if Hg is the only target analyte. It is critical that the bottles have tightly sealing caps to avoid diffusion of atmospheric Hg through the threads. Polyethylene sample bottles must not be used. Before samples are collected, consideration should be given to the type of data required (i.e., dissolved or total), so that appropriate preservation and pretreatment steps can be taken. An excess of BrCl should be confirmed either visually (presence of a yellow color) or with starch iodide indicating paper, using a separate sample aliquot, prior to sample processing or direct analysis to ensure the sample has been properly preserved.

26.31.3.2 Preservation
Samples are preserved by adding either 5 mL/L of pretested 12 N HCl or 5 mL/L BrCl solution to the sample bottle. If a sample is also being used for the determination of methyl mercury, it should be collected and preserved according to procedures in the method that will be used for determination of methyl mercury (e.g., HCl or H_2SO_4 solution). Preserved samples are stable for up to 90 days of the date of collection.

Samples that are acid-preserved may lose Hg to coagulated organic materials in the water or condensed on the walls. The best approach is to add BrCl directly to the sample bottle at least 24 h before analysis. If other Hg species are to be analyzed, these aliquots must be removed prior to the addition of BrCl. If BrCl cannot be added directly to the sample bottle, the bottle must be shaken vigorously prior to subsampling.

26.31.3.3 Storage
Sample bottles should be stored in clean (new) polyethylene bags until sample analysis.

26.31.3.4 Analytical Methods and Analytical Achievability

Limits of detection for inorganic mercury are 0.6 µg/L by ICP and 5 µg/L by flame AAS. The dithizone method can be used to detect mercury in a sample with >1 µg of mercury per 10 mL final volume. Alternatively, cold vapor AAS can also be used for the detection of mercury, with a detection limit of 0.05 µg/L (APHA et al. 1999). Multicomponent simultaneity analysis is possible with metals by AAS and ICP.

26.31.3.4.1 Hg Purging System.
Figure 26.18 shows the schematic diagram for the purging system. The system consists of the following:

Flowmeter/needle valve with a flow rate to the purge vessel at 350 ± 50 mL/min. Fluoropolymer fittings—connections between components requiring mobility are made with 3.2 mm OD fluoropolymer tubing because of its greater flexibility. Acid fume pretrap 10 cm long × 0.9 cm ID fluoropolymer tube containing 2–3 g of reagent grade, nonindicating, 8–14 mesh soda lime chunks, packed between wads of silanized glass wool. This trap is cleaned of Hg by placing on the output of a clean cold vapor generator (bubbler) and purging for 1 h with N_2 at 350 mL/min. Cold vapor generator (bubbler) 200 mL borosilicate glass (15 cm high × 5.0 cm diameter) with standard taper 24/40 neck, fitted with a sparging stopper having a coarse glass frit that extends to within 0.2 cm of the bubbler bottom.

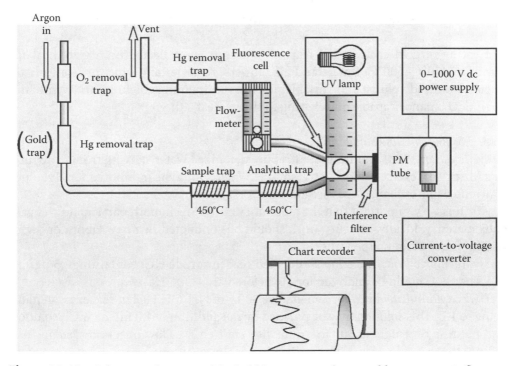

Figure 26.18 Schematic diagram of the bubbler, purge and trap, cold vapor atomic fluorescence spectrometer (CVAFS) system.

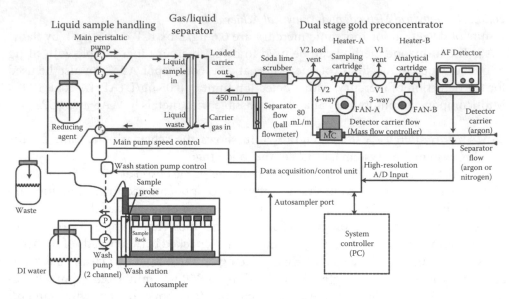

Figure 26.19 Schematic diagram of the flow injection, cold vapor atomic fluorescence spectrometer (CVAFS) system.

26.31.3.4.2 Dual Trap Hg⁰ Preconcentrating System. The system consists of gold-coated sand traps—10 cm long × 6.5 mm OD × 4 mm ID quartz tubing. The tube is filled with 3.4 cm of gold-coated 45/60 mesh quartz sand. The ends are plugged with quartz wool (Figure 26.19).

Heating of gold-coated sand traps—To desorb Hg collected on a trap, heat for 3 min to 450°C–500°C, timers and air blowers, recorder and pipettors—all plastic pneumatic fixed volume and variable pipettors in the range of 10 μL to 5.0 mL and analytical balance capable of weighing to the nearest 0.01 g.

26.31.3.5 Reagents and Standards

Reagent water: 18 MΩ minimum, ultrapure deionized water starting from a prepurified (distilled, reverse osmosis, etc.) source. Water should be monitored for Hg, especially after ion-exchange beds are changed.

Air: It is very important that the laboratory air be low in both particulate and gaseous mercury. Ideally, mercury work should be conducted in a new laboratory with mercury-free paint on the walls.

Hydrochloric acid: Trace metal–purified reagent-grade HCl containing <5 pg/mL Hg. The HCl should be analyzed for Hg before use.

Hydroxylamine hydrochloride: Dissolve 300 g of NH_2OH HCl in reagent water and bring to 1 L. This solution may be purified by the addition of 1.0 mL of $SnCl_2$ solution and purging overnight at 500 mL/min with Hg-free N_2. Flow injection systems may require the use of less $SnCl_2$ for purification of this solution.

Stannous chloride: Bring 200 g of $SnCl_2 \cdot 2H_2O$ and 100 mL conc. HCl to 1 L with reagent water. Purge overnight with mercury-free N_2 at 500 mL/min to remove all traces of Hg. Store tightly capped.

Bromine monochloride (BrCl): In a fume hood, dissolve 27 g of reagent-grade KBr in 2.5 L of low Hg HCl. Place a clean magnetic stir bar in the bottle and stir for ~1 h in the fume hood. Slowly add 38 g reagent-grade $KBrO_3$ to the acid while stirring. When all of the $KBrO_3$ has been added, the solution color should change from yellow to red to orange. Loosely cap the bottle, and allow to stir another hour before tightening the lid.

Caution: This process generates copious quantities of free halogens (Cl_2, Br_2, BrCl), which are released from the bottle. Add the $KBrO_3$ slowly in a fume hood.

Stock mercury standard: 1000 ppm aqueous Hg solution.

Secondary Hg standard: Add ~0.5 L of reagent water and 5 mL of BrCl solution to a 1.00 L Class A volumetric flask. Add 0.100 mL of the stock mercury standard to the flask and dilute to 1.00 L with reagent water. This solution contains 1 μg/mL (1.00 ppm) Hg. Transfer the solution to a fluoropolymer bottle and cap tightly.

Working Hg Standard A: Dilute 1 mL of the secondary Hg standard to 100 mL in a Class A volumetric flask with reagent water containing 0.5% by volume BrCl solution. This solution contains 10 ng/mL and should be replaced monthly, or longer if extended stability is demonstrated.

Working Hg Standard B: Dilute 0.10 mL of the secondary Hg standard to 1000 mL in a Class A volumetric flask with reagent water containing 0.5% by volume BrCl solution. This solution contains 0.10 ng/mL and should be replaced monthly, or longer if extended stability is demonstrated.

Initial precision and recovery (IPR) and ongoing precision and recovery (OPR) solutions: Using the working Hg standard A, prepare IPR and OPR solutions at a concentration of 5 ng/L Hg in reagent water. IPR/OPR solutions are prepared using the same amounts of reagents used for preparation of the calibration standards.

Nitrogen: Grade 4.5 (standard laboratory-grade) nitrogen that has been further purified by the removal of Hg using a gold-coated sand trap.

Argon: Grade 5 (ultrahigh-purity, GC-grade) argon that has been further purified by the removal of Hg using a gold-coated sand trap.

26.31.3.6 Protocol

26.31.3.6.1 Sample Preparation. Pour a 100 mL aliquot from a thoroughly shaken, acidified sample into a 125 mL fluoropolymer bottle. If BrCl was not added as a preservative, add the amount of BrCl solution, cap the bottle, and digest at room temperature for a 12 h minimum. For every 10 or fewer samples, pour 2 additional 100 mL aliquots from a selected sample, and process in the same manner as the samples. There must be a minimum of 2 MS/MSD pairs for each analytical batch of 20 samples.

26.31.3.6.2 Hg Reduction and Purging for the Bubbler System. Add 0.2–0.25 mL of NH_2OH solution to the BrCl-oxidized sample in the 125 mL sample bottle. Cap the bottle and swirl the sample. The yellow color will disappear, indicating the destruction of the BrCl. Allow the sample to react for 5 min with periodic swirling to be sure that no traces of halogens remain.

Connect a fresh trap to the bubbler, pour the reduced sample into the bubbler, add 0.5 mL of $SnCl_2$ solution, and purge the sample onto a gold trap with N_2 at 350 ± 50 mL/min for 20 min. When analyzing Hg samples, the recovery is quantitative, and organic interferents are destroyed. Thus, standards, bubbler blanks, and small amounts of high-level samples may be run directly in previously purged water. After very high samples, a small degree of carryover (<0.01%) may occur. Bubblers that contain such samples must be demonstrated to be clean prior to proceeding with low-level samples.

26.31.3.6.3   Hg Reduction and Purging for the Flow Injection System.   Add 0.2–0.25 mL of NH_2OH solution to the BrCl-oxidized sample in the 125 mL sample bottle or in the autosampler tube (the amount of NH_2OH required will be ~30% of the BrCl volume). Cap the bottle and swirl the sample. The yellow color will disappear, indicating the destruction of the BrCl. Allow the sample to react for 5 min with periodic swirling to be sure that no traces of halogens remain. Pour the sample solution into an autosampler vial and place the vial in the rack. Carryover may occur after analysis of a sample containing a high level of mercury. Samples run immediately following a sample that has been determined to result in carryover must be reanalyzed using a system demonstrated to be clean.

26.31.3.6.4   Desorption of Hg from the Gold Trap.   Remove the sample trap from the bubbler, place the Nichrome wire coil around the trap, and connect the trap into the analyzer train between the incoming Hg-free argon and the second gold-coated (analytical) sand trap. Pass argon through the sample and analytical traps at a flow rate of ~30 mL/min for ~2 min to drive off condensed water vapor. Apply power to the coil around the sample trap for 3 min to thermally desorb the Hg (as Hg^0) from the sample trap onto the analytical trap. After the 3 min desorption time, turn off the power to the Nichrome coil and cool the sample trap using the cooling fan. Turn on the chart recorder or other data acquisition device to start data collection, and apply power to the Nichrome wire coil around the analytical trap. Heat the analytical trap for 3 min (1 min beyond the point at which the peak returns to baseline). Stop data collection, turn off the power to the Nichrome coil, and cool the analytical trap to room temperature using the cooling fan. Place the next sample trap in line and proceed with analysis of the next sample. Peaks generated using this technique should be very sharp and almost symmetrical. Mercury elutes at ~1 min and has a width at half-height of about 5 s. Broad or asymmetrical peaks indicate a problem with the desorption train, such as improper gas flow rate, water vapor on the trap(s), or an analytical trap damaged by chemical fumes or overheating. Damage to an analytical trap is also indicated by a sharp peak, followed by a small, broad peak. If the analytical trap has been damaged, the trap and the fluoropolymer tubing downstream from it should be discarded because of the possibility of gold migration onto downstream surfaces. Gold-coated sand traps should be tracked by unique identifiers so that any trap producing poor results can be quickly recognized and discarded.

26.31.4 Data Analysis and Calculations

Separate procedures are provided for calculation of sample results using the bubbler system and the flow injection system, and for method blanks.

26.31.4.1 Calculations for the Bubbler System

Calculate the mean peak height or area for Hg in the bubbler blanks measured during system calibration or with the analytical batch (ABB; $n = 3$ minimum).

Calculate the concentration of Hg in ng/L (parts per trillion [ppt]) in each sample according to the following equation:

$$[\text{Hg}] \ (\text{ng/L}) = \frac{As - ABB}{CFm \times V}$$

where

As is peak height (or area) for Hg in sample
ABB is peak height (or area) for Hg in bubbler blank
CFm is mean calibration factor
V is volume of sample (L)

26.31.4.2 Calculations for the Flow Injection System

Calculate the mean peak height or area for Hg in the system blanks measured during system calibration or with each analytical batch (ASB; $n = 3$).

Calculate the concentration of Hg in ng/L in each sample according to the following equation:

$$[\text{Hg}] \ (\text{ng/L}) = \frac{As - ASB}{CFm \times V} \times \frac{V_{std}}{V_{sample}}.$$

where

As is peak height (or area) for Hg in sample
ASB is peak height (or area) for Hg in system blank
CFm is mean calibration factor
V_{std} is volume (mL) of used for standard sample – volume (mL) reagent used in standards
V_{sample} is volume (mL) of sample – volume (mL) reagent used in sample

Calculate the concentration of Hg in the method blanks (CMB), field blanks (CFB), or reagent blanks (CRB) in ng/L, using the equation and substituting the peak height or area resulting from the method blank, field blank, or reagent blank for *As*.

Determine the mean concentration of Hg in the method blanks associated with the analytical batch (a minimum of three). If a sample requires additional reagent(s) (e.g., BrCl), a corresponding method blank containing an identical amount of reagent must be analyzed. The concentration of Hg in the corresponding method blank may be subtracted from the concentration of Hg in the sample.

26.32 Study on Biogenic Methane Production in Different Habitats (Biogas)

Biogas is the gas produced by anaerobic microorganisms, primarily methane. It is also known as swamp gas, marsh gas, landfill gas, or digester gas produced through methanogenesis (biological production of methane) by a group of strictly anaerobic archaea called methanogens. It is an extremely important fuel source used in the generation of mechanical, electrical, and heat energy. It can be used as a source of fuel for domestic consumption and industry and can also be converted by microbial action or chemical means to methanol, which can be used as a fuel in internal combustion engines. However, methanogenesis can be an ecological problem as methane absorbs infrared radiation and thus is a greenhouse gas and its overproduction in nature may significantly promote future global warming.

In the biogas digesters, the biodegradable feedstock is converted into two useful products: gas and digestate. The digestate comprises manure and the remnants of the anaerobic microorganisms. One of the most promising way of methane production is through the bioconversion of waste material where methane is a normal by-product. The production of methane is carried out by a mixed microbial community. Of these, some microbes break the organic molecules and complex polymers and fatty acids into hydrogen, carbon dioxide, and alcohols (Figure 26.20).

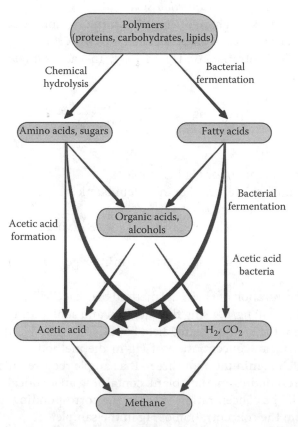

Figure 26.20   Processes used to generate methane as a fuel from polymeric waste material.

Methanogens utilize these fermentation products for methane production. Methanogenesis is commonly carried out by microbial populations involved in synergistic and mutualistic relationships. The evolution of methane from anaerobic digesters and other bioconversion processes occurs simultaneously with the evolution of carbon dioxide. The ratio of two gases, that is, methane and carbon dioxide, depends on the chemical composition of the substrate and the environmental conditions under which the bioconversion is carried out. Biogas is generally composed of 50%–75% methane, 25%–50% CO_2, 0%–10% nitrogen, 0%–1% hydrogen, 0%–3% H_2S, and 2% oxygen.

Two types of biogas plants are used for methane production: (a) floating-type biogas plant (Figure 26.21) and (b) fixed dome-type biogas plant (Figure 26.22). Biogas plant is mainly based on the design of the digester, an airtight tank in which anaerobic decomposition is carried out. It is well-shaped tank inside the ground, made up of concrete bricks and cement. There lies a cylindrical container above the digester to collect the gas, which is made up of steel. Sometimes dome of the digester acts as a storage tank for gas. Gas tank is connected with pipe having a valve, from which gas is taken out and used for cooking and other purposes. The slurry so prepared is fed in the digester through the inlet pipe and left for fermentation and the spent slurry is collected from the outlet pipe.

Several microorganisms are involved in anaerobic digestion and biogas production. The complex organic material is decomposed by hydrolysis, acidification, and methanogenesis. The biological production of methane occurs through a series of reactions involving several enzymes. During methanogenesis, electrons for CO_2 reduction generally come from H_2 but in certain methanogens few organic compounds such as alcohols can be oxidized to yield electrons for CO_2 reduction. Maximum biogas production results when the pH is between 6 and 7, temperature 35°C, C/N ratio is between 20 and 30, and solid water ratio is 1:1.

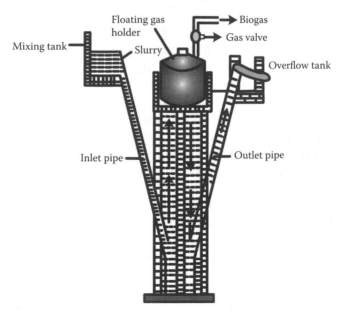

Figure 26.21   Floating gas holder-type biogas plant.

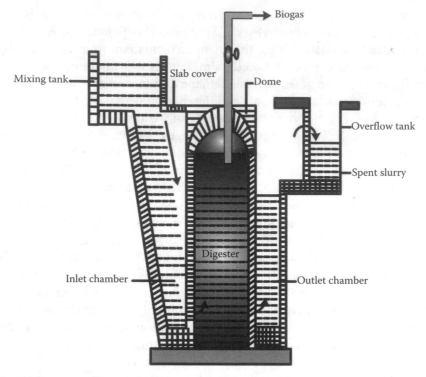

Figure 26.22   Fixed dome-type biogas plant.

26.32.1 Principle

Under anaerobic conditions, the organic materials are converted through microbiological reactions into gases (fuel) and organic fertilizer (sludge). The mixture of gases is composed of 63% methane, 30% CO_2, 4% nitrogen, and 1% hydrogen sulfide and traces of hydrogen, oxygen, and carbon monoxide. Methane is the main constituent of biogas. It is also referred to as biofuels, sewerage gas, Klar gas, sludge gas, will-o'-the wisp of marsh lands, fool's fire, gobar gas (cow dung gas), bioenergy, and fuel of the future. About 90% of energy of substrate is retained in methane. It is used in internal combustion engines to power water pumps and electric generators. It is also used as fuel in fuel-type refrigerators. Sludge is used as fertilizer. The most economical benefits are minimizing environmental pollution and meeting the demand of energy for various purposes. It is devoid of smell and burns with a blue flame without smoke (Figure 26.23).

26.32.2 Experimental Outline

In a biogas plant, fresh cattle dung, organic animal and agriculture waste, municipal sewage remain in the tank for several days where they undergo solubilization. Acidogenesis and methanogenesis result in CH_4 production. Methanogens are a unique group of bacteria playing a major role in the breakdown of substrate into gas form (Figure 26.24a and b).

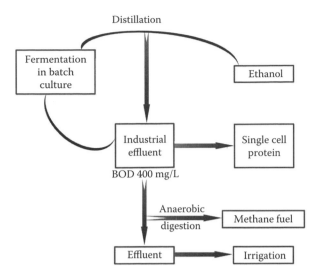

Figure 26.23 Production of single cell protein, ethanol, and biogas from the organic industrial waste.

26.32.3 Materials

Airtight cylindrical tank, feedstock (agriculture waste), water, microbes.

26.32.4 Protocol

The anaerobic digestion is a partial conversion of microorganisms of organic substrates into gases in the absence of air. The gas produced is collectively known as biogas (Figure 26.24a and b):

1. Feedstock is mixed with water, which is taken in an airtight cylindrical tank known as digester.
2. A digester is made up of concrete bricks and cement or steel. It has a side opening for adding organic materials for digestion. A cylindrical container above the digester is also maintained to collect the gas.
3. Fresh cattle dung is added into a charge pit leading to the digestion tank and remains in the tank for 50 days. Sufficient amount of gas is accumulated in gas tank in 50 days, which is used for household purposes. Digested sludge is removed from the base and is used as fertilizer.

26.32.5 Process

Anaerobic digestion is accomplished in three stages:

Solubilization: It is the initial stage, where feedstock is solubilized by water and microbial enzymes. The feedstock is suspended in water to make a slurry.

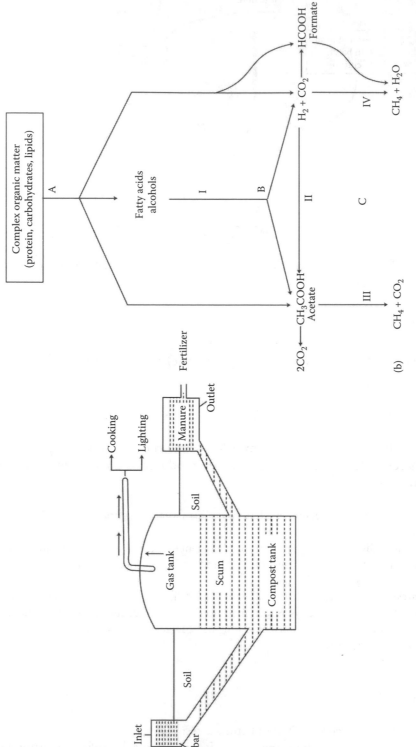

Figure 26.24 (a) Illustration of cattle dung gas plant. (b) Anaerobic digestion of organic matter and production of methane. A, Hydrolytic and fermentative bacteria; B, acidogenic bacteria (I, acetogenic dehydrogenation by proton-reducing acetogenic bacteria; II, acetogenic dehydrogenation by acetogenic bacteria); and C, methanogenesis by acetoclastic methanogens (i.e., acetate respiratory bacteria) (III; and hydrogen-oxidizing methanogens IV).

Acidogenesis: Facultative anaerobic bacteria attack the feedstock; H_2-producing acidogenic bacteria convert the sample organic material via oxidation/reduction reactions into acetate, H_2, and CO_2 and serve finally as food for the final stage.

Methanogenesis: This is the final stage of anaerobic digestion where acetate and $H_2 + CO_2$ are converted by CH_4-producing bacteria into CH_4, CO_2, H_2O, and other products.

26.32.5.1 Methanogens

Methanogens are a unique group of bacteria. They are obligate anaerobes and have slow growth rate. They play a major role in breakdown of substrate into gas form. In morphology, they are of various types such as cocci, bacilli, spirilli, and sarcinac.

All bacteria require H_2 and formate for growth and CH_4 production whereas *Methanosarcina barkeri* requires methanol, methylamine, and acetate for their growth.

26.32.5.2 Mechanism of CH_4 Formation

A detailed description and pathway of CH_4 formation are given as reactions:

$$CO + H_2O \rightarrow CO_2 + H_2$$

$$CO_2 + 4H_2 \rightarrow CH_4 + 2H_2Os$$

The preceding secondary reaction takes place in the presence of sufficient H_2.
Other reactions showing CH_4 formation from various substrates are given as follows:

$$(\text{Methanol}) \; 4CH_3OH \rightarrow 3CH_4 + CO_2 + 2H_2O$$

$$(\text{Formate}) \; 4HCOOH \rightarrow CH_4 + 3CO_2 + 2H_2O$$

$$(\text{Acetate}) \; CH_3CHOOH \rightarrow 12CH_4 + 12CO_2$$

26.32.5.3 Production from Different Feedstocks

Salvinia: This fern can serve as a good feedstock material for biogas production. Fermentation of *Salvinia* starts within 7–9 days on putting under water in a suitable container. Biogas yield is about 0.1–1/kg fresh weight for 4 weeks. Air-dried weed produces higher amount for 90 days and fermentation continues for 3 months. Thereafter, gas yield gradually declines.

Water hyacinth: This weed can be used for the production of gas for laboratory and domestic fuel. The decomposition rate is higher than cow dung for gas production. Water hyacinth is totally decomposed within 3 days in summer while cow dung takes 8 days. The ratios between total gas evolved by water hyacinth and cow dung under the identical conditions in summer and winter are 5:3 and 5:1, respectively.

Municipal sewage: Techniques have been evolved to make available sullage gas for cooking purpose or for industrial activities.

Suggested Readings

American Public Health Association (APHA). 1999. Standard methods for the examination of water and wastewater, American Water Works Association. Washington, DC: Water Environment Federation.

Anna, I. K., ed. 2011. *Microbial Bioremediation of Non-Metals: Current Research.* Norfolk, VA: Caister Academic Press.

Baker, M. B., Jr., P. F. Ffolliott, L. F. Debano, and D. G. Neary. 2004. *Riparian Areas of the Southwestern United States, Hydrology, Ecology, and Management.* Boca Raton, FL: Lewis Publishers, CRC Press.

Garte, S. J., ed. 1994. *Molecular Environmental Biology.* Lewis Publishers, Boca Raton, FL: CRC Press.

Nierenberg, W., ed. 2012. *Encyclopedia of Environmental Biology*, Vols. 1–3. London, U.K.: Elsevier.

Palmer, J. R. and J. N. Reeve. Structure and function of methanogen genes. In: *The Biochemistry of Archaea.* Kates, M., D. J. Kushner, A. T. Matheson, eds. Amsterdam, The Netherlands: Elsevier, 1993, pp. 497–534.

Technical Assistance Hydrology Project, Government of India & Government of the Netherlands. 1999. *Laboratory Manual on Standard Analytical Procedures for Water Analysis.* New Delhi, India.

Thauer, R. K. 1998. Biochemistry of methanogenesis: A tribute to Marjory Stephenson. *Microbiology* 144: 2377–2406.

Important Links

http://www.h2ou.com/h2wtrqual.htm
http://www.health.gov.sk.ca/water-testing-common-questions
http://www.articlesbase.com/non-fiction-articles/the-importance-of-water-testing-225983.html
http://www.dec.ny.gov/about/828.html
http://www.weatheroffice.gc.ca/analysis/index_e.html
http://water.epa.gov/lawsregs/lawsguidance/cwa/tmdl/airdeposition_index.cfm
http://www3.ag.purdue.edu/counties/marion/Pages/SoilSamplingTesting.aspx
http://www.ncbi.nlm.nih.gov/pmc/articles/PMC544252/
http://www.iisc.ernet.in/currsci/jul10/articles13.htm

27

Bioinformatics

27.1 Introduction

Bioinformatics derives knowledge from computer analysis of biological data, which deals with the study of methods of storing, retrieving, and analyzing biological data, such as nucleic acid (DNA/RNA) and protein sequence, structure, function, pathways, and genetic interactions. These can consist of the information stored in the genetic code, but also experimental results from various sources, patient statistics, and scientific literature. It generates new knowledge that is useful in such fields as drug design and development of new software tools to create that knowledge. Research in bioinformatics includes method development for storage, retrieval, and analysis of the data. It is a rapidly developing branch of biology and is highly interdisciplinary, using techniques and concepts from informatics, statistics, web technologies, artificial intelligence and soft computing, information and computation theory, structural biology, software engineering, data mining, image processing, modeling and simulation, discrete mathematics, control and system theory, circuit theory, and statistics, mathematics, chemistry, biochemistry, physics, and linguistics. It has many practical applications in different areas of biology and medicine. Commonly used software tools and technologies in this field include Java, XML, Perl, C, C++, Python, R, MySQL, SQL, CUDA, MATLAB®, and Microsoft Excel.

27.2 Development

The history of computing in biology goes back to the 1920s when scientists already thought about establishing biological laws solely from data analysis by induction (Lotka 1925). However, only the development of powerful computers and the availability of experimental data that can be readily treated by computation (e.g., DNA or amino acid sequences and three-dimensional structures of proteins) paved the way for the establishment of bioinformatics as an independent field today. The practical applications of bioinformatics are readily available through the World Wide Web, and are widely used in biological and medical research. As the field is rapidly evolving, the very definition of bioinformatics is still a matter of some debate.

The relationship between computer science and biology is a natural one for several reasons. First, the phenomenal rate of generation of biological data being produced provides challenges: massive amount of data has to be stored, analyzed, and made accessible. Second, the nature of the data is often such that a statistical method, and hence computation, is necessary. This applies in particular to the information on the building plans of proteins and of the temporal and spatial organization of their expression in the cell encoded by the DNA. Third, there is a strong analogy between the DNA sequence and a computer program (it can be shown that the DNA represents a Turing Machine).

Analyses in bioinformatics focus on three types of datasets: genome sequences, macromolecular structures, and functional genomics experiments (e.g., expression data, yeast two-hybrid screens), but bioinformatics analysis is also applied to various other data, for example, taxonomy trees, relationship data from metabolic pathways, the text of scientific papers, and patient statistics. Common activities in bioinformatics include mapping and analyzing DNA and protein sequences, aligning different DNA and protein sequences to compare them, and creating and viewing 3D models of protein structures. There are two fundamental ways of modeling a biological system (e.g., living cell) both coming under bioinformatic approaches (Figure 27.1).

Bioinformatics has a large impact on biological research. Giant research projects such as the human genome project would have been meaningless without the bioinformatics component. The goal of sequencing projects, for example, is not to corroborate or refute a hypothesis, but to provide raw data for later analysis. Once the raw data are available, hypotheses may be formulated and tested in silica. In this manner, computer experiments may answer biological questions, which cannot be tackled by traditional approaches. This has led to the foundation of dedicated bioinformatics research groups as well as to a different work practice in the average bioscience laboratory where the computer has become an essential research tool.

Three key areas are the organization of biological knowledge in databases, sequence data analysis, and structural bioinformatics.

27.2.1 Organization of Biological Knowledge in Databases

Biological raw data are stored in public databanks (such as Gene bank or EMBL for primary DNA sequences). The data can be submitted and accessed via the World Wide Web. Protein sequence databanks like trEMBL provide the most likely translation of all coding sequences in the EMIR databank. Sequence data are prominent, but also other data are stored, for example, yeast two-hybrid screens, expression arrays, systematic gene knockout experiments, and metabolic pathways.

The stored data need to be accessed in a meaningful way, and often contents of several databanks or databases have to be accessed simultaneously and correlated with each other. Special languages have been developed to facilitate this task (such as the Sequence Retrieval System [SRS] and the Entrez System). An unsolved problem is the optimal design of interoperating database systems. Database provides additional functionality such as access to sequence homology searches and links to other databases and analysis results. For example, SWISSPROT contains verified protein sequences and more annotations describing the function of a protein. Protein 3D structures are stored in specific databases (e.g., the Protein Data Bank now primarily created and developed by the Research Organism–specific databases have been

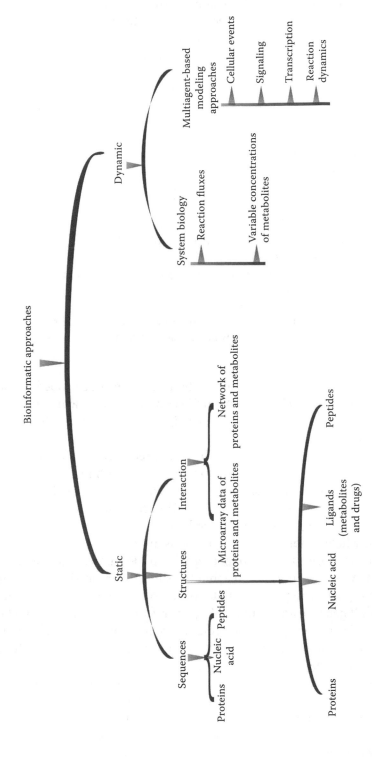

Figure 27.1 Details of bioinformatic approaches.

developed). Also, databases of scientific literature (such as PUBMED, MEDLINE) provide additional functionality, for example, they can search for similar articles based on word-usage analysis. Text recognition systems are being developed that extract automatically knowledge about protein function from the abstracts of scientific articles, notably on protein–protein interactions.

27.2.2 Sequence Data Analysis

The primary data of sequencing projects are DNA sequences. They are really valuable only through their annotation. Several layers of analysis with bioinformatic tools are necessary to arrive from a raw DNA sequence to an annotated protein sequence. For such analysis, one should follow the following guidelines:

1. Establish the correct order of contig sequences to obtain continuous sequence. A contig (from *contiguous*) is a set of overlapping DNA segments that together represent a consensus region of DNA.
2. Find the translation and transcription initiation sites, promoter sites and define open reading frames (ORF).
3. Find splice sites, introns, and exons.
4. Translate the DNA sequence into a protein sequence, searching all six frames.
5. Compare the DNA sequence to know protein sequences in order to verify exons etc., with homologous sequences.

Some completely automated annotation systems have been developed. GeneQuiz is such a system created in 1993 for automated genome analysis, which integrated different methods, a set of databases daily updated, and a user-friendly display using the World Wide Web.

The proteins sequences are further analyzed to predict function. The function can often be inferred if a sequence of a homologous protein with known function can be found. Homology searches are the predominant bioinformatics application, and very efficient search methods have been developed. The often difficult distinction between orthologous and paralogous sequences facilitates the functional annotation in the comparison of whole genomes. Several methods are known to detect glycosylation, myristoylation, and other sites to predict signal peptides in the amino acid sequence, which gives the valuable information about the subcellular location of a protein.

The ultimate goal of sequence annotation is to arrive at a complete functional description of all genes of an organism. However, function is an ill-defined concept. The simplified idea of "one gene–one protein" and/or "one structure–one function" cannot be taken into account as proteins have multiple functions depending on different contexts (e.g., subcellular location and the presence of cofactors). Currently, work on ontology is under way to explicitly define vocabulary that can be applied to all organisms as knowledge of gene and role of protein expression in cells is accumulating and changing. Families of similar sequences contain information on sequence evolution in the form of specific conservation pattern at all sequence positions. Multiple sequence alignments are useful for the following:

1. Building sequence profiles or hidden Markov models, to perform more sensitive homology searches. A sequence profile contains information about the variability of every sequence position improving structure prediction methods (secondary structure prediction). Sequence profile searches have become readily available through the introduction of PsiBLAST that uses a position-specific scoring matrices derived during the search. This tool is used to detect distant evolutionary relationships.
2. Studying evolutionary aspects, by the construction of phylogenetic trees from the pair-wise differences between sequences; for example, the classification with 70S, 30S RNAs established the separate kingdom of archaea.
3. Determining active site residues, and residues specific for subfamilies.
4. Predicting protein–protein interactions.
5. Analyzing single-nucleotide polymorphisms (SNPs) to hunt for genetic sources of diseases.

Several complete genomes of microorganisms and a few of eukaryotes are available. A wealth of additional information can be obtained by analysis of entire genome sequences. The complete genomic sequence contains not only all protein sequences but also sequences regulating gene expression. Comparisons of the genomes of genetically close organisms reveal genes responsible for specific properties of the organisms (e.g., infectivity). Protein interactions can be predicted from conservation of gene order or operon organization in different genomes. Also, the detection of gene fusion and gene fission (viz. one protein is split into two in another gnome) events helps to deduce protein interactions.

27.2.3 Structural Bioinformatics

The combination of the information from a protein structure with the tools of structural bioinformatics provides a much deeper understanding and new valuable insights in biochemical and molecular biology research, both in academia and industry. By using the tools of structural bioinformatics, one may answer questions like the use of structural information to increase the stability of protein at higher temperatures or different pH, to use structural information in drug discovery, and to make an amino acid sequence alignment. Tools of structural bioinformatics can give an idea about the protein 3D structure from a sequence alignment, and the best way to plan an amino acid replacement in the protein of interest. Advances in high-throughput techniques, in structural biology, have resulted in thousands of three-dimensional structures of proteins and their complexes with various ligands, including substrate analogs and inhibitors. It deals with generalizations about macromolecular 3D structure such as comparisons of overall folds and local motifs, principles of molecular folding, evolution, and binding interactions, and structure/function relationships, working both from experimentally solved structures and from computational models. Membrane proteins are targets of ~50% of drugs, and mutations in membrane proteins may result in diseases ranging from diabetes to cystic fibrosis. Computer simulations allow membrane proteins to "come alive," that is, one can simulate the motions of membrane proteins and use this to explore the relationship between (static) structure and dynamic function. This is relevant to a number of areas ranging from biomedicine to nanotechnology. It can also be called as computational structural biology.

Some experiments in bioinformatics currently in use are described herein.

27.3 Nucleotide Sequence Databases

27.3.1 Principle

To explore NCBI sequence repository and retrieve nucleotide sequence information.

27.3.2 Materials

Hardware/system requirements

1. Operating system: Windows/Linux
2. Broadband Internet connection

Software/other requirements
Connect to NCBI website

27.3.3 Protocol

1. Open an Internet browser and visit http://www.ncbi.nlm.nih.gov/
2. Select search database as "Nucleotide" and enter gene name.
3. Click "Search."
4. Click on the relevant record and download nucleotide sequence in Fasta format.
5. Save Fasta file for further analysis.

27.4 Protein Sequence Databases

27.4.1 Principle

To explore protein sequence repository and retrieve protein sequence information.

27.4.2 Materials

Hardware/system requirements

1. Operating system: Windows/Linux
2. Broadband Internet connection

Software/other requirements

Connect to UniProt website

27.4.3 Protocol

1. Open an Internet browser and visit http://www.uniprot.org/
2. Type your enzyme name in the query box and click "Search."
3. Select the appropriate enzyme based on the desired source.
4. Retrieve protein sequence in Fasta format.
5. Save Fasta file for further analysis.

27.5 Biomolecular Structure Databases

27.5.1 Principle

To explore protein structure repository and retrieve protein structure.

27.5.2 Materials

Hardware/system requirements

1. Operating system: Windows/Linux
2. Broadband Internet connection

Software/other requirements

Java runtime environment (JRE) installed in a system

27.5.3 Protocol

1. Open Internet browser and visit http://www.rcsb.org/pdb
2. Enter protein name and click "Search."
3. Select relevant PDB record and click to open PDB report.
4. View protein structure.
5. Download PDB file (.txt) and save it for further analysis.

27.6 Pair-Wise Sequence Alignment

27.6.1 Principle

To check similarity between two biological sequences.

27.6.2 Materials

Hardware/system requirements

1. Operating system: Windows/Linux
2. Broadband Internet connection

Software/other requirements
Connect to EBI

27.6.3 Protocol

1. Open an Internet browser and visit http://www.ebi.ac.uk/Tools/emboss/align/
2. Paste given sequence "A" in sequence 1 box and given sequence "B" in sequence 2 box.
3. Click "Run" and view the result.
4. Save the output file.
5. Compute similarity by changing method to "water" for local alignment.
6. Save the output file.
7. Prepare a report about alignment results with two methods and compare results (global versus local alignment).

27.7 Detection of CpG Island for Nucleotide Sequence

27.7.1 Principle

Detection of regions of genomic sequences that are rich in the CpG pattern is important because such regions are resistant to methylation and tend to be associated with genes that are frequently switched on. Regions rich in the CpG pattern are known as CpG islands. This exercise will help in detection of CpG island in given nucleotide sequence.

27.7.2 Materials

Hardware/system requirements

1. Operating system: Windows/Linux
2. Broadband Internet connection

Software/other requirements
JRE installed in the system

27.7.3 Protocol

1. Open an Internet browser and visit http://www.ebi.ac.uk/Tools/emboss/cpgplot/
2. Paste given nucleotide sequence in box.

3. Click "Run" and view the result.
4. Save the output file.
5. Prepare a report about detected CpG island in given.

27.8 Detection of Gene Structure of DNA Sequence

27.8.1 Principle

Eukaryotic genomic region consists of coding (exon) and noncoding region (intron). By analyzing DNA sequence and corresponding protein sequence in conjugation, one can understand underlying gene structure.

27.8.2 Materials

Hardware/system requirements

1. Operating system: Windows/Linux
2. Broadband Internet connection

Software/other requirements

JRE installed in system

27.8.3 Protocol

1. Open Internet browser and visit http://www.ebi.ac.uk/Tools/Wise2/
2. Paste given nucleotide sequence and protein sequence in respective boxes.
3. Click "Run" and view the result.
4. Save the output file.
5. Prepare a report about detected exons.

27.9 Translation of Nucleotide Sequence to Protein Sequence

27.9.1 Principle

DNA sequence is converted into protein by a biological process known as translation. Computationally nucleotide sequence can be converted into corresponding peptide sequence.

27.9.2 Materials

Hardware/system requirements

1. Operating system: Windows/Linux
2. Broadband Internet connection

Software/other requirements
JRE installed in the system

27.9.3 Protocol

1. Open an Internet browser and visit http://www.ebi.ac.uk/Tools/emboss/transeq/
2. Paste given nucleotide sequence.
3. Click "Run" and view the result.
4. Save the output file.
5. Rerun analysis by translating nucleotide sequence in all six reading frames.

27.10 Reports of Various Chemical and Physical Properties of Protein Sequence

27.10.1 Principle

Understanding of various physicochemical properties of protein helps in understanding wet-lab results and also helps in planning experiments.

27.10.2 Materials

Hardware/system requirements

1. Operating system: Windows/Linux
2. Broadband Internet connection

Software/other requirements
JRE installed in the system

27.10.3 Protocol

1. Open an Internet browser and visit http://expasy.org/tools/protparam.html
2. Paste the given protein sequence.
3. Click Compute parameter > Submit.
4. Note down important parameters computed by the tool.

27.11 Database Search "Blast"

27.11.1 Principle

To search for similar or homologous sequences in sequence databases:

27.11.2 Materials

Hardware/system requirements

1. Operating system: Windows/Linux
2. Broadband Internet connection

Software/other requirements
JRE installed in the system

27.11.3 Protocol

1. Open an Internet browser and visit http://www.ncbi.nlm.nih.gov/protein
2. Type enzyme name and click "Search."
3. A list of sequence links along with its source in brackets will appear.
4. Choose the appropriate enzyme based on the desired source.
5. The selected enzyme's sequence will appear.
6. Select "Run Blast."
7. Blast window will be opened; just click Blast at the bottom, and it will run automatically.
8. The result will appear as three parts in the window.
9. Analyze the result and prepare a report.

27.12 Phylogenetic Analysis

27.12.1 Principle

To determine the phylogenetic relationship between the given sequences.

27.12.2 Materials

Hardware/system requirements

1. Operating system: Windows/Linux
2. Broadband Internet connection

Software/other requirements
JRE installed in the system

27.12.3 Protocol

1. Open an Internet browser and go to the link http://expasy.org/tools/
2. Scroll down to Phylogenetic Analysis and click on Phylogeny.fr from the list given.

3. Select Phylogeny Analysis > "One Click" from the menu.
4. Select and copy the protein sequences given one by one and paste them to the Phylogeny.fr window.
5. Once all sequences are pasted, click "Submit" button.
6. Phylogenetic computation will start and result would be displayed once computation is over.
7. Download phylogenetic tree in any format (PDF/PNG.) and save it for interpretation.
8. Study the phylogenetic tree and prepare a short report based on your observation.

27.13 Molecular Pathways Database

27.13.1 Principle

To study pathways in which a gene/enzyme of interest is known to participate.

27.13.2 Materials

Hardware/system requirements

1. Operating system: Windows/Linux
2. Broadband Internet connection

Software/other requirements
JRE installed in the system

27.13.3 Protocol

1. Open an Internet browser and visit http://www.ncbi.nlm.nih.gov/gene
2. Search given gene name and retrieve Entrez Gene id (e.g., xxx).
3. Visit http://www.genome.jp/kegg/genes.html
4. Enter hsa:xxx (i.e., three-letter code for the organism and Entrez Gene ID) and click "Entry."
5. Display a detailed pathways report.
6. Save pathways report for further reference.

27.14 Proteomics

27.14.1 Predicting Subcellular Location of a Protein

27.14.1.1 Principle
The cellular functions are controlled by well coordinately temporal and spatial expression of proteins. The subcellular location of novel protein can be regarded as a critical

attribute in functional annotation of novel protein. Bioinformatics tool "TargetP" utilizes pattern found in N-terminus sequence, and discriminates between proteins destined for the mitochondrion, the chloroplast, the secretory pathway, and other locations.

27.14.1.2 Materials
Amino acid sequence from N-terminus of protein (at least first 130 amino acids from N-terminus are required). Internet connection with access to the following portals:

1. Target P server: http://www.cbs.dtu.dk/services/TargetP/
2. Protein sequence database "Uniprot": http://www.uniprot.org/

27.14.1.3 Protocol
1. Open browser (like Internet Explorer, Firefox).
2. Browse http://www.uniprot.org/
3. Type HLA_STAAU in "Query" text box and click "Search" (HLA_STAAU is Uniprot name for "alpha-hemolysin" isolated from *Staphylococcus aureus*).
4. Locate "Sequence" section in the result page, and click on "FASTA," copy Fasta sequence displayed in the result page.
5. Open notepad and paste Fasta sequence in a text editor. Save the file as "AlphaTox. fasta."
6. Browse http://www.cbs.dtu.dk/services/TargetP/
7. Locate "SUBMISSION" section.
8. Click "Browse" and select file "AlphaTox.fasta" created in Step 5.
9. Click "Submit" and wait for server-prediction results.
10. Note down character under "Loc" (interpretation help can be found at http://www.cbs.dtu.dk/services/TargetP-1.1/output.php).

27.14.1.4 Results and Observations
1. What is subcellular location of this protein?
2. Check with other prediction server like (http://pprowler.itee.uq.edu.au/pprowler_webapp_1-2/), does it also predict same subcellular location for the "AlphaTox.fasta"?

27.14.2 Exploring Protein–Protein Interactions

27.14.2.1 Principle
Proteins are one of the most functionally important biomolecules, which participates in various metabolic/nonmetabolic pathways and are thereby responsible for carrying out various cellular functions. A protein brings about functional effects by interacting with other proteins in a process. It is very important to get information about functional partners or interactants of a particular protein for complete understanding of functional role of any protein.

27.14.2.2 Materials
Internet connection with access to following portals:
STRING server: http://string-db.org/

27.14.2.3 Protocol

1. Open browser (like Internet Explorer, Firefox).
2. Browse http://string-db.org/
3. Type "TP53" in "protein name": text box and click on "GO!" and then click on "Continue ->."
4. Navigate down to "Info & Parameters ...."
5. Deselect all "Active Prediction Methods ..." except "Experiments" to view only experimentally reported protein–protein interaction for TP53.
6. Click on "Update Parameters."
7. Study interaction network and note down "Predicted Functional Partners."
8. Try to change various parameters and note their effect on interaction network.

27.14.2.4 Results and Observations

1. What can you infer about the role of TP53 on the basis of its main interaction partners?
2. Which parameter would be useful in detecting significant functional/interaction partner(s) of a protein of interest (TP53)?

27.15 Metabolomics

27.15.1 Exploring Human Metabolites

27.15.1.1 Principle

Metabolomics experiments help in profiling of metabolites present in sample (blood, urine, CSF, etc.). Metabolite concentration in the body can be regarded as a reflection of health status of an individual. Bioinformatics resources like Human Metabolome Database (HMDB) can be very useful for researchers, who wish to refer to information about metabolites available in the public domain.

27.15.1.2 Materials

Internet connection with access to following portals:
     HMDB server: http://www.hmdb.ca/

27.15.1.3 Protocol

1. Open browser (like Internet Explorer, Firefox).
2. Browse http://www.hmdb.ca/
3. Type "Dopamine" in "Search" text box and click on "Search" button.
4. Click on HMDB00073.
5. Navigate down to "Normal Concentrations" section.
     a. Note down concentration ("Value") in blood.
6. Navigate down to "Abnormal Concentrations" section.
     a. Note down concentration ("Value") in blood.
     b. Note down "Condition" (e.g., Alzheimer's disease [AD]).

27.15.1.4 Results and Observations

1. What difference did you noticed in dopamine concentration, in blood of normal versus AD?
2. Does dopamine concentration decreases in AD?

27.15.2 Metabolite Profiling

27.15.2.1 Principle

The concentration of metabolites is altered in case of physiological abnormalities or diseased conditions. Metabolite profiling is one of the first analytical steps in which we compare concentration of metabolites from biofluid samples against the normal range reported in the literature.

27.15.2.2 Materials

Internet connection with access to the following portals:
    MSEA server: http://www.msea.ca/MSEA/

27.15.2.3 Protocol

1. Open browser (like Internet Explorer, Firefox).
2. Browse http://www.msea.ca/MSEA/
3. Click Enrichment Analysis > Single Sample Profiling (SSP).
4. Click "Submit" (You can run an analysis for your data also, however, for running this sample protocol you can go with default data).
5. Click "Next."
6. Note down metabolites that deviate from the normal range (i.e., Comparison → H/L).
7. Click Next.
8. Select "Disease-associated metabolite sets (Blood)."
9. Click Submit.
10. Note down metabolite sets, which are enriched.

27.15.2.4 Result and Observation

Which is the topmost enriched metabolite set? (Check its detail to get an idea about metabolite(s) that significantly differed in the sample against normal range.)

27.16 Genomics

27.16.1 Meta-Analysis of Gene Expression

27.16.1.1 Principle

Gene expression studies use high-throughput array-based methods to estimate expression profiles in two or more biological conditions (e.g., cancer, precancer, normal). Such studies are performed by various research groups; at times, results from similar studies are conflicting. Meta-analysis is a method that can be used to detect genes that shows consistent behavior across studies performed independently by various research groups.

27.16.1.2 Materials

Internet connection with access to the following portals:

1. Rank product server: http://bioinformatics.biol.rug.nl/websoftware/rank/
2. Sample data: http://bioinformatics.biol.rug.nl/websoftware/rank/mydata.txt

27.16.1.3 Protocol

1. Open browser (like Internet Explorer, Firefox).
2. Browse http://bioinformatics.biol.rug.nl/websoftware/rank/rank_start.php.
3. Click "Browse" and select "mydata.txt" downloaded as a sample file (alternatively you can also upload your data).
4. Click "Upload file(s)."
5. Click "Proceed."
6. Note down the number of significantly up-/down-regulated genes.

27.16.1.4 Results and Observations

1. What could be reasons for very low number of significant genes detected in meta-analysis?
2. What would be the effect of changing FDR cutoff from 0.01 (default) to 0.001?

27.16.2 Annotating Gene List Obtained through Expression Studies

27.16.2.1 Principle

Array-based high-throughput studies result in the list of genes significantly expressed in conditions of interest (usually disease or abnormality) in comparison to background expression in normal biological condition, such list of genes is often known as gene list. Functional annotation of the gene list is one of the most crucial steps in getting insights into biological themes (pathways, processes, etc.), which are affected in diseased state. DAVID is one of the web-based tools that facilitate functional annotation of gene list.

27.16.2.2 Materials

Internet connection with access to the following portals:
    DAVID tool: http://david.abcc.ncifcrf.gov/

27.16.2.3 Protocol

1. Open browser (like Internet Explorer, Firefox).
2. Browse http://david.abcc.ncifcrf.gov/

3. Click Functional Annotation.
4. Step 1: Copy following list (list of probe ids from Affymetrix experiment):
   1007_s_at
   1053_at
   117_at
   121_at
   1255_g_at
   1294_at
   1316_at
   1320_at
   1405_i_at
   1431_at
   1438_at
   1487_at
   1494_f_at
   1598_g_at
5. Paste them under "A: Paste a list."
6. Step 2: Select Identifier: choose from the list "AFFYMETRIX_3PRIME_IVT_ID."
7. Step 3: List Type—Select "Gene List."
8. Click "Submit List."
9. Click "Functional Annotation Chart."
10. Explore the result page.

27.16.2.4 Results and Observations
1. Which genes are involved in "aromatase activity"?
2. Which is the most significant "Term" detected in this analysis?

27.17 Bioinformatics Approaches for DNA-Based Signatures of Species/Populations/Breeds/Races/Variety/Strains

27.17.1 Introduction

Molecular characterization of genetic resources has been adding objectivity and rationality in decision making for conservation. Plant, animal, fish, and microbial genetic resources are being characterized by various molecular markers, predominantly by microsatellite, AFLP, and SNP covering both nuclear genome as well as mitochondrial genome. These molecular markers have in-built "molecular clock" entrained with evolutionary timescale having "pictures" or "signatures" of speciation and differentiation of dynamic germplasm in evolutionary pace and scale.

Bioinformatics has not only revolutionized the germplasm characterization, but had been proven as an indispensable tool for molecular identification of species. Bioinformatics has become the most powerful tool of taxonomy right from microbial metagenome analysis of hitherto uncultured microbes, plant, animal, and fish species identification.

27.17.2 DNA Bar Coding of Species

DNA bar coding is a taxonomic method that uses a short genetic marker in an organism's mitochondrial DNA (mtDNA) to identify it as belonging to a particular species. It is based on a relatively simple concept: most eukaryote cells contain mitochondria and mtDNA have a relatively fast mutation rate, which results in significant variance in mtDNA sequences between species and, in principle, a comparatively small variance within species. A 648 bp region of the cytochrome *c* oxidase subunit I gene (COI) was initially proposed as a potential "bar code."

27.17.3 Origin of Species Bar Code

The use of the nucleotide sequence variations to investigate evolutionary relationships is not a new concept. Carl Woese used sequence differences in ribosomal RNA (rRNA) to discover archaea, which in turn led to the redrawing of the evolutionary tree, and molecular markers (e.g., allozymes, rDNA, and mtDNAvage). They have been successfully used in molecular systematics for decades. DNA bar coding provides a standardized method for this process via the use of a short DNA sequence from a particular region of the genome to provide a "bar code" for identifying species. In 2003, Paul D.N. Hebert from the University of Guelph, Ontario, Canada proposed the compilation of a public library of DNA bar codes that would be linked to named specimens. This library would provide a new master key for identifying species whose power will rise with increased taxon coverage and with faster, cheaper sequencing.

27.17.4 Examples of Species Bar Codes

27.17.4.1 Identifying Birds by Species Bar Code

In an effort to find a correspondence between traditional species boundaries established by taxonomy and those inferred by DNA bar coding, Hebert and coworkers sequenced DNA bar codes of 260 of the 667 bird species that breed in North America (Hebert et al. 2004a). They found that every single one of the 260 species had a different COI sequence. One hundred and thirty species were represented by two or more specimens; in all of these species, COI sequences were either identical or were most similar to sequences of the same species. COI variations between species averaged 7.93%, whereas variation within species averaged 0.43%. In four cases, there were deep intraspecific divergences, indicating possible new species. Three out of these four polytypic species are already split into two by some taxonomists. Hebert et al.'s (2004a) results reinforce these views and strengthen the case for DNA bar coding. Hebert et al. (2004a) also proposed a standard sequence threshold to define new species, and this threshold, the so-called "bar-coding gap," was defined as 10 times the mean intraspecific variation for the group under study.

27.17.4.2 Delimiting Cryptic Species by DNA Bar Code

The next major study into the efficacy of DNA bar coding was focused on the neotropical skipper butterfly, *Astraptes fulgerator* at the Area Conservacion de Guanacaste (ACG) in northwestern Costa Rica. This species was already known as a cryptic species complex, due to subtle morphological differences, as well as an unusually large variety of caterpillar food plants. However, several years would have been required for taxonomists to delimit the species completely. Hebert et al. (2004b) sequenced the COI gene of 484 specimens from the ACG. This sample included "at least 20 individuals reared from each species of food plant, extremes and intermediates of adult and caterpillar color variation, and representatives" from the three major ecosystems where *A. fulgerator* is found. Hebert et al. (2004b) concluded that *A. fulgerator* consists of 10 different species in northwestern Costa Rica. This highlights that the results of DNA bar-coding analyses can be dependent upon the choice of analytical methods so the process of delimiting cryptic species using DNA bar codes can be as subjective as any other form of taxonomy.

27.17.4.3 Identifying Flowering Plants by Species DNA Bar Code

Kress et al. (2005) suggested that the use of the COI sequence "is not appropriate for most species of plants because of a much slower rate of cytochrome *c* oxidase I gene evolution in higher plants than in animals." A series of experiments were then conducted to find a more suitable region of the genome for use in the DNA bar coding of flowering plants.

Three criteria were set for the appropriate genetic loci:

1. Significant species-level genetic variability and divergence
2. An appropriately short sequence length so as to facilitate DNA extraction and amplification
3. The presence of conserved flanking sites for developing universal primers

At the conclusion of these experiments, Kress et al. (2005) proposed the nuclear internal transcribed spacer region and the plastid trnH–psbA intergenic spacer as a potential DNA bar code for flowering plants. These results suggest that DNA bar coding, rather than being a "master key" may be a "master keyring," with different kingdoms of life requiring different keys.

27.17.4.4 Genus DNA Bar Codes of Microbes

27.17.4.4.1 Strain Identification of Fungi: Example, Puccinia graminis tritici Ug99:
P. graminis, the causal agent of stem rust, has caused serious disease of small cereal grains (wheat, barley, oat, and rye) worldwide. *P. graminis* is the first sequenced representative of the rust fungi (Uredinales), which are obligate plant pathogens. The rust fungi comprise >7000 species and are one of the most destructive groups of plant pathogens. Stem rust of wheat has been a serious problem wherever wheat is grown and has caused major epidemics in North America. In 1999, a new highly virulent race TTKS (Ug99) of *P. graminis* was identified in Uganda, and since then has spread, causing a widening epidemic in Kenya and Ethiopia.

Bioinformatics can play a very critical role in identification of species as well as strains and also its dynamics across globe. The plethora of data both from the host and parasite generated by using latest molecular or biotechnological tools can easily be analyzed by bioinformatics tools.

27.17.4.4.2 Bacterial Genus Identification by rDNA Fingerprinting: The technique of rDNA fingerprinting of microbes has become the gold standard as far as genus of microbes are concerned. The genus-specific PCR primers are available for plus minus assay of microbial genus identification (Figures 27.2 and 27.3).

27.17.4.4.3 Bacterial Species Identification by Gene-Specific DNA Signatures. The low rate of 16S rRNA evolution is responsible for the failure of this molecule to provide multiple diagnostic sites for closely related but ecologically distinct taxa. Rates of evolutionary substitution in protein-coding genes are one order of magnitude greater than those of 16S rRNA genes. Thus, some pairs of ecologically distinct taxa may have had time to accumulate neutral sequence divergence at rapidly evolving loci but not reflected at the 16S rRNA level. The *tuf* gene belongs to a large transcriptional unit, the str operon. The EF-Tu protein is encoded by the *tuf* gene in eubacteria. EF-Tu is a GTP-binding protein playing a central role in protein synthesis. It loads the aminoacyl tRNA molecule onto the ribosome during the translation process. The ubiquitous distribution of the gene encoding elongation factor Tu (EF-Tu) may render this gene a valuable phylogenetic marker for eubacteria.

Species-specific signature can be developed in Lactobacilli by downloading all available *tuf* gene sequences of *Lactobacillus casei, Lactobacillus paracasei,* and *Lactobacillus rhamnosus* from GenBank and aligning using ClustalW program. After completing *in silico*, restriction endonuclease analysis was performed with

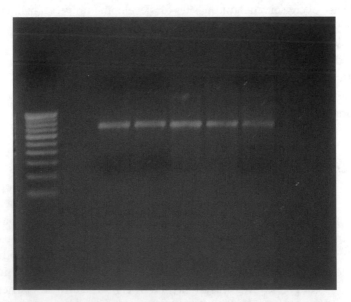

Figure 27.2　Amplified PCR products of 16S rRNA gene of *Lactobacilli.*

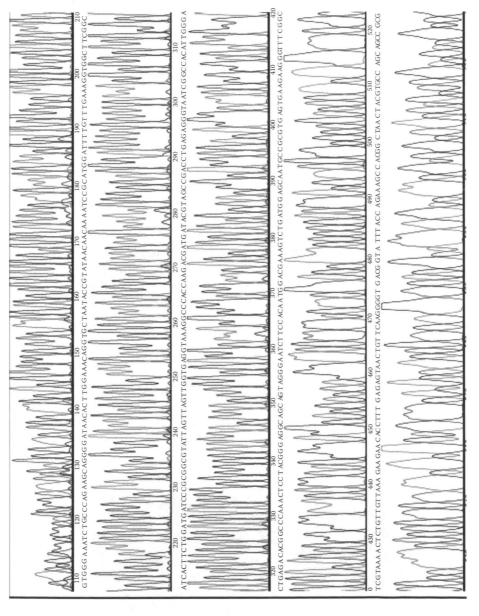

Figure 27.3 **(See color insert.)** *Lactobacillus* identification by rDNA fingerprinting.

Cleaver software with the *tuf* gene sequence of these three species. *Bsh*1236I, *Ksp*AI, and *Mun*I were the restriction enzymes found suitable for species differentiation.

Five *Lactobacillus* strains including three strains obtained from the National Collection of Dairy Cultures (NCDC), NDRI, Karnal and two isolates from churpi cheese made from yak milk, i.e., NCDC 66 (*L. paracasei*), NCDC 19, NCDC 24 (*L. rhamnosus*) were used in this study. *Tuf* gene was amplified using universal primer KEU-1 5'-AAY ATG ATI ACI GGI GCI GCI CAR ATG GA-3' and KEU-2 5'-AYR TTI TCI CCI GGC ATI ACC AT-3'. Restriction digestion was performed in total 30 µL reaction volume with 10 µL of PCR product by 10U of restriction enzyme (Fermentas) at 37°C for 8 h. Digested fragments were separated by agarose (2.5% w/v) gel electrophoresis, at 100 V for 1 h. The obtained result clearly shows that each enzyme can be individually used to differentiate three closely related *Lactobacillus* species, but out of them restriction enzyme *Ksp*AI was found to be best as digested fragment can be analyzed through modest laboratory equipments like agarose gel electrophoresis (Figure 27.4).

27.17.5 DNA-Based Signature of Domestic Species

mDNA markers have been proved to be successful in many species of domestic animals, being used especially for meat identification, poaching of wild animals, adulteration of dairy milk, and dairy products (like cheese) of various domestic animal species.

27.17.6 DNA-Based Signature of Domestic Animal Breeds

The prevalent markers used for the breeds are almost STR but very recently the SNP-based chip has proven its accuracy for breed signature along with details of admixture as well as very powerful for parentage and pedigree.

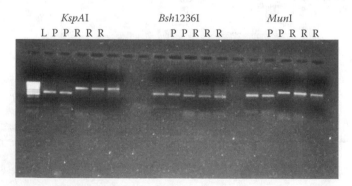

Figure 27.4   L—100 base ladder, P—*Lactobacills paracasei*, R—*Lactobacillus rhamnosus*. *Ksp*AI—*L. paracasei* 542,158 bp, *L. rhamnosa* 701 bp; *Bsh*1236I—*L. paracasei* 547,153 bp, *L. rhamnosa* 153 bp; *Mun* I—*L. Paracasei* 594,106 bp, *L. rhamnosa* 701 bp.

27.17.6.1 STR-Based Signatures of Breeds

A question has generally been asked at various scientific fora with regard to molecular characterization of breeds as to whether a livestock breed can be identified from a sample of blood, semen, hair, blood spot, carcass, etc. Various attempts have been made in the last couple of years by the molecular geneticists globally to answer this question. Some studies have succeeded in developing a technology for breed certification and breed-specific genetic/DNA signature in different breeds of cattle in Spain, Portugal, and France; horses in Norway; sheep in Spain; and camel in Kenya. The degree of accuracy of certification of a breed in these studies was very high ranging between 95% and 99%.

Three methods viz. (i) frequency method (Paetkau et al. 1995), (ii) Bayesian method (Rannala and Mountain 1997), and (iii) distance methods (Goldstein et al. 1995) have been used for developing breed-specific signatures. The Bayesian method has been reported to be more accurate with microsatellite data to the extent of >99% confidence limits (Bustamante et al. 2003, Corander et al. 2003).

Few attempts have been made to develop genetic signatures of some breeds of livestock in the recent past. For cases of doubtful breed identity where it becomes difficult to assign an individual to a particular breed due to individual being an admixture of breeds, the studies have been made to develop breed hybrid index.

27.17.6.2 SNP Chip-Based DNA Signature of Breeds

In Japan, Japanese Black and Holstein cattle are appreciated as popular sources of meat, and imported beef from Australia and the United States is also in demand, in the meat industry. Since the BSE outbreak, the problem of false sales has arisen: imported beef has sometimes been mislabeled as domestic beef due to consumer concerns. A method is needed to correctly discriminate between Japanese and imported cattle for food safety. The SNP 50K based chip can discriminate markers between Japanese and US cattle. There is a report where five US-specific markers (BISNP7, BISNP15, BISNP21, BISNP23, and BISNP26) have been developed with allelic frequencies that ranged from 0.102 (BISNP15) to 0.250 (BISNP7) and averaged 0.184. The combined use of the five markers would permit discrimination between Japanese and US cattle with a probability of identification 0.858 indicating that the potential of the bovine 50K SNP array as a powerful tool for developing breed identification markers (Figure 27.5). These markers would contribute to the prevention of falsified beef displays in Japan (Suekawa et al. 2010, Sasazaki et al. 2011).

27.17.6.3 Development of Breed-Specific Signatures/Profiles

In the United Kingdom, Signer et al. (2000) developed such test using minisatellite probe pCMS12 to differentiate breed assignment among three breeds of pig, viz. Chinese, Meishan, Large White and other European breeds. The linear discrimination analysis revealed that the DNA profiles were breed-specific.

In Finland, Primmer et al. (2000) have assigned a disputed fish to a specific population out of four suspected fish populations using seven microsatellite loci by Bayesian method with confidence limit of 99%.

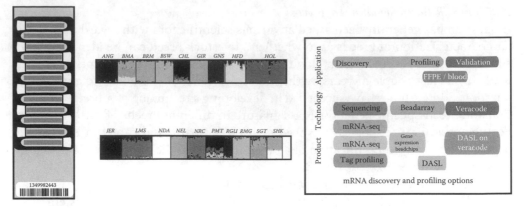

Figure 27.5 50K SNP chip of illumine and US cattle breed signature showing pure and admixture breeds. (Courtesy of United States Department of Agriculture, Washington, DC.)

In Spain, Arranz et al. (2001) have developed individual breed assignment test by using the Bayesian method with 99.63% accuracy among 5 Spanish sheep breed viz. Churra, Latxa, Castellana, Rasa-Aragonesa, and Merino using 18 microsatellite markers.

In Norway, Bjornstad and Roed (2001) reported individual breed assignment using 26 microsatellite loci in 6 breeds of horses, Fjord, Nordland/Lyngen, Dole, Trotter, Icelandic horse, and Shetland pony with >95% confidence limit.

In European countries, Canon et al. (2001) reported individual breed assignment to their population of origin with confidence limit of 99% for 18 local breeds of cattle of different countries: Alistana, Astruriana, Asturiana Valles, Sayaguesa, Tudanca, Avilena Negra-Iberica, Bruna del Pirineus, Morucha, Pirenaica, Retinta of Spain; Alentejana, Barrosa, Maronesa, Mertolenga, Mirandesa of Portugal; and Aubrac, Gasconne, Salers of France.

In Kenya, Mburu et al. (2003) have reported breed assignment in 4 breeds using 14 microsatellite loci of camel viz. Somali, Turkana, Rendille, Gabbra, using the maximum likelihood method up to 48% confidence limit. The reasons for low accuracy in breed assignment in Kenyan camel was attributed to weak genetic differentiation and gene flow between populations. Interestingly, the assignment accuracy was increased considerably when the four populations were grouped as two separate genetic entities. It was also suggested that the results of this study did not support the present classification of indigenous Kenyan dromedaries population into four distinct breeds based on sociogeographic criteria. On the contrary, the study further revealed that instead of four there were just two separate genetic entities Somali, and the other three populations in one group (Gabbra–Rendille–Turkana).

In Finland, Koskinen (2003) has assigned breeds of domestic dog using microsatellite with 100% accuracy and has refuted that DNA-based breed identification is feasible at the individual level of resolution.

The earlier findings indicated the possibilities of developing DNA-based test for identification of livestock and poultry breeds. Further, the clustering of different genetic populations (breeds) would also be feasible, in case there are no significant population differences. In such circumstances, only one breed signature of that cluster may be developed.

27.17.6.4 Development of Breed Hybrid Indices/Profiles

The concept of breed hybrid index has been introduced for cases where it becomes difficult to assign an individual to a particular breed due to the individual being an admixture of breeds belonging to adjoining areas. In India, such a situation is usually prevalent under field conditions due to interbreeding among the animals of different breeds of adjoining areas and poor commitment of the farmers in maintaining the purity of the indigenous breeds.

Campton and Utter (1985) developed a hybrid index, which can be regarded as a way of visualizing the relative assignment probabilities in an assignment test involving two parental assignment methods using microsatellites. It requires three samples, i.e., a sample of each of the two possible parental populations and a sample of the group of suspected "hybrids." The hybrid index value ranges between 0 and 1. The score of individuals from the parental populations is close to either 0 or 1, and scores of hybrids are intermediate. It is not necessarily beneficial to use all available loci, as loci with alleles appearing at high frequency in both parental populations tend to level out hybrid index values of individuals. Instead, it is recommended by trial and error to identify the set of loci that provides the best separation between the parental populations and then apply these loci to the suspected "hybrid" sample.

Campton and Utter (1985) developed their hybrid index for analyzing hybridization between cut throat (*Oncorhynchus clarkii*, Salmonidae) and rainbow trout (*Oncorhynchus mykiss*, Salmonidae), and they used allozyme loci for which the two species were almost fixed for alternate alleles.

Hansen et al. (2001) found that hybrid index statistic is also useful for microsatellite loci, and they applied it successfully in a study of interbreeding between wild and domesticated brown trout, where the parental populations were relatively weakly differentiated (FST = 0.04).

The hybrid and assignment indices can be calculated by first calculating assignment likelihood using the available softwares like Assignment Calculator, GeneClass, or Arlequin and then importing the data into a spreadsheet where the final calculations can be done. In the proposed study, the exotic alpine × beetal cross/admixture populations will be taken up as a model to test the hybrid breed/population assignment in terms of degree of admixture.

27.18 DNA-Based Signature of Plant Variety (Example, Basmati Rice)

Basmati rice has a typical pandan-like (*Pandanus amaryllifolius* leaf) flavor caused by the aroma compound 2-acetyl-1-pyrroline.

Difficulty in differentiating genuine traditional basmati from pretenders and the significant price difference between them has led fraudulent traders to adulterate traditional basmati. To protect the interests of consumers and trade, a PCR-based assay similar to DNA fingerprinting in humans allows for the detection of

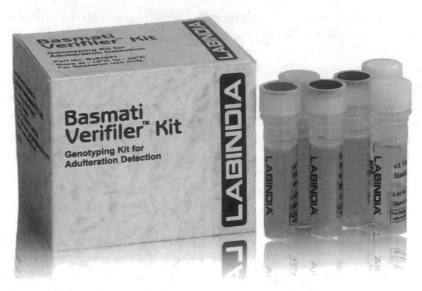

Figure 27.6   Basmati Verifiler™ Kit.

adulterated and nonbasmati strains. Its detection limit for adulteration is from 1% upward with an error rate of ±1.5%. Exporters of basmati rice use "purity certificates" based on DNA tests for their basmati rice consignments. It was developed at the Centre for DNA Fingerprinting and Diagnostics, Labindia, an Indian company that has released kits to detect basmati adulteration: World's First Single-tube, Multiplex (coamplify eight microsatellite loci) Microsatellite Assay-based Kit for Basmati Authentication.

The Basmati Verifiler™ Kit is the world's first product for establishing the authenticity of basmati rice samples via a molecular assay. The kit uses a PCR amplification technique based on simple sequence repeats (SSR) that provide the single-most discriminating assay for basmati genotyping (Figure 27.6).

27.19 DNA-Based Bar-Coded Signature of Fishes

Ward et al. (2005) described in a paper the potential of cox1 sequencing, or "bar coding," in to identification of fish species. In this study, 207 species of fish, mostly Australian marine fish, were sequenced (bar coded) for a 655 bp region of the mitochondrial cytochrome oxidase subunit I gene (*cox1*). Most species were represented by multiple specimens, and 754 sequences were generated. The GC content of the 143 species of teleosts was higher than the 61 species of sharks and rays (47.1% versus 42.2%), largely due to a higher GC content of codon position 3 in the former (41.1% versus 29.9%). Rays had higher GC than sharks (44.7% versus 41.0%), again largely due to higher GC in the third codon position in the former (36.3% versus 26.8%). Average within-species, genus, family, order, and class Kimura two-parameter (K2P) distances were 0.39%, 9.93%, 15.46%, 22.18%, and 23.27%, respectively. All species could be differentiated by their cox1 sequence, although single individuals of each of two species had haplotypes characteristic of a congener. Although DNA bar coding aims to develop species identification systems, some phylogenetic signal was apparent in the data. In the neighbor-joining tree for all 754 sequences, 4 major clusters were apparent: chimaerids, rays, sharks, and teleosts. Species within genera invariably clustered and generally so did genera within families. Three taxonomic groups—dogfishes of the genus *Squalus*, flatheads of the family Platycephalidae, and tunas of the genus *Thunnus*—were examined more closely. The clades revealed after bootstrapping generally corresponded well with expectations. Individuals from operational taxonomic units designated as *Squalus* species B through F formed individual clades, supporting morphological evidence for each of these being separate species.

27.20 Bioinformatics Tool and Protocol for SNP/STR Signatures

Advances in genome analysis technology are providing an unprecedented amount of information about animals, bacterial and viral organisms, and holding great potential for pathogen detection and identification. In this section, a rational approach to the development and application of nucleic acid signatures is described based on SNP and STR nucleotides.

Regardless of the origin of the SNPs (e.g., sequencing and public databases), once SNPs from a target organism and its nearest neighbors have been collected, it is necessary to identify those SNPs that will be useful for species and strain identification. The approach that has been taken is to use a database of SNP markers to enable phylogenetic analysis to identify evolutionary clades and the SNPs that define them. The need for large data storage capability, which facilitates data accessibility, automated SNP prediction (with reduction in manual intervention), signature delineation, and facilitated complex query capability, has been recognized. Many databases exist as local resources, although some universal databases housing eukaryotic SNP data have been established (e.g., dbSNP). Such global databases have not been developed for microbial SNP data. Each database created for SNP discovery and phylogenetic analysis will have different content and different structure that are determined by the use of the data. There is no single correct way to design a database but essential content is necessary not only to allow different polymorphism databases to communicate but also to provide essential information for analysis of the data. Three essential core elements have been defined and include

1. A unique SNP identifier (allele)
2. The data source (e.g., experimental or computational)
3. The sequence flanking the allele and the allele(s)

Many databases have been created for the storage and analysis of eukaryotic SNP data, some are comprehensive or genome-wide, and others are specialized or locus-specific. Both types of databases are essential. The comprehensive database will provide a genome-wide view of polymorphism, ideal for strain typing and identification. The locus-specific database will provide a more in-depth view of polymorphisms at a particular locus. A database should incorporate accurate information that can be used for downstream analyses and have the ability to integrate with other databases. Some additional information associated with SNPs should be implemented in the databases. A database and its associated pipeline should be able to process and store data from a variety of sources, not only from a sequencing machine but also from external sequence databases (e.g., GenBank, dbEST). The database should track the organism and project to which a SNP belongs along with genome-, gene-, and exon-specific information related to a SNP. A downstream analysis requires not only flanking sequences but also a reference sequence. Other information useful for quality assurance purposes and general data analysis include the algorithm by which a SNP was discovered and whether it was validated experimentally or not validated but computationally predicted and the method by which it was validated (e.g., genotyping assay or sequencing). The type of SNP should also be included (e.g., homozygous or heterozygous) along with the average allele frequency. Useful information, such as the position of the SNP relative to its reference sequence, contig, or amplicon, and whether the SNP is silent or pathogenic should be incorporated. To meet the needs of signature development, a relational database has been created to store information related to SNP discovery and downstream assay development. The information specific to SNP discovery and assay design is stored logically in database tables or entities enabling complex queries on SNPs and related data. Specifically, the SNP table includes, in addition to the SNP site alleles, the 5′ and 3′ flanking sequences for assay design. Information related to the gene, exon, and project are stored to facilitate downstream analysis, such as population studies. Algorithm-specific rank values and method are included, which enable the investigator to assess the actual quality of each SNP. The SNP table is the central entity in the database. Associated with each SNP is a name where each SNP can have more than one name. Each SNP can also be associated with one or more reference sequences. Reference sequences have multiple purposes including

- Serving as a template for PCR primer design
- Providing flanking sequence around a SNP
- Being included in a Phrap assembly to ensure an accurate assembly

Reference sequences also provide a starting point for functional annotation. The reference sequence has associated with it a name, GenBank accession or GI number, description, and sequence. Amplicons are sequences used for SNP prediction. Associated with an amplicon is information, such as the name and description of each amplicon, primers used for its amplification, and its expected size. Even though this database was designed for higher eukaryotes and their viruses, the data relationships will remain the same for prokaryotic SNP data. The SNP marker database serves as the repository of information required for downstream signature development and assay design activities.

27.20.1 BioEdit

BioEdit is a mouse-driven, easy-to-use sequence alignment editor and sequence analysis, and intended to supply a single program that can handle most simple sequence and alignment editing and manipulation functions that researchers are likely to do on a daily basis, as well as a few basic sequence analyses. For example, alignment of different nucleotide sequence of various bacterial strains shown step-by-step are File→New Alignment→Import→Accessary Applications→ClustalW Alignment→Multiple Alignment and to see the Alignment Result View→View Mode→Identity/Similarity (Figures 27.7 and 27.8).

27.20.2 GeneClass2

The effectiveness of SNPs for the assignment of various breeds of cattle and buffalo has already been investigated by analyzing nu, erpus SNPs. Breed assignment has been performed by comparing the Bayesian and frequentist methods implemented

Figure 27.7 BioEdit is a mouse-driven, easy-to-use, sequence alignment editor.

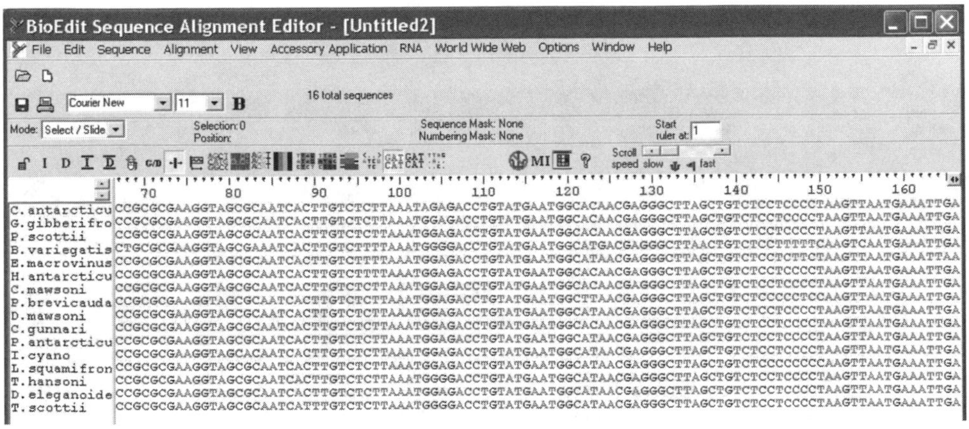

Figure 27.8 Nucleotide sequence data (16 different microbial strains) import in the main window.

in the Structure 2.2 and GeneClass2 software programs. The use of SNPs for the reallocation of known individuals to their breeds of origin and the assignment of unknown individuals has already been tested; example given with GeneClass2 in buffalo having reference and unknown data of buffalo breed. Steps: (1) Download the GeneClass2 software (freely available at http://www.montpellier.inra.fr/URLB/geneclass/geneclass.html). (2) Preparation of data files for reference and unknown samples. (3) Open the main window of the software (Figure 27.9) and import both files. (4) Choice of the parameters like Computational goal, Criteria for computation, Probability computation, and Selection Criteria. (5) By clicking on the start button we can see the result (Figure 27.10), and finally interpretation of the result can be drawn (Figure 27.11).

27.20.3 Cleaver

Cleaver is an application for identifying restriction endonuclease recognition sites that occur in some taxa (Jarman 2006). Differences in DNA fragment restriction patterns among taxa are the basis for many diagnostic assays for taxonomic identification, and are used in some procedures for removing the DNA of some taxa from pools of DNA from mixed sources. Cleaver analyzes restriction digestion of groups of orthologous DNA sequences simultaneously to allow identification of differences in restriction pattern among the fragments derived from different taxa. Cleaver is freely available without registration from its website (http://cleaver.sourceforge.net/) and can be copied, modified, and redistributed under the terms of the GNU general public license version 2 (http://www.gnu.org/licences/gpl). The program can be run as a script for computers that have Python 2.3 and necessary extra modules installed. This allows it to run on Gnu/Linux, Unix, MacOSX, and Windows platforms. Standalone executable versions for Windows and MacOSX operating systems are available. The protocol for using the software is shown in Figures 27.12 through 27.14.

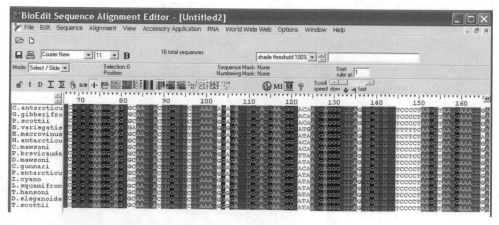

Figure 27.9 Alignment of all sequences showing nucleotide differences.

Figure 27.10 GeneClass2 software programs. http://www.montpellier.inra.fr/CBGP/softwares/

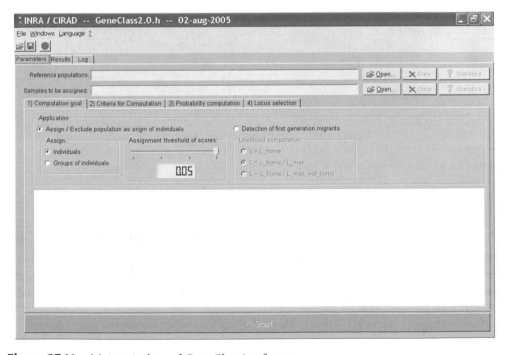

Figure 27.11 Main window of GeneClass2 software.

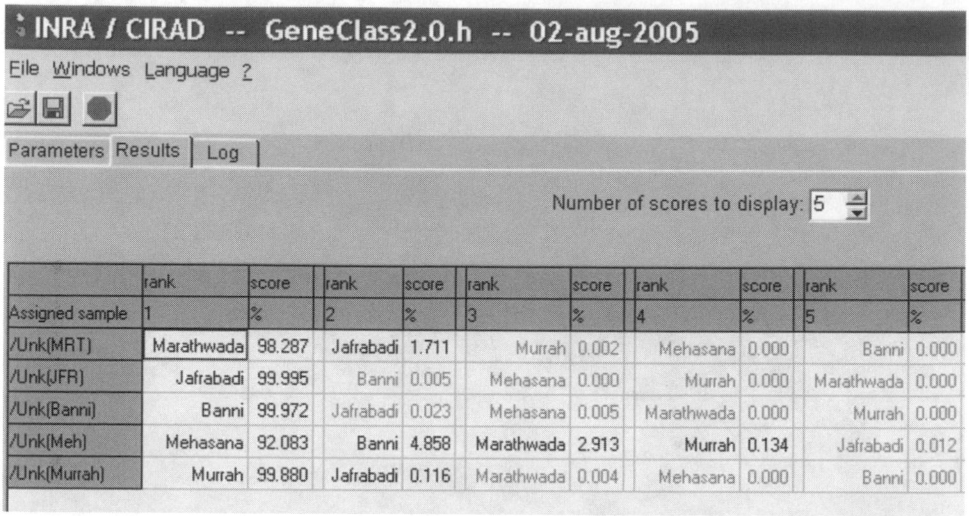

Figure 27.12 Identification of five unknown breeds of buffalo with reference data.

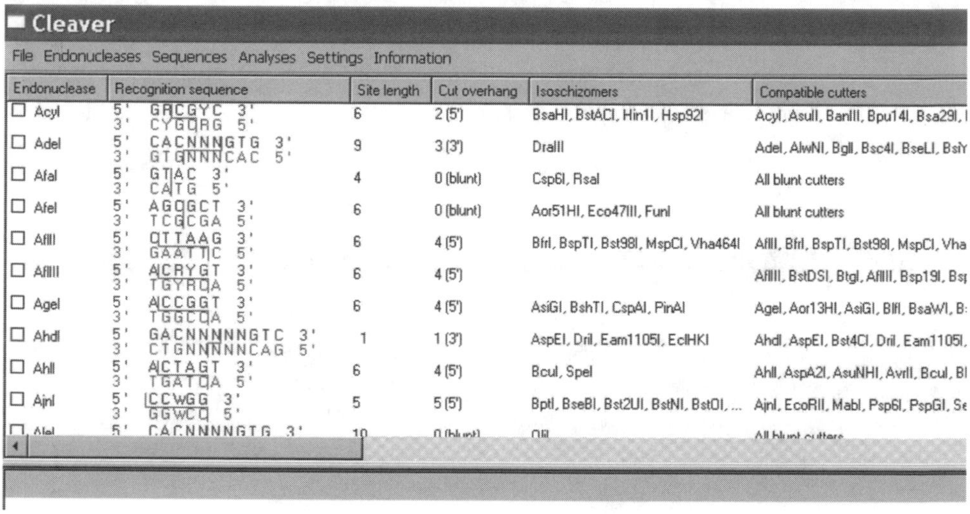

Figure 27.13 Main window of Cleaver software.

27.21 FastPCR

The FastPCR is an integrated tool for PCR primers or probe design, *in silico* PCR, oligonucleotide assembly and analyses, alignment, and repeat searching (Figure 27.15). The software utilizes combinations of normal and degenerated primers for all tools and for the melting temperature calculation is based on the nearest neighbor thermodynamic parameters.

The "*in silico*" (virtual) is used for PCR primers or probe searching or *in silico* PCR against whole genome(s) or a list of chromosome—prediction of probable PCR products and search of potential mismatching location of the specified primers or

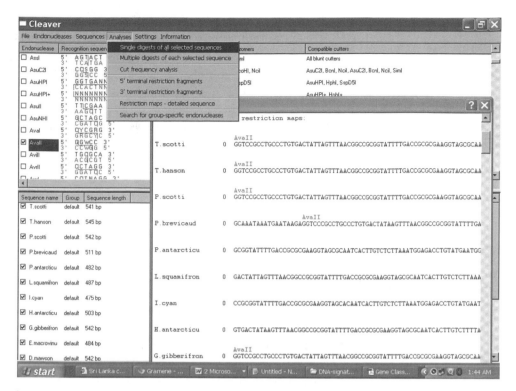

Figure 27.14   Restriction map analysis of variable sequences of different bacterial genomes using Cleaver software.

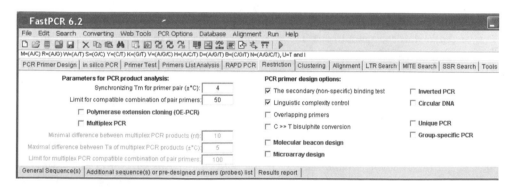

Figure 27.15   Main window of FastPCR software. For SSR search or any other analysis, we just need to prepare data file in notepad file and import in the main window. As per our need, we can import the data and analyze by clicking on Run, SSR search, or Primer list analysis options seeing the main window.

probes, comprehensive primer test, the melting temperature calculation for standard and degenerate oligonucleotides, primer PCR efficiency, primer's linguistic complexity, and dilution and resuspension calculator.

Primers (probes) are analyzed for all primer secondary structures including G-quadruplexes detection, hairpins, self-dimers, and cross-dimers in primer pairs.

FastPCR has the capacity to handle long sequences and sets of nucleic acid or protein sequences, and it allowed the individual task and parameters for each given sequences and joining several different tasks for a single run. It also allows sequence editing and databases analysis. Efficient and complete detection of various types of repeats developed (for DNA-based signature) and applied to the program with a visualization.

The program includes various bioinformatics tools for analysis of sequences with GC or AT skew, CG content and purine–pyrimidine skew, the linguistic sequence complexity, generation random DNA sequence, restriction analysis and supports the clustering of sequences and consensus sequence generation and sequences similarity and conservancy analysis.

Suggested Readings

Arranz, J. J., Y. Bayon, and F. S. Primitivo. 2001. Differentiation among Spanish sheep breeds using microsatellites. *Genet Sel Evol* 33: 529–542.

Baxevanis, A. D. 2001. *Bioinformatics: A Practical Guide to the Analysis of Genes and Proteins*, 2nd edn. New York: Wiley-Interscience.

Bjornstad, G. and K. H. Roed. 2001. Breeds demarcation and potential for breed allocation of horses assessed by microsatellite markers. *Anim Genet* 32: 59–65.

Bourne, P. E. 2003. *Structural Bioinformatics* (Methods of Biochemical Analysis). New York: Wiley-Liss.

Bustamante, C. D., R. Nielsen, and. D. L. Hartl. 2003. Maximum likelihood and Bayesian methods for estimating the distribution of selective effects among classes of mutations using DNA polymorphism data. *Theor Popul Biol* 63: 91–103.

Campton, D. E. and F. M. Utter. 1985. Natural hybridization between steelhead trout (*Salmo gairdneri*) and coastal cutthroat trout (*Salmo clarki clarki*) in two Puget Sound streams. *Can J Fish Aquat Sci* 42: 110–119.

Canon, J., P. Alexandrino, I. Bessa et al. 2001. Genetic diversity measures of local European beef cattle breeds for conservation purposes. *Genet Sel Evol* 33: 311–332.

Corander, J., P. Waldmann, and M. J. Sillanpaa. 2003. Bayesian analysis of genetic differentiation between populations. *Genetics* 163: 367–374.

Durbin R. 1998. *Biological Sequence Analysis: Probabilistic Models of Proteins and Nucleic Acids*. Cambridge, U.K.: Cambridge University Press.

Gibas, C. 2001. *Developing Bioinformatics Computer Skills*. Sebastopol, CA: O'Reilly Media.

Goldstein, D. B., A. R. Linares, L. L. Cavalli-Sforza et al. 1995. Genetic absolute dating based on microsatellites an origin of modern humans. *Proc Natl Acad Sci USA* 92: 6723–6727.

Hansen, M. M., D. E. Ruzzante, E. E. Nielsen et al. 2001. Microsatellite polymorphism in domesticated and wild brown trout *(Salmo trutta)*. *Ecol Appl* 11: 148–160.

Hebert, P. D. N., E. H. Penton, J. M. Burns et al. 2004a. Ten species in one: DNA barcoding reveals cryptic species in the neotropical skipper butterfly *Astraptes fulgerator*. *Proc Natl Acad Sci USA* 101: 14812–14817.

Hebert, P. D. N., M. Y. Stoeckle, T. S. Zemlak et al. 2004b. Identification of birds through DNA barcodes. *PLoS Biol* 2: 1657–1663.

Higgins, D. 2000. *Bioinformatics: Sequence, Structure and Databanks: A Practical Approach*. Oxford, U.K.: Oxford University Press.

Jarman, S. N. 2006. Cleaver: Software for identifying taxon specific restriction endonuclease recognition sites. *Bioinformatics* Advance Access published http://bioinformatics. oxfordjournals.org/content/early/2006/06/20/bioinformatics.btl330.full.pdf

Koskinen, M. T. 2003. Individual assignment using microsatellite DNA reveals unambiguous breed identification in the domestic dog. *Anim Genet* 34: 297–301.

Kress, W. J., K. J. Wurdack, E. A. Zimmer et al. 2005. Use of DNA barcodes to identify flowering plants. *Proc Natl Acad Sci USA* 102: 8369–8374.

Lotka, A. J. 1925. *Elements of Physical Biology*. Baltimore, MD: Williams & Wilkins.

Mburu, D. N., J. W. Ochieng, S. G. Kuria et al. 2003. Genetic diversity and relationships of indigenous Kenyan camel (*Camelus dromedaries*) populations: Implications for their classification. *Anim Genet* 34: 26–32.

Paetkau, D., W. Calvert, I. Stirling et al. 1995. Microsatellite analysis of population structure in Canadian polar bears. *Mol Ecol* 4: 347–354.

Primmer, C. R., M. T. Koskinen, and J. Piironen. 2000. The one that did not get away: Individual assignment using microsatellite data detects a case of fishing competition fraud. *Proc R Soc Lond B* 267: 1699–1704.

Rannala, B. and J. L. Mountain. 1997. Detecting immigration by using multi locus genotypes. *Proc Natl Acad Sci USA* 94: 9197–9221.

Sasazaki, S., D. Hosokawa, R. Ishihara et al. 2011. Development of discrimination markers between Japanese domestic and imported beef. *Anim Sci J* 82: 67–72

Signer, E. N., Y. E. Dubrova, and A. J. Jeffreys. 2000. Are DNA profiles breed specific? A pilot study in pigs. *Anim Genet* 31: 273–276

Suekawa, Y., H. Aihara, M. Araki et al. 2010. Development of breed identification markers based on a bovine 50K SNP array. *Meat Sci* 85: 285–288

Van Halden, J. 2011. *Statistics Applied to Bioinformatics*. Oxford, U.K.: Oxford University Press.

Visser, B., L. Hrselmen, and Z. A. Pretorius. 2008. Genetic comparison of Ug99 with selected South African races of *Puccinia graminis* f. sp. *tritici*. *Mol Plant Pathol* 10: 213–222.

Ward, R. D., T. S. Zemlak, B. H. Innes et al. 2005. DNA barcoding Australia's fish species. *Philos Trans R Soc Lond B Biol Sci* 360: 1847–1857.

Important Links

http://bioinformatics.uams.edu/useful_links.html
http://www.chem.ac.ru/Chemistry/Databases/MAIN.ru.html
http://www.ncbi.nlm.nih.gov/
http://www.cbil.upenn.edu/
http://www.expasy.org/
http://www.ebi.ac.uk/
http://imtech.res.in/raghava/www.html
http://bioinformatics.uams.edu/useful_links.html
http://www.ornl.gov/sci/techresources/Human_Genome/glossary/
http://www.genomicglossaries.com/content/proteomics.asp
http://www.phrap.org/

Appendix A

A.1 Enzyme Stocks and Buffers

A.1.1 Enzyme Stocks

Lysozyme (10 mg/mL): Dissolve 10 mg/mL solid lysozyme in 10 mM Tris–HCl (pH 8.0) immediately before use. Make sure that the pH of the Tris solution is 8.0 before dissolving the protein. Lysozyme will not work efficiently if the pH of the solution is less than 8.0.

Lyticase (67 mg/mL): Dissolve 67 mg/mL (900 units/mL) in 0.01 M sodium phosphate containing 50% glycerol just before use.

Pancreatic DNase I (1 mg/mL): Dissolve 2 mg crude pancreatic DNase I (Sigma or equivalent) in 1 mL of 10 mM Tris–HCl (pH 7.5), 150 mM NaCl, 1 mM $MgCl_2$.

When the DNase I is dissolved, add 1 mL of glycerol to the solution and mix by gently inverting the closed tube several times. Take care to avoid creating bubbles and foam. Store the solution in aliquots at –20°C.

Pancreatic RNase (1 mg/mL): Dissolve 2 mg crude pancreatic RNase I (Sigma or equivalent) in 2 mL of TE (pH 7.6).

Proteinase K (20 mg/mL): Dissolve 20 mg/mL lyophilized powder in sterile 50 mM Tris (pH 8.0), 1.5 mM calcium acetate. Divide the stock solution into small aliquots and store at –20°C. Each aliquot can be thawed and refrozen several times but should then be discarded. Unlike much crude preparations of protease (e.g., pronase), proteinase K need not be self-digested before use.

Trypsin: Prepare bovine trypsin solution 250 µg/mL in 200 mM ammonium bicarbonate (pH 8.9) (Sequencer grade; Boehringer Mannheim). Store solution in aliquots at –20°C.

Zymolyase 5000 (2 mg/mL): Dissolve 2 mg/mL in 0.01 M sodium phosphate containing 50% glycerol just before use.

A.2 Enzyme Dilution Buffers

A.2.1 DNase I Dilution Buffer

10 mM Tris–HCl (pH 7.8)
150 mM NaCl
1 mM $MgCl_2$

A.2.2 Polymerase Dilution Buffer

50 mM Tris–HCl (pH 8.1)
1 mM dithiothreitol (DTT)
0.1 mM ethylene diamine tetra-acetic acid (EDTA) (pH 8.0)
0.5 mg/mL bovine serum albumin
5% (v/v) glycerol
Prepare solution fresh for each use.

A.2.3 Sequenase Dilution Buffer

10 mM Tris–HCl (pH 7.5)
5 mM DTT
0.5 mg/mL bovine serum albumin
Store the solution at −20°C

A.2.4 Tag Dilution Buffer

25 mM Tris (pH 8.8)
0.01 mM EDTA (pH 8.0)
0.15% (v/v) Tween-20
0.15% (v/v) Nonidet P-40

A.3 Enzyme Reaction Buffers

Important: Whenever possible, use the 10× reaction buffer supplied by the manufacturer of the enzyme used. Otherwise, use the recipes given here.

A.3.1 10× Amplification Buffer

500 mM KCl
100 mM Tris–HCl (pH 8.3 at room temperature)
15 mM $MgCl_2$

Autoclave the 10× buffer for 10 min at 15 psi (1.05 kg/cm) on liquid cycle. Divide the sterile buffer into aliquots and store them at −20°C.

A.3.2 10× Bacteriophage T4 DNA Ligase Buffer

200 mM Tris–HCl (pH 7.6)
50 mM $MgCl_2$
50 mM DTT
0.5 mg/mL bovine serum albumin (Fraction V; Sigma) (optimal)

Divide the buffer in small aliquots and store at −20°C. Add ATP to the reaction to an appropriate concentration (e.g., 1 mM).

A.3.3 10× Bacteriophage T4 DNA Polymerase Buffer

330 mM Tris–acetate (pH 8.0)
660 mM potassium acetate
100 mM magnesium acetate
5 mM DTT

1 mg/mL bovine serum albumin (Fraction V; Sigma)
Divide the 10× stock into small aliquots and store frozen at −20°C.

A.3.4 10× Bacteriophage T4 Polynucleotide Kinase Buffer

700 mM Tris–HCl (pH 7.6)
100 mM $MgCl_2$
50 mM DTT

Divide the 10× stock into small aliquots and store frozen at −20°C.

A.3.5 5× BAL 31 Buffer

3 M NaCl
60 mM $CaCl_2$
100 mM Tris–HCl (pH 8.0)
1 mM EDTA (pH 8.0)

A.3.6 10× Dephosphorylation Buffer (for Use with CIP)

200 mM Tris–HCl (pH 3.8)
100 mM $MgCl_2$
10 mM $ZnCl_2$

A.3.7 1× EcoRI Methylase Buffer

50 mM NaCl
50 mM Tris–HCI (pH 8.0)
10 mM EDTA
80 μM S-adenosylmethionine

Store the buffer in small aliquots at −20°C.

A.3.8 10× Exonuclease III Buffer

660 mM Tris–HCl (pH 8.0)
66 mM MgCl$_2$
100 mM p-mercaptoethanol <!>
Add β-mercaptoethanol just before use.

A.3.9 10× Klenow Buffer

0.4 M potassium phosphate (pH 7.5)
66 mM MgCl$_2$
10 mM β-mercaptoethanol <!>

A.3.10 10× Linker Kinase Buffer

600 mM Tris–HCl (pH 7.6)
100 mM MgCl$_2$
100 mM DTT
2 mg/mL bovine serum albumin

Prepare fresh just before use.

A.3.11 Nuclease S1 Digestion Buffer

0.28 M NaCl
0.05 M sodium acetate (pH 4.5)
4.5 mM ZnSO$_4$·7H$_2$O

Store aliquots of nuclease S1 buffer at −20°C, and add nuclease S1 to a concentration of 500 units/mL just before use.

A.3.12 10× Proteinase K Buffer

100 mM Tris–HCl (pH 8.0)
50 mM EDTA (pH 8.0)
500 mM NaCl

A.3.13 10× Reverse Transcriptase Buffer

500 mM Tris–HCl (pH 8.3)
750 mM KCl
30 mM MgCl$_2$

A.3.14 RNase H Buffer

    20 mM Tris–HCl (pH 7.6)
    20 mM KCl
    0.1 mM EDTA (pH 8.0)
    0.1 mM DTT

Prepare fresh just before use.

A.3.15 5× Terminal Transferase Buffer

Most manufacturers supply a 5× reaction buffer, which typically contains

    500 mM potassium cacodylate (pH 7.2) <!>
    10 mM $CoCl_2 \cdot 6H_2O$
    1 mM DTT

5× terminal transferase (or tailing) buffer may be prepared according to the following method (Eschenfeldt and Berger, 1987):

1. Equilibrate 5 g of Chelex 100 (Bio-Rad) with 10 mL of 3 M potassium acetate at room temperature.
2. After 5 min, remove excess liquid by vacuum suction. Wash the Chelex three times with 10 mL of deionized H_2O.
3. Prepare a 1 M solution of potassium cacodylate, and equilibrate the cacodylate solution with the treated Chelex resin.
4. Recover the cacodylate solution by passing it through a Buchner funnel fitted with Whatman no. 1 filter paper.
5. Add H_2O, DTT, and cobalt chloride in order to the recovered cacodylate, making the final concentrations of 500 mM potassium cacodylate, 1 mM DTT, and 20 mM $CoCl_2$.

A.3.16 10× Universal KGB (Restriction Endonuclease) Buffer

    1 M potassium acetate
    250 mM Tris–acetate (pH 7.6)
    100 mM magnesium acetate tetrahydrate
    5 mM β-mercaptoethanol <!>
    0.1 mg/mL bovine serum albumin

Store the 10× buffer in aliquots at −20°C.

A.3.17 Hybridization Buffers

Alkaline transfer buffer (for alkaline transfer of DNA to nylon membranes)

    0.4 N NaOH <!>
    1 M NaCl

A.3.18 Church Buffer

1% (w/v) bovine serum albumin
1 mM EDTA
0.5 M phosphate buffer*
7% (w/v) SDS

Denaturation solution (for neutral transfer, double-stranded DNA targets only)

1.5 M NaCl
0.5 M NaOH <!>
HCl (2.5N)

Add 25 mL of concentrated HCl <!> (11.6 N) to 91 mL of sterile H_2O. Store the diluted solution at room temperature.

A.3.19 Hybridization Buffer with Formamide (for RNA)

50 mM PIPES (pH 6.8)
1 mM EDTA (pH 8.0)
0.4 M NaCl
80% (v/v) deionized formamide <!>

Use the disodium salt of PIPES to prepare the buffer, and adjust the pH to 6.4 with 1 N HCl.

A.3.20 Hybridization Buffer without Formamide (for RNA)

40 mM PIPES (pH 6.4)
0.1 mM EDTA (pH 8.0)
0.4 M NaCl

Use the disodium salt of PIPES to prepare the buffer, and adjust the pH to 6.4 with 1 N HCl.

A.3.21 Neutralization Buffer I (for Transfer of DNA to Uncharged Membranes)

1 M Tris–HCl (pH 7.4)
1.5 M NaCl

---

* 0.5 M phosphate buffer is 134 g of $Na_2HPO_4 \cdot 7H_2O$, 4 mL of 85% H_3PO_4 (concentrated phosphoric acid), H_2O to 1 L.

A.3.22 Neutralization Buffer II (for Alkaline Transfer
     of DNA to Nylon Membranes)

0.5 M Tris–HCl (pH 7.2)
1 M NaCl

A.3.23 Neutralizing Solution (for Neutral Transfer,
     Double-Stranded DNA Targets Only)

0.5 M Tris–HCl (pH 7.4)
1.5 M NaCl

A.3.24 Prehybridization Solution (for Dot, Slot, and Northern Hybridization)

0.5 M sodium phosphate (pH 7.2)*
7% (w/v) SDS
1 mM EDTA (pH 7.0)

A.3.25 Prehybridization/Hybridization Solution (for Plaque/Colony Lifts)

50% (v/v) formamide (optional) <!>
6 × SSC (or 6 × SSPE)
0.05 × bovine lacto transfer technique optimizer (BLOTTO)

A.3.26 Prehybridization/Hybridization Solution (for
     Hybridization in Aqueous Buffer)

6 × SSC (or 6 × SSPE)
5 × Denhardt's reagent
0.5% (w/v) SDS
1 mg/mL poly (A)
100 µg/mL salmon sperm DNA

A.3.27 Prehybridization/Hybridization Solution (for
     Hybridization in Formamide Buffers)

6 × SSC (or 6 × SSPE)
5 × Denhardt's reagent
0.5% (w/v) SDS

---

* 0.5 M phosphate buffer is 134 g of $Na_2HPO_4 \cdot 7H_2O$, 4 mL of 85% H_3PO_4 (concentrated phosphoric acid), H_2O to 1 L.

1 µg/mL poly (A)
100 µg/mL salmon sperm DNA
50% (v/v) formamide <!>

After a thorough mixing, filter the solution through a 0.45 µm disposable cellulose acetate membrane (Schleicher and Schuell Uniflow syringe membrane or equivalent). To decrease background when hybridizing under conditions of reduced stringency (e.g., 20%–30% formamide <!>), it is important to use formamide that is as pure as possible.

A.3.28 Prehybridization/Hybridization Solution (for Hybridization in Phosphate SDS Buffer)

0.5 M phosphate buffer (pH 7.2)*
1 mM EDTA (pH 8.0)
7% (w/v) SDS
1% (w/v) bovine serum albumin

Use an electrophoresis-grade bovine serum albumin. No blocking agents or hybridization rate enhancers are required with this particular prehybridization/hybridization solution.

A.3.29 20× SSC

Dissolve 175.3 g of NaCl and 88.2 g of sodium citrate in 800 mL of H_2O. Adjust the pH to 7.0 with a few drops of 14 N solution of HCl <!>. Adjust the volume to 1 L with H_2O. Dispense into aliquots. Sterilize by autoclaving. The final concentrations of the ingredients are 3.0 M NaCl and 0.3 M sodium acetate.

A.3.30 20× SSPE

Dissolve 175.3 g of NaCl, 27.6 g of $NaH_2PO_4 \cdot H_2O$, and 7.4 g of EDTA in 800 mL H_2O. Adjust the pH to 7.4 with NaOH <!> (~6.5 mL of a 10 N solution). Adjust the volume to 1 L with H_2O. Dispense into aliquots. Sterilize by autoclaving. The final concentrations of the ingredients are 3.0 M NaCl, 0.2 M $Na_2HPO_4 \cdot 7H_2O$, and 0.02 M EDTA.

A.3.31 Blocking Agents

Blocking agents prevents ligands from sticking to surfaces. They are used in molecular cloning to stop nonspecific binding of probes in Southern, Northern,

---

* 0.5 M phosphate buffer is 134 g of $Na_2HPO_4 \cdot 7H_2O$, 4 mL of 85% H_3PO_4 <!> (concentrated phosphoric acid), H_2O to 1 L.

and Western blotting. If left to their own devices, these probes would bind tightly and nonspecifically to the supporting nitrocellulose or nylon membrane. Without blocking agents, it would be impossible to detect anything but the strongest target macromolecules. No one knows for sure what causes nonspecific binding of probes. Hydrophobic patches, lignin impurities, excessively high concentrations of probe, overbaking or underbaking of nitrocellulose filters, and homopolymeric sequences in nucleic acid probes have all been blamed from time to time, together with a host of less likely culprits. Whatever the cause, the solution is generally simple. Treat the filters with a blocking solution containing a cocktail of substances that will compete with the probe for nonspecific binding sites on the solid support. Blocking agents work by brute force. They are used in high concentrations and generally consist of a cocktail of high molecular weight polymers (heparin, polyvinylpyrrolidone, nucleic acids), proteins (bovine serum albumin, nonfat dried milk), and detergents (SDS or Nonidet P-40). The following recommendations apply only to nylon and nitrocellulose filters. Charged nylon filters should be treated as described by the individual manufacturer.

A.3.32 Blocking Agents Used for Nucleic Acid Hybridization

Two blocking agents in common use in nucleic acid hybridization are Denhardt's reagent (Denhardt 1966) and BLOTTO (Johnson et al. 1984). Usually the filters carrying the immobilized target molecules are incubated with the blocking agents for an hour or two before the probe is added. In most cases, background hybridization is completely suppressed when filters are incubated with a blocking agent consisting of 6× SSC or SSPE containing 5× Denhardt's reagent, 1.0% SDS, and 100 mg/mL denatured, sheared salmon sperm DNA. This mixture should be used whenever the ratio of signal to noise is expressed to in the following, for example, when carrying out Northern analysis of low-abundance RNAs or Southern analysis of single-copy sequences of mammalian DNA. However, in most other circumstances (Grunstein–Hogness hybridization, Benton–Davis hybridization, Southern hybridization of abundant DNA sequences, etc.), a less expensive alternative is 6× SSC or SSPE containing 0.25%–0.5% nonfat dried milk (BLOTTO) (Johnson et al. 1984).

Blocking agents are usually included in both prehybridization and hybridization solutions when nitrocellulose filters are used. However, when the target nucleic acid is immobilized on nylon membranes, the blocking agents are often omitted from the hybridization solution. This is because high concentrations of protein are believed to interfere with the annealing of the probe to its target. Quenching of the hybridization signal by blocking agents is particularly noticeable when oligonucleotides are used as probes. This problem can often be solved by carrying out the hybridization step in a solution containing high concentrations of SDS (6%–7%), sodium phosphate (0.4 M), bovine serum albumin (1%), and EDTA (0.02 M) (Church and Gilbert 1984).

Heparin is sometimes used instead of Denhardt's solution or BLOTTO when hybridization is carried out in the presence of the accelerator, dextran sulfate. It is used at a concentration of 500 µg/mL in hybridization solutions containing dextran sulfate. In hybridization solutions that do not contain dextran sulfate, heparin is used at a concentration of 50 µg/mL (Singh and Jones 1984). Heparin (Sigma from porcine intestinal mucosa or equivalent) is dissolved at a concentration of 50 mg/mL in 4× SSPE or SSC and stored at 4°C.

A.3.33 Denhardt's Reagent

Denhardt's reagent is used for Northern hybridization, single-copy Southern hybridization, and hybridization involving DNA immobilized nylon membranes.

Denhardt's reagent is usually made up as a 50× stock solution, which is filtered and stored at –20°C. The stock solution is diluted tenfold into prehybridization buffer (usually 6× SSC or 6× SSPE containing 1.0% SDS and 100 µg/mL denatured salmon sperm DNA) in H_2O (Denhardt 1966):

1% (w/v) Ficoll 400
1% (w/v) polyvinylpyrrolidone
1% (w/v) bovine serum albumin (Sigma, fraction V)

A.3.34 BLOTTO <!>

BLOTTO is used for (1) Grunstein–Hogness hybridization, (2) Benton–Davis hybridization, and (3) all Southern hybridizations other than single-copy dot blots and slot blots.

1× BLOTTO is 5% (w/v) nonfat dried milk dissolved in H_2O containing 0.02% sodium azide <!>. 1× BLOTTO is stored at 4°C and is diluted 10–25 fold into prehybridization buffer before use. BLOTTO should not be used in combination with high concentrations of SDS, which will cause the milk proteins to precipitate. If background hybridization is a problem, Nonidet P-40 (a nonionic, non-denaturing detergent) may be added to a final concentration of 1% (v/v).

BLOTTO may contain high levels of RNase and should be treated with diethylpyrocarbonate (Siegel and Bresnick 1986) or heated overnight to 72°C (Monstein et al. 1992) when used in Northern hybridizations and when RNA is used as a probe. BLOTTO is not as effective as Denhardt's solution when the target DNA is immobilized on nylon filters.

A.3.35 Blocking Agents Used for Western Blotting

The best and least expensive blocking reagent is nonfat dried milk (Johnson et al. 1984). It is easy to use and is compatible with all of the common immunological detection system. The only time nonfat dried milk should not be used is when Western blots are probed for proteins that may be present in milk.

A.3.36 Blocking Buffer (TNT Buffer Containing a Blocking Agent)

10 mM Tris–HCl (pH 8.0)
150 mM NaCl
0.05% (v/v) Tween-20

Blocking agent (1% (w/v) gelatin, 3% (w/v) bovine serum albumin, or 5% (w/v) nonfat dried milk.

Opinion about which of the blocking agents is best varies from laboratory to laboratory. Blocking buffer can be stored at 4°C and reused several times. Sodium azide should be added to a final concentration of 0.05% (w/v) to inhibit the growth of microorganisms.

A.4 Extraction/Lysis Buffer and Solutions

A.4.1 Alkaline Lysis Solution I (Plasmid Preparation)

50 mM glucose
25 mM Tris–HCl (pH 8.0)
10 mM EDTA (pH 8.0)

Prepare solution I from standard stocks in batches of ~100 mL, autoclave for 15 min at 15 psi (1.05 kg/cm^2) on liquid cycle, and store at 4°C.

A.4.2 Alkaline Lysis Solution II (Plasmid Preparation)

0.2 N NaOH (freshly diluted from a 10 N stock) <!>
1% (w/v) SDS

Prepare solution II fresh and use at room temperature.

A.4.3 Alkaline Lysis Solution III (Plasmid Preparation)

5 M potassium acetate 60.0 mL
Glacial acetic acid <!> 11.5 mL
H$_2$O 28.5 mL

The resulting solution is 3 M with respect to potassium and 5 M with respect to acetate. Store the solution at 4°C and transfer it to an ice bucket just before use.

A.4.4 STET

10 mM Tris–HCl (pH 8.0)
0.1 M NaCl

1 mM EDTA (pH 8.0)
5% (v/v) Triton X-100

Make sure that the pH of STET is 8.0 after all ingredients are added. There is no need to sterilize STET before use.

A.5 Electrophoresis and Gel-Loading Buffers

A.5.1 Commonly Used Electrophoresis Buffers

TBE is usually made and stored as a 5× or 10× stock solutions. The pH of the concentrated stock buffer should be ~8.3. Dilute the concentrated stock buffer just before use and make the gel solution and the electrophoresis buffer from the same concentrated stock solutions. However, 5× stock solutions are more stable because the solutes do not precipitate during storage. Passing the 5× or 10× buffer stocks through a 0.22 μm filter can prevent or delay formation of precipitate.

Use Tris–glycine buffers for SDS–polyacrylamide gels. Table A.1 describes some of the common buffers used in electrophoresis.

A.5.2 Specialized Electrophoresis Buffers

10× alkaline agarose gel electrophoresis buffer

500 mM NaOH <!>
10 mM EDTA

TABLE A.1 Buffer Solutions for Various Types of Electrophoresis

Buffer	Working Solutions	Stock Solutions (L)
TAE	1×	50
	40 mM Tris–acetate	242 g Tris base
	1 mM EDTA	57.1 mL of glacial acetic acid
TBE	0.5×	5 ×
	45 mM Tris borate	54 g Tris base
		27.5 g boric acid
	1 mM EDTA	20 mL of 0.5 M EDTA (pH 8.0)
TPE	1×	10 ×
	90 mM Tris–phosphate	108 g Tris base
		15.5 mL phosphoric acid (85% 1.679 g/mL)
	2 mM EDTA	40 mL 0.5 M EDTA (pH 8.0)
Tris–glycine	1×	5 ×
	25 mM Tris–HCl	15.1 g Tris base
		94 g glycine (electrophoresis grade)
	0.1% SDS	50 mL 10% SDS (electrophoresis grade)

Add 50 mL of 10 N NaOH and 20 mL of 0.5 M EDTA (pH 8.0) to 800 mL of H_2O and then adjust the final volume to 1 L. Dilute the 10× alkaline agarose gel electrophoresis buffer with H_2O to generate a 1× working solution immediately before use. Use the same stock of 10× alkaline agarose gel electrophoresis buffer to prepare the alkaline agarose gel and the 1× working solution of alkaline electrophoresis buffer.

A.5.3 10× BPTE Electrophoresis Buffer

100 mM PIPES
300 mM Bis–Tris
10 mM EDTA

The final pH of the 10× buffer is ~6.5. The 10× buffer can be made by adding 3 g of PIPES (free acid), 5 g of Bis–Tris (free base), and 2 mL of 0.5 M EDTA to 90 mL of distilled H_2O and then treating the solution with diethylpyrocarbonate.

A.5.4 10× 3(N-Morpholino) Propanesulfonic Acid Electrophoresis Buffer

0.2 M 3(N-morpholino) propanesulfonic acid (MOPS) (pH 7.0) <!>
20 mM sodium acetate
10 mM EDTA (pH 8.0)

Dissolve 41.8 g of MOPS in 700 mL of sterile DEPC-treated H_2O <!>. Adjust the pH to 7.0 with 2 N NaOH. Add 20 mL of DEPC-treated 1 M sodium acetate and 20 mL of DEPC-treated 0.5 M EDTA (pH 8.0). Adjust the volume of the solution to 1 L with DEPC-treated H_2O. Sterilize the solution by passing it through a 0.45 μM Millipore filter, and store it at room temperature and protect from light. The buffer becomes yellow with age and/or if it is exposed to light or is autoclaved. Straw-colored buffer works well, but darker buffer does not.

A.5.5 3(N-Morpholino) Propanesulfonic Acid

MOPS is one of the buffers developed by Robert Good's laboratories in the 1970s to facilitate isolation of chloroplasts and other organelles (Good and Izawa 1972; Ferguson et al. 1980). In molecular cloning, MOPS is a component of buffers used for the electrophoresis of RNA through agarose gels (Lehrach et al. 1977, Goldberg 1980).

A.5.6 TAFE Gel Electrophoresis Buffer

20 mM Tris–acetate (pH 8.2)
0.5 mM EDTA

Use acetic acid to adjust the pH of the Tris solution to 8.2, and use the free acid of EDTA, not the sodium salt. Concentrated solutions of TAFE buffer can also be purchased (e.g., from Beckman). *Important*: The TAFE gel electrophoresis buffer must be cooled to 14°C before use.

A.5.7 Gel-Loading Buffers (Table A.2)

6× alkaline gel-loading buffer

> 300 mM NaOH
> 6 mM EDTA
> 18% (w/v) ficoll (type 400, Pharmacia)
> 0.15% (w/v) bromocresol green
> 0.25% (w/v) xylene cyanol

A.5.8 Bromophenol Blue Solution (0.4%, w/v)

Dissolve 4 mg of solid bromophenol blue in 1 mL of sterile H_2O. Store the solution at room temperature.

A.5.9 Bromophenol Blue Sucrose Solution

> 0.25% (w/v) bromophenol blue
> 40% (w/v) sucrose

A.5.10 Cresol Red Solution (10 mM)

Dissolve 4 mg of the sodium salt of cresol red (Aldrich) in 1 mL of sterile H_2O. Store the solution at room temperature.

TABLE A.2   6× Gel-Loading Buffers

Buffer Type	6× Buffer	Storage Temperature
TAE	0.25% (w/v) bromophenol blue, 0.25% (w/v) xylene cyanol, 40% (w/v) sucrose in distilled water	4°C
TBE	0.25% (w/v) bromophenol blue, 0.25% (w/v) xylene cyanol, 15% (w/v) Ficoll (type 400 Pharmacia) in distilled water	Room temperature
TPE	0.25% (w/v) bromophenol blue, 0.25% (w/v) xylene cyanol, 30% glycerol in distilled water	4°C
Tris–glycine	0.25% (w/v) bromophenol, 40% (w/v) sucrose in distilled water	4°C

A.5.11   10× Formaldehyde Gel-Loading Buffer

50% (v/v) glycerol (diluted in DEPC-treated H_2O) <!>
10 mM EDTA (pH 8.0)
0.25% (w/v) bromophenol blue
0.25% (w/v) xylene cyanol

A.5.12   Formamide-Loading Buffer

80% (w/v) deionized formamide <!>
10 mM EDTA (pH 8.0)
1 mg/mL xylene cyanol
1 mg/mL bromophenol blue

A.5.13   RNA Gel-Loading Buffer

95% (v/v) deionized formamide <!>
0.02% (w/v) bromophenol blue
0.025% (w/v) xylem cyanol
5 mM EDTA (pH 8.0)
0.025% (w/v) SDS

A.5.14   2× SDS Gel-Loading Buffer

100 mM Tris–HCl (pH 6.8)
4% (w/v) SDS (electrophoresis grade)
0.2% (w/v) bromophenol blue
20% (v/v) glycerol
200 mM DTT or β-mercaptoethanol <!>

1× and 2× SDS gel-loading buffer lacking thiol reagents can be stored at room temperature. Add the thiol reagents from 1 M (DTT) or 14 M (β-mercaptoethanol) stocks before the buffer is used.

A.5.15   2.5× SDS–EDTA Dye Mix

0.4% (v/v) SDS
30 mM EDTA
0.25% bromophenol blue
0.25% xylene cyanol
20% (w/v) sucrose

A.5.16 Special Buffers and Solutions

A.5.16.1 Elution Buffer
    50 mM Tris–HCl (pH 8.1–8.2)
    1.4 M NaCl
    15% (v/v) ethanol

A.5.16.2 KOH/Methanol Solution
This solution is for cleaning the glass plates used to cast sequencing gels. It is prepared by dissolving 5 g of KOH pellets in 100 mL of methanol <!>. Store the solution at room temperature in a tightly capped glass bottle.

A.5.16.3 λ Annealing Buffer
    100 mM in Tris–HCl (pH 7.6)
    10 mM $MgCl_2$

A.5.16.4 LB (Luria Broth) Freezing Buffer
    36 mM K_2HPO_4 (anhydrous)
    13.2 mM KH_2PO_4
    1.7 mM sodium citrate
    0.4 mM $MgSO_4 \cdot 7H_2O$
    6.8 mM ammonium sulfate
    4.4% (v/v) glycerol in LB broth

LB freezing buffer (Zimmer and Verrinder Gibbins 1997) is best made by mixing the previous solutions in 100 mL of LB to the specified concentrations. Measure 95.6 mL of the resulting solution into a fresh container and then add 4.4 mL of glycerol. Mix the solution well and then filter sterilize by passing it through a 0.45 µm disposable Nalgene filter. Store the sterile medium at a controlled room temperature (15°C–25°C).

A.5.16.5 $MgCl_2$–$CaCl_2$ Solution
    80 mM $MgCl_2$
    20 mM $CaCl_2$
    P3 buffer
    3 M potassium acetate (pH 5.5)

A.5.16.6 PEG–$MgCl_2$ solution
    40% (w/v) polyethylene glycol (PEG 8000)
    30 mM $MgCl_2$

Dissolve 40 g of PEG 8000 in a final volume of 100 mL of 30 mM $MgCl_2$. Filter sterilize the solution through a 0.22 µm filter, and store at room temperature.

A.5.16.7 QM Buffer
    750 mM NaCl
    50 mM MOPS (pH 7.0) <!>

15% (v/v) isopropanol
0.15 (v/v) triton X-100

A.5.16.8 Radioactive Ink <!>

Radioactive ink is made by mixing a small amount of ^{32}P with waterproof black drawing ink. It is convenient to make the ink in three grades: very hot (>2000 cps on a handheld mini monitor), hot (>500 cps on a handheld mini monitor), and cool (>50 cps on a handheld mini monitor). Use a fiber-tip pen to apply ink of the desired activity to the pieces or tape. Attach radioactive-warning tape to the pen, and store it in an appropriate place.

A.5.16.9 Sephacryl Equilibration Buffer
50 mM Tris–HCl (pH 8.0) 5 mM EDTA
0.5 M NaCl

A.5.16.10 SM and SM plus gelatin (per liter)
NaCl 5.8 g
$MgSO_4 \cdot 7H_2O$ 2 g
1 M Tris–HCl (pH 7.5) 50 mL
2% (w/v) gelatin solution 5 mL
H_2O to 1 L

Sterilize the buffer by autoclaving for 20 min at 15 psi (1.05 kg/cm^2) on liquid cycle. After the solution has cooled, dispense 50 mL aliquots into sterile containers. SM may be stored indefinitely at room temperature. Discard each aliquot after use to minimize the chance of contamination.

A 5.16.11 Sorbitol Buffer
1 M sorbitol
0.1 M EDTA (pH 7.5)
STE
10 mM Tris–HCl (pH 8.0)
0.1 M NaCl
1 mM EDTA (pH 8.0)

Sterilize by autoclaving for 15 min at 15 psi (1.05 kg/cm^2) on liquid cycle. Store the sterile solution at 4°C.

A.5.16.12 10× TEN Buffer
0.1 M Tris–HCl (pH 8.0)
0.01 M EDTA (pH 8.0)
1 M NaCl

A.5.16.13 TES
10 mM Tris–HCl (pH 7.5)
1 mM EDTA (pH 7.5)
0.1% (w/v) SDS

A.5.16.14 Triton/SDS Solution
  10 mM Tris–HCl (pH 8.0)
  2% (v/v) Triton X-100
  1% (w/v) SDS
  100 mM NaCl
  1 mM EDTA (pH 8.0)

Filter sterilize the solution through a 0.22 μm filter, and store at room temperature.

A.5.16.15 Tris–Sucrose
  50 mM Tris–HCl (pH 8.0)
  10% (w/v) sucrose

Filter sterilize the solution through a 0.22 μm filter, and store at room temperature.
  Solutions containing sucrose should not be autoclaved since the sugar tends to carbonize at high temperature.

A.5.16.16 Wash Buffer
  50 mM MOPS–KOH (pH 7.5–7.6) <!>
  0.75 M NaCl
  15% (v/v) ethanol

Adjust the pH of a MOPS/NaCl solution before adding the ethanol.

A.5.16.17 Yeast Resuspension Buffer
  50 mM Tris–HCl (pH 7.4)
  20 mM EDTA (pH 7.5)

A.5.17 Preparation of Organic Reagents Phenol

Mostly commercial liquefied phenol is clear and colorless and can be used in molecular cloning without redistillation. Occasionally, liquefied phenol is pink or yellow, which should not be used. Crystalline phenol is not recommended because it must be redistilled at 160°C to remove oxidation products, such as quinones that are known to break down phosphodiester bonds or cause cross-linking of RNA and DNA.

A.5.18 Equilibration of Phenol

Before use, phenol must be equilibrated to a pH of >7.8 because the DNA partitions into the organic phase at acidic pH. Always wear gloves, a full-face protector, and a lab coat while performing experiments using phenol:

1. Store liquefied phenol at –20°C. As needed, take out phenol from the freezer, allow it to warm to room temperature, and then melt at 68°C. Add hydroxyquinoline to a final concentration of 0.1%. This compound is an antioxidant, a partial inhibitor of

RNase, and a weak chelator of metal ions (Kirby 1956). In addition, its yellow color provides a convenient way to identify the organic phase.

2. To the melted phenol, add an equal volume of buffer (usually 0.5 M Tris–HCl [pH 8.0] at room temperature). Stir the mixture on a magnetic stirrer for 15 min. Turn off the stirrer, and when the two phases have separated, aspirate, as much as possible, the upper (aqueous) phase using a glass pipette attached to a vacuum line equipped with appropriate traps.

3. Add an equal volume of 0.1 M Tris–HCl (pH 8.0) to the phenol. Stir the mixture on a magnetic stirrer for 15 min. Turn off the stirrer and remove the upper aqueous phase as described in Step 2. Repeat the extractions until the pH of the phenolic phase is >7.8 (measure with pH strip).

4. After the phenol is equilibrated and the final aqueous phase has been removed, add 0.1 volume of 0.1 M Tris–HCl (pH 8.0) containing 0.2% β-mercaptoethanol <!>. The phenol solution may be stored in this form under 100 mM Tris–HCl (pH 8.0) in a light-tight bottle at 4°C for periods of up to 1 month.

A.5.19 Phenol:Chloroform:Isoamyl Alcohol (24:24:1)

A mixture consisting of equal parts of equilibrated phenol and chloroform:isoamyl alcohol <!> (24:1) is frequently used to remove proteins from preparations of nucleic acids. The chloroform denatures proteins and facilitates the separation of the aqueous and organic phases, and the isoamyl alcohol requires treatment before use. The phenol/chloroform/isoamyl alcohol mixture may be stored under 100 mM Tris–HCl (pH 8.0) in a bottle at 4°C for periods of up to 1 month.

A.5.20 Deionization of Formamide

Reagent grade formamide is sufficiently pure, which can be safely used without further treatment. However, if there is any yellow color in formamide <!>, deionize the formamide by steering on a magnetic stirrer with Dowex XG18 mixed-bed resin for 1 h and filter it twice through Whatman no. 1 paper. Store deionized formamide in small aliquots under nitrogen at −70°C.

A.5.21 Deionization of Glyoxal

Commercial stock solutions of glyoxal (40% or 6 M) contain various hydrated forms of glyoxal as well as oxidation products such as glyoxylic acid, formic acid, and other compounds that can degrade RNA. These contaminants must be removed by treatment with a mixed-bed resin such as Bio-Rad AG-510-X8 until the indicator dye in the resin is exhausted. The following steps are taken to deionize the glyoxal:

1. Immediately before use, mix the glyoxal with an equal volume mixed-bed ion-exchange resin (Bio-Rad AG-510-X8). Alternatively, pass the glyoxal through a small column of mixed-bed resin and then proceed to Step 3.

2. Separate the deionized material from the resin by filtration (e.g., through a uniflow plus filter Schleicher and Schuell).
3. Monitor the pH of the glyoxal by mixing 200 µL of glyoxal with 2 µL of a 10 mg/mL solution of bromocresol green in H_2O, and observe the color change. Bromocresol green is yellow at pH <4.8 and blue-green at pH >5.2.
4. Repeat the deionization process (Steps 1–2) until the pH of the glyoxal is >5.5.

Deionized glyoxal can be stored indefinitely at –20°C under nitrogen in tightly sealed microfuge tubes. Use each aliquot only once and then discard.

A.5.22 Chemical Stock Solution

Acrylamide solution (45% w/v)

Acrylamide (DNA-sequencing grade) <!>	434 g
N,N' methylenebisacrylamide	16 g
H_2O to	600 mL

Heat the solution to 37°C to dissolve the chemicals. Adjust the volume to 1 L with distilled water. Filter the solution through a nitrocellulose filter 0.45 µm pore size, and store the filtered solution in dark bottles at room temperature.

A.5.23 Actinomycin D (5 mg/mL)

Dissolve actinomycin D <!> in methanol <!> at a concentration of 5 mg/mL. Store the stock solution at –20°C in the dark.

A.5.24 Adenosine Diphosphate (1 mM)

Dissolve solid adenosine diphosphate in sterile 25 mM Tris–HCl (pH 8.0). Store small aliquots (~20 µL) of the solution at –20°C.

A.5.25 Ammonium Acetate (10 M)

Dissolve 770 g of ammonium acetate in 800 mL of distilled H_2O. Adjust the volume to 1 L with H_2O. Filter sterilize the solution through a 0.22 µm filter. Store the solution in tightly sealed bottles at 4°C or at room temperature. Ammonium acetate decomposes in hot H_2O and solutions containing it should not be autoclaved.

A.5.26 Ammonium Persulfate (10% w/v)

Ammonium persulfate <!>	1 g
H_2O to	10 mL

Dissolve 1 g ammonium persulfate in 10 mL of H_2O and store at 4°C. Ammonium persulfate decays slowly in solution; therefore, replace the stock solution after every 2–3 weeks. Ammonium persulfate is used as a catalyst for the copolymerization of acrylamide and bisacrylamide gels. The polymerization reaction is driven by free radicals generated by an oxidoreduction reaction in which a diamine (TEMED) is used as to adjust the catalyst (Chrambach and Rodbard 1971).

A.5.27 ATP (10 mM)

Dissolve an appropriate amount of solid ATP in 25 mM Tris–HCl (pH 8.0). Store the ATP solution in small aliquots at −20°C.

A.5.28 Calcium Chloride (2.5 M)

Dissolve 11 g of $CaCl_2 \cdot 6H_2O$ in a final volume of 20 mL of distilled H_2O. Filter sterilize the solution through a 0.22 μm membrane filter. Store in 1 mL aliquots at 4°C.

A.5.29 Coomassie Staining Solution

Take 45 mL methanol, 10 mL glacial acetic acid <!>, and 45 mL water in a measuring cylinder. Properly cover the mouth of the measuring cylinder by parafilm. Mix by inverting four to five times. Weigh 0.3 g of Coomassie brilliant blue R-250 in a conical flask (capacity 250 mL). Add the solvent in the conical flask containing Coomassie brilliant blue R-250. Mix by swirling the conical flask. Coomassie brilliant blue R-250 is highly soluble in this solvent. It will dissolve in almost no time. Filtration of the solution is not required.

A.5.30 Deoxyribonucleoside Triphosphates

Dissolve each deoxyribonucleoside triphosphates (dNTP) in H_2O at an approximate concentration of 100 mM. Use 0.05 M Tris base and a micropipette to adjust the pH of each of the solutions to 7.0. Dilute an aliquot of the neutralized dNTP appropriately, and read the optical density at the wavelengths given in the Table A.3. Calculate the

TABLE A.3   Extinction Coefficient (E) and λ Max of Each dNTP

Base	Wavelength (nm)	Extinction Coefficient (E) $(m^{-1} cm^{-1})$
A	259	154×10^4
G	253	137×10^4
C	271	9.10×10^3
T	267	9.60×10^3

actual concentration of each dNTP. Dilute the solutions with H_2O to a final concentration of 50 mM dNTP. Store each separately at –70°C in small aliquots.

For a cuvette with a path length of 1 cm, absorbance = EM. A 100 mL stock solution of each dNTP is commercially available.

For polymerase chain reactions (PCRs), adjust the dNTP solution to pH 8.0 with 2 N NaOH. Commercially available solutions of PCR-grade dNTPs require no adjustment.

A.5.31 Dimethyl Sulfoxide (DMSO) <!>

Purchase a high grade of DMSO (high purity grade). Divide the contents of fresh bottle into 1 mL aliquots in sterile tubes. Close the tubes tightly and store at –20°C. Use each aliquot only once and then discard.

A.5.32 Dithiothreitol (1 M)

Dissolve 3.09 g of DTT in 20 mL of 0.01 sodium acetate (pH 5.2) and sterilize by filtration. Dispense into 1 mL aliquots and store at –20°C. Under these conditions, DTT is stable to oxidation by air.

A.5.33 EDTA (0.5 M, pH 8.0)

Add 186.1 g of disodium EDTA-2H_2O to 800 mL of H_2O. Stir vigorously on a magnetic stirrer. Adjust the pH to 8.0 with NaOH (~20 g of NaOH pellets <!>). Dispense into aliquots and sterilize by autoclaving. The disodium salt of EDTA will not go into solution until the pH of the solution is adjusted to –8.0 by the addition of NaOH.

A.5.34 EGTA (0.5 M, pH 8.0)

EGTA is ethylene glycol bis(β-aminoethyl ether)-N,N,N',N'-tetra acetic acid. A solution of EGTA is made up essentially as described for EDTA previously and sterilized by either autoclaving or by membrane filtration. Store the sterile solution at room temperature.

A.5.35 Ethidium Bromide (10 mg/mL) <!>

Add 1 g of ethidium bromide to 100 mL of H_2O. Stir on a magnetic stirrer for several hours to ensure that the dye has dissolved. Wrap the container in aluminum foil or transfer the solution to a dark bottle and store at room temperature.

A.5.36 Gelatin (20% w/v)

Add 2 g of gelatin to a total volume of 100 mL of H_2O and autoclave the solution for 15 min at 15 psi (1.05 kg/cm²) on liquid cycle.

A.5.37 Glycerol (10% v/v)

Dilute 1 volume of extra pure grade glycerol in 9 volumes of sterile deionize water. Sterilize the solution by passing it through 0.22 μm membrane filter. Store in 200 mL aliquots at 4°C.

A.5.38 IPTG (20% w/v, 0.8 M)

IPTG is isopropylthio-β-D-galactoside. Make a 20% solution of IPTG by dissolving 2 g of IPTG in 8 mL of distilled H_2O. Adjust the volume of the solution to 10 mL with H_2O and sterilize by passing it through a 0.22 μm membrane filter. Dispense the solution into 1 mL aliquots and store at −20°C.

A.5.39 KCl (4 N,N,N',N' M)

Dissolve an appropriate amount of solid KCl in H_2O, autoclave for 20 min on liquid cycle and store at room temperature. Ideally, this solution should be divided into small (~100 μL) aliquots in sterile tubes and each aliquot thereafter should be used.

A.5.40 Lithium Chloride (LiCl, 5 M)

Dissolve 21.2 g of LiCl in a final volume of 90 mL of H_2O. Adjust the volume of the solution by passing it through a 0.22 μm membrane filter or by autoclaving for 15 min at 15 psi (1.05 kg/cm²) on liquid cycle. Store the solution at 4°C.

A.5.41 MgCl₂·6H₂O (1 N,N,N',N' M)

Dissolve 203.3 g of $MgCl_2 \cdot 6H_2O$ in 800 mL of distilled H_2O. Adjust the volume to 1 L with distilled H_2O. Dispense into aliquots and sterilize by autoclaving. $MgCl_2$ is extremely hygroscopic.

A.5.42 MgSO₄ (1 N,N,N',N' M)

Dissolve 12 g of $MgSO_4$ in a final volume of 100 mL of distilled H_2O. Sterilize by autoclaving or filter sterilization. Store at room temperature.

A.5.43 Maltose (20% w/v)

Dissolve 20 g of maltose in a final volume of 100 mL of distilled H_2O and sterilize by passing it through a 0.22 μm membrane filter. Store the sterile solution at room temperature.

A.5.44 NaOH (10 N) <!>

The preparation of 10 N NaOH involves a highly exothermic reaction, which can cause breakage of glass containers. Prepare this solution with extreme care in plastic beakers. To 800 mL of distilled H_2O, slowly add 400 g of NaOH pellets, stirring continuously. As an added precaution, place the beaker on ice. When the pellets have dissolved completely, adjust the volume to 1 L with distilled H_2O. Store the solution in a plastic container at room temperature. Sterilization is not necessary.

A.5.45 NaCl (Sodium Chloride, 5 M)

Dissolve 292 g of NaCl in 800 mL of distilled H_2O. Adjust the volume to 1 L with distilled H_2O. Dispense into aliquots and sterilize by autoclaving. Store the NaCl solution at room temperature.

A.5.46 Polyethylene Glycol (PEG) 8000 <!>

Working concentrations of PEG range from 13% to 14% (w/v). Prepare the appropriate concentration by dissolving PEG 8000 in sterile H_2O, warm it if necessary. Sterilize the solution by passing it through 0.22 μm membrane filter. Store the solution at room temperature.

PEG is a straight-chain polymer of a simple repeating unit $H(OCH_2 CH_2)n OH$. PEG is available in a range of molecular weights whose names reflect the number (n) of repeating units in each molecule. In PEG 400, for example, $n = 8$–9, whereas in PEG 4000, n range from 68 to 84, PEG induces macromolecular crowding of solutes in aqueous solution (Zimmerman and Minton 1993) and has a range of uses in molecular cloning, including those given in the following sections.

A.5.46.1 Precipitation of DNA Molecules according to Their Size

The concentration of PEG required for precipitation is in inverse proportion to the size of the DNA fragments (Lis and Schleif 1975, Ogata and Gilbert 1977, Lis 1980), precipitation and purification of bacteriophage particles (Yamamoto et al. 1970), and increasing the efficiency of reassociation of complementary chains of nucleic acid during hybridization, blunt-end ligation of DNA molecules, and end-labeling of DNA with bacteriophage T4 polynucleotide kinase and fusion of cultured cells with bacterial protoplasts (Schaffner 1980, Rassoulzadegan et al. 1982).

A.5.46.2 Potassium Acetate (5 N,N,N',N' M) <!>

5M potassium acetate	6 mL
Glacial acetic acid	11.5 mL
Distilled H$_2$O	28.5 mL

The resulting solution is 3 M with respect to potassium and 5 M with respect to acetate. Store the buffer at room temperature.

A.6 Sodium Lauryl Sulfate (SDS) (20% w/v) <!>

Dissolve 200 g of electrophoresis-grade SDS in 90 mL of distilled H$_2$O, heat to 68°C, and stir with a magnetic stirrer to assist dissolution. If necessary, adjust the pH to 7.2 by adding a few drops of concentrated HCl. Adjust the volume to 1 L with distilled H$_2$O. Store at room temperature. Do not autoclave.

A.6.1 Sodium Acetate (3 M, pH 5.2 and pH 7.0)

Dissolve 408.3 g of sodium acetate 3H$_2$O in 800 mL of H$_2$O. Adjust the pH to 5.2 with glacial acetic acid <!> or adjust the pH to 7.0 with dilute acetic acid. Adjust the volume to 1 L with H$_2$O. Dispense into aliquots and sterilize by autoclaving.

A.6.2 Spermidine (1 M)

Dissolve 1.45 g of spermidine (free-base form) in 10 mL or deionized H$_2$O and sterilize by passing it through a 0.22 μm membrane filter. Store the solution in small aliquots at −20°C. Make a fresh stock solution of this reagent every month.

A.6.3 SYBR Gold-Staining Solution <!>

SYBR gold (molecular probes) is supplied as a stock solution of unknown concentration in dimethyl sulfoxide <!>. Agarose gels are stained in a working solution of SYBR gold, which is a 1:10,000 dilution of SYBR gold daily, and stored in the dark at regulated room temperature.

A.6.4 Trichloroacetic Acid (100% Solution) <!>

Add 277 mL of H$_2$O to 500 g of trichloroacetic acid (TCA). The resulting solution will contain 100% (w/v) TCA.

A.6.5 X-Gal Solution (2% w/v)

X-gal is 5-bromo-4-chloro-3-indoly1–β-D-galactoside. Make a stock solution by dissolving X-gal in dimethylformamide <!> at a concentration of 20 mg/mL solution. Use a glass or polypropylene tube. Wrap the tube containing the solution in aluminum foil to prevent damage by light and store at –20°C. It is not necessary to sterilize X-gal solutions by filtration.

<!> Use with care and caution.

References

Chrambach, A. and D. Rodbard. 1971. Polyacrylamide gel electrophoresis. *Science* 172: 440–451.

Church, G. M. and W. Gilbert.1984. Genomic sequencing. *Proc Natl Acad Sci USA* 81: 1991–1995.

Denhardt, D. T. 1966. A membrane-filter technique for the detection of complementary DNA. *Biochem Biophys Res Commun* 23: 641–646.

Eschenfeldt, W. H. and S. L. Berger. 1987. Purification of large double stranded cDNA fragments. In *Guide to Molecular Cloning Techniques*, eds. S. L. Berger and A. R. Kimmel. Orlando, FL: Academic Press, pp. 335–337.

Eschenfeldt, W. H., R. S. Puskas, and S. L. Berger. 1987. Homopolymeric tailing. In *Guide to Molecular Cloning Techniques*, eds. S. L. Berger and A. R. Kimmel. Orlando, FL: Academic Press, pp. 337–342.

Ferguson, W. J., K. L. Braumschweiger, W. R. Braumscheiger et al. 1980. Hydrogen ion buffers for biological research. *Anal Biochem* 104: 300–310.

Goldberg, D. A. 1980. Isolation and partial characterization of the *Drosophila* alcohol dehydrogenase gene. *Proc Natl Acad Sci USA* 77: 5794–5798.

Good, N.E. and S. Izawa. 1972. Hydrogen ion buffers. *Methods Enzymol.* 24: 53–68.

Johnson, D. A., J. W. Gautsch, J. R. Sportsman et al. 1984. Improved technique utilizing nonfat dry milk for analysis of proteins and nucleic acids transferred to nitrocellulose. *Gene Anal Technol* 1: 3–8.

Kirby, K. S. 1956. A new method for the isolation of ribonucleic acids from mammalian tissues. *Biochem J* 64: 405–408.

Lehrach, H., D. Diamond, J. M. Wozney et al. 1977. RNA molecular weight determinations by gel electrophoresis under denaturing conditions, a critical reexamination. *Biochemistry* 16: 4743–4751.

Lis, J. T. 1980. Fractionation of DNA fragments by polyethylene glycol induced precipitation. *Methods Enzymol* 65: 347–353.

Lis, J. T. and R. Schleif. 1975. Size fractionation of double-stranded DNA by precipitation with polyethylene glycol. *Nucl Acids Res* 2: 283–289.

Monstein, H. J., T. Geijer, and G. Y. Bakalkin. 1992. BLOTTO-MF, an inexpensive and reliable hybridization solution in northern blot analysis using complimentary RNA probes. *Biotechniques* 13: 842–844.

Ogata, R. and Gilbert, W. 1977. Contacts between the lac repressor and the thymines in the lac operator. *Proc Natl Acad Sci USA* 74: 4973–4976.

Rassoulzadegan, M., B. Binetruy, and F. Cuzin. 1982. High frequency of gene transfer after fusion between bacteria and eukaryotic cells. *Nature* 295: 257–259.

Schaffner, W. 1980. Direct transfer of cloned genes from bacteria to mammalian cells. *Proc Natl Acad Sci USA* 77: 2163–2167.

Siegel, L. I. and E. Bresnick. 1986. Northern hybridization analysis of RNA using diethylpyrocarbonate-treated nonfat milk. *Anal Biochem* 159: 82–87.

Singh, L. and K. W. Jones. 1984. The use of heparin as a simple cost-effective means of controlling background in nucleic acid hybridization procedures. *Nucl Acids Res* 12: 5627–5638.

Wood, N. E. and S. Izawa.1972. H ion buffers. *Methods Enzymol* 24: 53–68.

Yamamoto, K. R., B. M. Alberts, R. Benzinger et al. 1970. Rapid bacteriophage sedimentation in the presence of polyethylene glycol and its application to large-scale virus purification. *Virology* 40: 734–744.

Zimmer, R. and A. M. Verrinder Gibbins. 1997. Construction and characterization of a large-fragment chicken bacterial artificial chromosome library. *Genomics* 42: 217–226.

Zimmerman, S. B. and A. P. Minton. 1993. Macromolecular crowding: biochemical, biophysical, and physiological consequences. *Annu Rev Biophys Biomol Struct* 22: 27–65.

Appendix B

B.1 Common Media Used in Microbiological Laboratories

Most of the media are available in dehydrated form from commercial manufacturers. Compositions, methods of preparation, and usage are found in their manuals, which are supplied at no extra cost.

B.1.1 Ammonium Medium (for Nitrosomonas)

$(NH_4)_2SO_4$ 2.0 g, $MgSO_4 \cdot 7H_2O$ 0.5 g, $FeSO_4 \cdot 7H_2O$ 0.03 g, NaCl 0.3 g, $MgCO_3$ 10.0 g, K_2HPO_4 1.0 g, distilled water to 1 L.

Medium is inoculated as soon as the medium is made up. Sterilization is not necessary for the isolation of *Nitrosomonas* from soil. The pH should be adjusted to 7.3. Sterilization is desirable for storage or other reasons, adjust the pH aseptically after sterilization with sterile 1 N HCl.

B.1.2 Bile Esculin Slants (Streptococcal Differentiation)

Heart infusion agar 4.0 g, esculin 1.0 g, ferric chloride 0.5 g, distilled water to 1 L.

Dispense into sterile 15×125 mm screw-capped tubes, sterilize in autoclave at 121°C for 15 min, and slant during cooling.

B.1.3 Blood Agar

Trypticase soy agar powder 40 g, distilled water to 1 L, final pH of 7.3, defibrinated sheep or rabbit blood 50 mL.

Liquefy and sterilize 1 L of Trypticase soy agar in a large Erlenmeyer flask. While the TSA is being sterilized, warm up 50 mL of defibrinated blood to 50°C. After cooling the TSA to 50°C, aseptically transfer the blood to the flask and mix by gently rotating the flask (cold blood may cause lumpiness). Pour 10–12 mL of the mixture into sterile Petri plates. If bubbles form on the surface of the medium, flame the surface gently with a Bunsen burner before the medium solidifies.

B.1.4 Bromothymol Blue Carbohydrate Broths

B.1.4.1 Make Up Stock Indicator Solution
Bromothymol blue 8 g, 95% ethyl alcohol 250 mL, distilled water 250 mL.
   Indicator is dissolved first in alcohol and then water is added.

B.1.4.2 Make Up Broth
Sugar base (lactose, sucrose, glucose, etc.) 5 g, tryptone 10 g, yeast extract 5 g, indicator solution 2 mL, distilled water 1 L final pH 7.0.

B.1.5 Chlorobium Agar Shake Deeps (Photosynthetic Bacteria)

The incompatibility of the ingredients of this medium at autoclave temperatures requires that the medium be made up of five different portions that can be sterilized separately and mixed at a lower temperature. Ideally, the medium is made up the day it is to be used and kept liquid in a water bath until the inoculations are made.

Solutions A (Prepare in 2000 mL flask)
NH_4Cl 1.0 g, KH_2PO_4 1.0 g, $MgCl_2$ 0.5 g, agar 15.0 g, water 700.0 mL

Solutions B
$NaHCO_3$ 2.0 g, water 100.0 mL

Solutions C
$Na_2S \cdot 9H_2O$ 1.0 g, water 100.0 mL

Solutions D
$FeCl_3 \cdot 6H_2O$ 5.0 g, water 100.0 mL

Solutions E
$(1\ N\ H_3PO_4)$

After sterilizing all five solutions in the autoclave at 121°C for 15 min, cool them to 50°C and pour solutions B, C, and D into solution A flask. Use solution E to adjust the pH of the medium to 7.3.
   Dispense aseptically to sterile 16 mm dia. soft glass test tube (10 mL per tube). Keep the tubes of media liquefied in a 50°C water bath until inoculations are completed.

B.1.6 Chlorobium Enrichment Medium

It is the best to prepare it without sterilization just before it is to be used. It is unstable in nature and also precludes storage of the medium for any length of time in the refrigerator. Ingredients are incompatible at high temperature.

NH_4Cl 1.0 g, KH_2PO_4 1.0 g, $MgCl_2$ 0.5 g, $NaHCO_3$ 2.0 g, $Na_2S \cdot 9H_2O$ 1.0 g
$FeCl_3 \cdot 6H_2O$ 0.0005 g, water 1 L

B.1.7 Chromatium Agar Shake Deeps

This medium is similar in composition to that used for Chlorobium. Thus, the problems are similar. The medium must be made up in three different portions that can be sterilized separately and mixed at a lower temperature. It is best to make up the medium the same day it is to be used. It must be kept liquid in a water bath until the inoculations are made.

Solutions A (Prepare in 2000 mL flask)
NH_4Cl 1.0 g, KH_2PO_4 1.0, $MgCl_2$ 0.5 g, agar 20.0 g, water 780.0 mL

Solutions B
$NaHCO_3$ 2.0 g, water 100.0 mL

Solutions C
$Na_2S \cdot 9H_2O$ 1.0 g, water 100.0 mL

After sterilizing each of three solutions in the autoclave at 121°C for 15 min, cool them to 50°C and pour solutions B and C into flask A.

Dispense aseptically to sterile 16 mm dia. soft glass test tube (10 mL per tube). Keep the tubes of media liquefied in a 50°C water bath until inoculations are completed.

B.1.8 Chromatium Enrichment Medium

This medium is similar in composition to the enrichment medium used for Chlorobium. Thus, the problems are similar. It is best to prepare this medium without sterilization and use it immediately; it does not store well, even in the refrigerator, because of its unstable nature.

NH_4Cl 1.0 g, KH_2PO_4 1.0 g, $MgCl_2$ 0.05 g, $NaHCO_3$ 2.0 g, $Na_2S \cdot 9H_2O$ 1.0 g, Water 1 L

B.1.9 Deca-Strength Phage Broth for Bacteriophage

Peptone 100 mg, yeast extract 50 g, NaCl 25 g, K_2HPO_4 80 g, distilled water 1 L. Final pH 7.6.

B.1.10 Glucose-Minimal Salts Agar (Media Used for Bacterial Genetics)

This medium is made from glucose, agar, and Vogel–Bonner medium E (50×).

B.1.11 Vogel–Bonner Medium E (50×)

Distilled water 45°C 670 mL, $MgSO_4 \cdot 7H_2O$ 10 g, citric acid monohydrate 100 g, K_2HPO_4 (anhydrous) 500 g, sodium ammonium phosphate ($NaHN_4PO_4 \cdot 4H_2O$) 175 g.

Add salt in the order indicated to warm water (45°C) in a 2 L beaker or flask placed on a magnetic stirring hot plate. Allow each salt to dissolve completely before adding the next. Adjust the volume to 1 L. Distribute into two 1 L glass bottles. Autoclave, loosely capped, for 20 min at 121°C.

B.1.12 Plates of Glucose-Minimal Salts Agar

Agar 15 g, distilled water 930 mL, 50× V–B salts 20 mL, 40% glucose 50 mL.

Add 15 g of agar to 930 mL of distilled water in 2 L flask. Autoclave for 20 min using slow exhaust. When the solution has cooled slightly, add 20 mL of sterile 50× V–B salts and 50 mL of sterile 40% glucose. For mixing, a large magnetic stir bar can be added to the flask before autoclaving. After all the ingredients have been added, the solutions should be stirred thoroughly. Pour 30 mL into each Petri plate. *Important*: The 50× V–B salts and 40% glucose should be autoclave separately.

B.1.13 Glucose Peptone Acid Agar

Glucose 10 g, peptone 5 g, monopotassium phosphate 1 g, magnesium sulfate ($MgSO_4 \cdot 7H_2O$) 0.5 g, agar 15 g, water 1 L.

While still liquid after sterilization, add sufficient sulfuric acid to bring the pH down to 4.0.

B.1.14 Glycerol Yeast Extract Agar

Glycerol 5 mL, yeast extract 2 g, dipotassium phosphate 1 g, agar 15 g, water 1 L.

B.1.15 M Endo MF Broth (Bacterial Count with Membrane Filter)

This medium is extremely hygroscopic in the dehydrated form and oxidizes to cause deterioration of the medium after the bottle has been opened. Once a bottle has been opened, it should be dated and discarded after 1 year. If the medium becomes hardened within that time, it should be discarded. Storage of the bottle inside a larger bottle that contains silica gel will extend shelf life. It is best to make up the medium the day it is to be used. It should not be stored over 96 h prior to use:

1. Into a 250 mL screw-capped Erlenmeyer flask, place the following:
   a. Distilled water 50 mL, 95% ethyl alcohol 2 mL, dehydrated medium (Endo MF broth) 4.8 g.
   b. Shake the previous mixture by swirling the flask until the medium is dissolved and then add another 50 mL of distilled water.
2. Cap the flask loosely and immerse it into a pan of boiling water. As soon as the medium begins to simmer, remove the flask from the water bath. Do not boil the medium any further.

3. Cool the medium to 45°C, and adjust the pH to between 7.1 and 7.3.
4. If the medium be stored for a few days, place it in the refrigerator at 2°C–10°C, with screw-cap tightened securely.

B.1.16 Milk Salt Agar (15% NaCl)

Prepare three separate beakers of the following ingredients:

1. Beaker containing 200 g of sodium chloride.
2. Large beaker (2000 mL size) containing 50 g of skim milk powder in 500 mL of distilled water.
3. Glycerol–peptone agar medium.
   $MgSO_4 \cdot 7H_2O$ 5.0 g, $MgNO_3 \cdot 6H_2O$ 1.0 g, $FeCl_3 \cdot 7H_2O$ 0.025 g, Difco proteose–peptone #3 5.0 g, Glycerol 10.0 g, agar 30.0 g, distilled water 500.0 mL

Sterilize the previous three beakers separately. The milk solution should be sterilized at 113°C–115°C (8 lb pressure) in autoclave for 20 min. The salt and glycerol–peptone agar can be sterilized at conventional pressure and temperature. After the milk solution has cooled to 55°C, add the sterile salt, which should also be cooled down to a moderate temperature. If the salt is too hot, coagulation may occur. Combine the milk salt and glycerol–peptone agar solutions by gently swirling with a glass rod. Dispense aseptically into Petri plates.

B.1.17 Minimal Agar (Bacterial Conjugation Experiments)

Glucose 2 g, $(NH_4)_2SO_4$ 1 g, K_2HPO_4 7 g, $MgSO_4$ 0.5 g, agar 15 g, distilled water 1 L.

B.1.18 Nitrite Medium (for Nitrobacteria)

$NaNO_2$ 1.0 g, $MgSO_4 \cdot 7H_2O$ 0.5 g, $FeSO_4 \cdot 7H_2O$ 0.03 g, NaCl 0.3 g, Na_2CO_3 1.0 g, K_2HPO_4 1.0 g, water 1 L.

The medium is inoculated as soon as the medium is made up. Sterilization is not necessary for the isolation of Nitrobacteria from soil. The pH should be adjusted to 7.3. Sterilization is desirable for storage or other reasons, adjust the pH aseptically after sterilization with sterile 1 N HCl.

B.1.19 Nitrogen-Free Glucose Agar (Free Living Nitrogen Fixer Microorganisms)

Add 15 g of agar to the basal salts portion of the previous recipe without nitrate, bring to a boil, and sterilize in the autoclave at 121°C for 15 min. One gram glucose is dissolved in 100 mL of water and sterilized separately in similar manner mix the two solutions aseptically, and dispense into sterile Petri plates.

B.1.20 Nitrogen-Free Medium (Azotobacter)

K_2HPO_4 1.0 g, $MgSO_4 \cdot 7H_2O$ 0.2 g, $FeSO_4 \cdot 7H_2O$ 0.05 g, $CaCl_2 \cdot 2H_2O$ 0.01 g, $Na_2MoO_4 \cdot 2H_2O$ 0.001 g, glucose* 10.0 g, distilled water 1 L.
   Adjust pH to 7.2.
   The medium is inoculated as soon as the medium is made up. Sterilization is not necessary for the isolation of *Azotobacter* from soil. The pH should be adjusted to 7.2. Sterilization is desirable for storage or other reasons, adjust the pH aseptically after sterilization with sterile 1 N HCl.

B.1.21 Phage Growth Medium (Isolation of Bacteriophages)

KH_2PO_4 1.5 g, Na_2HPO_4 3.0 g, NH_4Cl 1.0 g, $MgSO_4 \cdot 4H_2O$ 0.2 g, glycerol 10.0 g, acid-hydrolyzed casein 5.0 g, DL-tryptophan 0.01 g, gelatin 0.02 g, Tween-80 0.2 g, distilled water 1 L.
   Sterilize in autoclave at 121°C for 20 min.

B.1.22 Phage-Lysing Medium (Lytic Phage)

Add sufficient sodium cyanide (NaCN) to the previous growth medium to bring the concentration up to 0.02 M. For 1 L of lysing medium, this will amount to about 1 g (actually 0.98 g) of NaCN. When an equal amount of this lysing medium is added to the growth medium during the last 6 h of incubation, the concentration of NaCN in the combined medium is 0.01 M.

B.1.23 Russell Double Sugar Agar (Gram-Negative Intestinal Pathogens)

Beef extract 1 g, proteose peptone no. 3 (Difco) 12 g, lactose 10 g, dextrose 1 g, sodium chloride 5 g, agar 15 g, phenol red (Difco) 0.025 g, distilled water 1 L.
   Final pH 7.5 at 25°C.
   Dissolve ingredients in water, and bring to boiling. Cool to 50°C–60°C and dispense about 8 mL per tube (16 mm dia. tubes). Slant tubes to cool. Butt depth should be about ½″.

B.1.24 Skim Milk Agar

Skim milk powder 100 g, agar 15 g, distilled water 1 L.
   Dissolve the 15 g of agar into 700 mL of distilled water by boiling. Pour into a flask and sterilize at 121°C, 15 lb pressure.

---

* Sterilize separately.

In a separate container, dissolve the 100 g of skim milk powder into 300 mL of water heated to 50°C. Sterilize this milk solution at 113°C–115°C (8 lb pressure) for 20 min.

After the two solutions have been sterilized, cool to 55°C and combine in one flask, swirling gently to avoid bubbles. Dispense into sterile Petri plates.

B.1.25 Sodium Chloride (6.5%) Tolerance Broth (Streptococci Differentiation)

Heart infusion broth 25 g, NaCl 60 g, indicator (1.6 g bromocresol purple) 1 mL.

In 100 mL 95% ethanol, dextrose 1 g, distilled water 1 L.

Add all reagents together up to 1000 mL (final volumes). Dispense in 15×125 mm screw-capped tubes and sterilize in autoclave for 15 min at 121°C.

A positive reaction is recorded when the indicator changes from purple to yellow or when growth is obvious even though the indicator does not change.

B.1.26 Sodium Hippurate Broth (Streptococci Differentiation)

Heart infusion broth 25 g, sodium hippurate 10 g, distilled water 1 L.

Sterilize in autoclave at 121°C for 15 min after dispensing in 15×15 mm screw-capped. Tighten caps to prevent evaporation.

B.1.27 Soft Nutrient Agar (for Bacteriophage)

Dehydrated nutrient broth 8 g, agar 7 g, distilled water 1 L.

Sterilize in autoclave at 121°C for 20 min.

B.1.28 Spirit Blue Agar (Bacterial Hydrolysis)

This medium is used to detect lipase production by bacteria. Lipolytic bacteria cause the medium to change from pale lavender to deep blue.

Spirit blue agar (Difco) 35 g, lipase reagent (Difco) 35 mL, distilled water 1 L.

Dissolve the spirit blue agar in 1 L of water by boiling. Sterilize in autoclave for 15 min at 15 psi (121°C). Cool to 55°C and slowly add the 35 mL of lipase reagent, agitating to obtain even distribution. Dispense into sterile Petri plates.

B.1.29 Streptomycin Agar (Replica Plating Mutant Isolation)

To 1 L sterile liquid nutrient agar (50°C), aseptically add 100 mg of streptomycin sulfate. Pour directly into sterile Petri plates.

B.1.30 Top Agar (Bacterial Mutagenicity Test)

Tubes containing 2 mL of top agar are made up just prior to using from bottles of top agar base and his/bio stock solution.

B.1.31 His/Bio Stock Solution

D-Biotin (F.W.247.3) 30.9 mg, L-histidine HCL (F.W.191.7) 24.0 mg, distilled water 250 mL.

Dissolve by heating the water to the boiling point. This can be done in a microwave oven. Sterilize by filtration through 0.22 μm membrane filter or autoclave for 20 min at 121°C. Store in a glass bottle at 4°C.

B.1.32 Top Agar Base

Agar 6 g, sodium chloride (NaCl) 5 g, distilled water 1 L.

The agar may be dissolved in a steam bath or microwave oven or by autoclaving briefly. Mix thoroughly and transfer 100 mL aliquots to 250 mL glass bottles with screw caps. Autoclave for 20 min with loosened caps. Slow exhaust. Cool the agar and tighten the caps.

Just before using, add 10 mL of the his/bio stock solution to bottles of 100 mL of liquefied top agar base (45°C). After thoroughly mixing, distribute, aseptically, 2 mL of this mixture to sterile tubes (13 mm × 100 mm). Hold tubes at 45°C until used.

B.1.33 Tryptone Agar

Tryptone 10 g, agar 15 g, distilled water 1 L.

B.1.34 Tryptone Broth

Tryptone 10 g, distilled water 1 L.

B.1.35 Tryptone Yeast Extract Agar

Tryptone 10 g, yeast extract 5 g, dipotassium phosphate 3 g, sucrose 50 g, agar 15 g, water 1 L, pH 7.4.

B.1.36 Nutrient Broth/Agar (Routine Bacteriological Test)

Peptone 5 g, beef extract 3 g, sodium chloride 5 g, agar 15 g, water 1 L, pH 7.2.

Prepare plates/slants/broth (without agar) after routine steam sterilization by autoclaving at 121°C.

Appendix C

TABLE C.1   MPN Determination from Multiple Tube Tests

Number of Tubes Giving Positive Reaction Out of			MPN Index per 100 mL	95% Confidence Limits	
3 of 10 mL Each	3 of 1 mL Each	3 of 0.1 mL Each		Lower	Upper
0	0	1	3	<0.5	9
0	1	0	3	<0.5	13
1	0	0	4	<0.5	20
1	0	1	7	1	21
1	1	0	7	1	23
1	1	1	11	3	36
1	2	0	11	3	36
2	0	0	9	1	36
2	0	1	14	3	37
2	1	0	15	3	44
2	1	1	20	7	89
2	2	0	21	4	47
2	2	1	28	10	150
3	0	0	23	4	120
3	0	1	39	7	130
3	0	0	64	15	380
3	1	0	43	7	210
3	1	1	75	14	230
3	1	2	120	30	380
3	2	0	93	15	380
3	2	1	150	30	440
3	2	2	210	35	470
3	3	0	240	36	1300
3	3	1	460	71	2400
3	3	2	100	150	4800

Source: The American Public Health Association, Inc., *Standard Methods for the Examination of Water and Wastewater*, 12th edn., New York, 1999, p. 608.

Appendix D

TABLE D.1   Temperature Conversion Table (Centigrade to Fahrenheit)

°C	0	1	2	3	4	5	6	7	8	9
−50	**−58.0**	**−59.8**	**−61.6**	**−63.4**	**−65.2**	**−67.0**	**−68.8**	**−70.6**	**−72.2**	**−74.2**
−40	−40.0	−41.8	−43.6	−45.4	−47.2	−49.0	−50.8	−52.6	−54.4	−56.2
−30	−22.0	−22.8	−25.6	−27.4	−29.2	−31.0	−32.8	−34.6	−36.4	−38.2
−20	−4.0	−5.8	−7.6	−9.4	−11.2	−13.0	−14.8	−16.6	−18.4	−20.2
−10	+14.0	+12.2	+10.4	+8.6	+6.8	+5.0	+3.2	+1.4	−0.4	−2.2
−0	+32.0	+30.2	+28.4	+26.6	+24.8	+23.0	+21.2	+19.4	+17.6	+15.8
0	**32.0**	**33.8**	**35.36**	**37.4**	**39.2**	**41.0**	**42.8**	**44.6**	**46.4**	**48.2**
10	50.0	51.8	53.6	55.4	57.2	59.0	60.8	62.6	64.4	66.2
20	68.0	69.8	71.6	73.4	75.2	77.0	78.8	80.6	82.4	84.2
30	86.0	87.8	89.6	91.4	93.2	95.0	96.8	98.6	100.4	102.2
40	104.0	105.8	107.6	109.4	11.2	113.0	114.8	116.6	118.4	120.2
50	122.0	123.8	125.6	127.4	129.2	131.0	132.8	134.6	136.4	138.2
60	**140.0**	**141.8**	**143.6**	**145.4**	**147.2**	**149.0**	**150.8**	**152.6**	**154.4**	**156.2**
70	158.0	159.8	161.6	163.4	165.2	167.0	168.0	170.6	172.4	174.2
80	176.0	177.8	179.6	181.4	183.2	185.0	186.8	188.6	190.4	192.2
90	194.0	195.8	197.6	199.4	201.2	203.0	204.8	206.6	208.4	210.2
100	212.0	213.8	215.6	217.4	219.2	221.0	222.8	224.6	226.4	228.2
110	**230.0**	**231.8**	**233.6**	**235.4**	**237.2**	**239.0**	**240.8**	**242.6**	**244.4**	**246.2**
120	248.0	249.8	251.6	253.4	255.2	257.0	258.8	260.6	262.4	264.2
130	266.0	267.8	269.6	271.4	273.2	275.0	276.8	278.6	280.4	282.2
140	284.0	285.8	287.6	289.4	291.2	293.0	294.8	296.6	298.4	300.2
150	302.0	303.8	305.6	307.4	309.2	311.0	312.8	314.6	316.4	318.2
160	**320.0**	**321.8**	**323.6**	**325.4**	**327.2**	**329.0**	**330.8**	**332.6**	**334.4**	**336.2**
170	338.0	339.8	341.6	343.4	345.2	347.0	348.8	350.6	352.4	354.2
180	356.0	357.8	359.6	361.4	363.2	365.0	366.8	368.6	370.4	372.2
190	374.0	375.8	377.6	379.4	381.2	383.0	384.8	386.6	388.4	390.2
200	392.0	393.8	395.6	397.4	399.2	401.0	402.8	404.6	406.4	408.2
210	**410.0**	**411.8**	**413.6**	**415.4**	**417.2**	**419.0**	**420.8**	**422.6**	**424.4**	**426.2**
220	428.0	429.8	431.6	433.4	435.2	437.0	438.8	440.6	442.4	444.2
230	446.0	447.8	449.6	451.4	453.2	455.0	456.8	458.6	460.4	462.2
240	464.0	465.8	467.6	469.4	471.2	473.0	474.8	476.6	478.4	480.2
250	482.0	483.8	485.6	487.4	489.2	491.0	492.8	494.6	496.4	498.2

Note: $°F = °C \times 9/5 + 32$; $°C = °F - 32 \times 5/9$.

Appendix E

TABLE E.1   Autoclave Steam Pressure and Corresponding Temperatures

Steam Pressure (lb/in.²)	Temperature °C	°F	Steam Pressure (lb/in.²)	Temperature °C	°F	Steam Pressure (lb/in.²)	Temperature °C	°F
0	100.0	212.0						
1	101.9	215.4	11	116.4	241.5	21	126.9	260.4
2	103.6	218.5	12	117.6	243.7	22	127.8	262.0
3	105.3	221.5	13	118.8	245.8	23	128.7	263.7
4	106.9	224.4	14	119.9	247.8	24	129.6	265.3
5	108.4	227.1	15	121.0	249.8	25	130.4	266.7
6	109.8	229.6	16	122.0	251.6	26	131.3	268.3
7	111.3	232.3	17	123.0	253.4	27	132.1	269.8
8	112.6	234.7	18	124.1	255.4	28	132.9	271.2
9	113.9	237.0	19	125.0	257.0	29	133.7	272.7
10	115.2	239.4	20	126.0	258.8	30	134.5	274.1

Note: Figures are for steam pressure only, and the presence of any air in the autoclave invalidates temperature reading from Table D.1.

Appendix F

TABLE F.1   Autoclave Temperatures Related to the Presence of Air

Gauge Pressure (lb)	Pure Steam Complete Air Discharge		Two-Thirds Air Discharge 20 in. Vacuum		One-Half Air Discharge 15 in. Vacuum		One-Third Air Discharge 10 in. Vacuum		No Air Discharge	
	°C	°F	°C	°F	°C	°F	°C	°F	°C	°F
5	109	228	100	212	94	202	90	193	72	162
10	115	240	109	228	105	220	100	212	90	193
15	121	250	115	240	112	234	109	228	100	212
20	126	259	121	250	118	245	115	240	109	228
25	130	267	126	259	124	254	121	250	115	240
30	135	275	130	267	128	263	126	259	121	250

Appendix G

TABLE G.1 Significance of Zones of Inhibition in Kirby–Bauer Method of Antimicrobial Sensitivity Testing

Antimicrobial Agent	Disc Potency	Resistant, R (mm)	Intermediate, I (mm)	Sensitive, S (mm)
Amikacin	30 mcg	<14	15–16	>17
Amoxicillin/clavulanic acid	30 mcg			
Staphylococci		<19	14–17	>20
Other Gram-positive organisms		<13	14–17	>18
Ampicillin	75 mcg			
Gram-negative enterics		<13	14–16	>17
Staphylococci		<28		>29
Enterococci		<16		>17
Streptococci (not *Streptococcus pneumoniae*)		<21	22–29	>30
Haemophilus spp.		<18	19–21	>22
Listeria monocytogenes		<19		>20
Azlocillin (*Pseudomonas aeruginosa*)	75 mcg	<17		>18
Carbenicillin (*P. aeruginosa*)	100 mcg	<13	14–16	>17
Other Gram-negative organisms		<19	20–22	>23
Cefaclor	30 mcg	<14	15–17	>18
Cephalothin	30 mcg	<14	15–17	>18
Chloramphenicol	30 mcg	<12	13–17	>18
S. pneumoniae		<20		>21
Clarithromycin	15 mcg	<13	14–17	>18
S. pneumoniae		<16	17–20	>21
Clindamycin	2 mcg	<14	15–20	>21
S. pneumoniae		<20		>21
Erythromycin	15 mcg	<13	14–22	>23
S. pneumoniae		<15	16–20	>21
Gentamicin	10 mcg	<12	13–14	>15
Imipenem	10 mcg	<13	14–15	>16
Haemophilus spp.				>16
Kanamycin	30 mcg	<13	14–17	>18
Lomefloxacin	10 mcg	<18	19–21	>22
Loracarbef	30 mcg	<14	15–17	>18
Mezlocillin (*P. aeruginosa*)	75 mcg	<15		>16
Other Gram-negative organisms		<17	18–20	>19

(continued)

TABLE G.1 (continued)   Significance of Zones of Inhibition in Kirby–Bauer Method of Antimicrobial Sensitivity Testing

Antimicrobial Agent	Disc Potency	Resistant, R (mm)	Intermediate, I (mm)	Sensitive, S (mm)
Minocycline	30 mcg	<14	15–18	>19
Moxalactam	30 mcg	<14	15–22	>23
Nafcillin	1 mcg	<10	11–12	>13
Nalidixic acid	30 mcg	<13	14–18	>19
Netilmicin	30 mcg	<12	13–14	>15
Norfloxacin	10 mcg	<12	13–16	>17
Ofloxacin	5 mcg	<12	13–15	>16
Penicillin G (Staphylococci)	10 units	<28		>29
Enterococci		<14		>15
Streptococci (not *S. pneumoniae*)		<19	20–27	>28
Neisseria gonorrhoeae		<26	27–46	>47
L. monocytogenes		<19		>20
Piperacillin/tazobactam	100\10 mcg			
Staphylococci		<17		>18
P. aeruginosa		<17		>18
Other Gram-negative organisms		<14	15–19	>20
Rifampin	5 mcg	<16	17–19	>20
Haemophilus spp.		<16	17–19	>20
S. pneumoniae		<16	17–18	>19
Streptomycin	10 mcg	<11	12–14	>15
Sulfisoxazole	300 mcg	<12	13–16	>17
Tetracycline	30 mcg	<14	15–18	>19
S. pneumoniae		<17	18–21	>22
Tobramycin	10 mcg	<12	13–14	>15
Trimethoprim/sulfamethoxazole	1.25/23.75	<10	11–15	>16
Vancomycin	30 mcg	<14	15–16	>17

Glossary

2D chromatography: It is a type of chromatographic technique in which the injected sample is separated by passing through two different separation stages.

2D difference gel electrophoresis: Methods of comparing two or more samples on the same 2D gel electrophoresis.

2D gel electrophoresis: Electrophoresis in which a second perpendicular electrophoretic transport is performed on the separate components resulting from the first electrophoresis.

Abiotic factors: Nonliving; moisture, soil, nutrients, fire, wind, temperature, climate.

Abortive infection: When a virus infects a cell (or host), but cannot complete the full replication cycle, that is, a nonproductive infection.

Absorbance: Refers to the mathematical quantity of light.

Absorption: Refers to the physical process of absorbing light.

Acetogenic bacterium: Prokaryotic organism that uses carbonate as a terminal electron acceptor and produces acetic acid as a waste product.

Acetylene-block assay: Estimates denitrification by determining release of nitrous oxide (N_2O) from acetylene-treated soil.

Acetylene-reduction assay: Estimates nitrogenase activity by measuring the rate of acetylene reduced to ethylene.

Acid-fast stain: A staining property of *Mycobacterium* species where cells are stained with hot carbolfuchsin and will not decolorize with acid alcohol.

Acid soil: Soil with a pH value <6.6.

Acidity: Ability of a substance that reacts with a base.

Acidophile: Organism that grows best under acid conditions (down to a pH of 1).

Actinomycete: Nontaxonomic term applied to a group of high G + C base composition, Gram-positive bacteria that have a superficial resemblance to fungi.

Actinorhizae: Associations between actinomycetes and plant roots.

Activated sludge: Sludge particles produced in raw or settled wastewater (primary effluent) by the growth of organisms (including zoogleal bacteria) in aeration tanks in the presence of dissolved oxygen.

Activation energy: Amount of energy required to bring all molecules in 1 mol of a substance to their reactive state at a given temperature.

Active carrier: An individual who has an overt clinical case of a disease and who can transmit the infection to others.

Active site: Region of an enzyme where substrates bind.

Active transport: When a cell uses energy to move a substance into or out of the cell. It requires a carrier protein and the input of energy.

Adapt: To change in order to survive better in an environment.

Adaptation: A body structure/function that allows an organism to survive more easily—process of getting such a structure.

Adenosine triphosphate (ATP): Common energy-donating molecule in biochemical reactions. Also an important compound in transfer of phosphate groups.

Adjuvant: Material added to an antigen to increase its immunogenicity.

Adsorption: The process of interaction between the solute and the surface of an adsorbent.

Adsorption chromatography: In which one solvent is immobile (by adsorption on a solid support matrix) and another is mobile. It is the most common applications of chromatography.

Aerobe: Microbe that grows in the presence of oxygen. A strict aerobe grows only under such a condition.

Aerobic anoxygenic photosynthesis: Photosynthetic process in which electron donors such as organic matter or sulfide, which do not result in oxygen evolution, are used under aerobic conditions.

Aerobic respiration: Reactions that release energy from food using O_2.

Aerotolerant anaerobes: Microbes that grow under both aerobic and anaerobic conditions, but do not shift from one mode of metabolism to another as conditions change. They obtain energy exclusively by fermentation.

Affinity chromatography: A method of separating biochemical mixtures based on a highly specific interaction such as that between antigen and antibody, enzyme and substrate, or receptor and ligand.

Aflatoxin: A polypeptide secondary fungal metabolite that can cause cancer.

Agar: Complex polysaccharide derived from certain marine algae that is a gelling agent for solid or semisolid microbiological media.

Agarose: Nonsulfated linear polymer consisting of alternating residues of D-galactose and 3,6-anhydro-L-galactose. Agarose is extracted from seaweed, and agarose gels are used as the resolving medium in electrophoresis.

Agarose gel electrophoresis: Technique used for the separation of DNA fragments ranging from 50 bp to several megabases (millions of bases) using a specialized apparatus.

Agglutinates: The visible aggregates or clumps formed by an agglutination reaction.

Agglutination reaction: The formation of an insoluble immune complex by the cross-linking of cells or particles.

Aggregate: A clustered mass of solid, fluffy, or pelletized individual cells that can clog the pores of filters or other fermentation apparatus.

Airborne transmission: The type of infectious organism transmission in which the pathogen is truly suspended in the air and travels over a meter or more from the source to the host.

Akinetes: Specialized, nonmotile, dormant, thick-walled resting cells formed by some cyanobacteria.

Alcoholic fermentation: A fermentation process that produces ethanol and CO_2 from sugars.

Alga (plural, algae): Phototrophic eukaryotic aquatic microorganism.

Aliphatic: Organic compound in which the main carbon structure is a straight chain.

Alkaline soil: Soil having a pH value >7.3.

Alkalinity: Ability of a solution to neutralize acids to the equivalence point of carbonate or bicarbonate.

Alkalophile: Organism that grows best under alkaline conditions (up to a pH of 10.5).

Alkane: Straight chain or branched organic structure that lacks double bonds.

Alkene: Straight chain or branched organic structure that contains at least one double bond.

Allochthonous flora: Organisms that are not indigenous to the soil but enter soil by precipitation, diseased tissues, manure, and sewage.

Allosteric site: Site on the enzyme other than the active site to which a nonsubstrate compound binds.

Allotype: Allelic variants of antigenic determinant(s) found on antibody chains of some, but not all, members of a species, which are inherited as simple Mendelian traits.

Alpha hemolysis: A greenish zone of partial clearing around a bacterial colony growing on blood agar.

Alpha particle: A positively charged particle that consists of two protons and two neutrons bound together.

Alpha proteobacteria: One of the five subgroups of proteobacteria, each with distinctive 16S rRNA sequences.

Alternate current: The direction of current flowing in a circuit is constantly being reversed back and forth.

Alternative complement pathway: An antibody-independent pathway of complement activation that includes the C3–C9 components of the classical pathway and several other serum protein factors (e.g., factor B and properdin).

Alveolar macrophage: A vigorously phagocytic macrophage located on the epithelial surface of the lung alveoli where it ingests inhaled particulate matter and microorganisms.

Amensalism (antagonism): Production of a substance by one organism that is inhibitory to one or more other organisms. The terms antibiosis and allelopathy also describe cases of chemical inhibition.

Ames test: A test that uses a special *Salmonella* strain to test chemicals for mutagenicity and potential carcinogenicity.

Amino acid activation: The initial stage of protein synthesis in which amino acids are attached to transfer RNA molecules.

Amino acids: A class of 20 hydrocarbon molecules that combine to form proteins in living things.

Amino group: An $-NH_2$ group attached to a carbon skeleton as in the amines and amino acids.

Aminoacyl or acceptor site (A site): The site on the ribosome that contains an aminoacyl-tRNA at the beginning of the elongation cycle during protein synthesis; the growing peptide chain is transferred to the aminoacyl-tRNA and lengthens by an amino acid.

Aminoglycoside antibiotics: A group of antibiotics synthesized by *Streptomyces* and *Micromonospora*, which contain a cyclohexane ring and amino sugars; all aminoglycoside antibiotics bind to the small ribosomal subunit and inhibit protein synthesis.

Ammonia oxidation: Test drawn during manufacturing process to evaluate the ammonia oxidation rate for the nitrifiers.

Ammonification: Liberation of ammonium (ammonia) from organic nitrogenous compounds by the action of microorganisms.

Amoeba (plural, amoebae): Protozoa that can alter their cell shape, usually by the extrusion of one or more pseudopodia.

Amoeboid movement: Moving by means of cytoplasmic flow and the formation of pseudopodia (temporary cytoplasmic protrusions of the cytoplasm).

Amphibolic pathways: Metabolic pathways that function both catabolically and anabolically.

Amphitrichous: A cell with a single flagellum at each end.

Amphotericin B: An antibiotic from a strain of *Streptomyces nodosus* that is used to treat systemic fungal infections; it is also used topically to treat candidiasis.

Amplifier: A device for increasing the power of a signal.

Anabolism: Metabolic processes involved in the synthesis of cell constituents from simpler molecules. An anabolic process usually requires energy.

Anaerobe: Microbes grow in the absence of air or oxygen. Some anaerobic organisms are killed by brief exposure to oxygen, whereas oxygen may just retard or stop the growth of others.

Anaerobic respiration: Reactions that release energy from food without using O_2.

Analogue: A voltage or current signal that is a continuous function of the measured parameter.

Analogue meter: Scale and pointer meter capable of indicating a continuous range of values.

Analytical ultracentrifuge: Sample being spun can be monitored in real time through an optical detection system, using ultraviolet light absorption and/or interference optical refractive index sensitive system.

Anamorph: Asexual stage of fungal reproduction in which cells are formed by the process of mitosis.

Anemia: When there are too few blood cells.

Anergy: A state of unresponsiveness to antigens. Absence of the ability to generate a sensitivity reaction to substances that are expected to be antigenic.

Angular aperture: It is the apparent angle of the lens aperture as seen from the focal point.

Anion: A particle having a greater number of electrons than protons; also called a negative ion.

Anion exchange capacity: Sum total of exchangeable anions that a soil can adsorb. Expressed as centimoles of negative charge per kilogram of soil.

Annotation: The process of determining the location of specific genes in a genome map after it has been produced by nucleic acid sequencing.

Anode: An electrode through which electric current flows into a polarized electrical device.

Anoxic: Literally "without oxygen." An adjective describing a microbial habitat devoid of oxygen.

Anoxygenic photosynthesis: Type of photosynthesis in green and purple bacteria in which oxygen is not produced.

Antagonist: Biological agent that reduces the number of disease-producing activities of a pathogen.

Anterior: The front end of an organism.

Antheridium: A male gamete-producing organ, which may be unicellular or multicellular.

Anthrax: An infectious disease of animals caused by ingesting *Bacillus anthracis* spores. Can also occur in humans and is sometimes called woolsorter's disease.

Anthropogenic: Derived from human activities.

Antibiosis: Inhibition or lysis of an organism mediated by metabolic products of the antagonist; these products include lytic agents, enzymes, volatile compounds, and other toxic substances.

Antibiotic: Organic substance produced by one species of organism that in low concentrations will kill or inhibit growth of certain other organisms.

Antibody: Protein that is produced by animals in response to the presence of an antigen and that can combine specifically with that antigen.

Antibody-dependent cell-mediated cytotoxicity (ADCC): The killing of antibody-coated target cells by cells with Fc receptors that recognize the Fc region of the bound antibody. Most ADCC is mediated by NK cells that have the Fc receptor or CD16 on their surface.

Anticodon triplet: The base triplet on a tRNA that is complementary to the triplet codon on mRNA.

Antifoam agent: A chemical added to the fermentation broth to reduce surface tension and counteract the foaming (bubbles) that can be caused by mixing, sparging, or stirring.

Antigen: A substance, usually macromolecular, that induces specific antibody formation.

Antigen-binding fragment (Fab): "Fragment antigen binding." A monovalent antigen-binding fragment of an immunoglobulin molecule that consists of one light chain and part of one heavy chain, linked by interchain disulfide bonds.

Antigenic drift: A small change in the antigenic character of an organism that allows it to avoid attack by the immune system.

Antigenic shift: A major change in the antigenic character of an organism that alters it to an antigenic strain unrecognized by host immune mechanisms.

Antigen-presenting cells (APCs): APCs are cells that take in protein antigens, process them, and present antigen fragments to B cells and T cells in conjunction with class II MHC molecules so that the cells are activated.

Antimetabolite: A compound that blocks metabolic pathway function by competitively inhibiting a key enzyme's use of a metabolite because it closely resembles the normal enzyme substrate.

Antimicrobial agent: An agent that kills microorganisms or inhibits their growth.

Antisense RNA: A single-stranded RNA with a base sequence complementary to a segment of another RNA molecule that can specifically bind to the target RNA and inhibit its activity.

Antiseptic: Agent that kills or inhibits microbial growth but is not harmful to human tissue.

Aorta: Largest artery in the body that takes blood away from the heart.

Aplanospore: A nonflagellated, nonmotile spore that is involved in asexual reproduction.

Apoenzyme: The protein part of an enzyme that also has a nonprotein component.

Apoptosis: Programmed cell death. Apoptosis is a physiological suicide mechanism that preserves homeostasis and occurs during normal tissue turnover.

Aporepressor: An inactive form of the repressor protein, which becomes the active repressor.

Apothecium: Open ascoma of fungi in the phylum Ascomycota.

Arbuscular mycorrhiza (AM): Mycorrhizal type that forms highly branched arbuscules within root cortical cells.

Arbuscule: Special "tree-shaped" structure formed within root cortical cells by arbuscular mycorrhizal fungi.

Archaea: Evolutionarily distinct group (domain) of prokaryotes consisting of the methanogens, most extreme halophiles and hyperthermophiles, and *Thermoplasma*.

Archaebacteria: Older term for the Archaea.

Aromatic: Organic compounds that contain a benzene ring, or a ring with similar chemical characteristics.

Arthritis: Inflammation of joints.

Arthroconidium: A thallic conidium released by the fragmentation or lysis of the hypha. It is not notably larger than the parental hypha, and separation occurs at a septum.

Arthropod: Invertebrate with jointed body and limbs (includes insects, arachnids, and crustaceans).

Arthrospore: A spore resulting from the fragmentation of a hypha.

Artificially acquired active immunity: The type of immunity that results from immunizing an animal with a vaccine. The immunized animal now produces its own antibodies and activated lymphocytes.

Artificially acquired passive immunity: The type of immunity that results from introducing into animal antibodies that have been produced either in another animal or by *in vitro* methods. Immunity is only temporary.

Ascending chromatography: The solvent in chromatography moving against the gravitational force.

Ascocarp: A multicellular structure in ascomycetes lined with specialized cells called asci in which nuclear fusion and meiosis produce ascospores.

Ascogenous hypha: A specialized hypha that gives rise to one or more asci.

Ascoma (plural, ascomata): Fungal fruiting body that contains ascospores; also termed an ascocarp.

Ascospore: Spores resulting from karyogamy and meiosis that are formed within an ascus.

Ascus (plural, asci): Saclike cell of the sexual state formed by fungi in the phylum Ascomycota containing ascospores.

Aseptic: Sterile; free from bacteria, viruses, and contaminants such as foreign DNA.

Aseptic technique: Manipulating sterile instruments or culture media in such a way as to maintain sterility.

Assimilatory nitrate reduction: Conversion of nitrate to reduced forms of nitrogen, generally ammonium, for the synthesis of amino acids and proteins.

Associative dinitrogen fixation: Close interaction between a free-living diazotrophic organism and a higher plant that results in an enhanced rate of dinitrogen fixation.

Associative symbiosis: Close but relatively casual interaction between two dissimilar organisms or biological systems. The association may be mutually beneficial but is not required for accomplishment of a particular function.

Atom: It is the basic unit of matter that consists of a dense central nucleus surrounded by a cloud of negatively charged electrons.

Atomic absorption spectroscopy (AAS): A spectroanalytical procedure for the qualitative and quantitative determination of chemical elements employing the absorption of optical radiation (light) by free atoms in the gaseous state.

Attenuated: Weakened (attenuated) bacteria or viruses often used as vaccines; they can no longer produce disease but still stimulate a strong immune response similar to the natural microbes.

Autogenous infection: An infection that results from a patient's own microbiota, regardless of whether the infecting organism became part of the patient's microbiota subsequent to admission to a clinical care facility.

Autoimmune disease: A disease produced by the immune system attacking self-antigens.

Autoimmunity: Autoimmunity is a condition characterized by the presence of serum autoantibodies and self-reactive lymphocytes. It may be benign or pathogenic.

Autolysins: Enzymes that partially digest peptidoglycan in growing bacteria so that the peptidoglycan can be enlarged.

Autolysis: Spontaneous lysis.

Autoradiography: Detecting radioactivity in a sample, such as a cell or gel, by placing it in contact with a photographic film.

Autotroph: Organism that uses carbon dioxide as the sole carbon source.

Autotrophic nitrification: Oxidation of ammonium to nitrate through the combined action of two chemoautotrophic organisms, one forming nitrite from ammonium and the other oxidizing nitrite to nitrate.

Auxotroph: A mutated prototroph that lacks the ability to synthesize an essential nutrient and therefore must obtain it or a precursor from its surroundings.

Axenic: Literally "without strangers." A system in which all biological populations are defined, such as a pure culture.

Axial filament: The organ of motility in spirochetes. It is made of axial fibrils or periplasmic flagella that extend from each end of the protoplasmic cylinder and overlap in the middle of the cell. The outer sheath lies outside the axial filament.

Axon: Takes a message (impulse) out of nerve cell body to next nerve cell, muscle, or gland.

B cell: Also known as B lymphocyte. A type of lymphocyte derived from bone marrow stem cells that matures into an immunologically competent cell under the influence of the bursa of Fabricius in the chicken and bone marrow in nonavian species.

Bacillus: Bacterium with an elongated, rod shape.

Backflushing: A column switching technique in which a four-way valve placed between the injector and the column allows the mobile phase to flow in either direction.

Bacterial artificial chromosome (BAC): A cloning vector constructed from the *Escherichia coli* F-factor plasmid that is used to clone foreign DNA fragments in *E. coli*.

Bacterial capsule: A compact layer of polysaccharide exterior to the cell wall in some bacteria.

Bacterial photosynthesis: A light-dependent, anaerobic mode of metabolism.

Bactericide: An agent that kills bacteria.

Bacteriochlorophyll: Light-absorbing pigment found in green sulfur and purple sulfur bacteria.

Bacteriocin: Agent produced by certain bacteria that inhibits or kills closely related isolates and species.

Bacteriophage: Virus that infects bacteria; sometimes used as a vector.

Bacteroid: Altered form of cells of certain bacteria. Refers particularly to the swollen, irregular vacuolated cells of rhizobia in nodules of legumes.

Bandwidth: Also known as spectral bandwidth or bandpass; this relates to the physical size of the slit from which the light passes out from the monochromator.

Base composition: Proportion of the total bases consisting of guanine plus cytosine or thymine plus adenine base pairs. Usually expressed as a guanine + cytosine $(G + C)$ value, for example, 60% $G + C$.

Base pair: Two bases on different strands of nucleic acid that join together. In DNA, cytosine (C) always pairs with guanine (G), and adenine (A) always links to thymine (T). In RNA molecules, adenine joins to uracil (U).

Baseline: The baseline is the line drawn by the data system when the only signal from the detector is from the mobile phase.

Basidiospore: Spore resulting from karyogamy and meiosis that usually is formed on a basidium.

Basidium (plural, basidia): Clublike cell of the sexual state formed by fungi in the phylum Basidiomycota.

Batch culture: Large-scale cell culture in which cell inoculum is cultured to a maximum density in a tank or airlift fermentor, harvested, and processed as a batch.

B cell antigen receptor (BCR): A transmembrane immunoglobulin complex on the surface of a B cell that binds an antigen and stimulates the B cell.

Beer–Lambert law (Beer's law): The linear relationship between absorbance and concentration of an absorbing species.

Beta hemolysis: A zone of complete clearing around a bacterial colony growing on blood agar. The zone does not change significantly in color.

Beta particle: High-speed electron or positron, especially one emitted in radioactive decay.

Betaproteobacteria: One of the five subgroups of proteobacteria, each with distinctive 16S rRNA sequences. Members of this subgroup are similar to the alphaproteobacteria metabolically but tend to use substances that diffuse from organic matter decomposition in anaerobic zones.

Bilateral symmetry: Object can only be cut one way to create two matching pieces—mirror images.

Bile: Chemical made by the liver and released by the gallbladder into the small intestine—breaks down fat.

Binary fission: Division of one cell into two cells by the formation of a septum. It is the most common form of cell division in bacteria.

Binocular microscope: Microscope with a head that has two eyepiece lenses.

Binomial nomenclature: System of having two names, genus and specific epithet, for each organism.

Bioaccumulation: Intracellular accumulation of environmental pollutants such as organic materials by living organisms.

Bioactivity: Ability of a molecule to function correctly after it has been delivered to the active site of the body (*in vivo*).

Bioaugmentation: The addition to the environment of microorganisms that can metabolize and grow on specific organic compounds.

Bioavailability: Measure of the true rate and the total amount of drug that reaches the target tissue after administration.

Biochemical oxygen demand (BOD): Amount of dissolved oxygen consumed in 5 days by biological processes breaking down organic matter.

Biodegradable: Substance capable of being decomposed by biological processes.

Biodegradation: The breakdown of organic substances by microorganisms.

Biofilm: Microbial cells encased in an adhesive, usually a polysaccharide material, and attached to a surface.

Biogeochemical cycling: The oxidation and reduction of substances carried out by living organisms and/or abiotic processes that result in the cycling of elements within and between different parts of the ecosystem (soil, aquatic environment, and atmosphere).

Bioinsecticide: A pathogen that is used to kill or disable unwanted insect pests. Bacteria, fungi, or viruses are used, either directly or after manipulation, to control insect populations.

Biologic: A therapeutic agent derived from living things.

Biologic transmission: A type of vector-borne transmission in which a pathogen goes through some morphological or physiological change within the vector.

Biology: The study of life.

Bioluminescence: The production of light by living cells, often through the oxidation of molecules by the enzyme luciferase.

Biomagnification: Increase in the concentration of a chemical substance as it progresses to higher trophic levels of a food chain.

Biopharmaceutical: Therapeutic product created through the genetic manipulation of living things, including (but not limited to) proteins and monoclonal antibodies, peptides, and other molecules that are not chemically synthesized, along with gene therapies, cell therapies, and engineered tissues.

Bioprocessing: The use of organisms or biologically derived macromolecules to carry out enzymatic reactions or to manufacture products.

Bioreactor: A vessel capable of supporting a cell culture in which a biological transformation takes place (also called a fermentor or reactor).

Bioremediation: The process by which living organisms act to degrade or transform hazardous organic contaminants.

Biosynthesis: Production of needed cellular constituents from other, usually simpler, molecules.

Biotechnology: Use of living organisms to carry out defined physiochemical processes having industrial or other practical application.

Blood pressure: How hard the heart is pushing the blood against the artery walls.

Breathing: Physically moving air into and out of the body.

Bright-field illumination: Lighting that shines through a transparent specimen that appears dark against a bright background; also called as transmitted light illumination opposite of dark-field illumination.

Brittle bone disease: Genetic disease that causes bones to break easily.

Bronchitis: Inflammation of the bronchi in the lungs.

Broth: The contents of a microbial bioreactor cells, nutrients, waste, and so on.

Bruise: When small vessels break and blood pools under the skin.

Budding: Asexual reproduction (usually for yeast) beginning as a protuberance from the parent cell that grows and detaches to form a smaller, daughter cell.

Buffer: Measure of the ability of the solution to resist pH change when a strong acid or base is added.

Bursa: Cover on joints.

Bursitis: Inflammation of joint cover (bursa).

Burst size: The number of phages released by a host cell during the lytic life cycle.

Calorie: Unit used to measure the amount of energy in food.

Calvin cycle: Biochemical route of carbon dioxide fixation in many autotrophic organisms.

Capacitance: The ability of a capacitor to store energy in an electric field.

Capillary: The smallest blood vessel where diffusion happens.

Capillary gel electrophoresis: Bundle of extremely narrow gel-filled tubes used to significantly speed the electrophoresis process of separating biological material.

Capsid: A protein shell comprising the main structural unit of a virus particle.

Capsule: Compact layer of polysaccharide exterior to the cell wall in some bacteria.

Carbohydrate: Molecule made of carbon, hydrogen, and oxygen used for energy and to build cell structures.

Carbon cycle: Sequence where carbon dioxide is converted to organic forms by photosynthesis or chemosynthesis, recycled through the biosphere, with partial incorporation into sediments, and ultimately returned to its original state through respiration or combustion.

Carbon fixation: Conversion of carbon dioxide or other single-carbon compounds to organic forms such as carbohydrates.

Carbon–nitrogen (C/N) ratio: Ratio of the mass of organic carbon to the mass of nitrogen in soil or organic material.

Carboxyl group: A −COOH group attached to a carbon skeleton as in the carboxylic acids and fatty acids.

Carcinogen: Substance that causes the initiation of tumor formation, frequently a mutagen.

Cardiac muscle: Branched and striped muscle that the heart is made of.

Cardiovascular: Word used to refer to the system of the heart and blood vessels.

Carnivore: Animal whose body is designed to digest and use other animals for food.

Carrier: Term used most often in affinity chromatography, which refers to the support that is used to carry the active ligand, usually by a covalent bond. It can also refer to the support in other chromatographic modes.

Carrier ampholyte(s): Small soluble molecules that are capable of acting as an acid or a base in standard 2D gel electrophoresis.

Cartilage: Protective covering on ends of bones.

Cascade effects: A series of events that result from one initial cause.

Catabolism: Biochemical processes involved in the breakdown of organic compounds, usually leading to the production of energy.

Catabolite repression: Transcription-level inhibition of a variety of inducible enzymes by glucose or other readily used carbon source.

Catabolites: Waste products of catabolism, by which organisms convert substances into excreted compounds.

Catalyst: Substance that promotes a chemical reaction by lowering the activation energy without itself being changed in the end. Enzymes are a type of catalyst.

Cathode: An electrode through which electric current flows out of a polarized electrical device.

Cation: A particle having a lesser number of electrons than protons; also called a positive ion.

Cation exchange capacity (CEC): Sum of exchangeable cations that a soil can adsorb at a specific pH. Expressed as centimoles of positive charge per kilogram of soil (cmolc/kg).

Cell: Fundamental unit of living matter.

Cell constant: The ratio of distance between conductance-titration electrodes to the area of the electrodes, measured from the determined resistance of a solution of known specific conductance.

Cell cycle: The sequence of events in a cell's growth–division cycle between the end of one division and the end of the next.

Cell theory: All cells come from other cells; living things are built of cells and cells run the organism.

Cell wall: Layer or structure that lies outside the cytoplasmic membrane; it supports and protects the membrane and gives the cell shape.

Cell-mediated immunity: The type of immunity that results from T cells coming into close contact with foreign cells or infected cells to destroy them; it can be transferred to a nonimmune individual by the transfer of cells.

Cellular respiration: The chemical reactions in a cell that release energy from food.

Cellular slime molds: Slime molds with a vegetative phase consisting of amoeboid cells that aggregate to form a multicellular pseudoplasmodium; they belong to the division Acrasiomycota.

Cellulitis: A diffuse spreading infection of subcutaneous skin tissue caused by streptococci, staphylococci, or other organisms. The tissue is inflamed with edema, redness, pain, and interference with function.

Cellulose: Glucose polysaccharide (with beta-1,4-linkage) that is the main component of plant cell walls. Most abundant polysaccharide on earth.

Central nervous system: Brain and spinal cord.

Cephalosporin: A group of b-lactam antibiotics derived from the fungus *Cephalosporium*, which share the 7-aminocephalosporanic acid nucleus.

Cerebellum: Part of brain that coordinates movement and filters sensory input.

Chemical oxygen demand (COD): The amount of oxygen in mg/L required to oxidize both organic and oxidizable inorganic compounds.

Chemoautotroph: Organism that obtains energy from the oxidation of chemical, generally inorganic, compounds and carbon from carbon dioxide.

Chemoheterotroph: Organism that obtains energy and carbon from the oxidation of organic compounds.

Chemolithotroph: Organism that obtains energy from the oxidation of inorganic compounds and uses inorganic compounds as electron donors.

Chemoorganotroph: Organism that obtains energy and electrons (reducing power) from the oxidation of organic compounds.

Chemostat: Continuous culture device usually controlled by the concentration of limiting nutrient and dilution rate.

Chemotrophs: Organisms that obtain energy from the oxidation of chemical compounds.

Chitin: A tough, resistant, nitrogen-containing polysaccharide forming the walls of certain fungi, the exoskeleton of arthropods, and the epidermal cuticle of other surface structures of certain protists and animals.

Chlamydiae: Members of the genus *Chlamydia*: Gram-negative, coccoid cells that reproduce only within the cytoplasmic vesicles of host cells using a life cycle that alternates between elementary bodies and reticulate bodies.

Chlamydospore: Thick-walled resting structure that forms from the cell wall of a fungal hypha.

Chloramines: Compounds formed by the reaction of hypochlorous acid (or aqueous chlorine) with ammonia.

Chloramphenicol: A broad-spectrum antibiotic that is produced by *Streptomyces venezuelae* or synthetically; it binds to the large ribosomal subunit and inhibits the peptidyl transferase reaction.

Chlorophyll: Green pigment required for photosynthesis.

Chloroplast: Chlorophyll-containing organelle of photosynthetic eukaryotes.

Chromatin: The DNA-containing portion of the eukaryotic nucleus; the DNA is almost always complexed with histones.

Chromatography: Any technique used to separate different species of molecules (or ions) by subjecting them to two different carrier phases: mobile and stationary phases.

Chromogen: A colorless substrate that is acted on by an enzyme to produce a colored end product.

Chromogenic: Producing color; a chromogenic colony is a pigmented colony.

Chromophore group: A chemical group with double bonds that absorbs visible light and gives dye its color.

Chromosome: Genetic element carrying information essential to cellular metabolism.

Chyme: Liquefied food in digestive system.

Chytrid: Fungal organism in the phylum Chytridiomycota that consists of a spherical cell from which short thin filamentous branches (rhizoids) grow that resemble fine roots.

Cilia: Tiny hairlike structures for locomotion (microbes) or sweeping material through a tube (multicell).

Classification: Process of grouping based on common traits.

Clay: Soil particle <0.002 mm in diameter.

Clone: Population of cells all descended from a single cell.

Cloning vector: DNA molecule that is able to bring about the replication of foreign DNA fragments.

Coagulants: Chemicals that cause very fine particles to clump (floc) together into larger particles.

Coagulase: An enzyme that induces blood clotting; it is characteristically produced by pathogenic staphylococci.

Coccus: Spherical bacterial cells.

Codon: A sequence of three nucleotides in mRNA that directs the incorporation of an amino acid during protein synthesis or signals the start or stop of translation.

Coenocytic: Fungal hypha without crosswalls (septa), so that the nuclei present in the cytoplasm are free-floating and mobile.

Coenzyme: Low molecular weight chemical that participates in an enzymatic reaction by accepting and donating electrons or functional groups.

Cofactor: The nonprotein component of an enzyme; it is required for catalytic activity.

Cohesion: Force holding a solid or liquid together, owing to attraction between like molecules.

Colicin: A plasmid-encoded protein that is produced by enteric bacteria and binds to specific receptors on the cell envelope of sensitive target bacteria, where it may cause lysis or attack specific intracellular sites such as ribosomes.

Coliform: Gram-negative, nonspore-forming facultative rod that ferments lactose with gas formation within 48 h at 35°C.

Colloid fraction: Organic and inorganic matter with very small particle size and a correspondingly large surface area per unit of mass.

Colloids: Very small, finely divided solids (particles that do not dissolve) that remain dispersed in a liquid for a long time due to their small size and electrical charge.

Colonization: Establishment of a community of microorganisms at a specific site or ecosystem.

Colony: Clone of bacterial cells on a solid medium that is visible to the naked eye.

Colony forming units (CFU): The number of microorganisms that can form colonies when cultured using spread plates or pour plates, an indication of the number of viable microorganisms in a sample.

Colorless sulfur bacteria: A diverse group of nonphotosynthetic proteobacteria that can oxidize reduced sulfur compounds such as hydrogen sulfide.

Combinatorial biology: Introduction of genes from one microorganism into another microorganism to synthesize a new product or a modified product, especially in relation to antibiotic synthesis.

Community: All organisms that occupy a common habitat and interact with one another.

Compact bone: Strong outer layer of bone.

Competent: In a genetic sense, the ability to take up DNA.

Competition: Rivalry between two or more species for a limiting factor in the environment that usually results in reduced growth of participating organisms.

Competitive exclusion principle: Two competing organisms overlap in resource use, which leads to the exclusion of one of the organisms.

Complement system: A group of plasma proteins that plays a major role in an animal's defensive immune response.

Complementary: In reference to base pairing, the ability of two polynucleotide sequences to form a double-stranded helix by hydrogen bonding between bases in the two sequences.

Complementary DNA (cDNA): A DNA copy of an RNA molecule (e.g., a DNA copy of an mRNA).

Complex medium: Medium whose precise chemical composition is unknown. Also called undefined medium.

Complex viruses: Viruses with capsids having a complex symmetry that is neither icosahedral nor helical.

Contraction: When muscle fiber shortens in order to cause motion/movement.

Control setup: The setup in an experiment that is used as a normal comparison.

Culture: Population of microorganisms cultivated in an artificial growth medium.

Cyanobacterium: Prokaryotic, oxygenic phototrophic bacterium containing chlorophyll *a* and phycobilins, formerly the "blue-green algae."

Cyst: Resting stage formed by some bacteria, nematodes, and protozoa in which the whole cell is surrounded by a protective layer; not the same as endospore.

Cytochrome: Iron-containing porphyrin ring (e.g., heme) complexed with proteins that act as electron carriers in an electron transport chain.

Cytokine: A general term for nonantibody proteins, released by a cell in response to inducing stimuli, which are mediators that influence other cells.

Cytoplasm: Cellular contents inside the cell membrane, excluding the nucleus.

Cytoplasmic membrane: Selectively permeable membrane surrounding the cytoplasm.

Dalton: The unit of molecular weight, equal to the weight of a hydrogen atom.

Dark-field microscopy: Technique used to enhance the contrast in unstained specimens.

Dark-field plate: Circular iris that sits on the base of the microscope above the light source and reflects the light horizontally to the specimen.

Decay: The disintegration of the nucleus of an unstable nuclide by the spontaneous emission of charged particles, photons, or both.

Decomposer: Heterotrophic organism that breaks down organic compounds.

Decomposition: Chemical breakdown of a compound into simpler compounds, often accomplished by microbial metabolism.

Defined medium: Medium whose exact chemical composition is quantitatively known.

Degassing: Process of removing dissolved gas from the mobile phase before or during use.

Degradation: Process whereby a compound is usually transformed into simpler compounds.

Denaturation: Process where double-stranded DNA unwinds and dissociates into two single strands.

Denaturing gradient gel electrophoresis (DGGE): A technique for screening of single nucleotide polymorphisms (SNPs).

Dendrite: Picks up impulse for nerve and carries it to nerve cell body.

Denitrification: Reduction of nitrate or nitrite to molecular nitrogen or nitrogen oxides by microbial activity (dissimilatory nitrate reduction) or by chemical reactions involving nitrite (chemical denitrification).

Density: A physical property of matter, as each element and compound has a unique density associated with it.

Density gradient centrifugation: Sedimentation based on a spatial variation in density over an area.

Deoxyribonucleic acid (DNA): Polymer of nucleotides connected via a phosphate-deoxyribose sugar backbone; the genetic material of the cell.

Dependent variable: The part of the experiment that changes depending on what you do and what you are measuring as your results.

Descending chromatography: The solvent in chromatography moving toward the gravitational force.

Detector: An electronic device that quantitatively discerns the presence of the separated components as they elute.

Diatom: Alga with siliceous cell walls that persist as a skeleton after death.

Diazotroph: Organism that can use dinitrogen as its sole nitrogen source, that is, capable of N_2 fixation.

Differential centrifugation: Sedimentation based on differences in size and density of particles.

Differential medium: Culture medium with an indicator, such as a dye, which allows various chemical reactions to be distinguished during growth.

Diffusion: Movement of molecules from where there is a lot to where there is less; does not cost cell any energy.

Digestion: The breakdown of food to get out nutrients/energy.

Digital: An output signal that represents the size of an input in the form of a series of discrete quantities.

Dikaryon: Two nuclei present in the same hyphal compartment; they constitute a homokaryon when both nuclei are genetically the same or a heterokaryon when each nucleus is genetically different from the other.

Dilution plate count method: Method for estimating the viable numbers of microorganisms in a sample.

Dinitrogen fixation: Conversion of molecular dinitrogen (N_2) to ammonia and subsequently to organic combinations or to forms useful in biological processes.

Diploid: In eukaryotes, an organism or cell with two chromosome complements, one derived from each haploid gamete.

Direct count: Method of estimating the total number of microorganisms in a given mass of soil by direct microscopic examination.

Disinfectant: Agent that kills microorganisms.

Dissimilatory nitrate reduction to ammonium (DNRA): Use of nitrate by organisms as an alternate electron acceptor in the absence of oxygen resulting in the reduction of nitrate to ammonium.

Dissolved oxygen (DO): A measure of the oxygen dissolved in water expressed in milligrams per liter.

Dissolved solids: Chemical substances either organic or inorganic that are dissolved in a waste stream and constitute the residue when a sample is evaporated to dryness.

DNA fingerprinting: Molecular genetic techniques to assess possible differences among DNA in a sample.

DNA library: Collection of cloned DNA fragments that in total contain genes from the entire genome of an organism; also called a gene library.

DNA vaccine: A nucleic acid vaccine; genes coding for specific antigenic proteins are injected to produce those antigens and trigger an immune response.

Domain: Highest level of biological classification, superseding kingdoms. The three domains of biological organisms are the Bacteria, the Archaea, and the Eukarya.

Dominant gene: Gene for trait that is stronger and more likely to be seen.

Dominant trait: The trait that is stronger. Trait will be seen in organism even if it only has one copy of the dominant gene (e.g., Ss, SS).

Double beam: Light source is split in two. One beam illuminates the reference cell holder and the other illuminates the sample.

Double helix: Spiral shape—like a spiral staircase with handrails on each side; the shape DNA has.

Doubling time: Time needed for a population to double in number or biomass.

Downstream processing: Bioprocessing steps following fermentation and/or cell culture; a sequence of separation and purification activities needed to obtain the required product at the necessary level of purity.

dpm: Disintegrations per minute of the isotope in terms of efficiency.

Dynamic equilibrium: The balance every living thing or system must keep in order to stay alive.

Ecology: Science that studies the interrelations among organisms and between organisms and their environment.

Ecosystem: Community of organisms and the environment in which they live.

Ectomycorrhiza: Mycorrhizal type in which the fungal mycelia extend inward, between root cortical cells, to form a network (Hartig net) and outward into the surrounding soil.

Efficacy: The ability of a substance to produce a desired clinical effect; its strength and effectiveness.

Efficiency: Measurement of the sharpness of an eluting band or peak.

Effluent: Wastewater or other liquid—raw (untreated) and partially or completely treated—flowing from a reservoir, basin, treatment process, or treatment plant.

Electromotive force: The maximum potential difference between two electrodes of a galvanic or voltaic cell.

Electrode: An electrical conductor used to make contact with a nonmetallic part of a circuit.

Electrolyte: Any substance containing free ions that make the substance electrically conductive.

Electromagnetic field: Physical field produced by moving electrically charged objects.

Electromagnetic interference (EMI): Disturbance that affects an electrical circuit due to either electromagnetic induction or electromagnetic radiation emitted from an external source.

Electromagnetic radiation (EMR): Form of energy that exhibits wavelike behavior as it travels through space.

Electromagnetic technique: Active method that uses an electromagnetic (EM) signal to detect variations in subsurface conductivity.

Electron: Part of an atom that carries a positive charge and orbits around the outside of the atom.

Electron acceptor: Substance that accepts electrons during an oxidation–reduction reaction. An electron acceptor is an oxidant.

Electron beam: Beam of electrons to illuminate the specimen and produce a magnified image.

Electron capture detector (ECD): Very sensitive detector used in GLC/GC for analysis of halogenated compounds.

Electron donor: Substance that donates electrons in an oxidation–reduction reaction. An electron donor is a reductant.

Electron microscope: Type of microscope that uses electrons rather than light to create an image of the target.

Electron spin resonance (ESR): A sophisticated spectroscopic technique that detects free radicals or inorganic complexes in chemical and biological systems.

Electron transport chain: Final sequence of reactions in biological oxidations composed of a series of oxidizing agents arranged in order of increasing strength and terminating in oxygen.

Electrophoresis: Separation technique based on the motion of colloidal particles in an electric field.

Element: A collection of atoms of one type; each square of the periodic table shows one element.

Eluent: The mobile phase in a chromatographic separation.

Elution order: The order in which the components pass out of the column in a separation process.

Eluviation: Removal of soil material from a layer of soil as a suspension.

Embden–Meyerhof–Parnas pathway (Embden–Meyerhof pathway; EMP pathway): A biochemical pathway that degrades glucose to pyruvate.

Embryo: Fertilized egg (zygote) that is implanted into the uterus.

Emission flame spectroscopy: Used for the solution containing the relevant substance to be analyzed that is drawn into the burner and dispersed into the flame as a fine spray.

Emission spectrum: The spectrum of frequencies of EMR emitted by the element's atoms or the compound's molecules when they are returned to a lower energy state.

Emulsion: A liquid mixture of two or more liquid substances not normally dissolved in one another, one liquid held in suspension in the other.

Endergonic reaction: Chemical reaction that proceeds with the consumption of energy.

Endoenzyme: Enzyme that operates along the internal portions of a polymer.

Endogenous: Growing or developing from a cell or organism or arising from causes within the organism.

Endogenous respiration: A reduced level of respiration (breathing) in which organisms break down compounds within their own cells to produce the oxygen they need.

Endomycorrhiza: Mycorrhizal association with intracellular penetration of the host root cortical cells by the fungus as well as outward extension into the surrounding soil.

Endonuclease: A restriction enzyme that breaks up nucleic acid molecules at specific sites along their length.

Endophyte: Organism growing within a plant. The association may be symbiotic or parasitic.

Endoplasmic reticulum: An extensive array of internal membranes in eukaryotes.

Endospore: Differentiated cell formed within the cells of certain Gram-positive bacteria and extremely resistant to heat and other harmful agents.

Endotoxin: A poison in the form of a fat/sugar complex (lipopolysaccharide) that forms a part of the cell wall of some types of bacteria.

Enrichment culture: Technique in which environmental (including nutritional) conditions are controlled to favor the development of a specific organism or group of organisms.

Enteric bacteria: General term for a group of bacteria that inhabit the intestinal tract of humans and other animals.

Entner–Doudoroff pathway (ED pathway): A pathway that converts glucose to pyruvate and glyceraldehyde-3-phosphate by producing 6-phosphogluconate and then dehydrating it.

Envelope: A lipid membrane enveloping a virus particle.

Enzyme: Functional proteins within or derived from a living organism that functions as a catalyst to promote specific reactions.

Enzyme-linked immunosorbent assay (ELISA): A test to measure the concentration of antigens or antibodies.

Episome: Plasmid that replicates by inserting itself into the bacterial chromosome.

Epithelium (epithelial): The layer(s) of cells between an organism or its tissues or organs and their environment.

Epitope: The region of an antigen to which the variable region of an antibody binds.

Ericoid mycorrhiza: Type of mycorrhiza found on plants in the Ericales.

Essential amino acid: Amino acids that the body cannot make using basic atoms.

Eukaryote: Organism having a unit membrane-bound nucleus and usually other organelles.

Eutrophic: Having high concentrations of nutrients optimal, or nearly so, for plant or animal growth.

Evolution: Making of new species through the gradual change of old species using natural selection.

Exclusion chromatography (size exclusion chromatography): A chromatographic method in which molecules in solution are separated by their size and, in some cases, molecular weight.

Excretion: Removing cell waste from the body.

Excretory system: Group of organs that removes cellular waste from the body.

Exergonic reaction: Chemical reaction that proceeds with the liberation of energy.

Exoenzyme: Enzyme that acts at the end of a polymer cleaving off monomers and dimers and sometimes larger chain fragments.

Exogenous: Developing from outside; originating externally.

Exponential growth: Period of sustained growth of a microorganism in which the cell number constantly doubles within a fixed time period.

Exponential phase: Period during the growth cycle of a population in which growth increases at an exponential rate; referred to as logarithmic phase.

Express: To translate a cell's genetic information, stored in its DNA (gene), into a specific protein.

Expression system: Organism chosen to manufacture (by expression) a protein of interest through recombinant DNA technology.

Expression vector: A way of delivering foreign genes to a host, creating a recombinant organism that will express the desired protein.

Extracellular: Outside the cell.

Exudate: Low molecular weight metabolites that leak from plant roots into soil.

Eyepiece: Referred to as an ocular; the eyepiece is the lens nearest to your eye.

Eyepiece reticles: Sometimes also called "eyepiece micrometers." They are clear circular glass inserts with a scale inscribed on them.

Facultative organism: Organism that can carry out both options of a mutually exclusive process (e.g., aerobic and anaerobic metabolism).

Fat: Molecule made of one glycerol and two fatty acids.

Fat soluble vitamin: Vitamins that get stored in fat.

Feces: Waste made up of the part of food that is not digested.

Feedback inhibition: Inhibition by an end product of the biosynthetic pathway involved in its synthesis.

Fermentation: Process by which microorganisms break down complex organic substances generally in the absence of oxygen to produce alcohol and carbon dioxide.

Fermentor: A bioreactor used to grow bacteria or yeasts in liquid culture.

Fertilization: Fusion of a sperm and an egg.

Fertilizer: Any organic or inorganic material of natural or synthetic origin (other than liming materials) added to a soil to supply one or more elements essential to plant growth.

Fetus: An embryo that has developed an umbilical cord.

Filamentous: In the form of very long rods; many times longer than wide (for bacteria) and in the form of long branching strands (for fungi).

Fimbria (plural, fimbriae): Short filamentous structure on a bacterial cell.

Fine focus: A knob used to fine-tune the focus of a specimen in conjunction with the coarse focus.

Fission: Type of cell division in which overall cell growth is followed by formation of a crosswall that typically divides the fully grown cell into two similar or identical cells.

Flagella: Long hairlike tail used for swimming by some microbes and sperm.

Flagellate: Protozoan that moves by means of one to several flagella.

Flagellum (plural, flagella): Whiplike tubular structure attached to a microbial cell responsible for motility.

Flame ionization detector (FID): A common type of detector used in gas chromatography and supercritical fluid chromatography.

Flame photometer: Uses a flame that evaporates the solvent and also sublimates and atomizes the metal and then excites a valence electron to an upper energy state.

Flame photometric detector (FPD): Sulfur and phosphorus containing hydrocarbons burn in a flame, producing chemiluminescent species that are monitored at selective wavelengths.

Floc: A fluffy aggregate that resembles a woolly cloud.

Fluorescent confocal microscopes: Usually laser-based; designed to provide high-resolution images without interference from out-of-focus portions of the sample.

Flow rate (*F*): The volumetric rate of flow of mobile phase through an LC column.

Fluorescence: The ability to emit light of a certain wavelength when activated by light of another wavelength.

Fluorescence detectors: Fluorescence detectors project a specified wavelength of light into the sample, causing the component of interest to fluoresce, and the emitted light is detected.

Fluorescence illumination: Provides cooler and brighter light than tungsten. This is beneficial when viewing slides for long periods of time or observing live specimens, such as protozoa.

Fluorescence microscope: An optical microscope used to study properties of organic or inorganic substances using the phenomena of fluorescence and phosphorescence instead of, or in addition to, reflection and absorption.

Fluorescence spectroscopy: A type of electromagnetic spectroscopy that analyzes fluorescence from a sample.

Fluorescent: Able to emit light of a certain wavelength when activated by light of a shorter wavelength.

Fluorescent antibody: Antiserum conjugated with a fluorescent dye, such as fluorescein or rhodamine.

Fluxes: Rate of emission, sorption, or deposition of a material from one pool to another.

Focal length: Of an optical system, a measure of how strongly the system converges or diverges light.

Focal planes: Defined as the planes, perpendicular to the optic axis, which pass through the front.

Focus: The ability to achieve a clear image, typically achieved by moving either the eyepiece tubes or the stage.

Food chain: Movement of nutrients from one life form to another as a result of the different feeding habits and dietary requirements of organisms in an ecosystem.

Food web: Diagram of the interconnections of nutrient flow through a food chain.

Free energy: Intrinsic energy contained in a given substance that is available to do work, particularly with respect to chemical transformations; designated D G.

Fruiting body: Macroscopic reproductive structure produced by some fungi, such as mushrooms, and some bacteria, including myxobacteria.

Frustule: Siliceous wall and protoplast of a diatom.

Fulvic acid: Yellow organic material that remains in solution after removal of humic acid by acidification.

Fungi: Nonphototrophic, eukaryotic microorganisms that contain rigid cell walls. Kingdom of multicellular decomposers; rarely do locomotion.

Fungistasis: Suppression of germination of fungal spores or other resting structures in natural soils as a result of competition for available nutrients, presence of inhibitory compounds, or both.

Fusiform: Spindle-shaped; tapered at both ends.

Gamete: In eukaryotes, the haploid cell analogous to sperm and egg, which results from meiosis.

Gamma particle: Denoted as γ EMR of high frequency (very short wavelength) having energies in a range from 10,000 (10^4) to 10 million (10^7) eV.

Gas chromatography: Chromatographic technique in which the stationary phase is a solid or an immobile liquid, and the mobile phase is gaseous. The gaseous samples are separated based on their differential adsorption to the stationary phase.

Gas ionization: The condition of being dissociated into ions (as by heat or radiation or chemical reaction or electrical discharge).

Gas liquid chromatography (GLC): The mobile phase is a carrier gas, usually an inert gas such as helium or an unreactive gas such as nitrogen.

Gas vacuole: A subcellular organelle, found only in prokaryotes, which consists of clusters of hollow, cylindrical, gas-filled vesicles (gas vesicles).

Gas–solid chromatography (GSC): A subclassification of gas chromatography where the mobile phase is a gas and the stationary phase is a solid.

Geiger–Müller counter: An instrument that detects and measures the intensity of radiation.

Gel: Inert polymer, usually made of agarose or polyacrylamide, that separates macromolecules such as nucleic acids or proteins during electrophoresis.

Gene: A segment of DNA that carries the instructions for one trait.

Gene cloning: Isolation of a desired gene from one organism and its incorporation into a suitable vector for the production of large amounts of the gene.

Gene expression: An important stage of viral replication at which viral genetic information is expressed: one of the major control points in replication.

Gene probe: A strand of nucleic acid that can be labeled and hybridized to a complementary molecule from a mixture of other nucleic acids.

Generation time: Time needed for a population to double in number or biomass.

Genetic code: Information for the synthesis of proteins contained in the nucleotide sequence of a DNA molecule (or in certain viruses, of an RNA molecule).

Genetic engineering: Altering the genetic structure of an organism through technological means rather than traditional breeding.

Genome: Complete set of genes present in an organism.

Genome replication: The stage of viral replication at which a viral genome is copied to form new progeny genomes.

Genotype: The possible gene combinations from the cross of two parents.

Genus (plural, genera): The first name of the scientific name (binomial); the taxon between family and species.

Germ cell: The "sex cells" in higher animals and plants that carry only half of the organism's genetic material and can combine to develop into new living things.

Glass electrode: A type of ion-selective electrode made of a doped glass membrane that is sensitive to a specific ion.

Glass probe: Thermistors for use as precision sensing elements where curve matched; interchangeability is required for precise temperature control and precision temperature indication.

Glucose: Chemical used by living things for energy.

Glycolysis: Reactions of the Embden–Meyerhof (glycolytic) pathway in which glucose is oxidized to pyruvate.

Glycosidase: Enzyme that hydrolyzes a glucosidic linkage between two sugar molecules.

Glycosylation: Adding one or more carbohydrate molecules onto a protein (a glycoprotein) after it has been built by the ribosome; a posttranslational modification.

GMPs: Good manufacturing practices required by FDA regulations.

Golgi body: A cell organelle consisting of stacked membranes where posttranslational modifications of proteins are performed; also called Golgi apparatus.

Gram stain: Differential stain that divides bacteria into two groups.

Green (sulfur) bacteria: Anoxygenic phototrophs containing chlorosomes and bacteriochlorophyll c, c_s, d, or e and light-harvesting chlorophyll.

Groundwater: Portion of the water below the surface of the ground at a pressure equal to or greater than atmospheric.

Growth: In microbiology, an increase in both cell number and cellular constituents.

Growth factor: Organic compound necessary for growth because it is an essential cell component or precursor of such components and cannot be synthesized by the organism itself.

Growth hormone: A protein produced in the pituitary gland to control cell growth.

Growth rate: The rate at which growth occurs, usually expressed as the generation time.

Growth rate constant: Slope of $\log_{10}$ of the number of cells per unit volume plotted against time.

Growth yield coefficient: Quantity of biomass carbon formed per unit of substrate carbon consumed.

Guard column: A small column placed between the injector and the analytical column.

Habitat: Place where an organism lives.

Hemagglutination inhibition: An assay used for certain types of viruses that are able to agglutinate red blood cells.

Hemocytometer: A device originally designed for the counting of blood cells.

Half-life: The period of time it takes for the amount of a substance undergoing decay to decrease by half.

Halogen lamp: Provides the brightest illumination. The best microscope has halogen lighting, and stereomicroscope with top lighting also uses halogen lighting.

Halophile: Organism requiring or tolerating a saline environment.

Halotolerant: An organism capable of growing in the presence of NaCl but not requiring it.

Hanging drop method: One of the commonest methods to demonstrate bacterial motility.

Haploid: In eukaryotes, an organism or cell containing one chromosome complement and the same number of chromosomes as the gametes.

Hapten: A substance not inducing antibody formation but able to combine with a specific antibody.

Heart attack: When blood flow is blocked to a part of the heart and the heart muscle begins to die.

Heavy metals: Those metals that have densities >5.0 Mg/m^3.

Hemoglobin: A protein in red blood cells that holds onto the oxygen during transport.

Hemorrhoids: Varicose veins of the anus.

Herbivore: Animal whose body is designed to digest and use plants for food.

Heredity: Passing of traits from generation to generation.

Heterocyst: Differentiated cyanobacterial cell that carries out dinitrogen fixation.

Heterofermentation: Any fermentation in which there is more than one major end product.

Heterokaryon: Hypha that contains at least two genetically dissimilar nuclei.

Heterolactic fermentation: A type of lactic acid fermentation in which sugars (e.g., lactose, glucose) are fermented to a range of products.

Heterothallic: Hyphae that are incompatible with each other, each requiring contact with another hypha of compatible mating type that, upon fusion, forms a dikaryon or a diploid.

Heterotroph: Organism capable of deriving carbon and energy for growth and cell synthesis from organic compounds.

Heterotrophic nitrification: Biochemical oxidation of ammonium to nitrite and nitrate by heterotrophic microorganisms.

Hexose monophosphate pathway: A metabolic pathway present in a wide range of prokaryotic and eukaryotic microorganisms as well as in plants and animals.

High blood pressure: When the force of blood against the blood vessel walls is too much.

High-performance liquid chromatography (HPLC): A chromatographic technique that can separate a mixture of compounds and is used in biochemistry and analytical chemistry to identify, quantify, and purify the individual components of the mixture.

High-performance thin-layer chromatography (HPTLC): An enhanced form of thin-layer chromatography used to increase the resolution achieved and to allow more accurate quantitative measurements.

Holomorph: Whole fungus consisting of all sexual and asexual stages in its life cycle.

Homeostasis: The balance every living thing or system must keep in order to stay alive (same as dynamic equilibrium).

Homofermentation: Any fermentation in which there is only one major end product.

Homokaryon: Fungal hypha in which all nuclei are genetically identical.

Homolactic fermentation: A type of lactic acid fermentation in which sugars (e.g., glucose, lactose) are converted entirely, or almost entirely, to lactic acid.

Homothallic: Hyphae that are self-compatible in that sexual reproduction occurs in the same organism by meiosis and genetic recombination.

Horizontal axis: The line across the bottom of a graph.

Hormone: A protein released by an endocrine gland to travel in the blood and act on tissues at another location in the body.

Host: Organism capable of supporting the growth of a virus or other parasite.

Humic acid: Dark-colored organic material extracted from soil by various reagents (e.g., dilute alkali) that is precipitated by acid (pH 1–2).

Humic substances: Series of relatively high molecular weight, brown-to-black substances formed by secondary synthesis reactions.

Humification: Process whereby the carbon of organic residues is transformed and converted to humic substances through biochemical and chemical processes.

Humus: Total of the organic compounds in soil exclusive of undecayed plant and animal tissues, their "partial decomposition" products, and the soil biomass.

Hybrid: Organism with a mixed genotype.

Hybridization: Natural formation or artificial construction of a duplex nucleic acid molecule by complementary base pairing between two nucleic acid strands derived from different sources.

Hybridoma: An immortalized cell line (usually derived by fusing B lymphocyte cells with myeloma tumor cells) that secretes desirable antibodies.

Hydrocarbon: Any chemical compound containing only carbon and hydrogen elements.

Hydrogen bond: Chemical bond between a hydrogen atom of one molecule and two unshared electrons of another molecule.

Hydrogen-oxidizing bacterium: Facultative lithotrophs that, in the absence of an oxidizable organic source, oxidize H_2 for energy and synthesize carbohydrates with carbon dioxide as their source of carbon.

Hydrolysis: Using water to break down molecules.

Hygroscopic water: Water adsorbed by a dry soil from an atmosphere of high relative humidity.

Hymenium: Layer of hyphae that are fertile in producing asci (fungi in the phylum Ascomycota) or basidia (fungi in the phylum Basidiomycota) from the process of meiosis.

Hyperparasite: Parasite that feeds on another parasite.

Hyperthermophile: A prokaryote having a growth temperature optimum of 80°C or higher.

Hypha (plural, hyphae): Long and often branched tubular filament that constitutes the vegetative body of many fungi and fungus-like organisms.

Illumination system: The light source on light microscopes, typically mounted under the stage.

Illuviation: Deposition of soil material removed from one horizon to another in the soil.

Image: An artifact, for example, a 2D picture, that has a similar appearance to some subject, usually a physical object or a person.

Immersion oil: A special oil used with the 100× objective in order to concentrate the light and increase the resolution of the image.

Immobilization: Conversion of an element from the inorganic to the organic form in microbial or plant biomass.

Immortalize: To alter cells (either chemically or genetically) so that they can reproduce indefinitely.

Immunity: The ability of a human or animal body to resist infection by microorganisms or their harmful products such as toxins.

Immunoblot (western blot): Detection of proteins immobilized on a filter by complementary reaction with a specific antibody.

Immunoelectrophoresis: A method for separation and characterization of proteins based on electrophoresis and reaction with antibodies.

Immunofluorescence: A technique used for light microscopy with a fluorescence microscope and is used primarily on biological samples.

Immunogen: Substance that is capable of eliciting immune response.

Immunoglobulin: Antibody.

In vitro: Performed in the laboratory rather than in a living organism.

In vivo: In the body; in a living organism.

Incident ray: A ray of light that strikes a surface.

Independent variable: What you change at the beginning of the experiment to cause some effect.

Inducible enzyme: Enzyme synthesized (induced) in response to the presence of an external substance (the inducer).

Infection: Growth of an organism within another living organism.

Infection thread: Cellulosic tube in a root hair through which rhizobia can travel to reach and infect root cells.

Infrared (IR): The portion of the electromagnetic spectrum with wavelengths from about 0.75 μm to 1 mm.

Infrared spectroscopy: Deals with the infrared region of the electromagnetic spectrum, that is, light with a longer wavelength and lower frequency than visible light.

Inhibition: Prevention of growth or function.

Injector: A mechanism for accurately injecting a predetermined amount of sample into the mobile phase stream.

Inoculate: To introduce cells into a culture medium.

Inoculum: Material used to introduce a microorganism into a suitable situation for growth.

Insertion: Genetic mutation in which one or more nucleotides are added to DNA.

Insertion sequence (IS element): Simplest type of transposable element.

Integration: Process by which a DNA molecule becomes incorporated into another genome.

Interference: A phenomenon in which two waves superpose to form a resultant wave of greater or lower amplitude.

Interference microscope: Two separate light beams with much greater lateral separation than that used in phase contrast microscopy or in differential interference microscopy (DIC).

Interferon: A cytokine that inhibits virus reproduction.

Interneuron: Nerves of the brain and spinal cord between sensory and motor nerve.

Intersection: The place where two lines on a graph cross.

Intracellular: Inside the cell.

Ion: An atom or group of atoms in which the number of electrons is different from the number of protons.

Ion chamber: A gas-filled enclosure containing positive and negative electrodes that measures the amount of radiation passing through the enclosure according to the degree of ionization caused by the radiation.

Ion-exchange chromatography (or ion chromatography): A process that allows the separation of ions and polar molecules based on their charge.

Ionization (or ionizing) radiation: Radiation composed of particles that individually have sufficient energy to remove an electron from an atom or molecule.

Iris diaphragm: Found on high-power microscopes under the stage; the diaphragm is, typically, a five-hole disc with each hole having a different diameter.

Isoelectric focusing: Sample is "focused" both separating and concentrating the protein or peptide at its isoelectric point.

Isocratic: Constant composition mobile phase used in liquid chromatography.

Isoenzyme (isozyme): When two different enzymes catalyze the same reaction(s), they are isoenzymes of each other.

Isolation: Any procedure in which an organism present in a particular sample or environment is obtained in pure culture.

Isomorphous substitution: Substitution in a crystalline clay sheet of one atom by a similarly sized atom of lower valence.

Isopycnic centrifugation: Also known as density gradient centrifugation or equilibrium sedimentation; is a technique used to separate molecules on the basis of buoyant density.

Isotachophoresis (ITP): A nonlinear electrophoretic technique used in the separation of a variety of ionic compounds, ranging from small molecules like metal ions to large molecules like proteins.

Isotope: A chemical element having the same atomic number as another, but having a different atomic mass.

Karyogamy: Fusion in a cell of haploid (N) nuclei to form a diploid ($2N$).

Kilodalton (kDa): Unified atomic mass unit or dalton is a unit that is used for indicating mass on an atomic or molecular scale.

Kinetic measurements: The absorbance of a sample at a given wavelength is measured over a specified time.

Koch's postulates: Set of laws formulated by Robert Koch to prove that an organism is the causal agent of disease.

Lag phase: Period after inoculation of fresh growth medium during which population numbers do not increase.

Lamella (plural, lamellae): A thin-layer, platelike arrangement or membrane.

Laryngitis: Inflammation of the larynx.

Latent infection: Viruses that are able to downregulate their gene expression can establish a truly latent state. Latent virus infections typically persist for the entire life of the host.

Leaching: Removal of valuable metals from ores by microbial action.

Lectins: Plant proteins with a high affinity for specific sugar residues.

Light-emitting diode (LED): Produces light at a single wavelength, without the need for a monochromator.

Leghemoglobin: Iron-containing, red pigment(s) produced in root nodules during the symbiotic association between rhizobia and leguminous plants.

Lenses: Standard achromatic lenses help prevent color distortion.

Lichen: Fungus and an alga or a cyanobacterium living in symbiotic association.

Ligand: Molecule, ion, or group bound to the central atom in a chelate or a coordination compound.

Ligase: An enzyme that causes fragments of DNA or RNA to link together; used with restriction enzymes to create recombinant DNA.

Light compensation point: Where the rate of photosynthesis is lower than the rate of respiration, usually about 1% of the light intensity of sunlight.

Light microscopes: Any microscope that uses a source of light to create an image of the specimen and essentially includes all compound and stereomicroscopes.

Line graph: A picture of data where data points are connected.

Linear accelerators (LINAC): An apparatus for accelerating charged subatomic particles; used in radiotherapy, physics research, and the production of radionuclides.

Linearity: Measurement process that ensures accurate quantitation.

Lipopolysaccharide (LPS): Complex lipid structure containing unusual sugars and fatty acids found in many Gram-negative bacteria.

Lithotroph: Organism that uses an inorganic substrate such as ammonia or hydrogen as an electron donor in energy metabolism. There are two types of lithotrophs: chemolithotroph and photolithotroph.

Litter: Surface layer of the forest floor consisting of freshly fallen leaves, needles, twigs, stems, bark, and fruits.

Locomotion: Movement of organism from place to place.

Lophotrichous: Having a tuft of polar flagella.

Luxury uptake: The absorption by plants of nutrients in excess of their need for growth.

Lymphocytes: White blood cells that produce antibodies.

Lysis: Rupture of a cell, resulting in loss of cell contents.

Lysogeny: An association where a prokaryote contains a prophage, and the virus genome is replicated in synchrony with the host chromosome.

Lysosomes: Cell organelles containing enzymes responsible for degrading proteins and other materials ingested by the cell.

Macrofauna: Soil animals that are >1000 μm in length (e.g., vertebrates, earthworms, and large arthropods).

Macrokinetics: Movement of whole cells and their media within a bioreactor.

Macromolecule: Large molecule formed from the connection of a number of small molecules.

Macronutrient: A substance required in large amounts for growth, usually attaining a concentration of >500 mg/kg in mature plants. Usually refers to N, P, K, Ca, Mg, and S.

Magnification: The essence of a microscope is its ability to magnify a specimen.

Manure: Excreta of animals, with or without an admixture of bedding or litter, fresh, or at various stages of decomposition or composting.

Marrow: Tissue at the center of long bones that makes blood cells.

Mass flow (nutrient): Movement of solutes associated with net movement of water.

Matric potential: Portion of the total soil water potential due to the attractive forces between water and soil solids as represented through adsorption and capillarity.

Maturation: The stage of viral replication at which a virus particle becomes infectious.

Mechanical digestion: Using physical structures like teeth to shred or liquefy food.

Mechanical stage: A flat mechanism that sits on top of the stage and allows the viewer to move a specimen small distances—a task that is otherwise difficult at higher magnifications.

Media: A preparation made for the growth, storage, maintenance, or transport of microorganisms or other cells.

Medium (plural, media): Any liquid or solid material prepared for the growth, maintenance, or storage of microorganisms.

Medulla: Part of the brain that controls involuntary functions (e.g., heart rate); part of the brainstem.

Meiosis: In eukaryotes, reduction division, the process by which the change from diploid to haploid occurs.

Membrane: Any thin sheet or layer.

Menstruation: The monthly release of unfertilized egg and uterine lining.

Mesofauna: Soil animals between 200 and 1000 μm in length, including nematodes, oligochaete worms, smaller insect larvae, and small arthropods.

Mesophile: Organism whose optimum temperature for growth falls in an intermediate range of approximately 15°C–40°C.

Messenger RNA (mRNA): RNA molecule transcribed from DNA, which contains the information to direct the synthesis of a particular protein.

Metabolism: All biochemical reactions in a cell, both anabolic and catabolic.

Metabolites: Chemical by-products of metabolism, the chemical process of life.

Metamorphosis: When an organism looks completely different at different life stages and even has different organs.

Methanogenesis: Biological production of methane.

Methanogenic bacterium (methanogen): Methane-producing prokaryote; member of the Archaea.

Methanotroph: Organism capable of oxidizing methane.

Metric: Measurement system used by scientists.

Microaerophile: Organism that requires a low concentration of oxygen for growth.

Microaggregate: Clustering of clay packets stabilized by organic matter and precipitated inorganic materials.

Microanalysis: The chemical identification and quantitative analysis of very small amount of chemical substances.

Microbial biomass: Total mass of microorganism alive in a given volume or mass of soil.

Microbial population: Total number of living microorganisms in a given volume or mass of soil.

Microbiology: The study of microscopic life such as bacteria, viruses, and yeast.

Microbore: Columns with smaller than usual internal diameters (<2 mm) used in HPLC.

Microcarrier: A microscopic particle that supports cell attachment and growth in suspension culture.

Microencapsulated: Surrounded by a thin, protective layer of biodegradable substance referred to as a microsphere.

Microenvironment: Immediate physical and chemical surroundings of a microorganism.

Microfauna: Protozoa, nematodes, and arthropods generally <200 μm long.

Microflora: Bacteria (including actinomycetes), fungi, algae, and viruses.

Microhabitat: Clusters of microaggregates with associated water within which microbes function.

Microheterogeneity: Slight differences in the amino acid sequence of a protein.

Microinjection: Manually using tiny needles to inject microscopic material (such as DNA) directly into cells or cell nuclei.

Microkinetics: Movement of chemicals into, out of, and within the cell.

Micrometer: One-millionth of a meter or 10^{-6} m; the unit usually used for measuring microorganisms.

Micronutrient: Chemical element necessary for growth found in small amounts, usually <100 mg/kg in a plant. These elements consist of B, Cl, Cu, Fe, Mn, Mo, and Zn.

Microorganism: A microbe; a living thing too small to be seen by the naked eye.

Microscope: An instrument used to see objects that are too small for the naked eye.

Microtubules: Cellular organelles common in microorganisms: thin tubes that make structures involved in cellular movement.

Mineralization: Conversion of an element from an organic form to an inorganic state as a result of microbial decomposition.

Mirror: Allows to direct ambient light up through the hole in the stage and illuminate the specimen.

Mitochondria: Cell organelles; seat of respiration and electron transport referred to as the powerhouse of a cell.

Mitosis: Highly ordered process by which the nucleus divides in eukaryotes.

Mobile phase: Solvent that moves the solute through the column.

Mold: A filamentous fungus.

Molecule: The result of two or more atoms combining by chemical bonding.

Monera: Single-celled kingdom.

Monoclonal antibody (MAb): A highly specific, purified antibody that recognizes only a single antigen.

Monocular microscope: A compound microscope with a single eyepiece.

Monokaryon: Fungal hypha in which compartments contain one nucleus.

Most probable number (MPN): Method for estimating microbial numbers in soil based on extinction dilutions.

Motility: Movement of a cell under its own power.

mRNA: Messenger RNA; translated on ribosomes to produce proteins.

Mucigel: Gelatinous material at the surface of roots grown in normal nonsterile soil.

Mucilage: Gelatinous secretions and exudates produced by plant roots and many microorganisms.

Mucus: Wet slimy material that keeps organs and tissues moist and helps protect them.

Mulch: Any material, such as straw, sawdust, leaves, plastic film, and loose soil, that is spread upon the surface of the soil to protect the soil and plant roots from the effects of raindrops, soil crusting, freezing, or evaporation.

Multiwavelength measurement: Sample absorbance is recorded at multiple wavelengths.

Multicellular: Organisms composed of more than one cell.

Municipal solid waste: Combined consumer and commercial waste generated within a defined geographic area.

Mushroom: Large, sometimes edible, fruiting body produced by some fungi.

Mutagen: An agent (chemicals, radiation) that causes mutations in DNA.

Mutant: Organism, population, gene, or chromosome that differs from the corresponding wild type by one or more base pairs.

Mutation: Heritable change in the base sequence of the DNA of an organism.

Mutualism: Interaction between organisms where both organisms benefit from the association.

Mycelium (plural, mycelia): Mass of hyphae that form the vegetative body of many fungal organisms.

Mycobacterium: A genus of aerobic bacteria found in soil and water that are capable of biodegrading multi-ring compounds such as PAHs.

Mycophagous: Organisms that consume fungi, such as mycophagous nematodes.

Mycoplasma: Group of bacteria without cell walls that do not revert to walled forms; phylogenetically related to clostridia.

Mycorrhiza: Literally "fungus root." The symbiotic association between specific fungi with the fine roots of higher plants.

Mycorrhizosphere: Unique microbial community that forms around a mycorrhiza.

Mycovirus: Virus that infects fungi.

Myeloma: Lymphocytic cancer; a malignancy normally found in bone marrow.

N-acetylglucosamine: Sugar derivatives in the peptidoglycan layer of bacterial cell walls.

N-acetylmuramic acid: Sugar derivatives in the peptidoglycan layer.

Natural selection: A process where better-adapted individual organisms survive and reproduce and less well-adapted individual organisms die more and reproduce less.

Necrosis: Damage of living tissues because of infection or injury.

Necrotrophic: Nutritional mechanism by which an organism produces a battery of hydrolytic enzymes to kill and break down host cells and then absorb nutritional compounds from the dead organic matter.

Nematode: Multicellular eukaryote defined as an unsegmented, usually microscopic, roundworm.

Nerve cell: Structure used to build all parts of the nervous system—neuron.

Neutralism: Lack of interaction between two organisms in the same habitat.

Neutralization: Blocking of viral infection by antibodies; also, an assay that measures this process.

Neutron: Part of an atom's nucleus—carries no charge (neutral).

Niche: Functional role of a given organism within its habitat.

Nicotinamide adenine dinucleotide (NAD$^+$): Important coenzyme, functioning as a hydrogen and electron carrier in a wide range of redox reactions; the oxidized form of the coenzyme is written NAD$^+$, the reduced form as NADH.

Nicotinamide adenine dinucleotide phosphate (NADP$^+$): Important coenzyme, functioning as a hydrogen and electron carrier in a wide range of redox reactions; the oxidized form of the coenzyme is written NADP$^+$, the reduced form as NADPH.

Nitrate reduction (biological): Process whereby nitrate is reduced by plants and microorganisms to ammonium for cell synthesis (nitrate assimilation, assimilatory nitrate reduction) or to various lower oxidation states (N_2, N_2O, NO) by bacteria using nitrate as the terminal electron acceptor in anaerobic respiration.

Nitrification: Biological oxidation of ammonium to nitrite and nitrate or a biologically induced increase in the oxidation state of nitrogen.

Nitrifying bacteria: Chemolithotrophs capable of carrying out the transformations from NH_3 to NO_2^- or NO_2^- to NO_3^-.

Nitrogen cycle: Sequence of biochemical changes wherein nitrogen is used by a living organism, transformed upon the death and decomposition of the organism, and converted ultimately to its original state of oxidation.

Nitrogenase: Specific enzyme system required for biological N_2 fixation.

Nodulins: Unique proteins produced in root hairs or nodules in response to rhizobial infection.

Nomenclature: System of naming organisms.

Nonessential amino acids: Amino acids that the body can make using basic atoms.

Nonpolar: Possessing hydrophobic (water repelling) characteristics and not easily dissolved in water.

Normal-phase chromatography: A mode of chromatography carried out with a polar stationary phase and a nonpolar mobile phase.

Northern blot: Hybridization of single-stranded nucleic acid (DNA or RNA) to RNA fragments immobilized on a filter.

Nuclear magnetic resonance (NMR): A physical phenomenon in which magnetic nuclei in a magnetic field absorb and reemit EMR.

Nuclear reactor: A device used to generate power, in which nuclear fission takes place as a controlled chain reaction, producing heat energy that is generally used to drive turbines and provide electric power.

Nucleic acid: Polymer of nucleotides.

Nucleoid: Aggregated mass of DNA that makes up the chromosome of prokaryotic cells.

Nucleophilic compound: Chemical that attracts or is drawn to electron-deficient regions in other chemicals; reducing agents act as nucleophilic compounds.

Nucleoside: Nucleotide without the phosphate group.

Nucleotide: Monomeric unit of nucleic acid; consisting of a sugar (pentose), a phosphate, and a nitrogenous base.

Nucleus: Membrane-enclosed structure containing the genetic material (DNA) organized in chromosomes.

Numerical aperture (NA): A measure of the diameter of the aperture compared to the focal length of a lens and, ultimately, of the resolving power of a microscope.

Nutrient: A chemical that the body needs to live and grow.

Objective lens: The lens closest to the specimen that first receives the rays from the specimen (the object) and forms the image in the focal plane of the eyepiece.

Obligate: Adjective referring to an environmental factor (e.g., oxygen) that is always required for growth.

Ocularmeter: An ocular micrometer serves as a scale or rule. This is simply a disc of glass upon which equally spaced divisions are etched.

Oligonucleotide: Short nucleic acid chain, either obtained from an organism or synthesized chemically.

Oligotroph: Microorganism specifically adapted to grow under low nutrient supply.

Omnivore: Animal whose body is designed to digest and use plants and animals for food.

Oncogene: A gene that, when expressed as a protein, can lead cells to become cancerous.

Oogonium: Specialized sexual structure formed as a female gametangium by fungus-like organisms in the phylum Oomycota.

Oospore: Thick-walled spore formed within an oogonium by fungus-like organisms in the phylum Oomycota.

Operon: Cluster of genes whose expression is controlled by a single operator; typical in prokaryotic cells.

Organ: Structures made of tissues; it helps carry out one or more life processes.

Organ system: A group of organs that work together to carry out one or more of the life processes.

Organelle: A structurally discrete component that performs a certain function inside a eukaryotic cell.

Organic soil: Soil that contains a high percentage (>200 or >120–180 g/kg if saturated with water) of organic carbon.

Organism: A single, autonomous living thing. Bacteria and yeasts are organisms.

Organotroph: Organism that obtains reducing equivalents (stored electrons) from organic substrates.

Osmosis: Diffusion of water through a membrane from a region of low solute concentration to one of higher concentration.

Osmotic potential: Portion of total soil water potential due to the presence of solutes in soil water.

Ovulation: Release of egg from ovary.

Oxic: Containing oxygen; aerobic.

Oxidation: Process by which a compound gives up electrons, acting as an electron donor, and becomes oxidized.

Oxidation–reduction (redox) reaction: Coupled pair of reactions in which one compound becomes oxidized, while another becomes reduced and takes up the electrons released in the oxidation reaction.

Oxidative phosphorylation: Synthesis of ATP involving a membrane-associated electron transport chain and the creation of a proton-motive force.

Oxygen electrode: Also referred to as Clark electrode; an electrode that measures oxygen on a catalytic platinum surface using the net reaction.

Oxygenic photosynthesis: Use of light energy to synthesize ATP and NADPH by noncyclic photophosphorylation with the production of oxygen from water.

Paper chromatography: A technique that involves placing a small dot or line of sample solution onto a strip of chromatography paper.

Parasexual cycle: Nuclear cycle in which genes of haploid nuclei recombine without meiosis.

Parasitism: Feeding by one organism on the cells of a second organism, which is usually larger than the first.

Particle density: Density of the soil particles; the dry mass of the particles being divided by the solid (not bulk) volume of the particles, in contrast with bulk density.

Particle size (d_p): Effective diameter of a particle measured by sedimentation, sieving, or micrometric methods.

Partition chromatography: A category of chromatography techniques in which the solute equilibrates between the mobile phase and the stationary liquid.

Partition coefficient (K_d): A measure of how a compound distributes between two immiscible solvent phases.

Pasteurization: Process using mild heat to reduce microbial numbers in heat-sensitive materials.

Pathogen: An organism that causes disease in some other organism.

Pathogenicity: Ability of a parasite to inflict damage on the host.

Peak: When the detector registers the presence of a compound, the normal baseline signal it sends to the data system changes, resulting in a deflection from the baseline called a peak.

Peak identification: Performed by comparing the sample chromatogram to a chromatogram of a standard solution separated under the same conditions.

Peak shape: The profile of a chromatographic peak.

Peat: Unconsolidated soil material consisting largely of undecomposed, or only slightly decomposed, organic matter accumulated under conditions of excessive moisture.

Pectin: Important component of the plant cell walls containing chains of galacturonic acid that is often esterified with a methyl group.

Pellicle: Relatively rigid layer of proteinaceous elements just beneath the cell membrane in many protozoa and algae.

Peptides: Proteins consisting of fewer than 40 amino acids.

Peptidoglycan: Rigid layer of cell walls of bacteria; a thin sheet composed of *N*-acetylglucosamine, *N*-acetylmuramic acid, and a few amino acids. Also called murein.

Peribacteroid membrane: Plant-derived membrane surrounding one to several rhizobia within host cells of legume nodules.

Periplasmic space: Area between the cell membrane and the cell wall in Gram-negative bacteria containing certain enzymes involved in nutrition.

Perithecium: Flask-shaped ascocarp open at the tip; containing asci of fungi in the phylum Ascomycota.

Peritrichous flagellation: Having flagella attached to many places on the cell surface.

Permanent wilting point: Greatest water content of a soil at which indicator plants, growing in that soil, wilt and fail to recover when placed in a humid chamber.

Permeable: A membrane that lets anything in.

pH: Negative logarithm of the hydrogen ion activity.

pH meter: An electronic instrument used for measuring the pH (acidity or alkalinity).

Phagotrophic: Form of feeding where animals, such as protozoans, engulf particulate nutrients, such as bacterial cells or detritus.

Phase contrast: A contrast-enhancing microscopic technique shifts the light phase wavelength, thereby causing the light deviated by the specimen to appear dark on a light background.

Phenotype: Observable properties of an organism.

Phosphobacterium: Bacterium that is especially good at solubilizing the insoluble inorganic phosphate in soil.

Phosphodiester bond: Type of covalent bond linking nucleotides together in a polynucleotide.

Phospholipid: Lipids containing a substituted phosphate group and two fatty acid chains on a glycerol backbone.

Phosphorus cycle: Sequence of transformations undergone by phosphorus where it is transformed between soluble and insoluble, and organic and inorganic forms.

Photic zone: Uppermost layer of a body of water or soil that receives enough sunlight to permit the occurrence of photosynthesis.

Photoautotroph: Organism that is able to use light as its sole source of energy and carbon dioxide as sole carbon source.

Photocell: Resistor whose resistance decreases with increasing incident light intensity. It can also be referred to as a photoconductor.

Photodiode array (PDA): PDA detectors are UV/Vis detectors that record the absorbance of light at different wavelengths simultaneously.

Photoheterotroph: Organism that is able to use light as a source of energy and organic materials as carbon source.

Photomultiplier: An electronic sensing device used to detect electromagnetic energy of a wide range of frequencies and intensity levels.

Photophosphorylation: Synthesis of high-energy phosphate bonds, as ATP, using light energy.

Photosynthesis: Process of using light energy to synthesize carbohydrates from carbon dioxide.

Phototaxis: Movement toward light.

Phototroph: Organism that uses light as the energy source to drive the electron flow from the electron donors, such as water, hydrogen, or sulfide.

Phycobilin: Water-soluble pigment that occurs in cyanobacteria and functions as the light-harvesting pigments for photosystem II.

Phylogeny: Ordering of species into higher taxa and the construction of evolutionary trees based on evolutionary (genetic) relationships.

Phytoremediation: Use of plants to remediate contaminated soil or groundwater.

Phytostabilization: Use of soil amendments and plants to reduce bioavailability and off-site migration of contaminants.

Pilot plant: A medium-scale bioprocessing facility used as an intermediate in scaling up processes from the laboratory to commercial production.

Pilus (plural pili): Fimbria-like structure that is present on fertile cells and is involved in DNA transfer during conjugation. Sometimes called sex pilus.

Placenta: The structure that an embryo forms, attaching it to the uterus—allows the baby to be nourished and get O_2.

Plant growth-promoting rhizobacteria (PGPR): Broad group of soil bacteria that exert beneficial effects on plant growth usually as root colonizers.

Plaque: Localized area of lysis or cell inhibition caused by viral infection on a lawn of cells.

Plasmid: Covalently closed, circular piece of DNA that, as an extrachromosomal genetic element, is not essential for growth.

Plasmogamy: Fusion of the contents of two cells, including cytoplasm and nuclei.

Plastid: Specialized cell organelles containing pigments or protein materials.

Plate count: Number of colonies formed on a solid culture medium when uniformly inoculated with a known amount of soil, generally as a dilute soil suspension.

Polar: Possessing hydrophilic characteristics and generally water soluble.

Polar flagellation: Condition of having flagella attached at one end or both ends of the cell.

Polyacrylamide gel electrophoresis (PAGE or SDS PAGE): Electrophoresis in which a polyacrylamide gel is used as the diffusion medium.

Poly-beta-hydroxybutyrate (PHB): Common storage material of prokaryotic cells consisting of beta-hydroxybutyrate or other beta-alkanoic acids.

Polyclonal antiserum: Mixture of antibodies to a variety of antigens or to a variety of determinants on a single antigen.

Polymer: Large molecule formed by polymerization of monomeric units.

Polymerase: An enzyme that catalyzes the production of nucleic acid molecules.

Polymerase chain reaction (PCR): A method of duplicating genes exponentially.

Polysaccharide: Long chain of monosaccharides (sugars) linked by glycosidic bonds.

Polysome: Strings of ribosomes attached by strands of mRNA.

Pore space: Portion of soil bulk volume occupied by soil pores.

Porin: A protein channel in the LPS layer of Gram-negative bacteria.

Porosity: Volume of pores in a soil sample (nonsolid volume) divided by the bulk volume of the sample.

Posttranslational modifications: Protein processing done by the Golgi bodies after proteins have been constructed by ribosomes.

Posterior: The back end of an organism.

Potentiometer: An instrument for measuring the potential (voltage) in a circuit.

Pour plate: Method for performing a plate count of microorganisms.

Preparative chromatography: The process of using liquid chromatography to isolate a sufficient amount of material for other experimental or functional purposes.

Primary producer: Organism that adds biomass to the ecosystem by synthesizing organic molecules from carbon dioxide and simple inorganic nutrients.

Primer: Molecule (usually a polynucleotide) to which DNA polymerase can attach the first nucleotide during DNA replication.

Prions: Resembling viruses; these pathogens are composed only of protein, with no detectable nucleic acid.

Prokaryote: Organism lacking a unit membrane-bound nucleus and other organelles, usually having its DNA in a single circular molecule.

Promoter: Site on DNA where the RNA polymerase binds and begins transcription.

Propagule: Cell unit capable of developing into a complete organism.

Prophage: State of the genome of a temperate virus when it is replicating in synchrony with that of the host, typically integrated into the host genome.

Prosthetic group: Tightly bound, nonprotein portion of an enzyme; not the same as coenzyme.

Protein: Macromolecules whose structures are coded in an organism's DNA. Each is a chain of more than 40 amino acids folded back upon itself in a particular way.

Proteolytic: Capable of lysing (denaturing or breaking down) proteins.

Proton: Part of an atom's nucleus that carries a positive charge.

Proton-motive force (PMF): Energized state of a membrane created by expulsion of protons through action of an electron transport chain.

Protoplast: Cell from which the wall has been removed.

Protozoan (plural, protozoa): Unicellular eukaryotic microorganisms that move by either protoplasmic flow (amoebae), flagella (flagellates) or, cilia (ciliates).

Pseudomonad: Member of the genus *Pseudomonas*; a large group of Gram-negative, obligately respiratory (never fermentative) bacteria.

Pseudopodium (plural, pseudopodia): Protrusion of an amoeboid cell formed by the extrusion or streaming of the cytoplasm (but still enclosed in the membrane) for the purpose of movement or feeding.

Psychrophile: Organism that is able to grow at low temperatures (0°C) and showing a growth temperature optimum <15°C. Not able to grow above 20°C.

Pulsed field gel electrophoresis: Electrophoresis in which the direction of the electric field is changed periodically.

Pure culture: Population of microorganisms composed of a single strain. Such cultures are obtained through selective laboratory procedures and are rarely found in a natural environment.

Radial symmetry: Can be cut in many identical pieces like a pie.

Radioactive decay: Spontaneous decomposition of the nuclei of the atoms of radioactive substances. Measured as the proportion of the atoms in a radionuclide that decomposes per unit of time.

Radioactive fallout: Dissemination of radioactive substances through the atmosphere, and deposition on the environment generally causes radiation injury.

Radioactive tracer: A radioactive isotope replacing a stable chemical element in a compound and so able to be followed or tracked through one or more reactions or systems.

Radioimmunoassay (RIA): A technique in radiology used to determine the concentration of an antigen, antibody, or other protein in the serum.

Radioisotope: An isotope of an element that undergoes spontaneous decay with the release of radioactive particles.

Radiotherapy: The treatment of disease with ionizing radiation.

Reaction center: A photosynthetic complex containing chlorophyll (or bacteriochlorophyll) and other components, within which occurs the initial electron transfer reactions of photophosphorylation.

Reannealing: Process where two complementary single strands of DNA automatically hybridize back into a single, double-stranded molecule upon cooling.

Recalcitrant: Resistant to microbial attack.

Receptor: Structure that picks up information/stimuli.

Recessive gene: Gene for trait that is weaker and less likely to be seen.

Recessive trait: The trait that is weaker must have two copies, the recessive gene, in order to see trait.

Recombinant: Containing genetic material from another organism.

Recombinant DNA: DNA molecule containing DNA originating from two or more sources.

Recombination: Process by which genetic elements in two separate genomes are brought together in one unit.

Reducing equivalent (power): Electrons stored in reduced electron carriers such as NADH, NADPH, and $FADH_2$.

Reduction: Process by which a compound accepts electrons.

Reduction potential: Inherent tendency of a compound to act as an electron donor or an electron acceptor. Measured in millivolts.

Reductive dechlorination: Removal of Cl as Cl⁻ from an organic compound by reducing the carbon atom from C–Cl to C–H.

Reference electrode: It is an electrode that has a stable and well-known electrode potential.

Reflection: The change in direction of a wave front at an interface between two different media so that the wave front returns into the medium from which it originated.

Refraction: The change in direction of a wave due to a change in its speed. It is essentially a surface phenomenon.

Refractive index: It is expressed as a ratio of the speed of light in vacuum relative to that in the considered medium.

Relative centrifugal force (RCF): It is the measurement of the force applied to a sample within a centrifuge.

Replication: Conversion of one double-stranded DNA molecule into two identical double-stranded DNA molecules.

Repression: Process by which the synthesis of an enzyme is inhibited by the presence of an external substance (the repressor).

Resistance: It is the property of a component that restricts the flow of electric current.

Resolution: The ability of a lens to distinguish the fine details of the specimens being viewed.

Resolving power: The ability of an imaging device to separate points of an object that are located at a small angular distance.

Respiration: Cellular process of breaking down food for energy.

Respiratory system: Group of organs that bring in O_2 and remove CO_2.

Restriction enzyme: Bacterial enzyme that cuts DNA molecules at the location of particular sequences of base pairs.

Restriction fragment length polymorphism (RFLP): Method to identify differences between similar genes from different organisms.

Retina: A translucent filmy piece of tissue lining the back of the eyeball.

Retrovirus: Virus containing single-stranded RNA as its genetic material and producing a cDNA by the action of the enzyme reverse transcriptase.

Reverse transcription: Process of copying information found in RNA into DNA.

Reversed-phase chromatography (RPC): The most common HPLC mode. Uses hydrophobic packings such as octadecyl or octylsilane phases bonded to silica or neutral polymeric beads. The mobile phase used is usually water and a water miscible organic solvent such as methanol or acetonitrile.

R_f value retention factor: Defined as the ratio of the distance traveled by the substance to the distance traveled by the solvent.

Rhizobacteria: Bacteria that aggressively colonize roots.

Rhizobia: Bacteria capable of living symbiotically in roots of leguminous plants.

Rhizoid: Rootlike structure that helps to hold an organism to a substrate.

Rhizomorph: Mass of fungal hyphae organized into long, thick strands usually with a darkly pigmented outer rind and containing specialized tissues for absorption and water transport.

Rhizoplane: Plant root surfaces and usually strongly adhering soil particles.

Rhizosphere: Zone of soil immediately adjacent to plant roots in which the kinds, numbers, or activities of microorganisms differ from that of the bulk soil.

Ribonucleic acid (RNA): Polymer of nucleotides connected via a phosphate-ribose backbone, involved in protein synthesis.

Ribosomal RNA (rRNA): Types of RNA found in the ribosome; some participate actively in the process of protein synthesis.

Ribosome: Cell organelles that translate RNA to build proteins.

Roller bottle: A container with large growth surfaces in which cells can be grown in a confluent monolayer.

Root nodule: Specialized structure occurring on roots, especially of leguminous plants, in which bacteria fix dinitrogen and make it available for the plant.

Rotar: Rotating part of a mechanical device.

Saline soil: Soil containing sufficient soluble salt to adversely affect the growth of most crop plants.

Sample capacity: The amount of sample that can be injected into an LC column without overload.

Sampling rate: The frequency with which the detector checks the flow cell. Sampling rate is often called time constant.

Sand: Soil particle between 0.05 and 2.0 mm in diameter.

Sanitization: Elimination of pathogenic or deleterious organisms, insect larvae, intestinal parasites, and weed seeds.

Saprophyte: Nonparasitic nutritional mechanism by which an organism obtains its food exclusively from the degradation of nonliving organic material.

Saturated calomel electrode (SCE): It is a reference electrode based on the reaction between elemental mercury and mercury(I) chloride.

Scale-up: To take a manufacturing process from the laboratory scale to a scale at which it is commercially feasible.

Scanning electron microscope (SEM): A type of electron microscope that scans the surface of a sample using a beam of electrons.

Scintillation counter: A measuring instrument for counting individual ionizing events.

Sclerotium: Modified fungal hyphae that form a compact, hard vegetative resting structure with a thick pigmented outer rind.

Secondary metabolite: Product of intermediary metabolism released from a cell, such as an antibiotic.

Seed stock: The initial inoculum or the cells placed in growth medium from which other cells will grow.

Selective medium: Medium that allows the growth of certain types of microorganisms in preference to others.

Selectively permeable: A membrane that only lets certain things through.

Sensitivity: The sensitivity setting is the line between normal background noise and a true peak.

Sensory nerve: Carries information to the brain about the environment.

Septum (plural, septa): Crosswall (partition) dividing a parent cell into two daughter cells during binary fission or occurring between adjacent cells in hyphae.

Sequence: The precise order of bases in a nucleic acid or amino acids in a protein.

Serial dilution: Series of stepwise dilutions (usually in sterile water) performed to reduce the populations of microorganisms in a sample to manageable numbers.

Serology: Study of antigen–antibody reactions *in vitro*.

Serum: The watery portion of an animal or plant fluid (such as blood) remaining after coagulation.

Sheath: Tubular structure formed around a chain of cells or around a bundle of filaments.

Shorter wavelength: Shorter than the 240 nm that has so far been reached.

Siderochromes: Compounds produced by microorganisms that are involved with the uptake of iron by those microorganisms.

Silt: Soil particle with a diameter between 0.002 and 0.05 mm.

Single beam: All of the light passes through the sample holder.

Single wavelength measurement: Also known as fixed wavelength.

SIP: Steam in place or sterilize in place.

Site: In ecology, area described or defined by its biotic, climatic, and soil conditions as related to its capacity to produce vegetation.

Site-directed mutagenesis: Insertion of a different nucleotide at a specific site in a molecule using recombinant DNA methodology.

Skeletal muscle: Muscle found on bones; striped not branched.

Slide: A flat, rectangular, glass plate on which a specimen may be placed.

Slime layer: Diffuse layer of polysaccharide exterior to the cell wall in some bacteria.

Slime mold: Nonphototrophic eukaryotic microorganism lacking cell walls, which aggregate to form fruiting structures (cellular slime molds) or simply masses of protoplasm (acellular slime molds).

Smooth muscle: Muscle found in organs other than the heart.

Soil: Unconsolidated mineral or material on the immediate surface of the earth that serves as a natural medium for the growth of land plants.

Soil aggregate: Unit of soil structure generally <10 mm in diameter and formed by natural forces and substances derived from root exudates and microbial products that cement smaller particles into larger units.

Soil atmosphere: Gases occupying the pore space in soil.

Soil extract: Solution separated from a soil suspension or from a soil by filtration, centrifugation, suction, or pressure.

Soil horizon: Layer of soil or soil material approximately parallel to the land surface and differing from adjacent genetically related layers in physical, chemical, and biological properties or characteristics such as color, structure, texture, consistency, kinds and number of organisms present, and degree of acidity or alkalinity.

Soil microbiology: Branch of soil science concerned with soil-inhabiting microorganisms and their functions and activities.

Soil organic matter (SOM): Organic fraction of the soil exclusive of undecayed plant and animal residues. Often synonymous with humus.

Soil population: All the organisms living in the soil, including plants and animals.

Soil pore: That part of the bulk volume of soil not occupied by soil particles.

Soil quality: Continued capacity of soil to function as a vital living system to sustain biological productivity, maintain the quality of the environment, and promote plant, animal, and human health.

Soil salinity: Amount of soluble salts in a soil.

Soil texture: Relative proportions of the various soil separates in a soil.

Solarization: Method to control pathogens and weeds where moistened soil in hot climates is covered with transparent polyethylene plastic sheets, thereby trapping incoming radiation.

Somatic cell: In higher organisms, a cell that carries the full genetic makeup of an organism.

Southern blot: Hybridization of single-stranded nucleic acid (DNA or RNA) to DNA fragments immobilized on a filter.

Sparge: A sparger is the component of a fermenter that sprays air into the broth.

Species: The smallest grouping of organisms.

Specific activity: Amount of enzyme activity units per mass of protein.

Spherical aberration: An optical effect observed in an optical device (lens, mirror, etc.) that occurs due to the increased refraction of light rays.

Spirillum (plural, spirilla): Bacterium with a spiral shape that is relatively rigid.

Split beam: Light from the source is split into two paths with approximately 70% of the energy from the monochromator passing through the sample and the rest going to a separate feedback detector.

Sporangiospore: Spore formed within a sporangium by fungi in the phylum Zygomycota.

Sporangium: Fungal structure that converts its cytoplasm into a variable number of sporangiospores; formed by fungi in the phylum Zygomycota.

Spores: Specialized reproductive cell.

Spread plate: Method of performing a plate count of microorganisms.

Stable isotope: One that does not transmute into another element with emission of corpuscular or EMRs.

Stage: The platform on which slides and specimens are placed for viewing.

Stage micrometer: Is simply a microscope slide with a scale etched on the surface.

Staining: It is an auxiliary technique used in microscopy to enhance contrast in the microscopic image.

Stationary phase: A solid or a liquid supported on a solid.

Sterilization: Rendering an object or substance free of viable microbes.

Stimuli: A change in the environment that organisms can react to.

Storage polysaccharide: Energy reserve deposited in the cell when there is an excess of carbon available.

Strain: A population of cells that all descend from a single cell.

Stroke: When blood flow is cut off from part of the brain and nerve cells begin to die; caused by broken or blocked blood vessels.

Structural polysaccharide: Polysaccharide that serves primarily as a structural element in cell walls and coats and intercellular spaces and connective tissue where it gives shape, elasticity, or rigidity to plant or animal tissues and protection and support to unicellular organisms.

Substage: The part of the microscope below the stage, including the illumination system.

Substrate: Reactive material; the substance on which an enzyme acts.

Substrate-level phosphorylation: Synthesis of high-energy phosphate bonds through reaction of inorganic phosphate with an activated (usually) organic substrate.

Substratum: The solid surface on which a cell moves or on which cells grow.

Succession: Gradual process brought about by the change in the number of individuals of each species of a community and by the establishment of new species that gradually replace the original inhabitants.

Sulfur cycle: Sequence of transformations undergone by sulfur where it is taken up by living organisms, transformed upon death and decomposition of the organism, and converted ultimately to its original state of oxidation.

Supernatant: Material floating on the surface of a liquid mixture.

Support: The solid particles in a column.

Surface area: Area of the solid particles in a given quantity of soil or porous medium.

Surface soil: Uppermost part of the soil, ordinarily moved in tillage, or its equivalent in uncultivated soils ranging in depth from 7 to 20 cm.

Surfactant: Any substance that changes the nature of a surface, such as lowering the surface tension of water.

Suspension: Particles floating in a liquid medium or the mix of particles and the liquid itself.

Symbiosis: Living together in intimate association of two dissimilar organisms. The interactions between the organisms can be commensal or mutualistic.

Synergism: Association between organisms that is mutually beneficial. Both populations are capable of surviving in their natural environment on their own although, when formed, the association offers mutual advantages.

Systemic: Not localized in a particular place of the body; an infection disseminated widely through the body is said to be systemic.

Tailing: A peak with an asymmetrical factor of >1. An asymmetrical peak is the result of a component that is excessively retarded in eluting.

Taxon (plural, taxa): A group into which related organisms are classified.

Taxonomy: Study of scientific classification and nomenclature.

Teflon membrane: They are chemically stable and inert.

Teichoic acids: All wall, membrane, or capsular polymers containing glycerophosphate or ribitol phosphate residues.

Teleomorph: Sexual stage in reproduction in which cells are formed by the process of meiosis and genetic recombination.

Temperate virus: Virus that upon infection of a host does not necessarily cause lysis but whose genome may replicate in synchrony with that of the host.

Terminal electron acceptor: External oxidant (often oxygen) that accepts the electrons as they exit from the electron transport chain.

Thallus: Vegetative body that is not differentiated into tissue systems or organs.

Thermal conductivity detector (TCD): A filament temperature increases as analytes present in the carrier gas pass over it, causing the resistance to increase.

Thermophile: Organism whose optimum temperature for growth is between 45°C and 85°C.

Thin-layer chromatography (TLC): It is a type of chromatography that involves a stationary phase of a thin layer of adsorbent like silica gel, alumina, or cellulose on a flat, inert substrate.

T$_i$ plasmid: Conjugative tumor-inducing plasmid present in the bacterium *Agrobacterium tumefaciens* that can transfer genes into plants.

Tissue: A group of cells that have one or more jobs they do for organs.

Tissue culture: Growing plant or animal tissues outside of the body.

Titer: A measured sample.

Topsoil: Layer of soil moved in cultivation.

Total dissolved solids (TDS): It is a measure of the combined content of all inorganic and organic substances contained in a liquid in molecular, ionized, or microgranular suspended form.

Toxin: A chemical harmful to an organism.

Trace gas: Gas other than nitrogen and oxygen in the atmosphere; particularly those gases that are active in the chemistry or radiation balance of the atmosphere.

Tracer carbon dating: A naturally radioactive carbon isotope, with atomic mass 14 and half-life 5730 years, used in determining the age of ancient organic, geologic, or archeological specimens.

Trait: A genetically determined characteristic.

Transcription: Synthesis of an RNA molecule complementary to one of the two strands of a DNA double-stranded molecule.

Transduction: Transfer of host genetic information via a virus or bacteriophage particle.

Transfer RNA (tRNA): Type of RNA that carries amino acids to the ribosome during translation.

Transformation: Transfer of genetic information into living cells as free DNA.

Transgenic: Describes genetically modified plants or animals containing foreign genes inserted by means of recombinant DNA techniques.

Transgenics: The alteration of plant or animal DNA so that it contains a gene from another organism.

Translation: Synthesis of proteins using the genetic information in mRNA as a template.

Transmission electron microscope (TEM): It is a microscopy technique whereby a beam of electrons is transmitted through an ultrathin specimen, interacting with the specimen as it passes through.

Transmittance: It is the fraction of incident light (or other EMR) at a specified wavelength that passes through a sample.

Transposable element: Genetic element that can move (transpose) from one site on a chromosome to another.

Transposition: Movement of a piece of DNA around the chromosome, usually through the function of a transposable element.

Transposon: Transposable element which, in addition to genes involved in transposition, carries other genes; often confers selectable phenotypes such as antibiotic resistance.

Transposon mutagenesis: Insertion of a transposon into a gene; this inactivates the host gene leading to a mutant phenotype and also confers the phenotype associated with the transposon gene.

Tricarboxylic acid cycle (TCA cycle, citric acid cycle, Krebs cycle): Series of metabolic reactions by which pyruvate is oxidized completely to carbon dioxide, also forming NADH, which allows ATP production.

Trichome: Row of cells that have remained attached to one another following successive cell divisions. Trichomes are formed by many cyanobacteria and by species of *Beggiatoa*.

Trophic level: Describes the residence of nutrients in various organisms along a food chain ranging from the primary nutrient assimilating autotrophs to the predatory carnivorous animals.

Trypsin, tryptic digestion: Trypsin allows the growth of cells as independent microorganisms distinct from tissue culture by causing cell disaggregation.

Tryptic fragment analysis: Quantitating the resultant fragments caused by tryptic digestion.

Tungsten: It is the least expensive source of illumination. It is hotter and less bright than the other kinds of illumination.

Turbidostat: A turbidostat is designed to keep the organisms at a constant concentration. A turbidity sensor measures the concentration of organisms in the culture and adds additional medium when a preset value is exceeded.

Turbulent flow field: The state that results from mixing the contents of a fermenter or bioreactor to provide oxygen to the cells. That must be balanced against the shear that causes cell damage and death.

Turgor pressure: The pressure exerted on a plant cell wall by water passing into the cell by osmosis; also called hydrostatic pressure.

Ultracentrifuge: It is a centrifuge optimized for spinning a rotor at very high speeds, capable of generating acceleration as high as $20,00,000g$.

Ultraviolet: That portion of the electromagnetic spectrum below blue light (380 nm).

Ultraviolet/visible light (UV/Vis): The tunable or variable wavelength UV/Vis detector is the most popular form of detector.

Unicellular: Single-celled organisms.

Urea: Cell waste from breaking down proteins.

Urine: Metabolic waste filtered out by the kidneys.

Uronic acid: Class of acidic compounds of the general formula $HOOC(CHOH)_nCHO$ that contain both carboxylic and aldehydic groups, are oxidation products of sugars, and occur in many polysaccharides; especially in the hemicelluloses.

Uterus: The structure in a woman that holds a baby as it develops.

Vaccines: Preparations of antigens from killed or modified organisms that elicit immune response (production of antibodies) to protect a person or animal from the disease-causing agent.

Vacuolation: In cell and tissue culture; excess fluid, debris (aggregates), or gas (from sparging) can form inside a cell vacuole.

Varicose veins: When valves in veins break causing the blood to back up in the veins making them bulge out.

Vector: The plasmid, virus, or other vehicle used to carry a DNA sequence into the cell of another species.

Vegetative cell: Growing or feeding form of a microbial cell, as opposed to a resting form such as a spore.

Vein: Blood vessel that carries blood back to the heart.

Velocity: It is speed in a given direction.

Vertical axis: The line on a graph that goes up the side.

Vesicles: Spherical structures formed intracellularly by some arbuscular mycorrhizal fungi.

Vessel jacket: A temperature control method consisting of a double wall outside the main vessel wall. Liquid or steam flows through the jacket to heat (or cool) the fluid in the vessel.

Viability: Life and health; ability to grow and reproduce; a measure of the proportion of live cells in a population.

Viable: Alive; able to reproduce.

Viable but nonculturable: Organisms that are alive but cannot be cultured on laboratory media.

Viable count: Measurement of the concentration of live cells in a microbial population.

Vibrio: Curved, rod-shaped bacterial cell.

Villi: Fingerlike structures in the small intestine that absorb nutrients.

Virion: Virus particle; the virus nucleic acid surrounded by protein coat and, in some cases, other material.

Virulence: Degree of pathogenicity of a parasite.

Virus: Any of a large group of submicroscopic infective agents that typically contain a protein coat surrounding a nucleic acid core and are capable of growth only in a living cell.

Viscosity: Is a measure of the resistance of a fluid that is being deformed by either shear or tensile stress.

Visible light: EMR that is visible to the human eye and is responsible for the sense of sight.

Voltage: Electric potential energy per unit charge; measured in joules per coulomb (= volts).

Water content: Water contained in a material expressed as the mass of water per unit mass of oven-dry material.

Water-soluble vitamins: Vitamins that are not stored so you must eat every day.

Water-retention curve: Graph showing soil–water content as a function of increasingly negative soil water potential.

Wavelength scanning: Also known as spectral scanning or wave scan measurements.

Weathering: All physical and chemical changes produced in rock by atmospheric agents.

White rot fungus: Fungus that attacks lignin, along with cellulose and hemicellulose, leading to a marked lightening of the infected wood.

Wild type: Strain of microorganism isolated from nature. The usual or native form of a gene or organism.

Winogradsky column: Glass column with an anaerobic lower zone and an aerobic upper zone, which allows growth of microorganisms under conditions similar to those found in nutrient-rich water and sediment.

Xenobiotic: Compound foreign to biological systems. Often refers to human-made compounds that are resistant or recalcitrant to biodegradation and decomposition.

Xenon: Flash lamps provide a high-energy light source with a short warm-up time and long lamp life.

Xerophile: Organism adapted to grow at low water potential, that is, very dry habitats.

X-ray (x-radiation): A form of EMR having a wavelength in the range of 0.01–10 nm, corresponding to frequencies in the range 30 PHz to 30 EHz (3×10^{16}–3×10^{19} Hz) and energies in the range 120 eV to 120 keV.

X-ray diffraction (XRD): is a versatile, nondestructive technique that reveals detailed information about the chemical composition and crystallographic structure of natural and manufactured materials.

Yeast: Fungus whose thallus consists of single cells that multiply by budding or fission.

Yolk: Part of fertilized egg that nourishes embryo (animals that lay eggs).

Zoospore: An asexual spore formed by some fungi that usually can move in an aqueous environment via one or more flagella.

Zygospore: Thick-walled resting spore resulting from fusion of two gametangia of fungi in the phylum Zygomycota.

Zygote: In eukaryotes; the single diploid cell resulting from the union (fusion) of two haploid gametes.

Index